The Dictionary of
CELL AND MOLECULAR BIOLOGY

ELSEVIER *science &*
technology books

The Dictionary of Cell and Molecular Biology, Fifth Edition
J.M. Lackie

If you would like to purchase the electronic version of this book, please visit
http://store.elsevier.com/product.jsp?isbn=9780123849328

ELSEVIER

ACADEMIC
PRESS

The Dictionary of
CELL AND MOLECULAR
BIOLOGY

Fifth Edition

J. M. Lackie

Advisory Editors:

J. G. Coote
C. W. Lloyd

Contributors to earlier Editions:

S. E. Blackshaw*
C. T. Brett
A. S. G. Curtis
J. A. T. Dow**
J. G. Edwards
A. J. Lawrence
G. R. Moores

* 2nd Edition only
** Co-Editor, Editions 1–3

AMSTERDAM • BOSTON • HEIDELBERG • LONDON
NEW YORK • OXFORD • PARIS • SAN DIEGO
SAN FRANCISCO • SINGAPORE • SYDNEY • TOKYO

Academic Press is an imprint of Elsevier

Academic Press is an imprint of Elsevier
32 Jamestown Road, London NW1 7BY, UK
225 Wyman Street, Waltham, MA 02451, USA
525 B Street, Suite 1800, San Diego, CA 92101-4495, USA

Fifth edition 2013

Notice
No responsibility is assumed by the publisher for any injury and/or damage to persons or property as a
matter of products liability, negligence or otherwise, or from any use or operation of any methods, products,
instructions or ideas contained in the material herein. Because of rapid advances in the medical sciences,
in particular, independent verification of diagnoses and drug dosages should be made

British Library Cataloguing-in-Publication Data
A catalogue record for this book is available from the British Library

Library of Congress Cataloging-in-Publication Data
A catalog record for this book is available from the Library of Congress

ISBN: 978-0-12-384931-1

For information on all Academic Press publications
visit our website at elsevierdirect.com

Typeset by MPS Limited, Chennai, India
www.adi-mps.com

Printed and bound in United States of America

12 13 14 15 16   10 9 8 7 6 5 4 3 2 1

Contents

A Note Concerning Entries

The main entry word (the headword) is in bold **Arial** followed by synonyms set in italics in *Times New Roman* font. Within the definition itself words in **bold** are cross-references to other entries that will provide additional information. The absence of an emboldened cross-reference does not mean that there is not a headword entry, but it seemed unnecessary to cross-reference standard things (e.g. actin, microtubule). Where I have arbitrarily decided that the acronym is more likely to be sought than the full name, the acronym has the main definition (so, for example, 'bone morphogenetic proteins' have their main entry under 'BMPs'). There are different conventions for genes and proteins in botanical, zoological and microbiological practice (though these areas overlap, of course) and plant genes are capitalized in italics (eg. '*CONSTANS*'), whereas the plant protein is similarly capitalized but not italicized (e.g. 'SLEEPY'). However, proteins that were originally described in animal systems are in normal typeface (e.g. 'tubulin'), even though they occur in plants as well, although the genes are italicized (e.g. '*decapentaplegic*').

Although noun-adjectives are grammatically to be abjured, they are a convenient shorthand and are often used (e.g. a '*Xenopus* protein' rather than 'in *Xenopus* a protein'). Similarly, some definitions mention 'mutation in protein X' whereas to be correct one should say 'mutation in the gene encoding protein X'. I assume users have some basic biological knowledge and will not be confused by this.

Because they are so commonly the experimental organisms that have been used (and the ones for which full genomic sequences are available) I have almost always referred simply to *Drosophila, Arabidopsis, C. elegans* (not *Caenorhabditis elegans*), *S. cerevisiae* (not *Saccharomyces cerevisiae*), *S. pombe* (not *Schizosaccharomyces pombe*), *E. coli* (not *Escherichia coli*). Unless explicitly stated otherwise the amino acid (aa) count for mammalian proteins refers anthropocentrically to the human protein, in plants to that from *Arabidopsis*, in bacteria from *E. coli*. Generally I have given the amino acid count for the complete protein, including the signalling sequence that is removed from secreted proteins: where the processing is more dramatic, I have indicated the sizes of both precursor and active product.

Alphabeticization is according to fairly standard lexicographical practice and ignores hyphens and spaces (so 'co-culture' appears after 'Cockayne's syndrome' and before 'codanin', not before 'coacervate'). Numbers are treated as numbers, so the sequence is 'interleukin-1',' interleukin-2', 'interleukin-3', etc. even though the database I use would put 'interleukin 10' immediately after 'interleukin-1 converting enzyme' and before 'interleukins 2, 3, . . .etc. Some dictionaries would put 5-HT under 'F' (for 'five') but I have put such numeric entries at the very beginning, before 'A', where I think they are more likely to be found. Greek-letter prefixes are a problem and generally I have treated 'α' as 'alpha' (but it is probably worth checking under the non-prefixed name if an expected entry is not found).

J.M. Lackie
December 2011

Tables

Preface

Almost nobody is sad enough to read the prefaces of Dictionaries except perhaps other lexicographers and occasionally reviewers, but it is customary to produce a Preface and make an apologia. This dictionary is the lineal descendent of the Dictionary of Cell Biology, first published in 1989, that had a second edition in 1995 and was renamed as the Dictionary of Cell & Molecular Biology in 1999. All three were co-edited by Julian Dow, and it was Julian who arranged that the second and third editions were accessible through the Web: we made use of the abortive searches which we logged to guide us in adding new headwords to the 3rd Edition. In 2007, I produced the 4th Edition, more-or-less single-handedly rewriting many of the old entries and all the new ones, again guided by 'abortive searches'. This Edition has been produced in much the same way although without the benefit of knowing what people failed to find in the 4th Edition (which never made it online) but with help from John Coote and Clive Lloyd who scrutinised the microbiological and botanical entries respectively. The headword count indicates what has happened to cell biology ('molecular bioscience') in the intervening years. The headword count has inexorably risen from 4000, to 5000, to 7000, to 10,000 and in this edition to 12,000 (although ~1500 are 'signposts' to main entries). The expansion is not simply by accretion of the new — some 2400 headwords and definitions are completely new, more than 4000 definitions have been rewritten, some headwords have second definitions (~600), third definitions (~300), even sixth definitions (60) added, less than half are unchanged.

Many of the additions have been to extend the scope in plant cell biology, microbiology and bioinformatics but there are undoubtedly areas where more additions could be made. To some extent there is a bias of entries towards areas that I find interesting, although that is a fairly broad category, but the difficulty of defining where 'cell and molecular biology' begins and ends is no easier than it ever was. Some entries are there simply because practitioners of modern bioscience are often unfamiliar with vocabulary that the more classically trained zoologist or botanist would consider standard. In this edition I have removed a lot of the entries that related to specific pharmaceutical compounds while keeping some generic entries ('alpha blockers', 'NSAIDs', 'tetracycline antibiotics') and some that are commonly part of the experimentalist's toolkit and probably never used in the clinic. Diseases have only earned an entry if the molecular basis is known — but that is an expanding category.

The intention is that this edition does go online and so it is likely that people will be able to search the dictionary electronically. This does not mean that spelling is irrelevant and it is worth remarking that more that 50 recognisable misspellings of mitochondrion appeared in the abortive search file (and so did not produce a definition for the searcher). Computers are very literal-minded and it is often better to terminate the search-term before plural or adjectival suffixes — so searching for 'mitoch' produces 6 entries, all relating to mitochondria (mitochondrion, mitochondrial diseases, Mitochondrial Eve, mitochondrial trifunctional protein, mitochondrial neurogastrointestinal encephalopathy and mitochondrial DNA deletion syndrome) — but *not* 'mitochondria' which does not get an entry in its own right. Because it is electronically easy to search the whole text I have removed many of the synonyms, which in previous editions were in a different typeface after the main headword, and put them within the definition (and in many cases there were so many synonyms it was becoming very clumsy).

The apologia is obvious: there are undoubtedly mistakes, for which I take all the blame (John Coote and Clive Lloyd corrected some, but saw only definitions in their specialist areas) and some things that are out-of date simply because the subject moves so fast. Definitions that I wrote in early 2010 may no longer be correct by the time the Dictionary is published. Similarly, although I have checked all the weblinks in the final stages of finishing the manuscript, some will have disappeared. In general I have tried to reference full research articles for which there is free open access. When only the abstract is free, then that is what I have linked.

In writing definitions I have had frequent recourse to the various NCBI (National Center for Biotechnology Information) databases, especially the UniProt Protein Knowledge Base and OMIM (On-line Mendelian Inheritance in Man) and the failure to find a protein in these databases or in PubMed has led me to deem them 'obsolete', although I have left their entries on the unlikely offchance that somebody might find them when reading the older literature. For other things I have assumed that two independent sources are enough to validate a definition. Wikipedia is usually accurate (in this area anyway) but often provides far more information than required and I have not treated it as a source: a dictionary should provide a brief definition, not an essay.

Having admitted the (probable) existence of errors, I should be more than happy to correct them if they are drawn to my attention (john@lackie.nildram.co.uk): although the print version cannot easily be corrected, updating the electronic version should be possible.

John Lackie
January 2012

Preface to the Fourth Edition

Modern biology continues to evolve at an astonishing rate and the boundaries between the old sub-disciplines are becoming so blurred as to be almost irrelevant; we are all 'bioscientists' now. But there is a welcome trend towards integration across the continuum and 'translational research', the process of moving from a molecule directed against a cellular target to a therapeutic drug being administered to patients, has become a fashionable term. The Dictionary tries to serve the needs of bioscientists or clinicians who are unfamiliar with terminology from adjacent sub-specialities and is therefore extensive in its coverage rather than specialist. It is, however, a Dictionary and not an encyclopaedia; it does not have large chunks of text to cut and paste into an essay.

As before, the choice of new headwords has been determined to a significant extent by logging of abortive searches on the web-based version; people continue to misspell things in imaginative ways (for example, more than 50 recognisable but incorrect versions of 'mitochondrion' have been typed in to the search box!). The task of defining the new headwords has been greatly facilitated by the resources available on the Web: NCBI's Entrez PubMed has probably been the major source of information, supplemented by the OMIM database (On-line Medelian Inheritance in Man) and various other 'professional' databases. The definitions have been distilled from multiple sources – essentially the Dictionary summarises the searching that any web-connected scientist could do – and the aim has been to provide a succinct definition with sufficient cross-referencing to enable the context to be inferred.

Searches for TLAs (three-letter abbreviations) were also common and as a general principle I have defined those that came up in more than two Entrez PubMed abstracts, reasoning that this meant they were in sufficiently general use to justify being defined. It is noticable that some are very popular – thus there are 8 possible interpretations of CPA, DMP, CSP, etc. This is a problem when authors forget that others may not recognise which version they intend and fail to specify in full on first use.

The most difficult task, in some ways, was the revision of the older entries; some named proteins have apparently disappeared from the literature – but proving disappearance is almost impossible. Where appropriate I have drawn attention to such obsolescence but have retained the entries since a few rare souls do read the older literature and may find the definitions useful.

Thus, having tried to define things people failed to find and choosing not to delete older entries, the Dictionary has grown yet again: this edition has more than 10,000 headwords, some with multiple definitions. Once the paper version is out then the new edition will go onto the Web and again abortive searches will be logged to inform the inevitable 5th Edition. But feedback is always welcomed and since it would be extraordinary if some entries were less than perfect, do not hesitate to let me know. My apologies if you fail to find what you are looking for, or consider my attempt at a definition to be inaccurate or incomplete! I hope you do find it useful.

John Lackie
November 2006

Preface to the Third Edition

Although the title has changed we as Editors have talked of this as the Third Edition of the Dictionary of Cell Biology — and the change in title simply reflects the changes that have happened to Cell Biology in the last decade. In the Preface to the 1st Edition we commented that the boundaries of "cell biology" were difficult to define and this has certainly not changed — if anything the territory has grown! But, the molecular entries have continued to increase in number as more and more is known of the detail of cell structure and the complexity of signalling pathways. The new title reflects the content more accurately — though we have also added a number of entries that are neither cellular nor molecular.

Now that the Human Genome Project is nearing completion and we have the complete genome sequence of yeast and the nematode *Caenorhabditis elegans*, the big problem becomes that of assigning function to genes. Inevitably this will draw heavily on cell biology and many molecular biologists will find themselves entering new territory — and cell biologists they meet will need to talk the language of molecular biology. Cell biology encompasses an extremely wide range of experimental systems and has a very diverse vocabulary: this Dictionary should help to provide some guidance.

In preparing this volume, we have been much influenced by the usage of the Internet version of the 2nd Edition of the Dictionary of Cell Biology. Putting the Dictionary on the Net was an experiment — and one that has proved fascinating. For the first time, perhaps, it has been possible to monitor how people used the "book" and what they searched for. Approximately one-third of a million visits have been made to the site and it has been cross-referenced from various other web pages. We have maintained a log of abortive searches and also gave an e-mail address for feedback. Though we had put a tear-out sheet for comment and feedback in the first and second paper editions we had almost no response: by e-mail it was different!

Many abortive searches were a result of inability to spell or perhaps to type accurately — we can do little about that! But many searches were for things that we felt really should have been in — sometimes we had omitted to write an entry, sometimes the search was for something new, sometimes the search was for things that are part of the wider vocabulary of those trained in the older disciplines. Thus there were quite a lot of searches for fixatives well known to the histologist, for Latin names of species where the common name is well known, for syndromes that have emerged or been diagnosed (we suspect). We have tried to put in some entries to help on these aspects, tried to put in new things from the literature but, in the case of diseases, we have tended to put entries only where there is a known cell or molecular basis for the disease. This edition has more than 7000 entries and we have been more comprehensive with cross-referencing of synonyms and from the text: this Dictionary is almost double the size of the First edition! The entries from the first and second editions have been scrutinised and modified where necessary, particularly when there has been comment from readers, and many of the new entries are for words sought by on-line users. A variety of sources have been used and a brief list is appended.

We have always tried to provide short, clear definitions that are helpful to people with the widest range of backgrounds, and the huge volume of feedback we have received from the Internet edition suggests that these efforts are well received. Not only career cell biologists, but school teachers, high school students and journalists have all sent glowing testimonials. Our aim is to continue to develop this resource, both online and, given the speed at which the discipline is evolving, to produce another hard copy in due course. Meanwhile we hope that people will find the hard copy useful – it is much easier and quicker to search adjacent entries. We also hope that people will not hesitate to send new entries, suggest amendments and help in the evolution of this unusually interactive resource.

The online Dictionary is to be found on http://www.mblab.gla.ac.uk/dictionary

John Lackie
Julian Dow
January 1999

Preface to the Second Edition

In the preface to the first edition we commented that 'the subject was far from static', and we have not been disappointed. The last few years have seen rapid progress and an increasing reliance upon the powerful tools of molecular biology. In this second edition we have extended the coverage, particularly in molecular biology and neurobiology, though the sense of barely keeping up with the emergence of new names for proteins is ever more pressing. 'Cytokine' has overtaken 'Interleukin' as the numbers of cytokines have almost doubled and chemokines have arrived. The table of CD numbers has grown impressively and the G-proteins have proliferated; far more genes have been named and new proteins are legion. A measure of the rate of change is that we have had to include more than 1000 new entries; by the next revision we will have to start deleting entries — and anybody who scans the old literature (pre-1990!) will find names that have disappeared without trace. As before, the choice of entries partly reflects our own interests, but we have tried to be broadminded with our inclusions. This time Dr Susanna Blackshaw has written entries rather than acting as an external adviser, in order to increase the neurobiological content, and both editors have become much more involved with molecular biology. In the first edition we included a tear-off page for users to send us their neologisms and revisions but, disappointingly, had little response. We continue to hope that interested users of this book will provide suggestions for future improvements. Several people have helped with the revision of the manuscript itself and in particular we wish to thank Mrs Lynn Sanders and Mrs Joanne Noble for help with typing. Many colleagues who have not actually been directly involved have nevertheless been plagued with questions, and we are grateful for their patience. We hope that this new edition will be as useful as the first and that it will date no faster. On past experience, however, our rapidly growing subject will doubtless keep this project alive for years to come. We have taken the unusual step of making a searchable version of the Dictionary available on the Internet. Although undoubtedly less convenient than the paper version, this will carry incremental changes and updates — including those provided electronically by readers — until the third edition is published. Details are provided on the revision form at the back of the book: we hope you find this service useful.

John Lackie
Julian Dow
July 1994

Preface to the First Edition

The stimulus to write this dictionary came originally from our teaching of a two-year Cell Biology Honours course to undergraduates in the University of Glasgow. All too often students did not seen to know the meanings of terms we felt were commonplace in cell biology, or were unable, for example, to find out what compounds in general use were supposed to do. But before long it become obvious that although we all considered ourselves to be cell biologists, individually we were similarly ignorant in areas only slightly removed from our own — though collectively the knowledge was there. It was also clear that many of the things we considered relevant were not easy to find, and that an extensive reference library was needed. In that we have found the exercise of preparing the Dictionary informative ourselves, we feel that it may serve a useful purpose. An obvious problem was to decide upon the boundaries of the subject. We have not solved this problem: modern biology is a continuum and any attempt to subdivide it is bound to fail. 'Cell Biology' implies different things to zoologists, to biochemists, and indeed to each of the other species of biologists. There is no sensible way to set limits, nor would we wish to see our subject crammed into a well-defined niche. Inevitably, therefore the contents are somewhat idiosyncratic, reflecting our current teaching, reading, prejudices, and fancies. It may be of some interest to explain how we set about preparing the Dictionary. The list of entry words was complied largely from the index pages of several textbooks, and by scanning the subject indexes of cell-biological journals. To this were added entries for words we cross-referenced. The task of writing the basic entries was then divided amongst us roughly according to interests and expertise. We all wrote subsets of entries which were then complied and alphabetized before being edited by one of us. Marked copies were then sent out to a panel of colleagues who scrutinized entries in their own fields. All entries were looked at by one or more of this panel, and then the annotated entries were re-edited, corrections made on disc, and the files copy-edited for consistency of style. A very substantial amount of the handling of the compiled text and the preparation of the final discs was done by Dr A M Lackie who also acted as copy-editor. Glasgow is a major centre for Life Sciences, and we are fortunate in having many colleagues to whom we could turn for help. We are very grateful to them for the work which they put in and for the speed with which they checked the entries that we sent. Although we have tried hard to avoid errors and ambiguities, and to include everything that will be useful, we apologise at this stage for the mistakes and omissions, and emphasise that blame lies with the authors and not with our panel (though they have saved us from many embarrassments). Since there is no doubt Cell Biology is developing rapidly as a field, it is inevitable that usages will change, that new terms will become commonplace, that new proteins will be christened on gels, and that the dictionary will soon have omissions. Were the subject static this dictionary would not be worth compiling – and we cannot anticipate new words. Because the text is on disc, it will be relatively easy to update: please let us have your comments, suggestions for entries (preferably with a definition), and (perhaps) your neologisms. A sheet is included at the back of the dictionary for this purpose.

John Lackie
Julian Dow
1989

Numerical entries and A

2−5A/RNase L system An RNA cleavage pathway in which 2′−5′ linked oligoadenylates (2−5A) are produced from ATP by interferon-inducible synthetases. 2−5A activates pre-existing RNase L, resulting in the cleavage of RNAs within single-stranded regions. Activation of RNase L by 2−5A initiates an antiviral response. Abstract: http://www.ncbi.nlm.nih.gov/pubmed/9856285

3M syndrome A disorder (gloomy face syndrome, Le Merrer syndrome, dolichospondylic dysplasia, Yakut short stature syndrome) in which there is low birth weight dwarfism and facial abnormalities: caused by mutation in genes for **cullin 7** or **obscurin**-like protein-1.

5-HT See **serotonin**.

5q-minus syndrome A type of **myelodysplasia** associated with deletion of the long arm of chromosome 5 and therefore to multiple gene dysfunctions. Genes involved include those for aFGF, GM-CSF, IL-3, IL-4, IL-5, IL-9, both subunits of IL-12, the gene encoding IRF1 (interferon regulatory factor-1) and the receptors for M-CSF, PDGF, and FGF (FGFR4).

9E3 In chickens, CXC motif chemokine 8 (IL-8, CEF-4, embryo fibroblast protein 1, 103aa) constitutively produced by chicken cells infected with **Rous sarcoma virus**. May be an autocrine factor for fibroblast proliferation.

14-3-3 proteins Family of adapter proteins (ca 250aa) able to interact with a range of signalling molecules including c-Raf, Bcr, PI-3-kinase, polyoma middle T-antigen. Bind to phosphorylated serine residues in cdc25C and block its further activity, may bind Bad (death inducer) thereby blocking heteromeric interaction with Bcl-XL, and in plants bind and inhibit activity of phosphorylated nitrate reductase. Basic mode of action may be to block specific protein-protein interactions. There are seven highly conserved human 14-3-3 proteins (beta, gamma, epsilon, eta, sigma, theta, zeta) that are involved in cellular proliferation, checkpoint control and apoptosis: 14-3-3σ is **stratifin** (248aa). More than 200 14-3-3 target proteins have been identified, including proteins involved in mitogenic and cell survival signalling, cell cycle control and apoptotic cell death.

α **alpha** Words (proteins etc.) pre-fixed by 'α' are listed as though this was 'alpha'; alternatively look for main portion of word.

A Single letter code for alanine (in proteins) or adenosine (in nucleic acid or compounds such as ATP).

A4 protein See **amyloidogenic glycoprotein**.

A9 cells Established line of heteroploid mouse fibroblasts that are deficient in **HGPRT**.

A20 A stress response gene in endothelial cells that encodes a dual function enzyme (790aa) with de-ubiquitinating activity towards Lys63-linked poly-ubiquitin chains and Lys48 E3 ligase activity. The enzyme is an inhibitor of TNF signalling and acts by triggering degradation of RIP (**receptor interacting protein kinase**). Deficiency in A20 and TNF or TNF-R leads to spontaneous inflammation in mice.

A260 Spectrophotometric absorbance at 260 nm. The ratio of absorbance at 260 nm to that at 280 nm is often used as a quick assessment of the purity of nucleic acid samples since nucleic acids absorb strongly at 260 nm and proteins at 280 nm.

A431 cells A line of human epidermoid carcinoma cells.

A23187 A monocarboxylic acid extracted from *Streptomyces chartreusensis* that acts as a mobile-carrier calcium **ionophore**. See Table I3.

AAA family A diverse superfamily of ATPases associated with various cellular activities that have a characteristic AAA-domain. There are at least 300 known members in at least six families. AAA proteins are found in all organisms and are essential for cell cycle functions, vesicular transport, mitochondrial functions, peroxisome assembly and proteolysis. Mutations in proteins of the AAA-family cause a variety of human diseases. See also **NSF**, **PIKK** and **valosin-containing protein**. Family tree: http://jcs.biologists.org/content/114/9/1601.full

Aarskog-Scott syndrome An X-linked developmental disorder (faciogenital dysplasia) caused by mutations in **FYVE, RhoGEF and PH domain-containing protein-1**, FGD1.

The Dictionary of Cell and Molecular Biology. DOI: http://dx.doi.org/10.1016/B978-0-12-384931-1.00001-5

Aatll The restriction endonuclease Aatll (345aa) from *Acetobacter aceti*.

AA-tRNA See **aminoacyl tRNA**.

Ab Common abbreviation for antibody. See **immunoglobulins**.

abaecins Proline-rich basic antibacterial peptides (53aa in *Apis mellifera*) found in the **haemolymph** of bees. See **apidaecins**.

A band That portion of the **sarcomere** in which the thick myosin filaments are located. It is anisotropic in polarized light.

abaxial Located on the side away from the axis or facing away from the axis of stem or root. Typically the lower surface of leaves. *Cf.* **adaxial**.

ABC (1) Antigen binding cell or antigen binding capacity. (2) Avidin-biotin peroxidase complex. Used in visualizing antigen. Primary (antigen-specific) antibody is bound first, a second biotinylated anti-immunoglobulin antibody is then bound to the first antibody, then ABC complex that has excess biotin-binding capacity, is bound to the biotin on the second antibody and finally the peroxidase used to catalyse a colorimetric reaction generating brown staining. The method gives substantial signal enhancement. (3) See **ABC-exinuclease, ABC model of flower development, ABC proteins**.

ABCA1 An **ABC protein** (cholesterol efflux regulatory protein, 2261aa) that is important in the initial steps of the reverse cholesterol transport pathway; a cAMP-dependent and sulphonylurea-sensitive anion transporter. Mutations in ABCA1 cause Tangier disease. There is a homologous protein (1882aa) in *Arabidopsis* and a much smaller form (877aa) in *Dictyostelium*.

ABCD1 In ATPase binding cassette protein in the same family of transporter proteins such as **CFTR** and **MDR** proteins. See **adrenoleucodystrophy**.

ABC-excinuclease Enzyme complex, product of *uvrA*, *uvrB* and *uvrC* genes from *E. coli* that mediates incision and excision steps of DNA excision-repair. Enzyme has the ability to recognize distortion in DNA structure caused by, e.g., ultraviolet irradiation.

ABC model of flower development A model of flower development in angiosperms based on the observation of mutants with defects in floral organ development. The ABC model summarizes how the presence or absence of different classes of transcription factors in the different parts of the flower regulates the development of floral organs. Mutations in class A genes (*SQUAMOSA* (*SQUA*), *LIPLESS1* (*LIP1*), *LIPLESS2* (*LIP2*) in *Antirrhinum*, *APETALA2* (*AP2*), *APETALA1* (*AP1*) in *Arabidopsis*) affect sepals and petals. Mutations in

class B genes (*DEFICIENS* (*DEF*), *GLOBOSA* (*GLO*), *APETALA3* (*AP3*), *PISTILLATA* (*PI*)) affect petals and stamens, while those in class C (*PLENA* (*PLE*), *FARINELLI* (*FAR*), *AGAMOUS* (*AG*)) affect stamens and carpels. All three classes of genes are homeotic genes, which are translated into proteins.

ABC proteins A very large family of membrane proteins involved in active transport or regulation of ion channel function and that have an ATP binding cassette. They are found in both prokaryotes and eukaryotes and in humans important examples are P-glycoprotein (multidrug resistance transporter), cystic fibrosis transmembrane conductance regulator (**CFTR**), and **SUR**. In *Arabidopsis* there are at least 120 genes encoding ABC proteins, although the function of many remains undefined.

Abell-Kendall method A reference-standard method for estimating total serum cholesterol levels.

Abelson leukaemia virus *A-MuLV* A replication-defective retrovirus originating from the Moloney murine leukemia virus by acquisition of c-*abl*. The virus induces B-cell lymphoid leukemias within a few weeks. The v-*abl* product has tyrosine kinase activity.

abenzyme *abzyme* See **catalytic antibody**.

aberration Departure from normal; in microscopy two common forms of optical aberration cause problems, **spherical aberration** in which there is distortion of the image of the magnified object and **chromatic aberration** that leads to coloured fringes, a consequence of the unequal refraction of light of different wavelength.

abetalipoproteinaemia Autosomal recessive defect (acanthocytosis, Bassen-Kornweig syndrome) in which there is total absence of **apolipoprotein B** caused by a defect in in **microsomal triglyceride transfer protein**. Characteristic feature is presence of **acanthocytes**; later in life neurological disorders and retinitis pigmentosa develop and death is usually a consequence of cardiomyopathy.

abf1 (1) Activated B-cell factor 1 (musculin, class A basic helix-loop-helix protein 22, 206aa) a transcription repressor that may regulate antigen-dependent B-cell differentiation. (2) In *S. cerevisiae*, ARS-binding factor 1 (bidirectionally acting factor 1, DNA replication enhancer-binding protein OBF1, 731aa), a general regulatory factor that affects transcription of a wide range of genes. (3) In *Arabidopsis*, abscisic acid responsive elements-binding factor 1 (ABSCISIC ACID-INSENSITIVE 5-like protein 4, bZIP transcription factor 35, 392aa).

ABH Alpha-ketoglutarate-dependent dioxygenase (ABH1, DNA lyase ABH1, EC 4.2.99.18, alkylated DNA repair protein alkB homologue 1, EC 1.14.11.-, 389aa), a dioxygenase that repairs alkylated single-stranded DNA and RNA containing 3-methyl-cytosine by oxidative demethylation. There are a family of alkB enzymes alternatively named ABH1, ABH5, etc.

abiogenesis Spontaneous generation of life from non-living material. The ability to synthesise a small genome and provide sufficient purified proteins etc. to generate a viable life-form is getting closer. A completely synthesized genome has been introduced into a cell and shown to replicate.

abiotic stress The negative effect of non-biological (abiotic) factors on plant growth, often resulting in diminished productivity. Stressors include environmental factors such as wind, desiccation, heavy metals and salinity.

abl An oncogene, identified in Abelson murine leukaemia virus that encodes a non-receptor tyrosine kinase, abl (1130aa) that regulates cytoskeleton remodeling during cell differentiation, cell division and cell adhesion. c-Abl contains both a G-actin binding site and an independent F-actin site. See also **ABLV**.

ABLV The Abelson murine leukaemia virus, a mammalian **retrovirus**. Its transforming gene, *abl*, encodes a protein tyrosine kinase of the **src** family.

ABM paper Aminobenzyloxy methylcellulose paper: paper to which single-stranded nucleic acid can be covalently coupled.

ABO blood group system Probably the best known of the blood group systems, involves a single gene locus that codes for a fucosyl transferase. If the *H*-gene is expressed then fucose is added to the terminal galactose of the precursor oligosaccharide on the red cell surface and the *A*- or *B*-gene products, also glycosyl-transferases, can then add N-acetyl galactosamine or galactose to produce the A or B-antigens respectively. Antibodies to the ABO antigens occur naturally and make this an important set of antigens for blood transfusion. Transfusion of mismatched blood with surface red cell antigens that elicit a response leads to a transfusion reaction. The natural antibodies are usually IgM. See **Rhesus**, **Kell**, **Duffy** and **MN blood group antigens**. NCBI Details: http://www.ncbi.nlm.nih.gov/books/NBK2267/

abortive infection Viral infection of a cell in which the virus fails to replicate fully, or produces defective progeny. Since part of the viral replicative cycle occurs, its effect on the host can still be cytopathogenic.

abortive transformation Temporary transformation of a cell by a virus that fails to integrate into the host DNA.

ABP See **actin binding proteins**. Encyclopedia of Actin Binding Proteins: http://www.bms.ed.ac.uk/research/others/smaciver/Encyclop/encycloABP.htm

ABP1 An F-actin binding protein (592aa) from *S. cerevisiae* that regulates ARP2/3 complex-mediated actin assembly. Recruits the ARP2/3 complex to sides of pre-existing actin filaments, which may promote nucleation or stabilization of filament branches. Has an ADF domain and an SH3 domain through which it binds to **dynamin**. The ABP1 family includes drebrin-like protein A (447aa) from *Xenopus* and human drebrin-F (430aa).

ABP-50 Actin binding protein (Elongation factor 1-alpha, 50 kDa, 453aa) from *Dictyostelium* that crosslinks actin filaments into tight bundles and also promotes the GTP-dependent binding of aminoacyl-tRNA to the A-site of ribosomes during protein biosynthesis. Calcium insensitive; localized near cell periphery and in protrusions from moving cells.

ABP-67 Early name for **fimbrin**. In yeast encoded by *SAC6* gene, mutations in which lead to disruption of the actin **cytoskeleton**.

ABP-120 Actin binding protein (Dictyostelium gelation factor, 857aa) from *Dictyostelium*. A small rod-shaped molecule ($35-40$ nm long), dimeric, capable of cross-linking filaments. Has strong sequence similarities with **ABP-280**.

ABP-280 Actin binding protein (2647aa, 280 kDa) originally isolated from *Dictyostelium*, now identified as **filamin**.

abrin An AB toxin from seeds of *Abrus precatorius* (Indian licorice). The precursor (abrin-a, 528aa) is cleaved into three chains: abrin-a A chain (EC 3.2.2.22, rRNA N-glycosidase), a linker peptide and abrin-a B chain that has a lectin-like binding site for galactose and related residues in carbohydrate but, because it is monovalent, is not an agglutinin for erythrocytes. The A-chain N-glycosylates eukaryotic 28S rRNA and inhibits protein synthesis. Abrin-b, -c and -d are very similar. Agglutinin-1 (547aa) from *A. precatorius* inactivates 60S ribosomal subunits by removing adenine from position 4,324 of 28S rRNA.

abrineurin See **brain-derived neurotrophic factor**.

abscisic acid A growth-inhibiting plant hormone (dormin) found in vascular plants. Originally believed to be important in abscission (leaf fall), now known to be involved in a number of growth and developmental processes in plants, including, in some circumstances, growth promotion. Primary plant hormone that mediates responses to stress and signals through **cyclic ADP-ribose** as a second messenger.

absolute lethal concentration LC_{100} Lowest concentration of a substance that kills 100 % of test organisms or species under defined conditions. This value is dependent on the number of organisms used in its assessment.

absorption coefficient (1) Any of four different coefficients that indicate the ability of a substance to absorb electromagnetic radiation. Absorbance is defined as the logarithm of the ratio of incident and transmitted intensity and thus it is necessary to know the base of the logarithm used. Scattering and reflectance are generally ignored when dealing with solutions. (2) Ratio of the amount of a substance absorbed (uptake) to the administered quantity (intake).

absorption spectrum Spectrum of wavelengths of electromagnetic radiation (usually visible and UV light) absorbed by a substance. Absorption is determined by existence of atoms that can be excited from their ground state to an excited state by absorption of energy carried by a photon at that particular wavelength.

AB toxin Multi-subunit toxin in which there are two major components, an active (A) portion and a portion that is involved in binding (B) to the target cell. The A portion can be effective in the absence of the B subunit(s) if introduced directly into the cytoplasm. In the well-known examples, the A subunit has **ADP-ribosylating** activity. See: **cholera toxin**, **diphtheria toxin**, **pertussis toxin** also **colicins**, **Ricinus communis agglutinins**, **shiga toxin**, **tetanus toxin**, **VacA**.

ABTS Compound that will produce a water soluble, green colored product upon reaction with horseradish peroxidase; used in enzyme-linked immunoassays. It is light sensitive and must be kept in the dark both as a stock solution and as a working solution.

ACADs Acyl-coenzyme A dehydrogenases, a family of mitochondrial enzymes that catalyze the first dehydrogenation step in the beta-oxidation of fatty acyl-CoA derivatives. Different ACADs act on long, medium and short-chain fatty acids. ACAD8 (acyl-Coenzyme A dehydrogenase family, member 8, isobutyryl-CoA dehydrogenase, EC 1.3.99.-, 415aa) is responsible for a step in the breakdown of valine, the conversion of isobutyryl-CoA into succinyl-CoA.

Acanthamoeba Soil amoebae 20–30 μm in diameter that can be grown under **axenic** conditions and have been extensively used in biochemical studies of cell motility. They have been isolated from cultures of monkey kidney cells, and are pathogenic when injected into mice or monkeys.

acanthocyte Cell with projecting spikes; most commonly applied to erythrocytes where the condition may be caused naturally by **abetalipoproteinaemia** or experimentally by manipulating the lipid composition of the plasma membrane.

acanthoma A benign skin tumour composed of squamous or epidermal cells.

Acanthoscurria gomesiana Tarantula.

acanthosis nigricans A rare disease characterized by pigmentation and warty growths on the skin. Often associated with cancer of the stomach or uterus.

acanthosome (1) Spinous membranous **organelle** found in skin **fibroblasts** from **nude mice** as a result of chronic ultraviolet irradiation. (2) Sometimes used as a synonym for **coated vesicle** (should be avoided).

Acanthus Genus of spiny-leaved Mediterranean plants.

acapnia Medical condition (hypocapnia, hypocapnea, hypocarbia) in which there is a low concentration of carbon dioxide in the blood.

ACAT An enzyme (acyl coenzyme A:cholesterol acyltransferase, sterol o-acyltransferase) that catalyzes cholesterol ester formation from cholesterol and fatty acyl CoA substrates. Sterol esterification by ACAT or homologous enzymes is highly conserved in evolution. In human ACAT1 (EC 2.3.1.9, 427aa) is mitochondrial and ACAT2 (EC 2.3.1.26, 522aa) is located in the ER. The ACAT2 gene shows complementary overlapping with the 3-prime region of the TCP-1 gene in both mouse and human. These genes are encoded on opposite strands of DNA, as well as in opposite transcriptional orientation.

ACC The immediate precursor (1-aminocyclopropane-1-carboxylic acid) of the plant hormone ethylene in most vascular plants. Synthesized from S-adenosyl methionine by **ACC synthase**.

ACC synthase Enzymes (ACC methylthioadenosine lyase, EC 4.1.1.14: e.g. ACC1, 496aa; ACC6, 495aa) that catalyses conversion of S-adenosyl-methionine to **ACC**, first step in production of the plant hormone ethylene. In *Arabidopsis* there is a large family with different tissue distribution of the various isoenzymes.

accelerin Obsolete name for coagulation factor V.

accessory cells (1) Cells that interact, usually by physical contact, with T-lymphocytes and that are necessary for induction of an immune response. Include antigen presenting cells, antigen processing cells, etc. They are usually MHC Class II positive (see **histocompatibility antigens**). Monocytes, macrophages, dendritic cells, **Langerhans cells**, B-lymphocytes may all act as accessory cells. (2) In plants, a synonym for subsidiary cells, epidermal cells that surround the **guard cells** of the stomata and are morphologically distinct.

accessory pigments In photosynthesis, pigments that collect light at different wavelengths and transfer the energy to the primary system.

ACD See **acid-citrate-dextrose**.

ACE See **angiotensin** and **angiotensin-converting enzyme inhibitors**.

A cells α *cells* Cell of the endocrine pancreas (Islets of Langerhans) that form approximately 20% of the population; their opaque spherical granules may contain **glucagon**. See **B cells, D cells**.

acellular Not made of cells; commonest use is in reference to **acellular slime moulds** such as *Physarum* that are multinucleate syncytia.

acellular slime moulds Protozoa of the Order Eumycetozoida (also termed true slime moulds). Have a multinucleate plasmodial phase in the life cycle.

acentric Descriptive of pieces of **chromosome** that lack a **centromere**.

acervulus A small asexual fungal fruiting body produced by fungi of the order Melanconiales (Deuteromycota, Coelomycetes). They consist of a mat of hyphae which give rise to short-stalked conidiophores and emerge through the epidermis of infected plants and release conidiospores.

Acetabularia Giant single-celled **alga** of the Order Dasycycladaceae. The plant is 3–5 cm long when mature and consists of rhizoids at the base of a stalk, at the other end of which is a cap that has a shape characteristic of each species. The giant cell has a single nucleus, located at the tip of one of its rhizoids, which can easily be removed by cutting off that rhizoid. Nuclei can also be transplanted from one cell to another.

acetaminophen See **paracetamol**.

Acetobacter Genus of aerobic bacilli that will use ethanol as a substrate to produce acetic acid – thus will convert wine to vinegar.

acetocarmine A solution of carmine, a basic dye prepared from the insect *Coccus cacti*, in 45% glacial acetic acid; used for the staining of plant chromosomes by the squash method.

acetosyringone A phenolic inducer of the virulence genes (*vir* genes) of *Agrobacterium tumefaciens*, produced by wounding of plant tissue. Acetosyringone can act as a chemical attractant *in vitro* and thus may act as chemotactic agent in nature. Webpage: http://www.ndsu.nodak.edu/instruct/mcclean/plsc731/transgenic/transgenic2.htm

acetyl CoA Acetylated form of coenzyme A that is a carrier for acyl groups, particularly in the **tricarboxylic acid cycle**.

acetylation Addition, either chemically or enzymatically, of acetyl groups.

acetylcholine *Ach* Acetyl ester of choline. Perhaps the best characterized **neurotransmitter**, particularly at neuromuscular junctions. ACh can be either excitatory or inhibitory, and its receptors are classified as **nicotinic** or **muscarinic**, according to their pharmacology. In **chemical synapses** ACh is rapidly broken down by **acetylcholine esterases**, thereby ensuring the transience of the signal.

acetylcholine esterase An enzyme (EC 3.1.1.7, 614aa), found in the **synaptic clefts** of cholinergic synapses, that cleaves the **neurotransmitter** acetylcholine into its constituents, acetate and choline, thus limiting the size and duration of the postsynaptic potential. Many nerve gases and insecticides are potent acetylcholine esterase inhibitors, and thus prolong the time-course of postsynaptic potentials. Drugs that inhibit the breakdown of acetycholine in the CNS may slow the rate of decline in Alzheimer's disease.

acetylcholine receptor See **nicotinic acetylcholine receptor, muscarinic acetylcholine receptor**.

acetylglucosaminyltransferase gene See **LARGE**.

acetylsalicylic acid See **aspirin**.

Ach See **acetylcholine**.

A chain Shorter of the two polypeptide chains of insulin (21 residues compared to 30 in the B chain). Many other heterodimeric proteins have their smaller chain designated the A chain, so the term cannot be used without qualification. Other A chains: **abrin, activin**, C1q (see **complement**), **diphtheria toxin, inhibin, laminin**, mistletoe lectin, **PDGF**, relaxin, **ricin, tPA**.

achaete-scute complex A group of basic helix-loop-helix transcription factors (achaete, scute, lethal of scute, and asense) in *Drosophila* that are involved in regulation of nervous system development.

achene A type of simple, dry, one-seeded fruit, produced by many flowering plants. Does not split (dehisce) when mature.

achondrogenesis See **hypochondrogenesis**.

achondroplasia Failure of endochondral ossification responsible for a form of dwarfism (chondrodystrophia fetalis); caused by an **autosomal dominant mutation** in the gene for the fibroblast growth factor receptor-3 (in 98% of cases, the mutation is a G380R substitution, resulting from a G-to-A point mutation at nucleotide 1138). Relatively high incidence (1:20,000 live births), mostly (90%) new mutations. See **pseudoachondroplasia**.

achromatopsia See **colour blindness**.

Achyla Genus of aquatic fungi with a branched coenocytic mycelium.

aciclovir *formerly* acyclovir An antiviral drug that is effective against both type 1 and 2 herpes virus and varicella-zoster virus. The drug is a nucleoside analogue (hydroxyethoxymethyl-guanine) that is converted to the monophosphate by viral **thymidine kinase** and to the triphosphate by cellular enzymes. The triphosphate competitively inhibits viral DNA polymerase and can be incorporated into viral DNA where it blocks further replication.

acid fast cells Bacteria that are resistant to decolourization by acids during Gram staining procedures. They have high levels of mycolic acid in the cell wall e.g. *Mycobacteria*. Alternative staining methods are the **Ziehl-Neelsen** or **Kinyoun** methods.

acid growth theory The hypothesis that the growth-promoting effect of auxin is due to excretion of hydrogen ions that loosen bonds in the cell wall, allowing the cell to expand.

acid hydrolases Hydrolytic enzymes (EC 3.) that have a low pH optimum. The name usually refers to the **phosphatases**, **glycosidases**, **nucleases** and **lipases** found in the **lysosome**. They are secreted during **phagocytosis**, but are considered to operate as intracellular digestive enzymes.

acid phosphatase Enzyme (EC 3.1.3.2, 423aa) with acidic pH optimum, that catalyses cleavage of inorganic phosphate from a variety of substrates. Found particularly in **lysosomes** and **secretory vesicles**. Can be localized histochemically using various forms of the **Gomori procedure**.

acid protease Rather imprecise term for a proteolytic enzyme with an acid pH optimum, characteristically found in lysosomes. See **peptidases**.

acid secreting cells Large specialized cells of the epithelial lining of the stomach (parietal or oxyntic cells) that secrete 0.1N HCl, by means of H^+ antiport ATPases on the luminal cell surface.

acid-citrate-dextrose Citric acid/sodium citrate buffered glucose solution used as an anticoagulant for blood (citrate complexes calcium).

acidic FGF See **fibroblast growth factor**.

acidobacteria A phylum of Gram-negative bacteria, abundant in soils and sewage sludge.

acidocalcisome Electron-dense acidic organelles, 100–200 nm diameter, rich in calcium and polyphosphate. First identified in trypanosomes but subsequently in many eukaryotes. Brief abstract: http://www.ncbi.nlm.nih.gov/pubmed/15738951

acidophilic (1) Easily stained with acid dyes; (2) Flourishing in an acidic environment.

acidophils One class of cells found in the pars distalis of the **adenohypophysis**.

acidosis Condition in which blood or tissues are more acidic than normal. In clinical practice usually taken as being when arterial blood pH falls below 7.35. Respiratory acidosis arises through failure of the lungs to remove carbon dioxide, metabolic acidosis through over-production of acidic compounds—which can happen transiently as in lactic acid accumulation during anaerobic exercise.

acidosome Non-lysosomal vesicle in which receptor-ligand complexes dissociate because of the acid pH.

acinar cells Epithelial **secretory cells** arranged as a ball of cells around the lumen of a gland (as in the pancreas).

Acinetobacter A Gram-negative bacterium commonly found in soil and water, but also found on the skin of healthy people. *Acinetobacter*, particularly *Acinetobacter baumannii*, is an important opportunistic, nosocomial (hospital acquired) pathogen, and is very resistant to antimicrobials; relatively few antibiotics are effective.

acinus Small sac or cavity surrounded by **secretory cells**.

acoelomate Animal without a **coelom**. The Acoelomate Phyla include sponges, coelenterates and lower worms such as nematodes and platyhelminths.

aconitase Enzyme (EC 4.2.1.3) of the **tricarboxylic acid cycle** that catalyses isomerisation of citrate/isocitrate. Isoforms are found both in mitochondrial matrix (ACO2, 778aa) and cytoplasm (ACO1, 889aa). ACO1 is also an **iron responsive element** (IRE)-binding protein involved in the control of iron metabolism.

ACP Small acidic proteins (acyl carrier proteins) associated with fatty acid synthesis in many pro- and eukaryotic organisms. They are functional only when modified by attachment of the prosthetic group, $4'$-phosphopantetheine ($4'$-PP), which is transferred from CoA to the hydroxyl group of a specific serine residue. In eukaryotes they are part of the multi-protein complex, **fatty acid synthase** (EC 2.3.1.85), in prokaryotes all the functions are present in a single protein (2504aa).

acquired immune deficiency syndrome
See **AIDS**.

acquired immunity Classically, the reaction of an organism to a new antigenic challenge and the retention of a memory of this, as opposed to innate immunity. In modern terms, the **clonal selection** and expansion of a population of immune cells in response to a specific antigenic stimulus and the persistence of this clone.

Acrasidae Order of **Protozoa** also known as the cellular slime moulds. They normally exist as free-living phagocytic soil amoebae (vegetative cells), but when bacterial prey become scarce, they aggregate to form a pseudoplasmodium (*cf.* true **plasmodium** of Eumycetozoida), that is capable of directed motion. The grex, or slug, migrates until stimulated by environmental conditions to form a fruiting body or sorocarp. The slug cells differentiate into elongated stalk cells and spores, where the cells are surrounded by a cellulose capsule. The spores are released from the sporangium at the tip of the stalk and, in favourable conditions, an **amoeba** emerges from the capsule, feeds, divides and so establishes a new population. They can be cultured in the laboratory and are widely used in studies of cell-cell adhesion, cellular **differentiation**, **chemotaxis** and **pattern formation**. The commonest species studied are *Dictyostelium discoideum*, *D. minutum*, and *Polysphondylium violaceum*.

acrasin Name originally given to the **chemotactic** factor produced by cellular slime moulds (**Acrasidae**): now known to be **cyclic AMP** for *Dictyostelium discoideum*.

acridine orange A fluorescent vital dye that intercalates into nucleic acids. The nuclei of stained cells fluoresce green; cytoplasmic RNA fluoresces orange. Acridine orange also stains acid mucopolysaccharides, and is widely used as a pH-sensitive dye in studies of acid secretion. Probably carcinogenic.

acridines Heterocyclic compounds with a pyridine nucleus. Usually fluorescent and reactive with double stranded DNA as intercalating agents at very low concentrations. Hence dsDNA can be detected on gels by fluorescence after acridine staining. Mutagenic (causing frame-shift mutations), cytostatic (and hence antimicrobial). They also affect RNA synthesis and have been used for cell marking.

acritarchs Small organic structures found as fossils and presumed to be remnants of simple unicellular organisms (e.g. bacteria, dinoflagellates, marine algae). Found in sedimentary rocks from as early as the Precambrian.

acrocentric See **metacentric**.

acrodermatitis enteropathica A disorder in which there is intermittent simultaneous occurrence of diarrhoea and dermatitis with failure to thrive, caused by a mutation in the intestinal zinc-specific transporter SLC39A4.

acrolein Inflammable liquid with a sharp, disagreeable odour that will readily polymerize to form a plastic solid. Used in a qualitative test for glycerol. Acrolein is principally used as a chemical intermediate in the production of acrylic acid and its esters but also directly as an aquatic herbicide and algicide in irrigation canals. Can be produced by fires.

acromegaly Enlargement, as an adult, of the extremities of the body as a result of the overproduction of growth hormone (**somatotropin**), e.g. by a pituitary tumour.

acromesomelic dysplasia A group of disorders in which there is disproportionate shortening of skeletal elements. Various forms are recognised, caused by mutation in genes for the C-type **natriuretic peptide** receptor, or for cartilage-derived morphogenetic protein-1 (growth/differentiation factor 5).

acropetal Describing transport or differentiation occurring from the base towards the apex.

acrophase The time at which the peak of a rhythm occurs. Originally referred to the phase angle of the peak of a cosine wave fitted to the raw data of a rhythm. Dictionary of circadian physiology: http://www.circadian.org/dictionary.html

acroplaxome A marginal ring containing 10 nm-thick filaments of keratin 5 and F-actin that anchors the developing **acrosome** to the nuclear envelope. The ring is closely associated with the leading edge of the acrosome and to the nuclear envelope during the elongation of the spermatid head. Full description: http://www.molbiolcell.org/cgi/content/full/14/11/4628

acrosin Serine peptidase (EC 3.4.21.10, 402aa) stored in the **acrosome** of a sperm as an inactive precursor (proacrosin), activated by zona pellucida glycoproteins during the acrosome reaction and involved in penetration by the sperm of the outer layers of the egg. Activation involves the removal of a C-terminal proline-rich segment and the cleavage into light and heavy chains. Acrosin binding protein (543aa) is found in testis.

acrosomal process A long process actively protruded from the acrosomal region of the spermatozoon following contact with the egg and that assists penetration of the gelatinous capsule. See **acrosome**, **acrosin**.

acrosome Vesicle at the extreme anterior end of the spermatozoan, derived from the **lysosome**.

ACT (**1**) *ACT1* and *ACT2* are actin genes from yeast; ACT1 is the essential (conventional) actin, 375aa, 89% homologous in sequence with mouse cytoplasmic actin; *ACT2* encodes a 391aa protein 47% identical to yeast actin that is required for vegetative growth. Divergence from conventional actin by ACT2 is in regions associated with actin polymerization, DNAase I and myosin binding. (**2**) Act2 is human **macrophage inflammatory protein 1β**

(CCL4). **(3)** Artemisinin-based combination therapy. See **artemisinin, lumefantrine**.

ACT domain A domain, named after bacterial aspartate kinase, chorismate mutase and TyrA (prephenate dehydrogenase), that serves as an amino acid-binding site in feedback-regulated amino acid metabolic enzymes and is also found in several transcription factors. In *Arabidopsis* there are several proteins with ACT-domain repeats (ACR family) although their function is unknown at present. ACR proteins: http://www.biomedcentral.com/1471-2229/11/118

ActA Major surface protein (639aa) of *Listeria monocytogenes* that acts as the nucleating site for actin polymerization at one pole of the bacterial cell; assembly of the bundle of microfilaments pushes the bacterium through the cell – though the appearance is like a comet with a tail. ActA spans both the bacterial membrane and the peptidoglycan cell wall. Interacts with several mammalian proteins including the phosphoprotein **VASP**, actin and the **Arp2/3 complex**. A functionally similar protein, **IcsA**, is found in *Shigella*.

actagardine A type B **lantibiotic** (formerly gardimycin), produced by *Actinoplanes garbadinensis* (an actinobacterium) that has activitiy against Gram-positive bacteria and good antistreptococcal activity.

ACTH See **adrenocorticotrophin**.

actin A highly conserved protein of 42 kDa (377aa), very abundant in eukaryotic cells (8–14% total cell protein) and one of the major components of the **actomyosin** motor and the cortical microfilament meshwork. First isolated from **striated muscle** and often referred to as one of the muscle proteins. G-actin is the globular monomeric form of actin, 6.7×4.0 nm: it polymerizes to form filamentous F-actin. Various proteins (**profilin, thymosin beta-4, gelsolin**) stabilize G-actin (i.e., prevent it from polymerizing) and a range of other proteins associate with filamentous actin, stabilizing it, cross-linking it, moving along it or anchoring it to other components. There are various isoforms: actin alpha1 in skeletal muscle, actin alpha2 in smooth muscle, actin alphaC1 in cardiac muscle. Actins beta and gamma are found in non-muscle cells. See **actin binding proteins** also **centractin**.

actin binding proteins A diverse group of proteins that bind to **actin** and often link actin to other structures. Some bind to G-actin and prevent it from adding onto the linear F-actin filaments (e.g. **profilin**), or keep actin levels very low in the nuclear compartment (**exportin-6**) others act as nucleating centres for polymerization (e.g. **hisactophilin** and bacterial proteins such as **ActA**) or block the preferred assembly (barbed) end of filaments to stabilize their length (e.g. **capZ**). Some will cleave filaments (**gelsolin**), others link filaments to one

another as isotropic gels (**filamin**) or bundles that may be parallel or antiparallel (e.g. **villin**), yet others attach bundles or individual filaments to membranes or membrane-embedded adhesion proteins. Microfilaments may have regulatory proteins associated with them that prevent the attachment of myosin unless calcium is present (tropomyosin); filaments may be linked to other cytoskeletal elements through various scaffolding proteins (**IQGAPs**) or to a membrane-skeleton (e.g. **spectrin**). There are separate entries for many different actin binding proteins such as α-actinin, β-actinin, ABP-50, ABP-67, ABP-120, ABP-280, actobindin, actolinkin, actophorin, acumentin, ADF, adseverin, AFAP-110, aginactin, AIP, anillin, beta-CAP73, caldesmon, calicin, calpactins, ciboulot, coactosin, cofilin, comitin, connectin, coronin, cortactin, cortexillin, dematin, depactin, destrin, DNAase I, drebin, dynacortin, ectoderm-neural cortex-1, ezrin, fascin, fesselin, fimbrin, fodrin, formins, fragmin, gelactins, hisactophilin, histidine rich protein II, hsp90, IcsA, MAP2, MARCKS, nebulin, plastin, ponticulin, radixin, scruin, tau, tensin, transgelin, twinfilin. Encyclopedia of Actin Binding Proteins: http://www.bms.ed.ac.uk/research/others/smaciver/Encyclop/encycloABP.htm

actin depolymerizing factor An actin-binding protein of the ADF/**cofilin** family that will depolymerize microfilaments and bind G-actin, but does not cap filaments. The human form (ADF, destrin, 165aa) has 95% homology to chick ADF. ADF has a nuclear localization domain and the interaction with actin is regulated by phosphoinositides. Both ADF and cofilin are associated with **Hirano bodies**. The ADF domain (ADF-homology domain) is an actin-binding motif found in three phylogenetically distinct families of proteins, ADF/cofilins, **twinfilins** and **drebin**/ABP-1s.

actin meshwork Microfilaments inserted proximally into the plasma membrane and cross-linked by actin binding proteins to form a mechanically resistive network that may support protrusions such as **pseudopods** (sometimes referred to as the **cortical meshwork**).

actin-fragmin kinase Kinase (EC 2.7.11.1, 737aa) found in *Physarum polycephalum* that specifically phosphorylates threonine residues of actin that is in the EGTA-resistant 1:1 actin-fragmin complex. Regulates cytoskeletal dynamics.

actin-RPV See **centractin**.

actinfilin An actin-binding protein (642aa) of the BTB-Kelch protein family expressed predominantly in brain. Actinfilin has the same overall structure as **mayven, kelch**, and Enc-1, other actin-binding proteins. Self-associates through an amino-terminal POZ domain.

Actinia equina Common beadlet anemone; a **coelenterate**. See **equinatoxins**.

actinic keratosis Thickened area of skin as a result of excessive exposure to sunlight – particularly common in those with very fair skin. See **keratoses**.

Actinobacteria A subclass of the order **Actinomycetales**.

actinogelin An **actin binding protein** (115 kDa) from Ehrlich ascites cells that gelates and bundles **microfilaments**. Not in the current literature.

Actinomycetales Order of Gram-positive bacteria, widespread in soil, compost, and aquatic habitats. Most are saprophytic, but there are a few pathogens; some produce important antibiotics. Important genera include: *Actinomyces, Corynebacterium,* **Frankia**, *Mycobacterium, Streptomyces*.

actinomycins A class of polypeptide antibiotics isolated from soil bacteria of the genus *Streptomyces*. Actinomycin C is a mixture of antibiotics: actinomycins C1, C2 and actinomycin D. Actinomycin D binds to DNA at the transcription initiation complex and blocks the movement of **RNA polymerases** thereby preventing RNA synthesis in both pro- and eukaryotes. It is used as an investigative tool, being too toxic for therapeutic use.

actinomycosis A chronic granulomatous infection caused by various filamentous bacteria of the genus *Actinomyces*.

Actinophrys sol Species of **Heliozoa** often used in studies on microtubule stability: the **axopodia** are supported by a bundle of cross-linked microtubules arranged in a complex double-spiral pattern when viewed in cross-section.

Actinopterygii Teleost fishes, a Subclass of the Gnathostomata. Includes cod and herring. The fins of the Actinopterygii are webs of skin supported by bony or horny spines.

actinorhodin One of the four antibiotics produced by *Streptomyces coelicolor*, a blue-pigmented polyketide. Within the actinorhodin biosynthetic cluster are two co-transcribed genes, *actA* and *actII-ORF3*, encoding integral membrane proteins implicated in actinorhodin export. See undecylprodigiosin, **methylenomycin** and **CDA** (calcium-dependent antibiotic).

Actinosphaerium Genus that is a member of the Class **Heliozoa**, Order Actinophryida, Family Actinosphaeridae: Multinucleate cells, 80–200 μm in size; remarkable for long radial protruding axopodia that contain complex double spiral arrangements of many microtubules. It catches prey by protrusion and retraction of the axopodia. Similar to *Actinophrys*.

actinotrichium One of the aligned **collagen** fibres (actinotrichia, ca 2 μm diameter) that provide a guidance cue for **mesenchymal** cells in the developing fin of teleost fish.

action potential An electrical pulse that passes along the membranes of excitable cells, such as **neurons**, **muscle cells**, fertilized eggs and certain plant cells. The precise shape of action potentials varies, but action potentials always involve a large **depolarization** of the cell membrane, from its normal **resting potential** of -50 to -90 mV. In a neuron, action potentials can reach +30 mV, and last 1 ms. In muscles, action potentials can be much slower, lasting up to 1 s.

action spectrum The relationship between the frequency (wavelength) of a form of radiation, and its effectiveness in inducing a specific chemical or biological effect.

activated leukocyte cell adhesion molecule A cell adhesion molecule (ALCAM, CD166, 583aa) of the immunoglobulin superfamily, the human homologue of chicken DM-GRASP. Involved in neurite extension through heterophilic and homophilic interactions. May play a role in the binding of T- and B-cells to activated leukocytes. See **neurolin**.

activated macrophage A **macrophage** (mononuclear phagocyte) that has been stimulated by cytokines or endotoxin and that has greatly enhanced cytotoxic and bactericidal potential. Classically activated (M1) macrophages are proinflammatory and kill pathogens, and alternatively activated (M2) macrophages produce anti-inflammatory factors. Brief overview: http://www.rndsystems.com/cb_detail_objectname_SP04_MacrophageActivation.aspx.

activation A very general term, but in the context of fertilization the changes in the egg brought about by contact with the spermatozoon. Activation is the first stage in development and occurs independently of nuclear fusion. The first observable change is usually the cortical reaction that may involve elevation of the fertilization membrane; the net result is a block to further fusion and thus to polyspermy. In addition to the morphological changes, there are rapid changes in metabolic rate and an increase in protein synthesis from maternal mRNA.

activation energy The energy required to bring a system from the ground state to the level at which a reaction will proceed.

activation induced deaminase Enzyme (AID, cytidine aminohydrolase, EC 3.5.4.5, 198aa) that acts on single-stranded DNA during replication and deaminates cytosine to uracil (the mismatch is then processed by base-excision or mismatch repair systems); a mechanism for generating diversity in

B-cells for antibody production but strictly controlled in most cells. See **ADAR**.

activation loop A region on protein kinases of the **AGC kinase** class containing a conserved threonine residue (termed the activation loop site) phosphorylation of which is critical for their activity. In some cases autophosphorylation may occur. Analogous activation regions have been described in, for example, **histone acetyltransferase**.

activation tagging A method used extensively in plant molecular biology; a **T-DNA** tagging vector containing four transcriptional enhancers derived from the cauliflower mosaic virus is used to randomly insert this T-DNA into the plant genome through *Agrobacterium* infection and leads to the over-expression of genes near to the inserted T-DNA. Activation-tagging vectors that confer resistance to the antibiotic kanamycin or the herbicide glufosinate have been developed and the method has allowed the generation of a wide range of gain-of-function mutants.

active immunity Immunity resulting from the normal response to antigen. Only really used to contrast with **passive immunity** in which antibodies or sensitised lymphocytes are transferred from the reactive animal to the passive recipient.

active site The region of a protein that binds to a substrate molecule(s) and facilitates a specific chemical conversion. Produced by juxtaposition of amino acid residues as a consequence of the protein's **tertiary structure**.

active transport Transport up a gradient of **electrochemical potential** or, more precisely, unidirectional or vectorial transport produced within a membrane-bound protein complex by coupling an energy-yielding process to a transport process. In primary active transport systems the transport step is normally coupled to **ATP** hydrolysis within a single protein complex. In secondary active transport the movement of one species is coupled to the movement of another species down an electrochemical gradient established by primary active transport.

active zone Site of transmitter release on the presynaptic terminal of a chemical synapse. At the neuromuscular junction active zones are located directly across the synaptic cleft from clusters of **acetylcholine receptors**. Evidence from **conotoxin** binding studies suggests that presynaptic **calcium channels** are exclusively localized at active zones.

activin Dimeric growth factors of the TGFβ family with growth and differentiation-promoting effects on a range of cell types in addition to its original role (FSH-releasing) in gonadal sites. Composed of two **inhibin** beta chains; since there are two isoforms, A (426aa) and B (407aa), there are three forms of

activin, AA, BB and AB. The receptors are **activin receptor-like kinases**.

activin receptor-like kinases Receptors (ser/thr protein kinases) for **activin**. Ligand binding stimulates the formation of a receptor complex consisting of two type II and two type I transmembrane serine/threonine kinases. Type II receptors phosphorylate and activate type I receptors which autophosphorylate, then bind and activate SMAD transcriptional regulators. Activin receptor 1 (ALK1, ACVRL1, EC 2.7.11.30, 503aa) is mutated in **hereditary haemorrhagic telangiectasia type 2**. ALK2 (509aa) is defective in **fibrodysplasia ossificans progressiva**. ALK3 (CD292, bone morphogenetic protein receptor type-1A, 532aa) is a receptor for BMP-2 and BMP-4: defects cause a subset of juvenile polyposis syndrome cases. A number of other activin receptor-like kinases are known with variable specificty for TGFβ-family ligands.

activin response factor *ARF* A multiprotein complex induced in early *Xenopus* blastomeres by **activin**, Vg-1 and **TGFβ**, binds to activin-response element in the **mix**-2 homeobox gene. The ARF complex contains XMAD2, a *Xenopus* homologue of the *Drosophila* **Mad** gene product, and **FAST**-1, a winged-helix transcription factor.

actobindin A G-actin binding protein (88aa with two **beta-thymosin repeats**, binds two actin molecules) from *Acanthamoeba castellani* that blocks actin polymerization. A very similar protein is found in *Dictyostelium* and homologous proteins in *Drosophila* and *C. elegans* have three beta-thymosin repeats.

actolinkin Monomeric protein (20 kDa) from echinoderm eggs that was reported to link actin filaments by their barbed ends to the inner surface of the egg plasma membrane. Sole report:http://www.ncbi.nlm.nih.gov/pubmed?term=actolinkin

actomere Site of actin filament nucleation in sperm of some echinoderms in which the **acrosomal process** is protruded by rapid assembly of a parallel microfilament bundle.

actomyosin Generally: a motor system that is thought to be based on **actin** and **myosin**. The essence of the motor system is that myosin makes transient contact with the actin filaments and undergoes a conformational change before releasing contact. The hydrolysis of ATP is coupled to movement, through the requirement for ATP to restore the configuration of myosin prior to repeating the cycle. More specifically: a viscous solution formed when actin and myosin solutions are mixed at high salt concentrations. The viscosity diminishes if ATP is supplied and rises as the ATP is hydrolyzed. Extruded threads of actomyosin will contract in response to ATP.

actophorin An actin depolymerizing factor (ADF) (138aa in *Acanthamoeba castellanii*) that will sever actin filaments (longer filaments being more rapidly severed) and sequester G actin. A member of the ADF/cofilin family. Binds ADP-G-actin with higher affinity than ATP-actin and binding is very sensitive to the divalent cation present on the actin. Has high sequence homology with vertebrate **cofilin** and **destrin**, echinoderm **depactin** and some plant ADFs but lacks the nuclear localization sequence found in the vertebrate ADFs.

acumentin Protein that was named because it was thought to cap the pointed end of microfilaments, now known to be L-plastin (see **plastins**).

acuminate Describing a structure that is slenderly tapered with an included angle of less than 45°.

acute (1) Sharp or pointed. (2) Of diseases: coming rapidly to a crisis − not persistent (chronic).

acute inflammation Response of vertebrate body to insult or infection; characterized by redness (rubor), heat (calor), swelling (tumor), pain (dolor), and sometimes loss of function. Changes occur in local blood flow, and **leucocytes** (particularly **neutrophils** in the early stages) adhere to the walls of postcapillary venules (margination) and then move through the **endothelium** (diapedesis) towards the damaged tissue. Although acute inflammation is usually short-term and transient there are situations in which acute-type inflammation persists.

acute lymphoblastic leukaemia *ALL* See **leukaemia**.

acute myeloblastic leukaemia *AML* See **leukaemia**.

acute phase proteins Plasma proteins that increase in concentration during acute inflammation, particularly **C-reactive protein**, **orosomucoid** and **serum amyloid** A protein. Other proteins also increase in level, including mannose-binding protein, α-1-antitrypsin, α-2-macroglobulin, blood clotting factors (**Factors I-XII**), complement components, ferritin, ceruloplasmin and haptoglobin. The levels of negative acute phase proteins decrease.

acute phase reaction Response to acute inflammation involving the increased synthesis of various plasma proteins (**acute phase proteins**).

acute respiratory distress syndrome *ARDS* A severe inflammatory reaction in the lung that can lead to fluid accumulation and effectively drowning. There may also be multiple organ failure. ARDS was formerly known as adult respiratory distress syndrome to differentiate it from infant respiratory distress syndrome which is caused by a deficiency of lung surfactant. See **septic shock**.

acutely transforming virus Retrovirus that rapidly **transforms** cells, by virtue of possessing one or more oncogenes. Archetype: **Rous sarcoma virus**.

ACV (1) Aciclovir. (2) Alpha-aminoadipylcysteinyl-valine, precursor for isopenicillin synthesis.

ACV synthase Enzyme (alpha-aminoadipylcysteinyl-valine synthase, EC 6.3.2.26, 3770aa in *Emericella nidulans*) responsible for an early step in cephalosporin synthesis. ACV is acted upon by **isopenicillin N synthase** to produce isopenicillin N.

ACY1 A cytosolic, homodimeric, zinc-binding enzyme (aminoacylase 1, EC 3.5.14, 408aa) that catalyzes the hydrolysis of acylated L-amino acids to L-amino acids. Function may be the salvage of acylated amino acids. Expression of the gene is reportedly reduced or undetectable in small cell lung carcinoma lines and tumours.

acyclovir See **aciclovir**.

acyl An organic radical or functional group derived from an organic acid by the removal of the carboxylic hydroxyl group. The general formula is -COR (or RCO-), where R is an organic group and there is a double bond between the carbon and oxygen. Common examples are formyl (from formic acid), acetyl (from acetic acid), benzoyl (from benzoic acid) amongst others.

acyl homoserine lactone *AHL* An autoinducer involved in **quorum sensing** in some Gram-negative bacteria. AHL reaches a high concentration within the cell, and will trigger transcription, only if there are many cells nearby creating a high extracellular concentration of the autoinducer that will diffuse into the cell. First discovered as a regulatory mechanism for light emission by bioluminescent bacteria. Among Gram-positive bacteria, short peptides function as autoinducers.

acylation Introduction of an acyl (RCO-) group into a molecule: for example the formation of an ester between glycerol and fatty acid to form mono-, di-, or tri-acylglycerol, or the formation of an aminoacyl-tRNA during protein synthesis. Acyl-enzyme intermediates are transiently formed during covalent catalysis.

acyltransferase Enzymes of the class EC 2.3.1 that catalyse the transfer of acyl groups from a carrier such as acetylCoA to a reactant.

ADA See **adenosine deaminase**.

ADAM family A disintegrin and metalloprotease family of membrane-anchored peptidases that regulate cell behaviour by proteolytically modifying the cell surface and ECM. In some cases they are responsible for the proteolytic release of membrane-bound growth factors. Have both cell adhesion and protease activities. Members of the family include *C. elegans* MIG-17, **alpha-secretase**, **TACE**. A similar and related family are the ADAM metallopeptidases with thrombospondin Type 1 motif, **ADAMTS**.

ADAMTS ADAM metallopeptidases with a thrombospondin Type 1 motif, peptidases that are anchored to extracellular matrix through the binding of the thrombospondin motif to aggrecan. ADAMTS1 (967aa) cleaves aggrecan and may be involved in its turnover. Many others are known, all with generally similar properties but different substrate preferences.

adaptation A change in sensory or excitable cells upon repeated stimulation, that reduces their sensitivity to continued stimulation. Those cells that show rapid adaptation are known as phasic; those that adapt slowly are known as tonic. Can also be used in a more general sense for any system that changes responsiveness with time − for example by downregulation of receptors (**tachyphylaxis**) or through internal modulation of the signalling system, as in **bacterial chemotaxis**.

adaptins See **adaptor proteins**, **adaptor protein complex 3** and **clathrin-associated adaptor complexes**.

adaptor protein complex 3 An adaptor complex (AP-3) that has a similar heterotetrameric structure to the **clathrin-associated adaptor protein complexes** AP-1 and AP-2, but is not clathrin associated and has different adaptin subunits (delta subunit AP3D1, 1153aa, beta subunit AP4B1, 1094aa or AP3B2, 1082aa; a medium adaptin, mu-type, AP3M1, 418aa or AP3M2, 418aa and a small adaptin sigma-type AP3S1, 193aa or AP3S2, 193aa). AP-3 may be directly involved in trafficking to lysosomes. See **BLOCS**.

adaptor proteins A general term for proteins that link multi-molecular protein arrays. See **adaptor protein complex 3**, **clathrin-associated adaptor protein complexes**, **low-density lipoprotein receptor adaptor protein 1**, **signal-transducing adaptor proteins**, transmembrane adaptor proteins (**TRAPS**).

ADAR Small family of adenosine deaminases (adenosine deaminase acting on RNA, EC 3.5.4.-, ADAR1−3, editases) that edit adenosine residues to inosine in double-stranded RNA (dsRNA). Although this editing recodes and alters functions of several mammalian genes, its most common targets are non-coding repeat sequences, indicating the involvement of this editing system in currently unknown functions. ADAR1 (1226aa) is associated with the RNA surveillance protein HUPF1 (RENT1) in the supraspliceosome, ADAR2 (ADARB1, 741aa) is widely expressed in brain and other tissues and knock-out mice die young. Mice can be rescued by exonically-edited AMPA receptor suggesting that mRNA for

this receptor is major substrate for the enzyme. ADAR3 (739aa) may have a regulatory function. See **ADAT1**.

ADATs Human enzymes (adenosine deaminase, tRNA-specific; ADAT-1, 502aa; ADAT-2, 191aa; ADAT-3, 351aa) of the mammalian RNA-specific **adenosine deaminase** family (**ADARs**) that is involved in pre-mRNA editing of nuclear transcripts; ADAT1 deaminates adenosine-37 to inosine in eukaryotic tRNA(ala) and probably represents the human counterpart of the yeast protein Tad1p. ADAT-2 probably does the same to several tRNAs; ADAT-3 may be regulatory.

adaxial The surface of a plant organ such as a leaf or petal that during early development faced towards the axis. In the case of leaves the upper surface is usually adaxial. *Cf.* **abaxial**.

ADCC See **antibody dependent cell-mediated cytotoxicity**.

ADDA An unusual hydrophobic amino acid (3-amino-9-methoxy-2,6,8-trimethyl-10-phenyl-deca-4,6-dienoic acid) found in **microcystins** and **nodularins** and essential for their toxicity. Article: http://www.ncbi.nlm.nih.gov/pmc/articles/PMC168230/pdf/624086.pdf

Addison's disease Chronic insufficiency of the adrenal cortex as a result of tuberculosis or, specific **autoimmune** destruction of the **ACTH**-secreting cells and consequent underproduction of steroid hormones. Characterized by extreme weakness, wasting, low blood-pressure and pigmentation of the skin. See **polyendocrine syndrome**. Not to be confused with Addison's anaemia (**megaloblastic anaemia**). See **leukodystrophy**, which can have similar effects.

additive effect An effect that is simply the sum of the effects of separate exposures to two (or more) agents under the same conditions and there is no synergistic effect. Proving synergy requires demonstrating that the agents together produce an effect that is greater than the maximum either can produce alone.

addressins See **selectins**.

adducin Calmodulin-binding protein associated with the membrane skeleton of erythrocytes. A substrate for **protein kinase C**, it binds to **spectrin-actin** complexes (but only weakly to either alone) and promotes the assembly of spectrin onto spectrin-actin complexes unless micromolar calcium is present. A heterodimer of $\alpha\beta$ or $\alpha\gamma$ subunits (α, 737aa; β, 726aa; γ, 706aa), distinguishable from Band 4.1.

adductor muscle Large muscle of bivalve molluscs that is responsible for holding the two halves of the shell closed. Its unusual feature is its ability to maintain high tension with low energy expenditure by using a 'catch' mechanism, and the high content of **paramyosin**.

adenine One of the bases (6-aminopurine) found in **nucleic acids** and **nucleotides**. In DNA, it pairs with **thymine**.

adenine nucleotide translocator An abundant mitochondrial protein (ANT, ADP/ATP translocator, SLC25A4, 298aa) embedded asymmetrically in the inner membrane. The heterodimer forms a gated pore through which ADP is moved from the matrix into the cytoplasm.

adeno- Prefix that indicates an association with, or similarity to, glandular tissue.

adenocarcinoma Malignant neoplasia of a glandular epithelium, or **carcinoma** showing gland-like organization of cells.

adenofibroma A tumour composed of connective tissue and cells arranged in a gland-like manner, usually benign.

adenohypophysis Anterior lobe of the pituitary gland; responsible for secreting a number of hormones and containing a comparable number of cell types.

adenoid Generally, gland-like. Adenoids are lymphoid tissue in the nasopharynx.

adenoma Benign tumour of glandular epithelium.

adenomatous polyposis coli An autosomal dominant disorder (polyposis coli, familial adenomatous polyposis) characterized by the development of hundreds of adenomatous **polyps** in the large intestine, which have a tendency to progress to malignancy. The APC gene encodes a multidomain tumour suppressor (2843aa) that antagonises the **wnt** signalling pathway; mutations are associated with various tumours including gastric carcinomata. *Cf.* hereditary nonpolyposis colon cancer (see **GTBP**).

adenomyoma See **endometrioma**.

adenosine The **nucleoside** (9-β-D-ribofuranosyla-denine) formed by linking adenine to **ribose**.

adenosine deaminase An enzyme (ADA, EC 3.5.4.4, 363aa) which deaminates adenosine and 2-deoxyadenosine to inosine or 2'-deoxyinosine respectively. A rare autosomal recessive genetic defect in this enzyme is responsible for 20−30% of cases of **severe combined immunodeficiency disease** (SCID) and was the first candidate disease for gene replacement therapy. ADA deficiency causes an increase of dATP, which inhibits S-adenosylhomocysteine hydrolase, causing an increase in S-adenosylhomocysteine; both are particularly toxic to lymphocytes. See **adenosine deaminase complexing protein**.

adenosine deaminase complexing protein A protein (ADCP2, dipeptidyl peptidase IV, EC 3.4.14.5, conversion factor, T-cell activation antigen CD26, 766aa) that forms a complex with adenosine deaminase and generates the tissue-specific forms. It is a serine exopeptidase that cleaves X-proline dipeptides from the N terminus of polypeptides. It is an intrinsic membrane glycoprotein involved in the costimulatory signal essential for T-cell receptor-mediated T-cell activation. ADCP1 is known but there is some doubt whether it is actually involved in ADA-complexes.

adenosine diphosphate See **ADP**.

adenosine monophosphate See **AMP**, **cAMP**.

adenosine receptors Various adenosine receptors (purinergic receptors) have been identified, A1, A2A, A2B, and A3. All are seven membrane-spanning G-protein coupled receptors. There is considerable difference in properties of receptors from different species. A1 receptors (ADORA1, 326aa) and A3 (318aa) inhibit adenylyl cyclase whereas the A2A (412aa) and A2B (332aa) receptors stimulate adenylyl cyclase. Do not confuse with adrenoreceptors (**adrenergic receptors**).

adenosine triphosphate See **ATP**.

adenosquamous Description of a benign tumour of epithelial origin (**adenoma**) in which cells have flattened morphology, as opposed to being cuboidal or columnar.

adenoviral vector Vector used for gene transfer, usually replication defective due to a deletion in the E1 region (early genes). Many vectors also have deletions in E3 and E4. To produce infectious particles, plasmids containing the defective adenovirus genome and the gene to be expressed are introduced into cells constitutively expressing the E1A genes, such as human 293 cells. Have been used clinically for gene therapy (for **ornithine transcarbamylase deficiency**) but with untoward side effects in some cases and a fatality in one.

Adenoviridae Large group of viruses first isolated from cultures of adenoids. The **capsid** is an icosahedron of 240 hexons and 12 pentons and is in the form of a base and a fibre with a terminal swelling; the genome consists of a single, linear molecule of double-stranded DNA. They cause various respiratory infections in humans. Some of the avian, bovine, human and simian adenoviruses cause tumours in newborn rodents, generally hamsters. They can be classified into highly, weakly and non-oncogenic viruses from their ability to induce tumours *in vivo* though all of these groups will transform cultured cells. The viruses are named after their host species and sub-divided into many serological types e.g. Human Adenovirus type 3. See **adenoviral vector**.

adenylate cyclase Enzyme (adenylyl cyclase, EC 4.6.1.1) that produces cAMP (**cyclic AMP**) from **ATP** and acts as a signal-transducer coupling hormone binding to change of cytoplasmic cAMP levels. The name strictly refers to the catalytic moiety, but it is usually applied to the complex system that includes the hormone receptor and the GTP-binding modulator protein (see **GTP-binding proteins**). There are multiple tissue-restricted isoforms.

ADF (1) In chicken, the term actin depolymerizing factor has been applied to both **gelsolin** and **destrin**; **cofilin** is similar in function to destrin but the product of a different gene. Gelsolin (brevin, homogenin, 778aa) is a calcium-regulated, actin-modulating protein that binds to the barbed ends of actin monomers or filaments, preventing monomer exchange (end-blocking or capping). It can nucleate the assembly of monomers into filaments and sever pre-existing filaments at higher cvalcium concentrations. Destrin (165aa) severs actin filaments and binds to actin monomers. In *Arabidopsis* there are a number of ADFs (e.g. ADF-1, 139aa; ADF-10, 140aa). (2) See **ADF domain**. (3) **Adult T-cell leukaemia-derived factor**. (4) In *Drosophila* a transcription factor (ADH distal factor 1, 262aa) that regulates expression of alcohol dehydrogenase and other genes.

ADF domain An actin-binding module (ADF-homology domain) found in an extensive family of proteins with three phylogenetically distinct classes, ADF/**cofilins**, **twinfilins** and **drebin**/ABP-1s. Original description: http://www.ncbi.nlm.nih.gov/pmc/articles/PMC25446/

ADH (1) Antidiuretic hormone. See vasopressin. (2) Alcohol dehydrogenase (EC 1.1.1.1). In humans there are seven ADH isozymes: three in class-I (alpha, beta, and gamma, all 375aa), one in class-II (ADH4, pi, 380aa), one in class-III (ADH5, chi, 374aa), one in class-IV (ADH7, 386aa) and one in class-V (ADH6, 368aa). They form homo or heterodimers and bind zinc. They catalyse the conversion of an alcohol to an aldehyde or ketone. They are not restricted to animals and a number are known from plants (e.g. in *Arabidopsis*, ADH Class III, glutathione-dependent formaldehyde dehydrogenase, 379aa, important for formaldehyde detoxification), archaea and bacteria.

adherens junction Specialized cell-cell junction into which are inserted microfilaments (in which case also known as **zonula adherens**), or **intermediate filaments** (macula adherens or spot **desmosomes**).

adhesins General term for molecules involved in adhesion, but applied more specifically in bacteriology to various surface components that generally seem to behave as lectins, binding to surface carbohydrates. Mediate binding to eukaryotic cells and important in biofilm formation. Examples from *E. coli* include adhesin/invasin TibA autotransporter (989aa) that mediates adhesion to and invasion of human intestinal epithelial cells, fimbrial adhesin papG (335aa) that binds the Gal-α1-4Gal receptor found on epithelial cells of the urinary tract and AIDA-I autotransporter (1286aa) that mediates attachment to a range of cells and is a efficient initiator of biofilm formation.

adhesion plaque Another term for a **focal adhesion**, a discrete area of close contact between a cell and a non-cellular substratum, with cytoplasmic insertion of **microfilaments** and considerable electron-density adjacent to the contact area. On the cytoplasmic face are local concentrations of various proteins such as **vinculin** and **talin**.

adhesion site (1) In Gram-negative bacteria, a region where the outer membrane and the plasmalemma appear to fuse. May be important in export of proteins or viral entry. (2) Used rather generally of any region of a cell specialized for adhesion.

adiaconidium A large thick-walled conidium (20–70 μm diam.) produced by *Emmonsia crescens* in the lungs of humans and animals (see **adiaspiromycosis**).

Adiantum capillus-veneris A species of maidenhair fern, the source of a number of drugs.

adiaspiromycosis A rare, self-limited pulmonary infection (adiaspirosis, haplomycosis) caused by inhalation of the asexual conidia of the saprophytic soil fungus *Emmonsia crescens* (formerly *Chrysosporium parvum* var. *crescens*).

adipate The ionized form of adipic acid. Sodium adipate is used as an acidity regulator in food.

adipocere White or yellowish waxy substance formed post-mortem by the conversion of body fats to higher fatty acids.

adipocyte Mesenchymal cell in fat tissue that has large lipid-filled vesicles. There may be distinct types in white and **brown fat**. 3T3-L1 cells are often used as a model system; they can be induced to differentiate with **dexamethasone/insulin/IBMX** treatment.

adipofibroblasts Adipocytes from subcutaneous fat that have lost their fat globules and developed a fibroblastic appearance when grown in culture. Unlike skin fibroblasts they will take up fat from serum taken from obese donors, and probably retain a distinct differentiated state.

adipokines Peptide hormones produced by **adipocytes** and involved in metabolic regulation. The family includes **adiponectin**, **apelin**, **chemerin**, **leptin**, **omentin**, **resistin**, **vaspin** and **visfatin**.

adiponectin A protein (adipocyte complement-related 30 kDa protein, gelatin-binding protein, 244 aa) secreted only by **adipocytes** and an important adipokine involved in the control of fat metabolism and insulin sensitivity. Forms homotrimers and homomultimers. Antagonizes TNFα and inhibits endothelial NFκB signalling.

adipophilin The membrane-associated protein (adipose differentiation related protein, 437aa) that, together with **perilipin**, is constitutively associated with lipid droplets and plays a role in sustained fat storage and regulation of lipolysis. One of the PAT (perilipin, adipophilin, and TIP47) family of proteins.

adipose tissue Fibrous connective tissue with large numbers of fat-storing cells, **adipocytes**.

adipsin A **serine peptidase** (Adn, Factor D, C3 convertase activator, properdin factor D, EC 3.4.21.46, 253aa) with complement factor D activity (cleaves Factor B that is bound to C3b to produce the C3bBb convertase of the alternative pathway). Synthesized by **adipocytes**. Altered levels are characteristic of some genetic and acquired obesity syndromes.

adjuvant Additional components added to a system to enhance the action of its main component, typically to increase **immune response** to an **antigen**. See **Freund's adjuvant**.

ADM See **adrenomedullin**.

ADME Commonly used abbreviation for Absorption, Distribution, Metabolism and Excretion: the **pharmacokinetics** of a drug.

adn See **adipsin**.

A-DNA Right-handed double-helical **DNA** with approximately 11 residues per turn. Planes of base-pairs in the helix are tilted 20° away from perpendicular to the axis of the helix. Formed from **B-DNA** by dehydration.

adnexa Appendages; usually in reference to ovaries and Fallopian tubes but can, more generally, also be applied to accessory or adjoining organs.

adocia sulphate Adocia sulphates 1−6 are inhibitors of **kinesin**, isolated from sponge (*Haliclona* spp.). Bind to motor domain of kinesin, mimicking tubulin. Adocia sulphates-2 and -6 are the most active. Details: http://www.ncbi.nlm.nih.gov/pubmed/11674622

adoptive immunity Immunity acquired as a result of the transfer of lymphocytes from another animal.

ADP Adenosine diphosphate. Unless otherwise specified, the nucleotide 5'ADP, **adenosine** bearing a diphosphate (pyrophosphate) group in

ribose-O-phosphate ester linkage at position 5' of the ribose moiety. Adenosine 2'5' and 3'5'diphosphates also exist, the former as part of **NADP** and the latter in **coenzyme A**.

ADP-ribosylation A form of **post-translational** modification of protein structure involving the transfer to protein of the ADP-ribosyl moiety of **NAD**. Believed to play a part in normal cellular regulation as well as in the mode of action of several bacterial toxins.

ADP-ribosylation factor *ARF* Ubiquitous **GTP-binding protein** (181aa), N-myristoylated, stimulates **cholera toxin** ADP-ribosylation. Mediates binding of non-clathrin coated vesicles and AP1 (adaptor-protein 1) of **clathrin-coated vesicles** to Golgi membranes. At least six isoforms have been identified. ADP-ribosylation factor-binding proteins (Golgi-associated gamma-adaptin ear-containing ARF-binding proteins, GGA1-3, 639aa, 613aa & 723aa) mediate the ARF-dependent recruitment of clathrin to the *trans*-Golgi network and are important for protein sorting and trafficking between the *trans*-Golgi network and endosomes.

adrenal Endocrine gland adjacent to the kidney. Distinct regions (cortex and medulla) produce different ranges of hormones including **corticosteroids**, adrenaline and noradrenaline.

adrenaline *epinephrine* A hormone secreted (with **noradrenaline**) by the medulla of the **adrenal** gland, and by **neurons** of the **sympathetic nervous system** (adrenergic neurons), in response to stress. The effects are those of the classic 'fight or flight' response, including increased heart function, elevation in blood sugar levels, cutaneous vasoconstriction making the skin pale, and raising of hairs on the neck.

adrenergic neuron See **adrenaline**.

adrenergic receptors *adrenoreceptors* Receptors for **noradrenaline** and **adrenaline**. All are seven membrane spanning G-protein coupled receptors linked variously either to adenylate cyclase or phosphoinositide second messenger pathways. Three subgroups are usually recognized, the β-adrenergic receptors linked to G_s, the α1 linked to G_i, and the α2 linked to G_q. The β-adrenergic receptor gene is unusual in having no introns.

adrenocorticotropin A peptide hormone (ACTH, adrenocorticotrophin, 39aa) produced by the pituitary gland in response to stress (mediated by corticotrophin releasing factor, a 41aa peptide, from the hypothalamus). Stimulates the release of adrenal cortical hormones, mostly **glucocorticoids**. Derived from a larger precursor, **pro-opiomelanocortin**, by the action of an endopeptidase that also releases β-**lipotropin**. Alpha-**melanocyte-stimulating hormone** (α-MSH, melanotropin-alpha) consists of the N-terminal 13aa of ACTH. See also Table H2.

adrenoleucodystrophy Demyelinating disease (X-linked Schilder's disease) with childhood onset, due to a mutation in the **ABCD1** gene with the result that there is an apparent defect in peroxisomal beta oxidation and the accumulation of saturated very long chain fatty acids in all tissues of the body.

adrenomedullin A vasodilator peptide (ADM, 52aa) related to **calcitonin gene related peptide** (CGRP). In addition to vasodilatory effects it has been reported that adrenomedullin protects a variety of cells against oxidative stress induced by stressors such as hypoxia, ischemia/reperfusion, and hydrogen peroxide through the phosphatidylinositol 3-kinase (PI3K)-dependent pathway. Adrenomedullin-2 (intermedin, 148aa) may regulate gastrointestinal and cardiovascular activities through a cAMP-dependent pathway. The protective effects of adrenomedullin are in association with adrenomedullin binding protein-1 (AMBP-1, complement **factor H**, 1231aa). The receptors are **RAMPs** (receptor activity-modifying proteins). See **proadrenomedullin**. Article relating to AMBP-1: http://www.jbc.org/content/276/15/12292.long

adrenomyeloneuropathy See **leukodystrophy**.

adrenoreceptors See **adrenergic receptors**.

adriamycin Trade name for **daunorubicin**.

ADRP Adipose differentiation related protein. See **perilipins**.

adseverin An actin regulating protein (715aa) found in the adrenal medulla. Has microfilament severing, nucleating and capping activities similar to those of **gelsolin**, but does not crossreact immunologically. Has a phospholipid binding domain and its properties are regulated by phosphatidyl inositides and by calcium. May be identical to **scinderin**.

adsorption coefficient A constant, under defined conditions, that relates the binding of a molecule to a matrix as a function of the weight of matrix, for example in a column.

adult respiratory distress syndrome ARDS See **acute respiratory distress syndrome** and **septic shock**.

adult T-cell leukaemia-derived factor A homologue of **thioredoxin**. An autocrine growth factor produced by HTLV-1 or **EBV**-transformed cells that will induce expression of the interleukin 2 receptor-α (IL-2Rα).

advanced glycation endproducts See **AGE**.

adventitia Generally, the outer covering of an organ. Most commonly, the outer coat of the wall of

vein or **artery**, composed of loose **connective tissue** that is vascularized.

AEBPs Adipocyte enhancer-binding proteins. AEBP1 (aortic carboxypeptidase-like protein, 1158aa) has DNA-binding regions and may be a transcriptional repressor but probably does not have carboxypeptidase activity. It may positively regulate MAP-kinase activity in adipocytes, leading to enhanced adipocyte proliferation and reduced differentiation; also may enhance macrophage pro-inflammatory activity. AEBP2 (517aa) is a zinc-finger DNA-binding transcriptional repressor and a component of the **PRC2/EED-EZH1** complex.

Aedes Genus of mosquitos, several of which transmit diseases of man. A. aegypti is the vector of the **yellow fever virus**.

Aequorea victoria Hydrozoan jellyfish (a **coelenterate**) from which **green fluorescent protein** (GFP) was isolated. **Aequorin** can be isolated from A. victoria and A. forskaolea.

aequorin Protein (196aa) extracted from jellyfish (Aequorea victoria) that emits light (peak 470 nm) in proportion to the concentration of calcium ions. The blue light is transduced into green by green fluorescent protein by energy transfer. Can be used to measure calcium concentrations, but has to be microinjected into cells. Similar light-emitting molecules are found in other coelenterates (e.g. **berovin**, **mitrocomin**, **mnemiopsin**, **phialidin**). See also **bioluminescence**.

aerenchyma Type of **parenchyma** with large intercellular air spaces that gives buoyancy to aquatic plants and allow gas exchange.

aerobes Organisms that rely on oxygen.

aerobic respiration Controlled process by which carbohydrate is oxidized to carbon dioxide and water, using atmospheric oxygen, to yield energy.

aerolysin Channel-forming bacterial **exotoxin** (43aa) produced by Aeromonas hydrophila as a 50 kDa protoxin. Binds to specific receptor on target cells (probably glycophorin on human erythrocytes, but may be other proteins in other cells) and polymerizes to form a heptameric complex that inserts into the plasma membrane and has a pore of approximately 1.5 nm diameter with some properties similar to porin channels. Similar to Staphylococcal alpha toxin.

Aeromonas Genus of Gram-negative bacteria some species of which are pathogenic. Aeromonas salmonicida causes furunculosis in fish.

aerotaxis A taxis in response to oxygen (air).

aesculin A hydroxycoumarin extracted from the horse chestnut (Aesculus hippocastanum), used in homeopathic medicine to thin the blood, acting as an anticoagulant and possibly hypodiuretic. Has toxic effects at higher doses. Aesculin is also used as a additive in agar media for the isolation of glycopeptide-resistant enterococci.

aetiology Outmoded UK spelling of **etiology**.

AF2 (1) Activation function domain 2 of steroid receptors. (2) Antiflammin-2, a synthetic peptide inhibitor of PLA2. (3) A nematode FMRFamide-related neuropeptide. (4) A major African genotype (Af2) of **JC virus**.

afadin Protein (ALL1-fused gene from chromosome 6 protein, AF-6, 1824aa) containing a single **PDZ domain,** that forms a peripheral component of cell membranes at specialized sites of cell-cell junctions. The carboxyl termini of the Eph-related receptor tyrosine kinases EphA7, EphB2, EphB3, EphB5, and EphB6 interact with the PDZ domain. Binds **nectins**. It is fused with **MLL** in leukemias caused by translocations between chromosomes 6 and 11. Afadin-and α-actinin-binding protein (SSX2-interacting protein, 614aa) is part of an adhesion system that may connect the nectin-afadin and E-cadherin-catenin systems through α-actinin. Widely expressed, especially in brain.

afamin A vitamin E binding protein (alpha-albumin, 599aa) of the albumin family (the others being **albumin, alpha-fetoprotein** and vitamin D-binding protein **(Gc protein)**), present in small amounts in plasma (30 microgm/ml).

AFAP-110 An actin binding protein (actin filament-associated protein-110, 635aa) with a **PH domain;** has an alpha-helical N-terminal region capable of self-association through a leucine zipper interaction. AFAP-110 is a **src** substrate and phosphorylation regulates self association; a major function of AFAP-110 may be to relay signals from PKCα via activation of c-src leading to the formation of podosomes.

afferent Leading towards; afferent nerves lead towards the central nervous system, afferent lymphatics towards the lymph node. The opposite of efferent.

affinity An expression of the strength of interaction between two entities, e.g. between receptor and ligand or between enzyme and substrate. The affinity is usually characterized by the equilibrium constant **association constant** or **dissociation constant** for the binding, this being the concentration at which half the receptors are occupied.

affinity chromatography Chromatography in which the immobile phase (bed material) has a specific biological affinity for the substance to be separated or isolated, such as the affinity of an antibody

for its antigen, or an enzyme for a substrate analogue.

affinity labelling Labelling of the active site of an enzyme or the binding site of a receptor by means of a reactive substance that forms a covalent linkage once having bound. Linkage is often triggered by a change in conditions, for example in photo-affinity labelling as a result of illumination by light of an appropriate wavelength.

affinity maturation The change that occurs in the antibody producing system following exposure to a novel antigen; the antibody is of progressively higher affinity and there are memory lymphocytes capable of producing high affinity antibody upon re-exposure to the antigen. Affinity maturation occurs by the early selective differentiation of high affinity variants into antibody forming cells that persist in the bone marrow.

aflatoxins A group of highly toxic substances produced by the fungus *Aspergillus flavus*, and other species of *Aspergillus*, in stored grain or mouldy peanuts. They cause enlargement and death of liver cells if ingested, and may be carcinogenic.

AFLP Amplified fragment length polymorphism, a method for distinguishing closely-related organisms in which molecular markers are produced by high-stringency PCR-amplification of restriction fragments that are ligated to synthetic adapters and amplified using primers, complementary to the adapters, which carry selective nucleotides at their 3' ends. Technical description: http://www.keygene.com/services/technologies_AFLP.php

AFP See **alpha-fetoprotein**.

agammaglobulinaemia Sex-linked genetic defect that leads to the complete absence of immunoglobulins (IgG, IgM, and IgA) in the plasma as a result of the failure of pre-B cells to differentiate. Failure to produce a humoral antibody response leads to high incidence of opportunistic infections but cell-mediated immunity is unimpaired.

AGAMOUS A floral homeotic protein (252aa), a transcription factor involved in the control of organ identity during the early development of flowers. Acts as a C class cadastral protein by repressing the A class floral homeotic genes like *APETALA1*.

agar A **polysaccharide** complex extracted from seaweed (Rhodophyceae) and used as an inert support for the growth of cells, particularly bacteria and some cancer cell lines (e.g. sloppy agar).

agarose A galactan polymer purified from **agar** that forms a rigid gel with high free water content. Primarily used as an electrophoretic support for separation of macromolecules. Stabilized derivatives are used as macroporous supports in **affinity chromatography**. See **Sepharose**.

agatoxins Toxins from the American funnel web spider, *Agalenopsis aperta*. The μ-toxins are 36–38aa peptides that act on insect but not vertebrate voltage-sensitive sodium channels. The ω-agatoxins are more diverse and act on various calcium channels, mostly in neuronal cells, blocking release of neurotransmitters. Omega-agatoxin-Aa3a (76aa), for example, blocks N- and L-type calcium channels in both vertebrates and invertebrates.

AGC kinases A large sub-class of protein kinases including protein kinases A, G and C, protein kinase B (PKB)/akt, p70 and p90 ribosomal S6 kinases and phosphoinositide-dependent kinase-1 (PDK-1). All have an activation loop with a tyrosine phosphorylation site that is important for activity. Plant-specific AGC-VIII kinases (phototropins) are involved in a range of developmental processes, particularly by regulating auxin levels.

AGE Advanced glycation endproducts, modified proteins (glycation adducts) that arise as a result of the reaction of reducing sugars with proteins; analogous reactions occur during cooking (Maillard reaction products). They form at an accelerated rate in diabetes and contribute to the development of vascular disease. Receptor is **RAGE**.

age-related macular degeneration *ARMD* Damage in the central region of the retina, sometimes with accumulation of drusen (dry ARMD) or retinal neovascularization leading to detachment (wet ARMD). A common causes of visual impairment that can be caused by a variety of mutations or polymorphisms including those in genes for **fibulin**-6 (hemicentin), ABCR (retina-specific ABC transporter), fibulin-5, complement factor H, ERCC6 (involved in nucleotide excision repair), retinal homeobox RAXL1, serine peptidase 11, complement C3 and Toll-like receptor-4.

agenesis A general term for the failure of an organ or tissue to develop or to develop normally.

agglutination The formation of mult-component aggregates (agglutinates or flocs) of particles or cells. Usually much more rapid than **aggregation** and involves the addition of extrinsic agents such as **antibodies**, **lectins** or other bi- or poly-valent reagents.

agglutinins Agents causing **agglutination**, e.g. antibodies, **lectins, polylysine**.

agglutinogen The antigen (in the case of antibody) or ligand (in the case of lectin) with which an **agglutinin** reacts.

aggrecan The major structural **proteoglycan** of cartilage. It is a very large and complex macromolecule, comprising a core protein (2316aa) to which are linked around 100 chondroitin sulphate chains and several keratan sulphate chains, as well as O- and N-linked oligosaccharide chains. It binds to a link protein (around 40 kDa) and to hyaluronic acid forming large aggregates, hence its name.

aggregation The process of forming adhesions between particles such as cells. Aggregation is usually distinguished from **agglutination** by the slow nature of the process; not every encounter between the cells is effective in forming an adhesion.

aggresome See **sequestosome**.

aginactin In *Dictyostelium*, an agonist-regulated actin-filament barbed-end capping protein that blocks further addition of G-actin subunits, regulated by cAMP which is chemotactic for the slime mould. A heterodimer of alpha (CAP34, 281aa) and beta (CAP32, 272aa) subunits. Refolding of denatured aginactin is stimulated by heat shock cognate 70 kDa protein 1 (Hsc70) but Hsc70 does not influence the actin capping process.

agitoxins Peptide toxins from the yellow scorpion (*Leiurus quinquestriatus hebraeus*) that block voltage-regulated potassium channels. Closely related to **kaliotoxin**.

aglycone *aglycon, aglucon* The portion of a **glycoside** that remains when the sugar moiety is removed.

agmatinase In humans a mitochondrial enzyme (agmatine ureohydrolase, EC 3.5.3.11, 352aa) that degrades **agmatine** to **putrescine**. There is a similar enzyme in bacteria.

agmatine A metabolite (1-amino-4-guanidobutane) of arginine via arginine decarboxylase, metabolized to **putrescine** by **agmatinase**. Suppresses polyamine biosynthesis and polyamine uptake by cells by inducing **antizyme**. Binds to **imidazoline receptors** and α-2-adrenoreceptors but with different affinities and is thought to be the endogenous ligand for imidazoline (I1) receptors. On the basis of its distribution in the brain is proposed to be a neurotransmitter involved in behavioural and visceral control. An increasing range of biological activities are being described.

Agnatha A superclass of anguilliform (eel-shaped) chordates without jaws or pelvic fins. Lampreys and hagfishes.

agnoprotein A family of small, highly basic proteins (71 aa) encoded by neurotropic JC virus, SV40 and BK virus that regulate viral gene expression, inhibit DNA repair after DNA damage and interfere with DNA damage-induced cell cycle regulation. Agnoprotein is expressed during lytic infection of glial cells by JCV in progressive multifocal leukoencephalopathy (PML), and also in some JCV-associated human neural tumours, particularly medulloblastoma.

agonist (1) In neurobiology describing a neuron or muscle that aids the action of another. If the two effects oppose each other, they are antagonistic. (2) In pharmacology, a compound that acts on the same receptor, and with a similar effect, to the natural ligand.

agorins Major structural proteins of the membrane skeleton of P815 mastocytoma cells. Agorin I, 20 kDa; Agorin II, 40 kDa. Recent (2006) mention: http://jcb.rupress.org/content/174/6/851.long

agouti Central American rodent (*Dasyprocta* spp.) that has given its name to a grey flecked coat coloration in mice caused by alternate light and dark bands on individual hairs. The *agouti* gene codes for a 131aa secreted protein that regulates phaeomelanin synthesis in melanocytes but associated with the locus are genes important in embryonic development. Mice with dominant mutation at the Ay locus develop diabetes and obesity. The *agouti* gene product binds to **melanocortin** receptor one but does not antagonise α-**MSH** and has similar antiproliferative effects on melanoma cells in culture. The dark agouti (DA) rat has a high susceptibility to developing arthritis. The human agouti protein (132aa) is very similar.

agouti-related peptide A neuropeptide (AGRP, agouti related protein, 132aa), very similar to the **agouti** protein, produced by neurons in the arcuate nucleus of the brain. It is involved in reducing food intake by acting as an antagonist of the melanocortin types three and four receptors. Polymorphisms may be associated with obesity.

AGP Arabinoglycan-protein, a class of extracellular **proteoglycan**, found in many higher-plant tissues, and secreted by many suspension-cultured plant cells. Contains 90−98% **arabinogalactan** and 2−10% protein. Related to arabinogalactan II of the cell wall.

agranular vesicles Synaptic vesicles that do not have a granular appearance in EM; 40−50 nm in diameter, with membrane only 4−5 nm thick. Characteristic of peripheral cholinergic **synapses**; (see also **neurotransmitter**). Some are located very close to presynaptic membrane.

agranulocytosis Severe acute deficiency of **granulocytes** in blood.

agretope Portion of antigen that interacts with an **MHC** molecule.

agrin Secreted protein originally isolated from the synapse-rich electric organ of *Discopyge ommata* (Electric ray). The human homologue (2045aa) causes the aggregation of acetylcholine receptors and acetylcholine-esterase on the surface of muscle fibres of the **neuromuscular junction**. Has several EGF repeats and a protease inhibitor-like domain.

Agro-infiltration The use of vacuum infiltration or injection to transform plant cells with *Agrobacterium* harbouring the gene to be transferred.

Agrobacterium rhizogenes Bacterium that infects roots of dicotyledons, used to experimentally transfect root cultures.

Agrobacterium tumefaciens A Gram-negative, rod-shaped flagellated bacterium responsible for **crown gall** tumour in plants. Following infection part of the Ti plasmid, the T-DNA carrying tumour-inducing genes, is transferred to the plant by a process resembling conjugation and the presence of the bacterium is no longer necessary for the continued growth of the tumour. See **T-DNA**.

agropine One of the three **opines**.

AGS cells A line of human gastric **adenocarcinoma** cells.

AH receptor Cytoplasmic receptor for aryl hydrocarbons: see **AHR**.

AHNAKs A class of giant propeller-like proteins that interact with the β2 subunit of cardiac L-type Ca^{2+} channels although predominantly present in the nucleus. Ahnak1 (neuroblast differentiation-associated protein, desmoyokin, 5890aa), was first mistakenly though to be a desmosomal protein; a similar protein (Ahnak2, 5795aa) was identified later. **Dysferlin** is involved in the recruitment and stabilization of Ahnak to the sarcolemma. Ahnak and dysferlin are markers for **enlargeosomes**, a type of cytoplasmic vesicle.

AHR The aryl hydrocarbon receptor, a cytosolic protein (~ 800aa) encoded by the *Ahr* gene that binds a range of aryl hydrocarbons and dioxin and is then translocated to the nucleus where it forms a dimer with **ARNT** and binds to the **xenobiotic response element**. Has a basic helix-loop-helix motif. In development AHR plays an important role in the closure of the ductus venosus and the nuclear form of the activated AHR/aryl hydrocarbon nuclear translocator complex is responsible for alterations in immune, endocrine, reproductive, developmental, cardiovascular, and central nervous system functions.

Aicardi-Goutieres syndrome A genetically heterogeneous encephalopathy (Cree encephalitis, pseudo-TORCH syndrome, pseudotoxoplasmosis syndrome) caused by mutation in the *TREX1* gene encoding a 3−5 exonuclease or by mutations in the genes for subunits A, B, and C of ribonuclease H2 (RNAseH2).

AIDS Acquired Immune Deficiency Syndrome. A condition caused by infection with HIV (Human Immunodeficiency Virus, also called LAV or HTLV-3 in the early literature), resulting in a deficiency of **T-helper cells** and thus **immunosuppression**; as a result opportunistic infections are likely to occur and there is predisposition to certain types of tumour, particularly **Kaposi's sarcoma**.

AIF See **apoptosis inducing factor**.

AIFM1 A caspase-independent mitochondrial effector of apoptotic cell death (mitochondrial apoptosis-inducing factor 1, programmed cell death protein 8, 613aa). Binds to DNA in a sequence-independent manner and interacts with **XIAP**.

AIM (1) Genes absent in melanoma. AIM1 (beta/gamma crystallin domain-containing protein 1, 1723aa) may be a suppressor of malignant melanoma. AIM2 (344aa) is interferon inducible and represses NFκB transcriptional activity. See **PAAD domain** and **inflammasome**. (2) Melanoma antigen AIM1 (membrane-associated transporter protein, solute carrier family 45 member 2, SLC45A2, 530aa) may transport substances required for melanin biosynthesis. (3) See Aurora kinases. (4) In *S. cerevisiae*, AIM proteins are associated with altered inheritance rate of mitochondria (e.g. AIM2, altered inheritance rate of mitochondria protein 2, 246aa) and have diverse functions. (5) In *Arabidopsis*, AIM1 is a peroxisomal fatty acid beta-oxidation multifunctional protein (ABNORMAL INFLORESCENCE MERISTEM 1, 721aa) that is required for wound-induced jasmonate biosynthesis. (6) In many fungi the AIM14 subfamily of ferric reductases are probably cell surface metalloreductases (EC 1.16.1.-) involved in copper or iron homeostasis.

AIP (1) Aryl-hydrocarbon receptor-interacting protein (immunophilin homologue ARA9, 330aa) probably involved in downstream signalling. Defects are associated with various pituitary adenomas. (2) In *Arabidopsis* AIP1 (AKT1-INTERACTING 1, protein phosphatase 2C 3, EC 3.1.3.16, 442aa) negatively regulates potassium channel AKT1, whereas AIP2 (ABI3-interacting protein 2, 310aa) is an E3 ubiquitin-protein ligase that negatively regulates abscisic acid signalling. (3) Actin-interacting protein-1 is an actin binding protein from yeast, subsequently found elsewhere (606aa in human). Interacts with ADF/cofilin to produce bundles in *Dictyostelium* in response to osmotic stress. Aip1 and cofilin cooperate to disassemble actin filaments *in vitro* and are thought to promote rapid turnover of actin networks *in vivo*. (4) Alix/AIP1 (ALG-2-interacting protein X/apoptosis-linked-gene-2-interacting protein 1) is an adaptor protein involved in endosomal trafficking.

air-lift fermenter A fermenter in which circulation of the culture medium and aeration is achieved by injection of air into some lower part of the fermenter. Usually not suitable for animal cell production. Related to gas-lift systems where an inert gas is used to achieve circulation in anaerobic conditions.

AITR A TNF receptor superfamily member (activation-inducible TNFR family member, TNF receptor superfamily 18, TNFRSF18, glucocorticoid-induced TNFR family-related protein, GITR, 241aa) that is upregulated in human peripheral mononuclear cells in response to various signals, mainly after stimulation with anti-CD3/CD28 monoclonal antibodies or phorbol 12-myristate 13-acetate/ionomycin. AITR associates with TRAF1 (TNF receptor-associated factor 1), TRAF2, and TRAF3, and induces NFκB activation via TRAF2. Ligand is **AITRL**.

ajuba A protein (538aa) of the **zyxin/ajuba family**, that associates with the actin cytoskeleton, and is part of a signalling complex that regulates gene expression in response to environmental cues. Ajuba is involved in **Rac** activation during cell migration and activates **Aurora A** in order to commit to mitosis. Article: http://mcb.asm.org/cgi/content/full/25/10/4010?view=long&pmid=15870274

AKAP family A-kinase anchoring protein family, an extensive family of scaffolding proteins that can assemble and compartmentalize multiple signalling and structural molecules. AKAP1 (903aa) binds to type I and II regulatory subunits of protein kinase A and anchors them to the cytoplasmic face of the mitochondrial outer membrane. AKAP2 (859aa) binds to regulatory subunit RII of protein kinase A and may establish polarity in signalling systems. AKAP5 (AKAP79, 79 kDa, 427aa) binds and inhibits PKA, PKC and calcineurin and also binds to the β2-adrenergic receptor. AKAP17A (splicing factor, arginine/serine-rich 17A, 695aa) regulates splicing of some pre-mRNAs in a PKA-dependent manner. See **pericentrin**.

A kinase A class of cAMP-regulated serine/threonine kinases (PKA, EC 2.7.11.11) each a heterotetramer of two catalytic (e.g. PKA-catalytic subunit alpha, 351aa) and two regulatory subunits. The regulatory subunits bind cAMP and are released from the complex activating the catalytic subunits which act as monomers. A-kinase anchor protein (AKAP) binds to the regulatory subunits and localizes the holoenzyme. Various catalytic and regulatory subunits can be combined; some are more tissue-specific than others. There are four regulatory subunits, Iα, Iβ, IIα, and IIβ some constitutive, others inducible and with variable tissue distribution.

akirin See **fbi #2**.

Akt *PKB* Product of the normal gene homologue of v-*akt*, the transforming oncogene of AKT8 virus. A serine/threonine kinase (58 kDa) with SH2 and PH domains, activated by PI3kinase downstream of insulin and other growth factor receptors. AKT will phosphorylate **glycogen synthase kinase** (GSK3) and is involved in stimulation of Ras and control of cell survival. Three members of the Akt/PKB family have been identified, Akt/PKBα, (PKBα, EC

2.7.11.1, 480aa), AKT2 (PKBβ, 481aa) and AKT3 (PKB-γ, 479aa). Only AKT2 has been shown to be involved in human malignancy.

AKV A replication competent murine leukemia virus occurring endogenously in some mouse strains.

ALA synthase Enzyme (5-aminolevulinate synthase, EC 2.3.1.37, 640aa) responsible for the synthesis of 5-aminolevulinic acid, the rate-limiting first step in the haem biosynthetic pathway. The erythrocyte isoform (587aa) is distinct from that in other tissues and deficiencies cause **sideroblastic anaemia**.

aladin See **triple A syndrome**.

alae (1) Generally, flat, wing-like processes or projections, especially of bone. *Adj.* alar, alary. (2) In L1, dauer and adult *C. elegans*, longitudinally-oriented ridges that interrupt the cuticular annulations. Generated by lateral **seam cells**.

Alagille's syndrome Alagille syndrome-1 (arteriohepatic dysplasia) is caused by mutation in **jagged-1**, a ligand for the notch receptor or by mutation in the notch-2 gene.

alamethicin A polyene pore-forming **ionophore** that forms relatively nonspecific anion or cation transporting pores in plasma membranes or artificial lipid membranes. The pores formed by alamethicin are potential gradient-sensitive.

Aland Island eye disease An X-linked retinal disorder caused by mutation in the retina-specific calcium channel alpha-1-subunit gene. Defects in this protein can cause a form of **stationary night blindness**.

alanine Normally refers to L-α-alanine (Ala, A, 89Da) the aliphatic **amino acid** found in proteins. See Table A1. The isomer β-alanine is a component of the vitamin **pantothenic acid** and thus also of **coenzyme A**.

alanine aminotransferase Cytoplasmic enzymes (ALT, glutamic pyruvic transaminase, GPT1, EC 2.6.1.2, 496aa; GPT2, 523aa) in hepatocytes, whose increase in blood is highly indicative of liver damage and often taken as a possible sign of infection with non-A non-B hepatitis. Catalyses L-alanine and 2-oxoglutarate conversion into pyruvate and L-glutamate. In the brain it may protect against **glutamate** excitotoxicity.

alanine scan See **site-specific mutagenesis**.

alar cells *angular cells* Cells at the basal margins (angles) of a leaf.

alarmone A small signal molecule in bacteria that induces an alteration of metabolism as a response to stress. Many metabolic responses may be altered by a single alarmone. Guanosine-tetraphosphate (ppGpp) is an alarmone that regulates **stringent control**, an adaptive response of bacteria to amino acid starvation.

A layer The **S layer** in *Aeromonas* sp.

Albers-Schoenberg disease See **osteopetrosis**.

albinism Condition in which no **melanin** (or other pigment) is present in the hair, skin, and eyes. Can be caused by mutation in the gene for **tyrosinase** or reduced activity of the enzyme in milder forms. Other forms can be a result of mutation affecting P-protein (838aa) that may transport tyrosine into melanosomes and regulates their pH, or mutation affecting tyrosinase-related protein (537aa), a catalase that enhances melanin production, or solute carrier protein (SLC45A2). Ocular albinism involves only the eyes and is a result of mutation in G-protein coupled receptor GPR143.

albino (1) An organism deficient in **melanin** biosynthesis. Hair and skin are unpigmented and the retinal pigmented epithelium is transparent, making the eyes appear red. (2) In *Arabidopsis*, *ALBINO3* is a nuclear gene essential for chloroplast differentiation that encodes a chloroplast protein that is the translocase responsible for the Sec-independent insertion of light-harvesting chlorophyll-binding protein into the chloroplast membrane. Has similarities to **Oxa1** and **YidC**, proteins present in bacterial membranes and yeast mitochondria.

alboaggregin Alboaggregin A is a C-type lectin (a disulphide-linked heterotetramer of ∼130aa subunits) from the white-lipped tree viper (*Trimeresurus albolabris*); causes platelet aggregation through acting as a platelet GPIb agonist. Alboaggregin B (heterodimeric, 146aa and 156aa) also has platelet aggregating activity but is calcium-independent. Similar in action to **botrocetin**.

albolabrin See **disintegrin**.

Albright hereditary osteodystrophy A developmental disorder (AHO) often associated with pseudohypoparathyoidism, hypocalcaemia, and elevated parathyroid hormone levels. AHO, pseudohypoparathyroidism type Ia (PHP Ia) and pseudopseudohypoparathyroidism (PPHP) are all caused by mutations or imprinting defects in the gene coding for the alpha subunit of the G_s type of heterotrimeric G protein. See **progressive osseous heteroplasia**.

albumen *ovalbumin* Major protein (386aa) of the white of birds' eggs. See **albumin**.

albumin The term normally refers to serum albumins, the major protein components of the serum of vertebrates. They have a single polypeptide chain (609aa) with a multidomain structure containing multiple binding sites for many lipophilic metabolites, notably fatty acids and bile pigments. In the embryo their functions are fulfilled by **alpha-fetoproteins**.

The viability of analbuminaemic mutants (those deficient in albumin) raises serious questions about the biological role of albumin. See **afamin**, **albumen** and **Gc protein**.

Alcaligenes Widespread genus of Gram-negative aerobic bacilli found in the digestive tract of many vertebrates and on skin. Occasionally cause opportunistic infections.

ALCAM See **activated leukocyte cell adhesion molecule**.

Alcian blue Water-soluble copper phthalocyanin stain (Alcian blue 8GX) used to demonstrate acid **mucopolysaccharides**. By varying the ionic strength some differentiation of various types is possible. Similar dyes, Alcian green and Alcian yellow, have comparable properties.

aldolase A glycolytic enzyme (fructose-1,6-bisphosphate aldolase, EC 4.1.2.13, 363aa) that catalyzes the conversion of fructose-1,6-bisphosphate to glyceraldehyde 3-phosphate and dihydroxyacetone phosphate. Aldolase, like phosphofructokinase, enolase and hexokinase, binds actin filaments, probably in order to increase efficiency by locally concentrating enzyme and substrate. The glucose transporter **GLUT4**, is connected to actin via aldolase.

aldose reductase Enzyme (aldehyde reductase 1, EC 1.1.1.21, 316aa) that mediates conversion of glucose to sorbitol and the rate-limiting enzyme in the polyol pathway. Altered activity of aldose reductase is thought to play a part in the alterations in the vasculature seen as a complication of diabetes.

aldosterone A steroid hormone (mineralocorticoid) produced by the **adrenal** cortex, that controls salt and water balance in the kidney.

aldosterone secretion inhibiting factor A natriuretic factor (35aa) isolated from chromaffin cells that is an agonist at NPR receptors that inhibit aldosterone production. Closely related in structure to brain natriuretic peptide (BNP). Atrial natriuretic factor (ANF) also inhibits **renin** and **aldosterone** secretion. Original description: http://endo.endojournals.org/content/124/3/1591.short

aleurioconidium A type of **conidium**, often thick-walled and pigmented and only released by lysis or fracture of the supporting cell.

aleurone grain Membrane-bounded **storage granule** (aleurone body) within plant cells that usually contains protein. May be an **aleuroplast** or just a specialized **vacuole**.

aleuroplast A semi-autonomous organelle (**plastid**) within a plant cell that stores protein.

Aleutian disease A chronic, fatal disease of mink, caused by a parvovirus; recognised originally in mink homozygous for the *aleutian* gene that controls fur colour, but also affects raccoons, skunks and ferrets.

Alexander's disease Rare neurodegenerative disorder, usually fatal, caused by a *de novo* mutation in the glial fibrillary acidic protein (**GFAP**) gene. There is an abundance of protein aggregates (Rosenthal fibers) in astrocytes that contain the protein chaperones αB-crystallin and HSP27 as well as glial fibrillary acidic protein (GFAP).

***Alexandrium* spp.** Genus of **dinoflagellates** that produce toxins associated with shellfish poisoning.

algae A non-taxonomic term used to group several phyla of the lower plants, including the **Rhodophyta** (red algae), **Chlorophyta** (green algae), **Phaeophyta** (brown algae), and **Chrysophyta** (diatoms). Many algae are unicellular or consist of simple undifferentiated colonies, but red and brown algae are complex multicellular organisms, familiar to most people as seaweeds. Blue-green algae are a totally separate group of **prokaryotes**, more correctly known as Cyanophyta, or **Cyanobacteria**.

alginate Salts of alginic acids, occurring in the cell walls of some algae. Commercially important in food processing, swabs, some filters, fire-retardants amongst others. Calcium alginates form gels. Alginic acid is a linear polymer of mannuronic and glucuronic acids.

algorithm A process or set of rules by which a calculation or process can be carried out, usually now referring to calculations that will be done by a computer.

aliphatic Carbon compound in which the carbon chain is open (non-cyclic).

aliphatic amino acids The naturally occurring amino acids with aliphatic side chains are glycine, alanine, valine, leucine and isoleucine.

aliquot Small portion. It is common practice to subdivide a precious solution of reagent into aliquots that are used when needed, without handling the total sample.

aliskiren Potent and selective orally-available inhibitor of human **renin** designed by a combination of molecular modelling and crystallographic structure analysis.

ALK (1) In *S. cerevisiae*, serine/threonine-protein kinases (EC 2.7.11.1; ALK1, 760aa; ALK2, 676aa) **haspin**-like protein kinases involved in cell cycle regulation. (2) ALK tyrosine kinase receptor (EC 2.7.10.1, anaplastic lymphoma kinase, CD246, 1620aa), an orphan receptor. (3) See **activin receptor-like kinases**.

alkaline phosphatase Membrane-bound enzymes (EC 3.1.3.1) catalyzing cleavage of inorganic phosphate nonspecifically from a wide variety of phosphate esters, and having a high (>8) pH optimum. In humans there are three distinct forms: intestinal (ALPI, 528aa), placental (ALPP, 532aa), and liver/bone/kidney (tissue non-specific isozyme, ALPL, 524aa). Found in bacteria, fungi and animals but not in higher plants.

alkaloid A nitrogenous base. Usually refers to biologically active (toxic) molecules, produced as allelochemicals by plants to deter grazing. Examples: **ouabain**, **digitalis**.

alkaptonuria A recessive disorder in which deficiency of homogentisate 1,2-dioxygenase (EC 1.13.11.5, 445aa) leads to accumulation of **homogentisic acid** and deposition of brown pigment in skin, the sclera of the eye, connective tissue and joints (ochronosis). Urine blackens on standing.

alkylating agent A reagent that places an alkyl group, e.g. propyl, in place of a nucleophilic group in a molecule. See **alkylating drug**.

alkylating drug Cytotoxic drug which acts by damaging DNA in various ways: addition of alkyl groups to guanine leads to fragmentation during frustrated repair; nucleotides may be cross-bridged, blocking strand separation; miss-pairing leading to mutation may occur during replication. There are six groups of **alkylating agents**: nitrogen mustards; ethylenimes; alkylsulfonates; triazenes; piperazines; and nitrosureas. Common examples are **cyclophosphamide**, chlorambucil, busulphan and mustine.

ALL Acute lymphoblastic **leukaemia**.

all-or-nothing Of an action potential, meaning that action potentials once triggered are of a stereotyped size and shape, irrespective of the size of stimulus that triggered them. Digital signals are all-or-nothing (either 1 or 0).

allantoin A derivative of uric acid; occurs in allantoic fluid and in urine of certain mammals, and is also excreted by certain insects and gastropods. Bizarrely, is used in various cosmetics and in various oral hygiene preparations and eye drops.

allantois Outgrowth from the ventral side of the hindgut in embryos of reptiles, birds and mammals. Serves the embryo as a store for nitrogenous waste and in chick embryos fuses with the chorion to form the **chorioallantoic membrane** (CAM).

allatostatins Peptide hormones produced by the **corpora allata** of insects that reversibly inhibit the production of **juvenile hormone**. Similar peptides are found in other phyla. Allatostatin-4, smallest of the family, is DRLYSFGL-amide. Allatostatins may also be produced in other insect tissues, particularly mid-gut.

allele-specific oligonucleotide probes Short (15−20n) probes (ASO probes) intended for use under hybridization conditions in which the DNA duplex between probe and target is stable only if there is perfect base complementarity between them. Details: http://www.ncbi.nlm.nih.gov/bookshelf/br.fcgi?book=hmg&part=A457&rendertype=figure&id=A500

alleles *allelomorphs* Different forms or variants of a **gene** found at the same place, or **locus**, on a **chromosome**. Assumed to arise by **mutation**.

allelic disorders Clinical disorders caused by different mutations in a gene. Depending on the mutation the protein might not be produced (perhaps because of a premature termination signal) or might have reduced or (less often) enhanced activity or be uncontrolled (for example because of loss of a site for phosphorylation. An example is **Denys-Drash syndrome**.

allelic exclusion The process whereby one or more loci on one of the **chromosome** sets in a **diploid** cell is inactivated (or destroyed) so that the locus or loci is (are) not expressed in that cell or a clone founded by it. For example in mammals one of the X chromosome pairs of females is inactivated early in development (see **Lyon hypothesis**) so that individual cells express only one allelic form of the product of that locus. Since the choice of chromosome to be inactivated is random, different cells express one or other of the X chromosome products resulting in mosaicism. The process is also known to occur in **immunoglobulin** genes so that a clone expresses only one of the two possible allelic forms of immunoglobulin.

allelic imbalance A situation where one **allele** of a heterozygous gene pair is lost (loss of heterozygosity) or amplified; the molecular basis of aneuploidy and a frequent finding in tumours. The mechanisms leading to allelic imbalance are uncertain, but it is thought to result in dysregulation of oncogenes or tumour suppressor genes near the sites of imbalance.

allelochemical Substances effecting allelopathic reactions. See **allelopathy**.

allelomorph One of several alternative forms of a gene: commonly shortened to **allele**.

allelopathic See **allelopathy**.

allelopathy The deleterious interaction between two organisms or cell types that are **allogeneic** to each other (the term is often applied loosely to interactions between **xenogeneic** organisms). Allelopathy is seen between different species of plant, between various individual sponges, and between sponges and gorgonians.

allelotype Occurence of an **allele** in a population or an individual with a particular allele. The allelotype of a tumour, the expression of particular microsatellite markers or isoenzymes, can indicate whether it is of polyclonal or monoclonal origin and the extent to which there is development of **aneuploidy**.

Allen video enhanced contrast *AVEC-microscopy* A method for enhancing microscopic images pioneered by R. D. Allen. The digitized image has the background (an out-of-focus image of the same microscopic field with comparable unevenness of illumination etc.) subtracted, and the contrast expanded to utilize the potential contrast range. Interestingly, it is possible to produce images of objects that are below the theoretical limit of resolution – **microtubules** for example.

allergenic Describing a substance that will provoke an allergic response (see **allergy**). Commonly used to describe substances (allergens) that cause immediate type hypersensitivity reactions such as pollens or insect venoms.

allergens See **allergenic**.

allergic encephalitis See **experimental allergic encephalomyelitis**.

allergic rhinitis A response to inhaled allergens causing swelling of mucous membranes of the nose and upper respiratory tract. Hay fever is a common seasonal form, often a response to grass or tree pollen.

allergy In an animal, a **hypersensitivity** response to some antigen that has previously elicited an immune response in the individual, producing a large and immediate immune response. Allergies, for example to bee venom, are occasionally fatal in humans.

allicin An antibacterial compound, with a strong odour, produced when raw garlic (*Allium sativum*) is either crushed or somehow injured. The enzyme, alliinase, converts alliin in raw garlic to allicin. It is rapidly degraded and its medical efficacy is doubtful. Informative website: http://www.allicin.com/

Allium Genus that includes onions (*A. cepa*), leeks (*A. porrum*), and garlic (*A. sativum*).

allo-epitope See **allotope**.

alloantibody *alloserum* Antibody raised in one member of a species that recognizes genetic determinants in other individuals of the same species. Common in multiparous women and multiply-transfused individuals who tend to have alloantibodies to **MHC** or blood group antigens.

alloantigen An antigen that is characteristic of an individual, or some individuals, of a species, the product of different alleles (individuals are **allogeneic**). The antigens concerned are often of the **histocompatibility complex** and are referred to as alloantigens. The antigenic differences between individuals lead to an immune response to allografts and to graft rejection.

allochthonous Anything found at a site remote from that of its origin.

allogeneic Two or more individuals (or strains) are stated to be allogeneic to one another when the genes at one or more loci are not identical in sequence in each organism. Allogenicity is usually specified with reference to the locus or loci involved.

allograft Grafts between two or more individuals allogeneic at one or more loci (usually with reference to **histocompatibility** loci). As opposed to **autograft** and **xenograft**.

allolactose An isomer of lactose and the natural intracellular inducer of the **lac operon**.

allometric growth Pattern of growth such that the mass or size of any organ or part of a body can be expressed in relation to the total mass or size of the entire organism according to the allometric equation: $Y = ax^b$ where Y = mass of the organ, x = mass of the organism, b = growth coefficient of the organ, and a = a constant.

allometric scaling Adjustment of data to allow either for change in proportion between an organ or organs and other body-parts during the growth of an organism, or to allow for differences and make comparisons between species having dissimilar characteristics, for example, in size and shape.

allometry (1) Study of the relationship between the growth rates of different parts of an organism. (2) Change in the proportion of part of an organism as it grows.

allomone Compound produced by one organism that affects, detrimentally, the behaviour of a member of another species. If the benefit is to the recipient the substance is referred to as a **kairomone**, if both organisms benefit then it is a **synomone**.

allopatric speciation The accumulation of genetic differences in a geographically isolated subpopulation leading to the evolution of a new species.

allophycocyanin A **phycobilin** (APC; λ max = 650 nm) found in some Rhodophyceae and Cyanobacteria. It is used as a fluorochrome for conjugation to other molecules such as antibodies.

allopolyploidy Polyploid condition in which the contributing genomes are dissimilar. When the genomes are doubled fertility is restored and the organism is an amphidiploid. Common in plants but not animals.

allopurinol A **xanthine oxidase** inhibitor used in the treatment of gout.

alloserum See **alloantibody**.

allosomes One or more chromosomes that can be distinguished from **autosomes** by their morphology and behaviour. Synonyms: accessory chromosomes, heterochromosomes, sex chromosomes.

allospecific (1) In taxonomic terms allospecificity implies having the status of a distinct species, genetically distinct and isolated from other similar species, whereas there can be genetic exchange (successful interbreeding) between sub-species and the gene-pool is not closed. (2) In immunological usage it is shorthand for allele-specific; antigenically distinct and thus capable of being recognised by the immune system. Thus individuals, unless genetically identical, are allospecific and capable of rejecting grafts that differ in antigenic features (allotypic determinants).

allosteric Describing a binding site in a protein, usually an enzyme, to which a small molecule will bind and alter the properties of the protein probably by causing a conformational change. The allosteric binding site is spatially separated from the active site.

allosteric activator A compound that activates an enzyme through an allosteric interaction.

allotetraploidy Example of **allopolyploidy** in which the hybrid diploid genome (formed from two chromosome sets) doubles in chromosome number.

allotope *allotypic determinant* The structural region of an **antigen** that distinguishes it from another **allotype** of that antigen.

allotype Products of one or more **alleles** that can be detected as inherited variants of a particular molecule. Usually the usage is restricted to those **immunoglobulins** that can be separately detected antigenically. See also **idiotype**. In humans light chain allotypes are known as Km (Inv) allotypes and heavy chain allotypes as Gm allotypes.

alloxan A compound used to produce **diabetes mellitus** in experimental animals. Destroys pancreatic beta cells by a mechanism involving **superoxide** production.

allozyme Variant of an enzyme coded by a different allele. See **isoenzyme**.

aloisines Competitive inhibitors of ATP binding to the catalytic subunit of **cyclin-dependent kinases** and **GSK**-3. Inhibit cell proliferation by arresting cells in both G1 and G2. Abstract: http://www.ncbi.nlm.nih.gov/pubmed/12519061

alopecia Baldness – can take various forms; alopecia areata in which hair loss is patchy, alopecia universalis in which loss is complete. One form of the latter is caused by mutation in **hairless**.

alp gene cluster A gene cluster involved in polyketide antibiotic synthesis in *Streptomyces* that contains a β-ketoacyl synthase (alpA), a chain length factor (alpB), and an acyl carrier protein (alpC); there are also five regulatory genes *alpT*, *alpU*, and *alpV* that are predicted to encode proteins from the Streptomyces antibiotic regulatory protein (**SARP**) family and *alpW* and *alpZ* which encode proteins from the TetR transcriptional regulator family. The deduced product of *alpZ* shares homology with **γ-butyrolactone autoregulator receptors**.

alpha blockers Class of vasodilatory drugs (α-adrenoceptor blockers) that block the effect of noradrenaline, which is a vasoconstrictor, on peripheral blood vessels.

alpha complementation Complementation of assembly-incompetent mutants of *E. coli* β-galactosidase by a small (26aa) amino-terminal fragment of the lacZ product (the so-called alpha polypeptide) allows assembly of a functional tetramer that will convert **X-gal** to a blue product. By putting a polycloning site within the alpha-polypeptide gene fragment carried by a cloning vector, successful insertion of a sequence that is being cloned prevents complementation and restores the inactivity of β-gal. Colonies with inserts are therefore white and can be selected. This strategy requires *E. coli* host strains, such as DH5a, with mutations that are subject to alpha-complementation, but has the advantage that the vector is small, allowing correspondingly large inserts.

alpha factor Oligopeptide (WHWLQLKPGQPMY) mating pheromone of *S. cerevisiae*; exposure to this pheromone arrests yeast in G1 of the cell cycle and induces the shmoo phenotype. Binds to receptor coded by STE2.

alpha granules See **platelet**.

alpha satellite DNA A family of tandemly repeated sequences found at all normal human centromeres, usually chromosome-specific. FISH probes made of this type of DNA are mainly used to determine aneuploidy and to identify the origin of small marker chromosomes. There are two types of alpha-satellite in the human genome, both made up of approximately 171bp monomers. Alpha satellite probing: http://www.slh.wisc.edu/cytogenetics/procedures/fish/alphasat.dot

alpha-1-acid glycoprotein See **orosomucoid**.

alpha-1-antitrypsin A serine peptidase inhibitor (α-1-antiprotease, **serpinA1**, 418aa) present in plasma and an important inhibitor of neutrophil elastase; also inhibits plasmin and thrombin. Deficiency in the enzyme is associated with emphysema.

alpha-1-microglobulin A plasma protein, (AMBP, 352aa), one of the **lipocalin** superfamily, the product of a gene that encodes a precursor which is proteolytically cleaved into α-1-microglobulin and **bikunin**. Present in free form and in a complex with IgA; involved in regulation of the inflammatory process.

alpha-2-macroglobulin Large (1474aa) plasma antipeptidase with very broad spectrum of inhibitory activity against all classes of proteases. Apparently works by trapping the peptidase within a cage that closes when the peptidase-sensitive bait sequence is cleaved. The peptidase is still active against small substrates that can diffuse into the cage, and the conformational change that closes the trap alters the properties of the α-2-macroglobulin molecule so that it is rapidly removed from circulation. **Plasminogen activator** is one of the few peptidases against which α-2-macroglobulin is ineffective.

alpha-actinin A homodimer of antiparallel subunits (α-actinin-1, 892aa) that can link actin filaments end-to-end with opposite polarity. Originally described in the **Z-disc**, now known to occur in **stress-fibres** and at **focal adhesions**. The nonmuscle isoform (α-actinin-4, 911aa) contains **EF-hand** motifs. Alpha-actinin-2 (894aa) is found in all skeletal muscle fibres, whereas α-actinin-3 (901aa) expression is limited to a subset of type 2 (fast) fibres and the absence of expression is common in endurance atheletes and rare in elite sprinters. An actinin-like actin-binding domain has been found in the N-terminal region of many different actin-binding proteins (e.g. beta chain of spectrin (or fodrin), dystrophin, ABP-120, filamin and fimbrin.). See **calponin homology domain**.

alpha-aminoadipic semialdehyde synthase A bifunctional mitochondrial enzyme (LKR/SDH, 926aa) involved in degradation of lysine. The N-terminal portion has lysine-ketoglutarate reductase activity (EC 1.5.1.8), C-terminal has saccharopine dehydrogenase activity (EC 1.5.1.9).

alpha-amylase An endo-amylase enzyme (EC 3.2.1.1, 511aa) that rapidly breaks down starch to dextrins. In humans there are salivary and pancreatic isoforms.

alpha-B crystallinopathy A type of desmin-related **myopathy** characterized by myofibrillar degeneration that commences at the Z-disk. Mutations in α-**crystallin** B truncate the C-terminal domain of the protein that has the chaperone function.

alpha-cell See **A cells** of endocrine pancreas.

alpha-fetoprotein Protein (609aa) in plasma of vertebrate embryos which probably fulfil the function of **albumin** in the mature organism. Found in both glycosylated and nonglycosylated forms. Presence in the fluid of the **amniotic sac** is diagnostic of spina bifida in the human fetus.

alpha-glucosidase Enzyme (EC 3.2.1.20) that catalyzes the splitting of α-D-glucosyl residues from the non-reducing end of substrates to release α-glucose. In humans, lysosomal α-glucosidase (acid maltase, 952aa) is deficient in glycogen storage disease type 2. Neutral α-glucosidase AB (EC 3.2.1.84, α-glucosidase 2) is a heterodimer with a catalytic alpha subunit (GANAB, 944aa) and a beta subunit (glucosidase 2 subunit beta, PRKCSH, protein kinase C substrate 60.1 kDa protein heavy chain, 528aa) that has a regulatory function and is defective in polycystic liver disease. The enzyme is important in glycoprotein biosynthesis and cleaves sequentially the 2 innermost α-1,3-linked glucose residues from the high mannose oligosaccharide precursor of immature glycoproteins. There is also a neutral α-glucosidase C (EC 3.2.1.20, 914aa). In *E. coli* α-glucosidase (maltodextrin glucosidase, EC 3.2.1.20, 605aa) may regulate the intracellular level of maltotriose.

alpha-helix A particular helical folding of the polypeptide backbone in protein molecules (both fibrous and globular), in which the carbonyl oxygens are all hydrogen-bonded to amide nitrogen atoms three residues along the chain. The translation of amino acid residues along the long axis is 0.15 nm, and the rotation per residue, 100, so that there are 3.6 residues per turn.

alpha-neurotoxins Postsynaptic neurotoxins, many varieties of which are found in snake venoms. Two subclasses, short (four disulphides, 60−62aa) and long (five disulphides and 66−74aa). Examples include alpha-**bungarotoxin**, alpha-**cobratoxin**, **erabutoxins**.

alpha-sarcin A U2 ribonuclease toxin (EC 3.1.27.10, 177aa) from *Aspergillus giganteus*. Is generally cytotoxic and blocks protein synthesis by cleaving eukaryotic 28S rRNA. Like **restrictocin** has anti-tumour activity.

alpha-secretase See **secretases**.

alpha-synuclein See **synuclein**.

Alphavirus Genus of the **Togaviridae**.

alpomycin An antibiotic of the **angucycline** class.

Alport's syndrome Commonest of the hereditary nephropathies. Associated with nerve deafness and variable ocular disorders. The X-linked phenotype is the result of mutation in the gene for the alpha-5 chain of **basement membrane** collagen.

ALS See **amyotrophic lateral sclerosis**.

Alsever's solution A solution used for preserving red blood cells. 2.05% glucose, 0.42% sodium chloride, 0.8% tri-sodium citrate, adjusted to pH 6.1 with citric acid.

alsin The gene product (1657aa) mutated in three juvenile-onset neurodegenerative disorders including **amyotrophic lateral sclerosis** 2 (ALS2). Sequence motif searches within alsin predict the presence of Vps9, DH, and PH domains, implying that alsin may function as a guanine nucleotide exchange factor (**GEF**) for Rab5 and a member of the Rho GTPase family. It seems to control survival and growth of spinal motoneurons

ALT See **alanine aminotransferase**.

altenusin An antifungal penicillide isolated from *Alternaria* sp. Non-competitive, specific neutral sphingomyelinase (N-SMase) and strong pp60c-Src inhibitor. Inhibits c-fms receptor tyrosine kinase (CSF-1/m-CSF receptor tyrosine kinase) and exhibits anti-HIV-1 integrase activity. Dehydroaltenusin is a potent and selective myosin light-chain kinase inhibitor. Also reported to inhibit trypanothione reductase from *Trypanosoma cruzi*.

altered self hypothesis The hypothesis that the **T-cell** receptor in MHC-mediated phenomena recognizes a **syngeneic** MHC Class I or Class II molecule after modification by a virus or certain chemicals. See **MHC restriction**.

alternative oxidase pathway Pathway of mitochondrial electron transport in higher plants, particularly in fruits and seeds, that does not involve **cytochrome oxidase** and thus is resistant to cyanide.

alternative pathway See **complement**.

alternative splicing The process whereby different exons are combined to form variants of a protein. This occurs before transcribed **mRNA** leaves the nucleus and as a result a single gene can encode several different mRNA transcripts, usually in a cell- or tissue-specific fashion. It is estimated that ~80% of human genes are alternatively spliced: the commonest form is exon skipping, in which one or more exons are missing. Trans-splicing is the splicing of an identical short leader sequence, the spliced leader (SL), to the 5′ ends of multiple mRNAs. Alternative Splicing Analysis Tool: http://www.altanalyze.org/ Alternative Splicing Database Project: http://www.ebi.ac.uk/asd/

altronate The base of altronic acid, a six-carbon dicarboxylic acid that is a product of uronic acid metabolism in *E. coli* a pathway in which glucuronic and galacturonic acid are isomerized to form their corresponding keto-analogues, fructuronic and tagaturonic acid, respectively. In the presence of reduced NAD, fructuronic acid is reduced to D-mannonic acid whereas tagaturonic acid yields n-altronic acid. Altronate dehydratase (altronate hydrolase, EC 4.2.1.7, 495aa) converts D-altronate to 2-dehydro-3-deoxy-D-gluconate; altronate oxidoreductase (EC 1.1.1.58, tagaturonate dehydrogenase, 483aa) converts D-altronate to D-tagaturonate. Description of pathway: http://www.jbc.org/content/235/6/1559.long

Alu (1) Type II **restriction endonuclease**, isolated from *Arthrobacter luteus*. The recognition sequence is 5′- AG/CT-3′. (2) Alu sequences are highly repetitive sequences found in large numbers (100−500,000) in the human genome, and that are cleaved more than once within each sequence by the Alu endonuclease. The Alu sequences look like DNA copies of mRNA because they have a 3′ **poly-A tail** and flanking repeats. Cf. **long interspersed nucleotide element**.

alveolar cell Cell of the air sac of the lung.

alveolar cells Squamous epithelial cells that line the terminal dilations (alveoli) of the branched airways of the lung. A few (great alveolar cells) secrete lung surfactant.

alveolar macrophage Macrophage found in lung and that can be obtained by lung lavage; responsible for clearance of inhaled particles and lung surfactant. Metabolism slightly different from peritoneal macrophages (more oxidative metabolism), often have **multivesicular bodies** that may represent residual undigested lung surfactant.

Alveolata A monophyletic group of primarily single-celled eukaryotes with diverse modes of nutrition, including predation, photoautotrophy and intracellular parasitism. There are three main subgroups: ciliates, dinoflagellates and apicomplexans.

Alzheimer's disease A presenile dementia characterized cellularly by the appearance of unusual helical protein filaments in nerve cells (neurofibrillary tangles), and by degeneration in cortical regions of brain, especially frontal and temporal lobes. See also **senile plaques**. Some forms are associated with mutations in genes for **amyloid precursor protein** (APP) or **presenilins**. The apolipoprotein E4 allele (ApoE4) is associated with another form and various other loci may have risk-factor alleles.

amacrine cell A class of **neuron** of the middle layer of the **retina**, with processes parallel to the plane of the retina. They are thought to be involved in image processing.

Amanita phalloides Poisonous mushroom, the Death Cap; contains **amanitin** and **phalloidin**.

amanitin Group of cyclic peptide toxins (amatoxins) from *Amanita phalloides* (Death cap) and related species. Amanitin-α (amanitin-γ) is a cyclic octapeptide that is an inhibitor of **RNA polymerase** II in eukaryotes that blocks protein synthesis. The toxic effects are slow and usually fatal.

amassin Olfactomedin-family member (495aa) from the sea urchin *Strongylocentrotus purpuratus*.

Mediates a rapid cell-adhesion event resulting in a large aggregation of **coelomocytes**.

amastigote Stage in the life cycle of trypanosomatid protozoa; resembles the typical adult form of members of the genus *Leishmania*, in which the oval or round cell has a nucleus, kinetoplast, and basal body but lacks a flagellum.

amber codon See **termination codons**.

amber suppressor A mutation in a tRNA allele, altering it so that an amino acid is inserted at the **amber codon** and termination does not occur.

AMBP (1) See **bikunin**. (2) See **adrenomedullin**.

Ambystoma mexicanum Mexican axolotl (amphibian). A salamander that shows **neoteny**. The adult may retain the larval form, but can reproduce. The neotenous, aquatic axolotl will metamorphose into the terrestrial form if injected with thyroid or pituitary gland extract.

AMD See **age-related macular degeneration**.

amelia Congenital abnormality in which one or more limbs are completely absent. Tetra-amelia is a rare autosomal recessive human genetic disorder characterized by complete absence of all four limbs and other anomalies; a **WNT**3 mutation in tetra-amelia indicates that WNT3 is required at the earliest stages of human limb formation.

ameloblasts Columnar epithelial cells that secrete the enamel layer of teeth in mammals. Their apical surfaces are tapering (Tomes processes) and are embedded within the enamel matrix.

amelogenesis imperfecta Defective formation of dental enamel that can arise from a variety of mutations including those in genes for **amelogenin**, **enamelin**, FAM83H protein (1179aa), distal-less homeobox 3 (DLX3) protein (287aa), matrix metalloproteinase-20 or **kallikrein-4**.

amelogenins Extracellular matrix proteins (191aa and 206aa) of developing dental enamel; regulate form and size of hydroxyapatite crystallites during mineralisation. Hydrophobic and proline-rich, produced by **ameloblasts**.

ameroconidium A one-celled conidium.

Ames test One of a number of procedures used to test substances for likely ability to cause cancer that combines the use of animal tissue (usually liver-derived) to generate active metabolites of the substance with a test for mutagenesis in strains of *Salmonella typhimurium* engineered to be deficient in histidine synthesis, but capable of acquiring this ability by simple point or frame-shift mutations.

amethopterin See **aminopterin**.

AMH See **anti-Müllerian hormone**.

AMICA1 An immunoglobulin superfamily protein expressed on the surface of polymorphonuclear leucocytes and involved in adhesion to epithelia and transmigration across endothelium (adhesion molecule interacting with CXADR antigen 1, junctional adhesion molecule-like, 394aa).

amidation site A C-terminal consensus sequence, required for C-terminal amidation of peptides. Consensus is glycine, followed by two basic amino acids (arg or lys).

amiloride Drug that blocks sodium/proton **antiport** thereby inhibiting sodium reabsorption in renal epithelial cells; used clinically as a potassium-sparing diuretic. See **sodium channel #2**.

amiloride-sensitive sodium channels See **sodium channel**.

aminergic Generally a description of receptors that respond to amines. Term usually applied to neurons that release noradrenaline, dopamine or serotonin. *Cf.* adrenergic, cholinergic.

amino acid permease A widely distributed group of large integral membrane proteins, required for the entry of amino acids into cells. In *Arabidopsis*, for example, amino acid permease one (amino acid transporter AAP1, neutral amino acid transporter II, 485aa) is an amino acid-proton symporter with a broad specificity for histidine, glutamate and neutral amino acids.

amino acid receptors Ligand-gated **ion channels** with specific receptors for amino acid **neurotransmitters**. An extended protein superfamily that also includes subunits of the **nicotinic acetylcholine receptor**.

amino acid transmitters Amino acids released as neurotransmitter substances from nerve terminals and acting on postsynaptic receptors e.g. γ-aminobutyric acid (GABA) and glycine that are fast inhibitory transmitters in the mammalian central nervous system. Glutamate and aspartate mediate fast excitatory transmission. **Strychnine** (for glycine) and **bicuculline** (for GABA) are blocking agents for amino acid action.

amino acids Organic acids carrying amino groups. The L-forms of about 20 common amino acids are the components from which proteins are made. See Table A1, and Table C4 for the **codon** assignment.

amino-naphthalimide Inhibitor of **poly(ADP-ribose) polymerase** (PARP).

amino-sugar Monosaccharide in which an OH-group is replaced with an amino group; often acetylated. Common examples are D-galactosamine, D-glucosamine, neuraminic acid, muramic acid. Amino sugars are important constituents of bacterial

TABLE A1. Amino acids

Name	Abbreviation	Single letter	Side Chain	pKa*	M_r (Da)	Hydropathy index** (Kyte & Doolittle)	Codons
Alanine	ala	A	$-CH_3$		89.1	1.8	GC(X)
Arginine	arg	R	$-CH_2\ CH_2\ CH_2\ NH$ $(CN^+H_2)\ NH_2$	12	174.2	-4.5	CG(X) AGA AGG
Aspartic acid	asp	D	$-CH_2\ COO^-$	4.4	133.1	-3.5	GAU GAC
Asparagine	asn	N	$-CH_2\ CONH_2$		132.2	-3.5	AAU AAC
Cysteine	cys	C	$-CH_2\ SH$	8.5	121.2	2.5	UGU UGC
Glutamic acid	glu	E	$-CH_2\ CH_2\ COO^-$	4.4	147.2	-0.4	GG(X)
Glutamine	gln	Q	$-CH_2\ CH_2\ CONH_2$		146.2	-3.5	CAA CAG
Glycine	gly	G	$-H$		75.1	-3.5	GG(X)
Histidine	his	H	$-CH_2$ (imidazole ring, $_+HN=\!\!\!\diagdown\!\!\!\diagup NH$)	6.5	155.2	-3.2	CAU CAC
Iso-leucine	ile	I	$-CH\ (CH_3)\ CH_2\ CH_3$		131.2	4.5	AUU AUC AUA
Leucine	leu	L	$-CH_2\ CH\ (CH_3)_2$		131.2	3.8	CU(X) UUA UUG
Lysine	lys	K	$-CH_2\ CH_2\ CH_2\ CH_2\ NH_3^+$	10	146.2	-3.9	AAA AAG
Methionine	met	M	$-CH_2\ CH_2\ SCH_3$		149.2	1.9	AUG
Phenylalanine	phe	F	$-CH_2-$ (benzene ring)		165.2	2.8	UUU UUC
Proline	pro	P	$-NH_2^+\ -CH\ -COO^-$ $CH_2\quad CH_2$ CH_2		115.1	-1.6	CC(X)
Serine	ser	S	$-CH_2\ OH$		105.1	-0.8	UC(X)
Threonine	thr	T	$-CH\ (OH)\ CH_3$		119.1	-0.7	AC(X)
Tryptophan	trp	W	$-CH_2-$ (indole ring, N–H)		204.2	-0.9	UGG (UGA mitochondria
Tyrosine	tyr	Y	$-CH_2-$ (phenol ring) $-OH$	10	181.2	-1.3	UAU UAC
Valine	val	V	$-CH\ (CH_3)_2$		117.2	4.2	GU(X)

L-amino acids specified by the biological code for proteins.
*The value for side chain ionisation when the amino acid residue is present in a polypeptide.
**A measure of the tendency for the residue to be buried within the interior of a folded protein.

cell walls, some antibiotics, blood group substances, milk oligosaccharides, and chitin.

aminoacyl tRNA Complex of an **amino acid** to its **tRNA**, formed by the action of aminoacyl tRNA synthetase. Requires ATP, which forms the linkage between the two molecules.

aminoacyl tRNA synthetases Enzymes that attach an amino acid to its specific tRNA. An intermediate step is the formation of an activated amino acid complex with AMP; the AMP is released following attachment to the tRNA. One example among many is glycyl-tRNA synthetase (glycine-tRNA ligase, EC 6.1.1.14, 739aa).

aminoacylase Enzymes involved in hydrolysis of most N-acylated or N-acetylated amino acids. Aminoacylase-1 (EC 3.5.1.14, 408aa) hydrolyses N-acylated or N-acetylated amino acids except L-aspartate and mutations lead to a serious metabolic disorder. Aspartoacylase (EC 3.5.1.15, aminoacylase-2, 313aa) hydrolyzes N-acetyl-L-aspartic acid and deficiency leads to **Canavan's disease**.

aminoacylation Addition of an aminoacyl group (formed by the removal of hydroxyl group from α-carbonyl group of an α-amino acid) to a substrate, the best known of which is tRNA (forming an aminoacyl-tRNA).

aminocoumarin antibiotics A group of antibiotics, produced by various Streptomyces strains, that are active against Gram-positive bacteria by inhibiting DNA gyrase with a secondary target of topoisomerase IV (involved in chromosome decatenation). Abstract: http://www.ncbi.nlm.nih.gov/pubmed/15720250

aminocyclitol antibiotics Antibiotics produced by actinomycete that interfere with the 30S subunit of the bacterial ribosome. An example is Spectinomycin.

aminoglycoside 3'-phosphotransferase In *E. coli* an enzyme (kanamycin kinase, type I, 271aa) encoded by the kanamycin resistance transposon that confers resistance to kanamycin and structurally-related aminoglycoside antibiotics.

aminoglycoside antibiotics *oligosaccharide antibiotics* Group of antibiotics active against many aerobic Gram-negative and some Gram-positive bacteria. Composed of two or more amino sugars attached by a glycosidic linkage to a hexose nucleus; polycationic and highly polar compounds. Inhibit bacterial protein synthesis by binding to a site on the 30S ribosomal subunit thereby altering codon-anticodon recognition. Common examples are **streptomycin**, **gentamicin**, amikacin, **kanamycin**, **tobramycin**, netilmicin, neomycin, framycetin.

aminopeptidase Enzymes that remove the N-terminal amino acid from a protein or peptide, many are zinc metalloenzymes. Examples include leucylcystinyl aminopeptidase (cystinyl aminopeptidase, EC 3.4.11.3, 1025aa) that degrades peptide hormones such as oxytocin, vasopressin and angiotensin III and aminopeptidase N (EC 3.4.11.2, CD13, 967aa) that has a role in final digestion of peptides generated from hydrolysis of proteins by gastric and pancreatic proteases. Aminopeptidase Q is a zinc metallopeptidase (EC 3.4.11.-, 990aa) which may be important for placentation. Many other aminopeptidases are known with varying specificity, tissue distribution and undoubtedly different roles. See **endoplasmic reticulum aminopeptidase, leucine aminopeptidase, methionine aminopeptidase**.

aminophylline An inhibitor of cAMP **phosphodiesterase**.

aminopterin A **folic acid** analogue and inhibitor of **dihydrofolate reductase**, a component of **HAT medium**. A potent cytotoxic agent used in the treatment of acute **leukaemia**.

aminotransferases A family of enzymes (transaminases, EC 2.6.1.x) that transfer an amino group from an amino acid to an α-keto acid with pyridoxal phosphate as coenzyme. Glutamate oxaloacetate transaminase (EC 2.6.1.1, aspartate aminotransferase), which catalyses the transfer from glutamate to oxaloacetic acid (producing aspartic acid and a-ketoglutarate), exists in both mitochondrial (430aa) and cytosolic (413aa) forms. Branched chain aminotransferases (EC 2.6.1.42, BCAT1, 393aa; BCAT2, 392aa) act on amino acids with non-linear aliphatic sidechains (leucine, isoleucine and valine). Defects in branched-chain amino acid transamination can lead to **hypervalinaemia** and **hyperleucine-isoleucinaemia**.

amitosis An unusual form of nuclear division, in which the nucleus simply constricts, rather like a cell, without chromosome condensation or spindle formation. Partitioning of daughter chromosomes is haphazard. Observed in some Protozoa.

AML Acute myeloblastic **leukaemia**.

ammodytoxins Group II secretory phospholipases A2 (ammodytoxins A, B & C, EC 3.1.1.4, all 138aa) found in the venom of *Vipera ammodytes*. Act presynaptically to block acetylcholine release at peripheral nerve endings in the neuromuscular junction.

amnesic shellfish poisoning See **domoic acid**.

amniocentesis Sampling of the fluid in the **amniotic sac**. In humans this is carried out, between the 12th and 16th week of pregnancy, by inserting a needle through the abdominal wall into the uterus. By **karyotyping** the cells and determining the proteins present, it is possible to determine the sex of the foetus and whether it is suffering from certain congenital diseases such as **Down's syndrome** or **spina bifida**.

amniocyte Cell type found floating freely in the amnion sac; following **amniocentesis**, amniocytes can be subcultured and used for prenatal genetic diagnosis.

amnion The inner layer of the fluid-filled sac in which the embryos of terrestrial vertebrates develop. The sac is formed by the outgrowth of the extra-embryonic **ectoderm** and **mesoderm** as projecting folds. These folds fuse to form two epithelia separated by mesoderm and **coelom**. The outer layer is the **chorion**.

amniotes Tetrapod vertebrates with terrestrially-adapted eggs which may be laid or carried internally in the female. They include mammals, birds and reptiles. The egg is self-contained and provides a stable fluid environment in which the embryo is protected by a system of membranes that include the amnion, chorion, and allantois.

amniotic sac Sac, enclosing the embryo of amniote vertebrates, that provides a fluid environment to prevent dehydration during development of land-based animals. See **amnion**.

amoeba Genus of protozoa, but also an imprecise name given to several types of free-living unicellular phagocytic organism. Giant forms (e.g. *Amoeba proteus*) may be up to 2 mm long, and crawl over surfaces by protruding **pseudopods** (**amoeboid movement**). Amoebae exhibit great plasticity of form and conspicuous **cytoplasmic streaming**.

amoebiasis Dysentry caused by *Entamoeba histolytica*.

amoebocytes Phagocytic cells found circulating in the body cavity of coelomates (particularly annelids and molluscs), or crawling through the interstitial tissues of sponges. A fairly non-committal classification.

amoeboid movement Crawling movement of a cell brought about by the protrusion of **pseudopods** at the front of the cell (one or more may be seen in monopodial or polypodial amoebae, respectively). The pseudopods form distal anchorages with the surface.

AMP Adenosine monophosphate and, unless otherwise specified, 5'AMP, the nucleotide bearing a phosphate in ribose-O-phosphate ester linkage at position 5 of the ribose moiety. Both 2' and 3' derivatives also exist. See also **cyclic AMP**, (adenosine 3'5'-cyclic monophosphate).

AMP-PNP Non-hydrolysable analogue of ATP (5-adenylyl imidodiphosphonate) used in isolation of some motor proteins.

AMPA Synthetic agonist for one class of metabotropic **glutamate receptors**. See also **excitatory amino acid**.

amphetamine Drug of abuse that acts by increasing extraneuronal **dopamine** in the midbrain. Thought to displace dopamine in **synaptic vesicles**, leading to increased synaptic levels.

amphibolic Description of a pathway that functions not only in **catabolism**, but also to provide precursors for **anabolic** pathways.

amphimixis Sexual reproduction resulting in an individual having two parents. Invariably the case in most animals, with the exception of a few hermaphrodite organisms, but not uncommon in plants where a single individual may produce both male and female gametes (be monoecious) and be self-fertile.

Amphioxus Obsolete generic name for *Branchiostoma*. Name is descriptive — sharp at both ends.

amphipathic Of a molecule, having both **hydrophobic** and **hydrophilic** regions. Can apply equally to small molecules, such as phospholipids, and macromolecules such as proteins.

amphiphilic Having affinity for two different environments — for example a molecule with hydrophilic (polar) and lipophilic (non-polar) regions. Detergents are classic examples. Antonym of **amphipathic**.

amphiphysin Protein (695aa) of the nerve terminal that associates with synaptic vesicles and regulates exocytosis. Forms a heterodimer with **bridging integrator 1** (BIN1, amphiphysin-2, Myc box-dependent-interacting protein 1, 593aa) and interacts with **endophilin-B** and the AP-2 complex (adapter-related protein complex 2).

amphiploid *amphidiploids* A polyploid organism in which chromosomes come from two different species. Not uncommon in plant breeding.

amphiregulin A heparin-binding growth factor (Schwannoma-derived growth factor in rat, colorectum cell-derived growth factor in humans, 252aa) containing an **EGF-like domain**. Binds to the EGF receptor, though with lower affinity than EGF. See **HB-EGF**.

amphitrichous Having a flagellum at both ends of the cell.

amphitrophic Of organisms that can grow either photosynthetically or chemotrophically.

ampholyte Substance with **amphoteric** properties. Most commonly encountered as descriptive of the substances used in setting up electrofocusing columns or gels.

amphoteric Having both acidic and basic characteristics. This is true of proteins since they have both acidic and basic side groups (the charges of which balance at the **isoelectric point**).

amphotericin B Polyene antibiotic (Fungizone) from Streptomyces spp. Used as a fungicide, it is cytolytic by causing the formation of pores (5−10 molecules of amphotericin in association with cholesterol) that allow passage of small molecules through the plasma membrane and thus to cytolysis. Only acts on membranes containing sterols (preferentially ergosterol, hence selectivity for fungi). See also **filipin**.

amphoterin In rat, a heparin-binding protein (high mobility group protein 1, 215aa) that enhances nerve growth cone migration and neurite outgrowth in the developing brain. It is also a DNA binding protein that binds preferentially single-stranded DNA and is involved in V(D)J recombination by acting as a cofactor of the RAG complex. In humans high mobility group protein 1 (215aa) is similar and is also involved in V(D)J recombination. In humans, amphoterin-induced protein 1 (AMIGO-1, alivin-2, 493aa) promotes growth and fasciculation of neurites from cultured hippocampal neurons. AMIGO-2

(522aa) and AMIGO-3 (504aa) have similar properties.

amphotropic virus An virus that does not produce disease in its natural host, but will replicate in tissue culture cells of the host species and cells from other species. Amphotropic murine leukemia virus has been extensively used as a vector in experimental gene transfer.

ampicillin A semi-synthetic **penicillin** derivative with a broad range of antibacterial activity. Ampicillin resistance is often used as a marker for plasmid transfer in genetic engineering (e.g. **pBR322** is ampicillin resistant).

AMPK AMP-activated protein kinase, an enzyme that plays a key role in regulating energy homeostasis. AMPK-mediated phosphorylation switches cells from ATP consumption towards ATP production. AMPK is itself regulated by physiological stimuli which lead to its activation by AMPK kinases (AMPKK). Mammalian AMPKs are trimeric enzymes (EC 2.7.11.1) composed of catalytic a subunits (550aa or 552aa) and non-catalytic b (270aa or 272aa) and g (331aa or 569aa) subunits.

amplicon The DNA product of a **polymerase chain reaction**.

ampulla In anatomy, a small membranous vesicle. *Adj* ampullary.

amygdala Almond-shaped body in the lateral ventricle of the brain.

amylase See **alpha-amylase**; and **beta-amylase**.

amylin Natural hormone (islet amyloid polypeptide, IAPP, 37aa) produced by pancreatic **beta-cells** that moderates the glucose-lowering effects of insulin. Co-secreted with insulin, controls nutrient intake as well as nutrient influx to the blood by an inhibition of food intake, gastric emptying, and glucagon secretion. One of the **calcitonin family peptides**.

amyloid Glycoprotein deposited extracellularly in tissues in **amyloidosis**. Amyloid of immune origin (AIO) can be derived from immunoglobulin light chain produced by a single clone of plasma cells or the N-terminal part of lambda or kappa **L-chain**). Amyloid of unknown origin (AUO) is derived from **serum amyloid** A (SAA), one of the acute phase proteins that increases many-fold in inflammation. The polypeptides are organized as a beta pleated sheet making the material rather inert and insoluble. Minor protein components are also found. Should be distinguished from β-amyloid deposited in the brain and that is derived from **amyloid precursor protein**.

amyloid precursor protein A single-pass membrane protein (amyloidogenic glycoprotein, APP, 770aa) from which beta amyloid peptides (42–43aa), the principal component of amyloid fibrils, are derived (as well as a number of other cleavage products). There are multiple alternatively-spliced isoforms, some of which may be associated with predisposition to formation of senile plaques, characterisitic of individuals with **Alzheimer's disease**. The specific role of amyloid protein is unclear but it is thought that amyloid deposits may cause neurons to degenerate. Amyloid deposits also occur in brains of older **Down's Syndrome** patients. APP itself functions as a cell surface receptor and performs physiological functions on the surface of neurons relevant to neurite growth, neuronal adhesion and axonogenesis.

amyloidogenic glycoprotein See **amyloid precursor protein**.

amyloidosis Deposition of **amyloid**. A common complication of several diseases (leprosy, tuberculosis); often associated with perturbation of the immune system, although there may be immunosuppression or enhancement. In a rare inherited form the deposits are of **transthyretin**. The brain amyloid in **senile plaques** is derived from **amyloid precursor protein** and is distinct. See **familial primary localized cutaneous amyloidosis, Icelandic-type cerebroarterial amyloidosis**.

amylopectin Soluble and highly branched component of **starch** in which glucose chain is α-1,4 linked (α-1,6 at branch points). Along with amylose, forms the starch granules made by plants.

amyloplast A plant **plastid** involved in the synthesis and storage of starch. Found in many cell types, but particularly storage tissues. Characteristically has starch grains in the plastid **stroma**.

amylase A linear glucose polymer formed from α-D-glucopyranosyl units in α-1,4 linkage. Found both in starch (starch amylase) and glycogen (glycogen amylose). Insoluble in water and less resistant to hydrolysis than **amylopectin**.

amyotrophic lateral sclerosis Progressive degenerative disease (Lou Gehrig's disease, motor neuron disease) of motor neurons in the brain stem and spinal cord that leads to weakening of the voluntary muscles. A few cases are associated with mutations in the superoxide dismutase-1 gene. Susceptibility has also been associated with mutations in the genes encoding the heavy neurofilament subunit, peripherin, **dynactin** and **angiogenin**. Other variants are associated with mutation in genes for **alsin**, SAC domain-containing inositol phosphatase 3, **senataxin**, TAR DNA-binding protein and **VAMP**-associated protein B.

amyotrophy Wasting (atrophy) of muscle.

Anabaena A genus of **Cyanobacteria** that forms filamentous colonies with specialized cells (**heterocysts**), capable of nitrogen fixation. Ecologically important in wet tropical soils and forms symbiotic associations with the fern Azolla.

anabolic Of a process, route or reaction in which energy (often ATP-derived) is expended in order to synthesize more complex molecules. Tends to involve reductive steps as opposed to **catabolic** reactions or processes that are degradative, often oxidative, with attendant regeneration of ATP.

anabolic steroids Synthetic forms of male sex hormones (androgens) that promote tissue growth, especially of muscle.

anabolism Synthesis; opposite of **catabolism**.

anaemia US. *anemia* Reduced level of **haemoglobin** in blood for any of a variety of reasons including abnormalities of mature red cells, iron deficiency, haemolysis of erythrocytes, reduced **erythropoiesis** or haemorrhage (to name the most common). See separate entries for **aplastic anaemia**, **congenital dyserythropoietic anaemia**, **Diamond-Blackfan anaemia**, **erythroblastosis fetalis**, **Fanconi's anaemia**, **haemolytic anaemia**, **megaloblastic anaemia** (pernicious anaemia), **microangiopathic haemolytic anaemia**, **sickle cell anaemia**, **spherocytosis** and **warm antibody haemolytic anaemia**.

anaerobic The absence of air (specifically of free oxygen). Used to describe a biological habitat or an organism that has very low tolerance for oxygen.

anaerobic respiration Metabolic processes in which organic compounds are broken down to release energy in the absence of oxygen. Requires inorganic oxidizing agents or accumulation of reduced coenzymes.

anagenesis Progressive evolution of species through alterations in gene frequency in an entire population so that eventually the new population would be recognised as distinct from the ancestral species, as opposed to cladogenesis in which two species emerge.

analogous Of genes or gene products, performing a similar role in different organisms. *Cf.* **homologous**.

analysis of variance *ANOVA* A powerful statistical technique that distinguishes between variation due to specific causes and variation due to random factors.

analyte Substance or compound for which an analysis is being carried out.

anamnestic response Archaic term now replaced by such terms as **secondary immune response**, **immunological memory**.

anandamide Arachidonyl ethanolamide, an endogenous agonist for cannabinoid receptors.

anaphase The stage of **mitosis** or **meiosis** beginning with the separation of sister **chromatids** (or homologous **chromosomes**) followed by their movement towards the poles of the **spindle**.

anaphase-promoting complex *cyclosome* An unusually complicated E3 **ubiquitin** ligase, composed of 13 core subunits (total 1.5-MDa) and either of two loosely associated co-activators, Cdc20 and Cdh1, that is responsible for initiation of sister chromatid separation and the inactivation of cyclin-dependent kinases. Activated APC ubiquitinates **securin**, targeting it for degradation by the 26S proteasome and thereby relieving the inhibition of **separin**. The largest subunit, Apc1, serves as a scaffold that associates independently with two separable subcomplexes, one that contains Apc2 (**cullin**), Apc11 (RING), and Doc1/Apc10, and another that contains the three tetratricopeptide repeat (**TPR motif**)-containing subunits (Cdc27, Cdc16, and Cdc23). In S-phase the APC is inactivated by binding of cyclin A.

anaphylatoxin Originally used of an antigen that reacted with an **IgE** antibody thus precipitating reactions of anaphylaxis. Now restricted to defining a property of **complement** fragments C3a and C5a, both of which bind to the surfaces of **mast cells** and **basophils** and cause the release of inflammatory mediators.

anaphylaxis As opposed to **prophylaxis**. A system or treatment that leads to damaging effects on the organism. Now reserved for those inflammatory reactions resulting from combination of a soluble antigen with **IgE** bound to a **mast cell** that leads to degranulation of the mast cell and release of **histamine** and histamine-like substances, causing localized or global immune reponses. See **hypersensitivity**.

anaplasia Lack of differentiation, characteristic of some tumour cells.

Anaplasma phagocytophilum A Gram-negative bacterium (formerly *Ehrlichia phagocytophilum*) that is an intracellular parasite of neutrophils and causes anaplasmosis. Transmitted to humans by ticks, *Ixodes* spp.

anaplerotic Describing reactions that replenish **TCA cycle** intermediates and allow respiration to continue; for example, carboxylation of **phosphoenolpyruvate** in plants.

anastomosis Joining of two or more cell processes or multicellular tubules to form a branching system. Anastomosis of blood vessels allows alternative routes for blood flow.

anatoxins A group of low molecular weight neurotoxic alkaloids first described in the fresh-water cyanobacteria *Anabaena flos-aquae*, but subsequently found in other species. Anatoxin-a and homoanatoxin-a are secondary amines that bind and irreversibly activate nicotinic acetylcholine receptors; anatoxin-a(s) is the only natural organophosphate known and inactivates acetylcholine esterase in a similar fashion to synthetic organophosphate pesticides such as parathion and malathion. All cause muscle exhaustion by overstimulation and death through respiratory failure. Structure: http://www-cyanosite.bio.purdue.edu/cyanotox/toxins/anatoxin.html

ANCA Anti-neutrophil cytoplasmic antibodies seen in patients with a variety of inflammatory disorders including IBD (inflammatory bowel disease), Wegeners granulomatosis and hepatobiliary disorders. Two forms are recognized, peripheral ANCA (p-ANCA) where the antigen seems to reside at the periphery of the nucleus and cytoplasmic ANCA (c-ANCA) where the antigen is distributed throughout the cytoplasm of the neutrophil.

anchorage Attachment, not necessarily adhesive in character; the term ought to be more widely used since it avoids assumptions about mechanism.

anchorage dependence The necessity for attachment (and spreading) in order that a cell will grow and divide in culture. Loss of anchorage dependence seems to be associated with greater independence from external growth control and is probably one of the best correlates of **tumorigenic** events *in vivo*. Anchorage independence is usually detected by **cloning** cells in soft agarose; only anchorage-independent cells will grow and divide (as they will in suspension).

anchored PCR Variety of **polymerase chain reaction** in which only enough information is known to make a single primer. A known sequence is therefore added to the end of the DNA, perhaps by enzymic addition of a polynucleotide stretch or by ligation of a known piece of DNA. The PCR can then be performed with the gene-specific primer and the anchor primer.

ancovenin An inhibitor of angiotensin I converting enzyme isolated from the culture broth of a Streptomyces species; a 16aa **lantibiotic** containing unusual amino acids such as threo-beta-methyllanthionine, meso-lanthionine, and dehydroalanine. *Streptoverticillium cinnamoneum* produces a lantibiotic with similar properties, lanthiopeptin.

AND-34 See **BCAR-3**.

Andersen's syndrome An ion channel disorder (**channelopathy**), that has been linked to muscle abnormalities and developmental defects. The mutation affects an inwardly-rectifying potassium channel (Kir2.1, encoded by KCNJ2). N.B. Andersen's disease is **glycogen storage disease** Type IV.

Anderson's disease A disorder of severe fat malabsorption causing failure to thrive in infancy. Like chylomicron retention disease, it is caused by mutation in the gene for **sar1B**.

Androctonus mauretanicus mauretanicus Moroccan scorpion. See **kaliotoxin**.

androecious Describing a plant with only male sex organs.

androgen General term for any male sex hormone in vertebrates. Androgen insensitivity syndrome is caused by mutation in the androgen receptor. See **testicular feminization** and **Reifenstein syndrome**.

androgenesis (1) Development from a male cell. (2) Development of an egg after entry of male germ cell but without the participation of the nucleus of the egg.

androstenedione Precursor of testosterone and estrone produced in the testis or ovary from 17α-hydroxyprogesterone, or from dehydroepiandrosterone. Also secreted into the circualtion by the adrenal glands.

androsterone A steroid hormone that has weak androgenic activity.

anemia See **anaemia**.

anemone toxins Polypeptide toxins (mostly 30−45aa) from sea anemones (anthozoan coelenterates), most of which act on voltage-gated sodium channels (e.g. neurotoxin-1, 47aa, from *Anthopleura fuscoviridis*). Some, however, block voltage-regulated potassium channels (e.g. Potassium channel toxin ShK 35aa from *Stoichactis helianthus*).

anergy *Adj* anergic. Generally, a lack of energy. In Immunology, failure of lymphocytes that have been primed to respond to second exposure to the antigen. Consequence is a depression or lack of normal immunological function.

aneugenic Agents that induce changes in chromosome numbers, aneuploidy or polyploidy, rather than mutation. A range of chemicals and treatments (e.g. X irradiation) have been shown to be aneugenic although there are no universally accepted standard test methods as yet.

aneuploid Having a chromosome complement that is not an exact multiple of the haploid number. Chromosomes may be present in multiple copies (e.g. trisomy) or one of a homologous pair may be missing in a diploid cell.

ANF See **atrial natriuretic peptide**.

Angelman's syndrome Syndrome in which there is severe mental retardation and ataxic movement associated with absence of maternal 15q11q13, and the absence of the β3 subunit of **GABA receptor-A** or in some cases with defects in **genomic imprinting**. Absence of the paternally-derived region leads to clinically disting **Prader-Willi syndrome**.

angio-associated migratory cell protein A protein (AAMP, 434aa) that has a role in angiogenesis and cell migration. May act through the rhoA pathway.

angioedema A condition (formerly angioneurotic edema) in which large edomatous welts develop in the dermis and subcutaneous tissue. Can be caused in some cases by mutation in genes encoding complement C1 inhibitor or coagulation factor XII.

angiogenesis The process of **vascularization** of a tissue involving the development of new capillary blood vessels.

angiogenin Polypeptide (ribonuclease 5, EC 3.1.27.-, 147aa) that induces the proliferation of endothelial cells; one of the components of **tumour angiogenesis factor**. It has ribonucleolytic activity, although the biological relevance of this is unclear. It has also been suggested that angiogenin binds to an actin-like molecule present on the surface of endothelial cells.

angiokeratoma See **Fabry's disease**.

angioma A knot of distended blood vessels atypically and irregularly arranged. Most are not tumours but haematomas.

angiomatoid fibrous histiocytoma See **fibrous histiocytoma**.

angiomotin A protein (1094aa) that is important for maintainence of tight junctions (see **nadrin**), stimulates motility in endothelial cells and is found in tissues where angiogenesis is occurring. May be antagonised by **angiostatin**.

angiomyolipoma Rare, slow-growing benign lesions of kidney, composed of varying amounts of blood vessels, smooth muscle, and fat. A few cases are associated with **tuberous sclerosis**.

angioneurotic oedema See **angioedema**.

angiopathy Any disease of blood vessels. Microangiopathy affects small blood vessels and is, for example, a complication of diabetes, particularly in the vasculature of the retina. Macroangiopathy affects large blood vessels leading to coronary artery disease, cerebrovascular disease, and peripheral vascular disease. Hereditary angiopathy with nephropathy, aneurysms and muscle cramps (HANAC) is

caused by missense mutations in the collagen 4A1 gene.

angiopoietin Angiopoietin-1 (498aa) is the ligand for **Tie2**; angiopoietin-2 (496aa) is a natural antagonist. Angiopoietin-1, but not Ang-2, is chemotactic for endothelial cells: neither have effects on proliferation. Angiopoietin-3 (angiopoietin-related protein 1, 491aa) is one of a number of angiopoietin-related proteins, that have different tissue distributions.

angiopoietins Growth factors of the VEGF family that have their effect on blood vessels. Angiopoietin-1 (498aa) is the ligand for the endothelium-specific receptor tyrosine kinase **Tie2**; angiopoietin-2 (496aa) is a natural antagonist. Angiopoietin-3 (angiopoietin-related protein 1, 491aa) is secreted by various tissues. Angiopoietin-related protein-4 (406aa) is induced in endothelial cells by hypoxia and may have a protective function. It inhibits proliferation, migration, and tubule formation by endothelial cells and reduces vascular leakage and is involved in regulating glucose homeostasis, lipid metabolism, and insulin sensitivity. Various angiopoietin-like proteins are also known with miscellaneous tissue-specific functions. Not all are mitogenic for endothelial cells – for example angiopoietin-like seven (angiopoietin-like factor thee, cornea-derived transcript six protein, 346aa) reduces tumour growth and aberrant blood vessel formation in a mouse xenograft model and is a target gene of the WNT/-catenin signalling pathway.

Angiosperm Phylogeny Website A guide to the classification of angiosperms. Version 11, May 2011. The site has a useful glossary for the diverse terms used in plant systematics. Website: http://www.mobot.org/MOBOT/research/APweb/.

angiosperms The largest and most diverse group of land plants, differing from the gymnosperms in having flowers that produce seed. A defining characteristic of angiosperms is that the seed are enclosed within an ovule, in contrast to the naked-seeded gymnosperms.

angiostatin Potent angiogenesis inhibitor, a proteolytic fragment of plasminogen containing the first three or four kringle domains (K1-4). Mode of action unclear but it reportedly binds to **angiomotin**.

angiotensin A peptide hormone derived from angiotensinogen (485aa) that is released from the liver and cleaved in the circulation by **renin** to form the biologically inactive decapeptide angiotensin-1. This is in turn cleaved to form active angiotensin-2 (angiotensin 1-8) by angiotensin converting enzyme (ACE). Angiotensin-2 causes contraction of vascular smooth muscle, and thus raises blood pressure, and stimulates **aldosterone** release from the adrenal glands. Angiotensin-3 stimulates aldosterone release. Receptors for angiotensin are G-protein coupled: Type 1 (359aa) mediates the cardiovascular effects which may be counteracted by signalling through Type 2 (363aa). Angiotensin is finally broken down by angiotensinases such as prolylcarboxypeptidase (EC 3.4.16.2, 496aa).

angiotensin converting enzyme *ACE* Enzymes that will convert **angiotensin-1** into the active angiotensin-2: ACE1 (EC 3.4.15.1, 1306aa) is a dipeptidyl carboxypeptidase that removes the terminal His-Leu from angiotensin-1. ACE2 (EC 3.4.17.23, 805aa) will generate a nonapeptide (angiotensin 1–9) of unknown function and a heptapeptide (angiotensin 1-7, angiotensin-3) that has vasodilatory actions and is a ligand for the G-protein coupled receptor **mas1**. Genetic variations in ACE may be a cause of susceptibility to ischemic stroke and predisposition to microvascular complications of diabetes; mutations can cause **renal tubular dysgenesis**.

angiotensin II receptor antagonists Group of drugs that work by blocking binding of **angiotensin** II to its receptor and thus have effects similar to those of **angiotensin-converting enzyme inhibitors**; used to treat hypertension. Examples are candesartan, irbesartan, losartan and valsartan.

angiotensin-converting enzyme inhibitors *ACE inhibitors* Drugs that inhibit the enzymatic conversion of inactive **angiotensin**-1 to the active form (angiotensin-2); used in the treatment of hypertension and heart failure. Captopril and Enalapril are common examples.

angiotensinase See **angiotensin**.

angiotensinogen See **angiotensin**.

Angstrom unit Small unit of measurement (10^{-10} m) named after Swedish physicist and astronomer. Much used as a unit in early electron microscopy though since it is not in the approved mks system should probably be avoided (but nanometres, which are 10 times larger are sometimes less convenient).

angucycline antibiotics A large group of antibiotics (aromatic polyketides) with anti-tumour activity, particularly agains doxorubicin-resistant tumours. Examples include landomycins, moromycins, **pradimicin**.

anguilliform Eel-like in shape.

anhidrosis A disorder in which there is an absence of sweating. Can be caused by mutations in genes for NFκBIA or **IkappaB** kinase-gamma. Congenital insensitivity to pain with anhidrosis (CIPA) is caused by mutation in the neurotrophic tyrosine kinase-1 **NTRK1** gene.

anhydrobiosis Life without water, a form of cryptobiosis that occurs in situations of extreme desiccation. Some invertebrates (e.g. brine shrimps) and

some plants (e.g. the resurrection plant, *Craterostigma plantagineum*) show extreme ability to withstand desiccation.

Aniline Blue A triphenylmethane dye used for staining the polysaccharide callose, found in immature cross-walls and in sites modified by stress. Also stains the polysaccharide curdlan found on the surface of some Rhizobium species.

anillin An actin binding protein (1201aa) first identified in *Drosophila*, that is concentrated in the cleavage furrow in a pattern that resembles that of rhoA. Anillin has a conserved C-terminal domain that shares homology with **rhotekin**. May function as a scaffold protein to link rhoA with actin and myosin in the contractile ring.

animal pole The region of an animal oocyte that contains the nucleus (which is not centrally placed). The opposite pole is the **vegetal pole**, with the animal-vegetal axis between the poles passing through the nucleus. During **meiosis** of the **oocyte** the polar bodies are expelled at the animal pole. In many eggs there is also a graded distribution of substances along this axis, with pigment granules often concentrated in the animal half and yolk, where present, largely in the vegetal half.

animalised cells Cells of the 8–16 cell early blastula of sea urchins that have been shifted from vegetal to animal in their characteristics by manipulating the environmental conditions. See **animal pole** and **vegetal pole**.

anion exchanger Family of integral membrane proteins that perform the exchange of chloride and bicarbonate across the plasma membrane. Best known is **band III** of the red blood cell.

anionic detergents Detergents in which the hydrophilic funtion is fulfilled by an anionic grouping. **Fatty acids** are the best known natural products in this class, but it is doubtful if they have a specific detergent function in any biological system. The important synthetic species are aliphatic sulphate esters, e.g. sodium dodecyl sulphate (SDS or SLS).

aniridia Rare congenital absence or partial absence of the iris and other tissues in the eye; caused either by an autosomal dominant mutation in PAX6 (oculorhombin), an identifiable chromosome deletion of the short arm of chromosome 11, including band p13, or in sporadic cases by mutations in WT (Wilms tumour gene) and AN2 (aniridia 2 gene) or only in AN2.

anisogamy Mode of sexual reproduction in which the two gametes are of different sizes.

anisotropic Not the same in all directions.

ANK repeat Motif found in diverse proteins including **ankyrin** (hence the name), the **Notch** product, transcriptional regulators, cell cycle regulatory proteins and a toxin produced by the black widow spider. The motif (~33aa) is generally found as a tandem array of 2–7 repeats, though ankyrins contain 24 repeats. Their role is not established, but they may be involved in protein-protein binding.

ankylosing spondylitis Polyarthritis involving the spine (spondyloarthropathy, Marie-Strumpell spondylitis, Bechterew's syndrome), which may become more-or-less rigid. The disease seems to be associated with HLA-B27; those with this **histocompatibility antigen** are 300 times more likely to get the disease, 90% of sufferers have HLA-B27.

ankylosis Fusion of bones across a joint, a complication of **chronic inflammation**. See **ankylosing spondylitis**.

ankylostomiasis Infection of the small intestine by parasitic nematode (hookworms; *Ankylostoma duodenale* or *A. americanum*); can cause iron deficiency for those on an inadequate diet.

ankyrin Globular protein (ankyrin 1, 1881aa) with multiple **ANK repeats** that links **spectrin** and **Band III**) in the erythrocyte plasma membrane. Isoforms (ank-2, ankyrin-B, 3924aa; ankyrin G, ank-3, 4372aa) exist in other cell types.

anlage Region of the embryo from which a specific organ develops.

annealing (1) Toughening upon slow cooling. (2) Used in the context of DNA renaturation after temperature dissociation of the two strands. Rate of annealing is a function of complementarity. (3) Fusion of microtubules or microfilaments end-to-end.

Annelida Phylum of segmented (metameric) coelomate worms. Common earthworm (*Lumbricus terrestris*) is a familiar example of the phylum.

annellide A specialized conidiogenous cell that produces conidia in basipetal succession (i.e. with the oldest at the tip, the newest at the base); the tip of an annellide increases in length and becomes narrower as each subsequent conidium is formed.

annexins Group of calcium-binding proteins that interact with acidic membrane phospholipids. There are 12 mammalian annexin genes, now classified as ANXA 1-13 with A12 not assigned. Non-vertebrate annexins are classified as B (invertebrates), C (fungi and some unicellular organisms), D (plants) and E (protists). All the mammalian annexins have four conserved repeats of a 61aa domain that folds into five α-helices, except ANXA6, which has eight. Also known by several other names (e.g. lipocortins, endonexins), reflecting the history of their discovery in different contexts. See Table A2, **lipocortin**, **endonexin** I and II, **calpactin**, p70, and **calelectrin**.

annular vessels Xylem elements that are prevented from collapse by rings (annuli) of thickened cell wall.

annulate lamellae Cytoplasmic stacks of of narrow membrane cisternae with nuclear pore complexes found in the fertilized eggs of several mammalian species and unfertilized oocytes of others. Also a feature of virally-infected cells where they may be sites of viral coat-protein glycosylation and storage. Description: http://jcs.biologists.org/content/111/19/2841.full.pdf and http://www.ncbi.nlm.nih.gov/pmc/articles/PMC110484/

anoikis Apoptosis in normal epithelial and endothelial cells, important in the regulation of cell number in skin.

anomers The α- and β-forms of hexoses. Interconversion (**mutarotation**) is anomerisation and is promoted by mutarotases (aldose epimerases).

Anopheles Genus of mosquitos (order **Diptera**) which carry the *Plasmodium* parasites which causes malaria.

anorectic Describing a substance that suppresses hunger, hence anorexia, the lack of appetite.

anorexia nervosa A severe disturbance in eating behaviour, typically exhibited in late adolescence and early adulthood. Several loci have been identified that increase susceptibility, including polymorphisms in **brain derived neurotrophic factor**.

anosmia Condition of being unable to smell. Can be transiently induced by osmic acid.

anosmin A glycosylated peripheral membrane protein (adhesion molecule-like X-linked protein, Kallmann syndrome protein, 680aa) that has branch-promoting and guidance activity for developing neurons in the olfactory cortex and for gonadotropin-releasing hormone-secreting neurons in the hypothalamus. Defects cause **Kallmann's syndrome**.

ANOVA Statistical technique (analysis of variance) used to determine the source of variability in a set of data, separating variation due to specific cause and to random variation.

anoxia Total lack of oxygen, *cf.* **hypoxia**.

ANP See **atrial natriuretic peptide**.

ANP receptor Family of three receptors for **atrial natriuretic peptide**. ANP-A (EC 4.6.1.2, 1061aa) and ANP-B (1047aa) have intracellular **guanylate cyclase** activity and **protein kinase**-like domains. ANP-C (541aa) shares the extracellular ligand-binding and transmembrane domains, but lacks the functional intracellular domains, and is not thought to be involved in **signal transduction** but may be a clearance receptor.

TABLE A2. Vertebrate annexins

Name	Size	Synonyms	Expression
Annexin 1	346aa	Lipocortin 1, Calpactin 2, p35, Chromobindin 9	Ubiquitous
Annexin 2	339aa	Lipocortin 2, Calpactin 1, Protein I, p36, Chromobindin 8	Ubiquitous
Annexin 3	323aa	Lipocortin 3, Placental anticoagulant protein III, PAP-III, 35-alpha calcimedin	Neutrophils
Annexin 4	319aa	Lipocortin 4, Endonexin I, Protein II, Chromobindin 4, PAP-II, Carbohydrate-binding protein p33/p41	Ubiquitous
Annexin 5	320aa	Lipocortin 5, Endonexin 2, Vascular anticoagulant-alpha, VAC-a, Anchorin CII, PAP-I, Thromboplastin inhibitor	Ubiquitous
Annexin 6	673aa	Lipocortin 6, Protein III, Chromobindin 20, p68, p70, 67 kDa calelectrin, Calphobindin-II	Ubiquitous
Annexin 7	488aa	Synexin	Ubiquitous
Annexin 8	327aa	Vascular anticoagulant-b, VAC-b	Placenta & skin
Annexin 9	345aa	Annexin-31, Pemphaxin	Stratified squamous epithelium
Annexin 10	324aa	Annexin-14	Stomach
Annexin 11	505aa	56 kDa autoantigen, calcyclin-associated annexin 50	Ubiquitous
Annexin 13	316aa	Intestine-specific annexin	Small intestine

There are 12 mammalian annexin genes, now classified as ANXA 1-13 with A12 not assigned. Size is given for the human form.

ansamycins A family of antibiotics active against Gram-positive and some Gram-negative bacteria and with some antiviral activity. Examples include the **streptovaricins**, **geldanamycin** and **rifamycins**.

ANT See **adenine nucleotide translocator**.

antagonist Compound that inhibits the effect of a hormone or drug; the opposite of an **agonist**.

antennal complex Light-harvesting complexes (LHC) of protein and pigment, in or on photosynthetic membranes of bacteria, that are organized into arrays called antennae. They transfer photon energy to reaction centres.

antennapedia In *Drosophila* the product (378aa) of a homoeotic gene, a transcription factor that regulates segmental identity in the mesothorax and provides positional information on the anterior-posterior axis. Overexpression can cause antennas to be transformed into legs, hence the name.

anterograde transport Movement of material from the cell body of a **neuron** into axons and dendrites (retrograde axoplasmic transport also occurs).

antheridium A haploid structure or organ that produces the male gametes (antherozoids) of lower plants like mosses and ferns, also some algae and fungi. In many gymnosperms and all angiosperms, the antheridia have been reduced to a single generative cell within the pollen grain. *Cf.* **archegonium**.

Anthocerotophyta Hornworts, a group of bryophytes so-called because the sporophyte develops as a horn-like protrusion on the upper surface of the haploid gametophyte thallus.

anthocyanidin Molecule (an aglycone) that when complexed with a sugar moiety forms an **anthocyanin**.

anthocyanidin reductase An oxidoreductase of the flavonoid pathway (ANR, EC 1.3.1.77, leucoanthocyanidin reductase, 340aa) encoded by the BANYULS gene, involved in the biosynthesis of condensed tannins (flavonoid oligomers). Abstract: http://www.ncbi.nlm.nih.gov/pubmed/14725861

anthocyanins Flavonoid pigments stored in plant vacuoles and not directly involved in **photosynthesis**. Can mask the green of **chlorophyll**, and give the plant, especially their fruit and flowers, a red-purple colour. Used commercially as food colourants. Anthocyanins are powerful antioxidants, studied for their potential health-giving properties.

anthopleurins Peptide toxins (Anthopleurin A, B and C: 49aa, 49aa, 47aa) from the nematocysts of the sea anemone, *Anthopleura*. Affect sodium channel of nerve and muscle and increase the duration of the action potential.

anthozoa *actinozoa* Class of Cnidaria in which alternation of generations does not occur and the medusoid phase is entirely suppressed. Sea anemones, corals, sea pens. http://www.anthozoa.com/

anthracyclines Antibiotics derived from *Streptomyces peucetius*, used in cancer chemotherapy. An example is **daunorubicin**.

anthrax Highly contagious disease of man and domestic animals caused by *Bacillus anthracis*. Onset is rapid and the disease is often fatal. A variety of **anthrax toxins** are known.

anthrax toxins Two AB toxins produced by *Bacillus anthracis* that share a common binding (B) subunit, protective antigen (PA, 746aa) which is proteolytically cleaved to form PA63 which forms heptamers that bind either edema factor (EF, 800aa) or lethal factor (LF, 809aa) before being internalised by the target cell. Edema factor is a calmodulin-dependent adenylyl cyclase, lethal factor is an endopeptidase that acts on macrophages to induce a massive oxidative burst, the release of IL-1β and TNFα and cytolysis.

anti-idiotype antibody An antibody directed against the antigen-specific part of the sequence of an antibody or T-cell receptor. In principle an anti-idiotype antibody should inhibit a specific immune response.

anti-inflammatory drugs Drugs that inhibit the inflammatory response. There are two major classes, the **non-steroidal anti-inflammatory drugs** (NSAIDs) and the glucocorticoids.

anti-lymphocyte serum Immunoglobulins raised **xenogeneically** against lymphocyte populations. Referring particularly to antisera recognizing one or more **antigenic determinants** on T-cells. Of use in experimental **immunosuppression**.

anti-Müllerian hormone A dimeric glycoprotein hormone (AMH, Müllerian inhibiting substance, 560aa) belonging to the TGFb superfamily. Synthesized as a large precursor with a short signal sequence followed by the prepro hormone that forms homodimers. Prior to secretion, the mature hormone undergoes glycosylation and dimerization. High levels of AMH are produced by Sertoli cells during fetal and postnatal testicular development and suppress the development of Müllerian ducts that differentiate into the uterus and fallopian tubes in females. In the human female, AMH is produced by ovarian granulosa cells from 36 week of gestation to the menopause. The receptors are serine/threonine kinases.

anti-muscarinic drugs A subgroup of anticholinergic drugs that block the action of acetylcholine at **muscarinic** acetylcholine receptors. Effect is generally to relax smooth muscle of gut (antispasmodic) or airways (bronchodilatory). Example is **tolterodine**.

anti-oncogene See **tumour suppressor**.

anti-thymocyte globulin A polyclonal IgG fraction that selectively targets and destroys T-cells; used as an immunosuppressive agent.

antibiotic Originally, a substance produced by one microorganism that selectively inhibits the growth of another although now generally applied to any substance (often completely synthetic) that has an antibacterial or antifungal effect. There are separate entries for various classes of antibiotics that describe the mode of action and lead to specific examples: see **aminocoumarin, aminoglycoside, angucycline, aurovertins, beta-lactam, cephalosporins, fluoroquinolones, glutarimide**, glycopeptide antibiotics (see **vancomycin, macrolide, oxazolidinones, polyene, polypeptide, quinolone** and **tetracycline antibiotics**, also **lantibiotics**.

antibiotic resistance gene Gene that encodes an enzyme that degrades or excretes an antibiotic, thereby conferring resistance. Frequently found in cloning vectors like plasmids, and sometimes in natural populations of bacteria. Example: bacterial ampicillin resistance is conferred by expression of the **beta-lactamase** gene.

antibody General term for an **immunoglobulin**.

antibody dependent cell-mediated cytotoxicity *ADCC* Killing of target cells by lymphocytes or other leucocytes that carry antibody specific for the target cell, attached to their **Fc receptors**. The cell involved in the killing may be a passive carrier of the antibody.

antibody-directed drug therapy The targeting of a drug to particular cells by coupling it to an antibody directed against an appropriate cell surface marker.

antibody-induced lysis An imprecise term that should be avoided since the mechanism involved could be complement-based or due to natural killer cells. See **complement** and **natural killer cells**.

antibody-producing cell A **lymphocyte** of the B series synthesizing and releasing **immunoglobulin**. Equivalent to plasmacyte and **plasma cell**.

anticlinal Perpendicular to the nearest surface. Used particularly in botanical anatomy to describe the orientation of cell walls with respect to the surface of the plant or organ. If a cell divides anticlinally the daughter cells will be separated by an anticlinal wall and will both lie in the same plane with respect to the surface. *Cf.* **periclinal**.

anticoagulant Substance that inhibits the clotting of blood. The most commonly used are EDTA and citrate (both of which work by chelating calcium) and **heparin** (that interferes with **thrombin**, probably by potentiating **antithrombins**). Other compounds such as **warfarin** and dicoumarol act as anticoagulants *in vivo* by interfering with clotting factors.

anticodon Nucleotide triplet on **transfer RNA** that is complementary to the **codon** of the **messenger RNA**.

anticonvulsants Drugs used to prevent or reduce the severity and frequency of seizures that occur when electrical activity in the brain that controls motor systems becomes chaotic and paroxysmal. Commonest form of seizure is epilepsy but not all seizures cause convulsions, and not all convulsions are due to epileptic seizures. Commonly used drugs are carbamazepine, **phenytoin**, **valproate**, and **diazepam**.

antidepressants Drugs that relieve the symptoms of moderate to severe depression. Main classes are **tricyclic antidepressants**, **monoamine oxidase inhibitors**, SSRIs, and **lithium**.

antidiuretic Inhibiting the formation of urine; an antidiuretic drug. See **vasopressin** (antidiuretic hormone).

antidromic Running in the opposite direction: most common usage is in neurophysiology for the passage of an action potential in the opposite direction to that in which it would normally travel i.e. from the presynaptic region towards the cell body.

antiemetic drugs Drugs that stop vomiting and, to a lesser extent, nausea. They are used for motion sickness and for the side-effects of chemotherapy and some gastrointestinal disorders. Examples include **hyoscine**, **antihistamines**, **phenothiazines**, metoclopramide, and ondantseron.

antigen A substance that induces an **immune response**. Normally antigens have molecular weights greater than about 1 kDa. The **antigenic determinant** group is termed an **epitope** and the association of this with a carrier molecule (that may be part of the same molecule) makes it active as an antigen. Thus dinitrophenol-modified human serum albumin is antigenic to humans, dinitrophenol being the hapten. Usually antigens are foreign to the animal in which they produce immune reactions.

antigen presentation See **antigen presenting cell**.

antigen presenting cell A cell that carries on its surface antigen bound to MHC Class I or Class II molecules, and presents the antigen in this 'context' to T-cells. Antigen presenting cells (APCs) include macrophages, endothelium, **dendritic cells** and **Langerhans cells** of the skin. See also **MHC restriction**, **histocompatibility antigens**.

antigen processing Modification of an antigen by **accessory cells**. This usually involves endocytosis of the antigen and either minimal cleavage or unfolding. The processed antigen is then presented in modified form by the accessory cell.

antigen shift Abrupt change in surface antigens expressed by a species or variety of organisms. Usually seen in microorganisms where the change may allow escape from immune recognition. Antigenic drift is a more gradual change. See **antigenic variation**.

antigen-antibody complex The product of the reaction of **antigen** and **immunoglobulin**, an immune complex. If the antigen is polyvalent the complex may be insoluble; see also **glomerulonephritis**, **Arthus reaction**, Type III **hypersensitivity**. Immune complexes activate **complement** through the classical pathway.

antigenic determinant *epitope* That part of an antigenic molecule against which a particular immune response is directed. For instance a tetra- to pentapeptide sequence in a protein, a tri- to penta-glycoside sequence in a polysaccharide. See also **hapten**. In the animal most **antigens** will present several or even many antigenic determinants simultaneously.

antigenic variation The phenomenon of changes in surface **antigens** in parasitic populations of *Trypanosoma* and *Plasmodium* (and some other parasitic protozoa) in order to escape immunological defence mechanisms. At least 100 different surface proteins have been found to appear and disappear during antigenic variation in a clone of trypanosomes. Each antigen is encoded in a separate gene. Antigenic variation is also known to occur in free-living Protozoa and certain bacteria.

antihistamine A substance or drug which inhibits the actions of **histamine** by blocking its site of action. See **histamine receptors**.

antimere *Adj.* antimeric. A part on a bilaterally or radically symmetrical organism corresponding to a similar structure on the other side.

antimetabolite Drugs used in treatment of cancer which are incorporated into new nuclear material and prevent normal cell division. Common examples are **methotrexate**, cytoarabinose and **fluorouracil**.

antimicrobial peptides Peptides (AMPs) found in the haemolymph of insects. In *Drosophila* there are eight classes of AMPs that can be grouped into three families based on their main biological targets, Gram-positive bacteria (**defensin**), Gram-negative bacteria (**cecropins, drosocin, attacins, diptericin, MPAC**), or fungi (**drosomycin, metchnikowin**). *Drosophila* AMPs are synthesized by the fat body in response to infection, and secreted into the haemolymph.

antimitotic drugs Drugs that block mitosis; the term is often used of those which cause metaphase-arrest such as **colchicine** and the **vinca alkaloids**. Many anti-tumour drugs are antimitotic, blocking proliferation rather than being cytotoxic.

antimorph A **dominant negative** mutant expressing some agent that antagonizes a normal gene product.

antimuscarinic Term for a drug that acts on (inhibits) **muscarinic acetylcholine receptors**.

antimycin An antibiotic produced by Streptomyces that inhibits **QH2-cytochrome C-reductase**.

antioxidant Any substance that inhibits oxidation – usually because it is preferentially oxidised itself. Common examples are vitamin E (α-tocopherol) and vitamin C. Important for trapping free radicals generated during the **metabolic burst** and possibly for inhibiting ageing.

antipain Protease inhibitor that inhibits **papain, trypsin** and, to a lesser extent, **plasmin**. More specific for papain and trypsin than **leupeptin**.

antiparallel Having the opposite **polarity** (e.g. the two strands of a DNA molecule).

antiphospholipid syndrome An autoimmune disorder (familial lupus anticoagulant) in which there are antibodies directed against cellular phospholipid components. There may be up to 30 different auto-antibodies, including those against platelets, glycoproteins, coagulation factors, lamins, mitochondrial antigens, and cell surface markers. See **apolipoprotein H**.

antiplasmin Plasma protein (α-2-antiplasmin, serpin F2, 491aa) that inhibits plasmin (and Factors XIa, XIIa, **plasma kallikrein, thrombin** and **trypsin**) and therefore acts to regulate **fibrinolysis**.

antiplectic Pattern of **metachronal** coordination of the beating of **cilia**, in which the waves pass in the opposite direction to that of the active stroke.

antipodal cells Three cells of the **embryo sac** in angiosperms, found at the end of the embryo, away from the point of entry of the pollen tube.

antiport Transport of two different ions or molecules, in opposite directions, across a lipid bilayer. Energy may be required, as in the sodium pump; or it may not, as in Na^+/H^+ antiport.

antiproteases Substances (antiproteinases, antipeptidases) that inhibit proteolytic enzymes.

antipsychotic drugs A class of drugs (neuroleptic drugs, major tranquillizers) used to treat serious mental disorders. Most reduce dopamine levels in the CNS (e.g. **chlorpromazine**), but the atypical antipsychotics (e.g. **risperidone**) act on **serotonin**-based systems.

Antirrhinum majus A flowering plant (the common snapdragon), widely used as a model system for plant molecular genetics.

antisense In general the complementary strand of a coding sequence of DNA or of mRNA. Antisense RNA hybridizes with and inactivates mRNA.

antisepsis Processes, procedures or chemical treatments that kill or inhibit microorganisms in contrast to **asepsis** where microorganisms are excluded.

antiserum Serum containing **immunoglobulins** against specified **antigens**.

antispasmodic drugs Drugs that relax smooth muscle of the gut wall and relieve symptoms of indigestion, irritable bowel syndrome, and diverticular disease. Most are **antimuscarinic drugs**.

antitermination During transcription, failure of an **RNA polymerase** to recognize a termination signal: can be of significance in regulation of gene expression.

antithrombin A plasma glycoprotein (antithrombin III, serpinC1, 464aa) that inhibits serine peptidases involved in blood coagulation (thrombin and factors IXa, Xa and XIa). Its inhibitory activity is enhanced by **heparin**. Antithrombins I, II & IV are obsolete names used in early studies on coagulation.

antitoxin An **antibody** reacting with a toxin, e.g. anti-**cholera toxin** antibody.

antitrypsin A plasma inhibitor (α-1-antitrypsin, α-1-antiprotease, serpinA1, 418aa) of serine peptidases, particularly elastase, but it also has a moderate affinity for plasmin and thrombin.

antiviral drugs Drugs that inhibit virus infection. Two main categories have been developed so far;

those which inhibit or interfere with the replication of viral nucleic acid (nucleoside analogues such as **aciclovir** and **AZT**) and those that interfere with virus-specific enzymes such as proteases (e.g. saquinavir) or neuraminidases (e.g. Relenza, Tamiflu) that are important for processing of viral proteins to produce infective particles.

antizyme Repressor of **ornithine decarboxylase** (ODC). Antizyme (228aa) is a polyamine-inducible protein involved in feedback regulation of cellular polyamine levels. The N terminus of antizyme is not required for the interaction with ODC but is necessary to induce its degradation. Antizyme can be induced by IL-1 and it is inhibited by antizyme inhibitor 1 (448aa). The elaborate regulation of ODC activity in mammals still lacks a defined developmental role but an antizyme-like gene in *Drosophila*, gutfeeling (*guf*), is required for proper development of the embryonic peripheral nervous system.

antolefinin The human homologue of lin-37, a component (246aa) of the **DREAM complex**.

Antp See **antennapedia**.

antral Relating to an **antrum**.

antral mucosa Mucosa (pyloric mucosa) found in the gastric **antrum** which has coiled and branching antral glands that are lined by mucus cells interspersed with endocrine cells (chiefly G and D types), and a few parietal (**oxyntic**) cells.

antrum (1) A cavity or chamber, especially in bone. (2) The lower third of the stomach (pyloric antrum) that lies between the body of the stomach and the pyloric canal (3) In the ovary, the fluid-filled space within the follicle.

anucleate Having no nucleus.

anucleolate Literally, having no nucleoli. An anucleolate mutant of *Xenopus* (viable when **heterozygous**) is used in nuclear transplantation experiments because nuclei are of identifiable origin.

Anura Class of amphibians; the frogs and toads.

anxiolytic Drug that reduces anxiety, for example **benzodiazepines** and **barbiturates**.

AOAC The Association of Official Agricultural Chemists, a non-profit US scientific association that publishes standardised, chemical analysis methods. For example, AOAC method 991.43 is the official UK method for estimating dietary fibre in foodstuffs.

AP sites Sites in DNA (apurinic/apyrimidinic site) that have neither a purine nor a pyrimidine base, usually due to DNA damage but also as intermediates in base excision repair.

AP1 (1) A transcription factor, formed from a heterodimer of the products of the **proto-oncogenes** *fos* and *jun*. Binds the palindromic DNA sequence TGACTCA. (2) Adaptor protein found in the trans-Golgi network that links membrane proteins to clathrin (see **AP2**).

AP2 (1) Cis-acting transcription activator. Mutations in AP2-alpha (436aa) cause **branchiooculofacial syndrome**; AP2-beta (460aa) is mutated in **Char's syndrome**. (2) One of the multimeric adaptor proteins (APs; ca. 270 kDa) found in **clathrin**-associated complexes. AP2 is found at the plasma membrane and may bind preferentially to the cytoplasmic tail of the EGF receptor. Also associates with the EGF-R tyrosine kinase substrate eps15.

AP2/ERF superfamily A large gene family encoding transcription factors that have an AP2/ERF domain (60~70aa) that binds specifically to the 5′-GCCGCC-3′ motif. ERF is ethylene-responsive element-binding protein; ERF1A in *Arabidopsis* (268aa) is involved in the regulation of gene expression by stress factors and by components of stress signal transduction pathways.

AP3, AP4, AP5 Selective antagonists for **NMDA receptors**.

APA Amino pimelic acid, a low affinity rapidly dissociating competitive antagonist of **NMDA receptors**.

APAF-1 Protein (apoptosis protease activating factor 1, 1248aa) that binds to cytochrome c that has been released from mitochondria and links with **caspase**-9 which then activates caspase-3, initiating a cascade of events that end in apoptotic death of the cell. APAF1-interacting protein (242aa) inhibits APAF-1 and is anti-apoptotic.

apamin A small basic peptide (18aa) present in the venom of the honey bee (*Apis mellifera*). Blocks calcium-activated potassium channels and has an inhibitory action in the central nervous sytem.

APC (1) **Antigen presenting cell**. (2) **Adenomatous polyposis coli**. (3) **Allophycocyanin**.

ape-1 An excision repair endonuclease (apurinic-apyrimidinic endonuclease 1, APEX nuclease, EC. 4.2.99.18, 318aa) important in the cellular response to oxidative stress. It initiates repair of apurinic or apyrimidinic sites in DNA by catalyzing hydrolytic incision of the phosphodiester backbone immediately adjacent to the damage. See **SET complex**.

apelin A bioactive peptide that is the endogenous ligand for the G-protein coupled receptor **APJ**. The peptide is produced from a pre-proprotein of 77aa and exists in multiple molecular forms (ranging from 12aa to 36aa). Apelin partially suppresses cytokine production from mouse spleen and is involved in the regulation of blood pressure and blood flow.

Apert syndrome A developmental disorder (Apert-Crouzon disease) caused by mutation in the gene encoding **fibroblast growth factor** receptor-2.

APETALA A family of transcription factors (floral homeotic proteins) involved in control of the transition from vegetative growth to flower production in *Arabidopsis*. APETALA1 (AP1, agamous-like MADS-box protein AGL7, 256aa) promotes early floral meristem identity in synergy with LEAFY and is necessary for petal and sepal development. APETALA2 (AP2, 432aa) is a transcriptional activator involved in sepal and petal development, represses **AGAMOUS** and is important in seed development. APETALA3 (AP3, 232aa) forms a heterodimer with **PISTILLATA** (PI) that autoregulates AP1, AP3 and PI genes. See **ABC model of flower development**.

APH-1 A component (anterior pharynx defective, 265aa), along with **nicastrin** and **PEN-2** (presenilin enhancer) of the **presenilin**-dependent **gamma-secretase** complex. APH-1 apparently stabilizes the complex. Aph-1 is present at the cell surface, presumably in active gamma-secretase complexes, and interacts with the Notch receptor.

aphakia The absence of the lens of the eye. Can be caused by a homozygous nonsense mutation in the **forkhead** transcription factor, FOXE3.

aphid transmission factor A protein (159aa) encoded by a Cauliflower mosaic virus gene that is involved in transmission. Similar proteins are associated with other plant viruses.

aphidicolin Reversible inhibitor of eukaryotic **DNA polymerases**, a tetracyclic diterpenoid from Cephalosporium.

apical bud Bud at the tip of a growing plant; usually dominant, repressing the development of lower buds by the production of **auxin**.

apical cell In plants, the upper (chalazal) cell formed after the first division of the zygote, which will give rise to the majority of the embryo. *Cf.* **basal cell**.

apical dominance Growth-inhibiting effect exerted by actively-growing **apical bud** of higher-plant shoots, preventing the growth of buds further down the shoot. Thought to be mediated by the basipetal movement of **auxin** from the apical bud.

apical ectodermal ridge Ridge of tissue at the developing limb-bud of the vertebrate embryo, a transient structure critical to maintaining limb outgrowth.

apical meristem *eumeristem* The **meristem** at the tips of stems and roots. Composed of undifferentiated cells, many of which divide to add to the plant body but the central mass (the **quiescent centre**) remains inert and only becomes active if the meristem is damaged. The shoot apical meristem (SAM) can produce vegetative leaf buds or reproductive floral buds. The root apical meristem is covered by the root cap.

apical plasma membrane The cell membrane on the apical (inner or upper) surface of transporting epithelial cells. This region of the cell membrane is separated, in vertebrates, from the baso-lateral membrane by a ring of **tight junctions** that prevents free mixing of membrane proteins from these two domains.

Apicomplexa A large group of obligate parasitic protozoa (within the **Alveolata**) that are characterized by an apical complex of microtubules within the cell in the sporozoite and **merozoite** stages but lack flagella. The apical complex has the capacity to release **rhopteries**. The group includes *Plasmodium, Cryptosporidium, Babesia, Toxoplasma*.

apicoplast A relict, non-photosynthetic plastid found in most Apicomplexa, including malaria parasites such as *Plasmodium falciparum*.

apidaecins Proline-rich basic antibacterial peptides (140−280aa) found in the immune **haemolymph** of the honeybee.

apigenin A bioflavone (naringenin chalcone) considered to have a beneficial effect on human health as an antioxidant, radical scavenger and anti-inflammatory. Present in leafy plants and vegetables (e.g., parsley, celery) and said to have chemopreventive activity against UV-radiation.

Apis mellifera The honeybee, a Hymenopteran.

APJ (1) The G protein-coupled receptor for apelin (angiotensin receptor-like 1, 380aa) that is an alternative coreceptor with CD4 for HIV-1 infection. (2) In *S. cerevisiae*, APJ1 is a J domain-containing protein (528aa) that is thought to be a chaperone involved in protein folding and that interferes with propagation of the **PSI+** prion when overproduced.

aplanogamete Most dictionaries define an aplanogamete as a nonmotile gamete, found in certain lower algae but the term does not appear to be used in the literature. Instead (and perhaps sensibly) reference is made to non-motile male gametes as for example in red seaweed. Paper: http://www.ncbi.nlm.nih.gov/pmc/articles/PMC1690222/pdf/DKG4A3GCGVWKFM4D_266_1879.pdf

aplanospore A nonmotile spore, e.g those produced by fungi in the Zygomycotina and by certain algae (Chlorophyta and Chrysophyta).

aplasia Defective development of an organ or tissue so that it is totally or partially absent from the body.

aplastic anaemia Anaemia due to loss of most or all of the **haematopoietic** bone marrow. Usually all haematopoietic cells are equally diminished in number.

aplyronine A cytotoxic macrolide from the sea hare *Aplysia kurodai* that has actin-depolymerizing activity and anti-tumour activity. Isolation & structure: http://www.sciencedirect.com/science/article/pii/S0040402007001913

Aplysia Opisthobranch mollusc (sea hare) with reduced shell; favourite source of ganglia for neurophysiological study.

APM *ami-prophosmethyl* A phosphoric amide herbicide that binds tubulin, inhibiting microtubule polymerization. It depolymerises cortical microtubules, resulting in growth inhibition, and also perturbs the mitotic spindle, resulting in multinuclearity.

apoA etc. *apoA, apoB, apoC, apoE* See **apolipoprotein A**, **apolipoprotein B** amongst others.

APOBECs A family of evolutionary conserved proteins (apolipoprotein B mRNA editing enzyme, catalytic polypeptide-like) that are cytidine deaminases involved in foreign DNA clearance. By triggering C-to-U hypermutation in exogenous DNA cause its degradation. For example, can induce mutations in the first HIV DNA expressed as complementary DNA (cDNA) thereby destroying the coding and replicative capacity of the virus. However, HIV viral infectivity factor (**Vif**) interacts with APOBEC3G and triggers ubiquitination and degradation. APOBEC1 (Apolipoprotein B mRNA-editing enzyme 1, 236aa) is responsible for the postranscriptional editing of a CAA codon for Gln to a UAA codon for stop in the APOB mRNA.

apocarotenoids Compounds derived from carotenoids by oxidative cleavage (by **carotenoid cleavage dioxygenases**). They include vitamin A retinoids and **abscisic acid**.

apocrine Form of secretion in which the apical portion of the cell is shed, as in the secretion of fat by cells of the mammary gland. The fat droplet is surrounded by apical plasma membrane, and this has been used experimentally as a source of plasma membrane.

apocynin A selective inhibitor of NADPH-oxidase in activated polymorphonuclear (PMN) leukocytes, isolated from dogbane (Apocynum cannabinum). Prevents the generation of reactive oxygen species.

apoenzyme An enzyme without its **cofactor**. See **apoprotein**.

apogamy In plants with independent gametophytes (notably ferns), the term is used interchangeably with **apomixis**; it is also sometimes used to refer to the situation in which an embryo develops from a cell of the megagametophyte other than the egg cell.

apolipoprotein A Apolipoprotein(a) (4548aa) is the the main protein constituent of **lipoprotein(a)**. It has serine peptidase activity, inhibits tissue-type plasminogen activator and may be a ligand for **megalin**. Apo(a) is proteolytically cleaved to form mini-Lp(a) and Apo(a) fragments that accumulate in atherosclerotic lesions. Apolipoprotein A1 (ApoA1, 243aa) is the major apoprotein of high density lipoprotein (HDL) and defects in ApoA1 cause various deficiencies in HDL (including **Tangier disease**) (TGD), Iowa-type amyloid polyneuropathy-nephropathy and Type 8 amyloidosis. Apolipoprotein A2 (100aa) may stabilize HDL structure. ApoA4 is a component of chylomicrons and high-density lipoproteins. Defects in ApoA5 cause susceptibility to familial hypertriglyceridaemia and hyperlipoproteinaemia type 5. ApoA1 binding protein (288aa) interacts with both ApoA1 and ApoA2. Apolipoprotein L1 interacts with ApoA1.

apolipoprotein B The main apolipoprotein of chylomicrons and low density lipoproteins (LDL). There are two main forms in plasma, apoB48 synthesized by the gut, and apoB100 produced by the liver, although both are coded by the same gene differentially spliced. In familial hypobetalipoproteinemia (FHBL) there are very low plasma levels of apolipoprotein B although triglyceride levels are normal, unlike the situation in abetalipoproteinaemia (Bassen-Kornzweig syndrome, acanthocytosis) caused by deficiency in microsomal triglyceride transfer protein.

apolipoprotein C A family of apolipoproteins (C-I – C-IV) mostly found in VLDL (very low density lipoprotein). ApoC-II (101aa) is a necessary cofactor for the activation of lipoprotein lipase (EC 3.1.1.3) whereas ApoC-III (99aa) inhibits lipoprotein lipase.

apolipoprotein D A **lipocalin** (189aa) that is a component of HDL (high density lipoprotein), found in plasma as a homodimer, as a heterodimer with ApoA2 or as part of a macromolecular complex with lecithin-cholesterol acyltransferase. It is involved in the transport of various ligands.

apolipoprotein E An apolipoprotein (317aa) present in all plasma liproteins that mediates their binding, internalization, and catabolism. It is a ligand for the LDL (apo B/E) receptor. Different alleles of the ApoE gene are associated with variations in plasma cholesterol levels and the risk of coronary artery disease. The ApoE4 allele appears to be associated with late onset **Alzheimer's disease** although there seems to be no causative linkage. Defects in ApoE are a cause of **hyperlipoproteinemia** type III, **sea-blue histiocyte disease** and **lipoprotein glomerulopathy**.

apolipoprotein F A minor apolipoprotein (lipid transfer inhibitor protein, 326aa) mostly associated with low density lipoprotein and an inhibitor of cholesteryl ester transfer protein and cholesterol transport.

apolipoprotein H A multifunctional apolipoprotein (beta-2 glycoprotein 1, 345aa) that will bind negatively charged substances (e.g. heparin) and may inhibit the intrinsic blood coagulation cascade by binding to phospholipids on the surface of damaged cells. The apoH-phospholipid complex appears to the the autoantigen in some cases of **systemic lupus erythematosus** and in **antiphospholipid syndrome**.

apolipoprotein L A family of proteins (ApoL1–6, 337–433aa) that are important in cholesterol transport and are associated with ApoA lipoproteins.

apolipoproteins The protein component of plasma **lipoproteins**. See entries for apolipoprotein A, B and others.

apollo A $5'−3'$ exonuclease (DNA cross-link repair protein 1B, 532aa) important for telomere maintenance and protection during S-phase. Binds to **telomeric repeat binding factor 2** (TRF2) and is one of the accessory proteins in the **shelterin** complex. Defects in apollo may be a cause of Hoyeraal-Hreidarsson syndrome, a multisystem disorder affecting males.

apomixis In plants, reproduction without gamete fusion, i.e. asexually, the equivalent of **parthenogenesis** in animals. Seeds may form without fertilization (agamospermy) and in some fungi meiosis and gamete formation do not occur even though an **ascus** containing identical diploid spores is formed. Vegetative propagation from cuttings is not classed as apomixis. In angiosperms two forms are distinguished, gametophytic apomixis in which the embryo arises from an unfertilized egg cell (parthenogenetically) and sporophytic apomixis (adventitious embryony) where the embryo is formed from nucellus (nucellar embryony) or integumental tissue. See **apogamy**.

apomorphic A term used in cladistics for derived or advanced characteristics that arose relatively late in members of a group and therefore differ among them. These are useful in assessing genealogical links among taxa; the more recent the common ancestor, the more apomorphic traits are shared. Apomorphic characters for the primates, like humans, include a large brain size and binocular vision. Plesiomorphic characteristics are ancient homologies (e.g. pentadactyl limbs), symplesiomorphic characteristics are recently derived. Synaptomorphic characters are shared between taxa.

apomorphine An alkaloid of the **morphine** series that is an an agonist at dopamine D1 and D2 receptors but does not bind the opioid receptors.

apopain Protease (**caspase** 3, Yama protein, EC 3.4.22.56) responsible for the cleavage of poly(ADP-ribose) polymerase and necessary for apoptosis. Has two subunits derived from a common proenzyme identified as CPP32 (277aa). Will cleave and activate **sterol regulatory element binding proteins** (SREBPs), caspases 6, 7 and 9, and **huntingtin**.

apoplast One of the two compartments that make up a plant. The apoplast is external to the plasma membrane and includes cell walls, xylem vessels etc., through which water and solutes pass freely. The other compartment, the **symplast**, comprises the total cytoplasmic compartment (the **protoplasts** of cells in a plant are all connected through **plasmodesmata** and can be considered as a connected whole).

apoprotein The protein portion of a complex between a polypeptide and a second moiety of non-polypeptide nature. For example, **ferritin** lacking its ferric hydroxide core may be referred to as apoferritin.

apoptin A small proline-rich protein (121aa) derived from the chicken anaemia virus, induces apoptotic cell death selectively in cancer cells. Human apoptin-associating protein one (RING1 and YY1-binding protein, death effector domain-associated factor, 228aa) interacts with apoptin, **DEDD**, **FADD**, caspases eight and ten, the transcriptional repressor protein **YY1** and GAPB1 (a transcription factor capable of interacting with purine rich repeats).

apoptosis The most common form of physiological (as opposed to pathological) cell death. Apoptosis is an active process requiring metabolic activity by the dying cell; often characterized by shrinkage of the cell, cleavage of the DNA into fragments that give a so-called 'laddering pattern' on gels and by condensation and margination of chromatin. Often called programmed cell death, though this is not strictly accurate. Cells that die by apoptosis do not usually elicit the inflammatory responses that are associated with necrosis, though the reasons are not clear. See also: **apoptosis inducing factor**, **apoptosome**, **ced mutant**, **bcl-2**.

apoptosis inducing factor *AIF* A group of proteins involved in the induction of apoptosis. AIF1 (programmed cell death protein 8, 613aa) is a mitochondrial oxidoreductase that is translocated to the nucleus when apoptosis is induced; interacts with DNA by virtue of positive charges clustered on the AIF surface (i.e. independently of sequence) and with **XIAP**. AIF2 (373aa) is similar. AIF-like protein (AIF3, 605aa) does not translocate to the nucleus and induces apoptosis through a caspase-dependent pathway.

apoptosome Complex of approximately 700 kDa (occasionally 1.4 MDa) assembled by rapid oligomerization of mitochondrial cytochrome c bound to **Apaf-1** followed by a slower process of procaspase-9 recruitment and cleavage to form the p35/34 forms. **XIAP** binds to the caspase nine and inhibits further proteolytic activity.

apothecia The sexual fruiting bodies produced by various phyla of ascomycete fungi (discomycetes; Leotiomycetes and Pezizomycetes) of which *Sclerotinia sclerotiurum* is an example.

App(NH)p Non-hydrolysable analogue of ATP.

appetite-regulating hormone Peptide (growth hormone secretagogue, growth hormone-releasing peptide, motilin-related peptide, 117aa) that is cleaved to produce ghrelin-27 (27aa), ghrelin-28 (ghrelin) and obestatin (23aa). Ghrelin is the ligand for growth hormone secretagogue receptor type one (GHSR) and induces the release of growth hormone from the pituitary. Has an appetite-stimulating effect. Obestatin may be the ligand for GPR39 and may have the opposite (appetite-reducing) effect.

applagin See **disintegrin**.

apple domain A consensus sequence (~90aa) with six cysteines, that forms a characteristic, vaguely apple-shaped, pattern via disulphide bridges. Shared by **kallikrein** and coagulation factor XI, both **serine peptidases**.

appressorium A flattened outgrowth which attaches a fungal parasite to its plant host.

aprataxin One of the **histidine triad** (HIT) superfamily (forkhead-associated domain histidine triad-like protein, 356aa) involved in DNA repair. Interacts with several nucleolar proteins, including **nucleolin, nucleophosmin** and upstream binding factor-1 (UBF-1). Mutations in the gene cause ataxia with oculomotor apraxia type 1 (AOA1). Aprataxin and PNK-like factor (511aa) is involved in single-strand and double-strand DNA break repair.

apraxia The total or partial loss of the ability to perform coordinated movements (even though there is no obvious motor or sensory impairment).

APRIL One of the TNF superfamily of ligands (a proliferation-inducing ligand, TNF-related death ligand one, TNFSF13, CD256, 250aa) found at low level in normal tissues but abundantly expressed in tumour cells.

aprotinin Basic polypeptide (Trasylol, 100aa), originally isolated from bovine lung although now produced as a recombinant protein, that inhibits several serine peptidases (including **trypsin, chymotrypsin, kallikrein** and **pepsin**).

aptamer *aptamere* Double-stranded DNA or single-stranded RNA molecule that binds to a specific molecular target.

apterous Without wings.

APUD cells Acronym for amine-precursor uptake and decarboxylation cells: **paracrine** cells of which **argentaffin cells** are an example. Usage neither helpful nor memorable.

apurinic sites Sites in DNA (AP sites) from which purines have been lost by cleavage of the deoxyribose N-glycosidic linkage.

APV (1) Avian polyoma virus. (2) An antiviral protease inhibitor, amprenavir. (3) Avian pneumonvirus, an emerging disease in turkeys. (4) An NMDA antagonist, (DL-2-amino-5-phosphonopentanoic acid).

Apx (1) In *Xenopus* an apical plasma membrane protein (Apx, shroom1, 1420aa) that plays a role in the functional expression of the **amiloride-sensitive sodium channel**. Induces gamma-tubulin accumulation at cell-cell junctions. Has two domains (Apx/Shrm domains one and two) that are found in various other **shroom** proteins. The ASD1 domain mediates F-actin binding. (2) In *Arabidopsis*, Apx2 (cytosolic L-ascorbate peroxidase 2, EC 1.11.1.11, 251aa) has a key role in hydrogen peroxide removal. (3) In *C. elegans*, Apx1 (anterior pharynx in excess protein 1, 515aa) is a single-pass membrane protein that contributes to the establishment of the dorsal-ventral axis in the early embryo.

apyrase A multi-pass membrane protein (EC 3.6.1.5, ectonucleoside triphosphate diphosphohydrolase, ~520aa) that catalyses breakdown of ATP to AMP. There are tissue-specific isoforms and it may regulate purinergic neurotransmission and platelet aggregation (by hydrolysing ADP). Mosquitos produce an apyrase in their salivary glands (562aa) that facilitates taking a blood meal by preventing platelet aggregation and blood clotting. The enzyme is also found in plants (415aa in *Solanum tuberosum*). Apyrase from *Pisum sativum* (nucleoside-triphosphatase, NTPase, EC 3.6.1.15, 455aa) may be involved in RNA export from nuclei.

aquaglyceroporin A subclass of **aquaporins** that also promotes glycerol permeability, e.g. aquaporin-10 (aquaglyceroporin-10, small intestine aquaporin, 301aa). In *E. coli*, aquaglyceroporin (glycerol uptake facilitator protein, 281aa) is a glycerol transporter that produces only limited permeability to water and small uncharged compounds such as polyols.

aquaporins Integral membrane proteins, members of the **major intrinsic protein** (MIP) family, with two tandem repeats each containing three membrane-spanning domains and a pore-forming loop that greatly increases water permeability. They form homotetramers in the membrane. Aquaporin-1 (AQP1, CHIP28, 269aa) is found in various mammalian tissues but especially in kidney and erythrocytes. Aquaporin-2 is found in the renal collecting tubules. Many human aquaporins are known and some, for example aquaporin-9 (small solute channel 1, 295aa), form channels with broad specificity; the different aquaporins serve various physiological functions. Plant MIPs have been divided into five subfamilies; the **plasma membrane intrinsic proteins** (PIPs), the tonoplast intrinsic proteins (TIPs), the **nodulin-26** like intrinsic proteins (NIPs), the small basic intrinsic proteins (SIPs) and the **GlpF-like intrinsic protein** (GIPs). Prokaryotic examples are known, for example *E. coli* GLP, a glycerol-transporting channel protein which is paralogous to AQP3, AQP7 and AQP9. Transport Classification Database: http://www.tcdb. org/tcdb/index.php?tc=1.A.8

Aquifex aeolicus Thermophilic, hydrogen-oxidizing, microaerophilic, obligate chemolithoautotrophic bacterium, able to grow at a remarkable 96°C. Genome, approximately one-third that of *E. coli*, has been sequenced.

Aquificae A phylum of bacteria found in extreme environments such as hot springs.

AR-A014418 A cell-permeable inhibitor of **glycogen synthase kinase-3** (GSK-3) showing high potency (IC_{50} = 104 nM) and selectivity (does not significantly inhibit closely related kinases such as cdk2 or cdk5). Inhibition is competitive with respect to ATP (K_i = 38 nM). Details: http://hwmaint.jbc. org/cgi/content/abstract/278/46/45937

ara operon Operon involved in arabinose metabolism, especially the araBAD operon of *E. coli*. AraA encodes arabinose isomerase, which converts arabinose to ribulose; *AraB* encodes ribulokinase, which phosphorylates ribulose; *AraD* encodes ribulose-5-phosphate epimerase, which converts ribulose-5-phosphate to xylulose-5-phosphate which is then metabolized via the **pentose phosphate pathway**. Regulation of the operon is by the *AraC* gene product that binds to the operator, AraI, site and another AroO site, forming a loop in the DNA which represses transcription of the operon. In the presence of arabinose, AraC adopts an altered conformation which when bound at the AraI site helps to activate expression of the operon.

ara6 (1) A small GTPase (202aa) of the rab family, an endosomal marker in *Arabidopsis*. Ara7 is similar. (2) A chloroplast enzyme (ARA6, 349aa) involved in thiazole biosynthesis and may also have a role in DNA damage tolerance.

arabidiol A plant secondary metabolite (a triterpene) involved in plant defense, produced by the action of **PEN1** on 2,3-oxidosqualene.

Arabidopsis thaliana Mouse-ear cress (Thale cress, wall cress), one of the Brassicaceae. Much used as a model system for plant molecular biology, because is small and has a small genome (7×10^7 bp), now fully sequenced, and a short generation time (five to eight weeks). TAIR (The *Arabidopsis* Information Resource) is an online database/portal that can be used to search for information about the genome, source mutants amongst others. Website: www.Arabidopsis.org

arabinofuranosidases Glycosidases (α-arabinofuranosidases) that will break down polymers such as arabinan, arabinoxylan, and other polysaccharides, that are major components of plant cell wall hemicelluloses. Many microorganisms use these enzymes to generate soluble carbohydrates that can be used as carbon or energy sources.

arabinogalactans Plant cell-wall polysaccharides containing predominantly arabinose and galactose. Two main types are recognized: arabinogalactan 1, found in the pectin portion of angiosperms and containing α-1,4 linked galactan and α-arabinose sidechains; arabinogalactan II, a highly branched polymer containing β-1,3 and β-1,6 linked galactose and peripheral α-arabinose residues. Arabinogalactan II is found in large amounts on some gymnosperms, especially larches, and is related to **AGP**.

arabinoglycan-protein See AGP.

arabinose A **pentose** monosaccharide that occurs in both D- and L- configurations. D-arabinose is the 2-epimer of **D-ribose**, i.e. differs from D-ribose by having the opposite configuration at carbon 2. D-arabinose occurs *inter alia* in the polysaccharide arabinogalactan, a neutral **pectin** of the cell wall of plants, and in the metabolites **cytosine arabinoside** and adenine arabinoside.

arabinoxylan Polysaccharide with a backbone of **xylose** (β-1,4 linked) with side chains of **arabinose** (α-1,3 linked): constituent of **hemicellulose** of angiosperm cell wall.

araC See ara operon.

arachidonic acid An essential dietary component for mammals. The free acid (5,8,11,14 eicosatetraenoic acid) is the precursor for biosynthesis of the signalling molecules **prostaglandins**, thromboxanes, hydroxyeicosatetraenoic acid derivatives including **leukotrienes** and is thus of great biological significance. Within cells the acid is found in the esterified form as a major acyl component of membrane **phospholipids** (especially **phosphatidyl inositol**) and its release from phospholipids is thought to be the limiting step in the formation of its active metabolites.

arachnoid layer See meninges.

arboviruses Diverse group of single-stranded RNA viruses that have an envelope surrounding the capsid. Arthropod borne, hence the name, and multiply in both invertebrate and vertebrate host, causing e.g. yellow fever and encephalitis. The group is very heterogeneous and three major families are recognized: Togaviridae, Bunyaviridae, and Arenaviridae.

arbuscular (1) Having the characteristics of a dwarf tree or shrub of tree-like habit. (2) Characteristic of the much-branched haustorium formed within the host cells by some endophytic fungi in **vesicular-arbuscular mycorrhiza**.

Arcella A small amoeba of the Phylum Sarcodina that has a chitinous test (shell), dome-like on the top and concave on the bottom. Around 50–200 μm wide.

Archaea *formerly* Archaebacteria One of two domains of the prokaryotes (the other being **Bacteria**). The Archaea are subdivided into two phyla (Crenarchaeota and Euryarchaeota) in the current **Taxonomic Outline of Bacteria and Archaea**. They include extreme **halophiles, methanobacteria**, and sulphur-dependent extreme-**thermophiles**. Archaea differ from **Bacteria** in number of important features including ribosomal structure, the possession (in some cases) of **introns**, membrane composition and cell wall composition.

Archaebacteria See **Archaea**.

archaeocyte An amoeboid cell type of sponges (Porifera).

archegonium A multicellular female sex organ of liverworts, mosses, ferns and most gymnosperms, that contains the ovum. Archegonia in the simpler plants are typically located on the surface of the plant thallus, although in the hornworts they are embedded and in some mosses they are located on top of the leafy gametophore They are much-reduced and embedded within the developing ovule in gymnosperms. The term is not used for angiosperms or the gnetophytes.

archenteron Cavity in the **gastrula** that opens to the exterior at the blastopore.

archvillin See **supervillin**.

arcuate nucleus Region of the hypothalamus containing two major types of neurons. One type secretes **neuropeptide Y** and **agouti-related peptide** which act to increase appetite and decrease metabolism. The other type of neurons, called POMC/CART neurons, produce alpha-melanocyte-stimulating hormone (α-**MSH**), which inhibits eating.

ARD1 (1) Nuclear inhibitor of protein phosphatase 1 (NIPP-1, EC 3.1.4.-, 351aa) that binds but does not cleave RNA. May target protein phosphatase 1 (PP-1) to RNA-associated substrates and may also be involved in pre-mRNA splicing. (2) A subfamily of acetyltransferases found in *S. cerevisiae* (arrest-defective protein 1, 238aa) and various animal species (in humans, N-terminal acetyltransferase complex ARD1 subunit homolog A, EC 2.3.1.88, 235aa). When complexed with **NARG1**, displays alpha (N-terminal) acetyltransferase activity but on its own has epsilon (internal) acetyltransferase activity towards **HIF1A**, promoting its degradation. (3) ADP-ribosylation factor domain-containing protein 1 (tripartite motif-containing protein 23, RING finger protein 46, 574aa) a GTP-binding protein of unknown function. (4) In rice, an enzyme (acireductone dioxygenase 1, 1,2-dihydroxy-3-keto-5-methylthiopentene dioxygenase 1, 199aa) involved in methionine biosynthesis. (5) In *Archaeoprepona demophon* (One-spotted leafwing butterfly) an antifungal defensin (ARD1, 44aa).

ARDS See **acute respiratory distress syndrome**.

ARE (1) **Activin**-response element to which **ARF** binds. (2) **AU-rich element**.

area opaca In the early development of birds and reptiles, a whitish peripheral zone of **blastoderm** in contact with the yolk. The area pellucida is a central clear zone of blastoderm that does not have direct contact with the yolk; the area vasculosa is the region of extraembryonic blastoderm in which the blood vessels develop.

area pellucida See **area opaca**.

area vasculosa See **area opaca**.

arecoline An alkaloid isolated from *Areca catechu* (betel palm) and one of the major pharmacologically active components; a **muscarinic** acetylcholine receptor agonist. Effects on the central nervous system are similar to those of nicotine. Has also been used medicinally as an anthelminthic.

Arenaviridae Family of ssRNA viruses including **Lassa virus**, lymphocytic choriomeningitis virus, and the **Tacaribe** group of viruses; not all require arthropods for transmission, despite their inclusion in the **arbovirus** group.

areolar connective tissue Loose **connective tissue** of the sort found around many organs in vertebrates. Does not have marked anisotropy, nor a pronounced content of any particular matrix protein.

ARF (1) **ADP-ribosylation factor**. (2) **Auxin response factor**.

arfaptins ADP-ribosylating factor interacting proteins (arfaptin 1, 373aa; arfaptin 2, 341aa), cytosolic targets of **ADP-ribosylation** factor and **rac** GTPases.

arg An oncogene (Abelson-related gene, ABL2, v-abl Abelson murine leukemia viral **oncogene** homolog 2, related to *abl*, that encodes a **tyrosine kinase** (1182aa).

argentaffin cells Cells so-called because they will form cytoplasmic deposits of metallic silver from silver salts. Their characteristic histochemical behaviour arises from 5-HT (**serotonin**), which they secrete. Found chiefly in the epithelium of the gastrointestinal tract (though possibly of neural crest origin) their function is rather obscure, although there is a widely distributed family of such **paracrine** (local endocrine) cells (**APUD cells**).

argentation chromatography Modified form of standard thin layer chromatography in which the solid phase includes silver salts. Used for lipid analysis.

arginine An essential amino acid (Arg, R, 174Da); a major component of proteins and contains the guanido group that has a pK_a of greater than 12, so that it carries a permanent positive charge at physiological pH. See Table A1.

argonaute Eukaryotic translation initiation factors (argonaute 1–4, AGO1–4, ~857aa) that are required for RNA-mediated gene silencing. Argonaute proteins are concentrated in **GW bodies** and are components of RNA-induced silencing complexes (**RISC**), binding the single-stranded siRNA and miRNA. RNA is required for the integrity of GW bodies and RNase eliminates Argonaute two localization. In *Drosophila*, unlike in humans, both AGO1 and AGO2 have **slicer activity**. Prokaryotic Ago proteins have unknown function but are similar to eukaryotic proteins. RNA interference and related RNA silencing phenomena use short antisense guide RNA molecules to repress the expression of target genes. Contain amino-terminal **PAZ** (for PIWI/Argonaute/Zwille) domains and carboxy-terminal **PIWI** domains.

argyrophil cells Neuroendocrine cells that take up silver ions from a staining solution but require the addition of a reducing agent to precipitate metallic silver (unlike **argentaffin cells** which do not). **Carcinoids** of the foregut tend to be argyrophilic whereas those of the lower intestine tend to be argentaffinic.

ARIA Acetylcholine Receptor Inducing Activity, a polypeptide (**heregulin-1** residues 177–244) purified from chick brain that stimulates transcription of **acetylcholine receptor** subunits in nuclei that underlie the developing synapses in chick **myotubes**. ARIA, NDF, **heregulin**, and GGF are encoded by alternatively spliced transcripts of the same gene. ARIA activates **erbB** receptor tyrosine kinases.

ariadne In *Drosophila*, proteins (ariadne-1, 503aa; ariadne-2, 509aa) that may act as E3 ubiquitin-protein ligases. There is a human homologue (557aa).

ARID The AT-rich interaction domain (ARID) is an ancient DNA-binding domain conserved throughout the evolution of higher eukaryotes. The consensus sequence spans about 100aa, and structural studies identify the major groove contact site as a modified helix-turn-helix motif. Characteristic of a family that includes 15 distinct human proteins with important roles in development, tissue-specific gene expression and proliferation control.

arisostatins Arisostatins A and B are antibiotics of the tetrocarcin class, isolated from *Micromonospora*. Have activity against Gram-positive bacteria and antitumour activity.

Ark1 (1) In *S. cerevisiae* an actin-regulating serine/threonine kinase (Ark1, EC 2.7.11.1, 638aa) that regulates the organisation of the actin cytoskeleton and endocytosis; see **Prk1p**. (2) Beta-adrenergic receptor kinase (BARK, βARK1, 689aa) one of the G-protein-coupled receptor kinases (GRK2) that desensitises and downregulates beta adrenoreceptors (βARs) in the heart. (3) Aurora-related kinase 1: see **aurora kinases**.

armadillo In *Drosophila* a segment polarity protein of the **beta-catenin** family. The neural isoform interacts with neural cadherin, the cytoplasmic isoform is involved in **wingless** signalling. Present on the inner surface of cell membrane and links adherens junctions to the cytoskeleton. Has 13 **armadillo repeats.**

armadillo repeat Protein motif (42aa) originally described in the *Drosophila* **armadillo** protein. Usually found in multiple repeats that form a superhelix of helices with a positively charged groove. Mediates interactions with proteins such as **cadherins**, Tcf-family transcription factors and the tumour suppressor gene, APC (**adenomatous polyposis coli**).

ARMD See **age-related macular degeneration**.

ARNO ARF nucleotide-binding site opener, a guanine nucleotide exchange factor (**GEF**) for **ARFs**. See **cytohesins**.

ARNT (1) Aryl hydrocarbon receptor nuclear translocator (ARNT). A promiscuous bHLH-PAS (Per-ARNT-Sim) protein (789aa) that forms heterodimeric transcriptional regulator complexes with several other bHLH-PAS subunits to control a variety of biological pathways. In association with hypoxia-inducible factors is important in cellular adaptation to low oxygen environments; it is also a dimeric partner for the Ah receptor (**AHR**), and this complex is essential in regulating the adaptive metabolic response to polycyclic aromatic hydrocarbons. (2) ArnT, 4-amino-4-deoxy-l-arabinose transferase (undecaprenyl phosphate-alpha-4-amino-4-deoxy-L-arabinose arabinosyl transferase, EC 2.4.2.43, 550aa) An inner membrane protein that catalyses covalent addition of 4-amino-4-deoxy-L-arabinose (L-Ara4N) groups to lipid A in the outer membranes of bacteria such as *Salmonella typhimurium* and *E. coli*, the final step in the polymyxin-resistance pathway.

aromatase Microsomal enzyme complex (EC 1.14.14.1, cytochrome P450 19A1, 503aa) that converts testosterone to estradiol.

ARP-1 Nuclear receptor (apolipoprotein regulatory protein-1, COUP transcription factor 2, 414aa) that binds to response element with two core motifs, 5-RG(G/T)TCA, as do various receptors such as **COUP-TFI**, and **PPAR**.

Arp2/3 complex A stable assembly of two actin-related proteins (Arp2 and Arp3) with five other subunits; a central player in the cellular control of actin assembly. Arp2/3 caps the pointed end of actin filaments and nucleates actin polymerization with low efficiency. The GTPase **Cdc42** acts in concert with **WASP** family proteins to activate the Arp2/3 complex. See also **ActA**, **IcsA**.

arrestins Family of inhibitory proteins that bind to tyrosine-phosphorylated receptors, thereby blocking their interaction with G-proteins and effectively terminating the signalling. Arrestin (S antigen, 405aa) from retinal rods, competes with **transducin** for light-activated rhodopsin, thus inhibiting the response to light (adaptation). Arrestin-C (cone arrestin, 388aa) is similar. Immune responses to arrestin lead to autoimmune uveitis. Beta-arrestins (e.g. β-arrestin-1, 418aa; β-arrestin-2, 409aa) regulate agonist-mediated G-protein coupled receptor (GPCR) signalling by binding to the phosphorylated receptor and blocking its interaction with the relevant G-protein. See **Ark1** #2.

Arrhenius plot A plot of the logarithm of reaction rate against the reciprocal of absolute temperature. For a single stage reaction this gives a straight line from which the activation energy and the frequency factor can be determined. Often applied to data from complex biological systems when the form observed is frequently a series of linear portions with sudden changes of slope. Great caution must be observed in interpreting such slopes in terms of activation energies for single processes.

arrhythmogenic right ventricular dysplasia/cardiomyopathy *ARVD/ARVC* A range of disorders in which there is arrhythmia of the right ventricle of the heart and replacement of muscle with fibrous tissue. The mutated genes include variously those for TGFβ3, **ryanodine receptor 2**, **desmoplakin**, **plakophilin 2**, **desmoglein 2**, **desmocollin 2** and gamma **catenin**.

arrowheads Fanciful description given to the pattern of **myosin** molecules attached to a filament of **F-actin**. Easier to see if tannic acid is added to the fixative. The arrowheads indicate the polarity of the filament; the barbed (attachment) end is the site of major subunit addition.

ARRs *Arabidopsis* response regulators, a **two-component system** in plants, involved in histidine (His) to aspartate (Asp) phosphorelay signal transduction. The ARRs can be subdivided into three discrete groups, the A-type ARRs and B-type ARRs and a novel group of ARRs consisting of ARR22 and ARR24. The B-type ARRs are transcription factors acting as partially redundant positive regulators of cytokinin signal transduction. A-type ARRs are products of a family of 10 genes that are rapidly induced by cytokinin and are highly similar to bacterial two-component response regulators. They may be partially redundant negative regulators of cytokinin signalling but seem to have a more general function in the integration and coordination of several signalling processes, such as those of light, the circadian clock and control of stem cell number. Paper: http://www.biomedcentral.com/1471-2229/8/77

ARS (1) An autonomously replicating sequence, a DNA sequence originally isolated from *S. cerevisiae* that, when linked to a non-replicating sequence, can confer on the latter the ability to be replicated in a yeast cell. Transformations effected with the use of ARS occur at relatively high frequency but are unstable. Homologous recombination of the DNA of interest with the host cell chromosomes is not required for expression when ARS routes are used. (2) Arylsulphatase.

ARSACS An ataxia (autosomal recessive spastic ataxia of Charlevoix-Saguenay) caused by mutation in the gene encoding **sacsin**.

arsenite Arsenite, the trivalent form of arsenic, is a thiol-reactive oxidative stressor that is toxic, co-carcinogenic and known to inhibit protein synthesis. It activates multiple stress signalling pathways and has effects on gene control, such as the interruption of cell cycle control by initiating G2/M arrest. At low doses may have useful anti-tumour activity. It is used as a herbicide and pesticide.

Artemether A rapid-acting antimalarial drug of the same class as **artemisinin**.

Artemia salina Brine shrimp, a crustacean of the Order Anostraca.

artemin A protein ligand of the TGFβ family (neublastin, enovin, 220aa) for the **GDNF** receptor α3, important for survival of neurons from all peripheral ganglia and some neurons in the CNS.

artemis A protein (DNA cross-link repair protein 1C, 692aa) with single-strand-specific 5′-to-3′ exonuclease activity regulated by DNA-dependent protein kinase and involved in V(D)J recombination in the immune system and in DNA repair. Mutation can lead to various forms of **severe combined immunodeficiency syndrome**.

artemisinin Antimalarial drug extracted from the Chinese herb *Artemisia annua* (qinghaosu or sweet wormwood), used to treat uncomplicated falciparum malaria; a sesquiterpene lactone. More potent derivatives such as artemether and artesuna have been developed. Artemisinin is often used in combination therapy with, for example, **lumefantrine**.

arteriole Finest branch of an artery upstream of the capillary bed.

arteriosclerosis Imprecise term for various disorders of arteries, particularly hardening due to fibrosis or calcium deposition; often used as a synonym for **atherosclerosis**.

artery Blood vessel carrying blood away from the heart; walls have smooth muscle and are innervated by the **sympathetic nervous system**.

arthritis General term for inflammation of one or more joints. Many diseases may cause arthritis, although in most cases the cause of the inflammation is not understood. This is particularly true of **rheumatoid arthritis**, though knowledge of other forms is not much better.

Arthrobacter Genus of the **Actinobacteria**, obligate aerobic bacteria of irregular shape, found extensively in soil.

arthroconidium A type of thallic conidium released by either the splitting of a double septum or by the fragmentation or lysis of a disjunctor cell.

arthrogryposis A rare disorder characterized by multiple joint contractures. Can be a result of mutation in genes for β-tropomyosin, embryonic myosin heavy chain, isoforms of troponin T and troponin I although there are many other variants in which the genetic defects have not been identified. Arthrogryposis, renal dysfunction and cholestasis (ARC syndrome) is caused by mutations in the gene *VPS33B* that encodes a protein (617aa) involved in vacuolar sorting.

arthropathy Any disease affecting a joint – care should be taken not to confuse arthrosclerosis (stiffness of joints) with atherosclerosis.

arthropod The largest phylum of the animal kingdom, containing several million species. Arthropods are characterized by a rigid external skeleton, paired and jointed legs, and a haemocoel. The phylum Arthropoda includes the major classes Insecta, Crustacea, Myriapodia and Arachnida.

Arthus reaction A localized **inflammation** due to injection of **antigen** into an animal that has a high level of circulating **antibody** against that antigen. A haemorrhagic reaction with edema occurs due to the destruction of small blood vessels by thrombi. It may occur, as in Farmer's Lung, as a reaction to natural exposure to antigen.

articulins Membrane-associated protein complex of *Euglena*; two isoforms of 80 and 86 kDa, completely unlike spectrin, though functionally analogous. Have a core domain of 12-residue repeats, rich in valine and proline. May attach directly to membrane proteins. Articulins in protists: http://www.ncbi.nlm.nih.gov/pubmed/9788878

artifical selection Conventional genetic engineering, the selection of progeny by a plant or animal breeder, rather than by environmental factors (**natural selection**). Quite remarkable phenotypic alterations can be brought about – a classic example being the variety of breeds of dogs.

Artiodactyla Order of herbivorous even-toed mammals that includes antelopes, pig, cow, giraffe and hippopotamus.

ARVD See **arrhythmogenic right ventricular dysplasia/cardiomyopathy**.

aryl hydrocarbon receptor See **AHR**.

aryl sulphatase Aryl sulphatases A, B and C comprise a group of enzymes originally assayed by their ability to hydrolyze O-sulphate esters of aromatic substrates. Aryl sulphatase A (arylsulfatase A, EC 3.1.6.8, 507aa) is a lysosomal enzyme that hydrolyses cerebroside 3-sulphate and is deficient in metachromatic leukodystrophy. Aryl sulphatase B (N-acetylgalactosamine 4-sulphatase, EC 3.1.6.12, 533aa) is deficient in **Maroteaux-Lamy syndrome**. Aryl sulphatase C (steroid sulphatase, EC 3.1.6.2, 583aa) hydrolyzes estrogen sulphates. All three are deficient in **multiple sulphatase deficiency**. Many other aryl sulphatases (e.g. arylsulphatase E, ARSE, 594aa) are known.

arylation Addition of an aryl group to a substrate, often catalysed by palladium. Not a enzymatic reaction, although the removal of aryl groups (for example by **aryl sulphatase**) can be carried out in biological systems.

AS-2 See **adocia sulphate**.

AS-252424 Potent and selective inhibitor of phosphoinositide 3-kinase (PI(3)K) p110-γ, shown to be selective for class IB PI(3)K-mediated cellular effects, e.g. inhibits C5a-mediated phosphorylation of Akt in RAW 264.7 macrophages and neutrophil recruitment in a mouse model of peritonitis.

ASA-GSH cycle The ascorbate (AsA) and reduced glutathione (GSH) cycle that removes reactive oxygen species from the chloroplast, especially under stress conditions. Oxidative stress in plants: http://www.plantphysiol.org/content/155/1/93.full

asbestosis Fibrosis of the lung as a result of the chronic inhalation of asbestos fibres. The needle-like asbestos fibres are phagocytosed by alveolar macrophages but burst the phagosome and kill the macrophage and the cycle is repeated. **Mesothelioma**, a rare tumour of the mesothelial lining of the pleura, is associated with intense chronic exposure to asbestos dust, particularly that of crocidolite asbestos.

ASC Apoptosis-associated speck-like protein containing a CARD (TMS1, 195aa), an adaptor protein containing a caspase activation and recruitment (CARD) domain that links pathogen recognition by **PYRIN-domain** pathogen recognition receptors to the activation of downstream effectors such as caspase and NFκB. May be a component of the **inflammasome**.

Ascaris Genus of nematodes (Aschelminthes). *Ascaris suum* is the common roundworm of pigs; *Ascaris lumbricoides* causes ascariasis in Man.

Aschelminthes Cluster of invertebrate phyla of which the best known are nematodes and acanthocephala. All have a pseudocoelom and an unsegmented elongate body with terminal anus and a nonmuscular gut.

Aschoff bodies Small **granulomas** composed of **macrophages**, **lymphocytes** and multinucleate cells grouped around eosinophilic **hyaline** material derived from collagen. Characteristic of the **myocarditis** of rheumatic fever.

Ascidiacea Class of simple or compound **tunicates** that have a motile larva but sedentary adult form that filter-feeds. Sea squirts are the commonly known examples.

ascites tumour Tumour that grows in the peritoneal cavity as a suspension of cells (and is anchorage independent). **Hybridomas** are sometimes grown as ascites tumours, and the ascites fluid can then be used as the crude 'antiserum'.

ascocarp Fruiting body of the Ascomyce fungi.

Ascomycota Ascomycete fungi that produce spores, usually eight, in a structure known as an ascus. Includes yeasts and *Neurospora*. The taxonomy of the group is complex but given in detail on MycoNet. Subphyla are Taphrinomycotina (Schizosaccharomyces are a class), Saccharomycotina (Saccharomyces are a class) and Pezizomycotina. MycoNet: http://fieldmuseum.org/sites/default/files/Myconet_13a.pdf

ascorbic acid *Vitamin C* An essential dietary vitamin for humans and guinea pigs; deficiency causes scurvy. May act as a reducing agent in enzymic reactions, particularly those catalysed by **hydroxylases**. See also Table V1.

ascospore Diploid spore formed by ascomycete fungi, contained within an ascus.

ascothecium An ascocarp that is open and cup-shaped allowing spores to disperse readily. *Cf.* **cleistothecium**, **perithecium**.

ascus Elongated spore case containing 4 or 8 haploid sexual ascospores of ascomycete fungi (which include most yeasts).

asepsis State in which harmful microorganisms are absent. Aseptic technique aims to avoid contamination of sterile systems.

asexual Reproducing without a sexual process and thus without formation of gametes or reassortment of genetic characters.

ASF/SF-2 One of the SR-type splicing factors (see **SR proteins**) found in nuclear speckles and involved in pre-mRNA splicing. ASF/SF-2 (SRp30a, arginine/serine-rich splicing factor 1, 248aa) interacts with other spliceosomal proteins and plays a role in preventing exon skipping.

ASGP Membrane-associated mucin (ascites sialoglycoprotein, mucin-4, 2169aa) present on rat mammary carcinoma cells. ASGP-1 and ASGP-2 are generated from a single precursor; ASGP-2 acts as a membrane anchor for ASGP-1. Though ASGP is thought to have similar functions to **episialin** there is no sequence homology.

asialoglycoprotein The carbohydrate moiety of many vertebrate glycoproteins without the normal terminal residues of **sialic acid**. The asialo-derivatives of some plasma proteins are specifically bound by a receptor on the surface of liver parenchymal cells (the **scavenger receptor**).

ASIP (1) Agouti signalling protein (ASIP, 132aa) is the human homologue of mouse **agouti**, inhibits melanogenesis and the response of human melanocytes to α-melanotropin. (2) Atypical Protein Kinase C (aPKC) Isotype-Specific Interacting Protein (ASIP) specifically interacts with the atypical protein kinase C isozymes PKCλ and PKCζ. Overexpression of ASIP inhibits insulin-induced glucose uptake by specifically interfering with signals transmitted through PKCλ. ASIP is the mammalian homologue of *C. elegans* polarity protein PAR-3. ASIP and PAR-3 share three **PDZ domains**, and can both bind to aPKCs.

A-site Site on the **ribosome** to which aminoacyl tRNA attaches during the process of peptide synthesis. See also **P-site**.

ASK (1) *Arabidopsis* SHAGGY-related protein kinases, a family of kinases that have homology to mammalian **GSK3** and *Drosophila* SHAGGY. There are at least 10 ASK genes in the haploid *Arabidopsis* genome. (2) Apoptosis signal-regulating kinase (ASK1, mitogen-activated protein kinase kinase kinase 5, MAPKKK5, 1347aa), a mitogen-activated protein kinase kinase kinase that is upregulated under conditions of cellular stress, activates MAP2K4 and MAP2K6, which in turn phosphorylate c-Jun N-terminal kinase (**JNK**) and p38 MAP kinases, leading eventually to an apoptotic response.

Askenazy cells Abnormal thyroid epithelial cells (Hürthle cells, oxyphil cells, oncocytes) found in autoimmune *thyroiditis*. The cubical cells line small **acini** and have **eosinophilic** granular cytoplasm and often bizarre nuclear morphology. Hürthle cell tumours are associated with chromosomal abnormalities or mutations in the RAS gene, the PAX8/PPARG fusion gene, or the mitochondrial NADH-ubiquinone oxidoreductase 1a subcomplex-13 gene (NDUFA13; gene associated with retinoid- and interferon-induced mortality-19, GRIM19) gene.

asparaginase Enzyme (EC 3.5.1.1, 305aa) that hydrolyzes L-asparagine to L-aspartate and ammonia that is used as an anti-tumour agent especially against lymphosarcoma and lymphatic leukaemia in which the cells cannot synthesise asparagine.

asparagine The β-amide (β-asparagine, Asn, N, 132Da) of aspartic acid; the L-form is one of the 20 amino acids directly coded in proteins. Coded independently of aspartic acid. See Table A1.

aspartame Trademark for Asp-Phe Methyl Ester, an artificial sweetener.

aspartate L-aspartate (aspartic acid, Asp, D, 133Da) is one of the 20 amino acids directly coded in proteins; the free amino acid is an excitatory neurotransmitter. See Table A1.

aspartate beta-hydroxylase See **junctin**.

aspartate transaminase See SGOT.

aspartate transcarbamylase ATCase.   See CAD #1.

aspartic peptidase Relatively small family of endopeptidases of the class EC 3.4.23 (aspartic proteases, aspartic proteinases, aspartyl proteinases) with a pH optimum below five, the best known member being **pepsin** A. The catalytic centre is formed by two Asp residues that activate a water molecule, and this mediates the nucleophilic attack on the peptide bond. They have significant roles in human diseases (e.g. **renin** in hypertension, **cathepsin D** in metastasis of breast cancer, **beta-secretase** in Alzheimer's Disease).

aspartokinase Enzyme (aspartate kinase, EC 2.7.2.4) that phosphorylates L-aspartate to produce aspartyl phosphate, a step in biosynthesis of lysine, methionine and threonine. In chloroplasts and some bacteria there is a bifunctional aspartokinase/homoserine dehydrogenase two (aspartokinase, EC 2.7.2.4 and homoserine dehydrogenase, EC 1.1.1.3) with the activites in separate domains. In *E. coli*,

aspartokinase III (lysine-sensitive aspartokinase 3, 449aa) is allosterically inhibited by lysine, one of the end-products of the aspartic acid family branched pathway.

aspartyl protease See **aspartic peptidase**.

aspartylglycosaminuria A lysosomal storage disease caused by deficiency of the **threonine peptidase**, N-aspartyl-β-glucosaminidase (EC 3.5.1.26, 346aa) that cleaves the asparagine from the residual N-acetylglucosamines produced by breakdown of glycoproteins.

aspergillin (1) A name sometimes given to the black spore pigment produced by *Aspergillus niger*. (2) A name occasionally used (possibly misused) for toxins (ribonucleases) produced by *Aspergillus* sp., examples include **alpha-sarcin**, **mitogillin** and **restrictocin**.

Aspergillus *now* ***Emericella*** A genus of common ascomycete fungi consisting of several hundred mould species found in almost all oxygen-rich environments. Industrially important in production of organic acids, and a popular fungus for genetic study (esp. *A. niger*). Some species can cause infection (aspergillosis) in humans and other animals, some produce potent allergens, others mycotoxins, a group of fungal exotoxins including **aflatoxins** produced by *A. flavus*.

aspirin An analgesic, antipyretic and antinflammatory drug (acetyl salicylate). It is a potent **cyclooxygenase inhibitor** and blocks the formation of **prostaglandins** from **arachidonic acid**.

ASPM Protein (abnormal spindle-like microcephaly-associated protein, 3477aa) involved in regulating the mitotic spindle and coordinating mitosis, particularly during neurogenesis. ASPM is a major determinant of cerebral cortical size and mutations lead to microencephaly primary type five. It has 39 **IQ domains** in humans.

asporin A class I **small leucine-rich repeat proteoglycan** (periodontal ligament-associated protein one, PLAP1, 380aa) that negatively regulates periodontal ligament differentiation and mineralization. It has a stretch of aspartate residues at its N terminus that has homology with decorin and biglycan and is associated with skeletal tissues. An allele with an increased number of Asp repeats seems to be associated with susceptibility to osteoarthritis.

association constant K_a, K_{ass} The reciprocal of **dissociation constant**. A measure of the extent of a reversible association between two molecular species at equilibrium.

astacin A zinc-endopeptidase (EC 3.4.24.21, 251aa) from crayfish (*Astacus*), the prototype for the astacin family of metallo-endopeptidases. Family includes

BMP-1, meprin A, **stromelysin** one, and **thermolysin**.

astakine Prokineticin-like cytokine (104aa) from the Signal crayfish (*Pacifastacus leniusculus*) involved in haematopoiesis.

astaxanthin A naturally occurring red carotenoid pigment with antioxidant properties. Most crustaceans are tinted red by accumulated astaxanthin and the pink flesh of a healthy salmon is due to accumulated astaxanthin which is added to feed in fish farms to substitute for the astaxanthin in the diet of a wild salmon.

aster Star-shaped cluster of microtubules radiating from the polar **microtubule organizing centre** at the start of mitosis.

asthma Inflammatory disease of the airways involving marked eosinophil infiltration and remodelling of the airways. Attacks can be triggered by allergic responses, physical exertion, inhaled chemicals or stress and involve wheezing, breathlessness and coughing.

astrin A microtubule-associated protein (sperm-associated antigen five, mitotic spindle-associated protein p126, 1193aa) that localizes with mitotic spindles in M-phase and is essential for progression through mitosis, although astrin's function is unclear. Present in most tissues but highly expressed in the testis (see **ODF**).

astroblast An embryonic **astrocyte**.

astrocyte A glial cell (astroglial cell) found in vertebrate brain, named for its characteristic star-like shape. Astrocytes lend both mechanical and metabolic support for neurons, regulating the environment in which they function. See **oligodendrocytes**.

astrocytoma A neuro-ectodermal tumour (**glioma**) arising from **astrocytes**. Probably the commonest glioma, it has a tendency to anaplasia.

astroglia See **astrocytes**.

astrogliosis Hypertrophy of the **astroglia**, usually in response to injury.

Astropectinidae Family of echinoderms that includes many starfish species with long spines.

astrotactins Neuronal adhesion molecules (astrotactin-1, 1302aa; astrotactin-2, 1339aa) required for glial-guided migration of young postmitotic neuroblasts during development.

Astroviridae Small (28–30 nm) spherical non-enveloped viruses with 5- or 6-pointed star-shaped surface appearance in the electron microscope. The genome is a positive-sense single-stranded RNA. They can cause mild enteritis.

ASYMMETRIC LEAVES A protein (AS2) that is required for the development of normal leaf shape and for the repression of *KNOX* genes in the leaf. AS2 is a member of the recently identified, plant-specific LATERAL ORGAN BOUNDARIES (LOB)-domain gene family. Expression of AS2 at high levels represses of the KNOX homeobox genes *BREVIPEDICELLUS*, *KNAT2*, and *KNAT6* but not of the related *SHOOT MERISTEMLESS* gene.

asymmetrical dimethyl arginine Endogenous competitive inhibitor of nitric oxide synthase.

AT hook An AT-rich DNA-binding domain that occurs three times in mammalian high-mobility-group chromosomal proteins and in DNA-binding proteins from plants.

ATA See **aurintricarboxylic acid**.

ataxia Imbalance of muscle control. See **ataxia telangiectasia**.

ataxia telangiectasia A hereditary **autosomal** recessive disease (Louis Bar syndrome) in humans characterized by a high frequency of spontaneous chromosomal aberrations, neurological deterioration and susceptibility to various cancers. In part an immune deficiency disease and in part one of DNA repair; it is believed to be due to hypersensitivity to background ionizing radiation as a result of mutations in the *ATM* gene. Ataxia telangiectasia-like disorder is caused by mutation in the double-strand break repair protein MRE11A.

ataxia-oculomotor apraxia-1 An early-onset cerebellar ataxia caused in some cases by mutation in the gene encoding **aprataxin**.

ataxins A group of proteins that are defective in various forms of **spinocerebellar ataxia** (SCA) because of expanded numbers of trinucleotide (CAG) repeats (polyglutamine repeats). Ataxin one (815aa) binds RNA and may be involved in RNA metabolism; ataxin-1-like (brother of ataxin-1, BOAT, 689aa) will suppress the cytotoxicity of ataxin one. Ataxin three (Machado-Joseph disease protein 1, 376aa) represses transcription by interacting with transcriptional regulators and binding to histones. Ataxin seven (892aa) is a component of the STAGA transcription coactivator-HAT complex.

ATCase Enzyme (EC 2.1.3.2, aspartate transcarbamylase) that catalyses the first step in pyrimidine biosynthesis, condensation of aspartate and carbamyl phosphate. Positively allosterically regulated by ATP and negatively by CTP; classic example of an allosterically regulated enzyme. Bacterial ATCases exist in three forms: class A (ca 450–500 kDa), class B (ca 300 kDa) and class C (ca 100 kDa).

ATCC The American Type Culture Collection, a non-profit organisation linked to the UK's

Laboratory of the Government Chemist, repository of many eukaryotic cell lines (which may be purchased) and a source of information about cell lines. Comparable collections of microorganisms, protozoa etc. are kept. Website: www.atcc.org/

atelocollagen A highly purified pepsin-treated type I collagen from calf dermis which is of low immunogenicity because telopeptides are absent. It is used clinically for a wide range of purposes, including wound-healing, vessel prosthesis, as a bone cartilage substitute and more recently has been used as a complex with DNA or siRNA because these complexes are efficiently transduced into cells.

atelosteogenesis A disorder of bone and cartilage development caused by mutations in **filamin B** or the **diastrophic dysplasia** sulphate transporter.

ATF (1) Activating transcription factor (cyclic AMP-dependent transcription factor), a member of the **CREB**/ATF family of **bZip** transcription factors. Binds the cAMP response element (CRE). A number of ATF variants (ATF-1, 271aa; ATF-7, 494aa; etc.) have been described and are, in general, involved in mediating transcription in response to intracellular signalling. ATF-interacting protein (1270aa) couples transcription factors to the general transcription apparatus. (2) Artificial transcription factor. (3) Amino-terminal fragment.

ATG (1) See anti-thymocyte globulin. (2) ATG genes encode various proteins required for autophagy-related processes. These include cysteine peptidases ATG4A-D (autophagins 2, 1, 3 & 4; ~393aa), ubiquitin-like modifier-activating enzymes, serine kinases in yeast that are involved in transporting proteins/organelles destined for proteolytic degradation to the vacuole. Most Atg proteins (coded by the *ATG* genes) are co-localized at the pre-autophagosomal structure (PAS). ATG14 (beclin 1-associated autophagy-related key regulator, 492aa) has a role in autophagosome formation.

athanogene Apparently a neologism for a gene that has anti-apoptotic function. Has been applied particularly to **BAG-1**.

ATHB In *Arabidopsis* a family of transcription factors (homeobox leucine zipper proteins) some of which are induced by water deficit and by abscissic acid.

atheroma Degeneration of the walls of the arteries because of the deposition of fatty plaques in the **intima** of the vessel wall, scarring and obstruction of the lumen.

atherosclerosis Condition caused by the deposition of lipid in the wall of arteries in atheromatous plaques. Migration of smooth muscle cells from media to intima, smooth muscle cell proliferation, the formation of **foam cells** and extensive deposition

of extracellular matrix all contribute to the formation of the lesions that may ultimately occlude the vessel or, following loss of the endothelium, trigger the formation of thrombi. *Cf.* **arteriosclerosis** which is a more general term usually applied to arterial hardening through other causes.

AtKinesin-13A An internal-motor kinesin (794aa) from *Arabidopsis*, localized to Golgi stacks. Database entry: http://www.*Arabidopsis*.org/servlets/TairObject?accession=AASequence:1009118854&show AllNote=true

Atlas of Genetics and Cytogenetics in Oncology and Haematology A peer reviewed on-line journal and database with free internet access. Devoted to genes, cytogenetics, and clinical entities in cancer, and cancer-prone diseases. Homepage: http://atlasgeneticsoncology.org/index.html

atlastin A family of GTPases that tether membranes through formation of trans-homooligomers and mediate homotypic fusion of endoplasmic reticulum membranes. Have homology to guanylate binding protein-1 (GBP1), one of the **dynamin** family of large GTPases. Atlastin-1 (558aa) is expressed predominantly in the brain, interacts with **spastin** and is mutated in hereditary spastic paraplegia type 3A (SPG3A). Atlastin-2 (ADP-ribosylation-like factor 6-interacting protein 2, 583aa) and atlastin-3 (541aa) are closely related to atlastin-1, but are localized to the ER in non-brain tissue.

ATM A serine/threonine protein kinase (3056aa) of the PI3/PI4-kinase family encoded by the gene mutated in **ataxia telangiectasia** (AT). It is recruited to double-strand breaks in DNA by the MRN complex and then becomes enzymatically active. Not only involved in damage sensing but also in immune-cell generation of diversity, signal transduction and cell cycle control. It is defective in various leukemias.

atomic force microscopy A form of scanning probe microscopy, in which a microscopic probe is mechanically tracked over a surface of interest in a series of x-y scans, and the force encountered at each coordinate measured with piezoelectric sensors. This provides information about the chemical nature of a surface at the atomic level.

atonal In *Drosophila* a developmental protein (312aa) involved in neurogenesis. Required for the formation of chordotonal organs and photoreceptors. Forms a heterodimer with **daughterless**.

atopy Allergic (**hypersensitive**) response at a site remote from the stimulus (e.g. food-induced dermatitis).

ATP Adenosine 5′ triphosphate. Often referred to as the energy currency of the cell, a compound synthesized in cells from ADP by energy-yielding processes. Enzymic transfer of the terminal phosphate or pyrophosphate from ATP to a wide variety of substrates provides a means of transferring chemical free energy from metabolic to catabolic processes.

ATP binding site A consensus domain ('A' motif) found in a number of ATP or GTP-binding proteins, for example **ATP synthase, myosin heavy chain, helicases, thymidine kinase**, G-protein α-subunits, GTP-binding **elongation factors, Ras** family. Consensus is: (A or G)-XXXXGK-(S or T); this is thought to form a flexible loop (the P-loop) between α-helical and beta pleated sheet domains.

ATP synthase A proton-translocating **ATPase**, found in the inner membrane of **mitochondria, chloroplasts** and the plasmalemma of bacteria. It can be known as the F1/Fo or CF1/CFo ATPase, or as the class of F-type ATPases. In all these cases, the enzyme is driven in reverse by the large proton motive force generated by the **electron transport chain**, and thus synthesizes, rather than uses, **ATP**. See also **chemiosmosis, V-type ATPase, P-type ATPase**.

ATP-grasp proteins A superfamily of proteins that have an ATP-grasp motif, a two-domain ATP-binding fold. ATP-grasp domain-containing protein one is carnosine synthase one (EC 6.3.2.11, 827aa).

ATPase Any enzyme capable of releasing the terminal (γ) phosphate from ATP, yielding **ADP** and inorganic phosphate. The description could mislead, because in most cases the enzymic activity is not a straightforward hydrolysis, but is part of a coupled system for achieving an energy-requiring process, such as ion-pumping or the generation of motility.

ATR A serine/threonine protein kinase, of the **ATM** subfamily of PI3/PI4-kinases (EC 2.7.11.1, ataxia telangiectasia and Rad3-related protein, FRAP-related protein 1, 2644aa) which activates checkpoint signalling following DNA damage. Interacts with ATM and together they may regulate synaptic vesicle release, hence, perhaps the neurological symptoms of ataxia-telangiectasia.

atresia The pathological absence of an opening (e.g. anus) or passage (e.g. a portion of the GI tract).

atrial fibrillation See **fibrillation** #2.

atrial natriuretic factor Obsolete name for **atrial natriuretic peptide**.

atrial natriuretic peptide A polypetide hormone (ANP, atrial natriuretic factor, ANF, 28aa from 153aa precursor) released by the **atrium** of the heart in response to stretching caused by elevated blood pressure. ANP reduces blood pressure by stimulating the rapid excretion of sodium and water in the kidneys (reducing blood volume), by relaxing vascular

smooth muscle (causing vasodilation), and through actions on the brain and adrenal glands. See **natriuretic peptides** and **ANP receptor**.

atrichia Absence of hair. In some cases (atrichia with papular lesions) the defect is in **hairless**.

atrichoblast Root epidermal cell that will not differentiate into a root hair, in contrast to trichoblasts that do produce hairs.

atriopeptin Obsolete term originally used for the polypeptide hormones **atrial natriuretic peptide** (ANF, ANP) and **brain natriuretic peptide** (BNP).

atrioventricular septal defect A heart defect in which blood can bypass the lungs. There is considerable heterogeneity but can be caused by mutation in CRELD1 (cysteine-rich with EGF-like domain protein 1, 420aa) or **connexin 43** (heart connexin). Can be a complication of **Down's syndrome**.

atrium *Pl.* atria A cavity in the body, especially either of the two upper chambers of the heart in higher vertebrates.

atrophin Atrophin 1 (ATN1, dentatorubral-pallido-luysian atrophy protein, 1185aa) is a transcriptional corepressor. In **dentatorubral-pallidoluysian atrophy** the highly polymorphic region of polyglutamine repeats is expanded from the normal 7−23 copies to 49−75 repeats. Atrophin-1-related protein (arginine-glutamic acid dipeptide repeats protein, 1566aa) acts as a a transcriptional repressor during development and interacts with atrophin one, the interaction being stronger if there are more repeats. Atrophin-interacting proteins include **MAGI** (ATN interacting protein 1, 1455aa) and E3 ubiquitin-protein ligase Itchy homologue (ATN-interacting protein 4, 903aa).

atrophy Wasting away of tissue with loss of mass and/or function.

atropine An alkaloid, isolated from Deadly Nightshade, *Atropa belladonna*, that inhibits **muscarinic acetylcholine receptors**. Applied to the eye causes dilation of the pupil that is said to enhance the beauty of a woman, hence belladonna as the specific name of the plant from which the ancients extracted the drug.

attachment constriction See **centromere**.

attachment plaques Specialized structures at the ends of a chromosome by which it is attached to the nuclear envelope at **leptotene** stage of mitosis.

attacins Antibacterial proteins (~200−240aa) produced by insect haemocytes following bacterial challenge. May be basic or acidic and are fairly highly conserved between species. Induction of these

antimicrobial peptides in *Drosophila* involves NFκB elements.

attenuation In general, the reduction in the strength or intensity of a stimulus or signal. Viruses that have been passaged extensively may become attenuated (non-virulent), and can be used as a vaccine.

ATX See **autotaxin**.

A type particles Retrovirus-like particles found in cells. Non-infectious. The mouse genome contains around 1000 copies of homologous sequences.

A-type potassium channels *KCNA4* Rapidly inactivating voltage-gated potassium channels. They are activated by depolarization but only after a preceding hyperpolarization and are important for repetitive firing of cells at low frequencies. Include KCNA4 of the **shaker**-related subfamily, Shaw-related KCNC3 and KCNC4 and Shal-related KCND1, KCND2 and KCND3. Various Kv channel-interacting proteins (~250aa) modulate the activity of Kv4/D (Shal)-type channels. See **delayed rectifier channels**.

AU-rich element *ARE* A region in the 3′ untranslated region of an mRNA transcript to which proteins that affect stability bind. This has been shown for mRNA encoding oncoproteins, cytokines and transcription factors.

AUF-1 A protein (AU-rich element RNA-binding protein 1, heterogeneous nuclear ribonucleoprotein D0, 355aa) found in cytoplasmic mRNP granules containing untranslated mRNAs and involved in the post-transcriptional regulation of mRNA containing **AU-rich elements**. Also acts as a transcription factor.

AUG The **codon** in **messenger RNA** that specifies initiation of a polypeptide chain, or within a chain, incorporation of a **methionine** residue.

augmin A complex required in *Drosophila* for localizing gamma-tubulin to spindle microtubules and necessary for centrosome-independent microtubule generation within the spindle. It is formed by the interaction of the five Dgt proteins (Dgt2-6) and at least 3 other proteins. A human homologue has been found, the HAUS complex. Paper: http://www.ncbi.nlm.nih.gov/pubmed/18443220

Aurelia aurita Common jellyfish − transparent disc with four blue/purple horseshoe-shaped gonads clearly visible. Phylum Cnidaria; Class Scyphozoa.

aureolic acid The type compound of a family of anti-tumour antibiotics that includes mithramycin, chromomycin, and olivomycin. They interact with the DNA minor groove in high-GC-content regions with a requirement for magnesium ions.

aurintricarboxylic acid A general inhibitor of nucleases that also stimulates the tyrosine phosphorylation of **MAP kinases**, inhibits both major **calpain** isoforms. Potent and selective inhibitor of **SARS** coronavirus replication and apoptosis. Data-sheet: http://www.axxora.com/cell_cycle__dna__transaction-ALX-270-201/opfa.1.1.ALX-270-201.26.4.1.html

auristatins Synthetic members of the **dolastatin** class of tubulin polymerization inhibitors.

Aurora kinases A family of serine-threonine kinases (Aurora kinase A, 403aa; AURK-B, 344aa; AURK-C, 309aa) required for spindle assembly, centrosome maturation, chromosomal segregation and cytokinesis. Overexpression may cause genomic instability and is found in a wide range of tumours. Aurora-B (AIM-1) is a component of the **chromosomal passenger complex** that transfers from the inner centromere in early mitosis to the spindle midzone, equatorial cortex and midbody in late mitosis and cytokinesis.

aurosome Gold-containing secondary lysosome found in patients treated with gold complexes.

aurovertins A family of related antibiotics from the fungus *Calcarisporium arbuscula* that inhibit oxidative phosphorylation in mitochondria and in many bacterial species. Aurovertins B and D have identical biological properties and are more potent than aurovertin A. Inhibit the proton-pumping **F-type ATP synthase** by binding to β-subunits in its F1 catalytic sector.

Australia antigen *HBsAg* A viral envelope antigen of **hepatitis B** virus. Appearance of the antigen in serum is associated with a phase of high infectivity. First identified in the serum of an Australian aborigine.

autacoids Local hormones such as **histamine, serotonin, angiotensin, eicosanoids**.

autapomorphy A derived trait unique to any given taxon.

autapses Synapses formed by a neuron with itself, found in interneurons in the CNS. The functional significance of autaptic synapses is unclear.

autoantibody Antibody that reacts with an antigen that is a normal component of the body. Obviously this can lead to some problems, and autoimmunity has been proposed as a causative factor in a number of diseases such as rheumatoid arthritis. See also **systemic lupus erythematosus, Hashimoto's thyroiditis, myasthenia gravis**.

autocatalytic A compound that catalyses its own chemical transformation. More commonly a reaction that is catalysed by one of its products or an enzyme-catalysed reaction in which one of the products functions as an enzyme activator.

autoclave An apparatus for sterilization by steam at high pressure or the use of such an apparatus. A high temperature, e.g. 121°C for 15 minutes, is necessary to ensure killing of bacterial spores.

autocoid See **autacoid**.

autocrine Secretion of a substance, such as a **growth factor**, that stimulates the secretory cell itself. One route to independence of **growth control** is by autocrine growth factor production.

autofluorescence Property of a compound or material that will fluoresce in its own right — without the addition of an exogenous fluorophore. A common problem in fluorescence microscopy and in assays where the read-out is fluorescence.

autogamy Self-fertilisation, common in plants and also in some ciliate protozoa where gametic nuclei from a single micronucleus subsequently fuse to form the zygote nucleus.

autogenous Generated without external influence or input.

autograft Graft taken from one part of the body and placed in another site on the same individual.

autoimmune Adjective describing a situation in which the immune system responds to normal components of the body and auto-reactive T-cells or auto-antibodies are found. Several diseases are thought to have an autoimmune component although the autoantigen is not always obvious. Autoimmunity may be a failure to develop self-tolerance by the deletion of self-reactive clones of cells during development or because normal antigens get presented as though they were dangerous (associated with danger signals through the **toll-like receptors**). A third possibility is that bacterial or viral antigens mimic self-antigens and there is a cross-reaction.

autoimmune polyendocrinopathy syndrome A disorder characterized by the presence of two of three major clinical symptoms: Addison's disease, and/or hypoparathyroidism, and/or chronic mucocutaneous candidiasis. Type I is caused by mutation in the **autoimmune regulator** gene.

autoimmune regulator A transcriptional regulator (545aa) that binds to DNA as a dimer or as a tetramer. It promotes the expression of otherwise tissue-specific self-antigens in the thymus, which is important for self tolerance and the avoidance of autoimmune reactions.

autoinducer-2 A 5-carbon sugar that spontaneously cyclizes from 4,5-dihydroxy-2,3-pentanedione (DPD), a product of the LuxS enzyme in the catabolism of S-ribosylhomocysteine. It appears to be a universal signal molecule mediating interspecies communication among Gram-negative bacteria. See **quorum sensing**.

autologous Derived from an organism's own tissues or DNA. *Cf.* **heterologous, homologous**.

autolysis Spontaneous **lysis** (rupture) of cells or organelles produced by the release of internal hydrolyic enzymes. Normally associated with the release of lysosomal enzymes.

autonomic nervous system Neurons that are not under conscious control, comprising two antagonistic components, the **sympathetic** and **parasympathetic nervous systems**. Together, they control the heart, viscera, smooth muscle, amongst others.

autophagy Removal of cytoplasmic components, particularly membrane bounded organelles, by digesting them within **secondary lysosomes** (autophagic vacuoles). Particularly common in embryonic development and senescence. See **ATG #2**.

autophosphorylation Addition of a phosphate to a protein kinase (possibly affecting its activity) by virtue of its own enzymic activity.

autoradiography Technique in which a specimen containing radioactive atoms is overlaid with a photographic emulsion which, after an appropriate lapse of time, is developed, revealing the localization of radioactivity as a pattern of silver grains. Resolution is determined by the path length of the radiation, and so the low-energy β-emitting isotope, tritium, is usually used.

autoregulation Regulation of a gene encoding a transcription factor by its own gene product: a feedback process.

autoregulatory ribonucleoprotein complex A multi-subunit autoregulatory ribonucleoprotein complex (ARC) involved in translational control. It is composed of IGF2BP1 (insulin-like growth factor 2 mRNA-binding protein 1, a RNA-binding factor that affects mRNA nuclear export, localization, stability and translation.), PABP1 (**poly(a) binding protein 1**, PABPC1) and CSDE1 (cold shock domain-containing protein E1). Paper: http://onlinelibrary.wiley.com/doi/10.1111/j.1742-4658.2006.05556.x/full

autosomal dominant Gene located on an **autosome** that has a dominant effect — even though two copies of the gene exist, one of them normal. Often attributed to a **gain of function mutation**.

autosomal recessive Mutation carried on an **autosome** that is deleterious only in homozygotes.

autosomes Chromosomes other than the sex chromosomes.

autotaxin An enzyme originally associated with melanoma cells, known to stimulate motility in tumour cells, that turns out to be **lysophospholipase D** (ATX, EC 3.1.4.39, 863aa) and works by producing **lysophosphatidic acid**. Also stimulates migration of smooth muscle cells and neurite outgrowth.

autotroph Organisms that synthesize all their organic molecules from inorganic materials (carbon dioxide, salts amongst others). May be photo-autotrophs or chemo-autotrophs, depending upon the source of the energy. Also known as lithotrophic organisms.

auxesis Growth by increase in cell size rather than by increasing cell numbers.

auxilin A novel adaptin, a protein tyrosine phosphatase (EC 3.1.3.48, 913aa) that promotes uncoating of clathrin-coated vesicles.

auxin efflux carriers Membrane transport proteins that specifically convey auxin out of the plant cell. The best-characterized are proteins encoded by the **PIN multigene family**, located at the rootward or basal end of cells competent to transport auxin. This asymmetric localization explains the involvement of these proteins in maintaining cell polarity. Growth of the plant towards or away from environmental stimuli, such as light and gravity, is provided by the realignment of auxin efflux carriers. Bacteria and yeasts possess homologues of this transporter family and the AEC family may be distantly related to the bile acid: Na^+ symporter (BASS) family and the divalent anion: Na^+ symporter (DASS) family which include members found in animals. See **auxin influx carriers**. Article: http://www.biomedcentral.com/1471-2229/9/139 Transport carrier database: http://www.tcdb.org/tcdb/index.php?tc=2.A.69

auxin influx carriers Carrier proteins of the amino acid/polyamine transporter two family involved in proton-driven auxin influx. They facilitate acropetal (base to tip) auxin transport in conjunction with **auxin efflux carriers**.

auxin response factors Transcription factors (ARFs) that mediate responses to the phytohormone, **auxin**. ARF2 (859aa) promotes transitions between multiple stages of *Arabidopsis* development, ARF1 (665aa) acts in a partially redundant manner with ARF2. Many others are known. N.B. ARF is also an abbreviation for **ADP-ribosylation factor**.

AUXIN SIGNALING F-BOX A family of proteins that are components of a SCF (SKP1-cullin-F-box) protein ligase complex that mediates Aux/IAA proteins proteasomal degradation and auxin-regulated transcription. Involved in regulation of embryogenesis by auxin. An example is AUXIN SIGNALLING F-BOX-1 (AFB1, GRR1-like protein 1, 585aa); several others are known.

auxins A group of **plant growth substances** (often called phytohormones or plant hormones), the most common example being indole acetic acid (IAA).

IAA promotes the excretion of protons (hydrogen ions) that favour the breakage of intermolecular bonds. The resulting wall relaxation allows the cell to expand under turgor pressure.

auxotroph Mutant that differs from the wild-type in requiring a nutritional supplement for growth. A deficiency mutant.

auxotyping Method for strain-typing *Neisseria* by checking their requirements for specific nutrients in defined media.

available nutrients Nutrients in the extracellular environment that are accessible to uptake, not bound irreversibly to some other component or in insoluble complexes.

avascular necrosis of femoral head A disorder that usually leads to destruction of the hip joint, caused by mutation in the collagen COL2A1 gene (also mutated in **Perthe's disease**).

Avena sativa Cultivated oat.

averantin An intermediate in the biosynthetic pathway that generates **aflatoxins**.

avermectins Potent acaricides, insecticides and anthelminthics, originally isolated from *Streptomyces avermitilis*.

averufin An intermediate in the biosynthetic pathway that generates **aflatoxins**.

avian erythroblastosis virus See **avian leukaemia virus**.

avian flu A highly contagious strain of influenza that affects poultry and can be transmitted to humans; having switched species it could become extremely virulent (a similar phenomenon occurred with the 1918 strain of flu which caused more deaths than the preceding four years of war). The **H5N1** strain recently caused considerable concern.

avian leukaemia virus Group of C-type RNA tumour viruses (**Oncovirinae**) that cause various leukaemias and other tumours in birds. The acute leukaemia viruses, that are replication-defective and require helper viruses, include avian erythroblastosis (AEV), myeloblastosis (AMV) and myelocytomatosis viruses. AEV carries two transforming genes, v-*erbA* and v-*erbB*; the cellular homologue of the latter is the structural gene for the **epidermal growth factor** receptor. AMV carries v-*myb* and causes a myeloid leukaemia; avian myelocytomatosis virus carries v-*myc*. The avian lymphatic leukaemia viruses (ALV) are also **Retroviridae** but are replication-competent and induce neoplasia only after several months; they often occur in conjunction with replication-defective leukaemia viruses.

avian myeloblastosis virus *AMV* Retrovirus of the subfamily Oncornaviridae. Causes myelocytomatosis, osteopetrosis, lymphoid leukosis and nephroblastoma. May be a mixture of viruses.

avidin Biotin-binding protein (152aa) from egg-white. Binding is so strong as to be effectively irreversible – a diet of raw egg-white leads to biotin deficiency. Avidin affinity columns can be used to affinity-purify biotinylated proteins.

avidity Strength of binding, usually of a small molecule with multiple binding sites by a larger; particularly the binding of a complex antigen by an antibody. **Affinity** refers to simple receptor-ligand systems.

avirulent Organism or virus that does not cause infection or disease.

awn A bristle projecting from the end of a plant part, e.g. the beard of barley.

axenic A situation in which only one species is present. Thus an axenic culture is uncontaminated by organisms of other species, an axenic organism does not have commensal organisms in the gut and other things. Some organisms have obligate symbionts and cannot be grown axenically.

axial filaments The central filaments, of which there may be several hundred, of the periplasmic flagella of spirochaetes that rotate within the periplasmic space and cause the whole bacterium to flex like a corkscrew and thus to move. The central filaments are composed of at least three proteins (FlaB1, FlaB2 and FlaB3) that have significant homology with flagellin; the sheath is composed of FlaA protein (~43 kDa). Paper: http://jb.asm.org/cgi/content/full/182/23/6698

axil Member of the **axin** family. Interacts with **GSK3** and β-catenin. By enhancing phosphorylation and thus the subsequent degradation of β-catenin, inhibits axis formation in *Xenopus* embryos. Original description: http://www.ncbi.nlm.nih.gov/pubmed/9566905

axillary meristem A meristem in the angle between the stem and a leaf petiole that forms buds which, when released from apical dominance, can form a flower or a lateral shoot.

axins Negative regulators (axin-1, 862aa; axin-2, 843aa) of the **Wnt** signalling pathway: control dorsoventral patterning by down-regulating beta-**catenin** which ventralizes and by activating a JNK-pathway that dorsalizes. Axin interaction partner and dorsalization antagonist (AIDA, 306aa) inhibits the axin-mediated JNK activation.

AXL A proto-oncogene product, a receptor tyrosine kinase (EC 2.7.10.1, 887aa) for which gas 6 (See **gas genes**) is a ligand. Seems to function as a cell entry factor for filovirus.

axokinin Axonemal protein (56 kDa) that, when phosphorylated by a cAMP-dependent protein kinase, reactivates the **axoneme**. Probably a regulatory subunit of PKA – disappeared from the literature in 1986.

axolemma Plasma membrane of an axon.

axon Long process, usually single, of a **neuron**, that carries efferent (outgoing) **action potentials** from the cell body towards target cells. See **dendrite**.

axon hillock Tapering region between the cell body of a **neuron** and its axon. This region is responsible for summating the graded inputs from the **dendrites**, and producing **action potentials** if the threshold is exceeded.

axonal guidance General term for mechanisms that ensure correct projections by nerve cells in developing and regenerating nervous systems. Implies accurate navigation by **growth cones**, the highly motile tips of growing neuronal processes. See **growth cone collapse**.

axoneme The central microtubule complex of eukaryotic **cilia** and flagella with the characteristic '9 + 2' arrangement of tubules when seen in cross-section.

axonin Chick homologue of TAG-1. See **tax #2**.

axonogenesis The growth and differentiation of axonal processes by developing neurons. See **axon**.

axonotmesis A type of nerve injury in which the axon and myelin sheath are degraded but connective tissue remains: generally allows regeneration to occur.

axoplasm The **cytoplasm** of a **neuron**.

axopod Thin process (a few μm in diameter but up to 500 μm long), supported by a complex array of **microtubules**. Axopodia radiate from the bodies of **Heliozoa**.

axostyles Ribbon-like bundles of **microtubules** found in certain parasitic protozoa that may generate bending waves by **dynein**-mediated sliding of microtubules.

azacytidine The ribonucleoside of **5-azacytosine**.

azacytosine An analogue (5-azacytosine) of the pyrimidine base cytosine, in which carbon 5 is replaced by a nitrogen. In DNA, unlike cytosine, it cannot be methylated.

azaserine An analogue of **glutamine** that competitively inhibits various pathways in which glutamine is metabolized, hence an antibiotic and antitumour agent.

azathioprine An immunosuppressant drug that inhibits purine synthesis.

azide Usually the sodium salt NaN_3, an inhibitor of electron transport that blocks electron flow from cytochrome oxidase to oxygen. Frequently used to prevent growth of microorganisms in e.g. refrigerated antisera or chromatography columns.

azidothymidine See **AZT**.

azoospermia Absence of spermatozoa in the semen. Can be caused by deletion of ubiquitin-specific protease 9 (EC 3.1.2.15, 2555aa) or mutation in the gene encoding **synaptonemal complex** protein 3 (236aa).

azothioprine *azidothioprine* Immunosuppressive drug used to prevent graft rejection and to treat a variety of connective tissue disorders.

Azotobacter Genus of free-living, obligately aerobic, rod-shaped Gram-negative bacilli capable of fixing atmospheric nitrogen.

AZT An antiviral drug (azidothymidine, zidovudine.) derived from thymidine, used in treatment of AIDS. Blocks the enzyme that stimulates growth and multiplication of the human immunodeficiency virus.

azurin (1) Blue copper-containing protein from *Pseudomonas aeruginosa*. (2) Histochemical dye.

azurophil granules Primary lysosomal granules found in **neutrophil granulocytes**; contain a wide range of hydrolytic enzymes. Sometimes referred to as primary granules to distinguish them from the secondary or **specific granules**.

B

β *beta* Entry prefix is given as 'beta' (i.e. listed after 'bestrophins' and before 'betacellulin'); alternatively look for the main portion of the word.

B1 and B2 repeats B1 and B2 sequences are two major classes of short repeats dispersed throughout the mouse genome. They consist of 130 and 190 nucleotides and there are around 50,000 copies of each per haploid genome.

B7 Superfamily of co-stimulatory molecules that bind to CD28 or **CTLA-4** and regulate T-cell responses. They are part of the **immunoglobulin superfamily** and have an extracellular Ig variable-like (IgV) and constant-like (IgC) domains. B7-1 is CD80, B7-2 is CD86. Other members are ICOS-L, B7-H1, B7-DC. Several B7 homologues are expressed on cells other than professional antigen-presenting cells.

B12 See **vitamin B**.

B220 One isoform of the mouse CD45 antigen, predominantly expressed on B-cells. A receptor-type tyrosine-protein phosphatase C (EC 3.1.3.48, 1291aa).

Babes-Ernst granules Metachromatic intracellular deposits of polyphosphate found in *Corynebacterium diphtheriae* when the bacteria are grown on sub-optimal media. Stain reddish with methylene blue or toluidine blue.

Babesia Genus of protozoa that are found as parasites within red blood cells of mammals and are transmitted by ticks. Cause babesiosis (piroplasmosis).

BABY BOOM *BBM* A plant transcription factor (of the AP2/EREPB multigene family) that induces the ectopic formation of **somatic embryos** and cotyledon-like structures on *Arabidopsis* and *Brassica* seedlings.

baby hamster kidney cells See **BHK cells**.

Bac7 Proline-and arginine-rich antimicrobial peptide (**cathelicidin-3**, bactenecin-7, 190aa) isolated from bovine neutrophils. Bac-5 is similar. One of the **protegrin** family of peptides.

BAC library Library (bacterial artificial chromosome library) constructed in a vector with an **origin of replication** that allows its propagation in bacteria as an extra chromosome. Advantageous in constructing **genomic libraries** with relatively large DNA fragments (100–300 kb). See also **bacterial artificial chromosome**.

bacillaene An antibacterial polyketide from *Bacillus subtilis*.

Bacille Calmette-Guerin *BCG* An attenuated **mycobacterium** derived from *Mycobacterium tuberculosis*. The bacterium is used in tuberculosis vaccination. Extracts of the bacterium have remarkable powers in stimulation of lymphocytes and leucocytes and are used in **adjuvants**.

bacillibactin An iron-**siderophore** produced by many *Bacillus* sp.

bacillus Cylindrical (rod-shaped) bacterium. Bacilli are usually 0.5–1.0 μm long, 0.3–1 μm wide.

Bacillus amyloliquefaciens A plant-associated Gram-positive spore-forming bacterial pathogen that has potential as a biocontrol agent. It produces the antibiotics **fengycin**, **surfactin**, **iturin A**, the iron siderophore **bacillibactin** and the antibacterial polyketides **macrolactin**, **bacillaene** and **difficidin**.

Bacillus anthracis A soil-living Gram-positive, spore producing bacterium that produces **anthrax toxin**. Pathogenesis results from inhalation of spores which germinate in the body where cell growth is accompanied by toxin production.

Bacillus cereus A Gram-positive, facultatively aerobic spore forming bacterium. *B. cereus* food poisoning is caused by two distinct metabolites, the diarrheal type of illness by a large molecular weight protein; the vomiting (emetic) type of illness by a low molecular weight, heat-stable peptide.

Bacillus megaterium A Gram-positive, spore producing, rod-shaped bacterium found in the soil. It is one of the largest Eubacteria and is extensively used in biotechnology due to its size and cloning abilities. Enzymes produced by *B. megaterium* are used in production of synthetic penicillin, modification of corticosteroids and include several amino acid dehydrogenases.

Bacillus thuringiensis Soil-living Gram-positive, spore producing bacterium that produces a **delta-endotoxin** that is deadly to insects. Many strains exist, each with great specificity as to target Orders of insects. In general, the mode of action involves solubilization at the high pH within the target insect's gut, followed by proteolytic cleavage; the activated peptides form pores in the gut cell apical plasma membranes, causing lysis of the cells. The toxin has been genetically engineered into various plant species, 'GM plants', to confer insect

The Dictionary of Cell and Molecular Biology. DOI: http://dx.doi.org/10.1016/B978-0-12-384931-1.00002-7

resistance: many human consumers have reacted negatively although unaffected by the toxin.

bacitracins Branched cyclic peptides produced by strains of *Bacillus licheniformis*. Interfere with murein (**peptidoglycan**) synthesis in Gram-positive bacteria.

baclofen Skeletal muscle relaxant, a derivative of GABA that selectively binds GABA$_B$ receptor and inhibits release of other neurotransmitters in the CNS.

bacmid A baculovirus shuttle vector that can be propagated in both *E. coli* and in insect cells. Details: http://www.bioc.cam.ac.uk/baculovirus/info/Baculo_virus_system.php

bactenecins Highly cationic polypeptides (**cathelicidins**) found in lysosomal granules of bovine neutrophil granulocytes. They are thought to be involved in bacterial killing and occur in a third class of granules, the large granules, not found in the neutrophils of most species.

bacteraemia *US bacteremia* The presence of living bacteria in the circulating blood: usually implies the presence of small numbers of bacteria that are transiently present without causing clinical effects, in contrast to **septicaemia**.

Bacteria *formerly* Eubacteria One of the two major subdivisions of the prokaryotes that includes most Gram-positive bacteria, cyanobacteria, proteobacteria and mycoplasmas. Unlike the **Archaea** (*formerly* Archaebacteria) they have ester-linked lipids in the cytoplasmic membrane, **peptidoglycan** in the cell wall, and do not have **introns**. Bacteria are small with linear dimensions around 1 μm, do not have internal compartments, have circular DNA, and ribosomes of 70 S. Protein synthesis differs from that of eukaryotes so that it is possible to use anti-bacterial antibiotics that interfere with protein synthesis without affecting the eukaryotic host.

bacterial artificial chromosome *BAC* A vector used to construct a **genomic library** that has the sites necessary for the DNA to be handled and replicated as a bacterial chromosome. Like **YACs**, this allows clones to contain very large pieces of DNA (around 200 kb), so aiding rapid, low resolution **physical mapping.**

bacterial cell wall Bacterial cells walls are of two major types, those that retain the **Gram stain** (Gram-positive bacteria) and those that do not. Gram-positive bacteria have a wall approximately 50 nm thick containing **teichoic acid** and **peptidoglycan**, made up of repeating N-acetyl-glucosamine and N-acetyl-muramic acid. The wall of Gram-negative bacteria is separated from the cell membrane by the periplasmic space, is much thinner, contains a different peptidoglycan, and has an outer lipid bilayer (containing lipid A) resembling a membrane.

Some bacteria, notably mycoplasmas, do not produce a cell wall. **Archaea** generally have rigid walls but the peptidoglycan composition differs.

bacterial chemotaxis The response of bacteria to gradients of attractants or repellents. In a gradient of attractant the probability of deviating from a smooth forward path is reduced if the bacterium is moving up-gradient. Since the opposite is true if moving down-gradient, the effect is to bias displacement towards the source of attractant. Strictly should perhaps be considered a **klinokinesis** with adaptation.

bacterial flagella Thin filaments composed of **flagellin** subunits that are rotated by the basal motor assembly and act as propellors. In Gram-negative bacteria the motor consists of four rings, the MS-ring and C-ring are anchored in the plane of the cytoplasmic membrane and the L- and P-rings, which act as bushes, are linked respectively to the lipopolysaccharide and peptidoglycan layers of the cell wall. In Gram-positive bacteria, which lack an outer membrane, only the inner P, MS and C rings are present. The MS- and C-rings are linked by a series of Mot proteins which drive rotation of the flagellum. The associated Fli proteins function as a motor switch, reversing the direction of rotation of the flagellum in response to intracellular signals. If rotating anticlockwise (as viewed from the flagellar tip) the bacterium moves in a straight path, if clockwise the bacterium 'tumbles'. The direction of rotation is controlled through the bacterial chemotactic receptor system (see **bacterial chemotaxis**). Details: http://jb.asm.org/cgi/content/full/181/23/7149

bactericidal permeability-increasing proteins Secreted protein (BPI, 487aa) produced by cells of the myeloid series and part of the BPI/LBP/Plunc superfamily. The protein is cytotoxic for Gram-negative bacteria: the very basic N-terminal half has high affinity for lipopolysaccharide of the outer envelope. Several BPI-like proteins (e.g. BPIL1, bactericidal/permeability-increasing protein-like one, long palate, lung and nasal epithelium carcinoma-associated protein 2, 458aa) are associated with particular mucosal sites (BPIL1 with tonsils).

bacteriochlorophyll Varieties of **chlorophyll** (bacteriochlorophylls a, b, c, d, e and g) found in **photosynthetic bacteria** and differing from plant chlorophyll in the substituents around the tetrapyrrole nucleus of the molecule, and in the absorption spectra.

bacteriocide A substance that kills bacteria.

bacteriocins Exotoxins, often **plasmid** coded, produced by bacteria and which kill other bacteria (not eukaryotic cells). **Colicins** are produced by about 40% of *E. coli* strains: colicin E2 is a DNA-ase, colicin E3 an RNA-ase. See **lantibiotics**.

bacteriophages *phages* Viruses that infect bacteria. The bacteriophages that attack *E. coli* are termed coliphages, examples of these are lambda phage and the T-even phages, T2, T4 and T6. Basically, phages consist of a protein coat or **capsid** enclosing the genetic material, DNA or RNA, that is injected into the bacterium upon infection. In the case of virulent phages all synthesis of host DNA, RNA and proteins ceases and the phage genome is used to direct the synthesis of phage nucleic acids and proteins using the host's transcriptional and translational apparatus. These phage components then self-assemble to form new phage particles. The synthesis of a phage lysozyme leads to rupture of the bacterial cell wall releasing, typically 100–200 phage progeny. The temperate phages, such as **lambda**, may also show this lytic cycle when they infect a cell, but more frequently they induce **lysogeny**. The study of bacteriophages has been important for our understanding of gene structure and regulation. Lambda has been extensively used as a vector in recombinant DNA studies.

bacteriopheophytin-b One of the components of the bacterial photosynthetic **reaction centre** (see also **ubiquinone**).

bacteriorhodopsin A light-driven proton-pumping protein (248aa), similar to rhodopsin, found in 'purple patches' in the cytoplasmic membrane of the bacterium *Halobacterium halobium*. It is composed of seven transmembrane helices, and contains the light-absorbing **chromophore**, **retinal**. Light absorption maxima: 568 nm (light-adapted); 558 nm (dark-adapted). Each photon results in the movement of two protons from cytoplasmic to extracellular sides of the membrane. The resulting proton gradient is used (amongst other things) to drive synthesis of ATP by **chemiosmosis**.

bacteriostatic Adjective applied to substances that inhibit the growth of bacteria without necessarily killing them.

bacteroid Small, often irregularly rod shaped bacterium, e.g. those found in root nodules of nitrogen-fixing plants which form into a symbiosome, surrounded by plant cytoplasmic membrane, where nitrogen fixation takes place.

Bacteroidetes A bacterial phylum composed of three large classes of bacteria (Bacteroides, Flavobacteria, and Sphingobacteria). A few are opportunistic pathogens.

Baculoviridae A family of viruses that are invertebrate pathogens with a circular double-stranded genome ranging from 80–180 kbp. They can be divided into two genera: nucleopolyhedroviruses (NPV) and granuloviruses (GV). The virus may be sequestered in infected cells within inclusion bodies that may be crystalline and composed of a single viral protein (e.g. **polyhedrin**). There is specificity for host species, the commonest being lepidopteran larvae. Baculoviruses are used as eukaryotic **expression vectors** in insect or vertebrate cells to produce proteins that require post-translational modifications such as **glycosylation**, proteolytic cleavage and fatty acylation. See **EaA cells**, **Sf9 cells**. Virus taxonomy database: http://www.ncbi.nlm.nih.gov/ICTVdb/ICTVdB/00.006.htm

Bad A pro-apoptotic member of the **Bcl-2** family (Bcl2 antagonist of cell death, 168aa) that promotes cell death but is tightly regulated by survival factors. Several major signalling pathways influence cell death through their direct effects on the phosphorylation state of Bad.

BAF See **barrier-to-autointegration factor** and **BRG1-associated factors**.

BAF complex A multi-subunit chromatin remodelling complex in mammals that is functionally related to the **SWI/SNF** complex of yeast. It consists of 9 to 12 subunits, (**BRG1-associated factors**, BAFs), some of which are homologous to SWI/SNF subunits.

bafilomycin Microbial toxin, a macrolide antibiotic from *Streptomyces griseus* that is a specific inhibitor of the **V-type ATPase**; also blocks lysosomal cholesterol transport in macrophages.

bag cell neurons Cluster of electrically coupled neurons in the abdominal ganglion of *Aplysia* that are homogeneous, easily dissected out and release peptides that stimulate egg laying.

BAG family A family of proteins (**Bcl-2** associated athanogene products) that act as co-chaperones. BAG-1 (345aa) facilitates ubiquitin-proteasome-mediated protein degradation and potentiates the anti-apoptotic functions of **Bcl-2** to which it binds. Deregulated in a variety of malignancies. BAG-2 (211aa) inhibits the chaperone activity of Hsp70/HSC70 by promoting substrate release. BAG-3 (docking protein CAIR-1, 575aa) has similar effects. BAG4 (silencer of death domains, SODD, 457aa) associates with the TNF receptor 1 (TNFR1) but is displaced if TNF binds. BAG5 (447aa) inhibits **parkin** and enhances dopaminergic neuron degeneration. Large proline-rich protein BAG6 (1132aa) acts as a chaperone in various processes including apoptosis, insertion of tail-anchored (TA) membrane proteins to the endoplasmic reticulum membrane and regulation of chromatin.

bagassosis Respiratory disease, similar to **farmer's lung**, caused by inhalation of dust from mouldy sugar cane. A Type III **hypersensitivity** reaction to mould spores.

Bak One of the pro-apoptotic **Bcl-2** family, a mitochondrial membrane protein (Bcl-2 homologous antagonist/killer, 211aa). Loss of the interaction between Bak and the anti-apoptotic Bcl-2-family member Mcl1 allows interaction with p53, oligomerization of Bak and release of cytochrome c from mitochondria.

bakuchiol A compound isolated from seeds of *Psoralea corylifolia* (Chinese tree; used in traditional medicine). It is an antioxidant and inhibits various enzymes including tyrosine phosphatase 1B and DNA polymerase. Inhibits mitochondrial lipid peroxidation and the expression of inducible nitric oxide synthase. Shows antimicrobial and cytotoxic activity.

BAL See **dimercaprol**.

BALB/c Inbred strain of white (albino) mice. Used as a source for one of the various 3T3 cell lines.

Balbiani ring The largest **puffs** seen on the **polytene chromosomes** of Diptera are called Balbiani rings after the nineteenth century microscopist who first described polytene chromosomes.

Baller-Gerold syndrome A disorder of skeletal development that phenotypically overlaps with disorders such as **Saethre-Chotzen syndrome**. Caused by mutations in the RECQL4 gene that encodes a **RECQ helicase**.

balloon cell Nonspecific description of any cell with abundant clear cytoplasm. May arise through a variety of causes and includes some **carcinoid** cells, hepatocytes following some forms of toxic insult or viral infection, neurons or other cells in **storage diseases** and cells in some **melanomas**.

BALT See **bronchus associated lymphoid tissue**.

Bamforth-Lazarus syndrome See **hypothyroidism**.

BamH I Common Type II **restriction enzyme** (from *Bacillus amyloliquefaciens* H) that cuts the sequence G|GATCC. See Table R2.

BAN *British Approved Name*. Formal name for a medicinal substance, now generally identical to the recommended International Non-Proprietary Names (rINN). In general BANs have been used for the main entry and the old name is given as a synonym.

band 4.1 domain The band 4.1 domain (JEF domain) was first identified in the red blood cell protein band 4.1, and subsequently in **ezrin**, **radixin**, and **moesin** (ERM proteins) and other proteins, including the tumour suppressor **merlin**/schwannomin, **talin**, unconventional myosins VIIa and X, and **protein tyrosine phosphatases**. A structurally-related domain has also been demonstrated in the N-terminal region of two groups of tyrosine kinases: the **focal adhesion kinases** (FAK) and the **Janus kinases** (JAK). Additional proteins containing the 4.1/JEF (JAK, ERM, FAK) domain include plant kinesin-like calmodulin-binding proteins (KCBP).

Additional properties common to band 4.1/JEF domains of several proteins are binding of phosphoinositides and regulation by GTPases of the **rho** family.

band cells Immature **neutrophils** released from the bone-marrow reserve in response to acute demand.

band III A protein (911aa) of the human erythrocyte membrane, identified as the major anion transport/ exchange protein. Analogous proteins exist in other erythrocytes. A dimeric transmembrane glycoprotein, with binding sites for many cytoplasmic proteins, including **ankyrin**, on its cytoplasmic domain.

banding patterns Chromosomes stained with certain dyes, commonly quinacrine (Q banding) or Giemsa (G banding), show a pattern of transverse bands of light and heavy staining that is characteristic for the individual chromosome. The basis of the differential staining, which is the same in most tissues, is not understood: each band represents 5–10% of the length, about 10^7 base pairs, although this is not true for **polytene chromosomes** in *Drosophila* that show more than 4000 bands.

bandshift assay An assay (gel shift assay) for proteins, such as transcription factors, that band specific DNA sequences. A labelled oligonucleotide corresponding to the recognition sequence is incubated with an appropriate nuclear protein extract, and run on a non-denaturing acrylamide gel. Oligonucleotides that have been bound by proteins are retarded relative to those that are unbound.

bantam In *Drosophila*, a gene encoding a 21 nucleotide microRNA that promotes tissue growth and regulates the pro-apoptotic gene *hid*. Abstract: http:// www.ncbi.nlm.nih.gov/pubmed/12679032

BAP Synthetic plant growth regulator (6-benzylaminopurine), a cytokinin, that stimulates cell division.

BAPTA Calcium chelator with low affinity for magnesium. Absorption maximum shifts when calcium is bound so it can be used as an indicator of intracellular calcium concentration (though it will chelate calcium and therefore alter the situation. See **MAPTAM**.

bar peptidase In *S. cerevisiae* a peptidase (barrierpepsin, EC 3.4.23.35, extracellular 'barrier' protein, 587aa) that is excreted by yeast cells of mating type a and probably cleaves the **alpha factor** mating pheromone.

BAR proteins A superfamily of adaptor proteins (**BIN1/amphiphysin/Rvs** domain proteins) that apparently integrate signal transduction pathways that regulate membrane dynamics, the F-actin cytoskeleton, and nuclear processes. They bind to membranes but differ in their preferences for membrane curvature. BAR/N-BAR modules bind to membranes of high positive curvature, F-BAR modules bind to a different range of positive membrane curvatures and I-BAR modules bind to negatively curved membranes. Rvs in *S. cerevisiae* is Reduced viability upon starvation protein 167, 482aa) required for the formation of endocytic vesicles at the plasma membrane level.

Barakat's syndrome See **hypoparathyroidism**.

barbamide A chlorinated metabolite synthesized by a **PKS/NRPS** gene cluster in the cyanobacterium *Lyngbya majuscula*. It has molluscicidal activity.

barbital The BAN for barbitone or veronal (diethyl-malonyl-urea, 5,5-diethylbarbituric acid). Once widely used, as the sodium salt, as a sedative and hypnotic; also used in pH buffer solutions.

barbiturates Class of drugs that depress activity of the central nervous system, largely superceded by **benzodiazepines**.

BARD1 A protein (BRCA1-associated RING domain protein 1, EC 6.3.2.-, 777aa) that forms a heterodimer with BRCA1, mediates ubiquitin E3 ligase activity and is important for the cellular response to DNA damage. Also interacts with **CSTF**. May be a tumour suppressor.

Bardet-Biedl syndrome A genetically heterogeneous disorder with multiple variants of the syndrome known and deletions on various chromosomes. All are associated with the basal body and cilia of the cell and the mutations affect ciliary function. Known mutations affect products of ARL6 (ADP-ribosylation-like 6), MKKS (probably a chaperonin), TTC8 (tetratricopeptide repeat domain 8, required for ciliogenesis), TRIM32 (tripartite motif-containing protein 32), MKS1 (B9 domain-containing protein, associated with basal bodies and primary cilia), CEP290 (nephrocystin-6), C2ORF86 (a homologue of *Drosophila* planar cell polarity gene *fritz.*). The CCDC28B gene (coiled-coil domain-containing protein-28) modifies the expression of BBS phenotypes in patients who have mutations in other genes.

Barley yellow dwarf virus A plant virus that produces pale yellow or white lesions on barley leaves due to loss of chlorophyll.

barnase A ribonuclease (EC 3.1.27.-, 158aa) secreted by *Bacillus amyloliquefaciens*. It is inhibited intracellularly by barstar, a 90aa polypeptide.

barophiles Organisms that grow optimally at high hydrostatic pressure.

baroreceptor *baroceptor* In an organism, a receptor that is sensitive to pressure. There are baroreceptors in arteries, large veins and the right atrium of the heart that are sensitive to blood pressure.

Barr body Small dark-staining inactivated X chromosome seen in female (XX) cells. According to the **Lyon hypothesis**, which of the X chromosomes is inactivated is random.

barrier-to-autointegration factor A conserved protein (BAF, breakpoint cluster region protein 1, 89aa) essential in proliferating cells. BAF dimers bind dsDNA, histone H3, histone H1.1, **lamin** A and transcription regulators, plus **emerin** and other **LEM domain** nuclear proteins. Binding to emerin and lamin A is inhibited by phosphorylation at serine-4 of BAF and this is important in nuclear envelope disintegration at the start of mitosis. The association of viral DNA with chromatin requires the presence of BAF and emerin.

BARS In rats, cytosolic proteins (brefeldin A-ADP-ribosylated substrates) of 38 and 50 kDa (glyceralde-hyde-3-phosphate dehydrogenase and **C terminal binding protein one**) that become ADP-ribosylated following treatment of cells by **brefeldin A**.

barstar In *Bacillus amyloliquefaciens* an inhibitor (90aa) of the ribonuclease, **barnase**, that it releases.

Bartonella A genus of Gram-negative bacteria that are facultative intracellular parasites and opportunistic pathogens. They are transmitted by ticks, fleas mosquitoes amongst others.

Bartter's syndrome A set of disorders of salt reabsorption in the thick ascending loop of Henle. Can be caused by mutation in genes for kidney chloride channel B, SLC12A1 (sodium-potassium-chloride transporter-2) or the **ROMK** potassium channel (KCNJ1). Infantile Bartter's syndrome with sensorineural deafness is caused by mutation in **barttin** or in both the genes encoding chloride channels.

barttin The beta subunit (320aa) for the chloride channels CLCKNA and CLCKNB that are found in the basolateral membranes of renal tubules and in epithelia of the inner ear. Mutations can cause a form of **Bartter's syndrome**.

basal body Structure found at the base of eukaryotic **cilia** and **flagella** consisting of a continuation of the nine outer sets of axonemal microtubules but with the addition of a C-tubule to form a triplet (like the **centriole**). May be self-replicating and serves as a nucleating centre for axonemal assembly. Anchored in the cytoplasm by **rootlet system**. Synonymous with **kinetosome**.

basal cell (1) General term for a relatively undifferentiated cell in an epithelial sheet that divides to produce more specialized cells (act as a **stem cell**). In the stratified squamous epithelium of mammalian skin the basal cells of the epidermis (stratum basale) give rise by an unequal division to another basal cell and to cells that progress through the spinous, granular and horny layers, becoming progressively more keratinised, the outermost being shed as **squames**. In olfactory mucosa the basal cells give rise to olfactory and sustentacular cells. In the epithelium of epididymis their function is unclear, but they probably serve as stem cells. (**2**) In plants the larger of the two cells produced by the first (asymmetrical) division of the zygote. This cell (also called the suspensor) provides nutrients for the smaller apical cell which will form the embryo.

basal cell carcinoma Common **carcinoma** (BCC, rodent ulcer) derived from the basal cells of the epidermis. Often a consequence of exposure to sunlight and much more common in those with fair skin; rarely metastatic.

basal disc (**1**) Portion of the stalk of a cellular slime mould fruiting body that is attached to the substratum. (**2**) General name for the conical-shaped structure that anchors the stalk of a fungal fruiting body, a hydroid, or any other sessile organism, to the substratum.

basal ganglia Three large subcortical nuclei of the vertebrate brain: the putamen, the caudate nucleus and the globus pallidus. They participate in the control of movement along with the **cerebellum**, the corticospinal system and other descending motor systems. Lesions of the basal ganglia occur in a variety of motor disorders including **Parkinsonism** and **Huntington's chorea**.

basal lamina See **basement membrane**.

basal medium 199 The first tissue culture medium of defined composition, although for long term culture it is usually necessary to add fetal calf serum. Various modifications have been made (particularly to the salts used for buffering pH) but the basic formulation has wide applicability and is still used.

BASC complex A multiprotein complex (BRCA1-associated genome surveillance complex) composed of **BRCA1**, tumour suppressors and the DNA damage repair proteins MSH2, MSH6, MLH1, **ATM**, BLM and the **MRN complex** (MRE11, RAD50, and **nibrin**). It is involved in DNA repair following replication.

base analogues Purine and pyrimidine bases that can replace normal bases used in DNA synthesis and hence can be included in DNA, e.g. 5-bromouracil (replacing thymine) or 2-aminopurine replacing adenine. May be used for inducing mutations, including point mutations.

base excision repair A DNA repair mechanism in which DNA glycosylase excises a mutated base and creates an apurinic/apyrimidinic (AP) site. The AP site is recognised by AP endonuclease which cuts the strand upstream (5′) of the mutated base and polymerase 1 can then add a new (correct) base to the free 3′ end by pairing with the undamaged strand. Ligase then seals the join. Single bases are removed and replaced in 'short-patch repair'; up to 10 nucleotides can be replaced in 'long-patch repair'. See **excision repair**.

base-pairing The specific hydrogen-bonding between **purines** and **pyrimidines** in double-stranded nucleic acids. In DNA the pairs are **adenine** and **thymine**, and **guanine** and **cytosine**, while in RNA they are adenine and **uracil**, and guanine and cytosine. Base-pairing leads to the formation of a DNA double helix from two complementary single strands.

Basedow's disease Thyrotoxicosis. See **Graves' disease**.

basement membrane Extracellular matrix characteristically found under epithelial cells. There are two distinct layers: the basal lamina, immediately adjacent to the cells, is a product of the epithelial cells themselves and contains collagen type IV; the **reticular lamina** is produced by fibroblasts of the underlying **connective tissue** and contains fibrillar collagen.

basic leucine zipper *bZIP* Family of proteins having a basic region and a **leucine zipper**. The basic region is the DNA-binding domain and the leucine zipper is involved in protein-protein interactions to form homo- or heterodimers. Includes **AP-1**, **ATF** and **CREB** transcription factors.

basidiocarp *basidioma* The fruiting body of basidiomycete fungi. (**Basidiomycotina**).

Basidiomycetes See **Basidiomycotina**.

Basidiomycotina *Basidiomycetes* Subdivision or class of Eumycota (true fungi) in which the sexual spores (basidiospores) are formed on a basidium. Includes the Tiliomycetes comprising the rusts (Uredinales) and smuts (Ustilaginales), the Hymenomycetes and the Gasteromycetes.

basidiospore Spore of a basidiomycete fungus, usually uninucleate and haploid.

basidium Club-shaped organ involved in sexual reproduction in basidiomycete fungi (mushrooms, toadstools and others). Bears four haploid basidiospores at its tip.

basigin A protein of the immunoglobulin superfamily (collagenase stimulatory factor, extracellular matrix metalloproteinase inducer, EMMPRIN, leukocyte activation antigen M6, OK blood group antigen, CD147, 385aa) that is important in spermatogenesis, embryo implantation, neural network formation and tumour progression.

basilar membrane A thin layer of tissue covered with mesothelial cells that separates the cochlea from the scala tympani in the ear.

basionym In plant (but not animal) taxonomy a previous legitimate name for a species that has been changed as a result of changes in taxonomic knowledge. Thus *Vinca rosea* was the basionym for what is now called *Catharanthus roseus* (Madagascan periwinkle, or rosy periwinkle). In bacteriology a similar term, basonym, is used.

basket cells Fast-spiking inhibitory interneurons in the hippocampus that play an important part in the function of neuronal networks. They have many small dendritic branches that enclose the cell bodies of adjacent **Purkinje cells** in a basket-like array.

basolateral plasma membrane The plasma membrane of epithelial cells that is adjacent to the **basal lamina** or to the adjoining cells of the sheet. Differs both in protein and phospholipid composition from the **apical plasma membrane** from which it is isolated by **tight junctions**.

basonuclin Zinc finger transcription factors (basonuclin-1, 994aa; basonuclin-2, 1099aa) specific for squamous epithelium although they may have a role in the differentiation of spermatozoa and **oocytes**.

basophil Mammalian **granulocyte** with large heterochromatic basophilic granules that contain **histamine** bound to a protein and heparin-like mucopolysaccharide matrix. They are not phagocytic. Very similar to mast cells though it is not clear whether they have common lineage.

basophilia (1) Having an affinity for basic dyes. (2) Condition in which there is an excess of **basophils** in the blood.

bassoon A zinc-finger protein (3926aa) involved in orchestrating events at the presynaptic terminal along with others players such as **piccolo**. It is also involved in the cytoskeleton underlying the concentrations of vesicles at **ribbon synapses**. Forms a complex with ERC2 (see **CAST**) and **RIMS1**.

Bateman function Equation used in toxicology that expresses the build-up and decay in concentration of a substance (usually in plasma) based on first-order uptake and elimination in a one-compartment model.

Batesian mimicry A form of defensive colouration in which an animal is protected from predators because it resembles an unrelated but dangerous or unpalatable animal.

batrachotoxin *BTX* **Neurotoxin** from the Columbian poison frog *Phyllobates*. A steroidal alkaloid that affects sodium channels; batrachotoxin R is more effective than related batrachotoxin A.

batroxostatin See **disintegrin**.

Batten's disease Severe neurodegenerative disorder (juvenile neuronal ceroid lipofuscinosis, Spielmeyer-Vogt-Sjogren disease) which causes blindness, deafness, loss of muscle control and early death. A storage disease in which lipopigment accumulates in neurons and tissues because of mutations in the **CLN3** gene which encodes battenin.

battenin See **CLN3** and **Batten's disease**.

bavachin A flavonoid from *Psoralea corylifolia* (Babchi plant) used in Chinese traditional medicine. Inhibits acyl-coenzyme A: cholesterol acyltransferase, is a weak antioxidant and is said to stimulate bone formation.

Bax Protein (Bcl-2-like protein 4, 192aa) related to **Bcl-2**. Homodimers seem to promote apotosis in cultured cells, but heterodimers with Bcl-2 or Bcl-XL block cell death. Mice deficient in bax have selective hyperplasias. Bax seems to act as a tumour suppressor and is induced by **p53**, though is not solely responsible for p53-mediated apoptosis.

BAY 43-9006 A broad-spectrum kinase inhibitor and a potent inhibitor of Raf-1.

Bayer's patches Sites of adhesion between the outer and cytoplasmic membranes of Gram-negative bacteria. Also Bayer's junctions.

Bayesian statistics Statistical theory, based on Bayes' decision rule, that outlines a framework for producing decisions based on relative payoffs of different outcomes. Used in genetic counselling.

b-c1 complex A part of the **mitochondrial electron transport chain** that accepts electrons from **ubiquinone**, and passes them on to **cytochrome c**. The b/c1 complex consists of two cytochromes.

BCA (1) BCA-1 is a CXC chemokine (CXCL13, 109aa) that regulates B-cell migration in lymphoid tissues and binds to the chemokine receptor CXCR5. Also known as B-lymphocyte chemoattractant (BLC). (2) Benzethonium chloride assay (BCA), used for protein estimation. (3) Blackcurrant anthocyanin (BCA). (4) Beta-cyanoalanine (BCA).

BCAR3 A protein (breast cancer antiestrogen resistance-3, novel SH2-containing protein two, NSP2, 825aa) expressed in **tamoxifen**-resistant breast carcinoma cells. Like the mouse homologue (AND34, p130Cas-binding protein, 820aa) it is thought to be an adaptor protein that couples activated growth factor receptors to signalling molecules that regulate proliferation and cell migration.

BCECF Fluorescent dye used to monitor intracellular pH.

B-cell receptor A membrane-bound form of immunoglobulin that binds antigen. Each B-cell expresses one immunoglobulin and can recognise one antigen; the population of B-cells provides diversity. Following binding of antigen the B-cell receptor complex, which includes Ig-alpha (CD79a) and Ig-beta (CD79b) chains non-covalently associated with it, signals through a kinase cascade (**lyn, btk**, etc.) and stimulates proliferation and the production of **plasma cells** and memory B-cells.

B cells (1) Cells within discrete endocrine islands (**Islets of Langerhans**) embedded in the major exocrine tissue of vertebrate pancreas. The B or β-cells (originally distinguished by differential staining from A, C and D cells), are responsible for synthesis and secretion into the blood of insulin. (2) Casual term for **B-lymphocytes**.

BCG See **Bacille Calmette-Guerin**.

BCIP A substrate for **alkaline phosphatase** that forms a precipitate in the presence of nitroblue tetrazolium (NBT). Used in various colourimetric assays.

B chromosome Small **acentric** chromosome; part of the normal genome of some races and species of plants.

Bcl Diverse products of oncogenes rearranged by translocations to the immunoglobulin genes in human B-cell malignancies (B-cell lymphomas, hence 'Bcl'). Bcl-1 (295aa) interacts with cyclin-dependent kinases, Bcl-2 (239aa) is an integral protein of the inner mitochondrial membrane that inhibits **apoptosis** and is homologous with the *C. elegans* gene *ced-9*; see **ced mutant**. Bcl-3 (446aa) acts as an IκB homologue that interferes with NFκB signalling. Bcl-6 (706aa) is a transcriptional repressor.

Bcl-2 homology domains Domains within the Bcl-2 family of proteins (BH1, BH2 etc.) that are involved in protein-protein interactions. The Bcl2-homology domain-3 (BH3 domain) is the only BH domain in a sub-family of proteins (**Bim**, Bmf, **Bik, Bad, Bid, Puma**, Noxa and **Hrk**) that are pro-apoptotic through the activation of **Bax**-like relatives.

BCL10 B-cell lymphoma/leukemia 10 (mammalian CARD-containing adapter molecule E10, 233aa) an adapter protein that promotes apoptosis, pro-caspase-9 maturation and activation of NFκB via **NIK** and **IKK**.

Bcor Product (1755aa) of a widely expressed gene (BCL6 corepressor) that is mutated in patients with X-linked oculofaciocardiodental (OFCD) syndrome. It is a transcriptional corepressor that works in association with a complex of proteins capable of epigenetic modification of chromatin. These include Polycomb group (PcG) proteins, Skp-Cullin-F-box (SCF) ubiquitin ligase components and a Jumonji C (Jmjc) domain-containing histone demethylase.

bcr Region (breakpoint cluster region) on chromosome 22 involved in the **Philadelphia chromosome** translocation. *Bcr* is one of the two genes in the *bcr-abl* complex and encodes a GTPase-activator for rac1 and cdc42 (EC 7.11.1.1, 1271aa) that has serine/threonine kinase activity.

Bdellovibrio Small, motile spiral-shaped proteobacteria, common in soil and water environments, that attach to and then penetrate other Gram-negative bacteria where they replicate within the periplasmic space.

BDGF See **brain-derived neurotrophic factor**.

BDM An inhibitor of myosin ATPase.

B-DNA The structural form of **DNA** originally described by Crick and Watson. It is the form normally found in hydrated DNA and is strictly an average, approximate structure for a family of B forms. In B-DNA, the double helix is a right-handed helix with about 10 residues per turn and has a major and a minor groove. The planes of the base pairs are perpendicular to the helix axis.

BDNF See **brain-derived neurotrophic factor**.

beaded filaments Intermediate filaments (beaded-chain filaments) found in the lens fibre cells of the eye: composed of **filensin** and **phakinin**.

bean In humans a protein (259aa) that interacts with Nedd4 and is mutated in spinocerebellar ataxia-31.

Becatecarin Semi-synthetic water-soluble derivative of rebeccamycin. An antineoplastic compound that intercalates into DNA, inhibits topoisomerase II and triggers apoptosis.

Becker muscular dystrophy Benign X-linked muscular dystrophy with later onset and lower severity than **Duchenne muscular dystrophy**. **Dystrophin** is present but in a truncated form.

Becker's disease See **myotonia**.

Beckwith-Wiedemann syndrome Rare developmental disorder with growth abnormalities and predisposition to tumours (exomphalos-macroglossia-gigantism syndrome) that has a complex pattern of inheritance caused by mutation or deletion of imprinted genes within the chromosome 11p15.5 region. Specific genes involved include those for the **cyclin-dependent kinase inhibitor p57** KIP2, **H19**, and LIT1 (long-QT intronic transcript-1, an antisense transcript of part of the KCNQ1 gene).

beclin A coiled-coil myosin-like Bcl2-interacting protein (450aa) located in the Golgi that has a central role in autophagy and may have a role in antiviral host defense.

Becquerel (Bq) The Systeme Internationale (SI, MKS) unit of radioactivity, named after the discoverer of radioactivity, and equal to one disintegration per second. Supersedes the Curie (Ci). 1Ci = 37 **GBq**.

BED-type zinc finger domain A zinc-finger (50–60aa) named after the *Drosophila* proteins BEAF and DREF. Found in cellular regulatory factors and transposases from plants, animals and fungi. Diverse BED fingers are able to bind DNA and this may be the general function of this domain.

Bee1 (1) In *Arabidopsis* a bHLH transcription factor (brassinosteroid enhanced expression 1, bee1, 260aa) that is a positive regulator of brassinosteroid signalling. Bee2 and bee3 are similar. (2) In *S. cerevisiae* Bee1p/Las17p (633aa) is the homologue of the human Wiskott-Aldrich syndrome protein (**WASP**).

Beet western yellows virus A luteovirus that causes chlorosis and stunting of a wide range of dicotyledonous plants worldwide, not only members of the beet family.

Beggiatoales An Order of chemosynthetic sulphur-oxidizing gliding bacteria that occur mostly as filaments. Sulphur granules occur intracellularly.

beige mouse A mouse strain typified by beige hair and **lymphadenopathy**, reticulum cell neoplasms, and giant lysosomal granules in **leucocytes**. May be the murine equivalent of **Chediak-Higashi syndrome** of man.

Belousov-Zhabotinsky reaction An example of chemical oscillations in **dissipative structures**, giving rise to characteristic sustained spatial patterns (concentric rings or spirals) in concentrations of reactants. The reaction is set up by mixing 0.2 M malonic acid, 0.3 M sodium bromate, 0.3 M sulphuric acid and .005M ferroin (1,10-phenanthroline ferrous sulphate) and putting a thin layer (0.5–1.0 mm deep) in a petri dish. Patterns emerge.

belt desmosome Another name for the zonula adherens or **adherens junction**.

Bence-Jones protein Dimers of **immunoglobulin** light chains, normally produced by **myelomas**. Bence-Jones proteins are sufficiently small to be excreted by the kidney.

Benedict's solution Solution used in **Benedict's test** for glucose; the qualitative solution contains 0.07 M copper sulphate, 0.67 M sodium citrate, 0.94 M sodium carbonate. The quantitative solution differs slightly, having 0.072 M copper sulphate, 0.7 M sodium carbonate, 1.29 M potassium thiocyanate and very small amount of potassium ferrocyanide.

Benedict's test Test for glucose and other reducing disaccharides, involving the oxidation of the sugar by an alkaline copper sulphate solution (**Benedict's solution**), in the presence of sodium citrate, to give an deep red copper (I) oxide precipitate. Used in urine testing for diabetes.

benign tumour A clone of **neoplastic** cells that does not invade locally or **metastasise**, having lost **growth control** but not positional control. Usually surrounded by a fibrous capsule of compressed tissue.

benomyl A **benzimidazole** class fungicide and pesticide used on growing fruit and vegetables although there are concerns about the health risks.

benzamidine A potent inhibitor of **serine** endopeptidases such as thrombin and trypsin.

benzimidazole Benzimidazole and its derivatives are used in organic synthesis, vermicides and fungicides. An examples of a benzimidazole class fungicides is benomyl.

benzodiazepines Class of drugs that are anxiolytics or hypnotics. Enhance the inhibitory action of **GABA** by modulating **GABA$_A$ receptors**. Examples include diazepam (Valium), nitrazepam (Mogadon) and chlordiazepoxide (Librium).

benzopyrene Polycyclic aromatic compound, a potent mutagen and carcinogen.

Berardinelli-Seip congenital lipodystrophy See **lipodystrophy**.

Bergmann glia Cerebellar astrocytes (radial epithelial cells, Golgi epithelial cells) that have their cell bodies in the Purkinje cell layer and extend processes into the molecular layer. They express high densities of glutamate transporters and are important in the developing cerebellum and for modification of synapses.

Berk-Sharp technique *S1 mapping* A technique of genetic mapping in which **mRNA** is hybridized with **single stranded DNA** and the nonhybridized DNA then digested with **S1 nuclease**; the residual DNA that hybridized with the messenger is then characterized by restriction mapping.

Bernard-Soulier syndrome A bleeding disorder (benign macrothrombocytopenia) caused by deficiency in platelet membrane glycoprotein Ib alpha chain (CD42b, 626aa) which binds to the A1 domain of vonWillebrand Factor; platelets aggregate normally (*cf.* **Glanzmann's thrombasthenia**) but do not stick to collagen of sub-endothelial basement membrane.

berovin Calcium-activated Renilla-type luciferin from the coelenterate *Beroe*. See **aequorin**.

Best's carmine Histochemical stain that can be used to demonstrate the presence of glycogen, which stains deep red.

Best's disease A form of vitelliform **macular dystrophy** in which there is abnormal accumulation of lipofuscin within and beneath the retinal pigment epithelium, caused by mutation in the **bestrophin** gene.

bestatin An inhibitor of aminopeptidases and a potent, irreversible inhibitor of LTA4 hydrolase.

bestrophins A family of calcium-sensitive chloride channels. Bestrophin-1 (vitelliform macular dystrophy protein 2, 585aa) is mutated in **Best's disease**, and is normally localized to the basolateral plasma membrane of the pigmented retinal epithelium. Three other related proteins (bestrophin-2, 509aa; bestrophin-3, 668aa; bestrophin-4, 473aa) are also chloride channels but have distinct tissue distributions.

beta Prefix for many proteins and motifs. Headwords with a 'beta' prefix (that could be shown as the Greek letter) precede words such as betacellulin and betaine. Hyphens and spaces are ignored for alphabeticization purposes.

beta-1,3-glucanases (1) A diverse set of enzymes that hydrolyse 1,3-β glucosidic linkages in glucans. Important for remodelling complex carbohydrates of, for example, the cell wall in plants, for digesting complex carbohydrates, for defence against pathogens such as yeasts that have beta-glucans in their cell walls. (2) In *S. cerevisiae*, endo-1,3(4)-beta-glucanase one (endo-1,3-β-glucanase 1, EC 3.2.1.6, laminarinase-1, 1117aa) is involved in the dissolution of the mother-daughter septum during cell separation and like endo-1,3(4)-β -glucanase two (779aa) hydrolyses (1,3)- or (1,4)-linkages in β-D-glucans. (3) In *Bacillus subtilis* beta-glucanase (EC 3.2.1.73, lichenase, 242aa) hydrolyses (1, 4)-β-D-glucosidic linkages in β-D-glucans. (4) Enzymes of the pathogen response (PR-2) protein family that hydrolyse β-1,3-glucans, major structural components of fungal cell walls.

beta-2-microglobulin Immunoglobulin-like polypeptide (119aa), homologous with the constant region of Ig, that is found on the surfaces of most cells, associated non-covalently with Class I **histocompatibility antigens**.

beta-actin One of many actin isoforms but found (like gamma-actin) predominantly in non-muscle cells. Beta-actin is found particularly in regions where the actin meshwork is rapidly remodelling and is less common in microfilament bundles. See **actin**. Paper: http://www.ncbi.nlm.nih.gov/pmc/articles/PMC2706787/?tool = pubmed

beta-actinin See **capZ**.

beta-adrenoceptor blocking drugs Group of drugs (casually referred to as beta-blockers) which block β-adrenoreceptors. Used in the treatment of angina, hypertension, migraine, thyrotoxicosis and anxiety states. Some are relatively unselective (e.g. propranolol), others act primarily on β1-receptors (e.g. atenolol).

beta agonist Sympathomimetic drug that acts on β-2 adrenoreceptors and has a rapid bronchodilatory effect (acting on bronchial smooth muscle) if taken by inhalation. Used especially for asthma; salmeterol is an example.

beta-alpha-beta motif Protein motif comprising a beta strand-loop-helix-loop-strand arrangment, with the strands lying parallel.

beta-amylase A terminal amylase (EC.3.2.1.2) that cleaves maltose units from the non-reducing ends of starch, glycogen and related polysaccharides.

beta amyloid A fragment of **amyloid precursor protein** that is produced by the action of secretases (see **presenilins**). Aggregates of beta-amyloid accumulate as plaques in the brain in **Alzheimer's disease**.

beta arch Protein motif comprising two adjacent antiparallel beta strands joined by a coil and that are part of different sheets, usually forming a **beta sandwich**.

beta barrel Protein motif in which a series of (typically **amphipathic**) beta sheets is arranged around a central pore. Example: **voltage-gated ion channel**.

beta blockers See **beta-adrenoceptor blocking drugs**.

beta bulge Protein motif comprising a disruption of a **beta sheet**, usually by the insertion of a single residue.

beta-CAP73 A bovine actin binding protein (uveal autoantigen with coiled-coil domains and ankyrin repeats protein, 1401aa) that binds only the barbed ends of β-actin filaments (not filaments of the β-actin isoform). Has six predicted **ankyrin**-like repeats at the N-terminus. Regulates APAF1 expression.

beta carotene See **carotenes**.

beta-cells *β-cells* See **B cells**.

beta-COP See **coated vesicles**.

beta-defensins Small cationic peptides with antibacterial activity. See **defensins**.

beta-emitter A radionuclide whose decay is accompanied by the emission of β particles, most commonly negatively charged electrons. Many isotopes used in biology, such as ^3H, ^{14}C, ^{35}S, and ^{32}P are pure β emitters.

beta-fructosidase An enzyme (EC 3.2.1.80) that hydrolyses terminal, non-reducing (2,1)- and (2,6)-linked β-D-fructofuranose residues in fructans. Also known as exo-β-D-fructosidase, fructanase, invertase, sucrase. Substrates include sucrose, raffinose, inulin and levan. In *Arabidopsis* there are several fructosidases, secreted insoluble isoenzymes (β-fructofuranosidases, EC 3.2.1.26, cell wall invertases).

beta-galactosidase Enzyme (EC 3.2.1.23, 1024aa) that cleaves the terminal β-galactose from ganglioside substrates and other glycoconjugates; encoded by the LacZ gene in *E. coli*, and widely used as a **reporter gene**, because a variety of coloured or fluorescent compounds can be produced from appropriate substrates (typically **Xgal**, that produces a blue colour). LacZ is incorporated in many plasmid **vectors** to allow **blue-white colour selection**. The human gene is defective in GM1-gangliosidosis, a lysosomal storage disease.

beta-gal NONOate Glycosylated, cell permeable, nitric oxide donor.

beta-glucosidase A cytosolic enzyme (EC 3.2.1.21, 469aa) that catalyses the release of glucose by hydrolysis of the glycosidic link in various β-D-glucosides (compounds of the form R-β-D-glucose, where the group R may be alkyl, aryl, mono- or oligosaccharide). Favoured source: almonds, from which enzyme is known as emulsin. The lysosomal enzyme (glucosylceramidase, EC 3.2.1.45, 536aa) that catalyzes the breakdown of glucosylceramide to ceramide and glucose is defective in **Gaucher's disease**.

beta-glucuronidase Enzyme (EC 3.2.1.31, 651aa) that catalyses hydrolysis of a β-D-glucuronoside to D-glucuronate and the compound to which it was attached. Important in the degradation of dermatan and keratan sulphates. Often used as a marker enzyme for lysosomes.

beta hairpin Protein motif describing one possible arrangement of strands in a **beta sheet**. Strands are antiparallel and hydrogen-bonded, lying adjacent in the sheet.

beta helix Protein motif (β helix; solenoid) comprising a large right-handed coil (or super-helix), containing either two or three **beta sheets**.

beta interferon Recombinant version (165aa, nonglycosylated) of the endogenous biological compound, **interferon** -beta, produced in *E. coli* with an inserted human fibroblast-derived gene.

beta-lactam antibiotics A large group of bactericidal antibiotics that act by inhibiting bacterial cell wall synthesis and activating enzymes that destroy the cell wall. Examples are **penicillin**, **ampicillin**, **amoxicillin**.

beta-lactamase Enzyme (penicillin amido-beta-lactam hydrolase, EC 3.5.2.6, 377aa in *E. coli*) produced by some bacteria that makes them resistant to **beta-lactam antibiotics**. Competitively inhibited by **clavulanic acid**. Extended-spectrum beta lactamases

(ESBLs) will degrade the 'extended spectrum cephalosporins' (e.g. cefuroxime) and are being found more frequently: many are plasmid-encoded.

beta-oxidation The process whereby fatty acids are degraded in steps, losing two carbons as (acetyl)-CoA. Involves CoA ester formation, desaturation, hydroxylation and oxidation before each cleavage. See **omega-oxidation**.

beta pleated sheet Beta secondary structure in proteins consists of two almost fully extended polypeptide chains lying side by side, linked by interchain hydrogen bonds between peptide C=O and N-H groups. When multiple chains are involved, an extended sheet, the β-pleated sheet, is formed, which can consist of parallel or antiparallel sheets (where the chains run in the same or opposite directions) or mixed sheets.

beta prism Protein motif comprising three antiparallel **beta sheets** arranged in a triangular, prism shape. In the orthogonal prism, strands are orthogonal to the prism access; in the aligned prism, the strands and prism axis are parallel.

beta propeller Protein motif comprising 4−8 antiparallel **beta sheets** arranged like the blades of a propellor. Neuraminidase has a six-bladed beta propeller.

beta sandwich Protein motif comprising two **beta sheets** that pack together face to face, in a layered arrangement.

beta-secretase See **secretases**.

beta sheet See **beta pleated sheet**.

beta strand Region of polypeptide chain that forms part of a **beta sheet**.

beta-thymosin Beta-thymosin (44aa) is effectively an isolated WH2 domain (**WASP homology domain-2**) and beta-thymosin repeats are actin monomer-binding motifs found in many proteins that regulate the actin cytoskeleton. See **thymosin β4**.

beta trefoil Protein motif consisting of three **beta hairpins** forming a triangular shape.

beta turn Protein motif which consists of an abrupt 180° reversal in direction of a polypeptide chain. The turn is defined as being complete within four residues.

betacellulin One of the EGF family of growth factors, synthesized primarily as a transmembrane precursor (178aa), which is cleaved to produce the mature molecule (80aa) that binds to the EGF receptor. It is a potent mitogen for retinal pigment epithelial cells and for vascular smooth muscle cells.

betacyanin Red pigments of the **betalain** type, for example the red pigment found in beetroot.

betaine A derivative of glycine characterized by high water solubility. Can function as an osmotic agent in plant tissues. See **biogenic amines**.

betalain Nitrogen-containing red or yellow pigments functionally replacing **anthocyanins** in flowers and fruits of many Caryophyllales (ice plants, cacti, carnations). Also found in some higher fungi where their role is obscure. Are used for food colouring and have antioxidant and radical scavenging properties that provide protection in certain oxidative stress-related disorders.

betanidin Natural pigment (betanin, beetroot red, **betalain**) from beetroot.

Bethlem myopathy A rare myopathy that mainly affects the elbows and ankles caused by mutations in the **collagen** COL6A3 gene.

Bet v 1 The major birch (*Betula*) pollen allergen, a **PR protein** homologous to **Mal d 1**.

Betz cells Large pyramidal cell neurons (up to 100 μm diameter) with cell bodies in the primary motor cortex; axons project down the spinal cord.

BFA See **brefeldin A**.

bFGF Basic **fibroblast growth factor**.

BFU-E See **burst forming unit**-erythrocytic.

BGH (1) Bovine **growth hormone**, (bGH). (2) **Brunner's gland** hyperplasia. (3) *Blumeria graminis* f.sp. *hordei* (Bgh), a powdery mildew fungus.

BH3 (1) See **Bcl-2 homology domains**. (2) The borano (BH_3^-) group.

BHK cells A quasi-diploid established line of Syrian hamster cells (baby hamster kidney cells), descended from a clone (Clone 13) isolated by Stoker & McPherson from an unusually rapidly growing primary culture of new-born hamster kidney tissue. Usually described as fibroblastic, although smooth muscle-like in that they express the muscle intermediate filament protein **desmin**. Widely used as a viral host, in studies of oncogenic transformation and of cell physiology.

bHLH Basic **helix-loop-helix**; a class of transcription factors.

Biacore Proprietary name for an instrument that uses **surface plasmon resonance** to detect the binding of a substance to the surface of a flow chamber. Using this instrument it is possible to measure the on- and off-rates for the binding of a molecule to a defined surface, e.g. the binding of an antibody to the antigen-coated surface of the flow cell or of ligand to an immobilized receptor.

Bial's orcinol test See **orcinol**.

bialaphos A tripeptide antibiotic (phosphinothricyclalanylalanine, phosphinothricin tripeptide) produced by *Streptomyces hygroscopicus*, consists of two molecules of L-alanine and one molecule of the unusual amino acid phosphinothricin (PT). Used as a selective agent in transformation experiments in plant genetic engineering and as a herbicide.

bicoid An **egg-polarity gene** in *Drosophila*, concentrated at the anterior pole of the egg, and required for subsequent anterior structures. A **maternal-effect gene**.

bicuculline A specific blocking agent for the action of the amino acid neurotransmitter γ-aminobutyric acid (**GABA**) from *Dicentra cucullaria* and herbs of the genus *Corydalis*. See **GABA receptor**.

Bid A proapoptotic Bcl-2 family member (BH3-interacting domain death agonist, p22 BID, 195aa) that is cleaved into smaller fragments which induce cytochrome c release, production of ICE-like proteases and apoptosis. A substrate for the catalytic casein kinase two alpha subunit. DNA damage leads to **ATM**-mediated Bid phosphorylation, and this phosphorylation regulates a novel, pro-survival function of Bid important for S phase arrest. Forms heterodimers with the pro-apoptotic protein Bax or the anti-apoptotic protein Bcl-2. See **Bcl-2 homology domains**.

Bifidobacteria Genus of bacteria found as a normal part of the microflora of the lower intestine and thought to assist digestive processes. Are used as a probiotic for intestinal well-being.

big brain Product (696aa) of a neurogenic gene in *Drosophila* that is essential for differentiation of ectoderm and acts synergistically with Notch and Delta during the separation of neural and epidermal cell lineages. Member of the **major intrinsic protein**/aquaporin family.

Bigelowiella natans See **Chlorarachniophytes**.

biglycan A small proteoglycan (bone/cartilage proteoglycan I, 150–240 kDa) of the extracellular matrix. The core protein (368aa) is very similar to those of **decorin** and **fibromodulin**. All three have highly conserved sequences containing 10 internal homologous repeats of around 25aa with leucine-rich motifs. Biglycan has two glycosaminoglycan chains, either chondroitin sulphate or dermatan sulphate and N-linked oligosaccharides.

biguanides Drugs used in treating maturity onset diabetes. Appear to act by increasing peripheral utilization of glucose and are of particular value in obese diabetics, e.g. Metformin.

Bik BH3-only pro-apoptotic protein of Bcl-2 family (Bcl-2-interacting killer; Nbk/Bik, 160aa), targets the membrane of the endoplasmic reticulum. It is induced in human cells in response to several stress stimuli, including genotoxic stress (radiation, doxorubicin) and overexpression of **E1A** or **p53** but not by ER stress pathways resulting from protein malfolding. Endogenous cellular Bik regulates a **Bax**, **BAK**-dependent ER pathway that contributes to mitochondrial apoptosis. Loss of Nbk/Bik is common in clear-cell renal cell carcinoma.

bikunin A plasma protein (HI-30, urinary trypsin inhibitor, inter-alpha-(trypsin) inhibitor light chain, 147aa), one of the super-family of Kunitz-type protease inhibitors, derived from AMBP protein (352aa) that also gives rise to **alpha-1-microglobulin** (184aa) and a trypsin inhibitor, trypstatin (61aa). See **inter-alpha-inhibitor**.

bile pigments Pigments produced by the breakdown of haemoglobin. The main pigments are bilirubin (reddish-yellow) and its oxidation product biliverdin (green).

bile salts Sodium salts of the bile acids, a group of hydroxy steroid acids condensed with taurine or glycine, the commonest of which are the salts of taurocholic and glycocholic acids. They are powerful surfactants and are important in aiding absorption of fats from the intestine.

bilharzia See **schistosomiasis**.

bilicyanin A blue oxidation product of **bilirubin**.

Biliphytes An enigmatic eukaryotic group known only by SSU rDNA (DNA encoding small subunit ribosomal RNA) sequence and FISH-FITC images. They appear to have phycobiliproteins and there is some indication of **nucleomorphs**.

biliproteins See **phycobilins**.

bilirubin Red-brown pigment found in bile, formed by breakdown of haemoglobin.

biliverdin Green bile pigment formed by haemoglobin breakdown; can be converted into **bilirubin** by reduction.

Bim (1) Bcl-2 interacting mediator of death (198aa), a proapoptotic Bcl-2 homology domain 3 (BH3)-only protein. Critical for eliminating most effector T-cells following an acute T-cell response; the few that survive become memory cells. (2) In *Arabidopsis* a bHLH transcription factor (BIM1, 529aa) involved in positive brassinosteroid-signaling.

BIN Proteins of the **BAR protein** superfamily. BIN1 (**amphiphysin II**, bridging integrator 1, 593aa) inhibits the oncogenic properties of **myc**. BIN2 (546aa) is mostly found in haematopoietic tissues, lacks tumour suppressor functions and interacts with BIN1, but not amphiphysin. BIN3 (253aa) has only the BAR domain and is widely expressed.

bin mapping Also known as selective mapping: an approach to constructing a genetic linkage map of an organism such as a plant. In the first step a mapping population of usual size (N = 60–250) is used to construct a saturated framework map with markers placed on it with high precision, and a second step in which new markers are added to this map with lower precision using a selected subset of highly informative plants. The selection of this subset of plants is based on the number and position of recombinational crossover sites (or breakpoints) detected with the framework marker data in each plant. The breakpoints identified by the ensemble of the selected plants define a set of bins, i.e., chromosome fragments bounded by two adjacent breakpoints or by a distal breakpoint and the telomere, characteristic of each subset. Article: http://www.genetics.org/cgi/content/full/171/3/1305

binary fission Division of a cell into two daughter cells; nuclear division precedes division of the cell-body.

bindin Protein (~480aa) normally sequestered in the **acrosome** of a sea-urchin spermatozoon, and that through its specific **lectin**-like binding to the **vitelline membrane** of the egg confers species-specificity in fertilization.

binomial nomenclature The standard formal method of naming organisms by their genus and their species, usually said to have been introduced by Linnaeus. Note that the genus is capitalized and the species name is not and both should be italicized: thus humans are *Homo sapiens*.

binovular twins Non-identical twins, the products of two separate ova.

binucleate Having two nuclei.

bioaccumulation Accumulation of substances in living organisms because the rate of intake exceeds the capacity to excrete or metabolise the substance. Organisms at the top of a food chain can accumulate considerable amounts of some substances, most notorious of which was DDT.

bioactivation Metabolic conversion of a **xenobiotic** substance to a more toxic or active derivative.

bioassay An assay for the activity or potency of a substance that involves testing its activity on living material.

bioautography The use of cells to detect, by their attachment or other reaction, the presence of a particular substance, e.g. an adhesion protein on an electrophoretic gel.

bioavailability Relative amount of a drug (or other substance) that will reach the systemic circulation when administered by a route other than direct intravenous injection.

bioblasts When Altmann first observed mitochondria he considered them to be intracellular parasites and christened them bioblasts. Not current usage.

biochip (1) A silicon chip implanted into and functioning as part of a human body. (2) An array of proteins, oligonucleotides or other molecules, immobilised on a solid substratum (often a microscope slide) that can be probed with labelled reagents, mixtures or substances etc. to identify interactions or, in the case of oligonucleotides, sequence similarity through hybridization. Increasing miniturisation allows arrays to contain thousands of individual sites. See **gene chip**.

biocidal Capable of killing living organisms.

biocompatible Capable of remaining in contact with cells or tissues without causing adverse effects. This can be simply the absence of toxicity, in the case of a culture vessel for example, or more complex, as with materials that can be implanted into the body without exciting an inflammatory or thrombogenic response.

biodiversity The various genetic, taxonomic and ecosystem differences in the living organisms of a given area, environment, ecosystem, or indeed the whole planet.

BioEdit A biological sequence alignment editor written for the Windows operating system but now (2011) no longer maintained although still available. Homepage: http://www.mbio.ncsu.edu/bioedit/bioedit.html

bioengineering Rather imprecise category of activities that can range from bioreactor design, through prosthetic devices to environmental restoration. Cynically, describes any interface between biology and engineering that will attract students or funds.

biofilm A layer of bacteria, enclosed in a mucilaginous slime, attached to surfaces exposed to water or biological fluids. Multiple bacterial species may be present, as well as fungi, algae, protozoa, debris and corrosion products, and the behaviour of the bacteria in such films may be distinctly different from that exhibited in suspension culture. Plaque on teeth is a common example. See **quorum sensing**.

bioflavonoids Group of coloured phenolic pigments originally considered vitamins (Vitamins P, C2) but not shown to have any nutritional role. Responsible for the red/purple colours of many higher plants.

biogenic amines Amines found in both animals and plants that are frequently involved in signalling. There are several groups: ethanolamine derivatives include **choline, acetylcholine** and **muscarine**; catecholamines include **adrenaline, noradrenaline** and **dopamine**; polyamines include **spermine**; indolylalkylamines include tryptamine and **serotonin**; betaines include **carnitine**; polymethyline diamines include **cadaverine** and **putrescine**.

bioinformatics The discipline of using computers to collate and form datasets of interest to biologists. Usually used to refer to databases of DNA and protein sequences, and of mutations, disease and gene functions, in the context of genome projects.

biolistic A method for transfecting cells using a **gene gun**: an unattractive hybrid term for biological and ballistic.

biological oxygen demand The oxygen required to satisfy the biological demands of contaminated water. The biological demand is from microbial flora involved in digesting organic constitutents.

bioluminescence Light produced by a living organism. The best known system is firefly luciferase (an ATPase), which is used routinely as a sensitive ATP assay system. Many other organisms, particularly deep-sea organisms, produce light and even leucocytes emit a small amount of light when their oxidative metabolism is stimulated. Does not really differ from **chemiluminescence**, except that the light-emitting molecule occurs naturally and is not a synthetic compound like **luminol** or lucigenin.

biomarker (1) A biologically-derived substance, the presence of which in serum may be an indication of disease. (2) Any biological feature that is indicative of the status of the system whether that is an individual organism or an ecosystem.

biomaterials (1) Solid materials which are produced by living organisms, such as **chitin, fibroin** or bone. (2) Any materials which replace the function of living tissues or organs in humans.

biometry *Adj.* biometric Statistical methods applied to biological problems.

biomimetic Processes, substances, devices, or systems that imitate those found in biology. Biomimetics is an area of bioengineering in which new technology is based upon mechanisms, features, methods and accomplishments found in biological systems.

Bioperl A toolkit of perl modules useful in building bioinformatics solutions in the Perl programming language: the basis for a lot of the serious bioinformatics work at large genome centres. Homepage: http://www.bioperl.org

biopiracy A politically-charged term for the development by industrialised nations of materials native to developing countries, e.g. medicinal plants, without adequate compensation to their country of origin.

bioprospecting Investigating living organisms with the aim of discovering materials that can be exploited for commercial gain, often without recompense for the local inhabitants (**biopiracy**).

biopsy The removal of a small sample of tissue for diagnostic examination or the sample itself.

biopterin Growth factor for some protozoa; present in many tissues as the reduced form, tetrahydrobiopterin, where it acts as a coenzyme for hydroxylases. Tetrahydrobiopterin is an essential cofactor for NO production from L-arginine. Defects in biopterin synthesis can lead to hyperphenylalaninemia.

Biopython A set of freely available software tools for biological computation, written in Python programming language. Homepage: http://biopython. org/wiki/Main_Page

bioreactors Reaction vessel for producing a biological product by fermentation or cell culture, increasingly involving modified micro-organisms or cells that produce particular substances.

bioremediation Remediation of a contaminated environment through the use of biological agents, often genetically engineered or selected for the particular task, for example, the breakdown of organic molecules or accumulation and sequestration of a toxic material. Has proved more difficult than had perhaps been hoped.

biosecurity Methods adopted to prevent harmful effects brought about by other species. May involve the use of deliberately crippled strains of virus or micro-organism that are unlikely to survive outside laboratory conditions and/or containment at different levels of stringency.

biosynthesis Synthesis by a living system (as opposed to chemical synthesis).

bioterrorism The use of disease-carrying organisms or agricultural pests as a weapon in terrorism. Fears about this threat have led some Governments to impose increasingly stringent regulations on the experimental use of some organisms.

biotic stress The stress caused to a plant by other living organisms.

biotin *vitamin H* A prosthetic group for carboxylase enzymes. Important in fatty acid biosynthesis and catabolism and has found widespread use as a covalent label for macromolecules which may then be detected by high-affinity binding of labelled **avidin** or **streptavidin**. Essential **growth factor** for many cells.

biotinyl Acyl group derived from **biotin**. Addition of this moiety to another molecule (biotinylation), for example to a protein, will allow affinity purification or labelling with **avidin**.

biotrophic A mode of fungal infection of a plant in which fungi grow between the host cells and invade only a few of the cells to produce nutrient-absorbing structures termed **haustoria**. The infection site acts as a nutrient sink which disadvantages, but does not kill, the host.

bioweapon Any weapon based upon a biological agent, either a biologically-derived toxin (e.g. **botulinum toxin**) or an infectious agent (bacteria or viruses).

BiP Molecular chaperone (immunoglobulin heavy chain-binding protein, endoplasmic reticulum lumenal Ca^{2+}-binding protein grp78, 654aa) found in endoplasmic reticulum and related to Hsp70 family of **heat-shock proteins**.

bipolar cells A class of retinal **interneurons**, named after their morphology, that receive input from the photoreceptors and send it to the **ganglion cells**. Bipolar cells are **non-spiking neurons** ; their response to light is evenly graded, and shows **lateral inhibition**.

bipolar filaments Filaments that have opposite polarity at the two ends; classic example is the **thick filament** of striated muscle.

Birbeck granules Characteristic inclusion bodies seen by electron microscopy in **histiocytes** (Langerhans cells) of patients with histiocytosis X, a group of diseases with uncertain pathogenesis.

BIRCs A family of proteins (baculoviral IAP (inhibition of apoptosis) repeat-containing proteins) that are inhibitors of apoptosis. In humans, BIRC1 (neuronal apoptosis inhibitory protein, 1403aa) has three BIR domains (each ~70aa) and inhibits apotosis only in motor neurons, BIRC2 (c-IAP2, 618aa) interacts with TNF-receptor associated factors one and two (TRAF1 and TRAF2), BIRC3 (IAP1, 604aa) is similar but has different tissue distribution: both have **CARD domains** in addition to the BIR domains. BIRC4 is **XIAP0**, BIRC5 (apoptosis inhibitor four, survivin, 142aa) is a component of the **chromosomal passenger complex**, BIRC6 (ubiquitin-conjugating BIR domain enzyme apollon, 4829aa) is found only in some gliomas and ovarian tumour cell lines, BIRC7 (kidney IAP, melanoma IAP, 298aa) protects against apoptosis induced by TNF or agents such as etoposide. BIRC8 (236aa) is testis-specific. Article: http://genesdev.cshlp.org/content/13/3/239. full

birefringence A property exhibited by some materials that have a different refractive index according to the plane of polarization of the light. The effect is to rotate the plane of the refracted light so that, using crossed Nicholl prisms (polarizers set at right angles to give complete extinction), the birefringent material appears bright. The birefringence can arise through anisotropy of structure (form birefringence) or through orientation of molecules either because of mechanical stretching (stress birefringence) or because of alignment in flow (flow birefringence). A classic example, often used to demonstrate the effect, is a hair which, because of the orientation of the keratin, shows form birefringence.

Birk Barel syndrome A disorder in which there is mental retardation and facial dysmorphism caused by mutation in KCNK9 (TASK-3). See **TASKs**.

bisphosphoglycerate *BPG* A highly anionic compound that binds to deoxyhaemoglobin and allosterically alters the molecule, reducing its oxygen affinity and facilitating oxygen release in capillaries. Present in erythrocytes at about the same molar ratio as haemoglobin, the concentration being regulated by bisphosphoglycerate mutase (259aa, EC 5.4.2.4), which catalyses its production, and bisphosphoglycerate phosphatase (EC 3.1.3.13) which catalyzes its hydrolysis.

bisphosphonates Family of drugs used to prevent and treat **osteoporosis**. Stimulate apoptosis of osteoclasts.

Biston betularia Peppered moth; famous for the shift to the melanised form as industrial pollution turned trees black and gave the melanotic form a selective advantage - and for reversion to the lighter form following the Clean Air Act in Britain.

bithorax complex A group of **homeotic** mutations of *Drosophila* that map to the bithorax region on chromosome III. The mutations all cause the third thoracic segment to develop like the second thoracic segment to varying extents. The genes of the bithorax complex are thought to determine the differentiation of the posterior thoracic segments and the abdominal segments.

bitistatin A **disintegrin** (bitin, 83aa) found in the venom of the puff adder, *Bitis arietans*.

Bittner agent Earlier name, now superseded, for the mouse **mammary tumour virus**.

biuret reaction Formation of a purple colour when biuret (carbamoyl urea) or any compound with two or more peptide bonds (i.e. proteins) reacts with copper sulphate in alkaline solution. Used as a colourimetric test.

bivalent (1) Describing a molecule with two binding sites. (2) A term used of two homologous chromosomes when they are in synapsis during **meiosis**.

bivalirudin Synthetic 20aa peptide (Angiomax®) that is a specific and reversible direct thrombin inhibitor, binding both to the catalytic site and to the anion-binding exosite of circulating and clot-bound thrombin.

Björnstad's syndrome A disorder in which there is sensorineural deafness and twisted hair (pili torti) caused by mutation in the BCS1L gene that encodes a mitochondrial chaperone (419aa, a homologue of *S. cerevisiae* bcs1 protein) which is involved in the assembly of complex III of the mitochondrial respiratory chain.

BK channels Potassium channel (Maxi-K channel, slowpoke, slo, in *Drosophila*) activated by membrane depolarization or an increase in cytosolic Ca^{2+} that mediates export of K^+. The pore-forming alpha subunit (1236aa) is modulated by various modulatory beta subunits (encoded by KCNMB1, KCNMB2, KCNMB3, or KCNMB4) and the pore is an $\alpha_4\beta_4$ octamer.

bla gene Bacterial gene coding for **beta-lactamase**, important for antibiotic resistance.

black fever Synonym for **Rocky Mountain spotted fever**.

black membrane An artificial (phospho)lipid membrane formed by 'painting' a solution of phospholipid in organic solvent over a hole in a hydrophobic support immersed in water. Drainage of the solvent from the film produces diffraction colours until the thickness falls below the wavelength of light — it then appears to be black. The structure is an extended bimolecular leaflet and has been used experimentally to investigate ionophores.

black widow spider venom See **latrotoxins**.

blackhead An infectious disease (Infectious enterohepatitis, histomoniasis) caused by infection by the protozoon, *Histomonas meleagridis*. Mainly affects turkeys.

blackleg An acute infectious disease of cattle and sheep caused by *Clostridium chauvoei*.

blackwater fever An acute disease (haemoglobinuric fever) of tropical regions characterized by intravascular haemolysis, haemoglobinuria, and acute renal failure; classically seen in European expatriates chronically exposed to *Plasmodium falciparum*. Symptoms include fever, vomiting and passage of red or dark-brown urine.

BLAST Abbreviation for Basic Local Alignment Search Tool, a commonly-used web-based resource for identifying similarities in sequences of nucleotides or amino acids. The software is public domain and made available by the US Government-funded National Center for Biotechnology Information (NCBI). Link: http://blast.ncbi.nlm.nih.gov/Blast.cgi

blast cells Cells of a proliferative compartment in a cell lineage.

blast transformation The morphological and biochemical changes in B- and T-lymphocytes on exposure to **antigen** or to a **mitogen**. The cells appear to move from G0 to G1 stage of the cell cycle. They usually enlarge and proceed to S phase and mitosis later. The process probably involves receptor cross-linking on the plasma membrane.

blastema A group of cells in an organism that will develop into a new individual by asexual reproduction, or into an organized structure during regeneration.

blastic conidium See **conidium**.

blasticidin A peptidyl nucleoside antibiotic isolated from the culture broth of *Streptomyces griseochromogenes*. It inhibits protein synthesis in both prokaryotes and eukaryotes by interfering with peptide bond formation in the ribosome. Frequently used to select transfected cells carrying resistance genes Datasheet: http://www.cayla.com/support/datasheets/blastech.pdf

blasticidin resistance genes Various **blasticidin** resistance genes are known, one encoding an acetyl transferase (*bls*, 136aa), others (*bsr* and *BSD*) encoding deaminases (130aa and 140aa). The latter two are used as dominant selectable markers for transformation experiments in mammalian and plant cells. Datasheet: http://www.cayla.com/support/datasheets/blastech.pdf

blastocoel *US* blastocele The cavity formed within the mass of cells of the blastula of many animals during the later stages of cleavage.

blastocyst In mammalian development, cleavage produces a thin-walled hollow sphere, whose wall is the **trophoblast**, with the embryo proper being represented by a mass of cells at one side (inner cell mass). The blastocyst is formed before implantation and is equivalent to the **blastula**.

blastoderm In many eggs with a large amount of yolk, cell division (cleavage) is restricted to a superficial layer of the fertilized egg (meroblastic cleavage). This layer is termed the blastoderm. In birds it is a flat disc of cells at one pole of the egg and in insects an outer layer of cells surrounding the yolk mass.

blastoma *Pl.* blastomas or blastomata. Neoplasm composed of immature **blast cells**.

blastomere One of the cells produced as the result of cell division, cleavage, in the fertilized egg.

blastopore The opening formed by the invagination of cells during **gastrulation**. It is an opening from the **archenteron**, the primitive gut, to the exterior. In some animals this opening becomes the anus, whilst in others it closes up and the anus opens at the same spot or nearby. In some animals, e.g. chick, invagination occurs without a true blastopore

and the site at which the cells move in, the (**primitive streak**), may be termed a virtual blastopore.

blastula Stage of embryonic development of animals near the end of cleavage but before **gastrulation**. In animals where cleavage (cell division) involves the whole egg, the blastula usually consists of a hollow ball of cells.

***ble* gene** An antibiotic resistance gene in *Streptoalloteichus hindustanus*, *S. aureus* and *Klebsiella pneumoniae* that encodes a binding protein ($\sim$124aa), which stoichiometrically binds bleomycin-family antibiotics.

bleb Protrusion from the surface of a cell, usually approximately hemispherical; may be filled with fluid or supported by a meshwork of microfilaments.

blebbistatin A small molecule myosin inhibitor with high affinity and selectivity toward myosin II. It preferentially binds to the ATPase intermediate with ADP and phosphate bound at the active site, and it slows down phosphate release. Article: http://www.jbc.org/content/279/34/35557.full

bleomycin Any of a group of glycopeptide antibiotics from *Streptomyces verticillus*. Blocks cell division in G2: used to synchronize the division of cells in culture and as an antiproliferative agent in oncology.

Blepharisma Genus of ciliate protozoans of the order Heterotricha.

blepharoplast Alternative name for a **basal body**. An organelle derived from the **centriole** and giving rise to the **flagella**. Found chiefly in protozoa and algae.

BLIMP1 See **PRD1-BF1**.

blk B-lymphocyte kinase. A member of the **src-family** tyrosine kinases (505aa) involved in B-cell maturation; sustained activation of blk induces responses normally associated with the pre-B-cell receptor activation (See **preBCR**).

blocking antibody An antibody used in a reaction to prevent some other reaction taking place, for example one antibody competing with another for a cell surface receptor.

BLOCs Biogenesis of lysosome-related organelles complexes. Protein complexes involved in the biogenesis of specialised lysosome-related organelles such as melanosomes and platelet dense granules. Components of BLOC1 include **pallidin**, **muted**, **dysbindin**, **cappuccino**, **snapin** and others. BLOC-1 interacts with **adaptor protein-3** (AP-3) to form complex that affects the targeting of SNARE and non-SNARE cargoes. BLOC-2 is composed of **Hermansky-Pudlak syndrome** protein-3 (HPS3, 1004aa), HPS-5 (alpha-integrin-binding protein 63,

1129aa) and HPS-6 (ruby-eye protein homologue, 775aa) and interacts with alpha-integrin chains that have an aromatic residue before the first lysine of the conserved KXGFFKR motif.

blood The fluid that circulates throughout the vasculature, transporting oxygen, nutrients, hormones, waste products and heat. The cells of the blood (**erythrocytes**, **leucocytes** and **platelets**) are suspended in **plasma**. The fluid that remains after blood has clotted is serum, which differs from plasma in having lost fibrinogen and gained various platelet-released substances.

blood group antigens The set of cell surface antigens found chiefly, but not solely, on blood cells. More than fifteen different blood group systems are recognized in humans. There may be naturally occurring antibodies without immunization, especially in the case of the **ABO** system, and matching blood groups is important for safe transfusion. In most cases the antigenic determinant resides in the carbohydrate chains of membrane glycoproteins or glycolipids. See also **Rhesus**, **Duffy**, **Kell**, **Lewis** and **MN** blood groups. NCBI Blood group data: http://www.ncbi.nlm.nih.gov/books/NBK2264/

blood smear A thin air-dried film of blood on a microscope slide in which the cells are thinly spread and that can be stained to examine the blood cells (and any blood-borne parasites).

blood sugar Colloquial shorthand for the concentration of glucose in the blood.

blood vessels All the vessels lined with **endothelium** through which blood circulates.

blood-brain barrier The blood vessels of the brain (and the retina) are much more impermeable to large molecules (like antibodies) than blood vessels elsewhere in the body. This has important implications for the ability of the organism to mount an immune response in these tissues, although the basis for the difference in endothelial permeability is not well understood. Also prevents some substances from entering the brain from the blood, which has importance in drug treatment and toxicology.

blood-clotting factors See **Factors I-XII**.

Bloom's syndrome Rare disorder associated with genomic instability causing short stature, immunodeficiency and increased risk of all types of cancer. Caused by mutation in the gene encoding DNA **helicase** RecQ protein-like-3 (Bloom's syndrome protein, EC 3.6.1.12, 1417aa) that is involved in DNA replication and repair and is a component of the **BASC complex**.

blotting General term for the transfer of protein, RNA or DNA molecules from a relatively thick acrylamide or agarose **gel** to a paper-like membrane (usually nylon or nitrocellulose) by capillarity or an electric field, preserving the spatial arrangment. Once on the membrane, the molecules are immobilized, typically by baking or by ultraviolet irradiation, and can then be detected at high sensitivity by **hybridization** (in the case of DNA and RNA), or antibody labelling in the case of protein (immuno-blotting). RNA blots are called Northern blots; DNA blots, Southern; protein blots, Western. In Northwestern blotting protein is transferred but is probed with specific RNA. See also **dot** and **slot blots**.

blue naevus A non-malignant accumulation of highly-pigmented **melanocytes** deep in the **dermis**.

blue roses By some sophisticated genetic engineering Florigene managed to produce blue roses, a long-term ambition of plantsmen. This involved using **small interfering RNA** to switch off the endogenous **dihydroflavinol reductase gene** in a red rose, thereby blocking the **cyanidin** pathway, installing a pansy **delphinidin** gene and a new DFR gene to allow complete delphinidin synthesis and thus give a blue colour.

blue-green algae Group of prokaryotes that should now be referred to as **Cyanobacteria**.

blue-white colour selection Method for identifying bacterial clones containing plasmids with inserts. Many modern **vectors** have their **polycloning site** within a part of the LacZ gene encoding **beta-galactosidase**, which provides **alpha-complementation** in an appropriate mutant *E. coli* strain. This means that a re-ligated (empty) vector will produce blue colonies when grown on plates containing **IPTG** and **Xgal**, but colonies with a substantial insert in their plasmid's polycloning site are unable to produce functional beta-galactosidase, and so produce white colonies.

Bluescript Proprietary plasmid (pBluescript), sold by Stratagene. Very widely used.

Bluetongue virus An Orbivirus (a genus of the Reoviridae family) that causes serious disease (**bluetongue**) of sheep and milder disease in cattle and pigs. Transmitted by biting flies.

blunt end End of double stranded DNA that has been cut at the same site on both strands by a **restriction enzyme** that does not produce **sticky ends**.

Blym-1 An oncogene thought to be activated in Burkitt's lymphoma and chicken bursal lymphomas but that has disappeared from the current literature. Original description (1985): http://www.ncbi.nlm.nih.gov/pubmed/4015580

B-lymphocyte *B-cell* See **lymphocyte**.

BM-40 See **osteonectin**.

BMAA A neurotoxin found in cycad seeds, a non-protein amino acid (β-methylamino-L-alanine) believed to block the function of brain glutamate receptors. It is actually produced by cyanobacteria of the genus *Nostoc* that live on the plant's roots.

BMAL1 A bHLH transcription factor (brain and muscle ARNT-like protein 1, aryl hydrocarbon receptor nuclear translocator-like protein 1, 583aa) that is a component of the circadian clock oscillator. As a heterodimer with the **clock** protein it activates transcription of a number of proteins of the circadian clock. including **period**; transcription is inhibited in a feedback loop by period and by cryptochrome (CRY).

BMAPs Bovine **cathelicidins** (bovine myeloid antibacterial peptides). The active peptides (~27aa) are cleaved from larger precursors (~150aa) and are effective against Gram-negative and Gram-positive bacteria, including methicillin-resistant *Staphylococcus aureus*, and fungi.

B$_{max}$ Amount of drug required to saturate a population of receptors and a measure of the number of receptors present in the sample. Usually derived from **Scatchard plot** of binding data. Analogous to **V$_{max}$** in enzyme kinetics.

BME (1) Beta-mercaptoethanol (βME). (2) Engineered *Arabidopsis* lines that exhibit **beta-glucuronidase** expression in the micropylar end of the seed, named Blue Micropylar End, BME lines.

Bmf A pro-apoptotic protein (Bcl-2-modifying factor, 184aa). See **Bcl-2 homology domains**.

BMPs Bone morphogenetic proteins, multifunctional cytokines that are members of the **TGF** β superfamily. Activities are regulated by BMP-binding proteins **noggin** and **chordin**. Receptors are serine-threonine kinase receptors (TypesI and II) that link with **smad proteins**. *Drosophila* decapentaplegic (Dpp) is a homologue of mammalian BMPs. BMP2 is involved in regulating bone formation, BMP4 acts during development as a regulator of mesodermal induction and is over-expressed in fibrodysplasia ossificans. Follistatin inhibits BMP function in early *Xenopus* embryos. See also **osteogenin**.

BMR Basal metabolic rate, the number of calories required to maintain basic body functions at rest and in a thermal neutral zone.

BN-PAGE Blue-Native Polyacrylamide Gel-Electrophoresis, a modification of **polyacrylamide gel electrophoresis** (PAGE) in which the binding of Coomassie Blue (**Kenacid Blue**) to non-denatured (native) protein complexes confers negative charge which can be used to separate the proteins electrophoretically. A second dimension run using denaturing SDS can give further discrimination.

BNLF Proteins encoded by Epstein-Barr virus (EBV). BLNF-1 is a latent membrane protein (LMP-1, 386aa) that acts as a CD40 functional homologue to prevent apoptosis of infected B-lymphocytes. BNLF2a (60aa) is involved in viral evasion of HLA class I-restricted T-cell immunity. Associates with host TAP1 and TAP2 and prevents TAP-mediated peptide transport and subsequent loading. BLNF2b is uncharacterized.

BNP See **brain natriuretic peptide**.

BOAA Neurotoxic amino acid β−N-oxalylamino-L-alanine) found in the chickling pea (*Lathyrus sativa*) and responsible for **lathyrism**.

BODIPY Group of fluorescent dyes that can be conjugated to a range of molecules for use as probes in fluorescence microscopy. Some derivatives show a large fluorescent enhancement upon increasing the acidity of the solution and thus can be used in aqueous solution as fluorescent pH probes. Data sheet: http://pubs.acs.org/cgi-bin/abstract.cgi/joceah/2005/70/i10/abs/jo0503714.html

Bodonids Free-living kinetoplastids with an anteriorly directed dorsal flagellum, a posteriorly directed ventral flagellum and a simple feeding apparatus.

Bohr effect Decrease in oxygen affinity of **haemoglobin** when pH decreases or concentration of carbon dioxide increases.

Bollinger bodies Intracytoplasmic inclusion bodies in epithelial cells infected with fowl pox virus. They are aggregates of smaller bodies (Borrel bodies) which are the actual masses of the virus collected in the cells.

bombesin Tetradecapeptide **neurohormone** with both **paracrine** and **autocrine** effects first isolated from skin of fire-bellied toad (*Bombina bombina*); mammalian equivalent is **gastrin-releasing peptide** (GRP). Bombesin cross-reacts with GRP receptors. Both are **mitogenic** for Swiss 3T3 fibroblasts at nanomolar levels. Neuropeptides of this type are found in many tissues and at high levels in pulmonary (small cell carcinoma) and thyroid tumours. See **neuromedins**.

Bombyx mori Commercial silk moth.

bone marrow Tissue found in the centre of most bones, the major site of **haematopoesis** in adults and the most radiation-sensitive tissue of the body.

bone morphogenetic protein See **BMPs**.

bongkrekic acid An inhibitory ligand (531Da) of the mitochondrial adenine nucleotide trans-locator, also inhibits apoptosis by preventing **PARP** cleavage and **DEVD**ase activity. Highly toxic and can occur in fermented coconut contaminated by the bacterium *Burkholderia gladioli* pathovar *cocovenenans*.

booster response The enhanced response that occurs when an antigen is re-administered or when there is reinfection with a micro-organism if prior contact has elicited B-memory cells and helper T-cells.

bootstrap analysis A statistical method for testing the reliability of a dataset. It involves creating a set of pseudoreplicate datasets by re-sampling and is extensively used, for example, in phylogenetic analyses.

border cells (1) Cells of the root cap that have been shed but remain associated because they are enclosed in mucilage. (2) In *Drosophila*, a cluster of six to eight migratory cells found in the ovary and derived from the follicular epithelium. They are important for development of the micropyle. Are an important model system for studying control of migratory behaviour. Control of migration: http://jcb.rupress.org/content/192/3/513.long. (3) Neurons in the visual cortex that are sensitive to borders.

Bordetella pertussis A small, aerobic, Gram-negative bacillus, causative organism of whooping cough. Produces a variety of toxins including a dermonecrotising toxin, tracheal cytotoxin, an adenyl cyclase, **endotoxin**, and **pertussis toxin**, as well as surface components such as **fimbriae**, filamentous haemagglutinin and **pertactin.**

borealin One of the components of the **chromosomal passenger complex** (cell division cycle-associated protein 8, pluripotent embryonic stem cell-related gene 3 protein, Dasra-B, 280aa). Levels increase during G2/M phase and then reduce after exit from mitosis.

Borg Binder of rho GTPase: see **cdc42 effector proteins**.

Borna disease Virally-induced T-cell dependent immunopathological disorder of central nervous system. There are suggestions that Borna disease virus (a broadly distributed unclassified arthropod-borne virus that infects domestic animals and man) may be associated with some psychiatric disorders.

Borrel body See **Bollinger bodies.**

Borrelia burgdorferi A Gram-negative spirochaete, responsible for **Lyme disease**. Can be isolated from midgut of ticks (*Ixodes*) which transmit the disease.

Bos taurus Domestic cow.

Botox® Proprietary name for **Botulinum toxin** type A, injected into the skin as a temporary treatment to make lines on the face less apparent.

botrocetin A C-type lectin (venom coagglutinin, α, 133aa; β, 125aa) from *Bothrops jararaca* (pitviper) that induces binding of von Willebrand factor (vWF)

to platelet glycoprotein Ib (GPIb) and causes aggregation of blood platelets. There is a single **C-type lectin** domain in the α-subunit of the dimer.

Botrytis cinerea A non-specific **necrotrophic** fungal pathogen that attacks more than 200 plant species.

bottle cells The first cells to migrate inwards at the **blastopore** during amphibian **gastrulation**. The 'neck' of the bottle is at the outer surface of the embryo.

botulinolysin Cholesterol binding toxin from *Clostridium botulinum*.

botulinum toxin Neurotoxin (~1296aa, seven distinct serotypes, A–G) produced by certain strains of *Clostridium botulinum*. The bacterium produces the toxin as a complex with a haemagglutinin that prevents toxin inactivation in the gut. Proteolysis in the body results in cleavage into two fragments A and B; B binds to gangliosides and may stimulate the endocytosis of fragment A. The A subunit (EC 3.4.24.69) acts proteolytically on **SNAP-25** and blocks cholinergic synapses by inhibiting vesicle exocytosis. See **Botox, botulinus toxins C2 and C3, synaptobrevin, tetanus toxin.**

botulinus toxin C2 An AB toxin with binding subunit (721aa) and an enzymatic subunit (5431aa) that ADP-ribosylates monomeric G-actin and blocks the formation of microfilaments. Produced by C and D strains of *Clostridium botulinum*.

botulinus toxin C3 Toxin (mono-ADP-ribosyl-transferase C3, EC 2.4.2.-, 251aa) produced by C and D strains of *Clostridium botulinum*. An ADP-ribosyl transferase that inactivates **rho** and **rac**. Needs to be injected into cells and is a laboratory tool rather than a true toxin.

botulism Severe and often fatal poisoning due to eating food contaminated by *Clostridium botulinum* (see **botulinus toxin**).

boudicca A retrovirus-like **LTR retrotransposon** from *Schistosoma mansoni* similar to Ty3/gypsy retrotransposons.

Bouin's solution Bright yellow picric acid-based fixative that also contains formaldehyde and acetic acid. It has the advantage that specimens can be stored indefinitely and generally preserves nuclear morphology quite well.

boutons Small swellings in the terminal region of an axon or along the length of an axon (boutons *en passant*) where it makes contact with, for example, a muscle fibre. Synaptic vesicles are clustered in the boutons.

Bowman-Birk peptidase inhibitors Family of **serine peptidase** inhibitors found in seeds of leguminous plants and cereals.

box Casual term for a DNA sequence that is a characteristic feature of regions that bind regulatory proteins e.g. **homeobox**, **TATA box** and **CAAT box**.

Boyden chamber Simple chamber used to test for chemotaxis, especially of leucocytes. Consists of two compartments separated by a millipore filter (3–8 μm pore size); chemotactic factor is placed in one compartment and the gradient develops across the thickness of the filter (ca 150 μm). Cell movement into the filter is measured after an incubation period less than the time taken for the gradient to decay. See also **checkerboard assay**.

bozozok In zebrafish a **homeodomain** protein (dharma, nieuwkoid, 192aa) involved in dorsalization: expression is activated through beta-**catenin** signalling. Mutations cause defects in axial mesoderm and anterior neurectoderm and affect organizer formation. The dorsalization induced by bozozok is antagonised by another homeobox protein, ved, (ventrally expressed dharma/bozozok antagonist, 278aa).

BPG See **bisphosphoglycerate**.

BPI/LBP/Plunc super-family An extensive family of proteins that all have a single structural domain: the **lipopolysaccharide-binding protein** (LBP) / bactericidal permeability-increasing protein (BPI) / **cholesteryl ester transfer protein** (CETP) (LBP/BPI/ CETP) domain. The BPI-like proteins (PLUNCs) are expressed in oral and airway mucosa and may have a defensive role.

Bq See **Becquerel**.

brachydactyly Feature of a number of congenital abnormalities in which the fingers and toes are short as a result of premature closure of the epiphyses. Can be caused by mutation in genes for **Indian hedgehog**, BMP-receptor B1, growth/differentiation factor-5, receptor tyrosine kinase NTRKR2, homeobox protein HOXD13 and other currently unidentified genes.

Brachydanio rerio Now *Danio rerio*. See **zebrafish**.

brachyolmia A heterogeneous group of skeletal dysplasias, primarily affecting the spine. One form is the result of mutation in the **transient receptor potential channel**, TRPV4.

brachyury A transcriptional activator (protein T, 435aa) involved in regulating genes required for mesoderm formation and differentiation. Binds to a palindromic site (T-site) and is the founding member of the **T-box** family. The mouse mutant, in which it was originally identified, has a short tail. In humans, variations in T are associated with susceptibility to neural tube defects.

bracoviruses See **polydnaviruses**.

BRAD A web-based database of genetic data at the whole genome scale for important Brassica crops. Homepage: http://brassicadb.org.

Bradford assay Very commonly used assay for protein concentration based upon the absorbance shift in Coomassie Brilliant Blue G-250 (CBBG) when bound to arginine and aromatic residues. Standardisation of the assay with a protein of comparable arginine content is important and for some proteins, such as collagen, the results are very inaccurate.

brady- A prefix meaning slow. Bradycardia is the condition in which the heartbeat is unusually slow. *Cf.* **tachycardia**.

bradykinin Vasoactive nonapeptide (RPPGFSPFR) formed by action of peptidases on kininogens. Very similar to **kallidin** (which has the same sequence but with an additional N-terminal lysine). Bradykinin is a very potent vasodilator and increases permeability of post-capillary venules; it acts on endothelial cells to activate phospholipase A2. It is also spasmogenic for some smooth muscle and will cause pain.brahma

brahma In *Drosophila* a homeotic gene activator (1638aa), an ATP-dependent helicase that is part of the brahma complex, the equivalent of the yeast **SWI/SNF** chromatin-remodelling complex.

Brahma-related gene 1 The human brahma-related gene 1 (BRG1) encodes a protein (SMARCA4, 1647aa) that is an essential component of the SWI/SNF chromatin remodeling complexes necessary for normal mitotic growth and transcription regulation and implicated in multiple functions through its interaction with different proteins. Mutations of BRG1 have been found in multiple tumour cell lines.

brain natriuretic peptide *BNP* Brain peptide (134aa but cleaved into various ~25–30aa peptides) may function as a paracrine antifibrotic factor in the heart. Also plays a key role in cardiovascular homeostasis through natriuresis, diuresis, vasorelaxation, and inhibition of renin and aldosterone secretion. Specifically binds and stimulates the NPR1 receptor. See **natriuretic peptides**.

brain regions The central nervous system of mammals is complex and the terminology often confusing. In development the brain is generated from the most anterior portion of the neural tube and there are three main regions, fore, mid and hind brain. The lumen of the embryonic nervous system persists in the adult as the cerebral ventricles, filled with cerebrospinal fluid, which are connected to the central canal of the spinal cord. The forebrain develops to produce the cerebral hemispheres and basal ganglia and the diencephalon which forms the thalamus and hypothalamus. The cerebrum consists of two hemispheres, connected by the corpus callosum, the outer part being greatly expanded in man with the increased surface being thrown into fold (ridges are gyri, valleys are sulci). The outer layer (cerebral cortex) is responsible for so-called higher order functions such as memory, consciousness and abstract thought, the deeper layers (basal ganglia) include the caudate nucleus and putamen (collectively the striatum), amygdaloid nucleus and hippocampus. The hypothalamus controls endocrine function (hunger, thirst, emotion, behaviours, sleep), the thalamus coordinates sensory input and pain perception. The midbrain is relatively small and develops to form corpora quadrigemina and the cerebral peduncle. The hindbrain develops into two regions, the more anterior being the metencephalon, the region nearest the spinal cord being the myelencephalon. The metencephalon contains the cerebellum, responsible for sensory input and coordination of voluntary muscles, and the pons. The myelencephalon contains the medulla oblongata, which regulates blood pressure, heart rate and other basic involuntary functions, and dorsally the choroid plexus.

brain small vessel disease with haemorrhage A disorder caused by a defect in **collagen** type IV (COL4A1) that causes a predisposition to strokes.

brain-derived growth factor See **brain-derived neurotrophic factor**.

brain-derived neurotrophic factor Small basic protein (BDNF, BDGF, abrineurin, 247aa) originally purified from pig brain; a member of the family of **neurotrophins** that also includes **nerve growth factor** (NGF) and **neurotrophin-3**. In contrast to NGF, BDNF is predominantly (though not exclusively) localized in the central nervous system. It supports the survival of primary sensory neurons originating from the **neural crest** and ectodermal **placodes** that are not responsive to NGF. **Huntingtin** upregulates transcription of BDNF and vesicular transport of BDNF along microtubules.

branching factors See **strigolactones**.

branchiooculofacial syndrome A developmental defect that affects the bronchial system, face and eyes; caused by mutation in transcription factor **AP-2A**.

Branchiostoma Cephalochordates, of which *Branchiostoma* is one genus (*Amphioxus* is the obsolete generic name), are the most vertebrate-like of the invertebrates with a dorsal hollow nerve cord, notochord, postanal tail, and pharyngeal gill slits used for filter feeding.

BRAP See **bridging integrators**.

Brassica napus Oil-seed rape (*US.* canola). Source of edible oil (see **erucic acid**).

brassinolide See **brassinosteroids**.

brassinosteroids Plant hormones (polyoxygenated steroids) that have pronounced plant growth regulatory effects. Are released by mature cells in response to root environmental, pest, or disease stress. Commonest form is brassinolide.

brat In *Drosophila* a protein (brain tumour protein, 1037aa) involved in translation repression of **hunchback** mRNA, and probably other mRNAs in other tissues. Interacts with **nanos** and **pumilio** in determining posterior polarity. Acts as a growth suppressor in the larval brain.

brazzein A sweet protein (53aa) from the fruit of *Pentadiplandra brazzeana*. It is 2000-fold sweeter than sucrose on a molar basis and is defensin-like.

brca Genes (*brca1, brca2*) linked to familial breast cancer (breast-ovarian carcinoma syndrome). BRCA1 (1863aa) is a nuclear protein and a component of the tumour-suppressor complex (BRCA1-associated genome surveillance complex, **BASC**). BRCA2 (Fanconi anaemia group D1 protein, 3418aa) is involved in DNA double-strand break repair and homologous recombination. See **BRCT**, **BARD**.

BRCT domain A tandem repeat domain (BRCA C-terminal domain, 90-100aa) conserved across multiple organisms, involved in binding phosphoserine-proteins and found predominantly in proteins that are involved in cell cycle regulation and DNA repair. Picture and details: http://www-nmr.cabm.rutgers.edu/photogallery/proteins/htm/pagebrct.htm

Brdu Bromodeoxyuridine. See **BUdR**.

bread mould *US* bread mold Favourite mould for classroom study; grows 'spontaneously' on damp bread and can be any one of a number of common fungi, though often stated to be *Rhizopus nigricans* that has a cottonly growth pattern. Bluish-green to green moulds are usually *Penicillium* or *Aspergillus*. Black to brown-black molds can be *Aspergillus niger*, *Alternaria alternata*, *Cladosporium herbarum*, *Cladosporium sphaerospermum*, or *Stachybotrys chartarum* (a highly toxic mold). Reddish or pink molds are usually species of *Fusarium*, orange mould is likely to be *Neurospora* spp.

breakpoint cluster region See **bcr**.

brefeldin A A macrocyclic lactone synthesized from palmitic acid by several fungi including *Penicillium brefeldianum*. It was initially described as an antiviral antibiotic, but it was later found to inhibit protein secretion at an early stage, probably blocking secretion in a pre-Golgi compartment. Binds to the ARF 1/GDP/Sec7 complex and blocks

GEF activity at an early stage of the reaction, prior to guanine nucleotide release. A valuable tool for studying membrane traffic and the control of organelle structure.

brevetoxins Lipophilic 10- and 11-ring polyether toxins from the dinoflagellate *Ptychodiscus* (formerly *Gymnodiunium*) *brevis;* responsible for neurotoxic shellfish poisoning and associated with 'red tides'. Bind to the sodium channel of nerve and muscle making it hyper-excitable by altering the critical threshold level of depolarization required to generate an action potential.

Brevibacterium Genus of Gram-positive aerobic coryneform bacteria.

BREVIS RADIX A protein (BRX, 344aa) that regulates cell proliferation and elongation in the root. Levels are up-regulated by auxin and down-regulated by brassinolide; mutation leads to short primary roots with more laterals. Brevis radix-like one (BRXL1, 331aa) forms heterodimers with **BRX**.

BRG1-associated factors Proteins associated with the **Brahma-related gene one** (BRG1) product. BRG1-associated factor-1 (BAF57, SMARCE1, 411aa) is the 57 kDa subunit of the chromatin remodeling ATPase complex (**BAF complex**) homologous to the yeast **SWI/SNF complex**. Many different BAFs have been identified with particular roles in transcriptional activation during development. Article: http://www.ncbi.nlm.nih.gov/pmc/articles/PMC2366179/

BRI1 A plant protein (BRASSINOSTEROID INSENSITIVE 1, EC 2.7.10.1, EC 2.7.11.1, 1196aa) a **dual-specificity kinase** that regulates a signalling cascade activated by brassinosteroid binding that is involved in plant development, including expression of light- and stress-regulated genes, promotion of cell elongation, normal leaf and chloroplast senescence, and flowering. It phosphorylates BRI1-associated receptor kinase 1 (BAK1, SERK3). Bri1 suppressor protein 1 is a serine/threonine-protein phosphatase (BSU1, EC 3.1.3.16 793aa) although the name has also been applied to serine carboxypeptidase 24 (EC 3.4.16.6, 465aa) which does inhibit an early step in the signalling pathway.

bride of sevenless In *Drosophila* eye development, the ligand (boss, 896aa) for the **sevenless** tyrosine kinase receptor. Boss is expressed by the central R8 cell. It is unusual as a ligand for a tyrosine receptor kinase in that it is on the surface of another cell and has, in addition to a large extracellular domain, seven transmembrane segments and a C-terminal cytoplasmic tail.

bridging integrators A family of adaptor proteins with BAR domains (see **BAR proteins**). Bridging integrator 1 (BIN1, myc box-dependent-interacting protein 1, 593aa) is an **amphiphysin**-like protein that may be involved in regulation of synaptic vesicle endocytosis and may act as a tumour suppressor. It is mutated in some forms of myopathy. Bridging integrator 2 (BRAP1, breast cancer-associated protein 1, 565aa) is preferentially expressed in haematopoietic tissues. Bridging integrator 3 (253aa) has a role in the organisation of F-actin during cytokinesis.

Bright Transcription factor (B cell regulator of immunoglobulin heavy chain transcription, 593aa) that binds A+T-rich sequences in the intronic enhancer regions of the murine heavy chain locus and 5'-flanking sequences of some variable heavy chain promoters. Binds as a dimer.

bright-field microscopy See **light microscopy**.

Brilliant Cresyl Blue Dye (Brilliant Blue C) used in staining of bone marrow smears.

British Approved Name See **BAN**.

brittle cornea syndrome A connective tissue disorder (Ehlers-Danlos syndrome type VIB) in which the cornea is prone to rupture after minor trauma. Caused by mutation in the zinc finger protein ZNF469 (3925aa).

Brix domain A domain found in one archaean and five eukaryotic protein families which have a similar domain architecture with a central globular Brix domain. Proteins from the Imp4/Brix superfamily appear to be involved in ribosomal RNA processing.

Broca's area The left inferior convolution of the frontal lobe of the brain; damage to this region leads to impairment of speech production (Broca's aphasia), though not comprehension.

bromelain Cysteine endopeptidases (thiol protease) from pineapple (*Ananas comosus*). EC 3.4.22.32 is stem bromelain (212aa) and is distinct from EC 3.4.22.33, the fruit bromelain (351aa). The pineapple also produced a bromelain inhibitor (bromein, 251aa) which is cleaved to form various small inhibitory peptides.

bromocriptine An agonist for dopamine D2 receptors, used in the treatment of pituitary tumours and Parkinson's disease.

bromodomain Domain (~70aa) found in a variety of mammalian, invertebrate and yeast DNA-binding proteins, involved in binding to acetyllysines on histone tails.

bromophenol blue Dye used as pH indicator: changes from yellow to blue in the range 3.0–4.6.

bromophenol red Dye used as pH indicator: changes from yellow to red in range 5.2–6.8

bromothymol blue A pH sensitive dye, changes from yellow to blue in the pH 6.0–7.6 range.

bronchodilators Substances that dilate airways by relaxing the smooth muscle of the bronchial wall, thus relieving breathlessness caused acutely by asthma or chronically by obstructive pulmonary disease. Often administered by inhalation. Examples include **sympathomimetic drugs** (salbuterol), **antimuscarinic drugs**, and xanthines (theophylline).

bronchogenic carcinoma Lung cancer arising from the epithelium of the bronchial tract.

bronchus associated lymphoid tissue *BALT* Subset of mucosal associated lymphoid tissue found as lymphoid nodules in the lamina propria of the bronchus. Some mammalian species have it constitutively present (e.g. rabbits) but in others (e.g. humans) it can develop in response to infection.

bronzed diabetes See **haemochromatosis**.

brown adipose tissue Highly vascularised adipose tissue found in restricted locations in the body (in the inter-scapular region in the rat, for example). In hibernating animals and neonates, brown adipose tissue is important for regulating body temperature via non-shivering thermogenesis.

brown fat cells Brown fat is specialized for heat production and the **adipocytes** have many mitochondria in which an inner-membrane protein can act as an uncoupler of **oxidative phosphorylation** allowing rapid thermogenesis. See **brown adipose tissue**.

Brownian motion Random motion of small objects as a result of intermolecular collisions. First described by the 19th-century microscopist, Brown.

Brownian ratchet Mechanism proposed to explain protein translocation across membranes and force generation by polmerizing actin filaments. Relies upon asymmetry of *cis* and *trans* sides of the membrane or biased thermal motion as a result of polymerization. Still an hypothesis.

Brucella Genus of Gram-negative aerobic bacteria which occur as intracellular parasites or pathogens in man and other animals. *Brucella abortus* is responsible for spontaneous abortion in cattle and causes undulent fever (brucellosis), a persistent recurrent acute fever, in man.

Bruch's membrane The innermost layer of the **choroid** of the eye. There are five layers, the basal lamina of the retinal pigment epithelium, an inner collagenous zone, a central layer rich in elastin fibres, an outer collagenous zone and the basal lamina of the **choriocapillaris**.

Brugada's syndrome Sudden unexplained nocturnal death syndrome that can be caused by mutation in the gene encoding the pore-forming alpha subunit of **voltage-gated cardiac sodium channel-5** or a defect in the GPD1L-gene (glycerol-3-phosphatase dehydrogenase 1-like) which affects the properties of the ion channel.

Brunner's gland Small, branched, coiled tubular glands (duodenal glands) in the submucosa of the first part of the duodenum. Their secretion of alkaline mucus helps neutralize gastric acid from the stomach.

brush border The densely packed **microvilli** on the apical surface of, e.g. intestinal epithelial cells.

Bruton's agammaglobulinaemia Sex-linked recessive **agammaglobulinaemia** caused by a deficiency in **B-lymphocyte** function. See **btk**.

Bruton's tyrosine kinase See **btk**.

Brx (**1**) A-kinase anchor protein 13 (breast cancer nuclear receptor-binding auxiliary protein, lymphoid blast crisis oncogene, 2813aa), one of the **Dbl** family that increases estrogen receptor activity. It anchors cAMP-dependent protein kinase (PKA) and is a guanine nucleotide exchange factor (**GEF**) for rho. (**2**) A cysteine-rich hydrophobic domain 1 protein (CHIC1, brain X-linked protein, BRX, 224aa) a palmitoylated membrane protein preferentially expressed in brain and possibly mutated in an X-linked mental retardation syndrome. (**3**) Ribosome biogenesis protein BRX1 homologue (Brix domain-containing protein 2, 353aa) required for biogenesis of the 60S ribosomal subunit. (**4**) In *Arabidopsis*, a transcription regulator BREVIS RADIX (BRX, 344aa) that regulates cell proliferation and elongation in the root and shoot.

Bryological Glossary A multilingual glossary that defines many terms specific to bryophytes. The definitions cover English, French, German, Spanish, Portugese, Japanese, Latin and Russian although the core is the English version. For example the acuminate entry gives a definition together with: Latin − acuminatus, French − acuminé, German − langspitzig, zugespitzt, Portuguese − acuminado, Spanish − acuminado. Homepage: http://www.mobot.org/MOBOT/tropicos/most/Glossary/glosefr.html

Bryophyta Plant phylum that includes mosses and liverworts.

bryostatin General name for a group of compounds (complex lactones) isolated from bryozoans; activate protein kinase C (**PKC**), though after longer-term exposure cells down-regulate their PKC.

Bryozoa *Polyzoa* Phylum of invertebrates, 'moss animals', that are mainly marine and usually colonial.

BSE Bovine **spongiform encephalopathy**. A transmissible encephalopathy that affected large numbers of cattle in the UK during the 1990s and is widely believed to have arisen through consumption, by cattle, of feedstuff containing sheep tissues from animals with scrapie, although the agent is clearly different from that of scrapie. A link with 'new variant CJD' is strongly suspected.

BTB/POZ domain Structurally well-conserved domain (Bric-a-brac Tramtrack Broad/ Pox virus and Zinc finger complex; POZ/BTB domain, ~120aa) involved in protein-protein interactions and important in a wide range of cellular functions, including transcriptional regulation, cytoskeleton dynamics, ion channel assembly and gating, and targeting proteins for ubiquitination. Originally identified as a conserved motif present in the *Drosophila* bric-à-brac, tramtrack and broad complex transcription regulators and in many pox virus and zinc finger proteins.

BTC See **betacellulin**.

btk A **Tec** family tyrosine kinase (Bruton's tyrosine kinase, 659aa), defective in Bruton's agammaglobulinaemia. Mutations in btk lead to B-cell immunodeficiencies XLA in humans, Xid in mice although the exact mechanism by which btk regulates B-cell differentiation is unclear. Overexpression of btk enhances calcium influx following B-cell antigen receptor crosslinking. Btk interacts with other tyrosine kinases such as **fyn**, **lyn** and **hck**, which are activated upon stimulation of B- and T-cell receptors. **Itk** is the T-cell homologue of btk. **Sab** selectively binds the SH3 domain of btk. Btk interacts with membrane through **PH domain**, and **SHIP**, by reducing PIP3 levels, regulates this association.

BTX See **batrachotoxins**.

budding A type of cell division in fungi and in protozoa in which one of the daughter cells develops as a smaller protrusion from the other. Usually the position of the budding cell is defined by polarity in the mother cell. In some protozoa the budded daughter may lie within the cytoplasm of the other daughter.

BUdR Bromo-deoxyuridine (the deoxynucleoside of 5-bromo-uracil) an analogue of thymidine that induces point mutations because of its tendency to tautomerization: in the enol form it pairs with G instead of A. It is used as a mutagen, and also as a marker for DNA synthesis (the incorporation of BUdR can be recognized because the staining pattern differs: an even more sensitive method uses a monoclonal antibody staining procedure.)

buffer A system that acts to minimize the change in concentration of a specific chemical species in solution against addition or depletion of this species. pH buffers: weak acids or weak bases in aqueous solution. The working range is given by $pK_a \pm 1$. Metal ion buffers: a metal ion chelator e.g. **EDTA**, partially saturated by the metal ion acts, as a buffer for the metal ion.

buffy coat Thin yellow-white layer of leucocytes on top of the mass of red cells when whole blood is centrifuged.

bufotenine An indole alkaloid (mappine) with hallucinogenic effects, isolated from *Piptadenia* spp. (Mimosidae); first isolated from skin glands of toad (*Bufo* sp.).

Buggy Creek virus See **Chikungunya virus**.

bulb-type lectin domain A domain of ~115aa found in lectins from various bulbs (garlic, snowdrop etc.), generally binds mannose. The domain has an overall three dimensional fold very similar to that of **comitin** and curculin (a sweet tasting protein from *Curculigo latifolia*).

bulla *Pl.* bulli A blister or bleb. A circumscribed elevation above the skin containing clear fluid; larger than a vesicle.

bulliform cell An enlarged epidermal cell type found in longitudinal rows in the leaves of some grasses. May be responsible for rolling and unrolling of leaves in response to changes of water status.

bullous pemphigoid Form of pemphigoid (which also affects mucous membranes), in which blisters (bulli) form on the skin. Patients have circulating antibody (usually IgG) to **basement membrane** of **stratified epithelium**, although the antibody titre does not correlate with the severity of the disease.

bundle of His A bundle of small specialized conducting muscle fibres in the mammalian heart that is responsible for transmitting electrical impulses from atrium to ventricle.

bungarotoxins Toxins found in the venom of *Bungarus multicinctus*, the multi-banded krait. Alpha-bungarotoxin (74aa) causes virtually irreversible block of the vertebrate neuromuscular junction by binding (as a monomer) to each of the α-subunits of the postsynaptic nicotinic acetycholine receptors (nAChR). Has been much used in identifying, quantifying and localizing these receptors on muscle cells. Will also bind some neuronal nAChR. Beta-bungarotoxin is a two-chain phospholipase A2 neurotoxin that acts at the presynaptic site of motor nerve terminals and blocks transmitter release. Subunit A (EC 3.1.1.4, 147aa) is structurally homologous to other PLA2s, the B subunit (85aa) has homology with Kunitz-type serine peptidase inhibitors and **dendrotoxins**. Binds to a subtype of voltage-sensitive potassium channels. Kappa-bungarotoxin: (bungarotoxin 3.1, Toxin F, neuronal bungarotoxin, 87aa) has considerable homology with α-bungarotoxin; the homodimer is a potent antagonist for a subset of neuronal nAChR but is much less active against muscle receptors.

Büngner band A region of Schwann cells found at the end of a regenerating nerve. Cells may be organised as ordered columns along the endoneurial tube. The residual basal lamina of Schwann cells in this region can act as a guidance cue for the nerve growth cone.

Bunyaviridae A family of enveloped viruses infecting vertebrates and arthropods. The genome is monomeric and consists of three segments of circular; sometimes super-coiled, negative-sense and ambisense, single-stranded RNA (ssRNA). The virion is spherical or oval, 90–100 mm diameter. Some genera (e.g. **Hantavirus**) cause serious disease in humans. Virus taxonomy database: http://www.ncbi.nlm.nih.gov/ICTVdb/ICTVdB/00.011.htm

Burkholderia pseudomallei A pathogenic pseudomonad that causes **melioidosis**, a disease endemic in animals and humans in Southeast Asia. The closely related *B. mallei* causes glanders, a disease of horses occasionally transmitted to humans.

Burkitt's lymphoma Malignant tumour of **lymphoblasts** derived from B-lymphocytes. Most commonly affects children in tropical Africa: both **Epstein-Barr virus** and immune-suppression due to malarial infection are involved.

burr cells Triangular helmet-shaped cells found in blood, usually indicative of disorders of small blood vessels.

Bursa of Fabricius A **lymphoid tissue** found at the junction of the cloaca and the gut of birds giving rise to the so-called **B-lymphocyte** series.

bursicon Insect hormone: bursicon subunit alpha (cuticle-tanning hormone, 173aa) forms a heterodimer with partner of bursicon (bursicon subunit beta, 141aa) and activates a G protein-coupled receptor. It is produced by neurosecretory cells of the brain and affects many post-ecdysal processes such as cuticular tanning.

burst forming unit *BFU-E* A bone marrow **stem cell** lineage detected in culture by its mitotic response to **erythropoietin** and subsequent erythrocytic differentiation in about 12 mitotic cycles into erythrocytes.

Buruli ulcer A tropical ulcerative disease caused by infection with *Mycobacterium ulcerans*. Most of the tissue damage is due to the bacterial toxin, **mycolactone.**

butyric acid Acid from which butyrate ion is derived. Smells of rancid butter, hence the name.

butyrolactone autoregulator receptors A group of receptors that bind as homodimers to conserved DNA sequences but are released if the cognate ligand (γ-butyrolactone) binds, thereby allowing gene expression. They regulate polyketide antibiotic production and sometimes morphological differentiation in streptomycetes. See **alp gene cluster**. Article: http://mic.sgmjournals.org/content/153/6/1817.full.pdf

butyrophilins A sub-family of the immunoglobulin super-family (∼525aa), integral membrane glycoproteins of the secretory epithelium of mammary gland. that may have a receptor function by binding to xanthine dehydrogenase/oxidase or act as an internal receptor on the apical plasma membrane for cytoplasmic milk-fat droplets. The BTN gene family codes for seven proteins (BTN, BTN2A1, BTN2A2, BTN2A3, BTN3A1, BTN3A2, BTN3A3).

butyrylcholinesterase A family of enzymes (BChE, choline esterase II, pseudocholinesterase, EC 3.1.1.8, 602aa) produced mainly in the liver that hydrolyse esters e.g. **procaine** and **suxamethonium**. Occurs as a number of variants, dependent on four alleles, with variable degrees of cholinesterase function. The nature of the variants determines sensitivity to suxamethonium; the extreme is pseudocholinesterase deficiency. Is known to metabolize cocaine in humans.

BY2 cells Line of cells from tobacco (*Nicotiana tabacum*) that will grow in suspension culture.

byssinosis Asthma-like respiratory disease in people exposed to dust from vegetable fibres (cotton, jute, flax).

bystander effect Phenomenon in which mammalian cells irradiated in culture produce damage-response signals which are communicated to their un-irradiated neighbours. The mechanism is unclear but the implications are important in radiation biology.

bystander help Lymphokine-mediated nonspecific help by T-lymphocytes, stimulated by one antigen, to lymphocytes stimulated by other antigens. Somewhat controversial. Useful discussion: http://www.sciencedirect.com/science/article/pii/S1471490606002705

bystin A cytoplasmic protein (437aa) involved in processing of 20S pre-rRNA and biogenesis of 40S ribosomal subunits. Binds **trophinin, tastin** and cytokeratins.

bZip See **basic leucine zipper.**

C

c (1) Single letter code for cysteine (in proteins) or cytosine (in nucleic acids). (2) Prefix used to denote the normal cellular form of, for example, a gene such as *src* (c-*src*) that is also found as a viral gene (v-*src*).

C₀t curve A graphical method that can be used to indicate the complexity (or size) of DNA. The DNA is heated to make it single stranded, then allowed to cool. The kinetics of renaturation (reannealing) of the DNA can be followed spectroscopically and the rate is affected by the length of the DNA fragments and by the extent of repetition. Highly repetitive DNA anneals much faster than unique sequence DNA (single copies sequences). C_0 is the starting concentration of the specific DNA sequence in moles of nucleotides per liter and t is the reaction time in seconds. The C_0t value varies depending on the temperature of reassociation and the concentration of monovalent cations and it is usual to use fixed reference values: a reassociation temperature of 65°C and a Na⁺ concentration of 0.3 M NaCl. (**Cot1 DNA** has a C_0t value of 1.0). Detailed description: http://www.ncbi.nlm.nih.gov/books/NBK7567/

C1 First component of **complement**; actually three subcomponents, C1q (245aa), C1r (705aa) and C1s (688aa), that form a complex in the presence of calcium ions. C1q, the recognition subunit, has an unusual structure of collagen-like triple helices forming a stalk for its Ig-binding globular heads. Upon binding to immune complexes the C1 complex becomes an active endopeptidase that cleaves and activates **C4** and **C2**.

C1–C9 Proteins of the mammalian **complement** system. See individual numbered components.

C2 Second component of **complement**. A serine peptidase (C3/C5 convertase, EC 3.4.21.43, 752aa) cleaved by activated factor C1 into two fragments: C2b and C2a; C2a combines with complement factor 4b to generate the C3 or C5 convertase.

C2-8 Cell permeable compound that potently inhibits polyglutamine (polyQ) aggregation in Huntington's disease (HD) neurons. Suppresses neurodegeneration in *Drosophila in vivo*. Suspected to block the polymerization step of the polyQ aggregation. Article: http://www.ncbi.nlm.nih.gov/pmc/articles/PMC2034257/

C2-kinin Probably an artifact, a kinin-like fragment thought to be generated from **complement** C2 but more likely to be **bradykinin**.

C3 Third component of **complement** (1663aa). Both classical and alternate pathways converge at C3, which is cleaved to yield C3a, an **anaphylotoxin**, and C3b, which acts as an opsonin and is bound by **CR1**; C3b in turn can be proteolytically cleaved to iC3b (ligand for **CR3**) and C3dg by C3b-inactivator. C3b complexed with Factor B (to form C3bBb) will cleave C3 to give more C3b, although the C3bBb complex is unstable unless bound to **properdin** and a carbohydrate-rich surface. The C3b-C4b2a complex and C3bBb are both C5 convertases (cleave **C5**). Cobra venom factor (1642aa, found in the venom of *Naja kaouthia*, the monocled cobra) is homologous with C3b but the complex of cobra venom factor, properdin and Factor Bb is insensitive to C3b-inactivator.

C3 nephritic factor An autoantibody that binds to and stabilizes the C3 convertase enzyme of the alternate **complement** pathway (C3bBb) both in the presence and absence of serum regulatory proteins. Deficiencies are associated with type 2, dense-deposit membrano-proliferative glomerulonephritis and partial **lipodystrophy**. Article: www.ncbi.nlm.nih.gov/pubmed/3848661

C3 plants Plants that fix CO_2 in photosynthesis by the **Calvin-Benson cycle**. The enzyme responsible for CO_2 fixation is **ribulose bisphosphate carboxylase**, whose products are compounds containing three carbon atoms. C3 plants are typical of temperate climates. **Photorespiration** in these plants is high.

C3G Guanine nucleotide exchange factor (RAPGEF1, 1077aa) that activates **Rap1**. C3G is involved in signalling from **crk** to **JNK**.

C3HA An inbred strain of mice that originally was very suceptible to mammary tumours.

C4 Fourth component of **complement** (1744aa), although the third to be activated in the classical pathway. Becomes activated by cleavage (by C1) to C4b, which complexes with C2a to act as a C3 convertase, generating C3a and C3b. The C4b2a3b complex acts on C5 to continue the cascade. Human C4 is polymorphic at two loci, C4A and C4B.

C4 plants Plants found principally in hot climates whose initial fixation of CO_2 in photosynthesis is by the **HSK pathway**. The enzyme responsible is **PEP carboxylase**, whose products contain four carbon atoms. Subsequently the CO_2 is released and re-fixed by the **Calvin-Benson cycle**. The presence of the HSK pathway permits efficient photosynthesis at high light intensities and low CO_2 concentrations.

The Dictionary of Cell and Molecular Biology. DOI: http://dx.doi.org/10.1016/B978-0-12-384931-1.00003-9

Most species of this type have little or no **photorespiration**.

C5 Fifth component of **complement** (1676aa), which is cleaved by C5-convertase to form C5a, a 74aa anaphylotoxin and potent chemotactic factor for leucocytes, and C5b. C5a rapidly loses a terminal arginine to form $C5a_{desarg}$, which retains chemotactic but not anaphylotoxic activity. C5b combines with C6, C7, C8 and C9 to form a membranolytic complex.

C57BL Inbred strain of black mice developed around 1920 by Clarence Cook Little, who later founded the Jackson Laboratory.

C6, C7 Sixth (C6, 934aa) and seventh (C7, 843aa) components of the complement cascade. Contain EGF-like motifs. See **C5** and **C9**.

C8 Eighth component of complement, a heterotrimer of α (584aa), β (591aa) and γ (202aa) peptides. The alpha and gamma chains are disulphide linked.

C9 Ninth component of **complement** (559aa). Complexed with C5b, 6, 7 and 8 it forms a potent membranolytic complex (sometimes referred to as the membrane attack complex, MAC). Membranes that have bound the complex have toroidal pores any one of which may be enough to cause lysis. C9 is the pore-forming subunit of the complex.

c127 cells A nontumorigenic mouse epithelial cell line, derived from mammary gland, widely used in *in vitro* transformation assays due to its normal morphological appearance and its very low levels of spontaneous transformation

Ca^{2+} puffs Local intracellular signals generated by ligand-gated ion channels (inositol trisphosphate receptors, IP3Rs) that liberate Ca^{2+}. Puff sites represent pre-established, stable clusters of IP3Rs. Free article: http://www.ncbi.nlm.nih.gov/pmc/articles/PMC2897231/?tool = pubmed

CAAT box Nucleotide sequence in many eukaryotic promoters usually about 75bp upstream of the start of transcription. Binds **NF-1**.

cabin A nuclear **calcineurin**-binding protein (calcineurin inhibitor, 2220aa) with a negative regulatory role in the T-cell-receptor signalling pathway. It binds **MEF2B**, and probably all the MEF family of transcription factors and if intracellular calcium levels rise, activated calmodulin will displace the MEFs, allowing them to act. May also be required for replication-independent chromatin assembly.

cachectin Protein produced by macrophages that is responsible for the wasting (cachexia) associated with some tumours. Now known to be identical to tumour necrosis factor (**TNFα**). Has three 233aa subunits, all derived from a single highly-conserved gene.

Caco cells Cell line with epithelial morphology derived from a primary colonic carcinoma of a 72-year-old male Caucasian.

CAD (1) A multifunctional protein (2225aa) responsible for the first three steps of the *de novo* pyrimidine synthesis pathway: glutamine-dependent carbamoyl-phosphate synthase (EC 6.3.5.5), aspartate carbamoyltransferase (EC 2.1.3.2) and dihydro-orotase (EC 3.5.2.3). It is allosterically regulated and when phosphorylated loses feedback inhibition. **(2)** See **DFF**.

CADASIL Cerebral arteriopathy with subcortical infarcts and leukoencephalopathy (CADASIL), an autosomal dominant progressive disorder of the small arterial vessels of the brain leading to stroke and dementia. Caused by mutation in the *Notch3* gene.

cadaverine Substance formed by microbial action in decaying meat and fish by decarboxylation of lysine. The smell can be imagined. Like many of the other diamines (e.g. **putrescine**) has effects on cell proliferation and differentiation.

CadF (1) An adhesion molecule (Campylobacter adhesin to Fn, 326aa) of the OmpA family, produced by *Campylobacter* sp., that allows binding of the organism to the extracellular matrix. Abstract: http://www.ncbi.nlm.nih.gov/pubmed/16091041. **(2)** Cofilin/actin-depolymerizing factor homologue (see **twinstar**).

cadherins Integral membrane proteins involved in calcium dependent cell adhesion, usually tissue-specific. The cytoplasmic domain interacts with **catenins**. Cadherin-1 (E-cadherin, uvomorulin, CD324, 882aa) is found in in epithelia, cadherin-2 (N-cadherin, 906aa) in neural cells and cadherin-3 (CDH3, P-cadherin, 829aa) in placenta. Cadherin-4 (916aa) is retinal cadherin, cadherin-5 is vascular endothelial cadherin (784aa). The cadherin family includes **desmocollins**, **desmogleins** and **protocadherins**. See also **eplin**, **flamingo**, **nectin #3**, **nullo**, **otocadherin**, **plakophilins**. Mutations in cadherin-1 are associated with many carcinomas, defects in cadherin-3 cause **hypotrichosis with juvenile macular dystrophy** and **EEM syndrome**.

Caenorhabditis elegans Nematode much used in lineage studies since the number of nuclei is determined (there are 959 cells), and the nervous system is relatively simple. One of the first organisms to have its complete genome sequenced. The organism can be maintained axenically and there are mutants in behaviour, in muscle proteins, and in other features. Sperm are amoeboid and move by an unknown mechanism which does not seem to depend upon actin or tubulin.

caerulin Amphibian peptide hormone related to **gastrin** and **cholecystokinin**. A decapeptide that is one of the main constituents of the skin secretion of *Xenopus laevis*.

caeruloplasmin See **ceruloplasmin**.

caesium chloride *US.* cesium chloride Salt that yields aqueous solutions of high density. When equilibrium has been established between sedimentation and diffusion during ultracentrifugation, a linear density gradient is established in which macromolecules such as DNA band at a position corresponding to their own buoyant density.

CAF-1 (1) CCR4-NOT transcription complex subunit 7 (BTG1-binding factor 1, CCR4-associated factor 1, 285aa) an ubiquitous transcription factor involved in a wide range of processes. (2) See chromatin assembly factor.

caffeine A **xanthine** derivative that elevates cAMP levels in cells by inhibiting phosphodiesterases. In plants caffeine inhibits fusion of vesicles that form the new cross wall or cell plate. Only stubs of wall are formed at the cortex, resulting in an incompletely separated and therefore bi-nucleate cell.

caffeoyl CoA O-methyl-transferase An enzyme (CCOMT, EC 2.1.1.104, 121aa) that catalyses the formation of S-adenosylhomocysteine and feruloyl-CoA from S-adenosyl methionine and caffeoyl-CoA in the biosynthesis of **lignin**.

Caffey's disease A disorder (infantile cortical hyperostosis) caused by mutation in the alpha-1 collagen type I gene (COL1A1) which is unusual because there is an episode of massive subperiosteal new bone formation beginning before five months of age and that resolves before two years of age.

caged-ATP A derivative of ATP that is not biologically active until a photosensitive bond has been cleaved.

Cajal bodies Spherical structures (coiled bodies, $0.1-2.0\ \mu m$), varying in number from 1 to 5, found in the nucleus of proliferating cells like tumour cells, or metabolically active cells like neurons. Contain factors required for splicing (e.g. **coilin**), ribosome biogenesis and transcription and are dynamic structures.

CAK In *S. cerevisiae*, CDK-activating kinase (CAK1, EC 2.7.11.22, 368aa), in humans, cyclin-dependent kinase 7 (CAK1, 346aa). See **cyclin-dependent kinase activating kinase**.

calbindins A family of calcium-binding proteins with multiple **EF-hand** domains that includes calmodulin, parvalbumin, troponin C, and S100 protein. Calbindin 1 (261aa) buffers intracellular calcium levels and is found in primate striate cortex and other neuronal tissues. Calbindin 2 (calretinin, 271aa) is found in cerebellar Purkinje cells and is is abundant in auditory neurons. Calbindin 3 (intestinal calcium-binding protein, S100-G, 79aa) is found in the intestine and is induced by vitamin D.

calcein A calcium-chelating agent that fluoresces brightly in the presence of bound calcium. The acetomethoxy derivative can be transported into live cells and the reagent is useful as a viability test and for short-term marking of cells.

calcicludine Polypeptide toxin (60aa) from *Dendroaspis angusticeps* (Eastern green Mamba). Blocks most high-threshold calcium channels (L-, N- or P-type). Structurally homologous to Kunitz-type serine peptidase inhibitors and **dendrotoxins**.

calciferol See Table V1 **Vitamin D**.

calcimedins See Table A2 **Annexins**.

calcineurin A calmodulin-stimulated protein phosphatase (EC 3.1.3.16), the major calmodulin-binding protein in brain. and the only Ca^{2+}-dependent Ser/Thr protein phosphatase. There are two-subunits, the A subunit (521aa) has the catalytic activity, the B subunit (170aa) confers calcium sensitivity. Dephosphorylates **NFATc**, heat shock protein beta-1 (HSPB1) and the protein tyrosine phosphatase **slingshot** homologue-1 (SSH1). It is inhibited by binding of **immunophilin**-ligand complex (immunophilin alone does not bind) and is the target for the immunosuppressive drugs **ciclosporin A** and FK506, being a key enzyme in T-cell activation. Calcineurin-binding protein (cabin-1, 2220aa) may be a negative regulator of T-cell receptor (TCR) signalling by inhibiting calcineurin (see **cabin**). Review article: http://www.biosignaling.com/content/7/1/25

calcinosis Deposits of calcium salts, primarily hydroxyapatite crystals or amorphous calcium phosphate, in various tissues of the body.

calciosome A membrane compartment proposed to contain the intracellular calcium store released in response to hormonal activity and thought to be distinct from the ER. Now discredited.

calcipressins A family of endogenous **calcineurin** inhibitors with multiple alternatively spliced isoforms. Calcipressin 1 (regulator of calcineurin 1, Down's syndrome critical region protein 1, DSCR 1, myocyte-enriched calcineurin-interacting protein-1, MCIP 1, Adapt 78, 252aa) interacts with calcineurin and inhibits **NFAT**-mediated transcriptional activation. Calcipressin 1 has greater capacity to inhibit calcineurin when phosphorylated at the FLISPP motif, and this phosphorylation also controls the half-life of calcipressin 1 by accelerating its degradation. Calcipressin 2 (197aa) and calcipressin 3 (241aa) are similar but with different tissue distribution.

calciseptine Polypeptide toxin (60aa) from *Dendroaspis polylepis* (Black Mamba) that is a specific blocker of L-type calcium channels and will cause relaxation of smooth muscle and inhibition of cardiac muscle but has no effect on skeletal muscle.

calcitermin See **calgranulins**.

calcitonin A hormone produced by C-cells of the thyroid that causes a reduction of calcium ions in the blood. The precursor (141aa) is cleaved into calcitonin (32aa), which promotes incorporation of calcium and phosphate into bone and katacalcin (calcitonin carboxyl-terminal peptide, 21aa) which potently reduced plasma calcium levels.

calcitonin family peptides Family of small (32−51aa) highly homologous peptides that act through seven-transmembrane G-protein-coupled receptors. **Adrenomedullin** (ADM; 51aa) is a potent vasodilator and has receptors on astrocytes; **amylin** (37aa) is thought to regulate gastric emptying and carbohydrate metabolism; calcitonin (32aa) is involved in control of bone metabolism; **calcitonin-gene-related peptides** one and two (37aa) regulate neuromuscular junctions, antigen presentation, vascular tone and sensory neurotransmission. Receptors are themselves regulated by **RAMPs**.

calcitriol The form of vitamin D3 that is biologically active in intestinal transport and calcium resorption by bone.

calcium ATPase Usually refers to calcium-pumping ATPase such as that present in high concentration as an integral membrane protein of the **sarcoplasmic reticulum** of muscle (sarcoplasmic/endoplasmic reticulum calcium ATPase, SERCA, EC 3.6.3.8: SERCA1, 1001aa; SERCA2, 1042aa). This pump lowers the cytoplasmic calcium level and causes contraction to stop. Normal function of the pump seems to require a local phospholipid environment from which cholesterol is excluded. Other calcium pumping ATPases are found in other tissues, e.g. calcium-transporting ATPase type 2C member 1 (919aa) (deficient in Hailey-Hailey disease).

calcium binding proteins There are two main groups of calcium-binding proteins, those that are similar to **calmodulin**, and are called **EF-hand** proteins, and those that bind calcium and phospholipid (e.g. **lipocortin**) and that have been grouped under the generic name of **annexins**. Many other proteins will bind calcium, although the binding site usually has considerable homology with the calcium binding domains of calmodulin.

calcium blocker Any compound that blocks or inhibits transmembrane calcium ion movement. Most are potent vasodilators and some are anti-arrhythmic. Examples are nifedipine, verapamil and diltiazem.

calcium channel Membrane channel that is specific for calcium. Probably the best characterized is the voltage-gated channel of the sarcoplasmic reticulum which is ryanodine-sensitive. See **voltage-sensitive calcium channels**, **ryanodine receptor**.

calcium current Inflow of calcium ions through specific **calcium channels**. Critically important in release of transmitter substance from presynaptic terminals.

calcium dependent regulator protein Early name for **calmodulin**.

calcium phosphate precipitation Technique used for introducing DNA or chromosomes into cells; co-precipitation with calcium phosphate facilitates the uptake of DNA or chromosomes.

calcium pump A **transport protein** responsible for moving calcium out of the cytoplasm. See **calcium ATPase**.

calcium-sensing receptor In humans, a plasma membrane multi-pass G protein-coupled receptor (1078aa) that is expressed in the **parathyroid hormone**-producing cells of the parathyroid gland and the cells lining the kidney tubule. It regulates parathyroid hormone secretion and calcium homeostasis. Mutations can lead to **hypoparathyroidism**. In *Arabidopsis* a chloroplast calcium sensor (387aa) modulates cytoplasmic Ca^{2+} concentration and is crucial for proper stomatal regulation in response to elevated levels of external Ca^{2+}.

calcium/calmodulin-dependent kinases Widely distributed calcium-regulated kinases. Calcium/calmodulin-dependent kinase 1 (CaMK1, EC 2.7.11.17, 370aa) phosphorylates **synapsins**, **CREB** and **CFTR** and is involved in cellular processes such as transcriptional regulation, hormone production, translational regulation, regulation of actin filament organization and neurite outgrowth. It is involved in calcium-dependent activation of the ERK pathway. CaM Kinase II (EC 2.7.1.37) is a multi-subunit enzyme composed of four homologous subunits, α, β, γ, δ, all encoded by different genes and $\sim$500aa. The heteromultimeric holoenzyme (500−600 kDa) has 10 or 12 subunits with the ratio of α,β,γ and δ reflecting that present in the cell. It has been implicated in various neuronal functions including **synaptic plasticity**. It is highly concentrated in the postsynaptic region and undergoes autophosphorylation at several sites in a manner that depends on the frequency and duration of Ca^{2+} spikes. Constitutively active CaMKII produces dendritic exocytosis in the absence of calcium stimulus. CaMKII activation is the primary event leading to inactivation of both CSF (cytostatic factor) and MPF (maturation promoting factor; **cyclin**) in mammalian eggs. Other CaM kinases with restricted tissue distribution are also known, e.g. **DAP kinases**, **protein kinase IV**. Calcium/calmodulin-dependent protein kinase kinase 1 (CaMKK1, EC 2.7.11.17, 505aa) phosphorylates CaM kinases and is involved in regulating apoptosis.

calcivirus See **Caliciviridae**.

calcofluor A fluorochrome, extensively used in microscopy, that also exhibits antifungal activity and a high affinity for yeast cell wall chitin.

calcyclin Prolactin receptor associated protein (S100-A6, 90aa), one of the **S100** calcium-binding proteins, originally isolated from Erlich ascites tumour cells, but human and rat forms now identified. Regulated through the cell cycle. Binds to annexin II (p36) and to glyceraldehyde-3-phosphate dehydrogenase.

calcyphosin Cytoplasmic calcium binding protein (calcyphosine, 189aa) that contains an **EF-hand** motif involved in both Ca^{2+}-phosphatidylinositol and cyclic AMP signal cascades. There is a calcyphosin-like protein (208aa) with similar properties. Calcyphosin-2 (557aa) is widely distributed.

caldesmon Protein (793aa) originally isolated from smooth muscle but also found in non-muscle cells. Regulates actomyosin interactions in smooth muscle and nonmuscle cells by stimulating the binding of tropomyosin to actin filaments. Calcium-calmodulin binding to caldesmon causes its release from actin, though phosphorylation of caldesmon may also affect the link with actin. Caldesmon can block the effect of **gelsolin** on F-actin and will dissociate actin-gelsolin complexes and actin-profilin complexes.

calelectrin Protein originally isolated from *Torpedo* (electric eel). Now known to be **annexin** A6, (calphobindin-II, chromobindin-20, lipocortin VI, protein III, p68, p70, 673aa) that may regulate the release of Ca^{2+} from intracellular stores.

caleosins Proteins involved in storage lipid mobilization during seed germination and located on the surface of lipid bodies or associated with an ER-subdomain. For example, in *Arabidopsis* caleosin-1 (embryo-specific protein 1, 245aa). They are products of a large gene family found ubiquitously in higher plants and in several lipid-accumulating fungi. Characteristically have a well conserved EF-hand, a central hydrophobic region and a C-terminal region with several putative phosphorylation sites. See **oleosins**.

calexcitins In *C. elegans*, calcium and GTP-binding proteins (calexcitin-1, 204aa; calexcitin-2, 189aa) that interact with and activate the ryanodine receptor and activate the calcium ATPase. Calexcitins are high affinity substrates for PKCα and once phosphorylated move to membrane where they inactivate voltage-dependent K^+ channels. Have two EF-hand motifs, and some similarities with **ARFs** (ADP-ribosylating factors).

CALF A file format (Compact ALignment Format) for computational bioinformatics; a CALF file records the base qualities and mapping qualities of the aligned reads, and unaligned read data. Details and protocols: http://www.phrap.org/phredphrap/calf.pdf

calgizzarin Calcium-binding protein of the **S100** family (S100 calcium binding protein A11, 105aa), originally isolated from chicken gizzard, implicated in the regulation of cytoskeletal function through its calcium-dependent interaction with **annexin I**.

calgranulins Calcium binding myeloid-associated proteins of the **S100** family. Calgranulin-A (S100-A8, cystic fibrosis antigen, leukocyte L1 complex light chain, migration inhibitory factor-related protein-8, MRP-8, 93aa) is expressed by macrophages in chronic inflammation and has a range of other activities, including some antimicrobial properties. Calgranulin-B (S100-A9, migration inhibitory factor-related protein 14, 114aa) is expressed in inflammation and may inhibit protein kinases. Calgranulin-C (S100-A12, 92aa) is proteolytically cleaved to calcitermin (15aa), which possesses antifungal and some antibacterial activity.

calicin Actin binding protein (588aa) that contains three **kelch** repeats. Found in the post-acrosomal calyx region of vertebrate sperm.

Caliciviridae A family of non-enveloped viruses with a genome consisting of a single molecule of linear positive-sense, single-stranded RNA enclosed in a capsid with icosahedral symmetry. They usually infect a single species of vertebrate host. Genera include **noroviruses** and **lagoviruses**, a species of the latter having been deliberately (but prematurely) released to cause rabbit haemorrhagic fever as a biological control mechanism. N.B. often misspelt calcivirus. Virus database entry: http://www.ncbi.nlm.nih.gov/ICTVdb/ICTVdB/00.012.htm

calitoxins Small peptide toxins (79aa) from sea anemone *Calliactis parasitica* that lives commensally with hermit crabs. Acts on neuronal sodium channels.

callose A plant cell-wall polysaccharide (a β-(1–3)-**glucan**) found in phloem **sieve plates**, wounded tissue, pollen tubes, cotton fibres, and certain other specialized cells.

callus (1) In plants, undifferentiated tissue produced at a wound edge. Callus tissue can be grown *in vitro* and induced to differentiate by varying the ratio of the hormones **auxin** and **cytokinin** in the medium. (2) In animals, either a mass of new bony trabeculae and cartilaginous tissue formed by **osteoblasts** early in the healing of a bone fracture or a skin thickening caused by persistent irritation or friction.

calmegin A calcium binding membrane protein (610aa) of the **calreticulin** family, that probably plays a role in spermatogenesis.

calmidazolium Inhibitor of calmodulin-regulated enzymes; also blocks sodium channel and voltage-gated calcium channel.

calmodulin Ubiquitous and highly conserved calcium binding protein (149aa) with four **EF-hand** binding sites for calcium (three in yeast). Ancestor of **troponin** C, **leiotonin** C, and **parvalbumin**.

calnexin Calcium-binding lectin-like protein (MHCclass I antigen-binding protein p88, 592aa) of endoplasmic reticulum that couples glycosylation of newly synthesized proteins with their folding. Calnexin and **calreticulin** act together as chaperones for newly synthesized proteins and prevent ubiquitinylation and proteosomal degradation. Can be phosphorylated by **casein kinase** II.

calpactins Calcium-binding proteins from cytoplasm. Calpactin II is identical to lipocortin, and is one of the major targets for phosphorylation by pp60src. See **annexin**.

calpains Calcium-activated cytoplasmic thiol endopeptidases (formerly EC 3.4.22.17) with broad specificity, containing the **EF-hand** motif in the larger catalytic subunit. The smaller regulatory subunit (268aa) has a calmodulin-like domain. Calpain-1 (EC 3.4.22.52, 714aa) is activated by micromolar calcium, and is inhibited by **calpastatin**. Calpain-2 (EC 3.4.22.53, 700aa) requires millimolar calcium for activation. Many other tissue-specific calpains are now known e.g. calpain-9 (digestive tract-specific calpain, 690aa) and defects in some calpains are associated with disorders such as muscular dystrophy and diabetes.

calpastatin Cytoplasmic inhibitor (708aa) of some **calpains**. It consists of four repetitive sequences of 120–140aa (domains I, II, III and IV), and an N-terminal non-homologous sequence.

calpeptin A cell-permeable **calpain** inhibitor.

calphobindins Annexins V and VI (320aa, 673aa) found in placenta (see Table A2). Have substantial sequence homology with **lipocortin** and may function like **calelectrin**.

calphostin C One of a group of compounds isolated from *Cladosporium cladosporioides* that will inhibit **protein kinase** C with some specificity, though inhibits other classes of kinases if present in high concentration.

calponin homology domain A superfamily of actin-binding domains (CH domains) found in both cytoskeletal proteins and signal transduction proteins. Included are the actinin-type actin-binding domain (including **spectrin, fimbrin, ABP-280**) and the calponin-type. Usually present in pairs although **calponin** itself has only one.

calponins Calcium and calmodulin binding **troponin** T-like proteins involved in the regulation of smooth muscle contractility. Interact with F-actin and tropomyosin in a calcium sensitive manner and act as regulators of smooth muscle contraction (inhibit when not phosphorylated). Distinct from **caldesmon** and **myosin light chain kinase**, but have some antigenic cross-reactivity with cardiac Troponin-T. Three human isoforms have been identified: calponin-1 (calponin H1, basic calponin, 297aa), calponin-2 (calponin H2, neutral calponin, 309aa), calponin-3 (acidic calponin, 329aa).

calregulin See **calreticulin**.

calreticulin Ubiquitous **calcium-binding protein** (calregulin, high affinity calcium binding protein, HACBP, calreticulin-1, 417aa; calreticulin-2, 384aa) with lectin-like domains found in the **endoplasmic reticulum**. Act as chaperones for newly synthesized proteins, possibly in conjunction with **calnexin**. May be more selective in the proteins with which they associate than calnexin.

calretinin See **calbindin**.

calsarcin See **myozenin**.

calsenilin See **DREAM**.

calsequestrins Proteins that sequester calcium in the cisternae of sarcoplasmic reticulum (calsequestrin-1, calmitin, 390aa) is found in fast skeletal muscle, calsequestrin-2 (399aa) is the cardiac muscle form. Each molecule can bind between 18–50 calcium ions probably on a charged surface rather than a specific binding pocket.

calspectin Non-erythroid **spectrin**.

calspermin High affinity calcium/calmodulin binding protein (**calcium/calmodulin-dependent protein kinase** type IV, EC 2.7.11.17, 474aa) found in rat. May be involved in transcriptional regulation, regulation of microtubule dynamics, spermatogenesis and consolidation/retention of hippocampus-dependent long-term memory.

calstabin A **immunophilin** (FK506 binding protein 1B, 108aa) that acts as a channel-stabilizing protein for the cardiac **ryanodine receptor** channel complex with four molecules of calstabin per receptor. Depletion can cause an intracellular calcium leak and trigger a fatal cardiac arrhythmia. Article: http://www.pnas.org/content/102/27/9607.full

calsyntenins Type I transmembrane proteins of the cadherin superfamily. Calsyntenin-1 (Cst1, alcadein-alpha, Alzheimer-related cadherin-like protein, 981aa) is expressed in postsynaptic membranes of excitatory neurons, Cst2 (alcadein-gamma, 955aa) and Cst3 (alcadein-beta, 956aa) are associated with inhibitory GABAergic neurons. The calsyntenins interact with numerous other proteins including APBA2 (amyloid beta A4 precursor protein-binding family A member 2) and APP (Amyloid beta A4 precursor protein).

caltractin Calcium-binding (EF-hand) protein (centrin, 169aa) from *Chlamydomonas reinhardtii* that is a major component of the contractile striated rootlet system that links basal bodies to the nucleus. One of the **centrin** family.

caltrin In *Bos taurus* an inhibitor (peptide YY-2, 80aa) of calcium transport into spermatozoa although it binds to calmodulin, not directly to calcium. Inhibits the growth of microorganisms and may act as an antibiotic by permeabilizing the bacterial membrane. In rat, caltrin (serine protease inhibitor Kazal-type 3, 79aa) also inhibits calcium transport but its physiological function is to prevent the trypsin-catalyzed premature activation of zymogens within the pancreas. Not reported from humans.

caltropin Obsolete name for the calponin-caltropin complex.

calumenin A calcium-binding protein (crocalbin, 315aa) localized in the endoplasmic reticulum (ER) and involved in protein folding and sorting. Calumenin is one of the **CREC** superfamily.

calvarium One of the bones that makes up the vault of the skull (in humans these are the frontal, 2 parietals, occipital, and 2 temporals). Calvaria are often used in organ culture to investigate bone catabolism or synthesis.

calvasculin An actin bundling protein, (metastasin, placental calcium-binding protein, S100 calcium-binding protein A4, 101aa) member of the family of S100-related calcium-binding proteins, located on cytoskeletal elements of cultured mammary cells in a pattern which is identical to actin filaments stained with **phalloidin**. May be involved in the progression and metastasis of human colorectal neoplastic cells. Interacts, in a calcium-dependent manner, with **liprin-beta-1**.

Calvin-Benson cycle Metabolic pathway (Calvin cycle) responsible for photosynthetic CO_2 fixation in plants and bacteria. The enzyme that fixes CO_2 is **ribulose bisphosphate carboxylase** (RuDP carboxylase). The cycle is the only photosynthetic pathway in **C3 plants** and the secondary pathway in **C4 plants**. The enzymes of the pathway are present in the stroma of the chloroplast.

calycin superfamily A protein superfamily with obvious similarities of three-dimensional structure that are not easily discernible at the sequence level. There are three distinct families of ligand binding proteins within the superfamily: the **lipocalins**, the **fatty acid-binding proteins** (FABPs) and **avidins**. Article: ftp://salilab.org/jpo/www/PMDG/pmdg_2.html

calyculin A Non-protein toxin from marine sponge, *Discodermia calyx*; potent tumour promoter and an inhibitor of protein phosphatases of Types 1 and 2a.

CAM See **crassulacean acid metabolism** or **cell adhesion molecule**.

CaM kinase See **calcium/calmodulin-dependent kinases**.

cambium (1) In plants, a layer of cells (lateral meristem) which forms new cells on both sides and divides to thicken the plant during secondary growth. Located either in vascular tissue (vascular cambium), forming secondary xylem on the inside and secondary phloem on the outside, or in cork (cork cambium or phellogen). *Cf.* **apical meristems** that increase the length of the tissue during primary growth. (2) In metazoa, the inner region of the **periosteum** from which **osteoblasts** differentiate.

camera lucida Attachment for a microscope that permits both a view of the object and, simultaneously, of the viewer's hand and drawing implement, thus facilitating accurate drawing of the object of interest.

CAML A multi-pass membrane protein (calcium signal-modulating cyclophilin ligand, 296aa) involved in the mobilization of calcium as a result of the T-cell receptor /CD3 complex interaction. Binds to cyclophilin B and to one of the TNF receptor superfamily (TNFRSF13B).

cAMP See **cyclic AMP**.

campomelic dysplasia A disorder of the newborn with various skeletal and extraskeletal defects caused by haploinsufficiency of **sox9**.

camptothecin Cytotoxic plant alkaloid originally isolated from *Camptotheca acuminata* (Cancer tree, Tree of Life); inhibits DNA **topoisomerase** I.

Campylobacter Genus of Gram-negative microaerophilic motile bacteria with a single flagellum at one or both poles. Found in reproductive and intestinal tracts of mammals. Common cause of food poisoning and can also cause opportunistic infections, particularly in immunocompromised patients.

Camurati-Engelmann disease A disorder involving excessive bone growth (osteitis), particularly of the lower limbs. Some forms, but not all, are a result of mutation in the gene for **TGFβ** (transforming growth factor-beta).

canal cell (1) One of the short-lived cells present in the central cavity of the neck of the archegonium in mosses. (2) Cells of Schlemm's canal, irregular space or spaces in the sclerocorneal region of the eye that receive aqueous humor from the anterior chamber of the eye.

canaliculi In bone, channels that run through the calcified matrix between lacunae containing **osteocytes**. In liver, small channels between **hepatocytes** through which bile flows to the bile duct and thence to the intestinal lumen.

Canavan's disease A severe neurological disorder, fatal in early life, due to mutation in the gene encoding **aminoacylase-2** (aspartoacylase).

canavanine One of the non-protein amino acids (also known as secondary metabolites or anti-metabolites), that exist in plants especially legumes and their seeds. L-canavanine is an L-arginine antimetabolite shown to be a selective inhibitor of inducible **nitric oxide synthase** (iNOS).

cancellous bone *trabecular bone* Adult bone consisting of mineralized regularly-ordered parallel collagen fibres more loosely organized than the lamellar bone of the shaft of adult long bones. Found in the end of long bones.

cancer A general term for diseases caused by any type of malignant tumour.

cancer susceptibility gene See **tumour suppressor** gene.

Candida albicans A dimorphic fungus that is an opportunistic pathogen of humans (causing candidiasis).

canine distemper Paramyxovirus infection of dogs with secondary bacterial complications. Ferrets, foxes and mink are also susceptible.

Canis familiaris Dog.

canker (1) A plant disease in which there are well-defined necrotic lesions of a main root, stem or branch in which the tissues outside the xylem disintegrate. (2) Chronic eczema of the ear of dogs, often caused by mites.

cannabinoid Group of compounds, all derivatives of 2-(2-isopropyl-5-methylphenyl)-5-pentylresorcinol, found in cannabis. Most important members of the group are cannabidiol, cannabidol and various tetra-hydrocannabinols (THCs). Bind to the **cannabinoid receptors** and mimic actions of endogenous agonists **anandamide** and palmitoyl ethanolamine.

cannabinoid receptors G-protein-coupled receptors for **cannabinoids** (and endogenous agonists such as anandamide). CB1 receptors (472aa) are mostly found in brain and may mediate the psychotropic activities; act by inhibiting adenylate cyclase and inhibit L-type Ca^{2+} channel current. The CB2 receptors (360aa) are more peripheral and found extensively in the immune system; the ligand is 2-arachidonoyl-glycerol and the receptor inhibits adenylate cyclase and may function in inflammatory responses, nociceptive transmission and bone homeostasis. CB1 cannabinoid receptor-interacting protein 1 (164aa) has two isoforms, one of which will suppress CB1-mediated tonic inhibition of voltage-gated calcium channels.

canola Variety of oilseed rape (*Brassica napus*) grown extensively to produce oil for human consumption.

canonical Classical, archetypal or prototypic. For example, the canonical polyadenylation sequence is

AATAAA. A canonical pathway is the orthodox or accepted pathway.

cantharidin A pharmaceutical product obtained from the dried elytra of the Spanish fly (blister beetles) although can now be synthesized. Has potent vesicant properties and has been used topically in the treatment of warts and **molluscum contagiosum** but is considered dangerously toxic. Inhibits **protein phosphatase** 2A.

CAP (1) Catabolite gene activator protein. (2) Cyclase associated protein. (3) See **CAP proteins**. (4) See **cap binding protein**.

cap binding protein Protein (eukaryotic Initiation Factor 4E, eIF4E, 217aa) with affinity for cap structure at 5'-end of mRNA that probably assists, together with other initiation factors, in binding the mRNA to the 40S ribosomal subunit. Translation of mRNA *in vitro* is faster if it has a cap-binding protein.

CAP proteins Obsolete name for Fas-associated via death domain (**FADD**) proteins, cytotoxicity-dependent APO-1-associated proteins. CAP1 is **FADD**; CAP2 is hyperphosphorylated FADD; CAP3 is an intermediate of procaspase-8 processing, CAP4 is pro-FLICE; CAP5, CAP6 are cleaved prodomains of FLICE (**caspase 8**).

CAP3 A DNA sequence assembly programme said to produce fewer errors than **Phrap**. Link: http://pbil.univ-lyon1.fr/cap3.php

CAP-18 Cathelicidin antimicrobial peptide (18 kDa cationic antimicrobial protein, CAP-18, 170aa). It is cleaved into the antibacterial proteins FALL-39 (39aa) and LL-37 (37aa) that bind to bacterial lipopolysaccharide.

cap-dependent translation Protein synthesis (translation of mRNA) that is dependent on binding of the eIF4F complex (see **eukaryotic initiation factor**) to the 5'-cap of eukaryotic mRNA as opposed to the cap-independent mode of translation initiation involving the **internal ribosome entry site** (IRES).

capacitance flicker Brief closings of an **ion channel** during its open phases, observed during **patch clamp**; or rapid transition of an ion channel between open and closed states such that the individual channel openings cannot be distinguished properly due to the limited bandwidth of the patch clamp amplifier.

capacitation A process occurring in mammalian sperm after exposure to secretions in the female genital tract. Surface changes take place probably involved with the **acrosome** which are necessary before the sperm can fertilize an egg.

CapG A ubiquitous gelsolin-family actin modulating protein (macrophage capping protein, MCP, 348aa) that binds to the barbed ends of microfilaments but does not sever them. Has considerable sequence homology with **gelsolin**, and like gelsolin responds to calcium and to phosphoinositides. Found in cytoplasm, nucleus and melanosomes. Not the same as **capZ**, though functionally similar. The murine form is gCap39 (352aa).

capillary A small blood vessel linking an arteriole and a venule. The lumen may be formed within a single endothelial cell, and have a diameter less than that of an erythrocyte, which must deform to pass through. Blood flow through capillaries can be regulated by precapillary sphincters, and each capillary probably only carries blood for part of the time.

capillary electrophoresis High resolution electrophoretic method for separating compounds, such as peptides, by electro-osmotic flow through long (~ 100 cm), narrow ($<100\,\mu$m) silica columns. With sensitive detectors attomolar concentrations can be detected.

capnine Sulphonolipid, a sulphonic acid derivative of ceramide, isolated from the envelope of the Cytophaga/Flexibacter group of **Gram-negative** bacteria. The acetylated form of capnine seems to be necessary for gliding motility. Reported to block synthesis of antibacterial proteins such as **cathelicidins**. Poster: http://mpkb.org/home/publications/marshall_metagenomics_2007

capnophilic Organisms that grow best at concentrations of carbon dioxide higher than in air.

CAPON Carboxyl-terminal PDZ ligand of neuronal nitric oxide synthase protein (nitric oxide synthase 1 adaptor protein, 506aa) an adaptor protein involved in regulation of neuronal nitric-oxide synthesis. Forms a complex with NOS1 and **synapsins**.

capping (1) Movement of cross-linked cell-surface material to the posterior region of a moving cell, or to the perinuclear region. (2) The intracellular accumulation of intermediate filament protein in the pericentriolar region following microtubule disruption by colchicine. (3) The blocking of further addition of subunits by binding of a cap protein to the free end of a linear polymer such as actin. See also **cap binding protein, capZ**.

cappuccino (1) Actin-nucleation factor (1059aa) of the formin homology family that regulates the onset of ooplasmic streaming in *Drosophila*. Like **spire**, a maternal-effect locus that participates in pattern formation in both the anteroposterior and dorsoventral axes of the early embryo. (2) The mouse protein cappuccino (215aa) is defective in a model of **Herman-Pudlak syndrome** and the human homologue (217aa) is a component of the **BLOC-1** complex.

capsaicin Molecule in chilli peppers that makes them hot and will stimulate release of neurogenic peptides (Substance P, neurokinins) from sensory neurons. Acts on the **vanilloid receptor-1** (VR1). Can be used to desensitise nociceptors to which it binds and kills.

capsazepine Competitive **capsaicin** antagonist.

Capsicum Genus that includes red peppers, pimentos and green peppers. See **capsaicin**.

capsid A protein coat that covers the nucleoprotein core or nucleic acid of a virion. Commonly shows icosahedral symmetry and may itself be enclosed in an envelope (as in the **Togaviridae**). The capsid is built up of subunits (some integer multiple of 60, the number required to give strict icosahedral symmetry) that self-assemble in a pattern typical of a particular virus. The subunits are often packed, in smaller capsids, into 5- or 6-membered rings (pentamers or hexamers) that constitute the morphological unit (capsomere). The packing of subunits is not perfectly symmetrical in most cases and some units may have strained interactions and are said to have quasi-equivalence of bonding to adjacent units.

capsomeres See **capsid**.

capsule (1) Thick gel-like material attached to the wall of Gram-positive or Gram-negative bacteria, giving colonies a 'smooth' appearance. May contribute to pathogenicity by inhibiting phagocytosis. Mostly composed of very hydrophilic acidic polysaccharide, but considerable diversity exists. (2) A multicellular aggregate of **haemocytes** or **coelomocytes** that isolates a foreign body too large to be phagocytosed. In some insects the capsule is apparently acellular and composed of **melanin**. (3) Dense connective tissue sheath surrounding an organ.

capZ A heterodimeric microfilament capping protein (capping protein, cap32/34, α, 286aa; β, 277aa) originally found in *Dictyostelium* and *Acanthamoeba* that binds to the barbed ends of microfilaments and blocks further assembly; also binds to thin filaments in the Z-disc of striated muscle. Binding is not calcium sensitive and capZ does not sever filaments. Widely distributed in vertebrate cells, though in non-muscle cells is predominantly in the nucleus. Identical to β-actinin in chickens. CapZ-interacting protein (416aa) can be phosphorylated under conditions of stress and then binds to capZ limiting its ability to regulate actin filament dynamics.

carbachol Parasympathomimetic drug (carbamoyl choline) formed by substituting the acetyl of acetylcholine with a carbamoyl group; acts on both **muscarinic** and **nicotinic acetylcholine receptors** and is not hydrolyzed by **acetylcholine esterase**.

carbamoyl The radical, -CO-NH$_2$.

carbamoyl phosphate synthetase A mitochondrial enzyme (CPS-1, EC 6.3.4.16, 1500aaa) involved in the urea cycle of ureotelic animals, playing an important role in removing excess ammonia from the cell. Similar enzymic activity (glutamine-dependent carbamoyl-phosphate synthase, EC 6.3.5.5) present in one of the domains of **CAD protein**, responsible for a key step in pyrimidine biosynthesis.

carbamoylcholine See **carbachol**.

carbendazim A systemic benzimidazole fungicide used extensively in plant disease control although there are concerns about its endocrine disrupting effects. Probably works as a fungicide by interfering with spindle formation at mitosis.

carbohydrate-deficient-glycoprotein syndrome A set of glycosylation-system defects causing deficiencies in the synthesis and processing of N-linked glycans or oligosaccharides on glycoproteins. In Type 1 (Jaeken's syndrome, CDG1) phosphomannomutase-2 (EC 5.4.2.8, 246aa) is affected, in CDGIb mannosephosphate isomerase (EC 5.3.1.8, 423aa) and in CDGIc asparagine-linked glycosylation protein 6 (ALG6, EC 2.4.1.-, 507aa).

carbohydrates Very abundant compounds with the general formula $C_n(H_2O)_n$. The smallest are monosaccharides like glucose; polysaccharides (e.g. starch, cellulose, glycogen) can be large and indeterminate in length.

carbon monoxide See **carboxy-haemoglobin**.

carbon replica A surface replica of a specimen for examination in the electron microscope. The specimen is coated with a structureless carbon film by vacuum deposition, and the film and specimen are subsequently removed by dissolving in an appropriate solvent.

carbonic anhydrase Enzyme (carbonate dehydratase, EC 4.2.1.1, 260aa) that catalyses reversible hydration of carbon dioxide to carbonic acid. An intracellular enzyme of the erythrocyte, essential for the effective transport of carbon dioxide from the tissues to the lungs. Zinc is a co-factor. Defects in the enzyme cause autosomal recessive **osteopetrosis** type three (Guibaud-Vainsel syndrome) and defects in isoform four cause a type of **retinitis pigmentosa**.

carbonyl group Bivalent =C=O group found in aldehydes, ketones and carboxylic acids.

carboxy terminus See **c-terminus**.

carboxyglutamate An amino acid (γ-carboxyglutamate) found in some proteins, particularly those that bind calcium. Formed by post-translational carboxylation of glutamate.

carboxyhaemoglobin Haemoglobin co-ordinated with carbon monoxide. The affinity of haemoglobin for CO is higher than for O_2 and binding is almost irreversible, hence the toxicity of CO. Asphyxia occurs without cyanosis because carboxyhaemoglobin is pink.

carboxyl-terminal Src kinase homologous kinase A negative regulatory kinase of the src tyrosine kinase family, **csk** subfamily (CHK, csk Homologous Kinase, probably the same as megakaryocyte-associated tyrosine-protein kinase, protein kinase HYL, tyrosine-protein kinase CTK, 507aa) generally believed to inactivate src-family tyrosine kinases by phosphorylating their consensus C-terminal regulatory tyrosine. Has been reported to regulate the expression of the chemokine receptor, CXCR4. Expression is limited to brain and hematopoietic cells. CHK overexpression in neuroblastoma and astrocytoma cells inhibits their growth and proliferation and loss of CHK expression is associated with human brain tumours. Not to be confused with **checkpoint kinases** (Chks). Paper (2005): http://cancerres.aacrjournals.org/content/65/7/2840.long

carboxypeptidases Enzymes (particularly of pancreas) that remove the C-terminal amino acid from a protein or peptide. Carboxypeptidase A1, (EC 3.4.17.1, 419aa) will remove any amino acid; carboxypeptidase B (EC 3.4.17.2, 417aa) is specific for terminal lysine or arginine. Carboxypeptidase D (EC 3.4.17.22, metallocarboxypeptidase D, 1380aa) releases C-terminal Arg and Lys from polypeptides. Carboxypeptidase N (EC 3.4.17.3, 458aa) degrades vasoactive and inflammatory peptides containing C-terminal Arg or Lys. Many others are known and the nomenclature is complex and may be based on mechanism (metallo-carboxypeptidases, EC 3.4.17.-; serine carboxypeptidases, EC 3.4.16.- and cysteine carboxypeptidases, EC 3.4.18.-) or on substrate specificity (e.g. glutamate carboxypeptidase).

carboxysome Inclusion body (polyhedral body; 90−150 nm diameter) found in some Cyanobacteria and autotrophic bacteria; contains **ribulose bisphosphate carboxylase** (RUBISCO) and is involved in carbon dioxide fixation.

carcinoembryonic antigen Cell membrane proteins of the immunoglobulin superfamily that are normally restricted to fetal tissue and absent in adults. Involved in adhesion and are sometimes used as markers for tumours. Examples include carcinoembryonic antigen-related cell adhesion molecule 1 (biliary glycoprotein 1, CD66a, 526aa) which exists in multiple isoforms, some of which are secreted, and carcinoembryonic antigen-related cell adhesion molecule 19 (carcinoembryonic antigen-like 1, 300aa) which is ubiquitous.

carcinogen An agent capable of initiating development of malignant tumours. May be a chemical, a form of electromagnetic radiation, or an inert solid body.

carcinogenesis The generation of cancer from normal cells, correctly the formation of a **carcinoma** from epithelial cells, but often used synonymously with **transformation, tumorigenesis.**

carcinoid Intestinal tumour arising from specialized neuroendocrine cells with paracrine functions (APUD cells), also known as argentaffinoma. The primary tumour in the appendix is often clinically benign but hepatic secondaries may release large amounts of vasoactive amines to the systemic circulation.

carcinoma *Pl.* carcinomata Malignant neoplasia of an epithelial cell: by far the commonest type of tumour. Those arising from glandular tissue are often called **adenocarcinomas.** Carcinoma cells tend to be irregular with increased basophilic staining of the cytoplasm, have an increased nuclear/cytoplasmic ratio and polymorphic nuclei. See **basal cell carcinoma, Gorlin's syndrome, hepatocarcinoma, Merkel cell carcinoma, oat cell carcinoma, scirrhous carcinoma.**

carcinoma *in situ* Carcinoma that has not invaded or metastasised and remains in the tissue site of origin.

carcinomatosis *carcinosis* Cancer widely disseminated throughout the body and thus highly malignant.

carcinosarcoma A rare tumour with histological features of both carcinoma and sarcoma.

carcinosis See **carcinomatosis.**

CARD domain A protein motif (caspase activation and recruitment domain) composed of a bundle of six alpha-helices that is a feature of a wide range of proteins, particularly those involved in inflammation and apoptosis, where it is involved in the formation of multi-protein complexes through interaction with other CARD domains. There are 521 CARD domains in 492 proteins in the **SMART** database, many in the **MAGUK** family. Numerous CARD proteins are known, for example, CARD11 (CARD-containing MAGUK protein 1, 1154aa) which is involved in T-cell receptor mediated T-cell activation and CARD14 (1004aa) which activates NFκB via **BCL10** and **IKK.** CARD 8 is cardinal.

cardiac ankyrin repeat protein A protein (CARP, ankyrin repeat domain-containing protein 1, cytokine-inducible nuclear protein, 319aa) mainly expressed in activated vascular endothelial cells. May act as a nuclear transcription factor that negatively regulates the expression of cardiac genes. Expression is upregulated in wounds and CARP may be important in revascularisation. See **myopalladin.**

cardiac cell Strictly speaking any cell of or derived from the cardium of the heart, but often used loosely of heart cells.

cardiac glycoside Specific blockers of the Na^+/K^+ **pump** especially of heart muscle, e.g. **strophanthin.**

cardiac jelly Gelatinous extracellular material that lies between endocardium and myocardium in the embryo.

cardiac muscle Specialized striated but involuntary muscle, able to contract and expand indefinitely (for a lifetime anyway), found only in the walls of the heart. Responsible for the pumping activity of the vertebrate heart. The individual muscle cells are joined through a junctional complex known as the **intercalated disc,** and are not fused together into multinucleate structures as they are in **skeletal muscle.**

Cardif See **MAVS.**

cardinal A CARD-domain protein found in some **inflammasomes** (caspase recruitment domain-containing protein 8, apoptotic protein NDPP1, 431aa) and that inhibits NFκB activation.

cardioblast An embryonic mesodermal cell which will differentiate into heart tissue.

cardiofaciocutaneous syndrome A developmental disorder that can be caused by mutation in K-*ras* (see **ras**), BRAF (see **raf**), **MEK1** or MEK2, all of which feed into the RAS/ERK pathway.

cardiolipin A diphosphatidyl glycerol, purified from beef heart, that is also found in the membrane of *Treponema pallidum* and is the antigen detected by the (obsolete) Wasserman test for syphilis.

cardiomyopathy Any disease affecting the heart muscle. See **Danon's disease, dilated cardiomyopathy, hypertrophic cardiomyopathy, Naxos disease.**

cardiotoxins Active components of cobra (*Naja* spp) venom; basic polypeptides of 57−62aa with four disulphide bonds (cobramine A, cobramine B, cobra cytotoxin, gamma toxin, membrane-active polypeptide); cause skeletal and cardiac muscle contracture, interfere with neuromuscular and ganglionic transmission, depolarize nerve, muscle and blood cell membranes, thus causing haemolysis.

cardiotrophin-1 Cytokine (CT-1, 201aa) belonging to the **IL-6 cytokine** family. Binds to hepatocyte cell lines and induces synthesis of various **acute phase proteins,** is a potent cardiac survival factor and supports long-term survival of spinal motoneurons. Cardiotrophin-like cytokine factor 1 (B-cell-stimulating factor 3, novel neurotrophin-1, 225aa) stimulates B-cells.

Cardiovascular Gene Ontology Annotation Initiative A resource for the cardiovascular-research community with more than 4000 cardiovascular associated genes (2009). Homepage: http://www.ucl.ac.uk/cardiovasculargeneontology/

cardiovirus Genus of viruses belonging to the Family **Picornaviridae**, isolated mostly from rodents; cause encephalitis and myocarditis.

carditis Inflammation of the heart, including pericarditis, myocarditis and endocarditis, according to whether the enveloping outer membrane, the muscle or the inner lining is affected.

carmil In *Dictyostelium* a protein (capping protein, Arp2/3 and myosin I linker, leucine-rich repeat-containing protein p116, 1050aa) that acts as a scaffold protein in a motor complex comprising microfilament capping proteins, the Arp2/3 complex, type I myosins (myoB and myoC) and carmil itself. Human **RLTPR** is related.

Carney complex A multiple neoplasia syndrome affecting a range of tissues. Multiple endocrine glands may be involved (see **multiple endocrine neoplasia**) and there are some similarities to McCune-Albright syndrome (see **granins**). Carney complex type 1 (CNC1) is caused by mutation in the gene encoding protein kinase A regulatory subunit-1-alpha. The Carney complex variant associated with distal arthrogryposis is caused by mutation in the myosin heavy chain-8 (MYH8) gene.

carnitine Compound that transports long chain fatty acids across the inner mitochondrial membrane in the form of acyl-carnitine. Sometimes referred to as Vitamin Bt or Vitamin B7. See table V1.

carnitine palmitoyltransferase system System that regulates fatty acid oxidation/ketogenesis in the liver and is itself switched off by malonyl CoA. Carnitine palmitoyltransferase (EC 2.3.1.21) is a multi-pass membrane protein in the mitochondrial membrane that is responsible for catalyzing the coupling of carnitine to palmitoyl-CoA. CPT1 (liver isoform CPT1A, 773aa; muscle isoform CPT1B, 772aa; brain isoform, CPT1C, 803aa) is in the outer mitochondrial membrane, CPT2 (658aa) is in the inner mitochondrial membrane.

carnosine Dipeptide (β-Ala-His) found at millimolar concentration in vertebrate muscle. It has antioxidant properties.

Carnoy Histological fixative containing ethanol, chloroform and acetic acid. Better for nuclear structure than for cytoplasm.

caronte In chickens, one of the **cer/dan family** of **bone morphogenic protein** antagonists (272aa) that mediates the sonic hedgehog-dependent induction of left-specific genes in the lateral plate mesoderm.

carotenes Hydrocarbon **carotenoids** usually with nine conjugated double bonds. Beta-carotene is the precursor of Vitamin A, each molecule giving rise to two Vitamin A molecules.

carotenoid cleavage dioxygenases Enzymes that catalyze the cleavage of carotenoids. In *Arabidopsis*, the CCD enzyme family includes four CCDs and five 9-*cis* epoxycarotenoid dioxygenases (NCEDs). The NCEDs are involved in the biosynthesis of abscisic acid (e.g. NCED2, chloroplastic 9-*cis*-epoxycarotenoid dioxygenase, EC 1.13.11.51, 583aa). CCD1 (538aa) cleaves a variety of carotenoids symmetrically at both the 9−10 and 9′−10′ double bonds. Review: http://www.jstage.jst.go.jp/article/plantbiotechnology/26/4/26_351/_article

carotenoid pathway Carotenoids are natural pigments (at least 700 have been characterized) that are produced by either the widely distributed C40 pathway, in which two molecules of geranylgeranyldiphosphate (GGDP) are condensed to form phytoene or the C30 pathway, found only in a few bacteria such as *Staphylococcus* and *Heliobacterium*, in which two molecules of farnesyldiphosphate undergo condensation to form 4,4′-diapophytoene (dehydrosqualene). Diagram of the pathway: http://www.biomedcentral.com/1471-2229/6/13/figure/F1?highres = y

carotenoids Accessory lipophilic photosynthetic pigments in plants and bacteria, including **carotenes** and **xanthophylls**; red, orange or yellow, with broad absorption peaks at 450−480 nm. Act as secondary pigments of the **light-harvesting system**, passing energy to **chlorophyll** and as protective agents, preventing photoxidation of chlorophyll. Found in chloroplasts and also in plastids in some non-photosynthetic tissues, e.g. carrot root.

carotid body cells Cells derived from the neural crest, involved in sensing pH and oxygen tension of the blood.

carpogenic germination The process of formation of **apothecia** from quiescent **sclerotia**.

carrageenan *carrageenin* Sulphated cell wall polysaccharide found in certain red algae. Contains repeating sulphated disaccharides of galactose and (sometimes) anhydrogalactose. It is used commercially as an emulsifier and thickener in foods, and is also used to induce an inflammatory lesion when injected into experimental animals (probably activates **complement**).

carrier (1) In human genetics, a person heterozygous for a recessive disorder. (2) A non-radioactive compound added to a tracer quantity of the same compound, which is radiolabelled. (3) A molecule or molecular system which brings about the transport of a solute across a cell membrane either by active transport or facilitated diffusion. (4) An organism harbouring a parasite but showing no symptoms of disease, especially if infectious to others. **(5)** A more or less inert material used either as a diluent or vehicle (e.g. for the active ingredient of drug) or as a support (e.g. for cells in a bioreactor). **(6)** See **carrier proteins**.

carrier protein (1) Protein to which a specific ligand or hapten has been conjugated, used for raising an antibody. **(2)** Unlabeled protein added into an assay system at relatively high concentrations which distributes in the same manner as the labeled protein analyte that is present in very low concentrations. **(3)** Protein added to prevent nonspecific interaction of reagents with surfaces, sample components, and each other; albumin is often used for this purpose. **(4)** Protein found in cell membranes, which facilitates transport of small molecules across the membrane.

CART (1) The cocaine- and amphetamine-regulated transcript (CART) peptide: a hypothalamic neuropeptide (116aa) implicated in the control of appetite. **(2)** A paired homoebox transcription factor (ALX homeobox protein 1, cartilage homeoprotein 1, CART-1, 326aa), that may be involved in chondrocyte differentiation and may influence cervix development.

cartilage **Connective tissue** dominated by **extracellular matrix** containing **collagen** type II and large amounts of **proteoglycan**, particularly **chondroitin sulphate**. Cartilage is more flexible and compressible than bone and often serves as an early skeletal framework, becoming mineralised as the animal ages. Cartilage is produced by **chondrocytes** that come to lie in small lacunae surrounded by the matrix they have secreted.

cartilage matrix protein A major component (matrilin-1, 496aa) of the extracellular matrix of non-articular cartilage. It binds to collagen. *Cf.* **cartilage oligomeric matrix protein**.

cartilage oligomeric matrix protein A protein (**thrombospondin-5**, 757aa), present at high levels in the matrix surrounding **chondrocytes** and involved in interactions between chondrocytes and other matrix proteins. A disulphide-linked homopentamer that binds calcium ions. Mutations cause **pseudoachondroplasia** or epiphyseal dysplasia. *Cf.* **cartilage matrix protein**.

caryopsis A type of simple dry fruit, grains being the common example (e.g. those of wheat, rice).

caryotype See **karyotype**.

Cas See **p130Cas, Cas-family proteins**.

Cas-family proteins Large multidomain adapter molecules (Crk-associated substrate family) that transmit signals as intermediaries through interactions with signalling molecules such as **FAK** and other tyrosine kinases, as well as tyrosine

phosphatases. After Cas is tyrosine-phosphorylated, it acts as a docking protein for binding SH2 domains of Src-family kinases. See **p130cas**. Members include Human enhancer of filamentation 1 (**HEF1**), Src-interacting protein (Sin)/Efs, Cas-L (Crk-associated substrate lymphocyte type), Nedd9.

casamino acid Acid hydrolyzed **casein** used as a nitrogen source in some culture media.

casein Group of proteins isolated from milk. The α_s and β-caseins (185 and 226aa) have hydrophobic C-terminal domains that associate to give micellar polymers in divalent-cation rich medium. κ-casein (182aa) is a glycoprotein that stabilizes the micelles of α- and β-casein. See **casoxins**.

casein kinases A group of kinases operationally defined by their preferred phosphorylation of acidic proteins such as caseins. Casein kinase 1 (EC 2.7.11.1, multiple isoforms, ~415aa) is involved in **Wnt** signalling and is a central component of the circadian clock. May be a negative regulator of circadian rhythmicity by phosphorylating **period**. Casein kinase II (CKII) is a tetrameric complex of two catalytic subunits, alpha (337 or 391aa) and alpha (350aa) and two regulatory beta subunits (phosvitin, 215aa). CKII is also involved in Wnt signalling and regulates a broad range of transcription factors and in the brain has been associated with **long-term potentiation** by phosphorylating proteins important for neuronal plasticity. Casein kinase II phosphorylates exposed Ser or sometimes Thr residues, provided that an acidic residue is present three residues from the phosphate acceptor site. Consensus pattern: (S/T)-x-x-(D/E).

caseous necrosis *caseation* The development of a necrotic centre (with a cheesy appearance) in a tuberculous lesion.

CASK One of the **MAGUK** family (membrane-associated guanylate kinases) a calcium/calmodulin-dependent serine protein kinase (LIN2 homologue, EC 2.7.11.1, 926aa) a multidomain scaffolding protein with a role in synaptic transmembrane protein anchoring and ion channel trafficking. Contributes to neural development and regulation of gene expression via interaction with the transcription factor TRB1. CASK binds to **syndecan**-2, the actin-binding band 4.1 protein and presynaptic **neurexin**. The *Drosophila* homologue of CASK is CAKI or CAMGUK (898aa). LIN2 is *C. elegans* abnormal cell lineage protein 2 (961aa)

caskin CASK-interacting protein 1 (1431aa) that binds the CaM kinase domain of CASK and forms a complex with LIN7 proteins that are involved in establishing and maintaining the asymmetric distribution of channels and receptors at the plasma membrane of polarized cells.

TABLE C1. Human Caspases

Name	Synonyms	Substrate	
caspase-1*	EC 3.4.22.36, ICE, 404aa	pro-IL-1β	
caspase-2	EC 3.4.22.55, Ich-1$_L$, nedd2, 452aa		I
caspase-3	EC 3.4.22.56 CPP32, Yama, apopain, 277aa	PARP, PKCδ, actin, Gas2, PAK2, procaspases 6, 9, U1-SnRNP	E
caspase-4*	EC 3.4.22.57, Tx/Ich-2, ICE$_{rel}$-II, 377aa	pro-ICE	
caspase-5*	EC 3.4.22.58, ICE$_{rel}$-III, Ty, 434aa		
caspase-6	EC 3.4.22.59, Mch2, 293aa	Lamins A, C, B1	E
caspase-7	EC 3.4.22.60, Mch3, CMH-1, ICE-LAP3, 303aa	PARP	E
caspase-8	EC 3.4.22.61, Mch5, MACH, FLICE, 479aa	Procaspases 3, 4, 7, 9	I
caspase-9	EC 3.4.22.62, Mch6, ICE-LAP6, 416aa		I
caspase-10	EC 3.4.22.63, Mch4, 521aa	Procaspases-3, 7	I
caspase-11*		Procaspases-1, 3 and pro-IL-1beta	I
caspase-12*	341aa;	Inactive	
caspase-13	EC 3.4.22.-, 377aa		#
caspase-14	EC 3.4.22.-, 242aa	Involved in differentiation of skin	

*Involved in inflammation
#Only bovine form known.
I– 'Initiator' caspases; activate effector caspases.
E– 'Effector' caspases – cleave, degrade or activate other proteins.

casoxins Proteolytic fragments of alpha-s casein (casoxin-D, 7aa) or kappa casein (casoxin-C, casoxin-6, casoxin-A, casoxin-B, casoplatelin). Casoxins A, B, C & D have opioid antagonist activity. Casoxin C binds to the complement C3a receptor, casoplatelin inhibits platelet aggregation.

Casparian band Region of plant-cell wall specialized to act as a seal to prevent back-leakage of secreted material (analogous to **tight junction** between epithelial cells). Found particularly where root parenchymal cells secrete solutes into xylem vessels.

caspases Family of cysteine-aspartic acid peptidases involved in processing of **IL-1β** (caspase-1 is interleukin-1 converting enzyme) and in **apoptosis**. See **CARD domain** and Table C1.

caspofungin See **echinocandins**.

CASPR Contactin-associated proteins, of which there are several in human. Members of the neurexin superfamily and involved in transmembrane adhesion and signalling. Caspr-1 (neurexin-4, 1384aa) is involved in the formation of functionally distinct domains critical for saltatory conduction of nerve impulses in myelinated nerve fibres and in defining regions of axon-glia interaction. Other examples in human are contactin-associated protein-like proteins (caspr2, 1331aa; caspr3, 1288aa; caspr4, 1308aa; caspr 5, 1306aa) probably with similar roles.

cassette A pre-existing structure into which an insert can be moved. Used to refer to certain vectors. See **cassette mechanism**.

cassette mechanism Term used for genes such as the a and α genes that determine **mating-type** in yeast; either one or the other is active. In this **gene conversion** process, a double-stranded **nuclease** makes a cut at a specific point in the MAT (mating type) locus, the old gene is replaced with a copy of a silent gene from one or other flanking region, and the new copy becomes active. As the process involves replacing one ready-made construct with another in an active 'slot' it is called a cassette mechanism.

CAST (1) In mouse, a protein (CAZ-associated structural protein 1, ERC protein 2, 957aa) involved in the organization of the cytomatrix at the nerve terminals active zone (cytomatrix at the active zone, CAZ) which regulates neurotransmitter release.

Binds directly to other **CAZ proteins** such as **rim1**, **bassoon** and **piccolo**. Article: http://www.ncbi.nlm. nih.gov/pmc/articles/PMC2173811/. (2) The CAST family includes ELKS/Rab6-interacting/CAST family member 1 (Rab6-interacting protein 2, 1116aa) which is a regulatory subunit of the IKK complex and may also be involved in the organization of the cytomatrix at the nerve terminals active zone (CAZ). (3) CD3-epsilon-associated protein (CAST, DNA-directed RNA polymerase I subunit RPA34, 510aa). (4) In potato, calcium-binding protein CAST (199aa).

castanospermine Alkaloid inhibitor of α-**glucosidase I** isolated from seeds of *Castanospermum australe* (Blackbean). The effect is to leave N-linked oligosaccharides in their 'high-mannose', unmodified state.

Castleman's disease Disease characterized by lymph node swelling, hypergammaglobulinaemia, increased levels of **acute phase proteins** and increased numbers of platelets. Probably caused by excess **IL-6** production.

CAT See **chloramphenicol acetyltransferase**.

catabolin Obsolete name for interleukin-1 (**IL-1**).

catabolism The sum of all degradative processes, the opposite of anabolism.

catabolite Product of catabolism, the breakdown of complex molecules into simpler ones.

catabolite activator protein Protein (CAP, cAMP regulatory protein, 210aa) from *E. coli* that regulates the transcription of several catabolite-sensitive operons. For example, CAP and the **lactose repressor** protein act together to control lactose utilization. If glucose is present CAP will not bind to DNA and the lac genes are not expressed even if lactose is available. In the absence of glucose cAMP levels rise, CAP binds cAMP, undergoes a conformational change, and is then capable of binding to DNA and promoting transcription of derepressed genes in the presence of lactose.

catabolite repression Inducible enzyme systems in some microorganisms (such as the **lactose operon**) that are repressed when a more favoured carbon source, such as glucose, is available (see **catabolite activator protein**). Catabolite repression of the respiratory system is seen in yeast in high glucose concentrations, though the mechanism is different.

catalase Tetrameric haem enzyme (EC 1.11.1.6, 527aa) that breaks down hydrogen peroxide.

catalytic antibody *abzyme* Antibody raised against a transition-state analogue (e.g. a phosphate analogue of a carboxylic acid ester transition state) that can then catalyse the analogous chemical reaction, though not as effectively as a true enzyme. They can, however, be selected to catalyze a reaction for which there is no endogenous enzyme.

catalytic RNA Species of RNA that catalyse cleavage or transesterification of the phosphodiester linkage. Operates in the self-splicing of group I and group II introns and in the maturation of various tRNA species.

cataplerosis The net loss due to consumption or degradation of intermediates in a biochemical cycle; the opposite of anaplerosis.

cataract Opacity of the lens of the eye. Can be a result of environmental factors but there are various inherited forms, mostly due to mutations in genes encoding **crystallins** although there are forms caused variously by mutation in genes for heat-shock transcription factor-4, **phakinin**, SLC16A12 solute-carrier, the alpha-8 subunit of the gap junction protein, maf, chromatin-modifying protein 4B, the PITX3 homeobox protein, **connexin** 50 and **beaded filament** structural protein-2.

catch muscle See **adductor muscle**.

catechins Polyphenolic compounds found in a range of plants and, for example, in green tea, where there are four major ones, epicatechin (EC), epigallocatechin (EGC), epicatechin gallate (ECg) and epigallocatechin gallate (EGCg). Remarkable health benefits are said to be gained from these compounds.

catechol oxidase Any of a family of oxidoreductases (EC 1.10.3.1, polyphenol oxidase) that are involved in catalysing oxidation of mono and orthodiphenols to ortho-diquinones. Most are found in plants and contain copper; they are responsible for the brownish discolouration of, for example, potatoes and apples when cut open.

catechol-O-methyltransferase An enzyme (COMT, EC 2.1.1.6, 271aa) that catalyzes the O-methylation and inactivation of catecholamine neurotransmitters and catechol hormones. There are soluble and membrane-bound isoforms and two alleles with valine or methionine at position 158 and differeing in enzyme activity. A transmembrane form of the enzyme (catechol O-methyltransferase 2, LRTOMT2, 291aa) is required for auditory function.

catecholamine A type of **biogenic amine** derived from tyramine, characterized as alkylamino derivatives of o-dihydroxybenzene. Catecholamines include **adrenaline**, **noradrenaline** and **dopamine**, with roles as **hormones** and **neurotransmitters**.

catenate Two or more circular DNA molecules where one or more circles run through the enclosed space of another, like links in a chain.

catenins Proteins that link the cytoplasmic domain of adhesion molecules such as cadherins to the cytoskeleton. Alpha-catenin ($\alpha 1$, 906aa; $\alpha 2$, 956aa; $\alpha 3$, 895aa) is a **vinculin**-related protein involved in adherens junction-mediated intercellular adhesion, Beta-catenin (781aa) is a key downstream component of the canonical Wnt signalling pathway although the majority of β-catenin is localized to the cell membrane and is part of E-cadherin/catenin adhesion complex, binding E-cadherin and N-cadherin. Ubiquitination of β-catenin is greatly reduced in Wnt-expressing cells. See **dapper**. Beta-catenin-interacting protein 1 (81aa) blocks interaction with TCF transcription factors and negatively regulates Wnt signalling. Gamma-catenin (desmoplakin-3, **plakoglobin**, 745aa), associates with N-cadherin and E-cadherin and is a major component of desmosomes. Delta-catenin-1 (p120, p120ctn, cadherin-associated src substrate, 938aa) is involved in cell-cell adhesion complexes together with E-**cadherin**, α-, β- and γ-catenins. It binds to and inhibits the transcriptional repressor ZBTB33 which may activate the Wnt pathway. It is tyrosine-phosphorylated by ligand-activated EGF-, PDGF- and CSF1-receptors and loss of expression is associated with some invasive tumours. Delta-catenin-2 (neural plakophilin-related armadillo repeat protein, δ-**catenin**, neurojungin, CTNND2, 1225aa) functions as a transcriptional activator when bound to ZBTB33. Hemizygosity of the gene is associated with **cri-du-chat syndrome**.

caterpiller family Family of proteins with a nucleotide-binding domain and a leucine-rich region. Have a role in immunity, cell death and growth, and diseases. The caterpiller proteins are structurally similar to a subgroup of plant-disease-resistance (R) proteins and to the apoptotic protease activating factor 1 (APAF1) and are genetically linked to several human immunological disorders. They have both positive and negative immunoregulatory activity and may be intracellular sensors of pathogen products. An example is protein NLRC3 (CARD15-like protein, caterpiller protein 16.2, nucleotide-binding oligomerization domain protein, 1065aa) that may modulate T-cell activation. See **cryopyrin**.

CATH database An hierarchical domain classification of protein structures in the Protein Data Bank (PDB). Protein structures are classified using a combination of automated and manual procedures. CATH v3.4 is built from 104,238 PDB chains. Link to database: http://www.cathdb.info/

cathelicidins A group of cationic antimicrobial peptides mostly found in the secretory (azurophil) granules of neutrophils, although may be found elsewhere. They are synthesized as precursor propeptides of that contain a structurally varied C-terminal cationic region corresponding to the antimicrobial peptide, joined to a conserved cathelin-like propiece of approximately 100aa. The mature peptide is liberated by limited proteolysis by elastase or proteinase 3. Examples are LL-

37/hCAP18, CAP-18, SMAP-29, BMAP-27 and BMAP-28, protegrins and indolicidin.

cathelin A porcine **cathelicidin** (96aa) that was originally described as being a cysteine peptidase inhibitor though subsequent reports have suggested that this is due to contamination with **PLCPI**. Cathelin-like sequences are found upstream of several antimicrobial peptides (protegrins) and are expressed in the pro-peptides. See **cathelicidins**.

cathepsins A large family of peptidases, mostly lysosomal and usually requiring processing before they are active. Cathepsin A (lysosomal protective protein, carboxypeptidase C, EC 3.4.16.5, 480aa) is essential for the activity of beta-galactosidase and neuraminidase; it associates with these enzymes and exerts a protective function. Defects cause galactosialidosis (**Goldberg's syndrome**). Cathepsin B (EC 3.4.22.1, 339aa) is a cysteine endopeptidase cleaved into heavy (205aa) and light (47aa) chains which form a disulphide-linked heterodimer. Involved in intracellular degradation and turnover of proteins and possibly in tumour invasion and metastasis. Cathepsin C (EC 3.4.14.1, 463aa) is a dipeptidyl peptidase. Defects are a cause of **Papillon-Lefevre syndrome**, **Haim-Munk syndrome** and juvenile periodontitis. Cathepsin D (EC 3.4.23.5, 412aa) has pepsin-like specificity. Defects cause neuronal ceroid lipofuscinosis Type 10. Cathepsin E (EC 3.4.23.34, 401aa) is endosomal and may process antigenic peptides during MHC class II-mediated antigen presentation. Cathepsin F (EC 3.4.22.41, 484aa) is involved in intracellular protein degradation and has also been implicated in tumour invasion and metastasis. Cathepsin G (EC 3.4.21.20, 255aa) is a serine protease with trypsin- and chymotrypsin-like specificity. Cathepsin H (EC 3.4.22.16, 335aa) has aminopeptidase activity. Cathepsin K (cathepsin O, EC 3.4.22.38, 329aa) is involved in osteoclastic bone resorption and defects cause **pycnodysostosis**. Several other cathepsins (L, N, S, W, Z) with different substrate specificities are also known.

cationic proteins Proteins of azurophil granules of neutrophils, rich in arginine. A chymotrypsin-like peptidase found in azurophil granules is very cationic as is cathepsin G and neutrophil elastase. Eosinophil cationic protein (160aa) is particularly important because it damages **schistosomula** *in vitro*.

cationized ferritin Ferritin, treated with dimethyl propanediamine, and used to show, in the electron microscope, the distribution of negative charge on the surface of a cell. The amount of cationic ferritin binding is very approximately related to the surface charge.

catsper A family of proteins (cation channel sperm-associated protein 1, catsper-1, 780aa; catsper-2, 530aa; catsper-3, 398aa; catsper-4, 472aa, catsper-B,

1116aa) that associate variously as heterotetramers to form a voltage-gated calcium channel that plays a central role in calcium-dependent responses such as sperm hyperactivation, the acrosomal reaction and chemotaxis towards the oocyte.

caudate nucleus The most frontal of the **basal ganglia** in the brain. Damage to caudate neurons is characteristic of **Huntington's chorea** and other motor disorders.

Caudovirales An Order of viruses (see **virus taxonomy**), tailed bacteriophages, comprising three families: Myoviridae, Podoviridae and Siphoviridae. The genome is a single molecule of linear double-stranded DNA. The phage has a head with icosahedral symmetry and a tail with helical symmetry. Virus taxonomy site: http://www.ncbi.nlm.nih.gov/ICTVdb/ICTVdB/02.htm

Caulimoviridae DNA para-retroviruses that replicate in plants via a RNA intermediate evolved from **LTR retrotransposons** (probably of the Ty3/Gypsy family). The group can be subdivided into six genera: Badnavirus (e.g. Commelina yellow mottle virus), Caulimovirus (e.g. cauliflower mosaic virus), Cavemovirus (e.g. cassava vein mosaic virus), Petuvirus (e.g. petunia vein clearing virus), Soymovirus (e.g. soybean chlorotic mottle virus) and Tungrovirus (e.g. rice tungro bacilliform virus). Caulimoviruses may have genes encoding movement protein, the **aphid transmission factor** and the **inclusion body matrix protein** as well as genes necessary for the viral life cycle and transmission. Gypsy database: http://gydb.uv.es/index.php/Caulimoviridae

Caulobacter Genus of Gram-negative aerobic bacteria that have a stalk or holdfast. Found in soil and fresh water.

caulonema One of the two types of filamentous protonemata produced when bryophyte spores germinate. Caulonemata contain few chloroplasts and make angled cross-walls. Chloronemata (*sing.* chloronema) contain chloroplasts and make cross-walls at right angles to the main filament. See **protonema**.

caveola *Pl.* caveolae. Small invagination of the plasma membrane characteristic of many mammalian cells and associated with endocytosis. The membrane of caveolae contain integral membrane proteins, **caveolins** that interact with heterotrimeric G-proteins. Caveolar membranes are enriched in cholesterol and sphingolipids and may be the efflux route for newly synthesized lipids. **Clathrin** is not associated with caveolae.

caveolin scaffolding peptide-1 Peptide corresponding to the **caveolin**-1 scaffolding domain, the region that is essential for caveolin interaction with signalling molecules. The peptide will block the ability of noradrenaline and histamine to induce changes in the internal calcium ion concentrations of vascular smooth muscle cells by inhibiting the activation of phospholipase C-beta3 and MAPK.

caveolins Integral membrane proteins that are essential structural components of caveolae and serve as a scaffolding onto which signalling molecules, particularly G-protein coupled receptors, are assembled, and also act as negative regulators of signal transduction. There are three closely-related human isoforms (caveolin-1, 178aa; cav-2, 162aa; cav-3, 151aa) that are differentially expressed in various cell types. Defects in caveolin-1 cause a form of **lipodystrophy**; defects in caveolin-3 cause a whole range of disorders including forms of muscular dystrophy, cardiomyopathy and **long QT syndrome**.

Cavia porcellas Guinea pig.

CAZ proteins Proteins involved in the CAZ (cytomatrix at the active zone) complex that determines the site of synaptic vesicle fusion. The c-terminal region has a PDZ-binding motif that binds directly to **RIM** (a small G protein Rab-3A effector). The complex includes **bassoon**, **CAST**, **Munc-13** and **piccolo**. Database entry: http://www.ebi.ac.uk/interpro/DisplayIproEntry?ac = IPR019323

C banding Method (centromeric banding) of defining chromosome structure by staining with **Giemsa** and looking at the **banding pattern** in the heterochromatin of the **centromeric** regions. Giemsa banding (G banding) of the whole chromosome gives higher resolution. Q banding is done with quinacrine.

CBD A neurological disorder (corticobasal degeneration) caused by defects in tau (a **tauopathy**) with features similar to **Alzheimer's disease**, **Steele-Richardson-Olszewski syndrome** and **Parkinson's disease**.

CBF/DREB genes Widely distributed genes in higher plants that encode transcription factors which bind a DNA regulatory element and impart responsiveness to low temperatures and dehydration. The CBF gene family, belongs to the AP2/ERF (ethylene-responsive element-binding protein) superfamily.

CBHA (1) A synthesized histone deacetylase (**HDAC**) inhibitor, m-carboxycinnamic acid bishydroxamide. (2) **Cellobiohydrolase** A (CbhA).

cbl The product (EC 6.3.2.-, 906aa) of a proto-oncogene (Casitas B-lineage lymphoma proto-oncogene) originally identified in a murine retrovirus. Cbl serves as a substrate for receptor and non-receptor tyrosine kinases and as a multi-domain adaptor protein, binding to Grb2, **crk** and **p85** of PI-3-kinase. Acts as a negative regulator of tyrosine kinase signalling through its E3 ubiquitin ligase activity that ubiquitinates the receptors thereby promoting their degradation. This function is modulated through interactions with regulatory proteins including **CIN85** and PIX. The transforming mutant v-cbl lacks the ubiquitin ligase RING finger domain.

CBP See **CREB binding protein**.

CC10, CC16 See **Clara cell secretory protein**.

CCAAT box Consensus sequence for RNA polymerase, found at about -80 bases relative to the transcription start site. Less well conserved than the **TATA box**.

CCCP An **uncoupling agent** that dissipates proton gradients across membranes.

CCFDN Congenital cataracts facial dysmorphism and neuropathy syndrome, a developmental disorder caused by mutation in the **CTD** phosphatase.

CCK See **cholecystokinin**.

CCN family Family of proteins that includes cysteine-rich 61 (CYR61/CCN1, 381aa), connective tissue growth factor (**CTGF/CCN2**, 349aa), nephroblastoma overexpressed (**NOV/CCN3**, 357aa), and Wnt-induced secreted proteins (**WISP**) 1, 2 and 3. (CCN4, 367aa; CCN5, 250aa; CCN6, 354aa). CCN proteins play a role in cell differentiation and function. Each has four conserved cysteine rich modular domains that have sequence similarity, respectively, to the insulin-like growth factor binding protein, von Willebrand factor, thrombospondin repeat and the C-terminal cysteine knot domain of growth factors. Probably act mostly through binding and activating cell surface integrins.

CCNA2 Cyclin A2 (432aa) which accumulates steadily during G2 and is abruptly destroyed at mitosis.

CCR See **chemokine receptors** and **CCR4-NOT core complex**.

CCR4-NOT core complex In the nucleus a general transcription factor, and in the cytoplasm the major mRNA deadenylase involved in mRNA turnover. The complex in *S. cerevisiae* contains CCR4 (glucose-repressible alcohol dehydrogenase transcriptional effector, carbon catabolite repressor protein 4, 837aa), CAF1 (CCR4 associated factor, **POP2**), NOT1 (general negative regulator of transcription subunit 1, cdc39, 2108aa), NOT2, NOT3, NOT4, NOT5, CAF40 and CAF130. The NOT protein subcomplex negatively regulates the basal and activated transcription of many genes. There is a homologous CCR4-NOT complex in humans. N.B. CCR4 is also an abbreviation for the C-C chemokine receptor type 4.

CCR5 A member of the beta **chemokine receptor** family (CKR5, CD195, 352aa), a G-protein coupled receptor, present in various cells, especially macrophages, monocytes, and T-cells. It is the co-receptor that HIV uses to gain entry into macrophages and defective alleles of this gene have been associated with resistance to HIV infection. Ligands of this receptor include monocyte chemoattractant protein 2 (MCP-2), macrophage inflammatory protein 1 alpha (MIP-1α), macrophage inflammatory protein 1-beta (MIP-1β) and regulated on activation normal T expressed and secreted protein (RANTES).

CCSP **Clara cell** secretory protein (CCSP, CC16; secretoglobulin 1A1, **uteroglobin**, 91aa) is an abundant 16-kDa homodimeric protein secreted by non-ciliated secretory epithelial cells in the lung. It has an important protective role against the intrapulmonary inflammatory process. Clara cell 10-kDa protein (CC10) also has anti-inflammatory properties. Both proteins are markers of lung irritation.

CCT A protein complex (chaperonin containing T-complex polypeptide 1, formerly TCP-1 complex) that is composed of 8 related proteins (~ 60 kDa) encoded by independent and highly diverged genes, arranged as two stacked multimers, that mediates folding of various cytoskeletal proteins including actin and tubulin. The complex is found in the leading edge of fibroblasts and neurons. But see **CCT domain**. Article: http://www3.interscience.wiley.com/cgi-bin/fulltext/119235580/PDFSTART

CCT domain The CONSTANS, CO-like, and TOC1 domain, a highly conserved basic module of ~ 43aa, found near the C-terminus of plant proteins often involved in light signal transduction. There is a putative nuclear localization signal in the motif.

CCVs See **clathrin-coated vesicles**.

CD antigens Cell surface antigens, originally the 'cluster of differentiation' antigens found on different classes of lymphocytes but now no longer restricted to cells of the immune system. International workshops comparing molecules recognised by different monoclonal antibodies were the original basis of rationalizing the antigens being detected. By mid-2011 the numbering had reached CD363; earlier editions of this dictionary had a table listing the CD antigens but since various web-accessible versions are now available, this table has been omitted. List: http://www.immunologylink.com/cdantigen.html Wall chart: http://www.ebioscience.com/resources/literature-request.htm

CD2-associated protein An adapter protein (CAS ligand with multiple SH3 domains, 639aa) that links membrane proteins and the actin cytoskeleton. Involved in receptor clustering and cytoskeletal polarity in the junction between a T-cell and an antigen-presenting cell, may anchor the podocyte slit diaphragm to the actin cytoskeleton in renal glomerulus and required for cytokinesis. Mutations are associated with focal segmental glomerulosclerosis 3. See **p130Cas**, **Cas-family proteins**.

CD3 complex T-cell surface molecules of the immunoglobulin superfamily that mediate signalling from the **T-cell receptor** (TCR). The TCR/CD3 complex consists of either a TCR α/β or TCR γ/δ heterodimer coexpressed with the invariant CD3 gamma (CD3g, 182aa), delta (CD3d, 171aa), epsilon (CD3e, 207aa), zeta (CD247, 164aa), and (in the mouse) eta (CD247, 206aa) subunits. Defects in CD3delta cause severe combined immunodeficiency disease ($T^-B^+NK^+$ SCID), defects in other members of the complex can also cause immune dysfunction.

CDA An antibiotic (calcium-dependent antibiotic) produced by *Streptomyces coelicolor*, an acidic lipopeptide comprising an N-terminal 2,3-epoxyhexanoyl fatty acid side chain and several nonproteinogenic amino acid residues. It is effective against Gram-positive bacteria, producing transmembrane channels which conduct monovalent cations in the presence of calcium ions. Abstract: http://www.ncbi.nlm.nih.gov/pubmed/12445768

***cdc* genes** Cell division cycle genes, of which many have now been defined, especially in yeasts. See **cyclins**. The cyclin dependent kinases are also known as cdc2 kinases.

cdc42 Product of the cell division cycle-42 (*cdc42*) gene, one of the **rho** GTPase family (191aa) that regulates signalling pathways controlling diverse cellular functions including cell morphology, migration, endocytosis and cell cycle progression. Thought to induce filopodium formation by regulating actin polymerization at the cell cortex by activating the **Arp2/3 complex** in concert with **WASP** family proteins. The product of oncogene *Dbl* catalyzes the dissociation of GDP from this protein. Alternative splicing gives rise to at least two transcript variants. Cdc42-interacting protein 4 (Felic, Salt tolerant protein, thyroid receptor-interacting protein 10, TRIP-10, 601aa) is one of the **formin** binding family of proteins and is required for translocation of GLUT4 to the plasma membrane in response to insulin signalling. Also involved in reorganization of the actin cytoskeleton during endocytosis. Binds to lipids such as phosphatidylinositol 4,5-bisphosphate and phosphatidylserine and promotes membrane invagination and the formation of tubules. Also promotes cdc42-induced actin polymerization by recruiting WASL/N-WASP which in turn activates the Arp2/3 complex.

cdc42 effector proteins A family of proteins (CEPs, binders of Rho GTPases, Borgs) that act downstream of **cdc42** in regulating actin polymerization. CEP1 (Borg5, 210aa) induces membrane extensions in fibroblasts, CEP2 (Borg1, 210aa) induces pseudopod formation, as do the others (CEP3, Borg2, 254aa; CEP4, Borg4, 356aa; CEP5, Borg3, 148aa). Cdc42 small effector proteins (79aa and 84aa) act downstream of cdc42, inducing actin filament assembly and in activated T-cells, may be important for F-actin accumulation at the immunological synapse.

CDEP Human protein (chondrocyte-derived ezrin-like protein, FERM, RhoGEF and pleckstrin domain-containing protein 1, FARP1, 1045aa) containing the **ezrin**-like domain of the band 4.1 superfamily, **Dbl homology** (DH) and **pleckstrin homology** (PH) domains. CDEP mRNA is expressed not only in differentiated chondrocytes but also in various fetal and adult tissues.

CDGs A family of severe inherited diseases (congenital defects in glycosylation) caused by deficient N-glycosylation of proteins. They are characterized by under-glycosylated serum proteins and there are a range of multi-system defects. One form is caused by mutation in the gene for **GlcNAc-phosphotransferase**.

CDGSH iron sulphur domain proteins A family of proteins (CISD proteins, CDGSH iron sulphur domain proteins) that contain iron-sulphur (Fe_2-S_2) clusters and have a 39aa CDGSH domain. CDGSH1 (MitoNEET, 108aa) is located in the outer mitochondrial membrane and may be involved in the transport of iron into the mitochondrion and regulates the maximal capacity for electron transport and oxidative phosphorylation. CDGSH2 (MitoNEET-related 1, miner1, 135aa) is involved in calcium homeostasis in the endoplasmic reticulum and is mutated in one form of **Wolfram's syndrome**. CDGSH3 (mitonneet-related 2, miner2, 127aa) is mitochondrial.

cdk See **cyclin-dependent kinases**.

cDNA Complementary DNA; DNA that has been synthesized by viral **reverse transcriptase** from an mRNA template. The cDNA can be used, for example, as a probe to locate the gene or can be cloned in the double-stranded form. Unlike the genomic coding region the cDNA does not contain introns.

cDNA-AFLP cDNA-amplified fragment length polymorphism analysis. A PCR-based method in which cDNA is synthesized from total RNA or mRNA using random hexamers as primers. The fragments are then digested with two restriction enzymes and adapters are ligated to the ends of the fragments. The first amplification step amplifies only fragments that were digested by both restriction enzymes and have different adapters at the end. In later amplifications the mixture is fractionated into smaller subsets by selective PCR amplification using primers with one or more extra nucleotides as well as the adapter. The method allows detection of genes expressed at low levels.

CDP See **cytidine 5′diphosphate**.

CDR See **complementarity determining region**.

CDRP Calcium dependent regulator protein, an obsolete early name for **calmodulin**.

CDS (1) The enzyme CDP-DAG synthase (CDS-1, phosphatidate cytidylyltransferase 1, EC 2.7.7.41, 461aa; CDS-2, 445aa) that catalyzes the production of CDP-diacylglycerol, an important precursor for the synthesis of phosphatidylinositol (PtdIns), phosphatidylglycerol, and cardiolipin. (2) In *S. pombe*, a checkpoint kinase (cds1, EC 2.7.11.1, 460aa) responsible for blocking mitosis in the S phase.

cdt1 A licensing protein (cdc10-dependent transcript 1, 546aa) that is recruited first to the origin of DNA replication, followed by cell division cycle 6 (Cdc6) and mini-chromosome maintenance proteins (Mcms). *Drosophila* homologue is Double-parked (Dup). Cdt1 is present in cells in G1 phase where it is required for initiation of replication but once replication has been initiated Cdt1 is either exported out of the nucleus or degraded, thereby preventing another round of replication. Also inhibited by **geminin**.

CEA See **carcinoembryonic antigen**.

C-EBP A family of bZIP transcription factors (C/EBP, CCAAT-enhancer binding proteins: C/EBPalpha to zeta, ~250–360aa) that recognize two different motifs: the CCAAT homology common to many promoters and the enhanced core homology common to many enhancers. In *Drosophila*, C/EBP (slow border cell protein, 449aa) may be required for the expression of gene products mediating **border cell** migration.

cecropins Antimicrobial peptides 30–40aa, act by permeabilizing the membranes of bacteria. Originally isolated from haemolymph of *Hyalophora cecropia* (silkmoth) pupae, and subsequently in several other species of endopterygote insects. A structural and functional equivalent, Cecropin P1, has been isolated from the pig intestine although subsequently it appears that this is probably the product of an intestinal nematode parasite, *Ascaris suum*.

ced genes Genes identified in *Caenorhabditis elegans* after studies of developmental mutations in which cells did not die when expected. Implicated in the control of **apoptosis**. See **interleukin-1 converting enzyme**.

C.elegans See *Caenorhabditis elegans*.

celiac See **coeliac**.

cell An autonomous self-replicating unit (in principle) that may constitute an organism (in the case of unicellular organisms) or be a subunit of multicellular organisms in which individual cells may be more or less specialized (differentiated) for particular functions. All living organisms are composed of one or more cells. Implicit in this definition is that viruses are not living organisms – and since they cannot exist independently, this seems reasonable.

cell adhesion See **adhesins, cadherins, cell adhesion molecules** (CAMs), **contact sites A, DLVO theory, integrins, sorting out, uvomorulin** and various specialized junctions (**adherens junctions, desmosomes, focal adhesions, gap junction** and **zonula occludens**).

cell adhesion molecule *CAM* Although this could mean any molecule involved in cellular adhesive phenomena, it has acquired a more restricted sense; CAMs are molecules on the surface of animal tissue cells, antibodies (or Fab fragments) against which specifically inhibit some form of intercellular adhesion. Examples are LCAM (Liver Cell Adhesion Molecule) and NCAM (Neural Cell Adhesion Molecule), both named from the tissues in which they were first detected, although they actually have a wider tissue distribution. Others include **LeCAM, SynCAM, VCAM** and **activated leukocyte cell adhesion molecule** (ALCAM).

cell aggregation Adhesion between cells mediated by cellular adhesion mechanisms rather than by a cross-linking or agglutinating agent.

cell bank Collection of cells stored by freezing that can be used to re-initiate cell cultures or, potentially, in the case of stem cells, to restore normal function to the organism from which they were removed.

cell behaviour General term for activities of whole cells such as movement, adhesion and proliferation, by analogy with animal behaviour.

cell body Used in reference to **neurons**; the main part of the cell around the nucleus excluding long processes such as **axons** and **dendrites**.

cell centre Microtubule organizing centre (MTOC) of the cell, the pericentriolar region.

cell coat Imprecise term occasionally used for the **glycocalyx**.

cell culture General term referring to the maintenance of cell strains or lines in the laboratory. See Table C2.

cell cycle The sequence of events between mitotic divisions. The cycle is conventionally divided into G0, G1, (G standing for gap), S (synthesis phase during which the DNA is replicated), G2 and M (mitosis). Cells that will not divide again are considered to be in G0, and the transition from G0 to G1 is thought to commit the cell to completing the cycle and dividing.

cell death Cells die (non-accidentally) either when they have completed a fixed number of division cycles (around 60; the **Hayflick limit**) or at some earlier stage when programmed to do so, as in digit separation in vertebrate limb morphogenesis. Whether this is due to an accumulation of errors or a programmed limit is unclear; some transformed cells have undoubtedly escaped the limit. See **apoptosis**.

TABLE C2. Common cell lines

Name	Species	Tissue of origin	Cell type	Comment
3T3	Mouse	Whole embryo	Fib	Swiss or Balb/c types; very density dependent
A9	Mouse	From L929	Fib	HGPRT negative
B16	Mouse	Melanoma	Mel	High and low metastatic variants (F1 and F10)
BHK21	Hamster	Baby kidney	Fib	Syrian Hamster. Usually C13 (clone 13)
BSC-1	Monkey	Kidney	Fib	Derived from African Green monkey (*Cercopithecus aethiops*), often used in virus propagation
CHO	Hamster	Ovary	Epi	Chinese Hamster ovary
Daudi	Human	Burkitt lymphoma	Lym	
Don	Hamster	Lung	Fib	Chinese Hamster
EAT	Mouse	Ascites tumour	Fib	
GH1	Rat	Pituitary tumour	Fib	Secrete growth hormone
HeLa	Human	Cervical carcinoma	Epi	Established line
HEp2	Human	Laryngeal carcinoma	Epi	Probably HeLa now.
HL60	Human	Peripheral blood	Myl	Will differentiate to granulocytes or macrophages
L1210	Mouse	Ascites fluid	Lym	Grows in suspension; DBA/2 mouse
L929	Mouse	Connective tissue	Fib	Clone of L cell
MCIM	Mouse	Sarcoma	Fib	Methylcholanthrene-induced
MDCK	Dog	Kidney	Epi	Madin-Darby canine kidney
MOPC31C	Mouse	Plasmacytoma	Lym	Grows in suspension; secretes IgG
MRC5	Human	Embryonic lung	Fib	Diploid, susceptible to virus infection
P388D1	Mouse	—	Lym	Grows in suspension
PC12	Rat	Adrenal	Neur	Phaeochromocytoma; can be induced to produce neurites
PtK1	Potoroo	Female kidney	Epi	Small number of large chromosomes
PtK2	Potoroo	Male kidney	Epi	Cells stay flat during mitosis
Raji	Human	Burkitt lymphoma	Lym	Grows in suspension. EB virus undetectable
S180	Mouse	Sarcoma	Fib	Invasive; maintained *in vivo*
SV40-3T3	Mouse	From 3T3	Fib	Transformed by SV40 virus
U937	Human	Monocytic leukaemia	Myl	Will differentiate to macrophages
Vero	Monkey	Kidney	Fib	Virus studies
Wl38	Human	Embryonic lung	Fib	Diploid, finite division potential
WRC-256	Rat	Carcinoma	—	Walker carcinoma; many variants

Although there are a great many cell lines available through the cell culture repositories and from trade suppliers, there are a few "classic" lines that will be met fairly frequently. Many of these well known lines are listed above, but the table is not comprehensive.

Epi = epithelial; Fib = fibroblastic; Lym = lymphocytic; Mel = melanin containing; Myl = myeloid; Neur = neural.

cell division The separation of one cell into two daughter cells, involving both nuclear division (**mitosis**) and subsequent cytoplasmic division (**cytokinesis**).

cell electrophoresis Method for estimating the surface charge of a cell by looking at its rate of movement in an electrical field; almost all eukaryotic cells have a net negative surface charge. Measurement is complicated by the streaming potential at the wall of the chamber itself, and by the fact that the cell is surrounded by a layer of fluid (see **double layer**). The electrical potential measured (the zeta potential) is actually some distance away from the plasma membrane. One of the more useful approaches is to systematically vary the pH of the suspension fluid to determine the pK of the charged groups responsible (mostly carboxyl groups of sialic acid).

cell fate Of an embryonic parent (progenitor) cell or cell type, the range and distribution of differentiated tissues formed by its daughter cells. For example, cells of the **neural crest** differentiate to form (among other things) cells of the peripheral nervous system.

cell fractionation Strictly this should mean the separation of homogeneous sets from a heterogeneous population of cells (by a method such as **flow cytometry**), but the term is more frequently used to mean subcellular fractionation i.e. the separation of different parts of the cell by differential centrifugation, to give nuclear, mitochondrial, microsomal and soluble fractions.

cell free system Any system in which a normal cellular reaction is reconstituted in the absence of cells, for example, *in vitro* translation systems that will synthesize protein from mRNA using a lysate of rabbit reticulocytes or wheat-germ. Mostly superceded by systems in which pure enzymes are used, as in many molecular biological procedures.

cell fusion Fusion of two previously separate cells occurs naturally in fertilization and in the formation of vertebrate skeletal muscle, but can be induced artificially by the use of **Sendai virus** or fusogens such as polyethylene glycol. Fusion may be restricted to cytoplasm or nuclei may fuse as well. A cell formed by the fusion of dissimilar cells is often referred to as a **heterokaryon**.

cell growth Usually used to mean increase in the size of a population of cells though strictly should be reserved for an increase in cytoplasmic volume of an individual cell.

cell junctions Specialized junctions between cells. See **adherens junctions**, **desmosomes**, **tight junctions**, **gap junctions**.

cell line A cell line is a permanently established cell culture that will proliferate indefinitely given appropriate fresh medium and space. Lines differ from cell strains in that they have escaped the **Hayflick limit** and become immortalized. Some species, particularly rodents, give rise to lines relatively easily, whereas other species do not. No cell lines have been produced from avian tissues, and the establishment of cell lines from human tissue is difficult. Many cell biologists would consider that a cell line is by definition already abnormal and that it is on the way towards becoming the culture equivalent of a neoplastic cell.

cell lineage The lineage of a cell relates to its derivation from the undifferentiated tissues of the embryo. Committed embryonic progenitors give rise to a range of differentiated cells: in principle it should be possible to trace the ancestry (lineage) of any adult cell.

cell locomotion Movement of a cell from one place to another.

cell mediated immunity Immune response that involves effector T-lymphocytes (T-cells) and not the production of humoral antibody. Responsible for **allograft** rejection, delayed **hypersensitivity** and in defence against viral infection and intracellular protozoan parasites.

cell membrane Rather imprecise term usually intended to mean **plasma membrane**.

cell migration Implies movement of a population of cells from one place to another, as in the movement of neural crest cells during morphogenesis.

cell movement A more general term than cell locomotion, that can include shape-change, cytoplasmic streaming, amongst others.

cell plate In plants, the immature disk that is formed during cytokinesis by the **phragmoplast** and that separates the sister nuclei. It is made by Golgi-derived vesicles that fuse, forming a jelly-like disk that expands outwards (centrifugally) until it separates the mother cell into two compartments. Initially, the cell plate is rich in callose but matures over time with the addition of stronger cellulose microfibrils.

cell polarity (1) In epithelial cells the differentiation of apical and basal specializations. In many epithelia the apical and basolateral regions of plasma membrane differ in lipid and protein composition, and are isolated from one another by tight junctions. The apical membrane may, for example, be the only region where secretory vesicles fuse, or have a particular ionic pumping system. (2) A motile cell must have some internal polarity in order to move in one direction at a time: a region in which protrusion will occur (the front) must be defined. Locomotory polarity may be associated with the pericentriolar **microtubule organizing centre**, and can be perturbed by drugs that interfere with microtubule dynamics.

cell proliferation Increase in cell number by division.

cell recognition Interaction between cells that is possibly dependent upon specific adhesion. Since the mechanism is not entirely clear in most cases, the term should be used with caution.

cell renewal Replacement of cells, for example those in the skin, by the proliferative activity of basal stem cells.

cell sap Casual term effectively equivalent to the term 'cytosol'.

cell signalling Release by one cell of substances that transmit information to other cells.

cell sorting The process or processes whereby mixed populations of cells, e.g. in a reaggregate, separate out into two or more populations that usually occupy different parts of the same aggregate or separate into different aggregates. Cell sorting probably takes place in the development of certain organs. See **differential adhesion, flow cytometry**.

cell strain Cells adapted to culture, but with finite division potential. See **cell line**.

cell streaming See **cyclosis**.

cell synchronization A process of obtaining (either by selection, or imposition of a reversible blockade) a population of growing cells that are to a greater or lesser extent in phase with each other in the cycle of growth and division.

cell trafficking The movement of leucocytes through tissues, as for example the movement of lymphocytes from skin to lymph nodes, or from blood into tissues and back into the circulation.

cell wall Extracellular material serving a structural role. In plants the primary wall is pectin-rich, the secondary wall mostly composed of **cellulose**. In bacteria, cell wall structure is complex: the walls of **Gram-positive** and **Gram-negative** bacteria are distinctly different. Removal of the wall leaves a **protoplast** or **spheroplast**.

cell wall loosening In plant cells, the loosening of the chemical bonds that maintain the rigidity of the complex cell wall. The non-enzymatic protein **expansin** causes slippage between adjacent cellulose microfibrils, allowing the cell to expand. Enzymes such as pectin methylesterases and xyloglucan endotransglycosylases break other kinds of bonds during wall loosening.

cell-surface marker Any molecule characteristic of the plasma membrane of a cell or in some cases of a specific cell type. Enzymic marker for the plasma membrane, such as 5′-**nucleotidase** and Na^+/K^+ ATPase, have largely been superceded by antibodies directed against specific cell surface molecules that are characteristic of particular cell types (see **CD antigens**).

cellobiohydrolase An enzyme that hydrolyses 1,4-β-D-glucosidic linkages in cellulose, releasing cellobiose from the non-reducing end of the chain. Cellulose degradation requires, sequentially, endoglucanases which cut internal beta-1,4-glucosidic bonds; exocellobiohydrolases that cut cellobiose from the non-reducing end of the polymer and β-1,4-glucosidases which hydrolyze the cellobiose and other short oligosaccharides to glucose. Found in fungi such as *Aspergillus* (1,4-β-D-glucan cellobiohydrolase A, cellobiohydrolase D, EC 3.2.1.91, 452aa) and bacteria such as *Clostridium thermocellum* (Exoglucanase xynX, EC 3.2.1.91, 1,4-β-cellobiohydrolase, exocellobiohydrolase, 1087aa).

cellobiose Reducing **disaccharide** composed of two D-glucose moieties β-1,4 linked. The disaccharide subunit of cellulose, though not found as a free compound *in vivo*.

cellubrevin Vesicle-associated membrane protein 3 (VAMP-3, synaptobrevin-3, 100aa), a **SNARE** involved in vesicular transport from the late endosomes to the trans-Golgi network.

cellular engineering The use of techniques for constructing replacement parts of tissues for both fundamental investigation and as prosthetic devices. Often involves the interfacing of cells and non-living structures.

cellular immunity Immune response that involves enhanced activity by phagocytic cells and does not imply lymphocyte involvement. Since the term is easily confused with **cell mediated immunity** its use in this sense should be avoided.

cellular retinoic acid-binding protein One of the **calycin superfamily** of lipid-binding proteins (**fatty acid binding proteins**), which occur in invertebrates and vertebrates, and that acts as an initial receptor for **retinoic acid** and regulates its transport to the nucleus where there are retinoic acid receptors. There are several isoforms, CRABP-1 (137aa), CRABP-2 (138aa) and CRABP-4 (retinoid-binding protein 7, 134aa).

cellular slime mould See **Acrasidae**.

cellulases Enzymes that break down **cellulose**, and are involved in cell-wall breakdown in higher plants, especially during abscission. Produced in large amounts by certain fungi and bacteria. Degradation of cellulose **microfibrils** requires the concerted action of several cellulases. See **cellobiohydrolase**.

cellulitis Inflammation of the subcutaneous connective tissues (dermis), mostly affecting face or limbs. *Streptococcus pyogenes* is commonly the causative agent. Also known as erysipelas.

cellulose The major structural component of plant cell walls where it is found as microfibrils laid down in several orthogonal layers. Cellulose is a straight chain polysaccharide composed of β(1−4)-linked glucose subunits. Cellulose-synthesizing particles in the plasma polymerize multiple polyglucan chains that crystallize to form a microfibril of very high tensile strength.

cellulose synthase Family of processive glycosyl synthetases involved in production of **cellulose**; plant cellulose synthase (CESA) proteins are integral membrane proteins (~1000aa). The sequences all have considerable similarity but differ from genes found in *Acetobacter* and *Agrobacterium* sp. In *Arabidopsis* there are three different types of CESA proteins that contribute to the cellulose synthesizing complexes involved in making primary walls and another three in making secondary walls. *Arabidopsis* also has several cellulose synthase-like (*Csl*) genes. Review: http://genomebiology.com/2000/1/4/REVIEWS/3001/ref

cellulose synthesizing complex In higher plants, a cluster of cellulose-synthesizing enzymes forming a transmembranous, hexagonal rosette. Rosettes are at least 25 nm (perhaps 50 nm) in diameter and protrude into the cytoplasm. In algae, the complex takes the form of linear terminal bars.

cellulosome A large multienzyme complex used by many anaerobic bacteria for the efficient degradation of plant cell-wall polysaccharides such as cellulose. Consists of **scaffoldins** containing nine **cohesin** domains and a cellulose-binding domain, and at least 14 different enzymatic subunits, each containing a conserved duplicated sequence, or **dockerin** domain that is responsible for holding the complex on to the bacterial cell surface through a calcium-mediated protein-protein interaction between the dockerin module from the cellulosomal scaffold and a cohesin (Coh) module of cell-surface proteins located within the proteoglycan layer.

cenexin Outer dense fibre of sperm tails protein 2 (Odf2, 829aa) a major component of sperm tail **outer dense fibres**. Also involved as a scaffolding protein in the centriole.

cenocyte See **coenocyte**.

CENPs A large group of **centromere** associated proteins. CENP-A (140aa) is a histone H3 variant and CENP-A nucleosomes directly recruit a proximal CENP-A nucleosome associated complex (NAC) comprising three other proteins, CENP-M, CENP-N and CENP-T, along with CENP-U, CENP-C and CENP-H. CENP-K, CENP-L, CENP-O, CENP-P, CENP-Q, CENP-R and CENP-S assemble on the CENP-A NAC. CENP-E (2701aa) is a kinesin-related microtubule motor protein that is essential for chromosome movement during mitosis. CENP-F (3210aa) is a microtubule-binding protein required for kinetochore localization of dynein, **Lis1**, **NDE1** and NDE-L1, and regulates recycling of the plasma membrane by acting as a link between recycling vesicles and the microtubule network. In **CREST syndrome** there are autoantibodies to CENPs.

centaurins A large family of proteins involved in regulating cytoskeletal and membrane trafficking events. They are GTPase-activating proteins for the **ADP-ribosylation** factor family of small G-proteins (Arf-GAPs) and bind lipid signalling molecules from the phosphoinositide 3-kinase, phosphoinositide 4-P

5-kinase and phospholipase D pathways. Examples include centaurin-alpha-2 (Arf-GAP with dual PH domain-containing protein 2, 381aa), centaurin-beta-1 (ACAP-1, ARFGAP with coiled-coil, ANK repeat and PH domain-containing protein 1, 740aa), centaurin-delta-1 (ARAP-2, Arf-GAP, Rho-GAP domain, ANK repeat and PH domain-containing protein 2, PARX protein, 1704aa) and centaurin-gamma-3 (AGAP-3, Arf-GAP, GTPase, ANK repeat and PH domain-containing protein 3, CRAM-associated GTPase, 875aa). See **cytohesins**.

centractin Vertebrate actin-related proteins (centractin-alpha, Arp1A, actin-RPV, 376aa; Arp1B, 376aa) that have 54% sequence homology with muscle actin and 69% similarity with cytoplasmic actin. Centractins are associated with the vertebrate **centrosome** and are involved in microtubule-linked cytoplasmic movement as components of the **dynactin** complex.

central core disease A myopathy caused by mutation in the **ryanodine receptor-1** (RyR1) gene.

central lymphoid tissue See **lymphoid tissue**.

centrifugation The process of separating fractions of systems in a centrifuge. The most basic separation is to sediment a pellet at the bottom of the tube, leaving a supernatant at a given centrifugal force. In this case sedimentation is determined by size and density of the particles in the system amongst other factors. Density may be used as a basis for sedimentation in **density gradient** centrifugation. At very high g values molecules may be separated, i.e. **ultracentrifugation**. In continuous centrifugation the supernatant is removed continuously as it is formed.

centrins Acidic calcium-binding phosphoproteins involved in the duplication of centrosomes in higher eukaryotes, homologous to **caltractin**. Also found in striated flagellar roots of various algae and basal bodies of human sperm. There are three centrins in humans: centrin-1 (caltractin isoform 2, 172aa), centrin-2 (caltractin isoform-1, 172aa) and centrin-3 (167aa). A mutation in centrin causes genomic instability via increased chromosome loss in *Chlamydomonas reinhardtii*.

centriolar region See **pericentriolar region** or **centrosome**.

centriole Organelle of animal cells that is made up of two orthogonally arranged cylinders each with nine microtubule triplets composing the wall. Almost identical to **basal body** of cilium. The pericentriolar material, but not the centriole itself, is the major **microtubule organizing centre** of the cell. Centrioles divide prior to mitosis and the daughter centrioles and their associated pericentriolar material come to lie at the poles of the spindle. Various proteins are associated with the centriole: see **centrins**,

CENPs, **centriolin**. Plants have a similar microtubule organising centre but no centriole.

centriolin A centriolar protein (CEP110, 2325aa) that is involved in cell cycle progression and cytokinesis. In the late stage of cytokinesis centriolin anchors **exocyst** and **SNARE** complexes at the midbody allowing secretory vesicle-mediated abscission.

centroblast Stage of B-lymphocyte differentiation after antigen exposure and activation. Centroblasts are rapidly proliferating B-cells with little or no surface immunoglobulin. These cells undergo somatic mutation and class-switching of their immunoglobulin genes.

centrocyte Non-proliferating progeny of **centroblasts** that re-express surface immunoglobulin and are thought to be positively or negatively selected by their affinity for antigen.

centrolecithal Type of egg in which the yolk is in the centre.

centromere The region (attachment constriction) in eukaryote chromosomes where daughter chromatids are joined together. The **kinetochore**, to which the spindle chromosomes are attached, lies adjacent to the centromere. The centromere region generally consists of large arrays of repetitive DNA. See **CENP**.

centronuclear myopathy Muscle disorders (myotubular myopathies) characterized by slowly progressive muscular weakness and wasting. Can be caused by mutations in genes for **dynamin-2, bridging integrator 1** or **myotubularin**.

centrophilin Old name for **nuclear mitotic apparatus protein**.

centrosome The **microtubule organizing centre** which, in animal cells, surrounds the **centriole**, and which will divide to organize the two poles of the mitotic spindle. By directing the assembly of a cell's skeleton, this organelle controls division, motility and shape.

centrosphere Alternative (rare) name for **centrosome**.

Cepaea Genus of land snails. Two species, *C. hortensis* and *C. nemoralis* have been much studied as convenient examples of polymorphism in colour and banding pattern.

cephalosporins Group of broad-spectrum beta-lactam (tetracyclic triterpene) antibiotics isolated from culture filtrates of the fungus *Cephalosporium* sp. Inhibit bacterial cell wall synthesis in a similar way to penicillins, and are effective against Gram-positive bacteria. Various modifications have been made to produce new variants such as cephaloridine, cefalexin and cefprozil.

cer/dan family Secreted antagonists of **bone morphogenetic proteins** that are involved in complex morphogenetic processes such as the development of left-right asymmetry. They have a characteristic cysteine-knot motif. Examples include **caronte, cerberus, charon, dan, gremlin** and **sclerostin**.

ceramides A family of lipids composed of **sphingosine** and a fatty acid; found in **sphingomyelin** and **glycosphingolipids.**

cerberus Cysteine knot superfamily protein of the **cer/dan family** (DAN domain family member 4, 267aa), functioning as secreted-type **BMP** antagonist, consequently an antagonist of **Nodal, Wnt,** and BMP signalling. Cerberus function is required in the leading edge of the anterior dorsal-endoderm of the *Xenopus* embryo for correct induction and patterning of the neuroectoderm. Cerberus-short only inhibits **nodal**.

CERC An acronym for members of the **EF-hand** superfamily (CREC family) found in endoplasmic reticulum and Golgi: Cab-45, Erc-55, reticulocalbin and **calumenin**.

cercidosome Specialized organelle of trypanosomes, site of terminal oxidative metabolism.

Cercozoa Soil-dwelling amoeboflagellate protozoa, mostly heterotrophic and very abundant. Tend to form filopodia or reticulopodia. Some lineages have secondarily acquired chloroplasts (see **Chlorarachniophytes**). Image and details: http://tolweb.org/Cercozoa/121187#Introduction

cerebellar ataxia See **spinocerebellar ataxia.**

cerebellum Part of the vertebrate hindbrain, concerned primarily with somatic motor function, the control of muscle tone and the maintenance of balance. Important model for cell migration in developing mammalian brain owing to well-studied migratory pathway of the **granule cell** and to the existence of the neurological mutant mouse **weaver** in which granule cell migration fails.

cereblon A protein (442aa) that may modulate cell surface expression of KCNT1 (T-type potassium channel) and may be involved in memory and learning. Defects are the cause of autosomal recessive mental retardation type 2A. Has recently been shown to have **thalidomide**-binding properties.

cerebral amyloid angiopathy A disease of small blood vessels in the brain in which there are deposits of amyloid protein in the vessel walls. Often a consequence of ageing but there are some forms caused by mutation in, variously, genes for **amyloid precursor protein, cystatin C,** integral membrane protein 2B (266aa) or associated with high density lipoprotein deficiency caused by mutation in the ABCA1 gene.

cerebral cavernous malformations A set of disorders characterized by venous sinusoids and a predisposition to intracranial haemorrhage. Can be caused by mutation in genes for **KRIT1, malcavernin** or **programmed death 10**.

cerebral palsy A nonprogressive disorder of posture or movement, usually a consequence of perinatal damage but a few rare cases are caused by mutation in the gene encoding **glutamate decarboxylase-1**.

cerebroside Glycosphingolipid found in brain (11% of dry matter). **Sphingosine** core with fatty amide or hydroxy fatty amide and a single monosaccharide on the alcohol group (either glucose or galactose).

cerebrotendinous xanthomatosis A lipid-storage disorder caused by a deficiency of the mitochondrial sterol 27-hydroxylase, CYP27.

cereolysin A general name for a variety of cytolytic toxins released by *Bacillus cereus*. Cereolysin A is a phospholipase C (EC 3.1.4.3, 283aa), cereolysin B is a sphingomyelinase C (EC 3.1.4.12, 333aa) and both hydrolyse membrane components. Another cereolysin is a thiol-activated cytolysin (inactivated by oxygen, reactivated by thiol reduction) which binds to cholesterol in the plasma membrane and rearrangement of the toxin-cholesterol complexes in the membrane leads to altered permeability.

cernunnos A protein involved in the process of repairing double-strand breaks in DNA (non-homologous end-joining factor 1, 299aa) (see **NHEJ**) and and V(D)J recombination. Defects in cernunnos lead to a severe immunodeficiency condition associated with microcephaly and other developmental defects in humans. Cernunnos physically interacts with the XRCC4/DNA-LigaseIV complex. NHEJ is required for the generation of diversity in the immune system.

ceroid lipofuscinosis See **Batten's disease**.

cerulenin An antibiotic isolated from the culture broth of *Cephalosporium caerulens* that has a very broad range of antibacterial and antifungal activity and inhibits fatty acid and polyketide biosynthesis.

ceruloplasmin A blue copper-binding **dehydrogenase** protein (EC 1.16.3.1, 1065aa) with 6−7 cupric ions per molecule, found in serum (200−500 mg/ml). Apparently involved in copper detoxification and storage, and possibly also in mopping up excess oxygen radicals or **superoxide** anions. Mutation in the ceruloplasmin gene is associated with iron accumulation in tissues. See **Wilson's disease**.

cesium chloride See **caesium chloride**.

CFSE Fluorochrome that readily enters cells where it is de-esterified by non specific esterases and remains trapped. Used as a vital dye for cell tracking and lineage studies.

CFTR See **cystic fibrosis transmembrane conductance regulator**.

CFU-E Colony forming unit for **erythrocytes**.

CG island See **CpG island**.

CGD See **chronic granulomatous disease**.

CGH Comparative genomic hybridization, a method for analysing for copy number changes. DNA from subject tissue and from normal control tissue are labeled with different fluorescent tags, mixed and hybridized to normal metaphase chromosomes: using quantitative image analysis, regional differences in the fluorescence ratio of gains/losses vs. control DNA can be detected. The technique can also be used with DNA arrays rather than chromosomes.

cGMP See **cyclic GMP**.

CGRP See **calcitonin gene related peptide**.

CGT (1) In *Bacillus* sp. cyclomaltodextrin glucanotransferase (EC 2.4.1.19, 713aa) an enzyme that cyclizes part of a (1−4)-α-D-glucan chain. (2) In humans, ceramide UDP-galactosyltransferase (2-hydroxyacylsphingosine 1-β-galactosyltransferase, cerebroside synthase, EC 2.4.1.45, 541aa) an important enzyme in biosynthesis of the sphingolipids of the myelin sheath.

ch-TOG A human homologue of the **xmap215** microtubule stabilizer (cytoskeleton-associated protein 5, CKAP5, colonic and hepatic tumour overexpressed protein, 2032aa).

chaconine Toxic trisaccharide glycoalkaloid (α-chaconine) found in potatoes (*Solanum tuberosum*), where it is produced in increased amounts as a response to stress (and has insecticidal and fungicidal effect). Major alkaloid may be β-chaconine, a disaccharide breakdown product of α-chaconine. Related to **solanine**. Imparts an unpleasantly bitter taste. Has cytotoxic effects that have been investigated for tumour therapy and it has been reported that α-chaconine induces apoptosis of HT-29 cells through inhibition of **ERK** and, in turn, activation of **caspase-3**. Also has anti-cholinesterase activity. See **solanidine**.

chaeotropic See **chaotropic**.

chaetoglobosins Chaetoglobosin J is a fungal metabolite related to cytochalasins that will inhibit elongation at the barbed end of an actin microfilament. Chaetoglobosin A is produced by *Chaetomium globosum*. Chaetoglobosin K is a plant growth inhibitor and toxin from the fungus *Diplodia macrospora*.

Chaetognatha Phylum of hermaphrodite marine Coelomata with the body divided into three distinct regions: head, trunk and tail. Arrow-worms.

Chagas' disease South American trypanosomiasis caused by *Trypanosoma cruzi* and transmitted by blood-sucking reduviid bugs such as *Rhodnius*.

chalaza (1) The basal part of a plant ovule where nucellus and integument are joined. (2) One of two spirally twisted cords of dense albumen connecting the yolk to the shell membrane in a bird's egg.

chalcone Intermediate in biosynthesis of flavanones, **flavones** and **anthocyanidins** by plants. Synthetic derivatives have been shown to have anti-inflammatory properties.

chalcone isomerase Enzymes in the biosynthetic pathway for all classes of flavonoids. Chalcone isomerase 1 (chalcone-flavonone isomerase, EC 5.5.1.6, 246aa) is responsible for intramolecular cyclization of bicyclic chalcones into tricyclic (S)-flavanones. Many variants are known from different plants.

chalcone synthase An important enzyme (naringenin-chalcone synthase, EC 2.3.1.74, 395aa) in the flavonoid pathway. See **apigenin**.

chalone Cell-released tissue-specific inhibitor of cell proliferation thought to be responsible for regulating the size of a population of cells. Their existence has been doubted but **myostatin** appears to be an example.

Chanarin-Dorfman syndrome See **ichthyosiform erythroderma**.

Chang liver cells A human cell line with epithelial morphology, derived from non-malignant tissue but later shown to be a sub-line of **HeLa cells**. Extensively used in virology and biochemistry.

channel gating See **gating current**.

channel protein A protein that facilitates the diffusion of molecules/ions across lipid membranes by forming a hydrophilic pore. Most frequently multimeric with the pore formed by subunit interactions.

channel-forming ionophore An **ionophore** that makes an amphipathic pore with hydrophobic exterior and hydrophilic interior. Most known types are cation selective.

channelopathies Any disease that arises because of a defect in an ion channel which may be the result of alteration in the channel-forming subunit or in regulatory sub-units. Defects in channels for potassium, sodium, chloride and calcium ions are known, and a wide variety of diseases are the result. Many toxins from spiders, scorpions, snakes etc. act on ion channels.

channelopsin See **channelrhodopsin**.

channelrhodopsin Light-gated ion (proton) channels from *Volvox* (channelrhodopsin-1, 837aa; channelrhodopsin-2, 747aa), originally isolated from the eyespot of *Chlamydomonas reinhardtii* (channelopsin-1, archaeal-type opsin 1, 712aa) that can be used experimentally to activate cells by light-pulses. The channels open rapidly after absorption of a photon to generate a large permeability for monovalent and divalent cations. Recent paper: http://www.ncbi.nlm.nih.gov/pmc/articles/PMC2718366/

Chaos chaos Giant multinucleate fresh-water amoeba (up to 5 mm long) much used for studies on the mechanism of cell locomotion.

chaotropic Describing an agent that causes chaos, usually in the sense of disrupting or denaturing macromolecules. For example, iodide is often used in protein chemistry to break up and randomise **micelles**. In molecular biology guanidium isothiocyanate is used to provide a denaturing environment in which RNA can be extracted intact without exposure to **RNAases**.

chaperones Cytoplasmic proteins of both prokaryotes and eukaryotes (and organelles such as mitochondria) that bind to nascent or unfolded polypeptides and ensure correct folding or transport. Chaperone proteins do not covalently bind to their targets and do not form part of the finished product. Heat-shock proteins are an important subset of chaperones. Three major families are recognized, the **chaperonins** (groEL and hsp60), the hsp70 family and the **hsp90** family. Outside these major families are other proteins with similar functions including **nucleoplasmin**, secB, and T-cell receptor-associated protein.

chaperonins Subset of chaperone proteins found in prokaryotes, mitochondria and plastids - major example is prokaryotic GroEL (the eukaryotic equivalent of which is hsp60).

CHAPS Zwitterionic detergent used for membrane solubilization.

Char's syndrome A developmental disorder caused by missense mutations in the gene for transcription factor **AP2beta** which mainly affects **neural crest** cells.

Chara See **Characean algae**.

Characean algae Class of filamentous green algae (Charophyceae) exemplified by the genus *Chara*, in which the mitotic spindle is not surrounded by a nuclear envelope. Probably the closest relatives, among the algae, to higher plants. The giant internodal cells (up to 5 cm long) exhibit dramatic **cyclosis** and have been much used for studies on ion transport and cytoplasmic streaming.

Charcot-Leyden crystals Hexagonal bipyramidal crystals of **galectin-10** (eosinophil lysophospholipase, EC 3.1.1.5, 142aa) found in human tissues and secretions in association with increased numbers of peripheral blood or tissue eosinophils, a feature of parasitic infections and allergic processes.

Charcot-Marie-Tooth disease A clinically and genetically heterogeneous group of motor and sensory neuropathies affecting peripheral nerves, the most common inherited disorders of the peripheral nervous system. The affected genes can be for **connexin-32**, **early growth response-2**, frabin, ganglioside-induced differentiation-associated protein-1, **kinesin-like** protein, KIF1B, **LITAF**, **mitofusin-2**, myelin protein zero, **myotubularin-related** protein-2, myotubularin-related protein 13, neurofilament light polypeptide, **N-myc** downstream-regulated-1, **periaxin**, peripheral myelin protein-22, SH3 domain and tetratricopeptide repeats-containing protein 2. Other loci are known to be affected but the genes have not yet been identified.

CHARGE syndrome A complex set of abnormalities (**coloboma**, heart anomaly, **choanal atresia**, retardation of mental and somatic development, genital hypoplasia, ear abnormalities and/or deafness) caused by mutation in the gene encoding the **chromodomain helicase** DNA-binding protein-7 or the **semaphorin-3E** gene. The **PAX2** gene that is active in all the developmental primordia that are affected seems to be unaffected.

charon A Cerberus/Dan-family protein (243aa) that is a negative regulator of Nodal signalling during left-right patterning in zebrafish. Suppresses the dorsalizing activity of all three of the known zebrafish Nodal-related proteins (Cyclops, Squint and Southpaw).

chartins Obsolete name for microtubule-associated proteins (**MAPs**) of 64, 67 and 80 kDa, distinct from **tau protein** and isolated from neuroblastoma cells. They are regulated by **nerve growth factor** (NGF) and may influence microtubule distribution.

charybdotoxin Peptide (37aa) isolated from *Leiurus quinquestriatus hebraeus* (Yellow scorpion) venom that is a selective blocker of high conductance (maxi-K), different intermediate and small conductance calcium-activated potassium channels (SK channels), as well as a voltage-dependent potassium channel (Kv1.3/KCNA3).

CHD proteins Proteins (chromo-ATPase/helicase-DNA-binding proteins) that regulate ATP-dependent nucleosome assembly and mobilization through their conserved double **chromodomains**. Whether they act as activators or repressors depends upon associated proteins. For example, CHD1 (EC 3.6.4.12, 1710aa) is a component of the **SAGA complex**, CHD7 (2997aa) is mutated in **CHARGE syndrome**, CHD9 (peroxisomal proliferator-activated receptor A-interacting complex 320 kDa protein, 2897aa) is a transcriptional coactivator for PPARA.

***che* genes** Genes involved in specifying components of the **bacterial chemotaxis** response system. CheW (167aa) bridges CheA to the methyl-accepting chemotaxis proteins to allow regulated phosphotransfer to CheY (129aa) and CheB. CheA (EC 2.7.13.3, 654aa) is a histidine kinase activated by autophosphorylation in proportion to the degree of receptor methylation, and passes phosphate to CheB (a **two-component system**) or CheY. When phosphorylated or acetylated CheY exhibits enhanced binding to FliM (see *fli* **genes**) and reverses the direction of flagellar rotation (causes tumbling). CheB (EC 3.1.1.61, 349aa) is a methyl esterase activated by CheA phosphorylation, that demethylates (and upregulates) receptors that have been desensitized (methylated) by CheR (EC 2.1.1.80, 286aa). The degree of receptor methylation allows the bacterium to 'remember' chemical concentrations from the recent past, a few seconds, and compare them to those it is currently experiencing, thus 'knows' whether it is travelling up or down a gradient. Other proteins such as CheZ (214aa), CheC (CheY-P phosphatase, 219aa in *B. subtilis*) and CheD (chemoreceptor glutamine deamidase, EC 3.5.1.44, 166aa in *B. subtilis*) are involved in regulation. The details in *Bacillus subtilis* are slightly different from those in *E. coli* (given above), even though orthologous proteins are present. Comparative study of chemotaxis in *E. coli* and *B.subtilis*: http://www.plosbiology.org/article/info:doi/10.1371/journal.pbio.0020049

checkerboard assay Variant of the Boyden chamber assay for leucocyte chemotaxis introduced by Zigmond. By testing different concentrations of putative chemotactic factor in non-gradient conditions, it is possible to calculate the enhancement of movement expected due simply to chemokinesis and to compare this with the distances moved in positive and negative gradients. Good experimental design thus allows chemotaxis to be distinguished from chemokinesis.

checkpoint Any stage in the **cell cycle** at which the cycle can be halted and entry into the next phase postponed. Two major checkpoints are at the G1/S and G2/M boundaries. These are the points at which **cdc** proteins act.

checkpoint kinases Serine/threonine kinases (EC 2.7.11.1, chk1, 476aa; chk2, 543aa) that act downstream of **ATM** in response to detection of DNA damage. Chk 1 is an essential gene for normal cell division and chk 2 has been found to be mutated in many human tumours. Chk2, along with ATM, was found to regulate the transcription factor p53 by preventing its ubiquitination by the RING E3 ligase Mdm2. Chk2 is mutant in **Li-Fraumeni syndrome-2**.

Chediak-Higashi syndrome An autosomal recessive disorder, caused by mutation in the **lysosomal trafficking regulator** gene, characterized by the presence of giant lysosomal vesicles in phagocytes and in consequence poor bactericidal function. Some perturbation of microtubule dynamics seems to be involved. Reported from humans, albino Hereford cattle, mink, beige mice and killer whale.

chelation Binding of a metal ion by a larger molecule such as EDTA or protein (iron in haem is held as a chelate). The binding is strong but reversible and chelating agents can be used to buffer the free concentration of the ion in question.

chelerythrine An alkaloid from the greater celandine (*Chelidonium majus*), and named from the red colour of its salts. Chelerythrine chloride inhibits **PKC** by binding to the catalytic domain.

chemerin An **adipokine** (retinoic acid receptor responder protein 2, tazarotene-induced gene 2 protein, 163aa) whose expression is upregulated *in vitro* by the synthetic retinoid tazarotene. A secreted chemotactic protein for antigen-presenting cells which express chemokine-like receptor 1 (CMKLR1) and also potentiates insulin-stimulated glucose uptake.

chemical potential In biological systems the work required (in J mol^{-1}) to bring a molecule from a standard state (usually infinitely separated in a vacuum) to a specified concentration. More usually employed as chemical potential difference, the work required to bring one mole of a substance from a solution at one concentration to another at a different concentration, $\Delta m = RT.\ln (c_2/c_1)$. This definition is useful in studies of active transport; note that, for charged molecules, the electrical potential difference must also be considered (see **electrochemical potential**).

chemical synapse A nerve-nerve or nerve-muscle junction where the signal is transmitted by release from one membrane of a chemical transmitter that binds to a receptor in the second membrane. Importantly, signals only pass in one direction.

chemiluminescence Light emitted as a reaction proceeds. Can be used to assay ATP (using firefly luciferase) and the production of toxic oxygen species by activated phagocytes (using **luminol** or **lucigenin** as bystander substrates that release light when oxidized). See also **bioluminescence**.

chemiosmosis A theoretical mechanism (proposed by Mitchell) to explain energy transduction in the mitochondrion. As a general mechanism it is the coupling of one enzyme-catalysed reaction to another using the transmembrane flow of an intermediate species, e.g. cytochrome oxidase pumps protons across the mitochondrial inner membrane and ATP synthesis is 'driven' by re-entry of protons through the ATP-synthesizing protein complex. The alternative model is production of a chemical intermediate species, but no compound capable of coupling these reactions has ever been identified and chemiosmosis is generally accepted as being the correct model.

chemoattraction Non-committal description of cellular response to a diffusible chemical − not necessarily by a tactic response. Term preferable to 'chemotaxis' when the mechanism is unknown.

chemoautotroph Chemotrophic **autotroph** (chemotroph). Organism in which energy is obtained from endogenous light-independent reactions involving inorganic molecules.

chemodynesis Induction of cytoplasmic streaming in plant cells by chemicals rather than by light (photodynesis).

chemokine receptors Chemokine receptors are G-protein-linked **serpentine receptors** that, in addition to binding **chemokines**, are used as co-receptors for the binding of immunodeficiency viruses (**HIV, SIV, FIV**) to leucocytes. They are named according to the class of chemokine that they bind (see Table C3). CXCR4 (which binds chemokine CXC12) is a co-receptor for T-tropic viruses, CCR5 for macrophage-tropic (M-tropic) viruses. Individuals deficient in particular CCRs seem to be resistant to HIV-1 infection.

chemokines Small secreted proteins ($\sim 100aa$) that stimulate **chemotaxis** of leucocytes. Chemokines can be subdivided into classes on the basis of conserved cysteine residues. The α-chemokines (IL-8, NAP-2, Gro-α, Gro-γ, ENA-78 and GCP-2) have conserved CXC motif and are mainly chemotactic for neutrophils; the β-chemokines (MCP-1-5, MIP-1α, MIP-1β, eotaxin, RANTES) have adjacent cysteines (CC) and attract monocytes, eosinophils or basophils; the γ-chemokines have only one cysteine pair and are chemotactic for lymphocytes (lymphotactin), the δ-chemokines are structurally rather different being membrane-anchored, have a CXXXC motif and are restricted (so far) to brain (**fractalkine**). The receptors are **G-protein** coupled. See Table C3.

chemokinesis A response by a motile cell to a soluble chemical that involves an increase or decrease in speed (positive or negative orthokinesis) or of frequency of movement, or a change in the frequency or magnitude of turning behaviour (**klinokinesis**).

chemolithotroph Alternative name for a **chemoautotroph**.

chemoreceptor (1) A receptor that binds small molecules and triggers an intracellular signalling system. The term is not usually applied to hormone, neurotransmitter or cytokine receptors. (2) A cell or a group of cells specialized for responding to chemical substances in the environment.

chemorepellant A substance that repels cells and leads to their dispersion rather than their accumulation. The opposite of a chemoattractant.

chemostat Apparatus for maintaining a bacterial population in the exponential phase of growth by regulating the input of a rate-limiting nutrient and the removal of medium and cells.

chemosynthesis Synthesis of organic compounds by an organism using energy derived from oxidation

TABLE C3. Chemokines

Chemokine	Synonyms	Attracts	Receptor	Produced by:
CXC Family	α-chemokines			
(ELR)-positive				
CXCL1	NAP-1, MONAP, MDNCF, NAF, LAI, GCP, GRO1, GROα, MGSA-a, NAP-3	Neutrophils	CXCR1, CXCR2	Many cells, including monocytes, lymphocytes, fibroblasts, endothelial cells, mesangial cells.
CXCL2	MIP2-α, GROβ, GRO2, MGSA $-\beta$			
	Neutrophils	CXCR2	Activated	Monocytes, fibroblasts, epithelial and endothelial cells.
CXCL3	MIP-2β, GROγ	Neutrophils, basophils, endothelial cells	CXCR2	
CXCL5	SCYB5, ENA-78	Neutrophils, endothelial cells	CXCR2	
CXCL6	SCYB6, GCP-2, CKA-3		CXCR1, CXCR2	
CXCL7	PBP, CTAP-III, NAP-2, β-TG, low affinity platelet factor 4;	Fibroblasts (βTG), neutrophils (NAP-2)	CXCR2	Platelets
CXCL8	IL-8	Neutrophils	CXCR1, CXCR2	
(ELR)-negative				
CXCL4	Oncostatin A, platelet factor 4 (PF-4)	Neutrophils	CXCR3-B	Aggregated platelets, activated T-cells
CXCL10	interferon-inducible cytokine, γIP-10, IP-10, CRG-2, C7;	Monocytes	CXCR3	Keratinocytes, monocytes, T-cells, endothelial cells and fibroblasts
CXCL12	PBSF, Stromal cell-derived factor, SDF-1α,	Primordial germ cells	CXCR4 (fusin), CXCR7	Fibroblasts, bone marrow stromal cells.
CXCL13	BCA, B-lymphocyte chemoattractant (BLC)	B cells	CXCR5 (CD185)	
CXCL14	MIP-2γ, BRAK, BMAC	Neutrophils, myeloid cells	?	Breast and kidney
CC Family	β-chemokines			
CCL1	I-309, TCA3	Monocytes	CCR8	T-cells, mast cells
CCL2	MCP1, MCAF, JE, LDCF, GDCF, HC14, MARC	Monocytes, basophils	CCR2	Monocytes, T-cells, fibroblasts, endothelial cells, smooth muscle, some tumours. Upregulated by IFNγ
CCL3	MIP-1α, LD78β, pAT464, GOS19	Eosinophils	CCR1, CCR5	T-cells, B-cells, Langerhans cells, neutrophils, macrophages
CCL4	MIP-1β, HIMAP	Memory T cells	CCR1, CCR5	ACT-2, pAT744, hH400, hSISα, G26, HC21, MAD-5,
CCL5	RANTES, sisδ	Monocytes, memory T, eosinophils	CCR1, CCR2, CCR3	T-cells, macrophages

TABLE C3. (Continued)

Chemokine	Synonyms	Attracts	Receptor	Produced by:
CCL7	MCP3	Monocytes, T cells, NK cells, eosinophils, dendritic cells	CCR1, CCR2, CCR3	Fibroblasts
CCL8	MCP2	Monocytes	CCR3, CCR5	
CCL11	Eotaxin	Eosinophils	CCR3	
CCL13	MCP-4	Monocytes	CCR2, CCR3	
CCL14	HCC-1	T cells, monocytes, basophils, eosinophils	CCR1, CCR5	
CCL15	HCC-2, Lkn-1, MIP-1δ	T cells, monocytes, eosinophils, basophilsgranulocytes	CCR1	
CCL16	HCC-4, LEC, LCC-1	T cells, monocytes, eosinophils, basophils	CCR1, CCR2	
CCL17	TARC	T cells, dendritic cells, basophils	CCR4	Activated T-cells
CCL18	DC-CK1		?	
CCL19	MIP-3b, ELC	T cells, dendritic cells	CCR7	
CCL20	MIP-3a, LARC	T cells, B cells	CCR6	
CCL21	6Ckine, SLC	T cells, dendritic cells	CCR7	
CCL22	MDC, Monocyte-derived chemokine	T cells, dendritic cells, basophils	CCR4	Monocytes
CCL23	MPIF-1, CKb8	T cells, monocytes, granulocytes	CCR1	
CCL24	Eotaxin-2	Eosinophils, basophils, mast cells	CCR3	
CCL25	TECK	T cells, IgA$^+$ plasma cells	CCR9	
CCL27	CTACK	T cells	CCR10	
CCL28	MEC	T cells	CCR3, CCR10	
C Family	γ chemokines			
XCL1	Lymphotactin, SCM-1α	Lymphocytes	XCR1	
XCL2	SCM-1β	Lymphocytes	XCL1	
CX3C Family	δ chemokines			
CX$_3$CL1	Fractalkine, neurotactin	Monocytes, lymphocytes, neutrophils	CX$_3$CR1	Endothelial cells, brain.

Sources include: Michael R. Douglas, Karen E. Morrison, Michael Salmon and Christopher D. Buckley (2002), *Expert Reviews in Molecular Medicine* (www.ermm.cbcu.cam.ac.uk/02005318h.htm).

of inorganic molecules rather than light (see **chemotrophy** and **photosynthesis**).

chemotactic See **chemotaxis**.

chemotaxis A response of motile cells or organisms in which the direction of movement is affected by the gradient of a diffusible substance (chemotactic response). Differs from chemokinesis in that the gradient alters probability of motion in one direction only, rather than rate or frequency of random motion.

chemotherapy Treatment of a disease with drugs that are designed to kill the causative organism or, in the case of tumours, the abnormal cells.

chemotrophy Systems of metabolism in which energy is derived from endogenous chemical reactions rather than from food or light-energy, e.g. in deep-sea hot-spring organisms.

chemotropism Growth or possibly bending of an organism in response to an external chemical gradient. Sometimes used in error when the terms **chemotaxis** or **chemokinesis** should have been used. Do not confuse with chemotrophism (see **chemotrophy**).

chenodeoxycholic acid A bile acid (chenodesoxycholic acid) synthesized in the liver from cholesterol.

CHF See **chick heart fibroblasts**.

chi-squared χ-*squared* Common statistical test to determine whether the observed values of a variable are significantly different from those expected on the basis of a null hypothesis.

chiasma Junction points between non-sister **chromatids** at the first **diplotene** of **meiosis**, the consequence of a **crossing-over** event between maternal and paternally derived **chromatids**. A chiasma also serves a mechanical function and is essential for normal equatorial alignment at meiotic **metaphase** I in many species. Frequency of chiasmata is very variable between species.

chick heart fibroblasts The cells that emigrate from an explant of embryonic chick heart maintained in culture. Often considered as archetypal normal cell and were used as the 'normal' reference cells in the original studies of contact inhibition of locomotion by Abercrombie.

chickenpox A highly infectious 'childhood disease' caused by human herpes virus type 3 (varicella zoster, see **Herpesviridae**). The virus can remain latent in nervous tissue and may become reactivated as **shingles** under stress or immunodeficiency.

Chikungunya virus A positive-sense, single-stranded RNA virus, an Alphavirus of the Togaviridae family (Buggy Creek virus). Carried by *Aedes aegypti* and new strains apparently by *Aedes albopictus* (Tiger mosquito), the virus causes severe flu-like symptoms. The name comes from the Makonde word for 'that which bends up', a reference to the positions that victims take to relieve the joint pain. Cases have begun to appear in Europe (2007).

Chilomonas Genus of small cryptomonad flagellate protists, 20–40 μm long.

chimera Organism composed of two genetically distinct types of cells. Can be formed by the fusion of two early blastula stage embryos or by the reconstitution of the bone marrow in an irradiated recipient, or by somatic segregation. Since female mammals have one or the other X-chromosome more-or-less randomly inactivated they could also be considered chimeric.

chimerins *chimaerins* GTPase-activating proteins with high affinity for phorbol esters and **diacyl glycerol**. N-chimerin (alpha-chimerin, 459aa) may be important in neuronal signal-transduction mechanisms. Found in neurons in brain regions that are involved in learning and memory processes. Mutations in α-chimerin are responsible for Duane retraction syndrome 2. Beta-chimerin (Rho GTPase-activating protein 3, 468aa) is found mostly in the brain and pancreas and expression is much reduced in malignant gliomas.

Chinese hamster ovary cells Cell line (CHO cells) that is often used for growing viruses and for the production of therapeutic proteins. The genome was sequenced in early 2011.

CHIP (1) A co-chaperone and ubiquitin ligase (C-terminus of Hsp70-interacting protein, STIP1 homology and U box-containing protein 1, EC 6.3.2.-, E3 ubiquitin-protein ligase CHIP, 303aa) that interacts with **Hsp70** through an amino-terminal tetratricopeptide repeat (**TPR motif**). Important in protection against physiologic stress. (2) Aquaporin-1 (Aquaporin-CHIP, CHIP-28, 269aa), see **aquaporins**.

chiral stationary phase Stationary phase for liquid chromatography that has a chiral molecule attached and has differential binding characteristics for enantiomers of molecules passed through the column. Can therefore be used to separate racemic mixtures.

Chironomus Genus of flies (midges). Larvae live in fresh water and have been much studied because of the giant **polytene chromosomes** in the salivary glands; haemolymph contains haemoglobin in solution.

Chiroptera Bats. An Order of flying mammals, mainly insect or fruit-eating and nocturnal.

chitin Polymer (β-1,4 linked) of N-acetyl-D-glucosamine, extensively cross-linked; the major structural component of arthropod exoskeletons and fungal cell walls. Widely distributed in plants and fungi.

chitinase Enzyme (EC 3.2.1.14) that catalyses the hydrolysis of 1,4-β linkages of chitin. In humans there is a cytoplasmic acidic chitinase (476aa) that may be important in defence against nematodes, fungi and other pathogens and has a role in the inflammatory response. See also **chitotriosidase**. Plants have a range of chitinases that are important in defence against fungi (e.g. in *Orzya sativa*, Chitinase 1, Class I chitinase α, 323aa).

chitosan A polymer of 1,4-β-D-glucosamine and N-acetyl-D-glucosamine found in the cell wall of some fungi. Can be manufactured by deacetylation of chitin (from crustacean shells) and is sold as a health supplement, although of doubtful value. Is also used as a plant growth enhancer, and boosts the ability of plants to defend themselves against fungal infections.

chitosome Membrane-bound vesicular organelle (40–70 nm diameter) found in many fungi. Contains chitin synthetase that produces chitin microfibrils that are released and incorporated into the cell wall.

chitotriosidase In humans a lysosomal enzyme belonging to the glycosyl hydrolase 18 family, chitinase class II subfamily (chitinase 1, 466aa) that degrades chitin, chitotriose and chitobiose.

chk (1) See **chokh mutant** (2) **Checkpoint kinases**, CHK1 and CHK2. (3) **Choline kinases**. (4) **Carboxyl-terminal Src kinase homologous kinase**.

Chlamydiae A phylum of bacteria, all obligate intracellular parasites. The phylum includes the genus *Chlamydia*: *C. trachomatis* causes trachoma in man and and the sexually transmitted infection, chlamydia; *C. psittaci* causes economically important diseases of poultry. The genome is small, about one-third that of *E. coli*.

Chlamydomonas A genus of unicellular green algae, usually flagellated. Easily grown in the laboratory and have often been used in studies on flagellar function – a range of paralysed flagellar (pf) mutants have been isolated and studied extensively.

chlamydospore A type of survival spore produced by several groups of fungi (e.g. Zygomycota and mitosporic fungi), usually as a response to stress. May be formed at an apical or intercalary position and contains dense cytoplasm and storage compounds. The wall may be darkly pigmented.

chloragosome Cytoplasmic granule of unknown function found in the coelomocytes of annelids.

chloramphenicol An antibiotic from *Streptomycetes venezuelae* that inhibits protein synthesis in prokaryotes and in mitochondria and chloroplasts by acting on the 50S ribosomal subunit. It is relatively toxic but has a wide spectrum of activity against Gram-positive and Gram-negative cocci and bacilli (including anaerobes), Rickettsia, Mycoplasma, and Chlamydia.

chloramphenicol acetyltransferase Enzyme (CAT, EC 2.3.1.28, 219aa in *E. coli*) that inactivates the antibiotic **chloramphenicol** by acetylation. Widely used as a **reporter gene**.

Chlorarachniophytes Amoeboflagellate algae that acquired photosynthesis secondarily by engulfing a green alga and retaining its plastid (chloroplast) which is surrounded by four membranes with a **nucleomorph**, the reduced algal endosymbiont nucleus, between the plastid inner and outer membrane pairs. *Bigelowiella natans* is the model species. Article: http://www.pnas.org/content/100/13/7678.full

Chlorella Genus of green unicellular algae extensively used in studies of photosynthesis.

chlorenchyma Form of **parenchyma** tissue active in photosynthesis, in which the cells contain many **chloroplasts**; found especially in leaf **mesophyll**.

chlorhexidine Antiseptic and disinfectant that is used for dressing minor skin wounds or burns.

chloride channels Ion channels selective for chloride ions. Various types including **ligand-gated** Cl-channels at synapses (the **GABA**- and **glycine**-activated channels), as well as **voltage-gated** Cl-channels found in a variety of plant and animal cells. The voltage-gated channels (ClCs) are produced by a single protein with 10–12 transmembrane domains. The ClC channel family contains both chloride channels and proton-coupled anion transporters that exchange chloride or another anion for protons. ClC-1 (CLCN1, 988aa) is involved in setting and restoring the resting membrane potential of skeletal muscle and is defective in various kinds of **myotonia congenita**. See also **CFTR**, **MDR**. Chloride channels can be blocked by other anions (I^-, Br^-, or SO_4^{2-}), small compounds (e.g. niflumic acid) and some toxins (**chlorotoxin**, **picrotoxin**).

chloride current Flow of chloride ions through chloride-selective **ion channels**.

chloride intracellular channel proteins A family of proteins (CLIC proteins, ~240–250aa) expressed in a wide variety of tissues that may have diverse functions in addition to their role as chloride channels. Channel activity depends on the pH. They are expressed in the nucleus and in cytoplasmic locations. CLIC1 (241aa) is primarily expressed in the nucleus, CLIC4 (253aa) and CLIC5 (410aa) interact with the cortical actin cytoskeleton. Members of the family include **parchorin**, bovine **p64**.

Chlorobi A phylum of obligately anaerobic photoautotrophic bacteria, green sulphur bacteria, which use sulphite as an electron donor in photosynthesis. The 'model organism' is *Chlorobium tepidum*.

chlorocruorin A green respiratory pigment found in some Polychaeta. The prosthetic group is similar, but not identical, to reduced haematin.

Chloroflexi A phylum of bacteria that includes the class Chloroflexi, green non-sulphur bacteria which have green pigment in photosynthetic bodies called chlorosomes. It has recently been suggested that the phylum Thermomicrobia (a group of thermophilic green non-sulphur bacteria) should be considered as a class within the phylum Chloroflexi.

chloronema See **caulonema**.

chlorophyll The photosynthetic pigment, a porphyrin with an associated magnesium atom, closely related to bacteriochlorophylls. Chlorophyll a is ubiquitous, chlorophyll b is found in most plants, chlorophylls c1 and c2 are found in various algae and chlorophyll d is found in Cyanobacteria. Core antenna complexes bind chlorophyll a, while apoproteins of peripheral antenna complexes bind both chlorophyll a and chlorophyll b. During leaf senescence, chlorophyll (chl) is degraded to colourless linear tetrapyrroles, termed nonfluorescent chlorophyll catabolites. In two subsequent reactions catalyzed by **chlorophyllase** and **Mg dechelatase**, respectively, phytol and the central Mg atom are removed. Then the ring structure of pheophorbide (pheide) a is oxygenolytically opened by **pheophorbide a oxygenase** to produce red chlorophyll catabolite (RCC), which is rapidly converted to a primary fluorescent chlorophyll catabolite (pFCC).

chlorophyllases Enzymes that begin the degradation of **chlorophyll**. Chlorophyllase 1 (EC 3.1.1.14, 324aa) hydrolyses an ester bond in chlorophyll a to yield chlorophyllide and phytol and is induced by methyl jasmonate. Chlorophyllase 2 (318aa) is similar in function but constitutively present.

chlorophyllide a oxygenase The chloroplast enzyme (CAO, EC 1.13.12.14, 536aa) that is involved in chlorophyll b biosynthesis, acting specifically on non-esterified chlorophyllide a. Chlorophyll b synthesis is partly regulated on a transcriptional level by the expression of the CAO gene and partly by the stability of the CAO enzyme.

Chlorophyta Green algae. Division of algae containing photosynthetic pigments similar to those in higher plants and having a green colour. Includes unicellular forms, filaments and leaf-like thalluses (e.g. *Ulva*). Some members form **coenobia**, and the **Characean algae** have branched filaments.

chloroplast Photosynthetic organelle of higher plants. Lens-shaped and rather variable in size but approximately 5 μm long. Surrounded by a double membrane and contains circular DNA (though not enough to code for all proteins in the chloroplast). Like the mitochondrion, it is semi-autonomous. It resembles a cyanobacterium from which, on the endosymbiont hypothesis, it might be derived. The photosynthetic pigment, chlorophyll, is associated with the membrane of vesicles (thylakoids) that are stacked to form grana.

chloroquine Antimalarial drug that has the interesting property of increasing the pH within the **lysosome** when added to intact cells in culture. Chloroquine resistance seems to be due to enhanced ABC transporter activity (*Plasmodium falciparum* Chloroquine Resistance Transporter) that pumps chloroquine from the cell. Also used to treat some autoimmune diseases such as rheumatoid arthritis and SLE.

chlorosis Yellowing or bleaching of plant tissues due to the loss of chlorophyll or failure of chlorophyll synthesis. Symptomatic of many plant diseases, also of deficiencies of light or certain nutrients.

chlorosome Elongated membranous vesicles attached to the plasma membrane of green photosynthetic bacteria; contain the light-harvesting antenna complexes of bacteria in the sub-Order Chlorobiineae. Pigments include bacteriochlorophylls and carotenoids.

chlorotoxin A peptide (36aa) in the venom of the scorpion, *Leiurus quinquestriatus* that blocks small-conductance chloride channels of epithelial cells.

chlorpromazine Neuroleptic aliphatic phenothiazine, thought to act primarily as dopamine antagonist, but also antagonist to α-adrenergic, H1 histamine, muscarinic and serotonin receptors. Used clinically as an anti-emetic. Has been shown to alter fibroblast behaviour.

CHMPs Charged multivesicular body proteins (CHMPs), for example charged multivesicular body protein 1b (CHMP1.5, chromatin-modifying protein 1b, vacuolar protein sorting-associated protein 46-2, 199aa). Some (CHMP2, 3, 4 & 6) are components of the **ESCRT** complex, homologues of yeast Class E VPS proteins.

choanocytes Cells (collar cells) that line the radial canals of sponges. Have long flagella that are responsible for generating the feeding current.

choanoflagellates A group of flagellate protozoa that may be ancestors of sponges. Each has a single flagellum, surrounded by a ring of microvilli, forming a cylindrical or conical collar that capture prey drawn in by the water current generated by the flagellum. Usually sessile, being attached at the opposite pole to the flagellum, and in some cases are colonial. Resemble the **choanocytes** of sponges.

chocolate agar A non-selective, enriched growth medium used for growing fastidious respiratory bacteria, such as *Haemophilus influenzae*. It contains red blood cells, lysed by heating very slowly to 56°C, and does not contain chocolate.

chokh mutant Mutant zebrafish that lacks eyes from the earliest stages in development; the *chk* gene encodes the homeodomain-containing transcription factor, Rx3 (Retinal homeobox protein, 292aa) that is important in eye development.

cholangiocarcinoma Malignant tumours derived from the epithelium of the biliary duct system of the liver (**cholangiocytes**). Most (90%) are adenocarcinomas, and the remainder are squamous cell tumours.

cholangiocyte Epithelial cells that line the bile ducts. Form an important transporting epithelium actively involved in the absorption and secretion of water, ions, and solutes. Can give rise to **cholangiocarcinomas**.

cholate The sodium salt of cholic acid, that has strong detergent properties and can replace membrane lipids to generate soluble complexes of membrane proteins.

cholecalcin See **calbindin**.

cholecystokinin Polypeptide hormone (CCK, pancreozymin, 58aa and 33aa) secreted by I-cells of the duodenum. Stimulates secretion of digestive enzymes by the pancreas and contraction of the gall bladder. The C-terminal octapeptide is found in some dorsal root ganglion neurons where it presumably acts as a peptide neurotransmitter. The receptors are G-protein coupled (CCK-AR, 428aa; CCK-BR, 447aa).

cholera An acute bacterial infection by *Vibrio cholerae* that causes severe vomiting and diarrhoea leading to dehydration, often fatal. Many of the effect are the result of **cholera toxin** acting on intestinal epithelial cells and causing hypersecretion of chloride and bicarbonate followed by water.

cholera toxin An **AB toxin** from *Vibrio cholerae*, the bacterium responsible for cholera. The toxic A subunit (EC 2.4.2.36, 258aa) irreversibly activates adenyl cyclase by **ADP-ribosylation** of a G_s protein. The B subunit (124aa) has five identical monomers, binds to GM1 ganglioside and facilitates passage of the A subunit across the cell membrane.

cholesterol The major **sterol** of higher animals and an important component of cell membranes, in which it has a buffering effect on fluidity. Synthesized in the liver (see **HMG-CoA reductase**) and transported in the esterified form by plasma lipoproteins.

cholesterol binding toxins Family of pore-forming toxins ($\sim$500aa) from various genera of bacteria including *Streptococcus, Listeria, Bacillus* and *Clostridium*. Apparently bind to cholesterol and oligomerise to form a pore (20–50 nm depending on concentration of toxin): as a result cause cell lysis and are lethal. See **Streptolysin O**. Other examples include Pneumolysin from *S. pneumoniae*, Cereolysin O from *Bacillus cereus*, Thuringolysin O from *B. thuringiensis*, Tetanolysin from *Clostridium tetani*, Botulinolysin from *C. botulinum*, Perfringolysin O from *C. perfringens*, Listeriolysin O from *L. monocytogenes*.

cholesteryl ester transfer protein A phospholipid transfer protein (lipid transfer protein I, 493aa) that facilitates transport of phospholipids between lipoprotein particles. Deficiencies in the protein do not seem to have particularly deleterious effects. A member of the **BPI/LBP/Plunc superfamily**.

cholic acid A major bile acid with strong detergent properties. It can be used to solubilize membrane proteins.

choline A saturated amine that is esterified in the head group of phospholipids (phosphatidyl choline and sphingomyelin) and acetylated in the neurotransmitter acetylcholine. Otherwise a biological source of methyl groups.

choline acetyltransferase The enzyme that catalyzes the formation of acetylcholine from acetyl CoA and choline (choline acetylase, EC 2.3.1.6, 748aa). Defects cause congenital myasthenic syndrome with episodic apnea.

choline kinases Enzymes (EC 2.7.1.32, CHKa, 457aa; CHKb, 395aa) that catalyze the phosphorylation of choline to phosphocholine in the biosynthesis of phosphatidylcholine (PC).

cholinergic neurons Neurons in which acetylcholine is the neurotransmitter.

chondroblast Embryonic cartilage-producing cell.

chondrocalcin A calcium-binding peptide (246aa) cleaved from alpha-1 type II collagen by **ADAMTS3**, found in developing cartilage and in growth plate cartilage. It appears to play a role in enchondral ossification.

chondrocyte Differentiated cell responsible for secretion of extracellular matrix of **cartilage**.

chondrodysplasia A clinically and genetically diverse group of bone growth disorders. Mutations in various genes cause distinct forms and include those encoding cartilage-derived morphogenetic factor-1, **parathyroid hormone** receptor, collagen COL10A1, mitochondrial RNA-processing endoribonuclease, emopamil-binding protein and **arylsulfatase E**. Rhizomelic chondrodysplasia punctata can be caused by mutations in the PEX7 gene, which encodes the peroxisomal type two targeting signal (PTS2) receptor.

chondroitin sulphates Major components of the extracellular matrix and connective tissue of animals. They are repeating polymers of glucuronic acid and sulphated N-acetyl glucosamine residues that are highly hydrophilic and anionic. Found in association with proteins. 'Chondroitin' is sold as a dietary supplement to relieve osteoarthritis, but the evidence for efficacy is minimal.

chondronectin A 180 kDa homotrimeric protein (subunits 55 kDa) isolated from chick serum that specifically favours attachment of **chondrocytes** to Type II **collagen** if present with the appropriate cartilage **proteoglycan**. Structurally and chemically distinct from fibronectin and laminin. Not in recent literature. Original report: http://www.ncbi.nlm.nih.gov/pmc/articles/PMC1003606/pdf/annrheumd00420-0005.pdf

chondrosarcoma A malignant tumour (sarcoma) derived from chondrocytes.

CHOP A nuclear protein (C/EBP homologous protein 10, DNA damage-inducible transcript 3, GADD153, 169aa) that is a dominant-negative inhibitor of the transcription factors C/EBP and LAP (liver activator protein).

CHORD domain A zinc-binding 60aa domain (cysteine and histidine rich domain) with uniquely spaced cysteine and histidine residues, highly conserved from plants to mammals. CHORD domain-containing protein 1 (cysteine and histidine-rich domain-containing protein 1, chp1, 332aa) may regulate NOD1 (see **NOD proteins**) through an interaction with HSP90AA1. See **melusin**.

chordamesoderm Embryonic mesoderm that gives rise to the **notochord**.

chordin Dorsalizing factor (955aa) that binds to the ventralizing TGFβ family of bone morphogenetic proteins (BMPs) during embryogenesis and antagonises their effects during development. In adults, the differential distribution and regulation of chordin in normal and osteoarthritic cartilage and chondrocytes suggests an involvement in osteoarthritis. Also found in other tissues.

chordoma A rare slow-growing primary bone cancer in the skull and spinal column derived from **notochord** remnants.

chordotonal organs Insect sense organs that are sensitive to pressure, vibrations and sound.

chorea Involuntary repetitive jerky movements of the body. Seen in a number of neurological diseases including Sydenham's chorea (or Saint Vitus's dance) and Huntington's chorea.

chorioallantoic membrane (1) Protective membrane around the eggs of insects and fishes. **(2)** Extra-embryonic membrane surrounding the embryo of amniote vertebrates. The outer epithelial layer of the chorion is derived from the trophoblast, by the apposition of the **allantois** to the inner face of the **chorion**. The chorioallantoic membrane is highly vascularized, and is used experimentally as a site upon which to place pieces of tissue in order to test their invasive capacity.

choriocapillaris The vascular layer immediately adjacent to **Bruch's membrane** in the choroid of the eye. Development: http://www.nature.com/eye/journal/v24/n3/full/eye2009318a.html

choriocarcinoma Malignant tumour of trophoblast.

chorion (1) Protective membrane around the eggs of insects and fishes. **(2)** Extra-embryonic membrane surrounding the embryo of amniote vertebrates. The outer epithelial layer of the chorion is derived from the trophoblast, the inner from somatic mesoderm. See **chorioallantoic membrane**. **(3)** *Cf.* **choroid**.

chorionic gonadotrophin A heterodimeric glycoprotein hormone (chorionic gonadotropin) with an alpha subunit identical to that of **luteinizing hormone** (LH), **follicle-stimulating hormone** (FSH), and **thyroid-stimulating hormone** (TSH) and a unique beta subunit (165aa). Synthesized in the placenta it promotes the maintenance of the **corpus luteum** during the beginning of pregnancy causing it to secrete **progesterone**; will also stimulate **Leydig cells** to synthesize testosterone.

chorionic villus sampling Method for diagnosing human fetal abnormalities in the sixth to tenth week of gestation. Small pieces of fetally derived chorionic villi are removed for chromosomal analysis and, increasingly, for DNA-based testing for disease-associated alleles.

choroid Middle layer of the vertebrate eye, between **retina** and sclera. Well vascularized and also pigmented to throw light back onto the retina (the tapetum is an iridescent layer in the choroid of some eyes). Not to be confused with the **choroid plexus**.

choroid plexus A highly vascularized region of the roof of the ventricles of the vertebrate brain that secretes cerebrospinal fluid.

choroideremia A disorder (tapetochoroidal dystrophy) in which there is degeneration of the choriocapillaris, the retinal pigment epithelium, and the photoreceptors of the eye. It is caused by mutation in the gene encoding **rab** escort protein-1 (653aa) which is involved in membrane trafficking.

chp (1) Calcineurin homologous protein (calcium-binding protein p22, CHP, 195aa) required for constitutive membrane traffic. It inhibits GTPase-stimulated Na^+/H^+ exchange and calcineurin phosphatase activity. **(2) CHORD domain-containing protein 1**, chp1.

Christ-Siemens-Touraine syndrome See **hypohidrosis**.

Christmas disease Congenital deficiency of blood-clotting factor IX (first described in a patient called Stephen Christmas and reported in the Christmas issue of British Medical J., 1952). Inherited in similar sex-linked way to classical haemophilia.

chromaffin cells Cells of the chromaffin tissue in the adrenal medulla that produce either adrenaline or noradrenaline (not both). The **catecholamines** are associated with carrier proteins (chromogranins) in membrane vesicles (chromaffin granules). See **granins**. A chromaffinoma (**phaeochromocytoma**) is a tumour derived from chromaffin cells.

chromagen Any substance that can give rise to a coloured product when appropriately modified by, for example, an enzyme or the products of enzymic activity.

chromatic aberration Distortions in an optical system as a result of different wave-lengths (colours) being brought to a focus at slightly different points. Good microscope objectives are corrected for this at two wave-lengths (achromats) or at three wave-lengths (apochromats), as well as for **spherical aberration**.

chromatid Single chromosome containing only one DNA duplex. Two daughter chromatids become visible at mitotic metaphase, though they are present throughout G2.

chromatin Stainable material of interphase nucleus consisting of nucleic acid and associated histone protein packed into **nucleosomes**. Euchromatin is loosely packed and accessible to RNA polymerases, whereas heterochromatin is highly condensed and probably transcriptionally inactive.

chromatin assembly factor Heterotrimeric protein complex that couples DNA replication to histone deposition *in vitro*, but is not essential for yeast cell proliferation. Depletion of CAF-1 in human cell lines demonstrates that CAF-1 is, however, required for efficient progression through S-phase. Has been used as a marker of the proliferative state because the expression of both CAF-1A and CAF-1B, is massively down-regulated during quiescence in several cell lines. The CAF-1 complex comprises CAF1A (CHAF1A, 956aa), CAF1B (CHAF1B, 559aa) and CAF1C (histone-binding protein RBBP4, **retinoblastoma-binding protein** 4, RBBP4, 425aa) is responsible for assembling histone octamers. During mitosis CAF1B becomes hyperphosphorylated and is displaced into the cytosol and during G- through S and into G2 it is progressively dephosphorylated. N.B. CAF-1 is also the name for CCR4-NOT transcription complex subunit seven.

chromatin body Barr body; condensed X chromosome in female mammalian cells.

chromatography Techniques for separating molecules based on differential absorption and elution. A mixture is passed though a non-mobile matrix (paper in early methods, beads of resin or gel in more modern methods) and different elements are retained for differing times, sometimes very briefly because they are excluded from part of the matrix (size-exclusion chromatography), sometime until the ionic conditions are changed as eluent is passed down the column.

chromatophores (1) Pigment-containing cells of the dermis, particularly in teleosts and amphibians. By controlling the intracellular distribution of pigment granules the animal can blend with the background.; when dispersed the colour is obvious, when condensed the effect is minimised. **Melanocytes** and **melanophores** are melanin-containing chromatophores. (2) Term occasionally used for chloroplasts in the chromophyte algae.

Chromobacterium viscosum Source of lipase B which seems to be identical to *Burkholderia glumae* (*Pseudomonas glumae*) lipase (triacylglycerol hydrolase, EC 3.1.1.3, 358aa). *C. violaceum* produces a purple pigment, violacein, that has antibiotic-like properties. Genus related to *Neisseria*. Reference: http://www.uniprot.org/citations/7786905

chromobindin See **annexin**.

chromoblast An embryonic cell that will differentiate into a **chromatophore**.

chromobox proteins A family of highly conserved adapter proteins characterized by an N-terminal **chromodomain** and a C-terminal chromoshadow domain, separated by a hinge region. They are involved in heterochromatin packaging and are enriched at the centromeres and telomeres. Chromobox protein homologue 1 (CBX1, heterochromatin protein 1 homologue beta, 185aa) binds histone H3 tails that are methylated at Lys-9; chromobox protein homologue 2 (CBX2, 532aa) is a component of the **polycomb repressive complex-1** (PRC1). Chromobox protein homologue 3 (CBX3, 183aa) also binds methylated histone H3 and is involved in the formation of a functional kinetochore through interaction with MIS12 complex proteins. Chromobox protein homologue 4 (CBX4, E3 **SUMO**-protein ligase, 558aa) facilitates SUMO1 conjugation by UBE2I and is a component of PRC1. At least 4 other chromobox protein homologues are known.

chromocentre Condensed heterochromatic region of a chromosome that stains particularly strongly although in the polytene chromosomes of *Drosophila* the chromocentre is of under-replicated heterochromatin and stains lightly.

chromodomains Chromatin Organization Modifier Domains, domains of ~50−60aa found in many proteins that are involved in chromatin remodeling and regulation of the gene expression in eukaryotes. First recognised in the *Drosophila* chromatin proteins HP1 and Polycomb. They are implicated in the recognition of lysine-methylated histone tails and nucleic acids. CHD (for chromo-ATPase/helicase-DNA-binding) proteins regulate ATP-dependent nucleosome assembly and mobilization through their conserved double chromodomains. See **chromoshadow domain**. Chromodomain database: http://www.uib.no/aasland/chromo/Chromo-table.html

chromogranins See **chromaffin tissue** and **granins**.

chromomere Granular region of condensed **chromatin**. Used of chromosomes at leptotene and zygotene stages of meiosis, of the condensed regions at the base of loops on lampbrush chromosomes, and of condensed bands in polytene chromosomes of Diptera.

chromophore The part of a visibly coloured molecule responsible for light absorption over a range of wavelengths thus giving rise to the colour. By extension the term may be applied to UV or IR absorbing parts of molecules. Do not confuse with **chromatophores**.

chromophore-assisted laser inactivation A method of disrupting intracellular elements; the target protein is expressed as a fusion protein with a chromophore (e.g. GFP) and the cellular region of interest illuminated with a tightly-focussed laser beam. The local release of reactive oxygen species inactivates proteins within about 5 nm of the fusion protein.

chromoplast Plant chromatophore filled with red/orange or yellow carotenoid pigment. Responsible for colour of carrots and of many petals.

chromoshadow domain Domain (40–70aa) found in some **chromobox proteins** that mediates dimerization, transcription repression, and interaction with multiple nuclear proteins. Related to the **chromodomain**, and found in proteins that also have a classical chromo domain. Chromodomain-containing proteins can be divided into two classes depending on the presence, e.g. in *Drosophila* **HP1**, or absence, e.g. *Drosophila* **Polycomb** (Pc), of the chromoshadow domain.

chromosomal passenger complex A multiprotein complex that acts as a key regulator of mitosis, ensures correct chromosome alignment and segregation and is required for chromatin-induced microtubule stabilization and spindle assembly. The components are **Aurora B** kinase, inner centromere protein (INCENP), **borealin** and survivin (**BIRC5**).

chromosome A DNA duplex associated with proteins that is part of the DNA complement of the cell.

In eukaryotes the DNA is subdivided into variable numbers of linear chromosomes, presumably for convenience of handling, the proteins include histones; in prokaryotes the chromosome is circular and not associated with histones. In eukaryotes the chromosomes (which are replicated into two chromatids during S phase) become more tightly packed at mitosis and become aligned on the **metaphase plate**. Each chromosome has a characteristic length and banding pattern. See **C banding, G banding**.

chromosome condensation The tight packing of **chromosomes** in **metaphase**, in preparation for **nuclear division**.

chromosome map The sequence of genes along a chromosome, originally based upon the probability of recombination between loci (see **genetic linkage**), but direct mapping is possible with fluorescence *in situ* hybridisation (see **FISH**).

chromosome painting See **fluorescence *in situ* hybridization**.

chromosome segregation The orderly separation of one copy of each chromosome into each daughter cell at **mitosis**.

chromosome synapsis The close apposition of homologous chromosomes before cell division, or permanently in giant **polytene chromosomes**.

chromosome translocation The fusion of part of one chromosome onto part of another. Largely sporadic and random, there are some translocations at 'hot-spots' that occur often enough to be clinically significant. See **Philadelphia chromosome** and **spectral karyotyping**.

chromosome walking A procedure to find and sequence a gene whose approximate position in a chromosome is known by classical genetic linkage studies. Starting with the known sequence of a gene shown by classical genetics to be near to the novel gene, new clones are picked from a **genomic library** by **hybridization** with a short probe generated from the appropriate end of the known sequence. The new clones are then sequenced, new probes generated, and the process repeated until the gene of interest is reached.

chronaxie The shortest time required for excitation of a nerve when the electrical stimulus is twice the threshold intensity required to elicit a response if applied over a prolonged period.

chronic Persistent, long-lasting (as opposed to **acute**). Chronic inflammation is generally a response to a persistant antigenic stimulus.

chronic fatigue syndrome See **myalgic encephalomyelitis**.

chronic granulomatous disease Disease, usually fatal in childhood, in which the production of hydrogen peroxide by **phagocytes** does not occur. Catalase-negative bacteria are not killed and there is no luminol-enhanced **chemiluminescence** when the cells are tested. The absence of the oxygen-dependent killing mechanism is not itself fatal but seriously compromises the primary defence system. At least three separate lesions can cause the syndrome, the commonest being an X-linked defect in plasma membrane cytochrome (see **p91-phox**).

chronic lymphocytic leukaemia See **leukaemia**.

chronic myelogenous leukaemia See **leukaemia**.

chronic wasting disease A transmissible **spongiform encephalopathy** that affects cervids (deer).

chronospecies A species that has diverged from the ancestral form by small continuous changes so that it no longer resembles the original either morphologically or genetically. It represents a form of evolution that is distinct from speciation by divergence in which an isolated subset of a species changes its characteristics to the point at which it cannot interbreed with the other subset and is recognised as a separate species.

Chrysiogenetes A bacterial phylum containing only a single species, *Chrysiogenes arsenatis*, which uses arsenate as its electron donor and can survive in arsenic-contaminated environments. Some selenate-respiring bacteria may also fall within this phylum. Description of selenate-respiring forms: http://www. ncbi.nlm.nih.gov/pubmed/17435005

chrysolaminarin A modified **laminarin** found as a food reserve in Chrysophyceae.

Chrysophyceae A Class of eukaryotic algae in the division Heterokontophyta. Golden-brown in colour due to high levels of the **xanthophyll**, fucoxanthin. Mostly found in fresh water and are single-celled or colonial. Also called Chrysomonadida by protozoologists.

CHUK Kinase (conserved helix-loop-helix ubiquitous kinase, IκB kinase 1, IκBKA, EC 2.7.11.10, 745aa), a component of the **IkappaB**-kinase (**IKK**) core complex which consists of CHUK, IKBKB and IKBKG and is responsible for phosphorylation of IkappaB which dissociates from **NFκB**, allowing **NFκB** to move to the nucleus: the phosphorylated IκB is eventually degraded.

Churg-Strauss syndrome *allergic granulomatosis* **ANCA**-associated vasculitis, affecting mostly small blood vessels, in which there is no complement consumption and no deposition of immune complexes.

CHX (1) Cycloheximide. (2) Chlorhexidine.

chylomicron Colloidal fat globule found in blood or lymph; used to transport fat from the intestine to the liver or to adipose tissue. Has a very low density, a low protein and high triacylglyceride content.

chylomicron retention disease See **Anderson's disease**.

chymase Cathepsin G-like, serine peptidase, (EC 3.4.21.39, mast cell protease, 247aa) secreted by mast cells, that plays an important role in generating angiotensin II in response to injury of vascular tissues, and converts big endothelin 1 to the 31aa peptide **endothelin1**. **chymosin** Aspartic peptidase (EC 3.4.23.4, preprorennin, 381aa) from the abomasum (fourth stomach) of calf (*Bos taurus*) that has properties similar to **pepsin**. Will cleave casein to paracasein and is used in cheesemaking.

chymostatin Low molecular weight peptide-fatty acid compound of microbial origin that inhibits **chymotrypsins** and **papain**.

chymotrypsin A secreted serine peptidase from pancreas that preferentially hydrolyzes Phe, Tyr, or Trp peptide and ester bonds. The chymotrypsin-family of peptidases are all endopeptidases. In humans chymotrypsinogen B (EC 3.4.21.1), is cleaved into chymotrypsin B-chain A (13aa), -chain B (131aa) and -chain C (97aa). Chymotrypsin-C (EC 3.4.21., caldecrin, 268aa) has chymotrypsin-type peptidase activity and hypocalcaemic activity. Chymotrypsin-like elastase family member 3B (EC 3.4.21.70, 270aa) preferentially cleaves Ala-|-Xaa and does not hydrolyze elastin. Various chymotrypsin inhibitors are known from a variety of plants (e.g. subtilisin-chymotrypsin inhibitor-2A, 84aa, from barley) and from the common earthworm (*Lumbricus terrestris* chymotrypsin inhibitor, 86aa).

chytrid Member of the **Chytridiomycota**.

Chytridiomycota The only major group of true (chitin-walled) fungi that produce zoospores. Common as saprotrophs, facultative parasites and obligate parasites in moist soil and freshwater habitats. Chemotaxis of the rumen chytrid *Neocallimastix frontalis* has been extensively studied. *Batrachochytrium dendrobatidis* is capable of causing sporadic deaths in some amphibian populations and 100% mortality in others.

ciboulot G-actin binding protein (isoform A, 129aa; isoform C, 97aa) from *Drosophila*; plays a major role in axonal growth during brain metamorphosis. Has three thymosin-like (WH2) repeats and binds to G-actin in the same way as **profilin** and the complex can add to the barbed but not the pointed end of an F actin filament.

cicatrisation Contraction of fibrous tissue, formed at a wound site, by **fibroblasts**, thereby reducing the size of the wound but causing tissue distortion and disfigurement. Once thought to be due to contraction of collagen but now known to be due to cellular activity.

Cicer arietinum Chickpea, an important grain-legume in tropical and sub-tropical regions. Two main forms are cultivated, desi types (small seeds, angular shape, and coloured seeds with a high percentage of fibre) and kabuli types (large seeds, ram-head shape, beige coloured seeds with a low percentage of fibre). Not to be confused with chickling peas (*Lathyrus sativus*). Article: http://www.biomedcen-tral.com/1471-2229/8/106#B3

ciclosporin *cyclosporine* **BAN** for cyclosporine, a cyclic undecapeptide isolated from *Beauveria nivea* (*Tolypocladium inflatum*), peptide drug used as an immunosuppressive to prevent transplant rejection. It binds to cyclophilin in T-cells and the complex inhibits **calcineurin** and blocks the production of various cytokines, particularly interleukin-2, thus inhibiting cell-mediated immune responses. Can cause renal damage and the long-term consequences of suppressing immune function are not yet clear.

CIG Cold insoluble globulin, an obsolete early name for **fibronectin**.

ciguatoxin A large, heat stable, polyether toxin produced by certain strains of the dinoflagellate *Gambierdiscus toxicus*. The toxin can accumulate in large, predatory fish and is responsible for the poisoning syndrome known as ciguatera; activates neuromuscular sodium channels. More information: http://www.cdc.gov/nczved/divisions/dfbmd/diseases/marine_toxins/

ciliary body The circumferential tissue in the eye that includes the muscles that act on the eye lens to produce accommodation and the arterial circle of the iris. The ciliary body is covered by inner ciliary epithelium, continuous with the pigmented retinal epithelium, and the outer ciliary epithelium which secretes the aqueous humour.

ciliary ganglion Neural crest-derived ganglion acting as relay between parasympathetic neurons of the oculomotor nucleus in the midbrain and the muscles regulating the diameter of the pupil of the eye.

ciliary neurotrophic factor Neurotrophin (CNTF, 200aa) originally characterized as a survival factor for chick ciliary neurons *in vitro*. Subsequently shown to promote the survival of a variety of other neuronal cell types, and to promote the differentiation of bipotential **O-2A progenitor** cells to **type-2 astrocytes** *in vitro*. Now considered to be one of the **IL-6 cytokine family** since it acts through a receptor containing gp130.

ciliary proteome An open source of ciliary and basal body proteomics data. Ciliary Proteome Web Server at http://www.ciliaproteome.org/

ciliata Class of Protozoa all of which have cilia at some stage of the life cycle, and that usually have a meganucleus.

cilium *Pl.* cilia Motile appendage of eukaryotic cells that contains an **axoneme**, a bundle of microtubules arranged in a characteristic fashion with nine outer doublets and a central pair ('9 + 2' arrangement). Active sliding of doublets relative to one another generates curvature, and the asymmetric stroke of the cilium drives fluid in one direction (or the cell in the other direction).

CIN85 An adaptor protein (Cbl-interacting protein of 85 kDa, SH3 domain-containing kinase-binding protein 1, 665aa) involved in different cellular functions including the down-regulation of activated receptor tyrosine kinases and survival of neuronal cells. Cbl (ubiquitin ligase) binds to the activated receptor, further binding of the CIN85/**endophilin** complex leads to clathrin-mediated internalisation. Has three Src homology 3 (SH3) domains, a proline-rich region (PRR), and a coiled-coil domain.

CINC In rat Cinc1 is CXC chemokine-1 (growth-regulated alpha protein, 96aa) which has chemotactic activity for neutrophils and contributes to their activation during inflammation. Cinc2 (CXC-3, 100aa) is the homologue of MIP2, Cinc3 (CXC-2, 101aa) of MIP2α.

CINCA A rare congenital inflammatory disorder (chronic infantile neurologic cutaneous and articular syndrome, neonatal onset multisystem inflammatory disease, NOMID) caused by mutations in **cryopyrin**.

cinchocaine *dibucaine* Long-lasting local anaesthetic.

cingulin Rod-shaped dimeric protein (1197aa) found in cytoplasmic domain of vertebrate tight junctions. Contains globular and coiled-coil domains and interacts *in vitro* with several tight junction and cytoskeletal proteins, including the PDZ protein **ZO-1**. GEF-H1/Lfc, a guanine nucleotide exchange factor for RhoA, directly interacts with cingulin. Cingulin-like protein 1 (CGNL1, paracingulin, 1302aa) may anchor **tight junctions** of the apical junctional complex to the actin cytoskeleton.

cinnamycin See **duramycin**.

CIP/KIP A family (p21, p27, p57) of **cyclin-dependent kinase** inhibitors (CKI) targeting CDK2.

circadian rhythm Regular cycle of behaviour with a period of approximately 24 hours. In most animals the endogenous periodicity, which may be of longer or shorter duration, is entrained to 24 h by environmental cues (zeitgebers), often light (see

cryptochrome). The biochemical basis is complex and involves oscillations in transcription regulated by positive and negative inputs. See **period**, **clock** and **timeless**. Dictionary of circadian physiology: http://www.circadian.org/dictionary.html#N86

circular dichroism Differential absorption of right-hand and left-hand circularly polarized light resulting from molecular asymmetry involving a chromophore group. CD is used to study the conformation of proteins in solution. *Cf.* **linear dichroism**.

circular DNA DNA arranged as a closed circle. This brings serious topological problems for replication that are solved with **DNA topoisomerase**. Characteristic of prokaryotes but also found in mitochondria, chloroplasts and some viral genomes.

cirrus *Pl.* cirri Large motor organelle of hypotrich ciliates: formed from fused **cilia**.

CIS Cytokine-inducible SH2-containing protein (Protein G18, suppressor of cytokine signalling, 258aa) product of a gene activated by cytokine signals, part of the **SOCS** negative feedback system.

***cis*-activation** Activation induced by something on the same cell or on the same chromosome. For example, the activation of a gene by an activator binding upstream on the same chromosome rather than by a diffusible product. *Cis*-activation of adhesion molecules can occur through protein-protein contacts in the plane of the membrane but not if the two molecules are in different cells. *Cf.* **transactivation**.

***cis*-dominance** A term applied to the situation in which a gene or promoter affects only gene activity in the DNA duplex molecule in which it is placed, as opposed to *trans* effects when a gene or promoter on one DNA molecule can affect genes on another DNA molecule. *Cis*-dominance is seen only when the appropriate pair or set of genes are all *cis* to each other.

***cis*-golgi** See **Golgi apparatus**.

***cis*-regulatory modules** Short stretches of DNA that help regulate gene expression in higher eukaryotes. They may be up to 1 megabase away from the genes they regulate and can be located upstream, downstream, and even within their target genes.

***cis-trans* test** The complementation test with two or more interacting genes placed in *cis* and in *trans* relationships to each other. A double mutant genome is used in the *cis* test, the recombination product of the two single mutant genomes used in the *trans* test. If the wild type phenotype is restored by both *cis* and *trans* arrangements it is concluded that the two mutations are in different genes and hence that the phenotype is determined by more than one gene. If the *trans* test is negative and the *cis* positive this means that the two mutations are in the same gene. If both tests are negative then at least one of the

mutations must be dominant. Thus the double test provides a means of fine mapping of genes.

CISD See **CDGSH iron sulphur domain proteins**.

cisplatin Cytotoxic drug used in tumour chemotherapy. Binds to DNA and forms platinum-nitrogen bonds with adjacent guanines.

cisternae Membrane-bounded saccules of the smooth and rough endoplasmic reticulum and **Golgi apparatus**. Operationally might almost be considered as an extra-cytoplasmic compartment since substances in the cisternal space will eventually be released to the exterior.

cisternal stack The dispersed **Golgi** in plants.

cistron A genetic element defined by means of the *cis-trans* **complementation** test for functional allelism; broadly equivalent to the sequence of DNA that codes for one polypeptide chain, including adjacent control regions.

citric acid cycle Also known as **tricarboxylic acid cycle** or Krebs' cycle.

Citrobacter freundii A Gram-negative, facultatively anaerobic, rod-shaped member of the family Enterobacteriaceae; often the cause of significant opportunistic infections. *C. freundii* has also been associated with neonatal meningitis. Another species, *Citrobacter koseri* (formerly *C. diversus*) is responsible for urinary tract infections but can also cause meningitis.

citrulline An α-amino acid not found in proteins. L-citrulline is an intermediate in the urea cycle.

Citrus tristeza virus A virus of the family Closteroviridae, that affects citrus trees and has had a major economic impact.

CJD See **Creutzfeldt-Jakob disease**.

CKIs See **cyclin-dependent kinase inhibitors**.

CLA (1) Conjugated linoleic acid. (2) Cutaneous lymphocyte-associated antigen.

cladistics A method of classifying organisms into groups (taxa) based on 'recency of common descent', i.e. evolutionary relationship. Members of a clade possess shared derived characteristics.

cladogenesis See **anagenesis**.

cladogram A branching diagram (dendrogram) showing the relationships between groups of organisms determined by the methods of **cladistics**.

clamp connection In many **basidiomycete** fungi a short lateral branch of a binucleate cell develops. This is the developing clamp connection. One of the nuclei migrates into it. Both nuclei then undergo simultaneous mitosis so that one end of the cell contains two daughter nuclei from each of the parental nuclei. The nucleus in the branch and the two nuclei are separated off from the centre of the cell by **septa**. A single nucleus remains in the central region. The clamp connection then extends towards and fuses with the central section so that a binucleate cell is reformed.

clans of peptidases In the **MEROPS** peptidase database, 'clans' are composed of peptidases that can be grouped on the basis of evolutionary relationship. Many clans contain peptidases of only a single family (e.g. serine peptidases in Clan SB which contains subtilisin Carlsberg, S8 and sedolisin, S53) but others, designated 'P' have members with mixed catalytic sites (e.g. Clan PC contains both cysteine and serine peptidases).

Clara cells The main epithelial cell type in small airways; play an important physiological role in surfactant production, protection against environmental agents, regulation of inflammatory and immune responses in the respiratory system. See **CCSP**.

clarins A family of small integral membrane glycoproteins (clarin-1, Usher syndrome type-3 protein, 232aa; clarin-2, 232aa; clarin-3, transmembrane protein 12, Usher syndrome type-3A-like protein 1, 226aa) that may have a role in the excitory ribbon synapse junctions between hair cells and cochlear ganglion cells. Clarins belong to the hyperfamily of membrane proteins with 4 transmembrane domains (tetraspanins) that includes **connexins** and **claudins**. There is some sequence similarity to **stargazin**.

CLARP Caspase-like apoptosis regulatory protein (CASP8 and FADD-like apoptosis regulator, 480aa); does not have enzyme activity.

CLASP family A family of proteins (ORBIT/MAST/CLASP family) that are associated with the plus ends of microtubules, where they promote the addition of tubulin subunits to attached kinetochore fibres during mitosis and stabilize microtubules in the vicinity of the plasma membrane during interphase. Human examples are **CLIP**-associated proteins. In plants CLASP protein (1440aa) is involved in both cell division and cell expansion. CLASPs in plants: http://www.plantcell.org/content/19/9/2763.full

class switching Phenomenon that occurs during the maturation of an immune response in which, for example, B cells cease making IgM and begin making IgG that has the same antigen specificity. Switching between other immunoglobulin classes can occur.

clastogen Substance that causes chromosome breakage.

clastogenic Describing a mutagen (clastogen) that causes chromosome breakage. N.B. Clastogenesis has a different meaning in geology.

clathrin Protein composed of three heavy chains (clathrin heavy chain 1, 1675aa; clathrin heavy chain 2, 1640) and three light chains (248aa or 229aa), that forms the basketwork of 'triskelions' around a **coated vesicle**. There are two genes for light chains, each of which can generate two distinct transcripts by tissue-specific alternative splicing. See **coatamer**.

clathrin-associated adaptor protein complexes Adaptor protein complexes (AP-1, AP-2) involved in protein transport via clathrin-coated vesicles in different membrane traffic pathways. AP-1 is a heterotetramer composed of two large adaptins (gamma-type subunit AP1G1, 822aa and beta-type subunit AP1B1, 949aa), a medium adaptin (mu-type AP1M1, or AP1M2, 423aa) and a small adaptin (sigma-type AP1S1, 2 or 3, 158aa). The complex is involved in recruiting clathrin to the membrane, recognising sorting signals in the cytoplasmic domains of transmembrane cargo molecules and in recycling of coated vesicles. AP-2 is functionally similar to AP-1 but with different adaptins: two large adaptins (alpha-type subunit AP2A1, 977aa or AP2A2, 939aa and beta-type subunit AP2B1, 937aa), a medium adaptin (mu-type subunit AP2M1, 435aa) and a small adaptin (sigma-type subunit AP2S1, 142aa). N.B., there is a transcription factor complex **AP2** which is completely different.

clathrin-coated vesicles Class of **coated vesicles** important in the receptor-mediated endocytosis. They also mediate the transport of cargo from the *trans*-Golgi network to the endosomal/lysosomal compartment. *Cf.* **coatamer**.

claudins Integral membrane proteins with four transmembrane domains involved in tight junction structure in epithelial and endothelial cells, a family of 24 members displaying organ- and tissue-specific patterns of expression. Claudin-1 (senescence-associated epithelial membrane protein, 211aa) forms homopolymers and heteropolymers with some other claudins. It is also a co-receptor for hepatitis C entry into liver cells. See **paracellin-1**.

CLAVATA The CLAVATA (*CLV1* and *CLV3*) and *SHOOT MERISTEMLESS* (*STM*) genes specifically regulate shoot meristem development in *Arabidopsis*. CLAVATA1 is a serine/threonine receptor kinase (980aa) that probably binds the secreted peptide MCLV3 (12aa) derived from CLAVATA3 and activates a signal transduction cascade to restrict **WUSCHEL** expression. Article: http://dev.biologists.org/content/122/5/1567.full.pdf

clavulanic acid Compound isolated from *Streptomyces clavuligerusa* that acts as a competitive inhibitor of the beta-lactamases that confer resistance to beta-lactam antibiotics. Often used in combination with **amoxicillin**.

cleavage The early divisions of the fertilized egg to form blastomeres. The cleavage pattern is radial in some phyla, spiral in others.

cleavage factor Im complex Multi-protein complex involved in pre-mRNA 3′-processing. Components include NUDT21/CPSF5 and CPSF6 or CPSF7 (see CPSF.

cleft lip/palate-ectodermal dysplasia syndrome A developmental disorder (Zlotogora-Ogur syndrome, Margarita Island ectodermal dysplasia) caused by mutations in **nectin 1**.

cleistothecium An enclosed **ascocarp** within which are randomly dispersed asci. Spore dispersal requires intervention, as for example by wild pigs breaking open truffles (which are cleistothecia). Cleistothecia of *Aspergillus nidulans* have a dark red colouration and are surrounded by **Hülle cells**. *Cf.* **ascothecium, perithecium**.

Cleland's reagent See **dithiothreitol**.

CLI See **clusterin**.

CLIC proteins See **chloride intracellular channel proteins**.

climacteric A particular stage of fruit ripening, characterized by a surge of respiratory activity, and usually coinciding with full ripeness and flavour in the fruit. Its appearance is hastened by ethylene at low concentrations.

CLIPs Cytoplasmic linker proteins that link endocytic vesicles to microtubules. CLIP-1 (cytoplasmic linker protein-170, CAP-Gly domain-containing linker protein 1, Reed-Sternberg intermediate filament-associated protein, restin, 1438aa) is highly expressed in **Reed-Sternberg cells**. CLIP-1 binds to growing ends of microtubules that have bound plus end-binding proteins such as **EB1** and is part of the microtubule plus end tracking system that regulates microtubule dynamics. CLIP-2 (CAP-Gly domain-containing linker protein 2, CLIP-115, 1046aa) links microtubules to the dendritic lamellar body and may be involved in control of brain-specific organelle translocations. CLIP3 (CLIP-170 related 59 kDa protein, 547aa) is involved in *trans*-Golgi/endosome dynamics. CLIP-associating protein 1 (multiple asters homolog 1, Orbit homolog 1, 1538aa) is a microtubule plus-end tracking protein that promotes the stabilization of dynamic microtubules and is required for the polarization of the cytoplasmic microtubule arrays in migrating cells. It is part of the **CLASP family**. CLIP-associating protein 2 (1294aa) is similar.

CLL Chronic lymphocytic **leukaemia**.

CLN2 A yeast cyclin (545aa) that controls the G1/S (start) transition. See **FAR1**.

CLN3 (1) An evolutionarily conserved multi-pass membrane protein (CLN3, battenin, 438aa) that localizes in lysosomes and/or mitochondria and is mutated in **Batten's disease** (neuronal ceroid lipofuscinosis type 3). The *S. pombe* homologue of CLN3 regulates vacuole homeostasis but the function of battenin is uncertain. (2) In *S. cerevisiae*, a G1/S-specific cyclin (CLN3, 580aa) which may be an upstream activator of the G1 cyclins.

cloche Zebrafish mutation that affects differentiation of endothelial and haematopoietic cells and probably acts upstream of **flk-1**. Full paper: http://dev.biologists.org/content/126/12/2643.full.pdf

clock A gene involved in control of circadian rhythm. It encodes a basic helix-loop-helix transcription factor (circadian locomoter output cycles protein kaput, hCLOCK, EC 2.3.1.48, Class E basic helix-loop-helix protein 8, 846aa) that, as a heterodimer with aryl hydrocarbon receptor nuclear translocator-like protein 2, activates the transcription of *period* in a manner that can be regulated by input from cryptochromes. Period and timeless proteins block clock's ability to activate *period* and *timeless* promoters in a negative feedback loop.

clonal deletion One of the two main hypotheses advanced to explain the absence of autoimmune responses: now generally accepted. Clonal deletion is the programmed death of inappropriately stimulated, auto-reactive, clones of T-cells.

clonal selection The process whereby one or more clones, i.e. cells expressing a particular gene sequence, are selected by naturally-occurring processes from a mixed population. Generally the clonal selection is for general expansion by mitosis, particularly with reference to **B-lymphocytes** where selection with subsequent expansion of clones occurs as a result of antigenic stimulation only of those lymphocytes bearing the appropriate receptors.

clone A propagating population of organisms, either single cell or multicellular, derived from a single progenitor cell. Such organisms should be genetically identical, though mutation events may change this.

cloning The process whereby clones are established. The term covers various manipulations for isolating and establishing clones: in simple systems single cells may be isolated without precise knowledge of their genotype; in other systems (see **gene cloning**) the technique requires partial or complete selection of chosen genotypes; in plants the term refers to natural or artificial vegetative propagation.

cloning vector A plasmid **vector** that can be used to transfer DNA from one cell type to another. Cloning vectors are usually designed to have convenient restriction sites that can be cut to generate sticky ends to which the DNA that is to be cloned can be ligated easily.

Clonorchis sinensis Chinese liver fluke. Can infect man if inadequately cooked fish is eaten and can cause biliary obstruction as a result of liver infestation.

closed meristem Describing a type of root meristem in flowering plants in which there is a distinct meristem for the root cap, as opposed to the situation in open meristems where there is a common meristem for both cap and cortex. Abstract: http://www.jstor.org/pss/2432028

Clostridium Genus of Gram-positive anaerobic spore-forming bacilli commonly found in soil. Many species produce exotoxins of great potency, the best known being **botulinum toxin** from *C. botulinum*, and tetanus toxin from *C. tetani*. Among the toxins produced by *C. perfringens* are **perfringolysin** (theta toxin), an alpha-toxin (Phospholipase C), beta, epsilon and iota-toxins (act on vascular endothelium to cause increased vascular permeability), delta-toxin (a haemolysin), and kappa-toxin (a collagenase). *C. difficile* secretes an enterotoxin (toxin A, 2710aa) that is one of the peptidase C80 family.

CLSM Confocal laser scanning microscopy, the correct term for confocal microscopy which, in practice, depends upon a laser light source to work.

cluster of differentiation antigens See **CD antigens**.

cluster of orthologous group Classification system for genes of orthologous sequence, a derivative of the Bidirectional Best BLAST Hits (BDBH) system. A COG consists of genes from different species that are genome-specific best hits as identified by sequence similarity searches using BLAST The database has some limitations: the same COG number may be assigned to genes of similar yet distinct functions, the same gene may be assigned multiple COG numbers and the classification does not provide any indication of likely function. Database: http://www.ncbi.nlm.nih.gov/COG

clusterin A glycoprotein (aging-associated gene 4 protein, apolipoprotein J, complement cytolysis inhibitor, complement-associated protein SP-40, Ku70-binding protein 1, testosterone-repressed prostate message 2, 449aa) that is cleaved into clusterin alpha (222aa) and beta (205aa) chains. Clusterin is differentially regulated in several patho-physiological processes and invariably induced during apoptosis. Functions as extracellular chaperone.

c-maf One of a family of **b-Zip** transcription factors, product (373aa) of the cellular proto-oncogene homologous to v-*maf*, the musculoaponeurotic fibrosarcoma oncogene. Expressed during development of various organs and tissues, involved in a variety of developmental and cellular differentiation processes and in oncogenesis.

CMC (1) Carboxymethylcellulose, cellulose bearing negative charge and used in ion-exchange chromatography. Also used as a thickening agent. (2) Critical micelle concentration. (3) Cell-mediated cytotoxicity: term applied to the killing of cells by effector T-cells or **NK cells**.

CMD1 (1) In *S. cerevisiae*, the gene encoding calmodulin. (2) Chick homologue of **myoD** (298aa).

CML (1) Cell-mediated lympholysis: the lysis of target cells by T-cells. (2) Chronic myeloid **leukaemia**.

CMV See **cytomegalovirus**.

c-myc tag An **epitope tag** (EQKLISEEDL) derived from the c-myc protein. Supplier datasheet: http://www.biochem.boehringer-mannheim.com/prod_inf/manuals/epitope/epi_toc.htm

CN hydrolases A family of hydrolases involved in a wide variety of non-peptide carbon−nitrogen hydrolysis reactions in plants, animals and bacteria, producing important natural products such as auxin, biotin and precursors of antibiotics. The family includes **vanins**, choloylglycine hydrolase (conjugated bile acid hydrolase, CBAH, EC 3.5.1.24), penicillin acylase (EC 3.5.1.11) and acid ceramidase (EC 3.5.1.23). Database entry: http://pfam.sanger.ac.uk/family?PF02275

CNFs See **cytotoxic necrotising factors**.

Cnidaria Diverse Phylum of diploblastic animals (Coelenterata) with radial or biradial symmetry, that includes Classes Hydrozoa (freshwater polyps, small jellyfish), Scyphozoa (large jellyfish), and Anthozoa (sea anemones and stony corals). They possess a single cavity in the body, the enteron, which has a mouth but no anus. Characteristically the ectoderm has specialized stinging cells (cnidoblasts) containing **nematocysts**.

cnidoblast Developing form of **cnidocyte**.

cnidocyst See **nematocyst**.

cnidocyte Ectodermal cell of Cnidaria (coelenterates) specialized for defence or capturing prey. Each cell has a **nematocyst** that can be replaced once discharged.

CNP (1) C-type natriuretic peptide, the major **natriuretic peptide** in the brain, an important regulator of skeletal growth and that may have a role as a neuromodulator. (2) 2,3-cyclic nucleotide 3-phosphodiesterase, see **CNPase**.

CNPase A marker enzyme for oligodendrocytes (CNP, 2′, 3′-cyclic nucleotide 3′-phosphodiesterase, EC 3.1.4.37, 421aa). There are two isoforms, CNP1 and CNP2, both found abundantly in myelinating

TABLE C4. The codon assignments of the genetic code

First position (5' end)	Second position				Third position (3' end)
	U	C	A	G	
	Phe, F	Ser, S	Tyr, Y	Cys, C	U
U	Phe, F	Ser, S	Tyr, Y	Cys, C	C
	Leu, L	Ser, S	Stop: (ochre)	Stop: (opal)/(Trp)[a]	A
	Leu, L	Ser, S	Stop: (amber)	Trp, W	G
	Leu, L	Pro, P	His, H	Arg, R	U
C	Leu, L	Pro, P	His, H	Arg, R	C
	Leu, L	Pro, P	Gln, Q	Arg, R	A
	Leu, L	Pro, P	Gln, Q	Arg, R	G
	lle, I	Thr, T	Asn, N	Ser. S	U
A	lle, I	Thr, T	Asn, N	Ser, S	C
	lle 1: (Met)[a]	Thr, T	Lys, K	Arg, R: (stop)[a]	A
	Met, M (start)	Thr, T	Lys, K	Arg, R: (stop)[a]	G
	Val, V	Ala, A	Asp, D	Gly, G	U
G	Val, V	Ala, A	Asp, D	Gly, G	C
	Val, V	Ala, A	Glu, E	Gly, G	A
	Val, V: (Met)[b]	Ala, A	Glu, E	Gly, G	G

[a]Unusual codons used in human mitochondria.
[b]Normally codes for valine but can code for methionine to initiate translation from an mRNA chain.

cells and at much lower levels in non-myelinating cells.

CNQX **AMPA receptor** antagonist that does not affect **NMDA receptors**. See **glutamate receptors**.

CNS Standard abbreviation for the central nervous system, the brain and spinal cord.

CNTF See **ciliary neurotrophic factor**.

CNTs Concentrative nucleoside transporters, a family of sodium dependent transporters (SLC28), structurally unrelated to the equilibrative (sodium-independent) transporters (**ENTs**) although both types are important in nucleotide salvage pathways. There are three isoforms in humans, CNT1 (SLC28A1, 649aa), CNT2 (SLC28A2, 658aa) and CNT3 (SLC28A3, 691aa), differentially distributed between tissues. CNT1 operates preferentially on pyrimidines, CNT2 on purines and CNT3 on both.

coacervate Colloidal aggregation containing a mixture of organic compounds. One theory of the evolution of life is that the formation of coacervates in the primaeval soup was a step towards the development of cells.

coactosin Actin-binding protein (146aa) originally isolated from *Dictyostelium* but since found in humans (coactosin-like protein, 142aa). Binds 5-lipoxygenase and F-actin through different sites. Has homology to the ADF/**cofilin** family.

coagulase (1) An enzyme (plasminogen activator, fibrinolysin, EC 3.4.23.48, 312aa) produced by *Yersinia pestis* that converts human Glu-plasminogen to **plasmin**. (2) An enzyme (636aa) produced by some species of *Staphylococcus*; causes clot formation by reacting with prothrombin to form staphylo-thrombin which converts fibrinogen to fibrin.

coagulation factor General name for the blood clotting factors of plasma. See Table F1.

coat protein complex See **coated vesicles**.

coatamer See **coated vesicles**

coated pit First stage in the formation of a **coated vesicle**.

coated vesicles Vesicles formed as an invagination of the plasma membrane (a coated pit) and surrounded by a protein meshwork (composed of coatamers). There are three types of coated vesicles, those in which the coat is of **clathrin** with associated **adaptins**, those with coat protein complex I (COP-I) and those surrounded by COP-II. The COP-I coated vesicles are involved in retrograde transport from the *trans*-Golgi to the *cis*-Golgi and endoplasmic reticulum. The coat is composed of seven equimolar subunits (alpha (alpha-COP, 1224aa; see **xenin**), beta-COP (953aa), beta (906aa), gamma (874aa), delta (archain, 411aa), epsilon (308aa) and zeta (177aa or 210aa)) which are recruited in association with **ADP-ribosylation** factors (ARFs). COP-II vesicles are important for anterograde transport from ER to *cis*-Golgi and the coat consists of the essential proteins Sec23p, Sec24p, Sec13p, Sec31p, Sar1p and Sec16p. Sec24p and its two nonessential homologues Sfb2p and Sfb3p have been suggested to serve in cargo selection. **Sar1p** is involved in generating membrane curvature and vesicle formation. COPII coat proteins are required for direct capture of cargo and SNARE proteins into transport vesicles.

cobalamin See **vitamin B12**.

cobra venom factor See **C3**.

cobratoxin Polypeptide toxin (α-cobratoxin, long neurotoxin 1, neurotoxin 3, 71aa) from *Naja kaouthia*. One of the **α-neurotoxins** (curaremimetics), it binds to nicotinic **acetylcholine receptors** with high affinity.

cocaine An alkaloid derived from the coca plant (*Erythroxylum coca*) that acts to increase extra-neuronal **dopamine** in midbrain by binding to the uptake transporters for dopamine, norepinephrine and serotonin.

co-carcinogens Substances that, though not carcinogenic in their own right, potentiate the activity of a carcinogen. Strictly speaking they differ from **tumour promotors** in requiring to be present concurrently with the carcinogen.

cocci Bacteria with a spherical shape.

coccidiomycosis Infection with the fungus *Coccidioides immitis*. Responsible for chronic infection of cattle, sheep, dogs, cats and certain rodents.

coccidiosis Infection of animals and birds by protozoa of the genera *Eimeria* and *Isospora*, usually affecting the intestinal epithelium and causing enteritis.

Coccolithophores *Coccolithophorids* Single-celled algae, protists and phytoplankton belonging to the Haptophytes. They are spherical (15–100 μm in diameter) and are covered in calcium carbonate plates (platelets or scales) 2–25 μm across called coccoliths and are important microfossils, responsible for the major chalk strata of, for example, SE England. Coccolithophores are almost exclusively marine and are found in large numbers throughout the surface euphotic zone of the ocean where they can be responsible for enormous algal blooms that appear white because of the coccoliths. A well-studied and abundant example is *Emiliania huxleyi*. Emiliana homepage: http://www.noc.soton.ac.uk/soes/staff/tt/eh/

Coccomyxa A genus of algae, in the family Chlorococcaceae. A symbiotic partner in the lichen *Peltigera aphthosa*. Article: http://pcp.oxfordjournals.org/content/5/3/297.abstract

cocculin See **picrotoxin**.

cochlear hair cells Cells of the organ of Corti in the inner ear which have stereocilia (hairs: see **stereovillus**) on their apical surfaces. The mechanical displacement of the stereocilia caused by sound waves is converted into an electrical signal by opening membrane ion channels. Frequency is detected partly by the position of the responding cells in the cochlea. The inner hair cells are responsible for sound perception, the outer hair cells oscillate in response to electrical potential changes induced by the inner hair cells (see **prestin**) and improve the discrimination of different frequencies.

cochlin A secreted extracellular matrix protein (550aa) expressed in the cochlea and vestibule of the inner ear. Mutations cause a form of sensorineural deafness and may contribute to **Meniere's disease**.

Cockayne's syndrome A developmental disorder characterized by slow growth and abnormal development, caused by defects in the DNA **excision repair** system.

co-culture Growth of distinct cell types in a combined culture. In order to get some cells to grow at low (clonal) density it is sometimes helpful to grow them together with a **feeder layer** of **macrophages** or irradiated cells. The mixing of different cell types in culture is otherwise normally avoided, although it is possible that this could prove an informative approach to modelling interactions *in vivo*.

codanin A multipass membrane protein (1227aa) that may be involved in maintaining the integrity of the nuclear envelope. Reported to be cell cycle-regulated and active in the S phase. Mutations cause Type 1 congenital dyserythropoietic anaemia. The *Drosophila* homologue (protein vanaso, 1240aa) is the product of the gene *disks lost*. It is cytoplasmic, regulates cell proliferation and is essential for imaginal disk formation.

codocyte See **target cell**.

codominant Genes in which both alleles of a pair are fully expressed in the heterozygote as, for example, AB blood group in which both A and B antigens are present. Cf. **incomplete dominance**.

codon A triplet of bases in mRNA (and the DNA from which the message was transcribed) that is recognized by anticodons on transfer RNA and specifies an amino acid to be incorporated into a protein sequence. The code is degenerate, i.e. most amino acids have more than one codon. The stop-codon determines the end of a polypeptide. See Table C4.

codon-optimization A strategy that is intended to increase the expression of an engineered gene in an experimental system. The basis for the technique is that the multiple codons for some amino acids are not used with equal frequency and the high-frequency codons are not the same in prokaryotes and eukaryotes (for example). Thus in principle adjusting the codon usage to match the local pattern should optimize expression. Although the evidence for this is patchy there are a number of commercially available systems and algorithms for optimization. Controversial discussion: http://omicsomics.blogspot.com/2009/04/is-codon-optimization-bunk.html

coelenterate Animal of the Phylum **Cnidaria**. Mostly marine, diploblastic and with radial symmetry. Sea anemones and *Hydra* are well known examples.

coelenterazine An imidazolopyrazine derivative which, when oxidized by an appropriate luciferase enzyme, produces carbon dioxide, coelenteramide, and light. Luciferin is coelenterazine disulphate.

coeliac disease *US.* celiac disease. Gluten enteropathy: atrophy of **villi** in small intestine leads to impaired absorption of nutrients. Caused by sensitivity to **gluten** (protein of wheat and rye). Sufferers have serum antibodies to gluten and show delayed hypersensitivity to gluten; the risk factor is ten times greater in HLA-B8 positive individuals.

coelom Body cavity characteristic of most multicellular animals (all coelomates). Arises within the embryonic mesoderm, that is thereby subdivided into **somatic mesoderm** and **splanchnic mesoderm**, and is lined by the mesodermally-derived peritoneum. May be secondarily lost and it is unclear whether it evolved once or several times.

coenobium *Pl.* coenobia. Colony of cells formed by certain green algae, in which little or no specialization of the cells occurs. The cells are often embedded in a mucilaginous matrix. Examples: *Volvox, Pandorina.*

coenocytes Organism that is not subdivided into cells but has many nuclei within a mass of cytoplasm (a syncytium), as for example some fungi and algae, and the acellular slime mould *Physarum.*

coenzyme A term that can be used to mean either a low-molecular weight intermediate that transfers groups between reactions (e.g. NAD) or a catalytically active low-molecular weight component of an enzyme (e.g. haem). The former is sometimes considered a cofactor, in the latter case the coenzyme and apoenzyme together constitute the holoenzyme.

coenzyme A A derivative of adenosine triphosphate and pantothenic acid that can carry acyl groups (usually acetyl) as thioesters. Involved in many metabolic pathways, e.g. citric acid cycle and in fatty acid oxidation.

coenzyme M A compound (2-mercaptoethanesulphonic acid, 2-sulfanylethylsulfonate) involved in the formation of methane from carbon dioxide by methanogenic bacteria. See **methyl-coenzyme M reductase.**

coenzyme Q See **ubiquinone.**

cofactor A compound that is required for the biological activity of a protein. Organic co-factors include **flavin** and **haem**; inorganic cofactors are usually metal ions or iron-sulphur clusters, see **CDGSH iron sulphur domain proteins.** See **coenzyme.**

Coffin-Lowry syndrome A multisystem developmental disorder caused by mutations in the gene for ribosomal S6 kinase, a growth factor-regulated serine/threonine kinase.

cofilin A protein (166aa with muscle and non-muscle isoforms) that reversibly controls actin polymerization, in a pH-sensitive manner. Related to **destrin** and similar to **ADF** (actin depolymerizing factor).

COG See **cluster of orthologous group.**

cohesin domain Carbohydrate-binding domains found within **scaffoldins** that interact with **dockerin domains** and are responsible for maintaining the structural integrity and localisation of **cellulosomes**. Structure: http://www.ncbi.nlm.nih.gov/pubmed/9083107

cohesins Proteins that are highly conserved in eukaryotes and have close homologues in bacteria. They form a multicomponent cohesin complex that holds sister chromatids together until anaphase, is required for efficient repair of damaged DNA and has important functions in regulating gene expression in both proliferating and post-mitotic cells. Two cohesin core subunits, Smc1 and Smc3, are members of the structural maintenance of chromosomes family (see **SMC proteins**), connected by the Scc1 subunit (a kleisin) and a number of other cohesin-associated proteins. See **Cornelia de Lange syndrome**. Paper: http://genesdev.cshlp.org/content/22/22/3089.full.pdf+html

coil In describing the secondary structure of a protein, a motif that does not fit into standard classifications such as an **alpha-helix**.

coiled body An ubiquitous nuclear organelle (Cajal body) containing the mRNA splicing machinery (U1, U2, U4, U5, U6 and U7 snRNAs), the U3 and U8 snRNPs that are involved in pre-rRNA processing, together with nucleolar proteins such as **fibrillarin** and **coilin**. Ultrastructurally they appear to consist of a tangle of coiled threads and are spherical, between 0.5–1 μm in diameter. During mitosis, coiled bodies disassemble, coinciding with a mitotic-specific phosphorylation of p80 coilin.

coilin Protein (p80-coilin, 576aa) found in **coiled bodies**. A relatively short portion of the N-terminus seems to target the protein to the organelle.

co-isogenic A strain of animal that differs from others of the same inbred strain at only one locus.

Col-V A plasmid of *E. coli* that codes for **colicin V**, that confers resistance to complement-mediated killing, for a **siderophore** to scavenge iron, and for F-like pili that permit conjugation (see **sex pili**).

colanic acid An exopolysaccharide (M antigen), produced by many enterobacteria. It is a heteropolysaccharide containing a repeat unit with D-glucose, L-fucose, D-galactose, and D-glucuronate sugars that are nonstoichiometrically decorated with O-acetyl and pyruvate side chains. May serve as a protective barrier for pathogenic bacteria in the intestine.

colcemid Methylated derivative of **colchicine**.

colchicine Alkaloid (400 Da) isolated from the Autumn crocus (*Colchicum autumnale*) that blocks microtubule assembly by binding to the **tubulin** heterodimer (but not to tubulin). As a result of interfering with microtubule reassembly will block mitosis at **metaphase**.

cold agglutinins IgM antibodies that react with blood groups I and i in humans (precursors of the ABH and Lewis blood group substances) and agglutinate erythrocytes more actively below 32°C. Cause Raynaud's disease.

cold insoluble globulin *CIG* Name, now obsolete, originally given to fibronectin prepared from **cryoprecipitate**.

Coleoptera Beetles. An Order of Insecta with the the fore-wings (elytra) thickened and chitinized. One of the largest of insect groups – which led JBS Haldane to remark "that God (if he existed) must have had an inordinate fondness for beetles".

coleoptericin Inducible antibacterial peptide found in the haemolymph of a tenebrionid beetle following the injection of heat-killed bacteria. Peptide A (glycine-rich, 74aa) is active against Gram-negative bacteria; Peptides B and C are isoforms of a 43aa cysteine-rich peptide that has sequence homology with **defensins** and is active against Gram-positive bacteria. See **diptericins**, **cecropins**, **apidaecins**, **abaecins**.

coleoptile Closed hollow cylinder or sheath of leaf-like tissue surrounding and protecting the plumule (shoot axis and young leaves) in grass seedlings.

coleorhiza Closed hollow cylinder or sheath of leaf-like tissue surrounding and protecting the radicle (young root) in grass seedlings.

colforsin See **forskolin**.

colicins Bacterial exotoxins (**bacteriocins**) from coliforms that affect other bacteria. Colicins E2 and E3 are **AB toxins** with DNAase and RNAase activity respectively. Most other colicins are channel-forming transmembrane peptides. Coded on plasmids which can be transferred at **conjugation**.

coliform Gram-negative rod-shaped bacillus. **(1)** May be used loosely of any rod-shaped bacterium. **(2)** Any Gram-negative enteric bacillus. **(3)** More specifically, bacteria of the genera *Klebsiella* or *Escherichia*.

coliform test Test for **coliforms** in water samples. Commonly involves passage of a volume of water through a sterile membrane, trapping bacteria, followed by incubation of the membrane on eosin-methylene blue agar which is selective for coliforms.

colipase An essential cofactor for pancreatic triglyceride lipase that anchors the enzyme-colipase complex to the lipid-water interface. **Procolipase** (112aa) is processed to release **enterostatin** and the mature peptide which forms a 1:1 complex with lipase.

colistin A polypeptide antibiotic (polymyxin E) produced by *Bacillus polymyxa* var. *colistinus*. Act against most Gram-negative bacilli.

collagen Major structural proteins of extracellular matrix, unusual both in amino acid composition (very rich in glycine (30%), proline, **hydroxyproline**, lysine, and **hydroxylysine**; no tyrosine or tryptophan), structure (a triple helical arrangement of non-identical polypeptides giving a **tropocollagen** molecule, dimensions 300 nm × 0.5 nm), and resistance to peptidases. Most types are fibril-forming with characteristic quarter-stagger overlap between molecules producing an excellent tension-resisting fibrillar structure. Type I collagen is trimeric (one alpha-2(I) (COL1A2) and two alpha-1(I) (COL1A1) chains; 1366aa and 1464aa respectively) and forms the fibrils of tendon, ligaments and bones. Type II collagen (COL2A1, 1487aa) is a homotrimer of alpha 1(II) collagen (chondrocalcin) specific for cartilaginous tissues and defects are the cause of a variety of chondrodysplasias. Type III collagen is found in soft connective tissues and is composed of three identical alpha-1(III) (COL3A1, 1466aa) chains. Collagen type IV does not form fibrils and is characteristic of **basal lamina**. Collagen type V is fibril forming, Collagen type VI acts as a cell-binding protein. Many other forms are known, often with restricted tissue distribution, some glycosylated (glucose-galactose dimer on the hydroxylysine), and nearly all types can be crosslinked through lysine side-chains. Defects in post-translational processing of collagen can arise if there is vitamin C deficiency, causing **scurvy**. Mutation in various collagen genes underly a wide range of pathological conditions. See **FACIT collagens**.

collagenase Proteolytic enzyme capable of breaking native collagen. Once the initial cleavage is made, less specific peptidases will complete the degradation. Collagenases from mammalian cells (EC3.4.24.7) are metallo-enzymes and are collagen-type specific. May be released in latent (proenzyme) form into tissues and require activation by other peptidases before they will degrade fibrillar matrix. Bacterial collagenases (EC 3.4.24.3) are used in tissue disruption for cell harvesting.

collapsin response-mediator proteins
Family of cytosolic phosphoproteins important in neuronal morphogenesis. They are involved in the signal transduction of class 3 semaphorins (of which **collapsin** is one) that cause growth cone collapse and have a role in axon guidance, invasive growth and cell migration. Collapsin response-mediator protein-1 (CRMP-1, dihydropyrimidinase-related protein 1, unc-33-like phosphoprotein 3, 572aa) forms homotetramers or heterotetramers with other CRMPs. CRMP-2 enhances the advance of growth cones by regulating microtubule assembly and Numb-mediated endocytosis but when phosphorylated by rho kinase no longer binds to the tubulin dimer. In *C. elegans* unc-33 controls the guidance and outgrowth of neuronal cells. Although the CRMPs form tetramers like liver **dihydropyrimidinase** (DHPase) and share sequence similarity they lack the residues needed to bind the metal cofactor and do not seem to have enzyme activity.

collapsins A family of **semaphorins** from chick brain (e.g. collapsin-1, semaphorin-3A, 772aa; collapsin-5, semaphorin-3E, 785aa) that act as a repulsive cue in development and inhibit regeneration of mature neurons. Cause the collapse of the nerve growth cone at picomolar concentrations. Bind to **neuropilin**. See **collapsin response-mediator proteins**.

collar cell See **choanocyte**.

collectins Family of C-type **lectins** (collagenous lectins) important in the first-line defence against viruses and in opsonizing yeasts and bacteria. In humans there are three collectins: collectin-10 (collectin liver protein 1, 277aa) that binds: galactose > mannose = fucose > N-acetylglucosamine > N-acetylgalactosamine; collectin-11 (collectin kidney protein 1, 271aa) that binds to LPS and binds fucose > mannose but does not bind to glucose, N-acetylglucosamine and N-acetylgalactosamine; collectin-12 (collectin placenta protein 1, scavenger receptor class A member 4, 742aa) that promotes binding and phagocytosis of Gram-positive and Gram-negative bacteria and yeast. The family, which

is structurally similar to complement C1q, includes **surfactant proteins** A and D, **mannan-binding protein** and **conglutinin**.

collectrin A protein (transmembrane protein 27, 222aa) expressed in pancreatic beta cells and renal proximal tubular and collecting duct cells. Regulates the function of the SNARE complex and stimulates beta cell replication.

collenchymas Plant tissue in which the **primary cell walls** are thickened, especially at the cell corners. Acts as a supporting tissue in growing shoots, leaves and petioles. Often arranged in cortical 'ribs', as seen prominently in celery and rhubarb petioles. **Lignin** and **secondary walls** are absent; the cells are living and able to grow.

collenocytes Stellate cells with long thin processes that ramify through the inhalent canal system of sponges.

colliculus A small mound or elevation. The superior and inferior colliculi are elevations on the dorsal surface of the mid-brain.

colligative properties Properties that depend upon the numbers of molecules present in solution rather than their chemical characteristics.

collimating lens Lens that produces a non-divergent beam of light or other electromagnetic radiation. Simpler collimators involve slits. Essential in obtaining good illumination in microscopy and for many measuring instruments.

colloblast Specialized mushroom-shaped cells in the outer layer of the epidermis of ctenophores that contain adhesive which can be released to capture prey. More detail: http://www.ucmp.berkeley.edu/cnidaria/ctenophora.html

collodion Cellulose tetranitrate dissolved in a mixture of ethanol and ethoxyethane (1:7); the solution is used for coating materials and medically for sealing wounds and dressings.

collybistin A neurospecific rho-GTPase guanine nucleotide exchange factor (ARHGEF9, 516aa) of the **Dbl family** that is specific for Cdc42. Promotes the formation of **gephyrin** clusters and is negatively regulated by gephyrin. Mutations cause one form of **startle disease**.

colonization factors Bacterial surface features that facilitate adhesion to gut epithelial cells, for example the pili on enteropathogenic forms of *E. coli* that probably bind to GM1 gangliosides. Colonization factor antigens may be plasmid coded, are essential for pathogenicity and are strain-specific, for example K88 (diarrhoea in piglets), CFAI and CFAII on strains causing similar disease in man.

colony-forming unit (1) A stem cell from bone marrow that forms a colony, originally in the spleen of an irradiated mouse when the immune system is reconstituted by the injection of cells from another animal or now more commonly, in methylcellulose-thickened growth medium *in vitro*. CFU-S are pluripotent and form colonies in the spleen, CFU-E are partially committed and only produce cells of the erythroid lineage. CFU-G give rise to granulocytes, CFU-GM to granulocytes and macrophages, CFU-M to macrophages, CFU-GEMM to granulocytes, erythrocytes, macrophages and megakaryocytes. See **burst forming unit**. (2) A measure of the numbers of live organisms in a diluted sample put into culture (e.g. a nutrient agar plate for bacteria) or of viruses plated onto a culture of host cells.

colony-stimulating factor Cytokines involved in the maturation of various leucocyte, macrophage and monocyte lines. See **CSF-1**.

Colostrinin See **proline-rich polypeptide**.

colostrum The first milk secreted by an animal coming into lactation. May be especially rich in maternal lymphocytes and Ig and thus transfer immunity passively.

colour blindness The inability to distinguish the full spectrum of colours. Most often the defect is in the absorption spectrum of one of the three **opsins** in **retinal cones**. In protanopia, the defect is in the red-sensitive opsin (364aa), in deuteranopia (Daltonism) in the green-sensitive opsin (deutan, 364aa), in tritanopia the defect is in blue-sensitive opsin (348aa). Most of the red-green colour-vision defects (which are the commonest) arise as a result of unequal crossing-over between the red and green pigment genes. A defect in the alpha-subunit of the cone photoreceptor cGMP-gated cation channel can cause complete or partial colour blindness (achromatopsia) and although cones are present they are dysfuntional. In dichromatism only two of the three primary colours are recognised, probably the norm for most mammals other than primates.

columella *Pl.* columellae. (1) In general, a small 'column like' part, often forming the central axis of a structure. (2) In molluscs, the central portion of a gastropod shell. (3) The central sterile portion of the sporangium in various fungi, sometimes the central axis of fruits and cones. (4) The central portion of the root cap in higher plants. (5) The fleshy external end of the nasal septum in mammals.

columnar cells Cells of columnar epithelium in which the area in contact with the basal lamina is less than the lateral cell-cell contact (in contrast to cuboidal and squamous epithelia).

comb plates Large flat organelles formed by the fusion of many cilia. Vertical rows of comb-plates form the motile appendages of Ctenophores.

combination therapy Treatment of a disease with two or more drugs, classically used with several antibiotics in tuberculosis and now in cancer and AIDS.

combinatorial chemistry Method by which large numbers of compounds can be made, usually utilizing solid-phase synthesis. In the simplest form carrier beads would be treated separately so as to couple subunits A, B, & C, mixed, re-divided, and then subunits A, B & C added in the three reaction mixtures. Thus bead + AA, AB, AC, BA, BB, BC etc. would have been synthesized − though the products are mixed and deconvolution of an active mixture will be necessary to identify the active molecule. Using a relatively small number of reactions enormous diversity can rapidly be generated. Increasingly the term is used loosely for any procedure that generates highly diverse sets of compounds − the more recent tendency is to prefer high speed parallel synthesis in which each reaction chamber contains only one compound.

combined immunodeficiency Congenital immunodeficiency with thymic agenesis, lymphocyte depletion and hypogammaglobulinaemia: both cellular and humoral immune systems are affected, and life expectancy is low unless marrow transplantation is successful. See **severe combined immunodeficiency disease**.

combining site Any region of a molecule that binds or reacts with a given compound. Especially of the region of immunoglobulin that combines with the determinant of an appropriate antigen.

combretastatins Compounds originally derived from the bark of the African bush willow tree (*Combretum caffrum*), used in cancer treatment. They act by inhibiting tumour vascularisation by binding to tubulin in endothelial cells. Combretastatin A-4 is the most potent natural form but synthetic variants are emerging with slightly different efficacy. Details: http://www.chm.bris.ac.uk/motm/combretastatin/combv.htm

comC, comD, comE Operon involved in regulating competence for genetic transformation in *Streptococcus pneumoniae*. Com D, a transmembrane histidine kinase, binds **competence stimulating peptide** (CSP-1 or CSP-2, encoded by *comC1* and *comC2* respectively) and activates the response regulator ComE. Allelic variation in ComD determines the **pherotype** − whether CSP-1 or CSP-2 elicits a response.

comet assay Single cell gel (SCG) electrophoresis, a rapid and very sensitive fluorescent microscopic method to examine DNA damage and repair at individual cell level. Single cells embedded in agarose gel on a microscope slide, are lysed and DNA allowed to unwind before being subjected to an electophoretic field. Following staining with acridine orange the image is of a distinct head, comprising intact DNA and a tail, consisting of damaged or broken pieces of DNA. By adjusting the lysis and unwinding conditions the sensitivity can be adjusted to preferentially identify particular types of damage.

comitin In *Dictyostelium* an actin binding protein (185aa) with a bulb-type lectin domain that has high affinity for both G-actin and F-actin. Binds to vesicle membranes via mannose residues and links these membranes to the cytoskeleton.

COMMD A multifunctional protein (copper metabolism MURR1 domain-containing protein 1, 186aa) that inhibits NFκB and has a copper metabolism (MURR1) gene domain. Mutation may cause copper deficiency (**Menkes' disease**) or copper accumulation (**Wilson's disease**). There is a family of conserved proteins (COMM domain-containing 1−10) that form multimeric complexes with COMMD1.

committed cells Cells that have become committed to a particular pathway of differentiation (see **determination**). This happens at different stages in embryogenesis; generally this is an irreversible event in mammalian cells although reversing such committment would potentially allow regeneration of tissues.

common source epidemic Infection or intoxication of a large number of people from a contaminated single source such as food or water. Distinguished from a host to host infection where a single infected individual infects one or more susceptible people who in turn pass on the disease.

communicating junction See **gap junction**.

compaction Process that occurs during the morula stage of embryogenesis in which blastomeres increase their cell-cell contact area and develop gap junctions. In mice compaction occurs at the eight cell stage and after this the developmental fate of each cell becomes restricted.

companion cell Relatively small plant cell, metabolically very active and with little or no vacuole, found adjacent to a phloem **sieve tube** and originating with the latter from a common mother cell. Thought to be involved in translocation of sugars in and out of the sieve tube. Intermediary cells and transfer cells are variants.

compartment Conceptualized part of the body (organs, tissues, cells, or fluids) considered as an independent system for purposes of modelling and assessment of distribution and clearance of a substance or in development as a clonal territory. In the insect wing, for example, there are two compartments, anterior and posterior, each containing several clones, but clones do not cross the boundary. It seems from studies with homeotic mutants that cells

in different compartments are expressing different sets of genes. The evidence for such developmental compartments in vertebrates is sparse at present.

compartmental analysis Mathematical process leading to a model of transport of a substance in terms of **compartments** and rate constants for input and output.

competence stimulating peptide Pheromone (CSP-1, 41aa; CSP-2, 41aa) that is involved in a **quorum-sensing** mechanism that regulates competence for genetic transformation in *Streptococcus pneumoniae* and other species in that genus. Encoded by the *comC* gene. Receptor system is a **two-component** signal transduction system, ComD-ComE (TCS12). See **comC**.

competent cells (1) Bacterial cells with enhanced ability to take up exogenous DNA and thus to be transformed. Competence can arise naturally in some bacteria (*Pneumococcus, Bacillus* and *Haemophilus* spp); a similar state can be induced in *E. coli* by treatment with calcium chloride. Once competence has been induced the cells can be stored at low temperature in cryoprotectant and used when needed. (2) Cells capable of responding to an inducer in embryonic development.

competitive inhibition Inhibitor that occupies the active site of an enzyme or the binding site of a receptor and prevents the normal substrate or ligand from binding. At sufficiently high concentration of the normal ligand inhibition is lost: the K_m is altered by the competitive inhibitor, but the V_{max} remains the same.

complement A heat-labile system of enzymes in plasma associated with response to injury. Activation of the complement cascade occurs through two convergent pathways. In the classical pathway the formation of antibody/antigen complexes leads to binding of **C1**, the release of active esterase that activates **C4** and **C2** that in turn bind to the surface. The C42 complex splits **C3** to produce C3b, an opsonin, and C3a (anaphylatoxin). C423b acts on **C5** to release C5a (anaphylatoxin and chemotactic factor) leaving C5b that combines with C6789 to form a cytolytic membrane attack complex. In the alternate pathway C3 cleavage occurs without the involvement of C142, and can be activated by IgA, endotoxin, or polysaccharide-rich surfaces (e.g. yeast cell wall, zymosan). **Factor B** combines with C3b to form a C3 convertase that is stabilized by Factor P (**properdin**), generating a positive feedback loop. The alternate pathway is presumably the ancestral one upon which the sophistication of antibody recognition has been superimposed in the classical pathway. The enzymatic cascade amplifies the response, leads to the activation and recruitment of leucocytes, increases phagocytosis and induces killing directly. It is subject to various complex feedback controls that terminate the response (see **Factor H, Factor I**). A third pathway, the lectin-mediated pathway is similar to the classical pathway but is initiated by the binding of **mannose-binding lectin** and ficolins, instead of C1q (the recognition subunit of C1). There are separate entries for most of the components: see also **decay accelerating factor** and **complement regulatory protein**. Summary: http://d3jonline.tripod.com/07-Immunobiology/The_Complement_System.htm

complement cytolysis inhibitor See **clusterin**.

complement fixation Binding of **complement** as a result of its interaction with immune complexes (the classical pathway) or particular surfaces (alternative pathway).

complement receptors Receptors, mostly on myeloid cells, that bind different complement components or fragments. CR1 (complement receptor 1, C3b/C4b receptor, CD35, 2039aa) is involved in the phagocytosis of opsonized bacteria and the uptake of immune complexes. CR2 (complement C3d receptor, CD21, 1033aa) is also present on lymphocytes and is the site to which the **Epstein-Barr virus** binds. CR3 (CD11b/CD18, MAC-1, integrin alpha-M, 1152aa and integrin beta-2, 769aa) binds C3bi (iC3b). CR4 (integrin alpha-X, beta-2, CD11c/CD18: CD11c is 1163aa) binds C3dg, the fragment that remains when C3b is cleaved to C3bi, and fibrinogen.

complement regulatory protein In mouse a regulator of the complement system (crry, p65, 483aa) that acts as a cofactor for complement **factor I** and as a **decay accelerating factor**. In early embryonic development is important in maintaining feto-maternal tolerance.

complementarity determining region Hypervariable region within the antigen binding site of immunoglobulin molecules and T-cell antigen receptors. The sequence in this region determines which antigen (epitope) will bind. Also used to refer to the genomic sequence encoding the hypervariable regions.

complementary base pairs The bases in DNA that interact with one another and cause the two strands to be complementary: The pairings are guanine (G) with cytosine (c) and adenine (A) with thymine (T) (or uracil in RNA). The G-C pair has three hydrogen bonds, the A-T pair has only two.

complementary DNA See **cDNA**.

complementation The ability of a mutant chromosome to restore normal function to a cell that has a mutation in the homologous chromosome when a hybrid or heterokaryon is formed – the explanation being that the mutations are in different cistrons and

between the two a complete set of normal information is present.

complexins Cytosolic proteins involved in the regulation of neurotransmitter release, conserved **SNARE**-binding proteins that compete with alpha-**SNAP** for binding to **synaptobrevin**. Complexin I (synaphin-2, 134aa) is a marker of inhibitory synapses and is also involved in glucose-induced secretion of insulin by pancreatic beta-cells. Complexin II (synaphin-1, 134aa) mainly labels excitatory synapses and is also involved in mast cell exocytosis. Complexin-3 (158aa) and complexin-4 (160aa) regulate a late step in synaptic vesicle exocytosis.

Compton-North congenital myopathy A serious myopathy caused by mutation in **contactin-1** with secondary loss of beta2-syntrophin and alpha-dystrobrevin from the sarcolemma.

COMT Enzyme (catechol-O-methyltransferase, EC 2.1.1.6, 271aa) thought to functionally modulate dopaminergic neurons. Catalyses reaction between S-adenosyl methionine and catechol to form S-adenosyyl-L-homocysteine and guiacol and methylates catecholamines, thereby terminating their signalling activity. There are two functional polymorphisms in humans with high enzyme activity (COMT Val) and low enzyme activity (COMT Met) variants, although the significance is still unclear.

Con A See **concanavalin A**.

Con A binding sites See **Con A receptors**.

Con A receptors A common misuse of the term receptor. Con A binds to the mannose residues of many different glycoproteins and glycolipids and the binding is therefore not to a specific site. It could be argued that the receptor is Con A and cells have Con A ligands on their surfaces: certainly this would be less confusing.

conalbumin *ovotransferrin* Non-haem iron-binding protein (705aa) found in chicken plasma and egg white.

conantokins Class of small peptides (17–21aa) from cone shells (*Conus* spp.) that inhibit the **NMDA** class of glutamate receptors.

conarachin A seed storage protein (662aa) from *Arachis hypogaea* (Peanut).

concanamycin A *folimycin* Specific inhibitor of vacuolar H$^+$-ATPase, isolated from *Streptomyces* sp.; inhibits perforin-based cytotoxic activity, mostly due to accelerated degradation of perforin by an increase in the pH of lytic granules. Other concanamycins have been isolated with slightly different characteristics.

concanavalin A *Con A* A **lectin** isolated from the jack bean, *Canavalia ensiformis*. See Table L1 (Lectins).

concatamer Two or more identical linear molecular units covalently linked in tandem. Especially used of nucleic acid molecules and of units in artificial polymers.

condensation (1) Process of compression or increase in density. Chromosome condensation is a consequence of increased supercoiling that causes the chromosome to become shorter and thicker and thus visible in the light microscope. (2) A condensation reaction in chemistry is one in which two molecules combine to form a single larger molecule with the concomitant loss of a relatively small portion as water or similar small molecule. (3) The product of the process, for example water droplets that have formed on a cold surface exposed to water vapour.

condensin complex A multimolecular complex consisting of **kleisin**, structural maintenance of chromosomes proteins (**SMC proteins**) and three regulatory proteins (NCAPH/BRRN1 (Barren homolog protein 1), NCAPD2/CAPD2 and NCAPG). During interphase the complex is mostly cytoplasmic and to a minor extent, is associated with chromatin. Most of the complex becomes associated with chromatin during mitosis, providing structural support and rigidity. It dissociates from chomatin in late telophase.

condensing vacuole Vacuole (prozymogen granule) formed from the *cis* face of the Golgi apparatus by the fusion of smaller vacuoles. Within the condensing vacuole the contents are concentrated and may become semi-crystalline (**zymogen granules** or **secretory vesicles**).

conditional mutation A mutation that is only expressed under certain environmental conditions, for example temperature-sensitive mutants.

conditioned medium Cell culture medium that has already been partially used by cells. Although depleted of some components, it is enriched with cell-derived material, probably including small amounts of growth factors; such cell-conditioned medium will support the growth of cells at much lower density and, mixed with some fresh medium, is therefore useful in **cloning**.

cone cell See **retinal cone**.

cone dystrophy A retinal dystrophy in which there is progressive loss of retinal cones but the rods are unaffected. One form is caused by mutation in **guanylate cyclase-activating protein**, some other forms by mutation in the alpha or gamma subunits of cone-specific cGMP-phosphodiesterase or in the KCNV2 potassium channel. See **cone-rod dystrophy**.

cone-rod dystrophy A dystrophy in which there is loss of retinal cones followed by progressive loss of retinal rods. A range of mutations are known to be responsible, including those in the cone-rod homeobox-containing gene (*CRX* gene), the genes for a retina-specific ABC transporter (ABCA4), a membrane-associated phosphatidylinositol transfer protein (PITPNM3, 974aa), a guanylate cyclase (GUCY2D, EC 4.6.1.2, 1123aa) of the rod outer segment membrane, rab3A-interacting molecule-1 (RIM1, 1692aa), **semaphorin 4a**, the retina and anterior neural fold homeobox-like protein (RAXL1, 184aa). Other forms involve defects in **ADAM9**, **prominin-1**, **protocadherin-21** or the retinitis pigmentosa GTPase regulator-interacting protein (RPGRIP1, 1286aa). Loci for other forms of CORD are mapped. See **cone dystrophy**.

confluent culture A cell culture in which all the cells are in contact and thus the entire surface of the culture vessel is covered. It is also often used with the implication that the cells have also reached their maximum density, though confluence does not necessarily mean that division will cease or that the population will not increase in size.

confocal microscopy A system of (usually) **epifluorescence** light microscopy in which a fine laser beam of light is scanned over the object through the objective lens. The technique is particularly good at rejecting light from outside the plane of focus, and so produces higher effective resolution than is normally achieved. There are various ways of achieving the scanning, the original method involved mechanical movement of the stage, other methods involve spinning disks (Nipkow disk), acousto-optic deflectors or programmable arrays and achieve faster imaging (shorter exposure times). Detailed description: http://www.physics.emory.edu/~weeks/lab/papers/ebbe05.pdf

conformational change Alteration in the shape, usually the tertiary structure of a protein, as a result of alteration in the environment (pH, temperature, ionic strength) or the binding of a ligand (to a receptor) or binding of substrate (to an enzyme).

conformer A molecule that has the same structural formula but different conformation. Some conformers are energetically more favourable than others.

congenic Describing organisms that differ in **genotype** at (ideally) one specified locus. Strictly speaking these are conisogenics. Thus one homozygous strain can be spoken of as being congenic to another.

congenital dyserythropoietic anaemia A group of disorders affecting erythroid differentiation. One form is caused by mutation in the gene for **codanin 1**.

congenital insensitivity to pain A condition in which there is an absence of the sensation of pain, either because of mutation in the SCN9A gene encoding a **voltage-gated sodium channel** or damage to or absence of the relevant nerves as in congenital insensitivity to pain with anhidrosis (CIPA) where there is a defect in the neurotrophic tyrosine kinase-1 receptor (a **nerve growth factor** receptor).

congenital nephrosis See **alpha-fetoprotein**.

congenital severe combined immunodeficiency See **severe combined immunodeficiency disease**.

conglutinin Calcium-dependent lectin-like protein (371aa) present in bovine serum that causes **agglutination** of antibody-antigen-complement complexes; binds C3bi.

Congo red Naphthalene dye that is pH sensitive (blue-violet at pH 3, red at pH 5). Used as vital stain, also in staining for amyloid.

conidiophores A specialised stalk bearing **conidium** in the asexual fruiting body of Ascomycetes (Ascomycota).

conidium An asexual fungal spore, borne at the tip of a specialized **hypha** (conidiophore) rather than within a **sporangium**. In blastic conidiogenesis the spore is already evident before it separates from the conidiogenic hypha from which it is produced; in thallic conidiogenesis a cross-wall appears in the hypha delimiting a cell which then develops into a spore.

conjugate (1) Molecular species produced in a biological system by covalently linking two chemical moieties from different sources. (2) Material produced by attaching two or more substances together, e.g. a conjugate of an antibody with a fluorochrome or enzyme for use as a probe. (3) See **conjugation**.

conjugated trienols Oxidation products of alpha-farnesene probably involved in development of superficial scald in apple fruit, a postharvest disorder in which there is browning of the fruit skin.

conjugation Union between two gametes or between two cells leading to the transfer of genetic material. In eukaryotes the classic examples are in Paramecium and Spirogyra. Conjugation between bacteria involves an F$^+$ bacterium (with F-pili) attaching to an F$^-$; transfer of the F-plasmid then occurs through the sex pilus. In Hfr mutants the F-plasmid is integrated into the chromosome and so chromosomal material is transferred as well. Conjugation occurs in many Gram-negative bacteria (*Escherichia, Shigella, Salmonella, Pseudomonas* and *Streptomyces*).

Conn's syndrome Uncontrolled secretion of **aldosterone** usually by an adrenal adenoma.

connectin (1) Alpha-connectin is identical to **titin-1**, β-connectin to titin-2. Elastic connectin/titin molecules position the myosin filaments at the center of a sarcomere by linking them to the Z line. (2) *Drosophila* connectin (CON) is a cell surface protein of the leucine-rich repeat family (682aa) expressed on the surface of a subset of embryonic muscles and on the growth cones and axons of the motoneurons that innervate these muscles. It is attached to the cell surface via a GPI linkage and mediates homotypic cell-cell adhesion *in vitro*.

connective tissue Rather general term for mesodermally derived tissue that may be more or less specialized. Cartilage and bone are specialized connective tissue, as is blood, but the term is probably better reserved for the less specialized tissue that is rich in extracellular matrix (**collagen, proteoglycan,** etc.) and that surrounds other more highly ordered tissues and organs. See **areolar connective tissue**.

connective tissue diseases A group of diseases including rheumatoid arthritis, systemic lupus erythematosus, rheumatic fever, scleroderma and others, that are sometimes referred to as rheumatic diseases. They probably do not affect solely connective tissues but the diseases are linked in various ways and have interesting immunological features which suggest that they may be autoimmune in origin.

connective tissue growth factor A cysteine rich regulatory protein (CTGF, CCN-2, insulin-like growth factor-binding protein 8, 349aa) of the **CCN family**, secreted by vascular endothelial cells. CTGF stimulates the proliferation and differentiation of chondrocytes, induces angiogenesis, promotes adhesion of fibroblasts and epithelial cells, and binds to various other growth factors. Connective tissue growth factor-like (CTGFL/WISP-2, Wnt-1-induced signalling pathway protein-2, CCN5, 250aa) is expressed in primary osteoblasts and may modulate bone turnover. Inhibits osteocalcin production.

connective tissue-activating peptide III Cytokine (CXCL7, CTAPIII, low affinity platelet factor-4, 85aa) produced from **platelet basic protein**, that is chemotactic for leucocytes and acts as a **growth factor**. Various other active peptides such as neutrophil activating peptide are produced by cleavage.

connexin Generic term for proteins with four membrane-passing domains isolated from gap junctions. Currently 24 human genes for connexins have been identified, each with tissue- or cell-type-specific expression. Most organs and many cell types express more than one connexin. Connexin phosphorylation has been implicated in connexin assembly, gap junction turnover, and responses to tumour promoters and oncogenes. Connexin43 (Cx43, Gap junction alpha-1 protein, 382aa), the most widely expressed and abundant gap junction protein, can be phosphorylated at several different serine and tyrosine residues. Mutations in connexin genes are associated with peripheral neuropathies, cardiovascular diseases, dermatological diseases, hereditary deafness and cataract. See **connexon**.

connexon The functional unit of gap junctions. An assembly of six membrane-spanning proteins (**connexins**) having a water-filled gap in the centre. Two connexons in juxtaposed membranes link to form a continuous pore through both membranes.

conotoxins Toxins from cone shells (*Conus* spp). The α-conotoxins (13−18aa) are competitive inhibitors of nicotinic acetylcholine receptors. The μ-conotoxins are small (22aa) peptides that bind voltage-sensitive sodium channels in muscle, causing paralysis. The ω-conotoxins are similar in size and inhibit voltage-gated calcium channels, thereby blocking synaptic transmission. See **conantokins**.

Consed/Autofinish A software tool for viewing, editing, and finishing sequence assemblies created with **phrap**. Link: http://www.phrap.org/phredphrapconsed. html

Consense Part of the **PHYLIP** software package that reads a file of computer-readable phylogenetic trees and prints out a consensus tree. Description: http:// nebc.nerc.ac.uk/bioinformatics/docs/consense.html

consensus sequence Of a series of related DNA, RNA or protein sequences, the sequence that reflects the most common choice of base or amino acid at each position. Areas of particularly good agreement often represent conserved functional domains. The generation of consensus sequences has been subjected to intensive mathematical analysis.

conservative substitution In a gene product, a substitution of one amino acid with another having generally similar properties (size, hydrophobicity, etc.), such that the overall functioning is likely not to be seriously affected.

Conserved Domain Database Part of the **NCBI** (National Center for Biotechnology Information) Entrez database system that has annotated information about **conserved domains** within proteins. Link: http://www.ncbi.nlm.nih.gov/sites/entrez?db = cdd

conserved domains Functional units within a protein that have been used as building blocks in molecular evolution and recombined in various arrangements to make proteins with different functions.

CONSTANS *CO Arabidopsis* gene that promotes flowering in long days. Flowering is induced when *CONSTANS* messenger RNA expression coincides with the exposure of plants to light. A member of a family of 17 *CO*-like genes that are widely distributed in plants. Product is a nuclear zinc-finger protein (373aa).

constant region The C-terminal half of the light or the heavy chain of an immunoglobulin molecule. The amino acid sequence in this region is the same in all molecules of the same class or sub-class whereas the variable region is antigen-specific.

constitutive Constantly present, whether there is demand or not. Thus some enzymes are constitutively produced, whereas others are inducible.

constitutive transport element *CTE.* An RNA motif that has the ability to interact with intracellular RNA helicases and is essential for export of RNA from the nuclear compartment. Many retroviruses have such an element to ensure their unspliced mRNA is moved into the cytoplasm. See **Tap protein**.

constriction ring The equatorial ring of **microfilaments** that diminishes in diameter, probably both by contraction and disassembly, as **cytokinesis** proceeds.

ConSurf Database A database that provides evolutionary conservation profiles for proteins of known structure in the **Protein Data Bank** (PDB). It is often possible to identify key residues that comprise the functionally-important regions of the protein using this resource. Homepage: http://consurfdb.tau. ac.il/

contact activation pathway The process (formerly called the intrinsic pathway) of activation of blood clotting when collagen becomes exposed to platelets which bind via the Gp Ia/IIa receptor. Platelet adhesion is enhanced by recruitment of von Willebrand factor (vWF) and triggers the release of platelet granule contents which also play a part in initiating the complex cascade of events that culminate in cleavage of **Factor I** to form fibrin. See **tissue factor pathway** (formerly the 'extrinsic pathway').

contact following Behaviour shown by individual **slime mould** cells when they join a stream moving towards the aggregating centre. **Contact sites A** at front and rear of cell may be involved in *Dictyostelium*.

contact guidance Directed locomotory response of cells to an anisotropy of the environment, for example the tendency of fibroblasts to align along ridges or parallel to the alignment of collagen fibres in a stretched gel.

contact inhibition of growth/division See **density-dependent inhibition**.

contact inhibition of locomotion/movement Reaction in which the direction of motion of a cell is altered following collision with another cell. In heterologous contacts both cells may respond (mutual inhibition), or only one (non-reciprocal). Type I contact inhibition involves paralysis of the locomotory machinery, Type II is a consequence of adhesive preference for the substratum rather than the dorsal surface of the other cell.

contact inhibition of phagocytosis
Phenomenon described in sheets of kidney epithelial cells that, when confluent, lose their weak phagocytic activity, probably because of a failure of adhesion of particles to the dorsal surface in the absence of locomotory ruffles.

contact sensitivity Allergic response to contact with irritant, usually a **hypersensitivity**.

contact sites A Developmentally-regulated adhesion sites that appear on the ends of aggregation-competent *Dictyostelium discoideum* (see **Acrasidae**) at the stage when the starved cells begin to come together to form the **grex**. Originally detected by the use of Fab fragments of polyclonal antibodies, raised against aggregation-competent cells and adsorbed against vegetative cells, to block adhesion in EDTA-containing medium. (Cell-cell adhesion mediated by contact sites A, unlike that mediated by contact sites B, is not divalent cation-sensitive). The fact that a mutant deficient in csA behaves perfectly normally in culture is puzzling. Involved in linking the membrane to the actin cytoskeleton via **ponticulin**.

contact sites B See **contact sites A**.

contact-induced spreading The response in which contact between two **epithelial cells** leads to a stabilized contact and the increased spreading of the cells so that the area covered is greater than that covered by the two cells in isolation.

contactinhibin Plasma membrane **glycoprotein** of 60−70 kDa isolated from human diploid fibroblasts, which when immobilized on silica beads has been reported to reversibly inhibit the growth of cultured cells. Not in the UniProt database. Original report: http://jcb.rupress.org/content/111/6/2681.long

contactins A family of neuronal cell surface proteins of the immunoglobulin superfamily that mediate cell interactions during development. In humans contactin-1 (1018aa) is involved in the formation of paranodal axo-glial junctions through an interaction with CNTNAP1 (contactin-associated protein 1, **neurexin**-4) and in oligodendrocyte generation by acting as a ligand of NOTCH1. It is GPI-anchored. Contactin-2 (axonin-1, 1040aa) is involved in the initial growth and guidance of axons. Contactin-3 (brain-derived immunoglobulin superfamily protein-1, BIG-1, Plasmacytoma-associated neuronal glycoprotein, 1028aa), contactin-4 (BIG-2, 1026aa), contactin-5 (neural recognition molecule NB-2, 1100aa) and contactin-6 (neural recognition molecule NB-3,

1028aa) are thought to have similar roles. Several other contactins are known. In *Drosophila*, contactin (1390aa) is required for organization of septate junctions (the equivalent of vertebrate tight junctions). See **Caspr** (contactin-associated protens).

contig Colloquial term for a DNA sequence assembled from overlapping shorter sequences to form one large contigous sequence.

contractile ring See **constriction ring**.

contractile vacuole A specialized vacuole of eukaryotic cells, especially Protozoa, that fills with water from the cytoplasm and then discharges this externally by the opening of a permanent narrow neck or a transitory pore. Function is probably osmoregulatory.

contrapsin Trypsin inhibitor (serine protease inhibitor A3K, **serpin** A3K, 416aa) from rat and other rodents.

control element Generic term for a region of DNA, such as a **promoter** or **enhancer** adjacent to (or within) a gene that allows the regulation of gene expression by the binding of transcription factors.

control region General name for genomic DNA that, though binding of transcription factors to its promoters, enhancers and repressors, modulates the expression level of nearby genes.

controlled drugs Drugs that can only be prescribed under guidelines laid down in legislation. Usually drugs that have the potential to cause addiction and dependence.

convulxin *Cvx* Snake venom toxin, an octamer composed of four disulphide-linked $\alpha\beta$-heterodimers (158aa and 148aa) a **C-type lectin** from *Crotalus durissus terrificus* (South American rattlesnake), that activates platelets through the collagen receptor glycoprotein VI (GPVI)/Fc receptor gamma-chain (FcR γ-chain) complex leading to tyrosine phosphorylation and activation of the tyrosine kinase **Syk** and PLCγ2.

Coomassie Brilliant Blue Blue dye (Coomassie Brilliant Blue G-250, Brilliant Blue R, Acid Blue 90, Kenacid Blue) that binds nonspecifically to proteins, used in **Bradford method** for protein estimation and for detecting proteins on gels. Originally developed as a dye for wool; in acid solution has an absorbance shift from 465 nm to 595 nm when it binds to protein.

Coombs' test Diagnostic test to determine whether an individual's red cells are coated with autoantibodies or immune complexes. The erythrocytes are mixed with anti-human immunoglobulin and if antibody is present the red cells will agglutinate. The direct test involves the addition of anti-immunoglobulin antibody to washed erythrocytes;

the indirect test is for unbound antibodies against erythrocytes present in the serum.

cooperativity Phenomenon displayed by enzymes or receptors that have multiple binding sites. Binding of one ligand alters the affinity of the other site(s). Both positive and negative cooperativity are known; positive cooperativity gives rise to a sigmoidal binding curve. Cooperativity is often invoked to account for non-linearity of binding data, although it is by no means the only possible cause.

coordination complex Complex held together by coordinate (dipolar) bonds, covalent bonds in which the two shared electrons derive from only one of the two participants.

COP9 signalosome A conserved regulatory complex (constitutive photomorphogenic-9 complex), present in diverse eukaryotes, that acts as a regulator of the ubiquitin conjugation pathway by mediating the deneddylation of the cullin subunit of SCF-type E3 ubiquitin-protein ligase complexes. There are eight subunits (CSN1 to CSN8; total 450 kDa) with some homologies to the proteasome. Subunits include **Gps1** (491aa), **Jab1** (334aa), a coactivator of AP-1 transcription factor), CSN5 (product of a candidate oncogene in human breast cancer), TRIP15 (thyroid hormone receptor interactor-15), SGN3 (which has homology to the 26S proteasome S3 regulatory subunit) and others. Remarkably, human **GPS1** can substitute for FUS6 (the plant homologue) in the COP9 complex that represses photomorphogenesis in *Arabidopsis*. Inactivation of the COP9 signalosome will impair T-cell development.

COPI, COPII See **coatamer**.

copia-like retrotransposons A super-family of viral-like **LTR retrotransposons** (Ty1-copia group retrotransposons, Pseudoviridae) originally described in *Drosophila* where there are large numbers of closely-related sequences (20–60 depending on the strain). The copia element is about 5000 bp with identical terminal repeats and inverted repeats. Copia-like retrotransposons are ubiquitous in plants. In Ty1/copia elements, the pol gene encodes the protease, integrase, reverse transcriptase (RT), and RnaseH domains in that order from the 5′ end of the gene, the order of the domains being peculiar to this group. The Gypsy Database (GyDB) of mobile genetic elements: http://gydb.uv.es/index.php/Ty1/Copia

Coprinus Genus of fungi that have gills that autodigest once spores have been discharged giving rise to a black inky fluid.

copy number The number of molecules of a particular type on or in a cell or part of a cell. Usually applied to specific genes, or to plasmids within a bacterium.

copy-number polymorphisms *CNPs* A form of polymorphism that is becoming increasing recognised with the availability of high-density oligonucleotide microarrays for SNPs. Individuals differ by 11 CNPs on average, and copy number variation has been found in 70 different genes. Such polymorphisms occur in about 12% of human genomic DNA and variations may vary from about one kilobase to several megabases in size. Copy number polymorphism in the orthologous rat and human FcγRIIIb receptor genes is a determinant of susceptibility to immunologically mediated glomerulonephritis.

coracle A *Drosophila* **protein 4.1** homologue (1698aa), required during embryogenesis and localized to the cytoplasmic face of the septate junction in epithelial cells, where it interacts with the transmembrane protein **neurexin IV**.

CORD See **cone-rod dystrophy**.

cord blood Blood taken post-partum from the umbilical cord.

cord factor Glycolipid (trehalose-6,6′-dimycolate) found in the cell walls of Mycobacteria (causing them to grow in serpentine cords) and important in virulence, being toxic and inducing granulomatous reactions identical to those induced by the whole organism.

Cori's disease A glycogen storage disease (Type IIIa) caused by mutation in the gene encoding the glycogen debranching enzyme.

cork cambium A lateral meristem (bark cambium, pericambium or phellogen) found in the periderm of many vascular plants and responsible for secondary growth that replaces the epidermis in roots and stems.

corn (1) In Europe a generic term for various edible grains (barley, wheat, oats, rye etc.) whereas in the US specifically maize. (2) A localized overgrowth of the keratinized layer of the skin at a site of mechanical irritation. Often has a focal centre.

cornea Transparent tissue at the front of the eye. The cornea has a thin outer squamous epithelial covering and an endothelial layer next to the aqueous humour, but is largely composed of avascular collagen laid down in orthogonal arrays with a few fibroblasts. Transparency of the cornea depends on the regularity of spacing in the collagen fibrils.

cornea plana A corneal defect in which the radius of curvature is larger than normal. Can be a result of mutation in the gene for **keratocan**.

corneal endothelial dystrophy A degenerative disease of the cornea. Fuchs' endothelial corneal dystrophy is associated with mutation in the gene for the alpha-2 chain of collagen type VIII (COL8A2). Other forms include one caused by mutation in the

SLC4A11 gene, which encodes a sodium borate cotransporter.

Cornelia de Lange syndrome A developmental disorder caused by mutation in the NIPBL gene, which encodes a component of the **cohesin** complex, or mutations in other components of the complex.

corneocyte Cell of the **stratum corneum**, the outer layer of the skin, heavily keratinised and dead.

cornification The process of keratinization of the outer layers of the skin that eventually leads to death of keratinocytes (forming **corneocytes**) and ultimately to shedding of squames.

cornified epithelium Epithelium in which the cells have accumulated keratin and died. The outer layers of vertebrate skin, hair, nails, horn and hoof are all composed of cornified cells.

cornulin A protein (squamous epithelial heat shock protein 53, 495aa) specifically expressed in upper layers of cells in squamous epithelium. One of the **fused gene family** of proteins.

corona radiata The layer of cylindrical cells that surrounds the developing mammalian ovum.

coronal section A cross-section of the brain taken effectively where the edge of a crown would touch the head.

coronatine A phytotoxin produced by *Pseudomonas syringae* that functions partly as a mimic of methyl jasmonate.

Coronaviridae Family of single-stranded RNA viruses responsible for respiratory diseases. The outer envelope of the virus has club-shaped projections that radiate outwards and give a charateristic corona appearance to negatively stained virions. A coronavirus is responsible for **SARS**.

coronin F-Actin binding protein (p55, 445aa) of *Dictyostelium*. Associated with crown-shaped cell surface projections in growth phase cells. Accumulates at front of cells responding to a chemotactic gradient of cAMP. Amino-terminal domain has similarity to β subunits of heterotrimeric G-proteins; C-terminal has high α-helical content.

corpora allata Insect endocrine organs, located behind the brain, that secrete juvenile hormone. In some species paired and laterally placed, in others they fuse during development to form a single median structure, the corpus allatum.

corpus callosum Band of white matter at the base of the longitudinal fissure dividing the two cerebral hemispheres of the brain.

corpus luteum *Pl.* corpora lutea. Glandular body formed from the Graafian follicle in the ovary following release of the ovum. Secretes **progesterone**.

corpus striatum A subcortical part of the forebrain in from of the thalamus that has a striated appearance. Sometimes considered to consist of the striatum and the globus pallidus.

corralling The proposed confinement of membrane proteins within a diffusion barrier, thereby limiting long-range translational diffusion rates without affecting short range properties (e.g. rotation rates).

cortactin A p80/85 protein (amplaxin, 550aa) first identified as a substrate for src kinase. An F-actin binding protein that redistributes to membrane ruffles as a result of growth factor-induced **Rac1** activation. Has proline-rich and SH3 domains. Overexpression of cortactin increases cell motility and invasiveness. See **shanks**.

cortex (1) In plants, the outer part of stem or root, between the vascular system and the epidermis; composed of **parenchyma**. (2) Region of cytoplasm adjacent to the plasma membrane. (3) In histology, the outer part of organ, the interior sometimes being the medulla.

cortexillin Calcium-independent homodimeric actin binding (bundling) protein from *Dictyostelium* (cortexillin-1, 444aa; cortexillin-2, 441aa). The bundles are of anti-parallel filaments and associate into meshworks. Cortexillin activity is crucial for cytokinesis; cortexillin is enriched in the cleavage furrow and **dynacortin** is depleted. Article: http://www.nature.com/emboj/journal/v20/n14/full/7593868a.html

cortical granules In sea urchin eggs, specialised secretory granules that fuse with the egg membrane following fertilisation. Granule contents include (1) hyalin which forms a layer immediately surrounding the egg, (2) a colloid that raises the **fertilization membrane** by imbibing water, (3) a serine protease that destroys receptors for sperm on the vitelline membrane (4) a protein termed vitelline delaminase that may cleave the connection between the vitelline membrane and the oolemma, (5) a structural protein, that in the presence of H_2O_2 is polymerized onto the inner surface of the old vitelline membrane, now called the fertilization membrane and (6) ovoperoxidase, a haem-dependent peroxidase that functions to block polyspermy by interacting with the structural protein.

cortical layer See **cortical meshwork**.

cortical meshwork Sub-plasmalemmal layer of tangled microfilaments anchored to the plasma membrane by their barbed ends. This meshwork contributes to the mechanical properties of the cell surface and probably restricts the access of cytoplasmic vesicles to the plasma membrane.

cortical microtubule array Interphase microtubules associated with the plasma membrane in higher plant cells. Cortical microtubules, which are nucleated from dispersed initiation sites, self-organize to form this array of mainly parallel microtubules. Provides tracks for the movement of cellulose-synthesizing particles.

corticostatin Peptide (97aa) of the alpha-defensin family with anti-fungal and antiviral activity. It also inhibits corticotropin (ACTH) stimulated corticosterone production.

corticosteroids Steroid hormones produced in the adrenal cortex. Formed in response to **adrenocorticotrophin** (ACTH). Regulate both carbohydrate metabolism and salt/water balance. Glucocorticoids (e.g. cortisol, cortisone) predominantly affect the former and minerocorticoids (e.g. aldosterone) the latter.

corticotrophic *corticotropic* Describing something that stimulates the adrenal cortex. See **adrenocorticotrophin, corticotropin releasing hormone**.

corticotrophin *corticotrophin* See **adrenocorticotrophin**.

corticotrophin releasing factor See **corticotropin releasing hormone** and **adrenocorticotrophin**.

corticotropic See **corticotrophic**.

corticotropin releasing hormone-binding protein A secreted glycoprotein (CRH-BP, 322aa) that binds both **corticotropin releasing hormone** (CRH) and **urocortin** with high affinity and is structurally unrelated to the CRH receptors. It is an important modulator of CRH activity. CRH-BP orthologues have been identified in multiple invertebrate and vertebrate species and it is strongly conserved throughout evolution.

corticotropin-releasing hormone Key regulator of the hypothalamic-pituitary-adrenal axis, a peptide hormone (CRH, corticotropin-releasing factor, corticoliberin, 41aa) produced by the hypothalamus that stimulates corticotropic cells of the anterior lobe of the pituitary to produce ACTH (**adrenocorticotrophin**) and other biologically active substances (for example β-endorphin). Also produced by both the placenta and fetal membranes at term in man. Stressors cause a release of corticotrophin releasing hormone. The CRH hormone family has at least four ligands, two receptors (CRHR1, 444aa; CRHR2, 411aa, both G-protein coupled) and a binding protein.

cortisol The major adrenal glucocorticoid; stimulates conversion of proteins to carbohydrates, raises blood sugar levels and promotes glycogen storage in the liver.

cortisone Natural **glucocorticoid** (11-dehydroxycortisol) formed by 11β-hydroxysteroid

dehydrogenase action on hydrocortisone; inactive until converted into **hydrocortisone** in the liver.

cortistatin A neuropeptide (CST, 17aa) with high structural homology with **somatostatin** and able to bind to somatostatin receptors. Its mRNA is restricted to **GABA**-containing cells in the cerebral cortex and hippocampus. CST modulates the electrophysiology of the hippocampus and cerebral cortex of rats; hence, it may be modulating mnemonic processes.

Corynebacteria Genus of **Gram-positive** non-motile rod-like bacteria, often with a club-shaped appearance. Most are facultative anaerobes with some similarities to **mycobacteria** and **nocardiae**. *C. diphtheriae* (Klebs-Löffler bacillus) is the causative agent of diphtheria and produces a potent exotoxin, **diphtheria toxin**. *Corynebacterium minutissimum* is associated with a skin rash (erythrasma).

Corynebacterineae A sub-Order of bacteria that includes the Families Mycobacteriaceae, Nocardiaceae and Corynebacteriaceae.

COS cells Simian fibroblasts (CV-1 cells) transformed by **SV40** that is deficient in the origin of replication region. Express **large T-antigen** constitutively and if transfected with a vector containing a normal SV40 origin have all the other early viral genes necessary to generate multiple copies of the vector and thus to give very high levels of expression.

cos sites Sites on a **cosmid** vector that are required for integration into host DNA. The two cohesive ends, known as cos sites, are 12 nucleotides in length and the chromosome circularizes by means of these complementary cohesive ends.

cosegregation Of two genotypes, meaning that they tend to be inherited together, implying close linkage.

Cosmarium A very large genus of the Order Desmidiales of the **Chlorophyta**, with the cells deeply constricted to form two semicells. Each semicell contains a single, large chloroplast with two pyrenoids. The cells are slowly motile by mucilage secretion. Usually found in fresh water.

cosmid A type of **bacteriophage** lambda vector. Often used for construction of genomic libraries, because of their ability to carry relatively long pieces of DNA insert, compared with **plasmids**.

costa (1) Rod-shaped intracellular organelle lying below the undulating membrane of *Trichomonas*. Generates active bending associated with local loss of **birefringence** at the bending zone, probably as a result of conformational change in the longitudinal lamellae. Major protein approximately 90 kDa. (2) In *Drosophila*, a kinesin-like protein (costal 2, 1201aa) that is part of the **hedgehog signalling**

complex and may inhibit protein **cubitus interruptus** from activating hedgehog target genes. Vertebrate (*Xenopus, Danio*) homologues have been identified.

costamere Regular periodic sub-membranous arrays of **vinculin** in muscle cells; link sarcomeres to the membrane and are associated with links to extracellular matrix. See **dystrophin-associated protein complex**.

Cot curve Common mis-spelling of **C$_0$t** (C$_{(zero)}$t) which has become adopted by commercial suppliers of **Cot-1 DNA**.

Cot1 DNA DNA that is enriched for repetitive DNA sequences and has a **C$_0$t** value of unity. It is used to block non-specific hybridization in microarray screening. N.B. Should strictly perhaps be C$_0$t C$_{(zero)}$t since the enrichment of repetitive sequences is by their rapid reannealing and the C$_0$t value is a measure of the similarity but 'Cot1 DNA' is a registered trademark. Details: http://www.ncbi.nlm.nih.gov/books/NBK7567/box/A496/?report = objectonly

cotinine A metabolite of nicotine that persists in the body for two to four days after tobacco use and thus serum or urine levels of cotinine provide a more accurate marker of smoking or smoke inhalation than questionnaires.

co-translational transport Process whereby a protein is moved across a membrane as it is being synthesized. This process occurs during the translation of the message at membrane-associated **ribosomes** in **rough endoplasmic reticulum** during the synthesis of secreted proteins in eukaryotic cells.

cotton marker database A curated and integrated web-based relational database providing centralized access to all publicly available cotton microsatellites and single nucleotide polymorphisms. Link to database: http://www.cottonmarker.org/Downloads.shtml

Coturnix coturnix japonica Japanese quail. Used extensively in developmental biology because quail nuclei can easily be distinguished from those of the chicken and this facilitates grafting experiments for fate mapping.

cotyledon Modified leaf ('seed leaf'), found as part of the embryo in seeds, involved in either storage or absorption of food reserves. Dicotyledonous seeds contain two, monocotyledonous seeds only one. May appear above ground and show photosynthetic activity in the seedling.

Coulter counter Proprietary name for a particle counter used for bacteria or eukaryotic cells; works by detecting change in electrical conductance as fluid containing cells is drawn through a small aperture. (The cell, a non-conducting particle, alters the effective cross-section of the conductive channel).

coumarin Pleasant-smelling compound (O-hydroxy-cinnamic acid) found in many plants and released on wilting (probably a major component of the smell of fresh hay). Has anticoagulant activity by competing with Vitamin K. Coumarin derivatives have anti-inflammatory and antimetastatic properties and inhibit xanthine oxidase and the production of 5-HETE by neutrophils and macrophages. Various derivatives have these activities including esculentin, **esculin** (6,7-dihydroxycoumarin 6-O-D-glucoside), **fraxin**, **umbelliferone** (7-hydroxy coumarin) and **scopoletin** (6-methoxy-7-hydroxy-coumarin).

Councilman body An eosinophilic cytoplasmic inclusion body in liver cells that is a useful diagnostic indication of yellow fever infection.

counterion Ion of the opposite charge to that of an immobilised ionised molecule, for example the carboxyl residue of N-acetyl neuraminic acid on cell surface glycoptotein. The consequence is to alter the composition of the environment immediately adjacent to the cell surface. See **double layer**.

counter-transport Transport system across a membrane in which movement of a molecule in one direction is matched by the movement of a different molecule in the opposite direction. If both are charged then no potential gradient will develop, provided equal numbers of charges are moved in each direction. The opposite of **co-transport**.

counterstain Rather nonspecific stain used in conjunction with another histochemical reagent of greater specificity to provide contrast and reveal more of the general structure of the tissue. Light Green is used as a counterstain in the Mallory procedure, for example.

COUP-TFs Transcription factors of the steroid/thyroid hormone receptor family (chicken ovalbumin upstream promoter-transcription factors). Although the ligands are unknown, their regulatory role is clear. COUP-TFI (423aa) plays a critical role in glial cell development and central nervous system myelination; it is expressed in cells of oligodendrocyte lineage. COUP-TF2 (414aa) regulates apolipoprotein A-I gene expression. See ARP-1. COUP-TF-interacting protein 1 (B-cell lymphoma/leukemia 11A, 835aa) functions as a myeloid and B-cell proto-oncogene, COUP-TF-interacting protein-2 (894aa) is a tumour-suppressor protein aberrant in T-cell lymphomas.

coupled transporter (1) A membrane transport system in which movement of one molecule or ion down an electrochemical gradient is used to drive the movement of another molecule up-gradient. Common example is Na$^+$-coupled glucose transport that uses the considerable gradient in sodium ion concentrations, high outside and low within the cell, to drive glucose uptake. See **symport** and **antiport**.

(2) May be used loosely of a transport system that is controlled by, for example, G-proteins although the coupling is of a different kind.

coupling The linking of two independent processes by a common intermediate, e.g. the coupling of electron transport to oxidative phosphorylation or the ATP-ADP conversion to transport processes.

coupling factors Proteins responsible for coupling transmembrane potentials to ATP synthesis in **chloroplasts** and **mitochondria**. Include ATP-synthesising enzymes (F1 in mitochondrion), that can also act as ATP-ases.

Cowden disease A rare autosomal dominant multiple-hamartoma syndrome caused by germ-line mutations in **PTEN**.

COX (1) See **cyclo-oxygenase**. (2) COX1 is an alternative name for cytochrome c oxidase subunit 1 (EC 1.9.3.1) and COX is a prefixed abbreviation for many cytochrome oxidase subunits (e.g. COX-6B1, cytochrome c oxidase subunit VIb isoform 1).

Coxsackie viruses Enteroviruses of the **Picornaviridae** first isolated in Coxsackie, N.Y. Coxsackie A produces diffuse myositis, Coxsackie B produces focal areas of degeneration in brain and skeletal muscle. Similar to polioviruses in chemical and physical properties.

CPA (1) **Cyclophosphamide** (2) **cyclopiazonic acid** (3) **cyclopentyladenosine** (4) Carboxypeptidase A.

cpDNA Chloroplast DNA.

CPE (1) Cytopathic effect. (2) **Cytoplasmic polyadenylation element**. (3) *Clostridium perfringens* enterotoxin.

CpG island Region of genomic DNA rich in the dinucleotide C-G (CG island). Methylation of the C in the dinucleotide is maintained through cell divisions, and profoundly affects the degree of transcription of the nearby genes, and is important in developmental regulation of gene expression.There are around 30,000 CpG islands in a typical mammalian genome, and these tend to be undermethylated and upstream of housekeeping genes.

CPK See **creatine kinase**.

C polysaccharide A highly immunogenic polysaccharide (C-teichoic acid, C substance) released by pneumococci which contains galactosamine-6-phosphate and choline phosphate. It is mitogenic for T-cells. **C-reactive protein** is so called because it will precipitate this polysaccharide through an interaction with the choline phosphate haptenic determinants.

CPP32 Caspase-3. See Table C1 (Caspases).

C proteins Striated muscle thick filament-associated proteins (myosin-binding proteins, ~1141aa) that show up in the C-zone of the A-band as 43 nm transverse stripes. Immunoglobulin superfamily members, structurally related to various other myosin-binding proteins (**twitchin**, **titin**, **myosin light chain kinase**, **skelemin**, 86 kDa protein, **projectin**, **M-protein**).

CPS See **carbamoyl phosphate synthetase**.

CPSF Multi-protein complex (cleavage and polyadenylation specificity factor) comprising CPSF-1 (1443aa), CPSF-2 (782aa), CPSF-3 (684aa), CPSF-4 (269aa) and FIP1L1 (Pre-mRNA 3′-end-processing factor, 594aa) involved in mRNA polyadenylation. Complex binds the AAUAAA conserved sequence in pre-mRNA. CPSF5 (NUDT21, CFIM25, CPSF25, 227aa) and CPSF6 (551aa) or CPSF7 (471aa) are components of the **cleavage factor Im complex** that plays a key role in pre-mRNA 3′-processing. In *Arabidopsis* CPSF73-II (cleavage and polyadenylation specificity factor 73 kDa subunit II, EC 3.1.27.-, 613aa), one of a superfamily of zinc-dependent beta-lactamase fold proteins, may function as mRNA 3′-end-processing endonuclease and also be involved in the histone 3′-end pre-mRNA processing. Diagram: http://www.reactome.org/cgi-bin/eventbrowser?DB=gk_current&FOCUS_SPECIES=Homo%20sapiens&ID=71994&

CPT (1) Carnitine palmitoyltransferase. (2) Camptothecin.

CR1 See **complement receptors**. (Also CR2, CR3, CR4).

CRABP See **cellular retinoic acid-binding protein**.

cranial nerves Any of the ten to twelve paired nerves that have their origin in the brain of vertebrates (olfactory, optic, oculomotor, trochlear, trigeminal, abducens, facial, auditory / vestibulocochlear, glossopharyngeal, vagus, accessory and hypoglossal). Students often use mnemonics to remember the sequence: a polite one is 'Old Officers Often Trust The Army For A Glory Vague And Hypothetical'.

craniolenticulosutural dysplasia A disorder of craniofacial development caused by a mutation in the SEC23A gene (see **coated vesicles**) leading to abnormal endoplasmic reticulum-to-Golgi trafficking.

craniosynostosis A disorder in which the bones of the skull fuse earlier than normal. See **twist**.

crassulacean acid metabolism   *CAM* Physiological adaptation of certain succulent plants, in which CO_2 can be fixed (non-photosynthetically) at night into malic and other acids. During the day the CO_2 is regenerated and then fixed photosynthetically into the **Calvin-Benson cycle**. This adaptation permits the stomata to remain closed during the day, conserving water.

CRD domain The carbohydrate recognition domain found in C-type (calcium dependent) lectins; there are 32 highly conserved residues in all C-type carbohydrate-recognition domains. Various CRDs have been structurally analysed.

CRE (1) Gene of *E. coli* bacteriophage P1 (*cre*) that encodes a recombinase (343aa) that catalyzes site-specific recombination between two 34bp LoxP sites. One of the phage integrase family. Now used in vertebrate transgenics: see **lox-Cre system**. (2) Cyclic AMP response element, see **CREB #1**

C-reactive protein A protein of the **pentraxin** family (pentaxin 1, PTX1, 224aa) found in serum in various disease conditions particularly during the acute phase of immune response. C reactive protein is synthesized by hepatocytes and its production may be triggered by **prostaglandin** E1 or parogen. It consists of five polypeptide subunits forming a molecule of total molecular weight 105 kDa. It binds to polysaccharides present in a wide range of bacterial, fungal and other cell walls or cell surfaces and to **lecithin** and to phosphoryl- or choline-containing molecules. It is related in structure to serum **amyloid**. See also **acute phase proteins** and **C polysaccharide**.

creatine kinase Dimeric enzyme (creatine phosphokinase, CPK, EC 2.7.3.2, 381aa) that catalyses the formation of ATP from ADP and phosphocreatine in muscle. Different isoforms are found in skeletal muscle (MM), myocardium (MB) and brain (BB). In mitochondria the enzyme (CKMT1, 417aa; CKMT2, 419aa) is an octamer associated with the inner membrane.

creatine phosphate See **phosphocreatine** and **creatine kinase**.

CREB (1) **Basic leucine zipper** (bZip) transcription factor (Cyclic AMP response element binding factor, CREB-1, 341aa) involved in activating genes through cAMP; binds to CRE element TGANNTCA. Phosphorylation by cAMP-dependent protein kinase (PKA) at serine-119 is required for interaction with DNA and phosphorylation at serine-133 allows CREB to interact with CBP (**CREB binding protein**) leading to interaction with RNA polymerase II and with **MECT1**. Other forms of CREB are known (CREB3, 395aa; CREB5, 508aa). See **CREM**. (2) In *E. coli*, CreB is a transcriptional regulator in the two-component regulatory system CreC/CreB involved in catabolic regulation. (3) In *Aspergillus*, creB is ubiquitin carboxyl-terminal hydrolase (EC 3.4.19.12, 766aa) a component of the regulatory network controlling carbon source utilization through ubiquitination and deubiquitination involving CreA, CreB, CreC, CreD and AcrB. It deubiquitinates the CreA catabolic repressor.

CREB binding protein Transcriptional co-activator (CBP, EC 2.3.1.48, 2442aa) of **CREB** and of c-Myb. Only binds the phosphorylated form of CREB.

CREC family A family of low affinity, Ca^{2+}-binding, multiple EF-hand proteins that includes **reticulocalbins**, **calumenin** and other proteins of the endoplasmic reticulum where they regulate various activities. Sometimes 'CERC family' for Cab-45 (45 kDa calcium-binding protein, 362aa), Erc-55 (reticulocalbin-2), reticulocalbin and calumenin.

C-region The parts of the heavy or light chains of **immunoglobulin** molecules that are of constant sequence, in contrast to variable or V regions. The constancy of sequence is relative because there are several constant region genes and alleles thereof (see allotypes), but within one animal homozygous at the light and heavy chain constant region genes all immunoglobulin molecules of any one class have constant sequences in their C regions. The constant region sequences for the various different types of immunoglobulin, e.g. IgG, IgA, etc. will vary.

CREM A transcriptional regulator (cAMP response element modulator, 361aa) that binds the cAMP response element. Has a highly conserved leucine zipper dimerization domain and a basic DNA binding domain at its carboxyl terminus. There are multiple splice variants.

crenation Distortion of the erythrocyte membrane giving a spiky, echinocyte-like, morphology. Results from ATP depletion or an excess of lipid species in the external lipid layer of the membrane.

CREST syndrome A complex syndrome in which there is calcinosis, Reynaud's phenomenon, esophageal dysmotility, sclerodactyly and telangielactasia, a variant of **scleroderma**. There are probably both genetic and environmental factors involved. Most patients have autoantibodies to **CENPs** (kinetochore proteins).

Creutzfeldt-Jacob disease Rare fatal presenile dementia of humans, similar to **kuru** and other transmissible spongiform encephalopathies. Method of transmission unknown. Will induce a neurological disorder in goats 3–4 years after inoculation with CJD brain extract. A new variant, nvCJD, has recently been recognized and associated with **bovine spongiform encephalopathy**. See **prions**.

CRF See **corticotropin-releasing factor**.

cri-du-chat syndrome A severe developmental disorder caused by loss of part of the short arm of chromosome 5 in man. Results in severe congenital malformation and affected infants produce a curious mewling sound said to resemble the cry of a cat. Deletion of multiple genes, including the telomerase reverse transcriptase gene (see **hTERT**), is responsible for the phenotype.

CRIB cells A clone of MDBK cells that are resistant to bovine viral diarrhoea virus.

CRIB motif The cdc42/Rac interactive binding motif, a consensus sequence found in various signalling proteins (e.g. PAK kinases) and involved in the binding of Cdc42. The CRIB motif itself is insufficient for high-affinity binding to Cdc42 but requires the sequence segment C-terminal to the CRIB motif for enhanced affinity.

cribriform Perforated, sievelike.

cribrostatin A family of antibacterial and anti-neoplastic compounds isolated from the blue marine sponge *Cribrochalina* sp. A total synthesis of cribrostatin-6 has been published. Paper: http://jmm.sgmjournals.org/content/53/1/61.full

***Cricetulus griseus* Chinese hamster.** See **CHO** cells. The Syrian (Golden) hamster is *Mesocricetus auratus.*

Crigler-Najjar syndrome See **Gilbert's syndrome**.

Crimean-Congo haemorrhagic fever A tick-borne disease caused by a **Nairovirus**.

crinkled (1) In *Drosophila*, myosin VIIa (2167aa) an atypical myosin that is expressed in the setae, micro- and macrochaetae on the head, thorax and wing and is necessary for auditory transduction. (2) EDAR-associated death domain protein, see **ectodysplasin**.

crinophagy Digestion of the contents of secretory granules following their fusion with lysosomes.

crypto A growth factor (epidermal growth factor-like cripto protein CR1, teratocarcinoma-derived growth factor 1, 188aa) found in humans and many other vertebrates. During early embryogenesis, has a role as a coreceptor for the TGFβ subfamily of proteins, including **nodal**. Cripto-1 has also been shown to function as a ligand through a Nodal/Alk4-independent signalling pathway that involves binding to **glypican-1**. Expression levels are elevated in gastric and colorectal carcinomas. In *Xenopus*, interacts with **tomoregulin-1** during neural development. PMID: 16123806

CRISP See **cysteine-rich secretory proteins**.

critical concentration Concentration (dose) of a substance at which adverse functional changes, reversible or irreversible, occur in a cell or an organ.

critical point drying A method for preparing specimens for the scanning electron microscope that avoids the problems of shrinkage caused by normal drying procedures. Water in the specimen is replaced by an intermediate fluid, for example liquid carbon dioxide, avoiding setting up a liquid/gas interface, and then the second fluid is allowed to vaporise by raising the temperature above the critical point, the temperature at which the liquid state no longer occurs.

crk An oncogene, originally identified in a chicken **sarcoma**. Encodes two alternatively spliced adapter signalling proteins, CRKI (204aa) and CRKII (304aa) that have SH2 and SH3 domains (see **Grb2**) and may recruit cytoplasmic proteins to associate with receptor tyrosine kinases. Both CRKI and CRKII have been shown to activate kinase signalling and anchorage-independent growth *in vitro*. Crk-like protein (CRKL, 303aa) has similar properties.

CRM (1) **Cis-regulatory modules**. (2) See **CRM-1**. (3) CRM elements are plant centromere-specific Ty3/Gypsy **LTR retrotransposons**. The chromodomain is rather unusual in these elements.

CRM-1 Chromosomal region maintenance-1 (exportin-1, 1071aa), the main mediator, in many cell types, of export from the nucleus of cellular proteins that have a leucine-rich **nuclear export signal** (NES) and of RNAs. Inhibited by **leptomycin B**. In the nucleus, in association with RANBP3, binds co-operatively to the NES on its target protein and to Ran-GTPase in its active GTP-bound form. Docking of this complex to the nuclear pore complex is mediated through binding to **nucleoporins**.

CRMP (1) **Collapsin response-mediated protein**. (2) A haem-oxygenase (HO) enzyme inhibitor, chromium-mesoporphyrin (CrMP).

cro-protein Protein (76aa) synthesied by lambda bacteriophage in the lytic state. The cro-protein blocks the synthesis of the lambda repressor (that is produced in the lysogenic stage, and inhibits cro-protein synthesis). Production of the cro-protein in turn controls a set of genes associated with rapid virus multiplication.

crocalbin See **calumenin**.

crocin A yellow carotenoid pigment from *Gardenia jasminoides* and *Crocus sativus*; heat-stable and water-soluble but sensitive to oxidation.

Crohn's disease Inflammatory bowel disease that usually affects the terminal ileum and colon and seems to have both genetic and environmental causes; not well understood but generally considered to be autoimmune. Mutations in genes for CARD15 (caspase recruitment domain-containing protein 15) and **NOD2**) and polymorphism in the promoter for interleukin-6 are associated with susceptibility to Crohn's disease in some cases. There is also an association with variation in the ATG16L1 gene that encodes autophagy-16-like-1, a component of a large protein complex essential for **autophagy**.

cross hybridization Hybridization of a nucleic acid probe to a sequence that is similar, but not identical, to the target.

cross_match A general purpose software utility for comparing any two DNA sequence sets. It is slower but more sensitive than BLAST. Link: http://nbx8.nugo.org/bioinformatics/docs/cross_match.html

crossing over Recombination as a result of DNA exchange between homologous chromatids in meiosis, giving rise to **chiasmata**.

crossover Protein **motif** that describes the connection between strands in a parallel **beta sheet**. In principle, can be extended to the region between adjacent parallel **alpha-helices**.

croton oil Oil from the seeds of the tropical plant *Croton tiglium* (Euphorbiaceae), causes severe skin irritation and contains a potent **tumour promoter**, **phorbol ester**.

crotoxin Neurotoxin of Brazilian rattlesnake (*Crotalus durissus terrificus*) venom. A complex of an acidic non-enzymic protein (8400Da), probably responsible for membrane-binding activity, and a phospholipase A2 derived from a precursor, crotoxin acid chain (crotapotin, 138aa). The two components in combination are necessary to produce high neurotoxicity; neither does alone. Bind to specific proteins on presynaptic membranes and blocks acetylcholine release at the neuromuscular junction.

crown gall Gall, or tumour, found in many dicotyledonous plants, caused by *Agrobacterium tumefaciens*.

CRP (1) C-reactive protein. (2) calreticulin.

CRP-ductin The mouse homologue of **DMBT-1** (deleted in malignant brain tumours 1 protein, 2085aa).

crumbs *Drosophila* protein (crb, 2146aa) important in epithelial development. It determines the location of adherens junctions as a ring of adhesions around the cell through the SAC complex. The mammalian homologue (crumbs homologue-1, 1406) is important for photoreceptor morphogenesis, crb-2 (1285aa) is thought to be involved in determining cell polarity and crb-3 (120aa) regulates the morphogenesis of tight junctions.

CRY Cryptochrome, the blue-light receptor from *Arabidopsis*, flavoproteins (CRY-1, BLUE LIGHT UNINHIBITED-1, ELONGATED HYPOCOTYL-4, OUT OF PHASE-2, 681aa; CRY2, 612aa) that induce gene expression in response to blue light. Mutants lacking both CRY-1 and CRY-2 are deficient in phototropism. In humans CRY1 (586aa) in conjunction with CRY2 (593aa) regulates circadian rhythms by acting on **clock**-induced transcription.

cryofixation Fixation processes for microscopy, particularly scanning electron microscopy, carried out at low temperature to improve the quality of fixation. Often very low temperatures (liquid nitrogen or helium) and fast cooling (>10,000 degree/min) are used to prevent formation of ice crystals.

cryoglobulin Abnormal plasma globulin (IgG or IgM) that precipitates when serum is cooled. In cryoglobulinemia there is an excess of cryoglobulin that may be a result of a lymphoproliferative disorder, hepatitis C infection or an autoimmune disease.

cryomicroscopy Microscopy, either light or EM, of samples that have been prepared by rapid freezing and are maintained in a frozen glass-like state. Frozen sections can be prepared and examined rapidly and are therefore important in pathological examination of biopsy material.

cryoprecipitate The precipitate that forms when plasma is frozen and then thawed; particularly rich in **fibronectin** and blood-clotting Factor VIII.

cryoprotectant Substance that is used to protect from the effects of freezing, largely by preventing large ice-crystals from forming. The two commonly used for freezing cells are **DMSO** and glycerol.

cryopyrin Protein (NACHT, LRR and PYD domains-containing protein 3, NALP3, caterpiller protein 1.1, cold autoinflammatory syndrome 1 protein, PYRIN-containing APAF1-like protein-1, angiotensin/vasopressin receptor AII/AVP-like, 1034aa) that may be an upstream activator of NFκB signalling in response to stimuli such as bacterial RNA and infectious agents, leading to upregulation of IL-1β and IL-18 production. See **pyrin**. Mutations cause familial cold urticaria, **Muckle-Wells syndrome** and chronic infantile neurological cutaneous and articular syndrome.

crypt Deep pit that protrudes down into the connective tissue surrounding the small intestine. The epithelium at the base of the crypt is the site of stem cell proliferation and the differentiated cells move upwards and are shed three to five days later at the tips of the villi.

cryptic plasmid Plasmid that does not confer a phenotype and is only detected by direct observation of the DNA. Many do, however, have open reading frames.

cryptobiont An organism that lives hidden away or with all signs of life disguised as in dormancy.

cryptochrome See **CRY**.

cryptococcosis An infection, usually of the lung but more seriously a meningitis, by the encapsulated yeast Cryptococcus neoformans (*Filobasidiella neoformans*, formerly *Torula histolytica*). Often opportunistic in immunocompromised individuals.

cryptomonads Small algae (cryptophytes) in the **Hacrobia** grouping that have plastids with **nucleomorphs**. *Cryptomonas* is one genus. Some have trichocysts (ejectisomes) that will explosively release a coiled protein filament as a startle/escape reaction.

Cryptomonas Genus of flagellate protozoa with two slightly unequal flagella and a large chromatophore in some species.

cryptophycins Naturally occurring macrolides from the cyanobacterium *Nostoc* that are potent antimitotic agents, causing cell death at picomolar or low nanomolar concentrations. Bind to beta-tubulin (sharing a binding site with **dolastatin** 10, **hemiasterlin** and **phomopsin A**) and deactivate **bcl-2**, causing apoptosis.

Cryptosporidium Genus of obligate parasitic apicomplexan protozoa, responsible for many cases of intestinal upset caused by drinking infected water. Sporulated oocysts, once in the intestinal tract (usually the ileum) release sporozoites that penetrate intestinal epithelial cells and divide to form merozoites which can propagate and infect other cells. Eventually meronts differentiate into either macro- or microgametocytes, microgametocytes penetrate the macrogametocytes to form a zygote that undergoes meiotic division, producing sporozoites within a resistant oocyst that is then released in faeces, completing the cycle.

crystal violet Deepest blue of the **methyl violet dyes** (Gentian violet; methyl violet 10 B), used in **Gram stain**, for the metachromatic staining of amyloid, and as an enhancer for bloody fingerprints. Also used as a pH indicator, turning yellow below pH1.8. Was used topically for skin infections but as a suspected carcinogen this is no longer advisable. Stains database: http://stainsfile.info/StainsFile/dyes/42555.htm

crystallins Major proteins of the vertebrate lens present in high concentrations in the cytoplasm of lens fibre cells. Unlike most proteins they do not turn over during the lifetime of the individual. The alpha-crystallins (α-crystallin-A, CRYAA, 173aa; α-crystallin-B, CRYAB, 175aa) are members of the small heat shock protein (HSP20) family and contribute to the optical properties of the lens, but have additional functions in other tissues. The beta and gamma crystallins form a single family (all 170–250aa). Immunological cross-reactivity suggests that the sequences of crystallin subunits are relatively highly conserved in evolution. Defects in crystallins, either mutational or from environmental damage, cause cataracts.

CSAT Monoclonal antibody defining integral membrane protein of chick fibroblasts. Originally thought to recognize a trimeric complex, now thought to recognize two different β_1 integrins (with different α chains).

CSD See **chromoshadow domain**.

CSF-1 (1) See **MCSF**. (2) In *S. cerevisiae*, cold sensitive for fermentation protein 1 (2958aa), required for nutrient uptake at low temperatures.

CSIF Cytokine synthesis inhibiting factor, obsolete name for **interleukin-10**.

Csk Protein tyrosine kinase (c-src kinase, EC 2.7.10.2, 450aa) that phosphorylates tyrosine-504 in **lck** and other **src family** kinases, thereby allowing an inhibitory interaction with src kinase SH2 domain. (It is loss of the tyrosine residue phosphorylated by csk that makes v-src unregulated.) One of two kinases forming a subfamily, the other being **carboxyl-terminal Src kinase homologous kinase**.

CSP Ambigous abbreviation (often with a number suffix) for (1) **Caveolin scaffolding protein**. (2) **Competence stimulating peptide**. (3) **Cysteine string protein**. (4) **Caspase**. (5) **Chiral stationary phase**. (6) Circumsporozoite protein of *Plasmodium vivax*. (7) Common salivary protein, a secreted protein of the parotid gland. (8) Carotid sinus pressure. (9) **Ciclosporin**.

CST See **cortistatin**.

CSTF Cleavage stimulation factor, one of the factors required for polyadenylation and 3′-end cleavage of mammalian pre-mRNAs. The CSTF complex is composed of CSTF1 (431aa), CSTF2 (577aa) and CSTF3 (717aa). Interacts with **BARD1**.

c-strand A shorthand term for the complementary strand of a nucleic acid.

C-subfibre The third partial microtubule associated with the A- and B- tubules of the outer axonemal doublets in the **basal body** (and in the **centriole**) to form a triplet structure.

C substance See **C-polysaccharide**.

CT-1 See **cardiotrophin-1**.

CTAB Cationic detergent (cetyltrimethylammonium bromide) used for membrane solubilization.

CTACK Beta-chemokine (CCL27, cutaneous T-cell attracting chemokine, ESkine, 112aa) produced by keratinocytes that is important in regulating the migration of lymphocytes and keratinocyte precursor cells into skin. Upregulated during wound healing and in psoriasis and atopic dermatitis. Binds to CCR10. Also produced by a wider range of cells in a non-secreted form (by alternative splicing) and this form (PESKY) is targeted to the nucleus and has effects on the actin cytoskeleton.

CTAP III See **connective tissue-activating peptide III**.

CTCF A transcriptional regulator protein (CCCTC-binding factor, 727aa) with 11 highly conserved zinc finger domains; depending upon where it binds, can act as either transcriptional activator or repressor through histone acetylation or de-acetylation. Mutations in this gene have been associated with invasive breast cancers, prostate cancers, and **Wilms' tumours**.

CTD Protein domain (carboxy-terminal domain) unique to RNA polymerase-II that contains multiple repeats of the YSPTSPS sequence. The CTD plays an important part in organizing the various protein factors that regulate the processing of the 3' end of mRNAs made by Pol-II. Phosphorylation of the CTD allows interaction with CTD-interacting factor 1 (704aa) and the CTD small phosphatase-like protein (EC 3.1.3.16, 276aa) inhibits this interaction.

Ctenophora Phylum of biradially symmetrical triploblastic coelomates. Lack nematocysts and cilia though have comb plates (costae) arranged in eight rows. The comb jelly or sea gooseberry is the best known example.

C terminal binding proteins C terminal binding protein 1 (CTBP1, EC 1.1.1.-, 440aa) is involved in controlling the equilibrium between tubular and stacked structures in the Golgi, has dehydrogenase activity and is a transcriptional corepressor. CTBP2 (445aa) is an ubiquitous corepressor targeting diverse transcription regulators; **ribeye** is an alternatively-spliced isoform. CTBPs bind to the C-terminal portion of adenovirus E1A proteins and are important for adenoviral oncogenic properties.

c terminus The carboxy terminal end of a polypeptide or protein, the 1-carboxy function of the c-terminal amino acid that is not linked to another amino acid by a peptide bond.

CTF Large family of vertebrate nuclear protein transcription factors, (CCAAT box-binding transcription factors, TGGCA-binding proteins, ~400−600 aa), that bind to the palindrome TGGCAnnnTGCCA in a range of cellular promoters. Includes nuclear factor-1 (NF-1A, 509aa; NFl-B, 420; NF-1C, 508aa; NF-1X, 502aa.

CTGF See **connective tissue growth factor**.

CTL See **cytotoxic T-cells**.

CTLA-4 Type I transmembrane protein (CD152, 223aa) of the immunoglobulin superfamily. Found on activated T-cells and binds to CD80 on B-cells. Resembles CD28 but acts as a negative regulator of T-cell activation. Cytoplasmic domain interacts with SH2 domain of **Shp2** (protein tyrosine phosphatase) and possibly with PI3-kinase. CTLA-4-deficient mice develop a severe lymphoproliferative disorder.

co-transport In membrane transport describes tight coupling of the transport of one species (generally Na^+) to another (e.g. a sugar or amino acid). The transport of Na^+ from high to low concentration can provide the energy for transport of the second species up a concentration gradient. See secondary **active transport**.

CTX (1) **Cerebrotendinous xanthomatosis.** (2) **Ciguatoxin.** (3) Cross-linked C-telopeptide of collagen Type I, a marker for bone degradation. (4) **Cyclophosphamide.** (5) **Crotoxin** (Ctx), **cardiotoxin**, **conotoxin** or **cholera toxin**. (6) Bacterial genes (*CTX-M* genes) that encode extended-spectrum **beta-lactamases**.

C-type lectins One of two classes of **lectin** produced by animal cells, the other being the **S-type**. The C-type lectins require disulphide-linked cysteines and Ca^{2+} ions in order to bind to a specific carbohydrate (cf. S-type lectins). The carbohydrate recognition domain of C-type lectins consists of about 130aa which contains 18 invariant residues in a highly conserved pattern. These invariant residues include cysteines which probably form disulphide bonds. So far, all identified C-type lectins are extracellular proteins and include both integral membrane proteins, such as the **asialoglycoprotein** receptor, and soluble proteins.

C-type virus Originally C-type particles identified in mouse tumour tissue and later shown to be oncogenic RNA viruses (**Oncovirinae**) that bud form the plasma membrane of the host cell starting as a characteristic electron-dense crescent. Include feline leukaemia virus, murine leukaemia and sarcoma viruses.

CUB domain A widespread 110aa module (complement subcomponent C1r/C1s/embryonic sea urchin protein Uegf /bone morphogenetic protein 1 domain) found in functionally diverse, often developmentally regulated proteins, for which an antiparallel beta-barrel topology similar to that in immunoglobulin V domains has been predicted. CUB domains have been found in the dorso-ventral patterning protein **tolloid**, bone morphogenetic protein 1, a family of **spermadhesins**, **complement** subcomponents C1s/C1r and the neuronal recognition molecule A5. Acidic seminal fluid protein (aSFP) is built by a single CUB domain architecture. Not found in prokaryotes, plants and yeast.

cubilin The **intrinsic factor**-cobalamin receptor (cubilin, 3623aa) is a cotransporter which facilitates the uptake of lipoprotein, vitamins and iron bound to their respective transport proteins (intrinsic factor for vitamin B12 for example). Defects can cause megaloblastic anaemia.

cubitus interruptus A segment polarity gene in *Drosophila* important for specifying the anterior and posterior regions of each segment. The response to **hedgehog** signals in the anterior part of the segment requires **Smoothened** (Smo) and the transcription factor Cubitus interruptus (Ci) which is the *Drosophila* **Gli** homologue. Ci regulates the expression of the *decapentaplegic* (*dpp*) gene which encodes a TGFβ superfamily protein. Flybase entry: http://www.sdbonline.org/fly/segment/cubitus.htm

cuboidal epithelium Epithelium in which the cells are approximately square in vertical section, with the area in contact with the basal lamina comparable to the area of lateral cell-cell contact. In contrast to **columnar** and **squamous** epithelia.

cucurbitacins Feeding stimulants, glucosides, for diabroticite beetles, including corn rootworms and cucumber beetles. Have been used as anti-tumour agents because of their interference with **STAT**3 signalling. In fibroblasts interfere with **lysophosphatidic acid** signalling.

cuffing The accumulation of white cells in tissue immediately adjacent to a blood vessel in certain infections of the nervous system.

Culex pipiens Most widely-distributed species of mosquito. Salivary glands have giant **polytene chromosomes**.

cullins A family of proteins involved in cell cycle control in eukaryotes and components of ubiquitin-ligase complexes. Cullin-1 (cdc53, 776aa), cullin-2 (745aa) and cullin-3 (768aa) are components of multiple ECS (Elongin/cullin/SOCS-box protein) E3 ubiquitin-protein ligase complexes and the SCF ubiquitin protein ligase complex (see **skp**). Function is regulated by neddylation. Mutations in cullin-4B (913aa) are associated with Cabezas X-linked mental retardation syndrome. Cullin-5 (780aa) is the arginine vasopressin (AVP)-activated calcium-mobilizing receptor-1 (VACM1), a cell surface protein. The **3M syndrome** is caused by mutation in cullin-7 (1698aa). Cullin-9 (2517aa) is a cytoplasmic anchor protein in p53-associated protein complex.

cultivar A subspecific rank used in classifying cultivated plants; particular cultivars have distinct properties that can be maintained by vegetative propagation or judicious crossing. Shown as 'cv Name' following the Genus and species names.

culture To grow *in vitro*.

cumulative median lethal dose Estimate of the total administered amount of a substance that is associated with the death of 50% of animals when given repeatedly at doses which are fractions of the median lethal dose.

cumulus Cells surrounding the developing ovum in mammals, fancifully thought to resemble the eponymous clouds.

cupins A superfamily of functionally diverse proteins, found in Archaea, Eubacteria, and Eukaryota, that share a conserved beta-barrel fold. There are enzymic (e.g. microbial phosphomannose isomerases, cereal oxalate oxidases) and non-enzymic (e.g. AraC- type transcriptional regulators, auxin-binding proteins, seed storage proteins) members of the superfamily. In plants the superfamily includes **germin**-like proteins, **vicilins**, NEP-1-like proteins and legumins. Many are seed allergens. Article: http://mbe.oxfordjournals.org/cgi/content/full/18/4/593

Cuprophan A thin membrane of natural cellulose manufactured using the the cuproammonium process. Used for dialysis and ultrafiltration. Can be produced as hollow fibres. Commercial website: http://www.medicell.co.uk/Cuprophan.htm

curacin A An anti-mitotic compound from the marine cyanobacterium, *Lyngbya majuscula*, that binds to the same site on tubulin as colchicine.

curare Curare alkaloids (D-tubocurarine, alloferine, toxiferine and protocurarine) from *Strychnos toxifera* are the active ingredients of arrow poisons used by S. American Indians; they have muscle-relaxant properties because they block motor **endplate** transmission, acting as competitive antagonists for acetylcholine.

curcumin Principal pigment of turmeric, a polyphenol, with a natural yellow colour. Appears to have some anti-inflammatory and anti-oxidant actions and has been extensively used in traditional medicine.

CURL The compartment for uncoupling of receptors and ligands, the region where internalised receptor-ligand complexes are stripped of the ligand and recycled.

Currarino's syndrome A developmental disorder (sacral agenesis) affecting the sacral region caused by mutation in the *HLXB9* homeobox gene or mutations in *rheb* which is mapped to the same chromosomal region.

Cushing's syndrome A type of hypertensive disease in man due probably to the over-secretion of **cortisol** due in turn to excessive secretion of **adrenocorticotrophic hormone** (ACTH). Adrenal tumours are the usual primary cause.

cuticle In plants, the water-repelling waxy layer secreted by the epidermis of aerial parts.

cuticulins Components of the cuticle in *C. elegans* (cuticulin-1, 424aa; cuticulin-2, 231aa). Cuticulin-1 has a **ZP domain** which is probably involved in the formation of large polymeric aggregates. Abstract: http://www.ncbi.nlm.nih.gov/sites/entrez?db=pubmed&cmd=search&term=1864469

cutin Waxy hydrophobic substance deposited on the surface of plants. Composed of complex long-chain fatty esters and other fatty acid derivatives. Impregnates the outer wall of epidermal cells and also forms a separate layer, the cuticle, on the outer surface of the **epidermis**.

cutis laxa A connective tissue disorder characterized by loose, sagging, and inelastic skin. Can be a result of mutations in genes for **elastin, fibulin-4 or -5, pyrroline-5-carboxylate reductase, latent TGFβ-binding protein-4** or in the alpha-2 subunit of the **V-type ATPase**.

CV-1 cells Pseudodiploid cells (ATCC No CCL-70) derived from the kidney of a male African green monkey (*Cercopithecus aethiops*). Morphology is fibroblastic. Much used for transfection and for studies on viral infection.

C value paradox Comparison of the amount of DNA present in the haploid genome of different organisms (the C value) reveals two problems: the value can differ widely between two closely related species, and there seems to be far more DNA in higher organisms than could possibly be required to code for the modest increase in complexity.

CVF Cobra venom factor, see **C3**.

CVS See **chorionic villus sampling**.

CWCV domain A rare amino acid sequence (cys-trp-cys-val) found in extracellular and cell surface molecules, for example **testican**. May be involved in binding growth factors.

CWD See **chronic wasting disease**.

CXADR A single-pass transmembrane protein component (Coxsackievirus and adenovirus receptor, 365aa) of the apical junctional complex in epithelia and essential for tight junctional integrity (interacts with **ZO proteins**). May be a homophilic cell adhesion molecule. Recruits **MPDZ** (multiple PDZ domain protein) to intercellular contact sites. Probably involved in transepithelial migration of polymorphonuclear leukocytes through adhesive interactions with **AMICA1/JAML**. There are multiple isoforms, some of which are secreted.

cy3, cy5 Trade names for fluorescent cyanine dyes used for coupling to probes of various sorts.

cyanidin The cyanidin gene codes for an enzyme that modifies dihydrokaempferol, directing it into the cyanidin pigment pathway, which produces deep red, pink and lilac-mauve hues.

Cyanobacteria *Cyanophyta* Modern term for the blue-green algae, prokaryotic cells that use chlorophyll on intracytoplasmic membranes for photosynthesis. The blue-green colour is due to the presence of **phycobilins**. Found as single cells, colonies or simple filaments. Responsible in part for 'algal blooms' on lakes; some produce neurotoxins or

geosmin, a compound imparting an earthy flavour to water. In *Anabaena*, in which the cells are arranged as a filament, heterocysts capable of nitrogen-fixation occur at regular intervals. According to the **endosymbiont hypothesis** Cyanobacteria are the progenitors of **chloroplasts**.

cyanocobalamin Usual form of **Vitamin B12**. See table V1.

cyanogen bromide *CNBr* Agent that cleaves peptide bonds at methionine residues. The peptide fragments so generated can then, for example, be tested to locate particular activities.

Cyanophyta Blue-green algae See **Cyanobacteria**.

cybrid cell See **hybrid cells**.

Cycadophyta A Division of the plants, gymnosperms, characterized by a stout woody trunk with a crown of large, hard and stiff, evergreen leaves. Palm-like, but unrelated to palms. Now mainly tropical or sub-tropical although more widespread in the Jurassic.

cyclase associated protein Proteins (adenylyl cyclase-associated protein 1, CAP-1, 475aa; CAP-2, 477aa) that bind actin monomers and regulate filament dynamics. Implicated in a number of complex developmental and morphological processes, including mRNA localization and the establishment of cell polarity. CAP homologues are found in many eukaryotes.

cyclic ADP-ribose Second messenger (cADPR, adenosine 5′-cyclic diphosphoribose) synthesized by the multifunctional transmembrane ectoenzyme CD38 in various systems particularly platelets, microsomes and sea urchin eggs. Endogenous regulator of intracellular calcium. May act by regulating ryanodine receptor though other mechanisms are suggested.

cyclic AMP The 3′5′-cyclic ester of AMP (cAMP). The first second-messenger hormone signalling system to be characterized. Generated from ATP by the action of adenyl cyclase that is coupled to hormone receptors by **G-proteins** (**GTP-binding proteins**). cAMP activates a specific (cAMP dependent) protein kinase (PKA) and is inactivated by phosphodiesterase action giving 5′AMP. Also functions as an extracellular morphogen for some slime moulds.

cyclic GMP The 3′5′-cyclic ester of GMP (cGMP), a second-messenger in heart muscle and photoreceptors, generated by **guanylate cyclase**. See **ANP, nitric oxide**.

cyclic inositol phosphates The 1,2-cyclic derivatives of inositol phosphatide that are invariably formed during enzymic hydrolysis of phosphatidyl inositol species. Have been proposed as second messengers in hormone-activated pathways.

cyclic nucleotide phosphodiesterases

Enzymes (EC 3.1.4.-) often casually referred to simply as phosphodiesterases. Multiple isoenzymes are known. PDE1 (531aa) is calcium/calmodulin regulated, hydrolyses both cAMP and cGMP, and is important in CNS and vasorelaxation, PDE2 (941aa) is cGMP-stimulated and hydrolyzes cAMP. PDE3 (PDE3A, 1141; PDE3B, 1112aa) regulates vascular and airway dilation, platelet aggregation, cytokine production and lipolysis; PDE4 (PDE4A, 886aa; inhibited by rolipram) is important in control of airway smooth muscle and inflammatory mediator release but also has a role in CNS and in regulation of gastric acid secretion. PDE5 (875aa) and PDE6 (860aa) are cGMP-specific. PDE5 is involved in platelet aggregation; PDE6 is regulated by interaction with transducin in photoreceptors. PDE7 (482aa) is abundant in skeletal muscle and present in heart and kidney. PDE8 (829aa) is found in thyroid, PDE9 (593aa) is a high affinity cGMP-specific phosphodiesterase, PDE10 (779aa) is found mainly in the CNS. PDE11 (933aa) is ubiquitous and acts on both cAMP and cGMP. PDE12 cleaves $2',5'$-phosphodiester bond (see **2−5A/RNase L system**).

cyclic nucleotide-gated channels

In *Arabidopsis* a family of cyclic nucleotide-gated channels (CNGCs) involved in the control of growth processes and responses to abiotic and biotic stresses. May contribute to cellular cation homeostasis, including calcium and sodium, as well as to stress-related signal transduction. Examples are CNGC1 (716aa), CNGC2, (DEFENSE NO DEATH 1, 726aa). In human, CNGC1 (cGMP-gated cation channel alpha-1, 690aa) is the rod photoreceptor cGMP-gated channel involved in the photoreception, CNCG2 is involved in olfaction. Paper: http://www.biomedcentral.com/1471-2229/9/140/abstract

cyclic phosphorylation

Any process in which a phosphatide ester forms a cylic diester by linkage to a neighbouring hydroxyl group.

cyclic photophosphorylation

Process by which light energy absorbed by photosystem I in the chloroplast can be used to generate ATP without concomitant reduction of $NADP^+$ or other electron acceptors. Energised electrons are passed from PS-I to ferredoxin, and thence along a chain of electron carriers and back to the reaction centre of PS-I, generating ATP en route.

cyclic-dependent kinase

Family of serine/threonine kinases (cdks) including cdc28, cdc2 and p34cdc2 that are only active when they form a complex with cyclins. The complex is regulated by cdk-activating kinase. Cdk9 is atypical (see **cyclin-T1**).

cyclin-dependent kinase activating kinase

Kinase that activates cyclin-dependent kinases (cdks) by phosphorylation. In human cdk7 (346aa) is the catalytic subunit of the cdk-activating kinase (CAK) complex (cdk7, cyclin H and MAT1), a serine-threonine kinase that activates the cyclin-associated kinases cdk1, cdk2, cdk4 and cdk6 by threonine phosphorylation.

cyclin-dependent kinase inhibitors *CKIs*

Two classes of CKIs are known in mammals, the $p21^{CIP1/Waf1}$ class that includes $p27^{KIP1}$ and $p57^{KIP2}$ and that inhibit all G1/S **cyclin-dependent kinases** (cdks), and the p16INK4 class that bind and inhibit only Cdk4 and Cdk6. The $p21^{CIP1}$ inhibitor is transcriptionally regulated by p53 tumour suppressor, is important in G1 DNA-damage checkpoint, and its expression is associated with terminally differentiating tissues. Deletion of $p21^{CIP1}$ is non-lethal in mice; deletion of $p27^{KIP1}$ leads to relatively normal mice but with some proliferation disorders, deletion of $p57^{KIP2}$ causes fairly major developmental abnormalities similar to **Beckwith-Wiedemann syndrome**. See also **ICK1**, **Waf-1**.

cyclins

The proteins that change in concentration at different stages of the cell cycle. They interact with **cyclin-dependent kinases** (cdks) and products of the **cdc** genes. The G1/S transition is regulated by G1/S cyclins, entry into mitosis by G2/M cyclins. Cyclin A (A1, 465aa; A2, 432aa) interacts with cdk2 and **cdc2** and is essential at both G1/S and G2/M transitions; the G1/S transition also involves cyclin D (295aa) interacting with cdk4 and cdk6 and cyclin E (410aa) interacting with cdk2. Cyclin B1 (433aa) is essential for the G2/M transition and interacts with cdc2 to form a serine/threonine kinase holoenzyme complex (maturation promoting factor, MPF); it is ubiquitinated and destroyed abruptly at the end of mitosis. Cyclin B together with cdkK1 regulates the S to G2 progression. Various other cyclins are known: cyclin C (283aa) is a component of the mediator complex. Cyclin-F (786aa) is probably involved in control of the cell cycle during S phase and G2. Cyclin G1 (295aa) is associated with G2/M phase arrest in response to DNA damage. Cyclin-H (323aa) regulates cdk7, the catalytic subunit of the cdk-activating kinase complex. Cyclin-I (377a) does not change in concentration during the cell cycle, cyclin-J (372aa) has an unknown role but is phylogentically ancient. Cyclin-K (580aa) may play a role in transcriptional regulation, cyclin L1 is a transcriptional regulator involved in the pre-mRNA splicing process, cyclin-M (cyclin-related protein FAM58A, 248aa) may have a role in regulating the cell cycle; defects cause **STAR syndrome**. Cyclin-O (EC 3.2.2.-, uracil-DNA glycosylase 2, 350aa) will remove uracil residues accidentally incorporated into DNA or arising by deamination of cytosine. Cyclin-T1 (726aa) is the regulatory subunit of the CDK9/cyclin-T1 complex (positive transcription elongation factor B, P-TEFb), which is proposed to facilitate the transition from abortive to productive elongation by phosphorylating the **CTD** (carboxy-terminal domain) of the large subunit of RNA polymerase II.

Cyclin-Y (Cyclin fold protein 1, 341aa) is membrane associated.

cyclo-oxygenase *COX* Enzyme complex responsible for the synthesis of prostaglandins and thromboxanes from arachidonic acid; inhibited by aspirin-like drugs (non-steroidal anti-inflammatory drugs, **NSAIDs**), probably accounting for their anti-inflammatory effects. Two isoforms are known, COX-1 (prostaglandin G/H synthase 1, EC 1.14.99.1, 599aa) and COX-2 (604aa). Many **NSAIDs** inhibit both isoforms but selective COX-2 inhibitors such as celecoxib have the analgesic and anti-inflammatory activity without deleterious effects on the gastric mucosa. They do, however, have other side-effects and some have been withdrawn. Acetaminophen (paracetamol) has been postulated to act on a third isoform, COX-3, but this has remained elusive and there are alternative explanations. COX3: http://www.ncbi.nlm.nih.gov/pubmed/15705740

cyclodextrins Cyclic polymers of six, seven or eight α-1,4-linked D-glucose residues. The toroidal structure allows them to act as hydrophilic carriers of hydrophobic molecules.

cycloheximide Antibiotic (281Da) isolated from *Streptomyces griseus*. Blocks eukaryotic (but not prokaryotic) protein synthesis by preventing initiation and elongation on 80S ribosomes. Commonly used experimentally.

CYCLOIDEA Gene that regulates dorsoventral asymmetry in flowers. Encodes a transcription factor of the TCP family.

cyclolysin Protein (adenylate cyclase toxin, 1706aa), from *Bordetella pertussis* that is both an adenylate cyclase and a haemolysin; an important virulence factor of the organism. Article: http://cmr.asm.org/content/18/2/326.full.pdf

cyclopamine Teratogenic steroidal alkaloid produced by the skunk cabbage (*Veratrum californicum*) that causes developmental defects such as cyclopia (one eye in the middle of the face). Cyclopamine blocks activation of the Hedgehog response pathway, specifically the multipass transmembrane proteins **Smoothened** (Smo) and Patched (Ptch). Cyclopamine will block the oncogenic effects of mutations of Ptch in fibroblasts (see **Gorlin's syndrome**) and inhibits the growth of cells lacking Ptch function.

cyclopentyladenosine A specific agonist of the adenosine A(1) receptor.

cyclophilin Enzyme (EC 5.2.1.8, 754aa) with **PPIase** activity; binds the immunosuppressive drug **ciclosporin A**. See **immunophilin**. In *Arabidopsis* there are at least 14 cyclophilin-like proteins that are likely to be involved in a range of different activities. Some are induced by stress. *Arabidopsis*

cyclophilins: http://www.plantphysiol.org/content/134/4/1268.full

cyclophosphamide An alkylating agent and important immunosuppressant. Acts by alkylating SH and NH$_2$ groups especially the N7 of guanine.

cyclopiazonic acid A mycotoxin produced by *Penicillium cyclopium* that selectively inhibits the sarcoplasmic-endoplasmic reticulum Ca^{2+}-ATPase (SERCA).

cyclosis Cyclical streaming of the cytoplasm of plant cells, conspicuous in giant internodal cells of algae such as *Chara*, in pollen tubes and in stamen hairs of *Tradescantia*. Term also used to denote cyclical movement of food vacuoles from mouth to cytoproct in ciliate protozoa.

cyclosome See **anaphase-promoting complex**.

cyclosporin A See **ciclosporin**.

Cyclotella A genus of small drum-shaped diatoms with cells 3–5 μm in diameter. *Cyclotella meneghiniana* is perhaps the commonest species, widely used in growth experiments. Image: http://silicasecchidisk.conncoll.edu

cylindrospermopsin A protein synthesis inhibitor, a tricyclic alkaloid hepatotoxin originally isolated from the cyanobacterium *Cylindrospermopsis raciborskii* and subsequently found in other species of cyanobacteria. The mechanism of toxicity differs from that of microcystins.

CYP1A Cytochrome P450-1A, see **cytochrome P450**.

cypermethrin Synthetic pyrethroid used as an insecticide. Acts on sodium channels of neurons. Highly toxic to aquatic organisms, including fish. Usually a mixture of several isomers.

cypin A mammalian guanine deaminase (cytosolic PSD-95 interactor) that increases dendrite number when overexpressed. Cypin contains zinc binding, **collapsin response mediator protein** (CRMP) homology, and PSD-95, Discs large, zona occludens-1 binding domains. Binds tubulin via its CRMP homology domain to promote microtubule assembly; this interaction is blocked by **snapin**.

cypris Larval stage of Cirrepedia (barnacles) following nauplius stage.

cyst (1) A resting stage of many prokaryotes and eukaryotes in which a cell or several cells are surrounded with a protective wall of extracellular materials. (2) A pathological fluid-filled sac bounded by a cellular wall, often of epithelial origin, found on occasion in all species of multicellular animal. May result from a wide range of insults or be of embryological origin.

cystathionine γ-synthase In higher plants, the first enzyme (EC 2.5.1.48, 563aa) specific to methionine biosynthesis. It catalyses the reaction of succinyl-homoserine and cysteine to form cystathionine and succinate. Pathway of methionine biosynthesis in potato: http://www.biomedcentral.com/1471-2229/8/65

cystatins A large group of natural cysteine-peptidase inhibitors widely distributed both intra- and extra-cellularly. Family I is of **stefins**, Family II the cystatins, Family III the **kininogens**. Cystatin-1 (cystatin-SN, 141aa), cystatin-2 (cystatin-SA, 141), cystatin-4 (cystatin-S, 141) and cystatin-5 (cystatin-D, 142aa) are secreted in saliva. Cystatin 3 (cystatin C, CSTC, 120aa) is excreted in urine and is defective in **Icelandic-type cerebroarterial amyloidosis**. Cystatin-6 (cystatin-M, 149aa) is found in skin, cystatin-7 (cystatin-F, Leukocystatin, 145aa) is mainly found in peripheral blood cells and spleen. Cystatin-8 (cystatin-related epididymal spermatogenic protein, 142aa) is important in sperm development, cystatin-9 (159aa) may be involved in haematopoiesis and inflammation. Others are known and the various forms differ in their specific inhibitory capacity.

cysteine *Cys, C* The only amino acid to contain a thiol (SH) group. In intracellular enzymes the unique reactivity of this group is frequently exploited at the catalytic site (as in cysteine endopeptidases). In extracellular proteins found only as half-cystine in disulphide bridges or fatty acylated.

cysteine peptidase A clan of endopeptidases (thiol proteinase, thiol protease, subclass EC 3.4.22). All have a cysteine residue in the active site that can be irreversibly inhibited by sulphydryl reagents. Includes **cathepsins** and **papain**. Natural inhibitors are **alpha-2-macroglobulin** and **cystatins**.

cysteine phosphatases A large family of phosphatases that utilize a conserved 'CX5R' sequence motif to hydrolyze phosphoester bonds in proteins and in non-protein substrates. They can be divided into seven categories: (i) protein tyrosine phosphatases (PTPs); (ii) dual-specificity phosphatases (DSPs); (iii) Cdc25 phosphatases; (iv) myotubularin-related; and (iv) low molecular weight phosphatases; (v) inositol 4-phosphatases; and (vi) Sac1-domain phosphatases. Database: http://ptp.cshl.edu/

cysteine-rich PDZ-binding protein A protein (CRIPT, 101aa) involved in the cytoskeletal anchoring of DLG4 (disks large homologue four) in excitatory synapses.

cysteine-rich receptor-like protein kinases In *Arabidopsis* a sub-family of more than 40 receptor-like protein kinases that contain the DUF26 motif (Domain of Unknown Function 26) in their extracellular domains (e.g. cysteine-rich receptor-like protein kinase 1, cysteine-rich RLK1, EC 2.7.11.1, receptor-like kinase in flowers 2, 615aa). Considered to have important roles in the regulation of pathogen defence and programmed cell death they are transcriptionally induced by oxidative stress, pathogen attack and application of salicylic acid. Transcriptional regulation: http://www.biomedcentral.com/content/pdf/1471-2229-10-95.pdf

cysteine-rich secretory proteins A family of proteins (CRISPs) with 16 conserved cysteine residues. Cysteine-rich secretory protein-1 (CRISP-1, 249aa) is found in the male reproductive tract and may have a role in sperm-egg fusion; CRISP-2 (243aa) is found in testis and epididymis. CRISP-3 (245aa) is found in the salivary gland, pancreas and prostate, also in the specific granules of neutrophils. Some of the other members of the family have **LCCL** domains (e.g. CRISP-11, 497aa, which promotes matrix assembly). See **helothermine, resistin**.

cysteine string proteins Membrane-associated protein (CSP, DnaJ homolog subfamily C member 5, 198aa), composed of a string of palmitoylated cysteine residues, a linker domain, and an N-terminal J domain characteristic of the DnaJ/Hsp40 co-chaperone family. May regulate calcium-dependent neurotransmitter release at nerve endings. In *Drosophila*, cysteine string protein (249aa) is also thought to be important in presynaptic function.

cystic fibrosis Generalized abnormality of exocrine gland secretion that affects pancreas (blockage of the ducts leads to cyst formation and to a shortage of digestive enzymes), bowel, biliary tree, sweat glands and lungs. The production of abnormal mucus in the lung predisposes to respiratory infection, a major problem in children with the disorder. A fairly common (1 in 2000 live births in caucasians) **autosomal recessive** disease. See **cystic fibrosis transmembrane conductance regulator** (CFTR).

cystic fibrosis antigen Now known to be MIF related protein 8 (MRP-8). See **calgranulins**.

cystic fibrosis transmembrane conductance regulator *CFTR* Product of a gene frequently mutated in **cystic fibrosis**; a range of different mutations are known to occur, making preimplantation genetic diagnosis difficult. Gene encodes a chloride channel, an ABC protein (ABCC7, EC 3.6.3.49, cAMP-dependent chloride channel, 1480aa).

cystine The amino acid formed by linking two **cysteine** residues with a disulphide linkage between the two sulphydryl (SH) groups. The analogous compound present within proteins is termed two half cysteines.

cystinosin Lysosomal protein (367aa) responsible for cystine export. Defects lead to nephropathic cystinosis and mutations in the the cystinosin gene are the most common cause of inherited renal **Fanconi syndrome**. Highly conserved in mammals. Yeast Ers1 is a functional orthologue.

cystinosis An autosomal recessive lysosomal storage disorder caused by a defect in the lysosomal cystine carrier **cystinosin**.

cystoblast In *Drosophila*, a differentiating daughter cell derived from a germline stem cell as the first step in oogenesis. The cystoblast undergoes four rounds of synchronous divisions with incomplete cytokinesis to generate a syncytial cyst of 16 interconnected cystocytes, within the cyst one of the cystocytes differentiates into an oocyte. *Cf.* **gonioblast**.

cystocyte See **cystoblast**.

cytarabine See **cytosine arabinoside**.

cytidine Nucleoside consisting of D-ribose and the pyrimidine base cytosine. Cytidine 5'diphosphate (CDP) derived from CTP is important in phosphatide biosynthesis; activated choline is CDP-choline.

cytidylic acid Ribonucleotide of **cytosine**.

cytisine A natural alkaloid found in many of the Leguminosae family, and the main toxin of the common garden Laburnum. Has very high affinity for the α4β2-nicotinic ACh receptors.

cytoband See **cytogenetic band**.

cytocalbins Cytoskeleton-related calmodulin-binding proteins. Seemingly obsolete.

cytochalasins A group of fungal metabolites that inhibit the addition of G-actin to a nucleation site and therefore perturb labile microfilament arrays. Cytochalasin B from *Helminthosporium dematioideum* inhibits at around 1 mg/ml but at about 5 mg/ml begins to inhibit glucose transport. Cytochalasin D from *Metarrhizium anisopliae* is preferable because it only affects the microfilament system.

cytochemistry Branch of histochemistry associated with the localization of cellular components by specific staining methods, as for example the localization of acid phosphatases by the Gomori method. Immunocytochemistry involves the use of labelled antibodies as part of the staining procedure.

cytochrome b6-f complex A multi-subunit complex that mediates electron transfer between photosystem II (PSII) and photosystem I (PSI). The 4 large subunits are cytochrome b6 (petB, 222aa), subunit IV (PetD, 160aa), apocytochrome f (petA, 320aa) and the **Rieske protein** (petC), while the 4 small subunits (30−40aa) are PetG, PetL, PetM and PetN. The complex functions as a dimer. Inhibited by the quinone analogue **DBMIB**.

cytochrome oxidase An integral membrane protein of the inner mitochondrial membrane that is the last enzyme (EC 1.9.3.1.) of the electron transport chain. It accepts electrons from (i.e. oxidizes) cytochrome C and transfers the electrons to molecular oxygen. There are 13 subunits (COX1, COX2, etc.).

cytochrome P450 A large group of mixed-function oxidases (EC 1.14.14.1) of the cytochrome b type, involved, among other things, in steroid hydroxylation reactions in the adrenal cortex. In liver cells they are found in the microsomal fraction and can be induced for the detoxification of foreign substances, including drugs. Found in most animal cells and organelles, in plants, and in microorganisms. They catalyze a wide range of oxygenation steps in plant metabolism. The genes and proteins are named 'Cyp' suffixed by family, sub-family and specific identifiers. Functional genomics of cytochrome P450 in *Arabidopsis*: http://www-ibmp.u-strasbg.fr/ ~ CYPedia/

cytochromes Enzymes of the electron transport chain that are pigmented by virtue of their **haem** prosthetic groups. Very highly conserved in evolution.

cytogenetic band A chromosomal subregion (cytoband) visible microscopically after special staining. See **chromosome banding**.

cytogenetics The study of the chromosomal complement of cells, and of chromosomal abnormalities and their inheritance.

cytohesins A family of guanine nucleotide exchange factors (GEFs) for ADP-ribosylation factor (**ARF**) GTPases. Cytohesin-1 (PH, SEC7 and coiled-coil domain-containing protein-1, SEC7 homologue B2-1, 398aa) is ubiquitous, and acts on ARF1 and ARF5. Cytohesin-2 (ARF nucleotide-binding site opener, ARNO, 400aa) acts on ARF1, ARF3 and ARF6, cytohesin-3 (general receptor of phosphoinositides 1, Grp-1, 400aa) acts on ARF1 and binds the inositol head group of phosphotidylinositol 3,4,5-trisphosphate. Cytohesin-4 (394aa) is found mainly in leucocytes. Cytohesin-interacting protein (359aa) is a scaffolding and modulator protein that binds to cytohesin-1 and modifies the activation of ARF.

cytokeratins Generic name for the intermediate filament proteins of epithelial cells.

cytokine receptors Cytokine receptors can be classified into several sub-types: Type I recognize and respond to cytokines with four a-helical strands and are heterodimeric with different cytokine-binding and signalling subunits. The signalling subunits are often shared between receptors for different cytokines. Includes receptors for interleukins 2−7, 9, 11−13, 15, 21, 23 and 27, receptors for erythropoietin, GM-CSF, G-CSF, growth hormone, prolactin, oncostatin M and LIF. Type 2 receptors are only distantly related, are usually heterodimers or multi-mers with a high and a low affinity component and signal through JAKs. They include receptors for type1 and type 2 interferons and interleukins 10, 20, 22 and 28. Other cytokines (IL-1, IL-18) have receptors from the immunoglobulin superfamily or G-protein coupled chemokine receptors (IL-8). There are separate classes of receptors for the **TNF superfamily** and for TGFβ family.

cytokines Small proteins (in the range of 5–20 kDa) released by cells and that affect the behaviour of other cells. Not really different from hormones, but the term tends to be used as a convenient generic shorthand for **interleukins**, **lymphokines** and several related signalling molecules such as **TNF** and **interferons**. Generally growth factors would not be classified as cytokines, though TGF is an exception. Rather an imprecise term, though in very common usage. **Chemokines** are a subset of cytokines; see Table C3.

cytokinesis Process in which the cytoplasm of a cell is divided after nuclear division (mitosis) is complete. Cytokinesis in animal cells involves an actin-rich contractile ring that pinches the daughter cells apart. In plants, the microtubule-rich **phragmoplast** deposits the **cell plate**, a disk that grows outwards from the centre of the cell, eventually fusing with the mother cell's wall.

cytokinins Class of **plant growth substances** (plant hormones) active in promoting cell division. Also involved in cell growth and differentiation and in other physiological processes. Examples: **kinetin**, **zeatin**, benzyl adenine.

cytology The study of cells. Implies the use of light or electron microscopic methods for the study of morphology.

cytolysis Cell **lysis**.

cytolysosome Membrane-bounded region of cytoplasm that is subsequently digested.

Cytomegalovirus Probably the most widespread of the Herpetoviridae group. Infected cells enlarge and have a characteristic inclusion body (composed of virus particles) in the nucleus. Causes disease only *in utero* (leading to abortion or stillbirth or to various congenital defects), although can be opportunistic in the immunocompromised host. See **viperin**.

cyton (1) The region of a neuron containing the nucleus and most cellular organelles. (2) In Cestodes (e.g. tapeworms) the syncytial epithelium that forms the tegument has the nucleated, proximal cytoplasm, or cyton, sunk deep in the parenchyma. The cyton region contains Golgi complexes, mitochondria, rough ER, and other organelles involved in protein synthesis and packaging (3) In cercaria of *Schistosoma mansoni*, an aggregate of subtegumental

cells is found in a small, dorsoanterior area. These highly amorphous cells type, designated as cyton II, have a heterochromatic nucleus and a cytoplasm that is elaborated into coarse, tortuous processes.

cytonectin A 35 kDa adhesion protein, independent of divalent cations, expressed in a variety of organs and tissues, being evolutionarily conserved from human to avian species, overexpressed in Alzheimer's disease entorhinal cortex. There is only one literature reference (2002) but a patent application is filed and antibodies are commercially available. Sole paper: http://www.ncbi.nlm.nih.gov/pubmed/11895037

cytoneme A long thin projection from an epithelial cell, a filopodium.

cytophotometry Examination of a cell by measuring the light allowed through it following staining.

cytoplasm Substance contained within the plasma membrane excluding, in eukaryotes, the nucleus.

cytoplasmic bridge *plasmodesmata* Thin strand of cytoplasm linking cells as in higher plants, *Volvox*, between **nurse cells** and developing eggs, and between developing sperm cells. Unlike gap junctions, allows the transfer of large macromolecules.

cytoplasmic determinants Slightly imprecise term usually applied to non-randomly distributed factors in maternal (oocyte) cytoplasm that determine the fate of blastomeres derived from this region of the egg following cleavage. Thus blastomeres containing cytoplasm derived from the apical region (animal pole) of the echinoderm egg form ectodermal tissues. Can be used even more generally of any feature in the cytoplasm that determines how a process or activity proceeds.

cytoplasmic inheritance Inheritance of parental characters through a non-chromosomal means; thus mitochondrial DNA is cytoplasmically inherited since the information is not segregated at mitosis. In a broader sense the organization of a cell may be inherited through the continuity of structures from one generation to the next. It has often been speculated that the information for some structures may not be encoded in the genomic DNA, particularly in protozoa that have complex patterns of surface organelles. See maternal inheritance.

cytoplasmic polyadenylation element An uridine-rich sequence element (CPE, consensus sequence 5′-UUUUUAU-3′) within the mRNA 3′-UTR that recruits the **cytoplasmic polyadenylation element binding protein** (CPEB) to quiescent mRNA that has been stockpiled in the egg (and had the poly-A tail truncated). **Maskin** may repress translation of mRNA with a CPE.

cytoplasmic polyadenylation element binding factors RNA binding proteins (e.g. CPEB1, 566aa) that control polyadenylation-induced translation in germ cells and at postsynaptic sites of neurons. They bind to the **cytoplasmic polyadenylation element** and to **maskin**. In association with maskin repress translation of CPE-containing mRNA; repression is relieved by phosphorylation. Involved in the transport of CPE-containing mRNA to dendrites. CPEB recruits **CPSF** (cleavage and polyadenylation specificity factor), which physically interacts with both CPEB and another sequence in the mRNA, to the 3′ side of the CPE, called the **nuclear polyadenylation hexanucleotide**. This particular sequence (AAUAAA) is required for cytoplasmic polyadenylation. CPSF, once bound to the mRNA, recruits the enzyme poly A polymerase to the mRNA.

cytoplasmic streaming Bulk flow of the cytoplasm of cells. Most conspicuous in large cells such as amoebae, in vacuolated plant cells and the internodal cells of *Chara* where the rate of movement can be 50–60 μm/sec. Powered by the movement of myosin-coated organelles along actin filaments. See **cyclosis**.

cytoplast Fragment of cell with nucleus removed (in **karyoplast**); usually achieved by cytochalasin B treatment followed by mild centrifugation on a step gradient.

cytoproct Cell anus: region at posterior of a ciliate where exhausted food vacuoles are expelled.

cytorrhysis Process in which a plant cell wall collapses inward following water loss due to hyperosmotic stress. *Cf.* plasmolysis.

cytosine Pyrimidine base found in DNA and RNA. Pairs with guanine. Glycosylated base is **cytidine**.

cytosine arabinoside *cytarabine* Cytotoxic drug used in oncology (particularly **AML**) and against viral infections. Blocks DNA synthesis.

cytoskeleton General term for the internal components of animal cells which give them structural strength and motility (plant cells and bacteria use an extracellular **cell wall** for structural support instead). The major components of the cytoskeleton are the **microfilaments** (of **actin**), **microtubules** (of **tubulin**) and **intermediate filament** systems in cells.

cytosol That part of the cytoplasm that remains when organelles and internal membrane systems are removed.

cytosome (1) The body of a cell apart from its nucleus. (2) A multilamellar body found in cells of the lung.

cytostome A specialized region of various protozoans in which phagocytosis is likely to occur. Often there is a clear concentration of microtubules or/and microfilaments in the region of the cytostome. In ciliates there may be a specialized arrangement of cilia around the cytostome.

cytotactin See tenascin.

cytotoxic drugs A term that could be applied to any drug that kills cells but usually used to mean those drugs used to treat cancer. In practice this means targeting rapidly proliferating cells so there are usually side effects. They include alkylating drugs, cytotoxic antibiotics (e.g. doxorubicin, bleomycin), anti-metabolites and anti-mitotic drugs (eleutherobin, maytansine, taxanes, vinca alkaloids). Other examples are bakuchiol, camptothecin, cisplatin, equisetin, gemcitabine, hadacidin, hexacyclinic acid, kosinostatin, pentostatin, psoralidin, reticulol and rottlerin.

cytotoxic necrotising factors Toxins (CNF1, CNF2, 1014aa) produced by some strains of *E. coli*. Induce ruffling and stress fibre formation in fibroblasts and block cytokinesis by acting on p21 Rho. A similar factor (CNF-γ) is produced by *Yersinia pseudotuberculosis*. Effect on neutrophils: http://www.jleukbio.org/content/68/4/522.full

cytotoxic T-cells *CTLs*. Subset of T-lymphocytes (mostly CD8$^+$) responsible for lysing target cells and for killing virus-infected cells (in the context of Class I **histocompatibility antigens**).

cytotrophic Descriptive of any substance that promotes the growth or survival of cells. Not commonly used except in the tissue-specific case of **neurotrophic** factors.

cytotropic Having affinity for cells: not to be confused with cytotrophic.

cytotropism Movement of cells towards or away from other cells.

cytovillin See **ezrin**.

D

D Single letter code for aspartic acid.

D₂O Deuterium oxide. See **heavy water**.

DAB (**1**) See **diaminobenzidine**. (**2**) A phosphoprotein (DAB1, disabled homologue 1, 588aa) an adapter that functions in neural development, binds nonreceptor tyrosine kinases and interacts with **lipoprotein receptor-related protein-8**. Mutations in murine homologue are responsible for various behavioural changes ('scrambler' and 'yotari' mutants).

DABA See **diaminobenzoic acid**.

Dacapo In *Drosophila* a cyclin-dependent kinase inhibitor (245aa). Homologue of human p27(Kip1).

DAF (**1**) **Decay accelerating factor**. (**2**) Diaminofluorescein, a sensitive fluorescent probe for the real-time detection of nitric oxide (NO) *in vivo*. In the cell, DAF-2 reacts rapidly with NO in the presence of O_2 to form the highly fluorescent compound triazolofluorescein. (**3**) DAF-16 is a **forkhead** transcription factor.

DAG See **diacylglycerol, phosphatidyl inositol** and **dystroglycan**.

daidzein The aglycone of daidzin, an isoflavone phytoestrogen, a plant-derived nonsteroidal compound that possesses estrogen-like biological activity. Mainly found in legumes, such as soybeans and chickpeas, it has been found to have both weak estrogenic and weak anti-estrogenic effects.

DAL-1 A **band 4.1**-like protein (differentially expressed in adenocarcinoma of the lung, 4.1B, 1087aa) which is lost in approximately 60% of non-small cell lung carcinomas, and exhibits growth-suppressing properties in lung cancer cell lines.

Dalton The unit of atomic mass. *Abbrev.* Da.

Daltonism See **colour blindness**.

damaged DNA binding proteins Proteins (DDB1, p127, 1140aa; DDB2, p48, 427aa) that form a heterodimeric damaged DNA-binding protein complex involved in the initial recognition of UV-damaged DNA and that help recruit nucleotide **excision repair** factor. Mutations in the DDB2 gene are found in complementation group E of **xeroderma pigmentosum**.

DAN (**1**) A zinc finger protein (differential screening-selected gene aberrative in neuroblastoma, neuroblastoma suppressor of tumorigenicity 1, 180aa) a BMP-antagonist that regulates **BMP** activity spatially and temporally during patterning and partitioning of the medial otic tissue in ear development. One of a family of proteins (**cer/dan family**) which all have a C-terminal cystine knot-like (CTCK) domain. (**2**) In *Drosophila*, distal antenna (*dan*) and distal antenna-related (*danr*) genes, encode **pipsqueak motif** DNA-binding domain protein family members required for timely downregulation of **hunchback** in neuroblasts and for limiting the number of early-born neurons.

Dane particle The complete infective virion of **hepatitis B**, 42 nm diameter.

Danio rerio Formerly *Brachydanio rerio*, the **zebrafish**.

Danon's disease An X-linked lysosomal storage disease with characteristic intracytoplasmic vacuoles containing autophagic material and glycogen in skeletal and cardiac muscle. Defect is in **LAMP-2**, a lysosomal membrane structural protein, not an enzyme. There is cardiomyopathy, myopathy and variable mental retardation.

dansyl chloride A strongly fluorescent compound that will react with the terminal amino group of a protein. After acid hydrolysis of all the other peptide bonds, the terminal amino acid is identifiable as the dansylated residue.

DAP (**1**) See **death-associated proteins** and **DAP kinase**. (**2**) In humans, disks large-associated proteins (e.g. DAP-1, 977aa, which is part of the post-synaptic scaffold in neuronal cells): see **DLG**. There are several other disks-large associated proteins. (**3**) See **DNAX-activation proteins**. (**4**) A compound, diaminopimelate, that is a component of cell wall mucopeptide in some bacteria and as a source of lysine in all bacteria. (**5**) In *Arabidopsis*, the gene encoding a chloroplast enzyme (DAP-aminotransferase, LL-diaminopimelate aminotransferase, EC 2.6.1.83, 461aa) required for lysine biosynthesis. (**6**) In *S. pombe*, a cytochrome P450 regulator (dap1, 166aa) required for sterol biosynthesis.

DAP kinases A pro-apoptotic family of multidomain calcium/calmodulin (CaM)-dependent Ser/Thr protein kinases (death-associated protein kinases, DAPK1, 1430aa; DAPK2, 370aa; DAPK3, ZIP-kinase, 454aa) that are phosphorylated upon activation of the Ras-extracellular signal-regulated kinase (ERK) pathway. DAP-related apoptotic kinases (DRAK1, 414aa; DRAK2, 372aa) are highly expressed in placenta.

The Dictionary of Cell and Molecular Biology. DOI: http://dx.doi.org/10.1016/B978-0-12-384931-1.00004-0

DAPI stain Fluorochrome that binds to DNA and is used biochemically for detection of DNA and to stain the nucleus in fluorescence microscopy.

dapper In *Xenopus* a protein (dapper-1, 824aa) that positively regulates **dishevelled** signalling pathways during development. Impedes the degradation of beta-catenin, thereby enhancing the Wnt signalling pathway. Dapper-1A (frodo, 818aa) and dapper-1B (824aa) are restricted in their tissue distribution. There are homologues in human: dapper homologue 1 (hepatocellular carcinoma novel gene 3 protein, 836aa) and dapper homologue 3 (arginine-rich region 1 protein, 629aa) act in the dishevelled pathway; dapper homologue 2 (774aa) negatively regulates the **nodal** signalling pathway.

dapsone Drug related to the sulphonamides (diaminodiphenyl sulphone) that is used to treat **leprosy**. May act by inhibiting folate synthesis.

daptomycin A cyclic lipopeptide antibiotic active against Gram-positive bacteria that inserts into the bacterial cell membrane, causing rapid membrane depolarization, potassium ion efflux, followed by arrest of DNA, RNA and protein synthesis and eventually causing bacterial cell death. Article: http://jac.oxfordjournals.org/content/55/3/283.full

darcin In mouse, major urinary protein 20 (181aa) which stimulates female sexual attraction to male urinary scent. Promotes male aggressive behaviour. Binds most of the male pheromone, 2-sec-butyl-4,5-dihydrothiazole, in urine.

dardarin A mixed-lineage kinase (**MLK**) (leucine-rich repeat serine/threonine-protein kinase 2, EC 2.7.11.1, 2527aa) expressed in the brain but also, at lower levels in other tissues. Role unclear but mutations cause **Parkinson's disease** type 8 (PARK8).

Darier's disease A skin disorder (Darier-White disease, keratosis follicularis) caused by mutations in the gene encoding the sarcoplasmic reticulum calcium ATPase (SERCA2) which affects **desmosome** integrity.

dark current In the retina, a current caused by constant influx of sodium ions into the **rod outer segment** of retinal photoreceptors that depolarizes them to around -40 mV. It is blocked by light which causes hyperpolarization. The plasma membrane sodium channel is cGMP-gated and controlled through a cascade of amplification reactions initiated by photon capture by **rhodopsin** in the disc membrane.

dark field microscopy See **light microscopy**.

dark reaction The reactions in photosynthesis that occur after NADPH and ATP production, and that take place in the stroma of the chloroplast. By means of the reaction, CO_2 is incorporated into carbohydrate.

Darling's disease See **histoplasmosis**.

Darwinian fitness General measure of the fitness of an individual, the number of offspring of that individual that will survive to reproduce in the next generation. It is a measure of the survival of alleles of genes that contributed to the selective advantage. It can be calculated relative to the phenotype with the highest absolute reproductive success which normalises the values and gives the maximum Darwinian fitness a value of unity.

DASH complex A microtubule-binding subcomplex of the outer kinetochore that is essential for proper chromosome segregation. The complex is composed of ten different subunits (one of each) and is around 210 kDa. The subunits in *S. cerevisiae* are ASK1 (292aa), DAD1 (94aa), DAD2 (133aa), DAD3 (94aa), DAD4 (72aa0, DAM1 (343aa0, DUO1 (247aa), HSK3 (69aa), SPC19 (165aa) and SPC34 (295aa). The DASH complex mediates the formation and maintenance of bipolar kinetochore-microtubule attachments by forming closed rings around spindle microtubules and establishing interactions with proteins from the central kinetochore. N.B. See **defender against cell death**.

database A collection of records or of data, almost invariably stored in a computer system. The advantage of using a database is that all records with a particular attribute can be extracted easily. The database used for preparing this dictionary allows, for example, all headwords beginning with 'D' to be extracted (638) or all definitions that include the word database (68). Many web-based databases with useful data about genes and proteins are becoming available, together with tools that allow searching for similarities.

dATP The reduced form of **ATP** the deoxyribonucleotide that is incorporated into DNA, produced by the action of ribonucleotide reductase (EC 1.17.4.1).

Datura stramonium Jimson weed or thornapple (also called Loco Weed, Jamestown Weed, Angel's Trumpet; Zombie's Cucumber), the source of **scopolamine**.

Daudi B-lymphoblastoid cell line derived from a 16-year-old male black African with Burkitt's lymphoma. Have surface complement receptors and IgG and are **EBV** marker positive.

dauer larva Semidormant stage of larval development in nematodes (for example *C. elegans*), triggered by a pheromone: essentially a survival strategy.

daughterless A *Drosophila* DNA-binding protein (710aa) that forms heterodimers with **achaete-scute complex** proteins that act as transcriptional activators of neural cell fates and are involved in sex determination. Also interacts with **amos** and **atonal**.

daunomycin See **daunorubicin**.

daunorubicin *daunomycin* Cytotoxic anthracycline antibiotic used in cancer chemotherapy. Intercalates between base pairs in DNA causing uncoiling of the helix, ultimately inhibiting DNA synthesis and DNA-dependent RNA synthesis. Also inhibits **topoisomerase** II activity by stabilizing the DNA-topoisomerase II complex. Cytotoxic activity is cell cycle phase non-specific, although effects are maximal in S-phase. Produced by *Streptomyces peucetius*.

DAXX A ubiquitously expressed protein, death domain-associated protein 6 (Fas death domain-associated protein, 740aa) that acts as an adapter protein in a MDM2-DAXX-USP7 complex. Has a complex role in regulating stress-induced apoptosis. Binds to the Fas death domain and enhances Fas-mediated apoptosis and is a component of nuclear promyelocytic leukemia protein (PML) oncogenic domains **PODS**. Acts downstream of apoptosis signal-regulating kinase (**ASK1**).

DBA mice Oldest of all the strains of inbred mice. Various sub-strains (DBA1, DBA2) are used extensively in experimental work. DBA/1J mice are widely used as a model for rheumatoid arthritis: injection with typeII collagen induces a poly-arthritis. Details: http://www.informatics.jax.org/external/festing/mouse/docs/DBA.shtml

dbl Human oncogene (*dbl, MCF-2*) originally identified by transfection of NIH-3T3 cells with DNA from human diffuse B-cell lymphoma. The product (925aa) is cleaved into one of two forms, MCF2-transforming protein (528aa) and DBL-transforming protein (428aa), both of which are guanine nucleotide exchange factors (**GEFs**) for rho-family members. Both have the PH and **Dbl-homology domain**. The Dbl family of related proteins with similar functions includes Dbl itself, Dbs, Brx, Lfc, Lsc, Ect2, DRhoGEF2 and Vav.

Dbl homology domain A cytoskeletal modulation-type domain (DH domain, ~150aa) found in **Dbl**, Vav and the other Dbl-family proteins (Lfc, Lsc, Ect2, Dbs, Brx). DH domains are invariably located immediately N-terminal to a **PH domain** and the membrane localization and enzymatic activity of the DH domain requires the PH domain for normal function.

DBMIB Dibromothymoquinone, a quinone analogue that inhibits the **cytochrome b6f complex** by occupying the Q(o) site.

dbp5 (1) In *S. cerevisiae*, and many other organisms, an ATP-dependent RNA helicase (DBP5, DEAD box protein 5, helicase CA5/6, ribonucleic acid-trafficking protein 8, 482aa) associated with the nuclear pore complex and essential for mRNA export The human equivalent, ATP-dependent RNA helicase

DDX19B (EC 3.6.4.13, DEAD box RNA helicase DEAD5, 479aa) is also involved in mRNA export from the nucleus. (2) In humans, a protein (DBP-5, SON3, Bax antagonist selected in saccharomyces 1, BASS1, negative regulatory element-binding protein, 2426aa) that may be involved in pre-mRNA splicing and has a double-stranded RNA-binding domain. Represses hepatitis B virus (HBV) core promoter activity, transcription of HBV genes and production of HBV virions.

DCC The DCC (deleted in colorectal cancer) gene is deleted in 70% of colorectal cancers, encodes a protein of the immunoglobulin superfamily (1447aa) that is a receptor for **netrins** and is essential for axonal guidance in development. It attracts growth cones when netrin is bound but may repel axons in association with UNC-5. Seems to be a tumour suppressor. DCC is one of a family of receptors that includes *Drosophila* frazzled which binds **netrin** in association with one of the UNC-5 family of receptors.

D cells Cells (δ cells, delta cells) of the pancreas; about 5% of the cells present in primate pancreas with small argentaffin-positive granules. They produce somatostatin and have receptors for **gastrin** and acetylcholine. They are also found in the stomach and intestine.

DCF An inhibitor (2'-deoxycoformycin) of adenosine deaminase.

DCIP A commonly used electron acceptor dye (2,6-dichlorophenolindophenol, Tillman's reagent) which can accept electrons instead of P700. DCIP is blue in neutral solution and pink in acidic solution; the reduced form is colourless. Used in test kits for the detection of haemoglobin E.

DCL In plants the enzymes (Dicer-like, EC3.1.26.- 1) that process primary miRNA transcripts into an miRNA-miRNA* duplex. There are four Dicer-like (DCL) enzymes (DCL1, 1909aa; DCL2, 1388aa; DCL3, 1580aa; DCL4, 1702aa) encoded in *Arabidopsis*; DCL2 is involved in the biogenesis of viral short-interfering RNA (siRNA) and DCL3 in endogenous siRNA biogenesis such as retrotransposon siRNA. DCL4 may be involved in cleaving double-stranded RNA and processing **trans-acting siRNAs**.

DCMU An inhibitor (3-(3,4-dichlorophenyl)-1,1-dimethylurea, diuron) of photosynthetic electron chain transport, used agriculturally as a non-selective herbicide. Persists in the environment and breakdown products are also toxic.

DCPIP A blue dye that, when reduced by electron addition, becomes colourless. Often used in measurements of the electron transport chain in plants.

DCX See **doublecortin**.

DD-PCR See **differential display PCR**.

DDB complex See **damaged DNA binding proteins**.

ddNTP General name for dideoxy-nucleotide triphosphates used in **dideoxy sequencing**.

DDX A family of ATP-dependent RNA helicases for example, DDX1 (EC 3.6.4.13, DEAD box protein 1, 740aa), DDX50 (DEAD box protein 50, nucleolar protein Gu2, 737aa). DDX58 (DEAD box protein 58, retinoic acid-inducible gene 1 protein, 925aa) is involved in innate immune defense against viruses (see **TRIM** #2).

DE3 A T7 expression system developed under contract to the US Department of Energy (hence 'DE') that allows protein expression in competent bacterial cells to be put under the inducible control of IPTG. Competent cells, e.g. BL21(DE3), contain a DE3 lysogen which has the T7 RNA Polymerase under the control of the lacUV5 promoter – and are an all-purpose strain for high-level protein expression and easy induction with IPTG.

DEA See **dehydroepiandrosterone**.

deacetylase An enzyme that removes an acetyl group: one of the most active deacetylation reactions is the constant deacetylation (and reacetylation) of lysyl residues in histones (the half life of an acetyl group may be as low as 10 min). Acetylation (which removes a positive charge on the lysine ε-amino group) is thought to be increased in active genes, therefore deacetylation would be important in switching off genes. The main classes are **histone deacetylases** and **sirtuins**.

DEAD-box A four amino acid motif, -D-E-A-D- (asp-glu-ala-asp), similar to a Walker B-type ATP binding site (Mg^{2+}-binding aspartic acid). DEAD proteins have multiple DEAD motifs and several hundreds of these proteins can be identified in databases. Some are **DEAD-box helicases**, but not all have helicase activity and it may be that their role is to alter the conformation of RNA.

DEAD-box helicases Family of ATP-dependent DNA or RNA **helicases** with a **DEAD-box** motif that are important in RNA metabolism in both eukaryotes and prokaryotes and may have important functions in mediating microbial pathogenesis. Humans have 36 putative DEAD-box helicases including **p68**, a nuclear protein involved in cell growth. In *Drosophila* the helicase **vasa** is required for specification of posterior embryonic structures. See **DDX** and **DEAH-box helicases**).

DEAE- Group (diethyl-aminoethyl-) that is linked to cellulose or Sephadex to give a positive charge and thus to produce an anion exchange matrix for chromatography.

DEAH-box proteins One of the two major groups of RNA helicases, having a highly conserved motif (Asp-Glu-Ala-His; DEAH). In *S. cerevisiae* there are 7 members (Prp2p, Prp16p, Prp22p, and Prp43p involved in mRNA splicing, Dhr1p and Dhr2p involved in ribosome biogenesis, a seventh of uncertain function). See **DEAD-box helicases**.

deamination The removal of an amine group from a molecule, an important first step in the breakdown of amino acids. Deamination of nucleic acids is the spontaneous loss of the amino groups of cytosine (yielding uracil), methyl cytosine (yielding thymine), or of adenine (yielding hypoxanthine). It can be argued that the presence of thymine in DNA in place of the uracil of RNA stabilizes genetic information against this lesion, since repair enzymes would restore the GU base pair to GC.

DEAS See **dehydroepiandrosterone**.

death Simplistically, the complete and permanent cessation of life, although defining the exact point at which it occurs is difficult in complex multicellular organisms. See **apoptosis**, **cell death**, **programmed cell death**.

death domain Conserved domain (around 80aa) found in cytoplasmic portion of some **death receptors** (including the TNF-receptor), essential for generating signals that often lead to apoptosis.

death effector filaments Cytoplasmic structures that recruit caspases and trigger apoptosis. They are characteristic feature when apoptosis is triggered by the DED protein motif. See **NALP proteins**. Article: http://jcb.rupress.org/content/141/5/1243.long

death receptors Superfamily of **tumour necrosis factor** receptors (see Table T1), that trigger apoptotic cell death through interaction of various adapter proteins (FADD, TRADD etc.) with their cytoplasmic **death domains**. These adaptors then interact with **caspases** such as FLICE.

death-associated protein kinase See **DAP kinases**.

death-associated proteins Various proteins involved in inducing cell death. Death-associated protein-1 is a basic, proline-rich protein (DAP-1, 102aa) thought to mediate programmed cell death induced by interferon-gamma. DAP3 (398aa) is a component of the 28S mitochondrial ribosome subunit. DAP-6 is **DAXX**. Death-associated protein-like 1 (early epithelial differentiation-associated protein, 107aa) is expressed in hair follicles and may be a regulator of apoptosis. See **DAP kinases**.

death-effector domain A protein-association domain (DED) involved in the assembly of the death-inducing signalling complex (DISC). The seven standard DED-containing proteins are fas-associated death domain protein (FADD), Caspase-8 and 10, cellular FLICE-like inhibitory protein (c-FLIP), death effector domain containing DNA binding (**DEDD**), DEDD2 and phosphoprotein enriched in astrocytes 15-Kda (PEA-15). Abstract: http://www.ncbi.nlm.nih.gov/pubmed/18989622

death-inducing signalling complex The multi-protein complex (DISC) formed when a death receptor such as Fas trimerizes upon ligand binding and associates through the **death-effector domain** at the C-terminus with **FADD** and the caspase FLICE. The complex activates a cascade of effectors. A similar complex is associated with TNF/TNF receptor-1.

Debye-Hückel limiting law A method for estimating the activity of a molecule in solution, which makes some assumptions about the solute behaviour and takes account of ions present. The estimated activity is proportional to the concentration multiplied by the activity coefficient, gamma. The law is appropriate for dilute solutions of known ionic strength as is typical for many biological fluids.

Dec (1) Regulatory proteins (Dec1, BHLHB2, 424aa; Dec2, BHLH B3, 482aa), basic helix-loop-helix transcription-factor superfamily members, involved in the timing system underlying **circadian rhythms**. Transcripts of Dec2 and Dec1 show striking circadian oscillation in the suprachiasmatic nucleus, and Dec2 inhibits transcription from the Per1 (see **period**) promoter induced by Clock/Bmal1. Transcription of the Dec2 gene is regulated by several clock molecules and a negative-feedback loop. (2) Lymphocyte antigen 75 (DEC-205, C-type lectin domain family 13 member B, CD205, 1722aa) an endocytic receptor the directs captured antigens to a specialized antigen-processing compartment. (3) ADAM DEC1(A disintegrin and metalloproteinase domain-like protein decysin-1, 470aa) see **decysin**. (4) In humans, a protein (Deleted in esophageal cancer 1, 70aa) that may be a tumour suppressor. (5) In *Cochliobolus heterostrophus* (ascomycete fungus) decarboxylase DEC1 (253aa) that is required for **T-toxin** production. (6) In *Drosophila* a gene (*dec1*) the product of which (defective chorion-1 protein, 950aa) is essential for production of the egg shell. There are several different isoforms. (7) In *Lymnaea stagnalis* (Great pond snail) a set of proteins (DEC1, 919aa; DEC2, DEC3) thought to be involved in production of chitin.

decapentaplegic *Drosophila* gene (*dpp*) of which the product (588aa) acts as a morphogen during embryogenesis to pattern the dorsal/ventral axis. Related to **TGFβ** and acts in conjunction with **screw**. See **mothers against decapentaplegic**. Flybase entry: http://flybase.org/reports/FBgn0000490.html

decapping An important process in mRNA turnover and nonsense-mediated mRNA decay, the removal of the 7-methyl guanine cap 5′ structure

from oligoadenylated (but not polyadenylated) mRNA by the decapping complex. The complex consists of two proteins (DCP1A,582aa or DCP1B, 617aa and DCP2, 420aa), located in **GW bodies**. After mRNA has been decapped, further degradation is by the 5′ to 3′ exonuclease **XRN1**.

decapping activator complex A multi-subunit complex that activates **decapping** of mRNA consisting of the highly conserved heptameric Lsm1p−7p complex (made up of the seven **Like Sm proteins**, Lsm1p−Lsm7p) and its interacting partner, Pat1p (see **Pat1** #1). It activates decapping by an unknown mechanism and localizes with other decapping factors to the P-bodies in the cytoplasm. See **GW bodies**.

decay accelerating factor Plasma protein (CD55, 381aa) that regulates the alternative **complement** cascade by blocking the formation of the C3bBb complex (the C3 convertase of the alternative pathway). Present on cells that are exposed to complement as a GPI-anchored form which has a protective function. Defective glypiation in somatically-mutated erythrocyte stem cells causes **paroxysmal nocturnal haemoglobinuria**.

deconvolution Process in digital image handling whereby a composite image is formed using information from several separate images taken at different levels (focal planes). The final image can be rotated and viewed from different angles and has usually had noise filtered out so that the image is much clearer and sharper.

decorin A small proteoglycan (bone proteoglycan II, PG40, 90−140 kDa), of the extracellular matrix, so-called because it 'decorates' collagen fibres. The core protein (359aa, ∼42 kDa) is very similar to the core protein of **biglycan** and **fibromodulin**. All three have highly conserved sequences containing 10 internal homologous repeats of approximately 25aa with leucine-rich motifs. Decorin has one **glycosaminoglycan** chain, either chondroitin sulphate or dermatan sulphate and N-linked oligosaccharides.

decoy receptor In general, a non-functional receptor that competes for binding of a ligand and therefore reduces the effectiveness of the signal. In the TNFreceptor superfamily (see Table T1) the decoy receptors DcR1 and DcR2 compete with DR4 or DR5 receptors for binding to the ligand (**TRAIL**); decoy receptor 3 (DcR3) is a soluble receptor for **Fas ligand**.

dectins C-type lectin receptors (dectin-1, C-type lectin superfamily member 12, CLECSF12, dendritic cell-associated molecule, 247aa; dectin-2, CLECSF10, 209aa) which bind beta-1,3-linked and beta-1,6-linked glucans on cell walls of bacteria and fungi. Dectin-1 is necessary for the toll-like receptor-2 (TLR2)-mediated inflammatory response. They have cytoplasmic **ITAM** motifs which are phosphorylated when they bind.

decysin An **ADAM**-like protein (ADAM DEC1, A disintegrin and metalloproteinase domain-like protein decysin-1, 470aa) a peptidase that may be important in the control of the immune response and during pregnancy. Highly expressed in dendritic cells of the germinal centre.

DEDD Proteins involved in apoptosis. DEDD (Death effector domain-containing protein, 318aa) is a scaffold protein that directs caspase to certain substrates and facilitates their degradation during apoptosis. DEDD2 (DED-containing protein FLAME-3, FADD-like anti-apoptotic molecule 3, 326aa) may target caspases 8 and 10 to the nucleus.

dedifferentiation Loss of differentiated characteristics. In plants, most cells, including the highly differentiated haploid **microspores** (immature pollen cells) of angiosperms, can lose their differentiated features and give rise to a whole plant; in animals this is less certain, and there is still controversy as to whether the undifferentiated cells of the blastema that forms at the end of an amputated amphibian limb (for example) are derived by dedifferentiation, or by proliferation of uncommitted cells. Neither is it clear whether dedifferentiation in animal cells might just be the temporary loss of phenotypic characters, with retention of the **determination** to a particular cell type.

deep cells Cells (blastomeres) in the teleost blastula that lie between the outer cell layer and the yolk syncytial layer, and are the cells from which the embryo proper is constructed during gastrulation; much studied in the fish, *Fundulus heteroclitus*.

defective virus A virus genetically deficient in replication, but that may nevertheless be replicated when it co-infects a host cell in the presence of a wild-type helper virus. Most acute transforming **retroviruses** are defective, since their acquisition of oncogenes seems to be accompanied by deletion of essential viral genetic information.

defender against cell death Multi-pass membrane proteins that inhibit cell death (dolichyl diphospho-oligosaccharide protein glycosyltransferase subunit DAD1, oligosaccharyl transferase subunit DAD1, defender against cell death 1, EC 2.4.1.119, 113aa). In *Arabidopsis* DAD2 (115aa). N.B. DAD is used in other protein names: see **DASH complex**.

defensins Family of small (30−35aa) cysteine-rich cationic proteins found in vertebrate phagocytes (notably the azurophil granules of neutrophils) and active against bacteria, fungi and enveloped viruses. Alpha defensins are found primarily in neutrophils, macrophages and Paneth cells, β-defensins are found in a wider range of tissues. May constitute up to 5% of the total protein. Insect defensins have some sequence homology with the vertebrate forms. See **hepcidin antimicrobial peptide**.

Deferribacteres A bacterial phylum. Some species may be potential oral pathogens, others are found in the gut contents of deep-sea shrimps and mammals, in sewage sludge and a range of other unpleasant environments.

defined medium Cell culture medium in which all components are known. In practice this means that the serum (that is normally added to culture medium for animal cells) is replaced by insulin, transferrin and possibly specific growth factors such as **platelet-derived growth factor**.

definitive erythroblast Embryonic erythroblast found in the liver; smaller than primitive erythroblasts, they lose their nucleus at the end of the maturation cycle and produce erythrocytes with adult haemoglobin.

degeneracy The coding of a single amino acid by more than one base triplet (**codon**). Of the 64 possible codons, three are used for stop signals, leaving 61 for only 20 amino acids. Since all codons can be assigned to amino acids, it is clear that many amino acids must be coded by several different codons, in some cases as many as six. See Table C4.

degenerate PCR Polymerase chain reaction in which the primers are deliberately mixed (**degenerate primers**) to amplify a sequence that is imperfectly known (because translated from the peptide sequence and thus having uncertainty about codon usage). A powerful tool to find 'new' genes or gene families.

degenerate primer A single-stranded synthetic **oligonucleotide** designed to hybridize to DNA encoding a particular protein sequence. As the mapping of codons to amino acids is many-to-one, the oligonucleotide must be made as a mixture with several different bases at variable positions. The total number of different oligos in the resulting mixture is known as the degeneracy of the primer. Such primers are widely used in screening a **genomic library** or in degenerate **PCR**, to identify homologues of already known genes.

degenerins Products of *deg-1*, *mec-4* and *mec-10* genes in *C. elegans* which have homology with amiloride-sensitive sodium channels. Mutations cause neuronal degeneration, probably by disrupting ion fluxes. Deg-1 (degeneration of certain neurons protein 1, 778aa) and del-1 (664aa) are probably sodium channel subunits. Mec-4 (mechanosensory abnormality protein 4, 768aa) and mec-10 (mechanosensory abnormality protein 10, 724aa) are channel subunits required for touch sensitivity. A related protein, the product of unc-105 (degenerin-like protein unc-105, uncoordinated protein 105, 887aa) is an ion channel of the amiloride-sensitive sodium channel family which is permeable to small monovalent cations and may be mechanosensitive. There are several others including degenerin-like protein T28D9.7 (1069aa), unc-8 (777aa), asic-1 (795aa), asic-2 (545aa) which are all of the amiloride-sensitive sodium channel family.

degradosome Multienzyme complex (processome, RNA degradosome) in *E. coli* that contains exoribonuclease, polynucleotide phosphorylase (PNPase, EC 2.7.7.8, 711aa), endoribonuclease E (RNAase E, EC 3.1.26.12, 1061aa), enolase (EC 4.2.1.11, 432aa) and Rh1B (one of the DEAD-box family of ATP-dependent RNA helicases, EC 3.6.4.13, 421aa) or cold-shock DEAD box protein A (EC 3.6.4.13, 629aa). The degradosome has an essential role in RNA processing and decay. Article: http://www.pnas.org/content/104/5/1667.full

degranulation Release of secretory granule contents by fusion with the plasma membrane.

dehydration Removal of water, as in preparing a specimen for embedding or a histological section for clearing and mounting.

dehydrins A family of plant proteins induced in response to dehydration, cold stress or abscisic acid. They almost all have a run of seven contiguous serines in their central region. Intrinsically disordered proteins that act as chaperones to prevent protein aggregation. In *Arabidopsis* an example is dehydrin COR47 (265aa).

dehydroepiandrosterone A precursor (DHEA, DEA) of androstenedione from which testosterone and estrogens are produced. DHEA is produced from cholesterol by the adrenal glands, the gonads, adipose tissue, brain and in the skin. and is reversibly converted to the sulphated form (dehydroepiandrosterone sulphate; DHEAS) by sulphotransferase (EC 2.8.2.2) which is predominant in blood. Has an immunomodulatory role and the age-related decline in DHEA levels correlates with reduced immune competence. Reported to have tumour suppressive and antiproliferative effects in rodent tumours.

dehydrogenases Broad class of enzymes (oxidoreductases) that oxidize a substrate by transferring hydrogen to an acceptor that is usually either $NAD^+/NADP^+$ or a flavin enzyme, although other acceptors may be used. Most biological oxidations are of this type.

Deinococcus-Thermus A bacterial phylum composed of cocci: the Deinococcales include a single genus, *Deinococcus*, with several species that are resistant to radiation; the Thermales include several genera resistant to heat, notably *Thermus aquaticus*.

Deiters' cells (1) See **phalangeal cells**. (2) Deiters' neurons are large multipolar nerve cells, the predominant cells of the lateral vestibular nucleus.

Dejerine-Sottas neuropathy A severe degenerating early-onset neuropathy (**Charcot-Marie-Tooth disease** type 4F) that can be caused by mutations in the genes for MPZ (myelin protein zero, peripheral myelin protein), PMP22 (peripheral myelin protein 22), **periaxin**, EGR2 (early growth response-2) and possibly **connexin 32**.

Del elements Plant Ty3/Gypsy **LTR retrotransposons** of the chromovirus type.

delayed rectifier channels The potassium-selective **ion channels** of **axons**, so called because they change the potassium conductance with a delay after a voltage step. The name is used to denote any axon-like K channel. Various roles, e.g. regulation of pacemaker potentials, generation of bursts of **action potentials** or generation of long plateaux on action potentials. **A-type potassium channels** can be converted into delayed rectifier-type channels by alterations in membrane lipid; arachidonic acid and **anandamide** will change delayed-rectifier channels into A-type channels. Interconversion of channels: http://www.sciencemag.org/content/304/5668/265. abstract

delayed-type hypersensitivity See **hypersensitivity**.

deletion mutation A mutation in which one or more (sequential) nucleotides is lost from the genome. If the number lost is not divisible by three and is in a coding region, the result is a **frame-shift mutation**.

DELLA A family of nuclear growth repressors, first identified as **gibberellin** signalling components, that restrain the growth of plants. Later shown to mediate effects of other phytohormones. There are five distinct DELLAs encoded in the *Arabidopsis* genome, GAI (gibberellin insensitive), RGA (repressor of ga1-3), RGL1 (RGA-like 1), RGL2 and RGL3. RGA and GAI are negative regulators of gibberellin signalling. Gibberellin stimulates growth via 26S **proteasome**-dependent destruction of DELLAs, thus relieving DELLA-mediated growth restraint. See **GRAS family**.

delphilin A protein (GRID2IP, glutamate receptor, ionotropic, delta 2-interacting protein 1, 1211aa) a postsynaptic scaffolding protein at the parallel fiber-Purkinje cell synapse. It has two PDZ domains and a formin homology domain. Original report: http://www.jneurosci.org/content/22/3/803.full.pdf

delphinidin An anthocyanidin pigment that gives blue hues to flowers like violas and delphiniums. The so called delphinidin gene codes for an enzyme closely related to the **cyanidin** gene and modifies the anthocyanin **dihydrokaempferol** and directs pigment synthesis into the delphinidin pathway. Expression of this gene is responsible for the blue/violet colours of violas, delphiniums and grapes and the gene was engineered into roses to produce blue roses. Delphinidin, an active compound of red wine, inhibits endothelial cell apoptosis via the nitric oxide pathway and regulation of calcium homeostasis. Blue roses: www.physorg.com/pdf3581.pdf

Delta Product (433aa) of the serendipity locus in *Drosophila*, a transcriptional activator with 7 C_2H_2-type zinc fingers that controls **bicoid** gene expression during oogenesis. The human homologues (Delta-like proteins, DLL1, 723aa; DLL3, 618aa; DLL4, 685aa) are ligands in the Notch signalling pathway. N.B. There are delta subunits of many multi-component systems such as coatamers and elongation factors.

delta chains See **immunoglobulin**. The **heavy chains** (δ-chains) of mouse and human IgD immunoglobulins.

delta sleep-inducing peptide A natural somnogenic peptide (WAGGDASGE) found in neurons, peripheral organs, and plasma; induces mainly delta-type sleep in mammals. DSIP has effects in pain, adaptation to stress and epilepsy, also anti-ischemic effects. DSIP-immunoreactive peptide (TSC22 domain family protein 3, 134aa) protects T-cells from IL-2 deprivation-induced apoptosis through the inhibition of FOXO3A transcriptional activity.

delta virus Hepatitis D virus. A defective RNA virus requiring a **helper virus**, usually Hepatitis B virus, for replication. Delta virus infections may exacerbate the clinical effects of Hepatitis B.

delta-endotoxin A family of endotoxins that kill insects by making pores in the epithelial cell membrane of the insect midgut. One example, among many is (pesticidal crystal protein cry4Ba, 1136aa) produced by sporulating *Bacillus thuringiensis* subsp. *israelensis*. The size of the toxin varies and the toxic segment of the protein is located in the N-terminus.

dematin Actin microfilament bundling protein (erythrocyte membrane protein Band 4.9, 405aa) of the villin/gelsolin family. A substrate for PKC and PKA and bundling of actin is regulated by PKA-mediated phosphorylation.

demissine Toxic glycoalkaloid from *Solanum* spp. See **chaconine, solanine, potato glycoalkaloids**.

demyelinating diseases Diseases in which the myelin sheath of nerves is destroyed and that often have an autoimmune component. Examples are **multiple sclerosis**, acute disseminated encephalomyelitis (a complication of acute viral infection), **experimental allergic encephalomyelitis, Guillain-Barre syndrome**.

denaturation Reversible or irreversible loss of function in proteins and nucleic acids resulting from loss of higher order (secondary, tertiary or

quaternary structure) produced by non-physiological conditions of pH, temperature, salt or organic solvents.

dendrite *dendron* A long, branching outgrowth from a **neuron**, that carries electrical signals from synapses to the cell body; unlike an axon that carries electrical signals away from the cell body. This classical definition, however, lost some weight with the discovery of axo-axonal and dendro-dendritic synapses.

dendritic cells (1) In general, any cell with a branched morphology, but usually refers to cells involved in antigen presentation. (2) Follicular dendritic cells are found in germinal centres of spleen and lymph nodes and retain antigen for long periods. (3) Dendritic cells (accessory cells, antigen-presenting cells), positive for Class II histocompatibility antigens, are found in the red and white pulp of the spleen and lymph node cortex and are associated with stimulating T-cell proliferation. (4) Dendritic epidermal cells are T-cells found in epidermis and are involved in antigen recognition. They express predominantly $\gamma\delta$-TCR receptors. (5) Immature dendritic cells (**Langerhans cells**) are bone marrow-derived and are also found in the basal layers of the epidermis (but are distinct from dendritic epidermal cells). They are strongly MHC Class II positive and are responsible for **antigen processing** and **antigen presentation**. They migrate from skin to lymph nodes once they have encountered antigen and may be a subset of myeloid dendritic cells. (6) Myeloid dendritic cells (mDC) are most similar to monocytes: mDC-1 stimulate T-cells, mDC-2 are rarer and may be involved in combatting wound infection. (7) Plasmacytoid dendritic cells (pDC) resemble plasma cells but are able to produce high amounts of interferon-alpha and were formerly called interferon-producing cells (IPC). (8) **DOPA**-positive dendritic cells derived from neural crest are found in the basal part of epidermis.

dendritic lamellar body A membranous organelle in bulbous dendritic appendages of neurons linked by dendrodendritic gap junctions. Interacts with **CLIP-2**.

dendritic spines Wine-glass or mushroom-shaped protrusions, microfilament-rich, from dendrites that represent the principal site of termination of excitatory afferent neurons on interneurons, especially in the cortical regions. They contain post synaptic densities (PSD).

dendritic tree Characteristic (tree-like) pattern of outgrowths of neuronal **dendrites**.

Dendroaspis Genus of snakes. *Dendroaspis angusticeps* is the Eastern green mamba. *D. polylepis* is the Black mamba. See **calcicludine, calciseptine** and **dendrotoxins**.

Dendroaspis natriuretic peptide *DNP* **Natriuretic peptide**, 38aa, found in the venom of the snake *Dendroaspis angusticeps*. Binds to atrial natriuretic peptide receptor A but not to the ANP-receptors type B.

dendrogram A branching diagram, like a family tree, reflecting similarities or affinities of some sort between species.

dendron See **dendrite**.

dendrotoxins Polypeptides (57−60aa) isolated from *Dendroaspis* (snake) venom that are selective blockers of voltage-gated potassium channels in a variety of tissues and cell types. Have sequence similarity with Kunitz-type serine peptidase inhibitors. **Kalicludines** from *Anemonia sulcata* are structurally and functionally homologous to the dendrotoxins.

dengue Tropical disease caused by a flavivirus (one of the **arboviruses**), transmitted by mosquitoes. A more serious complication is dengue shock syndrome, a haemorrhagic fever probably caused by an immune complex hypersensitivity after re-exposure.

denitrifying bacteria Bacteria that break down nitrate and nitrite to gaseous nitrogen. Often found in soil and **biofilms** and may prove important in bioremediation since nitrate contamination of water is becoming a problem. Denitrification is not specific to any one phylogenetic group; the trait is found in about 50 genera of Proteobacteria and involves genes coding for NO_2-reductase (*nirK* and *nirS*) and N_2O reductase (*nosZ*).

dense bodies Areas of electron density associated with the thin filaments in smooth muscle cells. Some are associated with the plasma membrane, others are cytoplasmic.

density dependent inhibition of growth The phenomenon exhibited by most normal (**anchorage dependent**) animal cells in culture that stop dividing once a critical cell density is reached. The critical density is considerably higher for most cells than the density at which a monolayer is formed; for this reason, most cell behaviourists prefer the term density dependent inhibition of growth as this avoids any confusion with contact inhibition of locomotion, a totally different phenomenon that is contact dependent.

density gradient A column of liquid in which the density varies continually with position, usually as a consequence of variation of concentration of a solute. Such gradients may be established by progressive mixing of solutions of different density (as for example, sucrose gradients) or by centrifuge-induced redistribution of solute (as for **caesium chloride** gradients). Density gradients are widely used for centrifugal and gravity-induced separations of cells, organelles and macromolecules. The separations may exploit density differences between particles, or primarily differences in size, in which latter case the function of the gradient is chiefly to stabilize the liquid column against mixing.

density-enhanced phosphatase-1 A transmembrane protein (Dep-1, CD148, protein-tyrosine phosphatase-eta, HPTPε, EC 3.1.3.48, 1337aa) with eight extracellular FnIII domains and a single cytoplasmic tyrosine phosphatase domain. Dep-1 acts on several growth factor receptors, terminating signalling induced by receptor occupancy. Plays a role in cell adhesion, migration, proliferation and differentiation and may act as a tumour suppressor. Negatively regulates T-cell receptor (TCR) signalling. In many tumour cells Dep-1 is associated with a serine/threonine kinase that may regulate its activity.

Dent's disease A renal tubule disorder (nephrolithiasis) caused by mutation in the endosomal anion transport protein CLC-5 or in the *OCRL* gene which encodes a phosphatidylinositol 4,5-bisphosphate-5-phosphatase (EC 3.1.3.36, Lowe oculocerebrorenal syndrome protein, 901aa) involved in actin polymerization.

dentate nucleus Nerve cell mass, oval in shape, located in the centre of each of the cerebral hemispheres, the largest of the four deep cerebellar nuclei. The cell bodies of neurons involved in volitional movement are located in the dentate nucleus.

dentatorubral-pallidoluysian atrophy A neuropathology (Haw river syndrome, Naito-Oyanagi disease) caused by an expanded trinucleotide repeat (CAG repeat) in the gene encoding **atrophin-1**. The more repeats the earlier the onset and the more severe the effects.

denticle (1) Any small tooth-like structure. (2) The placoid scales of elasmobranchs.

dentin matrix protein A serine-rich acidic matrix protein (DMP-1, 513aa) present in the mineralized matrix of dentine that has a key role in mineralization. There are numerous potential phosphorylation sites, especially for kinases of the casein kinase II group. Oligomers of DMP-1 temporarily stabilize calcium phosphate nanoparticle precursors by sequestering them and preventing their further aggregation and precipitation. Binds to CD44 and RGD sequence-dependent **integrins**.

dentin sialophosphoprotein A protein (DSPP, 1301aa) that is cleaved to form dentin phosphoprotein (dentin phosphophoryn, DPP, 839aa) and dentin sialoprotein (447aa) the two major acidic matrix proteins of dentine. Mutations in the DSPP gene cause **dentinogenesis imperfecta**. See also **dentin matrix protein**.

dentine *dentin* The mineralized extracellular matrix from which teeth are formed, secreted by **odonto-blasts**. A structural biocomposite of needle-like crystals of hydroxyapatite embedded in a fibrous collagen matrix, very similar to compact bone.

dentinogenesis imperfecta A disorder restricted to teeth and distinct from **osteogenesis imperfecta**. Caused by mutation in the **dentin sialo-phosphoprotein** gene.

Denys-Drash syndrome A developmental defect affecting the urinogenital system, caused by mutation in the WT1 gene that encodes a zinc finger DNA-binding protein which can be a transcriptional activator or repressor depending on context. Meacham's syndrome and Frasier's syndrome are allelic disorders. See **Wilms' tumour**.

deoxycholate A bile salt formed by bacterial action from cholate; usually conjugated with glycine or taurine. The sodium salt is used as a detergent to make membrane proteins water soluble.

deoxyglucose Analogue of glucose (2-deoxyglucose) in which the hydroxyl on C-2 is replaced by a hydrogen atom. Since it is often taken up by cells but not further metabolized, it can be used to study glucose transport, and also to inhibit glucose utilization.

deoxyhaemoglobin Haemoglobin without bound oxygen.

deoxynojirimycin Deoxynojirimycin and nojirimycin act as a saccharide decoys, being imino sugars, and inhibit the ceramide glucosyltransferase (glucosylceramide synthase, EC 2.4.1.80, 394aa) that carries out the first step of glucosphingolipid synthesis. This blocks the production of cell surface glycoproteins. Produced by *Streptomyces* and *Bacillus* spp. and present in mulberry plants.

deoxyribonuclease *DNAase, DNase*. An **endonuclease** that hydrolyzes phosphodiester bonds and has a preference for DNA. DNAase I (EC 3.1.21.1, 282aa) yields di- and oligo-nucleotide $5'$ phosphates and binds to G-actin and the pointed ends of microfilaments, although the physiological relevance of the actin binding is unclear. DNAase II (360aa) yields $3'$ phosphates and hydrolyzes DNA under acidic conditions with a preference for double-stranded DNA. In chromatin, the sensitivity of DNA to digestion by DNAase I depends on its state of organization, transcriptionally active genes being much more sensitive than inactive genes. DNase gamma (liver and spleen DNase, 305aa) cleaves chromatin DNA to nucleosomal units and is involved in apoptotic cleavage of DNA (see **DNA laddering**).

deoxyribonucleic acid See **DNA**.

deoxyribonucleoside A purine or pyrimidine base N-glycosidically linked to 2-deoxy-D-ribofuranose. The phosphate esters of nucleosides are nucleotides.

deoxyribonucleotide A **deoxyribonucleoside**, ester linked to phosphate.

deoxyribose The sugar (2-deoxy-D-ribose) that, when linked by $3'-5'$ phosphodiester bonds, forms the backbone of DNA.

DEP domain A globular domain (Dishevelled, Egl-10, Pleckstrin homology domain, ~90aa), of unknown function present in prokaryotic and eukaryotic signalling proteins including those for which it is named. Mammalian regulators of G-protein signalling contain these domains. May target DEP domain-containing proteins to specific subcellular membranous sites. Domain structure: http://pawsonlab. mshri.on.ca/index.php?option=com_content&task= view&id=290&Itemid=64

Dep-1 See **density-enhanced phosphatase-1**

depactin Actin depolymerizing protein (150aa) originally isolated from eggs of the echinoderm *Asterias amurensis*. Of the ADF family, similar to **actophorin**.

dephosphorylation Removal of a phosphate group.

depolarization A positive shift in a cell's **resting potential** (that is normally negative), thus making it numerically smaller and less polarized, e.g. $-90\,\mathrm{mV}$ to $-50\,\mathrm{mV}$. The opposite of **hyperpolarization**. In the case of excitable cells, the resting potential is around $-70\,\mathrm{mV}$ and depolarisation (due to sodium ion influx) to below the threshold level will lead to the generation of an **action potential**.

depsipeptides Polypeptides that contain ester bonds as well as peptides. Naturally occurring depsipeptides are usually cyclic; they are common metabolic products of microorganisms and often have potent antibiotic activity (examples are **actinomycin**, enniatins, **valinomycin**).

depurination The spontaneous loss of purine residues from DNA. The N-glycosidic link between purine bases and deoxyribose in DNA has an appreciable rate of spontaneous cleavage *in vivo*, a lesion that must be enzymically repaired to ensure stability of the genetic information.

derepress To activate by suppressing a repressor; not an uncommon feature in gene regulation.

derivatization Modification of a molecule to change its solubility or other properties to enable analysis, for example by mass spectroscopy or chromatography, or to provide a label (e.g. fluorescent moiety) to facilitate identification and tracking. The molecule is said to have been derivatized.

dermal tissue The outer covering of plants, that includes the **epidermis** and periderm (non-living bark). *Cf.* **dermis**.

dermamyotome The embryonic region, derived from the dorsal portion of the **somite**, that gives rise to the **dermis** and axial musculature,

dermaseptins Closely related peptides (27−34aa) with broad-spectrum antibacterial activity produced by the skin of the South American frog, *Phyllomedusa sauvagei*. They are polycationic (Lys-rich), alpha-helical, and amphipathic, a structure that is believed to enable the peptides to interact with membrane bilayers, leading to permeation and disruption of the target cell. Dermaseptin S9 acts on both Gram-positive and Gram-negative bacteria. Dermaseptin O1 from *Phyllomedusa oreades* is active against *Trypanosoma cruzi*.

dermatan sulphate Glycosaminoglycan (15−40 kDa) typical of extracellular matrix of skin, blood vessels and heart. Repeating units of D-glucuronic acid-N-acetyl-D-galactosamine or L-iduronic acid-N-acetyl-D-galactosamine with 1−2 sulphates per unit. Broken down by L-iduronidase, but accumulates intra-lysosomally in **Hurler's disease** and **Hunter's syndrome**.

dermatitis herpetiformis A chronic itchy rash (Duhring's disease) associated with coeliac disease. There is IgA deposition in the dermis. The familial form is associated with HLA DQA1 and B1.

dermatofibrosarcoma protuberans A rare infiltrative skin tumour in which there is unregulated PDGF production. A chromosomal translocation generates a collagen-1/PDGF-B fusion protein that is processed to generate mitogenically-active PDGF-B.

dermatogens *protoderm* In plants, the primary meristem that gives rise to epidermis, a **histogen**.

dermatome (1) An area of skin innervated by a single spinal nerve. (2) An instrument used to produce thin slices of skin for grafting.

dermatomyositis One of a group of acquired muscle diseases called inflammatory myopathies. Probably auto-immune, with inflammation and weakness of muscles and often a purplish skin rash.

Dermatophagoides pteronyssinus House dust mite. Antigens extracted from mites and their faeces are a common cause of allergy to house dust in W European countries. Major allergen is Der p I, a cysteine proteinase.

dermatophyte A parasitic fungus which causes a skin disease in animals or man. Examples are ring worm and athlete's foot. The genera are *Epidermophyton, Microsporum,* or *Trichophyton*.

dermatopontin A tyrosine-rich acidic extracellular matrix protein (TRAMP, 201aa) that binds to cells and to dermatan sulphate proteoglycans, interacts with other ECM components, especially decorin, and regulates ECM formation and collagen fibrillogenesis. A molluscan homologue is a major shell matrix protein.

dermatosparaxis Recessive disorder of cattle in which a procollagen peptidase is absent. In consequence the amino- and carboxy-terminal peptides of procollagen are not removed, the **collagen** bundles are disordered, and the dermis is fragile. Similar to **Ehlers-Danlos syndrome** in humans.

dermcidin A protein (110aa) produced by eccrine sweat glands of the skin that has antimicrobial activity and limits infection by potential pathogens. Levels are apparently reduced in individuals with atopic dermatitis.

dermis Mesodermally derived **connective tissue** underlying the epithelium of the skin.

dermoid cyst A cystic teratoma that contains mature skin with hair follicles and sweat glands, usually benign. Many ovarian tumours are dermoid cysts.

DES See **diethylstilboestrol**.

DeSanctis-Cacchione syndrome A variant of **xeroderma pigmentosum** in which there is mutation in DNA excision repair protein ERCC-6 (ATP-dependent helicase ERCC6, Cockayne's syndrome protein, 1493aa) which is also defective in **Cockayne's syndrome** type B, cerebro-oculo-facio-skeletal syndrome type 1, susceptibility to age-related macular degeneration type 5 and UV-sensitive syndrome.

Descemet's membrane Not a membrane in a cell-biological sense but the basal lamina of the corneal endothelium which separates the cornea from the aqueous humour.

desensitization (1) Generally, a decrease in responsiveness following repeated exposure to a stimulus. (2) In immunology, reducing or abolishing the effects of a known allergen by giving gradually increasing doses until tolerance develops.

desert hedgehog One of the **hedgehog** family of morphogens (Dhh, 396aa), processed in much the same way as other members of the family. Essential for testicular development. Defects in Dhh have been associated with partial gonadal dysgenesis (PGD) accompanied by minifascicular polyneuropathy.

desferrioxamine Iron transporter from *Streptomyces pilosus* that chelates trivalent ions such as iron and aluminium. Used clinically to treat acute iron poisoning.

desmids Chlorophyte **algae** that are usually freshwater living and unicellular. Their cell wall often has elaborate ornamented shape.

desmin A protein (470aa) of Class III intermediate filaments, somewhat similar to **vimentin**, but characteristic of muscle cells. The filaments form a fibrous network connecting myofibrils to each other and to the plasma membrane from the periphery of the Z-line structures.

desmocalmin A protein (240 kDa) isolated (in 1985) from bovine desmosomes that binds calcium-calmodulin and cytokeratin-type intermediate filaments. Not mentioned in recent literature. Probably an obsolete synonym for a **desmocollin**.

desmocollins Glycoproteins (desmocollin-1, 894aa; desmocollin-2, 901aa; desmocollin-3, 896aa) of the cadherin superfamily of Ca^{2+}-dependent cell adhesion molecules isolated from **desmosomes** and involved in the interaction of plaque proteins and intermediate filaments. See **desmogleins** and **desmoplakin**.

desmogleins Like **desmocollins**, components of desmosome junctions. Single-pass type I membrane proteins (desmoglein-1, 1049aa; desmoglein-2, 1118aa; desmoglein-3, 999aa; desmoglein-4, 1040aa) with **cadherin** repeats.

desmoid tumour A fibrous neoplasm (aggressive fibromatosis, grade I fibrosarcoma) of connective tissue caused, in some cases, by mutation in the **adenomatous polyposis coli** (APC) gene or, in sporadic cases, by somatic mutation in the beta-**catenin** gene. Although non-malignant they are locally aggressive and tend to recur.

desmoplakin Cytoplasmic protein (desmoplakin, 2871aa) from the innermost portion of the desmosomal plaque. Involved in the organization of the desmosomal cadherin-plakoglobin complexes and the anchoring of intermediate filaments to the desmosomes. The desmoplakin-II isoform is characteristic of stratified epithelium. See **catenins**.

desmoplasia Growth of fibrous or connective tissue, for example the growth of dense fibrous tissue around some tumours.

desmosine Component of **elastin**, formed from four side chains of lysine and constituting a cross-linkage.

desmosome Specialized cell junction (macula adherens junction, spot desmosome) characteristic of epithelia into which intermediate filaments (tonofilaments of cytokeratin) are inserted. The gap between plasma membranes is of the order of 25–30 nm and is bridged by cadherins (**desmogleins** and **desmocollins**). **Desmoplakin** links the transmembrane cadherins to the cytoplasmic tonofilaments. Desmosomes are particularly conspicuous in tissues such as skin that have to withstand mechanical stress.

desmotubule Cylindrical membrane-lined channel through a **plasmodesma**, linking the cisternae of **endoplasmic reticulum** in the two cells.

desmoyokin A nuclear protein (neuroblast differentiation-associated protein AHNAK, 5890aa) that may be required for neuronal cell differentiation. It was originally identified as a desmosomal plaque protein on the basis of antibody reactivity but this seems to be coincidental and desmoyokin null mice do not show any phenotypic effect. Review: http://www.nature.com/jid/journal/v123/n4/full/5602494a.html

desorption The release of a substance from or through a surface: the greater the desorption, the less the substance will be retained on e.g. a chromatography column.

desoxy- See **deoxy-**.

desquamation Shedding of outer layer of skin (squames) or of cells from other epithelia.

destrin Widely distributed actin depolymerizing protein (165aa) that severs actin filaments and binds to actin monomers. One of the ADF family of proteins.

destruxins Cyclic **depsipeptide** fungal toxins that suppress the immune response in invertebrates.

desynapsis Separation of the paired homologous chromosomes at the **diplotene** stage of meiotic prophase I.

detergents Amphipathic, surface active, molecules with polar (water soluble) and non-polar (hydrophobic) domains. They bind strongly to hydrophobic molecules or molecular domains to confer water solubility. Examples include: sodium dodecyl sulphate, fatty acid salts, the Triton family, octyl glycoside.

determinate cleavage A type of embryonic cleavage in which each blastomere has a predetermined fate in the later embryo (in contrast to the situation in so-called 'regulating' embryos).

determination The commitment of a cell to a particular path of differentiation, even though there may be no morphological features that reveal this determination. Generally irreversible, but in the case of **imaginal discs** of *Drosophila* that are maintained by serial passage, **transdetermination** may occur.

detoxification reactions Reactions taking place generally in the liver or kidney in order to inactivate toxins, either by degradation or else by conjugation of residues to a hydrophilic moiety to promote excretion. Cytochrome P450 enzymes play an important part in oxidising toxins and other xenobiotics such as drugs as a first step in their inactivation and removal.

deuteranopia See **colour blindness**.

deuterated compound A compound in which hydrogen has been replaced, wholly or partially, by deuterium.

deuterium oxide Heavy water (D_2O), in which the hydrogen is replaced by deuterium. Will stabilize assembled microtubules.

Deuteromycetes Outmoded term for group now reclassified as **Deuteromycotina**. Includes fungi with no known sexual reproductive stages – the old Fungi Imperfecta.

Deuteromycotina Fungi (Fungi imperfecti, Deuteromycetes, deuteromycota) in which no sexual reproduction is known. Many appear to be Ascomycotina. Group includes many saprophytes, e.g. *Aspergillus*, *Penicillium* and plant parasites such as *Fusarium* and *Verticillium*.

deuterostome Embryonic developmental pattern in which the mouth does not form from the blastopore but from a second opening: includes echinoderms and chordates. Contrasts with morphogenesis in protostome phyla which include annelids, molluscs and arthropods. The two groups also differ in many aspects of early development including the pattern of early cleavage and the stage at which blastomeres become committed in differentiation; in deuterostomes the early blastomeres are equipotent whereas in protostomes there is earlier patterning and commitment to form particular cell lineages.

DEVD A derivatized peptide substrate for **caspase 3** with a benzyloxycarbonyl group (also known as BOC-DEVD or Z-DEVD) at the N-terminus to give improved cellular permeability. Can be conjugated with rhodamine or other fluorochromes for chromogenic assays. Fluoromethyl ketone (FMK)-derivatized peptides act as effective irreversible inhibitors.

devitrification Loss of transparency. Devitrification (opacity) of the lens of the eye occurs in cataract.

Devoret test Test for potential carcinogens based upon induction of prophage lambda in bacteria (*E. coli* K12 envA uvrB). There is a good correlation between ability of aflatoxins and benzanthracenes to induce lambda and their carcinogenicity in rodents. Description (1976): http://www.pnas.org/content/73/10/3700.full.pdf + html

dexamethasone Steroid analogue (**glucocorticoid**), used as an anti-inflammatory drug.

dexosome A small heat-stable membrane-bounded vesicle (exosome), 60–90 nm diameter, released by a dendritic cell and that carries antigen-loaded major histocompatibility complex class I and II molecules to naive dendritic cells.

dextrans High-molecular weight polysaccharides synthesied by some microorganisms. Consist of D-glucose linked by α-1,6 bonds (and a few α-1,3 and α-1,4 bonds). Dextran 75 (average molecular weight 75 kDa) has a colloid osmotic pressure similar to blood plasma, so dextran 75 solutions are used clinically as plasma expanders. They will also cause charge-shielding, and at the right concentrations, induce flocculation of red cells, a trick that is used in preparing leucocyte-rich plasma for white cell purification in the laboratory. Cross-linked dextran is the basis for **Sephadex**. Commercially derived from strains of *Leuconostoc mesenteroides*.

dextrins Low-MW mixtures of linear α-1,4-linked D-glucose polymers formed by hydrolytic degradation of starch or glycogen. Chain length can be variable.

dextrocardia *dexiocardia* A condition in which the heart is anatomically a mirror image of normal. The abdominal viscera may also be transposed (**situs inversus**). The incidence of left-handedness in people with dextrocardia is no more frequent than in the general population. See **left-right asymmetry**.

dextrorotatory Describing an optically active substance that rotates the plane of plane polarized light in a clockwise direction. Dextrorotatory compounds are often prefixed '(+)-' or 'd-'. The 'd-' prefix is based on the actual configuration of each enantiomer. Levorotation is the opposite (anti-clockwise rotation).

dextrose See **glucose**.

DFF Heterodimer of DNA fragmentation factor-α (DFFA, inhibitor of CAD, 331aa) and DFFB, (caspase-activated deoxyribonuclease, 338aa) that induces DNA fragmentation and chromatin condensation during apoptosis.

DFNA5 A gene that is mutated in autosomal dominant nonsyndromic sensorineural deafness; the product is a **gasdermin** expressed in the cochlea, (nonsyndromic hearing impairment protein 5, 496aa) related to **pejvakin**.

DFP A family of defense proteins (e.g. DFP-1, 168aa) expressed in insects following bacterial infection.

D gene segment A set of genes (diversity genes) in the complementarity-determining region (CDR3) that encodes immunoglobulin heavy chain. In humans there are at least 30 D gene segments, clustered between the IgVH locus and the JH genes. The various D genes differ in length, and are flanked at 5′ and 3′ ends by the conventional recombination signal sequences recognized by the **rag** recombinases (RAG1, RAG2). They can be inserted in either orientation, may use different reading frames and there may be multiple tandem insertions producing a longer D segment in the mature VDJ construct. T-cell receptors also have regions encoded by D genes in the β and δ chains. They contribute to the variability in the antigen-binding region of the heavy chain.

Dgt proteins Proteins involved in the *Drosophila* **augmin** complex (Dim gamma-tubulin proteins), necessary for localizing gamma-tubulin to spindle MTs but not to the centrosomes. Dgt6 (654aa) interacts with **Ndc80**, **Msps/XMAP215**, and gamma-tubulin to promote kinetochore-driven MT formation possibly mediating nucleation and/or initial stabilization of chromosome-induced MTs. Article: http://jcb.rupress.org/cgi/content/full/181/3/421

DH domain See **Dbl homology domain**.

DH5 Strain of *E. coli* K-12 that is disabled and non-colonising and can therefore be used in experimental work where escape would be undesirable. It is recombination (recA$^-$) and endonuclease (EndA$^-$, HsdR$^-$) deficient, highly transformable, and allows for selection by α-complementation.

DHA See **docosahexaenoic acid**.

DHAP (1) Dihydroxyacetone phosphate. (2) DHAP-AT is the enzyme (dihydroxyacetone phosphate acyltransferase, glycerone-phosphate O-acyltransferase, EC 2.3.1.42, 680aa) involved in the first step of **plasmalogen** biosynthesis. (3) 3,4-dihydroxyacetophenone or 3,4-**dihydroxyacetophynone**, one of the constituents of a traditional Chinese herbal medicine. (4) Initials of drugs used in combined chemotherapy for lymphoma (**dexamethasone**, cytarabine (**Ara C**), and **cisplatin**).

DHEA See **dehydroepiandrosterone**.

DHFR See **dihydrofolate reductase**.

Dhh See **desert hedgehog**.

DHT See **dihydrotestosterone**.

diabetes insipidus Rare form of **diabetes** in which the kidney tubules do not reabsorb enough water. The neurohypophyseal form is caused by mutation in the arginine vasopressin gene. The X-linked nephrogenic form is caused by mutation in the vasopressin V2 receptor, the autosomal nephrogenic type is caused by mutations in **aquaporin-2**. See **Wolfram's syndrome**.

diabetes mellitus Relative or absolute lack of **insulin** leading to uncontrolled carbohydrate metabolism. In juvenile onset (Type 1) diabetes there is autoimmune destruction of pancreatic B cells and the insulin deficiency tends to be almost total, whereas in adult onset (Type 2) diabetes (**NIDDM**) there seems to be no immunological component but an association with obesity. Maturity onset diabetes of the young

(MODY) occurs before 25 years of age and can be caused by mutations affecting various genes including those for hepatocyte nuclear factor-4alpha, glucokinase, hepatocyte nuclear factor-1alpha, insulin promoter factor-1, hepatic transcription factor-2, NeuroD transcription factor, the Krüppel-like transcription factor-11, carboxyl-ester lipase or the **PAX4** transcription factor.

diabetes related peptide See **amylin**.

diablo A kelch-like protein (623aa) in *Drosophila* that is thought to be a substrate-specific adapter of an E3 ubiquitin-protein ligase complex which mediates the ubiquitination and subsequent proteasomal degradation of target proteins. The mammalian homologue (**smac**/diablo) is a pro-apoptotic protein.

diabody A recombinant bispecific antibody (BsAb), constructed from heterogeneous single-chain antibodies. There is a five amino acid linker between VH and VL domains and the two antigen binding sites of the diabody molecule are located at opposite ends of the molecule at a distance of about 7 nm apart and pointing away from each other. See also **triabody**.

diacetoxyscirpenol Trichothecene mycotoxin produced by various species of fungi. Cytotoxic for human CFU-GM and BFU-E.

diacylglycerol Glycerol substituted on the 1 and 2 hydroxyl groups with long chain fatty acyl residues. DAG is a normal intermediate in the biosynthesis of phosphatidyl phospholipids and is released from them by phospholipase C activity. DAG from phosphatidyl inositol polyphosphates is important in signal transduction. Elevated levels of DAG in membranes activate protein kinase C by stabilizing its catalytically active complex with membrane-bound phosphatidylserine and calcium.

diacytosis Uncommon term for the discharge of an empty pinocytotic vesicle from a cell.

diad Anything with two-fold symmetry. A diad axis defines a plane where there is mirror-symmetry between the two halves of the structure.

diadinoxanthin A xanthophyll pigment from diatoms and dinoflagellates. Data sheet: http://epic.awi.de/Publications/Jef1997t.pdf

diakinesis The final stage of the first prophase of meiosis. The chromosomes condense to their greatest extent during this stage and normally the nucleolus disappears and the fragments of the nuclear envelope disperse.

diallyl trisulfide A natural compound derived from garlic and produced during the decomposition of allicin. Despite its reported lipid-lowering effects, the mechanisms of its actions are not yet clear.

dialysis Separation of molecules on the basis of size through a semi-permeable membrane. Molecules with dimensions greater than the pore diameter are retained inside the dialysis bag or tubing whereas small molecules and ions emerge in the dialysate outside the tubing.

diaminobenzidine Artificial substrate for **peroxidase**, producing a coloured (dark brown) reaction product. A potent carcinogen.

diaminobenzoic acid Compound (3,5-diaminobenzoic acid) used in fluorimetric determination of DNA content: gives fluorescent product when heated in acid solution with aldehydes. Assay method: http://www.ncbi.nlm.nih.gov/pubmed/2817360

diaminopimelic acid A diamino-carboxylic acid with two chiral centres, a constituent of the cell wall peptidoglycan of Bacteria, not known to occur in Archaea or Eukaryota.

Diamond-Blackfan anaemia A rare, progressive haematological disorder (congenital erythroid hypoplastic anaemia, erythrogenesis imperfecta, Aase-Smith syndrome II) which presents in early childhood and is often associated with other developmental defects. There is defective erythropoiesis and lack of nucleated erythrocytes in the bone marrow. Approximately 25% of cases of Diamond-Blackfan anemia are caused by mutation in the gene encoding ribosomal protein RPS19 but the disorder is heterogeneous and other forms are caused by defects in different ribosomal proteins (RPS24, RPS17, RPL35A, RPL5, RPL11, RPS7, RPS10 and RPS26).

diapedesis Archaic term for the emigration of leucocytes across the endothelium.

diaphanous In *Drosophila* a protein of the **formin** homology family (1091aa) required for cytokinesis in both mitosis and meiosis, possibly a component of the contractile ring or may control its function. Homologues in humans (diaphanous-related formin-1, DIAPH1, 1272aa; DIAPH2, 1101aa; DIAPH3, 1193aa) act in a rho-dependent manner to recruit **profilin** to the membrane and are required for the assembly of F-actin structures. The activity of diaphanous-releated formins is inhibited by an interaction between their N-terminal regulatory region and a conserved C-terminal segment termed the diaphanous autoinhibitory domain. Defects in DIAPH1 are the cause of a form of sensorineural deafness.

diaphanous-related formins See **diaphanous**.

diaphorase A ubiquitous class of flavin-bound enzymes capable of catalysing oxidation of NAD or NADPH in the presence of an electron acceptor other than oxygen – for example methylene blue, quinones or cytochromes. Examples in humans include diaphorase-1, NADH-cytochrome b5 reductase 3, EC 1.6.2.2, 301aa) involved in desaturation and elongation of fatty acids and cholesterol biosynthesis and DT-diaphorase (menadione reductase, NAD(P)H:quinone oxidoreductase 1, phylloquinone reductase, quinone reductase 1, EC 1.6.5.2, 274aa) which may be involved in detoxification pathways.

diaphysis Central portion of a long (limb) bone, composed of an outer wall of heavily mineralised bone and a central yellow (fatty) marrow-filled cavity. See **epiphysis**.

diastase Any enzyme that hydrolyses starch. Includes α-, β- and γ-amylases. Originally discovered as the enzyme produced during the germination of barley that is essential in the production of malt for brewing.

diastema (1) A thin yolk-free structure, which is considered to play an essential role in the induction of the cleavage furrow. (2) A gap in a row of teeth.

diastrophic dysplasia A form of short-limb dwarfism caused by mutation in a sulphate transporter (SLC26A2, 739aa) that may play a role in endochondral bone formation. See **atelosteogenesis**.

diatom Algae of the division Bacillariophyta; largely unicellular and characterized by having cell walls of hydrated silica embedded in an organic matrix. The cell walls are formed in two halves that fit together like the lid and base of a pillbox (or a petri dish) and often have elaborate patterns formed by pores. Diatoms are very abundant in marine and freshwater plankton. Deposits of the cell walls form diatomaceous or siliceous earths. Introduction to Bacillariophyta: http://www.ucmp.berkeley.edu/chromista/bacillariophyta.html

diatomic Describing a molecule composed of only two atoms which may be identical (e.g. O_2, molecular oxygen) or different (e.g. CO, carbon monoxide).

diatoxanthin A crystallizable xanthophyll found in diatoms.

diauxic growth See **diauxie**.

diauxie *diauxy* Adaptation of microorganisms to culture media that contain two different carbohydrates (e.g. glucose and lactose). In the first growth phase the sugar for which there are constitutive enzymes is utilized, then there is a brief pause while the enzyme systems for the second sugar are induced and synthesized.

diazo compound An organic compound that has two linked nitrogen atoms as a terminal functional group.

diazonium compounds Organic compounds that have a common functional $R\text{-}N_2^+ \ X^-$, where R is any organic residue such alkyl or aryl and X is an inorganic or organic anion. Diazotization is the process of forming diazonium compounds.

diazotization See **diazonium compounds**.

diazotroph An organism that is capable of nitrogen fixation. Assemblages of diazotroph species are often important in relatively impoverished soils such as salt marshes. The capacity to fix nitrogen occurs in a phylogenetically diverse range of bacterial species.

dibucaine See **cinchocaine**.

dibutyryl cyclic AMP *dbCAMP* An analogue of **cyclic AMP** that shares some of the pharmacological effects of this nucleotide, but is generally believed to enter cells more readily on account of its greater hydrophobicity.

dicentric Describing a chromosome that has two centromeres.

dicer A highly conserved ribonuclease (EC 3.1.26.-, 1922aa) that cleaves double-stranded RNA (dsRNA) or pre-miRNA and is required for formation of the RNA induced silencing complex (RISC).

dichlorobenzonitrile *dichlobenil* 2,6-dichloro-benzonitrile, an inhibitor of cellulose biosynthesis in higher plants.

dichlorophenoxyacetic acid A synthetic **auxin** (2,4-dichlorophenoxyacetic acid, 2,4-D), used as a selective herbicide that kills broad-leaved weeds but not grass crops. In much diluted form is used to promote cell division in tissue culture.

dichroism See **circular dichroism**.

dichromatism See **colour blindness**.

dicistronic mRNA An mRNA that has two cistrons, the downstream one being under the control of an **internal ribosome entry site**. There is some doubt concerning the internal initiation hypothesis and such mRNAs are not produced by eukaryotic cells. Dicistronic mRNAs produced by some plant and animal viruses are structurally dicistronic but functionally monocistronic; only the 5′ proximal cistron gets translated. Discussion: http://nar.oxfordjournals.org/content/33/20/6593.full#sec-8

dicistronic vectors *bicistronic vectors* A vector in which two internal ribosome entry sites independently initiate the translation of two proteins from a single RNA. See **dicistronic mRNA**. Examples of use: http://www.retrovirology.com/content/2/1/60, http://www.biomedcentral.com/1472-6750/7/74

Dicistroviridae A family of Group IV (positive-sense ssRNA) viruses that infect insects. The name reflects the dicistronic arrangement of the genome with internal ribosome entry sites. They resemble picornaviruses in some respects and were formerly classified as such. Abstract of review:http://www.ncbi.nlm.nih.gov/pubmed/19961327

Dick test An obsolete skin-test for immunity against the toxin of *Streptococcus pyogenes*, the organism that causes scarlet fever.

dickkopf A family of secreted proteins (220–350aa) that are important in vertebrate development, by locally inhibiting Wnt regulated processes such as antero-posterior axial patterning, limb development, somitogenesis and eye formation. Originally identified in *Xenopus* but human homologues (DKK1–4) have subsequently been identified. DKK1 (266aa) forms a ternary complex with the transmembrane protein **kremen** that promotes internalization of **low-density lipoprotein receptor-related proteins** 5 & 6 (LRP5/6).

diclofenac A non-steroidal anti-inflammatory drug. Responsible for the catastrophic decline in the numbers of Indian vultures (*Gyps indicus*) that, having eaten carcases contaminated by diclofenac, die of gout.

dicotyledonous Describing plants ('dicots') belonging to the large subclass of Angiosperms that have two seed-leaves (cotyledons). Includes the majority of herbaceous flowering plants and most deciduous woody plants of the temperate regions. *Cf.* **monocotyledonous**.

dictyBase A resource for the biology and genomics of *Dictyostelium discoideum*. Link: http://dictybase.org/

Dictyoglomi A bacterial phylum containing a single species, *Dictyoglomus thermophilum*, an extreme thermophile deriving energy from organic molecules (chemoorganotrophic). It produces a **xylanase**.

Dictyoptera Order of insects that includes the cockroaches and mantises.

dictyosome Organelle found in plant cells and functionally equivalent to the **Golgi apparatus** of animal cells.

dictyostatin A 22-member macrolactone first isolated from the marine sponge, *Corallistidae* sp. and subsequently synthesized. It stabilizes microtubules and will cause cell cycle arrest in G2/M at nanomolar concentrations. Shares many structural characteristics with discodermolide and is potentially an anti-tumour drug. Details: http://www.chem.wisc.edu/~burke/dictyostatin.htm

Dictyostelium A genus of the **Acrasidae**, the cellular slime moulds.

dictyotene Prolonged **diplotene** of meiosis: the stage at which oocyte nuclei remain during yolk production.

dideoxy sequencing The most popular method of DNA sequence determination (Sanger method *cf*. **Maxam-Gilbert method**). Starting with single-stranded template DNA, a short complementary primer is annealed, and extended by a DNA polymerase. The reaction is split into 4 tubes (called 'A, C, G or T') each containing a low concentration of the indicated dideoxy-nucleotide, in addition to the normal deoxynucleotides. Dideoxynucleotides, once incorporated, block further chain extension, and so each tube accumulates a mixture of chains of lengths determined by the template sequence. The 4 reactions are denatured and run out on an acrylamide sequencing gel in neighbouring lanes, and the sequence read up the gel according to the order of the bands. In modern automated methods the labelling of each the dideoxy-nucleotides is different and the mixture of products can be separated on a column and the products identifed spectrophotometrically.

dideoxynucleotide A nucleoside triphosphate that lacks hydroxyl groups at both its $2'$ and $3'$ carbons. The absence of the $3'$OH prevents normal DNA chain elongation which involves formation of a diester bond between the terminal $3'$OH of the chain with the $5'$-phosphate of the nucleotide being added. See **dideoxy sequencing**.

Didinium Fast moving carnivorous protozoan (of the Phylum Ciliophora) that feeds almost exclusively on live *Paramecium*.

DIDS An irreversible anion transport inhibitor, 4,4′-diisothiocyano-2,2′-disulphonic acid stilbene.

dieback A soilborne disease caused by two viruses from the family Tombusviridae. Susceptibility to dieback is common in lettuce (*Lactuca saliva* L.) but modern iceberg cultivars are resistant to this disease because of Tvr1, a single, dominant gene that provides durable resistance.

dieldrin One of the cyclodiene insecticides which acts on the $GABA_A$ receptor-chloride channel complex. Formerly used extensively on food crops and to control tsetse flies and other vectors of tropical diseases. Binds to soil where it has long half-life (years) in temperate regions and can bio-accumulate with toxic effects. Now generally banned and regarded as one of the most ecologically damaging toxins.

dielectric constant Relative permittivity when referring to the medium of a capacitor and independent of electric field strength. Under these conditions it is the ratio of the capacitance of the conductor to the capacitance it would have if the medium was replaced by vacuum. The dielectric constant (which is dimensionless) is a measure of the polarity of a solvent: water (very polar) has a dielectric constant of 80.10 at 20 °C while n-hexane (very non-polar) has a dielectric constant of 1.89 at 20 °C.

dielectrophoresis A phenomenon in which a force is exerted on a dielectric particle, even if uncharged, when it is subjected to a non-uniform electric field. It can be used to manipulate small particles such as cells or nanomaterials in 'lab-on-a-chip' systems. Explanatory video clip:http://www. youtube.com/watch?v = ngZZEubmAjw

Diels-Alder reaction Reaction used in organic synthesis of six-membered rings.

diencephalon In vertebrate central nervous system, the most rostral part of the **brain** stem, consisting of the thalamus, hypothalamus, subthalamus and epithalamus. It is a key relay zone for transmitting information about sensation and movement and also contains (in the hypothalamus) important control mechanisms for homeostatic integration.

diethylcarbamazine An anthelminthic drug that inhibits filarial arachidonic acid metabolism.

diethylstilbestrol An orally-active synthetic non-steroidal estrogen that was formerly used therapeutically but now recognised to be a potent teratogen. Exposure to diethylstilbestrol *in utero* can cause clear-cell adenocarcinoma of the vagina and cervix in adulthood and increases the risk of breast and testicular cancer.

difference threshold The change in a stimulus that is just detectable (the just noticeable difference or JND).

differential adhesion The differential adhesion hypothesis was advanced by Steinberg to explain the mechanism by which heterotypic cells in mixed aggregates sort out into isotypic territories. Quantitative differences in homo- and hetero-typic adhesion are supposed to be sufficient to account for the phenomenon without the need to postulate cell-type specific adhesion systems: fairly generally accepted, although tissue specific **cell adhesion molecules** are now known to exist.

differential display PCR Variation of the **polymerase chain reaction** used to identify differentially expressed genes; superceded by gene chip technologies. **mRNA** from two different tissue samples is reverse transcribed, then amplified using short, intentionally nonspecific primers. The array of bands obtained from a series of such amplifications is run on a high resolution gel, and compared with analogous arrays from different samples. Any bands unique to single samples are considered to be differentially expressed. Similar in aim to **subtractive hybridization**. See also **differential hybridization**.

differential hybridization Technique used to compare gene expression levels under different conditions by comparing two **cDNA** libraries. The two libraries are transcribed into RNA and each set of library products is labelled with a different fluorochrome. The two RNA samples are pooled and used to probe a **DNA array**; spectrophotometric analysis of the binding reveals whether binding is comparable or whether the products of one library bind preferentially (differentially) which indicated a higher level of that particular RNA species in that library. Alternative labelling strategies can be used.

differential interference contrast Method of image formation in the light microscope based on the method proposed by Nomarski (though strictly speaking all forms of optical microscopy rely to a greater or lesser extent on differential interference). The light beam is split by a **Wollaston prism** in the condenser, to form slightly divergent beams polarized at right angles. One passes through the specimen (and is retarded if the refractive index is greater), and one through the background nearby: the two are recombined in a second Wollaston prism in the objective and interfere to form an image. The image is spuriously 'three-dimensional' – the nucleus, for example, appears to stand out above the cell (or be hollowed out) because it has a higher refractive index than the cytoplasm. The Nomarski system has the advantage that there is no phase-halo, but the contrast is low and image formation with crowded cells is poor because the background does not differ from the specimen.

differential scanning calorimetry *DSC* Form of **thermal analysis** in which heat-flows to a sample and a standard at the same temperature are compared, as the temperature is changed. Applications: http://www.npl.co.uk/advanced-materials/measurement-techniques/thermal-analysis/differential-scanning-calorimetry

differential screening-selected gene aberrative in neuroblastoma See **DAN** #1.

differential stain (1) A histological stain which selectively stains some elements of a specimen more than others, either by giving them different colours, or different shades or intensities of the same colour. (2) Some stains bind reversibly and for best effect (greatest discrimination between different parts) the specimen is over-stained and then washed to remove stain; more will be released where binding is weakest.

differentially expressed in adenocarcinoma of the lung A protein (DAL-1, Band 4.1-like protein 3, 1087aa) lost in approximately 60% of non-small cell lung carcinomas. A growth regulator in the pathogenesis of meningiomas.

differentiation Process in development of a multicellular organism by which cells become specialized for particular functions. Requires that there is selective expression of portions of the genome; the fully differentiated state may be preceded by a stage in which the cell is already programmed for

differentiation but is not yet expressing the characteristic phenotype (**determination**).

differentiation antigen Any large structural macromolecule that can be detected by immune reagents and that is associated with the differentiation of a particular cell type or types. Many cells can be identified by their possession of a unique set of differentiation antigens. There is no implication that the antigens cause differentiation. See **CD antigens**.

difficidin An antibacterial polyketide from *Bacillus amyloliquefaciens* GA1.

Difflugia A small amoeba (200–250 μm long) of the Phylum Sarcodina. Possess a chitinous 'test' (shell) that is usually covered completely with sand grains. Feeds mainly on green algae.

diffraction Phenomenon that occurs when a wavetrain passes an obstacle: secondary waves are set up that interfere with the primary wave and give rise to bands of constructive and destructive interference. Around a point source of light, in consequence, is a series of concentric light and dark bands (coloured bands with white light), a diffraction pattern.

diffraction grating An optical component that has a periodic structure (often grooves or ridges) that splits and diffracts light into several beams travelling in different directions. Used in spectrophotometers to produce monochromatic light.

diffusion coefficient For the translational diffusion of solutes, diffusion is described by Fick's First Law, that states that the amount of a substance crossing a given area is proportional to the spatial gradient of concentration and the diffusion constant (D), that is related to molecular size and shape. A useful derived relationship is that the mean square distance moved (in three-dimensions) by molecules in time t is 6Dt. Diffusion coefficients of most ions are in the range of 0.6×10^{-9} to 2×10^{-9} m^2/s. For biological molecules the diffusion coefficients normally range from 10^{-11} to 10^{-10} m^2/s.

diffusion limitation The principle underlying the boundary layer hypothesis, that the proliferation of cells in culture is limited by the rate at which some essential component (almost certainly a growth factor) diffuses from the bulk medium into the layer immediately adjacent to the plasma membrane. By spreading out, a cell obtains a supra-threshold level of the factor and can divide; if unable to spread (because of crowding or poor adhesion) then the cell will remain in the G0 stage of the **cell cycle**.

diffusion potential Potential arising from different rates of diffusion of ions at the interface of two dissimilar fluids; a junction potential.

diffusion pump A type of pump in which a high speed jet of vapour sweeps gas molecules in the pump throat down into the bottom of the pump and out of the exhaust. Used to generate high vacuum.

digenetic Describing a pattern of reproduction involving alternation of sexual and asexual cycles in successive hosts. More specifically a parasite having two hosts, the classic example being Digenean trematodes (flukes).

DiGeorge syndrome See **thymic hypoplasia** and **velocardiofacial syndrome**.

digestive vacuole Intracellular vacuole into which lysosomal enzymes are discharged and digestion of the contents occurs. More commonly referred to as a **secondary lysosome**.

digitalis General term for pharmacologically active compounds from the foxglove (*Digitalis*). The active substances are the cardiac glycosides, digoxin, digitoxin, strophanthin and **ouabain**. Causes increased force of contraction of the heart, disturbance of rhythm and reduced beat frequency.

digitonin A glycoside obtained from *Digitalis purpurea* (foxglove), often used as a detergent to solubilize membrane proteins. The aglycone is digitogenin. Do not confuse with **digitalis**.

digitoxin A cardiac glycoside similar to **digoxin** but with longer-lasting effects.

diglyceride Generic term for any compound with two glyceryl residues. Formerly used for **diacyl glycerol** though this is inappropriate.

digoxin Cardiac glycoside from foxglove (*Digitalis lanata*), used to treat congestive heart failure and supraventricular arrhythmias. Inhibits Na^+/K^+-ATPase. Aglycone is **digoxigenin**.

digoxygenin Small molecule derived from foxgloves, that is used for labelling DNA or RNA probes, and subsequent detection by enzymes linked to anti-digoxygenin antibodies. Proprietary to Boehringer-Mannheim. See **digoxin**.

dihybrid The product of a cross between parents differing in two characters determined by single genes, each of which has two alleles. Heterozygous for two pairs of alleles.

dihydroflavanol reductase Plant enzyme (DFR, dihydrokaempferol 4-reductase, EC 1.1.1.219, 382aa) that modifies the precursor pigments for flower colour in the cyanidin, delphinidin and pelargonidin pathways. Precursor pigment molecules are colourless until modified by DFR, so any mutation that disrupts the DFR gene results in white flowers.

dihydrofolate reductase Enzyme (DHFR, EC 1.5.1.3, 187aa) involved in the biosynthesis of **folic acid** that transfers hydrogen from NADP to dihydrofolate, yielding tetrahydrofolic acid, the active form in humans, an essential vitamin cofactor in purine,

thymidine and methionine synthesis. Inhibitors (e.g. aminopterin and amethopterin, components of **HAT medium**) can be used as antimicrobial and anticancer drugs.

dihydrokaempferol The anthocyanin-precursor for all three primary plant pigments: **cyanidin, pelargonidin** and **delphinidin**.

dihydropyridines Specific blockers of some types of **calcium channel**, e.g. nifedipine and nitrenidine; among the most widely used drugs for the management of cardiovascular disease.

dihydropyrimidinase An enzyme (dihydropyrimidine amidohydrolase, hydantoinase, EC 3.5.2.2, 519aa) that catalyzes the second step in degradation of pyrimidines. Dihydropyrimidinase-related proteins are **collapsin response mediator proteins**, but lack enzyme activity.

dihydrotestosterone A potent metabolite of **testosterone**. Mediates many of the functional activities of testosterone (differentiation, growth-promotion) through the androgen receptor.

dihydroxyacetone A triose (glycerone) found in plants and used as a sunless tanning agent (a form of Maillard reaction occurs with proteins in the keratinized layer). Dihydroxyacetone phosphate is one of the products of the reduction of 1,3-bisphosphoglycerate by NADPH in the Calvin cycle, used in the synthesis of sedoheptulose 1,7-bisphosphate and fructose 1,6-bisphosphate. Also the product of the dehydrogenation of L-glycerol-3-phosphate.

dihydroxyacetophynone One of the constituents (Qingxintong) of a traditional Chinese herbal medicine, derived from *Ilex pubescens*. Inhibits platelet function.

dihydroxyphenylalanine The precursor of **dopamine** (L-DOPA, levodopa), made from L-tyrosine by tyrosine 3-mono-oxygenase. Used to treat **Parkinson's disease**.

dihydroxypurine See **xanthine**.

DiI Name ('di-i') used for fluorescent derivatives of indocarbocyanine iodide that have two long alkyl chains and are membrane soluble. Used as general stains for membranes, as specific probes for membrane fluidity measurements and as **vital dye**.

dikaryon Fungal hypha or mycelium in which there are two nuclei of different genetic constitution (and different mating type) in each cell (or hyphal segment). *Adj.* dikaryotic.

dikaryophase The period in the life cycle of an ascomycete or basidiomycete fungus in which the cells have two nuclei, i.e. between **plasmogamy** and **karyogamy**.

dilated cardiomyopathy A set of disorders in which the heart becomes weakened and enlarged. A wide range of causes includes myocarditis, coronary artery disease, systemic diseases, and myocardial toxins and only around 25% of cases are due to mutations. See **hypertrophic cardiomyopathy** and **phospholamban**.

dilution cloning Cloning by diluting the cell suspension to the point at which the probability of there being more than one cell in the inoculum volume is small. Inevitably on quite a few occasions there will not be any cells.

dimercaprol Drug (British Anti-Lewisite) used as an antidote to poisoning by heavy metals such as antimony, arsenic, bismuth, gold, mercury, thallium or lead, but not iron or cadmium. Not considered to be the drug of first choice because of toxic side-effects. Originally synthesized as an antidote against the vesicant arsenical war gases (Lewisite), based on the fact that arsenic products react with SH radicals.

dimethyl formamide Compound sometimes used as an alternative to **DMSO** for making stock solutions of compounds that are poorly water-soluble; the concentrated solution can then be diluted.

dimethyl sulphoxide See **DMSO**.

dimethylbenzene Organic compound with three isomers (ortho-, meta- and para-) that are found as a mixture in **xylene**.

diminuendo A mouse mutant that exhibits progressive loss of hearing and has hair cell anomalies. The defect is in the gene encoding a microRNA and there are 96 mRNA transcripts affected in the homozygous mutant.

dinitrophenol A small molecule used as an uncoupler of oxidative phosphorylation. Also used after reaction with various proteins to provide a strong and specific identified **haptenic** group.

dinoflagellates Common aquatic organisms of the order Dinoflagellida (for botanists Dinophyceae), now classed within the **Alveolata**; about half the species are photosynthetic. They are second only to diatoms as marine primary producers. They have two **flagella** lying in grooves in an often elaborately sculptured shell or **pellicle** that is formed from plates of cellulose deposited in membrane vesicles. The pellicle gives some dinoflagellates very bizarre shapes. Their chromosomes lack nucleosomes and centromeres and may have little or no associated protein and the group has sometimes been termed **mesokaryotic** (see **dinokaryon**). There is an extensive internal system of vacuoles (the vacuome) some of which are referred to as **pusules**. *Gymnodinium* and *Gonyaulax*, that cause 'red tides', produce toxins (e.g. **saxitoxin**) that can be accumulated by filter-feeding molluscs with potentially fatal results further up the food chain.

Another common genus is *Peridinium*. Images: http://tolweb.org/Dinoflagellates/2445

dinokaryon The nucleus of **dinoflagellates** which has unusual features. The DNA is not arranged as nucleosomes, there being very little basic protein associated with the DNA. The DNA content is very high and remains continuously condensed during both interphase and mitosis, the $3-6$ nm fibrils being packed in a highly ordered state. Mitosis is unusual (dinomitosis) with the nuclear envelope persisting and an extranuclear spindle with spindle microtubules penetrating the envelope.

dioecious Describing a plant population in which there are separate male and female plants. Androecious plants have only male organs, gynoecious plants are female. *Cf.* **monoecious**, **synoecious**.

dioptre Unit for the power of a lens, the reciprocal of focal length in metres.

dioxane A heterocyclic compound (1,4-dioxane), mainly used as a stabilizer for trichloroethane and as a laboratory solvent. Do not confuse with **dioxins**.

dioxins Heterocyclic compounds with two oxygen and four carbon atoms in the ring. Used casually as a generic name for chlorinated derivatives of dibenzo-p-dioxin, the most toxic of which is 2,3,7,8-tetrachlorodibenzo-p-dioxin. The polychlorinated dibenzodioxins have teratogenic, mutagenic and carcinogenic properties.

dioxygenase Any oxidoreductase system in EC 1.13.11; catalyses reactions in which two oxygen atoms (from O_2) are added to a substrate.

diphtheria The disease caused by the Gram-positive bacillus *Corynebacterium diphtheriae*. The pathological effects are largely due to **diphtheria toxin** which can cause myocarditis, polyneuritis, and other systemic effects.

diphtheria toxin An AB exotoxin (567aa) coded by β-corynephage of virulent *Corynebacterium diphtheriae* strains (that have a repressor of toxin production). The B subunit (342aa) binds to the heparin-binding EGF receptor on the surface of a target cell and facilitates the entry of the enzymically active A subunit (EC 2.4.2.36, 193aa) that ADP-ribosylates **elongation factor** 2 and blocks translation. The diphtheria toxin repressor (iron-dependent diphtheria tox regulatory element, 226aa) blocks expression of the toxin gene.

diphtheria toxoid Diphtheria toxin treated with formaldehyde so as to destroy toxicity without altering its capacity to act as antigen. Used for active immunization against diphtheria.

dipicolinic acid Dipicolinic acid (DPA, 2,6-pyridinedicarboxylic acid) and the Ca^{2+} complex of DPA (CaDPA) are major chemical components of bacterial endospores. DPA is a chelator of metal ions, reduces water availability within the endospore and intercalates between bases in DNA, thus contributing to heat resistance and chromosome stabilisation.

diplococcus Bacterial strain in which two spherical cells (cocci) are joined to form a pair like a dumb-bell or figure-of-eight.

Diplococcus pneumonia See ***Streptococcus pneumoniae***, the formal name for this organism.

diplohaplontic Organisms that show an alternation of generations; a haploid phase (**gametophyte**) exists in the life cycle between meiosis and fertilization (e.g. higher plants, many algae and fungi); the products of meiosis are spores that develop as haploid individuals from which haploid gametes develop and fuse to form a diploid zygote. See **diplontic**, **haplontic**.

diploid A diploid cell has its **chromosomes** in homologous pairs, and thus has two copies of each autosomal genetic **locus**. The diploid number (2n) equals twice the **haploid** number and is the characteristic number for most cells other than gametes.

Diplomonads An Order of flagellates, mostly parasitic. Best known member is *Giardia lamblia*.

diplonema A stage in **meiosis** (diplotene stage) at which the chromosomes are clearly visible as double structures. More commonly **diplotene**.

diplont Organisms in which only the zygote is diploid and the vegetative cells are haploid.

diplontic Organisms with a life cycle in which the products of meiosis behave directly as gametes, fusing to form a zygote from which the diploid, or sexually reproductive polyploid, adult organism will develop. Cf. **haplontic**, **diplohaplontic**.

diplophase The diploid phase of the life cycle: in most metazoa and higher plants, the majority of the lifespan. Cf. haplophase.

diplornavirus Taxonomically unsound proposal for a family of all double-stranded RNA viruses.

diplotene The final stage of the first prophase of meiosis. All four chromatids of a **tetrad** are fully visible and homologous chromosomes start to move away from one another except at **chiasmata**.

Diptera Order of insects with one pair of wings, the second pair being modified into balancing organs, the halteres; the mouthparts are modified for sucking or piercing. The insects show complete metamorphosis in that they have larval, pupal and imaginal (imago, adult) stages. The order includes the flies and mosquitoes; best known genera are ***Anopheles*** and ***Drosophila***.

diptericin An antimicrobial peptide (106aa) of the **attacin/sarcotoxin**-2 family. Originally isolated from the dipteran *Phormia terranovae* (blowfly) in which several variants occur, predominantly diptericin A (82aa).

directed evolution (1) An approach to developing molecules with specific desirable properties in which sequential rounds of synthesis are informed by data on the properties of the previous set. This approach has been applied, for example, to generating novel peptide ligands and enzymes with improved characteristics. (2) The creationist view that biological systems are so complex that there must have been an external director (aka deity). Evidence is unavailable for this belief which also masquerades as 'Intelligent design'.

DIRS A family of tyrosine recombinase retroelements (LTR retrotransposons) that differ from the two other YR-like families in having a conserved methyltransferase (MT) domain that is similar to those encoded by various bacteriophages. DIRS1 from *Dictyostelium discoideum* has inverted terminal repeats flanking an internal region containing genes for a putative Gag protein, a reverse transcriptase/ribonuclease H and a putative tyrosine recombinase. DIRS1-like elements have been found in metazoa including vertebrates. Journal article: http://mbe.oxfordjournals.org/content/21/4/746.long

disaccharide Sugar formed from two monosaccharide units linked by a glycosidic bond. The trehalose type are formed from two non-reducing sugars, the maltose type from two reducing sugars.

DISC See **death-inducing signalling complex** or **disrupted-in-schizophrenia 1**.

disc gel A gel support for electrophoresis in which there is a discontinuity in pH, or gel concentration, or buffer composition. Confusingly, nothing to do with shape.

discodermolide Anti-tumour drug (a polyhydroxylated alkatetraene lactone) that, like **taxol**, promotes formation of stable bundles of microtubules and competes with taxol for binding to polymerized tubulin. Isolated from the marine sponge, *Discodermia dissoluta*.

discoidin A lectin, isolated from the cellular slime mould *Dictyostelium discoideum* (see **Acrasidae**), that has a binding site for carbohydrate residues related to galactose. The lectin, that consists of two distinct species (discoidins I and II), is synthesized as the cells differentiate from vegetative to aggregation phase, and was originally thought to be involved in intercellular adhesion, but discoidin I is now thought to be involved in adhesion to the substratum by a mechanism resembling that of fibronectin.

Discomycetes A class of Ascomycete fungi. Includes the Lecanorales (lichen-forming fungi), and many saprophytic and mycorrhizal species, e.g. morels and truffles.

dishevelled *Drosophila* protein (623aa) involved in **wingless** (wg) signalling, possibly through the reception of the wg signal by target cells and subsequent redistribution of armadillo protein in response to that signal in embryos. This signal seems to be required to establish planar cell polarity and identity. The human homologues (Dvl-1, 695aa; Dvl-2, 736aa) participate in Wnt signalling by binding to the cytoplasmic C-terminus of **frizzled** family members and transducing the Wnt signal to down-stream effectors. See **axin, dapper** and **wnt**.

disintegration constant The constant (decay constant) that can be used to calculate the rate of decay of a radioactive element.

disintegrins Peptides found in the venoms of various snakes of the viper family, that inhibit the function of some **integrins** of the β_1 and β_3 classes. They were first identified as inhibitors of platelet aggregation and were subsequently shown to bind with high affinity to integrins and to block the interaction of integrins with **RGD**-containing proteins e.g. they block the binding of the platelet integrin $\alpha_{IIb}\beta_3$ to **fibrinogen**. Disintegrins are effective inhibitors at molar concentrations 500–2000 times lower than short RGDX peptides. They are cysteine-rich peptides (45–84aa), almost all with a conserved -RGD-sequence on a β-turn, presumed to be the site that binds to integrins. The assumption is that their biological role in the venom is to inhibit blood clotting. Found in many snake species, where they are called variously albolabrin, applagin, batroxostatin, bitistatin, echistatin, elegantin, flavoridin, halysin, kistrin, triflavin, and trigramin.

disjunction The separation of the two paired homologous chromosomes during meiotic anaphase.

disjunction mutant A mutant in which chromosomes are partitioned unequally between daughter cells at **meiosis**, as a result of nondisjunction.

dispase Trade name for a crude protease preparation used for disaggregating tissue in setting up primary cell cultures. Dispase gives less complete disaggregation than trypsin but survival of cells may be better.

dispermy Condition that arises if two spermatozoa fuse with a single ovum leading to diandric triploidy.

dispersion forces Forces of attraction between atoms or non-polar molecules that result from the formation of induced dipoles. Sometimes referred to as London dispersion forces. Important in the **DLVO** theory of colloid flocculation and thus in theories of cell adhesion.

disrupted-in-schizophrenia 1 The product of a susceptibility gene for mood disorders and schizophrenia, a protein (854aa) involved in the regulation of multiple aspects of embryonic and adult neurogenesis.

disseminated intravascular coagulation Complication of **septic shock** in which endotoxin (from Gram-negative bacteria) induces systemic clotting of the blood, probably indirectly through the effect of endotoxin on neutrophils. It may also develop in other situations where neutrophils become systemically hyperactivated.

dissipative structure A system maintained far from chemical/thermodynamic equilibrium, having the potential to form ordered structures. See **Belousov-Zhabotinsky reaction**.

dissociation Any process by which a tissue is separated into single cells. Enzymic dissociation with trypsin or other peptidases is often used.

dissociation constant In a chemical equilibrium of form $A + B = AB$, the equilibrium concentrations (strictly, activities) of the reactants are related such that $[A][B]/[AB] = $ a constant, K_d, the dissociation constant, that in this simplest case, has the dimensions of concentration. When A is H^+, this is the acid dissociation constant often designated K_a, and expressed as pK_a (the negative logarithm to base 10 of K_a).

distemper virus Paramyxovirus of the genus Morbillivirus. Commonest is the canine distemper virus that causes fever, vomiting and diarrhoea; variant that infects seals (Phocavirus) has caused significant mortality in recent years.

disulphide bond The -S-S- linkage. A linkage formed between the SH groups of two **cysteine** moieties either within or between peptide chains. Each cysteine then becomes a half-**cystine** residue. Disulphide linkages stabilize, but do not determine, secondary structure in proteins. They are easily disrupted by -SH groups in an exchange reaction and are not present in cytosolic proteins (cytosol has a high concentration of **glutathione** that has a free -SH residue).

DIT (1) In *S. cerevisiae* a spore wall maturation protein (DIT-1, 536aa) and a cytochrome (P450-DIT2, EC 1.14.14.-, 489aa) thought to catalyze the oxidation of tyrosine residues in the formation of LL-dityrosine. (2) In the nematode *Dirofilaria immitis* DIT33 (234aa) is an aspartyl peptidase inhibitor.

diterpene A fundamental class of natural products with about 5000 members known. The skeleton of every diterpene contains 20 carbon atoms although additional groups linked to the diterpene skeleton by an oxygen atom can increase the carbon atom count. Many have potent biological activity and diterpene acids are known to have substantial feeding deterrent and growth inhibiting effects on a variety of insect groups and are known to inhibit a variety of fungi. Diterpenes also form the basis for biologically important compounds such as **retinol**, retinal, and **phytol**.

dithioerythritol See **dithiothreitol**.

dithiothreitol A small-molecule redox reagent (Cleland's reagent) with two SH groups, used to protect sulphydryl groups from oxidation during protein purification procedures or to reduce disulphides to sulphydryl groups. Dithioerythritol is an epimer and has similar, though slightly weaker, effects.

diuretics Compounds (drugs) that produce diuresis. Usually subdivided into **thiazide diuretics** and the more potent **loop diuretics**. See also **potassium-sparing diuretics**.

diurnal Occurring during the day or repeating on a daily basis. Use of **circadian rhythm** for the latter avoids ambiguity.

divergence In evolutionary biology the process (divergent evolution) by which organs of different form and function arise from the same original structure. Cf. convergent evolution. See **homology**.

diversity gene segment See **D gene segment**.

diversity map A genetic map that identifies causal polymorphisms for important traits. The development of a diversity map for a particular species relies on the sequence polymorphisms in a small set of genotypes that have been chosen as representative of the genetic diversity that is exhibited in a 'core collection'.

division septum The cell wall that forms between daughter cells at the end of mitosis in plant cells or just before separation in bacteria.

dizygotic Twins arising as a result of the fertilization of two ova by two spermatozoa and thus genetically non-identical, in contrast to **monozygotic** twins.

DKK See **dickkopf**.

dlbcl Diffuse large B-cell lymphoma. See **large-cell lymphoma**.

DLGs A family of proteins that are homologues of the *Drosophila* protein 'disks large' (Dlg, 970aa) a tumour suppressor that is the prototype of the **MAGUKs**. In humans DLG1 (Disks large homologue-1, synapse-associated protein 97, SAP-97, 904aa) is a multidomain scaffolding protein required for normal development that recruits channels, receptors and signalling molecules to discrete plasma membrane domains. Other human DLG proteins are known: DLG2 (870aa) is required for perception of chronic pain through NMDA receptor signalling, DLG3 (817aa) is required for learning probably through a role in synaptic plasticity following

NMDA receptor signalling. DLG4 (724aa) is also involved in synaptic plasticity, DLG5 (1919aa) is important in maintenance of the structure of epithelial cells. Disks large associated proteins (DAPs) include disks large-associated protein 1 (DLGAP1, 977aa) which is a component of the postsynaptic scaffold in neuronal cells, DLGAP2 (1054aa) a synaptic adapter protein and DAP-5 (DLGAP5, 846aa) that is a potential cell cycle regulator.

DLL See **delta**.

D loop Structure (displacement loop) formed when an additional strand of DNA is taken up by a duplex so that one strand is displaced and sticks out like a D-shaped loop. Tends to happen in negatively supercoiled DNA, particularly in mitochondrial DNA as an intermediate during recombination.

DLVO theory Theory of colloid flocculation advanced independently by Derjaguin & Landau and by Vervey & Overbeek and subsequently applied to cell adhesion. There exist distances (primary and secondary minima) at which the forces of attraction exceed those of electrostatic repulsion; an adhesion will thus be formed. For cells there is quite good correlation between the calculated separations of primary and secondary minima and the cell separations in tight junctions (1−2 nm) and more general cell-cell appositions (12−20 nm) respectively, although it is clear that other factors (particularly **cell adhesion molecules**) also play an important part.

dlx A gene family (*DLX1-DLX6*) all containing a homeobox that is related to that of Distal-less (*Dll*), a gene expressed in the head and limbs of developing *Drosophila*. The product is a transcription factor (327aa) involved in the development of larval and adult appendages. In humans DLX1 (255aa) has a regulatory role in the development of the ventral forebrain and other DLX proteins are associated with the ventral forebrain and craniofacial development.

DM-GRASP See **activated leukocyte cell adhesion molecule**.

DMARD A Disease Modifying Anti-Rheumatic Drug, one used to treat rheumatoid arthritis rather than just relieving symptoms. Examples include gold, penicillamine, sulphasalazine and chloroquine, though none are as effective as would be desirable.

DMBT-1 The product (deleted in malignant brain tumour-1, glycoprotein 340, surfactant pulmonary-associated D-binding protein, hensin, salivary agglutinin, 2413aa) of a candidate tumour suppressor gene. May play roles in mucosal defense, cellular immune defense and epithelial differentiation. CRP-ductin is the mouse homologue.

DMEM Dulbecco Modified Eagle's Medium, a very commonly used tissue culture medium for mammalian cells.

dmf (1) **Dimethyl formamide**. (2) 3′,4′-dimethoxy-flavone, an antagonist of PCB126 nuclear receptor **AhR**.

dmp See **dentin matrix protein** (DMP-1).

DMSO Dimethyl sulphoxide. Much used as a solvent for substances that do not dissolve easily in water and that are to be applied to cells (for example cytochalasin B, formyl peptides), also as a cryoprotectant when freezing cells for storage. It is used clinically for the treatment of arthritis, although its efficacy is disputed.

DNA The genetic material (deoxyribonucleic acid) of all cells and many viruses. A polymer of **nucleotides**. The monomer consists of phosphorylated 2-deoxyribose N-glycosidically linked to one of four bases **adenine, cytosine, guanine or thymine**. These are linked together by 3′,5′-phosphodiester bridges. In the Watson-Crick double-helix model two complementary strands are wound in a right-handed helix and held together by hydrogen bonds between **complementary base pairs**. The sequence of bases encodes genetic information. Three major conformations exist **A-DNA, B-DNA** (that corresponds to the original Watson-Crick model) and **Z-DNA**.

DNA adduct DNA that has been modified by the covalent addition of another moiety. Most commonly a result of exposure to pro-oxidant species such as the hydroxyl free radical with formation of 8-hydroxyguanine.

DNA annealing The reformation of double-stranded DNA from thermally denatured DNA. The rate of reassociation depends upon the degree of repetition, and is slowest for unique sequences (this is the basis of the C_0t value; see **C_0t curve**).

DNA barcode Concept that a short sequence of DNA can be used as a species-specific identifier. For prokaryotes ribosomal 16S gene has been used: for eukaryotes one major project uses around 650 base pairs of mitochondrial cytochrome c oxidase. Although this is reasonably species-specific in Lepidoptera and birds, it may be less satisfactory for other phyla.

DNA binding proteins A general term for proteins that interact with DNA, although not those in which the interaction is of enzyme with substrate (i.e. excluding nucleases, polymerases, etc.). Binding may be sequence-specific, as in the case of **transcription factors**, in other cases the role is structural, particularly in the case of **histones**. See also **damaged DNA binding proteins**.

DNA chips A micro-array of DNA or polynucleotides on a solid support (often microscope-slide sized) that can be probed with labelled RNA or DNA that will hybridize. Chips can be designed to

look for expression of particular genes or for **SNPs**. Increasing miniturisation allows thousands of individual samples to be placed on a single chip.

DNA fingerprinting See **restriction fragment length polymorphism**.

DNA footprinting Technique for identifying the recognition site of DNA-binding proteins: see **footprinting**.

DNA fragmentation factor See **DFF**.

DNA glycosylase Class of enzymes involved in **DNA repair**. They recognize altered bases in DNA and catalyse their removal by cleaving the glycosidic bond between the base and the deoxyribose sugar. At least 20 such enzymes occur in cells. Examples include **uracil-DNA glycosylase, thymine-DNA glycosylase** and **formamidopyrimidine-DNA glycosylase**.

DNA gyrase A type II **topoisomerase** (EC 5.99.1.3) of *E. coli*, that is essential for DNA replication. There are two chains, an A chain (875aa) responsible for DNA breakage and rejoining, and a B chain (804aa) that catalyzes ATP hydrolysis, arranged as an A_2B_2 tetramer. Gyrase can induce or relax **supercoiling**, and catalyze the interconversion of other topological isomers of double-stranded DNA rings, including catenanes and knotted rings. The mechanochemical activity requires energy derived from ATP hydrolysis. Inhibited by **quinolone antibiotics**.

DNA helicase Mechanochemical enzymes (unwindase) in prokaryotes and eukaryotes that use the hydrolysis of ATP to unwind the DNA helix at the **replication fork**, to allow the resulting single strands to be copied. Two molecules of ATP are required for each nucleotide pair of the duplex. See **RECQ helicases, CHD proteins, RNA helicases**.

DNA hybridization See **hybridization**.

DNA iteron Repeated DNA sequence (~20 bp) found near the **origin of replication** of some plasmids and that bind plasmid-specific replication initiator protein. Mini-review: http://www2.hawaii.edu/~scallaha/SMCsite/MIcro671Links/10-PasmidIteron/ChattoraIteronRev.pdf

DNA ladder Term often applied to the molecular weight (base-pair) standards run in parallel with DNA samples on an electrophoretic gel. The 'rungs' represent different sizes of polynucleotides and calibrate the gel. See also **DNA laddering**.

DNA laddering The pattern of DNA fragmentation seen on a gel when DNA from apoptotic cells is examined. The DNA is fragmented into multiples of the 180 bp nucleosomal unit by an endonuclease, the DNase I family member, DNase gamma (endoG) (see **DFF**).

DNA library See **genomic library**.

DNA ligase Enzyme involved in DNA replication. The DNA ligase of *E. coli* seals nicks in one strand of double-stranded DNA, a reaction required for linking precursor fragments (**Okazaki fragments**) during discontinuous synthesis on the lagging strand. Nicks are breaks in the phosphodiester linkage that leave a free 3′-OH and 5′-phosphate. The ligase from phage T4 has the additional property of joining two DNA molecules having completely base-paired ends. In humans, DNA ligase 1 (LIG1, EC 6.5.1.1, 919aa) seals nicks in double-stranded DNA during replication, recombination and repair, LIG3 (1009aa) can correct defective DNA strand-break repairs, LIG4 (911aa) is involved in DNA non-homologous end joining (NHEJ). DNA ligases are crucial in joining DNA molecules and preparing radioactive probes (by nick translation) in recombinant DNA technology.

DNA markers (1) Genetic markers, for example polymorphisms such as **SNPs** and short tandem repeats that allow relationships to be established, genetic diversity to be estimated and so on. (2) Defined-length oligonucleotides used as size markers in gel electrophoresis (see **DNA ladder**.

DNA methylation Process by which methyl groups are added to certain nucleotides in genomic DNA. This affects gene expression, because methylated DNA is not easily transcribed. The degree of methylation is passed on to daughter strands at mitosis by maintenance DNA **methyltransferases**. Accordingly, DNA methylation is thought to play an important developmental role in sequentially restricting the transcribable genes available to distinct cell lineages (see genomic imprinting). Methylation may also allow targeting of specific regions by **methyl-CpG-binding proteins** which induce binding of other regulatory proteins such as histone deacetylases. In bacteria, methylation plays an important role in the restriction systems because many **restriction enzymes** cannot cut sequences with certain specific methylations.

DNA polymerase alpha-primase complex In *S. cerevisiae* DNA polymerase alpha (pol alpha subunit A, p180, 1468aa; subunit B, p74, 705aa) forms a four subunit complex with DNA primase (prim-2: p58, 528aa; p48, 409aa) and is the only enzyme able to start DNA synthesis *de novo*. The major role of the DNA polymerase alpha-primase complex (pol-prim) is in the initiation of DNA replication at chromosomal origins and in the discontinuous synthesis of **Okazaki fragments** on the lagging strand of the replication fork. Homologues in other eukaryotes are basically similar.

DNA polymerase delta complex A complex of DNA polymerase delta with proliferating cell nuclear antigen (**PCNA**) and replication factor C

that is involved in leading strand synthesis, completing Okazaki fragments initiated by the DNA polymerase alpha/primase complex and DNA repair. Pol d is one of the DNA polymerase B family that has two enzymatic activities: DNA synthesis (polymerase) and an exonucleolytic activity that degrades single stranded DNA in the 3′- to 5′-direction. It is a heterotetramer (subunits of 1107aa, 469aa, 466aa and 107aa).

DNA polymerases Enzymes (EC 2.7.7.7) involved in DNA template-directed synthesis of DNA from deoxyribonucleotide triphosphates. In prokaryotes there are five DNA polymerase classes: Pol-I and pol-II are involved in DNA repair, Pol-III is the main polymerase responsible for elongation, Pol-IV and Pol-V are **translesion synthesis** polymerases. There are at least 15 DNA polymerases in eukaryotes, with Pol-α apparently responsible for replication of nuclear DNA (see **DNA polymerase alpha-primase complex**), and Pol-γ for replication of mitochondrial DNA. Pol-β is implicated in repairing DNA, Retroviruses possess a unique DNA polymerase (**reverse transcriptase**) that uses an RNA template. DNA polymerases can be subdivided into seven different families: A, B, C, D, X, Y, and RT. See Table D1.

DNA primase Enzymes (RNA polymerases, RNA primase; EC 2.7.7.6) that catalyse the synthesis of short (~10 bases) RNA primers on single stranded (ss) DNA templates that are used by DNA polymerase to initiate the synthesis of **Okazaki fragments** on the lagging strand. Bacterial primases have three functional domains in the protein, a N-terminal 12 KDa fragment contains a zinc-binding motif, a central fragment of 37 KDa with conserved sequence motifs that are characteristic of primases, including the so-called 'RNA polymerase (RNAP)-basic' motif and a C-terminal domain of approximately 150 residues that interacts with the replicative helicase, DnaB, at the replication fork. Eukaryotic DNA primase is a heterodimer of large (p60) and small (p50) subunits that show little homology with prokaryotic primases.

DNA probe A short sequence of DNA that has been labelled isotopically or chemically and that can be used to detect a complementary nucleotide sequence. A diversity of labelling methods has been developed and DNA probes are increasingly being used for example in detection and identification of pathogens and in testing for genetic abnormalities in chromosomal DNA.

DNA profiling Forensic tool to compare samples of DNA. Generally preferred to the term 'DNA fingerprinting'.

DNA rearrangement Wholesale movement of sequences from one position to another in DNA,

TABLE D1. DNA Polymerases

Family	Distribution	Role	Examples
A	Prokaryotes, eukaryotes, viruses	replication and repair	[**T7 polymerase**], Pol γ, pol I.
B	Prokaryotes, eukaryotes	repair and replication	Pol II (bacteria), Pol B (archaea), and Pol α, δ, ε and ζ (eukaryota)
C*	Bacteria	replication	Pol-III
D	Archaea, Euryarchaeota	replication	
X	Eukaryotes	**base excision repair, non-homologous end-joining**	polB, pol σ, pol λ, pol μ, and [**TdT**]
Y	Prokaryotes, eukaryotes	translesion synthesis	Pol κ, DNA polymerase IV (EC 2.7.7.7)
RT	retroviruses, eukaryotes	reverse transcription	pol in viruses, telomerase in eukaryotes

*May be a subset of Family X.

such as occur somatically, for example in the generation of antibody diversity.

DNA renaturation See **DNA annealing**.

DNA repair Enzymic correction of errors in DNA structure and sequence that protects genetic information against environmental damage and replication errors. See **activation induced deaminase, ape-1, apollo, artemis, BASC complex, cernunnos, damaged DNA binding proteins, DNA glycosylase, DNA ligase, DNA polymerase, DNA polymerase delta complex, excision repair, KARP-1, Ku, lex A, mismatch repair, NHEJ, nucleotide-excision repair factor, PCNA, photolyase, poly(ADP-ribose) polymerase, rad proteins, rec proteins, RECQ helicases, SMC proteins, SOS system**; also diseases associated with deficiencies in error repair: **ataxia telangiectasia, Cockayne's syndrome, DeSanctis-Cacchione syndrome, Werner's syndrome, xeroderma pigmentosum**.

DNA replication The process whereby a copy of a DNA molecule is made, and thus the genetic information it contains is duplicated. The parental double stranded DNA molecule is replicated semi-conservatively, i.e. each copy contains one of the original strands paired with a newly synthesized strand that is complementary in terms of AT and GC base pairing. Though in this sense conceptually simple, mechanistically a complex process involving a number of enzymes.

DNA sequencing Determination of the nucleotide sequence of a length of DNA. Typically, this is performed by cloning the DNA of interest, so that enough can be prepared to allow the sequence to be determined, usually by the Sanger **dideoxy sequencing** method or the **Maxam-Gilbert method**. The resulting reactions are then run on a large sequencing gel, capable of resolving single nucleotide differences in chain length. Recently, **PCR**-based methods have obviated the need to clone the DNA under some conditions, and automated DNA sequencing using column chromatographic separation has become widely available. Eventually sequencing of whole genomes, even those of individuals, will probably become economically feasible and perhaps commonplace. Cost estimates (2011): http://blogs.nature.com/news/2011/01/600_genomes.html

DNA synthesis The linking together of nucleotides (as deoxyribonucleotide triphosphates) to form DNA. *In vivo*, most synthesis is **DNA replication**, but incorporation of precursors also occurs in repair. In the special case of retroviruses, DNA synthesis is directed by an RNA template (see **reverse transcriptase**). *In vitro* synthesis can be done using automated solid-phase methods and relatively long sequences (~200 bp) can be prepared.

DNA topoisomerase See topoisomerases.

DNA transfection Originally a term describing viral infection of animal cells by uptake of purified viral DNA rather than by intact virus particles, now much broader, see **transfection**.

DNA tumour virus Virus with DNA genome that can cause tumours in animals. Diverse, and found among the **Papillomaviridae, Polyomaviridae** (e.g. SV40), **Adenoviridae**. See **Epstein-Barr virus**.

DNA vaccine Vaccine in which the active principle is a DNA sequence that will be transiently expressed in host cells and generate antigens to stimulate an immune response. Are considered to have great potential especially in diseases for which it has been difficult to develop conventional vaccines (e.g. AIDS, malaria).

DNA virus A virus with a genome composed of double- or single-stranded DNA. Group I viruses possess double-stranded DNA and include, among other families, the **Adenoviridae, Herpesviridae, Papillomaviridae, Polyomaviridae, Poxviridae, Mimivirus** and many tailed bacteriophages. Group II viruses possess single-stranded DNA and include families such as the **Parvoviridae** and bacteriophage M13. See **DNA tumour virus**.

DNA-activated protein kinase A nuclear DNA-dependent serine/threonine protein kinase (DNA-dependent protein kinase, DNA-PK) that acts as a molecular sensor for DNA damage. Consists of a catalytic subunit (EC 2.7.11.1, 4128aa) and a dimer of the Ku autoantigen which directs it to DNA and activates the kinase. Phosphorylates various transcription factors, probably modulating their activity.

DNA-binding domain Domain in a protein that is responsible for binding to DNA, usually with sequence-specificity. Thus many transcription factors would be expected to have such domains. Classic examples are in **zinc finger proteins**, the **helix-turn-helix** proteins, and the **leucine zipper** proteins.

DNA-binding proteins Proteins that interact with DNA, typically to pack or modify the DNA e.g. histones, or to regulate gene expression, transcription factors. Among those proteins that recognize specific DNA sequences, there are a number of characteristic conserved 'motifs' (**DNA-binding domains**) believed to be essential for specificity.

DNA-PK See **DNA-activated protein kinase**.

dnaA **etc.** Genes in *E. coli* that are involved in coding for replication machinery. *dnaA* and *dnaP* produce proteins involved in replication at the chromosome origin; *dnaB, C* and *D* are involved in **primosome** formation; *dnaE* codes for subunits of polymerase II; *dnaF* for ribonucleotide reductase; *dnaG* codes for primase; *dnaH, Q, X* and *Z* for components of polymerase III; *dnaI* for protein involved at the replication fork; *dnaJ* and *dnaK* products (see **dnaJ, dnaK**) are necessary for survival at high temperature, also considered essential for phage lambda replication; *dnaL* and *M* are uncharacterized; dnaT protein interacts with *dnaC* product, *dnaW* codes adenylate kinase.

DNAase More commonly DNAse. See **deoxyribonuclease**.

Dnadist A program that uses nucleotide sequences to compute a distance matrix, under three different models of nucleotide substitution. The distance for each pair of species is an estimate of the divergence time between those two species. Details: http://cmgm.stanford.edu/phylip/dnadist.html

dnaJ In *E. coli*, a chaperone protein (HSP40, 376aa) that interacts with Hsp70-like **DnaK** protein and **GrpE** to disassemble a protein complex at the origins of replication of phage lambda and several plasmids. Participates in hyperosmotic and heat shock responses by preventing the aggregation of stress-denatured proteins and by disaggregating proteins, also in an autonomous, dnaK-independent fashion. Unfolded proteins bind initially to dnaJ which triggers more complex interactions required for efficient folding. The human homologue (396aa) is a co-chaperone of Hsc70 and has a role in protein import into mitochondria.

dnaK Bacterial molecular chaperone of the Hsp70 family (638aa). Interacts with **DnaJ** and **GrpE** in stress responses and in refolding of misfolded proteins. DnaK is itself a weak ATPase; ATP hydrolysis by DnaK is stimulated by its interaction with another co-chaperone, DnaJ and release of ADP is stimulated by GrpE.

DNase See **deoxyribonuclease**.

DNAX (1) In *E. coli*, *dnaX* is the gene that encodes DNA polymerase III subunit tau (EC 2.7.7.7, 643aa) which is cleaved to form DNA polymerase III subunit gamma (431aa). (2) See **DNAX-activation proteins**. (3) In humans, DNAX accessory molecule 1 (CD226, 336aa) is a receptor involved in intercellular adhesion, lymphocyte signalling, cytotoxicity and lymphokine secretion mediated by cytotoxic T-cells and NK cells.

DNAX-activation proteins Adapter proteins; DAP10 (DNAX-activation protein-10, haematopoietic cell signal transducer, 93aa) is a transmembrane protein that forms a homodimer with **NKG2** which is a receptor for the recognition of MHC class I HLA-E molecules by NK cells and some cytotoxic T-cells. DAP-12 (killer-activating receptor-associated protein, 113aa) is expressed by NK cells and B- and T-cells and has an immunoreceptor tyrosine-based activation motif (ITAM) that will bind **zap-70** and **syk** and thus activate NK cells, although it has been implicated in inhibitory signalling in mouse macrophages and dendritic cells. Defects cause **Nasu-Hakola disease**. See **TREM-2**. DAP12: http://www.ncbi.nlm.nih.gov/pubmed/17220916

dnd **genes** (1) A gene cluster found in *Streptomyces lividans* which sensitises its DNA to degradation during electrophoresis (the Dnd phenotype). The *dnd* gene cluster incorporates sulphur into the DNA backbone as a sequence-selective, stereospecific phosphorothioate modification. (2) The DND (Defence, No Death) loci of *Arabidopsis* regulate the extent of broad-spectrum disease resistance against a broad range of viral, bacterial, oomycete and fungal pathogens. Plants lacking a functional copy of the *DND1* or *DND2* gene are defective in hypersensitive response (HR) cell death but exhibit successful disease resistance. The *DND1* gene product (DND1, 726aa) encodes a cyclic nucleotide-gated ion channel. (3) The *dnd* gene in *Brachydanio* encodes a RNA-binding factor (dead end protein 1, 411aa) that positively regulates gene expression by prohibiting miRNA-mediated gene suppression. The mouse homologue (352aa) has a similar function.

DNP See **Dendroaspis natriuretic peptide**.

DNS Abbreviation for dansyl. See **dansyl chloride**.

docetaxel Taxane (Taxotere™) extracted from the English yew (*Taxus baccata*), slightly more potent than paclitaxel (**taxol**).

dockerin domains Domains (Doc domain) that bind the cellulose-degrading enzymes of the **cellulosome** to **cohesin** domains in the proteins of the bacterial cell wall. The Doc-I domain (65–70aa) which binds cohesin-I has a conserved fold of 42aa and two calcium-binding sites with sequence similarity to the **EF-hand** motif.

docking protein See **signal recognition particle-receptor**.

docosahexaenoic acid Any straight-chain fatty acid with 22 carbon atoms and six double bonds. The all-Z isomer is found in fish oils. It is a major omega-3 fatty acid in human brain, synapses, retina, and other neural tissues. See **protectin D1**.

dodecanoic acid *lauric acid* A saturated 12-carbon fatty acid, the sodium salt of which (sodium dodecyl sulphate, SDS or sodium lauryl sulphate) is extensively used as a detergent.

dodo A **peptidylprolyl isomerase** (166aa) that facilitates the degradation of the transcription factor CF2, which regulates expression of the *rhomboid* gene in *Drosophila* follicle cells. This is required to establish the dorsal/ventral polarity of the developing oocyte. Degradation is probably facilitated by isomerizing the prolyl peptide bond after MAPK-catalysed phosphorylation of the protein. Flybase entry: http://www.sdbonline.org/fly/torstoll/dodo1.htm

dok proteins A family of adaptor proteins that are 'downstream of tyrosine kinases' and are enzymatically inert docking molecules similar to the insulin receptor substrate family of proteins, with an amino-terminal pleckstrin homology (PH) domain, a central putative phosphotyrosine-binding (PTB) domain and numerous potential sites of tyrosine phosphorylation. When tyrosine-phosphorylated they link clustered receptors to other signalling molecules. DOK1 (docking protein-1, 481aa) appears to be a negative regulator of the insulin signalling pathway. DOK2 (412aa) may modulate proliferation induced by IL-4

and Bcr-Abl signalling. DOC7 (504aa) is thought to be an activator of the muscle-specific tyrosine-protein kinase receptor (**MuSK**) that plays an essential role in neuromuscular synaptogenesis. A number of other docking proteins have been identified.

dolastatins Family of peptides isolated from the marine nudibranch mollusc *Dolabella auricularia* (sea hare). Dolastatin 10 is a potent antimitotic penta-peptide that inhibits microtubule assembly. Complete synthesis is being used to generate variants. Dolastatin 11, a depsipeptide, binds to actin and stabilizes F-actin *in vitro*, like **phalloidin** and **jasplakinolide**, although acting at a different site. Tasidotin, a synthetic analogue of dolastatin 15, inhibits microtubule assembly and induces a G2-M block in treated tumour-derived cells. Lyngbyastatin 4 is a depsipeptide isolated from the marine cyanobacterium *Lyngbya confervoides* and is an analogue of dolastatin 13; it selectively inhibits elastase and chymotrypsin *in vitro*. See **doliculide**, **phomopsin A**.

dolichol Terpenoids with 13–24 isoprene units and a terminal phosphorylated hydroxyl group. Function as transmembrane carriers for glycosyl units in the biosynthesis of glycoproteins and glycolipids. The core oligosaccharide for N-glycosylation of proteins is constructed on a dolichol phosphate molecule prior to its donation to the nascent polypeptide chain.

doliculide An actin binding macrocyclic depsipeptide that stimulates actin assembly, extracted from the sea hare (*Dolabella auricularia*); competes with **phalloidin** for binding and has the same effects as **jasplakinolide**

dolipore septum A type of septum found in the Basidiomycetes that has channels which give cytoplasmic connectivity but restrict the exchange of organelles and nuclei. There is a pore cap (**parenthesome**) surrounding a septal swelling and septal pore. (Dolioform is an adjective describing a barrel-shaped object).

DOM (1) In *Drosophila* a transcription factor, Domina (Dom, 719aa) of the FKH/WH (forkhead/winged helix) family. (2) The *Drosophila* gene *dom* (domino) encodes a helicase (EC 3.6.4.-, 3198aa). (3) In humans the protein Dom3Z (Dom-3 homolog-Z, 396aa) may have pyrophosphohydrolase activity towards 5′ triphosphorylated RNA. (4) In *S. cerevisiae*, Dom34 (386aa) is involved in protein translation and is a member of the eukaryotic release factor 1 family.

domain Used to describe a part of a molecule or structure that shares common physicochemical features, e.g. hydrophobic, polar, globular, α-helical domains, or properties, e.g. DNA-binding domain, ATP-binding domain.

dominant Describing an allele that has an effect when present as a single copy in a heterozygous diploid organism or describing a phenotyic character due to a dominant gene. In dominant negative mutations the product suppresses the activity of the product of the normal allele by forming an inactive complex or by competing with the normal protein so that the overall activity is below a critical threshold level and function is abnormal (haplo-insufficiency). *Cf.* **apical dominance, recessive**.

domoic acid A tricarboxylic acid toxin similar to the **glutamate receptor** agonist **kainic acid**, originally isolated from the macroscopic red alga *Chondria armata*, (known locally in Japan as domoi), used as an antihelminthic in a traditional medicine. Acts preferentially upon a sub-class of ionotropic glutamate receptors found in nervous tissue. Has been identified as the cause of amnesic shellfish poisoning, the source of the toxin being the diatom *Pseudo-nitzschia* (previously *Nitzschia*) *pungens*. Details of poisoning: http://www.nwfsc.noaa.gov/hab/habs_toxins/marine_biotoxins/da/index.html

Donnai-Barrow syndrome An autosomal recessive disorder (facio-oculo-acoustico-renal syndrome), caused by mutation in the **lipoprotein receptor-related protein 2**.

Donnan equilibrium An equilibrium established between a charged, immobile colloid (such as clay, ion exchange resin or cytoplasm) and a solution of electrolyte. Ions of like charge to the colloid tend to be excluded and ions of opposite charge tend to be attracted; the colloid compartment is electrically polarized relative to the solution in the same direction as the colloid charges (a 'Donnan potential'); and the osmotic pressure is higher in the colloid compartment.

donor splice junction The junction between **exon** and an **intron** at the 5′ end of the intron. During **processing** of **hnRNA** the donor junction is spliced to the acceptor junction at the 3′ end of the intron and the intron is omitted. Introns invariably begin with GU and end in AG and mutational change in either can prevent normal splicing.

DOPA Precursor (L-DOPA, levodopa, 3-hydroxytyrosine) of the neurotransmitter dopamine, made from L-tyrosine by tyrosine 3-mono-oxygenase and used as a treatment for **Parkinsonism**.

DOPAC A metabolite (3,4-dihydroxyphenylacetic acid) of **dopamine** produced by the action of monoamine oxidase and which is further degraded by catechol-O-methyl transferase (COMT) to form homovanillic acid (HVA). Dopamine can also be degraded to 3-methoxytyramine (3-MT) by COMT and monoamine oxidase will catalyze degradation of 3-MT to HVA

dopamine A **catecholamine** neurotransmitter and hormone (153 Da), formed by decarboxylation of dihydroxyphenylalanine (DOPA). A precursor of **adrenaline** and **noradrenaline**. Dopamine released by dopaminergic neurons is taken up by the dopamine uptake transporter (sodium-dependent dopamine transporter, solute carrier family six members 3, 620aa) and this terminates the action. The uptake process is inhibited by **cocaine**. See **dopamine receptors**.

dopamine receptors Family of G **protein coupled** receptors for **dopamine** encoded by DRD genes. In humans, D1-like receptors (D1, 446aa; D5, 477aa) interact with G-proteins that are activators of adenylyl cyclase, D2-like (D2, 443aa; D3, 400aa; D4, 467aa) with G-proteins that are inhibitors of adenylyl cyclase. Most antipsychotic drugs are dopamine receptor antagonists and most neuroleptics were developed as D2 receptor antagonists. In *Drosophila* both D1 (511aa) and D2 (539aa) receptors activate adenylyl cyclase. Receptor subtypes: http://www.acnp.org/g4/GN401000014/CH014.html

doppel Prion-like protein (Dpl, prion protein 2, 176aa) related to the **prion** protein (PrP). In adults is found only in testis.

dormin See **abscisic acid**.

dorrigocins Glutarimide antibiotics with some antifungal activity. They will modify the morphology of ras-transformed NIH/3T3 cells from a transformed phenotype to a normal one by inhibiting the carboxyl methyltransferase involved in ras processing. Article: http://www.journalarchive.jst.go.jp/jnlpdf.php?cdjournal=antibiotics1968&cdvol=47&noissue=8&startpage=875&lang=en&from=jnlabstract

dorsal *Drosophila* polarity gene; homologue of the *rel* **proto-oncogene**. The product is a morphogenetic protein (999aa) that specifically binds to the kappa B-related consensus sequence in the enhancer region of zygotic genes that encode transcription factors such as snail, twist and **zerknuellt 1** and the morphogen **decapentaplegic**. The lateral or ventral identity of a cell depends upon the concentration of dorsal protein in its nucleus during the blastoderm stage. See **tube, pelle** and **toll**.

dorsal horn Region in the grey matter of the spinal cord, consisting of five zones (lamina I-V) where nociceptive information begins to be processed in the central nervous system. The dorsal (posterior) horn receives sensory input either from the skin, striated muscles or joints or from blood vessels and internal organs.

dorsal root ganglion Nodule on a dorsal root (the nerve bundle from the dorsal part of the spinal cord) that contains cell bodies of afferent spinal nerve neurons leading into the dorsal part of the spinal cord. Dorsal root ganglia from chick embryos are a classic source of neurites for cell culture.

dorsalin-1 Chicken protein (427aa) of the TGFβ family that stimulates **neural crest** differentiation, neural crest growth, bone growth and wound healing. The human homologue is probably growth differentiation factor 2 (**BMP9**).

dorsalization In development the series of events that define dorsal elements of the body. The Wnt signalling pathway plays a major role. Abstract: http://www.ncbi.nlm.nih.gov/pubmed/17681137

dosage compensation Genetic mechanisms that allow genes to be expressed at a similar level irrespective of the number of copies at which they are present. Usually invoked for genes that lie on sex chromosomes, that are thus present in different copy numbers in males and females.

dose-response curve Graph of the relation between dose (concentration of substance introduced into the system) and the effect (enzyme activity, membrane potential, mortality, etc.) that is being measured. Standard dose-response curves are similar to receptor-binding curves (sigmoidal), but exceptions are found when there is cooperativity or dual response modes. A dose-response curve with a standard slope has a **Hill coefficient** of unity.

dot blot Method for detecting a specific protein or message. A spot of solution is dotted onto nitrocellulose paper, a specific antibody or probe is allowed to bind and the presence of bound antibody/probe then shown by using a peroxidase-coupled second antibody, as in **Western blot** or by other visualization methods. See also **slot blot**.

double decomposition A reaction between two chemical substances that results in the exchange of a constituent from each compound and the two reactants generate two different products.

double diffusion The principle underlying the **Ouchterlony assay** in which antibody and antigen diffuse through a gel (usually agarose) from separate wells and form an opaque precipitin line where the antibody/antigen ratio is balanced.

double helix Conformation of a DNA molecule – like a twisted ladder.

double layer The zone adjacent to a charged particle in which the potential falls effectively to zero. An excess or deficiency of electrons on the surface (charge; not to be confused with the transmembrane potential) leads to an equivalent excess of ions of the opposite charge in the surrounding fluid. For most cells, that have negative charges, there will be an excess of cations immediately adjacent to the plasma membrane, and at physiological ionic strength the double layer is likely to be around 2–3 nm thick.

double minute An E3 ubiquitin-protein ligase (double minute 2 protein, oncoprotein Mdm2, p53-binding protein Mdm2, 491aa) that mediates ubiquitination of p53 and its degradation by the proteasome. See **mdm #2**

double minute chromosome Small paired extra-chromosomal bodies comprising circular DNA, associated with many tumours. There may be multiple copies and the resultant amplification of genes such as those for growth factors may be responsible for uncontrolled proliferation.

double mutant Organism in which there are two mutations, often necessary if the effect is to be seen because two parallel systems need to be affected, although in some cases one mutation can mask (suppress) the effect of the other.

double recessive An organism that is homozygous for a recessive allele and expresses the phenotype.

double strand break Serious form of damage to DNA in which both strands are cleaved. See **MRN complex**, **nibrin**.

double stranded RNA Form of RNA (dsRNA) in which there is base-pairing between complementary regions producing duplex structures similar to those of DNA. Found as the genetic material in some viruses. dsRNA introduced into cells can cause **RNA interference**, the sequence-specific degradation of mRNA or, sometimes, induction of interferon production.

doublecortin A developmentally expressed neuronal microtubule-associated protein (doublin, lissencephalin-X, 441aa) that seems to be required for initial steps of neuronal dispersion and cortex lamination during cerebral cortex development. Mutations in the human doublecortin gene result in abnormal neuronal migration, epilepsy, and mental retardation. A family of doublecortin like proteins (including the retinitis pigmentosum 1 (RP1) gene product and **doublecortin-like kinase 1**) has been identified, all with doublecortin-like (DCX) domains (usually in tandem). The domain (~80aa) is a microtubule-binding module and is involved in protein-protein interactions. Phylogenetically ancient with homologues in invertebrates and unicellular organisms. Doublecortin is widely used as a marker for newly generated neurons.

doublecortin-like kinase A serine/threonine-protein kinase (EC 2.7.11.1, 740aa) that may be involved in a calcium-signaling pathway controlling neuronal migration in the developing brain. Has two doublecortin domains.

doublet microtubules Microtubules of the axoneme. The outer nine sets are often referred to as doublet microtubules, although only one (the A tubule) is complete and has 13 protofilaments. The B-tubule has only 10 or 11 protofilaments, and shares the remainder with the A-tubule. A and B tubules differ in their stability and in the other proteins attached periodically to them; it is the **dynein** affixed to the A tubule attaching and detaching from the B tubule of the adjacent doublet that generates sliding movement in the **axoneme**.

doubletime A *Drosophila* ser/thr kinase (discs overgrown protein kinase, EC 2.7.11.1, 440aa) that forms a complex with **period** and by phosphorylation may influence its stability.

doubling time The time taken for a cell to complete the cell cycle.

Dounce homogenizer An apparatus, usually made of glass, with a tightly-fitting plunger (pestle) in a glass tube. Tissue is homogenised by shear-forces generated by rotation and gentle reciprocation of the plunger; by varying the clearance between pestle and wall the particle size in the homogenate can be altered.

Down's syndrome Formerly referred to as mongolism, because affected individuals supposedly had facial features typical of the eponymous ethnic group. It is caused by trisomy of chromosome 21 and the various effects are due to overproduction of products of triplicated genes. Common (1 in 700 live births) and incidence increases with maternal age. The cause is usually non-disjunction at meiosis but occasionally a translocation of fused chromosomes 21 and 14. See **Down's syndrome critical region**.

Down's syndrome critical region Region of chromosome 21 thought to be responsible for some, if not all, of the features of **Down's syndrome**. DSCR1 encodes **calcipressin** 1. Article on control of expression: http://www.jbc.org/content/280/33/29435.long.

down-regulation Reduction in the responsiveness of a cell to a stimulus following first exposure, often by a reduction in the number of receptors expressed on the surface (as a consequence of reduced recycling). The term is often used imprecisely.

downstream (1) Portions of DNA or RNA that are more remote from the initiation sites and that will therefore be translated or transcribed later. (2) Shorthand term for things that happen at a late stage in a sequence of reactions, for example in a signalling cascade.

doxorubicin Cytotoxic antibiotic from *Streptomyces peucetius*. Blocks **topoisomerase** and **reverse transcriptase** by intercalating into the DNA. Has been used in clinical oncology.

Doyne's honeycomb retinal dystrophy A disorder (malattia leventinese) caused by mutation in **fibulin-3**. Yellow-white deposits (drusen) accumulate beneath the retinal pigment epithelium.

DP-1, DP-2 (1) G protein-coupled receptors for prostaglandin D_2 with opposite effects. DP1 activation tends to ameliorate the pathology in asthma; DP2 is preferentially expressed on type 2 lymphocytes, eosinophils, and basophils and is thought to be important in the promotion of Th2-related inflammation. (2) Cell cycle-regulating transcription factors (DP-1 and DP-2) exist in humans and there are additional isoforms, DP-1alpha, 278aa; DP-1beta, 357aa). Form a heterodimer with **E2F** and regulate progression through the cycle. (3) A pneumococcal bacteriophage Dp-1 that produces lysin (EC 3.5.1.28, N-acetyl-muramoyl-L-alanine amidase, 296aa) that lyses bacterial cell walls.

DPA See **dipicolinic acid**.

dpe In *Drosophila*, a downstream promoter element that functions cooperatively with the initiator (Inr) for the binding of TFIID in the transcription of core promoters in the absence of a TATA box. Details: http://www.ncbi.nlm.nih.gov/pubmed/10848601

DPIP The dye dichlorophenol-indophenol (DCPIP) often used as an indicator for the activity of the electron transport system during the light-dependent reactions of photosynthesis, becoming colourless when reduced.

Dpl See **doppel**.

dpm (1) Disintegrations per minute, a measure of radioactivity. (2) Defects per million, in manufacturing processes.

DPN Diphosphopyridine nucleotide, an obsolete name for nicotinamide adenine dinucleotide (**NAD**).

Dpp See **decapentaplegic**.

DPPC Dipalmitoyl-phosphatidylcholine.

DR (1) In humans, DR1 (down-regulator of transcription 1, TATA-binding protein-associated phosphoprotein, 176aa). Forms a heterodimer with **DRAP1** that interacts with the TATA binding protein and functionally represses activated and basal transcription of class II genes. (2) Class II histocompatibility antigens, HLA-DR. (3) Various **death receptors** e.g. DR3 (death receptor 3), a member of the **Fas** gene family. The gene can be duplicated and duplication is more common in people with rheumatoid arthritis.

DRAK See **DAP kinase**.

DRAP (1) Dr1-associated protein (DR1-associated corepressor, negative co-factor 2-alpha, 205aa). The interaction of the DR1/DRAP1 heterodimer with TATA binding protein (TBP) inhibits the association of TBP with TFIIA and/or TFIIB and represses transcription of class II genes. (2) Down region aspartic protease (DRAP, beta-secretase 2, EC 3.4.23.45, 518aa) that proteolytically processes amyloid precursor protein (APP).

draxin Dorsal repulsive axon guidance protein (neucrin, 349aa) required for the development of spinal cord and forebrain commissures. Inhibits the stabilization of cytosolic beta-catenin and acts as an antagonist of Wnt signalling.

DRB Multiallelic locus in the Class II **MHC** DR region encoding β-chains. Over a hundred alleles have been reported at the DRB locus in humans, which is more polymorphic than most of these loci. Other loci are DQ and DP.

DRD See **dopamine receptors**.

DREAM A protein (downstream regulatory element antagonist modulator, calsenilin, potassium channel-interacting protein 3, 256aa) first identified as a Ca^{2+}-regulated transcriptional repressor belonging to the neuronal calcium sensor (NCS) family. It is preferentially expressed in the CNS and appears to have multiple functions in pain modulation, long-term potentiation, learning and memory. The effects may be a result of inhibition of the expression of NMDA receptors (see **glutamate receptors**). NMDA receptor modulation: http://www.ncbi.nlm.nih.gov/pubmed/20519532

DREAM complex A multi-protein complex originally described in *Drosophila* that can act as a transcription activator or repressor depending on the context. The complex contains multiple site-specific DNA-binding proteins such as *Drosophila* Myb–MuvB (MMB)/dREAM. Represses cell cycle-dependent (*cdc*) genes in quiescent cells. Other subunits include **retinoblastoma**-like proteins, histone-binding protein, RBBP4 (retinoblastoma-binding protein 4, **chromatin assembly factor** 1 subunit C, 425aa), LIN52 and LIN54. The complex dissociates in S phase when the LIN proteins form a subcomplex that binds to MYBL2. Mammalian homologues (e.g. **antolefinin**) have been described but not yet fully characterized. *Cf.* **LIN complex**. Description: http://genesdev.cshlp.org/content/21/22/2880.full.pdf

drebin A developmentally regulated F-actin binding brain protein which in the chicken has characteristic changes in expression related to developmental stage. Contains a single **ADF-H domain**. Same class as yeast ABP-1. Not in current usage.

dredd An effector of apoptosis in *Drosophila*, (caspase-8, EC 3.4.22.61, death-related ced-3/NEDD2-like protein, 494aa). Activated by **reaper**, **grim** and W (cell death protein W, wrinkled, 410aa). BG4 (239aa), the *Drosophila* homologue of **FADD** promotes cleavage of Dredd and is necessary and sufficient for enhancing Dredd-induced apoptosis.

D region See **D gene segment**.

drepanocyte Synonym for a sickle cell. (See **sickle cell anaemia**).

DRG See **dorsal root ganglion**.

Drickamer motif Either of the two highly conserved patterns of invariant amino acids found in the carbohdrate recognition domain of C-type and S-type lectins, as described by Drickamer. Usage obsolete although motifs still recognised.

drosha The nuclear RNase III enzyme (EC 3.1.26.3, 1374aa) that is the major nuclease involved in initiation of **microRNA** (miRNA) processing in the nucleus. A component of the **microprocessor complex**.

drosocin A cationic 19aa **antimicrobial peptide** secreted by *Drosophila* in response to bacterial infection. The peptide is glycosylated at Thr11 which appears to be important for its potent antimicrobial activity. Has sequence homology with **apidaecin** Ib.

drosomycin Antimicrobial peptide, 44aa with four intramolecular disulfide bridges, product of a toll-dependent immunity gene in *Drosophila* and other diptera. Has anti-fungal activity.

Drosophila A genus of small dipteran flies. The best known species is *D. melanogaster*, often called the fruit fly, but more correctly termed the vinegar fly. First investigated by T.H.Morgan and his group, it has been extensively used in genetic studies. More recently it has been used for studies of embryonic development.

drosulphakinins *Drosophila* homologues of the **gastrin** family of peptide hormones.

drug delivery The process of getting a drug into the appropriate compartment of the body: orally, if the drug can be absorbed through the gut, by inhalation, injection or by transdermal diffusion. If the target is superficial then topical application may be suitable and in this case absorption into deeper tissues may be undesirable.

DrugBank A database that combines detailed drug (i.e. chemical, pharmacological and pharmaceutical) data with comprehensive drug target (i.e. sequence, structure, and pathway) information. The database contained 6707 drug entries in September 2011. Link: http://www.drugbank.ca/

druse crystal (1) Crystals found in plant tissue, often composed of calcium oxalate, formed within the central vacuoles of parenchyma cells (idioblasts) in the cortex region just outside the phloem. (2) Extracellular deposits (drusen) that accumulate below the retinal pigment epithelium on **Bruch's membrane**.

DSB (1) **Double-strand breaks**. (2) Dsb proteins are thiol: disulphide interchange proteins (DsbA, 208aa; DsbB, 176aa; DsbC, 236aa; DsbD, 565aa; DsbG, 248aa) catalyze formation and isomerization of protein disulphide bonds in the periplasm of *E. coli* and other bacteria. DsbC is a disulphide isomerase and can convert aberrant disulphide bonds to correct ones. DsbG and DsbC are part of a periplasmic reducing system that controls the level of cysteine sulphenylation, and provides reducing equivalents to rescue oxidatively damaged secreted proteins.

dscam (1) An immunoglobulin superfamily molecule (2016aa) in *Drosophila* that is hypervariable (because of multiple alternative splicing options independently in three of the seven Ig domains) and confers adhesive specificity because only identical isoforms interact as cell adhesion molecules. Important in the developing nervous system because isoform 'recognition' allows repulsion between dendrites of a single neuron. (2) Down's syndrome cell adhesion molecule (2012aa) which is the human homologue of the *Drosophila* dscam molecule and also involved in neuronal self-avoidance. Article: http://www.ncbi.nih.gov/pmc/pubmed/15169762

DSIP See **delta sleep inducing peptide**.

DSPP See **dentin sialophosphoprotein**.

DST (1) In *Dictyostelium*, a family of serine/threonine-protein kinase (dst1, 737aa; dst2, 1142aa; dst3, 562aa; dst4, 485aa). (2) See **dystonin**. (3) In *S. cerevisiae*, transcription elongation factor S-II (Dst1), DNA strand transferase 1, pyrimidine pathway regulatory protein 2, 309aa) necessary for efficient RNA polymerase II transcription elongation.

D-TACC The homologue in *Drosophila* of the mammalian **TACC**s (transforming, acidic, coiled-coil-containing proteins). D-TACC (1226aa) is concentrated at centrosomes, interacts with microtubules and is essential for normal spindle function in the early embryo. Aurora A kinase activates D-TACC–**minispindles** complexes at centrosomes to stabilize microtubules. Original description: http://www.nature.com/emboj/journal/v19/n2/full/7592133a.html

DTE Dithioerythritol, see **dithiothreitol**.

DTH Delayed type **hypersensitivity**.

DTLET A synthetic agonist (deltakephalin, Tyr-D-Thr-Gly-Phe-Leu-Thr) for the G-protein coupled delta **opioid receptor**. Inhibits the release of Gonadotropin-Releasing Hormone (GnRH) from hypothalamic fragments containing the arcuate nucleus and the median eminence.

DTN See **dystrobrevin**.

DTNB A cell impermeable dithiol-oxidizing agent (Ellman's reagent, 5,5'-dithiobis-(2-nitrobenzoic acid)) that reacts with sulphydryl group on proteins and releases 5-sulphido-2-nitrobenzoic acid which absorbs strongly at 412 nm.

dual recognition hypothesis An outmoded hypothesis that is known to be incorrect now that the structure of the T-cell receptor has been determined. The proposal was that viral (and some chemical) antigens were recognized in association with **histocompatibility antigens** by separate receptors on the T-cell. The generation of cytotoxic T-cells was by association with Class I MHC antigens, of T-helper cells by association with Class II MHC antigens. See **altered self hypothesis**.

dual-specificity kinase A protein kinase that will phosphorylate target proteins either on serine/threonine residues or on tyrosine residues. Examples include SERK gene products such as mitogen-activated protein kinase kinase 4 (EC 2.7.12.2, MAPKK 4, JNK-activating kinase 1, SAPK/ERK kinase 1, 399aa) and CDC-like kinase 3 (EC 2.7.12.1, 638aa). Dual specificity phosphatases are also known (e.g. **MAPK phosphatase**).

Duane retraction syndrome A congenital disorder of eye movement one form of which (type 2) is caused by mutation in **chimerin-alpha**.

Duchenne muscular dystrophy A sex-linked **muscular dystrophy** confined to young males and to females with **Turner's syndrome**. It is characterized by degeneration and **necrosis** of skeletal muscle fibres which are not replaced by the satellite cells. Caused by mutation in **dystrophin**. The incidence of this disorder is about 1 in 4000 male births and of these a third are estimated to be new mutational events. See **Mdx mice**.

duct A tube formed of cells or lined with cells. A ductule is a small duct.

ductal cell carcinoma A tumour derived from cells of a duct. Common examples are mammary ductal carcinoma and pancreatic ductal carcinoma although it can occur in other glands.

ductin (1) In *Drosophila*, the **V-type ATPase** 16 kDa proteolipid subunit (vacuolar H$^+$ ATPase subunit 16-1, 159aa) the proton-conducting pore forming subunit of the integral membrane V0 complex of vacuolar ATPase. The name ductin reflects a (controversial) view that it may be a multifunctional transmembrane pore protein, also involved (for example) in gap junction formation. See also **connexin**. (2) See **CRP-ductin**.

ductless gland An **endocrine gland** that discharges directly into the blood.

dudulin Mouse ortholog of **STEAP**.

DUF26 motif Domain of Unknown Function 26, a characteristic of the cysteine-rich receptor-like kinases of *Arabidopsis* has four conserved cysteines that may form disulphide bridges. Details: http://www.biomedcentral.com/content/pdf/1471-2229-10-95.pdf

Duffy **Blood** **group** **system** determined by a single gene. The Duffy antigen/chemokine receptor (DARC) is also known as Fy glycoprotein (Fy, CD234, 336aa) found on erythrocytes and capillary endothelium: there are five phenotypes (Fy-a, Fy-b, Fy-o, Fy-x and Fy-y). The Fy(a-, b-) confers resistance to infection by *Plasmodium vivax* and *P. knowlesi* which use the receptor to enter red cells. DARC is the receptor for interleukin-8 and several other chemokines.

dumbbell (1) Dumbbell-type proteins have two domains linked by a long helical segment. Typical examples are calmodulin and troponin C. Paper: http://en.scientificcommons.org/42718177. (2) DNA dumbbells are stable, short segments of double-stranded DNA with closed nucleotide loops on each end making them resistant to exonucleases. Paper: http://nar.oxfordjournals.org/content/25/3/575.full. (3) In *S. cerevisiae*, dumbbell forming protein 4 (DDK kinase regulatory subunit DBF4, 704aa) is a regulatory subunit of the CDC7-DBF4 kinase that is involved in cell cycle regulation of premitotic and premeiotic chromosome replication and in chromosome segregation. (4) Also in *S. cerevisiae*, dumbbell former protein 8 (tethering factor for nuclear proteasome STS1, 319aa) targets proteasomes to the nucleus and facilitates the degradation of nuclear proteins. It is required for efficient chromosome segregation.

dumpy (1) A very large extracellular protein (2.5 MDa) required to maintain tension at epidermal-cuticle attachment sites in *Drosophila*. It has 308 EGF modules, interspersed with DPY modules, and terminating in a crosslinking ZP domain and membrane anchor sequence. Probably forms membrane-anchored fibres. Mutations in dumpy lead to wing blisters as do mutations in *piopio* and *papillote* which encode proteins with a full and a partial ZP domain, respectively. Abstract http://www.ncbi.nlm. nih.gov/pubmed/10837220?dopt = Abstract (2) In *C. elegans* there are a range of dumpy proteins: dumpy-2 (360aa) is cuticular collagen, dumpy-19 (693aa) is a multi-pass membrane protein required to orient neuroblasts QL and QR correctly on the anterior/posterior axis, dumpy-20 (440aa) may be involved in cuticle function and has two **BED-type zinc finger domains**, dumpy-31 is a zinc metalloprotease (nematode astacin 35, tollish protein 2, 592aa) involved in cuticular collagen maturation.

dunce *Drosophila* mutant (dnc) that is deficient in short-term memory. Gene codes for cAMP-specific 3′,5′-cyclic phosphodiesterase (EC 3.1.4.17, 1070aa) and mutation leads to elevated cAMP levels that in turn particularly affect the delayed rectifier potassium currents in neurons of brain centres associated with acquisition and retention. The effect of the mutation does also alter nerve terminal growth and synaptic plasticity. Comparable behavioural defects are associated with **rutabaga**.

Dunn chamber A special circular cell counting chamber slide. Cells are cultured on coverslips that are then inverted onto the slide and a temporally-stable gradient of potential chemoattractant is established across the annulus that separates inner and outer circular wells. Allows directed behaviour of slow-moving cells to be observed with time-lapse methodology. A more sophisticated version of the **orientation chamber** originally developed by Zigmond.

dunnisinoside An iridoid glycoside isolated from the leaves of *Dunnia sinensis*.

duplicon A chromosome-specific low copy-number repeat section of DNA which may vary from a few kilobases to hundreds of kilobases in length. Duplicated segments of genomic DNA can allow evolution of gene function and recombination between duplicated regions can result in deletion, further duplication, inversion etc. If the duplication includes a protein-coding region then there may be dosage effects. See **copy-number polymorphism**. Role in disease: http://genome.cshlp.org/content/10/5/597.full

Dupuytren's contracture Fibroma-like lesion of the palm of the hand that causes flexion contracture. Heritable and commoner in men. Never metastasises.

dura mater See **meninges**.

duramycins A group of Type B **lantibiotics** isolated from *Streptoverticillium* and *Streptomyces*, structurally similar to cinnamycin. All are potent inhibitors of phospholipase A2.

durotaxis The apparent preference of moving cells for a stiff substratum has been dubbed 'durotaxis', although it is a consequence of the physical properties of the substratum rather than being based upon gradient perception.

D-value A term used in microbiology, the decimal reduction time (D value) is the time required at a certain temperature to kill 90% of the organisms being studied.

dwarfin In *C. elegans*, a small family of proteins related to **Smad proteins**, involved in the TGFβ signalling pathway. They include dwarfin sma-2 (MAD protein homologue 1, 418aa), dwarfin sma-3 (393aa) and dwarfin sma-4 (570aa).

dyad Generally, any two entities regarded as some kind of unit. More specifically, half of a tetrad group of chromosomes that moves to one pole at the first meiotic division.

dyad symmetry element Two areas of a DNA strand whose base pair sequences are inverted repeats of each other (palindromic); one strand has the same base sequence (5′-to-3′) as its complementary strand

(5'-to-3'). In principle such dyad elements can lead to the formation of cruciform structures in DNA or hairpin loops in RNA. The former is energetically fairly unlikely, the latter are common. N.B. The term 'dyad repeat' is sometimes used of sequence duplications that favour the binding of dimeric proteins with DNA-binding regions: this does not imply palindromic sequences unless the second element is inverted.

dye coupling Measure of intercellular communication, usually through **gap junctions**. If a fluorescent dye (e.g. **lucifer yellow**) injected into one cell is seen to pass into a neighbouring cell, the presence of junctions at least able to pass solutes of that size can be inferred between the two. See also electrical coupling.

Dyggve-Melchior-Clausen syndrome A rare developmental disorder leading to short trunk dwarfism, microcephaly and psychomotor retardation; caused by a defect in **dymeclin**. The rough endoplasmic reticulum is dilated and there are enlarged and aberrant vacuoles. Mutations in the same gene cause Smith-McCort dysplasia.

dymeclin A protein (**Dyggve-Melchior-Clausen syndrome** protein, 669aa) that shuttles between the Golgi and the cytosol.

Dynabead™ Registered trademark for small (~1μm) spherical superparamagnetic particles which can be coated with various molecules (antibodies, lectins etc.) and used to purify the appropriate ligand by incubating the beads with the sample and then removing them magnetically. Have been used for cell separation. Invitrogen homepage: http://www.invitrogen.com/site/us/en/home/brands/Dynal/dynabeads_technology.html?cid = covinvggl89300000000665s&s_kwcid = TC|12178|dynabead||S|p|6948699259

dynacortin Actin binding protein (354aa) from *Dictyostelium* that cross-links actin filaments into parallel arrays. Is excluded from the cleavage furrow (see **cortexillin**). Original description: http://www.jbc.org/content/277/11/9088.long

dynactin The stable 20S multiprotein complex that is required for the cytoplasmic dynein-driven retrograde movement of vesicles and organelles along microtubules. Dynactin subunit 1 (p150-glued, 1278aa) binds directly to microtubules and cytoplasmic dynein. The other nine subunits are dynactin-2 (p50 dynamitin, 401aa), dynactin-3 (186aa), dynactin-4 (460aa), dynactin-5, (182aa), dynactin-6 (190aa) and **centractins**.

dynamic equilibrium An equilibrium state in which forward and reverse reaction rates are exactly balanced: changes in either rate will shift the relative concentrations of reactants or products. Most methods for analysing rate constants etc. assume that equilibrium has been reached but the dynamic nature of this position means that this may be a transient situation and may change, as is desirable in systems that must adapt to altered circumstances.

dynamic isomerism Tautomerism, a form of isomerism in which the two forms co-exist and interconvert dynamically.

dynamins A subfamily of GTP-binding proteins. Dynamin 1 (EC 3.6.5.5, 864aa), dynamin-2 (870aa) and dynamin-3 (869aa) are microtubule-associated mechanochemical proteins involved in producing microtubule bundles and able to bind and hydrolyze GTP. Probably involved in vesicular trafficking and receptor-mediated endocytosis. Dynamin-related proteins are found in *Arabidopsis* and are associated with cytokinesis and vesicular movement. There are tissue-specific and developmentally-regulated forms of dynamin in *Drosophila* (see **shibire**). Dynamin-binding protein (**tuba**, 1577aa) has 6 SH3 domains that are involved in linking dynamin with actin-regulating proteins.

dynamitin Subunit of the dynactin complex (p50, Jnm1p, 401aa) that modulates cytoplasmic dynein binding to an organelle, and plays a role in prometaphase chromosome alignment and spindle organizations. Overexpression will interfere with the **dynein**-dynactin interaction involved in vesicle transport and this has been used experimentally as a method for demonstrating the importance of such systems. **Immunophilins** link to dynein indirectly via dynamitin.

dynein regulatory complex A multiprotein complex (DRC) that regulates the activity of dynein in the ciliary axoneme and, by preventing all dyneins being active simultaneously (which would lock the system), converts the sliding of doublets into axonemal bending. Mutations in DRC subunits can suppress the paralysis caused by mutations in the radial spokes or the central microtubule pair. Only some components have actually been identified but there are thought to be at least seven. See **nexin** #1. Article: http://jcb.rupress.org/content/187/6/921.full.pdf+html

dyneins The motor ATPases involved with microtubule-associated movement. Axonemal dynein forms the side arms of the outer microtubule doublets in the ciliary axoneme and is responsible for the sliding of doublets and the bending of the cilium (see **dynein regulatory complex**). At least 14 human axonemal dynein heavy chains are known with different tissue distributions and there are several intermediate and light chain variants. Mutations in axonemal dynein are responsible for **primary ciliary dyskinesia** (see also **Kartagener's syndrome**). Cytoplasmic dynein is the multimeric complex responsible for intracellular retrograde motility of vesicles and organelles along microtubules (towards the minus ends). The complex, like myosin, is an ATPase; the force-producing power stroke is thought to occur on release of ADP. The homodimeric cytoplasmic dynein 1 complex consists of two catalytic heavy chains (HCs, 4646aa) and a number of non-catalytic subunits: intermediate chains (ICs, $\sim$630aa), light intermediate chains (LICs, 523aa) and light chains (LCs, $\sim$90aa) which vary. There are many permutations of cytoplasmic dynein components. In the older literature cytoplasmic dynein was referred to as MAP-IC.

dynorphin Opiate peptide derived from the hypothalamic precursor pro-dynorphin (that also contains the neoendorphin sequences). Contains the pentapeptide leu-**enkephalin** sequence. Its binding affinity is greater for the κ-type than for the μ-type **opioid receptor**.

dyrks A very conserved family of protein kinases (dual-specificity tyrosine phosphorylation-regulated kinases, EC 2.7.12.1) that autophosphorylate a tyrosine residue in their activation loop by an intramolecular mechanism and phosphorylate exogenous substrates on serine/threonine residues. Also have nuclear targeting signal, putative leucine zipper and a very conserved 13-histidine repeat sequence. Rat gene *dyrk* is a homologue of *Drosophila* minibrain (*mnb*) a gene involved in postembryonic neurogenesis; the human homologue maps to the Down's syndrome critical region on chromosome 21. There are several mammalian isoforms (in humans, DYRK1A, 763aa; DYRK1B, 629aa; DYRK2, 601aa; DYRK3, 588aa; DYRK4, 420aa).

dysautonomia A general term for any disease or malfunction of the autonomic nervous system. Familial dysautonomia (**hereditary sensory and autonomic neuropathy** type 3), is caused by mutations in the gene encoding IκB-kinase associated protein.

dysbindin A **dystrobrevin** binding protein (dystrobrevin-binding protein 1, Hermansky-Pudlak syndrome 7 protein, 351aa) that is a component of the **BLOC-1** complex. Defects give rise to **Hermansky-Pudlak syndrome type 7**. Ubiquitinated by **TRIM32** which leads to degradation. Dysbindin is thought to be involved in suceptibility to schizophrenia. The *Drosophila* homologue of dysbindin functions during synapse development, baseline neurotransmission, and synaptic homeostasis. In humans there are two dysbindin domain-containing proteins (DBNDD1, 158aa; DBNDD2, 259aa) and DNNDD2 may modulate the activity of casein kinase-1. Abstract: http://www.sciencemag.org/content/326/5956/1127.abstract

dyschromatosis symmetrical hereditaria An autosomal dominant disorder (reticulate acropigmentation of Dohi.) in which there are hyper- and hypo-pigmented areas on the dorsal parts of the extremities. Caused by mutation in the gene encoding double-stranded RNA-specific adenosine deaminase.

dysentery A severe diarrhoea caused by infection of the gut with *Shigella* (bacillary dysentery) or with *Entamoeba histolytica* (amoebic dysentery).

dysferlin Protein (2080aa), similar to **ferlin**, that associates with the plasma membrane in primary fibroblasts, skeletal and cardiac muscles and is the calcium ion sensor involved in the Ca^{2+}-triggered synaptic vesicle-plasma membrane fusion. Mutations in the dysferlin gene cause a form of limb girdle muscular dystrophy and Miyoshi myopathy. Related molecules, myoferlin and **otoferlin** are found in myoblasts and cochlea respectively.

dysgenic System of breeding or selection that is genetically deleterious or disadvantageous.

dysgerminoma A tumour derived from germ cells, usually in the ovary.

dyskeratosis congenita A rare multisystem disorder caused by defective telomere maintenance. Clinical features are highly variable and include bone marrow failure, predisposition to malignancy, and pulmonary and hepatic fibrosis. Caused by mutations in the genes for the RNA (TERC) or reverse transcriptase (TERT) components of **telomerase** or in **TIN2**.

dyskinetoplasty Absence of an organized **kinetoplast** (and of kinetoplast DNA) from a flagellate protozoan cell.

dysostosis A disorder of bone development, usually a defect in mineralization of cartilage (sometimes dyschondroplasia: see **endochondroma**). See **postaxial acrofacial dysostosis**.

dysplasia Literally 'wrong growth'. Usually used to denote early stage of carcinogenesis, marked by abnormal epithelial morphology.

dystonin Cytoskeletal linker protein (Bullous pemphigoid antigen 1, hemidesmosomal plaque protein, 7570aa) that anchors keratin-containing intermediate filaments to the inner plaque of hemidesmosomes.

dystrobrevins A family of widely expressed **dystrophin-associated proteins**. Alpha-dystrobrevin (dystrophin-related protein 3, 743aa) may be involved in the formation and stability of synapses as well as being involved in the clustering of nicotinic acetylcholine receptors. Beta-dystrobrevin (627aa) interacts with dystrophin short form DP71 and syntrophins 1 and 2 and binds dystrobrevin binding protein 1 (see **dysbindin**).

dystroglycan Component of the **dystrophin-associated protein complex** consisting of two proteins, α- and β-dystroglycans (cleavage products of dystrophin-associated glycoprotein, DAG1, 895aa). Beta-dystroglycan is a transmembrane protein that associates with **dystrophin** in the cytoplasm and α-dystroglycan, an extracellular glycoprotein, that binds to **agrin** and **neurexin**, thus linking actin through dystrophin and β-dystroglycan to the extracellular matrix. Dystrophin deficiency leads to a deficiency in the appearance of these proteins on the sarcolemma, even though they are not themselves defective. See **muscular dystrophy-dystroglycanopathy with brain and eye anomalies**.

dystrophic epidermolysis bullosa A group of diseases associated in all cases with mutations of the gene coding for type VII collagen. See **epidermolysis bullosa**.

dystrophin Skeletal muscle protein (3685aa) missing in **Duchenne muscular dystrophy** and defective in **Becker muscular dystrophy.** It is a component of the **dystrophin-associated protein complex** which links the cytoskeleton with the extracellular matrix. There are sequence homologies with nonmuscle α-actinin and with spectrin. In X-linked

dilated cardiomyopathy dystrophin is mutated but skeletal muscle is normal. See **utrophin.**

dystrophin-associated protein complex A protein complex (**syntrophin, dystrobrevin,** and **dystroglycan** isoforms) believed to provide a molecular link between the actin cytoskeleton and the extracellular matrix in muscle cells, thereby sustaining sarcolemmal integrity during muscle contraction. In muscle, defects lead to **muscular dystrophies**. Some of these functions are mediated by the **sarcoglycan** subcomplex. Also important for the clustering and anchoring of signalling proteins and ion and water channels.

dystrophy Literally a defect in nutrition but a general term for a condition in which there is degeneration of an organ or tissue. See **muscular dystrophy, corneal endothelial dystrophy.**

E

E Single letter code for glutamic acid.

E1 Adenoviral oncogenes that interact with the Rb **tumour suppressor** gene product. E1A (32 kDa, 289aa) releases E2F1 transcription factor from retinoblastoma protein thereby driving cell proliferation. Isoform early E1A 26 kDa protein stabilizes p53. E1B (large T antigen, 55 kDa, 495aa; small T-antigen, 19 kDa, 175aa) is anti-apoptotic and is essential for cell transformation by adenovirus and for the regulation of viral early gene transcription.

E1 enzymes See **ubiquitin conjugating enzymes**.

E1-E2-type ATPase A superfamily of ion-pumping ATPases (P-type ATPases), found in prokaryotes and eukaryotes. There are four main types: **(i)** Ca^{2+}-transporting ATPases (e.g. **sarcoplasmic-endoplasmic reticulum Ca^{2+}-ATPase**, SERCA), **(ii)** Na^+K^+-ATPases of plasma membrane and H^+K^+ ATPase of gastric mucosa, **(iii)** plasma membrane proton pumps of plants, fungi and lower eukaryotes, **(iv)** all bacterial P-type ATPases, except the Mg^{2+}-ATPase of *Salmonella typhimurium*. They are heterodimers with a catalytic alpha subunit that has multiple membrane-spanning domains and a beta subunit that stabilizes the alpha subunit and transports it to the membrane. The pumps are generally inhibited by ATP analogues such as vanadate, there are more selective inhibitors, e.g. ouabain for the Na^+K^+ ATPase, thapsigargin for the Ca^{2+}-ATPase).

E2 enzymes See **ubiquitin conjugating enzymes**. Diagram: http://www.ebi.ac.uk/interpro/potm/2004_12/Page2.htm

E2F Family of transcription factors originally identified through their role in transcriptional activation of the adenovirus E2 promoter, subsequently found to bind to promoters for various genes involved in the G1 and S phases of the cell cycle. E2F forms heterodimers with **DP-1** to produce an active transcriptional complex. E2F family members are regulated by interaction with **retinoblastoma** (Rb) proteins.

E3 ligase See **ubiquitin conjugating enzymes**. Webpage: http://www.ebi.ac.uk/interpro/potm/2004_12/Page2.htm

E5 **(1)** An oncogene from a **papillomavirus** that encodes a small protein (44aa) that binds and blocks the 16 kDa **proteolipid** of the **V-type ATPase**, producing abnormal intravesicular processing of growth factor receptors. Is also thought to modulate growth factor receptor function leading to a stimulation of growth factor signal transduction pathways.

(2) A venom allergen (226aa) from *Polistes exclamans* (Paper wasp).

E6 An oncogene from a papillomavirus that encodes a protein (158aa) that drives cell proliferation through an association with PDZ domain proteins and Rb (retinoblastoma) protein, and contribute to neoplastic progression. E6-mediated p53 degradation prevents the normal repair of chance mutations in the cellular genome. The *E7* product (~100aa) has similar effects.

E 64 An inhibitor of most cysteine peptidases in clan CA and some other peptidases but not a general inhibitor of cysteine peptidases.

EA-rosettes See **E-rosettes**.

EAA See **excitatory amino acid**.

EaA cells Insect cell line derived from haemocytes of the salt marsh caterpillar *Estigmene acrea*. An alternative line for baculovirus expression. See **Sf9 cells**.

EAC-rosettes See **E-rosettes**.

EACA See **epsilon-aminocaproic acid**.

Eadie-Hofstee plot Linear transformation of enzyme kinetic data in which the velocity of reaction (v) is plotted on the ordinate, v/S on the abscissa, S being the initial substrate concentration. The intercept on the ordinate is V_{max}, the slope is $-K_m$. Preferable to the **Lineweaver-Burke plot** because it gives equal weight to data points in any range of substrate concentration or reaction velocity.

EAE See **experimental allergic encephalomyelitis**.

EAP **(1)** EAP proteins in humans are involved in **ESCRT complexes** and are homologues of yeast Class E VPS proteins. Examples include EAP30 (ELL-associated protein of 30 kDa, vacuolar-sorting protein SNF8, 258aa) one of three components of the ESCRT-II complex. **(2)** A developmentally regulated embryonal protein in the chicken (EAP-300, 300 kDa) that is expressed by radial glia in various regions of the CNS. Probably identical to **paranemin**. Article: http://www.jbc.org/content/272/51/32489.long

early antigens Virus-coded cell surface antigens that appear soon after the infection of a cell by virus, but before virus replication has begun. Encoded by the so-called early genes.

early gene See **early antigens**.

The Dictionary of Cell and Molecular Biology. DOI: http://dx.doi.org/10.1016/B978-0-12-384931-1.00005-2

early growth response genes A family of cys₂-his₂-type zinc-finger proteins; transcription factors that are upregulated after a variety of stresses. The *Egr1* gene product (KROX-24, 543aa) acts upstream of TGFβ1 and activates the transcription of target genes whose products are required for mitogenesis and differentiation. Egr-2 (KROX-20, 476aa) binds to two sites in the promoter of HOXA4. Defects in Egr2 are a cause of Charcot-Marie-Tooth disease type 1D, Dejerine-Sottas syndrome and congenital hypomyelination neuropathy. Egr-3 (Zinc finger protein pilot, 387aa) is involved in muscle spindle development. Egr4 (486aa) is involved in mitogenesis and differentiation.

EAST (1) In chicken, epidermal growth factor receptor-associated protein with SH3 and TAM domain (signal transducing adapter molecule, STAM-2, 468aa). An EGF receptor substrate phosphorylated by src in response to EGF and PDGF. Enriched at focal adhesions and in some cells (MDCK) at cell-cell contacts so may play a role in EGF receptor-stimulated cytoskeletal reorganization. (2) *Drosophila* EAST protein (enhanced adult sensory threshold, 2342aa) may be involved in meiotic chromosome segregation. Has carboxypeptidase activity.

easter A *Drosophila* gene, that encodes a member of the trypsin family of serine peptidases that is present throughout the embryo as a zymogen but is activated in the ventral regions to promote normal embryonic polarity by acting on the **spätzle** ligand for the **Toll** receptor. Activation of easter is by proteolytic processing by the peptidases **nudel, gd** and **snake**.

EB1 A member of the RP/EB family of proteins (adenomatous polyposis coli-binding protein EB1, microtubule-associated protein, RP/EB family member 1, MAPRE1, end-binding protein 1, 268aa) that binds to plus-end of microtubules in interphase cells and during mitosis is associated with the centrosomes and spindle microtubules. Also binds to the APC protein (see **adenomatous polyposis coli**) and associates with components of the **dynactin** complex and the intermediate chain of cytoplasmic **dynein**.

EBBP Estrogen-responsive B box protein (tripartite motif-containing protein 16, 564aa) a cytoplasmic protein, more extensively expressed in the fetus: may be involved in keratinocyte differentiation.

Ebola virus Filovirus similar to **Marburg virus**, a single-stranded negative-sense RNA virus, that causes severe fever and bleeding, often fatal. Outbreaks have so far been mostly confined to Africa.

E box The CANNTG sequence motif, found in numerous promoters and enhancers; the binding site for the basic-helix-loop-helix transcription factors.

EBP50 A widely distributed protein (ERM-binding phosphoprotein 50 kDa, Na⁺/H⁺ exchanger regulatory factor, solute carrier family 9 isoform A3, SLC9A3, 834aa) involved in pH regulation to eliminate acids. Associates with **ezrin**, and with the **cystic fibrosis transmembrane conductance regulator** (CFTR) linking them to the cortical actin cytoskeleton. EBP50 has two PDZ domains; CFTR binds with high affinity to the first and **yes-associated protein** (YAP65) binds with high affinity to the second.

EBV See **Epstein-Barr virus**.

EC cells (1) **Embryonal carcinoma cells.** (2) Endocrine cells.

E classification Classification of enzymes based on the recommendations of the Committee on Enzyme Nomenclature of the International Union of Biochemistry. The first number indicates the broad type of enzyme (1 = oxidoreductase; 2 = transferase; 3 = hydrolase; 4 = lyase; 5 = isomerase; 6 = ligase (synthetase)). The second and third numbers indicate subsidiary groupings, and the last number, which is unique, is assigned arbitrarily in numerical order by the Committee.

EC number See **E classification** for enzymes.

EC₅₀ Effective concentration; concentration at which the substance concerned produces a specified effect in 50% of the organisms treated or 50% of the maximal effect.

eccrine Type of gland in which the secretory product is excreted from the cells.

ecdysis Moulting of the outer layers of the integument, as in arthropods. Regulated by **ecdysone** and **juvenile hormone**.

ecdysone Family of steroid hormones found in insects, crustaceans and plants. In insects, α-ecdysone stimulates moulting (ecdysis). The steadily maturing character of the moults is affected by progressively decreasing levels of **juvenile hormone**. β-ecdysone (ecdysterone) has a slightly different structure and is also found widely. Phytoecdysones are synthesized by some plants.

ECE See **endothelin converting enzymes**.

ECF See **eosinophil chemotactic peptide**.

Echinacea N American perennial plants of the genus *Echinacea* (e.g. purple coneflower). Herbal remedies prepared from these plants are extensively used and are claimed to boost the immune system.

echinocandins A class of antifungal agents, large lipopeptide molecules, that act on the fungal cell wall by way of noncompetitive inhibition of the synthesis of 1,3-beta-glucans. Examples include Caspofungin, Micafungin. Used in clinical treatment of candidiasis and aspergillosis.

echinocytes Erythrocytes (burr cells) that are shrunken and have a spiky appearance as a result of being placed in hypertonic medium.

Echinodermata Phylum of exclusively marine animals. The phylum is divided into five classes: the Asteroidea (starfish), the Echinoidea (sea urchins), the Ophiuroidea (brittle stars and basket stars), the Holothuroidea (the sea cucumbers) and the Crinoidea (sea lilies and feather stars).

Echinoidea Class of echinoderms (Echinodermata), commonly known as sea urchins.

echinoidin A multimeric lectin (147aa) specific for Gal-GalNAc and involved in defence against microorganisms from the coelomic fluid of the sea urchin *Anthocidaris crassispina*. The C-terminal sequence is highly homologous to C-terminal carbohydrate recognition portions of rat liver mannose-binding protein and several other hepatic lectins.

Echinosphaerium Previously *Actinosphaerium*. A **Heliozoan** protozoan. The organisms are multinucleate and have a starburst of radiating **axopodia**, the microtubules of which have been much studied.

echinosporin An antibiotic isolated from *Streptomyces echinosporus* and other *Streptomyces sp.* Has weak activity against Gram-positive and -negative microorganisms, inhibits the cell cycle at the G2/M phase and induces apoptosis.

echistatin **Disintegrin** found in the venom of the saw-scaled viper, *Echis carinatus*.

Echiuroidea A phylum of sedentary marine worm-like animals.

ECHO virus A group of picornaviruses (Enteric Cytopathic Human Orphan viruses) that are a common cause of enteric infections, particularly in children, and can occasionally cause aseptic meningitis.

ECL (1) Electrochemiluminescence: production of light during an electrochemical reaction, now being applied to various bioassay systems. (2) Enhanced chemiluminescence. Method for enhancing detection of proteins on blots. Involves the use of luminol that is oxidized by peroxidase-coupled antibody used to detect the protein of interest, and the light produced is then detected on film.

eclosion Emergence of an insect from its old cuticle at a moult, particularly from pupa to adult, but also from the egg.

ecm Common abbreviation for **extracellular matrix**.

E. coli See *Escherichia coli*.

EcoliWiki A database devoted to *Escherichia coli*. Link to database: http://ecoliwiki.net/colipedia/index.php/Welcome_to_EcoliWiki

EcoRI Probably the most commonly used type II **restriction endonuclease** (EC 3.1.21.4, 277aa) isolated from *E. coli*. It cuts the sequence GAATTC between G and A thus generating 5′ **sticky ends**. EcoRII (404aa) cuts the sequence CC(T/A)GG in front of the first C.

ecospecies A species that can be subdivided into sub-types (ecotypes) that are differentiated to suit various environments but can all interbreed.

ecotropic viral integration site 1 A gene (*Evi-1*) that encodes a transcriptional regulator, expressed at high level during the development of the mouse urinary system, Müllerian ducts, lung and heart, but at low level in most of the adult tissues; the human homologue (MDS1 and EVI1 complex locus protein EVI1, 1051aa) is expressed abundantly during development of various tissues.

ecotropic virus Generally, a virus that will only replicate in its original host species. Commonly refers to murine retroviruses, typified by the classical AKR murine leukaemia virus; other murine C-type viruses infect only heterologous species (xenotropic viruses), a third category (amphotropic viruses) infect cells of both original and heterologous species. The distinction is important when considering the risks of xenografts.

ecotype A population of a single species that is adapted to a particular set of environmental conditions. A subdivision of an **ecospecies**.

ECP See **eosinophil cationic protein**.

ectoderm The outer of the three germ layers of the embryo (the other two being mesoderm and endoderm). Ectoderm gives rise to epidermis and neural tissue.

ectoderm-neural cortex-1 A putative oncogene (*ENC-1*, p53-induced gene 10) that encodes an actin binding protein (589aa) of the **Kelch** family, that is an early and highly specific marker of neural induction in vertebrates. Expression of *ENC1* has been shown to induce the formation of neuronal processes and is involved in differentiation of neural crest cells. Also expressed in adipose tissue, where it appears to play a regulatory role early in adipocyte differentiation.

ectodermal dysplasia Ectodermal dysplasia, a disorder of teeth, hair, and eccrine sweat glands, is a feature of many diseases, mostly rare. The X-linked form of hypohidrotic ectodermal dysplasia is caused by mutation in the gene encoding **ectodysplasin-A** (EDA) and other forms by mutations in the EDA receptor (EDAR) or in the EDAR-associated death domain adaptor. McGrath's syndrome (ectodermal dysplasia syndrome) is caused by mutations in the **plakophilin-1** gene. See **desmosome**. Other forms may arise through mutations in **nectin-1**,

IKK-gamma, keratin-85, NFκBI, the products of the **Ellis-Van Creveld syndrome** genes (EVC and EVC2) or in unknown loci.

ectodysplasin A protein (EDA, ED1, 391aa in membrane, 232aa secreted form), probably of the TNFligand superfamily, expressed in keratinocytes, hair follicles, sweat glands, and in other adult and fetal tissues and involved in epithelial-mesenchymal signalling during morphogenesis. Defects in EDA cause ectodermal dysplasia type 1 (Christ-Siemens-Touraine syndrome, X-linked hypohidrotic ectodermal dysplasia). There are multiple isoforms and two receptors are known, one (TNFRSF-EDAR, 448aa) binds only isoform A1, the other, TNFRSF27 (X-linked ectodysplasin-A2 receptor, 297aa) only binds isoform A2. Both receptors activate NFκB, Jnk and cell death pathways. See **hypohidrosis**. The EDAR-associated death domain adaptor (protein crinkled homologue, 215aa) binds EDAR, TRAF1, TRAF2 and TRAF3 and mediates the activation of NFκB.

ectoenzyme Enzyme that is secreted from a cell or located on the outer surface of the plasma membrane and therefore able to act on extracellular substrates.

ectoglycosidase See **exoglycosidase**.

ectomycorrhiza Mycorrhiza (ectotrophic mycorrhiza) with a well-developed layer of fungal mycelium on the outside of the root interconnected with hyphae both within the root cortex and also ramifying through the soil. Fungus is often a basidiomycete. See **endomycorrhiza**, **vesicular-arbuscular mycorrhiza**.

ectopic Misplaced, not in the normal location.

ectoplasm Granule-free cytoplasm of amoeba lying immediately below the plasma membrane.

ectoplasmic tube contraction Model for amoeboid movement in which it was proposed that protrusion of a pseudopod is brought about by contraction of the sub-plasmalemmal region everywhere else in the cell thus squeezing the central cytoplasm forwards. See **frontal zone contraction theory**.

ectromelia (1) Congenital absence or gross shortening of long bones of limb or limbs. (2) Ectromelia virus (EV, mousepox) is a dsDNA virus of the poxvirus family that is a highly virulent natural pathogen of mice and has been used as a model for generalised infection and evasion of the immune response.

eczema A common form of atopic dermatitis, an inflammatory skin condition usually due to immune hypersensitivity to food or environmental allergens. A number of susceptibility loci have been identified.

ED1 (1) Antibody extensively used to identify rat monocytes/macrophages; marker antigen is a single-chain 90–110 kDa glycosylated protein, mostly on lysosomal membranes that is similar to human CD68. (2) See **ectodysplasin**.

ED$_{50}$ Median effective dose, that dose that produces a response in 50% of individuals or 50% of the maximal response.

edaphic Type of physical or chemical property of soil that influences plants growing on that soil.

edeines Pentapeptide amide antibiotics composed of four nonprotein amino acids, glycine, and polyamine produced by *Bacillus brevis* Vm4. They have antimicrobial and immunosuppressive activities and are universal inhibitors of translation.

edema See **oedema**.

EDGE Extraction of Differential Gene Expression, an open source software program for the analysis of DNA microarray experiments. Link to source: http://faculty.washington.edu/jstorey/edge

EDGs Endothelial differentiation G-protein-coupled receptors, a family of high affinity receptors for **sphingosine-1-phosphate** that are also low affinity receptors for the related lysophospholipid, sphingosylphosphorylcholine (SPC). EDG-1 (382aa), EDG-3 (378aa), EDG-5 (353aa) and EDG-8 (398aa) are very similar; EDG-6 (384aa) is more distantly related. The EDGs are related to lysophosphatidic acid (LPA) receptors.

Ediacara Extensive 'family' of ancient (600–540 million years old), pre-Cambrian soft-bodied animals, fossils of which were first described from the Ediacaran hills of South Australia. There are various bizarre forms and considerable uncertainty about the taxonomic postion of many; some appear to be of extinct phyla.

editosome Multiprotein complex (27S) involved in **RNA processing**. For example the editosome responsible for the postranscriptional editing of a CAA codon for Gln to a UAA codon for stop in the apolipoproteinB mRNA. Consists of a member of cytidine deaminase family of enzymes (apolipoprotein B mRNA editing catalytic polypeptide 1, APOBEC-1, 236aa) and a specificity factor (APOBEC1 complementation factor, ACF, 594aa) in addition to the target mRNA. This particular editosome is also involved in CGA (Arg) to UGA (Stop) editing in the NF1 mRNA.

Edman degradation The classic method for sequence determination of peptides using sequential cleavage of the N-terminal residue after reaction with Edman reagent (phenyl isothiocyanate). The N-terminal amino acid is removed as a phenylthiohydantoin derivative.

EDRF Endothelium-derived relaxation factor; see **nitric oxide**.

EDTA Ethylenediamine tetraacetic acid, often used as the disodium salt, a chelator of divalent cations; $\log_{10} K_{app}$ for calcium at pH7 is 7.27 (5.37 for magnesium) See **EGTA**.

EDTA-light chain Myosin regulatory light chains (153aa) from scallop muscle (two per pair of heavy chains), easily extracted by calcium chelation. Although the EDTA-light chains do not bind calcium they confer calcium sensitivity on the myosin heavy chains.

Edwards' syndrome Complex of abnormalities caused by trisomy 18.

EEA (1) Early endosome antigen 1 (EEA-1, zinc finger FYVE domain-containing protein 2, 1411aa) that binds phospholipid vesicles containing phosphatidylinositol 3-phosphate and is involved in endosomal trafficking. Autoantibodies to EEA1 have been reported in a patient with subacute cutaneous lupus erythematosus. (2) *Euonymus europaeus* agglutinin, an alpha-galactophilic lectin that binds the sugar moiety α-Gal (1,3)-β-Gal (1,4)-GlcNAc, particularly on endothelial cells.

EED A **polycomb group** (PcG) protein (WD protein associating with integrin cytoplasmic tails 1,WAIT-1, 441aa) that is a component of the **PRC2/EED-EZH** complexes that methylates Lys-9 and Lys-27 of histone H3 thereby repressing the target gene. Expression peaks at the G1/S phase boundary.

EEM syndrome A heterogeneous group of autosomal recessive disorders (ectodermal dysplasia with ectrodactyly and macular dystrophy, Albrectsen-Svendsen syndrome, Ohdo-Hirayama-Terawaki syndrome) in which there is abnormal development of ectodermal structures; caused by defects in **cadherin-3**.

EF-1 See **elongation factor**.

E face See **freeze fracture**.

EF-hand A very common calcium-binding motif. A 12aa loop with a 12aa α-helix at either end, providing octahedral coordination for the calcium ion. Members of the family include: **aequorin, α-actinin, calbindin, calcineurin, calcyphosin, calmodulin, calpain, calcyclin, diacylglycerol** kinase, **fimbrin,** myosin regulatory light chains, **oncomodulin, osteonectin, spectrin, troponin** C.

EF-Tu See **elongation factor**.

efferent Leading away from something. The opposite of **afferent**.

eflornithine An irreversible inhibitor of ornithine decarboxylase, used to slow hair growth and also in the treatment of African trypanosomiasis.

egasyn A liver carboxylesterase (esterase 22, EC 3.1.1.1, 567aa) involved in the detoxification of

xenobiotics and in the activation of ester and amide prodrugs. Found in the lumen of the endoplasmic reticulum.

EGF See **epidermal growth factor, EGF-like domain**.

EGF receptor Receptor tyrosine kinase (EC 2.7.10.1, erbB1, 1210aa) encoded by c-*erbB1*. One of Type I family of growth factor receptors that also includes TGFα receptor, heregulin receptor (HER1). Binds EGF and other members of the EGF family (TGFα, amphiregulin, betacellulin, heparin-binding EGF-like growth factor, GP30 and vaccinia virus growth factor). EGF receptor-associated protein with SH3 and TAM domains: see **EAST #1**.

EGF-like domain Region of 30–40aa containing 6 cysteines found originally in EGF, and subsequently in a range of proteins involved in cell signalling. Examples: **TGFα, amphiregulin, urokinase, tissue plasminogen activator, complement** C6-C9, **fibronectin, laminin** (each subunit at least 13 times), **nidogen, selectins**. It is also found in the *Drosophila* gene products: **Notch** (36 times) **Delta, Slit, Crumbs, Serrate**.

EGFP Enhanced **green fluorescent protein**, often used as a reporter gene.

egg-polarity gene A gene whose product distribution in the egg determines the anterior-posterior axis of subsequent development. Best characterized in *Drosophila*: see *bicoid*, **maternal-effect gene**.

eglin C A peptidase inhibitor (70aa) from leech (*Hirudo medicinalis*) but now available as a recombinant protein. A member of the potato chymotrypsin inhibitor family of serine peptidase inhibitors. In particular, inhibits neutrophil elastase and cathepsin G.

EGO complex A complex involved in conjunction with **TOR**, in the regulation of microautophagy in yeast.

Egr-1, Egr-2, Egr-3 See **early growth response genes**.

EGS See **external guide sequence**.

EGTA Like **EDTA** a chelator of divalent cations but with a higher affinity for calcium (log K_{app} 6.68 at pH 7) than magnesium (log K_{app} 1.61 at pH 7). Will also bind other divalent cations. Note: the 'apparent association constant', K_{app}, is used because protons compete for binding and the association constant varies according to pH. Thus, EGTA has $\log_{10} K_{app}$ for calcium of 2.7 at pH 5, 10.23 at pH 9.

EH domain A highly conserved motif (Eps15-homology domain, ~100aa) found in proteins that are primarily involved in regulating endocytosis and vesicle transport (e.g. **testilin, intersectin**). EH domains bind to proteins that contain the tripeptide asparagine-proline-phenylalanine (NPF) motif and there are often multiple EH domains. EH-domain proteins are found in many species ranging from yeast to mammals. Details: http://pawsonlab.mshri. on.ca/index.php?option = com_content&task = view& id = 214&Itemid = 64

Ehlers-Danlos syndrome The classical Ehlers-Danlos Syndrome (EDS-I & -II) is characterised by loose-jointedness and fragile, bruisable skin, and is due to defects in genes for collagen alpha-1(V), alpha-2(V) or alpha-1(I). EDS III is a benign form of classic EDS. Ehlers-Danlos syndrome type IV is an autosomal dominant disorder in which there is a defect in the gene for type III collagen. In Type VI there is a defect in the gene for lysyl hydroxylase. Other forms are also recognised with defects in various aspects of collagenous connective-tissue production. See **dermatosparaxis**.

Ehringhaus compensator Device used in **interference** or **polarization microscopy** to reduce the brightness of the object to zero in order to measure the phase retardation (optical path difference). The compensator consists of a birefringent crystal plate that can be tilted. An alternative to **Senarmont compensation** and has the advantage that it can be applied to retardations of more than one wavelength.

Ehrlich ascites A mouse adenocarcinoma-type cell line, originally derived from mammary carcinoma in 1905, and adapted to ascites form in 1932. Normally maintained by serial intraperitoneal passage. The cells have been extensively used as a model system.

Ehrlichia Genus of rickettsia that are the cause of emerging and serious tick-borne human zoonoses, and the cause of serious and fatal infections in companion animals and livestock.

EHS cells A line of mouse cells (Englebreth-Holm-Swarm sarcoma cells) that produce large amounts of basement membrane-type extracellular matrix (ecm), rich in **laminin**, collagen type IV, **nidogen** and heparan sulphate. Often used as a source of these ecm molecules.

eicosanoids Useful generic term for compounds derived from arachidonic acid. Includes **leukotrienes, prostacyclin, prostaglandins** and **thromboxanes**.

eIF-1, eIF-2 etc. See **eukaryotic initiation factor**.

Eimeria Coccidian protozoan. All coccidians are intracellular parasites of various vertebrates and invertebrates. *Eimeria tenella* infects chick intestinal epithelial cells and is of veterinary importance. The trophozoites invade host cells, proliferate as merozoites by schizogony which can then infect adjacent cells if released. Merozoites differentiate to male or female gamonts that fuse to form a zygote that undergoes division to form eight zoites that are retained within a zygocyst. If the zygocyst is ingested by a new host the zoites emerge, and reinfect the host as trophozoites.

Eisenberg algorithm An algorithm for calculating a **hydropathy plot**.

EJC See **exon junction complex**.

ejectosome (1) A cellular organelle that has the capacity to eject its contents. **Cryptomonads** have ejectosomes (extrusosomes, trichocysts) containing a coiled protein ribbon that, when ejected, propels the organism backwards as an escape mechanism. (2) Mycobacteria have actin-based ejectosomes that will expel the bacterium from its host cell without causing lysis.

EJP Prefix for proteins from *Erwinia sp.* (strain Ejp617) e.g. putative transport protein EJP617_24070 (562aa).

ektacytometry Method in which cells (usually erythrocytes) are exposed to increasing shear-stress and the laser diffraction pattern through the suspension is recorded; it goes from circular to elliptical as shear increases. From these measurements a deformability index for the cells can be derived.

Elaeis guineensis Oil palm, one of the family Palmaceae.

elafin An elastase-specific inhibitor (skin-derived antileukoproteinase, SKALP, WAP four-disulphide core domain protein 14, 117aa) found in the granular layer of skin, but not in the spinous or basal layers. Also inhibits proteinase 3. Gene expression is upregulated in psoriatic epidermis. Has two domains, one a transglutaminase substrate domain (cementoin moiety) that anchors it to ecm proteins, and one that is the elastase inhibitor.

elaioplast Unpigmented type of **plastid** modified as an oil-storage organelle.

ELAM-1 See **selectins**.

elastase Serine endopeptidase (formerly EC 3.4.4.7) that will digest **elastin, collagen** Type IV and a range of other proteins; inhibited by alpha-1-protease inhibitor of plasma. A range of elastases are known including leucocyte elastase from neutrophil granules (EC 3.4.21.37, elastase-2, 267aa) and several chymotrypsin-like elastases (elastase-1, EC 3.4.21.36, 258aa; elastase-2A, EC 3.4.21.71, 269aa and so on).

elasticoviscous Alternate form of the commoner term viscoelastic.

elastin Glycoprotein (786aa) randomly coiled and cross-linked to form the elastic fibres found in connective tissue. Like collagen, the amino acid composition is unusual with 30% of residues being glycine and with a high proline content. Cross-linking depends upon formation of **desmosine** from four lysine side groups. The mechanical properties of elastin are poorer in old animals. Elastin microfibril interface-located protein 1 (EMILIN1, 1016aa) may anchor smooth muscle cells to elastic fibres.

elastonectin Obsolete name originally given (1986) to a fibroblast-derived **elastin** -binding protein (120 kDa) found in extracellular matrix. Original paper: http://www.pnas.org/content/83/15/5517.short

ELAV proteins In *Drosophila* a RNA-binding protein (embryonic lethal abnormal visual protein, 483aa). The ELAV family of RNA-binding proteins is highly conserved in vertebrates and in humans, there are four ELAV-like proteins in humans: HuR (326aa) is ubiquitously involved in $3'$-UTR **ARE**-mediated myc stabilization. Binds avidly to the AU-rich element in Fos and IL-3 mRNAs. Hel-N1 (HuB, nervous system-specific RNA-binding protein Hel-N1, 359aa), HuC (paraneoplastic limbic encephalitis antigen 21, 367aa) and HuD (ELAV-like 4, human antigen-D, paraneoplastic encephalomyelitis antigen HuD, 380aa) are expressed in terminally differentiated neurons. See **AREs**.

electrical coupling A phenomenon that occurs between two cells that are in contact provided **gap junctions** are formed which allow the passage of electrical current (ionic coupling). Usually tested by impaling both cells with microelectrodes, injecting a current into one, and looking for a change in potential in the other. **Electrical synapses** are a specialized example but electrical coupling is not confined to excitable cells: many embryonic and adult **epithelia** are coupled, possibly to allow **metabolic cooperation**. See also **dye coupling**.

electrical synapse A connection between two electrically excitable cells, such as neurons or muscle cells, via arrays of **gap junctions**. This allows **electrical coupling** of the cells, and so an action potential in one cell moves directly into the other, without the 1 ms delay inherent in **chemical synapses**. Electrical synapses do not allow modulation of their connection, and so only occur in neuronal circuits where speed of conduction is paramount (e.g. the crayfish escape reflex). A few electrical synapses are rectifying, implying a more specialized property than a simple gap junction.

electrochemical potential Defined as the work done in bringing 1 mole of an ion from a standard state (infinitely separated) to a specified concentration and electrical potential. Measured in joules/mole. More commonly used to measure the electrochemical potential difference between two points (e.g. either side of a cell membrane), thus sidestepping the rather abstract concept of a standard state. If the molecule is uncharged or the electrical potential difference between two points is zero, the electrochemical potential reduces to the **chemical potential** difference of the species. At equilibrium, the

electrochemical potential difference (by definition) is zero; the situation can then be described by the **Nernst equation**.

electrochemiluminescence A light-emitting chemiluminescent reaction (ECL, electrogenerated chemiluminescence) that is preceded by an electrochemical reaction. This has the advantage, for assay systems, that the time and location of the light emission can be controlled. Thus it is possible to arrange that the electrochemical reaction will only occur if the components are physically adjacent, if, for example, a receptor linked to a magnetic bead has bound labelled ligand from solution. The beads are magnetically captured, electrically stimulated and the light emission is proportional to the binding of ligand. A common label is Ruthenium (II) tri-bipyridine, NHS ester.

electrodynamic forces London-Van der Waals forces: see **DLVO theory**.

electrofocusing Any technique whereby chemical species are concentrated using an applied electric field. See **isoelectric focusing**.

electrogenic pump Ion pump that generates net charge flow as a result of its activity. The sodium-potassium exchange pump transports two potassium ions inward across the cell membrane for each three sodium ions transported outward. This produces a net outward current that contributes to the internal negativity of the cell.

electrokinetic potential See **zeta potential**.

electrolyte A compound that dissociates into ions in solution.

electron microprobe A technique of elemental analysis in the electron microscope based on spectral analysis of the scattered X-ray emission from the specimen induced by the electron beam. Using this technique it is possible to obtain quantitative data on, for example, the calcium concentration in different parts of a cell, but it is necessary to use ultra-thin frozen sections.

electron microscopy Any form of microscopy in which the interactions of electrons with the specimens are used to provide information about the fine structure of that specimen. In transmission electron microscopy (TEM) the diffraction and adsorption of electrons as the electron beam passes normally through the specimen is imaged to provide information on the specimen. In **scanning electron microscopy** (SEM) an electron beam falls at a non-normal angle on the specimen and the image is derived from the scattered and reflected electrons. Secondary X-rays generated by the interaction of electrons with various elements in the specimen may be used for **electron microprobe** analysis. High voltage electron microscopy (HVEM) has two advantages, the increased voltage shortens the wavelength of the electrons (and therefore increases resolving power) but, more importantly for the biologist, the penetrating power of the beam is increased, and it becomes possible to look at thicker specimens. Thus it is possible, by using stereoscopic views (obtained with a tilting stage) to get a three-dimensional picture of the interior of a cell. See **scanning transmission electron microscopy** and *cf.* **atomic force microscopy, light microscopy.**

electron paramagnetic resonance Form of spectroscopy (EPR, electron spin resonance, ESR) in which the absorption of microwave energy by a specimen in a strong magnetic field is used to study atoms or molecules with unpaired electrons. Webpage: http://www.chem.queensu.ca/eprmmr/EPR_summary.htm

electron transport chain A series of compounds that transfer electrons to an eventual donor with concomitant energy conversion. One of the best studied is in the mitochondrial inner membrane, that takes NADH (from the **tricarboxylic acid cycle**) or FADH and transfers electrons via **ubiquinone**, cytochromes and various other compounds, to oxygen. Other electron transport chains are involved in **photosynthesis.**

electrophoresis Separation of molecules based on their mobility in an electric field. High resolution techniques normally use a gel support for the fluid phase. Examples of gels used are starch, acrylamide, agarose or mixtures of acrylamide and agarose. Frictional resistance produced by the support causes size, rather than charge alone, to become the major determinant of separation. The electrolyte may be continuous (a single buffer), or discontinuous, where a sample is stacked by means of a buffer discontinuity, before it enters the running gel/running buffer. The gel may be a single concentration or gradient in which pore size decreases with migration distance. In **SDS** gel electrophoresis of proteins or electrophoresis of polynucleotides, mobility depends primarily on size and is used to determined molecular weight. In pulse-field electrophoresis, two fields are applied alternately at right-angles to each other to minimize diffusion-mediated spread of large linear polymers. See also **electrofocusing, pulse-field electrophoresis.**

electrophoretogram Result of a zone electrophoresis separation or the analytical record of such a separation.

electroplax A stack of specialized muscle fibres found in electric eels, arranged in series. The fibres have lost the ability to contract; instead they generate extremely high voltages (ca 500 V) in response to nervous stimulation. They contain asymmetrically distributed **sodium-potassium ATPases, acetylcholine receptors** and **sodium gates** at extraordinarily high concentrations.

electroporation Method for temporarily permeabilising cell membranes so as to facilitate the entry of large or hydrophilic molecules (as in **transfection**). A brief (ca 1 msec) electric pulse is given with potential gradients of about 700 V/cm.

electroretinogram Record of electrical activity in the retina made with external electrodes.

electrospray mass spectroscopy Method of mass spectroscopy in which the sample is introduced as a fine spray from a highly charged needle so that each droplet has a strong charge. Solvent rapidly evaporates from the droplets leaving the free macromolecule. Beginning to be widely used because of its capacity to identify a wide range of compounds.

electrostatic forces The forces exerted by charges in close proximity: like charges repel, unlike charges attract. If two surfaces, such as those of animal cells, bear appreciable and approximately equal densities of charged groups on their surfaces there may be appreciable forces of repulsion between them. The range of these forces is determined in the main by the ionic strength of the intervening medium, forces being of minimal range at high ionic strength. The forces are effective over approximately twice the **double layer** thickness. See **DLVO theory.**

electrotonus The state of a nerve during the brief period that an electrical current is passing.

elegantin See **disintegrin**.

eleidin Clear substance found in stratum lucidum of skin, derived from keratohyalin and a precursor of keratin.

Elejalde's syndrome A developmental disorder (acrocephalopolydactylous dysplasia) that may be due to a defect in a fibroblast growth factor receptor gene. Elejalde's disease (neuroectodermal melanolysosomal disease) is distinct from Elejalde's syndrome and is caused by mutation in the MYO5A gene that encodes myosin heavy chain 12 (myoxin) (as in Griscelli's syndrome type 1).

elementary bodies (1) Inclusion bodies within cells, often of virus particles, although this term is more common in the older literature. (2) Infectious extracellular form of *Chlamydia*, consisting of electron-dense nuclear material and a few ribosomes surrounded by a rigid trilaminar wall. Once taken up by cells these reorganise into reticulate bodies.

elephantiasis Lymphatic filariasis, enlargement of the limbs, or of the scrotum, due to thickening of skin and blockage of lymphatic vessels by filarial nematode parasites, especially *Brugia malayia* and *Wuchereria bancrofti*.

eleutherobin Tricyclic compound (a diterpene glycoside) that, like taxol, will stabilize microtubule bundles by competing for the paclitaxel binding site. Originally isolated from a marine soft coral,

Eleutherobia aurea, also found in *Erythropodium caribaeorum*, an encrusting coral found in South Florida and the Caribbean. Synthetic routes for producing the compound have been devised.

eleutherosides Class of compounds, lignan glycosides, with anti-inflammatory and immunostimulatory activity isolated from the roots of *Eleutherococcus senticosus* (Siberian ginseng, a distant relative of Asian ginseng) and other medicinal herbs. Eleutherosides B and E have been most extensively studied.

ELF (1) Eph ligand family, see **ephrins**. (2) ETS-related transcription factors Elf-2, Elf3 etc. see **ets**.

elicitor In general any compound that induces a response in a system; more specifically, a substance that induces the formation of **phytoalexins** in higher plants. May be exogenous (often produced by potentially pathogenic microorganisms), or endogenous (possibly cell-wall degradation products).

elimination Disappearance of a substance from an organism, or a part of the organism, by processes of metabolism, secretion, or excretion. Rates of elimination are important in toxicology and pharmacology.

ELISA A very sensitive technique (enzyme-linked immuno-sorbent assay) for the detection of small amounts of protein or other antigenic substance. The basis of the method is the binding of the antigen by an antibody that is linked to the surface of a plate. Formation of an immune complex is detected by use of peroxidase coupled to antibody, the peroxidase being used to generate an amplifying colour reaction. Various ways of carrying out the assay are possible: if the aim is to detect antibody production from a myeloma clone, for example, then the antigen may be bound to the plate, and the formation of the antibody/antigen complex may be detected using peroxidase coupled to an anti-Ig antibody.

Elk proteins (1) Transcriptional activators of the **ets** family. Elk1 (ETS domain-containing protein Elk-1, p62 ternary complex factor, 428aa) is found in lung and testis. Binds to DNA at purine-rich sites. When phosphorylated by MAP kinase can form a ternary complex with the serum response factor and the ETS and SRF motifs of the fos serum response element. Others are elk3 (407aa) and elk4 (431aa). (2) Obsolete (and confusing) name for Eph-like kinases. See **Eph-related receptor tyrosine kinases**.

ELL See **elongation factor ELL**.

ellipsosome Membrane-bounded compartment containing cytochrome-like pigment and found in the retinal cones of some fish.

elliptocytosis A disorder in which erythrocytes are elliptical and there is some haemolytic anaemia. Caused by mutation in genes coding for components of the erythrocyte membrane skeleton. In the

Rhesus-linked form the defect is in Band 4.1, the rhesus-unlinked forms are defects in genes for alpha-spectrin, beta-spectrin or band 3.

Ellis-van Creveld syndrome A skeletal dysplasia caused by mutation in the *EVC* gene or the nonhomologous gene, *EVC2*. The *EVC* gene product is a single-pass membrane protein (992aa) found in developing vertebral bodies and other tissues, *EVC2* encodes limbin (1308aa) a multi-pass membrane protein involved in bone formation and skeletal development.

Ellman's reagent Reagent (DNTB, DTNB) used to estimate the number of free sulphydryl groups in peptides or proteins. Reacts to produce a coloured compound. Datasheet: http://www.interchim.com/interchim/bio/produits_uptima/tech_sheet/FT-UP01566%28DTNB%29.pdf

elongation factor *EF* Peptidyl transferase components of ribosomes that catalyse formation of the acyl bond between the incoming amino acid residue and the peptide chain. There are three classes of elongation factor: EF1α (462aa; EF-Tu, 394aa in prokaryotes) binds GTP and aminoacyl-tRNA, delivering it to the A site of ribosomes. EF-1β (225aa; EF-Ts, 283aa) helps in regeneration of GTP-EF-1a. EF-2 (858aa; EF-G, 704aa) binds GTP and peptidyl-tRNA and translocates it from the A site to the P site. Diphtheria toxin inhibits protein synthesis in eukaryotes by adding an ADP-ribosyl group to a modified histidine residue (diphthamide) in elongation factor 2.

elongation factor ELL An elongation factor (elongation factor 'eleven-nineteen lysine-rich leukemia', 621aa) that increases the rate at which RNA polymerase II transcribes DNA by suppressing transient pausing. ELL2 (640aa) and ELL3 (397aa) have similar effects. ELL-associated factor 1 (EAF1, 268aa) is a transcriptional transactivator of elongation by ELL and ELL2. Functionally similar to **elongin**.

elongins A set of transcription elongation factors, which act to increase the overall rate at which RNA polymerase II transcribes DNA. There are three subunits, Subunit A (elongin A, transcription elongation factor B polypeptide 3, SIII p110, 798aa) is transcriptionally active and its activity is strongly enhanced by binding to the dimeric complex of the SIII regulatory subunits B and C (elongin BC complex). Elongin B (SIII p18, 118aa); elongin C (transcription elongation factor B polypeptide 1, SIIIp15, 112aa). Elongin C has homology to Skp1. The von Hippel-Lindau (VHL) tumour suppressor protein will bind the elongin BC complex and prevent it from activating elongin A.

elutriation Separation of particles on the basis of their differential sedimentation rate.

EMA (1) Epithelial embrane antigen; see **episialin**. (2) In mouse, E2F-binding site modulating activity (EMA, 272aa) a transcriptional repressor that has some similarity with E2F but lacks the activation domain at the carboxy terminus. (3) Forkhead box protein 11-ema (Ectodermally-expressed mesendoderm antagonist, 373aa) a transcriptional activator in *Xenopus* that activates ectoderm and inhibits mesoderm and endoderm formation.

EMAP (1) Echinoderm microtubule-associated protein-like proteins (EMAP1, 815aa; EMAP2, 649aa; EMAP3, 896aa; EMAP5, 1969aa, EMAP6, 1958) that may affect the assembly dynamics of microtubules causing them to be slightly longer, but more dynamic. EMAP4 (restrictedly overexpressed proliferation-associated protein, Ropp 120, 981aa) is overexpressed during mitosis (in humans). All contain multiple **WD repeats**. (2) Endothelial monocyte-activating polypeptide II (EMAPII, 166aa) derived by cleavage from aminoacyl tRNA synthase complex-interacting multifunctional protein 1 (AIMP1, 312aa) a non-catalytic component of the **multisynthase complex**. EMAPII is an anti-angiogenic factor in tumour vascular development and may direct vascularisation of the developing lung.

EMB30 A protein (EMBRYO DEFECTIVE 30, GNOM, 1451aa) involved in pattern formation in plant development.

Embden-Meyerhof pathway The main pathway (Embden-Meyerhof-Parnas pathway, glycolysis) for anaerobic degradation of carbohydrate. Starch or glycogen is hydrolyzed to glucose-1-phosphate and then through a series of intermediates, yielding two ATP molecules per glucose, and producing either pyruvate (which feeds into the **tricarboxylic acid cycle**) or lactate.

embedding A technique used in preparing tissue for cutting thin sections for microscopical examination. The specimen is infiltrated by wax or plastic to provide mechanical support.

embolic gastrulation Gastrulation by invagination of part of the blastocyst wall, rather than overgrowth of the epiblast as happens with, for example, birds.

embolus A clot formed by platelets or leucocytes that blocks a blood vessel.

EMBOSS European Molecular Biology Open Software Suite (EMBOSS), a package of free Open Source software for molecular biology. Link: http://emboss.sourceforge.net/

embryo The developmental stages of an animal or, in some cases a plant, during which the developing tissue is effectively isolated from the environment by, for example, egg membranes, fetal membranes and various structures in plants.

embryo sac The female **gametophyte** in flowering plants (angiosperms) that develops within the ovule (megaspore) contained within an ovary at the base of the pistil of the flower. There are usually eight (haploid) cells in the female gametophyte: one egg, two synergids flanking the egg, two polar nuclei in the center of the embryo sac and three antipodal cells, at the opposite end of the embryo sac from the egg.

embryogenesis The processes leading to the development of an embryo from egg to completion of the embryonic stage.

embryoid In plants, an embryo-like structure that may subsequently grow into a plantlet; in animals, aggregates of cells (embryoid body) derived from embryonic stem cells, that will exhibit some differentiation *in vitro*.

embryoma A mass of rapidly growing cells (embryonal tumour) in embryonic tissue or, in the adult, derived from residual embryonic tissue. Embryomas may be benign or malignant, and examples include neuroblastomas and Wilms' tumour.

embryonal carcinoma cells Pluripotent cells of ectodermal origin, derived from **teratocarcinomas**.

embryonic induction The induction of differentiation in one tissue as a result of proximity to another tissue that may be a result of morphogenetic movements such as gastrulation. One of the best known examples is the induction of the neural tube in the ectoderm by the underlying chordamesoderm. Although the information to form the tube is present in the competent determined ectoderm, it must be elicited by the inducing tissue. In some cases it is known that cell-cell contact between epithelium and mesenchyme is necessary.

embryonic lethal abnormal visual proteins See **ELAV proteins**.

embryonic stem cell See **stem cell**.

Embryophyta The embryophytes (clade Embryophyta or Metaphyta) are the most familiar group of plants. They are often called land plants because they live primarily in terrestrial habitats, in contrast with the related green algae that are primarily aquatic. The embryophytes include trees, flowers, ferns, mosses, and various other green land plants. All are complex multicellular eukaryotes with specialized reproductive organs.

emergent A term applied to novel and unpredictable properties that can appear in a complex system. For example, it is argued that consciousness is an emergent property of brain neurophysiology.

Emericella nidulans New name for *Aspergillus nidulans*.

emerin A ubiquitous type II integral membrane protein (254aa) that forms part of a nuclear protein complex consisting of the barrier-to-autointegration factor (BAF), the nuclear lamina, nuclear actin and other associated proteins; apparently links A-type **lamins** to the inner nuclear envelope. Emerin is defective in some forms of X-linked **Emery-Dreifuss muscular dystrophy** (X-EDMD). In the heart emerin is associated with intercalated discs.

Emerson enhancement effect The effect on the rate of photosynthesis (in plants and algae) of illuminating simultaneously with far red light ($\lambda > 680$ nm) and light of shorter wavelength ($\lambda < 680$ nm). The effect is more than additive and provides evidence for the existence of the two photosystems I and II.

Emery-Dreifuss muscular dystrophy Form of X-linked muscular dystrophy, a degenerative myopathy characterized by weakness and atrophy of muscle without involvement of the nervous system. Can arise either from defects in **emerin** or nuclear **lamin** A.

emetine An alkaloid derived from ipecac root (*Cephaelis ipecacuanha*); used in the treatment of amebiasis and as an emetic. Inhibits protein synthesis at the translation stage by blocking translocation of peptidyl-tRNA from the A-site to the P-site on the ribosome. Emetine-resistant CHO cell lines have been extensively studied.

EMMPRIN See **basigin**.

emperipolesis Phenomenon in which lymphocytes are apparently phagocytosed by macrophages (histiocytes) in the lymph node; associated with massive lymphadenopathy, an inflammatory disorder of obscure aetiology. In the early literature lymphocytes and leucocytes were described as entering the cytoplasm of endothelial cells during their extravasation in post-capillary venules, a process also termed emperipolesis, although this was a misapprehension, and leucocytes were shown by ultrastructural studies to be moving between, not through, the endothelial cells.

EMT See **epithelial-mesenchymal transition**.

En/Spm A transposable element (enhancer/suppressor mutators) that causes mutable phenotypes, widespread in various distantly related plant species. Other similar transposons (of which the En/SPm system in maize is the prototype) are known (the CACTA family) and all carry common 28 bp terminal inverted repeats and subterminal repetitive regions and are known as the Tpn1 family. The suppressor affects genes which contain an En/Spm responsive transposable element in the transcribed sequences. The En/Spm encoded protein tnpA (transposase A) binds a defined cis element in the inserted transposon, repressing expression of the adjacent gene.

ENA-78 An alpha chemokine (epithelial derived neutrophil activating peptide-78; CXCL5, 114aa) that affects the growth, movement, or activation of cells involved in immune and inflammatory responses. It is chemotactic for activated T-cells, binds to CXCR2 and production is enhanced by $TNF\alpha$ in dermal fibroblasts and vein endothelial cells. It is induced by inflammatory cytokines in human colonic enterocyte cell lines. Mouse homologue is **LIX**.

enabled In *Drosophila* a protein (980aa) that forms part of a complex signalling network (together with Abl, trio and fra) that regulates axon guidance at the CNS midline. Required in part for robo-mediated repulsive axon guidance. May be involved in lamellipodial dynamics. See **mena**.

enamelin A secreted protein (1142aa) involved in the mineralization and structural organization of tooth enamel. Mutations cause some forms of **amelogenesis imperfecta**.

enamelysin Metallopeptidase (matrix metallopeptidase-20, MMP-20, 483aa) with substrate specificity for **amelogenin**. See Table M1.

enantiomer Either of a pair of stereoisomers of a chiral compound.

enaptin See **nesprins**.

Enc-1 See **ectoderm-neural cortex-1**.

encapsidate To envelop a virus in a protein shell (the **capsid**).

encephalitis Inflammation of the brain, caused by viral or bacterial infection. Japanese encephalitis is the major cause of viral encephalitis in Asia, caused by a flavivirus and transmitted by mosquitoes. See also **Aicardi-Goutieres syndrome, experimental allergic encephalomyelitis, Schilder's disease**.

encephalization The increased development of the head region, and in particular the brain, in the course of the evolution of an organism.

encephalopsin An extraretinal photoreceptor molecule (panopsin, opsin3, 402aa) primarily found in brain, that may play a role in non-visual photic processes such as the entrainment of circadian rhythm or the regulation of pineal melatonin production. Has highest homology to vertebrate retinal and pineal opsins. Encephalopsin is highly expressed in the preoptic area and paraventricular nucleus of the hypothalamus, is enriched in selected regions of the cerebral cortex, cerebellar Purkinje cells, a subset of striatal neurons, selected thalamic nuclei, and a subset of interneurons in the ventral horn of the spinal cord. See **opsin sub-families**.

encysted Describing something enclosed in a cyst or a sac.

end plate potential Depolarization of the sarcolemma as a result of acetylcholine release from the motoneuron causing an influx of sodium ions. The end plate potential (epp) is the sum of quantal **miniature end plate potentials**. Development of the end plate potential is blocked by curare.

End3 In *S. cerevisiae* an actin-regulatory protein (endocytosis protein 3, 349aa) that is defective in the endocytosis-deficient mutant, end3Delta. It is a component of the **Pan1** actin cytoskeleton-regulatory complex required for the internalization of endosomes during actin-coupled endocytosis. End3 regulates Pan1 function by preventing its phosphorylation by serine/threonine kinase **Prk1p**.

endarteritis Chronic inflammation of the arterial **intima**.

endergonic Describing a reaction that requires the input of energy.

endobrevin See **synaptobrevin**.

endocannabinoids Endogenous metabolites capable of activating the **cannabinoid receptors** (CB1 and CB2). Anandamide (arachidonylethanolamide) and 2-arachidonyl glycerol (2-AG) are the main endocannabinoids although both bind to both receptors.

endochondral Term for anything situated within, or occurring within, cartilage. Endochondral bone formation occurs on a cartilage scaffold.

endocrine gland Gland that secretes directly into blood and not through a duct. Examples are pituitary, thyroid, parathyroid, adrenal glands, ovary and testis, placenta and B cells of pancreas.

endocyte (1) In *Dictyostelium* cannibalistic phagosomes within the zygote giant cell that contain ingested amoebae. (2) In *Hydra*, engulfed 'nurse cells' found early in development. (3) Sometimes loosely (inaccurately) used of endodermally-derived cells.

endocytosis Uptake of material into a cell by the formation of a membrane-bound vesicle (an endocytotic vesicle).

endoderm A germ layer lying remote from the surface of the embryo that gives rise to internal tissues such as gut. *Cf.* **mesoderm** and **ectoderm**.

endodermis Single layer of cells surrounding the central stele (vascular tissue) in roots. The radial and transverse walls contain the hydrophobic **Casparian band**, that prevents water flow in or out of the stele through the **apoplast**. Also present in some stems.

endogenous Product or activity arising in the body or cell, as opposed to agents coming from outside.

endogenous pyrogen Fever-producing substance released by leucocytes (and Kuppfer cells in particular) that acts on the hypothalamic thermoregulatory centre. Now known to be interleukin 1 (**IL-1**).

endogenous retroviruses Viruses that are integrated into the genome and no longer able to produce infective virus. Human endogenous retroviruses (HERVs) comprise ~8% of the genome and proteins encoded by viral genes may be important in some autoimmune diseases, in particular multiple sclerosis. HERV genes also produce **syncytin** and there may be an association between HERVs and the HELLP syndrome and pre-eclampsia. The presence of endogenous retroviruses in xenogeneic tissue used clinically is a concern; porcine retroviruses (PERVs) may be present in heart valves derived from pigs.

endoglin Homodimeric membrane glycoprotein (CD105, 658aa) highly expressed on vascular endothelium. Forms an heteromeric complex with the signalling receptors for TGFβ. Its cytoplasmic domain interacts with the actin cytoskeleton via **zyxin-related protein 1** (ZRP-1). Defects in the endoglin gene cause **hereditary haemorrhagic telangiectasia type 1** (Osler-Rendu-Weber syndrome 1).

endoglycosidase Enzyme of the subclass EC 3.2 that has the ability to hydrolyze non-terminal glycosidic bonds in oligosaccharides or polysaccharides. Endoglycosidases F and H are often used as tools to determine the role of carbohydrate moieties on glycoproteins. Endo-F, the product of *Flavobacterium meningosepticum*, cleaves glycans of high mannose and complex type at the link to asparagine in the protein; Endo H is from *Streptomyces* spp. and is an endo-β-N-acetyl-glucosaminidase.

endolithic Growing within rock. Such habitats are important in arid areas (e.g. Arctic, Antarctic) and a range of organisms exploit the habitat, including algae, fungi and cyanobacteria. Bacteria within deep rock strata have been described and the whole field is of interest to astrobiologists.

endolymph Fluid that fills the membranous labyrinth of the inner ear.

endolyn Sialomucin core protein 24 (CD164, 197aa), present in membranes of endosomes and lysosomes, that regulates the adhesion of haematopoietic stem cells (CD34$^+$) to bone marrow stroma, stimulates proliferation of these cells and associates with the CXCR4 chemokine receptor. It regulates myoblast migration and promotes myoblast fusion into myotubes.

endometrioma Tumour (adenomyoma) of the endometrium consisting of glandular elements and a cellular connective tissue.

endometrium Mucous membrane that lines the uterus. The endometrium thickens during the menstrual cycle but if implantation of an embryo does not occur it reverts to its previous state and the excess tissue is shed at menstruation. If implantation does occur the endometrium becomes the decidua and is not shed until after parturition.

endomitosis Chromosome replication without mitosis, leading to polyploidy. Many rounds of endomitosis give rise to the giant **polytene chromosomes** of Dipteran salivary glands, though in this case the daughter chromosomes remain synapsed.

endomorphins Endogenous peptides (endomorphin-1, YPWF-NH$_2$; endomorphin-2, YPFF-NH$_2$) with high selective affinity for μ-opiate receptor. See **morphiceptin**.

endomycorrhiza *Pl. endomycorrhizae.* Plant-fungal symbiotic association (vesicular-arbuscular mycorrhiza, endotrophic mycorrhiza), common to many plant genera, in which fungi (usually zygomycota) penetrate roots and come to lie in close association with root cells. More common than **ectomycorrhiza**.

endomysium Connective tissue sheath surrounding individual muscle fibres.

endoneurium Connective tissue sheath surrounding individual nerve fibres in a nerve bundle.

endonexins (1) Calcium-dependent membrane-binding proteins. Endonexin 1 (**annexin A4**, chromobindin-4, lipocortin IV, 319aa) promotes membrane fusion and is involved in exocytosis. Endonexin-2 (annexin A5, lipocortin V, calphobindin I, placental anticoagulant protein I, thromboplastin inhibitor, vascular anticoagulant-alpha, anchorin CII, 320aa) can modify membrane properties and endonexin II forms voltage-gated divalent-cation selective channels. (2) Centromere protein R (Beta-3-endonexin, integrin beta-3-binding protein, nuclear receptor-interacting factor 3, 177aa) is a transcriptional coregulator that can activate or inhibit acording to the hormone receptor. See **CENP**.

endonuclease One of a large group of enzymes that cleave nucleic acids at positions within the chain. Some act on both RNA and DNA (e.g. S1 nuclease, EC.3.1.30.1, that is specific for single stranded molecules). **Ribonucleases** such as pancreatic, T1 etc. are specific for RNA, **Deoxyribonucleases** for DNA. Bacterial **restriction endonucleases** are crucial in recombinant DNA technology for their ability to cleave double-stranded DNA at highly specific sites.

endopeptidase An enzyme that cleaves protein at positions within the chain. Formally, the enzymes are peptidyl-peptide hydrolases, often referred to as **proteinases** or **proteolytic enzymes**.

endopeptidase 24.11 See NEP #1.

endophilin Family of proteins with a **BAR domain** involved in regulating clathrin-mediated endocytosis. Phosphorylation of endophilin by Rho-kinase inhibits their binding to **CIN85**, a key step in the internalisation of ligand-activated receptors such as EGF-R. Endophilin A1 (352aa) is implicated in synaptic vesicle endocytosis and interacts with synaptojanin. Endophilin B1 (Bax-interacting factor 1, SH3 domain-containing GRB2-like protein B1, 365aa) may be required for normal outer mitochondrial membrane dynamics. Endophilins were once erroneously thought to have lysophosphatidic acid acyl transferase (LPAAT) activity.

endoplasm Inner, granule-rich cytoplasm of amoeba.

endoplasmic reticulum Membrane system (ER) that ramifies through the cytoplasm. The membranes of the ER are separated by 50–200nm and the **cisternal** space thus enclosed constitutes a separate compartment. The Golgi region is composed of flattened sacs of membrane that together with ER and lysosomes constitute the GERL system. See also **smooth ER, rough ER**.

endoplasmic reticulum aminopeptidase An **aminopeptidase** (ERAP1, adipocyte-derived leucine aminopeptidase, EC 3.4.11.-, 941aa) thought to be important in metabolism of peptides such as **angiotensin** II.

endoplasmin Most abundant protein in microsomal preparations from mammalian cells (100-fold more concentrated in ER than elsewhere). The major chaperone of the ER, a glycoprotein (tumour rejection antigen gp96; TRA1, 803aa) with calcium-binding properties. Same as GRP94 (glucose-regulated protein-94). A member of the **hsp90** family of **heat-shock proteins**.

endoreduplication Replication of DNA in the absence of mitosis, leading to high ploidy levels. Common in plants, where endoreduplication seems to be associated with increased cell size.

endorepellin An anti-angiogenic and anti-tumour peptide (705aa) cleaved from **perlecan** that inhibits endothelial cell migration, collagen-induced endothelial tube morphogenesis and blood vessel growth in the chorioallantoic membrane.

endorphins A family of peptide hormones that bind to **opioid receptors**. Released in response to neurotransmitters and rapidly inactivated by peptidases. Physiological responses to endorphins include analgesia and sedation.

endosmosis Movement of water into a cell as a result of greater internal osmotic pressure. The water potential within the vascular sap of a plant cell must be lower than that in the bathing medium or sap of a neighbouring cell.

endosome (1) Endocytotic vesicle derived from the plasma membrane. More specifically an acidic non-lysosomal compartment in which receptor-ligand complexes dissociate. (2) A chromatinic body near the centre of a vesicular nucleus in some protozoa.

endosperm Tissue present in the seeds of angiosperms, external to and surrounding the **embryo**, that it provides with nourishment in the form of **starch** or other food reserves. Formed by the division of the **endosperm mother cell** after fertilization; may be absorbed by the embryo prior to seed maturation, or may persist in the mature seed. In plants, endosperm is a classic **syncytium**. The tissue may be initially liquid, containing multiple nuclei not separated by cell walls. Cellularization may occur later, when a branched phragmoplast forms between every nucleus to deposit an immature cell wall.

endosperm mother cell Cell of the higher plant embryo sac. Contains two 'polar nuclei', and fuses with the sperm cell from the pollen grain. Gives rise to the **endosperm**.

endospore (1) An asexual spore formed within a cell. (2) Inner part of the wall of a fungal spore.

endostatin A C-terminal fragment (183aa) of collagen18A (1754aa) that potently inhibits endothelial cell proliferation and angiogenesis. May inhibit angiogenesis by binding to the heparan sulphate proteoglycans involved in growth factor signalling.

endosymbiont hypothesis The hypothesis that semi-autonomous organelles such as mitochondria and chloroplasts were originally endosymbiotic bacteria or cyanobacteria. The arguments are convincing and although the hypothesis cannot be proven it is widely accepted.

endosymbiotic bacteria Bacteria that establish a symbiotic relationship within a eukaryotic cell. For example, the nitrogen-fixing bacteria of legume root nodules. See also **endosymbiont hypothesis**.

endothelial monocyte-activating polypeptide II A multifunctional polypeptide (aminoacyl tRNA synthase complex-interacting multifunctional protein 1, AIMP1, 312aa) that stimulates the catalytic activity of cytoplasmic arginyl-tRNA synthase. It is cleaved to produce a cytokine (endothelial monocyte-activating polypeptide 2 subfamily E, member 1, SCYE1, 166aa) that is proinflammatory, activates endothelial cells and is chemotactic for neutrophils and mononuclear phagocytes.

endothelin converting enzyme Integral membrane proteins belonging to the family of metalloproteinases involved in processing various neuropeptides. Endothelin converting enzyme 1 (ECE1, EC 3.4.24.71, 770aa) converts big-endothelin-1 (ET-1) to active ET-1. ECE-2 (883aa) does the same and is also involved in the processing of various neuroendocrine peptides, including neurotensin, angiotensin I, substance P, proenkephalin-derived peptides, and prodynorphin-derived peptides.

endothelin receptor There are two G-protein coupled receptors for **endothelins**, ET(A) (427aa) and ET(B) (442aa), present on vascular smooth muscle cells mediating vasoconstriction, and on endothelium mediating **nitric oxide** release. ET(A) binds ET-1 preferentially whereas ET(B) binds ET-1, ET-2 and ET-3 with equal affinity. Endothelin B receptor-like protein 1 (613aa) is an orphan receptor possibly with a role in the CNS where it forms a complex with **parkin-2**, STUB1 (E3 ubiquitin-protein ligase CHIP) and HSP70.

endothelins Group of peptide hormones (ET-1, ET-2, ET-3, all 21aa) released by endothelial cells. All have two disulphide bridges that hold them in a conical spiral shape. They are the most potent vasoconstrictor hormones known. Structurally related to the snake venom **sarafotoxins**. Pre-pro-endothelin-1 (203 residues) is cleaved to the biologically inactive big endothelin-1 (92aa) by **endothelin converting enzyme** which will further cleave big endothelin to form active endothelin-1. ET-1, the predominant form, is produced by endothelial cells, ET-2 and ET-3 by various tissues. In addition to their vasoconstrictive properties, endothelins have inotropic and **mitogenic** properties, influence salt and water balance, alter central and peripheral sympathetic activity and stimulate the **renin-angiotensin** -aldosterone system. Though ET-1 acting through **endothelin receptor** (A) is vasoconstrictive, it acts through ET (B) to induce the release of **nitric oxide** which is a vasodilator.

endothelioma A tumour, usually benign, derived from the epithelial lining of blood vessels or lymph channels (endothelium).

endothelium (1) Simple, generally squamous, epithelium lining blood vessels, lymphatics and other fluid-filled cavities (such as the anterior chamber of the eye). Mesodermally derived, unlike most epithelia. Modified in areas where there is lymphocyte traffic (see **high endothelial venule**). (2) In plants, radially elongated and metabolically active cells of the inner epidermis of the inner integument of the ovule.

endothelium-derived relaxation factor *EDRF.* See **nitric oxide**.

endotherm An animal that is able to maintain a body temperature above ambient by generating heat internally. In contrast to poikilothermic (so-called 'cold-blooded') animals.

endothermic Process or reaction that absorbs heat and thus requires a source of external energy in order to proceed.

endotoxin Heat-stable polysaccharide-like toxin bound to a bacterial cell. The term is used more specifically to refer to lipopolysaccharide (LPS) of the outer layer of the cell envelope of Gram-negative bacteria. There are three parts to the molecule, the **Lipid A** (six fatty acid chains linked to two glucosamine residues), the core oligosaccharide (branched chain of ten sugars), and a variable length polysaccharide side chain (up to 40 sugar units in smooth forms) that can be removed without affecting the toxicity (rough LPS). Some endotoxin is probably released into the medium and endotoxin is responsible for many of the virulent effects of Gram-negative bacteria.

endotrophic mycorrhiza An endomycorrhiza in which the fungal hyphae grow between and within the cells of the root cortex and connect with hyphae ramifying though the soil but which do not form a thick mantle on the surface of the root. Vesicular-arbuscular mycorrhizas and the mycorrhizas of orchids and of the Ericaceae are endotrophic.

endovanilloids Endogenous ligands of the transient receptor potential vanilloid type 1 (TRPV1) channels, one of the thermo-sensitive **TRP channels**. Include N-arachidonoyl-dopamine, N-oleoyl-dopamine and **anandamide**. See **vanilloid receptor**.

endplate The area of sarcolemma immediately below the synaptic region of the motor neuron in a neuromuscular junction.

enduracidin See **ramoplanin**.

Englebreth-Holm-Swarm sarcoma cells See **EHS cells**.

engrailed *Drosophila* gene that controls segmental polarity. It is the archetype for one of three subfamilies of **homeobox**-containing genes. The product (552aa) is a transcriptional regulator that represses activated promoters and is required for the development of the central nervous system.

enhanced chemiluminescence A sensitive method (ECL) for detecting specific proteins on gels by the use of a peroxidase-coupled antibody which oxidises luminol and produces light.

enhancement effect See **Emerson enhancement effect**.

enhancer A DNA **control element** frequently found 5′ to the start site of a gene, which when bound by a specific transcription factor, enhances the levels of expression of the gene, but is not sufficient alone to cause expression. **Promoters** differ in being sufficient to cause expression of the gene when bound; in practice, the two terms merge.

ENHANCER OF SHOOT REGENERATION 1 The *Arabidopsis* gene (*ESR1*) that is thought to be a key gene for commitment to *in vitro* shoot regeneration in tissue culture and regulates gene expression patterns in meristems. *ESR1* encodes a member of the ethylene responsive factor (ERF) family of transcription factors. The ESR1 protein (protein ENHANCER OF SHOOT REGENERATION 1, protein DORNROSCHEN, 328aa) is a transcriptional activator that upregulates various genes. Article: http://www.jstage.jst.go.jp/article/plantbiotechnology/26/4/26_385/_article

enhancer of zeste In *Drosophila*, a polycomb-group protein (histone-lysine N-methyltransferase E(z), EC 2.1.1.43, 760aa) involved in transcriptional repression and specifically required during the first 6 hours of embryogenesis. Homologues are found in plants and many metazoa: see **EZH**.

enhancer trap Technique for mapping gene expression patterns, classically in *Drosophila*. A **transposon** element carrying a **reporter gene** (usually **beta-galactosidase**), linked to a very weak promoter, is induced to 'jump' within the genome. If the P-element re-inserts within the sphere of influence of promoters and enhancers of some (random) gene, then the reporter gene is also expressed in a similar tissue-specific manner. Usually, many lines of flies carrying such random insertions are studied; if a line shows 'interesting' patterns of expression, it is then possible to clone the gene of interest.

enkephalins Natural **opiate** pentapeptides isolated originally from pig brain. Leu-enkephalin (YGGFL) and Met-enkephalin (YGGFM) bind particularly strongly to δ-type opiate receptors.

enlargeosomes Cytoplasmic organelles discharged by regulated exocytosis, identified by immunofluorescence of their membrane marker, **desmoyokin/Ahnak**. Best characterized in a PC12 clone, PC12-27, that is defective in classical neurosecretion. Their function remains unclear. Abstract: http://www.ncbi.nlm.nih.gov/pubmed/17488290

enols Tautomeric form of some ketones; any organic compounds with a hydroxyl group attached to a carbon that is linked to another carbon by a double bond. Loss of a proton generates an enolate anion.

enovin See **artemin**.

ensconsin Epithelial microtubule-associated protein (MAP7, 749aa) a microtubule-stabilizing protein that may be important during reorganization of microtubules during polarization and differentiation of epithelial cells.

entactin See **nidogen**.

Entamoeba Single celled eukaryotes that parasitize all classes of vertebrates, a few invertebrates and possibly other unicellular eukaryotes. All species have a simple life cycle consisting of an infective cyst stage and a multiplying trophozoite stage. *Entamoeba histolytica* is the only species that affects man and is the third leading cause of morbidity and mortality due to parasitic disease in humans. 'Non-pathogenic' entamoebiasis is now recognised as being due to infection with another species, *E. dispar*.

enteric Relating to the intestine.

Enterobacter Genus of enteropathic bacilli of the Klebsiella group. Not to be confused with the Family **Enterobacteria** of which they are members.

Enterobacteriaceae A large family of Gram-negative bacilli that inhabit the large intestine of mammals. Commonest is *Escherichia coli*; most are harmless commensals but others can cause intestinal disease (*Salmonella*, *Shigella*).

enterobactin See **enterochelin**.

enterochelin Iron-binding compound (enterobactin, a **siderophore**) of *E. coli* and *Salmonella* spp. A cyclic trimer of 2,3-dihydroxybenzoylserine.

enterochromaffin cells Neuroendocrine cells (Kulchitsky cells) found in the epithelia lining the lumen of the gastrointestinal tract. They produce and contain about 90% of the body's store of serotonin. See **enterochromaffin-like cells**.

enterochromaffin-like cells Distinct type of neuroendocrine cells (ECL-cells) found in the gastric mucosa underlying the epithelium, particularly in the acid-secreting regions of the stomach. Synthesize and secrete histamine in response to stimulation by the hormones gastrin and pituitary adenylyl cyclase-activating peptide. Unlike **enterochromaffin cells** do not produce serotonin.

Enterococcus Genus of Gram-positive cocci that occur singly, in pairs, or in short chains. They are facultative anaerobes and live mostly in the digestive tract. Most enterococcal infections of humans are due to *E. faecalis*.

enterocyte Epithelial cell of the intestinal wall.

enteron The body cavity of Cnidaria, corresponding to the **archenteron** of a gastrula.

enterostatin The N-terminal pentapeptide (VPDPR) cleaved from **procolipase**, suppresses fat intake after peripheral and central administration. Enterostatin alters 5-HT release in the brain, and 5-HT$_B$ receptor antagonists block the anorectic response to enterostatin. Release of enterostatin varies in a circadian fashion in some animals.

enterotoxins Group of bacterial **exotoxins** that act on the intestinal mucosa. By perturbing ion and water transport systems they induce diarrhoea. **Cholera toxin** is the best known example.

enterovirus A genus of **Picornaviridae** that preferentially replicate in the mammalian intestinal tract. It includes the **polioviruses** and **Coxsackie viruses**.

Entner-Doudoroff pathway Metabolic pathway for degradation of glucose in a wide variety of bacteria. Differs from the Embden-Meyerhoff pathway although end result is similar.

entomophthoromycosis A rare fungal disease of the tropics in which there is a chronic subcutaneous form of zygomycosis. It is caused by zygomycetes of the order Entomophthorales: *Conidiobolus coronatus*, *Conidiobolus incongruous*, and *Basidiobolus ranarum*.

entopic Developed or located in the normal anatomical location (opposite of ectopic).

entrainment The process whereby an endogenous clock-driven rhythm is synchronized with an external environmental rhythm. See **zeitgeber**.

ENTs Members of a family of integral membrane proteins (equilibrative nucleoside transporters) with 11 transmembrane domains. These sodium-independent nucleoside transporters are widely distributed in eukaryotes; typical inhibitors of mammalian ENTs are nitrobenzylmercaptopurine ribonucleoside, dilazep, and dipyridamole. Adenosine flux across cardiomyocyte membranes occurs mainly via equilibrative nucleoside transporters. PMAT (SLC29A4, ENT4, 530aa) is a Na$^+$-independent and membrane potential-sensitive transporter that transports monoamine neurotransmitters and the neurotoxin 1-methyl-4-phenylpyridinium (MPP$^+$) and may be a polyspecific organic cation transporter. See **CNTs** (concentrative nucleoside transporters).

enucleation Removal of the nucleus of a cell.

env Retroviral gene encoding viral envelope glycoproteins.

envelope (1) Lipoprotein outer layer of some viruses that is derived from plasma membrane of the host cell. (2) In bacteriology, the plasma membrane and cell wall complex of a bacterium.

envoplakin The transglutaminase cross-linked protein (2033aa) layer (cornified envelope) deposited under the plasma membrane of keratinocytes in outer layer of skin. Has sequence homology with **desmoplakin**, bullous pemphigoid antigen 1, and **plectin**.

ENY2 A component of the **SAGA** complex (enhancer of yellow 2 transcription factor homologue, 101aa) where it is part of a subcomplex that specifically deubiquitinates histones H2A and H2B.

enzyme induction An increase in enzyme secretion in response to an environmental signal. The classic example is the induction of β-galactosidase in *E. coli.*

enzyme nomenclature See **E classification.**

eosin A red dye used extensively in histology, for example in the standard 'H & E' (haematoxylin and eosin) stain used in routine pathology.

eosinophil Polymorphonuclear leucocyte (granulocyte) of the myeloid series, in which the granules stain red with eosin. Phagocytic, particularly associated with helminth infections and with hypersensitivity.

eosinophil cationic protein Arginine-rich protein (EC 3.1.27.-, ribonuclease 3, 160aa) in granules of eosinophils, that damages schistosomula *in vitro.* Not the same as the MBP (major basic protein) of the granules.

eosinophil chemotactic peptide Tetrapeptides (ECF, ECF of anaphylaxis: VGSE, AGSE) released by mast cells and said to attract and activate eosinophils. Not mentioned in current literature.

eosinophilia Condition in which there are unusually large numbers of **eosinophils** in the circulation, usually a consequence of helminth parasites or allergy.

eosinophilic (1) Having affinity for the red dye **eosin.** (2) Describing an inflammatory lesion characterized by large numbers of **eosinophils.**

eosinophilopoietin Obsolete name for a substance that stimulated eosinophil production from bone marrow. Now known to be **interleukin-5** (IL-5).

eotaxin Chemokine (CCL11, 97aa) originally thought to be eosinophil-specific but subsequently shown to attract IL-2- and IL-4-stimulated T-cells which also express the CCR3 receptor.

epalons Class of neuroactive steroids that are positive allosteric modulators of GABA via a neurosteroid site on the GABA$_A$ receptor/Cl$^-$ ion channel complex. Name derived from epiallopregnanolone, an endogenous metabolite of progesterone.

epaxial Describing muscles which are dorsal to the horizontal septum of the vertebra, cf. hypaxial muscles which lie ventral to the septum and constitute the majority of the muscullature including the diaphragm, the abdominal muscles, and all limb muscles.

ependymal cells Cells that line cavities in the central nervous system, considered to be a type of glial cell.

Eph-related receptor tyrosine kinases The largest known family of receptor tyrosine kinases, implicated in the control of axonal navigation and fasciculation and in vascular assembly. They are receptors for **ephrins** which are themselves cell-surface anchored and respond to binding to the receptor (signalling events occur in cells with receptors and in cells with ligands). Ephrin type-A receptor 1 (976aa) binds ephrin-A family ligands (GPI anchored), ephrin type-B receptor 2 (EPH-like kinase 5, renal carcinoma antigen NY-REN-47, tyrosine-protein kinase TYRO5, tyrosine-protein kinase receptor EPH-3, 1055aa) binds type B ephrins (integral membrane proteins). There are multiple forms of both type-A and type-B receptors.

ephedra A naturally occurring plant-derived substance (Ma huang) in which the principal active ingredient is **ephedrine**. Ephedra supplements are considered a health risk. Health risks of ephedra: http://nccam.nih.gov/news/alerts/ephedra/consumer-advisory.htm

ephedrine Alkaloid from plants of genus *Ephedra.* Structural analogue of epinephrine (**adrenaline**) the effects of which it mimics.

Ephemeroptera An order of insects (Mayflies) in which the adult life is very short and the mouthparts are reduced and functionless; the immature stages are active aquatic forms.

ephexins Rho guanine nucleotide exchange factors that link **ephrin** receptors (**Eph-related receptor tyrosine kinases**) with Rho GTPases and thus regulate actin remodeling in 'forward signalling'. Ephexin-1 (Eph-interacting exchange protein, 710aa) differentially activates RhoA, Rac1 and cdc2. Its DH and PH domains are both required to mediate interaction with the ephrin receptor, EPHA4. Other ephexins have specificity for different rho GTPases; ephexin-5 (RhoGEF-15, 841aa) acts on RhoA but not Rac1 or cdc42 and regulates vascular smooth muscle contractility.

ephrins Ligands of the **Eph-related receptor tyrosine kinases**, transmembrane molecules on other cells that respond to being bound by the receptor: both receptor and ligand are altered by the binding interaction and so signalling is bidirectional. Ephrins of class 'A' are GPI-anchored and bind to EphA receptors, ephrins of class B are transmembrane and bind to EphB receptors and to EphA4. Type A ephrins presumably change their association with other signalling molecules in the membrane, type B become tyrosine phosphorylated. Ephrin B1 (formerly Elk-L/Lerk2) is a ligand for the EphB2 receptor (formerly Nuk/Cek5/Sek3). ELF-2 (Eph ligand family 2, ephrin-B2 in mouse, 336aa) is another such ligand that binds to three closely related Eph family receptors, Elk, Cek10 (apparent ortholog of Sek-4 and HEK2), and Cek5 (apparent ortholog of Nuk/Sek-3).

epi- Prefix indicating something on, above or near. Epi-illumination is from above, epithelia cover (are on top of) other tissues.

epi-lipoxins Anti-inflammatory 5R-epimers of **lipoxins** formed *in vivo* in the presence of aspirin.

epibenthos Organisms living on the floor of a sea or lake.

epiblast The outer germinal layer of a metazoan embryo that gives rise to the ectoderm.

epiboly The process in early embryonic development in which a monolayer of dividing cells (blastoderm) spreads over the surface of a large yolk-filled egg (e.g. those of teleosts, reptiles and birds).

epicatechin Flavonoid found in chocolate, thought to be beneficial for blood vessel function.

epichromosomal Genetic material, for example an adenovirus used as a vector, that does not become integrated into the host chromosomes but proliferates in tandem. Use of such vectors avoids the risk of activating host genes in an inappropriate fashion.

epicotyl The first shoot of a plant embryo or seedling, above the point of insertion of the cotyledon(s). Can be relatively long in some seedlings showing **etiolation**.

epidemic An outbreak of infectious disease that spreads rapidly and affects a high proportion of susceptible people in a region as opposed to an endemic disease which is constantly present at low level. There is no universal set of criteria but complex rules for defining influenza epidemics are based upon time-series analysis of health services-based indicators collected on a weekly basis by a surveillance network. The theoretical requirements, in terms of infectivity, transmission rates, suceptibility etc., for epidemic spread are well-understood. Also used as an adjective. See **pandemic**. Influenza epidemics: http://www.ncbi.nlm.nih.gov/pubmed/9663521

epidermal cell (1) Cell of epidermis in animals. (2) Plant cell on the surface of a leaf or other young plant tissue, where bark is absent. The exposed surface is covered with a layer of cutin.

epidermal growth factor A mitogenic polypeptide (EGF, 53aa) initially isolated from male mouse submaxillary gland. The name refers to the early bioassay, but EGF is active on a variety of cell types, especially but not exclusively epithelial. A family of similar growth factors are now recognised. Human equivalent originally named **urogastrone** owing to its hormone activity. See **EGF receptor**.

epidermal hair See trichome.

EPIDERMAL PATTERNING FACTORs Morphogens (EPF1 and EPF2, ~120aa) that inhibit the development of stomata in the epidermis of leaves. They bind to the receptor TOO MANY MOUTHS.

epidermin See **lantibiotics**.

epidermis Outer epithelial layer of a plant or animal. May be a single layer that produces an extracellular material (as for example the cuticle of arthropods), or a complex stratified squamous epithelium, as in the case of many vertebrate species. In plants aerial tissues may be covered by a waxy cuticle that minimizes water-loss and in roots, the root tip is covered by a specialized root cap that aids penetration of the soil. During secondary growth of plants the epidermis becomes replaced by a cork layer produced by the cork cambium (**phellogen**).

epidermodysplasia verruciformis A rare dermatosis associated with a high risk of skin cancer in which there is an abnormal susceptibility to specific related human papillomavirus (HPV) genotypes. Caused by mutations in either of 2 adjacent genes on chromosome 17, those for transmembrane channel-like proteins 6 (TMC6, 805aa) and 8 (TMC8, 726aa).

epidermolysis bullosa A very rare condition in which the skin and internal body linings are very susceptible to blistering. Most cases of epidermolysis bullosa simplex (EBS) are associated with mutations of the genes coding for keratins 5 and 14 (intermediate filaments of basal keratinocytes). A range of other types (junctional EB, dystrophic EB, hemidesmosomal EB) arise from defects in dermal extracellular matrix proteins (collagen, laminin, plectin) or proteins that bind to them (e.g. integrins). Epidermolysis bullosa acquisita is different, being a chronic autoimmune condition with IgG autoantibodies against the noncollagenous (NC1) domain of type VII collagen.

epididymis Convoluted tubule connecting the vas efferens, that comes from the seminiferous tubules of the mammalian testis, to the vas deferens. Maturation and storage of sperm occur in the epididymis.

epifluorescence Method of fluorescence microscopy in which the excitatory light is transmitted through the objective onto the specimen rather than through the specimen; only reflected excitatory light needs to be filtered out rather than transmitted light which would be of much higher intensity.

epigen One of the epidermal growth factor (EGF) superfamily (152 aa), a highly mitogenic ligand for the EGF-receptors, ErbB1 and ErbB2 and a potent epithelial mitogen implicated in wound healing. The precursor is a transmembrane protein that is cleaved by the metallopeptidase ADAM17 to release the soluble bioactive ectodomain. Epigen processing: http://www.ncbi.nlm.nih.gov/pubmed/17169360

epigenesis The theory that development is a process of gradual increase in complexity as opposed to the preformationist view that supposed that mere increase in size was sufficient to produce adult from embryo. See **epigenetics**.

epigenetics The study of mechanisms involved in the production of phenotypic complexity in morphogenesis. According to the epigenetic view of differentiation, the cell makes a series of choices (some of which may have no obvious phenotypic expression, and are spoken of as **determination** events) that lead to the eventual differentiated state. Thus, selective gene repression or derepression at an early stage in differentiation will have a wide-ranging consequence in restricting the possible fate of the cell. See **epigenomics**.

epigenomics Epigenetic effects are mediated by either chemical modifications of the DNA itself (e.g. methylation) or by modifications of proteins that are closely associated with DNA (chromatin structure). Many of these epigenetic changes alter the whole genome and need to be considered holistically, rather than in isolation: epigenomics is the analysis of genome-wide consequences of epigenetic modifications. The Human Epigenome Project is a joint effort by an international collaboration that aims to identify, catalog and interpret genome-wide DNA methylation patterns of all human genes in all major tissues. Homepage of Epigenome project: http://www.epigenome.org/index.php

epiglycanin Very extensively glycosylated transmembrane glycoprotein originally found in mouse mammary carcinoma cells. The human orthologue (mucin 21, 566aa) is expressed by normal bronchial epithelial cells and by adenocarcinomas of the lung. It may reduce cell adhesion. First report of human orthologue: http://glycob.oxfordjournals.org/content/18/1/74.long

epilepsy A disorder in which there are recurrent seizures and sometimes loss of consciousness. The severity varies (grand mal, Jacksonian epilepsy, **MERRF syndrome myoclonic epilepsy of Unverricht and Lundborg**, petit mal, amongst others). See **doublecortin**.

epiligrin See **kalinin**.

epilithic Descriptor for an organism that grows on the exposed surface of a rock.

epilysin Matrix metallopeptidase 28 (520aa): may have a role in tissue homeostasis and repair.

epimers Diastereomeric molecules (stereoisomers that are non-superposable, non-mirror images of one another), that differ in configuration at only one stereogenic centre.

epimorphosis Pattern of regeneration in which proliferation precedes the development of a new part. Opposite of **morphallaxis**.

epinasty Asymmetrical growth of a leaf or stem that causes curvature of the structure.

epinemin Obsolete name for an intermediate filament-like protein (44.5 kDa monomer) associated with **vimentin** in non-neural cells. Last literature reference probably 1984.

epinephrine See **adrenaline**.

epiphysis (1) Region at the end(s) of long (limb) bones that is ossified separately and only becomes united with the main portion (diaphysis) of the bone once maturity is reached and no further growth in height will occur. The outer region of compact bone is relatively thin compared to that in the diaphysis and the remainder is composed of trabecular bone with red marrow. (2) In Echinoidea, one of the ossicles of Aristotle's lantern. (3) The pineal body (epiphysis cerebri). *Adj.* epiphysial.

epiphytotic A widespread outbreak of plant disease; the botanical equivalent of an epidemic.

epiplakin A member (5090aa) of the **plakin** family of cytolinker proteins originally identified as a human epidermal autoantigen. Somewhat atypical, having 65 plectin repeats but none of the other domains found in other plakins. Its function is unclear but it is probably involved in cross-linking intermediate filaments and linking them to other cytoskeletal elements. A knock-out mouse does not have major skin abnormalities. Reported to accelerate keratin bundling in proliferating keratinocytes during wound healing. Knock-out mouse: http://www.ncbi.nlm.nih.gov/pmc/articles/PMC1346887/?tool = pubmed

epiregulin A member of the epidermal growth factor family (secreted as 49aa from a 169aa membrane precursor) that is a ligand for ErbB1 and ErbB4 homodimers and all possible heterodimeric ErbB complexes although binds EGFR with low affinity. Stimulates the proliferation of fibroblasts, hepatocytes, smooth muscle cells, and keratinocytes but inhibits the growth of several tumour-derived epithelial cell lines. Original description: http://www.jbc.org/content/275/8/5748.full

episialin Heavily glycosylated membrane glycoprotein (mucin-1, polymorphic epithelial mucin, PEM, epithelial membrane antigen, EMA, epitectin, 1255aa) located in the apical domain of the plasma membrane of highly polarized epithelial cells where it may inhibit bacterial binding. A MUC1/GRB2/SOS1 complex is involved in ras signalling The extracellular domain may extend hundreds of nanometres beyond the plasma membrane; the increased expression in carcinoma cells may reduce the adhesion and mask antigenic properties of the cells. Similar functions are ascribed to **ASGP**, **epiglycanin** and **leukosialin**. Membrane mucins: www.springer.com/cda/content/document/.../9780896037205-c2.pdf

episodic ataxia A range of disorders in which there are intermittent episodes of ataxia. Some types have known causes, mostly mutations ion channels: In Type 1 the mutation is in the potassium channel gene KCNA1, in Type 2 mutation in the calcium ion channel gene CACNA1A, in Type 5 by mutation in CACNB4. Type 6 involves a mutation in the SLC1A3 glutamate-transporter gene.

episome Piece of hereditary material that can exist as free, autonomously replicating DNA or be attached to and integrated into the chromosome of the cell, in which case it replicates along with the chromosome. Examples of episomes include many **bacteriophages** such as lambda and the male sex factor of *E. coli*.

epistasis Non-reciprocal interaction of non-allelic genes – for example when the expression of one gene masks the expression of another. Thus a gene that blocks development of an organ will mask the effects of genes that would modify the form of that organ had it been developed.

epistasy See **epistasis**.

epitectin Synonym for **episialin** that has fallen out of use.

epithelial membrane antigen See **episialin**.

epithelial mesenchymal transition A process (EMT, epithelio-mesenchymal transition) in which epithelial cells assume a mesenchymal phenotype, a key event occurring during normal development and pathological processes including metastasis. Features include: the downregulation of cell adhesion molecules such as E-cadherin, the increased expression of matrix metallopeptidases to assist in the degradation of the basement membrane, the activation of the Rac/Rho/Cdc42 family leading to cytoskeletal rearrangement and the nuclear translocation of several transcription factors including β-catenin and the T-cell factor/lymphocyte enhancer factor 1 (TCF/LEF1) complex, **Snail1**, Snail2, and twist. The process is associated with stimulation by various growth factors such as IGF-1 and TGFβ-1. See **ZEB** and **interleukin-like EMT inducer**. Review: http://www.jci.org/articles/view/31200

epithelioid cells (1) Generally, a cell that has an appearance that is similar to that of epithelial cells. (2) More specifically, the very flattened macrophages found in granulomas (e.g. in tubercular lesions).

epithelioma A malignant tumour derived from epithelium, more commonly referred to as a carcinoma.

epithelium One of the simplest types of tissues. A sheet of cells, one or several layers thick, organized above a basal lamina (see **basement membrane**), and often specialized for mechanical protection or

active transport. Examples include skin, and the lining of lungs, gut and blood vessels.

epitope That part of an antigenic molecule to which the T-cell receptor responds or a site on a large molecule against which an antibody will be produced and to which the antibody will bind. See also **agretope**. Epitopes in proteins involve 4 or 5 amino acids (not necessarily linked in a simple sequence, it can depend upon tertiary structure) and in carbohydrates 3–5 sugars. There may be multiple epitopes on a large molecule and monoclonal antibodies directed against a single epitope can sometimes indicate, for example, important regions for interacting with other molecules. See **epitope mapping**, **phage display library**.

epitope library See **phage display library**.

epitope mapping The identification and definition of the **epitope** recognised by an antibody or T-cell receptor. Various methods can be used including synthetic peptides (where the sequence of the protein is known), phage display libraries, protein footprinting (using monoclonal antibody to protect the protein from proteolytic degradation), isolation and characterisation of the peptide bound to MHC, or expression cloning. Since some epitopes may involve glycosylation or other post-translational modifications to proteins the process is not necessarily straightforward.

epitope tag Short peptide sequence that constitutes an **epitope** for an existing antibody. Widely used in molecular biology to 'tag' transgenic proteins (as a translational fusion product) to follow their expression and fate by immunocytochemistry or Western blotting, but without having to raise antibodies against the specific protein. Example: **myc tag**. See also **flag tagging**.

epitrichium *periderm* *Adj.* epitrichial. The outer layer of the epidermis in an embryo or fetus, usually shed post-partum.

epizootic The veterinary equivalent of an epidemic disease.

eplin A LIM domain protein (epithelial protein lost in neoplasm, Lima-1, 759aa) that binds to actin monomers and filaments, increases the number and size of actin stress fibers and inhibits membrane ruffling. It links the cadherin-catenin complex to F-actin. There are two isoforms, an abundant alpha isoform and a less highly expressed beta isoform with different tissue distribution. Eplin interaction with cadherin-catenin: http://www.pnas.org/content/105/1/13.long

EPO See **erythropoietin**.

epothilones Compounds (epothilone A and B) isolated from myxobacterium *Sorangium cellulosum* Str 90. Cytotoxic to tumour cells as a result of inducing microtubule assembly and stabilization in a manner similar but not identical to that of **paclitaxel**. Mechanism of action: http://jco.ascopubs.org/content/22/10/2015.full

epp See **end plate potential**.

eppin Epididymal protease inhibitor (serine protease inhibitor-like with Kunitz and WAP domains 1, 133aa), expressed in epididymis and testis.

Eps15 An adaptor protein (896aa) that is involved in epidermal growth factor (EGF) receptor endocytosis and trafficking. It is phosphorylated by EGF receptor tyrosine kinase and the two may be brought together by ubiquitin (c-**cbl** ubiquitin-ligase activity is required for recruitment and co-localisation of EGFR and Eps15 in the endosomal compartment). See EH domain (Eps15 homology domain), **epsins**. Review: http://www.biosignaling.com/content/7/1/24

EPSC Excitatory postsynaptic current.

epsilon-aminocaproic acid An inhibitor (εACA, EACA, 6-aminohexanoic acid) of the plasmin-plasminogen system. Acts as a lysine side-chain mimic.

epsins A family of proteins involved in regulating endocytosis (**Eps15**-interacting proteins; epsin-1, 551aa; epsin-4, clathrin interactor 1, enthoprotin, 625aa) with an epsin N-terminal homology (ENTH) domain that binds phosphoinositides and a poorly structured C-terminal region that interacts with **ubiquitin** and the endocytic machinery, including **clathrin** and endocytic scaffolding proteins. Epsin 1 is an integral component of clathrin coats forming at the cell surface. A range of epsin-like proteins have been identified in animals, plants and fungi.

EPSP See **postsynaptic potential**.

Epstein-Barr virus Species of Herpetoviridae, that binds **CR2** and that causes infective mononucleosis (glandular fever) and, in the presence of other factors, tumours such as **Burkitt's lymphoma** and nasopharyngeal carcinoma. Review on EB virus entry into cells: http://jvi.asm.org/cgi/content/full/81/15/7825

equatorial plate Region of the mitotic spindle where chromosomes are aligned at metaphase: as its name suggests, it lies midway between the poles of the spindle.

equilibrative nucleoside transporters Integral membrane proteins with 11 transmembrane domains that transport nucleosides in a sodium-independent fashion. In humans there are four members of solute carrier family 29 (SLC29A1, 456aa, SLCA2, 456aa; SLCA3, 475aa; SLCA4, 530aa). ENT1 (SLC29A1) is mostly localized in the basolateral membrane in polarized MDCK cells. Defects in SLC29A3 are the cause of H syndrome

(hyperpigmentation, hypertrichosis, hepatosplenomegaly, heart anomalies and other problems beginning with 'H'). ENT4 is the plasma membrane monoamine transporter that transports various organic cations including monoamine neurotransmitters. Inhibitors of mammalian ENTs include the anticancer drugs dilazep and dipyridamole. See **concentrative nucleoside transporters**.

equilibrium constant *equilibrium dissociation constant, dissociation constant.* The ratio of the reverse and forward rate constants for a reaction of the type $A + B = AB$. At equilibrium the equilibrium constant (K) equals the product of the concentrations of reactants divided by the concentration of product, and has dimensions of concentration. $K =$ (concentration $A \times$ concentration B)/(concentration AB). The affinity (association) constant is the reciprocal of the equilibrium constant.

equilibrium dialysis Technique used to measure the binding of a small molecule ligand to a larger binding partner. The macromolecule is contained within a dialysis chamber and the diffusible ligand added to the exterior: once equilibrium is reached an excess of ligand inside the dialysis chamber is evidence of binding and it is possible to calculate the binding affinity from a measurement of the concentrations of ligand and that of the binding macromolecule.

equinatoxins Small peptide toxins (actinoporins, ~200aa) from *Actinia equina*. Form cation-selective pores of around 1 nm and causes cardiac stimulation and haemolysis.

equisetin Fungal metabolite with antibiotic and cytotoxic activity. Inhibitor of mitochondrial ATPases and HIV-1 integrase. Isolated from *Fusarium equiseti*.

equivalence The situation where two interacting molecular species are present in concentrations just sufficient to produce occupation of all binding sites. Only used to describe high avidity interactions, especially the antibody/antigen interaction.

ER See **endoplasmic reticulum**.

ERAB A mitochondrial enzyme (endoplasmic reticulum associated binding protein, EC 1.1.1.178, hydroxyacyl-CoA dehydrogenase II, 17-beta-hydroxysteroid dehydrogenase X, short-chain 3-hydroxyacyl-CoA dehydrogenase, amyloid beta-binding alcohol dehydrogenase, 261aa) that has a role in mitochondrial tRNA maturation, being part of mitochondrial ribonuclease P which cleaves tRNA molecules in their 5'-ends. By interacting with intracellular amyloid-beta, it may contribute to the neuronal dysfunction associated with Alzheimer's disease. Defects lead to various forms of mental retardation.

erabutoxins Polypeptide toxins (83aa) from venom of *Laticauda semifasciata* (Black-banded sea krait). Bind to nicotinic acetylcholine receptors and have an effect similar to curare.

erb Oncogenes, *erb A* and *erb B*, associated with avian erythroblastosis virus (an acute transforming retrovirus). The products of human cellular homologues of *erb B* (c-ErbB1, 1210aa; c-ErbB2, HER2/neu, 1255aa) are tyrosine kinase receptors for members of the epidermal growth factor family (see **EGF receptor**), c-ErbB3 (HER3, 1342aa) is activated by neuregulins and NTAK. c-ErbB4 (HER4, 1308aa) specifically binds and is activated by neuregulins but not by EGF, TGFα-A or amphiregulin. The products of *c-ErbA* are steroid hormone receptors.

ERBIN A protein (ErbB2 receptor-interacting protein, LAP2, densin-180-like protein, 1412aa) that binds to **p0071** and **Erb** B2; co-localized with **PAPIN** on the lateral membrane of epithelial cells. Acts as an adapter for the receptor ErbB2, and may restrict it to the basolateral membrane domain. Also found in hemidesmosomes.

erbstatin A compound isolated from *Streptomyces* MH435-hF3 that inhibits the autophosphorylation of the EGF receptor: erbstatin itself is unstable in serum but analogues are potent cell-permeable inhibitors of most protein kinases. **Tyrphostins** are derived from erbstatin.

ERC proteins See **CAST**.

ergastic substances Metabolically inert products of photosynthesis, such as starch grains and fat globules.

ergocalciferol Synonym for **calciferol**.

ergodic System or process in which the final state is independent of the initial state.

ergosterol A sterol found in ergot, yeasts and other fungi where it serves the same function as cholesterol. Does not occur in higher plants or animal cells, but is the most important precursor of vitamin D2 (provitamin D2, ercalciol) into which it is converted by the action of ultraviolet light on the skin.

ergot Fungal (*Claviceps purpurea*) infection of rye (*Secale cornutum*); the so-called ergot that replaces the grain of the rye is a dark, purplish sclerotium, from which the sexual stage, of the lifecycle will form after over-wintering. The fungus produces a mycotoxin (see **ergotamine**) that contaminates rye flour and causes ergotism (St. Anthony's Fire.). Various bizarre behavioural symptoms caused by ergot poisoning were ascribed, in the Middle Ages, to witchcraft. A serious toxin for domestic animals that are fed on cereals, particularly rye.

ergotamine An alkaloid produced by *Claviceps purpurea* infecting rye (see **ergot**). It is structurally similar to several neurotransmitters and is a vasoconstrictor. Can be used to treat migraine - it causes constriction of intracranial blood vessels by acting as an agonist at 5-HT_{1B} receptors and blocking effects of 5-HT_{1D} receptors. Side effects, however, arise through interaction with dopamine and noradrenaline receptors.

ERKs See **MAP kinases**.

ERM (1) Ezrin/radixin/moesin (ERM) proteins, involved in linking plasma membrane proteins to the cortical actin meshwork. E.g. erm-1 (566aa) from *C. elegans*. See **FERM domain**. (2) A member of the Ets family of transcription factors, (Erm, 510aa). (3) Genes (*ermA, ermB* etc.) coding for dimethyl- or methyl-transferases that confer drug resistance on many pathogenic bacteria. The transferases show absolute specificity for nucleotide A2058 in 23S rRNA; monomethylation at A2058 confers resistance to a subset of the macrolide, lincosamide, and **streptogramin** B (MLS(B)) group of antibiotics, dimethylation at A2058 confers high resistance to all MLS (B) and ketolide drugs.

E-rosetting Technique used to identify T-cells in leucocyte mixtures. Sheep erythrocytes (E) will bind to form a 'rosette' around T-cells because they express a LFA-3 homologue on their surfaces which binds to CD2. Erythrocytes from other species do not have the necessary surface ligand. EA-rosettes will form around B-cells which have Fc receptors that bind antibody (A) which is bound to antigen on the erythrocyte surface. Antibody-coated erythrocytes (EA) with bound complement (C) will bind to cells that have C3b or C3bi receptors (**CR1** or **CR3**) to form EAC-rosettes.

Errera's rule That the cell plate in plants adopts the same area-minimizing geometry as a soap film. This can be extended to explain asymmetrical divisions that give rise, for example, to guard cells. The cellular equivalent of the surface tension of soap films that drives their behaviour is probably dependent on actin microfilament meshworks. Recent developments: http://the-scientist.com/2011/08/01/plant-cells-and-soap-bubbles/

error-prone repair See **SOS system**.

erucic acid Trivial name for 22:1 fatty acid. Found in rape-seed (canola) oil.

ERV See endogenous retroviruses.

Erwinia chrysanthemi Phytopathogenic bacterium that causes soft-rot. Virulence factors include pectinases coded by *pelB, pelC, pelD, pelE, ogl, kduI* and *kdgT* that degrade the cell walls of the plant being attacked.

Eryf1 See **GATA transcription factors**.

erythraemia See erythrocytosis.

erythritol A polyol found in various fungi and algae, probably as a storage carbohydrate.

erythroblast Rather non-committal name for a nucleated cell of the bone marrow that gives rise to erythrocytes. See also **normoblast, burst forming unit, colony-forming unit, primitive** and **definitive erythroblasts**.

erythroblastosis fetalis Severe haemolytic disease of the neonate as a result of transplacental passage of maternal antibodies mainly directed against Rhesus blood group antigens.

erythrocyte A red blood cell. Cell that is specialized for oxygen transport, having a high concentration of **haemoglobin** in the cytoplasm (and little else). Biconcave, anucleate discs, very deformable, ca 7 μm diameter in mammals; nucleus contracted and chromatin condensed in other vertebrates.

erythrocyte ghost The membrane and cytoskeletal elements of the erythrocyte devoid of cytoplasmic contents, but preserving the original morphology. See Table E1.

erythrocyte membrane proteins A well-studied group of proteins easily isolated from erythrocyte ghosts. The nomenclature was based upon the position of bands on an SDS polyacrylamide gel with the largest, Band 1 (protein 1) being at the top. See Table E1.

erythrocyte sedimentation rate A standard haematological test that measures the rate of settling of erythrocytes in anti-coagulant treated blood. Sedimentation rates are more rapid in inflammation because fibrinogen levels are higher causing increased formation of rouleaux (stacked erythrocytes).

erythrocytosis An increase in the number of circulating differentiated red blood cells (erythraemia, polycythaemia. *US* erythremia, polycythemia). 'Erythrocytosis' implies that the cells are fully differentiated whereas in **polycythemia vera** this is not necessarily the case. Familial erythrocytosis-1 is caused by mutations in the **erythropoietin** receptor. Familial erythrocytosis-2 is caused by mutation in the **VHL tumour suppressor gene**.

erythrogenic toxin Toxin produced by strains of *Streptococcus pyogenes* that cause scarlet fever and that carry a lysogenic bacteriophage. Three antigenic variants of the toxin are known. It is a small protein (251aa), a **superantigen** that is complexed with hyaluronic acid and can intensify the effects of other toxins such as **endotoxin** and **streptolysin O**.

erythroid cell Cell that will give rise to erythrocytes.

TABLE E1. Erythrocyte membrane proteins

Band number[a]	MW (kDa)		Other name or function
1	240	2419aa	Spectrin α
2	220	2137aa	Spectrin β
2.1	200	1881aa	Ankyrin. Links band 3 to spectrin
3	93	911aa	Anion transporter (**SLC4A1**), **CD233**
4.1	82	864aa	Links spectrin to glycophorin
4.2	72	691aa	Stabilizes link between ankyrin and Band 3
4.5	46		Mixed nucleoside and glucose transporters
4.9	48	*405aa*	*Dematin:* bundles microfilaments
5	43	375aa	Actin; forms short oligomers. involved in gelation of spectrin and band 4.1
6	35	335aa	Glyceraldehyde 3-phosphate dehydrogenase
7	32	288aa	Stomatin (Band 7.2): missing in hereditary stomatocytosis

The mammalian erythrocyte ghost consists of a lipid bilayer linked to a cytoskeletal network. The proteins of the ghost vary across species, but there are some common patterns. Components are identified as far as possible by comparison with the proteins of the human erythrocyte ghost, after electrophoretic separation on SDS polyacrylamide gel, and numbered according to the Steck classification. (*J. Cell Biol.* 1974. 62, 1−29)

[a]These bands are visible when the gel is stained with a typical "protein" dye, e.g., Coomassie brilliant blue. Other bands are only detected when stained for carbohydrate with the Periodic Acid/Schiff reagent (PAS). Four bands are characterized:- PAS1, PAS2, PAS3 and PAS4. Of these PAS1 and PAS2 are the glycoprotein glycophorin (55kD) in different oligomeric states. PAS3 and PAS4 are minor components.

erythroid Krüppel-like factor Red cell-specific transcriptional activator (EKLF, 362aa) essential for establishing high levels of adult beta-globin expression.

erythroid transcription factor A general and rather unhelpful term for any transcription factor that binds to regulatory regions of genes expressed in erythroid cells. Important examples are GATA-1 (see **GATA transcription factors**) and **erythroid Krüppel-like factor**, but many are now known.

erythrokeratodermia variabilis A dermatosis caused by defects in **connexin-31** (gap junction beta-3 protein, GJB3, 270aa) which is also defective in a form of deafness.

erythroleukaemic cell Abnormal precursor (virally transformed) of mouse erythrocytes that can be grown in culture and induced to differentiate by treatment with, for example, DMSO. See **Friend murine erythroleukaemia cells**.

erythromelalgia Primary erythermalgia (erythermalgia, acromelalgia, Mitchell's disease, red neuralgia) is a disorder in which there is episodic redness and burning pain of the feet and lower legs caused by mutation in the SCN9A gene that codes for the alpha subunit of the Type 9 voltage-gated sodium channel (which is also defective in **congenital insensitivity to pain**). Secondary erythromelalgia is associated with thrombocythemia or myeloproliferative disorders.

erythromycin General name for a variety of widespectrum macrolide antibiotics isolated from *Saccharopolyspora erythraea* (formerly *Streptomyces erythraeus*). Inhibit protein synthesis by binding to the prokaryotic 50S ribosomal subunit and preventing translocation. A variety of proteins will confer resistance to erythromycin — either by degrading the antibiotic, by enhancing its export from the cell, or by causing modification to the RNA so that its affinity for erythromycin is reduced.

erythrophores **Chromatophores** that have red pigment.

erythropoiesis Process of production of erythrocytes in the marrow in adult mammals. A pluripotent stem cell (colony forming unit, CFU) produces, by a series of divisions, committed stem cells (**burst forming units** -erythrocytic, BFU-Es) that give rise to **CFU-Es**, cells that will divide only a few more times to produce mature erythrocytes. Each stem cell product can give rise to 211 mature red cells.

erythropoietin Glycoprotein hormone (EPO, 193aa) that regulates the process of **erythropoiesis** but also has cardioprotective and neuroprotective effects. EPO is mostly produced in the kidney in response to low oxygen levels which activate hypoxia-inducible transcription factors. Recombinant EPO was one of the first biopharmacuticals to be used therapeutically in patients. The receptor (508aa) is one of the type I cytokine receptor family; EPO binding leads to dimerization which triggers the JAK2/STAT5 signalling cascade. Defects in the receptor cause familial erythrocytosis-1.

erythropterin A red pterine pigment deposited in the epidermal cells or the cavities of the scales and setae of many insects.

ES cells See **stem cell**.

Esa1 (1) The catalytic component (445aa) of the **NuA4** histone acetyltransferase (HAT) multisubunit complex responsible for acetylation of histone H4 and H2A N-terminal tails in yeast. It is essential for cell cycle progression, gene-specific regulation and has been implicated in DNA repair. Almost all NuA4 subunits have clear homologues in higher eukaryotes, suggesting that the complex is conserved throughout evolution. (2) Epidermal surface antigen-1 (**flotillin-2**, 428aa).

Escherichia coli The archetypal bacterium for biochemists, used very extensively in experimental work. A rod-shaped Gram-negative bacillus ($0.5 \times 3-5\,\mu m$) abundant in the large intestine (colon) of mammals. Normally non-pathogenic, but the *E. coli* O157 strain, common in the intestines of cattle, has recently caused a number of deaths.

Escherichia coli haemolysins The α-haemolysin (HlyA, 1024aa) is a secreted virulence factor of the **RTX family** produced by many *E. coli* strains responsible for non-intestinal uropathogenic infections. The inactive precursor requires fatty acid acylation by accessory protein HlyC to be effective and will then form pores in a calcium-dependent manner that cause cytolysis. Activation of HlyA: http://www.medmicro.wisc.edu/labs/welch/publications/papers/200011_Lim_Hackett_JBiolChem.pdf

ESCRT complex A set of multi-protein complexes (endosomal sorting complex required for transport, ESCRT-0, I, II and III) that constitute a major pathway for the lysosomal degradation of mono-ubiquitinylated transmembrane proteins, and are critical for receptor downregulation, budding of the HIV virus and other normal and pathological cell processes. The ESCRT system is conserved from yeast to man. In *S. cerevisiae* a subset of the vacuolar protein sorting (VPS) proteins, the class E VPS proteins, are involved in the formation of **multivesicular bodies** and at least 17 different E-VPS proteins have been identified as being involved in forming ESCRT complexes. The homologues in humans are proteins like **TSG101**, **EAP30** and charged multivesicular body proteins (**CHMPs**). Overview: http://www.ncbi.nlm.nih.gov/pmc/articles/PMC1648078/

esculin A **coumarin** derivative extracted from the bark of flowering ash (*Fraxinus ornus*). Esculin is used in the manufacturing of pharmaceuticals and in diagnostic microbiology: Group D streptococci hydrolyze esculin to esculetin and dextrose. Esculetin reacts with an iron salt such as ferric citrate to form a blackish-brown coloured complex.

E selectin See **selectins**.

eserine An alkaloid (physostigmine) that has anticholinesterase activity, isolated from the Calabar bean (*Physostigma venenosum*).

E-site Site on the ribosome that binds deacylated tRNA after it leaves the **P-site** and prior to it leaving the ribosome. See also **A-site**.

ESKIMO1 A key gene involved in plant water economy as well as cold acclimation and salt tolerance. Mutations in *ESK1* cause altered expression of transcription factors and signalling components and of a set of stress-responsive genes. The set of 312 genes regulated by ESK1 overlap with sets of genes regulated by salt, osmotic and abscisic acid treatments rather than with genes regulated by cold acclimation or by the transcription factors CBF3 and ICE1, which have been shown to control genetic pathways for freezing tolerance. The wild-type *ESK1* gene encodes a 487aa protein and is a member of a large gene family of DUF231 domain proteins, mostly of unknown function. Recent paper: http://www.ncbi.nlm.nih.gov/pmc/articles/PMC3052256/?tool = pubmed

ESL **E-selectin** ligand-1 (ESL-1, Golgi apparatus protein 1, cysteine-rich fibroblast growth factor receptor, Golgi sialoglycoprotein MG-160, 1179aa).

espin An actin-bundling protein (ectoplasmic specialization protein, 854aa) that binds actin monomer via a WH2 domain. Important in regulating the organization, dimensions, dynamics and signalling capacities of the microvilli that mediate sensory transduction in various mechanosensory and chemosensory cells. Defects in espin cause a form of deafness. Espin-like protein (1005aa) lacks the WH2 domain although, like espin, it has 9 ankyrin repeats.

ESR See **erythrocyte sedimentation rate**. N.B ESR is also used for electron spin resonance (see **electron paramagnetic resonance**).

essential amino acids Those amino acids that cannot be synthesized by an organism and must therefore be present in the diet. The term is often applied anthropocentrically to those amino acids required by humans (Ileu, Leu, Lys, Met, Phe, Thr, Try and Val), though rats need two more (Arg and His).

essential fatty acids The three fatty acids required for growth in mammals, arachidonic, linolenic and linoleic acids. Only linoleic acid needs to be supplied in the diet; the other two can be made from it.

EST See **expressed sequence tag**.

established cell line See **cell line**.

esterase An enzyme that catalyses the hydrolysis of organic esters to release an alcohol or thiol and acid. The term could be applied to enzymes that hydrolyze carboxylate, phosphate and sulphate esters, but is more often restricted to the first class of substrate.

estradiol US name and **BAN** for oestradiol. Female sex hormone (272 Da) synthesized mainly in the ovary, but also in the placenta, testis, and possibly adrenal cortex. A potent **estrogen**.

estrogen A type of hormone (UK oestrogen now outmoded) that induces oestrus ('heat') in female animals. It controls changes in the uterus that precede ovulation, and is responsible for development of secondary sexual characteristics in pubescent girls. Some tumours are sensitive to estrogens. There are two main estrogen receptors, ER-alpha (595aa) and ER-beta (530aa) that are nuclear receptors, ligand-activated transcription factors which move from cytoplasm to nucleus once estrogen or estradiol is bound. Additional isoforms, generated by alternative mRNA splicing, have been identified in several tissues. There is also a G-protein coupled membrane receptor (chemoattractant receptor-like 2, flow-induced endothelial G-protein coupled receptor, 1375aa). See **estradiol**.

estrogen-responsive B box protein Protein (EBBP, tripartite motif-containing protein 16, 564aa) a retinoic acid receptor-beta2 transcriptional regulator. May regulate keratinocyte differentiation. Retinoid receptor association: http://www.ncbi.nlm.nih.gov/pubmed/19147277

ET See **endothelin**.

ethacrynic acid A loop diuretic (acts on the ascending limb of the loop of Henle and on the proximal and distal tubules); inhibits reabsorption of a much greater proportion of filtered sodium than most other diuretic agents. Inhibits glutathione S-transferase and signalling by NFκB. Used experimentally as a chloride pump blocker.

ethephon A compound that will release ethylene and is used as a plant growth hormone source.

ethidium bromide A fluorescent dye that intercalates into DNA and to some extent RNA. Intercalation into linear DNA is easier than into circular DNA and the addition of ethidium bromide to DNA prior to ultracentrifugation on a caesium chloride gradient was used to separate nuclear and mitochondrial or plasmid DNA for analytical purposes.

Because less intercalates into the circular DNA, the density remains higher. May be mutagenic.

ethmoid cells Small sinuses between the eyes, not cells in the cell-biological sense.

ethylene A gas at room temperature, the simplest alkene hydrocarbon. Acts as a plant growth substance (phytohormone, plant hormone), involved in promoting growth, **epinasty**, fruit ripening, senescence and breaking of dormancy. Its action is closely linked with that of **auxin**. Experimentally, ethylene inhibits the elongation of plant cells.

ethylmalonic encephalopathy A neurodevelopmental delay syndrome characterized by ethylmalonic and methylsuccinic aciduria caused by mutation in the *ETHE1* gene, which encodes a mitochondrial matrix protein (254aa), also found in the cytoplasm and which may function as a nuclear-cytoplasmic shuttling protein that binds transcription factor RelA.

etiolation Growth habit adopted by germinating seedlings in the dark. Involves rapid extension of shoot and/or hypocotyl and suppression of chlorophyll formation and leaf growth.

etiology The study of causation, commonly of the cause of disease.

etioplast Form of **plastid** present in plants grown in the dark. Lacks chlorophyll, but contains chlorophyll precursors and can develop into a functional chloroplast in the light.

etoposide Semi-synthetic lignan derivative synthesized from **podophyllotoxin**. Used as an anti-tumour drug; works by inhibiting topoisomerase II.

ets An oncogene originally found in E26 transforming retrovirus of chickens which will induce both myeloid and erythroid leukemias. The founder member of the ETS family of transcription factors (which includes Elfs and Elks). In humans c-ets1 (441aa) and c-ets2 (469aa) are nuclear proteins that regulate the initiation of transcription from a range of promoter and enhancer elements. ETS homologous factor (ETS domain-containing transcription factor, epithelium-specific Ets transcription factor 3, 300aa) is a transcriptional activator that may play a role in regulating epithelial cell differentiation and proliferation. ETS-related transcription factor Elf-1 (619aa) activates the LYN and BLK promoters and is one of several elf transcription factors (Elf2, 593aa; Elf3, 371aa; Elf4, 663aa; Elf5, 265aa). There are also ETS domain transcriptional repressors, for example ETS translocation variant 3 (PE1, mitogenic Ets transcriptional suppressor, 521aa) that is involved in growth arrest during terminal macrophage differentiation.

ETS domain DNA binding domain ($\sim$80aa), formed of three **alpha-helices** found in ets family of transcription factors. See **ets**. Database entry: http://

www.pdb.bnl.gov/scop/data/scop.1.001.004.003.009. html

Eubacteria At one time a name given to a subdivision of the prokaryotes (all except **Archaea**) now generally **Bacteria**. See also **Taxonomic Outline of Bacteria and Archaea**.

Eucaryote See **Eukaryota**.

euchromatin The chromosomal regions that are diffuse during interphase and condensed at the time of nuclear division. They show what is considered to be the normal pattern of staining (eu = true) as opposed to **heterochromatin**.

Eudorina Simple multicellular alga of the Order Volvocida — often quoted as illustrating the path to multicellularity. Small spherical or ovoid colonies of between 4–64 flagellated cells coexist within a gelatinous envelope. *Pandorina* and *Volvox* are similar though more complex.

Euglena *Euglena gracilis* and *E. viridis* are phytoflagellate protozoa of the algal order Euglenophyta (zoological order Euglenida). An elongate cell with two **flagella**, one emerging from a pocket at the anterior end, the organism exhibits positive **phototaxis**, determined by a photoreceptive spot on the basal part of the flagellum shaft being shielded by a carotenoid-containing stigma, 'eyespot' in the wall of the pocket.

euglenoid movement A type of movement shown typically by *Euglena*, that swims with a single flagellum moving in a screw-like fashion but also shows writhing of the body caused by sub-pellicular contractions of cytoplasmic filament networks which can produce co-ordinated peristalsis-like movements in some cases.

Eukaryota One of the major subdivisions of living organisms. Eukaryotic cells have linear DNA organised into chromosomes with nucleosomal structure involving histones, the nucleus separated from the cytoplasm by a two-membrane envelope and compartmentalisation of various functions in distinct cytoplasmic organelles. A range of other characteristics distinguish them from the prokaryotes (see **Bacteria** and **Archaea**).

eukaryotic initiation factor A multi-protein initiator factor complex (43S) that is involved in binding the small (40S) ribosomal subunit to mRNA, moving along the mRNA until a start site (AUG) is found and then binding the large (60S) subunit, a process that depends upon GTP hydrolysis. Once assembled the complete 80S riboosome proceeds to translate the mRNA into protein. eIF-1 is a low molecular weight factor critical for stringent AUG selection, eIF-2 has three subunits: alpha, beta (a guanine nucleotide exchange factor), and gamma and forms a ternary complex with Met-tRNAi and

GTP. eIF-3 contains at least 8 distinct polypeptides and plays a role in recycling of ribosomal subunits. eIF-4 is trimeric (eIF-4A, eIF-4E and eIF-4G) and associates with the 5′ cap of mRNA. eIF-5 interacts with the 40S ribosomal initiation complex and promotes the hydrolysis of bound GTP and subsequent release of eIF-3 from the 40S subunit. The 40S subunit is then able to interact with the 60S ribosomal subunit to form the functional 80S initiation complex.

eumelanin See **melanin**.

Eumycetozoida Order of Protozoa, includes true slime moulds (not the cellular slime moulds).

Eumycota Division of fungi having defined cell walls and forming hyphae. The other main group is the **Myxomycota**.

Euonymus lectin family A family of carbohydrate-binding proteins (158aa) from *Euonymus europaeus* (European spindle tree). There is a conserved structural domain (EUL domain) that is also found in some stress-induced plant proteins. Cloning: http://www.uniprot.org/citations/18451263

eupeptide bond Peptide bond formed between the α-carboxyl group of one amino acid and the α-amino group of another; the common peptide bond of proteins.

euploid A cell or an individual with a complete set of chromosomes.

euploidy Polyploidy in which the chromosome number is an integer multiple of the starting number.

Euplotes Genus of free living hypotrich Protozoa. Do not have cilia but may have undulating membranes for propulsion.

euryhaline Descriptive term for marine organisms that will tolerate a wide variation in salinity.

eurytopic Descriptive term for organisms that are able to survive in a wide range of environmental conditions.

eutely Phenomenon exibited by a few phyla, notably nematodes, where all individuals have the same number of cells (or nuclei in a coenobium).

Eutypa lata A fungal pathogen of grapes (*Vitis vinifera*) present in all grape growing areas and responsible for important economic losses.

Evans blue A diazo dye that binds to albumin and is commonly used to estimate blood volume (knowing the amount of dye injected and the concentration in a sample taken after the dye has distributed throughout the blood the total volume is easily calculated). Can also be used to demonstrate sites where there is leakage of plasma protein from blood vessels, for example in a site of inflammation.

even-skipped A **pair-rule gene** of *Drosophila*. Encodes a protein (eve, 376aa) that is involved in regulation of axonogenesis and is required for segmentation by restricting the products of **gap genes** to a precise periodic expression pattern. Flybase entry: http://flybase.org/reports/FBgn0000606.html

evening primrose oil An oil obtained from the seeds of *Oenothera biennis*, rich in gamma linolenic acid, the precursor for prostaglandin synthesis. Remarkable claims are made for the therapeutic value of this oil as a dietary supplement.

everninomicins Oligosaccharide antibiotics potently active against Gram-positive organisms and potentially useful to treat antibiotic resistant *S. aureus*. Several variants have been isolated from the fermentation broth of *Micromonospora carbonacea* var *africana*. Binds to the same site in the large 50S ribosomal subunit as **avilamycin**.

EVH1 domains A domain (Enabled/VASP homology 1 domain, WH1, RanBP1-WASP domain, ~115aa) found in multi-domain proteins implicated in a diverse range of signalling, nuclear transport and cytoskeletal events in species ranging from yeast to mammals. Many EVH1-containing proteins are involved in cytoskeletal organisation. EVH1 domains bind the proline-rich motif FPPPP with low-affinity. See **WH domains**.

Evi1 A transcriptional regulator (ecotropic virus integration site 1 protein homologue, MDS1 and EVI1 complex locus protein, 1051aa) involved in development, cell proliferation and differentiation. May also affect apoptosis through regulation of the JNK and TGFβ signalling. A chromosomal aberration involving EVI1 is a cause of chronic myelogenous leukemia.

evo-devo Evolutionary developmental biology (evo-devo), the study of developmental programs and patterns from an evolutionary perspective.

Ewing's sarcoma A highly malignant, metastatic, primitive small round cell tumour of bone and soft tissue. The tumour and various others are associated with chromosomal translocation of the EWS RNA-binding protein (Ewing sarcoma breakpoint region 1 protein, 656aa) to produce fusion proteins which are potent activators of transcription and are oncogenic. Abstract: http://www.ncbi.nlm.nih.gov/pubmed?term = 8084618.

exaptation Phenomenon in which a character or organ is used for a purpose other than the one for which it first evolved, for example the use of wings by penguins for swimming rather than flying.

excision repair Mechanism for the repair of environmental damage to one strand of **DNA** (loss of **purines** due to thermal fluctuations, formation of pyrimidine dimers by UV irradiation). The site of damage is recognized, excised by an **endonuclease**, the correct sequence is copied from the complementary strand by a **polymerase** and the ends of this correct sequence are joined to the rest of the strand by a **ligase**. The term is sometimes restricted to bacterial systems where the polymerase also acts as endonuclease. Deficiencies in the repair mechanism are responsible for **xeroderma pigmentosum, DeSanctis-Cacchione syndrome** and **Cockayne's syndrome**.

excitable cell A cell in which the membrane response to **depolarizations** is non-linear, causing amplification and propagation of the depolarisation (an **action potential**). Apart from neurons and muscle cells, electrical excitability can be observed in fertilized eggs, some plants and glandular tissue. Excitable cells contain **voltage-gated ion channels**.

excitation-contraction coupling Name given to the chain of processes coupling excitation of a muscle by the arrival of a nervous impulse at the **motor end plate** to the contraction of the filaments of the **sarcomere**. The crucial link is the release of calcium from the sarcoplasmic reticulum, and the analogy is often drawn between this and **stimulus-secretion coupling**, that also involves calcium release into the cytoplasm.

excitatory amino acid EAA The naturally occurring amino acids L-glutamate and L-aspartate and their synthetic analogues, notably **kainate, quisqualate**, and **NMDA**. They have the properties of excitatory neurotransmitters in the CNS, may be involved in long-term potentiation, and can act as **excitotoxins**. At least three classes of EAA receptor have been identified; the agonists of the N-type receptor are L-aspartate, NMDA, and ibotenate; the agonists of the Q-type receptor are L-glutamate and quisqualate; agonists of the K-type are L-glutamate and kainate. All three receptor types are found widely in the CNS, and particularly the telencephalon; N- and Q-type receptors tend to occur together, and may interact; their distribution is complementary to the K-type receptors. The ion fluxes through the Q and K receptors are relatively brief, whereas the flux through the N-type is longer, and carries a significant amount of calcium. Additionally the N-type receptor is blockaded by magnesium near the resting potential, and thus shows **voltage-gated ion channel** properties, leading to a regenerative response; this is why N-type receptors have been linked to long-term potentiation. Invertebrate glutamate receptors may have different properties.

excitatory synapse A synapse (either **chemical** or **electrical**) in which an action potential in the presynaptic cell increases the probability of an action potential occurring in the postsynaptic cell. See **inhibitory synapse**.

excitotoxin Class of substances that damage neurons through paroxysmal overactivity. The best known excitotoxins are the **excitatory amino acids**, that can produce lesions in the CNS similar to those of **Huntington's chorea** or **Alzheimer's disease**. Excitotoxicity is thought to contribute to neuronal cell death associated with stroke.

exein Protein exons: see **intein**.

exendin Group of peptide hormones, related to the glucagon family, found in the saliva of Gila monsters and have VIP/secretin-like biological activity. Exendin-1 (helospectin, 38aa) and exendin-2 (helodermin, 35aa) and exendin-4 (87aa) are from *Heloderma suspectum*. Exendin-3 (87aa) is from *H. horridum horridum*. They bind to the glucagon-like peptide-1 (GLP-1) receptor.

exergonic A biochemical reaction (exogonic reaction) in which there is a release of energy - a negative change in free energy that can be used to produce work. Such reactions will proceed spontaneously.

exfoliative toxins Epidermolytic toxins, (exfoliative toxin A, ET-A, 280aa; ET-B, 277aa) produced by some strains of *Staphylococcus aureus*. They have serine peptidase-like activity (EC = 3.4.21.-) and are of the S1B (chymotrypsin) family and are **superantigens**. They bind to **profilaggrin** and cause an impetigo-like condition called staphylococcal scalded skin syndrome (SSSS).

exine External part of pollen wall that is often elaborately sculptured in a fashion characteristic of the plant species. Contains **sporopollenin**. The term is also used for the outer part of a spore wall.

exobiology The study of putative living systems that, statistically, are likely to exist elsewhere in the universe.

exocrine Exocrine glands release their secreted products into ducts that open onto epithelial surfaces. See **endocrine**.

exocyst complex A multi-protein complex involved in spatial targeting and tethering of post-Golgi vesicles to the plasma membrane prior to their fusion and release of their contents to the exterior. In humans and yeast there are eight subunits (EXOC1-8, ranging from 725aa to 974aa) that interact with **centriolin**, rho-related GTP-binding proteins, GEFs and various other proteins. The yeast proteins are Sec3, Sec5, Sec6, Sec8, Sec10, Sec15, Exo70, and Exo84.

exocytosis Release of material from the cell by fusion of a membrane-bounded vesicle with the plasma membrane.

exocytotic vesicle Vesicle, for example a secretory vesicle or **zymogen granule**, that can fuse with the plasma membrane to release its contents.

exodus Beta-chemokines (exodus-1, CCL20, LARC, MIP3A, SCYA20, 96aa; exodus-2, CCL21, ECL, SLC, CKb9, TCA4, 6Ckine;,SCYA21, MGC34555, 134aa). Exodus-1 attracts lymphocytes but not monocytes, inhibits proliferation of myeloid progenitors in colony formation assays and may be involved in formation and function of the mucosal lymphoid tissues by attracting lymphocytes and dendritic cells towards epithelial cell. Binds CCR6. Exodus-2 inhibits haemopoiesis and is chemotactic *in vitro* for thymocytes and activated T-cells (particularly naive T-cells), but not for B-cells, macrophages, or neutrophils. May mediate homing of lymphocytes to secondary lymphoid organs. Binds to CCR7. Exodus-3 (CCL19, MIP3B, SCYA19, 98aa) also binds to CCR7 and is potently chemotactic for T-cells and B-cells but not for granulocytes and monocytes. It has an important role in trafficking of T-cells in thymus, and T-cell and B-cell migration to secondary lymphoid organs.

exoenzyme (1) An enzyme attached to the outer surface of a cell (an ectoenzyme) or released from the cell into the extracellular space. (2) An enzyme that only cleaves the terminal residue from a polymer (in contrast to an endoenzyme).

exogen (1) The stage of the life cycle of a hair follicle in which a hair exits the follicle. (2) An obsolete name for a plant that grows by means of a peripheral cambial layer. Basically, most dicotyledonous plants.

exoglycosidase Hydrolytic enzymes that cleave glycosidic bonds of terminal sugar moieties. Sequential exoglycosidase cleavage is used to sequence carbohydrates. Can also be used to refer to glycosidases that act on the exterior of a cell although the prefix ecto- is less ambiguous.

exogonic See **exergonic**.

exon In eukaryotes and Archaea, the sequences of DNA or of the RNA **primary transcript** that encode amino acid sequences in proteins. In the primary transcript neighbouring exons are separated by introns which are removed before the mature mRNA leaves the nucleus.

exon junction complex A complex of proteins that assembles on newly spliced RNA 20nt upstream of the exon-exon junction and is a central effector of messenger RNA functions. It is a dynamic structure consisting of a few core proteins and several more peripheral nuclear and cytoplasmic associated factors that join the complex only transiently either during EJC assembly or during subsequent mRNA metabolism. There are factors involved in mRNA export, cytoplasmic localization, and nonsense-mediated mRNA decay and it is claimed that the presence of an EJC enhances the translatability of mRNAs. Components include RNPS1 (RNA-binding protein with serine-rich domain 1, 305aa), Y14 (RNA-binding protein 8A, 174aa), SRm160, (serine/arginine repetitive matrix protein 1, 904aa), Aly/REF (THO complex subunit 4, 257aa) and Magoh (146aa).

exon shuffling Process by which the evolution of proteins with multifunctional domains could be accelerated. If exons each encoded individual functional domains, then **introns** would allow their recombination to form new functional proteins with minimal risk of damage to the sequences encoding the functional parts.

exon skipping Probably the commonest cause of alternative splicing in which, during RNA processing, one or more exons are omitted and the message is abbreviated accordingly. A recent estimate is that more than 1200 human genes exhibit exon skipping. Database: http://www.bioinfo.de/isb/2004/05/0021/main.html

exon trapping Technique for identifying regions of a genomic DNA fragment that are part of an expressed gene. The genomic sequence is cloned into an intron, flanked by two exons, in a specialized exon trapping vector, and the construct expressed through a strong promoter. If the genomic fragment contains an exon, it will be spliced into the resulting mRNA, changing its size and allowing its detection.

exonuclease Enzyme that digests the ends of a piece of DNA (*cf.* **endonuclease**). The nature of the digestion is usually specified (e.g. $5'$ or $3'$ exonuclease).

exopeptidase A peptidase of the class EC 3.4. that cleaves a peptide bond within three residues of the free N- or C-terminus of a peptide. Exopeptidases are further subdivided into aminopeptidases (EC 3.4.11), carboxypeptidases (EC 3.4.16-18), dipeptidyl- and tripeptidyl-peptidases (EC 3.4.14), peptidyl-dipeptidases (EC 3.4.15), dipeptidases (EC3.4.13) and acylaminoacyl-peptidases (EC3.4.19). Enzyme nomenclature: http://www.chem.qmul.ac.uk/iubmb/enzyme/

exosome (1) Antigen-presenting vesicle secreted by some **antigen-presenting cells**. Exosomes are membrane-bounded and enriched in MHC Class I and II proteins. Dendritic cells release exosomes that are sometimes referred to as dendritic cells are referred to as **dexosomes**. (2) The exosome (nuclear exosome) complex is involved in multiple RNA processing and degradation pathways and contains $3' \rightarrow 5'$ exoribonucleases. It is found in eukaryotes and archaea. The bacterial equivalent is the **degradosome**. (3) (*obsolete*) A DNA fragment taken up by a cell that does not become integrated with host DNA but nevertheless replicates.

exostosins Products of tumour suppressor genes (ext1, multiple exostoses protein 1, 746aa; ext2, 718aa) that are glycosyltransferases (EC 2.4.1.224, EC 2.4.1.225) required for the biosynthesis of heparansulphate. The two proteins form a hetero-oligomeric complex in the Golgi with greater enzymic activity than either alone. Exostosin-like 2 (EC 2.4.1.223, 330aa) is processed to release a soluble form from the membrane precursor. Exostosin-like 3 (919aa) is probably also a glycosyl transferase. Mutations in EXT1 or EXT2 are associated with multiple exostes (projections of bone capped by cartilage) and tumours. Heparan sulphate proteoglycan biosynthesis is important for signal transduction of Indian hedgehog (IHH), Wnt, and TGFβ. N.B. Ext is also used in naming of **extensin**. Role in chondrosarcoma: http://www.ncbi. nlm.nih.gov/pubmed/19179614

exothermic Process or reaction in which heat is produced — the opposite of endothermic.

exotoxins Toxins released from Gram-positive and Gram-negative bacteria in contrast to **endotoxins** that form part of the cell wall. Examples are **cholera, pertussis** and **diphtheria toxins**. Usually specific and highly toxic. See Table E2.

exp6 See exportins.

expansins A superfamily of plant cell wall-loosening proteins (all ca 250aa) that has been divided into two major groups, the expansin A (alpha-expansin) and expansin B (beta-expansin) families. In *Arabidopsis* there are 26 different α-expansin genes and 6 β-expansin genes. They facilitate cell expansion by selectively weakening the cell wall, although it appears that this is achieved through a non-enzymatic mechanism. Some β-expansins in grass pollen are group 1 pollen allergens.

experimental allergic encephalomyelitis An autoimmune disease (EAE) that can be induced in various experimental animals by the injection of homogenized brain or spinal cord in **Freund's adjuvant**. The antigen appears to be **myelin basic protein**, and the response is characterized by focal areas of lymphocyte and macrophage infiltration into the brain, associated with demyelination and destruction of the blood-brain barrier. Sometimes used as a model for demyelinating diseases, although whether this is entirely justifiable is not clear.

exponential phase A period during the microbial growth cycle when cells are growing and dividing at a maximum rate.

exportins Family of proteins within the **karyopherin** superfamily (that includes **importins**). Importins and exportins are both regulated by the small GTPase Ran, which is thought to be highly enriched in the nucleus in its GTP-bound form. Exportins interact with their substrates (proteins with **nuclear export signals**) in the nucleus in the presence of RanGTP and release them after GTP hydrolysis in the cytoplasm, causing disassembly of the export complex. Exportin 1 is the same as chromosome region maintenance-1 (**CRM1**, 1071aa). Exportin-6 (exp6, 1125aa) seems to be specific for

profilin-actin complexes and for maintaining the nucleus actin-free, exportin-T (962aa) mediates the nuclear export of aminoacylated tRNAs. **Leptomycin B** inhibits export by binding to exportins and preventing the binding interaction with the nuclear export signal.

expressed sequence tag A DNA sequence (500−800bp) derived from a reverse transcript of mRNA (cDNA) and thus can reasonably be supposed to have been translated into protein. The ESTs from a cDNA library provide an indication of the genes being expressed in the source cells at that time although some may be derived from mRNA that will be rapidly degraded and some will be fragments from a single mRNA. Thousands of ESTs are generated as part of **genome projects**. Approximately 8.3 million human ESTs were available in public databases in 2011, 1.5 million from *Arabidopsis*. EST database: http://www.ncbi.nlm.nih.gov/dbEST/

expression cloning Method of **gene cloning** based on **transfection** of a large number of cells with **cDNAs** in an **expression vector** (e.g. a cDNA library), then screening for a functional property (e.g. binding of a radiolabelled hormone to identify receptors, or induction of transforming activity for putative oncogenes).

expression profiling Shorthand for gene expression profiling, a procedure usually carried out using DNA microarrays to obtain an insight into tissue- and developmental-specific expression of genes and the response of gene expression to environmental stimuli.

expression vector A **vector** that results in the **expression** of inserted DNA sequences when propagated in a suitable host cell, i.e. the protein coded for by the DNA is synthesized by the host's system.

extended-spectrum beta-lactamases See **beta-lactamase**.

extensins Glycoproteins of the plant cell wall, characterized by having a high hydroxyproline content. Carbohydrate side-chains are composed of simple galactose residues and oligosaccharides containing 1−4 arabinose residues. Part of a larger class of hydroxyproline-rich glycoproteins **HGRPs**. At least 20 are known in *Arabidopsis*, for example, leucine-rich repeat extensin-like protein 1 (LRR/ EXTENSIN1, 744aa) that regulates cell wall formation and assembly in root hairs. Other extensins are involved in different plant tissues.

external guide sequence RNA oligonucleotides (EGSs) that serve as an RNA catalyst or ribozyme by directing bound mRNA to the ubiquitous cellular enzyme **RNAse P**. By virtue of their complementarity, they are mRNA-specific.

TABLE E2. Representative bacterial exotoxins

Name	Source	Target/mode of action
Aerolysin	*Aeromonas hydrophila*	Pore-forming
α-toxin	*Clostridium perfringens*	Phospholipase C
α-toxin	*Staphylococcus aureus*	Pore forming
Anthrax toxin	*Bacillus anthracis*	Three components, one a soluble adenyl cyclase
Bacteriocins	*Plasmid in E.coli*	Colicin E2 is a DNase, colicin E3 an RNase.
Botulinum toxins	*Clostridium botulinum*	Inhibits acetylcholine release
Botulinolysin	*Clostridium botulinum*	Cholesterol binding
Cereolysin	*Bacillus cereus.*	Cholesterol binding
Cholera toxin	*Vibrio cholerae*	ADP-ribosylation of G_s
Cytolethal distending toxin	*Campylobacter jejuni, Haemophilus ducre*	Genotoxin producing lesions in DNA, causes apoptosis
Cytotoxic necrotising factors	*Escherichia coli, Yersinia pseudotuberculosis*	Rho proteins
Diphtheria toxin	*Corynebacterium diphtheriae*	ADP-ribosylation of EF-2
δ-toxin	*Clostridium perfringens*	Cholesterol binding
β, ε and ι -toxins	*Clostridium perfringens*	Increase vascular permeability
Enterotoxin	*Staphyloccus aureus*	Neurotoxic
Enterotoxin	*Pseudomonas aeruginosa*	Causes diarrhoea
Erythrogenic toxin	*Streptococcus pyogenes*	Skin hypersensitivity (causes scarlet fever)
Exfoliatin	*Staphyloccus aureus*	Disrupts desmosomes
Haemolysins $\alpha, \beta, \chi, \delta$	*Staphyloccus aureus*	β is a sphingomyelinase C, γ is haemolytic, δ is a surfactant
Haemolysin	*Serratia marcescens*	Pore forming; different method to RTX toxins
Haemolysin	*Pseudomonas aeruginosa*	Toxic for macrophages
Heat-labile toxin	*Escherichia coli*	Similar to cholera toxin
Heat-stable enterotoxins	*Escherichia coli*	(1) Activates guanylate cyclase (2) Analogue of guanylin, acts via another receptor
Hyaluronidase	*Streptococcus pyogenes*	Digests hyaluronic acid in connective tissue
Kanagawa haemolysin	*Vibrio parahaemolyticus*	Haemolytic, cardiotoxic
κ-toxin	*Clostridium perfringens*	Collagenase
Leucocidin/alpha-toxin	*Staphyloccus aureus* and *Pseudomonas aeruginosa*	Lyses neutrophils and macrophages
Listeriolysin O	*Listeria monocytogenes*	Cholesterol binding
Perfringolysin (theta toxin)	*Clostridium perfringens*	Cholesterol binding
Pertussis toxin	*Bordetella pertussis*	ADP-ribosylates G_i
RTX family	Various Gram negative bacteria	Calcium-dependent pore forming toxins
Shiga toxin/verotoxin	*Shigella dysenteriae*	Blocks eukaryotic protein synthesis
Streptolysin D	*Streptococcus pyogenes*	Cholesterol binding
Streptolysin S	*Streptococcus pyogenes*	Membranolytic
Subtilysin	*Bacillus subtilis*	Haemolytic surfactant
Tetanolysin	*Clostridium tetani*	Cholesterol binding
Tetanus toxin	*Clostridium tetani*	Inhibits glycine release at synapse
Thuringolysin	*Bacillus thuringiensis*	Cholesterol binding
Toxic shock syndrome toxin	*Staphylococcus aureus*	Systemic shock
Toxin A	*Pseudomonas aeruginosa*	ADP-ribosylates EF-2

G_s, G_i: see GTP-binding proteins
EF-2: elongation factor 2

external transcribed spacers See **internal transcribed spacers**.

extinction coefficient Outmoded term for **absorption coefficient**.

extracellular matrix Any material produced by cells and secreted into the surrounding medium, but usually applied to the non-cellular portion of animal tissues. The extracellular matrix (ecm) of connective tissue is particularly extensive and the properties of the ecm determine the properties of the tissue. In broad terms there are three major components: fibrous elements (particularly **collagen**, **elastin**, or **reticulin**), link proteins (e.g. **fibronectin**, **laminin**), and space-filling molecules (usually **glycosaminoglycans**). The matrix may be mineralised to resist compression (as in bone) or dominated by tension-resisting fibres (as in tendon). The basal lamina of epithelial cells is another commonly encountered ecm. Although ecm is produced by cells, it has recently become clear that the ecm can influence the behaviour of cells quite markedly, an important factor to consider when growing cells *in vitro*: removing cells from their normal environment can have far-reaching effects.

extrachromosomal element Any heritable element not associated with the chromosome(s). It is usually a **plasmid** or the DNA of organelles such as mitochondria and chloroplasts.

extreme halophiles Organisms that can withstand extreme salt concentrations such as those found in salt lakes (and preserved food). There are six genera of Archaea that are classified as extreme halophiles. They are aerobic chemoheterotrophs which require complex energy and carbon sources (particularly, proteins and amino acids) and their minimum osmotic requirements exceed 1.5 M NaCl (8% wt/vol) although they will tolerate levels of up to 36% wt/vol).

extremophile An organism that requires an extreme enviroment in which to flourish – examples are **thermophiles** and **halophiles**.

extrinsic pathway See **tissue factor pathway**.

exudate The fluid which accumulates in tissues or in body cavities as a result of altered vascular permeability in the inflammatory response. Cells, particularly leucocytes, may also be present in large numbers. A peritoneal exudate induced in rabbits or mice by injection of saline with complement-activating glycogen is a source of large numbers of leucocytes, mostly neutrophils after a few hours but increasingly mononuclear cells at later stages.

Eya Class IV protein tyrosine phosphatases (EC 3.1.3.48, eyes absent homologues). Eya1 (592aa) specifically dephosphorylates Y142 of histone H2AX, promoting the recruitment of DNA repair complexes. Eya2 (538aa), Eya3 (573aa) and Eya4 (639aa) have similar activities. In *Drosophila* the eya protein (clift, 766aa) is required for the survival of eye progenitor cells at a critical stage in morphogenesis.

eyepiece graticule Grid (micrometer eyepiece, *US*. ocular micrometer) incorporated into the microscope eyepiece for measuring objects, often as a separate glass disc. May have any sort of scale and there are special type used in particle-size analysis consisting of a rectangular grid for selecting the particles and a series of graded circles for use in sizing. Need to be calibrated against a scale (micrometer slide) for each objective.

eyespot *stigma* An orange or red spot found near photoreceptive areas in motile cells of many algae, phytoflagellates and protozoa. Are assumed to help detect the direction of light in phototaxis. Pigments are carotenoids. See *Euglena*.

EZH Polycomb group (PcG) proteins (**enhancer of zeste** homologues) that are the catalytic component of the **polycomb repressive complexes** (PRC2/ EED-EZH complexes), histone-lysine N-methyltransferases (EC 2.1.1.43, EZH1, 747aa; EZH2, 746aa).

ezrin An actin binding protein (villin-2, cytovillin, p81, 586aa) that links the plasma membrane and microfilament bundles in the microvilli found in intestinal epithelium and elsewhere. See **FERM domain**.

F

F Single letter code for phenylalanine.

F1 hybrid First filial generation – product of crossing two dissimilar parents. If the parents are sufficiently dissimilar the hybrid may be sterile (for example in the crossing of horse and donkey to produce a mule) and the term F1 hybrid generally refers to such sterile hybrids which may, however, show desirable hybrid vigour.

F11 (1) In chickens, a neural cell recognition molecule (contactin-1, 1010aa) with **immunoglobulin** type C domains and **fibronectin** type III repeats. Like **neurofascin**, **TAG-1** and **fasciclin II**, thought to be associated with the process of **fasciculation**. (2) One of a family of proteins found in pox viruses. (3) Misleading abbreviation for blood coagulation factor XI.

Fab Fragment of immunoglobulin prepared by papain treatment. Fab fragments (45 kDa) consist of one light chain linked through a disulphide bond to a portion of the heavy chain, and contain one antigen binding site. They can be considered as univalent antibodies. Fab$_2$ fragments (90 kDa) are produced by pepsin treatment and consist of two linked Fab fragments. They are divalent but lack the complement-fixing (Fc) domain.

FABP See **fatty acid binding proteins**.

Fabry's disease X-linked **storage disease** (angiokeratoma) due to mutation in alpha-galactosidase (EC 3.2.1.22, also known as ceramide trihexosidase). The enzyme deficiency leads to inadequate breakdown of lipids, which accumulate to harmful levels in the eyes, kidneys, autonomic nervous system, and cardiovascular system.

facilitated diffusion A process (passive transport) by which substances are conveyed across cell membranes faster than would be possible by diffusion alone. This is generally achieved by proteins that provide a hydrophilic environment for polar molecules throughout their passage through the **plasma membrane**, acting as either shuttles or pores. See **symport**, **antiport**, **uniport**.

facilitation Greater effectiveness of synaptic transmission by successive presynaptic impulses, usually due to increased transmitter release.

facilitator neuron A neuron whose firing enhances the effect of a second neuron on a third. This allows the effects of neuronal activity to be modulated.

faciogenital dysplasia See **Aarskog-Scott syndrome**.

FACIT collagens Fibril Associated Collagens with Interrupted Triple helices; a set of collagens that do not form fibrils. Includes collagen types IX, XII, XIV, XIX and XXI.

FACS See **flow cytometry**.

FACT complex A stable heterodimeric complex of SSRP1 (structure-specific recognition protein 1, 709aa) and SUPT16H (chromatin-specific transcription elongation factor 140 kDa subunit, 1047aa) that facilitates chromatin transcription and is involved in multiple processes that require DNA as a template such as mRNA elongation, DNA replication and DNA repair. During transcription elongation the FACT complex acts as a histone chaperone that both destabilizes and restores nucleosomal structure.

F-actin Filamentous **actin**.

Factor B An important component of the alternative **complement** activation pathway (EC 3.4.21.47, C3/C5 convertase, properdin factor B, 764aa) that is cleaved by **factor D** into 2 fragments, Ba (234aa) and Bb (505aa). The latter has serine peptidase activity (S1 family) and combines with a cleaved fragment of C3 to form C3bBb, the C3 or C5 convertase. Ba inhibits the proliferation of preactivated B-lymphocytes. Defects in Factor B can cause an atypical form of haemolytic uremic syndrome.

Factor C (1) Limulus clotting factor C, a serine peptidase zymogen (EC 3.4.21.84, 1019aa) secreted into the haemolymph which initiates the coagulation cascade in horseshoe crabs (*Limulus polyphemus*, *Carcinoscorpius rotundicauda*). It is cleaved into four peptides, Factor C heavy chain (665aa), light chain (329aa), chain A (72aa) and chain B (257aa) and is activated by Gram-negative bacterial lipopolysaccharides and chymotrypsin. (2) In *S. cerevisiae* Replication Factor C (340aa) is a heteropentamer that is part of the ATP-dependent clamp loader (RFC and RFC-like) complexes for DNA clamps. (3) In humans, nuclear factor 1/C (508aa) is a CCAAT-box-binding transcription factor.

Factor D See **adipsin**.

Factor H A serum glycoprotein (complement factor H, CFH, beta-1H globulin, 1231aa) that is a cofactor in the inactivation of C3b by **factor I** and also increases the rate of dissociation of the C3bBb complex (C3 convertase) and the (C3b)NBB complex (C5 convertase) in the alternative complement pathway.

The Dictionary of Cell and Molecular Biology. DOI: http://dx.doi.org/10.1016/B978-0-12-384931-1.00006-4

TABLE F1. Blood clotting factors

Factor	Name	M_r (kDa)	Function
I	Fibrinogen	340	Cleaved to form fibrin
II	Prothrombin	70	Converted to thrombin by Factor X
III	Thromboplastin	—	Lipoprotein which acts with Vll to activate X
IV	Calcium Ions	—	Needed at various stages
V	Proaccelerin	—	Product accelerin, promotes thrombin production
Vll	Proconvertin	—	Activated by trauma to tissue
Vlll	Antihaemophilic factor	$>10^3$	Acts with IXa to activate X
IX	Christmas factor	55	See Vlll
X	Stuart factor	55	When activated converts II to thrombin
Xl	Thromboplastin antecedent	124	Converts IX to active form
Xll	Hagemann factor	76	Activated by surface contact
Xlll	Fibrin-stabilizing factor	350	Transglutaminase which cross-links fibrin

Deficiency of CFH allows uncontrolled activation of the alternate pathway which leads to renal damage, including membranoproliferative glomerulonephritis and atypical **haemolytic uremic syndrome**. Deficiency also increases susceptibility to meningococcal infections. Complement factor H-related protein 1, 330aa may be involved in complement regulation and can associate with lipoproteins. See **adrenomedullin**.

Factor I Complement factor I (EC 3.4.21.45, C3B/C4B inactivator, 583aa) that inactivates complement subcomponents C3b, iC3b and C4b by proteolytic cleavage. The precursor is cleaved to produce a heterodimer linked by disulphide bonds. Defects cause an atypical form of haemolytic uremic syndrome.

Factor P See **properdin**.

Factors I-XII Blood clotting factors, especially from humans. These factors form a cascade in which the activation of the first factor (they are numbered in approximately reverse order, so the last to be activated is factor I, fibrinogen) leads to enzymic attack on the next factor and so on, finally resulting in blood clotting. The proteolytic cascade provides amplification and rapid haemostasis and deficiencies cause various forms of **haemophilia**. Clotting can be triggered through the **contact activation pathway** (formerly called the intrinsic pathway), or the **tissue factor** pathway (formerly the extrinsic pathway) triggered by the trauma-induced production of tissue factor. See Table F1, **Hageman factor**, **protein-C**, **protein-S**, **von Willebrand factor**.

facultative heterochromatin That heterochromatin which is condensed in some cells and not in others, presumably representing stable differences in the activity of genes in different cells. The best known example results from the random inactivation of one of the pair of X chromosomes in the cells of female mammals, (**Lyonisation**).

FAD Flavin adenine dinucleotide, a prosthetic group of many flavin enzymes. See **flavin nucleotides**.

FADD Adaptor protein (Fas-associated via death domain, Mort1, 208aa) that links **death receptors** of the TNF receptor superfamily to **caspases** in the signalling pathway that leads to apoptotic cell death.

FAF1 Protein (Fas-associated factor 1, 650aa) involved in negative regulation of NFκB activation that interacts with the **pyrin domains** of several **PYPAFs**. Potentiates but cannot initiate FAS-induced apoptosis. FAS-associated factor 2 (UBX domain-containing protein 8, 445aa) may be involved in the translocation of misfolded proteins from the endoplasmic reticulum to the cytoplasm for degradation by the proteasome.

FAK See **focal adhesion kinase**.

Falconisation Trade name for the treatment of polystyrene to make it appropriate for use in cell culture by increasing its wettability and thus the ability of proteins and then cells to adhere. The main commercial process was probably corona discharge in air or other gas mixtures at low pressure. Treatment of polystyrene with sulphuric acid will produce the same effect. Now superceded by other methods.

Fallot's tetralogy See **tetralogy of Fallot**.

false positive A positive result in an assay that is due to something other than the effect of interest or the result of some other random factor. For example, inhibition of growth by general toxicity rather than inhibition of a growth-regulating pathway. To avoid such 'false positives' assays are often designed to give a positive readout if the inhibitor works.

familial adenomatous polyposis See **adenomatous polyposis coli**.

familial cold autoinflammatory syndrome
Rare systemic inflammatory diseases (familial cold urticaria) characterized by episodes of rash, arthralgia, fever and conjunctivitis after generalized exposure to cold. Type 1 is caused by mutation in **cryopyrin**, Type 2 is caused by mutation in NLRP12 (NACHT, LRR and PYD domains-containing protein 12).

familial haematuria Benign disorder in which there is persistent haematuria; caused by a defect in type IV collagen that thins the glomerular basement membrane.

familial Mediterranean fever An autoinflammatory disorder (FMF, familial paroxysmal polyserositis) caused by mutation in **pyrin**. Most affected individuals originate from Mediterranean or Middle-Eastern countries.

familial primary localized cutaneous amyloidosis A disorder associated with chronic skin itching and the deposition of epidermal keratin filament-associated amyloid material in the dermis, caused by a mutation in the **oncostatin M** receptor.

Fanconi's anaemia Defect in thymine-dimer excision from DNA predisposing to development of leukaemia. Caused by mutation in any one of the 13 Fanconi anaemia complementation group genes, *FANCA - FANCN*. The multisubunit nuclear FA complex is composed of FANCA, FANCB, FANCC, FANCE, FANCF, FANCG, FANCL/PHF9 and FANCM and is responsible for triggering monoubiquitination of the FANCD2 protein during S phase of the growth cycle and after exposure to DNA crosslinking agents. All bone-marrow derived cells are affected and there are diverse congenital malformations.

Fanconi's syndrome A transport disease (Fanconi-Bickel syndrome) caused by mutations in the *GLUT2* gene (which encodes solute carrier family 2, member 2, SLC2A2) which impairs the renal reabsorption of several substances (phosphate, glucose, amino acids).

far Western blot Form of Western blot in which protein/protein interactions are studied. Proteins are run on a gel and transferred to a membrane as in a normal **Western blot**. The proteins are then allowed to renature, incubated with a candidate protein, and the blot washed. Areas of the blot where the protein has adhered are then detected with an antibody.

FAR1 Yeast gene, induced by a factor, so called because it is a 'Factor ARrest' gene. Product is a cyclin-dependent protein kinase inhibitor (830aa) that causes cells to arrest in G1 phase, by interacting with the G1 cyclin, CLN2.

Farber's disease Lipogranulomatosis caused by deficiency of acid **ceramide** degrading enzyme (EC 3.5.1.23, N-acylsphingosine amidohydrolase, 395aa), a storage disease.

Farmer's lung Type III **hypersensitivity** response to *Micropolyspora faeni*, a thermophilic bacterium found in mouldy hay.

farnesyl transferase Enzyme of the endoplasmic reticulum (EC 2.5.1.21, squalene synthase, 417aa) that adds a farnesyl group to certain intracellular proteins. Also involved in biosynthesis of cholesterol, isoprenes, lipids, steroids and sterols. See **farnesylation**.

farnesylation The farnesyl group is the linear grouping of three isoprene units. It is post-translationally attached by **farnesyl transferase** to proteins that contain the C-terminal motif CAAX by cleavage and addition to the SH group of C; the free carboxylate group is also methylated. Believed to act as a membrane attachment device. See also **polyisoprenylation**.

Farr-type assay Method of radioimmunoassay in which free antigen remains soluble and antibody-antigen complexes are precipitated. Farr assays using PicoGreen DNA-binding dye have been used to detect levels of free anti-DNA in **systemic lupus erythematosus**. Article: http://onlinelibrary.wiley.com/doi/10.1046/j.1365-3083.2003.01261.x/pdf

Fas Cell surface transmembrane protein (Fas antigen, Fas receptor, CD95, TNFRSF6, 335aa), one of the TNF receptor superfamily, that binds **Fas ligand** and mediates apoptosis. May play a part in negative selection of autoreactive T-cells in the thymus.

Fas ligand The TNF superfamily ligand for **Fas** (TNFSF6, FASLG, 281aa), a type II transmembrane protein with three receptor-binding sites that induces trimerization of the receptor and induces apoptosis of the receptor-bearing cell. FASLG is expressed in activated splenocytes and thymocytes. A soluble N-terminal fragment (152aa) which has the receptor-binding extracellular domain is cleaved from the cell surface by the matrix metalloproteinase ADAM10 and acts as an antagonist. Defects in FASLG cause autoimmune lymphoproliferative syndrome type 1B (Canale-Smith syndrome).

Fas-associated factor 1 See **FAF1**.

Fas-associated via death domain See **FADD**.

fascicle Literally, a bundle. The tendency of **neurites** to grow together is bundle-formation (fasciculation) and involves **axon**-associated cell adhesion molecules such as **L1**, **F11**, **contactin**, **neurofascin** and transient axonal glycoprotein **TAG-1**. In insect nervous systems a related group of molecules, the **fasciclins**, are involved.

fasciclins Cell adhesion molecules of the **immunoglobulin superfamily** found in the central nervous system of insects. Involved with **fasciculation** of axons and probably in pathfinding during morphogenesis of the nervous system. The sequence of fasciclin-2 shows that it shares structural motifs with a variety of vertebrate **CAMs** such as **contactin**, **F11**, **neurofascin**, **TAG-1**, *Drosophila* **neuroglian**. In *Arabidopsis*, fasciclin-like arabinogalactan proteins (e.g. FLA1, 424aa; FLA18, 462aa) may be cell adhesion proteins.

fascicular cambium Form of **cambium** present in the vascular bundles of higher plants.

fasciculation (1) See **fascicle**. (2) A local involuntary muscle contraction of a small bundle (fascicle) of muscle fibres.

fascin Actin filament-bundling protein (493aa), the actin binding ability being regulated by phosphorylation; originally identified from sea-urchin eggs. An important structural component of microspikes, membrane ruffles, and stress fibres. Fascin-2 (492aa) is found in retinal photoreceptors and is defective in one form of retinitis pigmentosa, fascin-3 (498aa) is testis-specific.

FAST (1) Forkhead activin signal transducer, a winged helix transcription factor of the **forkhead** family, the DNA-binding component (FAST1, 365aa) of the **ARF** complex, binds to **activin** response element in *mix* gene promoter. (2) Fas-activated serine/threonine phosphoprotein (FAST1, 847aa) is the founding member of the FAST kinase domain-containing protein (FASTKD) family.

fast twitch muscle See **slow muscle**.

FASTA A bioinformatics resource, available on the web, that allows rapid DNA and protein sequence alignment and identifies matches in appropriate databases, making allowance for frameshifts and so on. Link: http://www.ebi.ac.uk/Tools/sss/fasta/

fasudil A selective inhibitor of **rho kinase** (ROCK1) that is a vasodilator and has some beneficial effects on memory.

fat cell See **adipocyte**.

fat droplets Microaggregates, mainly of triglycerides, visible within cells.

fatal familial insomnia A spongiform encephalopathy in which there is a mutation (D178N with methionine at position 129) in the **prion** protein gene; neuronal degeneration is limited to selected thalamic nuclei. A similar mutation (D178N) but with valine at position 129 is responsible for familial **Creutzfeldt-Jakob disease**.

fate map Diagram of an early embryo (usually a **blastula**) showing which tissues will arise from the cells in each region (i.e. their developmental fate). Fate maps are normally constructed by labelling small groups of cells in the blastula with vital dyes and seeing which tissues are stained when the embryo develops.

fats A term largely applied to storage lipids in animal tissues. The primary components are triglyceride esters of long-chain fatty acids.

fatty acid binding proteins Group of small cytosolic proteins (FABPs, ~130aa) that bind fatty acids and facilitate their intracellular transport. At least eight different types of human FABP occur, each with a specific tissue distribution and possibly with a distinct function.

fatty acid synthase In animal tissues a complex multifunctional enzyme consisting of two identical monomers. The FAS monomer (2511aa) catalyzes the formation of long-chain fatty acids from acetyl-CoA, malonyl-CoA and NADPH. The protein has 7 catalytic activities and an acyl carrier protein (ACP) role. The catalytic domains are: ACP-S-acetyltransferase (EC 2.3.1.38), ACP-S-malonyltransferase (EC 2.3.1.39), 3-oxoacyl-ACP synthase (EC 2.3.1.41), 3-oxoacyl-ACP reductase (EC 1.1.1.100), 3-hydroxypalmitoyl-ACP dehydratase (EC 4.2.1.61), enoyl-ACP reductase (EC 1.3.1.10) and oleoyl-ACP hydrolase (EC 3.1.2.14). In bacteria a range of different enzymes are involved, for example, in *E. coli*, cyclopropane-fatty-acyl-phospholipid synthase (EC 2.1.1.79, 382aa) and in plants there are some special enzymes involved in the synthesis of storage lipids, for example in *Arabidopsis*, very long-chain fatty acid condensing enzyme 18 (EC 2.3.1.119, 506aa). In *S. pombe*, the synthase is a heterodimer, the alpha subunit (1842aa) having acyl carrier, 3-oxoacyl-ACP reductase (EC 1.1.1.100) and 3-oxoacyl-ACP synthase (EC 2.3.1.41) activities, the beta (EC 2.3.1.86, 2073aa) having 3-hydroxypalmitoyl-ACP dehydratase (EC 4.2.1.61), enoyl-ACP reductase (EC 1.3.1.9), ACP-acetyltransferase (EC 2.3.1.38), ACP-malonyltransferase (EC 2.3.1.39) and S-acyl fatty acid synthase thioesterase (EC 3.1.2.14) activities.

fatty acids Chemically R-COOH where R is an aliphatic moiety. The common fatty acids of biological origin are linear chains with an even number of carbon atoms. Free fatty acids are present in living tissues at low concentrations. The esterified forms are important both as energy storage molecules and structural molecules. See **triglycerides**, **phospholipids**.

fatty streak Superficial fatty patch in the artery wall caused by the accumulation of cholesterol and cholesterol oleate in distended **foam cells**.

favism Haemolytic anaemia induced (probably by DOPA-quinone) in individuals who are glucose 6-phosphate dehydrogenase-deficient when they consume fava beans (from *Vicia fava*, the broad bean). G6PD deficiency is relatively common, probably because it confers some resistance to malaria.

FBI (1) Factor that binds to inducer of short transcripts protein 1 (FBI-1, zinc finger and BTB domain-containing protein 7A, 584aa), a protein that is involved in repressing T-cell instructive Notch signals and allowing early lymphoid progenitors to develop into the B-cell lineage. (2) In rats, fourteen-three-three beta interactant 1 (FBI-1, akirin-2, 201aa) a downstream effector of the toll-like receptor, TNF and IL-1β signalling pathways. (3) Bowman-Birk type proteinase inhibitor (FBI, 63aa) from *Vicia fava*.

F-box proteins Adapter proteins that are involved in associating proteins with the ubiquitin-driven proteolytic system. The F-box is a protein-protein interaction motif (~50aa) originally identified within *Neurospora crassa* negative regulator sulphur controller-2 but subsequently found in a wide variety of proteins including many cell cycle regulatory proteins, though various F-box proteins probably also play a part in regulation of transcription, signal transduction and development. See **SCF #2**.

Fc That portion of an immunoglobulin molecule (Fragment crystallizable) that binds to a cell when the antigen binding sites (**Fab**) of the antibody are occupied or the antibody is aggregated; the Fc portion is also important in **complement** activation. The Fc fragment can be separated from the Fab portions by pepsin. Fc moieties from different antibody classes and subclasses have different properties.

Fc receptors Receptors for the **Fc** portion of immunoglobulins. FcγRI (CD64, 374aa) is the receptor for IgG1, as are FcγRII-A (CD32, 317aa), FcγRII-B (CD32, 310aa), FcγRII-C (CDw32, 323aa) and FcγRIII (CD16A & B, 254aa & 233aa). The IgE Fc receptor (FcεR) is a tetramer of an alpha chain (257aa), a beta chain (244aa), and two disulfide linked gamma chains (86aa) with the gamma chains being involved in other classes of Fc receptor. FcαRI (CD89, 287aa) binds IgA. The distribution on cells varies and the consequences of receptor occupancy differ according to the sub-type. FcγRII-B has an inhibitory **ITIM** motif in the cytoplasmic domain. Fc receptor-like A (FcRX, 359aa) may be implicated in B-cell differentiation.

fccp A potent **uncoupling agent** of mitochondrial oxidative phosphorylation.

FCHo proteins F-BAR domain-containing Fer/Cip4 homology domain-only proteins (FCHO1, 889aa; FCHO2, 810aa) are nucleators of clathrin-mediated endocytosis, bind specifically to the plasma membrane and recruit the scaffold proteins **eps15** and intersectin, which in turn engage the adaptor complex AP2. Article: http://www.ncbi.nlm.nih.gov/pmc/articles/PMC2883440/?tool = pubmed

FCS See **fetal calf serum**.

Fechner's law *Weber-Fechner law* The proposition that sensory systems require a logarithmic change in the intensity of a stimulus to give a linear change in the perceived intensity. It is approximately true for a significant proportion of the sensitivity range of acoustic and photoreceptor systems and for weight.

Fechtner syndrome See **May-Hegglin anomaly**.

feedback regulation Control mechanism that uses the consequences of a process to regulate the rate at which the process occurs: if, for example, the products of a reaction inhibit the reaction from proceeding (or slow down the rate of the reaction), then there is negative feedback, something that is very common in metabolic pathways. Positive feedback is liable to lead to exponential increase and may be explosively dangerous in some cases. Other examples are the action of voltage-gated **sodium channels** in generating action potentials and the activation of blood clotting **factors V and VIII** by **thrombin**. Without damping, feedback can lead to resonance (hunting) and oscillation in the system.

feeder layer A layer of less fastidious cells used to condition the medium in order to culture some cell types, particularly at low density, as in cloning. The cells of the feeder layer are usually irradiated or otherwise treated so that they will not proliferate. In some cases the feeder layer may be producing growth factors or cytokines.

feedforward A control mechanism that is insensitive to the effects of a signal, the opposite of feedback system of control.

Feingold's syndrome A developmental disorder affecting various systems caused by mutation in the MYCN gene which encodes a transcription factor (N-*myc* proto-oncogene protein, 464aa) of the **myc** family.

feline immunodeficiency virus *FIV* Widespread lentivirus (retrovirus) that causes an immunodeficiency in domestic cats.The immunodeficiency may be due to failure to generate an IL-12-dependent Type I response. CXC-R4 and feline CD134 seems to be the surface receptors for viral binding; CD4 is not required in contrast with HIV infection and both CD4$^+$ and CD8$^+$ T-cells, B-cells and macrophages are infected.

feline sarcoma oncogene homologue See **fgr**.

Felty's syndrome A syndrome characterised by rheumatoid arthritis, splenomegaly and neutropenia. There are autoantibodies to elongation factor 1A-1 in many patients.

fengycins A family of antibiotics including the related plipastatin; decapeptides with a β-hydroxy fatty acid that have antifungal activity, although more specific for filamentous fungi. Useful as biocontrol agents agains plant pathogens. Article: http:// apsjournals.apsnet.org/doi/pdf/10.1094/MPMI-20-4-0430

fentanyl A powerful opioid analgesic resembling morphine in its action but approximately 80 times more potent as an analgesic. Agonist for mu **opiate receptors**.

FEP See **free erythrocyte protoporphyrin**.

ferlin A protein (2034aa) required for fusion of specialized vesicles with the sperm plasma membrane during spermiogenesis in *C. elegans*. Various mammalian homologues that mediate Ca^{2+}-dependent lipid-processing events have been identified. See **dysferlin**, **myoferlin** and **otoferlin**.

FERM domain Conserved domain of about 150aa found in Protein 4.1 (F), **ezrin** (E), **radixin** (R) and **moesin** (M) and a number of other cytoskeletal-associated proteins that link the cytoskeleton to proteins of the plasma membrane.

fermentation Breakdown of organic substances, especially by microorganisms such as bacteria and yeasts, yielding incompletely oxidized products. Some forms can take place in the absence of oxygen, in which case **ATP** is generated in reaction pathways in which organic compounds act as both donors and acceptors of electrons. Historically, the production of ethyl alcohol or acetic acid from glucose. Also applied to anaerobic **glycolysis** as in **lactate** formation in muscle.

fermitin See **kindlin**.

ferredoxins Low molecular weight iron-sulphur proteins that transfer electrons from one enzyme system to another without themselves having enzyme activity.

ferrichrome A cyclic hexapeptide (siderochrome) that forms a complex with iron atoms and is secreted by fungi to sequester and transport iron. The ferrichrome-iron complex is the ligand for a receptor found in fungi (in *S. cerevisiae*, siderophore iron transporter ARN1, ferrichrome permease, 627aa) and bacteria (ferrichrome ion receptor, FhuA, 747aa in *E. coli*). The latter also acts as a receptor for bacteriophages T1 and T5, and colicin M. Webpage: http://virtual-museum.soils.wisc.edu/ferrichrome/content.html

ferritin An iron storage protein of mammals, found in liver, spleen, and bone marrow, involved in iron detoxification and as a iron reserve. Morphologically a roughly spherical shell of 24 ferritin molecules (light chain, 175aa; heavy chain, 183aa) with a diameter of 12 nm having a central cavity into which the insoluble ferrous hydroxide/phosphate core is deposited. It is much used as an electron-dense label in electron microscopy. Cationised ferritin that has been treated with dimethyl propanediamine to give it a positive charge is used to show the distribution of negative charge on cell surfaces. Mutation in the iron-responsive element (IRE) of the ferritin light-chain gene is associated with **hyperferritinaemia-cataract syndrome**.

ferroportin A transmembrane iron transporter (SLC40A1, 571aa) expressed in macrophages, liver, spleen, kidney and duodenal enterocytes. Heterozygous mutations in the ferroportin gene result in an autosomal dominant form of iron overload disorder, type IV haemochromatosis. See **hepcidin**.

fertilin A heterodimeric protein of the **ADAM family** in the membrane of the sperm. The alpha subunit, ADAM1, in mouse 791aa, has regions with similarity to viral fusion peptides; ADAM1 in human is a pseudogene but ADAM20 (726aa) probably substitutes. The beta subunit (ADAM2, 735aa) is involved in integrin binding and mediating sperm-egg adhesion, the integrin being on the egg. See **zona pellucida** glycoprotein ZP2.

fertilisin Obsolete name term for an agglutinin from sea urchin egg, originally described by Lillie in 1913. Similar substances have been postulated to play a role in mammalian sperm-egg interactions. See **bindin**, **fertilin**, **speract**.

fertilization The essential process in sexual reproduction, involving the union of two specialized **haploid** cells, the gametes, to give a diploid cell, the zygote, which then develops to form a new organism. See *in vitro* **fertilization**.

fertilization membrane Membrane formed on the inner surface of the **vitelline membrane** of some eggs, notably of the sea urchin, following entry of the sperm. A wave of depolarization spreads over the egg surface and intracellular calcium levels rise, triggering the fusion of **cortical granules** that contain a complex variety of molecules that between them inhibit polyspermy. There is no human morphological equivalent, but similar events trigger changes in the **zona pellucida** and plasma membrane that prevent polyspermy.

ferulic acid Phenolic compound present in the plant cell wall that may be involved in cross-linking polysaccharide.

fes An oncogene, originally identified in avian and feline sarcomas. The *fps/fes* proto-oncogene is abundantly expressed in myeloid cells, and the Fps/Fes cytoplasmic protein-tyrosine kinase (822aa) is implicated in signalling downstream from haematopoietic cytokines, including interleukin-3 (IL-3), granulocyte-macrophage colony-stimulating factor (GM-CSF), and erythropoietin (EPO). Somatic mutations in sequences encoding the Fps/Fes kinase domain have been identified in human colorectal carcinomas.

fesselin A proline-rich actin binding protein (996aa), similar to **synaptopodin**, but isolated from turkey gizzard smooth muscle.

festuclavine An ergot alkaloid produced by *Aspergillus fumigatus*. Derivatives reportedly inhibit nucleoside uptake by human lymphoid leukemia Molt 4B cells.

fetal calf serum *FCS* Expensive component of standard culture media for many types of animal tissue cells.

fetuin A carrier protein in plasma, present in higher concentrations in childhood, being substituted by albumin in adulthood. Human fetuin is alpha-2-Heremans–Schmid glycoprotein (α2-HS-glycoprotein, AHSG, 367aa) which is cleaved to produce A and B-chains (282aa and 27aa). Fetuin B (382aa) is similar. AHSG promotes endocytosis, has opsonic properties, and has a high affinity for calcium. Bovine fetuin (359aa) is a major component of **fetal calf serum**.

fetus *foetus* In mammals, the later developmental stages *in utero* after the embryonic phase. In humans, an unborn infant from the eighth week of development onwards.

Feulgen reaction A cytochemical staining procedure for DNA: mild acid hydrolysis makes the aldehyde group of deoxyribose available to react with Schiff's reagent (a pararosaniline, rosaniline and magenta III mixture, decolourized by sulphurous acid) to give a purple colour. The reaction can be used quantitatively for spectrophotometric determination of the amount of DNA present See **periodic acid-Schiff reaction**.

FFA (1) Free fatty acids. (2) (Occasionally) fundus fluorescein angiography, an ophthalmological technique used to examine the vasculature of the retina.

F-factor Plasmid that confers the ability to conjugate (i.e. fertility) on bacterial cells, and carries the *tra* genes (transfer genes); first described in *E. coli*.

FGD See **FYVE, RhoGEF and PH domain-containing proteins**.

FGF See **fibroblast growth factor**.

fgf oncogenes Oncogenes (*K-fgf/hst, fgf-5, fgf-3*) encoding growth factors of the fibroblast growth factor (FGF) family that transform cells through an autocrine mechanism.

fgr An oncogene (feline sarcoma oncogene homologue) originally identified in Gardner-Rasheed feline sarcoma virus, one of the c-src family of cytoplasmic tyrosine kinases. The product of the homologous human c-*fgr* gene (tyrosine kinase fgr, 529aa) is activated in human B-cells following infection with Epstein-Barr virus. Regulates cell migration through effects on a signalling pathway involving **FAK**/Pyk2 and leading to activation of Rac and the Rho inhibitor p190RhoGAP.

FHA1 In plants, a transcriptional activator (in *Nicotiana tabacum*, 209aa).

FHF complex The FTS/Hook/FHIP complex (see **hook** #2). The FHF complex may function to promote vesicle trafficking and/or fusion via the homotypic vesicular protein sorting complex (the HOPS complex).

FHIT See **fragile histidine triad protein**.

FIAU Nucleoside analogue (2'-deoxy-2'-fluoro-5-iodo-1-beta-D-arabinofuranosyluracil) often used in labelled form (with ^{3}H, ^{14}C, ^{125}I or ^{18}F) in imaging studies to locate the site of expression of viral thymidine kinase.

Fibonacci series Named after the early medieval mathematician, a mathematical series produced by addition of the two preceding numbers, e.g. 0, 1, 1, 2, 3, 5, 8, 13, 21 etc. The series describes many patterns in nature, such as the numbers of petals, the pattern (phyllotaxy) of leaves around a stem or the numbers of spirals on the sunflower's seed head. This seems to be related to the maximization of space during the formation of each unit in the series with the result that leaves, for example, are maximally exposed to light.

fibrates A group of cholesterol-lowering drugs that are agonists for **peroxisome proliferator-activated receptor** alpha (PPARα).

fibre cell Greatly elongated type of plant cell with very thick lignified wall. Usually dead at maturity, this cell type is specialized for the provision of mechanical strength. Fibre cells and **sclereids** together make up the tissue known as **sclerenchyma**.

fibrillar centres Location of the nucleolar ribosomal chromatin at telophase: as the nucleolus becomes active the ribosomal chromatin and associated ribonucleoprotein transcripts compose the more peripherally located dense fibrillar component.

fibrillarin Highly conserved small nucleolar protein (snoRNP, rRNA 2'-O-methyltransferase, EC 2.1.1.-, 321aa) that associates with U3-snoRNP and is found in the **coiled body** of the nucleolus where it catalyzes the site-specific 2′-hydroxyl methylation of ribose moieties in pre-ribosomal RNA. The N terminus contains a glycine and arginine-rich domain (GAR domain). Yeast homologue is NOP1 (327aa). Expression of fibrillarin (and **nucleolin**) is greater in rapidly proliferating cells and in the early stages of lymphocyte activation. Autoantibodies to fibrillarin are found in some patients with scleroderma, systematic sclerosis, **CREST** syndrome and other connective tissue diseases.

fibrillation (1) In certain neurological diseases, twitching of individual muscle fibres, or bundles of fibres (fasciculation). (2) Unco-ordinated contraction of regions of the heart, as in atrial fibrillation or ventricular fibrillation.

fibrillin The major protein of extracellular 10−12 nm microfibrils found in both elastic and nonelastic connective tissue throughout the body. Mutations in fibrillin-1 (2871aa) are responsible for **Marfan's syndrome**. Mutations in the fibrillin-2 (2912aa) cause congenital contractural arachnodactyly (CCA; Beal's syndrome; distal arthrogryposis Type 9). Fibrillin-3 (2809aa) is also known. See **tropoelastin**.

fibrin The main protein component of a blood clot. Monomeric fibrin (fibrin A, FGA, 831aa and FGB, 447aa) is produced from **fibrinogen** by proteolytic removal of the highly charged (aspartate- and glutamate-rich) **fibrinopeptides** (16aa and 14aa) by thrombin, in the presence of calcium ions. The monomer readily polymerizes to form long insoluble fibres (23 nm periodicity; half-staggered) that are stabilized by covalent crosslinking (by Factor XIII, plasma transglutaminase).

fibrinogen Soluble plasma protein (340 kDa; 46 nm long), composed of 6 peptide chains, 2 each of $A\alpha$ (866aa), $B\beta$ (491aa) and γ (453aa) and present at about 2−3 mg/ml. See **fibrin**. Defects in fibrinogen genes can cause congenital **afibrinogenaemia** and a type of **amyloidosis**.

fibrinolysis Solubilization of fibrin in blood clots, chiefly by the proteolytic action of plasmin.

fibrinopeptides Very negatively-charged peptide fragments cleaved from **fibrinogen** by thrombin. Two peptides, A (16aa) and B (14aa) are produced from fibrinogen A and B respectively.

fibroadenoma A benign and usually encapsulated tumour with both glandular (adenomatous) and stromal (fibrous) elements.

Fibrobacteres A phylum which includes many of the bacteria found in the rumen and which digest cellulosic material.

fibroblast Resident cell of connective tissue, mesodermally derived, that secretes fibrillar procollagen, fibronectin and collagenase. Many cells in culture

adopt a fibroblast-like (fibroblastic) appearance but this does not mean that they are fibroblasts.

fibroblast growth factor A family of structurally related **growth factors** for mesodermal and neuroectodermal cells. Members include acidic FGF (FGF1, a-FGF, heparin-binding growth factor, HBGF 1, 155aa) and basic FGF (FGF2, b-FGF, HBGF 2, 288aa), the two founder members, but many more are now known (See Table F2). Both aFGF and bFGF lack a signal sequence and the pathway of release is unclear. In addition to their growth promoting activity FGFs play an important part in developmental signalling. Fibroblast growth factor-binding proteins acts as carrier proteins that release FGFs from storage in extracellular matrix and enhance the mitogenic activity of FGFs. Several are known (FGF-BP1, 234aa; FGF-BP2, 223aa etc.) with different distributions and binding affinities.

fibroblast growth factor receptor Family of **receptor tyrosine kinases** for **fibroblast growth factor**. The basic fibroblast growth factor receptor 1 (FGFR-1, c-fgr, CD331, 822aa) is also the receptor for FGF23 in the presence of **klotho**. Defects in FGFR1 cause Pfeiffer's syndrome, idiopathic hypogonadotropic hypogonadism, Kallmann's syndrome type 2 and other developmental abnormalities; a range of abnormalities are associated with defects in the other members of the family. Fibroblast growth factor receptor 2 (keratinocyte growth factor receptor, 822aa) binds both FGF1 and FGF2.

fibrocystin A large receptor-like protein (polyductin, tigmin, 4074aa), localized predominantly at the apical domain of polarized epithelial cells, that is mutated in autosomal recessive **polycystic kidney disease**. Fibrocystin-L (4243aa) is expressed ubiquitously but its function is unclear.

fibrodysplasia ossificans progressiva A rare disorder of intermittent progressive ectopic ossification caused by mutation in the gene for **activin receptor-like kinase-2.**

fibroid A benign tumour (leiomyoma) of smooth-muscle origin, usually in the uterus or occasionally in the gastrointestinal tract.

fibroin Structural protein of silk, one of the first to be studied with X-ray diffraction. It has a repeat sequence GSGAGA and is unusual in that it consists almost entirely of stacked antiparallel **beta pleated sheets**. The protomer is a disulphide-linked heavy and light chain and a p25 glycoprotein in molar ratios of 6:6:1 forming a complex of approximately 2.3 MDa.

fibroma A benign tumour composed of fibrous tissue and arising from mesenchymal cells.

fibromodulin A small proteoglycan (376aa) of the extracellular matrix. The core protein has a mass of

TABLE F2. Human fibroblast growth factors (FGFs)

Name	Synonym	Size*	Function
FGF1	AcidicFGF, hepatocyte growth factor 1	155	Mitogenic, angiogenic
FGF2	Basic FGF, HBGF2	288	Mitogenic, angiogenic
FGF3	HBGF3, proto-oncogene Int-2	239	Ear development?**
FGF4	Heparin secretory-transforming protein 1	206	Mitogenic
FGF5	HBGF5	268	Inhibitor of hair elongation
FGF6	HBGF6, heparin secretory-transforming protein 2	208	Strong mitogen
FGF7	Keratinocyte growth factor	197	Keratinocyte growth
FGF8	Androgen-induced growth factor	233	Autocrine mitogen
FGF9	Glia-activating factor	209	Glial cell growth?
FGF10	Keratinocyte growth factor 2	208	Wound healing?
FGF11	FHF-3	225	Development of nervous system?
FGF12	FGF homologous factor 1, FHF-1, myocyte-activating factor	243	Development of nervous system
FGF13	FHF-2	245	Development of nervous system
FGF14	FHF-4	247	Development of nervous system
FGF16	—	207	Hepatocellular proliferation
FGF17	—	216	Brain induction and patterning
FGF18	zFGF5	207	Mitogenic in liver and intestine
FGF19	—	216	Suppression of bile acid biosynthesis
FGF20	—	211	Neurotrophic factor
FGF21	—	209	Stimulates glucose uptake in differentiated adipocytes
FGF22	—	170	Hair development?
FGF23	Phosphatonin, tumour-derived hypophosphataemia-inducing factor	251	Regulates phosphate homeostasis

FGF15 (218aa) only described from mouse.
*Size is the number of amino acids.
**The '?' indicates that the function is only probable.

around 42 kDa and is very similar to the core protein of **biglycan, decorin** and lumican. All have highly conserved sequences containing 11 internal homologous leucine-rich repeats (~25aa). Fibromodulin has four keratan sulphate chains attached to N-linked oligosaccharides.

fibromyoma See **fibroid**.

fibronectin Glycoprotein of high molecular weight (2 chains each of 2386aa linked by disulphide bonds) that occurs in insoluble fibrillar form in extracellular matrix of animal tissues, and in soluble form in plasma, the latter previously known as cold-insoluble globulin. The various slightly different forms of fibronectin appear to be generated by tissue-specific differential splicing of fibronectin mRNA, transcribed from a single gene. Fibronectins have multiple domains that confer the ability to interact with many extracellular substances such as collagen, fibrin and heparin, and also with specific membrane receptors on responsive cells. Notable is the **RGD** domain recognized by **integrins**, and two repeats of the **EGF-like domain**. The Fibronectin Type III domain (FnIII), about 90aa, of which there are 15–17 per molecule, is a common motif in many cell surface proteins. Interaction of a cell's fibronectin receptors (members of the **integrin** family) with fibronectin adsorbed to a surface results in adhesion and spreading of the cell. Defects in fibronectin are the cause of glomerulopathy with fibronectin deposits type 2.

fibroplasia The production of fibrous tissue, e.g. in wound healing. See **retrolental fibroplasia**.

fibrosarcoma Malignant tumour derived from a connective tissue fibroblast.

fibrosis Deposition of avascular collagen-rich matrix (**fibrous tissue**) in a wound, usually as a consequence of slow fibrinolysis or extensive tissue damage as in sites of chronic inflammation.

fibrous histiocytoma A benign tumour, usually in the skin of the extremities, in which there are large numbers of spindle-shaped, fibroblast-like cells, probably derived from **histiocytes**. Malignant forms are the commonest type of soft-tissue sarcoma in adults. In children, angiomatoid fibrous histiocytoma is associated with somatic fusions of several genes, including ATF1 (cAMP-dependent transcription factor-1), EWS (**Ewing's sarcoma** gene) and **CREB1**.

fibrous lamina Alternative name for the **nuclear lamina**, the region lying just inside the inner nuclear membrane.

fibrous plaque Thickened area of arterial **intima** with accumulation of smooth muscle cells and fibrous tissue (collagen etc.) produced by the fat-laden smooth muscle cells. Below the thickening may be free extracellular lipid and debris that, if much necrosis is also present, is referred to as an **atheroma**.

fibrous tissue Although most connective tissue has fibrillar elements, the term usually refers to tissue laid down at a wound site — well-vascularized at first (**granulation tissue**) but later avascular and dominated by collagen-rich extracellular matrix, forming a scar. Excessive contraction and hyperplasia leads to formation of a **keloid**.

fibulin Family of conserved calcium-binding, cysteine-rich glycoproteins found in the extracellular matrix and in plasma. Fibulin-1 (703aa with multiple spliced isoforms) is incorporated into fibronectin-rich fibres in extracellular matrix and may be important in cell adhesion and motility. Binds fibrinogen and may be a tumour suppressor. Mutations in the gene cause synpolydactyly-2. Fibulin-2 (1184aa) binds to fibronectin and laminin in a calcium-dependent manner and is important in various developmental processes. Fibulin-3 (EGF-containing fibulin-like extracellular matrix protein-1, EFEMP1, 493aa) binds to the EGF receptor and induces signalling; it may regulate cell adhesion and motility during development and mutations are associated with Doyne's honeycomb retinal dystrophy. Fibulin-4 (EFEMP2, 443aa) is defective in a form of cutis laxa, fibulin-5 (developmental arteries and neural crest EGF-like, DANCE, 448aa) promotes integrin-mediated adhesion of endothelial cells and is defective in age-related macular degeneration-3. It is also defective in other forms of cutis laxa. Fibulin-6 (hemicentin-1, 5635aa) has 44 immunoglobulin-like domains and is defective in age-related macular degeneration type 1. Fibulin-7 (439aa) is important in tooth development.

ficin Cysteine endopeptidase (ficain, EC 3.4.22.3) that selectively cleaves at Lys-, Ala-, Tyr-, Gly-, Asn-, Leu- or Val-. Similar to papain. Commercial ficin is purified from the latex of the fig tree, *Ficus glabatra* or *Ficus carica*.

Fick's law Equations that describe the process of diffusion. The first law applies in steady-state and states that flux (J) is proportional to the concentration gradient (dC/dx), times the **diffusion constant** (diffusion coefficient) D for the molecule in that particular medium: $J = -D.dC/dx$ The second law applies in more realistic non-steady state conditions and requires the solution of a partial differential equation.

ficolins Family of molecules structurally very similar to **collectins**, **conglutinin**, and **surfactant proteins A and D** that can activate the complement system by binding to sugar residues on pathogens. When they bind to markers on microbial surfaces they trigger the cleavage of complement C4 by the serine peptidase **MASP2**. Ficolin-1 (collagen/fibrinogen domain-containing protein 1, ficolin-A, M-ficolin, 326aa) binds GlcNAc moieties and is secreted but also found on the surface of monocytes. Ficolin-2 (L-ficolin. 313aa) is secreted, ficolin-3 (Hakata antigen, 299aa) binds to a wider range of sugars (GalNAc, GlcNAc, D-fucose and lipopolysaccharides from *Salmonella* spp.). They have also been found in insects and ascidians.

Ficoll™ Synthetic branched co-polymer of sucrose and epichlorhydrin. Ficoll solutions have high viscosity and low osmotic pressures. Often used for preparing density gradients for cell separations. Ficoll-paque is the proprietary name for premixed Ficoll and diatrizoate (Hypaque) with a density of 1.077 g/cm^3 used as a cushion for separating lymphocytes (which do not pass through the Ficoll-Paque layer) from other blood cells in a one-step centrifugation method.

field ion microscope Type of microscopy in which the specimen is 'illuminated' with ions, often gallium ions, that are focused electrostatically. The ions remove components of the specimen, lower atomic masses first. These are imaged and provide information on elemental distribution with a resolution of perhaps 30 nm.

FIH-1 See **hypoxia-inducible factor**.

filaggrins Basic protein components of **kerato-hyalin granules** of the suprabasal cells of the skin. The precursor protein (filament-aggregating protein, 4061aa) is highly phosphorylated but when dephosphorylated can be proteolytically cleaved into 10–12 individual filaggrin peptides (~324aa) which cause aggregation of the keratin cytoskeleton. Aggregation collapses the cells into flattened squames. Mutations in the gene encoding filaggrin (FLG) have been identified as the cause of ichthyosis vulgaris (IV) and shown to be major predisposing factors for atopic dermatitis (AD) but not for asthma.

filamentous phage Single-stranded DNA bacteriophage of the genus Inoviridae. Examples that infect *E. coli* are M13, f1.

filaments See **axial filaments, beaded filaments, intermediate filaments, microfilaments, thick filaments** and **thin filaments**.

filamins A family of proteins that cross-link **F-actin** to form an isotropic network; the binding does not require Ca^{2+}. Filamin A (filamin 1; ABP280, 2647aa) is widely expressed and interacts with a range of proteins including integrins and transmembrane receptor complexes. Mutations in the X-linked gene can cause a range of developmental disorders. Filamin B (filamin beta; ABP 276/278, 2602aa) is ubiquitously expressed and connects the actin cytoskeleton to the plasma membrane. Various skeletal disorders result from mutation in the FLNB gene including atelosteogenesis. Filamin C (filamin 2, ABP280A, 2725aa) is muscle specific but functionally similar. It interacts with sarcoglycans and with the muscular dystrophy Ky protein. Filamin-binding LIM protein 1 (373aa) is migfilin. Filamin-A-interacting protein 1 (1213aa) controls the start of neocortical cell migration from the ventricular zone, filamin A-interacting protein 1-like (1135aa) inhibits endothelial proliferation and migration and is up-regulated by the angiogenesis inhibitors endostatin and fumagillin.

Filaria Nematodes, for example *Wuchereria bancrofti* and *Brugia malayi*, that cause filariasis and elephantiasis. Transmitted by insects. Can be treated with anthelminthics, or with antibiotics that kill the essential symbiotic bacteria, *Wolbachia* spp.

filensin Protein (beaded filament structural protein 1, 665aa) of the intermediate filament family found in lens fibre cells. Binds to **vimentin** and co-assembles with **phakinin** to form the lens-specific intermediate filament system referred to as **beaded filaments**. Does not assemble on its own to form filaments. Defects cause a form of juvenile-onset cataract.

filiform apparatus A complex of cell wall invaginations in **synergids**, cells of the plant embryo.

filiform papillae Curved tapering cone-shaped body on the tongue of rodents, of which the epithelial cell columns have been investigated in detail.

filipin Polyene antibiotic from *Streptomyces filipinensis*. Polymers of filipin associated with cholesterol in the cell membrane form pores which lead to cytolysis (as does **amphotericin B**)

filopodium *Pl.* filopodia. A thin protrusion from a cell, usually supported by microfilaments; may be functionally the linear equivalent of the leading lamella.

Filoviridae Family of single-stranded RNA viruses, similar in some respect to rhabdoviruses. Marburg and Ebola viruses are the only two genera known at present. Filovirus infections seem to cause intrinsic activation of the clotting cascade leading to haemorrhagic complications and high mortality. Morphologically, virions are very long filaments (up to 14 μm, 70 nm thick), sometimes branched. The RNA is contained within a nucleocapsid that is surrounded by a cell-derived envelope.

fimbria *Pl.* fimbriae. Protein filaments composed of **fimbrillin**, 3–10 nm in diameter and up to several micrometers long, protruding from the surface of Gram-positive and Gram-negative bacteria. They are important virulence factors involved in adhesion to target epithelia and there may be several hundred fimbriae per cell. *Cf.* **pilus**.

fimbrillin Major subunit protein of bacterial **fimbriae**. In *Porphyromonas* (*Bacteroides*) *gingivalis* fimbrillins are 347aa and the fimbriae bind to periodontal tissues via fibronectin and **statherin**. S-fimbrillin (181aa) in *E. coli* forms polar filaments 0.5-1.5 μm long, 100–300 per cell, involved in adhesion to epithelial surfaces. Binds to alpha-sialic acid-(2–3)-beta-Gal containing receptors and is one of the mannose-resistant haemagglutination fimbrial proteins. MatB (meningitis-associated and temperature regulated, 195aa) is a fimbrillin that is expressed in pathogenic *E. coli* strains at low temperatures. There is a family of fimbrial proteins involved in bacterial adhesion some encoded on plasmids (e.g. F17b-G fimbrial adhesin, 343aa).

fimbrin A highly conserved calcium-dependent actin binding protein. In humans there are tissue-specific isoforms referred to as **plastins**. In *S. pombe* fimbrin (ABP-67, 614aa) is associated with actin structures involved in the development and maintenance of cell polarity. In *Arabidopsis* fimbrin-1 (687aa) is ubiquitously expressed and cross-links actin filaments in a calcium independent manner, regulates the actin cytoskeleton, stabilizes F-actin and prevents depolymerization mediated by profilin. N.B. In *E. coli* fimbrin-like protein is completely different and is one of the fimbrial family of proteins that form fimbriae.

fingerprinting The basic principle of the technique is to digest a large molecule with a sequence-specific hydrolase to produce moderate size fragments that can then be run on an electrophoresis gel. Provided the hydrolase only cleaves at specific sites (e.g. between particular amino acids or bases) then the fragments should be characteristic of that molecule. The technique can be used to distinguish strains of virus or to differentiate between similar but non-identical proteins (peptide mapping). Not to be confused with **footprinting**.

FIO (1) In *S. pombe*, an iron transport multicopper oxidase (Fio-1, 622aa), an essential component of copper-dependent iron transport. Binds 4 copper ions. (2) In physiology, the fraction of inspired oxygen, the proportion of oxygen in air taken into the lungs.

Firmicutes A phylum of Gram-positive bacteria with low G + C levels in their DNA. They can be cocci or rod-shaped forms. Are further sub-divided into the anaerobic Clostridia, the Bacilli, which are obligate or facultative aerobes, and the **Mollicutes**, (mycoplasmas) although the latter are placed in a separate phylum (Tenericutes) by some authors.

FISH analysis See **fluorescence** *in situ* **hybridization**.

fish-eye disease A disorder of lipoprotein metabolism caused by mutation in the **LCAT** gene.

fission yeast See *Schizosaccharomyces pombe*.

FITC Fluorescein isothiocyanate, a reagent used to conjugate fluorescein to protein. FITC-labelled antibodies are extensively used for fluorescence microscopy: the fluorophore, when illuminated with UV, emits a yellow-green light.

FIV See **feline immunodeficiency virus**.

fixation Any chemical or physical treatment of cellular material that tends to result in its insolubilization, thus making it suitable for various types of processing for microscopy, such as **embedding** or staining. Typically, fixation involves protein denaturation.

fizzy Originally described in *C. elegans*, a WD repeat-containing protein (fizzy-1, 507aa). Homologues in plants (FIZZY-RELATED 1, FZR1, cell cycle switch protein CCS52A2, 475aa; FZR2, 483aa; FZR3, 481aa) and humans (CDC20-like protein 1, 496aa) are a regulatory subunit of the anaphase-promoting complex (APC) ubiquitin ligase.

FK506 Immunosuppressive drug (tacrolimus) that acts in a very similar way to ciclosporin, binding to an immunophilin and affecting calcineurin-mediated activation of the transcription factor **NFAT** in T-cells. See **FKBP**.

FKBP FK506 binding proteins, a family of small intracellular proteins (e.g. FKBP1A, EC 5.2.1.8, 108aa) that bind the immunosuppressive drug FK506 (tacrolimus), thus are immunophilins. Like **cyclophilin** have peptidyl-prolyl isomerase activity, but are not structurally similar. FKBP1 may modulate ryanodine receptor isoform-1.

FKHR A transcription factor (forkhead homologue in rhabdomyosarcoma; forkhead receptor; Foxo1, 655aa), a member of the hepatocyte nuclear factor 3/forkhead homeotic gene family, a nuclear hormone receptor (NR) intermediary protein. FKHR interacts with both steroid and nonsteroid NRs and can act as either a coactivator or corepressor, depending on the receptor type. Acts as a regulator of cell responses to oxidative stress. Expression is regulated by **KRIT1** (Krev interaction trapped protein 1, 736aa) a negative regulator of angiogenesis. FKHRL1 (forkhead transcription factor in rhabdomyosarcoma like-1, Foxo3, 673aa) triggers apoptosis in the absence of survival factors, including neuronal cell death upon oxidative stress. In the presence of survival factors it is phosphorylated by AKT1/PKB, interacts with 14-3-3 proteins and is retained in the cytoplasm. Survival factor withdrawal induces dephosphorylation allowing it to move to the nucleus and activate transcription of target genes. See **foxo**.

flag tagging Molecular biology technique, in which the gene encoding a protein of interest is engineered to include an octapeptide (N-DYKDDDDK-C) for which there is a good antibody. Because the octapeptide is hydrophilic it is likely to be accessible on the surface of the protein. The fate of the protein in a transfected cell or transgenic organism can then be followed easily. Other tags include the **myc**, **green fluorescent protein** or haemagglutinin epitopes. See also **epitope tag**.

flagella and basal body proteome See **ciliary proteome**. Ciliary Proteome Web Server: http://www.ciliaproteome.org/

flagellin Subunit protein (498aa in *E. coli*) of the **bacterial flagellum**.

flagellum *Pl.* flagella. Long thin projection from a cell used in movement. In eukaryotes flagella (like **cilia**) have a characteristic axial '9 + 2' microtubular array (**axoneme**) and bends are generated along the length of the flagellum by restricted sliding of the nine outer doublets. (See also **hispid flagella**). In prokaryotes the flagellum is made of polymerized **flagellin** and is rotated by the basal motor.

flame cells Specialized excretory cells found in Platyhelminthes (flatworms). The basal nucleated cell body has a distal cylindrical extension that surrounds an extracellular cavity lined by cilia. Mode of action unclear.

flamingo In *Drosophila* a protein involved in the **planar cell polarity pathway** (protocadherin-like wing polarity protein stan, flamingo, starry night, 3579aa). It mediates homophilic cell adhesion. There are vertebrate homologues, also of the cadherin family.

flanking sequence Short DNA sequences bordering a transcription unit. Often these do not code for proteins.

FLAP (1) Activator (5-lipoxygenase activating protein, 161aa) of ALOX5 (5-lipoxygenase), the enzyme responsible for the production of 5-HPETE from arachidonic acid, the first step in **leukotriene** synthesis. Anchors ALOX5 to the membrane. (2) Flap endonuclease 1 (DNase I, flap structure-specific endonuclease 1, maturation factor 1, 380aa) an enzyme with 5′-flap endonuclease and 5′−3′ exonuclease activities involved in DNA replication and repair. During DNA replication it cleaves the 5′-overhanging flap structure that is generated when DNA polymerase encounters the 5′-end of a downstream **Okazaki fragment**.

flare streaming Phenomenon described in isolated cytoplasm of giant amoeba when the medium contains Ca^{2+} and ATP. A loop of cytoplasm flows outward and then returns to the main mass − the appearance is reminiscent of flares visible around the eclipsed sun.

flat revertant Variant of a malignant-transformed animal tissue cell in which the characteristic high **saturation density** and piled-up morphology have reverted to the flatter morphology associated with non-transformed cells.

flavan Parent ring compound on which flavanols, flavanones, flavones, flavonols and flavonoids are based. Should be distinguished from flavin which shares the yellow colour but not structure.

flavin Group of variously substituted derivatives of 7,8-dimethylisoalloxazine. Yellow coloured. The flavin group is found in FAD, FADH and flavoproteins. Not to be confused with **flavan** and **flavones**.

flavin adenine dinucleotide See **flavin nucleotides**.

flavin nucleotides General term for flavin-adenine dinucleotide (FAD) or flavin mononucleotide (FMN). Act as prosthetic groups (covalently linked cofactors) for flavin enzymes.

Flaviviridae A family of enveloped RNA viruses with spherical virions 40−60 nm in diameter having surface projections of small spikes surrounded by a prominent fringe. The Flavivirus genus includes yellow fever virus (the source of the name); other species cause dengue haemorrhagic fever, Japanese encephalitis, tick-borne encephalitis and West Nile fever in humans as well as fevers in a range of animals. Other genera are Pestivirus (with species that cause bovine diarrhoea and swine fever) and Hepacivirus (the type species of which is **hepatitis C virus**).

flavodoxin Electron-transfer proteins, widely distributed in anaerobic bacteria, photosynthetic bacteria and cyanobacteria, that contain flavin mononucleotide as the prosthetic group. In *E. coli*, flavodoxin is reduced by the FAD-containing protein NADPH: ferredoxin (flavodoxin) oxidoreductase; flavodoxins serve as electron donors in the reductive activation of anaerobic ribonucleotide reductase, biotin synthase, pyruvate formate lyase, and cobalamin-dependent methionine synthase. Can substitute functionally for **ferredoxin**. Flavodoxins have been isolated from prokaryotes, cyanobacteria, and some eukaryotic algae.

flavones Specifically the compound and more generally a group of hydroxylated derivatives. Flavone glycosides occur widely as yellow pigments in angiosperms.

flavonoids A large group of secondary metabolites (glycosides) of bryophytes and vascular plants. Some are pigments found in the vacuole, others may be phytoalexins. Term can also be used more generally for any flavone, isoflavone or their derivatives.

flavoproteins Enzymes or proteins that have a **flavin nucleotide** as a coenzyme or prosthetic group. Oxidoreductases or electron carriers in the terminal portion of the electron transport chain.

flavoridin See **disintegrin**.

Flemming-without-acetic An excellent cytoplasmic fixative that contains chromic acid and osmium tetroxide.

fli genes Genes that encode flagellar motor switch proteins (*FliG, FliM, FliN* and *FliY*) in bacteria. FliG (331aa), FliN (137aa) and FliM (334aa) form a switch complex at the base of the basal body which interacts with the CheY and CheZ chemotaxis proteins, in addition to contacting components of the motor that determine the direction of flagellar rotation. FliY in *E. coli* is a cystine-binding periplasmic protein (266aa) involved in transport but in other microbes is associated with flagellar control.

FLICE FADD-like interleukin converting enzyme, **caspase** 8; see Table C1.

flip-flop A term used to describe the coordinated exchange of two phospholipid molecules from opposite sides of a lipid bilayer membrane. Now used to mean the passage of a phospholipid species from one lamella of a lipid bilayer membrane to the other.

flippase (1) See flp-frp recombinase. (2) In *E. coli* an enzyme, undecaprenyl phosphate-aminoarabinose flippase, a heterodimer of ArnE (111aa) and ArnF (128aa) that translocates 4-amino-4-deoxy-L-arabinose-phosphoundecaprenol from the cytoplasmic to the periplasmic side of the inner membrane. Similar enzymes are found in many bacteria.

FLIPR Machine (fluorescence imaging plate reader) for fluorescence imaging using a laser that is capable of illuminating a 96-well plate and a means of simultaneously reading each well thus enabling rapid measurements on a large number of samples. Used in high throughput screening.

FLIPs FLICE-inhibitory proteins, a family of proteins that inhibit the **caspase**, FLICE, and thus protect cells from apoptotic death. Viral FLIPs (v-FLIPs) contain two **death-effector domains** that interact with **FADD** and have been shown to be produced by various herpes viruses and molluscipox virus. The human cellular homologues of v-FLIPs are termed cellular FLICE-inhibitory protein (c-FLIP, FLAME-1, I-FLICE, CASP8 and FADD-like apoptosis regulator, Casper, CASH, MRIT, CLARP, usurpin, 480aa) and inhibit TNFRSF6 mediated apoptosis. Article: http://mcb.asm.org/cgi/content/full/21/24/8247

flk-1 One of the tyrosine kinase receptors (VEGFR-2, KDR, 1356aa) for **VEGF**, binds VEGF-121 and VEGF-C. See **flt** and **cloche**.

floral formula A way of describing a flower although there are various ways of writing the formula. For example the formula K4C(4)A8G(4) where K stands for calyx, C for corolla, A for androecium and G for gynoecium. More sophisticated formulas indicate whether there is fusion of parts, whether the flower is actinomorphic (symmetric, *) or zygomorphic (asymmetric, z) and so on. Extended description: http://employees.csbsju.edu/SSAUPE/biol308/Lecture/floral_form.htm

Floral Genome Project A project that aims 'to investigate the origin, conservation, and diversification of the genetic architecture of the flower, and develop conceptual and real tools for evolutionary functional genomics in plants'. It is supported by the US National Science Foundation. Link to homepage: http://fgp.bio.psu.edu/

florigen A postulated plant growth substance (hormone) that induces flowering but that remained elusive for many years. Now clear that in *Arabidopsis* it is a transcription factor (**CONSTANS**, 373aa) that regulates the *FLOWERING LOCUS T* (*FT*) gene. Homologues in other plants have been identified.

flotillins Integral membrane protein markers of detergent-resistant lipid microdomains (**caveolae**) involved in the scaffolding of large heteromeric complexes that signal across the plasma membrane. Flotillin 1 (reggie-2, reg2, 427aa) and flotillin-2 (reggie-1, epidermal surface antigen-1, 428aa) co-purify with caveolin. They have an evolutionarily conserved domain called the **prohibitin** homology (PHB) domain, and reggies/flotillins have been included within the SPFH (stomatin-prohibitin-flotillin-HflC/K) protein superfamily. Reggie/flotillin homologues are highly conserved among metazoans but are absent in plants, fungi and bacteria.

flow cytometry Slightly imprecise but common term for the use of the Fluorescence Activated Cell Sorter (FACS) or more often just an analyser. Cells are labelled with fluorescent dye and then passed, in suspending medium, through a narrow dropping nozzle so that each cell is in a small droplet. A laser-based detector system is used to excite fluorescence, and droplets with positively fluorescent cells are given an electric charge. Charged and uncharged droplets are separated as they fall between charged plates, and so collect in different tubes. The machine can be used either as an analytical tool, counting the number of labelled cells in a population, or to separate the cells for subsequent growth of the selected population. Further sophistication can be built into the system by using a second laser system at right angles to the first to look at a second fluorescent label, or to gauge cell size on the basis of light-scatter. The great strength of the system is that it looks at large numbers of individual cells, and makes possible the separation of populations with, for example, particular surface properties.

flow-mediated dilation Dilation of blood vessels in response to increase flow that is mediated by endothelium-derived factors such as nitric oxide. Normally expressed as the percentage maximum change in vessel diameter from baseline and assessed using Doppler methods following release of a temporary restriction of flow in the brachial artery by a cuff.

FLOWERING LOCUS A set of genes involved in regulating flowering. *FLOWERING LOCUS T* (*FT*) encodes **florigen**, *FLOWERING LOCUS C* encodes a MADS box protein (196aa) that is a transcription factor that regulates flowering time in the late-flowering phenotype of *Arabidopsis* by repressing '*SUPPRESSOR OF OVEREXPRESSION OF CONSTANT*'. See **CONSTANS**.

FLOWERING PROMOTING FACTOR-1 A protein (FPF1, 110aa) involved in promoting flowering in *Arabidopsis*. It is expressed in apical meristems immediately after photoperiodic induction of flowering in long-day plants. FPF1 modulates the acquisition of competence to flower in the apical meristem and overexpression leads to early flowering in *Arabidopsis*. It may have different effects in other plants.

flox See **flavodoxin**.

flp recombinase Yeast system for DNA rearrangement. In the presence of site-specific recombinase Flp (Protein Able, 423aa) a stretch of DNA flanked by matching FLP recognition target (FRT) sites is excised, and the ends rejoined. An example of a **cassette mechanism**. Important in **plasmid partitioning**.

FLRF-amide See **FMRFamide-related peptides**.

flt Fms-related tyrosine kinases of the src family, receptors for vascular endothelial growth factor isoforms. Flt-1 is VEGF-R1, (fms-related tyrosine kinase 1, 1338aa), flt-3 (Flk2; stem cell tyrosine kinase, STK-1, CD135, 993aa) has a ligand (FL cytokine, FLT3LG, 235aa) that stimulates the proliferation of early haematopoietic cells and synergizes well with a number of other colony stimulating factors and interleukins. Flt-4 is VEGFR-3 (1298aa) and binds only VEGF-C. Flt-2 does not appear in PubMed. See **VEGF**.

fluctuation analysis Method originally introduced in 1943 by Luria & Delbruck to estimate mutation rates in bacterial populations (see **fluctuation test**) but that can be applied to a variety of other situations, for example to determine how many ion channels contribute to the transmembrane current. On the assumption that each channel is either open or shut, the noise in the recorded current can be considered to arise from the statistical fluctuation in the number of channels open, and the magnitude of the fluctuation gives an estimate of the conductance of a single channel. The program Fluctuation AnaLysis CalculatOR (FALCOR) is a web tool designed for use with Luria–Delbrück fluctuation analysis to calculate the frequency and rate from various mutation assays in bacteria and yeast. Link to FALCOR: http://www.mitochondria.org/protocols/FALCOR.html.

fluctuation test Test devised by Luria & Delbruck to determine whether genetic variation in a bacterial population arises spontaneously or adaptively. In the original version the statistical variance in the number of bacteriophage-resistant cells in separate cultures of bacteriophage-sensitive cells was compared with variance in replicate samples from bulk culture. The greater variance in the isolated populations indicates that mutation occurs spontaneously before challenge with phage. (The proportion of resistant cells depends upon when after inoculation of the isolates the mutation arises – which will be very different in separate populations).

fluid bilayer model Generally accepted model for membranes in cells. In its original form, the model held that proteins floated in a sea of phospholipids arranged as a bilayer with a central hydrophobic domain. Although it is now recognized that some proteins are restrained by interactions with cytoskeletal elements, and that the phospholipid annulus around a protein may contain only specific types of lipid, the model is still considered broadly correct.

fluorescein Fluorophore commonly used in microscopy. Fluorescein di-acetate can be used as a vital stain, or can be conjugated to proteins (particularly antibodies) using isothiocyanate (**FITC**). Excitation is at 365 nm, and the emitted light is green-yellow (450–490 nm). The emission spectrum is pH-sensitive and fluorescein can therefore be used to measure pH in intracellular compartments.

fluorescence The emission of one or more photons by a molecule or atom activated by the absorption of a quantum of electro-magnetic radiation. Typically the emission, that is of longer wavelength than the excitatory radiation, occurs within 10^{-8} seconds : phosphorescence is a phenomenon with a longer or much longer delay in re-radiation. Note that gamma-rays, X-rays, UV, visible light and IR radiations may all stimulate fluorescence.

fluorescence activated cell sorter See **flow cytometry**.

fluorescence energy transfer Transfer of energy (fluorescence resonant energy transfer) from one fluorochrome to another. The emission wavelength of the fluorochrome excited by the incident light must approximately match the excitation wavelength of the second fluorochrome. If light at the second emission wavelength is detected, it implies that the two fluorochromes were physically within a few nanometres. Used as a technique to probe protein or cell interactions.

fluorescence in situ hybridization Technique of directly mapping the position of a gene or DNA clone within a genome by **in situ** **hybridization** to **metaphase** spreads, in which condensed chromosomes are distinguishable by light microscopy. The DNA probe is labelled with a fluorophore, and the hybridization sites visualized as spots of light by **epifluorescence**. Frequently, several probes can be used at one time, to mark specific chromosomes with different coloured fluorophores 'chromosome painting'.

fluorescence microscopy Any type of light microscopy in which intrinsic or applied reagents are visualized. Intrinsic fluorescence is often referred to as auto-fluorescence. The applied reagents typically include fluorescently-labelled proteins that are reactive with sites in the specimen. In particular, fluorescently-labelled antibodies are widely used to detect particular antigens in biological specimens. Fluorescence speckle microscopy is used to study the dynamics of protein assemblages such as microtubules using low levels of fluorescently-labelled protomers *in vivo*. The polymer incorporates some labelled protomer and develops a speckled appearance; individual speckles can be tracked in time and space.

fluorescence recovery after photobleaching *FRAP* The recovery of fluorescence on part of the surface of a cell (for example) after a fluorochrome label has been bleached by excitatory light. The bleached patch that starts off as a dark area will gradually recover fluorescence. The recovery is due to the re-population of the area by unbleached molecules and diffusion of bleached molecules to other areas. The rate and extent of recovery are a measure of the fluidity of the membrane and the proportion of labelled molecules that are free to exchange with adjacent areas. The technique is usually applied to cell surface fluidity or viscosity measurements, but is also applicable to other structures.

fluorescence speckle microscopy See **fluorescence microscopy**.

fluorexon Synonym for **calcein**.

fluoride The fluoride ion, F^-. Low levels of fluoride in drinking water markedly decrease the incidence of dental caries, probably because bacterial metabolism is much more sensitive to low fluoride levels. It has been claimed that fluoridation of drinking water, despite vehement protests by a minority of people, has been one of the most successful public health measures ever taken.

fluorite objective Microscope objective corrected for **spherical** and **chromatic aberration** at two wavelengths. Better than an ordinary objective corrected at one wavelength but inferior to (but much cheaper than) a planapochromatic objective.

fluorochromes Those molecules that are fluorescent when appropriately excited; fluorochromes such as fluorescein or tetramethyl rhodamine are usually used in their isothiocyanate forms (**FITC**, **TRITC**).

fluorography A method used to visualize substances present in gels, blots, etc. that involves incorporating fluor or scintillant into the gel which produces light when excited by radioactively labelled molecules that are being separated.

fluoroquinolones A group of broad-spectrum antibiotics related to nalidixic acid that inhibit DNA gyrase and topoisomerase.

Flybase The definitive web-based resource for *Drosophila* genes and genomes. Homepage: http://flybase.org/

flying-spot microscope A type of light microscope in which the object is scanned in two dimensions by a light spot formed by a cathode-ray tube. Transmitted energy is collected by a photomultiplier and an image, suitable for electronic analysis, is reconstructed using the timing circuits driving the cathode-ray tube. Analogous to the scanning electron microscope.

FMD (1) **Foot and mouth disease**. (2) **Flow-mediated dilation**. (3) **Fibromuscular dysplasia**.

f-met-leu-phe See **formyl peptides**.

fMLP See **formyl peptides**.

FMN Flavine adenine nucleotide. See **flavine nucleotides**.

Fmoc Abbreviation for the fluorenylmethyloxycarbonyl group. Fmoc chemistry is important in peptide synthesis: the bulky Fmoc group protects the amino group of the Fmoc-amino acid that is being added to the growing peptide; once the peptide bond has been formed the terminal amino group is 'deprotected' by treatment with a mild base, piperidine, and the next Fmoc-amino acid residue can be added.

FMRFamide-related peptides A group of neuropeptides, derived from a 113aa precursor, that have wide-ranging physiological effects, including the modulation of morphine-induced analgesia, elevation of arterial blood pressure, and increased somatostatin secretion from the pancreas. They include neuropeptide SF (NPSF, 11aa), neuropeptide FF (NPFF, 8aa) and neuropeptide AF (NPAF, 18aa). Many FMRF amides and FLRF-amides are important neuropeptides in invertebrates; all have the same C-terminal RF-amide sequence. Receptors are G-protein coupled.

fMRI A technique (functional magnetic resonance imaging) that provides high resolution, noninvasive imaging of neural activity detected by a blood oxygen level-dependent signal (haemoglobin is diamagnetic when oxygenated but paramagnetic when deoxygenated) based on the increase in blood flow to the local vasculature that accompanies neural activity in the brain.

fms An oncogene, originally identified in a feline sarcoma. The product of human c-*fms* is the colony stimulating factor-1 (MCSF)-receptor tyrosine kinase (CD115, 972aa), also a receptor for IL-34.

FmtA A protein (397aa) from *Staphylococcus aureus* that affects the methicillin resistance level, autolysis in the presence of Triton X-100 as well as the cell wall structure. FmtA has a low binding affinity for beta-lactams, and may be essential to the beta-lactam resistance mechanism. FmtA binds to peptidoglycan *in vitro* and is classed as a penicillin-binding protein. Article: http://www.jbc.org/content/282/48/35143.long

Fnr (1) Fumarate and nitrate reductase regulatory protein (250aa), that activates a number of operons in *E. coli* during anaerobic growth and is a global transcriptional regulator. Fnr senses oxygen via an N-terminal iron-sulphur cluster. In anaerobic conditions, Fnr is able to bind to specific DNA targets at promoters and modulate transcription. In aerobic conditions, Fnr is converted to a form unable to bind these targets. Article: http://nar.oxfordjournals.org/content/35/1/269.full. (2) Ferredoxin-NADP$^+$ reductase (FNR, EC 1.18.1.2, 360aa); an FAD-containing enzyme that catalyzes electron transfer between NADP(H) and ferredoxin. In *Arabidopsis*, the leaf chloroplast isoforms (360aa, 369aa) regulate the relative amounts of cyclic and non-cyclic electron flow to meet the demands of the plant for

ATP and reducing power. In *E. coli* the same enzyme (flavodoxin reductase, 248aa) protects against superoxide radicals due to methyl viologen in the presence of oxygen.

foam cells Lipid-laden macrophages and, to a lesser extent smooth muscle cells, found in **fatty streaks** on the arterial wall.

focal adhesion kinase Non-receptor protein-tyrosine kinase (FAK, pp125FAK, 1052aa) which is found at **focal adhesions** and is thought to mediate the adhesion or spreading processes. Activated by tyrosine-phosphorylation in response to integrin clustering induced by cell adhesion and by various other signals.

focal adhesions Areas of close apposition, and thus presumably anchorage points, of the plasma membrane of a fibroblast (for example) to the substratum over which it is moving. Usually $1 \times 0.2\ \mu m$ with the long axis parallel to the direction of movement; always associated with a cytoplasmic microfilament bundle that is attached via several proteins to the plasma membrane at an area of high protein concentration (this is noticeably electron-dense in electron micrographs). Focal adhesions tend to be characteristic of slow-moving cells and absent from cells with invasive potential such as leucocytes. See **focal adhesion kinase**.

focal segmental glomerulosclerosis A feature of several renal disorders in which there is localized scarring of glomeruli. Can be caused by mutations in non-muscle alpha-actinin-4 (911aa), transient receptor potential channel C6, a non-selective calcium permeant cation channel (931aa) or haploinsufficiency of CD2-associated protein, an adapter protein (Cas ligand with multiple SH3 domains, 639aa).

focus Group of (frequently **neoplastic**) cells, identifiable by distinctive morphology or histology.

fodrin Non-erythroid isoform of **spectrin** found in brain; a tetrameric protein (α, 2472aa; β, 2364aa).

foetus, foetal UK spelling for **fetus, fetal**, now probably obsolete and not used in this dictionary.

FOG1 A transcription regulator (zinc finger protein ZFPM1, friend of GATA1, 1006aa) that plays an essential role in erythroid and megakaryocytic cell differentiation.

folate Molecule (tetrahydrofolate, vitamin B9) that acts as a carrier of one-carbon units in intermediary metabolism. It contains residues of p-aminobenzoate, **glutamate**, and a substituted **pteridine**. The latter cannot be synthesized by mammals, which must obtain tetrahydrofolate as a vitamin or from intestinal microorganisms. One-carbon units are carried at three different levels of oxidation, as methyl-, methylene- or formimino- groups. Important biosyntheses

dependent on tetrahydrofolate include those of **methionine, thymine** and **purines**. Analogues of dihydrofolate, such as **aminopterin** and **methotrexate** block the action of tetrahydrofolate by inhibiting its regeneration from dihydrofolate.

folic acid Pteridine derivative that is abundant in liver and green plants, and is a growth factor for some bacteria. The biochemically active form is tetrahydrofolate (see **folate**).

Folin-Ciocalteau reagent A reagent used in the **Lowry assay** for proteins. A solution of copper tartarate with added detergent (sodium dodecyl sulphate) mixed with Folin's phenol reagent (phosphotungstic phosphomolybdic acid) which forms a blue colour with most proteins.

follicle Generally a small sac or vesicle. In plants, a fruit formed from a single carpel, that splits to release its seeds. In zoology it is used in various compund forms: the hair follicle is an invagination of the epidermis into the dermis surrounding the hair root; the ovarian follicle consists of an oocyte surrounded by one or more layers of **ovarian granulosa cells**. As the ovarian follicle develops a cavity forms and it is then termed a Graafian follicle.

follicle-stimulating hormone Pituitary hormone (FSH; follitropin) that induces development of ovarian follicles and stimulates the release of estrogens. The alpha subunit (92 aa) is the same as that of **luteinising hormone, thyroid stimulating hormone** and human **chorionic gonadotrophin**; the beta subunit (111 aa) which interacts with the G-protein coupled receptor (695aa) is specific to FSH.

follicular dendritic cells Cells of uncertain (mesenchymal or haematopoietic) lineage found in germinal centres. These cells present native antigens to potential memory cells, and only B-cells with high affinity B-cell receptors (BCR) bind. These bound B lymphocytes survive, whereas non-binding B-cells undergo apoptotic cell death.

follistatin An activin antagonist (activin-binding protein, 344aa) that inhibits the secretion of follicle stimulating hormone but not one of the inhibin family. Follistatin-like 1 (follistatin related protein, 308aa) may modulate the action of some growth factors on cell proliferation and differentiation. Follistatin-like-3 (263aa) binds and antagonizes members of the TGFb family, such as activin and BMP2. Follistatin-related protein 4 (842aa) and follistatin-related protein 5 (847aa) are also reported.

Fong's disease See **nail-patella syndrome**.

Fontana stain A stain (Fontana-Masson stain) used for melanin. Melanin granules reduce ammoniacal silver nitrate to produce a black deposit. Argentaffin, chromaffin, and some other lipochrome pigments do the same.

foot and mouth disease A highly infectious disease of cloven-hoofed animals caused by a picornavirus, of the genus Aphthovirus. Endemic in many countries but the UK is virus free because of a strict control policy involving culling of infected animals.

footprinting A technique used to identify the binding site of, for example, a protein on a nucleic acid sequence. The basic principle is to carry out a very limited hydrolysis of the DNA with or without the protein complexed and then to compare the digestion products. If a cleavage site is masked by the bound protein then the pattern of fragments when protein is present will be different and it is possible to work out, by a series of such procedures, exactly where the protein binds.

Foraminifera Group of Rhizopod Protozoa that secrete a test 'shell' and have slender pseudopods that extend beyond the test and unite to form networks. *Allogromia* is a genus within this group. Extensive remains of Foraminiferan tests are found in sedimentary rocks from the Ordovician to the present.

Forbes' disease A glycogen storage disease caused by mutation in the gene encoding the **glycogen debranching enzyme**.

foreign body giant cell Syncytium formed by the fusion of macrophages in response to an indigestible particle too large to be phagocytosed (e.g. talc, silica or asbestos fibres). There may be as many as 100 nuclei randomly distributed: similar cells but with the nuclei more peripherally located (**Langhans cells**) are found at the centre of tuberculous lesions.

forisomes P-protein bodies (about 1–3 µm wide and 10–30 µm long) found in the cytoplasm of phloem sieve tubes in Fabaceae (legumes). They change shape anisotropically in response to alterations in calcium concentration or pH and can exert mechanical force in both contraction and expansion. They act as valves in pores of the sieve plate by changing shape. The shape change is ATP-independent and the volume-change can be considerable (up to six-fold). May have potential in biomimetic devices. The proteins involved (~684aa) have been sequenced (late 2009). Abstract: http://www.nrc.org/procs/Nanotech2004v1/1/W31.01

forkhead Originally the product of the *Drosophila* homeobox gene *fork head*, a transcription factor. Subsequently many related transcription factors have been identified, and are now reclassified as forkhead box proteins (FOX proteins) with sub-classes ranging from FOXA to FOXS. All are winged-helix transcription factors containing the forkhead box motif that binds DNA, and are involved in cell growth, proliferation, differentiation, longevity and embryonic development. See **FAST** #1 (forkhead activin

signal transducer), **FKHR** (forkhead homologue in rhabdomyosarcoma) and **foxo**, also **HNF**.

forkhead activin signal transducer See **FAST** #1.

formaldehyde Commonly used fixative and antibacterial agent. As a fixative it is cheap and tends to cause less denaturation of proteins than does glutaraldehyde, particularly if used in a well-buffered solution (buffered formalin, formal saline). Old formaldehyde solutions usually contain cross-linking contaminants, and it is therefore often preferable to used a formaldehyde-generating agent such as paraformaldehyde. Formalin fumes, particularly in conjunction with HCl vapour, are potently carcinogenic.

formamidopyrimidine-DNA glycosylase A dual function enzyme (EC 3.2.2.23, DNA-(apurinic or apyrimidinic site) lyase mutM, EC 4.2.99.18, 269aa) involved in base excision repair of DNA damaged by oxidation or by mutagenic agents. Acts as DNA glycosylase that recognizes and removes damaged bases, particularly oxidized purines, such as 7,8-dihydro-8-oxoguanine (8-oxoG) and also has AP (apurinic/apyrimidinic) lyase activity and introduces nicks in the DNA strand. The human homologue is OGG1 (345aa).

formiminotransferase-cyclodeaminase A bifunctional enzyme (E.C. 2.1.2.5-E.C. 4.3.1.4, 541aa) involved in the histidine-degradation pathway which exhibits specificity for polyglutamylated folate substrates. The enzyme transfers the formimino group of formiminoglutamate to tetrahydrofolate and then catalyses the cyclodeamination of the formimino group. It is the autoantigen (**LC-1**) in autoimmune chronic active hepatitis type 2.

formins Family of conserved multidomain proteins in all eukaryotes, that regulate actin dynamics by accelerating nucleation rate, altering filament barbed end elongation/depolymerization rates and antagonising the binding of capping proteins. They interact with diverse signalling molecules and cytoskeletal proteins. There are three formin homology domains (FH1, FH2 and FH3), although not all are invariably present. The proline-rich FH1 domain mediates interactions with profilin, SH3-domain proteins, and WW-domain proteins. The FH2 domain is required for the self-association of formin proteins. The FH3 domain is less well conserved and may be important for determining intracellular localisation of formin family proteins. Mammalian formins are mDia1 and mDia2, yeast formins Bni1p and Cdc12p. Formins were originally identified as isoforms encoded by alternatively spliced products of the *ld* (limb deformity) locus of the mouse. Mutations in *ld* lead to disruption in pattern formation, small size, fusion of distal bones and digits of limbs and renal aplasia. Actin polymerization in plant cells is mainly controlled by formins. See **diaphanous**.

formyl peptides Informal term for small peptides with a formylated N-terminal methionine and usually a hydrophobic amino acid at the carboxy-terminal end (fMetLeuPhe was the first common example). These peptides stimulate the motor and secretory activities of leucocytes, particularly neutrophils and monocytes. Formyl peptide receptors are G-protein coupled: the high affinity human receptor (350aa) activates a phosphatidylinositol-calcium second messenger system and similar responses are triggered by the low affinity receptors (N-formyl peptide receptors 2 & 3, 351aa & 353aa). Leucocytes show chemotaxis towards formyl peptides but the term chemotactic peptides understates the range of activities the molecules will trigger. Thought to be synthetic analogues of bacterial signal sequences - though this is unproven. The leucocytes of many animals (e.g. pig, cow, chicken) do not respond.

fornix (1) Any arch-like structure. (2) A major nerve-fibre tract that connects the hippocampus to the septal nuclei and mammillary bodies in the brain.

forskolin Diterpene (colforsin) from the roots of *Coleus forskohlii* that stimulates adenylate cyclase and is often used in conjunction with inhibitors of phosphodiesterase to artificially increase intracellular levels of cAMP.

Forssman antigen A glycolipid heterophile antigen (globopentosylceramide) present on tissue cells of many species. It was first described for sheep red cells, and is not present on human, rabbit, rat, porcine or bovine cells. Antibody to the antigen are commonly found in plasma and may be involved in **Guillain-Barre syndrome**.

fortilin See TCTP (translationally controlled tumour protein).

forward scatter Scattering of electromagnetic waves (light, radio etc.) by particles significantly larger than the wavelength, in a direction that is within 90° of the direction of propagation of the incident beam. In **flow cytometry** the forward scatter is roughly proportional to the diameter of the cell and orthogonal scatter (side scatter) is proportional to the granularity: thus neutrophil granulocytes have higher side scatter than agranular lymphocytes. Dead cells have lower forward-scatter and higher side-scatter than living cells.

fos A leucine zipper protein (380 aa) encoded by the oncogene, v-*fos*, carried by the Finkel-Biskis-Jenkins and Finkel-Biskis-Reilly murine osteogenic sarcoma retroviruses and by the normal c-*fos* gene. Fos dimerises with Jun to form the AP-1 transcription factor. The fos family includes fos, fosB (338aa), fos-related antigen-1 (FRA1, fos-like antigen-1, Fosl-1, 271aa) and fos-related antigen-2 (FRA2, 326aa) all of which can dimerise with members of the jun family. Fra-1 lacks the transactivation domains so will suppress AP-1 activation.

fosmids An f-factor **cosmid**, used as a bacterially-propagated phagemid vector system for cloning genomic inserts approximately 40–50 kilobases in size. Largely superceded by **bacterial artificial chromosome** and P1 bacteriophage (PAC) vector sytems.

Fouchet's stain A histological method that uses the oxidizing action of Fouchet's reagent to convert bile pigment to green biliverdin. Van Gieson's stain is often used as a counterstain.

founder cell Cell that gives rise to tissue by clonal expansion. For most mammalian tissues there are considerably more than two founder cells, as can be determined by forming chimeras from genetically distinguishable embryos, but single founder cells have been found for the intestine and germ line in *C. elegans*.

four helix bundle Common protein motif in which four **alpha-helices** bundle closely together to form a hydrophobic core. Database entry: http://www.cathdb.info/cathnode/1.20.120

Fourier analysis Loosely, the use of Fourier transformations to convert a time-based signal to a frequency spectrum and back, allowing any periodic property of the signal to be identified.

fovea Small pit or depression on the surface of a structure or organ; the fovea centralis is the most cone-rich region of the retina with maximum acuity and colour sensitivity.

fox Prefix for proteins of the **forkhead** family of which many are now recognised, ranging from FoxA to FoxS. FoxP2 is involved in brain development and mutations result in disruption of neural pathways essential for human speech, although this does not equate with it being 'the gene for speech'. Mutation in the mouse Foxp3 gene gives rise to the 'scurfy' mouse (see **polyendocrine syndrome**). See **foxo**.

Fox-1 Gene encoding an RNA-binding protein (ataxin-2-binding protein 1, 397aa) that regulates alternative splicing events by binding to 5'-UGCAUGU-3' elements. Regulates alternative splicing of tissue-specific exons and of differentially spliced exons during erythropoiesis. A family of related genes is being identified. See **NeuN**.

foxo Sub-family of **forkhead** transcription factors (forkhead box sub-group O, foxhead 'other') that are negatively regulated by protein kinase B (PKB) in response to signalling by insulin and insulin-like growth factor in *C. elegans* (Foxo homologue, Daf-16) and mammals (human foxo1 is **FKHR**). dFOXO, the *Drosophila* homologue has also been described. Phosphorylated foxo is retained in the cytoplasm and cannot act as a transcription factor, dephosphorylation allows it to move to the nucleus and activate genes involved in cell death or cell cycle arrest. Foxo factors have been implicated in stress resistance and longevity.

F plasmid Fertility plasmid containing genes that allow transfer of the plasmid from a donor to a recipient in *E. coli* in the process of conjugation.

FPLC Fast protein liquid chromatography. Chromatographic method for protein purification that is much less commonly used now that recombinant proteins can be purified by affinity methods.

FPPs (1) Farnesyl pyrophosphate synthetase (FPPS, 419aa) a key enzyme in isoprenoid biosynthesis which catalyzes the formation of farnesyl diphosphate. There are two enzymically active domains: one a dimethylallyltranstransferase (EC 2.5.1.1), the other a geranyltranstransferase (EC 2.5.1.10). (2) Filament-like plant proteins (FPPs) a family of possible **lamin** functional homologues identified in *Arabidopsis*, tomato, and rice. There are four novel unique sequence motifs and two clusters of long coiled-coil domains separated by a noncoiled-coil linker. The *Arabidopsis* homologue of the FPP family binds in a yeast two-hybrid assay to MAF1, a nuclear envelope-associated plant protein.

fps See *fes*.

fra-1 Fos-related antigen-1 (271aa), one of the AP-1 (activator protein 1) family of transcription factors and, due to the lack of transactivation domain Fra-1, able to suppress activation of AP-1.

frabin See **FYVE, RhoGEF, and PH domain-containing protein-4**.

fractalkine Membrane-bound **chemokine** with CX3C motif (FK, CX3CL1, SCYD1, neurotactin, 397aa). Chemokine domain (76aa) is bound to membrane through mucin-like stalk (241aa) or can be released as a 95 kDa glycoprotein. Predominantly expressed in brain and is up-regulated on capillary vessels and microglia in LPS-induced inflammation and **EAE**. Soluble active fragments (76aa) generated by TNFα converting enzyme (TACE) are chemotactic for T-cells, natural killer cells and monocytes. The receptor is G-protein coupled (CX3CR1, 355aa).

fraction I protein See ribulose bisphosphate carboxylase/oxidase (**RUBISCO**).

fractionation A term used to describe any method for separating and purifying biological molecules. See also **cell fractionation**.

fragile histidine triad protein An AP3A (diadenosine 5′,5‴-P(1),P(3)-triphosphate) hydrolase, FHIT, EC 3.6.1.29, 147aa) thought to be involved in regulation of DNA replication, signalling stress responses and may be a tumour suppressor. The conserved histidine triad residues are required for enzymatic activity. FHIT is inactivated in various tumours and FHIT is a target for **src**-kinase. Histidine triad nucleotide-binding protein 5 (337aa) is an mRNA decapping enzyme.

fragile X syndrome Most frequent cause of mental retardation. There is an expanded **trinucleotide repeat**, CGG, in the 5′-untranslated region of the *fra(X)* gene (there may be more than 200 repeats).

fragilysin Metallopeptidases (EC 3.4.24.74) of the M10 family, produced by some (~10%) pathogenic strains of *Bacteroides fragilis* that can cause diarrhoea. The fragilysin enterotoxin acts by proteolytically damaging the intestinal epithelium, altering the barrier function, possibly by degrading the tight junction proteins. Article: http://www.mendeley.com/research/bacteroides-fragilis-toxin-fragilysin-disrupts-paracellular-barrier-epithelial-cells/

fragmentin See **granzymes**.

fragmin An actin-binding protein (371aa) from the slime mould *Physarum polycephalum*, that has calcium-sensitive severing and capping properties. See **actin-fragmin kinase**.

framboesia See **yaws**.

frame-shift mutation Insertion or deletion of a number of bases not divisible by three in an open reading frame in a DNA sequence. Such mutations usually result in the generation, downstream, of nonsense, chain-termination codons.

Frankenstein food A foodstuff (Frankenfood) made or derived from plants or animals that have been genetically modified by methods other than conventional breeding techniques. Usage is indicative of the style of the debate over GM crops.

Frankia Genus of **Actinomycetales** capable of nitrogen fixation, both independently and in symbiotic association with roots of certain non-leguminous plants, notably alder.

FRAP (1) See **fluorescence recovery after photobleaching**. (2) The human gene *FRAP1* encodes mTOR.

FRAS1 An extracellular matrix protein (4007aa) expressed in many adult tissues, with highest levels in kidney, pancreas, and thalamus. FRAS1-related extracellular matrix protein 1 (FREM1, QBRICK, 2179aa) is involved in epidermal differentiation and is required for epidermal adhesion during embryonic development FRAS1-related extracellular matrix protein 2 (ECM3 homologue, 3169aa) is required to maintain the integrity of the skin epithelium and renal epithelia. FRAS1-related extracellular matrix protein 3 (2135aa) may play a role in cell adhesion. Defects in FRAS1 or FREM2 cause **Fraser's syndrome**.

Fraser's syndrome A developmental disorder (cryptophthalmos with other malformations), caused by mutation in the **FRAS1** gene or in the **FREM2** gene.

Frasier's syndrome See **Denys-Drash syndrome**.

Frat The proto-oncogene *Frat1* (frequently rearranged in advanced T-cell lymphomas) was originally identified as a common site of proviral insertion in transplanted tumours of Moloney murine leukemia virus (M-MuLV)-infected Emu-Pim1 transgenic mice. The human homologues (Frat1, 279aa; GSK-3-binding protein Frat2, 233aa) are critical components of the **Wnt** signal transduction pathway that stabilize beta-catenin.

frataxin A protein (EC 1.16.3.1, 210 aa) that is involved in mitochondrial iron metabolism and is the product of the *X25* gene that is mutated in Friedreich's ataxia. Frataxin catalyzes the oxidation of Fe^{2+} to Fe^{3+} and will form a homopolymer that binds up to 10 iron atoms per protein molecule. Important in resisting oxidative stress. The bacterial frataxin orthologue, CyaY (106aa), is an iron-dependent inhibitor of iron-sulphur cluster formation. Abstract: http://www.ncbi.nlm.nih.gov/pubmed/19305405

fraxin Coumarinic glucoside from *Fraxinus excelsior* (Common Ash) that has anti-inflammatory and antimetastatic properties, the former probably because of its inhibitory effect on **5-HETE** production.

frazzled *Drosophila* orthologue of **DCC**, binds **netrin** and may present it to other netrin receptors of the UNC-5 family.

fredericamycin An antibiotic produced by a soil isolate of *Streptomyces griseus. In vitro*, fredericamycin A exhibits antibacterial, antifungal, and cytotoxic activities; *in vivo*, it has antitumour activity against P388 mouse leukaemia and some others. It inhibits protein and RNA synthesis in *Bacillus subtilis* and P388 cells.

free energy A thermodynamic term (Gibbs free energy, G) used to describe the energy that may be extracted from a system at constant temperature and pressure. In biological systems the most important relationship is: $\Delta G = -RTln(K_{eq})$, where K_{eq} is an equilibrium constant.

free erythrocyte protoporphyrin An analytical measurement used to assess the long-term effects of high level lead exposure. Free protoporphyrin (uncomplexed with haem) accumulates because lead inhibits incorporation of iron into haem.

free radical Highly reactive and usually short-lived molecular fragment with one or more unpaired electrons.

freeze cleavage See **freeze fracture**.

freeze drying Method commonly adopted to produce a dry and stable form of biological material that has not been seriously denatured. By freezing the specimen, often with liquid nitrogen, and then subliming water from the specimen under vacuum, proteins are left in reasonably native form, and can usually be rehydrated to an active state. Since the freeze-dried material will store without refrigeration for long periods, it is a convenient method for holding back-up or reference material, or for the distribution of antibiotics, vaccines amongst others.

freeze etching The process that occurs if a freeze fractured specimen is left for any length of time before shadowing. Water sublimes from the specimen thereby etching (lowering) those surfaces that are not protected by a lipid bilayer. Some etching will take place following any freeze cleavage process; in deep etching the ice surface is substantially lowered to reveal considerable detail of, for example, cytoplasmic filament systems.

freeze fracture Method of specimen preparation for the electron microscope in which rapidly frozen tissue is cracked to produce a fracture plane through the specimen. The surface of the fracture plane is then shadowed by heavy metal vapour, strengthened by a carbon film, and the underlying specimen is digested away, leaving a replica that can be picked up on a grid and examined in the transmission electron microscope. The great advantage of the method is that the fracture plane tends to pass along the centre of lipid bilayers, and it is therefore possible to get *en face* views of membranes that reveal the pattern of integral membrane proteins. The E-face is the outer lamella of the plasma membrane viewed as if from within the cell, the P-face the inner lamella viewed from outside the cell. Fracture planes also often pass along lines of weakness such as the interface between cytoplasm and membrane, so that outer and inner membrane surfaces can be viewed. Further information about the structure can be revealed by **freeze etching.** Extremely rapid freezing followed by deep etching has allowed the structure of the cytoplasm to be studied without the artefacts that might be introduced by fixation.

FREM See **FRAS1**.

French flag problem The French flag (tricolor) is used to illustrate a problem in the determination of pattern in a tissue, that of specifying three sharp bands of cells with discrete properties that do not have blurred edges using, for example, a gradient of a diffusible morphogen.

French press Hydraulic pressure system used to force a suspension of cells or organelles at very high pressure (140 MPa) through a small orifice; the shearing forces and abrupt pressure drop causes disruption of membrane-bound organelles.

frequenin Synaptic calcium-binding protein (neuronal calcium sensor, NCS-1, 190aa), originally found in *Drosophila* (187aa), involved in the regulation of neurotransmission in the central and peripheral nervous systems from insects to vertebrates. Regulates G protein-coupled receptor phosphorylation in a calcium dependent manner. Highly conserved and homologous to **recoverin** and visinin.

frequently rearranged in advanced T-cell lymphomas See **Frat**.

Freund's adjuvant A water-in-oil emulsion used experimentally for stimulating a vigorous immune response to an antigen that is in the aqueous phase. Complete Freund's adjuvant contains heat-killed tubercle bacilli; these are omitted from Freund's incomplete adjuvant. Unsuitable for use in humans because it elicits a severe granulomatous reaction.

Friedreich's ataxia Autosomal recessive disorder caused by trinucleotide (GAA) repeats that, unlike those in **Huntington's chorea** and **fragile X syndrome**, are within an intron of the gene that codes for frataxin, the protein deficient in the disease. The intra-intronic repeat may interfere with hnRNA processing and thus lead to a deficiency in frataxin production.

Friend helper virus Mouse (lymphoid) leukaemia virus present in stocks of Friend virus, that was believed at one time to assist its replication. Molecular cloning of Friend virus has since shown that it is non-defective. See **Friend murine leukaemia virus**.

Friend murine erythroleukaemia cells Lines of mouse erythroblasts transformed by the Friend virus, that can be induced to differentiate terminally, producing haemoglobin, by various agents such as dimethyl sulphoxide.

Friend murine leukaemia virus Murine leukaemia virus isolated by Charlotte Friend in 1956 whilst attempting to transmit the Erlich ascites tumour by cell-free extracts. Causes an unusual erythroblastosis-like leukaemia, in which anaemia is accompanied by large numbers of nucleated red cells in blood. Does not carry a host-derived oncogene, but seems to induce tumours by proviral insertion into specific regions of host genome.

Friend spleen focus-forming virus Defective (replication-incompetent) virus found in certain strains of **Friend helper virus**, detected by its ability to form foci in spleens of mice, and believed to be responsible in those strains for the production of a leukaemia associated with polycythemia rather than anaemia.

fringe In *Drosophila*, a glycosyltransferase (EC 2.4.1.222, O-fucosylpeptide 3-beta-N-acetylglucosaminyltransferase, 412aa) that regulates the location-specific expression of the **Notch** ligands **serrate** and **delta** in the developing wing. In humans the same enzyme (beta-1,3-N-acetylglucosaminyltransferase lunatic fringe, 379aa) is a mediator of somite segmentation and patterning and is defective in a form of spondylocostal dysostosis. The glycosyl transferase radical fringe (331aa) initiates the elongation of O-linked fucose residues attached to EGF-like repeats in the extracellular domain of Notch molecules and may be involved in limb formation and neurogenesis. Another transferase involved in modifying notch ligands is manic fringe (321aa).

frizzled A family of G protein-coupled receptors, originally described in *Drosophila*, mostly coupled to the beta-catenin canonical signalling pathway and encoded by tissue-polarity genes. In humans there are at least 8 members (e.g. frizzled-1, 647aa; frizzled-8, 694aa). Downstream signalling seems to involve Jnk/SAPK-like kinases, **Rho factor** A and the product of the gene **dishevelled** (dsh). Secreted frizzled-related proteins (e.g. secreted frizzled-related protein 1, 314aa) modulate Wnt signalling by competing with membrane-bound frizzled receptors for the binding of secreted Wnt ligands.

frontal zone contraction theory Old model proposed to account for the movement of giant amoebae in which cytoplasmic contraction at the front of the leading pseudopod (fountain zone) pulls viscoelastic cytoplasm forward in the centre of the cell and forms a tube of more rigid cytoplasm immediately below the plasma membrane behind the active region. The peripheral contracted cytoplasm relaxes into a weaker gel at the rear and is pulled forward in its turn. Contrasts with the **ectoplasmic tube contraction** model. An increased understanding of the dynamics of the cytoskeleton renders this of mild historical interest.

frontotemporal dementia A diverse group of adult-onset behavioural disturbances caused by degeneration of the frontal lobe of the brain and may extend back to the temporal lobe. Can be caused by mutation in the gene for tau or presenilin-1 (see Pick's disease). Ubiquitin-positive frontotemporal dementia is caused by mutation in the gene for progranulin. Other loci have also been identified.

frozen stock An approach to avoiding changes that occur in cell lines with repeated sub-culturing. Aliquots of an early passage of the cells are stored either in liquid nitrogen or at −70°C, usually in the presence of a cryoprotectant such as DMSO or glycerol, so that cells of comparable passage number can be used. The method also allows strains to be stored for long periods. Similar methods are used for storing semen for artificial insemination or eggs for *in vitro* fertilization.

FRT Flp recombinase target: see **flp recombinase**.

fructose Fruit sugar, a 6-carbon sugar (hexose) abundant in plants. Fructose has its reducing group (carbonyl) at C2, and thus is a ketose, in contrast to glucose that has its carbonyl at C1 and thus is an aldose. Sucrose, common table sugar, is the non-reducing disaccharide formed by an a-linkage from C1 of glucose to C2 of fructose (latter in furanose form). Fructose is a component of polysaccharides such as inulin, levan.

frustules The cell wall, largely composed of silica, of a diatom (Bacillariophyceae) consisting of two halves, the hypotheca fitting inside the epitheca, rather like the two halves of a petri-dish. The frustule is covered in delicate markings and intricate designs, which are a useful test of the resolving power of a microscope.

FSH See **follicle-stimulating hormone**.

F-spondin See **spondins**.

FTIR An analytical technique (Fourier Transform Infrared Spectroscopy) used to identify (generally) organic materials. This technique is based upon the absorption of various infrared light wavelengths by the material of interest.

FtsZ Filamentous temperature sensitive protein (383aa) involved in cell division in bacteria. It is, like tubulin, a GTP-binding protein with GTPase activity. Similar proteins are found in mitochondria and chloroplasts. Assembly, bundling and stability of FtsZ protofilaments is important for the formation and functioning of the cytokinetic Z-ring, which defines the cell-division plane, during bacterial division. FtsZ interacts with cell division protein FtsA (480aa) which may be a component of the septum and helps connect the FtsZ ring to the cytoplasmic membrane.

F-type ATPase Multi-subunit proton-transporting ATPase (F1F0 ATPase), related to the **V-type ATPase**. Found in the inner membrane of **mitochondria** and chloroplasts, and in bacterial **plasma membranes**. Normally driven in reverse by **chemiosmosis** to make **ATP**, and so also known as ATP synthase.

ftz See **fushi tarazu**.

Fuchs' endothelial dystrophy See **corneal endothelial dystrophy**.

fuchsin Synthetic rosaniline dye. Used as a red dye (in **Schiff's reagent**) and as an anti-fungal agent.

fucose L-fucose (6-deoxy-L-galactose) is a constituent of N-glycan chains of glycoproteins; it is the only common L-form of sugar involved. D-fucose

is usually encountered as a synthetic galactose analogue.

fucosyl transferase An enzyme catalysing the transfer of fucosyl residues from the nucleotide sugar GDP-fucose to glycan chains. Fucosyltransferase 1 (galactoside 2-alpha-L-fucosyltransferase 1, EC 2.4.1.69, 365aa) is important for synthesis of a soluble precursor oligosaccharide (H antigen) which is an essential substrate for the final step in the synthesis of soluble A and B blood group antigens. Fucosyltransferase 2 (343aa) is encoded by a different gene but catalyzes the same reaction. Fucosyltransferase 3 (galactoside 3(4)-L-fucosyltransferase, EC 2.4.1.65, 361aa) is involved in synthesis of Lewis blood group antigens. Fucosylglycoprotein 3-alpha-galactosyltransferase (glycoprotein-fucosylgalactoside alpha-N-acetylgalactosaminyltransferase, EC 2.4.1.40, glycoprotein-fucosylgalactoside alpha-galactosyltransferase, EC 2.4.1.37, 354aa) is responsible for the synthesis of the ABO blood group antigen. Various other fucosyl transferases are known.

fucoxanthin Carotenoid pigment of certain brown algae (**Phaeophyta**) and bacteria: absorbs light at 500−580 nm.

Fucus Genus of brown algae in the Class Phaeophyceae, common on the inter-tidal zone of northern seas.

Fugu rubripes Japanese Puffer fish. Famous (or notorious) for the poison (tetrodotoxin) found in lethal amounts in the poison gland and at low levels elsewhere. Also of interest and utility because of the very low levels of repetitive DNA found in the genome.

Fujian flu (1) An virulent strain of avian influenza (H5N1 subtype of influenza A) first decribed in the province of Fujian, China. (2) The human H3N2 subtype of the Influenza A virus (A/Fujian (H3N2)), responsible for outbreaks in 2003−2004 and still prevalent.

fukutin A secreted protein (461aa), possibly a glycosyl transferase, important for the normal glycosylation of **dystroglycan** and its localisation and function. Fukutin-related protein (495aa) is also involved in the glycosylation. Glycosylation is important for assembly of the **dystrophin-associated protein complex** and mutation in fukutin leads to Fukuyama **muscular dystrophy**, limb-girdle muscular dystrophy type 2I (where the mutation is in fukutin-related protein). See muscular **dystrophy-dystroglycanopathy with brain and eye anomalies**.

Fukuyama muscular dystrophy A form of progressive **muscular dystrophy** caused by mutation in the **fukutin** gene. The pathological changes are similar to those in Duchenne dystrophy. See muscular dystrophy-dystroglycanopathy with brain and eye anomalies.

fumagillin Naturally secreted antibiotic from *Aspergillus fumigatus* that inhibits endothelial cell proliferation by binding to **methionine aminopeptidase 2** and is therefore anti-angiogenic. Originally used against fungal *Nosema apis* infections in honeybees.

fumarate A dicarboxylic acid intermediate in the Krebs' cycle (tricarboxylic acid cycle). Can be derived from aspartate, phenylalanine and tyrosine for input to the Krebs' cycle.

fumonisins A group of economically important mycotoxins that contaminate maize-based food and feed products worldwide. Fumonisin B1, produced by *Fusarium nygamai* and probably the most toxic, inhibits sphingolipid biosynthesis.

functional cloning Strategy for cloning a desired gene that is based on some property (antigenicity, ligand binding, etc.) of the expressed gene (expression cloning). A cDNA library in a eukaryotic **expression vector** is transfected into a large number of cells. Cells expressing the protein of interest are then cloned, the plasmid recovered and the gene sequenced.

Fundulus heteroclitus The killifish. A teleost much used for the study of early embryonic development because the egg and embryo are transparent.

fura-2 Cell permeable fluorescent indicator that exhibits a spectral shift when it chelates calcium ions; the acetoxymethyl ester (fura-2 AM) is hydrolysed within the cell by non-specific esterases and trapped in the cytosol. See **quin-2**.

furan One of a class of heterocyclic aromatic compounds characterized by five-membered ring structure consisting of four CH_2 groups and one oxygen atom. The simplest furan compound is furan itself; a clear, volatile and mildly toxic liquid

furanocoumarins Toxic compounds found primarily in species of the Apiaceae (Umbelliferae) and Rutacea (citrus family) where they are protective against pests. They can cause skin lesions in humans and some are photoactive. Webpage: http://www.life.illinois.edu/berenbaum/newpage1.htm

furanone Class of compounds with furan-like rings of 4 carbon and one oxygen atom. A wide variety of furanones with chemical structures similar to the N-acylhomoserine lactones are produced in nature. Butenolides (2(5H)-furanones) have been isolated from *Streptomyces* spp. and are also produced by marine algae, by sponges, fungi, and ascidians. Some furanones are insect sex pheromones, others are important artificial flavoring compounds or produced during cooking or fermentation. Ascorbic acid is a furanone. Naturally occuring furanones may play a role in inhibiting bacterial infections and biofilm formation by interfering with quorum sensing.

furanose Any monosaccharide with a furanoid ring of 4 carbons and one oxygen.

furfural A viscous, colorless liquid (furfuraldehyde) that has a pleasant aromatic odor; turns dark brown or black upon exposure to air. It is the aldehyde of pyromucic acid, a derivative of **furan**. Used as a solvent and a feedstock in industrial-scale organic synthesis. Also used as a fungicide and nematicide. Datasheet: http://www.chemicalland21.com/arokorhi/industrialchem/solalc/FURFURAL.htm

furin **Subtilisin**-like eukaryotic endopeptidase (EC 3.4.21.75, 794aa) a **kexin** with substrate specificity for consensus sequence Arg-X-Lys/Arg-Arg at the cleavage site. Furin is known to activate the haemagglutinin of fowl plague virus and will cleave the HIV envelope glycoprotein (gp160) into two portions, gp120 and gp41, a necessary step in making the virus fusion-competent. See **notch**.

furunculosis Disease of fish caused by *Aeromonas salmonicida*. Major problem in fish farms.

Fusarium A large genus of filamentous fungi widely distributed in soil and in association with plants. Most are harmless saprophytes but some produce mycotoxins that can cause disease in plants or animals consuming e.g. infected grain. The mycotoxins include zearalenone, **diacetoxyscirpenol**, T-2 toxin, neosolaniol monoacetate, deoxynivalenol, nivalenol, **fumonisin B1**, fumonisin B2, moniliformin, fusarenon-X, HT-2 tioxin and beta-zearalenol.

fused In *Drosophila* a serine/threonine-protein kinase (EC 2.7.11.1, 805aa) maternally required for correct patterning in the posterior part of each embryonic metamere. There are mammalian homologues: see **SUFU**.

fused gene family A family of genes that encode proteins associated with keratin intermediate filaments and that are partially cross-linked to the cell envelope in keratinocytes. The proteins include **profilaggrin, trichohyalin, repetin, hornerin**, the profilaggrin-related protein and a protein encoded by c1orf10. The Ca^{2+} binding EF-hand domain has significant homology with that of **S100** proteins, and the family probably arose through fusion between S100-type proteins and proteins involved in cornification.

fushi tarazu (Japanese for 'too few segments') A **pair-rule gene** of *Drosophila* encoding a protein (410aa) that is required in alternating segment primordia, specifying the correct number of segments. May play a role in determining neuronal identity and be directly involved in specifying identity of individual neurons. Flies lacking *ftz* exhibit embryos with half the usual number of body segments

fusidic acid A steroid antibiotic related to **cephalosporin P**, isolated from fermentation broth of the fungus *Fusidium coccineum*. Blocks translocation of the elongation factor G (EF-G) from the ribosome during bacterial protein synthesis.

fusiform Tapered at both ends, like a spindle - though the current rarity of spindles makes this a somewhat unhelpful description.

fusin The chemokine receptor (CXCR4, 352aa) for which CXCL12 (stromal cell-derived factor-1) is the ligand. It is the co-receptor, together with CD4 for lymphotropic strains of HIV. **FIV** will infect CD4$^-$ cells so the chemokine receptor may be the original binding site and CD4 the co-receptor, rather than the converse. See Table C3.

fusion protein Protein formed by expression of a hybrid gene made by combining two gene sequences. Typically this is accomplished by cloning a cDNA into an **expression vector** in-frame with an existing gene, perhaps encoding e.g. beta-galactosidase. See **GST-fusion protein**.

Fusobacteria A bacterial phylum that includes the family Fusobacteriaceae within which is *Fusobacterium nucleatum* a filamentous, anaerobic, Gram-negative bacterium involved in several human diseases, including periodontal diseases and Lemierre's syndrome. Pathology: http://jmm.sgmjournals.org/content/53/10/1029.long

fusome A membrane and cytoskeletal organelle specific to the germline in *Drosophila*; spheroid throughout stem cells and **gonialblasts**, but branches extensively throughout interconnected secondary spermatogonia.

futile cycles Any sequence of enzyme-catalysed reactions in which the forward and reverse processes (catalysed by different enzymes) are constitutively active. Frequently used to describe the cycle of phosphorylation and dephosphorylation of phosphatidyl inositol derivatives in cell membranes.

FVB Inbred strain of mice used extensively in transgenic research because of its defined background, good reproductive performance, and prominent pronuclei, which facilitate microinjection of genomic material. Derived from Swiss mice and named because of sensitivity to Friend leukaemia virus B.

Fx (1) A homodimeric NADP(H)-binding protein (GDP-L-fucose synthase, tissue-specific transplantation-antigen 3, TSTA3, EC 1.1.1.271, 321aa) first isolated from erythrocytes and shown to be the enzyme responsible for the last steps of GDP-L-fucose synthesis from GDP-D-mannose in both procaryotic and eucaryotic cells. Upregulation of Fx expression, leading to an increase in GDP-L-fucose levels, may be a marker for hepatocellular carcinoma. (2) Obsolete name for **thymosin β4**.

fyn A non-receptor **tyrosine kinase** (p59-Fyn, 537aa), related to src and implicated in both brain development and adult brain function. Plays a role in T-cell signal transduction in concert with lck, and excess fyn activity in the brain is associated with conditions such as Alzheimer's and Parkinson's diseases.

FYVE A phosphatidylinositol 3-phosphate-binding **zinc-finger** domain (75–80aa), with two zinc binding centers. The FYVE domain is found in eukaryotic proteins that are involved in membrane trafficking and phosphoinositide metabolism and was named after the four proteins in which it was first found (Fab1, YOTB/ZK632.12, Vac1, and EEA1)

FYVE, RhoGEF and PH domain-containing proteins

A family of proteins (FGDs) with zinc fingers, GDP-exchange properties and **PH domains** involved in regulating actin-based motility. FGD1 (faciogenital dysplasia 1 protein, 961aa) catalyses GTP/GDP exchange on the rho-family GTPase, **cdc42** and is mutated in **Aarskog-Scott syndrome**. FGD2 (655aa) and FGD3 (725aa) may do the same. FGD4 (FGD1-related F-actin-binding protein, frabin, 766aa) has an additional actin-binding domain and when overexpressed *in vitro*, induces microspike formation FGD4 is thought to be is important in myelination of the peripheral nervous system; defects are the cause of **Charcot-Marie-Tooth disease** type 4H. FGD5 (1462aa) and FGD6 (1430aa) also act on cdc42.

G

G Single letter code for glycine, the simplest amino acid, or guanosine in nucleic acids or compounds such as GTP.

G0 Phase of the cell cycle in which non-proliferating cells are considered to exist. Entry into **G1** phase is a prelude to S-phase and eventually division.

G1 Phase of the eukaryotic **cell cycle** between the end of cell division and the start of DNA synthesis, **S phase**. G stands for gap.

G2 Phase of the eukaryotic **cell cycle** between the end of DNA synthesis and the start of cell division.

gab Adapter proteins (gab1, grb2-associated binder-1, 694aa; gab2, 676aa; gab3, 586aa; gab4, 574aa) that bind to **Grb-2**. Tyrosine phosphorylation of gab proteins by receptor tyrosine kinases mediates interaction with several proteins that contain SH2 domains. Gab-1 has some similarity to IRS-1 (insulin receptor substrate-1). As a substrate for the EGF-receptor and various others, may integrate growth-factor signals.

GABA Fast inhibitory **neurotransmitter** (gamma-aminobutyric acid) in the mammalian **central nervous system**; prevalent in higher regions of the **neuraxis**. Also mediates peripheral inhibition in crustaceans and in the leech *Hirudo medicinalis*.

GABA receptor Member of a family of receptors for neurotransmitters that includes the **glycine receptor** and the **nicotinic acetylcholine receptor**. Opened by γ-amino butyric acid (**GABA**). There are two main classes; **ionotropic** GABA$_A$ receptors and **metabotropic** GABA$_B$ receptor. The GABA$_A$ receptor is a ligand-gated chloride channel, pentameric, usually with two alpha (456aa), and two beta (474aa) subunits together with a fifth subunit, of which there are various types presumably conferring specific properties in particular neurons. The channel is blocked by **bicuculline** and **picrotoxin**, binds benzodiazepines, and its properties can be modified by phosphorylation. A related receptor, found in the retina and formerly called the GABA$_C$ receptor is now considered to be a member of the rho (ρ) subfamily of the GABA$_A$ receptors (GABA$_A$-ρ). The metabotropic GABA$_B$ receptor is negatively coupled to adenylate cyclase through a G$_o$-protein and thus acts indirectly on N-type calcium channels. It is found in the brain and differs in agonist specificity The inhibitory effects of GABA$_B$ are due to a reduction in catecholamine release.

GABI A collaborative network of different plant genomic research projects (Genomanalyse im biologischen System Pflanze). The projects are funded by the German Federal Ministry of Education and Research (BMBF) and the GABI database has information about commercial important plant genomes. Link to Database: http://www.gabi.de

G-actin Globular **actin**, the protomer for the assembly of **F-actin**.

gadd genes A set of genes (growth arrest- and DNA damage-inducible genes) activated when DNA is damaged. Gadd-34 (protein phosphatase 1 regulatory subunit 15A, PPP1R15A, MyD116 homologue, 674aa) recruits **protein phosphatase-1** to dephosphorylate the translation initiation factor eIF-2A/EIF2S1, thereby reversing the shut-off of protein synthesis initiated by stress-inducible kinases and facilitating recovery of cells from stress. May promote apoptosis by inducing p53 phosphorylation. Gadd-45α (165aa) which is strongly induced by X-rays binds to **proliferating cell nuclear antigen**. Gadd-45β (160aa) and gadd-45γ (159aa) mediate activation of stress-responsive MEKK4. Gadd-153 (DNA damage-inducible transcript 3 protein, C/EBP-homologous protein, CHOP, 169aa) acts as an inhibitor of the transcription factors C/EBP and LAP.

gadolinium *Gd* Lanthanide element. The trivalent ion blocks current through T-type voltage gated calcium channels and stretch- activated ion channels in a concentration-dependent manner and is used as an investigative tool.

GAG See **glycosaminoglycan**.

gag protein In HTLV-1 the *gag* gene encodes a polyprotein (429aa) which is cleaved into three proteins, matrix protein p19, capsid protein p24 and nucleocapsid protein p15-gag. The matrix protein targets Gag, Gag-Pro and Gag-Pro-Pol polyproteins to the plasma membrane, the capsid protein p24 forms the conical core of the virus that encapsulates the genomic RNA-nucleocapsid complex and nucleocapsid protein p15 is involved in the packaging and encapsidation of two copies of the genome. Other retroviruses have similar polyproteins. In Moloney murine leukemia virus there are four peptides produced from the polyprotein, in HIV there are six, but the basic functions are similar.

GAI Gibberellic acid insensitive. See **DELLA** and **GRAS family**.

The Dictionary of Cell and Molecular Biology. DOI: http://dx.doi.org/10.1016/B978-0-12-384931-1.00007-6

gain of function mutation Mutation that results in higher than normal levels of activity of the gene product, for example by deletion of a regulatory phosphorylation site on the protein. Examples are **oncogenic** mutations in genes involved in growth control.

GAL promoter Inducible promoter region of the yeast operon that encodes, among other things, the enzyme **beta-galactosidase** which will act on **Xgal** to produce a blue colour. Both yeast and bacterial forms of the promoter are used experimentally. The heterodimeric transcription factor that binds to the promoter (**GAL4**) is used in **yeast two-hybrid screening**.

GAL4 Yeast transcription factor that binds the UASG promoter domain. Often used in reporter gene constructs and in **yeast two-hybrid screening**. The GAL4 **enhancer trap**, used classically in *Drosophila*, has GAL4 as the **reporter gene**. Webpage: http://www.fly-trap.org/flytrap/html/docs/egal4.html

galactans Polymers of galactose, mostly found in plants. May be branched or unbranched.

galactocerebroside Surface antigen (GalC, galactosylceramide) characteristic of newly differentiated **oligodendrocytes**. GalC antibody is used to identify this glial cell type in cultures of rat optic nerve and brain.

galactosaemia A disorder in which the enzyme galactose-1-phosphate uridyl transferase (GALT; EC 2.7.7.12, 379aa), that converts galactose-1-phosphate into glucose-1-phosphate, is absent.

galactosamine An amino sugar found in some glycolipids, chondroitin sulphate and dermatan sulphate and the terminal carbohydrate that constitutes blood group A antigen.

galactose Hexose identical to glucose except that orientation of -H and -OH on carbon 4 are exchanged. A component of **cerebrosides** and **gangliosides**, and glycoproteins. **Lactose**, the disaccharide of milk, consists of galactose joined to glucose by a β-glycosidic link.

galactose binding protein A bacterial periplasmic protein (332aa), most studied in *E. coli*, that acts both as a sensory element in the detection of galactose as a chemotactic signal, and in the uptake of the sugar.

galactosyl transferase Enzyme catalysing the transfer of galactose units from the sugar-nucleotide, uridine diphospho-galactose (UDP-galactose) to an acceptor, commonly N-acetyl-glucosamine in a glycan chain, forming a glycosidic bond involving C1 of galactose. A range of different transferases are known. The transfer is to ceramide in the case of cerebroside synthase (UDP-galactose ceramide galactosyltransferase, EC 2.4.1.45, 541aa).

Galadriel elements Plant Ty3/Gypsy **LTR retrotransposons** of the chromovirus type.

galanin Neuropeptide (30aa from a 123aa precursor) found throughout the central and peripheral nervous system. Regulates gut motility and the activity of endocrine pancreas. The receptors are G-protein coupled; GALR1 (349aa) inhibits adenylyl cyclase, GALR2 (387aa) activates the phospholipase C/protein kinase C pathway, GALR3 (368aa) couples to a G protein of the G_i/G_o class but has a different profile to GALR1. Galanin-like peptide (GALP, 60aa) binds to all the galanin receptors but has a role in the regulation of energy homeostasis and reproduction.

galantamine *galanthamine* Tertiary amine compound originally derived from flowers (daffodils and snowdrops), now synthesized. A specific, competitive, and reversible **acetylcholine esterase inhibitor** shown to have mild cognitive and global benefits for patients with Alzheimer's disease.

galaptins See **galectin**.

galectins Family of conserved S-type beta-galactoside-binding lectins found in a range of metazoan phyla and involved in many biological processes such as morphogenesis, control of cell death, immunological response, and cancer. Galectins 1−3 are homodimeric, the remainder are monomeric. In humans, galectin-1 (galaptin, 135aa) mediates cell-cell and cell-substratum adhesion and plays a role in immune regulation. Galectin-2 (132aa) binds to lymphotoxin-alpha. Galectin-3 (IgE binding protein, Mac-2, e-BP, 250aa) is usually considered proinflammatory but can induce apoptosis in T-cells. It is part of the AGE-receptor (**RAGE**) complex. Galectin-4 (323aa) is monomeric but has two carbohydrate binding sites: it is restricted to small intestine, colon, and rectum. Galectin-7 (136aa) is confined to keratinocytes, galectin 8 is prostate carcinoma tumour antigen-1(PCTA1, 316aa). Galectin-9 (355aa) may have a role in thymocyte-epithelial interactions, galectin-10 (142aa) is the **Charcot-Leyden crystal** protein. Other galectins are known and although most are described from metazoa there are also galectins in fungi.

gallic acid Phenolic acid, commonly found in flowering plants, usually esterified with tannins. Reported to have anti-fungal and anti-viral properties, to act as a antioxidant and helps to protect cells against oxidative damage. Gallic acid is said to be cytotoxic for cancer cells, but not normal cells. Also used for making dyes and inks.

gallidermin A Type A **lantibiotic** from *Staphylococcus gallinarum*, a tetracyclic polypeptide similar to **epidermin**, and effective against Propionobacteria.

GALT See **gut-associated lymphoid tissue**.

galvanotaxis The directed movement of cells induced by an applied voltage. This movement is almost always directed toward the cathode, occurs at fields around 100 mV/mm, and is argued to be involved in cell guidance during morphogenesis, and in the repair of wounds. The term galvanotropism is used for neurons, since the cell body remains stationary and the neurites grow toward the cathode. Note that these processes involve cell locomotion, and are distinct from **cell electrophoresis**.

galvanotropism See **galvanotaxis**.

gambogic acid A natural product isolated from the resin of *Garcinia hurburyi* (gamboge tree), a potent inducer of apoptosis acting as an antagonist of anti-apoptotic Bcl-2 family proteins.

gamete Specialized haploid cell produced by meiosis and involved in sexual reproduction. Male gametes are usually small and motile (spermatozoa), whereas female gametes (oocytes) are larger and non-motile.

gametocyte (1) A cell that divides to produce gametes. (2) The sexual reproductive stage of the malaria parasite that develops within erythrocytes.

gametogenesis Process leading to the production of **gametes**.

gametophores *gametangiophore* In mosses and ferns (Archegoniata) the region that bears the sex organs (gametangia), the female **archegonia** and the male **antheridia**. In Bryopsida the leafy moss plant is called a gametophore and is the adult form of the haploid gametophyte which develops from the **protonema**.

gametophyte The haploid stage of the life cycle of plants; the major vegetative stage for simple plants like liverworts. In pteridophytes and higher plants (gymnosperms and angiosperms) the gametophyte phase is restricted to the ovary and the main part of the plant is the diploid sporophyte. In gymnosperms, the female gametophyte consists of around 2000 nuclei and forms **archegonia** which produce the egg cells for fertilization. In flowering plants, the megagametophyte (**embryo sac**) is even more reduced and may have only eight nuclei.

gamma-aminobutyric acid See **GABA**.

gamma-delta cells Lineage of T-cells possessing the γδ form of the T-cell receptor. Appear early in development and constitute around 5% of mature T-cells in peripheral lymphoid organs. May be the predominant form at epithelial surfaces. Most have neither CD4 nor CD8.

gamma-glutamyltransferase Heterodimeric enzyme (γ-glutamyl transpeptidase, EC 2.3.2.2, 662aa), highly glycosylated, attached to external surface of cell membrane; transfers γ-L-glutamyl residue (usually from glutathione) to the amino group of an amino acid. Elevated plasma levels are used as a diagnostic marker of hepatic disorders and pancreatitis. Type member of the threonine (T) peptidase family with Thr391 at the active site. The *E. coli* enzyme (non-glycosylated) is soluble and localized in the periplasmic space.

gamma-haemolysin See **gamma-toxin**

gamma-secretase A multi-protein complex of **presenilin, nicastrin, APH-1** and **PEN-2** together with regulatory subunits. See **secretases**.

gamma-toxin Two-component toxins produced by *Staphylococcus aureus*, forming a protein family with **leucocidins** and **Staphylococcal toxin** alpha. Two active toxins (AB and CB) can be formed combining one of the class-S components, HlgA (309aa) or HlgC (315aa), with the class-F component HlgB (325aa). Gamma-haemolysins form cation-selective pores (hetero-oligomers formed by three or four copies of each component) with marked similarities to those formed by alpha-toxin in terms of conductance, nonlinearity of the current-voltage curve, and channel stability in the open state. Article: http://www.ncbi.nlm.nih.gov/pmc/articles/PMC96427/

gamma-tubulin complex See **gamma-TuRC**.

gamma-TuRC The gamma-tubulin ring complex (γ-TuRC) is a multi-protein complex of approximately 2.2 MDa that nucleates microtubules at the centrosome. It consists of 1014 γ-tubulin molecules and at least six additional proteins. Each ring is approximately the diameter of a microtubule (25 nm) and is located at the negative (minus) end of the tubule at the centrosome. Abstract: http://www.ncbi.nlm.nih.gov/pubmed/10854328

Gammaretrovirus A genus of the retroviridae family. Many species contain oncogenes and cause sarcomas and leukemias. Examples include murine leukemia virus, the feline leukemia virus, the feline sarcoma virus, and the avian reticuloendotheliosis viruses. Endogenous retroviruses are closely related to exogenous gammaretroviruses. Many of the gammaretroviruses share a conserved RNA structural element called a core encapsidation signal. Database entry: http://www.expasy.org/viralzone/all_by_species/67.html

gammopathy A disorder (monoclonal gammopathy, monoclonal gammopathy of unknown significance, MGUS), often asymptomatic, in which there is **paraprotein** (M protein) in serum and urine. In some cases may progress to a B-cell malignancy or myeloma.

gamocyte Rarely used term for a phase of the life cycle of protozoa such as *Plasmodium* that develops from the trophozoite and later gives rise to gametes.

gamone A pheromone released by a gamete or hypha and that is attractive to another appropriate gamete or hypha in sexual reproduction. The glycoprotein gamone 1 (blepharmone, 305aa) is produced by mating type I cells of the ciliate *Blepharisma japonicum*, Type II produce gamone 2 (blepharismone). See **sirenin**.

ganglion (1) A cluster of **neurons** that in vertebrates is an appendage of the central nervous system. In invertebrates, the majority of neurons are organized as separate ganglia. (2) A ganglion or ganglion cyst, is a swelling on or around joints and tendons in the hand or foot.

ganglion cell A type of **interneuron** that conveys information from the retinal **bipolar**, horizontal and **amacrine cells** to the brain.

ganglioside A **glycosphingolipid** that contains one or more residues of N-acetyl or other neuraminic acid derivatives. Gangliosides are found in highest concentration in cells of the nervous system, where they can constitute as much as 5% of the lipid. GM gangliosides have a single sialic acid, GD forms have two.

gangliosidoses Diseases, such as **Tay-Sachs**, caused by inherited deficiency in enzymes necessary for the breakdown of gangliosides. Cause gross pathological changes in the nervous system, with devastating neurological symptoms.

gankyrin A proto-oncoprotein (p28-GANK, 226aa) that is involved in negative regulation of the tumour suppressors **retinoblastoma** and **p53**. Acts as a chaperone during the assembly of the 26S proteasome. It has seven **ankyrin** repeats.

GAP See **GTPase-activating protein, growth-associated proteins**.

gap **gene** Segmentation genes involved in specifying relatively coarse subdivisions of the arthropod embryo. Their products are transcription factors that are expressed under the control of maternal effect genes such as *bicoid* and *nanos*, and regulate **pair-rule genes**. In *Drosophila*, a number of these genes have been identified, including: *knirps, krüppel, hunchback, giant, huckebein* and *tailless*. The gap genes *orthodenticle* and *buttonhead* are required for the development of the head. See *eve*.

gap junction A junction (communicating junction) between two cells consisting of many pores that allow the passage of molecules up to about 900 Da. Each pore is formed by an hexagonal array (connexon) of six transmembrane proteins (**connexins**) in each plasma membrane: when mated together the pores open, allowing communication and the interchange of metabolites between cells. **Electrical synapses** are gap junctions, and **metabolic cooperation** depends upon the formation of gap junctions.

GAPDH See **glyceraldehyde-3-phosphate dehydrogenase**.

GAPT Growth factor receptor-bound protein 2 (Grb2)-binding adaptor protein, transmembrane. See **LAT adaptor**.

gar (1) Gar1 (217aa) is a small nucleolar protein required for ribosome biogenesis and telomere maintenance. It forms part of the H/ACA small nucleolar ribonucleoprotein (H/ACA snoRNP) complex. (2) In *S. pombe* gar2 is a nucleolar protein (500aa) containing a glycine- and arginine-rich (GAR) domain, required for 18S rRNA and 40S ribosomal subunit accumulation and assembly of the pre-ribosomal particles. The functional homologue of NSR1 from *S. cerevisiae*, and structurally related to **nucleolin** from vertebrates. (3) In *C. elegans*, G-protein-linked acetylcholine receptors (gar1, 713aa; gar2, 627aa; gar3, 611aa) of the muscarinic type. (4) Suppressor mutations (gar1 and gar2) that act semi-dominantly to restore gibberelin responsiveness to **gibberellin**-insensitive (gai) mutant of *Arabidopsis*.

garpin Leucine-rich repeat-containing protein 32 (glycoprotein A repetitions predominant, GARP, 662aa).

Gardner's syndrome A variant of familial **adenomatous polyposis coli**.

gargoyle cells Fibroblasts with large deposits of mucopolysaccharide, commonly found in storage diseases such as **Hurler's disease**.

gargoylism Old name for **Hurler's syndrome**.

GARP (1) A superfamily of plant transcription factors defined by G2 in maize, the *Arabidopsis* RESPONSE REGULATOR-B (ARR-B) proteins and the PHOSPHATE STARVATION RESPONSE1 (PSR1) protein of *Chlamydomonas*. Involved in a variety of key cellular functions, including regulation of transcription, phosphotransfer signalling, and differentiation. (2) In *E. coli* a galactarate transporter (garP, D-galactarate permease, 444aa) a multipass protein of the inner membrane involved in uptake of D-galactarate. (3) See **garpin**. (4) Glutamic acid-rich protein (cyclic nucleotide-gated cation channel beta-1, 1251aa) part of a non-selective cation channel important in both visual and olfactory signal transduction.

gas gangrene A condition caused by *Clostridium perfringens* infection; extracellular collagenase breaks down the collagen network.

gas **genes** (1) Growth arrest specific genes, associated with cellular quiescence. Their products have diverse roles. Gas1 protein (345aa) blocks entry to S phase. The human Gas2 gene product (313aa) is cleaved during apoptosis and the cleaved form induces dramatic rearrangements of the actin cytoskeleton. It has a calponin homology actin-binding domain and a Gas2-related microtubule-binding domain. Gas2-like protein 1 (681aa) is involved in the cross-linking of microtubules and microfilaments as are Gas2-like proteins 2 and 3 (880aa and 694aa). Gas3/peripheral myelin protein 22 (PMP22, 160aa) is a component of peripheral nerve myelin, and is defective in some peripheral neuropathies in humans. The *gas5* gene encodes multiple small nucleolar RNAs. Gas6 (721aa) is a secreted ligand for tyrosine-protein kinase receptors **AXL**, **TYRO3** and **MER**. Gas7 (476aa) may promote maturation and morphological differentiation of cerebellar neurons. Gas8 (gas11, 478aa) is a cytoskeletal linker which binds microtubules and probably functions in axonemal and non-axonemal dynein regulation. (2) In *S. cerevisiae*, GAS1 (1,3-beta-glucanosyltransferase GAS1, EC 2.4.1.-, glycolipid-anchored surface protein 1, glycoprotein GP115, 559aa) is involved in the elongation of 1,3-beta-glucan chains in cell wall biosynthesis and morphogenesis. There are other glycolipid-anchored surface proteins, GAS2, GAS3 etc. with similar activities.

GAS motif DNA motif (IFN-gamma activation site) in the FcεRI promoter and in various other interferon-gamma regulated genes.

gas vacuole A prokaryotic cellular organelle consisting of cylindrical vesicles around 75×300 nm, often in clusters. The wall of the gas vacuole, which is permeable to gases but not to water, is formed from a monolayer of a single protein. Gas vacuoles are found mainly in planktonic cyanobacteria and their prime function is to make the bacterium buoyant.

GASC1 Gene (gene amplified in squamous cell carcinoma 1, JMJD2C) that belongs to the JMJD2 subfamily of the jumonji family (that contain two **jumonji domains**). Product (1056aa) is a lysine-specific histone trimethyl demethylase.

gasdermins A family of proteins expressed specifically in cells at advanced stages of differentiation. Gasdermin-1 (gasdermin-A, 445aa) induces apoptosis and is expressed predominantly in the gastrointestinal tract. It is suppressed in gastric cancer. Gasdermin-B (411aa) may be important for maintaining the final differentiated state of epithelial cells. Gasdermin-C is **MLZE**. Other members of the gasdermin family include **pejvakin** and non-syndromic hearing impairment protein 5 (**DFNA5**).

gastric carcinoma Cancer of the stomach, probably the second most common carcinoma. There are various inherited cancer predisposition syndromes, including hereditary nonpolyposis colon cancer (See **GTBP**), **adenomatous polyposis coli**, **Peutz-Jeghers syndrome**, **Cowden disease**, and the **Li-Fraumeni syndrome**. Another form is caused by germline mutations in the E-cadherin gene. Only a minority of cases have a clear hereditable component.

gastric inhibitory polypeptide Peptide hormone (glucose-dependent insulinotropic polypeptide, GIP, 42aa) of the incretin family secreted by intestinal K cells that stimulates insulin release and inhibits the release of gastric acid and pepsin. The receptor is G-protein coupled. See **GIP**.

gastrin Peptide **hormones** secreted by cells of the gastric mucosa (G-cells) and D cells of pancreas in response to mechanical stress or high pH and that stimulate secretion of protons and pancreatic enzymes. The precursor (progastrin, 101aa) is cleaved into six peptides: gastrin-71, gastrin-52, big gastrin, gastrin-34, gastrin (gastrin-17, 17aa), gastrin-14 and gastrin-6. Some forms are sulphated (which increases their half-life). Pancreatic tumours overproduce the hormones in **Zollinger-Ellison syndrome**. Gastrin is competitively inhibited by **cholecystokinin**.

gastrin-releasing peptide A regulatory peptide (27aa cleaved from a 148aa precursor) that stimulates the release of gastrin, pancreatic polypeptide, glucagon, gastric inhibitory peptide and insulin. Can be cleaved to produce neuromedin-C (10aa) which is the mammalian equivalent of **bombesin**. It elicits gastrin release, causes broncho-constriction and vasodilation in the respiratory tract, and stimulates the growth and mitogenesis of cells in culture. The receptor (384aa) is G-protein-coupled.

gastrocnemius One of the two large muscles in the calf of the leg. The frog gastrocnemius muscle was a favourite physiological preparation used to demonstrate control of muscle contraction.

gastrocoel See **archenteron**.

gastrodermis The layer of cells that lines the body cavity of Cnidarians and is responsible for digestion of prey.

gastrointestinal stromal tumours Mesenchymal tumours derived from the interstitial cells of Cajal, the pacemaker cells that regulate peristalsis of the gut. They are caused by mutation in *kit* or in platelet derived growth factor receptor A (PDGFRA) genes.

gastroparesis Disorder in which paralysis of the stomach muscles delays the passage of food through the stomach. Often associated with Type 1 diabetes.

Gastropoda Class of the Phylum Mollusca; snails, slugs, limpets and conches.

gastrula Embryonic stage of an animal when **gastrulation** occurs; follows **blastula** stage.

gastrulation A complex and coordinated series of cellular movements that occurs at the end of **cleavage** phase of embryonic development in most animals. The details of these movements, gastrulation, vary from species to species, but usually result in the formation of the three primary germ layers, **ectoderm**, **mesoderm** and **endoderm**.

GATA transcription factors A family of zinc-finger transcription factors involved in vertebrate embryonic development. They bind to the GATA nucleotide motif. GATA1 (erythroid transcription factor, eryf1, NF-E1 DNA-binding protein, GF-1, 413aa) probably serves as a general switch factor for erythroid development in conjunction with the transcriptional regulator friend of GATA1 (FOG1, zinc finger protein ZFPM1, 1006aa) with which the GATA factors heterodimerize. Mutations in this gene have been associated with X-linked dyserythropoietic anaemia and thrombocytopenia. GATA2 (endothelial transcription factor, 480aa) regulates endothelin-1 gene expression in endothelial cells. GATA-3 (trans-acting T-cell-specific transcription factor, 443aa) binds to the enhancer of the T-cell receptor alpha and delta genes. Haplo-insufficiency in GATA-3 leads to HDR syndrome (hypoparathyroidism, sensorineural deafness and renal anomaly syndrome). GATA4 (442aa) acts as a transcriptional activator of ANF and mutations have been associated with cardiac septal defects. GATA5 (397aa) may be important in smooth muscle diversity. GATA6 (595aa) has an expression pattern similar to GATA4 but is also found in lung and liver. GATA-like protein-1 (GATA-type zinc finger protein 1, GLP1, 271aa) is a transcriptional repressor of GATA6 and is important in germ cell development. Many GATA transcription factors have been identified in *Arabidopsis*.

gated ion channel Transmembrane proteins, that allow a flux of ions to pass only under defined circumstances, a feature of excitable cells. Channels may be either **voltage-gated**, such as the **sodium channel** of neurons, or **ligand-gated** such as the **acetylcholine receptor** of cholinergic synapses. Channels tend to be relatively ion-specific and allow fluxes of typically 1000 ions to pass in around 1 ms; they are thus much faster at moving ions across a membrane than transport **ATPases**.

gating currents Small currents in the membrane just prior to the increase in ionic permeability, caused by the movement of charged particles within the membrane. So called because they open the 'gates' for current flow through ion channels.

Gaucher's disease A **lysosomal disease**, most common in Ashkenazi Jews, in which the defect is in glucocerebrosidase (**beta-glucosidase**, EC 3.2.1.45), leading to the intracellular accumulation of glucosylceramide (GlcCer) in cells of the mononuclear phagocyte series. GlcCer-laden cells are referred to as Gaucher cells and found in a range of tissues. Associated with hepatosplenomegaly (enlargement of liver and spleen) and, in severe early onset forms of the disease, with neurological dysfunction.

gax The product of one of the family of growth arrest genes (*gas genes* and *gadd* genes), a homeobox protein (growth arrest-specific homeobox, mesenchyme homeobox 2, MOX2, 304aa) involved in induction of mesoderm, somitogenesis and myogenic and sclerotomal differentiation.

G banding *Giemsa banding* **Banding pattern** observed in spreads of metaphase chromosomes, treated briefly with protease then stained with **Giemsa.** Can be used to identify the separate chromosomes. The deeply-staining G bands do not coincide with the pattern of quinacrine bands (Q bands).

gbb The *Drosophila* orthologue (glass bottom boat, Protein 60A, 455aa) of mammalian BMP5,6,7 and 8, involved in the regulation of lipid metabolism and control of growth. Loss-of-function mutant larvae are almost transparent because of the underdevelopment of fat body. The receptor is **Sax**. Abstract: http://stke.sciencemag.org/cgi/content/abstract/sigtrans;3/104/ec7

GbpA A protein (GlcNAc-binding protein A, 485aa) secreted by various bacteria, including *Vibrio cholerae*, that may promote attachment to epithelial cell surfaces in the intestine and to chitin. Article: http://jb.asm.org/content/191/22/6911.long

gCAP39 See capG.

GC box DNA binding motif (consensus sequence 5′-G/T G/A GGCG G/T G/A G/A C/T-3′) recognized by many mammalian general transcription factors.

G-cell Cell of the stomach mucosa that secretes **gastrin** in response to **gastrin releasing peptide** from the vagus nerve.

GCIP (1) Grap2 and cyclin-D-interacting protein, a helix-loop-helix protein (cyclin-D1-binding protein 1, human homologue of maid, 360aa) that is a putative tumour suppressor. Interacts with **p29** and with **GRB2**-related adaptor protein 2 (Grap2) and may negatively regulate cell cycle progression. (2) In *Rana pipiens*, guanylyl cyclase inhibitory protein (206aa) found in retina.

GCN3 (1) In *S. cerevisiae*, the alpha subunit of the translation initiation factor eIF2B (305aa), the guanine-nucleotide exchange factor for eIF2; activity subsequently regulated by phosphorylated eIF2; first identified as a positive regulator of **GCN4** expression. (2) In *Arabidopsis*, GCN20-type ATP-binding cassette protein GCN3 (ABC transporter F family member 3, 715aa)

GCN4 (1) In *S. cerevisiae* general control protein GCN4 (amino acid biosynthesis regulatory protein, 281aa), a basic leucine zipper protein, a transcription factor responsible for the activation of more than 30 genes required for amino acid or for purine biosynthesise in response to amino acid or purine starvation Gcn4 is regulated at the level of transcription, translation and protein stability, with its half-life ranging from approximately two minutes during growth in rich medium to ten minutes under amino acid starvation conditions. (2) In *Arabidopsis*, ABC transporter F family member 4 (GCN20-type ATP-binding cassette protein GCN4, 724aa).

GCN5 family A family of histone acetyltransferases (Gcn5-related N-acetyltransferases, GNAT, EC 2.3.1.48, 568aa in *Arabidopsis*, 837aa in *H. sapiens*) that acetylate histone H3.

GCP (1) **Granulocyte chemotactic protein 2** (GCP2). (2) **Glutamate carboxypeptidase II**.

Gc protein A plasma protein (group-specific component, Gc-globulin, vitamin D-binding protein, 474aa) of the **albumin** family, also found in cerebrospinal fluid and urine and on the surface of many cell types It is involved in sterol transport and scavenges G-actin. Highly polymorphic.

gd A *Drosophila* serine protease (serine protease gd, EC 3.4.21.-, gastrulation defective, 528aa) involved in activation of **easter**.

GDAP (1) The Genomic Disulfide Analysis Program, that provides web access to computationally predicted protein disulphide bonds for over one hundred bacterial and archaeal genomes. Link to website: http://www.doe-mbi.ucla.edu/Services/GDAP (2) Ganglioside-induced Differentiation-Associated Protein (358aa) that may be involved in a signal transduction pathway responsible for ganglioside-induced neurite differentiation.

GDF (1) **GDI displacement factor**. (2) **Growth (and) differentiation factor**

GDGF Glioma-derived growth factor, originally derived from glioma cells, now known as glia-derived neurotrophic factor (**GDNF**).

GDI GTP dissociation inhibitor: protein that inhibits GDP/GTP exchange on **ras**-like GTPases thereby maintaing them in the active GTP-bound form. GDIs specific for particular families of GTPases have been found (e.g. rab-GDI, rho-GDI).

gDNA Genomic DNA.

GDNF A disulphide-linked homodimeric neurotrophic factor (astrocyte-derived trophic factor 1, formerly glioma-derived growth factor, monomers of 211aa) that enhances survival and morphological differentiation of dopaminergic neurons. One of the cysteine-knot superfamily of growth factors, structurally related to **artemin**, **neurturin** and **persephin**. The GDNF-receptor (GFRα-1, 465aa) is one of a family of GPI-linked receptors: GFRα-2 is the receptor for neurturin, GFRα-3 for artemin and GFRα-4 for persephin. Forms a heterodimeric

(alpha/beta) receptor complex with the tyrosine kinase **ret** as the beta subunit.

GDP Guanosine diphosphate. Phosphorylation gives **GTP**.

Gea (1) In *S. cerevisiae*, guanine nucleotide exchange factors (Gea1, 1408aa; Gea2, 1459aa) for **ARF**. Contain a **Sec7**-like domain and are similar to **ARNO** and **cytohesin**. (2) In *Arabidopsis*, small proteins (GEA1 – GEA6) homologous to the early methionine labelled (Em) proteins of wheat. For example Gea1 (152aa), found in seeds where it may protect against the effects of desiccation.

GEFs Family of proteins (guanine nucleotide exchange factors) that facilitate the exchange of bound GDP for GTP on small G-proteins such as ras and rho and thus activate them. Act in the opposite way to **GTPase activating proteins** (GAPs) which promote the hydrolysis of bound GTP, thereby switching the G-protein to the inactive form. Family includes **cytohesin**, **ARNO**, Gea-1 and 2, kalirin and yeast **Sec7**.

gel Jelly-like material formed by the coagulation of a colloidal liquid. Many gels have a fibrous matrix and fluid-filled interstices: gels are viscoelastic rather than simply viscous and can resist some mechanical stress without deformation. Examples are the gels formed by large molecules such as collagen (and gelatin), agarose, acrylamide and starch.

gel electrophoresis See **electrophoresis**.

gel filtration An important method for separating molecules according to molecular size by percolating the solution through beads of solvent-permeated polymer that has pores of similar size to the solvent molecules. Unlike a continous filter that retards flow according to molecular size, separation is achieved because molecules that can enter the beads take a longer path than those that cannot (i.e. are retarded). Typical gels for protein separation are made from polyacrylamide, or from flexible (Sephadex) or rigid (agarose, Sepharose) sugar polymers. The size separation range is determined by the degree of cross-linking of the gel.

gel mobility shift See **gel retardation assay**.

gel retardation assay Test (mobility shift assay) for interaction between molecules by looking for a change in gel electrophoretic mobility. For example, to assay for levels of a transcription factor, cell extracts are incubated with a radiolabelled oligonucleotide corresponding to the recognition sequence of the transcription factor, and run on an agarose gel. Most of the radiolabel will run quickly through the gel, but radioactive label is retarded if it interacts with the transcription factor.

gelatin Heat-denatured collagen.

geldanamycin A naturally-occurring benzoquinone ansamycin antibiotic produced by *Streptomyces hygroscopicus*. Geldanamycin binds to heat shock protein 90 (**Hsp90**) and inhibits essential ATPase activity of Hsp90 causing inactivation, destabilization, and degradation of Hsp90 client proteins. Has anti-tumour activity, inhibits pp60src tyrosine kinase and c-*myc* gene expression in murine lymphoblastoma cells and has been shown to have a range of other effects.

geleophysic dysplasia A mucopolysaccharidosis caused by mutation in the **ADAMTS2** gene. Affected children appear to have happy faces (gelios = happy).

gelonin A type I ribosome-inactivating toxin (316aa) isolated from *Gelonium multiflorum* (*Suregada multiflora*, False lime). It is a RNA-N-glycosidase that depurinates RNA in ribosomes, thus inhibiting protein synthesis. Widely used to construct immunotoxins.

gelsolin Actin-binding protein (brevin, actin depolymerising factor, 782aa) that nucleates actin polymerization, but at high calcium ion concentrations (10^{-6} M) causes severing of filaments.

geminin Protein (209aa) that inhibits the assembly of the pre-replication complex by binding to **Cdt1** and preventing it from recruiting the minichromosome maintenance proteins to chromatin. It is ubiquitinated by the anaphase-promoting complex and degraded during the mitotic phase of the cell cycle. Also directly interacts with Six3 and Hox homeodomain proteins during embryogenesis and inhibits their functions. Geminin coiled-coil domain-containing protein 1 (334aa) promotes the initiation of chromosomal DNA replication by mediating TOPBP1- (DNA topoisomerase 2-binding protein 1) and CDK2- (cyclin-dependent kinase-2) dependent recruitment of CDC45L (cell division control protein 45 homologue) onto replication origins.

geminivirus A large family of plant viruses with circular, single-stranded DNA genomes that replicate through double-stranded intermediates. Have recently emerged as leading plant pathogens that cause severe crop losses worldwide, infecting a broad range of plant species.

gemins See **Gems** and **SMN complex**.

gemma *Pl.* gemmae. (1) Small multicellular structure involved in vegetative reproduction in algae, pteridophytes and bryophytes. (2) Same as chlamydospore. (3) A bud that will give rise to a new individual.

Gemmata obscuriglobus A member of the Planctomyces group of stalked bacteria that show extensive cell compartmentalisation. In *Gemmata* the nucleoid is surrounded by a nuclear envelope, a feature that blurs the distinction between prokaryotic and eukaryotic organisms.

Gemmatimonadetes A bacterial phylum with a single family, first identified from activated sludge in a sewage treatment system. They are Gram-negative pigmented rod-shaped aerobes, the type species being *Gemmatimonas aurantiaca*.

Gems Nuclear inclusions (gemini of the coiled bodies), resembling Cajal bodies in number and size. They are enriched in **SMC complexes**.

gene Originally defined as the physical unit of heredity but the meaning has changed with increasing knowledge. It is probably best defined as the unit of inheritance that occupies a specific locus on a chromosome, the existence of which can be confirmed by the occurrence of different allelic forms. In this sense it is almost an abstract concept and the gene for a disorder is actually a mutant allele. Alternatively it could be defined as the set of DNA sequences (**exons**) that are required to produce a single polypeptide although regulatory sequences, promoter sites and introns are an essential part of the locus although they do not encode the protein sequence. The second definition is also problematic since many proteins have alternatively spliced forms and some loci encode only RNA.

gene amplification Selective replication of DNA sequence within a cell, producing multiple extra copies of that sequence. The best-known example occurs during the maturation of the oocyte of *Xenopus*, where the set (normally 500 copies) of ribosomal RNA genes is replicated some 4000 times to give about 2 million copies. For many genes even a small increase in the normal complement (as in trisomy) can be deleterious.

gene annotation The process of adding comments, cross-links, information regarding alternative splicing, pseudogenes, promoter regions, the chromosomal location, homologies, etc. to gene entries in databases.

gene chip An array of oligonucleotides immobilized on a surface that can be used to screen an RNA sample (after reverse transcription) and thus a method for rapidly determining which genes are being expressed in the cell or tissue from which the RNA originated. There are two alternatives for the immobilized oligonucleotides: either a random set of defined sequences or known probes for genes (probes for cytokines, adhesion molecules and others). In both cases the position on the chip defines the sequence that hybridizes.

gene cloning The insertion of a DNA sequence into a **vector** that can then be propagated in a host organism, generating a large number of copies of the sequence.

gene conversion A phenomenon in which alleles are segregated in a 3:1 not 2:2 ratio in meiosis. May be a result of **DNA polymerase** switching templates and copying from the other homologous sequence, or a result of mismatch repair (nucleotides being removed from one strand and replaced by repair synthesis using the other strand as template).

gene dosage Number of copies of a particular **gene** locus in the genome; in most cases either one or two. An excess of copies can cause problems.

gene duplication A class of DNA rearrangement that generates a supernumerary copy of a gene in the genome. This would allow each gene to evolve independently to produce distinct functions. Such a set of evolutionarily-related genes can be called a 'gene family'.

gene expression The full use of the information in a gene via **transcription** and **translation** leading (usually) to production of a protein and hence the appearance of the **phenotype** determined by that gene. Gene expression is assumed to be controlled at various points in the sequence leading to protein synthesis and this control is thought to be the major determinant of cellular **differentiation** in eukaryotes.

gene family A set of genes (multigene family) coding for diverse proteins which, by virtue of their high degree of sequence similarity, are believed to have evolved from a single ancestral gene. An example is the immunoglobulin family where the characteristic features of the constant-domains are found in various cell surface receptors.

Gene Index An extensive collection of resources for handling genomic information and searching for genes in a range of plant, animal and microbial species. It includes Resourcerer, a microarray-resource annotation and cross-reference database built using the analysis of expressed sequence tags and GCOD (GeneChip Oncology Database) a collection of publicly available microarray gene expression data related to human cancers. Link to website: http://compbio.dfci.harvard.edu/tgi/

gene neighbour method Computational method of trying to deduce protein interactions based upon chromosomal location: if two genes are always found adjacent in several genomes then the assumption is that they are likely to be involved in a common function. Though obviously relevant in prokaryotes where there are operons, the principle seems to extend to eukaryotes. See **Rosetta Stone method** and **phylogenetic profile**.

Gene Ontology Consortium A bioinformatics initiative supported by the US National Human Genome Research Institute (NHGRI) that aims to standardize the representation and nomenclature of gene and gene product attributes across species and databases. It provides a controlled vocabulary of terms for describing gene product characteristics and gene product annotation data. The project started as a collaboration between **FlyBase** (*Drosophila*), the **Saccharomyces Genome Database** (SGD) and the **Mouse Genome Database** (MGD) but has subsequently been extended to include, among others, the **Cardiovascular Gene Ontology Annotation Initiative**, **dictyBase** (for *Dictyostelium discoideum*), **EcoliWiki** (for *E. coli*), **GeneDB** (databases for *S. pombe, Plasmodium falciparum, Leishmania major* and *Trypanosoma brucei*), **Gramene** (for grains, including rice), the **Rat Genome Database** (RGD), the **Reactome**, the *Arabidopsis* Information Resource (**TAIR**), **WormBase** (for *C. elegans*), the Zebrafish Information Network (**ZFIN**) for *Danio rerio*, and databases on several bacterial species. Link: http://www.geneontology.org/

gene regulatory protein Any protein that interacts with DNA sequences of a gene and controls its transcription.

gene therapy Treatment of a disease caused by malfunction of a gene by stable **transfection** of the cells of the organism with the normal gene. Although an attractive prospect, the technical problems of designing safe vectors and transfecting sufficient cells in the patient have made it virtually impossible for most diseases.

gene transfer General term for the insertion of foreign genes into a cell or organism. Synonymous with **transfection**.

GeneDB Database of genomic information for 35 genomes (in March 2011) including *Schizosaccharomyces pombe*, various parasites, bacteria, parasite vectors etc. Link: http://www.genedb.org/Homepage

GeneDoc Windows-based software for carrying out multiple sequence alignments, a way of investigating the relationship between structure and function in biomolecules. Link: http://www.nrbsc.org/gfx/genedoc/

general import pore *GIP* See **TOM complex**.

general insertion protein See **TOM complex**.

general receptor for phosphoinositides-1 See **cytohesins**.

generation time Time taken for a cell population to double in numbers, and thus equivalent to the average length of the cell cycle.

genetic burden See **genetic load**.

genetic code Relationship between the sequence of bases in **nucleic acid** and the order of amino acids in the polypeptide synthesized from it. A sequence of three nucleic acid **bases** (a triplet) acts as a 'codeword' (**codon**) for one amino acid. See Table C4.

genetic drift Random change in allele frequency within a population. If the population is isolated and the process continues for long enough may lead to speciation.

genetic engineering General term covering the use of various experimental techniques to produce molecules of DNA containing new genes or novel combinations of genes, usually for insertion into a host cell for cloning.

genetic linkage The term refers to the fact that certain genes tend to be inherited together, because they are on the same chromosome; the closer they are physically, the less likely is a recombination event separating them. Thus parental combinations of characters are found more frequently in offspring than non-parental. Linkage is measured by the percentage recombination between loci, unlinked genes showing 50% recombination. See **linkage equilibrium, linkage disequilibrium, synteny**.

genetic load In general terms the decrease in fitness of a population (as a result of selection acting on phenotypes) due to deleterious mutations in the population gene pool (genetic burden). More specifically, the average number of **recessive lethal mutations**, in the **heterozygous** state, estimated to be present in the genome of an individual in a population.

genetic locus The position of a gene in a linkage map or on a chromosome.

genetic recombination Formation of new combinations of alleles in offspring (viruses, cells or organisms) as a result of exchange of DNA sequences between molecules. It occurs naturally, as in **crossing-over** between homologous chromosomes in meiosis or experimentally, as a result of **genetic engineering** techniques.

genetic transformation Genetic change brought about by the introduction of exogenous DNA into a cell. See **transformation, germ-line transformation, transfection**.

genetically modified organism *GMO* In general usage, an organism that has been altered by modern methods of genetic engineering rather than by traditional selective breeding.

geneticin An antibiotic (antibiotic G418) used as a selection agent in transfection. It is toxic to bacteria, yeast and mammalian cells unless a bacterial geneticin-resistance gene (which can be expressed in eukaryotes) is present. The vector has the resistance gene and only positive transfectants survive.

genistein Phytoestrogen from soybeans that is an inhibitor of protein tyrosine kinases. Competes at ATP binding site and will inhibit other kinases to some extent. A potent vasorelaxant, probably through effects on **CFTR** and **NKCC1**.

genome The total set of genes carried by an individual or cell.

Genome Database for Rosaceae *GDR* A curated and integrated web-based relational database containing data of the genetically anchored peach physical map, annotated EST databases of apple, peach, almond, cherry, rose, raspberry and strawberry, Rosaceae maps and markers and all publicly available Rosaceae sequences. Link: http://www.bioinfo.wsu.edu/gdr/about_gdr.php

genome project Coordinated programme to completely sequence the genomic DNA of an organism. Usually, genomic sequencing is combined with several associated ventures; the **physical mapping** of the genome (to allow the genome to be sequenced); the sequencing of **expressed sequence tags** (to aid in the identification of transcribed sequences in the genomic DNA sequence); for non-human organisms, a programme of systematic mutagenesis, and genetic mapping of the mutants (to infer function of novel genes by reverse genetics); and an overarching computer database resource, to manage and give access to the data.

genomic clone Clone of a portion of DNA obtained from genomic DNA rather than of DNA produced by reverse transcription from mRNA. It will therefore contain introns as well as exons – but may not necessarily be expressed or complete.

genomic imprinting An epigenetic phenomenon in which there is parent-specific expression or repression of genes or chromosomes in offspring. Imprinted genes can be silenced by methylation (see **DNA methylation**) often of clusters of genes under the influence of **imprinting control regions** (ICRs) that are set differently during gametogenesis in male and female germlines. There are an increasing number of recognized chromosomal imprinting events in pathological conditions: e.g. preferential transmission of paternal or maternal predisposition to diabetes or atopy, preferential retention of paternal alleles in **rhabdomyosarcoma, osteosarcoma, retinoblastoma** and **Wilms' tumour**, preferential translocation to the paternal chromosome 9 of a portion of maternal chromosome 22 to form the **Philadelphia chromosome** of chronic myeloid leukaemia. See also **Prader-Willi syndrome, Angelman's syndrome**.

genomic instability Abnormally high rates (possibly accelerating rates) of genetic change occurring serially and spontaneously in cell-populations, as they continue to proliferate. A potentially serious feature of rapidly proliferating tumours. May well be a consequence of mutation or dysfunction of normal DNA damage-repair mechanisms.

genomic island Mobile genetic elements of length from 10 to 200 kb that have been integrated into an organism's genome. They allow lateral transfer of genes between a range of phylogenetically distinct organisms and can therefore facilitate the rapid spread of a trait (for example, multidrug-resistance in *Salmonella* based upon an antibiotic resistance gene cluster in Salmonella genomic island 1). Can sometimes be recognised because of atypical G + C content.

genomic library Type of DNA library in which the cloned DNA is from an organism's genomic DNA. As genome sizes are relatively large compared to individual **cDNAs**, a different set of **vectors** is usually employed in addition to **plasmid** and **phage**; see **bacterial** and **yeast artificial chromosomes**, **cosmid**.

genotoxin Toxin that works by causing damage to a gene or interfering with its function.

genotype The genetic constitution of an organism or cell, as distinct from its expressed features or **phenotype**.

gentamicin A group of aminoglycoside antibiotics produced by *Micromonospora* spp. Members include the closely related gentamycins C1, C2 and C1a, together with gentamycin A. They inhibit protein synthesis on 70S ribosomes by binding to the 23S core protein of the small subunit, that is responsible for binding mRNA. Mode of action similar to that of **kanamycin**, **neomycin**, paromomycin, spectinomycin and streptomycin. Active against strains of the bacterium *Pseudomonas aeruginosa*.

geotaxis See **gravitaxis**. The prefix gravi- is preferable since the gravitational fields used as cues need not necessarily be the Earth's.

geotropism See **gravitropism**.

gephyrin Peripheral membrane protein (736aa) of the cytoplasmic face of the glycinergic synapses in the spinal cord. Appears at developing postsynaptic sites before the **glycine receptor** and may therefore be important in clustering. Thought to interact with the glycine receptors, with microtubules and indirectly with the actin cytoskeleton. Catalyzes two steps in the biosynthesis of the **molybdenum cofactor**. Negatively regulates **collybistin**. See **startle disease**.

geranyl Prenyl group that can be post-translationally added to proteins generally at a CAAX motif (where X is usually leucine) and serves as a membrane-attachment device; geranyl-geranyl prenylation is more common. Geranyl pyrophosphate is an intermediate in cholesterol synthesis and in production of the geranyl-geranyl group. See **farnesylation**.

geranyl transferase Enzyme (farnesyl pyrophosphate synthetase, EC 2.5.1.10, 419aa) that catalyzes the formation of farnesyl diphosphate, a substrate for the post-translational transfer of geranyl-geranyl residue to protein and important in synthesis of sterols, dolichols, carotenoids, and ubiquinones. Geranylgeranyl pyrophosphate synthase (dimethylallyltranstransferase, EC 2.5.1.1, farnesyl diphosphate synthase, EC 2.5.1.29, geranylgeranyl diphosphate synthase, EC 2.5.1.10, 300aa) catalyzes the formation of geranylgeranyl pyrophosphate, an important precursor of carotenoids and geranylated proteins.

GERL The Golgi-endoplasmic reticulum-lysosome system. See individual entries for each of these membranous compartments of the trans-Golgi network.

germ cell Cell specialized to produce **haploid** gametes. The germ cell line is often formed very early in embryonic development and undergoes fewer cells divisions than somatic cells, reducing the chances of copy-errors. In some animals the only cells to retain the full complement of chromosomes. See **Weismann's germ plasm theory**.

germ cell nuclear factor A nuclear orphan receptor (GCNF, retinoid receptor-related testis-specific receptor, 480aa) that functions as a transcriptional repressor during gametogenesis and is transiently expressed in mammalian carcinoma cells during retinoic acid (RA) induced neuronal differentiation.

germ layers The main divisions of tissue types in multicellular organisms. Diploblastic organisms (e.g. coelenterates) have two layers, ectoderm and endoderm (the latter sometimes referred to as gastrodermis); triploblastic organisms (all higher animal groups) have mesoderm between these two layers. Germ layers become distinguishable during late blastula/early gastrula stages of embryogenesis, and each gives rise to a characteristic set of tissues, the ectoderm to external epithelia and to the nervous system for example, although some tissues contain elements derived from two layers.

germ-line therapy Gene therapy that introduces a new gene into the reproductive cells of the body and can therefore be inherited. Generally considered an inappropriate and potentially risky procedure at present. See **germ-line transformation**.

germ-line transformation Micro-injection of foreign DNA into an early embryo, so that it becomes incorporated into the **germ-line** of the individual, and thus stably inherited in subsequent generations of **transgenic** organisms. Typically, the DNA would be a **reporter gene** or cDNA in a **vector** such as a **transposon**, that might also carry a visible **marker gene** (such as eye or coat colour), so that successful transformation could readily be detected.

germin A protein marker of the onset of growth in germinating wheat, later shown to be associated with the cell wall and more recently to be an oxalate oxidase (EC 1.2.3.4, 224aa). Produces developmental and stress-related release of hydrogen peroxide in the apoplast. See **germin like proteins**.

germin like proteins Class of proteins (GLPs) with sequence and structural similarity to the cereal **germins** but mostly without oxalate oxidase activity. Germins and germin like proteins are developmentally regulated glycoproteins characterized by a beta-barrel core structure, a signal peptide, and are associated with the cell wall. GLPs are found in organisms ranging from myxomycetes, bryophytes, pteridophytes, gymnosperms and angiosperms and have diverse activities. Some may be associated with defence reactions.

germinal center kinases A subgroup of the Ste20 family of kinases that includes **MINK1**, Traf2- and Nck-interacting kinase (TNIK), **SOK1**. Germinal centre kinase (mitogen-activated protein kinase kinase kinase kinase 2, MAP4K2, EC 2.7.11.1, 820aa) specifically activates the SAPK pathway, is activated by TNFα and is highly expressed in the **germinal centre**. Germinal centre kinase-like kinase (HGK, hepatocyte progenitor kinase-like kinase, MAP4K4, 1239aa) activates **Jnk** and is upstream of **mTOR**.

germinal centre An aggregation of lymphocytes, mainly B-cells (centrocytes) and proliferating B-cells (centroblasts), that develops from a primary follicle in response to antigenic stimulation. **Antigen presenting cells** are also conspicuously present. May be sites at which B-memory cells are produced with receptors which recognize antigens in the complexes.

germlings Germinating asexual spores of *Neurospora crassa* that are attracted to each other to fuse and form hyphae called conidial anastomosis tubes (CATs).

gerodermia osteodysplastica A disorder characterized by wrinkly skin and osteoporosis caused by loss-of-function mutations in a **golgin** (rab6-interacting golgin, SCYL1BP1, 394aa) which is highly expressed in skin and osteoblasts.

Gerstmann-Straussler-Scheinker syndrome A familial **spongiform encephalopathy**. Transgenic mice with a mutant form of the **PrP** gene from patients with this syndrome develop degenerative brain disease that is similar, but not identical, to that caused by **scrapie**.

GF-1 See **GATA transcription factors**.

GFAP See **glial fibrillary acidic protein**.

GFP See **green fluorescent protein**.

GFR Glial cell line-derived neurotrophic factor (GDNF)-family receptor. See **GDNF**.

GGT See **gamma-glutamyltransferase**.

ghosts See **erythrocyte ghost**.

ghrelin See **appetite-regulating hormone**.

GHRH Growth hormone releasing hormone. See **growth hormone**.

GI (1) Common abbreviation for gastro-intestinal. (2) G$_i$: See **GTP-binding protein**. (3) Glycaemic index.

giant axonal neuropathy A severe neuropathy affecting peripheral nerves and the central nervous system caused by mutation in the gene encoding **gigaxonin**.

giant axons Extraordinarily large unmyelinated axons found in invertebrates. Some, like the squid giant axon, can approach 1 mm diameter. Large axons have high conduction speeds; the giant axons are invariably involved in panic or escape responses, and may (e.g. crayfish) have **electrical synapses** to further increase speed. Vertebrate axons with high conduction velocites are much narrower: they have a **myelin sheath**, allowing **saltatory conduction**.

giantism Excessive growth (gigantism) of the body, usually due to overproduction of growth hormone by the pituitary. May be a consequence of a benign pituitary tumour or, occasionally, multiple endocrine neoplasia type 1 (MEN-1), McCune-Albright syndrome (MAS), neurofibromatosis, or **Carney complex**.

Giardia Genus of flagellate protozoans, found as intestinal parasites of vertebrates. The human intestinal parasite is *Giardia lamblia* which can cause acute or chronic diarrhoea and a characteristic sulphurous breath.

giardins Proteins (alpha, beta and gamma forms with sizes ranging from 273aa to 311aa) involved in the attachment of *Giardia lamblia* to the intestinal mucosa and in the cytoskeletal disassembly and reassembly that marks the transition from infectious trophozoite to transmissible cyst.

gibberellic acids A major group of plant growth regulators or phytohormones, diterpenoid compounds with **gibberellin** activity in plants. At least 70 related gibberellic acids have been described and designated as a series GA1, GA2 etc. Gibberellic acid-regulated responses include the promotion of stem elongation, mobilization of food reserves in seeds, and other processes; its absence results in the dwarfism of some plant varieties. Responses are mediated by **DELLA**-domain-containing proteins including GAI, RGA and RGL1-3. See **SLEEPY**.

Giemsa A Romanovsky-type stain that is often used to stain blood films that are suspected to contain protozoan parasites. Contains both basic and acidic dyes and will therefore differentiate acid and basic granules in granulocytes. See **G banding**.

GIGANTEA In *Arabidopsis* a protein (1183aa) involved in regulation of circadian rhythm and photoperiodic flowering. Regulates **CONSTANS** in the long-day flowering pathway.

gigaxonin A cytoskeletal protein (kelch-like protein 16, 597aa) that is important in neurofilament networks. One of the cytoskeletal BTB/kelch (Broad-Complex, Tramtrack and Bric-a-brac) repeat family that is a substrate-specific adapter of an E3 ubiquitin-protein ligase. It controls protein degradation including that of **tubulin-folding cofactor** B. Mutations lead to **giant axonal neuropathy**.

Gilbert's syndrome Benign disorder (hyperbilirubinemia 1) causing mild jaundice. Caused by mutation in the gene for UDP-glucuronosyltransferase (UGT1A1, EC 2.4.1.17, 533aa) which is important for glucuronidation, a mechanism involved in the elimination of many lipophilic xenobiotics and endogenous substances such as bilirubin. More severe deficiencies of the enzyme lead to **Crigler-Najjar syndrome**.

Gilles de la Tourette syndrome *Tourette syndrome* A neurologic disorder characterised by motor and vocal tics and behavioural abnormalities. May be caused by defects in **slit and NTRK-like protein-1** although other genes are also implicated.

GILT Gamma-interferon inducible lysosomal thiolreductase (261aa) that cleaves disulphide bonds in proteins by reduction, facilitating subsequent lysosomal degradation and proteasomal MHC class II-restricted antigen processing.

Ginkgo biloba Ornamental tree originally native to China. Sole surviving member of the family Ginkgoales. Source of various bioactive compounds.

GINS complex Protein complex (go ichi ni san complex) that allows the MCM (minichromosome maintenance) helicase to interact with key regulatory proteins in the large replisome progression complexes (RPCs) that are assembled at eukaryotic DNA replication forks during initiation of S phase. The four subunits, Sld5 (GINS4, 223aa), Psf1 (Partner of Sld5-1, GINS1, 196aa), Psf2 (GINS2, 185aa) and Psf3 (GINS3, 216aa), are ubiquitous and evolutionarily conserved in eukaryotes. See **minichromosome maintenance proteins**.

GIP An over-used acronym: (**1**) **Gastric inhibitory polypeptide**. (**2**) **Glucose-dependent insulinotropic polypeptide**. (**3**) **General insertion protein**. (**4**)

Gip: Light-regulated inhibitory G-protein from washed microvilli of the photoreceptors of *Octopus*. (**5**) GTPase inhibitory protein that interferes with the binding of **GAP** to ras, or enhances nucleotide exchange (a **GEF**). (**6**) In *Drosophila*, Copia protein (Gag-int-pol protein, 1409aa) that is cleaved into Copia VLP protein and Copia protease. (**7**) The *gip2* oncogene encodes GTPase-deficient alpha-subunits of G_s or G_i-2 proteins and has been identified in tumours of the ovary and adrenal cortex. It will induce neoplastic transformation of Rat-1 cells but not NIH 3T3 cells and appears to be a tissue-selective oncogene.

girdin A ubiquitous protein (girders of actin filament, hook-related protein 1, coiled-coil domain-containing protein 88A, Akt phosphorylation enhancer, 1871aa) that is an important modulator of Akt-mTOR signalling. It is important for dynamic regulation of the actin cytoskeleton in cell motility and may also be involved in vesicle sorting in the early endosome.

GIRKs A family of potassium channels found in mammals and birds (G-protein-gated inward rectifying potassium channels), homo or hetero-oligomers of the various subunits. GIRK1 (Kir3.1, KCNJ3, 501aa) and GIRK4 (Kir3.4, 419aa) are found mainly in the atrium of the heart, GIRK1, GIRK2 (Kir3.2, 423aa), and GIRK3 (Kir3.3, 393aa) are found in neural tissue. A point mutation in the GIRK2 gene is the cause of the neurological and reproductive defects observed in the **weaver** mutant mouse.

GITR Glucocorticoid-induced tumour necrosis factor receptor (TNFRSF18, CD357, 241aa), the receptor for TNFSF18. Involved in interactions between activated T-cells and endothelial cells and in the regulation of T-cell receptor-mediated cell death.

GJA, GJB etc. Gap junction proteins: for example, GJA1 is gap junction alpha-1 (connexin-43). See **connexins**.

GL1 A transcription activator (trichome differentiation protein GL1, GLABROUS 1, 228aa) involved in epidermal cell fate specification in leaves. Regulates the production of a signal that induces the differentiation of hair (**trichome**) precursor cells on leaf primordia.

GLA (**1**) Gamma linolenic acid. (**2**) Protein found in bone (bone Gla protein or **osteocalcin**) and matrix (**matrix Gla protein**), important in calcium metabolism and skeletal development. (**3**) The gla-domain is a calcium-binding motif, found in Gla proteins and other vitamin K dependent proteins, consisting of 9-12 residues of gamma-carboxyglutamic acid (Gla) distributed through around 45aa. (**4**) Galactosidase-alpha (EC 3.2.1.22, GLA), defective in **Fabry's disease**.

gland Organ specialized for secretion by the infolding of an epithelial sheet. The secretory epithelial cells may either be arranged as an acinus with a duct or as a tubule. Glands from which release occurs to a free epithelial surface are exocrine; those that release product to the circulatory system are endocrine glands.

glanders A contagious bacterial disease of equiids due to infection by *Burkholderia pseudomallei* (formerly *Actinobacillus mallei, Malleomyces mallei*). See **melioidosis**.

glandular fever Self-limiting disorder of lymphoid tissue caused by infection with **Epstein-Barr virus** (infectious mononucleosis). Characterized by the appearance of many large lymphoblasts in the circulation.

Glanzmann's thrombasthenia Platelet dysfunction in which aggregation is deficient. Can be caused by mutation in the gene for platelet glycoprotein IIb (integrin α_{2B}, CD41, 1039aa) or for platelet glycoprotein IIIa (integrin β_3, CD61, 788aa). The integrin heterodimer α_{2B}/β_3 is a receptor for fibronectin, fibrinogen, plasminogen, prothrombin, thrombospondin and vitronectin. See Table I.1.

glaucoma A range of eye disorders in which the intra-ocular pressure rises, causing damage to optic nerve fibres. Major cause of blindness after the age of 45. Mutations in various genes increase susceptibility or cause different forms, including those for **myocilin**, **optineurin**, WD repeat-containing protein-36, and **OPA1**.

GlcNAc-phosphotransferase A multisubunit glycosyl transferase (DPAGT1, UDP-N-acetylglucosamine-dolichyl-phosphate N-acetylglucosaminephosphotransferase, EC 2.7.8.15, 408aa) that catalyzes the initial step in the synthesis of dolichol-oligosaccharides. Defects in DPAGT1 cause one type of glycosylation defect (CDG1J). (See **CDGs**).

GLD-1 A **STAR protein** (germline defective-1, 463aa) from *C. elegans*, essential for oogenesis. A translational repressor that acts through regulatory elements in the 3′ untranslated region of the sex-determining gene *tra-2*.

GLGF A motif, the Gly-Leu-Gly-Phe loop, found in the **PDZ** domain (also called DHR or GLGF domain) of diverse membrane-associated proteins.

GLI **oncogenes** Glioma-associated oncogenes that encode zinc-finger transcription factors of the Kruppel family. GLI1 (1106aa), GLI2 (1586aa) and GLI3 (1580aa) mediate **sonic hedgehog** signalling; defects in GLI2 are the cause of **holoprosencephaly** type 9, defects in GLI3 are found in **Pallister-Hall syndrome** and various forms of polysyndactyly. See **hedgehog signalling complex**.

gliadins Group of proline-rich storage proteins (ca 250–300aa) found in cereal seeds. Associate with **glutenin** to form **gluten**.

glial cells Specialized non-neuronal cells (neuroglia) that surround **neurons**, providing mechanical and physical support, and electrical insulation between neurons. Microglia are a myeloid-series subset; the macroglia are subdivided into **astrocytes**, **oligodendrocytes**, **ependymal cells** and **radial glial cells** in the central nervous system and **Schwann cells** in the peripheral system.

glial fibrillary acidic protein A type III intermediate filament protein (GFAP, 432aa), characteristic of **astrocytes**. Mutations in the gene cause **Alexander's disease**.

glial filaments Intermediate filaments of glial cells, made of **glial fibrillary acidic protein**.

glibenclamide See **sulphonylureas**.

glicentin See **glucagon**.

gliding motility Mode of cell motility exhibited by, for example, gregarines. There are no obvious motile appendages and the motor mechanism is poorly understood, although actin and two class XIV unconventional myosins have been cloned from *Gregarina polymorpha* and are thought to be involved. Abstract: http://onlinelibrary.wiley.com/doi/10.1002/cm.10178/abstract

glioblastoma Highly malignant brain tumour derived from glial cells. See **gliomas**.

gliomas Neuroectodermal tumours of neuroglial origin: include astrocytomas, oligodendroglioma, and ependymoma derived from **astrocytes**, **oligodendrocytes** and **ependymal cells** respectively. All infiltrate the adjacent brain tissue, but they do not metastasise. Glioblastoma is an aggressive form of glioma and may be associated with a viral infection.

gliostatin See **PD-ECGF**.

globin The polypeptide moiety of haemoglobin. In the adult human most haemoglobin molecules have two α (141aa) and two β (146aa) globin chains although a small percentage (~3%) have delta chains (147aa), not beta. Other variants are found in fetal development. Other members of the globin family are found in nematodes (*C. elegans*), annelids (e.g. *Lumbricus terrestris* (common earthworm)), molluscs (e.g. *Aplysia limacina*) and insects (e.g. globin CTT-III or erythrocruorin III in *Chironomus thummi thummi* (midge))).

globoside Major neutral glycosphingolipid found in kidney and erythrocytes.

globular protein Any protein that adopts a compact morphology is termed globular. Generally applied to proteins in free solution, but may also be used for compact folded proteins within membranes.

globus pallidus See **basal ganglia**.

glomerulonephritis Inflammatory response in the kidney glomerulus that often arises because immune complexes cannot pass through the basement membrane of the fenestrated epithelium where plasma filtration occurs. Circulating neutrophils are trapped on the accumulated adhesive immune-complexes (that also activate complement). Immune complex tends to be irregularly distributed in contrast to the picture in **Goodpasture's syndrome**. In Berger's nephropathy (IgA disease, see **Henoch-Schönlein purpura**), the immune complexes involve IgA.

glomulin A ubiquitous protein (FK506-binding protein-associated protein, 594aa) that is essential for normal development of the vasculature and is associated with **glomus tumours**. Binding to the **immunophilin** is inhibited by FK506 and rapamycin and glomulin may be the natural ligand. May function as an membrane anchoring protein.

glomus tumour Benign cutaneous neoplasms with rounded glomus cells (derived from perivascular smooth muscle) around distended vein-like channels. Caused by mutations in the **glomulin** gene.

gloomy face syndrome See **3M syndrome**.

GLP-1 (1) **Glucagon-like peptide 1**. (2) **Germinlike protein**-1. (3) A **GARP** homologue found in the protozoan *Giardia lamblia*, an important transcriptional activator. (4) One of the glutaredoxin-like proteins, a subgroup of glutaredoxins with a serine replacing the second cysteine in the CxxC-motif of the active site.

GlpF-like intrinsic protein A subset of the **aquaporins**. GlpF is the glycerol uptake facilitator protein (aquaglyceroporin, 281aa in *E. coli*) that transports glycerol across the cytoplasmic membrane, with limited permeability to water and small uncharged compounds.

GLPs Glycerol-transporting channel proteins, a subset of the **aquaporins** that includes AQP3, AQP7, and AQP9, several nematode paralogues, a yeast paralogue, and *E. coli* GLP (GlpF, glycerol uptake facilitator protein, 281aa).

glucagon The glucagon precursor (180aa) can be cleaved by prohormone convertase into glicentin (69aa), glicentin-related polypeptide (30aa), oxyntomodulin (37aa), glucagon (29aa), glucagon-like peptide 1 (GLP-1, 37aa), GLP-1(7−37), GLP-1(7−36) and glucagon-like peptide 2 (33aa) depending upon the tissue. Glucagon is the counter-hormone to insulin and is secreted by the **A cells** of the Islets of Langerhans in the pancreas in response to a fall in blood sugar levels. The insulin to glucagon ratio in the liver determines the rates of gluconeogenesis and glycogenolysis. GLP-1 and GLP-2 are produced by **intestinal L cells**: GLP-1 is a potent stimulator of glucose-dependent insulin release, GLP-2 stimulates intestinal growth and up-regulates villus height in the small intestine. Oxyntomodulin significantly reduces food intake. Glicentin may modulate gastric acid secretion. Structurally related peptides include gastric inhibitory polypeptide, **secretin**, **vasoactive intestinal peptide**, **growth hormone releasing factor**, **PACAP** and **exendins**. The glucagon receptor (477aa) signals through adenylate cyclase and also mediates an increase in intracellular calcium. See also Table H2. Dedicated website: http://www.glucagon.com/

glucans Glucose-containing polysaccharides, including **cellulose**, **callose**, **laminarin**, **starch**, and **glycogen**.

glucocorticoid receptor interacting protein See **GRIP** #2.

glucocorticoids Steroid hormones (both natural and synthetic) that promote **gluconeogenesis** and the formation of glycogen at the expense of lipid and protein synthesis. They also have important anti-inflammatory activity. Type compound is hydrocortisone (cortisol), other common examples are cortisone, prednisone, prednisolone, dexamethasone, betamethasone; *Cf.* **mineralocorticoids**, see also Table H3. The glucocorticoid receptor (nuclear receptor subfamily-3, group-C, member-1, NR3C1, 777aa) is a ubiquitous ligand-activated transcription factor. The receptor is cytoplasmic and complexed with Hsp90, Hsp70, FK binding protein and other proteins until activated by binding of the corticosteroid, upon which it dissociates from the binding proteins and moves to the nucleus where it binds to glucocorticoid response elements (GRE) and modulates other transcription factors.

glucomannan Hemicellulosic plant cell-wall polysaccharide containing glucose and mannose linked by $\beta(1-4)$-glycosidic bonds. May contain some side-chains of galactose, in which case it may be termed galactoglucomannan. A major polysaccharide of gymnosperm wood (softwood).

gluconeogenesis Synthesis of glucose from non-carbohydrate precursors, such as pyruvate, amino acids and glycerol. Takes place largely in liver, and serves to maintain blood glucose under conditions of starvation or intense exercise.

glucosamine An amino-sugar (2-amino-2-deoxy-glucose), a component of **chitin**, **heparan sulphate**, **chondroitin sulphate**, and many complex polysaccharides. Usually found as β-D-N-acetylglucosamine.

glucosaminoglycan See **glycosaminoglycan**.

glucose *dextrose* Six-carbon sugar (aldohexose) widely distributed in plants and animals. Breakdown of glucose (**glycolysis**) is a major energy source for metabolic processes. In green plants, glucose is a major product of photosynthesis, and is stored as the polymer **starch**. In animals it is obtained chiefly from dietary di- and polysaccharides, but also by **gluconeogenesis**, and is stored as **glycogen**. Storage polymer in microorganisms is **dextran**.

glucose transporter Generic name for any protein that transports glucose. In bacteria these may be **ABC proteins**, in mammals they belong to a family of 12-transmembrane integral transporters. GLUT1 (SLC2A1, 492aa) is ubiquitously expressed, with particularly high levels in human erythrocytes and in the endothelial cells lining the blood vessels of the brain. GLUT2 (SLC2A2, 524aa) is a low-affinity transporter, present in liver, intestine, kidney, and pancreatic beta cells. GLUT3 (496aa) is expressed primarily in neurons. GLUT4 (509aa) is insulin-responsive transporter, restricted to striated muscle and adipose tissue, and defects may be associated with some forms of **diabetes**. GLUT5 (501aa) is a fructose transporter. A number of others are known.

glucose-1-phosphate Product of glycogen breakdown by phosphorylase. Converted to glucose-6-phosphate by phosphoglucomutase.

glucose-6-phosphate Phosphomonoester of glucose that is formed by transfer of phosphate from ATP, catalysed by the enzyme **hexokinase**. It is an intermediate both of the glycolytic pathway (next converted to fructose-6-phosphate), and of the NADPH-generating **pentose phosphate pathway**.

glucose-6-phosphate dehydrogenase Ubiquitous enzyme (EC 1.1.1.49, 515aa), present in bacteria and all eukaryotic cell types, that catalyses the conversion of D-glucose-6-phosphate to D-glucono-1,5-lactone 6-phosphate, using NADP as a cofactor and generating NADPH. The first step in the pentose pathway. See **hexose monophosphate shunt**. The enzyme is highly polymorphic and X-linked: in heterozygotic females X-inactivation means a cell expresses only one allele (in accord with the **Lyon hypothesis**) and this can be used to determine whether a single cell is likely to have given rise to a clone of cells, as in a tumour.

glucose-dependent insulinotropic polypeptide See **gastric inhibitory polypeptide**.

glucose-regulated protein See **endoplasmin**.

glucosinolate Compounds in which a glucose residue is linked by a thioglucoside bond to an amino acid derivative. About 120 different glucosinolates are known to occur naturally in plants. Some act as endogenous pesticides in brassicas such as broccoli and Brussels sprout, and are responsible for the bitter or sharp taste of many common foods such as mustard, horseradish, cabbage and Brussels sprouts.

glucosylation Transfer of glucose residues, usually from the nucleotide-sugar derivative UDP-glucose, either to other sugars, as in the synthesis of complex carbohydrates, to various small molecules to produce glucosides or as a post-translational modification of a protein, for example the transfer of glucose to a galactose residue on the hydroxylysine of collagen. Glucosylation of certain long-lived proteins by a non-enzymic reaction with free glucose may contribute to ageing. See **AGE**.

glucuronic acid Uronic acid (GA, GlcA) formed by oxidation of OH group of glucose in position 6. D-glucuronic acid is widely distributed in plants and animals as a subunit of various oligosaccharides.

glucuronidase An enzyme that catalyses the hydrolysis of polysaccharides containing glucuronic acid. Alpha glucuronidase (EC 3.2.1.139, 841aa in *Aspergillus*) is involved in the hydrolysis of alpha-glucuronide-containing **xylan**. Beta-glucuronidase (EC 3.2.1.31, 651aa in human, 603aa in *E. coli*) is a tetrameric glycoprotein that hydrolyses β-D-glucuronic acid residues from the non-reducing end of glycosaminoglycans, particularly dermatan and keratan sulphates. It is often used as a reporter gene (see **GUS reporter system**). Mutations in the gene lead to mucopolysaccharidosis type VII (Sly's syndrome), a lysosomal storage disease. See **egasyn**.

glucuronoxylan Hemicellulosic plant cell-wall polysaccharide containing glucuronic acid and xylose as its main constituents. Has a β(1-4)-xylan backbone, with 4-O-methylglucuronic acid side-chains. Arabinose and acetyl side-chains may also be present. Major polysaccharide of angiosperm wood (hardwood).

glufosinate A non-selective chemical herbicide that can be used only on crops tolerant to it − notoriously, those genetically engineered for such resistance. It is a natural compound isolated from two species of *Streptomyces* that inhibits the activity of glutamine synthetase, which is necessary for the production of glutamine and for ammonia detoxification.

GLUT1, GLUT2, etc. See **glucose transporters**.

glutamate An **excitatory amino acid**, the major fast excitatory neurotransmitter in the mammalian central nervous system and the excitatory neuromuscular transmitter in arthropod skeletal muscles. Excess glutamate in food causes Chinese Restaurant Syndrome. See **glutamate receptor**.

glutamate carboxypeptidase II A zinc metalloenzyme (GCP2, EC 3.4.17.21, prostate-specific membrane antigen, 750aa) with folate hydrolase and N-acetylated-alpha-linked-acidic dipeptidase (NAALADase) activity.

Important for folate uptake in the intestine and for hydrolysis of **NAAG** neuropeptide in the brain.

glutamate decarboxylases The enzymes responsible for synthesis of the neurotransmitter, **GABA**, (glutamate decarboxylase 1, GAD67, EC 4.1.1.15, 594aa; glutamate decarboxylase 2, GAD65, 585aa). Defects in GAD1 are the cause of one form of cerebral palsy. In *E. coli* the glutamate decarboxylase system helps to maintain a near-neutral intracellular pH when cells are exposed to extremely acidic conditions. The enzyme is also found in plants.

glutamate pyruvate transaminase See **alanine aminotransferase**.

glutamate receptors Members of a superfamily of **amino acid receptors**, implicated in many important brain functions including **long-term potentiation** (LTP). There are at least four major pharmacologically-distinct glutamate-gated **ion channel** subtypes, named after their most selective agonists: N-methyl-D-aspartate (**NMDA**), kainate (KA), **quisqualate/ AMPA** and L-2-amino-4-phosphobutyrate (APB). A fifth subtype, for which the agonist is trans-1-amino-cyclopentane 1,3 dicarboxylate (APCD), is a G-protein-coupled receptor. NMDA-type receptors are implicated in memory and learning, neuronal cell death, ischemia and epilepsy. See glutamate receptor interacting proteins (**GRIP #1**).

glutamic acid One of the 20 α-amino acids (Glu, E, 147Da) commonly found in proteins. Plays a central role in amino acid metabolism, acting as precursor of **glutamine**, proline and **arginine**. Also acts as amino group donor in synthesis by transamination of alanine from pyruvate, and aspartic acid from oxaloacetate. Glutamate is also a neurotransmitter (see **glutamate receptors**); the product of its decarboxylation is the inhibitory neurotransmitter **GABA**.

glutamic peptidases A subset of peptidases, fungal endopeptidases formerly termed 'pepstatin-insensitive carboxyl proteinases'.

glutamine One of the 20 amino acids (Gln, Q, 146 Da) commonly found (and directly coded for) in proteins. It is the amide at the γ-carboxyl of the amino acid **glutamate**. Glutamine can participate in covalent cross-linking reactions between proteins, by forming peptide-like bonds by a transamidation reaction with lysine residues. This reaction, catalysed by clotting factor XIII stabilizes the aggregates of fibrin formed during blood-clotting. Media for culture of animal cells contain some 10 times more glutamine than other amino acids, the excess presumably acting as a carbon source.

glutaraldehyde A dialdehyde used as a fixative, especially for electron microscopy. By its interaction with amino groups (and others) it forms cross-links between proteins.

glutaredoxins A ubiquitous family of proteins (thiol-transferases, all ~100aa) in eukaryotes, prokaryotes and viruses which catalyze the reduction of disulphide bonds in their substrate proteins, becoming oxidised themselves; they are then reduced non-enzymatically by **glutathione** and the oxidized glutathione is regenerated by **glutathione reductase**.

glutarimide antibiotics Antibiotics that block peptide chain elongation by interacting with the 60S ribosomal subunit, for example **cycloheximide** which is commonly used experimentally.

glutathione The tripeptide γ-glutamylcysteinylglycine. It contains an unusual peptide linkage between the γ-carboxyl group of the glutamate side chain and the amine group of cysteine. The concentration of glutathione in animal cells is around 5 mM and its sulphydryl group is kept largely in the reduced state. This allows it to act as a sulphydryl buffer, reducing any disulphide bonds formed within cytoplasmic proteins to cysteines. Hence, few, if any, cytoplasmic proteins contain disulphide bonds. Glutathione is also important as a cofactor for the enzyme **glutathione peroxidase**, in the uptake of amino acids and participates in **leukotriene** synthesis. See **glutaredoxins**.

glutathione peroxidase A detoxifying enzyme (EC 1.11.1.9) that eliminates hydrogen peroxide and organic peroxides. Glutathione is an essential cofactor for the enzyme and its reaction involves the oxidation of glutathione (GSH) to glutathione disulphide (GSSG). The GSSG is then reduced to GSH by **glutathione reductase**. Glutathione peroxidase, (GPX), has a selenocysteine residue in its active site. There are multiple forms of the enzyme, for example: cytoplasmic GPX1 (203aa) and GPX2 (190aa), extracellular GPX3 (226aa), phospholipid hydroperoxide glutathione peroxidase, GPX4 (197aa) which protects cells against membrane lipid peroxidation. GPX5 (221aa) is secreted specifically in the epididymis. GPX6 (232aa) is found in mitochondria in *Arabidopsis*, where there are other GPX forms as well.

glutathione reductase A homodimeric enzyme (EC 1.8.1.7) in which each subunit (522aa) can be divided into 4 domains; domains 1 and 2 bind FAD and NADPH, respectively, and domain 4 forms the interface. It catalyses the NADP-dependent reduction of glutathione disulphide (GSSG) to glutathione (GSH). This maintains a GSH:GSSG ratio in the cytoplasm of around 500:1. There are mitochondrial and cytoplasmic isoforms.

glutathione S transferase A superfamily of enzymes (EC 2.5.1.18) found in eukaryotes and prokaryotes, that will couple **glutathione** to a xenobiotic as the first step in removal. They occur in cytosolic, mitochondrial, and microsomal forms (see

MAPEG proteins). The GST superfamily includes, for example, dichloromethane dehalogenase (EC 4.5.1.3, 288aa) from *Methylobacterium extorquens*, mitochondrial glutathione-dependent dehydroascorbate reductase 1 (EC 2.5.1.18, DHAR1, 213aa) from *Arabidopsis*, glutathione S-transferase theta-2 (GSTT2, 244aa) from human. The mammalian GST family consists of cytosolic dimeric isoenzymes subdivided into at least six classes: alpha, mu, pi, theta, zeta and omega, all ~200aa. The GST gene is often used as a fusion with a gene of interest that is being expressed in a bacterial system. The fusion construct can be purified easily from lysate by passage down a glutathione affinity column, the purified construct then being eluted with glutathione. The GST can then be cleaved proteolytically from the protein of interest though often the complete fusion protein can be used.

glutathione synthase A family of eukaryotic enzymes (EC 6.3.2.3, 474aa) that catalyse synthesis of glutathione from ATP, γ-L-glutamyl-L-cysteine and glycine. GSS deficiency can cause a severe disorder (5-oxoprolinuria or pyroglutamic aciduria); a mild form affects erythrocytes.

glutelin Group of proteins found in seeds of cereals. In rice the glutelins are around 500aa. A smaller form (glutelin-2, 223aa) is found in maize.

gluten Protein-rich fraction from cereal grains, especially wheat. When hydrated forms a sticky mass responsible for the mechanical properties of bread dough. **Glutelins** and **gliadin** form a substantial component.

glycaemic *US.* glycemic. Relating to the concentration of glucose in the blood.

glycation The non-enzymic coupling of sugar moieties to proteins. *Cf.* **glycosylation**, see **AGE**.

glycemic index A ranking of foods based on their overall effect on blood glucose levels. Slowly absorbed foods have a low GI rating, whilst foods that are more quickly absorbed will have a higher rating.

glyceraldehyde-3-phosphate Three-carbon intermediate of the glycolytic pathway formed by the cleavage of fructose 1,6-bisphosphate, catalysed by the enzyme **aldolase**. Also involved in reversible interchange between **glycolysis** and the **pentose phosphate pathway**.

glyceraldehyde-3-phosphate dehydrogenase Glycolyticenzyme (GAPD, GAPDH, G3PD, EC 1.2.1.12, 335aa) that catalyses the reversible oxidative phosphorylation of glyceraldehyde-3-phosphate. Has been shown to interact with various elements of the **cytoskeleton**, and with the trinucleotide repeat in the **huntingtin** gene. Also has nitrosylase activity (EC 2.6.99.-) which is probably responsible for its involvement with transcription, RNA transport, DNA replication and apoptosis through cysteine-mediated S-nitrosylation of nuclear target proteins such as sirtuin and histone deacetylases. Article: http://www.nature.com/ncb/journal/v12/n11/full/ncb2114.html

glycerination Permeabilisation of the plasma membrane of cells by incubating in aqueous glycerol at low temperature. The technique was first applied to muscle which, once glycerinated, can be made to contract by adding exogenous ATP and calcium.

glycerol A metabolic intermediate, but primarily of interest as the central structural component of the major classes of biological lipids, triglycerides and phosphatidyl phospholipids. Also used as a **cryoprotectant**.

glycine The simplest amino acid (Gly, G, 75.1 Da). It is a common residue in proteins, especially collagen and elastin, and is not optically active. It is also a major inhibitory **neurotransmitter** in spinal cord and brainstem of vertebrate **central nervous system**.

glycine receptor Chloride-channel forming receptor composed of four alpha (457aa) and one beta subunit (497aa) with a central pore. Binding of glycine, β-alanine or taurine increases ion flux in neurons that is blocked by strychnine which binds to the alpha subunits. Mutation in the alpha or beta subunit genes cause **startle disease**.

glycine-rich proteins (1) Plant proteins that are implicated in several independent physiological processes. For example, glycine-rich RNA-binding protein 8 (*Arabidopsis*, 169aa) may have a role in RNA transcription or processing during stress. Glycine-rich cell wall structural protein 2 (*Oryza*, glycine-rich protein 1, 185aa) is responsible for plasticity of the cell wall. (2) Cysteine and glycine-rich protein 2 in humans (CRP2, smooth muscle cell LIM protein, 193aa) is down-regulated in response to PDGF-BB or cell injury and seems to have a role in the development of the embryonic vascular system.

glycocalyx An old and probably obsolete term that referred to the region, seen by electron microscopy, external to the outer dense line of the **plasma membrane** that appears to be rich in glycosidic compounds such as proteoglycans and glycoproteins. Since these molecules are often integral membrane proteins and may be denatured by the processes of fixation for electron microscopy, it might be better to avoid the term or to refer to membrane glycoproteins or to proteoglycans associated with the cell surface.

glycocholate Anion of the bile acid, glycholic acid. Usually found in bile as the sodium salt. Has powerful detergent properties.

glycoconjugate Any biological macromolecule containing a carbohydrate moiety — thus a generic term to cover **glycolipids**, **glycoproteins** and **proteoglycans**.

glycodelin A **lipocalin** (progestagen-dependent endometrial protein, pregnancy-associated endometrial alpha-2-globulin, placental protein-14, 180aa) found in two differentially glycosylated forms, glycodelin-A in amniotic fluid with contraceptive and immunosuppressive activities and glycodelin-S in seminal plasma. It is highly homologous to beta-**lactoglobulin** and may be angiogenic. Article: http://edrv.endojournals.org/cgi/reprint/23/4/401

glycogen Branched polymer of D-glucose (mostly $\alpha(1-4)$-linked, but some $\alpha(1-6)$ at branch points). Size range very variable, up to 105 glucose units. Major short-term storage polymer of animal cells, and is particularly abundant in the liver and to a lesser extent in muscle. In the electron microscope glycogen has a characteristic 'asterisk/star' appearance.

glycogen debranching enzyme A monomeric protein (1515aa) with two catalytic activities, both required for the breakdown of branched regions of glycogen. There are two catalytic sites, one with amylo-1,6-glucosidase (EC 3.2.1.33) activity, the other with 4-alpha-glucanotransferase (EC 2.4.1.25) activity. Deficiencies cause glycogen storage disease III (Forbes' disease). The glycogen debranching enzyme from *E. coli* (EC 3.2.1.-, 657aa) is simpler and although it hydrolyzes alpha-1,6-glucosidic linkages, has little activity against native glycogen; a **glycogen phosphorylase** is needed as well.

glycogen phosphorylase The enzyme (EC 2.4.1.1) that sequentially removes glycosyl residues from unbranched glycogen, producing glucose-1-phosphate. Complete hydrolysis of branched glycogen also requires a **glycogen debranching enzyme**. Activity is regulated allosterically and by **phosphorylase kinase**. There are tissue-specific isozymes in muscle (GPMM, 842aa), liver (GPLL, 847aa) and brain (GPBB, 862aa). Mutations in the liver enzyme lead to glycogen storage disease VI (**Hers' disease**), in the muscle enzyme to **McArdle's disease**.

glycogen storage diseases A range of **storage diseases** caused by mutations in genes required for glycogen metabolism or its regulation. They include: **Pompe's disease**, **Danon's disease**, **McArdle's disease** and **Hers' disease**.

glycogen synthase kinase A serine-threonine protein kinase (GSK3, EC 2.7.11.26) involved in the control of glycogen metabolism but also important as a negative regulator in various intracellular signalling pathways. There are two human isoforms, GSK3α (483aa) and GSK3β (420aa). Levels of GSK are rapidly reduced by insulin and various growth factors, releasing a constraint on protein synthesis.

GSK3 itself is regulated by **PKB**. In *Drosophila* the homologue (shaggy, 1067aa) is involved in the specification of cell fate as a negative regulator of the Wnt/beta-catenin pathway and in *Xenopus* the homologue is involved in regulation of dorso-ventral patterning. In *S. pombe* gsk3 (387aa) interacts with cdc14 which is thought to play a role in the initiation and completion of mitosis. GSK3 is identical to Tau protein kinase I and phosphorylation of tau is known to occur in the neurofibrillary tangles of Alzheimer's disease.

glycogenin In eukaryotes, a self-glucosylating protein that primes glycogen granule synthesis in skeletal muscle and heart (glycogenin-1, EC2.4.1.186, 350aa) and in liver (glycogenin-2, 501aa). Glycogenin binds glucose from UDP-glucose, then more glucose units are added forming an alpha-1,4-glycan of around 10 residues attached to Tyr-195. Further addition of glucose subunits is then carried out by glycogen synthetase. Defects cause a form of glycogen storage disease.

glycogenoses See **glycogen storage diseases**.

glycolic acid Hydroxyacetic acid; found in young plants and green fruits. Glycolate is formed from ribulose-1,5-bisphosphate in a seemingly wasteful side reaction of photosynthesis, known as **photorespiration**.

glycolipid Oligosaccharides covalently attached to lipid as in the glycosphingolipids (GSL) found in plasma membranes of all animal and some plant cells. The lipid part of GSLs is sphingosine in which the amino group is acylated by a fatty chain, forming a **ceramide**. Most of the oligosaccharide chains belong to one of four series, the ganglio-, globo-, lacto-type 1 and lacto-type 2 series. Blood group antigens are GSLs.

glycolipid transfer protein See **phospholipid transfer protein**.

glycolysis The conversion of a monosaccharide (generally glucose) to pyruvate via the glycolytic pathway (i.e. the **Embden-Meyerhof pathway**) in the cytosol. Generates ATP without consuming oxygen and is thus anaerobic.

glycomics The study and analysis of the carbohydrate moieties of complex macromolecules (glycolipids and glycoproteins) found within an organism, by analogy with proteomics.

glycophorins Sialylated transmembrane glycoproteins of the erythrocyte. Glycophorin A (131aa) is highly O-glycosylated with terminal sialic acid residues. The peptide chain has the **MN blood group antigens** at its N-terminus. Glycophorin B (CD235b, 91aa) has Ss and U antigens of the MNS blood group system. Glycophorin C (CD236, glycoprotein beta, glycoconnectin, 128aa) is a minor constituent and carries the Gerbich blood group antigens. The

Plasmodium falciparum protein PfEBP-2 (erythrocyte binding protein 2; baebl; EBA-140) binds to glycophorin C. Glycophorin D is a ubiquitous alternatively spliced form of glycophorin C. Glycophorin E (78aa), is a minor sialoglycoprotein.

glycoprotein Protein with covalently attached sugar units, either bonded via the OH group of serine or threonine (O-glycosylated) or through the amide NH_2 of asparagine (N-glycosylated). Includes most secreted proteins (serum albumin is the major exception) and proteins exposed at the outer surface of the plasma membrane. Sugar residues found include: mannose, N-acetyl glucosamine, N-acetyl galactosamine, galactose, fucose and sialic acid.

glycoprotein-2 A GPI-anchored membrane-associated protein (pancreatic secretory granule membrane major glycoprotein GP2, ZAP75, 537aa) that has a **ZP domain** and interacts with **syncollin**.

glycosaminoglycan attachment site In **proteoglycans** the **glycosaminoglycan** sidechains are attached to a core protein through a xyloside residue which is linked to a serine residue. The serine is sometimes in a consensus motif (S-G-X-G) but not all such motifs are glycosylated. Article: http://www.jbc.org/content/276/16/13411.full

glycosaminoglycans Polysaccharides (GAGs) that are covalently linked to core proteins to form **proteoglycans** (formerly mucopolysaccharides). The side chains are usually attached about every 12th residue (but see **glycosaminoglycan attachment site**) and enormous variability is possible by permutation of the different side chains and the number attached. The polysaccharides are of repeating disaccharide units (more than 100) of amino sugars, at least one of which has a negatively charged side-group (carboxylate or sulphate). The common GAGs are, chondroitin sulphate (D-glucuronic acid- N-acetyl-D-galactosamine-4 or -6-sulphate), dermatan sulphate (D-glucuronic acid- or L-iduronic acid-N-acetyl-D-galactosamine), keratan sulphate (D-galactose- N-acetyl-D-glucosamine-sulphate), and heparan sulphate (D-glucuronic acid- or L-iduronic acid-N-acetyl-D-glucosamine). Proteoglycans are non-covalently attached by link proteins to hyaluronate (D-glucuronic acid-N-acetyl-D-glucosamine: MW up to 10 million Da) to produce a large hydrated space-filling polymer found in **extracellular matrix**. The extent of sulphation is variable and there is tremendous diversity.

glycosidase *glycosylase* General and imprecise term for an enzyme that degrades linkage between sugar subunits of a polysaccharide. Any of the EC 3.2 class of hydrolases that cleave glycosidic bonds. They may distinguish between α and β links, for example, but are not very substrate specific. See **endoglycosidase**.

glycoside Molecule in which a sugar group (glycone) is bonded to a non-sugar (aglycone) through its anomeric carbon to another group via an O-glycosidic bond or an S-glycosidic bond (thioglycosides). Glycosides are often storage forms of molecules that are inactive until the glycosidic bond is hydrolysed. Glycosides can be classified either on the basis of the sugar and its linkage or on the nature of the aglycone.

glycosidic bond Bond between anomeric carbon of a sugar and the group to which it is attached (which may be in another sugar, as in polysaccharides, or in non-sugar molecules).

glycosome A **peroxisome** containing glycolytic enzymes in protozoa of the Kinetoplastida (e.g. trypanosomes).

glycosphingolipids Glycolipids of the cell membrane in which the lipid is **sphingosine** with the amino group acylated by a fatty chain, forming a **ceramide**. Depending upon the glycosylation they can be subdivided into cerebrosides (with a single oligosaccharide side chain), gangliosides, which have one or more sialic acid residues and globosides which have more than one side chain and the sugars are usually a combination of N-acetylgalactosamine, D-glucose or D-galactose. Many blood group antigens are GSLs.

glycosyl phosphatidyl inositol See **GPI-anchor**.

glycosyl transferase Class of ectoenzymes (EC 2.4) that catalyse the transfer of a sugar (monosaccharide) unit from a sugar nucleotide derivative to a sugar or amino acid acceptor. Various sugars, including galactose, glucose, N-acetylglucosamine, N-acetylneuraminic acid, mannose and fucose may be transferred.

glycosylation The enzymic addition of sugar moieties to lipids or proteins. Proteins can be glycosylated in two ways: in N-linked glycosylation the linkage is to the amide nitrogen of asparagine, in O-glycosylation the sugar is attached to the hydroxy side group of serine or threonine. *Cf.* **glycation**.

glycosylation disorders A family of severe inherited diseases caused by a defect in protein N-glycosylation. Serum proteins are under-glycosylated and there are a wide variety of clinical features, including disorders of the nervous system development, psychomotor retardation, dysmorphic features, hypotonia, coagulation disorders, and immunodeficiency.

glycyrrhizin The main sweet tasting compound in liquorice root, a triterpenoid saponin glycoside, 30−50 times sweeter than sucrose.

glyoxalines See **imidazoles**.

glyoxisome Organelle found in plant cells, containing the enzymes of the **glyoxylate cycle**. Also contains catalase and enzymes for **beta-oxidation** of fatty acids. Together with the **peroxisome** makes up the class of organelles known as **microbodies**.

glyoxylate cycle Metabolic pathway present in bacteria and in the **glyoxisome** of plants, in which two acetyl-CoA molecules are converted to a 4-carbon dicarboxylic acid, initially succinate. Includes two enzymes not found elsewhere, isocitrate lyase and malate synthase. Permits net synthesis of carbohydrates from lipid, and hence is prominent in those seeds in which lipid is the principal food reserve.

glyoxysome See **peroxisome**.

glyphosate A broad spectrum, non-selective systemic herbicide, effective in killing all plant types including grasses, perennials and woody plants. It is inactivated when it comes into contact with soil and, although an organophosphate, is not a cholinesterase inhibitor.

glypiation See **GPI-anchor**.

glypicians A family of cell surface proteoglycans with heparan sulphate side-chains. Glypican-1 (558aa) is GPI-anchored on the external face of the plasma membrane and is released as a secreted form. Glypican-3 is thought to bind **insulin-like growth factor-2** (IGF-2) and modulate its function: defects in glypican-3 are responsible for **Simpson-Golabi-Behmel syndrome** and glypican-3 is upregulated in some hepatocellular carcinomas. Glypican-6 (555aa) may be a coreceptor for growth factors, extracellular matrix proteins, proteases and anti-proteases.

GM-CSF A cytokine (granulocyte-macrophage colony stimulating factor, GMCSF, CSF2, 144aa) that stimulates the formation of granulocyte or macrophage colonies from myeloid stem cells isolated from bone marrow. The receptors is a heterodimer of a low-affinity alpha subunit (CSF2RA, CD116, 400aa) and a beta subunit (CSF2RB, CD131, 897aa) which raises affinity and is common to the GMCSF, IL-3 and IL-5 receptors. A soluble form of the alpha subunit is secreted.

Gm-types Genetically determined allotypic antigens found on IgG of some individuals.

GM130 Golgin subfamily A member 2 (130 kDa cis-Golgi matrix protein, golgin-95, 1002aa). See **golgins**.

GMO See **genetically modified organism**.

GNAS locus The locus that encodes the alpha subunit of trimeric **G proteins** (Gnas, adenylate cyclase-stimulating G alpha protein, 394aa), but a complex locus that is **genomically imprinted** and encodes 5 main transcripts: Gs-alpha (Gnas), XLas (extra large alpha-s protein, 1037aa), NESP55 (neuroendocrine secretory protein 55, 245aa) which has no sequence overlap with Gnas, the A/B transcript (which uses the alternative first exon A/B) and an antisense GNAS transcript. The latter two transcripts are ubiquitously expressed but non-coding. XLas isoforms are paternally derived, the Gnas isoforms are biallelically derived and the Nesp55 isoforms are maternally derived. Mutations can cause pseudohypoparathyroidism, Albright hereditary osteodystrophy, pseudopseudohypoparathyroidism, McCune-Albright syndrome, progressive osseus heteroplasia, polyostotic fibrous dysplasia of bone and some pituitary tumours. Article: http://content.karger.com/produktedb/produkte.asp?typ=fulltext&file=83895

Gnetophyta Three genera of woody plants grouped in the gymnosperms. There are about 70 extant species of Gnetophyta, subdivided into three families, Gnetaceae, Welwitschiaceae and Ephedraceae.

GNOM See **EMB30**.

gnotobiotic Organism or environment completely or almost completely depleted of all organisms or all other organisms. Animals that are SPF (specific pathogen free) are gnotobiotic.

GnRH See **gonadotropin-releasing hormone**.

GNRP Guanine nucleotide releasing protein, an old name for a type of protein that facilitated binding of GTP by ras-like GTPases, resulting in activation of the ras signal, now usually referred to as exchange factors (**GEFs**). Examples include **sos**, RasGRP, and C3G.

G$_o$ A specific class (o for other) of signal-transducing heterotrimeric **GTP-binding proteins** (G proteins) expressed in high levels in mammalian brain. Other G-proteins are involved in stimulation (G$_s$) or inhibition (G$_i$). Do not confuse with G0 (G-zero), the resting stage of the cell cycle.

goblet cell (1) Cell of the epithelial lining of small intestine that secretes mucus and has a very well-developed Golgi apparatus. (2) Cell type characteristic of larval lepidopteran midgut, containing a potent H$^+$-ATPase, and thought to be involved in maintenance of ion and pH gradients.

gold nanoparticles Nano-scale gold particles (AuNPs) of variable size from 1.9 nm upwards, that can be used for cell imaging, diagnostics (as a contrast agent), targeted delivery of adsorbed drugs, and as a stain for molecules to which they have been conjugated. Supplier datasheet: http://www.nanoprobes.com/Applic.html

Goldberg's syndrome A lysosomal storage disease (galactosialidosis) associated with a combined deficiency of beta-galactosidase and neuraminidase, secondary to a defect in protective protein (cathepsin A)

which is required for normal processing of the two enzymes.

Goldberg-Hogness box See **TATA box**.

Goldberg-Shpritzen megacolon syndrome
An autosomal recessive trait with complex developmental abnormalities, often associated with **Hirschsprung's disease** and caused by mutation in the KIF-1 binding protein (621aa) which is required for organization of axonal microtubules, and axonal outgrowth and maintenance during development of the peripheral and central nervous system. It regulates mitochondrial transport by modulating the kinesin family motor, KIF1B.

Golden rice
Rice that has been genetically engineered by the insertion of genes that produce high levels of beta-carotene, which is converted to Vitamin A within the body, and should help to prevent blindness due to vitamin deficiency. The carotene makes the rice yellow in colour, hence the name.

Goldmann equation
Equation (Goldmann constant field equation) that describes the electrical potential across a membrane in terms of the distributions and relative permeabilities of the main permeant ions (typically sodium, potassium and chloride). Assumes that the electrical field across the membrane is constant, and that there are no **active transport** processes; but nonetheless gives a reasonable approximation to real membranes.

Golgi apparatus
Also known as the Golgi body, Golgi vesicles; in plants, the dictyosome; in flagellate protozoa, the parabasal body. Intracellular stack of flattened membrane-enclosed disks in which glycosylation and packaging of secreted proteins takes place; part of the **GERL** complex. Vesicles from endoplasmic reticulum fuse with the *cis*-Golgi region (the inner concave face) and progress through the vesicular stack to the *trans*-Golgi, where they are directed towards the plasma membrane or to lysosomes.

Golgi cells
Inhibitory interneurons in the granular layer of the cerebellum, the first identified example of a neuronal inhibitory feedback network. They receive excitatory input from mossy fibres and synapse onto the soma of granule cells and unipolar brush cells.

golgins
Coiled-coil proteins associated with the Golgi apparatus, necessary for tethering events in membrane fusion and as structural supports for Golgi cisternae. They localize to particular Golgi subdomains via their C termini and are important is sorting. Golgins such as GM130 (golgin-95), golgin-45 and p115 bind to Rab GTPases via their coiled-coil domains. Article: http://jcb.rupress.org/content/183/4/607.long

golli proteins
Alternatively spliced products of the **myelin basic protein** gene, found in the nervous and immune systems. In the immune system golli proteins can act as negative regulators of T-cell signalling pathways; in the nervous system they are important during development of neurons and oligodendrocytes. Abstract: http://www.cell.com/immunity/abstract/S1074-7613%2806%2900265-2

Gomori procedure
Cytochemical staining procedure used to localize acid phosphatases. Depends upon the production of phosphate ions from organic phosphoesters such as β-glycerophosphate. The phosphate in the presence of lead ions causes the formation of a precipitate of lead salt that is converted to the brown electron-dense sulphide of lead by the action of yellow ammonium sulphide. Gomori trichrome stain is used for connective tissues.

gonadotrophin-releasing hormone
A decapeptide hormone (gonadoliberin I, gonadorelin, GnRH-I, luliberin I, luteinizing hormone-releasing hormone I) produced by hypothalamic neurons, derived from progonadoliberin-1 (92aa): stimulates the secretion of gonadotrophins (both luteinizing and follicle-stimulating hormones). The receptor is G-protein coupled, Ca^{2+}-dependent and mutations in the receptor lead to hypogonadotropic hypogonadism.

gonadotrophins
US. gonadotropins. Group of glycoprotein hormones from the anterior lobe of the pituitary gland which are heterodimers of a common alpha chain (116aa) and a unique beta chain which confers biological specificity to **thyroid-stimulating hormone**, **follicle-stimulating hormone**, **luteinizing hormone** and **chorionic gonadotrophin** (beta chain 165aa). Chorionic gonadotrophin stimulates the ovaries to synthesize the steroids that are essential for the maintenance of pregnancy.

gonadotropins See **gonadotrophins**.

gonialblast
In *Drosophila* spermatogenesis the germline stem cell divides, one daughter becomes a gonialblast, while the other remains a stem cell. Each gonialblast executes four divisions as secondary spermatogonia, which exit the mitotic cycle and enter a meiotic and differentiation program as a clone of 16 spermatocytes. *Cf.* **cystoblast**

Goniomonads
Colourless flagellates, marine or freshwater, laterally flattened, part of the cryptophyte group but a distinct lineage. The type species is *Goniomonas truncata* (formerly *Monas truncata*).

Gonium
A genus of colonial green algae, members of the Volvocales. Have, variously, 4−16 cells arranged as a flat plate. Image: http://protist.i.hosei.ac.jp/pdb/Images/Chlorophyta/Gonium/index.html

gonococcus
Casual name for the Gram-negative diplococcus, *Neisseria gonorrhoeae*, which causes gonorrhoea.

gonosome Collective name for the reproductive zooids of a colonial animal such as a hydroid.

Gonyaulax Genus of **dinoflagellates**. Responsible for red tides and associated shellfish poisoning due to **saxitoxin**. Some species are bioluminescent.

Goodpasture's syndrome An autoimmune disease in which there is deposition of immune complexes, especially in the glomerulus, which induce an inflammatory response. The antigen is the alpha-3 chain of type IV collagen.

gooseberry A **segment-polarity gene** of *Drosophila* encoding a homeobox protein (449aa) that has a **paired box domain**.

goosecoid A homeodomain transcription factor (243aa) expressed in the dorsal lip of the *Xenopus* blastopore; may be a key factor in specifying these cells as the organizer of the embryo (**Spemann's organizer**). Name derived from *Drosophila* homedomain proteins to which its homoeodomain is similar: **gooseberry** and **bicoid**. Homologous genes have been identified in human, mouse, zebrafish, and chick. Goosecoid-2 (205aa) is a similar paired box homeobox protein; the human gene is located in the region deleted in DiGeorge syndrome (**thymic hypoplasia**) and **velocardiofacial syndrome**.

Gorlin's syndrome A rare autosomal dominant disorder (naevoid basal cell carcinoma) caused by mutations in the ***patched*** (*PTCH1* or *PTCH2*) genes or the homologue of the *Drosophila* gene *suppressor of fused* (*SUFU*) gene that encodes a component of the **sonic hedgehog**/patched signalling pathway.

GOS (1) In *S. cerevisiae* a SNARE (GOS1, 223aa) involved in retrograde transport within the Golgi complex. A similar protein (vesicle transport protein GOS15, BET1-like protein, 111aa) is found in humans. (2) Various proteins designated GOS1, GOS9 etc. from rice. GOS2 is translation initiation factor 1 (protein translation factor SUI1 homolog, protein eIF1, 115aa). (3) GOS19-1 is an obsolete name for human **macrophage inflammatory protein 1α**.

Gossypium Genus of plants that includes cotton.

gossypol A natural phenol derived from the cotton plant, probably present for anti-insect defence. It inhibits several dehydrogenase enzymes, has some contraceptive and anti-malarial activity.

gout Recurrent acute arthritis of peripheral joints caused by the accumulation of monosodium urate crystals which cause damage to phagosome membranes in neutrophils, leading to the release of contents which exacerbate the inflammatory response. Colchicine may be effective as a treatment because it inhibits lysosome-phagosome fusion.

gp- Generic prefix for a glycoprotein, usually followed by a number which relates to the (approximate) molecular weight in kDa.

gp41 Envelope glycoprotein of **HIV** (41 kDa, 345aa), derived from envelope glycoprotein gp160 (encoded by *env*) by cleavage into gp41 and gp120 which remain associated. The N-terminal part of gp41 is thought to mediate fusion between viral and cellular membranes.

gp100 Melanosomal matrix protein (661aa) involved in melanin synthesis, one of the **melanoma-associated antigens** often used in diagnostic testing.

gp130 Signalling subunit (918aa) of receptors for the **IL-6-family** of cytokines. Degradation of gp130 is regulated through a phosphorylation-dephosphorylation mechanism in which protein phosphatase-2A is crucially involved.

GPBB See **glycogen phosphorylase**.

GPI-anchor A glycosyl phosphatidyl inositol (PI) moiety (a pigtail) post-translationally added to a protein as a means of anchoring it to membranes. The phosphatidyl inositol is linked through glucosamine and mannose to a phosphoryl ethanolamine residue bound to the C-terminal amino acid of the protein. The addition (glypiation) is synchronized with the removal of a large C-terminal polypeptide sequence that is usually hydrophobic and could itself have formed a membrane anchor. The surface proteins of many unicellular protozoa very commonly have this modification, the best known being the variable surface glycoprotein of trypanosomes and of **malaria** parasites, but examples are probably present in all eukaryotic plasma membranes.

GpIb Heterodimeric glycoprotein (α, 626aa; β, 206aa) of platelet membranes that, together with Gp5 (560aa) and Gp9 (177aa) forms the receptor for von Willebrand factor and is important for thrombus formation. Defects in any of the components causes **Bernard-Soulier syndrome**.

GPIHBP1 A cell surface protein on the luminal face of the capillary endothelium (glycosylphosphatidylinositol-anchored high density lipoprotein-binding protein 1, 184aa) that binds high density lipoproteins and plays an important part in the lipolytic processing of chylomicrons. GPIHBP1-deficient mice accumulate chylomicrons in the plasma even on low-fat diets. Article: http://www.ncbi.nlm.nih.gov/pmc/articles/PMC1913910/?tool=pubmed

G-protein (1) See **GTP-binding proteins**. (2) The spike glycoprotein of vesicular stomatitis virus. This has been an important protein for investigation of membrane transport in eukaryotic cells.

G-protein receptor kinases See **GRKs**.

G-protein-coupled receptors Cell surface receptors (GPCRs) that are coupled to heterotrimeric **G-proteins** (GTP-binding proteins). All G-protein coupled receptors seem to have seven membrane-spanning domains (are **serpentine receptors**), and have been divided into 2 subclasses: those in which the binding site is in the extracellular domain e.g. receptors for glycoprotein hormones, such as **thyroid stimulating hormone** (TSH) and **follicle-stimulating hormone** (FSH), and those in which the ligand-binding site is likely to be in the plane of the 7 transmembrane domains, e.g. **rhodopsin**, and receptors for small **neurotransmitters** and **hormones**, e.g. **muscarinic acetylcholine receptor**.

GPS (1) G-protein pathway suppressors, which suppress **ras** and **MAPK**-mediated signalling and interfere with **Jnk** activity. Gps1 is actually COP9 signalosome complex subunit 1 (491aa) an essential part of the complex that regulates ubiquitin conjugation. Gps2 (327aa) suppresses G-protein- and mitogen-activated protein kinase-mediated signal transduction. (2) The GPS domain (G-protein coupled receptor proteolytic site, ~50aa) contains a cleavage site in **latrophilin**. The domain is found in many otherwise unrelated cell surface receptors but there is no evidence that it is a cleavage site in any of the other receptors.

GPT See **alanine aminotransferase**.

Graafian follicle Final stage in the differentiation of follicles in the mammalian ovary. Consists of a spherical fluid-filled blister on the surface of the ovary that bursts at ovulation to release the oocyte.

gradient perception Problem faced by a cell that is to respond directionally to a gradient of, for example, a diffusible attractant chemical. In a spatial mechanism the cell would compare receptor occupancy at different sites on the cell surface; a temporal mechanism would involve comparison of concentrations at different times, the cell moving randomly between readings. In pseudospatial sensing, the cell would detect the gradient as a consequence of positive feedback to protrusive activity if receptor occupancy increased with time as the protrusion moved up-gradient. Relatively few cell types have been unambiguously shown to detect gradients.

graft-versus-host disease A complication that can occur when an allogeneic graft containing lymphocytes (bone marrow or other tissues) is made. The immune cells of the graft attack the host: acute GVH disease is characterized by selective damage to the liver, skin, mucosa and the gastrointestinal tract, chronic GVH disease affects a broader range of tissues.

Gramene Genomic database for grains, including rice and wheat. Link to database: http://www.gramene.org/

Gram stain A staining method in which a heat-fixed bacterial smear is stained with crystal violet (methyl violet), treated with 3% iodine/potassium iodide solution, washed with alcohol and counter-stained. The method differentiates bacteria into two main classes, **Gram-positive**, which retain the stain, and **Gram-negative**, which do not. Certain bacteria, notably **mycobacteria**, that have walls with high lipid content, show acid-fast staining – the stain resists decoloration in strong acid.

Gram-negative bacteria Bacteria with thin **peptidoglycan** walls bounded by an outer membrane containing **endotoxin** (lipopolysaccharide). See **Gram stain**.

Gram-positive bacteria Bacteria with thick cell walls containing **teichoic** and **lipoteichoic acid** complexed to the **peptidoglycan**. See **Gram stain**.

gramicidins Antibiotics isolated from *Bacillus brevis*. Gramicidins A, B and C make up 80%, 6%, and 14% of a mixture referred to as gramicidin D. They are linear pentadecapeptides that will form a dimeric monovalent cation-specific pore across membranes. It is thought on the basis of the kinetics of pore formation and loss that two molecules form a membrane-spanning helix that constitutes a pore lined with polar residues. Gramicidin S is a cyclic decapeptide used as a topical antibiotic.

grammotoxin Toxin (omega-theraphotoxin-Gr1a, 36aa) from the Chilean pink tarantulas, *Grammostola spatulata*, that inhibits non-L-type (P/Q and N-type) **voltage-sensitive calcium channels**, thus resembling ω-**conotoxins** and ω-**agatoxins** in its effect, although it binds to a site associated with gating and is not a pore-blocker. Binds to some potassium channels (Kv2.1) with lower affinity.

graniferous Term for grain-producing plants.

granins Family of related acidic proteins (chromogranins and secretogranins) found in many endocrine cell secretory vesicles. Chromogranin A (pituitary secretory protein I, 457aa) is cleaved into 13 peptides including vasostatin-1 (76aa) and -2 (113aa) which are negatively inotropic and pancreastatin (48aa) which inhibits glucose induced insulin release from the pancreas. Chromogranin B (secretogranin 1, 677aa) is cleaved into at least two major peptides, GAWK peptide (74aa) and CCB peptide (60aa). Chromogranin C (secretogranin II, 617aa) is cleaved to produce secretoneurin which is chemotactic for monocytes and eosinophils and regulates endothelial cell proliferation. In mouse, secretogranin V (neuroendocrine protein 7B2, 212aa) is a specific chaperone for prohormone convertase-2.

granular component of nucleolus Area of nucleolus that appears granular in the electron microscope and contains 15 nm diameter particles that are maturing ribosomes. In contrast to the pale-staining and **fibrillar** areas.

granulation tissue Highly vascularized tissue that replaces the initial fibrin clot in a wound. Vascularization is by ingrowth of capillary endothelium from the surrounding vasculature. The tissue is also rich in fibroblasts (that will eventually produce the **fibrous tissue**) and leucocytes.

granule cells Various small neurons found in various parts of the brain. Cerebellar granule cells, are glutamatergic and receive excitatory input from mossy fibers. Olfactory bulb granule cells are GABAergic and axonless; granule cells in the dentate gyrus have glutamatergic axons. Granule cells of layer four in the cerebral cortex receive inputs from the thalamus.

granulins See **progranulin**.

granulocyte Leucocyte with conspicuous cytoplasmic granules. In humans the granulocytes are also classified as polymorphonuclear leucocytes and are subdivided according to the staining properties of the granules into **eosinophils, basophils** and **neutrophils** (using a **Romanovsky-type stain**); some invertebrate blood cells are also referred to, not very helpfully, as granulocytes.

granulocyte chemotactic protein 2 An alpha-chemokine (GCP-2, CXCL6, chemokine alpha 3, CKA-3, SCYB6, small inducible cytokine B6 precursor, 114aa) originally isolated from cytokine-stimulated osteosarcoma cells. Complements the activity of IL-8 as neutrophil chemoattractant and activator in humans but is a major neutrophil chemokine in the mouse. Receptors are the high affinity IL8 receptors CXCR1 and CXCR2.

granulocyte/macrophage colony stimulating factor See **GM-CSF**.

granulocytopenia Low granulocyte numbers in circulating blood.

granuloma Chronic inflammatory nodular lesion characterized by large numbers of cells of various types (macrophages, lymphocytes, fibroblasts, giant cells), some degrading and some repairing the tissues. There may be a surrounding cuff of lymphocytes. Granulomas may be a result of infection (e.g. in tuberculosis and leprosy), persistent non-degradable foreign objects or various autoimmune diseases. See **chronic granulomatous disease**.

granulopoiesis The production of **granulocytes** in the bone marrow.

granulysin A **saposin**-like protein (lymphokine LAG-2, T-cell activation protein 519, 145aa) present in cytotoxic granules of cytolytic T-cells and natural killer (NK) cells and released upon antigen stimulation. Kills a wide range of intracellular pathogens

(Gram-positive and Gram-negative bacteria, mycobacteria, fungi, and parasites).

granum *Pl.* grana. Stack of **thylakoids** in the chloroplast, containing the **light-harvesting system** and the enzymes responsible for the **light-dependent reactions** of photosynthesis.

granzymes Family of serine endopeptidases found in cytotoxic T-cells and NK cells that are involved in **perforin**-dependent cell killing. Granzyme A (cytotoxic T-lymphocyte-associated serine esterase-3, CTLA3, fragmentin-1, EC 3.4.21.78, 262aa) is necessary for target cell lysis in cell-mediated immune responses. Granzyme A cleaves SET (inhibitor of granzyme A-activated DNase), disrupting its binding to nucleoside diphosphate kinase A and releasing an inhibition. Granzyme B (fragmentin-2; CTLA1; EC 3.4.21.79, 247aa) is similar but enters target cells in a perforin-independent manner by binding to cell surface cation-independent **mannose-6-phosphate receptor**. Various other granzymes are known.

GRAS A composite regulatory element (gonadotropin releasing hormone (GnRH) receptor activating sequence) that interacts with multiple classes of transcription factors including **Smads**, **AP-1** and a **forkhead** DNA binding protein. It is involved in controlling expression of the receptor which is upregulated by **activin** and downregulated by **follistatin**. But see **GRAS family**.

GRAS family Family of of plant regulatory proteins named after **GAI** (gibberellin insensitive), RGA (repressor of ga1-3), and SCR (SCARECROW), the first three of its members isolated. SCR plays a significant role in the radial patterning of both roots and shoots. The **DELLA** proteins are a sub-family of the GRAS family. The *Arabidopsis* genome encodes at least 33 GRAS protein family members, although the role of only a few is currently known. But see **GRAS**.

GRASP (1) Golgi reassembly stacking protein (Golgi membrane-associated protein p59, GRASP65, 440aa) involved in the postmitotic assembly of Golgi stacks from mitotic Golgi fragments. GRASP2 (Golgi phosphoprotein 6, 55 kDa, 452aa) is similar and forms a rab2 effector complex with **Golgin** 45 in the medial Golgi essential for normal protein transport and Golgi structure. (2) GRIP1-associated protein 1 (GRASP1, 841aa) interacts with **GRIP**1, GRIP2 and AMPA receptors. (3) GRASP1, (**general receptor for phosphoinositides-1**-associated scaffold protein, 395aa) has a role in intracellular trafficking and contributes to the macromolecular organization of group 1 metabotropic glutamate receptors (mGluRs) at synapses. It is localized at post-synaptic membranes. (4) See **ATP-grasp**.

Graves' disease Autoimmune disease (hyperthyroidism) characterized by goitre, exophthalmia and

thyrotoxicosis, caused by antibodies to the thyrotropin receptor resulting in constitutive activation of the receptor and increased levels of thyroid hormone. In Caucasians is associated with HLA-B8 and DR3. Variations in several genes, including the CTLA4 gene, the vitamin D receptor gene, and the vitamin D binding protein gene may contribute to susceptibility to the disease.

gravitaxis Directed locomotory response to gravity.

gravitropism Directional growth of a plant organ in response to a gravitational field - roots grow downwards, shoots grow upwards. Achieved by differential growth on the sides of the root or shoot. A gravitation field is thought to be sensed by sedimentation of **statoliths** (starch grains) in root caps.

gray platelet syndrome A disorder, probably genetically heterogeneous, in which platelets lack platelet-specific alpha-granule proteins and have few or no alpha-granules. **Weibel-Palade bodies** in endothelial cells, which have the same contents as alpha granules, are normal indicating that the disorder is restricted to megakaryocytes. Mutations in **GATA-1** also cause a similar abnormality in platelet morphology.

Grb-2 An adapter protein (growth factor receptor bound protein 2, 217aa) that links the cytoplasmic domain of growth factor receptor to **sos** and to **shc** through its SH2 and SH3 domains, and thus is important in the assembly of the signalling complex. Homologous to yeast sem-5 and *Drosophila* drk. Grb2-associated binder-1 (gab-1, 694aa) may integrate signals from different protein-tyrosine kinase receptors. See **gab**. Grb2-related adaptor proteins (GRAP1, 217aa; GRAP2, GADS, 330aa) link other growth factor receptors to the ras signalling pathway. GRAP2 may be leukocyte-specific.

Greek key A protein **motif** that involves five or six strands to form a stereotypical Greek key beta-barrel which can have multiple possible topologies. The twisted arrangement is notionally similar to a pattern often seen on Greek vases. Abstract: http://www.ncbi.nlm.nih.gov/pubmed/10861931

green algae See **Chlorophyta**.

green fluorescent protein Protein (GFP, 238aa) from luminous jellyfish, *Aequorea victoria*. Excited by blue light (max 395 nm: as produced from **aequorin** luminescence), it emits green light (509 nm). The gene has been cloned and mutagenised to give brighter fluorescence and different colour variants. These have become valuable transgenic tools in **flag tagging** and as **reporter genes** and **cell lineage** markers.

gregarine movement Peculiar **gliding motility** shown by gregarines (Protozoa).

Greig's cephalopolysyndactyly syndrome A disorder caused by mutations in *GLI-3* the glioma-related oncogene that encodes a zinc-finger DNA-binding transcription factor from the GLI-Kruppel gene family. The shape of the skull is abnormal and there is syndactyly (fusion of digits).

gremlin A **BMP** antagonist (cysteine knot superfamily 1, DAN domain family member 2, down-regulated in Mos-transformed cells protein, increased in high glucose protein 2, 184aa) that down-regulates BMP4 signalling in a dose-dependent manner and is fairly widely expressed. Required for early limb outgrowth and patterning in maintaining the FGF4-sonic hedgehog feedback loop. Gremlin transgenics show bone fractures and reduced bone mineral density by 20−30%, compared with controls. Gremlin-2 (168aa) inhibits BMP2 and BMP4 in a dose-dependent manner. Gremlin-3 (189aa) seems to play a role in specification of the left-right axis. Other BMP antagonists are **noggin,** chordin, cerberus, and **DAN**.

grex The multicellular aggregate formed by cellular slime moulds (**Acrasidae**): the slug-like grex migrates, showing positive phototaxis and negative gravitaxis, until culmination (the formation of a fruiting body) takes place. Coordination of the activities of the hundreds of thousands of individual amoebae that compose the grex may involve pulses of cyclic AMP in *Dictyostelium discoideum*, a species in which cAMP is the chemotactic factor for aggregation.

grey crescent A region near the equator of the surface in the fertilized egg of various amphibia, often of greyish colour, that appears to contain special morphogenetic properties.

GRF See **growth hormone releasing hormone**.

Grim (1) In *Drosophila* an activator of apoptosis (138aa) that acts on the effector, **dredd**. Like **reaper** and **hid** the effect is blocked by caspase inhibitors and *Drosophila* homologues of mammalian **IAPs**. (2) In humans, GRIM19 (gene associated with retinoic and interferon-induced mortality 19 protein, 144aa) is an accessory subunit of the mitochondrial membrane respiratory chain NADH dehydrogenase and is also a regulator of cell-death pathways.

GRIP (1) Glutamate receptor-interacting proteins (GRIP-1, 1118aa; GRIP-2, 1043aa) are found in postsynaptic terminals, contain 7 **PDZ** domains and are involved as scaffolds for the assembly of signalling complexes. GRIP1-associated protein 1 (GRASP-1, 841aa) interacts with GRIP1, GRIP2 and AMPA receptors. (2) In the mouse, glucocorticoid receptor interacting protein (GRIP-1, 1462aa) is a transcriptional coactivator probably the orthologue of human protein transcription intermediary factor 2 (TIF2) and partially homologous to steroid receptor coactivator 1 (SRC-1). GRIP1 interacts with the hormone binding domains (HBDs) of all five steroid receptors in a hormone-dependent manner and also with HBDs of class II nuclear receptors, including thyroid receptor alpha, vitamin D receptor, retinoic acid receptor alpha, and retinoid X receptor alpha. (3) In *Arabidopsis*, GRIP (788aa) is a Golgi matrix protein that tethers vesicles to Golgi membranes. (4) The GRIP domain is a feature of various proteins involved in maintaining Golgi structure, for example GRIP and coiled-coil domain-containing protein 1 (Golgi coiled-coil protein 1, 775aa), RANBP2-like and GRIP domain-containing protein 1/2 (1748aa) and some of the **golgins**.

Griscelli's syndrome A range of disorders in which there are disturbances of skin pigmentation and in some cases neurological or immunological impairments. Type 1 is cause by mutation in myosin 5A which interacts with **melanophilin**, which is defective in Type 3. Type 2 is associated with immunodeficiency and is caused by mutation in rab27A, product of the ras-related gene in megakaryocytes.

griseofulvin Polyketide antibiotic from *Penicillium griseofulvum*. Used therapeutically as an antifungal. Blocks microtubule assembly and thus mitosis.

GRKs The family of G-protein receptor kinases (EC 2.7.11.16) that phosphorylate agonist-occupied G-protein-coupled receptors and down-regulate their activity. GRK-1 is **rhodopsin kinase**; many others are known. See **arrestins, Ark1**.

gRNA Small RNA molecules (guide RNA, 60−80 nucleotides) that are found in the **editosome**. Guide RNAs are complementary to edited portions of the mature mRNA and contain poly-U tails that donate the U's added during editing.

gro See **melanoma growth-stimulatory activity** protein.

groEL See **chaperonins**.

Gronblad-Strandberg syndrome See **pseudoxanthoma elasticum**.

groucho In *Drosophila* a transcriptional co-repressor (730aa) that regulates transcription when recruited to specific target DNA by hairy-related bHLH proteins but does not act on repressor regions of **even-skipped, kruppel**, or **knirps** transcription factors. Maternally required for neurogenesis; interacts with Notch and Delta. Homologous repressors (groucho-1, 535aa; groucho-2, 761aa) have been found in zebrafish and there are related mammalian proteins (e.g. in human, transducin-like enhancer protein 1, enhancer of split groucho-like protein 1, 770aa) which inhibits NFκB-regulated gene expression and Wnt signalling.

ground meristem Partly differentiated meristematic tissue (primary meristem) derived from the apical meristem and giving rise to relatively unspecialised plant tissues (ground tissues) such as **parenchyma** and **collenchyma**.

ground tissue Plant tissues other than those of the vascular system and the dermal tissues. Composed of relatively undifferentiated cells.

ground-glass cells Hepatitis B infected liver cells. The Hepatitis B surface antigen appears as fine granules either diffusely spread through-out the cytoplasm or concentrated in the cytoplasm peripheral to the sinusoid space and when stained by **orcein** gives the 'ground glass' appearance to the cells.

group specific antigen See **gag protein**.

growth (and) differentiation factors Family of proteins (GDFs), part of the TGFβ superfamily, involved in growth and differentiation. GDF1 (372aa) may mediate cell differentiation events during embryonic development. GDF2 (429aa) is bone morphogenetic protein-9 (BMP9). Mis-sense mutations of GDF-3 (364aa) cause ocular and skeletal anomalies. GDF-5 (also known as bone morphogenetic protein-14 and cartilage-derived morphogenetic protein-1, 501aa) is important in bone repair; GDF6 (455aa) is also known as BMP13; GDF7 (450aa) is BMP12; GDF-8 (375aa) is **myostatin**; GDF-9 (454aa) is required for ovarian folliculogenesis; GDF11 (407aa) is BMP11, important in anterior/posterior patterning of the axial skeleton.GDF-15 (308aa) is macrophage inhibitory cytokine-1 (MIC-1).

growth arrest specific genes See **gas** genes.

growth cone A specialized region at the tip of a growing **neurite** that is responsible for sensing the local environment and moving toward the neuron's target cell. Growth cones are hand-shaped, with several long **filopodia** that differentially adhere to surfaces in the embryo. Growth cones can be sensitive to several guidance cues, for example, surface adhesiveness, growth factors, neurotransmitters and electric fields (**galvanotropism**).

growth cone collapse Loss of motile activity and cessation of advance by **growth cones.** There are specific molecules that inhibit the motility of particular growth cones, and that are important in establishing correct pathways in developing nervous systems.

growth control When applied to cells usually means control of growth of the population, i.e. of the rate of division rather than of the size of an individual cell.

growth factor receptor bound protein See **Grb-2**.

growth factors A diverse set of proteins that regulate the division and differentiation of cells, for example EGF (**epidermal growth factor**), PDGF (**platelet-derived growth factor**), FGF (**fibroblast**

growth factor). Growth factors could be considered cytokines and some cytokines are growth factors. Hormones such as insulin and somatomedin are also growth factors; the status of NGF (**nerve growth factor**) is more uncertain. Perturbation of growth factor production or of the response to growth factor is important in neoplastic transformation. Other growth factors (for which there are separate entries) include **activin, amphiregulin, betacellulin, bone morphogenetic proteins, brain-derived neurotrophic factor**, the **CCN family, CSF-1, CTGF, epiregulin, GDNF, growth (and) differentiation factors, HB-EGF, hepatocyte GF, heregulins, insulin-like GFs, keratinocyte GF, midkine, neuregulin, neurotrophins, nodal, oncomodulin, p185, schwannoma-derived GF, transforming growth factor** and **vascular endothelial GF**.

growth hormone Polypeptide (somatotropin, 217aa) produced by the anterior pituitary that stimulates liver to produce IGF-1 (**insulin-like growth factor-1**). Important for growth in childhood (deficiencies cause pituitary dwarfism) but continues to be produced in adulthood after growth has finished, when it is important for maintaining bone mineral density. The receptor is G-protein coupled (638aa) and is cleaved to produce a soluble secreted form, growth hormone-binding protein, that acts as a reservoir of growth hormone in plasma. Multimers of hormone bind to membrane-forms of the receptor and cause dimerization, activating a signalling cascade. Mutations in the receptor lead to **Laron dwarfism**.

growth substances See **plant growth substances**.

growth-associated proteins Group of developmentally regulated polypeptides (GAP-43, B-50, pp46, F1) thought to be critical for the formation of neural circuitry although all turn out to be **neuromodulin** (238aa) from different regions of the brain. Neuromodulin is a substrate for phosphorylation by protein kinase C. Other proteins that regulate neurite outgrowth include brain acid-soluble protein (BASP1, neuronal axonal membrane protein NAP-22, 227aa) which is associated with the membranes of growth cones.

growth-hormone releasing hormone Hypothalamic hormone (growth hormone-releasing factor, growth hormone-regulating hormone, somatocrinin, somatorelin, somatoliberin, active fragment 44aa) that acts on the adenohypophysis to stimulate the secretion of **growth hormone** (somatotropin). Release is inhibited by **somatostatin**. The receptor is G-protein coupled (423aa). See Table H2.

growth-rate dependent control One of the two regulatory systems used by *E. coli* to to adjust rRNA output. **Stringent control** regulates rRNA production according to amino acid availability; growth-rate dependent control regulates according to the steady-state growth rate. Growth-rate dependent control ensures rRNA synthesis relative to total cell protein is proportional to the square of the steady-state growth rate. Article: http://www.pnas.org/content/92/4/1117.full.pdf

growth/differentiation factors See **growth (and) differentiation factors**.

GRP (1) See **endoplasmin** (glucose-regulated protein-94). (2) See **gastrin-releasing peptide**. (3) GRP-1 is **cytohesin-3**, a guanine nucleotide exchange factor with **PH domain** that binds PtdIns (3,4,5)P3 and also has domain with homology to yeast Sec7. See **cytohesin, ARNO**, Gea-1 and Gea-2, other members of the family. (4) See **BiP** (glucose-regulated protein-78). (5) GrpE is a bacterial chaperone of the **Hsp70** family that works in conjunction with **dnaJ** and dnaK. Dimeric GrpE is the co-chaperone for dnaK, and acts as a nucleotide exchange factor, stimulating the rate of ADP release from dnaK 5000-fold.

G$_s$ See **GTP-binding protein**.

GSE complex In *S. cerevisiae* a GTPase complex required for intracellular sorting of GAP1p (general amino acids permease) out of the endosome to the plasma membrane to maintain adequate amino acid levels for protein synthesis. Abstract: http://www.nature.com/ncb/journal/v8/n7/full/ncb0706-648.html

GSG domain A domain (GRP33 / Sam68 / GLD-1 domain, ~200aa) found in a family of RNA-binding proteins that mediates dimerisation and RNA binding of several regulatory proteins such as **sam68** and **quaking**. Also called STAR, signal transduction and activator of RNA. PMID: 14718919

GSK See **glycogen synthase kinase**.

GSS (1) **Glutathione synthetase**. (2) **Gerstmann-Straussler-Scheinker** disease (3) Genome Survey Sequences (GSS) database, a database with DNA sequences based on genomic rather than expressed (cDNA) sequences. Link to database: http://www.ncbi.nlm.nih.gov/dbGSS/

GST See **glutathione S transferase**.

GTBP Protein (G/T binding protein, MutS homologue-6, MSH6, 1068aa) that is important in mismatch recognition in human cells (see **mismatch repair**). The heterodimer of GTBP with hMSH2 (one of the **MSH** family) binds mismatches (G paired with T) as a first step in excision repair. Absence of either protein predisposes to tumours and mutations in GTBP are responsible for some cases of hereditary nonpolyposis colon cancer. The GTBP gene is located near that for hMSH2 and GTBP can be considered one of the MSH family.

GTP Guanosine 5'-triphosphate. Like **ATP** a source of phosphorylating potential, but is separately synthesized and takes part in a limited, distinct set of energy-requiring processes. Synthesis is by a substrate-linked phosphorylation involving succinyl coenzyme A, part of the tricarboxylic acid cycle. GTP is required in protein synthesis, the assembly of microtubules, and for the activation of regulatory G-proteins **GTP-binding proteins**.

GTP-binding proteins There are two main classes of GTP-binding proteins (G-proteins): **(1)** the heterotrimeric G-proteins that associate with receptors of the seven transmembrane domain superfamily and are involved in signal transduction, and the small cytoplasmic G-proteins. The $G\alpha$ subunit ($\sim$350aa) of the heterotrimeric G-proteins has the nucleotide binding site and dissociates from the $\beta\gamma$ subunits (β, $\sim$340aa: γ, $\sim$70aa) when GTP is bound, and in this state will interact with various second messenger systems, either inhibiting (G_i) or stimulating (G_s). The $G\alpha$ subunit has slow GTPase activity and once the GTP is hydrolyzed it reassociates with the $\beta\gamma$ subunits. There is less diversity among the $\beta\gamma$ subunits, but they may have direct activating effects in their own right. Most $\beta\gamma$ subunits are post-translationally modified by myristoylation or isoprenylation that may alter their association with membranes. Stimulatory G-proteins are permanently activated by **cholera toxin**, inhibitory ones by **pertussis toxin**. **Transducin** was one of the first of the heterotrimeric G-proteins to be identified. **(2)** The small G-proteins are a diverse group of monomeric GTPases that include **ras**, **rab**, **rac** and **rho** and that play an important part in regulating many intracellular processes including cytoskeletal organization and secretion. Their GTPase activity is regulated by activators (GAPs) and inhibitors (GIPs) that determine the duration of the active state. See also **GEFs**, **ras-like GTPases**, Table G1.

GTPase-activating protein A protein that stimulates the GTPase activity of small **GTP-binding protein**, usually specific to a particular member of the family (e.g. a rho-GAP). Hydrolysis of the bound GTP to GDP inactivates the protein: this can be reversed by the action of **GEFs**, GTP-exchange factors that stimulate release of GDP and allow binding of GTP. Reactivation can be blocked by **GDIs** (guanosine nucleotide dissociation inhibitors) that prevent release of GDP. GAPs may be regulated by phospholipids and by phosphorylation on a tyrosine residue by growth factor receptors (PDGF-R, EGF-R). They also have **SH2** and **SH3** domains through which other interactions can occur. The **neurofibromatosis** type 1 gene (NF1) codes for a protein homologous to GAP.

Guam dementia An amyotrophic lateral sclerosis/parkinsonism dementia complex (ALS/PDC) that has an extremely high rate of incidence among the Chamorro people of Guam. May be caused by the toxin **BMAA**.

guanidinium chloride Chloride salt of guanidinium $(C(NH_2)_3)^+$ (guanidine hydrochloride), a powerful chaotropic agent that is used to denature proteins and to dissociate nucleic acid and protein in RNA isolation procedures.

guanine One of the constituent bases of nucleic acids, nucleosides and nucleotides, 2-amino 6-hydroxy purine.

guanine deaminase An ubiquitous enzyme (EC 3.5.4.3, guanine aminohydrolase, p51-nedasin, 454aa) that catalyzes the hydrolytic deamination of guanine, producing xanthine and ammonia. The alternatively spliced form in neurons is **cypin**.

guanosine The nucleoside formed by linking ribose to guanine.

guanosine 5′-triphosphate See **GTP**.

guanylate cyclase Enzyme (EC 4.6.1.2) that catalyzes the synthesis of the second messenger guanosine 3′,5′- cyclic monophosphate (cyclic GMP, cGMP) from guanosine 5′-triphosphate. Two soluble forms of guanylate cyclase are known, heterodimers of highly related subunits (alpha, 690aa or 732aa; beta, 619aa) that are directly stimulated by nitric oxide. The plasma membrane form of guanylate cyclase is an integral membrane protein with an extracellular receptor for peptide hormones, a transmembrane domain, a **protein kinase**-like domain, and a guanylate cyclase domain. Examples: sea urchin receptors for **speract** and **resact**; **atrial natriuretic peptide** receptors. Intestinal guanylate cyclase (guanylyl cyclase C, 1073aa) is the receptor for *E. coli* heat-stable enterotoxin. See **guanylate cyclase-activating protein**.

guanylate cyclase-activating proteins Proteins that activate **guanylate cyclase** and increase the production of cGMP. GCAP1 is a lipid-anchored calcium-binding protein (201aa) of the calmodulin superfamily, found only in photoreceptor inner segments, that stimulates synthesis of cGMP when calcium levels are low, the key step in recovery of the dark state of rod photoreceptors. Mutations cause **cone dystrophy**. Guanylate cyclase activator 2A (115aa) is cleaved to produce HMW-guanylin (94aa) and guanylin (15aa), the latter being the endogenous activator of intestinal guanylate cyclase, binding to the same region of the enzyme as heat-stable enterotoxins. Guanylate cyclase activator 2B (112aa) is cleaved into guanylate cyclase C-activating peptide 2 (GCAP-II, 24aa) and uroguanylin (16aa) and may be important in regulating kidney function.

guanylin See **guanylate cyclase-activating proteins**.

guard cells Plant cells occurring in pairs in the **epidermis**, flanking (guarding) each **stoma**. Changes in **turgor** in the guard cells cause the stoma to open and close.

TABLE G1. The Properties of Heterotrimeric G Protein Subunits

Alpha subunits:

α	MW (kDa)	Signal detector	Effector	Toxin	Comments
α_s	44–45	α-adrenergic, glucagon and many other receptors	Activates adenylate cyclase	CT	Stimulatory. At least four splice variants known, and relative concentration varies from one cell type to another.
α_{i1}	40.4	β-adrenergic, muscarinic cholinergic, opiate and many other receptors	Inhibits adenylate cyclase and PLC (phosphoinositde hydrolysis).	PT	Inhibitory. The sequence for $G_{i\alpha 1}$, and $G_{i\alpha 2}$ were derived from two different cDNA clones. The relative importance of the two forms, in the cell, is not yet clear
α_{i2}	40.5		Inhibits adenylate cyclase, PLC, involved in regulating K^+-channels, Ca^{2+} channels.	PT	
α_{i3}	40.5		Probably same as $\alpha_{i1,2}$	PT	
α_{olf}	44	Seven-membrane spanning	Activates adenylate cyclase	CT	In olfactory neurons
α_{gust}	40.3	Taste sensors	Unclear	Both	Component of gustducin, restricted to tongue.
α_z	40.9	Unknown	Inhibits adenylate cyclase	None	In brain and neuronal cells
α_q	42.2		Activates PLCβ (PI hydrolysis)	None	Widely expressed except in T-cells.
α_{11-14}	42–45		Activation PLCβ	None	Widely distributed though with variations between tissues
α_o	39	Unknown	Inhibits neuronal calcium channels	PT	'o' stands for other G protein. It was first detected in large amounts in brain. Splice variants known.
α_{t1}	40	Rhodopsin	cGMP-PDE	Both	Subunit of Transducin, found in the rod cells of the retina
α_{t2}	40.4	—	cGMP-PDE	Both	Found in the cone cells of the retina: probably the α subunit of the cone's analogue of Transducin

Beta and gamma subunits:

β	MW (kDa)	Tissue distribution
β1	37	Beta subunit in transducin
β2	38	Widely distributed
β3	37	Wide, particularly cone cells of retina
β4	37	Wide, especially brain, eye. lung, heart and testis
γ1	8.4	Rod cells of retina
γ2	7.8	Brain
γ3	8.3	Brain and retina
γ4		Unknown
γ5	7.3	Wide
γ7	7.5	Brain

Based partly upon G-protein linked receptor Facts Book (Academic Press) S. Watson and S Arkinstall (1994).

guard mother cell Small cell with dense cytoplasm that divides symmetrically to form a pair of **guard cells** that flank the stomatal pore.

Guarnieri body Acidophilic inclusion body found in cells infected with **vaccinia** virus; composed of viral particles and proteins, it is the location of virus replication and assembly.

Guibaud-Vainsel syndrome A form of **osteopetrosis** (marble brain disease), caused by mutation in the gene encoding carbonic anhydrase II.

guidance See **contact guidance**.

guide RNA *gRNA* (1) Any RNA sequence involved in sequence-specific recognition of target RNA (see **RNA interference**). (2) Small RNA molecules that hybridize to specific mRNAs and direct their **RNA editing**.

Guillain-Barre syndrome *Landry-G-B syndrome* Acute inflammatory demyelinating polyneuropathy, probably autoimmune and sometimes associated with prior *Campylobacter jejuni* infection. There are antibodies to gangliosides of peripheral nerves. In rare cases there is mutation in the peripheral myelin protein (PMP22; **gas**-3, growth arrest-specific-3 gene). Causes a temporary paralysis, particularly of the extremities.

Gunther's disease See **porphyria**.

gurken In *Drosophila*, a TGFα-like ligand (295aa) for the EGF-receptor torpedo, important for determining the anterior-posterior and dorsal-ventral axes of the egg. The gurken-torpedo signal is the first to specify dorsal fate and in turn induces the production of the ligand **spätzle** that binds to **toll** and activates the transcription factor dorsal. FlyBase entry: http://www.sdbonline.org/fly/torstoll/gurken1.htm

GUS reporter system A reporter gene system using the *E. coli* beta-glucuronidase gene which is detected either histochemically, using 5-bromo-4-chloro-3-indolyl glucuronide (X-Gluc) which is hydrolysed to give a blue colour, spectrophotometrically, using p-nitrophenyl β-D-glucuronide or fluorimetrically using 4-methylumbelliferyl-β-D-glucuronide.

gustducin Taste-cell specific **GTP-binding protein**. Has a novel Gα subunit (354aa) that resembles **transducin** more than any other Gα.

gut-associated lymphoid tissue *GALT* Mucosal lymphoid tissue associated with the gut (Peyer's patches, tonsils, adenoids, mesenteric lymph nodes and the appendix).

GVH See **graft-versus-host disease**.

GW182 RNA-binding protein (1962aa) initially shown to associate with a specific subset of mRNAs and to reside within discrete cytoplasmic foci named **GW bodies**. Required for miRNA-dependent repression of translation and for siRNA-dependent endonucleolytic cleavage of complementary mRNAs by **argonaute** family proteins. Originally identified as the target of an autoantibody in serum from a patient with a sensory ataxic polyneuropathy. GW182 has multiple ($\sim$60) glycine (G)-tryptophan(W) repeats.

GW bodies *P bodies* Small, generally spherical, cytoplasmic domains that vary in number and size in various mammalian cell types and appear to be the site for mRNA processing, storage and degradation. Several proteins co-locate in these processing granules, including **GW128** and **argonaute** 1 and 2 which are associated with **siRNA** operations. May well be equivalent to maternal granules in eggs which are accumulations of long-lived maternal mRNA.

Gymnospermae One of the two major division of seed-bearing vascular plants of which the most common members are conifers. See **Angiospermae**.

gynoecious Describing a plant with only female sex organs.

gynoecium The female reproductive organs of a flower.

gypsy-like retrotransposons One of the two most commonly found classes of **LTR retrotransposons** (Ty3/gypsy-like retrotransposons). All mouse LTR retrotransposons are gypsy-like. In rice 17% of the rice genome consists of LTR retrotransposons, two-thirds being gypsy-like or copia-like, with more of the former. The Gypsy element of *Drosophila melanogaster* can act as a retrovirus and horizontally transfer itself from one Drosophila species to another. The Ty3/gypsy retrotransposons can be subdivided into two main groups, one containing all those with a chromodomain at the C-terminal end of their integrases (chromoviruses), the second with the non-chromoviral Ty3/Gypsy LTR retrotransposons and retroviruses of plants and animals.

gyrate atrophy A progressive form of night blindness with sharply demarcated areas of retinal atrophy. It is caused by deficiency in ornithine-aminotransferase (EC 2.6.1.13) and can be partly treated by dietary restriction of arginine.

gyrin See **synaptogyrins**.

gyrus Any of the ridge-like folds of the cerebral cortex.

H

H Single letter code for histidine.

H2 antigen An antigen of the H2 region of the **major histocompatibility complex** of mice, the equivalent of the human HLA antigens. Divided into Class I and Class II antigens.

H2 blocker Antagonist of the histamine type 2 (H2) receptor. Drugs of this type block gastric acid secretion and are therefore clinically useful in treating duodenal ulcers.

H2 complex See **H2 antigens** and **histocompatibility antigens**.

H5N1 Strain of avian influenza that it is feared may give rise to a strain that can become highly infectious (transmissible) in man and potentially cause a major pandemic. The fears are partly based upon the fact that a similar avian flu strain gave rise to the 1918 flu pandemic. The strain type derives from the main antigenic determinants of influenza A and B viruses, the haemagglutinin (H or HA, of which there are 16 subtypes) and neuraminidase (N or NA, of which there are 9 subtypes) both of which are transmembrane glycoproteins of the viral envelope.

H19 gene Gene in mouse and human that has a differentially methylated domain (DMD); this domain is a methylation-sensitive insulator that blocks access of the insulin-like growth factor 2 **IGF2** gene to shared enhancers on the maternal allele and inactivates H19 expression on the methylated paternal allele (see **imprinting centre 1**). The imprinted H19 gene produces a noncoding RNA of unknown function. Hypermethylation of H19 in humans is associated with **Beckwith-Wiedemann syndrome** and hypomethylation of the H19 promoter with Silver-Russell syndrome.

H89 Isoquinoline **PKA** inhibitor, also reported to inhibit **Rho-kinases (ROCKs)**.

Ha-*ras* Harvey-*ras*; see *ras*.

HA-tag An epitope tag often added to recombinant proteins to aid purification or localisation. The tag is amino acid residues 97−115 of the influenza virus haemagglutinin (HA) protein.

habituated culture A plant tissue culture that can grow independently of exogenously added auxin.

HACBP High affinity Ca^{2+}-binding protein. See **calreticulin**.

Hacrobia A proposed grouping of Cryptophytes (cryptomonads) and Haptophytes. Webpage: http://tolweb.org/Hacrobia/124797

haem *US*. heme. Compounds of iron complexed in a porphyrin (tetrapyrrole) ring that differ in side chain composition. Haems are the prosthetic groups of **cytochromes** and are found in most oxygen carrier proteins. Haem- and haemato- (*US* heme- and hemato-) are prefixes indicating something related to blood. Entries are under the UK English form.

haem oxygenases A vascular enzyme, haem oxygenase-1 (HO-1, HSP32, EC 1.14.99.3, 288aa) an inducible heat-shock/stress protein that metabolize haem to form carbon monoxide, iron and biliverdin. The CO inhibits nitric oxide synthase and may be a neurotransmitter. HO-2 (316aa) is constitutive and present in endothelial cells, adventitial nerves of blood vessels and neurons in autonomic ganglia. HO-3 (290aa) is similar to HO-2 and only reported from at.

haemagglutination Agglutination of red blood cells, often used to test for the presence of antibodies directed against red-cell surface antigens or carbohydrate-binding proteins or viruses in a solution. Requires that the agglutinin (haemagglutinin) has at least two binding sites.

haemangioblast Earliest mesodermal precursor of both blood and vascular endothelial cells. In human embryos a marker, **flk-1**, is detectable early in week 4. Haematopoiesis in vertebrates occurs in two phases, the early phase from primitive haemangioblasts in the extra-embryonic mesoderm of the yolk sac, producing mostly erythroid and some myeloid lineages. The second phase involves definitive haemangioblasts from intra-embryonic mesoderm and produces a wider range, including lymphoid lineages.

haemangioma A benign tumour (angioma) produced by proliferation of capillary endothelial cells. Cutaneous haemangiomas include capillary naevi, port-wine stains, vin rose patches, cavernous haemangiomata, Campbell de Morgan spots and telangiectasias.

Haemanthus katherinae The African blood lily, chiefly known because of classic time-lapse studies done on mitosis in endosperm cells.

haematochrome A red or orange pigment (β-carotene) accumulated in the cells of some green algae, probably for protective purposes. Algae, for

The Dictionary of Cell and Molecular Biology. DOI: http://dx.doi.org/10.1016/B978-0-12-384931-1.00008-8

example *Trentepohlia*, have haematochrome and are often photosynthetic symbionts in lichens, hence the bright red colouration of these lichens. Haematochrome in *Chlamydomonas nivalis* is responsible for 'red snow', snow that is pink or red and seen occasionally on Scottish hills.

haematocrit Relative volume of blood occupied by erythrocytes. An average figure for humans is 45 ml per cent, i.e. a packed red cell volume of 45 ml in 100 ml of blood.

haematoma A swelling in which there has been leakage of blood into connective tissue. Includes bruises (ecchymoses) and petechiae (haematomas less than 3 mm in diameter).

haematopoiesis Production of blood cells involving both proliferation and differentiation from stem cells. In adult mammals usually occurs in bone marrow. See **haemangioblast**.

haematopoietic cell kinase A protein tyrosine kinase of the **src family** (Hck, p59-HCK/p60-HCK, 526aa) found in lymphoid and myeloid cells and that is bound to B-cell receptors in unstimulated B-cells. May couple the Fc receptor to the activation of the respiratory burst and regulate neutrophil degranulation. Deletion of hck and src or hck and fgr leads to severe developmental anomalies and impaired immunity in mice.

haematopoietic stem cell A **stem cell** involved in production of blood cells (haematopoiesis).

haematoxylin Basophilic stain (Natural Black 1, C.I. 75290) that gives a blue colour (to the nucleus of a cell for example), commonly used in conjunction with eosin that stains the cytoplasm pink/red (the H&E staining of routine histopathology). Various modifications of haematoxylin have been developed, often used with aluminium or iron salts as mordants. See **Heidenhain's iron haematoxylin**.

haemochromatosis An iron-overload disorder in which **haemosiderin** is deposited in excess in the organs of the body, giving rise to cirrhosis of the liver, enlargement of the spleen, diabetes (bronzed diabetes) and skin pigmentation. Classic haemochromatosis (HFE) is usually caused by mutation in the HFE gene which encodes HFE protein (348aa) that binds to the transferrin receptor and reduces its affinity for iron-loaded transferrin. Haemochromatosis type 2A is caused by mutation affecting **hemojuvelin**, HFE2B by mutation in hepcidin antimicrobial peptide. Type 3 is caused by mutation in the gene encoding transferrin receptor-2, type 4 by mutation in the **ferroportin** gene.

haemocoel The body cavity derived from the blood vessels that replaces the coelom in arthropods and molluscs. It is a secondary cavity, not generated from the **archenteron**.

haemocyanin Blue, oxygen-transporting, copper-containing protein found freely dissolved in the blood (haemolymph) of molluscs and crustacea. A very large protein with 20–40 subunits and molecular weight of 2–8 million Daltons, and having a characteristic cuboidal appearance under the electron microscope. Prior to the introduction of immunogold techniques, it was used for electron-microscopic localization by coupling to antibody. In *Panulirus interruptus* (California spiny lobster) a heterohexamer of haemocyanin A (657aa), B (657aa) and C (661aa). Keyhole limpet haemocyanin (KLH) is widely used as a carrier in the production of antibodies.

haemocytes Blood cells, associated with a haemocoel, particularly those of insects and crustacea. Despite the name they are more leucocyte-like, being phagocytic and involved in defence and clotting of haemolymph, and not involved in transport of oxygen.

haemocytoblast A pluripotential stem cell of the haematopoietic tissue from which monopotential stem cells of various lineages arise.

haemocytometer A special glass slide used for counting blood cells etc. under the microscope. A grid of lines is engraved on the bottom of a shallow rectangular trough and a stiff coverslip is placed over the trough so that the grid demarcates known volumes. Cell suspension can then be introduced and the number of cells per square counted. Various grid patterns are available (Neubauer and Fuchs-Rosenthal being common). A similar device on a smaller scale (Helber cell) is used for bacterial counting.

haemodialysis The dialysis of blood against a physiological saline as an artificial substitute for kidney function. Small molecules are removed, not all of which are unwanted.

haemoglobin Tetrameric globular oxygen-carrying protein of vertebrates and some invertebrates. There are two α (142aa) and two β (147aa) chains (very similar to myoglobin) in adult humans; the haem moiety (an iron-containing substituted porphyrin) is held non-covalently in a non-polar crevice in each peptide chain. Fetal haemoglobin has gamma chains (147aa) instead of beta chains and has higher oxygen affinity. Adult haemoglobin A2 (HbA2), which is only ~3.5% of total, has delta (147aa) chains instaed of beta.

haemoglobinaemia The presence of haemoglobin in the blood as a result of erythrocyte lysis.

haemoglobinopathies Disorders due to abnormalities in haemoglobin. See **sickle cell anaemia** and **thalassaemia**.

haemolymph Circulating body fluid of invertebrates such as insects that have a **haemocoel**. Cells in the haemolymph are usually referred to as **haemocytes**.

haemolysins Bacterial **exotoxins** that lyse erythrocytes. See **cyclolysin**, **Kanagawa haemolysins**, **leucocidin**, **perfringolysin O**, **streptococcal toxins**, **streptolysin O**, **streptolysin S**, **tetanolysin** and **thiol-activated haemolysins**.

haemolysis Leakage of haemoglobin from erythrocytes due to membrane damage.

haemolytic anaemia Anaemia resulting from reduced red-cell survival time, either because of an intrinsic defect in the erythrocyte (hereditary **spherocytosis** or ellipsocytosis, enzyme defects, **haemoglobinopathies**), or through the action of an extrinsic agent such as autoantibody (autoimmune haemolytic anaemia), *Plasmodium*, **haemolysins** or mechanical damage to erythrocytes.

haemolytic disease of the newborn See **erythroblastosis fetalis**.

haemolytic plaque assay A method used to detect individual cells secreting antibody *in vitro*. Sheep red cells (treated so as to bind the antibody) are mixed with the cell suspension to be assayed in a thin layer of agarose, and incubated. Cells that are producing antibody are revealed, when complement is added, by an area of haemolysis surrounding them.

haemolytic uremic syndrome A disorder of the microvasculature associated with mutations in **complement factor H**. There are distorted erythrocytes, 'burr cells' causing microangiopathic haemolytic anaemia, thrombocytopenia, and acute renal failure. Can also be caused by infection with enterohaemorrhagic *E. coli* OH157:H7.

haemonectin A 60 kDa protein (fetuin-A, alpha-2-HS-glycoprotein) found in the bone marrow matrix of rabbits specifically aiding adhesion of granulocyte-lineage cells. The sequence shows 60–70% similarity with that of **fetuin** from other mammal species.

haemopexin A single-chain haem-binding plasma β1-glycoprotein (462aa) that transports haem to the liver for breakdown and iron recovery; unlike **haptoglobin** does not bind haemoglobin. Present at around 1 mg/ml in plasma. Structurally related to **vitronectin** and some **collagenases** which have haemopexin-like domains (the **pexin family**).

haemophagocytic lymphohistiocytosis A rare disorder in which activated lymphocytes and macrophages infiltrate several organs and there is erythrocyte phagocytosis causing anaemia. The molecular defect in Type 1 is unknown; other forms are a result of mutations affecting **perforin**, unc13D homologue (1090aa, involved in cytotoxic granule exocytosis in lymphocytes), **syntaxin-11** or syntaxin-binding protein-2.

haemophilia Disease in which there is a deficiency in blood-clotting. The classical X-linked form (haemophilia A) is a deficiency in factor VIII, haemophilia B (**Christmas disease**) is a factor IX deficiency, **von Willebrand disease** is a deficiency of vWF.

Haemophilus influenzae A non-motile Gram-negative coccobacillus (**Pfeiffer's bacillus**) sometimes associated with influenza virus infections, causes pneumonia and meningitis. *H. influenzae* type b (Hib) is a major cause of lower respiratory tract infections in children.

haemopoiesis See **haematopoiesis**.

haemorrhage Blood loss from a damaged blood vessel, implicitly substantial.

haemorrhagic Related to or causing haemorrhage (bleeding).

haemosiderin An intracellular yellow-brown iron compound (Iron(III) oxide/iron hydroxide), usually in macrophages, derived from breakdown of haemoglobin via ferritin. Not in itself harmful, but a sign of bleeding and subsequent haemolysis, or of iron-overload. See **haemochromatosis**.

Hageman factor *Not Hagemann factor* Blood-clotting factor XII, a β-globulin (615aa), that is activated to form Factor XIIa by contact with surfaces. Factor XIIa activates factor XI, generates **plasmin** from plasminogen and **kallikrein** from prekallikrein and triggers the complement cascade. Hageman factor is important both in clotting and activation of the inflammatory process. Dr John Hageman was the first person to be identified with a deficiency in the factor. Mutations cause hereditary angioedema (HAE) type III.

Hailey-Hailey disease A blistering disease (familial chronic benign pemphigus) caused by a defect in epidermal desmosomes and adherens junctions, even though they appear morphologically normal. There is a mutation affecting ATP2C1 (calcium-transporting ATPase).

Haim-Munk syndrome A disorder (Cochin Jewish disorder) characterized by palmoplantar keratosis, onychogryphosis and periodontitis caused by a defect in **cathepsin** C.

hair cells (1) Cells found in the epithelial lining of the labyrinth of the inner ear. The hairs are **stereovilli** up to 25 μm long that restrict the plane in which deformation of the apical membrane of the cell can be brought about by movement of fluid or by sound. Movement of the single **stereocilium** transduces mechanical movements into electrical receptor potentials. **(2)** In plants hairs are thin-walled cells that cover many surfaces. They are convenient for studying cytoplasmic streaming and for observing mitosis. *Tradescantia* stamens are a common source.

hairless **(1)** A transcriptional corepressor (1189aa) for the thyroid hormone receptor, product of the human homologue of the mouse hairless gene. It apparently regulates hair growth and mutations lead to **alopecia** universalis congenita and congenital atrichia. **(2)** In *Drosophila* (where hairs are completely different) a potent antagonist (1077aa) of neurogenic gene activity during sensory organ development during which a pair of cells differentiate to form a hair shaft (trichogen) and socket (tormogen) and act as mechanoreceptors. Suppressor of hairless (594aa) is a transcriptional regulator that plays a central role in Notch signalling. There is a human homologue, also involved in notch signalling and unconnected with hair development.

hairpin Protein motif formed by two adjacent regions of a polypeptide chain that lie antiparallel and alongside each other. Depending on whether the polypeptide is in **alpha-helix** or **beta strand** configuration, can be described as alpha hairpin or beta hairpin, respectively. Webpages: http://swissmodel.expasy.org/course/text/chapter2.htm

hairy One of the *Drosophila* **pair-rule** genes that encodes a bHLH transcription factor of the **hairy/ enhancer of split/deadpan** (HES) family of proteins. *Drosophila* HES family proteins are key repressors in the developmental processes of segmentation, neurogenesis, and sex determination. All have a highly conserved **bHLH domain**, an adjacent Orange domain, which confers specificity among family members, and a C-terminal tetrapeptide motif, WRPW, which has been shown to be necessary and sufficient for the recruitment of the corepressor, **groucho**.

hairy cell leukaemia A rare disorder in which there are hairy cells (abnormal white blood cells with hair-like projections) in the blood. The proliferating cells are B-cells which may accumulate in bone marrow and inhibit production of other blood cell types. Several variants are recognised.

hairy/enhancer of split/deadpan Family of basic helix-loop-helix (bHLH) proteins that function as transcriptional repressors implicated in preventing tissue-specific determination of stem cells. See **hairy**.

Hakata antigen See **ficolin**.

half-life $t_{1/2}$ The period over which the activity or concentration of a specified chemical or element falls to half its original activity or concentration. Typically applied to the half-life of radioactive atoms but also applicable to any other situation where the population is of molecules of diminishing concentration or activity.

Haller cell Air-filled cavities in the head, pneumatized infraorbital ethmoid cells, not cells in the cell-biological sense.

Hallervorden-Spatz disease See **infantile neuroaxonal dystrophy**.

halobacteria Bacteria that live in conditions of high salinity (**halophiles**).

Halobacterium halobium Photosynthetic (halophilic) archaean that has patches of purple membrane containing the pigment **bacteriorhodopsin**. Has been extensively used as an experimental species and the genome (2,571,010 bp) has been sequenced.

halophile Literally, salt-loving (halophilic): used to describe an organism that tolerates saline conditions, in extreme cases in concentrations considerably in excess of those found in normal sea water such as salt lakes. Some Archaea (e.g. *Halobacterium halobium*) are notable for their ability to survive extremes of salinity (**extreme halophiles**).

halophyte Plant that grows in or tolerates salt-rich environments.

halorhodopsin Light-driven chloride ion pump of halobacteria (276aa in some strains of *Halobacterium halobium*), a retinylidene protein very similar to **bacteriorhodopsin**. The cells accumulate potassium and chloride ions to counter the dehydrating osmotic effects of high extracellular salt (sodium chloride).

halysin See **disintegrin**.

hamartin Protein (1164aa) encoded by tumour suppressor gene TSC1 that negatively regulates mTORC1 signalling; interacts with the **tuberin** protein. See **tuberous sclerosis**.

hamartoma Tumour-like but non-neoplastic overgrowth of tissue that is disordered in structure. Examples are haemangiomas (that include the vascular naevus or birthmark) and the pigmented naevus (mole).

hammerhead ribozyme A **ribozyme** in which there are three helical regions radiating from a central core.

HAMP See **hepcidin antimicrobial peptide**.

hanatoxins Peptide toxins (HaTx1 and HaTx2, 35 aa) from the venom of the Chilean tarantula, *Grammostola spatulata*. Inhibit the Kv2.1 voltage-gated K^+ channel and are unrelated in primary sequence to other K^+ channel inhibitors.

Hand-Schüller-Christian disease A chronic disorder which is histologically a type of **Langerhans cell histiocytosis**, with onset between three and five years. There is lipid accumulation and histiocytic granulomas in bone, particularly in the skull, the skin, and viscera.

hanging-drop preparation A preparation in which the specimen is suspended in a drop of medium on a coverslip that is inverted over a cavity ground into a microscope slide and sealed at the edges to prevent evaporation. Now probably only of historical interest except in crystallography where hanging drops are sometimes used to try to obtain crystals for X-ray diffraction.

Hanks' balanced salts solution A physiological saline (HBSS) that can be sterilized by autoclaving and was devised by Hanks. It is phosphate-buffered to pH 7.0−7.2, contains bicarbonate and (usually) phenol red as a pH indicator. Suitable for mammalian and avian cells in temporary culture but is not a growth medium.

Hansenula See *Pichia*.

Hanta virus Viruses of the bunyaviridae family, first identified near the Hantaan River in Korea. The genome consists of three negative-sense, single-stranded RNA segments. Cause haemorrhagic fevers although pathogenicity varies with the serotype from asymptomatic infection to highly fatal disease. The viruses persist in rodents and infection is transmitted by aerosols.

haplodiploid A reproductive system in which females are diploid and males are haploid, as in insects such as bees.

haploid Describes a nucleus, cell or organism possessing a single set of unpaired chromosomes. **Gametes** are haploid.

haploinsufficiency Situation in which a single copy of a gene is insufficient to allow normal functioning. The heterozygote is therefore affected.

haplontic Organisms in which meiosis occurs in the zygote, giving rise to four haploid cells (e.g. many algae and protozoa), only the zygote is diploid and this may form a resistant spore. *Cf.* **diplontic**, **diplohaplontic**.

haplotype One of the two sets of alleles in a diploid organism. Also used to refer to the set of alleles on one chromosome or a part of a chromosome, i.e. one set of alleles of linked genes. Its main current usage is in connection with the linked genes of the **major histocompatibility complex**.

hapten Could be considered an isolated **epitope**: although a hapten (by definition) has an antibody directed against it, the hapten alone will not induce an immune response if injected into an animal, it must be conjugated to a carrier (usually a protein). The hapten constitutes a single antigenic determinant; perhaps the best known example is dinitrophenol (DNP) that can be conjugated to BSA and against which anti-DNP antibodies are produced (antibodies to the BSA can be adsorbed out). Because the hapten is monovalent, immune complex formation will be blocked if the soluble hapten is present as well as the hapten-carrier conjugate (assuming there is more than one hapten per carrier then an immune precipitate can be formed). Competitive inhibition by the soluble small molecule is sometimes referred to as haptenic inhibition, and this term has carried over into lectin-mediated haemagglutination where monosaccharides are added to try to block haemagglutination: the blocking sugar defines the specificity of the lectin.

haptenic inhibition See **hapten**.

haptoglobin An **acute phase protein**, acid α2-plasma glycoprotein (406aa then cleaved into alpha and beta chains) that binds to oxyhaemoglobin that is free in the plasma, and the complex is then removed in the liver. Tetrameric (2α, 2β subunits): the existence of two different α chains (α1, 83aa; α2, 142aa) in humans means that haptoglobins can exist in three variants in heterozygotes. It is homologous to serine peptidases but has no enzymic activity.

haptonema Filament extending between the paired flagella of certain unicellular algae (**haptophytes**). Supported by six or seven microtubules (not in an axoneme-like array) and apparently used for capturing prey, in a manner analogous to the axopodia of **heliozoa**.

Haptophytes A phylum of algae, typically with two slightly unequal flagella, both of which are smooth, and a unique organelle called a **haptonema**. Grouped with **cryptomonads** in the **Hacrobia** super-Phylum. Examples include coccolithophores (the Order **Coccolithales** within the Class Prymnesiophyceae).

haptotaxis Strictly speaking, a directed response of cells in a gradient of adhesion, but often loosely applied to situations where an adhesion gradient is thought to exist and local trapping of cells seems to occur.

Hardy-Weinberg law Mathematical formula that gives the relationship between gene frequencies and genotype frequencies in a population. If genotypes are AA, Aa, aa and the frequency of alleles A,a are respectively p,q, then $p + q = 1$ and AA:Aa:aa is p^2:$2pq$:q^2. Deviation from this equilibrium distribution suggests adverse survival characteristics of organisms with one of the alleles.

harlequin ichthyosis See **ichthyosis**.

harmonin Actin bundling and PDZ domain-containing scaffold protein (552aa) that forms complexes with CDH23 (cadherin-23; **otocadherin**) through two of the PDZ domains and the two proteins are co-expressed in the stereocilia of **hair cells** and in retinal photoreceptors. Also interacts with SANS protein. Mutations in the harmonin gene are the cause of Usher syndrome type 1C (USH1C), a rare, autosomal recessive syndrome of congenital deafness and progressive blindness. Harmonin-interacting, ankyrin repeat-containing protein (HARP, 417aa) is expressed in kidney and small intestine.

Hartig net A hyphal network, typical for ectomycorrhiza, that extends into the root, penetrating between epidermal and cortical cells. It is the site of nutrient exchange between the fungus and the host plant.

Hartnup disease Amino acid transport defect that can be caused by mutations in the SLC6A19 (solute carrier family-6) gene; leads to excessive loss of monoamino monocarboxylic acids (cystine, lysine, ornithine, arginine) in the urine, and poor absorption in the gut. See **iminoglycinuria**.

Harvey sarcoma virus See *ras*.

Hashimoto's thyroiditis Autoimmune disease in which there is destruction of the thyroid by autoantibodies usually directed against **thyroglobulin**, thyroid peroxidase (EC 1.11.1.7, 933aa) and **thyroid-stimulating hormone** (TSH) receptors. Hashimoto's encephalopathy is associated with Hashimoto's thyroiditis but alpha-enolase (EC 4.2.1.11) is the autoantigen.

haspins A family of atypical eukaryotic protein serine/threonine kinases, conserved in animals, fungi, and plants. Haspin (germ cell-specific gene 2 protein, haploid germ cell-specific nuclear protein kinase, 798aa) is mainly associated with chromosomes and phosphorylates histone H3 at threonine-3 during mitosis. It appears to be required for protection of cohesion at mitotic centromeres; yeast homologues (**Alk1** and Alk2), are also implicated in regulation of mitosis. Article: http://www.pnas.org/content/106/48/20198.long

Hassell's corpuscle Spherical or ovoid bodies 20−50 μm in diameter in the medulla of the thymus (thymic corpuscles), composed of flattened concentrically-arranged whorls of keratinised or hyaline cells surrounding dead cells in the core.

HASTY The plant orthologue of **exportin 5** (1202aa), involved in export of miRNA from the nucleus to the cytoplasm.

HAT medium A selective growth medium for animal tissue cells that contains hypoxanthine, the folate antagonist aminopterin (amethopterin) and thymine. Used for selection of hybrid somatic cell lines, as in the production of monoclonal antibodies. In HAT medium, cells are forced to use these exogenous bases, via the salvage pathways, as their sole source of purines and pyrimidines. Parental cells lacking enzymes such as **HGPRT** or **thymidine kinase** (TK) can be eliminated whilst hybrids grow.

Hatch-Slack-Kortshak pathway *Hatch-Slack pathway*. Metabolic pathway responsible for primary CO_2 fixation in **C4 plant** photosynthesis. The enzymes that are found in **mesophyll** chloroplasts include **PEP carboxylase**, that adds CO_2 to phosphoenolpyruvate to give the 4-carbon compound, oxaloacetate. Four-carbon compounds are transferred to bundle-sheath chloroplasts, where the CO_2 is liberated and re-fixed by the **Calvin-Benson cycle**. The HSK pathway permits efficient photosynthesis under conditions of high light intensity and low CO_2 concentration, avoiding the non-productive effects of photorespiration.

HAUS The human homologue of the **augmin** complex, formed of 8 subunits, an evolutionary conserved multisubunit protein complex that regulates centrosome and spindle integrity. Disruption of HAUS alters **nuclear mitotic apparatus protein** localization. Abstract: http://www.ncbi.nlm.nih.gov/pubmed/19427217

haustorium A projection from a cell or tissue of a fungus or higher plant, that penetrates another plant and absorbs nutrients from it. In fungi it is a **hyphal** projection that penetrates into the cytoplasm of a host plant cell; in parasitic angiosperms it is a modified root.

Haversian canals Small channels found in compact bone, They run along the length of the long bone and provide major blood-vessel supply to the osteocytes. Each Haversian canal is surrounded by concentric layers of bone, the whole forming an osteon.

HAX1 A protein of the bcl2 family (HS1-associated protein X1, 279aa) that potentiates Galpha13-mediated cell migration, is involved in clathrin-mediated endocytosis, may inhibit caspases and regulate intracellular calcium levels. Interacts with the haematopoietic and lymphoid-restricted intracellular protein (**HS1**) and with **cortactin**. Mutations in the *HAX1* gene lead to recessive severe congenital neutropenia (**Kostmann's syndrome**).

Hayflick limit See **cell death, cell line**.

HB-EGF A growth factor for smooth muscle cells (heparin-binding epidermal growth factor, 208aa) that, like **amphiregulin**, has a long N-terminal extension that seems to confer the ability to bind to heparin and also to other connective tissue macromolecules (glycosaminoglycans) and cell surface molecules such as CD44. Synthesized as a type I transmembrane protein (proHB-EGF, the binding site for **diptheria toxin**); the ectodomain is cleaved from the cell surface by a metallopeptidase and the soluble portion binds EGF receptors, HER-1 and HER-4. The soluble form can, however, become immobilized so that the effective local concentration may be much higher and the effects may differ from those of soluble growth factors. *Cf.* **HBGF**.

HBGF   Heparin-binding growth factor. See **fibroblast growth factor** and **midkine**.

HBV   Hepatitis B virus.

HCC1   An RNA binding protein (hepatocellular carcinoma protein 1, transcription coactivator CAPER, RNA binding motif protein 39, 530aa) a transcriptional coactivator for steroid nuclear receptors. Concentrated in nuclear speckles where it co-localizes with core spliceosomal proteins.

HCG   Human **chorionic gonadotrophin**.

H-chain   Heavy chain of immunoglobulin; see **IgG**, **IgM**, and others.

hck   See **haematopoietic cell kinase**.

HCP   See **histidine-rich calcium-binding protein**.

HCT   **(1)** Haematocrit (Hct). **(2)** Human colon cancer cell line, of which there are many variants, HCT-15, HCT-116 amongst others.

HCV   Hepatitis C virus.

HCV core protein   Hepatitis C core protein, the first 191aa of the viral precursor polyprotein that is cotranslationally inserted into the membrane of the endoplasmic reticulum; a viral structural protein that is considered to influence multiple cellular processes, blocks the activity of caspase-activated DNase and thus inhibits apoptotic cell death. An 'alternate reading frame' (ARF) overlaps with the core protein-encoding sequence and encodes the unstable ARF protein (ARFP, F-protein, 162aa).

HDAC   See **histone deacetylases**.

HDL   High density **lipoprotein**.

hdm2   A ubiquitin **E3 ligase** (human double minute 2, 491aa) that is the principal negative regulator of **p53**; inhibits p53 transcriptional activity and subjects it to degradation by an E3 ligase activity. Overexpression causes increased susceptibility to tumours. See **mdm** #2.

HDR syndrome   Hypoparathyroidism, sensorineural deafness and renal anomaly syndrome. See **GATA**.

Heaf test   A tuberculin test in which tuberculin is injected intradermally with a multiple puncture apparatus. A positive reaction indicates the presence of T-cell reactivity to mycobacterial products. Superceded by the **Mantoux test**.

heart muscle   See **cardiac muscle**.

heat shock factors   A ubiquitous family of DNA-binding proteins in eukaryotes (HSP1, 529aa; HSP2, 536aa, and others) that bind heat shock promoter elements (HSE) and activate transcription. Under normal conditions are retained in the cytoplasm bound to heat shock proteins.

heat-shock proteins   Families of proteins (hsps) conserved through pro- and eukaryotes, induced in cells as a result of a variety of environmental stresses, though some hsps are constitutively expressed. Some serve to stabilize proteins in abnormal configurations, play a role in folding and unfolding of proteins and the assembly of oligomeric complexes, and may act as **chaperonins**. Hsp90 (HSP 90-alpha, 732aa; HSP 90-beta, 724aa) complexes with inactive steroid hormone receptor and is displaced upon ligand binding. Four major subclasses are recognized: hsp90, hsp70, hsp60 and small hsps. Hsps have been suggested to act as major immunogens in many infections. See **BiP**, **endoplasmin**.

heavy chain   The larger polypeptide in a multimeric protein. The difference can vary and immunoglobulin heavy chain is of 50 kDa, the light chain of 22 kDa, whereas the myosin heavy chain at 220 kDa is far larger than myosin light chains ($\sim$20 kDa). Heavy chain diseases are rare B-cell disorders in which aberrant immunoglobulin heavy chains (of various kinds) are produced.

heavy water   Water in which the heavy isotope of hydrogen is present (deuterium oxide, D_2O). Used by cell biologists to stabilize **microtubules**.

HECT   A catalytic domain (Homologous to E6-AP C Terminus, $\sim$350aa) that characterizes a class of ubiquitin E3 ligases that have a direct role in catalysis during ubiquitination. See **ubiquitin conjugating enzymes**.

hedgehog   A protein morphogen that activates an essential cellular pathway (hedgehog pathway) required during the embryogenesis of various organisms ranging from *Drosophila* to mammals. In *Drosophila*, hedgehog establishes the anterior-posterior axis of the embryonic segments and patterns the larval imaginal disks. Binds to the **patched** receptor, which functions in association with **smoothened**, to activate the transcription of target genes *wingless* and *decapentaplegic*. The protein precursor (471aa) is autocatalytically cleaved to a 173aa N-terminal portion that remains lipid-anchored to cells (by cholesterol added by the C-terminal portion) and has signalling activity and a non-signalling C-terminal portion (214aa). Mammalian homologues are also important in morphogenesis (see **sonic hedgehog**, **Indian hedgehog** and **desert hedgehog**). The basic pathway seems to be the same in mammals and homologues of all the various components have been identified. In humans, hedgehog-interacting protein (700aa) modulates hedgehog-family signalling in several cell types. Useful overview: http://www.abcam.com/index.html?pageconfig=resource&rid=10858&pid=10039

hedgehog signalling complex A multi-subunit complex downstream of **smoothened** in *Drosophila*. The complex is microtubule-associated and has components that include the kinesin-related protein Costal2 (see **costa** #2), the serine/threonine protein kinase **fused**, and the zinc finger transcriptional activator **cubitus interruptus** (Ci). The mammalian homologues of Ci are the **Gli** family of transcription factors. See **hedgehog**.

HEF1 A docking protein of the **Cas family** (human enhancer of filamentation 1, neural precursor cell expressed, developmentally downregulated-9, NEDD-9, Cas-L, 834aa) which has a coordinating role in tyrosine-kinase-based signalling related to cell adhesion and involved in both integrin and growth factor signalling pathways. Isoforms p105 and p115 are predominantly cytoplasmic and associate with focal adhesions while p55 associates with mitotic spindle.

hefutoxin See **kappa toxin** #2.

Heidenhain's Azan A particularly beautiful but extremely time-consuming trichrome staining method that, properly carried out, results in chromatin, erythrocytes and neuroglia being stained red, mucus blue, collagen sharp blue and cytoplasmic granules red, yellow or blue. Now very rarely used.

Heidenhain's iron haematoxylin One of many haematoxylin-based staining solutions and one that is particularly good for photography or automatic image processing because of the intensity of black staining that can be achieved. Requires differentiation in iron alum and thus the intensity of staining can be adjusted according to the specimen. Sections stained with Heidenhain are usually counterstained with, for example, eosin or orange G.

Heidenhain's Susa Good general-purpose histological fixative but has the disadvantage of containing mercuric chloride.

HEK-293 cells Human embryonic kidney cell line transformed by sheared adenovirus type 5 DNA; were first described in 1977. They may have originated from a rare neuronal cell in the kidney since they stain strongly and specifically with antibodies to several neurofilament proteins and are a poor model for kidney cells. Easily transfected and used extensively in biotechnology.

HeLa cells An established line of human epithelial cells derived from a cervical carcinoma in the patient Henrietta Lacks. The malignant phenotype is probably a result of mutation in **hTERT**

(telomerase). They proliferate extremely well in culture and have contaminated a number of cell lines. Article: http://www.ncbi.nlm.nih.gov/pubmed?term=PMID%3A%2020450767; Lines affected by cross-contamination: http://onlinelibrary.wiley.com/doi/10.1002/ijc.25242/full

Helber cell See **haemocytometer**.

helicase See **DNA helicase**.

Helicobacter pylori S-shaped or curved Gram-negative bacteria $(0.5-0.9 \times 3.0 \, \mu m)$, non-spore forming, can be flagellate; found in human stomach. Was originally named *Campylobacter pyloridis*. Infection with *H. pylori* is now considered to be a major predisposing cause of gastric ulcers and antibiotic therapy is increasingly used.

helicoidal cell wall Type of plant cell wall in which each wall layer contains parallel **microfibrils**, but in which the orientation of the microfibrils changes by a fixed angle from one layer to the next to give a 'rotating-ply' texture. When sectioned this multi-layered structure has the appearance of bow-shaped arcs. A similar architecture of fibrillar material is seen in some insect exoskeletons.

Heliozoa Amoeboid **Protozoa**, Order Heliozoida. They are generally free-floating, spherical cells with many straight, slender microtubule-supported **pseudopods** radiating from the cell body like a sunburst. These modified pseudopods are termed **axopodia**. Genera include *Actinophrys* and *Echinosphaerium*.

Helisoma trivolvis Pulmonate mollusc whose relatively simple nervous system contains large identifiable cells and is consequently, like *Hirudo* and *Aplysia*, a favourite organism for studying neural mechanisms at the cellular level; and in particular for studying isolated neurons in culture.

helix-coil transition See **random coil**.

helix-destabilizing protein (1) In *E. coli*, a single-stranded DNA-binding protein (178aa) involved in DNA replication. Bind cooperatively to single-stranded DNA, preventing the reformation of the duplex and extending the DNA backbone, thus making the exposed bases more accessible for base-pairing. Essential for replication of the chromosomes and single-stranded DNA phages, also for repair. (2) In humans, single-strand RNA-binding protein (heterogeneous nuclear ribonucleoprotein A1, 372aa) a component of the spliceosome involved in the packaging of pre-mRNA into hnRNP particles and in transport of poly(A) mRNA from the nucleus to the cytoplasm.

helix-loop-helix A motif (HLH, 40–50aa) associated with transcription factors, allowing them to recognize and bind to specific DNA sequences. Two alpha-helices are separated by a loop. Sequence-specific binding requires dimerization. HLH proteins with an additional basic region of about 15aa next to the HLH domain are referred to as basic HLH (bHLH) proteins and bind the **E box** motif. Examples: myoblast MyoD1, c-myc, *Drosophila* daughterless, hairy, twist, scute, achaete, asense. Not the same as **helix-turn-helix**.

helix-turn-helix A motif associated with transcription factors, allowing them to bind to and recognize specific DNA sequences. There is a core 22aa segment that is useful for identifying proteins as being in this class. Two amphipathic alpha-helices are separated by a short sequence with a beta pleated sheet. One helix lies across the major groove of the DNA, while the recognition helix enters the major groove and interacts with specific bases. An example in *Drosophila* is the homeotic gene product **fushi tarazu**, that binds to the sequence TCAATTAAATGA. Not the same as **helix-loop-helix**. Minireview: http://www.jbc.org/content/264/4/1903.full.pdf+html

helospectin See **exendin**.

helothermine Protein toxin (242aa) from venom of the Mexican beaded lizard, *Heloderma horridum horridum*. Probably acts by inhibiting the ryanodine-sensitive calcium release channel of sarcoplasmic reticulum. One of a family of **cysteine rich secretory proteins**.

helper factors An obsolete name for factors apparently produced by helper T-cells that acted specifically or nonspecifically to transfer T-cell help to other classes of lymphocytes. The existence of specific T-cell helper factor is uncertain and although many cytokines could be described as such, the term is better avoided.

helper T-cell See **T-helper cells**.

helper virus A virus that will allow the replication of a co-infecting defective virus by producing the necessary protein.

helveticin A bacteriocin (333aa) produced by *Lactobacillus helveticus* that inhibits the growth of closely related *Lactobacillus* species.

hema-, hemo-, hemato US form of UK English haema-, haemo-, haemato-. The UK form is used for headwords (except for **hemojuvelin**).

heme See **haem**.

hemiasterlin A tripeptide derived from marine sponges (including *Hemiasterella minor*) that will cause depolymerization of microtubules; binds to the vinca-alkaloid binding-site on beta tubulin and prevents polymerization. See **phomopsin A**.

hemibiotrophic Describing a fungal lifestyle in which **biotrophic** and **necrotrophic** developmental stages are sequentially established. An example is the behaviour of *Colletotrichum destructivum*, the causal agent of tobacco anthracnose, a disease that causes spots and cankers to develop all over the plant.

hemicellulose Class of plant cell-wall polysaccharide that cannot be extracted from the wall by hot water or chelating agents, but can be extracted by aqueous alkali. Includes **xylan, glucuronoxylan, arabinoxylan, arabinogalactan II, glucomannan, xyloglucan**, and galactomannan. Part of the cell-wall matrix.

hemicentins Extracellular matrix proteins (hemicentin1, fibulin-6, 5635aa; hemicentin-2, 5065aa), members of the immunoglobulin superfamily, expressed in skin fibroblasts and retinal pigment epithelium. A mutation affecting hemicentin-1 has been implicated in familial age-related maculopathy (**AMD**).

hemicyst Blebs or blisters formed by a confluent monolayer of epithelial cells in culture as a result of active fluid accumulation between cell sheet and substratum.

hemidesmosomal epidermolysis bullosa A rare disease in which there is intraepidermal blistering at the most basal aspect of the lower cell layer. Can arise from a disorder of **plectin** and is associated with muscular dystrophy or from a defect of the $\alpha_6\beta_4$ integrin receptor and associated with pyloric atresia. See **epidermolysis bullosa**.

hemidesmosome Specialized junction between an epithelial cell and its basal lamina. Although morphologically similar to half a desmosome (into which intermediate cytokeratin filaments are also inserted) integrins are involved rather than **cadherins**.

hemimetabolous Of an insect, a species without any marked change in body-plan from larval to adult, apart from the development of wings. Examples: grasshoppers and crickets. (*cf.* holometabolous.)

Hemiptera An order of Insecta with two pairs of wings and mouthparts adapted for piercing and sucking; Many feed on plant juices. Include bugs, cicadas, aphids, plant lice, scale insects, leaf hoppers and cochineal insects.

hemizygote Nucleus, cell or organism that has only one of a normally diploid set of genes. In mammals the male is hemizygous for the X chromosome.

hemojuvelin A glycosylphosphatidylinositol-linked protein of the repulsive guidance molecule family (426aa) involved in iron metabolism, a co-receptor for the **bone morphogenetic proteins** two and four. Enhances BMP-induced **hepcidin** expression and binds **neogenin**, a receptor involved in a variety of cellular signalling processes and regulates shedding of hemojuvelin. Mutations affecting hemojuvelin are frequently the cause of juvenile **haemachromatosis**.

hemorrhage See haemorrhage.

HEN1 (1) In humans, helix-loop-helix protein 1, (133aa) a DNA-binding protein, possibly involved in cell differentiation in the developing nervous system. (2) In *C. elegans*, hesitation behaviour protein 1 (117aa). (3) In *Drosophila*, a methyl transferase (391aa), product of the *Pimet* gene that methylates RNA and is involved in posttranscriptional gene silencing. (4) In *Arabidopsis*, a methyl transferase (942aa) that stabilizes the miRNA-miRNA* duplex by adding methyl groups to the 3' ends.

Henderson-Hasselbach equation Equation of the form: $pH = pK_a - \log(\text{concentration }(A^-)/\text{concentration }(HA))$ used for the calculation of the pH of solutions where the ratio $[A^-]/[HA]$ is known and HA and A^- are respectively the protonated and deprotonated forms of an acid and pK_a is the acid dissociation constant.

Henoch-Schönlein purpura An inflammatory disorder, thought to be an immunoglobulin A (IgA) mediated autoimmune phenomenon, characterized by a generalized vasculitis involving the small vessels of the skin. It is the most common vasculitis in children. In adults the predominant feature may be glomerulonephritis.

Hensen's node Thickening of the avian blastoderm at the cephalic end of the primitive streak (primitive knot), the avian equivalent of Spemann's organizer in amphibia. Presumptive notochord cells become concentrated in this region. Is a source of morphogens in the developing embryo and ciliary activity in this region generates a L−R asymmetry.

Hep2 cells Line established from human laryngeal carcinoma from a 56 year old Caucasian male but probably contaminated with HeLa cells. Extensively used in viral studies being susceptible to arboviruses and measles virus.

Hepadnaviridae A family of viruses that cause liver disease. The genome consists of partially double-stranded, partially single stranded circular DNA that replicates via an RNA intermediate which is reverse-transcribed into DNA. The type species of the genus Orthoheptadnavirus is **hepatitis B** virus; related viruses have been isolated from subhuman primates and from various rodents. The Aviheptadnavirus genus infect birds, the type species being duck hepatitis virus. Virus taxonomy database: http://www.ncbi.nlm.nih.gov/ICTVdb/ICTVdB/00.030.htm

heparan sulphate A glycosaminoglycan (ca 50 kDa but variable) that is attached to core protein in membrane-associated **proteoglycans** such as **syndecans** and **glypicans** and matrix proteins such as **perlecan**, **agrin** and collagen XVIII. Composed largely of glucuronic acid (GlcA) linked to N-acetylglucosamine (GlcNAc) but with some N and O-linked sulphated sugars; closely related to **heparin**. Heparan sulphate is important for binding various growth factors (see **FGF**) and adhesion molecules such as **NCAM** have heparan sulphate binding sites.

heparanase Lysosomal enzyme (EC 3.2.1.-, 543aa), an endoglycosidase, that will cleave heparan sulphate into characteristic large molecular weight fragments. Degradation of extracellular matrix heparan sulphate by heparanase is a key step in the extravasation of tumour cells and migrating leukocytes, and also in processes such as angiogenesis, wound healing, and smooth muscle proliferation.

heparin Sulphated mucopolysaccharide (3−40 kDa), found in granules of mast cells, that inhibits the action of **thrombin** on **fibrinogen** by potentiating anti-thrombins, thereby interfering with the blood-clotting cascade. Platelet factor IV will neutralise heparin. Commonest disaccharide is 2-O-sulphated iduronic acid linked to N-sulphated glucosamine.

heparin-binding growth factor See **fibroblast growth factor**.

hepatic lipase An important enzyme in triglyceride metabolism (hepatic triglyceride lipase, EC 3.1.1.3, 499aa) that catalyzes the hydrolysis of phospholipids, mono-, di-, and triglycerides, and acyl-CoA thioesters. It is an important enzyme in HDL metabolism. It is bound to and acts at the endothelial surfaces of hepatic tissues. Mutations affecting hepatic lipase result in abnormally triglyceride-rich low and high density lipoproteins. See **lipoprotein lipase**.

hepatic stellate cells See Ito cells.

hepatitis Inflammation of the liver. Can be caused by **hepatitis viruses**. .

hepatitis A Small (27 nm diameter) single-stranded RNA virus with some resemblance to **enteroviruses** such as polio. Causes infectious hepatitis.

hepatitis B The virus responsible for serum hepatitis but also associated with hepatocellular carcinoma. Can persist in asymptomatic carriers. The virion (**Dane particle**) is 42 nm diameter, with an outer sheath enclosing inner 27 nm core particle containing the circular viral DNA. Aggregates of the envelope proteins are found in plasma and are referred to as hepatitis B surface antigen (HBsAg; previously called Australia antigen).

hepatitis C An enveloped RNA virus (HCV) responsible for a fairly high proportion of cases of hepatitis of the non-A, non-B, type, but not all. It is a member of the **Flaviviridae** with a particle size of approximately 50 nm diameter and a positive-sense RNA genome. Chronic HCV infection can cause hepatocellular carcinoma.

hepatitis D A small circular RNA (single-stranded, negative sense) virus (delta virus) that is replication defective and cannot propagate in the absence of another virus. In humans, hepatitis D virus infection only occurs in the presence of **hepatitis B** infection.

hepatitis E virus Virus responsible for enterically transmitted non-A, non-B hepatitis worldwide. A spherical, non-enveloped, single stranded RNA virus.

hepatitis non-A, non-B Hepatitis caused by a virus that is neither Hepatitis A or B and has no antigenic cross-reaction with either; often, but not always, **hepatitis C virus**.

hepatitis viruses See **hepatitis A, B, C, D, E, non-A non-B**. Hepatitis F may be responsible for a common infection in the Far East. Hepatitis G (HGV) is an RNA virus, a member of the Flavivirus family similar to hepatitis C. There is already a candidate nonA, nonE virus which may be Hepatitis H.

hepato- A prefix denoting something associated with the liver.

hepatocarcinoma Malignant tumour (hepatocellular carcinoma, hepatoma) derived from hepatocytes. Associated with **hepatitis B** or **hepatitis C** in 80−90% of cases.

hepatocyte Epithelial cell of liver. Often considered the paradigm for an unspecialised animal cell. Blood is directly exposed to hepatocytes through fenestrated endothelium, and hepatocytes have receptors for sub-terminal N-acetyl-galactosamine residues on asialo-glycoproteins of plasma.

hepatocyte growth factor Polypeptide mitogen (HGF, scatter factor, hepatopoietin A, lung fibroblast-derived mitogen) that stimulates proliferation in hepatocytes but also in a number of other cells. HGF is synthesized as a single chain precursor (728aa) that is proteolytically cleaved to give a heavy chain (463aa) and a light chain (234aa) linked by a single disulphide bond. It contains multiple copies of the **kringle** domain. Both the single chain precursor and the two-chain forms of HGF are biologically active and HGF is generally isolated as a mixture of the two. HGF also alters cell motility and is now known to be identical to **scatter factor**. The HGF receptor (1390aa) is the *met* oncogene product, a receptor tyrosine kinase.

hepatocyte growth factor-like protein See **macrophage stimulating protein**.

hepatocyte nuclear factor See **HNF**.

hepatolenticular degeneration See **Wilson's disease**.

hepatoma See **hepatocarcinoma**.

hepatoma transmembrane kinase A receptor tyrosine kinase (htk, EphB4, 987aa) of the **EPH** class that is expressed abundantly in placenta and in a range of primary tissues and malignant cell lines. The ligand is **Htk-L**.

hepatopancreas Digestive gland of crustaceans with functions approximately analogous to liver and pancreas of vertebrates − enzyme secretion, food absorption and storage.

Hepatophyta Liverworts.

hepcidin antimicrobial peptide A peptide hormone (HAMP, LEAP1, liver-expressed antimicrobial peptide, 25aa) produced in the liver from an 84aa precursor: regulates duodenal iron absorption and iron trafficking in the reticuloendothelial system, also has antimicrobial properties. Hepcidin inhibits the cellular efflux of iron by binding to and inducing the degradation of **ferroportin**, the sole iron exporter in iron-transporting cells. The active peptide has a unique 17-residue stretch with 8 cysteines forming 4 disulphide bridges. HAMP is most active against Gram-positive bacteria, but also inhibits the growth of certain yeast and Gram-negative species.

HEPES Very commonly used buffer (4-(2-hydroxyethyl)-1-piperazine-ethane-sulphonic acid) for tissue culture medium. Its pK_a of 7.5 makes it ideal for most cell culture work. Since related compounds are molluscicides it may be unsuitable for some invertebrate cultures. One of the series of zwitterionic buffers described by Good.

HepG2 cells Cell line derived from hepatic carcinoma. Epithelial in morphology; produce a variety of proteins such as prothrombin, alpha-fetoprotein, C3 activator and fibrinogen.

hephaestin A transmembrane ferroxidase (1158aa) that acts in conjunction with **ferroportin** and that has been implicated in duodenal iron and copper export. Hephaestin has approximately 50% sequence identity with the plasma multicopper ferroxidase, **ceruloplasmin**. Hephaestin-like protein 1 (1159aa) has similar properties.

heptad repeats Tandem repeat sequence in which a group of seven amino acids occurs many times in a protein sequence. Most coiled-coil sequences contain heptad repeats.

HER Family of receptors (HER-2, erbB2; HER-3, erbB3; HER-4, erbB4) of the EGF-receptor family of receptor **tyrosine kinases**. Ligands are **neuregulins**. HER-3 binds heregulin. Overexpression of HER-2, human homologue of erbB2, correlates with poor prognosis in breast carcinoma.

heraclenin A furanocoumarin extracted from the plant *Opopanax chironium* (sweet myrrh) that is an inhibitor of T-cell-receptor-mediated proliferation in human primary T-cells; induces DNA fragmentation at the G2/M phases of the cell cycle. Despite a close structural similarity to **imperatorin**, induces apoptosis in a mechanistically different way. Abstract: http://www.ncbi.nlm.nih.gov/pubmed/15104479

herbimycin A Tyrosine kinase inhibitor from *Streptomyces hygroscopicus*.

herculin Product of the muscle regulatory gene *Myf-6*. Also known as MRF4 (muscle regulatory factor-4).

hereditary angioedema A disorder in which there are episodes of subcutateous edema in the respiratory and GI tracts. In some cases a result of mutation in the complement C1 inhibitor gene leading to overproduction of **C2-kinin**, in other cases by mutation in mutation in the gene encoding coagulation factor XII (**Hageman Factor**).

hereditary haemorrhagic telangiectasia An autosomal dominant vascular dysplasia (Osler-Rendu-Weber disease) characterised by thinning of blood vessel walls and thus a predisposition to bleeding. Malformations in the arteriovenous system occur in skin, mucosa, and viscera. Type 1 is caused by mutation in the gene for **endoglin**, Type 2 by mutation in **activin receptor-like kinase-1**.

hereditary lymphedema A complex of disorders in which there is edema caused by blockage in lymphatic drainage (lymphedema). Type 1A (Milroy's disease) is caused by mutation in the FLT4 gene which encodes vascular endothelial growth factor receptor-3. Type II (Meige lymphedema) is caused by mutation in the forkhead family transcription factor gene MFH1 (FOXC2).

hereditary nonpolyposis colon cancer See GTBP.

hereditary sensory and autonomic neuropathies A heterogeneous group of disorders in which there is sensory dysfunction. HSAN1 is caused by mutation in the gene that encodes serine palmitoyltransferase-1 (EC 2.3.1.50, 473aa), the key enzyme in sphingolipid biosynthesis, HSAN2 by mutation in **WNK1**, HSAN3 (see **dysautonomia**) by mutation in the **IkB** kinase-associated protein. HSAN4 is caused by mutation in a **neurotrophin** receptor, HSAN5 by mutation in the **nerve growth factor** beta subunit.

hereditary spastic paraplegias See **spastic paraplegias**.

heregulins Neuregulin-1 (heregulin-1, HRG1, HRGalpha, beta, neu differentiation factor; 222aa) is the ligand for **HER3**. See **ARIA**.

hERG Gene (human ether-a-go-go related gene) encoding the pore-forming subunit of cardiac IKr, the channel that is one of the two responsible for repolarization of the cardiac action potential (the outward rectifying delayed current). Mutations are associated with **long QT2 syndrome** and **short QT syndrome**.

Hermansky-Pudlak syndrome A genetically heterogeneous disorder in which there is abnormal vesicle trafficking to lysosomes and related organelles, such as melanosomes and platelet dense granules. In mice, at least 16 loci are associated with HPS, including sandy (*sdy*). The *sdy* mutant mouse expresses no **dysbindin** protein owing to a deletion in the gene *Dtnbp1* and mutation of the human ortholog *DTNBP1* causes a novel form of HPS called HPS-7. See **BLOCs**.

Herpesvirales An Order of viruses (see **virus taxonomy**). The genome is not segmented and contains a single molecule of linear double-stranded DNA, 120−220 thousand nucleotides long.

Herpesviridae A family of large DNA viruses in the Order Herpesvirales: Herpes simplex (human herpes virus-1, HHV-1, HHV-2; HSV) causes coldsores and genital herpes; Varicella-zoster (human herpes virus 3 (HHV-3)) causes chicken-pox and shingles; **Epstein-Barr virus** (EBV, HHV-4) causes glandular fever, cytomegalovirus (HHV-5) causes congenital abnormalities and is an opportunistic pathogen; Roseolovirus (HHV-6) causes **Sixth disease** and HHV-8 is Kaposi's sarcoma-associated herpesvirus. Herpes simplex type 2 and EBV are associated with human tumours (cervical carcinoma for the former and **Burkitt's lymphoma** and nasopharyngeal carcinoma in the case of EBV). Varicella establishes a lifelong latent infection of sensory neurons in human dorsal root ganglia and has a tendency to resurgence (shingles) if the immune system is suppressed.

Herring bodies Granules within neurosecretory axons in the posterior lobe of the pituitary gland containing either **vasopressin** or **oxytocin**.

Hers' disease Glycogen storage disease VI.

herstatin An alternatively spliced **HER2** product (419aa) with growth-inhibitory properties in experimental systems. Article: http://www.neoplasia.com/pdf/manuscript/v10i07/neo08314.pdf

HES proteins See *hairy*.

HETE A family of hydroxyeicosenoic acid (C20) derivatives (hydroxytetraeicosenoic acid) of arachidonic acid produced by the action of lipoxygenase. Potent pharmacological agents with diverse actions. See also **HPETE**.

heteroagglutination (1) The adhesion of spermatozoa to one another by the action of a substance produced by the ova of another species. (2) The adhesion of erythrocytes to one another when blood of different groups is mixed. *Cf.* iso-agglutination. See **agglutinin**.

heterochromatin The chromosomal regions that are condensed during interphase and at the time of nuclear division. They show what is considered an abnormal pattern of staining as opposed to **euchromatin**. Can be subdivided into constitutive regions (present in all cells) and facultative heterochromatin (present in some cells only). The inactive X chromosome of female mammals is an example of facultative heterochromatin.

heterochrony Ancestral shifts in developmental timing that gave rise to morphological changes. Thought to provide an explanation for the evolution of new forms.

heteroclitic antibody An antibody that was produced in response to an antigen but turns out to have higher affinity for an antigen that was not present in the original immunisation, presumably because of molecular mimicry.

heterocyst Specialized cell type found at regular intervals along the filaments of certain **Cyanobacteria**; site of nitrogen fixation formed under conditions of nitrogen limitation.

heterodimer A dimer in which the two subunits are different. One of the best known examples is tubulin that is found as an α-tubulin/β-tubulin dimer. Heterodimers are relatively common, and it may be that the arrangement has the advantage that, for example, several different binding subunits may interact with a conserved signalling subunit.

heteroduplex Double-stranded nucleic acid in which the two strands are different, either of different heritable origin, formed *in vitro* by annealing similar DNA strands with some complementary sequences, or formed of mRNA and the corresponding DNA strand.

heterogamy (1) Situation in which gametes of different sizes are produced by different mating types or sexes. (2) The condition in a flowering plant species of having two or more types of flowers.

heterogenous nuclear RNA Originally identified as a class of RNA (hnRNA), found in the nucleus but not the nucleolus, which is rapidly labelled and with a very wide range of sizes, 2–40 kilobases. It represents the primary transcripts of **RNA polymerase** II and includes precursors of all **messenger RNAs** from which introns are removed by splicing.

heterokaryon Cell that contains two or more genetically different nuclei. Found naturally in many fungi and produced experimentally by cell fusion techniques, e.g. **hybridoma**. See **heterokaryosis**.

heterokaryosis The coexistence of many dissimilar nuclei in, for example, the multinucleate cells of arbuscular mycorrhizal fungi (Glomeromycota). The opposite state is homokaryosis.

heterokonts The phylum Heterokontophyta (stramenopiles) comprises a range of classes, many being diatoms (Bacillariophyceae) but including large algae (Chrysophyceae, Phaeophyceae) and oomycetes. The group are characteriesd by a motile stage in the life cycle, the motile forms having two different flagella, the anterior flagellum with lateral bristles or mastigonemes, the other flagellum is a smooth whiplash-type. Heterokont chloroplasts have two pairs of membranes and are thought to be derived from symbiotic eukaryotic red algae; they have chlorophyll a and chlorophyll c, and the accessory pigment fucoxanthin.

Heteroloboseans A phenotypically diverse group of heterotrophic amoeboflagellates that have paddle-shaped mitochondrial cristae, eruptive pseudopodia and a flagellar apparatus consisting of parallel basal bodies (e.g. *Acrasis, Percolomonas, Tetramitus* and *Psalteriomonas*).

heterologous Derived from the tissues or DNA of a different species. *Cf.* **autologous, homologous**.

heterophile antibody An antibody raised against an antigen from one species that also reacts against antigens from other species. Also used of systems such as the **Forssman antigen** where antibody against antigens from a variety of species is present without immunization.

heteroplasia *heteroplasmy* The replacement of normal tissue by invasive (malignant) tissue or by inappropriate normal tissue as a result of abnormal cellular differentiation.

heterosis Hybrid vigour, the superiority of a heterozygotic organism over the homozygote.

heterospecific Ab An antibody that binds, sometimes more strongly, to an antigen other than that which was used as an immunogen. Can also be used of artificially produced antibody in which the two antigen binding sites are for different antigens.

heterospory Condition in vascular plants where the spores are of different sizes, the smaller producing male prothalli, the larger female prothalli.

heterothallic Situation in some fungi and algae in which there are two mating types and the individual thallus is self-sterile even if hermaphrodite.

heterotopia A condition in which an organ or part of an organ is found in an abnormal position. In X-linked periventricular heterotopia neuronal cell bodies are mislocated as a result of mutation in the gene encoding **filamin-A**. Autosomal recessive periventricular heterotopia with microcephaly (ARPHM) is caused by mutation in the gene encoding ADP ribosylation factor (ARF) guanine nucleotide exchange protein-2 (ARFGEF2 gene). Both form of heterotopia can be associated with seizures and epilepsy. **Lissencephaly** type I is also a heterotopia.

heterotrophy An organism that requires carbon compounds from other plant or animal sources and cannot synthesize them itself – not an **autotroph**.

heterotypic Of different types. Thus heterotypic adhesion would be between dissimilar cells, in contrast to homotypic adhesion between cells of the same type.

heterozygosity index Measure of the number of gene loci for which an individual is heterozygous. A value of zero indicates complete identity at all loci (as in some highly inbred strains of mice). Low heterozygosities for allozyme loci can indicate that there have been episodes in which the population was very small (population bottleneck) whereas a high heterozygosity indicates a lot of genetic variability in the population.

heterozygote Nucleus, cell or organism with different **alleles** of one or more specific genes. A heterozygous organism will produce unlike gametes and thus will not breed true.

HEV (1) **High endothelial venule.** (2) See **hevein**.

hevein Protein (Hev b 6.02, 43aa) found in the rubber tree, *Hevea brasiliensis*, has a **CRD domain**, which is known to bind chitin and GlcNAc-containing oligosaccharides. The antifungal activity of hevein-like proteins has been associated with their chitin-binding activities. Hevein is a major IgE-bindingallergen in natural rubber latex (as used in gloves). A hevein-like protein is found in *Arabidopsis*.

hexitol Sugar alcohol with six carbon atoms. Natural examples are sorbitol, mannitol.

hexokinase Enzyme (HK, EC 2.7.1.1) that catalyses the transfer of phosphate from ATP to glucose to form glucose-6-phosphate, the first reaction in the metabolism of glucose via the glycolytic pathway (**glycolysis**). It is normally cytoplasmic or associated through a conserved domain with **porin** in the outer mitochondrial membrane and is allosterically inhibited by G-6-P. There are four major glucose-phosphorylating isoenzymes in vertebrates with different tissue distributions. HK1 (917aa) is a red-cell isoform, HK2 (917aa) is the major hexokinase expressed in skeletal muscle, HK3 (923aa) is in white blood cells and HK4 (glucokinase, 465aa) is expressed only in mammalian liver and pancreatic islet beta cells. HK1 is defective in a rare form of nonspherocytic haemolytic anaemia. HK4 is defective in maturity-onset diabetes of the young type two and hyperinsulinaemic hypoglycaemia type three.

hexon Subunit of a hexameric structure or with hexameric symmetry, in particular the arrangement of most of the capsomers of **Adenoviridae** – one capsomer surrounded by six others to form the hexon.

hexosaminidase Enzyme (EC 3.2.1.52, N-acetyl β-hexosaminidase) involved in the metabolism of **gangliosides**. The alpha subunit (529aa) is deficient in **Tay-Sachs disease** and the beta subunit (556aa) in Sandhoff's disease.

hexose Monosaccharide containing six carbon atoms, e.g. **glucose, galactose, mannose**.

hexose monophosphate shunt See **pentose phosphate pathway**.

Hey proteins Products of Hes-related repressor genes. See *hairy*.

Heymann nephritis Rat model of human membranous nephropathy; an autoimmune disease in which the antigen is megalin (**lipoprotein receptor-related protein-2**).

HFE See **haemochromatosis**.

Hfr High frequency, in the sense of bacterial **conjugation**.

HGF See **hepatocyte growth factor**.

HGPRT Enzyme (hypoxanthine-guanine phosphoribosyl transferase, EC 2.4.2.8, 218aa) that catalyses the first step in the pathway for salvage of the purines hypoxanthine and guanine. The phosphoribosyl moiety is transferred from an activated precursor, 5-phosphoribosyl 1-pyrophosphate. Since animal cells can synthesize purines *de novo*, HGPRT-mutants can be selected by their resistance to toxic purine analogues. A genetic lesion in HGPRT in humans underlies the **Lesch-Nyhan syndrome**. See **HAT medium**.

HHH syndrome A very rare autosomal recessive metabolic disorder (hyperornithinaemia-hyperammonemia-homocitrullinuria syndrome) caused by a deficiency of the mitochondrial ornithine transporter, one of the urea cycle components. Caused by mutation in the SLC25A15 gene on chromosome 13 (13q14).

Hib Bacterium of the genus *Haemophilus* that can cause meningitis in young children.

hid *Drosophila* gene (head involution defective) that is involved in positive regulation of apoptosis (like **reaper** and **grim**). See **bantam**.

hidden Markov models Graphical models, originally developed for the study of machine learning and speech recognition, that have been applied to analysis of gene sequences and phylogenetics. They describe a probablility distribution over an infinite number of sequences.

HIF-1 See **hypoxia-inducible factor**. A transcription factor (hypoxia-inducible factor-1, HIF-1A, 826aa) involved in responses to environmental stress (particularly, but not exclusively, hypoxia) by regulating the expression of genes that are involved in glucose supply, growth, metabolism, redox reactions and blood supply. Also essential for myeloid cell activation in response to inflammatory stimuli.

high density lipoprotein See **lipoproteins**.

high endothelial venule Venules (HEV) in which the endothelial cells are cuboidal rather than squamous. Found particularly in lymph nodes where there is considerable extravasation of lymphocytes as part of normal traffic. Morphologically similar endothelium is found associated with some chronic inflammatory lesions. Express particular adhesion molecules accounting for preferential lymphocyte adhesion.

high mannose oligosaccharide A subset of the oligosaccharide chains (e.g. $(Glc)_3(Man)_9$ $(GlcNAc)_2$) that are added post-translationally to certain asparagine residues of secreted or membrane proteins in eukaryotic cells (N-glycans); contain 5–9 mannose residues, but lack the sialic acid-terminated antennae of the so-called complex type. They are subsequently trimmed by glucosidases that remove the glucose residues and mannosidases that remove all but three mannose residues and then terminal sugars such as N-acetylglucosamine, galactose, sialic acid and fucose may be added to generate complex glycans. **Deoxynojirimycin** inhibits the trimming reactions.

high-energy bond Chemical bonds that release more than 25 kJ/mol on hydrolysis: their importance is that the energy can be used to transfer the hydrolyzed residue to another compound. The risk in using the term is that students may think the bond itself is different in some way, whereas it is the compound that matters. Hydrolysis of creatine phosphate yields 42.7 kJ/mol; of phosphoenolpyruvate, 53.2; ATP to ADP, 30.5: the latter is important because it shows that energetically the hydrolysis of creatine phosphate will suffice to reconstitute ATP, hence the use of creatine phosphate in muscle.

high-mobility group proteins Small, non-histone, nuclear proteins involved in the regulation of inducible gene transcription, integration of retroviruses into chromosomes, and the induction of neoplastic transformation and metastatic progression. They bind preferentially to the minor groove of many AT-rich promoter and enhancer elements through have conserved motifs (AT-hooks). Lack of HMGA1 (215aa) is reported to cause insulin resistance and diabetes; HMGIY (107aa) has been suggested as a potential oncogene. HMG2 (208aa) is a component of the **SET complex**.

high-voltage electron microscopy *HVEM* See **electron microscopy**.

Hill coefficient A measure of cooperativity in a binding process. A Hill coefficient of one indicates independent binding, a value of greater than one shows positive cooperativity – binding of one ligand facilitates binding of subsequent ligands at other sites on the multimeric receptor complex. Worked out originally for the binding of oxygen to haemoglobin (Hill coefficient of 2.8). See **Hill plot**.

Hill plot Graphical method for estimating the **Hill coefficient** for a reaction or the binding of ligands: the initial reaction velocity or fractional binding saturation is plotted against substrate concentration.

Hill reaction Reaction, first demonstrated by Robert Hill in 1939, in which illuminated chloroplasts evolve oxygen when incubated in the presence of an artificial electron acceptor (e.g. ferricyanide). The reaction is a property of **photosystem II**.

Hind Type II **restriction endonucleases** isolated from *Haemophilus influenzae* Rd. HindII, the first such enzyme to be isolated, cleaves 'GTPyPuAc' between the unspecified pyrimidine (Py) and purine (Pu) residues to generate **blunt ends**. HindIII cleaves the sequence 'AAGCTT' between the two As, generating **sticky ends**.

hinge region Flexible region of a polypeptide chain – for example, as in immunoglobulins between Fab and Fc regions, and in myosin the S2 portion of heavy meromyosin.

HIP-10 The protein product of the PRPF40A gene (huntingtin-interacting protein 10, pre-mRNA-processing factor 40 homologue A, formin-binding protein 3 or 11, huntingtin-interacting protein A, Fas ligand-associated factor 1, 957aa) that binds to **WASL** and suppresses its translocation from the nucleus to the cytoplasm.

hippo See **Salvador-Warts-Hippo pathway**.

hippocalcin A neuronal calcium binding protein (193aa) related to **recoverin**. Found in the brain, originally in pyramidal cells of the hippocampus. May be involved in the calcium-dependent regulation of rhodopsin phosphorylation.

hippocampus Area of mammalian forebrain that is important for short term memory and spatial navigation. It is also the site of long-term synaptic plasticity (see **long-term potentiation**) which is exhibited by defined synaptic pathways in the hippocampus.

Hirano bodies Paracrystalline intracellular inclusions, composed of microfilament-associated proteins including actin, alpha-actinin, and vinculin, found in the brain of patients with neurodegenerative disorders.

Hirschsprung's disease Aganglionic megacolon, a congenital malformation caused by absence of ganglion cells in myenteric and submucosal neural plexuses of gut. In some cases defect is due to mutation in RET receptor tyrosine kinase, in others to mutations in **endothelin**-3, the endothelin-beta receptor or in **GDGF**. When associated with **Waardenburg's syndrome**, the defect is due to mutation in Sox10. See **Mowat-Wilson syndrome**.

hirudin An anticoagulant (65aa) present in the saliva of the leech, *Hirudo medicinalis*, that prevents blood clotting by inhibiting the action of thrombin on fibrinogen.

Hirudo medicinalis The medicinal leech. The **central nervous system** of this annelid contains a relatively small number of large, identifiable cells. This has made the leech, like the molluscs *Aplysia* and *Helisoma*, a chosen preparation for studying nervous system mechanisms at the cellular level. Related species of leeches are the organisms of choice for cellular and molecular genetic studies of early development, since the early embryos also contain identifiable cells.

his tag A short sequence of histidine residues (usually six) engineered onto the N- or C-terminal of an expressed protein that can be used as an epitope tag for detection with antibody or for purification on a nickel affinity column. The tag is often arranged with a convenient peptidase-sensitve linker so that it can be removed from the purified protein.

hisactophilin A histidine-rich actin binding protein (118aa) from *Dictyostelium*. Promotes F-actin polymerization and binds to microfilament bundles. It is very pH sensitive as a result of having 31 histidine residues and may act as an intracellular pH sensor that links chemotactic signals to the microfilament system. Two similar proteins (hisactophilin-2, 118aa and hisactophilin-3, 119aa) are known. Structure (though not sequence) very similar to **FGF** and **IL1**.

hispid flagella Eukaryotic flagella with two rows of stiff protrusions (mastigonemes) at right-angles to the long axis of the shaft. In hispid flagella, the normal relationship between the direction of flagellar wave-propagation and the direction of movement is reversed; a proximal to distal wave pulls the organism forward.

histamine A biogenic amine formed by decarboxylation of histidine that mediates a range of responses through **histamine receptors**. Histamine is stored in **mast cells** and basophils and its release is responsible for some of the characteristic features of inflammation and, in extreme cases, anaphylaxis (see Type I **hypersensitivity**). Also present in some venoms.

histamine receptors G protein-coupled receptors of two types. In peripheral tissues, particularly skin and airways, the H1 subclass (487aa) mediate smooth muscle contraction and increased capillary permeability; they also induce catecholamine release from adrenal medulla, as well as mediating neurotransmission in the central nervous system. Antihistamine drugs for allergy act as antagonists at the H1 receptors. The H2 subclass (359aa) mediate gastric acid secretion and are the target for anti-ulcer drugs (H2 blockers).

histatins Salivary proteins (51−57aa) secreted by parotid and sub-mandibular glands, that are precursors of the protective proteinaceous layer on tooth surfaces (enamel pellicle). Multiple proteolytic products are found in saliva. Histatins have antibacterial and antifungal activities, particularly against *Candida albicans*, and inhibit matrix metallopeptidases MMP2 and MMP9. They counteract dietary tannins by causing them to precipitate and are produced in larger amounts in animals with a tannin-rich diet.

histidine An amino acid (His, H, 155 Da) with an imidazole side chain with a pK_a of 6−7. Acts as a proton donor or acceptor and has high potential reactivity and diversity of chemical function. Forms part of the catalytic site of many enzymes. See Table A1.

histidine ammonia-lyase An enzyme (histidase, EC 4.3.1.3, 657aa) that converts histidine into ammonia and urocanic acid. Mutations in the gene cause histidinaemia.

histidine rich protein II A pH sensitive actin binding protein from Plasmodium falciparum; also binds phosphatidylinositol 4,5-bisphosphate. Similar to **hisactophilin**.

histidine triad A motif (HIT) found in a superfamily of nucleotide hydrolases and transferases that act on the alpha-phosphate of ribonucleotides. The motif is HϖHϖH$\varpi\varpi$ (where H is histidine and ϖ (phi) is a hydrophobic amino acid). Histidine triad nucleotide-binding protein 1 (HINT1, 127aa) hydrolyzes adenosine 5′-monophosphoramidate substrates. It was originally thought to be a PKC inhibitor but this does not seem to be the case. HINT2 (163aa) is probably involved in steroid biosynthesis. HINT5 (337aa) is an mRNA decapping enzyme. See **aprataxin** and **fragile histidine triad**.

histidine-aspartate phosphorelay A two-component signal transduction mechanism implicated in a wide variety of cellular responses to environmental stimuli. The relay system involves a sensor histidine kinase (HK), a phosphotransfer intermediate (HPt), and a response regulator (RR). They were originally described in bacterial systems but subsequently found in plants (see **ARRs**) where they sense signals from external and/or internal stimuli such as ethylene, cytokinin, and osmolarity. In *S. pombe* a His-Asp phosphorelay system is involved in oxidative stress responses.

histidine-rich calcium-binding protein Major protein (699aa) found in the lumen of **sarcoplasmic reticulum**, where it may play a role in sequestering calcium ions. Highly acidic, with multiple repeats of highly conserved domains, believed to be responsible for calcium binding. Binds low-density lipoprotein with high affinity.

histidine-rich glycoprotein A protein (HRG, 525aa) present in platelets and abundant in plasma. It binds to heparin and to plasminogen, and inhibits fibrinolysis. May mediate the contact activation phase of the intrinsic blood coagulation cascade. Defects in HRG cause a form of thrombophilia (a tendency to thrombosis).

histiocytes Long-lived resident **macrophages** found within tissues.

histiocytosis X See **Langerhans' cell histiocytosis**.

histoblasts Population of small diploid epithelial cells in Dipteran larvae that do not form typical **imaginal discs**, yet resemble them in some ways.

histochemistry Study of the chemical composition of tissues by means of specific staining reactions.

histocompatibility See **histocompatibility antigens**.

histocompatibility antigens A set of plasmalemmal glycoproteins that are crucial for T-cell recognition of antigens. Particularly the HLA system in humans and the H2 system in mice. The major histocompatibility (MHC) antigens are responsible for rapid (e.g. seven days in the mouse) graft rejection and other immune phenomena. The minor histocompatibility antigens are involved in much slower rejection phenomena. Only histocompatible animals (monozygotic twins and highly inbred strains of mice) will accept grafts from other individuals without immunosuppressive intervention. There are two classes of major histocompatibility antigens encoded by the **major histocompatibility complex**: (1) Class I; histocompatibility antigens composed of two glycosylated subunits, a heavy chain of 44 kDa and β2-microglobulin (12 kDa). The heavy chain may be coded by K, D or L genes of mouse H2 and A, B or C genes of human HLA complex. Class I antigens are important in T-cell killing and are recognized in conjunction with the foreign cell surface antigens (**MHC restriction**). (2) Class II antigens such as human HLA-DR and murine H2 Ia antigens are heterodimeric with α (254aa) and β (266aa) chains. Found mostly on B-lymphocytes, macrophages and accessory cells. The response of T-helper cells requires that the foreign antigen is presented in conjunction with the appropriate Class II antigens.

histogen An outmoded concept that there were discrete meristems (dermatogen, periblem, plerome) at the shoot or root tip that gave rise exclusively to particular tissues in that organ (epidermis, cortex, stele + pith).

histogenesis The process of formation of a tissue, involving differentiation, morphogenesis, and other processes such as angiogenesis, growth control, cellular infiltration and so on.

histoma *histioma* A benign tumour in which the tissue morphology is relatively normal.

histone acetylation Modification of histone by the addition by **histone acetyl transferases** (HATs, EC 2.3.1.-) of acetyl groups to N-terminal lysine residues. Effect is to reduce the affinity between histones and DNA making it easier for RNA polymerase and transcription factors to access the promoter region. In most cases, histone acetylation enhances transcription, an effect reversed by **histone deacetylases** (HDACs). Acetylation, particularly of histone H4, has also been proposed to play an important role in replication-dependent nucleosome assembly.

histone acetyltransferase Enzymes that catalyse the addition of an acetyl group to the N-terminal lysine of histone. See **histone acetylation**. A number of enzymes have been identified, for example, histone acetyltransferase KAT2B (EC 2.3.1.48, 832aa) that acetylates the core histones (H3 and H4) and nucleosome core particles. It inhibits cell-cycle progression, counteracts the mitogenic activity of the adenoviral oncoprotein E1A and is part of a multiprotein chromatin remodelling complex. The HAT1 gene encodes the catalytic subunit of histone acetyltransferase type B (EC 2.3.1.48, 419aa) that acetylates histones H2A and H4 and may be involved in telomeric silencing.

histone deacetylases A family of 11 enzymes in humans (HDACs; EC 3.5.1.-) that are involved in the control of gene expression by removing acetyl groups from histones (**see histone acetylation**). The HDAC transcriptional repressor complex consists of HDAC1 (482aa), HDAC2 (488aa) and **retinoblastoma binding proteins** 4 and 7 (RBBP4 and RBBP7); other multi-component repressor complexes are known.

histone methyltransferase Enzymes that catalyze the transfer of one to three methyl groups from S-adenosyl methionine to lysine and arginine residues of histones. Methylated histones bind DNA more tightly and transcription is inhibited. Euchromatic histone methyltransferase-1 (histone-lysine N-methyltransferase, EC 2.1.1.43, 1298aa) is a component of the E2F6 transcriptional repressor complex, has specificity for core histone H3 and is defective in some cases of chromosome 9q subtelomeric deletion syndrome. See **Jumonji**.

TABLE H1. Classes of histones

Class	Average M_r (kDa)	% Arginine	% Lysine
H1	23	1.5	29
H2A	14	8	11
H2B	14	5	16
H3	15	13.5	10
H4	11	14	11

histones Proteins found in the nuclei of all eukaryotic cells where they are complexed to DNA in **chromatin** and **chromosomes**. They are of relatively low molecular weight and are basic, having a very high arginine/lysine content. They are highly conserved and can be grouped into five major classes. Two copies of H2A, H2B, H3 and H4 bind to about 200 base pairs of DNA to form the repeating structure of chromatin, the **nucleosome**, with H1 binding to the linker sequence. They may act as nonspecific repressors of gene transcription. See **histone acetylation**, **histone methyltransferase** and Table H1.

histoplasmosis A disease (Darling's disease) of animals and man due to infection by the soil fungus *Histoplasma capsulatum*. Affects the lungs in man where they are phagocytosed by alveolar macrophages but not killed. A culture filtrate (histoplasmin) is intradermally injected in a skin test for the disease.

histotope A site on an MHC Class I or Class II antigen (see **histocompatibility antigen**) recognized by a T-cell.

HIV Human immunodeficiency virus, a lentivirus, previously known as HTLV-III, human lymphotrophic virus type III, and also referred to as LAV, lymphadenopathy-associated virus; the retrovirus that causes acquired immunodeficiency syndrome (**AIDS**) in humans, by killing $CD4^+$-lymphocytes (T-helper cells). There are two species of HIV: HIV-1 is the virus that was initially discovered, is more virulent, more infective and is the cause of the majority of HIV infections globally. HIV-2 has lower infectivity and is largely confined to West Africa. Macrophage (M-tropic) strains of HIV-1 use CD4 and the β-chemokine receptor CCR5 for entry into cells and are able to replicate in macrophages and $CD4^+$ T-cells. T-tropic isolates replicate in primary $CD4^+$ T-cells as well as in macrophages and use the α-chemokine receptor, CXCR4, as a co-receptor for entry.

HKT-transporters A sub-family of the trkH potassium transporter family. HKT1 in *Arabidopsis* is a sodium transporter (506aa) important for salt tolerance and Na^+ exclusion from leaves.

HL60 cells Human promyelocytic cell line established from a patient with acute myeloid leukaemia. HL60 cells have an amplified c-*myc* proto-oncogene and can be induced to differentiate into neutrophil or eosinophil-like cells by various treatments. Have some, but not all, of the features of normal blood cells.

HLA The **histocompatibility antigens**, human leucocyte antigens.

Hmc Periplasmic high molecular mass cytochrome c, the product of the first gene in a large operon (hmc operon) of *Desulfovibrio vulgaris*. HmcA has sixteen haem moieties attached to a single polypeptide chain, is associated with a membrane-bound redox complex, and is involved in electron transfer from the periplasmic oxidation of hydrogen to the cytoplasmic reduction of sulphate. Article: http://www.jbc.org/content/277/49/47907.full

HMEC (1) Human microvascular endothelial cells. (2) *(Rarely)* Human mammary epithelial cells.

HMG (1) See **high-mobility group proteins**. (2) The HMG box is a domain (79aa) found in high-mobility group proteins, and shared by **sox** and **sry** proteins, that is involved in binding linear DNA in a sequence-specific manner, causing it to bend through a large angle. Binding of HMG box proteins influences the binding of other transcription factors. (3) See **HMG-CoA reductase**.

HMG-CoA reductase Integral membrane protein (3-hydroxy-3-methylglutaryl-CoA reductase, EC 1.1.1.34, 888aa) found in endoplasmic reticulum and peroxisomes. It is the rate-limiting enzyme in cholesterol biosynthesis, catalysing the reaction between hydroxy-methyl-glutaryl-CoA and two molecules of NADPH to produce **mevalonate**. When sterols and nonsterol end products of mevalonate metabolism accumulate in cells the enzyme is ubiquitinated and rapidly degraded. It is the target for cholesterol-lowering drugs (see **fibrates**).

HMM Heavy meromyosin, the soluble fragment of myosin produced by trypsin proteolysis, that retains the ATPase activity and that will bind to F-actin to produce a characteristic **arrowhead** pattern (unless ATP is present, in which case it detaches). Papain cleavage of HMM yields S1 and S2 subfragments, the former having the ATPase activity.

hMSH Human **mismatch repair** genes. See **msh** and **Muir-Torre syndrome**.

HNF Family of transcription factors enriched in liver. Heterozygous mutations in hepatocyte nuclear factor (HNF)-1α (641aa) and HNF-1β (557aa) result in maturity-onset diabetes of the young. Mutations in HNF4A (474aa) have a similar effect. HNF3A (forkhead box protein A1, 472aa) is involved in embryonic development, establishment of tissue-specific gene expression and regulation of gene expression in differentiated tissues.

hnRNA See **heterogenous nuclear RNA**.

HNT (1) Human neurotrophin. See **IgLON** and **neurotrophin**. (2) Line of human neuroteratoma cells (hNT).

Hodgkin's disease A human lymphoma (Hodgkin's lymphoma) that appears to originate in a particular lymph node and later spreads to the spleen, liver and bone marrow. Giant cells, the Sternberg-Reed cells, with mirror-image nuclei are diagnostic. There is immunological depletion, caused perhaps by the excessive growth of neoplastic histiocytes, and this is often the indirect cause of death. Four types of the disease are recognized depending on the relative predominance of various neoplastic derivatives of the lymphoid series.

Hoechst 33258 A fluorescent dye (emission at ~ 460 nm) that binds DNA and can be used to visualize chromosomes and to monitor animal cell cultures for contamination by microorganisms such as mycoplasma.

Hogness box See **TATA box**.

holandry Inheritance of characters borne on the male chromosome and therefore only expressed in the male.

Holliday junction A four-stranded DNA junction, the cross-over region at meiosis that results in homologous genetic recombination, first proposed by Holliday in 1964 as a structural intermediate in a mechanistic model. The actual conformation of a DNA crossover was speculated to be a four-way-junction with separate DNA helices, or with stacked helices in either a parallel or an antiparallel orientation of the helices. Crystal structures have confirmed the antiparallel stacked-X conformation and that it can switch to the open-X conformation in order to slide. Products of the *msh* genes (MSH4 and MSH5) associate with Holliday junctions but do not engage in repair.

holoblastic Describing eggs that exhibit total cleavage, for example those of sea urchins, and mammals.

holocentric Description of a chromosome in which the centromere is diffuse rather than discrete.

holocrine Form of secretion in which the whole cell is shed from the gland, usually after becoming packed with the main secretory substance. In mammals, sebaceous glands are one of the few examples.

holoenzyme The complete enzyme complex composed of the protein portion (**apoenzyme**) and cofactor or coenzyme.

hololena toxin Toxins from the funnel-web spider *Hololena curta* that irreversibly block insect presynaptic sodium channels. They are μ-agatoxins (curtatoxins) heterodimers of Mu-AGTX-Hc1a (36aa), Mu-AGTX-Hc1b (38aa) or Mu-AGTX-Hc1c (38aa) subunits.

holometabolous Of an insect, a species with a marked change in body-plan from larva to pupa to adult. Examples: flies and wasps. (*cf.* hemimetabolous.)

holoprosencephaly A developmental disorder in which the prosencephalon (embryonic forebrain) fails to develop normally; severe forms may cause cyclopia. Various mutations can cause the disorder; affected genes include the *SIX3* homeobox gene, **sonic hedgehog**, *TGIF* (encodes TGFβ-induced factor), *ZIC2* (encodes the zinc-finger protein of the cerebellum-2), *PTCH1* (see **patched**) and the *GLI2* **oncogene**.

homeobox Conserved DNA sequence originally detected by DNA−DNA hybridization in many of the genes that give rise to **homeotic mutants** and segmentation mutants in *Drosophila*. The homeobox consists of about 180 nucleotides coding for a sequence of 60aa, sometimes termed the homeodomain, of which about 80−90% are identical in the various homeodomains identified from *Drosophila*. Homeoboxes have also been detected in the genomes of vertebrates, with about 75% amino acid homology and a similar sequence has been found in the *MAT* gene of yeast. The homeodomain is involved in binding to DNA. Three subfamilies of homeobox-containing proteins can be identified, based on the archetypal *Drosophila* genes *engrailed, antennapedia* and *paired*.

homeodomain See **homeobox**.

homeostasis *homoeostasis* The tendency towards a relatively constant state. A variety of homoeostatic mechanisms operate to keep the properties of the internal environment of organisms within fairly well-defined limits.

homeotic genes A highly conserved set of genes (*hox* genes) that act as master switches to regulate the development of organ systems, appendages, body segments, etc. They have a **homeobox** region and their products are transcription factors. To the surprise of zoologists the homeoboxes for insect systems (e.g. eye) can be replaced by the vertebrate equivalent even though the actual structures have always been considered analogous rather than homologous. Mutations in homeotic genes can cause major developmental abnormalities (homeotic mutants) in which one body part, organ or tissue, is transformed into another part normally associated with another segment. Examples are the **antennapedia** and **bithorax complex** mutants of *Drosophila* and **synpolydactyly** in humans. Interestingly, linear order within the genome maps to order of expression in the embryo. This may be required for the **transcriptional silencing** of certain homeotic genes (see **Polycomb**). The homeobox genes are clustered and activated in a 3′ to 5′ direction; those activated early in development and responsible for morphologically anterior systems are at the 3′ end. Vertebrates have four duplicate sets (paralogues) of the ten ancient homeotic genes, known as *Hoxa, Hoxb, Hoxc* and *Hoxd*.

homeotic mutant See **homeotic genes**.

homer In mammals there are three genes (*homer-1*, *-2*, and *-3*) that encode at least six homer proteins (~350aa) through differential splicing. Homer proteins are thought to be postsynaptic density scaffolding proteins involved in the clustering of a subset of metabotropic glutamate receptors. A single homologous gene is found in *Drosophila* where it is required for the function of neural networks controlling locomotor activity and behavioural plasticity. See also **GRIP** and **postsynaptic protein**.

homing-endonuclease gene A selfish gene found in fungi, plants, and bacteria that has the ability to create a second copy of itself in individuals that have only one. They are a type of restriction enzyme typically encoded by introns (inteins) and the gene encoding the endonuclease is located within the recognition sequence which the enzyme cuts, thus preventing the cut in HEG$^+$ DNA. In heterozygotes the HEG$^-$ strand is cut and repaired using the HEG$^+$ strand as template, thereby introducing a second copy of the gene.

homocysteine An amino acid, the oxidized form of cysteine, that is an intermediate in methionine metabolism. Epidemiological studies have shown that an elevated level of plasma homocysteine (strongly influenced by dietary and genetic factors) is associated with a higher risk of coronary heart disease, stroke and peripheral vascular disease (though this does not imply causality).

homocystinuria Recessive condition in which the enzyme (cystathione synthetase, EC 4.2.1.22) that converts homocysteine and serine into cystathione, a precursor of cysteine, is missing and there are elevated levels of homocysteine in blood and urine. There are major effects in connective tissue, circulation and nervous system.

homoeo- See **homeo-**.

homogalacturonan A major component of plant primary cell walls, a linear α-1,4-linked GalA homopolymer with, typically, around 100 subunits. Article: http://mplant.oxfordjournals.org/content/2/5/851.full

homogentisic acid An intermediate in the metabolism of tyrosine and phenylalanine. See **alkaptonuria**.

homograft Outmoded term for a graft from one individual of a species to another of the same species. Includes **allogeneic** grafts (allografts) between genetically dissimilar individuals, and syngeneic grafts between identical individuals (twins).

homokaryon A multinucleated cell (syncytium) in which all the nuclei are identical. *Cf.* **heterokaryon**.

homologous (1) The state of having the same relative position, proportion, value, or structure. *Cf.* **heterologous, autologous**. (2) Derived from the tissues or DNA of a member of the same species. (3) Of genes, similar in sequence, *cf.* **analogous, paralogous**. (4) Homologous **recombination** involves exchange of homologous loci and is the way to generate **knockouts** in **transgenic** mice. (5) Homologous chromosomes are identical with respect to genetic loci, and the maternal and paternal homologues form a pair (synapse) during mitosis.

homosporous A species which produces only one type of spore.

homozygote Nucleus, cell or organism with identical **alleles** of one or more specific genes.

Hoodia gordonii A succulent plant used in South Africa for its appetite suppressant properties; the active constituent may be an oxypregnane steroidal glycoside but many commercial preparations lack activity.

hook (1) Basal portion of bacterial flagellum, to which is distally attached the **flagellin** filament. Proximally the hook is attached to the rotating spindle of the motor. In some bacteria (Myxobacteria) the rotation of the hook itself (without an attached flagellum) may directly cause forward gliding movement. (2) In *Drosophila* a gene encoding a large homodimeric protein (679aa) that stabilizes organelles of the endocytic pathway, probably by acting as a cytoskeletal linker. The human homologue hook1 (728aa) is required for spermatid differentiation and is a component of the FTS/Hook/FHIP complex (FHF complex). Hook2 (719aa) and hook3 (718aa) have similar involvement in the FHF complex. (3) See hook-related proteins. (4) See **AT hook**.

hook-related proteins Family of proteins from which **hook proteins** are derived. There are several conserved domains, including a unique C-terminal HkRP domain. The central region of each protein is comprised of an extensive coiled-coil domain, and the N-terminus contains a putative microtubule-binding domain. Hook-related protein 1 (girdin, 1871aa) is a key modulator of the AKT-mTOR signalling pathway that controls the correct neuron positioning, dendritic development and synapse formation during adult neurogenesis. Hook-related protein 2 (daple, 2028aa) is a negative regulator of the canonical Wnt signalling pathway. Hook-related protein 3 (coiled-coil domain-containing protein 88B, 1476aa) is a third member of the family.

Hopp Wood An algorithm for calculating a **hydropathy plot**.

horizontal transmission Transmission of a disease between individuals of the same generation, as opposed to vertical transmission from mother to offspring.

hormesis A dose response phenomenon in which low doses stimulate, but high doses cause inhibition, The dose-response curve may be J-shaped or an inverted U-shape, the latter being observed, for example, with the effect of chemotactic peptides on neutrophil adhesion.

hormone A substance secreted by specialized cells that affects the metabolism or behaviour of other cells possessing functional receptors for the hormone. Hormones may be hydrophilic, like **insulin**, in which case the receptors are on the cell surface, or lipophilic, like the **steroids**, where the receptor can be intracellular. Endocrine hormones are released into blood, exocrine hormones into specialised ducts. Although neurotransmitters, growth factors and cytokines all act on more-or-less remote cells that have the appropriate receptors they are not generally classed as hormones. See Tables H2 and H3.

TABLE H2. Polypeptide hormones and growth factors of vertebrates

Name	M_r (Da)	Residues	Source	Actions
Polypeptide hormones				
Adrenocorticotrophin (ACTH)*	4.5 k	39	Anterior pituitary ACTH-MSH family	Stimulates glucocortoid production from adrenals.
Amylin (Islet Activating Polypeptide)		37	Co-secreted with insulin by pancreatic β cells	Moderates action of insulin
Angiotensin II*		8	Formed from angiotensinogen, *via* angiotensin I	Acts on adrenal gland to stimulate aldosterone release, elevates blood pressure, mitogen for vascular smooth muscle.
Atrial natriuretic peptide (ANP, ANF)		21–28	Atria of heart, brain, adrenal glands	Acts on kidney to produce natriuresis & diuresis, relaxes vascular smooth muscle, inhibits catecholamine release from adrenal medulla.
Bombesin*		14	Skin, gut P-cells, nerves	Acts on CNS, gut smooth muscle, pancreas, pituitary kidney, heart; mitogenic *in vitro*.
Bradykinin*		9	Formed in plasma	Dilates blood vessels, increases capillary permeability.
Caerulein	1352	10	Amphibian skin	Hypotensive, stimulates gastric secretion.
Calcitonin*	4.5 k	32	Thyroid, parathyroid	Opposes parathyroid hormone, hypocalcaemic, hypophosphataemic.
Corticotrophin releasing factor (CRH)*		41	Hypothalamus	Stimulates release of corticotrophin.
Endorphins*		15–31	Pituitary, brain;	Opiates.
Endothelins ET-1, ET-2, ET-3		21	Endothelium	Vasoconstrictors, also inotropic and mitogenic.
Enkephalins*		5	Adrenal medulla, brain, gut.	Opiates.
Exendins (helospectin, helodermin)		39		Glucagon-related, from Gila monster.

TABLE H2. (Continued)

Name	M_r (Da)	Residues	Source	Actions
Follicle stimulating hormone (FSH)	30 k		Anterior pituitary	Acts on gonads.
Gastric inhibitory peptide (GIP)		43	Gut	Inhibits gastric acid secretion, stimulates intestinal secretion, stimulates insulin & glucagon secretion.
Gastrin*		17 or 34	Gut G-cells	Stimulates acid secretion, muscle contraction.
Ghrelin		28	Stomach	Ligand for growth hormone secretagogue receptor. Stimulates appetite.
Glucagon (HGF)*	3550	29	A cells of pancreas	Hyperglycaemic.
Glucagon-like peptide-1 (an incretin)		36, 37	Intestinal epithelial cells	Insulinotropic
Gonadotrophin-releasing hormones		10	Hypothalamus	Stimulate release of LH and FSH
Growth hormone (somatotropin, GH)*	21.5 k	191	Pituitary	Regulates organismal growth.
Growth hormone releasing factor (somatoliberin)	44 k		Hypothalamus	Stimulates release of growth hormone.
Hepcidin antimicrobial peptide		84	Liver	Regulates duodenal iron absorption and trafficking.
Human chorionic gonadotrophin (hCG)	30 k		Anterior pituitary, placenta	Maintains corpus luteum during pregnancy.
Hypothalamic thyrotropic hormone *releasing factor (TRF, TRH)	362	3	Hypothalamus	Stimulates release of thyrotropin from anterior pituitary.
Kallidin		10	Formed in plasma	Dilates blood vessels, increases capillary permeability.
Kisspeptins		54		Regulate reproduction via the hypothalamic-releasing hormones.
Lipotropins*		50–90	Anterior pituitary	Stimulate lipid breakdown.
Luteinizing hormone (LH)	30 k		Anterior pituitary	Acts on gonads.
Motilin*		22	Gut enterochromaffin-2	Increases contractile response of stomach muscle cells.
Neurotensin*		13	Gut, hypothalamus	Effects on smooth muscle tone; may also be a neurotransmitter.
Oxytocin*	1007	8	Posterior pituitary	Uterine contraction, lactation.
Pancreastatin		49	Pancreas and gut	Inhibits insulin release.
Pancreatic polypeptide*		36	PP-cells of pancreas;	Released on feeding; alters gut muscle tone, gut mucosa.
Pancreozymin-cholecystokinin (PZ-CCK)*		8, 33 or 39	Gall bladder, brain, gut I-cells	Secretion of enzymes & electrolytes, secretion of insulin, glucagon, pancreatic polypeptide, contraction of gall bladder.
Parathyroid hormone (PTH)	9.5 k	84	Parathyroid glands	Raises kidney cAMP & blood Ca by stimulating bone release.
Peptide YY		36	Gut	Inhibit pancreatic secretion and release of cholecystokinin.
Placental lactogen		191	Placenta	Promotes lactation.

TABLE H2. (Continued)

Name	M_r (Da)	Residues	Source	Actions
Prolactin	23 k	199	Pituitary	Promote mammary growth, lactation.
Renin	40 k	gp	Kidney	Cleaves angiotensinogen to angiotensin I in plasma.
Secretin*		27	Gut S-cells	Stimulates alkali secretion from pancreas: decreases gastric acid secretion.
Somatostatin (SRIF)*		14	Hypothalamus, D-cells of pancreas	Inhibits release of several hormones, including growth hormone.
Substance P*		11	Gut enterochromaffin-1 cells, nerves	Contracts gut musculature, decreases blood pressure.
Thyroid stimulating hormone (TSH)	30 k		Anterior pituitary	Stimulates release of thyroid hormones.
Vasoactive intestinal peptide (VIP)*		28	Lung, gut H-cells, nervous tissue	Vasodilation, bronchodilation, stimulates insulin, glucagon, prolactin secretion.
Vasopressin (antidiuretic hormone, ADH)*		8	Posterior pituitary	Antidiuretic vasopressor.
Growth factors				
Activin	26 k	115		TGF family; 2 inhibinβ subunits.
Amphiregulin	35−45 k	84		Heparin-binding growth factor.
Amphoterin	30 k	215		Enhances nerve growth cone migration.
Brain-derived neurotrophic factor (BDNF)	28 k	247	Neurotrophin-family	Neurotrophic in CNS.
Epidermal growth factor (urogastrone, EGF)		53	Mouse submaxillary gland, urine	Stimulates epidermal growth, formation of GI tract.
Fibroblast growth factors (FGFs) See Table F2	130 k	59 (a), 288 (b)	Brain, pituitary	Stimulates proliferation of fibroblasts, adrenal cells, chondrocytes, endothelia.
Glial-derived neurotrophic factor (GDNF; formerly Glioma-derived growth factor)	30	211	Glial cells	Specific for midbrain dopamine neurons.
GM-CSF (MG1, CSFalpha)	23 k	141	Many tissues	Promotes granulocyte or macrophage colony formation.
G-CSF (murine; CSFβ in human)	25/30 k	42	Many tissues	Promotes differentiation in myeloleukaemic cells.
Heregulins (Neuregulin)	44 k	640		Growth factors of the [EGF] family.
Inhibin		426	TGFβ family	Inhibits FSH synthesis & secretion.
Insulin*	6 k	21 + 30	β cells of pancreas	Hypoglycaemic, growth factor.
Insulin-like growth factor I (IGF I, somatomedin C)		70	Liver, kidney	Growth factor, released into plasma.
Insulin-like growth factor II (IGF II, multiplication stimulating activity, MSA)		67	Cultured hepatocytes	Mitogen for several cell types.
M-CSF (CSF-1)	40−70 k	-	Mouse tissues	Promotes macrophage differentiation.
MultiCSF (IL3)	23−30 k	-	Mouse primed T-cells	Proliferation of various haematopoietic stem cells.

TABLE H2. (Continued)

Name	M_r (Da)	Residues	Source	Actions
Midkine	13 k			TGFβ family.
Nerve growth factor (NGFβ)	130 k	118	Salivary gland, snake venom	Tropic and trophic effects, mainly on sensory and sympathetic neurons. Peripheral nerve targets.
Platelet-derived growth factor (PDGF)	3 k		Platelets	Stimulate fibroblast proliferation, wound healing.
Somatomedin A		50–80	Liver, kidney	Stimulates growth of peripheral nervous system.
Stem cell factor (Steel factor)	35 k dimer			Haematopoietic growth factor.
Transforming growth factor (TGF)α	5–20 k		Carcinomata	Acts via EGF receptor.
TGFß	2×25 k	112	Tumour cells	Multifunctional role in tissue damage. Promotes or inhibits proliferation depending on cell type.
Thrombopoietin	19 k			Regulates production of platelets.
Vascular endothelial growth factor (VEGF)	45 k (dimer)	165		Important in angiogenesis.

*Neuropeptide. Other neuropeptides include calcitonin gene related peptide, neurokinin, neuromedin, neuropeptide Y, proctolin, and carnosine.

TABLE H3. Steroid hormones

Name	M_r (Da)	Distribution	Actions
Female sex hormones			
β-Estradiol	272	Ovary, placenta, testis	Estrogen
Progesterone	314	Corpus luteum, adrenal, testis, placenta	Regulates pregnancy, menstrual cycle
Male sex hormones			
Testosterone, Dihydrotestosterone	288	Testes	Male secondary sexual characteristics, anabolic effects
Glucocorticoids			
Cortisol	362	Adrenal cortex	Gluconeogenesis, anti-inflammatory
Cortisone	360	Adrenal cortex	Glucocorticoid, anti-inflammatory
Corticosterone	346	Adrenal cortex	
Mineralocorticoids			
Aldosterone	360	Adrenal cortex	Stimulates resorption of Na^+ from kidneys
Thyroid hormones			
Thyroxine (T4)	777	Thyroid	Controls basal metabolic rate
Triiodothyronine (T3)	651	Thyroid	
Other			
Vitamin D (calciferol)	384	Skin (+ sunlight)	Calcium and phosphate metabolism
Ecdysone	481	—	Moulting hormone of insects and nematodes.

hormone-sensitive lipase An enzyme (EC 3.1.1.79, 1076aa) involved in the mobilization of fatty acids from triglyceride stores in adipocytes and in steroidogenic tissues it converts cholesteryl esters to free cholesterol for steroid hormone production.

hornerin A protein (2850aa) of the **fused gene family**. Probably involved in cornification (keratinization) of epidemal cells.

horseradish peroxidase A large enzyme (EC 1.11.1.7, 353aa) isolated from horeseradish (*Armoracia rusticana*), frequently used in conjunction with diaminobenzidine (which it oxidises to produce a brown precipitate) as an intracellular marker to identify cells both at light- and electron-microscopic levels.

host-range mutant A mutant of phage or animal virus that grows normally in one of its host cells, but has lost the ability to grow in cells of a second host type.

host-versus-graft reaction The normal lymphocyte-mediated reactions of a host against allogeneic or xenogeneic cells acquired as a graft or otherwise, which lead to damage or/and destruction of the grafted cells. The reaction involves CD4$^+$ T-cells that respond to differences in MHC Class I antigens. The opposite of **graft-versus-host response**. The common basis of graft rejection.

hotspot A region of DNA that is particularly prone to mutation or transposition.

housekeeping genes Genes that code for proteins or RNAs that are important for all cells and are thus constitutively active. Term used by contrast with luxury proteins, those that are only produced by differentiated cells. Housekeeping genes are often used as internal controls in RT-PCR analyses to assess gene expression since they are assumed to be expressed in all tissues at comparable levels (which may be an unwarranted assumption). A subset are essential genes, those absolutely required for survival. There is a database of essential genes, DEG 6.8 (Nov. 2011) that contains 12995 essential genes from various prokaryotes and eukaryotes. Link to database: http://tubic.tju.edu.cn/deg or http://www.essentialgene.org.

housekeeping proteins Products of **housekeeping genes**.

hox **genes** Homeobox-containing genes of vertebrates.

HP1 In *Drosophila*, heterochromatin protein 1 (206aa), a structural component of heterochromatin, involved in gene repression by binding histone H3 tails methylated at 'Lys-9'. Human homologues (e.g. chromobox protein homolog 1, heterochromatin protein 1 homologue beta, 185aa) are also involved in gene silencing: they are released from chromatin at M-phase of mitosis. There are heterochromatin protein 1-binding proteins 3 (e.g. HP1-BP74, 553aa).

HPETE Intermediate in leukotriene synthesis (5-HPETE, 5-hydroperoxy-eicosatetraenoic acid). Generated from arachidonic acid by 5-lipoxygenase and the starting point for the formation of 5-HETE or of leukotriene A$_4$. Other related enzymes add the hydroperoxy group in different positions.

hpf (1) High power field (of a microscope). (2) Hours post-fertilization. (3) Human plasma fibrinogen. (4) HPF-1 cells, a line of human fibroblasts. (5) High-pressure freezing, a technique used in cryofixation. (6) High pass filter.

HPGDS Haematopoietic PGD synthase (EC 5.3.99.2, 199aa) a bifunctional enzyme responsible for the production of PGD$_2$, a prostaglandin involved in smooth muscle contraction/relaxation and a potent inhibitor of platelet aggregation and also for the conjugation of glutathione to a wide range of aryl halides and organic isothiocyanates.

HPLC Chromatographic method (high pressure liquid chromatography) in which the sample is forced at high pressure (400–1000 atm) through a tightly-packed column of finely-divided particles that present a very large surface area. Solvent pumped through the column elutes compounds at different rates according to their properties, eluants are detected as they leave the column, usually spectrophotometrically. Because HPLC gives gives good separation very rapidly (but is expensive), manufacturers tend to speak of 'high performance liquid chromatography' as an encouragement to purchasers. A wide range of stationary phases, solvents etc. are available and the method can be scaled up or down. Datasheet: http://www.isco.com/WebProductFiles/Applications/105/Application_Notes/HPLC_System_Configuration.pdf

HPRT See **HGPRT**.

HPV Human **papillomavirus**; causes warts.

HRG1 See **heregulins**, **ARIA**.

HRGP Class of plant glycoproteins (hydroxyproline-rich glycoproteins) and proteoglycans rich in hydroxyproline, that includes **AGP**, **extensin** and certain **lectins**. Found in the cell wall and are produced in response to injury.

Hrk A pro-apoptotic Bcl-2 homology domain 3 (BH3)-peptide (harakiri, death protein 5, 91aa) that has been found in brain tissues of AIDS patients, particularly in macrophages. Interacts with the death-repressor proteins Bcl-2 and Bcl-X(L) and expression of Hrk in mammalian cells induces rapid cell death.

HS1 (1) Haematopoietic lineage cell-specific protein (486aa) a leukocyte-specific homologue of **cortactin** that is phosphorylated by **lck** in response to T-cell receptor ligation. Involved in triggering the formation of an F-actin scaffold at the **immunological synapse**. HS1-binding protein-1 (HS1-BP1, HAX1, 279aa) promotes cell survival, potentiates GNA13-mediated cell migration and is involved in the clathrin-mediated endocytosis pathway. Defects lead to **Kostmann's syndrome**. HS1-BP3 (392aa) associates with **SH3 domains** of HS1 and may modulate IL-2 signalling. (2) An adapter protein (**14-3-3 protein** theta, 14-3-3 protein tau, HS1, 245aa). (3) Heteroscorpine-1 (95aa) a toxin with antibacterial properties from *Heterometrus laoticus* (Thai giant scorpion).

HSA (1) Human serum **albumin**. (2) An adhesin (Hsa, 2178aa) of *Streptococcus gordonii* that binds sialic acid moieties on platelet membrane glycoprotein Ibα. (3) Prefix for the name of microRNA (**miRNA**) derived from *Homo sapiens*. Link to miRNA database: http://www.mirbase.org/cgi-bin/mirna_summary.pl?org=hsa

HSC See **hedgehog signalling complex**.

HSCR See **Hirschsprung's disease**.

HSET A motor protein (kinesin-like 2, 673aa) of the **kinesin-14** family, a microtubule-dependent motor required for bipolar spindle formation. May contribute to movement of early endocytic vesicles. Like the *Drosophila* protein **Ncd**, it moves towards the minus end of the microtubule (like cytoplasmic dynein).

HSK pathway See **Hatch-Slack-Kortshak pathway**.

HSP32 See **haem oxygenases**.

hsp60, hsp70, hsp90 See **heat-shock proteins**.

HSR (1) Hypersensitivity reaction. (2) Heat shock response. (3) Homogeneously staining region (of a chromosome).

hst (1) Product of a human oncogene fibroblast growth factor-4 (heparin secretory-transforming protein 1, 206aa). (2) In *S. cerevisiae*, a family of NAD-dependent histone deacetylases involved in telomeric silencing (e.g. HST1, 503aa), members of the **sirtuin** family.

HT1080 cells Line of human fibrosarcoma cells derived from a 35-year-old male patient.

hTERT The catalytic subunit of telomerase (human telomerase reverse transcriptase, 1132aa), the ribonucleoprotein which *in vitro* recognizes a single-stranded G-rich telomere primer and adds multiple telomeric repeats to its 3′ end by using an RNA template. TERT has similarity to reverse transcriptases and may represent a universal subunit for telomerases. Ectopic expression of telomerase in normal human cells extends their replicative life span although does not necessarily transform them. Over-expression is associated with some tumour cells although it may not be the only change involved in progression.

Htk-L Membrane-anchored ligand (Ephrin B2, 333aa) for **EPH** class receptor tyrosine kinases such as EPHA4, EPHB4 (**hepatoma transmembrane kinase**) and EPHA3.

HTLV A family of retroviruses (human T-cell leukaemia/lymphoma viruses). HTLV-1 (adult T-cell lymphoma virus type 1) causes leukaemia and sometimes a mild immunodeficiency. In addition to *gag*, *pol* and *env*, the virus carries a coding sequence *pX* that does not seem to have a normal genomic homologue and is not a conventional oncogene. The protein product of the pX region is a short-lived nuclear protein (around 40 kDa). T-cells transformed with HTLV-1 continue to proliferate and are independent of **interleukin-2**. It is associated with a progressive demyelinating upper motor neurone disease, tropical spastic paraparesis. Type II (HTLV-2) was originally isolated from a T-cell line from a patient with hairy cell leukaemia and has about 70% genomic homology with HTLV-1. HTLV-III was the name originally given to **HIV**, although this usage is obsolete, and the name is now applied to simian T-lymphotropic virus-3.

HTRF Assay methodology (homogeneous time-resolved fluorescence) used in high throughput screening. An absorbing fluorochrome is coupled to one component, the emitting fluorochrome with slow-release characteristics coupled to the other and if the two components are in proximity (because they bind) then **fluorescence energy transfer** between emitting and absorbing fluorochromes gives a signal which is analysed in a **time-resolved fluorescence** system. No separation step is required – all the reagents are mixed together and inhibitors of binding will reduce the output signal.

HU proteins Small (~90aa) homodimeric proteins, found in a range of prokaryotes, architectural non-sequence specific DNA binding proteins associated with DNA supercoiling and chromosome condensation (histone-like proteins) and initiation of DNA replication.

HuH7 cells A line of human hepatoblastoma cells with an epithelioid morphology.

Hülle cells Thick-walled cells with characteristic thin-walled pores, usually associated with **cleistothecia** of *Aspergillus* where they may act as nurse cells.

human Generally ignored at the beginning of a noun-phrase; 'human chorionic gonadotrophin' is defined under 'chorionic gonadotophin'. Eukaryotic proteins are the human type unless otherwise specified.

human double minute 2 See **hdm2**.

human endogenous retroviruses See **endogenous retroviruses**.

human enhancer of filamentation 1 A multifunctional docking protein (HEF1, neural precursor cell expressed, developmentally downregulated-9, NEDD-9, Cas-L, 835aa) of the **Cas family** a key protein coordinating tyrosine-kinase-based signalling related to cell adhesion. There are four isoforms: p115, p105, p65, and p55.

Human Genome Diversity Project The plan to analyse and compare variations in DNA samples from hundreds of different ethnic groups, in order to understand the origins and migrations of human populations and the genetic basis of differing susceptibility to disease. Opposition from some groups has caused the project considerable difficulty. It is not connected to the Human Genome Project.

human immunodeficiency virus See **HIV**.

humanized antibody Usually a mouse monoclonal antibody directed against a target of particular therapeutic value that has been modified to have all regions except the antigen-binding portion substituted by human immunoglobulin domains. The procedure should make the antibody minimally antigenic when administered to patients though some anti-idiotype antibodies may still be made to the antigen binding site.

humic acid A complex mixture of partially decomposed and transformed organic materials. There are several subclasses of humic acids, (tannins, lignins, fulvic acids) and their properties are complex.

humoral immune responses Immune responses mediated by antibody. *Cf.* **cell mediated immunity**.

hunchback Key regulatory gene in the early segmentation gene hierarchy of *Drosophila*. Codes for a transcription factor (758aa) of the Cys2-His2 **zinc finger** type. See **nanos**.

Hunter's syndrome Recessive **mucopolysaccharidosis**, X-linked, in which dermatan and heparan sulphates are not degraded. Because two lysosomal enzymes (heparan sulphate sulphatase, EC 3.1.6.13 and α-iduronidase, EC 3.2.1.76) are involved in the breakdown of these glycosaminoglycans, fibroblasts from Hunter's syndrome will complement the fibroblasts from **Hurler's disease** patients in culture; by recapture of lysosomal enzymes from the medium, both types of cells in mixed culture become competent to digest glycosaminoglycans.

huntingtin Protein product (3142aa) of the *IT15* gene that is widely expressed and required for normal development. There are variable numbers of polyglutamine repeats in Huntington's chorea (44 in the commonest form of the disease) which increase the interaction of huntingtin with huntingtin-associated protein-1 (HAP-1, 671aa) which is enriched in the brain and may be associated with pathology. Huntingtin-interacting protein (HIP-1, 1037aa) is membrane-associated and has similarity to cytoskeleton proteins. It has a role in clathrin-mediated endocytosis and trafficking. Various other huntingtin interacting proteins have been identified.

Huntington's chorea Maturity onset disease characterized by progressive loss of neuronal functioning. Caused by unstable amplification of a trinucleotide $(CAG)_n$ repeat within the coding region of a gene encoding **huntingtin**.

HuP proteins (1) Proteins involved in the hydrogen uptake system in bacteria such as *Rhizobium leguminosarum* which produce hydrogen as a byproduct of the activity of nitrogenase. (2) (*less commonly*) Human homologues of paired-box homeobox (*Pax*) genes. (3) A bacterial gene *hup* encodes the bacterial histone-like DNA-binding protein, HU (94aa in *Nostoc sp.*) which is essential for heterocyst differentiation.

HuR An RNA-binding protein (Hu-antigen R, ELAV-like protein 1, 326aa) that regulates the stability and/or the translation of mRNAs such as those for myc, fos, and IL-3.

Hurler's syndrome An autosomal recessive **storage disease** in which the deficiency is in a-L-iduronidase (EC 3.2.1.76, 653aa) an enzyme involved in degradation of heparan and dermatan sulphates. Extensive deposits of mucopolysaccharide are found in **gargoyle cells**, and in neurons. There are three different clinical syndromes associated with the enzyme deficiency: Hurler's (mucopolysaccharidosis IH, MPS IH), Scheie's (MPS IS), and Hurler-Scheie (MPS IH/S) syndromes. See **Hunter's syndrome**.

Hürthle cells See **Askenazy cells**.

HUT-78 cells A cell line derived from a male patient with **Sezary syndrome**. Have features of mature T-cells of the helper/inducer phenotype and release IL-2.

Hutchinson-Gilford progeria syndrome See **progeria** and **Werner's syndrome**.

HUVEC Endothelial cells (human umbilical vein endothelial cells) derived from the large vein in the umbilical cord which is usually discarded together with the placenta after childbirth. The cells can be removed as a fairly pure suspension by mild enzymatic treatment of the vein followed by some mechanical distraction and will grow relatively easily in culture, retaining their differentiated characteristics for several passages.

HVEM High-voltage **electron microscopy**.

hyaline Clear, transparent, granule-free; as for example hyaline cartilage and the hyaline zone at the front of a moving amoeba.

hyaluronic acid Polymer composed of repeating dimeric units of glucuronic acid and N-acetyl-glucosamine. May be of extremely high molecular weight (3−4 million daltons) and forms the core of complex proteoglycan aggregates found in extracellular matrix. The hydrated volume is very large (1000x greater than the dry volume) and solutions are very viscous.

hyaluronidases Enzymes (EC 3.2.1.35) that degrade **hyaluronic acid**. Hyaluronidase 1 (435aa) is lysosomal and defective in mucopolysaccharidosis type 9; hyaluronidase 2 (473aa) is GPI-anchored and hydrolyzes high molecular weight hyaluronic acid to produce an intermediate-sized product which is further hydrolysed by other hyaluronidases. Both may have a role in tumour progression. Other types are hyaluronidase 3 (417aa), hyaluronidase 4 (481aa) and hyaluronidase PH-20 (509aa). The latter is lipid-anchored on the sperm surface and is involved in sperm-egg adhesion.

hybrid antibody Artificially-produced antibody made by fusing **hybridomas** producing two different antibodies; the hybrid cells produce three different antibodies, only one of which is a **heterophile antibody** with two different antigen-binding sites. Can also be prepared chemically from two antibodies.

hybrid cells Any cell type containing components from one or more genomes, other than zygotes and their derivatives. Hybrid cells may be formed by **cell fusion** or by **transfection**. See **heterokaryon**. Cybrid cells have the nucleus from one species in the enucleated cytoplasm of an egg from another species.

hybrid dysgenesis Genetic phenomenon, in which two strains of organism produce offspring at anomalously low rates. Example: the P-M system of *Drosophila*, in which P-strain males (containing multiple **P elements**) mated to M-strain females produce sterile hybrids.

hybrid specific amplification A simple method designed to isolate the common fraction of two DNA samples while avoiding the background due to repeated sequences. The method is based on the **suppression PCR** principle, associated with a **Cot1 DNA** pre-hybridization step. Description of method:

http://www.ncbi.nlm.nih.gov/pmc/articles/PMC55899/?tool = pubmed

hybridization (1) The production of an organism by mating individuals of different species. Most animal hybrids are sterile (e.g. mules), but plant hybrids can be fertile and may show hybrid vigour (**heterosis**). (2) Hybridization of nucleic acids occurs when single-stranded **nucleic acids** are allowed to interact so that complexes, or hybrids, are formed by molecules with sufficiently similar, complementary sequences. By this means the degree of sequence identity can be assessed and specific sequences detected (See C_0t **curve**). The hybridization can be carried out in solution or with one component immobilized on a gel or, most commonly, nitrocellulose paper. Hybrids are detected by various means: visualization in the electron microscope; by radioactively labelling one component and removing non-complexed DNA; or by washing or digestion with an enzyme that attacks single-stranded nucleic acids and finally estimating the radioactivity bound. Hybridizations are done in all combinations: DNA-DNA (DNA can be rendered single-stranded by heat denaturation), DNA-RNA or RNA-RNA. *In situ* hybridizations involve hybridizing a labelled nucleic acid (often labelled with a fluorescent dye) to suitably prepared cells or histological sections. This is used particularly to look for specific **transcription** or localization of genes to specific chromosomes (**FISH analysis**).

hybridoma A cell hybrid in which a tumour cell forms one of the original source cells. In practice, confined to hybrids between T- or B-lymphocytes and appropriate **myeloma** cell lines.

hydatid cyst Large cyst in the viscera of sheep, cattle or man following ingestion of eggs of the tapeworm *Echinococcus granulosus*, a cestode. In the normal host (carnivores such as dog) the eggs develop into intestinal worms but in abnormal hosts the larvae penetrate through the wall of the intestine and migrate to liver and other organs. The cyst contains many scolices that can form further cysts if liberated: it is thus a metastasising parasite. Do not confuse with **hydatidiform mole**.

hydatidiform mole Abnormal conceptus in which an embryo is absent and there is excessive proliferation of placental villi in the uterus. In most cases the tissue is diploid XX with both X chromosomes being of paternal origin and thus it is thought that it may arise by fertilization of a dead ovum. May occasionally become invasive though the metastases regress following removal of the mole. Mutations in the NALP7 gene (see **NALP proteins**) cause recurrent hydatidiform moles. Not to be confused with **hydatid cyst**.

Hydra Genus of freshwater coelenterates (**Cnidaria**). They are small, solitary and only exist in the polyp form, which is a radially-symmetrical cylinder that is attached to the substratum at one end and has a mouth surrounded by tentacles at the other. They have considerable powers of regeneration and have been used in studies on positional information in morphogenesis.

hydraulic motor By altering the internal osmotic pressure within a cell, water will enter and a considerable expansion of the compartment will occur. This has been used as a motor device in plants (turgor pressure), in eversion of **nematocysts**, and possibly in the production of other cellular protrusions.

hydrocortisone Moderately potent **corticosteroid** (cortisol, 17-hydroxy-corticosterone) with both mineralocorticoid and glucocorticoid activity, produced by cells of the zona reticularis in the adrenal gland. Has potent anti-inflammatory effects.

hydrogel A gel in which water forms the bulk phase.

hydrogen peroxide Hydrogen peroxide (H_2O_2) is produced by vertebrate phagocytes and is used in bacterial killing (the **myeloperoxidase**-halide system).

hydrogenosome Membrane-bounded organelle found in certain anaerobic trichomonad and some ciliate protozoa: contains hydrogenase and produces hydrogen from glycolysis.

hydrolase One of a class of enzymes (EC Class 3) catalysing hydrolysis of a variety of bonds, such as esters, glycosides, peptides.

hydrolytic enzymes See **hydrolase**.

hydropathy plot Approximate way of deducing the higher order structure of a protein, based on the principle that $20-30$ consistently hydrophobic residues are necessary to make a membrane-spanning α-helix. For each amino acid residue, a weighted average of the hydrophobicity of the residue and its immediate neighbours are calculated and graphically displayed as a hydropathy plot, with hydrophobic domains plotted as positive numbers. There are several formulae for the calculation, that differ in the calculation of the moving average, the window size, and the scoring system for hydrophobicity of individual residues (e.g. Kyte-Doolittle, Hopp Wood, Eisenberg).

hydrophilic group A polar group or one that can take part in hydrogen bond formation, e.g. OH, COOH, NH_2. Confers water solubility, or in lipids and macromolecules causes part of the structure to make close contact with the aqueous phase.

hydrophobic bonding Interaction driven by the exclusion of non-polar residues from water. It is an important determinant of protein conformation and of lipid structures, and is considered to be a consequence of maximizing polar interactions rather than a positive interaction between apolar residues.

hydrophobins Class of surface active proteins produced by filamentous fungi that have a role in hyphal development and in sporulation. Make fungal structures, such as spores, hydrophobic. In *Aspergillus fumigatus*, hydrophobin (rodlet protein, 159aa) is involved in resistance to environmental stress.

hydropic Generally used as an adjective to describe one of the early signs of cellular degeneration in response to injury, the accumulation of water in the cell, causing swelling.

hydroxyapatite The calcium phosphate mineral, $Ca_{10}(PO_4)_6(OH)_2$, found both in rocks of nonorganic origin and as a component of bone and dentine. Used as column packing for chromatography, particularly for separating double-stranded DNA from mixtures containing single-stranded DNA.

hydroxylysine Post-translationally hydroxylated lysine is found in **collagen** and commonly has galactose and then glucose added sequentially by glycosyl transferases. The extent of glycosylation varies with the collagen type.

hydroxymethylbilane synthase See **uroporphyrinogen I synthetase**.

hydroxyproline A post-translationally modified proline residue. In collagen specific proline residues on the amino side of a glycine residue become hydroxylated at C4, before the polypeptides become helical, by the activity of **prolyl hydroxylase** (which requires ascorbic acid to keep the iron residue in a reduced state). The presence of hydroxyproline is essential to produce stable triple-helical tropocollagen, hence the problems caused by ascorbate deficiency in scurvy. This unusual amino acid is also present in considerable amounts in the major glycoprotein of primary plant cell walls (see **HRGP**).

hydroxyproline-rich glycoprotein See **HRGP**.

hydroxytetraecosanoic acid See **HETE**.

hydroxytryptamine See **serotonin**.

hydroxyurea Inhibitor of DNA synthesis (but not repair), used as an antimetabolite in treating cancer.

hydrozoa A class of **Cnidaria**, in which there is alternation of generations; the hydroid phase is usually colonial, and produces the free-swimming medusoid phase by budding.

hygromycin B Aminoglycoside antibiotic from *Streptomyces hygroscopicus* that is toxic for both pro- and eukaryotic cells. Inhibits peptide chain elongation by yeast polysomes by preventing elongation factor EF2-dependent translocation. Used as a selection agent in transfection — the hygromycin-B-phosphotransferase gene, *hph*, in the vector confers resistance.

hyoscine An alkaloid (scopolamine), present in *Datura meteloides*, that has a sedative effect on the central nervous system.

Hypaque(TM) Compound (sodium diatrizoate) used in contrast media for radiology and in FicollPaque™ for blood cell separation by centrifugation (see **Ficoll**).

hypaxial See **epaxial**.

hyper-IgD syndrome An autosomal recessive disorder caused by mutations in the gene encoding **mevalonate kinase**. There are recurrent episodes of fever.

hyperaccumulator General term for an organism that accumulates high levels of a substance from the environment. Usually used in reference to plants that absorb large amounts of heavy metals from the soil.

hyperadrenalism Overactivity of the adrenal glands. Symptoms depend upon the cell type affected and the hormone being over-produced. See **Cushing's syndrome**.

hypercalcaemia Condition in which calcium levels in the blood are above normal levels (9−10.5 mg/dL) usually as a result of **hyperparathyroidism** or malignancy.

hypercholanemia A disorder associated with a mutation in tight junction protein 2 (see **ZO-1**) in which there are elevated serum bile acid concentrations.

hypercholesterolaemia High serum levels of cholesterol often simply as a result of high dietary intake. In some cases can be caused by a defect in lipoprotein metabolism or, for example, defects in the **low density lipoprotein receptor** (familial hypercholesterolaemia) or the **low-density lipoprotein receptor adaptor protein**. Other causes include mutation in the LDL receptor-binding domain of **apolipoprotein B-100** and mutation in the gene encoding proprotein convertase subtilisin/kexin type 9 (PCSK9).

hyperchromasia Histopathological term for cells in which the nuclei appear to be smudged or opaque and are more darkly stained. Usually an indication of pre-cancerous behaviour.

hyperCKemia A disorder caused by defects in caveolin-3 in which serum creatine kinase levels are persistently high.

hyperexcitability Can be used in the context of behaviour, but physiologically, a state in which the threshold for neuronal firing is reduced and there are spontaneous action potentials. This may exhibit itself as hyper-algesia (increased sensitivity to pain) and

there are clinical conditions in which this maybe due to autoimmune dysfunction.

hyperferritinaemia-cataract syndrome An autosomal dominant disorder caused by mutation in the iron-responsive element (IRE) in the 5′-noncoding region of the ferritin light chain gene. There is congenital nuclear **cataract** (because ferritin accumulates in lens cells) and elevated serum **ferritin** levels.

hyperforin A natural phloroglucinol purified from *Hypericum perforatum* (St. John's Wort). Often used as an antidepressant, apparently working by blocking serotonin re-uptake (see **SSRIs**). Found to be inhibitory to *Staphylococcus aureus* multiresistant to conventional antibiotics as well as against other Gram-positive bacteria and reported to promote apoptosis of B-cell chronic lymphocytic leukemia cells.

hypergammaglobulinaemia A condition in which the concentration of immunoglobulins in the blood exceeds normal limits. May be because there is continuous antigenic stimulation in chronic infections, a result of auto-immune disease, or an abnormal proliferation of B-cells as in **Waldenstrom's macroglobulinaemia** or in **myelomatosis**. See **hyper-IgD syndrome**.

hyperglycaemia An excess of plasma glucose (usually more than 11 mmol/L) that can arise through a deficiency in insulin production. See **diabetes mellitus, diabetes insipidus**.

hyperimmune serum Serum prepared from animals that have recently received repeated injections or applications of a chosen antigen; thus the serum should contain a very high concentration of polyclonal antibodies against that antigen.

hyperinsulinism A condition in which there is oversecretion of insulin leading to hypoglycaemia, usually caused by a pancreatic tumour. Inherited forms can arise through mutation in genes for the SUR1 or Kir6.2 subunits of the inwardly rectifying potassium channel of pancreatic beta cells, or mutation in genes for glucokinase (EC 2.7.1.1), 3-hydroxyacyl-CoA dehydrogenase (HADH; EC 1.1.1.35), the insulin receptor or glutamate dehydrogenase (GDH; EC 1.4.1.3). There is also a form associated with enteropathy and deafness caused by a 122 kb deletion on chromosome 11.

hyperkalaemia An excessive level of potassium in the blood. Most cases of hyperkalemia are caused by disorders that reduce potassium excretion by the kidney, either because of dysfunction or a deficiency of aldosterone (as in **Addison's disease**), the hormone that regulates potassium excretion. See **pseudohypoaldosteronism**.

hyperkalaemic periodic paralysis See **voltage-gated sodium channels** and **periodic paralysis**.

hyperkeratosis Overgrowth of the horny layer of the skin, a feature of bullous ichthyosiform erythroderma (epidermolytic hyperkeratosis) caused by mutations in **keratin** genes, of palmoplantar hyperkeratoma and various **ichthyoses**.

hyperlipidaemia Condition in which plasma lipid concentrations are above normal: various forms are recognised according to the lipid type affected. In Type I for example there is a deficiency of lipoprotein lipase (LPL) or altered apolipoprotein C2 with high levels of chylomicrons and extremely elevated triglycerides. In Type IIa (familial **hypercholesterolemia**) **low-density lipoprotein** (LDL) levels are raised because of a defect in the LDL receptor recycling system and in Type IIb both LDL and VLDL levels are affected. In some cases there is an increased risk of cardiovascular disease.

hyperlipoproteinaemia The same as **hyperlipidaemia**.

hypermastigote Large multi-flagellate symbiotic protozoan found in the gut of termites and wood-eating cockroaches. Most bizarre example of the group is *Mixotricha paradoxica* that actually has few flagella and is propelled by spirochaetes (bacteria) that are attached to special bracket-like regions of the cell wall.

hypermetropia A visual defect (hyperopia, long-sightedness) in which the image forms behind the plane of the retina.

hyperopia See **hypermetropia**.

hyperosmotic Of a liquid, having a higher osmotic pressure (usually than the physiological level, 275–295 mOsmol/kg).

hyperostosis corticalis generalisata See **van Buchem's disease**.

hyperoxaluria A disorder caused by mutation in the gene encoding liver peroxisomal alanine-glyoxylate aminotransferase (EC 2.6.1.44, 392aa) in which there is high urinary oxalate excretion.

hyperparathyroidism Condition in which there is overproduction of **parathyroid hormone**, usually because of hyperplasia of the gland, leading to high blood calcium levels (**hypercalcemia**) and low blood phosphate levels. Can arise through mutation in the **parafibromin** (HRPT2) gene or in the MEN1 gene (mutations cause **multiple endocrine neoplasia type I**).

hyperphosphataemia Abnormal concentrations of phosphate in the blood (above the normal levels of 1.0–1.5 mmol/L), often associated with **hypocalcaemia**. See **hypoparathyroidism**.

hyperplasia Increase in the size of a tissue as a result of enhanced cell division. Once the stimulus (wound healing, mechanical stress, hormonal overproduction) is removed the division rate returns to normal (whereas in neoplasia proliferation continues in the absence of a stimulus).

hyperpolarization A negative shift in a cell's **resting potential** (which is normally negative), thus making it numerically larger i.e. more polarized. The opposite of **depolarization**.

hypersensitivity (1) In immunology, a state of excessive and potentially damaging immune responsiveness as a result of previous exposure to antigen. If the hypersensitivity is of the immediate type (antibody-mediated), then the response occurs in minutes; in delayed hypersensitivity the response takes much longer (about 24 hours) and is mediated by primed T-cells. Hypersensitivity responses are not simply divisible into the two types, and it is now more common to subdivide immediate responses into types I, II, and III, the delayed response being of type IV. Type I responses involve antigen reacting with IgE fixed to cells (usually mast cells) and are characterized by histamine release; anaphylactic responses and urticaria are of this type. In type II responses circulating antibody reacts with cell surface or cell-bound antigen, and if complement fixation occurs, cytolysis may follow. In type III reactions immune complexes are formed in solution and lead to damage (serum sickness, glomerulonephritis, **Arthus reaction**). Delayed-type responses of Type IV involve primed lymphocytes reacting with antigen and lead to formation of a lymphocyte-macrophage granuloma without involvement of circulating antibody. (2) In botany, a hypersensitive response is an active response of plant cells to pathogenic attack in which the cell undergoes rapid necrosis and dies. Associated with the production of **phytoalexins**, lignin, and sometimes **callose**. The response is thought to prevent a potential pathogen from spreading through the tissues.

hypertelorism A condition in which there is an unusually large separation of organs. Most commonly encountered in orbital hypertelorism (an increased distance between the eyes) which is a feature of some developmental abnormalities.

hyperthermophile Members of the Archaea that live and thrive in temperatures above 60°C, sometimes above 100°C (cf. thermophiles, which have a tolerance ceiling of about 60°C).

hyperthyroidism See **Graves' disease**.

hypertonic Of a fluid, sufficiently concentrated to cause osmotic shrinkage of cells immersed in it. Note that a mildly **hyperosmotic** solution is not necessarily hypertonic for viable cells that are capable of regulating their volumes by **active transport**. See **hypotonic, isotonic**.

hypertriglyceridaemia Abnormally high levels of plasma triglycerides, a feature of most **hyperlidaemias**. Susceptibility has been associated with mutation in the apolipoprotein A5 gene and the lipase I gene, and with polymorphism in the RP1 retinitis pigmentosa gene.

hypertrophic cardiomyopathy Abnormal thickening of the muscle of the left ventricle. The condition can be caused by mutation in various genes including those for myosin heavy chain-six or seven, myosin light chain two or three, cardiac troponin T2 or I, tropomyosin 1, cardiac myosin-binding protein C, **titin**, cardiac muscle alpha actin, cysteine-rich protein-3, myosin light chain kinase-2, caveolin-3 and with mutations in genes for mitochondrial tRNAs for glycine and isoleucine.

hypertrophy Increase in size of a tissue or organ as a result of cell growth, rather than an increase of cell number (**hyperplasia**), though often both processes occur.

hyperuricaemia An abnormally high level of uric acid in the blood. See **uromodulin**.

hypervalinemia Abnormally high levels of valine in plasma and urine, probably as a result of a defect in a branched-chain amino-acid aminotransferase (EC 2.6.1.42, BCAT1, 386aa; BCAT2, 392aa) involved in leucine, isoleucine and valine catabolism.

hypervariable region Those regions of the heavy or light chains of immunoglobulins in which there is considerable sequence diversity in a single individual. These regions specify the antigen affinity of each antibody.

hypha Filament of fungal tissue that may or may not be separated into a file of cells by cross-walls (septa). It is the main growth form of filamentous fungi, and is characterized by growth at the tip followed by lateral branching.

hypoadrenalism Underactivity of the adrenal glands. See **triple-A syndrome**.

hypoblast The innermost germinal layer in the embryo of a metazoan animal that gives rise to the endoderm and sometimes also to the mesoderm.

hypocalcaemia The condition in which calcium levels in blood are below normal.

hypochlorhydria Condition in which there is reduced secretion of hydrochloric acid by gastric acid-secreting cells. Can be a feature of **Menetrier's disease**.

hypochondrogenesis A disorder in which there is abnormal bone growth. Can be caused by mutation in the gene for collagen type II or in the DTDST gene that encodes a sulphate transporter (solute carrier family 26 member 2, SLC26A2, 739aa) needed for normal sulphation of cartilage proteoglycans. A more severe disorder (achondrogenesis) is caused by mutation in the gene for TRIP11 (thyroid hormone receptor interactor 11, 1979aa).

hypocotyl The filamentous axis that grows out of the plant embryo, bearing at its tip the seed-leaves, or cotyledons. It usually elongates rapidly in the dark and its growth is inhibited once exposed to light.

hypocretin See **orexin**.

hypogammaglobulinaemia A condition in which the immunoglobulin level in plasma is below the normal range. Congenital, chronic and transient types are known. See **hypergammaglobulinaemia, Bruton's disease**.

hypogeal *hypogaeous* Describing: (1) An organism that lives below the ground surface. (2) A plant that germinates with the cotyledons remaining in the soil.

hypogonadism Condition in which the gonads do not produce normal levels of reproductive homones. Can be caused by mutations in several genes, including those for the gonadotropin-releasing hormone receptor, the G protein-coupled receptor-54, the nasal embryonic LHRH factor and the fibroblast growth factor receptor-1. See also **Kallmann's syndrome**. Hypogonadism is also a feature of disorders such as **testicular feminization**.

hypohidrosis Abnormally low production of sweat, sometimes restricted to parts of the body. The X-linked form (Christ-Siemens-Touraine syndrome) is a result of mutation in the **ectodysplasin-A** gene. Autosomal forms arise from mutation in the ectodysplasin anhidrotic receptor (EDAR) gene or in the EDAR-associated death domain gene (*EDARADD* gene).

hypomagnesaemia with hypocalcuria and nephrocalcinosis A disorder of kidney function in which tight junction proteins (**claudins**) are defective. In the standard form the mutation is in **paracellin-1** (claudin 16), and a related form with severe ocular involvement is caused by mutations in the claudin-19 gene.

hypomorph An individual in which the level of some gene product is reduced because of a mutation.

HYPONASTIC LEAVES A double-stranded RNA-binding domain (dsRBD)-containing plant protein (HYL1, 419aa) that assists the release of the miRNA-miRNA* duplex formed from the primary miRNA transcript by dicer-like-1 (DCL1).

hyponatraemia A decreased level of sodium in the blood, usually due to increased vasopressin levels but can be a result of adrenal insufficiency, congenital adrenal hyperplasia or hypothyroidism.

hypoparathyroidism A disorder in which there is underproduction of **parathyroid hormone** or defects in the parathyroid signalling system. There is **hypocalcaemia** and **hyperphosphataemia.** Can be a result of mutation in genes for the **calcium-sensing receptor,** the parathyroid hormone itself, or for the GCM2 transcription factor (important in parathyroid differentiation and a homologue of the *Drosophila* 'glial cells missing' (*gcm*) gene). There is also an X-linked form, possibly a defect in SOX3. Hypoparathyroidism-retardation-dysmorphism syndrome (Sanjad-Sakati syndrome) is caused by mutation in the gene encoding tubulin-specific chaperone E which is also affected in Kenny-Caffey syndrome. Hypoparathyroidism, sensorineural deafness, and renal disease (HDR) syndrome (Barakat's syndrome) is a result of haploinsufficiency of the transcriptional enhancer GATA3. Hypoparathyroidism can also be a feature of **autoimmune polyendocrinopathy syndrome**.

hypophosphatasia A disorder in which there is defective bone mineralization caused by mutation in the tissue-nonspecific isoenzyme of **alkaline phosphatase**.

hypopituitarism Any condition in which the activity of the pituitary gland is low. This can arise from mutations in pituitary-specific transcription factor (PIT1), the paired-like homeodomain transcription factor (prophet of PIT1, PROP1), the Rathke pouch homeobox (HESX1) and the LIM homeobox-3 (LHX3). X-linked panhypopituitarism can be a result of duplications in the *SOX3* gene. In Simmonds' disease there is a failure of the anterior lobe of the pituitary to produce any one or more of its six hormones (ACTH, TSH, FSH, LH, GH, and prolactin).

hypothalamo-pituitary-adrenal axis A neuroendocrine system involving the hypothalamus, the pituitary gland and the adrenal glands that is responsible for controlling stress reactions and the levels of cortisol and other stress related hormones.

hypothalamus The region of the brain involved in control of the autonomic nervous system and that links the nervous and endocrine systems.

hypothyroidism The condition in which there are low levels of thyroid hormone. Iodine deficiency is the commonest cause (see **goitre**) but the condition can also be caused by **Hashimoto's thyroiditis** or a deficiency of hypothalamic or pituitary hormones. **Pendred's syndrome** (goitre with deafness) is caused by mutation in an anion transporter (pendrin) gene. Nongoitrous hypothyroidism can be caused by

mutation in the thyroid-stimulating hormone receptor gene or in the *PAX8* gene. Athyroidal hypothyroidism (Bamforth-Lazarus syndrome) is caused by mutations in the gene encoding thyroid transcription factor-2. Other forms can be caused by mutations in genes for iodotyrosine deiodinase, thyroid oxidase-2, thyroid peroxidase, the sodium-iodide symporter and defects in thyroglobulin synthesis.

hypotonia-cystinuria syndrome A disorder in which muscle tone is poor (hypotonia), there is cystinuria and various neurological and developmental abnormalities. A deletion on chromosome 2 disrupts the *SLC3A1* gene (an amino acid carrier) and the *PREPL* gene that encodes a prolyl oligopeptidase.

hypotonic Of a fluid, having a concentration that will cause osmotic shrinkage of cells immersed in it. Not necessarily hypo-osmotic.

hypotrichosis with juvenile macular dystrophy A rare autosomal recessive disorder caused by defects in **cadherin-3**.

hypovolaemia An abnormal reduction of the volume of blood or blood plasma leading to a form of clinical shock in which blood pressure falls.

hypoxanthine Purine base present in inosine monophosphate (IMP) from which adenosine monophosphate (AMP) and guanosine monophosphate (GMP) are made. The product of deamination of adenine, 6-hydroxy-purine.

hypoxanthine guanine phosphoribosyl transferase See HGPRT.

hypoxia Condition in which there is an abnormally low level of oxygen in the blood and tissues.

hypoxia-inducible factor Basic helix-loop-helix-PAS domain transcription factors (HIF1α, 826aa, HIF3α, 669aa) that are master regulators of the response to hypoxia and other environmental stressors. Form heterodimers with **ARNT** and affect genes involved in energy metabolism, angiogenesis, and apoptosis. Also essential for myeloid cell activation in response to inflammatory stimuli. They are normally short-lived and are targeted for proteasomal degradation by the activity of a **prolyl hydroxylase** that acts as an oxygen sensor and is inhibited by hypoxia. An inhibitor, factor-inhibiting HIF-1 (FIH-1, 349aa), regulates HIF activity and plays a role in modulation of the hypoxic response by notch signalling.

H zone Central portion of the A-band of the **sarcomere**, the region that is not penetrated by thin (actin) filaments when the muscle is only partially contracted. The **M-line** is in the centre of the H zone.

I

I Single letter code for isoleucine.

I-309 A CC-**chemokine** (CCL1, 96aa) secreted by activated T-cells that is chemotactic for monocytes but not neutrophils. It is a ligand for CCR8.

Ia antigen (1) Antigens coded for by the I-region of the MHC complex. Most are MHC Class II molecules composed of α (365aa) and β (266aa) polypeptide chains. (2) Ia antigen-associated invariant chain (p33, CD74, 296aa) forms a nonameric complex ($\alpha_3\beta_3\gamma_3$) with peptide-free class II alpha/beta heterodimers and directs transport of the complex from the endoplasmic reticulum to compartments where peptide loading of class II takes place. Serves as cell surface receptor for MIF. (3) Protein Ia is a porin expressed on the outer surface of *Neisseria meningitidis* where it is a major antigen.

IAA See **indole acetic acid**.

IAPP Islet amyloid peptide. See **amylin**.

iatrogenic Descriptor for a disease caused by attempts at therapy.

I-band The isotropic band of the sarcomere of striated muscle, where only thin filaments are found. Unlike the A-band, the I-band can vary in width depending upon the state of contraction of the muscle when fixed.

IBA RESPONSE-5 In plants, a putative dual-specificity protein phosphatase (IBR5, 257aa), part of a SCF protein ligase complex that when mutated alters response to **indole-3-butyric acid** (IBA). Dephosphorylates and inactivates MAP kinase MPK12.

IBD (1) Inflammatory bowel disease. See **Crohn's disease** and ulcerative colitis. (2) Identity by descent. Two genes at a locus have identity by descent (IBD) if they were both inherited from a common ancestor. Important in linkage studies for disease susceptibility.

iberiotoxin Peptide toxin (37aa) from *Mesobuthus tamulus* (Eastern Indian scorpion) that is selective for the high-conductance calcium-activated K^+ channels (maxi-K channels). Similar (and highly homologous in sequence) to **charybdotoxin** but more selective.

IBMPFD See **inclusion body myopathy with Paget's disease and frontotemporal dementia, Paget's disease of bone**.

IBMX Isobutylmethylxanthine, a compound often used experimentally to inhibit cAMP phosphodiesterase and raise intracellular cAMP levels.

ibotenate An **excitotoxin** from *Amanita sp.*, that acts on the NMDA class of **glutamate receptors**.

ibuprofen Non-steroidal anti-inflammatory drug (NSAID) that inhibits **cycloxygenases** I and II (COX-1, COX-2). Related NSAIDs are ketoprofen, flurbiprofen and naproxen.

iC3b Inactivated C3b (C3bi). See **complement**.

IC$_{50}$ Concentration of an inhibitor at which 50% inhibition of the response is seen; should only be used of *in vitro* test systems. Needs to be used with caution because ED$_{50}$ (for example) is the concentration that causes an effect in 50% of tests - not necessarily the same thing.

ICAD (1) See **DFF**. (2) A family of staphylococcal intercellular adhesion proteins (in *S. aureus*: poly-beta-1,6-N-acetyl-D-glucosamine synthesis protein IcaD, biofilm polysaccharide intercellular adhesin synthesis protein IcaD, 101aa) multi-pass membrane proteins important for formation of biofilms.

ICAMs Intercellular adhesion molecules. Membrane glycoproteins of the immunoglobulin superfamily that are ligands for beta-2 integrins. ICAM-1 (CD54, 532aa) is the ligand for LFA-1 (CD11a/CD18: β2-integrin) and to a lesser extent Mac-1 (CD11b/CD18) and is expressed on the luminal surface of endothelial cells. ICAM-1 expression is upregulated on endothelium in response to IL1 or TNF treatment and is upregulated on activated T- and B-cells. Not only is it an important ligand for leucocyte adhesion, but for binding of rhinovirus and of *Plasmodium falciparum*-infected erythrocytes. ICAM-2 (CD102, 275aa) is constitutively expressed and is not upregulated by inflammatory cytokines. ICAM-3 (CD50, 547aa) is involved in activation of kinase signalling in the immune response and is a ligand for LFA-1. ICAM-4 (CD242, formerly LW blood group antigen, 271aa) is an erythrocyte-specific ligand for CD11c/CD18. ICAM-5 (telencephalin; 924aa) seems to play a role in neuronal targeting in the developing forebrain.

ICAP See **integrin cytoplasmic domain-associated protein**.

iccosomes Immune-complex coated bodies, formed when antigen is injected into an immune animal, and found in follicular **dendritic cells**. May serve as a reservoir of antigen to maintain B-cell memory.

The Dictionary of Cell and Molecular Biology. DOI: http://dx.doi.org/10.1016/B978-0-12-384931-1.00009-X

ICE **(1)** Integrative and conjugative elements, a diverse group of mobile genetic elements that can carry genes encoding a variety of properties into many bacterial hosts. They are transferred between cells by conjugation but unlike conjugative plasmids, they do not replicate autonomously and become integrated into the host chromosome. They share a common core set of genes that are required for integration, excision, transfer and regulation but can carry a variety of other genes such as those conferring antibiotic resistance or heavy metal resistances Article: http://genomebiology.com/2009/10/6/R65. **(2)** See **interleukin 1 converting enzyme**.

ice nucleation proteins Outer membrane proteins produced by some Gram-negative bacteria that promote the nucleation of ice, apparently by aligning water molecules along repeated domains of 48aa. Consist of 16aa repeats containing the conserved octamer AGYGSTxT. Bacteria use them to nucleate ice and cause freezing damage to various plants with which they are associated, thus releasing nutrients for growth. Now finding commercial use in snow-making at ski resorts.

Icelandic-type cerebroarterial amyloidosis An autosomal dominant form of **cerebral amyloid angiopathy** caused by mutation in the gene encoding **cystatin** C.

I-cell disease A human disease (mucolipidosis II) in which the lysosomes lack hydrolases but high concentrations of these enzymes are found in extracellular fluids. Caused by mutation in the gene for **stealth protein** GNPTAB (GlcNAc-phosphotransferase, EC 2.7.8.17, 1256aa) that catalyzes the initial step in the addition of the lysosome recognition marker (mannose-6-phosphate) to lysosomal hydrolases so that they are not directed into the lysosomes but are released.

ICF syndrome Immunodeficiency, centromeric instability and facial anomalies syndrome, a rare autosomal recessive disease caused by mutations in the DNA methyltransferase gene DNMT3B (EC 2.1.1.37, 853aa).

ichnoviruses See **polydnaviruses**.

ichthyin See **NIPA1 gene**.

ichthyosiform erythroderma Skin disorders in which there is abnormal proliferation and hyperkeratinisation, often accompanied by fragility and a tendency for blistering and infection. Bullous ichthyosiform erythroderma (epidermolytic hyperkeratosis) is caused by point mutations in keratin genes (KRT1 and KRT10). Nonbullous autosomal recessive ichthyoses can be caused by mutation in the transglutaminase-1 gene (which also causes lamellar ichthyosis type 1), the 12R-lipoxygenase gene, the lipoxygenase-3 gene and a rare form,

Chanarin-Dorfman syndrome, by mutation in the ABHD5 gene that encodes a lysophosphatidic acid acyltransferase (349aa).

ichthyosis Skin disorders in which there is abnormal scaling. Ichthyosis vulgaris is caused by mutation in the gene for **filaggrin**. Lamellar ichthyosis type 1, an autosomal recessive disorder, is caused by mutation in the gene for keratinocyte transglutaminase. Lamellar ichthyosis type 2 and congenital ichthyosis with harlequin fetus (harlequin ichthyosis) are caused by mutations in the gene for an ABC-transporter, ABCA12. Other types of lamellar ichthyosis are a result of mutations in genes for cytochrome P450 (CYP4F22) or at a locus on chromosome 17. X-linked ichthyosis (XLI), is caused by mutation in **arylsulphatase** C or **multiple sulphatase deficiency**. Ichthyosis with leukocyte vacuoles, alopecia and sclerosing cholangitis is caused by mutation in the gene for **claudin-1**. See **ichthyosiform erythroderma** and **NIPA1**.

icilin A synthetic **TRP channel** super-agonist (331Da) that is nearly 200-fold more potent than menthol. Icilin induces sensations of intense cold when applied orally in humans.

ICK1 Inhibitor (191aa) of cyclin-dependent kinase identified in *Arabidopsis*. Has some limited similarity with mammalian p27Kip1 kinase inhibitor. Inhibits plant cdc2 kinase but not human p34cdc2.

ICM **(1)** Inner cell mass of the mammalian blastocyst. **(2)** Intermediate cell mass (ICM), a major site of zebrafish primitive haematopoiesis. **(3)** Ischemic cardiomyopathy.

IcsA Actin nucleating protein (outer membrane protein IcsA autotransporter, 1102aa) in the outer membrane of virulent strains of *Shigella flexneri*. Like **ActA**, IcsA is responsible for unipolar assembly of an F-actin bundle that pushes the bacterium through the cytoplasm. Activation of the CDC42 effector N-WASP by the IcsA protein promotes actin nucleation by **Arp2/3** complex. The precursor is cleaved into the outer membrane protein IcsA (706aa) and outer membrane protein IcsA translocator (344aa).

ICSH Interstitial cell stimulating hormone. Name given to luteinizing hormone in males, where it stimulates the production of testosterone by Leydig cells.

ICSI Intracytoplasmic sperm injection, a method used for *in vitro* fertilization.

ICTVdB The Universal Virus Database of the International Committee on Taxonomy of Viruses. See **virus taxonomy**. Link to database: http://www.ncbi.nlm.nih.gov/ICTVdb/canintro1.htm

ID$_{50}$ The dose of an infectious organism required to produce infection in 50% of subjects.

idazoxan An imidazoline alpha-2-adrenoreceptor antagonist, used as an experimental tool.

IDDM Insulin-dependent diabetes melitus. See **diabetes**.

idioblast See **sclereid**. Literally, a unique cell, a clearly distinct, specialised and/or differentiated cell embedded within a tissue of less specialised cells.

ideogram *karyogram* Picture or diagram of the chromosome complement of a cell that is arranged to show the general morphology including relative sizes, positions of centromeres and so on. Often prepared as a photo-montage.

idiomorphs Alleles, responsible for defining mating types in self-incompatible ascomycetes, that have marked sequence dissimilarities. Although in any given species one idiomorph is distinct from the other, each individual idiomorph is highly conserved in DNA sequence within families and at the level of protein function among families.

idiopathic Applied to disease of unknown origin or peculiar to the individual.

idiophase In a growing culture of micro-organisms, the slow-growing or nongrowing productive phase, following the trophophase or growth phase, in which secondary metabolites are produced that play no role in the growth of the microorganisms.

idiotope An antigenic determinant (**epitope**) unique to a single clone of cells and located in the variable region of the immunoglobulin product of that clone or in the T-cell receptor. The idiotope forms part of the antigen-binding site. Any single immunoglobulin may have more than one idiotope. Idiotopes are also associated with the antigen-binding sites of T-cell receptors and are the epitopes to which an anti-**idiotype** antibody or T-cell binds.

idiotype The set of **idiotopes** that characterise the antigen combining site of a particular immunoglobulin. Anti-idiotype antibodies combine with the sequences specific to the binding site, may block immunological reactions, and may resemble the epitope to which the first antibody reacts. See **network theory** #2.

IDL See **lipoproteins**.

I-domain Binding domain of around 200aa in the N-terminal part of a subunits of **integrins**. The I-domain has intrinsic ligand-binding activity that is divalent cation-dependent.

Id proteins Helix-loop-helix proteins (inhibitor of DNA binding proteins, ID1, 155aa; ID2, 134aa etc.) that do not have a basic DNA-binding domain but form heterodimers with other HLH proteins, thereby inhibiting DNA binding. They regulate tissue-specific transcription within several cell lineages and

play a part in cell growth, senescence, differentiation, and angiogenesis.

iduronic acid A uronic acid, derived from the sugar idose, and bearing one terminal carboxyl group. With N-acetyl-galactosamine-4-sulphate, a component of dermatan sulphate. See **iduronidase**.

iduronidase An enzyme (α-1-iduronidase; EC 3.2.1.76, 653aa) that hydrolyses the bonds between iduronic acid and N-acetylgalactosamine-4-sulphate; a lysosomal enzyme absent in **Hurler's disease**.

IEF See **isoelectric focusing**.

iejimalides A group of macrolides (iejimalides A-D), isolated from a tunicate, *Eudistoma*, that inhibit V-ATPase activity, cause growth inhibition in a variety of cancer cell lines at nanomolar concentrations and depolymerize the microfilament network. Abstract: http://www.ncbi.nlm.nih.gov/pubmed/20039309

IFN See **interferons**.

IFs See **initiation factors**; **eukaryotic initiation factors** (eIFs).

IgA Major class of immunoglobulin of external secretions in mammals, also found in serum. In secretions, found as a dimer (400 kDa) joined by a short J-chain and linked to a secretory piece or transport piece. In serum found as a monomer (170 kDa). IgAs are the main means of providing local immunity against infections in the gut or respiratory tract and may act by reducing the binding between an IgA-coated microorganism and a host epithelial cell. Present in human colostrum but not transferred across the placenta. Have α heavy chains.

IgD An immunoglobulin (184 kDa) present at a low level in plasma ($3-400$ µg/ml) but is a major immunoglobulin on the surface of B-cells where it may play a role in antigen recognition. Its structure resembles that of IgG but the heavy chains are of the δ type.

IgE Class of immunoglobulin (188 kDa) associated with immediate-type **hypersensitivity** reactions and helminth infections. Present in very low amounts in serum and mostly bound to mast cells and basophils that have an IgE-specific Fc-receptor (FcϵR). IgE has a high carbohydrate content and is also present in external secretions. Heavy chain of ϵ-type.

IGF See insulin-like growth factor.

IgG The classical immunoglobulin class also known as 7S IgG (150 kDa). Composed of two identical light and two identical heavy chains, the constant region sequence of the heavy chains being of the γ type. The molecule can be described in another way as being composed of two **Fab** and an **Fc** fragment. The Fabs include the antigen combining sites; the

Fc region consists of the remaining constant sequence domains of the heavy chains and contains cell-binding and complement-binding sites. IgGs act on pathogens by agglutinating them, by opsonizing them, by activating complement-mediated reactions against cellular pathogens and by neutralising toxins. In some mammals, including man, they can pass across the placenta to the fetus as maternal antibodies, unlike other Ig classes. In humans four main subclasses are known, IgG2 differs from the rest in not being transferred across the placenta and IgG4 does not fix complement. IgG is present at 8−16 mg/ml in serum.

IGIF Interferon-inducing factor, an obsolete name for **IL-18**.

IgLON A family of GPI-anchored cell adhesion molecules of the immunoglobulin superfamily involved in nueronal guidance. The original members were **LAMP** (limbic system-associated membrane protein, IgLON3, 338aa), **OBCAM** (opioid-binding cell adhesion molecule, IgLON1, 345aa) and **neurotrimin** (IgLON2, 338aa). Another member is NEGR1 (IgLON4, 354aa).

IgM An IgM molecule (macroglobulin, 970 kDa) is built up from five IgG type monomers joined together, with the assistance of J chains, to form a cyclic pentamer. IgM binds complement and a single IgM molecule bound to a cell surface can lyse that cell. IgM is usually produced first in an immune response before IgG. The human red cell isoantibodies are IgM antibodies. Heavy chain (μ chain) is rather larger than the heavy chains of other immunoglobulins.

IGS (1) Intergenic spacer, most commonly applied to the region between ribosomal genes in eukaryotes and prokaryotes. (2) In *Petunia*, isoeugenol synthase 1 (323aa) that catalyzes the synthesis of the phenylpropene, isoeugenol, an aromatic constituent of spices.

IgSF See **immunoglobulin superfamily**.

IgX An immunoglobulin class found in Amphibia.

IHC Immunohistochemistry. See **immunocytochemistry**.

IHF See **integration host factor**.

Ihh See **Indian hedgehog**.

IK See **immunoconglutinin**.

IkappaB *IκB* Endogenous inhibitors of **NFκB** that act by binding to the p65 subunit and preventing it moving into the nucleus. (IκBα, 317aa; IκBβ, 356aa; IκBγ (NEMO, NFκB essential modulator), 419aa; IκBε, 500aa). IκB-kinases (**IKK**) phosphorylate IκBs, targeting them for destruction.

IKK Serine/threonine kinases that phosphorylate the NFκB inhibitor **IkappaB**. The IκB-kinase (IKK)

core complex consists of CHUK (IκB kinase-1, conserved helix-loop-helix ubiquitous kinase, EC 2.7.11.10, 745aa), IκBKB (756aa) and IκBKG (419aa); probably four alpha/CHUK-beta/IκBKB dimers associated with four regulatory gamma/IκBKG subunits. The IKK core complex seems to associate with regulatory or adapter proteins to form a IKK-signalosome holocomplex. X-linked anhidrotic ectodermal dysplasia with immunodeficiency is caused by mutations in the gene encoding IKK-gamma.

IL-1, IL-2, etc See **interleukins** and individual interleukin-n entries.

IL-2 inducible T-cell kinase Tec-family protein tyrosine kinase (Itk, Lyk, 620aa) involved in T-cell activation; interacts with SLP-76 (SH2 domain-containing leukocyte protein of 76 kD) adaptor protein and **THEMIS**. Defects in ITK cause a rare immunodeficiency characterized by extreme susceptibility to infection with Epstein-Barr virus. Abstract: http://stke.sciencemag.org/cgi/content/abstract/stke.3962007pe39

IL-6 cytokine family Family of cytokines that all act through receptors sharing a common gp130 subunit (918aa). Includes **interleukin-6, interleukin-11, ciliary neurotrophic factor, LIF, oncostatin M** and **cardiotrophin-1**.

ILEI A cytokine (interleukin-like EMT inducer, FAM3C, 227aa) that promotes epithelial to mesenchymal transition (EMT), tumour formation, and late events in metastasis in epithelial cells.

ILK See **integrin-linked kinase**.

ImageJ A Java-based image processing program, freely available as open-source. Link: http://rsb.info.nih.gov/ij/docs/index.html

imaginal disc Epithelial infoldings in the larvae of holometabolous insects (e.g. Lepidoptera, Diptera) that rapidly develop into adult appendages (legs, antennae, wings etc.) during metamorphosis from larval to adult form. By implanting discs into the haemocoele of an adult insect their differentiation can be blocked, though their **determination** remains unchanged although occasionally **transdetermination** occurs. The hierarchy of transdetermination has been studied in great detail in *Drosophila*.

Imd pathway The immune deficiency pathway, an innate immune response system in invertebrates, homologous to the vertebrate tumour necrosis factor receptor (TNFR) signalling pathways (the other main system involves **toll-like receptors** which are found in both vertebrates and invertebrates). The Imd pathway is usually activated by peptidoglycan recognition proteins that bind diaminopimelic acid (DAP)-type peptidoglycan present in the cell wall of Gram-negative bacteria and many Gram-positive bacteria and downstream involves the **JNK** signalling cascade and a NFκB transcription factor, Relish (Rel-p110).

IME2 Meiosis induction protein kinase (IME2/SME1, EC 2.7.11.1, 645aa) essential for the initiation of meiosis and sporulation in *S. cerevisiae*. See **RIMs.**

imidazoles A group of chemically-related heterocyclic compounds (glyoxalines), active against fungi, a range of bacteria, and in some cases as anthelminthics. Examples are ketoconazole, tiabdazole.

imidazoline Heterocyclic compound such as agmatine and idazoxan, derived from imidazole. Imidazoline receptors have been controversial but are thought to have a variety of physiological functions and two main sub-types, I1 and I2, are recognised: The I1 receptor (nischarin, 1504aa) mediates the sympatho-inhibitory actions, for example lowering blood pressure, via diacylglycerol and arachidonic acid. The I2 receptors bind monoamine oxidase. The I3 receptor is found on pancreatic beta cells but its signalling mechanism is unknown.

imine Any organic compound with one or more imino groups.

imino acid Organic acid containing an imino group in place of two hydrogen atoms. Also applied to cyclic alkylamino derivatives of aliphatic carboxylic acids, such as proline although these should properly be referred to as azacycloalkane carboxylic acids.

iminoglycinuria A defect in amino acid transport leading to abnormal excretion of glycine, proline and hydroxyproline in the urine: more seriously, absorption in the intestine may be inadequate. See **Hartnup disease.**

immediate early gene Class of genes (IEGs) whose expression is low or undetectable in quiescent cells, but whose transcription is activated within minutes after extracellular stimulation such as addition of a **growth factor**. c-*fos* and c-*myc* protooncogenes were among the first IEGs to be identified. Many IEGs encode transcription factors and therefore have a regulatory function.

immortalization Escape from the normal limitation on growth of a finite number of division cycles (the Hayflick limit), by variants in animal cell cultures, and cells in some tumours. Immortalization in culture may be spontaneous, as happens particularly readily in mouse cells, or induced by mutagens or by **transfection** of certain oncogenes.

immotile cilia syndrome See **primary ciliary dyskinesia.**

immune complex Multimolecular antibody-antigen complexes that may be soluble or insoluble depending upon their size and whether or not complement is present. Immune complexes can be filtered from plasma in the kidney, and the deposition of the complexes gives rise to **glomerulonephritis** probably because of the trapping of neutrophils via their Fc receptors. The presence of immune complexes in body fluids is a feature of **Arthus** type hypersensitivity and serum sickness for example.

immune response Alteration in the reactivity of an organism's immune system in response to an antigen; in vertebrates, this may involve antibody production, induction of cell-mediated immunity, complement activation or development of immunological tolerance.

immune system In mammals the immune system has two major elements, an innate non-specific mechanism involving phagocytosis of foreign objects, including pathogens, or inhibition of attachment to mucosal surfaces and a superimposed highly specific system with **immunological memory** that involves recognition of **epitopes** expressed by pathogens (the humoral immune system involving **antibody**) or on the surface of infected or altered cells (**cell mediated immunity**). There are also recognition mechanisms based on the **toll-like receptor system** and on carbohydrate recognition (**collectins, C-reactive protein, ficolins, mannan-binding lectin**) and the anti-viral **interferon** system (see **oligoadenylate synthetases, RIG-1**). In invertebrates there are two major systems, one involving toll-like receptors as in vertebrates, the other the **Imd system**. Mutation affecting any of the components is likely to be deleterious since the current status of the immune system derives from a long history of competition between pathogens and reluctant hosts.

immunity A state in which the body responds specifically to antigen and/or in which a protective response is mounted against a pathogenic agent. May be innate or may be induced by infection or vaccination, or by the passive transfer of antibodies or of immunocompetent cells. Use of the term should not be restricted to vertebrates and invertebrate immune responses, though less sophisticated, are important for resisting pathogens. Plants also have a range of defence mechanisms against pathogens that could be considered immune systems.

immunization The induction of immunity by deliberate eliciting a primary immune response to antigens associated with disease, while avoiding the disease itself. Classically done by exposure to live attenuated bacteria or virus (vaccination) or purified products (such as bacterial toxins). DNA vaccines are designed to induce the production of disease-specific antigens and are being developed for some diseases for which it has been difficult to produce vaccines.

immunoadsorbent Any insoluble material, e.g. cellulose, with either an antigen or an antibody bound to it and that will bind its corresponding anti-body or antigen thus removing it from a solution.

immunoblotting See **blotting**.

immunoconglutinin *IK* Antibodies that react with complement components or their breakdown products. Usually directed against C3b or C4. High levels of IgG immunoconglutinins are found in plasma from patients with systemic lupus erythematosus.

immunocytochemistry Techniques for staining cells or tissues using antibodies against the appropriate antigen. Although in principle the first antibody could be labelled, it is more common (and improves the visualization) to use a second antibody directed against the first (an anti-IgG). This second antibody is conjugated either with fluorochromes, or appropriate enzymes for colorimetric reactions, or gold beads (for electron microscopy), or with the biotin-avidin system, so that the location of the primary antibody, and thus the antigen, can be recognized.

immunodeficiency disease Heterogeneous group of diseases resulting in impaired efficiency of the immune response. The cause may be congenital (genetic/primary immune deficiency) or be acquired as a result of disease e.g. **AIDS**. Genetic defects may be in a single arm of the immune response e.g. lack of a particular complement component, or may affect the whole system e.g. **severe combined immunodeficiency disease** (SCID). See **Bloom's syndrome, cernunnos, feline immunodeficiency virus, Griscelli's syndrome, HIV, HTLV, ICF syndrome, IL-2 inducible T-cell kinase, neutrophil immunodeficiency syndrome, Nijmegen breakage syndrome, Omenn syndrome, reticular dysgenesis, thymic hypoplasia, Wiskott-Aldrich syndrome.**

immunoelectrophoresis Any form of **electrophoresis** in which the molecules separated by electrophoresis are recognized by precipitation with an antibody. Usually done in non-denaturing agarose gels where epitopes are preserved.

immunofluorescence A test or technique in which one or other component of an immunological reaction is made fluorescent by coupling with a **fluorochrome** such as fluorescein, phycoerythrin or rhodamine so that the occurrence of the reaction can be detected as a fluorescing antigen-antibody complex. Used in microscopy to localize small amounts of antigen or specific antibody. A wide variety of fluorochromes allows several different antigens to be simultaneously visualized, often in an aesthetically very attractive manner. In indirect immunofluorescence the first antibody, that is directed against the antigen to be localized, is used unlabelled, and the location of the first antibody is then detected by use of a fluorescently labelled anti-IgG (against IgGs of the species in which the first antibody was raised). The advantage is that there is some amplification,

and a well-characterized goat anti-rabbit IgG antibody can, for example, be used against a scarce specific antibody raised in rabbits. The same technique can be used for ultrastructural localization of the first antibody by substituting peroxidase or gold-labelled second antibody. Useful table of fluorochromes at http://flowcyt.salk.edu/fluo.html and at http://www.sciencegateway.org/resources/fae1.htm

immunogenicity The property of being able to evoke an immune response within an organism. Immunogenicity depends partly upon the size of the substance in question, and partly upon how unlike host molecules it is. Highly conserved proteins tend to have rather low immunogenicity.

immunoglobulin superfamily *IgSF* A large group of proteins with immunoglobulin-like domains. Most are involved with cell surface recognition events. Sequence homology suggests that Igs, MHC molecules, some **cell adhesion** molecules and cytokines receptors share close homology, and thus belong to a **multigene family.**

immunoglobulins See **IgA, IgD, IgE, IgG,** and **IgM**.

immunological memory The systems responsible for the situation where reactions to a second or subsequent exposure to an antigen are more extensive than those seen on first exposure (but see also **immunological tolerance**. The memory is best explained by clonal expansion and persistence of clones of memory cells following the first exposure to antigen. The long-lasting immune memory is humoral and resides in B-lymphocytes, although it appears that persistence of the antigen may be essential. T-cell memory is shorter (see **T-memory cells**).

immunological network The hypothesis advanced by Jerne that the entire specific immune system within an animal is made up of a series of interacting molecules and cell surface receptors, based on the idea that every antibody combining-site carries its own marker antigens or **idiotypes** and that these in turn may be recognized by another set of antibody combining-sites and so on.

immunological surveillance The hypothesis that lymphocyte traffic ensures that all or nearly all parts of the vertebrate body are surveyed by visiting lymphocytes in order to detect any altered-self material, e.g. mutant cells.

immunological synapse The name given to the region where a CD4$^+$ T-cell is in contact with an antigen-presenting cell (APC). The site has T-cell receptors, co-receptors, signalling and adhesion molecules, all stabilized by cytoskeletal and adapter molecules such as ezrin, F-actin, and CD43. Activation of the T-cells requires a few hours, after which the T-cell return to the circulation and the APCs undergo apoptosis.

immunological tolerance Specific unresponsiveness to antigen. Self-tolerance is a process occurring normally early in life due to suppression of self-reactive lymphocyte clones. Tolerance to foreign antigens can be induced in adult life by exposure to antigens under conditions in which specific clones are suppressed. Tolerance is not the same as immunological unresponsiveness, since the latter may be very nonspecific as in immunodeficiency states.

immunophilin Generic term for an intracellular protein that binds immunosuppressive drugs such as **ciclosporin**, **FK 506** or **rapamycin**. Both **cyclophilin** and the receptor for FK506 are peptidyl prolyl *cis-trans* isomerases (rotamases) although the immunosuppressive effect is probably through an interaction with **calcineurin**.

immunoprecipitation The precipitation of a multivalent antigen by a bivalent antibody, resulting in the formation of a large complex. The antibody and antigen must be soluble. Precipitation usually occurs when there is near equivalence between antibody and antigen concentrations.

immunoreceptor tyrosine-based activation motif See ITAM, also ITIM.

immunoregulation The various processes by which antibodies may regulate immune responses. At a simple level, secreted antibody neutralises the antigen with which it reacts thus preventing further antigenic stimulation of the antibody-producing clone. At a more complex level, anti-idiotype antibodies can be shown to develop against the first antibodies in some cases, and perhaps further anti-idiotype antibodies against them. This is the main concept of the **immunological network** theory.

immunostimulatory complexes See ISCOMs.

immunosuppression Reduction in immune responsiveness that occurs when T- and/or B-clones of lymphocytes are depleted in size or suppressed in their reactivity, expansion or differentiation. It may arise from activation of specific or nonspecific T-suppressor lymphocytes of either T- or B-clones, or by drugs that have generalized effects on most or all T- or B-lymphocytes. **Ciclosporin** A and FK506 act on T-cells, as does antilymphocyte serum; alkylating agents such as cyclophosphamide are less specific in their action and damage DNA replication, while base analogues interfering with guanine metabolism act in a similar way. See **immunophilin**.

immunotoxins A toxin conjugated to either an immunoglobulin or to a Fab fragment directed against a specified antigen thereby targeting it to a specific site or cell type.

imperatorin A furanocoumarin biosynthesized from umbelliferone that can be isolated from various plants such as *Opopanax chironium*. It inhibits T cell-receptor-mediated proliferation in human primary T-cells. Abstract: http://www.ncbi.nlm.nih.gov/pubmed/15104479

importins Proteins of the karyopherin family that bind nuclear localization signals (NLS) on proteins destined for the nucleus and that, in conjunction with ran and pp15 are involved in transport. Importins α and β form a heterodimer in the cytoplasm that corresponds to the 'NLS-receptor': both have multiple repeated arm-domains. Importin α binds importin β through a NLS- like region on importin β; importin β binds ran-GTP and also several nucleoporins. Importin dissociates from its cargo once in the nucleus and is then free to recycle to the cytoplasm. Since importins α and β recycle from the nucleus to the cytoplasm at different rates they presumably dissociate at some stage in the cycle. There are multiple importins with different cargo specificities. *Cf.* **exportins**.

imprinting See **genomic imprinting**.

imprinting centre 1 *IC1* An **imprinting control region** that regulates whether the imprinted gene (usually a cluster of genes) is active or inactive. The imprinting centre on chromosome 11 regulates expression of **H19** and **IGF2** alleles, that on chromosome 15 regulates the Prader-Willi/Angelman syndrome region. There are at least two different types of imprinting centre. The first controls higher order chromatin structure, and partitions genes into expressed or repressed domains. The second controls repressive chromatin modifications through a noncoding RNA.

imprinting control regions A chromosomal region that determines whether imprinted genes are expressed or not according to its methylation, which depends upon the parent from which the gene derived. The region is a regulated transcriptional insulator that binds the transcriptional repressor zinc-finger protein CTCF (727aa) which preferentially interacts with unmethylated DNA.

IMPs (1) Insulin-like growth factor 2 mRNA-binding proteins (IMP1, 577aa; IMP2, 599aa; IMP3, 579aa). RNA-binding proteins that regulate mRNA nuclear export, localization, translation and stability. Bind to the 5'-UTR of the insulin-like growth factor 2 (IGF2) mRNAs with different selectivity. Their expression seems to be temporally and spatially regulated during development. (2) Intramembrane protease-3 (IMP3, signal peptide peptidase-like-2A, 520aa), a **presenilin**-like protease. (3) U3 small nucleolar ribonucleoprotein proteins (e.g. IMP3, Interacting with MPP10 protein 3, 184aa) that is involved in 18S ribosomal pre-RNA processing in mammals and yeast. (4) Inositol monophosphatases (e.g. IMP2,

EC 3.1.3.25, 288aa). **(5) Importins** (e.g. Imp4, Ran-binding protein 4, 1081aa).

in ovo In the egg.

in silico Term used (rather jokingly) for experiments done using a computer database (i.e. on a silicon chip). It is now possible to match a small sequence of nucleotides with a full-length gene by running a search on a database; often it is then possible to order the appropriate cDNA in a vector ready for use. Derived by analogy with *in vitro* and *in vivo*.

in situ Literally, in place. Used particularly in the context of *in situ* **hybridization**.

***in situ* hybridization** (1) Technique for revealing patterns of gene expression in a tissue. The tissue is fixed and prepared (usually by sectioning) on a slide, and a labelled DNA or RNA probe is hybridized to the sample. It binds only to complementary mRNA sequences, and so only stains cells that are transcribing the gene in question (it was usually assumed that this meant that the mRNA was then translated, but this assumption now seems less justifiable). (2) Use of the technique for identifying the position of a gene on a chromosome by hybridizing a probe to spread chromosomes. Widely used in *Drosophila* salivary gland polytene chromosome 'squashes', but now also possible with a variety of eukaryotic nuclei. FISH analysis is the use of **fluorescence *in situ* hybridization**.

in vitro Literally, in glass; general term for cells in culture as opposed to in a multicellular organism (***in vivo***). Modern tissue culture techniques rarely use glass vessels, but the term persists.

***in vitro* fertilization** *IVF* Fertilisation of an ovum outside the body (in culture). Usually several eggs are used and the zygotes produced by sperm-egg fusion are allowed to undergo several cell divisions *in vitro* before the embryo of choice is implanted into the uterus. The opportunity therefore exists for pre-implantation genetic diagnosis to identify embryos with abnormal genetic characteristics or select female embryos to avoid X-linked disorders. More recently intracytoplasmic sperm injection (with a micro-pipette) has been used, rather than relying upon normal sperm-egg fusion, especially if the donor's sperm cell count is low.

in vivo Literally, in life; used of cells in their natural multicellular environment or of experiments done on intact organisms rather than on isolated cells in culture (***in vitro***).

inactivation The process by which, for example, voltage gated sodium channels that have been activated (opened) by **depolarization** subsequently close during the depolarization. Distinguished from activation by its slower kinetics.

inbred strain Any strain of animal or plant obtained by a breeding strategy that tends to lead to homozygosity. Such breeding strategies include brother-sister mating and back-crossing of offspring with parents. See also **congenic**.

inclusion bodies Nuclear or cytoplasmic structures with characteristic staining properties, usually found at the site of virus multiplication. Semicrystalline arrays of **virions, capsids,** or other viral components.

inclusion body matrix protein A protein encoded by Caulimovirus (IBMp, transactivator/viroplasmin protein, TAV, 520aa) that enhances the translation of downstream ORFs on polycistronic mRNAs derived from cauliflower mosaic virus.

inclusion body myopathy with Paget's disease and frontotemporal dementia A disorder in which there is lower-body motor neuron degeneration and skeletal disorganization resembling **Paget's disease of bone** caused by mutation in **valosin-containing protein**.

incomplete dominance Condition (co-dominance) in which members of the F1 generation are intermediate in character between the two parents; the heterozygote is phenotypically different from both parents.

incomplete metamorphosis In insects a pattern of development in which there is a more or less gradual change from the immature to the mature state and there is no pupal stage. Usually the young resemble the parents, but lack wings and mature sexual organs.

incretins Hormones, for example glucagon-like peptide-1 and glucose-dependent insulinotropic polypeptide, that stimulate insulin secretion from pancreatic beta-cells. They are produced in the intestine and stimulate the insulin release required after eating.

incubation time The lag time between infection by a pathogen and the first appearance of symptoms.

indels Insertion/deletion points in a DNA sequence. Indel polymorphisms are relatively common.

index case The first or original case of a disease. The term is used in the epidemiology of infectious disease and in genetics it is synonymous with the proband or propositus.

Indian hedgehog One of the **hedgehog family** (Ihh, 411aa), actively involved in endochondral bone formation. Binds to the **patched** receptor, which functions in association with **smoothened**, to activate the transcription of target genes. Post-translationally cleaved into N-terminal (175aa) and C-terminal (209aa) portions: the N-terminal portion remains lipid-anchored to the cell surface, the

C-terminal portion has no signalling activity but is a cholesterol transferase that modifies the N-terminal portion. Brachydactyly type A1 and acrocapitofemoral dysplasia are caused by mutations in the Indian hedgehog gene.

indicator species A species whose presence or absence indicates particular conditions in a habitat or that are particularly susceptible to the effects of some environmental factor. Patterns of gene expression in such species may provide sensitive indicators of low levels of pollutants.

indirect immunofluorescence See **immunofluorescence**.

indirubin-3′-monoxime The active ingredient of Danggui Longhui Wan, a mixture of plants used in traditional Chinese medicine. It is a selective inhibitor of cyclin-dependent kinases (CDKs) and inhibits the proliferation of a large range of cells, mainly through arresting the cells in the G2/M phase of the cell cycle. Powerful inhibitor of GSK-3β (IC$_{50}$ ~100 nM) and tau phosphorylation *in vitro* and *in vivo* at Alzheimer's disease-specific sites. Abstract: http://www.ncbi.nlm.nih.gov/pubmed/10559866

indole acetic acid *IAA* The most common naturally occurring **auxin**. Promotes growth in excised plant organs, induces adventitious roots, inhibits axillary bud growth, regulates gravitropism.

indole-3-butyric acid *IBA* A precursor of auxin but active in its own right. Used to promote rooting.

indoleamine dioxygenase An enzyme (Ido, EC 1.13.11.52, 403aa) which converts tryptophan to N-formyl-kynurenine and is induced in vascular endothelium by proinflammatory cytokines. Implicated in regulation of blood pressure during infection with malaria and in mice with LPS-induced endotoxaemia. Abstract: http://stke.sciencemag.org/cgi/content/abstract/sigtrans;3/113/ec78

indolicidin Bovine **cathelicidin** (13aa).

indometacin *indomethacin* Non-steroidal anti-inflammatory drug that blocks the production of **arachidonic acid** metabolites by inhibiting **cyclo-oxygenase**. Indometacin is the British Approved Name (BAN) for this drug.

indomethacin See **indometacin**.

inducer cells Cells that induce other nearby cells to differentiate in specified pathways. A distinction can, and perhaps should, be made between those cells that evoke a predetermined pathway of differentiation in the target cells and those cells that can actually induce new and unexpected differentiations.

induction See **embryonic induction** or **enzyme induction**.

infantile neuroaxonal dystrophy A neurodegenerative disease caused in many cases by mutation in the gene for **phospholipase A2-G6**.

infantile paralysis Old synonym for **poliomyelitis**.

infarct An area of tissue that has become necrotic as a result of anoxia, usually as a result of thrombotic occlusion of the blood supply (infarction).

infectious hepatitis See **hepatitis A**.

infectious jaundice See *Leptospira*.

infectious mononucleosis See **glandular fever**.

inflammasome A multi-protein complex (>700 kDa) that activates **caspases** 1 and 5 which are involved in the processing and secretion of pro-inflammatory cytokines IL-1β and IL-18. The NALP1 inflammasome is composed of NALP1 (see **NALP proteins**), ASC, caspase 1 and 5. The NALP2/3 inflammasomes contain, in addition to NALP2 or NALP3, the caspase recruitment domain (CARD)-containing protein **cardinal**, ASC and caspase 1. Other forms of inflammasome include one formed as a result of recognition of cytosolic dsDNA by **AIM2** which then forms a caspase-1-activating inflammasome with ASC.

inflammation Response to injury. **Acute inflammation** is dominated by vascular changes and by neutrophil leucocytes in the early stages, mononuclear phagocytes later on. Leucocytes adhere locally and emigrate into the tissue between the endothelial cells lining the post-capillary venules. Plasma exudation from vessels may lead to tissue swelling, but the early vascular changes are independent of and not essential for the later cellular response. In chronic inflammation, where the stimulus is persistent, the characteristic cells are **macrophages** and **lymphocytes**.

inflammatory bowel disease A group of chronic relapsing intestinal inflammatory disorders (IBD) generally subdivided into **Crohn's disease** and **ulcerative colitis**.

influenza virus Member of the **Orthomyxoviridae** that causes influenza in humans. There are three types of influenza virus; Type A causes the world-wide epidemics (pandemics) of influenza and can infect other mammals and birds; Type B only affects humans; Type C causes only a mild infection. Types A and B virus evolve continuously, resulting in changes in the antigenicity of their spike proteins, preventing the development of prolonged immunity to infection. The spike proteins, external haemagglutinin (HA) and **neuraminidase** have been studied as models of membrane glycoproteins. See **H5N1**.

informosome Obsolete name for a cytoplasmic complex of mRNA and non-ribosomal protein.

infusoria A archaic collective term for various minute aquatic organisms found in freshwater pond water, probably some of the first specimens ever observed under a microscope. Strictly speaking the term relates to ciliate protozoa. Usage seems to be current in the (translated) Russian literature.

INGs A family of tumour suppressors (inhibitors of growth) that are functionally similar to **p53**. ING1 (p33ING1, 422aa) is associated with squamous cell carcinomas of the head and neck. ING2 (p32, p33ING2, 280aa) is a component of a mSin3A-like corepressor complex (see **sin3**), which is probably involved in deacetylation of nucleosomal histones and has its activity modulated by phosphoinositide binding. ING3 (p47ING3, 418aa) is part of the **NuA4 histone acetyltransferase complex**. ING4 (p29ING4, 249aa) is part of the HBO1 complex which has a histone H4-specific acetyltransferase activity. This complex may also have ING5 (p28ING5, 240aa) as a component.

inhibin Polypeptide hormone secreted by the hypophysis that selectively suppresses the secretion of pituitary FSH (**follicle-stimulating hormone**). The molecule is a heterodimer (alpha, 366aa and either betaA, 426aa or betaB, 407aa) and is one of the **TGFβ** family. The β subunit is shared with **activin** which opposes many of the effects of inhibin. Inhibin-binding protein (immunoglobulin superfamily member 1, 1336aa) seems to be a co-receptor in inhibin signalling although the high affinity receptor for inhibin is the activin receptor type 2.

inhibition constant K_i The equilibrium dissociation constant for the reaction between enzyme and inhibitor. $K_i = [E]*[I]/[EI]$ where [E], [I], and [EI] are concentrations of enzyme, inhibitor and the complex respectively.

inhibitor of apoptosis proteins Products (IAPs) of a family of evolutionarily conserved genes that inhibit apoptosis, originally identified as baculovirus genes that allow survival of host cell whilst the virus replicates. Several human homologues have subsequently been identified e.g. X-linked **XIAP**. See **BIRCs** (Baculoviral IAP repeat-containing proteins).

inhibitory post-synaptic potential *ipsp* See **post-synaptic potential**.

inhibitory synapse A synapse in which an action potential in the presynaptic cell reduces the probability of an action potential occurring in the postsynaptic cell. The most common inhibitory neurotransmitter is **GABA**; this opens channels in the postsynaptic cell which tend to stabilize its resting potential, thus rendering it less likely to fire. See **excitatory synapse**.

initial cell Actively dividing plant cell in a **meristem**. At each division one daughter cell remains in the meristem as a new initial cell, and the other is added to the growing plant body. Animal equivalent is a stem cell (which would be ambiguous in a plant).

initiation codon The start codon, 5'-AUG in mRNA, at which polypeptide synthesis is started. It is recognized by formylmethionyl-tRNA in bacteria and by methionyl-tRNA in eukaryotes.

initiation complex Complex between mRNA, 30S ribosomal subunit and formyl-methionyl-tRNA or methionyl-tRNA that requires GTP and **initiation factors** to function.

initiation factors The set of catalytic proteins required, in addition to mRNA and ribosomes, for protein synthesis to begin. In bacteria three distinct proteins have been identified: IF-1 (72aa), IF-2 (890aa) and IF-3 (180aa). IFs 1 and 2 enhance the binding of initiator tRNA to the **initiation complex**. In Archaea there are five initiation factors (in *Methanocaldococcus jannaschii*, aIF1, 102aa; aIF2a, 266aa; aIF2b, 143aa; aIF2g, 411aa; aIF6, 228aa). There is a greater diversity of **eukaryotic initiation factors** (eIFs).

INK4 A family of tumour suppressors (inhibitors of CDK4: p15, INK4b, 138aa; p16, INK4a, 156aa; p18, INK4c, 168aa; p19, INK4d, 166aa), that inhibit the D-type cyclin-dependent kinases (CDK4 and CDK6). p16 has been found to be mutated more frequently than p53 in various cancer cells. *Cf.* **CIP/KIP**.

inner cell mass A group of cells in the mammalian blastocyst that give rise to the embryo and are potentially capable of forming all tissues, embryonic and extra-embryonic, except the trophoblast.

inner sheath The material that encases the two central microtubules of the ciliary axoneme.

innexins Highly conserved protein components of **gap junctions** in protostomes like *Drosophila*, *C. elegans*, and *Hirudo*, distantly related to the vertebrate **pannexins** but considered to be part of the same multi-gene family.

inoculum Cells added to start a culture or, in the case of viruses, viruses added to infect a culture of cells. Also a term for biological material injected into an animal to induce immunity (a vaccine).

inosine The 'fifth base' of nucleic acids, hypoxanthine attached to a ribose ring. Important because it fails to form specific pair bonds with the other bases. In **transfer RNAs**, this property is used in the **anticodon** to allow matching of a single tRNA to several codons. **PCR** performed with primers containing inosine tolerates a limited degree of mismatch between primer and template, useful when trying to clone homologous protein by using degenerate primers.

inositol A cyclic hexahydric alcohol with six possible isomers. The biologically active form is myo-inositol.

inositol 3-kinase See **PI3-kinase**.

inositol phosphates Virtually all of the possible phosphorylated states of inositol have been reported to occur in living tissues. The hexaphosphate, **phytic acid**, is abundant in many plant tissues and is a powerful calcium chelator. See **PI3-kinase**.

inositol phosphoglycans A family of putative second messengers of insulin, released outside cells by hydrolysis of membrane-bound glycosylphosphatidylinositols. There are two sub-families, IPG-A and IPG-P; IPG-A inhibits PKA from bovine heart, decreases phosphoenolypyruvate carboxykinase mRNA levels in rat hepatoma cells, and stimulates lipogenesis and inhibits leptin release in rat adipocytes. IPG-P stimulates bovine heart pyruvate dehydrogenase phosphatase.

inositol trisphosphate Inositol 1,4,5 trisphosphate (IP3, InsP3) an important second messenger. It is derived from the membrane phospholipid phosphatidyl inositol bisphosphate by the action of a specific phospholipase C enzyme (PLC-γ). Binds to and activates a calcium channel in the endoplasmic reticulum.

inositol-requiring enzyme 1a A conserved dual endoribonuclease (EC 3.1.26.-) /protein kinase (serine/threonine kinase, EC 2.7.11.1) involved in the endoplasmic reticulum stress response in yeast, flies, and worms. The protein (IRE1alpha, 977aa) has a kinase domain that is activated by trans-autophosphorylation triggered by unfolded proteins in the luman of the ER; the kinase activity is required for activation of the endoribonuclease domain. Once activated the endoribonuclease domain splices **XBP1** mRNA to produce an unfolded-protein response transcriptional activator and triggering growth arrest and apoptosis. Article: http://www.ncbi.nlm.nih.gov/pmc/articles/PMC3006623/?tool = pubmed

inotropic A substance or process that alters the rate of heartbeat. Adrenaline, for example has a positive inotropic effect and increases rate of beating.

insect defensins See **defensins**.

insertin Protein fragment (351aa) derived by proteolysis from **tensin**, found in chicken gizzard smooth muscle. Binds to the barbed ends of actin filaments and apparently allows insertion of further monomers.

insertion sequence Mobile nucleotide sequences (IS elements) that occur naturally in the genomes of bacterial populations. When inserted into bacterial DNA, they inactivate the gene concerned; when they are removed the gene regains its activity. Many IS elements have, however, been shown to activate the expression of neighbouring genes. Closely related to transposons and range in size from a few hundred to a few thousand bases, but are usually less than 1500 bases. All carry genes encoding a transposase, directing insertion, and possess short inverted nucleotide repeats at their ends that target the host DNA. The database for insertion sequences: http://www-is.biotoul.fr/

insertional mutagenesis Generally, mutagenesis of DNA by the insertion of one or more bases but can also cover oncogenesis by insertion of a retrovirus adjacent to a cellular proto-oncogene or a strategy of mutagenesis with **transposons**. In the latter case the progeny of a round of transposition are screened by **PCR**, with transposon- and gene-specific primers, to identify those in which the transposon is close to the gene of interest. As PCR can only produce products up to 1−2 kb, a large fraction of progeny identified as positive by PCR will have a transposon close enough to the gene to inactivate or otherwise alter its pattern of expression.

inside-out patch See **patch clamp**.

inside-out vesicle Small closed vesicles (IOV) surrounded by a bilayer membrane, produced by mechanical disruption of cell membranes, with the cytoplasmic face exposed. Some vesicles are right-side out (ROV).

insig proteins Transmembrane proteins of the endoplasmic reticulum, products of insulin-induced genes insig-1 (277aa) and insig-2 (225aa) required for sterol-mediated inhibition of the proteolytic processing (by **SCAP**) of sterol regulatory element-binding proteins (**SREBPs**) to their active nuclear forms. Thought to play a central role in cholesterol homeostasis.

instructive theory Theory of antibody production, now considered untenable, in which antigen acted as template for the production of specific antibody as opposed to the **clonal selection** theory in which pre-existing variation occurs and appropriate clones are selectively expanded.

insufflation The action of blowing gas, air, vapour or powder into a body cavity, usually the lungs.

insulin A polypeptide hormone (proinsulin, 110aa), highly conserved and found in both vertebrates and invertebrates. Post-translationally cleaved into A (21aa) and B (30aa) chains which remain linked by two disulphide bonds, by removal of intervening C peptide (31aa). Secreted by the **B cells** of the pancreas in response to high blood sugar levels, it induces hypoglycaemia. Defective secretion of insulin is the cause of **diabetes mellitus**. Insulin is also a mitogen, has sequence homologies with other **growth factors**, and is a frequent addition to cell culture media for demanding cell types. The insulin receptor (CD220, 1382aa) is a receptor tyrosine kinase, post-translationally cleaved to alpha and beta subunits which form a tetrameric complex (two alpha chains with ligand-binding activity, two beta chains with kinase activity). Insulin binding triggers association with **IRS1** and activation of a PI3kinase signalling system.

insulin sensitizer Category of drug (thiazolidine-diones (TZDs), e.g rosiglitazone) that enhance insulin action in muscle, fat and other tissues. Require the presence of insulin in order to work.

insulin-induced genes See **insig proteins**.

insulin-like growth factor Polypeptide growth factors (IGF I, somatomedin C, 195aa; IGF II, somato-medin A, 180aa) with considerable sequence similarity to insulin and more potently mitogenic. The IGFII gene is imprinted and only the maternal allele is normally expressed; epigenetic changes cause Silver-Russell syndrome (see **H19 gene**). Increased expression is associated with the **Beckwith-Wiedemann syndrome**. The IGFI receptor (CD221, 1367aa) is a receptor tyrosine kinase similar to the insulin receptor and also binds insulin. IGF2R is also the cation-independent mannose-6-phosphate receptor (CD222, 2491aa) and it is not clear whether it is involved in signalling. See **IMPs #1**.

***int* oncogenes** Oncogenes first identified as targets for insertional activation by the mouse mammary tumour virus (MMTV) in mammary carcinomas. *Int1*, *Int2* and *Int3* are unrelated genes; the similarity in nomenclature is simply that they are targets for MMTV insertion mutation (insertion sites 1, 2 and 3). Int1 is a homologue of *Drosophila* **wingless** and a member of the **Wnt** family (Wnt-1); The *Int2* product is fibroblast growth factor-3; the *Int3* product is **Notch4**.

integral membrane protein A protein that is firmly anchored in a membrane (unlike a peripheral membrane protein). Most is known about the integral proteins of the plasma membrane, where important examples include hormone receptors, ion channels, and transport proteins. An integral protein need not cross the entire membrane; those that do are referred to as transmembrane proteins.

integrase (1) An enzyme of the bacteriophage lambda (λ) that catalyses the integration of phage DNA into the host DNA. (2) Integrases (INTs) in **retrotransposons** (retroelements) are zinc finger endonucleases that make targeted double-stranded breaks in genomic DNA and stimulate homologous recombination and gene targeting. The INT domain is usually, but not invariably, the C-terminal region of the pol polyprotein.

integration Incorporation of the genetic material of a virus into the host genome.

integration host factor *IHF* A heterodimeric DNA-binding protein belonging to the bacterial histone-like protein family that is found in a broad range of prokaryotes (in *E. coli*, IhfA, 99aa; IhfB, 94aa) and that functions in processes that involve high order protein-complexes such as in replication where it binds to oriC, transcriptional regulation where it interacts with RNA polymerase, and site-specific recombination. IHF's primary function is structural: it binds to DNA in a sequence specific manner and introduces a sharp bend ($>160°$) in the DNA that facilitates interaction between components in a nucleoprotein complex. It is also important in the lysogenic life cycle of bacteriophage lambda, as it is required for recombination, which inserts lambda DNA into the *E. coli* chromosome, and for the synthesis of int and cI repressor, phage proteins necessary for DNA insertion and repression, respectively.

integrator complex A multiprotein complex involved in the transcription of small nuclear RNAs (snRNA) U1 and U2 and in their 3′-box-dependent processing. The complex is evolutionarily conserved in metazoans and directly interacts with the C-terminal domain of the RNA polymerase II largest subunit. Components include INTS1 (a nuclear membrane transmembrane protein, 2190aa), INTS2, INTS3, INTS4, INTS5, INTS6, INTS7, INTS8, INTS9/RC74, INTS10, CPSF3L/INTS11 and INTS12. Two of the integrator subunits display similarities to the subunits of the cleavage and polyadenylation specificity factor (**CPSF**) complex.

integrator gene In the Britten & Davidson model for the coordinate expression of unlinked genes in eukaryotes, sensor elements respond to changing conditions by switching on appropriate integrator genes, which then produce transcription factors that activate appropriate subsets of structural genes.

integrin cytoplasmic domain-associated protein-1 Protein (alternatively spliced isoforms ICAP1alpha, 200aa; ICAP1beta, 150aa) that interacts with the conserved NPXY (asn-pro-X-tyr) sequence motif found in the C-terminal region of beta-1 **integrin** and may recruit them to focal contacts during integrin-dependent cell adhesion. It interacts with **KRIT1** in regulating vascular morphogenesis. The activity of ICAP1 is regulated by calcium/calmodulin-dependent protein kinase II (CAMK2).

integrin-linked kinase A multi-functional cytoplasmic serine/threonine kinase (ILK, 452aa) that associates with the cytoplasmic domain of beta integrins and probably mediates inside-out integrin signalling. Implicated in regulating processes such as cell proliferation, survival, migration and invasion.

integrins Superfamily of cell surface proteins that are involved in binding to extracellular matrix components or other cells. Most are heterodimeric with a β subunit of ($\sim$790aa) that is conserved through the superfamily, and a more variable α subunit of $\sim$1100aa. The first examples described were **fibronectin** and **vitronectin** receptors of fibroblasts, which bind to an RGD (Arg-Gly-Asp) sequence in the ligand protein, though the context of the RGD motif seems important and there is also a divalent cation-dependence. Subsequently the platelet IIb/IIIa surface glycoprotein (fibronectin and fibrinogen receptor) and the **LFA-1** class of leucocyte surface protein were recognized as integrins, together with the **VLA proteins**. The requirement for the RGD sequence in the ligand does not seem to be invariable. Signalling is bidirectional; ligand-binding can activate **integrin-linked kinase** and trigger intracellular events (outside-in signalling) and the activity of the ligand-binding site can be modified by the cell thereby altering interactions with the environment (inside-out signalling). See Table I1.

integron Class of prokaryotic DNA element composed of a DNA integrase gene adjacent to a recombination site, at which one or more genes can be found inserted. Frequently, antibiotic resistance genes are found inserted in integron sites in samples of resistant bacteria.

inteins Intervening sequences (introns) in proteins, by analogy with those in mRNA, the polypeptide chain is removed by proteolytic processing and splicing to produce the mature protein. Some intein polypeptides are site-specific endonucleases as well as protein splicing catalysts. Endonuclease action results in insertion of the intein nucleic acid in a target site. About 90% of inteins contain domains whose amino acid sequences are about 34% similar to those of the HO site-specific endonuclease involved in the **cassette mechanism**.

inter-alpha-inhibitor A plasma serine peptidase inhibitor (inter-alpha-(trypsin) inhibitor) that inhibits trypsin, plasmin, and lysosomal granulocytic elastase; assembled from one or two heavy chains (H1, 911aa; H2, 946aa; H3, 890aa) and one light chain, **bikunin**. The heavy chains are encoded by separate

TABLE I1. Vertebrate integrins

Subunits		Ligand	Binding Site	Synonyms
β_1	α_1	Collagens, Laminin		VLA-1
	α_2	Collagens, Laminin	-DGEA-	VLA-2, Platelet glycoprotein Ia/IIa
	α_3	Collagens, Fibronectin, Laminin	-RGD-?	VLA-, chick integrin
	α_4	Fibronectin (alternatively spliced domain), VCAM-1	-EILDV-	VLA-4
	α_5	Fibronectin (RGD)	-RGD-	VLA-5 'Fibronectin receptor', Platelet glycoprotein Ic/IIa
	α_6	Laminin		VLA-6, Platelet glycoprotein Ic'/IIa
	α_7	Laminin		
	α_8	Fibronectin		
	α_9	Tenascin C, osteopontin		
	α_V	Vitronectin, Fibronectin	-RGD-	
β_2	α_L	ICAM-1, ICAM-2		LFA-1, CD11a
	α_M	C3bi, Fibrinogen, Factor X, ICAM-1		Mac-1, Mo-1, CR3, CD11b
	α_X	Fibrinogen, C3bi?	-GPRP-	gp150,95, CR4, CD11c
	α_D	VCAM-1		
β_3	αIIb	Fibronectin, Fibrinogen, Vitronectin, Thrombospondin,	-RGD-	Platelet glycoprotein IIb/IIIa
		von Willebrand Factor	-KQAGDV-	

TABLE I1. (Continued)

Subunits		Ligand	Binding Site	Synonyms
	α_V	Collagen, Fibronectin, Fibrinogen, Osteopontin, Thrombospondin, Vitronectin, von Willebrand Factor	-RGD-	"Vitronectin receptor"
β_4	α_6	Laminin-5 > laminin-1		
β_5	α_V	Vitronectin	-RGD-	
β_6	α_V	Fibronectin	-RGD-	
β_7	α_4	Fibronectin (alternatively spliced domain), VCAM-1	-EILDV-(In fibronectin only)	LPAM-2
	α_{IEL}	Expressed on intraepithelial lymphocytes		Doesn't bind MAdCAM-1, VCAM-1, or fibronectin
β_8	α_V	Vitronectin		

Known permutations of alpha and beta subunits

	beta1	beta2	beta3	beta4	beta5	beta6	beta7	beta8
alpha1	Yes	–	–	–	–	–	–	–
alpha2	Yes	–	–	–	–	–	–	–
alpha3	Yes	–	–	–	–	–	–	–
alpha4	Yes	–	–	–	–	–	Yes	–
alpha5	Yes	–	–	–	–	–	–	–
alpha6	Yes	–	–	Yes	–	–	–	–
alpha7	Yes	–	–	–	–	–	–	–
alpha8	Yes	–	–	–	–	–	–	–
alpha9	Yes	–	–	–	–	–	–	–
alphaD	–	Yes	–	–	–	–	–	–
alphaL	–	Yes	–	–	–	–	–	–
alphaM	–	Yes	–	–	–	–	–	–
alphaV	Yes	–	Yes	–	Yes	Yes	–	Yes
alphaX	–	Yes	–	–	–	–	–	–
alphaIIb	–	–	Yes	–	–	–	–	–
alphaIELb	–	–	–	–	–	–	Yes	–

genes (ITH1-3). The product of a fourth gene (H4, 930aa) has sequence similarities and a polymorphism in this gene is associated with susceptibility to hypercholesterolaemia. Heavy chain H5 (1313aa) may be a tumour suppressor.

interactome The complete set of molecular interactions going on within cells. Unlike the genome or, to a lesser extent the proteome, the interactome has a temporal aspect - not all interactions take place at one time or even in one cell's lifetime.

intercalated disc An electron-dense junctional complex, at the end-to-end contacts of cardiac muscle cells, that contains **gap junctions** and **desmosomes**. Most of the disc is formed of a convoluted type of **adherens junction** into which the actin filaments of the terminal sarcomeres insert (they are therefore equivalent to half Z-bands); desmosomes are also present. The lateral portion of the stepped disc contains gap junctions that couple the cells electrically and thus coordinate the contraction.

intercalation Insertion into a pre-existing structure; e.g. nucleotide sequences into DNA or RNA, alternatively, molecules into structures such as membranes.

intercellular Between cells: can be used either in the sense of connections between cells (as in intercellular junctions), or as an antonym for intracellular.

intercellular adhesion molecule See ICAM.

interdigitating cells Cells found particularly in thymus-dependent regions of lymph nodes; they have dendritic morphology and **accessory cell** function.

interference diffraction patterns The patterns arising from the recombination of beams of light or other waves after they have been split and one set of rays have undergone a phase retardation relative to the other. Such patterns formed by simple objects give information on the correctness of the focus and the presence or absence of optical defects.

interference microscopy Although all image formation depends on interference, the term is generally restricted to systems in which contrast comes from the recombination of a reference beam with light that has been retarded by passing through the object. Because the phase retardation is a consequence of the difference in refractive index between specimen and medium, and because the refractive increment is almost the same for all biological molecules, it is possible to measure the amount of dry mass per unit area of the specimen by measuring the phase retardation. Quantification of the phase retardation is usually done by using a compensator to reduce the bright object to darkness (see **Senarmont** and **Ehringhaus compensators**). Two major optical systems have been used – the **Jamin-Lebedeff** system and the **Mach-Zehnder** system. These instruments are often referred to as interferometers, since they are designed for measuring phase retardation. Although their use has passed out of fashion, it may be that they will be employed more frequently in future in conjunction with image analysing systems.

interference reflection microscopy An optical technique for visualising the topography of the side of a cell in contact with a planar substratum, and for providing information on the separation of the plasmalemma from the substratum. Interference between the reflections from the substratum-medium interface and the reflections from the plasmalemma-medium interface generate the image.

interferon regulatory factors A family of transcription factors that are activated by the binding of interferons to cell surface receptors and are induced by interferon. For example IRF1 (325aa) binds to the upstream regulatory region of type I IFN and IFN-inducible MHC class I genes (the interferon consensus sequence) and activates those genes. IRF9 (ISGF3, 393aa) forms a complex with dimerized phosphorylated STAT1/STAT2, enters the nucleus and binds to the ISRE and activates the transcription of interferon stimulated genes. Virally encoded IRFs, which may interfere with cellular IRFs, have also been identified.

interferon stimulated response element A conserved **response element** (ISRE) present upstream of some, but not all, IFN-alpha/beta-responsive genes. In many cases, the binding site for **interferon regulatory factors**.

interferon-inducing factor See IGIF.

interferons A family of mammalian glycoproteins (IFNs) that mediate non-specific antiviral activity. Type 1 interferons include IFNα (leucocyte interferon: there are many variants, all ~189aa), IFNβ (fibroblast interferon, 187aa), IFNδ, IFNω (195aa) and IFNκ (207aa). Type 2 interferon, IFNγ (166aa) is produced by immune cells after antigen stimulation and is usually classed as a cytokine. There is no homology between types 1 and 2. Type 3 interferon (IFNλ) is a collective term referring to IL28A, IL28B, and IL29. Receptors for Type 1 interferons signal through the JAK/STAT pathway, receptors for IFNγ are type II cytokine receptors; two receptors bind the IFNγ homodimer.

intergenic spacer Non-transcribed sequences that separate 18S-5.8S-28S ribosomal DNA sequences. They are useful as sensitive markers of evolutionary distance and molecular taxonomy because they are not subject to any selection pressure (*cf.* **internal transcribed spacers** which must fold in such a way that nucleases can cut the transcript at the appropriate points).

intergenic suppression The situation where a primary gene and the gene that suppresses it do not lie in the same chromosomal locus. *Cf.* **intragenic suppression**.

interleukin A variety of substances produced by leucocytes (though not necessarily exclusively) and that function during inflammatory responses. (This is the definition recommended by the IUIS-WHO Nomenclature Committee). Interleukins are of the larger class of T-cell products, **lymphokines**. Now more frequently considered as **cytokines**. The pace of new cytokine discovery now seems to have stalled at IL-35 but this may be a temporary lull. Generally speaking the information given for the individual members relate to the human form which is often, but not always, similar to mouse equivalent.

interleukin-1 Cytokines (IL-1α (LAF; MCF), 159aa; IL-1β (IFNβ-inducing factor, OAF, catabolin), 153aa) secreted by **macrophages** or **accessory**

cells, although they lack classical signal sequences, and involved in the activation of both T- and B-lymphocytes in response to antigens or mitogens, as well as affecting a wide range of other cell types. The products of the IL-1α and IL-1β genes both bind to the same receptor (p80, CD121a, 569aa), a member of the immunoglobulin superfamily expressed predominantly on T-cells and cells of mesenchymal origin. There is an endogenous receptor antagonist, IL-1RA (152aa) that binds to the receptor but does not elicit effects. IL-1α, IL-1β and IL-1RA are remarkably different in sequence though similar in binding properties. See also **catabolin**, **endogenous pyrogen**.

interleukin-1 converting enzyme Cytoplasmic cysteine endopeptidase (ICE, caspase-1, EC 3.4.22.36, 404aa) that is uniquely responsible for cleaving proIL-1β (269aa) into mature IL-1β (153aa); the active cytokine is then released by a non-standard mechanism (there is no signal sequence and it does not pass through the Golgi). The enzyme seems to be composed of two non-identical subunits derived from a single proenzyme. ICE has some homology with ced-9 of *C. elegans*, a caspase involved in apoptosis.

interleukin-2 Cytokine (IL-2, T-cell growth factor (TCGF), thymocyte stimulating factor (TSF), 153aa) released by activated T-cells that causes activation, stimulates and sustains growth of other T-cells independently of the antigen. Blocking production or release of IL-2 would block the production of an immune response. The receptor is a non-covalent dimer of an alpha and a beta subunit that has three forms, high affinity heterodimer, an intermediate affinity monomer (beta subunit, CD122, 551aa), and a low affinity monomer (alpha subunit, CD25, 272aa).

interleukin-3 Product (IL-3, HCGF, multipotential colony-stimulating factor, MCGF, multi-SCF, 152aa) of mitogen-activated T-cells: **colony-stimulating factor** for bone-marrow stem cells and mast cells; controls the production, differentiation, and function of granulocytes and the monocyte-macrophage lineage. Species specific — the human form is ineffective in mouse systems and *vice versa*. The receptor is a heterodimer with a ligand-specific alpha chain (IL3Rα, CD123, 378aa) and a beta chain (CD131, 897aa) that is also a subunit of receptors for IL-5 and GMCSF.

interleukin-4 Cytokine (IL-4, B-cell stimulating factor, BSF-1, 153aa) that provides B-cell help, stimulates IgG1 and IgE production, regulates T-helper subsets. An alternatively-spliced variant lacking exon 2 is an antagonist of the complete molecule. Species specific — the human form is ineffective in mouse systems and *vice versa*. The IL-4 receptor alpha subunit (CD124, 825aa) binds IL-4 (or IL-13) and then forms a functional receptor by interacting with the IL-2 receptor subunit gamma. The IL-4/

IL-13 responses are involved in regulating IgE production and, chemokine and mucus production at sites of allergic inflammmation. Soluble IL-4R inhibits IL-4-mediated cell proliferation and IL-5 up-regulation by T-cells.

interleukin-5 A B-cell growth and differentiation factor (IL-5, B-cell differentiation factor, eosinophil differentiation factor, T-cell replacing factor, 134aa) that also stimulates eosinophil precursor proliferation and differentiation. The receptor is a heterodimer of the IL-5-binding alpha subunit (CD125, 420aa) and a beta subunit (897aa) shared with IL-3 and GM-CSF receptors.

interleukin-6 Cytokine (IL-6; BCSF; IFN-β2, 212aa) that is co-induced with interferon from fibroblasts, a B-cell differentiation factor, a hybridoma growth factor, an inducer of acute phase proteins, and a colony stimulating factor acting on mouse bone marrow. The receptor is a heterodimer of a ligand-binding alpha chain (CD126, 468aa) and a signal-transducing component (gp130, CD130) which is shared with receptors for other members of the **IL-6 cytokine family**.

interleukin-7 Single-chain cytokine (IL-7, lymphopoietin 1, 177aa) that is a haematopoietic growth factor that stimulates the proliferation of lymphoid progenitors and important in certain stages of B-cell maturation. Produced by monocytes, T-cells and NK cells. The receptor is a heterodimer of the IL-7-binding alpha subunit (CD127, 459aa) and the common IL-2Rgamma subunit. The receptor is expressed on activated T-cells, although a soluble form is also found. Defects in IL-7R cause autosomal recessive **severe combined immunodeficiency disease** (T⁻/B⁺/NK⁺ SCID) and variations are associated with susceptibility to multiple sclerosis type 3.

interleukin-8 One of the first **chemokines** to be isolated; one of the C-X-C family (IL-8; neutrophil activating protein, NAP-1, CXCL8, 99aa). The precursor is cleaved to form a variety of active peptides, the most common being IL-8$_{(1-77)}$. Secreted by a variety of cells and potently chemokinetic and chemotactic for neutrophils, basophils and T-cell but not monocytes. The receptor (CD181/CD182, 350aa/360aa) is G-protein coupled.

interleukin-9 Cytokine (IL-9, mast cell growth factor (MCGF), 144aa) produced by T-cells, particularly when mitogen stimulated, that stimulates the proliferation of erythroid precursor cells (**BFU-E**). May act synergistically with erythropoietin and together with IL-3 promotes mast cell growth. Supports IL-2 independent and IL-4 independent growth of helper T-cells. Receptor (CD129, 521aa) is a type I cytokine receptor

interleukin-10 Cytokine (IL-10, cytokine synthesis inhibiting factor (CSIF), TGIF, 178aa) produced by

Th2 helper T-cells, some B-cells and LPS-activated monocytes. Regulates cytokine production by a range of other cells and is an inhibitor of immune responses. The IL-10 family of cytokines (IL-10, IL-19, IL-20, IL-22, IL-24, IL-26, IL-28, and IL-29) has a diverse set of activities, even though there are shared receptors.

interleukin-11 Pleiotropic cytokine (IL-11, adipogenesis inhibitory factor, megakaryocyte CSF, 199aa) originally isolated from primate bone marrow **stromal cell** line. Stimulates T-cell-dependent B-cell maturation, megakaryopoiesis, various stages of myeloid differentiation. Receptor shares gp130 subunit with other members of **IL-6 cytokine family**.

interleukin-12 Heterodimeric cytokine (NK stimulatory factor; cytotoxic lymphocyte maturation factor, IL-12/p70, 219aa and 328aa) that enhances the lytic activity of **NK** cells, promotes Th1 responses, induces **interferon-γ** production and stimulates the proliferation of activated T-cells and NK cells. Is secreted by human B-lymphoblastoid cells (NC-37). May play a role in controlling immunoglobulin isotype selection and is known to inhibit IgE production. The receptor is a heterodimer of IL-12RB1 (662aa) and IL-12RB2 (862aa) that signals through the JAK/STAT pathway. IL-12RB1 forms a heterodimer with IL-23R The interleukin-12 family of cytokines includes IL-12, IL-23, IL-27 and IL-35.

interleukin-13 Cytokine (IL-13, NC30, p600, 132aa) with anti-inflammatory activity. Produced by activated T-cells; inhibits IL-6 production by monocytes and also the production of other pro-inflammatory cytokines such as TNFα, IL-1, IL-8. Stimulates B-cells and the production of IL1-RA. Gene is located in cluster of genes on human chromosome 5q that also has IL-4 gene (see **5q minus syndrome**).

interleukin-14 Protein no longer considered a cytokine and renamed alpha-**taxilin**.

interleukin-15 Cytokine (162aa) of the IL-2 family that has effects very similar to IL-2 but in addition potently chemotactic for lymphocytes. Levels are elevated in the rheumatoid joint. Receptor shares β and γ subunits with IL-2 receptor but has unique α-subunit.

interleukin-16 Cytokine (IL-16, lymphocyte chemoattractant factor, LCF, 121aa) cleaved from a much larger inactive pro-IL-16 (1332aa) by caspase-3. Secreted mainly from CD8$^+$ cells and will induce migratory responses in CD4$^+$ cells (lymphocytes, monocytes and eosinophils). May bind to CC-CKR-5 and contribute to the blocking of **HIV** internalisation.

interleukin-17 A family of cytokines, the first of which (IL-17A, cytotoxic T-lymphocyte-associated serine esterase 8, CTLA-8, 155aa) is a pro-inflammatory T-cell product that acts on receptors on a range of cells to activate NFκB. Induces expression of IL-6, IL-8 and ICAM-1 in fibroblasts and enhances T-cell proliferation stimulated by sub-optimal levels of **PHA**. Receptors (IL-17R-A, CD217, 866aa) are Type I transmembrane proteins, though soluble forms are also found. IL-17B, -C, -D, -E and -F all have similar structures but differ in their effects and the receptors (IL17R A-E) to which they bind with differing affinities (e.g. IL-17R-B, 502aa, is the receptor for IL-17B and IL-17E).

interleukin-18 A pro-inflammatory cytokine (interferon gamma inducing factor; IGIF, IL-1γ, 157aa) that has sequence homology with IL-1β and IL-1RA. Augments natural killer cell activity in spleen cells and has a role in angiogenesis. Precursor (193aa) is processed by caspase-1.

interleukin-19 Melanoma differentiation associated protein-like protein (177aa). One of the IL-10 family of cytokines; produced by activated monocytes and B-cells; induces IL-6 and TNF production by monocytes, IL-4, IL-5, IL-10 and IL-13 production by activated T-cells; shares receptor with IL-20 and Il-24.

interleukin-20 Member of IL-10 family of cytokines (176aa); produced by monocytes and keratinocytes; is an autocrine factor for keratinocyte function, differentiation and proliferation. The receptor is a type 2 **cytokine receptor** and the IL-20RA (553aa)/IL-20RB (311aa) dimer is a receptor for IL-19, IL-20 and IL-24. The IL-20RA/IL-10RB dimer is a receptor for IL-26.

interleukin-21 An immunoregulatory IL-2-like cytokine (155aa), produced by activated CD4$^+$ T-cells; signals through a specific IL-21R and the IL-2R gamma-chain. Influences the function of T-cells, NK cells, and B cells. N.B. IL-22 was mistakenly named IL-21 in some early reports.

interleukin-22 A cytokine (IL-10-related T cell-derived inducible factor, IL-TIF, 179aa) produced by activated T-cells. IL-22 is a ligand for CRF2-4, a member of the class II cytokine receptor family that, together with a member of the interferon receptor family (IL-22R), enables IL-22 signalling. Cell lines have been identified that respond to IL-22 by activation of STATs 1, 3, and 5, but were unresponsive to IL-10. Unlike IL-10, IL-22 does not inhibit the production of proinflammatory cytokines by monocytes in response to LPS, but it has modest inhibitory effects on IL-4 production from Th2 T-cells.

interleukin-23 A heterodimer of IL-12p40 and a different subunit, p19 (189aa). The heterodimeric receptor is composed of IL-12RB1 and IL-23R, activates the Jak-STAT signalling cascade, stimulates memory rather than naive T-cells and promotes production of proinflammatory cytokines. IL-23 induces autoimmune inflammation and thus may be responsible for autoimmune inflammatory diseases.

interleukin-24 Cytokine (IL-24, suppression of tumorigenicity 16 protein (ST16), melanoma differentiation-associated gene-7, FISP, 206aa) of the IL-10 family. Involved in megakaryocyte differentiation. Inhibits melanoma cell proliferation. Induces IL-6 and TNF production by monocytes.

interleukin-25 A T-helper-2 (Th2) produced cytokine of the IL-17 family (IL-25, IL-17E, 177aa) that has a role in allergic inflammation. Induces activation of NFκB and stimulates production of IL-8. Upregulates IgG and IgE production, eosinophil levels and inflammatory response. Effects are mediated through induction of IL-4, IL-5 and IL-13.

interleukin-26 Cytokine (IL-26, AK155, 171aa) of the IL-10 family that induces secretion of IL-8 and IL-10. May play a role in local mechanisms of mucosal immunity and seems to have a proinflammatory function. Activates STAT1 and STAT3, MAPK1/3 (ERK1/2), JUN and AKT, induces expression of SOCS3, TNFα and IL-8. The IL-26 receptor complex is highly specific for IL-26 although the individual subunits of the IL-26 receptor complex are components in receptor complexes for other class II cytokines.

interleukin-27 One of the IL-6/IL-12 family of long type I cytokines (IL-27, IL-17D), a heterodimer of EBI3 (EBV-induced gene 3, a 34 kDa glycoprotein related to the p40 subunit of IL-12 and IL-23, 229aa), and p28 (28 kDa glycoprotein related to the IL12 p35, 243aa). Has both anti- and proinflammatory properties and can regulate T helper cell development, suppress T-cell proliferation, stimulate cytotoxic T-cell activity, induce isotype switching in B-cells. Expressed by monocytes, endothelial cells and dendritic cells.

interleukin-28A Ligand (interferon λ2, 200aa) for class II cytokine receptor, distantly related to members of the IL-10 family and type I IFN family. Like interferon I has antiviral activity and up-regulates MHC class I antigen expression. IL-28A, IL-28B (IFNλ-4, 193aa) and IL-29 all signal through the same heterodimeric receptor complex that is composed of the IL-10 receptor β (IL-10 Rβ) and a novel IL-28 receptor β (IFN-IR).

interleukin-28B See **interleukin-28A**.

interleukin-29 Cytokine (IL-29; IFNλ1, 200aa) similar to **interleukin-28A**; signals through the same receptor.

interleukin-30 The p28 subunit of **interleukin-27**: not in the UniProt database and only of historical interest.

interleukin-31 A member of the alpha-helical family of cytokines, (IL-31, 164aa) that activates STAT3 and possibly STAT1 and STAT5 through the IL-31 heterodimeric receptor composed of IL-31RA and **oncostatin M**-receptor. IL-31 may function in skin immunity and is constitutively expressed by keratinocytes.

interleukin-32 An inflammatory cytokine (IL-32, natural killer cells protein 4, TNF-inducing factor, 234aa) that induces TNFα, IL-8 and MIP-2 production. Apparently unrelated to any other cytokines but activates typical cytokine signal pathways. At least four isoforms (α, β, γ, δ) are reported.

interleukin-33 A cytokine (IL-33, nuclear factor from high endothelial venules, NFHEV, interleukin-1 family member 11, 270aa) expressed at high level in high endothelial venules found in tonsils, Peyer's patches and mesenteric lymph nodes. The receptor is IL-1 receptor-like-1 (ST2), activates NFκB and MAP kinases and leads to expression of Th2-associated cytokines.

interleukin-34 A cytokine (242aa) that promotes the differentiation and viability of monocytes and macrophages and the ligand for colony-stimulating factor-1 receptor.

interleukin-35 An inhibitory cytoine of the IL-12 family that may be specifically produced by regulatory T-cells. It is a heterodimer of IL-27beta (encoded by Epstein-Barr-virus-induced gene 3 (Ebi3)) and IL-12alpha (p35). IL-35 will expand $CD4^+CD25^+$ Treg cells, suppress the proliferation of $CD4^+CD25^-$ effector cells and inhibits Th17 cell polarization. Article: http://www.ncbi.nlm.nih.gov/pmc/articles/PMC2631363/?tool = pubmed

interleukin-like EMT inducer Secreted cytokine (ILEI, 227aa) that is the product of the FAM3C gene, one of the FAM3 (FAM3AD) family that have no sequence homology to known genes. ILEI promotes epithelial to mesenchymal transition and altered expression in breast and colonic carcinomas is predictive of metastasis formation. Expression is stimulated by TGFβ.

intermediary cells A type of **companion cell**.

intermediate filaments A class of cytoplasmic filaments of animal cells so named originally because their diameter (nominally 10 nm) in muscle cells was intermediate between thick and thin filaments. Unlike microfilaments and microtubules, the protein subunits of intermediate filaments show considerable diversity and tissue specificity. See **cytokeratins**, **desmin**, **glial fibrillary acidic protein**, **neurofilament** proteins, **nestin** and **vimentin**; see also Table I2.

intermediate-density lipoproteins See **lipoproteins**.

intermedin See **adrenomedullin**.

TABLE I2. Intermediate filaments and sequence-related proteins

Name of protein	Sequence homology group	M_r (kDa)	Cell type
Cytokeratins: epithelial keratins (~20), trichocytic (hair) keratins (~13)	I (acidic)	40–60	Epithelial
	II (neutral-basic)	50–70	
Vimentin	III	53	Many, especially mesenchymal
Desmin	III	52	Muscle
Glial fibrillary acidic protein	III	51	Glial cells. Astrocytes
Peripherin	III	57–58	Co-expressed with neurofilaments in periperal neurons.
Neurofilament polypeptides (L, M & H)	IV	57–150	Neurons (vertebrates)
		60–120	Neurons (invertebrates)
α-Internexin		68	CNS neurons
Synemin (3 alternatively spliced isoforms)		230	Smooth muscle, co-localised with desmin
Syncoilin		53	Striated and cardiac muscle; associates with α-dystrobrevin
Nuclear Lamins	V	60–70	All eukaryotic cells
Nestin	VI	200	Developing rat brain

intermembrane space Region between the two membranes of mitochondria and chloroplasts. On the **endosymbiont hypothesis**, this space would represent the original phagosome.

internal bias Applied to the motile behaviour of crawling cells that, in the short term, show **persistence** and do not behave as true random walkers. Any intrinsic regulation of the random motile behaviour of the cell could be considered as internal bias.

internal membranes General term for intracellular membrane systems such as endoplasmic reticulum. Not particularly helpful, but has the advantage of being non-committal.

internal ribosome entry site A region within a **dicistronic mRNA** that allows ribosome binding and the initiation of translation that is distinct from the normal end-dependent recruitment of ribosomes to the start site. IRESes were originally described in picornaviral mRNAs and there is some controversy as to whether they occur elsewhere or whether the dicistronic mRNA is processed to produce a monocistronic form by splicing. Review: http://genesdev.cshlp.org/content/15/13/1593.full.pdf + html

internal transcribed spacers DNA sequences that separate the eukaryotic 18S, 5.8S and 26S rRNA genes that are found as repeat units arranged in tandem arrays in the nucleolar organizing regions. Each repeat unit consists of a transcribed region (having genes for 18S, 5.8S and 26S rRNAs and the external transcribed spacers i.e. ETS1 and ETS2) and a non-transcribed spacer (**intergenic spacer**, NTS) region. In the transcribed region, internal transcribed spacers (ITS) are found on either side of 5.8S rRNA gene and are described as ITS1 and ITS2. The ITS region is highly conserved intraspecifically but variable between different species so is often used in taxonomy.

internalin Surface proteins (InlA, InlB) that mediate entry of *Listeria monocytogenes* into epithelial cells that express E-cadherin or L-CAM. There appears to be an diverse internalin multigene family (Inl C, D, E, F) although not all the products are involved in bacterial entry into cells. Article: http://www.plospathogens.org/article/info%3Adoi%2F10.1371%2Fjournal.ppat.1000900

International Cucurbit Genomics Initiative A genomics database for melons, cucumbers and watermelons. Link to database: http://www.icugi.org/

International HapMap Project A partnership of scientists and funding agencies from many nations that aims to develop a public resource that will help researchers find genes associated with human disease and response to pharmaceuticals. Homepage: http://hapmap.ncbi.nlm.nih.gov/

International normalized ratio A system established by the World Health Organization (WHO) for reporting the results of blood coagulation (clotting) tests. It is not a test but a mathematical calculation that corrects for the variability in **prothrombin time (PT)** results attributable to the variable sensitivities (**international sensitivity index**, ISI) of the **thromboplastin reagents** used by laboratories.

International sensitivity index Value assigned to each batch of thromboplastin reagent after comparing each batch to a 'working reference' reagent preparation. This 'working reference' has been calibrated against internationally accepted standard reference preparations which have an ISI value of 1. The more sensitive the thromboplastin reagent, the longer the resulting **prothrombin time** (PT); its ISI will be less than 1. See **international normalized ratio**.

interneurons Neurons that connect only with other neurons, and not with either sensory cells or muscles. They are thus involved in the intermediate processing of signals.

internexin Neuronal intermediate filament protein (α-internexin, 499aa). Subunit of Type IV filaments found in neurons of CNS.

interphase The stage of the cell or nucleus when it is not in mitosis, hence comprising most of the cell cycle.

intersectin An adapter protein (SH3 domain-containing protein 1A, 1721aa) that may indirectly link endocytic membrane traffic and the actin assembly machinery. Has two **EH domains**. Interacts with dynamin, CDC42, SNAP25 and SNAP23 and binds clathrin-associated proteins and other components of the endocytic machinery.

interstitial cell stimulating hormone See **ICSH**.

interstitial cells (1) Cells lying between but distinct from other cells in a tissue, a good example being the interstitial cells in *Hydra* that serve as stem cells. (2) Cells lying between the testis tubules of vertebrates and that are responsible for the secretion of testosterone.

intervening sequence Alternative but uncommon name for an **intron**.

intestinal calcium-binding protein Calcium-binding protein (ICaBP, protein S100-G, calbindin-D9k, vitamin D-dependent calcium-binding protein, 79aa) with two **EF-hand** motifs, induced by vitamin D3. See **S100**.

intestinal epithelium The endodermally-derived epithelium of the intestine, but usually implies the absorptive epithelium of small intestine. The apical surfaces of these cells have **microvilli** (which increases the absorptive surface, and probably also provides a larger surface area for enzyme activity). The lateral sub-apical regions have well developed intercellular **junctions**.

intestinal L cells Endocrine cells in the mucosa of distal ileum and colon that sense nutrients and secrete glucagon-like peptides (GLP-1 and GLP-2) and **peptide YY** in response. Also have **toll-like receptors** and are important in immune responsiveness. Distinct from the murine cell line referred to as L cells.

intima Inner layer of blood vessel wall, comprising an endothelial monolayer on the luminal face with a subcellular elastic extracellular matrix containing a few smooth muscle cells. Below the intima is the **media**, then the **adventitia**. The term may be applied to other organs.

intimin Bacterial protein (**adhesin**) located in the outer membrane of enteric Gram-negative pathogens that mediates adhesion between bacterium and mammalian cell. Intimin binds to the **translocated intimin receptor**, a bacterially-produced protein that is transferred to the mammalian host cell. Intimins are important in pathology of attaching and effacing pathogens such as enterohaemorrhagic *E. coli* O157:H7. Intimin has four Ig-like domains and a C-terminal lectin-like domain that is similar to that in **invasin**.

intine Inner layer of the wall of a pollen grain, resembling a **primary cell wall** in structure and composition. Also used for the inner wall layer of a **spore**.

intragenic suppression The situation where a primary gene and the mutated gene that suppresses it lie within the same locus.

intramembranous particles Particles (or complementary pits) seen in freeze fractured membranes. The cleavage plane is through the centre of the bilayer, and the particles are usually assumed to represent **integral membrane proteins** (or polymers of such proteins).

intrinsic factor A glycoprotein (gastric intrinsic factor, transcobalamin III, 417aa) normally secreted by the epithelium of the stomach and that binds vitamin B12; the intrinsic factor/B12 complex is selectively absorbed by the distal ileum, though only the vitamin is taken into the cell.

intrinsic pathway See **tissue factor pathway**.

introgression The process of introduction of genes from one species into the gene pool of another; in conventional plant genetic engineering (plant breeding) by backcrossing an interspecific hybrid with one of its parents. An introgression line (IL) in plant molecular biology is a line of a crop species that contains genetic material derived from a similar species, for example a wild relative. An example of collection of ILs (called IL-Library) is one in which chromosome fragments from *Solanum pennellii* (a wild variety of tomato) introgressed in *Solanum lycopersicum* (the cultivated tomato).

intron A non-coding sequence of DNA within a eukaryotic gene (an intervening sequence, *cf.* **exon**), that is transcribed into hnRNA but is then removed by RNA splicing in the nucleus, leaving a mature mRNA that is then translated in the cytoplasm. Introns are poorly conserved and of variable length, but the regions at the ends are self-complementary, allowing a hairpin structure to form naturally in the hnRNA; this is the cue for removal by RNA splicing. Introns are thought to play an important role in allowing rapid evolution of proteins by **exon shuffling**. Genes may contain as many as 80 introns.

intrusive growth A process in which cell extensions invade between other cells and extend the territory occupied by a cell, although the cell body remains in its original location. In animals the classic example is the extension of axonal and dendritic processes, in plants it is seen in the extension of fibres (as in phloem bast fibres), sclereids, lactiferous tubules and the growth of pollen tubes.

intussusception The insertion of new material into the thickness of an existing cell wall or other structure.

inulin A polysaccharide of variable molecular weight (around 5 kDa), that is a polymer of fructofuranose. Widely used as a marker of extracellular space, an indicator of blood volume in insects (by measuring the dilution of the radiolabel), and in food for diabetics.

invadopodia Actin-rich protrusions, very similar to **podosomes**, that are used by tumour cells to degrade extracellular matrix (ECM) and invade tissues. **Supervillin** reportedly potentiates the function of invadopodia. Invadopodia are said to differ from podosomes in having Nck1 or Grb2.

invariant chain Polypeptide chain of invariant sequence (Ii, CD74, MHC Class II gamma chain, 296aa) that masks the peptide binding groove of MHC class II molecules inside antigen presenting cells until antigen is encountered in the lysosomal compartment. Three invariant chains trimerize and associate with three class II alpha/beta MHC heterodimers; the nine-membered complex is then moved through the ER system, the invariant chain is released and degraded and the antigen-binding groove of the MHC dimer is exposed.

invasins Bacterial proteins that promote bacterial penetration into mammalian cells. The invasin produced by *Yersinia pseudotuberculosis* (985aa) seems to bind to the fibronectin receptor (α_5-β_1 integrin) at a site close to the fibronectin binding site, though the invasin does not have an RGD sequence. Unlike simple adhesins, the invasins trigger cytoskeletal rearrangements (via integrins) and an active process of internalization. See **internalins** and **thrombospondin-related anonymous protein**.

invasion A term that should be used with caution; although most cell biologists would follow Abercrombie in meaning the movement of one cell type into a territory normally occupied by a different cell type (see **invasion index**), some pathologists might not agree.

invasion index An index devised by Abercrombie & Heaysman as a means to estimate the invasiveness of cells *in vitro*. The index is derived from measurements on confronted explants of the cells and embryonic chick heart fibroblasts growing in tissue culture: it is the ratio of the estimated movement, had the cells not been hindered, and the actual movement in the zone in which collision occurs.

inverse agonist *reverse antagonist* Any ligand that binds to receptors and reduces the proportion in the active form. Has the opposite effects to an agonist and may actually reduce the background level of activity. Not the same as a **partial agonist**.

inversin A protein (nephrocystin-2, 1065aa) that interacts with the **anaphase-promoting complex** subunit-2, has a role in primary cilia function and is involved in the cell cycle. Mutations lead to reversal of left-right polarity (**situs inversus**) and cyst formation in the kidneys (nephronophthisis) but is distinct from **Kartagener's syndrome**. See **planar cell polarity pathway**.

inversion heterozygote Individual in which one chromosome contains an inversion whereas the homologous chromosome does not.

invertase (1) Enzyme (sucrase, EC 3.2.1.26, 581aa in *S. pombe*) that catalyses the hydrolysis of sucrose to glucose and fructose, so-called because the sugar solution changes from dextro-rotatory to laevo-rotatory during the course of the reaction. (2) Generally a name for an enzyme that catalyses certain molecular rearrangements. DNA invertases are a class of **resolvase**.

involucrin Marker protein (585aa) for **keratinocyte** differentiation first appearing as a cytoplasmic protein in the upper spinous layer of the **epidermis** but becomes cross-linked by transglutaminase. Together with **trichohyalin** and **loricrin** forms the scaffold for the cell envelope.

involuntary muscle A muscle that is not under conscious control, as for example, the heart.

involution (1) Restoration of the normal size of an organ. (2) Infolding of the edges of a sheet of cells, as in some developmental processes, notably gastrulation.

ion channel A transmembrane pore that provides a hydrophilic channel for ions to cross a lipid bilayer down their electrochemical gradients. Some degree of ion specificity is usually observed, and typically a million ions per second may flow. Channels may be permanently open, like the potassium leak channel; or they may be **voltage-gated**, like the **sodium channel** or **ligand-gated** like the acetylcholine receptor.

ion exchange Interchange of ions of similar charge between a solution and a solid phase (an ion exchange resin).

ion-exchange chromatography Separation of molecules by absorption and desorption from charged polymers. An important technique for protein purification. For small molecules the support is usually polystyrene, but for macromolecules, cellulose, acrylamide or agarose supports give less nonspecific absorption and denaturation. Typical charged residues are CM (carboxymethyl) or DEAE (diethylaminoethyl).

ion-selective electrode An electrode half-cell, with a semi-permeable membrane that is permeable only to a single ion. The electrical potential measured between this and a reference half-cell (e.g. a calomel electrode) is thus the **Nernst potential** for the ion. Given that the solution filling the ion selective electrode is known, the activity (rather than concentration) of the ion in the unknown solution can be measured. Commercial ion-selective electrodes frequently use a hydrophobic membrane containing an **ionophore**, such as **valinomycin** (for potassium) or **monensin** (for sodium). A pH electrode is made with a thin membrane of pH sensitive (i.e. proton permeable) glass.

ionic coupling See **electrical coupling**.

ionizing radiation Radiation capable of ionizing, either directly or indirectly, the substances it passes through. Alpha and β radiation are far more effective at producing ionization (and therefore are more likely to cause tissue or cell damage) than γ radiation or neutrons.

ionomycin Antibiotic (a diacidic polyether) that acts as a potent and selective calcium ionophore; more effective than **A23187**. Ionomycin binds Ca^{2+} in the 7.0–9.5 pH range. Ionomycin induces apoptosis in immature B cell lines, e.g. in Burkitt's lymphoma cells, and in cultured embryonic rat cortical neurons.

ionophore A molecule that allows ions to cross lipid bilayers. There are two classes: carriers and channels. Carriers, like **valinomycin**, form cage-like structures around specific ions and diffuse freely through the hydrophobic regions of the bilayer. Channels, like **gramicidin**, form continuous aqueous pores through the bilayer, allowing ions to diffuse through. See **ion channels** and Table I3.

ionotropic receptors Ligand-gated ion channels involved in fast inhibitory or excitatory neurotransmission. Examples include the **nicotinic acetylcholine receptors** of the neuromuscular junction, some **glutamate receptors** in the CNS and **GABA receptors**. Cf. **metabotropic receptors**.

iontophoresis Movement of ions as a result of an applied electric field. For example the delivery of a charged molecule from the end of a micropipette without hydraulic flow.

IP-10 A cytokine (interferon-inducible protein-10, CXCL10, 10 kDa, 98aa) that binds to CXCR3 and selectively chemoattracts Th1 lymphocytes and monocytes, and inhibits cytokine-stimulated haematopoietic progenitor cell proliferation. Additionally, it is angiostatic and mitogenic for vascular smooth muscle cells

IP3 See **inositol trisphosphate**.

IPG See **inositol phosphoglycans**.

IPNS Non-haem iron-dependent oxidase (isopenicillin N synthase, EC1.21.3.1, 331aa in *Penicillium notatum*) that catalyses the formation of isopenicillin N from alpha-aminoadipylcysteinyl-valine (ACV).

ipomeamarone One of the furanoterpenoids (ipomoeamarone) produced in sweet potato (*Ipomoea batatas*) that has become infected with the black rot fungus, *Ceratocystis fumbriata*. A phytoalexin, hepatotoxic.

iporin A rab1-interacting protein (RUN and SH3 domain-containing protein 2, 1516aa) The RUN domain is present in several proteins that are linked to the functions of GTPases in the Rap and Rab families. Iporin is ubiquitously expressed and immunofluorescence staining displays a punctuate cytosolic distribution. Also interacts with another rab1 interacting partner, the **GM130** protein.

IPS-1 See **MAVS**.

IPSP See **post-synaptic potential**.

ipt oncogene Plant oncogene from *Agrobacterium tumefaciens* that encodes **isopentenyl transferase** under the control of the transcriptional regulator Ros.

IPTG A compound (isopropyl β-D-thiogalactoside) used experimentally to trigger gene expression that

TABLE I3. Ionophores

	M_r (Da)	Ion selectivity	Comments
Neutral:			
CA 1001	685	Ca	Selective for Ca
Cryptate 211	288	Li > Na > K≈Rb≈Cs;Ca > Sr≈Ba	Amino ether; one of a substantial family
Enniatin A	681	K > Rb≈Na > Cs >> Li	Cyclic hexadepsipeptide
Enniatin B	639	Rb > K > Cs > Na >> Li; Ca > Ba > Sr > Mg	Cyclic hexadepsipeptide
Monactin	750	NH4 > K > Rb > Cs > Na > Ba	Macrotetralide
Narasin	765	K > Na	4-methylsalinomycin
Nonactin	736	NH4 > K≈Rb > Cs > Na	Macrotetralide, product of Actinomyces strains.
Salinomycin	773	K > Na	Polyether antibiotic, used as coccidiostat
Valinomycin	1110	Rb > K > Cs > Ag >> NH4 > Na > Li	Depsipeptide, uncoupler
Carboxylic:			
A23187	523	Li > Na > K; Mn > Ca > Mg > Sr > Ba	Predominantly selective for divalent cations
Ionomycin	747	Mn > Ca > Mg >> Sr > Ba	Diacidic polyether
Monensin	670	Na >> K > Rb > Li > Cs	Blocks transport through Golgi
Nigericin	724	K > Rb > Na > Cs >> Li	
X-537A (lasalocid)	590	Cs > Rb≈K > Na > Li; Ba > Sr > Ca > Mg	Macrotetralide
Channel-forming			
Alamethicin	–	K > Rb > Cs > Na	Peptide; voltage dependent
Gramicidin A	≈1700	H > Cs≈Rb > NH4 > K > Na > Li	Peptide; works as a dimer
Monazomycin	≈1422	Cs > Rb > K > Na > Li	Polyene-like; voltage sensitive
Miscellaneous			
Nystatin	926	Cation selective	Activates Na/K ATPase
Palytoxin	2670	Na	Converts Na/K ATPase into channel; highly toxic

Many uncouplers such as FCCP (carbonyl cyanide-trifluoro-methoxyphenylhydrazone) also act as ionophores. Amphotericin, and filipin may be anion-specific ionophores.

is under the control of **gal promoter**, particularly used in expression systems for producing protein.

IQ motif A basic amphiphilic helix motif, usually present in 1−7 copies, that is a crucial determinant for **calmodulin** binding. Found, for example in the C-terminal region of voltage-gated calcium channel subunits, and in myosin molecules as the binding sites for light chains, often calmodulin.

IQGAPs A conserved group of scaffolding proteins (IQ motif-containing GTPase activating proteins, p195; IQGAP1, 1657aa; IQGAP2, 1575aa) typically present as multiple isoforms and that bind to various proteins including actin, catenin, rho family GTPases and calmodulin. Have a **calponin homology domain** in the amino terminal region that confers actin binding capacity. Have been implicated in cytokinesis, cell-cell adhesion, transcription, cytoskeletal architecture, and signalling pathways.

IRAK Part of the kinase cascade (IL-1 receptor associated kinases: IRAK1, 712aa; IRAK2, 625aa; IRAK3, 596aa; IRAK4, 460aa), that mediates **toll-like receptor** (TLR) signalling and eventually leads to NFκB translocation to the nucleus and altered gene expression. Has homology with **pelle**. IRAK associates with the IL-1R once **interleukin-1** (IL-1) binds. Interleukin-1 receptor-associated kinase 1-binding protein 1 (260aa) is part of the IRAK1-dependent TNFRSF1A signalling pathway.

IRE See **iron-responsive element**.

I-region (1) The inducible gene region of the genome of *E. coli* involved in the lactose operon. (2) Region of the murine genome coding the I-A and I-E immune response (Ia) antigens, polymorphic MHC Class II proteins expressed mainly on B-cells and antigen presenting cells.

IRFs See **interferon regulatory factors**.

Ir genes Immune response genes, located within the MHC of vertebrates. Originally recognized as controlling the level of immune response to various synthetic polypeptides, they are now also recognized as mapping within the regions controlling T-cell help and suppression (**I-region**). Encode Ir-associated antigens.

iridocyte A cell of the **tapetum** in the eye that is filled with iridescent crytals of guanine and reflects light.

iridoid A monoterpene secondary metabolite found in a wide variety of plants and in some animals. Often intermediates in the biosynthesis of alkaloids.

iridoid glycosides Constitutively expressed plant metabolites (e.g. aucubin, catalpol) known to affect both herbivores and pathogens. Often intermediates in the biosynthesis of alkaloids.

iridoviruses A genus of dsDNA viruses of insects in the family Iridioviridae; the crystalline array of virus particles in the cytoplasm of epidermal cells gives infected insects an irridescent appearance. Other genera are Chloriridovirus, Lymphocystivirus, Megalocytivirus, and Ranavirus, mostly infecting invertebrates or cold-blooded vertebrates.

IRL-2500 An **endothelin** ET(B) receptor antagonist.

iron regulatory proteins See **iron responsive element binding protein**.

iron-responsive element A translational-control sequence in the 5′UTR of **ferritin** mRNA and the 3′UTR of **transferrin** receptor mRNA recognised by **iron responsive element binding proteins** (IRE-BPs). Important in the regulation of iron metabolism. Mutations in the IRE of L-ferritin mRNA lead to increased plasma ferritin levels and eventually lead to cataract (hyperferritinaemia-cataract syndrome).

iron responsive element binding proteins Proteins that bind to **iron-responsive elements** (IREs), stem-loop structures found, for example, in the 5′-UTR of ferritin. IREBP1 **cytoplasmic aconitate hydratase, aconitase, EC 4.2.1.3,** 889aa) is an iron sensor that acts as an aconitase when iron levels are high and as a mRNA binding protein that regulates uptake, sequestration and utilization of iron when cellular iron levels are low. IREBP2 (Iron regulatory protein 2, 963aa) has only the mRNA binding properties. IRE/IRP interactions stabilize transferrin receptor 1 (TfR1) mRNA against endonucleolytic cleavage, and inhibit translation of ferritin: as a results there is increased uptake of circulating transferrin-bound iron. N.B. Sometimes referred to as iron regulatory proteins (IRPs) but IRP is also used for **Int-1-like proteins**. See **Wnt**.

IRS Multi-site docking proteins (insulin receptor substrate, IRS-1, 1242aa; IRS-2, 1338; IRS-4, 1257aa) that bind to the cytoplasmic region of ligand-occupied receptors such as those for insulin, IGF2, growth hormone, several interleukins (IL-4, IL-9, IL-13), and other cytokines, become tyrosine-phosphorylated and are then bound by the SH2 domains of e.g. p85 of **PI-3-kinase**, **Grb-2** and **PTP-2**. Association with **14-3-3 proteins** may, however, interrupt the association between the insulin receptor and IRS and regulate insulin sensitivity. Mutations in IRS-1 and IRS-2 are associated with increased susceptibility to Type 2 diabetes.

Is element See **insertion sequence**.

ischaemia *US*. ischemia. Inadequate blood flow leading to hypoxia in the tissue.

ISCOMS Small cage-like structures (immunostimulatory complexes) that make it possible to present viral proteins to the immune system in an array, much as they would appear on the virus. Produced by mixing the viral protein with Quill A, a substance isolated from the Amazonian oak (*Quillaia saponaria*), in the presence of detergent. ISCOMS are being used successfully as adjuvants in vaccines.

ISI See **International sensitivity index**.

islet amyloid peptide See **amylin**.

islet cell autoantigen An autoantigen (p69, 483aa) associated with insulin-dependent diabetes mellitus. It binds to the small GTPase rab2 and has an extended BAR domain, that interacts with membrane phospholipids. ICA69 is enriched in the Golgi complex and probably has a role in regulating secretory granule transport and neurotransmitter secretion.

islet cells Cells of the **Islets of Langerhans** within the pancreas. See **A cells, B cells, D cells**.

Islets of Langerhans Groups of cells found within the pancreas: A cells and B cells secrete **insulin** and **glucagon**. See also **D cells**.

iso-agglutination (1) The agglutination of spermatozoa by the action of a substance produced by the ova of the same species. (2) The adhesion of erythrocytes to one another within the same bloodgroup. *Cf.* **hetero-agglutination**.

isoantibody Antibody made in response to antigen from another individual of the same species.

isobavachalcone A **chalcone** from *Angelica keiskei* and *Psoralea corylifolia* that inhibits platelet aggregation, Akt signalling, induction of Epstein-Barr virus early antigen (EBV-EA) and matrix metallopeptidase-2. Has DNA strand-cleaving activity.

isochors Long stretches (>300 kb) of relatively GC- or AT-rich sequences of DNA associated with R and G chromosome bands respectively. Each of the ~3200 isochores in the human genome has homogeneous level of GC but some have <40% GC and are gene poor, others have >52%GC and are gene rich. There is a web-based tool (GC-profile) for visualizing and analyzing the variation of GC content in genomic sequences. Homepage: http://tubic. tju.edu.cn/GC-Profile/

isocitrate An intermediate in the **tricarboxylic acid cycle** (citric acid cycle).

isodesmosine A rare amino acid found in **elastin**, formed by condensation of four molecules of lysine into a pyridinium ring. Isodesmosine and **desmosine** are involved in the intramolecular crosslinks between elastin chains.

isoelectric focusing A high-resolution method for separating molecules, especially proteins, that carry both positive and negative charges, essentially **electrophoresis** in a stabilized pH gradient. Molecules migrate to the pH corresponding to their **isoelectric point**. The gradient is produced by electrophoresis of amphiphiles, heterogenous molecules giving a continuum of isoelectric points. Resolution is determined by the number of amphiphile species and the evenness of distribution of their isoelectric points.

isoelectric point The pH at which a protein carries no net charge. Below the isoelectric point proteins carry a net positive charge; above it a net negative charge. Due to a preponderance of weakly acid residues in almost all proteins, they are nearly all negatively charged at neutral pH. The isoelectric point is of significance in protein purification because it is the pH at which solubility is often minimal, and at which mobility in an **isoelectric focusing** system is zero (and therefore the point at which the protein will accumulate).

isoenzymes Variants of enzymes (isozymes) that catalyse the same reaction, but owing to differences in amino acid sequence can be distinguished by techniques such as electrophoresis or isoelectric focusing. Different tissues often have different isoenzymes. The sequence differences generally confer different enzyme-kinetic parameters that can sometimes be interpreted as fine-tuning to the specific requirements of the cell types in which a particular isoenzyme is found. Strictly speaking allozymes are enzymes from different alleles of the same gene and isozymes are from different genes, but isozyme tends to be used interchangeably for both.

isoflavone A class of plant-derived compounds related to flavonoids and that are **phytoestrogens**. The major dietary source is soy bean in which the major isoflavones are **genistein** and **daidzein**. They are said to have nutritional benefits.

isoform A protein having the same function and similar (or identical sequence), but the product of either of a different gene or of alternative splicing and (usually) tissue-specific. Rather stronger in implication than 'homologous'.

isogamy Situation in which the two gametes (isogametes) that fuse to form the zygote are morphologically identical.

isohaemagglutinins Natural antibodies that react against normal antigens of other members of the same species.

isokont Descriptor for flagella of equal length or identical morphology.

isolecithal Describing eggs that have relatively little yolk, uniformly distributed. *Cf.* **centrolecithal, telolecithal**.

isoleucine Hydrophobic amino acid (Ileu, I, 131 Da). See Table A1.

isoline (1) A line on a map, chart, or graph connecting points of equal value. (2) In plant breeding a set of genetically similar (isogenic) plant lines that carry, for example, different specific genes for resistance to a particular pathogen.

isolog *isologue* Term that has been used to indicate similarity rather than identity between genetic sequences; unclear whether this will become standard usage. *Cf.* **homologue**, paralogue amongst others.

isomastigote A protozoan having two or four flagella of equal length at one pole.

isomers Alternative stereochemical forms of molecules containing the same atoms.

isometric tension Tension generated in a muscle without contraction occurring: cross-bridges are being re-formed with the same site on the thin filament, and the tension (in striated muscle) is proportional to the overlap between thick and thin filaments.

isopentenyl transferase An enzyme (isopentenyl-diphosphate:tRNA isopentenyltransferase, EC 2.5.1.75, 467aa) that transfers a dimethylallyl group onto the adenine at position 37 of both cytosolic and mitochondrial tRNAs. In *Agrobacterium tumefaciens* an enzyme (EC 2.5.1.-; trans-zeatin producing protein, 243aa) involved in biosynthesis of the plant growth hormone **cytokinin**. Overexpression of the isopentenyltransferase gene (ipt) from the Ti-plasmid of *A. tumefaciens* increases cytokinin levels and leads to the generation of shoots from transformed plant cells. See *ipt* **oncogene**.

isopeptide bond A peptide bond other than a **eupeptide bond**, e.g. bond formed between β-carboxyl of aspartic acid or with the ε-amino group of lysine.

isoprenoids Large family of molecules that include **carotenoids**, phytoids, prenols, steroids, **terpenoids** and tocopherols. May form only a portion of a molecule being attached to a nonisoprenoid portion. Isoprenoids are synthesized from the diphosphate of isopentyl alcohol (isopentenyl diphosphate). Isoprenylation (prenylation) is a post-translational modification of some proteins, e.g. the addition of geranyl-geranyl- or farnesyl- moieties, which will facilitate membrane-association.

isoprostanes Class of **prostaglandin**-like compounds produced by peroxidation of lipoproteins; thought to play a causative role in atherogenesis. 8-Isoprostane is considered a marker of oxidative stress.

isopycnic Having equal density: thus in equilibrium density gradient centrifugation a particle (molecule) will cease to move when it reaches a level at which it is isopycnic with the medium.

isosbestic Wavelength at which the absorption coefficients of equimolar solutions of two different substances are identical.

isoschizomers Restriction enzymes that share the same recognition sequences.

isosmotic Having the same osmotic pressure.

isothiocyanate Any of a group of sulphur-containing compounds, some of which are produced by cabbages, and other cruciferous vegetables and act as herbicides or fungicides. Allyl isothiocyanate is also called mustard oil. General formula, $-N{=}C{=}S$.

isotonic (1) Of a fluid, having a concentration that will not cause osmotic volume changes of cells immersed in it. Note that an isotonic solution is not necessarily isosmotic. See **hypotonic**, **hypertonic**. (2) When applied to muscle contractions, a contraction in which the tension remains constant. Since the contractile force is proportional to the overlap of the filaments, and the overlap is varying, the numbers of active cross-bridges must be changing.

isotropic Describing an environment in which there are no vectorial or axial cues and the properties are the same at all points.

isotype (1) Applied to a set of macromolecules sharing some features in common. In immunology

isotype describes the class, subclass, light chain type and subtype of an immunoglobulin. See **isotype switching**. (2) Antigenic determinant that is uniquely present in individuals of a single species. (3) A conventionalised method for the graphical display of statistical data.

isotype switching The switch of immunoglobulin isotype that occurs, for example, when the immune response progresses (IgM to IgG). The switch from IgM to IgG involves only the constant region of the heavy chains (from μ to γ), the light chain and variable regions of the heavy chain remaining the same, and involves the switch regions, upstream (on the 5′ side) of the constant region genes, at which recombination occurs. Similarly, IgM and IgD with the same variable region of the heavy chain, but with different heavy chain constant regions (μ and δ), seem to coexist on the surface of some lymphocytes.

isotypic variation Variability of antigens common to all members of a species, for example the five classes of immunoglobulins found in humans. See **idiotype** and **allotype**.

isozyme See **isoenzyme**.

ISRE See **interferon stimulated response element**.

ISSR The region (inter-simple sequence repeat) of the genome that lies between **microsatellite** loci.

IstR Antitoxin RNAs (inhibitor of SOS-induced toxicity by RNA, IstR-1, 75nt; IstR-2, 140nt), a non-coding RNA first identified in *E. coli* where it counters the toxic protein TisB. IstR-1 is expressed throughout growth, whereas istR-2 is induced by mitomycin C. IstR-1 binds tisAB mRNA and promotes RNase III-dependent cleavage, thereby inactivating the mRNA for translation. SOS induction also leads to depletion of the IstR-1 pool, concomitant with accumulation of tisAB mRNA. Under such conditions, TisB exerts its toxic effect, slowing down growth. Article: http://www.mendeley.com/research/small-rna-istr-inhibits-synthesis-sosinduced-toxic-peptide/

I-TAC A CXC chemokine (CXCL11, interferon-inducible T-cell alpha chemoattractant, small inducible cytokine B11, IP9, interferon-gamma-inducible protein 9, 94aa) that is induced by IFNγ and IFNβ. Induction by IFNγ is enhanced by TNFα in monocytes, dermal fibroblasts and endothelial cells, and by IL-1 in astrocytes. I-TAC has potent chemoattractant activity for interleukin-2 activated T-cells. May be important in CNS diseases that involve T-cell recruitment and mRNA is expressed by basal layer keratinocytes in a variety of skin disorders. Binds to CXCR3 and induces calcium release in activated T-cells.

ITAM A motif of four amino acids (immunoreceptor tyrosine-based activation motif, YxxL) duplicated in the cytoplasmic domain of immunoglobulin (Ig) superfamily molecules. The tyrosines become phosphorylated following receptor occupancy and can then bind **zap-70** and **Syk**, initiating T-cell activation. ITAMs are found in T- and B-cell receptors, Fc receptors, NFAM1 (NFAT activating protein with ITAM motif-1) and **dectin-1**. *Cf.* **ITIMs**.

ITIM A short motif (immunoreceptor tyrosine-based inhibitory motif) with a tyrosine residue that can be phosphorylated by src-family kinases and then recruits, via SH2 domain interactions, phosphatases such as **shp-1** that inhibit immune responses. ITIMs are found in killer inhibitory receptors, (**KIRs**) and CD22.

ITK See **IL-2 inducible T-cell kinase**.

Ito cells Hepatic stellate cells that become activated in liver fibrosis due to intoxication or hepatotoxic compounds such as carbon tetrachloride. Activation is associated with expression of a sodium/calcium exchanger.

ITS See **internal transcribed spacers**.

iturins Lipopeptide antibiotics from *Bacillus amyloliquefaciens* and *Bacillus subtilis*. The iturin family that includes iturin A, mycosubtilin, and bacillomycin, are heptapeptides with a β-amino fatty acids and have strong antifungal activity. Bacteria producing these antibiotics have useful plant-protective effects. Article: http://apsjournals.apsnet.org/doi/pdf/10.1094/MPMI-20-4-0430

IUPAC The International Union of Pure and Applied Chemistry.

IVET A promoter-trapping technique (*in vivo* expression technology) extensively used in functional genomics. It was the first practical strategy described for selecting bacterial genes expressed preferentially during infection of an animal host; genes are detected because they are highly expressed in host tissues or cell culture infection models, but poorly expressed on laboratory media.

IVF See *in vitro* **fertilization**.

IVS Standard nomenclature for description of sequence variations in intronic nucleotides; the number of the preceding exon, a plus sign, and the position in the intron (intervening sequence), e.g. IVS2 + 1G > T denotes the G to T substitution at nucleotide +1 of intron 2.

J

J (1) The joule, SI unit of energy. (2) Used in the single letter code for amino acids to represent trimethyl lysine, e.g. in calmodulin.

J774.2 cells Mouse (Balb/c) monocyte/macrophage cells with surface receptors for IgG and complement.

JAB (1) Cytokine-inducible inhibitor of **JAKs**, JAK-binding protein. See **SOCS**. (2) Jun activation domain-binding protein-1 (JAB1, COP9 signalosome complex subunit 5, 334aa) the protease subunit of the **COP9 signalosome** complex.

jacalin A D-galactose-specific lectin from *Artocarpus integer* (Jack fruit, *Artocarpus integrifolia*) that binds the T-antigen (Thomsen-Friedenreich-antigen) and is a potent and selective stimulant of distinct T- and B-cell functions. Shows a unique ability to specifically recognize IgA-1 from human serum. It is a tetramer of four alpha chains (133aa) associated with two or four beta chains (20aa).

JACKDAW A plant zinc finger protein (503aa) that regulates tissue boundaries and asymmetric cell division by delimiting SHORT-ROOT movement. Rapidly up-regulates SCARECROW. Counteracted by **MAGPIE**. Article: http://dev.biologists.org/content/137/9/1523

Jacobson's organ See **vomeronasal organ**.

Jaeken's syndrome See **carbohydrate deficient-glycoprotein syndrome**.

jagged-1 Single pass membrane proteins that are ligands of **Notch** receptors (jagged-1, CD339, 1218aa; jagged-2, 1238aa) originally identified in rat. In humans mutation in jagged-1 is associated with **tetralogy of Fallot** and with **Alagille syndrome-1**.

JAK Family of intracellular tyrosine kinases (Janus kinases or, possibly, 'just another kinase') that associate with cytokine receptors (particularly but not exclusively interferon receptors) and are involved in the JAK-STAT signalling cascade. Lack SH2 and SH3 domains. JAK1 (EC 2.7.10.2, 1154aa) has a FERM domain that mediates interaction with JAKMIP1 (see below). JAK2 (1132aa) is involved in downstream signalling from cytokines and from cytokine-family receptors for growth hormone, prolactin, leptin, erythropoietin, granulocyte-macrophage colony-stimulating factor and thrombopoietin. Chromosomal aberrations involving JAK2 are found in various forms of leukemia; mutations are associated with polycythemia vera, essential thrombocythemia, myelofibrosis and acute myelogenous leukemia. JAK3 (124aa) is involved in the interleukin-2 and interleukin-4 signalling pathways

and is mutated in $T^-B^+NK^-$ **severe combined immunodeficiency disease**. JAKMIP1 (Janus kinase and microtubule-interacting protein 1, 626aa) may play a role in the microtubule-dependent transport of $GABA_B$ receptors: other JAKMIPs are known.

jamaicamides Compounds (jamaicamides A–C) from the marine cyanobacterium, *Lyngbya majuscula*. Jamaicamide A is a lipopeptide with potent sodium channel blocking activity. The biosynthesis of these compounds involves a gene cluster, that for jamaicamides being 57 kbp in length, with 17 ORFs that encode for proteins ranging in length from 80 to 3936aa that are arranged in order of their utilisation in the synthesis. Article: http://www.biomedcentral.com/1471-2180/9/247

Jamin-Lebedeff system Interference microscopy in which object and reference beams are split and later recombined by birefringent calcite plates, but pass through the same optical components (in contrast to the **Mach-Zehnder system**).

JAMM motif A conserved motif in metallopeptidases (JAB1/MPN/Mov34 metalloenzyme motif) originally identified in the CSN5 subunit of the COP9 signalosome and the regulatory particle number 11 (Rpn11) subunit of the proteasome. Eukaryotic proteins with a JAMM motif hydrolyse iso-peptide linkages involving ubiquitin and ubiquitin-like proteins; JAMM-proteins are found in Archaea and Bacteria. (JAB1 is Jun activation domain-binding protein 1, mov34 is the murine 26S proteasome non-ATPase regulatory subunit 7, the MPN domain is named for Mpr1, Pad1 N-terminal: Mpr1 is Rpn11 in *S. cerevisiae*, Pad1 is the *S. pombe* homologue).

JAMs Junctional adhesion molecules, members of the immunoglobulin superfamily. Jam-1 (JAM-A, platelet adhesion molecule 1, CD321, 299aa) is involved in epithelial tight junction formation in association with **PARD3**, has a role in regulating monocyte transmigration and is involved in platelet activation. It is a ligand of the integrin, LFA1, and a reovirus receptor. Jam-2 (JAM-B, vascular endothelial junction-associated molecule, CD322, 298aa) interacts with T-cells but not other leukocytes and may be involved in lymphocyte homing. Jam-3 (JAM-C, 310aa) may be involved in adhesion other than at tight junctions. It is a counter-receptor for the leukocyte integrin Mac-1 and promotes neutrophil transendothelial migration *in vitro* and *in vivo*. Jam4 (immunoglobulin superfamily member 5, 407aa)

The Dictionary of Cell and Molecular Biology. DOI: http://dx.doi.org/10.1016/B978-0-12-384931-1.00010-6

mediates calcium-independent homophilic cell adhesion and interacts with **MAGI-1** at tight junctions. N.B. There is some confusion in nomenclature with JAM-C being Jam-2 in some papers.

janiemycin See **ramoplanin**.

Janus kinase See **JAK**.

jasmonate Group of organic compounds that occur in plants and are thought to control processes such as growth and fruit ripening and to aid the plant's defences against disease and insect attack. They are derived from alpha-linolenic acid or hexadecatrienoic acid. The prohormone jasmonic acid (JA) is conjugated to amino acids such as isoleucine to form the active hormone jasmonoyl-isoleucine (JA-Ile). Abstract: http://stke.sciencemag.org/cgi/content/abstract/sigtrans;3/109/cm3

jasplakinolide A cyclic peptide (~710 Da) isolated from the marine sponge, *Jaspis johnstoni*. Behaves like **phalloidin** in stabilising F-actin and binds to the same or similar site, but permeates into cells readily and so can be used to assay actin dependent processes in living cells. Although it stabilizes actin filaments *in vitro*, it can disrupt actin filaments *in vivo* and induce polymerization of monomeric actin into amorphous masses.

JAZ proteins (1) A family of transcriptional regulators (e.g. jasmonate ZIM-domain protein-1, 253aa) that repress methyl jasmonate-regulated transcription in *Arabidopsis*. JAZ proteins are normally bound to transcription factors and inhibit their activity. (2) In *Xenopus*, Just another zinc finger protein (524aa) that binds preferentially to dsRNA, but also to RNA-DNA hybrids.

JAZF1 A zinc-finger transcription factor (JAZF1, juxtaposed with another zinc finger gene-1, TAK1-interacting protein 27, 243aa). A chromosomal aberration involving JAZF1 may be a cause of endometrial stromal tumours, an oncogene (JAZF1-SUZ12) is the product of a chromosomal translocation that brings together the N-terminal portion of JAZF1 and the C-terminus part of **SUZ12** (suppressor of zeste-12).

J cells (1) A line of rat pancreatic cells (AR 4-2 J cell line). (2) A derivative of BHK-tk⁻ cells (thymidine kinase negative), J1.1-2 cells, that lack gD receptors (gD is a herpesvirus envelope glycoprotein that binds to the potential host cell entry receptors TNFRSF14/HVEM, PVRL1 and PVRL1) http://www.ncbi.nlm.nih.gov/pubmed/9811737. (3) A line of rat cells derived from pigmented retinal epithelium, RPE-J cells. (4) A subcloned cell line (PC-J) which was isolated from a metastatic human prostate cell line, PC-3. (5) A feline T-lymphocyte cell line, FeT-J. (6) See **J774.2 cells**.

J chain Polypeptide chain (J-piece, joining chain, 137aa), found in IgA and in IgM, that joins heavy chains (H chains) to each other to form dimers (in IgA) and pentamers (in IgM). Disulphide bonds are formed between the J chain and H chains near the Fc ends of the heavy chains. Despite the similar name, it is not identical with the **J region** nor coded for by the **J gene**.

JC virus A human **retrovirus** (John Cunningham virus) similar to **polyoma** virus, but which has only recently been found associated with any human cancer. The causative agent of progressive multifocal leukoencephalopathy in immunocompromised patients.

J domain A sequence that is similar to the initial 73aa of the *E. coli* protein **DnaJ** and characteristic of **DnaJ**-like proteins. The domain is involved in regulating the ATPase activity of heat shock protein 70 (Hsp70) by stimulating adenosine triphosphate (ATP) hydrolysis. Article: http://www.ncbi.nlm.nih.gov/pmc/articles/PMC312864/

jelly roll Complex protein topology in which 4 **Greek key** motifs form an 8-stranded **beta sandwich**. So called because the overall structure resembles a jelly roll (US name for a swiss roll). Article: http://www.chm.bris.ac.uk/org/woolfson/papers/paper06.pdf

Jensen's syndrome See **Mohr-Tranebjaerg syndrome**.

Jervell and Lange-Nielsen syndrome A disorder characterized by congenital deafness, **long QT**, ventricular arhythmias and a high risk of sudden death. Caused by mutation in KCNQ1 or KCNE1, components of the delayed rectifier potassium channel.

jervine Steroidal alkaloid produced by the skunk cabbage (*Veratrum californicum*); has similar effects to **cyclopamine**.

J genes Genes coding for the joining segment (joining region) which links the V (variable regions) to the C (constant) regions of immunoglobulin light and heavy chains. During lymphoid development recombination between the V gene, one of the 27 D genes and one of the 6 J genes produces a rearranged VDJ gene (or VJ in light chains) to which a constant region is added to produce a complete heavy or light chain.

JH domains Domains (JAK homology domains) found in **JAKs**. JH1 is the the catalytic protein tyrosine kinase domain, the JH2 domain is a catalytically inactive pseudokinase but essential for normal kinase activity and may regulate JH1 activity. The JH3 and JH4 domains have homology with Src-homology-2 (SH2) domains and bind protein tyrosine phosphatase-2A. JH4–JH7 constitute a **FERM domain**.

Jijoye cells Human lymphablastic cell line, CD23 positive used as a model for B-lymphoctes. Derived from tumour tissue of a boy with **Burkitt's lymphoma**.

jimpy Mouse mutant with reduced life-span due to a recessive sex-linked defect in **PLP** gene. Has a severe CNS **myelin** deficiency associated with complex abnormalities affecting all glial populations. Jimpy-J4, the most severe of the jimpy mutants has virtually no PLP protein. See **Pelizaeus-Merzbacher disease**.

jinggangmycin An antifungal antibiotic (validamycin A), a trehalase inhibitor, from *Streptomyces hygroscopicus*. Used extensively as a fungicide for the control of sheath blight disease of rice.

jingzhaotoxins Toxins (theraphotoxins) (34–80aa) in the venom of the Chinese earth tiger tarantula (*Chilobrachys guangxiensis*). Inhibit voltage-gated sodium channels in a similar fashion to **hanatoxin1** and **SGTx1**. Some are **knottins**.

JM109 Strain of competent *E.coli* K12 cells said to be an ideal host for many molecular biology applications.

JNK Family of kinases (Jun kinases, c-jun N-terminal kinases, stress-activated protein kinases SAPKs) involved in intracellular signalling cascades. JNKs are distantly related to **ERKs** and are activated by dual phosphorylation on tyrosine and threonine residues. In addition to **c-Jun** will also phosphorylate p53. JNK1 (mitogen-activated protein kinase-8, EC 2.7.11.24, 427aa), JNK2 (MAPK-9, 424aa) and JNK3 (MAPK10, 464aa) are activated by environmental stress and pro-inflammatory cytokines and will phosphorylate a number of transcription factors, primarily components of AP-1 such as c-Jun. JNK3 is specific to a subset of neurons in the CNS. JNK-interacting protein 1 (JIP1, 711aa) is one of a group of scaffold proteins that aggregate specific components of the MAPK cascade to form a functional JNK signalling module.

Jnm1p See **dynamitin**.

Job's syndrome A disorder (hyperimmunoglobulin E recurrent infection syndrome) caused by mutation in **STAT3**. Originally (and erroneously) thought to be due to a defect in neutrophil chemotaxis.

joining chain See **J chain**.

Jones-Mote hypersensitivity Form of delayed type hypersensitivity, mediated by CD4$^+$ T-cells, characterised by infiltration of the skin by basophils.

jouberin A protein (Abelson helper integration site 1 protein homologue, 1196aa) that interacts with **nephrocystin-1**; downregulation of jouberin expression is important in early differentiation of

haematopoietic cells. Mutation is associated with one form of **Joubert's syndrome**.

Joubert's syndrome A genetically heterogeneous disorder with neurological, retinal, renal and other dysfunctions. Several of the mutations affect genes involved in the function of primary cilia. Affected proteins include inositol polyphosphate-5-phosphatase (INPP5E), TMEM216 (a transmembrane protein involved in ciliogenesis), **nephrocystins** 1 and 6, **meckelin**, RPGRIP1L (1315aa, similar to RPGRIP1, a protein present at the photoreceptor connecting cilium and mutated in **Leber's congenital amaurosis type VI**, ADP-ribosylation factor-like-2, coiled-coil and C2 domain-containing protein 2A (1532aa) and the product of an open reading frame on chromosome 10 (CXORF5).

J piece See **J chain**.

J region See **J genes**.

jumonji A protein (Jumonji/ARID domain-containing protein 2, 1246aa) that regulates histone methyltransferase complexes and plays an essential role in embryonic development. The **ARID** domain binds DNA and there are two conserved jumonji domains (JmjC and JmjN). The JmjC proteins, of which several are known in plants (*Arabidopsis* encodes 21 JmjC domain proteins), animals and yeast, are histone demethylases.

jumping gene Populist term for **transposon**.

jun (1) Protein (V-jun avian sarcoma virus 17 oncogene homologue, p39, 331aa) that dimerises with **fos** via a **zipper** motif to form the transcription factor AP1. JunB (347aa) is an NFκB-regulated jun-like protein that can also dimerise with fos and competes with c-jun. See **JNK**. (2) Jun a 1 (367aa) is a major pollen allergen from juniper (*Juniperus ashei*), jun v 1 is from *Juniperus virginiana* (Eastern red cedar).

junction A component (225aa) of a quaternary complex (junctin, calsequestrin, triadin, and the ryanodine receptor) in junctional sarcoplasmic reticulum that may be required for normal Ca^{2+} release. Triadin and junctin share similar structures. Junctin-1 (cardiac junctin) is an isoform of aspartate beta-hydroxylase (758a) with only the first 239aa; junctin-2 (210aa) is more widely distributed; junctate is a similar short isoform (313aa). They all lack the catalytic domain of aspartate beta-hydroxylase (EC 1.14.11.16) which catalyzes the post-translational hydroxylation of aspartic acid or asparagine residues within epidermal growth factor-like domains of numerous proteins. Article: http://www.ncbi.nlm.nih.gov/pmc/articles/PMC2777989/?tool=pubmed

junction potential Potential difference at the boundary between dissimilar solutions; arises from differences in diffusion constants between ions.

junctional adhesion molecule See **JAM**.

junctional basal lamina Specialized region of the extracellular matrix surrounding a muscle cell, at the **neuromuscular junction**. May be responsible for localization of **acetylcholine receptors** in the synaptic region, and also binds acetylcholinesterase to this region.

junctional epidermolysis bullosa A collection of diseases characterized by blistering within the lamina lucida. Primary subtypes include a lethal subtype termed Herlitz or JEB letalis (a severe defect in laminin 5), a nonlethal subtype termed JEB mitis, and a generalized benign type termed generalized atrophic benign EB (GABEB). Mutations in genes coding for laminin 5 subunits (α3 chain, laminin β3 chain, laminin γ2 chain), collagen XVII (BP180), α6 integrin, and β4 integrin have been demonstrated. See **epidermolysis bullosa**.

junctions See **adherens junction, desmosome, gap junction, zonula occludens.**

junk DNA Genomic DNA that serves, as yet, no known function.

Jurkat cells Human T-lymphocyte line derived from a 14-year old boy with acute T-cell leukaemia.

Much used for studies of IL-2 production *in vitro* but, although a convenient model system, they are not identical to real T-cells, particularly in their activation behaviour.

juvenile hormone A insect hormone which affects the balance between mature and juvenile attributes of certain tissues at each moult. In particular, the **imaginal discs** of many larval insects only develop into adult wings, sexual organs or limbs when blood juvenile hormone levels fall below a threshold level. There is a complex interaction between juvenile hormone and **ecdysone**. Synthetic analogues of JH include farnesol and methoprene, which have been tested for insecticide potential (known, with diflubenzuron, as Insect Growth Regulators, IGRs; see also **chitin**).

juvenile polyposis An autosomal dominant condition that predisposes to various types of tumours. Can be a result of mutation in genes for either **Smad4** or bone morphogenetic protein receptor-1A (**ALK3**).

juxtacrine activation Activation of target cells by membrane-anchored growth factors; also used for activation of leucocytes by **PAF** bound to endothelial cell surface.

K

K Single letter code for lysine.

K562 cells A line of blast cells (lymphoblasts) established from a female Caucasian patient with chronic myelogenous leukaemia in terminal blast crisis. They are highly undifferentiated and of the granulocytic series. Recent studies have shown the K562 blasts are multipotential, haematopoietic malignant cells that spontaneously differentiate into recognisable progenitors of the erythrocyte, granulocyte and monocytic series.

K_a (1) Acid **dissociation constant**. Often encountered as pK_a (i.e. $-\log_{10}K_a$). (2) Association constant (K_{ass}). The equilibrium constant for association, the reciprocal of K_d, with dimensions of litres/mole. Better to use K_d, thereby removing any ambiguity.

KaiABC A clock protein complex originally identified in the cyanobacterium *Synechococcus elongatus* PCC 7942, that is essential in maintaining circadian rhythmicity in cyanobacteria. It acts as a promoter-non-specific transcription repressor. The Kai proteins do not share any homology with any known eukaryotic clock genes. The KaiABC complex is composed of a homodimer of KaiA (circadian clock protein, 284aa), a homodimer of KaiB (102aa) and a homohexamer of KaiC (circadian clock protein kinase, EC 2.7.11.1, 519aa). Kai C is a serine/threonine kinase that autophosphorylates (positively and negatively regulated by KaiA and KaiB respectively) and accumulates at a particular phase of the circadian cycle: phosphorylation correlates with clock speed. First description: http://www.sciencemag.org/content/281/5382/1519.abstract

kainate An agonist for the K-type **glutamate receptor**. It can act as an **excitotoxin** producing symptoms similar to those of **Huntington's chorea**, and is also used as an anthelminthic drug. Originally isolated from the red alga *Digenea simplex* (Kainin-sou).

kaiso A transcriptional regulator (zinc finger- and BTB domain-containg protein 33, 672aa) that interacts with delta-**catenin**-2. Induced in vascular endothelium by wounding.

KAL See **anosmin**.

kala-azar Visceral leishmaniasis, a disease caused by infection with the protozoan *Leishmania donovani* spread by the bite of infected sand flies; characterized by enlargement of the liver and spleen, anaemia, wasting and fever.

kalicludines Nematocyst toxins (58−59aa) isolated from the sea anemone, *Anemonia sulcata*, that are structurally homologous to **dendrotoxins** and to Kunitz-type peptidase inhibitors. Block voltage-sensitive potassium channels (Kv1.2) and inhibit typsin. Article: http://www.jbc.org/cgi/content/full/270/42/25121

kalinin The major glycoprotein (laminin 5, epiligrin, nicein) of the basal lamina that underlies epidermal keratinocytes. Like other laminin types, kalinin is composed of three different polypeptide chains, alpha (3333aa), beta (1172aa) and gamma (1193aa), which are disulphide-linked into a cruciform molecule with one long (170 nm) and three short arms with globules at each end. Laminin-5 is thought to be involved in cell adhesion (via integrin α_3/β_1 in focal adhesions and integrin α_6/β_4 in hemidesmosomes), in signal transduction via tyrosine phosphorylation of focal adhesion kinase and in differentiation of keratinocytes. Kalinin is absent in patients with lethal junctional **epidermolysis bullosa**.

kaliocin See **lactoferrin**.

kaliotoxin Toxin from the scorpion, *Androctonus mauretanicus m.*, (kaliotoxin-1, 38aa) that blocks high conductance Ca^{2+}-activated K^+ channels (KCa). Related toxins are found in other scorpion species. Closely related to **charybdotoxin**, **noxiustoxin**, **iberiotoxin** and **agitoxins**.

kalirin A neuronal Rho GEF of the **Dbl** family (P-CIP10, huntingtin-associated protein-interacting protein, DUO, EC 2.7.11.1, 2985aa), a serine/threonine protein kinase with Dbl and pleckstrin-homology domains, involved in regulating actin organization, endocytosis, exocytosis and free radical production. There are numerous kalirin isoforms with different functional domains and tissue locations. Genetic variation in kalirin is associated with susceptibility to coronary heart disease.

kaliseptine A nematocyst toxin from the sea anemone, *Anemonia sulcata* (36aa) that blocks voltage-sensitive potassium channels (Kv1/KCNA). Although it binds to the same site as **dendrotoxins** and **kalicludines** there is no sequence homology.

kallidin Decapeptide (lysyl-bradykinin) derived from kininogen-1 in kidney. Like **bradykinin**, an inflammatory mediator (a **kinin**); causes dilation of renal blood vessels and increased water excretion.

kallikreins Plasma serine endopeptidases normally present as inactive prekallikreins which are activated by **Hagemann factor**. Act on **kininogens** to produce **kinins** and plasminogen to produce plasmin.

The Dictionary of Cell and Molecular Biology. DOI: http://dx.doi.org/10.1016/B978-0-12-384931-1.00011-8

Kallikrein 1 (tissue kallikrein, EC 3.4.21.35, 262aa) generates **kallidin** from kininogen-1, as does kallikrein 2 (glandular kallikrein-1, tissue kallikrein-2, EC 3.4.21.35, 261aa). Kallikrein 3 (semenogelase, EC 3.4.21.77, 261aa) is prostate-specific antigen. Kallikrein 4 (enamel matrix serine peptidase-1, 254aa) is defective in **amelogenesis imperfecta**. Kallikrein 5 (stratum corneum tryptic enzyme, 293aa) may be involved in desquamation, kallikrein-6 (neurosin, protease M, 244aa) degrades alpha-synuclein and prevents its polymerization and may be involved in the pathogenesis of Parkinson's disease. Kallikrein-7 (EC 3.4.21.117, stratum corneum chymotryptic enzyme, 253aa) is involved in shedding of squames and possibly in generation of inflammatory cytokines, kallikrein 8 (260aa) is the human homologue of mouse **neuropsin**. Several other kallikreins are known. Plasma kallikrein (EC 3.4.21.34, Fletcher factor, kininogenin, 638aa) is the main form in plasma and converts kininogen to bradykinin: it is processed into a catalytic light chain and a heavy chain that binds kininogen through its four apple domains.The gamma subunit of nerve growth factor is a kallikrein involved in the processing of NGF.

Kallmann's syndrome A syndrome characterized by hypogonadotropic hypogonadism and anosmia, caused by a defect of migration and targeting of gonadotropin-releasing hormone-secreting neurons and olfactory axons during embryonic development. The gene responsible for the X-linked form of the disease encodes **anosmin**. Other forms are caused variously by defects in fibroblast growth factor receptor-1, **prokineticin** receptor-2, prokineticin-2, chromodomain-helicase-DNA-binding protein 7 (see **CHD proteins**) or fibroblast growth factor-8.

Kanagawa haemolysins Thermostable bacterial exotoxins (three variants, all 189aa) produced by *Vibrio parahaemolyticus*. The mechanism of haemolysis is unclear but involves cation leakage. The Kanagawa phenomenon is an *in vitro* haemolytic plaque assay, now rarely used. See Table E2.

K antigen Capsular polysaccharide antigens of Gram-negative bacteria, often used to define strain-types (**serotypes**), e.g. *E. coli* K12. In *Klebsiella*, many serotypes (77) are known and there is a correlation between K-antigen serotype and pathogenicity.

Kaposi's sarcoma A sarcoma of spindle-shaped cells mixed with angiomatous tissue caused by human herpesvirus 8 and associated with long-term immunosuppression. Usually classed as an angioblastic tumour.

kappa chain See **L-chain**.

kappa particle Gram-negative bacterial endosymbiont, *Caedobacter taeniospiralis*, that confers the 'killer' trait on infected *Paramecium* which are resistant to the toxin liberated by infected forms. Killing activity is associated with the **induction** of defective phage in the endosymbiont, leading to the release of R-bodies, coded for by the phage genome and apparently of mis-assembled phage-coat protein.

kappa toxin (1) Exotoxin (κ-toxin) produced by *Clostridium perfringens*; a collagenase that presumably aids tissue infiltration. (2) Kappa-hefutoxins, isolated from the venom of the scorpions *Heterometrus fulvipes* and *H. spinifer*, block the voltage-gated K^+-channels, Kv1.3 and Kv1.2, and slow the activation kinetics of Kv1.3 currents, and are part of the short scorpion toxin superfamily.

kaptin Actin-associated protein 2E4 (436aa) that binds to F-actin in an ATP-dependent manner. Associated with platelet activation and also found in lamellipodia and the tips of the stereocilia of the inner ear.

Kar3 A **kinesin-14** motor of *S. cerevisiae* (729aa) essential for yeast nuclear fusion during mating and required for mitosis and karyogamy. Like **NCD**, differs from kinesin in that it moves towards the minus end of the microtubule (like cytoplasmic dynein). Kar3 forms a heterodimer with either Cik1 or **Vik1**, both of which are noncatalytic polypeptides. Kar3Cik1 depolymerizes microtubules from the plus end and promotes minus-end-directed microtubule gliding. The interaction with Vik1 is distinct and also important in regulating interactions with microtubules.

Karak syndrome A form of neurodegeneration with brain iron accumulation caused by mutation in the gene for **phospholipase A2−G6**. See also **infantile neuroaxonal dystrophy**.

KARP-1 A fragment of a protein (Ku86 autoantigen related protein-1, 156aa) expressed from the human Ku86 autoantigen locus that appears to play a role in mammalian DNA double-strand break repair. KARP-1-binding protein (centrosomal protein of 170 kDa, Cep170, 1584aa) is associated with spindle microtubules during mitosis and interacts with polo-like kinase 1 (PLK1).

Kartagener's syndrome Condition (situs inversus) in which the normal left/right asymmetry of the viscera is reversed. Associated with a **dynein** defect (dynein is absent or dysfunctional in some cases) and with **immotile cilia syndrome**.

karyogamy The formation of a zygote by fusion of two gametic nuclei, usually immediately after cytoplasmic fusion, although in some fungi there may be a prolonged binucleate stage (dikaryophase). See **plasmogamy**.

karyogram See **idiogram**.

karyokinesis Division of the nucleus, whereas cytokinesis is the division of the whole cell.

karyopherins Superfamily of proteins that are nucleocytoplasmic shuttling receptors (**importins**

and **exportins**), which bind to transport signals on the cargoes, and by means of interactions with **nuclear pore complex** (NPC) proteins (nucleoporins), direct cargo translocation through the NPC.

karyoplast A nucleus isolated from a eukaryotic cell surrounded by a very thin layer of cytoplasm and a plasma membrane. The remainder of the cell is a cytoplast.

karyorrhexis Degeneration of the nucleus of a cell. There is contraction of the chromatin into small pieces, with obliteration of the nuclear boundary.

karyosome (1) An aggregation of chromatin in the interphase nucleus, not the nucleolus. (2) A single, compact cluster of meiotic chromosomes that forms after the completion of recombination in prophase I: seen in *Drosophila* and also humans. Phosphorylation of barrier-to-autointegration factor (BAF) by the nucleosomal histone kinase-1 has a critical function in karyosome formation. Article: http://www.ncbi. nlm.nih.gov/pmc/articles/PMC2099182/

karyotype The complete set of chromosomes of a cell or organism. Used especially for the display prepared from photographs of mitotic chromosomes arranged in homologous pairs.

kassinin A tachykinin (**neuromedin**) of amphibian origin (12aa) from *Kassina senegalensis* (Senegal running frog) that binds preferentially to the mammalian tachykinin-2 (NK-2) receptor.

kasugamycin *KSM* An aminoglycoside antibiotic that inhibits the initial step of protein synthesis. Sensitivity to KSM is in part due to interaction with KsgA (kasugamycin dimethyltransferase, a rRNA adenine dimethyltransferase, EC 2.1.1.-, 273aa) which methylates adenosine residues at positions 1518 and 1519 of the 16S rRNA.

Katablepharids Heterotrophic flagellates, oval or cylindrically ovate, with one anterior and one posterior flagellum emerging from a shallow subapical groove. They are distantly related to **cryptomonads** and **goniomonads**. The type species is *Katablepharis phoenikoston*. Details and images: http://tolweb.org/Katablepharids/2413#Characteristics

katanin An ATP-dependent microtubule-severing enzyme, similar to **spastin**. Consists of a subunit termed P60 (katanin A1, EC 3.6.4.3, 491aa) that breaks the lattice of the microtubule and another subunit termed P80 (katanin B1, 655aa) which contains 6 WD repeats and may serve as a targeting subunit. In *Arabidopsis*, katanin p60 (BOTERO 1, ECTOPIC ROOT HAIR 3, FAT ROOT, FRAGILE FIBER 2, 523aa) may be required for the correct orientation of cortical microtubule arrays during cellular elongation.

katanosin B Katanosin B and plusbacin A(3) are naturally occurring cyclic depsipeptide antibiotics from strains of *Cytophaga* and *Pseudomonas* respectively, containing a lactone linkage. Antibacterial activity is due to blocking of transglycosylation in bacterial cell wall peptidoglycan synthesis via a mechanism differing from that of vancomycin.

Kawasaki's disease An acute, self-limited vasculitis of infants and children. A functional polymorphism of the gene for inositol-1,4,5-trisphosphate 3-kinase C (ITPKC) is significantly associated with susceptibility.

kazal proteins Family of **serine peptidase** inhibitors. Includes seminal **acrosin** inhibitors, pancreatic secretory trypsin inhibitor (PSTI) and Bdellin B-3 from leech.

kazrin Component of the cornified envelope of keratinocytes (775aa). May link adherens junctions and desmosomes.

KB cells A line of cells originally described as 'epidermoid carcinoma established from the mouth of a Caucasian man in 1954'; however, probably a **HeLa** subclone.

KC Mouse chemokine (CXCL1, platelet-derived growth factor-inducible protein KC, 96aa) processed to gnerate KC (5−72) which is chemotactic for neutrophils.

K$_{cat}$ Catalytic constant of an enzyme, also referred to as the turnover number. Represents the number of reactions catalysed per unit time by each active site.

K cells (1) See **killer cells**. (2) A subpopulation of gut cells that secrete glucose-dependent insulinotropic polypeptide.

K$_d$ An equilibrium constant for dissociation. Thus, for the reaction: $A + B = C$, at equilibrium $K_d = [A][B]/[C]$. Dimension: moles per litre in this case. K_d is the reciprocal of K_a. In general the concept of K_d is more readily understood than that of K_a; for example, in considering the conversion of A to C by the binding of ligand B, the $K_d = [B]$ when $[A] = [C]$. Thus K_d is equal to the ligand concentration which produces half-maximal conversion (response).

KDEL Single letter code for the C-terminal amino acid consensus, in animals and many plants, for proteins targeted to the **endoplasmic reticulum**. Other variants in some plants and other Phyla include HDEL, DDEL, ADEL and SDEL. The KDEL receptors (ER lumen protein retaining receptors, ~212aa) are multi-pass membrane proteins with seven hydrophobic domains.

KDR See **VEGF**.

Kearns-Sayre syndrome A complex disorder involving ophthalmoplegia, pigmentary degeneration of the retina, and cardiomyopathy, caused by various mitochondrial deletions.

Kelch proteins The kelch family of proteins is defined by a ~50aa repeat (Kelch domain) that has been shown to associate with actin. *Drosophila* kelch (1477aa) is required to maintain ring canal organization during oogenesis.

Kell Blood group system. The K antigen is relatively uncommon (9%) but after the **Rhesus** antigens is the next most likely cause of haemolytic disease of the newborn. Kell glycoprotein is a transmembrane, single-pass protein that carries the Kell antigens. It is an endothelin-3-converting enzyme. NCBI details: http://www.ncbi.nlm.nih.gov/books/NBK2270/

keloid A bulging scar, resulting from the overgrowth of granulation tissue at the site of a healed skin injury which is then slowly replaced by collagen type 1. The scar grows continuously and invasively beyond the confines of the original wound. The tendency to produce keloids seems to be heritable and involves the expression of the TGFβ1 gene by the neovascular endothelial cells.

Kenacid blue See **Coomassie Brilliant Blue**.

kendrin See **pericentrin**.

Kenny-Caffey syndrome See hypoparathyroidism.

K$_{eq}$ The equilibrium constant for a reversible reaction. $K_{eq} = [AB]/[A][B]$.

keratan sulphate A **glycosaminoglycan** with a repeating disaccharide unit of galactose β1−4 linked to N-acetyl glucosamine and either or both of which can be sulphated. The polysaccharide chain is linked to various core proteins including **aggrecan, fibromodulin, keratocan, lumican, mimecan, osteoadherin** and **proline arginine-rich end leucine-rich repeat protein** (PRELP). Keratan sulphate I (KSI) is N-linked to asparagine residues via N-acetylglucosamine and KSII is O-linked to specific serine or threonine residues via N-acetyl galactosamine. Both are found in a range of tissues.

keratinizing epithelium An epithelium such as vertebrate epidermis in which a keratin-rich layer is formed from intracellular **cytokeratins** as the outermost cells die.

keratinocyte Skin cell, of the keratinized layer of epidermis: its characteristic **intermediate filament** protein is **cytokeratin**.

keratinocyte growth factor A growth factor (KGF, fibroblast growth factor 7, 194aa) mitogenic for epithelial cells but not fibroblasts or endothelial cells. It is a ligand for fibroblast growth factor receptor-2.

keratins Group of highly insoluble fibrous proteins (of high α-helical content) which are found as constituents of the outer layer of vertebrate skin and of skin-related structures such as hair, wool, hoof and horn, claws, beaks and feathers. Extracellular keratins are derived from **cytokeratins**, a large and diverse group of **intermediate filament** proteins.

keratocan Core protein (352aa) to which keratan sulphate glycosaminoglycan chains are attached, especially in the cornea. Defects cause abnormal curvature of the cornea (autosomal recessive cornea plana).

keratohyalin granules Granules found in live cells of **keratinizing epithelia** and which contribute to the **keratin** content of the dead cornified cells. Some, but not all, contain sulphur-rich keratin.

keratoma A callus, not a tumour.

keratoses Benign but precancerous lesions of skin (actinic keratoses) associated with ultraviolet irradiation.

ketogenesis Production of **ketone bodies**. Occurs in mitochondria, mostly in liver.

ketoglutarate An intermediate of the **tricarboxylic acid cycle**, also formed by deamination of **glutamate**.

ketone body A misleading term for compounds, acetoacetate, β-hydroxybutyrate and acetone that accumulate in the body following starvation, in diabetes mellitus and in some disorders of carbohydrate metabolism.

ketosynthase domain A diverse set of domains in fungal **polyketide** synthases that are responsible for a catalytic step in chain elongation. Fungal polyketide synthases (PKSs) are modular enzymes responsible for the biosynthesis of several mycotoxins, each typically consisting of a single set of domains which function iteratively during the assembly of the polyketide chain. The ketosynthase domain acts collaboratively with an acyl carrier protein (ACP) domain. The ACP of one module then engages the KS domain of the next module to facilitate chain transfer.

ketothiolase deficiency A disorder of isoleucine catabolism (alpha-methylacetoacetic aciduria, 3-oxothiolase deficiency) caused by mutation in the mitochondrial acetyl-CoA acetyltransferase-1 gene (**ACAT1**).

kettin A very large (4250aa in *C. elegans*, 18141aa in *Drosophila*) actin-binding protein of the **connectin/titin** family with many immunoglobulin-like (Ig) repeats (31 in *C. elegans*, 53 in *Drosophila*), which is associated with the thin filaments in invertebrate muscles. Appears to be an important regulator of myofibrillar organization and provides mechanical stability to the myofibrils during contraction. May also have a role in chromosome condensation and chromosome segregation during mitosis. See **projectin, sallimus**.

KEULE A SNARE-interacting protein (666aa) that regulates vesicle trafficking involved in cytokinesis and root hair development in *Arabidopsis* (see **KNOLLE**).

Keutel's syndrome Abnormal (ectopic) calcification of cartilage as a result of mutation in the gene for **matrix Gla protein.**

kexins Family of eukaryotic subtilisin-like peptidases. Examples include kexin, a yeast serine peptidase of the S8 family (EC 3.4.21.61, 814aa) that processes the precursors of alpha-factors and killer toxin; the mammalian kexin-like **proprotein convertases,** and **furin.** There are prokaryotic kexin-like peptidases.

keyhole limpet haemocyanin Haemocyanins (3125aa and 3421aa) from the keyhole limpet (*Megathura crenulata*) often used as a carrier for the production of antibodies.

KGF See **keratinocyte growth factor.**

KH domain An evolutionarily conserved sequence (K homology domain, ~70 aa) first identified in the human heterogeneous nuclear ribonucleoprotein (hnRNP) K and subsequently in a variety of nucleic acid-binding proteins, e.g. the **AU-rich element** RNA-binding protein KSRP. The KH domain binds RNA in a sequence-specific fashion and may be present in multiple copies.

Ki antigen Component (PA28gamma, 254aa) of the **proteasome** activator protein **PA28** or 11S regulator, the nuclear autoantigen (Ki) recognised by the monoclonal antibody Ki-67. During interphase, the antigen is located only in the nucleus, but in mitosis most of the protein is relocated to the surface of the chromosomes. The fraction of Ki-67-positive tumour cells (the Ki-67 labeling index) is often correlated with the clinical course of the disease. Autoantibodies against a variety of nuclear antigens including Ki antigen are found in patients with **systemic lupus erythematosus** (SLE).

KIAA genes Uncharacterized human genes isolated at the Kazusa DNA Research Institute, Japan and made available in the Human Unidentified Gene-Encoded (HUGE) protein database. Links: http://www.kazusa.or.jp/huge

KID domain One of two domains (the kinase-inducible domain) forming the **CREB** transactivation domain; the KID domain and a glutamine-rich constitutive activator (Q2) synergize to stimulate target gene expression in response to cAMP. Ser-133 phosphorylation of CREB within the KID domain promotes target gene activation via complex formation with the **KIX domain.**

KIF (1) **Keratin** intermediate filaments. (2) **Kinesin** superfamily proteins (KIFs). (3) **Ki antigen** (KiF).

killer cells (1) Mammalian cells which can lyse antibody-coated target cells. They have a receptor for the Fc portion of IgG, and are probably of the mononuclear phagocyte lineage, though some may be lymphocytes. Not to be confused with **cytotoxic T-cells** (CTL) which recognize targets by other means and are clearly a subset of T-lymphocytes: this confusion exists in the early literature. (2) (NK cells) Natural killer cells are CD3-negative large granular lymphocytes, mediating cytolytic reactions that do not require expression of Class I or II major **histocompatibility antigens** on the target cell. (3) (LAK cells) Lymphokine-activated killer cells are NK cells activated by **interleukin-2**.

killer plasmid Linear double-stranded DNA plasmids (k1 and k2) found in some strains of the yeast *Kluyveromyces marxianus* as multiple cytoplasmic copies. Cells with plasmids secrete a glycoprotein toxin.

kilobase *kb, kbp* One thousand base pairs of DNA. Strictly should probably be kbp (kilobase pairs) but usually truncated.

KIN10/11 Protein kinases (SNF1-related protein kinases, EC 2.7.11.1, KIN10, 535aa; KIN11, 512aa) from *Arabidopsis* that trigger gene expression reprogramming in response to stress. The catalytic subunit of a trimeric SNF1-related protein kinase (SnRK) complex. Homologues in yeast are **Snf1 kinase** and in mammals, AMPK.

kinase (1) Historically, a protein (enzyme) that speeds up a process. (2) Usually a shorthand term for a **protein kinase,** an enzyme that catalyses the transfer of phosphate from ATP to a protein substrate (onto serine/threonine or tyrosine residues). Many protein kinases have separate entries (e.g. **protein kinase A**). (3) Sometimes an abbreviation for phosphokinase, which leaves some ambiguity since phosphokinases can act on substrates other than proteins but in many cases the intention is to describe a **protein kinase.** It is then helpful to specify the substrate, e.g. creatine phosphokinase (**creatine kinase**), phosphatidyl-inositol kinase (see **PI-3-kinase**). Kinase database: http://kinase.com/

Kindler's syndrome A dermatological disorder (hereditary acrokeratotic poikiloderma) caused by defects in **kindlin-1.**

kindlins A family of proteins associated with cell adhesion and involved in two-way signalling that occurs at attachment sites (inside-out signalling in which **talin** activates integrins and outside-in signalling through integrins that enables adhesion and spreading). They bind to beta-integrin cytoplasmic tails directly and cooperate with talin in integrin activation. Kindlin-1 (kindlerin, fermitin family homologue 1, 677aa) is localized in basal epidermal keratinocytes and defects are associated with **Kindler's**

syndrome. Kindlin-2 (mitogen induced gene 1, MIG1: see **migfilin**) is required for the stabilization and maturation of focal adhesions and stress fibres in activated fibroblasts and myofibroblasts. Kindlin-3 (MIG2-like protein, 667aa) is associated with activation of platelet and leucocyte integrins and deficiency is associated with severe bleeding. Fermitin-1 and -2 are *Drosophila* proteins (708aa and 715aa) involved in activation of integrins and in maintaining muscle integrity. Article: http://www.ncbi.nlm.nih.gov/pmc/articles/PMC2603460/?tool=pubmed

kinectin Integral membrane protein (1357aa) of the endoplasmic reticulum and probably other membrane compartments; binds to **kinesin** and accumulates in integrin-based adhesion complexes as the membrane anchor for kinesin-driven vesicle movement. Kinectin has extensive α-helical coiled-coil regions and, like the myosin tail with which it has sequence and structural similarities, may form a very long molecule, possibly 100 nm in length when fully extended.

kinesin Cytoplasmic protein that is responsible for moving vesicles and particles towards the distal (plus) end of microtubules. Differs from cytoplasmic **dynein** (MAP1C) in the direction in which it moves and its relative insensitivity to vanadate. It has two heavy chains and two light chains (in kinesin-1, 963aa and 573aa). There are many kinesin-like proteins that form the kinesin superfamily (KIFs) that transport membranous organelles and macromolecules along microtubules, well-studied in axons and dendrites. Kinesin-14 proteins (formerly C-terminal motor proteins) have a C-terminal motor domain that moves towards the minus end of microtubules. There are at least four members of the group (**Ncd**, **KAR3**, CgCHO2, AtKCBP).

kinesis Alteration in the movement of a cell, without any directional bias. Thus speed may increase or decrease (**orthokinesis**) or there may be an alteration in turning behaviour (**klinokinesis**). See **chemokinesis**.

kinetin A **cytokinin** (6-furfurylaminopurine) added to plant tissue culture media to promote cell division and hence callus formation. By modifying the ratio of this cytokinin to auxin, callus can be induced to generate shoots. It also enhances germination and fruit set, reduces apical dominance and breaks lateral bud dormancy. Obtained by heat-treatment of DNA and for a long time not thought to occur naturally. Has been used in various cosmetic treatments claimed to have anti-ageing properties.

kinetochore Multilayered structure, a pair of which develop on the mitotic chromosome, adjacent to the **centromere**, and to which spindle microtubules attach – but not at the end normally associated with a **microtubule organizing centre**. See centromere associated proteins, (**CENP**) and **septin 7**.

kinetodesma *Pl.* kinetodesmata. An array of longitudinally-oriented cytoplasmic fibrils with cross-striations that connect the ciliary kinetosomes of ciliates and may have a role in coordinating ciliary activity. The kinetodesmal fibrils arise independently from kinetosomes and each fibril has a limited length of around 5–6 kinetosomal intervals. The basal bodies of the cilia are located to the right of the associated kinetodesma when a ciliate is seen in normal orientation. The protein composition is complex and a family of epiplasmins (200–400aa) have been identified. (NB. The definition in the earlier editions of this dictionary has been so extensively plagiarised that this definition has been rewritten). Articles: http://www.ncbi.nlm.nih.gov/pmc/articles/PMC2223830/pdf/583.pdf, http://www.uniprot.org/citations/16427359

kinetoplast Mass of mitochondrial DNA, usually adjacent to the flagellar **basal body**, in flagellate protozoa.

kinetosome **Basal body** of cilium: used mostly of ciliates.

kinety A row of **kinetosomes** and associated **kinetodesma** in a ciliate protozoan.

kingdom In modern taxonomy, any of the six major groupings used in the classification of living organisms: Monera, Archaea, Protista, Plantae, Animalia, Fungi. Many alternative schemata have been proposed and Monera are often considered to include Archaea, making only five kingdoms.

kininogen Inactive precursor in plasma from which **kinin** is produced by proteolytic cleavage. The high-molecular-weight form of kininogen (Williams-Fitzgerald-Flaujeac factor; 644aa) is cleaved to produce: kininogen-1 heavy chain (362aa), T-kinin (Ile-Ser-Bradykinin, 14aa), bradykinin (kallidin I, 10aa), Lysyl-bradykinin (kallidin II, 9aa), kininogen-1 light chain (255aa) and low molecular weight growth-promoting factor (4aa). It is enzymically inactive but is a cofactor for the activation of kallikrein and Hageman factor. Low-molecular-weight kininogen, like the high molecular weight form, is an inhibitor of **cysteine peptidases**.

kinins Inflammatory mediators that cause dilation of blood vessels and altered vascular permeability. Kinins are small peptides produced from **kininogen** by **kallikrein**, and are broken down by kininases. Act on phospholipase and increase arachidonic acid release and thus prostaglandin (PGE_2) production. See **bradykinin, kallidin, C2-kinin**.

Kinyoun method A method of staining acid-fast microorganisms, specifically mycobacteria. It uses carbolfuchsin as a primary stain, followed by decolourization with an acid-alcohol solution and methylene blue as a counterstain. Acid-fast organisms stain red and nonacid-fast organisms blue.

KIR (1) Sets of receptors on killer cells (killer cell inhibitory receptors, killer cell immunoglobulin-like receptors). KIR2D receptors (CD158) bind HLA-C alleles and inhibit the activity of NK cells, KIR3D bind to HLA-B alleles. They are distinct from the CD94/NKG2 KIR C-type lectin that is HLA-E specific. Between them these receptors confer self-tolerance by active signalling in the killer cell. (2) Kir channels are a superfamily of inwardly-rectifying potassium channels.

Kirsten sarcoma virus A murine sarcoma-inducing retrovirus, generated by passaging a murine erythroblastosis virus in newborn rats. Source of the Ki-*ras* oncogene: *KRAS1* is a pseudogene but *KRAS2* is mutated in 17 to 25% of all human tumours.

KIS A serine/threonine kinase (kinase interacting with stathmin, 419aa), expressed in all adult tissues, that phosphorylates cyclin-dependent kinase inhibitor **p27** (**Kip1**) and **stathmin**. May be involved in trafficking or processing of RNA

KiSS1 A metastasis suppressor gene that inhibits metastasis of human melanomas and breast carcinomas. The human *KiSS1* gene encodes a C-terminally amidated active peptide (**kisspeptin**, 145aa) that is the ligand of a G-protein-coupled receptor.

kisspeptin Product of the *KiSS* gene that is cleaved into: metastin (kisspeptin-54, 54aa), kisspeptin-14, kisspeptin-13 and kisspeptin-10. Metastin, a metastasis suppressor, is the endogenous ligand of the G-protein coupled receptor GPR54. Kisspeptin-10 regulates invasive activity of the trophoblast and also acts as a regulator of reproductive maturity via the hypothalamic **gonadotrophin-releasing hormone** (GnRH) system. Mutations of the GPR54 gene are linked to absence of puberty onset and hypogonadotrophic hypogonadism in humans. Article: http://endo.endojournals.org/content/151/7/3247.long

kistrin Naturally occurring inhibitor (68aa) of platelet aggregation found in the venom of Malayan pit viper *Agkistrodon rhodostoma*. Kistrin has an RGD site that competes for the platelet IIb/IIIa **integrin** and is therefore one of the **disintegrins**.

kit An oncogene, identified in Hardy-Zuckerman 4 feline sarcoma virus, that encodes a receptor tyrosine kinase (mast cell growth factor receptor, 976aa) which binds **stem cell factor**. See **W locus**.

KIX domain A highly conserved domain (~90aa) in CREB binding protein (CBP) and p300 that binds to the phosphorylated **KID domain** of **CREB**. The KIX domain of CBP also recognizes the transactivation domains of other nuclear factors, including Myb, Jun, cubitus interruptus, and HTLV-1 virally encoded Tax protein. Database entry: http://www.expasy.ch/cgi-bin/nicedoc.pl?PDOC50952

Kjellin's syndrome See **spastizin**.

Klebs-Löffler bacillus See *Corynebacterium diphtheriae*.

Klebsiella Genus of Gram-negative bacteria, non-motile and rod-like, facultatively anaerobic, associated with respiratory, intestinal and urinogenital tracts of mammals. *K. pneumoniae* is associated with pneumonia in humans and is a common **nosocomial infection**.

Klein-Waardenburg syndrome See **Waardenburg's syndrome**.

kleisin In *C. elegans* a protein (821aa) associated with condensing chromosomes at metaphase. The human homologue, kleisin-beta (condensin-2 complex subunit H2, 605aa) is a component of the **condensin complex**. The complex also contains structural maintenance of chromosomes proteins (**SMC proteins**) and three regulatory proteins, NCAPH/BRRN1 (Barren homolog protein 1), NCAPD2/CAPD2 and NCAPG. See also **cohesins**.

Klenow fragment Larger part of the bacterial DNA polymerase I (928aa) that remains after treatment with **subtilisin**; retains some but not all exonuclease and polymerase activity.

kleptoplastid A plastid stolen from another algal cell, for example those found in the dinoflagellates *Amphidinium* spp. and *Dinophysis* spp.

KLF6 See **zf9**.

KLH See **keyhole limpet haemocyanin**.

Klinefelter's syndrome Human genetic abnormality in which the individual, phenotypically apparently male, has three sex chromosomes (XXY).

klinokinesis Kinesis in which the frequency or magnitude of turning behaviour is altered. Bacterial chemotaxis can be considered as an adaptive klinokinesis; the probability of turning is a function of the change in concentration of the substance eliciting the response.

Klippel-Feil syndrome A congenital shortening of the neck. The autosomal dominant form is caused by mutation in the GDF6 gene encoding growth and differentiation factor 6 (cartilage-derived morphogenetic protein-2, 455aa).

klotho A membrane protein (EC 3.2.1.31, 1012aa) that is processed to produce a secreted peptide with weak glycosidase activity towards glucuronylated steroids. The klotho peptide is involved in calcium and phosphorus homeostasis by inhibiting the synthesis of active vitamin D, is essential for the specific interaction between FGF23 and FGFR1 and may be an anti-aging hormone by virtue of inhibiiton of insulin/IGF1 signalling.

K_m *Michaelis constant* A kinetic parameter used to characterize an enzyme; defined as the concentration of substrate that permits half-maximal rate of reaction. An analogous constant **K_a** is used to describe binding reactions, in which case it is the concentration at which half the receptors are occupied.

km-fibres Parallel overlapping bundles of microtubules (each ribbon with ~20 microtubules) running longitudinally below the cortex of heterotrich ciliates such as *Stentor*, to the right of a row of basal bodies (**kinety**) to which they are attached. They are contractile (by sliding of microtubules relative to one another) and lie adjacent to the myonemes (M fibres) which are also contractile and may act as an antagonistic contractile system. It is not clear how these filamentous structures are related to the **kinetodesmata** of *Paramoecium*. Article: http://jcb.rupress.org/content/57/3/704.full.pdf

KNAT2 A transcription factor in the TALE/KNOX homeobox family (homeobox protein knotted-1-like 2, Protein ATK1, 310aa) that is expressed in shoot apices of seedlings, in the receptacle and developing pistil of flowers and in axillary buds of inflorescence stems. It may maintain cells in an undifferentiated, meristematic state.

Kniest's syndrome A **chondrodysplasia** caused by mutations in the collagen type II gene (COL2A1).

knirps A *Drosophila* **gap gene**, asymmetric distribution of which is essential for normal expression of striped patterns of **pair-rule genes** and thus abdominal segmentation. Encodes a steroid/thyroid orphan nuclear receptor, a transcriptional repressor (nuclear receptor subfamily 0 group A member 1, 429aa). There is a knirps-related protein (nuclear receptor subfamily 0 group A member 2, 647aa).

Knobloch's syndrome An autosomal recessive disorder in which there is myopia, vitreoretinal degeneration with retinal detachment, macular abnormalities and abnormalities of cranial development. One form is caused by a mutation in **collagen XVIII (COL18A)**.

knock-in A transgenic animal to which a new gene has been added (and often another similar gene deleted) rather than eliminated (knocked-out).

knockout Informal term for a transgenic organism in which the function of a particular gene has been completely eliminated (a 'null allele'). See also **homologous recombination, transposon**. Transgenic mice (knockout mice) often show disappointingly little phenotypic change, usually because there are alternative mechanisms or because the right challenge is not being made (some genes are probably unnecessary for the survival of a well-fed laboratory mouse in very well-regulated surroundings).

KNOLLE A cytokinesis-specific syntaxin-related molecule (310aa) in *Arabidopsis* that is required for the fusion of vesicles at the cell plate, the immature cross-wall. It interacts with SNAP33 and/or NPSN11 to form a t-SNARE complex and with KEULE.

knottins A structural superfamily of proteins that have a minimum of three disulphide linkages arranged as a disulphide through disulphide knot in which one disulphide bridge crosses the macrocycle formed by the two other disulphides and the interconnecting backbone. Examples include **agouti-related peptide**, **conotoxin** and various antimicrobial peptides. In November 2010 the knottin database contained 33 families, 161 structures, 1994 sequences. Link to database: http://knottin.cbs.cnrs.fr/

KNOX proteins Homeodomain proteins that are normally expressed only in apical meristems. Some phytohormones such as gibberellins (GAs) and cytokinins (CKs) are potential targets of the KNOTTED1-like homeobox (KNOX) protein.

KNUCKLES A zinc-finger transcriptional repressor (161aa) of cellular proliferation involved in regulation of patterning along the proximo-distal axis of the developing *Arabidopsis* gynoecium.

Koch's postulates The criteria, first advanced by Robert Koch in 1890, by which the causative agent of a disease can be unambiguously identified. For an organism to be accepted as the causative agent it must be (i) present in all cases, (ii) isolatable in pure culture, (iii) inoculation with the pure isolated organism should cause the disease and (iv) the organism should be observable in the experimentally infected host.

Kohler illumination The recommended type of optical microscope illumination in which the image of the lamp filament is focused in the lower focal plane of the substage condenser. As opposed to collimated illumination in which the light-emitting surface is imaged in the object. Collimated illumination requires even intensity across the light-emitting surface but is preferable for certain types of microscopy. Kohler illumination gives even illumination on the object even if there are irregularities in the brightness of the light emitting surface.

koilocytes Large cells with cleared cytoplasm and pyknotic nuclei with inconspicuous nucleoli. Koilocytosis is induced by human papilloma virus infection of the superficial epithelial cells of the uterine cervix.

Kostmann's syndrome Autosomal recessive disease characterized by profound **neutropenia**. It appears that bone marrow precursor cells fail to respond to the endogenous (normal) levels of functional G-CSF though they will respond to pharmacologic doses of G-CSF and the G-CSF receptors seem normal. Caused by mutation in the **HAX1 gene** that encodes HS1-binding protein 1.

Kozak consensus Consensus for **translational** start site of an **mRNA**. Although the trinucleotide ATG (coding for methionine) is generally considered as the start site, statistical analysis of a large number of mRNAs reveals several conserved residues around this sequence. In eukaryotes, RNNMTGG; in prokaryotes, MAYCATG (where R = A/G; Y = C/T; M = C/A; W = A/T; N = A/T/C/G). Recent article: http://www.ncbi.nlm.nih.gov/pmc/articles/PMC2241899/

KRAB Subset of **zinc finger**-type transcription factors that have a Kruppel-associated box (a domain of ~75aa).

Krabbe's disease A lysosomal storage disease affecting the white matter of the central and peripheral nervous system caused by deficiency in galactosylceramidase (galactocerebrosidase) that degrades a major lipid in myelin. An atypical form is caused by deficiency in **saposin A**.

kranz anatomy An arrangement of leaf tissue found in **C4 plants** such as maize and sugarcane, a sheath of tightly packed cells around the vascular bundles. The effect is to provide a site in which CO_2 can be concentrated around RuBisCO, thereby reducing photorespiration.

K-ras Kirsten-*ras*; see also **Kirsten sarcoma virus**.

Krebs cycle Tricarboxylic acid cycle or citric acid cycle.

kremen Proteins that are receptors for **dickkopf**. Kremen-1 (dickkopf receptor-1, kringle domain-containing transmembrane protein 1, 473aa) cooperates with dickkopf to block Wnt/beta-catenin signalling. Kremen-2 (462aa) is similar.

Krev1 Ras-related protein Rap-1A (184aa) that competes with ras for ras-GAPs and raf and can reverse the mitogenic effect of ras.

kringle Triple-looped, disulphide-linked protein domain (~80aa), found in some serine peptidases and other plasma proteins, including plasminogen (5 copies), tissue plasminogen activator (2 copies), thrombin (2 copies), hepatocyte growth factor (4 copies), apolipoprotein A (38 copies). Resemble the eponymous Scandinavian pastry.

KRIT A protein (KREV interaction trapped 1, cerebral cavernous malformations 1 protein, 736aa) that interacts with RAP1A (see **Krev1**) and is involved in determining endothelial cell shape and function by affecting cytoskeletal structure. Mutations lead to abnormal endothelial tube formation and the formation of **cerebral cavernous malformations** (CCM1). KRIT1 also interacts with **malcaverin**, a protein that is defective in CCM2.

krüppel Gap gene of *Drosophila*, encoding a zinc-finger transcription factor. See **KRAB** and **Krüppel-like factors**.

Krüppel-like factors A subfamily of transcription factors characterized by the presence of a conserved DNA-binding domain comprising three **Krüppel**-like zinc fingers. Various tissue-specific forms have been identified, for example, KLF1 (**erythroid Krüppel-like factor**, 362aa), KLF2 (lung krüppel-like factor, 355aa). KLF17 (389aa) acts as a negative regulator of epithelial-mesenchymal transition and metastasis in breast cancer.

KTX See **kaliotoxin**.

Ku A heterodimeric protein (732aa and 609aa) with ATP-dependent DNA unwinding activity (DNA helicase II) involved in the non-homologous end joining (**NHEJ**) repair pathway for rejoining DNA double-strand breaks. See **KARP-1**.

Kulchitsky cells See **enterochromaffin cells**.

Kunitz domain Domain (50−60aa) that characterises a family of serine protease inhibitors that act on peptidases of the S1 family, mostly confined to metazoa. The type example for this family is **aprotinin** (bovine pancreatic trypsin inhibitor); other examples include trypstatin, a rat mast cell inhibitor of trypsin, and tissue factor pathway inhibitor precursor. The domain can act alone. Kunitz-type serine protease inhibitors SPINT1 (478aa) and SPINT2 (252aa) inhibit the peptidase that activates **hepatocyte growth factor**.

Kuppfer cell Specialized macrophage of the liver sinusoids; responsible for the removal of particulate matter from the circulating blood (particularly old erythrocytes).

Kurloff cells Mononuclear cells with monocyte/lymphocyte properties, found in the blood and organs of guinea pigs and capybaras. They contain one large oval, elongated, or rounded cytoplasmic inclusion body, ranging from 111 m in diameter. Article: http://www.jwildlifedis.org/cgi/content/full/41/2/431

kuru Degenerative disease of the central nervous system found in members of the Fore tribe of New Guinea: a **spongiform encephalopathy**.

kwashiorkor Form of severe malnutrition of children in the tropics. Generally considered to be due to protein deficiency though it could be due to deficiency in a single essential amino acid. *Cf.* **marasmus**.

Kyasanur forest disease A tick-borne viral haemorrhagic fever, endemic to South Asia. The virus is one of the genus Flavivirus.

kynurenine pathway The major route of L-tryptophan catabolism, activated by IFNγ and IFNα, resulting in the production of nicotinamide adenine dinucleotide and other neuroactive intermediates, in particular **quinolinic acid**.

kyphoscoliosis peptidase A transglutaminase-like peptidase (561aa) that may specifically degrade **filamin C** and is required for normal muscle growth and stabilization of the neuromuscular junction. Kyphoscoliosis is a spinal deformity.

Kyte-Doolittle An algorithm for calculating a **hydropathy plot** for a protein sequence.

L

L Single letter code for leucine.

L1 (1) Neural adhesion molecule (CD171, 1257aa) of the **immunoglobulin superfamily** with binding domains similar to **fibronectin**. The purified molecule, immobilized on a culture dish, is a potent substrate for neurite outgrowth. *In vivo* is important in the development of the nervous system being involved in neuron-neuron adhesion, neurite fasciculation and outgrowth of neurites. Binds **axonin**. See also **neuroglian** and **NCAM**. (2) A major capsid protein (531aa) of Human papillomavirus type 16.

L2 (1) Inner lipoyl domain (L2) of the dihydrolipoyl acetyltransferase. (2) Rat lung epithelial L2 cells. (3) Reovirus lambda2 core spike (*L2*) gene. (4) A minor capsid protein (473aa) of Human papillomavirus type 16. Interacts with major capsid protein L1. (5) *Toxocara canis* L2 excretory/secretory antigen.

L32 Ribosomal protein (188aa) that forms part of the 50 S ribosomal subunit, found in both prokaryotes and eukaryotes. Ribosomal protein L32 of yeast binds to and regulates the splicing and the translation of the transcript of its own gene.

L929 cells Line of fibroblasts originally cloned from areolar and adipose tissue of an adult C3H mouse.

LA-PF4 Synonym of **connective tissue-activating peptide III**.

La protein Protein (La autoantigen, La ribonucleoprotein domain family 3, LARP3, 408aa) transiently bound to unprocessed cellular precursor RNAs that have been produced by polymerase III; stabilizes nascent pre-tRNAs from nuclease degradation, influences the pathway of pre-tRNA maturation, and assists correct folding of certain pre-tRNAs. Mainly located in the nucleus. A nuclear autoantigen in **Sjogren's syndrome, systemic lupus erythematosus**, and neonatal lupus. La-related protein (LARP1) has a similar RNA-binding domain and a second conserved RNA-binding domain, the LARP1 domain. Other LARPs have been found and, for example, LARP7 is a negative transcriptional regulator of polymerase II genes.

lac operon See **lactose operon**.

laccase A group of multi-copper proteins (EC.1.10.3.2, urishiol oxidases) of low specificity. Act on both O- and P-quinols, and often also on aminophenols and phenylenediamine. First found in the lac (lacquer) tree but now known to be widespread. In *Arabidopsis*, laccase-1 (581aa) is involved in lignin degradation and detoxification of lignin-derived products

lactacystin A microbial metabolite (376 Da) isolated from *Streptomyces*. Widely used as a selective and irreversible inhibitor of the 20S **proteasome**.

lactadherin Mucin-associated milk glycoprotein (milk-fat globule-EGF factor 8, 387aa) important in intestinal epithelial homeostasis and the promotion of mucosal healing. Promotes VEGF-dependent neovascularization, contributes to phagocytic removal of apoptotic cells in many tissues and is a ligand for the $\alpha_v\beta_3$ and $\alpha_v\beta_5$ integrins; also binds phosphatidylserine-enriched surfaces. Has been used as a tumour marker because it is expressed in most breast cancer cells. Protective for suckling young because it binds to rotavirus and inhibits replication. The precursor is cleaved into a short form (186aa) and a smaller fragment, medin (50aa) which is the main constituent of aortic medial amyloid.

lactalbumin A milk protein. Alpha-lactalbumin (lactose synthase B protein, lysozyme-like protein 7, 142aa) is the regulatory subunit of **lactose synthetase**.

lactase The enzyme (EC 3.2.1.23, 1927aa) required for digestion of lactose: most adults have lost this enzyme and may be lactose intolerant as a result. Adult Caucasians, however, maintain lactase, apparently due to a noncoding variation in the MCM6 (minichromosome maintenance-6) gene upstream of the lactase gene, which enhances activation of the lactase promoter.

lactate The terminal product of anaerobic glycolysis. Accumulation of lactate in tissues is responsible for the so-called oxygen debt.

lactate dehydrogenase The enzyme (LDH, EC 1.1.2.3, 332aa) that catalyses the formation and removal of lactate. The appearance of LDH in cell culture medium is often used as an indication of cell death and the release of cytoplasmic constituents.

lacticin A two-peptide **lantibiotic** produced by *Lactococcus lactis*. Both peptides, LtnA1 (59aa) and LtnA2 (65aa), must interact to produce antibiotic activity. LtnA1 interacts specifically with lipid II in the outer leaflet of the bacterial cytoplasmic membrane and then recruits LtnA2 to generate a three-component complex which inhibits cell wall biosynthesis and forms a potassium-permeable pore.

Lactobacillus Genus of Gram-positive anaerobic or facultatively aerobic bacilli; the product of their

The Dictionary of Cell and Molecular Biology. DOI: http://dx.doi.org/10.1016/B978-0-12-384931-1.00012-X

glucose fermentation is lactate. Important in production of cheese, yoghurt, sauerkraut and silage.

lactoferrin Iron-binding protein (lactotransferrin, 710aa) of very high affinity for iron (K_d 10^{-19} at pH 6.4, 26-fold greater than that of **transferrin**), found in milk and in the specific granules of neutrophil leucocytes. By sequestering iron may have a bacteriostatic effect. Cleaved into: kaliocin-1 (31aa), lactoferroxin-A (6aa), lactoferroxin-B (5aa) and lactoferroxin-C (7aa). The lactoferroxins have opioid antagonist activity.

lactoferroxin See **lactoferrin**.

lactogenic hormone See **prolactin**.

lactoglobulin A globular protein (178aa) present in most milk (except human); constitutes 50–60% of bovine whey protein. Can be allergenic in humans.

lactoperoxidase Peroxidase enzyme (EC 1.11.1.7, 712aa) from milk. Used experimentally to generate active iodine as a non-permeant radiolabel for membrane proteins. Has homology to myeloperoxidase and eosinophil peroxidase.

lactose The major sugar in human and bovine milk. Conversion of lactose to lactic acid by *Lactobacillus* is important in the production of yoghurt and cheese.

lactose carrier protein The proton symport protein (lactose permease, 417aa), product of the *LacY* gene, coded for in the **lactose operon** and responsible for the uptake of lactose by *E. coli*.

lactose operon Group of adjacent and coordinately controlled genes concerned with the metabolism of lactose in *E. coli*. The lac operon was the first example of a group of genes under the control of an **operator** region to which a **lactose repressor** binds. When the bacteria are transferred to lactose-containing medium, allolactose (which forms by transglycosylation when lactose is present in the cell) binds to the repressor, inhibits the binding of the repressor to the operator, and allows transcription of mRNA for enzymes involved in galactose metabolism and transport across the membrane (beta-galactosidase, galactoside permease, and thiogalactoside transacetylase). *LacZ* codes for beta-galactosidase, *LacY* for the permease, *LacA* for the transacetylase.

lactose repressor Protein (tetramer of 360aa subunits) that normally binds with very high affinity to the **operator** region of the **lactose operon** and inhibits transcription of the downstream genes by blocking access of the polymerase to the promoter region. When the lactose repressor binds allolactose, its binding to the operator is reduced and the gene set is derepressed.

lactose synthase The enzyme responsible for the synthesis of lactose from glucose and UDP-galactose. The catalytic A chain (EC 2.4.1.22, 398aa) is derived from β-1,4-galactosyltransferase 1 by processing; the regulatory B subunit is α-**lactalbumin**. The Golgi complex form catalyzes the production of lactose in the lactating mammary gland, the cell surface form functions as a recognition molecule. Mutations cause a disorder of glycosylation which affects multiple systems.

lacuna *Pl.* lacunae. Small cavity or depression, for example, the space in bone where an **osteoblast** is found.

LacZ See **lactose operon**.

LAD syndrome Leucocyte adhesion deficiency syndrome. See **LFA-1**.

laddering A regular (ladder-like) pattern of oligonucleotide sizes on electrophoretic gels characteristic of apoptotic cells. It is a consequence of the cleavage of the DNA strand between **nucleosome** beads by **endonucleases**.

laf-3 An *Arabidopsis* mutant (long after far-red 3) deficient in a protein (583aa and 576aa isoforms) that may regulate nucleo-cytoplasmic trafficking of intermediates involved in phytochrome A signal transduction. The mutants show defective hypocotyl elongation in response to far-red light.

laforin A dual specificity phosphatase (EC 3.1.3.16, EC 3.1.3.48, 331aa) that may be involved in the control of glycogen metabolism and acts as a scaffold protein to facilitate protein phosphatase 1 regulatory subunit 3C (317aa) ubiquitination by **malin**. Defects can cause progressive **myoclonic epilepsy** type 2.

lag phase The interval after which cells in a microbial population will start to grow after inoculation into fresh medium.

LAG-1 (1) Longevity-assurance protein-1. (2) Lymphocyte activation gene-1 protein.

LAL test See *Limulus polyphemus*.

LALF peptide A cyclic peptide (Limulus Anti-LPS Factor peptide), based on amino acids 31–52 of Limulus anti-LPS factor, that binds **lipid A** and neutralizes LPS *in vivo* and *in vitro*.

LAMB2 An important protein constituent (laminin beta2, formerly s-laminin, 1798aa) of certain kidney and muscle basement membranes. Defects in the gene are responsible for **Pierson's syndrome**, an autosomal recessive disorder, usually fatal in early infancy.

lambda bacteriophage λ *phage* Bacterial DNA virus, first isolated from *E. coli*. Its structure is similar to that of the **T even phages**. It shows a **lytic**

cycle and a **lysogenic** cycle, and studies on the control of these alternative cycles have been very important for the understanding of the regulation of gene **transcription**. It is used as a cloning vector, accommodating fragments of DNA up to 15 kilobase pairs long. For larger pieces, the **cosmid** vector was constructed from its ends.

lambda chain See **L-chain**.

lamellar phase See **phospholipid bilayer**.

lamellipodium Flattened projection from the surface of a cell, often associated with locomotion of fibroblasts.

lamina (1) Flat sheet; as in **basal lamina**. (2) See **nuclear lamina**. (3) A program, LAMINA (Leaf shApe deterMINAtion), for the automated analysis of images of leaves. Link: http://www.biomedcentral.com/1471-2229/8/82#B20

lamina propria Fibrous layer of connective tissue underlying the basal lamina (**basement membrane**) of an epithelium. May contain smooth muscle cells and lymphoid tissue in addition to fibroblasts and extracellular matrix.

laminarin Storage polysaccharide of *Laminaria* and other brown algae; made up of $\beta(1-3)$-glucan with some $\beta(1-3)$ linkages.

laminin Important component of the basal lamina, first identified in matrix produced in culture by mouse **EHS cells**, and composed of an A chain (3075aa) and two B chains (1786aa) although different tissue-specific forms of laminin occur. In laminin from placenta the A chain is replaced with merosin, in laminin found near the neuromuscular junction the B1 chain is replaced by laminin B2 (**LAMB2**, formerly s-laminin (synapse laminin)). Laminin induces adhesion and spreading of many cell types and promotes the outgrowth of neurites in culture. The three chains A, B1 and B2 have been renamed alpha, beta, and gamma respectively and sub-types of each are known. **Kalinin** is laminin 5. See **netrins, neurexins**.

lamins Proteins that form the nuclear lamina, a polymeric structure intercalated between chromatin and the inner nuclear envelope in metazoa. A-type lamins (lamin A/C, 664aa) have C-terminal sequences homologous to the head and tail domains of **keratins**. B-type lamins (lamins B1, 586aa; B2, 600aa) are expressed ubiquitously and have a role in spindle assembly. Mutations in A-type lamins are associated with Emery-Dreifuss **muscular dystrophy, cardiomyopathy, lipodystrophy** and **progeria**. See **emerin**.

lampbrush chromosomes Large chromosomes (as long as 1 μm), actually meiotic **bivalents**, seen during prophase of the extended meiosis in the oocytes of some Amphibia. Segments of DNA form loops in pairs along the sides of the sister chromosomes, giving them a brush-like appearance. These loops are not permanent structures but are formed by the unwinding of **chromomeres** and represent sites of very active RNA synthesis.

LAMPs Lysosomal-associated membrane proteins, heavily glycosylated but to a variable extent. Lamp-1 (LEP100, LGP120, CD107a, 417aa) and Lamp-2 (410aa) may protect lysosomal and plasma membranes from attack by lysosomal enzymes and have been implicated in metastasis. Defects in Lamp-2 cause **Danon's disease**.

Landry-Guillain-Barre syndrome See **Guillain-Barre syndrome**.

Langerhans See **Islets of Langerhans** and **Langerhans' cells**.

Langerhans' cell histiocytosis A disorder (histiocytosis X) in which there is abnormal proliferation of **Langerhans cells**, high levels of soluble **RANKL** and **IL-17A** and multiple granulomas. Also used by some pathologists to describe a set of syndromes including, in decreasing severity: Letterer-Siwe disease, **Hand-Schuller-Christian disease** and eosinophilic granuloma of bone. The classification and causes of these disease are confused at present.

Langerhans' cells Immature **dendritic cells** derived from bone marrow, strongly MHC Class II positive and weakly phagocytic, found in the basal layers of the epidermis where they serve as **accessory cells**, responsible for **antigen processing** and **antigen presentation**. Having been exposed to antigen they migrate to the lymph nodes. Part of the immune surveillance system, their location means that they are readily exposed to antigens that penetrate the dermal barrier.

Langhans' giant cells Multinucleate cells formed by fusion of epithelioid macrophages and associated with the central part of early tubercular lesions. Similar to **foreign body giant cells**, but with the nuclei peripherally located.

Langmuir trough A device for studying the properties of lipid monolayers at an air/water interface. A moveable barrier connected to a balance allows measurement of surface pressure.

Langmuir-Blodgett film In biophysics, an ordered monolayer of molecules produced on the surface of water. An **amphipathic** molecule is floated at low concentration on the surface of the water, and steadily compressed into an ordered surface by moving a barrier across the surface (usually done in a **Langmuir trough**).

lanthanum *La* Lanthanum salts are used as a negative stain in electron microscopy, and as calcium-channel blockers.

lanthionine A non-protein amino acid composed of two alanine residues crosslinked on their β-carbon atoms by a thioether linkage. Found in human hair, lactalbumin, and feathers, also in bacterial cell walls. Are a component of gene encoded peptide antibiotics, **lantibiotics**.

lanthiopeptin See **ancovenin**.

lantibiotics Polycyclic thioether peptide antibiotics containing the unusual amino acids lanthionine or methyllanthionine. Produced by a large number of Gram-positive bacteria, such as *Streptococcus* and *Streptomyces*, to attack other Gram-positive bacteria: thus considered bacteriocins. The prototype lantibiotic, **nisin**, inhibits peptidoglycan synthesis and forms pores through specific interaction with the cell wall precursor lipid II. The flexible amphiphilic type-A lantibiotics (e.g. nisin and epidermin) act primarily by pore formation in the bacterial membrane. The rather rigid and globular type-B lantibiotics inhibit enzyme functions through interaction with the respective substrates: **mersacidin** and actagardine inhibit the cell wall biosynthesis by complexing lipid II, whereas the cinnamycin-like peptides inhibit phospholipases by binding phosphoethanolamine. Other examples are **ancovenin**, **subtilin**. Article: http://www.iptonline.com/articles/public/IPT_27_p22nonprint.pdf

LAP (1) A transcription factor (liver activator protein, IL6-dependent DNA-binding protein, CCAAT/enhancer-binding protein beta, CEBPB, 345aa) that stimulates transcription of genes for **acute phase proteins** and is important in regulation of genes affected by cyclin D1. The LAP gene is unusual in having no introns. (2) See **erbin** (LAP2). (3) Leukemia associated protein (E3 ubiquitin-protein ligase LAP, 206aa) of viruses including Myxoma. (4) The LAP (LRR and PDZ) protein family includes leucine-rich repeat-containing protein 7 (densin-180, LAP1, 1537aa) that may be involved in the organization of synaptic cell-cell contacts. (5) Scribble homologue (LAP4, 1630aa) is a scaffold protein involved in apical/basal polarity in epithelia, an activator of Rac GTPase activity. (6) Leucine aminopeptidase (e.g. cytosol aminopeptidase 3, LAP3, EC 3.4.11.1, 519aa) that cleaves unsubstituted N-terminal amino acids from various cytoplasmic peptides. Similar enzymes are found in mitochondria, in chloroplasts, in secreted form, and as bacterial enzymes.

LAPF One of the **phafins**, an adapter protein (lysosome-associated apoptosis-inducing protein containing PH and FYVE domains, 279aa) that recruits phosphorylated p53 to lysosomes.

LAR Leucocyte antigen-related protein (LAR-PTP, EC 3.1.3.48, 1501aa in rat), is the prototype of a family of transmembrane protein tyrosine phosphatases with extracellular domains composed of Ig and fibronectin type III (FnIII) domains and two cytoplasmic catalytic domains, one active, one inactive. LAR-family phosphatases (LAR, PTPδ, PTPσ) play a role in axon guidance, mammary gland development, regulation of insulin action and glucose homeostasis. Interacts with **GRIP1**, **PPFIA1**, PPFIA2 and PPFIA3. See **liprins**.

LARGE A glycosyltransferase (acetylglucosaminyltransferase-like 1A, 756aa) that is required for normal glycosylation of dystroglycan; defects in either Large or dystroglycan cause abnormal neuronal migration. There are mutations in Large in both the myodystrophy mouse and congenital muscular dystrophy type 1D (MDC1D). See **muscular dystrophy-dystroglycanopathy with brain and eye anomalies**.

large T-antigen See **T-antigen**.

large-cell lymphoma Highly malignant group of tumours arising from transformed lymphocytic precursors. Diffuse large cell lymphomas (DLCL) are mostly of B-cell origin and are the commonest of the non-Hodgkin's lymphomas; mutations in p53 are common. Less common are anaplastic large-cell lymphomas that generally arise from T-cells. Article: http://emedicine.medscape.com/article/202969-overview

Laron dwarfism Human growth defect in which the growth hormone receptor is mutated and cells do not respond to growth hormone. A Laron syndrome-like phenotype, associated with immunodeficiency, is caused by a postreceptor defect, a mutation in the STAT5B (signal transducer and activator of transcription 5B) gene.

Larsen's syndrome A skeletal disorder caused in some cases by mutation in the gene for **filamin B**.

Las17p *S. cerevisiae* homologue of human Wiskott-Aldrich syndrome protein (**WASP**); interacts with the **Arp2/3** complex.

LASPs See **LIM and SH3 proteins**.

LASS1 The human homologue (LAG1, UOG-1, 350aa) of the yeast longevity assurance gene product, a membrane protein of the ER involved in sphingolipid synthesis. In *S. cerevisiae* the LAG genes, a subgroup of the homeobox gene family, encode proteins differentially expressed during the replicative life span that are necessary for N-stearoyl-sphinganine synthesis and play a role in determining yeast longevity, although the mechanism of action is unclear. All Lag1 homologues contain a highly conserved stretch of 52aa known as the Lag1p motif.

Lassa virus Virulent and highly transmissible member of the **Arenaviridae** whose normal host is a rodent (*Mastomys natalensis*); first recorded from Nigeria. Causes an acute haemorrhagic fever.

LAT (1) The LAT adapter (linker for activation of T-cells, 262aa) is a transmembrane protein essential for the transmission of T-cell receptor (TCR)-mediated signalling. The intracytoplasmic domain contains nine tyrosines which, when phosphorylated by ZAP-70 upon receptor aggregation, recruit SH2 domain-containing cytosolic enzymes and adapters. LAT is tyrosine phosphorylated in human platelets in response to collagen, collagen-related peptide (CRP), and FcγRIIA cross-linking. (2) L-type amino acid transporters (e.g. LAT1, CD98, SLC7A5, 507aa) involved in sodium-independent, high-affinity transport of large neutral amino acids such as phenylalanine, tyrosine, leucine, arginine and tryptophan, when associated with 4F2hc/CD98 heavy subunit encoded by SLC3A2. (3) In *Solanum tuberosum* (potato) an anther-specific protein LAT52 (161aa).

late gene Gene expressed relatively late after infection of a host cell by a virus, usually encoding structural proteins for the viral coat.

latency (1) In electrophysiology: the time between onset of a stimulus and peak of the ensuing **action potential**. (2) Of an infection, a period in which the infection is present in the host without producing overt symptoms. See **latent virus**.

latent membrane protein Oncogenic **Epstein-Barr virus** (EBV)-encoded latent membrane proteins (LMP1, 360–400aa depending upon viral strain; LMP2, 497aa) interact with various intracellular signalling systems to induce proliferation or block apoptosis. LMP1 acts as a constitutively active TNF receptor that upregulates anti-apoptotic genes. It is a potential target for immunotherapy of some cases of Hodgkin's disease, nasopharyngeal carcinomas, EBV-associated natural killer (NK)/T lymphomas, and chronic active EBV infection. LMP2 signals through **lyn** and up-regulates the survivin gene through the NFκB pathway.

latent TGFβ-binding proteins A family of proteins that bind TGFβ and may be involved in the assembly, secretion and targeting of TGFβ1 to sites at which it is stored and/or activated. LTBP1, (1721aa) and LTBP3 (1303aa) may have a structural roile in extracellular matrix, LTBP2 (1821aa) may be involved in elastic-fibre organization and assembly. LTBP4 (1624aa) is mutated in a severe form of **cutis laxa** (Urban-Rifkin-Davis syndrome).

latent virus Virus integrated within host genome but inactive: may be reactivated by stress such as ultraviolet irradiation.

lateral diffusion Diffusion in two dimensions, usually referring to movement in the plane of the membrane, such as the motion of fluorescently labelled lipids or proteins measured by the technique of **fluorescent recovery after photobleaching** (FRAP).

lateral inhibition A simple form of information processing. The classic example is found in the eye, whereby ganglion cells are stimulated if photoreceptors in a well defined field are illuminated, but their response is inhibited if neighbouring photoreceptors are excited (an 'on field/off surround' cell), or *vice versa* an 'off field/on surround' cell. The effect of lateral inhibition is to produce edge- or boundary-sensitive cells, and to reduce the amount of information that is sent to higher centres; a form of peripheral processing.

***LATERAL ORGAN BOUNDARIES* family** A large, plant-specific family of DNA-binding transcription factors, some of which have been implicated in a variety of developmental processes. The *Arabidopsis LOB* gene is expressed at the boundaries of lateral organs during vegetative and reproductive plant development. There are 43 proteins encoded in the *Arabidopsis* genome that contain a LOB domain (which has motifs resembling a zinc-finger and a leucine zipper). Article: http://nar.oxfordjournals.org/content/35/19/6663.full

latex Milky fluid that exudes from cells and vessels (laticifers) when many plants are cut. It is a watery solution containing many different substances including terpenoids (which form rubber), alkaloids (e.g. opium alkaloids), sugar, starch and others.

lathyrism Disorder of collagen cross-linking as a result of copper sequestration by nitriles. (Lysyl oxidase is a copper-containing metalloenzyme). Can be caused by eating seeds of the sweet pea (*Lathyrus odoratus*; odoratism). In animals a neurological disease, caused by eating chickling peas (*Lathyrus sativus*) which contain the neurotoxic amino acid beta-N-oxalylamino-L-alanine (BOAA), a glutamate analogue.

latrophilin Calcium-independent alpha-latrotoxin receptor 1 (latrophilin-1, lectomedin-2, 1474aa) which has high affinity for α-latrotoxin. Probably regulates exocytosis. Other latrophilins (latrophilin-2, 1459aa; latrophilin-3, 1447aa) are similar. It is proteolytically cleaved into 2 subunits, an extracellular subunit and a seven-transmembrane subunit (see **GPS** #2).

latrotoxins Toxins in the venom of Black Widow spiders (*Latrodectus* spp.). There are five insecticidal toxins, α, β, γ, δ, and ε-latroinsectotoxins, one toxin affecting crustaceans, α-latrocrustatoxin, and a vertebrate-specific neurotoxin, α-latrotoxin (1401aa). They all act by forming pores that allow calcium influx which triggers neurotransmitter release.

latrunculin A macrolide inhibitor of actin polymerization (binds G actin rendering it assembly incompetent), now often used in preference to cytochalasin because of greater potency. Latrunculins A and B were isolated from the Red Sea sponge *Negombata* *magnifica* (Demospongiae, Latrunculiidae), although chemical syntheses have been developed. The compound is found within vesicles in archaeocytes and choanocytes and may be for defensive purposes.

laulimalide A natural product from a marine sponge (*Cacospongia mycofijiensis*); a microtubule-stabilizing agent that binds to tubulin at a site distinct from that of the **taxanes**.

LAX **gene family** A family of plant genes encoding auxin influx/permease sequences (MtLAXs). Article: http://www.nature.com/cr/journal/v12/n3/abs/7290131a.html

lazy leucocyte syndrome A rare human complaint in which neutrophils display poor locomotion towards sites of infection. Thought to be due to a defect in the cytoplasmic actomyosin system leading to impaired movement and reduced deformability that affect release from the bone marrow and emigration into tissues, but sufficiently rare that this has not been explored in detail.

LB medium A growth medium for *E. coli*, typically contains 10 g of tryptone and 5 g of yeast extract per litre. LB medium was originally developed by Bertani to optimize *Shigella* growth and plaque formation. The abbreviation, according to Bertani, was intended to stand for lysogeny broth.

Lbc Lymphoid blast crisis oncogene product (A-kinase anchor protein 13, AKAP-13, p47, 2813aa) that acts as an adapter protein to selectively couple G alpha-13 and Rho. One of the Dbl-like family.

LBD In *Arabidopsis*, a large family of proteins containing LOB domains (see *LATERAL ORGAN BOUNDARIES* **family**). E.g. LOB domain-containing protein 1 (LBD1, ASYMMETRIC LEAVES 2-like protein 8, 190aa).

LBP (1) Serum protein (lipopolysaccharide binding protein, 481aa) that binds lipid A of lipopolysaccharide; levels rise in patients with severe Gram-negative sepsis and it may function as a recognition molecule in conjunction with CD14. (2) Transcription factor LBP-1 (upstream-binding protein 1, 540aa) a transcriptional activator.

LC-1 (1) Liver cytosol antigen type 1, subsequently identified as **formiminotransferase cyclodeaminase**. An autoantibody to the antigen is a marker for childhood autoimmune chronic active hepatitis type 2. (2) LC is often an abbreviation for light chain.

LCAT An enzyme (lecithin:cholesterol acyltransferase, EC 2.3.1.43, 416aa) bound to high-density lipoproteins (HDLs) and low-density lipoproteins in the plasma that catalyzes the formation of cholesterol esters in lipoproteins. There are two autosomal recessive disorders caused by mutations of the LCAT gene; in familial LCAT deficiency

(complete LCAT deficiency) all activity is lost, in **fish-eye disease** and **Norum disease** there is a partial defect.

LCCL domain A conserved protein module (~100aa) found in various proteins, named for three of them, Limulus factor C, **cochlin**) and late gestation lung protein Lgl1. May be involved in LPS binding. Database entry: http://supfam.cs.bris.ac.uk/SUPERFAMILY/cgi-bin/scop.cgi?sunid = 69848

L cells Cell line established by Earle in 1940 from mouse connective tissue. L929 cells are a subclone of this original line.

L-chain *light chain* Although **light chains** are found in many multimeric proteins, L-chain usually refers to the light chains of immunoglobulins. These are of 22 kDa and of one of two types, kappa (κ) or lambda (λ). A single immunoglobulin has identical light chains (2 κ or 2 λ). Light chains have one variable and one constant region. There are **isotype** variants of both κ and λ.

lck Lymphocyte-specific protein tyrosine kinase (p56-Lck, 509aa), a non-receptor protein-tyrosine kinase of the **src-family** that is involved in transduction of **T-cell receptor**-mediated activation by tyrosine phosphorylation of **STAT5**. Interacts with **zap70** and **fyn**.

LCM An anchor polypeptide in cyanobacteria (~980aa), involved in photosynthesis.

LCR See **locus control region**.

LD78 Chemokines (CCL3, ~92aa) that probably play an inhibitory role in haematopoiesis. LD78α binds to CCR1, CCR4 and CCR5; has inflammatory and chemokinetic properties. LD78β is chemotactic for lymphocytes and monocytes and is a ligand for CCR1, CCR3 and CCR5; it down-regulates the expression of CCR5 in monocytes and macrophages and is a potent HIV-1 inhibitor (acting through CCR5), neutrophil activator (through CCR1) and eosinophil activator (through CCR1 and CCR3).

LDH See **lactate dehydrogenase**.

LDL See **low density lipoprotein**.

LDL-receptor See **low density lipoprotein receptor**.

LE body A globular mass of nuclear material composed of DNA and antinuclear antibody; derived from lymphocyte nuclei and associated with lesions of **systemic lupus erythematosus**. Original description: http://www.ncbi.nlm.nih.gov/pmc/articles/PMC2136812/pdf/575.pdf

LE cell Phagocyte that has ingested nuclear material (LE bodies) from another cell: characteristic of **systemic lupus erythematosus**.

leader sequence In bacteria the upstream untranslated region (5'UTR) of mRNA that has the ribosome binding motif (Shine-Dalgarno sequence). It may have control elements such as the **iron-responsive element** or other binding regions that are sensitive to the levels of particular metabolites. In some cases the leader sequence is actually transcribed into a leader peptide that contains several residues of the amino acid being regulated. If the translation of the leader peptide region is restricted by shortage of the amino acid then the operon for synthesis of that amino acid is activated, if the leader peptide is translated then further transcription is prematurely terminated and the operon remains inactive.

leading lamella The front of a crawling cell, such as a fibroblast, supported by a meshwork of microfilaments which excludes most cytoplasmic granules. Protrusion of the leading lamella and the formation of new adhesions is an essential part of the locomotory process.

leaf explant tissue culture A common form of plant tissue culture in which a sterile piece of leaf is placed on agar in growth medium with the addition of cytokinin and auxin to promote callus formation.

leaf spot Any of several plant diseases characterized by the appearance of dark spots on the leaves.

LEAFY A transcription factor (420aa) that interacts with **APETALA1** in flower development and regulates expression of the ABC classes of floral homeotic genes.

LEAFY COTYLEDON A family of plant transcription factors involved in regulating development and seed maturation.

leaky mutation Mutation in which sub-normal function exists – for example if a mutation leads to instability in a protein rather than its complete absence or there is reduced expression of a gene.

Leber's congenital amaurosis A group of autosomal recessive retinal dystrophies originally described by Leber in 1869. They are the most common cause of congenital visual impairment. Currently 14 different types are recognised with identified mutations in retinal guanylate cyclase (Leber's congenital amaurosis-1, LCA1), a retinal pigmented epithelium-specific protein of 65 kDa (RPE65, LCA2), spermatogenesis associated protein-7 (SPATA7, LCA3), arylhydrocarbon receptor interacting protein-like 1 (AIPL1, LCA4), **lebercilin** (LCA5), retinitis pigmentosa GTPase regulator-interacting protein 1 (RPGRIP1, LCA6), cone-rod homeobox-containing transcription factor (CRX, LCA7), Drosophila crumbs homologue 1 (CRB1, LCA8), 290 kDa centrosomal protein (CEP290, LCA10), inosine-5-monophosphate dehydrogenase (LCA11), retinol dehydrogenase-12 (LCA13), lecithin retinol acyltransferase (LCA14). Loci for the others are known but the mutated genes have not yet been identified. See **Leber's optic atrophy**.

Leber's optic atrophy A disorder (Leber's hereditary optic neuritis, hereditary optic atrophy, Leber's disease) in which there is acute or subacute loss of central vision, usually in middle-age. The disease is associated with mutations in multiple genes encoded by the mitochondrial genome and therefore maternally inherited. See **Leber's congenital amaurosis**.

lebercilin A protein (697aa) that may be involved in minus end-directed microtubule transport and that interacts with many proteins associated with centrosomal or ciliary functions. Mutated in **Leber's congenital amaurosis** Type 5 (LCA5). Lebercilin-like protein (670aa) has been identified although its function has not been investigated. Article: http://hmg.oxfordjournals.org/content/8/1/51.long

LEC A chemokine (liver-expressed chemokine, CCL16, novel CC chemokine (NCC)-4, SCYA16, LCC-1, HCC4, LMC, Ck-beta-12, monotactin-1, SCYL4, 120aa) chemotactic for monocytes and a functional ligand for CCR1, CCR2 and CCR5. It is constitutively expressed by hepatocytes.

LECAM L-selectin. See **selectins**.

lecithin Phospholipids from egg yolk (usually hen's eggs). A mixture of phosphatidyl choline and phosphatidyl ethanolamine, but usually refers to phosphatidyl choline.

lecithinase Phospholipase C.

lecticans A term infrequently used for a family of chondroitin sulphate proteoglycans that have a hyaluronan-binding and C-type lectin domain and act as cross-linkers in the extracellular matrix. In humans includes **aggrecan**, **versican**, neurocan and brevican. Webpage: http://www.imperial.ac.uk/research/animallectins/ctld/mammals/Groups/GroupI.html

lectin Proteins obtained particularly from the seeds of leguminous plants, but also from many other plant and animal sources, that have binding sites for specific mono- or oligosaccharides. Named originally for the ability of some to selectively agglutinate human red blood cells of particular blood groups. Lectins such as **concanavalin A** and **wheat germ agglutinin** are widely used as analytical and preparative agents in the study of glycoproteins. They are classified according to the carbohydrate-recognition domain (CRD) of which there are two main types, **S-type lectins** and **C-type lectins**. See **selectin** and Table L1.

left-right asymmetry The body plan of vertebrates has only partial bilateral symmetry, and there are various left/right asymmetries. In humans the reverse pattern (e.g. of heart and viscera) is **situs inversus** due to a **dynein** defect but a number of

TABLE L1. Lectins

Some common plant-derived lectins

Many plant-derived lectins have been characterised and only a small sample of commonly encountered ones are listed below.

Source	Abbreviation	Sugar specificity
Bandieraea simplicifolia	BSL1	α-D-gal > α-D-GalNAc
Concanavalla ensiformis (Jack bean)	ConA	α-D-Man > α-D-Glc > α-D-GlcNAc
Dolichos biflorus	DBA	α-D-GalNAc
Lens culinaris (lentil)	LCA	α-D-Man > α-D-Glc > α-D-GlcNAc
Phaseolus vulgaris (red kidney bean)	PHA	β-D-Gal (1−4)-D-GlcNAc
Arachis hypogaea (peanut)	PNA	β-D-Gal (1−3)-D-GalNAc
Pisum sativum (garden pea)	PSA	α-D-Man > α-D-Glc
Ricinus communis (castor bean)	RCA1	β-D-Gal > α-D-Gal
Sophora japonica	SJA	β-D-GalNAc > β-D-Gal > α-D-Gal
Glycine max (soybean)	SBA	α-D-GalNAc > β-D-GalNAc
Ulex europaeus (common gorse)	UEA1	α-L-fucosyl
Triticum vulgaris (wheat germ)	WGA	β-D-GlcNAc(1−4) GlcNAc > β-D-GlcNAc(1−4)-β-D-GlcNAc

Animal lectins

Sub-family	Example
C-type Lectins	
Endocytic receptors	macrophage mannose receptor
Collectins	tetranectin
Selectins	ELAM-1
Lymphocyte lectins	CD69
Proteoglycans	versican core protein
Miscellaneous	endothelial cell scavenger receptor
Viral lectins	gp22−24 of vaccinia virus
Snake venom	botrocetin
Invertebrate lectins	echinoidin
S-type Lectins	
Galectins 1−10	

Gal = galactose; GalNAc = galactosamine; Glc = glucose; GlcNAc = glucosamine; Man = mannose.

other genes are known to be implicated in the specification of the L-R axis. These include *nodal*, *lefty*, *rotatin*, the gene for the homeobox transcription factor ptx2 and **ZIC3**. An autosomal form of visceral heterotaxy (HTX2) can be caused by mutation in the *CFC1* gene that encodes the cryptic protein (223aa), a co-receptor for nodal, or in the type IIB **activin** A receptor.

lefty Proteins of the **TGFβ** family (lefty-1 and lefty-2, both 366aa) involved in the determination of **left-right asymmetry**. Involved in regulation of **nodal**. Mutations in the gene for lefty-2 cause left-right axis malformations in humans. In mice the equivalent genes are expressed only on the lefthand side of the embryo.

Legg-Calve-Perthes disease See **Perthes' disease**.

leghaemoglobin Form of haemoglobin (in *Vicia faba*, 144aa) found in the nitrogen-fixing root-nodules of legumes. Binds oxygen, and thus protects the nitrogen-fixing enzyme, **nitrogenase**, that is oxygen sensitive. Similar haemoglobins are found in plants that do not have symbiotic nitrogen fixing bacteria and are referred to as non-symbiotic haemoglobins.

Legionella Genus of Gram-negative asporogenous bacteria. Most species are pathogenic in humans, causing pneumonia-like disease, e.g. Legionnaires' disease, named after an outbreak in Philadelphia amongst members of an American Legion reunion.

legumins Plant storage proteins that occur in the seeds of many dicot families (e.g. legumin A, 517aa in *Pisum sativum*); a compound similar to legumin is also produced in monocots. Typically have two sub-units, one acidic, one basic, derived by proteolysis from a precursor. The quaternary structure involves six acidic and six basic polypeptides all linked by disulphide bonds. Often up-regulated by biotic stress. See **cupins**.

Leidig cells See **Leydig cells**.

Leigh's syndrome A progressive neurodegenera-tive disorder with considerable genetic heterogene-ity; can be caused by mutations in both nuclear- and mitochondrial-encoded genes for components of energy-generating systems, including mitochondrial respiratory chain complexes I–V and the pyruvate dehydrogenase complex.

leiomodins Actin-binding proteins of the **tropomo-dulin** family. Leiomodin-1 (600aa) is expressed in smooth muscle, some striated muscle, eye and thryoid and is the thyroid-associated ophthalmopathy autoantigen. Leiomodin-2 (cardiac leiomodin, 547aa) may block the pointed end of actin filaments. Leiomodin-3 (560aa) is the fetal form.

leiomyoma Benign tumour of smooth muscle in which parallel arrays of smooth muscle cells form bundles which are arranged in a whorled pattern. The amount of fibrous connective tissue is very vari-able. Leiomyoma of the uterus (fibroid) is the com-monest form.

Leishman stain Romanovsky-type stain; a mix-ture of basic and acid dyes that differentially stains various classes of leucocytes in blood smears.

Leishmania A genus of protozoan parasites (of the Order Trypanosomatida) that live intracellularly in macrophages. Various forms of the disease are known, depending upon the species of parasite: in particular visceral leishmaniasis (**kala-azar**), and mucocutaneous leishmaniasis.

LEKTI A serine protease inhibitor (lymphoepithelial Kazal-type-related inhibitor, 1064aa), the main inhibitor of kallikrein-7 of the cornified layer of the skin. Probably important for the anti-inflammatory and antimicrobial protection of mucous epithelia. Mutations lead to **Netherton's syndrome**.

LEM domain The LEM domain (lamina-associated polypeptide-**emerin**-MAN1) is a motif ($\sim$40aa) shared by a group of **lamin**-interacting proteins in the inner nuclear membrane (INM) and in the nucle-oplasm that mediates binding to **barrier-to-autoin-tegration factor**. LEMD3 (LEM domain-containing protein 3, MAN1, 911aa) is a repressor of TGFβ, activin, and BMP signalling; mutations cause Buschke-Ollendorff syndrome, a skeletal dysplasia, and melor-heostosis, a rare mesenchymal dysplasia. Database entry: http://supfam.cs.bris.ac.uk/SUPERFAMILY/cgi-bin/scop.cgi?sunid = 63451

lengsin A lens-specific member of the glutamine synthetase superfamily (509aa) but enzymatically inactive. Binds intermediate filament proteins such as **vimentin** and **phakinin** and may have a structural role.

lenticels A small patch of the periderm in which intercellular spaces are present allowing some gas exchange between the internal tissues of the stem and the atmosphere.

Lentisphaerae A bacterial phylum that has been suggested to be a member of a super-phylum together with the Planctomycetes, Verrucomicrobia and Chlamydiae. *Lentisphaera araneosa* is a marine bacterium. Abstract: http://www.ncbi.nlm.nih.gov/pubmed/16704931

Lentivirinae Subfamily of the **Retroviridae**, non-oncogenic retroviruses that cause 'slow diseases' that are characterized by horizontal transmission, long incubation periods and chronic progressive phases. Visna virus is in this group, and there are similarities between visna, and equine infectious anaemia virus.

lentoid (1) Lens-shaped or lens-like. (2) Spherical cluster of retinal cells, formed by aggregation *in vitro*, that has a core of lens-like cells inside which accumulate proteins characteristic of normal lens. The cells concerned derive from retinal glial cells. *In vivo* may form ectopically in the eye through defects in **Wnt** signalling.

LEOPARD syndrome A syndrome in which there are multiple lentigines, electrocardiographic conduction abnormalities, ocular hypertelorism, pul-monic stenosis, abnormal genitalia, retardation of growth and sensorineural deafness. Can be caused by mutations in the *PTPN11* gene or the *RAF1* gene (see **Noonan's syndrome**).

Lepore haemoglobin Variant haemoglobin in a rare form of **thalassemia**: there is a composite δ-β chain as a result of an unequal **crossing-over** event. The composite chain is functional but synthesized at reduced rate. Anti-Lepore haemoglobins have the converse, the N-terminus of beta-globin and the C-terminus of delta-globin.

leprosy Disease (Hansen's disease) caused by *Mycobacterium leprae*, an obligate intracellular parasite that survives lysosomal enzyme attack by possessing a waxy coat. Leprosy is a chronic disease associated with depressed cellular (but not humoral) immunity; the bacterium requires a lower temperature than 37°C, and thrives particularly in peripheral Schwann cells (leading to destruction of peripheral nerves and a failure to respond to pain) and macrophages. Only humans and the nine-banded armadillo are susceptible. There are now dapsone-resistant strains.

leptin Product (167aa) of the ob (obesity) locus. Found in plasma of mouse and man: reduces food uptake and increases energy expenditure. Mutations in mouse lead to obesity but this seems only very rarely to be the case in man. The receptor is a single-pass membrane protein of the type I cytokine receptor family (1165aa) and signals through JAK2/STAT3. In rat the receptor is mutated in the fa/fa strain which show profound obesity of early onset.

leptine Toxic glycoalkaloid from *Solanum sp.* (esp. wild potato). Aglycone is acetylleptinidine; different trisaccharides are added to produce Leptine I and II. See **chaconine**.

leptinine Glycoalkaloid from Solanaceae. Similar to **chaconine**. Aglycone is leptinidine.

leptocyte An abnormally thin erythrocyte, large in diameter and with a central pigmented area and a peripheral ring of haemoglobin. Found in some types of anaemia.

leptomycin B Antibiotic first identified in *Streptomyces sp.* as an unsaturated, branched-chain fatty acid with antifungal activity. Has a range of inhibitory activities but the key one is inhibition of the export of proteins from the nucleus. Binds to **exportin-1** in its conserved central region at a critical cysteine residue and prevents formation of the complex between the exportin and the nuclear export signal of cargo proteins.

leptonema See **leptotene**.

Leptospira Genus of aerobic **spirochaete** bacteria with both saprophytic and pathogenic species. Pathogenic *Leptospira* spp., such as *L. interrogans*, *L. borgpetersenii*, *L. weilii* and *L. kirschner*, are the causative agents of leptospirosis (Weil's disease), a mild chronic infection in rats and many domestic animals. The bacteria are excreted continuously in the urine and contact with infected urine or water can result in infection of humans via cuts or breaks in the skin. Leptospirosis, in which the organism localises in the kidneys causing renal failure, is an occupational hazard for sewerage and farm workers.

leptotene A Classical term for the first stage of **prophase** I of **meiosis**, during which the chromosomes condense and become visible.

Lesch-Nyhan syndrome A sex-linked recessive inherited disease in humans that results from mutation in the gene for the purine salvage enzyme **HGPRT**, located on the X chromosome. Results in severe mental retardation and distressing behavioural abnormalities, such as compulsive self-mutilation.

let-7 A **microRNA** (lethal-7, 21 nucleotides), one of the first to be described. In *C. elegans* let-7 is required for timing of cell fate determination; loss of function leads to early death, over-expression to premature expression of adult traits. Temporal upregulation of let-7 miRNA in the seam cells is required for their terminal differentiation. Let-7 complementary sites are upstream regulatory sites in genes such as that for let-60 (ras orthologue in *C. elegans*) that bind the microRNA. In humans, various let-7 homologues map to regions deleted in human cancers.

lethal mutation Mutation that eventually results in the death of an organism carrying the mutation.

LETS Original name (now obsolete) for a large extracellular transformation/trypsin sensitive cell-surface protein that was altered on transformation *in vitro*: now known to be **fibronectin**.

Letterer-Siwe disease See **Langerhans cell histiocytosis**.

Leu enkephalin See **enkephalins**.

Leu-phyllolitorin A bombesin-related peptide, Leu8-**phyllolitorin**, that increases branching of developing airways and augments thymidine incorporation in cultured lung buds. Central injection of Leu8-phyllolitorin has been shown to produce hypothermia in animals exposed to a cold environment. See **phyllolitorin**.

leuc- or leuco- *US* leuk- or leuko-. Prefix meaning white. E.g. **leucocyte**. The c and k forms are interchangeable although in the UK leucocyte and leukaemia are probably the common forms.

leucine The most abundant amino acid found in proteins (Leu, L, 131 Da). Confers hydrophobicity and has a structural rather than a chemical role. See Table A1.

leucine aminopeptidase An **exopeptidase** (EC 3.4.11.3) that removes neutral amino acid residues, not just leucine, from the N-terminus of proteins. In *C. elegans*, leucine aminopeptidase 1, 491aa, in *Arabidopsis*, chloroplast leucine aminopeptidase 3, 583aa. Related enzymes are found in bacteria.

leucine zipper Motif found in certain **DNA-binding proteins**. In a region of around 35aa, every seventh is a leucine. This facilitates dimerisation of two such proteins to form a functional transcription factor. Examples of proteins containing leucine zippers are products of the **proto-oncogenes** *myc*, *fos* and *jun*. See also **AP-1**.

leucine-responsive regulatory proteins Family of bacterial transcriptional regulators that control a large variety of genes, including those coding for cell appendages and other potential virulence factors. In *E. coli*, LRP (164aa) mediates a global response to leucine and activates several operons.

leucine-rich repeat *LRR* Short motif (around 24aa) with 5−7 leucines generally at positions 2, 5, 7, 12, 21, 24. Forms an amphipathic region and is probably involved in protein-protein interactions.

leucinopine An analogue of **nopaline** found in crown gall tumours induced by *Agrobacterium tumefasciens* that do not synthesize octopine or nopaline.

leucocidin *Panton-Valentine leucocidin* Distinct exotoxins from staphylococcal and streptococcal species of bacteria that cause killing or lysis of myeloid but not lymphoid cells. Only a small number of strains produce the toxin but Panton-Valentine leucocidin (PVL)-producing *Staphylococcus aureus* are an emerging problem. The active haemolysin has two subunits, S and F, (slow and fast on chromatographic columns) that are inactive alone but interact to form a pore; the F subunit is shared with gamma-haemolysin. The PVL genes are carried by two phages, namely phiPVL and phiSLT.

leucocyte *US.* leukocyte. Generic term for a white blood cell. See **basophil, eosinophil, lymphocyte, monocyte, neutrophil**.

leucocytopenia See **leucopenia**.

leucocytosis An excess of **leucocytes** in the circulation.

leucoderma See **vitiligo**.

leucopenia *leucocytopenia* An abnormally low number of white cells in the blood.

leucoplast Colourless **plastid**, that may be an **etioplast** or a storage plastid (**amyloplast, elaioplast** or **proteinoplast**).

leucosulfakinin Cockroach **peptide hormones** (10−11aa), that affect gut motility. Related to **gastrin**.

leukaemia *US.* leukemia. Malignant neoplasia of **leucocytes**. Several different types are recognized according to the stem cell that has been affected, and several virus-induced leukaemias are known. Both acute and chronic forms occur: (**1**) In acute lymphoblastic leukaemia (ALL) there are large numbers of primitive lymphocytes (high nuclear/cytoplasmic ratio characteristic of dividing cells and few specific surface antigens expressed); tends to be common in the young. (**2**) In acute myeloblastic leukaemia (AML), which is more common in adults, the proliferating cells are of the **myeloid** haematopoietic series and the cells appearing in the blood are primitive **granulocytes** or **monocytes**. (**3**) In chronic lymphocytic leukaemia (CLL) there are excessive numbers of normal, mature lymphocytes in the circulation, usually B-cells. A disease of middle or old age. (**4**) Chronic myelogenous leukaemia (CML) is commonest in middle-aged or elderly people, characterized by excessive numbers of circulating leucocytes of the myeloid series, most commonly neutrophils (or precursors), but occasionally eosinophils or basophils. (**5**) Hairy cell leukaemia is a rare CLL in which the proliferating B-cells have hair-like cytoplasmic projections on their surfaces. (**6**) Virally-induces leukaemias include those caused by **Abelson leukaemia virus, avian leukaemia virus,** feline leukaemia virus, **Friend murine leukaemia virus, HTLV** and **Moloney murine leukaemia virus**.

leukaemia inhibitory factor Polypeptide **growth factor** or **cytokine** (LIF, 202aa) with a wide range of activities including the induction of terminal differentiation in leukaemic cells. Regulates growth and differentiation of primordial germ cells and embryonic stem cells but has effects on peripheral neurons, osteoblasts, adipocytes and various cells of the myeloid lineage. Given to adult animals induces weight loss, behavioural disorders and bone abnormalities. Many of the effects of LIF *in vitro* can be mimicked by **interleukin-6, oncostatin M** and **ciliary neurotrophic factor**, all of which interact indirectly with gp130, a shared tranducer subunit. N.B., not the same as leukocyte inhibitory factor (leukocyte migration inhibitory factor), an operational definition of an activity and probably not a distinct factor.

leukemia See **leukaemia**.

leuko- See **leuco-**.

leukodystrophy A set of disorders in which there is a defect in myelination causing progressive degeneration of the white matter of the brain. Adult-onset autosomal dominant leukodystrophy is associated with duplication of the **lamin B1** gene. Adrenoleukodystrophy (Siemerling-Creutzfedt disease, Bronze Schilder disease) is an X-linked disorder in which there is a defect in peroxisomal beta oxidation caused by mutation in the ATPase binding cassette protein (ABCD1) gene. Long-chain fatty acids accumulate in tissues. Neonatal adrenoleukodystrophy is a result of mutation in the peroxisome receptor gene (*PTS1R, PEX5*) or **peroxin** genes. Metachromatic leukodystrophies are a group of disorders caused by deficiencies in **aryl sulphatase A** or **saposin B** or by **multiple sulphatase deficiency**. In adrenomyeloneuropathy there are muscular and neural side effects.

leukosialin Widely distributed membrane-associated mucin, the major sialoglycoprotein (CD43, sialophorin, 400aa) of thymocytes and mature T-cells.

Transmembrane protein with extensive O-linked glycosylation (75–85 oligosaccharides on the 239aa extracellular domain). Extends at least 45 nm beyond plasma membrane. Similar but not homologous to **episialin**.

leukosis The correct term for an excess of leucocytes in the circulation and other parts of the body rather than leucocytosis.

leukotrienes A family of hydroxyeicosatetraenoic (HETE) acid derivatives (LTA_4, LTB_4, LTC_4, LTD_4, LTE_4). LTA_4 and LTB_4 are modified lipids; leukotrienes C, D and E have the lipid conjugated to glutathione (LTC_4) or cysteine (LTD_4, LTE_4) to form the peptidyl leukotrienes. A mixture of the latter (LTC_4, LTD_4, LTE_4) constitute SRS-A, the **slow reacting substance of anaphylaxis**, that has potent bronchoconstrictive effects. LTB_4 is a potent neutrophil chemotactic factor.

leupeptin Family of modified-tripeptide peptidase inhibitors. Commonest is N-acetyl-Leu-Leu-argininal.

levodopa The L-amino acid precursor of dopamine that increases dopamine levels in the basal ganglia thereby improving mobility in Parkinsonism. Usually given together with a dopa-decarboxylase inhibitor (**carbidopa**) to prevent peripheral effects of dopamine.

Lewis blood group Surface antigens on red cells, terminal fucose residues on the H antigen of the A, B, H system added by fucosyl-transferases (FUT2, FUT3). There are only three phenotypes: Le(a^-b^-); Le(a^+b^-); and Le(a^-b^+).

Lewy body Hyaline eosinophilic concentrically-laminated inclusions found in the **substantia nigra** and locus ceruleus of patients with **Parkinson's disease** and Lewy body dementia. Composed of **alpha-synuclein** associated with ubiquitin, neurofilament protein, synphilin and alpha B crystallin.

lex A In bacteria a repressor protein (EC 3.4.21.88, 202aa) for genes involved in the SOS system of DNA repair. In the presence of single-stranded DNA, recA interacts with lexA causing an autocatalytic cleavage which disrupts the DNA-binding part of lexA derepressing the SOS regulon.

Leydig cell Interstitial cells of the mammalian testis, involved in synthesis of testosterone.

LFA-1 Lymphocyte function-related antigen-1, a heterodimeric lymphocyte plasma-membrane protein (α_L, CD11a, 1170aa; β, CD18, 95 kDa, 769aa) that binds **ICAM-1**, particularly involved in cytotoxic T-cell killing. One of the **integrin** superfamily of adhesion molecules. Deficiency of LFA-1 in leucocyte adhesion deficiency (LAD) syndrome leads to severe impairment of normal defences and poor survival prospects. The related surface adhesion molecules which share a common β subunit (sometimes

referred to as the LFA-1 class of adhesion molecules) are Mac-1 (α_M, CD11b, 1152aa) and p150,95 (α_X, CD11c, 150 kDa, 1163aa) and are defective in severe forms of LAD if the β subunit is missing. Mac-1 (also known as Mo-1 in earlier literature) is the complement C3bi receptor (CR3) and is present on mononuclear phagocytes and on neutrophils; p150,95 is less well characterized, but is particularly abundant on macrophages. Integrin $\alpha_D\beta_2$ (CD11d, 1162aa) is a receptor for ICAM3 and VCAM1 and may be involved in clearing lipoproteins from atherosclerotic plaques and in phagocytosis of blood-borne particles.

LFA-3 Lymphocyte function-related antigen-3 (CD58, 250aa), the ligand for the CD2 adhesion receptor that is expressed on cytolytic T-cells. LFA-3 is expressed on endothelial cells at low levels. The CD2/LFA-3 complex is an adhesion mechanism distinct from the **LFA-1/ICAM-1** system, and binding of erythrocyte LFA-3 to T-cell CD2 is the basis of **E-rosetting**.

Lfc Murine oncoprotein of the Dbl-related family (LBC's first cousin, lymphoid blast crisis-like 1, rho guanine nucleotide exchange factor 2, rhobin, 985aa). Contains a Dbl-homology domain in tandem with a **PH domain** and is similar to Lsc, Lbc, **Tiam-1** and Dbl.

L-forms Bacteria lacking cell walls, a phenomenon usually induced by inhibition of cell-wall synthesis, sometimes by mutation.

LG3 peptide A peptide (195aa) cleaved from **perlecan** that has anti-angiogenic properties that require binding of calcium ions for full activity.

LGP2 Cytoplasmic RNA helicase (EC 3.6.4.13, 678aa) related to **RIG1** and **melanoma differentiation-associated protein 5** and involved in the innate immune defense against viruses. The interaction of LGP2 with intracellular dsRNA produced during viral replication, triggers a transduction cascade leading to the expression of antiviral cytokines.

LH See **luteinizing hormone**.

LHRF See **luteinizing hormone releasing factor**.

LHX A family of homeobox proteins that have a **LIM domain**. At least eight LHX transcription factors are known in humans (~350–400aa), involved in control of differentiation of various tissues.

Li-Fraumeni syndrome A predisposition to tumours caused by mutation in the **p53** gene or in **checkpoint kinase-2**.

lichen A large group of symbiotic associations between fungi and a photosynthetic partner, usually a green alga but sometimes a cyanobacterium. Several genera of algae and of fungi are involved and the associations are so stable and of such varied but distinct types that the lichens have been classified into genera and species. A variety of incompatibility phenomena are often manifest between individual lichens. Confined to terrestrial habitats and often used as indicators of pollution status of the environment.

lichen planus Rare skin disorder in which there is marked hyperkeratosis and extensive infiltration of lymphocytes into the lower epidermis.

Liddle's disease A type of salt-sensitive human hypertension caused by mutation in the beta or gamma subunit of the multi-subunit epithelial **amiloride-sensitive sodium channel** (ENaC).

LIF See **leukaemia inhibitory factor**. Also used for leukocyte inhibitory factor.

ligand Any molecule that binds to another; in normal usage a soluble molecule such as a hormone or neurotransmitter, that binds to a receptor. The decision as to which is the ligand and which the receptor is often a little arbitrary when the broader sense of receptor is used (where there is no implication of transduction of signal). In these cases it is probably a good rule to consider the ligand to be the smaller of the two – thus in a lectin-sugar interaction, the sugar would be the ligand (even though it is attached to a much larger molecule, recognition is of the saccharide).

ligand-gated ion channel A transmembrane **ion channel** the permeability of which is increased by the binding of a specific **ligand**, typically a neurotransmitter at a **chemical synapse**. The permeability change is often drastic; such channels let through effectively no ions when shut, but allow passage at up to 10^7 ions s^{-1} when a ligand is bound. The receptors for both **acetylcholine** and **GABA** have been found to share considerable sequence homology, implying that there may be a family of structurally related ligand-gated ion channels.

ligand-induced endocytosis The formation of coated pits and then **coated vesicles** as a consequence of the interaction of ligand with receptors, which then interact with **clathrin** and associated proteins (coatomers) on the cytoplasmic face of the plasma membrane and come together to form a pit. Not all coated vesicle uptake of receptors requires receptor occupancy.

ligase amplification reaction Method for detecting small quantities of a target DNA, with utility similar to **PCR**. It relies on DNA ligase to join adjacent synthetic oligonucleotides after they have bound the target DNA. Their small size means that they are destabilised by single base mismatches, and so form a sensitive test for the presence of mutations in the target sequence.

ligases Major class of **enzymes** that catalyse the linking together of two molecules (synthetases, category 6 in the **E classification**) e.g. DNA ligases that link two fragments of DNA by forming a **phosphodiester** bond.

ligatin A peripheral membrane protein (584aa) that binds and localizes glycoproteins to the external cell surface and within endosomes (a trafficking receptor for phosphoglycoproteins).

LIGHT A type II transmembrane protein of the TNF superfamily (TNFSF14, 240aa) produced by activated T cells. Forms a membrane-anchored homotrimeric complex that will bind the lymphotoxin β receptor (LTβR) and the herpes simplex virus entry mediator. Name apparently based on 'homologous to lymphotoxin, exhibits inducible expression, competes with herpesvirus glycoprotein D for herpes virus entry mediator on T cells' and justifiably abbreviated.

light chain Nonspecific term used of the smaller subunits of several multimeric proteins, for example **immunoglobulin, myosin, dynein, clathrin**. See also **L-chain**.

Light Green A stain often used for counterstaining cytoplasm following iron haematoxylin; a component of Masson's trichrome stain.

light harvesting complex A large family of proteins, accessories to the core light-gathering phostosystems I and II. LHCI and LHCII are associated with PSI and PSII respectively. They improve photosynthesis at low light energy levels.

light microscopy In the simplest form of light microscopy, bright-field microscopy, the image is formed mainly by absorption of light by the specimen, although there is a minor contribution from diffraction. By using various dyes (histological stains) different parts of the specimen have their light absorption increased, aiding identification. In interference methods the image is formed as a result of constructive or destructive interference between light that has passed through the specimen, which has a different refractive index, with a reference beam. Phase contrast microscopy is the commonest of the interference methods and is routinely used to observe live cells. More complex interference methods allow quantitative estimation of the extent of the refractive index differences between various part of the specimen. Dark-field microscopy (dark ground microscopy) is used to image small objects that diffract light and appear bright against a dark background. Polarisation microscopy involves illumination with plane polarised light, and an analyser that is rotated to give total extinction of light that has not had the plane of polarisation rotated and the background is therefore dark. In these circumstances only **birefringent** objects appear bright. **Fluorescence microscopy** involves illumination with ultraviolet

TABLE L2. Types of light microscopy

Method	Physical parameter detected
A. With axial illumination	
*Without Spatial filtration**	
I. Bright field	Absorption by specimen (may be operated in Visible, UV or IR. regions of the spectrum and in quantitative microspectrophotometric modes)
II. Interference:	
Transmitted	Path difference arising in specimen, qualitative or quantitative
Reflected (interference reflection = IRM)	Path difference in films 1 to 10 wavelengths thick next to substrate. For cell contacts.
III. Fluorescence	Natural fluorescence or that of probes applied to system.
IV. Dark-field	Refractive index discontinuities revealed by scattered light.
V. Polarisation	Birefringent and/or dichroic properties.
With Spatial filtration	
VI. Confocal scanning microscopy	Contrast and resolution enhanced by selection of light paths modified by the object at the back focal plane of the objective. Bright-field or fluorescence modes. Usually combined with video processing of the image.
VII. Phase contrast	Path differences revealed as contrast differences non-quantitatively and non-regularly, using phase plate at back focal plane.
VIII. Differential interference contrast (DIC) = Nomarski	Path difference gradients revealed as contrast or colour differences.
IX. Out-of-focus phase contrast	Path differences revealed as diffraction patterns.
B. With anaxial illumination	
With Spatial filtration	
X. Hoffman modulation contrast	Path differences
XI Single side-band edge enhancement (SSEE microscopy)	Path differences from first order diffractions.

NOTES. Nearly all systems can be run in the epi (incident) illumination mode. Video (image) processing can enhance contrast and resolution in images by the application of simple algorithms to expand the grey scale, reduce noise and subtract background. More complex processing is possible, including the extraction of further information by Fourier transforms.
*Spatial filtration.** This is the application of methods to remove those ray paths which have not interacted with the object. This is done at the back focal plane of the objective. It can also be applied to select or remove ray paths that have interacted in some specified way with the object.

light at a wavelength that excites fluorescence, usually in the visible spectrum, in dyes used to stain the specimen: a system of filters blocks the exciting light. The illumination is often through the objective (epi-illumination) rather than through a condenser. See also **confocal microscopy, fluorescence speckle microscopy, interference microscopy, interference reflection, nanovid microscopy, ratio-imaging fluorescence microscopy, total internal reflection fluorescence microscopy. Electron microscopy** has a separate entry. See Table L2.

light scattering A phenomenon exhibited by suspensions of particles in fluid. The extent of the scattering is related, in a complex fashion, to the size and shape of the particles. See **nephelometer**.

light-dependent reaction (1) Any reaction that can be activated by light. (2) Usually, the reaction taking place in the chloroplast in which the absorption of a photon leads to the formation of ATP and NADPH.

light-harvesting system Set of photosynthetic pigment molecules that absorb light and channel the energy to the photosynthetic **reaction centre**, where the light reactions of **photosynthesis** occur. In higher plants, contains **chlorophyll** and **carotenoids**, and is present in two slightly different forms in **photosystems I and II**.

Lightcycler Tradename for a **real time PCR** machine used to quantify specific DNA sequences, or, after reverse transcription, mRNA levels. It uses fluorescent detection during the PCR reaction to quantify the DNA at each cycle.

lignin A complex polymer of phenylpropanoid subunits, laid down in the walls of plant cells such as **xylem** vessels and **sclerenchyma**. Imparts considerable strength to the wall, and also protects it against degradation by microorganisms. Lignification also waterproofs the hollow tubes (tracheary elements) of the water-transporting xylem system. Lignin is laid down as a defence reaction against pathogenic attack, as part of the **hypersensitive response** of plants. Dicotyledonous angiosperm lignins contain two major monomer species, termed guaiacyl (G) and syringyl (S) units linked through at least five different dimer bonding patterns. See **shikimic acid pathway**, **phenylalanine ammonia lyase**, **protolignin**, **phenylpropanoid pathway**, **caffeoyl CoA O-methyl-transferase**.

lignostilbene alphabeta-dioxygenase Bacterial and fungal oxidoreductase enzymes (EC 1.13.11.43, ~500aa) thought to be involved in lignin biodegradation. One product is vanillin.

Like Sm proteins A subfamily of proteins with homology to **Sm proteins** that form heptameric complexes (Lsm1p-7p in the cytoplasm; Lsm2p-8p in the nucleus) involved in RNA decapping (localized in the cytoplasm and the nucleus respectively). Lsm1p-7p is highly conserved in eukaryotes and is involved in RNA **decapping**; the Lsm2p-8p complex binds to U6 snRNA and functions in RNA splicing. Article: http://ukpmc.ac.uk/articles/PMC2553750

LIM and SH3 proteins Proteins (LASPs) with a **LIM domain** and an **SH3 domain**, ubiquitously expressed and involved in cytoskeletal architecture and the organization of focal adhesions. LASP-1 (261aa) is involved in neuronal differentiation and plays a role in the migration and proliferation of certain cancer cells. that is overexpressed in breast and ovarian cancer. LASP2 (LIM/nebulette, 270aa) is a splice variant of **nebulin**.

LIM domain Zinc-binding domain found in proteins required for developmental decisions. Contain 60aa conserved, cysteine-rich, repeats that will interact with a second LIM domain. Named after first 3 genes in group: Lin-11 (*C. elegans* − required for asymmetric division of blast cells), Isl-1 (mammalian insulin-gene binding enhancer protein), mec-3 (*C. elegans* − required for differentiation of a set of sensory neurons). LIMS1 (LIM and senescent cell antigen-like domains-1, Particularly interesting new Cys-His protein, PINCH1, 314aa) has 5 LIM domains and is an obligate partner of **integrin-linked kinase** both of which are necessary for proper control of cell shape change, motility, and survival. See **LIMD1**.

LIM kinases Serine/threonine kinases with two LIM motifs and a C-terminal protein kinase domain. LIMK1 (647aa) has actin-binding properties and

regulates microfilament dynamics by phosphorylating ADF/cofilin family members. LIMK2 (638aa) will phosphorylate myelin basic protein and histone *in vitro*.

LIM mineralization protein A protein (PDZ and LIM domain protein 7, enigma, 457aa) that may act as an adapter that localizes LIM-binding proteins to actin filaments. Involved in osteogenesis.

limatin Actin-binding LIM protein 1 (778aa) that may be involved in development of the retina and in axon guidance.

limb bud An outpushing of mesenchyme surrounded by a simple epithelium that will develop to form the limb in a vertebrate embryo. The distal region is referred to as the progress zone. There has been extensive study of positional information within the limb-bud that determines, for example, the proximal-distal pattern of bone development and the anterior-posterior specification of digits.

limbic system Those regions of the central nervous system responsible for autonomic functions and emotions. Includes hippocampus, amygdaloid nucleus and portions of the mid-brain.

LIMD1 A protein (LIM domain-containing protein 1, 676aa), one of the **zyxin/ajuba family** that acts as a transcriptional regulator suppressing the expression of most genes with E2F1-responsive elements although is usually located in the cytoplasm. May act as a tumour suppressor. Article: http://www.pnas.org/content/101/47/16531.long

LIME A transmembrane adaptor (**TRAP**) required for B-cell receptor (BCR)-mediated B-cell activation. LIME (Lck-interacting transmembrane adaptor, 295aa) is expressed in mouse splenic B-cells. Upon BCR cross-linking, LIME is tyrosine phosphorylated by **lyn** and associated with lyn, **Grb2**, **PLC-gamma2**, and **PI3K**.

limit of resolution See **resolving power**.

limitin A murine cytokine (interferon alpha-11, interferon-zeta, 182aa) of the interferon alpha/beta family. Suppresses B lymphopoiesis through ligation of the interferon-α/β (IFNα/β) receptor, activation of **Tyk2** and the up-regulation and nuclear translocation of **Daxx**.

limonene The major component of the oil extracted from citrus rind; D-limonene is characteristic of oranges, L-limonene of lemons. Used for a wide variety of purposes as a solvent and additive to cleaning solutions.

Limulus polyphemus Now renamed *Xiphosura*, though *Limulus* is still in common usage. The king crab or horseshoe crab, found on the Atlantic coast of North America. It is more closely related to the arachnids than the crustacea, and horseshoe crabs are the only surviving representatives of the subclass Xiphosura. Its compound eyes have been widely used in studies on visual systems, but it is probably better known from the Limulus-amoebocyte lysate (LAL) test; LAL is very sensitive to small amounts of **endotoxin**, clotting rapidly to form a gel, and the test is used clinically to test for septicaemia.

LIN complex A multiprotein complex in human cells that is required for transcriptional activation of G2/M genes. It consists of five different proteins RbAp48, LIN9, LIN37, LIN52 and LIN54 and is related to the **DREAM complex** in *Drosophila* and the DRM complex in *C. elegans*. *Cf.* **LINC complex**.

LIN-2 See CASK.

lin-4 One of the first **microRNA** species to be described. In *C. elegans* lin-4 negatively regulates protein lin-28 (abnormal cell lineage protein 28, 227aa) that controls the choice of stage specific cell fates. Degradation of the lin28mRNA depends upon a lin-4 complementary element (LCE) in the 3'-UTR. Lin28 may negatively regulate the larval to adult transition by suppressing the microRNA **let-7**.

LINC complex A multi-protein complex of **Sun proteins** and **nesprins** that links the nucleoskeleton and cytoskeleton across the nuclear envelope. N.B. not the same as the **LIN complex**.

lincomycin A macrolide antibiotic active against Gram-positive bacteria, isolated from *Streptomyces lincolnensis*. Blocks protein synthesis by binding to the 50S subunit of the ribosome and interfering with the peptidyl transferase reaction.

LINE See **long interspersed nucleotide element**.

line probe assay A PCR-based assay used to detect the presence of specific nucleic acid in a sample using oligonucleotide probes immobilized as lines on nitrocellulose paper. A positive outcome is the appearance of a coloured line. Commercially-available kits are used, for example, to detect the presence of *Mycobacterium tuberculosis* in sputum and whether it is a drug-resistant strain, but it is potentially applicable to a wide range of diagnostic challenges.

linear dichroism A spectroscopic technique that uses linearly polarized light (polarized in only one direction): the linear dichroism is the difference in absorption of light polarized parallel and perpendicular to an orientation axis. The technique is used to study the functionality and structure of molecules but in LD experiments the molecules need to have a preferential orientation otherwise the LD = 0. See **circular dichroism**.

Lineweaver-Burke plot A plot of 1/v against 1/S for an enzyme-catalyzed reaction, where v is the initial rate and S the substrate concentration. From the equation: $1/v = 1/V_{max} \cdot (1 + K_m/S)$ the parameters

V_{max} and K_m can be determined. The equation over-weights the contribution of the least accurate points and other methods of analysis are preferred; see **Eadie-Hofstee plot**.

lining epithelium An epithelium lining a duct, cavity or vessel, that is not particularly specialized for secretion or as a mechanical barrier. Not a precise classification.

linkage Tendency for certain genes tend to be inherited together, because they are on the same chromosome. Thus parental combinations of characters are found more frequently in offspring than non-parental. Linkage is measured by the percentage recombination between loci.

linkage disequilibrium The occurrence of some alleles together, more (or less) often than would be expected. Random assortment produces linkage equilibrium (although slowly if the two loci are close) but a disequilibrium will only persist if there is some selective advantage in the association.

linoleic acid An essential **fatty acid** (9, 12, octadecadienoic acid); occurs as a glyceride component in many fats and oils.

linolenic acid An 18-carbon fatty acid with three double bonds (9, 12, 15, octadecatrienoic acid) and α- and γ-isomers. Essential dietary component for mammals. See **fatty acids**.

lipaemia Presence in the blood of an abnormally large amount of lipid.

lipases Enzymes that break down mono-, di- or triglycerides to release fatty acids and glycerol. Calcium ions are usually required. Triglyceride lipases (EC 3.1.1.3) hydrolyze the ester bond of triglycerides.

lipid A The lipid associated with polysaccharide in the **lipopolysaccharide** (LPS) of Gram-negative bacterial cell walls.

lipid bilayer See **phospholipid bilayer**.

Lipid II A membrane-anchored cell-wall precursor that is essential for bacterial cell-wall biosynthesis. Consists of a GlcNAc-MurNAc- pentapeptide subunit linked by a pyrophosphate to a polyisoprenoid anchor 11 subunits long. The target molecule for at least four different classes of antibiotic, including **vancomycin** and several **lantibiotics**.

lipid transfer protein See **cholesteryl ester transfer protein**, **microsomal triglyceride transfer protein** and **phospholipid transfer protein**.

lipidoses Storage diseases in which the missing enzyme is one that degrades sphingolipids (sphingomyelin, ceramides, gangliosides). In **Tay-Sachs disease** the lesion is in hexosiminidase A, an enzyme that degrades ganglioside Gm2; in **Gaucher's disease**, glucocerebrosidase; in **Niemann-Pick** disease, sphingomyelinase.

lipids Biological molecules soluble in apolar solvents, but only very slightly soluble in water. They are an heterogenous group (being defined only on the basis of solubility) and include fats, waxes and terpenes. See Table L3.

lipins A family of homologous proteins (lipins1, 2 and 3). Lipin-1 (phosphatidate phosphatase LPIN1, EC 3.1.3.4, 890aa) catalyzes the dephosphorylation of PA to yield diacylglycerol and inorganic phosphate. Lipin acts as a nuclear transcriptional coactivator to modulate lipid metabolism gene expression and is required for normal adipose tissue development. Mutations in the lipin-1 gene are associated with autosomal recessive recurrent myoglobinuria and in the mouse with fatty liver dystrophy. Mutations in lipin-2 (896aa) are found in patients with **Majeed syndrome**. Lipin-3 (851aa) has similar functions.

lipoamide The functional form of lipoic acid in which the carboxyl group is attached to protein by an amide linkage to a lysine amino group.

lipocalins Members of the calycin superfamily of carrier proteins that transport small, hydrophobic molecules, such as retinol, porphyrins, odorants. Characterized by two orthogonally-stranded **beta sheets** which protect the cargo from the aqueous environment. Examples: α-1-microglobulin, lipocalin-1 (176aa), **purpurin**, **orosomucoid**. Neutrophil-gelatinase-associated lipocalin (NGAL, 198aa) is an iron-trafficking protein has been used as a biomarker in the detection of acute renal failure. Many tissue-specific lipocalins are known.

lipocortin The name given to a calcium-binding protein that was thought to secreted by macrophages and that acted as an inhibitor of phospholipase A2 enzymes. Lipocortin-1 is **annexin A1** and does not seem to be secreted. Several different lipocortins have been described although their function has been contested.

lipocyte Liver cell that stores lipid.

lipodystrophy A set of disorders in which the distribution of adipose tissue in the body is altered and in extreme cases there may be a near-absence of adipocytes. Mutation in genes for various proteins are known to be involved including those for AGPAT2 (lysophosphatidic acid acyltransferase, LPAAT, 1-acyl-sn-glycerol-3-phosphate acetyltransferase, EC 2.3.1.51), **seipin**, **lamin** A/C, and **PPARγ**. In some cases of partial lipodystrophy, there is an association with C3 nephritic factor. Berardinelli-Seip congenital lipodystrophy is caused by mutation in the gene for **caveolin-1**.

lipofectamine Proprietary liposome preparation (lipofectin), formulated from cationic lipids, for lipid-mediated **transfection** of cultured cells.

lipofuscin Brown pigment characteristic of ageing. Found in lysosomes and is the product of peroxidation of unsaturated fatty acids and symptomatic, perhaps, of membrane damage rather than being deleterious in its own right.

lipoic acid Regarded as a coenzyme in the oxoglutarate dehydrogenase complex of the **tricarboxylic acid cycle**, an organosulphur compound (6,8-dithiooctanoic acid, thioctic acid). Involved generally in oxidative decarboxylations of α-keto acids. A growth factor for some organisms.

lipoid congenital adrenal hyperplasia A severe form of congenital adrenal hyperplasia in which the synthesis of steroid hormones is blocked due to a defect in **steroidogenic acute regulatory protein**. Affected individuals are all phenotypic females.

lipolysis The breakdown of fat into fatty acids and glycerol.

lipoma A benign tumour composed of fatty tissue.

lipoma-preferred partner A protein (LIM domain-containing preferred translocation partner in lipoma, LPP, 612aa) with considerable homology to **zyxin** and containing three **LIM domains**. It shuttles between the cytoplasm and the nucleus and is found at cell-to-cell contacts and focal adhesions. May be involved in maintaining cell shape and in signalling. In some benign lipomas the LPP gene from chromosome 3 becomes fused to the DNA-binding molecule, HMGIC (high mobility group 1C).

lipomodulin Obsolete name for **lipocortin** isolated from neutrophils.

lipophilic Having an affinity for lipids, and thus hydrophobic.

lipophilin See **proteolipid protein**.

lipophorin A family of high-density lipoproteins (600−700 kDa) from insect **haemolymph**, that transport diacyl glycerols, hormones, morphogens and other relatively hydrophobic molecules. The molecule comprises (in *Drosophila*) heavy (apolipophorin-1, 2645aa) and light (apolipophorin-2, 681aa) subunits derived from a single precursor, the remainder of the molecular weight being accounted for by the high lipid content (40−50%, depending on insect species). Lipophorin forms large aggregates during the haemolymph clotting process. The receptor has five functional domains with similarity to vertebrate very low density lipoprotein receptor.

TABLE L3. Lipids

(i) FATTY ACIDS These are the most important feature of the majority of biological lipids. They occur free in trace quantities and are important metabolic intermediates. They are esterified in the majority of biological lipids. Compounds are included either because they are common components of biological lipids or are used in synthetic "model" analogues of these lipids. General formula R-COOH. Branched chain compounds are widespread, but are not found in mammalian lipids. All the examples given are straight-chain compounds.

Saturated fatty acids			**Unsaturated fatty acids**		
Number of carbon atoms	Name	M_r (Da)	Designation*	Name	M_r (Da)
2	Acetic	60	Mono-unsaturated acids.		
3	Propionic	74.1	16:1 (*cis* 9)	palmitoleic	254.2
4	Butyric	88.1	18:1 (*cis* 9)	oleic	282.5
5	Valeric	102.1	18:1 (*trans* 9)	elaidic	282.5
6	Hexanoic (caproic)	116.2	18:1 (*cis* 11)	*cis*-vaccenic	282.5
7	Heptanoic	130.2	18:1 (*trans* 11)	*trans*-vaccenic	282 5
8	Octanoic (caprylic)	144.2	Poly-unsaturated acids (all *cis* double bonds).		
9	Nonanoic (pelargonic)	158.2	18:2 (9, 12)	linoleic	280.4
10	Decanoic (capric)	172.2	18:3 (9, 12, 15)	α-linolenic	278.4
11	Undecanoic	186.3	18:3 (6, 9, 12)	γ-linolenic	278.4
12	Lauric	200.3	20:4 (5, 8, 11, 14)	arachidonic (eicosenoic)	304.5
13	Tridecanoic	214.4			
14	Myristic	228.4	22:6 (4, 7, 10, 13, 16, 19)	dodecosahexaenoic acid	328.6
15	Pentadecanoic	242.4			
16	Palmitic	256.4			
17	Margaric	270.7			
18	Stearic	284.5			
20	Eicosanoic (Arachidic)	312.5			
22	Docosanoic (Behenic)	340.6			

*Number of carbon atoms: number of double bonds (position and configuration of bonds)

(ii) ACYL GLYCEROLS Glycerol esters of fatty acids. Acyl glycerols are the parent compounds of many structural and storage lipids. Diglycerides (DG) may be considered as the parent compounds of the major family of phosphatidyl phospholipids. Triglycerides (TG) are important storage lipids.

Diglycerides Present as trace components of membranes. They are important metabolites and second messengers in signal-response coupling.

(a) OH

CH₂– CH–CH₂

O   O

C=O C=O

R₁   R₂

i.e.

OH

(a) This carbon is asymmetric. See below under phosphatidic acid.

(iii) SPHINGOLIPIDS

Important and widespread classes of phospholipids and glycolipids.

TABLE L3. (Continued)

The parent alcohol is SPHINGOSINE:

$$CH_3(CH_2)_{12}CH=CH-\underset{\underset{NH}{|}}{\overset{\overset{OH}{|}}{CH}}-CH-CH_2-X$$

$$\underset{Y}{\overset{|}{NH}}$$

where X = OH and the primary amino-group is free.

SPHINGOSINE is normally substituted at X and Y. When Y is a long-chain unsaturated fatty acyl group the derivative is a CERAMIDE. When the CERAMIDE carries uncharged sugars as the X substituent this is a CEREBROSIDE and where the sugars include sialic acid it is a GANGLIOSIDE.

(iv) PHOSPHOLIPIDS

In animal cell membranes the major class of phospholipids are the phosphatidyl phospholipids for which phosphatidic acid can be considered as the simplest example. These are diacylglycerol (DG) derivatives and in most cases DG is the immediate metabolic precursor.

Outline structure (see diglyceride):

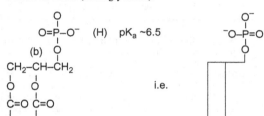

(b) As in diglycerides, this carbon atom is asymmetric. The biologically important configuration is *syn*.
R1 is usually saturated and R2 is unsaturated in animal cell membranes.

Phosphatidyl phospholipids. Derived from phosphatidic acid by esterification of the phosphate group.

Base (substituent)	Phospholipid class	Abbreviation	Ionic status
None	Phosphatidic acid	PA	Anionic
Choline	Phosphatidyl choline	PC	Neutral
Ethanolamine	Phosphatidyl ethanolamine	PE	Neutral
Glycerol	Phosphatidyl glycerol	PG	Anionic
Inositol	Phosphatidyl inositol (Ptdyl. Ins.)	PI	Anionic
Inositol 4-monophosphate	Phosphatidyl inositol 4-phosphate		
(Ptdyl. Ins. 4-phosphate)	PIP	Anionic	
Inositol 4,5 diphosphate	Phosphatidyl inositol 4,5-diphosphate		
(Ptdyl. Ins. 4,5 bisphosphate)	PIP2	Anionic	
Phosphatidyl glycerol	Diphosphatidyl glycerol (Cardiolipin)		Anionic
Serine	Phosphatidyl serine	PS	Anionic

Sphingomyelin. (SM) is an analogue of phosphatidyl choline in which the diacylglycerol component is replaced by a CERAMIDE. Common variants of these structures are:-

Ether phospholipids in which the diacylglycerol structure is modified so that one or both acyl groups are replaced by ether groups.

$$\underset{\underset{\underset{R}{|}}{\overset{\overset{|}{O}}{C=O}}}{\overset{CH-}{|}} \quad becomes \quad \underset{\underset{R}{|}}{\overset{\overset{\overset{CH-}{|}}{O}}{|}}$$

TABLE L3. (Continued)

Plasmalogens in which the 1-acyl group is replaced by a 1-alkenyl group. Plasmalogens are abundant lipid components of many membranes.

$$
\begin{array}{ccc}
\text{CH}- & & \text{CH}- \\
| & & | \\
\text{O} & \text{becomes} & \text{O} \\
| & & | \\
\text{C=O} & & \text{CH} \\
| & & \parallel \\
\text{R} & & \text{CH} \\
& & | \\
\end{array}
$$

Phosphonolipids In which the ester linkage between the base (choline or ethanolamine) is replaced by a P-C (phosphono) linkage.

$$
\begin{array}{ccc}
\quad\; \text{O} & & \quad\; \text{O} \\
\quad\; \parallel & & \quad\; \parallel \\
-\text{O}-\text{P}-\text{O}-\text{CH}- & \text{becomes} & -\text{O}-\text{P}-\text{CH}- \\
\quad\; | & & \quad\; | \\
\quad\; \text{O}^- & & \quad\; \text{O}^-
\end{array}
$$

Lysophospholipids. Derivatives of phosphatidyl phospholipids in which one of the acyl groups has been removed (enzymically). Lysophosphatidyl choline (lysolecithin) is a common, but trace component of membranes.

(P)-choline   (P)-choline

OH

(**v**) **STEROLS**. Of this large class of compounds only one member, CHOLESTEROL, is an important structural lipid. It is the single most abundant lipid in the plasma membrane of many animal cell types.

---

lipopolysaccharide *LPS* The major constituent of the cell walls of Gram-negative bacteria. Highly immunogenic and stimulates the production of endogenous pyrogen **interleukin-1** and **tumour necrosis factor** (TNF) through binding to the CD14/Toll-like receptor 4 (TLR4)/Lymphocyte antigen 96 (MD2) receptor complex. Has three parts: variable polysaccharide (**O-antigen**) side chains, core polysaccharides, and **lipid A**.

lipopolysaccharide binding protein An **acute phase protein** (452aa) of the **BPI/LBP/Plunc superfamily** that binds **lipopolysaccharide** and interacts with CD14 LPS-receptor complex, triggering the upregulation of various inflammatory genes.

lipoprotein glomerulopathy A rare kidney disorder caused by mutation in the ApoE gene.

lipoproteins An important class of serum proteins in which a lipid core with a surface coat of phospholipid monolayer is packaged with specific proteins (**apolipoproteins**). Important in lipid transport, especially cholesterol transport. Classified according to density: chylomicrons, large low density particles; very low density (VLDL); low density (LDL), intermediate density (IDL), high density (HDL) and very high density (VHDL) species. VLDL has a density of 0.94–1.006 g/ml and only about 10% protein: VHDL has a density greater than 1.21 g/ml, a protein content about 57%, the remainder being lipid. Intermediate-density lipoproteins are formed by the degradation of VLDL and are composed mainly of triacylglycerols and cholesterol esters: they are further degraded to low density lipoprotein (LDL) in the liver. IDL and LDL are rapidly cleared from plasma by receptor-mediated endocytosis (see **low density lipoprotein receptor**). Lipoprotein(a) is formed from **apolipoprotein(a)**.

liposomes Artificially formed single or multi-layer spherical lipid bilayer structures. Made from solutions of lipids etc. in organic solvents dispersed in aqueous media. Under appropriate conditions liposomes form spontaneously. Often used as models of the plasma membrane. May also be used experimentally and therapeutically for delivering drugs etc. to cells since liposomes can fuse with a plasma membrane and deliver their contents to the interior of the cell (see **lipofectamine**). Vary in size from submicron diameters to (in a few record-breaking cases) centimetres.

lipoteichoic acid Compounds formed from **teichoic acid** linked to glycolipid and a major component of the walls of most Gram-positive bacteria. The lipoteichoic acid of streptococci may function as an **adhesin** and will stimulate production of inflammatory cytokines through toll-like receptor 2, CD14 and CD36.

lipotropin Polypeptide hormone (lipotropic hormone, LPH, adipokinetic hormone, β form, 91aa; γ form has residues 1–58) derived from **pro-opiomelanocortin** in the pituitary hypophysis. Acts through melanocortin receptors on the adenylyl cyclase system to stimulate lipolysis, steroidogenesis and melanin production. Part of the **ACTH** group of hormones. **Endorphins** are produced by proteolysis of LPH.

lipovitellin The predominant lipoprotein found in the yolk of egg-laying animals, both vertebrate and invertebrate, and is involved in lipid and metal storage; lipovitellin-1 (1096aa) and lipovitellin-2 (238aa) are formed from the precursor **vitellogenin**.

lipoxins Bioactive **eicosanoids**, generated by **lipoxygenases** acting on arachidonic acid, that activate human monocytes and inhibit neutrophils as the inflammatory response begins to resolve. LXA_4 and LXB_4 act through a G protein-coupled receptor and inhibit neutrophil chemotaxis, transmigration, superoxide generation and NFκB activation. Cysteinyl-lipoxins LXC_4, LXD_4 and LXE_4 are antagonists for the **leukotriene** receptors. **Epi-lipoxins**, **resolvins** and **protectins** are analogous.

lipoxygenase Enzyme (5-lipoxygenase, 5-LO, EC 1.13.11.34, 674aa) that catalyses the addition of a hydroperoxy group to the 5-position of arachidonic acid, the first step in **leukotriene** synthesis.

liprins Family of proteins that recruit and anchor **LAR** family phosphatases. The C-terminal portion of liprins binds to the membrane-distal phosphatase domain of LAR and the N-terminal region may be involved in dimerisation. Some liprins are widely distributed, others are more tissue specific. May affect LAR distribution in the cell, in particular bringing LAR to regions of contact between cell and extracellular matrix. See **PPFIA**.

Lis1 Product (lissencephaly protein-1, platelet-activating factor acetylhydrolase IB subunit alpha, 410aa) of a gene that is deleted or mutated in patients with **lissencephaly**. Required for activation of Rho GTPases and actin polymerization at the leading edge of locomoting cerebellar neurons and postmigratory hippocampal neurons. Interacts with nuclear distribution gene E homologue-like 1 **NDE1**.

lissencephaly A severe human neuronal-migration defect. See **Lis1**.

Listeria monocytogenes Rod-shaped Gram-positive bacterium. Widespread and able to grow at an unusually wide range of temperatures (0–45°C). Normally **saprophytic** but is an opportunistic parasite, in that it can survive within cells (particularly leucocytes) and can be transmitted transplacentally. It has caused a number of serious outbreaks of food poisoning with a high mortality rate in recent years.

listeriolysin O **Cholesterol binding toxin** (529aa) from *Listeria monocytogenes*. A major virulence factor required for the escape of bacteria from phagosomes into the host cell's cytoplasm.

LITAF Lipopolysaccharide-induced tumour necrosis factor-alpha factor (p53-induced gene 7 protein, 161aa) that may regulate the transcription of specific genes including that for TNFα. Defects in LITAF cause Charcot-Marie-Tooth disease type 1C.

lithium *Li* The lightest of the alkali metals, although the cation has the largest hydration shell. Lithium salts are used as an antidepressant and are thought to act by inhibiting the regeneration of **inositol** from IP3 and thus reducing the efficiency of the **phosphatidyl inositol** signalling pathways.

lithostathine A secreted protein (lithostathine-1-alpha, islet cells regeneration factor, pancreatic stone protein, regenerating protein 1 alpha, 166aa) that may inhibit spontaneous calcium carbonate precipitation and thus inhibit the formation of calcareous deposits. May be associated with neuronal sprouting in brain, and with brain and pancreas regeneration. See **REG #1**.

lithotroph Cell or organism that depends upon inorganic compounds as electron donors for energy production.

litorin A nonapeptide fronm *Litoria aurea* (Green and golden bell frog) that mimics **bombesin** in its mitogenic effects, and has a carboxy-terminal octapeptide in common with bombesin.

liver cells Usually implies **hepatocytes**, even though other cell types are found in the liver (**Kupffer cells**, **Ito cells** and **lipocytes** for example). Hepatocytes are relatively unspecialised epithelial cells and are the biochemist's 'typical animal cell'.

liver cytosol antigen See **formiminotransferase-cyclodeaminase**.

liver X receptor See **LXR**.

liver-expressed chemokine See **LEC**.

LIX (1) A murine neutrophil-chemoattractant CXC chemokine (lipopolysaccharide-induced CXC chemokine; CXCL5; SCYB5; AMCF-2, 132aa), homologue of **ENA-78**. GCP-2(1–78) and GCP-2(9–78) are produced by proteolytic cleavage after secretion from fibroblasts and epithelial cells. (2) A family of proteins expressed in developing limbs. LIX (281aa) is found transiently in the developing hind-limb of chickens; homologues are found in human (282aa), as is a LIX1-like protein (337aa).

Ljungan virus Picornavirus first isolated from bank voles (*Clethrionomys glareolus*) in which it causes myocarditis and diabetes-like symptoms. Can occasionally infect humans and may be involved in various human diseases, although this is still contentious. Website: http://www.ljunganvirus.org/

LKB1 A serine/threonine kinase (STK11, 433aa) originally discovered as the product of the gene mutated in the autosomal dominant human disorder, Peutz-Jeghers syndrome (PJS). People with PJS develop benign polyps in the gastrointestinal tract but also have a 15-fold increased risk of developing malignant tumours in other tissues. Interacts with **p53** and deficiency is associated with defective apoptosis. Activates **NUAKs**.

LKR (1) Lysine-alpha-ketoglutarate reductase, see **alpha-aminoadipic semialdehyde synthase**. (2) LKR-13, a lung adenocarcinoma cell line derived from K-ras (LA1) mice. (3) The G-protein coupled leucokinin receptor (Lkr, 542aa) from *Drosophila*.

LL-37 An antibacterial protein (37aa) derived from **cathelicidin**. Binds to bacterial lipopolysaccharides.

LMM Light meromyosin, the rod-like portion of the myosin heavy chain (predominantly α-helical) that is involved in lateral interactions with other LMM to form the thick filament of striated muscle, and that is separated from heavy meromyosin (HMM) by cleavage with trypsin.

LMP (1) **Latent membrane protein**. (2) Low molecular weight proteins (LMP), components of the **proteasome**. (3) **LIM mineralization protein-1**.

LMPTP Class II protein tyrosine phosphatase (low molecular weight phosphotyrosine protein phosphatase, EC 3.1.3.48, 158aa) which also acts as a general cytosolic acid phosphatase (EC 3.1.3.2) acting on low-MW aryl phosphates.

L-*myc* Relative of the *myc* **proto-oncogene** overexpressed in lung **carcinoma**. The product, Class E basic helix-loop-helix protein 38 (364aa) binds DNA when dimerized with another bHLH protein such as **Max**.

LNCaP cells A line of androgen-sensitive prostate cancer cells derived from a lymph node metastatic lesion of human prostatic adenocarcinoma; widely used in the study of prostate cancer.

lobomycosis A chronic mycosis of the skin caused by the fungus *Lacazia loboi* (formerly *Loboa loboi*).

lobopodia Hemispherical protrusions from the front of a moving tissue cell.

local circuit theory A generally accepted model for neuronal conduction, by which depolarization of a small region of a neuronal plasma membrane produces transmembrane currents in the neighbouring regions, tending to depolarize them. As the **sodium channels** are **voltage- gated**, the depolarization causes further channels to open, thus propagating the action potential.

locomotion Term used by some authors to distinguish movement of cells from place-to-place from movements such as flattening, shape-change, **cytokinesis** amongst others.

locus *Pl*. loci. (1) Generally, a place or site, a location. (2) In genetics, the site in a linkage map or on a chromosome where the gene for a particular trait is located. Any one of the alleles of a gene may be present at this site.

locus coeruleus A dense cluster of neurons in the dorso-rostral pons region of the brain. It is the major source of norepinephrine with neuronal projections throughout most of the CNS, including the cerebral cortex, hippocampus, thalamus, midbrain, brainstem, cerebellum, and spinal cord. Considered to be a key brain centre for anxiety and fear.

locus control region Region of DNA which contains the **promoters** and **enhancers** that regulate the expression of a particular gene. Often taken to be a single region 0−2 kb upstream of the transcriptional **start site**, although there are probably few genes where things are that simple.

Lod score Logarithm of the odds score: a statistical parameter that quantifies the probability that there is linkage between traits or markers. For non-X-linked genetic disorders a Lod score of +3 (1000:1) is usually taken to indicate linkage.

Loeffler's medium A culture medium containing coagulated horse serum and other complex tissue extracts used to culture *Corynebacterium diphtheriae* in diagnostic bacteriology.

logarithm of the odds score See **Lod score**.

logP $logK_{ow}$. The octanol-water **partition coefficient**, serves as a quantitative indicator of lipophilicity, an important parameter in pharmacokinetics and for predicting distribution of a chemical in the environment.

LOH See **loss of heterozygosity**.

lomasome Membranous structure, often containing internal membranes, located between the plasma membrane and cell wall of plant cells. Included in the more general term, **paramural body**.

long interspersed nuclear elements *LINEs* Autonomously replicating **transposons** of 6−7 kb that make up a significant part (ca. 15%) of the mammalian genome. LINE1 has two open reading frames, one of which encodes a reverse transcriptase that is also used by non-autonomous **short interspersed nuclear elements** (SINEs). LINE2 is apparently more ancient and is no longer being replicated. LINEs do not have **long terminal repeats**, unlike viral retrotransposons. See **mammalian-wide interspersed repeats**.

long QT syndrome A group of disorders due to defects in ion channels (channelopathies) which increase the risk of sudden heart failure. The QT interval is defined from the characteristics of the electrocardiogram and represents the time taken for electrical activation and inactivation of the ventricles. Various channels are affected in different type of the syndrome including the KQT-like voltage-gated **potassium channel-1** (KCNQ1) (also mutated in **Jervell and Lange-Nielsen syndrome**); an inwardly rectifying potassium channel, KCNH2 (the human homologue of *Drosophila* ether-a-go-go); voltage-gated sodium channel (SCN5A and SCN4B); a delayed rectifier potassium channel and a voltage-dependent L-type calcium channel). Other forms are the result of defects in **ankyrin-2, caveolin-3** or **AKAP-9**.

long-term potentiation Increase in the strength of transmission at a **synapse** with repetitive use that lasts for more than a few minutes. As a form of long-term **synaptic plasticity** it is important as a possible cellular basis of learning and memory storage. It has been studied most extensively at excitatory synapses onto principal neurons of the **hippocampus** where it was first demonstrated. Selective inhibition of **NMDA receptor** channels has been shown to block LTP, and to block spatial learning.

long-terminal repeat *LTR* Identical DNA sequences, several hundred nucleotides long, found at either end of **transposons** and DNA that is reverse transcribed from **retroviral** RNA. They are thought to have an essential role in integrating the transposon or provirus into the host DNA. LTRs have inverted repeats, that is, sequences close to either end are identical when read in opposite directions. In proviruses the upstream LTR acts as a promoter and enhancer and the downstream LTR as a polyadenylation site.

longevity-assurance protein-1 *LAG-1* See **LASS1**.

loop of Henle The hairpin-loop part of the kidney tubule, the region that connects the proximal convoluted tubule to the distal convoluted tubule in the renal medulla. Different parts of the loop are permeable to water or to ions or are the site of active transport and the ultrafiltrate formed in the glomerulus is modified by retrieval of ions and water.

lophocytes Cells found beneath the dermal membrane of a few species of sponges. Have been postulated to constitute a primitive nervous system though this is uncertain.

lophotrichous Descriptor for a cell with flagella arranged as a tuft at one end.

lorica Shell or test secreted by a protozoan; often vase-shaped.

loricrin Major protein of the **keratinocyte** cell envelope (312aa). Some glutamines and lysines are cross-linked to other loricrin molecules by trans-glutaminases. Mutations in the gene for loricrin are responsible for the inherited skin diseases Vohwinkel's syndrome and progressive symmetric erythrokeratoderma.

loss of function mutation Mutation that causes the loss of function in a protein or system. Much

more common than gain of function mutations since there are more ways to disrupt things than to add additional functionality.

loss of heterozygosity *LOH* Situation in which heterozygosity is lost in some tissues, probably as a result of mitotic recombination so that one daughter cell obtains two identical alleles and the other gets two mutant alleles or a major deletion on the homologous chromosome so that only one allele remains. It is fairly common in various tumours and indicates tumour progression.

Lou Gehrig's disease See **amyotrophic lateral sclerosis**.

Louis Bar syndrome See **ataxia telangiectasia**.

LOV domains Photosensitive domains (light-oxygen-voltage-sensing domains) of the blue-light sensitive protein YtvA (261aa) from *Bacillus subtilis* also found in **phototropin** from higher plants. LOV domains are a special class of **PAS domain** and each LOV domain noncovalently binds a flavin mononucleotide as a chromophore. Article: http://www.pnas.org/content/100/10/5938.full

low-affinity platelet factor IV See **connective tissue-activating peptide III**.

low density lipoprotein See **lipoproteins**.

low density lipoprotein receptor The cell-surface receptor (LDL-receptor, 860aa) that mediates the endocytosis of LDL by cells. Genetic defects in LDL-receptors lead to abnormal serum levels of LDL and hypercholesterolaemia. The oxidized low-density lipoprotein receptor 1 (273aa) mediates the recognition, internalization and degradation of oxidatively modified low density lipoprotein (oxLDL) by vascular endothelial cells. It has lectin-like properties (see **collectins**) and is also important as a receptor for advanced glycation end (AGE) products, activated platelets, monocytes, apoptotic cells and both Gram-negative and Gram-positive bacteria. It can trigger inflammatory responses and is pro-atherogenic.

low density lipoprotein receptor-related proteins A family of receptors (LRPs) with a diverse range of functions. LRP1 (CD91; apolipoprotein E receptor; α-2-macroglobulin receptor, 4544aa) is involved in endocytosis, phagocytosis of apoptotic cells, clearance of triglyceride-rich lipoproteins from plasma and the uptake of apolipoprotein E-containing lipoprotein particles by neurons. It mediates the endocytosis and degradation of secreted amyloid precursor protein (APP) and APP-generated beta-amyloid. LRP1B (4599aa) is also involved in clearance. LRP2 (megalin; glycoprotein-330) binds lipoprotein lipase, apolipoprotein E-enriched beta-VLDL and apolipoprotein J/clusterin. It is the target antigen of **Heymann nephritis** and is mutated in

Donnai-Barrow syndrome. LRP3 (770aa) is widely expressed but does not bind lipoproteins and its function is unclear. LRP4 (1905aa) is involved in the negative regulation of the canonical Wnt signalling pathway, by antagonizing LRP6-mediated activation of this pathway. LRP5 (LRP7, 1615aa) is part of the Wnt-Fzd-LRP5-LRP6 complex that triggers beta-catenin signalling. Mutations are associated with osteoporosis-pseudoglioma syndrome, exudative vitreoretinopathy and various disorders of bone mineralisation. LRP6 (1613aa) is a co-receptor for Wnt with LRP5 and mutations are associated with early coronary artery disease. LRP8 (apolipoprotein E receptor 2, 963aa) is the receptor for reelin and apolipoprotein E (apoE)-containing ligands. (See **DAB1**). LDL receptor-related protein associated protein 1 (LRPAP1, α-2-macroglobulin receptor-associated protein (MRAP), 357aa) forms a complex with the α-2-macroglobulin receptor heavy and light chains.

low-density lipoprotein receptor adaptor protein 1 A transmembrane adapter protein (LDLRAP1, 308aa) that interacts with the cytoplasmic tail of the low density lipoprotein receptor (LDLR), phospholipids, and components of the endocytic machinery (**clathrin** or AP2) and is required for for efficient binding of LDL to the LDL-receptor and internalization of the complex. Mutation in the encoding gene (ARH) causes autosomal recessive **familial hypercholesterolaemia**.

low-density lipoprotein receptor-related proteins A large family of cell surface proteins involved in endocytosis and of signalling pathways such as the Wnt pathway. LRP-1 is the α-2-macroglobulin receptor (apolipoprotein E receptor, CD91, 4544aa precursor) involved in endocytosis and in phagocytosis of apoptotic cells. LRP-5 (1615aa) and LRP-6 (1613aa) are components of the Wnt-Fzd-LRP5-LRP6 complex that triggers beta-catenin signalling through inducing aggregation of receptor-ligand complexes into signalsomes (see **dickkopf**). LRP-8 (963aa) is the receptor for **reelin** and apolipoprotein E. Many others are known although their functions are not always understood in detail.

Lowry assay One of the most commonly used assays for protein content − the paper describing it is said to be the most frequently cited in the biological literature. Depends upon the interaction of **Folin-Ciocalteau reagent** with tyrosine or phenylalanine. Proteins that are deficient in these amino acids (e.g. collagen) will be underestimated and for this reason the Bradford assay is often preferred.

lox Site in bacteriophage P1 DNA that is recognized by the **cre** recombinase. Now used in vertebrate transgenics: see **lox-Cre system**.

lox-Cre system Site-specific recombination system from *E. coli* bacteriophage P1. Now used in transgenic animals to produce conditional mutants. If two **lox** sites are introduced into a transgene, the intervening DNA is spliced out if active **cre recombinase** is expressed.

LPA See **lysophosphatidic acid**.

LPL The enzyme (lipoprotein lipase, EC 3.1.1.34, 475aa) that hydrolyses triglycerides of circulating chylomicrons and very low density lipoproteins. Familial lipoprotein lipase deficiency is characterized by increased plasma trigylceride levels after digestion of dietary fat because there is delayed clearance of chylomicrons.

LPS See **lipopolysaccharide**.

L-ring See **S-ring**.

LRK1 The homologue in *C. elegans* of the familial Parkinsonism gene PARK8/LRRK2, a kinase (leucine-rich repeat serine/threonine-protein kinase 1, PARK8-related kinase, EC 2.7.11.1, 2393aa) that antagonizes the effects of **pink-1** both in the regulation of axon guidance and the stress response. It is required for the polarized localization of synaptic vesicle proteins.

LRP (1) **Leucine-responsive regulatory proteins**. (2) **Low density lipoprotein receptor-related proteins**. (3) LRP16 is a protein (MacroD1, 325aa) with a single macro domain that has a role in estrogen signalling. It amplifies the transactivation function of the androgen receptor, to which it binds via the macro domain, and is upregulated in breast cancer MCF-7 cells.

LRR See **leucine-rich repeat**.

LSAMP A lipid-anchored adhesion molecule (limbic system-associated membrane protein, Iglon family member 3, 338aa) that mediates selective neuronal growth and axon targeting, particularly in the limbic region of the brain.

Lsc One of the Dbl-like oncoproteins. See **Lfc**.

LST8 A **WD-repeat protein** (326aa) that is a subunit of both mTORC1 and mTORC2 (see **TORC**). Resembles G-protein beta subunits and regulates the kinase activity of TORC2.

LTA$_4$, LTB$_4$, LTC$_4$, etc. See **leukotrienes**.

LTP See **long-term potentiation**.

LTR retrotransposons Retrotransposons with long terminal repeat sequences of very varied length. They integrate into the genome by means of an integrase that has tight sequence specificity. They are subdivided into **copia-like** (Ty1/copia-like), **gypsy-like** (Ty3/gypsy-like) and **Pao-BEL-like** groups. The copia-like and gypsy-like superfamilies are found at high copy-number in all eukaryotes. Pao-BEL like elements have so far only been found in animals. LTR-type retrotransposons account for about 8% of the human genome and approximately 10% of the mouse genome. The Gypsy Database of mobile genetic elements: http://gydb.uv.es/index.php/LTR_retroelements

L-type channels A class of **voltage-sensitive calcium channels**. L-type channels are found in neurons, neuroendocrine cells, smooth, cardiac and striated muscle, are involved in neurotransmitter release at some synapses and inactivate relatively slowly. They are activated at membrane potentials more positive than -30 mV. The long-lasting properties and possible role in long-term potentiation were the reason for them being designated L-type. They are insensitive to ω-**conotoxin** but inhibited by dihydropyridines, benzodiazepines and phenylalkylamines. Mutation can cause hypokalemic **periodic paralysis**.

lucifer yellow Bright yellow fluorescent molecule (similar to fluorescein), widely used by microinjection in developmental biology and neuroscience to study the outline of cells, in cell lineage studies, or as an indicator of **dye coupling** between cells.

luciferase Enzyme (luciferin 4-monooxygenase, EC 1.13.12.7, 548aa) from *Luciola lateralis* (firefly) that catalyses the production of light in the reaction between luciferin and ATP. Used by the male firefly for producing light to attract females, and used in the laboratory in a **chemiluminescence** bioassay for ATP and as a reporter gene.

luciferins Small-molecule substrates for **luciferase** that catalyses an oxidative reaction leading to bioluminescence.

lucigenin Compound used as a bystander substrate in assaying the **metabolic burst** of leucocytes by **chemiluminescence**. When oxidized by **superoxide** it emits light.

Lucké carcinoma A renal carcinoma, caused by a herpesvirus, in frogs; it aroused interest because its abnormal growth appears to be dependent on a restricted temperature range. Nuclei from these cells give rise to normal frogs if transplanted into enucleated eggs, giving support to epigenetic theories of neoplasia.

luliberin See **gonadotropin-releasing hormone**.

lumen A cavity or space within a tube or sac.

lumican Keratan sulphate proteoglycan (338aa) in cornea and other tissues. In adult cartilage lumican exists predominantly in a glycoprotein form lacking keratan sulphate, whereas the juvenile form of the molecule is a proteoglycan.

lumicolchicine A derivative of **colchicine** produced by exposure to ultraviolet light and that does not inhibit tubulin polymerization, although it has many of the nonspecific effects of colchicine.

luminol Compound used as a bystander substrate (like lucigenin) in assaying the **metabolic burst** of leucocytes by **chemiluminescence**. When oxidized by the myeloperoxidase/hydrogen peroxide system, it emits light. Also used forensically to detect traces of blood.

luminometer Laboratory instrument used to measure light emission, for example in measurement of ATP levels by interaction with luciferase which produces light.

lumirhodopsin Altered form of **rhodopsin** produced as a result of illumination.

lumisome Subcellular membrane-enclosed vesicle that is the site of bioluminescence in some marine **coelenterates**.

lupus erythematosus Skin disease (discoid lupus erythematosus) in which there are red scaly patches, especially over the nose and cheeks. May be a symptom of **systemic lupus erythematosus**.

lusitropic Anything that relaxes the heart, the opposite of **inotropic**.

lutein An orange-red carotenoid pigment (xanthophyll) found in many plants and animals and an important anti-oxidant. It is found in the macula where it may have light-filtering and anti-oxidant functions.

luteinizing hormone A **gonadotropin** composed of an α-chain (92aa) common to other gonadotropins, and a hormone-specific β-chain (141aa). Acts with **follicle-stimulating hormone** to stimulate sex hormone release. Luteinizing hormone releasing factor is a decapeptide (see **gonadotropin-releasing hormone**).

luteotrophic hormone See **prolactin**.

luxury protein A term sometimes used to describe those proteins that are produced specifically for the function of differentiated cells and are not required for general cell maintenance (the so-called 'housekeeping' proteins).

LXR Ligand-activated transcription factors (liver X receptor alpha, oxysterols receptor LXRα, 447aa; LXRβ, 460aa) of the nuclear receptor subfamily 1, that induce genes controlling cholesterol homeostasis and lipogenesis. Expressed in liver, intestine and adrenal gland. They form a heterodimers with retinoid-X receptor (RXR) in which RXR provides the DNA-binding function, LXR the ligand-binding. Ligand may be oxysterol metabolites of cholesterol.

Ly-6/uPAR proteins A family of proteins with one or more repeat units of the Ly-6/uPAR domain (lymphocyte antigen-6/**urinary plasminogen**

activator) a domain in which there is a distinct disulphide bonding pattern between 8 or 10 cysteine residues. There are two subfamilies, one comprises GPI-anchored glycoprotein receptors with 10 cysteine residues, the other includes the secreted single-domain snake and frog cytotoxins and **slurp**. Article: http://www.ncbi.nlm.nih.gov/pmc/articles/PMC2144295/pdf/10211827.pdf

lyases Enzymes of the EC Class 4 (see **E classification**) that catalyse the non-hydrolytic removal of a group from a substrate with the resulting formation of a double-bond; or the reverse reaction, in which case the enzyme is acting as a synthetase. Include decarboxylases, aldolases and dehydratases.

Lyb antigen Surface antigens of mouse B-cells. Largely superceded by use of **CD antigen** markers.

lycopene A linear, unsaturated hydrocarbon **carotenoid** (536 Da); the major red pigment in some fruit. Extracted and used as an antioxidant.

Lycopodiophyta Clubmosses.

Lyme disease Disease caused by *Borrelia burgdorferi*, a tick-borne spirochaete.

lymph Fluid found in the lymphatic vessels that drain tissues of the fluid that filters across the blood vessel walls from blood. Lymph carries lymphocytes that have entered the lymph nodes from the blood.

lymph node Small organ (lymph gland) made up of a loose meshwork of reticular tissue in which are enmeshed large numbers of lymphocytes, macrophages, and accessory cells. Recirculating lymphocytes leave the blood through the specialized high endothelial venules of the lymph node and pass through the node before being returned to the blood through the lymphatic system. Because the lymph nodes act as drainage points for tissue fluids, they are also regions in which foreign antigens present in the tissue fluid are most likely to begin to elicit an immune response.

lymphadenitis Inflammation of **lymph nodes**.

lymphoblast Often referred to as a blast cell. Usually the suffix -blast is for cells that give rise to differentiated cells but a lymphoblast is a differentiated product of a T or B-lymphocyte that re-enters the mitotic cycle as a result of an antigenic stimulus.

lymphocyte White cells of the blood that are derived from stem cells of the lymphoid series. Two main classes, T- and B-lymphocytes (**T-cells** and **B-cells**), are recognized, the latter responsible (when activated) for production of antibody, the former subdivided into subsets (**T-helper cells, T-regulatory cells, cytotoxic T-cells, gamma-delta cells, Th17 cells, T-memory cells**), and responsible both for cell-mediated immunity and for stimulating B-lymphocytes.

lymphocyte activation The change in morphology and behaviour of lymphocytes exposed to a mitogen or to an antigen to which they have been primed. The result is the production of **lymphoblasts**, cells that are actively engaged in protein synthesis and that divide to form effector populations. Sometimes referred to as lymphocyte transformation but this is not like the transformation associated with oncogenic viruses, and 'activation' is therefore preferable.

lymphocyte activation gene-1 protein A CC chemokine (LAG-1, CCL4, 92aa) of the MIP-1β family that binds to both CCR5 and to CCR8.

lymphocyte function-related antigen See **LFA-1, LFA-3**.

lymphocyte transformation See **lymphocyte activation**.

lymphocytic leukaemia See **leukaemia**.

lymphocytopenia A condition in which the number of lymphocytes in the blood is abnormally low (below normal levels of $1-5 \times 10^9$/litre). Cf. **lymphocytosis**.

lymphocytosis An abnormally high blood lymphocyte count.

lymphoid cell Cells derived from stem cells of the lymphoid lineage: large and small lymphocytes, plasma cells.

lymphoid enhancer-binding factor 1 A protein of the TCF/LEF family (T-cell-specific transcription factor 1-alpha, 399aa) involved in the Wnt signalling pathway. May play a role in hair cell differentiation and follicle morphogenesis (see **nude mice**).

lymphoid tissue Tissue that is particularly rich in lymphocytes and accessory cells such as macrophages and reticular cells. The primary lymphoid tissues are the **thymus** and bone marrow which produce lymphocytes; the secondary tissues (misleadingly sometimes referred to as peripheral lymphoid tissues), are **lymph nodes**, spleen, **Peyer's patches**, pharyngeal tonsils, adenoids, and (in birds) the **Bursa of Fabricius**. See also **BALT, GALT, MALT, SALT** (bronchus, gut, mucosal and skin-associated lymphoid tissues).

lymphokine Increasingly uncommon term for a subset of the **cytokines**, those produced by leucocytes. Examples are **interleukins, interferon** γ, lymphotoxin (tumour necrosis factor β), granulocyte-monocyte colony-stimulating factor (**GM-CSF**).

lymphoma Malignant neoplastic disorder of lymphoreticular tissue that produces a distinct tumour mass, not a leukaemia (in which the cells are circulating). Includes tumours derived both from the lymphoid lineage and from mononuclear phagocytes; lymphomas arise commonly (but not invariably) in lymph nodes, spleen, or other areas rich in lymphoid tissue. Lymphomas are subclassified as **Hodgkin's disease**, and Non-Hodgkin's lymphomas (e.g. **Burkitt's lymphoma**, large-cell lymphoma, histiocytic lymphoma).

lymphosarcoma Synonym for a **lymphoma**.

lymphotoxin Cytotoxic product of T-cells: the term is usually restricted to **tumour necrosis factor β** which is also known as lymphotoxin.

lymphotropic Having an affinity for lymphocytes – as, for example, some forms of HIV.

lyn Non-receptor tyrosine **kinase** (512aa), related to src. Plays a critical role in B-cell development and intracellular signalling. Lyn-deficient mice exhibit splenomegaly, elevated serum IgM, production of autoantibody and, later, glomerulonephritis.

lyngbyastatin See **dolastatins**.

Lyon hypothesis Hypothesis, first advanced by Lyon, concerning the random inactivation of one of the two X chromosomes of the cells of female mammals. In consequence females are chimaeric for the products of the X chromosomes, a situation that has been exploited in females who are heterotypic for isozymes of glucose-6-phosphate dehydrogenase as a means to confirm the monoclonal origin of papillomas and of atherosclerotic plaques.

lyophilic Characteristic of a material that readily forms a colloidal suspension. Molecules of the solvent form a shell around the particles; if the solvent is water then 'hydrophilic'.

lyophilization Now generally restricted to mean freeze drying – removal of water by sublimation under vacuum.

lyotropic series A listing of anions and cations in order of their effect on protein solubility (tendency to cause salting out). Salting out occurs when ions outcompete the protein for water molecules to form a hydration shell.

lysergic acid diethylamide Hallucinogenic compound (LSD) related to **ergot** which affects G-protein coupled receptors, including dopamine receptors, adrenoreceptors and particularly the serotonin 5-HT2$_A$ receptor.

lysigeny A mechanism for the formation of aerenchyma in which cells die to create the gas space. Lysigenous aerenchyma is found in many important crop species, including barley, wheat, rice and maize. See **schizogeny**.

lysine Amino acid (Lys, K, 146 Da) the only carrier of a side-chain primary amino group in proteins. Has

important structural and chemical roles in proteins. See Table A1.

lysis Rupture of cell membranes and loss of cytoplasm.

lysogeny The ability of some phages to survive in a bacterium as a result of the integration of their DNA into the host **chromosome**. The integrated DNA is termed a prophage. A regulator gene produces a **repressor protein** that suppresses the lytic activity of the phage, but various environmental factors, such as ultraviolet irradiation, may prevent synthesis of the repressor, leading to normal phage development and lysis of the bacterium. The best known example of a lysogenic bacteriophage is the **lambda bacteriophage**.

lysophosphatides Mono-acyl derivatives of diacyl phospholipids that are present in membranes as a result of cyclic deacylation and reacylation of phospholipids. Membranolytic in high concentrations, and fusogenic at concentrations that are just sublytic. May have important modulatory roles. See **lysophosphatidic acid**.

lysophosphatidic acid A signalling phospholipid with a single acyl chain that will, for example, induce neurite retraction and the formation of retraction fibres in young cortical neurons by actin rearrangement. Also implicated in various inflammatory signalling cascades. The lysophosphatidic acid receptor-1 (LPAR1, endothelial differentiation gene 2, EDG2, 364aa) is G-protein coupled and expressed in various adult organs. LPAR2 (EDG4, 351aa) has a different tissue distribution and at least three other similar receptors are known. Receptors: http://arjournals.annualreviews.org/doi/abs/10.1146/annurev.pharmtox.010909.105753

lysophospholipase D Enzyme (EC 3.1.4.39, 863aa) found in mammalian plasma and serum, that hydrolyzes lysoPE and lysoPC with different fatty acyl groups to the corresponding lysophosphoric acids. See **autotaxin**.

lysosomal diseases Diseases (also known as storage diseases) in which a deficiency of a particular lysosomal enzyme leads to accumulation of the undigested substrate for that enzyme within cells. Not immediately fatal, but within a few years lead to serious neurological and skeletal disorders and eventually to death. See the following diseases or syndromes: **Batten's, cystinosis, Danon's, Fabry's, Farber's, Gaucher's, glycogen storage diseases, Hunter's, Hurler's, Krabbe's, lipidoses, Morquio-Brailsford, mucopolysaccharidoses, Niemann-Pick, Pompe's, Sanfilipo, Salla, Sandhoff's, Scheie, Sly's, Tay-Sachs**. See also **I-cell disease**.

lysosomal enzymes A range of degradative enzymes, most of which operate best at acid pH. The best known marker enzymes are **acid**

phosphatase and **beta-glucuronidase**, but many others are known.

lysosomal trafficking regulator A protein (LYST, beige homologue, 3801aa) that appears to be required for sorting endosomal resident proteins into late multivesicular endosomes by a mechanism involving microtubules. The gene is mutated in **Chediak-Higashi syndrome** and the beige mouse. See **neurobeachin**.

lysosome Membrane-bounded cytoplasmic organelle containing a variety of hydrolytic enzymes that can be released into a phagosome or to the exterior. Release of lysosomal enzymes in a dead cell leads to autolysis (and is the reason for hanging game, to tenderise the muscle), but it is misleading to refer to lysosomes as 'suicide bags', since this is certainly not their normal function. Part of the GERL complex or *trans*-Golgi network. Secondary lysosomes are phagocytic vesicles with which primary lysosomes have fused. They often contain undigested material.

lysosome-associated membrane glyco-proteins See **LAMPs**.

lysosome-phagosome fusion A process that occurs after the internalisation of a primary phagosome. Fusion of the membranes leads to the release of lysosomal enzymes into the phagosome. Some species of intracellular parasite evade immune responses by interfering with this process.

lysosome-related organelles A family of organelles that includes melanosomes, platelet dense bodies, and cytotoxic T-cell granules. These organelles are all affected in **Hermansky-Pudlak syndrome**.

lysosomotropic Having affinity for and thus accumulating in **lysosomes**.

lysozyme Glycosidase (muramidase, EC 3.2.1.17, 148aa) that hydrolyzes the bond between N-acetyl-muramic acid and N-acetyl-glucosamine, thus cleaving an important polymer of the cell wall of many bacteria. Present in tears, saliva, and in the **lysosomes** of phagocytic cells, it is an important antibacterial defence, particularly against Gram-positive bacteria.

lysyl oxidase Extracellular copper-containing metalloenzyme (EC 1.4.3.13, 417aa) that deaminates lysine and hydroxylysine residues in collagen or elastin to form aldehydes, these then interact with each other or with other lysyl side chains to form cross-links. See **lathyrism**.

Lyt antigen A set of surface glycoproteins on mouse T-cells. Possession of Lyt 1 partly defines a T-helper cell, and of Lyt 2 and 3 suppressor and cytotoxic cells respectively. Formerly known as Ly antigens but this is largely superceded by **CD antigen** nomenclature. See also **Lyb antigen**.

lytic Describing something that causes lysis.

lytic complex The large multi-protein (2000 kDa) cytolytic complex (membrane attack complex) formed from complement C5b6789. See **complement**.

lytic infection The normal cycle of infection of a cell by a **virus** or **bacteriophage**, in which mature virus or phage particles are produced and the cell is then lysed releasing virus particles which infect other cells. *Cf.* **lysogeny**.

lytic vacuole Vacuole that contains hydrolytic enzymes, found in plant cells and analogous to the **lysosome** of animal cells but differing in morphology, function, enzyme content and mode of origin.

LZTR-1 A **BTB/kelch** protein (leucine-zipper-like transcriptional regulator 1, 840aa), deleted in the majority of **DiGeorge syndrome** patients, believed to act as a transcriptional regulator.

M

M Single letter code for methionine.

m119 See **Mig**.

Mab See **monoclonal** antibody.

MAC (1) Membrane attack complex. See **complement** and **C9**. (2) In plants, the MAC complex probably regulates defense responses through transcriptional control and is essential for plant innate immunity. See **PRL #2**.

Mac-1 The $\alpha_M\beta_2$ **integrin** of leucocytes (CR3, CD11b/CD18). See **LFA-1**.

MacArdle's disease See **glycogen storage diseases**.

MacConkey's agar Agar-based selective medium used for isolation of Gran-negative bacteria from faeces, etc. The presence of crystal violet and bile salts inhibit growth of Gram-positive bacteria. Contains lactose and neutral red as an indicator so that lactose fermenting bacteria producing an acidic end-product will form red-pink colonies.

macedocin An anticlostridial bacteriocin **lantibiotic** (22aa) produced by *Streptococcus macedonicus*, a natural cheese isolate. Inhibits a broad spectrum of lactic acid bacteria, as well as several food spoilage and pathogenic bacteria.

Mach-Zehnder system Interferometric system of microscopy in which the original light beam is divided by a semi-transparent mirror: object and reference beams pass through separate optical systems and are recombined by a second semi-transparent mirror. Interference fringes are displaced if the optical path difference for the reference beam is greater, and this can be compensated with a wedge-shaped auxiliary object. The position of the wedge allows the phase-retardation of the object to be measured. The Mach-Zehnder system was used in a microscope designed by Leitz.

Machupo virus A member of the **Arenaviridae** that may cause a severe haemorrhagic fever in humans. The natural hosts are rodents and transmission from human to human is not common.

macro domain A conserved protein domain found in prokaryotes and eukaryotes and some positive-strand RNA viruses; may have various roles, amongst them the binding of ADP-ribose. See **LRP16**.

macrocytes Abnormally large red blood cells, numerous in **megaloblastic anaemia** (pernicious anaemia).

microglia A general term for glial cells other than **microglia**; **oligodendrocytes** and **astrocytes**.

macroglobulin Globulin such as IgM that has a high molecular weight: 400 kDa in the case of **IgM**, 725 kDa in the case of **alpha-2-macroglobulin**.

macrolactin An antibacterial polyketide from a *Bacillus subtilis* isolate that inhibits a number of multidrug-resistant Gram-positive bacterial pathogens. Abstract: http://aac.asm.org/cgi/content/short/50/5/1701

macrolide antibiotics Broad-spectrum bacteriostatic antibiotics active against most aerobic and anaerobic Gram-positive cocci and Gram-negative anaerobes. Inhibit bacterial protein synthesis by blocking the 50S ribosomal subunit. Examples are erythromycin, lincomycin, azithromycin, and clindamycin.

macromolecule Non-specific and rather imprecise term covering proteins, nucleic acids and polysaccharides, but probably not phospholipids.

macronucleus The larger nucleus (or sometimes nuclei) in ciliate protozoans. Derived from the micronucleus by the formation of **polytene chromosomes**. The DNA in the macronucleus is actively transcribed but the macronucleus degenerates before conjugation.

macrophage Relatively long-lived phagocytic cell of mammalian tissues, derived from blood **monocyte**. Macrophages from different sites have distinctly different properties. Main types are peritoneal and alveolar macrophages, tissue macrophages (**histiocytes**), **Kuppfer cells** of the liver, and **osteoclasts**. In response to foreign materials may become stimulated or activated. Macrophages play an important role in killing of some bacteria, protozoa, and tumour cells, release substances that stimulate other cells of the immune system, and are involved in **antigen presentation**. May further differentiate within chronic inflammatory lesions to epithelioid cells or may fuse to form **foreign body giant cells** or **Langhans' giant cells**. Classically activated (M1) macrophages are proinflammatory and kill pathogens, and alternatively activated (M2) macrophages produce anti-inflammatory factors.

macrophage activation syndrome A serious disorder in which there is excessive activation of well-differentiated macrophages. It is associated with **Still's disease** and may be autoimmune disease-associated reactive **haemophagocytic lymphohistiocytosis**.

The Dictionary of Cell and Molecular Biology. DOI: http://dx.doi.org/10.1016/B978-0-12-384931-1.00013-1

macrophage colony-stimulating factor See MCSF.

macrophage inflammatory proteins A range of chemokines, originally identified in mice but with human counterparts (given here). MIP1 is a C-C **chemokine** now recognized to exist in various forms, MIP1α (CCL3, 92aa) and MIP1β (CCL4, SISα, TY-5, L2G25B, 464.1, GOS 19-1, 92aa), MIP1-gamma (CCL9) and MIP1-delta (CCL15). MIP2δ is chemotactic for T-cells and monocytes, but not for neutrophils, eosinophils, or B-cells. Has pro-inflammatory and chemokinetic properties. MIP2α is a CXC chemokine (CXCL1,Groβ, 107aa), MIP-2G, CXCL14, 111aa). MIP3 (CCL23, 120aa) is chemotactic for monocytes, resting T-cells, and neutrophils, MIP3α (CCL20; LARC, Exodus, 96aa) is mainly a lymphocyte chemoattractant, as is MIP3β (CCL19, 98aa). MIP4 (CCL18, 89aa) attracts lymphocytes but not monocytes or granulocytes. The nomemclature is complex with many having multiple synonyms.

macrophage inhibition factor *MIF* macrophage migration inhibition factor Substance with a complex history, partly because of doubts concerning the assay systems that were used to define the phenomenon: originally considered to be a mixture of lymphokines (including a 14 kDa glycoprotein) produced by activated T-cells that reduced macrophage mobility and probably increased macrophage-macrophage adhesion. More recently it has been considered a pro-inflammatory cytokine (glycosylation-inhibiting factor, L-dopachrome isomerase/tautomerase, EC 5.3.3.12, 115aa) involved in the innate immune response to bacterial pathogens. Whether the enzymic activity is relevant remains unclear.

macrophage stimulating protein A serum protein (hepatocyte growth factor-like protein, 711aa) belonging to the plasminogen-related growth factor family and the ligand for **ron** receptor tyrosine kinase. A pleiotropic growth factor, also inhibits NO production by macrophages in response to IFNγ and LPS-induced production of various inflammatory mediators. Hepatocyte growth factor-like protein homologue (brain rescue factor 1, 715aa) may be a pseudogene product.

macrosialin A membrane protein (CD68, 354aa) with a short isoform in the plasma membrane and a longer isoform in membranes of endosomes. Highly expressed by blood monocytes and tissue macrophages. Binds to tissue- and organ-specific lectins or selectins, allowing homing of macrophage subsets to particular sites. Belongs to the same family as **LAMP**-1 and -2.

MACS (1) Membrane-anchored C peptides; peptides derived from the C-terminal heptad repeat domain of HIV-1 gp41. (2) Medium-chain acyl-CoA synthetases; enzymes that catalyze the ligation of medium chain fatty acids with CoA. (3) Magnetic-activated cell separation or magnetic assisted cell sorting. (4) Membrane-associated adenylyl cyclases (mACs). (5) Mammalian artificial chromosomes (MACs).

macula *Pl.* maculae. An anatomically distinct area or spot (macule) on the skin (hence immaculate, unspotted).

macula adherens A spot **desmosome**.

macula densa A group of close-packed epithelial cells in the distal convoluted tubule of the kidney that control renin release by producing prostaglandin in response to changes in the sodium concentration of the fluid passing through the tubule.

macula lutea The central region of the retina surrounding the **fovea**. Macular degeneration is an ophthalmological disorder affecting this region (see **age-related macular degeneration** and **macular dystrophy**).

macular corneal dystrophy A disorder in which small areas of opacity develop in the cornea as a result of mutations in the gene for corneal N-acetylglucosamine-6-sulphotransferase (EC 2.8.2.-, 395aa) involved in extracellular matrix biosynthesis.

macular dystrophy A group of ophthalmological disorders in which there is patchy (macular) degeneration of retinal cells or other parts of the visual system. The juvenile-onset form (Best's disease) is caused by mutation in the **bestrophin** gene. The adult form is usually due to mutation in the **peripherin 2** gene. See also **Stargardt's disease-3**.

MAD (1) Mitotic arrest deficient proteins (mitotic spindle assembly checkpoint proteins), components of the spindle-assembly checkpoint that prevents the onset of anaphase until all chromosomes are properly aligned at the metaphase plate. Examples include Mad1 (mitotic arrest deficient 1-like protein 1, 718aa), the human homologue of the MAD1 protein of *S. cerevisiae* and Mad2 (MAD2L1, 205aa). (2) Basic **helix-loop-helix** leucine zipper transcription factors involved in **Dpp** signalling, MAD (mothers against dpp) contains a specific DNA-binding activity that activates an enhancer in a *Drosophila* wing-patterning gene, *vg* (vestigial). The homologues in mammals are more usually referred to as **Smad proteins** (MAD-related proteins). Smad1 is mothers against decapentaplegic homologue 1 (mad homologue-1, 465aa).

Madin-Darby canine kidney cells Line of canine epithelial cells (MDCK cells) that grow readily in culture and form confluent monolayers with relatively low trans-monolayer permeability (varies between clones). Often used as a general model for epithelial cells.

madindolines Non-cytotoxic indole alkaloids that bind competitively and noncovalently to the extracellular domain of the membrane glycoprotein gp130, the common subunit of receptors for the **IL6 family** of cytokines, and block activation of the JAK/STAT signal transduction pathway. Synthetic analogues are potent IL-6 inhibitors. Supplier factsheet: http://www.axxora.com/indexOP.php?PID = ALX-350-328&fromnewsletter = 1

MADS-box A conserved motif found in a family of transcription factors, originally in MCM1 (minichromosome maintenance 1) from *S. cerevisiae*, AGAMOUS from *Arabidopsis*, DEFICIENS from *Antirrhinum majus* and SRF (serum response factor) from *Homo sapiens* although more than 100 are now described. In plants, MADS-box genes include various homeotic genes involved in the determination of floral organ identity (see **ABC Model of Flower Development**). But see **mothers against decapentaplegic**. DNA binding by MADS-box transcription factors: a molecular mechanism for differential DNA bending. http://mcb.asm.org/cgi/content/abstract/17/5/2876. MADS-box Gene Home Page http://www.biologie.uni-hamburg.de/b-online/e28_2/home.htm

Madurella A genus of ascomycete fungi that can be a cause of chronic inflammatory infection of the foot (Madura foot).

MAF (1) Imprecise term for a cytokine (lymphokine), macrophage activating factor, that activates macrophages. Main example is interferon gamma. (2) See c-maf. Maf-A, Maf-B, Maf-F, Maf-G, Maf-K are homologues of the maf transcription factor.

MAG (1) A cell adhesion molecule (myelin-associated glycoprotein, 626aa) of the **immunoglobulin superfamily** that plays a part in regulating neurite outgrowth. See nogo and **oligodendrocyte-myelin glycoprotein**. (2) An allergen from *Dermatophagoides farinae* (American house dust mite). (3) Synonym for glycerol-3-phosphate acyltransferase 3 (EC 2.3.1.15, 434aa). (4) In *S. cerevisiae*, a RING-finger protein, MAG2.

magainins Cationic peptides of about 20aa with antimicrobial activity, originally found in amphibian skin. Probably have membrane insertion and lytic properties. Sequence related to **melittin**.

MAGE antigens See **melanoma associated antigens**.

MAGI A family of **MAGUK** proteins (membrane-associated guanylate kinase inverted) with an inverted arrangement of protein-protein interaction domains. MAGI-1 (membrane-associated guanylate kinase, WW and PDZ domain-containing protein 1, 1491aa) is present at adherens and tight junctions, where it acts as a structural and signalling scaffold regulating various cellular and signalling processes, and associates with **megalin** and **synaptopodin**, MAGI-2 (**atrophin**-1-interacting protein-1, AIP1, 1455aa) co-immunoprecipitates with **stargazin** and

interacts with **PTEN**. MAGI-3 (1506aa) has similar functions.

Magnaporthe A genus of ascomycete fungi, many species of which are cereal pathogens, e.g. *Magnaporthe grisea*, rice blast fungus.

magnesium An essential divalent cation (Mg^{2+}). The major biological role is as the chelated ion in ATP and presumably other triphosphonucleotides. The Mg^{2+}/ATP complex is the sole biologically active form of ATP. The other essential role of Mg^{2+} is as the central ion of chlorophyll. Cellular concentration is less than 5 mM, serum concentration approx. 1 mM.

magnetosome Enveloped compartment in magnetotactic bacteria containing magnetite particles. Aquatic species which favour low oxygen concentrations apparently use this organelle to detect the vertical component of the Earth's magnetic field, and swim downwards.

magnetotaxis Tactic response to magnetic field. See **magnetosome**.

magnocellular neuron Large electrically-excitable neurosecretory cell in the magnocellular region of the hypothalamus. Perhaps the first class of neuron from the central nervous system shown to be sensitive to **nerve growth factor** (previously thought only to act at the periphery). Some secrete **oxytocin**, others **vasopressin**.

Magnoliophyta Flowering plants.

mago nashi A *Drosophila* protein (147aa) involved in the bidirectional intercellular signaling between the posterior follicle cells and oocyte to establish spatial coordinates that induce axis formation. Interacts with **tsu** and is part of the **exon junction complex**. There are two homologues (146 & 148aa) in human and others in *Arabidopsis* and yeast.

MAGPIE In *Arabidopsis* a zinc-finger transcription factor (506aa) that regulates tissue boundaries and asymmetric cell division. Interacts with **JACKDAW**, **SCARECROW** and **SHORT ROOT**.

MAGUKs A family of proteins (membrane-associated guanylate kinases) involved in the assembly of multiprotein complexes on the inner surface of the plasma membrane at regions of cell-cell contact. There are four subfamilies, **DLG**-like, ZO-1-like (see **ZO proteins**), p55-like, and LIN2-like (see **CASK**). Many contain caspase recruitment domains (see **CARD**). See also **MAGI**.

maitotoxin Toxin from the dinoflagellate, *Gambierdiscus toxicus*. Activates L-type voltage-sensitive calcium channels (**VSCC**) and mobilizes intracellular calcium stores. Very potent (LD_{50} of 50 ng).

maize streak virus One of the Geminiviridae, endemic in sub-Saharan Africa. It is transmitted by an African leafhopper, *Cicadulina mbila*. Transgenic maize strains resistant to the virus have recently (2007) been developed. Article: http://onlinelibrary.wiley.com/ doi/10.1111/j.1467-7652.2007.00279.x/pdf

Majeed syndrome A rare disorder in which there are recurrent episodes of fever and inflammation in the bones and skin; caused by mutation in the **lipin 2** gene.

major facilitator superfamily domain-containing proteins See **MFSDs**.

major histocompatibility antigen See **histocompatibility antigens**.

major histocompatibility complex *MHC* The set of gene loci specifying major histocompatibility antigens, e.g. *HLA* in man, *H-2* in mice, *RLA* in rabbits, *RT-1* in rats, *DLA* in dogs, *SLA* in pigs, amongst others. In humans, MHC Class I encodes peptide-binding proteins, heterodimers of an alpha chain and a common beta-2-microglobulin. The peptide-binding complex presents antigen fragments to cytotoxic T-cells. Class I loci also encode antigen-processing molecules such as **TAP** and **tapasin**. MHC Class II genes encode heterodimeric peptide-binding molecules in antigen-presenting cells, MHC Class III genes encode some complement components and cytokines.

major intrinsic proteins *MIP* Family of structurally related proteins with 6 transmembrane segments, associated with gap junctions or vacuoles that all form transmembrane channels some of which are aquaporins, other transport glycerol. The MIP of lens fibres (aquaporin-0, MIP-26, 263aa) is found in lens fibre gap junctions and may be responsible for regulating the osmolarity of the developing lens. In its native form it is an aquaporin but as the lens develops it is proteolytically truncated and MIPs from adjacent cells interact as an adhesive junction. Mutations can cause cataracts. Other members: **nodulin**-26 (soybean), **tonoplast intrinsic protein** (TIP) found in plant storage vacuoles, *Drosophila* neurogenic protein **big brain**. Page in Transporter Classification Database: http://www.tcdb.org/search/ result.php?tc = 1.A.8

major urinary proteins Proteins in the urine of rodents that bind pheromones. MUP-1 (180aa in mouse) is one of the **lipocalin** family. MUP-4 (178aa) is expressed in the vomeronasal organ and may bind pheromones that have been carried by urinary MUPs. MUP-20 (181aa) is a pheromone (darcin) that stimulates female sexual attraction to male urinary scent and promotes aggressive behaviour in males. Abstract: http://www.uniprot.org/citations/ 18064011

majusculamide C A microfilament-depolymerizing cyclic depsipeptide from the cyanobacterium *Lyngbya*

majuscula, an analogue of the natural antitumour agent **dolastatin**. Majusculamides A and B are lipodipeptides. Abstract: http://pubs.acs.org/doi/abs/10.1021/ jo00176a004

Mal d 1 A major apple (*Malus domestica*) allergen responsible for the symptoms of oral allergy syndrome. It is a pathogenesis-related (PR) protein (159aa) of the PR-10 family.

malabsorption syndrome A variety of conditions in which digestion and absorption in the small intestine are impaired. Multiple causes including lymphoma, amyloid and other infiltrations, **Crohn's disease**, gluten-sensitive enteropathy and the sprue syndrome in which the villi atrophy for unknown reasons.

malacia General pathological term for softening of any organ or tissue.

malaria See *Plasmodium*.

Malassez cells Cells found in the periodontal ligament as 'epithelial rests of Malassez'. Malassez cells retain the major characteristics of epithelial cells throughout their differentiation from the root sheath epithelium into the rests of Malassez.

malate The ion from malic acid, a component of the citric acid cycle.

malcavernin A protein (cerebral cavernous malformations 2 protein, 444aa) that may be a scaffold protein for MAP2K3-MAP3K3 signalling and seems to modulate MAP3K3-dependent p38 activation induced by hyperosmotic shock. It interacts with **KRIT1**, the CCM3 gene product (protocadherin-10) and the cytoplasmic domain of integrins. Mutations cause a type of cerebral cavernous malformation, vascular abnormalities of the CNS.

MALDI-TOF Common abbreviation for matrix assisted laser-desorption ionisation (MALDI), a method for generating molecular ions, combined with time of flight (TOF) mass spectroscopy. An increasingly important method for analysing biological samples.

maleate The ion from maleic acid, often used in biological buffers.

malignant In general, describing any disease that can become life-threatening if untreated. As applied to tumours means that the primary tumour has the capacity to show **metastatic spread** (metastasise) and implies loss of both **growth control** and positional control.

malignant hyperthermia Channelopathy of calcium channels, a rare but often fatal genetic condition during anesthesia. Type 1 is caused by mutation in the **ryanodine** receptor gene, type 5 by mutation in the CACNA1S gene that encodes a subunit of the voltage-dependent **L-type calcium channel**. Other types are known but the genes involved are not yet identified.

malin E3 ubiquitin-protein ligase NHLRC1 (EC 6.3.2.-, 395aa) which in complex with **laforin** and HSP70 suppresses the cellular toxicity of misfolded proteins by promoting their degradation. Defects in NHLRC1 are a cause of progressive **myoclonic epilepsy** type 2.

Mallory's one-step stain A modified trichrome method that can give good tissue differentiation using a relatively brief procedure.

malonate The ion from malonic acid, HOOC. CH_2COOH. Malonate is a competitive inhibitor for succinate dehydrogenase in the **tricarboxylic acid cycle**. Malonyl-SCoA is an important precursor for fatty acid synthesis.

Malpighian layer (1) In mammalian skin, the innermost layer of the epidermis. (2) A palisade layer of the seed, cells with thickened walls in the outer epidermis (exotesta).

Malpighian tubule Blind-ending tubule opening into the lower intestine of insects and responsible for fluid excretion − the arthropod equivalent of the kidney.

MALT See **mucosal associated lymphoid tissue**.

maltase Lysosomal enzyme (EC 3.2.1.20, alpha-glucosidase, 1857aa) that hydrolyzes maltose (and the glucose trimer maltotriose) to glucose, during the enzymic breakdown of starch. Also secreted by the cells of intestinal villi. Mutations cause **glycogen storage disease type II** (Pompe's disease).

maltoporins A family of bacterial outer membrane proteins (maltose-inducible porin, Lambda receptor protein, 446aa) that form homotrimeric complexes involved in the transport of maltose and maltodextrins. Entry in Transporter Classification Database: http://www.tcdb.org/search/result.php?tc = 1.B.3

maltose Disaccharide intermediate of the breakdown of starch, glucose-α(1−4)-glucose. Fermentable substrate in brewing.

maltose-binding protein Protein (396aa) of the bacterial (*E. coli*) periplasm that links with **MCP-II** and is involved in the chemotactic response to maltose; probably derived from a similar protein that links with the maltose transmembrane transport system. Article: http://jb.asm.org/content/179/24/7687. long

MAML proteins A family of three co-transcriptional regulators (mastermind-like proteins) essential for **Notch** signalling. They have distinct tissue-specific distributions. MAML1 (1016aa) localizes to nuclear bodies and interacts with **MEF2C**. MAML2

(1173aa) is widely expressed. MAML3 (1133aa) acts selectively on different notch promoters. Mastermind itself is a neurogenic *Drosophila* protein (1594 aa).

mammalian expression vector In molecular biology, a **vector** that will produce large amounts of eukaryotic protein (taxonomy notwithstanding, not necessarily a protein from a mammal).

mammalian-wide interspersed repeat *MIR* The second most common type of interspersed repeat in primates, 260 bp tRNA-derived **short interspersed nucleotide elements** (SINEs). There may be as many as 300,000 copies constituting 1−2% of the total DNA. See **Alu**, **long interspersed nucleotide elements**.

mammamodulin Protein (52−55 kDa) that was reportedly expressed by hormone-independent mammary tumour cells and that affected morphology, motility, growth and hormone-receptor expression. There have been no recent reports nor entries in databases and the term apparently only survives in dictionary entries that are verbatim copies of the definition in the 3rd Edition of this dictionary.

mammary gland Milk-producing gland of female mammals. An adapted sweat gland, it is made up of milk-producing alveolar cells, surrounded by contractile myoepithelial cells, together with considerable numbers of fat cells. Milk production is hormonally controlled.

mammary tumour virus *MMTV* **Retrovirus** that induces mammary carcinoma in mice. (mouse mammary tumour virus, Bittner agent). Isolated from highly inbred strains that had very high incidence of the tumours, after the discovery that the disease was transmitted in milk by nursing mothers. Endogenous provirus is present in germ-line of all inbred mice. Transcription of the provirus is regulated by a viral promoter that increases transcription in response to glucocorticoid hormones. May transform by proviral insertion activating the cellular *int-1* oncogene.

mammary-derived growth inhibitor Heart-type **fatty acid binding protein** (MDGI, H-FABP, FABP3, 133aa) involved in intracellular transport of long-chain fatty acids, that inhibits proliferation of mammary carcinoma cells. Overexpression or knockout of the gene in mice causes no overt phenotypic change.

mandibuloacral dysplasia A progeria-like disorder associated with partial **lipodystrophy**. Can be caused by mutation in genes for lamin A/C or a zinc metallopeptidase (ZMPSTE24, prenyl protein-specific endoprotease 1, 475aa). See **progeria**.

Manduca sexta A species of Lepidopteran insect, also known as the tobacco hornworm moth. The caterpillars, which are very large, are used in studies of ion transport, moulting, and as a system for transgenic gene expression (see **baculovirus**).

manganese   *Mn* An essential trace element. Present in cells as concentrations of around 0.01 mM. Activates a wide range of enzymes, e.g. **pyruvate carboxylase** and one family of **superoxide dismutases**. Resembles **magnesium** and may replace it in many enzymes when it can modify substrate specificities. The addition of manganese salts to buffer solutions will often make cells very adhesive.

mannan   Mannose-containing polysaccharides found in plants as storage material, in association with cellulose as hemicellulose. In yeasts a wall constituent.

mannan-binding protein   An acute phase protein (MBP, mannan-binding lectin, MBL, collectin-1, 248aa), a calcium-dependent lectin, structurally related to **complement** C1, that binds specific carbohydrates (**mannans**) on the surface of various microorganisms including bacteria, yeasts, parasitic protozoa, and viruses; activates the complement cascade through **MASP** and promotes phagocytosis. Deficiency is associated with frequent infections in childhood. See **collectins**.

mannitol   Hexitol related to D-mannose. Found in plants, particularly fungi and seaweeds.

mannopeptimycins   Mannopeptimycins alpha, beta, gamma, delta, and epsilon are cyclic glycopeptide antibiotics produced by *Streptomyces hygroscopicus* LL-AC98. Inhibit cell wall biosynthesis through lipid II binding. Article: http://aac.asm.org/content/47/1/62.long

mannoprotein   Glycoproteins in which there are many mannose residues. Mannoproteins are a major component of the cell wall of *S. cerevisiae*, comprising about 40% of its mass. Mannoproteins are believed to be determinants of cell wall permeability, and some are essential for developmental events such as mating and transition to hyphal growth.

mannose   Hexose identical to D-glucose except that the orientation of the -H and -OH on carbon 2 are interchanged (i.e. the 2-epimer of glucose). Found as constituent of polysaccharides and glycoproteins. Mannose-6-phosphate is formed by phosphorylation in the **Golgi complex** of certain mannose residues on N-glycan chains of lysosomal enzymes. Believed to function as targeting signal that causes entry of these enzymes to the lysosomes.

mannose-6-phosphate receptor   There are two unrelated receptors, both required for transport of phosphorylated lysosomal enzymes from the Golgi complex and the cell surface to lysosomes. The cation-dependent mannose 6-phosphate receptor is a transmembrane protein (46 kDa, 277aa) and functions as a dimer. The extracellular portion is homologous to that of the cation-independent receptor (CD222, 2491aa) which is also the receptor for insulin-like growth factor II (IGF2R) and has a role in the intracellular trafficking of lysosomal enzymes, the activation of **TGFβ** and in degradation of IGF2. Mannose-6-phosphate receptor-binding protein 1 (perilipin-3, 434aa) is required for the transport of both mannose 6-phosphate receptors from endosomes to the *trans*-Golgi network.

mannosidases   Enzymes catalysing hydrolysis of the glycosidic bond between mannose residues and a variety of hydroxyl-containing groups. Alpha-mannosidases (EC 3.2.1.113) in rough endoplasmic reticulum and *cis*-Golgi are responsible for progressively removing four mannose residues during the synthesis of the complex-type N-linked glycan chains of glycoproteins. They are inhibited by **deoxymannojirimycin**. Beta-mannosidase (EC 3.2.1.25) is a lysosomal enzyme that carries out the final exoglycosidase step in degradation of N-linked oligosaccharide moieties of glycoproteins. Defects in mannosidases are responsible for some forms of **lysosomal diseases** (storage diseases).

mantle-cell lymphoma   A subtype of B-cell non-Hodgkin's lymphoma derived from CD5$^+$ antigen-naive B-cells from within the mantle zone that surrounds normal germinal centre follicles.

Mantoux test   Test for tuberculin reactivity in which tuberculin PPD (purified protein derivative) is injected intracutaneously. The injection site is examined after 2−3 days, a positive reaction, indicating current or previous infection with *Mycobacterium tuberculosis* (in an uninoculated individual), is an oedomatous and reddened area caused by T-cell reactivity.

MAP kinases   Serine-threonine kinases (mitogen-activated protein kinases, MAPKs, EC 2.7.11.24) that are activated when quiescent cells are treated with mitogens, and are involved in regulation of the cell cycle. There is a signalling cascade in which mitogen activates MAPkinasekinasekinase (MAPKKK, MAP3K), which activates MAPkinasekinase (MAPKK, MAP2K), which activates MAPkinase. Confusingly, MAPkinases also phosphorylate **microtubule associated proteins** (MAPs). MAP kinases can be subdivided into six groups: **i.** extracellular signal-regulated kinases, (externally regulated kinases, classical MAP kinases) ERK1 (MAPK3, 379aa) and ERK2 (MAPK1, 360aa); **ii.** c-Jun N-terminal kinases (**JNKs**; MAPK8, MAPK9, MAPK 10), also known as stress-activated protein kinases (SAPKs); **iii.** p38 isoforms (MAPK11, MAPK12/ ERK6), MAPK13, MAPK14); **iv.** ERK5 (MAPK7) which is critical for endothelial function and maintenance of blood vessel integrity; **v.** ERK3/4 atypical MAPKs. ERK3 (MAPK6) and ERK4 (MAPK4); **vi.** ERK7/8. (MAPK15). This classification is unlikely to remain definitive. See **MAPKK, MAPKKK**.

MAP1C (1) See **microtubule-associated proteins**. (2) In *Arabidopsis* a mitochondrial and chloroplastic enzyme (methionine aminopeptidase 1C, EC 3.4.11.18, 344aa) that removes the amino-terminal methionine from nascent proteins.

MAP4Ks Mitogen-activated protein kinase kinase kinase kinases. For example MAP4K1 (MAPK/ERK kinase kinase kinase 1, EC 2.7.11.1, haematopoietic progenitor kinase, 833aa), which may be involved in the response to environmental stress, upstream of the JUN N-terminal pathway. May also be involved in haematopoietic lineage decisions and growth regulation. See **MAP kinase, MAPKK, MAPKKK**.

MAPEG proteins A family of transmembrane proteins (Membrane-Associated Proteins in Eicosanoid and Glutathione metabolism) with diverse functions. They include 5-lipoxygenase-activating protein (gene FLAP), leukotriene C4 synthase (EC 2.5.1.37), microsomal **glutathione S-transferase** II (EC 2.5.1.18) (GST-II) and prostaglandin E synthase.

MAPK phosphatases Group of dual-specificity phosphatases that dephosphorylate the tyrosine and threonine residues of the activation-loop of **MAP kinases**. For example, dual specificity protein phosphatase 1 (DUSP1, EC 3.1.3.16, EC 3.1.3.48, 367aa) acts on Erk2 and is induced by oxidative stress and heat shock. DUSP22 (JNK-stimulatory phosphatase-1, 184aa) dephosphorylates and deactivates p38 and stress-activated protein kinase/c-Jun N-terminal kinase (SAPK/JNK). There are tissue-specific variants, and forms with different MAPK preference activated in response to different signals.

MAPKKKs Family of kinases (MAP kinase kinase kinases, MAP3Ks) that regulate the activity of the kinases (MAP2Ks) that phosphorylate MAP kinases (MAPKs). Lie at the top of signalling cascades and are activated by mitogens. For example, mitogen-activated protein kinase kinase kinase 1 (MAP3K1, MAPK/ERK kinase kinase 1, MEKK1, EC 2.7.11.25, 1512aa) is activated by autophosphorylation following oligomerization that occurs after binding MAP4K2 or TRAF2. See **ASK1, MAP4Ks**.

MAPKKs MAP kinase kinases (MAP2Ks, MAP/ERK kinases, MEKs), dual specificity kinases that phosphorylate MAP kinase on both the tyrosine and threonine residues, in that order. They are activated by phosphorylation by **MAPKKK**. For example, dual specificity mitogen-activated protein kinase kinase 1 (MAP2K1, EC 2.7.12.2, 393aa) activates ERK1 and ERK2. See **MAP kinases**.

MapMan A software tool that displays large datasets (e.g. gene expression data) onto diagrams of metabolic pathways or other processes and thus enables easier interpretation of results. Incorporates some statistical tools. Originally built for *Arabidopsis*

thaliana but now extended to other species. Link: http://www.gabipd.org/projects/MapMan/

mappine See **bufotenine**.

MapPop A software program that facilitates the selection of optimal subsets from mapping populations for the purposes of **bin mapping**. Download available from: http://www.sciencecentral.com/site/466271

MAPs See **microtubule-associated proteins** and mitogen activated protein kinases (**MAP kinases**).

MAPTAM Compound that readily enters cells where it is converted to 5-methyl **BAPTA**, an indicator of calcium concentration.

Marburg virus One of the family **Filoviridae** that causes Marburg disease, a severe haemorrhagic fever developed in many people who work with African green monkeys. Related to **Ebola virus**.

Marchantiophyta Liverworts.

MARCKS Protein (myristoylated alanine-rich protein kinase C-substrate, 332aa) that is membrane associated through the myristoyl residue when phosphorylated and cytoplasmic otherwise. An F-actin crosslinking protein when membrane associated, binds calmodulin and **synapsin**. Implicated in macrophage activation, neurosecretion and growth-factor dependent mitogenesis. Plays a role in insulin-dependent endothelial signalling to PIP2, and is a critical determinant of actin assembly and directed cell movement in the vascular endothelium. MARCKS-related protein (195aa) may couple the protein kinase C and calmodulin signal transduction systems. Abstract: http://www.ncbi.nlm.nih.gov/pubmed/21097841

Marek's disease Infectious cancer of the lymphoid system (lymphomatosis) in chickens, caused by a contagious Herpesvirus. An effective vaccine is now available.

marenostrin See **pyrin**.

Marfan's syndrome Dominant disorder of connective tissue in which limbs are excessively long and loose-jointed. Due to mutation in the fibrillin-1 gene that causes collagen fibril-assembly disorder that can be mimicked in mice by aminonitriles that interfere with crosslinking.

margaratoxin Charybdotoxin-related peptide toxin (39aa) from scorpion *Centruroides margaritatus*. Blocks mammalian voltage-gated potassium channels (Kv1.3) in neural tissues and lymphocytes. Similar to **charybdotoxin, kaliotoxin** and **noxiustoxin**.

marginal band A bundle of equatorially-located microtubules that stabilize the biconvex shape of platelets and avian erythrocytes. They are unusual in that they do not derive from the centrosomal **MTOC**.

margination Adhesion of leucocytes to the endothelial lining of blood vessels, particularly postcapillary venules; often, but not always, a prelude to leaving the circulation and entering the tissues.

mariner Group of **transposons** with broad phylogenetic distribution (arthropods, nematodes, planaria, humans). Mariner elements consist of a transposase gene flanked by short inverted repeats.

Marinesco-Sjogren syndrome A disorder in which there is cerebellar ataxia, cataracts, and retarded somatic and mental maturation caused by disruption of the SIL1 gene that encodes a nucleotide exchange factor (SIL1, BiP-associated protein, 461aa) required for protein translocation and folding in the endoplasmic reticulum.

marker gene Gene that confers some readily-detectable phenotype on cells carrying the gene, either in culture, or in transgenic or chimeric organisms. Gene could be an enzymic **reporter gene**, a selectable marker conferring antibiotic resistance, or a cell membrane protein with a characteristic **epitope**.

marker rescue The restoration of gene function by replacing a defective gene with a normal one by recombination. Most common application is in co-infection of cells with a mutant phage that is unable to replicate and a wild-type phage; recombination between the phages repairs the replication defect and the recombinant derivative of the mutant phage can reproduce — it has been rescued.

Markov process A **stochastic** process in which the probability of an event in the future is not affected by the past history of events.

Maroteaux-Lamy syndrome Mucopolysaccharidosis type VI; deficiency of the lysosomal enzyme arylsulphatase B; resembles **Hurler's disease** in some respects.

MARPS (1) A conserved family of 'muscle ankyrin repeat proteins' induced by stress/strain injury signals and can associate with the elastic region of **titin**/connectin. See **cardiac ankyrin repeat protein**. (2) Microtubule associated repetitive proteins from membrane skeleton of *Trypanosoma brucei*; probably stabilize microtubules — there are tandem 38aa repeats with tubulin-binding capacity. Not in the recent literature.

Marshall's syndrome A disorder in which there is a distinctive facial appearance, eye abnormalities, hearing loss, early-onset arthritis and short stature; caused by mutation in the gene for collagen XI (COL11A1). Has features in common with **Stickler's syndrome** but is considered distinct.

marvel domain A domain with a four transmembrane-helix architecture that has been identified in myelin and lymphocyte protein (MAL), physins, gyrins and occludin families. The MAL gene encodes a proteolipid protein (Myelin and lymphocyte protein, MAL protein, T-lymphocyte maturation-associated protein, 153aa) involved in vesicular trafficking to the apical plasma membrane. MAL2 (176aa) is involved in polarized transport within epithelial cells, particularly the transport of endosomes to the apical surface. MARVEL domain-containing protein 2 (MarvelD2, tricellulin, 558aa) has a role in the formation of the epithelial barrier that separates the endolymphatic and perilymphatic spaces of the organ of Corti; defects in marvelD2 lead to a form of deafness. MarvelD3 co-localises with **occludin** at tight junctions in intestinal and corneal epithelial cells. N.B. There are various unrelated Mal proteins (Molybdenum cofactor sulphurase in *Drosophila*, an important allergen (**Mal d 1** in *Malus domestica*, a maltose permease in *Saccharomyces*).

mas Proto-oncogene (*mas1*) originally identified in a human epidermoid carcinoma cell line. The *mas1* gene product is a G-protein-coupled receptor (325aa) for the peptide angiotensin 1-7 that is a functional antagonist of the angiotensin-2 type 1 receptor. There is a mas-related G-protein coupled receptor (MRG, 378aa).

masked messenger RNA Long-lived and stable mRNA found originally in the oocytes of echinoderms and constituting a store of maternal information for protein synthesis that is unmasked (derepressed) during the early stages of morphogenesis. In these early stages the rate of cell division is so rapid that transcription from the embryonic genome cannot occur. Undoubtedly not restricted to oocytes, and the term can be applied to any mRNA which is present in inactive form.

maskin In *Xenopus* oocytes, transforming acidic coiled-coil-containing protein 3 (cytoplasmic polyadenylation element-binding protein-associated factor, 931aa) that outcompetes eif4g to bind eif4e, thereby preventing translation and 'masking' the mRNA. See **cytoplasmic polyadenylation element binding factor** (CPEB).

MASP Serine peptidases (EC 3.4.21.104, MASP-1, 699aa; MASP-2; 686aa) in the **complement** system, activated by the binding of mannan to **mannan-binding lectin** and probably responsible for the proteolytic cleavage and activation of C2 and C4. They are similar to the **C1q**-associated proteases, C1r and C1s. All four have a C1r/C1s-like domain, an **EGF-like domain** and a second C1r/C1s-like domain, two complement control protein (CCP) domains and a serine protease domain.

maspardin A cytosolic protein (33-kDa acidic cluster protein, 308aa) that is a negative regulatory factor in CD4-dependent T-cell activation. Mutated in a form of **spastic paraplegia** (Mast syndrome;

SPG21). The name derives from 'mast syndrome, spastic paraplegia, autosomal recessive, with dementia'.

maspin Serpin-B5 (375aa) expressed in normal mammary epithelium and down-regulated in mammary tumours.

mass spectrometry *MS* Widely used method for determining relative mass and abundance of ionised molecules or fragments of molecules (generated e.g. by **MALDI**) that are accelerated *in vacuo* by electrical means and their trajectories within the beam of charged particles modified by the application of magnetic fields; particles are separated on the basis of mass/charge ratio. The spectrum of peaks on the detector system indicates the composition of the mixture and is diagnostic of the starting material.

Masson's trichrome stain Trichrome stains are used particularly for connective tissue. Masson's trichrome method uses haemalum, acid fuchsin and methyl blue and has the effect of staining nuclei blue-black, cytoplasm red and collagen blue.

mast cell Resident cell of connective tissue that contains many granules rich in **histamine** and heparan sulphate. Release of histamine from mast cells is responsible for the immediate reddening of the skin in a weal-and-flare response. Very similar to **basophils** and possibly derived from the same stem cells. Two types of mast cells are now recognized, those from connective tissue and a distinct set of mucosal mast cells; the activities of the latter are T-cell dependent.

mastermind-like proteins See **MAML proteins**.

mastigonemes Lateral projections from eukaryotic flagella. May be stiff and alter the hydrodynamics of flagellar propulsion, or flexible and alter the effective diameter of the flagellum (flimmer filaments).

mastocytoma A benign nodular skin tumour that is infiltrated by **mast cells**. *Cf.* **mastocytosis**.

mastocytosis A variety of disorders (mast cell disease) in which tissue is infiltrated by large numbers of mast cells.

mastoparans Basic peptides (~14aa) from wasp venoms. Mastoparan-X from *Vespa xanthoptera* (Japanese yellow hornet) causes mast cell degranulation and activates G proteins that couple to phospholipase C. Mastoparan-V1 (15aa) from *Vespula vulgaris* (common wasp) is chemotactic for neutrophils and has potent antimicrobial activity. Analogous to **melittin** in honey bee venom.

maternal antibody Any antibody transferred from a mammalian mother transplacentally or, in some mammals, in the colostrum. See **IgG**.

maternal inheritance Inheritance through the maternal cell line, e.g. through the oocyte and eggs. Mitochondrial genes are maternally inherited and various other non-Mendelian forms of inheritance may also appear as maternal inheritance.

maternal mRNA Messenger RNA found in oocytes and early embryos that is derived from the maternal genome during oogenesis. See **masked messenger RNA**.

maternal-effect gene Gene, usually required for early embryonic development, whose product is secreted into the egg by the mother. The phenotype is thus determined by the mother's, rather than the egg's, genotype, *cf.* **zygotic-effect gene**. See also **egg-polarity gene**.

mating-type genes Genes that, in *S. cerevisiae*, specify into which of the two mating types (a and α) a particular cell falls. Only unlike mating type haploids will fuse. The interest derives from the way in which mating type is switched: the existing gene is removed and a new gene, derived from a (silent) master copy elsewhere in the genome is spliced in. Later this gene will in its turn be replaced by a new copy of the old gene, also derived from a silent 'master'. The a- and α-genes code for pheromones that affect cells of the opposite mating type. Similar mating-type genes are known from other yeasts, and the switching mechanism (**cassette mechanism**) may be used more generally.

Matrigel Proprietary name for gel-forming matrix material derived from **EHS cells**. Cells grown on Matrigel often show morphological characteristics distinct from those seen on a solid tissue-culture substratum – and are probably in a more normal environment both chemically and physically.

matrilysin Matrix metalloproteinase (MMP-7, EC 3.4.24.23, 267aa) that is thought to play a part in cellular invasion of tissues by digesting extracellular matrix and more generally in tissue remodelling. Matrilysin-2 is MMP-26. See Table M1.

matriosome See **striosome**.

matrix Ground substance in which things are embedded or that fills a space (as for example the space within the mitochondrion). Most common usage is for a loose meshwork within which cells are embedded (e.g. extracellular matrix), although it may also be used of filters or absorbent material.

matrix attachment region Region (scaffold/matrix attachment region) of chromatin sequence that binds directly to nuclear matrix. Found in intergenic DNA, especially flanking the 5' ends of genes or clusters of genes. Many have a sequence motif called the MAR/SAR recognition signature sequence that binds to MAR binding proteins.

TABLE M1. Matrix metallopeptidases

Enzyme	Designation	Main substrate
Interstitial collagenase	MMP-1	Fibrillar collagens
Gelatinase A	MMP-2	Progelatinase A
Stromelysin-1	MMP-3	Non-fibrillar collagen, Ln, Fn
Matrilysin	MMP-7	Ln, Fn, non-fibrillar collagen
Neutrophil collagenase	MMP-8	Fibrillar collagens I, II & III
Gelatinase B	MMP-9	Interleukin-8 precursor, Metastasis-suppressor KiSS-1 precursor.
Stromelysin-2	MMP-10	Ln, Fn, non-fibrillar collagen
Stromelysin-3	MMP-11	Serpin
Metalloelastase	MMP-12	Elastin
Collagenase-3	MMP-13	Fibrillar collagens
MT-MMP	MMP-14	Pro-gelatinase-A
Membrane-type matrix metallopeptidase-3	MMP-16	Pro-gelatinase-A
Collagenase 4 (Xenopus)	MMP-18	α-1-proteinase inhibitor, type I collagen α-1
RASI-1, RASI-6	MMP-19	Aggrecan core protein
Enamelysin	MMP-20	Amelogenin
Xenopus MMP-21	MMP-21	?
MMP-27 (*Homo sapiens*)	MMP-22	?
MIFR protein (*Homo sapiens*)	MMP-23B	Synthetic
Matrilysin-2	MMP-26	Fibrinogen, vitronectin, matrilysin, α-1-proteinase inhibitor
Epilysin	MMP-28	?

For more detailed information consult the Merops peptidase database http://merops.sanger.ac.uk/
Fn = fibronectin; Ln = laminin.

matrix attachment region-binding proteins
Proteins (MARBPs) involved in the attachment of chromatin to the nuclear matrix. In metazoa the function is carried out by scaffold attachment factors (SAFB1, 915aa; SAFB2, 995aa) and scaffold/matrix-associated region-1-binding proteins (BANP, 519aa). In *Arabidopsis* the nuclear envelope-associated protein, MAR-binding filament-like protein 1 (MFP1, 726aa) binds to DNA and interacts with MFP1 attachment factor 1 (MAF1, WPP domain-containing protein 1, WPP1, 155aa) which regulates mitotic activity in roots. WPP1 localizes preferentially to the outer nuclear membrane near nuclear pore complexes and is redistributed to the cell plate during cytokinesis. It interacts with farnesyl-diphosphate farnesyltransferase (squalene synthase, EC 2.5.1.21, 410aa), WIPs (See **WIP** #2) and WPP domain-interacting tail-anchored protein 1 and 2 (703 and 627aa). The **WPP domain** is required for the nuclear envelope localization.

matrix Gla protein A protein (103aa) associated with the organic matrix of bone and cartilage. Thought to act as an inhibitor of bone formation. Requires vitamin K-dependent gamma-carboxylation by glutamate carboxylase for its function. Secreted by vascular smooth muscle cells and chondrocytes and upregulated by Vitamin D. Comparable in function to bone Gla protein (**osteocalcin**). Mutations are associated with **Keutel's syndrome**.

matrix metallopeptidases A family of enzymes (MMPs, matrix metalloproteinases) that degrade proteins of the extracellular matrix such as collagen and elastin, some of which are resistant to other peptidases. MMP1 is collagenase 1, MMP2 is gelatinase A, MMP3 is stromelysin 1, MMP7 is **matrilysin** and so on. See Table M1, also **tissue inhibitors of metalloproteinases**.

matrix metalloproteinases *MMP* See **matrix metallopeptidases**.

matrix proteins Proteins of the outer layer of the cell wall of Gram-negative bacteria.

maturation-promoting factor See **cyclin**.

Mauthner neuron Large neuron in the **mesencephalon** of fishes and amphibians. A rare example of an individually identifiable neuron in a vertebrate nervous system.

MAVS Mitochondrial membrane protein (mitochondrial antiviral signalling protein, IPS-1, Cardif; VISA, 540aa) that is a **RIG1** adaptor and activates NFκB and IRF3. Hepatitis C virus NS3-4A protease cleaves MAVS, causing its relocation from the mitochondrial membrane to the cytosolic fraction, resulting in disruption of signalling to the antiviral immune response. Contains an amino-terminal CARD domain and a carboxyl-terminal mitochondrial transmembrane sequence that localizes to the mitochondrial membrane.

Max Myc associated factor X, a basic helix-loop-helix transcription factor (160aa): forms homodimers that interact with CACGTG motif of DNA repressively, but will form heterodimers with **Myc protein** that bind the same motif with greater affinity and activate the downstream gene.

Maxam-Gilbert method A method of DNA sequencing, based on the controlled degradation of a DNA fragment in a set of independent, nucleotide-specific reactions. The resulting fragments have characteristic sizes depending on the sequence of the template that can be resolved on a sequencing gel. Although no longer the main protocol, Maxam-Gilbert sequencing still has advantages, e.g. for oligonucleotides or covalently modified DNA. See also **dideoxy sequencing**.

maxiprep Slang, denoting a large scale purification of **plasmid** from a bacterial culture. Usually used to describe preparations from 100−500 ml culture. See also **miniprep, midiprep, megaprep**.

May-Hegglin anomaly A disorder involving thrombocytopenia, enlargement of platelets, and characteristic leukocyte inclusions (Dohle bodies), caused by mutation in the gene encoding nonmuscle myosin heavy chain-9. Mutation in this gene also cause Fechtner and Sebastian syndromes.

maytansine A cytotoxic antibiotic isolated from the Ethiopian shrub, *Maytenus serrata*. It blocks microtubule assembly by binding to the same site on tubulin as **rhizoxin**.

mayven An actin binding protein (kelch-like 2, 593aa) expressed predominantly in the CNS, has six **kelch** repeats. Thought to have a role in the formation of processes from oligodendrocyte precursor (O2-A) cells. Associates with the SH3 domain of **fyn**. See **actinfilin**.

mazF/mazE system A type II toxin / antitoxin system in *Staphylococcus aureus*. MazF (120aa), the toxin, is a sequence-specific endoribonuclease that cleaves mRNA, thus inhibiting protein synthesis and inducing bacterial stasis. MazE (56aa) binds to the mazF toxin and counteracts its endoribonuclease activity. MazG also binds mazF and further inhibits its activity. Similar systems exist in *E. coli* and other bacteria. The toxin is stable, whereas the antitoxin is labile under conditions of stress and the toxin can then exert its toxic effect to inhibit cell growth. Article: http://www.ncbi.nlm.nih.gov/pmc/articles/PMC2168618/

M-band Central region of the **A-band** of the **sarcomere** in striated muscle.

MBC (1) Metastatic breast cancer or male breast cancer. (2) Minimum (minimal) bactericidal concentration. (3) **Carbendazim**, methyl benzimidazol-2-ylcarbamate. (4) Microbial biomass carbon.

MBF (1) In *S. pombe*, a protein complex (Mlu1-binding factor) that contains the proteins Res1p and Res2p and binds to the Mlu1 cell-cycle box (**MCB**) element in DNA, activating the transcription of genes required for S phase. Abstract: http://www.nature.com/ncb/journal/v3/n12/abs/ncb1201-1043.html. (2) In *S. cerevisiae*, a transcriptional coactivator (multiprotein-bridging factor 1, suppressor of frameshift mutations protein 13, 151aa) that recruits TATA box binding protein to promoters that have bound GCN4 activator protein. A homologue is found in *Arabidopsis*.

MBL See **mannan-binding lectin**.

MBP (1) See **myelin basic protein**. (2) See **mannan-binding protein**. (3) Myrosinase-binding proteins. The predicted plant proteins MBP1 (642aa) and MBP2 are similar to lectins and plant aggregating factors. (4) In *S. cerevisiae* a transcription factor (MBP1, 833aa) that binds to MCB elements (Mlu I cell cycle box) found in the promoter of most DNA synthesis genes.

MCAF See **monocyte chemotactic and activating factor**.

MCAK Mitotic centromere-associated kinesin (kinesin-like protein 6, 725aa). Promotes ATP-dependent removal of tubulin dimers from microtubules and regulates the turnover of microtubules at the kinetochore.

McArdle's disease See **glycogen storage diseases**.

mcb (1) Metaplastic carcinoma of the breast, a tumour with a mixture of epithelial and mesenchymal elements. (2) Monochlorobimane. (3) Multicolour banding (of chromosomes). (4) **MluI Cell-cycle Boxes**. (5) Monochlorobenzene.

McCune-Albright syndrome See **granins**.

M cells Cells (microfold cells) of the follicle-associated epithelium of **Peyer's patches** that take up antigenic materials from the gut and transfer them to dendritic cells and T-cells lying below. They lack microvilli on their apical surface, but have broader microfolds.

MCF7 cells Line of human estrogen receptor-positive breast adenocarcinoma cells derived from a metastatic site.

MCH (1) **Melanin-concentrating hormone**. (2) Mean erythrocyte haemoglobin content.

M-channels Voltage-sensitive **potassium channels** inactivated by **acetylcholine**, responsible for the M current, a slowly activating and deactivating potassium conductance which plays a critical role in determining the subthreshold electrical excitability of neurons. ACh acting at **muscarinic acetylcholinesterase receptors** produces an internal messenger that turns off this class of K channel. They are products of KCNQ2−5 genes: KCNQ2 encodes the channel subunit Kv7.2 (872aa) which forms a multimer with KNCQ3 (Kv7.3, 872aa). KCNQ4 (Kv7.4, 695aa) is found in cochlear hair cells, KCNQ5 (Kv7.5, 932aa) is widely expressed and forms multimers with KCNQ3.

MCHC Mean corpuscular haemoglobin concentration; the amount of haemoglobin per erythrocyte, a standard measure in haematology (32 to 36 g/dl).

MCL A software package (Markov CLustering) that uses an efficient algorithm for detecting protein families on the basis of sequence data. Link to webpage: http://micans.org/mcl/

MCM (1) **Mini-chromosome maintenance proteins** (2) Modified Chee's medium, sometimes used for hepatocytes. (3) Microglia-conditioned medium. (4) **Methylmalonyl-CoA mutase**. (5) See **MCM-41**.

MCM-41 A silicate (Mobile Crystalline Material-41) produced by a templating mechanism that has arrays of non-intersecting hexagonal channels. By changing the length of the template molecule, the width of the channels can be controlled to be within 2 to 10 nm. The walls of the channels are amorphous SiO_2; used as a selective adsorption agent and molecular sieve.

MCP (1) See **methyl-accepting chemotaxis proteins**. (2) See **monocyte chemotactic and activating factor**, MCP-1. (3) See **gCAP39**. (4) Major capsid protein (MCP) of many viruses.

MCPA A herbicide. Current view is that MCPA is not genotoxic *in vivo* which is consistent with its lack of carcinogenicity in rats and mice. Review: http://mutage.oxfordjournals.org/content/20/1/3.full. pdf + html

MCS See **polycloning site**.

MCSF Cytokines (macrophage colony-stimulating factors: M-CSFα, 256aa; M-CSFβ, 554aa; M-CSFγ, 438aa) that play an important role in the activation and proliferation of **microglial cells** both *in vitro* and in injured neural tissue, are important in adipocyte hyperplasia, in osteoclast differentiation and in one of the early events in atherosclerosis, monocyte to macrophage differentiation in the arterial intima. The receptor for MCSF (**fms**) is expressed on pluripotent precursors and on mature osteoclasts and macrophages. Mutation in MCSF lead to osteoclast deficiency and **osteopetrosis**.

mct-1 Malignant T cell-amplified sequence 1 (multiple copies in a T-cell malignancy, 181aa) an anti-oncogene induced by DNA damaging agents, that plays a role in cell cycle regulation. It is over-expressed in T-cell lymphoid cell lines and in non-Hodgkin's lymphoma cell lines as well as in some primary large B-cell lymphomas; causes malignant transformation of murine fibroblasts.

M-current See **M-channels**.

MCV (1) Mean cell volume or mean corpuscular volume, a standard parameter in haematological analyses. (2) Measles-containing vaccine. (3) **Molluscum contagiosum virus**.

MDCK cells **Madin-Darby canine kidney** cells.

MDGI See **mammary-derived growth inhibitor**.

Mdm (1) Mdm1 is a nuclear protein (714aa). (2) Mdm2 (double minute 2 protein, p53-binding protein Mdm2, 491aa) is an E3 ubiquitin-protein ligase that targets **p53** and the retinoblastoma gene product for destruction by the proteasome. Mdm4 (490aa), has similar properties. (Mdm because originally described as 'mouse double minute'). (3) In *S. cerevisiae*, MDM10 (mitochondrial distribution and morphology protein 10, 493aa) connects the endoplasmic reticulum and mitochondria and is involved in the **TOM complex**. MDM12 (271aa) is similar. MDM20 (796aa) is a subunit of the N-terminal acetyltransferase B complex that acetylates the N-terminal methionine residues of all proteins beginning with Met-Asp or Met-Glu.

Mdr (1) See **multidrug transporter**. (2) Multi-drug resistant.

MDSC See **myeloid-derived suppressor cells**.

Mdx mouse Mouse mutant deficient in **dystrophin** and a model system for **Duchenne muscular dystrophy**.

ME See **myalgic encephalomyelitis**.

Meacham's syndrome See **Denys-Drash syndrome**.

mean residence time In pharmacokinetics, the average time a drug molecule remains in the body or an organ after rapid intravenous injection.

measles virus A morbillivirus of the **Paramyxoviridae** that causes the childhood disease measles and can cause subacute sclerosing panencephalitis in adults.

mec element A type of Staphylococcal cassette chromosome (SCC) element that encodes methicillin-resistance. Methicillin resistance strains containing an intact mec element in their chromosomes are defective in adhesion to fibrinogen and fibronectin. SCCmec elements are classified into different types based on the characteristics of the *mec* and *ccr* (cassette chromosome recombinase) gene complexes and are further classified into subtypes according to their 'junkyard DNA' region. Link to webpage: http://www.staphylococcus.net/

mechanoreceptor A sense organ or cell specialized to respond to mechanical stimulation.

Meckel syndrome A developmental disorder (Meckel-Gruber syndrome) in which there are renal cysts and developmental anomalies of the central nervous system. Meckel syndrome type 1 is caused by mutation in MKS1 (559aa) involved in centrosome migration to the apical cell surface during early ciliogenesis and that interacts with **meckelin**. Mutation in MKS1 can also cause a form of **Bardet-Biedl syndrome**. Type 2 is caused by mutation in transmembrane protein 216 (138aa), Type 3 is caused by mutation in meckelin. Type 4 is caused by mutation in the gene for nephrocystin-6 which is also associated with Joubert syndrome-5, Leber congenital amaurosis and **Senior-Loken syndrome**.

meckelin Protein (995aa) that is mutated in **Meckel syndrome** type 3 and is found in the primary cilium and at the plasma membrane in ciliated cells. Appears to mediate a fundamental developmental stage of ciliary formation and epithelial morphogenesis.

mecp2 See **methyl-CpG-binding proteins**.

MECT (1) Mucoepidermoid carcinoma-translocated 1 (MECT1, CREB-regulated transcription coactivator 1, TORC-1, 634aa), a transcriptional coactivator for CREB1 which activates transcription through both consensus and variant cAMP response element (CRE) sites. A fusion protein consisting of the N-terminus of CRTC1 joined to the C-terminus of MAML2 is found in some mucoepidermoid carcinomas. (2) In *Arabidopsis*, a chloroplast enzyme (MECT, 2-C-methyl-D-erythritol 4-phosphate cytidylyltransferase, EC 2.7.7.60, 302aa) involved in biosynthesis of pigments and gibberellins.

Medaka *Oryzias latipes*, the Japanese killifish, used as an experimental organism in studying vertebrate development. NCBI database entry: http://www.ncbi.nlm.nih.gov/UniGene/UGOrg.cgi?TAXID = 8090

media Avascular middle layer of the artery wall (tunica media), composed of alternating layers of elastic fibres and smooth muscle cells.

mediator complex A multi-protein scaffold complex composed of at least 21 subunits assembled as three structurally distinct submodules. It binds to promoters by association with regulatory proteins and forms the basis for the assembly of a functional preinitiation complex with RNA polymerase II and the general transcription factors. Originally described in *S. cerevisiae* but homologues are found in mammals. Can act as both a positive and negative regulator of transcription. One component is **cyclin C**.

Medicago truncatula Barrel Medic or Barrel Clover, a small legume native to the Mediterranean region that is used as a model organism for legumes, having a compact genome, simple genetics, short generation time, relatively high transformation, and large collections of mutants and ecotypes. Medicago HapMap: http://medicagohapmap.org/

medicinal leech The freshwater leech (*Hirudo medicinalis*) formerly used for blood-letting. The large size and simple nervous system have made it a favourite animal for some neurophysiological studies. See **hirudin**.

medin See **lactadherin**.

medium Shorthand for culture medium or growth medium, the nutrient solution in which cells or organs are grown.

medulla The inner region of any tissue or organ, surrounded by the cortex. In wood, medullary rays are radial sheets or ribbons, primarily parenchyma, extending vertically through the tree across and perpendicular to the growth rings and involved in the radial transmission of sap.

medulla oblongata Region of the brain where the spinal cord tapers into the brain stem. Neurons in this region regulate some very basic functions such as respiration.

medulloblastoma A highly malignant brain tumour that is thought to arise from cerebellar **granule cells**. Can be caused by mutation in the *SUFU* (suppressor of fused) or *BRCA2* genes or, in sporadic forms, somatic mutations in various genes, including *patched* (see **Gorlin syndrome**), beta-catenin and **APC**.

MEF (1) Mouse embryonic fibroblasts. (2) Myocyte-enhancer factors (MEFs), a group of transcription factors of the **MADS** superfamily. MEF2A (serum response factor-like protein 1, 507aa) binds specifically to the MEF2 element found in many muscle-specific genes but is also involved in the activation of numerous growth factor- and stress-induced genes and promotes cell survival during nervous system development. MEF2B (365aa) also binds the MEF element and is expressed in skeletal and cardiac muscle and in brain. MEF2C (473aa) is additionally expressed in various myeloid cells where its transactivation activity is enhanced by **LPS** acting through the MAP kinase **p38**. The result of activation is increased c-*jun* transcription. MEF2D (521aa) is expressed in myotubes and undifferentiated myoblasts.

megakaryocyte *myeloplax* Giant polyploid cell of bone marrow. A single megarkaryocyte gives rise to 3–4,000 platelets.

megalin See **lipoprotein receptor-related protein** 2.

megaloblast An embryonic cell that will give rise to erythroblasts by mitotic division within the blood vessels. See **megaloblastic anaemia**.

megaloblastic anaemia *pernicious anaemia* A heterogeneous group of disorders that have common morphological characteristics. Erythrocytes are larger than normal, neutrophils can be hypersegmented, and megakaryocytes are abnormal. **Megaloblasts** are found in the bone marrow, and often macrocytes in the peripheral blood. The usual cause is a deficiency in vitamin B12 or folic acid.

megaprep Slang, denoting a medium scale purification of **plasmid** from a bacterial culture. Usually used to describe preparations from more than 500 ml culture medium. See also **miniprep**, **midiprep**, **maxiprep**.

megaspore A haploid spore, produced by a plant sporophyte (megasporocyte), that develops into a female **gametophyte**.

megasporocyte The megaspore mother cell that undergoes meiotic divisions to produce a haploid megaspore within the plant ovule. The megaspore gives rise to the female **gametophyte**.

Meige lymphedema See **hereditary lymphedema**.

meiocytes Cell that will undergo meiosis; a little-used term.

meiosis A specialized form of nuclear division in which there are two successive nuclear divisions (meiosis I and II) without any chromosome replication between them. Each division can be divided into 4 phases similar to those of mitosis (pro-, meta-, ana- and telophase). Meiosis reduces the starting number of 4n chromosomes in the parent cell to n in each of the 4 daughter cells. Each cell receives only one of each homologous chromosome pair, with the maternal and paternal chromosomes being distributed randomly between the cells. This is vital for the segregation of genes. During the prophase of meiosis I (classically divided into stages: **leptotene**, **zygotene**, **pachytene**, **diplotene** and **diakinesis**), homologous chromosomes (each already replicated to form

two chromatids) pair to form **bivalents**, thus allowing **crossing-over**, the physical exchange of chromatid segments. This results in the **recombination** of genes. Meiosis occurs during the formation of **gametes** in animals, which are thus haploid and fertilization gives a diploid egg. In plants meiosis leads to the formation of the spore by the sporophyte generation.

meiospore Haploid spore formed after a meiotic division.

meiotic spindle The meiotic equivalent of the **mitotic spindle**.

MEKK See **MAPKKK**.

MEKs See **MAPKK**.

Mel-14 Antibody that reacts with L-selectin (CD62L). Blocks lymphocyte binding to **HEV** both *in vitro* and *in vivo*.

melanin Group of pigments, high molecular weight polymers of indole quinone, found in feathers, cuttle ink, human skin, hair and eyes, some neurons and a few other locations. Melanins can be of various colours, black/brown, yellow, red and violet. In mammalian skin, and in the pigmented retinal epithelium of the eye, the two main forms are eumelanin (black/brown) and pheomelanin (red). Neuromelanin (dark brown) is present in some neurons, for example in the substantia nigra. See **melanocytes**.

melanin-concentrating hormone *MCH* A hypothalamic neuropeptide (19aa) involved in the regulation of energy balance. The prohormone (165aa) is processed to produce MCH, and neuropeptides NEI and NGE. Two G-protein coupled receptors (MCHR1 & 2) are known (see **SLC**). Antagonists of MCH1R diminish food intake in rodents and have anxiolytic and antidepressant properties.

melanocortin A group of pituitary peptide hormones derived from **pro-opiomelanocortin** that include **adrenocorticotropin** (ACTH) and the alpha, beta and gamma melanocyte-stimulating hormones (MSH). Melanocortin-1 receptor (MC1R) binds **melanocyte stimulating hormone**. Melanocortin-2 receptor (MC2R, corticotropin receptor) binds **ACTH**. Mutations in the MC2R gene result in glucocorticoid deficiency-1. MC3R recognizes the core heptapeptide sequence of melanocortins and, like MC4R is expressed primarily in the brain and mutations in both are associated with obesity. MC5R appears to be involved in regulating exocrine glands. Melanocortin-2 receptor accessory protein (MRAP, fat tissue-specific low molecular weight protein, 172aa) interacts with MC2R and is essential for its function.

melanocyte Cells that synthesize melanin pigments. The pigments are stored in melanosomes (**chromatophores**) that can be redistributed in the cytoplasm to change pigment patterns in fish, amphibians and reptiles.

melanocyte-stimulating hormone A releasing hormone derived from **pro-opiomelanocortin** and produced in the mammalian hypophysis and related structures in lower vertebrates. Made up of α-MSH (melanotropin-alpha, 13aa, identical to amino acids 1–13 of **ACTH**, β-MSH (melanotropin-beta, 22aa) and γ-MSH (12aa). It is the ligand for the melanocortin-1 receptor and mutations in this receptor are associated with red hair and fair skin. In animal models αMSH is important in regulating food intake by binding and activating the brain melanocortin-4 receptor.

melanocytes Neural crest-derived cells of the epidermis and retinal epithelium, that synthesize and store **melanin** pigments in **melanosomes**. Skin colouration varies according to the level of melanin production rather than differences in melanocyte numbers. *Cf.* **melanophore**.

melanoma Neoplasia derived from **melanocytes**; benign forms are moles, but often are highly malignant. Generally the cells contain melanin granules and for this reason they have been used in studies on metastasis because the secondary tumours are easily located in lung.

melanoma differentiation-associated proteins Proteins associated with melanomas. Melanoma differentiation-associated protein 5 (Mda5) is a cytoplasmic RNA-helicase (interferon-induced helicase C domain-containing protein 1, EC 3.6.4.13, 1025aa) that is a receptor for dsRNA and plays a part in anti-viral responses (see **RIG-1**). Mda-6 is cyclin-dependent kinase inhibitor 1 (p21), Mda-7 protein (206aa) is IL24 and is upregulated in terminally-differentiated melanoma cells Mda-9 is syntenin-1 (see **syndecans**). In addition there are a various proteins referred to as **melanoma-associated antigens**.

melanoma growth-stimulatory activity Chemokine of the CXC subfamily (MGSA, neutrophil-activating protein 3, NAP-3, gro, CXCL1, 107aa). Potent **mitogen**. Activates, and is chemotactic for, neutrophils. N-terminal processed forms GROα(4-73), GROα(5-73) and GROα(6-73) are produced by proteolytic cleavage after secretion from peripheral blood monocytes and are ~30-fold more potent.

melanoma-associated antigens Antigens associated with differentiated melanoma cells. They include Melan-A (melanoma antigen recognised by T-cells, MART-1, 118aa) which is involved in melanosome biogenesis, melanoma-associated antigen MUC18 (CD146, 646aa), melanoma-associated ME20 antigen (pMel17, p100, silver locus protein homologue, 661aa), melanoma-associated antigen ME491 (CD63, granulophysin, 238aa),

melanotransferrin (melanoma-associated antigen p97, CD228, 738aa) involved in iron uptake. A further group are the MAGE antigens, a superfamily of proteins associated with melanomas and other tumours and part of a larger 'cancer/testis antigen' family. The MAGE superfamily includes five families: MAGE-A, MAGE-B, MAGE-C, MAGE-D, and **necdin**. The MAGE domain (~200aa) may mediate interaction with **neurotrophin**. See also **melanoma differentiation-associated proteins**.

melanophilin A member of the **rab** family (synaptotagmin-like protein 2a, 600aa) that interacts with **myosin5A** in melanosome transport. Defects in melanophilin cause one form of **Griscelli's syndrome**.

melanophore See **chromatophores**.

melanopsin A photopigment (opsin 4, 534 aa), initially cloned from cultured photosensitive dermal melanophores derived from *Xenopus* embryos, found in a subpopulation of mammalian retinal ganglion cells that are intrinsically photosensitive. Important for photo-entrainment of behaviour; more closely related to opsin proteins found in invertebrates. Gq-coupled. See **opsin sub-families**.

melanosome Membrane-bounded organelle found in melanocytes; when **melanin** synthesis is active their internal structure is characteristic, containing melanofilaments that have a periodicity of around 9 nm and are arranged in parallel arrays. Mature melanosomes, in which the filamentous structure is masked by the dense accumulation of melanin, are transferred to keratinocytes in the skin. Also found in **pigmented retinal epithelium** and in some cells of the connective tissue. Melanosomes, unlike **chromatophores**, cannot be rapidly relocated within the cell.

melanotropin See **melanocyte-stimulating hormone**.

MELAS syndrome A genetically heterogeneous set of disorders (mitochondrial myopathy, encephalopathy, lactic acidosis, and stroke-like episodes syndrome) caused by mutations in various mitochondrial tRNA genes, as is **MERRF syndrome**.

melatonin A hormone (N-acetyl 5-methoxytryptamine) secreted by the pineal gland. In lower vertebrates causes aggregation of pigment in melanophores, and thus lightens skin. In humans believed to play a role in establishment of circadian rhythms, hence its use to try to ameliorate the effects of jet-lag. The receptors are inhibitory G- protein coupled: melatonin receptor 1A (MTR1A; 350aa) is of high affinity and is predominantly found in the suprachiasmatic nucleus (location of the circadian clock), MTR1B (362aa) is found in retina and to a lesser extent in brain. The melatonin-related receptor (617aa) does not bind melatonin but forms heterodimers with MTR1A and B.

Meleda disease A dermatological disorder caused by mutation in the gene for secreted Ly6/PLAUR domain-containing protein 1 (**SLURP1**).

melibiose Disaccharide of galactose and glucose. Found in exudates and nectar of some plants; used as a pheromone by the desert locust and some other some insects. Added to a variety of cosmetic preparations although no evidence exists for any beneficial effects.

melioidosis A fatal infectious disease (Whitmore's disease), predominantly in the tropics, caused by the bacterium *Burkholderia* (formerly *Pseudomonas*) *pseudomallei*. Affects lymph nodes and viscera; clinically and pathologically similar to glanders. Has been seen as potential bioterror agent.

melittin The major component (26aa) of bee venom, responsible for the pain of the sting. Can lyse cell membranes and activate phospholipase A2 enzymes; it has a very high affinity for **calmodulin** but the biological relevance of this is unclear.

mellitose See **raffinose**.

melting curve Graph of the melting (denaturation and strand separation, detected by change in absorbance) of DNA as a function of temperature, from which can be calculated the temperature (T_m) at which half the molecules have undergone thermal denaturation. Melting curve analysis can distinguish products of the same length but different GC/AT ratio and products with the same length and GC content, but differing in their GC distribution along the sequence will have very different melting curves. Small sequence differences may even be observable.

melusin A protein (integrin beta-1-binding protein 2, 347aa) that may be involved in muscle development. Involved in heart hypertrophy in response to mechanical overload.

membrane Generally, a sheet or skin. In cell biology the term is usually taken to mean a modified lipid bilayer with integral and peripheral proteins, such as the plasma membrane and in eukaryotes other intracellular membrane systems such as endoplasmic reticulum. Because this usage is so general, it is advisable to avoid other uses where possible, particularly in histology or ultrastructure.

membrane attack complex See **complement**.

membrane capacitance The electrical capacitance of a cellular membrane. Plasma membranes are excellent insulators and dielectrics: capacitance is the measure of the quantity of charge that must be moved across unit area of the membrane to produce unit change in membrane potential, and is measured in Farads. Most plasma membranes have a capacitance around 1 microfarad cm^{-2}.

membrane depolarization See **depolarization**.

membrane fluidity The property exhibited by biological membranes within their physiological temperature range. Membranes are viscous 2-dimensional fluids; embedded proteins and the constituent phospholipids are more-or-less free to diffuse in the plane of the membrane.

membrane fracture See **freeze fracture**.

membrane potential More correctly, transmembrane potential difference: the electrical potential difference across a plasma membrane. See **resting potential**, **action potential**.

membrane protein A protein with regions permanently attached to a membrane (peripheral membrane protein), or inserted into a membrane (integral membrane protein). In some cases integral membrane proteins may have regions exposed on either side (transmembrane proteins). Insertion into a membrane implies hydrophobic domains in the protein. All **transport proteins** are integral membrane proteins.

membrane recycling The process whereby membrane is internalised, fuses with an internal membranous compartment, and is then re-incorporated into the plasma membrane. In cells that are actively secreting by an exocrine method (in which secretory granules fuse with the plasma membrane), it is obviously essential to have some way of reducing the area of the plasma membrane. The membrane can then be used to form new secretory vesicles. The converse is true for phagocytic cells.

membrane transport The transfer of a substance from one side of a plasma membrane to the other, in a specific direction, and at a rate faster than diffusion alone. See **active transport**.

membrane vesicles Closed unilamellar shells formed from membranes either in physiological transport processes or else when membranes are mechanically disrupted. They form spontaneously when membrane is broken because the free ends of a lipid bilayer are highly unstable.

membrane zippering See **zippering**.

memory cells See **immunological memory**.

mena Protein of the EVL family (Ena-VASP like, 541aa) that binds to **Tes** and is a component of focal adhesions. The binding to Tes blocks interaction with other EVL family members and it has been suggested that overproduction allows metastatic migration. Mice deficient in mena exhibit subtle defects in forebrain commissure formation. Mouse mena/VASP double mutants die perinatally and display defects in neurulation, craniofacial structures, and the formation of several fibre tracts in the CNS and peripheral nervous system. Homologous to *Drosophila* **enabled**.

menadione *vitamin K3* Synthetic naphthoquinone derivative with properties similar to those of Vitamin K. See also Table V1.

menaquinone Class of naphthaquinones (vitamins K) produced by bacteria in the intestine and important in blood-clotting.

Mendelian inheritance Inheritance of characters according to the classical laws formulated by Gregor Mendel, which give the classic ratios of segregation in the F2 generation. In sexually-reproducing organisms, any process of heredity explicable in terms of chromosomal segregation, independent assortment and homologous exchange. *Cf.* **maternal inheritance**.

Menetrier's disease A hyperproliferative disorder of the stomach (giant hypertrophic gastritis) caused by overexpression of **transforming growth factor** alpha (TGFα) in the stomach, but not elsewhere. TGFα is a ligand for the EGF-receptor.

Meniere's disease An autosomal dominant disorder characterized by hearing loss associated with episodic vertigo. There are autoantibodies to **cochlin** in some patients and in other cases there may be mutations in cochlin.

menin Nuclear protein (610aa) that binds to DNA through its nuclear localization signals, product of the MEN1 gene. Defects in this gene lead to **multiple endocrine neoplasia** type 1 and MEN1 is thought to be a tumour supressor that blocks the transcription factor JunD.

meninges Three layers of tissue surrounding the brain and spinal cord, the outermost dura mater, the arachnoid layer (arachnoid mater) and the inner pia mater.

meningitis Inflammation of the **meninges** of the brain and spinal cord. It can be caused by viral infections, by lymphocytic infiltrations and by various bacteria, but the most serious form is due to infection by *Neisseria meningitidis* with rapidly fatal consequences in up to 70% of untreated cases.

meningococcus *Neisseria meningitides* Gram-negative non-motile pyogenic coccus that is responsible for epidemic bacterial **meningitis**.

meniscocyte Obsolete name for a sickle-shaped erythrocyte.

Menkes' disease X-linked human defect in copper metabolism (kinky hair disease, steely hair disease, copper transport disease, incorrectly Menke's disease) caused by mutation in the gene encoding the alpha polypeptide of Cu^{2+}-transporting ATPase, resulting in an inability to absorb copper from the gut. Copper is required in synthesis of various elements of connective tissue so the disease is serious and lethal at an early age.

menthol A camphor compound extracted from common (wild) mint (*Mentha arvensis*): The L-isomer is the chief constituent of peppermint oil. It is used as an antiseptic, local analgesic and as a flavouring, such as vanillin.

meprins Members of the astacin family of zinc metalloendopeptidases, capable of hydrolyzing a wide variety of peptide and protein substrates including a variety of signalling peptides such as bradykinin and TGFα. Meprin 1 (endopeptidase 2, EC 3.4.24.18, alpha subunit, 746aa; beta, 701aa) is found as meprin A, a heterotetramer of 2 α and 2 β subunits and is secreted and meprin B which contains only beta subunits and is plasma-membrane associated. Inhibited by actinonin, a naturally occurring antibacterial agent.

mercaptans *thiols* Any of a class of organic compounds with a sulphydryl group bonded to a carbon atom. Low-molecular-weight mercaptans have very disagreeable odours; methyl mercaptan is produced as a decay product of animal and vegetable matter; allyl mercaptan is released when onions are cut; butanethiol (butyl mercaptan) derivatives are present in skunk secretions.

mercaptoethanol A pungent water-soluble thiol, not of biological origin. Used in biochemistry to cleave disulphide bonds in proteins or to protect sulphydryl groups from oxidation.

MERF Methionine enkephalin-Arg-Phe, a general opioid-receptor agonist. See **encephalins**.

meristem Group of actively dividing plant cells, found as apical meristems at the tips of roots and shoots and as lateral meristems in vascular tissue (vascular **cambium**) and in cork tissue (**phellogen**). Also found in young leaves, and at the bases of internodes in grasses. Consists of small non-photosynthetic cells, with primary walls and relatively little vacuole.

Merkel cell carcinoma Rare and highly malignant skin tumour (primary neuroendocrine carcinoma of the skin) that arises from neuroendocrine cells with features of epithelial differentiation.

merlin Product (schwannomin, neurofibromin 2, 595aa) of the **neurofibromatosis** 2 (NF2) tumour suppressor gene, a diverged member of the ERM family of membrane-associated proteins. Merlin is defective or absent in schwannomas and meningiomas and 80% of merlin mutants have significantly reduced cell adhesiveness. Probably regulates the **Salvador-Warts-Hippo pathway**. Phosphorylation blocks the tumour suppressor activity which can be restored by the action of protein phosphatase 1.

MeroCaM Calcium-sensitive fluorophore that can be used to measure calcium levels within live cells.

merocrine Commonest mode of secretion in which a secretory vesicle fuses with the plasma membrane and releases its contents to the exterior.

merogony (1) Development of only a portion of an ovum. (2) **Schizogony** resulting in the production of merozoites.

meromyosin Fragments of myosin-II formed by trypsin digestion. Heavy meromyosin (HMM) has the hinge region and ATPase activity, light meromyosin (LMM) is mostly a-helical and is the portion normally laterally associated with other LMM to form the thick filament itself. HMM can be proteolytically further digested by pappain to produce the S1 region that has the ATPase activity and the S2 hinge region.

meront The asexual developmental stage of certain protozoa, especially nonsporozoa, that gives rise to merozoites.

MEROPS Database of peptidases classified into 'Families' on the basis of their catalytic site (see Table P1) or '**Clans of peptidases**' on the basis of evolutionary relationships. The database provides sequence data as well as links to structural information and also lists inhibitors. It provides far more information than this dictionary. Link to database: http://merops.sanger.ac.uk/

merosin See **laminin**.

merotomy Partial cutting: used in reference to experiments in which protozoa are enucleated and the behaviour of the residual cytoplasm is studied.

merozoite Stage in the life-cycle of the malaria parasite (*Plasmodium*): formed during the asexual division of the schizont. Merozoites are released and invade other cells.

merozygote *merodiploid* A bacterium that is in part haploid and in part diploid because it has acquired exogenous genetic material e.g. during transduction or conjugation.

MERRF syndrome A disorder (myoclonic epilepsy associated with ragged-red fibres) that can result from mutation in various mitochondrial genes including those for some tRNAs (for lysine, leucine, histidine, serine and phenylalanine) and for the NADH-ubiquitone oxidoreductase subunit of Complex I.

mersacidin A Type B **lantibiotic** (20aa) that inhibits cell wall biosynthesis.

mesangial cells Cells found within the glomerular lobules of mammalian kidney, where they serve as structural supports, may regulate blood flow, are phagocytic, and may act as **accessory cells**, presenting antigen in immune responses.

mesencephalon Region of the brain below the thalamus and above the pons developed from the middle of the three cerebral vesicles of the embryonic nervous system. Includes the superior and inferior colliculi and cerebral peduncles.

mesenchyme Embryonic tissue of mesodermal origin. *Adj.* mesenchymal.

mesocotyl In monocots, the tubular, white, stem-like tissue that connects the seed and the tip of the plant (coleoptile) that penetrates the soil. Expansion of the mesocotyl pushes the coleoptile above the soil surface.

mesoderm Middle of the three **germ layers**; gives rise to the musculo-skeletal, blood vascular, and urinogenital systems, to connective tissue (including that of dermis) and contributes to some glands.

mesokaryotic Description applied to the unusual nuclei (dinokaryons) of Dinoflagellates, in which the chromosomes lack histone and are attached to the nuclear membrane. This arrangement was once considered to be an intermediate between the nucleoid region of prokaryotes and the true nuclei of eukaryotes but this is no longer considered appropriate and the term should probably be allowed to fall into historical oblivion.

mesolimbic pathway One of the dopaminergic pathways in the brain, involved in modulating behavioural responses to stimuli that activate feelings of reward.

mesomere (1) A **blastomere** that is intermediate in size between a micromere and a macromere. (2) The middle zone of the mesoderm from which excretory tissue will develop. (3) The median series of bones supporting the pectoral fin of sarcopterygian fish.

mesopelagic Descriptive term for the intermediate regions between the surface and depths of the ocean.

mesophase Arrangement of phospholipids in water where the liquid-crystalline phospholipids form multi-layered parallel-plate structures (smectic mesophase), each layer being a bilayer, the layers separated by aqueous medium.

mesophile Organism that thrives at moderate temperatures (between, say, $20-40°C$).

mesophyll Tissue found in the interior of leaves, made up of photosynthetic (parenchyma) cells, also called **chlorenchyma** cells. Consists of relatively large, highly vacuolated cells, with many **chloroplasts**. Includes **palisade parenchyma** and spongy mesophyll.

mesosecrin Obsolete name for a glycoprotein (46 kDa) secreted by mesothelial cells (including endothelium). Cloning revealed it to be identical to plasminogen activator inhibitor-1 (PAI-1).

mesosome Invagination of the plasma membrane in some bacterial cells, observed in electron micrographs, sometimes with additional membranous lamellae inside. May have respiratory or photosynthetic functions, but some microbiologists consider them artifacts of chemical fixation.

mesothelin A cell surface GPI-linked glycoprotein (630aa), a differentiation antigen that is strongly expressed in normal mesothelial cells, mesotheliomas, nonmucinous ovarian carcinomas, and some other malignancies. First described as the antigenic target of the monoclonal antibody K1. It is proteolytically processed to produce megakaryocyte-potentiating factor (MPF, 250aa) which is secreted and potentiates megakaryocyte colony formation *in vitro*.

mesothelioma Malignant tumour of the **mesothelium**, usually of lung; frequently caused by exposure to asbestos fibres, particularly those of crocidolite, the fibres of which are thin and straight and penetrate to the deep layers of the lung. Because of their shape, the fibres puncture the macrophage phagosome and are released, leading to a chronic inflammatory state that is thought to contribute to development of the tumour.

mesothelium Simple squamous epithelium of mesodermal origin. It lines the peritoneal, pericardial and pleural cavities and the synovial space of joints. The cells may be phagocytic.

mesotocin A diuretic hormone cleaved from (mesotocin-neurophysin MT, 125aa in *Bufo japonicus*) produced by magnocellular preoptic neurons in the hypothalamus in amphibians, reptiles and birds. Mesotocin itself is 9aa and binds to a G-protein coupled receptor (389aa).

messenger RNA *mRNA* Single-stranded RNA molecule that specifies the amino acid sequence of one or more polypeptide chains. This information is translated during protein synthesis when ribosomes bind to the mRNA. In prokaryotes, mRNA is the primary transcript from a DNA sequence and protein synthesis starts while the mRNA is still being synthesized. Prokaryote mRNAs are usually very short-lived (average $t_{1/2}$ 5 mins). In contrast, in eukaryotes the primary transcripts (**hnRNA**) are synthesized in the nucleus and they are extensively processed to give the mRNA that is exported to the cytoplasm where protein synthesis takes place. This processing includes the addition of a $5'$–$5'$-linked 7-methyl-guanylate 'cap' at the $5'$ end and a sequence of adenylate groups at the $3'$ end, the **poly-A tail**, as well as the removal of any **introns** and the splicing together of **exons**; only 10% of hnRNA leaves the nucleus. Eukaryote mRNAs are comparatively long-lived with $t_{1/2}$ ranging from 30 mins to 24 hours.

met (1) Product of the c-*met* oncogene, a receptor tyrosine kinase (hepatocyte growth factor receptor, scatter factor receptor, EC 2.7.10.1, 1390aa), a disulphide-linked heterodimer formed of an alpha chain (50 kDa) and a beta chain (145 kDa). Overexpressed in a significant percentage of human cancers. (2) In *S. cerevisiae* MET proteins are involved in biosynthesis of methionine and other sulphur-containing compounds such as glutathione. Some (e.g. MET4, 672aa) are involved in transcriptional activation of the genes, others (e.g. MET30, 640aa) are negative regulators. (3) See **met repressor-operator complex**.

met repressor-operator complex Repressor protein, 104aa, product of the *metJ* gene, which regulates methionine biosynthesis in *E. coli*. Dimeric molecules bind to adjacent sites 8 base-pairs apart on the DNA; sequence recognition is by interaction between antiparallel beta-strands of protein and the major groove of the B-form DNA duplex.

met-enkephalin See **enkephalins**.

metabolic burst *respiratory burst* Response of phagocytes to particles (particularly if **opsonized**), and to agonists such as **formyl peptides** and **phorbol esters**; an enhanced uptake of oxygen leads to the production, by an NADH-dependent system, of hydrogen peroxide, superoxide anions, and hydroxyl radicals, (reactive oxygen species, ROS) all of which play a part in bactericidal activity. Defects in the metabolic burst, as in **chronic granulomatous disease**, predispose to infection particularly with **catalase**-positive bacteria, and are usually fatal in childhood.

metabolic cooperation *metabolic coupling* Transfer between tissue cells in contact of low molecular weight metabolites such as nucleotides and amino acids. Transfer is via channels constituted by the **connexons** of gap junctions, and does not involve exchange with the extracellular medium. First observed in cultures of animal cells in which radiolabelled purines were transferred from wild-type cells to mutants unable to utilize exogenous purines.

metabolic half life The time required for 50% of a substance to be metabolized in the body.

metabolic syndrome A group of metabolic risk factors in one person. They include: abdominal obesity, high levels of triglycerides, low HDL-cholesterol and high LDL-cholesterol, elevated blood pressure, insulin resistance, high fibrinogen or plasminogen activator inhibitor−1 levels in the blood and elevated C-reactive protein levels. Metabolic syndrome is associated with high risk of coronary heart disease.

metabolism Sum of the chemical changes that occur in living organisms. Often subdivided into anabolism (synthesis) and catabolism (breakdown).

metabolome All the low molecular weight molecules present in cells in a particular physiological or developmental state. Unlike the genome it has a temporal component and the complexity of the data is considerable. Metabolomics is the study of information gleaned from endogenous metabolic profiling. Metabonomics is the study of dynamic changes in the metabolome.

metabonomics See **metabolome**.

metabotropic Type of neurotransmitter receptor which affects cell activity but not through changes in ion channel properties (*cf.* **ionotropic**).

metacentric Descriptive of a chromosome that has its centromere (**kinetochore**) at or near the middle of the chromosome. In sub-metacentric chromosomes the two arms (p and q) are unequal. *Cf.* telocentric chromosomes where the centromere is near one end and acrocentric chromosomes in which the short (p) arm is hardly distinguishable.

metachromasia The situation where a stain when applied to cells or tissues gives a colour different from that of the stain solution (metachromatic staining).

metachromatic See **metachromasia** .

metachromatic leukodystrophy See **leukodystrophy**.

metachronal rhythm See **metachronism**.

metachronism Type of synchrony (metachronal rhythm) found in the beating of cilia. A metachronal process is one that happens at a later time, and the synchronization is such that the active stroke of an adjacent cilium is slightly delayed so as to minimize the hydrodynamic interference; coordination is by visco-mechanical coupling. Different patterns of metachronal synchronization are recognized: in symplectic m. the wave of activity in the field passes in the same direction as the active stroke of the individual cilium; in antiplectic m. the opposite is true. In dexioplectic and laeoplectic m. the wave of activity in the field is normal to the beat axis. Symplectic and antiplectic m. are considered orthoplectic, the other forms as diaplectic.

metacyclic The final developmental stage of a trypanosome in the tsetse fly, the infective form of the parasite that is transmitted by the tsetse fly when it takes a blood meal from a mammalian host.

metafemales Human females with four X chromosomes in addition to 44 autosomes.

metagon RNA particle found in *Paramecium*, where it behaves as mRNA, and that can behave like a virus if ingested by the protozoan *Didinium*. Metagon RNA hybridizes specifically with DNA from paramecia bearing an M gene.

metalloenzyme An enzyme that contains a bound metal ion as part of its structure. The metal may be required for enzymic activity, either participating directly in catalysis, or stabilizing the active conformation of the protein.

metallopeptidases *metalloproteases, metalloproteinases* The most diverse group of peptidases, with more than 30 families identified. A divalent cation, usually zinc, activates the water molecule and is held in place by amino acid ligands, usually three in number. See **matrix metalloproteinases**.

metalloprotein A protein that contains a bound metal ion as part of its structure.

metalloproteinase See **metallopeptidase**.

metallothioneins Small cysteine-rich metal-binding (chelating) proteins found in the cytoplasm of eukaryotes. Synthesis can be induced by glucocorticoids and by heavy metals such as zinc, cadmium, copper and mercury. Metallothioneins probably serve a protective function and each metallothionein can bind 7−12 metal ions by interaction with the cysteines. Human metallothioneins are classified into various types: Type 1 has nine sub-types (e.g. metallothionein 1A, 61aa), Type 2 (61aa) binds zinc, Type 3 (68aa) is abundant in a subset of astrocytes in the normal human brain, Type 4 (62aa) binds zinc and copper and may regulate zinc metabolism during the differentiation of stratified epithelia. Metallothionein gene promoters are used in studies of gene expression. Database entry: http://www.uniprot.org/docs/metallo.txt

metamere Unit of **segmentation** or metamerism.

metameric In chemistry, having two or more constitutional isomers. In biology, having a segmented body form (metamerism).

metamorphosis Change of body form, for example in the development of the adult frog from the tadpole or the butterfly from the caterpillar.

metaphase Classically the second phase of **mitosis** or one of the divisions of **meiosis**. In this phase the **chromosomes** are well condensed and aligned along the **metaphase plate**, making it an ideal time to examine the chromosomes in a cytological preparation (metaphase spread) that flattens the nuclei.

metaphase plate The plane of the **spindle** approximately equidistant from the two poles along which the chromosomes are lined up during **mitosis** or **meiosis**. Also termed the equator.

metaphysis *Pl.* metaphyses. The region of a bone where growth occurs, part way between the shaft (diaphysis) and the end (epiphysis).

metaplasia Change from one differentiated phenotype to another, for example the change of simple or transitional epithelium to a stratified squamous form as a result of chronic damage.

metastasis Development of secondary tumour(s) at a site remote from the primary (metastatic spread); a hallmark of malignant cells. Involves local invasion (in most cases), passive transport, lodgement, and proliferation at a remote site.

metastasis associated proteins Components of the **nucleosome-remodelling and histone-deacetylase complex**, MTA1 (715aa) and MTA2 (p53 target protein in deacetylase complex, 668aa) may be involved in the regulation of gene expression by covalent modification of histone proteins. MTA3 also has a role in maintenance of the normal epithelial architecture by repressing **snail** transcription in a histone deacetylase-dependent manner, and regulating E-cadherin levels.

metatropic dysplasia A severe spondyloepimetaphyseal dysplasia (metatropic dwarfism) caused by defects in **vanilloid receptor** TrpV4. The characteristics are short limbs with limitation and enlargement of joints and usually severe kyphoscoliosis.

metavinculin See **vinculin**.

metchnikowin In *Drosophila* a proline-rich **antimicrobial peptide** (26aa), with antibacterial and antifungal properties. Induction of metchnikowin gene expression can be mediated either by the **toll** pathway or by the *imd* gene product.

methaemoglobin An oxidized non-functional form of haemoglobin containing ferric iron that is produced by the action of oxidizing poisons.

Methanobacterium A genus of strictly anaerobic Archaea that reduce CO_2 using molecular hydrogen, H_2, to give methane (methanogenesis). Methanobacteria and other methanogens are found in the anaerobic sediment at the bottom of ponds and marshes (hence marsh gas as the common name for methane), associated with deep-sea hydrothermal vents and as part of the microflora of the rumen in cattle and other herbivorous mammals.

methanochondrion A structure of involuted plasma membrane found in many methanogenic bacteria and thought to be an organelle of methane formation.

methanotroph Bacteria (Methylococcaceae and Methylocystaceae) that can use methane as their sole source of energy. Can be found in soil at landfill sites and in lake sediments where methane is being generated.

methicillin See **meticillin**.

methionine An amino acid (Met, M, 149 Da) that contains the $-SCH_3$ group that can act as a methyl donor (see **S-adenosyl methionine**). Common in proteins but at low frequency. The met-x linkage is subject to specific cleavage by cyanogen bromide. See also **formyl peptides**, and Table A1.

methionine adenosyltransferase The enzyme (EC 2.5.1.6) that catalyzes the formation of S-adenosylmethionine from methionine and ATP. Defects in isoform 1 (395aa) cause a form of hypermethioninaemia.

methionine aminopeptidase A mitochondrial metallopeptidase (Peptidase M 1, MAP1, EC 3.4.11.18, 478aa) that removes N-terminal methionine from proteins, facilitating further modification. Methionine aminopeptidase 2 (eukaryotic initiation factor-2 (eIF-2)-associated protein; p67) is responsible for processing proteins containing N-terminal Met-Val and Met-Thr sequences *in vivo* and is the target for the anti-angiogenic drug **fumagillin**. Knockout of the gene causes an early gastrulation defect; targeted deletion specifically in the haemangioblast lineage results in abnormal vascular development and embryonic lethality.

methionine puddle A term used to describe a region of a protein surface composed of a cluster of methionine side chains. Proposed as the active hydrophobic site of calmodulin and also of signal recognition particle (SRP). The concept is of a highly fluid hydrophobic patch.

methotrexate An analogue of dihydrofolate that inhibits **dihydrofolate reductase** and kills rapidly growing cells. Therapeutic agent for leukaemias, but with a low therapeutic ratio.

methyl- $-CH_3$ Specific reference to the methyl group is made when macromolecules are modified after synthesis by enzymic addition of methyl groups. The group is transferred to nucleic acids and proteins. See also **methyl transferase** and **DNA methylation**.

methyl violet dyes Mixtures of tetramethyl, pentamethyl and hexamethy pararosanilin (the latter being **crystal violet**). All are deep blue except at very acid pH and increasing methylation makes the colour darker. Metachromatic and used to demonstrate presence of **amyloid**. Gentian violet is an imprecise synonym.

methyl-accepting chemotaxis proteins Chemotactic-signal transducer proteins of the inner cytoplasmic face of the bacterial plasma membrane (MCPs) with which the receptors of the outer face interact. Attractants increase the level of methylation while repellents decrease the level of methylation (methyl groups are added by the methyltransferase

CheR and removed by the methylesterase CheB); methylation is part of the **adaptation** process. They are not directly connected to the flagellar motor system used in chemotaxis. Four different MCPs are known in *E. coli*, each with a separate set of receptors. MCP1 (Tsr, 551aa) is the receptor for the attractant L-serine and related amino acids and is also responsible for negative chemotaxis away from a wide range of repellents, including leucine, indole, and weak acids; MCP2 (Tar, 553aa) is the receptor for the attractant L-aspartate and related amino acids. It mediates positive taxis to maltose via an interaction with the periplasmic maltose binding protein and negative taxis to cobalt and nickel. MCP3 (Trg, 546aa) mediates taxis to the sugars ribose and galactose via periplasmic ribose- or galactose-binding proteins. MCP4 (Tap, 533aa) mediates taxis toward dipeptides via periplasmic dipeptide-binding protein.

methyl-coenzyme M reductase The enzyme (MCR, EC 2.8.4.1) responsible for microbial formation of methane. In *Methanothermobacter marburgensis* (*Methanobacterium thermoautotrophicum*) a hexamer composed of 2α (550aa), 2β (443aa), and 2γ (249aa) subunits that non-covalently binds two nickel porphinoid coenzymes.

methyl-CpG-binding proteins Proteins that bind to methylated CpG-rich regions of DNA, act as transcriptional repressors and contribute to silencing (see **DNA methylation** and **genomic imprinting**). MeCP1 (MBD1, 605aa) is widely expressed and upregulated by interferon. MeCP2 (MBD2, 486aa) is defective in some forms of **Angelman syndrome**, **Rett's syndrome** and various other neurological disorders. MeCP3 (MBD3, 291aa) does not bind DNA by itself but is involved in recruitment of histone deacetylases and DNA methyltransferases. MeCP4 (MBD4, 580aa) is a mismatch-specific DNA N-glycosylase involved in DNA repair. MeCP5 (MBD5, 1494 aa) is defective in some forms of mental retardation. MeCP6 (MBD6, 1003aa) is also known.

methylcholanthrene Carcinogenic polycyclic hydrocarbons formed during incomplete combustion of organic material. Experimentally 3-methylcholanthrene is used as a mutagen.

methylene blue Water-soluble dye (Swiss blue, Basic Blue 9, tetramethythionine chloride) that can be reduced to a colourless form and can be oxidized by atmospheric oxygen. Used as a stain in bacteriology and histology.

methylenomycin An antibiotic produced by *Streptomyces coelicolor*. Unlike the others (actinorhodin, undecylprodigiosin and CDA), it is plasmid encoded. It is effective against most Gram-positive and some Gram-negative bacteria. Article: http://mic.sgmjournals.org/cgi/reprint/95/1/96?view = long&pmid = 822125

methylmalonyl-CoA mutase A mitochondrial adenosylcobalamin-requiring enzyme (MUT, EC.5.4.99.2, 750aa) that catalyzes the rearrangement of methylmalonyl-CoA to succinyl-CoA and is involved in the degradation of several amino acids, odd-chain fatty acids and cholesterol. Defects are associated with methylmalonic acidemia, a disorder of organic acid metabolism that is frequently fatal.

methylotroph A diverse group of organisms that utilize methanol as an energy source. Examples include the bacterium *Methylococcus capsulatus* and the yeast *Pichia pastoris*.

methyltransferases Enzymes that transfer a methyl group from S-adenosyl methionine to a substrate. DNA (cytosine-5)-methyltransferase (EC 2.1.1.37, DNMT1, 1616aa) has a strong preference for hemi-methylated DNA, and at S-phase associates with DNA replication sites, preserving the pattern of methylation that is important in epigenetic inheritance. Other forms (DNMT3A, 912aa; DNMT3B, 853aa) are known; DNMT2 (tRNA (cytosine-5-)-methyltransferase, EC 2.1.1.29, 391aa) specifically methylates tRNA for asparagine. DNMTL (DNA (cytosine-5)-methyltransferase 3-like, 387aa) is catalytically inactive but required for activity of DNMT3A and 3B. Histone-specific methyltransferases are important in transcriptional regulation (see **histone methyltransferase**) and some (e.g. **EZH**) are part of the **polycomb repressive complex**). Methyltransferases are also important in bacterial chemotaxis where the **methyl-accepting chemotaxis proteins** become methylated in the course of adaptation.

methylxanthines Naturally occurring purine alkaloids such as theobromine, theophylline and caffeine (trimethyl-xanthine). They inhibit cAMP phosphodiesterase and thus cause an increase in the intracellular cAMP concentration. Found in various beverages, coffee, tea and others.

meticillin Synthetic penicillinase-resistant beta-lactam antibiotic (formerly methicillin). No longer used therapeutically but the defining antibiotic for **meticillin-resistant *Staphylococcus aureus*** (MRSA).

meticillin-resistant *Staphylococcus aureus* *MRSA* A strain of *Staphylococcus* that is resistant to meticillin (methicillin) and other beta-lactam antibiotics, an increasing problem, particularly in hospitals, suggesting that bacteria are beginning to win the arms race against antibiotics. Many are now also resistant to the newer antibiotics such as **vancomycin**.

metorphamide Amidated **opioid** octapeptide from bovine brain. Derived by proteolytic cleavage from proenkephalin.

metrizoate The sodium salt of metrizoate (3-acetimido-5-(N-methyl-acetamido)-triiodobenzoate, TN

Isopaque) is used to produce solutions with high densities suitable for cell density gradient separations. Is radio-opaque and used in diagnostic radiology. See **Ficoll**.

metzincins A clan of multidomain zinc endopeptidases that share a common core architecture characterized by a long zinc-binding consensus motif and a methionine-containing Met-turn. They are active in unspecific protein degradation such as digestion of intake proteins and tissue development, maintenance, and remodeling, but they are also involved in highly specific cleavage events to activate or inactivate themselves or other (pro)enzymes and bioactive peptides. Metzincins are subdivided into families: **astacins**, **ADAMs**, serralysins, matrix metalloproteinases, snapalysins, leishmanolysins, and pappalysins. Article: http://www.jbc.org/content/284/23/15353.long

mevalonate The product of the reaction between hydroxy-methyl-glutaryl-CoA and two molecules of NADPH, catalysed by **HMG-CoA reductase**, the starting material for the synthesis of cholesterol, other sterols and e.g. geranylgeraniol groups for post-tranlational modification of proteins.

mevalonate kinase An enzyme (EC 2.7.1.36, 396aa) involved in synthesis of isoprenoids (terpenoids) and sterols. Mutation in the gene can lead to mevalonic aciduria and **hyper-IgD syndrome**. In **Zellweger's syndrome** and **neonatal adrenoleukodystrophy** peroxisomal mevalonate kinase activity is reduced.

mevalonic acid Key intermediate in polyprenyl biosynthesis and thus cholesterol synthesis. Derived from hydroxymethylglutaryl-CoA (HMG-CoA) – a reaction inhibited by **statins**.

mevinolin A fungal metabolite from *Aspergillus terreus* that is a potent competitive inhibitor of **HMG-CoA-reductase**.

mex In *C. elegans,* proteins involved in developing asymmetry during embryogenesis. Mex-5 and mex-6 are zinc-finger transcription factors that, when phosphorylated by **par1**, are localized to the anterior cytoplasm of the zygote. Mex-3 (415aa) has RNA-binding activity.

MFSDs A superfamily of transporters (major facilitator superfamily domain-containing proteins) with 10–12 membrane-spanning domains found in bacteria, archaea and eukaryotes. They act as permeases (uniporters, symporters or antiporters) for a wide range of small charged and uncharged molecules. At least 18 families are recognised with diverse roles. For example MFSD2A (543aa) has a role in thermogenesis via the beta-adrenergic signalling pathway, MFSD10 (455aa) confers cellular resistance to apoptosis induced by indometacin and diclofenac possibly by acting as an efflux pump. Defects in

MFSD8 are the cause of a form of ceroid lipofuscinosis.

Mg dechelatase The term applied to the substance responsible for removal of magnesium from chlorophyll after the first step in degradation by chlorophyllase. It may be a low molecular weight metal-chelating substance rather than an enzyme.

MG132 A potent and reversible inhibitor (carbobenzoxy-L-leucyl-L-leucyl-L-leucinal) of the **proteasome** system at nanomolar levels; at higher concentrations also activates c-Jun N-terminal kinase (**JNK1**) and inhibits NFκB activation ($IC_{50} = 3\ \mu M$). Prevents β-secretase cleavage.

MGF Mast cell growth factor (stem cell factor, 273aa), ligand for **kit**. Stimulates mast cell proliferation and proliferation of both myeloid and lymphoid haematopoietic progenitors in bone marrow culture. Acts synergistically with other cytokines.

MGP See **matrix Gla protein**.

MGSA See **melanoma growth-stimulatory activity**.

MHC See **major histocompatibility complex**.

MHC restriction Restriction on interaction between cells of the immune system because of the requirement to recognize foreign antigen in association with MHC antigens (major histocompatibility antigens). Thus, cytotoxic T-cells will only kill virally-infected cells that have the same Class I antigens as themselves, whereas helper T-cells respond to foreign antigen associated with Class II antigens.

MHCP (1) Methylhydroxy chalcone polymer, an active component of cinnamon, that will increase glucose metabolism of adipocytes roughly 20-fold *in vitro*. (2) MHC-Peptide Interaction Database (MHCP), a curated database for sequence-structure-function information on MHC-Peptide interactions. See also **SYFPEITHI**. Link to database: http://www.hsls.pitt.edu/guides/genetics/obrc/immunology/URL1152465729/info

mib In *Danio rerio* an E3 ubiquitin-protein ligase (mindbomb, 1030aa) that mediates ubiquitination of **Delta** receptors, which act as ligands of **Notch**. Mutants have a disorganized brain and neural tube. *Drosophila* and human homologues have similar enzyme activity. Human mindbomb homologue 1 (MIB1, DAPK-interacting protein-1, DIP1; 1006aa) regulates the cellular levels of **death-associated protein kinase-1** (Dapk1). MIB2 (1013aa) is **skeletrophin**.

micelle One of the possible ways in which amphipathic molecules may be arranged; a spherical structure in which all the hydrophobic portions of the molecules are inwardly directed, leaving the hydrophilic portions in contact with the surrounding aqueous phase. The converse arrangement will be found if the major phase is hydrophobic.

Michaelis constant See K_m and **Michaelis-Menten equation**.

Michaelis-Menten equation Equation derived from a simple kinetic model of enzyme action that successfully accounts for the hyperbolic (adsorption-isotherm) relationship between substrate concentration S and reaction rate V. $V = V_{max} \times S/(S + K_m)$, where K_m is the Michaelis constant and V_{max} is maximum rate approached at very high substrate concentrations.

microaerophiles Organisms that grow well at low-oxygen concentrations but are unable to survive normal oxygen concentrations. Examples are *Borrelia burgdorferi, Helicobacter pylori, Lactobacillus* spp. and *Treponema pallidum*.

microangiopathic haemolytic anaemia A subgroup of haemolytic anaemia in which there are small fragemnts of red cells (microcytes) in the circulation that have been generated mechanically by being forced through a fibrin meshwork. Can be a result of **disseminated intravascular coagulation**.

microarray A small membrane or glass slide containing samples of biological material (DNA, protein etc.) arranged in a regular pattern and used as an analytical tool. Arrays are becoming progressively smaller as analytical techniques improve.

microbicide Substance that kills microbes; rather an imprecise term.

microbody See **peroxisome**.

microcarrier Microcarriers are small solid or in some cases immiscible liquid spheres, on which cells may be grown in suspension culture. They provide a means of obtaining large yields of cells in small volumes. The cells must exhibit anchorage dependence of growth and the dimensions of the carrier bead may be important in controlling growth rate. The term is imprecise and has other potential meanings.

microcephalin A protein (MCPH1, 835aa) implicated in chromosome condensation and DNA damage-induced responses. Mutations cause premature chromosome condensation with microcephaly and mental retardation (PCC syndrome) and also microcephaly Type 1 and it is thought that microcephalin may be important in neurogenesis, particularly regulating the size of the cerebral cortex. Contains three **BRCT domains**. The *Drosophila* homologue (779aa) is associated with development of **mushroom bodies**. Abstract: http://www.ncbi.nlm.nih.gov/pubmed/16311745?dopt = Abstract

microchimerism The presence of two genetically distinct populations of cells in an individual or in an organ, one population being at a low concentration. May arise through transfer of cells between mother and fetus, between twins *in utero*, or as a result of blood transfusions and transplantation.

microcinematography The making of films using a microscope and cine camera.

Micrococcus Genus of Gram-positive aerobic bacteria, cells around $1-2\mu m$ in diameter. *M. lysodeikticus* (now *M. luteus*) was commonly used as the source of bacterial cell wall suspension on which lysozyme activity was measured by a decrease in turbidity.

microcolliculus *Pl.* microcolliculi. Broad swelling ($0.5\mu m$) on the dorsal surface of a moving epidermal cell in culture, that moves rearward as the cell moves forward (as do ruffles on fibroblasts).

microcystins Toxic 7aa cyclic peptides that inhibit liver function. The amino acid composition of the individual microcystins or **nodularins** may vary, but the novel hydrophobic amino acid, 3-amino-9-methoxy-10-phenyl-2,6,8 trimethyl deca,4,6 dienoic acid (ADDA) is essential to its pharmacological activity. Potent inhibitors of the serine/threonine protein phosphatases; bind to the same site as **okadaic acid**. The microcystins are generated via a mixed polyketide synthase/non-ribosomal peptide synthetase (PKS/NRPS) gene cluster, expression of which is positively correlated with increased light intensity and with iron starvation.

microcysts Spherical resting structures formed by the genus *Sporocytophaga*, a member of the Cytophaga group, which are long, slender Gram-negative rods that move by gliding. Many cytophagas digest polysaccharides such as cellulose and chitin. Some Cytophaga species are fish pathogens and can cause serious problems in the fish farming industry.

microcytes Abnormally small red blood cells. See **microangiopathic haemolytic anaemia**.

microdialysis Dialysis on a small scale, giving microlitre range samples. Used for example in studies of *in vivo* release of transmitters in brain tissue.

microelectrode An electrode, with tip dimensions small enough (less than $1\mu m$) to allow non-destructive puncturing of the plasma membrane. This allows the intracellular recording of **resting** and **action potentials**, the measurement of intracellular ion and pH levels (using **ion-selective electrodes),** or **microinjection**. Microelectrodes are generally pulled from glass capillaries, and filled with conducting solutions of potassium chloride or potassium acetate to maximize conductivity near the tip. Electrical contact, if required, is usually made with a silver chloride-coated silver wires.

microfibril Basic structural unit of the plant cell wall, made of cellulose in higher plants and most algae, **chitin** in some fungi, and **mannan** or **xylan** in a few algae. Higher plant microfibrils are about 10 nm in diameter, and extremely long in relation to their width, a paracrystalline array. Microfibrils have great tensile strength and, set amongst other matrix components, provide the main structural element of the plant cell wall. Each microfibril is extruded from a large cellulose-synthesizing rosette, a hexagonal transmembrane complex that moves along the plasma membrane, powered by the polymerization of glucose and aligned by cortical microtubules.

microfilament Cytoplasmic filament, 5–7 nm thick, of **F-actin** that can be decorated with **HMM**; may be laterally associated with other proteins (tropomyosin, α-actinin) in some cases, and may be anchored to the membrane. Microfilaments are conspicuous in **adherens junctions**.

microfilaria A stage in the life cycle of parasitic nematodes in the family Onchocercidae (filarial nematodes such as *Onchocerca volvulus*, *Brugia malayi*, and *Wuchereria bancrofti* that cause diseases in humans). Motile embryos (150–300µm long) are the infective forms in the insect vector. The release of microfilariae by the adult female may be periodic, thereby increasing the chances of being transmitted by blood-sucking insects that bite at particular times of the day.

microglial cell Small glial cells of mesodermal origin, with scanty cytoplasm and small spiny processes. Distributed throughout grey and white matter. Derive from monocytes and invade neural tissue just before birth; capable of enlarging to become macrophages.

microglobulin Any small globular plasma protein. See **alpha-1-microglobulin, beta-2-microglobulin**.

microinjection The insertion of a substance into a cell through a **microelectrode**. Typical applications include the injection of drugs, histochemical markers (such as **horseradish peroxidase** or lucifer yellow) and RNA or DNA in molecular biological studies. To extrude the substances through the very fine electrode tips, either hydrostatic pressure (pressure injection) or electric currents (ionophoresis) is employed.

micromere One of the small cells which are formed in the upper or animal hemisphere of a fertilised egg during **holoblastic** cleavage.

micrometer eyepiece See **eyepiece graticule**.

micronucleus The smaller nucleus in ciliate protozoans, fully active in inheritance and passed after meiosis to conjugating pairs. Gives rise to the macronucleus or macronuclei. Genes in the micronucleus are not actively transcribed.

microperoxidase Part of a cytochrome c molecule that retains haem group and has **peroxidase** activity.

microperoxisome See **peroxisomes**.

microphyll A leaf, not necessarily a small one, that is supplied by a single unbranched vascular strand.

micropinocytosis Pinocytosis of small vesicles (around 100 nm in diameter). Does not depend upon microfilaments and is therefore insensitive to cytochalasin. May involve **dynamin 2**. Article: http://jcs.biologists.org/cgi/content/short/jcs.010686v1

microplate reader Analytical instrument for spectroscopic analysis of multi-well plates, commonly 96-well but increasingly having more (but smaller) wells per plate to economise on reagents. Different instruments will analyse radioactivity, optical absorbance, fluorescence or luminescence.

micropore filters *Millipore filter* Filters made of a meshwork of cellulose acetate or nitrate and with defined pore size. They can be autoclaved, and the smaller pore sizes (0.22 µm, 0.45 µm) are used for sterilising heat-labile solutions by filtering out microorganisms. Larger pore-size filters are used in setting up **Boyden chambers**. They are about 150 µm thick and should be distinguished from **Nucleopore filters**. Millipore is a trade name for micropore filters.

microprobe See **electron microprobe**.

microprocessor complex A RNA- and haem-binding multi-protein complex that is involved in the initial step of **microRNA** (miRNA) biogenesis. Composed of haem-free or haem-bound DiGeorge syndrome critical region 8 protein (DGCR8, 773 aa) and the RNAse **drosha**; cleaves long primary miRNAs (pri-miRNAs) to the intermediate 60–70 nt precursor miRNAs (pre-miRNAs) that are then further processed by **dicer**.

micropyle (1) Small hole or aperture in the protective tissue surrounding a plant ovule, through which the pollen tube enters at fertilization. Develops into a small hole in the seed coat through which, in many cases water enters at germination. (2) Perforation in the shell (chorion) of an insect's egg through which the sperm enters at fertilization.

microRNA *miRNA* Small noncoding RNA molecules (usually ∼22 nt) generated by the **microprocessor complex** and **dicer** from precursors with a characteristic hairpin secondary structure. Many miRNA sequences are evolutionarily conserved and they have important regulatory functions. They are associated with proteins such as **argonaute** in **RISCs** and induce the sequence-specific degradation of mRNAs. The first examples were **lin-4** and **let-7** that control developmental timing in *C. elegans* but subsequently homologues were discovered in many species including humans where, for example, let-7 represses the HMGA2 oncogene. MicroRNAs are functionally similar to **small interfering RNAs** (siRNAs) but differ in being derived from endogenous transcripts that can form local hairpin structures, which give rise to only a single miRNA molecule. (Natural siRNAs come from long exogenous or endogenous double-stranded RNA molecules from which numerous siRNAs from both strands of the dsRNA are generated). The naming conventions involve a three or four letter prefix to designate the species, (e.g. hsa-miR-101 from *Homo sapiens*) and there is a searchable database of published miRNA sequences with annotation. The mature sequences are designated 'miR' in the database, whereas the precursor hairpins are labelled 'mir'. Aberrant miRNA expression is observed in several human malignancies. MicroRNA database: http://www.mirbase.org/

microsatellites Short sequences of di- or trinucleotide repeats of very variable length distributed widely throughout the genome. Using PCR primers to the unique sequences upstream and downstream of a microsatellite their location and polymorphism can be determined and the technique is extensively used in investigating genetic associations with disease. See **satellite DNA**.

microscopic polyangiitis Necrotizing **vasculitis**, with few or no immune deposits, affecting small vessels (capillaries, venules, arterioles).

microscopy The visualisation of small objects: see **light microscopy** or **electron microscopy**. See also **atomic force m.**, **confocal m.**, **fluorescence speckle m.**, **interference m.**, **interference reflection**, **nanovid m.**, **ratio-imaging fluorescence m.**, **scanning ion conductance m.**, **scanning probe m.**, **stimulated emission depletion m.**, **total internal reflection fluorescence m.**

microsequencing Term generally, though rather imprecisely, used for the sequencing of very small amounts of protein – often as a prelude to producing an oligonucleotide probe, screening a cDNA library and cloning.

microsomal fraction See **microsomes**.

microsomal triglyceride transfer protein *MTTP* A heterodimeric protein that catalyses the transport of triglyceride, cholesteryl ester, and phospholipid from phospholipid surfaces. The smaller subunit is **protein disulphide isomerase** (EC 5.3.4.1, 504aa), the larger (894aa) is a product of one of the vitellogenin gene family. The large subunit is defective in **abetalipoproteinaemia**.

microsomes Heterogenous set of vesicles 20–200 nm in diameter formed from the **endoplasmic reticulum** when cells are disrupted. By differential centrifugation a microsomal fraction can be purified from a cell homogenate .

microspikes Projections from the leading edge of some cells, particularly, but not exclusively, nerve **growth cones**. They are usually about 100 nm diameter, 5–10 μm long, and are supported by loosely bundled microfilaments. They are referred to by some authors as filopodia. Functionally a sort of linear version of a ruffle on a leading lamella.

microspore (1) A haploid spore produced by a plant sporophyte that develops into a male gametophyte. In seed plants, it corresponds to the developing pollen grain at the uninucleate stage. (2) The smaller of the spores of a heterosporous species. (3) See Microsporidia.

Microsporidia Obligate intracellular parasites, considered to be extremely reduced fungi, ranging in size from 1.5 to 2.0 μm. There are eight genera of microsporidia that can infect humans: more common are *Encephalitozoon* spp., *Septata intestinalis* and *Enterocytozoon bieneusi*; the less common are *Brachiola* spp, *Microsporidium* spp., *Nosema* spp., *Pleistophora* spp., *Trachipleistophora* spp., and *Vittaforma* spp. Microsporidia multiply extensively with the host cell cytoplasm; the life cycle includes repeated divisions by binary fission (merogony) or multiple fission (schizogony) and spore production (sporogony); spores can then infect other cells.

microtome A device used for cutting thin sections from an embedded specimen, either for light- or electron-microscopy.

microtrabecular network Complex network arrangement seen using the high-voltage electron microscope to look at the cytoplasm of cells prepared by very rapid freezing. The suggestion was that most cytoplasmic proteins are in fact loosely associated with one another in this fibrillar network and are separate from the aqueous phase that contains only small molecules in true solution. If it exists, then it must certainly be very labile in cells where there is cytoplasmic flow and rapid organelle movement; now considered artefactual by most microscopists.

microtubule Cytoplasmic tubule, 25 nm outside diameter with a 5 nm thick wall. Made of **tubulin** heterodimers packed in a three-start helix (or of 13 protofilaments, looked at another way), and associated with various other proteins (**MAPs**, **dynein**, **kinesin**). Microtubules of the ciliary **axoneme** are more permanent than cytoplasmic and spindle microtubules.

microtubule organizing centre *MTOC* A rather amorphous region of cytoplasm which determines the number and pattern of microtubules in the cytoplasm and is the location from which microtubules radiate. In animal cells the **pericentriolar region** is the major organising centre for cytoplasmic and spindle microtubules. Basal bodies are organising centres for axonemal microtubules. The MTOC has gamma-tubulin (see **gamma-tubulin complex**) which serves as a marker. Activity of MTOCs can be regulated, but the mechanism is unclear.

microtubule-associated proteins A group of proteins (MAPs) that form part of the electron-lucent zone around a microtubule. MAP1A (2805aa), 1B (2468aa), 2A (1827aa) and 2B (1827aa) are associated with brain microtubules and form projections. MAP4 (1152aa) co-purifies with MAPs 1 and 2. Microtubule associated protein 2 is associated with axons rather than dendrites. MAP1C is in a separate class, being a motor molecule (cytoplasmic dynein 1 heavy chain, 4646aa; CD2HC, 4307aa, associated with 6 or 7 light and intermediate chains of about 350–500aa) the two-headed cytoplasmic equivalent of ciliary dynein. It is responsible for retrograde transport (transport towards the centrosome). MAP7 is **ensconsin**. MAP9 (aster-associated protein, 647aa) is involved in spindle assembly. In *Drosophila* microtubule-associated protein futsch (5495aa) is necessary for dendritic and axonal organization and growth at the neuromuscular junction through the regulation of the synaptic microtubule cytoskeleton; jupiter (208aa) binds to all MTs. In *Arabidopsis* there are various MAPs (e.g. MAP65-1, 587aa; MAP65-6, 608aa) that have diverse roles in regulating microtubule arrays. Microtubule-associated protein TORTIFOLIA1 (CONVOLUTA, 864aa) is a plant-specific MAP that regulates the orientation of cortical microtubules and the direction of organ growth. See also **EMAP**. N.B. **MAP kinases** (mitogen-activated protein kinases) can, confusingly, phosphorylate microtubule-associated proteins.

microvillus *Pl.* microvilli. Projection from the apical surface of an epithelial cell that is supported by a central core of microfilaments associated with bundling proteins such as **villin** and **fimbrin**. In the intestinal **brush border** the microvilli presumably increase absorptive surface area, whereas the stereovilli (**stereocilia**) of the cochlea have a distinct mechanical role in sensory transduction.

Microviridae A diverse group of single-stranded DNA bacteriophages, also known as φX phage group or isometric ssDNA phages.

Mid1p *S. pombe* homologue of **anillin**. Mid2p is another anillin homologue but is not orthologous with Mid1p and influences **septin** ring organization at the site of cell division. Overproduction of Mid2p depolarizes cell growth and affects the organization of both the septin and actin cytoskeletons.

midbody Dense structure formed during **cytokinesis** at the cleavage furrow. It consists of remnants of **spindle fibres** and other amorphous material and disappears before cell division is completed.

midbrain See **mesencephalon**.

middle lamella First part of the plant **cell wall** to be formed, laid down in the **phragmoplast** during cell division as the **cell plate**. Subsequently makes up the central part of the double cell wall that separates two adjacent cells, cementing together the two primary walls. Rich in **pectin**, and relatively poor in **cellulose**.

midiprep Slang, denoting a medium scale purification of **plasmid** from a bacterial culture. Usually used to describe preparations from 10–100 ml culture. See also **miniprep, maxiprep, megaprep**.

midkine Heparin-binding growth factor (neurite growth-promoting factor 2, NEGF2, 123aa) of the TGFβ; superfamily; has 50% sequence identity with heparin-binding growth-associated molecule (HB-GAM). Structurally unrelated to fibroblast growth factor (FGF). Midkine was originally described as associated with tooth morphogenesis induced by epithelial-mesenchyme interactions. **Nucleolin** binds midkine. Review: http://www.ncbi.nlm.nih.gov/pubmed/20431264

MIF See **macrophage inhibition factor**, or **migration inhibitory factor**.

Mig (1) In *S. cerevisiae* MIG1 is a regulatory protein (CAT4, 504aa) that is involved in glucose repression of the *SUC, GAL* and *MAL* genes as well as of the *CAT8* gene. MIG2 (382aa) has a similar role. (2) In *S. cerevisiae*, multicopy inhibitor of growth protein 3 (MIG3) is a transcriptional repressor involved in response to toxic agents. (3) In humans, mitogen-inducible gene products: Mig2 (fermitin family homologue 2, **kindlin**-2, 680aa) recruits **migfilin** to cell-ECM focal adhesion sites. Mig2B is kindlin-3. Mig6 (ERBB receptor feedback inhibitor 1, 462aa) is a negative regulator of EGFR signalling in skin morphogenesis. (4) A cytokine, (monokine induced by interferon-gamma, MIG, CXCL9, 125aa). (5) In *C. elegans*, ADAM family mig-17 (EC 3.4.24.-, abnormal cell migration protein 17, 509aa) a metallopeptidase secreted from muscle cells and required for the migration of distal tip cells.

migfilin A **LIM-domain** containing protein (filamin-binding LIM protein 1, FBLP1, 373aa) that localizes to cell-matrix adhesions, associates with actin filaments, and is essential for cell shape modulation. Migfilin interacts with the cell-matrix adhesion protein Mig-2 (mitogen inducible gene-2, see **kindlin**) and **filamin** through its C- and N-terminal domains, respectively.

migraine A condition characterized by painful headaches, together with nausea and/or photophobia and phonophobia. Probably caused by malfunction in the serotonin system and the **triptans**, serotonin 5-HT1B/1D receptor agonists, are helpful. Familial hemiplegic migraine is usually caused by mutations in a gene coding for the P/Q-type calcium channel alpha subunit; less commonly by mutations in the Na$^+$/K$^+$-ATPase gene, ATP1A2 or a sodium channel alpha-subunit coding gene, SCNA1.

migrastatin A macroketone compound isolated from *Streptomyces platensis*, closely related to the **dorrigocins**. Migrastatin and its analogues may inhibit metastasis, possibly by targeting **fascin**.

migration inhibitory factor Factor that inhibits macrophage movement (macrophage migration inhibitory factor). Originally defined on basis of inhibition of emigration of mononuclear cells from capillary (haematocrit) tubes; more recently a name given to a proinflammatory cytokine (EC 5.3.2.1, glycosylation-inhibiting factor, 115aa) involved in innate immunity to pathogens. It has phenylpyruvate tautomerase and dopachrome tautomerase activity but it is unclear whther this is physiologically relevant. See **macrophage inhibitory factor**.

mil (1) Avian homologue of v-*raf-1* product (RAF proto-oncogene serine/threonine-protein kinase, 647aa). (2) In humans, Mil1 (Bcl-2-like protein 13, 485aa) a single-pass mitochondrial membrane protein that may promote the activation of caspase-3 and apoptosis.

Miller unit A standardized amount of β-galactosidase activity, a 'Miller unit', named after Jeffrey Miller who published the first protocol (1972) for a beta-galactosidase (β-gal) assay using o-nitrophenyl-β-D-galactoside (ONPG) as a substrate (Miller assays).

Miller-Dieker lissencephaly syndrome A disorder caused by a deletion of several genes on chromosome 17, one of which is the LIS1 gene (see **lissencephaly**). The features include developmental defects in the brain, growth retardation, and abnormalities in a range of other organs. See **postaxial acrofacial dysostosis** (Miller's syndrome).

Millipore filter Trade name for a well-known brand of **micropore filters**.

Milroy's disease See **hereditary lymphedema type IA**.

MIM A WH2 domain-containing protein (missing in metastasis, metastasis suppressor 1, 356aa) that binds ATP-actin monomers more tightly than ADP-actin monomers; interacts directly with the SH3 domain of **cortactin**. MIM appears to regulate cell motility by modulating different **Arp2/3** activators.

mimecan One of the small leucine-rich proteoglycans (SLRPs), a secreted glycoprotein (osteoglycin, 298aa) that induces bone formation in conjunction with TGFβ.

mimitin A mitochondrial protein (myc-induced mitochondrial protein, NADH dehydrogenase ubiquinone 1 alpha subcomplex assembly factor 2, 169aa) upregulated by proinflammatory cytokines. Mimitin interacts with a microtubular protein (MAP1S). Acts as a molecular chaperone for assembly of mitochondrial complex I. Article: http://www.biomedcentral.com/1471-2121/10/23

mimivirus An exceptionally large double-stranded DNA virus found in *Acanthamoeba polyphaga*. The particle size of 400 nm makes it dimensionally comparable to mycoplasma. The genome is 1.2 megabases in size and there are 1262 putative open reading frames, only 10% of which resemble proteins of known function.

Mimosa pudica The 'sensitive plant' whose leaflets fold inwards very rapidly when touched. A more vigorous stimulus causes the whole leaf to droop, and the stimulus can be transmitted to neighbouring leaves.

mimosine A non-protein amino acid found in leaves, pods and seeds of tropical legumes of the genus *Leucaena*. It is toxic to animals and is an extremely effective inhibitor of DNA replication in mammalian cells, preventing the formation of replication forks in cells approaching the G/S boundary. It is an iron/zinc chelator and may inhibit iron-dependent ribonucleotide reductase and the transcription of the cytoplasmic serine hydroxymethyltransferase gene (SHMT).

mimotope Compound that mimics the structure of a conformational **epitope** and that will elicit an identical antibody response (whereas a mimetic would not have the same antigenicity). Mostly used of peptides from **phage display libraries**; potentially useful as vaccines.

MIN6 cells A line of cells derived from *in vivo* immortalized insulin-secreting pancreatic beta cells.

Minamata disease Classic and infamous case of poisoning by methylmercury in humans who ate fish and shellfish contaminated with waste water (in 1956) from a chemical plant in Minamata, South West Kyushu, Japan. The disease is characterised by sensory and motor disturbances and fetal damage in pregnant women.

minD An ATPase (septum site-determining protein minD, cell division inhibitor minD, 270aa) required for the correct placement of the division site in *E. coli*. In the presence of ATP, MinD binds to the membrane and recruits MinC (231aa), forming a complex that can destabilize the cytokinetic **Z-ring**. MinE (88aa), which is also recruited to the membrane by MinD, displaces MinC and stimulates the MinD ATPase, resulting in the oscillation of the Min proteins between the poles of the cell to destabilize **ftsZ** filaments that have formed before they mature into polar Z rings. Article: http://jb.asm.org/content/187/2/629

mindin See **spondins**.

mineralocorticoid Natural or synthetic corticosteroid that acts on water and electrolyte balance by promoting retention of sodium ions and excretion of potassium ions in the kidney. Aldosterone is the most potent natural example and is produced in the outer layer of the adrenal cortex.

mini-chromosome maintenance proteins

Family of conserved proteins (MCM proteins) involved in the initiation and elongation of DNA replication forks in archaea and eukaryotes. MCM proteins can form hexameric complexes that possess ATP-dependent DNA unwinding activity. In eukaryotes MCM2-7 is a heterohexameric **helicase**, composed of six related subunits, that assembles at the DNA replication fork. MCM-2 is often used as a marker for proliferation competence; MCM-4 and other MCMs are phosphorylated during the cell cycle, at least in part by cyclin-dependent kinases. MCM-7 is upregulated in a variety of tumours including neuroblastoma, prostate, cervical and hypopharyngeal carcinomas.

miniature end plate potential See **postsynaptic potential**.

miniature inverted-repeats transposable elements *MITEs* Class III **transposons**. Recurring motifs consisting of almost identical sequences of about 400 base pairs flanked by characteristic inverted repeats of about 15 base pairs. They are too small to encode proteins. Thousands of copies have been identified in the genomes of *C. elegans* and of rice and known to occur in humans. Database: http://gydb.uv.es/index.php/Intro

minicell Spherical fragment of a bacterium produced by abnormal fission and not containing a bacterial chromosome.

minichromosome (1) A chromatin structure consisting of viral DNA complexed with histones resembling a small chromosome found in virus-infected eukaryotic cells. (2) A plasmid that contains a chromosomal **origin of replication**. (3) See **mini-chromosome maintenance proteins**.

minimal inhibitory concentration *MIC* The smallest concentration of an antimicrobial substance that will inhibit growth of a test organism.

minimal medium The simplest tissue culture medium (minimal essential medium, MEM) that will support the proliferation of normal cells.

miniprep Slang, denoting a small scale purification of plasmid from a bacterial culture. Usually used to describe preparations from 1–10 ml culture. See also **midiprep**, **maxiprep**, **megaprep**.

minisatellite Class of highly repetitive **satellite DNA** (variable number tandem repeats, VNTRs), comprising variable (typically 10–20) repeats of short (e.g. 64 bases) DNA sequences. The high level of polymorphism of such minisatellites make them very useful in genomic mapping.

minispindles In *Drosophila* a gene (*msps*) that encodes a protein (2042aa) involved in the structural integrity of the mitotic spindle. Mutation results in the formation of one or more small additional spindles in diploid cells. Original description: http://jcb.rupress.org/content/146/5/1005.abstract

minK (1) Widely expressed protein (voltage-gated potassium channel subfamily E member 1, KCNE1, 129aa) that forms delayed rectifier **potassium channels** by aggregation with other membrane proteins. A variety of channels have been shown to have minK associated with them and minK seems to be important in regulating structure and activity of the channel. (2) See **misshapen-like kinase 1**.

minor allele frequency A measure used to assess the informativeness of SNP loci and is related to expected heterozygosity where the number of alleles is two, as is usually the case for SNPs. The occurrence of a triallelic SNP is very rare. SNPs with MAF values $>/=0.05$ or 0.10 are considered common and useful for most applications, whereas SNPs with MAF values $>/=0.30$ are the most informative and transferable across various genotypes.

minute mutant A class of recessive lethal mutants of *Drosophila* The heterozygotes grow more slowly, are smaller and less fertile than the wild type flies. There are about 40 loci that produce minute mutants. See **mdm**.

Min proteins See **MinD**.

MIP (1) See **major intrinsic protein**. (2) See **macrophage inflammatory protein**. Macrophage inflammatory protein is known to have various subclasses, MIP-1α, MIP-1β, MIP-2. See Table C3 (chemokines). (3) Major intrinsic proteins: MIP26 is aquaporin-0 (263aa) found in lens fibres. (4) M9 region interaction protein (transportin-1, importin beta-2, karyopherin beta-2, 898aa) a nuclear transport receptor. (5) Mitochondrial intermediate peptidase (MIP, EC 3.4.24.59, 713aa) that processes proteins that are imported into the mitochondrion. (6) In *Drosophila*, myoinhibitory-like protein (211aa), the precursor for a number of different allostatins (e.g. Drostatin-B1, MIP-1, 9aa) that may have a role in gut motility. (7) In *Legionella*, outer membrane protein MIP (macrophage infectivity potentiator, peptidyl-prolyl *cis-trans* isomerase, rotamase, EC 5.2.1.8, 233aa) a virulence factor. (8) In *S. pombe*, a WD repeat-containing protein (mip1, 1313aa) that facilitates the functioning of the meiotic regulator mei2. (9) In *S. cerevisiae*, an RNA-binding protein, MIP6 (659aa) that interacts with the mRNA export factor MEX67. (10) Microtubule-interacting protein associated with TRAF3 (MIP-T3, 691aa) that inhibits IL13 signalling.

MIR (1) See **mammalian-wide interspersed repeat**. (2) The mir-17–92 polycistron is a cluster of microRNAs located in a region of DNA that is amplified in human B-cell lymphomas; levels of the primary or mature miRNAs derived from this locus are often substantially increased in these cancers making the **polycistron** a potential oncogene. (3) See **microRNA**.

miracidium The ciliated first-stage larva of a trematode, 150–180 μm in length by 70–80 μm in width, the free-living stage responsible for infecting the intermediate host; in the case of *Schistosoma sp.* the aquatic snail.

miranda *Drosophila* gene, the product of which (830aa) co-localises with prospero in mitotic neuroblasts and apparently directs prospero exclusively to the ganglion mother cell (GMC) during the asymmetric division that gives rise to another neuroblast and the GMC. Anchors prospero selectively to the basal side of the cell cortex during mitosis and releases it after cytokinesis. Miranda has two **leucine zipper** motifs and eight consensus sites for PKC phosphorylation.

miRBase A searchable database of published miRNA sequences and annotation of the sequences. Release 18 (Nov. 2011) has 18,226 entries. Link to database: http://www.mirbase.org/

miRNA See **microRNA**.

MIS12 In *S. pombe*, a component (259aa) of the **NMS complex** that is required for correct segregation of chromosomes, for maintaining the inner centromere structure and kinetochore function. The human homologue (215aa) has a comparable role and is part of the MIS12 complex that consists of MIS12 together with DSN1 (kinetochore-associated protein DSN1 homologue, 356aa), NSL1 (kinetochore-associated protein NSL1 homologue, 281aa) and PMF1 (polyamine-modulated factor 1, 205aa).

mismatch repair A DNA repair system that detects and replaces wrongly paired (mismatched) bases in newly replicated DNA or in DNA that has been damaged in some way. In long patch repair tracts of a few kilobases are removed and replaced using the correct (parental) strand as a template. In short patch repair only around 10 nucleotides are involved and this is the method used when there is damage rather than inaccurate synthesis. The gap is filled by DNA polymerase. The *E. coli* mismatch correction enzyme is encoded by three *mut* genes *mutH*, *mutL* and *mutS*, that is directed to the newly synthesized strand and removes a segment of that strand including the incorrect nucleotide. The mismatch repair system is highly conserved and in eukaryotes the **MLH genes** are homologous to the

mutL gene of *E. coli; msh* genes are homologues of the *mutS* genes. *PMS* genes are homologues of the yeast mismatch repair enzymes (postmeiotic segregation increased genes). Mutations in mismatch repair systems are associated with various malignancies, for example mismatch repair cancer syndrome (**Turcot's syndrome**) and **Muir-Torre syndrome**.

missense mutation A mutation that alters a **codon** for a particular amino acid to one specifying a different amino acid.

misshapen-like kinase 1 A serine/threonine kinase (MINK1, misshapen/NIK-related kinase, MEKKK6, MAP4K6, 1332aa) one of the **germinal centre kinase** (GCK) family, that may play a role in the response to environmental stress upstream of the JUN N-terminal pathway. May play a role in development of the brain.

MITEs See **miniature inverted-repeats transposable elements**.

mitochondrial diseases Illnesses, frequently neurological, which can be ascribed to defects in mitochondrial function. If the defect is in the mitochondrial rather than the nuclear genome unusual patterns of inheritance can be observed. Examples include various myopathies, **Bjornstad's syndrome**, **Kearns-Sayre syndrome**, **Leber's optic atrophy**, **Leigh's syndrome**, **MELAS syndrome**, **MERRF syndrome**, **Navajo neurohepatopathy** and **sideroblastic anaemia with spinocerebellar ataxia**. See also **mitochondrial DNA deletion syndrome**.

mitochondrial DNA deletion syndrome A diverse set of disorders caused by loss or mutation of mitochondrial enzymes. The hepatocerebral form can be a result of mutation in the nuclear gene for mitochondrial deoxyguanosine kinase, the MPV17 gene (involved in mitochondria homeostasis) or the C10ORF2 gene that encodes **twinkle**. The myopathic form is caused by mutation in the nuclear-encoded mitochondrial thymidine kinase gene, the encephalomyopathic form is caused by mutation in the succinyl-CoA synthase gene.

Mitochondrial Eve Purportedly the most recent common matrilineal ancestor of all living humans since all mitochondria are supposed to derive from the egg — although there has been some recent doubt cast upon this. 'Mitochondrial Eve' was a member of a population that existed about 150,000 years ago and produced multiple lineages, but all others have disappeared.

mitochondrial neurogastrointestinal encephalopathy A progressive multi-system disorder that can be caused by mutations in the genes encoding either thymidine phosphorylase (**gliostatin**) or DNA polymerase gamma. There are multiple deletions of mitochondrial DNA.

mitochondrial trifunctional protein A mitochondrial multi-enzyme complex (TFP) composed of 4 hydroxyacyl-CoA dehydrogenase-alpha (HADHA, 763aa) and 4 hydroacyl-CoA dehydrogenase-beta (HADHB, 474aa) subunits, that catalyzes the last three steps in the long-chain fatty acid beta-oxidation pathway in mitochondria. TFP deficiency leads to a wide clinical spectrum of disease ranging from severe neonatal/infantile cardiomyopathy and early death to mild chronic progressive sensorimotor polyneuropathy with episodic rhabdomyolysis.

mitochondrion *Pl.* mitochondria. Highly pleiomorphic organelle of eukaryotic cells that varies from short rod-like structures present in high number to long branched structures. Contains DNA and **mitoribosomes**. Has a double membrane and the inner membrane may contain numerous folds (cristae). The inner fluid phase has most of the enzymes of the **tricarboxylic acid cycle** and some of the urea cycle. The inner membrane contains the components of the **electron transport chain**. Major function is to regenerate ATP by oxidative phosphorylation (see **chemiosmotic hypothesis**).

mitofusin Transmembrane GTPases of the outer mitochondrial membrane (mitofusin-1, 741aa; mitofusin-2, 757aa) that mediate mitochondrial fusion. Form homomultimers and heteromultimers and if overexpressed induce the formation of mitochondrial networks. Upregulation of mitofusin-2 inhibits angiotensin II-induced myocardial hypertrophy. Defects in mitofusin-2 cause a form of **Charcot-Marie-Tooth disease**. Abstract: http://www.ncbi.nlm.nih.gov/pubmed/21106870

mitogen activated kinases *MEKs, MAPKs* See **MAP kinases**.

mitogen inducible genes See **kindlin**.

mitogenesis The process of stimulating transit through the cell cycle especially as applied to lymphocytes. Concanavalin A is a mitogen for T-cells; the best mitogen for B-cells is Cowan strain *Staphylococcus aureus*.

mitogenic Causing re-entry of cells into the cell cycle, not just into mitosis.

mitogillin See **restrictocin**.

mitomycin C Aziridine antibiotic isolated from *Streptomyces caespitosus*. Inhibits DNA synthesis by cross-linking the strands and is used as an antineoplastic agent. Most active in late G1 and early S phase. Mitomycin-treated cells are sometimes used as feeder layers.

mitoplasts Isolated mitochondria without their outer membranes. They have finger-like processes, and retain the capacity for oxidative phosphorylation.

mitoribosomes Mitochondrial ribosomes that more closely resemble prokaryotic ribosomes than cytoplasmic ribosomes of the cells in which they are found, though they are even smaller (55S) and have fewer proteins than bacterial ribosomes.

mitosis The usual process of nuclear division in the somatic cells of **eukaryotes**. Mitosis is classically divided into four stages. The chromosomes are actually replicated prior to mitosis during the **S phase** of the **cell cycle**. During the first stage, prophase, the chromosomes condense and become visible as double strands (each strand being termed a chromatid) and the **nuclear envelope** breaks down. At the same time the mitotic **spindle** forms by the polymerization of **microtubules** and the chromosomes are attached to spindle fibres at their kinetochores. In metaphase the chromosomes align in a central plane, the metaphase plate, perpendicular to the long axis of the spindle. During anaphase the paired chromatids are apparently pulled to opposite poles of the spindle by means of the spindle fibre microtubules attached to the kinetochore, though the actual mechanism for this movement is still controversial. This separation of chromatids is completed during telophase, when they can be regarded as chromosomes proper. The chromosomes now lengthen and become diffuse and new nuclear envelopes form round the two sets of chromosomes. This is usually followed by cell division (cytokinesis) in which the cytoplasm is also divided to give two daughter cells. Mitosis ensures that each daughter cell has a **diploid** set of chromosomes that is identical to that of the parent cell.

mitotic death Cell death that occurs after mitosis even though the fatal injury may have occurred much earlier. Often observed following radiation damage which can fragment chromosomes.

mitotic index The fraction of cells in a sample that are in mitosis. It is a measure of the relative length of the mitotic phase of the **cell cycle**.

mitotic recombination The process of somatic **crossing-over** can occur between **homologous chromosomes** during mitosis, a very rare event because the chromosomes do not normally pair. When it occurs it can lead to new combinations of previously linked genes. Although infrequent, mitotic recombination has been utilized for genetic analysis in *Aspergillus* and in studies on developmental **compartments** in *Drosophila* where the frequency of mitotic recombination can be increased by X-irradiation.

mitotic shake-off method A method of collecting cells in **mitosis**, so that the chromosomes can be examined and the karyotype determined. Many cultured cells round up during mitosis and so become less firmly attached to the culture substratum. Cells in mitosis thus can be removed into suspension by gentle shaking of the culture vessel, leaving the non-mitotic cells still attached. The number of cells that are in mitosis is usually increased by using a drug, such as **colcemid**, that blocks mitosis at **metaphase**.

mitotic spindle The microtubule-based structure that forms as the nuclear envelope disappears in the earliest stages of **mitosis**. It has the shape of two cones apposed at their bases. The poles are the sites of the duplicated centrioles (in animals) and act as microtubule organising centres. The spindle-pole fibres (microtubules) interdigitate with microtubules of opposite polarity from the other pole and with those from the kinetochores of chromosomes that become aligned (at metaphase) at the equatorial plane. See also **spindle pole body**.

mitrocomin Calcium-activated photoprotein (198aa) from coelenterate *Halistaura*. See **aequorin**.

mitsugumins Mitsugumin-23 ia a transmembrane protein (transmembrane protein 109, 243aa) with three putative transmembrane segments that forms homotrimers. It is preferentially expressed in excitable tissues, including striated muscle and brain, where it may act as counter-ion channel synchronising functions including the release of calcium from intracellular stores. Mitsugumin-29 is synaptophysin-like protein 2, (264aa in mouse). Tripartite motif-containing protein 72 is mitsugumin-53 (477aa), a muscle-specific protein that is important for cell membrane repair.

mix Homeobox genes (*mix1*, *mix2*), expressed in prospective mesoderm and endoderm after mid-blastula stage, that respond to TGFβ (**transforming growth factor**-β) superfamily signals including activin, TGFβ, Vg-1 and BMP-4, but not to non-TGF inducers. Human MIXL1 (232aa) is important in proper axial mesendoderm morphogenesis and endoderm formation. In adults, restricted to lymphocyte progenitors and secondary lymph tissues.

mixed lymphocyte reaction A mitogenic reaction in T-cells when allogeneic (i.e. mixed) lymphocytes are brought together, provided they are mismatched in histocompatibility loci. Once used as a test for possible graft compatibility in human grafting. Now known that a negative reaction is a poor predictor of graft acceptance.

mixotroph An organism the combines two or more fundamental methods of nutrition.

MKK A MAP/ERK kinase kinase. See **MAPKK**.

MKPs See **MAPK phosphatase**.

MKS1 (1) See **Meckel syndrome**. (2) In *S. cerevisiae* is a negative regulator of the ras-cAMP pathway (584aa) involved in transcriptional regulation of galactose-inducible genes. (3) In *Arabidopsis* MAP kinase 4 substrate 1 (222aa) a regulator of plant defense responses.

MLCK See **myosin light chain kinase**.

MLEE Form of two-dimensional **electrophoresis** (multi-locus enzyme electrophoresis) used to distinguish **polymorphisms** between strains or populations.

***MLH* genes** Eukaryotic genes involved in DNA **mismatch repair**. MLH1 (MutL protein homologue 1, 756aa) heterodimerizes with PMS2 to form MutL alpha which binds to MutS alpha or MutS beta (see *MSH* genes) that has bound to a mismatch site. Several other MLH proteins are known in yeast and mammalian systems.

M-line Central part of the **A-band** of striated muscle (and of the **M-band**): contains M-line protein (**myomesin**), **creatine kinase**, and glycogen phosphorylase b. Involved in controlling the spacing between thick filaments.

MLKs A family of serine/threonine kinases (mixed lineage kinases) with the features of mitogen-activated protein kinase kinase kinases (MAPKKKs). The MLK family (MLK1, MAP3K9, EC 2.7.11.25, 1104aa; MLK2/MST, MAP3K10, 954aa; MLK3/SPRK/PTK, MAP3K11, 847aa; MLK4, 1036aa) regulate signalling by the **JNK** and **p38** MAPK pathways. In *Drosophila* the MLK homologue, slipper, regulates JNK to control dorsal closure during embryonic morphogenesis.

MLL MLL1 is a histone-lysine N-methyltransferase (EC 2.1.1.43, 3969aa), product of the mixed lineage leukaemia gene, a DNA-binding protein that positively regulates the expression of target genes, including many homeobox (*HOX*) genes. MLL is cleaved by **taspase-1** to produce fragments that heterodimerize to stabilize the complex and define its intranuclear destination. Chromosomal aberrations involving MLL are a cause of acute leukemias. Homologous to *Drosophila* trithorax. A number of other MLLs are known, all of which have methyl transferase activity and association with myeloid malignancies.

MLR See **mixed lymphocyte reaction**.

MLST A nucleotide sequence-based approach (multilocus sequence typing) for the unambiguous characterisation of isolates of bacteria and other organisms using the sequences of internal fragments of (usually) seven house-keeping genes. It is based upon the concept of multilocus enzyme electrophoresis (**MLEE**) adapted so that alleles at each locus are defined directly, by nucleotide sequencing, rather than indirectly from the electrophoretic moblity of their gene products.

MluI cell-cycle boxes A hexameric nucleotide sequence (MCB element) found near the start site of yeast genes expressed at G1/S that may play a role in controlling entry into the cell division cycle; cut by the Type II restriction enzyme MluI.

MLV (1) **Murine leukaemia virus**. (2) Modified-live virus. (3) Multilamellar liposomes or multilamellar vesicles.

MLVA A subtyping method (multilocus variable-number tandem repeat analysis) used to distinguish closely related bacterial isolates for investigation of disease outbreaks and for studying phylogenetic patterns among isolates. More sensitive than multilocus sequencing typing (MLST) in the context of Public Health investigations. Article: http://www.biomedcentral.com/1471-2180/9/278

MLZE A cytoplasmic protein (melanoma-derived leucine zipper-containing extranuclear factor, **gasdermin** C, 508aa) related to **pejvakin** and of unknown function, although upregulated in melanoma. The leucine zipper may be non-functional.

MMR (1) Measles, mumps, and rubella vaccine, a childhood vaccine. Mistakenly claimed to be associated with autism and the resulting media-scare led to a reduction in immunisation, an increase in measles infection and probably avoidable deaths. (2) **Mismatch repair**. (3) In many bacteria, multidrug resistance protein, an efflux pump.

MMTV Mouse **mammary tumour virus**.

MN blood group antigens A pair of blood group antigens governed by genes that segregate independently of the ABO locus. The alleles are co-dominant and there are three types MM, NN, and MN. **Glycophorin** has M or N activity and this is associated with oligosaccharides attached to the amino terminal portion of the molecule. M-type glycophorin differs from N-type in amino acid residues 1 and 5, although the antigenic determinants are associated with the carbohydrate side chains. The MN phenotype is a genetic marker for essential arterial hypertension. Abstract: http://eurheartj.oxfordjournals.org/content/16/9/1269.short

mnemiopsin Calcium-activated photoprotein from the coelenterate *Mnemiopsis*. See **aequorin**.

MNNG A potent mutagen and carcinogen (N-methyl-N′-nitro-N-nitrosoguanidine), an alkylating agent used experimentally to induce mutations.

Mnt Protein (Max binding protein, myc antagonist, ROX, 582aa) that interacts with **Max** and functions as a transcriptional repressor. Not a member of the **Myc** or **Mad** families. Binds to same site on Max as does **Sin3**.

Mo-1 See **LFA-1**.

mobile genetic elements See **transposons**.

mobile ion carrier See **ionophore**.

modification enzyme Bacterial enzyme that introduces minor bases into DNA or RNA or, more commonly is part of a restriction/modification system that alters, by methylation, bases already incorporated. Serves to alter the sequence so that **restriction enzymes** no longer bind to their recognition site in the host DNA.

moesin Membrane-organising extension spike protein (577aa), a member of the **FERM domain** (protein 4.1, **ezrin**, **radixin**, moesin) protein family, all members of which have been shown to serve as cytoskeletal adaptor molecules. Found in all tissues. Phosphorylation on Thr-558 is essential for the formation of microvilli-like structures.

Mog Myelin oligodendrocyte glycoprotein (247aa), one of the Ig superfamily of membrane proteins, a minor component of the myelin sheath that mediates homophilic cell-cell adhesion.

Mohr-Tranebjaerg syndrome A progressive form of deafness (dystonia-deafness syndrome) caused by mutation in the TIMM8A (DDP) gene, the product of which is a mitochondrial intermembrane chaperone (mitochondrial import inner membrane translocase subunit Tim8 A, 97aa) that participates in the import and insertion of some multi-pass transmembrane proteins into the mitochondrial inner membrane (see **general import pore**). Mutation in TIMM8A also causes Jensen's syndrome (opticoacoustic nerve atrophy with dementia), a generalized degenerative disease of the CNS.

molecular clock (1) The rate of fixation of mutations in DNA and thus a measure of the rate of genetic diversification. (2) A biological system capable of maintaining a timing rhythm or pulse. All such clocks are thought to be entrained by a natural oscillator such as the **diurnal** rhythm.

Molisch reaction A colorimetric test for carbohydrates using alpha-naphthol.

Mollicutes Sub-group of the **Firmicutes**, mycoplasmas. Although they lack cell walls and so do not respond to Gram staining, they are considered Gram-positive and lack the second membrane found in Gram-negative bacteria.

molluscan catch muscle Muscle responsible for holding closed the two halves of the shell of bivalves. Specialized to maintain tension with low expenditure of ATP. Rich in **paramyosin**.

molluscum bodies Intracellular inclusions of poxviruses found in cells of human epidermis; harmless, but contagious. Associated with skin lesions (**molluscum contagiosum**).

Molluscum contagiosum virus Poxvirus that causes a benign viral disease of the skin. The virus is large, 240−320 nm diameter. See **molluscum bodies, Poxviridae**.

Moloney murine leukaemia virus Replication-competent retrovirus that causes leukaemia in mice, isolated by Moloney from cell-free extracts of a transplantable mouse sarcoma.

Moloney murine sarcoma virus Replication-defective retrovirus, source of the oncogene v-*mos*, responsible for inducing fibrosarcomas *in vivo*, and transforming cells in culture.

MOLT-4 cells A stable T-cell leukaemia line that will grow in suspension culture. Derived from a human male with acute lymphoblastic leukaemia.

molybdenum cofactor A coordination complex formed between **molybdopterin** and an oxide of molybdenum, an essential cofactor for sulphite oxidase, xanthine dehydrogenase and aldehyde oxidase. Molybdenum cofactor deficiency can arise through mutation in either of the two enzymes involved in its synthesis (molybdenum cofactor biosynthesis protein 1, 636aa; MOCS2, 188aa) or in the **gephyrin** gene. Deficiency causes severe neurological damage and **xanthinuria**.

molybdopterin A pyranopterin (394 Da) that binds molybdenum or tungsten and is a cofactor for various enzymes. See **molybdenum cofactor**.

MOM (1) The mitochondrial outer membrane; MOM is the prefix for many mitochondrial outer membrane proteins, in conjunction with the molecular mass in kiloDaltons e.g. MOM19 and MOM72, proteins of 19 and 72 kDa respectively. (2) In *C. elegans*, mom2 (362aa) is the ligand for members of the **frizzled** family of receptors and mom4 (536aa) is MAP kinase kinase kinase, a component of the Wnt signalling pathway.

Mona (1) In mouse, an adapter protein (GRB-2-related monocytic adapter protein, Grap2, Gads, 322aa) that links **Slp-76** and **LAT** upon T-cell receptor activation and, in platelets, collagen receptor activation. Platelet activation by thrombin results in rapid induction of Mona expression. (2) Multicentric osteolysis, nodulosis and arthropathy (MONA) a bone disorder caused by mutation in matrix metallopeptidase 2 (neutrophil gelatinase).

monastrol Membrane-permeable drug that reversibly inhibits activity of mitotic **kinesin** (Eg5). Treatment of cells with monastrol results in the formation of monopolar spindles.

monellin Basic non-glycosylated heterodimeric protein (44aa and 50aa) from the serendipity berry (*Dioscoreophyllum cumminsii*) that has intensely sweet taste (1000-fold sweeter than sugar. See **thaumatin**.

monensin A sodium **ionophore** (671 Da) from *Streptomyces cinnamonensis*. Has antibiotic properties, and is used as a feed additive in chickens. Also used in **ion-selective electrodes**.

Monera An obsolete taxonomic group, the prokaryotes, now subdivided into Archaea and Bacteria.

monoacyl-glycerol lipase See **monoglyceride lipase**.

monoamine neurotransmitters See **biogenic amines**.

monoamine oxidase inhibitors Class of drugs formerly used to treat severe depression. Now rarely used except in patients unresponsive to **tricyclic antidepressants** or **SSRIs** because of their dangerous interactions with foods containing tyramine and with **sympathomimetic** drugs.

monoamine oxidases Enzymes (EC 1.4.3.4, 527aa, 520aa) that catalyze breakdown of several **biogenic amines**, such as serotonin, adrenaline, noradrenaline, dopamine. See **monoamine oxidase inhibitors**.

monocentric chromosome Chromosome with a single **centromere**, as is the case for most chromosomes.

monocistronic RNA A messenger RNA that gives a single polypeptide chain when **translated**. Virtually all eukaryotic mRNAs are monocistronic, but some bacterial mRNAs are polycistronic especially those transcribed from **operons**.

monoclonal Used of a cell line whether within the body or in culture to indicate that it has a single clonal origin. Monoclonal antibodies (Mabs) are produced by a single clone of **hybridoma** cells, and are therefore a single species of antibody molecule.

monocotyledonous Describing plants in which the developing plant has only one **cotyledon**. Grasses are perhaps the commonest examples of the Class (which also contains palms, lilies and orchids). *Cf.* **dicotyledonous**.

monocyte Mononuclear phagocyte circulating in blood that will later emigrate into tissue and differentiate into a **macrophage**.

monocyte chemoattractant protein-1 See **monocyte chemotactic and activating factor**.

monocyte chemotactic and activating factor Cytokine (MCAF, MCP-1, CCL2, 76aa), co-induced with **interleukin-8** on stimulation of endothelial cells, fibroblasts or monocytes by various inflammatory mediators: activates and is chemotactic for **monocytes**. A **chemokine**.

monoecious Describing a plant that has both male and female reproductive units; these can occur simultaneously or sequentially (protoandrous plants function first as males and then change to females;

monoglyceride lipase The enzyme (monoacylglycerol lipase, EC 3.1.1.23, 303aa) that converts monoacylglycerides to free fatty acids and glycerol. It hydrolyses the endocannabinoid 2-arachidonoyl glycerol, an agonist of the cannabinoid (CB1) receptor, to arachidonic acid and glycerol. Levels are reported to be elevated in aggressive tumour cell lines.

protogynous plants do things in the reverse order). *Cf.* **dioecious, synoecious**.

monokines An obsolete term for **cytokines** derived from monocytic cells (macrophages).

monolayer (1) A single layer of any molecule, but most commonly applied to polar lipids. Can be formed at an air/water interface in experimental systems. The term should not be used to describe one layer of a lipid bilayer, for which the term 'leaflet' is generally used. (2) A single layer of cells in culture. Most 'normal' cells will not grow on top of one another and proliferation ceases when the culture substratum is covered by a monolayer. This non-random distribution is generated by contact inhibition of locomotion, a phenomenon in which colliding cells change direction rather than move over one another. Of the theories why some (but by no means all) types of cells stop growing when a monolayer is formed, present evidence favours limitation by supply of growth factors from the medium, rather than any inhibitory effect of contact on growth; the monolayer may become confluent (all cells in contact) long before proliferation ceases.

Mononegavirales An Order of viruses (see **virus taxonomy**). They have a non-segmented, negative sense RNA genome. The order includes four families: Bornaviridae (Borna disease virus), Rhabdoviridae (e.g. Rabies virus), Filoviridae (e.g. Marburg virus, Ebola virus) and Paramyxoviridae (e.g. Newcastle disease virus, measles virus)

mononuclear cells Cells of the blood other than erythrocytes and **polymorphonulear leucocytes** (neutrophils, basophils and eosinophils). Mononuclear phagocytes are monocytes and their differentiated products, macrophages.

monopodial Adjective describing an **amoeba** that has only one **pseudopod** (as opposed to polypodial forms).

monosaccharide A simple sugar that cannot be hydrolyzed to smaller units. Empirical formula is $(CH_2O)_n$ and range from trioses ($n = 3$) to heptoses ($n = 7$).

monosome (1) A single ribosome attached to a strand of mRNA. (2) A ribosome that has dissociated from a polysome. (3) Chromosome in an aneuploid set that does not have a homologue (see **monosomy**).

monosomy Situation in a normally **diploid** cell or organism in which one or more of the **homologous chromosome** pairs is represented by only one chromosome of the pair. For example, sex determination in grasshoppers depends on the fact that females are XX and males XO; that is, males have only one sex chromosome and are monosomic for the X chromosome.

MOPC Cell lines (e.g. MOPC-21, MOPC-315) derived from a murine myeloma (plasmacytoma) originally induced by intraperitoneal injections of mineral oil in highly inbred BALB/CJ mice.

MOPS A buffer (morpholino-propane sulphonic acid), a synthetic zwitterionic compound with a pKa of 7.2, that is non-toxic and has a low temperature coefficient. Widely used in biochemical studies, largely as a replacement for phosphate buffers.

Moraxella *Moraxella catarrhalis* is a Gram-negative, aerobic, oxidase-positive diplococcus that was described for the first time in 1896. The organism has also been known as *Micrococcus catarrhalis*, *Neisseria catarrhalis*, and *Branhamella catarrhalis*. It is commensal in the upper respiratory tract and is a common cause of otitis media and sinusitis and an occasional cause of laryngitis and pneumonia. Article: http://cmr.asm.org/content/15/1/125.full

Morbilli virus Genus of viruses (of the **Paramyxoviridae**). Type species is **measles virus**; other species include canine distemper virus (CDV) and the related seal virus (phocine distemper virus, PDV).

morphallaxis Regenerative process in which part of an organism is transformed directly into a new organism without replication at the cut surface.

morphiceptin A tetrapeptide amide fragment (YPFP-NH$_2$) of a milk protein, α-casein, that is a mu-selective **opioid receptor** ligand. May occur naturally.

morphine An **opioid** alkaloid, isolated from opium, with a complex ring structure. It is a powerful analgesic with important medical uses, but is highly addictive. Functions by occupying the receptor sites for the natural neurotransmitter peptides, **endorphins** and **enkephalins**, but is stable to the peptidases that inactivate these compounds.

morphogen Diffusible substance that carries information relating, for example, to position in the embryo, and thus determines the differentiation that cells perceiving this information will undergo.

morphogenesis 'Shape formation': the processes that are responsible for producing the complex shapes of adults from the simple ball of cells that derives from division of the fertilized egg.

morphogenetic movements Movements of cells or of groups of cells in the course of development. The invagination of cells in gastrulation is one of the most dramatic of morphogenetic movements; another much-studied example is the migration of neural crest cells.

morphometry Method that involves measurement of shape. A variety of methods exist to enable one to examine, for example, the distribution of objects in a 2-D section of a cell and then to use this to predict the shapes and the distribution of these objects in three dimensions.

Morquio-Brailsford disease A rare storage disease (**lysosomal disease**) characterized by dwarfism, kyphosis and skeletal defects in the hip joint. The defect is in degradation of keratan sulphate with deposits in tissues of glycosaminoglycans due to N-acetylgalactosamine-6-sulphatase deficiency. Keratin sulphate is excreted in large amounts in urine Occurs in two forms depending on which gene is mutated: in type A there is a deficiency of galactosamine-6-sulphate sulphatase; in Type B a deficiency of β-galactosidase.

mortalin Member of the **HSP70** family (75 kD glucose regulated protein, GRP75, MOT-1, MOT-2, 679aa) located in mitochondria, implicated in the control of cell proliferation and cellular aging. Interacts with **frataxin**.

morula Stage of development in holoblastic embryos. The morula stage is usually likened to a spherical raspberry, a cluster of blastomeres without a cavity.

mos An oncogene (Moloney murine sarcoma viral oncogene homologue), originally identified in mouse **sarcoma**, encoding a serine/threonine **protein kinase** (346aa in humans). Normal c-*mos* is expressed only in the germ cells of both testis and ovary. Overexpression of c-*mos* proto-oncogene product stimulates activity of **jun**.

mosaic egg A type of egg in which there appears to be a firmly committed fate map from the outset in distinction to 'regulating' embryos. The distinction is, however, only based upon the timing of differentiative events, and within a few divisions the regulating embryo also becomes a mosaic of determined cells. The classic example of a mosaic embryo is in molluscs where the polar lobe provides a clear morphological marker. Removal of a single blastomere in a mosaic embryo leads to the formation of an embryo deficient in particular tissues: in regulating embryos it is possible to remove early blastomeres and the remainder will compensate.

mosaicism A condition in which an individual is composed of a mixture of cells that are karyotypically or genotypically distinct.

motheaten A mouse strain in which the gene for **Shp1** is defective, leading to immunosuppression. The homozygous me/me mice have a very short life expectancy.

mothers against decapentaplegic In *Drosophila*, a protein (mothers against dpp, 455aa) from the **Smad protein** family. A mutation in the gene *MAD* in the mother represses the gene *decapentaplegic* in the embryo.

motif A small structural element that is recognisable in several proteins, e.g. **alpha-helix**. Useful guide: http://www.brc.dcs.gla.ac.uk/~drg/courses/bioinformatics_mscIT/slides/slides5/sld001.htm

motilin Peptide (22aa) derived from a 115aa precursor, found in duodenum, pituitary and pineal that stimulates intestinal motility. Apparently unrelated to other hormones. The receptor is G-protein coupled.

motogen Term proposed for substances that stimulate cell motility — by analogy with those that stimulate cell division (**mitogens**). **Scatter factor** (hepatocyte growth factor) is an example, though it seems likely that factors may be motogens for some cells and mitogens for others and may be motogens, mitogens or both depending upon the local conditions in which the cell is operating.

motoneuron A **neuron** (motor neuron) that connects functionally to a **muscle fibre**.

motor end plate A **neuromuscular junction**.

motor neuron See **motoneuron**.

motor neuron disease See **amyotrophic lateral sclerosis**.

motor protein Proteins that bind ATP and are able to move on a suitable substrate with concomitant ATP hydrolysis. Most eukaryotic motor proteins move by binding to a specific site on either actin filaments (myosin) or on microtubules (dynein, kinesin). They are normally elongated molecules with two active binding sites although some kinesin analogues have a single site. The distal end of the molecule normally binds adaptor proteins that enable them to make stable interactions with membranous vesicles or with filamentous structures, which then constitute the 'cargo' to be moved along the substrate filament.

Mott cells Plasma cells containing large eosinophilic inclusions (Mott bodies, Russell bodies) derived from rough endoplasmic reticulum and filled with IgM; found in lymphoid tissues in multiple myeloma and African **trypanosomiasis**.

mouse double minute See **Mdm**.

Mouse Genome Database An international database resource for the laboratory mouse, providing integrated genetic, genomic, and biological data. Link: http://www.informatics.jax.org/

movement protein A protein found in various plant viruses, involved in intra- and inter-cellular movement of virus. (1) In bean yellow dwarf virus (a ssDNA virus in the family Geminiviridae, genus *Mastrevirus*) it is a small membrane protein (V2, 92aa) in the host membrane that interacts with capsid protein. One of the mastrevirus movement protein family. (2) Movement protein BC1 (293aa) from Abutilon mosaic virus (Geminiviridae, *Begomovirus*) transports the DNA-nuclear shuttle protein complex to the host cell periphery and facilitates movement to the exterior. It is one of the begomovirus movement protein BC1 family. (3) Movement protein (MP, 303aa) from Odontoglossum ringspot virus (ssRNA positive-strand viruses, Virgaviridae, *Tobamovirus*) is involved in transport of the virus from the initially infected cells to adjacent cells, possibly by modifying the function of the plasmodesmata. Binds to RNA and single-stranded DNA and is one of the tobamoviruses movement protein family. (4) Movement protein (one of four fragments derived from RNA2 polyprotein, 1046aa) from Cowpea mosaic virus (a ssRNA positive-strand virus, Picornavirales, Secoviridae, Comovirinae, *Comovirus*) is assembled into tubules that allow the transport of virions from cell to cell through plasmodesmata. It is one of the the comoviridae genome polyprotein M family. (5) Various other 'movement proteins' are found in different viruses; they are not motor proteins.

Moviol A cheap and easily prepared semi-permanent mounting medium for immunofluorescence microscopy. It contains glycerol and Mowiol® a polyvinyl alcohol. Datasheet: http://www.fluoresbrite.com/SiteData/docs/TDS%20777/ba9fc00865bea9c61762776e7062aadd/TDS%20777.pdf

Mowat-Wilson syndrome A developmental disorder (Hirschsprung disease-mental retardation syndrome) caused by mutation in the *ZEB2* gene. There is mental retardation, delayed motor development, epilepsy, and a wide spectrum of clinically heterogeneous features. Not always associated with **Hirschprung's disease**.

MPDZ Multiple PDZ domain protein, a component (2042aa) of the NMDA-receptor signalling complex that may play a role in control of AMPA-receptor potentiation and synaptic plasticity in excitatory synapses.

MPF See **cyclin**.

M-phase Mitotic phase of cell cycle of eukaryotic cells, as distinct from the remainder, which is known as interphase (and that can be further subdivided as G1, S and G2). Beginning of M is signalled by separation of centrioles, where present, and by the condensation of chromatin into chromosomes. M-phase ends with the establishment of nuclear membranes

around the two daughter nuclei, normally followed immediately by cell division (**cytokinesis**).

M-phase promoting factor Protein whose levels rise rapidly just before, and fall away just after, **mitosis**. Now known to be the **Cyclin B2-Cdc2** complex.

MPM-2 A monoclonal antibody raised against HeLa cells which recognises various mitosis-specific phosphoproteins.

MPO See **myeloperoxidase**.

M-protein (1) Galactoside carrier in *E. coli*. (2) Cell surface antigen of *Brucella*. (3) Structural protein in the M-line of striated muscle (**myomesin**). (4) **Streptococcal M-protein**. (5) See **paraprotein**. (6) Matrix protein of various viruses (348aa in Sendai virus), important in virion assembly and budding. Forms a shell at the inner face of the plasma membrane and concentrates the HN and F glycoproteins.

MPSS Massively Parallel Signature Sequencing, an approach to analysing the level of expression of virtually all genes in a sample by counting the number of individual mRNA molecules produced from each gene without the genes being identified and characterized. The sensitivity is at a level of a few molecules of mRNA per cell. Tagged PCR products produced from cDNA are amplified, attached to microbeads, further amplified and a sequence signature of $\sim 16-20$ bp is identified from each bead; approximately 1,000,000 sequence signatures are obtained per experiment. The level of expression of any single gene is calculated by dividing the number of signatures from that gene by the total number of signatures for all mRNAs present in the dataset. Article: http://bfgp.oxfordjournals.org/cgi/content/abstract/1/1/95; method in more detail: http://www.ncbi.nlm.nih.gov/projects/genome/probe/doc/TechMPSS.shtml

MPTP Compound that causes dopaminergic neuronal degeneration, used to treat mice to produce a model for Parkinson's disease 'MPTP mouse model'. Inbred mouse strains differ remarkably in their susceptibility to MPTP, indicating a genetic element.

MPV (1) Mean platelet volume, a determinant of platelet function, and a risk factor for atherothrombosis. (2) Mouse parvovirus-1.

MRC-5 cells Cell line established from normal human male fetal lung tissue. Will double 50 to 60 times before showing senescence. Often used as 'normal' cells.

MreB In *E. coli*, rod shape-determining protein MreB (347aa) involved in cell shape. MreB forms spiral-shaped bands around the inside of the cell beneath the cytoplasmic membrane. Inactivation of the *mreB* gene causes the cell to become coccus-shaped and coccus-shaped bacteria lack this gene. May also be a negative regulator of peptidoglycan glycosyltransferase 3 (FtsI, EC 2.4.1.129, penicillin-binding protein 3, 588aa) that is involved in septum formation.

MRF-4 Member of the **MyoD** family of muscle regulatory proteins (muscle-specific regulatory factor 4, myf-6, 242aa).Involved in muscle differentiation and will induce fibroblasts to differentiate into myoblasts. See **herculin**.

MRI See f**MRI, nuclear magnetic resonance**.

M-ring See **S-ring** and **bacterial flagella**.

MRN complex A multi-protein complex (MRE11-RAD50-NBN complex) involved in recruiting **ATM** to double strand breaks in DNA and additionally is important for DNA recombination, maintenance of telomere integrity and meiosis. The complex has single-strand endonuclease activity and double-strand-specific $3'-5'$ exonuclease activity, provided by MRE11A (708aa). RAD50 (DNA repair protein RAD50, 1312aa) may bind DNA ends and hold them together for repair. There are two heterodimers of RAD50 and MRE11A associated with a single **nibrin** (NBN). The MRN complex is a sub-component of the **BASC complex**.

mRNA See **messenger RNA**.

MRP See **calgranulins**.

MRSA See **meticillin-resistant** *Staphylococcus aureus*.

MS2 (1) Type of F-specific coliphage (RNA bacteriophage). (2) Occasionally used as an abbreviation for MS-MS, mass-spectroscopy (MS) followed by a second MS analysis of the fragments from the first analysis. (3) Disintegrin and metalloproteinase domain-containing protein 8 (ADAM 8, CD156a, cell surface antigen MS2, 824aa).

MSC (1) Mesenchymal stem cells. (2) Bone marrow stromal cells, slightly confusing because there can be bone marrow mesenchymal stem cells.

MSCL (1) In *E. coli*, the large-conductance mechanosensitive channel (MscL, 136aa) that opens in response to stretch forces in the membrane lipid bilayer that can be caused by osmotic swelling. (2) Mesyl Chloride (MsCl), methanesulphonyl chloride.

MSDS (1) Musculoskeletal disorders (MSDs). (2) Material safety data sheet. (3) Mean square displacements, of atoms, particles amongst others. (4) Membrane-spanning domains. (5) In *S. cerevisiae*, cell cycle serine/threonine-protein kinase CDC5/MSD2 (705aa).

msh (1) *Msh* genes encode various proteins involved in mismatch repair. They are the human homologues of the *MutS* genes in *S. cerevisiae* which are themselves homologues of the *E. coli* MutHLS mismatch repair system that involves the *MutH, MutL, MutS,* and *MutU* genes. In humans there are five MSH proteins (MSH2, 934aa; MSH3, 1137aa; MSH4, 936aa; MSH5, 834aa; MSH6, **GTBP**, 1360aa) which form various heterodimers and are involved in different mismatch repair processes. They interact with the **MLH** and **PMS** proteins. Mutations in the *MSH2* gene result in **hereditary nonpolyposis colorectal cancer-1** and an *MSH3* frameshift mutation has been observed in an endometrial carcinoma. MSH4-MSH5 heterodimers bind uniquely to **Holliday junctions**. (2) The MSH homeobox family includes the *Drosophila* muscle segmentation homeobox protein (515aa) and in humans, **MSX1**. Other members of the family are known in a range of metazoan phyla. (3) See **melanocyte-stimulating hormone**.

MSI (1) Microsatellite instability (see **satellite DNA**). (2) Amphipathic antimicrobial peptides, MSI-78 and MSI-594 derived from **magainin**-2 and **melittin**, respectively. (3) In *Arabidopsis*, WD-40 repeat-containing proteins (MSI-1, medicis, 424aa; MSI-2, 415aa etc.) a histone-binding components of several complexes which regulate chromatin metabolism. (4) In *S. cerevisiae*, heat shock protein homologue SSE1 (chaperone protein MSI3, 693aa). (5) In humans, RNA-binding protein musashi homologues (MSI-1, 362aa; MSI-2, 328aa). Musashi homologues regulate the expression of target mRNAs at the translational level.

msps Protein (mini spindles, 2049aa) involved in the *Drosophila* **augmin** complex. Homologues are found in other organisms, XMAP215 (*Xenopus* microtubule associated protein, 2065aa) and human CKAP5 (cytoskeleton-associated protein 5, colonic and hepatic tumour over-expressed gene protein, Ch-TOG, 2032aa) as well as in various fungi and plants.

MSS4 (1) Mammalian suppressor of SEC4 (rab interacting factor, ras-specific guanine-releasing factor, RASGRF3, 123aa) that may have a role in vesicular transport. See TCTP (translationally controlled tumour protein). (2) In *S. cerevisiae*, MSS4 encodes a phosphatidylinositol-4-phosphate 5-kinase (779aa) that synthesizes phosphatidylinositol (4,5)-bisphosphate.

Msx (1) Homeobox genes generally expressed in areas of cell proliferation and in association with multipotent progenitor cells. For example, in humans MSX-1 (hox7, 297aa) is a transcriptional repressor with a role in limb-pattern formation, craniofacial development and odontogenesis.

MT See **microtubule**.

MT2 cells A line of lymphoid cells transformed by human T-cell leukemia virus (HTLV-I), derived from an adult T-cell leukaemia.

MTA (1) **Metastasis associated proteins**. (2) **Metallothionein A**.

MTOC See **microtubule organizing centre.**

mTOR The mammalian target of rapamycin. See **raptor, rictor, TOR** and **TORC.**

MTS (1) **Microtubules** (MTs). (2) **Metallothioneins.** (3) Mitochondrial targeting sequence. (4) See **MTS assay.**

MTS assay A cell viability assay in which MTS is chemically reduced by live cells into formazan; the intensity of the colour is a good indication of the number of live cells. Often used in place of the **MTT** assay.

MTT assay A assay for cell viability assay which depends upon a mitochondrial dehydrogenase acting upon MTT to produce dark-blue formazan from MTT (3-(4,5-dimethylthiazol-2-yl)- 2,5-diphenyltetrazolium bromide). Only live cells go blue. An alternative to the **MTS assay.**

MTTP See **microsomal triglyceride transfer protein.**

MUC-1 See **episialin, epitectin.**

mucilage Sticky mixture of carbohydrates in plants.

Muckle-Wells syndrome A periodic fever syndrome (urticaria-deafness-amyloidosis syndrome) caused by mutation in **cryopyrin.**

mucocyst Small membrane-bounded vesicular organelle in **pellicle** of ciliate protozoans that will discharge a mucus-like secretion.

mucolipidosis A **lysosomal disease** (storage disease) which differs from a **mucopolysaccharidosis** in that excessive sugars are not found in the urine. Type I (sialidosis) is caused by mutation in the gene for neuraminidase (EC 3.2.1.18), Type II (I-cell disease) by complete absence of the αβ subunits of **GlcNAc-phosphotransferase**, whereas Type III (classic pseudo-Hurler polydystrophy) is caused by severely reduced levels of the αβ subunits. The gamma subunits are normal in both forms but are mutated in mucolipidosis III gamma. Mucolipidosis IV (sialolipidosis) is caused by mutation in the gene for **mucolipin-1** and lysosomal hydrolases are normal. See also **Goldberg's syndrome.**

mucolipins Mucolipin 1 (580aa), defective in mucolipidosis IV, is a cation channel that probably has a role in the endocytic pathway and in the control of membrane trafficking of proteins and lipids. Mucolipin-2 (566aa) and mucolipin-3 (553aa) are other members of the **polycystin** sub-family of transient receptor potential Ca^{2+} channels.

mucopeptide Synonym for **peptidoglycan.**

mucopolysaccharide The polysaccharide components of proteoglycans, now more usually known as **glycosaminoglycans.**

mucopolysaccharidoses Lysosomal diseases (storage diseases) in which there is an inability to break down glycosaminoglycans. Mucopolysaccharidosis I is **Hurler syndrome,** mucopolysaccharidosis II is **Hunter syndrome,** mucopolysaccharidosis III is **Sanfilippo syndrome.** Type IVA (Morquio syndrome A) is due to a deficiency in galactosamine-6-sulphatase (EC 3.1.6.4) and the milder Morquio IVB due to beta-galactosidase deficiency: both are needed for the breakdown of keratan sulphate. Mucopolysaccharidosis VI is **Maroteaux-Lamy syndrome,** mucopolysaccharidosis VII is **Sly syndrome,** mucopolysaccharidosis IX is caused by mutation in the hyaluronidase I gene. See **mucolipidosis.**

mucosal associated lymphoid tissue MALT Lymphoid tissue and lymphoid aggregates associated with mucosal surfaces; includes bronchus associated lymphoid tissue (BALT) and gut associated lymphoid tissue (GALT).

mucous gland A type of **merocrine** gland that produces a thick (mucopolysaccharide-rich) secretion (as opposed to a **serous gland**).

mucous membrane mucosa An endodermally-derived epithelium that lines body cavities that are exposed to the external environment and internal organs. In many cases there are specialised cells that secrete a thick (mucous) material from the apical surface.

mucus Viscous solution secreted by various membranes; rich in glycoprotein.

Muir-Torre syndrome An autosomal, dominantly inherited disorder characterized by sebaceous neoplasms and visceral malignancies; the lesions are characterised by loss of hMSH2 expression (see **msh #1**).

Müller cell Supporting cell of the neural retina. Cell body and nucleus lie in the middle of the inner nuclear region, their bases form the internal and external limiting membranes.

multi-locus enzyme electrophoresis See **MLEE.**

multicopy inhibition Inhibition of translation of the transcript of a **transposase** gene by a multicopy **plasmid** with suitable inhibitory gene. The plasmid inhibits transposition events in the host bacterium.

multidrug transporter Mdr, P-glycoprotein Closely related family (**ABC proteins**) of integral membrane glycoproteins in both eukaryotes and prokaryotes that export a variety of small compounds, including drugs, from the cytoplasm.

multienzyme complex Cluster of distinct enzymes catalysing consecutive reactions of a metabolic pathway that remain physically associated through purification procedures. Multifunctional enzymes, found in eukaryotes, differ in that the several enzymic activities are associated with different domains of a single polypeptide.

multigene family See **gene family**.

multinet growth hypothesis The idea that cellulose microfibrils are initially wrapped around the plant cell in a flat helix whose alignment changes as the cell expands and newer layers of microfibrils are deposited upon the plasma membrane. The hypothesis predicts that layers of microfibrils are passively realigned from transverse through oblique to longitudinal as the cell grows but this does not fully explain various wall patterns. So named because each layer of cellulose microfibrils was (wrongly) thought to form an interwoven net.

multiple cloning site See **polycloning site**.

multiple endocrine neoplasia A disorder in which there is a high frequency of peptic ulcer disease and primary endocrine abnormalities involving the pituitary, parathyroid, and pancreas. Type I (MEN1) is caused by mutation in the gene encoding **menin** which is also defective in inherited forms of **Zollinger-Ellison syndrome**. MEN2A and MEN2B are caused by mutation in the *ret* protooncogene, and MEN4 is caused by mutation in the cyclin-dependent kinase inhibitor 1B (CDKN1B, p27/KIP1) gene. See also **giantism**.

multiple epiphyseal dysplasia A generalized skeletal dysplasia. In multiple epiphyseal dysplasia with myopia and conductive deafness (EDMMD) there is a mutation in the **collagen** COL2A1 gene.

multiple isomorphous replacement Method of solving the phase problem in X-ray crystallography of proteins. Protein crystals consists of an array of geometrically identical unit cells arranged in three dimensions, each unit cell containing one or more identical asymmetric units. By substitution of a heavy atom at a small number of sites in each molecule it may be possible to produce isomorphous crystals with identical geometry and molecular structure. Diffraction patterns from the unlabelled protein and two or more isomorphous derivatives can then be used to calculate the phases of the unlabelled crystal and, with the amplitude data, the molecular structure deduced.

multiple myeloma See **myeloma cell**.

multiple myeloma oncogene-1 Interferon regulatory factor 4. See **IRFs**.

multiple paragangliomata Tumours (glomus tumours) derived from paraganglia (chromaffin bodies) located throughout the body. Those with chromaffin type cells with endocrine activity and are usually referred to as **phaeochromocytomas**. Familial paragangliomas (PGL1-4) are caused by mutations in the genes for various subunits of succinate-ubiquinone oxidoreductase.

multiple sclerosis Neurodegenerative disease characterized by the gradual accumulation of focal plaques of demyelination particularly in the periventricular areas of the brain but not affecting peripheral nerves. Onset usually in middle age with intermittent progression. Susceptibility has been associated with mutations affecting tyrosine phosphatase (CD45), the IL7-receptor and CD24. An autoimmune component has been suggested and there is some association with particular HLA haplotypes.

multiple sulphatase deficiency A lysosomal disease caused by mutation in the **sulphatase-modifying factor-1** gene.

multipotent cell Progenitor or precursor cell that can give rise to diverse cell types, but not all, in response to appropriate environmental cues. *Cf.* **totipotent**.

multipotential colony-stimulating factor See **interleukin-3**.

multisynthase complex A multi-protein complex composed of a bifunctional glutamyl-prolyl-tRNA synthase, the monospecific isoleucyl, leucyl, glutaminyl, methionyl, lysyl, arginyl and aspartyl-tRNA synthases, and three auxiliary proteins, EEF1E1/p18 (eukaryotic translation elongation factor 1 epsilon-1, 174aa) and aminoacyl tRNA synthase complex-interacting multifunctional proteins 1 and 2 (AIMP1/p43, 312aa; AIMP2/p38, 320aa).

multivesicular body Secondary **lysosome** around 50 nm diameter containing intraluminal vesicles that are generated by invagination and scission from the limiting membrane of the endosome. MVBs are involved in degradation of membrane proteins, such as stimulated growth factor receptors, lysosomal enzymes and lipids and the MVB pathway appears to require the sequential function of **ESCRT** complexes ESCRT-O, -I,-II and -III.

munc13 Homologues of *C. elegans* UNC-13 protein (munc13-1, 1703aa; munc13-2, 1591aa; munc13-3, 2214aa; munc13-4, 1090aa) that are involved in vesicle maturation during exocytosis as a target of the diacylglycerol second messenger pathway; munc13-4 is particularly involved in lymphocyte granule exocytosis. Interact with **syntaxins**. The MHD1 (munc homology domain-1) and MHD2 domains mediate localization on recycling endosomes and lysosomes. Defects in munc13-4 (UNC13D) are the cause of haemophagocytic lymphohistiocytosis type 3.

MUP (1) 4-methylumbelliferylphosphate, a fluorogenic substrate used in phosphatase assays. (2) Motor unit potential, action potentials occurring in muscle units. (3) **Major urinary protein**. (4) See **mup genes**. (5) In *S. cerevisiae*, permeases for methionine (high affinity permease MUP1, 574aa). MUP3 (546aa) is a low affinity permease. (6) In human, mitochondrial thiamine pyrophosphate carrier (mitochondrial uncoupling protein 1, MUP1, SLC25A19, 320aa). **(7)** In *C. elegans*, the gene (*mup-2*) for troponin T. **(8)** *Mup* genes in bacteria including *S. aureus*, are a cluster of genes encoding type I polyketide synthases and monofunctional enzymes that are involved in conversion of the product of the polyketide synthase into the active antibiotic, **mupirocin**. *MupM* and *MupR* encode isoleucyl-tRNA synthetases (EC 6.1.1.5).

mupirocin A polyketide-derived antibiotic from *Pseudomonas fluorescens* NCIMB10586, a mixture of pseudomonic acids (PA) that target isoleucyl-tRNA synthase and as a result inhibit cell wall synthesis.

muramic acid Subunit of **peptidoglycan** (**murein**) of bacterial cell walls.

muramidase See **lysozyme**.

muramyl dipeptide Fragment of **peptidoglycan** from cell wall of mycobacteria that is used as an **adjuvant**.

murein See **peptidoglycan**.

murine leukaemia virus A group of Type C **Retroviridae** infecting mice and causing in some strains lymphatic **leukaemia** after a long latent period. Nearly all are replication competent and v-*onc* negative. See also **Abelson leukaemia virus**, **Moloney murine leukaemia virus**, **Friend murine leukaemia virus**.

muscarine Toxin (alkaloid) from the mushroom *Amanita muscaria* (Fly Agaric) that binds to **muscarinic acetylcholine receptors**.

muscarinic acetylcholine receptor Distinct from the **nicotinic acetylcholine receptor** in having no intrinsic ion channel; seven membrane-spanning G-protein-coupled receptors. In humans five subtypes of muscarinic receptors have been identified (M1, 460aa; M2, 466aa; M3, 490aa; M4, 479aa; M5, 532aa). Primary transducing effect is adenylate cyclase inhibition or modification of phosphatidyl inositol turnover.

muscimol A psychoactive alkaloid in many mushrooms of the genus *Amanita*. A potent, selective agonist of the $GABA_A$ receptor.

muscle Tissue specialized for contraction. See also **muscle cell**, **twitch muscle**, **smooth muscle**, **catch muscle**, **cardiac muscle**.

MUSCLE A piece of software that can be used to align sequences (MUltiple Sequence Comparison by Log-Expectation, MUSCLE). Link: http://www.ebi.ac.uk/Tools/msa/muscle/

muscle cell Cell of muscle tissue; in striated (skeletal) muscle it is a **syncytium** formed by the fusion of embryonic **myoblasts**, in cardiac muscle a cell

linked to the others by specialized junctional complexes (**intercalated discs**), in smooth muscle a single cell with large amounts of actin and myosin capable of contracting to a small fraction of its resting length.

muscle fibre Component of a skeletal muscle comprising a single syncytial cell that contains **myofibrils**.

muscle spindle A specialized muscle fibre found in tetrapod vertebrates. A bundle of muscle fibres that is innervated by sensory neurons. Stretching the muscle causes the neurons to fire; the muscle spindle thus functions as a stretch receptor.

muscle-eye-brain disease See **muscular dystrophy-dystroglycanopathy with brain and eye anomalies**.

muscular dystrophy A group of diseases in which there is progressive degeneration and/or loss of muscle fibres, usually without nervous system involvement. **Duchenne muscular dystrophy** is the commonest form. See also **Becker muscular dystrophy, Emery-Dreifuss MD, Fukuyama MD, myotonic dystrophy**. Limb girdle MD generally causes weakness in the shoulder and pelvic girdles and can be dominant or recessive and caused by mutations in genes for **myotilin, lamin A/C, caveolin-3, calpain-3, dysferlin, sarcoglycans, telethonin**, tripartite motif-containing protein-32, **fukutin**-related protein, **titin** and protein O-mannosyltransferases-1 (see **muscular dystrophy-dystroglycanopathy with brain and eye anomalies**). Other forms of muscular dystrophy can arise through mutations in **laminin** alpha-2 (merosin-deficient MD type 1A), **selenoprotein N**, collagen type VI (Ullrich congenital MD, **Bethlem myopathy**). Myofibrillar myopathy (MFM) refers to a group of morphologically homogeneous, but genetically heterogeneous chronic neuromuscular disorders caused by mutations in the genes for alpha-B-**crystallin**, LIM domain-binding protein 3, **filamin C, myotilin** or **desmin**. Patient-group website: http://www.muscular-dystrophy.org/

muscular dystrophy-dystroglycanopathy with brain and eye anomalies A set of disorders involving severe early-onset muscle weakness, mental retardation and pathological change in the eye (muscle-eye-brain disease is one variant), caused by mutations affecting glycosylation of **dystroglycan**. The enzymes involved include protein O-mannosyltransferase-1 (POMT1, EC 2.4.1.109, 747aa) that transfers mannosyl residues to the hydroxyl group of serine or threonine residues. Both POMT1 and POMT2 (750aa) are necessary for enzyme activity. Defects in either can cause **Walker-Warburg syndrome**, mutation in POMT1 causes limb-girdle **muscular dystrophy** type 2K. Other forms involve mutations in **fukutin**, fukutin associated protein and an acetylglucosaminyltransferase encoded by the **LARGE** gene. The hypoglycosylation of **dystroglycan** interferes with its interactions with other proteins such as laminin, neurexin, and agrin (see **dystrophin-associated protein complex**).

musculin See abf1.

mushroom bodies A pair of conspicuous structures (corpora pedunculata) in the insect brain thought to be crucial for olfactory associated learning. They have a roughly hemispherical calyx, a protuberance joined to the rest of the brain by a central nerve tract or peduncle. See **microcephalin**. Article: http://learnmem.cshlp.org/content/5/1/11.long

MuSK Muscle-specific receptor tyrosine kinase (869aa) involved in agrin signalling and regulation of the formation of the neuromuscular junction. Mice lacking MuSK fail to form neuromuscular junctions. Mutations in humans can lead to a form of myasthenic syndrome.

mut (1) See **mismatch repair**. (2) See **methylmalonyl-CoA mutase**.

mutagens Agents that cause an increase in the rate of mutation; includes X-rays, ultraviolet irradiation (260 nm), and various chemicals. The **Ames test** is often used to test for mutagenicity of compounds.

mutarotation Change in optical rotation with time as an optical isomer in solution converts into other optical isomers.

mutation A change in the DNA sequence of an organism, which may arise in any of a variety of different ways. See **conditional mutation, deletion m., dominant, frame-shift m., gain of function m., leaky m., lethal m., loss of function m., missense m., neutral m., nonsense m., point m., silent m., somatic m., suppressor m., temperature-sensitive m.**, also **mutation rate**.

mutation rate The frequency with which a particular mutation appears in a population or the frequency with which any mutation appears in the whole genome of a population. Normally the context makes the precise use clear. See **fluctuation analysis**.

muted The human homologue (187aa) of a mouse protein, mutated in **Hermansky-Pudlak syndrome** (see **BLOC-1**). Fibroblasts derived from the muted (muted brown) mouse strain exhibit reduced levels of **pallidin**, suggesting that the absence of the muted protein destabilizes pallidin.

mutein Protein with altered amino acid sequence – usually enough to alter properties. Uncommon usage.

MVB See **multivesicular body**.

MVP A component (major vault protein, lung resistance-related protein, 893aa) of the **vault** particle present in most normal tissues but more highly expressed in epithelial cells with secretory and excretory functions. Overexpressed in some multi-drug resistant tumours.

Mx proteins A family of GTPases (70–100 kDa) found in interferon-treated cells. Mx1 (MxA; p78; interferon induced 78 kDa protein, 662aa) is found in the nucleus and the mouse homologue confers resistance to influenza A virus by blocking transcription of the viral RNA genome. Other Mx proteins are cytoplasmic and are related to dynamin; Mx proteins are involved in the innate antiviral response of fish.

myalgic encephalomyelitis *ME* A long-term post-viral syndrome with chronic fatigue and muscle pain on exercise (chronic fatigue syndrome). Although dismissed for many years as a psychosomatic construct it has become recognised as a real phenomenon, although diagnostic markers are poor.

myasthenia gravis A disease characterized by progressive easy fatigue of certain voluntary muscle groups on repeated use. Muscles of the face or upper trunk are especially likely to be affected. In most, and perhaps all cases, due to the development of autoantibodies against the **acetylcholine receptor** in neuromuscular junctions. Immunization of mice or rats with this receptor protein leads to a disease with the features of myasthenia. Congenital myasthenic syndrome with episodic apnea (formerly familial infantile myasthenia gravis) is caused by mutation in the **choline acetyltransferase** gene.

myasthenic syndrome Muscular weakness that can be a consequence of defects in components of the neuromuscular junction (congenital myasthenic syndrome), autoimmune disorders (**myasthenia gravis**), or mutations in calcium channels at the neuromuscular junction (Lambert-Eaton myasthenic syndrome, LEMS). See **rapsyn**.

myb A transcriptional activator (640aa), product of the c-*myb* oncogene, that is important in the control of proliferation and differentiation of haematopoietic progenitor cells. Originally identified in avian myeloblastosis. Myb-like protein 2 (700aa) is part of the LIN complex. Related proteins are found in *Arabidopsis*.

myc A **proto-oncogene**, identified in several avian tumours. The myc proteins are transcription factors with a C-terminal basic helix-loop-helix-zipper domain. Myc-**Max** heterodimers specifically bind the sequence CACGTG with higher affinity than homodimers of either and activate transcription. Myc in association with Miz1 (Myc-interacting zinc finger protein-1) can repress transcription. N-myc is found in neuroblastomas. See **Burkitt's lymphoma**.

myc tag Epitope tag frequently expressed as a translational fusion with a transgenic protein of interest. As there are good antibodies to the myc epitope, this allows localization of the fusion gene product by **immunocytochemistry** or **Western blot**, or its immunoaffinity purification.

myceliogenic germination The formation of an infective mycelium from the resting **sclerotia** of fungal pathogens such as *Sclerotinia sclerotiorum*.

mycelium Mass of **hyphae** that constitutes the vegetative part of a fungus (the conspicuous part in most cases is the fruiting body). Similar, though smaller, structures are found with some saprophytic bacteria such as *Nocardia*.

mycetocytes Cells containing symbiotic microorganisms, found in the **mycetome** of insect, ticks and mites.

mycetome A specialised organ in some species of insects, ticks and mites, that is the site for intracellular symbionts such as bacteria, fungi and rickettsiae.

mycobacteria Bacteria of the Order Actinomycetales that have unusual cell walls that are resistant to digestion, being waxy, very hydrophobic, and rich in lipid, especially esterified **mycolic acids.** Staining properties differ from those of Gram-negative and Gram-positive organisms, being acid-fast. Many are intracellular parasites, causing serious diseases such as **Buruli ulcer, leprosy** (*Mycobacterium leprae*) and **tuberculosis** (*M. tuberculosis*; *M. bovis* causes tuberculosis in cattle and humans; the attenuated strain is **Bacille Calmette-Guerin** (BCG), used for immunization). Mycobacterial cell walls have strong immunostimulating (**adjuvant**) properties due to muramyl dipeptide (MDP).

mycolactone A family of macrolide toxins produced by *Mycobacterium ulcerans* that have immunosuppressive effects and inhibit production of macrophage inflammatory protein (MIP) 1α, MIP-1β, RANTES, interferon-γ-inducible protein 10, and monocyte chemoattractant protein 1, but not IL-12, TNFα or IL-6. See **Buruli ulcer.**

mycolic acids Saturated fatty acids found in the cell walls of **mycobacteria**, *Nocardia* and **corynebacteria**. Chain lengths can be as high as 80, and the mycolic acids are found in waxes and in glycolipids.

mycophenolic acid Antibacterial and antitumour compound from *Penicillium brevicompactum*. Inhibits *de novo* nucleotide synthesis.

mycoplasma Prokaryotic microorganisms (Mollicutes) lacking cell walls, and therefore resistant to many antibiotics. Parasitic for animals and plants. Formerly known as pleuro-pneumonia-like organisms (PPLO). *Mycoplasma pneumoniae* is a causative agent of pneumonia in humans and some domestic animals. They are troublesome contaminants of animal cell cultures, in which they may grow attached to or close to cell surfaces, subtly altering properties of the cells, but escaping detection unless specifically monitored. Similar organisms, spiroplasms, cause various diseases in plants.

mycorrhiza Fungi associated with roots of higher plants: relationship is mutually beneficial and in some cases essential to survival of the higher plant. See **vesicular-arbuscular mycorrhiza, endomycorrhiza**.

mycosides Complex glycolipids found in cell walls of **mycobacteria**. Non-toxic, non-immunogenic molecules that influence the form of the colony and the susceptibility of the bacteria to bacteriophages.

mycosis Any disease caused by a fungus. But see **mycosis fungoides**.

mycosis fungoides A human disease in which a frequent secondary feature is fungal infection of lesions in the skin. Recognized as a tumour of T-cells that accumulate in the dermis and epidermis and cause loss of the epidermis.

myd (1) Common abbreviation for **myotonic dystrophy.** (2) A gene (*myd*) that is involved in the determination of muscle cells (committment to the myogenic lineage), not the same as *MyoD*. (3) In the Large(myd) mouse, **dystroglycan** is incompletely glycosylated and thus cannot bind its extracellular ligands, causing a muscular dystrophy that is usually lethal in early adulthood. The Large(myd) mutation alters the composition and organization of the sarcolemma of fast-twitch skeletal muscle fibres. (4) Myeloid differentiation factor (MyD-88) is a key adaptor protein that plays a major role in the innate immune pathway.

myelin The main constituent of the **myelin sheath** of nerve axons.

myelin basic protein Major component (304aa) of the myelin sheath in mammalian CNS. An alternatively spliced isoform (18 kDa) is used as an antigen to induce **experimental allergic encephalomyelitis**, possibly a model for some neurodegenerative disorders. The 'shiverer' (shi) mouse has a mutation in the mouse MBP gene. See also **myelin proteolipid protein**.

myelin figures Structures that form spontaneously when bilayer-forming phospholipids (e.g. egg lecithin) are added to water. They are reminiscent of the concentric layer structure of myelin.

myelin proteolipid protein See **proteolipid protein** #1.

myelin sheath An insulating layer surrounding vertebrate peripheral **neurons**, that dramatically increases the speed of conduction. It is formed by specialized **Schwann cells**, that can wrap around neurons up to 50 times. The exposed areas are called **nodes of Ranvier**: they contain very high densities of **sodium channels**, and **action potentials** jump from one node to the next, without involving the intermediate axon, a process known as saltatory conduction.

myeloblastin See p29 #10.

myeloblasts Cells of the bone marrow that divide to produce **myelocytes** and the myeloid series of cells.

myelocytes The bone marrow cells that give rise to **myeloid cells**. They are derived from **myeloblasts**.

myelodysplasia A group of disorders in which the bone marrow malfunctions and fails to produce normal numbers of blood cells. Various categories are recognised: refractory anaemia with or without ring sideroblasts (RA or RARS); refractory cytopenia with multilineage dysplasia (RCMD); 5q$^-$ syndrome; refractory anaemia with excess blasts (RAEB); unclassified (none of the previous types). Myelodysplasia and leukaemia with monosomy 7 is due to loss of one copy of chromosome 7, the key missing gene possibly being EZH2 (enhancer of zeste homologue 2, 746aa: see **polycomb repressive complex-2**). Alpha-thalassemia myelodysplasia syndrome is caused by mutation in the chromatin-remodeling factor ATRX (X-linked helicase 2, 2492aa). Thiamine-responsive myelodysplasia is caused by a defect in a thiamine transporter (SLC19A2, 497aa).

myelodysplastic syndrome See **myelodysplasia**.

myeloid cells One of the two classes of marrow-derived blood cells; includes **megakaryocytes**, erythrocyte-precursors, **mononuclear phagocytes**, and all the **polymorphonuclear leucocytes**. That all these are ultimately derived from one stem cell lineage is shown by the occurrence of the **Philadelphia chromosome** in these, but not **lymphoid** cells. Most authors tend, however, to restrict the term 'myeloid' to mononuclear phagocytes and granulocytes and commonly distinguish a separate erythroid lineage.

myeloid nuclear differentiation antigen A protein (MNDA, 407aa) that may act as a transcriptional activator/repressor in the myeloid lineage and probably has a role in the response to interferon. Reduced levels are found in some myelodysplasias. Has a DAPIN (pyrin) domain.

myeloid-derived suppressor cells MDSC A heterogeneous population of cells that expand during cancer, inflammation and infection, and suppress T-cell responses. They infiltrate tumours, induce tumour-specific T-cell tolerance and facilitate tumour growth and metastasis. Article: http://www.ncbi.nlm.nih.gov/pmc/articles/PMC2828349/?tool = pubmed

myeloma cell Neoplastic **plasma cell**. The proliferating plasma cells often replace all the others within the marrow, leading to immune deficiency, and frequently there is destruction of the bone cortex. Because they are monoclonal in origin they secrete a monoclonal immunoglobulin (a Bence-

Jones protein) which is excreted in the urine. Myeloma cell lines are used for producing **hybridomas** in raising monoclonal antibodies.

myeloperoxidase A metallo-enzyme containing iron (EC 1.11.1.7, 745aa: cleaved into 89 kDa myeloperoxidase, 84 kDa myeloperoxidase, myeloperoxidase light chain and myeloperoxidase heavy chain), found in the lysosomal (azurophil) granules of myeloid cells, particularly macrophages and neutrophils; responsible for generating potent bacteriocidal activity by the hydrolysis of hydrogen peroxide (produced in the **metabolic burst**) in the presence of halide ions. Deficiency of myeloperoxidase is not fatal, and the enzyme is reportedly entirely absent in chickens.

myeloplax *megakaryocyte* A giant cell of bone marrow and other haematopoietic organs that gives rise to the blood-platelets. May be multinucleated.

myf *myogenic factor* A group of transcription factors involved in differentiation of muscle. Myf3 is myoblast determination protein 1 (MyoD1, 320aa), myf5 (255aa) is structurally related to **MyoD1**. Muscle-specific regulatory factor 4 (myf6, 242aa) when defective causes a mild centronuclear myopathy. Murine myf6 is herculin (242aa).

myo-inositol 'Muscle sugar' – an obsolete name for inositol.

myoblast Cell that by fusion with other myoblasts gives rise to **myotubes** that eventually develop into skeletal muscle fibres. The term is sometimes used for all the cells recognisable as immediate precursors of skeletal muscle fibres. Alternatively, the term is reserved for those post-mitotic cells capable of fusion, others being referred to as presumptive myoblasts.

myobrevin See **synaptobrevin**.

myocarditis Inflammation of heart muscle, usually due to bacterial or viral infection.

myocardium Middle and thickest layer of the wall of the heart, composed of cardiac muscle.

myocilin A cytoskeletal protein (trabecular meshwork glucocorticoid-inducible response protein, TIGR, 504aa) expressed in many ocular tissues, including the **trabecular meshwork**, but also in various other tissues. Mutations in the *MYOC* gene encoding myocilin are responsible for primary open-angle glaucoma (POAG). Interacts with **flotillin-1**.

myoclonic Descriptive of a type of congenital tremor involving jerky spasms of muscles. See **myoclonic epilepsy of Lafora**, **myoclonic epilepsy of Unverricht and Lundborg**.

myoclonic epilepsy of Lafora A disorder that can be caused by mutation in genes for **laforin** or **malin**.

myoclonic epilepsy of Unverricht and Lundborg A convulsive disorder caused by mutation in the **stefin B** (cystatin B) gene. A similar disorder, progressive myoclonic epilepsy type 1B (EPM1B) is caused by mutation in the **prickle 1** gene that encodes REST-interacting LIM domain protein (831aa), probably a nuclear receptor.

myocyte-enhancer factor 2 (1) A family of transcriptional activators in the **MADS** superfamily that bind to the MEF2 locus of many muscle-specific genes. In humans, MEF-2A (serum response factor-like protein 1, 507aa) mediates cellular functions not only in skeletal and cardiac muscle development, but also in neuronal differentiation and survival. Mutations in MEF2A are responsible for autosomal dominant coronary artery disease with acute myocardial infarction. MEF2B (365aa) may be involved in muscle-specific and/or growth factor-related transcription. MEF2C (473aa) controls cardiac morphogenesis and myogenesis, is involved in vascular development and plays an essential role in hippocampal-dependent learning and memory by suppressing the number of excitatory synapses. Also involved in development of platelets and B-cells. Defects in MEF2C are the cause of mental retardation-stereotypic movements-epilepsy. MEF2D (521aa) is involved in development of skeletal muscle, cardiac muscle and neuronal tissue (where it regulates neuronal apoptosis). MEF2-activating motif and SAP domain-containing transcriptional regulator (415aa) is a transcriptional coactivator of MEF2C. (2) In *S. cerevisiae*, MEF2 is mitochondrial ribosome-releasing factor 2 (819aa), a GTPase that mediates the disassembly of ribosomes from mRNA.

myoD1 A transcriptional activator (myoblast determination protein 1, Class C basic helix-loop-helix protein 1, 320aa) involved in muscle differentiation (myogenic factor) and that will induce fibroblasts to differentiate into myoblasts. Activates muscle-specific promoters but efficient DNA binding requires dimerization with another bHLH protein. Interacts with and is inhibited by **twist**. myoD family inhibitor (myogenic repressor I-mf, 246aa) associates with myoD family members, and masks the nuclear localization signal so they do not move to the nucleus and myogenesis is repressed. myoD family inhibitor domain-containing protein (246aa) modulates the activity of both cellular and viral promoters.

myoepithelial cell Cell (basket cell, basal cell) found between epithelium of exocrine glands (e.g. salivary, sweat, mammary, mucous) and their basement membranes, which resembles a smooth muscle cell, and is thought to be contractile.

myoferlin Calcium/phospholipid-binding protein (2061aa) that plays a role in the plasmalemma repair mechanism of endothelial cells. Highly expressed in myoblasts undergoing fusion and is concentrated at the membrane sites of both myoblast-myoblast and myoblast-myotube fusions. See **ferlin** and **dysferlin**.

myofibril Long cylindrical organelle of striated muscle, composed of regular arrays of thick and thin filaments, and constituting the contractile apparatus.

myofibroblasts Histological term for fibroblast-like cells that contain substantial arrays of actin microfilaments, myosin and other muscle proteins arranged in such a way as to suggest that they produce contractile forces. Are commonly described as occurring in granulation tissue (formed during wound healing) and in certain forms of arterial thickening where they are found in the intima. Behave in much the same way as smooth muscle cells and have markers characteristic of these cells.

myogenesis The developmental sequence of events leading to the formation of adult muscle that occurs in the animal and in cultured cells. In vertebrate skeletal muscle the main events are: the fusion of **myoblasts** to form **myotubes** that increase in size by further fusion to them of myoblasts, the formation of **myofibrils** within their cytoplasm, and the establishment of functional **neuromuscular junctions** with **motoneurons**. At this stage they can be regarded as mature muscle fibres.

myogenin A basic HLH transcription factor of the **myoD** family (myogenic factor 4, 224aa).

myoglobin Protein (153aa) found in red skeletal and cardiac muscle. The first protein for which the **tertiary structure** was determined by **X-ray diffraction**, by J.C. Kendrew's group working on sperm whale myoglobin. A single polypeptide chain with a **haem** group bonded via its ferric iron to two histidine residues. Myoglobin has a higher affinity for oxygen than **haemoglobin** at all partial pressures. In capillaries oxygen is effectively removed from haemoglobin and diffuses into muscle fibres where it binds to myoglobin which acts as an oxygen store. Muscles that are active most of the time tend to have a lot of myoglobin.

myomesin A major protein of the **M-line** of the **sarcomere** that has fibronectin-type III-like and immunoglobulin C2-like motifs and that may link the intermediate filament cytoskeleton to the M-disk in striated muscle and integrins to the cytoskeleton in non-muscle cells. Myomesin 1 (190 kDa titin-associated protein, 190 kDa connectin-associated protein, 1685aa) binds myosin, titin, and light meromyosin (skelemin is the murine homologue). There are other isoforms (myomesin-2, 1465aa; myomesin-3, 1437aa).

myoneme Contractile organelle of ciliate protozoans; referred to as M-bands in *Stentor*, where they are composed of 8−10 nm tubular fibrils. The **spasmoneme** of peritrich ciliates was originally called a myoneme.

myopalladin A protein (1320aa) present in the Z-line of the sarcomere that links **nebulin** or **nebulette** to alpha-actinin in skeletal and cardiac muscle respectively. It colocalizes with **cardiac ankyrin repeat protein** in the I band.

myopathic carnitine deficiency See **systemic carnitine deficiency**.

myopathy A disorder in which muscle fibres do not function properly and there is muscle weakness. There are separate entries for: **alpha-B crystallinopathy, Bethlem myopathy, centronuclear myopathy, Compton-North congenital myopathy, Danon's disease, dilated cardiomyopathy, dermatomyositis, hypertrophic cardiomyopathy, inclusion body myopathy with Paget disease and frontotemporal dementia, ischaemic cardiomyopathy, MELAS syndrome, myosclerosis** and **nemaline myopathy**.

myopodin See **synaptopodin**.

myosclerosis A chronic inflammation of skeletal muscle (myosclerotic myopathy, congenital myosclerosis of Lowenthal) caused by mutations in **collagen** type VI (COL6A2).

myosin A family of motor ATPases that interact with F-actin filaments (thin filaments in muscle, microfilaments in the cytoplasm of non-muscle cells). An increasing number of different myosins are being described. The best known myosin, that of the thick filaments of striated muscle, is myosin II, a hexamer of two **myosin heavy chains** and two pairs of **myosin light chains** (see also **meromyosin**). In skeletal muscle there are seven heavy chain isoforms, two developmental, three in adult skeletal muscle, one also expressed in cardiac muscle, one expressed primarily in extrinsic eye muscles. Mutations in the gene for myosin heavy chain 9 lead to the **May-Heggelin anomaly**. Most other myosins do not self-assemble into filaments and although their N-terminal motor domain is conserved, their C-terminal tails are very variable and in some cases (e.g. myosin I) interact with 'cargo' vesicles which are transported along actin filaments.

myosin heavy chain The large subunit of striated muscle **myosin** (~2000aa) that has the motor ATPase activity. The head region of the molecule binds actin, changes shape, hydrolyses ATP in order to release the actin-myosin linkage and revert to its former shape. Each cycle consumes an ATP and slides the two filaments relative to one another by a single step. In the thick filament the head is attached through a flexible hinge-region to the tail which interacts laterally with other myosin tails to form the backbone of the filament. **Heavy meromyosin** is a subfragment of the heavy chain of myosin II. See also **myosin light chains**.

myosin light chain kinase Calmodulin-regulated kinase that phosphorylates myosin II light chains, probably regulating their function: molecular weight varies according to source, 1914aa in smooth muscle, 596aa in skeletal and cardiac muscle

myosin light chains Small subunit proteins (175aa and 194aa) of the hexameric thick filament **myosin** II, all with sequence homology to **calmodulin**, but not all with calcium-binding activity. There are two pairs of different light chains in each hexamer. Various light chains are characteristic of different muscle types. Some probably regulate the ATPase activity of the heavy chain directly when they bind calcium, others indirectly when phosphorylated by **myosin light chain kinase**.

myosin-binding proteins A family of thick filament-associated proteins. MyBPC is located in the crossbridge region in a fixed molar ratio with myosin heavy chains. MyBPC2 (1142aa) is found in fast skeletal muscle, MyBPC1 (1141aa) in slow muscle, MyBPC3 (1273aa) in cardiac muscle. PKA phosphorylation of MyBPC3 accelerates the kinetics of crossbridge cycling and regulates myocyte power output. Mutations can cause **hypertrophic cardiomyopathy.** Myosin binding protein H (MyBPH, 477aa) and myosin-binding protein H-like (354aa) belong to the immunoglobulin superfamily. See **obscurin**.

myostatin One of the TGFβ superfamily (growth/differentiation factor-8a, 375aa), a negative regulator of skeletal muscle growth. Mutation of the myostatin gene in mice, cattle, and humans causes a massively developed skeletal musculature, characterized by muscle hypertrophy and hyperplasia.

myotilin A cytoskeletal protein (titin immunoglobulon domain protein, TTID, 498aa) that co-localizes and interacts with **alpha-actinin** in the sarcomeric I-bands. The C-terminal portion contains two Ig-like domains homologous to **titin**. Myotilin is expressed in skeletal and cardiac muscle, and defects in the myotilin gene can cause one form of myofibrillar myopathy (see **muscular dystrophy**).

myotonia A skeletal muscle disorder in which there is delayed relaxation of muscle after voluntary contraction. There is no atrophy or hypertrophy of the muscle. Myotonia congenita (Thomsen's disease, Becker's disease) are the result of mutations in the muscle **chloride channel**, CLCN1 (988aa), which regulates electric excitability. Potassium-aggravated myotonia is caused by a mutation in the SCN4A gene that encodes the alpha subunit (1836aa) of a **voltage-gated sodium channel** and that is also mutated in paramyotonia congenita and hyperkalemic **periodic paralysis**.

myotonic dystrophy An autosomal dominant disorder (dystrophia myotonica) in which there is progressive muscle weakening and wasting, the most common form of **muscular dystrophy** in adults. Dystrophia myotonica 1 (DM1) is caused by an amplified trinucleotide repeat in the 3'-untranslated region of the dystrophia myotonica protein kinase gene (the kinase, 624aa, has homology with cAMP-dependent serine-threonine kinases). More than 50 repeats will cause pathological effects; in severe cases there may be >1000 repeats. DM2 (Ricker's syndrome) is caused by expansion of a CTG repeat in intron 1 of the zinc finger protein-9 gene (pathogenic alleles contain from 75 to 11,000 repeats).

myotoxins Small basic proteins (42–45aa) in rattlesnake venom. Induce rapid necrosis of muscle.

myotube Elongated multinucleate cells (three or more nuclei) that contain some peripherally located **myofibrils**. They are formed *in vivo* or *in vitro* by the fusion of **myoblasts** and eventually develop into mature muscle fibres that have peripherally located nuclei and most of their cytoplasm filled with myofibrils. There is no very clear distinction between myotubes and muscle fibres proper.

myotubularin A dual-specificity phosphatase (myotubularin 1, EC 3.1.3.48, 603aa) that acts on both phosphotyrosine and phosphoserine. May be involved in a signal transduction pathway necessary for late myogenesis, but is ubiquitously distributed. Mutations in myotubularin 1 lead to X-linked myotubular myopathy-1. Myotubularin-related protein 2 (643aa) is a phosphatase that acts on lipids with a phosphoinositol headgroup; it is mutated in a recessive form of **Charcot-Marie-Tooth** neuropathy. There are at least 13 myotubularin-related genes although some may encode catalytically inactive proteins.

myozenins A family of striated muscle-specific **calcineurin**-interacting proteins that are involved in linking Z-disk proteins (alpha-actinin, gamma-filamin, TCAP/telethonin, LDB3/ZASP) and localizing calcineurin signalling to the sarcomere. Myozenin 1 (calsarcin 2, 299aa) is expressed only in adult fast skeletal muscle, myozenin 2 (calsarcin-1, 264aa) is expressed specifically in adult cardiac and slow-twitch skeletal muscle. Myozenin 3 (calsarcin 3, 251aa) is restricted to skeletal muscle.

myristic acid The myristoyl group is a relatively uncommon C_{14} fatty acyl residues of phospholipids in biological membranes (see Table L3) but is found as an N-terminal modification (myristoylation) of a large number of membrane-associated proteins and some cytoplasmic proteins. It is a common modification of viral proteins. In all known examples, the myristoyl residue is attached to the amino group of N-terminal glycine by myristoyl CoA:protein N-myristoyl transferase. The specificity of the myristoyl transferase enzymes is extremely high with respect to the fatty acyl residue. For many proteins, the addition of the myristoyl group is essential for membrane association. There is some evidence that myristoylated proteins do not interact with free lipid bilayer, but require a specific receptor protein in the target membrane.

myrosinase Myrosinase 1 (EC 3.2.1.147, sinigrinase 1, thioglucosidase 1, beta-glucosidase 38, 541aa) an enzyme in *Arabidopsis* that degrades **glucosinolates** to glucose, sulphate and any of the products: thiocyanates, isothiocyanates, nitriles, epithionitriles or oxazolidine-2-thiones. These toxic degradation products can deter insect herbivores. Seems to function in abscisic acid (ABA) and methyl jasmonate (MeJA) signalling in guard cells. Hydrolyzes sinigrin. Myrosinase 2 (547aa in *Arabidopsis*) is comparable and either one will suffice. Myrosinase-binding protein (642aa) belongs to the jacalin family.

MYST family A family (MOZ, YbF2, Sas2, Tip60-like family) of histone acetyltransferases (HATs). In mammals, the MYST family is large and divergent and MYST family proteins are involved in a wide range of cell functions ranging from transcription activation and silencing, apoptosis, cell cycle progression, DNA replication and repair. In *Arabidopsis* there are three other classes of HATs (the GCN5-family, p300/CBP and TAFII250 (transcription initiation factor TFIID 250 kDa subunit) classes) in addition to the two closely related MYST family proteins HAM1 and HAM2.

myxamoebae In the Myxomycetes, such as *Physarum*, each spore on germination produces two amoeboid cells, myxamoebae, which then transform into flagellated cells.

Myxobacteria Group of Gram-negative bacteria, found mainly in soil. They are non-flagellated with flexible cell walls. They show a gliding motility, moving over solid surfaces leaving a layer of slime (myxo = slime). At some stage in their growth the cells of this group swarm together and form fruiting bodies and spores in a fashion similar to the **slime moulds** but at a much smaller scale.

myxoma virus A poxvirus (see **Poxviridae**) that causes myxomatosis. Originally isolated from a species of wild rabbit, *Sylvilagus*, in Brazil, in which it causes a mild non-fatal disease, it was found to be 99% fatal in the European rabbit *Oryctylagus*. It causes the characteristic, subcutaneous gelatinous swellings, 'myxomata' and usually kills in 2–5 days. It has been used to control rabbit populations in Australia and Britain, but there are signs that they have developed immunity.

myxomycota The slime moulds. Amoeboid heterotrophic organisms, formerly classed as fungi but now as Protista. May be free-living phagocytes that feed on bacteria or may be parasitic within plant cells. Includes Acrasiomycetes (cellular slime moulds, e.g. *Dictyostelium*) and Myxomycetes (acellular slime moulds, e.g. *Physarum*).

Myxoviridae Obsolete taxon of single-stranded RNA viruses now divided into **Orthomyxoviridae** and **Paramyxoviridae**.

N

N (1) Nitrogen. (2) Single-letter code for **asparagine**. (3) Abbreviation for amino, as in N-terminal (the end of a peptide chain in which the amino (-NH₂) group is free).

N51 Mouse homologue of **melanoma growth stimulatory activity** protein.

NAA See **naphthalene acetic acid**.

NAAG N-acetyl-L-aspartyl-L-glutamate. One of the three most prevalent peptide neurotransmitters in the central nervous system; suppresses glutamate signalling by activating presynaptic metabotropic glutamate receptors (mGluR₃). It is removed by specific NAAG peptidases (**glutamate carboxypeptidase II**). See **ZJ-43**.

Nabothian cysts Benign cysts on the cervix of the womb composed of endocervical columnar cells covered by squamous metaplasia and filled with mucus.

NAC See **nascent polypeptide associated complex**.

NAC proteins A superfamily of plant-specific transcription factors that are involved in a variety of developmental events as well as in biotic and abiotic stress responses. Named for three genes (NAM (no apical meristem), ATAF (*Arabidopsis* transcription activation factor) and CUC (cup-shaped cotyledon)) and characteristically have a NAC domain. There are at least 135 NAC genes in *Arabidopsis*. Review: http://www.ncbi.nlm.nih.gov/pubmed/20600702?dopt=AbstractPlus&holding=f1000,f1000m,isrctn

N-acetyl glucosamine A sugar unit (2-acetamido glucose) found in glycoproteins and various polysaccharides such as **chitin**, bacterial **peptidoglycan** and in **hyaluronic acid**.

N-acetyl muramic acid Sugar unit of bacterial peptidoglycan, consisting of N-acetyl glucosamine bearing an ether-linked lactyl residue on carbon 3. The repeating unit of the cell wall polysaccharide is N-acetyl muramic acid linked to N-acetyl glucosamine via a β(1-4)-glycosidic bond that can be cleaved by **lysozyme**.

N-acetyl neuraminic acid A 9-carbon sugar, structurally a condensation product of N-acetyl mannosamine and pyruvate. Also known as sialic acid, but more correctly is a member of the family of **sialic acids**. Found in **glycolipids**, especially **gangliosides**, and in **glycoproteins**, and therefore in the **plasma membrane** of animal cells, to the outer surface of which it contributes negative charge by virtue of its carboxylate group.

NAD Coenzyme (nicotinamide adenine dinucleotide, NAD⁺, formerly DPN) in which the nicotine ring undergoes cyclic reduction to NADH and oxidation to NAD. Acts as a diffusible substrate for dehydrogenases amongst others. NADH⁺ is one source of reducing equivalents for the electron transport chain and the source of ADP-ribose (see **ADP-ribosylation**). Nicotinamide adenine dinucleotide phosphate (NADP, formerly TPN) is an analogue of NAD mostly generated by the **hexose monophosphate shunt**, important in many biosynthetic pathways and the **metabolic burst** in neutrophils. Nicotinic acid adenine dinucleotide phosphate (NAADP) second messenger that triggers the release of calcium ions.

NADP See **NAD**.

NADPH oxidases Enzymes (NOX1, 564aa; NOX2, cytochrome b-245 heavy chain, p91phox, 570aa) that catalyze the transfer of electrons from NADPH to molecular oxygen to generate reactive oxygen species (ROS). NOX3 (Mitogenic oxidase 2, MOX2, 568aa) has a role in the biogenesis of otoconia/otoliths, crystalline structures of the inner ear involved in the perception of gravity. NOX4 (renal NADPH-oxidase, 578aa) may function as an oxygen sensor regulating the KCNK3/TASK-1 potassium channel and HIF1A activity in the kidney. NOX5 (765aa) is calcium-dependent and functions as a calcium-dependent proton channel. Large NOX 1 (dual oxidase 1, thyroid oxidase 1, 1551aa) and large NOX2 (dual oxidase 2, p138 thyroid oxidase, 1548aa) produce hydrogen peroxide and are important in thyroid hormone synthesis and lactoperoxidase-mediated antimicrobial defense at the surface of mucosa. The NADPH oxidase activator, (NOXA1, p67phox-like factor, p51-nox, 476aa) activates NOX1 and NOX2. The NADPH oxidase organiser (NOXO1, p41-NOX, Nox-organizing protein 1, SH3 and PX domain-containing protein 5, 376aa) potentiates the superoxide-generating activity of NOX1 and NOX3 and targets NOX-activators to NOX and to different subcellular compartments. Nox-organizing protein 2 (neutrophil cytosol factor 1, p47phox, 390aa) is defective in one form of chronic granulomatous disease. See **phox**.

nadrin A neuron-specific rho GTPase-activating protein (neuron-associated developmentally-regulated protein, RhoGAP17, 846aa in mouse) involved in regulated exocytosis and in the maintenance of tight junctions by regulating the activity of CDC42. The

The Dictionary of Cell and Molecular Biology. DOI: http://dx.doi.org/10.1016/B978-0-12-384931-1.00014-3

homologue in humans is RICH-1 (RhoGAP interacting with CIP4 homologue protein 1, 881aa) has a **BAR domain** that interacts with **angiomotin**, so that it is recruited to tight junctions.

Naegleria A genus of soil-dwelling amoebae that can cause opportunistic infections. *Naegleria gruberi* is normally an amoeboid protozoan but can transform into a swimming form with two **flagella** in fluid-phase. *Naegleria fowleri* is found in warm water and can cause primary amoebic meningoencephalitis.

naevus *US.* nevus. Tumour-like but non-neoplastic hamartoma of skin. A vascular naevus is a localized capillary-rich area of the skin ('strawberry birthmark'; sometimes the much more extensive 'port-wine stain'). A mole (benign melanoma) is a pigmented naevus, a cluster of melanocytes containing melanin. A blue naevus is an unusual mole in which the melanocytes are deep in the **dermis**.

nagarse Old name for a broad-specificity serine peptidase (nagarase, subtilisin BPN', alkaline protease, EC.3.4.21.62, 382aa) from *Bacillus amyloliquefaciens*.

Nagler's reaction Standard method for identifying *Clostridium perfringens*. When the bacterium is grown on agar containing egg yolk, an opalescent halo is formed around colonies that produce α-toxin (lecithinase).

Na⁺/H⁺ exchanger regulatory factor See EBP50.

nail-patella syndrome A disorder (Turner-Kieser syndrome, Fong disease) in which there is dysplasia of the nails and the patellae are absent; caused by mutation in the LIM homeobox transcription factor gene (LMX1B).

Nairovirus A genus of the **Bunyaviridae** with a genome consisting of circular, negative-sense single stranded RNA. One of the group is responsible for **Crimean-Congo haemorrhagic fever**.

Naja kaouthia Asian cobra (one of the Elapidae). See alpha-**cobratoxin**.

nalidixic acid Synthetic antibiotic that interferes with **DNA gyrase** and inhibits prokaryotic replication. Often used in selective media.

NALP proteins A subfamily of the caterpiller protein family, cytoplasmic proteins that are implicated in the activation of proinflammatory caspases and are involved in **inflammasomes**. NALP1 (NACHT, LRR and PYD domains-containing protein 1, caspase recruitment domain-containing protein 7, death effector filament-forming ced-4-like apoptosis protein, 1473aa) will form cytoplasmic **death effector filaments**, activates various caspases, stimulates apoptosis and binds ATP. NALP 7 (980aa) may play

a role in cell proliferation and is a negative regulator of IL-1β; mutations are associated with recurrent **hydatidiform mole**.

Namalwa cells Line of human B-lymphocytes grown in suspension and used to produce interferon (stimulated by Sendai virus infection). Derived from patient with **Burkitt's lymphoma**.

nanobacteria A group of small (<200 nm) bacteria that were claimed to be responsible for producing glomerular aggregates of apatite in culture medium. Subsequently shown to be non-living but self-propagating mineral-fetuin complexes. Article: http://www.ncbi.nlm.nih.gov/pubmed/18282102

nanog A homeobox transcription regulator (305aa) involved in inner cell mass and embryonic stem cell proliferation and self-renewal.

nanopore A microscopic pore or opening, strictly speaking one between 10^{-7} and 10^{-9} m.

nanos Maternal RNA-binding protein in *Drosophila* (401aa) that is required for germ cell proliferation and self-renewal. Forms a complex with **pumilio** and **brat** that regulates translation and mRNA stability.

nanovid microscopy Technique of bright-field light microscopy using electronic contrast enhancement and maximum numerical aperture.

NAP-1 See **interleukin-8**.

NAP-3 See **melanoma growth stimulatory activity**.

naphthalene acetic acid *NAA* A synthetic auxin, often used in plant physiology and in plant tissue culture media because it is more stable than **IAA**.

naphthylamine Potent carcinogen; used in production of aniline dyes, one of the first chemicals to be associated with a tumour (bladder cancer). The compound itself is not directly carcinogenic; a metabolite produced by hydroxylation (1-hydroxy-2-aminonaphthalene) is detoxified in the liver by conjugation with glucuronic acid, but reactivated by a glucuronidase in the bladder.

napins Small, basic, water-soluble seed storage proteins from *Brassica napus* (oilseed rape). Napin (180aa) is processed into a mature form with two polypeptide chains (3.8 and 8.4 kDa) linked by two disulphide bridges. Interacts with calmodulin and has antifungal properties. A number of napins are produced,

napthoquinones Plant pigments derived from napthoquinone.

NARG1 NMDA receptor-regulated protein 1 (N-terminal acetyltransferase, tubedown-1, gastric cancer antigen Ga19, 866aa) that forms a complex with **ARD1** that has alpha (N-terminal) acetyltransferase activity which may be important for vascular, hematopoietic and neuronal growth and development. It regulates retinal neovascularization.

naringenin A flavonoid that can act as an anti-oxidant, free radical scavenger, anti-inflammatory, carbohydrate metabolism promoter, and immune system modulator. It is the predominant flavanone in grapefruit.

NASBA Nucleic acid sequence-based amplification, a technology used for the continuous amplification of nucleic acids in a single mixture at one temperature. Normally applied to amplify single-stranded target RNA, producing RNA amplicons, although it can be modified to amplify DNA.

nascent polypeptide associated complex *NAC* A highly conserved heterodimeric complex (NACA, 215aa; basic transcription factor 3b, BTF3, 206aa), a peripheral component of cytoplasmic ribosomes, that interacts with nascent polypeptides as they emerge; it may also act as a transcriptional coactivator and binds to nucleic acids. NAC prevents targeting of nascent polypeptide chains that lack a **signal sequence** to the ER, the opposite function to that carried out by the **signal recognition particle**. Interacts with alpha-**taxilin**. Mutations in NAC cause severe embryonically lethal phenotypes in mice, *Drosophila*, and *C. elegans*.

naso-pharyngeal carcinoma Carcinoma, highly prevalent in Southern China, associated with infection by **Epstein-Barr virus** and probably exposure to inhaled co-carcinogens, although additional risk factors are known.

nastic movement Non-directional movement of part of a plant in response to external stimulus. The tips of growing shoots of plants that twine around supports show nastic movement. See **epinasty**.

Nasu-Hakola disease A recessive disorder (polycystic lipomembranous osteodysplasia with sclerosing leukoencephalopathy) in which there are psychotic symptoms that soon progress to presenile dementia and bone cysts in wrists and ankles. Can be caused by loss-of-function mutations in the TYROBP gene that encodes DNAX-activation protein 12 (DAP12, TYRO protein tyrosine kinase-binding protein, 113aa).

natriuretic peptides Family of four peptides all sharing significant sequence and structural homology, that stimulate secretion of sodium in the urine (natriuresis). All act through guanylyl cyclase. The mammalian members are atrial natriuretic peptide (ANP, 153aa), B-type natriuretic peptide (BNP, 134aa), C-type natriuretic peptide (CNP, 126aa) and possibly osteocrin/musclin. C-type natriuretic peptide is cleaved into three shorter peptides (CNP-22, CNP-29 and CNP-53), causes vasodilation and is important in control of blood pressure, possibly antagonising ANP. It also regulates the proliferation and differentiation of chondrocytes in the cartilaginous growth plate of long bones. It is produced by endothelial and renal cells and is considered an autocrine regulator of endothelium as well as a neuropeptide. Effects may oppose those of **atrial natriuretic peptide**. Receptors are guanylate cyclases: Atrial natriuretic peptide receptor 1 (NPR1, EC 4.6.1.2, 1061aa) binds ANP and BNP; NPR-B (NPR2, EC 4.6.1.2, 1047aa) is the receptor for CNP; NPR-C (NPR3, 541aa) is the atrial natriuretic peptide clearance receptor. Mutations in NPR2 cause a skeletal dysplasia (Maroteaux-type **acromesomelic dysplasia**). See also **dendroaspis natriuretic peptide** and **plant natriuretic peptides**.

natural killer cells *NK cells* See **killer cells**.

natural selection The generally accepted mechanism whereby genotype-environment interactions occurring at the phenotypic level lead to differential reproductive success of individuals and hence to modification of the gene pool of a population.

nauplius The typical first larval stage of Crustacea; approximately egg-shaped, unsegmented, and with three pairs of appendages and a median eye.

NaV channels See **sodium channels**.

Navajo neurohepatopathy An autosomal recessive multisystem disorder prevalent in the Navajo population of the southwestern United States, caused by a mutation in the MPV17 gene that encodes a protein (176aa) involved in mitochondrial homeostasis.

Naxos disease An autosomal recessive disorder (palmoplantar keratoderma with arrhythmogenic right ventricular cardiomyopathy and wooly hair) first reported in families on the island of Naxos and due to mutation in **plakoglobin**. Similar disorders are linked to mutations in **desmoplakin**.

NB-LRR genes Nucleotide-binding site/leucine-rich repeat genes. The largest group of plant disease resistance genes (more than 600 in rice), also mediate pathogen recognition in animals. Two major sub-classes are the TIR-NB-LRR genes that encode proteins with similarity to **Toll** and **Interleukin 1** receptors, while the non-TIR class typically contains a coiled-coil (CC) domain in the N-terminal region. As might be expected, there is strong divergence of the genes in response to differing selection pressures on various plant genera.

Nbk Natural born killer, a proapoptotic **BH3**-only protein; see **Bik**.

NBT See **nitroblue tetrazolium**.

N-cadherin See **cadherins**.

NCAM Neural cell adhesion molecule. One of the first of the **cell adhesion molecules** (CAMs) to be isolated from chick brain. A member of the **immunoglobulin superfamily**, as is NgCAM (neural-glial CAM). Initially defined by adhesion-blocking antiserum. Thought to be important in divalent-cation independent (**L1**) intercellular adhesion of neural and some embryonic cells. The human homologues (NCAM1, CD56, 858aa; NCAM2, 837aa) have similar roles in neuronal adhesion and fasciculation. See also **neuroglian**.

NCAP (1) Non-cell-autonomous proteins, proteins exchanged between plant cells via plasmodesmata. (2) N-acetyl-4-cystaminylphenol; used topically for the treatment of hyperpigmentation. (3) N-terminal residue of a protein.

NCBI The National Center for Biotechnology Information, an important US organisation that manages resources such as BLAST, OMIM and PubMed as well as many others devoted to aspects of cell and molecular biology (nucleotide sequences, ESTs, SNPs and others). Access is free and unrestricted and of inestimable value to the community. Homepage: http://www.ncbi.nlm.nih.gov/guide/

NCD Protein of the **kinesin-14 protein** family that differs from kinesin in that it moves towards the minus end of the microtubule (like cytoplasmic dynein). Implicated in spindle organization. In *Drosophila*, the protein product of the *ncd* gene (claret segregational, 700aa) is required for normal chromosomal segregation in meiosis in females, and in early mitotic divisions of the embryo. See also **Kar3**.

Nck Small adaptor proteins (Nck1, 377aa; Nck2, 380aa) with SH2 and SH3 domains. Similar to **Crk** and **Grb2**. The Nck family of adaptors function to link tyrosine phosphorylation induced by extracellular signals with downstream regulators of actin dynamics. Nck1 has a role in the DNA damage response, and interacts with **WIP**.

N-chimaerin A phorbol ester/diacyl glycerol binding protein (459aa) found in brain. A GTPase-activating protein for **rac**.

N-CoR Nuclear receptor corepressors, proteins involved in transcriptional repression by thyroid hormone and retinoic acid receptors (NCoR1, 2440aa; NCoR2, silencing mediator of retinoic acid and thyroid hormone receptor, SMRT, 2525aa). The binding of the corepressor leads to the assembly of a larger complex that may, for example, contain **Sin3** and **histone deacetylases** (HDAC). NCoR1 and NCoR2 are paralogs and possess similar molecular architectures and mechanistic strategies, but exhibit distinct molecular and biological properties. See **RIP140**.

NCP1 (1) Family of cell-adhesion molecules that includes **neurexin** IV, contactin associated protein

(**Caspr**) and **paranodin**, present at the vertebrate axo-glial synaptic junction. (2) Neutrophil cationic protein (25aa in pigs).

NCS family Neuronal calcium sensor family of proteins. See **frequenin**, **visinin**.

Ndc80 A kinetochore protein (kinetochore-associated protein 2, highly expressed in cancer protein, Hec1, retinoblastoma-associated protein HEC, 642aa) that is a component of the kinetochore-associated NDC80 complex, required for chromosome segregation and spindle checkpoint activity. Interacts with **Dgt proteins** in the **augmin/HAUS** complex. Homologues are found in yeasts and most metazoa.

NDE1 A protein (nuclear distribution protein nudE homolog 1, 346aa) that is required for centrosome duplication and for the formation and function of the mitotic spindle. It controls the orientation of the mitotic spindle during division of cortical neuronal progenitors which determines whether both progeny have proliferative capacity or whether one is postmitotic. Nuclear distribution gene E homologue-like 1 (Ndel1, 345aa) is homologous to Nde1 and may function as both a cysteine protease and a centrosomal structural protein interacting with dynein and the dynein-regulator Lis1. Binds **disrupted-in-schizophrenia 1**. See **CENP**.

Ndk Nucleoside diphosphate kinase, an enzyme (EC 2.7.4.6) that generates nucleoside triphosphates or their deoxy derivatives by terminal phosphotransfer from ATP or GTP. There are multiple isoforms (at least seven) in humans.

nearest neighbour analysis Statistical method that can be used to analyse spatial distributions (e.g. of organisms in an environment) according to whether they are clustered, random or regular. It has also been used, for example, to determine the frequency of pairs of adjacent bases in DNA, revealing a deficiency of the pair CG in most eukaryotes.

nebulette A protein (actin-binding Z-disk protein, 1014aa) found in Z-discs of cardiac muscle and dense bodies in nonmuscle cells. It has sequence and structural homology with **nebulin** and has 23 nebulin repeats. See **myopallidin**.

nebulin Family of large matrix proteins (6669aa) found in the **N-line** of the **sarcomere** of striated muscle; binds and stabilizes F-actin. Consist of many (more than 200) repeats of conserved actin-binding motifs. There are many tissue and development-stage specific isoforms. Mutations are associated with a form of **nemaline myopathy**. Nebulin-related anchoring protein (NRAP, 1730aa) is found in the myotendinous junction in skeletal muscle and the intercalated disc in cardiac muscle. See **nebulette**.

necdin A growth suppressor (321aa) expressed predominantly in postmitotic neurons and implicated in terminal differentiation; functionally similar to the retinoblastoma protein, repressing the activity of cell-cycle-promoting proteins. Necdin-like proteins are a family within the **MAGE superfamily**. Necdin is one of the chromosome 15 products disrupted in **Prader-Willi syndrome** and only the paternal allele is expressed (see **genomic imprinting**).

necrosis Death of some or all cells in a tissue as a result of injury, infection or loss of blood supply. Necrosis, unlike **apoptosis** is likely to elicit an inflammatory response.

necrotising fasciitis A rare infection of the deeper layers of skin and subcutaneous tissues caused by a variety of different bacteria (e.g. *Streptococcus pyogenes*, *Staphylococcus aureus*, *Clostridium perfringens*) often in immunocompromised individuals. There is extensive damage to tissue.

necrotrophic A mode of fungal infection in which host cells are killed, in part by the plant's own hypersensitive responses. Examples include the fungi *Botrytis cinerea* and *Sclerotinia sclerotiorum*. *Cf.* **biotrophic**.

nectin (1) Another name for SAM (substrate adhesion molecule) e.g. fibronectin. (2) A protein forming the stalk of mitochondrial ATPase. (3) Calcium-independent immunoglobulin-like adhesion molecule that interacts with other nectin molecules to cause cell-cell adhesion and interacts, through its cytoplasmic domain-associated protein **afadin**, with **catenin** which interacts in turn with the cytoplasmic domain of **cadherin**. Nectins are receptors for various viruses, including poliovirus. Nectin-like protein 1 is **synCAM 3**.

nectinepsin An extracellular matrix protein (359aa) of the **pexin family**, isolated from quail neuroretina. There is only one literature reference. Sole reference: http://www.jbc.org/content/271/42/26220.full

Nedd (1) Nedd proteins (neural-precursor-cell-expressed and developmentally down regulated proteins) are a diverse group involved in various developmental processes. Nedd1 (660aa) is required for mitosis progression and promotes the nucleation of microtubules from the spindle. Nedd4 (1319aa) is an E3 ubiquitin-protein ligase. Nedd4 family-interacting proteins (NDFIP1, 221aa; NDFIP2, 336aa) activate HECT domain-containing E3 ubiquitin-protein ligases. See **bean**. Nedd8 is an ubiquitin homologue (81aa) that activates **cullin** complexes by providing a recognition site for an ubiquitin-conjugating enzyme. Neddylation, the conjugation of Nedd8 to the conserved lysines of cullins, is essential for *in vivo* cullin-organized E3 activities. Deneddylation, removal of the Nedd8 moiety, requires the isopeptidase activity of the COP9 signalosome (CSN). Neddylated Cul1 and Cul3 are unstable. Nedd8 ultimate buster 1 (negative regulator of ubiquitin-like proteins 1,618aa) down-regulates Nedd8 activity. (2) Nedd2 is **caspase 2**. See Table C1.

Needleman-Wunsch algorithm An algorithm commonly used in bioinformatics to align protein or nucleotide sequences.

nef HIV protein (206aa) that is important for pathogenesis, enhances infectivity, is required for optimal virus replication and alters numerous pathways of T-cell function. Has a proline-rich sequence that interacts with **hck**-SH3 domain and activates kinase activity of hck; blocks MAP3K5 thereby inhibiting apoptosis. Nef itself has a sorting signal (ENTSLL) that functions as an endocytosis marker and it has some amino acid homology with alpha-scorpion toxins that bind to potassium channels.

negative feedback A phenomenon in which the products of a process can act at an earlier stage in the process to inhibit their own formation. The term was first used widely in conjunction with electrical amplifiers where negative feedback was applied to limit distortion of the signal by the amplification mechanism. Tends to stabilize the process. *Cf.* **positive feedback**.

negative regulation Negative feedback in biological systems, usually mediated by allosteric regulatory enzymes.

negative staining Microscopic technique in which the object stands out against a dark background of stain. For electron microscopy the sample is suspended in a solution of an electron-dense stain such as sodium phosphotungstate and then sprayed onto a support grid. The stain dries as structureless solid and fills all crevices in the sample. When examined in the electron microscope the sample appears as a light object against a dark background. Quite fine structural detail can be observed using negative staining and it has been used extensively to study the structure of viruses and other particulate samples.

negative-stranded RNA virus Class V **viruses** that have an RNA genome that is complementary to the mRNA, the positive strand. They also carry the virus-specific **RNA polymerase** necessary for the synthesis of the mRNA. Includes **Rhabdoviridae**, **Paramyxoviridae** and Myoviridae (e.g. the T-even phages).

NEGR1 See **IgLON**.

Negri body Acidophilic cytoplasmic inclusion (mass of **nucleocapsids**) characteristic of rabies virus infection.

NEI A neuropeptide (neuropeptide-glutamic acid-isoleucine, 13aa) derived from pro-melanin concentrating hormone by the action of prohormone convertase-2 in neurons.

Neisseria Gram-negative non-motile pyogenic cocci. Two species are serious pathogens, *N. meningitidis* (see **meningitis**) and *N. gonorrhoeae*. The latter associates specifically with urinogenital epithelium through surface **pili**. Both species seem to evade the normal consequences of attack by phagocytes by the possession of a capsule that mediates resistance to both phagocytosis and complement-mediated killing.

nek kinases A family of serine/threonine-protein kinases (never in mitosis A-related kinases, NimA-like protein kinases) involved in mitotic or meiotic regulation. Nek1 (1258aa) may also have tyrosine kinase activity and is implicated in the control of meiosis. Nek2 (445aa) is an integral component of the mitotic spindle-assembly checkpoint which is necessary for proper chromosome segregation during metaphase-anaphase transition. Several others are known.

nelin See **nexilin**.

nemaline Fibrous or thread-like.

nemaline bodies Sarcoplasmic inclusions composed largely of alpha-actinin and actin, probably derived from the Z-disc. See **nemaline myopathies**

nemaline myopathies A set of clinically and genetically heterogeneous disorders characterized by abnormal thread- or rod-like (nemaline) structures (nemaline bodies) and progressive weakening in muscle. Caused by mutations in genes encoding various muscle proteins (tropomyosin-3, **nebulin**, alpha-actin-1, beta-tropomyosin, troponin T1, cofilin-2).

nematocyst *cnidocyst* Stinging mechanism used for defence and prey capture by *Hydra* and other members of the Cnidaria (Coelenterata). It is located within a specialized cell, the **nematocyte** and consists of a capsule containing a coiled tube. When the nematocyte is triggered, the wall of the capsule changes its water permeability and the inrush of water causes the tube to invert explosively ejecting the nematocyst from the cell. The tube is commonly armed with barbs and may also contain toxin.

nematocyte *cnidoblast* Stinging cells found in *Hydra*, used for capturing prey and for defence. There are four major types, containing different sorts of **nematocysts**: stenoteles (60%), desmonemes, holotrichous isorhizas, and atrichous isorhizas. They differentiate from interstitial cells and are almost all found in the tentacles.

Nematoda A class of the phylum, Aschelminthes. Unsegmented worms with an elongate rounded body pointed at both ends, a mouth and alimentary canal, and a simple nervous system. The sexes are separate and larvae resemble the adults; many species are of economic importance as pests; most are free-living but some are parasitic. Best known example in cell biology is **Caenorhabditis elegans** which, because of the determinate number of cells, has proved a valuable experimental organism especially now that its genome is sequenced and the effects of many mutations analysed. *C. elegans* has an unusual amoeboid spermatozoon that is actively motile yet appears to lack both actin and tubulin.

nematosome Electron-dense cytoplasmic inclusion (0.3–0.5 μm diameter) found in cytotrophoblast cells of the placenta and in some neurons.

NEMO See IκB.

N-end rule The N-end rule holds that the *in vivo* half-life of a protein is determined by the N-terminal residues. The rule is conserved between prokaryotes and eukaryotes. See **N-recognins**.

neoblasts Totipotent stem cells responsible for regeneration and tissue renewal in planarians.

neogenin A receptor of the immunoglobulin superfamily (1461aa), closely related to the axon guidance receptor DCC (deleted in colorectal cancer). Binds **netrin** and other repulsive guidance molecules and also **hemojuvelin**. Netrin-1-neogenin interaction is positively chemotactic for axons; the interaction between neogenin and RGMA protein induces repulsion. May also regulate diverse developmental processes, including neural tube and mammary gland formation, myogenesis and angiogenesis. See also **protogenin**.

neomycins Aminoglycoside antibiotics from *Streptomyces fradiae*; bind to duplex RNA with high affinity. Neomycin resistance is often used as a selective marker for transfected mammalian cells with an aminoglycoside phosphotransferase (*neo*) gene incorporated in the plasmid.

neoplasia Literally new growth, usually refers to abnormal new growth, which may be benign or malignant. Unlike **hyperplasia**, neoplastic proliferation persists even in the absence of the original stimulus. Neoplasia can occur in tumours, where there is an actual swelling, but there are other proliferative disorders, such as **leukaemias**, all colloquially referred to as 'cancer'. The cells in **benign tumours** do not spread and such tumours are not life-threatening unless the growth pressure causes damage. Malignant neoplasia involves the loss of both growth control and positional control and malignant cells will invade territory normally occupied by other cells (see **contact inhibition of locomotion**) and may form secondary tumours (**metastases**) at sites well away from the primary tumour.

neopterin A pteridine derivative, the D-erythro enantiomer of **biopterin**, produced by human monocytes/macrophages upon stimulation with interferon-gamma. Increased neopterin concentrations in human serum and urine indicate activation of cell-mediated (Th1-type) immune responses.

neotenin See **juvenile hormone**.

neoteny *paedomorphosis* The persistence in the reproductively-mature adult of characters usually associated with the immature organism.

neoxanthin A **xanthophyll carotenoid** pigment, found in higher plant chloroplasts as part of the **light-harvesting system**.

NEP (1) Cell surface zinc endopeptidase (neutral endopeptidase, EC 3.4.24.11, neprilysin, enkephalinase, common acute lymphocytic leukemia antigen (CALLA), CD10, 750aa), that hydrolyzes regulatory peptides such as **ANP**. Spontaneously hypertensive hamsters have elevated levels of NEP in two organs that contribute appreciably to vascular resistance, skeletal muscle and kidney. In humans, an important cell surface marker in acute lymphocytic leukemia. (2) A necrosis and ethylene inducing protein (NEP-1, 253aa) capable of triggering plant cell death, that was purified from culture filtrates of *Fusarium oxysporum*. NEP1-like proteins (NLPs) are a novel family of microbial elicitors of plant necrosis. (3) In *C. elegans*, a protein (231aa) involved in 40S ribosomal subunit biogenesis.

nephelometer Instrument for measuring turbidity in which light scattered orthogonally to the incident beam is measured (nephelometry). Light scattering depends upon the number and size of particles in suspension.

nephrin A single-pass membrane protein (1241aa) involved in the development or function of the kidney glomerular filtration barrier. May anchor the **podocyte** slit diaphragm to the actin cytoskeleton. Mutations in the nephrin gene are associated with congenital nephrotic syndrome of the Finnish type (NPHS1). Nephrin-like protein 3 (kin of IRRE-like protein 2, 708aa) interacts with **podocin**.

nephroblastoma See **Wilms' tumour**.

nephrocystins Proteins originally identified as being associated with various forms of **nephronophthisis**, disorders in which there is a defect in **primary cilia**. Nephrocystin-1 (732aa) may play a role in the control of epithelial cell polarity and interacts with **p130Cas** (BCAR1), proline-rich tyrosine kinase-2 (PTK2B), and **tensin** in embryonic kidney and testis. Nephrocystin-4 (nephroretinin, 1426aa) is involved in similar signalling complexes; the proteins interact with one another and and colocalize with alpha-tubulin especially in primary cilia and the microtubule organizing centre.

Nephrocystin-2 (**inversin**, 1065aa) is required for normal renal development and establishment of left-right axis. Probably acts as a molecular switch between different **Wnt** signalling pathways. Nephrocystin-3 (1330aa) may be important for mechano-sensing in the kidney. Nephrocystin-5 (IQ motif-containing protein B1, IQCB1; p53- and DNA damage-regulated IQ motif protein, PIQ, 598aa) interacts with calmodulin and the retinitis pigmentosa GTPase regulator and mutations are associated with **Senior-Loken syndrome**. Nephrocystin-6 (centrosomal protein of 290 kDa, 2479aa) is required for the correct localization of ciliary and phototransduction proteins in retinal photoreceptor cells; mutations are associated with **Meckel syndrome Type 4** and various other sensory-system defects.

nephron The structural and functional unit of the vertebrate kidney. It is made up of the glomerulus, Bowman's capsule and the convoluted tubule.

nephronophthisis A group of cystic kidney disorders that lead to kidney failure but are associated with various non-renal dysfunctions, all caused by defects in **primary cilia**. See **nephrocystins**.

nephropathia epidemica A disease caused by Puumala virus (genus *Hantavirus*, family Bunyaviridae). The vector and natural reservoir of Puumala virus is a small rodent, the bank vole (*Clethrionomys glareolus*).

neprilysin See **NEP** #1.

Nernst equation A basic equation of biophysics that describes the relationship between the equilibrium potential difference across a **semipermeable membrane**, and the equilibrium distribution of the ionic permeant species. It is described by: $E = (RT/zF) \cdot \ln([C1]/[C2])$, where E is the potential on side 2 relative to side 1 (in volts), R is the gas constant ($8.314 \ J \ K^{-1} \ mol^{-1}$), T is the absolute temperature, z is the charge on the permeant ion, F is the Faraday constant ($96,500 \ °C \ mol^{-1}$) and [C1] and [C2] are the concentrations (more correctly activities) of the ions on sides 1 and 2 of the membrane. It can be seen that this equation is a solution of the more general equation of **electrochemical potential**, for the special case of equilibrium. The equation described the voltage generated by ion-selective electrodes, like the laboratory pH electrode; and approximates the behaviour of the resting plasma membrane (see **resting potential**).

Nernst potential See **Nernst equation** and **ion-selective electrodes**.

nerve cell See **neuron**.

nerve growth cone See **growth cone**.

nerve growth factor *NGF* A peptide (118aa) with both chemotropic and chemotrophic properties for **sympathetic** and **sensory neurons**. There are three types of subunits, alpha, beta and gamma, which interact to form a 7S, 130 kDa complex. Found in a variety of peripheral tissues, NGF attracts **neurites** to the tissues by chemotropism, where they form synapses. The successful neurons are then 'protected' from neuronal death by continuing supplies of NGF. It is also found at exceptionally high levels in snake venom and male mouse submaxillary salivary glands, from which it is commercially extracted. NGF was the first of a family of nerve tropic factors to be discovered. Amino acids 1–81 show homology with proinsulin. Besides its peripheral actions, NGF selectively enhances the growth of **cholinergic neurons** that project to the forebrain and that degenerate in **Alzheimer's disease**. The receptor is TNF receptor superfamily-16 (TNFRSF16) and binds other **neurotrophins**.

nesfatin See **nucleobindin**

nesidioblast Precursor cell of pancreatic **B cells**. In nesidioblastosis (familial hyperinsulinemic hypoglycemia) these cells fail to mature properly and promiscuously secrete various hormones. Caused either by mutation in the gene that encodes the regulatory **sulphonylurea receptor** (SUR1) subunit of the inwardly rectifying potassium channel (Kir6.2) of the beta cell or mutations in the channel itself.

nesprins A family of alpha-actinin type actin-binding proteins. Nesprin-1 (nuclear envelope spectrin repeat protein 1, synaptic nuclear envelope protein 1, SYNE1, enaptin, 8797aa) forms a linking network between organelles and the actin cytoskeleton and also forms part of the LINC complex that links the nucleoskeleton with the cytoskelton. Nesprin-2 (6885aa) is similar in function and different isoforms are found in various cell compartments. Nesprin-3 (975aa), an outer nuclear membrane protein, lacks an actin-binding domain and associates with the cytoskeletal linker protein **plectin**. Nesprin-4 (404aa) promotes kinesin-dependent apical migration of the centrosome and Golgi apparatus and basal localization of the nucleus in secretory epithelia. Mutations in nesprin-1 cause a form of **spinocerebellar ataxia**.

nested PCR Variety of **polymerase chain reaction**, in which specificity is improved by using two sets of primers (nested primers) sequentially. An initial PCR is performed with the 'outer' primer pairs, then a small aliquot is used as a template for a second round of PCR with the 'inner' primer pair.

nestin Large (1621aa) class VI intermediate filament protein found in stem cells of developing brain, later replaced by neurofilament proteins. Functionally similar to other intermediate filament proteins but the sequence is very different.

Netherton's syndrome A disorder of hair, 'bamboo hair', trichorrhexis nodosa and skin (ichthyosiform erythroderma) caused by mutation in the gene for **LEKTI** (lymphoepithelial Kazal-type-related inhibitor).

netrins A conserved family of soluble laminin-related proteins (450–650aa) that are **chemotropic** for embryonic commissural neurons: netrin 1 is secreted by the floorplate, whereas netrin-2 is distributed ventrally except for the floorplate. The netrins are homologous to the product of unc-6, a gene identified in studies of neuronal development of *C. elegans*. Receptors include members of the Deleted in Colorectal Cancer (**DCC**) protein family (see *Drosophila* **frazzled**), and members of the UNC-5 family. UNC-5 tyrosine phosphorylation is known to be important for netrin to induce cell migration and axonal repulsion. Several other netrins are known and some, e.g. netrin-4 (beta-netrin, 628aa), are important in kidney and vascular development as well as in the nervous system.

netropsin Basic peptide antibiotic with antitumour and antiviral activity isolated from *Streptomyces netropsis*. Binds selectively in minor groove of **B-DNA** at AT-rich regions and will induce A to B transition.

network theory (1) In general, theories about the properties of networks of which signalling pathways and the Internet are classic examples. An important property of a network is its robustness to perturbation – alternative pathways can be used to compensate for damage or deletion. (2) In immunology, a theory proposed by Jerne (1974) that the immune system is controlled by a network of interactions between antigen binding sites (paratopes) each of which is capable of binding an epitope on an external antigen and also an idiotope, with a shape resembling the epitope, present on another immunoglobulin molecule.

neu See **erb**.

NeuN Antigenic marker (neuronal nuclei protein) frequently used to identify mature neurons, now shown to be fox-3, one of the **Fox-1** family of splicing factors. Article: http://www.jbc.org/content/284/45/31052.long

neurabin Neuron-specific actin-binding proteins (neurabin-1, 1095aa; neurabin-2, spinophilin, 817aa) that tether protein phosphatase-1 to regions of actin-rich postsynaptic density and will cause F actin bundling. Neurabin-2 is particularly associated with dendritic spines.

neural cell adhesion molecule See **NCAM**.

neural crest A group of embryonic cells that separate from the **neural plate** during neurulation and migrate to give several different lineages of adult cells: the spinal and autonomic ganglia, the **glial cells** of the peripheral nervous system, and non-neuronal cells, such as **chromaffin cells, melanocytes,** and some haematopoietic cells.

neural induction In vertebrates the formation of the nervous system from the **ectoderm** of the early embryo as a result of a signal from the underlying **mesoderm** of the archenteron roof; also known as primary neural induction. The mechanism of neural induction is not yet clear but in *Xenopus* neural induction results from the combined inhibition of **BMP** receptor-regulated serine/threonine kinases and activation of receptor tyrosine kinases that signal through MAPK and phosphorylate **Smad1** in the linker region, further inhibiting Smad1 transcriptional activity. NCBI summary article: http://www.ncbi.nlm.nih.gov/books/NBK10823/

neural plate A region of embryonic ectodermal cells, called neuroectoderm, that lies directly above the **notochord**. During **neurulation**, the cells change shape, so as to produce an infolding of the neural plate (the neural fold) that then seals to form the neural tube.

neural retina Layer of nerve cells in the vertebrate retina, embryologically part of the brain. The incoming light passes through nerve-fibres and intermediary nerve cells of the neural retina, before encountering the light-sensitive rods and cones at the interface between neural retina and the pigmented retinal epithelium.

neural tube See **neural plate, neurulation.**

neuraminic acid See **N-acetyl neuraminic acid.**

neuraminidase Enzyme (sialidase, EC3.2.1.18) catalysing cleavage of neuraminic acid residues from oligosaccharide chains of glycoproteins and glycolipids. Since these residues are usually terminal, neuraminidases are generally exo-enzymes, although an endoneuraminidase is known. For use as a laboratory reagent, common sources are from bacteria such as *Vibrio* or *Clostridium*. A neuraminidase is one of the transmembrane proteins of the envelope of influenza virus and a target for antiviral drugs.

neurapraxia Nerve injury in which the capacity to conduct impulses is lost but the nerve does not degenerate. *Cf.* **neurotmesis, axonotmesis.**

neuraxin (1) In rat, microtubule-associated protein 1B (2459aa) that is cleaved to produce MAP1 light chain LC1. May have a role in neurite extension. (2) Formerly a C-terminal fragment of microtubule-associated protein 5 (MAP5).

neuregulins A family of growth and differentiation factors that are related to epidermal growth factor (EGF) and are ligands for the **ErbB** family of tyrosine kinase transmembrane receptors. Induce growth and differentiation of epithelial, glial and muscle cells in culture. Gene disruption is lethal during embryogenesis with heart malformation and defects

in Schwann cells and neural ganglia. Neuregulin-1 (heregulin-1, 222aa) is produced from a membrane-bound precursor (640aa) and is the ligand for ErbB3 and ErbB4. There are multiple isoforms of neuregulin 1 with diverse functions. Type II neuregulin has a different precursor (850aa) but the secreted product (293aa) binds to the same receptors as neuregulin-1. Neuregulin-3 (precursor, 720aa, secreted form 359aa) binds only to ErbB4 and may be a survival factor for oligodendrocytes. Neuregulin-4 (61aa) is a low affinity ligand for ErbB4.

neurexins A large family of highly variable cell-surface molecules, related to laminin, agrin and the alpha-**latrotoxin** receptor, that may function in synaptic transmission and/or synapse formation. Each of the three vertebrate neurexin genes encodes two major neurexin variants, alpha- and beta-neurexins, that are composed of distinct extracellular domains linked to identical intracellular sequences (e.g. neurexin 1α, 1477aa; neurexin 1-β, 442aa). Alpha-neurexins regulate pre-synaptic N- and P/Q-type Ca^{2+} channels. Beta-neurexin binds to pre-synaptic **neuroligin**. Neurexin IV (contactin-associated protein 1, caspr, p190, 1384aa) seems to play a role in the formation of functionally distinct domains essential for saltatory conduction of nerve impulses in myelinated nerves. In *Drosophila*, it is a component (1284aa) of septate junctions and binds to **contactin** and **coracle**. See also **NCP, paranodin.**

neurite A process growing out of a neuron. As it is hard to distinguish a **dendrite** from an **axon** in culture, the term neurite is used for both.

neurobeachin A **lysosomal-trafficking regulator** (2946aa) that binds to type II regulatory subunits of protein kinase A and anchors/targets them to the membrane. Found in brain and at low levels in a number of other tissues. Neurobeachin-like protein 1 (2694aa) is a candidate for involvement in **amyotrophic lateral sclerosis.**

neuroblast Cells arising by division of precursor cells in neural ectoderm (**neurectoderm**) that subsequently differentiate to become neurons.

neuroblastoma Malignant tumour derived from primitive ganglion cells. Mainly a tumour of childhood. Commonest sites are adrenal medulla and retroperitoneal tissue. The cells may partially differentiate into cells having the appearance of immature neurons.

neurocalcin A dimeric calcium-binding protein (neurocalcin-delta, 193aa) from neurons that belongs to a family with **recoverin**, visin, VILIP and **hippocalcin**. The interaction with F-actin requires calcium. Neurocalcin is myristoylated, and also binds clathrin and tubulin. Abundant in CNS. The *Drosophila* homologue (190aa) is involved in the calcium-dependent inhibition of rhodopsin phosphorylation.

neuroectoderm Ectoderm on the dorsal surface of the early vertebrate embryo that gives rise to the cells (neurons and glia) of the nervous system. Also called the **neural plate**.

neuroendocrine cell See **neurohormone**.

neuroepithelium See **neuroectoderm**.

neurofascin An axonal member of the L1 subgroup of the immunoglobulin superfamily, implicated in neurite extension during embryonic development and in cell adhesion. The canonical isoform is 1347aa but there are at least 13 alternatively-spliced isoforms. Binds **ankyrin**.

neurofibrillary tangle A characteristic pathological feature of the brain of patients with **Alzheimer's disease**, tangles of coarse **neurofibrils** within large neurons of the cerebral cortex. Whether this causes neuronal degeneration or is a secondary consequence remains contentious.

neurofibrils Filaments found in neurons; not necessarily **neurofilaments** in all cases, and in the older literature 'fibrils' are composed of both microtubules and neurofilaments. Originally used by light microscopists to describe much larger fibrils seen particularly well with silver-staining methods.

neurofibromatosis Tumours of neuronal sheath. Type 1 neurofibromatosis, the commonest form, is associated with the the von Recklinghausen neurofibromatosis locus that encodes the NF 1 protein, **neurofibromin, a GTPase activating protein** for the ras proteins. Neurofibromatosis type 2 is caused by inactivating mutations in the *NF2* (**neurofibromin-2**) tumour suppressor gene, the product of which is **merlin**. The Type 2 form is characterized by tumours of the eighth cranial nerve (usually bilateral), meningiomas of the brain, and schwannomas of the dorsal roots of the spinal cord, and has few of the hallmarks of the peripheral (Type 1) form of **neurofibromatosis**. Other variants of the disease are recognised. Mutations in the *SPRED1* gene cause neurofibromatosis 1-like syndrome.

neurofibromin A GTPase-activating protein (2839aa) that activates ras protein; mutations in the gene lead to **neurofibromatosis** Type 1. Neurofibromin-2 is **merlin**, and mutations cause **neurofibromatosis** Type 2.

neurofilament The class of **intermediate filaments** found in axons of nerve cells. In vertebrates assembled from three distinct protein subunits (NF-L 543aa; NF-M, 916aa; NF-H, 1026aa). These proteins, if introduced into fibroblasts, will incorporate into the vimentin filament system. Mutations in NF-L are associated with one form of **Charcot-Marie-Tooth disease**. *Cf.* **neurofibrils**.

neurogenesis Differentiation of the nervous system from the **ectoderm** of the early embryo. There are major differences between neurogenesis in vertebrates and invertebrates.

neurogenins Family of bHLH transcription factors involved in specifying neuronal differentiation. neurogenin-1 (neurogenic differentiation factor 3, 237aa) is related to *Drosophila* atonal and is expressed in and required for specification of dopaminergic progenitor cells; it inhibits the differentiation of neural stem cells into astrocytes. Neurogenin 2 (atonal homolog 4, 272aa) and neurogenin 3 (214aa) control distinct phases of neurogenesis involved in differentiation of different classes of sensory neurons. Mutation in the neurogenin 3 gene, which is also involved in early differentiation of pancreatic endocrine cells, is associated with congenital malabsorptive diarrhoea.

neuroglia See **glial cells**.

neuroglian Protein isolated from *Drosophila* nervous system (1302aa) that is a member of the **immunoglobulin superfamily**, the invertebrate homologue of L1-CAM with strong sequence homology to mouse **NCAM** and **L1**. Two different forms of neuroglian arise by differential splicing. These have identical extracellular domains but differ in the size of the cytoplasmic domains: the long form is restricted to neurons in central and peripheral nervous systems of embryos and larvae and may play a role in neural and glial cell adhesion in the developing embryo. The short isoform may be a more general cell adhesion molecule involved in other tissues and imaginal disk morphogenesis. Vital for embryonic development and essential for formation of septate junctions.

neurogranin A postsynaptic substrate (78aa) for protein kinase C involved in synaptic development and remodeling. Binds calmodulin at low Ca^{2+} levels and the binding is affected by PKC phosphorylation. Neurogranin apparently enhances **long-term potentiation** by promoting calcium-mediated signalling; knock-out mice have learning difficulties.

neurohaemal organs Organs specialised for the release of products of neurosecretory cells into the circulating blood or body fluid. Term usually applied to invertebrate organs such as the corpora cardiaca of insects.

neurohormone A hormone secreted by specialized **neurons** (neuroendocrine cells); e.g. releasing hormones.

neurokinin See **tachykinins**.

neuroleptic drugs Literally 'nerve-seizing': used of drugs that antagonise the effects of **dopamine**.

neuroligins A family of postsynaptic transmembrane proteins, ligands for beta-**neurexins**. Neuroligin 1 (832aa) and neuroligin-2 (550aa) interact with SAP90/PSD95, a multidomain scaffolding protein thought to anchor proteins to postsynaptic sites. Neuroligin 3 (gliotactin homologue, 848aa) may also be involved in glia-glia or glia-neuron interactions in the developing peripheral nervous system. Neuroligin 4 (816aa) exists in X- and Y-linked forms and mutations in the X-linked form are associated with some forms of autism.

neurolin A growth-associated cell surface glycoprotein of the immunoglobulin superfamily of cell adhesion molecules from goldfish and zebrafish (CD166 homologue, DM-GRASP homologue, activated leukocyte cell adhesion molecule A, 564aa) which has been shown to be involved in axonal pathfinding in the goldfish retina and suggested to function as a receptor for axon guidance molecules.

neuromedins Family of neuropeptides. Four classes are recognised: **kassinin**-like, **bombesin**-like, **neurotensin**-like and neuromedin U (NMU). NMU (174aa) stimulates smooth muscle contractions of specific regions of the gastrointestinal tract. The NMU receptors are G-protein coupled: NMUR1 (426aa) is abundantly expressed in peripheral tissues, NMUR2 (415aa) in specific regions of brain, particularly ventromedial hypothalamus where levels of NMU are reduced following fasting.

neuromelanin See **melanin**.

neuromeres Alternate swellings and constrictions seen along the **neuraxis** at early stages of **neural tube** development, thought to be evidence of intrinsic segmentation in the central nervous system. Neuromeres or segments in the hindbrain region are called **rhombomeres** and have been shown to be lineage-restriction units, each constructing a defined piece of hindbrain.

neuromodulation Alteration in the effectiveness of **voltage-gated** or **ligand-gated ion channels** by changing the characteristics of current flow through the channels. The mechanism is thought to involve **second messenger** systems.

neuromodulin Protein (growth-associated protein 43, pp46, 243aa) associated with actively growing axons, especially in the **growth cone**. Binds **calmodulin**, is phosphorylated by **protein kinase C** especially in forms of synaptic plasticity.

neuromuscular junction A **chemical synapse** between a motoneuron and a muscle fibre. Also known as a motor end plate.

neuron *neurone, nerve cell* An **excitable cell** specialized for the transmission of electrical signals over long distances. Neurons receive input from sensory cells or other neurons, and send output to muscles or other neurons. Neurons with sensory input are called sensory neurons, neurons with muscle outputs are called motoneurons; neurons that connect only with other neurons are called interneurons. Neurons connect with each other via **synapses**. Neurons can be very long and a single **axon** can be several metres in length. Although signals are usually sent via **action potentials**, some neurons are **non-spiking**.

neuronal calcium sensor-1 See **frequenin**.

neuronal ceroid lipofuscinoses A group of progressive childhood neurological disorders caused by defects in lysosomes. Variants are caused by mutations in palmitoyl-protein thioesterase-1, lysosomal tripeptidyl peptidase (EC 3.4.14.9), battenin (see **Batten's disease**), a soluble lysosomal glycoprotein (CLN5), a voltage-gated potassium channel in the brain, **major facilitator superfamily domain-containing protein-8** and **cathepsin D**. Other forms are known although the genes have not yet been identified.

neuronal plasticity Ability of nerve cells to change their properties e.g. by sprouting new processes, making new synapses or altering the strength of existing synapses. See **long-term potentiation** and **synaptic plasticity**.

Neuropeptides AF, FF, SF See **FMRFamide-related peptides**.

Neuropeptide Y Peptide neurotransmitter (36aa) found in adrenals, heart and brain. Potent stimulator of feeding and regulates secretion of gonadotrophin-releasing hormone. **Leptin** inhibits NPY gene expression and release. Receptors are G-protein coupled.

neuropeptides Peptides with direct synaptic effects (peptide neurotransmitters) or indirect modulatory effects on the nervous system (peptide neuromodulators). See **neuropeptide Y, FMRFamide-related peptides**, Table N1.

neurophysins Carrier proteins (90−97aa) that transport neurohypophysial hormones along axons, from the hypothalamus to the posterior lobe of the pituitary. Neurophysin 1 (94aa) binds **oxytocin**, neurophysin 2 (93aa) binds **vasopressin**. Both the hormone and the carrier are part of the same precursor polypeptide. Comparable and related carrier/hormone sets are found in a wide range of phyla.

neuropil *neuropile* A network of axons, dendrites and synapses within the central nervous system of vertebrates or within the brain and the central portion of segmental ganglia of arthropods.

neuropilin Neuropilin-1 (NRP1, CD304, 923aa) is a receptor for two unrelated ligands with disparate

TABLE N1. Neurotransmitters

Transmitter	Peripheral nervous system	Central nervous system
Noradrenaline	Some postganglionic sympathetic neurons	Diverse pathways especially in arousal and blood pressure control
Dopamine	Sympathetic ganglia	Diverse; perturbed in Parkinsonism and schizophrenia
Serotonin	Neurons in myenteric plexus	Distribution very similar to that of noradrenergic neurons. Lysergic acid (LSD) may antagonise
Acetylcholine	Neuromuscular junctions (nmj). All postganglionic parasympathetic and most postganglionic sympathetic neurons	Widely distributed, usually excitatory. Possibly antagonises dopaminergic neurons
GABA	Inhibitory at nmj of arthropods	Inhibitory in many pathways
Glutamate	Excitatory at nmj of arthropods	Widely distributed; excitatory
Glycine	—	Diverse; particularly in grey matter of spinal cord
Aspartate	Locust nmj	—
Neuropeptides	Diverse actions in both peripheral and central nervous systems; see Table H2	
Histamine	—	Minor role
Purines	Particularly neurons controlling blood vessels	Mostly inhibitory
Octopamine	Invertebrate nmj	—
Substance P	Sensory neurons of vertebrates	Sensory neurons
RF-amides	Invertebrates	

Other substances known, or proposed to have neurotransmitter function are: adrenaline, agmatine, β-alanine, cholecystokinin, taurine, proctolin, and cysteine.

activities, vascular endothelial growth factor-165 (**VEGF**165) and **semaphorins/collapsins**, mediators of neuronal guidance. The membrane-bound iso-form is a VEGF receptor, the soluble form binds and sequesters VEGF. Neuropilin 2 (931aa) has similar properties but a subtly different tissue distribution. Neuropilin and tolloid like-1 (NETO1, 533aa) is a retina and brain-specific transmembrane protein with homology to both neuropilin and *Drosophila* **tolloid**. It is involved in the development and/or maintenance of neuronal circuitry and is an accessory subunit of the neuronal N-methyl-D-aspartate **glutamate receptor**. NETO2 (525aa) is an accessory subunit of kainate-sensitive glutamate receptors,

neuropore The anterior and posterior openings of the neural tube of the early embryo; usually close at very specific times in development.

neuropsin A secreted serine peptidase (opsin-5, ovasin, kallikrein-8, EC 3.4.21.118, 260aa) with a role in neuronal plasticity; its expression has been shown to be upregulated in response to injury to the CNS. Has also been implicated in learning and memory and the type II splice form of neu-ropsin is only found in hominoid species (humans and apes). Neuropsin has 25–30% amino acid identity with all known opsins. It is expressed in the eye, brain, testis and spinal cord. It is also involved in skin desquamation and keratinocyte proliferation. See **opsin sub-families, neurosin**.

neurosecretory cells Cells that have properties both of electrical activity, carrying impulses, and a secretory function, releasing hormones into the bloodstream. In a sense, they are behaving in the same way as any chemically-signalling neuron, except that the target is the blood (and remote tissues), not another nerve or postsynaptic region.

neurosin An arginine-specific serine endopeptidase (kallikrein 6, protease M, 244aa), expressed by neurons and glial cells. Expression is similar, but not identical, to that of **neuropsin**, especially following injury to the CNS. Degrades amyloid precursor protein, myelin basic protein, gelatin, casein and extracellular matrix proteins such as fibronectin, laminin, vitronectin and collagen. By degrading alpha-**synuclein** it prevents its polymerization, so may be involved in the pathogenesis of Parkinson's disease. See **neuropsin**.

Neurospora An Ascomycete fungus, haploid and grows as a **mycelium**. There are two mating types, and fusion of nuclei of two opposite types leads to meiosis followed by mitosis. The resulting eight nuclei generate eight ascospores, arranged linearly in an ordered fashion in a pod-like **ascus** so that the various products of meiotic division can be identified and isolated. Because of this, *Neurospora crassa* is one of the classic organisms for genetic research; studies on biochemical mutants led Beadle & Tatum to propose the seminal 'one gene-one enzyme' hypothesis.

neurosteroids Steroids synthesized in the brain that have effects on neuronal excitability. The **epalons** may regulate type A **GABA receptors** by allosteric potentiation.

neurotactin (1) See **fractalkine**. (2) *Drosophila* neurotactin (846aa) is a transmembrane receptor, one of the cholinesterase-homologous protein family, involved in cell adhesion and axon fasciculation.

neurotensin A hormone (13aa) derived from a precursor (170aa) from which **neuromedins** (large neuromedin N, 125aa; neuromedin N, 5aa) are also generated. May have an endocrine or paracrine role in the regulation of fat metabolism and causes contraction of smooth muscle.

neurotmesis A nerve injury in which both the nerve and the nerve sheath are disrupted. Does not usually allow regeneration. *Cf.* **axonotmesis**.

neurotoxin A substance, often exquisitely toxic, that inhibits neuronal function. Neurotoxins act typically against the **sodium channel** (e.g. **TTX**) or block or enhance **synaptic transmission** (**curare**, **bungarotoxin**).

neurotransmitter A substance released from the presynaptic terminal of a **chemical synapse** in response to depolarization by an action potential, diffuses across the synaptic cleft, and binds a ligand-gated ion channel on the postsynaptic cell. This alters the resting potential of the postsynaptic cell, and thus its excitability. Examples: **acetylcholine**, **GABA**, **noradrenaline**, **serotonin**, **dopamine**. See Table N1.

neurotrimin A GPI-anchored cell adhesion molecule (IgLON family member 2, 344aa), expressed at high levels during brain development.

neurotrophic Involved in the nutrition (or maintenance) of neural tissue. Classic example is **nerve growth factor**, see **neurotrophins**. *Cf.* **neurotropic**.

neurotrophins Molecules with closely related structures that are known to support the survival of different classes of embryonic neurons. NT-3 (hippocampal-derived neurotrophic factor, NGF-2, 257aa) shows strong similarities to **nerve growth factor** and **brain-derived neurotrophic factor** (including strictly conserved domains that contain 6 cysteine residues) but has different neuronal specificity and regional expression. The NT3 receptor is TrkC tyrosine kinase (839aa) but it is also bound by TNFRSF16. NT4 (neurotrophin-5, 210aa) is a survival factor for peripheral sensory sympathetic neurons. Novel neurotrophin-1 (cardiotrophin-like cytokine factor 1, 225aa) is a cytokine that stimulates B-cells. See (NGF), **brain-derived neurotrophic factor** (BDNF), **GDGF** and **ciliary neurotrophic factor.**

neurotropic Having an affinity for, or growing towards, neural tissue. Rabies virus, which localizes in neurons, is referred to as neurotropic; can also be used to refer to chemicals. Not to be confused with **neurotrophic**.

neurotubules Neuronal **microtubules**.

neurturin A potent neurotrophic factor (197aa), closely related to glial cell line-derived neurotrophic factor (**GDNF**), the two forming a distinct TGFβ subfamily (TRNs, TGFβ-related neurotrophins). The receptor is a tyrosine kinase receptor RET which has a GPI-linked coreceptor. Neurturin will support the survival of sympathetic neurons in culture. Defects in neurturin are a cause of **Hirschsprung's disease**.

neurula The stage in vertebrate embryogenesis during which the neural plate closes to form the central nervous system.

neurulation The embryonic formation of the **neural tube** by closure of the neural plate, directed by the underlying notochord.

neutral mutation A mutation that has no selective advantage or disadvantage. Considerable controversy surrounds the question of whether such mutations can exist.

neutral protease Peptidase that is optimally active at neutral pH: may be from any of several classes of peptidase and the category is not recognised in **MEROPS**. Eukaryotic example is **calpain**; a range of bacterial enzymes were originally designated neutral proteases, e.g. neutral protease from *Staphylococcus hyicus*, hyicolysin, a metallopeptidase.

neutropenia Condition in which the number of **neutrophils** circulating in the blood is below normal. (Normal levels are $4.3 - 10.8 \times 10^9$ cells per litre.) See **Kostmann's syndrome**, and **Chediak-Higashi syndrome**.

neutrophil Commonest (2500–7500/mm^3) blood leucocyte (neutrophil granulocyte, polymorphonuclear leucocyte, PMN or PMNL); a short-lived phagocytic cell of the **myeloid** series, which is responsible for the primary cellular response to an acute inflammatory episode, and for general tissue homeostasis by removal of damaged material. Adheres to endothelium (**margination**) and then migrates into tissue, possibly responding to chemotactic signals. Contain **specific** and **azurophil granules**.

neutrophil immunodeficiency syndrome A rare disorder in which neutrophil function is severely impaired leading to recurrent severe bacterial infections; caused by mutation in the *rac2* gene, which is part of the NADPH oxidase complex.

neutrophil-activating protein See **interleukin-8** and **melanoma growth-stimulatory activity**.

nevus See **naevus**.

Newcastle Disease virus A paramyxovirus (avian paramyxovirus-1) that causes the disease, fowl-pest, in poultry; the lethality of different strains varies and highly virulent strains can cause major mortality in all species of birds.

nexilin An F-actin binding protein (nelin, 675aa) involved in regulating cell migration through association with the actin cytoskeleton. Has an essential role in the maintenance of Z-disk and sarcomere integrity. Localized at cell-matrix adherens junction. Defects in nexilin cause a type of dilated cardiomyopathy.

nexin (1) The linkage between adjacent microtubule doublets of the **axoneme** (between the A tubule of one and the B-tubule of the other), important for the transformation of interdoublet sliding into axonemal bending. There is a repeat at 96 nm intervals. It has been suggested recently that nexin is the multiprotein **dynein regulatory complex**. Article: http://www.ncbi.nlm.nih.gov/pmc/articles/PMC2806320/?tool=pubmed#__secid3611834. (2) See **protease nexin-1**. (3) See **sorting nexins**.

Nezelof's syndrome Congenital T-cell deficiency associated with thymic hypoplasia and distinct from Bruton type agammaglobulinemia and from **severe combined immunodeficiency disease** (SCID). The molecular defect is unknowm.

NF-1 (1) Nuclear factor-1; CCAAT-binding transcription factor. Family of dimeric transcription factors ($\sim$500aa); there are four NF1 genes, NF1-A, -B, -C (**CTF/NF1**) and -X that give rise to multiple isoforms by alternative splicing in many tissues. Essential for adenovirus DNA replication and the

transcription of many cellular genes. (2) Do not confuse with **neurofibromatosis** Type 1 (NF1).

NF-B Nuclear factor-B, more commonly **NFκ-B**. Not to be confused with **NF1-B**.

NF-E1 See **GATA transcription factors**.

NFAR Conserved protein (nuclear factor associated with dsRNA, interleukin enhancer-binding factor 3, ILF3, NFAT-90, 894aa) that may facilitate double-stranded RNA-regulated gene expression at the level of post-transcription. There are multiple alternatively spliced variants.

NFAT A family of transcription factors (nuclear factor of activated T-cells, NFATc1 - NFATc5, mostly $\sim$940aa) involved in regulation of IL-2, IL-4 and IFNγ gene transcription (in concert with other transcription factors). NFAT is cytoplasmic until dephosphorylated by **calcineurin**, a step that is inhibited by **ciclosporin** and **FK506**, then translocates to the nucleus. NFATc also controls the expression of the endogenous calcineurin inhibitory proteins (calcipressins), thereby forming a negative feedback loop.

NFκB A transcription factor originally found to switch on transcription of genes for the kappa class of immunoglobulins in B-cells, but subsequently shown to activate the transcription of more than 20 genes in a variety of cells and tissues. NFκB is found in the cytoplasm in an inactive form, bound to the protein IκB. A variety of stimuli, such as tumour necrosis factor, phorbol esters and bacterial lipopolysaccharide activate it, by releasing it from IκB, allowing it to enter the nucleus and bind to DNA. NFκB then forms homo- or heterodimeric complexes with Rel-like domain-containing proteins RELA/p65, RELB, NFκB1/p105, NFκB1/p50, REL and NFκB2/p52; and the heterodimeric p65–p50 complex appears to be most abundant. The dimerisation and DNA binding activity are located in N-terminal regions of 300 amino acid that are similar to regions in the Rel and dorsal transcription factors. NFκB p100 subunit (900aa) is processed to produce the p52 form; the p105 subunit (968aa) is processed to produce p50.

NFκB-inducing kinase Mitogen-activated protein kinase kinase kinase 14 (NIK, MAP3K14, 947aa) a lymphotoxin beta-activated serine/threonine kinase exclusively involved in the activation of NFκB and its transcriptional activity. Nik-related protein kinase (NRK, 1582aa) may phosphorylate **cofilin-1** and induce actin polymerization in the late stages of embryogenesis. Involved in the TNFα-induced signalling pathway.

Ngaro A family of tyrosine recombinase retroelements found in zebrafish, fungi and echinoderms.

NgCAM In chickens, neural-glial cell adhesion molecule (1266aa) one of the immunoglobulin superfamily and the L1/neurofascin/NgCAM family. See **CAM**.

NGF See **nerve growth factor**.

N-glycanase Enzyme that cleaves asparagine-linked oligosaccharides from glycoproteins. N-glycanase 1 (EC 3.5.1.52, 654aa) specifically deglycosylates denatured N-linked glycoproteins in the cytoplasm and assists their proteasome-mediated degradation.

N-glycans Oligosaccharides, based on the common core pentasaccharide $Man_3GlcNAc_2$, that are linked to a protein backbone via an amide bond to asparagine residues in an Asn-X-Ser/Thr motif, where X can be any amino acid, except Pro. Glycosylation with N-glycans (N-glycosylation) occurs co-translationally; the completed oligosaccharide is transferred from the dolichol precursor to the Asn of the target glycoprotein by oligosaccharyltransferase (OST) in the cisternal space of the ER where trimming and further modification may also take place. In the Golgi, high mannose N-glycans can be converted to a variety of complex and hybrid forms. Database entry: http://prosite.expasy.org/cgi-bin/prosite/nicedoc.pl?PDOC00001

NHE A family of ion exchangers that includes six isoforms (e.g. NHE1, Solute carrier family 9 member 1, 815aa) that function in an electroneutral exchange of intracellular H^+ for extracellular sodium ions to regulate intracellular pH. The various isoforms have different tissue distributions.

NHE-RF Cytoplasmic scaffolding phosphoprotein (Na^+-H^+ exchanger regulatory factor, ezrin-radixin-moesin binding protein 50, solute carrier family 9 isoform A3 regulatory factor 1, SLC9A3R1, 358aa) involved in **protein kinase A** (PKA) mediated regulation of ion transport and recruiting **PTEN** to the cytoplasmic region of the PDGF receptor, inhibiting PI3kinase signalling.

NHEJ Non-homologous end-joining, the primary mammalian DNA repair mechanism that occurs through recognition and repair of double-strand breaks by a variety of proteins that process and rejoin DNA termini by direct ligation. See **synapsis** #2. Proteins known to play a role in NHEJ include the DNA-dependent protein kinase catalytic subunit (DNA-PKcs), the **Ku** heterodimer, XRCC4, **cernunnos** and DNA ligase IV. Originally thought to be restricted to eukaryotes but now known to occur in prokaryotes (see **DNA ligase D**).

niacin See **Vitamin B**.

nibrin Protein (cell cycle regulatory protein p95, Nijmegen breakage syndrome protein 1, 754aa) defective in **Nijmegen breakage syndrome**. Has two modules found in cell cycle checkpoint proteins and a **forkhead**-associated domain. It is the p95

protein component of the double strand break repair complex (**MRN complex**).

nicalin A single-pass membrane protein found in ER (**nicastrin**-like protein, 563aa) that may antagonize **nodal** signalling during mesodermal patterning and interacts with nodal modulators.1

nicastrin Transmembrane glycoprotein (709aa) that interacts with β-**amyloid precursor protein** (βAPP) and with **presenilins**. Part of the gamma-secretase complex involved, together with presenilins, in processing of βAPP to amyloid-β-peptide and in processing of **notch**. Also found in *Arabidopsis*.

nick A point in a double-stranded DNA molecule where there is no **phosphodiester bond** between adjacent nucleotides of one strand, typically through damage or enzyme action.

nick translation A technique used to radioactively label DNA. *E. coli* DNA polymerase I will add a nucleotide, copying the complementary strand, to the free 3′-OH group at a nick, at the same time its exonuclease activity removes the 5′-terminus. The enzyme then adds a nucleotide at the new 3′-OH and removes the new 5′-terminus. In this way one strand of the DNA is replaced starting at a nick, which effectively moves along the strand. Nick translation refers to this translation or movement and not to protein synthesis. In practice, DNA is mixed with trace amounts of **DNAase** I to generate nicks, **DNA polymerase** I and labelled nucleotides. Because the nicks are generated randomly the DNA preparation can be uniformly labelled and to a high degree of specific activity.

nicotinamide adenine dinucleotide See **NAD**.

nicotinamide adenine dinucleotide phosphate See **NAD**.

nicotine A plant alkaloid from tobacco; blocks transmission at nicotinic synapses. See **nicotinic acetylcholine receptor**.

nicotinic acetylcholine receptor *nAChR* Integral membrane protein of the postsynaptic membrane to which **acetylcholine** binds. The receptor contains an integral **ion channel**; as a result of binding of acetylcholine, ion channels in the subsynaptic membrane are opened. At the **neuromuscular junction**, the nicotinic acetylcholine receptor initiates muscle contraction. Currently the best characterized ion channel protein: made of a hetero-pentamer of related subunits, although a homo-pentamer is functional in insects. Structural studies show that the acetylcholine binding site and the ionic channel are part of the same macromolecular unit. The nAChR mediates rapid transduction events (1 ms) whereas receptors activating **G-protein**-coupled receptors operate on slower time scales (millisecond to second range).

nicotinic acid A precursor of **NAD**, a product of the oxidation of nicotine.

NIDDM See **diabetes**.

nidogen A dumbbell shaped 150 kDa sulphated glycoprotein, found in all basement membranes, consisting of three globular regions, G1–G3. G1 and G2 are connected by a thread-like structure, whereas that between G2 and G3 is rod-like. The nidogen G2 region binds to collagen IV and **perlecan**. The connecting rod has 5–6 **EGF-like domains** of cysteine-rich repeats, one of which has an RGD sequence for cellular interaction. Also interacts with laminin and ablation of the high-affinity nidogen-binding site of the laminin gamma1 chain is lethal. Oddly, the nidogen G2 beta-barrel domain has structural similarity to green fluorescent protein. There are two isoforms, nidogen-1 (entactin, 1247aa) and nidogen-2 (osteonidogen, 1375aa) with broadly similar properties.

Nidovirales An Order of viruses (see **virus taxonomy**).

Niemann-Pick disease Severe lysosomal storage disease. Types A & B are caused by deficiency in **sphingomyelinase**; excess sphingomyelin is stored in 'foam' cells (macrophages) in spleen, bone-marrow and lymphoid tissue. Niemann-Pick disease Type C1 and Type D are due to a defect in Niemann-Pick C1 protein (1278aa) with sequence similarity to the morphogen receptor 'patched', the putative sterol-sensing regions of **SREBP** and **HMGCoA reductase** and that has a critical role in regulating of intracellular cholesterol trafficking. Niemann-Pick disease type C2 protein (epididymal secretory protein E1, 51aa) is also involved in the regulation of the lipid composition of sperm membranes. Cf. **Pick's disease**.

nif (1) Nif genes are a complex set of genes in nitrogen fixing bacteria that code for the proteins required for **nitrogen fixation**, particularly the **nitrogenase**. Present as an operon in **Klebsiella** and carried on plasmid in **Rhizobium**. (2) In *S. pombe* Nif1 (Nim1-interacting factor 1, 681aa) is a negative regulator of mitosis. (3) In humans, NIF3-like protein 1 (NIF3L1, 377aa) is involved in alternative splicing; NIF3L1-binding protein 1 (Ngg1-interacting factor 3-like protein 1-binding protein 1, THO complex subunit 7 homolog, 204aa) is a component of the spliceosome, part of the THO subcomplex of the TREX complex.

nifedipine A calcium channel blocker (346 Da) used experimentally and therapeutically as a coronary vasodilator.

niflumic acid A rather nonspecific inhibitor of calcium-regulated chloride channels.

nigericin An ionophore capable of acting as a carrier for K^+ or Rb^+ or as an exchange carrier for H^+ with K^+. Originally used as an antibiotic and derived from *Streptomyces hygroscopicus*. Has been used in investigating chemiosmosis and other transport systems.

NIH 3T3 cells Very widely used mouse fibroblast cell line; 3T3 cells have been derived from different mouse strains and it is therefore important to define the particular cell line. NIH strain were from the National Institute of Health in the USA. Cf. Swiss 3T3, Balb/c 3T3.

Nijmegen breakage syndrome *NBS* Autosomal recessive chromosomal instability syndrome characterized by microencephaly, growth retardation, immunodeficiency, and predisposition to tumours. Cells from patients are hypersensitive to ionizing radiation in the same way as cells from **ataxia telangiectasia**. See **nibrin**.

NIK A **MAP kinase** kinase kinase (NFκB-inducing kinase, MAP3K14, 947aa) involved in CD3/CD28 activation of IL-2 transcription. Splenic T-cells from aly/aly mice (defective in NIK) have a severe impairment in IL-2 and GM-CSF but not TNF secretion in response to CD3/CD28. Apparently activates the CD28 responsive element (CD28RE) of the IL-2 promoter and strongly synergizes with c-rel in this activity. NIK and IκB-binding protein (NIBP, trafficking protein particle complex subunit 9, Tularik gene 1 protein, 1148aa) activates NFκB by phosphorylating the **IKK** complex; defects can cause a form of mental retardation. There is a NIK- and IκB-binding protein homolog (1320aa) in *Drosophila*.

ninein A protein (glycogen synthase kinase 3 beta-interacting protein, 2090aa) found in the centrosome and required for the positioning and anchorage of the minus-end of microtubules in epithelial cells. Autoantibodies against ninein are found in autoimmune disease such as **CREST syndrome**. Ninein-like protein (1382aa) is involved in microtubule organization of interphase cells.

ninhydrin Pale yellow compound used to detect amino acids and proteins (compounds containing free amino or imino groups) with which it forms a deeply coloured purple-blue compound.

NIPA1 Product of the NIPA1 gene (non-imprinted gene in Prader-Willi syndrome/Angelman syndrome chromosome region 1, ichthyin, 329aa) a Mg^{2+} transporter (transports other divalent cations less efficiently). It is defective in a form of **spastic paraplegia** and in **ichthyosis**. Like the *Drosophila* orthologue, spichthyin, seems to inhibit **bone morphogenetic protein** (BMP) signalling (as do **spastin** and **spartin**) by promoting the internalization and degradation of BMP receptors. Article: http://hmg.oxfordjournals.org/content/18/20/3805.long

Nipah virus A virus of the Paramyxoviridae family that causes encephalitis, frequently fatal. It is one of two viruses in the genus *Henipavirus* (the other being *Hendravirus*).

nischarin Imidazoline receptor 1 (1504aa) that acts either as the functional imidazoline-1 receptor (I1R) candidate or as a membrane-associated mediator of the I1R signalling.

nisin A post-translationally modified **lantibiotic** (34aa), widely used as a food preservative. Forms aqueous pores in bacterial plasma membrane. Gallidermin and epidermin possess the same putative lipid II binding motif as nisin although both are shorter (22aa).

Nissl granules Discrete clumps of material seen by phase contrast microscopy in the perikaryon of some neurons, particularly motor neurons. They are basophilic and contain much RNA, and are regions very rich in rough endoplasmic reticulum. Their reaction following damage to neurons is characteristic; they disperse through the cytoplasm giving a general basophilia to the whole cell body.

Nitella Characean alga that has giant, multinucleate internodal cells. These show **cytoplasmic streaming** at rates of several micrometres per second and have been used as models for motile phenomena in cells, and in studies on ionic movement.

nitric oxide Gas (NO) produced from L-arginine by the enzyme **nitric oxide synthase**. Acts as an intracellular and intercellular messenger (endothelium-derived relaxation factor) in a wide range of processes, in the vascular and nervous systems. The intracellular 'receptor' is a soluble (cytoplasmic) **guanylate cyclase**. In the immune system, large amounts can be generated as a cytotoxic attack mechanism. NO signalling is phylogenetically widespread, suggesting it is an ancient mechanism.

nitric oxide synthase Enzyme (NO synthase, NOS, EC 1.14.13.39) that produces the vasorelaxant **nitric oxide** (endothelium-derived relaxation factor) from L-arginine. There are several isoforms, two constitutive and calmodulin dependent (NOS1, 1434aa; endothelial NOS, NOS3, 1203aa), and an inducible calcium-independent form (iNOS, NOS2, 1153aa). The *Drosophila* homologue (1349aa) generates NO which is a second messenger in diverse signalling pathways.

nitroblue tetrazolium A yellow dye that can be taken up by phagocytosing neutrophils and reduced to deep-blue insoluble formazan if the **metabolic burst** is normal. Reduction does not take place in **chronic granulomatous disease**.

nitrocellulose paper Paper with a high nonspecific absorbing power for biological macromolecules. Very important as a receptor in **blotting** methods. Bands are transferred from a chromatogram or electropherogram either by blotting on

nitrocellulose sheets or by electrophoretic transfer. The replica can then be used for sensitive analytical detection methods.

nitrogen fixation The incorporation of atmospheric nitrogen into ammonia by various bacteria, catalysed by **nitrogenase**. This is an essential stage in the nitrogen cycle and is the ultimate source of all nitrogen in living organisms. In the sea, the main nitrogen fixers are **Cyanobacteria**. There are several free-living bacteria in soil that fix nitrogen including species of *Azotobacter, Clostridium* and *Klebsiella. Rhizobium* only fixes nitrogen when in symbiotic association, in root nodules, with leguminous plants. The oxygen-sensitive nitrogenase is protected by plant-produced leghaemoglobin and the plant obtains fixed nitrogen from the bacteria. See *Frankia*.

nitrogen mustards A series of tertiary amine compounds having vesicant (blistering) properties similar to those of mustard gas. They can alkylate compounds such as DNA and derivatives have been used for cancer chemotherapy.

nitrogenase Oxygen-labile enzymes (EC 1.18.6.1) found in nitrogen-fixing bacteria that reduce nitrogen to ammonia (also ethylene to acetylene). See **nitrogen fixation**.

nitrosamines Molecules with a N−N=O group (N-nitrosamines): many are carcinogens or suspected carcinogens.

Nitrospira A bacterial phylum. The genus *Nitrospira* within the phylum are nitrite-oxidizing bacteria, important in marine habitats and in sewage sludge.

NK cells See **killer cells**.

NKCC1 Electroneutral cation-coupled chloride cotransporter (Na^+-K^+-$2Cl^-$ co-transporter-1, SLC12A2, 1212aa) found in kidney; regulated by **SPAK** and **OSR1** and inhibited by **thiazide diuretics**.

NKG2 Type II integral membrane proteins (NKG2-A/NKG2-B, CD159a, 233aa; NKG2-C, CD159c, 231aa; NKG2-D, killer cell lectin-like receptor subfamily K member 1, NK cell receptor D, CD314, 216aa; NKG2-E, 240aa; NKG2-F, 158aa) that are receptors for the recognition of MHC class I HLA-E molecules by NK cells and some cytotoxic T-cells.

NLA Product of the NLA (nitrogen limitation adaptation) gene, a RING-type ubiquitin ligase localized to the nuclear speckles, where it interacts with the *Arabidopsis* ubiquitin conjugase 8. NLA is a positive regulator for the development of the adaptability of *Arabidopsis* to nitrogen limitation. Article: http://www.ncbi.nlm.nih.gov/pubmed/17355433?dopt=AbstractPlus&holding=f1000,f1000m,isrctn

N-lines Regions in the sarcomere of striated muscle. The N1 line is in the I-band near the Z-disc, the N2

line is at the end of the A-band. The N-lines may represent the location of proteins such as **nebulin** that contribute to the stability of the sarcomere.

NM23H1 A tumour suppressor, product of the NME1 gene, a granzyme A-activated DNase (metastasis inhibition factor nm23, nucleoside diphosphate kinase A, EC 2.7.4.6, 152aa) that is part of the **SET complex**. It is inhibited by SET until released by granzyme A when it relocates to the nucleus, generates single-strand breaks and induces apoptosis. Important in the synthesis of nucleoside triphosphates other than ATP. Also has serine/threonine-specific protein kinase, geranyl and farnesyl pyrophosphate kinase, histidine protein kinase and $3'-5'$ exonuclease activities. Involved in cell proliferation, differentiation and development, signal transduction, G protein-coupled receptor endocytosis, and gene expression. Required for neural development including neural patterning and cell fate determination.

NMDA A powerful agonist (N-methyl-D-aspartic acid) for a subclass of **glutamate receptors** (NMDA receptors). See **excitatory amino acids**. Overactivation of NMDA receptors on the surface of brain cells post ischaemia, activates **SREBP-1**, which subsequently causes cell death.

N-methyl-D-aspartate See **NMDA**.

NMRI Nuclear **magnetic resonance imaging**.

NMS complex A multi-protein complex (Ndc80-MIND-Spc7 complex) in *S. pombe* important in kinetochore function during late meiotic prophase and throughout the mitotic cell cycle, required for correct segregation of chromosomes and for maintaining the inner centromere structure. The complex consists of **mis12** and the kinetochore proteins mis13 (329aa), mis14 (210aa), ndc80 (624aa), nnf1 (145aa), nuf2 (441aa), spc7 (1364aa), spc24 (198aa) and spc25 (238aa). Homologous complexes are found in other eukaryotes. See **MinD**.

N-myc See **myc**.

NO See **nitric oxide**.

NO synthase See **nitric oxide synthase**.

Nocardia Genus of Gram-positive bacteria that form a **mycelium** that may fragment into rod- or coccoid-shaped cells. They are very common **saprophytes** in soil but some are opportunistic pathogens of humans, causing nocardiosis.

nociceptin See **orphanin FQ**.

nociception Detection of pain. Nociceptors (nocireceptors, pain receptors) are sensory nerve-ending that send signals that cause pain in response to certain stimuli. See **capsaicin**.

nocistatin See **orphanin FQ**.

nocodazole Microtubule disrupting compound that binds to the tubulin heterodimer rendering it assembly-incompetent.

Noctiluca A bioluminescent dinoflagellate. Responsible for many instances of marine phosphorescence.

NOD mice A strain of mice (non-obese diabetic mice) that have unique histocompatibility antigens; pancreatic B cells are destroyed by an autoimmune response.

NOD proteins A family of cytoplasmic proteins (nucleotide-binding oligomerization domain proteins, NODs, caspase recruitment domain proteins, CARDs) that are pathogen recognition receptors. They bind breakdown products of bacterial peptidoglycan and through receptor-interacting protein 2 (RIP2) stimulate **NFκB** activation. See **CARD domain**. NOD1 (CARD4, 953aa) which binds a N-acetylglucosamine-N-acetylmuramic acid tripeptide from Gram-negative peptidoglycan has structural similarity to a class of disease-resistance plant proteins that induce localized cell death at the site of pathogen invasion. NOD2 (CARD15, Inflammatory bowel disease protein 1, 1040aa) appears to have a function in the adaptive immune system in T-cell differentiation, independently of its role as a pattern recognition receptor in the innate immune system.

nodal A protein (347aa) of the TGFβ superfamily, originally identified in mouse (354aa) but with homologues in most vertebrates that is expressed in the epiblast and visceral endoderm of the mammalian embryo. Nodal signals induce mesoderm and endoderm. See **cerberus**. Nodal modulators (NOMO1, 1222aa; NOMO2, 1267aa; NOMO3, 1222aa) may antagonise nodal signalling; they interact with **nicalin**.

node A knob, knot, protuberance, or swelling (e.g. lymph node). In networks, a connection point. In plants a point on a plant stem at which one or more leaves are attached.

node of Ranvier A region of exposed neuronal plasma membrane in a myelinated axon. Nodes contain very high concentrations of **voltage-gated ion channels**, and are the site of propagation of action potentials by saltatory conduction.

nodularins Hepatotoxic cyclic pentapeptides similar to microcystins. The amino acid composition may vary but the hydrophobic amino acid, ADDA, is essential for activity. Like **microcystin** binds to the same site on serine/threonine phosphatases as **okadaic acid** and inhibit activity.

nodulins *Nod factors* The products of genes that are induced during the formation and function of a root nodule, resulting from the infection of a legume's roots by rhizobial bacteria. Some (e.g. Nod22) may protect against oxidative stress, others (e.g. nodulin-26) are **aquaporins**, others may be enzymes (e.g. nodulin-35 which is a uricase-2 isozyme, EC 1.7.3.3, 309aa).

noelins A sub-family of extracellular proteins with proposed roles in neural and neural crest development, members of the **olfactomedin** family. Noelin 1 (olfactomedin 1, 485aa) regulates the production of neural crest cells by the neural tubeis. Noelin 3 (olfactomedin-3, optimedin, 478aa) appears to be a downstream target regulated by Pax6 in eye development. Noelin-2 is olfactomedin-2 (454aa). See **pancortin**.

noggin Dorsalising factor (232aa) homologous to that produced by **Spemann's organizer** region of the amphibian embryo. Binds and inhibits **bone morphogenetic protein** and essential for cartilage morphogenesis and joint formation. Mutations are associated with bone or joint fusion abnormalities (synostoses). See **chordin**.

nogo A neuronal multi-pass membrane protein, expressed by oligodendrocytes, involved in diverse processes that include axonal fasciculation and apoptosis. Like **MAG** and **OMgp**, nogo can cause growth cone collapse and inhibit neurite outgrowth *in vitro*. There are three splice variants, nogo-A (reticulon 4, 1192aa), nogo-B and nogo-C, the latter two lacking the axon-inhibiting extracellular domain of 66aa. Nogo may blocking axonal regeneration in the CNS. Nogo, MAG and OMgp lack sequence homologies but all bind to the Nogo receptor (NgR, 473aa), a GPI-linked cell surface molecule widely expressed in brain which, in turn, binds p75 (neurotrophin receptor) and activates **RhoA**.

nojirimycin See **deoxynojirimycin**.

Nomarski differential interference contrast See **differential interference contrast**.

NompA A transmembrane protein in *Drosophila* (no-mechanoreceptor-potential-A, 1549aa, various isoforms) with a large extracellular segment containing a **ZP domain**; thought to be involved in sound-detection and required to connect mechanosensory dendrites to sensory structures. Abstract: http://www. cell.com/neuron/abstract/S0896-6273%2801% 2900215-X

non-coding DNA DNA that does not code for part of a polypeptide chain or RNA.This includes **introns** and **pseudogenes**. In eukaryotes the majority of the DNA is non-coding. The non-coding strand is the so-called nonsense strand, as opposed to the sense strand which is actually translated into mRNA. See **non-coding RNA**.

non-coding RNA Any **RNA** that does not encode a protein. Historically this meant mostly **ribosomal RNA** and **tRNAs** but more recently the importance of a range of different non-coding RNAs has been recognised, and the category includes **microRNAs**, **catalytic RNA** (ribozyme), piwi-interacting RNA (see **piwi**), small nuclear RNAs and small nucleolar RNAs, small interfering RNA (**siRNA**) and long non-coding RNAs.

non-competitive inhibitor A compound that binds at a site other than the substrate-binding site and inhibits the catalytic activity of an enzyme.

non-cyclic photophosphorylation Process by which light energy absorbed by **photosystems I and II** in chloroplasts is used to generate ATP (and also NADPH). Involves photolysis of water by photosystem II, passage of electrons along the photosynthetic electron transport chain with concomitant phosphorylation of ADP, and reduction of $NADP^+$ using energy derived from photosystem I.

non-disjunction Failure of homologous chromosomes or sister **chromatids** to separate at meiosis or mitosis respectively. It results in aneuploid cells. Non-disjunction of the X chromosome in *Drosophila* allowed Bridges to confirm the theory of chromosomal inheritance.

non-equivalence Term used in cell determination for cells that will give rise to the same sorts of differentiated tissues but that have different positional values (e.g. cells of fore-limb and hind-limb buds).

non-histone chromosomal proteins A very heterogeneous group of proteins found in chromatin, mostly acidic in contrast to the other major group, the basic histones. Includes DNA polymerases, transcription factors, regulator proteins, amongst others.

non-Hodgkin's lymphoma See **lymphoma**.

non-homologous end-joining See **NHEJ**.

non-ionic detergent Detergent in which the hydrophilic head-group is uncharged. In practice hydrophilicity is usually conferred by -OH groups. Examples are the polyoxyethylene p-t-octyl phenols known as Tritons, and octyl glucoside. Non-ionic detergents can be used to solubilize intrinsic membrane proteins with less tendency to denature them than charged detergents. They do not usually cause disassembly of structures such as microfilaments and microtubules that depend on protein-protein interactions.

non-LTR retrotransposons The non-LTR class of retrotransposons integrate into the genome using a mechanism in which an endonuclease nicks the chromosome and DNA synthesis is initiated using the 3′ hydroxyl of the broken strand of target DNA as the primer for reverse transcription. There are two main subdivisions, autonomous and non-autonomous non-LTR retroelements. The autonomous non-LTR retroelements can be subdivided further into R2 elements and **long interspersed nuclear elements** (LINEs).

Non autonomous non-LTR retroelements are **short interspersed nuclear elements** (SINEs). They are probably much more ancient than the **LTR transposons**. Article: http://jvi.asm.org/cgi/content/full/78/6/2967

non-Mendelian inheritance In eukaryotes, patterns of gene transmission not explicable in terms of segregation, independent assortment and linkage. May be due to **cytoplasmic inheritance**, **gene conversion**, meiotic drive, and so on.

non-Newtonian fluid A fluid in which the viscosity varies depending upon the shear stress. The effect can arise because of alignment of non-spherical molecules as flow is established or because of suspended deformable particles, as in blood.

non-reciprocal contact inhibition Collision behaviour between different cell types in which one cell shows contact inhibition of locomotion, and the other does not. An example is the interaction between sarcoma cells and fibroblasts (the former not being inhibited).

non-spiking neuron A neuron that can convey information without generating action potentials. As passive electrical potentials are attenuated over distances greater than the space constant for a neuron (typically 1 mm), this implies that most non-spiking neurons are involved in signalling over relatively short distances. Typical examples are invertebrate stretch receptors and **interneurons** in the central nervous system.

non-steroidal anti-inflammatory drugs See NSAIDs.

non-transcribed spacer See **intergenic spacers** and **internal transcribed spacers**.

Nonidet Trade name for non-ionic detergents, usually octyl- or nonyl- phenoxy-polyethoxy-ethanols.

nonpermissive cell Originally a cell of a tissue type or species that does not permit replication of a particular virus. Early stages of the virus cycle may be possible in such a cell and in the case of tumour viruses the cell may become **transformed**. Now used in a more general sense, of agents and treatments other than viruses.

nonpolar group *hydrophobic group* Group in which the electronic charge density is essentially uniform, and that cannot therefore interact with other groups by forming hydrogen bonds, or by strong dipole-dipole interactions. In an aqueous environment, nonpolar groups tend to cluster together, providing a major force for the folding of macromolecules and formation of membranes. Clusters are formed chiefly because they cause a smaller increase in water structure (decrease in entropy) than dispersed groups. (Nonpolar groups interact with each

other only by the relatively weak London-van der Waals forces).

nonreceptor protein-tyrosine kinase See **tyrosine kinases**.

nonsense codon *nonsense triplet* See **termination codons** and Table C4.

nonsense mutation Mutation in coding DNA producing a **nonsense codon** that prevents the protein from being synthesized.

nonsense strand See **non-coding DNA**.

nonsense-mediated mRNA decay A eukaryotic mRNA surveillance mechanism that detects and degrades mRNAs that have premature termination codons (PTCs). A substantial proportion of alternative splicing events will generate mRNAs with PTCs; skipping the exon with the PTC allows the mRNA to survive. The proteins involved in nonsense-mediated decay are highly conserved across species from plants to humans. See **tsu**. Article: http://www.ncbi.nlm.nih.gov/pmc/articles/PMC395752/?tool=pubmed

Noonan's syndrome A set of developmental disorders, a type of dwarfism, affecting both males and females although it was originally described as the male version of Turner's syndrome. Various mutations cause different types of the syndrome: affected genes include those for the nonreceptor protein tyrosine phosphatase **Shp2**, **raf1**, Kirsten-**ras**, N-ras, the leucine-rich repeat protein SHOC2, cbl and son of sevenless homologue 1 (**SOS1**). Neurofibromatosis-Noonan syndrome is a result of mutation in the **neurofibromin** gene. See **LEOPARD syndrome**.

nopaline An **opine**. The gene for nopaline synthase is carried on the T-DNA of the **Ti plasmid**.

noradrenaline Catecholamine neurohormone (norepinephrine, arterenol), the neurotransmitter of most of the sympathetic nervous system (of so-called adrenergic neurons): binds more strongly to α-adrenergic receptor than β-adrenergic receptor. Stored and released from **chromaffin cells** of the adrenal medulla.

norepinephrine See **noradrenaline**.

norleucine *Nle, Ahx* Non-protein amino acid. Formyl-Norleucyl-Leucyl-Phenylalanine has been used as a substitute for fMLP in studies on neutrophil chemotaxis since it is not so susceptible to oxidation.

normoblast Nucleated cell of the **myeloid cell** series found in bone marrow, that give rise to red blood cells.

normocyte Erythrocyte of normal size and shape.

Norovirus Genus of viruses of the family Caliciviridae. Noroviruses contain a positive strand RNA genome of approximately 7.5 kb. Causes intestinal illness. Norwalk virus is in this genus.

norrin Norrie disease protein (X-linked exudative vitreoretinopathy 2 protein, 133aa) that activates the canonical Wnt signalling pathway and plays a central role in retinal vascularization. Interacts with FZD4 (see **frizzled**). Abstract: http://www.ncbi.nlm.nih.gov/pubmed/17955262?dopt=Abstract

norsolorinic acid An intermediate in the biosynthetic pathway that generates **aflatoxins**.

Northern blot See **blotting**.

Northwestern blot See **blotting**.

Norum's disease A disorder of lipoprotein metabolism cause by deficiency in the lecithin:cholesterol acyltransferase **LCAT** gene (which is also mutated in **fish-eye disease**).

Norwalk virus See *Norovirus*.

nosocomial infections Hospital-acquired infections: commonest are due to *Staphylococcus aureus*, *Pseudomonas aeruginosa*, *E. coli*, *Klebsiella pneumoniae*, *Serratia marcescens*, and *Proteus mirabilis*. The strains responsible are often antibiotic-resistant.

Nostoc Common genus of fresh-water nitrogen-fixing **cyanobacteria** that form colonies of intertwined filaments in a gelatinous sheath.

Notch Family of large transmembrane receptor proteins (350 kDa) that mediate developmental cell-fate decisions; Notch contains 36 repeats of the **EGF-like domain**. Notch precursor is cleaved by a **furin**-like convertase and activation, following ligand-binding, involves further proteolysis by **TACE** to release the extracellular portion which allows a further proteolytic step that releases an intracellular portion that is a transcriptional activator. Mammalian Notch gene mutations have been associated with leukaemia, breast cancer, stroke and dementia (see **CADASIL**). Notch proteins in humans (notch 1; 2555aa; notch2, 2471aa; notch 3 (2321aa; notch 4, 2003aa) are receptors for the membrane-bound ligands Jagged and **Delta**. In *Drosophila*, notch (2703aa) provides positive and negative signals in regulating the differentiation of the central and peripheral nervous system and eye, wing disk, oogenesis, segmental appendages such as antennas and legs, and muscles; binds transmembrane ligands encoded by Ser (**serrate** protein) and Dl (**Delta** protein). In *Drosophila* wing development, the notch receptor is activated at the dorsal/ventral boundary and is important in growth and patterning. **Fringe** (fng) is also involved in Notch signalling, encoding a pioneer protein. Notch signalling is modulated by proteins such as **dishevelled**.

notexins Notexins Np and Ns are phospholipase A2 isoforms found in the venom of *Notexis scutatus scutatus* (tiger snake) that block acetylcholine release at the neuromuscular junction.

notochord An axial mesodermal tissue found in embryonic stages of all chordates and protochordates, often regressing as maturity is approached. Typically a rod-shaped mass of vacuolated cells immediately below the nerve cord, and may provide mechanical strength to the embryo.

notoplate Region of the **neural plate** overlying the **notochord**.

nov A proto-oncogene (nephroblastoma overexpressed, CCN3) first identified in an avain nephroblastoma. The homologous product in humans is an angiogenic inducer of the **CCN family** (insulin-like growth factor-binding protein 9, 357aa), broadly expressed in derivatives of all three germ layers during mammalian development. Aberrant expression is associated with vascular injury and a broad range of tumours.

nova A family of RNA-binding proteins. Nova-1 (neuro-oncological ventral antigen 1, 510aa) may regulate RNA splicing or metabolism in a specific subset of developing neuron (see **NPTB**). Nova-2 (492aa) is similar.

novobiocin An aminocoumarin antibiotic obtained from *Streptomyces niveus* and other *Streptomyces* species, used clinically chiefly against staphylococci and other Gram-positive organisms. Acts as an inhibitor of prokaryotic DNA **gyrase** and eukaryotic type II **topoisomerase** enzymes, interferes with *in vitro* chromatin assembly using purified histones, DNA and **nucleoplasmin**.

NOX See **NADPH oxidases**.

Noxa See **Bcl-2 homology domains**.

noxiustoxin **Charybdotoxin**-related peptide toxin (39aa) from scorpion *Centruroides noxius*. Blocks mammalian voltage-gated potassium channels and high-conductance calcium-activated potassium channels.

NP-40 Non-ionic detergent (Nonidet P40) useful for the isolation and purification of functional membrane proteins. Apparently no longer available commercially but an equivalent, IGEPAL CA-630 (octylphenoxy)polyethoxyethanol, is said to be chemically indistinguishable.

NPAF See **FMRFamide-related peptides**.

NPC (1) **Niemann-Pick disease** type C (NPC), a storage disease, or Niemann-Pick C1 protein. (2) **Nuclear pore complex**. (3) Nasopharyngeal carcinoma. (4) Neural precursor cells.

NPFF See **FMRFamide-related peptides**.

NPL4 A component of the ubiquitin degradation system (nuclear protein localization protein 4 homologue, 608aa) that forms a complex with **UFD1L** and **valosin containing protein** (VCP) that is involved in exporting misfolded proteins from the ER to the cytoplasm for degradation and for the regulation of spindle disassembly at the end of mitosis. The complex of NPL4, UFD11 and VCP is necessary for the formation of a closed nuclear envelope.

NPRAP Delta-catenin-2. See **catenins**.

N-protein (1) Nucleoprotein (nucleocapsid protein, 370aa) of viruses such as Borna disease virus that encapsidates the genome, protecting it from nucleases. (2) Anti-terminator protein of the **lambda bacteriophage** and other phages that plays a key role in the early stages of infection. During the early phase, only two genes *N* and *cro* are transcribed, by transcription of the DNA in opposite directions. N-protein binds to sites on the DNA (nut sites for N-utilisation), prevents rho-dependent termination and allows transcription of the genes. (3) Obsolete name for **GTP-binding proteins** (G-proteins); should be avoided because of confusion with N-protein of bacteriophages.

NPSF See **FMRFamide-related peptides**.

NPTB Neural polypyrimidine tract-binding protein (polypyrimidine tract-binding protein 2, 531aa) an RNA-binding protein which binds to polypyrimidine tracts in introns and mediates negative regulation of exon splicing. Interacts with **nova-1**.

Nramps Integral membrane proteins (natural resistance associated macrophage protein-1, SLC11A1, 550aa; nramp-2, 568aa) expressed only in macrophages/monocytes and PMNs. Localized to endosomal/lysosomal compartment and rapidly recruited to the phagosome membrane following phagocytosis. Mutations in Nramp1 impair macrophage killing of intracellular parasites such as *Mycobacterium tuberculosis* and are associated with onset of rheumatoid arthritis. Nramp2 is very similar to Nramp1 but expressed in more tissues and is known to be an iron transporter. Yeast homologues Smf1 and Smf2 transport divalent cations.

N-recognins A class of E3 ligases (see **ubiquitin conjugating enzymes**) that bind to particular N-terminal residues of a substrate protein that mark that protein for degradation (N-recognin-1, E3 ubiquitin-protein ligase UBR1, EC 6.3.2.-, 1749aa; N-recognin-2, 1755aa). See **N-end rule**.

Nrf2 A transcription factor (nuclear factor erythroid 2-related factor 2, 605aa) that binds to antioxidant response (ARE) elements and up-regulates genes in response to oxidative stress. A cytosolic inhibitor of Nrf2 (kelch-like ECH-associated protein 1,624aa) retains Nrf2 in the cytosol and targets it for ubiquitination and degradation, thereby repressing antioxidant reponses.

NRG (1) See **neuregulin**. (2) *Drosophila* neuroglian (Nrg). (3) In *S. cerevisiae*, a transcriptional repressor (NRG1, zinc finger protein MSS1, 231aa) involved in regulation of glucose repression.

NRK cells A line of normal rat (*Rattus norvegicus*) kidney cells that grow adherently and exhibit epithelial morphology.

NS1 (1) Viral non-structural protein, NS1, a nuclear, dimeric protein that is highly expressed in infected cells and has dsRNA-binding activity. Various biochemical functions, such as ATP binding, ATPase, site-specific DNA binding and nicking, and helicase activities, have been assigned to the protein NS1 and it also suppresses immune response by inhibiting production of inflammatory cytokines. Various other non-structural proteins of viruses are NS2, amongst others.

NSAIDs Non-steroidal anti-inflammatory drugs, a category that includes aspirin, ibuprofen and a wide range of derivatives. Mostly act on the production of early low molecular weight mediators of the acute inflammatory response (particularly **COX-1** and others). Are particularly good at inhibiting swelling but often have undesirable side-effects including gastric irritation.

NSF Homotetrameric protein (N-ethyl-maleimide sensitive factor, EC 3.6.4.6, 744aa) involved, together with three **SNAPs**, in mediating vesicle traffic between medial and trans-Golgi compartments. One of the **AAA family**.

NSMCEs Proteins (non-structural maintenance of chromosomes elements) that interact with **SMC proteins**. Non-structural maintenance of chromosomes element 1 homolog (NSMCE1, 266aa) and NSMCE2 (E3 SUMO-protein ligase component of the SMC5/6 complex, 247aa) interact with the SMC5/SMC6 complex at double-strand breaks in DNA.

NSO cells Murine **myeloma** cell line (plasmacytoma).

NSP1 (1) In *S. cerevisiae*, an essential component of the nuclear pore complex (nucleoporin NSP1, nucleoskeletal-like protein, p110, 823aa) which mediates nuclear import and export. (2) NSP is often used for viral non-structural proteins.

NT2 cells Human teratocarcinoma cell line with properties similar to those of progenitor cells in the central nervous system (CNS). Can differentiate into all three major lineages, neurons, astrocytes, and oligodendrocytes.

NT3 Neurotrophin-3.

NTAL A transmembrane adaptor protein (non-T cell activation linker, 243aa), a **TRAP**, found in mature B-cells and phosphorylated following immunoreceptor engagement. Phosphorylated NTAL recruits **Grb2**.

N-terminal kinases See **JNK**. N-terminal kinase-like protein (SCY1-like protein 1, 808aa) regulates COPI-mediated retrograde traffic but does not seem to have kinase activity.

ntk (1) Non-receptor tyrosine kinase: includes **src**, **yes** and **fyn**. (2) Old name for a nervous tissue and T-lymphocyte kinase. (3) In *Nicotiana tabacum*, shaggy-related protein kinase NtK1 (409aa) that may mediate extracellular signals to regulate transcription in differentiating cells.

NTR domain A domain type implicated in inhibition of zinc metalloproteinases of the metzincin family. Found in **netrins**, complement proteins C3, C4, C5, secreted **frizzled**-related proteins, type I procollagen C proteinase enhancer proteins (**PCOLCE proteins**) and **tissue inhibitors of metalloproteinases** (TIMPs). Article: http://www.jbc.org/content/278/28/25982.full

NTRK Receptor tyrosine kinases that bind **neurotrophins**. NTRK1 (neurotrophic tyrosine kinase receptor, 790aa) is the high affinity receptor for **nerve growth factor**; mutations are associated with sensory neuropathy type IV. Somatic rearrangements of NTRK1 have constitutive tyrosine kinase activity, see **trk**. NTRK3 (839aa) is the receptor for neurotrophin-3. See **SLIT and NTRK-like proteins**.

N-type channels A class of **voltage-sensitive calcium channels**. Restricted to neurons and neuroendocrine cells where they are involved in regulation of neurotransmitter or neurohormone release. Require substantial depolarization to become activated and become inactivated in a time-dependent fashion. Potently inhibited by ω-conotoxin.

NuA4 histone acetyltransferase complex A multi-protein complex involved in transcriptional activation of select genes by acetylation of nucleosomal histones H4 and H2A. The complex has a catalytic subunit KAT5/TIP60 (a histone acetyltransferase, EC 2.3.1.48, 513aa) and many other associated proteins including **ING3** and EPC1 (enhancer of polycomb homolog 1, 836aa). The NuA4 complex interacts with MYC and the adenovirus E1A protein. TIP60, EPC1, and ING3 together constitute a minimal HAT complex termed Piccolo NuA4. NuA4 may have a role in DNA repair when directly recruited to sites of DNA damage.

NUAKs A family of AMPK-related kinases (NUAK family **SNF1**-like kinase, NUAK1, EC 2.7.11.1, 661aa; NUAK2, 628aa) activated by the tumour suppressor **LKB1**. Involved in tolerance to glucose starvation.

nucellus The inner portion of the plant ovule consisting of diploid maternal tissue. It is surrounded by the integuments (one in gymnosperms, two in angiosperms) that will form the seed-coat. It has the function of a **megasporangium** and during ovule development contains a megasporocyte (megaspore mother cell) which gives rise to haploid cells by meiosis; three of the four haploid cells degenerate, leaving a single megaspore surrounded by the nucellar tissue.

nuclear actin binding protein Nuclear protein, dimer of 34 kDa subunits. Binds actin with Kd of around 25 mM. Has not appeared in recent literature but could be what is now termed **gCap39** or **myopodin**.

nuclear envelope Membrane system that surrounds the nucleus of eukaryotic cells. Consists of inner and outer membranes separated by perinuclear space and perforated by nuclear pores. The term should be used in preference to the term 'nuclear membrane' which is potentially very confusing. The plant NE appears to differ from the metazoan and yeast NE in both function and composition. Higher plant cells lack centrosomal structures and instead use the NE as the major site of microtubule nucleation during mitosis (see **matrix attachment region-binding proteins**. Plants do not have **lamins** or identified **nucleoporins**.

nuclear export signal A leucine-rich motif (e.g. LQLPPLERLTL, the NES in rev protein of HIV-1) in proteins destined for export from the nucleus by **exportins** such as **CRM-1**.

nuclear factor 1 See **NF-1**.

nuclear lamina A fibrous protein network lining the inner surface of the **nuclear envelope.** The extent to which this system also provides a scaffold within the nucleus is controversial. Proteins of the lamina in metazoa and yeast are **lamins** A, B and C, which have sequence homology to proteins of **intermediate filaments**.

nuclear localization signal *NLS* In eukaryotes, a peptide signal sequence that identifies a protein as being destined for the nucleus (see **importins**). Frequently the signal sequence is a collection of basic amino acids downstream of a helix-breaking proline; e.g. SV40 T (Pro-Lys-Lys-Lys-Arg-Lys-Val). *Cf.* **nuclear export signal**.

nuclear magnetic resonance *NMR* Biophysical technique which allows the spectroscopy or imaging of molecules containing at least one paramagnetic atom (e.g. ^{13}C, ^{31}P). Although non-invasive, the scale of the equipment needed to generate the radio-frequency electromagnetic and magnetic fields, and the computer power needed to analyse the results, are non-trivial. Widely used as a diagnostic and investigative technology in medical practice (functional magnetic resonance imaging, **fMRI**).

nuclear matrix Protein latticework filling the nucleus that anchors **DNA replication** and **transcription** complexes. In plants the nuclear matrix constituent protein (NMCP, 1042aa) may function as a **lamin** homologue. Identified in carrot, celery and *Arabidopsis*; contains a central domain with long α-helices exhibiting heptad repeats of apolar residues and terminal domains that are predominantly nonhelical. It also contains potential **nuclear localisation sequence** motifs. NMCP1 shares a weak sequence similarity to myosin, tropomyosin and IF proteins and fractionates with the nuclear matrix.

nuclear membrane See **nuclear envelope**.

nuclear mitotic apparatus protein A microtubule-binding protein (Numa1, formerly centrophilin, 2115aa) that is an abundant component of interphase nuclei and important in mitotic spindle assembly and maintenance, tethering microtubules to spindle poles. NuMA and its invertebrate homologues have a similar tethering role at the cell cortex, and mediate asymmetric divisions during development. May be a component of the nuclear matrix.

nuclear polyadenylation hexanucleotide Sequence of mRNA (AAUAAA) that is required for polyadenylation in the nucleus. It is also necessary for cytoplasmic polyadenylation. See **cytoplasmic polyadenylation element**.

nuclear pore Openings in the nuclear envelope, ca10 nm diameter, through which molecules such as nuclear proteins (synthesized in the cytoplasm) and mRNA must pass. Pores are generated by a large protein assembly, the nuclear pore complex. See **nucleoporins**.

nuclear receptor Generic term for a receptor in the nucleus that binds diffusible signal molecules such as steroid hormones.

nuclear RNA The nucleus contains RNA that has just been synthesized, but in addition there is some that seems not to be released, or is only released after further processing, the heterogenous nuclear RNA (**hnRNA**) and small RNA molecules associated with protein to form **snRNPs** (small nuclear ribonucleoproteins).

nuclear run-on An assay (sometimes called a nuclear run-off) used to identify genes that are being transcribed at a particular instant; nuclei are rapidly isolated from cells, and incubated briefly with labelled nucleotides. This gives a population of labelled RNAs that were being transcribed immediately before isolation. These can be studied directly; or (more commonly) used as a probe to identify corresponding cDNAs.

nuclear transplantation Experimental approach in study of nucleo-cytoplasmic interactions, in which a nucleus is transferred from one cell to the cytoplasm (which may be anucleate) of a second.

nuclear transport Passage of molecules in and out of the nucleus, presumably via nuclear pores. Passage of proteins into the nucleus may depend on possession of a **nuclear localization sequence**. See **exportins, importins**.

nuclear transport factor A protein (Ntf2, placental protein 15, PP15, 127aa) that facilitates protein transport into the nucleus. Interacts with the **nucleoporin** p62 and with **Ran**.

nuclease An enzyme (EC 3.1.33.1) capable of cleaving the phosphodiester bonds between nucleotide subunits of nucleic acids.

nucleation A general term used in polymerization or assembly reactions where the first steps are energetically less favoured than the continuation of growth. Polymerization is much faster if a preformed seed is used to nucleate growth. (e.g. microtubule growth is nucleated from the **microtubule organizing centre**, although the nature of this nucleation is not known).

nucleic acids Linear polymers of nucleotides, linked by $3',5'$ phosphodiester linkages. In DNA, deoxyribonucleic acid, the sugar group is deoxyribose, and the bases of the nucleotides adenine, guanine, thymine and cytosine. RNA, ribonucleic acid, has ribose as the sugar, and uracil replaces thymine. DNA functions as a stable repository of genetic information in the form of base sequence. RNA has a similar function in some viruses but has a much more diverse range of functions: an informational intermediate (mRNA), a transporter of amino acids (tRNA), in a structural capacity, as an enzyme (**ribozyme** or, as has more recently emerged, involved in regulation of message stability (small interfering RNA, siRNA, see **RNA interference**).

nucleobindin Originally identified as a 55 kDa DNA- and calcium-binding leucine zipper protein that enhanced anti-DNA antibody production when added to cultures of autoimmune MRL/lpr (lupus-prone) mouse spleen cells. Two related forms, nucleobindin-1 (461aa) and nucleobindin-2 (420aa) have subsequently been identified. Nucleobindin-1 is the major calcium-binding protein of the Golgi and may have a role in calcium homeostasis and bone mineralization. Nesfatin, produced by cleavage of NEFA/nucleobindin-2, is expressed in the appetite-control hypothalamic nuclei in rats and is involved in the regulation of feeding behaviour. Excess nesfatin-1 in the brain leads to a loss of appetite, satiety, and a drop in body fat and weight whereas a lack of nesfatin-1 has the opposite effects. Nesfatin is associated with melanocortin signalling. Further detail: http://www.copewithcytokines.de/cope.cgi?key=Nucleobindin

nucleocapsid The coat (**capsid**) of a virus plus the enclosed nucleic acid genome.

nucleocytoplasmic transport See **nuclear transport**.

nucleoid Region of bacterial cell that contains the DNA.

nucleolar organizer Loop of DNA that has multiple copies of rRNA genes. See **nucleolus**.

nucleolin A major nucleolar protein (710aa) associated with intranucleolar chromatin and pre-ribosomal particles. It induces chromatin decondensation by binding to histone H1 and may play a role in pre-rRNA transcription and ribosome assembly. In the cytoplasm it is found in mRNP granules that contain untranslated mRNAs.

nucleolus A small dense body within the nucleus of eukaryotic cells, visible by phase contrast and interference microscopy in live cells throughout interphase. Contains RNA and protein, and is the site of synthesis of ribosomal RNA. The nucleolus surrounds a region of one or more chromosomes (the nucleolar organizer) in which are repeated copies of the DNA coding for ribosomal RNA.

nucleomorph Small, reduced eukaryotic nuclei found in certain plastids in cryptomonads and chlorarachniophyte algae that have acquired plastids secondarily by engulfing respectively a red or green alga. The nucleomorph is found between the inner and outer pairs of membranes surrounding the chloroplast and is thought to be a remnant of the engulfed algal nucleus. The nucleomorph genome of the cryptomonad *Guillardia theta* has been completely sequenced: it has only three chromosomes and a total size of 551 kb. Article: http://jhered.oxfordjournals.org/content/100/5/582.full

nucleophosmin A nucleolar protein (nucleolar phosphoprotein B23, numatrin, 294aa) found as a decamer formed by two pentameric rings associated in a head-to-head fashion and involved in diverse cellular processes such as ribosome biogenesis, centrosome duplication, protein chaperoning, histone assembly, cell proliferation, and regulation of tumour suppressors TP53/p53 and ARF. Chromosomal aberrations involving the nucleophosmin gene NPM1 are associated with forms of non-Hodgkin's lymphoma, acute promyelocytic leukaemia and myelodysplastic syndrome. One of the **nucleoplasmin** family.

nucleoplasm By analogy with cytoplasm, that part of the nuclear contents other than the nucleolus and chromosomes.

nucleoplasmins Chaperone proteins (chromatin decondensation proteins) that bind to core histones and transfer DNA to them in an ATP-dependent reaction during the assembly of regular nucleosomal arrays (nucleosomes). Nucleoplasmin-2 (214aa) is probably involved in sperm DNA decondensation during fertilization. Nucleoplasmin-3 (178aa) may act as a chaperone. See **nucleophosmin**.

Nucleopore filter Filter of defined pore size made by etching a polycarbonate filter that has been bombarded by neutrons, the extent of etching determining the pore size. Very thin, with neat circular holes going right through the membrane, not a complex meshwork like micropore filters.

nucleoporins A large family of proteins that make up the nuclear pore complex that regulates the traffic of proteins and nucleic acids into and out of the nucleus. Many contain N-acetyl-glucosamine residues. Nomenclature is NUPnn where nn is the size in kDa. For example, **pericentrin** is NUP75.

nucleoproteins (1) Structures containing both nucleic acid and protein. Examples are chromatin, ribosomes, certain virus particles. (2) Sometimes, proteins that associate with nucleic acid.

nucleoside Purine or pyrimidine base linked glycosidically to ribose or deoxyribose, but lacking the phosphate residues that would make it a nucleotide. Ribonucleosides are adenosine, guanosine, cytidine and uridine. Deoxyribosides are deoxyadenosine, deoxyguanosine, deoxycytidine and deoxythymidine (the latter is almost universally referred to as thymidine).

nucleoskeletal DNA DNA that is proposed to exist mostly to maintain nuclear volume and not for coding protein.

nucleosome Repeating units of organization of chromatin fibres in chromosomes, consisting of around 200 base pairs of DNA, and two molecules each of the **histones** H2A, H2B, H3 and H4. Most of the DNA (around 140 base pairs) is believed to be wound around a core formed by the histones, the remainder joins adjacent nucleosomes, thus forming a structure reminiscent of a string of beads.

nucleosome remodeling and histone deacetylase complex *NuRD* complex A multiprotein complex that promotes transcriptional repression by histone deacetylation and nucleosome remodeling. It consists of a core **histone deacetylase** (HDAC) complex together with **metastasis associated proteins, methyl-CpG-binding domain proteins** and chromodomain-helicase-DNA-binding proteins.

nucleosome remodeling factor complex *NURF* complex A multi-subunit complex consisting of **SMARCA1**, BPTF (bromodomain and PHD finger-containing transcription factor, 3046aa) and the **retinoblastoma**-binding proteins, RBBP4 and RBBP7 that uses the energy of ATP hydrolysis to catalyze nucleosome sliding, thereby altering chromatin structure and regulating transcription.

nucleotidase Enzyme (5'-nucleotidase, EC 3.1.3.5) that cleaves the 5' monoester linkage of nucleotides, and converts them to the corresponding nucleoside.

nucleotide binding fold Protein motif consisting of a fold or pocket with certain conserved residues, required for the binding of nucleotides.

nucleotide-binding site/leucine-rich repeat genes See NB-LRR genes.

nucleotide-excision repair factor Multi-protein complexes (NEFs) involved in nucleotide-**excision repair**; possess DNA damage-recognition and endodeoxynuclease activities. In yeast Rad14 and Rad1-Rad10 form one subassembly called NEF1, the Rad4-Rad23 complex is named NEF2, Rad2 and TFIIH constitute NEF3, and the Rad7-Rad16 complex is called NEF4.

nucleotides Phosphate esters of **nucleosides**. The metabolic precursors of nucleic acids are monoesters with phosphate on carbon 5 of the pentose (known

as 5' to distinguish sugar from base numbering). However many other structures, such as adenosine 3'5'-cyclic monophosphate (cAMP), and molecules with two or three phosphates are also known as nucleotides. See Table N2.

nucleus (1) The major organelle of eukaryotic cells, in which the chromosomes are separated from the cytoplasm by the **nuclear envelope**. (2) In neuroanatomy a distinct region of the CNS, for example the **suprachiasmatic nucleus**.

nucleus accumbens A region of the brain involved in functions ranging from motivation and reward to feeding and drug addiction.

nudE See NDE1.

nude mice Strains of athymic mice bearing the recessive allele nu/nu (a mutated form of the **forkhead** homologue 11 gene) which lack body hair and all or most of the T-cell population. Show no rejection of either allografts or xenografts; nu/nu alleles on some backgrounds have near normal numbers of T-cells. Knockout mice in which FOXN1 gene is deleted also show the 'nude' phenotype.

TABLE N2. Nucleotides

Phosphate esters of nucleosides, which are themselves conjugates between the biological bases and sugars, either ribose or 2-deoxyribose.

Nucleosides are derived from the bases by the addition of a sugar in the position indicated (H).

Adenine   Cytosine   Guanine   Uracil   Thymine

The structure of nucleotides, exemplified by adenine derivatives is:

tri   di   mono

phosphonucleotide

TABLE N2. (Continued)

The phosphate may also be a cyclic diester involving two hydroxyl groups of the sugar. eg. 3′,5′ cyclic AMP

3′,5′ Cyclic AMP

Ribonucleotides are precursors of RNA and also common metabolic intermediates and regulators; Examples of the shorthand nomenclature are given.

	Adenine	Cytosine	Guanine	Uracil
Mononucleotide	AMP	CMP	GMP	UMP
Dinucleotide	ADP	CDP	GDP	UDP
Trinucleotide	ATP	CTP	GTP	UTP
Cyclic nucleotide	3′,5′ cyclic AMP		3′,5′ cyclic GMP	

Deoxyribonucleotides, required for the synthesis of DNA, are made by the biological reduction of the corresponding ribose dinucleotides and the deoxyribonucleotides are phosphorylated to give the triphosphonucleotides. dTMP is made by methylation of dUMP, which is then phosphorylated to give dTTP.

	Adenine	Cytosine	Guanine	Thymidine
Dinucleotide	dADP	dCDP	dGDP	
Trinucleotide	dATP	dCTP	dGTP	dTTP

Nucleotides occur as part of other biological molecules, e.g. NAD is the ADP-ribose derivative of nicotinamide. Nucleotide adducts are important intermediates in anabolic processes. CDP derivatives occur in the biosynthesis of lipids. UDP and TDP derivatives are important in sugar metabolism.

nudel A *Drosophila* serine peptidase (EC 3.4.21.-, 2616aa). Nudel, **pipe** and **windbeutel** together trigger the protease cascade within the extra-embryonic perivitelline compartment which induces dorsoventral polarity of the embryo. See **easter**.

null cell Lymphocytes lacking typical markers of T- or B-cells and capable of lysing a variety of tumour or virus-infected cells without obvious antigenic stimulation, also effect **antibody-dependent cell lysis**, and in humans carry CD16 marker.

null mutant Mutation in which there is no gene product.

nullo In *Drosophila* a protein (213aa) that stabilizes the actin-myosin network stability during cellularization of the initially syncytial embryo. May increase actin-actin interactions or membrane-to-cytoskeleton attachments.

NuMA See **nuclear mitotic apparatus protein**.

Numb In *Drosophila* a nuclear protein (556aa) required for determination of cell fate during sensory organ formation in embryos. The mammalian homologue (mNumb, 603aa) is membrane-associated, has multiple splicing isoforms, and segregates to only one of the daughter cells of a dividing neural precursor. Important in neuronal differentiation in the central nervous system. The signalling pathway mediated by Numb is antagonistic to that mediated by **Notch**; both play essential roles in enabling the two daughters to adopt different fates after a wide variety of asymmetric cell divisions. The cytoplasmic ligands of numb are **E3 ligases**.

numerical aperture *N.A.* An important parameter describing the properties of a lens. The N.A. is the product of the refractive index of the medium (1 for air, 1.5 for immersion oil) and the sine of the angle, i, the semi-angle of the cone formed by joining objects to the perimeter of the lens. The larger the value of N.A., the better the resolving power of the lens; most objectives have their N.A. value engraved on the barrel and this should be quoted when describing an optical system. (Resolving power also depends upon the wavelength of light being used).

NUPs See **nucleoporins**

NURF complex See **nucleosome remodeling factor complex**.

nurse cells Cells accessory to egg and/or sperm formation in a wide variety of organisms. Usually thought to synthesize special substances and to export these to the developing gamete.

NusA In *E. coli* a transcription elongation protein (495aa) involved in the termination and antitermination of transcription through a binding interaction with the core enzyme of the DNA-dependent RNA polymerase.

nutlins Small selective inhibitors (cis-imidazoline derivatives) of murine double minute-2 (Mdm2)-ligase binding to p53 that will drive tumour cells into apoptosis.

NXFs Nuclear RNA export factors. NXF1 (Tip-associated protein, mRNA export factor TAP, 619aa) is involved in the export of mRNA from the nucleus to the cytoplasm in association with **NXT**. NXF2 (TAP-like protein 2, 626aa) has a similar function as do a number of other NXFs.

NXT Proteins that stimulate the export of nuclear export signal-containing proteins and various RNA species. NXT1 (NTF2-related export protein, p15, 140aa) forms a heterodimer with **NXF1** through which it interacts with **nucleoporins**. NXT2 (142aa) has a similar role.

Nycodenz® Proprietary name for a dense non-ionic substance, solutions of which will form density gradients for separation procedures if centrifuged. Solutions are heat-stable and can be sterilized by autoclaving. A density of 2.1g/ml can be achieved.

nyctalopin A **small leucine-rich repeat proteoglycan** (481aa) that is defective in an X-linked form of **stationary night blindness**.

nyctinasty The **nastic movement** of higher plants in response to the onset of darkness. A circadian rhythm.

nystatin A polyene antibiotic active against fungi. The name is derived from "New York State Health Department" where it was discovered as a product of *Streptomyces noursei*. Exhibits selectivity for Na+ and increases the activity of the Na+-K+ pump.

NZB, NZW mice *New Zealand Black, NZ White* Inbred strains of mice which develop spontaneous auto-immune diseases. There is evidence for an underlying retroviral aetiology and a NZB virus has been isolated.

O

O-2A progenitor cells Bipotential progenitor cells in rat optic nerve that give rise initially to oligodendrocytes and then to type-2 astrocytes. Production of type-2 astrocytes from O-2A progenitor cells *in vitro* is triggered by **ciliary neurotrophic factor** (CNTF).

O-antigens Tetra- and penta-saccharide repeat units of the cell walls of Gram-negative bacteria. They are a component of **lipopolysaccharide**. Full-length O-antigen makes the cell surface smooth and hydrophilic and more difficult to phagocytose. Rough strains of bacteria have shorter O-antigen and more hydrophobic surfaces.

O-glycans Glycans linked to protein through the hydroxyl group of serine or threonine. O-Linked glycosylation is a post-translational modification, usually to fairly large proteins, does not require a consensus sequence and no oligosaccharide precursor is required for protein transfer. The commonest O-linked glycans contain an initial GalNAc residue. Cf. **N-glycans**.

OAT2 (1) The transporter (organic anion transporter 2, SLC22A7, 548aa) that mediates sodium-independent movement of a range of organic anions (e.g. prostaglandin E2, glutarate, L-ascorbic acid, salicylate) in liver and kidney. (2) Ornithine aminotransferase 2 (EC 2.6.1.13, 396aa in *S. aureus*) a bacterial enzyme that catalyzes the interconversion of ornithine to glutamate semialdehyde.

oat cell carcinoma Form of carcinoma of the lung in which the cells are small, spindle-shaped and dark-staining. May derive from argyrophilic **APUD** cells of the mucosa and certainly tends to be associated with endocrine symptoms.

OBCAM A cell-adhesion molecule (OPCML, **IgLON** family member 1, 345aa), that binds opioids in the presence of acidic lipids. Probably GPI-linked. Defects in OPCML are a cause of susceptibility to ovarian cancer.

obelin Calcium-activated photoprotein (195aa) in the photocyte of the colonial hydroid coelenterate, *Obelia geniculata*.

obestatin Peptide (23aa) that has appetite-suppressing activity. A product of the cleavage of **appetite-regulating hormone** (117aa), the other two products being **ghrelin-27** and ghrelin-28 which are appetite stimulators. The receptor is G-protein coupled (GPR39).

obscurin Giant multidomain muscle protein (EC 2.7.11.1, 7968aa) that apparently plays a role in spatial positioning of contractile proteins and the structural integration and stabilization of myofibrils, especially at the stage of myosin filament incorporation and A-band assembly. Is similar to **myosin-binding protein C** and **titin** with which it co-assembles. Obscurin-like protein 1 (OBSL1, 1401aa) is widely expressed, particularly in heart. Defects in OBSL1 are the cause of **3M syndrome** type 2.

OCA (1) A multi-pass membrane protein of the melanosome (OCA2, P protein, melanocyte-specific transporter protein, 838aa) that has a role in regulating melanosome pH and the post-translational processing of tyrosinase. Genetic variations in OCA2 are associated with skin/hair/eye pigmentation variability type 1 (SHEP1) and mutations lead to oculocutaneous albinism type 2. The mouse equivalent is mutated in the 'pink-eyed dilution' (p) mouse. (2) In *S. cerevisiae*, OCA1 is a putative tyrosine-protein phosphatase (EC 3.1.3.48, 238aa) required for protection against superoxide stress. OCA4 (oxidant-induced cell-cycle arrest protein 4, 362aa) is required for replication of Brome mosaic virus. (3) In *S. pombe* OCA3 (overexpression-mediated cell cycle arrest protein 3, 282aa) may be involved in cell cycle regulation.

occludens junction Tight junction. See **zonula occludens**.

occludin A four-pass integral plasma-membrane protein (522aa), a functional component of the **zonula occludens**. The predicted structure has two extracellular loops and N- and C-terminal cytoplasmic domains, the latter interacting with **ZO proteins**.

occlusion bodies Large proteinaceous structures $(0.3 \times 0.5\mu m)$ formed late in the infection of cells by **baculovirus**; major protein is **polyhedrin** embedded within which are virions.

ocellus *Pl.* ocelli. A simple eye or eyespot found in some invertebrates.

ochratoxins Mycotoxins produced by some *Aspergillus* and *Penicillium* species. Dietary exposure can cause renal damage and may be carcinogenic.

ochre codon See **termination codons**.

OCIF See **osteoprotegerin**.

oct (1) A family of transcription factors (octamer-binding transcription factors) that act as RNA

The Dictionary of Cell and Molecular Biology. DOI: http://dx.doi.org/10.1016/B978-0-12-384931-1.00015-5

Polymerase II promoters. They have a **POU domain** and are **leucine zipper** proteins that bind to **octamer** sequences. Oct4 is required to maintain the pluripotency and self-renewal of embryonic stem cells. **(2)** A family of organic cation transporters (for example, OctN1, solute carrier family 22, member 4, SLC22A4, 551aa) is a proton/organic cation transporter involved in active secretion of cationic compounds, including xenobiotics, across the renal epithelial brush-border membrane.

octamer **(1)** An assembly of eight histone proteins (2 each of H2A, H2B, H3 and H4) that forms the core of the nucleosome. **(2)** Eight-base sequence motif (octamer motif) common in eukaryotic promoters. Consensus is ATTTGCAT; binds various transcription factors (see **oct #1**).

octanoylation The post-translational modification (lipoylation) of a protein by formation of an ester or amide of octanoic acid. **Ghrelin** is modified with an O-linked octanoyl side group on its third serine residue and this modification is crucial for its physiological effects.

octopaline An **opine**.

octopamine A **biogenic amine** found in both vertebrates and invertebrates (identified first in the salivary gland of *Octopus*). Octopamine can have properties both of a hormone and a neurotransmitter, and acts as an adrenergic agonist.

octyl glucoside A biological detergent used to solubilize membrane proteins. Easily removed from hydrophobic proteins.

ocular coloboma A congenital abnormality in which there is defective closure of the optic cup; can be caused by mutation in the PAX6 homoeobox gene.

oculodentodigital dysplasia A developmental disorder (oculodentoosseous dysplasia, ODDD) caused by mutation in **connexin-43**.

OD600 Optical density at 600 nm; provides a reasonable estimate of cell numbers present in a growing bacterial or yeast culture.

ODC **(1)** **Ornithine decarboxylase**. **(2)** Oxygen (oxyhaemoglobin) dissociation curve.

ODF **(1)** **Osteoclast differentiation factor**. **(2)** **Outer dense fibres**. **(3)** Orientation distribution function, a formal mathematical method for describing orientation of materials (fibres, crystals and others).

odontoblasts Columnar cells derived from the dental papilla after **ameloblasts** have differentiated, and that give rise to the dentine matrix that underlies the enamel of a tooth.

odontogenic epithelial cells Epithelial cells that will give rise to teeth.

odorant receptors A diverse set of G-protein coupled receptors (>1000) in the olfactory receptor cells located in the upper part of the nasal epithelium in mammals. In *Drosophila* there are at least 16 genes for G-protein coupled receptors that are differentially expressed on sensory neurons in the antenna. Webpage: http://www.sdbonline.org/fly/aignfam/odorcpt1.htm#dafka3

odoratism See **lathyrism**.

oedema *US*. edema. Swelling of tissue: can result from increased permeability of vascular endothelium or increased blood pressure, for example as a result of being at high altitude without adequate acclimatisation.

Oedogonium The type genus of the Oedogoniaceae; freshwater green algae with long unbranched filaments.

oestrogen *US*. estrogen. See **estrogen**.

oestrone See **estrogen**.

OFAGE Orthogonal field alternation gel electrophoresis. Method in which macromolecules are electrophoresed in a gel using electric fields applied alternately at right-angles to each other.

Oguchi disease A form of **stationary night blindness** in which there is abnormally slow dark adaptation caused by mutation in genes either for **arrestin** or **rhodopsin kinase**.

oil body Small droplets (0.2 to 1.5 μm in diameter), containing mostly triacylglycerol, that are surrounded by a phospholipid/**oleosin** annulus. Found in oil-rich seeds.

okadaic acid A toxin first isolated from the sponge *Halichondria okadai*, although produced by various species of dinoflagellates; a complex lipophilic polyether. It is a potent inhibitor of serine-threonine-specific protein **phosphatases** 1 and 2A and can act as a tumour promoter. **Microcystins** and **nodularins** bind to the same site on the phosphatase.

Okazaki fragments Short fragments of newly synthesized DNA strands produced during DNA replication. All the known **DNA polymerases** synthesize DNA in the $5'$ to $3'$ direction. However as the strands separate, replication forks will be moving along one parental strand in the $3'$ to $5'$ direction and $5'$ to $3'$ on the other parental strand. On the former, the leading strand, DNA can be synthesized continuously in the $5'$ to $3'$ direction. On the other, the lagging strand, DNA synthesis can only occur when a stretch of single stranded DNA has been exposed and proceeds in the direction opposite to the movement of the replication fork (still $5'$ to $3'$).

It is thus discontinuous and the series of fragments are then covalently linked by **ligases** to give a continuous strand. Such fragments were first observed by Okazaki using pulse-labelling with radioactive thymidine. In eukaryotes, Okazaki fragments are typically a few hundred nucleotides long, whereas in prokaryotes they may contain several thousands of nucleotides.

oleic acid See **fatty acids** and Table L3.

oleosins A large family of proteins (140 – 450aa) that form a hydrophilic shell around **oil body** inclusions in plant cells. Oleosin has three distinct domains; the N- and C-terminal domains, which are amphipathic, and a central extremely hydrophobic domain with long stretches of nonpolar amino acids. May have a structural role in stabilizing the lipid body during dessication of the seed.

oleosome Plant spherosome rich in lipid that serves as a storage granule in seeds and fruits. There are none of the enzymes characteristic of lysosomes.

olfactomedins A family of proteins with a conserved protein motif through which extracellular protein-protein interactions occur. Although they share sequence similarity they have diversified roles in many important biological processes. Includes **amassin**, **myocilin**, **noelins**, which enhances neural crest generation in chick, **pancortins**, **photomedins** and **tiarin**.

olfactory ensheathing cells *OECs* Specialised glial cells found in the outer nerve layer of the olfactory bulb, enclosing the terminal non-myelinated axonal branches of olfactory neurons. They provide guidance cues for developing neurites and support continual regeneration of olfactory receptor neurons throughout life. Transplantation of OECs to damaged spinal cord has some promise for improving neuronal regrowth. Diagram: http://www.nature.com/nrn/journal/v2/n5/fig_tab/nrn0501_369a_F2.html

olfactory epithelium The **epithelium** lining the nose. Has the **odorant receptors** responsible for the sense of smell.

olfactory neuron Sensory neuron from the lining of the nose. They are some of the only neurons that continue to divide and differentiate throughout an organism's life.

oligoadenylate synthetases A family of interferon-induced enzymes ($2' – 5'$ oligoadenylate synthetases, EC 2.7.7.-, ranging from 400 – 1087aa) activated by double-stranded RNA, that polymerize ATP which then activates a latent endoribonuclease that degrades viral and cellular RNAs. There are four human genes, OAS1, OAS2, OAS3 and OAS-like gene (OASL). Mutations cause greater susceptibility to viral infection.

oligodendrocyte Neuroglial cell of the central nervous system in vertebrates whose function is to myelinate CNS axons.

oligodendrocyte-myelin glycoprotein An integral membrane protein (OMgp, myelin-oligodendrocyte glycoprotein, MOG, 440aa) expressed in oligodendrocytes and outer myelin lamellae; like **myelin associated glycoprotein** (MAG) and **nogo**-A binds to the nogo receptor, causes growth cone collapse and inhibits neurite outgrowth *in vitro*.

oligodontia-colorectal cancer syndrome See **axins**.

oligomycin A bacterial toxin that inhibits oxidative phosphorylation by binding to the d-subunit of the **F-type ATP synthase** (oligomycin sensitivity-conferral protein).

oligonucleotide Linear sequence of up to 20 nucleotides joined by phosphodiester bonds. Above this length the term polynucleotide begins to be used.

oligopeptide A peptide composed of a small number of amino acids as opposed to a polypeptide. Peptides from three to about 40aa might be classed as such.

oligophrenin See **p21-activated kinases**.

oligosaccharide A saccharide composed of a small number of sugars, either O- or N-linked to the next sugar. The boundary between oligosaccharides and polysaccharides is arbitrary and undefined.

oligosaccharin An oligosaccharide derived from the plant cell wall that in small quantities induces a physiological response in a nearby cell of the same or a different plant, and thus acts as a molecular signal. Sometimes considered to be a plant hormone or plant growth substance. The best authenticated examples are involved in host-pathogen interactions and in the control of plant cell expansion.

oligotroph Organism that can grow in an environment poor in nutrients.

olomoucine Purine derivative that is an inhibitor of cyclin-dependent kinase **cdk5** but inactive against cdk4 and cdk6. A similar compound, **roscovitine**, is somewhat more potent.

omega fatty acid Type of fatty acid found in unsaturated fat. Two forms are found, omega-3 and omega-6. Generally regarded as beneficial in reducing risk of heart disease.

omega-oxidation Minor metabolic pathway in the ER for medium chain-length fatty acids. The omega carbon in a fatty acid is the carbon furthest in the alkyl chain from the carboxylic acid: this carbon is progressively oxidized first to an alcohol and then to a carboxylic acid, creating a molecule with a carboxylic acid on both ends. The dicarboxylic acid product can enter the beta-oxidation pathway to be shortened at both ends of the molecule at the same time.

Omenn syndrome A **severe combined immunodeficiency disease** with hypereosinophilia, caused by mutation in the **RAG genes** or in **artemis**.

omentin An **adipokine** that enhances insulin-mediated glucose-uptake in the visceral adipose tissues and activates **Akt/PKB**. Omentin 1 (intelectin-1, endothelial lectin HL-1, intestinal lactoferrin receptor, galactofuranose-binding lectin, 313aa) and omentin 2 (intelectin-2) expression levels are decreased with obesity and associated with insulin resistance.

OMgp See **oligodendrocyte-myelin glycoprotein**.

OMIM The Online Mendelian Inheritance in Man database that is a catalogue of more than 12000 human genes and genetic disorders, authored and edited by Dr. V. A. McKusick and his colleagues at Johns Hopkins, and made freely available by NCBI, the National Center for Biotechnology Information. An invaluable resource that provides much more information than the brief Dictionary entries. Link: http://www.ncbi.nlm.nih.gov/sites/entrez?db = OMIM&itool = toolbar

ommatidium *Pl.* ommatidia. Single facet of an insect compound eye. Composed of a set of photoreceptor cells, overlain by a crystalline lens.

Onchocerca Genus of filarial nematode parasites that cause river blindness.

oncocytes See **Askenazy cells**.

oncogen Uncommon synonym for **carcinogen**.

oncogene Mutated and/or overexpressed version of a normal gene of animal cells (the **proto-oncogene**) that in a dominant fashion can release the cell from normal restraints on growth, and thus alone, or in concert with other changes, convert a cell into a tumour cell. Viral versions of cellular genes (prefixed v-, e.g. v-*src* cf. the normal cellular version, c-*src*) are responsible for transformation by retroviruses. There are separate entries for *abl, akt, arg, Axl, bcl, brx,c-maf, cbl, crk, dbl, E1,E5,* **ectoderm-neural cortex-1**, *erb, ets,* **feline sarcoma oncogene homologue,** *fes, fms, fos, fyn,GASC1,GLI* oncogenes, *gip2, hst, int* oncogenes, *JAZF1, jun,kit, maf, mas,met, mos,* **multiple myeloma oncogene-1,** *myb, neu, nov, pim, raf, ral, ras, rel, ret, ros,sea,sis, ski, src* and the **src-family of kinases,** *tre, trk* **and** *yes.* See Table O1.

oncomodulin See **parvalbumin**.

OncoMouse™ Registered proprietary name for transgenic mice, genetically-engineered to contain the activated v-Ha-*ras* oncogene fused to a mouse zeta-globin promoter. The patenting of these animals was a matter of considerable debate. Because they are predisposed to papillomas they have been used extensively in screening tumour promoters and non-genotoxic carcinogens and for assessing antitumour and antiproliferative agents.

oncoprotein 18 See **stathmin**.

oncostatin M Multifunctional cytokine (252aa) of the **IL6 cytokine family.** Produced by activated T-cells; inhibits tumour cell growth and induces IL-6 production by endothelial cells via the tyrosine kinase p62yes. Oncostatin binds to the common gp130 subunit of the IL6 family, in conjunction with either the **leukaemia inhibitory factor** receptor subunit or an specific 979aa subunit (OSM receptor beta, OSMRβ). Mutations in OSMRβ cause **familial primary localized cutaneous amyloidosis**.

Oncovirinae The sub-family of retroviruses (**Retroviridae**) that can cause tumours. No longer considered to be an appropriate class since the genera of the sub-family are unrelated.

ontogeny The total of the stages of an organism's life history. Once believed to recapitulate phylogeny, though this is an outmoded conceit.

oöcyst (1) In some Protozoa, the cyst formed around two conjugating gametes, generally with a thick protective wall that facilitates survival. (2) In **Sporozoa**, the passive phase from which sporozoites are released.

oocyte The developing female gamete before maturation and release.

oocyte expression Technique whereby the cellular translational machinery of an oocyte (typically *Xenopus*) is utilized to generate functional protein from microinjected mRNA or to produce protein encoded by an introduced expression vector.

oogenesis The process of egg formation.

oogonium Female sexual structure in certain algae and fungi, containing one or more gametes. After fertilization the oogonium contains the oöspore.

oomycetes Fungus-like organisms in which the mycelium is non-septate, i.e. lacks cross-walls, and the nuclei are diploid. Sexual reproduction is oogamous. Differ from fungi in several ways, e.g. having a cellulose cell wall, and are now classified as being in the kingdom Protoctista and are related to heterokont, biflagellate, golden-brown algae.They are important plant pathogens (*Phytophthora, Pythium,* downy mildews, white blister rusts).

TABLE O1. Oncogenes and tumour viruses

Acronym	Virus	Species	Tumour origin	Comments
abl	Abelson leukaemia	mouse	Chronic myelogenous leukaemia	TyrPK(src)
akt	AKT8	human		Ser/Thr kinase (PKB)
arg	Abelson leukaemia	mouse		TyrPK
bcl		human	B cell lymphoma	Involved in apoptosis
crk		chicken	Sarcoma	Adaptor signalling proteins
dbl		human	Diffuse B-cell lymphoma	Encodes a GEF
E1A	Adenovirus			Interacts with Rb product
E1B	Adenovirus			Interacts with p53
E5,6,7	Papilloma			E5 blocks processing of growth factor receptors, E6 & E7 drive proliferation
erbA	Erythroblastosis	chicken		Homology to human glucocorticoid receptor
erbB	Erythroblastosis	chicken		TyrPK EGF/TGFc receptor
ets	E26 myeloblastosis	chicken		Nuclear
fes(fps)[a]	Snyder-Theilen sarcoma Gardner-Arnstein sarcoma	cat		TyrPK (src)
fgr	Gardner-Rasheed sarcoma	cat		TyrPK(src)
fms	McDonough sarcoma	cat		TyrPK CSF-1 receptor
fps(fes)[a]	Fujinami sarcoma	chicken		TyrPK(src)
fos	FBJ osteosarcoma	mouse		Nuclear, TR
gip2		rat	Ovarian and adrenal	GTPase-defective Gs & Gi
hst	NVT	human	Stomach tumour	FGF homologue
intl	NVT	mouse	MMTV-induced carcinoma	Nuclear, TR
int2	NVT	mouse	MMTV-induced carcinoma	FGF homologue
ipt	Agrobacterium	plants		Isopentenyl transferase
jun	ASV17 sarcoma	chicken		Nuclear, TR
kit	Hardy-Zuckerman 4 feline sarcoma	cat	Sarcoma	PTK receptor for SCF
mas	NVT	human	Epidermoid carcinoma	Potentiates response to angiotensin II?
met	NVT	mouse	Osteosarcoma	TyrPK GFR L?
mil (raf)	Mill Hill 2 acute leukaemia	chicken		Ser/ThrPK
mos	Moloney sarcoma	mouse		Ser/ThrPK
myb	Myeloblastosis	chicken	Leukaemia	Nuclear, TR
myc	MC29 Myelocytomatosis	chicken	Lymphomas	Nuclear TR?

TABLE O1. (Continued)

Acronym	Virus	Species	Tumour origin	Comments
N-myc	NVT	human	Neuroblastomas	Nuclear
neu(ErbB2)	NVT	rat	Neuroblastoma	TyrPK GFR L?
pim-1, -2, -3		human	Prostatic carcinoma	Ser/ThrPK
ral(mil)^b	3611 sarcoma	mouse		Ser/ThrPK
Ha-ras	Harvey murine sarcoma	rat	Bladder, mammary and skin carcinomas	GTP-binding
Ki-ras	Kirsten murine sarcoma	rat	Lung, colon carcinomas	GTP-binding
N-ras	NVT	human	Neuroblastomas, Leukaemias	GTP-binding
ral			Small GTPase	
rel	Reticuloendotheliosis	turkey		
ret		human	Thyroid, multiple endocrine neoplasia	RTK for GDNF
ros	UR2	chicken		TyrPK GFR L?
sea	S13 avian erythroblastosis homologue			RTK for hepatocyte growth factor
sis	Simian sarcoma	monkey		One chain of PDGF
src	Rous sarcoma	chicken		TyrPK
ski	SKV770	chicken		Nuclear
tre		human	Transfected 3T3	Deubiquitinating enzyme
trk	NVT	human	Colon carcinoma	RTK for NGF
yes	Y73, Esh sarcoma	chicken		TyrPK(src)

^a fps/fes are species equivalents.
^b mil/raf are species equivalents.
GFR L = From sequence, a growth-factor receptor for unknown ligand
MMTV = Mouse mammary tumour virus
NVT = Isolated from non-retroviral tumour. In most cases detected by transfection of 3T3 cells
RTK = receptor tyrosine kinase, Ser/ThrPK = Serine, Threonine protein kinase; TyrPK = Tyrosine protein kinase
TR = Transcriptional regulator.
See separate entries for cbl, fgf oncogenes.

oöspore A zygote with food reserves and a thick protective wall, formed from a fertilized oösphere in some algae and the Oomycetes.

ootid An immature female gamete that will develop into an ovum.

OPA1 Optic atrophy protein 1, 960aa a single-pass protein of the inner mitochondrial membrane that is cleaved to produce a mitochondrial dynamin-like 120 kDa protein (873 or 766aa), a GTPase that is required for mitochondrial fusion and regulation of apoptosis.

opal See **termination codons**.

Opalina A genus of parasitic protozoa found in the guts of frogs and toads. They look superficially like ciliates, but are classified in a separate group.

OPC (1) Oropharyngeal candidiasis, an opportunistic infection in immunosuppressed patients. (2) Oral and pharyngeal cancer. (3) Oligodendrocyte precursor cell, cells that give rise to oligodendrocytes and possibly neurons and astrocytes. They express most neurotransmitter receptors. (4) Organophosphorus compounds. (5) OPC-14523 (OPC), a compound with high affinity for sigma and 5-HT1A receptors that shows 'antidepressant-like' effects in animal models of depression. (6) See **OBCAM**.

open meristem See **closed meristem**.

open reading frame *ORF* A possible **reading frame** of DNA which is capable of being translated into protein, i.e. is not punctuated by stop codons. (This capacity does not indicate that the ORF is actually translated.)

operator The site on DNA to which a specific **repressor protein** binds and prevents the initiation of transcription at the adjacent **promoter**.

operon Groups of bacterial genes with a common **promotor**, that are controlled as a unit and produce mRNA as a single, **polycistronic** messenger. An operon consists of two or more structural genes, which usually code for proteins with related metabolic functions, and associated control elements that regulate the transcription of the structural genes. The first described example was the **lac operon**.

OPG See **osteoprotegerin**.

opiates See **opioids**.

opines Carbon compounds produced by crown galls and hairy roots induced by *Agrobacterium tumefaciens* and *A. rhizogenes* respectively. They are utilized as nutritional sources by *Agrobacterium* strains that induced the growth and are, in some cases, chemoattractants. Chemotactic activity seems to be specific for plasmids carrying the relevant opine synthase gene. **Octopaline**, **nopaline**, mannopine and agrocinopines A and B are examples.

opioids Naturally occuring basic (alkaloid) molecules with a complex fused ring structure. Group includes the opiates such as morphine, found in the opium poppy (*Papaver somniferum*) and synthetically modified derivatives. Effective pain-killers but tend to be addictive and are drugs of abuse. Weak opioids include codeine and dihydrocodeine; strong opioids include **morphine**, diamorphine, fentanyl and phenazocine. There are four distinct classes of G-protein-coupled opioid receptors, δ, μ and κ together with an orphan receptor, the ligand for which is **orphanin FQ**; the endogenous ligands are **enkephalins** and **endorphins**.

opportunistic infection An infection caused by an organism (opportunistic pathogen) that is normally non-pathogenic, usually as a result of deficiencies in the immune system.

opsins General term for the apoproteins of the **rhodopsin** family. They are G-protein coupled receptors and many variants are known: (1) the vertebrate visual (transducin-coupled) and non-visual opsin subfamily, (2) the **encephalopsin**/tmt-opsin subfamily, (3) the Gq-coupled opsin/**melanopsin** subfamily, (4) the Go-coupled opsin subfamily, (5) the **neuropsin** subfamily, (6) the **peropsin** subfamily and (7) the retinal photoisomerase subfamily. See **colour blindness**.

opsonin Substance that binds to the surface of a particle (opsonization) and enhances the uptake of the particle by a phagocyte. Probably the most important opsonins in mammals derive from **complement** (C3b or C3bi) or immunoglobulins (which are bound through the **Fc receptor**).

optic atrophy An autosomal dominant disorder in which there is progressive loss of vision during childhood; Type I is caused by mutation in **OPA1**.

optic nerve A bundle of neurons that connects the retinal photoreceptor system to the **optic tectum** of the midbrain. Embryologically, a CNS tract rather than a peripheral nerve. Popular experimental preparation for studies of regeneration of **retino-tectal connections** in lower vertebrates and also for studies of glial cell lineage in CNS. The optic nerve arises in front of the photoreceptors and where it passes out of the eye there is a blind spot, the optic disc.

optic tectum A region of the midbrain in which input from the optic nerve is processed. Because the retinally derived neurons of the optic nerve 'map' onto the optic tectum in a defined way, the question of how this specificity is determined has been a long-standing problem in cell biology. Although there is some evidence for adhesion gradients and for some adhesion specificity, the problem is unresolved.

optical diffraction A technique used to obtain information about repeating patterns. Diffraction of visible light can be used to calculate spacings in the object.

optical isomers Isomers (stereoisomers) differing only in the spatial arrangement of groups around a central atom. Optical isomers rotate the plane of polarized light in different directions. For most biological molecules in which the possibility of optical isomerism exists, only one of the isomers is functional, although there are some exceptions. For example both glucose and mannose are used and D-amino acids are found in the cell walls of some bacteria (whereas L-forms are universal in proteins).

optical tweezers A technique (laser tweezers, optical trap) that involves focusing a beam of light onto a microscopic particle. The particle is trapped as a result of the forces (pico-newtons) exerted by radiation pressure and with a laser beam it is possible to move small organelles around under the microscope or to measure the forces that motor molecules are exerting by measuring the force needed to oppose their activity.

optimedin See **noelins**.

optineurin A protein (huntingtin-interacting protein 7, transcription factor IIIA-interacting protein, 577aa) with an important role in the maintenance of the Golgi complex, in membrane trafficking and in exocytosis. It links myosin VI to the Golgi complex and interacts with rab8. It is neuroprotective in the eye and optic nerve and defects cause one form of primary open angle glaucoma.

Orai proteins A family of proteins (Orai1, calcium release-activated calcium channel protein 1, CRACM1, 301aa; Orai2, CRACM2, 254aa; Orai3, CRACM3, 295aa) that are subunits of the store-operated calcium channels involved in calcium signalling in lymphocytes. Orai1 is defective in a severe combined immunodeficiency disease (SCID). See **STIM proteins**.

orange domain A helix-loop-helix domain (60 – 100aa), found in specific DNA- binding proteins that act as transcription factors. A functional domain for the **Hairy**/E(SPL) proteins.

orbit In *Drosophila*, a **CLASP family protein** (orbit, CLIP-associating protein, misexpression suppressor of ras 7, protein multiple asters, protein chromosome bows, 1491aa) that binds to plus ends of microtubules and promotes their stabilization.

Orbivirus Genus of **Reoviridae** that infects a wide range of vertebrates and insects. The genome is double-stranded RNA and the capsomeres are a characteristic doughnut-shape. Best known example is Bluetongue virus.

orcein A purple dye originally extracted from lichens, formerly used as a food colouring and as a microscopical stain. Contains a variety of phenazones: hydroxy-orceins, amino-orceins and amino-orceinimines and is the standard histological stain for elastic fibres and for **ground-glass cells** (hepatitis B infected liver cells).

orcinol The compound used to produce Bial's reagent (an acidified alcoholic solution of orcinol with ferric chloride). Bial's orcinol test is used to distinguish pentoses (which produce a green to deep blue colour) from hexoses which give a muddy brown/gray colour.

ORCs Proteins that form the **origin of replication complex (ORC)**. The ORC has six highly conserved protein subunits and in human, ORC is cell cycle-dependently regulated: it is sequentially assembled at the exit from anaphase of mitosis and disassembled as cells enter S phase. Origin recognition complex subunit 1 (ORC1, replication control protein 1, 861aa); ORC2, 577aa; ORC3, 711aa; ORC4, 436aa; ORC5, 435aa; ORC6, 252aa.

orexin *hypocretin* Orexin-A and -B are potent orexigenic (appetite-stimulating) peptides (hypocretin-1, 33aa; hypocretin-2, 28aa) that are derived from the same precursor peptide (131aa). They play an important role in the regulation of food intake and sleep-wakefulness; mutations in orexin are associated with narcolepsy. Orexin receptor-1 (425aa) is a fairly

selective excitatory receptor for orexin-A, exclusively coupled to the Gq subclass of heterotrimeric G proteins, which activate phospholipase C mediated signalling. Orexin receptor 2 (444aa) binds both orexins with comparable affinity.

orf A skin disease of sheep and goats, caused by a parapox virus. Can infect those who work with sheep.

organ culture Culture *in vitro* of pieces of tissue (as opposed to single cells) in such a way as to maintain some normal spatial relationships between cells and some normal function. *Cf.* **tissue culture**.

organelle A structurally discrete component of a cell. Usually restricted to membrane-bounded structures: a ribosome would not be considered an organelle by many cell biologists.

organizing centre See **microtubule organizing centre**.

organogenesis The process of formation of specific organs in a plant or animal involving morphogenesis and differentiation.

orientation chamber Chamber designed by Zigmond in which to test the ability of cells (particularly **neutrophils**) to orient in a gradient of chemoattractant. The chamber is similar to a haemocytometer, but with a depth of only ca. 20μm. The gradient is set up by diffusion from one well to the other, and the orientation of cells towards the well containing chemoattractant is scored on the basis of their morphology or by filming their movement. A variation on this is the **Dunn chamber**.

origin of replication Regions of DNA where the pre-replication complex assembles and from where replication begins. The origin recognition complex (ORC) recruits **cdc6** and **cdt1** during interphase; phosphorylation by **cyclin-dependent kinase** releases these and allows **DNA helicase** to begin the process of replication.

ornithine decarboxylase The enzyme (EC 4.1.1.17, 461aa) that converts ornithine to putrescine (dibasic amine) by decarboxylation. Rate-limiting in the synthesis of the polyamines spermidine and spermine that regulate DNA synthesis.

ornithine transcarbamylase A nuclear-encoded mitochondrial matrix enzyme (EC 2.1.3.3, 354aa) that catalyzes the second step of the urea cycle in mammals. Mutations are the most common cause of inherited urea cycle disorders.

ornithine-aminotransferase The mitochondrial enzyme (OAT, EC 2.6.1.13, 439aa in *Homo*) that catalyses the conversion of ornithine into glutamate 5-semialdehyde. Mutations cause hyperornithinemia with gyrate atrophy of choroid and retina, a slowly progressive blinding disorder. In plants delta-OAR

(475aa) converts ornithine to pyrroline-5-carboxylate (P5C) and is essential for arginine catabolism (arginine and Arg-rich proteins are important storage forms of organic nitrogen in many plants). Research article: http://www.biomedcentral.com/1471-2229/8/40

orosomucoid An **acute phase protein** (α-1-seromucoid, α-1-acid glycoprotein-1, orosomucoid-1, OMD1, 201aa; OMD2, α-1-acid glycoprotein 2, 201aa) present at 0.6-1.2 mg/ml of plasma and highly glycosylated; levels are increased in inflammation, pregnancy, and various diseases. May modulate the activity of the immune system during the acute-phase reaction.

orotic acid *orotate* Intermediate in the de novo synthesis of pyrimidines. Linked glycosidically to ribose 5′-phosphate, orotate forms the pyrimidine nucleotide orotidylate, that on decarboxylation at position five of the pyrimidine ring yields the major nucleotide uridylate (uridine 5′-phosphate).

orphan receptor A receptor for which no ligand has been identified.

orphanin FQ *nociceptin* Endogenous ligand for the fourth opioid receptor (opioid-like orphan receptor), a 17aa peptide resembling **dynorphin**. Cleaved from a prepro-protein (176aa) together with neuropeptide 1 (30aa) and neuropeptide 2 (17aa). The same precursor protein also gives rise to nocistatin (17a) that is reported to antagonize several effects of nociceptin by acting on a different receptor.

orthodromic Conduction of impulses in the normal direction down a nerve fibre, from the cell body towards the distal pre-synaptic region, as opposed to the antidromic direction.

orthogonal Describing an arrangement of elements (fibres, cells) at approximately right angles to one another. Confluent fibroblasts often become organized into such arrays; other examples are the packing of collagen fibres in the cornea, and cellulose fibrils in the plant cell wall.

orthograde transport Axonal transport from the cell body of the neuron towards the synaptic terminal. Opposite of retrograde transport and dependent on a different mechanochemical protein **(kinesin)** interacting with microtubules.

orthokinesis Kinesis in which the speed or frequency of movement is increased (positive orthokinesis) or decreased.

orthologous genes Genes related by common phylogenetic descent and usually with a similar organisation. *Cf.* **paralogous genes**.

Orthomyxoviridae A family of viruses (Class V) with a genome consisting of a negative strand of RNA (10-14 kb) that is present as several (6 − 8) separate segments each of which acts as a template for a single mRNA. The **nucleocapsid** is helical and has a viral-specific RNA polymerase for the synthesis of the mRNAs. They leave cells by budding out of the plasma membrane and are enveloped with two classes of viral spike protein in the envelope. One has **haemagglutinin** activity (H) and the other acts as a **neuraminidase** (N) and both are important in the invasion of cells by the virus. The major viruses of this group are the influenza viruses which are classified on the basis of the haemagglutinin and neuraminidase (e.g. **H5N1**).

Orthopoxviridae See **Poxviridae**.

orthotropism A **tropism** in which a plant part (e.g. a shoot or root), becomes aligned directly towards (positive-) or away from (negative-) the source of the orientating stimulus. Most seedling shoots are negatively orthogravitropic and positively orthophototropic.

Oryza sativa Rice An important cereal foodstuff which has been the target of much plant breeding and genomics. The two major cultivated subspecies are indica and japonica. See **Golden rice, Oryzabase, Rice Annotation Project Database, RetrOryza, RICD** (Rice Indica cDNA Database), **Rice Genome Annotation Project**.

Oryzabase A comprehensive rice science database established in 2000 by a rice researchers committee in Japan. It has information about genetic resources, chromosome maps, and other aspects of rice genomics. See also **RetrOryza** and **Rice Annotation Project Database**. Link to database: http://www.nig.ac.jp/labs/PlantGen/english/oryzabase-e/index.html

Os- A prefix for microRNAs, enzymes etc. from *Oryza sativa* (rice). Thus OsDCLs are Dicer-like (DCL) proteins from rice.

oscar (1) Osteoclast-associated immunoglobulin-like receptor (282aa); regulated osteoclastogenesis. (2) See **oskar**.

Oscillatoria princeps Large cyanobacterium that exhibits gliding movements, possibly involving the activity of helically arranged cytoplasmic fibrils of 6 − 9 nm diameter.

oscillin Soluble protein (glucosamine-6-phosphate isomerase 1, EC 3.5.99.6, 289aa) from mammalian sperm that is involved with the oscillations in calcium concentration that occur in the egg following fertilization. Has sequence similarity with prokaryote hexose phosphate isomerase.

OSCP Oligomycin sensitivity-conferral protein. See **oligomycin**.

osculum A large pore on the surface of a sponge through which water is exhaled. See **ostium**.

oskar Product of an **egg-polarity gene** in *Drosophila*, a maternal effect protein (606aa) that is concentrated at the posterior pole of the egg, and required for subsequent posterior structures. Directs localization of the posterior determinants **nanos** and **staufen**.

Osler-Rendu-Weber disease See **hereditary haemorrhagic telangiectasia** type 1.

OSMED A skeletal dysplasia (otospondylomegaepiphyseal dysplasia) with severe hearing loss caused by defects in **collagen** type XI (COL11A2).

osmiophilic Having an affinity for **osmium tetroxide**.

osmium tetroxide OsO_4 Used as a post-fixative/stain in electron microscopy. Membranes in particular are osmiophilic, i.e. bind osmium tetroxide.

osmole The amount of a solute, dissolved in water, that produces a solution with the same osmotic pressure as would be produced by dissolving one mole of an ideal non-ionized solute. Sea water is approximately 1000 milliosmolar, mammalian isotonic saline is about 290 milliosmolar.

osmoreceptors Cells specialized to react to osmotic changes in their environment and in mammals involved in the regulation of secretion of antidiuretic hormone by the neurohypophysis.

osmoregulation Processes by which a cell regulates its internal osmotic pressure. These may include water transport, ion accumulation or loss, synthesis of osmotically active substances such as glycerol in the alga *Dunaliella*, activation of membrane ATPases amongst others.

osmosis The movement of solvent through a membrane impermeable to solute, in order to balance the chemical potential due to the concentration differences on each side of the membrane. Frequently mis-used in general writing.

osmotic pressure See **osmosis**. The pressure required to prevent osmotic flow across a semipermeable membrane separating two solutions of different solute concentration. Equal to the pressure that can be set up by osmotic flow in this system.

osmotic shock Passage of solvent from a hypotonic solution into a membrane-bound structure due to osmosis, causing rupture of the membrane. A method of lysing cells or organelles.

OSR1 A serine/threonine-protein kinase (oxidative stress-responsive kinase-1, 527aa) that regulates downstream kinases such as **p21-activated kinase-1** in response to environmental stress. Interacts with some, but not all, chloride channels of the SLC12A class. See SPAK and **WNK**.

osseous heteroplasia A rare autosomal dominant disorder in which bone forms in the dermis and other inappropriate sites. It is caused by paternally inherited inactivating mutations of the GNAS1 gene that encodes the alpha subunit of heterotrimeric Gs, a gene also mutated in **Albright hereditary osteodystrophy**, pseudohypoparathyroidism Ia, and pseudopseudohypoparathyroidism.

osteo- A prefix indicating bone-related.

osteoadherin See **osteomodulin**

osteoarthritis Disease of joints due to mechanical trauma. There is major disturbance in homeostasis of extracellular matrix with cartilage degradation (involving **matrix metalloproteinases**) and loss of normal joint function. Unlike **rheumatoid arthritis** there is no autoimmune element and, though there is inflammation, it is generally considered to be secondary rather than causative.

osteoblast Mesodermal cell that gives rise to bone.

osteocalcin Polypeptide (bone γ-carboxyglutamic acid protein, BGP, bone gla protein, 50aa) found in the extracellular matrix of bone. Binds **hydroxyapatite**. Has limited homology of its leader sequence with that of other vitamin K-dependent proteins such as prothrombin, Factors IX and X, and Protein C.

osteoclast Large multinucleate cell formed from differentiated **macrophage**, responsible for breakdown of bone.

osteoclast differentiation factor A member of the tumour-necrosis factor (TNF) superfamily (TNFSF11, ODF, receptor activator of NFκB ligand, RANKL, osteoprotegerin ligand, 317aa), a key cytokine involved in the differentiation of the immune system and the regulation of immunity as well as in bone metabolism. In particular, RANKL-deficient mice show defects in the early differentiation of T-cells, suggesting that RANKL is a novel regulator of early thymocyte development. RANKL-RANK signalling is essential for osteoclast development and plays a major role in pathological bone destruction. **Osteoprotegrin** and anti-RANKL antibody act as a specific inhibitors of RANKL and have therapeutic value in osteoporosis and rheumatoid arthritis.

osteocyte Osteoblast that is embedded in bony tissue and which is relatively inactive.

osteogenesis Production of bone.

osteogenesis imperfecta Heterogenous group of human genetic disorders that affect connective tissue in bone, cartilage and tendon. Bones are very brittle and fracture-prone. Type I is a dominantly inherited, generalized connective tissue disorder caused by mutation in either of the collagen Type 1A genes (COL1A1 and COL1A2); other types also involve mutations in one or the other of these genes but have phenotypic differences.

osteogenin Bone morphogenetic protein-3 (**BMP3**, 472aa), associated with extracellular matrix of adult and fetal cartilage; negatively regulates bone density. See **bone morphogenetic protein**.

osteoglycin See **mimecan**.

osteoid Uncalcified bone matrix, the product of osteoblasts. Consists mainly of collagen, but has **osteonectin** present.

osteoma Benign tumour of bone.

osteomalacia Softening of bone caused by vitamin D deficiency: adult equivalent of rickets.

osteomodulin A keratan sulphate proteoglycan found in mineralised extracellular matrix (osteoadherin, 421aa) that binds osteoblasts via the integrin $\alpha_v\beta_3$ and also binds hydroxyapatite. Has 12 leucine-rich repeats and is important in mineralisation.

osteonectin Calcium-binding protein of bone (basement membrane protein BM-40, secreted protein acidic and rich in cysteine, SPARC, 303aa), may regulate cell growth through interactions with the extracellular matrix and cytokines. Binds to both collagen and **hydroxyapatite**.

osteopetrosis *Albers-Schoenberg disease* The formation of abnormally dense and brittle bone, as opposed to **osteoporosis**. Type 1 is caused by defects in **low density lipoprotein receptor related protein** 5 that mediates wnt signalling in association with **axin**, Type 2 by mutation in chloride channel 7; other forms can be caused by mutation in any one of the following: the TCIRG1 subunit of the vacuolar proton pump, the osteopetrosis-associated transmembrane protein 1 (334aa), the **osteoprotegerin** ligand, the pleckstrin homology domain-containing protein M1 (926aa), **carbonic anhydrase II**, TNFRSF11A (see **Paget's disease of bone**) or **MCSF**.

osteopontin Bone-specific sialoprotein (bone sialoprotein 1, secreted phosphoprotein 1, nephropontin, uropontin, 314aa) that links cells and the hydroxyapatite of mineralised matrix; has an **RGD** sequence through which it is bound by integrin $\alpha_v\beta_3$, also binds CD44. Found mostly in calcified bone and is produced by osteoblasts. Synthesis is stimulated by **calcitriol** (Vitamin D3)

osteoporosis Loss of bone mineral density associated with low levels of **estrogen** in older women and in some cases by variations in type 1 collagen.

osteoporosis-pseudoglioma syndrome See **LDL receptor**-related protein-5.

osteoprotegerin A soluble decoy receptor (TNFRSF11B, osteoclastogenesis inhibitory factor, OCIF, OPG, 401aa) for **osteoclast differentiation factor/RANKL** that inhibits both differentiation and function of osteoclasts and thus acts as an inhibitor of bone destruction. OPG-deficient ($OPG^{-/-}$) mice exhibit severe osteoporosis and deficiency of OPG in human has been shown to result in juvenile **Paget's disease of bone**.

osteosarcoma Malignant tumour of bone (probably neoplasia of **osteocytes**).

osterix A zinc finger transcription factor (431aa) that is essential for osteoblast differentiation and bone formation.

ostium Generally, a mouth-like aperture. In sponges, an inhalant opening on the surface; in arthropods, an aperture in the wall of the heart by which blood enters the heart from the pericardial cavity; in mammals, the internal aperture of a Fallopian tube.

OTC (1) Ornithine transcarbamylase. (2) 'Over the counter' drugs that do not require a doctor's prescription.

otocadherin A **cadherin** (cadherin 23, 3354aa), that is involved, together with **protocadherin 15** (PCDH15), in forming the tip links at the tips of stereocilia of the hair cells of the inner ear that gate the mechanoelectrical channel. The extracellular domain has 27 repeats with significant homology to the cadherin ectodomain and otocadherin also interacts with **harmonin**. Mutations cause deafness (**Usher syndrome 1D**).

otoferlin A transmembrane protein (1997aa) that acts as the calcium ion sensor involved in synaptic vesicle-plasma membrane fusion and in the control of neurotransmitter release; interacts with **SNAP-2** and **syntaxin-1**. Otoferlin expression in mouse hair cells correlates with afferent synaptogenesis. Mutation can cause deafness in humans. See **ferlin**.

ouabain A plant alkaloid (strophanthin G) from *Strophantus gratus*, that specifically binds to and inhibits the **sodium-potassium ATPase**. Related to **digitalis**.

Ouchterlony assay Immunological test for antigen-antibody reactions in which diffusion of soluble antigen and antibody in a gel leads to precipitation of an antigen-antibody complex, visible as a whitish band. Because there is radial diffusion of the reagents, a very wide range of ratios of antigen to antibody concentration develop and precipitation will probably occur somewhere in the gel even when no care is taken with quantitation of the system.

outer dense fibres *ODF* Filamentous structures located on the outside of the axoneme in the midpiece and principal piece of the mammalian sperm tail that may help to maintain the passive elastic structures and elastic recoil of the sperm tail. The proteins composing the fibres include Odf1 (ODF27,

241aa) that utilizes its leucine zipper to associate with Odf2 (**cenexin**, 829aa), Odf3 (254aa) and Odf4 (257aa). Odf5 is **rhophilin 1**.

outron A sequence found at the 5′ end of pre-mRNAs that are to be trans-spliced: contains an intron-like sequence, followed by a splice acceptor.

outside-out patch See **patch clamp**.

oval cells Stem cells of the adult liver that can differentiate into both hepatocytes and bile duct epithelial cells. In normal adult liver oval cells are quiescent, existing in low numbers, and proliferate following severe, prolonged liver trauma. There is some evidence implicating oval cells in the development of hepatocellular carcinoma.

ovalbumin A major protein constituent of egg white. A phosphoprotein (386aa) with one N-linked oligosaccharide chain. Synthesis is stimulated by estrogen. The gene, of which there is only one in the chicken genome, has eight exons and is of 7.8 kbase; it was one of the first genes to be studied in this sort of detail.

ovalocytosis Hereditary disorder of erythrocytes relatively common in areas where malaria is endemic. Not only are the erythrocytes more rigid, but there is a mutation in **band III**, the anion transporter. Other forms are caused by defects in various elements of the erythrocyte membrane skeleton.

ovarian follicle See **follicle**.

ovarian granulosa cells Cells that form the multilayered cumulus oophorus that surrounds the oocyte in the Graafian **follicle**. They secrete sex steroids and growth factors.

overlap index A measure of the extent to which a population of cells in culture forms multilayers. The predicted amount of overlapping is calculated knowing the cell density, the projected area of the nucleus (usually), and assuming a Poisson distribution. The actual overlap is measured on fixed and stained preparations and the ratio of actual/predicted is derived. A value of 1 implies a random distribution with no constraint on overlapping; normal fibroblasts may have values as low as 0.05. Although a useful measure it does not unambiguously indicate the reason for the effect, which may be **contact inhibition of locomotion** or differential adhesion of cells between substratum and other cells.

overlapping In cell locomotion, the situation in which the **leading lamella** of one cell moves actively over the dorsal surface of another cell – should be distinguished from **underlapping**.

overlapping genes Different genes whose nucleotide coding sequences overlap to some extent. The common nucleotide sequence is read in two or three different reading frames thus specifying different polypeptides.

oviductin A component of the vitelline membrane that is responsible for converting the outer egg membrane into a fertilizable form as it passes down the oviduct. It is an enzyme (ovochymase-2, oviductal protease, EC 3.4.21.120, 1004aa in *Xenopus*) that selectively hydrolyzes the envelope glycoprotein gp43. The homologous mammalian enzyme is rather smaller.

ovocleidin A major protein (142aa) of the calcified layer of the avian eggshell.

ovomucin A glycoprotein that constitutes $2-4\%$ of the protein of egg albumin; a trypsin inhibitor.

ovomucoid Egg-white protein (210aa) produced in tubular gland cells in the epithelium of the chicken oviduct in response to progesterone or oestrogen. It is the dominant allergen of egg-white.

ovum *Pl.* ova. An egg cell.

owl-eye cells Enlarged cells infected with **Cytomegalovirus** that contain large inclusion bodies surrounded by a halo, hence the name.

Oxa1 Translocase (435aa) involved in the insertion of proteins into the inner mitochondrial membrane. Similar to **Albino3** and **YidC**.

oxacladiellanes Class of compounds that includes **sarcodictyin A** and **eleutherobin** that stabilize microtubules, and the **valvidones** and **eleuthosides** that are anti-inflammatory.

oxalic acid A naturally occurring dicarboxylic acid, found in plants. Toxic to higher animals by virtue of its calcium binding properties; it causes the precipitation of calcium oxalate in the kidneys, prevents calcium uptake in the gut, and is not metabolized.

oxaloacetate Metabolic intermediate. Couples with acetyl CoA to form citrate, i.e. the entry point of the **tricarboxylic acid cycle**. Formed from aspartic acid by transamination.

oxazolidinones Antibiotics that are effective against many Gram-positive bacteria that have developed resistance to the older antibiotics. They interfere with the assembly of the two bacterial ribosomal subunits (30S and 50S). An example is linezolid.

oxidation The process in which a compound donates electrons to an oxidizing agent. Also combination with oxygen or removal of hydrogen in reactions where there is no overt passage of electrons from one species to another.

oxidation-reduction potential See **redox potential**.

oxidative metabolism The controlled oxidation of compounds to produce energy for metabolic processes. Respiration in the biochemical sense.

oxidative phosphorylation The phosphorylation of ADP to produce ATP, coupled to the **respiratory chain**.

oxidative stress A condition of increased oxidant production in animal cells; reactive oxygen species including superoxide, hydroxyl radicals and singlet oxygen are potently bacteriostatic but also potentially very damaging.

oxidoreductase Any oxidase that uses molecular oxygen as the electron acceptor. See **dehydrogenases**.

oxygen electrode An electrode that is designed as a sensitive detector of oxygen consumption; involves a PTFE (Teflon) membrane that is permeable to oxygen; underneath this membrane is a compartment containing a saturated KCl solution and two electrodes, a platinum cathode and a silver anode. A fixed polarising voltage is applied between the electrodes and the resulting current (approx. 1 microamp) is proportional to oxygen concentration.

oxygen radical Any oxygen species that carries an unpaired electron (except free oxygen). Examples are $\cdot$OH, the hydroxyl radical and O_2^-, the superoxide anion. These radicals are very powerful oxidizing agents and cause structural damage to proteins and nucleic acids. They mediate the damaging effects of ionizing radiation.

oxygen-dependent killing One of the most important bactericidal mechanisms of mammalian phagocytes involves the production of various toxic oxygen species (hydrogen peroxide, superoxide, singlet oxygen, hydroxyl radicals) through the **metabolic burst**. Although anaerobic killing is possible, the oxygen-dependent mechanism is crucial for normal resistance to infection, and a defect in this system is usually fatal within the first decade of life (**chronic granulomatous disease**). See **myeloperoxidase, chemiluminescence**.

oxygenases Enzymes (Subclass EC 1.13) catalysing the incorporation of the oxygen of molecular oxygen into organic substrates; they differ from those in EC 1.14 in that a second hydrogen donor is not required. Sub-subclasses are EC 1.13.11, when two atoms of oxygen are incorporated (dioxygenases); EC 1.13.12, when only one oxygen atom is used (mono-oxygenases), and EC 1.13.99, for other cases. Both types are used by bacteria in degradation of aromatic compounds. Dioxygenases all contain iron, e.g. tryp-2,3 dioxygenase. Examples of mono-oxygenases are the enzymes that hydroxylate proline and lysine of collagen, using α-ketoglutarate.

oxylipins Secreted, hormone-like lipogenic molecules derived from linolenic acid. Examples include signalling compounds such as **jasmonates**, antimicrobial and antifungal compounds such as leaf aldehydes or divinyl ethers, and a plant-specific blend of volatiles including leaf alcohols.

oxyntic cell Cell (parietal cell) of the gastric epithelium that secretes hydrochloric acid.

oxyntomodulin A peptide (37aa) that consists of the 29aa sequence of **glucagon** followed by an 8aa carboxyterminal extension. Produced by cleavage of proglucagon by prohormone convertases. Oxyntomodulin inhibits meal-stimulated gastric acid secretion in rodents and has been shown to inhibit food intake following icv administration in rats. No specific receptor has been identified although it does act on glucagon and **GLP-1** receptors.

oxyphil cells See **Askenazy cells**.

OxyR A bacterial peroxide sensor and transcription regulator, which can sense the presence of reactive oxygen species and induce an antioxidant system. In *E. coli*, OxyR (305aa) is activated by hydrogen peroxide via the formation of a disulfide bond between two conserved cysteine residues (C199 and C208) and then activates the expression of a regulon of hydrogen peroxide-inducible genes.

oxysome Multimolecular array that acts as a unit in **oxidative phosphorylation**.

oxytocin *ocytocin* A peptide hormone (9aa) produced by cleavage of a precursor (oxytocin-neurophysin 1, 125aa): **neurophysin** specifically binds oxytocin. Induces smooth muscle contraction in uterus and mammary glands. Related to **vasopressin**. The receptor (389aa) is G-protein coupled.

Oxyuranus scutelatus scutelatus The Australian taipan snake. See **taicatoxin**.

P

p (1) Single letter code for proline. (2) Prefix, usually of a number, indicating a protein (sometimes a phosphoprotein), e.g. p53. So many proteins have been assigned a 'pNN' name that it is difficult to be comprehensive in the entries that follow. Many viral proteins are designated as pNN, also many peroxidases in plants. The number is often the (apparent) molecular weight in kilodaltons based upon mobility in gel electrophoresis. (3) Designating a region of a chromosome (see p region.

p0071 A member of the **armadillo repeat** protein family (plakophilin-4, 1211aa), most closely related to p120(ctn) (**catenin**) and the **plakophilins**. Depending upon the cell type may be localized in adherens junctions and desmosomes and may colocalize with desmoplakin. Interacts with PDZ domain-containing protein 4 which may play a role in the regulation of tight junction formation. See **PAPIN, ERBIN**.

p14 (1) An isoform (173aa) of cyclin-dependent kinase inhibitor 2A, p14ARF (ARF being 'alternative reading frame', not **ADP ribosylation factor**). See **INK4**. (2) The p14 gene in *Schistosoma mansoni* encodes one of a family of eggshell proteins and is expressed only in vitelline cells of mature female worms in response to a male stimulus.

p15 (1) Activated RNA polymerase II transcription cofactor 4 (p15, 127aa) that mediates transcriptional activation of class II genes. (2) Binding partner of the export receptor **TAP**. (3) One of three tumour suppressors p15 (INK4b), encoded by the INK4/ARF locus (the others are ARF and p16 (INK4a)). (4) PCNA-associated factor (p15(PAF), 111aa) which binds **proliferating cell nuclear antigen** (PCNA) and is overexpressed in several types of tumour.

p16 INK4a See **p15**.

p17 A viral matrix (structural) protein in HIV (HIV-1 p17, 231aa), derived by cleavage from the gag-pol polyprotein, that also acts as a cytokine which promotes proliferation, proinflammatory cytokine release and HIV-1 replication in preactivated, but not resting, human T-cells.

p18 (1) Ink4c See **Ink4**. (2) A haploinsufficient tumour suppressor encoded by v-*maf* and a key factor for ATM/ATR-mediated p53 activation. (3) A surface protein produced by the fish pathogen *Flavobacterium psychrophilum*. (4) N-terminal truncated version of **Bax** (p18 Bax.). (5) An antimicrobial peptide, P18

(KWKLFKKIPKFLHLAKKF-NH$_2$) designed from a **cecropin** A-**magainin** 2 hybrid.

p19 (1) Cyclin-dependent kinase 4 inhibitor D (p19**INK4D**, 166aa). (2) A line of murine embryonal carcinoma stem cells (P19 cells). (3) One subunit of the cytokine **IL-23**. (4) p19arf is a tumour suppressor that blocks the ubiquitin-ligase activity of mdm2 from targeting and inactivating **p53**. Absence of p19 therefore leads to decreased p53 activity and facilitates the formation of tumours. The CDK inhibitor Cdkn2a (p16INK4a) also controls the p53 pathway by generating an alternative transcript that encodes Cdkn2a (p14ARF) in humans or Cdkn2a (p19ARF) in mice.

p20 (1) A truncated C/EBPbeta isoform (p20 CCAAT enhancer-binding protein beta). (2) *Bacillus thuringiensis* helper (chaperone) protein P20, important in production of vegetative insecticidal proteins (VIPs). (3) *Xenopus* p20 protein influences the stability of MeCP2 (**methyl-CpG-binding protein 2**) in the human homologue cause **Rett's syndrome**. (4) A cardiac heat shock protein, p20. (5) *E. coli* p20 is a thioredoxin-dependent thiol peroxidase. (6) A cyclin-dependent kinase, Cdk5/p20. (7) One of the subunits of **ARP2/3** (p20-ARC, 168aa). (8) A protein (p20-CGGBP, CGG-binding protein1, 167aa) that binds to the unmethylated former of the trinucleotide repeat in the 5′UTR that is a feature of **fragile X syndrome**.

p21 A **cyclin dependent kinase** (CDK) inhibitor (Waf1, Cip1, CDKN1A, 21 kDa, 164aa) that inhibits multiple cdks; transcriptionally regulated by **p53**. See **p21-associated kinaseS** (PAK),

p21-activated kinases *PAKs* Serine/threonine protein kinases of the **STE20** subfamily that form an activated complex with GTP-bound ras-like proteins (P21, cdc2 and rac1). Activated, autophosphorylated PAK1 (545aa) acts on a variety of targets, shows highly specific binding to the SH3 domains of phospholipase Cγ and of adapter protein **Nck**, and regulates various morphological and cytoskeletal changes. **Caspase**-mediated cleavage of PAK2 (507aa) generates a constitutively-active catalytic fragment (34 kDa) and induces apoptosis in **Jurkat** cells. Its activity is apparently essential for the formation of apoptotic bodies. PAK3 (oligophrenin 3) has three tissue-restricted isoforms and mutations are associated with a form of X-linked mental retardation.

p22 phox See **phox**.

p23 (1) A member of the p24 family (p23, p24, p25) of cargo receptors important for vesicular trafficking

The Dictionary of Cell and Molecular Biology. DOI: http://dx.doi.org/10.1016/B978-0-12-384931-1.00016-7

between Golgi complex and ER. (2) Prostaglandin E synthase 3 (EC 5.3.99.3, telomerase-binding protein p23, Hsp90 co-chaperone, 160aa) a co-chaperone that disrupts receptor-mediated transcriptional activation by promoting disassembly of transcriptional regulatory complexes. (3) In *Arabidopsis*, peroxidase 23 (EC 1.11.1.7, 349aa). (4) See **pancreatitis-associated protein-1**.

p24 (1) A family of small and abundant Class I transmembrane proteins found in all eukaryotes, major components of COPI- and COPII-coated vesicles and implicated in cargo selectivity of ER to Golgi transport. There are four subfamilies (α, β, γ and δ). Overview: http://mbe.oxfordjournals.org/content/26/8/1707.full.pdf + html (2) One of the tubulin polymerization-promoting proteins (p24, p25-alpha, 219aa). (3) A capsid protein (p24, 192aa) in *Orgyia pseudotsugata* multicapsid polyhedrosis virus. (4) There are a family of baculoviridae p24 proteins.

p25 (1) A transmembrane protein (p25) that has a role in localization of protein tyrosine phosphatase TC48 to the ER. **p23** and p25 are members of a family of putative cargo receptors involved in vesicular trafficking between Golgi complex and ER. (2) Potent proteolytic fragment of **p35** (cdk5). (3) P25 and P28 proteins are essential for *Plasmodium* parasites to infect mosquitoes. (4) P25 (fibrohexamerin, 220aa) is one of three polypeptide components of the **fibroin** synthesized in the larval silk gland of silkworm.

p26 (1) An abundantly expressed small heat shock protein (192aa) from *Artemia* (brine shrimp) (2) One of a family of capsid proteins in Baculoviridae.

p27kip1 A member of the Cip1/Kip1 family of cyclin-dependent kinase inhibitors.

p28 (1) See **p25** definition 3. (2) One of the subunits of **IL-27** (243aa). (3) A poxvirus **RING** protein (E3 ubiquitin-protein ligase) gene, p28, encoding an anti-apoptotic factor. (4) Subunit (p28) of eukaryotic initiation factor 3 (eIF3k). (5) Replicase proteins p28 and p65 of mouse hepatitis virus.

p29 (1) See **BAG-1**. (2) Papain-like protease, p29 from the N-terminal portions of the *Cryphonectria hypovirus* 1 (CHV1)-EP713-encoded open reading frame that shares similarities with the potyvirus-encoded suppressor of RNA silencing HC-Pro and can suppress RNA silencing in the natural host, the chestnut blight fungus *Cryphonectria parasitica*. (3) Subline (P29) of the mouse Lewis lung cancer line. (4) Type IV collagen-binding protein (p29) of ML-SN2 (murine *fyn* cDNA-transfected clone). (5) An adhesin from *Mycoplasma fermentans*. (6) Stable and active **calpain** cleavage product of the cdk5 neuronal activator p39 (cf. **p25**). (7) Antigen (P29) localized in the dense granules of *Toxoplasma gondii*. (8) Estrogen-receptor-related protein, p29. (9) **Synaptogyrin** (p29) is a synaptic vesicle protein

that is uniformly distributed in the nervous system. (10) An autoantigen, in **Wegener's granulomatosis** (myeloblastin, EC 3.4.21.76, p29, neutrophil proteinase 4, 256aa). Degrades elastin, fibronectin, laminin, vitronectin, and collagen types I, III, and IV. (11) An ABC transporter (ATP-binding protein p29, 245aa) in *Mycoplasma genitalium*.

p30 (1) Any one of a number of viral proteins. HTLV-I p30 interferes with **toll like receptor** (TLR)-4 signalling. (2) A major surface protein, P30 (336aa), from *Toxoplasma gondii*. (3) Terminal organelle protein (P30, 274aa) of *Mycoplasma pneumoniae*, an adhesin that has a role in gliding motility that is distinct from its requirement for adherence. (4) Serine-arginine-rich protein p30 directs alternative splicing of glucocorticoid receptor pre-mRNA to glucocorticoid receptor beta in neutrophils.

P32 (1) Beta-emitting radioisotope of phosphorus (^{32}P), often used for labelling biomolecules. (2) Protein from *Spiroplasma citri*, (32 kDa deduced from the electrophoretic mobility). The P32-encoding gene (714 bp) is carried by a large plasmid of 35.3 kbp present in transmissible strains and missing in non-transmissible strains. (3) A cofactor of splicing factor ASF/SF-2 (4) Antigenic protein of African swine fever virus. (5) A mitochondrial matrix protein that binds the capsid of Rubella virus. (6) An adhesin (294aa) from *Mycoplasma gallisepticum*.

p33 (1) Tomato bushy stunt virus replication protein p33, also found in other tombus viruses. (2) A tumour suppressor p33 (ING1b). (3) Complement and kininogen binding protein gC1qR/p33 (gC1qR), distinct from **calreticulin**. (4) Eukaryotic translation initiation factor 3 subunit G (translation initiation factor eIF3 p33 subunit, TIF35, 274aa) that binds to mRNA and rRNA.

p34 (1) General transcription factor IIH subunit 3 (TFIIH p34, 308aa) (2) A cyclin-dependent kinase, (p34 kinase, cdc2, cdk1, 297aa). See **cyclin**. (3) Zinc finger CCCH domain-containing protein 12D (ZC3H12D, p34, 312aa), probably a ribonuclease and possibly a tumour suppressor. (4) Alpha- and gamma-adaptin-binding protein p34 (315aa). (5) A rickettsial protein (300aa) belonging to the cation diffusion facilitator (CDF) transporter family. (6) See **Arp 2/3 complex**.

p35 (1) One of two **cyclin-dependent kinase** 5 activators cleaved from cdk5 activator 1 (p35, 307aa and p25, tau protein kinase II 23 kDa subunit, 209aa). (2) **IL-12** p35, the alpha subunit of IL-12 p70. (3) *Baculovirus* caspase inhibitor, p35 protein. (4) P35 surface antigen of *Toxoplasma gondii*. (5) **Uroplakin**3b.

p36 (1) A member of a family of secreted proteins distributed throughout the genus *Mycobacterium*. The central domain of these proteins contains several amino acid PGLTS repeats, which differ considerably between species. P36, also called exported repetitive protein (Erp) in *M. tuberculosis*, has been shown to be associated with virulence. **(2)** A subunit of **eukaryotic initiation factor 3**. **(3)** One of the immunodominant sperm antigens identified by antibodies eluted from the spermatozoa of infertile men. **(4)** See **BAG-1**. **(5)** See **endonexin**.

p37 **(1)** A major structural protein of African swine fever virus, p37 protein. **(2)** A membrane lipoprotein of *Mycoplasma hyorhinis* (high affinity transport system protein p37, 409aa). There are similar proteins in other mycobacteria. **(3)** Isoform 4 of the AU-rich element RNA-binding protein 1 (AUF1) which, like the p40 isoform, has AU-rich element (**ARE**)-mRNA-destabilizing activity. **(4)** One of two RNA binding proteins (p34 and p37) from *Trypanosoma brucei*. **(5)** A 37-kDa platelet-agglutinating protein, p37, probably prethrombin-2. **(6)** *Vaccinia* virus major envelope protein P37. **(7)** The 37-kDa protein (P37) of *Borrelia burgdorferi* elicits an early IgM response in **Lyme disease**. **(8)** A **nucleoporin**, NUC37. **(9)** Long tail fiber protein p37 of bacteriophage T4 (1026aa) that recognizes the bacterial receptor.

p38 **(1)** Serine/threonine protein kinases (p38MAPK) activated by MAP kinase kinase (MKK6b) that acts in the signalling cascade downstream of various inflammatory cytokines such as IL-1 and TNFα. There are four isoforms of p38 MAP kinase: p38-α (MAPK14), -β (MAPK11), -γ (MAPK12 or ERK6) and -δ (MAPK13 or SAPK4). Homologous to the yeast HOG protein. See MEF2. Diagram of pathway: http://www.biocarta.com/pathfiles/m_p38mapkPathway.asp **(2)** Multi-tRNA synthetase cofactor (312aa), one of three auxiliary proteins (EEF1E1/p18, AIMP2/p38 and AIMP1/p43) associated with the complex.

p39 **(1)** An activator of cdk5, like **p35**. See **p29** definition 6. **(2)** Bacterioferritin (BFR) or P39 proteins of *Brucella spp.* that are T dominant antigens. **(3)** A subunit of **eukaryotic initiation factor 3**. **(4)** Transcription factor AP-1; see **jun**.

p40 **(1)** Subunit of both **IL-12** and **IL-23**. **(2)** A subunit (363aa) of ribonuclease P. **(3)** A 40 kDa erythrocyte membrane protein (LanC-like protein 1, 399aa) **(4)** A structural protein in some polyhedrosis viruses. (348aa in *Bombyx mori* nuclear polyhedrosis virus). **(5)** A 40 kDa Rab9 effector protein (372aa) required for endosome to trans-Golgi network (TGN) transport; interacts with PtdIns 5-P/PtdIns 3,5-P2-producing kinase, PIK**Fyve**. **(6)** Nucleolysin TIA-1 isoform p40 (T-cell-restricted intracellular antigen-1, TIA-1, 386aa) involved in alternative pre-RNA splicing.

p41 **(1)** The MHC II-associated chaperone molecule of **invariant chain** (inhibitory p41 Ii) that may regulate stability and activity of cathepsin L in antigen-presenting cells. **(2)** A putative regulatory

component of the mammalian **Arp2/3 complex**, p41-Arc. **(3)** Human herpesvirus 6 (HHV-6) early protein, p41. **(4)** Polypeptide p41 of a Norwalk-like virus is a nucleic acid-independent nucleoside triphosphatase. **(5)** In the Archaean, *Sulfolobus acidocaldarius*, a probable signal recognition particle protein (docking protein, 369aa). **(6)** **NADPH oxidase** organiser 1, p41-NOX (376aa). **(7)** In *Borrelia burgdorferi* the subunit (flagellin, P41, 336aa) that forms the core of the flagellum.

p42 **(1)** One of the **MAP kinases**, p42/MAP-kinase is erk2, p44MAP-kinase is erk1. Often collectively referred to as p42/p44MAP kinase or p44/p42 MAPK. **(2)** Nucleoporin Nup43 (p42, 380aa), part of the nuclear pore complex. **(3)** In *S. cerevisiae*, GU4 nucleic-binding protein 1 (ARC1, P42, 376aa) that binds specifically to G4 quadruplex nucleic acid structures. **(4)** The protein p42.3 (tumour specificity and mitosis phase-dependent expression protein, 394aa) is found in 5-month fetal tissues and in some tumours. **(5)** In *Drosophila*, transcription initiation factor **TFIID** subunit 9 (p42, Protein enhancer of yellow 1, 278aa).

p43 **(1)** Former name for aminoacyl-tRNA synthetase-interacting multi-functional protein (AIMP1, 312aa), a auxiliary factor associated with a macromolecular tRNA synthetase complex; also has cytokine activity and acts on endothelial and immune cells to control angiogenesis and inflammation. **(2)** Mitochondrial triiodothyronine receptor. **(3)** Subunit (p43) of telomerase from the ciliate *Euplotes aediculatus*. **(4)** Human placental isoferritin is composed of a 43 kDa subunit (p43) and ferritin light chains. **(5)** In *Xenopus*, a 5S RNA binding protein (p43, thesaurin-B, 365aa) which is a major constituent of oocytes and comprises part of a 42S ribonucleoprotein storage particle.

p44 **(1)** A **MAP kinase**, also called erk1. See **p42**. **(2)** The 44-kDa major outer membrane proteins (P44s) of *Anaplasma phagocytophilum*, the causative agent of human granulocytic anaplasmosis. **(3)** Splice variant of **arrestin**, p44. **(4)** Interferon-induced hepatitis C-associated microtubule aggregate protein, p44 (444aa). **(5)** A subunit of **eukaryotic initiation factor-3**. **(6)** Subunit 9 of the 26S **proteasome**, p44.5. **(7)** A subunit (395aa) of the transcription factor IIH (TFIIH). **(8)** Methylosome protein 50 (p44/Mep50, WD repeat-containing protein 77, androgen receptor cofactor p44, 342aa) a non-catalytic component of the 20S **PRMT5**-containing methyltransferase complex.

p45 **(1)** A component (373aa) of the NF-E2 transcription factor complex essential for regulating erythroid and megakaryocytic maturation and differentiation. **(2)** **F-Box protein** p45(SKP2) (424aa) the substrate-specific receptor of ubiquitin-protein ligase involved in the degradation of p27(Kip1). **(3)** Nucleoporin p58/p45 (nucleoporin-like protein 1, 599aa), part of the nuclear pore complex. (**4**) **Caspase 1** precursor (404aa). (**5**) **Golgin**-45 (p45 basic leucine-zipper nuclear factor, 400aa), required for normal Golgi structure and for protein transport from the ER to the cell surface via the Golgi. (**6**) 26S protease regulatory subunit 8 (proteasome subunit p45, p45/SUG, 404aa), an AAA ATPase.

p46 (**1**) An isoform of **Shc**, p46Shc. (**2**) An isoform of cyclin-dependent kinase-11 (cdk11p46) derived from a larger isoform (cdk11p110). (**3**) A 46-kDa glucose 6-phosphate translocase (P46), one of the components of the glucose-6-phosphatase enzyme complex of the ER. (**4**) Isoform of **JNK/SAPK**, p46JNK. (**5**) One of two subunits of mouse DNA primase; p46 is the catalytic subunit capable of RNA primer synthesis, the role of p54 is unclear, although it has a nuclear localization signal. (**6**) Natural cytotoxicity triggering receptor 1 (NKp46, CD335, 304aa) of natural killer cells. (**7**) Polymerase delta-interacting protein 3 (46 kDa DNA polymerase delta interaction protein, 421aa), involved in regulation of translation. (**8**) Thioredoxin-like protein p46 (endoplasmic reticulum resident protein 46, 432aa) a protein disulphide isomerase.

p47 (**1**) See **phox**. (**2**) A family of GTPases (p47 GTPases, immunity-related GTPases (IRG) family) that are essential, interferon-inducible resistance factors in mice, active against a broad spectrum of intracellular pathogens. (**3**) Constitutively expressed heat shock protein, p47. (**4**) Synonym for **pleckstrin**. (**5**) NSFL1 cofactor p47 (370aa) that reduces the ATPase activity of **valosin-containing protein**. (**6**) Eukaryotic translation initiation factor 3 subunit F (eIF3 p47, 357aa).

p48 (**1**) One of two splice variants of visual arrestin p44 and p48 (S-arrestin, retinal S-antigen, 48 kDa protein, rod photoreceptor arrestin, 405aa). (**2**) Nucleotide excision repair protein p48 (427aa) encoded by the DDB2 gene. (**3**) A DNA binding subunit of the transcription factor PTF1 (pancreas specific transcription factor 1a, 328aa). (**4**) Surface lipoprotein P48 of *Mycoplasma bovis*. (**5**) Norwalk virus nonstructural protein p48 that may disrupt intracellular protein trafficking in infected cells. (**6**) Cytokine receptor-like factor 3 (type I cytokine receptor-like factor p48, 442aa). (**7**) The e-subunit (p48) of mammalian initiation factor 3 (eIF3e, 445aa). (**8**) The most conserved subunit of mammalian DNA polymerase alpha-primase. See **p49** #4. (**9**) Retinoblastoma-binding protein p48 (histone-binding protein RBBP4, **chromatin assembly factor-1**, 425aa) a component of several complexes which regulate chromatin metabolism.

p49 (**1**) A 49-kDa protein (p49/STRAP, serum response factor binding protein 1, 429aa) that specifically interacts with an acidic amino acid motif in the N-terminus of **GLUT4**. (**2**) A member of the **p35** family, p49, that inhibits mammalian and insect **caspases**. p49 will block apoptosis triggered by treatment with **Fas ligand** (FasL), **TRAIL**, or ultraviolet radiation but not apoptosis induced by **cisplatin**. (**3**) An HLA-G1-specific inhibitory receptor (p49) present on NK cells from placenta but undetectable in peripheral-blood NK cells. (**4**) A subunit of the DNA polymerase-alpha/primase complex, PRIM1 (sometimes p48, 420aa), the others being p180 (POLA), p68 (POLA2) and the primase p58 (PRIM2A). (**5**) A structural protein (438aa), p49, in African swine fever virus. (**6**) A protein of unknown function (469aa), in *Streptomyces lividans*. (**7**) In mouse, tubulin-tyrosine ligase-like protein 1 (p49, tubulin polyglutamylase complex subunit 3, 423aa) responsible for post-translationally modifying alpha and beta tubulin.

p50 (**1**) A subunit of **NFκB** derived by cleavage from the p105 subunit. (**2**) DNA polymerase delta subunit p50 (EC 2.7.7.7, 469aa) (**3**) Major capsid protein P50 (467aa) from Invertebrate iridescent virus 6. (**4**) Activating signal cointegrator 1 complex subunit 1 (Trip4 complex subunit p50, 400aa) that enhances NFκB, SRF and AP1 transactivation. (**5**) See **dynamitin**. (**6**) A cytosolic protein **secernin-1** (p50, 414aa) that regulates exocytosis in mast cells. (**7**) A spindle body protein, spindolin (spheroidin, p50, 341aa) from *Choristoneura biennis* (spruce budworm) entomopoxvirus. (**8**) **Proteasome** subunit P50 (26S protease regulatory subunit 6A, 439aa).

p51 (**1**) A p53 homologue, p51/p63/p73L/p40/KET. (**2**) Subunit, p51, (440aa derived from the 1447aa gag-pol polyprotein) of **HIV-1** reverse transcriptase. The other subunit is p66. (**3**) Major antigenic 51-kDa protein in *Neorickettsia risticii*, P51. (**4**) The large subunit (p51, 365aa) of the neonatal IgG-Fc receptor. (**5**) Guanine deaminase (EC 3.5.4.3, p51-nedasin, 454aa) that catalyzes the deamination of guanine to produce xanthine.

p52 (**1**) A transcription factor (p52, 454aa) derived from NFκB2 (p100) when **IKKalpha** activates a non-canonical NFκB pathway important in inducing genes involved in adaptive immunity. (**2**) Isoform of **Shc**. (**3**) A transcriptional coactivator (see **p75**. (**4**) Polypeptide 4 (462aa) of general transcription factor IIIH (TFIIH).

p53 An important tumour suppressor (393aa), one of a family that is mutated in many tumours ($\sim$60%) and if only one p53 gene is functional (Li-Fraumeni syndrome) there is a predisposition to tumours. In unstressed cells p53 levels are kept low by the action of **hmdm2** ubiquitin ligase but in response to stress p53 protein binds DNA and stimulates **p21** production: p21 interacts with cdk2 and the complex inhibits progression through the cell cycle. The p53 family consists of **p53**, **p63**, and p73, with multiple isoforms and splice variants of each.

p54 (**1**) An arginine-rich 54 kDa nuclear protein (p54, 484aa) that may be involved in pre-mRNA splicing. **(2)** An envelope protein (176aa) of African swine fever virus. **(3)** A synonym of the peptidyl-prolyl *cis-trans* isomerase **FKBP5**. **(4)** A 54 kDa **nucleoporin** (507aa). **(5)** A 54 kDa nuclear RNA- and DNA-binding protein (p54(nrb), non-POU domain-containing octamer-binding protein (nono), 471aa) that binds the conventional octamer sequence in double stranded DNA and binds single-stranded DNA and RNA at a completely different site. **(6)** Protein product of the cellular proto-oncogene c-*ets1* (441aa), a transcription factor. **(7)** Transcriptional regulatory protein p54 (487aa), a transcriptional repressor for zinc finger transcription factors EGR1 and EGR2. **(8)** ATP-dependent RNA helicase p54 (483aa) that may be involved in mRNA decapping. **(9)** See **p56 #1**.

p55 **(1)** Membrane-associated guanylate kinase (**MAGUK**) protein (erythrocyte protein p55, 466aa) that is found in the **stereocilia** of outer hair cells of the inner ear and is an essential regulator of neutrophil polarity. There are a subfamily of MAGUK-p55 proteins that are homologues of the *Drosophila* **discs large** (**dlg**) proteins. **(2)** TNF-receptor alpha, p55 (TNFRSF1A, 455aa). **(3)** *Drosophila* ortholog (chromatin assembly factor, p55/dCAF-1) of retinoblastoma protein RbAp46/RbAp48. **(4)** A subunit of the 26S **proteasome** (456aa). **(5)** See **coronin**. **(6)** Phosphatidylinositol 3-kinase 55 kDa regulatory subunit gamma (461aa): forms a heterodimer with **p110**. **(7)** A protein disulfide-isomerase, the beta subunit (508aa) of prolyl 4-hydroxylase (EC 1.14.11.2). **(8)** A 55 kDa actin-bundling protein, fascin (493aa).

p56 **(1)** Viral stress-inducible murine protein, P56, that inhibits initiation of translation by binding to the 'e' subunit of eukaryotic initiation factor 3 (eIF3). P54 has similar effects. **(2)** **Src-family kinase**, p56(lck) and other related kinases also designated p56. See **lck**. **(3)** **Borna disease** virus surface glycoprotein (p56), implicated in viral entry involving receptor-mediated endocytosis. **(4)** The cortical granules of the eggs of mice, rats, hamsters, cows, and pigs contain a pair of proteins designated p62/p56. **(5)** Docking protein 2 (p56(**dok-2**), 412aa) that acts downstream of receptor or non-receptor tyrosine kinases. **(6)** Cyclin-dependent kinase-like 2 (493aa). **(7)** Coiled-coil domain-containing protein 91 (p56 accessory protein, 441aa) involved in regulating membrane traffic through the trans-Golgi network.

p57 **(1)** See **neuromodulin**. **(2)** An oxypregnane steroidal glycoside, P57AS3 (P57), the only reported active constituent from *Hoodia gordonii*. **(3)** An actin-binding protein p57 is mammalian **coronin**. **(4)** p57/kip2 is a **cyclin-dependent kinase** inhibitor of the p21CIP1, p27KIP1 family. Deficiency of this inhibitor casues major developmental defects in mice similar to those seen in **Beckwith-Wiedemann syndrome**. Inhibits G1/S phase cdks. The gene is maternally imprinted in human and mouse.

p58 **(1)** A cyclin-dependent protein kinase (**PITSLRE/ CDK11(p58)** 795aa). **(2)** p58/ERGIC-53 is a calcium-dependent animal lectin that acts as a cargo receptor, binding to a set of glycoproteins in the endoplasmic reticulum (ER) and transporting them to the Golgi complex. **(3)** Protein kinase inhibitor of 58 kDa (P58(IPK), DnaJ homolog subfamily C member 3, 504aa) is a cellular inhibitor of the mammalian double-stranded RNA-activated protein kinase (PKR). **(4)** Insulin receptor substrate p53/p58 (IRSp53) is involved in cytoskeletal dynamics. **(5)** The large p58 subunit (509aa) of human DNA primase is important for primer initiation (see **DNA polymerase alpha-primase complex**). **(6)** A family of MHC class-I-specific NK receptors (**KIR** family) (e.g. p58.1, CD158a, 348aa). Bind HLA-C alleles and inhibit the activity of NK cells thus preventing cell lysis. **(7)** XP-C repair-complementing complex 58 kDa protein (UV excision repair protein RAD23 homolog B, 409aa). **(8)** Protein disulphide-isomerase A3 (EC 5.3.4.1, 58 kDa glucose-regulated protein, 505aa). **(9)** **Nucleoporin** p58/p45 (599aa). **(10)** 26S **proteasome** non-ATPase regulatory subunit 3 (proteasome subunit p58, 534aa).

p59 **(1)** **Src family** kinases, p59-fyn (537aa) and p59-hck (526aa). **(2)** p59, **oligoadenylate synthetase**-like gene (OASL) product (514aa) that binds double-stranded RNA and DNA. **(3)** p59(scr), a protein predominantly expressed in the testis and developmentally regulated during spermatogenesis. **(4)** p59 is the human homologue of **GRASP55** (Golgi reassembly-stacking protein 2, 452aa). **(5)** **FKBP4** (HSP-binding immunophilin, 459aa), a component of the steroid receptor complex.

p60 **(1)** *Listeria monocytogenes* protein p60 that affects haemolytic activity and the uptake of bacteria by macrophages. **(2)** **Katanin** p60, a microtubule-severing protein. **(3)** Type 1 TNF receptor, p60 (TNFRSF1A, 455aa). The Type 2 receptor is p80. **(4)** **Chromatin assembly factor** 1 p60 subunit (CAF-1 p60, 468aa in *S. cerevisiae*). **(5)** Tyrosine kinase p60 (c-**src**) (pp60-c-src, 536aa). **(6)** **Caveolin** isoform, cav-p60. **(7)** Early T-cell activation antigen p60 (CD69, 199aa).

p61 **(1)** Proto-oncogene tyrosine-protein kinase Yes (p61-yes, 543aa). **(2)** Immunodominant antigen, P61, from *Nocardia brasiliensis*. **(3)** A capsid protein (238aa) of *Citrus tristeza virus*.

p62 **(1)** Subunit 4 (460aa) of **dynactin**. **(2)** Signalling adaptor (p62, **DOK1** 481aa) that appears to be a negative regulator of the insulin signalling pathway. **(3)** Polyubiquitin-binding protein p62/SQSTM1, the **sequestosome 1** (SQSTM1) gene product (440aa), an adapter protein which may regulate the activation of NFκB1 by TNFα, NGF and IL-1. **(4)** Subunit of the protein kinase C zeta (PKCζ)-p62-Kvβ (beta-subunit of delayed rectifier K$^+$ channel) complex, a Kv channel-modulating complex. (5) TFIIH basal transcription factor complex p62 subunit (548aa).

p63 (1) A transcription factor (680aa) of the **p53 family**; complex cross-talk between **Notch** and p63 is involved in the balance between keratinocyte self-renewal and differentiation. Mutations in p63 are involved in at least five distinct malformation syndromes. (2) Cytoskeleton-associated protein 4 (CKAP-4, 602aa), a membrane protein. (3) Mitogen-activated protein kinase 4 (p63-MAPK, 587aa). (4) UV radiation resistance-associated gene protein (p63, 699aa) that forms a complex with **beclin-1** and PI3kinase.

p64 (1) A chloride channel (p64, CLIC4, 253aa) of intracellular membranes; present in regulated secretory vesicles. See **parchorin**. (2) **Interleukin 2**-receptor gamma chain (p64, CD132, 369aa), common to several IL-receptors. (3) A major envelope glycoprotein in polyhedrosis viruses. (4) Transcription factor p64 (**myc** proto-oncogene product).

p65 (1) RelA/p65, a subunit of **NFκB**. (2) Proline-rich P65 protein of *Mycoplasma pneumoniae* (423aa). (3) Serine/threonine-protein kinase **WNK1**. (4) **Synaptotagmin**-1. (5) Golgi reassembly-stacking protein of 65 kDa (GRASP-65, 440aa). (6) In mouse, complement regulatory protein Crry (483aa), a cofactor for complement factor I. (7) See **plastins**.

p66 (1) An accessory subunit (POLD3, 466aa) in the **DNA polymerase delta complex**. (2) A subunit (548aa) of **eukaryotic initiation factor-3**. (3) An isoform of the protein encoded by *shc*. (4) Transcriptional repressor p66-beta (p66/p68, 593aa) (5) In *Physarum polycephalum*, a 66 kDa stress protein (601aa).

p67 (1) Methionine aminopeptidase 2 (EC 3.4.11.18, 478aa), an enzyme that removes the amino-terminal methionine from nascent proteins. Also binds eukaryotic initiation factor-2 (eIF-2). (2) CD33, a sialic acid-binding immunoglobulin-like lectin, SIGLEC3 (gp67, 364aa). (3) See **phox**. (4) In *Arabidopsis*, chloroplastic RNA-binding protein P67 (688aa) involved in chloroplast RNA processing. (5) **Syntaxin**-binding protein 1 (594aa). (6) In *S. cerevisiae*, a nuclear localization sequence-binding protein (414aa). (7) In guinea pig, zona pellucida sperm-binding protein 3 receptor (533aa).

p68 (1) One of the **DEAD-box helicases**, DEAD box protein 5 (RNA helicase p68, 614aa). (2) **Annexin** A6 (673aa). (3) See **p66**. (4) Interferon-inducible RNA-dependent protein kinase (eukaryotic translation initiation factor 2-alpha kinase 2, p68 kinase, 551aa). (5) Transforming protein p68/c-ets-1 (485aa) in chicken. See **ets**.

P69 (1) Pancreatic islet cell autoantigen 1 (483aa). (2) An **oligoadenylate synthetase**, p69 OAS/p71 OAS, 719aa). (3) Outer membrane protein P.69 of

Bordetella pertussis (677aa) derived from the **pertactin** autotransporter (p93, 910aa). (4) ABC transport system permease protein p69 (543aa) of *Mycoplasma genitalium*. (5) See **tramtrack**. (6) In turnip yellow mosaic virus a 69 kDa protein (628aa) that acts as a suppressor of RNA-mediated gene silencing.

p70 (1) **Annexin VI**. (2) p70 **S6 kinase** (525aa). (3) The heterodimeric form of **IL-12**, p70. (4) Ubiquitin-associated and SH3 domain-containing protein B, UBASH3B (suppressor of T-cell receptor signalling 1, STS1, 650aa). (5) A subunit of the p70/p80 thyroid autoantigen (lupus autoantigen, Ku antigen, **Ki antigen**). (6) Interleukin-2 receptor subunit beta (p70-75, CD122, 551aa).

p71 (1) A transcriptional activator, zinc finger DNA-binding protein p52/p71 (642aa). (2) See **p69** #2.

p72 One of the **DEAD-box** RNA helicases (DDX17, RNA-dependent helicase p72, 650aa).

p73 (1) One of the **p53** family of transcription factors. Overexpression of p73 occurs in malignant myeloproliferations; hypermethylation is common in malignant lymphoproliferative disorders, especially acute lymphoblastic leukemia (ALL) and non-Hodgkin's lymphomas. See **E1A** oncogene. (2) The Rho GTPase-activating protein-24 (p73 rhoGAP, filamin A-associated rhoGAP, FILGAP, 748aa).

p74 A protein (645aa) in *Autographa californica* nuclear polyhedrosis virus (AcMNPV) that is essential for virulence and is one of the baculoviridae p74 family.

p75 (1) A sialic acid-binding immunoglobulin-like lectin (adhesion inhibitory receptor molecule-1, SIGLEC7, 467aa), found mostly in natural killer cells. (2) Transcriptional coactivator p75/p52 (lens epithelium-derived growth factor, LEDGF, 530aa), an alternatively-spliced isoform of **p52**. (3) Low-affinity nerve growth factor receptor **TNFRSF16**, CD271, low affinity neurotrophin receptor p75NTR, 427aa). (4) KH type-splicing regulatory protein (710aa) that binds to the dendritic targeting element and may play a role in mRNA trafficking. (5) A-kinase anchor protein 5 (**AKAP5**) in *Bos taurus*.

p76 (1) A member of the transmembrane 9 superfamily of proteins, TM9SF2 (663aa), that may function as an endosomal ion channel or small molecule transporter. (2) Phospholipase B domain-containing protein 2 (589aa) that may function as a phospholipase. (3) Protein p76 IgBP (874aa), a cleavage product from adenosine monophosphate-protein transferase and cysteine protease ibpA (p120, 4095aa) of *Haemophilus somnus*, involved in virulence.

p77 A serine peptidase (251aa) from *Tribolium castaneum* (Red flour beetle).

p78 (1) A component of the **mediator complex** (ARC92, MED25, 747aa). (2) Microsphere protein 1, (MCRS1, cell cycle-regulated factor p78, 462aa) that modulates the activity of **DAXX** by recruiting it to the nucleolus and may be an inhibitor of TERT telomerase activity. (3) Interferon induced proteins (myxovirus resistance 1 & 2, MX1, MX2, 662aa, 715aa) that are members of the large GTPase family and have antiviral activity. (4) Serine/threonine-protein kinase p78 (MAP/microtubule affinity-regulating kinase 3, STK10, 776aa) that phosphorylates microtubule-associated proteins and cdc25C.

p80 (1) Type I **interleukin-1** receptor (IL-1R1, CD121a, 569aa). (2) **IL-12** p80, a homodimer of p40 subunits. (3) p80 **coilin** (576aa), a nuclear autoantigen that accumulates in **Cajal bodies**. (4) Type 2 TNF receptor, p80 (TNFRSF1B, CD120b, 461aa). (5) Katanin p80 WD40-containing subunit B1 (p80 katanin, 655aa) part of a complex that severs microtubules in an ATP-dependent manner. (6) TFIIH basal transcription factor complex 80 kDa subunit (760aa), an ATP-dependent 5′-3′ DNA helicase. (7) WD repeat-containing protein 48 (677aa), that regulates deubiquitinating complexes. (8) Telomerase protein component 1 (p80 telomerase homologue, 2627aa). (9) An endosomal membrane protein in *Dictyostelium* (p80, 530aa), one of the SLC31A transporter family. (10) Capsid protein p80 (691aa) of *Autographa californica* nuclear polyhedrosis virus.

p81 (1) Ezrin (p81, cytovilli, villin-2, 586aa). (2) Catalase-peroxidase (EC 1.11.1.6, EC 1.11.1.7, 737aa) of *Mycobacterium vanbaalenii*.

p82 (1) Eukaryotic translation initiation factor 3 subunit B (p82, eIF-3-eta, 712aa) in *Arabidopsis*. (2) A-kinase anchor protein 4 (AKAP4, 849aa in mouse).

p83 Rho-related BTB domain-containing protein 2 (deleted in breast cancer 2 gene protein, 727aa).

p84 (1) A regulatory subunit of **PI3Kgamma**, present in human, mouse, chicken, frog, and fugu genomes. Broadly expressed in cells of the murine immune system. (2) **STAT1beta** (750aa) that mediates signalling by interferons. (3) A tyrosine-phosphorylated protein (SIRP-alpha-1, SHP substrate 1, SHPS1, MYD1, macrophage fusion receptor, CD172a, 503aa) that is dephosphorylated by **shp** and is the receptor for CD47. (4) A nuclear matrix protein (p84, THO complex subunit 1, 657aa) a component of the THO subcomplex of the **TREX complex**.

p85 (1) Regulatory subunit (724aa) of phosphatidylinositol 3-kinase (see p110). (2) Transcription initiation factor TFIID 85 kDa subunit (TAFII-80, 704aa) in *Drosophila*. (3) Pluronic P85; pluronic block copolymers are potent sensitizers of multi-drug resistant (MDR) cancer cells. (4) **Cortactin** (563aa) in chicken, a substrate for src kinase. (5) Extracellular matrix protein 1 (secretory component p85, 540aa)

involved in endochondral bone formation as a negative regulator of mineralization.

p87 (1) A regulatory subunit of **PI3Kgamma** (p87 (PIKAP), PI3K-γ adapter protein, 503aa), functionally homologous to **p101** in many ways: binds to both p110-γ and G$_{β/γ}$ and mediates activation of p110-γ downstream of G protein-coupled receptors. Highly expressed in heart. (2) A mitochondrial inner membrane protein (mitofilin, p87/89, 758aa).

p90 Calnexin (major histocompatibility complex class I antigen-binding protein p88, 592aa).

p91 (1) See **phox**. (2) P24 **oleosin** isoform B (223aa) of *Glycine max* (soybean).

p95 See **nibrin**.

p100 See **NFκB**.

p101 Regulatory subunit of PI3Kgamma that binds to G$_{β-γ}$ and recruits the catalytic p110-γ subunit to the plasma membrane. See **PI-3-kinases**.

p107 Protein (107 kDa, 936aa) with many similarities to the **retinoblastoma** gene product. Binds to **E2F** and is found in the **cyclin**/E2F complex together with p33cdk2. Acts as a tumour suppressor in the context of activated H-**ras**.

p110 Catalytic subunit of phosphatidyl inositol 3-kinase (**PI-3-kinase**) of which various isoforms (α, β and δ) are coupled to receptor tyrosine kinases. The activity of the p110 is regulated by various subunits of which there are at least 7 (p85α, p85β, p55-γ and their splicing variants). Upon growth factor stimulation, p110 is recruited to the membrane and activated via the interaction of SH2 domains on the regulatory subunit and phosphotyrosine motifs on the stimulated RTKs. The PI3K-γ isoform is activated by receptor-stimulated G proteins and recruited to the membrane by **p101**.

p120 (1) One of the **catenins** (p120ctn; p120-catenin, catenin delta-1, 968aa) that regulates adherens junction stability in cultured cells. There is evidence that p120 affects NFκB activation and immune homeostasis in part through regulation of **rho** GTPases. Binds and inhibits the transcriptional repressor ZBTB33, which may lead to activation of the **Wnt** signalling pathway. (2) See **p76** #3.

p125 (1) A protein (1000aa) that interacts with Sec23p, is only expressed in mammals and exhibits sequence homology with phosphatidic acid-preferring phospholipase A1 although does not have enzyme activity. Appears to be essential for localisation of COPII-**coated vesicles** to ER exit sites. (2) Xeroderma pigmentosum group C-complementing protein (940aa), involved in DNA excision repair. (3) DNA polymerase delta catalytic subunit (EC 2.7.7.7, 1107aa). (4) p125(FAK): see **focal adhesion kinase**.

p130cas An adaptor protein (breast cancer anti-estrogen resistance protein 1, BCAR1, 870aa) that, once phosphorylated by **src family** kinases, can act as docking protein for proteins with **SH2** domains; a key mediator of focal adhesion turnover and cell migration. Related proteins are Sin/Efs and Nedd9 (a melanoma metastasis gene). Founder member of the **Cas-family proteins** that serve as docking proteins in integrin-mediated signal transduction.

p150 (1) A protease (1301aa) in Rubella virus derived from non-structural polyprotein **p200**. (2) Subunit (p150) of **dynactin** (dynactin 1) that is only present in the dynactin members of the CAP-Gly family of proteins. Also referred to as p150(Glued). (3) The hVPS34/p150 phosphatidylinositol (PtdIns) 3-kinase complex that regulates late endosomal phosphatidylinositol signalling. (4) Leukocyte beta2 integrin, CD11c/CD18 (p150,95/CR4). (5) p150(Sal2), a vertebrate homologue of the *Drosophila* homeotic transcription factor Spalt. (6) See **PAF** #2.

p185 (1) The *Erb-B2* gene product **HER-2**, a 185-kDa transmembrane receptor tyrosine kinase (EC 2.7.10.1, 1255aa) of the epidermal growth factor (EGF) receptor family, also called p185/neu or c-erbB-2. (2) See **Philadelphia chromosome**.

P388.D1 cells Clonal derivative of P388 cells, a mouse macrophage line that produces a large amount of interleukin-1.

p400 protein (1) Inositol 1,4,5-triphosphate receptor (InsP3-R) type 1 (inositol 1,4,5-trisphosphate-binding protein P400, 2749aa in mouse) that binds InsP3 and also has calcium channel activity. Involved in release of calcium from the ER. (2) Adenovirus E1A-associated p400 (3160aa) is one of the **SWI2/SNF2** family of chromatin remodeling proteins, a component of the NuA4 histone acetyl-transferase complex.

P680 Form of chlorophyll that has its absorption maximum at 680 nm. See **photosystem II**.

P700 Form of chlorophyll that has its absorption maximum at 700 nm. See **photosystem I**.

PA28 Proteasome activator protein (11S regulator, 254aa) composed of two homologous subunits (alpha and beta) and a separate but related protein termed **Ki antigen** or PA28gamma.

PAAD domain A conserved domain (PYRIN, AIM, ASC and Death domain) found in more than 35 human proteins that are involved in apoptosis and inflammatory signalling pathways. Belongs to the death domain superfamily and is associated with protein-protein interactions. See **pyrin**, **AIM**, **ASC** and **death domain**. http://www.ncbi.nlm.nih.gov/pubmed/16678172

PABA Compound (p-aminobenzoic acid) present in yeast as an intermediate in the synthesis of **folic acid**. Sometimes called Vitamin BX, although is not a true vitamin. Used as an UV-blocking ingredient in sunscreen.

PABPs See **poly(A) binding proteins**.

PAC (1) Proteasome assembly chaperone 1 (PAC1, Down syndrome critical region protein 2, 288aa). (2) Polyaluminum chloride, used in gel column chromatography. (3) Protein antigen c (PAc, 1565aa) of *Streptococcus mutans*, an adhesin. (4) A ligand-mimetic anti-alphaIIb/beta3 monoclonal antibody, PAC-1. (5) Photoactivated adenylyl cyclase (PAC), a blue-light photoreceptor that mediates photomovement in *Euglena gracilis*. (6) In *S. cerevisiae*, a nuclear distribution protein (PAC1, 494aa) with a role in positioning the mitotic spindle at the bud neck during cell division; targets cytoplasmic dynein to microtubule plus ends. Also a protein (PAC2, 518aa) involved in the assembly of alpha-tubulin. (7) A dual specificity protein phosphatase PAC-1 (EC 3.1.3.48, EC 3.1.3.16, 314aa) that acts on MAP kinases ERK1 and ERK2. (8) The PAC cloning vector is a P1-derived artificial chromosome vector that will accept large DNA fragments.

PACAP Pituitary adenylate cyclase-activating peptide (176aa). It is cleaved into PACAP-related peptide (PRP-48, 48aa), pituitary adenylate cyclase-activating polypeptide 27 (PACAP-27, 27aa) and PACAP-38. Member of the **secretin** superfamily of neuropeptides expressed in both the brain and peripheral nervous system, with neurotrophic and neurodevelopmental effects *in vivo*. Promotes the differentiation of **PC12 cells**. Receptor is G-protein coupled.

pachynema Rare synonym for **pachytene**.

pachytene Classical term for the third stage of prophase I of meiosis, during which the homologous chromosomes are closely paired and **crossing-over** takes place.

pacifastin A heterodimeric serine protease inhibitor isolated from the haemolymph of *Pacifastacus leniusculus* (Signal crayfish). Related inhibitors have subsequently been found in locust haemolymph. The two subunits are separately encoded and are covalently linked. The heavy chain of pacifastin (977aa) is related to transferrin, the light chain of pacifastin (420aa) is the proteinase inhibitory subunit.

Pacinian corpuscles In vertebrates, pressure receptors in the skin. The nerve ending is surrounded by multiple concentric layers of connective tissue that provide mechanical resistance to deformation.

paclitaxel See **taxol**.

pacsins A family of cytoplasmic adaptor phospho-proteins (protein kinase C and casein kinase substrate in neurons) involved in vesicle trafficking. They interact with **dynamin, synaptojanin** and N-WASP (see **WASP-family proteins**). Pacsin 1 (434aa) interacts with **huntingtin** in a repeat length-sensitive manner, interacting more strongly with mutant huntingtin. Pacsin 2 (486aa) and pacsin 3 (424aa) are more ubiquitously expressed and are not restricted to neurons. They do not interact with huntingtin.

pactamycin Antibiotic, isolated from *Streptomyces pactum* that inhibits translation in pro- and eukaryotes by preventing release of initiation factors from the 30S initiation complex. The binding site is distinct from that of tetracycline and that of hygromycin B. Research article: http://www.mrc-lmb.cam. ac.uk/ribo/homepage/pdf/brodersen_cell2000.pdf

pad-1 (1) In *C. elegans*, patterning defective protein 1 (2417aa) that may be involved in protein traffic between late Golgi and early endosomes and is essential for cell patterning during gastrulation. (2) In *S. pombe*, the 26S proteasome regulatory subunit rpn11: see **JAMM motif**. (3) In *Arabidopsis*, proteasome subunit alpha type-7-A. (4) In *S. cerevisiae*, phenylacrylic acid decarboxylase (EC 4.1.1.-, 242aa) an enzyme that confers resistance to cinnamic acid; there is a similar enzyme in *E. coli* (probable aromatic acid decarboxylase, 197aa)

PADGEM *P-selectin* See **selectins**.

paedogenesis Sexual reproduction by immature or larval forms. See **neoteny**.

paedomorphosis Synonym for **neoteny**.

PAF (1) **Platelet activating factor**. (2) The PAF protein complex is a transcriptional regulatory complex that associates with the RNA polymerase II subunit POLR2A and with a **histone methyltransferase** complex throughout the transcription cycle. The components include **parafibromin**, Leo1 RNA polymerase II associated factor, SH2 domain-binding protein 1 (p150) and PAF1 (pancreatic differentiation protein 2).

PAGE See **polyacrylamide gel electrophoresis**.

Paget's disease Breast carcinoma characterized by large cells with clear cytoplasm in the skin of the nipple. See **Paget's disease of bone.**

Paget's disease of bone A chronic disease (osteitis deformans) in which there is progressive enlargement and softening of bones, particularly of the skull and of the lower limbs due to activated osteoclasts. Genetically heterogeneous; some cases (PDB2) are caused by mutation in the TNFRSF11A gene which encodes **RANK**, a protein essential in **osteoclast** formation. PDB Type 3 is caused by mutation in the SQSTM1 gene (see **sequestosome**), the product of which is associated with the RANK pathway. Juvenile Paget's disease can result from

osteoprotegerin deficiency caused by homozygous or compound heterozygous mutation in the TNFRSF11B gene. Inclusion-body myopathy with Paget's disease and frontotemporal dementia (IBMPFD) is a disease of muscle, bone, and brain caused by mutations in the gene encoding **valosin-containing protein** (VCP). See **Paget's disease**.

pagoda cells Ganglion cells, from the central nervous system of a leech, with a spontaneous firing pattern that can look a little like a pagoda on an oscilloscope.

PAI See **plasminogen activator inhibitor**.

pair-rule genes Segmentation genes in *Drosophila*, expressed sequentially between gap genes and **segment-polarity genes**. There are about eight pair-rule genes expressed only in alternate segments (odd or even) of the developing embryo. Loss-of-function mutants thus lack alternate segments. Examples: *even-skipped* (*eve*), *fushi tarazu* (*ftz*), *hairy*. See *Pax* **genes**.

paired Developmentally regulated gene in *Drosophila* that contains the paired box domain. The product (613aa) is homologous to the human **Pax3** protein (479aa). Research article: http://www.imls.uzh.ch/research/noll/publ/Dev_2001_128_395_405.pdf

paired box domain A conserved domain (124aa) found in several developmentally-regulated proteins in *Drosophila* (e.g. paired, gooseberry, Pox) and *Pax* homeobox gene family products in mouse and human. Database entry: http://pfam.sanger.ac.uk/family?acc = PF00292

PAKs See **p21-activated kinases**.

PAL (1) **Phenylalanine ammonia lyase**. (2) Pyothorax-associated lymphoma, a rare B-cell non-Hodgkin's lymphoma. (3) Peptidoglycan-associated lipoprotein, a highly conserved structural outer membrane protein among Gram-negative bacteria. See **tol-pal proteins**. (4) The PAL motif is found in the C-terminal region of the aspartyl peptidases **gamma-secretase** and **signal peptide peptidase**.

palindromic sequence Nucleic acid sequence that is identical to its complementary strand when each is read in the correct direction (e.g. TGGCCA). Palindromic sequences are often the recognition sites for **restriction enzymes**. Degenerate palindromes with internal mismatching can lead to loops or hairpins being formed (as in tRNA).

palisade parenchyma Tissue found in the upper layers of the leaf **mesophyll**, consisting of regularly-shaped elongated parenchyma cells, orientated perpendicular to the leaf surface, which are active in **photosynthesis**.

paladin An actin-associated proline-rich protein (1383aa) that binds to vasodilator-stimulated phosphoprotein (**VASP**), **alpha-actinin**, **ezrin**, and **profilin** and required for organization of normal actin cytoskeleton and thus in regulating cell shape and motility. Genetic variations are associated with susceptibility to pancreatic cancer type 1. See **pallidin**.

pallidin Protein (172aa) encoded by a gene that is mutated in the pallid mouse strain. There is a human homologue. Involved in the development of lysosome-related organelles, such as melanosomes and platelet-dense granules, and may be more generally involved in intracellular vesicle trafficking as part of the **BLOC** complex. Has no homology to any other known protein and no recognizable functional motifs. Pallidin bind to **syntaxin** 13, F-actin and **muted**. *Cf.* **palladin**.

Pallister-Hall syndrome A pleiotropic autosomal dominant developmental disorder with variable consequences, often perinatally lethal. Caused by mutation in the *gli3* oncogene. NCBI datasheet: http://www.ncbi.nlm.nih.gov/bookshelf/br.fcgi?book = gene&part = phs

Pallister-Killian syndrome Rare disorder with multiple congenital abnormalities, seizures and mental retardation. Cause is an extra metacentric chromosome 12p (tetrasomy 12p) only in skin fibroblasts — so that the body is a tissue-specific mosaic.

palmdelphin A ubiquitous protein (paralemmin-like protein, 551aa) abundant in cardiac and skeletal muscle. It is a cytosolic isoform of **paralemmin-1** and may associate with endomembranes or cytoskeleton-linked structures. Abstract: http://www.ncbi.nlm.nih.gov/pubmed/16323283

palmitic acid One of the most widely distributed of fatty acids, n-hexadecanoic acid. The palmitoyl residue is one of the common acyl residues of membrane phospholipids. It is also found as a thioester attached to cysteine residues on some membrane proteins. Palmitoylated proteins are often transmembrane proteins with the modified residue on the cytoplasmic face of the membrane. The specificity of the transferase for the acyl residue is not high and both stearoyl and oleoyl residues can replace the palmitoyl residue. (*cf.* myristoylation).

palmitoylation See **palmitic acid**.

palmitvaccenic acid A fatty acid (11,12-hexadecenoic acid) that is a marker for arbuscular mycorrhizal fungi.

palynology The study of the past occurrence and abundance of plant species by an analysis of pollen grains and other spores that have been preserved in peat and sedimentary deposits of known age. The characteristic morphology and toughness of pollen makes the technique possible.

palytoxin *PTX* Linear peptide (2670 Da) from corals of *Palythoa spp.* that binds to Na^+/K^+ ATPase at a site overlapping that of **ouabain** and converts it into a channel. Extremely toxic and said to be the most potent animal-derived toxin. It is a complex molecule with 64 stereocenters and a backbone of 115 contiguous carbon atoms but has been synthesized. There are suggestions that the coral is simply concentrating the toxin made by the dinoflagellate *Ostreopis siamensis*.

Pam3 A lipopeptide from the outer cell membrane of bacteria that binds to **toll-like receptor-2** and causes a marked inflammatory response. A synthetic analogue (Pam3-Cys-Ala-Gly) has been used experimentally to stimulate cytokine production from monocytes. Research article: http://www.ncbi.nlm.nih.gov/pmc/articles/PMC1363802/pdf/immunology00051-0082.pdf

PAMP (1) A pathogen-associated molecular pattern, a set of molecular structures (epitopes) not shared with the host but shared by related pathogens and relatively invariant. PAMP-triggered immunity (PTI) inhibits pathogen infection of plants. Abstract: http://www.cell.com/current-biology/abstract/S0960-9822%2808%2901030-0 (2) A plasmid carrying an ampicillin-resistance gene.

Pan1 A yeast actin cytoskeleton-associated protein (1480aa) required for the internalization of endosomes during actin-coupled endocytosis. It activates the **Arp2/3 complex**. The PAN1 actin cytoskeleton-regulatory complex is composed of at least END3 (endocytosis protein 3, 349aa), PAN1, and SLA1 (1244aa).

pan-b A gene in *E. coli* that encodes the first enzyme of the **pantothenate** biosynthesis pathway, ketopantoate hydroxymethyltransferase (3-methyl-2-oxobutanoate hydroxymethyltransferase, EC 2.1.2.11, 264aa).

pancortin In mouse, a neuron-specific **olfactomedin**-related glycoprotein (**noelin**, neuronal olfactomedin-related ER localized protein, olfactomedin-1, 485aa) which may have a role in regulating the production of neural crest cells There are alternatively spliced variants.

pancreastatin Peptide hormone to (49 aa) that inhibits insulin release from the pancreas. Derived from **chromogranin A** by proteolytic processing in several peptide hormone-producing cells, such as pancreatic islet cells and gut endocrine cells.

pancreatic acinar cells Cells of the pancreas that secrete digestive enzymes; the archetypal secretory cell upon which much of the early work on the sequence of events in the secretory process was done.

pancreatic peptide Peptide (PP, pancreatic hormone) synthesized in **islets of Langherhans**, that acts as a regulator of pancreatic and gastrointestinal functions. Produced as a larger propeptide, which is enzymatically cleaved to yield the mature active peptide (36aa).

pancreatic triglyceride lipase Lipolytic enzyme (PNLIP, EC 3.1.1.3, 465aa) that hydrolyses ester linkages of triglycerides; cofactor is **colipase**. Similar hepatic and gastric/lingual isozymes exist. Plays a key role in dietary fat absorption by hydrolysing dietary long chain triacyl-glycerol to free fatty acids and monoacylglycerols in the intestinal lumen.

pancreatitis-associated protein I A secreted protein (PAP I, HIP, p23, Reg2, islet neogenesis-associated protein, 175aa) initially characterized as being overexpressed in acute pancreatitis, now also associated with a number of inflammatory diseases, such as **Crohn's disease**. Encoded by the hepatocarcinoma-intestine-pancreas (HIP) gene. May be involved in the control of bacterial proliferation.

pancreozymin See **cholecystokinin**.

pancytopenia Simultaneous decrease in the numbers of all blood cells: can be caused by aplastic anaemia, hypersplenism, or tumours of the marrow.

pandemic An epidemic of an infectious disease that spreads through human populations across a large region. The World Health Organization (WHO) definition requires that the disease is new to a population, causes serious illness and is spread easily and sustainably among humans.

Pandorina A genus of colonial green algae in which the cells are held together in a gelatinous matrix. More complex than ***Eudorina***, less complex than ***Volvox***. Image: http://www.hib.no/avd_al/naturfag/plankton/english/plankton/plankton-algae/green_algae/pandorina.html

Paneth cells Coarsely granular secretory cells found in the basal regions of crypts in the small intestine and most abundant in the distal small intestine. Secrete alpha-defensin microbicidal peptides as mediators of innate enteric immunity.

panicle A branched inflorescence, a compound raceme, for example that found in grasses.

panmictic Describing a population in which there is random mating.

pannexins **Innexin** homologues found in vertebrates, a highly conserved family in worms, molluscs, insects and mammals. Both innexins and pannexins are predicted to have four transmembrane regions, two extracellular loops, one intracellular loop and intracellular N and C termini. In humans three pannexins are known (Pannexin 1, 426aa; PANX2, 633aa; PANX3, 392aa); pannexin 1 forms mechanosensitive ATP-permeable channels in erythrocytes, others are structural components of gap junctions.

panning Method in which cells are added to a dish with a particular surface coat or a layer of other cells and the non-adherent cells are then washed off. Those that remain are expressing particular surface adhesive properties and can be cloned, or in the case of an expression library, the identity of the adhesion molecule can be determined.

pannus (1) Vascularized granulation tissue rich in fibroblasts, lymphocytes, and macrophages, derived from synovial tissue; overgrows the bearing surface of the joint in rheumatoid arthritis and is associated with the breakdown of the articular surface. (2) Granulation tissue that invades the cornea from the conjunctiva in response to inflammation.

pantetheinase An ubiquitous enzyme (EC 3.5.1.92, 513aa) that will hydrolyse pantotheine to pantothenic acid (vitamin B5) and cysteamine, a potent antioxidant. The vanin-1 gene product is a GPI-anchored pantetheinase, an ectoenzyme.

pantetheine An intermediate (N-pantothenylcysteamine) in the biosynthesis of CoA and a growth factor for *Lactobacillus*. Pantethine is the disulphide dimer of pantetheine and is sold as a health food supplement.

PANTHER A database (Protein ANalysis Through Evolutionary Relationships) that classifies proteins on the basis of function. Version 7.0, released in May 2010, contains 6594 protein families, divided into 62,972 functionally distinct protein subfamilies. Link: http://www.pantherdb.org/

P antigens The determinants of the P blood group system; there are three antigens, the globosides Pk (globotriosylceramide, CD77) and P (globoside) and the paragloboside, P1, all consisting of monosaccharides added sequentially to lactosyl-ceramide. There are five phenotypes. The Donath-Landsteiner antibody reacts with P antigen and is a 'cold IgG' that binds at low temperatures but elutes at 37°C; its binding causes **paroxysmal cold haemoglobinuria**. P and Pk act as 'receptors' for the binding of various pathogenic bacteria and P antigen also binds parvovirus B19. Blood group antigen database: http://www.ncbi.nlm.nih.gov/gv/mhc/xslcgi.cgi?cmd=bgmut/systems_info&system=p

Panton-Valentine leucocidin See **leucocidin**.

pantonematic flagella Eukaryotic **flagella** without **mastigonemes**; *cf.* **hispid flagella**.

pantophysin Ubiquitously expressed **synaptophysin** homologue (259aa) found in cells of non-neuroendocrine origin. May be a marker for small cytoplasmic transport vesicles.

pantothenate kinase A set of regulatory enzymes involved in CoA biosynthesis (EC 2.7.1.33; PANK1, 598aa; PANK2, 570aa; PANK3, 370aa; PANK4, 773aa), phosphorylate **pantothenate** (vitamin B5). Mutation in the PANK2 gene which encodes the mitochondrial enzyme (570aa) causes pantothenate kinase-associated neurodegeneration (Hallervorden-Spatz disease: see **infantile neuroaxonal dystrophy**).

pantothenic acid Vitamin of the B2 group. See Table V1.

Pao/BEL family A family of **LTR retrotransposons** (semotiviruses), so far found only in metazoan genomes. This family was originally characterized with the discovery of element sequences such as Pao (widely distributed in various animal phyla), Bel (from insects), and Tas (from nematodes and cnidarians) and the various 'Cer'-like sequences in *C. elegans*. They can be divided into five lineages called Pao, Sinbad (from *Schistosoma mansoni,*), Bel, Tas and Suzu (from echinoderms and vertebrates). The Gypsy Database of mobile genetic elements. http://gydb.uv.es/index.php/Bel/Pao

PAP technique (1) Colloquial abbreviation for **Papanicolaou's stain**, (2) Peroxidase-antiperoxidase method for obtaining an enhanced peroxidase reaction to indicate antibody binding to antigen. In the first stage the material, e.g. a section, is reacted with a specific antiserum (say rat) against the antigen. In the next stage a large excess of (say) rabbit anti-rat immunoglobulin is applied so that only one of the binding sites is bound to the first antibody. Then a rat antiperoxidase antiserum is bound to the second antibody's unfilled sites and finally peroxidase is added and binds to the third antiserum before the peroxidase is used to develop a colour reaction.

Pap test *Pap smear* A diagnostic procedure in which a smear of cervical cells are stained with **Papanicolaou's stain** to identify pre-cancerous changes, cancer, infection or inflammation.

papain A cysteine peptidase (EC 3.4.22.2, 345aa) from *Carica papaya* (pawpaw) that is thermostable and active even in the presence of denaturing agents. Although it will cleave a variety of peptide bonds there is greatest activity one residue towards the C-terminus from a phenylalanine.

Papanicolaou's stain *PAP stain* A complex stain for detecting malignant cells in cervical smears. Contains in separate staining stages: (a) haematoxylin, (b) Orange-G phosphotungstic acid and (c) Light green, Bismarck Brown, Eosin and phosphotungstic acid.

papaverine An opium alkaloid that acts as a smooth muscle relaxant, probably by blocking membrane calcium channels and inhibiting phosphodiesterase.

paper chromatography Separation method in which filter paper is used as the support. Not a very sensitive method, but historically important as one of the first methods available for separating natural compounds.

papilla (1) A small nipple-like projection occurring in various animal tissues and organs. (2) A small blunt hair on plants.

papilloma Benign tumour of epithelium. Warts (caused by papilloma virus) are the most familiar example, and each is a clone derived from a single infected cell.

Papillomaviridae A family of viruses with a genome consisting of a single molecule of circular, supercoiled, double-stranded DNA 5300–8000 nucleotides long. (Originally a member of the Papovaviridae family that was subsequently subdivided into two new families, Papillomaviridae and **Polyomaviridae**.) The papillomaviruses cause papillomas (warts) in their hosts and are usually confined to the keratinocytes of specific epithelia but can be oncogenic. Human papilloma virus 1 (HPV1) tends to infect the soles of the feet, HPV2 more commonly affects the hands. Virus database: http://www.ncbi.nlm.nih.gov/ICTVdb/ICTVdB/00.099.htm

PAPIN Plakophilin-related armadillo-repeat protein-interacting protein (2642aa) that is diffusely distributed on the plasma membrane of epithelial cells. Has six **PDZ domains** and interacts with **p0071**, a **catenin**-related protein. See **plakophilins**.

Papovaviridae An obsolete taxonomic family of oncogenic DNA viruses that was subdivided into the **Papillomaviridae** and the **Polyomaviridae**.

pappalysins Metalloendopeptidases (EC 3.4.24.79, pappalysin-1, PAPP-A, pregnancy-associated plasma protein A, 1627aa) that cleave insulin-like growth factor (IGF) binding protein-4 (IGFBP-4) and and IGFBP-5, causing a dramatic reduction in affinity for IGF-I and -II and making growth factor available. They are also involved in IGF-dependent degradation of IGFBP-2. Pappalysin-2 (1791aa) may be a IGFBP-5 proteinase in many tissues.

PAR (1) See **protease-activated receptor**. (2) In *C. elegans* a set of proteins involved in partitioning, the development of asymmetry during embryogenesis. Par1 and par4 are serine/threonine-protein kinases (1192aa and 617aa) that phosphorylate the asymmetry effectors mex-5 and mex-6 and thereby restrict them to the anterior cytoplasm of the zygote. Par5 is a 14-3-3-like protein (248aa). Other partitioning defective proteins (PAR3, PAR6) affect tight junction formation (see **PARDs**).

parabiosis Surgical linkage of two organisms so that their circulatory systems interconnect.

paracellin Paracellin-1 (claudin-16, 305aa) is one of the **claudin** family expressed at tight junctions of renal epithelial cells of the thick ascending limb of the loop of Henle; mutations in the paracellin-1 gene cause familial hypomagnesaemia with hypercalciuria and nephrocalcinosis (FHHNC) with severe renal Mg^{2+} wasting. Renal tubular dysplasia is an autosomal recessively inherited disorder in Japanese black cattle that is due to deletion mutations in the claudin-16 gene.

paracentric Descriptor for a portion of a chromosome that does not include the **centromere**. A paracentric inversion is one in which a paracentric portion of a chromosome has been rotated through 180° and re-inserted in the same location. If no genes have been lost this may not be particularly deleterious.

Paracentrotus lividus The purple sea urchin, often used in developmental biology. Web resource: http://www.eol.org/pages/599658

paracortex Mid-cortical region of a lymph node, an area that is particularly depleted of T-cells in thymectomised animals, and is referred to as the thymus-dependent area.

paracrine Form of signalling in which the target cell is close to the signal-releasing cell. Neurotransmitters and neurohormones are usually considered to fall into this category. Cf. **endocrine, exocrine, juxtacrine**.

paradominance A condition in which heterozygous individuals are phenotypically normal unless there is loss of heterozygosity in some lineages during development, so that some cells become homozygous or hemizygous for the mutation, making the individual a mosaic of normal and abnormal cells. A few rare disorders, particularly of skin, probably arise through paradominant inheritance.

parafibromin A tumour suppressor protein (cdc73 homologue, 531aa) that is a component of the **PAF** protein complex. It is encoded by the HRPT2 gene which is mutated in **hyperparathyroidism-2**.

paraganglia *chromaffin bodies* Small groups of neural-crest derived chromaphil cells associated with various ganglia. Can give rise to **multiple paragangliomata** or **phaeochromocytomas**; in the former the cells are of the non-chromaffin (chemosensory) type, in the latter the cells secrete hormone.

paragloboside See **P antigen**.

parainfluenza virus Species of the **Paramyxoviridae**; there are four types: Type 1 is also known as Sendai virus or Haemagglutinating Virus of Japan (HVJ), and the inactivated form is used to bring about cell fusion. Types 2–4 cause mild respiratory infections in humans.

parakeratosis Condition in which there is retention of nuclei in the **stratum corneum** of the epidermis. This is a normal finding on mucous membranes and also occurs in **psoriasis**.

paralemmin A phosphoprotein (387aa) lipid-anchored in membranes and possibly involved in regulating cell shape. It is post-translationally modified by prenylation and palmitoylation and the lipid anchor is necessary for function. Paralemmin-2 (379aa) has a different tissue distribution. See **palmdelphin**.

paralogous genes Genes (paralogues) that arise by duplication of an ancestral gene and divergence of function. Cf. **orthologous genes**.

Paramecium *Paramoecium* Genus of ciliate protozoans. The 'slipper animalcule' is cigar-shaped, covered in rows of cilia and about 250 µm long. Free-swimming, common in freshwater ponds: feeds on bacteria and other particles. Reproduces asexually by binary fission, and sexually by conjugation involving the exchange of micronuclei. See **kappa particle**.

paramural body Membranous structure located between the plasma membrane and cell wall of plant cells. If it contains internal membranes, it may be called a **lomasome**; if not, it may be termed a plasmalemmasome.

paramutation An allele-dependent transfer of epigenetic information, which results in the heritable silencing of one allele by another. In maize the mop1 (mediator of paramutation 1) gene is required for paramutation; the gene product is an RNA-dependent RNA polymerase. A similar modification of the mouse *Kit* gene in the progeny of heterozygotes with the null mutant *Kit(tm1Alf)* (a lacZ insertion) has been reported. Even the homozygous wild-type offspring maintain, to a variable extent, the white spots characteristic of *Kit* mutant animals.

paramylon Storage polysaccharide of *Euglena* and related algae, present as a discrete granule in the cytoplasm and consisting of β(1-3)-glucan.

paramyosin Protein (200−220 kDa, ∼850aa) that forms a core in the thick filaments of invertebrate muscles. The molecule is rather like the rod part of myosin and has a two-chain coiled-coil a-helical structure, 130 × 2 nm. Paramyosin is present in particularly high concentration in the **catch muscle** of bivalve molluscs, where it forms the almost crystalline core of the thick filaments.

Paramyxoviridae Class V viruses of vertebrates. The genome consists of a single negative strand of RNA as one piece. The helical nucleocapsid has a virus-specific RNA polymerase (transcriptase) associated with it. They are enveloped viruses: main members are **Newcastle Disease virus, measles virus,** and the **parainfluenza viruses.**

paranemic Topological term for the joint that is made by wrapping one circle around another without cutting either circle. The two circles can always be pulled apart. See **plectonemic.**

paranemin Developmentally-regulated intermediate filament protein (1748aa) in chickens, associated with **desmin** and **vimentin** filaments. Contains the rod domain characteristic of all cytoplasmic intermediate filament proteins and coassembles with desmin in muscle cells. **EAP-300** and IF-associated protein (IFAPa-400) are highly homologous to paranemin and it has significant homology with human **nestin** and frog tanabin.

paranode Region flanking the **nodes of Ranvier** in myelinated fibres where glial cells closely appose and form specialized septate-like junctions with axons. These junctions contain a *Drosophila* **neurexin** IV-related protein, Caspr/**Paranodin** (NCP1).

paranodin In rodents, one of the **NCP** family of transmembrane neuronal cell adhesion molecules found at synaptic junctions and highly enriched in paranodal regions of myelinated axons. In rat, contactin-associated protein 1 (Caspr1, neurexin-4, p190, 1381aa). See **CASPR.**

paraoxonase Human serum proteins (EC 3.1.1.2, 45 kDa glycoprotein) located on high density **lipoprotein** (HDL) and implicated in detoxification of organophosphates and possibly preventing the oxidation of LDL. There are isoforms with different substrate specificities and multiple polymorphisms in the PON1 and PON2 genes. Mice lacking serum paraoxonase are susceptible to organophosphate toxicity and to atherosclerosis.

parapatric speciation Formation of a new species without geographical separation, usually assumed to arise through subdivision of an environmental niche with reproductive isolation of those members of the parent species that occupy the new niche. *Cf.* **allopatric** and **sympatric speciation.**

paraplegin A nuclear-encoded mitochondrial ATP-dependent metallopeptidase (EC 3.4.24.-, 795aa) mutated in **hereditary spastic paraplegia** Type 7 (SPG7). The mutation causes a mitochondrial dysfunction that appears to disrupt axonal transport.

paraprotein *M-protein* Protein found in plasma of patients with monoclonal **gammopathy**: can be composed of whole immunoglobulin molecules (IgG, IgM, IgA) or their constituent subunits (heavy or light chains) and derived from a single clone of cells.

parasegment In development of *Drosophila*, the genetic boundaries between developing segments are thought to lie along the middle of each visible segment. To distinguish them from the segments in everyday use, these compartments are called 'parasegments'.

parasitaemia Infection of a host by a parasite or the level of infection by the parasite, depending upon context.

parasite-derived neurotrophic factor *PDNF* A trans-sialidase on the surface of *Trypanosoma cruzi* that is a substrate and an activator of *Akt*. PDNF increases the expression of *Akt* and suppresses the transcription of genes that encode proapoptotic factors. The result is that infected cells are protected from apoptosis induced by oxidative stress and the proinflammatory cytokines. Abstract: http://stke.sciencemag.org/cgi/content/abstract/2/97/ra74

parasympathetic nervous system One of the two divisions of the vertebrate **autonomic nervous sytem.** Parasympathetic nerves emerge cranially as pre-ganglionic fibres from oculomotor, facial, glossopharyngeal and vagus nerves, and from the sacral region of the spinal cord. Most neurons are cholinergic and responses are mediated by **muscarinic acetylcholine receptors.** The parasympathetic system innervates, for example, salivary glands, thoracic and abdominal viscera, bladder and genitalia. *Cf.* **sympathetic nervous system.**

parathormone See **parathyroid hormone.**

parathyroid hormone A peptide hormone (parathyrin, parathormone, 84aa) that stimulates osteoclasts to increase blood calcium levels, the opposite effect to **calcitonin.** Mutations lead to **hypoparathyroidism.** Parathyroid hormone-related protein (PTHRP, 141aa), regulates endochondral bone development and epithelial-mesenchymal interactions. The G-protein coupled parathyroid hormone receptor 1 (PTHR1, 585aa) binds both parathyroid hormone and PTHRP.

paratope (1) In immune network theory, an **idiotope**; an antigenic site of an antibody that is responsible for that antibody binding to an antigenic determinant (**epitope**). (2) The site on a ligand molecule to which a cell-surface receptor binds.

paratyphoid Enteric fever due to infection by *Salmonella* spp other than *S. typhi*; usually *Salmonella enterica* serovar Paratyphi. The disease is acquired through ingestion of heavily contaminated food and water and is similar to, but milder than, typhoid fever.

paraxial Lying along an axis – commonest use is in reference to paraxial mesoderm, the mesoderm that forms somites as opposed to the axial mesoderm that forms notochord.

parchorin Parchorin, (p64, CLIC4, 253aa) and the related chloride intracellular channel (CLIC) proteins are thought to be auto-inserting, self-assembling intracellular anion channels involved in a wide variety of fundamental cellular events including regulated secretion, cell division and apoptosis.

pardaxins Polypeptides (33aa) from the toxin gland of *Pardachirus marmoratus* (Red sea flatfish) that form an eight-subunit voltage-dependent pore that will induce neurotransmitter release. May act as shark repellants.

PARDs (1) Partitioning defective proteins, first described in *C.elegans* where PAR3 (partitioning defective protein 3, 1379aa) and PAR6 (309aa) interact and are essential for apicobasal and anterior-posterior asymmetries associated with cell adhesion and gastrulation during the first few cell cycles of embryogenesis (see **PAR** #2). In humans partitioning defective 3 homologue (PARD3, 1356aa) is an adapter protein involved in asymmetrical cell division and cell polarization processes. There is a partitioning defective 3-like protein (PARD3B, 1205aa) and three forms of partitioning defective 6 (PARD6A, PARD6alpha, 346aa; PARD6B, 372aa; and PARD6G, 376aa). PARD3 in association with JAM1 is important in the formation of epithelial tight junctions but PARD3 in association with PARD6B may block the PARD3/Jam-1 interaction. The PARD6-PARD3 complex links GTP-bound Rho small GTPases to atypical protein kinase C proteins. (2) In *E. coli* parD protein (83aa) is involved in **plasmid partitioning**.

parenchyma (1) Type of unspecialised cell making up the ground tissue of plants. The cells are large and usually highly vacuolated, with thin, unlignified walls. They are often photosynthetic, in which case they may be termed **chlorenchyma**. (2) The functional component of an organ, as opposed to the connective tissue (stroma).

parenteral Administration of a substance to an animal by any route other than the alimentary canal.

parenthesome Structure shaped rather like a parenthesis '(', found on either side of pores in the septum of a basidiomycete fungus. More logically called septal pore caps.

parfocal Describing microscope objectives mounted in such a way that changing objectives does not cause the specimen to go out of focus.

parietal cell See **oxyntic cell**.

parkin Gene mutated in an unusual form of **Parkinson's disease** (autosomal recessive juvenile parkinsonism). Gene is large (500 kb), very active in the **substantia nigra**, and codes for an E3 ubiquitin-protein ligase (465aa). The parkin coregulated gene (*PACRG*) encodes a protein of 296aa, probably also linked to the ubiquitin/proteosome system. p53-associated parkin-like cytoplasmic protein (PARC, **cullin-9**, 2517aa) anchors **p53** in the cytoplasm of lung carcinoma cells. See

synphilin. The parkin-associated endothelin receptor-like receptor (613aa) is an orphan G protein-coupled receptor.

Parkinson's disease A disease (paralysis agitans) characterised by tremor and associated deficiency of L-dopa (dihydroxyphenylalanine) production in the substantia nigra of the brain and the progressive loss of dopaminergic neurons. Autosomal dominant Parkinson's disease can be caused by mutation or triplication of the alpha-**synuclein** gene. Other forms are a result of mutation in **ubiquitin carboxyl-terminal esterase L1**, **dardarin**, **parkin**, a serine peptidase (HTRA2), **synphilin-1**, or the *DJ1* oncogene. Mitochondrial mutations may also cause or contribute to Parkinson's disease and other loci have also been implicated.

PARL See **rhomboid**.

paroral membrane In ciliates a compound ciliary organelle lying along the right side of the oral area; not a membrane in the normal sense.

paroxysmal cold haemoglobinuria An antibody-induced anaemia (Donath-Landsteiner syndrome) caused by a cold-reacting polyclonal immunoglobulin G (IgG) known as the Donath-Landsteiner autoantibody. Antibody-binding occurs at temperatures below normal body temperature and complement-dependent lysis follows after warming. Autoantibody formation may be a result of infection with a microorganism-derived antigen that induces antibodies that cross-react with the P antigen in the erythrocyte membrane.

paroxysmal extreme pain disorder A disorder in which there are paroxysms of rectal, ocular, or submandibular pain. Caused by mutation in the alpha subunit of the **voltage-gated sodium channel-9** (SCN9A).

paroxysmal nocturnal haemoglobinuria Disease in which there is haemolysis by complement as a result of deficiency in **decay accelerating factor** on the surface of red cells derived from somatically mutated haematopoeitic stem cells. The severity depends upon the proportion of mutated cells in the circulation. *Cf.* **paroxysmal cold haemoglobinuria**.

PARP See **poly(ADP-ribose)polymerase**.

parthenocarpy The formation of seedless fruit because fertilization does not occur. Can be spontaneously in some plants, e.g. banana, and in other plants can be induced by application of **auxin**. *Cf.* **stenospermocarpy**.

parthenogenesis Development of an ovum without fusion of its nucleus with a male pronucleus to form a zygote. See **apomixis**. In plants haploid parthenogenesis, in which the resulting plant is haploid, can occur; diploid parthenogenesis is usually a result of incomplete meiosis.

parthenolide A sesquiterpene lactone, the primary bioactive compound in feverfew (*Tanacetum parthenium*); an inhibitor of NFκB with anti-inflammatory and anti-tumour activity.

partial agonist Agonist for a receptor population that is unable to produce a maximal response even if all the receptors are occupied.

particle gun An apparatus used to transfect cells using gold microparticles (0.6 μm in diameter) that are coated with plasmid DNA.

partition coefficient Equilibrium constant for the partitioning of a molecule between hydrophobic (oil) and hydrophilic (water) phases. A measure of the affinity of the molecule for hydrophobic environments, and thus, for example, a rough guide to the ease with which a molecule will cross the plasma membrane.

parvalbumins Calcium binding proteins (110aa), found in muscle, with sequence homology to **calmodulin** and **troponin C** but with only two **EF-hand** calcium binding sites. Parvalbumin-beta is also known as oncomodulin.

parvins A family of actin binding proteins from the α-actinin super-family that form a complex with **integrin-linked kinase**. Alpha-parvin (actopaxin, 372aa) and beta-parvin (affixin, 364aa) are located at focal contacts, some cell -cell adhesion junctions, ruffling membranes and the nucleus. Gamma-parvin (331aa) is specifically expressed in several lymphoid and monocytic cell lines and lacks the nuclear localization signal present in alpha and beta-parvin.

Parvoviridae Class II viruses. The genome of these simple viruses is single stranded DNA and they have an icosahedral nucleocapsid. The autonomous parvoviruses have a negative strand DNA and include viruses of vertebrates and arthropods. The defective Adeno-associated viruses cannot replicate in the absence of helper adenoviruses and have both positive and negative stranded genomes, but packaged in separate virions.

parvulins A subfamily of the **peptidyl prolyl** *cis-trans* **isomerases** (PPIase), highly conserved in all metazoans. The human parvulin PIN1 (EC 5.2.1.8, 163aa) is a nuclear protein implicated in the regulation of mitosis through interaction with CDC25 and **polo-like kinase-1**. Phosphorylation of PIN1's **WW domain** (at Ser-16) by PKA abolishes the interactions between Pin1 and its target proteins. Parvulin 14 (PIN4, 131aa) is preferentially located in the mitochondrial matrix.

PAS (1) See **periodic acid-Schiff reaction**. (2) p-aminosalicylic acid. (3) Pre-autophagosomal structure, from which the autophagosome is thought to originate. (4) See **PAS domain**. (5) See **PAS genes**. (6) PAS kinase (PAS domain-containing serine/threonine-protein kinase, PASK, 1323aa) is a conserved nutrient-responsive protein kinase.

PAS domain A ubiquitous protein module (Per-Arnt-Sim domain) with a common three-dimensional fold involved in a wide range of regulatory and sensory functions. See PAS #6, ARNT (aryl hydrocarbon receptor nuclear translocator) and **period**.

PAS genes Genes in *S. cerevisiae* associated with peroxisomes also called *PEX* genes. The product of *PAS1* (peroxisomal ATPase PEX1, Peroxin-1, 1043aa) is part of the peroxisomal protein import machinery. Together with PEX6, mediates the ATP-dependent relocation and recycling of the **peroxisomal targeting signal-1** (PTS1) import receptor PEX5 from the peroxisomal membrane to the cytosol. See **peroxins**.

pasin Name given to cytoplasmic adapter proteins that interact with the sodium-potassium ATPase. They do not appear in recent literature: pasin-2 is probably **moesin**.

passage Term that derives originally from maintenance of, for example, a parasite by serially infecting host animals, passaging the parasite each time. Subsequently also used to describe the subculture of cells in culture, and the passage number, the number of sub-culturing events, is not equivalent to cell division number. With increasing passage number mutations accumulate and selection pressures will alter the characteristics of the culture.

passage cells In plants, cells of the endodermis opposite the protoxylem that remain thin-walled and retain their Casparian band, or a similar short cell in the exodermis.

passive immunity Immunity acquired by the transfer from another animal of antibody or sensitised lymphocytes. Passive transfer of antibody from mother to offspring is important for immune defence during the perinatal period.

passive transport The movement of a substance, usually across a plasma membrane, by a mechanism that does not require metabolic energy. See **active transport, transport protein, facilitated diffusion, ion channels**.

Pasteur effect Decrease in the rate of carbohydrate breakdown that occurs in yeast and other cells when switched from anaerobic to aerobic conditions. A consequence of the relatively slow flux of material through the biochemical pathways of respiration compared with those of fermentation because more ATP is produced via respiration than via fermentation of sugars.

Pasteurella pestis Old name for *Yersinia pestis*.

pasteurization The use of precisely controlled heart to reduce the microbial load in milk and other heat-sensitive liquids and to increase their shelf-life.

PAT family Family of proteins (**perilipin, adipophilin**, and **TIP47**) associated exclusively with lipid droplets and involved in regulating lipid deposition and mobilisation. They are not restricted to the lipid droplet surface but also pervade the droplet core. Research article: http://www.jbc.org/content/280/28/26330.full

Pat1 (1) In *S. cerevisiae*, DNA topoisomerase 2-associated protein PAT1 (796aa), a component of the Like Sm protein (Lsm) complex in the **decapping activator complex** that is involved in RNA processing. The human homologue (PAT1-like protein 1, 770aa) is required for processing body (P body) formation. Abstract: http://www.ncbi.nlm.nih.gov/pubmed/17936923 (2) Also in *S. cerevisiae*, the gene (PAT1, PXA2) that encodes peroxisomal long-chain fatty acid import protein 1 (Peroxisomal ABC transporter 2, 853aa). (3) In *Arabidopsis*, Scarecrow-like transcription factor PAT1 (PHYTOCHROME A SIGNAL TRANSDUCTION 1, 490aa), one of the **GRAS family**.

patatins A group of storage glycoproteins found in potatoes that also have lipase activity, possibly as a defence mechanism for the plant. The patatin domain has subsequently been found in bacteria and animals. An extensive family of human patatin domain-containing proteins have been identified: patatin-like phospholipase domain-containing protein 1 (PNPLA1, 532aa) is a lipid hydrolase, PNPLA2 (EC 3.1.1.3, adipose triglyceride lipase, calcium-independent phospholipase A2, desnutrin, pigment epithelium-derived factor, transport-secretion protein 2, 504aa) catalyzes the initial step in triglyceride hydrolysis in adipocyte and non-adipocyte lipid droplets. PNPLA6 (neuropathy target esterase, EC 3.1.1.5, 1366aa) is defective in a form of spastic paraplegia. Research article: http://mic.sgmjournals.org/cgi/content/full/150/3/522

patch clamp A specialized and powerful variant of the **voltage clamp** method, in which a patch electrode of relatively large tip diameter (5 μm) is pressed tightly against the plasma membrane of a cell, forming an electrically tight, 'gigohm' seal. The current flowing through individual **ion channels** can then be measured. Different variants on this technique allow different surfaces of the plasma membrane to be exposed to the bathing medium: the contact just described is a 'cell-attached patch'. If the electrode is pulled away, leaving just a small disc of plasma membrane occluding the tip of the electrode, it is called an 'inside-out patch'. If suction is applied to a cell-attached patch, bursting the plasma membrane under the electrode, a 'whole cell patch' (similar to an intracellular recording) is formed. If the electrode is withdrawn from the whole-cell patch, the membrane fragments adhering to the electrode reform a seal across the tip, forming an 'outside-out patch'.

patched A *Drosophila* segmentation polarity protein (1286aa) that is the receptor for **hedgehog** and acts in conjunction with **smoothened**. The human homologue (PTCH1, 1447aa) is the receptor for sonic hedgehog (SHH), indian hedgehog (IHH) and desert hedgehog (DHH) and seems to have a tumour suppressor function. A second gene, PTCH2, has high homology and somatic mutations are associated with **basal cell carcinoma**. See **Gorlin syndrome**.

patching Passive process in which integral membrane components become clustered following cross-linking by an external or internal polyvalent ligand. See **capping**.

path analysis A statistical approach to evaluating the influence of different variables on, for example, the probablility of tumorigenesis. It can be thought of as a form of multiple regression focusing on causality and has been applied with some success to evaluating the relevance of biomarkers for occupational exposure to carcinogens. Abstract: http://www.ncbi.nlm.nih.gov/pubmed/17548684?dopt=Abstract

pathogenesis-related proteins Plant proteins (PR proteins) upregulated in response to pathogens and stress. They include anti-fungal β-1,3-glucanases and chitinases.

pathogenicity island Region of bacterial chromosome of foreign origin that contains clusters of virulence-associated genes.

patj In humans a protein associated to tight junctions (InaD-like protein, 1801aa), a scaffolding protein that may regulate protein targeting, cell polarity and integrity of tight junctions. The *Drosophila* homologue is a membrane-associated protein (871aa) involved in establishing cell polarity as a component of the **SAC complex**. It is restricted to the sub-apical domain of epithelial cells and probably participates in the assembly, positioning and maintenance of adherens junctions.

pattern formation One of the classic problems in developmental biology is the way in which complex patterns are formed from an apparently uniform field of cells. Various hypotheses have been put forward, and there is now evidence for the existence of gradients of diffusible substances (morphogens) specifying the differentiative pathway that should be followed according to the concentration of the **morphogen** around the cell. See *Pax* genes.

patulin A mycotoxin (a polyketide lactone) produced by certain species of *Penicillium*, *Aspergillus* and *Byssochlamys* growing on fruit, particularly apples, pears and grapes. It is a carcinogen resistant to low pH and tolerant of high temperature. Causes breaks in DNA and inhibits aminoacyl-tRNA synthetase.

Pauly test A colorimetric reaction for identification of imidazole compounds using diazotized sulphanilic acid which reacts with histidine to give a red colour and with tyrosine to give an orange colour.

PAUP* Phylogenetic Analysis Using Parsimony. Software package used for phylogenetic tree reconstruction, a process in which the ancestral relationships among a group of organisms are inferred from their DNA sequences. Link: http://paup.csit.fsu.edu/

pavementing Term used to describe the **margination** of leucocytes on the endothelium near a site of damage.

pawn Mutant of *Paramecium* that, like the chesspiece, can only move forward and is unable to reverse to escape noxious stimuli. Defect is apparently in the voltage-sensitive calcium channel of the ciliary membrane.

Pax genes A conserved family of vertebrate genes (paired domain genes, paired box genes) encoding transcription factors with a DNA-binding domain similar to that of pair-rule genes of *Drosophila* and important in pattern formation. Pax1 (Hup48) is involved in development of the vertebral column (mutated in the mouse strain, undulated). Pax 2 is important in nephrogenesis (mutated in **renal-coloboma syndrome**), Pax3 (Hup2) is involved in neurogenesis (mutated in **Waardenburg's syndrome**). Pax4 and Pax6 are involved in pancreatic development, Pax 5 in B-cell differentiation. Pax6 is involved in development of the eye, Pax7 is required for neural crest formation and Pax8 is essential for the formation of thyroxine-producing follicular cells.

paxillin Cytoskeletal protein (591aa) that localizes, like **talin**, to focal adhesions, to dense plaques in smooth muscle, and to the myotendonous and neuromuscular junctions of skeletal muscle. There are isoforms with different affinities for **vinculin** and **focal adhesion kinase**.

PAZ domain PIWI/Argonaute/Zwille domains. See **piwi** and **argonaute**. Database entry: http://pfam.sanger.ac.uk/family?acc = PF02170

PBMC A mixture of **monocytes** and **lymphocytes** (peripheral blood mononuclear cells), blood leucocytes from which **granulocytes** have been separated and removed.

P bodies See **GW bodies**.

PBP (1) **Platelet basic protein**. (2) **Penicillin-binding protein**. (3) Nuclear receptor coactivator PBP (**peroxisome proliferator-activated receptor** (PPAR)-binding protein) functions as a coactivator for PPARs and other nuclear receptors. (4) Pheromone-binding proteins (PBPs) located in the antennae of male moths.

pBR322 Plasmid that is one of the most commonly used *E. coli* cloning vectors.

PC12 A rat **phaeochromocytoma** cell line from adrenal medulla. Widely used in the study of stimulus-secretion coupling, and because it differentiates to resemble sympathetic neurons on application of nerve growth factor.

PCA (1) Principal component analysis; a widely used technique in exploratory data analysis. The first principal component accounts for most of the variability in a sample, each successive component for lesser and lesser sources of variability. (2) p-chloroamphetamine. (3) Primary cutaneous aspergillosis. (4) Prostate cancer antigen-1 (pca-1).

PCAF A histone acetyltransferase (EC 2.3.1.48, p300/CREB-binding protein-associated factor, 832aa) which promotes transcriptional activation.

PCC syndrome An autosomal recessive disorder caused by mutation in **microcephalin** and characterized by premature chromosome condensation in early G2 leading to a high frequency (>10%) of prophase-like cells in lymphocytes, fibroblasts, and lymphoblast cell lines. In affected individuals there is microcephaly and mental retardation. Research article: http://www.ncbi.nlm.nih.gov/pmc/articles/PMC379095/?tool = pubmed

PCD (1) Programmed cell death (see **apoptosis**. (2) **Primary ciliary dyskinesia**. (3) Premature centromere division. (4) Mutant mice (pcd mice) in which there is Purkinje cell degeneration, often used as a model for neurodegenerative disorders.

pcDNA Expression vectors used experimentally. pcDNA3.1$^{(+)}$ and pcDNA3.1$^{(-)}$ are 5.4 kb vectors derived from pcDNA3 and designed for high-level stable and transient expression in mammalian hosts. They contain Human cytomegalovirus immediate-early (CMV) promoter, multiple cloning sites in the forward (+) and reverse (−) orientations and a neomycin resistance gene for selection of stable cell lines.

pCEF-4 See **9E3**.

pcg body In embryonic cells, polycomb group (PcG) proteins reside in about 50 to 100 nuclear foci termed PcG bodies, which may be concentrated areas of transcriptional repression. See **polycomb repressive complexes**.

PCGF (1) The genes that encode a family of transcriptional repressors (polycomb group RING finger proteins) involved in the **polycomb repressive complex-1** (PRC1). *PCGF1* encodes polycomb group RING finger protein 1 (nervous system polycomb-1, 259aa). At least six others are known. (2) A yeast plasmid vector for the construction of GFP (green fluorescent protein) fusion proteins in budding yeast.

pCMBS An organomercurial sulphydryl-reactive compound (p-chloromercuriphenylsulphonic acid) that inhibits water movement through **aquaporin-1**.

PCMT Enzyme (protein-L-isoaspartate (D-aspartate) O-methyltransferase, EC 2. 1.1.77, 227aa) that catalyses the methyl esterification of the free alpha-carboxyl group of abnormal L-isoaspartyl residues, which occur spontaneously in protein and peptide substrates as a consequence of molecular ageing.

pCMV Usually part of the name of an expression vector that contains the cytomegalovirus (CMV) promoter. e.g. pCMV-Tag from Stratagene.

PCNA *cyclin* Commonly used marker (proliferating cell nuclear antigen) for proliferating cells, a well-conserved nuclear protein (261aa) that associates as a trimer, and as a trimer interacts with DNA polymerases δ and ε for which it is an auxiliary factor for DNA repair and replication. Transcription of PCNA is modulated by **p53**. PCNA-associated factor (p15PAF, 115aa) may be involved in protecting cells from UV-induced cell death.

PCOLCE proteins Extracellular matrix proteins (procollagen C-proteinase enhancer proteins; PCOLCE1, 449aa; PCOLCE2, 415aa) that enhance the activities of procollagen C-proteinases by binding to the C-propeptide of procollagen I. They have three structural modules, two **CUB domains** followed by a C-terminal netrin-like **NTR domain**. Sequence conservation between PCOLCE proteins from different organisms suggests a conserved binding surface for other protein partners. Research article: http://www.jbc.org/content/278/28/25982.full

PCP pathway See **planar cell polarity pathway**.

PCR See **polymerase chain reaction**.

PCR *in situ* hybridization A technique for detecting rare mRNA or viral transcripts in a tissue. Tissue sections are subjected to **PCR**, usually in a temperature-cycling oven, before detection of the (hugely amplified) transcript.

PCTA-1 A secreted protein (prostate carcinoma tumour antigen-1, **galectin**-8, 317aa) that is highly expressed in prostate cancer and, because it is secreted, may be a useful serum marker. See **STEAP**.

PD-1 See **programmed cell death**.

PD 98059 A flavone derivative (2′-amino-3′-methoxyflavone) that is a selective mitogen-activated protein kinase kinase-1 (MEK-1) inhibitor. Article: http://www.cellsignal.com/pdf/9900.pdf

PD 184352 CI-1040 Potent and selective non-competitive inhibitor of **MEK1** that has potent anti-proliferative effects.

PD-ECGF A dimeric enzyme (thymidine phosphorylase, EC 2.4.2.4) that is also a cytokine (platelet-derived endothelial cell growth factor, gliostatin, 482aa). Produced by platelets, fibroblasts and smooth muscle cells. Stimulates endothelial proliferation *in vitro* and angiogenesis *in vivo*. Also promotes survival and differentiation of neurons.

PDE Phosphodiesterase. Any enzyme (in EC 3.1 class) that catalyses the hydrolysis of one of the two ester linkages in a phosphodiester. PDE-I (EC 3.1.4.1) catalyses removal of 5′-nucleotides from the 3′-end of an oligonucleotide. PDE-II (EC 3.1.16.1) catalyses removal of 3′-nucleotides from the 5′-end of a nucleic acid. Often the name is used loosely when cAMP-phosphodiesterase is meant. See **cyclic nucleotide phosphodiesterases**.

PDGF See **platelet-derived growth factor**.

PDK (1) **Pyruvate dehydrogenase kinase**. (2) **Phosphoinositide-dependent kinase**-1.

P domain See **trefoil motif**.

pds1 (1) In *S. cerevisiae*, the gene that encodes **securin**. (2) In *Zea mays*, the gene that encodes phytoene dehydrogenase, a chloroplast enzyme (EC 1.14.99.-, 571aa) involved in carotenoid biosynthesis.

PDX-1 (1) Transcription factor (pancreas-duodenum homeobox-1, insulin promoter factor 1, somatostatin transcription factor-1, 283aa) that plays a central role in regulating insulin gene transcription and differentiation of both exocrine and endocrine pancreas. (2) In fungi, and other plants, pyridoxine biosynthesis protein PDX1 (343aa) involved in synthesis of vitamin B6 and also with an indirect role in resistance to singlet oxygen-generating photosensitizers.

PDZ domains Domains (80−90aa) found in various intracellular signalling proteins associated with the plasma membrane; named for the postsynaptic density, disc-large, ZO-1 proteins in which they were first described. May mediate formation of membrane-bound macromolecular complexes, for example of receptors and channels, by homotypic interaction, also of cell-cell junctions. Usually bind to short linear C-terminal sequences in the protein with which they interact. Laboratory homepage: http://sysbio.harvard.edu/csb/macbeath/research/pdz.html

PE See **phosphatidyl ethanolamine**.

PEA-15 An acidic serine-phosphorylated protein (phosphoprotein enriched in astrocytes, 15 kDa, 130aa) highly expressed in the CNS, where it blocks ras-mediated inhibition of integrin activation and modulates the ERK/MAPkinase cascade. Has a protective role against cytokine-induced apoptosis. PEA-15 is phosphorylated in astrocytes by CaMKII and protein kinase C in response to endothelin which may determine whether it influences proliferation or apoptosis.

peanut agglutinin Lectin (PNA, 273aa) from *Arachis hypogaea* that binds to membrane **glycoproteins** containing β-D-gal (1-3) D-galNAc; used to investigate differential adhesiveness in developing systems.

pectin Class of plant cell wall polysaccharide, soluble in hot aqueous solutions of chelating agents or in hot dilute acid. Includes polysaccharides rich in galacturonic acid, rhamnose, arabinose and galactose, e.g. the polygalacturonans, rhamnogalacturonans, and some arabinans, galactans and arabinogalactans. Prominent in the **middle-lamella** and **primary cell wall**. Pectins provide the gelling agents for fruit preserves (jams). Experimentally, they can be hardened by addition of calcium.

PEDF A natural extracellular component of the retina (pigment epithelium-derived factor, serpin-F1, 418aa), a non-inhibitory serpin, that is a potent inhibitor of angiogenesis, induces extensive neuronal differentiation in retinoblastoma cells and inhibits endothelial cell injury *in vitro*.

pedicels See **podocytes**.

pedin A peptide of 13aa that stimulates foot formation in *Hydra*. The precursor protein, thypedin (1089aa) contains 13 copies of the peptide.

PEG See **polyethylene glycol**.

pegylation Covalent coupling of **polyethylene glycol** (PEG) to a molecule. Pegylated proteins have improved stability, biological half-life, water solubility, and immunological characteristics following injection. The addition of 40 to 50 kDa of PEG is sufficient to increase the size of a small molecule to such an extent that it is less readily excreted through the kidneys and persists in the body for longer. In addition, as they are more or less surrounded by the attached PEGs, pegylated proteins are less rapidly degraded.

PEI Polyethylenimine.

PEITC Phenethyl isothiocyanate, a compound with tumour-inhibiting activity.

pejvakin A **gasdermin** (352aa) essential in the activity of auditory pathway neurons and mutated in a form of autosomal recessive neuronal deafness. It has a nuclear localization signal and a zinc-binding motif.

pelargonidin One of the three primary plant pigments of the anthocyanin class. See **cyanidin**, **delphinidin**.

pelB Pectin lyase B(EC 4.2.2.10, 378aa in *Aspergillus*) a virulence factor of *Erwinia chrysanthemi*, a pectinase that degrades cell walls of plants. The 18 residue N-terminal leader sequence of pelB is used in various vector constructs.

P element A class of *Drosophila* **transposon**, widely used as a vector for reporter genes, for efficient germline transformation, and for **enhancer trap** or **insertional mutagenesis** studies. It encodes P transposase (751aa) which is required for insertion and excision (see **P(GAL4)**).

Pelizaeus-Merzbacher disease An X-linked dysmyelinating disease resulting from defects in PLP (proteolipid protein) gene. Mouse model is jimpy. Pelizaeus-Merzbacher-like disorders (PMLD) are described, some caused by mutation in the GJA12 gene that encodes **connexin 46**.

pellagra Chronic disease due to a deficiency of vitamin B3 (niacin) or of tryptophan. Often a consequence of a diet consisting predominantly of maize where the nicotinic acid is in bound form and there is a lack of the tryptophan precursor of nicotinic acid.

pelle *Drosophila* serine/threonine protein kinase (EC 2.7.11.1, 501aa) that is involved in the activation of **dorsal** (NFκB homologue), the signalling pathway that establishes embryonic dorso-ventral polarity. The IL-1 receptor-associated kinase 1 (Pelle-like protein kinase, 710aa) has similarities.

pellicle (1) In general, a thin skin or film. (2) The outer covering of a **protozoan**: the plasma membrane plus underlying reinforcing structures, for example the membrane-bounded spaces (alveoli) just below the plasma membrane in ciliates. (3) In dentistry, a protein film that forms rapidly on the surface of a clean tooth.

Pelomyxa Genus of giant amoebae, usually 500–800 μm but occasionally larger; multinucleated; found in fresh water.

PEM Polymorphic epithelial mucin. See **episialin**.

pemphigus A group of autoimmune dermatological diseases characterized by the production of bullae (blisters). Pemphigus vulgaris, and pemphigus foliaceus are caused by antibodies against **desmogleins** 3 and 1 respectively. See **bullous pemphigoid** and **Hailey-Hailey disease**.

PEN1 (1) Pentacyclic triterpene synthase 1 (arabidiol synthase, EC 5.4.99.-, 766aa), a plant enzyme that converts oxidosqualene to **arabidiol**. (2) In potato, a syntaxin, (PEN1, 239aa). (3) See **penaeidin-1**. (4) *Cf.* **PEN-2** (presenilin enhancer-2).

PEN-2 One of the four components (presenilin enhancer-2, 101aa) of **gamma-secretase**; catalyzes the intramembrane cleavage of integral membrane proteins such as Notch receptors, APP and **presenilin**, conferring gamma-secretase activity on the latter.

penaeidins A family of small (5–6 kDa) antimicrobial peptides (e.g. pen-1, 50aa) originally identified in the haemolymph of the Pacific white shrimp, *Litopenaeus vannamei*. Activity is predominantly directed against Gram-positive bacteria. There are multiple classes, based on primary structure, that vary in target specificity and effectiveness. All, regardless of class or species, are composed of two different domains: an unconstrained proline-rich domain of variable length and relatively low conservation, and a well-conserved disulphide bond-stabilized cysteine-rich domain. The C-terminal cysteine-rich domain of penaeidins resembles a motif found in several chitin-binding proteins isolated from plants and confers chitin-binding capacity. A curated database (Penbase) of all penaeidins has been established. Database: http://www.penbase. immunaqua.com/

Pendred's syndrome An autosomal recessive disease characterized by goitre and congenital sensorineural deafness. The Pendred's syndrome gene (PDS gene) encodes pendrin (sodium-independent chloride/iodide transporter, solute carrier family 26 member 4, 780aa).

pendrin See **Pendred's syndrome**.

Penelope retrotransposons A family of retrotransposons (PLEs) described in many animal genomes. They have LTRs (long terminal repeats) that may be in either direct or inverted orientations flanking a coding region with RT (reverse transcriptase) and EN (endonuclease) domains. The RT domain is closer to telomerase RTs (**TERTs**) than to any other characterized RTs. The EN domain has similarity to bacterial repair endonuclease UvrC. PLE transposons can retain introns during transposition and seem to be an ancient class of retroelements. Abstract: http://www.ncbi.nlm.nih. gov/pubmed/17345670?dopt = Abstract

penetrance The proportion of individuals with a specific genotype who express that character in the phenotype.

penguin A *Drosophila* protein (737aa) that has a **pumilio** homology domain.

penicillamine Product (dimethyl cysteine) of acid hydrolysis of **penicillin** that chelates heavy metals (lead, copper, mercury) and assists in their excretion in cases of poisoning. Also used in treatment of rheumatoid arthritis although its mode of action as an anti-rheumatic drug is not clear.

penicillin Probably the best known of the **beta-lactam antibiotics**, derived from the mould *Penicillium notatum*. It blocks the cross-linking reaction in **peptidoglycan** synthesis, and therefore destroys the bacterial cell wall making the bacterium very susceptible to damage. See **penicillin-binding protein**.

penicillin-binding protein Proteins (PBPs) that catalyze both polymerization of glycan chains (glycosyltransferases) and cross-linking of penta-peptidic bridges (transpeptidases) during the biosynthesis of the peptidoglycan bacterial cell wall. PBPs are the targets for beta-lactam antibiotics and thus play key roles in drug-resistance mechanisms. Altered penicillin-binding protein 2X (PBP 2X), for example, is essential to the development of penicillin and cephalosporin resistance in *Streptococcus pneumoniae*.

Pennisetum glaucum Pearl millet, a staple food and fodder crop in sub-Saharan Africa and the Indian subcontinent.

pentose phosphate pathway Alternative metabolic route (pentose shunt, hexose monophosphate pathway, phosphogluconate oxidative pathway) to the **Embden-Meyerhof pathway** for breakdown of glucose. Diverges from the E-M pathway when **glucose-6-phosphate** is oxidized to ribose 5-phosphate by the enzyme glucose-6-phosphate dehydrogenase (EC: 1.1.1.49). This step reduces **NADP** to NADPH, generating a source of reducing power in cells for use in reductive biosyntheses. In plants, part of the pathway functions in the formation of hexoses from CO_2 in photosynthesis. Also important as source of pentoses, e.g. for nucleic acid biosynthesis. It is the main metabolic pathway in activated neutrophils, rendering them relatively insensitive to inhibitors of oxidative phosphorylation. Congenital deficiency of the first enzyme in the shunt produces a sensitivity to infection similar to that seen in **chronic granulomatous disease**.

pentoses Sugars (monosaccharides) with five carbon atoms. Include **ribose** and **deoxyribose** of nucleic acids, and many others such as the aldoses **arabinose** and **xylose**, and the ketoses **ribulose** and **xylulose**.

pentraxins Family of proteins that share a discoid arrangement of five non-covalently linked subunits and the pentraxin domain (pentaxin domain). Neural petraxin 1 (432aa) may mediate uptake of degraded synaptic material and binds the snake toxin taipoxin. Neural pentraxin 2 (431aa) may be involved in long-term plasticity in the nervous system. Pentraxin 3 (pentaxin 3, pentraxin-related protein, 381aa) is involved in the regulation of innate resistance to pathogens, inflammatory reactions, possibly clearance of self-components and female fertility. Pentraxin 4 (478aa) is widely distributed and binds calcium. The short pentraxins include serum amyloid P and C reactive protein (**CRP**). The long pentraxins include pentraxin 3 and neuronal pentraxins. Database entry: http://smart.embl-heidelberg.de/smart/do_annotation. pl?DOMAIN = PTX

PEP See **phosphoenolpyruvate**.

PEP carboxylase Enzyme (EC 4.1.1.31) responsible for the primary fixation of CO_2 in **C4 plants**. Carboxylates PEP (**phosphoenolpyruvate**) to give oxaloacetate. Also important in **crassulacean acid metabolism**, since it is responsible for CO_2 fixation in the dark.

pepducin Synthetic lipidated peptides that act on the inside cell surface by blocking signalling from G-protein-coupled receptors. Pepducins based on the cleaved portion of the **thrombin receptor** PAR1 can act as antagonists and may have potential as chemotherapeutic agents in invasive breast carcinoma in which PAR1 is upregulated. Abstract: http://cancerres.aacrjournals.org/content/69/15/6223.short

peplomers Glycoproteins of the outer viral envelope; particularly large and conspicuous in Coronavirus and responsible for the 'sun-burst' appearance.

Pepper EST A database of EST/cDNA sequences from 21 libraries constructed from different tissues and from leaf tissue of *Capsicum annuum* cv. Bukang in stressful conditions. (In late 2009 the database has 116,412 ESTs). Link: http://genepool.kribb.re.kr/pepper/

pepsin Aspartic peptidase (EC 3.4.23.1, formerly EC 3.4.4.1) from stomach of vertebrates; 'acid protease' or 'carboxyl proteinase' in the older literature. Cleaves preferentially between two hydrophobic amino acids (e.g. F-L, F-Y), and will attack most proteins except protamines, keratin and highly glycosylated proteins. Enzymatically active pepsin (326aa) is autocatalytically cleaved from the inactive zymogen, pepsinogen, at acid pH in the presence of HCl. One of the peptides cleaved off in this process is a pepsin inhibitor and has to be further degraded to allow the pepsin to have full activity. Pepsin is the Type example of peptidase family A1.

pepsinogen See **pepsin**.

pepstatin A hexapeptide from *Streptomyces* spp. that inhibits pepsin and other aspartic peptidases.

peptidase Alternative name for a **protease** or proteinase; there is a move to phase out those terms and use the general term 'peptidase'. Peptidases can be grouped into clans and families. Clans are groups of families for which there is evidence of common ancestry. Families are grouped by their catalytic type, aspartic; **cysteine**; **glutamic**; **metallopeptidases**; **serine**; **threonine**; and unknown-type peptidases. A recent initiative, designed to rationalise the classification, is the MEROPS database. Peptidase database: http://merops.sanger.ac.uk/

peptide bond The amide linkage between the carboxyl group of one amino acid and the amino group of another. The linkage does not allow free rotation and can occur in *cis* or *trans* configuration, the latter the most common in natural peptides, except for links to the amino group of proline, which are always *cis*.

peptide histidine methionine One of the **secretin family** of neuropeptides. Human analogue (PHM, 27aa, N-terminal histidine and C-terminal methionine) of peptide histidine-isoleucine (PHI), generated from the same precursor as **vasoactive intestinal peptide**, prepro-VIP. Has vasodilatory activity.

peptide map A shorthand term for the pattern of fragments, separated on a gel, produced by the action of a peptidase (protease) on a protein. The pattern is diagnostic of the protein and a map can be produced from a single band on a gel; the method is being superceded by mass-spectroscopy.

peptide neurotransmitter See **neuropeptides**.

peptide nucleic acid *PNA* Synthetic **nucleic acid** mimic, in which the sugar-phosphate backbone is replaced by a peptide-like polyamide. Instead of 5′ and 3′ ends, PNAs have N and C termini. Their resistance to both **nucleases** and **peptidases**, and their ability to bind closely to complementary DNA or RNA sequences, made them promising candidates in **antisense** and **gene therapy** technologies but this has not really been realised. Webpage: http://www.highveld.com/pages/pna.html

peptide receptor Specific receptor for **peptide neurotransmitters**.

peptide YY Gut-derived peptide hormone of the NPY family (PYY, 97aa) that inhibits exocrine pancreatic secretion, has a vasoconstrictory action, inhibits jejunal and colonic mobility and has anorectic properties. It is cleaved into a 34aa fragment (PYY 3-36). Acts as an antagonist of the neuropeptide Y2 receptor (Y2R).

peptidoglycan *murein* Cross-linked polysaccharide-peptide complex of indefinite size found in the inner cell wall of all bacteria (50–90% of the wall in Gram positive, 10% in Gram negative). Consists of chains of approximately 20 residues of β(1-4)-linked N-acetyl glucosamine and N-acetyl muramic acid cross-linked by small peptides (4–10 residues).

peptidomimetics In general any compound that mimics the properties of a peptide; in practice often a synthetic peptide with some non-natural amino acids in the sequence that confer greater stability or resistance to degradation.

peptidyl prolyl *cis-trans*-isomerases Enzymes (PPIases, EC 5.2.1.8) that catalyze the *cis-trans* isomerization of prolyl bonds thereby affecting tertiary structure of some proteins. Three distinct classes of PPIases have been identified: **cyclophilins**, FK506-binding proteins (**FKBPs**) and **parvulins**.

The isomerase activity is apparently irrelevant to the immunosuppressive action of the **immunophilins.**

peptidyl transferase The enzymic activity (EC 2.3.2.12) of the large subunit of a ribosome, the catalysis of the formation of a **peptide bond** between the carboxy-terminus of the nascent chain and the amino group of an arriving tRNA-associated amino acid.

peptidyl-arginine deiminase Enzyme (PAD, EC 3.5.3.15, 663aa) responsible for formation of protein-bound **citrulline** from arginine, a major amino acid in the inner root sheath and medulla of the hair follicle. Substrate is **trichohyalin** and post-synthetic modification of trichohyalin by PAD alters its properties so that it is able to act as a rigid matrix component.

peptoid Oligomer composed of N-substituted gly-cines, a specific subclass of **peptidomimetics** in which sidechains are appended to nitrogen atoms along the molecule's backbone, rather than to the α-carbons (as they are in amino acids).

peptones Mixture of partial degradation products of proteins; sometimes used in culture media for micro-organisms (peptone broth).

PER kinase Enzymes that phosphorylate the clock protein **period**. Candidates are *Drosophila* **double-time** and casein kinase II.

Percoll Trademark for colloidal silica coated with polyvinylpyrrolidone that is used for density gradi-ents. Inert and will form a good gradient rapidly when centrifuged. Useful for the separation of cells, viruses, and subcellular organelles.

perforins Perforins 1 and 2 form tubular transmem-brane complexes (16 nm diameter) at the sites of tar-get cell lysis by **NK cells** and **cytotoxic T-cells**. Perforin-1 (cytolysin, 555aa) is defective in many cases of familial haemophagocytic lymphohistiocy-tosis (FHLH), a heterogeneous autosomal recessive disorder characterized by hyperactivation of mono-cytes/macrophages.

perfringolysin O Cholesterol binding toxin (theta toxin; θ-toxin, 500aa) from *Clostridium perfringens*. Shares with other thiol-activated haemolysins a highly conserved sequence (ECTGLAWEWWR) near the C-terminus.

periaxin Protein, localized to the plasma membrane of **Schwann cells,** that plays an important role in the myelination of the peripheral nerve. Two isoforms exist coded by a single gene. L-periaxin (1461aa) is localized to plasma membrane of Schwann cells, S-periaxin (147aa) is diffusely cytoplasmic. Both possess **PDZ-domains**. Mutations in the gene for periaxin are a cause of **Dejerine-Sottas syndrome**.

peribacteroid membrane Membrane derived from the plasma membrane of a plant cell and that surrounds the nitrogen-fixing bacteroids in legume root nodules. Has a high lipid content and may regulate the passage of material from the plant cell cytoplasm to the symbiotic bacterial cell. The idea that it restricts **leghaemoglobin** to the peribacteroid space seems untenable since leghaemoglobin is found in the cytoplasm of some cells.

pericanalicular dense bodies Obsolete name for electron-dense membrane-bounded cytoplasmic organelles (lysosomes) found near the canaliculi in liver cells.

pericarp That part of a fruit that is produced by thickening of the ovary wall. Composed of three layers, epicarp (skin), mesocarp (often fleshy) and endocarp (membranous or stony in the case of e.g. plum).

pericentric inversion Chromosomal inversion in which the region that is inverted includes the kinetochore.

pericentrin Conserved coiled-coil protein (Pcnt, NUP75, 3336aa) of the nucleoporin family found in the pericentriolar region where it is involved in orga-nization of microtubules during meiosis and mitosis; concentration highest at metaphase, lowest at telo-phase. A similar but larger human protein, kendrin, has been identified. **AKAP9** (also known as AKAP350, CG-NAP or hyperion) and pericentrin share a well conserved 90aa domain near their C-termini.

pericentriolar region An amorphous region of electron-dense material surrounding the centriole in animal cells: the major **microtubule organizing centre** of the cell.

perichondrium The fibrous connective tissue sur-rounding cartilage. The outer layer is fibroblast-rich, the inner layer contains mostly undifferentiated chondroblasts and chondrocytes. It becomes vascu-larised and becomes the periostium in mature bone.

pericline In botanical anatomy periclines are planes that are parallel to the outer surface. For a dome-like surface, e.g. the tip of a growing shoot, two kinds of periclines can be distinguished, meridional (longitu-dinal) and latitudinal (transverse). For an organ with bilateral symmetry longitudinal and transverse peri-clines can be distinguished. A periclinal cell wall is parallel to a nearby surface, usually the outer surface of the plant. Anticlines are trajectories perpendicular to periclines. Usage in geology is slightly different.

pericycle A cylinder of parenchyma cells that lies just inside the endodermis and forms the outermost part of the stele in roots. In dicotyledonous plants the pericycle divides to give rise to lateral roots.

pericyte Cell associated with the walls of small blood vessels: not a smooth muscle cell, nor an endothelial cell.

periderm The outer cork layer of a plant that replaces the epidermis of primary tissues. Cells have their walls impregnated with **cutin** and **suberin**.

peridinin Accessory pigment (carotenoid), part of the light-harvesting complex in dinoflagellates (Dinophyceae).

peridinium *peridium* General term for the outer wall of the fruiting body of a fungus.

perikaryon Cell body surrounding the nucleus of a neuron - does not include axonal and dendritic processes.

perilipins Proteins (perilipin 1, 522aa; perilipin 2 (adipophilin, adipose differentiation-related protein), 437aa; perilipin 4, 1357aa; perilipin 5, 463aa) that coat lipid droplets in adipocytes. Perilipin phosphorylation by **PKA** is apparently essential for the translocation of **hormone-sensitive lipase** from the cytosol to the lipid droplet, a key event in stimulated lipolysis. Perilipin 3 (434aa) is required for the transport of mannose 6-phosphate receptors from endosomes to the trans-Golgi network.

perimysium The connective tissue sheath that binds muscle fibres into bundles.

perinuclear space Gap 10–40 nm wide between the two membranes of the **nuclear envelope**.

period *per Drosophila* gene regulating circadian rhythm. Expressed in **CNS**, **Malpighian tubules**, and a number of other tissues. Per contains a structural **PAS domain**, a nuclear localization sequence and a cytoplasmic localization domain that restricts it to the cytoplasm in the absence of Tim (product of *timeless*) with which it forms a heterodimer. The mammalian homologues are PER1 (rigui, 1290aa) and PER2 (1255aa) both expressed with circadian periodicity in the suprachiasmatic nucleus. See **dec**.

periodic acid-Schiff reaction *PAS* A method for staining carbohydrates: adjacent hydroxyl groups are oxidized to form aldehydes by periodic acid (HIO_4) and these aldehyde groups react with Schiff's reagent (basic fuchsin decolourised by sulphurous acid) to give a purple colour. Used in histochemistry and in staining gels on which glycoproteins have been run.

periodic fever An autosomal dominant disorder (familial hibernian fever) in which there are recurrent attacks of fever, pain, skin lesions and myalgia. Caused by mutation in the TNFα receptor Type 1 (TNFRSF1A).

periodic paralysis A set of disorders in which there are episodes of partial paralysis of muscles.

The hyperkalemic form is caused by mutation in the **voltage-gated sodium channel** gene SCN4A (see **paramyotonia congenita**), the hypokalemic form by mutations in the gene encoding a subunit of the **L-type calcium channel** (CACNL1A3) or in the SCN4A gene. The thyrotoxic form can be caused by mutation in the voltage-regulated potassium channel, KCNE3. See also **Andersen's syndrome**.

peripheral lymphoid tissue Secondary lymphoid tissue, not necessarily located peripherally. See **lymphoid tissue**.

peripheral membrane protein Membrane proteins that are bound to the surface of the membrane and not embedded in the hydrophobic region. Usually soluble and were originally thought to bind to integral proteins by ionic and other weak forces (and could therefore be removed by high ionic strength buffers). However, it is now clear that some are covalently linked to molecules that are part of the membrane bilayer (see **glypiation**), and that there are others that fit the original definition but are perhaps more appropriately considered proteins of the cytoskeleton (e.g. Band 4.1 and **spectrin**) or extracellular matrix (e.g. **fibronectin**).

peripherin (1) Type III intermediate filament protein (470aa) co-expressed with **neurofilament** triplet proteins. (2) A highly conserved photoreceptor-specific glycoprotein (346aa) found on the rim region of rod outer segment disk membranes. Thought to be essential for assembly, orientation and physical stability of outer segment disks. Defects can cause retinitis pigmentosa or macular dystrophy.

periphilin A protein (gastric cancer antigen Ga50, 458aa) that is incorporated into the cornified cell envelope, colocalized with **periplakin** in differentiated keratinocytes.

periplakin One of the **spectraplakin** superfamily (1756aa), a component of the cornified envelope of keratinocytes that may link the cornified envelope to desmosomes and intermediate filaments. Can form homodimers or heterodimers with **envoplakin**.

periplasmic binding proteins Transport proteins (bacterial solute-binding proteins) located within the **periplasmic space**. Some act as receptors for bacterial chemotaxis, interacting with **MCPs**, others in transport of small molecules (e.g. periplasmic dipeptide transport protein of *E. coli*, 535aa) or ions (e.g. molybdate-binding periplasmic protein, 257aa).

periplasmic space Structureless region between the plasma membrane and the cell wall of Gram-negative bacteria.

periseptal annulus Organelle associated with cell division in Gram-negative bacteria. There are two circumferential zones of cell envelope in which membranous elements of the envelope are closely associated with **murein**. The annuli appear early in division and in the region between them, the periseptal compartment, the division septum is formed.

perisperm A diploid maternal tissue derived from the **nucellus**, a storage organ in some seeds.

perithecium A flask-shaped **ascocarp** with a pore or ostiole at the top through which the ascospores are discharged. *Cf.* **ascothecium, cleistothecium.**

peritoneal exudates A term most commonly used to describe the fluid drained from the peritoneal cavity some time after the injection of an irritant solution. For example, a standard method for obtaining rabbit neutrophil leucocytes is to inject intraperitoneally saline with glycogen (to activate complement) and drain off the leucocyte-rich peritoneal exudate some hours later.

peritrichous Descriptor for bacteria that have flagella distributed uniformly over the surface of the cell.

periventricular heterotopia An X-linked neurological disorder in which neurons fail to migrate into the cerebral cortex, caused by mutation in the gene for **filamin A**. Filamin B may partly compensate by forming homodimers instead of the normal filamin A/B heterodimers. An autosomal recessive form is caused by mutation in ADP ribosylation factor guanine nucleotide exchange protein-2.

periviscerokinins A family of small peptides (9−13 aa) found in the central nervous system and at neurohaemal release sites in insects, mostly of fairly primitive orders (cockroaches, locusts) although structurally similar peptides have been identified in the *Drosophila* genome.

perivitelline space The gap between the plasma membrane of the ovum and the **zona pellucida**.

PERK Protein kinase-like endoplasmic reticulum kinase (eukaryotic translation initiation factor 2-alpha kinase, EC 2.7.11.1, 1116aa) that inactivates EIF2 by phosphorylation, part of the unfolded protein response (UPR)-induced G1 growth arrest due to the loss of cyclin D1.

perlecan Proteoglycan (heparan sulphate proteoglycan 2) found in all basement membranes. The core protein (4391aa) is 80 nm long with 5−7 variable-length globular domains and has two or three heparan sulphate chains about 40−60 kDa located in the N-terminal end of the molecule. May occasionally contain both heparan sulphate and chondroitin/dermatan sulphate chains. It contains domains homologous to **LDL receptor**, **laminin**, neuronal cell adhesion molecule (**N-CAM**) and **epidermal growth factor** (EGF). Binds to **nidogen** and plays an important role in maintaining the structural integrity of the basement membrane. In man, non-functional mutations of perlecan cause a lethal chondrodysplasia;

partially functional mutations of perlecan also cause **Schwartz-Jampel syndrome**. The core protein is cleaved into **endorepellin** (705aa) and **LG3 peptide** (195aa).

permease General term for a membrane protein that increases the permeability of the plasma membrane to a particular molecule, by a process not requiring metabolic energy. See **facilitated diffusion**.

permissive cells Cells of a type or species in which a particular virus can complete its replication cycle.

permissive temperature Of a temperature-sensitive mutation, a temperature at which the mutated gene product behaves normally, and so the cell or organism survives as if wild-type (compare with restrictive temperature, at which the gene product takes on a mutant phenotype).

pernicious anaemia See **megaloblastic anaemia**.

peropsin A rhodopsin homologue (RRH, 337aa) localized to the microvilli of the retinal pigment epithelium that surround the photoreceptor outer segments. May act as a retinal isomerase. See **opsin**.

peroxidases A family of enzymes (EC 1.11.1.n) that catalyse reduction of hydrogen peroxide by a substrate that loses two hydrogen atoms. Many are haem enzymes (but not, for example, glutathione peroxidase). Haloperoxidases generate reactive halogen species that are important in intracellular killing (see **metabolic burst**). Within cells, may be localized in peroxisomes. Coloured reaction-products allow detection of the enzyme with high sensitivity, so peroxidase-coupled antibodies are widely used in microscopy and **ELISA**. **Lactoperoxidase** is used in the catalytic surface-labelling of cells by radioactive iodine. There are 73 peroxidase genes in *Arabidopsis*.

peroxins Products of *PEX* genes, of which at least 29 are known, that are involved in **peroxisome** biogenesis. For example, peroxin 1 (peroxisome biogenesis disorder protein 1, 1283aa) is an AAA ATPase that is required for stability of PEX5 and for protein import into the peroxisome matrix. Peroxin-2 (RING finger protein 72, 305aa) is defective in **Zellweger's syndrome**, peroxin-3 (373aa) is responsible for the assembly of membrane vesicles before the matrix proteins are translocated. See **peroxisome biogenesis disorders**.

peroxiredoxins A family of peroxide detoxifying enzymes (e.g. 2-cysteine peroxiredoxin A, BAS1, EC 1.11.1.15, 266aa) that have an important function in the chloroplast. Levels of 2CPA are regulated by the redox-sensitive transcription factor Rap2.4. See **sestrins**.

peroxisomal targeting sequence A consensus sequence on cytoplasmic proteins destined for accumulation in **peroxisomes**. The PTS-1 sequence involves a C-terminal serine-lysine-leucine (SKL) sequence. Examples include firefly (*Photinus pyralis*) luciferase and urate oxidase. The PTS1 receptor is encoded by the PEX5 gene. Some peroxisomal matrix proteins have a different, and less conserved sequence, the PTS-2 motif, which is recognized by the PTS2 receptor, encoded by the PEX7 gene.

peroxisome Organelle (0.5–1.5 μm) bounded by a single membrane and containing peroxidase and catalase; found in all major groups of eukaryotes and probably having a single evolutionary origin, based upon a common set of proteins implicated in peroxisome biogenesis and maintenance. The enzymatic content varies substantially across species: peroxisomes in trypanosomatid species contain enzymes involved in glycolytic reactions and are known as glycosomes, in some plants they have enzymes of the glyoxylate cycle and are called glyoxysomes. In filamentous fungi, a particular type of peroxisome is called the **Woronin body.** Microperoxisomes (microbodies) are of 150–250 nm diameter. Peroxisomes are important in lipid metabolism: see **peroxisome biogenesis disorders, peroxins**, and peroxisome proliferator-activated receptors **PPAR**. Review: http://rstb. royalsocietypublishing.org/content/365/1541/765.full

peroxisome biogenesis disorders Fatal autosomal recessive diseases in which peroxisomes are defective or deficient. There are 12 complementation groups, most of which have been linked to specific gene mutations. Examples are Zellweger's syndrome and neonatal adrenoleukodystrophy. See **peroxins**.

persephin A factor (156aa) of the TGFβ family that has neurotrophic activity for mesencephalic dopaminergic and motor neurons. It is the ligand for **GDNF** family receptor alpha-4 (GFRα-4).

persistence (1) The tendency of a cell to continue moving in one direction: an internal bias on the random walk behaviour that cells exhibit in isotropic environments. (2) Descriptive of viruses that persist in cells, animals, plants or populations for long periods, often in a non-replicating form. Persistence is achieved by strategies such as integration into host DNA, immunological suppression, or mutation into forms with slow replication.

pertactin An outer membrane protein (Omp) and virulence factor from *Bordetella* species that is involved in binding to host cells through an RGD sequence. The periplasmic protein (pertactin autotransporter, P93, 910aa) is itself responsible for passage to the outer membrane where it is cleaved into Omp P.69 (677aa) and pertactin translocator (199aa). Recombinant P.69 is a component of modern whooping cough vaccines.

Perthes' disease *Legg-Calve-Perthes disease* Avascular necrosis of the femoral head caused by mutation in the gene (COL2A1) encoding type II collagen.

pertussis toxin An **AB toxin**, that is a major virulence factor for *Bordetella pertussis*. The active (A) subunit (269aa) is an NAD-dependent ADP-ribosylase that acts on inhibitory **GTP-binding proteins**; the binding (B) subunit is a pentamer with subunits S2 (226aa), S3 (227aa), S5 (133aa), and two copies of S4 (152aa); S5 binds the subunits together.

PESKY Non-secreted isoform of the murine beta chemokine **CTACK** (CCL27) that has a nuclear localization signal.

PEST sequence Amino acid motif (Pro-Glu-Ser-Thr) that is found in many short-lived eukaryotic proteins and plays a role in their degradation. Research paper: http://www.biomedcentral.com/1471-2091/3/29

pestiviruses Genus of the family **Flaviviridae** responsible for swine fever and bovine diarrhoea.

pET expression system System that uses the pET vector to produce large amounts of protein in a bacterial expression system. The pET vector (plasmid) contains a ampicillin resistance marker, the lacI gene, the T7 transcription promoter, the lac operator region 3′ to the **T7 promoter**, and a polylinker region. There are two origins of replication - one which enables the production of a single stranded vector under appropriate conditions, and the other the conventional origin of replication. The gene for the protein of interest is cloned into the polylinker region, a bacterial host expressing T7 polymerase under lac control is transfected and the system activated by IPTG (which displaces repressor from the lac operon of the vector and the T7 polymerase).

petite mutants A class of yeast mutants, most studied in *S. cerevisiae*. Mutants grow slowly and rely on anaerobic respiration: mitochondria, although present, have reduced cristae and are functionally defective (termed promitochondria). There are three types of petite mutant: (i) Segregational mutants that show Mendelian behaviour and result from mutations in mitochondrial genes located in the nucleus; (ii) Neutral petites, which are recessive genotypes and result from the complete absence of mitochondrial DNA; (iii) Suppressive petites, in which most of the mitochondrial DNA is lost (60–99%), though what remains is often amplified.

Peutz-Jeghers syndrome An autosomal dominant disorder, caused by mutation in the serine-threonine kinase 11 gene (**LKB1**) and associated with an increased risk of developing malignant tumours.

PEV (1) Position-effect variegation, a heterochromatin-associated gene silencing phenomenon. (2) Porcine enterovirus 1 (PEV-1).

PEVK domains Domains first identified in verte-brate titin that bind actin and and associated with the elasticity of the protein.

pexin family A family of proteins that contain a similar haemopexin repeat domain. Includes **haemo-pexin, vitronectin**, and most matrix metallopepti-dases. The haemopexin domain is involved in protein binding.

Peyer's patches Lymphoid organs located in the submucosal tissue of the mammalian gut containing very high proportions of IgA-secreting precursor cells. The patches have B- and T-dependent regions and germinal centres. A specialized epithelium lies between the patch and the intestine. Involved in gut-associated immunity.

PF 4 Platelet factor 4, a cytokine (CXCL4, SCYB4, 101aa).

P-face See **freeze fracture**.

Pfam An open-access database with a large collec-tion of multiple sequence alignments and hidden Markov models covering many common protein domains and families. Can be used, for example, to view the domain organisation of proteins. In November 2011 Pfam 26 included 13,672 families. Link: http://pfam.sanger.ac.uk/

PFG Pulsed field gradient. PFG-NMR is a well-established method for the determination of diffusion coefficients, which are indicative of molecular size and shape.

PFK See **phosphofructokinase**.

Pfr The form of **phytochrome** that absorbs light in the far red region, 730 nm, and is thus converted to **Pr**. It slowly and spontaneously converts to Pr in the dark.

PGA (1) Poly (gamma-glutamic acid) (γ-PGA), a naturally occurring biodegradable polymer produced by *Bacillus subtilis*. (2) Phenylglyoxylic acid, (3) Penicillin G acylase, (EC 3.5.1.11, 846aa) an enzyme in the semi-synthetic production of β-lactam antibiotics. (4) 3-phosphoglyceric acid.

P(GAL4) Synthetic **P element** of *Drosophila*, com-prising **long-terminal repeats** flanking a mini-**white** gene to mark flies carrying the P element by their red eye colour, **Bluescript** to allow plasmid rescue of DNA flanking the genomic insertion site, and a gene encoding the yeast transcription factor **GAL4**, downstream of a weak (permissive) promoter. Although itself unable to move within the genome, P(GAL4) mobilization can be induced by crossing in a source of transposase. Patterns of expression of neighbouring genes can be detected (see **enhancer trapping**) by crossing in a reporter gene (e.g. **lacZ**, **green fluorescent protein** (GFP)) under the control of the UAS promoter recognized by GAL4.

PGC Peroxisome proliferator-activated receptor gamma coactivator-1. See **PPAR**.

PGD Preimplantation genetic diagnosis, a technique for screening embryos that have been fertilized *in vitro*. The range of tests available is increasing rapidly, raising some ethical concerns.

pGEM Proprietary name for cloning vectors with multiple restriction sites. Various modified forms are sold to facilitate particular cloning requirements.

pGEX Proprietary name for a family of expression vectors that have an expanded multiple cloning site that facilitates the unidirectional cloning of cDNA inserts obtained from libraries. Proteins are expressed as fusion proteins with **glutathione S-transferase** (GST) that facilitates purification on Glutathione Sepharose™ 4B.

PGK See **phosphoglycerate kinase**.

pGLO Plasmid that encodes the gene for **GFP** and a gene for resistance to the antibiotic ampicillin. The expression of the gene for GFP can be switched on in transformed cells by adding the sugar arabinose to the medium.

P glycoprotein See **multidrug transporter**.

PGP (1) P glycoprotein (Pgp). See **multi-drug trans-porter**. (2) Protein gene product 9.5 (PGP 9.5) is an ubiquitin hydrolase that has been used as a marker for neural and neuroendocrine cells although the speci-ficty is questionable. Article: http://www.nature.com/modpathol/journal/v16/n10/full/3880873a.html

PGs Prostaglandins (PGA, PGB, PGD, PGE, PGF, PGG, PGH, PGI). PGA is **prostaglandin** A, etc. PGI is more commonly known as **prostacyclin**.

P granules Ribonucloprotein granules found in germ cells, and their precursors, in the nematode *C. elegans*. P granules are segregated during early embryogenesis into those blastomeres that eventually produce the germ line. All the known protein components of P granules contain putative RNA-binding motifs and the granules are probably involved in post-transcriptional regulation. Research article: http://www.andrologyjournal.org/cgi/rapidpdf/jandrol.109.008292v1

pH A logarithmic scale ($-\log_{10}$ [H^+]) for the mea-surement of the acidity or alkalinity of an aqueous solution. Neutrality corresponds to pH 7, whereas a 1 molar solution of a strong acid would approach pH 0, and a 1 molar solution of a strong alkali would approach pH 14.

PH domain Pleckstrin homology domain (~120aa) found in various intracellular signalling cascade proteins (e.g. **pleckstrin, tec family kinases**). Seem to be involved in interactions with phospholipids, particularly PIP3. At one stage it was suggested that they were involved in the interaction with heterotrimeric G-proteins.

PH-30 A sperm surface transmembrane protein involved in sperm-egg fusion. In many mammals exists as a heterodimer composed of alpha (ADAM1) and beta subunits. In human, fertilin subunit alpha is a pseudogene.The α-subunit has some similarities to viral fusion proteins, and the β-subunit (PH-30, beta-fertilin, ADAM2, 735aa) has an **ADAM** domain (a disintegrin and metalloprotease domain) similar to that in soluble integrin ligands (**disintegrins**). The metallopeptidase part of the domain is post-translationally removed.

PHA (1) **Phytohaemagglutinin**. (2) *Pha-2* is the *C. elegans* homologue of the vertebrate homeobox gene *Hex*. (3) Polyhydroxyalkanoates; the organically-produced basis for biodegradable plastics.

phaeochromocytoma *US.* pheochromocytoma. A normally benign neoplasia (**neuroblastoma**) of the **chromaffin tissue** of the adrenal medulla. In culture, the cells secrete enormous quantities of **catecholamines**, and can be induced to form neuron-like cells on addition of (for example) cyclic AMP or nerve growth factor. Excessive production of adrenaline and noradrenaline leads to secondary hypertension, sometimes paroxysmal.

phaeomelanin See **melanin**.

Phaeophyta *US.* Pheophyta. Brown algae. Division of algae, generally brown in colour, with multicellular, branched thalluses. Includes large seaweeds such as *Laminaria* and *Fucus*. The brown colour is due to the **xanthophylls**, fucoxanthin and lutein. Many have **laminarin** as a food reserve and alginic acid as a wall component.

phaeophytin *pheophytin* Chlorophyll from which the metal ion (magnesium) has been removed and two protons substituted.

phafins A family of proteins that contain both **PH** (pleckstrin homology) and **FYVE** domains. Phafin-1 (279aa) may induce apoptosis through the lysosomal-mitochondrial pathway. Phafin-2, 249aa. See **LAPF** and **Aarskog-Scott syndrome**.

phage See **bacteriophage**.

phage conversion Alteration of the phenotype of a bacterial host by **lysogeny** of a bacteriophage. An important example is conversion of a non-toxin producing strain of *Corynebacterium diphtheriae* to a diphtheria toxin-producing strain by lysogeny with phage beta.

phage display library A set of bacteriophages with diverse inserts that are expressed as translational fusion proteins with a phage coat protein. This makes them easy to screen with antibodies. Widely used to identify epitopes recognized by some particular antibody (an 'epitope library'); commercial phage display libraries randomly encoding all possible six or seven residue peptides can be screened with the antibody, and the inserts of bound phage sequenced to build up a picture of the binding profile of the antibody.

phage integrase family See **integrase, resolvase, cre recombinase**

phage typing A method of identifying bacteria on the basis of their susceptibility to a range of bacteriophages; confusion may arise if the bacteria carry plasmids encoding **restriction endonucleases.**

phagemid Bacteriophage whose genome contains a **plasmid** that can be excised by co-infection of the host with a 'helper phage'. Useful as **vectors** for library production, as the library can be amplified and screened as phage, but the inserts of selected **plaques** can readily be prepared as plasmids without subcloning. An example of a commercial phagemid is λZap, from which pBluescript can be excised with helper phage.

phagocyte A cell that is capable of phagocytosis. The main mammalian phagocytes are **neutrophils** and **macrophages**.

phagocytic vesicle Membrane-bounded vesicle enclosing a particle internalised by a phagocyte. The primary phagocytic vesicle (phagosome) will subsequently fuse with **lysosomes** to form a secondary phagosome in which digestion will occur.

phagocytosis Uptake of particulate material by a cell (endocytosis). See **opsonization, phagocyte**.

phakinin Eye-lens specific protein (beaded filament structural protein 2415aa) that co-assembles with **filensin** (3 phakinin molecules per filensin) to form beaded-chain **intermediate filaments**. Phakinin has very strong sequence homology with **cytokeratins** but lacks the rod domains that are involved in filament formation. It is effectively a tail-less intermediate filament protein. Mutations can cause autosomal dominant juvenile-onset **cataract**.

phalangeal cells Epithelial cells of the organ of Corti (in the inner ear). The outer phalangeal cells are also called Deiter's cells.

phalloidin Cyclic peptide (789 Da) from the Death Cap (*Amanita phalloides*) that binds to, and stabilizes, F-actin. Fluorescent derivatives are used to stain actin in fixed and permeabilized cells, although there is some uptake by live cells.

phallotoxins Toxic compounds (bicyclic peptides) produced by *Amanita phalloides*. Bind to F-actin and inhibit depolymerisation; hepatotoxicity is the primary cause of problems. **Phalloidin** is a phallotoxin.

phantom genes DNA sequences formerly thought to contain functional genes (often on the basis of computational methods), but have never been shown to be expressed.

PHAP-I Acidic leucine-rich nuclear phosphoprotein 32 family member A (pp32, putative HLA class II-associated protein 1, mapmodulin, 249aa) A potent heat-stable protein phosphatase 2A inhibitor and a component of the **SET complex**. Upregulated levels increase apoptosis of tumour cells.

pharate Of an insect, having its new cuticle formed beneath its present cuticle, and thus ready for its next moult.

pharmacodynamics The study of how drugs affect the body: contrast with **pharmacokinetics**.

pharmacogenetics The study of genetic causes of individual variations in drug response; the term is often used interchangeably with **pharmacogenomics**.

pharmacogenomics Pharmacogenomics involves genome-wide analysis of the genetic determinants of drug efficacy and toxicity rather than individual genetic differences (polymorphisms). Often, however, used interchangeably with **pharmacogenetics**.

pharmacokinetics The study of what the body does to drugs, in contrast to **pharmacodynamics**. The pharmacokinetics of a drug relate to the rate and extent of its uptake (absorption), its transformation as a result of metabolism, the kinetics of distribution of the drug and its metabolites in the tissues, and the rate and route of elimination (excretion) from the body. Commonly abbreviated as **ADME** (absorption, distribution, metabolism and excretion).

pharming (1) The commercial production of substances from transgenic plants or animals for medical (pharmaceutical) use. (2) The covert redirection of computer users from legitimate websites to counterfeit sites in order to gain confidential information.

phase contrast microscopy See **light microscopy**.

phase separation The separation of fluid phases that contain different concentrations of common components. Occurs with partially miscible solvents used in many biochemical separation methods. Temperature-dependent phase separation occurs with some detergent solutions. With reference to membranes means the segregation of lipid components into 'domains' that have different chemical composition.

phase variation Alteration in the expression of surface antigens by bacteria. For example, *Salmonella* can express either of two forms of **flagellin**, H1 and H2, that are coded by different genes. Their expression is controlled by inversion of a segment of chromosome containing the promoter for the H2 gene, which if functional promotes expression of H2 and a repressor of the H1 gene. Inversion removes expression of H2 and also relieves repression of the H1 gene. This occurs about every 1000 bacterial divisions, and is under the control of another gene, *hin*, that is within the invertable sequence.

phaseolin Vacuolar storage proteins of the 7S class, and the major trimeric seed storage protein found in the bean, *Phaseolus vulgaris*.

phaseollin A **phytoalexin** produced by *Phaseolus* (bean) plants in response to pathogens or other stress. Not a misprint for **phaseolin**.

phaseolotoxin A phytotoxic compound produced by *Pseudomonas syringae* that has antimicrobial effects and inhibits ornithine carbamoyltransferase. Synthesis involves a gene cluster.

phasic See **adaptation**.

phasing of nucleosomes A non-random arrangement of **nucleosomes** on DNA, in which, at certain segments of the genome, nucleosomes are positioned in the same way relative to the nucleotide sequence in all cells. Most nucleosomes are arranged randomly, but phasing has been detected in some genes.

phasmid *phagemid* A type of cloning vector developed as a hybrid of the filamentous phage M13 and plasmids to produce a vector that can grow as a plasmid, and also be packaged as single stranded DNA in viral particles. Hybrid phage/plasmid formed by integration of plasmid containing the att site, and lambda phage, mediated by phage integrase site-specific recombination.

PHB (1) Poly(beta-3-hydroxybutyrate), a biodegradable polymer that can be produced by bacteria. (2) The prohibitin homology domain: see prohibitin.

PHC Polyhomeotic-like proteins (PHC1, early development regulatory protein 1, 1004aa; PHC2, 858aa; PHC3 983aa) components of the **polycomb repressive complex-1**.

PHD-type See **ubiquitin conjugating enzymes**.

phelloderm Tissue in the bark of tree roots and shoots containing parenchyma-like cells. Produced by cell division in the **phellogen**.

phellogen **Meristematic** tissue in plants, giving rise to cork (phellem) and phelloderm cells. Also termed 'cork cambium'.

phenocopy An environmentally produced phenotype simulating the effect of a particular genotype.

phenol red Dye used as pH indicator: changes from yellow to red in range 6.8–8.4. Very commonly used in tissue culture medium (which turns acid as it becomes exhausted) though it can interfere with luminescence assays.

phenology The study of periodic plant and animal life cycle events and the effects of seasonal variations in climate.

phenome Phenotypic equivalent of the genome – the sum or matrix of all phenotypic characters that can be expressed. See **phenotype**.

phenothiazines A group of antipsychotic drugs, thought to act by blocking dopaminergic transmission in the brain. Examples are **chlorpromazine** and **trifluoperazine**.

phenotype The characteristics displayed by an organism under a particular set of environmental factors, regardless of the actual genotype of the organism.

phenylalanine An essential amino acid (Phe, F, 165 Da) with an aromatic side chain. See Table A1.

phenylalanine ammonia lyase A family of enzymes (e.g. PAL1, EC 4.3.1.24, 725aa) involved in the synthesis of **lignin** and other phenolic compounds from phenylalanine (see **phenylpropanoid pathway**). Used as an enzymic marker for lignification and other developmental processes in plant cells. Commonly activated as part of a plant's response to disease.

phenylethylamine Backbone for compounds which have important physiological functions within the body as neurotransmitters in the central nervous system and hormones in the blood circulation as well as alkaloids found in substances such as chocolate.

phenylketonuria *PKU* Congenital absence of phenylalanine hydroxylase, an enzyme that converts phenylalanine into tyrosine. Phenylalanine accumulates in blood and seriously impairs early neuronal development but dietary control can prevent the problem. Incidence highest in Caucasians.

phenylmethylsulphonyl fluoride *PMSF* A broad-spectrum serine peptidase inhibitor that works by sulphonylating the active-site histidine.

phenylpropanoid pathway The plant phenylpropanoid pathway is responsible for the synthesis of a wide variety of secondary metabolic compounds, including **lignins**, salicylates, coumarins, hydroxycinnamic amides, flavonoid phytoalexins, pigments, UV light protectants, and antioxidants. Phenylpropanoids affect qualities such as texture, flavour, colour, and processing characteristics. Phenylpropanoids are derived from cinnamic acid, which is formed from phenylalanine by **phenylalanine ammonia-lyase** (PAL), the branch point enzyme between the primary **shikimic acid pathway** and secondary (phenylpropanoid) metabolism. Diagram of pathway to lignin: http://www.mm.helsinki.

fi/MMSBL/english/research/Gerberalab/lignin_phenylpropanoid.html; Full pathway: http://www.biomedcentral.com/1471-2229/6/26/figure/F1

phenylthiourea An inhibitor (PTU, phenylthiocarbamide, PTC) of **tyrosinase** and melanin synthesis, widely used in zebrafish research to suppress pigmentation in developing embryos/fry. There is a common human polymorphism in the ability to taste PTC.

pheomelanin See **melanin**.

pheophorbide a oxygenase The regulated enzyme (EC 1.14.12.20, accelerated cell death 1, lethal leaf-spot 1 homologue, 537aa) that catalyzes the key reaction of chlorophyll catabolism during senescence. Cleaves pheophorbide a (pheide a) to a primary fluorescent catabolite (pFCC) which is then degraded by red chlorophyll catabolite reductase (RCCR). Research article: http://www.botany.unibe.ch/nutr/abstr_repr/PNAS_100_15259.pdf

pheophytin See **phaeophytin**.

pheresis Procedure in which the blood is filtered, particular elements (cells or plasma) separated and the remainder returned to the donor's circulation. In leukapheresis, white cells are removed; in plasmapheresis, plasma is removed and all cellular components returned.

pheromone A volatile hormone or behaviour modifying agent. Normally used to describe sex attractants (e.g. bombesin for the moth *Bombyx*) but includes volatile aggression stimulating agents (e.g. isoamyl acetate in honey bees).

pherotype Strain-specific variation of *Streptococcus pneumoniae* in the response to the quorum-sensing pheromone **competence stimulating peptide** (CSP). The pherotypes are determined by whether the response is to CSP-1 or CSP-2. See **comCDE**.

PHF1 A transcriptional repressor (PHD finger protein 1, polycomb-like protein 1, 567aa) that may promote methylation of histone H3 by the **PRC2/EED-EZH2** complex.

phi X-174 *φ X-174* Bacteriophage of *E. coli* with a single-stranded DNA genome and an icosahedral shell. This was the first DNA phage to be fully sequenced: the genome consists of 10 genes, some of which are **overlapping genes**.

phialidin Calcium-activated photoprotein (clytin, 198aa) from the coelenterate *Clytia gregaria* (*Phialidium gregarium*). Emission peak 470 nm (blue). See **aequorin**.

phialospores microconidia, spermatia A type of spore produced by various ascomycetes. Phialospores arise at the apex of specialised cells (phialides) by the formation of a cross wall, and each contains a single nucleus.

Philadelphia chromosome Characteristic chromosomal abnormality of chronic myelogenous **leukaemia** in which there is a reciprocal translocation between chromosomes 9 and 22. The bcr, 'breakpoint cluster region' of chromosome 22 forms a fusion protein (p185, bcr/abl) with part of the *abl* gene of chromosome 9 that has unregulated tyrosine kinase activity.

phloem Tissue forming part of the plant vascular system, responsible for the transport of organic materials, especially sucrose, from the leaves to the rest of the plant. Consists of **sieve tubes, companion cells, fibre cells** and **parenchyma**.

phoA Alkaline phosphatase (EC 3.1.3.1, 471aa) from *E. coli*.

Phoma medicaginis The causal agent of spring black stem and leaf spot in alfalfa (*Medicago sativa*), a necrotrophic fungus of the phylum Ascomycota.

phomopsin A A peptide from the fungus *Phomopsis leptostromiformis* that acts on tubulin to block microtubule assembly. Shares a binding site (distinct from the **vinca alkaloid** binding site) on beta-tubulin with **cryptophycin** 1, cryptophycin 52, **dolastatin 10** and **hemiasterlin** (the **rhizoxin/maytansine site**). Has anti-tumour properties.

phorbol esters Polycyclic compounds isolated from croton oil (from the seeds of *Croton tigliumin*), in which two hydroxyl groups on neighbouring carbon atoms are esterified to fatty acids. The commonest of these derivatives is phorbol myristoyl acetate (PMA). Potent co-carcinogens or tumour promotors, they are diacyl glycerol analogues and activate protein kinase C irreversibly.

phormicin Insect **defensin** (94aa) produced by the blowfly, *Phormia terranovae* and responsible for the anti Gram-positive activity of immune haemolymph.

phosducin Protein (246aa) that inhibits Gs-GTPase activity by binding the $G_{\beta\gamma}$ subunit of heterotrimeric G-proteins and making them unavailable for signalling. Isolated from bovine brain and found in retina, pineal gland and many other tissues. Activity of phosducin is inhibited if phosphorylated by a cAMP-dependent protein kinase. May be defective in **retinitis pigmentosa** and **Usher syndrome type II**. A number of phosducin-like proteins that regulate G protein function have been identified.

phosmid Cosmid-phage hybrid vector which has the phage 1 origin of replication. Was developed to facilitate restriction enzyme mapping.

phosphatases Enzymes that hydrolyze phosphomonoesters. Acid phosphatases are specific for the single-charged phosphate group and alkaline phosphatases for the double-charged group. These specificities do not overlap. The phosphatases comprise a very wide range of enzymes including broad- and narrow-specificity members. Phosphoprotein phosphatases specifically de-phosphorylate a particular protein and are essential if phosphorylation is to be used as a reversible control system; they are also specific for phosphoserine/threonine or phosphotyrosine residues within the target protein.

phosphatides The family of phospholipids based on 1, 2 diacyl 3-phosphoglyceric acid. See **phospholipids**.

phosphatidic acid PA The simplest of the **phospholipids**, diacyl glycerol 3-phosphate, a glycerol backbone with acyl groups derived from long-chain fatty acids covalently linked to carbons 1 and 2 and a phosphate group on C3. The acyl group on C1 is usually a saturated fatty acid, that on C2, unsaturated. It is present in low concentrations in membranes, is an intermediate in the synthesis of diacyl glycerol and the immediate precursor of most of the phosphatidyl phospholipids (except phosphatidyl inositol) and of triacyl glycerols. See **lysophosphatidic acid**.

phosphatidyl choline PC The major phospholipid of most mammalian cell membranes where the 1-acyl residue is normally saturated and the 2-acyl residue unsaturated. Choline is attached to phosphatidic acid by a phosphodiester linkage. Major synthetic route is from diacyl glycerol and CDP-choline. Forms monolayers at an air/water interface, and forms bilayer structures (**liposomes**) if dispersed in aqueous medium. A zwitterion over a wide pH range. Readily hydrolyzed in dilute alkali.

phosphatidyl ethanolamine PE A major structural phospholipid in mammalian systems. Tends to be more abundant than phosphatidyl choline in the internal membranes of the cell, and is an abundant component of prokaryotic membranes. Ethanolamine is attached to phosphatidic acid by a phosphodiester linkage. Synthesized from diacyl glycerol and CDP-ethanolamine.

phosphatidyl inositol PI An important minor phospholipid in eukaryotes, involved in signal transduction processes. Contains myo-inositol linked through the 1-hydroxyl group to phosphatidic acid. The 4-phosphate (PIP) and 4,5 bisphosphate derivatives (PIP2) are formed and broken down in membranes by the action of specific kinases and phosphatases (futile cycles). Signal-sensitive phospholipase C enzymes remove the inositol moiety, in particular from 1,4,5 trisphosphate (PIP2) as inositol 1,4,5-triphosphate ($InsP_3$: IP3). Both the diacyl glycerol and inositol phosphate products act as **second messengers**.

phosphatidyl inositol-3-kinase See **PI-3-kinases**.

phosphatidyl serine *PS* An important minor species of phospholipid in membranes with serine attached to phosphatidic acid by a phosphodiester linkage. Synthesis is from phosphatidyl ethanolamine by exchange of ethanolamine for serine. Distribution is asymmetric, as the molecule is only present on the cytoplasmic side of cellular membranes. It is negatively charged at physiological pH and interacts with divalent cations; involved in calcium-dependent interactions of proteins with membranes (e.g. protein kinase C).

phosphocreatine *creatine phosphate* A compound that serves as an immediate energy reserve for muscle, present at about 20 mM in striated muscle. It is synthesized and broken down by creatine phosphokinase to buffer ATP concentration.

phosphodiester bond An imprecise term for the linkage of two parts of a molecule through a phosphate group. Examples are found in RNA, DNA, phospholipids, cyclic nucleotides, nucleotide diphosphates and triphosphates.

phosphodiesterase An enzyme that cleaves phosphodiesters to give a phosphomonoester and a free hydroxyl group. For example PDE-1A (calcium/calmodulin-dependent 3′,5′-cyclic nucleotide phosphodiesterase 1A, EC 3.1.4.17, 535aa) catalyses the hydrolysis of cyclic nucleotide second messengers (with a higher affinity for cGMP than for cAMP). Other examples include RNAase, DNAase, phospholipases C and D, but in casual usage the cAMP-phosphodiesterase is usually meant (see **cyclic nucleotide phosphodiesterases**).

phosphodiesterase inhibitors Compounds that inhibit the breakdown of the important intracellular second messenger cyclic AMP, although strictly speaking the cAMP-phosphodiesterase (EC 3.1.4.17) is only one of a number of such enzymes. Allowing cAMP to accumulate potentiates the action of the sympathetic nervous system. Most commonly encountered examples are the xanthines such as theophylline.

phosphoenolpyruvate *PEP* An important metabolic intermediate. The enol (less stable) form of pyruvic acid is trapped as its phosphate ester, giving the molecule a high phosphate transfer potential. Formed from 2-phosphoglycerate by the action of enolase.

phosphofructokinase The pacemaker enzyme (6-phosphofructo-1-kinase, EC 2.7.1.11) of glycolysis. Converts fructose 6-phosphate to fructose 1,6-bisphosphate. A tetrameric allosteric enzyme that is sensitive to the ATP/ADP ratio. In humans there are muscle (M), liver (L), and platelet (P) isoenzymes, encoded by different genes. Deficiency of the muscle form leads to glycogen storage disease type VII.

phosphoglycerate The molecules 2-phosphoglycerate and 3-phosphoglycerate which are intermediates in glycolysis. 3-phosphoglycerate is the precursor for synthesis of phosphatidic acid and diacyl glycerol, hence of phosphatidyl phospholipids.

phosphoglycerate kinase An X-linked enzyme (PGK, EC 2.7.2.3, 417aa) that plays a key role in the glycolytic pathway catalysing the phosphorylation of 3-phospho-D-glycerate to 3-phospho-hydroxypyruvate. PGK-1 also acts as a polymerase alpha cofactor protein (primer recognition protein).

phosphoglycerate mutase A widely distributed enzyme (254aa) that interconverts 3- and 2-phosphoglycerate with 2,3-bisphosphoglycerate as the primer of the reaction. Can also act as phosphoglycerate synthase (EC 5.4.2.4) and phosphoglycerate phosphatase (EC 3.1.3.13), but with a reduced activity. There is a muscle-specific isozyme (M) and another (B) elsewhere; deficiency of the muscle isoform leads to myopathy similar to that caused by deficiency in **phosphofructokinase**.

phosphoinositide-dependent kinase A protein kinase (EC 2.7.1.37, 556aa) that is critical for the activation, in a PIP2 or PIP3-dependent manner (see **phosphatidyl inositol**), of many downstream protein kinases in the **AGC kinase** superfamily, through phosphorylation of the activation loop site on these enzymes.

phosphokinase See **kinase**.

phospholamban Integral membrane homopentameric protein (subunits of 52aa) that is the endogenous regulator of the sarcoplasmic reticulum calcium ATPase (**SERCA**). Phosphorylation by protein kinase A and dephosphorylation by protein phosphatase 1 modulate the inhibitory activity of phospholamban. A mutation in the human phospholamban gene, deleting arginine 14, results in lethal, hereditary cardiomyopathy. See **sarcolipin**.

phospholemman A small transmembrane protein (PLM, FXYD domain-containing ion transport regulator 1, 92aa) that, like **phospholamban**, interacts with P-type ATPases and regulate ion transport in cardiac cells and other tissues. Induces a hyperpolarization-activated chloride current when expressed experimentally in *Xenopus* oocytes. Phospholemman-like protein (FXYD3, 67aa, mammary tumour 8 kDa protein) is overexpressed in many mammary tumours.

phospholipase A class of enzyme that hydrolyzes ester bonds in phospholipids. There are two types: aliphatic esterases (phospholipase A1, A2 and B) that release fatty acids, and phosphodiesterases (types C and D) that release diacyl glycerol or phosphatidic acid respectively. Type A2 is widely distributed in venoms and digestive secretions. Types A1, A2 and C (the latter specific for phosphatidyl inositol) are present in all mammalian tissues. Type C is also found as a highly toxic secretion product of pathogenic bacteria (e.g. *Clostridium perfringens* alpha-toxin). Type B attacks monoacyl phospholipids and is poorly characterized. Phosphatidylinositol bisphosphate-specific phospholipase C is important in generating **diacylglycerol** and **inositol trisphosphate**, both **second messengers**. See **phospholipase A**, **phospholipase D**.

phospholipase A An aliphatic esterase type of **phospholipase**. Phosphatidic acid-selective phospholipase A1 (lipase H, 451aa) generates **lysophosphatidic acid** and is mutated in autosomal recessive hypotrichosis. PLA1-alpha (456aa) acts specifically on phosphatidylserine. Phospholipases A2 (PLA2, EC 3.1.1.4) are diverse with secreted, cytosolic and lipoprotein-associated classes. Group I PLA2s include a pancreatic form (PLA2G1B, EC 3.1.1.4, 148aa) and are also found in the venom of cobras and kraits; Group II are extracellular and require calcium, examples are found in synovial fluid (PLA2G2A) and in the venom of rattlesnakes and vipers); Type III is found in bee and lizard venom; Group IV are cytosolic and will generate arachidonic acid as the starting point for **eicosanoid** synthesis and are important in the inflammatory response. Other groups have differing substrate specificities or tissue distributions. Lipoprotein-associated PLA2s (Group VII, EC 3.1.1.47, lp-PLA2) hydrolyse **platelet activating factor**. PLA2G6 (806aa) is a calcium-independent PLA2 (see **infantile neuroaxonal dystrophy** and **Karak syndrome**). A PLA2 receptor (PLA2R1, 1463aa) may be involved in internalising secreted forms of the enzyme for degradation. Phospholipase A2-activating protein (PLAP, 795aa) regulates inflammatory processes. See also **patatins**.

phospholipase D A widely distributed family of phospholipases that play a central signalling function in eukaryotic cells. Phosphatidylcholine-specific phospholipases D (EC 3.1.4.4: PLD1, 1074aa; PLD2, 931aa; PLD3, 490aa; PLD4, 506aa; PLD5, 536aa but inactive; PLD6, 252aa) hydrolyse PC to phosphatidic acid and choline and may be important in signal transduction, membrane trafficking, and the regulation of mitosis. Glycosylphosphatidylinositol-specific phospholipase D (GPI-PLD, EC 3.1.4.50, 840aa) is abundant in serum and can remove GPI anchors from membrane proteins. A specific PE-specific phospholipase D type enzyme (N-acyl-PE-hydrolyzing phospholipase D, 393aa) is important in the biosynthesis of N-acylethanolamines such as **anandamide**.

phospholipid The major structural lipid of most cellular membranes (except the chloroplast, which has galactolipids). Contain phosphate, usually as a diester. Examples include phosphatidyl phospholipids, plasmalogens and sphingomyelins. See Table L3.

phospholipid bilayer A lamellar organization of phospholipids that are packed as a bilayer (7 nm thick) with hydrophobic acyl tails inwardly directed and polar head groups on the outside surfaces (see black lipid membrane). It is this bilayer that forms the basis of membranes in cells, though in most cellular membranes a very substantial proportion of the area may be occupied by integral proteins, some of which may be surrounded by a restricted set of phospholipids. The bilayer is fluid in two-dimensions and membrane proteins can move (see **patching**, **fluorescence recovery after photobleaching**). The lipid composition of inner and outer leaflets differ although flip-flop can occur, causing redistribution. The triple-layered appearance of membranes seen in electron microscopy 'unit membrane' is thought to arise because **osmium tetroxide** binds to the polar regions leaving a central, unstained, hydrophobic region.

phospholipid transfer protein Cytoplasmic proteins (e.g. lipid transfer protein II, 476aa) that bind phospholipids and facilitate their transfer between cellular membranes. May also cause net transfer from the site of synthesis. Phosphatidylinositol transfer protein (PIPTP, 271aa) is specific for PI. Glycolipid transfer protein (209aa) is a soluble protein that selectively accelerates the transfer of glycosphingolipids and glycoglycerolipids. See **cholesteryl ester transfer protein**.

phosphomannose See **mannose-6-phosphate**.

phosphoprotein Protein that has phosphate groups esterified to serine, threonine or tyrosine (S,T or Y). The addition, by a **protein kinase**, of a phosphate group, or its removal by a **protein phosphatase** usually regulates the function.

phosphoramidite Nucleotide derivative used in oligonucleotide synthesis.

phosphorescence (1) Emission of light following absorption of radiation. Emitted light is of longer wavelength than the exciting radiation and is a result of decay of electrons from the triplet to the ground state. Lasts longer than fluorescence (electron decay from singlet to ground state) and occurs after a longer delay. (2) Popularly misused as a term for biological luminescence, e.g. by fireflies.

phosphorimaging Method for detecting radioactivity using 'phosphor' compounds that emit visible light when exposed to radiation; used in the same way as autoradiography (e.g. for detecting labelled bands on gels) but is much more sensitive.

phosphorylase See **glycogen phosphorylase**.

phosphorylase kinase The enzyme (EC 2.7.1.38) that regulates the activity of **glycogen phosphorylase** and glycogen synthetase by addition of phosphate groups. A large and complex enzyme, consisting of 4 copies of an α-β-γ-δ tetramer with several isoforms of the α, β and γ subunits. The alpha (1235aa) and beta (1093aa) subunits have regulatory functions, the gamma subunit (406aa) has the catalytic activity, the delta subunit is **calmodulin**. Activity is regulated by phosphorylation. Integrates the hormonal and calcium signals in muscle. Mutations can cause **glycogen storage disease**.

phosphotransferase An enzyme of the EC 2.7 class that transfers a phosphate group from a donor to an acceptor. Very important in metabolism.

phosphotyrosine Strictly speaking, tyrosine phosphate, but normally refers to the phosphate ester of a protein tyrosine residue. Present in very small amounts in tissues, but important in systems that regulate cell proliferation and thus in studies of malignancy. The *src* gene product (pp60src) was one of the first kinases shown to phosphorylate at a tyrosine residue.

phosvitin An egg-yolk storage protein, a cleavage fragment (47aa in ducks) of **vitellogenin**. It is very highly phosphorylated and may be important for binding and storing cations such as calcium and iron.

photoadduct Generally, any compound formed between two reacting molecules as a result of exposure to light. Most commonly used in the context of covalent modifications of DNA as a result of UV-irradiation, sometimes for therapeutic reasons, as with **psoralens** in **photodynamic therapy**. See **photoaffinity labelling**.

photoaffinity labelling A technique for covalently attaching a label or marker molecule onto another molecule such as a protein. The label, which is often fluorescent or radioactive, contains a group that becomes chemically reactive when illuminated (usually with ultraviolet light) and will form a covalent linkage with an appropriate group on the molecule to be labelled: proximity is essential. The most important class of photoreactive groups used are the aryl azides, which form short-lived but highly reactive nitrenes when illuminated.

photobleaching Light-induced change in a **chromophore**, resulting in the loss of its absorption of light of a particular wavelength. A problem in fluorescence microscopy where prolonged illumination leads to progressive fading of the emitted light because less of the exciting wavelength is being absorbed.

photolithography Originally a form of lithography in which light-sensitive plates or stones were exposed to a photographic image, now more widely applied to any process in which selective masking generates light patterns which cause chemical transformations on exposed areas of a photosensitive surface, an approach that has been used in semiconductor manufacture and in production of complex substrata for cell behavioural studies, especially to demonstrate **contact guidance**.

photolyase Family of ubiqitous enzymes (DNA photolyases) found in bacteria, archaebacteria and eukaryotes that can repair UV-induced DNA damage. The protein (between 454 to 614aa) is associated with two prosthetic groups, FADH and a light-harvesting cofactor, MTHF (5,10-methenyltetrahydofolyl polyglutamate). Light is needed for the repair step. An example in *E. coli* is deoxyribodipyrimidine photolyase (EC 4.1.99.3, 472aa). The human homologues are **cryptochromes** which do not, apparently, have the repair capacity.

photolysis Light-induced cleavage of a chemical bond, as in the process of photosynthesis.

photomedins **Olfactomedin-family** proteins of the retina; photomedin-1 (olfactomedin-like protein 2A, 652aa) is selectively expressed in the outer segment of photoreceptor cells and photomedin-2 (750aa) in all retinal neurons. Photomedins preferentially bind to chondroitin sulphate-E and heparin.

photoperiodism Events triggered by duration of illumination or pattern of light/dark cycles: often the wavelength of the illuminating light is important, as for example in control of circadian rhythm in plants. See **phytochromes**.

photophosphorylation The synthesis of ATP that takes place during photosynthesis. In non-cyclic photophosphorylation the photolysis of water produces electrons that generate a **proton motive force** which is used to produce ATP, the electrons finally being used to reduce $NADP^+$ to NADPH. When the cellular ratio of reduced to non-reduced NADP is high, **cyclic photophosphorylation** occurs and the electrons pass down an electron transport system and generate additional ATP, but no NADPH.

photopigment Any pigment involved in **photosynthesis** in plants. Includes **chlorophyll**, **carotenoids** and **phycobilins**.

photoreceptor A specialized cell type in a multicellular organism that is sensitive to light. This definition excludes single-celled organisms, but includes non-eye receptors, such as snake infra-red (heat) detectors or photosensitive pineal gland cells. See **retinal rods**, **retinal cones**.

photorespiration Increased respiration that occurs in photosynthetic cells in the light, due to the ability of **RuDP carboxylase** to react with oxygen as well as carbon dioxide. Reduces the photosynthetic efficiency of **C3 plants**.

photosynthesis Process by which green plants, algae, and some bacteria (green and purple bacteria) absorb light energy and use it to synthesize organic compounds (initially carbohydrates). In green plants occurs in **chloroplasts** that contain the photosynthetic pigments. Occurs by slightly different processes in **C3**

and **C4 plants**. See also **Z scheme of photosynthesis** and contrast with **chemosynthesis**.

photosynthetic unit Group of photosynthetic pigment molecules (**chlorophylls** and **carotenoids**) that supply light to one **reaction centre** in photosystems I or II.

photosystem I Photosynthetic system in **chloroplasts** in which light of up to 700 nm is absorbed and its energy used to bring about charge separation in the **thylakoid** membrane. The electrons are passed to ferredoxin and then used to reduce $NADP^+$ to NADPH (non-cyclic electron flow) or to provide energy for the phosphorylation of ADP to ATP (cyclic photophosphorylation).

photosystem II Photosynthetic system in **chloroplasts** in which light of up to 680 nm is absorbed and its energy used to split water molecules, giving rise to a high energy reductant, Q^-, and oxygen. The reductant is the starting point for an electron transport chain that leads to **photosystem I** and that is coupled to the phosphorylation of ADP to ATP.

phototaxis Movement of a cell or organism towards (positive phototaxis) or away from (negative p.) a source of light.

phototransduction The transformation by photoreceptors (e.g. **retinal rods** and **cones**) of light energy into an electrical potential change.

phototrophic Describing any organism that can utilize light as a source of energy.

phototropins Proteins that function as blue light photoreceptors for phototropism, chloroplast relocation, stomatal opening and leaf flattening in *Arabidopsis*. Serine/threonine protein kinases (EC 2.7.11.1; phototropin-1, 996aa; phototropin-2, 915aa) that are autophosphorylated as a result of irradiation with blue light. Required for blue-light mediated mRNA destabilization.

phototropism Movement or growth of part of an organism (e.g. a plant shoot) towards (positive phototropism) a source of light, without overall movement of the whole organism.

phox (**1**) The NADPH oxidase system in phagocytes which generates large quantities of microbicidal superoxide and other oxidants upon activation (the respiratory or **metabolic burst**). The membrane-associated portion is a heterodimer of p22phox (CYBA, 195aa) and p91phox (CYBB, 570aa), the A and B subunits of cytochrome b245. There are cytosolic components, all with SH3 domains: p40phox (339aa) associates primarily with p67phox (526aa) to form a complex with p47phox (390aa). The cytosolic components only associate with the integral membrane component following activation. Deletion or mutation in the cytosolic proteins can cause forms of chronic granulomatous disease although the

commonest form involves mutation in p91. (**2**) Paired mesoderm homeobox proteins (e.g. PHOX2B, 314aa).

phragmoplast The cytokinetic apparatus that helps deposit the new cross-wall in higher plant cells. It can form out of the central spindle at the end of mitosis and is therefore composed of two opposed circlets of microtubules, actin filaments and associated membranes. Phragmoplasts can also develop *de novo*, as in endosperm, between non-sister nuclei without any spindle remnant. Vesicles are directed to the midline to form the cell plate. This causes the structure to expand out centrifugally (contrast with animal cytokinesis) until it contacts that part of the cortex predicted before mitosis by the preprophase band

phragmosome In large plant cells, the cytoplasmic raft that develops across the vacuole and in which mitosis and cytokinesis take place. It occupies the plane described by the cortical preprophase band and, like this structure, predicts the plane of cell division.

phrap A program for assembling shotgun DNA sequence data. Link: http://www.phrap.org/phredphrap-consed.html

phred Software that reads DNA sequencing trace files, calls bases, and assigns a quality value to each called base. The quality value is a log-transformed error probability, specifically $Q = -10 \log_{10}(Pe)$ where Q and Pe are respectively the quality value and error probability of a particular base call. Link: http://www.phrap.org/phredphrapconsed.html

phycobilins Photosynthetic pigments (biliproteins) found in certain algae, especially red algae (Rhodophyta) and **cyanobacteria**.

phycobiliprotein Subunits of the major accessory light-harvesting complexes (LHC) of most cyanobacteria and red alga and present in the thylakoid lumen of cryptophytes. Phycobilins, the chromophore groups, are covalently linked to phycobiliproteins. Examples include allophycocyanin, phycocyanin, phycoerythrin and phycoerythrocyanin.

phycobilisome An accessory light energy harvesting structure in **cyanobacteria**. They have cores of allophycocyanin with radiating rods composed of discs of **phycocyanin** and **phycoerythrin**. Linker polypeptides attach the core to the **thylakoid** membranes. These structures, 20–70 nm across, contain the pigments named above that transfer light energy to chlorophyll a. The pigments are extracted and used as fluorochromes for labelling various probe reagents.

phycocyanin Blue **phycobilin** (λmax = 617 nm) found in some algae, and especially in **cyanobacteria**.

phycoerythrin Red **phycobilins** (λmax = 575 nm) found in some algae, especially red algae (**Rhodophyta**).

phycoerythrocyanin A blue phycobilin (λmax = 575 nm) found in Cyanobacteria and red algae

Phycomycetes An obsolete taxonomic class of fungi with hyphae that are usually non-septate (without cross walls). Now classified as **zygomycetes**.

phycoplast The cytokinetic apparatus responsible for cytoplasmic separation in some algae. It contains microtubules parallel to the division plane. The cell may divide by a combination of phycoplast-dependent wall formation in the centre of the cell, completed by in-furrowing of the cortex. Contrast with the phragmoplast of higher plants in which the microtubules are perpendicular to the division plane.

PHYLIP A software package (the PHYLogeny Inference Package) for generating unrooted consensus trees. Link to source: http://evolution.genetics.washington.edu/phylip.html

phyllolitorin Bombesin-like peptide sub-family, originally identified from the skin of the South American frog *Phyllomedusa sauvagei*. The precursor (90aa) has a signal peptide sequence, an amino-terminal extension peptide, the phyllolitorin peptide (9aa) and a carboxy-terminal extension peptide. See **Leu-phyllolitorin**.

phyllotaxy In plants, the arrangement of leaves on a stem. Examples include spiral patterning, where leaves arise in a rotary pattern around the axis, and opposite, where pairs of leaves arise at each node. The pattern arises at the shoot apical meristem as a regular series of bud initials.

phylogenetic profile Computational method of trying to deduce functional interactions between proteins based upon the presence of both the proteins in some genomes and neither in other genomes: the argument is that it is unlikely their joint presence would be invariable if they did not interact. See **Rosetta Stone method** and **gene neighbour method**.

physaliphorous cells Cells of chordoma (tumour derived from notochordal remnants) that appear vacuolated because they contain large intracytoplasmic droplets of mucoid material.

Physarum A member of the Myxomycetes (acellular slime moulds). Normally exists as a multinucleate diploid **plasmodium** that may be many centimetres across, but if starved and stimulated by light will produce a fruiting body and spores that later germinate to produce haploid amoeboid cells, myxamoebae, which may transform into flagellated swarm cells. Either of these cell types may fuse to produce a diploid zygote that forms the plasmodium by synchronous nuclear division without cell division.

Easily grown in the laboratory and much used for studies on cytoplasmic streaming and on the cell cycle (because they show synchronous DNA synthesis and nuclear division).

Physcomitrella patens A moss; the first non-vascular plant to have its complete genome sequenced. Web portal: http://moss.nibb.ac.jp/

physical mapping The process of assembling genomic DNA clones that completely cover a genetic locus. In **genome projects**, this is an essential prerequisite for sequencing; in **positional cloning**, it assists in designing a strategy to identify the gene of interest. The procedure is to screen candidate clones for a series of characteristic marker sequences, based either on **satellite DNA**, or on **PCR**-derived sequence tagged sites. Clones that share particular markers are assumed to overlap in that region, and computer analysis is used to identify the smallest set of clones that completely cover the region.

physins Sub-group of the **tetraspan vesicle membrane proteins**. Mammalian physins include **synaptophysin**, **synaptoporin**, **pantophysin**, and mitsugumin29.

phytanic acid A branched-chain fatty acid (3,7,11,15-tetramethyl hexadecanoic acid), a derivative of the phytol side-chain of chlorophyll. In humans it comes from consumption of dairy products, ruminant animal fats, and certain fish and accumulates in **Refsum's syndrome** and **Zellweger's syndrome**. See **peroxisomal biogenesis disorders**.

phytic acid Inositol hexaphosphate, found in plant cells, especially in seeds, where it acts as a storage compound for phosphate groups.

phytoalexins Toxic compounds produced by higher plants in response to attack by pathogens and to other stresses. Sometimes referred to as plant antibiotics, but are nonspecific, having a general fungicidal and bactericidal action. Production is triggered by **elicitors**. Examples: **ipomeamarone**, **pisatin**, **phaseollin**.

phytochelatins Metallothionein-type peptides of plants that bind heavy metals such as cadmium, zinc, lead, mercury and copper. General form is (γ-glutamyl-cysteinyl)$_n$-glycine where n is from 2–11. Involved in the detoxification of heavy metals and the homeostasis of non-essential metals. Synthesized by enzymes such as glutathione gamma-glutamylcysteinyltransferase 1 (EC 2.3.2.15, phytochelatin synthase 1, cadmium tolerance protein, 485aa).

phytochrome Plant pigment protein that absorbs red light and then initiates physiological responses governing light-sensitive processes such as germination, growth and flowering. Exists in two forms, **Pr** and **Pfr**, that are interconverted by light.

phytochrome-interacting factors *PIFs*
Transcription factors that activate genes involved in light-sensitive development. For example PIF1 (basic helix-loop-helix protein 15, 478aa) inhibits chlorophyll biosynthesis and seed germination in the dark but is broken down by light, relieving the inhibition. PIF3 (524aa) acts positively in the **phytochrome** signalling pathway. PIF7 (366aa) is expressed in flowers and acts negatively in the phytochrome B signalling pathway under prolonged red light. PIFs interact with **DELLA** proteins.

phytoestrogen A diverse group of plant-derived non-steroidal substances, mostly coumestans, prenylated flavonoids and isoflavones, that have weak **estrogen**-like properties. Their potential effects on humans have caused some concern although no serious impact has been demonstrated.

phytohaemagglutinin *PHA* Sometimes used as synonym for **lectins** in general, but more usually refers to lectin from seeds of the red kidney bean *Phaseolus vulgaris*. Binds to oligosaccharide containing N-acetyl galactosyl residues. Binds to both B- and T-lymphocytes, but acts as a **mitogen** only for T-cells.

phytohormones See **plant growth substances**.

phytol A long-chain fatty alcohol (C20) forming part of **chlorophyll**, attached to the protoporphyrin ring by an ester linkage.

phytoremediation The use of plants to decontaminate soil, for example by absorbing pollutants such as heavy metals.

phytosterols Steroid alcohols (plant sterols) that occur naturally in plants as structural components of cell membranes (a role fulfilled in animal cells by cholesterol). Whether they have beneficial effects in the human diet is disputed.

phytotoxins Any plant-derived toxin, usually present to deter herbivores or pests although some have proved to be useful pharmacological agents. Examples include aristolochic acids, pyrrolizidine alkaloids, beta-**carotene**, **coumarin**, alkenylbenzenes, **ephedrine**, kavalactones, anisatin, St. John's wort ingredients (**hyperforin**), cyanogenic glycosides, **picrotoxin**, **solanine** and **chaconine**, thujone, and glycyrrhizinic acid. See **Fusarium mycotoxins**.

phytozome A website tool for green plant comparative genomics that allows searches of the genomes of a range of plants (25 in version 7, late 2011), ranging from Aquilegia coerulea to *Zea mays*. Link to resource: http://www.phytozome.com/index.php

PI See **phosphatidyl inositol**.

pi protein Polypeptide (π protein, replication initiation protein, 305aa) that is required for the initiation of DNA replication in the R6K antibiotic-resistance plasmid; the *pir* gene is carried on the plasmid, of which there are 12−18 copy equivalents in the *E. coli* chromosome. Research article: http://www.pnas.org/content/76/3/1150.full.pdf

PI-103 Potent, cell permeable and ATP-competitive selective inhibitor of DNA-PK, phosphoinositide 3-kinase (PI-3-K), the rapamycin-sensitive (mTORC1) and rapamycin-insensitive (mTORC2) complexes of the protein kinase mTOR. Paper: http://www.ncbi.nlm.nih.gov/pubmed/16697955

PI-3-kinases Lipid kinases (phosphatidyl inositol-3-kinases; PI kinases) that phosphorylate phosphatidylinositol phosphate on the 3 position. They are key enzymes in the signalling cascade downstream of many receptors, particularly receptor tyrosine kinases such as PDGF-receptor (in the case of Class Ia). Classical form has p85 regulatory subunit (of which there are vaious isoforms) and p110 enzymatic subunit. The p85 adaptor associates with the cytoplasmic domain of various **growth factor receptors** through SH2 domains that bind to phosphotyrosine residues in the ligated (phosphorylated) receptor and with the catalytic subunit. An increasing family is being identified: the PI3K-γ isoform is activated by receptor-stimulated **G proteins** and recruited to the membrane by **p101**; others are regulated by calcium. Most, but not all, are inhibited by **wortmannin**.

pia mater See **meninges**.

PIAS A family of nuclear proteins (protein inhibitor of activated STAT; PIAS1 651aa; PIAS2, 621aa; PIAS3, 628aa; PIAS4, 410aa) that act as E3-type small ubiquitin-like modifier (SUMO) ligases, stabilizing the interaction between the SUMO-conjugating enzyme UBE2I and the substrate, and as a SUMO-tethering factor. Play a crucial role as transcriptional coregulators in various cellular pathways, including the STAT pathway, the p53 pathway, the Wnt pathway and the steroid hormone signalling pathway. Also involved in gene silencing.

piccolo (1) A **CAZ protein** (aczonin, 5183aa) that may act as a scaffolding protein at the presynaptic side of synaptic junctions, possibly regulating synaptic vesicle trafficking. Interacts with various proteins in the presynaptic region including RIM (**retinoblastoma**-interacting myosin-like), prenylated **rab** acceptor protein 1, profilin and **bassoon**. (2) See **NuA4 histone acetyltransferase complex**.

Pichia A genus (formerly *Hansenula* or *Hyphopichia*) in the family Saccharomycetaceae but evolutionarily distant from *S. cerevisiae*. Pichia pastoris, a methylotrophic yeast that can use methanol as the sole carbon source, has been developed into a heterologous protein expression system for production of recombinant protein therapeutics. *Pichia canadensis* (*Hansenula wingei*) has been used for studies on mating type.

Pick's disease Rare neurodegenerative disease similar in clinical symptoms to **Alzheimer's disease**. Affects mostly frontal and temporal lobes. Caused by mutations in the gene encoding **tau** or, in some cases, **presenilin-1**. See **Niemann-Pick disease**.

PICK1 A neuronal protein (protein interacting with C kinase-1, 415aa) that is phosphorylated by protein kinase C (**PKC**) and interacts with the GTP-bound forms of ADP-ribosylation factor-1. It binds AMPA receptors and acid-sensing ion channels and may regulate Golgi-to-endoplasmic reticulum vesicle transport. Acts as an adapter protein involved in clustering of various receptors containing PDZ recognition sequences.

Picornavirales An Order of viruses (see **virus taxonomy**). There are five Families: Dicistroviridae, Iflaviridae, Marnaviridae, Picornaviridae and Secoviridae.

Picornaviridae A family of viruses (Class IV viruses) within the Order **Picornavirales**. They have a single positive strand of RNA and an icosahedral capsid. There are two main classes: enteroviruses, which infect the gut and include poliovirus, and the rhinoviruses that infect the upper respiratory tract (common cold virus, Coxsackie A and B, Foot-and-Mouth disease virus and hepatitis A).

picrotoxin *cocculin* Toxic plant alkaloid, found primarily in the fruit 'Fish-berries' of *Anamirta cocculus*, an East Indian woody vine. Acts as a non-competitive antagonist of GABA-A receptors and, since GABA is an inhibitory neurotransmitter, infusion of picrotoxin has a stimulative effect, sometimes used as an antidote for barbiturate poisoning.

Pierson's syndrome Rare autosomal recessive disease (microcoria-congenital nephrosis syndrome) due to mutation in the gene encoding laminin beta-2 (**LAMB2**).

pigment cells Cells that contain pigment: see **melanocytes**, **chromatophores**.

pigmentation Multiple genes influence normal human skin, hair, and/or eye pigmentation. Variation arises from differences in the number of **melanosomes** produced, the type of melanin synthesized (black-brown eumelanin or red-yellow pheomelanin), and the size and shape of the melanosomes. For example, variants of the **OCA2** gene play a role in determining blue versus non-blue eye colour, and blond versus brown hair (SHEP1, skin, hair, eye-pigmentation 1). The SHEP2 association (red hair and fair skin) is determined by variation in the **melanocortin 1** receptor (MC1R). Many other associations are known, each affected by particular

genetic polymorphisms: SHEP3, **tyrosinase;** SHEP4, SLC24A5 (potassium-dependent sodium/calcium exchanger); SHEP5, melanoma antigen **AIM-1**; SHEP6, SLC24A4; SHEP7, **kit** ligand expression; SHEP9, agouti signalling protein **ASIP**; SHEP10, **two-pore channel 2**.

pigmented retinal epithelium *retinal pigmented epithelium, RPE* Layer of unusual phagocytic epithelial cells lying below the photoreceptors of the vertebrate eye. The dorsal surface of the PRE cell is closely apposed to the ends of the rods, and as discs are shed from the rod outer segment they are internalised and digested by the PRE. Do not have **desmosomes** or **cytokeratins** in some species.

pigtail See **GPI-anchor**.

PIIF Proteinase-inhibitor inducing factor. A serine peptidase inhibitor (wound-induced proteinase inhibitor 1, 111aa) produced by tomato plants in response to attack by insects. It inhibits the proteinase that the insect secretes to digest plant tissues. May be mobile within the plant, thus inducing inhibitor formation away from the site of original attack.

pikachurin A protein (agrin-like protein, EGF-like, fibronectin type-III and laminin G-like domain-containing protein, 1017aa) involved in the formation of ribbon synapses in retinal photoreceptors and in visual perception. Interacts with alpha-**dystroglycan**.

PIKK A family of phosphatidylinositol 3-kinase-related protein kinases (PIKK). RUVBL1 and RUVBL2 (**AAA family** proteins RuvB-like 1 and 2) associate with PIKK family members, control PIKK abundance at least at the mRNA level, and are involved in various cellular processes, including transcription, RNA modification, DNA repair, and telomere maintenance. Abstract: http://stke.sciencemag.org/cgi/content/abstract/sigtrans;3/116/ra27

pilin (1) General term for the protein subunit of a **pilus**. (2) Protein subunit (121aa) of F-pili, **sex pili** coded for by the F-plasmid.

pilomatrixoma A firm, circumscribed tumour (epithelioma calcificans of Malherbe), that is caused in some cases by mutation in the beta-catenin gene.

pilus *Pl.* pili. Hair-like projection from surface of some bacteria.; specialized **sex-pili** are involved in conjugation with other bacteria. Major constituent is a protein, **pilin**. The term is often used interchangeably with fimbria although some authors reserve pilus for projections involved in conjugation and fimbria for those involved in adhesion to target epithelia.

pim Family of oncogenes encoding serine/threonine **protein kinases**. Pim-1 (404aa) may affect the structure or silencing of chromatin by phosphorylating heterochromatin protein 1 gamma (HP1γ), has a role in cytokine signal transduction and affects cell proliferation and survival. It is implicated in multiple human cancers, including prostate cancer, acute myeloid leukemia and other haematopoietic malignancies. Pim-2 (334aa) is 53% identical to Pim-1 at the amino acid level and shares substrate preference; pim-3 (326aa), is aberrantly expressed in human pancreatic cancer and phosphorylates **bad**.

PIN multigene family (1) In plants a family of genes encoding **auxin efflux carriers** (MtPINs) *Cf.* auxin influx/permease sequences (MtLAXs). PIN1 (auxin efflux carrier component 1, PIN-FORMED protein, 622aa) is involved in setting up the auxin gradient which is required for organogenesis. The ARF-GEF protein GNOM is required for the correct recycling of PIN1 between the plasma membrane and endosomal compartments. Mutations lead to the formation of pin-shaped inflorescences (hence the name) and abnormalities in lateral organs. (2) See **parvulins**.

PIN-1 See **parvulins** and **PIN multigene family**.

pinacocyte Flattened polygonal cell that lines ostia and forms the epidermis of sponges. Capable of synthesizing collagen.

pinitol Naturally-occurring compound found in certain plants, trees and foods, such as soy. It is claimed to have an insulin-like activity and to cause insulin sensitisation although double-blind clinical trials do not show it to have any detectable effects (neither toxic nor therapeutic).

pink1 A mitochondrial outer membrane serine/threonine-protein kinase (PTEN-induced putative kinase 1, BRPK, EC 2.7.11.1, 581aa) that protects against mitochondrial dysfunction during cellular stress. It has a mitochondrial targeting sequence and interacts with **parkin**. Defects in pink1 are the cause of autosomal recessive early-onset Parkinson's disease Type 6. A homologue is found in *C. elegans* (641aa) where it acts antagonistically to **LRK-1** in stress response and neurite outgrowth. Research article: http://www.jbc.org/content/284/24/16482.long

pinocytosis Uptake of fluid-filled vesicles (pinosomes, phagocytotic vesicles), usually less than 150 nm diameter, into cells (endocytosis). Macropinocytosis and micro-pinocytosis are distinct processes, the latter being energy independent and involving the formation of receptor-ligand clusters on the outside of the plasma membrane, and clathrin on the cytoplasmic face. Micropinocytotic vesicles are around 70 nm diameter.

Pinophyta Conifers.

pinosome See **pinocytosis**.

PINX1 A microtubule-binding protein (Pin2-interacting protein X1, TRF1-interacting protein, 1328aa) involved in the **shelterin** complex that binds telomeric repeat binding factor 1 (TRF1), mediates TRF1 and **TERT** accumulation in the nucleolus and enhances TRF1 binding to telomeres. Overexpression of PINX1 causes shortening of telomeres; depletion has the opposite effect and increases tumorigenicity in animal model systems. Has **telomerase**-inhibiting activity and is potentially a tumour suppressor. In *S. pombe* PinX1-related protein 1 (284aa) is involved in rRNA-processing and inhibits telomerase.

pioneer species Species that colonise 'bare' environments, for example after fire or pollution has destroyed the previous flora and fauna. Most attention has been paid to plants but bacterial and animal pioneers must also exist.

piopio A *Drosophila* wing-blister gene that encodes a protein (462aa) required for the microtubule organizing activity of the apical junctions. The protein interacts with dumpy and papillote.

PIP2 See **phosphatidyl inositol**.

pipe In *Drosophila*, a heparan sulphate 2-O-sulphotransferase (EC 2.8.2.-, 514aa) involved in dorsoventral axis patterning in early embryos through a protease cascade that activates **easter**. See **nudel**.

piperazines A class of compounds with a core piperazine group; many have useful pharmacological properties. Piperazine itself is an anthelminthic.

PIPES One of the Good buffers (1,4-piperazinediethanesulfonic acid); $pK_a(20°C) = 6.8$.

pipsqueak motif A DNA-binding motif (~ 50aa), structurally related to the helix-turn-helix domain, found in a family of transcription factors. The family includes proteins from fungi, sea urchins, nematodes, insects, and vertebrates. Expression pattern of pipsqueak family: http://onlinelibrary.wiley.com/doi/10.1002/dvdy.20046/full

piroplasm Class of Protista, Phylum Apicomplexa (Sporozoa or Telosporidea), which includes the tick-transmitted parasite, *Babesia*.

Pisaster Echinoderm of the Class Asteroidea, a starfish.

pisatin **Phytoalexin** produced by peas.

PIST A protein (Golgi-associated PDZ and coiled-coil motif-containing protein, PDZ protein interacting specifically with TC10, 462aa) involved in intracellular protein trafficking and degradation. Overexpression causes intracellular retention and lysosomal degradation of **CFTR**.

PISTILLATA A transcription factor (208aa) involved in the genetic control of flower development by interacting with floral homeotic proteins **APETALA1** and **SEPELLATA3**.

pit Region of the plant cell wall in which the **secondary wall** is interrupted, exposing the underlying **primary cell wall**. One or more **plasmodesmata** are usually present in the primary wall, communicating with the other half of a pit pair. May be simple or bordered; in the latter case, the secondary wall overarches the pit field. Do not confuse with **coated pits**.

pit1 Pituitary-specific positive transcription factor 1 (291aa) that activates transcription of growth hormone and prolactin genes in the pituitary.

PITSLRE kinase family A family of p34Cdc2-related protein kinases (CDK11 family kinases), named according to the single amino acid code of an important regulatory region; generated by alternative splicing and promoter utilization from three duplicated and tandemly linked genes on human chromosome 1. Their function has been related to cell-cycle regulation, splicing and apoptosis. Homologues are found in non-mammalian species. In humans, PITSLRE serine/threonine-protein kinase CDC2L1 (cell division protein kinase 11B, EC 2.7.11.22, 795aa).

pituicytes Dominant intrinsic cells of the neural lobe of the hypophysis. Have long branching processes and resemble **neuroglia**: secrete **antidiuretic hormone**.

Pitx genes A family of genes (*PTX1, 2 & 3*) that encode paired-like homeodomain transcription factors expressed in normal pituitary and aberrantly expressed in various pituitary adenomas. PITX1 (314aa) may play a role in the development of anterior structures. PITX2 (solurshin, 317aa) mutations are associated with **Rieger's syndrome**, and the PTX2C isoform is involved in left–right asymmetry of the developing embryo. PITX3 (302aa) defects are associated with various eye abnormalities.

piwi Family of *Drosophila* genes that play an essential role in stem cell self-renewal, gametogenesis and RNA interference in diverse organisms ranging from *Arabidopsis* to human (*hiwi* genes). The piwi domain is a highly conserved motif within **argonaute** proteins, that has been shown to adopt an RNase H fold critical for the endonuclease cleavage activity of **RISC**. Piwi interacting RNAs (piRNAs) are germ cell-specific RNAs required for male germ cell development.

pK$_a$ See **association constant**.

PKA, B etc. See **protein kinase A, protein kinase B**, and so on.

PKB See Akt.

PKI (1) Protein kinase inhibitor peptide. (2) See Ki antigen. (3) A Her-2/Her-1 inhibitor (PKI-166). (4) The negative log dissociation constant of a competitive inhibitor (pK$_i$). (5) Potato-kallikrein-inhibitors (PKI).

PKN See **protein kinase N**.

PKR See **protein kinase R**.

PKS/NRPS Enzymes (polyketide synthase/non-ribosomal peptide synthetases) encoded in gene clusters and involved in natural product synthesis in many microorganisms. See **microcystins** and **jamaicamides**.

PKU See **phenylketonuria**.

PLA2 See **phospholipase A**.

placental calcium-binding protein See **S100**.

placode Area of thickened ectoderm in the embryo from which a nerve ganglion, or a sense organ will develop.

plakalbumin Fragment of **ovalbumin** produced by **subtilisin** cleavage: more soluble than ovalbumin itself.

plakins A family of giant cytoskeleton binding proteins. One member is bullous pemphigoid antigen 1 (Bpag1)/dystonin, which has neuronal and muscle isoforms with actin-binding and microtubule-binding domains at either end separated by a plakin domain and several spectrin repeats. In epithelial cells plakins connect intermediate filaments to desmosomes and hemidesmosomes. See **epiplakin, spectraplakin**.

plakoglobin See **catenins**.

plakophilins Members of the **armadillo** family (arm-repeat) family of proteins, found in various cell types, both as an architectural component in desmosomes and dispersed in cytoplasmic particles. Plakophilin 1 (PKP1) has two isoforms, PKP1a (726aa) and PKP1b (747aa) with PKP1b restricted to the nucleus. Mutations in PKP1 cause ectodermal dysplasia-skin fragility syndrome. Plakophilin-2 (837aa and 881aa isoforms) links cadherins to intermediate filaments; defects are associated with arrhythmogenic right ventricular dysplasia/cardiomyopathy (ARVC). Plakophilin-3 (PKP3, 797aa) interacts with **plakoglobin, desmoplakin** and the epithelial keratin 18 and can bind all three **desmogleins, desmocollin**-3a and -3b, and possibly also desmocollin-1a and -2a. Plakophilin 4 (1149aa) is **p0071**.

planapochromat Expensive microscope objective that is corrected for **spherical aberration** and **chromatic aberration** at three wavelengths.

planar cell polarity pathway *PCP pathway* A non-canonical **Wnt** signalling pathway in *Drosophila* that determines planar cell polarity, perpendicular to the apical/basal polarity. It leads to cells being polarized in the plane of the epithelial sheet The main gene products involved are **frizzled**, van gogh/strabismus (see **vang-like protein**), **prickle, dishevelled, flamingo**, and diego. Homologous proteins are used in vertebrates during regulation of cell movements in gastrulation and various other morphogenetic movements. Inhibition of the signalling pathway disrupts endothelial cell growth, polarity, and migration, but this inhibition is reversed in vertebrates if the pathway is activated downstream by Daam-1 (disheveled-associated activator of morphogenesis 1, 1078aa), diversin (ankyrin repeat domain-containing protein 6, 727aa) or inversin. Webpage: http://rgd.mcw.edu/wg/pathway/the_planar_cell_polarity_wnt_signaling_pathway

Planctomycetes A bacterial phylum: examples are found in samples of brackish, and marine and fresh water. They appear to be rather distantly related to other bacteria and are closest to the Verrucomicrobia and Chlamydiae. Research article: http://www.pnas.org/content/100/14/8298.long

plant growth substances Substances that, at low concentration, influence plant growth and differentiation. Formerly referred to as plant hormones or phytohormones, these terms are now suspect because some aspects of the 'hormone concept', notably action at a distance from the site of synthesis, do not necessarily apply in plants. Also known as 'plant growth regulators'. The major classes are **abscisic acid**, **auxin**, **cytokinin**, **ethylene** and **gibberellin**; others include steroid and phenol derivatives.

plant growth-promoting rhizobacteria Bacteria that colonize plant roots and promote plant growth. Some may stimulate induced systemic resistance to various pathogens.

plant natriuretic peptides A class of systemically mobile molecules found in xylem and distantly related to **expansins**. They are recognized by anti-human atrial natriuretic polypeptide rabbit antibodies. PNP-A (130aa) from *Arabidopsis* has a role in the regulation of ion and solute homeostasis.

plantaricin C A **lantibiotic** produced by *Lactobacillus plantarum* LL441 with Type A and Type B lantibiotic properties.

planula Free-living larval form of a hydrozoan cnidarian. Has an outer layer of ciliated ectoderm and an inner mass of endoderm cells, is flattened and bilaterally symmetrical.

plaque assay (1) Assay for virus in which a dilute solution of the virus is applied to a culture dish containing a layer of the host cells; convective spread is prevented by making the medium very viscous. After incubation the 'plaques', areas in which cells have been killed (or transformed), can be recognized, and the number of infective virus particles in the original suspension estimated. (2) Assay for cells producing antibody against erythrocytes or against antigen that has been bound to the erythrocytes. The cell is surrounded by a clear plaque of haemolysis. Basic principle behind the assay is the same as for the virus plaque assay.

plaque-forming cell Antibody-secreting cell detected in a **plaque assay**.

plaque-forming unit *pfu* Number of Ig-producing cells or infectious virus particles per unit volume. Of a virus like bacteriophage λ, the number of viable viral particles, established by counting the number of plaques (see **plaque assay**) formed by serial dilution of the library. For example, a cDNA library might have a titre of 50,000 pfu/μl of library.

plasma Acellular fluid in which blood cells are suspended. Serum obtained by defibrinating plasma (plasma-derived serum) lacks platelet-released factors and is less suitable to support the growth of cells in culture.

plasma cell A terminally differentiated antibody-forming, and usually antibody-secreting, cell of the B-cell lineage.

plasma kallikrein See **kallikrein**.

plasma membrane The external, limiting **phospholipid bilayer** membrane of cells. See **transmembrane proteins**.

plasma membrane intrinsic proteins *PIPs* A subset of **aquaporins** in plants.

plasmacytoma Malignant tumour of **plasma cells**, very similar to a **myeloma** (plasmacytomas usually develop into multiple myeloma). Can easily be induced in rodents by the injection of complete **Freund's adjuvant**. Plasmacytoma cells are fused with primed lymphocytes in the production of monoclonal antibodies.

plasmal reaction A rare histochemical staining procedure that involves the reaction of aldehydes, derived from **plasmalogens** by treatment with mercuric chloride, and **Schiff's reagent**.

plasmalemma Archaic name for the plasma membrane of a cell (the term often included the cortical cytoplasmic region). Adjectival derivative (plasmalemmal) still current.

plasmalemmasome See **paramural body**.

plasmalogens A group of glycerol-based phospholipids in which the aliphatic side chains are not attached by ester linkages. More than 60% of the phosphatidylethanolamine in the brain is in the plasmalogen form, possibly because it allows closer packing in the lipid bilayer and greater impermeability. In bacteria plasmalogens are only found in anaerobes. Plasmalogen synthesis can be defective in some forms of **Zellweger's syndrome**. Blogpage: http://schaechter.asmblog.org/schaechter/2010/08/plasmalogens-have-evolved-twice.html

plasmid *episome* A small, autonomously replicating, piece of cytoplasmic DNA that can be transferred from one organism to another. They can be linear or circular DNA molecules and are found in both pro- and eukaryotes. 'Stringent' plasmids occur at low copy number in cells, 'relaxed' plasmids at high copy number, *ca.* 10−30. Plasmids can become incorporated into the genome of the host, or can remain independent. An example is the **F-factor** of *E. coli*. May transfer genes, and plasmids carrying antibiotic-resistant genes can spread this trait rapidly through the population. Widely used in genetic engineering as vectors of genes (**cloning vectors**).

plasmid partitioning The process whereby newly replicated plasmids are distributed uniformly to daughter cells during cell division. Important for low copy-number plasmids. Plasmid partitioning in yeast requires the proteins REP1 (partitioning protein REP1, Protein Baker, trans-acting factor B, 373aa), REP2 (Protein Charlie, trans-acting factor C, 296aa) and a cis-acting locus STB (REP3). Protein Able is the site-specific recombinase **FLP**. Equivalent proteins in bacteria are ParA, an ATPase essential for plasmid movement during partition and ParB a protein that can bind tightly to the partition site, ParS and is required for capture of the plasmid at the cell centre prior to partition.

plasmid prep Generic term for the isolation of **recombinant** plasmids from liquid bacterial culture, usually by alkaline/detergent lysis, selective precipitation of other components, and affinity purification of plasmid. As this is the most exciting thing most molecular biologists ever do, there is an informal shorthand for the scale of the preparation based on the size of the overnight culture: see **miniprep**, **midiprep**, **maxiprep** and **megaprep**.

plasmin Trypsin-like **serine endopeptidase** of the peptidase S1 family (fibrinolysin, fibrinase, thrombolysin, E.C. 3.4.21.7) that is responsible for digesting **fibrin** in blood clots and also acts on activated **Hagemann factor** and complement. Generated from **plasminogen** by the action of **plasminogen activator** which cleaves plasminogen to produce the heavy chain (561aa), the light chain (230aa), and **angiostatin**.

plasminogen Inactive precursor (810aa) of **plasmin**, 200 mg/l in blood plasma. Has multiple copies of the **kringle** domain.

plasminogen activator *PA* **Serine peptidase** that acts on **plasminogen** to generate **plasmin**. There are two forms, **urinary PA** (urokinase, uPA, EC 3.4.21.73, 431aa) and tissue PA (tPA, EC 3.4.21.68, 562aa) which are similar to **streptokinase** and **urokinase**. Polymorphisms in uPA are associated with susceptibility to late-onset Alzheimer's disease. Plasminogen activator activity has been implicated in cell invasiveness, and the uPA receptor (**uPAR**, CD87), controls matrix degradation in tissue remodelling and is important for normal cell migration and tumour cell metastasis. See **SLURP**.

plasminogen activator inhibitors Plasminogen activator inhibitor-1 (PAI-1, mesosecrin, 402aa) is a **serpin**. PAI-2 (415aa) is secreted by the placenta and only present in significant amounts during pregnancy. See **protease nexin-1**.

plasmodesma *Pl.* plasmodesmata. Narrow tube of cytoplasm penetrating the plant cell wall, linking the protoplasts of two adjacent cells. A desmotubule runs down the centre of the tube, which is lined by plasma membrane.

plasmodium (1) Multinucleate mass of protoplasm bounded only by a plasma membrane; the main vegetative form of acellular slime moulds (e.g. *Physarum*). (2) See *Plasmodium*.

Plasmodium Genus of parasitic protozoa that cause malaria. *Plasmodium vivax* causes the tertian type, *P. malariae* the quartan type and *P. falciparum* the quotidian or irregular type of disease. The names refer to the frequency of fevers which occur when the merozoites are released from the erythrocytes. The life-cycle is complex and involves an intermediate host, the female mosquito (*Anopheles*) that infects vertebrate host when taking a blood meal. Predominant form of the organism in humans is the intracellular parasite (the **merozoite**) in the erythrocyte, where it undergoes a form of multiple cell division termed **schizogony**. As a result the erythrocyte bursts and the progeny infect other erythrocytes. Eventually some cells develop into gametes that, when ingested by a female mosquito, will fuse in her gut to form a zygote (ookinete). Multiple cell division within the resultant oocyte, attached to the gut wall, gives rise to infective sporozoites; these migrate to the salivary glands and are ejected with the saliva the next time the mosquito takes a blood-meal.

plasmogamy Fusion of cytoplasm that occurs when protoplasts or gametes fuse. In most organisms the latter is followed more or less immediately by karyogamy (fusion of nuclei); in some fungi it may result in the formation of a heterokaryon.

plasmolysis Process by which the plant cell protoplast shrinks, so that the plasma membrane becomes partly detached from the wall. Occurs in solutions of high osmotic potential, due to water moving out of the protoplast by osmosis. The shrunken protoplast can remain attached to the cell wall by numerous thin cytoplasmic filaments called Hechtian threads or strands.

plastid Type of plant cell organelle, surrounded by a double membrane and often containing elaborate internal membrane systems. Partially autonomous, containing some DNA, RNA and ribosomes, and reproducing itself by binary fission. Includes **amyloplasts**, **chloroplasts**, **chromoplasts**, **etioplasts**, **leucoplasts**, **proteinoplasts**, and **elaioplasts**. Develop from **proplastids**.

PLASTID MOVEMENT IMPAIRED 15 A protein in *Arabidopsis* (574aa) that is involved in the chloroplast avoidance response to high intensity blue light in which chloroplasts relocate to the anticlinal side of exposed cells.

plastins A family of actin-bundling and cross-linking proteins. Plastin-1 (intestinal-plastin, I-plastin, 629aa) is found in intestinal microvilli (human homologue of avian fimbrin). Plastin-2 (lymphocyte cytosolic protein-1, L-plastin, p65, 627aa) is expressed in leukocytes, transformed fibroblasts, and various tumour cell lines. Plastin 3 (T-plastin, 630aa) is found in intestinal microvilli, hair cell stereocilia, and fibroblast filopodia.

plastochron The time interval between the generation of successive primordia by the apical meristem. E.g., the time taken to form two leaf initials.

plastochron index *P.I.* A measure of plant age on a morphological, rather than a temporal time scale, useful because plants of very different temporal age may be at the same stage of development because of differences in growth conditions or genotype. The leaf plastochron (leaf age) index is a way of measuring the age of leaves. Details: http://biomath.biology. usu.edu/Fipse/Labs/Plantstruc/plaschron.htm

plastocyanin An electron-carrying protein present in chloroplasts, forming part of the electron transport chain and associated with **photosystem I**. Contains two copper atoms per molecule. The major isoform in *Arabidopsis* chloroplasts is 167aa (DNA-damage-repair/toleration protein DRT112) but there is a minor isoform, 171aa.

plastoglobuli Globules found in plastids, containing principally lipid, including **plastoquinone**.

plastoquinone A **quinone** present in chloroplasts, forming part of the photosynthetic electron transport chain. Closely associated with **photosystem II**. May be stored in **plastoglobuli**. Webpage:http://www.cyberlipid.org/vitk/plas0002.htm

plate count Number of bacterial colonies that grow on a nutrient agar plate under defined conditions. Used to estimate numbers of viable cells in a population and bacterial contamination of water amongst others.

platelet Anucleate discoid cell (3 μm diameter) found in large numbers in blood; important for blood coagulation and for haemostasis. Platelet α-granules contain lysosomal enzymes, beta-thromboglobulin, platelet-derived growth factor, fibrinogen, platelet factor-4 and thrombospondin; dense granules contain ADP (a potent platelet aggregating factor), and **serotonin** (a vasoactive amine). They also release **platelet-derived growth factor** which presumably contributes to later repair processes by stimulating fibroblast proliferation.

platelet activating factor A phospholipid (PAF, PAFacether, 1-0-hexadecyl-2-acetyl-sn-glycero-3-phosphorylcholine) that is chemotactic and has potent inflammatory, smooth-muscle contractile and hypotensive activity. The receptor (342aa) is a rhodopsin-type GTP-binding protein that acts through an inositol 1,4,5-trisphosphate /calcium signalling pathway. PAF is degraded by PAF acetylhydrolase (EC 3.1.1.47, Group-VIIA phospholipase A2, 441aa).

platelet basic protein Protein (PBP, 94aa) in platelet alpha granules that is derived from a 128aa precursor and further processed by N-terminal cleavage to yield connective tissue activating peptide-III (85aa, CTAP-III), **beta-thromboglobulin** (81aa), and neutrophil activating peptide-2 (NAP-2, 70aa).

platelet factor 3 Phospholipid-rich particles (40% protein, 42% phospholipids, 13% cholesterol and 5% triacylglycerols) derived from platelets that contribute to the blood clotting cascade by forming a complex (thromboplastin) with other plasma proteins and activating **prothrombin**.

platelet factor 4 Platelet-released cytokine (CXCL4, 101aa) released from alpha granules following platelet aggregation. Neutralizes the anticoagulant effect of heparin, is chemotactic for neutrophils and monocytes and inhibits endothelial cell proliferation.

platelet-derived growth factor *PDGF* The major mitogen in serum for growth in culture of cells of connective tissue origin and important for wound healing. It is a heterodimer of PDGF-A and PDGF-B (211aa and 241aa) linked by disulphide bonds. The B chain is almost identical in sequence to p28sis, the transforming protein of simian sarcoma virus, that can transform only those cells that express receptors for PDGF, suggesting that transformation is caused by **autocrine** stimulation. Receptor is a **tyrosine kinase**.

platyspondyly The condition in which one or more vertebrae are flattened; a feature of **Morquio-Brailsford disease**, **spondyloepiphyseal dysplasia** tarda and **Kniest's syndrome**.

PLCPI Stefin type peptidase inhibitor (porcine leucocyte cysteine protease inhibitor, stefin-D1, 103aa) that co-purifies with **cathelin**. Inhibits papain and cathepsins L and S by forming a tight complex.

PLD (1) **Phospholipase D**. (2) Pegylated liposomal doxorubicin. (3) PLD-118 is a novel, oral antifungal drug, formerly BAY 10-8888, a synthetic derivative of the naturally occurring beta-amino acid cispentacin. (4) The Protein Ligand Database (PLD) is a publicly available web-based database now available as Binding MOAD which had 16,948 protein-ligand structures in 2010. Link to database: http://www.BindingMOAD.org.

pleckstrin A platelet protein (platelet P47, platelet and leukocyte C kinase substrate, pleckstrin-1, 350aa) that is the main target for phosphorylation by PKC. Pleckstrin-2 (353aa) is bound to the cell membrane and may be involved in cytoskeletal rearrangement and the formation of lamellipodia. Pleckstrin homology domains (**PH domains**) are found in a number of proteins (including **FYVE, RhoGEF and PH domain-containing proteins**.

plectin Abundant linker protein (hemidesmosomal protein 1, 4684aa) that links intermediate filaments with microtubules and microfilaments. Has an actin-binding domain with homology to that of the **dystrophins** and the C-terminal region has homology to **desmoplakin** and binds vimentin, desmin, GFAP, cytokeratins, lamin B and integrins. There are 33 plectin repeats. See **spectraplakin**. Defects in plectin are the cause of some forms of epidermolysis bullosa simplex.

plectonemic One of two topologically distinct forms of supercoiling of DNA: plectonemic supercoiling is found in DNA molecules freely suspended in solution; solenoidal supercoiling is a characteristic of DNA molecules wrapped around histones. Plectonemic associations of DNA molecules cannot be disrupted by deproteination, whereas paranemic joints can.

pleiotrophin A secreted heparin binding protein (heparin-binding neurite outgrowth-promoting factor 1, osteoblast-specific factor 1, 168aa) that stimulates mitogenesis and angiogenesis and neurite and glial process outgrowth guidance activities *in vitro*. One of a family that includes **midkine**.

pleiotropic Having multiple effects. For example, the cyclic-AMP concentration in a cell will have a variety of effects because the cAMP acts to control a protein kinase that in turn affects a variety of proteins.

pleomorphic Having more than one body shape during the life cycle or having the ability to change shape or to adopt a variety of shapes.

plesiomorphic See **apomorphic**.

PLETHORA A small family of transcriptional activators in plants. PLETHORA1 (AP2-like ethylene-responsive transcription factor, 574aa) binds to promoters involved in stress responses and is essential for root **quiescent centre** specification. Modulates the polar transport of auxin in the root by regulating the distribution of auxin transporters (PIN proteins). PLETHORA2 (568aa) has similar activity

Pleurobrachia Small free-swimming marine organism, member of the Phylum Ctenophora. Roughly spherical and transparent with most of the body made up from transparent jelly-like material. The animal has two long tentacles for catching prey, and swims

by means of eight rows of **comb plates** (made of fused cilia) that run along the body.

pleuropneumonia-like organism *PPLO* See **mycoplasma**.

plexins Transmembrane receptors of **semaphorins**. Subtype-specific functions of the majority of the nine members of the mammalian plexin family are largely unknown. Nomenclature is idiosyncratic (plexin 4 is SEX, plexin 5 is SEP, plexin 2 is OCT) Plexin-A1 (NOV, 1896aa) is the receptor for semaphorin 3A; binding of which leads to local translation of RhoA in the nerve growth cone and subsequent collapse. Plexin domain-containing protein 1 (PLXDC1, tumour endothelial marker 3 or 7, 500aa) may interact with cortactin in is found in endothelial cells in tumours but not in normal tissue. PLXDC2 (tumour endothelial marker 7-related protein, 529aa) may be involved in tumour angiogenesis.

PLGA (1) A biodegradeable and biocompatible polymer (poly(lactic-co-glycolic acid)) used in therapeutic devices. (2) Pediatric Low Grade Astrocytoma. (3) Polymorphous low-grade adenocarcinoma, a rare salivary gland tumour.

PLGF Growth factor (placental growth factor, 149aa) related to **PDGF/VEGF** involved in angiogenesis and endothelial cell growth.

P-light chain A **myosin light chain** that is regulated by phosphorylation (by **myosin light chain kinase**) rather than by binding calcium.

PLL In *Arabidopsis* a family of protein phosphatases. PLL-1 (protein phosphatase 2C 29, POLTERGEIST-LIKE 1, 783aa) is involved in the regulation of pedicel length and **CLAVATA** signalling pathways. Other members of the family have different tissue distributions.

P-loop See **ATP binding site**.

pluripotent stem cell See **stem cell**.

pluronic block copolymers Synthetic copolymers of ethylene oxide and propylene oxide that are amphiphilic and will form micelles that enclose drug molecules for effective delivery. Pluronic is a registered trade mark. Various sizes are available with different properties. Datasheet: http://www2.basf.us/performancechemical/pdfs/Pluronic_NF_Grades.pdf

plusbacin A See **katanosin B**.

pluteus Free-swimming ciliated larval stage of some echinoderms.

PM1 (1) Particulate matter (PM) in air with a diameter less than 1 μm. PM2.5, PM10, etc are particles of greater diameter. (2) Bacterial strain PM1 will rapidly and completely biodegrade the petrol additive methyl tertiary butyl ether (MTBE) in groundwater. (3) A transformed CD4$^+$ T-cell clone derived from the Hut78 T-cell line. (4) Monoclonal antibody (precursor marker 1: PM1) that labels most neuroepithelial cells in day 4 embryonic chick retinal sections. (5) A lignin-degrading basidiomycete, strain PM1 (CECT 2971).

PMA Phorbol myristate acetate, a **phorbol ester**. In the earlier literature often (ungrammatically) TPA, tumour promotor activity.

PMAT Plasma membrane monoamine transporter, see **ENTs**.

PMF (1) **Proton motive force**. (2) Potential of mean force (PMF), a concept used in molecular dynamics, the energy associated with the probability of being in a particular state.

PML body Nuclear structure (promyelocytic leukaemia body, Kremer body) containing multimers of PML protein and various other nucleoproteins including the Nijmegen breakage syndrome protein (p95/Nbs1, **nibrin**). PML protein (RING finger protein 71, tripartite motif-containing protein 19, 882aa) is probably a transcription factor. Sumoylated forms localize to the PML nuclear bodies.

PMN Polymorphonuclear leucocyte (PMNL), an idle shorthand for an eosinophil, **basophil** or **neutrophil granulocyte**, but usually intended to mean the latter.

PMS In *S. cerevisiae*, PMS1 (postmeiotic segregation protein 1, 873aa) is required for **mismatch repair**. Forms a heterodimer with **MLH1**. The human homologue (932aa) is similar. Defects in PMS1 are the cause of hereditary non-polyposis colorectal cancer type 3. PMS2 is mismatch repair endonuclease (PMS1 protein homolog 2, 862aa). See **MLH** and **mismatch repair**.

PMSF See **phenylmethylsulphonyl fluoride**.

PNA See **peptide nucleic acid**.

pneumococci See *Streptococcus*.

Pneumocystis jiroveci Organism (formerly *Pneumocystis carinii*), related to ustomycetous yeasts, that is a common cause of pneumonia in immunocompromised patients.

pneumocyte Cells that line the alveoli of the lung. Type I pneumocytes are squamous. Type II pneumocytes are smaller, roughly cuboidal cells, usually found at the alveolar septal junctions, responsible for secreting surfactant and will replicate to replace damaged Type I pneumocytes.

pneumolysin Cholesterol binding toxin (471aa) from *Streptococcus pneumoniae*.

PNMT Terminal enzyme (phenylethanolamine N-methyl transferase, EC 2.1.1.28, 282aa) in the catecholamine biosynthetic pathway; converts noradrenaline to adrenaline.

PNPLA6 See **patatins**.

podocalyxin A single-pass transmembrane protein (528aa) heavily glycosylated and found on the apical membrane of rat renal glomerular epithelial cells (**podocytes**) and also in endothelial, haematopoietic, and tumour cells. Function as an anti-adhesive molecule that maintains an open filtration pathway between neighboring foot processes in the podocyte. Colocalizes with actin filaments, **ezrin** and SLC9A3R1 in a punctate pattern at the apical cell surface where microvilli form.

podocin A protein (383aa) that has a role in the regulation of glomerular permeability, probably acting as a linker between the plasma membrane and the cytoskeleton. Interacts with **nephrin** and nephrin-like protein-3.

podocytes Cells of the visceral epithelium that closely invest the network of glomerular capillaries in the kidney. Most of the cell body is not in contact with the **basal lamina**, but is separated from it by trabeculae that branch to give rise to club-shaped protrusions, known as pedicels, interdigitating with similar processes on adjacent cells. The complex interdigitation of these cells produces thin filtration slits that seem to be bridged by a layer of material (of unknown composition), that acts as a filter for large macromolecules. See **nephrin**, **podocalyxin**, and **P-cadherin**.

podophyllotoxin Glucoside toxin (414 Da), derived from the roots of *Podophyllum peltatum* (American Mayapple), that binds to tubulin and prevents microtubule assembly. **Etoposide** is a derivative of podophyllotoxin.

podoplanin A type I membrane glycoprotein (aggrus, glycoprotein 36, 162aa) that is a marker for normal lymphatic endothelial cells (PA2.26 antigen bound by MAb D2-40), although expression is upregulated in various carcinomas. May be involved in cell migration and organization of the actin cytoskeleton and is localized in actin-rich microvilli, filopodia, lamellipodia and ruffles. It is extensively O-glycosylated and glycosylation is necessary for its effect on platelet aggregation.

podosomes Punctate substratum-adhesion complexes in osteoclasts. Contain **vinculin**, **talin**, **fimbrin** and **F-actin**. Podosomes form a broad ring of contacts with the underlying bone and the enclosed area below the cell is then absorbed.

poikilocytosis Irregularity of red cell shape.

point mutation Mutation that causes the replacement of a single base pair with another pair.

pokeweed mitogen Any of the **lectins** derived from the pokeweed, *Phytolacca americana*, all of which will stimulate T-cells. Binds β-D-acetylglucosamine.

pol genes Genes coding for **DNA polymerases** of which there are three in *E. coli*, *polA*, *polB*, and *polC* coding for polymerases I, II, and III respectively. *Pol* genes in **oncogenic** retroviruses code for **reverse transcriptase**.

polar body In animals each meiotic division of the oocyte leads to the formation of one large cell (the egg) and a small polar body as the other cell. Polar body formation is a consequence of the very eccentric position of the nucleus and the spindle.

polar granules Granules containing a basic protein found in insect eggs that induce the formation of germ cells into which they become incorporated.

polar lobe In some molluscs a clear protrusion close to the vegetal pole of the fertilized cell prior to the first cleavage, that becomes associated with only one of the daughter cells. Removal of the first polar lobe, or of any polar lobe that forms at a subsequent mitosis, leads to defects in the embryo; it seems that the polar lobe contains special morphogenetic factors.

polar plasm Differentiated cytoplasm associated with the animal or vegetal pole of an oocyte, egg or early embryo.

polarity Literally 'having poles' (like a magnet), but used to describe cells that have one or more axes of symmetry. In epithelial cells, the polarity meant is between apical and basolateral regions; in moving cells, having a distinct front and rear. Some cells seem to show multiple axes of polarity (which will hinder forward movement).

polarization microscopy See **light microscopy**.

pole cell A cell at or near the animal or vegetal pole of an embryo.

pole fibres Microtubules inserted into the pole regions of the mitotic spindle (each pole is the product of the division of the centrioles and constitutes a **microtubule organizing centre**.

polehole *Drosophila* homologue of the *raf* oncogene.

poliovirus A member of the enterovirus group of **Picornaviridae** that causes poliomyelitis, an acute inflammation of the central nervous system. Poliovirus receptor-related proteins are single-pass membrane proteins, nectins. Poliovirus receptor-related protein-1, PVRL1, is nectin-1 (CD111, 517aa), PVRL3 is nectin-3 (CD113, 549aa). They are involved in cell-cell adhesion through heterophilic trans-interactions with other **nectins** or nectin-like proteins (see **syncams**).

pollen mother cell A diploid plant cell that forms four **microspores** by meiosis; the microspores give rise to pollen grains in seed plants.

pollen tube A tubular outgrowth produced when a pollen grain germinates. In angiosperms, after the pollen has attached to the stigma, the tube grows tdown to the embryo sac to deliver the male gamete(s) and complete fertilization. In *Arabidopsis* growth of the pollen tube is guided by gradients of **GABA**.

polo Founding member (576aa) of the family of **polo-like kinases** (Plks), identified in a *Drosophila* screen for mutants affecting spindle pole behaviour.

polo-like kinases A conserved family of serine/threonine kinases, characterized by the presence of a C-terminal domain termed the Polo-box domain (PBD) in addition to the N-terminal kinase domain, with many members throughout various species. Multiple Plks are present in mammalian cells (Plk1, 603aa; Plk2/Snk (serum inducible kinase), 685aa; Plk3/Fnk/Prk, (cytokine-inducible kinase, proliferation-related kinase), 646aa; Plk4/Sak, 970aa) and *Xenopus* (Plx1-3), whereas in other species only one member has been identified, like Polo in *Drosophila*, Cdc5 in budding yeast and Plo1 in fission yeast. Required for various stages of mitosis particularly the G2/M transition. Plk1 associates with spindle poles up to metaphase, but relocalizes to the equatorial plane, as anaphase proceeds. Article: http://www.ncbi.nlm.nih.gov/pmc/articles/PMC2133970/pdf/jc13561701.pdf

poly(A) binding proteins *PABPs* Proteins that bind to the 3′ poly(A) tail of mRNA and probably involved in cytoplasmic processes such as pre-mRNA splicing; *in vivo* may bind to other RNA sequences. PABP1 (636aa) is involved in nuclear events associated with the formation and transport of mRNP to the cytoplasm. PABP2 (polyadenylate-binding nuclear protein 1, 306aa) stimulates **polyA polymerase** (PAPOLA) that adds the poly(A) tail to pre-mRNA. It is also involved in nucleocytoplasmic trafficking and nonsense-mediated decay of mRNA. PABP3 (631aa) is testis-specific, PABP4 (644aa) is upregulated in activated T-cells, PABP5 (382aa) is found in fetal brain and some adult tissues. PABPs are components of the autoregulatory ribonucleoprotein complex (ARC). There are PABP-interacting proteins (PAIP1, 480aa) that are involved in initiating or repressing (PAIP2, 127aa) translation. See **cytoplasmic polyadenylation element**, **maskin**. Article: http://nar.oxfordjournals.org/content/33/22/7074.full

poly(ADP-ribose)polymerase An abundant nuclear protein (PARP1, EC 2.4.2.30, 1014aa) activated by DNA nicks and involved in the **base excision repair** pathway. PolyADP ribosylation, brought about by ADP-ribosyl protein ligase, is a post-transcriptional modification of proteins including **p53**. Other members of the PARP family may regulate gene transcription by polyadenylation of histones and gene silencing. PARP knockout mice show defects in fibroblast proliferation and impaired capacity to handle radiation-induced damage. One of the earliest proteins cleaved by **caspase 3** in apoptosis.

poly A polymerases Enzymes (polynucleotide adenylyltransferases, EC 2.7.7.19, PAPOLA, 745aa; PAPOLB, 636aa) that add the poly(A) tail to the 3′ end of pre-mRNA. See **poly(ADP-ribose)polymerase**.

poly-A tail Polyadenylic acid sequence of varying length found at the 3′ end of most eukaryotic mRNAs though not those of histones. The poly-A tail is added post-transcriptionally to the primary transcript by **poly A polymerases** as part of the nuclear processing of RNA yielding **hnRNAs** with 60–200 adenylate residues in the tail. In the cytoplasm the poly-A tail on mRNAs is gradually reduced in length. The function of the poly-A tail is not entirely clear but it is the basis of a useful technique for the isolation of eukaryotic mRNAs. The technique uses an **affinity chromatography** column with oligo(U) or oligo(dT) immobilized on a solid support. If cytoplasmic RNA is applied to such a column, poly-A-rich RNA (mRNA) will be retained. See **poly(A) binding proteins**.

poly(glycerol sebacate) A biodegradable elastomer that has been used as a support material for neural reconstruction.

poly-immunoglobulin receptor One of the immunoglobulin superfamily (764aa) found on the basolateral surface of various glandular epithelia where it binds polymeric IgA and IgM; the complex is transported across the cell and secreted at the apical surface with concurrent proteolysis that releases the secretory piece from the transmembrane portion.

poly(lactide-co-glycolide) A biomaterial used as a support matrix for tissue engineering. It is biodegradable and its properties can be adjusted by varying the proportions of the glycolic acid and lactic acid monomers.

poly-Q diseases A group of diseases (polyglutamine disorders), including Huntingdon's disease, caused by expansion of polyQ-encoding repeats within otherwise unrelated genes. Mutant protein accumulates as insoluble aggregates in neuronal cells, causing their death.

poly-unsaturated fatty acid See **PUFA**.

polyacrylamide gel electrophoresis *PAGE* Analytical and separative technique in which molecules, particularly proteins, are separated by their different electrophoretic mobilities in a hydrated gel. The gel suppresses convective mixing of the fluid phase through which the electrophoresis takes place, and contributes molecular sieving. Commonly carried out in the presence of the anionic detergent sodium dodecylsulphate (SDS). SDS denatures proteins so that non-covalently associating subunit polypeptides migrate independently, and by binding to the proteins confers a net negative charge roughly proportional to the chain weight. See also **SDS-PAGE**.

polyadenylic acid Polynucleotide chain consisting entirely of residues of adenylic acid (i.e. the base sequence is AAAA....AAAA). Polyadenylic chains of various lengths are found at the 3′ end of most eukaryotic mRNAs, the **poly-A tail**.

polyamine Compounds that are polycationic at physiological pH, and can bind and interact with various other molecules within the cell. In particular interact with DNA but also may modulate ion channels and act as growth factors. **Spermine** has four positive charges, **spermidine** has three. The precursor of both, **putrescine**, has two.

polyanion Macromolecule carrying many negative charges. The commonest in cell-biological systems is nucleic acid.

polycaprolactone A biocompatible polymer used as a support matrix in tissue engineering. It biodegrades rather slowly and is also used as a suture material.

polycation Macromolecule with many positively charged groups. At physiological pH the most commonly used in cell biology is poly-L-lysine; this is often used to coat surfaces thereby increasing the adhesion of cells (which have net negative surface charge). See also **cationized ferritin**.

polycistronic mRNA A single **mRNA** molecule that is the product of the **transcription** of several tandemly arranged genes (a polycistron); typically the mRNA transcribed from an **operon** in prokaryotes. More recently it has emerged that a set of miRNAs are coordinately produced from a polycistron (see **mir-17-92 polycistron**) and in trypanosomatids gene regulation seems to occur by processing poly-cistronic transcripts and through differential stability of the resulting mRNAs.

polyclonal antibody An antibody produced by several clones of B-lymphocytes as would be the case in a whole animal. Usually refers to antibodies raised in immunized animals, whereas a **monoclonal** antibody is the product of a single clone of B-cells, usually maintained *in vitro*.

polyclonal compartment An area or volume in an animal that has a defined boundary but is composed of the progeny of several cells. Examples are found close to the mid-line of the wing of *Drosophila*.

polycloning site Region of a phage or plasmid vector (multiple cloning site, MCS, polylinker) that has been engineered to contain a series of **restriction sites** that are usually unique within the entire vector. This makes it particularly easy to insert or excise (subclone) DNA fragments.

polycomb *Drosophila* gene, that when mutated leads to extra sex combs on the legs of male flies, suggesting that the posterior legs have become anterior legs. There are at least 10 genes in the Polycomb group; they are thought to act by **transcriptional silencing** of **homeotic genes**. See **polycomb repressive complexes**.

polycomb repressive complexes Multi-protein complexes involving products of the polycomb group (PcG) genes which were initially identified as regulators of homeotic genes in *Drosophila* and are implicated in regulation of stem cell self-renewal and in cancer development. The PcG complexes are mostly associated with heterochromatin, where they inhibit gene expression by histone modification. There are numerous human polycomb group homologues and there are at least two distinct sets of complexes, PRC1 which includes at least one paralog each of **PCGF**, Ring1 (E3 ubiquitin ligase), **PHC**, and **chromobox** proteins and PRC2 which includes **EED**, **EZH** and **SUZ12**. There are two PRC2 complexes, PRC2/EED-EZH1 which comprises **EED**, **EZH1**, **SUZ12**, **retinoblastoma binding protein-4** (RBBP4) and **AEBP2**. The PRC2/EED-EZH2 complex is composed of EED, EZH2, RBBP4, RBBP7 and SUZ12 and methylates lysine residues on histone H3, leading to transcriptional repression of the affected target gene. The PRC2/EED-EZH2 complex may also associate with **histone deacetylase**, HDAC1. It seems that covalent modification of histone by PRC2 may 'mark' the site for PRC1 ubiquitinylation.

polycystic kidney disease An autosomal dominant renal disorder in adults characterised by renal and hepatic cysts and intracranial aneurysm. Can be caused by mutation in **polycystins**. An infantile form is caused by mutation in the gene for **fibrocystin**.

polycystic ovary syndrome A metabolic syndrome (Stein-Leventhal syndrome) with many other symptoms; ovarian cysts arise through incomplete follicular development or failure of ovulation. Associated with insulin resistance and consequent hyperinsulinaemia and (frequently) hyperlipidaemia and obesity. The genetic lesion is undefined at present.

polycystins A family of transmembrane proteins with a variety of roles; at least six have been identified in humans. Polycystin-1 (4280aa) and polycystin-2 (968aa) interact to form a non-selective cation channel *in vitro*, although polycystin-2 on its own will form a calcium permeable cation channel. The polycystins may be important for normal tubulogenesis in the kidney and dysfunction of these proteins causes autosomal dominant **polycystic kidney**

disease. They also mediate mechanosensing in the primary cilia (monocilia) of kidney cells. Similar proteins (for example polycystic kidney disease protein 1-like 1, 2849aa) are found in other tissues and there are homologues in *C. elegans* (location of vulva defective 1, 3330aa, which also localizes to cilia and the ciliary base and is involved in male mating behaviour). Website: http://www.humpath.com/polycystins

polycythaemia *US* polycythemia. An increase in the haemoglobin content of the blood, either as a result of decreased plasma volume or an increase in red cell numbers (erythrocytosis). In polycythaemia vera (Vaquez-Osler disease) there is increased production of erythrocytes as a result of somatic mutation in the **JAK2** gene in a single haematopoietic stem cell. Familial erythrocytosis can be caused by mutation in the erythropoietin receptor gene or the **VHL tumour suppressor gene**.

polydnaviruses Viruses (PDVs) that have been described in thousands of parasitoid wasp species; have a segmented DNA genome in viral particles and an integrated form that persists as a provirus in the wasp genome. Two genera of phylogenetically unrelated PDVs exist, the bracoviruses (BVs) and the ichnoviruses (IVs), associated with braconid and ichneumonid wasps, respectively.

polyductin See **fibrocystin**.

polyendocrine syndrome An autoimmune disorder (the antigen to which the response is mounted is in the **B cells** of the pancreas) in which there is involvement of several organ systems. Autoimmune polyendocrinopathy syndrome type 1 is caused by mutation in the autoimmune regulator gene (AIRE), product of which (545aa) is probably a transcription factor. In type 2 (Schmidt's syndrome) there is **Addison's disease** and autoimmune thyroid disease and/or insulin-dependent diabetes mellitus. In Type 3 there is autoimmune thyroid disease and other autoimmune disorders but not Addison's disease. X-linked polyendocrinopathy, immune dysfunction and diarrhoea (XPID) is caused by mutation in **FOXP3**.

polyene antibiotics Group of structurally-related antibiotics produced by *Streptomyces* spp. Interact with sterols in eukaryotic membranes. Examples are **amphotericin B** and **nystatin**.

polyethylene glycol *PEG* A hydrophilic polymer that interacts with cell membranes and promotes fusion of cells to produce viable hybrids. Often used in producing **hybridomas**. See **pegylation**.

polyethylenimine Synthetic polymers (PEIs) with a high cationic charge density which function as transfection reagents based on their ability to compact DNA or RNA into complexes.

polygalacturonan Plant cell wall polysaccharide consisting predominantly of galacturonic acid. May also contain some rhamnose, arabinose and galactose. Those with significant amounts of rhamnose are termed **rhamnogalacturonans**. Found in the **pectin** fraction of the wall.

polygalacturonase Enzyme (EC 3.2.1.15) that degrades **polygalacturonan** by hydrolysis of the glycosidic bonds that link galacturonic acid residues. Important in fruit ripening and in fungal and bacterial attack on plants. In *Arabidopsis* polygalacturonase ADPG1 (pectinase ADPG1, 431aa) is involved in cell separation in pod shatter and anther dehiscence; ADPG2 (433aa) is involved in a wider range of similar cell separation events. Other polygalacturonases are known. Polygalacturonase inhibitors (PGIP1, PGIP2, both 330aa) inhibit fungal polygalacturonase and are important for resistance to phytopathogenic fungi. Transgenic *Arabidopsis* plants over-expressing PGIPs exhibit enhanced resistance to *Botrytis cinerea*.

polygenic Describing something that is controlled or caused by the action of many genes. Thus many of the major non-infectious diseases (for example arthritis, cardiovascular disease, asthma, diabetes) are likely to be caused by the interaction of many genes; no single gene mutation is responsible, rather the coincidence of polymorphic variants that together contribute risk factors that predispose an individual to the disease. Polygenic inheritance is also referred to as quantitative or multifactorial inheritance.

polyglycolide A biocompatible polymer used mostly as biodegradable suture material, but also as a support matrix in tissue engineering. Degrades relatively quickly.

polyhedron Major protein of the crystalline matrix of viral polyhedral bodies (occlusion bodies) that form within baculovirus-infected cells. In *Autographa californica* nuclear polyhedrosis virus polyhedrin (major occlusion protein, 245aa) protects the virus from the outside environment until it is ingested by the host.

polyhydroxybutyrate A polymer that is produced by some micro-organisms (e.g. *Alcaligenes eutrophus*, *Bacillus megaterium*) in response to physiological stress. It is relatively resistant to hydrolytic degradation and is biocompatible, making it potentially useful in tissue reconstruction.

polyisoprenylation See **geranylation**.

polyketides A very diverse class of compounds which make up a significant proportion of the antibiotics synthesized by actinomycetes. Their biosynthesis is catalysed by multifunctional enzymes, polyketide synthases, which are related to the fatty acid synthases. See **PKS/NRPS**. Article: http://www.plosbiology.org/article/info:doi%2F10.1371%2Fjournal.pbio.0020035

polylactic acid A biodegradable polymer that can be produced from natural products. Has been used as a plastic for various applications and has been used in various tissue-engineering applications (sutures, stents, support films and others).

polylinker See **polycloning site**

polylysine A polycationic polymer of **lysine** (20–30 residues) used to mediate adhesion of living cells to synthetic culture substrates, or of fixed cells to glass slides (for observation by fluorescence microscopy, for example).

polymer A macromolecule made of repeating (monomer) units or **protomers**. Polymer formation (polymerization) often requires **nucleation** and will only occur above a certain critical concentration.

polymerase chain reaction *PCR* A method used for *in vitro* amplification of DNA. Two synthetic oligonucleotide primers, which are complementary to two regions of the target DNA to be amplified (one for each strand), are added to the target DNA sample, in the presence of excess deoxynucleotides and **Taq polymerase**, a heat-stable DNA polymerase. In a series (typically 30) of temperature cycles, the target DNA is repeatedly denatured (at around 90°C), annealed to the primers (typically at 50–60°C) and a daughter strand extended from the primers (72°C). As the daughter strands themselves act as templates for subsequent cycles, DNA fragments matching both primers are amplified exponentially, rather than linearly. The original DNA need thus be neither pure nor abundant, and the PCR reaction has accordingly become widely used not only in research, but in clinical diagnostics and forensic science. See entries for: **AFLP**, **anchored PCR**, **cDNA-AFLP**, **degenerate PCR**, **differential display PCR**, **hybrid specific amplification**, **ligase amplification reaction**, **MPSS**, **nested PCR**, **PCR *in situ* hybridization**, **RAPD**, **rapid amplification of DNA ends**, **real time PCR**, **RT-PCR**, **suppression PCR**, **suppression subtractive hybridization**, **vectorette method**.

polymeric immunoglobulin receptor See **secretory component**.

polymorphic epithelial mucin See **episialin**.

polymorphism (1) The existence, in a population, of two or more alleles of a gene, where the frequency of the rarer alleles is greater than can be explained by recurrent mutation alone (typically greater than 1%). HLA alleles of the **major histocompatibility complex** are very polymorphic. (2) The differentiation of various parts of the units of colonial animals into different types of unit specialized for different purposes, e.g. as in the colonial hydroid *Obelia*.

polymorphonuclear leucocyte *PMNL, PMN* Mammalian blood leucocyte (granulocyte) of myeloid series in distinction to mononuclear leucocytes: see **neutrophil, eosinophil, basophil**.

polymyxins Group of cyclic peptide antibiotics produced by *Bacillus* spp. Molecular weights are around 1–2 kDa. Act against many Gram-negative bacteria, working apparently by increasing membrane permeability. See **colistin**.

polynucleotide Linear sequences of **nucleotides**, in which the 5′-linked phosphate on one sugar group is linked to the 3′ position on the adjacent sugars. In the polynucleotide DNA the sugar is **deoxyribose** and in RNA, **ribose**. They may be double-stranded or single-stranded with varying amounts of internal folding.

polyol Any polyhydric alcohol. Common examples are inositol, mannitol and sorbitol.

Polyomaviridae A family of DNA tumour viruses (originally a genus in the obsolete family Papovaviridae). The genus *Polyoma* (the type species of which is simian virus 40, SV40) was isolated from mice, in which it causes no obvious disease, but when injected at high titre into baby rodents, including mice, causes tumours of a wide variety of histological types (hence poly-oma). *In vitro*, infected mouse cells are permissive for virus replication, and thus are killed, whilst hamster cells undergo abortive infection, and at a low frequency become transformed.

polyp (1) Growth, usually benign, protruding from a mucous membrane. (2) The sessile stage of the Cnidarian (**coelenterate**) life-cycle; the cylindrical body is attached to the substratum at its lower end, and has a mouth surrounded by tentacles bearing **nematocysts** at the upper end; *Hydra* and the feeding-polyps of the colonial *Obelia* are examples.

polypeptide Chains of α-**amino acids** joined by peptide bonds. Distinction between peptides, oligopeptides and polypeptides is arbitrarily by length; a polypeptide is perhaps more than 10 residues.

polypeptide antibiotics Bactericidal antibiotics (**bacitracin, colistin, polymyxins**) with activity against Gram-negative aerobic bacilli including *Pseudomonas aeruginosa*. Act by disrupting the bacterial cell membrane.

polyphemusin A small peptide (18aa) isolated from the American horseshoe crab (*Limulus polyphemus*) that inhibits the growth of Gram-negative and Gram-positive bacteria. Related to tachyplesin-1 (17aa with a 77aa precursor) from *Tachypleus tridentatus* (Japanese horseshoe crab). Synthetic derivatives of polyphemusin bind to chemokine receptors on lymphocytes and can block chemotaxis and binding of HIV to its co-receptor.

polyphenism An adaptation in which a genome is associated with discrete alternative phenotypes in different environments, for example the solitary and migratory forms of the locust.

polyploidy Of a nucleus, cell or organism that has more than two **haploid** sets of **chromosomes**. A cell with three haploid sets (3n) is termed triploid, four sets (4n) tetraploid and so on.

polypodial Describing an amoeba with several pseudopods.

polyprotein Protein that, after synthesis, is cleaved to produce several functionally distinct polypeptides. Some viruses produce such proteins, and some polypeptide hormones seem to be cleaved from a single precursor polyprotein (**pro-opiomelanocortin**, for example).

polypyrimidine tract-binding proteins Heterogenous nuclear ribonucleoproteins (PTBPs) that bind to the polypyrimidine tract of introns and are involved in pre-mRNA splicing. PTBP1 (531aa) is widely distributed, PTBP2 (531aa) is neuron-specific and the level of PTBP2 is controlled by a **miRNA** (microRNA124A1) that targets PTBP1 and leads to its degradation: in the absence of PTBP1 the cell expresses PTBP2. **IMP1** competes with PTBP1 binding to IGF2 leader 3 mRNA.

polyribosome *polysome* Functional unit of protein synthesis consisting of several **ribosomes** attached along the length of a single molecule of mRNA.

polysaccharide Polymers of (arbitrarily) more than about ten monosaccharide residues linked glycosidically in branched or unbranched chains.

polysialic acid A post-translationally added moiety that may regulate cell-cell interactions by conferring negative charge (and electrostatic repulsion) at the cell surface when added to glycoproteins. Thus the low PSA form of **NCAM** is thought to promote cell-cell contact and enhance **fasciculation** whereas NCAM with a high PSA content is thought to prevent close membrane-membrane apposition.

polysome See **polyribosome**.

polysomy Situation in which all chromosomes are present, and some are present in greater than the diploid number, for example, trisomy 21 (**Down's syndrome**).

polyspermy Penetration of more than one spermatozoon into an ovum at time of fertilization. Occurs as normal event in very yolky eggs (e.g. bird), but then only one sperm fuses with the egg nucleus. Many eggs have mechanisms to block polyspermy.

Polysphondylium A genus of **Acrasidae**, the cellular slime moulds.

polytene chromosomes Giant chromosomes produced by the successive replication of homologous pairs of chromosomes, joined together (synapsed) without chromosome separation or nuclear division. They consist of many (up to 1000) identical chromosomes (strictly chromatids) running parallel and in strict register. The chromosomes remain visible during interphase and are found in some ciliates, ovule cells in angiosperms, and in larval Dipteran tissue. The best known polytene chromosomes are those of the salivary gland of the larvae of *Drosophila melanogaster* which appear as a series of dense bands interspersed by light interbands, in a pattern characteristic for each chromosome. The bands, of which there are about 5,000 in *D.melanogaster*, contain most of the DNA (ca 95%) of the chromosomes, and each band roughly represents one gene. The banding pattern of polytene chromosomes provides a visible map to compare with the linkage map determined by genetic studies. Some segments of polytene chromosome show chromosome **puffs**, areas of high transcription.

polytropic virus See **xenotropic virus**.

polyuridylic acid Homopolymer of uridylic acid. Historically, was used as an artificial mRNA in cell-free **translation** systems, where it coded for polyphenylalanine; thus began the deciphering of the genetic code.

polyvinylidene fluoride *PVDF* A polymer that is very non-reactive and can be used in contact with biological materials and in medical applications. Also used in sensors.

polyvinylpyrrolidone *PVP* Polymer used to bind phenols in plant homogenates, and hence to protect other molecules, especially enzymes, from inactivation by phenols. Also occasionally used to produce viscous media for gradient centrifugation.

POMC See **pro-opiomelanocortin**.

POMC/CART neurons See **arcuate nucleus**.

Pompe's disease Severe glycogen **storage disease** (glycogen storage disease Type II) caused by deficiency in lysosomal α(1-4)-glucosidase (EC 3.2.1.20, acid maltase, 952aa), responsible for glycogen hydrolysis. Even though the non-lysosomal glycogenolytic system is normal, glycogen still accumulates in the lysosomes.

Ponceau red Dye (Ponceau S; Fast Ponceau 2B) used to stain proteins.

ponticulin Developmentally-regulated multi-pass membrane glycoprotein (143aa) from *Dictyostelium* that regulates actin binding and nucleation. It is unusual in having both transmembrane domains and a glycolipid anchor. Preferentially located at actin-rich regions such as sites of cell adhesion together with gp80 (contact site A protein). Many ponticulin-like proteins have been described in *Dictyostelium*. Phospholipid vesicles containing ponticulin have been used to form solid supported and tethered bilayer lipid membranes. Article: http://www.jbc.org/content/278/4/2614.long

POP2 (1) In *S. cerevisiae* POP2 protein (poly(A) ribonuclease POP2, EC 3.1.13.4, CCR4-associated factor 1, CAF1, 433aa) is part of a global transcription regulatory complex and is required for expression of many genes. It is a nuclease of the **DEDD superfamily**, probably the catalytic component of the **CCR4-NOT core complex** that mediates 3′ to 5′ mRNA deadenylation involved in mRNA turn-over. (2) In *S. pombe* there is a WD repeat-containing protein pop2 (proteolysis factor sud1, 703aa) involved in maintenance of ploidy through proteasome dependent degradation of CDK inhibitor rum1 and S-phase initiator cdc18.

popliteal pterygium syndrome A disorder phenotypically similar to **van der Woude's syndrome**, caused by mutation in the **IRF-6** gene.

population diffusion coefficient Coefficient that describes the tendency of a population of motile cells to diffuse through the environment. Its use presupposes that the cells move in a random-walk. Can also be applied to populations of free-living motile organisms.

porins Transmembrane matrix proteins (outer membrane channel proteins, 200–500aa) found in the outer membranes of Gram-negative bacteria. Associate as trimers to form channels (1 nm diameter, ca 10^5 per bacterium) through which hydrophilic molecules of up to 600 Da can pass. Similar porins are also found in outer mitochondrial membranes (VDAC, voltage-dependent anion-selective channel, 283aa). Multiple genes coding for VDAC homologues have been discovered in eukaryotic genomes, but their function is unclear.

porocyte Type of cell in asconoid sponges (small, simple sponges with a tube-shaped body) through which water enters the spongocoel.

porphobilinogen deaminase See **uroporphyrinogen I synthetase**.

porphyria Any of a group of disorders in which there is excessive excretion of porphyrins or their precursors. Most of the mutations are in enzymes of the haem synthesis pathway: **uroporphyrinogen I synthetase** in the case of acute intermittent porphyria, coproporphyrinogen oxidase (EC 1.3.3.3) in coproporphyria, ferrochelatase (EC 4.99.1.1) in erythropoietic porphyria, protoporphyrinogen oxidase (EC 1.3.3.4) in porphyria variagata. Gunther's disease (congenital erythropoietic porphyria) is caused by mutation in the uroporphyrinogen III synthase gene (EC 4.2.1.75) and porphyria cutanea tarda due to deficiency of uroporphyrinogen decarboxylase (uroporphyrinogen III cosynthase, EC 4.1.1.37). Deposition of porphyrins in the skin can lead to photosensitivity and dermatitis.

porphyrins Pigments derived from porphin (a 20-carbon heterocyclic compound): all are chelates with metals (Fe, Mg, Co, Zn, Cu, Ni). Constituents of haemoglobin, chlorophyll, cytochromes.

***Porphyromonas gingivalis* collagenase** Peptidase (proteinase C, 330aa) with collagenase activity from a Gram-negative anerobe associated with periodontal lesions (Strain W83 has been fully sequenced). Catalytic site of unknown character, hence in U-family of peptidases. Inhibited by EDTA and thiol blocking agents.

POSH A scaffold protein (plenty of SH3s, SH3 domain-containing RING finger protein 1, SH3RF1, 888aa) for the Jun N-terminal kinase (**JNK**) signal transduction pathway that may also act as an E3 ubiquitin-protein ligase. It has four SH3 domains. In mammals, important in pathways leading to apoptosis, in *Drosophila* the homologue (838aa) is part of the Imd pathway of immunity. Article: http://www.ncbi.nlm.nih. gov/pmc/articles/PMC1371032/?tool = pubmed

position effect Effect on the expression of a gene depending upon its position relative to other genes on the chromosome. Moving (transposing) a gene from an inactive region to an active region can alter expression markedly − sometimes with unfortunate consequences as with the **Philadelphia chromosome** abnormality that leads to chronic myelogenous leukemia.

positional cloning Identification of a gene based on its location in the genome. Typically, this will result from **linkage** analysis based on a mutation in the target gene, followed by a **chromosome walk** from the nearest known sequence.

positional information The instructions that are interpreted by cells to determine their differentiation in respect of their position relative to other parts of the organism, e.g. digit formation in the limb bud of vertebrates.

positive control Mechanism for gene regulation that requires that a regulatory protein must interact with some region of the gene before transcription can be activated (as opposed to removal of a restraint).

positive feedback See **feedback**.

positive strand RNA viruses Class IV and VI viruses that have a single-stranded RNA **genome** that can act as mRNA (plus strand) and in which the virus RNA is itself infectious. Includes **Picornaviridae**, **Togaviridae** and **Retroviridae**.

post-transcriptional gene silencing Inactivation of a gene by destruction of the mRNA, similar to quelling in fungi and **RNA interference** (RNAi) in animals.

post-translational modification Changes that occur to proteins after peptide bond formation has occurred. Examples include glycosylation, acylation, limited proteolysis, phosphorylation, isoprenylation.

postaxial acrofacial dysostosis A rare autosomal recessive disorder (Miller's syndrome) caused by mutations in the gene encoding dihydroorotate dehydrogenase (DHODH, EC 1.3.3.1, 395aa) an enzyme that catalyzes the fourth step in *de novo* pyrimidine biosynthesis.

postcapillary venule That portion of the blood circulation immediately downstream of the capillary network; the region having the lowest wall-shear stress, and the most common site of leucocytic margination and endothelial transmigration (diapedesis).

postsynaptic cell In a chemical **synapse**, the cell that receives a signal (binds neurotransmitter) from the presynaptic cell and responds with depolarization. In an electrical synapse, the postsynaptic cell would just be downstream, but since many electrical synapses are **rectifying**, one of the two cells involved will always be postsynaptic.

postsynaptic potential In a synapse, a change in the **resting potential** of a postsynaptic cell following stimulation of the presynaptic cell. For example, in a cholinergic synapse, the release of acetylcholine from the presynaptic cell causes channels to open in the postsynaptic cell. Each channel opening causes a small depolarization, known as a **miniature end plate potential** (mepp); these sum to produce an excitatory postsynaptic potential (epsp). Inhibitory neurotransmitter receptors such as **GABA receptors** and **glycine receptors**, cause a hyperpolarization (an inhibitory post-synaptic potential, ipsp) which reduces the probability of generating an action potential.

postsynaptic protein A conserved peripheral membrane protein (rapsyn, 412aa) closely associated with the cytoplasmic portion of nicotinic acetylcholine receptors, anchoring them in the postsynaptic membrane.There is a family of post-synaptic density proteins (PSD proteins) involved in the clustering of receptors on the post-synaptic cell including PSD-95 protein (synapse-associated protein 90, 723aa), a membrane-associated guanylate kinase (**MAGUK**) responsible for the clustering of NMDA receptors and K^+ channels (see **cypin**). See also **AKAP79**, **neuroligins, neurabin**

POT1 A **telombin family** protein (protection of telomeres 1, 634aa) that is a part of the **shelterin** complex and binds specifically to the single-stranded

G-overhang strand of telomeric DNA. *Cf.* **telomere repeat binding factors** which bind to double-stranded DNA.

potassium channels A diverse set of ion channels that are selective for potassium ions, encoded by more than 80 genes and regulated by an even greater number of other gene products. There are four main classes: **(i)** calcium-activated channels (**BK** and **SK** subtypes), **(ii)** inwardly-rectifying channels (**ROMK** (Kir1.1), G-protein coupled receptor-regulated (**GIRKs**, Kir3.x), and ATP-sensitive (Kir6.x, associated with sulphonyl urea receptors), **(iii)** tandem-pore domain channels (**TWIK, TRAAK, TREK** and **TASK**) and **(iv)** voltage-gated channels (**hERG**, (Kv11.1); KvLQT1 (Kv7.1). See **A-type channels**), **delayed rectifier channels, DLGs, minK, shaker, weaver.** Disorders are associated with **Andersen syndrome, Bartter's syndrome, episodic ataxia-1, hyperinsulinism, Jervell and Lange-Nielsen syndrome, long QT syndrome, nesidioblastosis, neuronal ceroid lipofuscinosis-6, short QT syndrome** and **spinocerebellar ataxia-13.** Because blocking potassium channels will prevent restoration of the **resting potential**, toxins that affect the channels will cause paralysis and are common in venoms etc. Examples include **agitoxin, ammodytoxins, anemone toxins, apamin, bungarotoxin, charybdotoxin, iberiotoxin, kaliotoxin, margaratoxin, noxiustoxin, scyllatoxin, SGTx1, ShK toxin.**

potato blight Destructive disease of the potato caused by either of the parasitic fungi *Alternaria solani* (early blight) or *Phytophthora infestans* (late blight). Late blight was responsible for the Irish potato famine of the 1840s.

potato glycoalkaloids A variety of defensive glycoalkaloids produced by potatoes, and other members of the Solanaceae, in response to stress. They are active against pathogens such as viruses, bacteria, fungi and insects. The major alkaloids are **chaconine** and **solanine**, but a range of others are known including: leptinine I & II, leptine I & II, commmersonine, demissine, α-solamargine, α-solasonine, β-solamarine, α-solamarine and α-tomatine. All are likely to be toxic to mammals and many act by inhibition of acetylcholine esterase.

potato lectin Lectin (chitin-binding lectin 1, 323aa) from the potato, *Solanum tuberosum*. Binds to branched or linear N-acetyllactosamine-containing glycosphingolipids and also to lactosylceramide.

potency In toxicology, an expression of relative toxicity of an agent as compared to a given or implied standard or reference, in pharmacology the relationship between the therapeutic effect and the dose.

potentiation (1) Increase in quantal release at a synapse following repetitive stimulation. Whereas **facilitation** at synapses lasts a few hundred milliseconds, potentiation may last minutes to hours. (2) Phenomenon in which a substance or physical agent, at a concentration or dose that does not itself have an effect, enhances the effect or response to another substance or physical agent. Sometimes referred to as 'priming'.

potocytosis Transport of small molecules across membrane using **caveolae** rather than **coated vesicles**.

Potter's syndrome A condition (renal adysplasia, renal agenesis) in which kidney development is abnormal or, in severe cases, absent. In some cases there is mutation in the *ret* protooncogene or in the **uroplakin** IIIA gene.

POU domain A conserved protein domain (~150aa), composed of a 20aa **homeobox** domain and a larger POU-specific domain, found in some transcription factors. Named POU (Pit-Oct-Unc) after 3 such proteins: Pit-1 that regulates expression of certain pituitary genes, Oct-1 and 2, that bind an octamer sequence in the promoters of histone H2A and some immunoglobulin genes, and Unc-86, a transcription factor involved in nematode sensory neuron development. Database entry: http://smart.embl-heidelberg.de/smart/do_annotation.pl?DOMAIN = POU

Poxviridae A family of Class I viruses with an double-stranded DNA genome that codes for more than 30 polypeptides. They are the largest viruses (~200 nm × 300 nm) and the envelope is complex, consisting of many layers, and includes lipids and enzymes, amongst which is a DNA-dependent RNA polymerase. Unlike other DNA viruses they multiply in the cytoplasm of the host cell. Four genera cause human disease: Orthopox (**variola, vaccinia**, cowpox, monkeypox and, until eradicated, smallpox); Parapox: (orf, pseudocowpox, bovine papular stomatitis virus); Yatapox (tanapox, **yaba** virus); Molluscipox (**molluscum contagiosum virus**).

POZ domain A protein-protein interaction domain (poxvirus zinc finger domain, ~120aa) that characterizes a family of transcription factors that have an N-terminal BTB/POZ domain and zinc fingers at their C terminus. Involved in the control of growth arrest and differentiation in several types of mesenchymal cell. See **BTB/POZ domain**.

pp- (1) Prefix, usually to a number, identifying a phosphoprotein. The number is generally the molecular weight in kDa. (2) PP1 is **protein phosphatase-1**. (3) PP2 is a Myb-related protein (421aa) in the moss *Physcomitrella patens*. (4) PP2 is paralytic peptide 2 (23aa) from *Manduca sexta*. (5) pp46: see neuromodulin. (6) pp60 is the phosphoprotein (60 kDa, 526aa) encoded by the **src** oncogene. A protein tyrosine kinase; see src family.

PPAR Nuclear hormone receptor family (peroxisome proliferator-activated receptors) with three known forms, PPARα, γ, δ with two isoforms of PPARγ, PPARγ1 and PPARγ2. PPARα (468aa) and PPARδ (441aa) regulate β-oxidative degradation of fatty acids, PPARγ (505aa) promotes lipid storage by regulating adipocyte differentiation.They are implicated in metabolic disorders predisposing to atherosclerosis and inflammation. PPARα-deficient mice show prolonged response to inflammatory stimuli. There is a transcriptional coactivator for PPAR and for other nuclear receptors, PPAR binding protein (PPARBP, 1581aa). PPARγ coactivator-1 (PGC, 910aa) is a transcriptional coactivator that is involved in various aspects of energy metabolism including regulating the expression of **uncoupling proteins**. Other coactivators have been identified such as oxidative stress-associated Src activator (FAM120A, 1118aa) and the PPARα-interacting cofactor complex (PRIC285, 2649aa).

PPD Protein (purified protein derivative) purified from the culture supernatant of tubercle bacteria (*Mycobacterium tuberculosis*) and used as a test antigen in **Heaf** and **Mantoux tests**.

PPFIA Members of the LAR protein-tyrosine phosphatase-interacting protein (**liprin**) family (PPFIA1, protein tyrosine phosphatase receptor type f polypeptide-interacting protein alpha 1, LAR-interacting protein 1, liprin alpha1, 1202aa). Binds to the intracellular membrane-distal phosphatase domain of tyrosine phosphatase LAR, and appears to localize LAR to cell focal adhesions. This interaction may regulate the disassembly of focal adhesions. Alternatively spliced transcript variants have been described. PPFIA2 (1257aa) is closely related to PPFIA1; expression is downregulated by androgens in a prostate cancer cell line. PPFIA3 (1194aa) and PPFIA4 (701aa) have also been identified.

PPI See **proton pump inhibitor**.

PPIase See **peptidyl-prolyl** *cis-trans* **isomerase**.

PPLO Pleuropneumonia-like organisms. See **mycoplasma**.

PPND A neurological disorder (pallido-ponto-nigral degeneration) caused by defects in **tau protein** in which there are ocular motility abnormalities, dystonia and urinary incontinence, besides progressive parkinsonism and dementia.

P-proteins (1) Proteins found in large amounts in phloem sieve tubes. There are two major proteins: PP1, the phloem filament protein (appears as thin strands when seen in the electron microscope), which contains structural motifs in common with cysteine proteinase inhibitors, and PP2. The latter has lectin activity and RNA-binding properties. PP2 is widely distributed through the vascular plants (even in the absence of PP1) and though it has conserved sequence motifs is polymorphic in size

between species. The PP2 superfamily includes genes encoding enzymes involved in lignin synthesis such as hinokiresinol synthase. See **forisomes**. N.B. PP1 is more commonly **protein phosphatase** 1. Article: http://www.ncbi.nlm.nih.gov/pubmed/12529520 (**2**) A protein (melanocyte-specific transporter protein, 838aa) involved in regulating the pH of melanosomes and defective in a form of oculocutaneous albinism. (**3**) Mitochondrial ribonuclease P proteins (MRPP1, 403aa; MRPP3, 583aa) function in mitochondrial tRNA maturation.

PQ401 Potent inhibitor of insulin-like growth factor I receptor (IGF-IR) signalling and breast cancer cell growth in culture and *in vivo*.

Pr The form of **phytochrome** that absorbs light in the red region (660 nm), and is thus converted to **Pfr**. In the dark the equilibrium between Pr and Pfr favours Pr, which is therefore more abundant.

Prader-Willi syndrome Syndrome in which there is an absence of paternal chromosome 15q11q13, a region containing the imprinted **SNRPN** gene, the **necdin** gene and possibly others. Short stature, obesity and mild mental retardation are features of the syndrome. Uniparental disomy leads to differences between this and **Angelman syndrome** where it is the equivalent maternal region that is deleted. See **genomic imprinting**.

pradimicin A mannose-binding antifungal antibiotic of the angucycline type, isolated from the actinomycete, *Actinomadura hibisca*, that will induce apoptosis-like cell death in *S. cerevisiae* probably as a result of binding to cell-wall mannans. It may also have some anti-viral activity.

pRb Protein product (928aa) of the retinoblastoma gene.

PRC (**1**) See **polycomb repressive complex**. (**2**) Bacterial protease (**Prc protease**). (**3**) Progesterone receptor isoform, PRc. (**4**) Plasma renin concentration. (**5**) See **PRC-barrel domain**.

PRC2/EED-EZH complexes See **polycomb repressive complexes**.

PRC-barrel domain Superfamily of protein domains, the PRC-barrels, approximately 80aa, widely represented in bacteria, archaea and plants. This domain is also present at the carboxyl terminus of the pan-bacterial protein RimM, which is involved in ribosomal maturation and processing of 16S rRNA. Prototype is the PRC-barrel identified in the H subunit of the purple bacterial photosynthetic reaction center (PRC-H).

Prc protease A bacterial peptidase, in *E. coli* (EC 3.4.21.102, tail-specific protease, 682aa) exhibits specificity *in vitro* for proteins with nonpolar carboxyl termini and may be involved in protection of the bacterium from thermal and osmotic stresses.

PRD1-BF1 A transcriptional repressor (positive regulatory domain I-binding factor 1, B lymphocyte-induced maturation protein 1, BLIMP1, 789aa) that is involved in determining whether immature B-cells differentiate into plasma cells rather than memory cells.

p region The centromere divides each chromosome into two regions: the smaller one, which is the p region, and the bigger one, the q region. Loci are identified, for example, as 1p35, (Chromosome 1, p-region, sub-region 35).

PRELP Proline arginine-rich end leucine-rich repeat protein. A connective tissue glycoprotein (PRELP, prolargin, 382aa) of the leucine-rich-repeat (LRR) family with some similarities to **fibromodulin** and **lumican**. Abundantly expressed in juvenile and adult, but not neonatal,cartilage. Binds heparin and heparan sulphate and collagens type I and type II through its leucine-rich repeat domain. May anchor basement membranes to the underlying connective tissue.

pre-pro-protein A pre-protein is a precursor form with a signal sequence that specifies its insertion into or through membranes. A pro-protein is one that is inactive; the full function is only present when an inhibitory sequence has been removed by proteolysis. A pre-pro-protein has both sequences still present and usually only accumulate as products of *in vitro* protein synthesis.

pre-prophase band Band of microtubules 1–3 μm wide that appears just below the plasma membrane of a plant cell before the start of mitosis. The position of the pre-prophase band determines the plane of cytokinesis and of the cell plate that will eventually separate the two cells.

preBCR Receptor (preB cell antigen receptor) on immature B-cells that contains the immunoglobulin mu heavy chain (Ig mu) and signals to the preB cell that heavy chain rearrangement has been successful, a process termed heavy chain selection. In association with the **B-cell receptor** signals through tyrosine kinases including **blk**.

prebiotics Non-digestible food ingredients believed to beneficial to health because they selectively stimulate the growth of beneficial bacteria residing in the gut. Many prebiotics stimulate the growth of bifidobacteria and lactobacilli *in vivo* and specific strains from these genera have been shown to suppress bacterial infections including those caused by ingestion of *Salmonella typhimurium*. This is, however, a contentious field and evidence exists to the contrary. Article: http://www.biomedcentral.com/1471-2180/9/245

precipitin Obsolete term for an antibody that forms a precipitating complex (a precipitin line) with an appropriate multivalent antigen.

prednisone Synthetic steroid (1,4-pregnadiene-17a, 21-diol-3,11,20-trione) that acts as a glucocorticoid, with powerful anti-inflammatory and anti-allergic activity.

pregnanediol A steroid derived from **progesterone**.

pregnenolone A steroid prohormone, synthesized from cholesterol, a precursor for the synthesis of progesterone, mineralocorticoids, glucocorticoids, androgens, and estrogens.

preimplantation genetic diagnosis See **PGD**.

preleptonema Rarely-used term that designates an extra stage in the prophase of meiosis I. Usually included in **leptotene**.

prenylation See **isoprenoids**.

preprophase Rarely used term to designate an extra stage of mitosis, normally included as part of prophase.

preprophase band In higher plants, the band of microtubules (PPB), actin filaments and associated membranous components that forms in the G2 phase of the cell cycle and predicts the plane of cell division. This cortical ring is attached to the plasma membrane and is connected with the nucleus, which usually lies within this division plane. The band disappears by metaphase. In cytokinesis, the phragmoplast expands outwards from between the sister nuclei to insert the cell plate at the cortical site formerly marked by the PPB.

presecretory granules Vesicles near the maturation face of the Golgi. Also known as Golgi condensing vacuoles.

presenilins Multi-pass transmembrane proteins (PS1 and PS2), found in Golgi. Mutations in genes for PS1 (467aa) are associated with 25% of early onset **Alzheimer's disease** and altered amyloid β protein (β-**amyloid precursor protein**) processing. PS1 is a functional homologue of SEL-12 (444aa), a protein found in *C. elegans* that facilitates signalling mediated by **notch**, and the expression of PS1 seems to be essential for the spatio-temporal expression of Notch1 and Dll1 (Delta-like gene 1) during embryogenesis. In *Drosophila*, presenilin homologues are also involved in notch signalling. PS1 and PS2 (448aa) are also similar to *C. elegans* Spe-1, a gene involved in protein trafficking in the Golgi during spermatogenesis. Presenilin enhancer 2 (PEN2, 101aa) is a component of the **gamma-secretase** complex. Presenilin-associated rhomboid-like protein (PARL, 379aa) seems to be involved in suppression of apoptosis in lymphocytes and neurons.

prespore cells Cells in the rear portion of the migrating slug (grex) of a cellular slime mould, which will later differentiate into spore cells. Can be recognized as having different proteins by immuno-cytochemical methods. See also **Acrasidae**.

prestalk cells Cells at the front of the migrating grex of cellular slime moulds that will form the stalk upon which the **sorocarp** containing the spores is borne. See **prespore cells**.

prestin A motor protein (solute carrier family 26 member 5, 744aa), a bidirectional voltage-to-force transducer, that converts auditory stimuli to length changes in the outer cochlear hair cells and mediates sound amplification in the ear. Rapid conformational change in the prestin molecule occurs following anion (bicarbonate and chloride ions) redistribution when membrane ion channels open.

presynaptic cell In a chemical synapse, the cell that releases neurotransmitter that will stimulate the **postsynaptic cell**. In an electrically-synapsed system, the cell that has the first action potential, but since electrical synapses are usually rectifying, one of the two cells involved is always presynaptic.

prezygonema Rare term for an extra stage in the prophase of meiosis I. Usually included in **zygotene**.

PRH See **proline-rich homeodomain**.

Pribnow box See **promoter**.

prickle In *Drosophila* a polarity-determining protein in the planar cell polarity pathway (protein spiny legs, 1299aa). It regulates cell movement through its association with **disheveled**. In humans, prickle-like protein 1 (REST/NRSF-interacting LIM domain protein 1, 831aa) is necessary for nuclear localization of the **REST corepressor** and is mutated in **myoclonic epilepsy of Unverricht and Lundborg 1B**. Prickle 2 (844aa) is coexpressed with prickle 1 in brain, eye, and testis. (NRSF is neuron-restrictive silencer factor.)

prickle cell Large flattened polygonal cells of the stratum germinosum of the epidermis (just above the basal stem cells), that appear in the light microscope to have fine spines projecting from their surfaces; these terminate in desmosomes that link the cells together, and have many tonofilaments of **cytokeratin** within them.

primary cell culture Of animal cells, the cells taken from a tissue source and their progeny grown in culture before subdivision and transfer to a subculture.

primary cell wall A plant cell wall that is still able to expand, permitting cell growth. Growth is normally prevented when a **secondary wall** has formed. Primary cell walls contain more **pectin** than secondary walls, and no lignin is present until a secondary wall has been laid at the plasma membrane. (N.B. The secondary wall is internal, not external, to the primary wall.)

primary cilia Cilia or ciliary rudiments that act as mechanosensory organelles in a range of tissues. By analogy with electrical systems, cilia act as motors, primary cilia as generators. A range of diseases are associated with dysfunction of primary cilia, including **Kartagener's syndrome, polycystic kidney disease, nephronophthisis, Bardet-Biedl syndrome** and **Meckel syndrome**. See **primary ciliary dyskinesia**.

primary ciliary dyskinesia Disorder (immotile cilia syndrome.) in which ciliary function is abnormal as a result of a defect in **dynein** or other axonemal components. Can result in **Kartagener syndrome**, and **situs inversus**.

primary immune response The immune response to the first challenge by a particular antigen. Usually less extensive than the secondary immune response, being slower and shorter-lived with smaller amounts of lower affinity antibody being produced.

primary lateral sclerosis A progressive paralytic neurodegenerative disorder, can be caused by mutations in the gene encoding alsin and an allelic disorder with juvenile **amyotrophic lateral sclerosis-2**.

primary lymphoid tissue See **lymphoid tissues**.

primary lysosome A **lysosome** before it has fused with a vesicle or vacuole.

primary meristem Synonym for an **apical meristem**.

primary oocyte The enlarging ovum before maturity is reached, as opposed to the secondary oocyte or polar body.

primary pigmented nodular adrenocortical disease A rare adrenal defect caused by mutations in PDE11A (dual 3',5'-cyclic-AMP and -GMP phosphodiesterase 11A). The adrenal glands are of normal size but have multiple small yellow-to-dark brown nodules. Causes ACTH-independent **Cushing syndrome**

primary spermatocyte A stage in the differentiation of the male germ cells. Spermatogonia differentiate into primary spermatocytes, showing a considerable increase in size in doing so; primary spermatocytes divide into secondary spermatocytes.

primary structure The lowest level of structural organisation in a macromolecule, in proteins, the amino acid sequence, in nucleic acids the sequence of bases. Secondary structure of a protein is the folding of the peptide chain determined by interactions between amino acids of the chain (into alpha-helical coils or beta-pleated sheets), the tertiary structure is the way in which the helices or sheets are folded or arranged to give the three-dimensional structure of the protein. Quaternary structure refers to the arrangement of protomers in a multimeric protein. Comparable hierarchical levels of organisation are seen in nucleic acids, for example clover-leaf secondary structure in tRNA.

primary transcript RNA transcript immediately after transcription in the nucleus, before **RNA splicing** or polyadenylation to form the mature **mRNA**.

primary tumour The mass of tumour cells at the original site of the neoplastic event - from the primary tumour **metastasis** will lead to the establishment of secondary tumours.

primase See **DNA primase**.

primer extension Technique for finding the transcriptional start site of a gene. **mRNAs** cannot be relied upon to be complete at the 5′ end, so a labelled antisense oligonucleotide **primer** is designed to complement the putative mRNA near its 5′ end, and used to prime a **reverse transcription** reaction. The products are run on a sequencing gel, and the lengths of products allow the putative start sites to be deduced.

priming Treatment that does not in itself elicit a response from a system but that induces a greater capacity to respond to a second stimulus. See **potentiation**.

primitive erythroblast Large cell with euchromatic nucleus found in mammalian embryos. In the mouse, the cells are located in the yolk sac and are responsible for early production of erythrocytes with fetal haemoglobin.

primitive streak Thickened elongated region of cells in early mammalian and avian embryos that marks the location of embryonic axis. Hensen's node is at the end of the primitive streak until the cellular movements of gastrulation cause it to regress caudally.

primordial germ cells Germ cells at the earliest stage of development. Since germ cells may originate in the embryo at some distance from the gonads they then have to migrate to the gonadal primordia, a process that may involve chemotaxis or, more probably, random movement with trapping.

primosome Complex of proteins involved in the synthesis of the RNA primer sequences used in DNA replication. Main components are **primase** and **DNA helicase** that move as a unit with the **replication fork**.

P-ring See **S-ring**.

prions Suggested, and generally accepted, as the causative agents of several infectious diseases (transmissible **spongiform encephalopathies**) such as **scrapie** (in sheep), kuru, **Creutzfeldt-Jakob disease** and **Gerstmann-Straussler-Scheinker syndrome** in man. Prions (proteinaceous infective particles) apparently contain no nucleic acid, only prion protein (PrP, 253aa). Normal PrP (PrPc) may be neuroprotective but is conformationally altered in the encephalopathies to PrPsc. See **doppel** and **shadoo**. A number of yeast prion proteins have been identified, for example **[PSI +]**. Yeast prions: http://www.landesbioscience.com/journals/prion/HinesPRI5-3.pdf

pristane A saturated terpenoid alkane (2,6,10,14-tetramethylpentadecane) extracted from shark liver. Will induce a lupus-like syndrome in non-autoimmune mice and a form of experimental arthritis.

PRK1 See **protein kinase N** and **polo-like kinases**.

Prk1p In *S. cerevisiae* an actin-regulating serine/threonine kinase (EC 2.7.11.1, p53-regulating kinase 1, 810aa) that is localized to cortical actin patches, which may be sites of endocytosis. Involved in regulating the **Sla1p/End3p/Pan1** complex. See **Ark1**.

PRL (1) **Protein tyrosine phosphatases** modified by farnesylation and found in regenerating liver (PRL-1, 173aa; PRL-2, 167aa; PRL-3, 173aa; PTP4A1, PTP4A2, and PTP4A3, respectively). PRLs stimulate progression from G1 into S phase during mitosis and their expression is induced by DNA-damaging stimuli in a p53-dependent manner. They enhance cell proliferation, cell motility and invasive activity, and promote cancer metastasis. http://www.ncbi.nlm.nih.gov/pubmed/18471976 (2) In *Arabidopsis*, protein pleiotropic regulatory locus 1 (PRL1, 486aa) a pleiotropic regulator of glucose, stress and hormone responses and a component of the MAC complex. (3) **Prolactin**.

PRMTs A ubiquitous family of arginine methyltransferases that catalyse the transfer of one or more methyl groups to the guanidino nitrogens of arginine. PMRT1 (Interferon receptor 1-bound protein 4, 361aa). PRMT7 (histone-arginine N-methyltransferase PRMT7, EC 2.1.1.125, **myelin basic protein**-arginine N-methyltransferase PRMT7, EC 2.1.1.126, 692aa) is a component of the 20S methyltransferase complex (the methylosome) that modifies specific arginines to dimethylarginines in several spliceosomal Sm proteins, targeting them to the survival of motor neurons (SMN) complex. Histone methylation is involved in **genomic imprinting**. At least 10 human forms are known, other have been identified in yeasts and higher plants.

PRND See **doppel**.

pro-drug Compound that is pharmacologically inactive (or relatively inactive) but is metabolised to the active form of the drug once in the body.

pro-enzyme Enzyme that does not have full (or any) function until an inhibitory sequence has been removed by limited proteolysis. See also **zymogen**.

pro-opiomelanocortin Polyprotein (POMC, 241aa) produced by the anterior pituitary that is cleaved by serine peptidases to yield **adrenocorticotrophin**, α, β and γ **melanocyte-stimulating hormones**, lipotropic hormones, β-**endorphin**, and other fragments. See **melanocortin**.

PRO40 A WW domain protein (1316aa) that controls fruiting body formation in the ascomycete *Sordaria macrospora*. PRO40 is associated with **Woronin bodies**. See **SO**. Article: http://www.ncbi.nlm.nih.gov/pmc/articles/PMC1899833/

proacrosin Proenzyme (zymogen) of **acrosin** (E.C. 3.4.21.10) activated by zona pellucida glycoproteins into beta-acrosin during the acrosome reaction. Activation involves the removal of a C-terminal segment rich in proline residues and the cleavage of the Arg23-Val24 bond, leading to the formation of the light (23aa) and heavy (301aa) chains.

proadrenomedullin The precursor polypeptide proteolytically cleaved to produce **adrenomedullin** and proadrenomedullin N terminal 20 peptide (PAMP) which is hypotensive, angiogenic and has some antimicrobial capability. The receptor is MrgX2, one of the family of Mas-related genes or sensory neuron-specific G protein-coupled receptors.

proanthocyanidins Class of flavonoid complexes (oligomeric proanthocyanidins; OPCs) found in grape seeds and skin; act as antioxidants, they are the the main precursors of the blue-violet and red pigments in plants. See **anthocyanidin**.

probe General term for a piece of DNA or RNA, corresponding to a gene or sequence of interest, that has been labelled either radioactively, or with some other detectable molecule, such as **biotin**, **digoxygenin** or **fluorescein**. Stretches of DNA or RNA with complementary sequences will hybridize, so probes can be used to label viral **plaques**, bacterial colonies or bands on a gel that contain the gene of interest. See also **Northern blots, Southern blots**.

procambium Plant **meristem** that gives rise to the primary vascular system.

procaryote See **prokaryote**.

procentriole The forming centriole, initially as a ring of 9 singlet microtubules adjacent to the existing centriole at the start of S-phase. In *C. elegans* a serine/threonine kinase, ZYG-1 (Zygote defective protein 1, 709aa) is essential for centrosome duplication. Multiple procentrioles are present in some cells as a structure called the blepharoplast.

procolipase Precursor of **colipase**, the protein cofactor for pancreatic lipase. The N-terminal pentapeptide is cleaved off as **enterostatin**. Procolipase$^{(-/-)}$ knockout mice have a severely reduced fat digestion and fat uptake.

procollagen Triple-helical trimer of collagen molecules in which the terminal extension peptides are linked by disulphide bridges; the terminal peptides are later removed by specific proteases (**procollagen peptidases**) to produce a **tropocollagen** molecule.

procollagen peptidases Enzymes (EC 3.4.24.19) of the peptidase family M12 (astacin family) that remove the terminal extension peptides of **procollagen**; deficiency of these enzymes leads to **dermatosparaxis** or **Ehlers-Danlos syndrome**. Activity is increased by Ca^{2+} and by an enhancer glycoprotein.

proctolin Bioactive neuropeptide (RYLPT) that modulates interneuronal and neuromuscular synaptic transmission in a wide variety of arthropods; orphan G protein-coupled receptor CG6986 of *Drosophila* is probably the proctolin receptor.

prodigiosin See *Serratia marcescens*.

profilaggrin A major protein component (400 kDa) of the **keratohyalin granules** of mammalian epidermis. It contains 10 to 12 tandemly repeated **filaggrin** units and is processed into the intermediate filament-associated protein filaggrin by specific dephosphorylation and proteolysis during terminal differentiation of the epidermal cells. One of the **fused gene family**.

profilin A ubiquitous actin-binding protein (140aa) that forms a complex with G-actin rendering it incompetent to nucleate F-actin formation. The profilin-actin complex seems to interact with inositol phospholipids that may regulate the availability of nucleation-competent G-actin. There are two isoforms in mammals (profilin-1 and -2) and several more in plants. Profilin is the allergen in white birch pollen that causes hay fever-like symptoms.

progenitor cell See **stem cell**.

progeria Accelerated ageing syndrome in which most of the characteristic stages of human senescence are compressed into less than a decade. Hutchinson-Gilford progeria syndrome is caused by mutation in the **lamin** A gene. See **mandibuloacral dysplasia, Werner's syndrome**.

progestagens Hormones (progestogens, gestagens) that produce effects similar to those of the natural hormone **progesterone**. The synthetic forms are sometimes called progestins.

progesterone Hormone (luteohormone, 314 Da) produced in the **corpus luteum**, an antagonist of **estrogens**. Promotes proliferation of uterine mucosa and the implantation of the blastocyst; prevents further follicular development. See Table H3.

programmed cell death A form of cell death, best documented in development, in which activation of the death mechanism requires protein synthesis (see **programmed death proteins**). Morphologically, the cell appears to die by **apoptosis** though this is not necessarily the case. Presumably requires some form of genetic code that determines that certain cells are to die at specific stages and specific sites during development. Classic example is the death of cells in the spaces between the developing digits of vertebrates, thus dividing them.

programmed death proteins Proteins involved in **programmed cell death**. Programmed death-1 (PD-1, PCD1, CD279, 288aa) induces cell death when it binds one of the ligands (programmed cell death 1 ligand 1, CD274, 290aa; PDL2, 182aa), membrane proteins of the immunoglobulin superfamily that are homologues of B7. PD-1 is a T-cell receptor of the B7 family; polymorphism in a regulatory region is associated with susceptibility to systemic **lupus erythematosus**. Programmed cell death protein 2 (PCD2, 344aa) may be a DNA-binding protein. PCD4 (469aa) inhibits translation initiation and cap-dependent translation possibly by hindering the interaction between EIF4A1 and EIF4G. It also inhibits the helicase activity of EIF4A and modulates the activation of JUN kinase. PCD6 (191aa) is a calcium-binding protein required for T-cell receptor-, Fas-, and glucocorticoid-induced cell death. PCD7 (485aa) interacts with RNA-binding protein 40 that is involved in pre-mRNA splicing. PCD10 (PCD10, 212aa) is defective in **cerebral cavernous malformations** Type 3 and seems to be involved in vascular morphogenesis. Various PCD-interacting proteins are known.

progranulin A growth factor (proepithelin, acrogranin, 593aa) expressed in many epithelial cells, macrophages and monocyte-derived dendritic cells and from which **granulins** are derived. Granulins are cysteine-rich polypeptides, some of which have growth modulatory activity. The tertiary structure is similar, but not identical, to that of EGF, although granulins do not bind to EGF receptors. They have cytokine-like activity and may be involved in inflammation, wound repair, and tissue remodeling. Granulin-4 promotes proliferation of A431 epithelial cells and is antagonised by granulin-3. Mutations in the progranulin gene (PGRN) have been shown to cause familial frontotemporal lobar degeneration with ubiquitin-positive inclusions (FTDL-U). Progranulin mediates proteolytic cleavage of **TDP-43**.

progress zone An undifferentiated population of mesenchyme cells beneath the **apical ectodermal ridge** of the chick limb bud from which the successive parts of the limb are laid down in a proximodistal sequence.

progressive external ophthalmoplegia A set of disorders in which there is adult onset of weakness in muscles controlling the eyes. The cause is multiple mitochondrial DNA deletions with defects variously in genes for nuclear-encoded DNA polymerase-gamma, adenine nucleotide translocator, and **twinkle**.

progressive supranuclear palsy See Steele-Richardson-Olszewski syndrome.

prohibitin A highly conserved membrane protein (prohibitin-1, 272aa) with multiple functions in the nucleus and the mitochondria. Prohibitin is involved in mitochondrial biogenesis and function, and is a potential tumour suppressor that represses the activity of E2F transcription factors while enhancing p53-mediated transcription. Prohibitin-2 (299aa) acts as a mediator of transcriptional repression by nuclear hormone receptors via recruitment of histone deacetylases and is an estrogen receptor-selective coregulator. The prohibitin homology domain **PHB domain** (~150aa) is found in various proteins that are linked to lipid rafts including **flotillin, stomatin**, and various bacterial proteins.

prohormone A protein hormone before processing to remove parts of its sequence and thus make it active.

proinsulin See **insulin**

projectin Projectin and **kettin** are **titin**-like proteins mainly responsible for the high passive stiffness of insect indirect flight muscles. Projectin is very large (1 MDa, 6658aa) and has its amino-terminus embedded in the Z-bands of the sarcomere with an adjacent elastic region, possibly the **PEVK**-like domain. A member of the functionally and structurally heterogeneous family of **myosin light chain kinases**, its location is different in synchronous and asynchronous muscles. See **sallimus**.

Prokaryotes Unicellular organisms, **Eubacteria** and **Archaea**, characterized by the possession of one or more simple naked DNA chromosomes, usually circular, without a nuclear membrane and possessing a very small range of organelles. *Cf.* **Eukaryota**.

prokineticins Small cysteine-rich secreted proteins with cytokine-type activity. Prokineticin 1 in humans (PK1, endocrine gland vascular endothelial growth factor, mambakine, 105aa) specifically acts on capillary endothelial cells from endocrine glands and induces proliferation, migration, and fenestration. May be complementary to **VEGF**. Prokineticin 2 (PK2, Bv8, 129aa) may be the output signalling molecule from the suprachiasmatic nucleus regulating circadian rhythms. Both PK1 and PK2 induce contraction of gastrointestinal muscle. Prokineticins are members of a family that includes a nontoxic protein purified from the venom of the black mamba (*Dendroaspis polylepis polylepis*) and **astakine**. The prokineticin receptors are G-protein coupled. Mutations in the PK2 gene cause some forms of **Kallmann's syndrome**.

prolactin Pituitary lactogenic hormone (lactotropin, mammotropin, luteotropic hormone, LTH, luteotropin, 199aa) that promotes milk production from the mammary glands. It is synthesized on ER-bound ribosomes as preprolactin (227aa) with an N-terminal signal peptide. The conversion of preprolactin to prolactin has been much used as an assay for membrane insertion. Prolactin in serum is found in three forms, big-big prolactin (macroprolactin), big prolactin and little prolactin, the latter having had some additional residues removed. Macroprolactin is an inactive complex of prolactin with anti-prolactin autoantibody. The receptor (598aa) is a cytokine-type receptor that signals through the JAK/STAT pathway.

prolactin-releasing peptide A releasing hormone (PrRP) originally reported to act in the anterior lobe of the pituitary gland to stimulate **prolactin** release but subsequently shown to have various other effects. In the central nervous system PrRP (87aa) inhibits food intake, stimulates sympathetic tone, and activates stress hormone secretion. Receptor is G protein-coupled receptor 10 (GPR10).

prolamellar body The disorganized membrane aggregations in chloroplasts that have been deprived of light (**etioplasts**).

prolamine proteins Plant storage proteins that form granules within specialised areas of ER (prolamine protein bodies; PPBs). Prolamine mRNA specifically locates to restricted areas of ER, apparently through a microfilament-based mechanism, and prolamines are retained, even though they lack a lumenal retention signal, by BiP-mediated folding and aggregation into PPBs.

proliferating cell nuclear antigen See PCNA.

proliferative unit A term applied to cells of the epidermis that derive from a single basal cell, a column of cells running outwards from the stem cell.

proliferin A family of prolactin-like hormones (e.g. proliferin-1, prolactin-2C2, mitogen-regulated protein-1, 224aa in mouse), associated with the serum-triggered induction of cell division and involved in angiogenesis of the uterus and placenta and probably fetal growth.

proline One of the 20 amino acids directly coded for in proteins (Pro, P, 115 Da). Structure differs from all the others, in that its side chain is bonded to the nitrogen of the α-amino group, as well as the α-carbon. This makes the amino group a secondary amine, and so proline is described as an imino acid. Has strong influence on secondary structure of proteins and is much more abundant in collagens than in other proteins, occurring especially in the sequence glycine-proline-**hydroxyproline**. A proline-rich region seems to characterize the binding site of **SH3** domains. See Table A1.

proline arginine-rich end leucine-rich repeat protein See PRELP.

proline-rich homeodomain protein A transcriptional repressor (haematopoietically-expressed homeobox protein, HHEX, PRH/Hex, 270aa) that may have a role in haematopoietic differentiation. Acts early in development by enhancing canonical **Wnt**-signalling and later by inhibiting NODAL-signalling. It is thought to form octamers and controls gene expression at both transcriptional and translational levels.

proline-rich polypeptide A peptide isolated from ovine colostrum that has a regulatory effect on the immune response. A nonapeptide fragment (VESYVPLFP) seems to have full activity. Colostrinin$^{(TM)}$ is a complex mixture of proline-rich polypeptides derived from colostrum and claimed to have beneficial effects in Alzheimer's disease.

proline-rich proteins A heterogeneous group (PRPs) that includes collagens, complement 1q, and salivary PRPs, that have unusual tertiary structure and salivary PRPs that may protect agains dietary tannins. Protein domains that bind proline-rich motifs (PRMs) are frequently involved in signalling events. See **proline-rich homeodomain protein**.

prolyl hydroxylases Enzymes that are responsible for post-translational modification of proteins by hydroxylation of proline to **hydroxyproline** with iron at the active site and reducing agent, usually ascorbic acid, as an essential cofactor. A cytoplasmic prolyl 4-hydroxylase (hypoxia-inducible factor prolyl hydroxylase 2, Egl nine homolog 1, 426aa) and a transmembrane form (502aa) catalyze the post-translational formation of 4-hydroxyproline in **hypoxia-inducible factor** alpha proteins. Prolyl 4-hydroxylase-1 (EC 1.14.11.2, leucine- and proline-enriched proteoglycan 1, leprecan-1) is a heterotetrameric enzyme (alpha subunits ca 543aa; the beta subunit is a protein disulphide-isomerase, EC 5.3.4.1) responsible for hydroxyproline formation in procollagen. Prolyl 3-hydroxylase 1 (EC 1.14.11.7, 736aa) is a basement membrane-associated chondroitin sulphate proteoglycan responsible for hydroxylation of proline in basal lamina procollagen. There are various other prolyl hydroxylases.

promastigote Stage in the life cycle of certain trypanosomatid protozoa (e.g. *Leishmania*) that resembles the typical adult form of members of the genus *Leptomonas*. The cell is elongate or pear-shaped with a central nucleus and at the anterior end a **kinetoplast** and a basal body from which arises a single long, slender flagellum.

prometaphase Rarely-used term that designates an extra stage in mitosis, starting with the breakdown of the nuclear envelope. Usually lumped in with **metaphase**.

prominin A stem cell marker, prominin-1 (CD133, 865aa), a pentaspan membrane protein found on membrane protrusions of the apical surface of neuroepithelial cells and on haematopoietic stem cells. It binds cholesterol-rich plasma membrane microdomains and is important in retinal development. There are tissue-specific splice variants of prominin-1; prominin-2 (834aa) has also been identified.

promoter A region of DNA to which RNA polymerase binds before initiating the transcription of DNA into RNA. The nucleotide at which transcription starts is designated +1 and nucleotides are numbered from this with negative numbers indicating upstream nucleotides and positive downstream nucleotides. Most bacterial promoters contain two **consensus sequences** that seem to be essential for the binding of the polymerase. The first, the Pribnow box, is at about -10 and has the consensus sequence 5′-TATAAT-3′. The second, the -35 sequence, is centred about -35 and has the consensus sequence 5′-TTGACA-3′. Most factors that regulate gene transcription do so by binding at or near the promoter and affecting the initiation of transcription. Much less is known about eukaryote promoters; each of the three RNA polymerases has a different promoter. RNA polymerase I recognizes a single promoter for the precursor of rRNA. RNA polymerase II, that transcribes all genes coding for polypeptides, recognizes many thousands of promoters. Most have the Goldberg-Hogness or **TATA box** that is centred around position -25 and has the consensus sequence 5′-TATAAAA-3′. Several promoters have a CAAT box around -90 with the consensus sequence 5′-GGCCAATCT-3′. There is increasing evidence that all promoters for genes for **'housekeeping' proteins** contain multiple copies of a GC-rich element that includes the sequence 5′-GGGCGG-3′. Transcription by polymerase II is also affected by more distant elements known as enhancers. RNA polymerase III synthesizes 5S ribosomal RNA, all tRNAs, and a number of small RNAs. The promoter for RNA polymerase III is located within the gene either as a single sequence, as in the 5S RNA gene, or as two blocks, as in all tRNA genes.

promoter insertion Activation of a gene by the nearby **integration** of a virus. The **long-terminal repeat** acts as a promoter for the host gene. A form of **insertional mutagenesis**.

promoter trap A method used to identify microbial promoters that are active in a specified niche, for instance, during the interaction of a microorganism with its host. The promoter trap consists of a promoterless essential growth factor or antibiotic resistance gene and a linked reporter gene in a vector with a multiple cloning site at the 5′ end. Random DNA fragment are cloned into the multiple cloning site; survival indicates that the promoter is specifically active under the conditions being experienced by the host. Constitutive promoters can be excluded by looking at expression of the reporter gene. The original technique has been modified in various ways and is a powerful approach for functional genomics.

promyelocytes Cells of the bone marrow that derive from myeloblasts and will give rise to myelocytes; precursors of **myeloid cells** and neutrophil granulocytes.

pronase Mixture of proteolytic enzymes from *Streptomyces griseus*. At least four enzymes are present, including trypsin and chymotrypsin-like peptidases.

pronucleus Haploid nucleus resulting from meiosis. In animals the female pronucleus is the nucleus of the ovum before fusion with the male pronucleus. The male pronucleus is the sperm nucleus after it has entered the ovum at fertilization but before fusion with the female pronucleus. In plants the pronuclei are the two male nuclei found in the pollen tube.

properdin Component (Factor P) of the alternative pathway for **complement** activation: complexes with C3b and stabilizes the alternative pathway C3 convertase (C3bBbP) that cleaves C3. Mutation in the properdin gene causes increased susceptibility to infection by *Neisseria* species.

prophage The genome of a **lysogenic** bacteriophage when it is integrated into the chromosome of the host bacterium. The prophage is replicated as part of the host chromosome.

prophase Classical term for the first phase of mitosis or of one of the divisions of meiosis. During this phase the chromosomes condense and become visible in the microscope.

propidium iodide A fluorescent stain (668.4 Da) for DNA and also for detecting dead cells which are permeable and therefore stain. Excitation at 488 nm stimulates fluorescence at 562–588 nm.

Propionibacterium Genus of Gram-positive anaerobic bacteria that will ferment glucose to propionic acid or acetic acid. *P. acnes* contributes to the skin condition, acne.

proplastid Small, colourless **plastid** precursor, capable of division. It can develop into a chloroplast or other form of plastid, and has little internal structure. Found in cambial and other young cells.

propositus See **index case**.

proprotein convertases Peptidases (**kexins**) that process proproteins. Proprotein convertase-1 (PC1, neuroendocrine convertase 1, EC 3.4.21.93, 753aa) and PC2 (EC 3.4.21.94, 638aa) act on **proopiomelanocortin**, renin, enkephalin, dynorphin, somatostatin and insulin. PC4 (proprotein convertase subtilisin/kexin-type 4, PCSK4, 755aa) is involved in processing pro-IGF2 (pro-**insulin-like growth factor II**) to produce the intermediate IGF2 (1−102) that is further cleaved to the active form by another peptidase. PC5 (PCSK5, 913aa) mediates posttranslational endoproteolytic processing of several integrin alpha subunits. The gene encoding the proprotein convertase subtilisin/kexin type 9 (PCSK9, 692aa) is linked to familial hypercholesterolemia and the product is involved in hepatocyte-specific low-density lipoprotein receptor degradation. Proprotein convertase subtilisin/kexin type 1 inhibitor (ProSAAS, 260aa) is cleaved to produce seven different peptides (e.g. big-SAAS, 26aa; big PEN-LEN, 40aa) that act as endogenous inhibitors.

prorenin Inactive precursor of **renin** although it may have an active role; when complexed to the renin receptor it has enzyme activity despite retaining the propeptide. Article: http://ndt.oxfordjournals.org/content/22/5/1288.full

prosaposin A glycoprotein (sphingolipid activator protein, sulphated glycoprotein-1, SGP-1, proactivator polypeptide 524aa) produced in high concentration in the testis, spleen, and brain from which **saposins** are proteolytically cleaved. There are two alternatively-spliced forms, one (65 kDa) destined for lysosomes (independently of the mannose-6-phosphate receptor, probably involving **sortilin**), one (70 kDa) secreted. Inhibitors of sphingolipid biosynthesis, such as **fumonisin B1**, inhibit the trafficking of prosaposin to lysosomes.

Prosite Searchable database of conserved protein domains. Useful in inferring likely function of novel proteins. Link: http://www.expasy.org/prosite/

prosome (1) Obsolete name for a **proteasome**. N.B. the Mondofacto Dictionary definition, plagiarised from earlier editions of this dictionary, is misleading. Letter to Nature: http://www.nature.com/nature/journal/v331/n6152/abs/331192a0.html (2) The anterior region of a copepod comprising the cephalosome (head region) and metasome (thorax) and excluding the urosome (tail).

prospero *Drosophila* gene, product of which is asymmetrically distributed in the division of neural stem cells (neuroblasts) and is not present in one daughter (pluripotent neuroblast) but is retained in the ganglion mother cell (which has more restricted developmental potential). Prospero protein (1703aa), is a homeobox protein, once released from interaction with **miranda**, translocates to the nucleus and causes differential gene expression. The human homologue (737aa) is also involved in the regulation of neuronal development.

prospherosome Proposed stage in the development of **spherosomes** in plant cells. There is an accumulation of lipid in the prospherosome that is mobilized at a later stage.

prostacyclin Unstable **prostaglandin** (PGI2) released by mast cells and endothelium, a potent inhibitor of platelet aggregation; also causes vasodilation and increased vascular permeability. Release is enhanced by **bradykinin**.

prostaglandins *PGs* Group of compounds derived from arachidonic acid by the action of **cyclo-oxygenases** that produces cyclic endoperoxides (PGG2 and PGH2) that can give rise to **prostacyclin** (PGI2) or **thromboxanes** as well as prostaglandins. Were originally purified from prostate (hence the name), but are now known to be ubiquitous in tissues. PGs have a variety of important roles in regulating cellular activities, especially in the inflammatory response where they may act as vasodilators in the vascular system, cause vasoconstriction or vasodilation together with bronchodilation in the lung, and act as hyperalgesics. Prostaglandins are rapidly degraded in the lungs, and will not therefore persist in the circulation. Prostaglandin E2 (PGE2) acts on **adenylate cyclase** to enhance the production of **cyclic AMP**.

prostanoids Collective term for **prostaglandins**, **prostacyclins** and **thromboxanes**: slightly narrower than **eicosanoids**.

prosthetic group A tightly bound non-polypeptide structure required for the activity of an enzyme or other protein, e.g. the **haem** of **haemoglobin**. They differ from coenzymes in being permanently bound to the apoprotein.

protachykinin The protein precursor of several peptide hormones, **substance P** (11aa), neurokinin A (substance K, neuromedin L, 10aa), neuropeptide K (36aa), neuropeptide gamma (19aa) and C-terminal-flanking peptide (16aa). See **tachykinins**.

protamine Highly basic (arginine-rich) protein that replaces **histone** in sperm heads, enabling DNA to pack in an extremely compacted form, e.g. clupein, iridin (4 kDa). Most mammals have only protamine P1 (50aa), but humans, mice and horses also have P2 protamines (57aa). See also **transition proteins**.

protanopia See **colour blindness**.

protease See also **peptidase**, the preferred modern term. The term was normally reserved for endopeptidases that have very broad specificity and would cleave most proteins into small fragments. These are usually the digestive enzymes, e.g. **trypsin**, **pepsin** etc., or enzymes of plant origin (e.g. ficin, papain) or bacterial origin (e.g. pronase, proteinase K). Proteases are widely used for peptide mapping and for structural studies. See Table P1.

protease activated receptors See **thrombin receptors**.

TABLE P1. Peptidases

Family	Feature	Inhibitors	Examples
Aspartic (formerly acid- or carboxyl- proteinases)	Two Asp at active site	Pepstatin	Pepsin
Cysteine	CysSH at active site	Iodoacetate	Papain, caspase 1, cathepsin B, bromelain
Glutamic	Glu136 at active site	1,2-epoxy-3-(*p*-nitrophenoxy)propane	Scytalidoglutamic peptidase, endopeptidases from fungi
Metallo	Metal ion, often zinc	o-phenanthroline, EDTA	See Table M1
Serine	Serine at active site	Organic phosphate esters (DFP, PMSF)	Trypsin, chymotrypsin, thrombin, plasmin, elastase, subtilisin
Threonine	Thr at active site	L-azaserine,	γ-glutamyltransferase
Unknown	Unknown catalytic site	Often unknown	Collagenase from Porphyromonas gingivalis

Proteolytic enzymes, now properly referred to as peptidases, the older names (proteases; proteinases) being deprecated, can be divided into 'mechanistic' sets (families) according to their mode of action. Most inhibitors tend to be specific for one set alone (the important exception being the plasma inhibitor **alpha-2-macroglobulin**). Alternatively, peptidases can be classified simply according to whether they act on terminal amino acids (exopeptidases; aminopeptidases act at the N-terminal, carboxypeptidases at the C-terminus) or on peptide bonds within the chain (endopeptidases). An important web-based classification scheme has recently been developed and should be consulted for more information (the **MEROPS** peptidase database, http://merops.sanger .ac.uk/ This also lists inhibitors). See also **clans of peptidases**.

protease inhibitors Any inhibitor of an enzyme (peptidase) that breaks down proteins, but commonly used as shorthand for drugs, for example, indinavir, that inhibit the action of the protease involved in producing mature virus particles and that are used in combination therapy for AIDS.

protease M See **neurosin**.

protease nexin-1 A serine protease inhibitor (PN-1, glial derived neurite promoting factor, glia-derived nexin, serpinE2, 397aa) that is a physiological inhibitor of thrombin, plasmin, and plasminogen activators, but when associated with glycosaminoglycans mainly inhibits thrombin. It is neuroprotective *in vitro* and highly expressed in a developmentally regulated manner in the nervous system.

proteasome *proteosome* A multimeric complex responsible for degradation of nuclear and cyto-plasmic proteins, found in all eukaryotes and in Archaea. The basic 20S form has 28 protein subunits ($\alpha1-\alpha7$, $\beta1-\beta7$)$_2$ arranged in four stacked rings (alpha-beta-beta-alpha) with the active sites directed to the interior ($\sim$5 nm diameter) channel. The 26S proteasome consists of a 20S core particle and two 19S regulatory caps each consisting of 19 individual proteins with a 10-protein base that binds directly to the alpha ring of the 20S core particle, and a 9-protein lid where polyubiquitin is bound Cells of

the immune system, in response to inflammatory cytokines, particularly interferon gamma, produce a proteasome (the immunoproteasome) with alterna-tive 11S subunits (PA28 complex, 11S regulator) with different substrate preferences, not involving ubiquitin. The 11S complex is composed of 2 homol-ogous subunits called PA28-alpha (Reg1α, 249aa) and -beta (Reg1β, 239aa), which form a hexameric ring. This complex is expressed constitutively in anti-gen-presenting cells. A third subunit, PA28gamma (254aa), is the **Ki antigen**. Nomenclature is complex: see Table P2.

protectin Obsolete name for a GPI-anchored **com-plement** regulatory glycoprotein (CD59) which pre-vents the formation of the membrane attack complex. See the unrelated **protectins**.

protectins A family of compounds derived from omega-3 fatty acids (eicosapentaenoic acid and docosahexaenoic acid), analogous to **lipoxins** and **resol-vins**. Protectin D1 (PD1, neuroprotectin D1, 10,17S-docosatriene) blocks T-cell migration *in vivo*, inhibits TNFα and interferon-gamma secretion, and promotes apoptosis mediated by raft clustering. See **protectin**.

protegrins A family of five porcine **cathelicidins**.

protein A linear polymer of amino acids joined by peptide bonds in a specific sequence.

TABLE P2. Proteasome subunits

20S CORE SUBUNITS

	Synonyms (*S. cerevisiae*)	Human equivalent	Size (kDa)
$\alpha 1$	C7/PRS2	iota	24
$\alpha 2$	Y7	C3	27
$\alpha 3$	Y13	C9	29
$\alpha 4$	PRE6	C6	28
$\alpha 5$	PUP2	zeta	29
$\alpha 6$	PRE5	C2	26
$\alpha 7$	C1/PRS1	C8	31
$\beta 1$	PRE3	LMP2	16
$\beta 2$	PUP1	MECL1	25
$\beta 3$	PUP3	C10	23
$\beta 4$	C11/PRE1	C7	22
$\beta 5$	PRE2	LMP7	23
$\beta 6$	C5/PRS3	C5	25
$\beta 7$	PRE4	N3/beta	26

19S CAP SUBUNITS

	Synonyms (yeast)	Human ortholog	Size (kDa)
Rpt1	Yta3/Cim5	S7/Mss1	52
Rpt2	Yta5	S4	49
Rpt3	Yta2/Ynt1	S6/Tbp7	48
Rpt4	Sug2/Crl13	S10b	49
Rpt5	Yta1	S6/Tbp1	48
Rpt6	Sug1/Cim3	S8/Trip1	50
Rpn1	Hrd2/Nas1	S2-Trap2	109
Rpn2	Sen3	S1	104
Rpn3	Sun2	S3	60
Rpn4	Son1/Ufd5		60
Rpn5			52
Rpn6	NAS5	S9	50
Rpn7		S10	49
Rpn8	NAS3	S12	38
Rpn9	RPN8		46
Rpn10	Mcb1/Sun1	S5a	30
Rpn11	Mpr1	Poh1	34
Rpn12	Nin1	S14	32

The 26S proteasome consists of a 20S protease core that is capped at one or both ends by the 19S regulatory particle. The 20S core particle has two copies each of seven different α and seven different β subunits arranged into four stacked rings ($\alpha_7\beta_7\beta_7\alpha_7$).

In addition there are a number of proteasome interacting proteins which further regulate its function.

Based upon http://biochemie.web.med.uni-muenchen.de/feldmann/proteasome_units.html

protein 4.1 An abundant protein (864aa), originally identified in human erythrocytes, that stabilizes the **spectrin**/actin cytoskeleton. A member of the **ERM** family with characteristic **FERM domain**. Multiple tissue-specific isoforms are generated by alternative pre-mRNA splicing, differential use of two translation initiation sites, and post-translational modifications. Isoforms include erythrocyte 4.1 (4.1R, EPB41), a neuronal form (4.1N, EPB41L1), a widely distributed general form from a different gene (4.1G, EPB41L2), a brain form (4.1B, EPBL3), DAL1, **differentially expressed in adenocarcinoma of the lung**) is another variant. Brain protein 4.1 (**synapsin I**) is the best characterized of the nonerythroid forms of protein 4.1. Mutations in 4.1R can cause hereditary elliptocytosis. Protein 4.1 can be found in nucleoplasm and centrosomes at interphase, in the mitotic spindle during mitosis, in perichromatin during telophase, as well as in the midbody during cytokinesis.

protein 4.2 A major protein (EPB42, 691aa) of the erythrocyte membrane skeleton. It has homology with transglutaminase but has no enzymatic cross-linking activity. Defects cause the Japanese type of recessive spherocytic **elliptocytosis** and some forms of recessive haemolytic anaemia.

protein A Protein (508aa) obtained from *Staphylococcus aureus* that binds the Fc portion of immunoglobulin molecules without interfering with antigen binding. Widely used in purification of immunoglobulins, and in antigen detection, e.g. by **immunoprecipitation**. A very effective B-cell mitogen.

protein B (1) Cell surface protein of Group B streptococci that, like protein A, will bind Fc region of immunoglobulin – but preferentially IgA. (2) A transferrin-binding protein in *Neisseria meningitidis* serogroup B (3) Centromere protein B (CENPB, 599aa) interacts with centromeric heterochromatin and binds to the CENP-B box in centromeric alpha-satellite DNA (alphoid DNA).

protein C Vitamin K-dependent glycoprotein (blood coagulation factor XIV, 461aa) that is the zymogen of a serine endopeptidase (activated protein C, EC 3.4.21.69) found in plasma. Activated protein C (from which a 42aa peptide has been proteolytically removed by thrombin-thrombomodulin), in combination with **protein S**, will hydrolyze blood-clotting Factors Va and VIIIa, thereby inhibiting blood coagulation. Heterozygous mutation in the gene leads to autosomal dominant hereditary thrombophilia; homozygous mutation causes the more severe autosomal recessive form.

Protein Data Bank The PDB archive contains information about experimentally-determined structures of proteins, nucleic acids, and complex assemblies. In May 2012 there were 81,227 structures. See also **Proteopedia**. US: http://www.rcsb.org/pdb, http://www.wwpdb.org/, Europe: http://www.ebi.ac.uk/pdbe/, Japan: http://www.pdbj.org/

protein disulphide isomerase A family of enzymes (EC 5.3.4.1) that rearrange disulphide bonds in proteins but have other roles. For example, in humans protein disulphide isomerase A1 (cellular thyroid hormone-binding protein, 508aa) acts as a chaperone at high concentration and is a subunit of various multi-subunit enzymes such as **prolyl 4-hydroxylase** and **microsomal triacylglycerol transfer protein**. Protein disulphide-isomerase A3 (p58, endoplasmic reticulum resident protein 60, 505aa) is a subunit of the TAP complex (peptide loading complex). Protein disulphide-isomerase A6 (440aa) may function as a chaperone that inhibits aggregation of misfolded proteins. Many enzymes of this class have been identified in plants

protein engineering A colloquial term for the use of recombinant DNA technology to produce proteins with desired modifications in the primary sequence. See **site-specific mutagenesis**.

protein G Protein (60aa) from Group C and G streptococci that binds the Fc portion of IgG; less species-specific than **protein A**.

protein kinase An enzyme that catalyses the transfer of phosphate from **ATP** to hydroxyl side chains on proteins, causing changes in function. Most phosphate on proteins of animal cells is on **serine** residues, less on **threonine**, with a very small amount on **tyrosine** residues. Tyrosine kinases phosphorylate proteins on tyrosine and are important in signalling (see **receptor tyrosine kinases**); serine/threonine kinases phosphorylate serine or threonine. Both PKA and PKG phosphorylate exposed serine or threonine residues near at least two consecutive N-terminal basic residues, with a consensus pattern: [RK](2)-x-[ST]. See **protein kinase A, C, G** and others. Link to PPSearch: http://www.ebi.ac.uk/2can/tutorials/function/PPSearch.html

protein kinase IV A calcium/calmodulin-dependent serine/threonine kinase (brain CaM kinase IV, 503aa) found in brain, T-cells and postmeiotic male germ cells. Present in nucleus where it phosphorylates and activates **CREB** and CREM-tau. See **calspermin** and **reticalmin**. May be important in preventing apoptosis during T-cell development and during activation of T-cells in response to mitogens.

protein kinase A A family of serine/threonine protein kinases (PKA, cAMP-dependent protein kinase, EC 2.7.11.1) whose activity is dependent on the level of **cyclic AMP** (cAMP) in the cell. Consists of two regulatory and two catalytic subunits; cAMP binds to the regulatory subunits causing conformational change and activation of the catalytic subunits. An important regulatory enzyme having pleiotropic effects because of the diversity of substrates.

protein kinase B See **Akt**.

protein kinase C Family of protein serine/threonine kinases (PKCs, EC 2.7.11.1) activated by phospholipids that play an important part in intracellular signalling. The classical PKCs (α, $\beta1$, $\beta2$, γ) are also calcium dependent and can be activated by diacyl glycerol, one of the products of phospholipase C activity or, non-physiologically, by phorbol esters. A growing set of non-classical calcium independent isoforms are known. The catalytic domain is highly conserved and specific properties are conferred by a variety of regulatory domains including a pseudo-substrate region which is displaced upon activation. The specific physiological substrates for these enzymes are not yet well defined. Protein kinase C tends to phosphorylate serine or threonine residues near a C-terminal basic residue, with the consensus pattern: ST-x-RK.

protein kinase G cGMP dependent serine/threonine protein kinase (PKG, 671aa), member of the **AGC kinase** family, that regulates **p21-activated kinase**. PKG is important for relaxation of vascular smooth muscle and inhibition of platelet aggregation. Type I is soluble, type II is membrane-bound.

protein kinase inhibitor peptide *PKI* Endogenous thermostable peptides (76−78aa) that are competitive inhibitors of cAMP-dependent protein kinase (PKA) activity. Distinct PKI isoforms (PKIα, PKIβ, PKIγ have been identified and each isoform is expressed in the brain.

protein kinase M-zeta *PKMζ* The independent catalytic domain of the atypical **PKC** zeta isoform, that is necessary and sufficient for the maintenance of hippocampal **long term potentiation** (LTP) and the persistence of memory in *Drosophila*. PKMζ is expressed in rat forebrain and is apparently important for synaptic plasticity. http://www.ncbi.nlm.nih.gov/pubmed/12857744

protein kinase N Serine/threonine kinases regulated by **rho**-dependent phosphorylation. Kinase activity resides in the C-terminal region (which has high homology with the catalytic domain of PKC) and there is some sequence homology with **rhophilin**. PKN1 (PRK-1, 942aa) is ubiquitous, is activated by phospholipids and arachidonic acid, binds to rho-GTP and possibly regulates cytoskeletal changes. PKN2 (PRK-2, PAK-2, 984aa) is activated by cardiolipin and acidic phospholipids, binds **rac** and rho, which will activate its kinase activity and interacts with SH3 domain of **Nck** and PLCγ. PKN3 (PRK-3, 889aa) contributes to invasiveness in prostatic carcinoma. See **polo-like kinases**.

protein kinase R *PKR* RNA-activated serine/threonine kinase (eukaryotic translation initiation factor 2-alpha kinase 2, 551aa), dsRNA-dependent; expression is induced by interferon. A member of a family of evolutionary conserved dsRNA binding molecules that includes *E. coli* RNase III, *Drosophila* staufen, and *Xenopus* 4F.1. Interaction with dsRNA structures of greater than 35bp causes PKR to autophosphorylate, subsequently to catalyze the phosphorylation of substrate targets and cause inhibition of protein synthesis in the cell. Contains an N-terminal RNA-binding domain and a C-terminal kinase domain. Plays a key role in interferon-mediated host defense against viral infection, and is implicated in cellular transformation and apoptosis.

protein L (1) A protein (719aa) from *Peptostreptococcus magnus* that has an affinity for immunoglobulin kappa light chains (but not lambda chains) from various species. (2) Centromeric protein L is a component of the centromere complex and required for proper kinetochore function.

protein phosphatases Protein phosphatase types 1 (PP1) and 2A (PP2A) represent two major families of serine/threonine protein phosphatases implicated in the regulation of many cellular processes, including cell growth and apoptosis in mammalian cells. Both types are oligomeric complexes comprising a catalytic structure (PP1c or PP2AC) containing the enzymatic activity and at least one more interacting subunit. See **protein tyrosine phosphatases**.

protein S A vitamin K-dependent plasma protein (635aa) that inhibits blood clotting by serving as a nonenzymatic cofactor for activated **protein C** which inactivates coagulation Factors Va and VIIIa. In plasma $\sim$60% is complexed with complement C4b-binding protein and only 40% is active. It binds to phosphatidylserine-positive apoptotic cells in a calcium-dependent manner and facilitates the clearance of early apoptotic cells. Mutation leads to a form of thrombophilia.

protein sequencing There are two major methods, Edman degradation and mass spectroscopy, the latter becoming more popular as instrumentation becomes more readily available. In the Edman degradation method peptides of no more than about 50 residues are affixed to a solid support and the N-terminal sequences sequentially reacted with **Edman reagent**, removed and identified. Larger proteins can be proteolytically fragmented in order to make analysis possible. In mass spectroscopic methods the protein is proteolytically cleaved and small peptides are identified by their characteristic mass, often using **electrospray** methods to introduce them into the spectrometer. With sufficient data, and by comparison with database information on proteins of known sequence, it is possible to work out the full sequence using overlapping peptides from different proteolytic digestions.

protein tyrosine phosphatase *PTP* A subset of the **cysteine phosphatases** (EC 3.1.3.48) that specifically cleave the phosphate from a tyrosine residue in a protein, thus reversing the action of a **tyrosine kinase**. PTPs are divided into four distinct classes: Class I PTPs, are the largest group and include receptor and non-receptor types and dual specificity phosphatases (DSPs) (see **MAPK phosphatases, slingshot, PTEN** and myotubularin. Class II has only one member, low-molecular-weight phosphotyrosine phosphatase (**LMPTP**). Class III contains cdc25 A, B and C and Class IV has four members (See **Eya**). Database: http://ptp.cshl.edu/

protein Z (**1**) A vitamin K-dependent plasma protein (400aa) that promotes the association of thrombin with phospholipid surfaces. (**2**) Major storage protein (**serpin-Z 4**, 399aa) of barley endosperm. (**3**) In many viruses a RING-finger protein (90−100aa) involved in virion assembly and budding and inhibits viral transcription and RNA synthesis by interacting with the viral polymerase L.

protein zero *Pφ* The major **glycoprotein** of peripheral nerve **myelin**, an integral transmembrane protein (248aa), synthesized by **Schwann cells**. Mutations are associated with **Charcot-Marie-Tooth disease type 1, Dejerine-Sottas syndrome** and congenital hypomyelinating neuropathy.

proteinase inhibitors (**1**) A variety of endogenous inhibitors that regulate the activity of peptidases including one that inhibits neutrophil lysosomal elastase and cathepsin G (human seminal plasma inhibitor-I, HUSI-I, 138aa) and in various mucous secretions (secretory leukocyte protease inhibitor). Proteinase inhibitor 8 (374aa) is a **serpin**, an endogenous **furin** inhibitor released from human platelets. See also **alpha-1-antitrypsin, alpha-2-macroglobulin, PIIF**. (**2**) 'Protease inhibitors' have become the casual name for a class of antiviral drugs.

proteinase K A serine peptidase (*Tritirachium* alkaline proteinase, endopeptidase K, EC 3.4.21.64, 384aa) isolated from the filtrate of the fungus *Tritirachium album* (*Engyodontium album*) that will hydrolyze keratin at aromatic and hydrophobic residues. Often used to remove protein contaminants from DNA preparations. Supplier webpage: http://www.worthington-biochem.com/PROK/default.html

proteinase-inhibitor inducing factor See **PIIF**.

proteinoplast *proteoplast* Form of **plastid** adapted as a protein storage organelle; the protein may be crystalline.

Proteobacteria A major bacterial phylum that includes various pathogenic genera but also **myxobacteria**. All are Gram-negative, many are motile. Can be divided into five classes: Alphaproteobacteria (e.g. *Rhodobacter, Rhizobium, Rickettsia*), Betaproteobacteria (e.g. *Neisseria, Spirillum*), Gammaproteobacteria (e.g. *Pseudomonas, Vibrio, Escherichia*), Deltaproteobacteria (e.g.

Bdellovibrio, Myxococcus) and Epsilonproteobacteria (e.g. *Helicobacter*).

proteoglycan A high molecular weight complex of protein and polysaccharide, characteristic of structural tissues of vertebrates, such as bone and **cartilage**, but also present on cell surfaces. Important in determining viscoelastic properties of joints and other structures subject to mechanical deformation. **Glycosaminoglycans** (GAGs), the polysaccharide units in proteoglycans, are polymers of acidic disaccharides containing derivatives of the amino sugars glucosamine or galactosamine.

proteoheparan sulphate A **proteoglycan** containing as its **glycosaminoglycan** heparan sulphate whose constituent N-acetyl glucosamine is often sulphated. Hence highly negatively charged. **Syndecan** is one example.

proteolipid protein (**1**) The major protein (~50%) of the myelin sheath of neurons in the CNS. A highly conserved multi-pass membrane protein. Proteolipid protein 1 (PLP1, lipophilin, 277aa) has an alternatively spliced isoform, DM20, and another form, PLP2 (152aa), has also been described in mouse. Cellular function obscure but mutations lethal e.g. **jimpy** mouse and **Pelizaeus-Merzbacher disease** of man. (**2**) Obsolete term for hydrophobic integral membrane proteins.

proteolysis Cleavage of proteins by peptidases (proteases). Limited proteolysis occurs where proteins are functionally modified (activated in the case of zymogens) by highly specific peptidases.

proteolytic enzyme See **peptidase** and **protease**.

proteome (**1**) All the proteins encoded by the genome of an organism. Though all proteins are coded for within the genome, not all are expressed in every cell and their differential expression, temporally and spatially, is key to understanding how cells and organisms work. Alternative splicing and post-translational modification (e.g. **glycosylation, phosphorylation, prenylation, sumoylation**) or proteolytic processing adds further complexity. (**2**) In a narrower sense, the set of proteins present in a cell at a particular time and under specific environmental conditions.

proteomics The study of **proteomes**, by analogy with genomics.

Proteopedia A free, collaborative 3D encyclopedia of proteins and other molecules. Its stated aim is 'To collect, organize and disseminate structural and functional knowledge about protein, RNA, DNA, and other macromolecules, and their assemblies and interactions with small molecules, in a manner that is relevant and broadly accessible to students and scientists'. Link: http://proteopedia.org/wiki/index.php/Main_Page

proteosome See **proteasome**.

Proteus (1) Genus of highly motile Gram-negative bacteria. They are found largely in soil but are also found in the intestine of humans. They are opportunistic pathogens; *P. mirabilis* is a major cause of urinary tract infections. (2) An urodele amphibian, a blind albino cave dweller with external gills.

Proteus syndrome A highly variable, severe disorder of asymmetric and disproportionate overgrowth of body parts with features that overlap with other overgrowth syndromes. It may be a form of somatic mosaicism but the suggestion that it is associated with **PTEN** mutations seems to have been excluded.

prothallus Independent gametophyte phase of horsetail or fern.

prothrombin See **thrombin**.

prothrombin time *PT* The time until a fibrin clot forms, measured in seconds, after a specific volume of **thromboplastin** reagent is added to the sample of citrated blood plasma. Used in calculating the **international normalised ratio** (INR).

protirelin See **thyrotropic-releasing hormone**.

Protista The kingdom of eukaryotic unicellular organisms. It includes the **Protozoa**, unicellular eukaryotic algae and some fungi (myxomycetes, acrasiales and oomycetes).

proto-oncogene The normal, cellular equivalent of an oncogene; thus usually a gene involved in the signalling or regulation of cell growth. In general, cellular proto-oncogenes are prefixed with a 'c', rather than their abnormal viral counterparts, that are prefixed with a 'v', e.g. c-*myc* and v-*myc*.

protocadherins A large subfamily of the **cadherins**, calcium-dependent cell adhesion molecules found in a range of species and originally thought to be an ancestral form. Protocadherin 1 (PCDH1, cadherin-like protein 1, 1060aa) is highly expressed in the brain and neuro-glial cells. Protocadherin 11 exists in two forms, one encoded on the X-chromosome, (PCDH11X, 1347aa), the other on the homologous region of the Y chromosome (PCDH11Y, 1340aa). Protocadherin 15 (1955aa) is expressed in stereocilia of the inner ear and in retinal photoreceptors. Together with cadherin 23 (**otocadherin**) forms tip links, extracellular filaments that connect the stereocilia and are thought to gate the mechanoelectrical transduction channel. Mutations cause **Usher Syndrome 1F**. Defects in PCDH19 are the cause of female-restricted epilepsy with mental retardation. Many of the protocadherins seem to be associated with the formation of specific neuronal connections in the CNS. In *Drosophila* there is a protocadherin-like wing polarity protein (stan, 3579aa).

protochlorophyllide Precursor of chlorophyll, found in **proplastids** and **etioplasts**. Lacks the phytol side chain of chlorophyll.

protofilaments The longitudinal rows, usually 13, of tubulin heterodimers that form the microtubule. N.B. this is only one way of describing the structure which is a complex polymer: it can also be described as being an imperfect helix with one turn of the helix containing 13 heterodimers.

protogenin A receptor of the **DCC/neogenin** family (protein Shen-Dan, 1150aa). May play a role in anteroposterior axis elongation.

protolignin An immature form of **lignin** that can be extracted from the plant cell wall with ethanol or dioxane.

protolysosome A **primary lysosome**.

protomers Subunits from which a larger structure is built. Thus the tubulin **heterodimer** is the protomer for microtubule assembly, G-actin the protomer for F-actin. Because it avoids the difficulty that arises with, for example, dimers that serve as subunits for assembly, it is a useful term that deserves wider currency.

proton ATPase An ion pump that actively transports hydrogen ions across lipid bilayers in exchange for ATP. Major groups are the **F-type ATPases**, that run in reverse to synthesize ATP in bacterial, mitochondrial and chloroplast membranes ('ATP synthase'); and the **V-type ATPases** found in intracellular vesicles with an acidic lumen, and on certain epithelial cells (e.g. kidney intercalated cells). Gastric H^+/K^+ ATPase is a proton-ATPase.

proton motive force *PMF* The proton gradient across the plasma membrane of a prokaryote or the inner mitochondrial membrane that provides the coupling between oxidation and ATP synthesis. In bacteria used to drive the flagellar motor. See **chemiosmosis**.

protonema A thread-like chain of cells that forms the earliest stage (the haploid phase) of a bryophyte. It grows by apical elongation and eventually gives rise to the more leaf-like gametophore. See **caulonema**.

protonophore **Ionophore** that carries protons. Many **uncoupling agents** are protonophores.

protophloem Primary phloem, the first phloem to be produced; characteristically matures while the organ is elongating.

protoplast A bacterial cell deprived of its cell wall, for example by growth in an isotonic medium in the presence of antibiotics that block synthesis of the wall **peptidoglycan**. Alternatively, a plant cell similarly deprived by enzymic treatment.

protoporphyrin Porphyrin ring structure lacking metal ions. The most abundant is protoporphyrin IX, the immediate precursor of **haem**.

protostome Invertebrate phylum in which the mouth forms from the embryonic blastopore. Major protostome phyla are Annelida, Mollusca and Arthropoda. See **deuterostome**.

prototroph An organism able to grow in unsupplemented medium. *Cf.* **auxotroph**.

protoxylem The first-formed primary xylem with narrow tracheary elements that have annular or helical thickening. Becomes stretched and crushed as the organ elongates.

Protozoa A very diverse group comprising some 50,000 eukaryotic organisms that consist of one cell. Because most of them are motile and heterotrophic, the Protozoa were originally regarded as a phylum of the animal kingdom. However it is now clear that they have only one common characteristic, that they are not multi-cellular, and Protozoa are now usually classed as a Sub-Kingdom of the Kingdom **Protista**. On this classification the Protozoa are grouped into several phyla, the main ones being the Sarcomastigophora (flagellates, heliozoans and amoeboid-like protozoa), the Ciliophora (ciliates) and the Apicomplexa (sporozoan parasites such as *Plasmodium*).

protruding A protein (zinc finger FYVE domain-containing protein-27, ZFYVE27, 404aa) that promotes neurite formation by directed membrane trafficking through interaction with the GDP-bound form of Rab11 when phosphorylated in response to nerve growth factor. Also interacts with **spastin** and the chaperone FKBP38. Mutation is associated with autosomal dominant **hereditary spastic paraplegia** Type 33. Article: http://www.sciencemag.org/cgi/content/full/314/5800/818

provacuoles In plant cells small vacuoles budded directly from the rough endoplasmic reticulum that fuse with other provacuoles to form vacuoles. Since vacuoles may contain hydrolytic enzymes, it is therefore possible to consider them as analogues of primary lysosomes in animal cells.

provirus The genome of a virus when it is integrated into the host cell DNA. In the case of the retroviruses, their RNA genome has first to be transcribed to DNA by **reverse transcriptase**. The genes of the provirus may be transcribed and expressed, or the provirus may be maintained in a latent condition. The integration of **oncogenic** viruses can lead to cell **transformation**.

prozymogen granule See **condensing vacuole**.

PRP (1) Platelet-rich plasma. (2) Proline-rich polypeptide. (3) Proline-rich proteins. (4) Prion proteins. (5) See Prp3, Prp19.

Prp3 A U4/U6-associated splicing factor, (pre-mRNA-splicing factor 3, 683aa).

Prp19 An essential splicing factor (precursor RNA processing-19) and a member of the U-box family of E3 ubiquitin ligases. The Prp19-associated complex consists of at least eight protein components and is involved in **spliceosome** activation by specifying the interaction of U5 and U6 with pre-mRNA. See **Prp3**; do not confuse with **PrP** (prion protein) or proline rich proteins **PRPs**.

PS See **phosphatidyl serine**.

PS2 (1) Presenilin-2. (2) Trefoil factor 1 (pS2/TFF1).

PSA Antigen (prostate specific antigen, semenogelase, EC 3.4.21.77, 261aa) in serum that is used as a marker for prostatic hyperplasia/carcinoma. A serine endopeptidase of the **kallikrein** family. See **PSCA**.

PSCA A cell-surface antigen (prostate stem cell antigen, 123aa) associated with prostatic carcinoma; a member of the Thy-1/Ly-6 family of GPI-anchored surface proteins, expressed primarily in basal cells of normal prostate tissue, suggesting that it is a potential stem-cell marker. Expression is up-regulated in cancer epithelia and is detected in 80% of prostate cancer. See **STEAP**.

PSD-proteins See **postsynaptic protein**.

P-selectin See **selectins**.

pseudoachondroplasia A growth disorder caused by mutation in **cartilage oligomeric matrix protein**.

pseudogenes Non-functional DNA sequences that are very similar to the sequences of known genes. Examples are those found in the β-like globin gene cluster. Some probably result from gene duplications that become non-functional because of the loss of promoters, accumulation of stop codons, mutations that prevent correct processing amongst others. Some pseudogenes contain a **poly-A tail** suggesting that a mRNA, at some point, was copied into DNA that was then integrated into the genome.

pseudohypoaldosteronism Excessive salt secretion in infancy despite high levels of **aldosterone**. Can be caused by mutation in the alpha, beta or gamma subunits of the epithelial sodium channel (ENaC) or in the mineralocorticoid receptor gene. **Hyperkalemia** that occurs despite normal renal glomerular filtration (Gordon hyperkalemia-hypertension) is caused by a defect in WNK1 or WNK4, kinases that are apparently activated by hyperosmotic stress. See **SPAK**.

pseudohypoparathyroidism See **Albright hereditary osteodystrophy**, **granins**.

pseudoknots A topological structure, composed of non-nested double-stranded stems connected by single-stranded loops into which RNA can fold. mRNA pseudoknots have a stimulatory function in programmed ribosomal frameshifting.

Pseudomonas Genus of Gram-negative bacteria. They are rod-shaped and motile, possessing one or more polar **flagella**. Several species produce characteristic water-soluble fluorescent pigments. They are found in soil and water. *P. syringae* is a plant pathogen responsible for superficial frost damage in plants by triggering ice nucleation. It produces a range of phytotoxins (see **coronatine, syringomycin, syringopeptin, tabtoxin,** and **phaseolotoxin**). Activation of phytotoxin synthesis is controlled by diverse environmental factors. *P. tolaasii* is the causal agent of bacterial blotch on cultivated mushrooms. *P. aeruginosa*, normally a soil bacterium, is an opportunistic pathogen of humans who are immunocompromised and is a major cause of **nosocomial infection**. It can also infect the wounds of victims with severe burns, causing the formation of blue pus. *P. alcaligenes* is a Gram-negative aerobic bacterium of the *P. aeruginosa* group that will degrade polycyclic aromatic hydrocarbon and is used in bioremediation.

pseudomurein A major cell wall component of some Archaea that has the same role as bacterial peptidoglycan although is chemically distinct.

Pseudonaja textilis textilis Australian brown snake. See **textilotoxin**.

pseudopod Blunt-ended projection from a cell – a feature of cells that have an amoeboid pattern of movement.

pseudopseudohypoparathyroidism See **Albright hereditary osteodystrophy**.

pseudopterosins Class of natural compounds (diterpene-pentose glycosides) isolated from the soft coral *Pseudopterogorgonia elisabethae*, and that interfere with arachidonic acid metabolism. Have anti-inflammatory and analgesic properties.

pseudospatial gradient sensing Mechanism for sensing a gradient of a diffusible chemical in which the cell sends protrusions out at random; up-gradient protrusions are stabilized by positive feedback (because receptor occupancy is rising with time) and others are transitory because of adaptation. Possibly the mechanism by which neutrophils sense chemotactic gradients.

pseudostratified Describing a simple epithelium (in which all cells are in contact with the basal lamina) that appears multi-layered because some cells have very thin basal strands linking them to the basal lamina. Images of epithelia: http://www.mhhe.com/biosci/ap/histology_mh/simpleep.html

pseudouridine Unusual nucleotide (5-β-D-ribofuranosyluracil) found in some tRNA: glycosidic bond is associated with position 5′ of **uracil**, not position 1′.

pseudovirus Virus-like particle composed of a viral coat protein enclosing an unrelated DNA sequence. Pseudoviruses are potentially useful as a means of delivering DNA into cells for therapeutic purposes or to induce antigen production and thus act as a vaccine.

pseudoxanthoma elasticum A disorder of elastic tissue (Gronblad-Strandberg syndrome) usually caused by mutation in the gene for the ATP-binding cassette C6 protein (ABCC6, 1503aa).

PSGL-1 An extended mucin-like transmembrane glycoprotein (P-selectin glycoprotein ligand, CD162, 412aa) expressed as a homodimer on neutrophils, monocytes and most lymphocytes. See **PADGEM**.

psi (1) In *S. pombe* a protein (379aa) required for nuclear migration during mitosis and for the normal initiation of translation. (2) In *S. cerevisiae*, [PSI$^+$] is a prion protein induced by [PSI$^+$] induction protein 2 (PIN2, 282aa) and is the model for a range of other yeast prion proteins. Yeast prions: http://www.landesbioscience.com/journals/prion/HinesPRI5-3.pdf

Psilotum nudum A spore-producing vascular plant (Whisk fern), an epiphyte occasionally found as a terrestial plant in rocky crevices in sandy soils. Related to ferns but may not be as primitive as formerly thought.

P-site The peptidyl-tRNA binding site on the **ribosome**, the one to which the growing chain is attached; the incoming **aminoacyl-tRNA** attaches to the A-site.

PSM A cell-surface antigen (prostate-specific membrane antigen, 750aa) overexpressed in prostate cancer, a type II transmembrane protein with hydrolase activity and 85% identity to a rat neuropeptidase, also expressed in the small intestine and the brain. May have a role in neuropeptide catabolism in the brain. See **STEAP**.

psoralens Drugs capable of forming photoadducts with nucleic acids if ultraviolet-irradiated.

psoralidin A cytotoxic drug isolated from the medicinal herb, *Psoralea corylifolia* (bakuchi). It inhibits tyrosine phosphatase 1B.

psoriasis Chronic inflammatory skin disease characterized by epidermal hyperplasia. Lesions may be limited or widespread and in the latter case the disease can be life-threatening. Unlike many chronic inflammatory conditions seems to be T-cell mediated. There is a fairly strong genetic predisposition.

psychrophile Organism that grows best at low temperatures.

psychrophilic Growing best at a relatively low temperature. In the case of micro-organisms, having a temperature optimum below 20°C. Such organisms are common in the marine environment.

psyllium Common name for several members of the plant genus *Plantago*, the seeds of which are used commercially for the production of mucilage. Seed husks are an ingredient in high-fibre breakfast cereals that are claimed to have cholesterol-lowering properties.

PTB domain Domain (phosphotyrosine-binding domain, ~100−150aa) that is present in many proteins downstream from receptors that are tyrosine-phosphorylated when they bind ligand (e.g. **shc**, **IRS-1** and **dok proteins**).

P-Tef A yeast expression vector. Datasheet: http://www.dualsystems.com/fileadmin/pdf/Vector%20maps/Yeast_expression_vectors/pTEF-MF.pdf

PTEN A dual-specificity protein phosphatase, dephosphorylating tyrosine-, serine- and threonine-phosphorylated proteins and also acts as a lipid phosphatase, removing phosphate from the 3 position of inositol tris-phosphate (phosphatase and tensin homolog deleted on chromosome ten, mutated in multiple advanced cancers, MMAC1/PTEN, TEP-1, EC 3.1.3.16, EC 3.1.3.48, EC 3.1.3.67, 403aa), a key modulator of the AKT-mTOR signalling pathway and an important tumour suppressor. Somatic mutations in PTEN occur in multiple tumours, most markedly glioblastomas. Germ-line mutations in PTEN are responsible for Cowden disease (CD), a rare autosomal dominant multiple-hamartoma syndrome.

pteridine Nitrogen-containing compound composed of two six-membered rings (pyrazine and pyrimidine rings). Structural component of **folic acid** and **riboflavin** and parent compound of pterins such as xanthopterin, a yellow pigment found in the wings of some butterflies.

Pteridophyta Division of the plant kingdom (the vascular cryptogams) that includes ferns, whisk ferns, horsetails and clubmosses.

Pteridospermae See **seed fern**.

PTGS See **post-transcriptional gene silencing**.

PtK2 cells Cell line from *Potorous tridactylis* (potoroo or kangaroo rat) kidney. Often used in studies on mitosis because there are only a few large chromosomes and the cells remain flattened during mitosis.

PTP See **protein tyrosine phosphatase**.

ptRNA The precursor form of tRNA, the 5′-leader sequence of which is removed by **ribonuclease P** (RNAse P).

PTS (1) **Phosphotransferase** system. (2) Platinum monosulphide (PtS).

PTX (1) Paclitaxel. (**taxol**). (2) **Pertussis toxin**. (3) **Pentoxifylline**. (4) Pectenotoxins, dinoflagellate toxins from *Dinophysis sp.* that accumulate in shellfish such as *Pecten*. (5) **Palytoxin**, a toxin from the dinoflagellate *Ostreopsis sp.* (6) **Picrotoxin**. (7) Homeobox transcription factor, Ptx-2 (see **left-right asymmetry**).

ptyalin Obsolete name for salivary alpha-amylase.

P-type ATPase See **E1-E2-type ATPase** *cf.* **F-type ATPase**, **V-type ATPase**.

P-type calcium channels A subclass of **P/Q voltage-sensitive calcium channels** (CaV2.1, CACNA1A) of the 'high-voltage activated' (HVA) group. Purkinje cells in the cerebellum contain predominantly P-type VSCC (hence P-type), the **Q-type** being responsible for a prominent calcium current in cerebellar granule cells. Activation requires a substantial depolarization and inactivation is slow. The channel is a multi-subunit complex of alpha-1 (2505aa), which forms the pore, alpha-2, beta and delta subunits in a 1:1:1:1 ratio. Mutations are associated with **episodic ataxia-2**, familial hemiplegic **migraine**, **spinocerebellar ataxia-6** and idiopathic generalized **epilepsy**. They are inhibited by funnel toxin and omega-**agatoxin**.

PU box Purine rich sequence 5′-GAGGAA-3′, recognized by transcription factors such as Ets and Sp-1.

PubMed An invaluable service provided by the National Center for Biotechnology Information (NCBI) at the National Library of Medicine (NLM), located at the U.S. National Institutes of Health (NIH). Pubmed provides access to the peer-reviewed biomedical literature and can be searched. For almost all articles the abstract is freely available and in many cases the full text of an article can be accessed via PubMed There are also links to other NCBI databases such as **OMIM**. If it doesn't appear somewhere in PubMed, it probably isn't important. Link: http://www.ncbi.nlm.nih.gov/sites/entrez

pUC9 *E. coli* **phagemid vector**, derived from pUC19 and M13MP9. pUC18 and pUC19 are commonly used plasmid cloning vectors, double-stranded circles, 2686 base pairs in length. pUC18 and pUC19 are identical except that they contain multiple cloning sites (MCS) arranged in opposite orientations. Datasheet: http://vectordb.atcg.com/vectordb/vector_descrip/PUC9.html

Puccinia graminis A basidiomycete fungus that causes stem rust, an economically important disease of cereal crops. Most cereal cultivars have been bred for resistance but a new variant, Ug99, is causing a spreading epidemic.

puckered A *Drosophila* gene that encodes a phosphatase (476aa) that down-regulates the n-Jun-N terminal kinase pathway.

PUFA A fatty acid with multiple unsaturated bonds (a poly-unsaturated fatty acid). Increasing the ratio of unsaturated to saturated fatty acids can alter the behaviour of cells, probably by altering physical characteristics of membranes and thus influencing the behaviour of integral membrane proteins. Whether this is true for whole organisms is less clear.

puffs Expanded areas of a **polytene chromosome**. At these areas the chromatin becomes less condensed and the fibres unwind, though they remain continuous with the fibres in the chromosome axis. A puff usually involves unwinding at a single band, though they can include many bands as in **Balbiani rings**. Puffs represent sites of active RNA transcription. The pattern of puffing observed in the larvae of *Drosophila*, in different cells, and at different times in development provides possibly the best evidence that differentiation is controlled at the level of transcription.

pulchellin A highly toxic type 2 ribosome-inactivating protein (a heterodimeric glycoprotein, an AB ptoxin) isolated from seeds of *Abrus pulchellus tenuiflorus*.

pull-down assay A form of affinity purification in which a 'bait' protein (e.g. either **GST**-tagged so that it can be purified on a glutathione column or directly coupled to a matrix) is used to capture proteins (prey) with which it interacts. Often used to confirm results of **yeast two-hybrid screening** studies.

pullulan An extracellular glucan, produced from starch by *Aureobasidium pullulans*, a chain of maltotriose units linked by alpha-1,4- and alpha-1,6-glucosidic bonds. Used to produce edible films.

pullulanase An enzyme (EC 3.2.1.41, pullulan-6-glucanohydrolase, 1090aa) that will hydrolyse (1−6)-alpha-D-glucosidic linkages in pullulan, amylopectin and glycogen. It is an extracellular, cell surface-anchored lipoprotein produced by Gram-negative bacteria of the genus *Klebsiella*. A similar enzyme is found in chloroplasts where it may be important in starch breakdown and synthesis.

pulse-chase An experimental protocol used to determine cellular pathways, such as precursor-product relationships. A sample (organism, cell or cellular organelle), is exposed for a relatively brief time to a radioactively labelled molecule, the pulse. The labelled compound is then replaced with an excess of the unlabelled molecule, the chase (cold chase). The sample is then examined at various later times to determine the fate or location of radioactivity incorporated during the pulse.

pulse-field electrophoresis A method used for high-resolution electrophoretic separation of very large (megabase) fragments of DNA. Electric fields 100° apart (the angle may vary) are applied to the separation gel alternately. The continuous change of direction prevents the molecules aligning in the electric field and greatly improves resolution on the axis between the two fields.

Puma (1) In humans, a p53 up-regulated modulator of apoptosis, (193aa) one of the Bcl-2 family (see **Bcl-2 homology domains**. (2) In *Parascaris univalens* a 227 kDa spindle- and centromere-associated protein (1955aa) that is involved in the organization of the spindle apparatus and its interaction with the centromeres. (3) In rodents, PUMA-G is a G-protein coupled receptor for nicotinic acid (niacin) and (D)-beta-hydroxybutyrate.

pumilio In *Drosophila* a maternal protein that has sequence-specific RNA-binding properties and regulates translation and mRNA stability. It is required for abdominal development and to support proliferation and self-renewal of germ cells. Homologues in humans (pumilo-1, 1186aa; PUM-2, 1066aa) may be important in stem cell proliferation and self-renewal and regulate microRNA-dependent gene silencing. Pumilio-family RNA binding domains (Puf repeats or Pumilio homology domains) mediate sequence specific RNA binding and usually occur as a tandem repeat of eight domains.

punctin Proteins of the extracellular matrix, deposited in a punctate manner, hence the name. They are **ADAM**-related glycoproteins (punctin 1, ADAMTSL1, 497aa; punctin 2, ADAMTSL3, 1690aa) with thrombospondin repeats, but without the metalloprotease and disintegrin-like domains.

punctuated equilibrium A view of the evolutionary process that holds that there were long periods of stasis interrupted by relatively short periods of rapid change and speciation.

Punnett's square The checkerboard (matrix) method used to determine the types of zygotes produced by fusion of gametes from parents of defined genotype.

puratrophin-1 A protein (Purkinje cell atrophy-associated protein 1, pleckstrin homology domain-containing family G member 4, 1191aa) that may be involved in intracellular signalling and cytoskeletal dynamics in the Golgi apparatus. Defects were initially thought to be the cause of spinocerebellar ataxia-31 but this has subsequently been shown to be incorrect (see **bean**).

purine A heterocyclic compound with a fused pyrimidine/imidazole ring. Planar and aromatic in character. The parent compound for the purine bases of nucleic acids (adenine, guanine, hypoxanthine).

purinergic receptors Receptors for which the ligands are purine nucleotides (e.g. ATP).

Purkinje cell A class of **neuron** in the **cerebellum**; the only neurons that convey signals away from the cerebellum. See **Purkinje fibres**.

Purkinje fibres Specialized cardiac muscle cells that conduct electrical impulses through the heart and are involved in regulating the beat.

puromycin An antibiotic used as an experimental tool that acts as an **aminoacyl tRNA** analogue. Binds to the A-site on the **ribosome**, forms a peptide linkage with the growing chain and then causes premature termination.

purple membrane Plasma membrane of *Halobacterium* and *Halococcus*, that contains a protein-bound carotenoid pigment that absorbs light and uses the energy to translocate protons from the cytoplasm to the exterior. The proton gradient then provides energy for ATP synthesis. The binding protein is called **bacteriorhodopsin**, or purple membrane protein.

purpurin (1) A secretory retinol-binding protein (196aa) in developing chicken retinas. (2) A highly active photodynamic therapy sensitiser, purpurin-18. (3) An anthraquinone constituent from madder (*Rubia tinctorum*) root, reportedly antimutagenic.

pus A viscous fluid, usually yellow-white, around a persistent infected object or as a result of local infection with bacteria that are resistant to phagocytic killing. Mostly formed from dead neutrophils. A greenish colour can be due to myeloperoxidase and blue pus is found in certain infections of *Pseudomonas aeruginosa* that produce the blue pigment, pyocyanin.

pustules Intracellular vacuoles found in **dinoflagellates** arising from ducts that open at the flagellar bases. The function is unknown.

putrescine A dibasic amine associated with putrifying tissue. Associates strongly with DNA. Has been suggested as a growth factor for mammalian cells in culture. Metabolic precursor of the polyamines **spermine** and **spermidine**.

Puumala virus See **nephropathia epidemica**.

PVDF See **polyvinylidene fluoride**.

Pvf *Drosophila* homologues of mammalian growth factors (PDGF- and VEGF-related factors). Pvf1, Pvf2 and Pvf3 are ligands for a single receptor (PVR) encoded by stasis. PVF1/PVR signalling is involved in apoptosis and in some developmental signalling.

PVP Polyvinyl pyrrolidone. Water-soluble white compound that when dissolved makes a very viscous solution.

Pvr (1) A receptor tyrosine kinase (1215aa) in *Drosophila* that is related to platelet-derived growth factor (PDGF) and vascular endothelial growth factor (VEGF) receptors and binds **Pvfs**. (2) In humans, the poliovirus receptor (nectin-like protein 5, CD155, 417aa) that mediates NK cell adhesion and triggers NK cell effector functions

pycnidium *Pl.* pycnidia. An asexual fruiting body produced by mitosporic fungi in the form order Sphaeropsidales (Deuteromycota, Coelomycetes). An internal cavity is lined with **conidiophores** and an opening develops at maturity allowing spores (pycnidiospores) to be dispersed.

pycnodysostosis An osteochondrodysplasia caused by a defect in **cathepsin K**.

PYK2 A calcium-dependent proline-rich **tyrosine kinase** (focal adhesion kinase 2, FAK2, 1009aa), activated by tyrosine phosphorylation and associated with **focal adhesion** proteins. Has been linked to proliferative and migratory responses in a variety of mesenchymal and epithelial cell types, is activated in neurones following NMDA receptor stimulation via PKC and is involved in hippocampal **long-term potentiation**.

pyknosis Contraction of nuclear contents to a deep-staining irregular mass; a sign of cell death.

PyMOL A molecular viewer and 3D molecular editor intended for visualization of 3D chemical structures including atomic resolution X-ray crystal structures of proteins, nucleic acids and carbohydrates, as well as small molecule structures. It is based on open-source software. Link to homepage: http://www.pymol.org/

pyocin-S1 One of the colicin/pyosin nuclease family (killer protein, 618aa) from *Pseudomonas aeruginosa* that kills nonimmune *P. aeruginosa* strains via a specific receptor. It degrades chromosomal DNA and inhibits lipid synthesis in sensitive cells. Article: http://ukpmc.ac.uk/classic/articlerender. cgi?artid = 1264615

pyocyanin Blue-green phenazine pigment produced by *Pseudomonas aeruginosa*; has antibiotic properties.

PYPAFs Proteins (PYRIN-containing apoptotic protease-activating factor 1-like proteins; **NALPs**) involved in inflammatory signalling by regulating NFκB activation and cytokine processing, and implicated in autoimmune and inflammatory disorders.

pyramidal cells Commonest nerve cells of the cerebral cortex.

pyranose Sugar structure in which the carbonyl carbon is condensed with a hydroxyl group (i.e. in a hemi-acetal link), forming a ring of five carbons and one oxygen. Most hexoses exist in this form, although in sucrose, fructose is found with the smaller (four carbon) furanose ring.

pyranoside Compound containing a pyrenose ring of five carbons and one oxygen atom.

pyrenoid Small body found within some chloroplasts, that may contain protein. In green algae may be involved in starch synthesis.

pyridoxal phosphate The coenzyme derivative of vitamin B6. Forms **Schiff's bases** of substrate amino acids during catalysis of transamination, decarboxylation and racemisation reactions.

pyrimidine A heterocyclic 6-membered ring, planar and aromatic in character. The parent compound of the pyrimidine bases of nucleic acid (uracil, thymine and cytosine).

pyrin Protein (marenostrin, 781aa) that probably controls the inflammatory response in myelomonocytic cells at the level of the cytoskeleton. The cytoplasmic isoform is associated with microtubules and with the filamentous actin of perinuclear filaments and peripheral lamellar ruffles. A second isoform is found in the nucleus. Pyrin is defective in **familial Mediterranean fever**. The pyrin domain (PYD, DAPIN domain) is a subfamily of the death domain (DD) superfamily, involved in protein-protein interactions and found in a range of signalling molecules involved in the development of innate immunity against intracellular pathogens through activation of inflammatory mediator pathways. ASC links pathogen recognition by PYD-containing proteins (Caterpiller proteins, **NALPs**, **NOD proteins**, **PYPAFs**) to the activation of downstream effectors. See **cryopyrin**, **FAF1**, **inflammasomes**, **myeloid nuclear differentiation antigen**, **PAAD domain**.

Pyrococcus furiosus An extremophile Archaean with an optimum growth temperature of 100°C. The DNA polymerase (Pfu DNA polymerase) is thermostable and an important reagent in PCR.

pyrogen Substance or agent that produces fever. The major endogenous pyrogen in mammals is **interleukin-1**.

pyroninophilic cells Cells that stain strongly with methyl green pyronin and have bright red cytoplasm indicative of large amounts of RNA, implying very active protein synthesis. **Plasma cells** are very pyroninophilic for example.

pyrophosphate Two phosphate groups linked by esterification. Released in many of the synthetic steps involving nucleotide triphosphates (e.g. protein and nucleic acid elongation). Rapid cleavage by enzymes that have high substrate affinity ensures that these reactions are essentially irreversible.

pyrosequencing A DNA sequencing technique based on the detection of released pyrophosphate (PPi) during DNA synthesis. The unknown sequence is used as a template and a cascade of enzymatic reactions generate light in proportion to the number of incorporated nucleotides. Nucleotides are added separately allowing the sequence to be deduced (only the complementary base will be incorporated, triggering pyrophosphate production and subsequent activation of the luciferase system to produce light.)

Pyrrhophyta An alternative name for **dinoflagellates**.

pyrrole ring A heterocyclic ring structure, found in many important biological pigments and structures that involve an activated metal ion, e.g. chlorophyll, haem.

pyrroline-5-carboxylate reductase The mitochondrial enzyme (P5CR-1, EC 1.5.1.2, 319aa) catalyzes the last step in proline biosynthesis and is defective in **cutis laxa**. The cytoplasmic form (P5CR2, 320aa) is similar.

pyruvate Important 3-carbon intermediate in many metabolic pathways, particularly of glucose metabolism and the synthesis of many amino acids. **Pyruvate carboxylase** catalyses the formation of 4-carbon oxaloacetate from pyruvate, CO_2 and ATP in **gluconeogenesis**. Pyruvate dehydrogenase (PDH, EC 1.2.1.51) uses pyruvate to produce acetylCoA. Phosphorylation of PDH by pyruvate dehydrogenase kinase (PDK, EC 2.7.11.2) results in inactivation and is an important control in energy metabolism. There are multiple tissue-specific PDK isozymes. Pyruvate dehydrogenase phosphatase deficiency can cause lactic acidosis.

pyruvate carboxylase A nuclear-encoded mitochondrial enzyme (EC 6.4.1.1, a homotetramer of 1178aa subunits) enzyme that catalyses the formation of oxaloacetate from pyruvate, CO_2 and ATP in **gluconeogenesis**. It is an important regulatory enzyme in gluconeogenesis, lipogenesis, and neurotransmitter synthesis.

Pythium Genus of cellulose-walled fungi in the Oomycota that are best known as pathogens of crop plants causing *Pythium* blight, damping-off and other seedling diseases, and also progressively destroy the root tips of older plants. A few *Pythium* species are, however, aggressive parasites of other fungi and may be useful as biological control agents.

Q

Q Single letter code for glutamine.

Q banding See **banding patterns** and **quinacrine**.

Q beta *Qβ* An RNA virus (bacteriophage) that infects *E. coli*. Genome circular, single-stranded and acts both as template for replication of a complementary strand and as messenger RNA.

Q box *Q motif* The Q motif is characteristic of and unique to the DEAD box family of helicases and has an invariant glutamine (Q). The consensus is G-F-c-c-P-T-P-I-Q, where c is a charged sidechain residue. Research article: http://www.nature.com/emboj/journal/v23/n13/full/7600272a.html

Q enzyme A rare synonym for 1,4-α-glucan branching enzyme (glycogen-branching enzyme, EC 2.4.1.18, 702aa). See **aminopeptidase Q**.

Q fever Typhus-like illness caused by rickettsia, *Coxiella burneti*. Mainly a disease of domestic animals but can be caught by man.

q region See **p-region**.

Q-FISH Quantitative **fluorescent** *in situ* **hybridization**.

Q-type channels A class of **voltage-sensitive calcium channels** with the pore-forming subunit being of the alpha-1A type (CACNA1A, 2506aa). May be identical or very similar to **P-type channels**. Inhibited by neurotransmitters that act through G-protein-coupled receptors, high concentrations of ω-**conotoxin** and ω-**agatoxin**. Mutations are associated with familial migraine and some forms of ataxia.

Q10 Ratio of the velocity of reaction at one temperature and that at a temperature 10°C lower. Usually around 2 for biological reactions.

QH2-cytochrome c reductase Membrane-bound complex (EC 1.10.2.2, ubiquinol-cytochrome-c reductase) in the mitochondrial inner membrane, responsible for electron transfer from reduced coenzyme Q to cytochrome c. Contains cytochromes b and c1, and iron-sulphur proteins.

qmf1, qmf2, qmf3 Quail homologues of **MyoD**, **myogenin** and **myf-5**, respectively.

QSAR See **structure-activity relationship**.

QT interval An important parameter in analysing cardiac activity, the time that elapses between the start of the Q wave and the end of the T wave in the ventricles of the heart during a single heartbeat. See **long QT**, **short QT**.

QTL analysis Quantitative trait locus analysis. A statistical approach to linking quantitative phenotypic data with genotypic data (e.g. SNPs or minisatellite markers) in an attempt to explain the genetic basis of variation in complex (polygenic) traits. Article: http://www.nature.com/scitable/topicpage/quantitative-trait-locus-qtl-analysis-53904

quail Small galliform bird. Quail embryos are often use in developmental studies because quail cells can be distinguished from chicken cells, yet the two are sufficiently closely related that it is possible to graft embryonic tissue from one to the other. *Coturnix coturnix coturnix* is the European quail.

quaking A family of KH-type RNA-binding proteins (**STAR proteins**) originally identified in the mouse quaking mutant which exhibits body tremor and severe dysmyelination. The murine protein (QK1, 341aa) plays a central role in myelinization. The human homologue is very similar. In *Arabidopsis* there are a number of related proteins with KH domains, but their function is unclear.

quantal mitosis A controversial concept in cellular differentiation proposed by H. Holtzer and defined by him as a mitosis 'that yields daughter cells with metabolic options very different from those of the mother cell as opposed to proliferative mitoses in which the daughter cells are identical to the mother cell'. Implicit in this is the idea that the changes in cell **determination** that occur during development take place at these special quantal mitoses.

quantasome Smallest structural unit of photosynthesis, a particulate component of the **thylakoid** membrane containing chlorophyll and cytochromes.

quantitative character A phenotypic trait displaying continuous variation, therefore likely to be polygenic. See **QTL analysis**.

quantitative structure-activity relationship *QSAR* See **structure-activity relationship**.

quantitative trait loci See **QTL analysis**.

quantum dot A small particle of semiconductor material, typically a few nanometers in diameter; the small size makes quantum mechanical effects more significant than in the (macroscopic) bulk material. Have been used for intravital staining and have the

The Dictionary of Cell and Molecular Biology. DOI: http://dx.doi.org/10.1016/B978-0-12-384931-1.00017-9

advantage that they are available in a wide range of colours.

quantum yield *quantum requirement* A measure of photosynthetic efficiency expressed in moles of photons absorbed per mole of CO_2 fixed or O_2 evolved. Theoretically, φ-max is 0.125, meaning that 8 moles of photons are required to reduce 1 mole of CO_2 in the absence of photorespiration. Research article: http://www.life.illinois.edu/delucia/Singass%20et%20al.pdf

quasi-equivalence Term used to describe the way in which subunits pack into a quasi-crystalline array as, for example, in viral coat assembly. There is usually some strain in the packing.

quaternary structure See **primary structure**.

Quellung reaction Swelling of the capsule surrounding a bacterium as a result of interaction with anticapsular antibody; consequently the capsule becomes more refractile and conspicuous.

quercetin Mutagenic flavonol pigment found in many plants. Inhibits **F-type ATPases**.

queuine tRNA-ribosyl transferase An enzyme (EC 2.4.2.29, guanine insertion enzyme, tRNA-guanine transglycosylase, 403aa) that catalyses the exchange of queuine for guanine in **tRNA**-guanine at the wobble position of tRNAs with GUN anticodons (tRNA-Asp, -Asn, -His and -Tyr), thereby forming the hypermodified nucleoside **queuosine**.

queuosine A modified guanosine derivative (nucleoside Q) found in the first anticodon position of tRNAs for histidine, aspartic acid, asparagine and tyrosine in bacteria and eukaryotes. (The first anticodon in tRNA pairs with the third wobble position of a codon.)

quiescence In cells, the state of not dividing (see **G0**); in neurons, the state of not firing.

quiescent centre A distinct region of mitotically inactive cells in the centre of the root meristem. In *Arabidopsis* the quiescent centre inhibits differentiation of surrounding cells and is a niche of stem cells (pluripotent cells). The transcription factor **SCARECROW** (SCR) is required for specification of the QC.

quiescent stem cell See **stem cell**.

quin2 A fluorescent calcium indicator. Resembles the chelator **EGTA** in ability to bind calcium much more tightly than magnesium. Binding of calcium causes large changes in ultraviolet absorption and fluorescence.

quinacrine A fluorescent dye that intercalates into DNA helices. Chromosomes stained with quinacrine show typical banding patterns of fluorescence at specific locations, **Q bands**, that can be used to recognize chromosomes and their abnormalities.

quinate dehydrogenase A plant enzyme (quinate:NAD oxidoreductase, EC 1.1.1.24, 329aa in *Aspergillus nidulans*) converting hydroquinic acid (from the **shikimic acid pathway**) to quinic acid. The enzyme is activated by a calcium- and calmodulin-dependent phosphorylation.

quinine An alkaloid isolated from cinchona bark. (*Cinchona* is a genus of S American trees and shrubs). Used as an antimalarial drug although resistance is now widespread. It is believed to act by inhibiting the formation of haemozoin from haem so that toxic levels of haem accumulate in the infected erythrocyte.

quinolinic acid One of the end products of the **kynurenine pathway**; thought to play a role in the pathogenesis of several major neuroinflammatory diseases; an endogenous NMDA receptor agonist and neurotoxin.

quinolone antibiotics A class of antibiotics (quinolones and fluoroquinolones) that inhibit the activity of **DNA gyrase**. The older quinolones, nalidixic acid and cinoxacin, are active principally against Enterobacteriaceae but the newer fluoroquinolones have a broader spectrum of activity, for example ciprofloxacin, a fluorinated derivative of nalidixic acid which is more soluble than the parent compound allowing it to reach therapeutic levels in the blood.

quinone reductase Enzymes that reduce quinones to phenols usually using NADH or NADPH as a source of reductant. Quinone reductase 1 (EC 1.6.5.2, 274aa) is involved in detoxification pathways as well as in some biosynthetic processes.

quinupristin A **streptogramin** antibiotic produced by streptomycetes; active agains a broad spectrum of bacteria.

quisqualate An agonist of the Q-type **excitatory amino acid** receptor. See **kainate**.

quorum sensing Phenomenon whereby single bacteria are able to sense the population density of bacteria in the immediate neighbourhood. Depends upon the accumulation of species-specific signalling molecules (e.g. N-acyl-L-homoserine lactones), although species that commonly live together may share the same signalling molecule. Cell density controls specific gene transcription and significant behavioural changes can be triggered when critical population densities are reached.

R

R Single-letter code for **arginine**.

R17 bacteriophage Bacteriophage with RNA genome that codes for **RNA synthetase** and for the coat protein, a protein to which the RNA is attached and that is involved in attachment to the bacterium.

rab genes (1) One of the three main groups of **ras**-like genes specifying a family of small GTP-binding proteins (the others are **ras** and **rho**). Rab proteins are involved in vesicular traffic and seem to control translocation from donor to acceptor membranes (see Table R1). The Rab escort protein-1 (REP1) is component A of RAB geranylgeranyl transferase, a heterodimeric enzyme (A and B subunits) that attaches a geranylgeranyl moiety to rab1A and rab3A: mutations cause **choroideremia**. Rim1-alpha, a Rab3a-interacting molecule, is involved in hippocampal late-phase long-term potentiation. (2) In plants, a gene family 'responsive to abscisic acid' that encode proteins such as dehydrin Rab18 (186aa), although there are also rab-like proteins related to the ras-family.

rabaptin A protein (rab GTPase-binding effector protein 1, rabaptin-5, 862aa) that acts as a linker between gamma-adaptin, rab4A and rab5A and involved in endocytic membrane fusion. See **adaptor proteins**.

rabies virus Species of the **Rhabdoviridae** that causes rabies in humans. The virus infects the cells in the brain, causing a fatal encephalomyelitis.

rabin A human protein (rab-3A-interacting protein, rabin-3, 476aa) that also interacts with SSX2 protein. See **SSX proteins**.

rabphilins Cytosolic proteins expressed in neurons and neuroendocrine cells that bind with high affinity to members of the rab3 family of GTPases when they have GTP bound, and to actin indirectly through alpha-actinin. There is a rab3A binding domain at the amino terminus of the protein and two C2 domains that bind to phospholipids in a Ca^{2+}-dependent manner at the carboxyl terminus. Rabphilin-3A (exophilin-1, 694aa) is probably involved with rab-3A in synaptic vesicle traffic. Rabphilin-3A-like protein (no C2 domains protein, 315aa) has a rab-binding domain and is involved in the late steps of regulated exocytosis, both in endocrine and exocrine cells. It may act as a rab3B effector protein in epithelial cells. Rabphilin-11 (WD repeat-containing protein 44, 913aa) is the downstream effector for rab11. Research article: http://www.jbc.org/content/280/41/34974.full

rac Members of the **ras** superfamily of small GTP-binding proteins, involved in regulating the actin cytoskeleton. The activated form of rac (GTP-rac) seems to induce membrane ruffling (whereas **rho** acts on stress fibres). Rac may be activated by specific **GAPs** such as **bcr** and **N-chimaerin**. Rac1 (p21-Rac1, ras-related C3 botulinum toxin substrate 1, ras-like protein TC25, cell migration-inducing gene 5 protein, 192aa); rac2 and rac3 have similar properties but different tissue distributions with rac2 restricted to haematopoietic tuissue; defects in rac2 are the cause of **neutrophil immunodeficiency syndrome**. See **RAC/ROP small GTPases**.

RAC/ROP small GTPases Rho-family plant-specific ras-related G proteins that interact with a range of receptors and are involved in signalling systems that regulate actin dynamics, production of reactive oxygen species, proteolysis and gene expression. Rac1 (RAC-ROP-like G-protein, 218aa in *Hordeum vulgare*) is only one of a set of similar proteins. Abstract of article: http://dx.doi.org/10.1016/j.tplants.2006.04.003

Rac1 immune complex An innate immune system in rice (*Oryza sativa*) in which the small GTPase, Rac1, plays a key role. Rac1 interacts with RACK1A. (See **RACKs**)

RACE See **rapid amplification of DNA ends**.

raceme A flower cluster arranged in a linear array with the lowest flowers usually opening earliest.

racemic mixture *racemate* A mixture containing equimolar amounts of two enantiomers (D- and L-forms) of a chiral molecule.

rachitic See **rickets**.

RACKs Proteins (receptors for activated C kinase), usually anchored to specific areas of the cell, which selectively bind activated protein kinase C and thus control the regions of the cell on which it acts. They have multiple WD40-repeats. RACK1 (guanine nucleotide-binding protein beta-2-like 1, 317aa) is implicated in numerous signalling pathways. In plants RACK genes are regulators of development and RACK1 proteins regulate innate immunity by interacting with multiple proteins in the Rac1 immune complex in rice.

rad (1) Abbreviation for **radian**. (2) Unit of radiation 1 rad = 0.01 Gy. (3) rad1 is a *S. pombe* checkpoint control gene important in both DNA damage-dependent and replication-dependent cycle control; various rad genes are of comparable function in other

The Dictionary of Cell and Molecular Biology. DOI: http://dx.doi.org/10.1016/B978-0-12-384931-1.00018-0

TABLE R1. Human rab proteins

Protein	size (aa)	Controls
rab1	205	Vesicle traffic from ER to Golgi
rab2	212	Cesicle traffic from ER to Golgi
rab3A	220	Late step in synaptic vesicle fusion
rab3b	219	Vesicles associated with polymeric immunoglobulin receptor
rab4	213	Trafficking of cardiac beta adrenergic receptors
rab5	215	Early endocytic pathway.
rab6	208	Dynein-mediated vesicle movement in Golgi
rab7	207	Vesicles directed to endosomal/lysosomal compartments
rab8	207	Localization of apical proteins in intestinal epithelia
rab9A	201	Transport of mannose 6-phosphate receptors
rab9b	201	Transport between endosomes and trans-Golgi
rab10	200	GLUT4 trafficking to the cell surface
rab11	216	Neurite formation and receptor recycling
rab12	244	Membrane traffic in Sertoli cells
rab13	203	Endocytic recycling of [occludin]
rab14	215	Apical targeting pathway in epithelia
rab15	212	Early endosomal membranes
rab18	206	Apical endocytosis/recycling
rab21	225	Integrin internalization and recycling
rab23	237	Defects cause of acrocephalopolysyndactyly type 2
rab24	203	Autophagy-related processes?
rab25	213	Promotes invasive migration
rab27A	221	Cytotoxic granule exocytosis in lymphocytes
rab27B	218	Targeting of uroplakins?
rab31 (rab22B)	194	Protein transport?
rab32	225	An A-kinase anchoring protein
rab35	201	A fast recycling pathway
rab41 (rab43)	212	Protein transport?

The ras-related rab proteins are involved in vesicular traffic and the number of identified variants is increasing rapidly. This table is not exhaustive.
? indicates that the function is uncertain

organisms (Hrad1 from man, Mrad1 from mouse, RAD17 from *S. cerevisiae*). (4) See **rad proteins**.

rad proteins Yeast proteins coded by genes originally identified as being particularly sensitive to X-rays, many of which are involved in DNA repair and required for spontaneous and induced mitotic recombination, meiotic recombination and mating-type switching. Human homologues of many of these proteins were subsequently identified. Rad21 (double-strand-break repair protein, 631aa) is a component of the **cohesin** complex that is proteolytically cleaved at the metaphase-anaphase transition by **separin**, allowing sister chromatids to separate. Rad23 (398aa in *S. cerevisiae*) is important in nucleotide **excision repair**; the N-terminal domain has similarity with **ubiquitin** and links rad23 to the **proteasome**. It binds to rad14 and TFIIH and forms a stable complex with rad4. Rad51 (400aa in *S. cerevisiae*, 339 in humans) is the functional homologue of **recA** and promotes ATP-dependent homologous pairing and DNA strand exchange; binds Rad52. The role of these proteins in DNA repair means that mutations are associated with tumour susceptibility.

radial cleavage Cleavage pattern, in holoblastic eggs, characteristic of the **deuterostomes**, in which the spindle axes are parallel or at right angles to the polar axis of the oocyte. *Cf.* **spiral cleavage**.

radial glial cell A type of glial cell, organized as parallel fibres joining the inner and outer surfaces of the developing cortex. They are thought to play a role in **neuronal guidance** in development. See **contact guidance**.

radial spoke The structure that links the outer microtubule doublet of the ciliary axoneme with the sheath that surrounds the central pair of microtubules and restricts the sliding of doublets relative to one another; digestion of the radial spokes will allow sliding apart of the doublets. The spokes are arranged periodically along the axoneme every 29 nm, have a stalk about 32 nm long and a bulbous region adjacent to the sheath. Among the 18 spoke proteins (RSPs) identified in *Chlamydomonas* so far, at least 12 have apparent homologues in humans. Many of them have domains associated with signal transduction, suggesting that the spoke stalk is both a scaffold for signalling molecules and itself a transducer of signals.

radiation inactivation An old method in which high energy particles (e.g. electrons) were used to inactivate lyophilised proteins and give an indication of the molecular weight (proportional to the target size).

radicicol A macrocyclic antifungal compound, structurally unrelated to geldanamycin, but that competes for the same binding site on Hsp90. Inhibits src tyrosine kinase.

radioimmunoassay *RIA* Any system for testing antigen-antibody reactions in which use is made of radioactive labelling of antigen or antibody to detect the extent of the reaction. A standard approach is to spike the sample with a known amount of radio-labelled analyte and estimate the concentration of analyte in the sample by measuring how much radio-label is bound. Analyte in the sample competes with the labelled material for binding. A radioimmuno-precipitation assay (RIPA) is used to confirm the presence of antibodies using **protein A** to precipitate immune complexes that have bound added radiolabelled antigenic fragments.

radioisotope Form of a chemical element with unstable neutron number, so that it undergoes spontaneous nuclear disintegration. Major use in biology is to trace the fate of atoms or molecules that follow the same metabolic pathway or enzymic fate as the normal stable isotope, but that can be detected with high sensitivity by their emission of radiation. Also used to locate the position of the radioactive metabolite, as in **autoradiography**, and to measure relative

rates of synthesis of compounds from radioactive precursors.

Radiolaria Subclass of the **Sarcodina**. Marine protozoans with silicaceous exoskeleton and radiating filopodia. Database: http://www.radiolaria.org/

radixin Barbed-end capping actin-binding protein (583aa) found in **adherens junctions** and in the cleavage furrow of many cells. Has a **FERM domain**.

raf Serine/threonine protein kinase implicated in signal-reponse transduction pathways involving tyrosine-kinases. Originally identified as a viral oncogene (v-*raf*) in a murine sarcoma. Raf-1 (c-raf, EC 2.7.11.1, 648aa) is part of the ras-dependent signalling pathway and defects are the cause of one form of **Noonan syndrome**. Various tissue-restricted forms are known, A-raf (606aa) in the urinogenital system, B-raf (766aa) in neural tissue and testis. Defects in B-*raf* are a cause of **cardiofaciocutaneous syndrome**. The avian *mil* oncogene is the avian homologue of murine *raf*.

raffinose *mellitose* A non-reducing trisaccharide found in sugar beet and many seeds, consisting of the disaccharide **sucrose** bearing a D-galactosyl residue linked $\alpha(1\text{-}6)$ to its glucose group.

rag (1) The RAG complex is a multiprotein complex that mediates the DNA cleavage phase during V(D)J recombination in the production of immunoglobulin and T-cell receptor diversity. Rag1 (V(D)J recombination-activating protein 1, 1043aa) has domains with endonuclease activity and with E3 ubiquitin-protein ligase activity. Rag2 (527aa) is another component. RAG1-activating protein 1 (stromal cell protein, 221aa) is thought to activate Rag1. Defects in components of the complex lead to **severe combined immunodeficiency disease** and **Omenn syndrome**. (2) Members of a ras-related GTP-binding protein family (ragA, 313aa; ragB, 374a; ragC, 399aa; ragD, 400aa) that are involved in the RCC1/Ran-GTPase pathway. Bind GTP but do not seem to have GTPase activity.

RAGE (1) Receptor of Advanced Glycation Endproducts, an immunoglobulin superfamily member (237aa) that binds advanced glycation endproducts (**AGE**), **amphoterin**, amyloid β peptide, and members of the **S100** family. May be important in triggering inflammatory response. (2) Renal tumour antigen 1 (RAGE, MAPK/MAK/MRK overlapping kinase, EC 2.7.11.22, 419aa).

RAIDD A dual-domain adapter protein (receptor-interacting protein RIP-associated ICH-1/CED-3-homologous protein with a death domain, death domain-containing protein CRADD, 199aa) that mediates the recruitment of caspase-2 to **tumour necrosis factor** receptor-1 (TNF-R1) signalling complex through **RIP kinase**. May have an additional function in cell differentiation and RAIDD-deficiency may be embryonically lethal.

Raji cells A line of **EBV**-transformed lymphocytes with surface Fc receptors that grows in suspension; derived in 1963 from a Nigerian boy with **Burkitt's lymphoma**. Article: http://www.nature.com/leu/journal/v19/n1/full/2403534a.html; Details: http://www.lgcstandards-atcc.org/attachments/17457.pdf

rak A src-family tyrosine kinase (FRK (fyn-related kinase)/RAK, 505aa) found in nucleus, originally isolated from breast cancer cells. Has N-terminal **SH2** and **SH3** domains and has similarities with src. Binds to **Rb** protein and leads to growth suppression. Formerly GTK (gut tyrosine kinase)/Bsk (beta-cell src-homology kinase) /IYK (intestinal tyrosine kinase).

ral Oncogene related to **ras**. Protein product (ralA, 206aa; ralB 206aa) is a multifunctional small GTPase involved in tumorigenesis and in controlling intracellular membrane trafficking. It is mainly activated by factors downstream of ras.

Raman spectroscopy Method for measuring the Raman spectrum, the plot of Raman scattering of light that produces weak radiation at frequencies not present in the incident radiation. The spectrum is characteristic of the compound and independent of the wavelength of the incident light.

RAM domain (1) An N-terminal region (RBP-Jkappa-associated module) in notch essential for interaction with other proteins in the signalling pathway. (2) A domain associated with transcriptional regulation of amino-acid metabolism in prokaryotes. The RAM domain binds small ligands. Archaea and Bacteria also have proteins with a RAM domain, but lacking the DNA-binding domain, called standalone RAM-domain (SARD) proteins (headless Lrp/AsnC proteins, demi-FFRPs) with uncertain function. Article: http://www.ncbi.nlm.nih.gov/pmc/articles/PMC2242884/

ramoplanin A peptide antibiotic that works by sequestration of lipid intermediates for peptidoglycan biosynthesis, making them unavailable to the late-stage peptidoglycan biosynthesis enzymes. Ramoplanin is structurally related to two cell wall active lipodepsipeptide antibiotics, janiemycin, and enduracidin, and is functionally related to members of the **lantibiotic** class of antimicrobial peptides (mersacidin, actagardine, **nisin**, and epidermin) and glycopeptide antibiotics (**vancomycin** and **teicoplanin**).

RAMPs (1) Type I transmembrane proteins (receptor activity modifying proteins) that determine receptor phenotype of various G-protein-coupled receptors. If the calcitonin receptor-like receptor is transported to the membrane by RAMP1 (148aa) it behaves as a **calcitonin gene related peptide**-receptor; if associated

with RAMP2 (175aa) its glycosylation pattern is different and it acts as an **adrenomedullin**-receptor. RAMP3 (148aa) has similar effects to RAMP1. (2) Retinoic acid-regulated nuclear matrix-associated protein (RAMP, denticleless protein homologue, 730aa) that is a substrate-specific adapter of a DCX (DDB1-CUL4-X-box) E3 ubiquitin-protein ligase complex required for cell cycle control and the DNA damage response.

ramus A physically and physiologically independent individual plant.

ran (1) Small G-protein (216aa) required, together with **importins** α and β and pp15, for protein transport into the nucleus. The only known nucleotide exchange factor for ran is nuclear (RCC1), whereas the only known activating factor is cytoplasmic. This would provide a mechanism for vectorial transport. Ran-GTP binds importin and may cause dissociation of the transport complex. GTP-loaded Ran also induces the assembly of microtubules into aster-like and spindle-like structures in *Xenopus* egg extract. (2) In *Arabidopsis* a copper-transporting ATPase (RAN1, EC 3.6.3.4, RESPONSIVE TO ANTAGONIST 1, 1001aa) involved in copper import into the cell and essential for ethylene signalling. N.B. plants also have ran G-proteins (ran1, 221aa; ran2, 221aa.

ran-binding proteins Proteins that bind the small G protein **ran**. Ran-binding protein 1 (ranbp1, 210aa) inhibits GTP exchange on ran whereas ran-binding protein 2 (358 kDa nucleoporin, 3224aa) is an E3 SUMO-protein ligase which facilitates SUMO1 and SUMO2 conjugation by the conjugating enzyme UBE2I and is involved in transport factor (Ran-GTP, karyopherin)-mediated protein import. Ran-binding protein 3 (567aa) acts as a cofactor for **exportin**-mediated nuclear export. Others, such as ran-binding protein 9 (729aa) and ranbp10 (620aa) act as adapters or are **importins**.

Rana pipiens Common European frog.

ranatensin Subfamily of **bombesin**-like peptides from the skin of *Rana pipiens*, of which **neuromedin**-B (NMB) is the mammalian form. See **phyllolitorin**.

random amplification of polymorphic DNA See **RAPD**.

random coil A term originally invented by polymer chemists to describe a disordered tangle of a linear polymer chain with curved sections. In DNA parlance the random coil refers to the structure that results from melting or other forms of separation of the double helix, ie. helix-coil transition.

random priming Method of labelling a DNA probe for use in hybridization. Double-stranded DNA is denatured to form a single stranded template. Random oligonucleotide primers (usually hexamers) are allowed to anneal, nucleotides and DNA polymerase added, and new DNA fragments synthesized in the presence of trace amounts of radioactive or non-radioactive label. The result is a population of short, labelled DNA molecules of indeterminate length that represent the whole length of the template DNA.

random walk A description of the path followed by a cell or particle when there is no bias in movement. The direction of movement at any instant is not influenced by the direction of travel in the preceding period. If changes of direction are very frequent, then the displacement will be small, unless the speed is very great, and the object will appear to vibrate on the spot. Although the behaviour of moving cells in a uniform environment can be described as a random walk in the long term, this is not true in the short term because of **persistence**.

RANK A TNF superfamily receptor (TNFRSF11A, 616aa) that activates NFκB. Ligand is **RANKL**. See **Paget's disease of bone**.

RANKL Ligand for RANK (receptor activator of NFκB), and the decoy receptor **osteoprotegerin** (OPG); part of the regulatory system for osteoclast development and function.

RANTES Cytokine of the C-C subfamily (regulated upon activation normal T-expressed and secreted, CCL5, 91aa) produced by T-cells, and chemotactic for monocytes, memory T-cells and eosinophils. Uniquely among the **chemokines**, it is down-regulated when the secreting cells are activated.

RAP (1) An oncogene (*rap*) related to *ras* that encodes small G-proteins of the ras family (rap1B, 184aa; rap2A, 183aa) that are key regulators of cell adhesion and cell migration in a number of systems. Rap seems to antagonise **ras** activity. (2) N-RAP (**nebulin**-related anchoring protein, 1175aa) is muscle-specific and concentrated in myofibril precursors during sarcomere assembly and at intercalated disks in adult heart. (3) *Bacillus subtilis* has several Rap proteins that are regulators antagonised by Phr signalling peptides that are imported into the cell. The processes regulated by many of these Rap proteins and Phr peptides are unknown but rapK-phrK regulates the expression of a number of genes activated by the response regulator ComA such as those involved in competence development and the production of several secreted products. (4) In *S. cerevisiae* a DNA binding protein (RAP1, 827aa) homologous to human **TRF** involved in regulating telomere structure. (5) Rhoptry-associated protein 1 (rap1) of *Plasmodium falciparum*, a potential component of an anti-malarial vaccine. (6) Rap2.4 is a redox-sensitive transcription factor that regulates expression of 2-cysteine **peroxiredoxin** in the chloroplast. (7) RAP74 is the large subunit of the transcription initiation factor (**TFIIF**). (8) In *Arabidopsis* a transcription factor (MYC2,

JASMONATE INSENSITIVE 1, 623aa). (9) See **receptor-associated protein** (RAP).

rapamycin Immunosuppressive macrolide antibiotic with structural similarity to **FK506**; inhibits T- and B-cell proliferation but at a much later stage than FK506, despite binding to the same **immunophilin**. Inhibits TOR (target of rapamycin) in the Ras/MAP kinase signalling pathway.

RAPD Random amplification of polymorphic DNA, a variant of the **polymerase chain reaction** used to identify differentially expressed genes. mRNA from two different tissue samples is reverse transcribed, then amplified using short, intentionally nonspecific primers. The array of bands obtained from a series of such amplifications is run on a high resolution gel and compared with analogous arrays from different samples. Any bands unique to single samples are considered to be differentially expressed; they can be purified from the gel, and sequenced and used to clone the full-length cDNA. Similar in aim to **subtractive hybridization**. See also **differential display PCR**.

raphide crystal A needle-shaped crystal, usually of calcium oxalate, found in the vacuole of some plant cells.

raphidosome Rod-shaped particle found in bacterial cells near the DNA-rich region.

rapid amplification of DNA ends *RACE* Techniques, based on the **polymerase chain reaction**, for amplifying either the 5′ end (5′ RACE) or 3′ end (3′ RACE) of a cDNA molecule, given that some of the sequence in the middle is already known. The two procedures differ slightly; in the more straightforward 3′ RACE, first strand cDNA is prepared by reverse transcription with an oligo-dT primer (to match the poly-A tail), from an mRNA population believed to contain the target. PCR then proceeds with a gene-specific, forward-facing primer and an oligo-dT reverse facing primer. 5′ RACE is an example of **anchored PCR**; the first-strand cDNA population is tailed with a known sequence, either by homopolymer tailing (e.g. with dA) or by ligation of a known sequence. PCR then proceeds as before with a primer specific for the gene, and one specific for the added tail.

rapsyn See **postsynaptic protein**.

raptor Binding partner (regulatory associated protein of mTOR, 1335aa) of the mammalian target of rapamycin (mTOR); the raptor-mTOR complex is a key component of a nutrient-sensitive signalling pathway that regulates cell size by controlling the accumulation of cellular mass. The raptor-mTOR complex phosphorylates the rapamycin-sensitive forms of S6K1, while the distinct **rictor**-mTOR complex phosphorylates the rapamycin-resistant mutants of S6K1.

ras One of the family of oncogenes involved in cellular transformation by murine **sarcoma viruses** but acquired when the Harvey virus was passaged through rats where it caused a sarcoma. The gene product, p21ras is a small GTP-binding protein that acts as a regulatory G-protein. See **ras-like GTPases**. The K-*ras*, H-*ras* and N-*ras* genes encode the human cellular homologues of the Kirsten and Harvey murine sarcoma virus oncogenes and the neuroblastoma oncogene respectively. The products (189aa) are functionally similar although antigenically different.

ras-like GTPases Family of small G-proteins (rac, rab, rag, ran, rad, rheb, rho, gem/kir, ric, rin, rit, Ypt). The **rab** subfamily is required for membrane traffic in eukaryotic cells, **ral** has been associated with growth factor-induced DNA synthesis and oncogenic transformation. **Ran** is highly conserved and found in the nucleus. **Rho** and **rac** are involved in cytoskeletal control. Rin, ric and rit lack prenylation sequences and are well conserved between *Drosophila* and man; rin is confined to neuronal cells. Ypts are the yeast homologues, of which 11 are known. The rad subfamily – rad (ras associated with diabetes, 308aa), gem (immediate early gene expressed in mitogen-stimulated T-cells, kir, 296aa) – bind calmodulin in a calcium-dependent manner via a C-terminal extension. and also have various serine phosphorylation sites so their activity may be regulated by kinases including CaMKII, PKA, PKC and CKII. **Rheb** is a ras homologue enriched in brain. The **rag** GTPases are linked to the **mTOR** system, **ral** has been associated with oncogenic transformation. NFκB inhibitor-interacting ras-like protein 1 (192aa) is atypical and acts as a potent regulator of NFκB activity by preventing the degradation of NFκB inhibitor beta.

Rat Genome Database A genomic and genetic database for the rat (*Rattus norweigicus*). Link: http://rgd.mcw.edu/

ratio-imaging fluorescence microscopy A method of measurement of intracellular pH or intracellular calcium levels, using a fluorescent probe molecule (see **fura-2**), in which the two different excitation wavelengths are used, and the emitted light levels compared. If emission at one wavelength is sensitive to the intracellular ion level, and emission at the other wavelength is not, then standardisation for intracellular probe concentration, efficiency of light collection, inactivation of probe and thickness of cytoplasm can all be performed automatically.

Rauber's cells Cells from the caudolateral deep part of the avian blastoderm (Rauber's sickle) or from the thin layer of trophoblast (Rauber's layer) that covers the inner cell mass in the early mammalian embryo.

Rauwolfia serpentina Indian snake-root. Source of various pharmacologically-active compounds, including the alkaloids **reserpine**, serpentine, sarpagine and ajmalicine, that have been used in traditional Ayurvedic medicine.

RAW 264.7 cells Murine monocyte/macrophage line derived from ascitic tumour induced with **Abelson leukaemia virus** in Balb/C mice. Passaging protocol: http://www.signaling-gateway.org/data/cgi-bin/ProtocolFile.cgi/afcs_PP00000159.pdf?pid = PP00000159

Rb Tumour suppressor gene encoding a nuclear protein that, if inactivated, enormously raises the chances of development of cancer, classically **retinoblastoma**, but also other **sarcomas** and **carcinomas**.

rbc Red blood cell or erythrocyte.

RBD (1) Ras-binding domain, originally described in **Raf** kinase. (2) Receptor-binding domain, a rather general term for a domain in a viral protein or a toxin that binds to a receptor on the surface of a target cell.

RBL-1 cells Rat basophilic leukaemia cell line: shows wide variation but can be used as a model for basophils.

R body A protein structure, refractile in the optical microscope, found in various bacteria, probably related to plasmid presence. Found both in free-living pseudomonads and in various bacteria endosymbiotic in *Paramecium*. Has toxic activity against *Paramecium* and confers killer characteristics on *Paramecium* that ingest bacteria containing the structure.

RBS (1) **Rutherford backscattering spectrometry**. (2) The rbs operon encodes the genes responsible for ribose utilization in bacteria. (3) Ribosome binding site.

RC3 See **neurogranin**.

rDNA DNA that codes for ribosomal RNA.

RDP Ribosomal Database Project, a database of aligned and annotated rRNA gene sequences. Database: http://rdp.cme.msu.edu/

RDW (1) Red cell distribution width, a standard parameter in haematology, actually a measure of deviation of the mean corpuscular volume (MCV): (Standard deviation of MCV / mean MCV) 100. (2) See **RDW rats**.

RDW rats Strain of dwarf hypothyroid rats isolated from Wistar-Imamichi rats. There is a missense point mutation of the **thyroglobulin** (*Tg*) gene.

RE1-silencing transcription factor A transcriptional repressor (REST, neuron-restrictive silencer factor, 1069aa) of neuronal genes in non-neuronal tissues that binds to the neuron-restrictive silencer element (NRSE) and acts in conjunction with the **REST corepressor** or the **Sin3 corepressor**. Reported to blocks the expression of an miRNA (miR-21) that prevents embryonic stem cells from reproducing themselves and causes them to differentiate into specific cell types. Article: http://www.molecularstation.com/science-news/2008/03/re1-silencing-transcription-factor-rest-proteins-dual-role/

reaction centre The site in the chloroplast that receives the energy trapped by chlorophyll and accessory pigments, and initiates the electron transfer process.

reactive oxygen species *ROS* Oxygen-containing radicals or reactive ions such as superoxide, singlet oxygen and hydroxyl radicals, the product of the **respiratory burst** in phagocytes and responsible for bacterial killing as well as incidental damage to surrounding tissue.

Reactome A free online curated resource of core biological pathways and reactions developed through collaboration between Cold Spring Harbor Laboratory, The European Bioinformatics Institute, and The Gene Ontology Consortium. Link: http://www.reactome.org/

reading frame One of the three possible ways of reading a nucleotide sequence. As the genetic code is read in non-overlapping triplets (**codons**) there are three possible ways of **translation** of a sequence of nucleotides into a protein, each with a different starting point. For example, given the nucleotide sequence: AGCAGCAGC, the three reading frames are: AGC AGC AGC, GCA GCA, CAG CAG.

readthrough Transcription or translation that continues beyond the normal termination signals in DNA or mRNA respectively.

reagin Reaginic antibodies; an outdated name for IgE.

real time PCR *RT-PCR* A method in which the rate of accumulation of PCR products is measured in real time using a fluorescent marker. The signal increases in direct proportion to the amount of PCR product which allows, from the kinetics, an estimation of the original concentration of the target. Rather misleadingly often called RT-PCR bringing the risk of confusion with reverse-transcriptase PCR.

reannealing Renaturation of a DNA sample that has been dissociated by heating. In reannealing the two strands that recombine to form a double-stranded molecule are from the same source. Differences in the rate of reannealing led to the early recognition of repetitive sequences – which rapidly recombine (have low values on the C_0t **curve**).

reaper Regulator of apoptosis (65aa) in *Drosophila*, acts on the effector caspase, **dredd**. Has no known homology with vertebrate proteins but reaper-induced apoptosis is blocked by **caspase** inhibitors and human **inhibitor of apoptosis proteins** (IAPs). See also **grim** and **hid**.

rebeccamycin Weak topoisomerase I inhibitor, structurally similar to staurosporine but without kinase-inhibitory properties.

rec proteins Proteins encoded by recombination (*rec*) genes and required for genomic repair and recombination in all organisms. RecA (339aa) aligns a single strand of DNA with a duplex DNA and mediates a DNA strand switch; recB, C and D are subunits of the bipolar DNA helicase that employs two single-stranded DNA motors of opposite polarity to drive translocation and unwinding of duplex DNA and thus facilitates loading of **RecA** protein onto ssDNA produced by its helicase/nuclease activity. This process is essential for RecBCD-mediated homologous recombination. As part of the process a loop of single stranded DNA (D-loop) is generated. Rec-8 (human homologue, 547aa) is a **cohesin**. Many other rec proteins are known, often suffixed by a number (e.g. rec104 from *S. cerevisiae*, a potential transcriptional regulator of a number of early meiotic genes (182aa)).

recapitulation The outmoded theory that the stages of development (ontogeny) recapitulated the evolutionary stages through which an organism had passed (phylogeny).

receiver cell Cells in the photosynthetic tissues of plants into which the solutes from xylem are pumped.

receptor In general terms, a relatively large molecule, often membrane-bound, that binds to, or responds to something more mobile (the ligand), with high specificity. Examples: **acetylcholine receptor, photoreceptors, nuclear receptors**. In a few unusual cases both receptor and ligand are membrane bound (e.g. **plexins** and their ligands, **semaphorins**). The term is occasionally misused: Concanavalin A receptors are really ligands for a soluble receptor (lectin).

receptor activator of NFκB *RANK* See **osteoclast differentiation factor**.

receptor downregulation A phenomenon observed in many cells: following stimulation with a ligand the number of receptors on the cell surface for that ligand diminishes because internalisation exceeds replenishment. Often used very loosely, thus destroying the utility of the term.

receptor interacting protein kinases A serine/threonine protein kinase (RIP kinase, RIP1, RIPK1, 671aa) that plays a key role in TNFα-induced IκB kinase (**IKK**) activation and subsequent activation of transcription factor NFκB. The **death domain** (DD) of RIP interacts with the DD of

TRADD (TNFR1-associated death domain protein) in two different ways: one that subsequently recruits **CRADD** (apoptosis/inflammation) and another that recruits NFκB (survival/proliferation). Multiple isoforms (RIP1, 2, 3, & 4) have been described. See **RAIDD** and RIP-like kinase (**RICK**).

receptor interacting protein-1 A name conferred on a disparate set of proteins and thus extrememely ambiguous, even when prefixed to indicate which is meant. (**1**) NRIP1 (nuclear receptor-interacting protein 1, RIP140, 1158aa) associates with a transcriptional activation domain of steroid receptors. (See **N-CoR**). Nuclear receptor coactivator 2 is **glucocorticoid receptor-interacting protein 1** (GRIP1) but this is also ambiguous. (**2**) **GRIPs** (glutamate receptor-interacting proteins, e.g. GRIP1, 1128aa) are scaffold proteins for neurotransmitter receptors. (**3**) Other RIP1s interact specifically with the C-terminal tail of the angiotensin II type 2 receptor (angiotensin II type 2 receptor-interacting protein 1, ATIP1) or with the leucine-rich repeat Ig domain-containing **Nogo** receptor, or the cannabinoid receptor 1 (CRIP-1, 164aa). (**4**) TRIP1 is eukaryotic translation initiation factor 3 subunit 2 (325aa) or TGFβ receptor interacting protein 1 in *Clonorchis sinensis* (TRIP1, 327aa). TRIP1 (406aa) is the thyroid-hormone receptor associated protein 1 (TRIP1, 406aa). (**5**) Another RIP1 is part of the NFκB signalling network. (**6**) See **receptor interacting protein kinases**.

receptor potential The transmembrane potential difference of a sensory cell. Such cells are not excitable (do not generate an action potential), but their response to stimulation is a gradual change in their **resting potential**.

receptor protein tyrosine phosphatase *RPTP* A family of **protein tyrosine phosphatases** (receptor-type tyrosine-protein phosphatases, RPTPs, EC 3.1.3.48).

receptor tyrosine kinases See **tyrosine kinases**.

receptor-associated protein See **low density lipoprotein receptor-related protein**. See also **rap** and **TRAFs**.

receptor-mediated endocytosis Endocytosis of molecules by means of a specific receptor protein that normally resides in a coated pit, but may enter this structure after complex formation occurs. The structure then forms a coated vesicle that delivers its contents to the endosome whence it may enter the cytoplasm or the lysosomal compartment. Many bacterial toxins and viruses enter cells by this route.

receptors for activated C Kinase See **RACKs**.

receptosome Synonym for **endosome**.

recessive An **allele** or **mutation** that is only expressed phenotypically when it is present in the homozygous form. In the heterozygote it is obscured by dominant alleles.

recombinant DNA Spliced DNA formed from two or more different sources that have been cleaved by **restriction enzymes** and joined by **ligases**.

recombinant protein Protein product from a gene that has been cloned and introduced into an appropriate expression system.

recombinase Enzymes that mediate **site-specific recombination** in prokaryotes. See **resolvases**, **tyrosine recombinase retroelements** and **flp- recombinase**.

recombination The creation, by a process of intermolecular exchange, of chromosomes combining genetic information from different sources, typically two genomes of a given species. Site-specific, homologous, transpositional and non-homologous (illegitimate) types of recombination are known. Recombination can be intragenic, between two alleles of a gene (**cistron**), or intergenic where there is information exchange between non-allelic genes. See **site-specific recombination** and **homologous recombination**.

recombination nodule Transient structures (about 90 nm diameter) associated with the **synaptonemal complex** during pachytene. The number of nodules corresponds to the number of genetic exchanges and so it seems probable that they are the site of recombination events.

recon Unit of genetic **recombination**, the smallest section of a chromosome that is capable of recombination. Until intragenic recombination was shown it was thought to represent a gene, it may be no more than a single base pair. More details: http://www.microbiologyprocedure.com/genetics/fine-structure-of-gene/modern-definition-of-gene.htm

recoverin Calcium-binding protein (cancer-associated retinopathy protein, p26, 200aa) containing 4 **EF-hand** motifs that inhibits rhodopsin kinase and prevents premature phosphorylation of rhodopsin until the opening of cGMP-gated ion channels causes a decrease in intracellular calcium levels, signalling completion of the light response. At one time was wrongly thought to be an activator of photoreceptor guanylate cyclase. When it binds calcium ions a conformational change causes a myristoyl group to be displaced from a binding pocket thus altering its physicochemical properties and allowing interaction with other proteins. Found in serum of patients with cancer-associated retinopathy. A member of a family of neuronal calcium sensors that includes **visinin**, **hippocalcin**, **neurocalcin**, S-modulin, visinin-like protein, and **frequenin**. Entry in signalling gateway database: http://www.signaling-gateway.org/molecule/query?afcsid = A004096

RECQ helicases In *E. coli* an ATP-dependent DNA helicase (recQ, EC3.6.4.12, 609aa) involved in **mismatch repair**. Five human homologues, RECQ-like helicases, are known: the RECQL2 gene is defective in Werner's syndrome, RECQL3 gene in **Bloom's syndrome** and RECQL4 in **Rothmund-Thomson syndrome**. Mutations in RECQL1 and RECQL5 are not associated with known disorders.

rectifying synapse An **electrical synapse** at which current flow can only occur in one direction.

red blood cell See **erythrocyte**.

red chlorophyll catabolite reductase The constitutively expressed chloroplast enzyme (RCCR, EC 1.3.1.80, accelerated cell death protein 2, 319aa) that degrades primary fluorescent catabolite produced by the action of **pheophorbide a oxygenase** on pheophorbide in the linked series of reactions involved in chlorophyll breakdown. (See **chlorophyll**).

red drop effect Experimental observation that the photosynthetic efficiency of monochromatic light is greatly reduced above 680 nm, even though chlorophyll absorbs well up to 700 nm. Led to the discovery of the two light reactions of photosynthesis; see **photosystems I and II**.

redox potential The reducing/oxidising power (oxidation/reduction potential) of a system measured by the potential at a hydrogen electrode.

Reed-Sternberg cell Giant histiocytic cells, a common feature of **Hodgkin's disease**.

reeler An autosomal recessive mouse mutant, with ataxia and a reeling gait. The defect is in **reelin** which leads to major developmental defects in neuronal architecture in the brain.

reelin Protein product (3461aa) of mouse **reeler** gene, an extracellular matrix serine peptidase produced by pioneer neurons and important in cortical neuronal migration. Regulates microtubule function in neurons and modulates cell adhesion by enzymic activity. It is secreted by Cajal-Retzius cells in the forebrain and by granule neurons in the cerebellum. The human homologue (3460aa) is deficient in a form of lissencephaly.

REEP Receptor expression-enhancing protein. A multipass membrane protein (e.g. REEP1, 201aa; REEP2, 252aa) that promotes functional expression of G-protein coupled odorant receptors on the cell surface. Mutations of REEP1 are associated with a form of **spastic paraplegia** (SPG31).

refractile Adjective usually used in describing granules within cells that scatter (refract) light. Not to be confused with refractory.

refractory period Most commonly used in reference to the interval (typically 1 ms) after the passage of an **action potential** during which an axon is incapable of responding to another. This is caused by inactivation of the sodium channels after opening. The maximum frequency at which neurons can fire is thus limited to a few hundred Hertz. An analogous refractory period occurs in individuals of *Dictyostelium discoideum*, which are insensitive to extracellular cyclic AMP immediately after a pulse of cAMP has been secreted. The term can be applied to any system where a similar insensitive period follows stimulation.

Refsum's disease A rare recessive lipid disorder in which there is accumulation of the branched chain fatty acid, phytanic acid, because of deficient alpha-oxidation of (14)C-phytanic acid to pristanic acid, the normal mechanism of degradation. Clinical features include retinitis pigmentosa, chronic polyneuropathy, and cerebellar signs. The adult form can be caused by mutation in either the gene encoding phytanoyl-CoA hydroxylase or that encoding **peroxin**-7. The infantile form is caused by mutation in peroxins 1, 2 or 3.

REG (1) A family of small secreted C-lectins implicated in regeneration, proliferation, and differentiation of the pancreas, liver, and gastrointestinal mucosa. Reg1 (lithostathine, pancreatic stone protein, REG1A, REG1B, 166aa) is encoded by the *reg1* gene (regenerating gene 1) and may be associated with neuronal sprouting in brain, and with brain and pancreas regeneration. Others (REG2, 3, 4) are associated with other tissues and may be stress proteins. (2) In *S. cerevisiae* REG2 (338aa) is a regulatory subunit that binds to type-1 protein phosphatase. (3) Components of the 11S **proteasome** activator that stimulates peptidase activity and enhances the processing of antigens for presentation. REGα (PA28α, 249aa) forms a heptamer in solution with a 2–3 nm cone-shaped pore, but will preferentially form a heteromeric complex with REGβ (PA28β, 239aa) and REGγ (PA28γ, Ki antigen, 254aa). (4) See **flotillins**.

regeneration Processes of repair or replacement of missing structures.

regulatory sequence DNA sequence (control element) to which regulatory molecules such as **promotors** or **enhancers** bind, thereby altering the expression of the adjacent gene.

regulatory T-cell See **T-regulatory cells**.

regulon A situation in which two or more spatially separated genes are regulated in a coordinated fashion by a common regulator molecule. The term is also applied to sets of coexpressed genes in eukayotes.

Reifenstein's syndrome Partial androgen insensitivity, a type of male pseudohermaphroditism caused by mutation in the androgen receptor gene on the X-chromosome.

Reina elements Plant Ty3/Gypsy LTR retrotransposons of the chromovirus type.

rel Protein product of the *rel* oncogene of avian reticuloendotheliosis virus that forms homo- or heterodimeric complexes with other rel-homology domain (RHD)-containing proteins (RELA/p65, RELB, NFκB1/p105, NFκB1/p50, REL and NFκB2/p52). The dimers bind at kappa-B sites in the DNA of their target genes and different dimer combinations act as transcriptional activators or repressors. v-rel (503aa) is a truncated and mutated form of c-rel (619aa) and transforms cells by increasing the expression of genes regulated by rel/NFκB proteins and inhibits apoptosis.

rel homology domain Domain (RHD, ~300aa) found in a family of eukaryotic transcription factors, including NFκB, Dorsal, Relish, NFAT, among others. Phosphorylation of the domain may regulate its activity. Has two immunoglobulin-like beta-barrel subdomains that grip the DNA in the major groove. See **rel**.

relaxase Part of the **relaxosome** nucleoprotein complex that facilitates bacterial plasmid conjugation. Mobilization protein A (mobA, DNA strand transferase, 709aa) has two domains: DNA relaxase (EC 5.99.1.2, DNA nickase) and DNA primase (EC 2.7.7.-) responsible for locally unwinding DNA and cleaving one of the DNA strands. MobA forms a complex with mobB (148aa) and mobC (94aa).

relaxation time Time taken for a system to return to the resting or ground state or a new equilibrium state following perturbation. Often used in context of receptor systems that have a **refractory period** after responding and then relax to a competent state. Can be used more precisely to mean the time for a system to change from its original equilibrium value to 1/e of this original value. In magnetic resonance imaging the spin-lattice and spin-spin relaxation times are important time constants.

relaxin Polypeptide hormone structurally similarity to **insulin**, produced as a prohormone (prorelaxin-H2, 185aa) by the corpus luteum and found in the blood of pregnant animals. Acts to cause muscle relaxation during parturition, but is a multi-functional endocrine and paracrine factor that is important in several organs, including the normal and diseased cardiovascular system. Among other things it regulates matrix metalloproteinases (MMPs) and thus the properties of extracellular matrix. Human relaxin has an A chain of 24aa and a B chain of 29aa linked by two disulphide bonds. Prorelaxin-H1 is restricted to the prostate. Relaxin-3 may play a role in neuropeptide signalling processes. The receptors are G-protein-coupled. See **relaxin-like factor**.

relaxin-like factor A hormone (insulin-like peptide 3, 131aa) produced by **Leydig cells** that has a role in regulating the onset of puberty in males and in control of the normal positioning of the gonads in the male fetus. It binds to the **relaxin** receptor.

relaxosome Complex multi-subunit structure forming at the plasmid origin of replication during bacterial conjugation. The relaxosome introduces a site- and strand-specific nick from which the physical transfer of a single strand of DNA is initiated. Some components are encoded by the tra genes but there is one host-encoded protein, integration host factor. TraI is a DNA helicase I (EC 3.6.4.12, 1756aa), TraJ (228aa) regulates the expression of transfer genes, TraK lipoprotein (242aa) is involved in pilus formation, TraY (131aa) is required for strand-specific nicking at oriT, the transfer origin. There are other components involved in the process. See relaxase. Research article: http://www.jbc.org/content/270/47/28381.full

relay cell An interneurone of the central nervous system, particularly one linking afferent and efferent neurones of a reflex arc.

release factor (1) Proteins that recognise termination codons and trigger the release of the polypeptide from the ribosome-mRNA complex. In *E. coli* there are three release factors (peptide chain release factors 1, 2 and 3; 360aa, 365aa and 529aa). RF-1 is specific for UAG/UAA codons and RF-2 for UGA/UAA; RF-3 is not codon-specific. In eukaryotes the release factor (eukaryotic peptide chain release factor, ERF1, 437aa) forms a heterodimer with a GTP-binding subunit (ERF3B, 628aa). It is also a component of the mRNA surveillance **SURF complex**. Mitochondrial release factors are in the same family as the prokaryotic ones. (2) Releasing factors are a set of hormones secreted by the hypothalamus that stimulate the pituitary gland to release other hormones.

RELT One of the TNF receptor superfamily (receptor expressed in lymphoid tissues, TNFRSF19L, 430aa). The ligand is not one of the known TNF superfamily.

renal cell carcinoma The commonest kidney tumour, arising from proximal tubule cells.

renal-coloboma syndrome *papillorenal syndrome* An autosomal dominant disorder characterized by both ocular and renal anomalies caused by heterozygous mutation in the **PAX2** gene.

renaturation The conversion of denatured protein or DNA to its native configuration. This rarely occurs for most proteins. DNA renaturation is also termed **reannealing**.

Renilla reniformis A soft coral, the sea pansy, source of Renilla luciferase (36 kDa monomeric protein that does not require post-translational modification for activity. Often included in vectors as a reporter gene. Therefore, like firefly luciferase, the enyzme may function as a genetic reporter immediately following translation and is often used as an internal control reporter in combination with experimental reporter vectors. Technical bulletin: http://www.promega.com/tbs/tb550/tb550.html

renin An **aspartic peptidase** (angiotensinogenase, EC 3.4.23.15, 406aa) released from the walls of afferent arterioles in the kidney when blood flow is reduced, plasma sodium levels drop, or plasma volume diminishes. Catalyses splitting of **angiotensin** I from **angiotensinogen**, an α2-globulin of plasma. Renin inhibitors are used to treat hypertension. The renin/prorenin receptor (350aa) is expressed as a membrane protein in various tissues and may mediate renin-dependent cellular responses by activating ERK1 and ERK2.

rennet A preparation containing the enzyme chymosin (rennin) that cleaves κ-casein and causes milk to curdle. Used in the production of cheese; originally prepared from the mucous membrane of the stomach of calves but now generally made by engineered bacteria.

Reoviridae A family of viruses (Class III), with a segmented double-stranded RNA genome. Each of the 8–10 segments encodes a different polypeptide and only one strand of the RNA (minus strand) acts as template for **mRNA** (plus strand). They have an icosahedral capsid, and the **virion** includes all the enzymes needed to synthesize mRNA. The viruses originally included in this group (Respiratory, Enteric, Orphan viruses) did not seem to cause any disease in humans but several pathogenic viruses are now classed as reoviruses including **orbivirus**, a tick-borne virus that causes Colorado tick fever, and **rotavirus**. Virus database: http://www.ncbi.nlm.nih.gov/ICTVdb/ICTVdB/00.060.htm

repair nucleases Class of enzymes involved in DNA repair. It includes **endonucleases** that recognize a site of damage or an incorrect base pairing and cut it out, and exonucleases that remove neighbouring nucleotides on one strand. These are then replaced by a **DNA polymerase**. See **cernunnos, damaged DNA binding proteins, DNA ligase, DNA polymerase, excision repair, mismatch repair, NHEJ** (non-homologous end-joining) and the **SOS system**.

repetin Protein (784aa), product of one of the **fused gene family**, involved in cornified cell envelope formation. Contains EF hands of the S100 type and internal tandem repeats typical of cell envelope precursor proteins.

repetitive DNA Nucleotide sequences in DNA that are present in the genome as numerous copies. Originally identified by the value on the C_0t **curve** derived from kinetic studies of DNA renaturation. These sequences are not thought to code for polypeptides. One class of repetitive DNA, termed highly repetitive DNA, is found as short sequences, 5–100 nucleotides, repeated thousands of times in a single long stretch. It typically comprises 3–10% of the genomic DNA and is predominantly **satellite DNA**. Another class, which comprises 25–40% of the DNA and termed moderately repetitive DNA, usually consists of sequences about 150–300 nucleotides in length dispersed evenly throughout the genome, and includes **Alu** sequences and **transposons**.

replica methods Methods in the preparation of specimens for transmission electron microscopy. The specimen (for example, a piece of **freeze fractured** tissue) is shadowed with metal and coated with carbon and then the tissue is digested away. The replica is then picked up on a grid and it is the replica that is examined in the microscope.

replica plating Technique for testing the genetic characteristics of bacterial colonies. A dilute suspension of bacteria is first spread, in a petri dish, on agar containing a medium expected to support the growth of all bacteria, the master plate. Each bacterial cell in the suspension is expected to give rise to a colony. A sterile velvet pad, the same size as the petri dish, is then pressed onto it, picking up a sample of each colony. The bacteria can then be 'stamped' onto new sterile petri dishes, plates, in the identical arrangement. The media in the new plates can be made up to lack specific nutritional requirements or to contain antibiotics. Thus colonies can be identified that cannot grow without specific nutrients or that are antibiotic-resistant, and cells with mutations in particular genes can be isolated.

replicase Generic (and rather unhelpful) term for an enzyme that duplicates a polynucleotide sequence (either RNA or DNA). The term is more usefully restricted to the enzyme involved in the replication of certain viral RNA molecules (e.g. Replicase polyprotein 1a, 4382aa of human SARS coronavirus which is cleaved into 11 peptides).

replication Copying, but usually the production of daughter strands of nucleic acid from the parental template.

replication factors Proteins involved in eukaryotic DNA replication. Replication factor A (replication protein A) is a heterotrimeric single-stranded DNA-binding protein (subunits p70, 616aa; p32, 270aa; p14, 121aa) that associates, together with the kinase **ATM**, at sites where homologous regions of DNA interact during meiotic prophase and at breaks associated with meiotic recombination after **synapsis**. It is essential for replication of SV40 virus.

Replication factor C (activator 1) in *S. cerevisiae* has five-subunits (RFC1, 861aa; RFC2, 353aa; RFC3, 340aa; RFC4, 323aa; RFC5, 354aa.) and is an ATP-dependent clamp-loader important for assembling the components of the DNA replicating machinery at the replication fork.

replication foci *replication factories* Small regions (~150 nm diameter) within the nucleus that form transiently during S phase and where DNA synthesis occurs. There are high local concentrations of the appropriate enzymes.

replication fork Point at which DNA strands are separated in preparation for **replication**. Replication forks thus move along the DNA as replication proceeds.

replicative intermediate Intermediate stage(s) in the replication of a RNA virus; a copy of the original RNA strand, or of a single-strand copy of the first replicative intermediate. Essentially an amplification strategy.

replicons Tandem regions of replication in a chromosome, each about 30 μm long, derived from an **origin of replication**. By definition a replicon must contain an origin of replication.

replisome Complex of proteins involved in the replication (elongation) of DNA that moves along as the new complementary strand is synthesized. On this basis a minimum content would be DNA polymerase III and a **primosome**, but the components are actually far more complex. An RNA-replisome has been proposed as a putative ancestor of the ribosome.

reporter gene A gene that encodes an easily assayed product (e.g. **CAT**) that is coupled to the upstream sequence of another gene and transfected into cells. The reporter gene can then be used to see which factors activate response elements in the upstream region of the gene of interest. A commonly used reporter is green fluorescent protein which can easily be detected microscopically.

repressor protein A protein that binds to an **operator** of a gene preventing the **transcription** of the gene. The binding affinity of repressors for the operator may be affected by other molecules. Inducers bind to repressors and decrease their binding to the operator, while co-repressors increase the binding. The archetype of repressor proteins is the lactose repressor protein that acts on the **lac operon** and for which the inducers are β-galactosides such as **lactose**; it is a polypeptide of 360aa that is active as a tetramer. Other examples are the λ repressor protein of lambda bacteriophage that prevents the transcription of the genes required for the lytic cycle leading to **lysogeny**, and the cro protein, also of lambda, which represses the transcription of the λ repressor protein establishing the lytic cycle. Both of these are active as dimers and have a common structural feature, the **helix-turn-helix** motif, that is thought to bind to DNA with the helices fitting into adjacent major grooves.

reproductive cloning Cloning of an organism, by intracytoplasmic nuclear transfer into an enucleated oocyte *in vitro*, with the intention that full development should occur; in the case of mammals this means that the blastocyst must successfully implant in a surrogate mother and the fetus must come to term. In most countries this is illegal for humans, although the technique has been used in sheep and a few other species. Strictly speaking the embryo is chimeric since only nuclear genes have been transferred and mitochondrial genes are derived from the egg into which the nucleus was transplanted.

repulsive guidance molecule *repellant guidance molecule* A GPI-linked protein involved in neuronal differentiation, apoptosis and repulsive axon guidance. Binds to **neogenin**. Three RGM isoforms (RGM-A, 450aa; RGM-B, 437aa; RGM-C, **hemojuvelin**, 426aa), a-RGMc) exist in vertebrates. Other repellant molecules include draxin (dorsal repulsive axon guidance protein, neucrin, 349aa) and homologues of **slit**. See also **netrin**.

RER See **rough endoplasmic reticulum.**

resact A peptide hormone (14aa) from the sea urchin *Arbacia punctulata* affecting sperm motility and metabolism. Receptor is a plasma-membrane **guanylate cyclase** (986aa) and resact is a chemoattractant for sperm.

resealed ghosts Membrane shells formed by lysis of erythrocytes that are resealed by adjusting the cation composition of the medium. Relatively impermeable, although more permeable than the original membrane.

reserpine Alkaloid derived from *Rauwolfia serpentina* or *R. vomitoria*; blocks the packaging of noradrenaline into presynaptic vesicles. Useful experimental tool to determine the involvement of sympathetic innervation.

residual body (1) **Secondary lysosomes** containing material that cannot be digested. (2) The surplus cytoplasm shed by spermatids during their differentiation to spermatozoa. Usually the cytoplasm from several spermatids connected by **cytoplasmic bridges**. (3) Surplus cytoplasm containing pigment and left over after production of merozoites during schizogony of **malaria** parasites.

resilin Amorphous rubber-like protein (pro-resilin, 620aa) found in insect cuticle: similar to **elastin**, though there is no fibre formation. One of the most elastic materials known and very important in insect flight. The first exon of the resilin gene has been expressed in a bacterial system and the cross-linked product has a resilience (recovery after deformation) greater than that of unfilled synthetic polybutadiene, a high resilience rubber.

resiniferatoxin Highly potent analogue of **capsaicin** from *Euphorbia resinifera* (flowering cactus) that is an agonist at **vanilloid receptor-1**. Research article: http://www.molecularpain.com/content/4/1/3

resistin An **adipokine** (cysteine-rich secreted protein FIZZ3 (found in inflammatory zone 3), 108aa produced by adipocytes; seems to suppress the ability of insulin to stimulate glucose uptake into adipose cells. FIZZ2 (resistin-like protein beta, 111aa) probably has similar functions.

resolution Complete return to normal structure and function: used, for example, of an inflammatory lesion, or of a disease. See also **resolving power**.

resolvase A enzyme complex found in all living organisms that will sort out (resolve) the tangle produced by a recombination event between two DNA duplexes. They fall into two main families, the phage **integrases** and the site-specific **recombinases**. In *E. coli* resolvase (protein D, 268aa) is one of the 'phage' integrase family and cleaves at the rfsF site. Tyrosine recombinases (xerC, 298aa; xer D, 298aa) are other members of the phage integrase family and form a heterotetrameric complex that catalyzes two consecutive pairs of strand exchanges. Transposon gamma-delta resolvase (183aa) catalyzes the site-specific recombination of a transposon and is one of the site-specific recombinase resolvase family. The **Holliday junction** resolvase (EC 3.1.22.4, 120aa) is one of the rusA family of endonucleases that resolves Holliday junction intermediates made during homologous genetic recombination and DNA repair. They seem to be related to the Type II restriction endonucleases. http://nar.oxfordjournals.org/cgi/reprint/28/22/4540.pdf

resolving power (1) The resolution (resolving power) of an optical system defines the closest proximity of two objects that can be seen as two distinct regions of the image. This limit depends upon the **Numerical Aperture** (N.A.) of the optical system, the contrast step between objects and background and the shape of the objects. The often quoted Airy limit applies only to self-luminous discs. (2) In genetics, the smallest map distance measurable by an experiment involving a certain number of classified recombinant progeny.

resolvins Family of endogenous 17R-hydroxy docosanoids (atypical eicosanoids analogous to **lipoxins**) biosynthesised from omega-3 fatty acids, especially if aspirin is present, that exert an anti-inflammatory effect. The resolvin E series derive from eicosapentaenoic acid, the D series from docosahexaenoic acid. Article: http://jem.rupress.org/content/196/8/1025.full

resonance energy transfer See **fluorescence energy transfer**.

resorufin Pink fluorescent dye; a caged form of resorufin (non-fluorescent unless activated - released —

by irradiation with UV) coupled to G-actin and microinjected has been used as a marker for microfilaments in the leading lamella of moving cells.

respiration Term used by physiologists to describe the process of breathing and by biochemists to describe the intracellular oxidation of substrates coupled with production of ATP and oxidized coenzymes (NAD^+ and FAD). This form of respiration may be anaerobic as in glycolysis, or aerobic in the case of oxidations operating via the **tricarboxylic acid cycle** and the **electron transport chain**.

respiratory burst See **metabolic burst**.

respiratory burst oxidase homologues Plant respiratory burst oxidase homologues (Rboh) are NADPH oxidases homologous to those of the human neutrophil **phox** system. For example, RBHO-A (EC 1.6.3.-, EC 1.11.1.-, 902aa) is a calcium-dependent enzyme that generates superoxide. Others produce reactive oxygen species important in the response to pathogens.

respiratory chain The mitochondrial **electron transport chain**.

respiratory enzyme complex The enzymes that make up the **respiratory chain**: NADH-Q reductase, succinate-Q reductase, cytochrome reductase, cytochrome C and cytochrome oxidase.

respiratory quotient *RQ* Molar ratio of carbon dioxide production to oxygen consumption. Varies according to the main substances being oxidised (RQ is 1 for pure carbohydrate but ~ 0.7 for fats).

respiratory syncytial virus An enveloped ssRNA virus of the **Paramyxoviridae** family responsible for common respiratory tract infections.

response elements The recognition sites of certain transcription factors, e.g. **CREB**, ISRE. Most are located within 1 kb of the transcriptional start site.

REST corepressor An essential component (coREST1, 426aa) of a BHC histone deacetylase complex that represses transcription of neuron-specific genes in non-neuronal cells. The BHC complex acts by deacetylating and demethylating specific sites on histones. There are several releated corepressors (REST corepressors 2 and 3, 523aa and 495aa)

resting potential The electrical potential of the inside of a cell, relative to its surroundings. Almost all animal cells are negative inside; resting potentials are in the range -20 to -100 mV, -70 mV typical. Resting potentials reflect the action of the **sodium pump** only indirectly; they are mainly caused by the subsequent diffusion of potassium out of the cell through potassium leak channels. The resting potential is thus close to the **Nernst potential** for potassium. See **action potential**.

restriction endonuclease See restriction enzymes.

restriction enzymes Class of bacterial enzymes (restriction endonucleases) that cut DNA at specific sites. In bacteria their function is to destroy foreign DNA, such as that of **bacteriophages** (host DNA is specifically modified and protected by methylation at these sites). Type I restriction endonucleases occur as a complex with the methylase and a polypeptide that binds to the recognition site on DNA. They are often not very specific and cut at a remote site. Type II restriction endonucleases are the classic experimental tools. They have very specific recognition and cutting sites. The recognition sites are short, 4–8 nucleotides, and are usually **palindromic sequences**. Because both strands have the same sequence running in opposite directions the enzymes make double-stranded breaks, which, if the site of cleavage is off-centre, generates fragments with short single-stranded tails; these can hybridize to the tails of other fragments and are called sticky ends. They are generally named according to the bacterium from which they were isolated (first letter of genus name and the first two letters of the specific name). The bacterial strain is identified next and multiple enzymes are given Roman numerals. For example the two enzymes isolated from the R strain of *E. coli* are designated Eco RI and Eco RII. The more commonly used restriction endonucleases are shown in Table R2. Type III restriction enzymes recognize two separate non-palindromic sequences that are inversely oriented and cut DNA about 20–30 base pairs after the recognition site. Restriction enzyme database: http://rebase.neb.com/rebase/rebase.html

restriction fragment length polymorphism *RFLP* Technique (DNA fingerprinting) that allows familial relationships to be established by comparing the characteristic polymorphic patterns that are obtained when certain regions of genomic DNA are amplified (typically by PCR) and cut with certain restriction enzymes. In principle, an individual can be identified unambiguously by RFLP (hence the use of RFLP in forensic analysis of blood, hair or semen). Similarly, if a polymorphism can be identified close to the locus of a genetic defect, it provides a valuable marker for tracing the inheritance of the defect.

restriction fragments The fragments of DNA generated by digesting DNA with a specific restriction endonuclease. Each of the fragments ends in a site recognized by that specific enzyme.

restriction map Map of DNA showing the position of sites recognized and cut by various **restriction endonucleases**.

restriction nucleases See restriction endonucleases.

TABLE R2. Recognition sequences of various type II restriction endonucleases

Enzyme	Bacterium from which enzyme is derived	Recognition sequence[a]
Aha III	*Aphanothece halophytica*	↓ TTT|AAA
Alu I	*Arthrobacter luteus*	↓ AG|CT
Ava I	*Anabaena variabilis*	↓ C PyC|GPuG
Bam HI	*Bacillus amyloliquefaciens* H	↓ m G GA|TCC
Bst EII	*Bacillus stearothermophilus* ET	↓ G GTNACC
Cla I	*Caryphanon latum*	↓ AT C|GAT
Dde I	*Desulfovibrio desulfuricans*	↓ C TNAG
Eco RI	*Escherichia coli*	↓ m G AA|TTC
Eco RII	*Escherichia coli*	↓ m CC(AT)TT
Eco RV	*Escherichia coli*	↓ GAT|ATC
Hae II	*Haemophilus aegyptius*	↓ PuGC|CG Py
Hae III	*Haemophilus aegyptius*	↓m GG|CC
Hha I	*Haemophilus haemolyticus*	m ↓ GC|GC
Hin dl II	*Haemophilus influenzae* R_d	m ↓ AA G|CTT
Hin f I	*Haemophilus influenzae* R_f	↓ G ANTC
Hpa I	*Haemophilus parainfluenzae*	↓ GTT|AAC
Hpa ll	*Haemophilus parainfluenzae*	↓ m CC|GG

TABLE R2. (Continued)

Enzyme	Bacterium from which enzyme is derived	Recognition sequence[a]
Kpn I	*Klebsiella pneumoniae* OK8	↓ GGT|AC C
Mbo I	*Moraxella bovis*	↓ GA|TC
Msp I	*Moraxella* species	↓ C C|GG
Mst I	*Microcoelus* species	↓ TGC|GCA
Pst I	*Providencia stuartii*	↓ CTG|CA G
Pvu I	*Proteus vulgaris*	↓ CGA|T CG
Sac I	*Streptomyces achromogenes*	↓ GAG|CT C
Sma I	*Serratia marcescens*	↓ CCC|GGG
Xba I	*Xanthomonas badrii*	↓ T CT|AGA
Xho I	*Xanthomonas holcicola*	↓ C TC|GAG

[a] 5'-XXX X\|XXXX-3' (↓: Cleavage site \|: Axis of symmetry).
Pu = Purine, ie. A or G are recognized. Py = Pyrimidine, ie. C or T are recognized. N = Any base m. X = base methylated by corresponding methylase, where known, to give N6-methyladenosine or 5-methylcytosin.

restriction point A point, late in **G1** of the cell cycle, after which the cell must, normally, proceed through to division at its standard rate. See **checkpoint**.

restriction site Any site in DNA that can be cut by a **restriction enzyme**.

restrictive temperature See **permissive temperature**.

restrictocin A ribonuclease (mitogillin, EC 3.1.27.-, 176aa) produced by *Aspergillus restrictus* and *A. fumigatus*. It cleaves 28S rRNA in eukaryotic ribosomes and inhibits protein synthesis. Has some antitumour activity and can cause an allergic reaction. One of the ribonuclease U2 family (aspergillins; see **alpha-sarcin**).

resveratrol A fungicidal phenol (trans-3,5,4'-trihydroxystilbene) with antioxidant properties found in grape skins and other plant tissues.

ret A human oncogene (rearranged during transfection, *ret*), encoding a receptor **tyrosine kinase** (EC 2.7.10.1, 1114aa) the ligands for which are members of the glial-cell-line-derived neurotrophic factor (**GDNF**) family. Gain of function mutations cause hereditary medullary thyroid carcinoma, loss of function mutations gene are associated with **Hirschsprung's disease**. *Ret* mutations are also associated with multiple endocrine neoplasia.

retention signal The sequence of amino acids on proteins that indicates that they are to be retained in the secretory processing system, for example the **Golgi apparatus**, and not passed on and released. Can be applied to any similar situation in which sorting of macromolecules into different compartments occurs. The carboxy-terminal tetrapeptide KDEL sequence is an important retrieval sequence for retention in the mammalian ER (HDEL in yeast).

reticular dysgenesis One of the rarest and most severe forms of combined immunodeficiency caused by mutation in the mitochondrial adenylate kinase-2 gene. There are no granulocytes or lymphocytes in the blood and the condition leads to fatal infection very quickly.

reticular fibres Fine fibres of **reticulin** (Type III collagen) found in extracellular matrix, particularly in lymph nodes, spleen, liver, kidneys and muscles.

reticular lamina The lower region of extracellular matrix underlying an epithelial monolayer, separated from the basal surface of the epithelial cells by the basal lamina. The reticular lamina contains fibrillar elements (collagen, elastin etc.) and is probably secreted by fibroblasts of the underlying connective tissues. The reticular lamina and the basal lamina together form what older textbooks referred to as the **basement membrane**.

reticulin Constituent protein of **reticular fibres**: collagen Type III.

reticulocalbins Low-affinity calcium-binding proteins with 6 EF-hand motifs (but only 4 functional calcium-binding sites) found in the lumen of the endoplasmic reticulum. Include: reticulocalbin-1 (331aa), reticulocalbin-2 (calcium-binding protein ERC-55, E6-binding protein, 317aa), reticulocalbin-3 (EF-hand calcium-binding protein RLP49, 328aa). See **CREC family**.

reticulocyte Immature red blood cells found in the bone marrow, and in very small numbers in the circulation.

reticulocyte lysate Cell lysate produced from **reticulocytes**; used as an *in vitro* translation sytem.

reticuloendothelial system The phagocytic system of the body, including the fixed macrophages of tissues, liver and spleen. Rather old-fashioned term that is coming back into use; mononuclear phagocyte system is probably better when only phagocytes are meant.

reticulol An antibiotic isolated from the culture broth of a strain of *Streptoverticillium sp.* that inhibits topoisomerase I and cAMP-phosphodiesterase.

reticulum cells Cells of the **reticuloendothelial system**, found particularly in lymph nodes, bone marrow, and spleen. In lymph nodes they are stromal cells and probably not reticuloendothelial cells in the current sense of that term.

retina Light-sensitive layer of the eye. In vertebrates, looking from outside, there are four major cell layers: (i) the outer neural retina, which contains neurons (ganglion cells, **amacrine cells**, **bipolar cells**) as well as blood vessels; (ii) the photoreceptor layer, a single layer of rods and cones; (iii) the **pigmented retinal epithelium** (PRE or RPE); (iv) the choroid, composed of connective tissue, fibroblasts, and including a well-vascularised layer, the choriocapillaris, underlying the basal lamina of the PRE. Behind the choroid is the sclera, a thick organ capsule. See **retinal rods**, **retinal cones**, **rhodopsin**. In molluscs (especially cephalopods such as the squid) the retina has the light-sensitive cells as the outer layer with the neural and supporting tissues below.

retinal Aldehyde of **retinoic acid** (vitamin A); complexed with opsin forms **rhodopsin**. Photosensitive component of all known visual systems. Absorption of light causes retinal to shift from the 11-*cis* form to the all-*trans* configuration, and through a complex cascade of reactions excites activity in the neurons synapsed with the rod cell.

retinal cone A light-sensitive cell type of the retina, that, unlike **retinal rods**, is differentially sensitive to particular wavelengths of light, and is important for colour vision. In the human eye there are three types of cones, sensitive respectively to long wavelengths (564–580 nm, yellow), medium-wavelengths (534–545 nm, green) and shorter wavelengths (420–440 nm, violet). Defects can lead to **colour blindness**. They are concentrated in the fovea and are important for high-acuity vision. The membrane infoldings of the outer segment are incompletely separated in cones.

retinal ganglion cell See **ganglion cell**.

retinal pigmented epithelial cell See **pigmented retinal epithelium** and **retina**.

retinal rod Major photoreceptor cell of the vertebrate retina, columnar cells (about 40 μm long, 1 μm diameter) having three distinct regions: a region adjacent to, and synapsed with, the neural layer of the retina contains the nucleus and other cytoplasmic organelles, below this is the inner segment, rich in mitochondria, connected through a thin neck to the outer segment which consists of a stack of discs (membrane infoldings) that are continually replenished near the inner segment and that are shed from the distal end and phagocytosed by the pigmented epithelium. The membranes of the discs are rich in **rhodopsin** which undergoes a conformational change when it absorbs light and triggers a cascade of amplification reactions that eventually produce a depolarization. The thin linking region betwen the inner and outer segments contains a **ciliary body**, There are about 125 million rods in a human eye, more in the peripheral areas; they are responsible for vision at low light levels.

retinitis pigmentosa Disease caused by abnormalities of rods, cones or pigmented retinal epithelial cells, with progressive loss of vision. Many different mutations can lead to the condition including defects in an oxygen-regulated photoreceptor protein ORP1 (2156aa), the homologue of *Drosophila* **crumbs**, the retinitis pigmentosa GTPase regulator (1020aa), **peripherin**, **tubby-like protein-1**, and **PIM1**-associated protein. Other forms are known and mutations in at least 35 genes are associated with the condition.

retino-tectal connection The developmental process in which nerve fibres from the developing retina are mapped on to the tectum of the brain. There seems to be a good positioning system in operation, and a variety of mechanisms probably operate, including control of the fasciculation of fibres in the optic nerve, and some specific recognition of the correct target area by the nerve growth cone.

retinoblastoma Malignant tumour of the retina, caused by mutation in gene encoding the tumour suppressor Rb, a nuclear protein (928aa) that is a negative regulator of cellular proliferation and interacts with transcription factors such as **E2F** to block transcription of growth-regulating genes. The *Rb* gene plays a role in normal development, not just that of the retina. See **retinoblastoma-binding proteins**.

retinoblastoma-binding proteins A diverse set of proteins that bind to the **retinoblastoma** gene product. Most regulate transcription in one way or another. See Table R3.

retinoic acid *vitamin A* The aldehyde (**retinal**) is involved in photoreception, but retinoic acid has other roles. There are cytoplasmic retinoic acid binding proteins and retinoic acid response elements that regulate gene transcription. See also Table V1.

retinoic acid-receptor-related orphan receptors Members of the nuclear hormone receptor family. Retinoid-related orphan receptor-alpha (RORα,

TABLE R3. Human retinoblastoma binding proteins (RBBPs)

	Synonyms	Size (aa)	Function
RBBP1	AT-rich interactive domain-containing protein 4A	1257	Interacts with the viral protein-binding domain of the retinoblastoma protein
RBBP1L1	RBBP1-like 1 Breast cancer-associated antigen BRCAA1	1312	Component of a Sin3A corepressor complex
RBBP2	Lysine-specific demethylase 5A, EC 1.14.11.-	1690	Demethylates lysine-4 in histone H3
RBBP2-H1	RBBP2-homolog 1	1544	Demethylates lysine-4 in histone H3
RBBP3	E2F1	437	Transcription factor
RBBP4	Chromatin assembly factor 1 subunit C, CAF-I p48, nucleosome-remodeling factor subunit RBAP48,	425	Regulates chromatin metabolism and assembly
RBBP5		538	Involved in methylation and dimethylation at Lys-4 of histone H3
RBBP6	E3 ubiquitin-protein ligase	1792	Promotes ubiquitination of YBX1*
RBBP7	Nucleosome-remodeling factor subunit RBAP46	425	Component of complexes which regulate chromatin metabolism
RBBP8	RIM, CtBP-interacting protein**	897	Possible tumour suppressor
RBBP9	Putative hydrolase RBBP9, RBBP10	186	Confers resistance to growth-inhibitory effects of TGFβ₁
RB1	Retinoblastoma-associated protein	928	Tumor suppressor, transcription repressor of E2F1 target genes

*YBX1: Y-box-binding protein 1: mediates regulation of pre-mRNA alternative splicing
**CtBP: C-terminal binding protein, corepressor of diverse transcription regulators

nuclear receptor subfamily 1 group F member 1, NR1F1, 556aa) regulates a number of genes involved in lipid metabolism and interacts with **hypoxia-inducible factor** 1. RORβ (NR1F2, 470aa) is required for normal postnatal development of rod and cone photoreceptor cells. RORγ (NR1F3, 518aa) is highly expressed in skeletal muscle; RORγ-deficient mice lack all lymph nodes and Peyer's patches.

retraction fibres Thin projections from crawling cells associated with areas where the cell body is becoming detached from the substratum, but **focal adhesions** persist. Usually contain a bundle of microfilaments that are under tension.

retrograde axonal transport The transport of vesicles from the synaptic region of an axon towards the cell body: involves the interaction of microtubule-associated protein-1C (cytoplasmic dynein) with microtubules.

retrolental fibroplasia Retinal detachment (retinopathy of prematurity) which is the consequence of overproduction of fibrous tissue linked to excessive vascularisation and scarring of the neonatal retina. Excessive oxygen is a risk factor but not the main cause.

retromer Multi-subunit complex that regulates retrieval of the cation-independent mannose 6-phosphate receptor from endosomes to the trans-Golgi network. See **sorting nexins**.

RetrOryza A database of the rice LTR-retrotransposons. Long terminal repeat (LTR)-retrotransposons comprise a significant portion of the rice genome. See **Oryzabase**. Database: http://www.retroryza.org/

retrotransposon A **transposable element** that replicates by producing an RNA transcript of itself, translating this back into a DNA copy with reverse transcriptase and inserting this DNA copy at a different site on the chromosome or in a different chromosome in the same cell. Retrotransposons account for a significant part of the overall genome (around 40–45% in humans, 50% in maize and more than 90% in wheat). Retroelements (retrotransposons and retroviruses) can currently be divided into four systems or groups; **LTR retrotransposons, non-LTR retrotransposons**, the **tyrosine recombinase (YR) retroelements** and the **Penelope retrotransposons**. LTR-retrotransposons only differ from retroviruses in the absence of the *env* gene that encodes the envelope proteins necessary for cell-to-cell transfer. See also **short interspersed nuclear elements, long interspersed nuclear elements** (SINEs, LINEs),

mammalian-wide interspersed repeats. The Gypsy Database (GyDB) of mobile genetic elements: http://gydb.uv.es/index.php/Intro

retroviral vector See **Retroviridae**. Retroviral vectors are used in the genetic modification of cells as a means of introducing foreign DNA into the genome and RVs encoding histochemical markers (**reporter genes**) are used in the study of cell lineages in vertebrates.

Retroviridae A family of viruses with a single-stranded RNA genome (Class VI). The viral **reverse transcriptase** generates a DNA replica within the host cell. The old classification into subfamilies (oncovirinae, lentivirinae, spumavirinae) is no longer considered appropriate and the retroviridae are currently separated into the sub-family Orthoretrovirinae with six genera, alpha-, beta-, gamma-, delta- and epsilon-retroviruses and lentivirus and the Spumaretrovirinae as a separate sub-family. See **retroviral vector**. Virology databases: http://www.ncbi.nlm.nih.gov/ICTVdb/ICTVdB/00.061.html & http://www.virology.net/Big_Virology/BVretro.html

Rett's syndrome Severe progressive neurological disorder that mainly affects baby girls, causing dyspraxia and impaired learning and communication. The gene associated with the disorder is on the X chromosome, and encodes methyl-CpG-binding protein 2 (MeCP2, 486aa), which regulates transcription of various other proteins. Lethal in males. An atypical form of Rett's syndrome can be caused by mutation in the cyclin-dependent kinase-like 5 (CDKL5) gene and a variant by mutation in the FOXG1 gene that encodes a forkhead-box transcription factor. Patient website: http://adam.about.com/encyclopedia/001536.htm

reverse antagonist See **inverse agonist**.

reverse genetics The technique of determining a gene's function by first sequencing it, then mutating it, and then trying to identify the nature of the change in the phenotype.

reverse passive haemagglutination A test for the presence of multivalent antigen that uses antibodies are bonded to the surface of red blood cells; if the antigen is present there is haemagglutination. Used as a test for e.g. **Hepatitis B** virus in serum.

reverse transcriptase RNA-directed DNA polymerase. Enzyme first discovered in retroviruses, that can construct double-stranded DNA molecules from the single-stranded RNA templates of their genomes. Reverse transcription now appears also to be involved in movement of certain mobile genetic elements, such as the Ty plasmid in yeast, in the replication of other viruses such as **Hepatitis B**, and possibly in the generation of mammalian **pseudogenes**. See **hTERT** and **telomerase**.

reversion Reversion of a mutation occurs when a second mutation restores the function that was lost as a result of the first mutation. The second mutation causes a change in the DNA that either reverses the original alteration or compensates for it.

Revesz's syndrome A developmental disorder characterised by characteristic bilateral exudative retinopathy. See **TIN2**.

Reynold's number A dimensionless constant that relates the inertial and viscous drag that act to hinder a body moving through fluid medium. For cells the Reynold's number is very small; viscous drag is dominant, and inertial resistance can be neglected.

RF (1) See **release factor**. (2) Retardation factor (Rf) in chromatographic separation, is the ratio of the distance travelled by the substance of interest to the distance simultaneously travelled by the mobile phase: always less than 1.

RFC (1) **Replication Factor** C. (2) Reduced-folate carrier, (solute carrier family 19 member 1, SLC19A1, 591aa) one of the major proteins mediating transmembrane folate transport.

RFLP See **restriction fragment length polymorphism**.

RGD A domain found in **fibronectin** and related proteins such as **disintegrins**; recognized by **integrins**. In most cases, the consensus is -R-G-D-S- (arginine-glycine-aspartic acid-serine).

RGS proteins Regulators of G-protein signalling proteins; a large and diverse family initially identified as GTPase activating proteins (GAPs) of heterotrimeric G-protein Gα-subunits, now recognised to have a broader range of activities. Characteristically have a RGS domain with a conserved stretch of 120aa responsible for direct binding to activated G-protein alpha subunits

RH factor (1) Rhesus factor: see **rhesus blood group**. (2) RH maps are radiation hybrid genetic maps produced by a technique that was originally developed for constructing long-range maps of mammalian chromosomes. It involved fragmenting chromosomes with X-irradiation and estimating the distance between genes on the basis of the probability of them being separated.

rhabdom The photosensitive portion of each ommatidium of the compound eye of arthropods. The rhabdom is formed by a cluster of rod-like cells that are tightly packed and have, on their inner faces, densely packed microvilli (forming the rhabdomeres) orthogonally disposed with respect to the long axis of the cells composing the rhabdom and containing rhodopsin in their membranes. In the majority of insects, the rhabdomeres of the neighbouring cells are interlaced to form a fused rhabdom. The orientation of the microvilli may be important for detecting the polarity of light.

rhabdomere See **rhabdom**.

rhabdomyosarcoma Malignant tumour (sarcoma) derived from satellite cells of striated muscle.

Rhabdoviridae Class V viruses with a single negative strand RNA genome (11,000–15,000 nucleotides long) and an associated virus-specific RNA polymerase. The capsid is bullet shaped 45–100 nm in diameter, 100–430 nm long, with surface projections (spikes 5–10 nm long and about 3 nm in diameter). The virus is enveloped by membrane formed when it buds out of the plasma membrane of infected cells. The budded membrane contains host lipids but only viral glycoproteins that form the spikes. The group includes **rabies virus**, **vesicular stomatitis virus** and a number of plant viruses. Virus database: http://www.virology.net/Big_Virology/BVRNArhabdo.html

rhamnogalacturonan Plant cell-wall polysaccharide consisting principally of rhamnose and galacturonic acid. Present as a major part of the pectin of the primary cell wall. Two types known: rhamnogalacturonan I (RG-I), the major component, which contains rhamnose, galacturonic acid, arabinose and galactose, and rhamnogalacturonan II, (RG-II), containing at least four different sugars in addition to galacturonic acid and rhamnose.

rhamnolipid Simple glycolipids, normally produced by *Pseudomonas aeruginosa*, composed of a fatty acid tail with either one or two rhamnose rings at the carboxyl end of the fatty acid. They have biosurfactant properties that are exploited in bioremediation of oil and heavy-metal contaminated soils. Rhamnolipids also have antifungal properties and are used as agricultural fungicides. Orthologs have been found in non-infectious *Burkholderia thailandensis*, and pathogenic *B. pseudomallei*.

rhamnose A sugar (6-deoxy L-mannose) found in plant glycosides.

rheb A ras-related GTP-binding protein (ras homologue enriched in brain, 184aa) that activates the protein kinase activity of **mTOR** through FK506 binding protein-8 (FKBP38). Despite the name the highest levels are found in skeletal and cardiac muscle. See **Currarino's syndrome**.

rheotaxis A **taxis** in response to the direction of flow of a fluid.

Rhesus blood group Human blood group system with allelic red cell antigens C, D and E. The D antigen is the strongest. Red cells from a Rhesus-positive fetus cross the placenta and can sensitise a Rhesus-negative mother, expecially at parturition. The mother's antibody may then, in a subsequent pregnancy, cause haemolytic disease of the newborn

if the fetus is Rhesus-positive. The disease can be prevented by giving anti-D IgG during the first 72 hours after parturition to mop up D^+ red cells in the maternal circulation.

rheumatic fever Disease involving inflammation of joints and damage to heart valves that follows streptococcal infection and is believed to be autoimmune, ie. antibodies to streptococcal components cross-react with host tissue antigens.

rheumatoid arthritis Chronic inflammatory disease in which there is destruction of joints. Generally considered to be an autoimmune disorder in which immune complexes are formed in joints and excite an inflammatory response (complex-mediated **hypersensitivity**), although the antigen(s) responsible is unclear. Cell-mediated (Type IV) hypersensitivity also occurs, and macrophages accumulate. This in turn leads to the destruction of the synovial lining (see **pannus**). For reasons that are also unclear, the pattern of joint damage tends to be bilaterally symmetrical. The most effective therapies involve blocking the effects of TNF either with antibodies or decoy receptors.

rheumatoid factor Complex of IgG and anti-IgG formed in joints in **rheumatoid arthritis**. Serum rheumatoid factors are more usually formed from IgM antibodies directed against IgG.

rhinovirus Picornaviridae that largely infect the upper respiratory tract. Include the common cold virus and foot-and-mouth disease virus.

Rhizobium Gram-negative bacterium that fixes nitrogen in association with roots of some higher plants, notably legumes. Forms root nodules, in which it is converted to the nitrogen-fixing **bacteroid** form. See *nif* **genes**.

rhizoid Portion of a cell or organism that serves as a basal anchor to the substratum.

rhizoplast Striated contractile structure attached to the basal region of the **cilium** in a variety of ciliates and flagellates. May regulate the flagellar beat pattern, and is sensitive to calcium concentration. Composed of a 20 kDa protein rather similar to **spasmin**.

Rhizopoda Phylum of protozoa that move with pseudopods. Includes single celled amoebae such as *Amoeba proteus*.

rhizopodin Cytostatic drug, a potent actin-binding macrolide, from the culture broth of the myxobacterium, *Myxococcus stipitatus*. Causes adherently growing L929 mouse fibroblasts and PtK2 potoroo kidney cells to produce long, narrow, branched extensions in the same way as **latrunculin** and apparently disrupts the actin cytoskeleton.

rhizoxin An antimitotic drug that binds to beta-tubulin at the maytansine-binding site and blocks microtubule assembly. It is produced by *Burkholderia rhizoxina*, an endosymbiont of the pathogenic plant fungus, *Rhizopus microsporus*.

rho See **ras-like GTPases**.

rho factors *ρ factors* A hexameric ring-shaped helicase (419aa) that uses the energy derived from ATP hydrolysis to dissociate RNA transcripts from the ternary elongation complex in *E. coli*. Mutations in rho may cause the RNA polymerase to read through from one **operon** to the next. Not to be confused with rho (see **ras-like GTPases**).

rho kinases *ROCKs* The best-characterized effectors of the small G-protein rhoA, activated by binding of rhoA. Rho-associated protein kinase 1 (ROCK-1, EC 2.7.11.1, 1354aa) is a serine/threonine kinase that phosphorylates and activates death-associated protein kinase 3 (DAPK3), which then regulates myosin light chain phosphatase. ROCK2 (1388aa) regulates the assembly of the actin cytoskeleton, promotes formation of stress fibres and focal adhesions and has a role in smooth muscle contraction.

rhodamines A group of triphenylmethane-derived dyes are referred to as rhodamines, lissamines and so on. Many are fluorescent and are used as fluorochromes in labelling proteins and membrane probes.

Rhodnius prolixus Reduviid blood-sucking bug, vector of *Trypanosoma cruzi* in South America. Much used by insect physiologists because the the transition from one instar to the next is triggered by a blood meal. The blood meal, which may substantially exceed the body weight of the insect, triggers complex changes in the mechanical properties of abdominal cuticle allowing the insect to swell, and activates excretory mechanisms. The haemolymph contains relatively few **haemocytes**, which made some of the classical parabiosis experiments much easier.

Rhodophyta *red algae* Division of algae, many of which have branching filamentous forms and red coloration. The latter is due to the presence of **phycoerythrin**. The food reserve is floridean (starch), found outside the plastid. The walls contain sulphated galactans such as **carrageenan** and **agar**.

rhodopsin Light-sensitive pigment (visual purple, opsin-2, 348aa) formed from **retinal** linked through a **Schiff's base** to **opsin**: rhodopsin is an integral G-protein coupled membrane protein comprising some 40% of the membrane.of the discs of **retinal rods**. The opsins in cones are related. See also **bacteriorhodopsin** and **opsin sub-families**.

rhodopsin kinase A G-protein dependent receptor kinase (GRK1, 563aa) that mediates adaptation of photoreceptors to light and protects against light-induced injury. Phototransduction initiated by light is quenched rapidly by RK-phosphorylation of photoexcited rhodopsin. The phosphorylated rhodopsin is then bound by cytosolic **arrestin**, and thereby uncoupled from the G protein **transducin**. Defects in GRK1 are a cause of one form of stationary night blindness. GRK4 specifically phosphorylates the activated forms of G protein-coupled receptors but only one of the four isoforms (GRK4-alpha, 578aa) can phosphorylate rhodopsin. See **GRKs**, **recoverin**.

Rhodospirillum rubrum A purple nonsulphur bacterium with a spiral shape; contains the pigment **bacteriochlorophyll** and under anaerobic conditions photosynthesises using organic compounds as electron donors for the reduction of carbon dioxide. The purple colour results from the presence of **carotenoids**, though the bacteria are often more red or brown.

rhomboid *rho* Members of the peptidase S54 family involved in intramembrane proteolysis that releases active polypeptides from their membrane anchors. In *Drosophila* rhomboid-1 (EC 3.4.21.105, 355aa) promotes the cleavage of the membrane-anchored TGFalpha-like growth factor Spitz, allowing it to activate the *Drosophila* EGF receptor. See **dodo**. In humans some members of the family do not have peptidase activity (rhomboid family member 1, p100hRh, rhomboid 5 homolog 1, 855aa; rhomboid 5 homolog 2, rhomboid veinlet-like protein 5, 856aa) but rhomboid-related proteins do have enzymic activity (EC 3.4.21.105, rhomboid-related protein 1, RHBDL1, 438aa; RHBDL-2, 303aa; RHBDL3, 404aa). Presenilins-associated rhomboid-like protein (mitochondria intramembrane cleaving protease, PARL, 379aa) is required for the control of apoptosis during postnatal growth. Yeast rhomboid proteins are involved in processing various mitochondrial proteins such as cytochrome c peroxidase.

rhombomere Neuromeres or segments in the hindbrain region that are of developmental significance. Shown to be lineage restriction units in that cells of adjacent rhombomeres do not mix with each other. Regulatory genes have been shown to be expressed in patterns in the developing hindbrain that relate to the neuromeric or **segmentation** pattern.

rhophilin A rho GTPase (643aa) that has sequence homology with the N terminal region of **protein kinase N** though possesses no catalytic activity. Has a **PDZ domain** at the carboxy terminus suggesting that it may act as an adaptor molecule. Rhophilin-associated tail protein 1 (Ropn1, ropporin, 212aa) is found in the sperm flagellum where it may form a complex with rhophilin. Ropn1-like protein (**AKAP**-associated sperm protein, 230aa) has homology with ropporin. Abstract: http://www.ncbi.nlm.nih.gov/pubmed/8571126

rhoptry Electron-opaque dense body found in the apical complex of parasitic protozoa of the Phylum Apicomplexa.

rhotekin A protein (rhotekin-1, 563aa) that mediates **rho** signalling to activate NFκB; overexpressed in many cancer-derived cell lines. Rhotekin-2 (pleckstrin homology domain-containing family K member 1, 609aa) may have a role in lymphopoiesis.

RIA See **radioimmunoassay**.

ribbon synapse Ultrastructurally distinct type of synapse found in a variety of sensory receptor cells such as retinal **photoreceptor** cells, **cochlear hair cells** and vestibular organ receptors, as well as in a non-sensory neuron, the retinal bipolar cell. Unlike most neurons, these cells do not use regenerative action potentials but release transmitter in response to small graded potential changes. Ribbon synapses have different exocytotic machinery from conventional synapses in containing dense bars or ribbons anchored to the presynaptic membrane covered with a layer of synaptic vesices. The ribbons have been proposed to shuttle synaptic vesicles to exocytotic sites. See **ribeye** and **bassoon**.

ribeye An isoform of **C terminal binding protein 2** that probably acts as a scaffold for **ribbon synapses** in the retina.

ribityl side chain Univalent radical derived from ribitol found, for example, in riboflavin.

ribityllumazine A precursor (6,7-dimethyl-8-ribityllumazine) for riboflavin.

riboflavin *vitamin B2* Ribose attached to a flavin moiety that becomes part of FAD and FMN. See also Table V1.

ribonuclease *RNAse, RNAase* Any enzyme that cleaves RNA. May act as endonucleases or exonucleases depending upon the type of enzyme. Generally recognize target by tertiary structure rather than sequence. Examples include RNAse 1 (ribonuclease A, EC 3.1.27.5, 156aa) a secreted pancreatic enzyme, extremely heat-stable and often used experimentally. RNAse 2 (eosinophil cationic protein, 161aa) is not secreted and is found in granules of eosinophils; it is selectively chemotactic for dendritic cells. Ribonuclease 3 (**drosha**, EC 3.1.26.3, 1374aa) is a double-stranded (ds) RNA-specific endoribonuclease that is involved in the initial step of microRNA (miRNA) biogenesis. Ribonuclease 4 (RNAse L, 741aa) functions in the interferon antiviral response. RNAse 5 is **angiogenin**. Ribonuclease 6 (ribonuclease T2, EC 3.1.27.-, 256aa) is a ubiquitous secreted form. Ribonuclease E is an RNAase involved in the formation of 5S ribosomal RNA from pre-rRNA, F is stimulated by interferons and cleaves viral and host RNAs and thus inhibits protein synthesis. Ribonuclease H1 (EC 3.1.26.4, 286aa) specifically degrades the RNA of RNA-DNA hybrids. RNAse P (EC 3.1.26.5) is a protein complex involving a RNA moiety and at least 8 protein subunits that generates mature tRNA molecules by cleaving their 5'-ends.

ribonucleic acid See **RNA**.

ribonucleoprotein *RNP* A protein / RNA complex. Besides **ribosomes** (with which RNP was originally almost synonymous), in eukaryotic cells both initial RNA transcripts in the nucleus (**hnRNA**) and cytoplasmic **mRNAs** exist as complexes with specific sets of proteins. Processing (splicing) of the former is carried out by small nuclear RNPs (**snRNPs**). Other examples are the **signal recognition particle** responsible for targeting proteins to endoplasmic reticulum and a complex involved in termination of transcription.

ribonucleotide reductase Enzyme (EC 1.17.4.1) that catalyzes the biosynthesis of deoxyribonucleotides, the precursors necessary for DNA synthesis, from the corresponding ribonucleotides. Heterodimer of a large (RRM1, 792aa) and a small subunit (RRM2, 389aa; RRM2B, 351aa).

ribophorins Glycoproteins of the rough endoplasmic reticulum that interact with ribosomes whilst co-translational insertion of membrane or secreted proteins is taking place. They are involved in posttranslational glycosylation and are also part of the regulatory subunit of the 26S **proteasome**. Ribophorin-1 (dolichyl-diphosphoolligosaccharide-protein glycosyltransferase subunit 1, EC 2.4.1.119, 607aa) catalyzes the transfer of a high mannose oligosaccharide to an asparagine residue as does ribophorin 2 (631aa).

riboprobe Somewhat casual term for an RNA segment used to probe for a complementary nucleotide sequence, either in the mRNA pool or in the DNA of a cell.

ribose A monosaccharide pentose of widespread occurrence in biological molecules, e.g. RNA.

ribose-binding protein Periplasmic binding protein of the bacterial solute-binding protein 2 family (296aa in *E. coli*) that interacts either with the high-affinity ribose transport system or with the methyl-accepting chemotaxis protein, **MCP** III (trg).

ribosomal protein Proteins present within the ribosomal subunits. In prokaryotes there are 34 proteins in the large (50S) subunit and 21 in the small (30S) subunit. In eukaryotes there are 50 in the large (60S) subunit and 33 in the small (40S) subunit) proteins. Mitochondrial ribosomes have 48 in the large (39S) subunit and 29 in the small (28S) subunit. The proteins of Archaean ribosomes resemble those of eukaryotes rather than of bacteria.

ribosomal RNA *rRNA* The RNA found in ribosomes. In bacteria the small 30S ribosomal subunit contains the 16S rRNA, the large 50S subunit contains 5S and 23S rRNAs. In Archaea, the ribosomal RNA is more similar to eukaryotic rRNA; the sequence of archaeal 16S rRNA has been important in constructing the phylogentic tree. The 18S rRNA in most eukaryotes is in the small ribosomal subunit, and the large subunit contains three rRNA species (5S, 5.8S and 28S rRNAs). Mammalian cells have 2 mitochondrial (12S and 16S) rRNAs. Ribosomal RNA database: http://www.arb-silva.de/ & http://rdp.cme.msu.edu/

ribosome A heterodimeric multi-subunit enzyme complex composed of **ribosomal proteins** and **ribosomal RNA**. Interacts with aminoacylated tRNAs, and mRNAs and translates protein coding sequences from messenger RNA. During protein elongation, the nascent protein is held at the P-site (peptidyl-tRNA complex), while aminoacyl-tRNAs bearing new aminoacids are bound at the A-site. Similar ribosomes are found in all living organisms, all composed of large and small subunits, as well as in chloroplasts and mitochondria. Differences are apparent between prokaryotic and eukaryotic ribosomes; ribosomes of Archaea resemble those of eukaryotes rather than those of bacteria.

riboswitch Genetic control elements that can be found in the 5'-untranslated region of certain messenger RNAs of prokaryotes. They bind specifically to small molecules, usually a metabolite, which causes an allosteric change in the mRNA and alters the expression of the gene. Riboswitches triggered by thiamine pyrophosphate (TPP) have been shown to control operons in *E. coli* but several eukaryotes (*Arabidopsis, Neurospora*) have RNA domains that conform to the consensus sequence and structure of the bacterial TPP riboswitch class. http://rnajournal.cshlp.org/content/9/6/644.full

ribotype A strain-specific characteristic based upon ribosomal RNA sequence polymorphisms. Ribotyping of bacterial strains, such as those in *Corynebacteria*, is important in clinical microbiology. Diptheria ribotype database: http://www.dipnet.org/home.php

ribozyme RNA with catalytic capacity – an enzyme made of nucleic acid not protein. Of particular interest because of the implications for self-replicating systems in the earliest stages of the evolution of (terrestrial) life.

ribulose 1,5-bisphosphate An intermediate in the **Calvin-Benson cycle** of photosynthesis.

ribulose bisphosphate carboxylase *RUBISCO* A multisubunit enzyme complex composed of 8 large (479aa) and 8 small (180aa), disulphide-linked (D-Ribulose 1,5-diphosphate carboxylase; EC 4.1.1.39) that is responsible for CO_2 fixation in photosynthesis. Carbon dioxide is combined with ribulose diphosphate to give two molecules of 3-phosphoglycerate, as part of the **Calvin-Benson cycle**. It is the sole

CO_2-fixing enzyme in C3 plants, and collaborates with **PEP carboxylase** in CO_2 fixation in C4 plants. In the presence of oxygen the products of the reaction are one molecule of phosphoglyceric acid and one molecule of phosphoglycolic acid. The latter is the initial substrate for photorespiration and this oxygenase function occurs in C3 plants where the enzyme is not protected from ambient oxygen; in C4 plants the enzyme acts exclusively as a carboxylase since it is protected from oxygen. Also known as Fraction 1 protein, the major protein of leaves.

ribulose diphosphate carboxylase *RuDPC, RuDP carboxylase*. See **ribulose bisphosphate carboxylase**, the recommended name.

RICD The Rice Indica cDNA Database, an online database for *Oryza sativa* L. *indica* subspecies, hyperlinked to the Rice Annotation Project Database (RAP-DB) and The TIGR **Rice Genome Annotation Resource**. Link: http://202.127.18.228/ricd/index.html

Rice Annotation Project Database *RAP-DB* An annotated resource for the rice genome. Version 6.1 was released in June 2009. http://rapdb.dna.affrc.go.jp/, http://rapdb.lab.nig.ac.jp/.and http://rice.plantbiology.msu.edu/

Rice yellow mottle virus *RYMV* A single-stranded RNA virus (Sobemovirus) encoding four open reading frames. Probably the most damaging pathogen of rice, widespread in Africa.

RICH-1 See **nadrin**.

ricin See *Ricinus communis* **agglutinins**.

Ricinus communis agglutinins Glycoprotein lectins isolated from seeds of the castor bean, *Ricinus communis*. Agglutinin I (RCA-I, 564aa) has binding specificity similar to ricin, but is much less toxic. Agglutinin-II (ricin) is an **AB toxin**; the A subunit (267aa) enzymatically inactivates ribosomes by depurinating A4324 in the 28S rRNA fragment of the eukaryotic 60S rRNA, and is disulphide-linked to the carbohydrate-binding B subunit (262aa) specific for β-galactosyl residues. See **sarcin/ricin loop**.

RICK Receptor-interacting serine/threonine-protein kinase 2 (RIP2, EC 2.7.11.1, RIP-like-interacting CLARP kinase, 540aa) that activates pro-caspases 1 and 8, potentiates caspase 8-mediated apoptosis and activates NFκB. Important in the innate and adaptive immune responses.

Ricker's syndrome See **myotonic dystrophy**.

rickets Disease (rachitis) due to deficiency of vitamin D that leads to defective ossification and softening of bones. *Adj.* rachitic.

Rickettsia Genus of Gram-negative bacteria responsible for a number of insect-borne diseases of man (including scrub typhus and **Rocky Mountain spotted fever**). Obligate intracellular parasites.

rictor The rapamycin-insensitive companion (1708aa) of **mTOR** that forms a complex which modulates the phosphorylation of PKCalpha and the actin cytoskeleton and regulates cell growth signalling pathways. Rictor shares homology with pianissimo from *D. discoideum* and STE20p from *S. pombe*. See **raptor**.

Rieger's syndrome An autosomal dominant disorder of morphogenesis in which there is abnormal development of the anterior segment of the eye. One form is caused by mutation in the paired homeobox transcription factor **ptx2**.

Rieske protein Iron-sulphur protein components of the cytochrome bc1 complex and **cytochrome b6-f complex**. The bacterial and mitochondrial bc1 complex catalyses the oxidoreduction of ubiquinol and cytochrome c, generating an electrochemical potential. In the chloroplast a component of the **cytochrome b6-f complex** (petC, EC 1.10.99.1, plastohydroquinone:plastocyanin oxidoreductase iron-sulphur protein, 229aa), an essential protein for photoautotrophism that confers resistance to photo-oxidative damage. The Rieske subunit acts by binding either a ubiquinol or plastoquinol anion, transferring an electron to the 2Fe-2S cluster, then releasing the electron to the cytochrome c or cytochrome f haem iron. The Rieske domain has a 2Fe-2S centre: two conserved cysteines coordinate one Fe ion while the other is coordinated by two conserved histidines.

rifamycin Antibiotic produced by *Streptomyces mediterranei* that acts by inhibiting prokaryotic, but not eukaryotic DNA-dependent RNA synthesis. Blocks initiation, but not elongation of transcripts. Rifampicin is a semi-synthetic variant.

RIG1 (1) Product of the retinoic acid-inducible gene 1 (RIG1, DDX58, 925aa) an RNA helicase of the DEAD/H box family responsible for the initial sensing of infection by RNA viruses. When RIG1 binds double-stranded viral RNA it forms multimers and activates NFκB and interferon regulatory factors through the adaptor **MAVS**. Multimer formation is blocked by **LGP2**. (2) Retinoic acid receptor responder protein 3 (RIG1, 164aa) may mediate some of the growth suppressive effects of retinoids.

rigor Stiffening of muscle as a result of high calcium levels and ATP depletion, so that actin-myosin links are made, but not broken.

RIMs (1) A family of proteins (rab-3-interacting molecule 1, regulating synaptic membrane exocytosis 1, rim1, 1692aa; rim2, 1411aa;) that regulate release of synaptic vesicles at synapses in the CNS during short-term plasticity. (2) Retinoblastoma-interacting myosin-like protein, RIM. (retinoblastoma-binding protein 8, 897aa) See **retinoblastoma**. (3) In *S. cerevisiae* RIM proteins are regulators of **IME2** (e.g. RIM20, regulator of IME2 protein, pH-response regulator protein palA, 661aa) cleaves the transcriptional repressor RIM101 in response to alkaline ambient pH).

RIN (1) A MADS-box transcription factor involved in ripening of fruit. (2) RPM1-interacting proteins (E3 ubiquitin protein ligases) in *Arabidopsis* involved in regulating plant resistance protein (RPM1 and RPS2)-dependent hypersensitive responses. (3) In humans, Ras interaction/interference protein 1 (RIN1, ras and rab interactor 1, 783aa) that interacts with the GTP-bound form of ras proteins and prevents the association between raf1 and Ras.

RING See E3 ligase and **RING finger motif**.

RING finger motif A zinc finger motif (40–60aa) found in various nuclear proteins and in some receptor-associated proteins. The RING (Really Interesting New Gene) finger, or $Cys_3HisCys_4$, family of zinc binding proteins play important roles in differentiation, oncogenesis, and signal transduction.

Ringer's solution Isotonic salt solution used for mammalian tissues; original version (for frog tissues) has been much modified and the term is often used loosely to mean any physiological saline.

rINNs Recommended International Non-Proprietary Names for medicinal substances. British Approved Names (BANs) have been harmonised with rINNs.

RIP140 Receptor-interacting protein 140 (nuclear receptor-interacting protein 1, 1158aa) a corepressor that modulates transcriptional activation by steroid receptors.

RIPA (1) HTH-type transcriptional repressor of iron proteins A (RipA, 331aa in *Corynebacterium glutamicum*), an AraC-type regulator of bacterial genes under conditions of iron limitation. (2) In *Dictyostelium discoideum*, ras-interacting protein A (ripA, 838aa) involved in integrating chemotaxis and signal relay pathways that are essential for aggregation. (3) Radioimmuno-precipitation assay. See **RIA**.

Ripk2 Receptor-interacting serine/threonine-protein kinase 2.

rippling muscle disease A rare disorder in which mechanical stimulation of skeletal muscle causes visible ripples to move over the muscle. One form is caused by a defect in **caveolin-3**.

RISCs RNA-induced silencing complexes involved in **RNA interference**: siRNAs incorporate into RISCs and provide guide functions for sequence-specific ribonucleolytic activity. Protein components of RISCs include **argonaute** family members, **dicer**, nucleases, and other factors.

ristocetin Mixture of the antibiotics ristocetin A and B: isolated from actinomycete, *Amycolatopsis* (*Nocardia*) *lurida*. Used experimentally to induce platelet aggregation.

rit 1 One of the ras-like small GTPases (ras-like without CAAX protein 1, 219aa) involved in the NGF signalling pathway.

R-loop Structure formed when RNA hybridizes to double-stranded DNA by displacing the identical DNA strand.

RLTPR A protein (RGD-, leucine-rich repeat, tropomodulin domain and proline-rich domain containing protein, 1435aa) that is expressed in most tissues. One of the **CARMIL** family.

RMI complex A complex that plays an important role in the processing of homologous recombination intermediates to limit DNA crossover formation in cells. The complex contains **topoisomerase** 3A (TOP3A) and RecQ-mediated genome instability proteins, RMI1 and RMI2. The RMI complex interacts with Bloom's syndrome protein (BLM).

RMP pathway A metabolic pathway (ribulose monophosphate pathway, allulose phosphate pathway) used by methylotrophic bacteria for the conversion of formaldehyde to hexose sugars amongst others. In the first stage ribulose-5-phosphate is condensed with HCHO.

RNA Ribonucleic acid. A linear polymer of $3'-5'$-linked ribose moieties with a purine (A or G) or pyrimidine (C or U) base attached to each in the 1 position. RNA has an informational role, a structural role and an enzymic role and is thus used in a more versatile way than either DNA or proteins. Considered by many to be the earliest macromolecule of living systems. RNA is transcribed from a DNA template and may be subsequently modified to include bases such as pseudouridine, ribothymidine and hypoxanthine. See **catalytic RNA**, **double stranded RNA**, **guide RNA**, **heterogeneous nuclear RNA**, **masked messenger RNA**, **messenger RNA**, **microRNA**, **non-coding RNA**, **nuclear RNA**, **RNA editing**, **RNA helicases**, **RNA interference**, **ribosomal RNA**, short hairpin RNA (shRNA), **small interfering RNA**, **small nuclear RNA**, **TAR**, transfer RNA (**tRNA**), **wobble hypothesis**.

RNA editing A process responsible for changes in the final sequence of mRNA that are not coded in the DNA template but alter the information content. Excludes mRNA **splicing** and modifications to tRNA. Various kinds of editing are known, the commonest being cytidine (C) to uridine (U) substitution, though polyadenylation of mitochondrial mRNA and guanosine (G) insertion in paramyxoviruses are other examples of editing. The process involves guide RNA (gRNA) in some cases but not all. Though the commonest examples are in organelles, the process

of editing does also occur in nuclear transcripts. See **ADARs**.

RNA helicases A large family of highly conserved enzymes (EC 3.6.4.13) that unwind nucleic acid helices using ATP as an energy source. RNA helicases are involved in transcription, splicing and translation. Examples include ATP-dependent RNA helicase A (DEAH box protein 9, 1270aa) that unwinds double-stranded DNA and RNA in a $3'$ to $5'$ direction, spliceosome RNA helicase BAT1 (428aa), part of the TREX complex specifically associated with spliced mRNA, **dicer**, the **DEAD-box helicases** and **RIG1**.

RNA interference An important mechanism of translational regulation in which mRNA is selectively degraded, following the binding of **small interfering RNA** (siRNA) and the formation of RISCs. It can be used experimentally to selectively delete a particular protein. See **microRNA**.

RNA plasmid (1) Double-stranded RNA found in yeasts, also known as killer factors. (2) In maize there is a family of autonomously replicating single- and double-stranded RNA plasmids in mitochondria. (3) Short hairpin RNA (shRNA) can be introduced into cells as part of an RNA plasmid to silence selected genes.

RNA polymerase mediator complex See **mediator complex**.

RNA polymerases Enzymes that polymerize ribonucleotides in accordance with the information present in DNA. Prokaryotes have a single enzyme for the three RNA types that is subject to stringent regulatory mechanisms. Eukaryotes have type I that synthesizes all **rRNA** except the 5S component, type II that synthesizes **mRNA** and **hnRNA** and type III that synthesizes **tRNA** and the 5S component of rRNA.

RNA primase See **DNA primase**.

RNA primer The short RNA sequence synthesized by **DNA primase**.

RNA processing Modifications of primary RNA trancripts that do not alter the information content, unlike **RNA editing**. Includes splicing, cleavage, base modification, capping and the addition of **poly-A tails**.

RNA recognition motif *RRM* A common motif (RNA-binding domain or ribonucleoprotein domain) found in a variety of RNA binding proteins, including various hnRNP proteins, proteins implicated in regulation of alternative splicing, and protein components of snRNPs. The domain is structurally flexible although characterized by the packing of two alpha-helices on a four-stranded beta-sheet. Database entry: http://pfam.sanger.ac.uk/family?acc = PF00076

RNA splicing See **alternative splicing**.

RNA thermometer An unusual class of mRNAs that regulate their translation initiation rate with temperature.

RNA tumour virus See **Retroviridae**.

RNAse protection assay Sensitive and quantitative alternative to **Northern blots** for the measurement of gene expression levels. Labelled antisense cRNA is transcribed from a DNA clone in an appropriate vector, hybridized with an mRNA sample, and single stranded RNA digested away with RNAse, then run out on a gel. The amount of labelled RNA surviving is directly proportional to the amount of target mRNA present in the sample.

RNAses *RNAases* See **ribonucleases**.

RNP See **ribonucleoprotein**.

RO-3306 An inhibitor of CDK1/cyclin B1 and CDK1/cyclin A that reversibly arrests human cells at the G2/M border of the cell cycle.

Robertsonian translocation A special type of non-reciprocal translocation that can occur in the five human acrocentric chromosome pairs, 13, 14, 15, 21, and 22. The two chromosomes break at their centromeres and the long arms fuse to form a single chromosome with a single centromere. The short arms, which have non-essential genes, form a reciprocal product but this is usually lost within a few cell divisions. Originally described in grasshoppers.

Robinow's syndrome A genetically heterogeneous developmental disorder some forms of which are caused by mutation in the ROR2 gene (neurotrophic tyrosine kinase receptor-2, NTRKR2) which can also cause **brachydactyly** type B. NTRKR2 is involved in the early formation of chondrocytes.

Robinow-Sorauf syndrome A phenotypic variant of **Saethre-Chotzen syndrome** caused by mutation in **twist**.

robo Receptor (roundabout), for **slit**. There are multiple roundabout homologues expressed in different tissues (e.g. roundabout homolog 1, 1651aa).

Rock See **rho kinases**.

Rocky Mountain spotted fever Acute infectious tick-borne rickettsial disease caused by *Rickettsia rickettsii*.

rod cell See **retinal rod**.

rod outer segment See **retinal rod**.

rodent ulcer See **basal cell carcinoma**.

rolling circle mechanism A mechanism of DNA replication in many viral DNAs, in bacterial F factors during mating, and of certain DNAs in gene amplification in eukaryotes. DNA synthesis starts with a cut in the + strand at the replication origin, the 5′ end rolls out and replication starts at the 3′ side of the cut around the intact circular DNA strand. Replication of the 5′ end (tail) takes place by the formation of **Okazaki fragments**.

rolling pebbles *Drosophila* proteins, one isoform, Rols7 (1900aa), is essential for myoblast fusion and overlaps with alpha-actinin and the N-terminus of D-titin/kettin/zormin in the Z-line of the sarcomeres.

Romanovsky-type stain Composite histological stains including methylene blue, Azure A or B and eosin, sometimes with other stains. Often used to stain blood films and allow differentiation of the various leucocyte classes. Examples are Giemsa, Wright's, and Leishman's stain.

ROMK An inwardly-rectifying ATP-dependent **potassium channel** (renal outer medullary potassium channel, Kir1.1, KCNJ1, 391aa) that transports potassium out of cells in the kidney. Mutations lead to a form of **Bartter's syndrome**.

ron A receptor tyrosine kinase (recepteur d'origine nantaise, CD136) for **macrophage stimulating protein**, a member of the receptor family that includes the proto-oncogene *met* and the avian oncogene *sea*. It is a trans-membrane heterodimer (one alpha- and one beta-chain disulphide-linked and derived from a 1400aa precursor).

root cap Tissue found at the apex of roots, overlying the root apical meristem and protecting it from friction as the root grows through the soil. The outer cells slough off as the root penetrates the soil. Secretes a polysaccharide mucilage as a lubricant.

root hair cell Root epidermal cell, part of which projects from the root surface as a thin tube, thus increasing the root surface area and promoting absorption of water and ions. It provides a mechanism for entry of a symbiotic bacterium that causes the tip of the hair to wrap around it, into a shepherd's crook configuration.

root nodule Globular structure formed on the roots of certain plants, notably legumes and alder, by symbiotic association between the plant and a nitrogen-fixing microorganism (*Rhizobium* in the case of legumes and *Frankia* in the case of alder and a variety of other plants).

rootlet system Microtubules associated with the base of the flagellum in ciliates and flagellates. Also associated with this region is the **rhizoplast**.

ros (1) An oncogene, *ros*, identified in bird **sarcoma** but highly-expressed in a variety of tumour cell lines, encoding a receptor **tyrosine kinase** that has high homology with the *Drosophila* **sevenless** protein. (2) Reactive oxygen species (ROS); see **metabolic burst**. (3) In *Agrobacterium* transcriptional regulator Ros is a prokaryotic zinc finger protein that regulates the plant *ipt* **oncogene**.

roscovitine A purine derivative that is a potent and selective inhibitor of **cyclin-dependent kinases**.

Rosetta Stone method A method in bioinformatics used to detect functional linkage between two proteins based on the principle that if two proteins A and B are found as separate sequences in some species but as a fused protein with both A and B regions in others then it is probable that the proteins are functionally associated. For example yeast genes Pur2 and Pur3 encode enzymes in the purine synthesis pathway: in *C. elegans* the single Ade 5,7,8 protein has sequences highly homologous with both Pur2 and Pur3. Knowing that both are found in a single protein makes it very probable that they are involved in a linked function. See **phylogenetic profile**, **gene neighbour method**.

rotamase Prokaryotic peptidyl-prolyl *cis-trans* isomerase, homologue of **immunophilins** but not inhibited by **cyclosporin**. Located in the periplasm. *E. coli* rotamase surA (EC 5.2.1.8, survival protein A, 428aa) is a chaperone involved in the correct folding and assembly of outer membrane proteins. In *Arabidopsis* rotamase CYP20-3 (peptidyl-prolyl *cis-trans* isomerase, 260aa) accelerates the folding of proteins in chloroplasts.

rotamer A rotational isomer – conformationally different by rotation at a single bond.

rotatin A protein (2226aa) involved in the genetic cascade that governs left-right specification and required for correct asymmetric expression of **nodal**, **lefty** and **PITX2**.

Rotavirus Genus of the Reoviridae having a double-layered capsid and 11 double stranded RNA molecules in the genome. They have a wheel-like appearance in the electron microscope, and cause acute diarrhoeal disease in their mammalian and avian hosts. A web-based tool for rotavirus genotype differentiation: http://rotac.regatools.be

rotenone An inhibitor of the **electron transport chain** that blocks transfer of reducing equivalents from NADH dehydrogenase to coenzyme Q. A very potent poison for fish and for insects.

Rothmund-Thomson syndrome A rare autosomal recessive dermatosis associated with multiple developmental defects, caused in some cases by mutation in the **RECQ-like helicase**, RECQL4.

Rotifera Small, unsegmented, pseudocoelomate animals (wheel animalcules) of the phylum Aschelminthes. Found in many freshwater environments and in moist soil, where they inhabit the thin films of water around soil particles. Move using cilia which may cause them to rotate, hence their name.

rottlerin A compound originally thought to be a selective protein kinase C delta (PKCδ) inhibitor but selectivity is poor and it is also an uncoupler of mitochondrial oxidative phosphorylation.

rough endoplasmic reticulum *RER* Membrane system of eukaryotes that forms sheets and tubules. Contains the receptor for the signal receptor particle and binds ribosomes engaged in translating mRNA for secreted proteins and the majority of transmembrane proteins. Also a site of membrane lipid synthesis. The membrane is very similar to the nuclear outer membrane. The lumen contains a number of proteins that possess the C-terminal signal **KDEL**.

rough microsome Small vesicles obtained by sonicating cells and that are derived from the rough endoplasmic reticulum. Have bound ribosomes and can be used to study protein synthesis.

rough strain Bacterial strains that have altered outer cell wall carbohydrate chains causing colonies on agar to change their appearance from smooth to dull. In Streptococci the smooth strains are virulent whereas the rough strains are not. This is partly because the rough strains are much more readily phagocytosed.

rouleaux Cylindrical masses of aggregated red blood cells. Horse blood will spontaneously form rouleaux, in other species it can be induced by reducing the repulsion forces between erythrocytes, for example by adding dextran of appropriate molecular weight.

Rous sarcoma virus *RSV* The virus responsible for the classic first cell-free transmission of a solid tumour, the chicken **sarcoma** reported by Rous in 1911. An avian **C-type** oncorna virus, original source of the *src* **gene**.

royalisin Insect **defensin** found in honeybee royal jelly.

R plasmid A conjugative plasmid (R factor; drug resistance factor) that confers resistance to one or more antibiotics or other poisonous compounds in a bacterium. Conjugation permits rapid spread through a population and R plasmids are a major problem in clinical medicine. For example, R100, a 94.3 kbp plasmid able to transfer between enteric bacteria, encodes resistance to sulphonamides, streptomycin, spectinomycin, fusidic acid, chloramphenicol, tetracycline and mercury.

RPMI 1640 Culture medium developed at Roswell Park Memorial Institute that utilizes a bicarbonate buffering system. Widely used for the culture of human normal and neoplastic leukocytes and when properly supplemented, will support growth of many types of cultured cells, including fresh PHA-stimulated human lymphocytes.

RPOTs A small family of nuclear-encoded T3/T7 phage-type DNA-dependent RNA polymerases (RPOTs, EC 2.7.7.6) responsible for transcription of mitochondrial and chloroplast genes in *Arabidopsis*. At least two nuclear-encoded RPOTs (RPOTm and RPOTmp) are located in mitochondria. RPOTm is important for normal pollen tube growth, female gametogenesis and embryo development. Abstract: http://www.ncbi.nlm.nih.gov/pubmed/20231244

RQ See **respiratory quotient**.

RRM See **RNA recognition motif**.

rRNA See **ribosomal RNA**.

RSC complex A chromatin-remodelling complex, related to the **SWI/SNF complex**, binds nucleosomes and naked DNA with comparable affinities. The RSC complex of *S. cerevisiae* is closely related to the SWI/SNF complex and they share conserved components. The RSC proteins Sth1, Rsc8/Swh3, Sfh1 and Rsc6 are homologues of the SWI/SNF proteins Swi2/Snf2, Swi3, Snf5 and Swp73 respectively.

RS domain See **SR proteins**.

R-spondin See **spondins**

RSV See **Rous sarcoma virus**.

RTE A region of an RNA molecule (RNA transport element) that is found in retrotransposons. RTE-directed mRNA export is mediated by a unknown cellular factor that act independently of the CRM1 nuclear export receptor, and is conserved among vertebrates.

RTF A plasmid that carries antibiotic resistance factors. There are two parts, one the RTF (resistance transfer factor) carrying genes for replication and transmission of the plasmid and the second consisting of one or more sequentially linked R determinants (resistance determinants).

RT-PCR A common form of PCR (reverse transcriptase polymerase chain reaction; reverse transcription PCR) in which the starting template is RNA, so an initial reverse transcriptase step is required to make a DNA template. Although some thermostable polymerases have appreciable reverse transcriptase activity it is more common to perform an explicit reverse transcription, inactivate the reverse transcriptase or purify the product, and proceed to a separate conventional PCR. Abbreviation ambiguous because also used sloppily for **real time PCR**.

RTX toxins A group of related cytolysins and cytotoxins produced by Gram-negative bacteria including *E. coli, Proteus vulgaris* (haemolysin), *Pasteurella haemolytica* (leukotoxin) and *Bordetella pertussis*

(adenylate cyclase-haemolysin). Characteristically contain a repeat domain (hence the designation, repeats in toxins) with glycine and aspartate-rich motifs repeated within the domain. All are produced in inactive pro-form that must be post-translationally modified to generate an active toxin and are calcium-dependent pore-forming toxins. See *E. coli* **haemolysin**. *Vibrio cholerae* RTX toxin causes the depolymerization of actin stress fibers, through the unique mechanism of covalent actin cross-linking. Article: http://www.ncbi.nlm.nih.gov/pmc/articles/PMC314022/

R-type channels A class of neuronal **voltage-sensitive calcium channels** of the 'high-voltage activated' (HVA) group that are blocked by nickel, and partially by omega-**agatoxin**-IIIA but are unaffected by dihydropyridines, phenylalkylamines and **conotoxins**. The alpha subunit is alpha-1E (2313aa), the beta and alpha-2/delta subunits regulate the activity. Thought to carry much of the current, stimulated by glutamate release in response to ischaemia, that induces neuronal death.

rubidium *Rb* One of the alkali earth metals, used to substitute for potassium in some ion flux experiments.

Rubinstein-Taybi syndrome A syndrome of mental retardation and morphological abnormalities that can be caused by mutation in the gene encoding CREB-binding protein or in the EP300 gene (see **p300**).

RUBISCO See **ribulose bisphosphate carboxylase**.

rubp See **ribulose bisphosphate carboxylase**.

RUDP carboxylase See **ribulose bisphosphate carboxylase**.

Ruffini's corpuscles Ovoid encapsulated sensory nerve ending in subcutaneous tissue. Probably mechanosensors.

ruffles Projections at the leading edge of a crawling cell. In time lapse films the active edge appears to ruffle. The protrusions are apparently supported by a microfilament meshwork, and can move centripetally over the dorsal surface of a cell in culture.

rugae Wrinkles. *Adj.* rugose, or if the wrinkles are relatively small, rugulose.

runt A protein (510aa) that regulates the expression of pair-rule genes in *Drosophila*. The runt domain is found in a range of transcription factors such as **Runx**.

Runx A family of transcription factors containing a **runt** domain. Chromosomal aberrations involving Runx1 (runt-related transcription factor-1, 453aa) are associated with a variety of disorder such as acute myelogenous leukaemia and some platelet disorders. Runx2 (521aa) is a global regulator of osteoblast differentiation, expressed several days before osteoblast genes in bone anlage and is inhibited by **twist**. Runx3 (acute myeloid leukemia 2 protein, oncogene AML-2 product, 415aa) binds to promoters of **lck**, IL-3 and GM-CSF and enhancers of the T-cell receptor genes.

Russell-Silver syndrome A growth retardation disorder (Silver-Russell dwarfism) caused by epigenetic changes of DNA hypomethylation at the telomeric **imprinting control region** (ICR1) on chromosome 11p15, involving the H19 and IGF2 genes.

rutabaga *Drosophila* memory mutant; gene codes for calcium/calmodulin-responsive adenylyl cyclase (2248aa); net result is elevated cAMP levels and a comparable behavioural defect to **dunce**.

ruthenium red A stain used in electron microscopy for acid mucopolysaccharides on cell surfaces. Also binds tightly to tubulin dimers and the **ryanodine receptor**.

rutin A phenolic compound produced as a defensive secondary metabolite by plants; it has generalized detrimental effects on herbivores and pathogens.

RXL motif *Cy motif* A hydrophobic region on some cyclin substrates that enhances the binding interaction with a hydrophobic region on the cyclin (although is neither necessary nor sufficient).

ryanodine Drug that blocks the release of calcium from the sarcoplasmic reticulum of skeletal muscle. Ryanodine-binding proteins have also been found in the CNS. The water-soluble plant extract ryania from the powdered stem of the tropical shrub *Ryania speciosa* has been used as an insecticide. The extract contains several structurally related ryanoids including ryanodine. See **ryanodine receptor**.

ryanodine receptor *RyR* Large transmembrane proteins that form tetrameric Ca^{2+} channels in association with 4 **calstabin** molecules which release calcium ions from the sarcoplasmic reticulum into the cytosol during muscle contraction. They are stimulated to transport Ca^{2+} into the cytosol by recognizing Ca^{2+} on the cytosolic side of the SR (calcium-induced calcium release) a positive feedback that leads to a rapid response. There are tissue-specific isoforms (skeletal muscle (RyR1, 5032aa) in skeletal muscle, RyR2 (4965aa) in cardiac muscle and RyR3 (4872aa) in brain. Show sequence similarity with $InsP_3$-gated calcium channels of the endoplasmic reticulum but are pharmacologically distinct. RyR1 mutations are associated with malignant hyperthermia.

ryk In atypical receptor protein tyrosine kinase (604aa) that is a coreceptor along with Frizzled for **Wnt** ligands and binds to **dishevelled**, through which it activates the canonical Wnt pathway.

S

S Single letter code for serine.

S1 (1) A soluble fragment (102 kDa) of **heavy mero-myosin** produced by papain cleavage: it retains the ATPase and actin-binding activity and motor function, and can be used to decorate actin filaments for identification by electron microscopy. (2) Ribosomal protein S1 plays a critical role in translation initiation and elongation in *E. coli* and is believed to stabilize mRNA on the ribosome. (3) See **S1 nuclease**.

S1 mapping See **Berk-Sharp technique** and **S1 nuclease**.

S1 nuclease (1) A single-strand specific nuclease (EC 3.1.30.1, 287aa), usually isolated from certain *Neurospora* and *Aspergillus* species, that degrades single-stranded nucleic acids and is more active against DNA than RNA. Used in the **Berk-Sharp technique** (S1 nuclease mapping) and to remove single-stranded extensions from DNA to produce blunt ends. (2) Ribonuclease S-1 (EC 3.1.27.1, 228aa) from *Pyrus pyrifolia* (Japanese pear) is an endonuclease and is involved in self-incompatibility. (3) See **pyocin-S1**.

S2 Fibrous fragment of heavy meromyosin (**HMM**). Links the **S1** head to the light meromyosin (**LMM**) region that lies in the body of the thick filament and acts as a flexible hinge.

S6 kinase A family of serine/threonine kinases, activated by **MAP kinase**, that phosphorylate ribosomal protein S6 to elevate protein production in cells stimulated by a **mitogen**. One example is S6 kinase alpha-1 (EC 2.7.11.1, MAP kinase-activated protein kinase 1a, 735aa) which may be involved in activating **CREB**.

S9 (1) Post-mitochondrial supernatant fraction of liver homogenate rich in drug-metabolising enzymes (cytochrome P450s), sometimes used in the **Ames test**. (2) A ribosomal protein found ubiquitously in the small subunit of ribosomes of prokaryotes and eukaryotes. (3) See **dermaseptins**.

S100 A large family of calcium-binding proteins containing an **EF-hand**. They also bind zinc at a different site and with higher affinity. S100A1 (94aa) is found particularly in heart but also other tissues. S100A4 (placental calcium-binding protein, Mts1, metastasin, p9Ka, pEL98, CAPL, calvasculin, Fsp-1, 101aa) is ubiquitously expressed and upregulated in various pathological conditions. S100A6 is **calcyclin**. S100A7 (101aa) is highly upregulated in psoriatic skin. S100A8 (**calgranulin A**) and S100A9

(calgranulin B) form heterodimeric **calprotectin**. S100A10 (97aa) is the light chain of **calpactin**, S100A12 (92aa) is calgranulin C. S100B (92aa) is involved with the regulation of protein phosphorylation in brain and has neurotrophic and mitogenic activity. Other members of the family include **calmodulin** and **troponin**.

S180 *sarcoma 180* Highly malignant mouse sarcoma cells, often passaged in **ascites** form. Used in some of the classical studies on **contact inhibition of locomotion**.

Sab Protein (SH3BP5, 455aa) that binds selectively to SH3 domain of **btk** and binds to and serves as a substrate for **JNK**.

Sabouraud's dextrose broth A culture medium used for moulds, yeasts and pathogenic fungi, particularly those associated with skin infections. It is acidic (pH 5.6) and selective for fungi. Informative supplier website: http://www.condalab.com/pdf/1205.pdf

SAC complex A multiprotein complex composed of **crumbs**, **patj** and **sdt** involved in determining polarity of epithelial cells.

saccade An eye movement in which the eyes jump from one point to another; the images are integrated centrally to give the impression that the whole field of view is being seen.

saccharomicins Saccharomicins A and B are heptadecaglycoside antibiotics isolated from the fermentation broth of the rare actinomycete *Saccharothrix espanaensis*. They are active both *in vitro* and *in vivo* against bacteria and yeast by disrupting membranes. http://www.ncbi.nlm.nih.gov/pmc/articles/PMC90028/

Saccharomyces Genus of Ascomycetes; yeasts. Normally haploid unicellular fungi that reproduce asexually by budding but have a sexual cycle in which cells of different mating types fuse to form a diploid **zygote**. Economically important in brewing and baking, and are also suitable eukaryotic cells for the processes of genetic engineering, and for the analysis of, for example, cell division cycle control by selecting for mutants (see **cdc genes**). *S. cerevisiae* is baker's yeast; *S. carlsbergensis* is now the major brewer's yeast. See also *Schizosaccharomyces pombe*.

Saccharomyces Genome Database A scientific database of the molecular biology and genetics of baker's yeast, *Saccharomyces cerevisiae*. Link to database: http://www.yeastgenome.org/

The Dictionary of Cell and Molecular Biology. DOI: http://dx.doi.org/10.1016/B978-0-12-384931-1.00019-2

saccharopine An intermediate in the aminoadipic pathway for the synthesis or degradation of lysine, synthesized from lysine and α-oxoglutarate in mammalian liver.

saccharopine dehydrogenase Cytoplasmic enzyme (EC 1.5.1.9, 429aa) that catalyzes the NAD^+-dependent cleavage of **saccharopine** to L-lysine and 2-oxoglutarate. In some organisms this enzyme is found as a bifunctional polypeptide with lysine ketoglutarate reductase, the first two linked enzymes of lysine catabolism. One of the AlaDH/PNT (alanine dehydrogenase/pyridine nucleotide transhydrogenase) family of enzymes.

sacsin A protein (4579aa) that is highly expressed in the central nervous system where it may act as a chaperone in protein folding. Mutations in the gene cause autosomal recessive spastic ataxia of Charlevoix-Saguenay (**ARSACS**), an early onset neurodegenerative disease.

sad1 (1) Constitutive membrane-bound component (514aa) of the yeast **spindle pole body** (SPB) that interact with Kms1. See **sun proteins**. (2) PremRNA-splicing factor (snRNP assembly-defective protein 1, 448aa) required for splicing of pre-mRNA in *S. cerevisiae*. (3) A serine/threonine kinase (BR serine/threonine-protein kinase 1, EC 2.7.11.1, 794aa) required for the differentiation of forebrain neurons. (4) SUN domain-containing protein 1 (812aa), a component of SUN-protein-containing complexes (LINC complexes) which link the nucleoskeleton and cytoskeleton. (5) *Arabidopsis SAD1* (supersensitive to ABA and drought) mutation increases plant sensitivity to drought stress and abscissic acid in seed germination. (6) Mycobacterial semialdehyde dehydrogenases, Sad1 and Sad2.

S-adenosyl methionine An activated derivative of **methionine** (S-(5′-deoxyadenosine-5′)-methionine), produced by **methionine adenosyltransferase**, that functions as a methyl group donor, in (for example) phospholipid methylation and bacterial **chemotaxis**.

Saethre-Chotzen syndrome An autosomal dominant syndrome in which there is fusion of bones (craniosynostosis) in the skull that are normally separate, caused by loss-of-function mutation in **twist-1**.

safranin A histological stain (safranin O, basic red 2), a mixture of dimethyl and trimethyl safranin, used as a counterstain. It stains nuclei red and metachromatically stains cartilage yellow.

SAGA (1) See **SAGA complex**. (2) In *Aspergillus fumigatus* a component of the actin cytoskeleton-regulatory complex (sagA, endocytosis protein 3, end3, 404aa) required for the internalization of endosomes.

SAGA complex A multisubunit **histone acetyltransferase** complex (Spt-Ada-Gcn5-acetyltransferase) originally described as being involved in transcriptional regulation in *S. cerevisiae*, required for RNA polymerase II-dependent transcription of several genes and that facilitates the binding of TATA-binding protein (TBP) during transcriptional activation. Components include **ataxin-7, ataxin-7-like protein 3**, enhancer of yellow 2 transcription factor homologue (**ENY2**), histone acetyltransferase GCN5L2, transcription initiation protein SPT3 homologue (SUPT3H), transcription factor II (TFIID) subunit TAF10, transformation/transcription domain-associated protein (**TRRAP**) and ubiquitin carboxyl-terminal hydrolase 22 (USP22).

SAGE *Serial Analysis of Gene Expression* A method for analysing overall gene expression patterns by generating short sequence tags (10–14 bp), each uniquely identifying a transcript, linking these together to produce a long serial molecule that can be cloned and sequenced and then quantifying the number of times a particular tag is observed, this being a measure of the expression level of the corresponding transcript. The necessary software is freely available. Link to Sagenet homepage: http://www.sagenet.org/

sagittal section Section through the median vertical longitudinal plane of an animal.

salicylic acid A naturally occuring antiseptic and anti-inflammatory compound (2-hydroxybenzoic acid) found especially in the bark of willow (*Salix*); the acetylated form is **aspirin** (acetylsalicylic acid). The ester, ethyl salicylate, is oil of wintergreen. In plants can act as a hormone and is a second messenger of oxidative stress signalling. It is involved in plant defence responses, senescence and regulation of flowering time.

salinosporamide A A potent proteasome inhibitor from the marine bacterium, *Salinispora tropica*. Currently in clinical trials for the treatment of multiple myeloma.

SALL proteins A family of zinc-finger transcription factors (SALL1, sal-like 1, 1324aa) involved in organogenesis. SALL1 defects cause **Townes-Brock syndrome**. It is the human homologue of the homeotic gene *spalt* (*sal*) of *Drosophila* required for the specification of posterior head and anterior tail regions in development. Other related proteins are known (SALL2, 1007; SALL3, 1300). SALL4 (1054aa) defects are associated with upper limb abnormalities.

Salla disease See **sialin**.

sallimus One of two proteins in insects (**projectin** and sallimus) that are functional homologues of vertebrate **titin**. The *Drosophila* gene *sallimus* (*sls*) encodes a protein of 2 MDa. The 5′ half of *sls* codes for zormin and **kettin**; both proteins contain Ig domains and can be expressed as separate isoforms. Abstract of article: http://www.ncbi.nlm.nih.gov/pubmed/17316686

Salmonella Genus of Enterobacteriaceae; motile, Gram-negative bacteria that, if invasive, cause enteric fevers (e.g. typhoid, caused by *S. typhi*), food-poisoning (usually *S. typhimurium* or *S. enteridis*, the latter notorious for contamination of poultry), and occasionally septicaemia in non-intestinal tissues.

SALT See **skin associated lymphoid tissue**.

saltatory Describing a pattern of movement in which there are distinct jumps (kangaroos move in a saltatory fashion). Intracellular particles can exhibit saltatory movements although the mechanism is unclear. Saltatory conduction is the rapid method by which nerve impulses move down a myelinated axon with excitation occurring only at **nodes of Ranvier**. This reduces the capacitance of the neuron, allowing much faster transmission. See **myelin**, **Schwann cells**. Saltatory replication is the sudden amplification (in generational terms) of a DNA sequence to generate many tandem copies, possibly the way in which **satellite DNA** arose.

salusins Bioactive peptides (salusin-alpha, 28aa; salusin-beta, 20aa) derived from **torsin 2A**. Salusin-alpha has mild hypotensive effects; salusin-beta stimulates arginine-vasopressin release from rat pituitary and causes rapid and profound hypotension and bradycardia. Informative supplier webpage: http://www.phoenixbiotech.net/Catalog%20Files/Salusin/Salusin.htm

salvador See **Salvador-Warts-Hippo pathway**.

Salvador-Warts-Hippo pathway A tumour-suppressor signalling pathway involved in tissue growth control and axis specification in *Drosophila*. The pathway is a ser/thr kinase cascade in which hippo (STE20-like kinase MST, 669aa), in complex with its regulatory scaffold protein salvador (608aa), phosphorylates and activates the ser/thr kinase warts (1105aa) that is complexed with its regulatory protein Mats (Mps one binder kinase activator-like 1, 219aa), which in turn phosphorylates and inactivates the **yorkie** oncoprotein. It also seems to be important in mammalian tumours by limiting activity of the yorkie oncoprotein; homologues of all the *Drosophila* components are found in humans. Research article: http://www.sciencedirect.com/science/article/pii/S0960982207020751

salvage pathways Metabolic pathways that allow synthesis of important intermediates from materials that would otherwise be waste products. An experimentally important pathway is that from hypoxanthine to nucleotides. See **HGPRT**.

SAM domain A protein interaction module (sterile alpha motif domain, ~70aa) found in diverse signal-transducing proteins including **EPH-related receptor tyrosine kinases**, serine/threonine protein kinases, cytoplasmic scaffolding and adaptor proteins, regulators of lipid metabolism, GTPases and transcription factors. SAM domains are known to form homo- and hetero-oligomers. Structural information: http://130.15.90.72/sam_domain.htm

sam68 Protein (src-associated in mitosis, 68 kDa, 443aa) that associates with and is tyrosine phosphorylated by src in a mitosis-specific manner. Has KH-type RNA-binding, SH2 and SH3 domains. Interacts with RNA, src-family kinases, grb2 and PLCγ. Inhibition of phosphorylation of sam68 by **radicicol** will block exit from mitosis.

Sandhoff's disease A **lysosomal disease** (storage disease) caused by mutation in the gene encoding the beta subunit of hexosaminidase (EC 3.2.1.52). GM2 gangliosides accumulate, particularly in neurons, leading to progressive neurodegeneration. Clinically indistinguishable from **Tay-Sachs disease**.

SANDO See **sensory ataxic neuropathy**.

Sanfilippo syndrome A **lysosomal disease** (mucopolysaccharidosis type III) characterized by severe central nervous system degeneration. There are four sub-types in which different enzymes are affected (heparan N-sulfatase in type A; alpha-N-acetylglucosaminidase in type B; acetyl CoA:alpha-glucosaminide acetyltransferase in type C; N-acetylglucosamine 6-sulphatase I in type D). Cross correction (complementation) of co-cultured fibroblasts from apparently clinically-identical patients can occur if a different enzyme is missing.

Sanger-Coulson method See **dideoxy sequencing**.

Sanjad-Sakati syndrome See **hypoparathyroidism**.

SANS Scaffold protein containing ankyrin repeats and SAM domain. A scaffold protein (461aa) found in the apical region of cochlear and vestibular hair cells but not in stereocilia. It interacts with **harmonin** and myosin 7A. Mutations cause a form of deafness (**Usher Syndrome 1G**).

S-antigen (1) An abundant protein (S-arrestin, 409aa) of the retina and pineal gland that elicits experimental autoimmune uveitis. See **arrestins**. (2) Soluble heat-stable antigens (250–650aa, very variable between isolates) from the surface of *Plasmodium falciparum* that are responsible for antigenic heterogeneity and are found in sera of some infected individuals.

SAP (1) See **serum amyloid P-component**. (2) **SLAM-associated protein**. (3) See **saposin**. (4)

Spliceosome-associated proteins of various sizes (e.g. SAP145, pre-mRNA-splicing factor SF3b 145 kDa subunit, 895aa). **(5)** SIT4-associating protein SAP155 in *S. cerevisiae*, a positive regulator of protein phosphatase SIT4. **(6) Histone deacetylase** complex subunit SAP30 (sin3-associated polypeptide p30, 220aa). **(7)** Receptor-type tyrosine-protein phosphatase H (EC 3.1.3.48, stomach cancer-associated protein tyrosine phosphatase 1, SAP-1, 1115aa). **(8)** See **sapintoxin D** (SAPD).

sapecin A **defensin** produced by the flesh fly, *Sarcophaga peregrina*.

sapintoxin D Fluorescent, highly potent phorbol ester (12-N-methylanthraniloylphorbol 13-acetate; SAPD) that is a selective activator of protein kinase C (PKC)-alpha.

SAPK4 (1) Stress-activated protein kinase 4, one of the MAP kinase family (EC 2.7.11.24, MAPK13, 365aa), closely related to p38 MAP kinase (MAPK11, SAPK2, 364aa), both of which can be activated by proinflammatory cytokines and cellular stress. Transcription factor ATF2, and **stathmin** are substrates of SAPK4. **(2)** SNF1-type serine-threonine protein kinase (SAPK4, EC 2.7.11.1, osmotic stress/abscisic acid-activated protein kinase, 4360aa) regulates stress-responsive gene expression in rice. There are many similar stress-responsive kinases in rice (SAPK1- SAPK10).

saponins Glycosidic surfactants produced by plant cells. Used to permeabilise membranes (being less harsh than, for example, Triton X-100) and as foaming agents in some beverages.

saporin A ribosome-inactivating protein (253aa), from seeds of the plant *Saponaria officinalis* (soapwort). Acts as an RNA-N-glycosidase, specifically depurinising the 28S RNA of ribosomes.

saposins Small (∼80aa), heat-stable glycoprotein activators of lysosomal glycosphingolipid hydrolases (saposins A, B, C and D) derived by proteolysis from a single precursor, prosaposin. Lysosomal degradation of several sphingolipids requires the presence of saposins and they are important for exchange of lipids from CD1 molecules to lysosomal antigens. Other functions include neuritogenic/neuroprotection effects and induction of membrane fusion. Saposins B, C, and D share common structural features including a lack of tryptophan, a single glycosylation sequence, the presence of three conserved disulphide bonds, and a common helical bundle motif. Saposin A contains an additional glycosylation site and a single tryptophan. Saposin-like proteins (SAPLIPs) have the same motif and similar location of six cysteines; they include surfactant protein B (SP-B), *Entamoeba histolytica* pore-forming peptide, **granulysin**, NK-lysin, acid sphingomyelinase and acyloxyacyl hydrolase. Deficiency of saposin A causes atypical **Krabbe's disease**, deficiency of

saposin B causes metachromatic **leukodystrophy**, deficiency of saposin C causes an atypical form of **Gaucher's disease**.

sapoviruses A genus of the Calicivirus family responsible for mild gastroenteritis in infants.

saprophyte Organism that feeds on complex organic materials, often the dead and decaying bodies of other organisms. Many fungi are saprophytic

SAR (1) See **structure-activity relationship** and **systemic acquired resistance. (2)** Secretion-associated and ras-related protein (sar1p). In yeast, a small GTPase (190aa) that controls the assembly of the **coat protein complex II** (COPII) that surrounds vesicles that are involved in export from the ER. The Sar1p-GTP complex initiates membrane curvature during vesicle biogenesis. The human homologue is SAR1 (SAR1A and SAR1B, both 198aa); mutations in SAR1B are associated with **Anderson disease** and chylomicron retention disease.

sarafotoxins Group of cardiotoxic venoms (SRTX, 21aa) from *Atractaspis engaddensis* (burrowing asp), structurally related to the **endothelins**.

sarcin/ricin loop A highly conserved sequence found in the RNA of all large ribosomal subunits. **Alpha-sarcin** and **ricin** both inactivate ribosomes by cleaving a single bond in the loop. This prevents the interaction with elongation factors and stops translation.

sarcodictyins Tricyclic compounds isolated from the Japanese soft coral *Bellonella albiflora*, that stabilize microtubule bundles.

Sarcodina A large phylum of aquatic protozoa that includes Amoebae, Foraminifera and Radiolaria.

sarcoglycans Transmembrane proteins that associate with **dystroglycan** in the **sarcolemma** and together with **dystrophin** and **syntrophins** link the contractile machinery to the extracellular matrix. The sarcoglycans are subdivided into the sarcoglycan alpha/epsilon and sarcoglycan beta/delta/gamma/zeta families. Alpha-sarcoglycan (50DAG, A2, adhalin, 387aa), β-sarcoglycan (43DAG, A3b, 318aa), γ-sarcoglycan (35DAG, A4, 294aa), δ-sarcoglycan (289aa), ε-sarcoglycan (437aa) and ζ-sarcoglycan (299aa). Defects in sarcoglycans have been shown to be associated with autosomally inherited **muscular dystrophy**.

sarcoidosis Disease (Besnier-Boeck disease) of unknown aetiology in which there are chronic inflammatory granulomatous lesions in lymph nodes and other organs with an accumulation of CD4[+] T-cells and a Th1 immune response.

sarcolemma Plasma membrane of a striated muscle fibre.

sarcolipin A proteolipid (31aa) that inhibits the cardiac sarcoplasmic reticulum Ca^{2+} ATPase (SERCA2a) by direct binding and is superinhibitory if it binds as a binary complex with **phospholamban** (PLN).

sarcoma A malignant tumour derived from connective tissue cells, e.g. osteosarcoma (from bone). Some sarcomas are of viral origin: see **Rous sarcoma virus**, and **src gene**. Less common than carcinomas but tend to have an earlier onset and a worse prognosis.

sarcoma virus Virus that causes tumours that originate from cells of connective tissue such as **fibroblasts**. See **Rous sarcoma virus** and **src gene**.

Sarcomastigophora Phylum of unicellular protozoa with pseudopodia or flagella or both.

sarcomere Repeating subunit from which the **myofibrils** of striated muscle are built. Has **A-** and **I-bands**, the I-band being subdivided by the **Z-disc**, and the A-band being split by the **M-line** and the H-zone.

sarcoplasm Cytoplasm of striated muscle fibre.

sarcoplasmic reticulum Endoplasmic reticulum of striated muscle, specialized for the sequestration of calcium ions that are released upon receipt of a signal relayed by the **T tubules** from the neuromuscular junction.

sarcoplasmic-endoplasmic reticulum Ca^{2+}-ATPase *SERCA* The calcium-ATPase pump (EC 3.6.3.8, 1043aa) of the sarcoplasmic reticulum (SR) that moves calcium ions from the sarcoplasm into the SR, reducing the calcium concentration around the myofibrils and allowing the muscle to relax. A different ATPase, the plasma-membrane Ca^{2+} ATPase (PMCA, EC 3.6.3.8, 1258aa), pumps calcium to the extracellular space. There are muscle type-specific and non-muscle isoforms.

sarcosine A natural amino acid (N-methyl glycine) found in muscle and other tissues; has a sweet taste and is used in toothpaste. Reported to be an endogenous antagonist of glycine transporter-1 and have some beneficial effects in treating schizophrenia.

sarcosyl *sarkosyl* A mild, biodegradable anionic surfactant (N-lauroyl-sarcosine) derived from fatty acids and **sarcosine**, used in preparing solubilised fractions of biological materials.

sarcotoxins Potent bactericidal proteins produced by insects in response to injury. They are cytotoxic to Gram-positive and Gram-negative bacteria. Sarcotoxin-1A (63aa) from *Sarcophaga peregrina* (Flesh fly) is a cecropin, as are the other sarcotoxin-1 variants; sarcotoxin-2 forms (e.g. sarcotoxin-2A, 294aa) are related to **attacins**.

sarin Nerve gas; inhibitor of acetylcholine esterase.

SARPs Streptomyces Antibiotic Regulatory Proteins. Pathway-specific regulatory proteins encoded within antibiotic biosynthetic-gene clusters in several antibiotic pathways in *Streptomyces*. They activate transcription by binding to a tandemly arrayed set of heptameric, direct repeats located around the -35 region of their cognate promoters. There are also pleiotropic regulatory genes. Review: http://mic.sgmjournals.org/cgi/reprint/142/6/1335?view=long&pmid=8704973

SARS Severe Acute Respiratory Syndrome. A highly infectious respiratory disease caused by a **coronavirus**, first recognised in the Far East in 2003 where it caused a major epidemic with a fatality rate of 11%.

satellite cell (1) Sparse population of mononucleate cells found in close contact with muscle fibres in vertebrate skeletal muscle. Seem normally to be inactive, but may be important in regeneration after damage. May be considered a quiescent stem cell. (2) An alternative name for glial cell.

satellite chromosome A small segment of a chromosome that is connected, usually to the short arm of an acrocentric chromosome, by a narrow neck (satellite stalk) which appears as an unstained gap in chromosome preparations. In humans chromosomes 13–15 and 21–22 have satellites. *Cf.* **satellite DNA**.

satellite DNA DNA containing highly repetitive sequences, that has a base composition (and thus density in ultracentrifugation) significantly different from normal DNA. Typically 10% of mammalian, and 50% of insect genomes are composed of satellites. As satellites are dispersed widely in the genome, are easily detectable (with a highly repetitive probe) and are frequently polymorphic in length, they are ideal markers for **linkage** studies of disease or inheritance, and for genomics. Minisatellites (variable number of tandem repeats, VNTRs) have core repeats of 9–80 bp, microsatellites (short tandem repeats, STRs, simple sequence repeats, SSRs) have repeats of only 2–6 bp. See also **Fragile X, Huntington's chorea**.

satellite virus A term used in plant virology for a virus associated functionally, at least for the purpose of its own replication, with another virus.

saturated fatty acids Fatty acids without double bonds. In eukaryotic membranes mostly stearic, palmitic and myristic acids, linear aliphatic chains with no double bonds. Prokaryotes have numerous branched-chain saturated fatty acids.

saturation The state in which all receptors or binding sites are effectively occupied all the time, can be said to occur in a simple binding equilibrium when the concentration of ligand is more than five times the K_d value, although strictly it will only be true at infinite ligand concentration.

saturation density The maximal population density achieved by a cell type grown under particular *in vitro* culture conditions. Although transformed cells generally grow to a higher saturation density than normal cells this is not necessarily the case. Many factors affect the final density achieved by a cell population; the critical factor may be availability of surface upon which to spread or the serum concentration in the medium. Population densities in culture never approach those found in whole organisms.

sauvagine Peptide (40aa) originally isolated from the skin of the frog, *Phyllomedusa sauvagei*, closely related to **corticotrophin releasing factor** and to urotensin I.

sax *saxophone* The receptor in *Drosophila* for **gbb**, the ortholog of the human Activin Receptor-Like Kinase-1 and -2. A serine/threonine protein kinase (570aa). Research article: http://www.genetics.org/content/183/2/563.long

saxitoxin *STX* **Neurotoxin** produced by the 'red tide' dinoflagellates, particularly *Alexandrium* (formerly *Gonyaulax*) *catenella*. It accumulates in shellfish and when ingested binds to the **sodium channel**, blocking the passage of action potentials. Its action closely resembles that of **tetrodotoxin**. The toxin was originally isolated from the clam, *Saxidomus giganteus* and is responsible for paralytic shellfish poisoning.

sbcc protein In *E. coli*, a nuclease (1048aa) that cleaves DNA hairpin structures. Forms a heterodimer with sbcD.

SBDS See **Shwachman-Diamond syndrome**.

SC-35 See **ASF/SF-2**.

scabies A contagious skin disease caused by the mite *Sarcoptes scabiei* which burrows in the horny layer of the skin and causes an inflammatory reaction.

scaffold-attachment regions Specific DNA sequences involved in interactions with the protein nuclear scaffold. Scaffold attached regions (SARs) are thought to be the bases of DNA loops and are sections of genome of variable length.

scaffold/radial loop model Model for chromatin organization in eukaryotic metaphase chromosomes. Involves a non-histone protein core that is coiled and to which the linear DNA molecule has an ordered series of attachment points every 30–90 kbp, with intervening DNA forming a loop that is supercoiled or folded. The 150–200 nm-diameter central core contains structural maintenance of chromosomes 2 (SMC2) protein and topoisomerase II.

scaffolding proteins In general, proteins that assist in the formation of large multi-molecular complexes. The term is applied, for example, to proteins that maintain the clustering of particular receptors at synapses (see, *inter alia*, **caveolins**, **flotillins**), to proteins involved in assembling the viral capsid, and to proteins involved in eukaryotic chromosome structure. See also **AKAP79, involucrin, titin**.

scaffoldins A family of proteins (CipA, cipB etc., ~2000aa) of the bacterial **cellulosome** that organize and position other protein subunits into the complex. The scaffoldins (scaA (cipBc), scaB, scaC, and scaD) can also serve as an attachment device for fastening the cellulosome to the cell surface and/or for its targeting to cellulose substrate.

SCAMPs See **secretory carrier-associated membrane proteins**.

scanning electron microscopy *SEM* Technique of electron microscopy in which the specimen is coated with heavy metal, and then scanned by an electron beam. The image is built up on a monitor screen, in the same way as the raster builds an image on a cathode ray tube. The resolution is not so great as with transmission electron microscopy, but preparation is easier (often by fixation followed by **critical point drying**), the depth of focus is relatively enormous, the surface of a specimen can be seen (though not the interior unless the specimen is cracked open), and the image is aesthetically pleasing.

scanning ion conductance microscopy A contact-free scanning technique which uses electrical resistance changes to detect the distance between the scanning tip, a glass microelectrode, and an insulator, which can be the surface of a live cell. Variants of the technique involve using current pulses to avoid electrode drift and other interference. Article: http://www.formatex.org/microscopy3/pdf/pp968-975.pdf

scanning probe microscopy *SPM* Methods for visualizing surfaces at microscopic scale that rely on moving a tiny probe over a surface (usually in an x-y scan), and recording some property of interest (current, force) at each coordinate. These techniques have the ability to resolve detail down to single atoms. See also **scanning tunnelling microscopy, scanning ion conductance microscopy** and **atomic force microscopy**.

scanning transmission electron microscopy *STEM* Method of electron microscopy in which image formation depends upon analysis of the pattern of energies of electrons that pass through the specimen. Has comparable resolving power to conventional transmission EM.

scanning tunnelling microscopy *STM* A form of ultra-high resolution microscopy of a surface in which a very small current is passed through a surface and is detected by a microprobe of atomic dimensions at its tip that scans the surface by use of a piezodrive. In the simplest form the current transferred to the probe is recorded as an indication of the contours of molecules on the surface above the local plane. In more complex forms feedback is used to hold the probe at a constant difference and the signal in the feedback loop indicates the contours of the molecule. Capable of resolving single atoms and known to work for non-conducting molecules as well as conducting ones.

SCAP SREBP cleavage-activating protein (1279aa), a protein involved in feedback inhibition of the sterol regulatory element-binding protein (**SREBP**) pathway. Sterols prevent movement of the SCAP/ SREBP complex from the endoplasmic reticulum to the Golgi, where proteolytic cleavage of SREBPs would release the transcription factor domain and activate genes for lipid biosynthesis.

SCAPER S phase cyclin A-associated protein in the endoplasmic reticulum. A zinc-finger regulatory protein (1399aa) that transiently maintains the **cyclinA2-CDK2** complex in the cytoplasm. Does not interact with cyclinA2 uncomplexed with CDK2, nor with other cyclin-CDC complexes.

scar A WASP-related protein, binds the p21 subunit of the **Arp 2/3** complex and is an endogenous activator of actin polymerization. See **WAVEs**.

scarb Gene family that encodes **scavenger receptors** of class B. These receptors bind ligands such as phospholipids, cholesterol ester, lipoproteins, phosphatidylserine and apoptotic cells and are located in **caveolae**. They facilitate the flux of free and esterified cholesterol between the cell surface and extracellular donors and acceptors. Scarb1 (CD36-like 1, 552aa), scarb2 (CD36L-2, 478aa).

SCARECROW Transcription factor (GRAS-family protein 20, 653aa) required for **quiescent centre** specification and maintenance of surrounding stem cells, and for the asymmetric cell division involved in radial pattern formation in roots. Also required for normal shoot gravitropism. There is a family of Scarecrow-like (SCL) putative transcription factors. See **GRAS family**.

SCARF Endothelial cell scavenger receptors. SCARF1 (scavenger receptor class F, member 1, acetyl LDL receptor, 830aa) is involved in the binding and degradation of acetylated low density lipoprotein but also in heterophilic adhesion. SCARF2 (scavenger receptor expressed by endothelial cells 2, 866aa) is probably an adhesion protein mediating homophilic and heterophilic interactions. Mutations in SCARF2 are associated with **Van Den Ende-Gupta syndrome**.

Scatchard plot A method for analysing data for freely reversible ligand/receptor binding interactions. The graphical plot is: (Bound ligand/Free ligand) against (Bound ligand); the slope gives the negative reciprocal of the binding affinity, the intercept on the x-axis the number of receptors (Bound/Free becomes zero at infinite ligand concentration). The Scatchard plot is preferable to the **Eadie-Hoffstee plot** for binding data because it is more dependent upon the values at high ligand concentration, the most reliable values. A non-linear Scatchard plot is often taken to indicate heterogeneity of receptors, although this is not the only explanation possible.

scatter factor A motility factor (**motogen**) isolated from conditioned medium in which human fibroblasts have been grown. It causes colonies of epithelial and endothelial cells, in culture, to separate into single cells that move apart i.e. they scatter. It has been shown to be identical to human **hepatocyte growth factor** (728aa), but it is not mitogenic for all cell types. Defects are responsible for a form of autosomal recessive deafness.

scavenger receptors Structurally diverse family of receptors on macrophages that are involved in the uptake of modified **LDL** and have been implicated in development of atherosclerotic lesions. Six classes are recognized with different binding preferences. Macrophage scavenger receptors Class A bind a wide range of ligands, including bacteria and scavenger receptors may be important in recognizing apoptotic cells. See **collectins** and **scarb**.

Scenedesmus A non-motile colonial alga, of the Order Chlorococcales, consisting of 2, 4 or 8 elongated cells, often with long spines on the terminal cells. Common in ponds and as planktonic forms in rivers and lakes.

S cells See **secretin**.

SCF (1) See **stem cell factor**. (2) SCF complexes are a class of E3 ubiquitin protein ligases (a multiprotein aggregate of Skp1p-cdc53p-F-box protein) that play a role in regulation of cell division. The **F-box** component of the complex gives substrate specificity for ubiquitinylation. See **skp** and cullin.

scFv Single-chain variable fragment. A fusion of the variable regions of the immunoglobulin heavy and light chains linked together with a short peptide linker and produced by recombinant methods in bacteria. Can be selected to be specific for antigens of choice by various selection methods.

***SCG10* gene** A neural-specific gene (superior cervical ganglion-10) that encodes a growth-associated protein (179aa) expressed early in the development of neuronal derivatives of the neural crest and that is associated with the membranous organelles that accumulate in growth cones. SCG10-like protein (SCLIP, stathmin-like 3, 180aa) forms a complex

(2 tubulins: 1 stathmin-like protein) thereby affecting microtubule dynamics. See **stathmin**.

Scheie syndrome Mucopolysaccharidosis (**lysosomal disease**) in which there is a defect in α-L-iduronidase. Fibroblasts from Scheie syndrome patients do not cross-correct fibroblasts from **Hurler's disease**; Hurler and Scheie syndromes represent phenotypes at the severe and mild ends of the clinical spectrum.

Schick test A test, introduced by Schick in 1913, to assess the degree of susceptibility or immunity of individuals to diphtheria by challenging with a small amount of diphtheria toxin injected intracutaneously into one forearm.

Schiff base The product of the reaction of a primary amine with an aldehyde or ketone, an imine. When an arylamine is used the Schiff base may form an intermediate in a staining reaction, e.g. for polysaccharides.

Schiff's reagent See **periodic acid-Schiff reaction**.

Schilder's disease A rare progressive demyelinating disorder (myelinoclastic diffuse sclerosis, diffuse sclerosis, encephalitis periaxialis) which usually begins in childhood, a variant of multiple sclerosis.

schistocytes Fragments of red blood cells found in the circulation.

schistosomiasis *bilharzia* Disease caused by trematode worms (flukes). Three main species, *Schistosoma haematobium*, *S. japonicum*, and *S. mansoni*, cause disease in man. Larval forms of the parasite live in freshwater snails; cercariae liberated from the snail burrow into skin, transform to the schistosomulum stage, and migrate to the urinary tract (*S. haematobium*), liver or intestine (*S. japonicum*, *S.mansoni*) where the adult worms develop. Eggs are shed into the urinary tract or the intestine and hatch to form miracidia which then infect snails, completing the life cycle. Adult worms cause substantial damage to tissue and seem to resist immune damage by mechanisms that are not fully understood.

schistosomulum *Pl. schistosomula.* See **schistosomiasis**.

schizocoel Coelom that is developed within the mass of mesoderm by splitting or cleavage. *Cf.* **enterocoel**.

schizogeny A mechanism of **aerenchyma** formation in plants in which development results in the cell separation. Schizogenous aerenchyma is common in wetland species like *Rumex* (dock) and is formed by cell separation, without the cells dying. See **lysigeny**.

schizogony The division of cells, especially of protozoans, in non-sexual stages of the life history of the organism.

Schizosaccharomyces pombe Species of fission yeast commonly used for studies on cell cycle control because there is a distinct G2 phase to the cycle. Only distantly related to the budding yeast *Saccharomyces cerevisiae*. A further advantage for experimental studies is that some mammalian introns are processed correctly.

Schmidt's syndrome See **autoimmune polyendocrinopathy syndrome**.

Schulman-Upshaw syndrome See **thrombotic thrombocytopenic purpura**.

Schultz-Charlton test Old test for scarlet fever in which antibody to **erythrogenic toxin** of *Streptococcus pyogenes* is injected subcutaneously.

Schwann cell A specialized glial cell that wraps around vertebrate **axons** providing extremely good electrical insulation. Separated by **nodes of Ranvier** about once every millimetre, at which the axon surface is exposed to the environment. See **saltatory conduction**, **myelin**.

Schwannoma-derived growth factor *SDGF* See **amphiregulin**.

Schwartz-Jampel syndrome A progressive growth disorder. Type 1 (SJS1) is caused by mutation in the gene encoding **perlecan**. The Silverman-Handmaker type of dyssegmental dysplasia is caused by a different mutation in the same gene and has a more severe phenotype. Neonatal Schwartz-Jampel syndrome type 2 (Stuve-Wiedemann syndrome) has a more severe phenotype caused by mutation in the **leukaemia inhibitory factor** receptor (LIFR) gene.

Schwartzmann reaction Mis-spelling of **Shwartzman** reaction.

SCID See **severe combined immunodeficiency disease**.

scinderin See **adseverin**.

scintillation counting Technique for measuring quantity of a radioactive isotope present in a sample. In biology, liquid scintillation counting is mainly used for β emitters such as ^{14}C, ^{35}S and ^{32}P and particularly for the low energy β emission of ^3H. Gamma emissions are often measured by counting the scintillations that they cause in a crystal. Autoradiographic images can be enhanced by using a screen of scintillant behind the film.

scintillation proximity assay Assay system in which antibody or receptor molecule is bound to a bead that will emit light when β emission from an isotope occurs in close proximity, i.e. from a radioactively-labelled ligand. Avoids the need for scintillant in order to measure the amount of bound isotope — and thus the amount of antigen or ligand present.

scintillons Cytoplasmic bodies (ca. 0.5 μm in diameter) found in the cortical region of dinoflagellates as an outpocket of the main cell vacuole. They contain **luciferase** and **luciferin** and are responsible for the production of luminescence, a brief (0.1 sec) blue flash (max 476 nm) when stimulated, usually by mechanical disturbance. Dinoflagellates produce most of the bioluminescence in the oceans.

SCIP POU-domain transcription factor (Oct-6, Tst-1, 451aa) expressed by promyelinating Schwann cells (where it represses expression of the myelin structural genes) and, in tissue culture, by oligodendrocyte progenitors.

scirrhous carcinoma Carcinoma having a hard structure because of excessive production of dense connective tissue.

sclereid Type of **sclerenchyma** cell that differs from the **fibre cell** by not being greatly elongated. Often occurs singly (an idioblast) or in small groups, giving rise to a gritty texture in, for instance, the pear fruit, where it is known as a 'stone cell'. May also occur in layers, e.g. in hard seed coats.

sclerenchyma Plant cell type with thick lignified walls, normally dead at maturity and specialized for structural strength. Includes **fibre cells**, that are greatly elongated, and **sclereids**, that are more isodiametric. Intermediate types exist.

scleritis An inflammatory disease that affects the conjunctiva, sclera, and episclera (the connective tissue between the conjunctiva and sclera). It is associated with underlying systemic diseases such as **Wegener's granulomatosis** or rheumatoid arthritis in about half of the cases.

scleroderma A chronic autoimmune connective tissue disorder (systematic sclerosis) in which there is immune activation, vascular damage, and fibrosis of the skin and major internal organs. There are some genetic predisposing factors and antibodies to **fibrillarin** are found in some cases. See **CREST syndrome, RNaseP.**

sclerosis Pathological hardening of tissue. But see **amyotrophic lateral sclerosis, Balo's concentric sclerosis, multiple sclerosis, Schilder's disease, scleroderma, tuberous sclerosis.**

sclerosteosis A rare autosomal recessive bone dysplasia affecting mainly the skull and mandible and leading to facial paralysis and hearing loss. Caused by mutation in the coding region of the gene for **sclerostin**. In **van Buchem's disease** there is down-regulation of sclerostin production.

sclerostin An antagonist (213aa) for **bone morphogenic protein** (BMP) that represses osteoblast differentiation and function but in turn is inactivated by **noggin**. Sclerostin has a cystine knot motif (residues 80–167) similar to that in the **cer/dan family**. The gene is mutated in **sclerosteosis** and downregulated in **van Buchem's disease**. Sclerostin-domain containing protein-1 (ectodin, 206aa) inhibits BMP2, BMP4, BMP6, and BMP7.

sclerotia Quiescent, multicellular aggregates of vegetative hyphae (several mm to several cm in size) which allow species of the fungus *Sclerotinia* to survive for long periods of time under adverse conditions. They have a black, melanized rind which is resistant to microbial invasion and physical insult. Sclerotia can undergo **myceliogenic** or **carpogenic germination** to produce infective hyphae or **apothecia**.

sclerotin Hard, dark-coloured cross-linked (tanned) protein found in the cuticle of insects and some other arthropods.

Sclerotinia sclerotiorum A nonspecific, ascomycete pathogen of plants. The fungus infects 64 families of plants, many of economic importance. About 90% of the life cycle of *Sclerotinia* species is spent in soil as **sclerotia**, which germinate under the right conditions to form a mycelium which can infect a host, or an **apothecium**. Resource page: http://www.sclerotia.org/

SCN (1) The suprachiasmatic nucleus of the hypothalamus. (2) The thiocyanate anion (SCN⁻). (3) See voltage-gated sodium channels.

SCO-spondin See **spondins**.

scoliosis An abnormal lateral curvature of the spine which may be idiopathic or a secondary feature of disorders including **Marfan's syndrome, dysautonomia, neurofibromatosis, Friedreich's ataxia** and **muscular dystrophies.**

scombrotoxin Causative agent of scombroid poisoning (histamine poisoning), caused by eating foods (spoiled fish, some cheeses) with high levels of histamine and possibly other vasoactive amines and compounds produced by bacteria.

scopolamine *hyoscine* An alkaloid found in thorn apple (*Datura stramonium*). Related to atropine both in effects and structure and acts as a **muscarinic acetylcholine receptor** antagonist.

scopoletin A naturally occurring fluorescent component (7-hydroxy-6-methoxycoumarin) of some plants that acts as a plant growth inhibitor. Said to lower blood pressure in hypertension and raise it in hypotension; also to be bacteriostatic and antiinflammatory. An acetylcholine esterase inhibitor.

scorpion toxins Polypeptide toxins (~7 kDa) with four disulphide bridges. The α-toxins are found in venom of Old World scorpions, β-toxins in those of the New World. Bind with high affinity to the voltage-sensitive **sodium channel** of nerve and muscle (α and β toxins bind to different sites).

scotophobin Peptide (15aa) isolated from brains of rats trained to avoid the dark that will transfer this aversion to naive animals. The original claim was treated with considerable scepticism and the topic remains contentious.

scramblases A family of cytoplasmic membrane-associated proteins (phospholipid scramblase-1, PLSCR1, 318aa; PLSCR2, 224aa; PLSCR3, 295aa; PLSCR4, 329aa) that mediate an ATP-independent Ca^{2+}-dependent transbilayer flip/flop of membrane lipids, causing the loss of the normal phospholipid asymmetry of the plasma membrane. They have some similarities with **tubby**. May have a role in initiating fibrin clot formation, activating mast cells, the recognition of apoptotic and injured cells and amplifying the interferon response. Research article: http://bioinformatics.oxfordjournals.org/content/25/2/159.long

scrapie A chronic neurological disease of sheep and goats, similar to other **spongiform encephalopathies** and much used as a model for studying the diseases. Controversy still surrounds the nature of the transmissible agent, although the idea of slow viruses has been overtaken by Prusiner's **prion** hypothesis, which is now fairly generally accepted. Atypical forms of the disease seem to be emerging in sheep of the genotypes that are resistant to the classical form.

screw A protein (400aa) in *Drosophila* that acts together with **decapentaplegic** in specifying dorsal cell fates in the embryo.

scRNP Small cytoplasmic ribonucleoprotein. See **small interfering RNA**.

scrub typhus See **shimamushi fever**.

scruin Actin-binding protein (alpha, 918aa; beta, 916aa) found associated with the acrosomal process of *Limulus polyphemus* sperm. Scruin holds the microfilaments of the core process in a strained configuration so that the process is coiled. The myosin binding sites on the microfilaments are blocked so HMM decoration is impossible, indicating that there is an unusual packing conformation; when the scruin-actin binding is released the process straightens, the conformation of the actin changes and myosin binding is possible.

scurfy A murine X-linked lymphoproliferative disease, similar to the **Wiskott-Aldrich syndrome** in humans, caused by mutation of the *foxp3* gene which affects the development and maintenance of $CD4^+$ **regulatory T-cells**.

scurvy Disease caused by vitamin C deficiency. The effects are due to a failure of the hydroxylation of proline residues in collagen synthesis, and the consequent failure of fibroblasts to produce mature collagen. See **hydroxyproline**.

scutellum Part of the embryo in seeds of the Poaceae (grasses). Can be considered equivalent to the cotyledon of other monocotyledenous seeds. During germination, absorbs degraded storage material from the endosperm and transfers it to the growing axis.

SCY cytokine superfamily A superfamily of small cytokines with chemokine activity. The SCYA family are CC-chemokines (CCL1–CCL28), the SCYB family are CXC chemokines (CXCL1–CXCL16), the SCYC family are the C-chemokines (XCL1 and XCL2). SCYD1 is **fractalkine** (neurotactin, CX3CL1); SCYE1 is the gene for **EMAP-II**. See Table C3.

scyllatoxin Toxin (31aa) from the scorpion *Leiurus quinquestriatus hebraeus* that specifically blocks low conductance calcium-dependent potassium channels (**SK channels**) that are also a target for **apamin**.

scyphozoa Jellyfish. A class within the phylum Cnidaria in which the polyp stage is inconspicuous or completely absent. Fuller description: http://www.mbl.edu/BiologicalBulletin/KEYS/INVERTS/3/Dscyphozoakeys.htm

scytalidopepsins Pepstatin-insensitive acid endopeptidases from the fungus *Scytalidium lignicolum*. Scytalidopepsin A (EC 3.4.23.31) is one of the sedolisin family of serine-carboxyl peptidases. Scytalidopepsin B (EC 3.4.23.32, scytalidoglutamic peptidase, 260aa) is the type peptidase of the glutamic (G1) peptidase family (See **MEROPS**).

SD sequence See **Shine-Dalgarno** region.

SDAM An autosomal dominant form of caudal dysgenesis (sacral defect with anterior meningocele) caused by mutations in the VANGL1 gene that encodes **vang-like protein 1**.

SDF-1 A chemokine (stromal cell derived factor-1, CXCL12, 93aa) that controls many aspects of stem cell function, including trafficking and proliferation. The receptor is **CXCR4**. Originally known as pre-B cell growth-stimulating factor and identical to human intercrine reduced in hepatomas (hIRH).

SDGF Schwannoma-derived growth factor. See **amphiregulin**.

SDH Most commonly, **(1) Succinate dehydrogenase**. The SDHA, SDHB, SDHC and SDHD genes encode the subunits of succinate dehydrogenase (succinate: ubiquinone oxidoreductase), a component of both the Krebs cycle and the mitochondrial respiratory chain. Less often: **(2) Sorbitol dehydrogenase**.

(3) Saccharopine dehydrogenase. **(4)** Serine dehydratase. **(5)** Shikimate 5-dehydrogenase.

SDS Sodium dodecyl sulphate (sodium lauryl sulphate). Anionic detergent that at millimolar concentrations will bind to and denature proteins, forming an SDS-protein complex. The amount of SDS bound is proportional to the molecular weight of the protein, and each SDS molecule, bound by its hydrophobic domain, contributes one negative charge to the protein thus swamping its intrinsic charge. This property is exploited in the separation of proteins by **SDS-PAGE**.

SDS-PAGE Polyacrylamide gel electrophoresis (PAGE) in which the charge on the proteins results from their binding of **SDS**. Since the charge is proportional to the surface area of the protein, and the resistance to movement proportional to diameter, small proteins migrate further.

sdt In *Drosophila* a protein (stardust, EC 2.7.4.8) with guanylate kinase activity that is a component of the SAC complex that determines epithelial polarity. There are multiple isoforms rannging from 731aa to 2020aa.

sea An oncogene (S13 avian erythroblastosis oncogene homologue) that encodes a member of the Met/**hepatocyte growth factor**/scatter factor family of **receptor tyrosine kinases**.

SEA0400 Potent and selective inhibitor of the Na^+/Ca^{2+} exchanger. Research article: http://www.ncbi.nlm.nih.gov/pmc/articles/PMC1575948/?tool = pubmed

sea-blue histiocyte disease A disorder caused by a defect in **apolipoprotein E**. There is splenomegaly, mild thrombocytopenia and many **histiocytes** in the bone marrow with cytoplasmic granules that stain bright blue (hence the name).

sea hare See *Aplysia*.

seam cells Specialised epithelial cells in *C. elegans* that lie along the apical midline of the hypodermis, at the extreme left and right sides between nose and tail. During postembryonic development, they can act as stem cells to produce neurons and are responsible for production of the cuticular alae in L1 stage, dauer larvae, and adults. See let-7. Description: http://www.wormatlas.org/handbook/anatomyintro/anatomyintro.htm

sebacic acid A naturally occurring dicarboxylic acid commercially produced from castor oil and used in production of various plastics. As a copolymer with glycerol (poly(glycerol sebacate)) has promising properties for tissue engineering. Website of commercial supplier: http://www.sebacicacid.com/

Sebastian syndrome See **May-Hegglin anomaly**.

SEC **(1)** Serpin-enzyme complex, see **serpins**. **(2)** Selenocysteine (Sec). **(3)** Size-exclusion **chromatography** (SEC). **(4)** The general secretory (sec) pathway in bacteria (see **sec-dependent transport**). **(5)** See **sec proteins**.

SecA A bacterial protein (901aa in *E. coli*) that couples ATP hydrolysis to the transfer of proteins into and across the cell membrane. Acts as a receptor for the preprotein-secB complex and as an ATP-driven molecular motor. Interacts with the secYEG preprotein conducting channel. See **sec-dependent transport**.

sec-dependent transport Pathway for the secretion of proteins across the inner membrane into the periplasm of Gram-negative bacteria using the general translocase SecYEG. The SecA protein recognises the signal sequence of proteins destined for export. Also used for the insertion of inner membrane proteins, in some cases in association with **YidC**. Other translocation mechanisms involve the **Tat** system and YidC. SecYEG is a trimeric complex where Y and E are related to sec61α and γ subunits (see **secA**).

sec proteins Proteins involved in **sec-dependent transport**. In *S. cerevisiae* Sec 1 (724aa) is involved in the final stage of protein secretion. Sec7 has a domain of around 200aa that is found in several **GEFs** for ADP ribosylation factors (**ARFs**); mutations lead to accumulation of Golgi cisternae and loss of secretory granules. Various other Sec proteins (Sec13, Sec16, Sec23, Sec24, Sec31) are components of the COPII **coat protein complex**. Sec61 is a conserved heterotrimeric protein-conducting channel (translocon) in eukaryotes (homologous to the SecY channel in eubacteria and archaea) that associates with the sub-complex Sec62/Sec63 and translocates proteins across cellular membranes and integrates proteins containing hydrophobic transmembrane segments into lipid bilayers. Sec61a (Ssh1, 476aa) is a multispanning membrane protein, Sec61b (SEB2, 88aa) kinetically facilitates cotranslational translocation and interacts with the 25-kD subunit of the signal peptidase complex (SPC25), Sec61γ (SSS1, 80aa) is the third component. Sec65 in *S. cerevisiae* is a subunit of the **signal recognition particle** similar to mammalian SRP19. See also **exocyst complex**.

secernins A subfamily of the peptidase C69 family. Secernin-1 (p50, 414aa) was originally isolated from bovine brain cytosol and shown to regulate exocytosis in mast cells. Secernin-2 (425aa) and secernin-3 (424aa) are known. Research article: http://www.molbiolcell.org/cgi/content/full/13/9/3344

second messenger The intracellular mediator of responses to signalling substances such as peptide hormones that do not cross the plasma membrane but bind to receptors on the cell surface. The second messenger may be the same for several different intercellular signalling molecules and may therefore integrate signals. Examples include cyclic AMP, cyclic GMP, IP3 and diacylglycerol.

secondary granules See **specific granules**.

secondary immune response The response of the immune system to the second or subsequent occasion on which it encounters a specific antigen.

secondary lymphoid tissue See **lymphoid tissue**.

secondary lysosome Term used to describe intracellular vacuoles formed by the fusion of lysosomes with organelles (**autosomes**) or with primary phagosomes. **Residual bodies** are the remnants of secondary lysosomes containing indigestible material.

secondary metabolite A product excreted by a microorganism as it enters **stationary phase**.

secondary phloem Phloem formed during secondary growth by the activity of a vascular **cambium** as opposed to primary phloem that is derived from procambium.

secondary product An end-product of plant cell metabolism, which accumulates in, or is secreted from, the cell. Includes **anthocyanins**, **alkaloids**, amongst others. Some are of major economic importance, e.g. as drugs. In contrast to a primary product that is involved in the vital metabolism of the plant.

secondary structure See **primary structure**.

secondary wall That part of the plant cell wall which is laid down on top of the **primary cell wall** after the wall has ceased to increase in surface area. Only occurs in certain cell types, e.g. tracheids, vessel elements and sclerenchyma. Differs from the primary wall both in composition and structure, and is often diagnostic for a particular cell type.

secondary xylem Xylem formed by the activity of a vascular **cambium** as a plant grows. Wood is composed largely of secondary xylem.

secretagogue Substance that induces secretion from cells; originally applied to peptides inducing gastric and pancreatic secretion.

secretases A family of peptidases involved in processing of membrane-bound precursor molecules. Alpha-secretase (**ADAM10**, EC 3.4.24.81, CD156c, 748aa) has broad proteolytic activity and cleaves **amyloid precursor protein** (APP) to the soluble non-amyloidogenic product sAPP alpha. It also processes heparin-binding epidermal growth-like factor, ephrin-A2 and TNF. Beta-secretase-1 (BACE-1, Beta-site APP cleaving enzyme 1, EC 3.4.23.46, 501aa) is the major beta-secretase *in vivo*, an aspartic peptidase that generates the N-terminus of the beta-amyloid protein from APP; further cleavage is then carried out by gamma-secretase. It is an integral membrane protein. Beta-secretase-2 (EC 3.4.23.45, theta-secretase, memapsin-1, 518aa) has similar activity. Gamma-secretase is a multi-protein complex that will cleave amyloid precursor protein (APP) and other type I transmembrane proteins such as **notch** and **E-cadherin**. There are four components, **presenilin**, **nicastrin**, **APH-1** and **PEN-2**, all required for proteolytic activity, and a fifth component, CD147, with a regulatory role.

secretin Peptide hormone (27aa derived from a 121aa precursor peptide) secreted by S cells in the duodenal mucosa. Stimulates secretion of pepsin by the pancreas and bile, but inhibits secretion of gastric acid. One of the secretin family of hormones that also includes **gastric inhibitory peptide, glucagon, glucagon like peptide-1, growth hormone releasing hormone, helodermin, peptide histidine methionine**, pituitary adenylate cyclase-activating polypeptide (**PACAP**) and **vasoactive intestinal peptide**. The receptor (440aa) is G-protein coupled.

secretion Release of synthesized product from cells. Release may be of membrane-bounded vesicles (merocrine secretion) or of vesicle content following fusion of the vesicle with the plasma membrane (apocrine secretion). In holocrine secretion whole cells are released.

secretion systems There are various different bacterial secretory systems. In the Type I secretion system (T1SS, TOSS) there is a contiguous channel that traverses the inner and outer membranes of Gram-negative bacteria. There are only three protein subunits: the ABC protein, membrane fusion protein (MFP), and outer membrane protein (OMP). The Type I secretion system transports various molecules including ions, drugs and proteins of various sizes (20–900 kDa). The Type II system (T2SS) in Gram-negative bacteria depends upon the Sec or Tat system for initial transport of proteins into the periplasm from where they pass through the outer membrane via a multimeric (12–14 subunits) complex of pore forming secretin proteins. In addition to the secretin proteins, 10–15 other inner and outer membrane proteins compose the full secretion apparatus, many with as yet unknown function. In some bacteria, certain proteins are shared between the pilus complex and the type II system. The Type III system (TTSS, T3SS) is essential for pathogenicity of many Gram-negative bacteria, and allows them to translocate proteins into eukaryotic host cells. The type III secretion apparatus (T3SA) is a multisubunit membrane-spanning macromolecular assembly comprising more than 20 different protein, some of which have sequence homology with flagellar proteins. The TTSS encoded in Salmonella Pathogenicity Island 2 (SPI2) is critical for adaptation to the intracellular environment within both phagocytic and epithelial cell types. The Type IV secretion system (T4SS, TFSS) is homologous to the conjugation machinery of bacteria and to archaeal flagella. It is capable of transporting both DNA and proteins. Discovered in *Agrobacterium tumefaciens* which uses this system to introduce the T-DNA portion of the **Ti plasmid** into the plant host, which in turn causes the development of a crown gall.

secretogranins See **granins**.

secretoneurin A polypeptide (33aa) derived by proteolytic cleavage from secretogranin II (617aa; see **granins**) that stimulates dopamine release from striatal neurons, and gonadotropin II secretion from the pituitary; inhibits serotonin and melatonin release from pinealocytes. Activates monocyte migration, and probably has a role in neurogenic inflammation. The receptor is G-protein coupled.

secretor A person who secretes ABO blood group substances into mucous secretions, e.g. saliva; at least 80% of humans are secretors.

secretory carrier-associated membrane proteins A family of integral membrane proteins (SCAMPs, **tetraspan vesicle membrane proteins**) of post-Golgi membranes that function as recycling carriers to the cell surface. At least five members of the family have been identified (SCAMP1, 338aa; SCAMP2, 329aa; SCAMP3, 347aa, SCAMP4, 229aa; SCAMP5, 235aa) with different tissue distributions and transport activities. In *Arabidopsis* similar proteins (e.g. SCAMP1, 282aa) are probably involved in membrane trafficking.

secretory cells Cells specialized for secretion, usually epithelial. Those that secrete proteins characteristically have well developed rough endoplasmic reticulum, whereas conspicuous smooth endoplasmic reticulum is typical of cells that secrete lipid or lipid-derived products (e.g. **steroids**). In plants secretory cells release substances such as mucins and latex.

secretory component *secretory piece* A secreted polypeptide chain (585aa) cleaved from the **polyimmunoglobulin receptor** (764aa). This receptor binds polymeric IgA and IgM at the basolateral surface of epithelial cells, the complex is then transported across the cell and is secreted at the apical surface. During the secretory process the secretory component is cleaved from the transmembrane segment of the receptor.

secretory proteins In eukaryotes, proteins synthesized on **rough endoplasmic reticulum** and destined for export. Nearly all proteins secreted from cells are glycosylated in the **Golgi apparatus**, although there are exceptions (e.g. **albumin**). In prokaryotes, secreted proteins may be synthesized on ribosomes associated with the plasma membrane or exported post-translation.

secretory vesicle Membrane-bounded vesicle derived from the **Golgi apparatus** and containing material that is to be released from the cell. The contents may be densely packed, often in an inactive precursor form (**zymogen**).

securin An inhibitor (pituitary tumour-transforming gene 1 protein, 202aa) of the anaphase activator **separin** (separase/Esp1p). It is ubiquitinated by activated **anaphase-promoting complex**; loss of securin leads to proteolytic cleavage of **cohesin**. Defects in securin may contribute to chromosomal instability.

sedimentation Settling of a component of a mixture under the influence of gravity or centrifugation so that the mixture separates into two or more phases or zones.

sedimentation coefficient The ratio of the velocity of sedimentation of a molecule to the centrifugal force required to produce this sedimentation. It is a constant for a particular species of molecule, and the value is given in **Svedberg units** (S) that are non-additive.

sedimentation test A standard blood test that involves measuring the rate of settling of erythrocytes in anti-coagulant treated blood. Erythrocyte sedimentation rates (ESR) are increased in inflammation.

sedlin An evolutionarily conserved protein (140 aa), a subunit of the Transport Protein Particle (TRAPP) complex, involved in targeting and fusion of endoplasmic reticulum (ER)-derived transport vesicles to the Golgi acceptor compartment. Encoded by the causative gene SEDL for spondyloepiphyseal dysplasia tarda, a progressive skeletal disorder. The intracellular chloride channel protein CLIC1 has been shown to associate with sedlin by yeast two-hybrid screening. Research article: http://www.ncbi.nlm.nih.gov/pmc/articles/PMC2871040/?tool = pubmed

sedoheptulose Seven-carbon sugar, whose phosphate derivatives are involved in the **pentose phosphate pathway** and the **Calvin-Benson cycle**.

seed fern *Pteridospermae* A group of extinct ferns that flourished in the Devonian period.

Segawa syndrome A childhood disorder (infantile Parkinsonism) in which there is severe motor retardation. The autosomal recessive form is caused by mutation in the **tyrosine hydroxylase** gene and an autosomal dominant form by mutation in the gene that encodes GTP cyclohydrolase I (EC 3.5.4.16), the rate-limiting enzyme in synthesis of tetrahydrobiopterin, an essential cofactor for phenylalanine, tyrosine, and tryptophan hydroxylases. Infantile parkinsonism-dystonia syndrome is caused by mutation in the SLC6A3 gene that encodes a dopamine transporter.

segment long-spacing collagen See **SLS collagen**.

segmentation Organization of the body into repeating units called segments, a common feature of several phyla, e.g. arthropods and annelids, although the segments arise by very different mechanisms. Segmentation also occurs during embryonic development in vertebrates, e.g. partition of the mesoderm into **somites**, and is a feature of early **CNS** development. See **rhombomeres, neuromeres**.

segmentation gene Genes required for the establishment of **segmentation** in the embryo. In *Drosophila* about 20 such genes are required, for example *gooseberry*.

Seip syndrome A rare autosomal recessive disease (Berardinelli-Seip congenital lipodystrophy type 2) characterised by an almost complete absence of adipose tissue and severe insulin resistance. It is caused by mutation in the gene encoding **seipin**. See **lipodystrophy**.

seipin A single-pass membrane protein (398aa) of the endoplasmic reticulum, highly expressed in the brain and testis. Seipin mutations are involved in various disorders including congenital **lipodystrophy** type 2, Silver syndrome and distal hereditary motor neuropathy type V.

seismonasty The movement of a plant part in response to vibration (a special case of the response to touch, the thigmonastic response). The folding of the leaflets of *Mimosa* is the classic example. Informative webpages: http://www.biologie.uni-hamburg.de/b-online/e32/32d.htm

selectins *addressins* A family of cell adhesion molecules that have selective carbohydrate-binding capacity (selective lectins, hence the name). They are integral membrane glycoproteins with an N-terminal, **C-type lectin** domain, followed by an EGF-like domain, a variable number of repeats of the short consensus sequence of complement regulatory proteins and a single transmembrane domain. The prefix letter is based on the site of their original identification, although they are not restricted to these locations. E-selectin (endothelial selectin, CD62E, endothelial leukocyte adhesion molecule-1, ELAM-1, 610aa) is upregulated on endothelial cells of post-capillary venules where it is important for neutrophil margination during acute inflammation; it binds sialylated Lewis X and a particular glycoform of ESL-1 that is present on myeloid cells. L-selectin (leucocyte-endothelial cell adhesion molecule, LECAM, CD62L, LAM-1, MEL-14 antigen, leu-8, 372aa) is also involved in leucocyte-endothelial adhesion, for lymphocyte homing to **high endothelial venules** and for implantation of the embryo to the uterine wall.

It binds to carbohydrates on CD34, CD162, GlyCam and MAdCAM. Once the leucocytes have migrated into tissue (diapedesis) the L-selectin is enzymatically removed by a membrane-associated metallopeptidase, a sheddase. P-selectin (platelet selectin CD62P, PADGEM, GMP-140, LECAM-3, 830aa) is rapidly upregulated in platelets and endothelial cells when activated, but only transiently expressed. It mediates rolling of neutrophils, platelets and some T-cell subsets to the endothelial lining of blood vessels. Although knock-out animals show defective cellular infiltration into inflammatory sites it is necessary to inhibit both P-selectin and E-selectin to see total blockade.

selective serotonin re-uptake inhibitors *SSRIs* A class of drugs that inhibit the uptake of serotonin into presynaptic cells. This increases the duration of the signal and also reduces serotonin production through a feedback mechanism. They are extensively prescribed for depression, anxiety disorders, and some personality disorders. An example is Prozac.

selector genes A group of genes that determines which part of a developmental pattern cells will be allocated within a developmental segment. *Antennapedia* is an example and the neural selector gene *cut*, that encodes a homeobox transcription factor, is required for the specification of the correct identity of external (bristle-type) sensory organs in *Drosophila*.

selenium *Se* An essential trace element that must be added as a supplement in serum-free culture media for most animal cells. Some plant species are Se tolerant and will accumulate very high concentrations of Se (accumulators), but most plants are Se non-accumulators and are Se-sensitive. See **selenocysteine** and **selenoprotein**. Review: http://arjournals.annualreviews.org/doi/abs/10.1146/annurev.arplant.51.1.401

selenocysteine *Sec* An unusual amino acid of proteins, the selenium analogue of **cysteine**, the selenium atom replacing sulphur. Involved in the catalytic mechanism of seleno-enzymes such as formate dehydrogenase of *E. coli*, and mammalian **glutathione peroxidase**. A special **opal suppressor** tRNA that recognizes certain UGA nonsense codons allow it to be genomically encoded.

selenoprotein A protein that contains **selenocysteine**. Selenoprotein N (590aa) contains a single selenocysteine residue and is a glycoprotein localized within the endoplasmic reticulum and found in brain, muscle, lung and placenta. Mutations in the gene lead to rigid-spine muscular dystrophy. Selenoprotein P (381aa) has multiple selenocysteine residues and may be involved in antioxidant activity and possibly in selenium transport. Mitochondrial capsule selenoprotein (116aa) is important for the maintenance and stabilization of the crescent structure of the sperm mitochondria. Selenoprotein W (87aa) may be involved in redox-related processes and may have a role in the myopathies of selenium deficiency. Selenoprotein Z (524aa) is thioredoxin reductase 2 (EC 1.6.4.5). Various other selenoproteins are known and many are involved in redox reactions.

self-antigens *autoantigens* Normal components of the body that can be antigens in an autoimmune response.

self-assembly The formation of higher-order structures from subunits (protomers) without any external source of information (priming structures or templates).

self-cloning Any system in which inappropriate cell types or organisms are eliminated because they possess some character that allows them to die or to remove themselves from the system. Thus a transfected cell with genetic material including a drug resistance marker will be self-cloning in the presence of the drug and non-transfected cells will die.

self-incompatibility The inability of pollen grains to fertilize flowers of the same plant or its close relatives. It is a mechanism to ensure out-breeding within some plant species, e.g. in the case of the **S gene complex** in Brassicas where the genes encoding for SI specificity in pistil (**SRK**) and pollen (**SCR**) are thought to be preserved because there is rarely or never any recombination.

self-replicating Literally, replication of a system by itself without outside intervention. In practice often taken to refer to systems that replicate without the contribution of any information from outside the system.

self-splicing Self-catalysed removal of group 5 **introns** from mRNA, mediated by six paired conserved regions.

sem-5 A cell-signalling gene of *C. elegans* that encodes a protein (228aa) with SH2 and SH3 domains; acts in vulval development and sex myoblast migration.

semaphorins Family of proteins that mediate neuronal guidance by inhibiting **nerve growth cone** movement. Both transmembrane and secreted proteins are included and many domains of the proteins are highly conserved between invertebrates and vertebrates. Most are around 750aa with a conserved 'sema' domain of up to 500aa extracellularly with a single immunoglobulin C2-type domain C-terminally to this. **Collapsin**, responsible for the collapse of nerve growth cones of chick sensory neurites in culture following contact with retinal axons was one of the first semaphorins described. Receptors are

plexins; see also **collapsin response-mediator proteins**. See Table S1.

semelparity The production of offspring only once in the lifetime of the organism. *Adj.* semelparous.

semiautonomous Describing systems or processes that are not wholly independent of other systems or processes.

semiconservative replication The system of replication of DNA found in all cells in which each daughter cell receives one old strand of DNA and one strand newly synthesized at the preceding **S phase**. The existence of semiconservative replication was demonstrated by the Meselson-Stahl experiment and implied the two- or multi-strandedness of DNA.

semigamy A rare type of facultative apomixis in plants controlled by an incompletely dominant autosomal gene. During semigamy, the sperm and egg cells undergo cellular fusion, but the sperm and egg nucleus fail to fuse in the embryo sac, giving rise to diploid, haploid, or chimeric embryos composed of sectors of paternal and maternal origin. Early report: http://jhered.oxfordjournals.org/content/71/2/117.extract

semipermeable membrane A membrane that is selectively permeable to only one (or a few) solutes. The potential developed across a membrane permeable to only one ionic species is given by the **Nernst equation** for the species: this is the basis for the operation of **ion-selective electrodes**.

Semliki forest virus Enveloped virus of the alphavirus group of **Togaviridae**. First isolated from mosquitoes in the Semliki Forest in Uganda; not known to cause any human illness although it causes a lethal encephalitis in rodents. The synthesis and export of its three spike glycoproteins, via the endoplasmic reticulum and Golgi complex, have been used as a model for the synthesis and export of plasma membrane proteins.

SEN1 See **senataxin**.

Senarmont compensation In interference microscopy, compensation for the phase difference introduced by the object, measured by introducing a quarter-wavelength plate and rotating the analyser: the angle of rotation is proportional to the optical path difference.

senataxin A protein (2,677aa), probably a helicase (EC 3.6.1.-), which may be involved in RNA maturation. Involved in DNA double-strand break response generated by oxidative stress. The gene is mutated in **spinocerebellar ataxia autosomal recessive Type 1** and in a rare autosomal dominant form of juvenile **amyotrophic lateral sclerosis**. Has homology to SEN1 of *S. cerevisiae*, a helicase required for endonucleolytic cleavage of introns from all families of precursor tRNAs. It is essential for vegetative growth of the yeast.

TABLE S1. Semaphorins

Class	New Name	Old name	Features of Class
Invertebrate			
Class 1	Sema-1a	G-Sema I, D-Sema I, T-sema I, Ce-Sema I	TM domain and short cytoplasmic tail
	Sema-1b	Sema 1b	
Class 2	Sema 2a	D-Sema II, Ce-Sema II, gSemaII	Secreted and have Ig domain
Vertebrate			
Class 3	Sema3A	C-Collapsin-1 (Coll-1), H-Sema III, M-SemD, R-Sema III, Sema-Z1a	Ig domain, short basic domain, secreted
	Sema3B	M-SemaA, H-SemaA, H-Sema V	
	Sema3C	M-SemE, C-Coll-3, H-Sema E	
	Sema3D	C-Coll-2, Sema-Z2	
	Sema3E	C-Coll-5, M-Sema H	
	Sema3F	H-Sema IV, M-Sema IV, H-Sema-3F	
Class 4	Sema4A	M-SemB	Ig domain, TM domain, short cytoplasmic domain
	Sema4B	M-SemC	
	Sema4C	M-sema F	
	Sema4D	CD100. M-Sema G, C-Coll-4	
	Sema4E	Sema-Z7	
	Sema4F	M-Sema W, R-Sema W, H-Sema W	
	Sema4G		
Class 5	Sema5A	M-SemF	7 thrombospondin repeats, TM and short cytoplasmic domain
	Sema5B	M-SemG	
Class 6	Sema6A	M-Sema Via	TM and cytoplasmic domain
	Sema6B	M-SemaVIb, R-Sema Z	
	Sema6C	M-Sema Y, R-Sema Y	
Class 7	Sema7A	H-Sema K1, H-Sema L, M-Sema L, M-Sema K1	Ig domain and GPI anchor
Viral			
Class V	SEMAVA	Vaccinia sema, Variola sema	Truncated sema domain
	SEMAVB	AHV sema	Sema domain + Ig domain

Based upon Recommendation of Semaphorin Nomenclature Committee. (1999) Cell 97: 551–552.
All have Sema domain – which is also found in some functionally unrelated proteins.

Sendai virus *Haemagglutinating virus of Japan (HVJ)* Parainfluenza virus type 1 (Paramyxoviridae). Can cause fatal pneumonia in mice, and may cause respiratory disease in humans. The ability of ultraviolet-inactivated virus to fuse mammalian cells has been extensively used in the study of **heterokaryons** and **hybrid cell** lines.

senescent cell antigen An antigen (62 kDa) that appears on the surface of senescent erythrocytes and is immunologically cross-reactive with isolated **band III**. Seems to be recognized by an autoantibody, and the immunoglobulin-coated erythrocyte is then removed from circulation by cells such as Kuppfer cells of the liver that have Fc receptors. Intracellular cleavage of intact band III by a calcium-activated peptidase, **calpain**, may reveal the antigen *in situ*. See **LIM kinases**.

senile plaque Characteristic feature of the brains of **Alzheimer's disease** patients and aged monkeys, consisting of a core of amyloid fibrils surrounded by dystrophic neurites. The principal component of amyloid fibrils in senile plaques is B/A4, a peptide of about 4 kDa that is derived from the larger **amyloid precursor protein** (APP). The B/A4 sequence is located near the C-terminus of APP.

Senior-Loken syndrome An autosomal recessive disease characterised by **nephronophthisis** and **Leber congenital amaurosis**. Various forms of the disorder are caused by mutations in genes encoding **nephrocystins**.

sensitization A state of increased responsiveness, usually referring to the state of an animal after primary challenge with an antigen. The term is frequently used in the context of **hypersensitivity**.

sensory ataxic neuropathy A clinically heterogeneous systemic disorder with variable features resulting from mitochondrial dysfunction. Sensory ataxic neuropathy, dysarthria, and ophthalmoparesis (SANDO) is caused by mutation in the nuclear gene that encodes DNA polymerase-gamma or by mutation in the gene for **twinkle**. Spinocerebellar ataxia with epilepsy (SCAE) is similar.

sensory neuron (1) A **neuron** that receives input from sensory cells. (2) Sensory cells such as cutaneous mechanoreceptors and muscle receptors.

SEPALLATA A MADS-box subfamily of transcription factors that interact with ABCD floral homeotic genes in specifying different floral whorls. SEPALLATA1 (SEP1, agamous-like MADS-box protein AGL2, 215aa) forms a heterodimer with AGAMOUS and interacts with other members of the family.

separin A caspase-like cysteine endopeptidase (separase, EC 3.4.22.49, 2120aa), involved in cleaving the **cohesin** complex that links sister chromatids until the metaphase to anaphase transition. Separin activity is inhibited until **securin** is ubiquitinated by the **anaphase-promoting complex** and proteolytically degraded. Cyclin-dependent kinase-1 (CDK1) phosphorylates securin and blocks ubiquitination by the APC. In *S. cerevisiae* separin is Esp1.

Sephacryl Trade-name for a covalently cross-linked allyl dextrose gel formed into beads. Used in **gel filtration** columns for separating molecules in the size range 5 kDa to 1.5 million Da.

Sephadex Trade-name for a cross-linked dextran gel in bead form used for **gel filtration** columns: by varying the degree of cross-linking the effective fractionation range of the gel can be altered.

Sepharose Trade-name for a gel of agarose in bead form from which charged polysaccharides have been removed. Used in **gel filtration** columns.

septate junction An intercellular junction found in invertebrate epithelia that is characterized by a ladder-like appearance in electron micrographs. Thought to provide structural strength and to provide a barrier to diffusion of solutes through the intercellular space. Occurs widely in transporting epithelia, and is (perhaps controversially) considered analogous to tight junctions (**zonula occludens**).

septic shock Condition of clinical shock caused by **endotoxin** in the blood. A serious complication of severe burns and abdominal wounds, frequently fatal. Part of the problem seems to be due to increased leucocyte adhesiveness, which leads to massive sequestration of neutrophils in the lung, increased vascular permeability, and acute (adult) respiratory distress syndrome.

septicaemia A potentially life-threatening infection in which many bacteria are present in the blood. Commonly referred to as blood poisoning. See **bacteraemia**.

septins Family of evolutionary conserved proteins (cytoskeletal GTPases, peanut-like proteins, cell division control-related proteins, ~350aa) that polymerize into hetero-oligomeric complexes that form filaments, and associate with cellular membranes, actin filaments and microtubules. GTPase activity is required for filament formation. They were first identified in *S. cerevisiae* where they are associated with cytokinesis and septum formation and form a ring of 10 nm filaments underlying the plasma membrane in the mother-bud neck. Homologous proteins, associated with cleavage furrows, are reported from *Drosophila*, amphibians and mammals but not from protozoa or plants, and they are also involved in membrane dynamics, vesicle trafficking, apoptosis, and cytoskeletal remodelling. There are at least 14 human septins with tissue-specific isoforms in many cases. A deficiency in septin-5 degradation may contribute to the development of early onset Parkinson's disease 2, septin-7 interacts with **CENP-E** at the kinetochore. Mutations in septin-9 are associated with hereditary neuralgic amyotrophy. Review: http://genomebiology.com/2003/4/11/236

septum Literally a separating wall. Mainly applied to the structure composed of plasmalemmae and cell wall material formed in cell division in prokaryotes and fungi. Also applied to the sealing layers in various packages of sterile fluids, or barriers through which injections, needles etc., may be passed.

Seqboot A data-analysis package that allows bootstrap, jackknife, or permutation resampling of molecular sequence, restriction site, gene frequency or character data. It can be used to generate multiple data sets that are resampled versions of the input data set. Link to webpage: http://cmgm.stanford.edu/phylip/seqboot.html

Sequenase™ Proprietary name for a genetically engineered form of T7 DNA polymerase (EC. 2.7.7.7) used in DNA sequencing. Affymetrix homepage: http://www.affymetrix.com/estore/index.jsp

sequence homology Strictly, refers to the situation where nucleic acid or protein sequences are similar because they have a common evolutionary origin. Often used loosely to indicate that sequences are very similar. Sequence similarity is observable; homology is an hypothesis based on observation.

sequestosome *aggresome* Protein aggregates (perinuclear inclusion bodies) composed of ubiquitin-linked proteins destined for degradation. Sequestosome-1 (ubiquitin-binding protein p62, 440aa) is an adapter protein which binds ubiquitin and may regulate the activation of NFκB1 by TNFα, nerve growth factor (NGF) and interleukin-1 and may regulate various signalling cascades. One form of **Paget's disease of bone** is caused by mutation in the gene. Overexpression of **parkin** or disruption of microtubules apparently blocks the formation of aggresomes.

sequon A consensus sequence of amino acids, as for example the tripeptide motif Asn-Xaa-Ser/Thr that is the site for N-linked glycosylation.

SER See **smooth endoplasmic reticulum**.

SERCA See **sarcoplasmic-endoplasmic reticulum Ca²⁺-ATPase**.

serglycin *proteoglycan 1* An intracellular **proteoglycan**, found particularly in the storage granules of connective tissue mast cells. The core protein consists of 153aa with 24 serine-glycine repeats between amino acids 89 and 137, hence the name. The serine-glycine repeats are the linkage sites for around 15 **glycosaminoglycan** chains that are either heparin or highly sulphated chondroitin sulphate. These negatively charged chains are thought to concentrate positively charged proteases, histamine and other molecules within the storage granules.

sericin A serine-rich protein (1186aa in *Bombyx mori*) found in silk. It forms a sticky coat on the fibroin threads and cements them together. It is finding a range of uses in biotechnological applications.

serinc Class of carrier proteins (serine incorporators, e.g., serinc-1, 453aa; serinc-2, 456aa) that facilitate the synthesis of serine-derived lipids, phosphatidylserine and sphingolipids. Serinc is a unique protein family that shows no amino acid homology to other proteins but is highly conserved among eukaryotes. The members contain 11 transmembrane domains. Research article: http://www.jbc.org/content/280/42/35776.long

serine One of the amino acids (Ser, S, 105 Da) found in proteins and that can be phosphorylated. See Table A1.

serine dehydratase A gluconeogenic enzyme (EC 4.2.1.13, SDH, 328aa), one of the beta-family of pyridoxal phosphate-dependent (PLP) enzymes, catalyzes the deamination of L-serine and L-threonine to yield pyruvate or 2-oxobutyrate.

serine hydroxymethyltransferase One of the alpha- class of pyridoxal phosphate enzymes (EC 2.1.2.1, SHMT1, 444 and 483aa isoforms; SHMT2, 494aa is mitochondrial), a catabolic enzyme involved in converting serine to glycine and a key enzyme in the formation and regulation of the folate one-carbon pool. *E. coli* SHMT (417aa) has little sequence similarity to the enzyme family. See **mimosine**.

serine peptidases Peptidases (serine proteases) that share a common reaction mechanism based on formation of an acyl-enzyme intermediate on a specific active serine residue. Most are inhibited by generic serine peptidase inhibitors (**serpins**) and irreversibly inactivated by a series of organophosphorus esters, such as di-isopropylfluorophosphate (DFP). They are, however, diverse in molecular structure and catalytic mechanisms and are not homologues of each other. Examples are **trypsin**, **chymotrypsin** and the bacterial enzyme **subtilisin**.

SERK genes Plant genes that encode **somatic embryogenesis** receptor-like kinases. SERK1 (EC 2.7.10.1, 625aa) is a dual specificity kinase that will phosphorylate serine/threonine- and tyrosine-containing substrates. It is involved in the **brassinolide** signalling pathway. SERK1 and SERK2 act redundantly and double mutants of SERK1 and SERK2 are completely male sterile. SERK3 (BRASSINOSTEROID INSENSITIVE 1-associated receptor kinase 1, 615aa) forms a heterodimer with **BRI1** which it can phosphorylate; this may change the equilibrium between plasma membrane-located BRI1 homodimers and endocytosed BRI1-BAK1 heterodimers. SERK4 and SERK5 are also known but their function is unclear.

serogroup A group of bacteria or other microorganisms that have a certain antigen in common.

serosa (1) A serous epithelium, having **serous glands** or cells, as opposed to a mucous membrane. (2) Thin infolding of the lining of the peritoneal cavity that forms the **omentum**.

serotonin A **neurotransmitter** and **hormone** (5-hydroxytryptamine, 5-HT, 176 Da), found in vertebrates, invertebrates and plants. It is important in the mammalian CNS and various psychoactive drugs (such as selective serotinin-reuptake inhibitors) are extensively used to treat depression. Serotonin is produced by enterochromaffin cells in the gut and is stored in platelets. Most receptors are G-protein coupled except for the 5-HT$_3$ receptor which is a ligand gated ion channel permeable to sodium, potassium, and calcium ions. In plants serotonin may be present at high levels in fruits.

serotype The genotype of a unicellular organism as defined by antisera directed against antigenic determinants expressed on the surface.

serous gland An exocrine gland that produces a watery, protein-rich secretion, as opposed to a carbohydrate-rich mucous secretion.

serpentine receptors See **G-protein-coupled receptors**.

serpins Superfamily of proteins, mostly **serine peptidase** inhibitors, that includes **alpha 1-antitrypsin**, complement C1 inhibitor, **plasminogen activator inhibitor 1**, **maspin**, **PEDF**, **protease nexin-1** and **vaspin**. The serpin-enzyme complex (SEC) formed from a serpin and a serine peptidase binds to a hepatocyte receptor (SEC receptor) that mediates catabolism of α-1-antitrypsin/elastase complexes and elevates α-1-antitrypsin synthesis.

serrate (1) The transmembrane ligand (1404aa) for **Notch**, contains 14 repeats of the **EGF-like domain**, expressed on dorsal cells of *Drosophila* wing, activates Notch on ventral cells and induces the expression of **delta** protein. Serrate protein expression is reciprocally induced by delta and modulated by **fringe**. (2) In plants, a zinc finger protein, (SERRATE, 720aa) that is required for proper processing of primary miRNAs to miRNAs, for pre-mRNA splicing and for the accumulation of the trans-acting small interfering RNA (ta-siRNA).

Serratia marcescens A Gram-negative bacterium that is very common in soil and water; most strains produce a characteristic pigment, prodigiosin. Opportunistic human pathogen, infecting mainly hospital patients. *Serratia marcescens* haemolysin (ShlA, 1608aa) is an unusual pore-forming toxin, that requires ShlB (557aa) for its secretion and activation. ShlA not only forms pores in erythrocytes but also in fibroblasts and epithelial cells. Haemolysin: http://iai.asm.org/cgi/reprint/72/1/611.pdf

Sertoli cell Tall columnar cells found in the mammalian testis closely associated with developing spermatocytes and spermatids. Probably provide appropriate microenvironment for sperm differentiation and phagocytose degenerate sperm.

serum Fluid that is left when blood clots; the cells are enmeshed in **fibrin** and the clot retracts because of the contraction of platelets. It differs from plasma in having lost various proteins involved in clot formation (**fibrinogen**, **prothrombin**, various blood-clotting factors such as **Hagemann factor**, **Factor VIII** etc.) and in containing various platelet-released factors, notably **platelet-derived growth factor**. For this reason serum is a better supplement for cell culture medium than defibrinated plasma (plasma-derived serum).

serum amyloid The fibrils deposited in tissues in secondary amyloidosis which are unrelated to immunoglobulin light chains (in contrast to the situation in primary amyloidosis) and are made of amyloid A protein (AA protein). This is derived from serum amyloid A (SAA) that is the apolipoprotein of a high-density lipoprotein and an acute phase protein. Partial proteolysis converts SAA into the **beta pleated sheet** configuration of the amyloid fibrils. Amyloid P protein is also found as a minor component of the fibrils (in both primary and secondary amyloidosis) and is derived from serum amyloid P a **pentraxin** with similarity to **C-reactive protein**. The physiological role remains obscure.

serum- and glucocorticoid-inducible kinases Serine/threonine protein kinases (SGKs, of which several isoforms have been identified) that are transcriptionally regulated by corticoids, serum, and cell volume. SGK1 (EC 2.7.11.1, 431aa) activates certain potassium, sodium, and chloride channels and plays an important role in the regulation of epithelial ion transport by inactivating (by phosphorylation) Nedd4-2, an E3 ubiquitin-protein ligase that targets the epithelial Na$^+$ channel (ENaC) and the excitatory amino acid transporter (EAAT-2) for degradation. SGK2 (427aa) is involved in the activation of potassium channels. SGK3 (496aa) also activates potassium channels and mediates IL-3-dependent survival signals. SGKs belong to the AGC Ser/Thr protein kinase family that includes **Akt**.

serum hepatitis See **hepatitis B**.

serum requirement The amount of serum that must be added to culture medium to permit growth of an animal cell in culture. Transformed cells frequently have less stringent serum requirements than their normal counterparts.

serum response element *SRE* DNA motif (20 bp) found (for example) in the c-fos **promoter**, which is bound by the **serum response factor**.

serum response factor Transcription factor (SRF, p67SRF) which interacts with Elk-1 (p62TCF) to bind the **serum response element** promoter motif found in many growth-related genes.

serum sickness A **hypersensitivity** response (Type III) to the injection of large amounts of antigen, as might happen when large amounts of antiserum are given in a passive immunization. The effects are caused by the presence of soluble immune complexes in the tissues.

sestrins A family of conserved proteins that accumulate in cells exposed to stress and whose expression is modulated by **p53**. They are involved in reducing **peroxiredoxins** which have been over-oxidised by hydrogen peroxide, thereby regenerating their protective anti-oxidant function. Sestrins contain a predicted redox-active domain homologous to bacterial **AhpD**. May also be regulators of cellular growth. In humans sestrin-2 (480aa) and sestrin-3 (492aa) are also known. Defects in sestrin-1 (SESN1, p53-regulated protein PA26, 492aa) may be involved in **heterotaxia**. In *Drosophila*, dSesn (sestrin homologue, 497aa) appears to be a negative feedback regulator of **TOR**. Weblog: http://anti-agingfirewalls.com/2010/03/07/sestrins-longevity-and-cancers/

SET complex An endoplasmic reticulum-associated complex, originally discovered as a granzyme A (GzmA) target in cells undergoing caspase-independent T cell-mediated death, that contains 3 DNases (the base excision repair endonuclease APE1, 5′-3′ exonuclease TREX1, and endonuclease NM23-H1). NM23-H1 and TREX1 are activated by GzmA cleavage of the inhibitor **SET protein** to cause single-stranded DNA damage Other components include a DNA binding protein (HMGB2) that preferentially binds to distorted or damaged DNA, and the PP2A inhibitor **pp32/PHAP-I**. The SET complex plays an important role in the early phase of the HIV-1 lifecycle by inhibiting autointegration.

SET protein A protein (suppressor of variegation, enhancer of zeste, and trithorax, 290aa) involved in apoptosis, transcription, nucleosome assembly and histone binding. It is a specific inhibitor of **protein phosphatase** 2A and modifies phosphorylation of histone H4 with effects on nucleosome structure. The SET domain is found in a number of proteins involved in embryonic development in plants and animals. For example, SETD8 (SET domain-containing protein 8, histone-lysine N-methyltransferase SETD8, EC 2.1.1.43, 393aa) specifically monomethylates Lys20 of histone H4, probably contributing to the maintenance of proper higher order structure of DNA during mitosis. See **SET complex**.

seven-membrane spanning receptors See **G-protein-coupled receptors**.

sevenless A *Drosophila* gene (*sev*) that is required for development of the R7 cell in each **ommatidium** in the eye. Gene product is a **receptor tyrosine kinase** (EC 2.7.10.1, 2554aa), related to the insulin receptor, for which the ligand is the product of the *bride of sevenless* gene. In the downstream signalling cascade **son-of-sevenless** plays an important part.

severe combined immunodeficiency disease A range of serious disorders of the immune system (severe congenital immunodeficiency disease, SCID). In all forms there is deficiency in T cell-mediated immune responses (T⁻) but B-cells and NK cells may or may not be affected. The commonest form is caused by mutation in the interleukin 2-receptor (IL2RG) gene (X-linked SCID; T^-, B^+, NK^-). Autosomal recessive SCID (T^-, B^+, NK^- SCID) is caused by mutation in the JAK3 gene; T^-, B^+, NK^+ SCID is caused by mutation in the IL7R gene, the CD45 gene, or the CD3D gene; T^-, B^-, NK^- SCID is caused by mutation in the ADA (**adenosine deaminase**) gene; T^-, B^-, NK^+ SCID is caused by mutation in the **RAG1 and 2** genes; T^-, B^-, NK^+ SCID combined with sensitivity to ionizing radiation is caused by mutation in the **artemis** gene. Gene therapy has been used to treat some patients with ADA deficiency.

severin A protein (362aa) of the villin/gelsolin family from *Dictyostelium* that binds to the barbed ends of F-actin microfilaments in a Ca^{2+}-dependent manner, blocking further assembly. Not, apparently, essential for movement.

sex chromatin Condensed chromatin of the inactivated X chromosome in female mammals (**Barr body**).

sex chromosome The chromosome that determines the sex of an animal. In humans the two sex chromosomes (X and Y) are dissimilar, the female has two X chromosomes (XX), and the male is heterogametic (XY); sperm are either X or Y. A portion of the X and Y chromosomes is similar and is known as the pseudoautosomal region. In many organisms, there is only one sex chromosome, and one sex is XX, the other X0. A different system operates in birds, some fish and crustaceans, and some insects; the the ovum determines the sex, the male being homogametic (ZZ) and the female ZW. There is apparently no homology between the XY and ZW systems.

sex-duction The transfer of genes from one bacterium to another by the process of conjugation. May involve one bacterium with an F′-plasmid, in which case the process is called F-duction.

sex hormone Hormone that is secreted by gonads, or that influences gonadal development. Examples are **estrogen, testosterone, gonadotrophins**.

sex-linked disorder A genetic defect most commonly seen in heterogametic individuals (males in mammals), the consequence of a mutation in a gene on the unpaired portion of the X chromosome so that there is only one copy of the allele. Only homozygous females will show the defect.

sex pili Fine filamentous projections (**pili**) on the surface of a bacterium that are important in conjugation. Often seem to be coded for by plasmids that confer conjugative potential on the host; in the case of the F-plasmid, the F-pili are 8–9 nm diameter and several microns long, composed of **pilin**. Whether the pili merely serve to establish and maintain adhesive contact between the partners in conjugation, or whether DNA is actually transferred through the central core of the pilus is still unresolved, although a simple adhesion role is more generally accepted.

Sezary cells See **Sezary syndrome**.

Sezary syndrome A cutaneous T-cell lymphoma, affecting $CD4^+$ cells, an advanced form of **mycosis fungoides**, in which skin all over the body is reddened, itchy, peeling, and painful. There may also be patches, plaques, or tumours on the skin. Cancerous T-cells (Sezary cells) are found in the blood and infiltrating the skin.

Sf9 cells Insect cell line derived from *Spodoptera frugiperda* much used for production of recombinant protein. The gene to be expressed is incorporated into a **baculovirus** vector which is used to infect the cells.

S gene complex Genes coding for molecular components of the pollen-stigma recognition system in the cabbage genus (*Brassica*). The gene products govern the **self-incompatibility** response and include a glycoprotein found on the stigma surface and a lectin on the pollen grain surface that binds to the stigma glycoprotein.

SGK See **serum- and glucocorticoid-inducible kinases**.

SGOT Enzyme (serum glutamic-oxaloacetic transaminase, aspartate transaminase-1, EC 2.6.1.1, 413aa) that reversibly catalyses the transfer of an amine group from glutamic acid to oxaloacetic acid, forming alpha-ketoglutaric acid and aspartic acid. High levels in serum are an indication of liver or heart damage. There are cytoplasmic, mitochondrial and chloroplastic isozymes.

SGPT Old name for serum glutamic-pyruvic transaminase (alanine aminotransferase-2, EC 2.6.1.2, 523aa).

sgRNA See **subgenomic RNA**.

SGS3 (1) In *Arabidopsis*, a protein (SUPPRESSOR OF GENE SILENCING 3, 626aa) required for post-transcriptional gene silencing and natural virus resistance. May bind nucleic acids and is essential for the biogenesis of **trans-acting siRNAs**. (2) In *Drosophila*, a protein (salivary glue protein, Sgs-3, 307aa) expressed in the salivary glands of mid- to late-instar larvae.

SGT-1 In *S. cerevisiae* a protein (SGT1, 395aa) involved in ubiquitination by associating with the SCF (Skp1p/Cdc53p/F box protein) ubiquitin ligase complex. (See **skp**). Required for both entry into S phase and kinetochore function. Also involved in the cyclic AMP (cAMP) pathway, possibly involved in the assembly or the conformational activation of specific multiprotein complexes. The human Sgt1 (suppressor of G2 allele of SKP1 homologue, 365aa, product of the SUGT1 gene) will work in yeast and is also thought to be involved in ubiquitinylation, whereas SGT1 (ecdysoneless homologue, 644aa) is a regulator of p53 stability and function. The *Drosophila* homologue (ecdysoneless, 684aa) is required for oocyte development. Plant homologues (643aa in *Arabidopsis*) are required for disease resistance mediated by nucleotide-binding site/leucine-rich repeat (NBS-LRR) proteins.

SGTx1 A peptide toxin (34aa) from the venom of the tarantula, *Scodra griseipes*. Inhibits Kv2.1 potassium channels in rat cerebellar granule neurons and is a gating-modifier. Homologous to **hanatoxin**.

SH domains Domains (Src homology domains) within proteins that, from their homology with **src**, are involved in the interaction with phosphorylated tyrosine residues on other proteins (SH2 domains) or with proline-rich sections of other proteins (SH3 domains). The SH1 domain has tyrosine kinase activity, the SH4 domain has myristoylation and membrane-localisation sites.

SH1, SH2, etc. See **SH domains**.

SH3BP4 A protein (SH3-domain binding protein 4, TTP, 963aa) involved in cargo-specific control of clathrin-mediated endocytosis, specifically controlling the internalization of the **transferrin** receptor (TfR). It interacts with endocytic proteins, including **clathrin, dynamin**, and the TfR, and localizes selectively to TfR-containing coated-pits and -vesicles. There are 3 Asn-Pro-Phe (NPF) motifs, an **SH3 domain**, a PXXP motif, a bipartite nuclear targeting signal, and a tyrosine phosphorylation site. See **TTP** which has alternative meanings.

shadoo A GPI-linked protein (151aa) expressed in the brain. The gene (SPRN, 'shadow of prion protein') is a paralog of PRNP that encodes the prion proteins (PrP) and is highly conserved from fish to mammals. Evolution of Vertebrate Genes Related to Prion and Shadoo Proteins – Clues from

Comparative Genomic Analysis; http://mbe.oxford-journals.org/cgi/content/full/21/12/2210

shadowing Procedure used in electron microscopy, in which a thin layer of material, usually heavy metal or carbon, is deposited onto a surface from one side, in such a way as to cast 'shadows'. Deposition is usually done by vaporizing the metal on an electrode under vacuum.

shaker *Drosophila* gene encoding a subunit of the Kv1.1 voltage-gated potassium channel (a heterotetramer that includes shaker, 655aa) involved in regulation of sleep; the legs of mutant flies shake when they are anaesthetised. Related genes, *shab*, *shal* and *shaw* are known in flies and humans. Mutations in the human gene can cause episodic ataxia Type 1. In *Arabidopsis* the homologue (potassium channel AKT6, 888aa) is a selective inward-rectifying potassium channel that plays an important role in pollen tube development.

SHANKs A family of adapter proteins (SH3 and multiple ankyrin repeat domains proteins), SHANK1 (somatostatin receptor-interacting protein, 2161aa), SHANK2 (cortactin-binding protein 1, 1253aa) and SHANK3 (proline-rich synapse-associated protein 2, 1741aa). They are important for the formation of postsynaptic densities in excitatory synapses and control dendritic spine morphology. Shank-interacting protein-like 1 (sharpin, SHANK-associated RH domain-interacting protein, 387aa) may be involved in development of the immune system and controlling inflammation; it is commonly up-regulated in multiple human cancer types

sharpin See **SHANKs**.

shc A family of proteins (SHC-transforming proteins, SH2 domain-containing-transforming proteins C, Shc1, 583aa; Shc2, 582aa; Shc3, 594aa; Shc4, 630aa) that couple activated growth factor receptors to signalling pathways in various tissues. Isoforms p46Shc and p52Shc of Shc1, once phosphorylated, couple activated receptor tyrosine kinases to **ras** via the recruitment of the **GRB-2**/sos complex. p66Shc is involved in signal transduction pathways that regulate the response to oxidative stress and life span. Overexpression of shc will transform fibroblasts. Shc SH2 domain-binding protein 1 (SHCBP1, 672aa) interacts with p52shc and may be part of a different downstream signalling pathway.

shear stress response element *SSRE* A response element proposed to activate gene expression in response to fluid shear stress. In the human PDGF-A promoter there is a GC-rich region near the TATA box that is required for shear-inducible reporter gene expression. There have been suggestions, however, that the signalling is through the **ras** pathway.

Sheldon-Hall syndrome See **arthrogryposis**.

shelterin A protein complex that protects chromosome ends from all aspects of the DNA damage response and regulates **telomere** maintenance. Components include **telomeric repeat binding factors** (TRF1 and TRF2), TRF1-interacting factors 1 and 2 (**TIN2** and **PINX1**), **TRF2-interacting protein**, protection of telomeres 1 (**POT1**), TIN2-interacting protein and accessory proteins such as **apollo**. There are other non-shelterin proteins at chromosome ends. Research article: http://genesdev.cshlp.org/content/19/18/2100.full

shibire *Drosophila* gene that encodes **dynamin** (EC 3.6.5.5, 877aa). Shibire is temperature sensitive and in affected flies synaptic vesicles are depleted at high temperatures but are restored in nerve terminals when endocytosis resumes at lower temperatures.

shiga toxin Bacterial AB toxin from *Shigella dysenteriae*. The B (binding) subunit (89aa) interacts with globotriaosylceramide in human intestinal microvilli. The A subunit (EC 3.2.2.22, 315aa) inactivates 60S ribosomal subunits thereby blocking eukaryotic protein synthesis. Shiga-like toxins (e.g. SLT-1 and SLT-2, verotoxin, of *E. coli*) are structurally-related toxins that cleave a single residue from the 28S rRNA subunit of ribosomes thus blocking interaction with elongation factors eEF-1 and eEF-2.

Shigella Genus of non-motile Gram-negative enterobacteria (Escherichiae group): cause dysentery. See **shiga toxin**.

shikimate 5-dehydrogenase A key enzyme (SDH, SKDH, EC 1.1.1.25, 272aa in *E. coli*) in the aromatic amino acid biosynthesis pathway (**shikimic acid pathway**), catalyzes the reversible reduction of 3-dehydroshikimate to shikimate.

shikimic acid pathway Metabolic pathway in plants and microorganisms, by which the aromatic amino acids (phenylalanine, tyrosine and tryptophan) are formed from phosphoenolpyruvate and erythrose-4-phosphate via shikimic acid. The aromatic amino acids in turn serve as precursors for the formation of lignin and other phenolic compounds in plants. Inhibitors of this pathway are used as herbicides. Diagram: http://4e.plantphys.net/article.php?ch = t&id = 23

shimamushi fever An acute fever (scrub typhus, flood fever, Japanese river fever, tsutsugamushi fever) caused by *Rickettsia tsutsugamushi*, transmitted by the bite of a larval mite (chigger), *Leptotrombidium akamushi*.

Shine-Dalgarno region A poly-purine sequence found in bacterial mRNA about 7 nucleotides in front of the **initiation codon**, AUG. The complete sequence is 5′-AGGAGG-3′ and almost all messengers contain at least half of this sequence. It is complementary to a highly-conserved sequence at the 3′ end of 16S ribosomal RNA, 3′-UCCUCC-5′, and it is thought to be involved in the binding of the mRNA to the ribosome.

shingles Disease in adults caused by *Varicella zoster* virus (Herpetoviridae), that in children causes chicken pox. Disease arises by reactivation (usually associated with a decline in cell-mediated immunity) of latent virus that persists in spinal or cranial sensory nerve ganglia.

SHIP Lipid phosphatase (SH2-containing inositol phosphatase, 1188aa) containing an SH2 domain; dephosphorylates 5′-inositol phosphate and thus modulates PI3-kinase signalling downstream of growth factor and insulin receptors. Important in regulation of mast cell degranulation and cytokine signal transduction in lymphoid and myeloid cells. Negative signalling through SHIP appears to inhibit the **ras** pathway by competition with **GRB-2** and **shc** for SH2 domain binding.

ShK toxin A toxin (35aa) from the sea anemone *Stichodactyla helianthus* that blocks the voltage-gated potassium channel Kv1.3 in T-cells.

shmoo Polarized morphological form of the yeast *S. cerevisiae* that has been exposed to mating pheromone. (either **alpha factor** or a-factor). The cytoskeleton and proteins involved in mating are localized to a cell-surface projection; the tips of the projections from the two cells eventually fuse.

shoot apical meristem The small number of cytoplasmically dense cells at the shoot tip, comprising the apical dome. Division of these cells generates other aerial primordia and all subapical tissue.

shootin A protein (631aa) involved in the generation of internal asymmetric signals required for neuronal polarization. Acts upstream of PI3K (phosphoinositide 3-kinase), being required for spatially localized PI3K activity.

SHOOTMERISTEMLESS A KNOX1 homeobox gene involved in maintaining the undifferentiated state of the **shoot apical meristem**.

Shope fibroma virus Poxvirus associated with the production of benign skin tumours in cottontail (but not domestic) rabbits.

Shope papilloma virus A member of the Papillomaviridae that produces **papillomas** (warts) in cottontail rabbits.

short hairpin RNA See shRNA.

short interfering RNA See **small interfering RNA**.

short interspersed nuclear elements *SINEs* **Retrotransposons** (short interspersed nucleotide elements) around 300 bp found in most animal genomes in very large numbers. They lack **long terminal repeats** and are nonautonomous, relying on the reverse

transcriptase of the **long interspersed nucleotide element** LINE1 to replicate. The best known is the human **Alu** repeat but most others are derived from tRNAs. See **mammalian-wide interspersed repeats** (MIRs).

short QT syndrome A disorder in which the QT interval in the cardiac cycle is short and there can be paroxysmal atrial fibrillation and sudden cardiac failure. There are various forms, all caused by mutations in potassium channels: in short QT syndrome-1 (SQT1) the KCNH2 gene product **hERG**, in SQT2 the KCNQ1 gene for the **shaker**-related subfamily of potassium channels, and in SQT3 the KCNJ2 gene product (Kir2.1). See **long QT syndrome**.

SHORT ROOT A transcription factor (531aa) of the **GRAS family**, required for specification of the **quiescent centre** and maintenance of surrounding stem cells, and for the asymmetric cell divisions involved in radial pattern formation in shoots and roots. Required for normal shoot gravitropism. Controls the transcription of **SCARECROW**.

short stop In *Drosophila* a **spectraplakin** homologue with many (>12) isoforms, including the short stop/Kakapo long isoform (5201aa) encoded by the gene *shot*. Shot isoforms are similar to **spectrin** and **dystrophin**, with an actin-binding domain followed by spectrin repeats. In short stop **plakin** repeats are inserted between the actin-binding domain and spectrin repeats of shot. Localized to **adherens junctions** of embryonic and follicular epithelia.

short tandem repeat See **satellite DNA**.

shot (1) A *Drosophila* **plakin** family member with actin binding and microtubule binding domains. In *Drosophila*, it is required for a wide range of processes, including axon extension, dendrite formation, axonal terminal arborization at the neuromuscular junction, tendon cell development, and adhesion of wing epithelium. See **short stop**. (2) Short stature homeobox protein 2 (Paired-related homeobox protein SHOT, 331aa) expressed during cranofacial development as well as in heart.

shotgun approach Casual term for any approach that analyses a large number of small samples rather than a single large one: a shotgun fires many small pellets over a wider area whereas a rifle fires a single large bullet. Generally used in the context of sequence analysis (shotgun sequencing) carried out by chopping DNA (or even the whole genome) into many small sections of random length, sequencing them all, some many times, and pasting them together using computer methods that recognise overlap, as opposed to working systematically from one end.

Shp (1) Protein tyrosine phosphatases with SH2 domains that are recruited to the ITIM motif of receptor tyrosine kinases and play an important role in the control of cytokine signalling. Shp-1 (haematopoietic cell phosphatase, PTP1C, PTPN6, 595aa) is important in regulating antigen responses in T-cells. See **motheaten**. Shp-2 (Syp, PTP2C, PTPN11, 593aa) functions downstream of several growth factor receptors and has a role in cell spreading and migration; the homozygous mutation in mice is lethal. Mutations can lead to **Noonan's syndrome** or LEOPARD syndrome. Shp-substrate 1 (signal regulatory protein alpha-1, SIRP-alpha-1, 503aa) is an Ig-like cell surface receptor for CD47 that may play a role in synaptogenesis. and is involved in the negative regulation of receptor tyrosine kinase-coupled cellular responses induced by cell adhesion, growth factors or insulin. CD47 binding prevents maturation of immature dendritic cells and inhibits cytokine production by mature dendritic cells. (2) Small heterodimer partner (shp, 257aa) an orphan nuclear receptor belonging to the nuclear receptor superfamily of transcription factors. May be a negative regulator of signalling pathways.

shRNA *short hairpin RNA* Small RNA molecules (50–100 bases) generated from plasmids that are processed by the cellular machinery to produce **small interfering RNA** (siRNA) and thus suppress the targeted gene. Research article: http://www.ncbi.nlm.nih.gov/pmc/articles/PMC1343552/?tool = pmcentrez

shrooms A family of actin-binding proteins with PDZ domains and the **Apx**/Shrm domain 2 (ASD2), involved in regulating cell shape during morphogenesis. Originally identified as being important in neural tube development in mice; mutations lead to the 'mushrooming' outward of the neural folds. Shroom 1 (852aa) may be involved in the assembly of microtubule arrays during cell elongation, Shroom 2 (1616aa) is involved in morphological changes in endothelial cells during spreading and may interact with gamma-tubulin. Shroom 3 (1986aa) interacts with microfilament bundles and also gamma-tubulin. Shroom 4 (1498aa) may regulate the spatial distribution of myosin II. A *Drosophila* homologue (1576aa) is involved in cell elongation by regulating the formation of microtubule arrays.

SH-SY5Y cells A dopaminergic neuroblastoma cell line, derived by multiple subcloning of SK-N-SH cells, from a four year-old girl.

shuttle flow See **cytoplasmic streaming**.

shuttle vector A **cloning vector** that will replicate in cells of more than one organism, e.g. *E. coli* and yeast. This combination allows DNA from yeast to be grown in *E. coli* and tested directly for **complementation** in yeast. Shuttle vectors are constructed so that they have the origins of replication of the various hosts.

Shwachman-Diamond syndrome A disorder (Shwachman-Bodian-Diamond syndrome) characterized by exocrine pancreatic insufficiency and various other abnormalities, caused by mutations in the SBDS gene. Heterozygous mutations in the SBDS gene have been associated with predisposition to aplastic anaemia. The gene product (250aa) may be involved in the biogenesis of the 60S ribosomal subunit and translational activation of ribosomes.

Shwartzman reaction Reaction that occurs when two injections of **endotoxin** are given to the same animal, particularly rabbits, 24 hours apart. In the local Shwartzman reaction the first injection is given intradermally, the second intravenously, and a haemorrhagic reaction develops at the dermal site. If both injections are intravenous the result is a generalized Shwartzman reaction, often accompanied by **disseminated intravascular coagulation**. The reaction depends upon the response of platelets and neutrophils to endotoxin.

sialic acid See **neuraminic acid**.

sialidase See **neuraminidase**.

sialidosis See **mucolipidosis**.

sialin A tetraspanin lysosomal protein (solute carrier family 17 member 5, 495aa) responsible for export of anionic substances, particularly sialic acid. Defects lead to Salla disease or infantile free sialic acid storage disease, both neurodegenerative conditions.

sialoglycoprotein Glycoprotein in which the N- or O-glycan chains include residues of **neuraminic acid**.

sialophorin See **leukosialin**.

sialyl Lewis x Sialylated form of CD15 (sLex, CD15s), the ligand for E-, P- and L-**selectins** and an important blood group antigen. Expressed on neutrophils, basophils and monocytes and only some lymphocytes. Also present on some **HEV**. Deficiency in sialyl Lewis-X will cause leucocyte adhesion deficiency Type II.

sialylate To add **sialic acid** to a glycoprotein or glycolipid, usually in a terminal position.

sialyltransferase A family of enzymes (EC 2.4.99.-, CMP-N-acetylneuraminate:acceptor N-acetylneuraminyl transferases) involved in the transfer of sialic acid (N-acetyl neuraminic acid) to oligosaccharide chains on glycoproteins or glycolipids. They are substrate-specific.

siamois A homeobox transcriptional activator (246aa) mediator of the dorsal **Wnt** signalling pathway, necessary for formation of the Spemann organizer and dorsoanterior development in *Xenopus*.

SIC1 Substrate and inhibitor of the cyclin-dependent protein kinase CDC28 (284aa) from *S. cerevisiae*. Interacts with the MAP kinase HOG1. See **Skp**.

sickle cell anaemia Disease common in ethnic groups originally from areas in which malaria is endemic. The cause is a point mutation in haemoglobin (valine instead of glutamic acid at position 6), and the altered haemoglobin (HbS) crystallizes readily at low oxygen tension. In consequence, erythrocytes from homozygotes change from the normal discoid shape to a sickled shape when the oxygen tension is low, and these sickled cells become trapped in capillaries or damaged in transit, leading to severe anaemia. In heterozygotes, the disadvantages of the abnormal haemoglobin are apparently outweighed by increased resistance to *Plasmodium falciparum* malaria, probably because parasitized cells tend to sickle and are then removed from circulation.

sideramines See **sideromycins**.

sideroblastic anaemia X-linked sideroblastic anaemia (XLSA) is caused by mutation in the gene encoding delta-aminolevulinate synthase-2 (**ALA synthase**), the enzyme that catalyzes the first committed step of haem biosynthesis. Sideroblastic anemia with spinocerebellar ataxia (ASAT) is an X-linked mitochondrial disease caused by mutation in ATP-binding cassette (ABC) transporter ABC7 (752aa) that is involved in iron homeostasis.

sideroblasts Nucleated red blood cells containing Pappenheimer bodies: small, deeply basophilic granules that contain ferric iron.

siderochromes See **ferrichromes**.

sideromycins *formerly* sideramines or siderochromes. Antibiotics covalently linked to siderophores that are actively transported into Gram-positive and Gram-negative bacteria. Examples include albomycin, a derivative of ferrichrome with a bound thioribosylpyrimidine moiety that inhibits seryl-t-RNA synthetase and salmycin, a ferrioxamine derivative with a bound aminodisaccharide, that is thought to inhibit protein synthesis. Research article: http://www.ncbi.nlm.nih.gov/pmc/articles/PMC2757582/?tool = pubmed

siderophilin See **transferrin**.

siderophilins Family of non-haem iron chelating proteins (about 80 kDa) found in vertebrates. Examples are **lactoferrin** and **transferrin**.

siderophores Natural iron-binding compounds that chelate ferric ions (which otherwise form insoluble colloidal hydroxides at neutral pH and are then inaccessible) and are then taken up together with the metal ion.

siderosis (1) A lung disease (a type of pneumonoconiosis) caused by the inhalation of metallic particles. (2) Excessive deposition of iron in tissues.

sieve plate The perforated end walls separating the component cells (sieve elements) that make up the phloem **sieve tubes** in angiosperms. The perforations permit the flow of water and dissolved organic solutes along the tube, and are lined with **callose**. The plates are readily blocked by further deposition of callose when the sieve tube is stressed or damaged.

sieve tube The structure within the phloem of higher plants that is responsible for transporting organic material (sucrose, raffinose, amino acids, etc.) from the photosynthetic tissues (e.g. leaves) to other parts of the plant. Made up of a column of cells (sieve elements) connected by **sieve plates**. In gymnosperms the sieve cells retain their nuclei and the conducting connections have less obvious pores than sieve plates.

sigma factors Prokaryotic **initiation factors** (σ factors) that promote attachment of DNA-dependent **RNA polymerase** to specific initiation sites. Every molecule of RNA polymerase has one sigma factor subunit which is released following attachment and different sigma factors are involved under different environmental conditions. For example, in *E. coli*, σ70 (RpoD, 613aa) is the primary sigma factor for housekeeping proteins, σ54 (RpoN, 477aa) is involved under conditions of nitrogen limitation, σ38 (RpoS, 111aa) under starvation conditions, amongst others.

sigma receptors Receptors originally identified by their ability to bind various psychoactive drugs but then shown to be non-opioid receptors. Sigma receptor 1 (sigma non-opioid intracellular receptor 1, 223aa) is widely expressed and involved in lipid transport from the ER, regulates various other receptors and potassium channels and has a role in calcium signalling. Considered to be involved in learning, memory and mood alteration. Sigma 2 receptors, unlike sigma-1 are not coupled to Gi/o proteins and sigma-2 receptor specific ligands induce apoptosis in a dose-dependent fashion.

signal peptidase complex *SPC* A multi-subunit serine peptidase (protease) complex located in the ER membrane that cleaves the **signal sequence** from proteins that are destined for export. Cleavage occurs as soon as the cleavage site of the translocating polypeptide is exposed in the lumen of the ER. Mammalian signal peptidase is a complex of five different polypeptide chains; SPC12 and SPC25 have substantial cytoplasmic domains and span the membrane twice, SPC18, SPC21, SPC22/23 are single-spanning membrane proteins mostly exposed to the lumen of the endoplasmic reticulum. See also **sec61**.

signal peptide See **signal sequence**.

signal peptide peptidase *SPP* Like **gamma-secretase**, an unusual aspartyl peptidase (presenilin-like protein 3, EC 3.4.23.-, 377aa) that catalyzes the intramembrane proteolysis of some signal peptides after they have been cleaved from a preprotein. A family of SPP-like proteins (SPPLs) of unknown function has been identified.

signal recognition particle *SRP* A complex between a 7S RNA and six proteins. The SRP binds to the nascent polypeptide chain of eukaryotic proteins with a **signal sequence** and halts further translation until the ribosome becomes associated with the rough endoplasmic reticulum. One of the SRP proteins (srp54) binds GTP and in association with 7S RNA and srp19 has GTPase activity.

signal recognition particle-receptor Receptor for the **signal recognition particle** (SRP) found in the membrane of the endoplasmic reticulum. Also known as docking protein. Heterodimeric, both protomers (α, 638aa; β, 271aa) having GTP-binding capacity, though dissimilar binding sites. Not until the complex of SRP, ribosome, message, and nascent polypeptide chain binds to the SRP-receptor is the block to further chain elongation released – and concurrently the SRP is released, leaving the ribosome attached to the ER membrane. **Co-translational transport** of the polypeptide delivers it into the lumen of the ER.

signal sequence (1) An N-terminal sequence ($\sim$20aa, usually hydrophobic but with some positive residues) that interacts with **signal recognition particle** and is present on proteins that are destined either to be secreted or to be membrane components. It is normally absent from the mature protein, having been removed from the growing peptide chain by the **signal peptidase complex** located on the cisternal face of the endoplasmic reticulum. (2) Any sequence that determines the organelle for which the protein is destined.

signal transduction *Signal-response coupling* The cascade of processes by which an extracellular signal (typically a hormone or neurotransmitter) interacts with a receptor at the cell surface, causing a change in the level of a **second messenger** (for example calcium or cyclic AMP) and ultimately effects a change in the cell's functioning (for example, triggering glucose uptake, or initiating cell division). Can also be applied to sensory signal transduction, e.g. of light at photoreceptors.

signal-response coupling See **signal transduction**.

signal-transducing adaptor proteins Adaptor molecules (signal-transducing adaptor molecule 1, STAM1, 540aa; STAM2, 525aa) that link cytokine and growth factor receptors and **STATs**. STAM-binding protein (EC 3.1.2.15, 424aa) is a zinc metallopeptidase that specifically cleaves 'Lys-63'-linked polyubiquitin chains. It interacts with the SH3 domain of STAM1 and potentiates various signalling pathways.

signet-ring cell An adipocyte with a large central fat-filled vacuole or an epithelial cell with a mucin-filled vesicle in which the nucleus is displaced to one side to give an appearance reminiscent of a signet ring.

silanization Modification (silanisation, silanizing) of hydroxyl groups on silica or glass surfaces with silane coupling agents e.g. (3-mercaptopropyl)-trimethoxysilane) to give the inactive -O-SiR3 grouping. Silanization can neutralize surface charges, thus eliminating non-specific binding. Often used to prepare surfaces for the binding of DNA fragments in microarrays.

silent gene A gene that is not phenotypically expressed. This can arise because of mutation of the gene itself (e.g. a nonsense codon causing premature termination of translation), a defect in some upstream control, or a mutation that renders the product inactive in some way.

silent mutation Mutations that have no effect on **phenotype** because they do not affect the activity of the product of the gene, usually because of codon ambiguity.

siliconization Non-covalent coating of surface with a layer of silicone oil making it less adhesive or reactive. See **silanizing**.

silicosis Inflammation of the lung caused by foreign bodies (inhaled particles of silica): leads to fibrosis but unlike **asbestosis** does not predispose to neoplasia.

siliqua A long dry dehiscent fruit of cruciferous plants, for example oilseed rape, divided into two compartments by a central septum to which seeds are attached.

Silver-Russell syndrome See **H19 gene**.

Silverman-Handmaker type dyssegmental dysplasia See **Schwartz-Jampel syndrome**.

Simian Virus 40 See **SV40**.

Simmonds' disease *hypopituitarism* The failure of the anterior lobe of the pituitary to produce any one or more of its six hormones (ACTH, TSH, FSH, LH, GH, and prolactin).

simocyclinone D8 A potent DNA gyrase inhibitor, an **aminocoumarin antibiotic** produced by *Streptomyces antibioticus* Tü 6040. The mode of action is atypical: it binds to the GyrA subunit of the enzyme and prevents binding to DNA. The producing organism is protected by the production of a specific efflux pump which prevents build-up of the metabolite. Research article: http://www.jic.ac.uk/STAFF/mark-buttner/pdfs/SimR_MolMicro_2009.pdf

simple epithelium An epithelial layer composed of a single layer of cells all of which are in contact with the basal lamina (see **basement membrane**). May be cuboidal, columnar, squamous or **pseudostratified**. Images http://www.mhhe.com/biosci/ap/histology_mh/simpleep.html

simple sequence length polymorphism *SSLP* Tandem repeat sequences that are of variable length in different individuals, i.e. exhibit polymorphism. Since the flanking sequences are the same in all individuals it is relatively easy to choose primers that will selectively amplify such repeat sequences and the length variation (diffence in repeat number) can be detected easily on the basis of the product size.

simple sequence repeats *SSRs* See **satellite DNA**.

Simpson-Golabi-Behmel syndrome *SGBS* An X-linked disorder characterized by excessive growth, similar to **Beckwith-Wiedemann syndrome**. Type 1 is caused by mutation in the **glycipan-3** gene which reduces the binding of insulin-like growth factor-2 which normally limits its availability. Type 2 has been associated with a mutation in the CXORF5 gene.

Sin3 Component of the multiprotein complex involved in repression of transcription. Sin3 is a corepressor (Sin3A, 1219aa; Sin3B, 1130aa) and forms a complex with Rpd3 histone deacetylase; the complex then interacts with DNA-binding proteins. Yeast SIN3 (1536aa) is involved in the repression of a diverse range of genes. Sin3 does not itself bind to DNA.

Sindbis virus Enveloped virus of the alphavirus group of **Togaviridae**. It is thought to be an infection of birds spread by fleas, and there is little evidence that it causes any serious infection in humans. The synthesis and export of the spike proteins, via the endoplasmic reticulum and Golgi complex, have been used as a model for the synthesis and export of plasma membrane proteins.

SINEs See **short interspersed nuclear elements.**

single cell protein Protein(s) produced by single cells of pure or mixed cultures of algae, yeasts, fungi or bacteria. Of possible commercial importance in providing food sources from biotechnological processes.

single nucleotide polymorphism See **SNP**.

single stranded DNA *ssDNA* DNA that consists of only one chain of nucleotides rather than the two **base-pairing** strands found in DNA in the double-helix form. **Parvoviridae** have a single-stranded DNA genome. Single-stranded DNA can be produced

experimentally by rapidly cooling heat-denatured DNA to prevent **renaturation**.

single-channel recording Variant of **patch clamp** technique in which the flow of ions through a single channel is recorded.

single-stranded conformational polymorphism, *SSCP* Technique for detecting point mutations in genes by amplifying a region of genomic DNA (using asymmetric **PCR**) and running the resulting product on a high quality gel. Single base substitutions can alter the secondary structure of the fragment in the gel, producing a visible shift in its mobility.

singlet oxygen $^{(1)}O_2$ An energised but uncharged form of oxygen (the diamagnetic form of molecular oxygen) that is produced in the **metabolic burst** of leucocytes and that can be toxic to cells. One of the so-called 'reactive oxygen species (ROS)'.

SipA An actin-binding protein from *Salmonella enterica*, (Salmonella invasion protein A, 685aa) that apparently stabilizes F-actin bundles by increasing polymerization and decreasing depolymerization. Salmonella force their way into nonphagocytic host intestinal cells to initiate infection. Uptake is triggered by delivery into the target cell of over thirty specialised effector proteins via two distinct type III secretion systems. SipA is one of the effectors of the Salmonella pathogenicity island-1 (SPI-1) type 3 secretion system. SipC is also involved in the process. *Salmonella* pathogenesis: http://www.ncbi.nlm.nih.gov/pubmed/19157959

sir2 Protein (silent information regulator-2 protein, 562aa) from *Saccharomyces cerevisiae*, original member of the **sirtuin** family to be identified. A NAD$^+$-dependent protein **histone deacetylase**.

sirenin Sexual pheromone, a bicyclic sesquiterpenediol, produced by female gametes of the water moulds, *Allomyces macrogynus*, and *A. arbuscula*. Male gametes respond chemotactically.

siRNA See **small interfering RNA.**

sirtuins Family of NAD-dependent protein **histone deacetylases** (HDACs, silent information regulator 2 (Sir2) enzymes) that includes **Sir2** and its mammalian orthologues: play an important role in epigenetic gene silencing, DNA recombination, cellular differentiation and metabolism, and the regulation of aging. Human sirtuin 1 (SIRT1, 747aa) and sirtuin-2 (389aa) have greatest homology with *S. cerevisiae* Sir2 protein; SIRT3 (399aa), SIRT4 (314aa) and SIRT5 (310aa) are mitochondrial and more closely resemble prokaryotic sirtuin sequences. SIRTs 1-5 are widely expressed in fetal and adult tissues. The class IV sirtuins include SIRT6 (355aa) which is chromatin-associated and involved in DNA repair,

SIRT7 (400aa) which is associated with active rRNA genes (rDNA) and histones.

sis (1) An oncogene, *sis*, originally identified in monkey **sarcoma**, encoding a B-chain of **PDGF**. Human c-*sis*/PDGF-B proto-oncogene has been shown to be overexpressed in a large percentage of human tumour cells. (2) See **macrophage inflammatory protein 1** α and β. (SIS). (3) Small intestinal submucosa: porcine SIS has been used as a cell-free, biocompatible biomaterial for surgical repair purposes.

sister chromatid One of the two **chromatids** making up a **bivalent**. Both are semi-conservative copies of the original chromatid.

SIT Transmembrane adaptor protein (**TRAP**), a disulphide-linked homodimer (SHP2-interacting transmembrane adapter protein, 196aa), that is expressed in lymphocytes. Interacts with the SH2-containing protein tyrosine phosphatase 2 (**SHP2**) via an immunoreceptor tyrosine-based inhibition motif (**ITIM**). Also interacts with **grb2**.

Site-1 protease *S1P* See **ski**.

site-directed mutagenesis See **site-specific mutagenesis**.

site-specific mutagenesis An *in vitro* technique (site-directed mutagenesis) in which an alteration is made at a specific site in a DNA molecule, which is then reintroduced into a cell. Various techniques are used; for the cell biologist, a very powerful approach to determining which parts of a protein or nucleotide sequence are critical to function. For proteins it is common to systematically substitute alanine for various residues (an alanine scan).

site-specific recombination A type of **recombination** that occurs between two specific short DNA sequences present in the same or in different molecules that are recognised by site-specific recombinases which bind, cleave, excise, exchange and rejoin the strands. Examples include the **lox-Cre system**, **resolvases**, Flp-Frt recombination, lambda integrase.

sitosterolemia A disorder caused by a defect in ABC transporters (ABCG8 or ABCG5) that normally cooperate to limit absorption of sterols in the intestine and promote sterol secretion in bile. Affected individuals have very high levels of the plant sterol, sitosterol, in the plasma which can cause **xanthomas** and increases the risk of coronary artery disease.

situs inversus Condition in which the normal asymmetry of the body (in respect of the circulatory system and intestinal coiling) is reversed to produce a mirror image of the normal situation. It occurs in approximately 50% of patients with primary ciliary dyskinesia (immotile cilia syndrome, Kartagener's syndrome) because ciliary function during development is necessary for the normal asymmetry.

SIV Simian immunodeficiency virus. Very similar to **HIV** and used extensively as an animal model.

Sixth disease A benign disease (roseola) of children under two years old caused by human herpes virus-6 and -7 (*Roseolovirus*). There is a transient rash following a brief fever.

Sjogren's syndrome One of the so-called connective tissue diseases that also include **rheumatoid arthritis**, **systemic lupus erythematosus** and **rheumatic fever**. Characterized by inflammation of conjunctiva and cornea. It is an autoimmune disease associated with HLA alleles DRB1*03 and DQB1*02. See **La protein**.

SK channels A subfamily of voltage-independent Ca^{2+}-activated **potassium channels** (small conductance calcium-activated potassium channels) with four members (SK1 (KCa2.1), 543aa; SK2 (KCa2.2), 579aa; SK3 (Kca2.3), 736aa; SK4 (KCa3.1), 427aa). The channel complex is composed of 4 subunits each of which binds to a calmodulin subunit giving the calcium sensitivity. They are thought to regulate neuronal excitability by contributing to **slow afterhyperpolarization**. Some are selectively inhibited by **apamin** and **scyllatoxin**, others by **charybdotoxin**.

skelemin The murine homologue of **myomesin 1** (1667aa).

skeletal muscle A rather nonspecific term usually applied to the striated muscle of vertebrates that is under voluntary control. The muscle fibres are syncytial and contain myofibrils, tandem arrays of **sarcomeres**.

skeletrophin A RING finger-dependent ubiquitin ligase (mindbomb homolog 2, MIB2, 1013aa), which mediates ubiquitination of **delta** receptors, that are ligands of **Notch** proteins. It is down-regulated in many melanomas.

ski (1) An oncogene, *ski* (Sloan-Kettering Institute proto-oncogene), identified in avian **carcinoma**, encoding a nuclear protein. C-ski is a regulating factor for fibroblast proliferation and an important co-repressor of **Smad3**. (2) **Subtilisin**/kexin isozyme-1 (SKI-1), otherwise known as Site-1 protease (S1P), a Golgi proteinase involved in activation of sterol-regulated element-binding proteins (**SREBPs**). (3) The homologue of *Drosophila* skinny hedgehog (SKI1, hedgehog acyltransferase, HHAT, EC 2.3.1.-, 493aa) that catalyses N-terminal palmitoylation of **sonic hedgehog**.

skin associated lymphoid tissue *SALT* The cells of the immune system associated with the dermis and epidermis. Includes **Langerhans cells** and resident phagocytes although some authors would include cutaneous nerve termini containing calcitonin gene-related peptide (CGRP): release of the peptide will stimulate mast cells to release cytokines such as IL-10 and TNFα. More of a concept than a morphological entity: *cf*. **gut-, bronchus-,** and **mucosal-associated lymphoid tissue**.

SKIP (1) Ski-interacting protein (nuclear receptor coactivator NCoA-62, SNW domain-containing protein 1, 536aa) involved in vitamin D-mediated transcription. See **ski**. (2) In *Arabidopsis*, a large family of F-box proteins (SKP1-interacting partners) that are components of SCF E3 ubiquitin ligase complexes. (3) The protein (SKIP, **SifA** and kinesin interacting protein, PH domain-containing family M member 2, 1019aa) that regulates kinesin, has a role in Golgi morphology and is necessary for *Salmonella* infectivity. The close proximity of the SCV (*Salmonella* containing vacuole) to the Golgi may facilitate interception of endocytic and exocytic transport vesicles to obtain nutrients. SifA binds SKIP and appears to be important for promoting bacterial replication. Efficient localisation of SifA to the SCV is mediated by the SPI-1 effector SipA. Research article: http://www.ncbi.nlm.nih.gov/pmc/articles/PMC2647982/

SKL Carboxyl-terminal amino acid sequence (serine-lysine-leucine, SKL), the consensus **peroxisomal targeting sequence 1** (PTS1) that directs a polypeptide to peroxisomes in plants, animals, and yeasts.

Skn-1 (1) In *C. elegans* a maternally and zygotically expressed transcription factor (skinhead-1, 623aa) that specifies the fate of certain blastomeres during early development. Binds DNA with high affinity as a monomer even though it has a basic region similar to basic-leucine zipper (bZIP) proteins that only bind as dimers. (2) In *S. cerevisiae*, beta-glucan synthesis-associated protein (SKN1, 771aa) required for synthesis of the major beta-glucans of the yeast cell wall.

skotomorphogenesis *etiolation* Development, growth and differentiation of a plant under conditions of darkness.

skp Components of the **SCF** complex, (S-phase kinase-associated proteins; p19, Skp1, 163aa; p45, Skp2, 424aa). The SCF (Skp1-cullin-F-box complex) ubiquitin protein ligase of *S. cerevisiae* triggers DNA replication by catalysing ubiquitinylation of the S phase cyclin-dependent kinase inhibitor, SIC1. The complex also regulates the function of NFκB. See **Sgt1**.

skyllocytosis Phagocytic process in *Allogromia* 'A marine rhizopod with which the general zoologist is usually acquainted'.

SL1 (1) Transcription factor SL1/TIF-IB complex, which is involved in the assembly of the preinitiation complex during RNA polymerase I-dependent transcription. The complex is composed of four proteins

including TATA-binding protein **TBP**. (2) Stem-loop structure 1 (SL1), a characteristic feature of some viral RNAs.

Sla1p In yeast, an actin regulatory protein (1244aa) that, together with **End3p** and Sla2p is important in maintaining a rapid turnover of F-actin in cortical patches. Binds both to activators of actin dynamics (**Las17p** and Pan1p) and to cargo proteins, such as the pheromone receptor Ste2p.

SLAMs Signalling lymphocyte-activation molecules. A family of Ig superfamily receptors on T- and B-cells (SLAM1 is CD150, 335aa). They recruit various cytoplasmic signalling molecules and co-activate immune responses. SLAM-associated proteins (SAPs) are adaptors with SH2 domains. SAP (128aa) is T-cell-specific and blocks recruitment of **Shp2**, it is mutated in X-linked lymphoproliferative disease (**XLP**).

SLAP (1) Src-like adapter proteins (SLA1, 281aa; SLA2, 261aa), that have SH2 and SH3 domains and interact with **zap-70**, **Syk**, **LAT**, and T-cell receptor (TCR)-zeta chain as negative regulators of immune responses. (2) Sarcolemmal-associated protein (sarcolemmal membrane-associated protein, 828aa) that forms a homodimer which interacts with myosin and is found in transverse tubules, near the junctional sarcoplasmic reticulum and along the Z- and M-lines in cardiomyocytes. May play a role during myoblast fusion.

S layer A paracrystalline array which completely covers the surfaces of many pathogenic bacteria (Archaea and Bacteria). Usually consist of a single (glyco-) protein species with molecular masses ranging from about 40 to 200 kDa that form lattices of oblique, tetragonal, or hexagonal architecture, and probably represent the earliest cell wall structures. About 10 nm thick.

SLC (1) Secondary lymphoid-tissue chemokine (CCL21, 134aa) that is selectively chemotactic *in vitro* for thymocytes and activated T-cells, and may mediate homing of lymphocytes to secondary lymphoid organs. (2) **Sodium-lithium countertransport**. (3) A G-protein coupled receptor, somatostatin-like receptor 1 (SLC-1), for which the ligand is **melanin-concentrating hormone** (MCH).

sleeping sickness See *Trypanosoma*.

SLEEPY In *Arabidopsis* an F-box protein (SLEEPY1, SLY1, 151aa) a subunit of a SCF E3 ubiquitin ligase complex that positively regulates **gibberellic acid** signalling. See also **SNEEZY**.

slicer activity The cleavage of target RNA by enzymes such as **argonaute** in **RNA interference**.

sliding filament model Generally accepted model for the way in which contraction occurs in the **sarcomere** of striated muscle, by the sliding of the **thick filaments** relative to the **thin filaments**.

slime moulds Two distinct groups of organisms, the cellular slime moulds **Acrasidae** that include *Dictyostelium*, and the **acellular slime moulds** or Myxomycetes that include *Physarum*.

slingshot A family of protein tyrosine phosphatases that are involved in control of actin dynamics (human slingshot homologue-1, SSH1, EC 3.1.3.48, EC 3.1.3.16, 1049aa). **Cofilin** is activated by dephosphorylation by slingshot (SSH) family phosphatases. SSH activity is strongly increased by binding to filamentous actin but is inhibited when phosphorylated itself. **Calcineurin** will dephosphorylate and activate slingshot; interaction with **14-3-3 proteins** inhibits phosphatase activity and also blocks recruitment to lamellipodia and slingshot stimulation by actin.

SLIP1 (1) A protein (SLBP (stem loop binding protein)-interacting protein 1, 222aa) involved in replication-dependent translation of histone mRNAs that end with a stem-loop rather than a poly-A tail. (2) In *Drosophila*, Slo-interacting protein 1 (767aa) that may selectively reduce calcium-activated potassium channel (**Slo**) currents (see **BK channels**).

slipped strand mispairing Mispairing of the complementary DNA strands of a single DNA double helix during replication. Slippage can be either forward (causing deletion) or backward (causing insertion). **Short tandem repeats** are thought to be particularly prone to slipped strand mispairing

slit Family of proteins first identified in *Drosophila* and subsequently in many vertebrates. Secreted diffusible proteins (Slit1,1534aa; Slit2, 1529aa) that act as chemorepellants for migrating neurons during development. Slit genes are expressed in the ventral midline of the neural tube and the effect of the repulsion is to force neurons to move in the rostral migratory stream and prevents axons from crossing between hemispheres. Slit3 (1532aa) is expressed more strongly in peripheral tissues. Receptor is **robo** (roundabout). See **slit and NTRK proteins**.

SLIT and NTRK-like proteins A family of proteins that affect neuronal migration by stimulating or inhibiting neuronal outgrowth. SLIT and NTRK-like protein-1 (SLIK1, 696aa) enhances neuronal dendrite outgrowth; defects may be a cause of **Gilles de la Tourette syndrome**. SLIK2 (845aa), SLIK3 (977aa), SLIK4 (837aa), SLIK5 (958aa) and SLIK6 (841aa) suppress neurite outgrowth. NTRK1 is the high-affinity nerve growth factor receptor.

SLK A microtubule-associated kinase (Ste20-like kinase, 1235aa), one of the germinal centre sub-family of **STE20**-like kinases, involved in turnover of **focal adhesions**.

SLM (1) RNA-binding proteins, (**STAR proteins**) (Sam68-like mammalian proteins, SLM1, 346aa; SLM2, 349aa) related to SAM68, that regulate splice site selection. They may function as adapters for Src kinases during mitosis. Expression of SLM2 is reported to be upregulated in patients with various kidney diseases. (2) In *S. cerevisiae*, phosphatidylinositol 4,5-bisphosphate-binding proteins (SLM1, TORC2 effector protein, 686aa; SLM2, 656aa); the heterodimer of SLM1-SLM2 is the effector of the TORC2- and calcineurin-signaling pathways. (3) In *S. cerevisiae* SLM3 is mitochondrial tRNA-specific 2-thiouridylase 1 (EC 2.8.1.-, 417aa), SLM4 (162aa) is a component of the GSE and EGO complexes. SLM5 is the gene for mitochondrial asparaginyl-tRNA synthetase (EC 6.1.1.22, 492aa) and SLM6 that for methyl methanesulphonate-sensitivity protein 1 (1407aa), involved in protection against replication-dependent DNA damage. (4) In *Silene latifolia* (white campion) SLM proteins are transcription factors. (e.g. SLM1, 248aa, SLM5, 257aa).

slot blot A **dot blot** in which samples are placed on a membrane through a series of rectangular slots in a template. This has some advantages because hybridization artefacts are usually circular.

slow after hyperpolarization Hyperpolarization of the presynaptic membrane that follows an action potential. It is generated by the activation of small-conductance calcium-activated potassium channels (**SK channels**). The hyperpolarization limits the firing frequency of repetitive action potentials (spike-frequency adaption) and is essential for normal neurotransmission.

slow muscle Striated muscle used for long-term activity (e.g. postural support). Depends on oxidative metabolism and has many mitochondria and abundant **myoglobin**.

slow reacting substance of anaphylaxis *SRS-A* Potent bronchoconstrictor and inflammatory agent released by mast cells; an important mediator of allergic bronchial asthma. A mixture of three **leukotrienes** (LTC_4 mainly, LTD_4 and LTE_4).

slow virus (1) Specifically one of the **Lentivirinae**. (2) Any virus causing a disease that has a very slow onset. Diseases such as sub-acute **spongiform encephalopathy**, Aleutian disease of mink, **scrapie**, **kuru**, and **Creutzfeldt-Jacob disease** may be caused by slow viruses although this hypothesis has fallen out of favour. See also **prion**.

SLP-76 An adapter protein required for T-cell receptor (TCR) signalling. See **TRAPS**. Interacts with **IL-2-inducible T-cell kinase**.

SLRPs See **small leucine-rich repeat proteoglycans**.

SLS collagen Segment long spacing collagen. Abnormal packing pattern of collagen molecules formed if ATP is added to acidic collagen solutions, in which lateral aggregates of molecules are produced. Each aggregate is 300 nm long, and the molecules are all in register. If SLS-aggregates are overlapped with a quarter-stagger, the 67 nm banding pattern of normal fibrils is reconstituted.

slug Snail-related zinc-finger transcription factor, slug (SNAI2, 268aa), is critical for the normal development of neural crest-derived cells and loss-of-function Slug mutations have been proven to cause piebaldism and **Waardenburg's syndrome** type 2.

slurps Secreted Ly6/PLAUR domain-containing proteins. Proteins with anti-tumour activity. Slurp 1 (103aa) is involved in maintaining the physiological and structural integrity of the keratinocyte layers of the skin, and defects are a cause of **Meleda disease**. Slurp 2 (97aa) is expressed in a wider range of tissues. Research article: http://www.ncbi.nlm.nih.gov/pmc/articles/PMC2144295/pdf/10211827.pdf

Sly's syndrome An autosomal recessive lysosomal storage disease (mucopolysaccharidosis type VII) caused by mutation in the gene encoding beta-glucuronidase.

Sm proteins (1) Sm proteins were originally isolated as the antigens targeted by anti-Sm antibodies in a patient (Stephanie Smith) with systemic lupus erythematosus. They are RNA-binding proteins associated with a set of uridine-rich small nuclear RNA (snRNA) molecules U1, U2, U4 and U5. The Sm core proteins (SmB, SmB', SmN, SmD1, SmD2, SmD3, SmE, SmF and SmG) combine with other proteins and snRNAs to form small nuclear ribonucleoproteins (snRNPs) which are components of the **spliceosome**. A related snRNA, U6, binds to a complex of seven **Like Sm proteins**. The seven Sm and eight LSm proteins are conserved in eukaryotes; related proteins are found in Archaea (Sm1 and Sm2) and in Bacteria (Hfq and YlxS homologues). (2) Sec1/Munc18-like proteins (SM proteins, syntaxin-in-binding proteins) that bind to the SNARE complexes formed between SNARE molecules on the surfaces of vesicles and on their target membranes and regulate the process of membrane fusion. See **Sec proteins**. http://www.sciencemag.org/content/323/5913/474.short. (3) See SMN complex.

smac In humans, the homologue of **diablo** (second mitochondria-derived activator of caspases, direct IAP binding protein with low isoelectric point; often smac/diablo, 239aa) promotes apoptosis by activating caspases in the cytochrome c/Apaf-1/caspase-9 pathway and opposes the effects of **inhibitor of apoptosis proteins** (IAPs). It is the mammalian functional homologue of *Drosophila* **Reaper**, **Grim** and **Hid**. Smac export from mitochondria into the cytosol is provoked by cytotoxic drugs and DNA damage but release can be reduced if **Bcl-2** is overexpressed.

Smad proteins Intracellular proteins (homologues of *Drosophila* **mothers against decapentaplegic**, MAD-related proteins and *C. elegans* **dwarfin**/sma

proteins) that mediate signalling from receptors for extracellular TGFβ-related factors. At least eight are identified. Smad2 (467aa) is essential for embryonic mesoderm formation and establishment of anterior-posterior patterning. Smad4 (552aa) is important in gastrulation. Smads1 (465aa) and 5 (465aa) are activated (serine/threonine phosphorylated) by **BMP** receptors, Smad2 and 3 (425aa) by **activin** and TGFβ receptors. Smads activated by occupied receptors form complexes with Smad4/DPC4 and move into the nucleus where they regulate gene expression. Interact with **forkhead activin signal transducer**. See **juvenile polyposis**.

small acid-soluble spore proteins *SASPs* DNA-binding proteins in the spores of some bacteria, thought to stabilize the DNA in the **A-DNA** configuration, so protecting it from cleavage by enzymes or UV light.

small basic intrinsic proteins *SIPs* A subset of the **aquaporins**. SIP1 (small basic intrinsic protein 1-1, 240aa) is found in the endoplasmic reticulum of *Arabidopsis*.

small cell carcinoma Common malignant neoplasm of bronchus. Cells of the tumour have endocrine-like characteristics and may secrete one or more of a wide range of hormones, especially regulatory peptides like **bombesin**.

small interfering RNA *siRNA* Small double-stranded RNA molecules (20–25 bp) that interfere with the expression of specific mRNAs (**RNA interference**). They are a powerful experimental method for investigating the function of particular genes and are being developed as potential therapeutic agents. Some will activate gene expression (small activating RNAs, saRNAs) and others will activate innate antiviral systems (see **RIG1**). *Cf.* **microRNAs**. See **transacting siRNAs**.

small leucine-rich repeat proteoglycans *SLRPs* A family of proteoglycans characterised by **leucine-rich repeats**. Includes **asporin**, **decorin**, **biglycan**, **osteomodulin**, **fibromodulin**, and osteoglycin (mimecan).

small nuclear RNA *snRNA* Abundant class of RNA found in the nucleus of eukaryotes, usually including those RNAs with sedimentation coefficients of 7S or less. They are about 100–300 nucleotides long. Although 5S rRNA and tRNA are of a similar size, they are not normally regarded as snRNAs. Most are found in complexes with proteins (see **ribonucleoprotein** particles, snRNPs) and at least some have a role in processing hnRNA.

small temporal RNAs *stRNAs* Early name for **microRNAs**.

smallpox See **variola virus**.

SMARC The **SWI/SNF complex**-related, matrix-associated, actin-dependent regulators of chromatin (SMARC), also called BRG1-associated factors (BAFs), are components of human SWI/SNF-like chromatin-remodelling protein complexes. The SMARCA4 gene encodes the transcription activator BRG1 (EC 3.6.1.-, BRG1-associated factor 190A, 1647aa) an ATP-dependent helicase that interacts with nuclear hormone receptors to potentiate transcriptional activation.

SMART A free web-based resource, a 'Simple Modular Architecture Research Tool', that allows the identification and annotation of protein domains. SMART-7 (2012) has models for 1009 protein domains from more than 2 million proteins in 1133 species. Link to web-pages: http://smart.embl-heidelberg.de/

SMC proteins A family of proteins (structural maintenance of chromosomes proteins) involved in chromosome cohesion during the cell cycle and in DNA repair. They are the central components of the **cohesin** complex which forms a large ring structure that traps sister chromatids until the complex is degraded at anaphase; the complex may also play a role in spindle pole assembly during mitosis. Cohesin complexes are composed of SMC1 (1233aa)/SMC3 (1217aa) heterodimers attached via their hinge domain, **RAD21** which links them and one **STAG protein** (STAG1, STAG2 or STAG3), which interacts with RAD21. Defects in SMCs are the cause of **Cornelia de Lange syndrome**. The SMC5 (1101aa)/SMC6 (1091aa) complex (in association with non-structural maintenance of chromosomes element 2 homologue, NSMCE2) is involved in maintaining the integrity of DNA at double-strand breaks and promoting repair by homologous recombination. Research article: http://www.nature.com/emboj/journal/v25/n14/full/7601218a.html

SMCT A Na^+-coupled transporter (sodium-coupled monocarboxylate transporter, SLC5A8, 610aa) for lactate, pyruvate and short-chain fatty acids. Resorbs lactate in the kidney and maintains blood lactate homeostasis.

SMG1C complex A mRNA surveillance complex that recognizes and degrades mRNAs containing premature translation termination codons. Components include a serine/threoinine kinase (SMG1, 3657aa), SMG8 (991aa) and SMG9 (512aa). The complex forms part of the larger **SURF complex**. Other SMG proteins are involved in different mRNA surveillance systems (nonsense-mediated mRNA decay) and in the telomerase complex.

Smith-Lemli-Opitz syndrome An autosomal recessive disorder in which there are multiple congenital malformations and mental retardation. Caused by mutations in the gene encoding sterol delta-7-reductase (DHCR7).

Smith-Magenis syndrome A set of complex developmental abnormalities generally caused by a 3.7-Mb interstitial deletion in chromosome 17p11.2, but occasionally by mutations in the RAI1 gene (retinoic acid inducible-1) which is located in this region. A milder phenotype is associated with duplication of the same chromosomal region.

Smith-McCort dysplasia A rare autosomal recessive osteochondrodysplasia caused by defects in **dymeclin**. Has features in common with **Dyggve-Melchior-Clausen syndrome** which is caused by mutation in the same gene.

Smith-Watermann alignment Algorithm for detecting sequence similarities when searching a genomic database.

SMMCI A disorder in which there is a solitary median maxillary central incisor (SMMCI) in both the deciduous and permanent dentition. Caused by a mutation in the **sonic hedgehog** gene.

SMN complex A large macromolecular complex (survival of motor neurons complex) found in all metazoan cells and involved in the production of spliceosomal small nuclear ribonucleoproteins (snRNPs). The complex is found in the cytoplasm and in the nucleus in **Gems** (gemini of the coiled bodies). The constituent proteins include SMN, **Gemin2**, Gemin3 (a DEAD-box RNA helicase), Gemin4, Gemin5 (a WD-repeat protein), Gemin6 and Gemin7. SMN protein (294aa) is the product of the survival of motor neurons (*SMN*) gene which is defective in **spinal muscular atrophy**. The complex binds to specific sequences in the snRNAs. SMN complex substrates include **Sm proteins, like Sm proteins,** RNA helicase A, fibrillarin and GAR1, the RNP proteins hnRNP U, Q and R, as well as p80-**coilin**. Research article: http://hmg.oxfordjournals.org/cgi/content/full/14/12/1605

SMO (1) See **smoothened**. (2) **Spermine oxidase**. (3) The SMO genetic locus in strains of the fungus *Magnaporthe grisea* directs the formation of correct cell shapes in asexual spores, infection structures and asci.

smooth endoplasmic reticulum *SER* An internal membrane structure of the eukaryotic cell, similar to the rough endoplasmic reticulum (RER) but without the ribosome-binding function. Tends to be tubular rather than sheet-like, may be separate from the RER or may be an extension of it. Abundant in cells concerned with lipid metabolism and proliferates in hepatocytes when animals are challenged with lipophilic drugs.

smooth microsome Fraction produced by ultracentrifugation of a cellular homogenate. It consists of membrane vesicles derived largely from the **smooth endoplasmic reticulum**.

smooth muscle Muscle tissue in vertebrates made up from long tapering cells that may be anything from 20–500μm long. Smooth muscle is generally involuntary, and differs from striated muscle in the much higher actin/myosin ratio, the absence of conspicuous sarcomeres, and the ability to contract to a much smaller fraction of its resting length. Smooth muscle cells are found particularly in blood vessel walls, surrounding the intestine (particularly the gizzard in birds), and in the uterus. The contractile system and its control resemble those of motile tissue cells (e.g. fibroblasts, leucocytes), and antibodies against smooth muscle myosin will cross-react with myosin from tissue cells, whereas antibodies against skeletal muscle myosin will not. See also **dense bodies**.

smooth strain See **rough strain**.

smoothelin Actin-binding protein (917aa with multiple isoforms) expressed abundantly in visceral (smoothelin-A) and vascular (smoothelin-B) smooth muscle and often used as a marker protein for contractile smooth muscle cells. Co-localizes with alpha-smooth muscle actin in stress fibres, posesses a **calponin homology domain** and is thought to have a role in the contractile process. Various smoothelin-like proteins are also known (e.g. smoothelin-like 1, CHASM, calponin homology-associated smooth muscle, 457aa) although they do not bind to F-actin *in vitro*.

smoothened In *Drosophila* a G-protein coupled receptor (smo, 1036aa) that associates with **patched** protein that has bound **sonic hedgehog** and transmits its activation signal to a microtubule-associated **hedgehog signalling complex**. It is thought to convert the Gli family of transcription factors from transcriptional repressors to transcriptional activators. Protein kinase A (PKA) and casein kinase I (CKI) regulate Smo cell-surface accumulation and activity. In the absence of hedgehog patched represses the constitutive signalling activity of smo through **fused**. The mammalian smoothened homologue (in humans, 787aa) is expressed on the primary cilium. See **cyclopamine**.

SMRT Nuclear receptor co-repressor (silencing mediator of retinoic acid and thyroid hormone receptors, 2525aa) that mediates the transcriptional repression activity of some nuclear receptors by promoting chromatin condensation, thus preventing access of the basal transcription. Distinct from **N-CoR**.

snail A zinc-finger transcription factor (264aa) involved in epithelial-mesenchymal transition; binds conserved E-box elements in the prostaglandin dehydrogenase (PGDH) promoter to repress transcription and downregulates E-**cadherin**. The snail family includes **slug** which suppresses several epithelial markers and adhesion molecules.

snake A *Drosophila* serine peptidase (EC 3.4.21.-, 435aa) involved in activating **easter**, a key component of the dorso-ventral patterning pathway in development.

SNAP (1) Originally, soluble **NSF** attachment (accessory) protein (SNAP), involved in the control of vesicle transport by binding together with NSF, to **SNAREs**. The name has morphed to become 'synaptosomal-associated proteins' with the size (kDa) as the suffix. SNAP23 (23 kDa, 211aa) is part of the high affinity receptor for the general membrane fusion machinery and an important regulator of transport vesicle docking and fusion. Binds **snapin**, multiple **syntaxins** and **synaptobrevins/**VAMPs. SNAP25 (206aa) is a t-SNARE involved in regulating neurotransmitter release. SNAP29 (258aa) and SNAP47 (464aa) are involved in various membrane trafficking activities. In *Arabidopsis* SNAP33 (SNAP25 homologous protein, 300aa) is ubiquitous and acts as a t-SNARE. (2) S-nitroso-N-acetyl penicillamine.

snapin A **SNARE**-associated protein (SNAP25-binding protein, 136aa) involved in synaptic vesicle docking and fusion and also a component of the **BLOC1** complex.

SNAREs Receptors for **SNAPs**. The neuronal receptor for vesicle-SNAPs, v-SNARE, is **synaptobrevin**, also known as VAMP-2. The target (t-SNARE) associated with the plasma membrane of the axonal terminal is **syntaxin**. The SNAP-SNARE complex is apparently responsible for regulating vesicle targeting: neurotoxins such as **tetanus toxin** and **botulinum toxin** selectively cleave SNAREs or SNAPs. See also **cellubrevin**.

SNEEZY *SNE* A plant F-box protein (**SLEEPY** (SLY1) homologue, 157aa) a component of a SCF-type E3 ligase complex that positively regulates the gibberellin signalling pathway. SNEEZY can replace SLY1 in the **gibberellic acid**-induced proteolysis of RGA, one of the **DELLA** proteins.

Snf1 kinase In *Saccharomyces* a multimeric complex required for transcriptional, metabolic, and developmental adaptations in response to glucose limitation. The complex is heterotrimeric with a catalytic alpha subunit SNF1 (carbon catabolite-derepressing protein kinase, EC 2.7.11.1, 633aa), one of the three related beta subunits SIP1, SIP2 or GAL83, and the regulatory gamma subunit SNF4. Homologous systems are found in plants (**KIN10/11**) and animals (SNF1-related kinase, SNRK).

SNIP (1) Smad nuclear-interacting protein 1 (396aa) that down-regulates NFκB signalling. (2) An enzyme (snRNA incomplete 3′ processing) involved in processing snRNA in *Trypanosoma brucei*. The central 158aa domain of SNIP is related to the exonuclease III (ExoIII) domain of the 3′→5′ proofreading epsilon subunit of *E. coli* DNA polymerase III. (3) SNAP-25-interacting protein (src kinase signalling inhibitor 1, 1055aa) is co-distributed with SNAP-25 in brain where it regulates dendritic spine morphology and is involved in calcium-dependent exocytosis.

snoRNA Small nucleolar ribonucleic acid; guide chemical modifications of ribosomal and transfer RNAs and other small nuclear RNAs (snRNAs). See **fibrillarin**.

SNP Single nucleotide polymorphism. Any of the variations in single nucleotides in a DNA sequence that contribute to human individuality; at least 10 million are known. Arbitrarily, SNPs are defined as variations that occur in 1% or more of the population. SNP-mapping is increasingly used in searching for genetic associations of disease and the **International HapMap Project** makes the data publicly available.

SnRKs SNF-related serine/threonine kinases (EC 2.7.11.1, 765aa). In *Arabidopsis* a superfamily of kinases (CDPK-SnRK superfamily) with at least 34 members from seven subclasses involved in a range of signalling activities. See **Snf1 kinase**. Research paper: http://www.plantphysiol.org/content/132/2/666.full

snRNA See **small nuclear RNA**.

snRNP Small nuclear ribonucleoprotein. See **small nuclear RNA**.

SNRPN One of the **Sm proteins** (small nuclear ribonucleoprotein polypeptide N, 240aa) an important trans-acting factors in constitutive pre-mRNA splicing, found predominantly in central neurons. The gene encoding the peptide is bicistronic and encodes 2 polypeptides, the SmN splicing factor and the SNRPN upstream reading frame (SNURF) polypeptide. Deletion of the paternal copy leads to **Prader-Willi syndrome**, that of the maternal copy to **Angelman syndrome**.

SO In *Neurospora crassa* the homologue of the WW domain protein **PRO40**. Said to contribute to the sealing efficiency of pores plugged by **Woronin bodies** after hyphal injury.

SOC media A cell growth medium used to ensure maximum transformation efficiency, it is 'Super Optimal Broth' with the 'B' in SOB changed to 'C', for catabolite repression, reflective of the added glucose. Details: http://openwetware.org/wiki/SOB

SOCS Family of proteins suppressor of cytokine signalling, SOCS 1-7 and CIS (cytokine-inducible SH2-containing protein, 258aa) with SH2 domains that are induced in response to cytokines and act as negative feedback regulator of **JAKs** in the intracellular signalling pathway. SOCS1 (JAK-binding protein, JAB, STAT-induced STAT inhibitor-1, SSI-1, 211aa) is induced by IL6 and is probably the substrate recognition component of an ECS (Elongin BC-CUL2/5-SOCS-box protein) E3 ubiquitin ligase complex which mediates the ubiquitination and subsequent proteasomal degradation of target proteins and seems to recognize JAK2. Other SOCs family members are induced by different cytokines and have slightly different targets (e.g. SOCS2, 198aa, is a negative regulator in the growth hormone/IGF1 signalling pathway).

sodium channel (1) Voltage-gated sodium channels (NaV1.1, etc.) are responsible for electrical excitability of neurons. They are multi-subunit transmembrane channels with an aqueous pore around 0.4 nm diameter, with a negatively charged region internally (the 'selectivity filter') that blocks the passage of anions. The alpha subunits form the pore itself, the beta subunits vary between tissues and regulate the sensitivity. The channel opens in response to a small depolarization of the cell (usually caused by an approaching action potential), by a multistep process. Around 1000 sodium ions pass in the next millisecond, before the channel spontaneously closes (an event with single-step kinetics). The channel is then refractory to further depolarizations until returned to near the **resting potential**. There are around 100 channels per mm^2 in unmyelinated axons; in myelinated axons, they are concentrated at the **nodes of Ranvier**. Mutations are associated with various forms of **epilepsy**, muscular disorders (**periodic paralysis**, **myotonia**), **long QT syndrome**, **Brugada's syndrome-1**, **erythromelalgia**, **congenital insensitivity to pain** and **paroxysmal extreme pain disorder**. Many toxins (e.g. **tetrodotoxin**) affect the channel. (2) Amiloride-sensitive sodium channels are a family of nonvoltage-gated sodium-permeable ion channels inhibited by the diuretic drug, amiloride. They are well characterized in epithelia where they constitute the rate-limiting step for sodium ion reabsorption. They are heterotrimers of highly homologous alpha, beta and gamma subunits which also have homology with **degenerins** from *C. elegans*. Gain of function mutations in the beta or gamma subunits is the cause of **Liddle's disease**. A delta subunit that will assemble with the beta and gamma subunits to form a functional channel is also known. See **Apx/Shrm domains**.

sodium cromoglicate A drug (*formerly* sodium cromoglycate) used prophylactically for allergic asthma. It acts on mast cells to inhibit release of bronchoconstrictors by indirectly blocking calcium ion influx.

sodium cromoglycate See **sodium cromoglicate**.

sodium dodecyl sulphate See **SDS**.

sodium lithium countertransport Erythrocyte sodium/lithium countertransport (SLC, Na/LiCT), sodium-stimulated lithium efflux from lithium-loaded erythrocytes, has been observed to be abnormal in several hypertension-related diseases and is increased in patients with diabetes mellitus. But see **solute carrier family proteins**.

sodium pump See **sodium-potassium ATPase**.

sodium-potassium ATPase A major transport protein, not a **sodium channel**, the sodium pump of the plasma membrane. A multi-subunit enzyme (EC 3.6.1.3), having a catalytic alpha subunit with multiple isoforms and a smaller beta subunit. The pump moves 3 sodium ions out of the cell, and 2 potassium ions in, for each ATP hydrolyzed. The sodium gradient established is used for several purposes (see **facilitated diffusion**, **action potential**), while the potassium gradient is dissipated through the potassium leak channel.

sodoku Rat-bite fever, an infection by *Streptobacillus moniliformis* or *Spirillum minus*, following the bite of a rat or contact with rat saliva.

soft agar Semi-solid agar used to gelate medium for culture of animal cells. Placed in such a medium, over a denser agar layer, the cells are denied access to a solid substratum on which to spread, so that only cells that do not show **anchorage-dependence** (usually transformed cells) are able to grow.

SOK1 (1) A Ste20 protein kinase (Sterile 20/oxidant stress-response kinase 1, EC 2.7.11.1, 426aa) of the germinal centre kinase (GCK) family, that is activated by oxidant stress and chemical anoxia. It localizes to Golgi apparatus where it appears to regulate protein transport events, cell adhesion, and cell polarization for locomotion. (2) In *S. cerevisiae* a suppressor (901aa) of a cyclic AMP-dependent protein kinase mutant.

sol-gel transformation Transition between more fluid cytoplasm (**endoplasm**) and stiffer gel-like **ectoplasm** proposed as a mechanism for amoeboid locomotion: since the endoplasm cannot really be considered a simple fluid and has viscoelastic properties like a gel, the term is misleading and the mechanism is not generally accepted.

solanidine The aglycone of **solanine** and **chaconine** that is produced by the action of solanidine UDP-glycosyltransferases. Solanine has a branched trisaccharide galactose-(glucose)- rhamnose moiety, chaconine a glucose-(rhamnose)-rhamnose moiety.

solanine A glycoalkaloid toxin found in species of the nightshade family (Solanaceae) that includes potatoes. It can occur naturally in the any part of the plant, including the leaves, fruit, and tubers. It is very toxic even in small quantities. Solanine has both fungicidal and pesticidal properties, and it is one of the plant's natural defenses. Has sedative and anticonvulsant properties, and has been used as a treatment for bronchial asthma. See **solanidine**, **chaconine**.

Solanum tuberosum The potato.

SolEST A database that integrates different EST datasets from both cultivated and wild Solanaceae species and from two species of the genus *Coffea*. Link to database: http://biosrv.cab.unina.it/solestdb

soluble tyrosine kinases An obsolete term for tyrosine kinases that are not membrane associated.

solurshin See **PITX genes**.

solute carrier family proteins *SLCs* Superfamily of proteins involved in the transport of molecules across membranes. 46 sub-classes are recognised, each with several members. For example, solute carrier family 1 (SLC1) comprises neuronal/epithelial high affinity glutamate/ neutral amino acid transporters, SLC2 family is of facilitated glucose transporters (GLUT proteins) and has 14 members. The superfamily does not include ATP Binding Cassette (ABC) transporters, ion channels or **aquaporins**.

somaclone Plants derived from somatic cells that have been grown in culture. Somaclonal variation is variation seen after repeated subculture.

somatic cell Usually any cell of a multicellular organism that will not contribute to the production of gametes, i.e. most cells of which an organism is made: not a **germ cell**. Notice, however, the alternative use in **somatic mesoderm**.

somatic cell genetics Method for identifying the chromosomal location of a particular gene without sexual crossing. Unstable **heterokaryons** are made between the cell of interest and another cell with identifiably different characteristics (or without the gene in question), and a series of clones isolated. By correlating retention of gene expression with the remaining chromosomes, it is possible to deduce which chromosome must carry the gene. Human-mouse heterokaryons have been extensively used in this sort of work.

somatic cell nuclear transfer Transfer of the nucleus of a somatic cell into an enucleated egg as a means of cloning.

somatic embryogenesis The process in plants (asexual embryogenesis) in which somatic cells differentiate into embryos and ultimately into plants via a series of characteristic morphological stages. Somatic cells in plants retain **totipotency**.

somatic hybrid Heterokaryon formed between two somatic cells, usually from different species. See **somatic cell genetics**.

somatic mesoderm That portion of the embryonic mesoderm that is associated with the body wall, and is divided from the splanchnic (visceral) mesoderm by the coelomic cavity.

somatic mutation Mutation that occurs in the somatic tissues of an organism, and that will not, therefore, be heritable, since it is not present in the **germ cells**. Some neoplasia is due to somatic mutation; a more conspicuous example is the reversion of some branches of variegated shrubs to the wild-type (completely green) phenotype. Somatic mutation is probably also important in generating diversity in V-gene regions of immunoglobulins.

somatic recombination One of the mechanisms used to generate diversity in antibody production is to rearrange the DNA in B-cells during their differentiation, a process that involves cutting and splicing the immunoglobulin genes. Somatic recombination via homologous crossing-over occurs at a low frequency in *Aspergillus*, *Drosophila* and *Saccharomyces* and in mammalian cells in culture. It may be detected through the production of homozygous patches or sectors after mitosis of cells heterozygous for suitable marker genes.

somatocrinin Growth hormone-releasing factor. See **growth hormone-releasing hormone**.

somatoliberin Growth hormone-releasing factor, somatotropin-releasing factor. See **growth-hormone releasing hormone**.

somatomedin Generic term for **insulin-like growth factors** (IGFs) produced in the liver and released in response to **somatotropin**. Somatomedins stimulate the growth of bone and muscle, and also influence calcium, phosphate, carbohydrate and lipid metabolism. Somatomedin A (180aa) is IGF-II, somatomedin C (153aa) is IGF-I. Somatomedin B is a serum factor of uncertain function derived by proteolytic cleavage from **vitronectin**. See **somatostatin**.

somatostatin Gastrointestinal and hypothalamic peptide hormone (two forms: 14 and 28aa); found in gastric mucosa, pancreatic islets, nerves of the gastrointestinal tract, in posterior pituitary and in the central nervous system. Inhibits gastric secretion and motility: in hypothalamus/pituitary inhibits **somatotropin** release. Somatostatin acts through five G protein-coupled receptor subtypes (SSTR1-5), displaying a tissue specific distribution with multiple subtypes present on many cells. A neuropeptide, cortistatin, strongly resembles somatostatin. See also Table H2.

somatotrope Cell in the anterior pituitary which secretes growth hormone.

somatotrophin *US.* somatropin. Hormone (human growth hormone, hGH, 191aa) released by anterior pituitary that stimulates release of **somatomedin**, thereby causing growth. See also Table H2.

somites Segmentally arranged blocks of mesoderm lying on either side of the notochord and neural tube during development of the vertebrate embryo. Somites are formed sequentially, starting at the head. Each somite will give rise to muscle (from the myotome region), spinal column (from the sclerotome), and dermis (from dermatome).

son-of-sevenless *Drosophila* ras-GRF (GDF releasing factor, 1956aa), mammalian homologues of which play an important part in intracellular signalling. See **sos**.

sonic hedgehog Secreted protein (Shh, 465aa) processed in the same way as other members of the **hedgehog** family, involved in organization and patterning of several vertebrate tissues during development. The zone of polarizing activity (**ZPA**) that determines anterior-posterior patterning of the limb expresses sonic hedgehog. Defects in Shh or the downstream signalling pathway are a cause of **holoprosencephaly 3**, **VACTERL syndrome** and **SMMCI**.

sorbitol *glucitol* The polyol (polyhydric alcohol) corresponding to glucose. Occurs naturally in some plants, is used as a growth substrate in some tests for bacteria, and is sometimes used to maintain the tonicity of low ionic strength media.

sorbitol dehydrogenase A member of the superfamily of medium chain dehydrogenases/reductases, (EC 1.1.1.14, SORD, 357aa), the second enzyme in the polyol pathway; oxidizes sorbitol to fructose in the presence of NAD^+.

sorbose A monosaccharide hexose: L-sorbose is an intermediate in the commercial synthesis of **ascorbic acid**.

soredium *Pl.* soredia. Specialised asexual reproductive products of lichens consisting of algal cells surrounded by fungal hypae; emerge from soralia on the lichen suface and are wind-dispersed.

Soret band A very strong absorption band in the blue region (414 nm) of the optical absorption spectrum of a haem protein.

Soret effect The mass diffusion of chemical species due to an imposed thermal gradient. A temperature gradient across a fluid or gaseous mixture generally leads to a net mass flux and the build-up of a concentration gradient. This effect is known as thermal diffusion or Ludwig-Soret-effect. The amplitude of the concentration gradient is determined by the Soret coefficient, S_T.

Sorghum bicolor An important cereal crop for food, fuel, and biofeedstock. Unlike rice it is a C4 plant.

sorocarp Fruiting body formed by some cellular **slime moulds**; has both stalk and spore-mass.

sortilin A sorting receptor (neurotensin receptor 3, 831aa) in the Golgi compartment and a clearance receptor on the cell surface. Required for protein transport from the Golgi to lysosomes and important for the sorting and trafficking of sphingolipid activator proteins (**saposins**).

sorting nexins A large family of proteins that characteristically have a phox-homology (PX) domain and are required for the endocytosis and sorting of transmembrane proteins. Sorting nexin-1 (SNX1, 522aa) is a component of the mammalian **retromer** complex. Human SNX1, -2, and -4 have been proposed to play a role in receptor trafficking and have been shown to bind to several receptor tyrosine kinases. SNX-6 (406aa) interacts with members of the TGFβ family of receptor serine-threonine kinases. Yeast Vps5p is the orthologue of mammalian sorting nexin-1.

sorting out Phenomenon observed to occur when mixed aggregates of dissimilar embryonic cell types are formed *in vitro*. The original aggregate sorts out so that similar cells come together into homotypic domains, usually with one cell type sorting out to form a central mass that is surrounded by the other cell type. Much controversy has arisen over the years as to the underlying mechanism, whether there is specificity in the adhesive interactions (which would imply tissue-specific receptor-ligand interactions), or whether it is sufficient to suppose that there are quantitative differences in homo- and hetero-typic adhesion (the **differential adhesion** hypothesis). With the exception perhaps of the main protagonists, most cell biologists consider that there are probably elements both of tissue specificity (**CAMs**) and of quantitative adhesive differences involved.

sorus A group of **sporangia** or spore cases, e.g. on the underside of fern leaves.

sos (1) Guanine-nucleotide releasing factors (sos1, 1333aa; sos2, 1332aa), the mammalian homologues of **son-of-sevenless**. The proline-rich region of sos binds to the **SH3** domain of **GRB2**. Has homology with CDC-25, the yeast GTP-releasing factor for ras. A family of related proteins are now known and include vav, **C3G**, Ost, NET1, Ect2, RCC1, tiam, RalGDS, and **Dbl**. (2) See **SOS system**.

SOS system The DNA repair system also known as error-prone repair in which apurinic DNA molecules are repaired by incorporation of a base that may be the wrong base but that permits replication. **RecA** protein is required for this type of repair. SOS genes function in control of the cell cycle in prokaryotes and eukaryotes. See **lexA**.

Sotos' syndrome A disorder of growth (cerebral gigantism) caused by mutation in the NSD1 gene (nuclear receptor-binding suppressor of variation,

enhancer of zeste, trithorax domain protein 1) that encodes a co-regulator of the androgen receptor (histone-lysine N-methyltransferase, H3 lysine-36 and H4 lysine-20 specific, EC 2.1.1.43, 2696aa). See **SET protein, SET complex**.

source-sink relationship In plants, the balance between source tissues that produce sugars and amino acids by photosynthesis and the sink tissues that import these products.

Southern blots See **blotting**. Originally developed by Dr Ed Southern, hence the name.

sox (1) SoxR: Redox sensory protein in *E. coli*. (2) SOX syndrome: Sialadenitis, osteoarthritis and xerostomia syndrome. (3) *Sox* genes (Sry-related HMG-box genes) are a family of genes involved in many developmental processes. Sox-2 regulates transcription of *FGF-4* gene, sox-3 is involved in neural tube closure and lens specification, sox-9 is related to **sry** and found in mouse testis, sox-10 is important in neural crest development.

soybean trypsin inhibitor *SBTI* A **Kunitz**-type trypsin inhibitor (181aa) that forms a stable, stoichiometric, enzymically-inactive complex with trypsin, chymotypsin, plasmin and related peptidases.

SP proteins (1) A family of Cys_2His_2 zinc finger transcription factors that bind with high affinity to GC-rich motifs and regulate the expression of a large number of genes involved in a variety of processes such as cell growth, apoptosis, differentiation and immune responses. Specificity protein 1 (Sp1, 785aa) has been linked to vascular endothelial growth factor (VEGF) expression in pancreatic cancer cells. Research article: http://cancerres.aacrjournals.org/content/64/18/6740.full.pdf + html (2) In *Bombyx mori*, a methionine-rich larval storage protein (747aa). (3) In *Macrozoarces americanus* (Ocean pout) an antifreeze protein (ice-structuring protein SP1, 63aa). (4) **Tunichrome** Sp-1 is a modified pentapeptide from the haemocytes of the ascidian *Styela plicata*. (5) In *Crotalus adamanteus* (Eastern diamondback rattlesnake) a venom serine peptidase (EC 3.4.21.-, 259aa).

Sp value The specificity score or profile determined, e.g. for PDZ domains. The score ranges from zero (least specific) at a given ligand position to one (most specific). It allows classification of the domains into specificity classes with preferences for particular ligands. Research article: http://www.plosbiology.org/article/info:doi/10.1371/journal.pbio.0060239

SPA, SPD See **surfactant proteins A and D**.

spacer DNA The DNA sequence between genes. In bacteria, only a few nucleotides long. In eukaryotes, can be extensive and include **repetitive DNA**, comprising the majority of the DNA of the genome. The term is used particularly for the spacer DNA between the many tandemly-repeated copies of the ribosomal RNA genes (see **intergenic spacer** and **internal transcribed spacers**).

SPAK A serine/threonine kinase (Ste20/SPS-1-related proline-, alanine-rich kinase, 547aa) that phosphorylates NKCC1 (Na^+-K^+-$2Cl^-$ co-transporter-1), leading to its activation. SPAK and **OSR1** (oxidative stress-responsive kinase-1) are phosphorylated and activated by **WNKs** (with no K (lysine) protein kinase-1).

SPARC A secreted protein, acidic and rich in cysteine (SPARC, 303aa) overexpressed in the fibroblasts of skin biopsy specimens obtained from patients with systemic sclerosis. See **osteonectin**.

sparsomycin An antibiotic from *Streptomyces sparsogenes*, that inhibits **peptidyl transferase** in both prokaryotes and eukaryotes by blocking the peptidyl transfer step on the 50S/60S ribosomal subunit. More detail: http://www.microbiologyprocedure. com/bacterial-cheomotherapy/sparsomycin.htm

spartin A cytosolic and membrane-associated protein (666aa) that may be implicated in endosomal trafficking. A frameshift mutation in the spartin gene is the cause of Troyer syndrome (a form of **spastic paraplegia**).

spasmin Protein (180aa in *Vorticella*) that forms the **spasmoneme**. Thought to change its shape when the calcium ion concentration rises, and to revert when the calcium concentration falls: the reversible shape change is used as a motor mechanism. Contraction does not require ATP, relaxation does, probably to pump calcium ions back into the smooth endoplasmic reticulum.

spasmoneme Contractile organelle found in *Vorticella* and related ciliate protozoans. Capable of shortening faster than any actin-myosin system, and of expanding actively. See **spasmin**.

spastic paraplegias A set of neurodegenerative disorders in which there is lower-limb spasticity (increased muscle tone with involuntary spasms or sustained contractions) and muscle weakness with a variety of different causes. Various subtypes are caused by defects in **atlastin**, kinesin-5A, **maspardin**, **myelin proteolipid protein**, **NIPA1**, **paraplegin**, **PNPLA6**, **protrudin**, receptor expression-enhancing protein-1, **spartin**, **spastin**, **spatacsin** and **strumpellin**; in other forms the defect is mapped but the defective protein as yet unidentified. Inheritance is most commonly autosomal dominant, but X-linked and autosomal recessive forms also occur.

spastin An ATPase (EC 3.6.4.3, 616aa) of the **AAA family**, related to the microtubule-severing protein **katanin**. It is involved in membrane traffic from the endoplasmic reticulum to the Golgi and may have a role in axon growth and the formation of axonal branches. Mutations in the encoding gene (SPG4)

are responsible for most cases of autosomal dominant **spastic paraplegia**.

spastizin Zinc finger FYVE domain-containing protein (ZFYVE26, 2539aa) found in a wide range of tissues. Mutations in the encoding gene cause a form of autosomal recessive **spastic paraplegia** (Kjellin's syndrome).

spatacsin A multi-pass membrane protein (spasticity with thin or atrophied corpus callosum syndrome protein, 2443 aa) expressed throughout the brain, and highly expressed in the cerebellum. The encoding gene is defective in autosomal recessive **spastic paraplegia** Type 11.

spatial sensing Mechanism of sensing a gradient in which the signal is compared at different points on the cell surface and cell movement directed accordingly. Translocation of all or part of the cell is not required. See **temporal** and pseudospatial gradient sensing.

spätzle *spaetzle* The protein (326aa) that is proteolytically processed (see **easter**) to produce the ligand (spätzle C-106) for **toll** in dorso-ventral polarity signalling in *Drosophila* development. In the adult it is involved in antifungal defense.

SPAZ domain A domain (serine- and proline-rich AZU-1 domain, 83aa) found in **TACC2** (AZU-1), TACC1, TACC3 and the *S. cerevisiae* gene product BCK1, a member of the MAPK kinase kinase family of serine/threonine kinases. It is thought that the domain may be a member of the Ig superfamily and as such may function as a protein-binding interface.

speciation Formation of new biological species. Usually considered to require isolation of a subpopulation of the ancestral species either geographically (classically seen with remote islands), by occupying a different niche (adaptive radiation), through acquisition of behavioural changes that restrict mating, so that distinct genetic variations accumulate and prevent further interbreeding, or by polyploidy that makes interbreeding with the original species impossible. See **sympatric speciation**.

species A group of individuals that are capable of interbreeding with each other but not with members of other groups. Usually a result of some form of isolation; in the early stages **speciation** can be reversed if reproductive isolation is lost or the niches change and overlap. Species are grouped into genera and divided into subspecies and varieties or cultivars. The name of the genus should be capitalised and the species name should not — thus *Homo sapiens*, genus *Homo*, species *sapiens* and both should be italicised. This nicety is increasingly neglected.

specific activity The number of activity units (whatever is appropriate) per unit of mass, volume or molarity. Perhaps most often encountered in the context of radiochemicals, the number of microcuries per micromole.

specific granules One of the two main classes of granules (secondary granules) found in neutrophil leucocytes: contain lactoferrin, lysozyme, Vitamin B12 binding protein and elastase. Are released more readily than the **azurophil** (primary) granules which have typical lysosmal contents.

spectinomycin A bacteriostatic aminocyclitol antibiotic produced by the soil bacterium, *Streptomyces spectabilis*; binds to the 30S subunit of the bacterial ribosome, thus inhibiting protein synthesis.

spectral karyotyping A technique that allows scientists to visualize all of the human chromosomes at one time by 'painting' each pair of chromosomes in a different fluorescent color; **translocations** show up very conspicuously because the affected chromosome is multi-coloured.

spectraplakins Large cytoskeletal linker proteins that bind cytoplasmic filaments (microfilaments, microtubules and intermediate filaments). They often have two **calponin homology domains**, a **plakin** domain, a series of **plectin** repeats, numerous **spectrin** repeats and a GAS2 domain: multiple isoforms are produced from each gene. The human spectraplakin (macrophin 1, trabeculin-alpha, actin crosslinking factor 7, 5373aa) is ubiquitously expressed. See **desmoplakin**, **envoplakin**, **epiplakin**, **kazrin**, **periplakin**, **plectin**. In *Drosophila* **short-stop** is a spectraplakin homologue.

spectrin Membrane-associated dimeric protein (alpha, 2419aa; beta, 2137aa) of erythrocytes. Forms a complex with **ankyrin**, **actin**, and probably other components of the 'membrane cytoskeleton', a meshwork of proteins underlying the plasma membrane, potentially restricting the lateral mobility of integral proteins and providing mechanical stability. Isoforms have been described from other tissues (**fodrin**, **TW-240/260 kDa protein**), where they are assumed to play a similar role. Contains the **EF-hand** motif.

spectrophotometry Quantitative measurement of the absorption of visible, ultraviolet or infrared light to determine the concentrations of substances in the sample.

Spemann's organizer Signalling region located on the dorsal lip of the blastopore in the early embryo, essential for defining the main body axis.

speract A peptide hormone, produced by proteolysis of sperm-activating peptide (296aa) in the the jelly coat of the eggs of the sea-urchins *Strongylocentrotus purpuratus* and *Hemicentrotus pulcherrimus*. Cleavage produces speract and speract-like decapeptides that stimulate sperm respiration and motility. Receptors are a plasma-membrane **guanylate cyclase** (EC 4.6.1.2, 1125aa) and a scavenger receptor (532aa).

speriolin Spermatogenesis and centriole-associated protein 1 (591aa) that interacts with CDC20 (**fizzy**) and forms a complex with CDC20, CDC27 and gamma tubulin.

spermatids The haploid products of the second meiotic division in spermatogenesis. Differentiate into mature spermatozoa.

spermatocytes Cells of the male reproductive system that undergo two meiotic divisions to give haploid **spermatids**.

spermatogenesis The process whereby primordial germ cells form mature spermatozoa.

spermatogonium Plant gonad cell that undergoes repeated mitoses, leading to the production of **spermatocytes**.

Spermatophyte Division of the plant kingdom, consisting of plants (phanerogams) that reproduce by means of seeds. The existent groups are cycads, Ginkgo (the sole member of the group), conifers, gnetophytes and angiosperms; others are known from the fossil record (e.g. seed ferns).

spermatozoon Mature sperm cell (male gamete).

spermidine A polybasic amine (polyamine), N-(3-aminopropyl)-1,4-butanediamine; see **spermine**.

spermine Polybasic amine (polyamine), N, N' bis (3-aminopropyl)-1,4-butanediamine. Found in human sperm, in ribosomes and in some viruses. Involved in nucleic acid packaging. Synthesis is regulated by **ornithine decarboxylase** which plays a key role in control of DNA replication.

spermine oxidase An inducible enzyme (SMO, EC 1.5.3.-, 555aa) that may play a direct role in the cellular response to the antitumour polyamine analogues. Oxidises both spermine and N(1)-acetylspermine, but not spermidine.

SPF Describing animals (specific pathogen free) that have been raised in carefully controlled conditions so that they are not infected with any known pathogens. Usually delivered by Caesarean section and raised in strict quarantine.

S phase The synthesis (S) phase of the cell cycle during which DNA replication occurs.

spherical aberration Deficiency in simple lenses in which the image is sharp in the centre but out-of-focus at the periphery of the field, more a problem when taking photographs than when observing directly. Lenses compensated for this defect are referred to as plan- lenses (e.g. **planapochromat**).

spheroblast *spheroplast* Bacterium or yeast that has been treated in such a way as to have lost the outer cell wall. This makes the cell osmotically fragile but makes it easier to get large molecules across the plasma membrane in e.g. transfection.

spherocytosis A condition in which erythrocytes lose their biconcave shape and become spherical. It occurs as cells age, and is also found in individuals with abnormal cytoskeletal proteins such as **spectrin**. Hereditary spherocytosis can cause haemolytic anaemia.

spheroplast See **spheroblast**.

spherosome Lysosome-like lipid body (oleosome) in plants that probably derives from the endoplasmic reticulum and is a site for lipid storage, particularly in oil-rich seeds.

sphingolipid Structural lipid of which the parent structure is sphingosine rather than glycerol. Synthesized in the Golgi complex.

sphingomyelin A **sphingolipid** in which the head group is phosphoryl choline. A close analogue of phosphatidyl choline. In many cells the concentration of sphingomyelin and phosphatidyl choline in the plasma membrane seems to bear a reciprocal relationship. See **lipidoses**.

sphingomyelinase A stress-activated enzyme (sphingomyelin phosphodiesterase-1, acid sphingomyelinase, EC 3.1.4.12, 629aa) that generates ceramide from **sphingomyelin**, which serves as a second messenger in initiating apoptosis. Deficiency in the enzyme leads to **Niemann-Pick disease**. Neutral sphingomyelinases (sphingomyelinase 2, 3 and 4; 423aa, 655aa and 827aa respectively) are found in various tissues and have similar capabilities. Sphingomyelinases are components of some spider venoms.

sphingosine A long-chain amino alcohol that bears an approximate similarity to glycerol with a hydrophobic chain attached to the 3-carbon. Forms the class of **sphingolipids** when it carries an acyl group joined by an amide link to the nitrogen. Forms **sphingomyelin** when phosphoryl choline is attached to the 1-hydroxyl group. Gives rise to the cerebroside and ganglioside classes of glycolipids when oligosaccharides are attached to the 1-hydroxyl group. Not found in the free form. See **sphingosine-1-phosphate**, **sphingosine kinase**.

sphingosine kinase The enzyme (sphinganine kinase, EC 2.7.1.91, SHPK1, 384aa) that catalyses the formation of the signalling molecule **sphingosine-1-phosphate**. Sphingosine kinase 2 (SHPK2, 654aa) is present in the nucleus in complexes with the histone deacetylases HDAC1 and HDAC2.

sphingosine-1-phosphate A signalling molecule released from activated platelets and by a number of other cell types in response to growth factors and cytokines. It acts extracellularly through G-protein coupled receptors (see **EDGs**) and intracellularly as a second messenger, reportedly inhibiting histone deacetylases (HDACs). The cellular effects include growth related effects, such as proliferation, differentiation, cell survival and apoptosis, and cytoskeletal effects, such as chemotaxis, aggregation, adhesion, morphological change and secretion.

SPHK See **sphingosine kinase**.

Spi-1 (1) Proto-oncogene encoding a transcription factor (PU1) that binds to purine-rich sequences (PU boxes) expressed in haematopoietic cells. (2) *Salmonella typhimurium* pathogenicity island 1 (SPI-1), a region that carries genes mediating invasion into intestinal epithelial cells and that induce cell death in murine macrophages. (3) Vaccinia virus host range/antiapoptosis genes, SPI-1 and SPI-2. (4) In *S. cerevisiae*, a cell wall protein (stationary phase-induced protein 1, 148aa) with a role in adaptation and resistance to cell wall stress. (5) In *S. pombe* a GTP-binding protein (216aa) involved in nucleocytoplasmic transport.

spichthyin See **NIPA1**.

spin labelling The technique of introducing a grouping with an unpaired electron to act as an electron spin resonance (ESR) reporter species. This is almost invariably a nitroxide compound (-N-O) in which the nitrogen forms part of a sterically hindered ring. See **electron paramagnetic resonance**.

spina bifida A neural tube defect in which the bones of the spinal canal fail to meet and fuse during development. Varies in severity depending upon the extent of the failure: mild forms (spina bifida occulta), where only the most posterior vertebrae are affected, may be almost asymptomatic. In more serious forms (myelomeningocele and meningocele) there can be severe disability. The addition of folic acid to the diet of women of child-bearing age significantly reduces the incidence of neural tube defects.

spinal cord Elongated, approximately cylindrical part of the central nervous system of vertebrates that lies in the vertebral canal, and from which the spinal nerves emerge.

spinal ganglion, dorsal root ganglion Enlargement of the dorsal root of the spinal cord containing cell bodies of afferent spinal neurons. Neural outgrowth from dorsal root ganglia has been studied extensively *in vitro*.

spinal muscular atrophy A group of autosomal recessive disorders with symmetrical muscle weakness and atrophy caused by degeneration of the anterior horn cells of the spinal cord. They are all a result of mutation in the telomeric copy of the survival of motor neuron gene (SMN1; see **SMN complex**). The severity and age of onset are affected by variable expression of the centromeric SMN2 gene.

spindle See **mitotic spindle, muscle spindle**.

spindle fibres Microtubules of the spindle that interdigitate at the equatorial plane with microtubules of the opposite polarity derived from the opposite pole **microtubule organizing centre**. Usually distinguished from **kinetochore** fibres that are microtubules that link the poles with the kinetochore, although these could be included in a broader use of the term.

spindle pole body In *S. cerevisiae*, the functional equivalent of the **centrosome** that organizes both the spindle and cytoplasmic microtubules throughout the cell cycle. It is anchored to the nuclear envelope. See **sad1**.

spinner culture Method for growing large numbers of cells in suspension by continuously rotating the culture vessel.

spinocerebellar ataxia A group of genetic disorders in which gait becomes uncoordinated, and there is slow and progressive loss of control of hands, speech, and eye movements. Identified defects are in genes for beta-III **spectrin**, FGF14, inositol 1,4,5-trisphosphate receptor, **nesprin 1**, **PKC, puratrophin-1, senataxin, twinkle**, twinky and the CABC1 gene that encodes a chaperone-like protein. Some forms are caused by expanded $(CAG)_n$ trinucleotide repeat in the **ataxin** or TATA box-binding protein genes, others by defects in voltage-dependent calcium or potassium channels or in the gene for the brain-specific regulatory subunit of protein phosphatase PP2A. Many other forms have yet to be linked to specific molecular defects.

spinophilin See **neurabin**.

spiral cleavage Pattern of early cleavage found in molluscs and annelids (both **mosaic eggs**). The animal pole blastomeres are rotated with respect to those of the vegetal pole. The handedness of the spiral twist shows **maternal inheritance**.

spire *Drosophila* protein (1020aa) of the Spir family (interacts with **cappuchino**). Spir proteins nucleate actin polymerization by binding four actin monomers to a cluster of four WASP-homology domain 2 (WH-2 domains) in the central region of the proteins. This mechanism is distinct from actin nucleation by the **Arp2/3** complex or by **formins**. There are two human spire homologues (756aa and 714aa).

spirillum *Pl.* spirilla. A fairly rigid helically twisted (corkscrew-shaped) bacterial cell often, but not necessarily, a member of the genus *Spirillum*. Common

examples are the spirochaetes Vibrio cholerae and Treponema pallidum, the causative agents of cholera and syphilis respectively.

Spirochaetes A phylum of Gram-negative bacteria, which have long, helically coiled cells. The phylum includes *Treponema pallidum*, the causative agent of syphilis, *Leptospira*, and *Borrelia burgdorferi*.

Spirogyra Genus of green filamentous algae found in freshwater ponds. Contain helically disposed ribbon-like chloroplasts.

Spiroplasma citri A plant-pathogenic mollicute phylogenetically related to Gram-positive bacteria. *Spiroplasma* cells are restricted to the phloem sieve tubes and are transmitted from plant to plant by the leafhopper vector *Circulifer haematoceps*.

Spirostomum Genus of large free-living ciliate protozoans with an elongated body.

splanchnic Relating to the viscera. See also **splanchnic mesoderm**.

splanchnic mesoderm That portion of the embryonic **mesoderm** that is associated with the inner (endodermal) part of the body in contrast to **somatic mesoderm** which is associated with the body wall. The two mesodermal regions are separated by the **coelom**.

splenocytes Vague term usually referring to lymphoid cells or mononuclear phagocytes of the spleen.

splice variants Proteins that are related but differ in their sequence as a result of **alternative splicing**. Alternative Splicing Database Project: http://www.ebi.ac.uk/asd/

spliceosomes The macromolecular RNA-protein complexes involved in intron removal and exon ligation as mRNA is processed. Components include U2, U5, and U6 snRNAs and Prp8 (see **Prp proteins**). Different sub-classes of spliceosome process particular classes of introns and the minor spliceosome, which contains U5, U11, U12, U4atac, and U6atac is involved in processing U12-dependent introns.

splicing See **alternative splicing** and **spliceosome**.

split gene See **introns**.

split ratio The fraction of the cells in a fully grown culture of cells that should be used to start a subsequent culture. Minimum may be dictated by inadequacies of the medium that result in poor growth of some cells at high dilution.

spokein Term formerly used for the constituent protein of the **radial spokes** of the ciliary **axoneme**. Since a number of complementary spoke mutants are known to occur in *Chlamydomonas*, and one mutant lacks 17 proteins, it seems likely that spokein is a complex mixture. The radial spokes are regularly repeating axonemal structures composed of at least 23 proteins and are required for normal axonemal motility. See **radial spoke proteins**.

spondins A sub-family of thrombospondin type I proteins involved in axonal growth and guidance by affecting adhesion. Spondin 1 (F-spondin, vascular smooth muscle cell growth-promoting factor (VSGP), 624aa) is a secreted extracellular matrix protein that interacts with a central sequence of amyloid precursor protein (APP) and prevents proteolytic processing by beta-secretase. Spondin 2 (M-spondin, mindin, 331aa) is a ligand for neutrophil and macrophage integrins and binding via mindin regulates the expression of Rho GTPases in dendritic cells. Mindin-deficient mice exhibit defective inflammatory and immune responses. The R-spondins regulate beta-**catenin** signalling. R-spondin 1 (RSPO1, 263aa) is expressed in enteroendocrine cells as well as in epithelial cells from various tissues. Mutations are associated with palmoplantar hyperkeratosis with squamous cell carcinoma of skin and sex reversal. R-spondin 2 (cristin 2, 243aa) apparently functions in a positive feedback loop to stimulate the WNT/beta-catenin cascade. RSPO3 (273aa) regulates kidney cell proliferation. RSPO4 and a RSPO4 splice variant induce epithelial proliferation in the gastrointestinal tract; mutations are associated with **anonychia** congenita. SCO-spondin (5147aa) is secreted by the **subcommissural organ** and is a constituent of Reissner's fibres.

spondyloepiphyseal dysplasia A subclass of chondrodysplasia in which cartilage is defective either because of mutation in genes encoding collagen IIA1 or II or those for other extracellular matrix components such as the chondroitin 6-sulphotransferase 3 gene or the genes encoding **aggrecan** or **sedlin**. The consequence is short stature and spinal abnormalities. See **pseudoachondroplasia**.

spondylometaphyseal dysplasia A type of bone dysplasia affecting the spine and **metaphyses** of bones leading to extreme dwarfism and usually a range of other problems. In some cases (Strudwick type) a result of mutation in the collagen COL2A1 gene, in other cases (Kozlowski type) by mutation in a **vanilloid receptor**. Other types are known but the molecular defect uncharacterised as yet.

spongiform encephalopathies A group of unusual diseases with very long incubation periods and a fatal progressive course in which there is a characteristic spongiform degeneration of brain cortex. The two main human diseases are **kuru** and **Creutzfeldt-Jakob disease**, other forms are scrapie in sheep and goats, bovine spongiform encephalopathy (BSE) and mink encephalopathy. Experimentally can be studied in mice although intracortical transfer of infective material (usually affected brain tissue

but without any nucleic acid present) is necessary. Controversy still surrounds the causative agent although general opinion now favours Prusiner's **prion** hypothesis. See also **fatal familial insomnia**, **Gerstmann-Straussler-Scheinker syndrome**.

spongin Collagenous protein (with homology to collagen IV) that forms the extracellular matrix of silicaceous sponges, a meshwork that links the silica-rich spicules. Research paper: http://mbe. oxfordjournals.org/cgi/content/full/23/12/2288?view= long&pmid=16945979

spongioblast Cell found in developing nervous system: gives rise to **astrocytes** and **oligodendrocytes**.

spongioblastoma Rare tumours of childhood and adolescence, a glioma in which cells resemble embryonic spongioblasts. Although sometimes described as a class of neuroepithelial tumours, are probably not a distinct entity but could be considered either ependymomas or neuroblastomas.

spongiocytes Lipid droplet-rich cells from the middle region of the cortex of the adrenal gland.

spongy parenchyma Tissue usually found in the lower part of the leaf **mesophyll**. Consists of irregularly-shaped, photosynthetic parenchyma cells, separated by large air spaces.

spontaneous transformation Transformation of a cultured cell that occurs without the deliberate addition of a transforming agent. Cells from some species, especially rodents, are particularly prone to such spontaneous transformation.

sporadic Of a **tumour** or genetic disease, a novel occurrence without any previous family history of the disease (*cf.* inherited). Examples of diseases with both sporadic and inherited forms are **retinoblastoma** and **Wilms' tumour**.

sporangium Spore case, within which asexual spores are produced. *Cf.* conidium.

spore Highly resistant dehydrated form of reproductive cell produced under conditions of environmental stress. Usually have very resistant cell walls (integument) and low metabolic rate until activated. Bacterial spores may survive quite extraordinary extremes of temperature, dehydration or chemical insult. Gives rise to a new individual without fusion with another cell.

sporocarp Multicellular structure in fungi, lichens, ferns or other plants, the site of spore formation.

sporophyte The spore-producing plant generation. The dominant generation in **pteridophytes** and higher plants, and alternates with the **gametophyte** generation.

sporopollenin Polymer of **carotenoids**, found in the exine of the pollen wall. Extremely resistant to chemical or enzymic degradation.

sporotrichosis A fungal infection, usually of the skin, caused by *Sporothrix schenckii*. Mainly an occupational disease of farmers, gardeners, and horticulturists.

Sporozoa Class of spore-forming parasitic protozoa without cilia, flagella or pseudopodia. See **Apicomplexa**.

sporozoite Infective stage of the life cycle of **Apicomplexa** such as *Plasmodium* and *Cryptosporidia*.

spot desmosome Macula adherens: see **desmosome**.

SPRED A family of membrane-associated proteins (sprouty-related EVH1 domain-containg proteins, SPRED1, 444aa; SPRED2, 418aa; SPRED3, 410aa) that inhibit growth factor mediated activation of MAP kinase, possibly in collaboration with **caveolin**. Mutations in SPRED1 cause neurofibromatosis 1-like syndrome. See **sprouty**. Structure: http://www.thesgc.org/structures?terms = SPRED

sprouting (1) Production of new processes (outgrowths) by nerve cells: e.g. by embryonic neurons undergoing primary differentiation; by adult neurons in response to nervous system damage; or by dissociated neurons redifferentiating in culture. (2) Casual term for the germination of a seed.

sprout In *Drosophila* a protein (SPRY, 589aa) that inhibits tracheal branching by antagonizing the BNL-FGF pathway. Mammalian homologues negatively regulate **fibroblast growth factor** signalling in a variety of systems. SPRY1 (sprouty homologue 1, 319aa) is downregulated in approximately 40% of prostate cancers. SPRY2 (315aa) and SPRY3 (288aa) also inhibit FGF signalling; SPRY4, 299aa) suppresses the insulin receptor and EGFR-transduced MAPK signalling pathway. See **SPRED**.

Spumavirinae Formerly a subfamily of the **Retroviridae**, now considered a single genus (*Spumavirus*) of the subfamily Spumaretrovirinae. Single-stranded enveloped RNA viruses that will induce the formation of syncytia in susceptible cell cultures; the polykaryons then undergo a characteristic foamy degeneration.

squalene A 30-carbon isoprenoid lipid found in large quantities in shark liver oil and in smaller amounts in olive oil, wheat germ oil, rice bran oil and yeast. A key intermediate in the biosynthesis of **cholesterol**.

squames Flat, keratinised, dead cells shed from the outermost layer of a squamous **stratified epithelium**.

squamous epithelium An epithelium in which the cells are flattened. May be a **simple epithelium** (e.g. **endothelium**) or a **stratified epithelium** (e.g. **epidermis**).

squamous-cell carcinoma Carcinoma that develops from the squamous layer of the epithelium. Slow growing but more likely to be metastatic than basal cell carcinomas.

squid giant axon Large axons, up to 1 mm in diameter, that innervate the mantle of the squid. Because of their large size, many of the pioneering investigations of the mechanisms underlying resting and action potentials in excitable cells were done on these fibres.

squidulin Squid calcium-binding protein (SCaBP, 149aa) with four **EF-hand** motifs from the optic lobe of squid (*Loligo pealeii*),

S region (1) The non-MHC gene in the midst of the murine H-2 **major histocompatibility complex** that codes for complement component C4. Sometimes confusingly known as the gene for the type III MHC product in mice. (2) The switch region, a region 5′ to each immunoglobulin heavy chain constant region, that is important in isotype switching. Research article: http://www.jbc.org/content/263/15/7397.full.pdf

SR proteins A family of highly conserved nuclear phosphoproteins required for constitutive pre-mRNA splicing and also influence **alternative splicing**. They are characterized by one or two N-terminal RNA-binding domains and a C-terminal SR domain enriched in serine-arginine dipeptides (RS domain). SR proteins influence splice site selection and are required at an early step in **spliceosome** assembly.

SRBC Sheep red blood cells, often used in immunological assays.

src family Family of protein tyrosine kinases of which src (p60-src, EC 2.7.10.2, 536aa) was the first example (see *src* **gene**). Includes fyn, yes, fgr, lyn, hck, lck, blk and yrk. All cells studied so far have at least one of these kinases which act in cellular control. Family members all have characteristic src-homology (**SH domain**) structure, a kinase domain (SH1), SH2 and SH3 domains, and a domain (SH4) which has myristoylation and membrane-localisation sites. Inter-domain interactions, themselves regulated by phosphorylation (see **csk**), regulate the activity of the kinase.

src gene The gene in Rous sarcoma virus that encodes an unregulated cytoplasmic tyrosine kinase (pp60-vsrc) and is responsible for the transforming activity. The normal c-src kinase is regulated and does not induce uncontrolled cell proliferation.

SRE (1) **Serum response element**. (2) Sterol regulatory element, see **SREBP**.

SREBP Family of transcription factors (sterol regulatory element-binding proteins, SREBP1, 1147aa; SREBP2, 1141aa) that bind to steroid response elements in the promoter regions of genes involved in the metabolism of cholesterol and fatty acids.

SRF See **serum response factor**.

S-ring The M and S rings of the bacterial flagellar motor are actually a single, double-flanged ring made from subunits of just one protein (FliF, 552aa) and do not, contrary to earlier ideas, contribute to the generation of torque. See **bacterial flagella**.

SRK A putative serine/threonine-protein kinase receptor (S-receptor kinase, EC 2.7.11.1, 849aa in *Brassica oleracea*) found mainly in the pistil and anther and involved in the **self-incompatibility** system, probably acting in combination with S-locus-specific glycoproteins that are kinase-activating ligands.

SRP See **signal recognition particle**.

SRS-A See **slow reacting substance of anaphylaxis**.

SRTX See **sarafotoxin**.

sry proteins Family of high-mobility group (**HMG**) proteins that bind to a subset of sequences recognized by C/EBP family of DNA-binding proteins. *Sry* (sex-related gene on Y) itself is the primary testis-determining gene on the Y chromosome. The protein product, Sry (testis determining factor, 204aa), binds to DNA and causes it to bend sharply thereby affecting the expression of other genes on the Y chromosome. **Sox-10** is related to Sry.

SSB (1) Single-stranded DNA binding (SSB) protein (helix-destabilizing protein, 178aa in *E. coli*) that binds selectively to single-stranded DNA intermediates during DNA replication, recombination and repair. (2) Single-strand breaks (SSB) in DNA. (3) Gene encoding Lupus La protein, (Sjogren syndrome type B antigen, 408aa) the target antigen of autoantibodies in sera of patients with **Sjogren's syndrome** and **systemic lupus erythematosus** (SLE).

SSCP See **single-stranded conformational polymorphism**.

ssDNA phage Single-strand DNA phages such as MS2, FX174, as opposed to double-stranded DNA phages or RNA phages.

SSEA (1) **Stage-specific embryonic antigen**. (2) SseA is a key *Salmonella* virulence determinant, a small (secretion system effector A, 104aa), basic pI protein that serves as a type III secretion system chaperone for SseB and SseD. (3) SseA, the translation product of the *E. coli* sseA gene, is a protein (rhodanese-like protein, EC 2.8.1.2, 281aa) with 3-mercaptopyruvate: cyanide sulphurtransferase activity *in vitro*.

SSH (1) See **suppression subtractive hybridization**. (2) **Slingshot** homologue. (3) Sec sixty-one protein homolog, one component (Ssh1, 490aa) of the **sec61** translocon complex in *Saccharomyces*.

SSI-1 See SOCS.

SSLP See **simple sequence length polymorphism**.

ssrA A bacterial gene that encodes a stable RNA (transfer-messenger RNA, tmRNA) for a peptide tag which is cotranslationally added to truncated polypeptides, targeting them for rapid proteolysis. This constitutes a general mechanism in bacteria to rescue stalled ribosomes, for example those that arrive at the end of an mRNA without a stop codon. Research article: www.ncbi.nlm.nih.gov/pmc/articles/PMC95270/

SSRE See **shear stress response element**.

SSRIs Antidepressant drugs (selective serotonin reuptake inhibitors) that inhibit reuptake of serotonin released in the brain, thereby prolonging its action as a neurotransmitter.

SSU rDNA Small subunit ribosomal DNA. The sequence of SSU rDNA has been extensively used in molecular taxonomy. The SSU rDNA (18S rDNA, 17S rDNA, 16S-like rDNA) sequences from ascomycetes are listed on Myconet. Myconet: http://fieldmuseum.org/explore/myconet

SSX proteins Products of the SSX gene family that has at least five functional and highly homologous members, SSX1 to SSX5. They are localized in the nucleus, are diffusely distributed and may act as transcriptional repressors. SSX2 (synovial sarcoma, X breakpoint 2, 188aa) interacts with **rabin**. Abstract: http://www.ncbi.nlm.nih.gov/pubmed/12007189

stable nuclear transformation A hereditable change in a cell experimentally induced by introduction of a new gene which becomes integrated into the host DNA.

stable transfection When **transfecting** animal cells, a clone of cells in which the **transgene** has been physically incorporated into the genome. It thus provides stable, long-term expression although is more difficult to produce.

stachyose Digalactosyl-sucrose, a compound involved in carbohydrate transport in the phloem of many plants, and also in carbohydrate storage in some seeds.

stacking gel An upper layer of weak gel at the top of a gel in which an electrophoretic separation is to be run. Because the upper stacking gel is weak all the large molecules move through it rapidly, accumulate at the very top of the separating gel and thus all start from the same level.

STAG proteins Components of the **cohesin** complex that interact directly with **RAD21**. STAG1 (stromal antigen-1, 1258aa), STAG2 (1231aa) and STAG3 (1225aa).

STAGA complex A human chromatin-acetylating transcription coactivator that interacts with pre-mRNA splicing and DNA damage-binding factors *in vivo*. Contains homologues of most yeast **SAGA complex** components. See **ataxin 7**. Article: http://www.ncbi.nlm.nih.gov/pubmed/11564863

stage-specific embryonic antigen A marker (SSEA1) for murine pluripotent stem cells that plays an important role in adhesion and migration of the cells in the preimplantation embryo. In humans it is the LewisX antigen (CD15, 3-fucosyl-N-acetyl-lactosamine) that can be used to define immature retinal progenitor cells (see **sialyl Lewis X**). Other SSEAa are cell surface antigens; SSEA-3 is the murine glycolipid GB5.

staggered cut Situation when the two strands of a DNA molecule are cut in different places, slightly offset so that the two halves have 'sticky ends', a short segment of single-stranded DNA overhanging the duplex.

stalked bacteria Bacteria (prosthecate bacteria) that have appendages that allow them to attach to surfaces in aquatic environments.

staminode A rudimentary, sterile or abortive stamen.

stanniocalcin Glycoprotein hormone (STC-1, 179aa in *Oncorhynchus keta*, Chum salmon), secreted by the corpuscle of Stannius, an endocrine gland in teleosts. Prevents hypercalcaemia and inhibits calcium uptake through gills. Mammalian homologues (STC1, 247aa; STC2, 302aa) have been identified and are involved in various physiological processes, such as calcium and phosphate homeostasis.

stanol Plant-derived sterol (24-alpha-ethylcholestanol), similar to cholesterol but poorly absorbed from the intestine. Thought to compete with cholesterol for binding sites and by decreasing dietary intake of cholesterol helps lower blood levels.

Staphylococcal cassette chromosome elements Genomic islands ubiquitously disseminated among staphylococci, that capture foreign DNA segments The staphylococcal cassette chromosome **mec elements** (SCCmec) carry meticillin-resistance. The SSCs have cassette chromosome recombinase genes (ccrA and ccrB) that encode polypeptides with partial homology to recombinases of the invertase/resolvase family that catalyze precise excision of the SCC and other genes carried within the element.

Staphylococcal scalded-skin syndrome See **exfoliatin**.

Staphylococcal toxins *Staphylococcus aureus* produces several membranolytic toxins, alpha-, beta-, gamma-, and delta-haemolysins and **leucocidin**. Alpha-toxin (319aa) is secreted as monomers which then associate to form heptameric oligomer with a pore of 1 nm: it preferentially attacks platelets and cultured monocytes. Beta-toxin is a Mg^{2+}-dependent **sphingomyelinase** C. The gamma-toxin locus expresses three proteins, two class S components (HlgA and HlgC) and one class F component (HlgB) which form S/F heterodimers, have potent proinflammatory effects and may be important in pathogenesis of toxic shock syndrome. Delta toxin is a small peptide (26aa) that is very amphipathic and surface active and has properties similar to **melittin**. Staphylococcal virulence regulator protein A (Multidrug export protein mepA, 451aa) is involved in the export of the toxins from the cell.

staphylococcins Bacteriocins produced by staphylococci.

Staphylococcus Genus of non-motile Gram-positive bacteria that are found in clusters, and that produce important exotoxins (see Table E2). *Staphylococcus aureus* (*S. pyogenes*) is pyogenic, an opportunistic pathogen, and responsible for a range of infections. It has **protein A** on the surface of the cell wall. Coagulase production correlates with virulence: hyaluronidase, lipase, and **staphylokinase** are released in addition to the toxins.

staphylokinase Enzyme (streptokinase, EC 3.4.99.22, 163aa) released by *Staphylococcus aureus* that acts as a **plasminogen activator**.

STAR proteins (1) A family of of RNA-binding proteins (signal transduction and activation of RNA); evolutionarily conserved from yeast to humans and important for a number of developmental decisions. Contain a conserved KH domain as well as two conserved domains called QUA1 and QUA2. Examples include mouse **quaking**, *C. elegans* germline defective-1 (**GLD-1**) and human RNA-binding protein T-Star (Sam68-like mammalian protein 2, 346aa). (2) An intestine-specific membrane-bound **guanylate cyclase** that binds **guanylin** and is the receptor for bacterial heat-stable enterotoxins such as the *E. coli* ST toxin, STa. (3) In *Drosophila*, a protein (star, 597aa) involved in EGF receptor signalling and photoreceptor development. Interacts with the EGF receptor **torpedo** in the eye. (4) StAR-related lipid transfer proteins contain a START domain and are involved in lipid transport. **Steroidogenic acute regulatory protein** is STAR1. For example, StAR-related lipid transfer proteins 3, 4 and 5 (STARD3, 445aa; STARD4, 205aa; STARD5, 213aa) are involved in intracellular lipid and sterol transport, STARD8 (1023aa) accelerates GTPase activity of rhoA and cdc42, but not rac1. StARD9 (4614aa) is found in the CNS, muscle and a few other tissues. Several others are known.

STAR syndrome A developmental disorder (syndactyly, telecanthus, anogenital and renal malformations) caused by mutation in the *FAM58A* gene that encodes **cyclin M**.

starch Storage carbohydrate of plants made from branched and unbranched polymers of glucose. It is made in choroplasts by photosynthesis and is stored in the form of starch grains. Starch grains occur in storage organelles termed amyloplasts, ready for use. The potato tuber is a rich source of stored starch. In cells of the root cap, starch grains form statoliths, involved in the perception of gravity.

Stargardt's disease A childhood disorder in which there is retinal degeneration in the central (macular) region, Various genetic defects can cause the disease including mutation in genes encoding the retina-specific ABC transporter-A4 or the cyclic-nucleotide gated ion channel beta-4. Another form is caused by mutation in the ELOVL4 gene encoding elongation of very long-chain fatty acids-like 4 (314aa) a photoreceptor-specific component of the fatty acid elongation system. Mutation in **peripherin** are also implicated in some forms.

stargazin A brain-specific murine protein (323aa), the gamma-2 subunit of a voltage-dependent calcium channel, structurally similar to **clarins**, mutated in the mouse mutant stargazin which has seizures characteristic of absence epilepsy. A related molecule (stargazin-like protein, 447aa) is found in *Drosophila*.

start codon See **initiation codon**.

start site Imprecise term for either a **transcriptional** or a **translational** start site.

startle disease *stiff man syndrome* A genetically heterogeneous disorder with neurologically-induced muscular rigidity and an exaggerated startle response (hyperexplexia). Can be caused by mutations in genes encoding the alpha-1 or beta subunits of the glycine receptor, the presynaptic glycine transporter-2 (SLC6A5), **gephyrin**, and **collybistin**.

statherin An acidic tyrosine-rich phosphoprotein (62aa) secreted mainly by salivary glands. Binds **fimbrillin**.

stathmin A coiled-coil cytosolic phosphoprotein (oncoprotein 18, stathmin-1, 149aa) that binds to two tubulin heterodimers and increases the rate of rapid 'catastrophic' disassembly of microtubules. Overexpressed in some tumours and probably regulated by phosphorylation, possibly by **SAPK4**. Stathmin-2 (superior cervical ganglion-10 protein, 179aa) is neuron-specific and may be important in neuronal differentiation, and in modulating membrane

interaction with the cytoskeleton during neurite outgrowth. Stathmin-3 (180aa) is also neuron-specific.

statins Lipid-lowering drugs that inhibit **HMG Co-A reductase**, a key enzyme in cholesterol biosynthesis.

stationary night blindness A retinal disorder in which there are reduced numbers of retinal rods, the photoreceptors used for low-light vision. X-linked forms can be caused by mutation in **nyctalopin** or the retina-specific calcium channel alpha-1-subunit. Autosomal forms can be caused by mutation in the metabotropic **glutamate** receptor-6, calcium-binding protein-4, **rhodopsin**, **cyclic nucleotide phosphodiesterase**-6B or in rod-specific **transducin**. The Oguchi-type in which dark adaptation is abnormally slow, is caused by mutation in **arrestin** or in rhodopsin kinase.

stationary phase The stage of a cell culture at which cell proliferation ceases, because of nutrient depletion, and before senescence and cell death begins to reduce numbers of live cells.

statocyst A sense organ used to perceive gravity and thus body orientation. It has a cavity enclosing a **statolith** that is lined with sensory cells.

statocyte A root-tip cell containing one or more **statoliths**, involved in the detection of gravity in geotropism.

statolith (1) A type of **amyloplast** found in root-tip cells of higher plants. It can sediment within the cell under the influence of gravity, and is thought to be involved in the detection of gravity in geotropism. (2) A solid particle found in the cavity of a **statocyst**. It stimulates sensory cells lining the cavity with which it comes in contact under the influence of gravity.

STATs A family of proteins (signal transducers and activators of transcription) that have **SH2** domains which bind phosphotyrosine residues in receptors, particularly cytokine-type receptors; they are then phosphorylated by **JAKs**, dimerize and translocate to the nucleus and act as transcription factors. Many STATs are known; some are relatively receptor-specific, others more promiscuous, so that a wide range of responses is possible with some STATs being activated by several different receptors, sometimes acting synergistically with other STATs. STAT1 (750aa) mediates signalling by interferons alpha and gamma whereas STAT2 (851aa) is only activated by Type 1 interferons. STAT3 (770aa) binds to the interleukin-6 (IL-6)-responsive elements identified in the promoters of various acute-phase protein genes, STAT4 (748aa) is involved in IL12 signalling. Several others are known in humans and in a range of other metazoa. Mutations in STAT1 affect innate immunity, mutations in STAT3 are associated with hyper-IgE syndrome (**Job's syndrome**). See **signal-transducing adaptor proteins**.

staufen In *Drosophila* a RNA binding protein (1026aa) a maternal effect protein that binds to the 3′ UTR of specific mRNAs and acts in their localization, particularly maintaining cellular asymmetry in the oocyte through interaction with microtubules. Mammalian homologues staufen 1 (577aa) and staufen 2 (570aa) are also involved in the localization of mRNA. Staufen2, a brain-specific isoform has been shown to shuttle between nucleus and cytoplasm and can enter the nucleolus.

staurosporine Inhibitor of PKC-like protein kinases, originally derived from *Streptomyces staurosporeus*. Has a rather broad inhibitory spectrum.

STE20 Evolutionarily conserved serine/threonine kinases (EC 2.7.11.1), subdivided into the **p21-activated kinase** (PAK) and **germinal centre** kinase (GCK) families, that regulate fundamental cellular processes including the cell cycle, apoptosis, and stress responses. In *S. cerevisiae* Ste20p (sterile 20 protein, 939aa) is a mitogen-activated protein kinase kinase kinase kinase (MAP4K) involved in the mating pathway. Human homologues (at least 28) have the kinase domain but linked to other domains that confer specificity.

stealth proteins A protein family that is conserved from bacteria to higher eukaryotes. In bacteria they help pathogens to avoid the host innate immune system and are involved in the biosynthesis of exopolysaccharides such as capsular polysaccharides of pathogenic streptocoocci. Stealth protein GNPTAB is involved in adding mannose residues to enzymes destined for lysosomes and is defective in human **I-cell disease**.

STEAP One of a small family of cell-surface antigens that are expressed in prostate cancer and are potential targets for antibody-mediated therapy and diagnosis. STEAP1 (six-transmembrane epithelial antigen of the prostate, EC 1.16.1.-, 339aa) is expressed at the cell-cell junctions of the secretory epithelium of prostate and strongly expressed in prostate cancer cells. It is a metalloreductase that can reduce Fe^{3+} to Fe^{2+} and Cu^{2+} to Cu^{+} using NAD^{+} as an acceptor. STEAP2 (490aa), STEAP3 (488aa) and STEAP4 (459aa) have similar properties but some are more ubiquitously expressed. See **papin, PSM, PTCA-1, PSCA**.

stearic acid N-octadecanoic acid. See **fatty acids**.

steatoblasts Cells that give rise to fat cells (adipocytes).

steel factor Murine equivalent of **stem cell factor**, the ligand (273aa) for kit.

Steele-Richardson-Olszewski syndrome A neurological disorder (progressive supranuclear palsy)

that can be caused by defects in **tau protein** although the more typical forms are linked to other loci. There are some features resembling **Parkinson's disease**.

stefins Family of **cysteine peptidase** inhibitors. Stefin A (cystatin A, keratolinin, 98aa) inhibits **cathepsins** D, B, H and L. Stefin B (cystatin B, 98aa) has anti-peptidase activity against cathepsins L, H and B. Mutations in stefin B are associated with **myoclonic epilepsy of Unverricht and Lundborg**.

stele *vascular cylinder* The vascular tissue (xylem and phloem) inside the cortex of roots and stems of vascular plants.

stem cell (1) Cell that gives rise to a lineage of cells (progenitor cell). (2) More commonly used of a cell that, upon division, produces dissimilar daughters, one replacing the original stem cell, the other differentiating further (e.g. stem cells in basal layers of skin, in haematopoietic tissue, and in meristems). Embryonic stem cells (ES cells) are totipotent, able to produce any cell type; following modification *in vitro* they can be used to produce chimeric embryos and thus transgenic animals. Pluripotent stem cells are more limited and can produce only some differentiated cells. Quiescent stem cells may have a repair function and for example, the satellite cells in the skeletal muscles of mammals are quiescent myoblasts that will proliferate after wounding and give rise to more muscle cells by fusion. Induced pluripotent stem cells (iPSCs) are differentiated cells that have been experimentally induced to revert to a pluripotent state. Originally this required transfection of genes with the risk that the iPSCs were potentially neoplastic; the hope is that eventually it will be possible to take cells from a patient, revert them to pluripotency and use them therapeutically.

stem cell factor A growth factor (mast cell growth factor, 273aa, 245aa when processed), the ligand for **kit**. Stimulates proliferation of myeloid and lymphoid series. See **steel factor**.

stem cell-derived tyrosine kinase The murine homologue of the human **ron** receptor tyrosine kinase (STK, macrophage-stimulating protein receptor, EC 2.7.10.1, CD136, 1378aa). The precursor is cleaved into alpha (281aa) and beta (1068aa) chains which form a disulphide-linked heterodimer. Expressed on macrophages. Ligand is macrophage-stimulating protein (**MSP**), a serum protein activated by the coagulation cascade.

stem-and-loop structure Term for the structure of tRNAs which has four base-paired 'stems' and three 'loops' (not base-paired), one of which contains the anticodon.

stenohaline Descriptive of an organism that is unable to tolerate a range of salinities Cf euryhaline.

stenospermocarpy One cause of seedlessness in, for example, grapes. Pollination and fertilization occur as normal, but the embryo and/or endosperm abort two to four weeks after fertilization; seed development ceases but the ovary wall pericarp continues to grow and forms berries. See **parthenocarpy**.

stent A device inserted into a blood vessel to keep it open, usually a small metal coil or mesh tube. Similar devices are used in the GI tract and ureter.

Stentor Genus of large multinucleate protozoa (up to 2 mm long) with a ring of apical cilia used in feeding on bacteria. Spirotrich ciliates. Usually attached to a surface in freshwater ponds etc. but can relocate. Sometimes called the 'trumpet animalcule' because of its shape.

stereocilium See **stereovillus**.

stereotaxis A system by which the precise 3-D coordinates of a target site are identified, allowing focussed delivery of radiation or precise insertion of a microelectrode.

stereovillus Microfilament bundle-supported projection, several microns long, from the apical surface of sensory epithelial cells (**hair cells**) in inner ear: like a **microvillus**, but larger. It is stiff and may act as a transducer directly, or merely restrict the movement of the single sensory cilium (which does have an axoneme). Also described on cells of pseudo-**stratified epithelium** of the epididymal duct. Sometimes called stereocilia but they do not have axonemes so this is potentially confusing.

sterigmatocystin An intermediate in the biosynthetic pathway that generates **aflatoxins**.

Sternberg-Reed cells See **Hodgkin's disease**.

steroid finger motif See **steroid receptor**.

steroid hormones A group of structurally related hormones, based on cholesterol from which they are synthesized. They control sex and growth characteristics, are highly lipophilic, so can readily cross the plasma membrane, and are unique in that their receptors are in the nucleus, rather than on the plasma membrane. The five major subgroups are **glucocorticoids**, **mineralocorticoids**, androgens, estrogens, and **progestagens**.

steroid receptors Receptors for **steroid hormones** that have a conserved domain (the 'steroid finger' motif) containing two C4-type **zinc fingers**. Type I are cytoplasmic until they bind a hormone and release a chaperone, then they move to the nucleus and act as transcription factors. Type II are nuclear transcription factors and other members of this type bind non-steroid ligands (thyroid hormone, vitamin A).

steroid response element DNA sequence in the promoter region of a gene that is recognized and bound by a **steroid receptor**. Informative web-pages: http://www.ultranet.com/~jkimball/BiologyPages/S/SteroidREs.html

steroidogenic acute regulatory protein A mitochondrial protein (START domain-containing protein 1, StAR1, 285aa) that mediates cholesterol transfer and promotes steroid hormone production in the ovary, testis and adrenal gland. Defective in congenital lipoid adrenal hyperplasia (lipoid CAH).

sterol regulatory element *SRE* See **sterol regulatory element-binding proteins**, **steroid response element**.

sterol regulatory element binding proteins *SREBPs* A family of membrane-bound transcription factors important in both cholesterol and fatty acid metabolism. SREBPs (SREBP1, 1147aa; SREBP2, 1142aa) regulate the expression of over 30 genes. SREBPs are regulated by proteolytic cleavage (see **insig-1**, **SCAP**), rapid degradation by the ubiquitin-proteasome pathway, and **sumoylation**.

sterols Molecules that have a 17-carbon steroid structure, but with additional alcohol groups and side chains. Commonest example is **cholesterol**.

Stickler's syndrome An autosomal dominant connective tissue disorder in which there are ocular, orofacial, auditory, and skeletal anomalies. Type 1 is caused by mutation in the collagen 2A1 gene and is morphologically similar to **Wagner's syndrome**, Type 2 is a result of mutation in collagen 11A1 and Type 3 by mutation in collagen 11A2. An autosomal recessive form can be caused by mutation in the COL9A1 gene.

sticky ends The short stretches of single-stranded DNA produced by cutting DNA with **restriction endonucleases** whose site of cleavage is not at the axis of symmetry. The cut generates two complementary sequences that will hybridize (stick) to one another or to the sequences on other DNA fragments produced by the same restriction endonuclease.

stilbene synthase Important enzymes (stilbene synthase 1, EC 2.3.1.95, 392aa in grape, *Vitis vinifera*; many others are known) in the stilbene pathway. The levels of stilbenes increases in grapevine in response to biotic and abiotic stress, but also during berry ripening. Resveratrol is a stilbene.

Still's disease A systemic inflammatory disorder (systemic-onset juvenile rheumatoid arthritis). Adult-onset Still's disease is probably just the adult version of the same disorder.

STIM proteins Single-pass transmembrane proteins localized predominantly in the membrane of the endoplasmic reticulum that appear to function as sensors of ER Ca^{2+} levels and redistribute into a punctate pattern when calcium levels are low. STIM1 (stromal interaction molecule 1, 685aa) communicates with and opens CRAC channels located in the plasma membrane by interacting with **Orai1**. STIM2 (746aa) can form heterodimers with STIM1 and may inhibit STIM1-mediated Ca^{2+} influx.

stimulated emission depletion microscopy A method of obtaining greater resolution in optical microscopy than is possible with standard confocal techniques. A small region of the specimen is illuminated with a very brief flash of light that stimulates fluorescence but is followed by illumination with a 'ring' of light at a wavelength that depletes the capacity to fluoresce. The net effect is to restrict the stimulation of fluorescence to an extremely small spot, smaller than that set by the point source of a confocal microscope and improving the resolution by an order of magnitude.

stimulus-secretion coupling A term used to describe the events that link receipt of a stimulus with the release of materials from membrane-bounded vesicles (the analogy is with excitation-contraction coupling in the control of muscle contraction). A classical example is the link between membrane depolarization at the presynaptic terminal and the release of neurotransmitter into the synaptic cleft.

sting cells Nematocysts of coelenterates.

stipe A stalk, especially of fungal fruiting bodies or of large brown algae.

stipule A small appendages at the base of a leafstalk in certain plants; usually paired.

STK (1) **Stem cell-derived tyrosine kinase**. (2) Serine/threonine protein kinase e.g. LKB1, p61. (3) An ambiguous abbreviation that can refer to **flt3**, **Aurora kinase B**, mitogen-activated protein kinase HOG1 in *Setosphaeria turcica* (Northern corn leaf blight disease fungus) and various serine/threonine kinases. (4) **Streptokinase**.

stochastic Random or probabilistic.

stoichiometry Ratio of the participating molecules in a reaction — in the case of an enzyme-substrate or receptor-ligand interaction should be a small integer.

Stoke's radius Stoke's law of viscosity defines the frictional coefficient for a particle moving through a fluid, a coefficient that depends upon the viscosity of the fluid and the radius of the particle. The apparent radius of a molecule sedimenting under centrifugal force calculated from this law (the Stoke's radius) is a feature of the tertiary structure and thus informative about the molecule in question.

stolon A creeping horizontal stem that forms roots. A characteristic method of spreading in plants such as the strawberry.

stoma *Pl.* stomata. Pore in the epidermis of leaves and some stems, which permits gas exchange through the epidermis. Can be open or closed, depending upon the physiological state of the plant. Flanked by stomatal **guard cells**.

STOMAGEN A mesophyll-derived factor (45aa) that promotes epidermal stomatal development in *Arabidopsis*, generated from a 102aa precursor. If overexpressed or externally applied it increases stomatal density. It may compete with **EPIDERMAL PATTERNING FACTOR**s for binding to the receptor TOO MANY MOUTHS.

stomatin An erythrocyte integral membrane protein (Band 7.2b, 288aa), defective in overhydrated **hereditary stomatocytosis**. Stomatin is found in most tissues and extensively throughout the animal kingdom, although its function is unclear. Stomatin-like 1 (STOML1, SLP1, 394aa), STOML2 (356aa) and STOML3 (287aa) are known in humans; STOML2 is overexpressed in a range of epithelial tumours. Similar genes (mec-2, unc-24, unc-1) are found in *C. elegans* and mutations in all of these affect the nervous system.

stomatocytosis A condition in which erythrocytes adopt an abnormal shape because of ion leakage across the plasma membrane. In one form (hereditary stomatocytosis-I, overhydrated stomatocytosis) there appears to be a defect in **stomatin**. A second form, stomatocytosis-II, in which there is no increase in fragility, has been reported. A different disorder, dehydrated hereditary stomatocytosis (xerocytosis; desiccytosis), is caused by excessive leakage of potassium from the erythrocytes which lose water and become more fragile.

stone cell See **sclereid**.

stonin A family of adapter proteins (stonin-1, stoned B-like factor, 735aa; stonin-2, stoned B, 905aa) that are involved in the endocytic machinery and may facilitate clathrin-coated vesicle uncoating. In *Drosophila* stoned B (1262aa) may mediate the retrieval of **synaptotagmin** from the plasma membrane.

STOP (1) A calmodulin-binding microtubule-stabilizing protein (stable tubulin-only protein, microtubule-associated protein 6, 813aa) that inhibits cold temperature-induced disassembly, particularly in neurites. A similar protein, MAP6 domain-containing protein 1 (199aa) probably has similar properties. (2) In *Arabidopsis* a protein, SENSITIVE TO PROTON RHIZOTOXICITY 1 (STOP1, 499aa) a zinc-finger transcription factor that is important in responding to major stress factors in acid soils such aluminum (Al^{3+}). (3) See short stop (**shot**).

stop codon   See **termination codons**.

stop transfer sequence   Amino acid sequence (membrane anchor sequence) that causes cessation of the co-translational transfer of a protein across a membrane and leaves the protein embedded in the membrane. Generally consists of long sequences of hydrophobic residues.

storage diseases   See **lysosomal diseases**.

storage granules   (1) Membrane-bounded vesicles containing condensed secretory materials (often in an inactive, zymogen, form). Otherwise known as zymogen granules or condensing vacuoles. (2) Granules found in plastids, or in cytoplasm; assumed to be 'food reserves', often of glycogen or other carbohydrate polymer.

storage pool disease   SPD Platelet defects in which the numbers and contents of granules are affected. In alpha-SPD (gray platelet syndrome) there are severe reductions in the alpha-granules and their contents, in delta-SPD the defect is only in the dense granules (delta-granules) and in a third form (alpha/delta-SPD) both types of granule are affected.

storage vacuole   In plants, vacuoles used to sequester products used for defence, or toxins that would harm the plant, or materials for later use. Contrast with lytic vacuoles.

STR   See **satellite DNA**

strain birefringence   See **birefringence**.

Strasburger cell   A nucleated cell in phloem tissue of gymnosperms, closely associated with a sieve cell.

stratified epithelium   An epithelium composed of multiple layers of cells, only the basal layer being in contact with the basal lamina (see **basement membrane**). The basal layer is of **stem cells** that divide to produce the cells of the upper layers; in skin, these become heavily keratinised before dying and being shed as squames. Stratified epithelia usually have a mechanical/protective role.

stratifin   An adapter molecule (14-3-3-sigma, epithelial cell marker protein 1, 248aa) found mainly in tissues enriched in stratified squamous keratinizing epithelium. Has a modulatory role in a range of signalling pathways.

stratum corneum   Outermost layer of skin, composed of clear, dead, scale-like cells with little remaining except **keratin**.

stratum granulosum   Layer of granular cells underlying the **stratum corneum** in the skin of vertebrates. The cells accumulate keratin and gradually become compressed to form the cornified cells of the outermost layer.

stratum lucidum   A thin, clear layer of dead cells in vertebrate skin, lying between the **stratum corneum** and **stratum granulosum** of thick skin, as on palms of the hands and the soles of the feet.

stratum Malpighii   One of the layers of the skin in vertebrates, lying between the proliferating cells of the basal layer (**stratum germinatum**) and the **stratum granulosum** where keratin deposition occurs. Also known as the prickle cell layer. Location of many of the disorders of the skin.

streptavidin   A protein (183aa) from *Streptomycetes avidinii* that has an **avidin** domain and binds **biotin** with high affinity. Forms a homotetramer that binds four biotin molecules. A useful experimental tool although its biological function is unclear.

Streptobacillus   A Gram-negative bacterium (*Streptobacillus moniliformis*) infective agent of rat-bite fever (**sodoku**).

streptococcal M-protein   Cell wall protein of streptococci used as a taxonomic marker for different strains of Group A streptococci (more than 100 serotypes are known). The M-protein confers anti-phagocytic properties on the cell and is present as hair-like **fimbriae** on the surface. M-protein is an important virulence factor, and antibodies directed against M-protein are essential for phagocytic killing of the bacteria.

streptococcal toxins   Group of haemolytic **exotoxins** released by *Streptococcus* spp. α-haemolysin: 26–39 kDa (four types), forms ring-like structures in membranes (see **streptolysin O**). Lipid target unclear; β-haemolysin: a hot-cold **haemolysin** with sphingomyelinase C activity; γ-haemolysin: complex of two proteins (29 and 26 kDa) that act synergistically, rabbit erythrocytes particularly sensitive; δ-toxin: heat-stable peptide (5 kDa) with high proportion of hydrophobic amino acids. Seems to act in a detergent-like manner (*cf.* **subtilysin**), but may form hydrophilic transmembrane pores by cooperative interaction with other δ-toxin molecules; leucocidin (Panton-Valentine leucocidin): two components F (fast migration on CM-cellulose column: 32 kDa), and S (slow: 38 kDa). Mode of action contentious. See also *Streptococcus*, **streptolysins O and S**, **erythrogenic toxin**. *Streptococcus pyogenes* releases a range of pyogenic (fever-inducing) exotoxins, several of which are superantigens with sequence homology to staphylococcal toxins (SPEA, erythrogenic toxin, 251aa; SPEC, 235aa). SPEB is a cysteine peptidase (streptopain, EC 3.4.22.10, 398aa).

streptococcins   Bacteriocins released by streptococci.

Streptococcus   Genus of Gram-positive cocci that grow in chains. Some species (*Strep. pyogenes* in particular) are responsible for important diseases in humans (pharyngitis, scarlet fever, rheumatic fever): *Strep. pneumoniae* (*Pneumococcus pneumoniae; Diplococcus pneumoniae;* Fraenkel's bacillus) is the main culprit in lobar- and broncho-pneumonia. *Streptococci* have anti-phagocytic components (hyaluronic acid-rich capsule and **M-protein**), and release various toxins (**streptolysins O and S, erythrogenic toxin**) and enzymes (**streptokinase, streptodornase**, hyaluronidase, and proteinase). The α-haemolytic streptococci (viridans streptococci) produce limited haemolysis on blood agar; include *S. mutans, S. salivarius, S. pneumoniae.* The β-haemolytic streptococci, of which *S. pyogenes* is the only species, though there are many serotypes, produce a broad zone of almost complete haemolysis on blood agar as a result of streptolysin O and S release. The γ-streptococci are non-haemolytic (e.g. *S. faecalis*).

streptodornase Mixture of four DNAases (EC 3.1.21.1) released by streptococci. By digesting DNA released from dead cells the enzyme reduces the viscosity of pus and allows the organism greater motility.

streptogramins Group of antibiotics that consist of mixtures of two structurally distinct compounds, type A and type B, which are separately bacteriostatic, but bactericidal in appropriate ratios. They act at the level of inhibition of translation through binding to the bacterial ribosome. An example is **quinupristin**.

streptokinase A non-enzymic **plasminogen activator** (440aa) released by *Streptococcus pyogenes.* Activates plasminogen by forming a complex. There are various forms (streptokinases A, C, G).

streptolydigin Antibiotic produced by *Streptomyces lydicus* that blocks peptide chain elongation by binding to bacterial RNA polymerase. Research paper: http://www.jbc.org/content/270/41/23930.long

streptolysin O Oxygen-labile thiol-activated haemolysin (571aa). Haemolysis is inhibited by cholesterol, and only cells with cholesterol in their membranes are susceptible. Toxin aggregates are linked to cholesterol to form a channel 30 nm diameter in the membrane, and non-osmotic lysis follows. Markedly inhibits neutrophil movement and stimulates secretion, but has little effect on monocytes. *Cf.* **streptolysin S**.

streptolysin S Oxygen-stable haemolysin of *Streptococcus pyogenes.* Thought to be a peptide of ca 28aa: causes zone of β-haemolysis around streptococcal colonies on blood agar. Like complement-mediated haemolysis it appears to act in a one-hit mechanism. Toxic to leucocytes, platelets, and several cell lines. Has sequence homologies to the **bacteriocin** class of antimicrobial peptides. A nine-gene operon is required for SLS production: this includes a candidate gene for a bacteriocin pre-propeptide, SagA, the likely SLS precursor and

candidate genes for chemical modification of the bacteriocin propeptide and self protection, an ABC transporter for export and maturation proteolysis of the leader peptide, and an internal terminator motif for differential transcription of structural gene and accessory gene mRNAs. *Cf.* **streptolysin O.** Article: http://www.ncbi.nlm.nih.gov/pmc/articles/PMC143243/?tool=pmcentrez

Streptomyces Genus of Gram-positive spore-forming bacteria that grow slowly in soil or water as a branching filamentous mycelium similar to that of fungi. Important as the source of many antibiotics, e.g. **streptomycin, tetracycline, chloramphenicol, macrolides.**

streptomycin A water-soluble **aminoglycoside** antibiotic from the bacterium *Streptomyces griseus.* Commonly used antibiotic in cell culture media: acts only on prokaryotes, and blocks transition from **initiation complex** to chain-elongating ribosome.

streptopain See **streptococcal exotoxins**.

streptovaricins Antibiotics of the ansamycin class, produced by various Actinomycetes, that block initiation of transcription in prokaryotes. (*Cf.* **rifamycins** and **rifampicin**).

streptozotocin Methyl nitroso-urea with a 2-substituted glucose, used as an antibiotic (effective against growing Gram-positive and Gram-negative organisms), and also to induce a form of diabetes in experimental animals (rapidly induces pancreatic **B cell** necrosis if given in high dose). By using multiple low doses in a particular strain of mice, it is possible to produce insulitis followed later by diabetes, a model for Type 1 (juvenile-onset) diabetes in humans.

stress related genes The strongly regulated biotic stress related genes in plants include several PR-proteins (endochitinase (TC60929), chitinase (Q7XB39), osmotin-like proteins (P93621), thaumatin (Q7XAU7), disease response protein (Q45W75), tumour related proteins (P93378)), three 1,3-β-glucanases and several proteinase inhibitors. Early induction of genes encoding chitinases and 1,3-β-glucanases is a typical response of plants towards fungal pathogens.

stress-fibres Bundle of microfilaments and other proteins found in fibroblasts, particularly slow-moving fibroblasts cultured on rigid substrata. Shown to be contractile; have a periodicity reminiscent of the **sarcomere**. Anchored at one end to a **focal adhesion**, although sometimes seem to stretch between two focal adhesions.

stress-induced proteins Alternative and preferable name for heat-shock proteins of eukaryotic cells, which emphasises that the same small group of proteins is stimulated both by heat and various other stresses.

stresscopin See **urocortin III**.

striated muscle Muscle in which the repeating units (**sarcomeres**) of the contractile **myofibrils** are arranged in register throughout the cell, resulting in transverse or oblique striations observable at the level of the light microscope, e.g. the voluntary (skeletal) and cardiac muscle of vertebrates.

strigolactones Carotenoid-derived plant hormones (sesquiterpenes) implicated in inhibition of shoot branching. Strigolactones act as a host-recognition signal for arbuscular mycorrhizal fungi, are released by the host plant and stimulate hyphal branching 'branching factors'. They are also seed germination stimulants for the parasitic weeds *Striga* and *Orobanche*.

string A *Drosophila* protein of the Cdc25 family (M-phase inducer phosphatase, EC 3.1.3.48, 479aa) that activates mitosis by hydrolyzing phosphotyrosine 15 of the cyclin dependent kinase Cdc2. String is activated by **tribbles** and is transcribed in a spatial pattern controlled by the anterior-posterior and dorsoventral patterning systems. *Cf.* **cysteine string protein**.

stringency In nucleic acid **hybridization**, the labelled **probe** is used to label matching sequences by base-pairing. Unbound probe is removed though a series of stringency washes. Low stringency washing (low temperature, high ionic strength), allows some mismatching of probe and target, and thus allows the detection of similar sequences at some cost in specificity. By contrast, high stringency conditions allow only closely matching sequence to remain base-paired.

stringent control One of the two regulatory mechanisms used by *E. coli* to adjust rRNA output to amino acid availability (*cf.* **growth-rate dependent control**). Stringent control is mediated by alarmone ppGpp.

striosome One of the components of the striatum region of the basal ganglia in the brain. The striatum is made of two parts, the matriosome and the striosome. Both receive input from the cortex (mostly frontal) and from dopaminergic (DA) neurons, but the striosome projects principally to DA neurons in the ventral tegmental area (VTA) and the substantia nigra (SN), the matriosome projects back to the frontal lobe.

strobila *Pl.* strobilae. A part or structure that buds to form a series of segments. In Scyphozoa, the sessile larval stage that produces medusoids by transverse fission. In Cestoda, the segmented body consisting of proglottides. *Cf.* **strobilus**.

strobilus *Pl.* strobili. (1) The cone-like reproductive structures of most gymnosperms and some pteridophytes. (2) An angiosperm inflorescence of cone-like appearance. *Cf.* **strobila**.

stroma (1) The soluble, aqueous phase within the chloroplast, containing water-soluble enzymes such as those of the **Calvin-Benson cycle**. The site of the **dark reaction** of photosynthesis. (2) Loose connective tissue with few cells.

stromal cell A non-committal name for a resident cell of loose fibrous connective tissue.

stromatolites Large rounded, multilayered fossils found in rocks dating from the Early Archaean (3.5mya), possibly before 'modern' cyanobacteria evolved. The layers are produced as a result of photosynthetic activity of bacterial colonies causing the precipitation of calcium carbonate that is combined with other sedimentary material trapped within the mucilage surrounding the colony. New layers of bacteria then develop on the outer surface. Modern equivalents are produced by cyanobacteria under certain circumstances e.g. on the beaches at Shark Bay in W. Australia.

stromelysins Matrix metallopeptidases (stromelysin-1, EC 3.4.24.17, MMP-3, 477aa; stromelysin-2, EC 3.4.24.22, MMP-10, 476aa; stromelysin-3, EC 3.4.24.-, MMP-11, 488aa) involved in breaking down the extracellular matrix.

stromovascular cells The non-adipocyte cells of adipose tissue, presumed to be mostly pre-adipocytes. Research article: http://jas.fass.org/content/82/2/429.full.pdf + html

stromule Dynamic tubular structures emerging from the surface of plastids. Their function is unclear but by increasing the plastid/cytoplasm contact area they may facilitate the exchange of metabolites or signals. Recent paper: http://www.biomedcentral.com/1471-2229/11/115/abstract

strong promoter Promoter that, when bound by a transcription factor, strongly activates expression of the associated gene. The term is widely used, but lacks precision.

Strongylocentrotus purpuratus Common sea urchin. Echinoderms are popular tools for developmental biology because the early embryo is transparent and cell movements can easily be observed. Sea urchin eggs are available in large numbers and were a convenient material for early biochemical studies on histones and mRNA. Because they are relatively large they have also been convenient for various electrophysiological studies.

strongyloidiasis Intestinal infestation (strongyloidosis) of man with the nematode worm, *Strongyloides stercoralis*.

strophanthin Mixture of glycosides from *Strophanthus kombe* (a tropical liana) with properties similar to those of digoxin and **ouabain** (strophanthin G). See **digitalis**.

STRP Short tandem repeat polymorphism. See **satellite DNA**.

structural bioinformatics The branch of bioinformatics associated with prediction of the secondary structure of proteins and nucleic acids. Much software has been developed to do this and there are a number of free packages available. The capabilities of such programs and the availability of software changes rapidly. See **PyMol**. RNA secondary structure: http://openwetware.org/wiki/Wikiomics: RNA_secondary_structure_prediction

structural gene A gene that codes for a product (e.g. an enzyme, structural protein, tRNA), as opposed to a gene that serves a regulatory role.

structure-activity relationship *SAR* The relationship between the strucure of a compound and its activity, usually done by systematically varying the structure and analysing binding or inhibitory activity. A more sophisticated analysis using complex computational methods to derive appropriate parameters of shape and electronic distribution produces a quantitative structure-activity relationship (QSAR). The hope is that eventually it will be possible to design better inhibitors (drugs) by this means, but rational design is not yet straightforward.

Strumpell's disease A **spastic paraplegia** caused by a defect in **strumpellin**.

strumpellin A component (1159aa) of the **WASH complex**, ubiquitously expressed and defective in some forms of **spastic paraplegia** (Strumpell's disease).

struvite A mineral composed of orthorhombic crystals of magnesium ammonium phosphate. Struvite stones in the kidney are composed of a struvite-carbonate-apatite matrix and are associated with urinary infections, particularly with urease-producing bacteria, including *Ureaplasma urealyticum* and *Proteus*.

strychnine **Alkaloid** obtained from the Indian tree *Strychnos nux-vomica*; specific blocking agent for the action of the amino acid transmitter glycine. Convulsive effects of strychnine are probably due to its blockage of inhibitory synapses onto spinal cord motoneurons.

Stuve-Wiedemann syndrome See **Schwartz-Jampel syndrome**.

STX (1) See **saxitoxin**. (2) Shiga toxin. See **shiga-like toxin**. (3) See **sialyltransferase**.

Stylonychia mytilus Large ciliate protozoan of the Order Hypotrichida, that has compound cilia (cirri) that can be used for walking or swimming.

S-type lectins *galectins* One of two classes of **lectin** produced by animal cells, the other being the **C-type lectins**. The carbohydrate binding activity of the S-type lectins requires their cysteines to have free thiols and, unlike the C-type lectins, does not need divalent cations. They mostly have molecular masses in the range 14–16 kDa and often form dimers and higher oligomers. The carbohydrate recognition domain contains a number of critically conserved amino acids and largely binds to β-galactosides. S-type lectins occur as cytoplasmic proteins and lack a signal sequence for secretion, yet do exist extracellularly.

subacute Description of a disease that progresses more rapidly than a chronic disease and more slowly than an acute one.

subacute sclerosing panencephalitis *SSPE* Chronic progressive illness seen in children a few years after measles infection, and involving demyelination of the cerebral cortex. Virus apparently persists in brain cells: usually considered a **slow virus** disease.

subcommissural organ A brain gland (SCO) present in all vertebrate phyla. It secretes glycoproteins into the cerebrospinal fluid, where some aggregate to form Reissner's fibres and others remain soluble. Among the secreted proteins is SCO-**spondin** which is probably involved in axonal growth and/or guidance. Research article: http://www.cerebrospinal-fluidresearch.com/content/5/1/3

suberin Fatty substance, containing long-chain fatty acids and fatty esters, found in the cell walls of cork cells (phellem) in higher plants. Also found in the **Casparian band**. Renders the cell wall impervious to water.

subgenomic RNA *sgRNA* Copies of viral RNA from positive-strand RNA viruses that are truncated at the 5′ ends, but have the same 3′ ends. The viral genome has multiple open reading frames with only a single 3′ end: the subgenomic RNA is a transcript of a downstream ORF which would not normally be expressed because the stop-site at the end of the upstream ORF terminates transcription. A special subgenomic promoter (RNA-promoter) between the ORFs permits this to happen. Helpful website: http://www.dias.kvl.dk/Plantvirology/evirusgenes/evirsubgenomic.html

substance K Neurokinin A (neuromedin L). See **tachykinins**.

substance P A **vasoactive intestinal peptide** (11aa) derived from **protachykinin-1** found in the brain, spinal ganglia and intestine of vertebrates.

Induces vasodilation, salivation and increased capillary permeability. The receptor is G-protein coupled (neurokinin 1 receptor, 407aa).

substantia nigra Area of darkly pigmented dopaminergic neurons in the ventral midbrain thought to control movement and damaged in **Parkinsonism**.

substrate (1) A substance that is acted upon by an enzyme. (2) A culture medium containing suitable compounds to permit growth of a species of bacterium. (3) A surface. See **substratum**.

substratum The solid surface over which a cell moves, or upon which a cell grows: should be used in this sense in preference to **substrate**, to avoid confusion.

subtilin Cationic pore-forming **lantibiotic** (56aa) produced by *Bacillus subtilis*. Acts preferentially on Gram-positive microorganisms but the producer cells have immunity mediated by the four genes *spaIFEG*. SpaFEG is an **ABC transporter**-2 subfamily member, a multidrug resistance protein. SpaI is a membrane-localised lipoprotein that may be a subtilin-intercepting protein. Mechanism of target recognition: http://aac.asm.org/content/52/2/612.full

subtilisin Extracellular serine endopeptidase (EC 3.4.21.62, ~380aa) produced by *Bacillus spp.*

subtilysin Haemolytic surfactant produced by *Bacillus subtilis*, a hexapeptide linked to a long-chain fatty acid.

subtractive hybridization *subtraction cloning* Technique used to identify genes expressed differentially between two tissue samples. A large excess of **mRNA** from one sample is hybridized to cDNA from the other, and the double stranded hybrids removed by physical means. Remaining cDNAs are those not represented as RNA in the first sample, and thus presumably expressed uniquely in the second. To improve specificity, the process is often repeated several times. See also **differential screening**.

subunits Components from which a structure is built; thus myosin has six subunits, microtubules are built of tubulin subunits. In some cases it may be more informative to speak of **protomers**.

succinate Intermediate of the **tricarboxylic acid cycle** and **glyoxylate cycle**.

succinate dehydrogenase (1) SDH (EC 1.3.5.1, succinate-coenzyme Q reductase, complex II in electron transport, succinate-ubiquinone oxidoreductase). An enzyme complex of 4 subunits located in the inner mitochondrial membrane. It has two main activities, as part of the citric acid cycle, oxidizing succinate into fumarate while passing electrons on to FAD and as complex II of the electron transport chain, which uses electrons freed

from succinate, to reduce ubiquinone to ubiquinol. (2) An enzyme (EC 1.3.99.1, fumarate reductase, fumarate dehydrogenase) that oxidises succinate to fumarate with the involvement of an acceptor. A bacterial enzyme or degraded form of mitochondrial SDH.

succinyl CoA An intermediate product in the **tricarboxylic acid cycle**, a combination of succinic acid and coenzyme A.

succinylcholine Cholinergic antagonist (suxamethonium chloride), two acetylcholine molecules linked by their acetyl groups. Binds to the acetylcholine receptor but unlike acetylcholine persists for long enough to cause the loss of electrical excitability and acts as a skeletal muscle relaxant.

succulents Plants adapted to dry conditions. Storage of water in leaves and other organs gives them their succulent or swollen appearance.

sucrase See **invertase**.

sucrose Table sugar, a non-reducing disaccharide, α-D-glucopyranosyl-β-D-fructofuranose.

Sudan stains Histochemical stains used for lipids.

sudden oak death A disease of oak trees caused by the fungus *Phytophthora ramorum*.

SUFU A negative regulator of the **sonic hedgehog/patched** and beta-catenin signalling pathways in vertebrates (suppressor of **fused**, 484aa), part of a corepressor complex that downregulates **GLI-1** (glioma-associated oncogene-1) transactivation and in the cytoplasm sequesters GLI1, GLI2 and GLI3, probably targeting them for proteosomal degradation. Defects in SUFU are a cause of **medulloblastoma**.

sugars See separate entries, and Table S2.

sulcus *Pl.* sulci. A groove or furrow e.g. on the surface of the cerebrum in mammals.

sulfonamides Synthetic bacteriostatic antibiotics derived from sulphanilamide (a red dye) with a wide spectrum of activity against most Gram-positive and many Gram-negative organisms. Sulfonamides inhibit multiplication of bacteria by acting as competitive inhibitors of p-aminobenzoic acid in the folic acid metabolism cycle. In UK formerly sulphonamides but sulfonamide is the BAN.

sulfur, sulfo- The British spelling, sulphur, sulpho- is used throughout except where the British Approved Name (BAN) for drugs is the 'sulf-' spelling.

sulphatase-modifying factor 1 The enzyme in the endoplasmic reticulum (SUMF1, EC 1.8.99.-, 374aa) that catalyses the post-translational formation of C-alpha-formylglycine (FGly) from cysteine at the active site of **sulphatases**. Mutations lead to **multiple sulphatase deficiency**.

TABLE S2. Sugars

Monosaccharides

PENTOSES

L-arabinose	D-ribose	D-xylose	2-deoxy-D-ribose MW 134 1

HEXOSES

D-fructose	D-galactose	D-glucose	D-mannose

Free amino-sugars are not found in structural oligosaccharides but N-acetyl aminohexoses are widely distributed. Most common are:

N-acetylgalactosamine   N-acetylglucosamine

Other common components of structural oligosaccharides are:

fucose   sialic acids (N-acetyl-neuraminic acid)

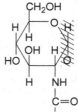

Hexose derivatives found in proteoglycans also include:

D-glucuronic acid	muramic acid

The list includes only the most common compounds found in metabolic pathways and in structural molecules. The structures are presented as Haworth models and it should be noted the configuration at the carbon which carries the carbonyl oxygen is not determined unless the hydroxyl-group takes part in a glycosidic linkage, which it always does in higher oligomers. The convention for depicting glycosidic linkages is:

glycosyl carbon → acceptor hydroxyl. Configuration not defined in free molecule.

Shaded regions have configurations that are not defined in free molecules.

Sulphated derivatives of N-acetyl aminohexoses are also widespread and include the 4- and 6- sulphate esters of N-acetyl glucosamine and N-acetyl galactosamine.

Disaccharides and Polysaccharides

These are fully specified by the residue names, sequence, bond-direction, and the position numbers of the carbon atoms giving rise to the linkage. The configuration around the glycosidic carbon is also specified as alpha or beta.

sulphatases Esterases (EC 3.1.6.-) that degrade sulphate esters such as heparan sulphate and dermatan sulphate. See, for example, **aryl sulphatase** also **multiple sulphatase deficiency**.

sulpholipids Any lipid containing sulphur although usually restricted to mean those in which the polar head group contains sulphate species. Sulpholipid-1 (SL-1) is an abundant sulphated glycolipid and potential virulence factor found in *Mycobacterium tuberculosis*. In plants they are important intermediates in the sulphur cycle.

sulphonylurea receptor The sulphonylurea receptor-1 (ABCC8, SUR-1, 1581aa) is an ABC transporter that regulates glucose-induced insulin secretion by controlling K^+-ATP channel activity of the pancreatic beta-cell membrane. They form a stoichiometric complex with 4 pore-forming inwardly-rectifying Kir6.2 molecules to produce the ATP-sensitive potassium (KATP) channels in neurons and neuroendocrine cells. SURs have two nucleotide-binding folds that sense changes in the metabolic status (ATP/ADP ratio) of the cell. Glibenclamide, a sulphonyl urea drug, binds to SUR and labelled forms of the drug can be used to show the distribution of the receptors.

sulphonylureas Group of drugs which act by augmenting insulin secretion. Used in the treatment of Type II diabetes. Examples include glibenclamide which is the sulphonylurea most commonly used in experimental studies. See **sulphonylurea receptor**.

sulphur-oxidising bacteria Aerobic chemolithotrophic bacteria that use sulphur or H_2S as an electron donor for oxygen to produce sulphate with release of energy.

sulphur-reducing bacteria Anaerobic bacteria (sulphate-reducing bacteria) that respire with sulphate or sulphur acting as electron acceptor instead of oxygen to produce H_2S.

sulphydryl reagents Compounds that bind to SH groups. Include p-chlormercuribenzoate, N-ethyl maleimide, iodoacetamide. Very important in studies of protein structure because they affect disuphide bonding.

SUMO Small ubiquitin-related modifier proteins See sumoylation.

sumoylation Post-transcriptional modification of a protein by the conjugation of SUMO (small ubiquitin-related modifier) proteins: stabilizes some proteins and may alter subcellular localization. Three different SUMO proteins are conjugated to proteins, SUMO-1, SUMO-2 and SUMO-3. The SUMO-2 and SUMO-3 genes are closely related, with 86% sequence identity while SUMO-1 is less closely related with about 50% sequence identity with SUMO-2 and SUMO-3. SUMO-1 (101aa) conjugates as a monomer, while SUMO-2 (95aa) and SUMO-3 (103aa) are conjugated to proteins as higher molecular weight polymers. Targets include p53, ran-Gap and various transcription factors, including C/EBP proteins and c-Myb.

Sun proteins A conserved family of proteins that have a Sun domain (Sad1/UNC-84 homology domain). They are important components of various complexes involved in chromosomal movement, positioning of nuclear pore complexes and linkage of the nucleoskeleton with the cytoskeleton (the LINC complex). Interact with **lamins** and with nesprins. They were originally cloned by homology with *C. elegans* protein UNC-84 that is involved in nuclear migration/positioning. In *S. cerevisiae* Sun 4 (septation protein SUN4, 420aa) is involved in the process of septum formation. The NSUN family (NOL1/NOP2/Sun domain family) are S-adenosyl-L-methionine-dependent methyltransferases, for example, Nsun2 (tRNA (cytosine-5-)-methyltransferase NSUN2, EC 2.1.1.29, 767aa).

superantigen Antigens, mostly of microbial origin, that activate all those T-lymphocytes that have a T-cell receptor with a particular Vβ sequence; as a consequence superantigens activate large numbers of T-cells and elicit an inappropriately strong immune response. Superantigens are presented on MHC Class II, but are not processed, and bind with high affinity but not in the normal peptide-binding groove of the MHC molecule. Presentation is not MHC-restricted. Staphylococcal enterotoxins are the best known superantigens and are pyrogenic, increase host susceptibility to endotoxic shock, suppress immunoglobulin production and stimulate CD4$^+$ T-cells. Superantigen diseases include *Staphylococcus aureus* food poisoning, **toxic shock syndrome**, and scarlet fever (produced by **erythrogenic toxin** of *Streptococcus pyogenes*). The *Mls* gene product in mice can act as a self-superantigen.

supercoiling In circular DNA or closed loops of DNA, twisting of the DNA about its own axis changes the number of turns of the double helix. If twisting is in the opposite direction to the turns of the double helix, i.e. anticlockwise, the DNA strands will either have to unwind or the whole structure will twist or supercoil - termed negative supercoiling. If twisting is in the same direction as the helix, clockwise, which winds the DNA up more tightly, positive supercoiling is generated. DNA that shows no supercoiling is said to be relaxed. Supercoiling in circular DNA can be detected by electrophoresis because supercoiled DNA migrates faster than relaxed DNA. Circular DNA is commonly negatively supercoiled and the DNA of eukaryotes largely exists as supercoils associated with protein in the **nucleosome**. The degree of supercoiling can be altered by **topoisomerases**.

superhelix A supercoil of a molecule, like DNA, that is already coiled.

SUPERMAN A transcriptional regulator (204aa) in *Arabidopsis* that acts indirectly to prevent the B class homeotic proteins **APETALA3** and **PISTILLATA** from acting during flower development.

superoxide Term used interchangeably for the superoxide anion $^{(\cdot)}O_2^-$, or the weak acid $HO_2^{(\cdot)}$. Superoxide is generated both by prokaryotes and eukaryotes, and is an important product of the **metabolic burst** of neutrophil leucocytes. A very active oxygen species, it can cause substantial damage, and may be responsible for the inactivation of plasma antiproteases that contributes to the pathogenesis of emphysema.

superoxide dismutase Any of a range of metalloenzymes (EC 1.15.1.1) that catalyse the breakdown of superoxide to produce hydrogen peroxide and oxygen. The eukaryotic form (SOD1) is a copper-zinc enzyme (154aa) and the prokaryotic (SodA, 206aa in *E. coli*) and mitochondrial (SOD2, 222aa in humans) form has manganese as a cation. Other metal ions (iron, nickel) are found in some prokaryotes. They play an important role in protecting tissues from superoxide-induced damage. Superoxide dismutase copper chaperone (274aa) delivers copper and zinc to SOD1. See **amyotrophic lateral sclerosis 1**.

supershift Phenomenon in band-shift assays where the reduction in mobility on a gel induced by a binding interaction with a protein is enhanced by the addition of an antibody to the protein (or another interacting protein). Net result is that the mobility of the band of interest is further decreased (shifted).

supervillin An actin binding protein (2214aa) of the **gelsolin/villin** family, that forms a high-affinity link between the actin cytoskeleton and the membrane. Isoform 1 (archvillin) is muscle-specific and is important in myogenesis and the assembly of myosin II. Isoform 2 (supervillin) is more ubiquitously expressed, somewhat smaller (1788aa), and may down-regulate focal adhesions by binding to thyroid receptor-interacting protein 6 (**TRIP6**).

suppression PCR A modification of **PCR**, competition (or suppression) hybridization in which there is selective suppression of the amplification of one or more closely related sequences, useful when the target is present as a rare sequence in a large excess of a closely related sequence. A 5′ extension is included in one (or both) primer(s) that will form a hairpin loop with internal sequence of the amplicon and is then refractory to further amplification. Sequences containing a perfect match to the 5′extension are not amplified (are suppressed) and only sequences containing mismatches or lacking the sequence is unaffected. Descriptive article: http://www.ncbi.nlm.nih.gov/pmc/articles/PMC1184225/

suppression subtractive hybridization *SSH* A method developed for cDNA comparisons, based on PCR suppression by inverted terminal repeats (PS-effect). In complex mixtures, the PS-effect allows precise amplification only of molecules that are flanked by different adapters at opposing termini (asymmetrically flanked molecules). This principle is used in Suppression Subtractive Hybridization (SSH), where the molecules of interest, those unique to the test sample, are driven to the asymmetrically flanked state and selectively amplified.

suppressor factor (1) Factors released by T-suppressor cells (**T-regulatory cells**). (2) See **suppressor mutation** and **termination codons**.

suppressor mutation Mutation that alleviates the effect of a primary mutation at a different locus. May be through almost any mechanism that can give a primary mutation, but perhaps the most interesting class are the **opal**, **amber** and ochre suppressors, where the anticodon of the tRNA is altered so that it misreads the termination codon and inserts an amino acid, preventing premature termination of the peptide chain.

suppressor T-cells See **T-regulatory cells**.

suprachiasmatic nucleus *SCN* A region in the hypothalamus, immediately above the optic chiasm. The SCN generates a circadian rhythm of neuronal activity, which regulates many different body functions. **Melanopsin**-containing ganglion cells in the retina have a direct connection to the SCN.

SUR See **sulphonylurea receptor**.

suramin A polysulphonated naphthylurea that uncouples G-proteins from receptors, inhibits phospholipase D and inhibits binding of EGF, PDGF to cell surface receptors. An antagonist of P_{2x} and P_{2y} purinergic receptors. Has been used in treatment of trypanosomiasis.

SURF complex A mRNA surveillance complex, composed of eukaryotic **release factors** 1 and 3 (ERF1, ERF3A or 3B), elongation factor (EEF2), regulator of nonsense transcripts 1 (UPF1/RENT1) and the **SMG1C complex** that includes the serine/threonine-protein kinase SMG1.

surface plasmon resonance Alteration in light reflectance as a result of binding of molecules to a surface from which total internal reflection is occurring. Used in the Biacore (Pharmacia Trademark) machine that detects the binding of ligand to surface-immobilised receptor or antibody.

surface potential The electrostatic potential due to surface charged groups and adsorbed ions at a surface. It is usually measured as the **zeta potential** at the Helmholtz slipping plane outside the surface.

surface-active compound Usually, in biological systems, a detergent-like molecule that is amphipathic and that will bind to the plasma membrane, or to a surface with which cells come in contact, altering its properties from hydrophobic to hydrophilic, or *vice versa*.

surfactant A **surface-active compound**; the best known example of which is the lung surfactant (see **surfactant proteins**).

surfactant proteins Pulmonary surfactant is mainly (90%) lipid associated with four surfactant-associated proteins: 2 collagenous, carbohydrate-binding glycoproteins (SP-A and SP-D of the **collectin** family) and 2 small hydrophobic proteins (SP-B and SP-C). Production of surfactants at parturition allows the lungs to fill with air because the alveolar surface becomes hydrophobic. SP-A is oligomeric, consisting of 18 protomers (248aa) with lectin-like domains that recognize glycoconjugates, lipids and protein determinants on both host cells and invading microorganisms. SP-B (381aa) is a disulphide-linked homodimer. SP-C (197aa) is defective in a form of pulmonary surfactant metabolism dysfunction and variations are associated with respiratory distress syndrome in premature infants. SP-D (375aa), an oligomeric complex of 4 homotrimers, contributes to the lung's defense against inhaled microorganisms. Binds maltose residues and other alpha-glucosyl moieties. It also occurs at the luminal surface of the gastric mucosa.

surfactin A lipopeptide biosurfactant (a cyclic heptapeptide coupled to an C13−C15 acyl residue) produced by surfactin synthase (a heterotrimeric enzyme, 3587aa, 3583aa and 1275aa) in *Bacillus subtilis*. Because it can be produced from cheap feedstock by biofermentation it is being investigated for use in bioremediation. It has antibiotic properties. Swarming motility protein swrC (1052aa) confers self-resistance to surfactant on *B. subtilis*.

surrogate marker A readily measured parameter that is associated with a disease state and that can be monitored as a substitute for more complex clinical signs that may be less easy to quantify.

survivin See **BIRCs**.

Sus scrofa Domestic pig.

sushi domains Domains (complement control protein (CCP) modules, short consensus repeats (SCR), ~60aa) identified in a wide variety of complement and adhesion proteins; involved in protein-protein and protein-ligand interactions. Contain four cysteines forming two disulphide bonds in a 1−3 and 2−4 pattern. See **sushi peptides**.

sushi peptides Peptides derived from the lipopolysaccharide (LPS)-binding domains of an LPS-sensitive serine peptidase, Factor C, from the horseshoe crab (*Carcinoscorpius rotundicauda*); have potent antibacterial properties. See **sushi domains**.

suspensor cell Plant cell linking the growing embryo to the wall of the embryo sac in developing seeds.

sustentacular Something that supports or maintains. Sustentacular cells (Sertoli cells) of the testis are involved in support and possibly nutrition of developing sperm.

suxamethonium See **succinylcholine**.

SUZ12 A **polycomb group** protein (suppressor of zeste 12 protein homologue, 739aa) that is a component of PRC2/EED-EZH complexes which methylate histone H3. See **zeste**, **EZH**.

SV2 (1) An integral membrane glycoprotein of synaptic vesicles, SV2 has several isoforms (A, 742aa; B and C) and is found in vertebrate neuronal and endocrine tissues, where it positively regulates vesicle fusion by maintaining the readily releasable pool of secretory vesicles. Interacts with synaptotagmin-1 in a calcium-dependent manner and binds adapter protein AP-2. Similar proteins have been found in invertebrates. Botulinum neurotoxin A binds to SV2 and is internalised. SV2 related protein (SVOP, 548aa) is probably an anion transporter (SLC22 family). (2) A line of immortal nontumorigenic human lung fibroblasts, MRC-5 SV2 cells.

SV3T3 Swiss 3T3 cells transformed with **SV40**.

SV40 Simian virus 40. A small DNA **tumour virus**, member of the **Papovaviridae**. Isolated from monkey cells, which were being used for the preparation of **poliovirus** vaccine, and originally named 'vacuolating agent' owing to a cytopathic effect observed in infected cells. Found to induce tumours in newborn hamsters. In culture, transforms the cells of many non- and semi-permissive species, including mouse and human. See also **T-antigen**.

S value See **Svedberg unit** and **sedimentation coefficient**.

Svedberg unit The unit named after a pioneer of the ultracentrifuge that is applied to the sedimentation coefficient of a particle in a high-speed or ultracentrifuge. The unit S is calculated as follows, $S = $ rate of sedimentation $\times 1/\rho^2 r$, where ρ is the speed of rotation in radians per second and r is the radius to a chosen point in the centrifuge tube. 1 Svedberg unit is defined as the time taken to travel unit distance for a particle under the influence of unit centrifugal acceleration, 10^{-13} seconds. The units are non-additive: a particle formed from two 5S particles will not have a sedimentation coefficient of 10S.

swainsonine Fungal indolizidine alkaloid that is a reversible inhibitor of lysosomal alpha-**mannosidase** and of the Golgi complex alpha-mannosidase II that is involved in processing the oligosaccharide chains of glycoproteins.

SWAM In mouse an antibacterial protein (SWAM1, single WAP motif protein 1, elafin-like protein 1, 80aa; SWAM2, 85aa).

swat A program for searching one or more DNA or protein query sequences, or a query profile, against a sequence database, using an efficient implementation of the Smith-Waterman or Needleman-Wunsch algorithms. Link to website: http://www.phrap.org/phred-phrapconsed.html

SWI/SNF complex The SWI/SNF (Switch/Sucrose non-fermentable) complex remodels nucleosome structure in an ATP-dependent manner. In yeast the SWI/SNF chromatin remodelling complex has 11 tightly associated polypeptides (SWI1, SWI2, SWI3, SNF5, SNF6, SNF11, SWP82, SWP73, SWP59, SWP61, and SWP29). SWP59 and SWP61, encoded by the ARP9 and ARP7 genes respectively, are members of the actin-related protein (ARP) family. The similarity of ARP7 and ARP9 proteins to the **heat-shock protein** and HSC family of ATPases suggests that chromatin remodelling may involve chaperone-like activities. See **BAF complex**, **RSC complex**.

Swiss 3T3 cells An immortal line of fibroblast-like cells established from whole trypsinised embryos of Swiss mice (not an inbred stock) under conditions that favour establishment of cells with low saturation density in culture.

switch regions The nucleotide sequences in heavy chain immunoglobulin genes located in the introns at the 5′ end of each CH locus concerned with DNA recombination events that lead to changes in the type of heavy chain produced by a B-cell, e.g. IgM to IgG switching. These regions are highly conserved. See **isotype switching**.

Sybr green A cyanine dye that binds to DNA forming a complex that absorbs at 488 nm and emits green light (λmax 522 nm). A much safer dye that has replaced ethidium bromide. Extensively used in quantitative PCR and in general gel staining.

Sydenham's chorea A condition, 'Saint Vitus' Dance' characterized by rapid, uncoordinated jerking movements (chorea). Caused by childhood infection with Group A beta-haemolytic streptococci (which also cause rheumatic fever) that leads to an autoimmune response that destroys cells in the basal ganglia.

SYFPEITHI A database comprising more than 7000 peptide sequences known to bind class I and class II MHC molecules. Link to database: http://www.syfpeithi.de/

syk A non-receptor tyrosine kinase (spleen tyrosine kinase, 630aa), an effector of the B-cell receptor signalling pathway. Contains two tandem SH2 domains through which it interacts with the **ITAM** motif. More widely distributed than **zap70** and important in signalling in both myeloid and lymphoid cells.

symbiont One of the partners in a symbiotic relationship.

symbiosis Living together for mutual benefit. See **symbiont, symbiotic algae**.

symbiotic algae Algae (often *Chlorella* spp.) that live intracellularly in animal cells (e.g. endoderm of *Hydra viridis*). The relationship is complex, because lysosomes do not fuse with the vacuoles containing the algae, and the growth rates of both cells are regulated to maintain the symbiosis. There is considerable strain-specificity. The term is imprecise, since there are many other symbiotic algae (as in lichens) where the relationship is different.

sympathetic nervous system One of the two divisions of the vertebrate **autonomic nervous system** (the other being the **parasympathetic nervous system**). The sympathetic preganglionic neurons have their cell bodies in the thoracic and lumbar regions of the spinal cord, and connect to the paravertebral chain of sympathetic ganglia. Innervate heart and blood vessels, sweat glands, viscera, and the adrenal medulla. Most sympathetic neurons, but not all, use noradrenaline as a post-ganglionic neurotransmitter.

sympathetic ophthalmia An autoimmune response in which there is inflammatory disease in the undamaged eye following perforating injury to the other eye.

sympathomimetic Mimicking the sympathetic nervous system. Sympathomimetic drugs mimic the stimulation of the sympathetic nervous system to produce, for example, tachycardia and increased output from the heart or bronchodilatation and vasodilatation.

sympatric speciation Development of a new species without (geographical) isolation. Controversial. See **speciation**.

Symphyta Symphyta (sawflies) are a sub-order of the Hymenoptera. Website: http://www.earthlife.net/insects/symphyta.html

symplast The intracellular compartment of plants, consisting of the cytosol of a large number of cells connected by **plasmodesmata**. See **apoplast**.

symplectic metachronism See **metachronism**.

symplesiomorphic See **apomorphic**.

symport A mechanism of transport across a membrane in which two different molecules move in the same direction. Often, one molecule can move up an electrochemical gradient because the movement of the other molecule is more favourable (see **facilitated diffusion**). Example: the sodium/glucose cotransport. See **antiport**, **uniport**.

synaphin See complexin.

synapomorph **Apomorphic** features possessed by two or more taxa in common. Phylogenic trees are based upon identifying groups united by synapomorphies.

synapse A connection between **excitable cells**, by which an excitation is conveyed from one to the other. (1) Chemical synapse: one in which an **action potential** causes the exocytosis of neurotransmitter from the presynaptic cell, which diffuses across the synaptic cleft and binds to **ligand-gated ion channels** on the postsynaptic cell. These ion channels then affect the resting potential of the postsynaptic cell. (2) Electrical synapse: one in which electrical connection is made directly through the cytoplasm, via **gap junctions**. (3) Rectifying synapse: one in which action potentials can only pass across the synapse in one direction (all chemical and some electrical synapses). (4) Excitatory synapse: one in which the firing of the presynaptic cell increases the probability of firing of the postsynaptic cell. (5) Inhibitory synapse: one in which the firing of the presynaptic cell reduces the probability of firing of the postsynaptic cell. (6) See **immunological synapse**.

synapsins Family of phosphoproteins that coat synaptic vesicles and bind to the cytoskeleton. Synapsin Ia and Ib (Brain protein 4.1, 705aa) are alternatively spliced variants as are synapsins IIa and IIb (582aa). Synapsin III (580aa) has a more restricted distribution. The synapsins are implicated in neurotransmitter release, axonogenesis and synaptogenesis. Can be phosphorylated by several **protein kinases** and interact with small G proteins such as **rab**. Synapsin 1 forms a ternary complex with nitric oxide synthase and **CAPON**. Mutations in synapsin 1 are associated with a form of X-linked epilepsy.

synapsis (1) The specific pairing of the chromatids of homologous chromosomes during **prophase** I of meiosis. It allows **crossing-over** to take place. (2) Process that brings the ends of double-strand breaks in DNA together, prior to end joining in **NHEJ**. Synapsis results in the autophosphorylation of DNA-PKcs, which is required to make the DNA ends available for ligation.

synaptic cleft The narrow space between the presynaptic cell and the postsynaptic cell in a chemical **synapse**, across which the **neurotransmitter** diffuses.

synaptic facilitation See **facilitation**.

synaptic plasticity Change in the properties of a synapse, usually in the context of learning and memory. Very few synapses provide simple 1:1 transfer of **action potentials**, and very small changes in the efficiency of a synapse (usually mediated by changes in either the pre- or postsynaptic membrane) can have profound influences on the electrical properties of a neuronal circuit. See also **neuronal plasticity**.

synaptic transmission The process of propagating a signal from one cell to another via a **synapse**.

synaptic vesicle Intracellular vesicles containing **neurotransmitter** found in the presynaptic terminals of chemical synapses.

synaptobrevins Small integral membrane proteins of synaptic vesicles that have v-SNARE coiled-coil homology domains and interact with target-**SNARES**. Two neuronal isoforms, VAMP1 (118aa) and VAMP2 (116aa) are known that bind **SNAPs** and **syntaxin**. Clostridial toxins encoding zinc endopeptidases, such as tetanus toxin and botulinum toxin, block synaptic release by degrading synaptobrevins. Other related proteins occur in other tissues serving a similar role in vesicle targeting through interactions with SNAREs. VAMP3 (100aa) is **cellubrevin** and involved in protein recycling. VAMP4 (141) is involved in removing an inhibitor (probably synaptotagmin-4) of calcium-triggered exocytosis during the maturation of secretory granules. VAMP5 (myobrevin, 102aa) is found in muscle. VAMP7 (synaptobrevin-like protein 1, 220aa) is ubiquitous and involved in endosome-lysosome transport. VAMP8 (endobrevin, 100aa) is associated with dense-granule secretion in platelets. VAMP-like proteins are found in a wide range of organisms, including protozoa, yeasts and *Arabidopsis*.

synaptogenesis Formation of a **synapse**.

synaptogyrin A family of proteins involved in synaptic plasticity (synaptogyrin 1, SYNGR1, 233aa; SYNGR2, (cellugyrin), 224aa, SYNGR3, 229aa; SYNGR4 234aa). Synaptogyrin-3 is involved in the positive regulation of dopamine transporter activity.

synaptojanins Inositol 5-phosphatases involved in vesicle recycling; the amino-terminal region has homology with yeast Sac1 (involved in phospholipid metabolism) and the C-terminal region has proline-rich sequences that probably interact with **SH3** domains of **amphiphysin** and **grb-2**. Synaptojanin-1 (EC 3.1.3.36, 1573aa) is involved in clathrin-mediated endocytosis in presynaptic terminals through an interaction with **dynamin**, **Eps15** and amphiphysin. Synaptojanin-2 (1496) has a much wider tissue distribution but a similar role in vesicle recycling. Synaptojanin-2-binding protein (145aa) is a mitochondrial outer-membrane protein.

synaptomorphic See **apomorphic**.

synaptonemal complex Structure, identified by electron microscopy, lying between chromosomes during **synapsis**; consists of two lateral plates closely apposed to the chromosomes and connected to a central plate by filaments. It appears to act as a scaffold, but is not apparently essential for **crossing-over**. In *S. cerevisiae*, a mutation eliminating the synaptonemal complex protein, Zip1 (875aa), increases the meiosis I nondisjunction rate of nonexchange chromosomes, suggesting it is a backup system for those chromosome where recombination has not occurred. Various

synaptonemal proteins have been identified (SYCP1, 976aa; SYCP2, 1,530aa; SYCP3, 236aa and synaptosomal central element proteins, Syce 1, 329aa and Syce2, 171aa). Research article: http://www.pnas.org/content/107/2/781.full.pdf + html

synaptophysin Abundant glycoprotein component (major synaptic vesicle protein p38, 313aa) of small synaptic vesicle with four transmembrane domains; both N- and C-termini are located cytoplasmically. The transmembrane organization and putative quaternary structure resemble that of **connexins**. Synaptophysin-like protein 2 (mitsugumin-29, 264aa) is involved in communication between the T-tubular and junctional sarcoplasmic reticulum (SR) membranes. See **synaptoporin**.

synaptopodin An actin-associated proline-rich protein (929aa) found in kidney podocytes and a subset of mature telencephalic dendritic spines of neurons; a regulator of rhoA signalling and cell migration. Blocks Smurf1-mediated ubiquitination of rhoA, thereby preventing the targeting of rhoA for proteasomal degradation. May be involved in synaptic plasticity in telencephalic neurons. Synaptopodin-2 (myopodin, 1093aa) is skeletal muscle-specific and localized to the Z-disc in mature muscle. Shuttles between the nucleus and the cytoplasm in a differentiation-dependent and stress-induced fashion.

synaptoporin *SPO* Putative channel protein (265aa) of synaptic vesicles, and a member of the **synaptophysin/connexin** superfamily. It has 58% amino acid identity to synaptophysin, with highly conserved transmembrane segments but a divergent cytoplasmic tail.

synaptosome A subcellular fraction prepared from tissues rich in chemical **synapses**, used in biochemical studies. Consists mainly of vesicles from presynaptic terminals.

synaptotagmins Calcium-binding synaptic vesicle proteins that bind acidic phospholipids and recognize the cytoplasmic domain of the **neurexins**. Function as Ca^{2+} sensors that facilitate SNARE-mediated membrane fusion. Synaptotagmin-1 (SYT1, p65, 422aa) is only one of a large family (at least 15 members) that probably differ subtly in their binding partners. SYT1 forms a complex with SV2B (synaptic vesicle glycoprotein 2B), **syntaxin 1** and **SNAP25**; SYT2 (419aa) interacts with stonin-2 and SCAMP5 (**secretory carrier-associated membrane protein-5**). Similar proteins are found in many metazoa.

SynCAMs Synaptic cell adhesion molecules. Immunoglobulin superfamily adhesion molecules of the **nectin** subfamily (for example, syncam-1, cell adhesion molecule 1, nectin-like protein 2, 442aa) involved in calcium-independent homo- and heterotypic cell adhesion, not restricted to neuronal cells. Syncam 1 is implicated in presynaptic differentiation along with β-**neurexin/neuroligin**. It also interacts with poliovirus receptor-related protein 3 (**nectin-3**) and is involved in a wide range of other activities, acting as a tumour suppressor, promoting natural killer cell cytotoxicity amongst other things.

synchronous cell population A culture of cells that all divide in synchrony. Particularly useful for certain studies of the cell cycle, cells can be made synchronous by depriving them of essential molecules, which are then restored. Synchronization breaks down after a few cycles, however, as individual cells have unique division rates.

syncoilin An intermediate filament-type III protein (482aa) found in striated and cardiac muscle where it binds to **desmin** and to α-**dystrobrevin** and thereby to the **dystrophin-associated protein complex**.

syncollin Protein (134aa) found within zymogen granules that is required for efficient regulated exocytosis. Normally exists as a doughnut-shaped homo-oligomer (possibly a hexamer) in close association with the luminal surface of the zymogen granule membrane.

syncytin Membrane proteins encoded by **human endogenous retroviruses** genes that have retained their original fusogenic properties and participate in trophoblast fusion (formation of the **syncytiotrophoblast**) during placental morphogenesis. Syncytin 1 (538aa), the product of a HERV-W gene, is cleaved into a surface protein (SU, gp50, 297aa) that mediates recognition of the type D mammalian retrovirus receptors SLC1A4 and SLC1A5, and a transmembrane protein (TM, gp24, 221aa) that acts a class I viral fusion protein. SU and TM form a heterodimer. Syncytin 2 (product of a HERV-FRD gene, 538aa) is similar.

syncytiotrophoblast Syncytial layer that forms the outermost fetal layer in the placenta and is thus the interface with maternal tissue. Has invasive capacity – though in a regulated manner. See **syncytin**.

syncytium An **epithelium** or tissue in which there is cytoplasmic continuity between the constituent cells. Striated muscle is the classic example.

syndactyly A relatively common congenital abnormality in which the webbing between finger or toes persists because of a failure in the normal developmental programmed death of inter-digital tissue. Can be caused by mutation variously in genes encoding **connexin-43**, homeobox HOXD13, **fibulin-1**, **low density lipoprotein receptor-related protein** 4 or regulatory elements of the sonic hedgehog system. Syndactyly is associated with several other disorders although the mutations responsible are unclear.

syndecans Integral membrane proteoglycans associated largely with epithelial cells. The core protein has an extracellular domain that is modified by addition of multiple heparan sulphate, chondroitin or dermatan sulphate chains and N-linked oligosaccharide. The heparan sulphate chains bind to proteins of the extracellular matrix, including collagens, fibronectin and **tenascin**. The cytoplasmic domain is thought to interact with actin filaments thus linking the cytoskeleton with fibrous elements of the extracellular matrix. Members of the family include syndecan-1 (CD138, 310aa), syndecan-2 (fibroglycan, heparan sulphate proteoglycan core protein, 201aa). Syndecan-3 (442aa) may have a role in signalling and binding of heparin-binding growth-associated molecule increases phosphorylation of c-src and **cortactin**. Syndecan-4 (amphiglycan, ryudocan core protein, 198aa) interacts with CUB domain-containing protein 1, the adaptor proteins syntenin-1 and -2 (syndecan binding protein, SDBP1, melanoma differentiation-associated protein 9, scaffold protein Pbp1, 298aa; SDBP2, 292aa) and various other proteins. Syndecans bind gp120 of HIV and mediate viral entry to the host cell. The *Drosophila* protein (399aa) is required for axonal and myotube guidance and is a necessary component in slit/robo signalling.

syne proteins See **nesprins**.

Synechocystis A freshwater cyanobacterium capable of both phototrophic growth in sunlight and heterotrophic growth in darkness. *Synechocystis sp.* PCC6803 is well studied and exhibits circadian rhythms.

Synedra A genus of freshwater diatoms. Details: http://craticula.ncl.ac.uk/EADiatomKey/html/Synedra.html

synemin A typeVI intermediate filament isolated from avian smooth muscle (1604aa in *Gallus gallus*), but a homologue (synemin, desmuslin, 1565aa) is also found in mammalian muscle. Forms heteropolymeric IFs with desmin and/or vimentin that link to adherens-type junctions. Three synemin isoforms, (339aa, 1251aa and 1563aa) are produced by alternative splicing and have different tissue distributions.

synergids Cells of the egg apparatus in plants; two cells at the apex of the embryo sac closely associated with the egg cell and directly involved with the process of fertilisation.

Synergistetes A phylum of bacteria; Gram-negative, rod-shaped obligate anaerobes. Some species have been implicated in periodontal disease, gastrointestinal infections and soft tissue infections.

synergy An interaction that produces a more-than-additive effect. Demonstrating a synergistic effect requires that the second compound produces an additional effect in the presence of a maximally-effective level of the other. This criterion is not always met.

synexin Annexin 8. See Table A2.

syngamy Fusion of two haploid gametic nuclei to form the diploid nucleus of the zygote.

syngeneic Organisms that are antigenically identical – monozygotic twins or highly inbred strains of animals. Cells or tissues introduced into a syngeneic host will not be rejected because they are histocompatible (do not differ in histocompatibility antigens).

synkaryon A somatic hybrid cell in which chromosomes from two different parental cells are enveloped in a single nucleus.

synoecious Describing the condition in which there are both male and female flowers in a single composite inflorescence or, in the case of bryophytes, having male and female organs together on a branch.

synomone See **allomone**.

synoviocytes Fibroblastic cells of the synovial membrane (lining) that produce synovial fluid and the extracellular matrix of the bearing surface of the joint.

synovium Connective tissue that forms the bearing surface of the joint and that is eroded in arthritis.

synphilin A neural protein (alpha-synuclein-interacting protein, synphilin-1, 919aa with alternatively spliced isoforms). Isoforms 1 and 2 form a heterodimer: isoform 2 inhibits E3 ubiquitin-protein ligase (SIAH1) and inhibits proteasomal degradation of target proteins. Can be ubiquitinylated by **parkin** to produce cytoplasmic inclusions resembling **Lewy bodies**. Defects may cause **Parkinson's disease.** See **synuclein**.

synpolydactyly See **syndactyly**.

syntaxins A large family of receptors (t-SNAREs) for intracellular transport vesicles that bind **synaptotagmin** and **taxilins** in a calcium-dependent fashion and interact with voltage dependent calcium and potassium channels. Syntaxins have a hydrophobic C-terminal region that anchors the protein on the cytoplasmic surface of cellular membranes, a central conserved region and a variable N-terminal cytoplasmic domain. Syntaxin 1 (288aa) is mainly expressed in brain tissue and is thought to function specifically in neurotransmitter release, whereas syntaxin 2, 3, and 4 are involved in more general vesicle trafficking. In *Arabidopsis*, syntaxin 121 (syntaxin-related protein At-Syr1, 346aa) functions in the secretory pathway and binds to SNAP33.

syntenic Describing genes that lie on the same chromosome. Some loci are syntenic in both human and mouse, others are not.

syntenin See **syndecans**.

synthetase Enzymes of Class 6 in the **E classification**; catalyse synthesis of molecules, their activity being coupled to the breakdown of a nucleotide triphosphate.

syntrophins A family of adapter proteins that interact with **dystrophin, dystrobrevin** and diacylglycerol kinase (DGK)-zeta at the plasma membrane of muscle and nerve and may link receptors to the actin cytoskeleton. Alpha-1-syntrophin-alpha (syntrophin-1, Pro-TGFα cytoplasmic domain-interacting protein 1, TACIP1, 505aa) is mainly expressed in skeletal muscle and heart, and is important in synapse formation. Beta-1-syntrophin (syntrophin-2, 538aa) and beta-2-syntrophin (syntrophin-3, 540aa) are ubiquitous and interact with the other members of the syntrophin family. Gamma-1-syntrophin (syntrophin-4, 517aa) is brain-specific but gamma-2-syntrophin (syntrophin-5, 539aa) is widely expressed. See **dystrophin-associated protein complex**.

synuclein Family of proteins (α-synuclein, 140aa; β-, 134aa; γ-, 127aa) with structural resemblance to apolipoproteins, abundant in neuronal cytosol and particularly in presynaptic terminals where they may be involved in the regulation of dopamine release. In **Alzheimer's disease** α-synuclein is a component of plaque amyloid; mutations in the α-synuclein gene are the cause of various forms of **Parkinson's disease** (types 1 and 4) and the protein accumulates in **Lewy bodies**. Interacts with **synphilin**.

syphilis A contagious sexually-transmitted disease caused by infection with the spirochaete *Treponema pallidum*. Can also be vertically transmitted from mother to fetus.

syringomycin A phytotoxic lipopeptide produced by *Pseudomonas syringae* that form pores in plasma membranes and is also a powerful surfactant. **Syringopeptin** has similar effects.

syringopeptin A phytotoxic cyclic lipodepsipeptide produced by *Pseudomonas syringae* that oligomerizes to form pores in cell membranes.

syringyl alcohol A phenylpropanoid alcohol (sinapyl alcohol), one of the three precursors of lignin.

systematic sclerosis See **scleroderma, fibrillarin**.

systemic Describing something with global rather than local effects: systemic insecticides, for example, become distributed throughout a plant, systemic diseases affect the whole body.

systemic acquired resistance *SAR* In plants, long-term resistance to pathogen attack, triggered by elicitors released by the pathogen (e.g. bacteria). In contrast to the hypersensitive response, which is a localized response involving cell death at the site of infection, SAR is displayed by the whole plant. SAR is associated with the expression and activity of a set of SAR genes which include the pathogenesis-related (PR) genes.

systemic carnitine deficiency A condition in which **carnitine** levels are low in non-muscle tissue. Mutations in the SLC22A5 gene, which encodes the sodium ion-dependent carnitine transporter (OctN2), cause a failure of reabsorption of carnitine in the kidney. Myopathic carnitine deficiency is caused by a different mutation and is restricted to skeletal muscle.

systemic lupus erythematosus *SLE* An autoimmune disease with antinuclear and other antibodies (often anti-RNA) in plasma. Immune complex deposition in the glomerular capillaries is a particular problem. Multiple genes affect susceptibility, mostly genes that regulate the immune suystem although resistance is associated with a polymorphism in the **Toll-like receptor-5** gene. See **lupus erythematosus**.

systemin An 18aa wound hormone released from sites on tomato leaves damaged mechanically or by insects. Through the **systemin receptor** regulates more than 20 defensive genes. Produced, by proteolytic cleavage, from prosystemin (200aa).

systemin receptor A leucine-rich repeat (LRR) receptor with serine/threonine-protein kinase activity in *Solanum peruvianum* (Peruvian tomato) (EC 2.7.11.1, brassinosteroid LRR receptor kinase, 1207aa). Almost identical to the brassinosteroid receptor in *Solanum lycopersicum* (tomato) although **brassinosteroid** and **systemin** bind to different sites. The receptor regulates an intracellular cascade including, depolarization of the plasma membrane, the opening of ion channels, an increase in intracellular Ca^{2+}, activation of a **MAP kinase** activity and a phospholipase A2 activity. As a consequence linolenic acid, is released from plant membranes and converted to **jasmonic acid**.

systems biology Fashionable term for an integrated approach to biology in which effects on the whole organism are studied as opposed to the reductionist approaches of modern molecular bioscience. This Dictionary deliberately contains definitions that span a wide range of biological sub-specialities since molecular cell biologists are generally well aware of the importance of putting their work into context, but may need a glossary for some of the more arcane terminologies.

syzygy (1) In some parasitic protozoa the pairing of gamonts prior to sexual fusion. (2) In gregarines the end-to-end attachment of the sporonts. (3) In some crinoids the fusion of organs or skeletal elements.

T

T Single letter code for threonine (in proteins) or thymine (in DNA).

T3, T4 See **thyroid hormones**.

T7 A lytic T-odd bacteriophage that infects *E.coli*, the prototype of a group of virulent phages that have an icosahedral head and short, stubby tail to which are attached 6 tail fibers The genome (39,937 bp) contains 56 genes that encode 59 known proteins. Genes are grouped into three classes: Class I genes, which enter the host cell and are transcribed first, moderate the transition in metabolism from host to phage. Class II genes are primarily responsibly for T7 DNA replication, and class III genes code for particle, maturation, and packaging proteins. The T7 promoter consists of a highly conserved 23 bp sequence that is selective for the phage T7 RNA polymerase; used in the **pET Expression System**.

T7 polymerase A DNA-directed RNA polymerase (EC 2.7.7.6, 883aa) in **T7** phage that catalyzes the synthesis of RNA in the 5′ to 3′ direction very specifically in the presence of a DNA template containing a T7 promoter. There is also a DNA-dependent DNA polymerase (EC 2.7.7.7, 704aa) composed of two subunits, one encoded by the phage and the other encoded by the host.

TAARs A family of G-protein-coupled receptors (trace amine-associated receptors) for trace amines (TAs) such as p-tyramine and beta-phenylethylamine, although most of them do not have known ligands. May be involved in pheromone-like signalling and are surprisingly different even between closely-related species (e.g. man and chimpanzee).

TA cloning Cloning strategy for **PCR** products that relies on the tendency of **Taq polymerase** to add an extra dA at the 3′ end of newly synthesized DNA strands, thus leaving a single base 3′ overhang. Vectors are accordingly prepared with single base dT 3′ overhangs, allowing ligation of **sticky ends**.

tabtoxin A phytotoxic compound (an unusual beta-lactam antibiotic) produced by *Pseudomonas syringae* that is potently antimicrobial and functions by inhibiting glutamine synthetase. Tabtoxin causes wildfire disease of tobacco. Tabtoxin resistance protein (TTR) is an acetyltransferase (EC 2.3.1.-, 177aa) that renders tabtoxin-producing pathogens tolerant to the toxin.

Tacaribe complex Group of 8 **Arenaviridae**, isolated from bats in S. America, that cause severe haemorrhagic disease in humans.

TACC proteins Transforming, acidic, coiled-coil-containing proteins, a family of proteins originally described in mammals although a *Drosophila* homologue is known (**D-TACC**, 1226aa). The function of the TACC proteins is unknown, but the genes encoding the known TACC proteins are all associated with genomic regions that are rearranged in certain cancers. TACC1 (taxin1, 805aa) is weakly concentrated at centrosomes during mitosis. TACC2 (anti zuai-1, AZU-1, 2948aa) is involved in microtubule-dependent coupling of the nucleus and the centrosome and may be a tumour suppressor that promotes correct tissue morphogenesis. Both TACC1 and TACC2 contain a **SPAZ domain** (Ser/Pro-rich AZU-1 domain). TACC3 (ERIC1, 838aa) also interacts with microtubules. It is up-regulated in various cancer cell lines. TACC2: http://www.ncbi.nlm.nih.gov/pubmed/20335520

TACE A zinc metallopeptidase (TNFα converting enzyme, ADAM17, EC 3.4.24.-, 827aa) of the **ADAM** family involved in processing of TNFα and releasing the soluble cytokine from the inactive membrane-bound precursor. Also mediates the cleavage and shedding of **fraktalkine** (CX3CL1). See **notch**.

tachycardia An abnormal increase in heart-rate. Ventricular tachycardia can progress to ventricular fibrillation and sudden death. Atrial tachycardia can occur without the ventricular rate being affected.

tachykinins A group of neuropeptide hormones derived from **protachykinin**, including **substance P**, substance K (neurokinin A) and neurokinin B in mammals, eledoisin from *Octopus* and physalaemin (amphibian). All have 10 or 11 residues with a common -FXGLM-NH$_2$ ending. Elicit a wide range of responses from neurons, smooth muscle, endothelium, exocrine glands and cells of the immune system; effects similar in many ways to **bradykinin** and **serotonin**. Abstract of review: http://pharmrev.aspetjournals.org/content/54/2/285.abstract

tachyphylaxis A decrease in the response to an agonist following repeated exposure. Can arise through a variety of mechanisms.

tachyplesin See **polyphemusin**.

Tachypleus tridentatus The Japanese horseshoe crab or Chinese horseshoe crab, one of the Limulidae family, similar in many respect to *Limulus polyphemus*.

tachyzoite An asexual stage of rapid growth in the tissue phase of certain coccidial infections such as *Toxoplasma gondii*. The proliferation occurs in parasitophorous vacuoles in the infected cells.

The Dictionary of Cell and Molecular Biology. DOI: http://dx.doi.org/10.1016/B978-0-12-384931-1.00020-9

tacrine A reversible acetylcholinesterase inhibitor used in the treatment of Alzheimer's disease (TN Cognex). It also inhibits histamine N-methyltransferase.

tacrolimus *FK506* An important immunosuppressant drug that inhibits T-cell activation. Originally derived from *Streptomyces tsukubaensis*. Binds to **FKBP-12**.

TAF (1) **TATA-binding protein**-associated factors. TAF genes encode components of the transcription factor IID (**TFIID**) complex, PCAF histone acetylase complex and TBP-free TAFII complex (TFTC). The TAF1 gene encodes TFIID subunit 1 (EC 2.7.11.1, p250, 1872aa), the largest component and core scaffold of the transcription factor complex and has domains with Ser/Thr kinase activity. TAF2 encodes subunit 2 (1199aa), TAF3 encodes sbunit 3 (140 kDa TATA box-binding protein-associated factor, TAF140, 929aa). There are at least 15 TAF genes encoding eponymous factors. (2) See **tumour angiogenesis factor**.

TAG-1 (1) Transient axonal glycoprotein. A surface glycoprotein (contactin-2, 1040aa) that is expressed transiently on commissural and **motoneurons** in developing vertebrate nervous system. TAG-1 (**axonin-1** in chicken, **tax-1** in rat) and **L1** have been shown to be on different segments of the same embryonic spinal axons. (2) Triacylglycerol lipase (EC 3.1.1.3, 499aa). (3) In *E. coli* a gene (*tag*) encoding DNA-3-methyladenine glycosylase 1 (EC 3.2.2.20, 187aa) that excises 3-methyladenine from DNA damaged by alkylating agents.

tagaturonate An intermediate (5-ketogluconate) in the galacturonate catabolism pathway. It is reportedly an activator of mannonate dehydratase. Tagaturonate reductase is an oxidoreductase (altronate oxidoreductase, EC 1.1.1.58, 483aa) that catalyzes the reaction: D-altronate + NAD$^+$ = D-tagaturonate + NADH. Article: http://mic.sgmjournals.org/cgi/reprint/144/11/3111.pdf

tagma *Pl*. tagmata. A distinct section of an arthropod, consisting of two or more adjoining segments, e.g. the thorax of an insect.

tagmosis The grouping or fusion of somites to form definite regions (tagmata) in a metameric animal.

taicatoxin Complex oligomeric protein toxin from *Oxyuranus scutelatus scutelatus* (taipan snake). Blocks high- but not low-threshold calcium channels of heart muscle. The oligomer contains a neurotoxin-like peptide (8 kDa), a phospholipase (EC 3.1.1.4, probably **taipoxin**, 16 kDa) and a serine-peptidase inhibitor (88aa).

taipoxin Heterotrimeric toxin from *Oxyuranus scutelatus scutelatus* (taipan snake). All three subunits (α, β, γ) have homology with pancreatic phospholipase A2. Binds to neuronal **pentraxins** and blocks transmission at the neuromuscular junction.

TAIR The Arabidopsis Information Resource, a comprehensive resource about *Arabidopsis* that incorporates information about the structure and organization of the *Arabidopsis* genome and the functions of its estimated 29,000 genes. Weblink: http://www.arabidopsis.org/help/quickstart.jsp

talin Protein (2541aa) that binds to **vinculin**, but not to actin, and is associated with the sub-plasmalemmal cytoskeleton. The amino-terminal head consists of a **FERM domain** that binds an NPxY motif within the cytoplasmic tail of most **integrin** beta subunits. Talin 2 (2542aa) has similar properties.

Talon resin Proprietary name for immobilized nickel-beads. Used to purify **his-tagged** recombinant proteins.

Tamiami virus Arenavirus of the **Tacaribe complex**.

tamoxifen Synthetic anti-estrogen used in chemotherapy of breast carcinoma. Probably has other effects, including inhibition of chloride channel conductance.

tanabin See **paranemin**.

tandem repeats Copies of genes repeated one after another along a chromosome: for example the 40S-rRNA genes in somatic cells of toads, of which there are about 500 copies.

tandem-pore domain channels Potassium 'leak' channels composed of dimers of alpha subunits each of which has two pore-forming domains. See **TWIK, TREK, TASKs** and **TRAAK**.

tangential longitudinal section A section of an approximately cylindrical organ taken longitudinally along a tangent at its surface.

Tangier disease A disorder (HDL deficiency Type 1) caused by mutation in the gene encoding ATP-binding cassette-1, a cholesterol efflux pump. As a result there are low levels of high density lipoprotein (HDL) in plasma, hypercholesterolaemia, enlargement of the liver, spleen and lymph nodes and peripheral neuropathy. Familial HDL deficiency is allelic to Tangier disease. See also **LXR**. NCBI entry: http://www.ncbi.nlm.nih.gov/books/NBK22201/ html

tankyrases Poly (ADP-ribose) polymerases (TRF1-interacting ankyrin-related ADP-ribose polymerases: tankyrase-1, TNKS1, PARP5A, 1327aa; tankyrase-2, PARP5B, TNKS2, 1166aa) that may regulate vesicle trafficking and modulate the subcellular distribution of SLC2A4/GLUT4-vesicles. Has PARP activity and can modify **telomeric repeat binding factor 1** (TRF1, TERF1), contributing to the regulation of telomere length. Tankyrase-1-binding protein (1729aa) binds to the ANK repeat domain of TNKS1 and TNKS2.

tannic acid Penta-(*m*-digalloyl)-glucose, or any soluble **tannin**; used in electron microscopy to enhance the contrast. Addition of tannic acid to fixatives greatly improves, for example, the image obtained of tubulin subunits in the microtubule, or the **HMM** decoration of microfilaments.

tannins Complex phenolic compounds found in the vacuoles of certain plant cells, e.g. in bark. They are strongly astringent, and are used in tanning and dyeing.

tanshinones The major bioactive compounds (diterpene quinones) of *Salvia miltiorrhiza* Bunge (Danshen) roots, which are used in Chinese traditional medicine. Danshen extracts contain diterpene quinone and phenolic acid derivatives, including tanshinone (I, IIA, and IIB), cryptotanshinone, isocryptotanshinone, miltirone, tanshinol (I and II), and salviol (7, 25). These compounds have antioxidant properties. Total synthesis: http://web.mit.edu/chemistry/rld/totalsyn/tanshinonei.html

Tap protein (1) Member (Tip-associated protein, TAP, 616aa) of the evolutionarily conserved nuclear RNA export factor (NXF) family of proteins that mediates the sequence nonspecific nuclear export of cellular mRNAs as well as the sequence-specific export of retroviral mRNAs bearing the **constitutive transport element** (CTE). Contains separate domains for binding to **nucleoporins** and **NXT1** and the functional heterodimer with NXT/p15 mediates the export process. (2) Antigen peptide transporters (APT1, Peptide transporter TAP1, ATP-binding cassette subfamily B member 2, ABCB2, 808aa; APT2, TAP2, ABCB3, 686aa) that are involved in the transport of antigens from the cytoplasm to the endoplasmic reticulum for association with MHC class I molecules. See **tapasin**. (3) In *Antirrhinum majus* TAP1 (107aa) is found in stamens, TAP2 (131aa) in the tapetum of the anthers.

tapasin An accessory protein (Tap-binding protein, 428aa) required for the interaction of MHC Class I with **TAPs** thus ensuring efficient peptide binding. Tapasin is related to the immunoglobulin superfamily and has an ER retention signal.

tapetum (1) Layer of reflective tissue just behind the pigmented retinal epithelium of many vertebrate eyes. May consist either of a layer of guanine crystals, or a layer of connective tissue. In bovine eyes reflects a blue-green iridescent colour. (2) Layer of cells in the sporangium of a vascular plant that nourishes the developing spores.

Taq polymerase A heat-stable **DNA polymerase** that is normally used in the **polymerase chain reaction**. It was isolated from *Thermus aquaticus*.

TAR Trans-activating-responsive region. Region of non-translated mRNA, a 59-base stem-loop structure in the 5′ untranslated region of all nascent HIV-1 transcripts which acts as a binding site for the **Tat**

transactivator protein of HIV-1. There is a direct correlation between Tat binding to TAR RNA and trans-activation. See **TAR-43**. http://www.ncbi.nlm.nih.gov/pmc/articles/PMC42493/pdf/pnas01484-0624.pdf

TAR syndrome A rare disorder(thrombocytopenia-absent radii) associated with a microdeletion (200 kbp) on 1q21.1 that is necessary but not sufficient to cause the phenotype; an additional modifier (mTAR) is thought to be required.

target cell *codocyte* An erythrocyte with increased surface area to volume ratio, so called because it resembles a target with a bullseye. Can arise as a result of liver disease, iron deficiency or **thalassemia**.

targeting signal Peptide sequence within a protein that determines where in the cell the protein will be located. Thus there are targeting signals for proteins that accumulate in the nucleus, others for ER, lysosomes and so on.

Tarui disease See **glycogen storage diseases**.

tasidotin See **dolastatins**.

TASKs A family of tandem-pore potassium channels (TWIK-related acid-sensitive K^+ channels) involved in cellular responses to changes in extracellular pH. TASK1 (KCNK3, 394aa) is pH-dependent, voltage-insensitive, and rectifies according to the potassium ion concentration on either side of the membrane (an outward rectifier when external potassium concentration is low, an inward rectifier when external potassium concentration is high). TASK2 (KCNK5, 499aa) is similar but loses outward rectification at high external K^+ concentrations. TASK3 (KCNK9, 374aa) is a background channel; defects cause **Birk Barel syndrome**. TASK4 (KCNK17, 332aa) produces rapidly activating and non-inactivating outward rectifier K^+ currents. TASK5 (KCNK15, 330aa) may need to be associated with another protein to form a channel. Related channels are found in *Drosophila* (TASK6, TASK7).

taspase-1 An endopeptidase (threonine aspartase 1, EC 3.4.25.-, 420aa) with a N-terminal threonine as the active site nucleophile, that is responsible for cleaving **MLL** and for proper HOX gene expression. The proenzyme is cleaved into alpha and beta chains. Abstract of paper: http://www.ncbi.nlm.nih.gov/pubmed/14636557

taspoglutide A glucagon-like peptide-1 (GLP-1) agonist used in treating Type 2 diabetes.

tastin A component (trophinin-associated protein, 778aa) of the cell adhesion complex that includes **trophinin** and **bystin**. The interaction is mediated by bystin.

TAT (1) Tyrosine aminotransferase (TAT, EC 2.6.1.5, 454aa). (2) See **tat protein**. (3) See **Tat system**.

Tat protein Transactivator protein (86aa) from lentiviruses, notably **HIV**, a sequence-specific RNA binding protein that recognizes the **TAR** region in viral transcripts and enhances transcription. It is also secreted and will induce endothelial cell migration and invasion *in vitro* and rapid angiogenesis *in vivo*. Peptides from this protein are potent neurotoxins, implying a possible route for HIV-mediated toxicity. Research article: http://www.biomedcentral.com/1471-2121/9/32

Tat system Translocase system (twin-arginine translocation system) of the inner bacterial membrane that transports fully-folded proteins with firmly bound co-factors and is the major pathway for the translocation of cofactor-containing enzymes participating in the respiratory and photosynthetic electron transport chains. Proteins transported via Tat contain a characteristic twin-arginine motif in their signal peptide. Independent of the **secYEG** system. The system is structurally and functionally related to the pH-dependent protein import pathway of the plant chloroplast thylakoid membrane and has three functionally distinct membrane-bound components TatA (89aa), TatB (171aa), and TatC (258aa) for *E. coli*. Research article: http://www.biomedcentral.com/1471-2180/9/114

TATA box *Goldberg-Hogness box* A consensus sequence found in the promoter region of most genes transcribed by eukaryotic **RNA polymerase** II. Found about 25 nucleotides before the site of initiation of transcription and has the consensus sequence: 5'-TATAAAA-3'. This sequence seems to be important in determining accurately the position at which transcription is initiated. See **TATA-binding protein**.

TATA-binding protein A component (TBP, 339aa) of TFIID and of **SL1/TIF-IB** complex, responsible for positioning the polymerase. Also involved in positioning RNA polymerase II in which case it binds directly to the **TATA box**. Spinocerebellar ataxia type 17, a neurodegenerative disorder in man, is caused by an expanded polymorphic polyglutamine-encoding trinucleotide repeat in the gene for TBP. Three TBP-associated proteins (TAFs) are bound to each other and to TBP in the SL1 transcription factor.

tau protein Protein (tau factor, microtubule-associated protein tau, 758aa) that co-purifies with **tubulin** through cycles of assembly and disassembly, and was the first microtubule associated protein to be characterized. Tau proteins are a family made by alternative splicing of a single gene and are found in all cells but particularly neurons. The C-terminus binds axonal microtubules while the N-terminus binds neural plasma membrane components. See **MAPs** and **tauopathy**.

tau tubulin kinases Serine/threonine kinases (EC 2.7.11.1) that will phosphorylate tau. Tau tubulin kinase 1 (TTBK1, brain-derived tau kinase, 1321aa), Tau tubulin kinase 2 (TTBK2, 1244aa). Phosphorylation may reduce binding to microtubules or to membrane components, depending upon the site. Mutations in TTBK2 are associated with **spinocerebellar ataxia 11**.

tauopathy Neuropathy in which there is deposition of an excess of the microtubule-binding protein **tau** (MAPT) in brain lesions. The paired helical filaments (PHFs) of neurofibrillary tangles in Alzheimer's disease are composed mainly of hyperphosphorylated tau protein. Defects in MAPT are a cause of various neurodegenerative disorders (pallido-ponto-nigral degeneration (**PPND**), corticobasal degeneration and atypical **Steele-Richardson-Olszewski syndrome**).

taurine Compound (2-aminoethanesulphonic acid) derived from cysteine by oxidation of the sulphydryl group and decarboxylation. Present in the cytoplasm of some cells (particularly neutrophils) at high concentration.

taurocholate Major bile salt (derived from taurocholic acid) with strong detergent activity. Formed by conjugation of taurine with cholate.

tautomerase An enzyme that catalyses a **tautomerism**, for example the keto-enol equilibrium. E.g. oxaloacetate tautomerase, EC 5.3.2.2. But some other enzymes may be called tautomerases, e.g. D-dopachrome decarboxylase (EC 4.1.1.84) which is also called phenylpyruvate tautomerase.

tautomerism Form of isomerism in which there are two or more arrangements usually of hydrogens bonded to oxygen. Keto-enol tautomerism is one common example. The balance between two coexisting tautomers may shift with time or as a result of changes in conditions.

tautomycin An antibiotic, isolated from *Streptomyces spiroverticillatus*, that is an inhibitor of Type 1 and Type 2a protein phosphatases.

tax (1) Tax-1 is a transcriptional enhancer (transactivating transcriptional regulatory protein 1, 353aa) encoded by human T-cell leukemia virus type 1 (HTLV-1) and involved in transcriptional regulation, cell cycle control, and transformation. Interacts with **TIP1** (Tax interacting protein 1). Tax2 (331aa) is similar but from HTLV-2, Tax3 (350) is from HTLV-3. (2) Tax-1 is an axonal surface glycoprotein (contactin-2, 1040aa), the human homologue of rat **TAG-1** and chicken **axonin**-1. It is GPI-linked to neuronal plasma membrane and involved in adhesion. There are 6 Ig-like and 4 fibronectin III-like domains. Will support neurite outgrowth *in vitro*. (3) Tax-4 (604aa) in *S. cerevisiae* is a positive regulator of INP51 (inositol-1,4,5-trisphosphate 5-phosphatase 1) activity and phosphatidylinositol 4,5-bisphosphate turnover.

tax-responsive element The portion of the transcriptional enhancer region in the LTR of HTLV-1 responsible for trans-activation by **Tax-1**. The enhancer region to which the transcription factor binds has three direct repeats of 21 bp; the tax-responsive element (TRE) of around 12 bp is distinct from the cAMP-responsive element (CRE) and the NFκB or AP-2 binding sites. Research article: http://www.ncbi.nlm.nih.gov/pmc/articles/PMC250893/pdf/jvirol00075-0034.pdf

taxane Generic name for a cytotoxic drug that inhibits cell growth by stabilising GDP-tubulin thereby blocking microtubule disassemble and mitosis. Best known is paclitaxel (**taxol**). The taxanes were originally isolated from yews (*Taxus* sp.).

taxilins Proteins that interact with the **syntaxins** that do not form SNARE complexes and are implicated in intracellular vesicle traffic and calcium-dependent exocytosis. Alpha-taxilin (formerly IL-14, 546aa) interacts with syntaxin-3 and the **nascent polypeptide-associated complex**. Beta-taxilin (684aa) interacts preferentially with syntaxin-1A, is expressed in muscle and is reproted to promote motor nerve regeneration. Gamma-taxilin (lipopolysaccharide-specific response protein 5, factor inhibiting ATF4-mediated transcription, 528aa) interacts with syntaxins 1A, 3A and 4A and also modulates osteoblast function by inhibiting the transcription factor ATF4. Research article: http://www.ncbi.nlm.nih.gov/pubmed/15184072

taxis A response in which the direction of movement is affected by an environmental cue. Should be clearly distinguished from a **kinesis**.

taxol Drug isolated from Pacific yew (*Taxus brevifolis*) that stabilizes microtubules by blocking disassembly: analogous in this respect to **phalloidin** that stabilizes microfilaments. Strictly speaking Taxol is the proprietary name for paclitaxel but taxol is commonly used in the cell-biological literature pre-dating the clinical use.

Taxonomic Outline of Bacteria and Archaea *TOBA* Probably the most widely accepted taxonomic scheme for Bacteria and Archaea, based on the phylogeny of the 16S rRNA gene. The most recent version is Release 7.7 of March 6, 2007. The Archaea are subdivided into two phyla (Crenarchaeota and Euryarchaeota), the Bacteria into 25 (**Acidobacteria, Actinobacteria, Aquificae, Bacteroidetes, Chlamydiae, Chlorobi, Chloroflexi, Chrysiogenetes, Cyanobacteria, Deferribacteres, Deinococcus-Thermus, Dictyoglomi, Fibrobacteres, Firmicutes, Fusobacteria, Gemmatimonadetes, Lentisphaerae, Nitrospira, Planctomycetes, Proteobacteria, Spirochaetes, Thermotogae, Thermodesulfobacteria, Thermomicrobia** and **Verrucomicrobia**). Other classifications have 3 additional Archaean phyla (Korarchaeota, Nanoarchaeota, Thaumarchaeota) and three additional Bacterial phyla: Caldiserica, **Synergistetes** and Tenericutes (which includes the class **Mollicutes**). The current classification is likely to evolve as more phyla are recognised. Weblink: http://www.taxonomicoutline.org/index.php/toba/index

Tay-Sachs disease An autosomal recessive lysosomal **storage disease (lipidosis)** in which hexosaminidase A (EC 3.2.1.52) is absent and the **ganglioside** GM2 accumulates in the brain causing cells to swell and die. Usually fatal in early childhood.

TB See **tuberculosis**.

TBC domains Domains (Tre-2/Bub2/Cdc16-domains) that are predicted to encode GTPase-activating proteins (GAPs) for rab family G proteins. The TRE17 gene (also referred to as Tre-2 and USP6) was originally identified by virtue of its ability to transform NIH 3T3 cells and its TBC domain actually functions to bind GDP-Arf6 and promote its plasma membrane localization. Research article: http://mcb.asm.org/cgi/content/full/24/22/9752

tBID A truncated form of cytosolic **BID** that translocates to mitochondria. Produced as a result of TNFR1/Fas engagement

T-box genes The T-box gene family codes for transcription factors (and putative transcription factors) that share a unique DNA-binding domain, the T-domain. In all metazoans studied from *C. elegans* to man, they are found as a small, highly conserved group of genes; mutations are associated with developmental defects. See **brachyury**. Database: http://www.genenames.org/genefamily/tbx.php

TBP (1) **TATA-binding protein**. (2) Thioredoxin binding protein-2 (TBP-2, thioredoxin-interacting protein, 392aa). (3) Transferrin-binding protein from Neisseria (TBP-2, 712aa). (4) TAT-binding protein, a regulatory component of the 26S proteasome e.g. in humans, TAT-binding protein 7 (TBP7, proteasome 26S subunit ATPase 4, 418aa).

TC10 A **rho**-family small GTPase (rhoQ, 205aa) involved in epithelial cell polarization, the production of filopodia and possibly CFTR trafficking to the plasma membrane.

TCA cycle See **tricarboxylic acid cycle**.

T-cell *T lymphocyte* A class of lymphocytes, so called because they are of thymic origin and have been through thymic processing. Involved primarily in cell-mediated immune reactions and in the control of B-cell development. They bear **T-cell receptors** (CD3) that bind antigen and lack Fc or C3b receptors. Numerous subsets are now recognised: **T-helper cells, T-regulatory cells, cytotoxic T-cells, gamma-delta cells, Th17 cells, T-memory cells**.

T-cell factors A family of transcription factors (TCFs, lymphoid enhancer-binding factors, LEFs) that are activated by the **wnt** signalling system and are repressed by **CREB binding protein**. Coactivator is beta-catenin. TCF1 (T-cell specific transcription factor, more correctly transcription factor 7, TCF7, 384aa) is found predominantly in T-cells but also in proliferating intestinal epithelial cells and in the basal epithelial cells of mammary gland. Transcription factor 1 (TCF1, 631aa) is hepatocyte nuclear factor 1-alpha (**HNF1α.** TCF4 (T-cell transcription factor 4, TCF7-like 2, 619aa) is constitutively activated in colorectal carcinoma and other glandular tumours.

T-cell growth factor See **interleukin-2**.

T-cell leukaemia/lymphoma viruses See **HTLV.**

T-cell receptor The receptor on the surface of **T-cells** that is responsible for binding antigen in association with proteins of the major histocompatibility complex **MHC.** The major T-cell receptor (TCR) is a heterodimer of immunoglobulin superfamily molecules (α and β, 42-44 kDa in mouse, 50–40 kDa in humans), each with variable and constant regions. A second heterodimer on CD3$^+$ cells has γ (35 kDa in mice, 55 kDa in humans) and δ (45 kDa in mice, 40 kDa in humans) chains and is not **MHC-restricted.** The γδ T-cell receptors (TCRs) are formed on very early T-cells in the thymus. Antigen binding to the TCR triggers activation of the cells as a first step in an immune response (see **ITAMs, lck, zap70**). The TCR forms a complex with various accessory proteins of the Ig superfamily but invariant (gamma, delta, epsilon and zeta components of the **CD3 complex**) and the co-receptor, CD4. Diagram: http://www-bioc.rice.edu/~mev/TCR2.html

TCTP An abundant protein (translationally controlled tumour protein, IgE-dependent histamine-releasing factor, (HRF), fortilin, 172aa) in different eukaryotic cell types. The sequence homology of TCTP between different species is very high, and it belongs to the **MSS4/DSS4** superfamily. TCTP is involved in both cell growth and human late allergy reaction, as well as having a calcium binding property.

T-cytoplasm Texas male-sterile cytoplasm, the cytoplasm in a Texas variety of maize (corn) which produces sterile pollen which simplifies plant breeding (avoids hand de-tasselling of the plants). Unfortunately T-cytoplasm has mitochondria that are susceptible to **T-toxin** produced by the southern corn blight fungus, a serious plant pathogen until the problem was recognised. Article: http://www.ncbi.nlm.nih.gov/pmc/articles/PMC542986/

TDG Enzyme (thymine-DNA glycosylase, EC 3.2.2.29, 410aa) responsible for repair of G/T mispairings. A enzyme with the same function is found in the Archaean, *Methanobacterium thermoformicicum*.

T-DNA (1) Transfer DNA (T-DNA), part of the Ti plasmid that is transferred to plant cells from *Agrobacterium tumefaciens*. Following its nuclear import, the single-stranded T-DNA is stripped of its escorting proteins, most likely converts to a double-stranded (ds) form, and integrates into the host genome. Several genes on the T-DNA have been identified, the most important for crown gall induction are concerned with **indole acetic acid** (IAA) synthesis and the synthesis of a **cytokinin**. The transfer of these genes from bacterium to plant induces cell division and explains the hormone independence of crown gall tissue. T-DNA is the most frequently used insertion element for gene tagging in *Arabidopsis*. See **activation tagging**. (2) Rarely: DNA coding for tRNA (tDNA).

TDP-43 A DNA and RNA-binding protein (TAR DNA binding protein-43, 43 kDa, 414aa) that binds to pyrimidine-rich motifs in the transcription-activating response region (**TAR**). It is involved in the regulation of **CFTR** splicing by binding to the UG repeated motifs in the polymorphic region near the 3′-splice site of exon 9 causing skipping of this exon. Defects cause **amyotrophic lateral sclerosis** type 10 and frontotemporal lobar degeneration, where it is found in ubiquitin-positive, tau-negative cytoplasmic inclusions in neurons.

TDT (1) Transmission disequilibrium test. A test for linkage between a genetic marker and a disease susceptibility locus. (2) **Terminal deoxynucleotidyl transferase** (TdT). (3) In plants, tonoplast dicarboxylate transporter (540aa) a carrier protein indirectly involved in the uptake of malate and fumarate to the vacuole.

Tdt See **terminal deoxynucleotidyl transferase**.

Tec kinases A family of non-receptor protein **tyrosine kinases** involved in signalling, mostly in haematopoietic cells. Includes Tec (EC 2.7.10.2, 631aa), found in T-cells, B-cells and liver cells; Bruton's tyrosine kinase (**Btk**), **IL2-inducible T-cell kinase** (Itk/Emt/Tsk), bone marrow tyrosine kinase gene in chromosome X protein (Bmx/Etk), and in *Drosophila* kinases Esrc29 and Rlk/Txk/Txl. Unlike **src family** kinases are not regulated by C-terminal phosphorylation. Have **PH** and **TH domains** in N-terminal region. Additional information: http://www.sigmaaldrich.com/life-science/your-favorite-gene-search/pathway-overviews/tec-kinase-signaling.html

tectorial membrane A layer of extracellular matrix rather than a lipid bilayer; the tectorial membrane covers the neuroepithelium of the cochlea and is in contact with the bundles of stereocilia on specialized sensory hair cells. Sound causes movement of the tectorial membrane which deforms the stereocilia and induces a neural signal. See **tectorins**.

tectorins Major protein components (alpha-tectorin, 2155aa; beta-tectorin, 329aa) of the **tectorial membrane** (a layer of extracellular matrix in the inner ear). Filamentous homodimers of α-tectorin, or heterodimers with β-tectorin are formed through their **ZP domain** interactions.

tegument Generally, an outer covering (integument). In viruses the region between the envelope and the **nucleocapsid** usually containing proteins involved in assisting replication and avoiding immune responses (especially in **Herpesviridae**).

teichoic acid Acidic polymers (glycerol or ribitol linked by phosphodiester bridges) found in cell wall of Gram-positive bacteria. May constitute 10−50% of wall dry weight and are cross-linked to peptidoglycan. Related to **lipoteichoic acid**.

teichuronic acid A long chain polysaccharide in the cell wall of Gram-positive bacteria, such as *Micrococcus luteus*, composed of almost a hundred disaccharide repeating units of D-glucose and N-acetyl-D-mannosaminuronic acid (ManNAcA) residues. It is important in protecting the microbe and interacting with the host cells.

teicoplanin Glycopeptide antibiotic used as a less toxic alternative to vancomycin for treating meticillin-resistant *Staph. aureus*. Actually a group of at least six glycopeptides, all with very similar activities, A2−1 to A2−5, and A3−1 being the main components. Work by inhibiting formation of cell walls in Gram-positive bacteria by interfering with formation of links in the cell wall (transglycosylation), acting on amino acyl-D-alanyl-D-alanine residues.

tektins Family of filamentous proteins (tektin-1, 418aa; tektin-2, 430aa, tektin-3, 490aa; tektin-4, 435aa; tektin-5, 485aa) associated with outer-doublet microtubules in ciliary and flagellar axonemes in all eukaryotes except higher plants. The filaments are 2 nm in diameter and are important for the structural integrity of the microtubules. Have homology with some intermediate filament proteins (keratins and lamins). Review: http://genomebiology.com/2008/9/7/229

telangiectasia Condition in which capillary vessels are dilated. See **ataxia telangiectasia** and **mastocytosis**.

telencephalin See **ICAMs**.

teleomorph Mycological term that describes the form of a fungus when reproducing sexually. The anamorph form describes the fungus when reproducing asexually and the holomorph is a description of the whole fungus, encompassing both forms of description.

teleost melanophores Large stellate cells found in the epidermis of fish. Cytoplasmic pigment granules (containing **melanin**) can be centrally located, or rapidly dispersed, using a microtubule-associated system. Altering the granule distribution changes the colour of the skin. **Chromatophores** containing other pigments extend the capacity to adopt protective colouration.

telethonin *T-cap* A muscle protein (167aa) that mediates the antiparallel assembly of **titin** at the sarcomeric Z-disc by interacting with **myozenin-1**

and -3 and the N-terminal Ig-like domains of 2 titin molecules. Defects in telethonin are a cause of familial hypertrophic **cardiomyopathy**, limb-girdle **muscular dystrophy** type 2G, and dilated cardiomyopathy type 1N.

teleutospore Thick-walled resting spore of rust and smut fungi. *Ustilago coicis* teleutospores can retain their ability to germinate even when maintained under dry conditions for about five years.

telocentric See **metacentric**.

telodendria Branched terminations of axons and axon collaterals, the distal ends of which are slightly enlarged to form synaptic bulbs.

telokin The C-terminal 155aa of smooth muscle **myosin light chain kinase** (MLCK) and also independently expressed. In some muscles may regulate the contractile state: when phosphorylated it inhibits myosin phosphatase, binds to unphosphorylated myosin, and keeps the muscle relaxed. Research article: http://ajplung.physiology.org/content/294/6/L1166.long

telolecithal Describing a type of egg which is relatively large, with yolk constituting most of the volume of the cell. Cytoplasm is concentrated at one pole. Typical of sharks, reptiles and birds.

telombin family A family of proteins associated with telomeres and that bind to single-stranded DNA. The original example was **pot1** (555aa) from Schizosaccharomyces pombe but subsequently found in many eukaryotes (human, 634aa; chicken, 778aa, Euplotes crassus, 460aa).

telomerase A DNA polymerase (telomerase reverse transcriptase, TERT, EC 2.7.7.49, 1132aa) that adds multiple telomeric repeats 'TTAGGG' to the **telomere**. In addition to the catalytic TERT, the enzyme complex also has an essential 159 residue RNA sequence, the telomerase RNA component (TERC) that provides a template for the replication of the G-rich telomere sequences (so that the enzyme could in fact be considered a reverse transcriptase). Overexpression of telomerase in cells can extend the replicative lifespan but does not transform them. Defects in TERC or TERT are the cause of short telomeres in some forms of aplastic anaemia and cases of increased susceptibility to adult-onset idiopathic pulmonary fibrosis. Heterozygous deletion of TERT is found in patients with **cri-du-chat syndrome**. See **telomeric repeat binding factors** and **shelterin**.

telomere The repetitive DNA found at the end of a chromosome that is protective during DNA replication. The telomere is shortened at each division cycle but new 6 bp repeats are added by **telomerase** although with age there seems to be progressive shortening. See **shelterin**.

Telomeric Repeat Amplification Protocol
assay *TRAP assay* An assay for **telomerase** activity in cells. In the first step, telomerase in the cell extract adds a number of telomeric repeats (TTAGGG) onto the 3′ end of a biotinylated oligonucleotide. In the second step, the extended products are amplified by PCR with a deoxynucleotide mix containing dCTP labeled with dinitrophenyl (DNP). This extenstion/amplification reaction generates a readily identifiable ladder of products with 6 base increments starting at 50 nucleotides. Descriptive article: http://www.clinchem.org/cgi/content/full/47/3/519

telomeric repeat binding factors Proteins (TRF-1, TERF-1, 439aa; TRF-2, 500aa) that co-localize with telomeres in interphase and at chromosome ends during mitosis. They bind as homodimers to the 'TTAGGG' tandem repeats and recruit other proteins to the **shelterin** complex that regulates telomere length. Defects in TRF2 can lead to exposure of the DNA ends and activation of the ATM (ataxia telangiectasia mutated)-mediated DNA damage response. **Tankyrases** can inhibit binding of TRFs to telomeres.

telomestatin A telomerase inhibitor, isolated from *Streptomyces anulatus* 3533- SV4, which is known to stabilize G-quadruplex structures at 3′ single-stranded telomeric overhangs (G-tails) and rapidly dissociates POT-1 (Protection of Telomere 1) and associated proteins such as telomeric repeat binding factor-2 (TRF2) from telomeres in cancer cells. Telomestatin has a large polycyclic planar structure resembling a G-quadruplex. Research article: http://cancerres.aacrjournals.org/cgi/content/full/66/14/6908

telopeptides Portions of the amino acid sequence of a protein that are removed in maturation of the protein. Best known examples are the N- and C-terminal telopeptides of procollagen that are involved in development of the quaternary structure and are then proteolytically removed by **procollagen peptidases**.

telophase The final stage of mitosis or meiosis, when chromosome separation is completed.

TEMED Compound   (N,N,N,N-Tetramethyl-Ethylenediamine) used in conjunction with ammonium persulphate to accelerate polymerization of acrylamide.

temperate phage A bacteriophage that integrates its DNA into that of the host (**lysogeny**) as opposed to virulent phages that lyse the host.

temperature-sensitive mutation *ts mutation* A type of conditional mutation in organism, somatic cell or virus that makes it possible to study genes whose total inactivation would be lethal. Such ts mutations can also make possible studies of the effect of reversible switching (by temperature changes) in expression of the mutated gene. The usual mechanism of temperature sensitivity is that the mutated gene codes for a protein with a temperature-dependent conformational instability, so that it possesses normal activity at one temperature (the permissive temperature), but is inactive at a second (non-permissive) temperature.

template A structure that in some direct physical process can cause the patterning of a second structure, usually complementary to it in some sense. In current biology almost exclusively used to refer to a nucleotide sequence that directs the synthesis of a sequence complementary to it by the rules of Watson-Crick base-pairing.

temporal sensing Mechanism of gradient sensing in which the value of some environmental property is compared with the value at some previous time, the cell having moved position between the two samplings. Initial movement is random; until the second observation is made the gradient cannot be detected. See **spatial** and **pseudospatial** sensing mechanisms. Bacterial chemotaxis (so called) is based on this mechanism.

tenascins Proteins of the extracellular matrix (240 kDa subunit: usually as a hexabrachion, a six-armed hexamer of more than 1000 kDa) that are involved in guidance of neurons, and various developmental processes. They are ligands for integrins. Tenascin C (cytotactin, myotendinous antigen, 2201aa) has 13 or 14 EGF-like repeats and up to 15 fibronectin (FN) type III domains and is found in tendons and embryonic extracellular matrix; Tenascin-R (1358aa) has 4.5 EGF-like repeats and 9 FN type III domains and is found mostly in the CNS; Tenascin-X (4289aa) is encoded by gene X in the major histocompatibility complex (MHC) class III gene region and has 18.5 EGF-like repeats and 29 -32 FN type III domains. Other variants have also been described: Tenascin Y in birds, similar to Tenascin X and Tenascin N from mice, similar to Tenascin W. Tenascin M1 (teneurin-1) is found only in fetal brain.

tensegrity The hypothesis that cells can behave like structures in which shape results from balancing tensile and hydrostatic forces.

tensins Actin-binding proteins present in some **focal adhesions** and in the submembranous cytoskeleton. Has an SH2 domain and can be **tyrosine phosphorylated**; interacts with **PI3-kinase** and **JNK**signalling pathways. Tensin-1 (1735aa) is ubiquitous, tensin-2 (tensin-like C1 domain-containing phosphatase, 1409aa) interferes with **Akt1** signalling, tensin-3 (tumour endothelial marker 6, 1445aa) has a more restricted tissue distribution, as does tensin-4 (715aa). See **PTEN**.

tenuin Sub-plasmalemmal protein (400 kDa) from rat **adherens junctions**, associated with membrane insertions of **microfilament** bundles, and membrane adjacent to circumferential microfilament bundles of epithelial cells. Reported in 1989 but has not reappeared in recent literature.

TEP1 (1) Telomerase protein component 1 (telomerase-associated protein 1, 2627aa) a component of the **telomerase** ribonucleoprotein complex. (2) In *S. cerevisiae* a phosphatidylinositol-3,4,5-trisphosphate 3-phosphatase (TEP1, EC 3.1.3.67, 434aa) which in humans is termed **PTEN** (TGFα-regulated and epithelial cell-enriched phosphatase, mutated in multiple advanced cancers 1, MMAC1, 403aa).

teratocarcinoma Malignant tumour (teratoma) thought to originate from primordial germ cells or misplaced blastomeres that contains tissues derived from all three embryonic layers, e.g. bone, muscle, cartilage, nerve, tooth-buds and various glands. Accompanied by undifferentiated, pluripotent epithelial cells known as embryonal carcinoma cells.

teratogen An agent capable of disrupting developmental processes. A notorious example is **thalidomide** but many mutagens can have teratogenic effects, the consequences being partly affected by genotype and by the developmental stage at which exposure occurs.

teratoma See **teratocarcinoma**.

terminal bar Obsolete name for **zonula occludens** (tight junction).

terminal buttons The small swellings at the end of an axon that release neurotransmitters; the presynaptic region.

terminal cisternae Regions of the **sarcoplasmic reticulum** adjacent to **T tubules**, and from which calcium is released when striated muscle is activated.

terminal deoxynucleotidyl transferase *TdT* An intranuclear DNA polymerase (EC 2.7.7.31, 509aa) that catalyzes the template-independent addition of deoxynucleotides to the 3′-hydroxyl terminus of oligonucleotide primers. Such terminal additions at the junctions of rearranging V(D) J gene segments greatly contribute to antigen-receptor diversity. TdT is expressed only on immature lymphocytes and acute lymphoblastic leukaemia cells and has been identified in several vertebrate species, where it is highly conserved.

terminal uridylyl transferase An enzyme (TUT1, EC 2.7.7.52, 874aa) that specifically catalyzes the uridylylation of **U6 snRNA** and is essential for cell proliferation. It is phosphorylated when there is DNA damage.

terminal web The cytoplasmic region at the base of microvilli in intestinal epithelial cells, a region rich in microfilaments from the microvillar core and from **adherens junctions**, in myosin, and in other proteins characteristic of an actomyosin motor system.

termination codons The three codons, UAA known as **ochre**, UAG as **amber** and UGA as **opal**, that do not code for an amino acid but act as signals for the termination of protein synthesis; sometimes called nonsense codons or stop codons. They are not recognised by any tRNA and termination is catalysed by protein release factors. There are two release factors in *E. coli*; RF1 recognizes UAA and UAG, RF2 recognizes UAA and UGA. Eukaryotes have a single GTP-requiring factor, eRF. An ochre mutation changes a codon to UAA, an ochre suppressor codes for an altered tRNA that recognize the UAA and carries the same amino acid as specified by the original codon or a neutral substitute amino acid. (Amber and opal mutations and suppressors are similar). Ochre suppressors will suppress **amber codons**.

terpenes Polymers of 5-carbon isoprene units, very abundant in plants although their function is unclear. Terpenoids (sometimes classed as terpenes) are modified terpenes. Common examples of plant terpenoids are citral, menthol, camphor and the cannabinoids. In animals **dolichol**, an important carrier species in the formation of glycoproteins, is a terpenoid. Squalene, an intermediate in the synthesis of cholesterol, is a terpene. Many terpenoids are synthesized from mevalonate (see **HMG-CoA reductase pathway**), but in plants and many bacteria a second **terpenoid pathway** operates.

terpenoid pathway Many terpenoids are synthesized through the **mevalonate pathway** but a second alternative pathway operates in plants and many bacteria (the 2-C-methyl-D-erythritol 4-phosphate/1-deoxy-D-xylulose 5-phosphate pathway, MEP/DOXP pathway), is responsible for synthesis of a wide variety of monoterpenes, diterpenes, and carotenoids. Diagram of pathway: http://www.genome.jp/kegg/pathway/map/map00900.html. Research article: http://www.ncbi.nlm.nih.gov/pmc/articles/PMC26933/pdf/pq008251.pdf

TERT See **telomerase**.

tertiary structure The third level of structural organization in a macromolecule. The primary structure of a protein (for example) is the amino acid sequence, the secondary structure is the folding of the peptide chain (e.g. alpha-helical or beta-pleated), the tertiary structure is the way in which the helices or sheets are folded or arranged to give the three-dimensional structure of the protein. Quaternary structure refers to the arrangement of promoters in a multimeric protein.

tes A putative tumour suppressor gene, encoding **testin**.

testa Outer covering of a seed, also called the seed-coat; derived from the integument of the ovary.

testicans A family of secreted calcium-binding proteoglycans (SPARC/osteonectin, CWCV and kazal-like domain proteoglycans, SPOCK1, 439aa; SPOCK2, 424aa; SPOCK3, 436aa) with chondroitin sulphate and heparan sulphate O-linked side chains, found in the extracellular matrix of brain where they may act as inhibitors of metallopeptidases. They are related to proteins involved in adhesion, migration, and cell proliferation and may be important in neural development. See **osteonectin, kazal proteins, CWCV domain.**

testicular feminization Complete androgen insensitivity syndrome. The process that can occur in genetic males if the receptors for male sex hormones are defective. They develop as females and are unresponsive to male hormones.

testilin A protein (EH domain-containing protein 1, PAST homolog 1, 534aa) highly expressed in testis and involved in membrane trafficking of recycling endosomes.

testin A scaffold protein (421aa) found in **focal adhesions** that may play a role in cell adhesion, cell spreading and in the reorganization of the actin cytoskeleton. Depending upon its conformation may bind actin, **Mena**, and vasodilator-stimulated phosphoprotein (**VASP**) or alternatively, **alpha-actinin**, paxillin, and **zyxin**. It is the product of a putative tumour suppressor gene (*tes*) that maps to human chromosome 7q31.1; when expressed, negatively regulates proliferation of T47D breast carcinoma cells. Research article: http://jcb.rupress.org/content/161/1/33.long

testosterone Male sex hormone (androgen) secreted by the interstitial cells of the testis of mammals and responsible for triggering the development of sperm and of many secondary sexual characteristics.

tetanolysin Thiol-activated haemolysin (50 kDa) released by the bacterium *Clostridium tetani*, one of the **cholesterol binding toxins**. Tetanolysin apparently forms water-filled pores (20−50 nm) that may vary in size, depending on the tetanolysin concentration utilized.

tetanospasmin See **tetanus toxin**.

tetanus *lock-jaw* Disease caused by the bacterium *Clostridium tetani*, spores of which persist in soil, but can proliferate anaerobically in an infected wound. Disease entirely due to the **tetanus toxin**, released by bacterial autolysis.

tetanus antitoxin Antibody to **tetanus toxin**, usually from horses hyper-immunized with *Clostridium tetani* exotoxin. Can cause serum sickness, an immune response to the horse proteins.

tetanus toxin An **AB toxin** (tetanospasmin, EC 3.4.24.68, 1315aa) released by *Clostridium tetani*; becomes active when cleaved proteolytically to heavy (100 kDa) and light (50 kDa) chains held together by disulphide bond. The heavy chain binds to disialogangliosides (GD2 and GD1b), and part of the peptide (the amino-terminal B-fragment) forms a pore through which the light A-chain, a zinc

endopeptidase, enters the cytoplasm. The endopeptidase specifically attacks **synaptobrevin** and blocks the release of neurotranmitters: it is the toxin that is responsible for the pathology of tetanus. See also **botulinum toxin**.

tetherin A protein (bone marrow stromal antigen 2, CD317, 180aa) that attaches the budding HIV virus particle to the host cell. The molecule has two lipid domains, one attaches to the virus, one to the cell. An artificial molecule with the same physical properties but no sequence similarity has comparable effects. Degradation of CD317 is important for virion budding and **vpu** targets CD317 for proteasomal degradation.

tetracycline antibiotics A group of closely related bacteriostatic antibiotics, with similar antibacterial spectrum and toxicity. They act by binding to the 30S subunit of the bacterial ribosome and inhibiting protein synthesis. Effective against many streptococci, Gram-negative bacilli, rickettsiae, spirochetes, Mycoplasma, and Chlamydia. Tetracycline itself is produced by *Streptomyces aureofasciens*.

tetrad Four homologous chromatids paired together during first meiotic prophase. More generally, any group of four objects.

tetraethylammonium ion *TEA* A monovalent cation widely used in neurophysiology as a specific blocker of potassium channels. It is similar in size to the hydrated potassium ion but gets stuck (reversibly) in the channels.

tetrahydrobiopterin See **biopterin**.

tetrahydrocannabinol *THC* A **cannabinoid** and one of the more psychoactive components of cannabis.

Tetrahymena Genus of ciliate protozoa frequently used in studies on ciliary axonemes, self-splicing RNA and telomere replication.

tetralogy of Fallot A congenital heart disease in which there is an incomplete septum between left and right ventricles and the baby is cyanotic (blue-baby). Can be a result of mutation in the human homologue of **jagged-1** or the CSX gene that encodes a cardiac-specific homeobox.

TETRAN See **TPO**.

tetranectin A plasma glycoprotein (ca. 10 mg/l), a homotetramer of 181aa subunits each with a **C-type lectin**-like domain, thought to regulate proteolytic processes via its binding to plasminogen kringle 4 and indirect activation of plasminogen. Decreased plasma levels of tetranectin correlate with cancer progression.

tetraploid Nucleus, cell or organism that has four copies of the normal **haploid** chromosome set.

tetraspan vesicle membrane proteins Ubiquitous and abundant family of widely expressed integral membrane proteins (tetraspanins; TVPs) that associate extensively with one another and with other membrane proteins to form specific membrane microdomains. They are characterized by four transmembrane regions and cytoplasmically located end domains. TVP-containing vesicles shuttle between various membranous compartments and are localized in biosynthetic and endocytotic pathways. There are three families of TVPs: **physins**, gyrins (e.g. **synaptogyrin**) and **secretory carrier-associated membrane proteins** (SCAMPs). Examples are CD9 (Tspan29) and CD81 (Tspan28). See **tetraspanin web**.

tetraspanin web A network of **tetraspan vesicle membrane proteins** (tetraspanins), **ADAMs** and **integrins**, that facilitates cellular interactions and cell fusion. The partners vary between cell types and a single tetraspanin type can interact with several different proteins. CD81 (tetraspanin-28, 2326aa) localizes in **immune synapses** of both B- and T-cells; both CD9 (Tspan29, 228aa) and CD81 are involved in egg-sperm fusion. Review article: http://physiologyonline. physiology.org/cgi/content/full/20/4/218

tetrocarcin Originally discovered as an antibiotic active against Gram-positive bacteria. Tetrocarcin A (TC-A) is an inhibitor of the anti-apoptotic function of **Bcl-2**.

tetrodotoxin *TTX* A potent **neurotoxin** (319 Da) from the Japanese puffer fish (*Fugu rubripes*). It binds to the voltage-activated sodium channel, blocking the passage of action potentials. It is actually a bacterial toxin and is accumulated in several fish as well as a diverse range of invertebrate and vertebrate phyla (where it has often been given other names, e.g. tarichatoxin in newts and maculotoxin in the blue-ringed octopus). The action of the toxin is similar to that of **saxitoxin**.

tetrose General term for a monosaccharide with 4 carbon atoms.

Teunissen-Cremers syndrome See **ankylosis**.

Texas red Rhodamine-type fluorophore well suited for excitation by the 568 nm spectral line of Ar−Kr mixed-gas lasers or the 594 nm spectral line of the orange He−Ne lasers; emits at about 620 nm, with very little spectral overlap with fluorescein, often conjugated to antibodies for use in FACS analysis.

textilinin Kunitz-type serine peptidase inhibitor (Txln, 59aa) from the venom of *Pseudonaja textilis* that has effects on plasmin that are kinetically distinct from those of **aprotinin** (with which it has ~50% sequence homology). Two distinct forms have been isolated. The effect is to slow fibrinolysis.

textilotoxin The toxin in the venom of *Pseudonaja textilis* (Australian common brown snake) that blocks neuromuscular transmission. All six subunits

(~120aa) of the toxin have some phospholipase A2 activity. See **textilinin**.

TFG (1) See **TRK-fused gene**. (2) *Trigonella foenumgraecum* (fenugreek; TFG), leaf extracts of which exert analgesic, anti-inflammatory and anti-pyretic effects in various experimental models. (3) Total ginkgo flavone glycosides (TFG) used in Chinese herbal medicine. (4) TFG1 and TFG2 encode the two larger subunits of the **TFIIF** complex (5) Often a misprint for **TGF**.

TFII The set of general transcription factors that are involved in the activity of **RNA polymerase II**. TFII-A (two subunits in yeast and three (TFIIA-α and -β from one gene (376aa before cleavage), TFIIA-γ (109aa) from another gene) in humans) binds directly to **TATA-binding protein** (TBP) and stabilizes its binding to DNA. TFII-B (316aa) also binds directly to TBP, and further stabilizes TBP binding to the TATA element. It is also involved, cooperatively with TFII-F (517aa) with recruiting RNA polymerase II. TFIID is a multi-component transcription factor that has a DNA-binding subunit (TATA-binding protein, TBP, 339aa) and several TBP-associated factors (or TAFs); it acts to nucleate the transcription complex, which interact directly with TFIIB. TFIIE (439aa) regulates initiation of transcription.TFIIF is a heterodimer: TFIIF-alpha (RNA polymerase II-associating protein 74; RAP74; EC 2.7.11.1) is the large subunit, TFIIF-beta is smaller (26 kDa), also known as RAP30. Helps recruit RNA polymerase II to the initiation complex and promotes translation elongation. TFIIH has ten subunits, some with helicase and ATPase activities, others that phosphorylate serine residues in the polymerase, or are associated with cell cycle control. TFIIH is also involved in nucleotide excision repair. Other components (TFIIG, TFIIJ) also associate with the complex. TFIIX is a general name for various TBP accessory proteins.

TFIII The general transcription factor for genes that are transcribed by eukaryotic **RNA polymerase** III. TFIIIA is a zinc-finger protein (365aa) that binds the promoters of the 5 S RNA genes and then forms a complex with TFIIIC, which can then interact with TFIIIB. The TFIIIB complex has two activities, alpha and beta. The TFIIIB-alpha complex is composed of TBP (**TATA binding protein**), BDP1 (TFIIIB component B'', 2624aa), and a sub-complex containing both BRF2 (TFIIIB 50 kDa subunit, 419aa) and at least four stably associated proteins. The TFIIIB-beta activity complex is composed of TBP, BDP1 (TFIIIB 90 kDa subunit, 677aa) and BRF1; both TFIIIB α and β are necessary for transcription. TFIIIC is a multisubunit complex that initiates transcription complex assembly on tRNA. It includes TFIIICα (2109aa), TFIIICβ (911aa), TFIIICγ (886aa) that directly binds tRNA, TFIIICδ (EC = 2.3.1.48, 822aa) that has histone acetyltransferase activity, TFIIICε (519aa) and TFIIIC subunit 6 (213aa).

TFIIX A general name for various accessory proteins involved in the binding of RNA Polymerase II to DNA in association with **TBP**.

TFP (1) **Trifluoperazine**. (2) **Mitochondrial trifunctional protein** (TFP). (3) Type IV pili (Tfp) of various bacteria.

TGEs Regions (tra-2 and GLI elements) of the 3′ untranslated region of the sex-determining gene *tra-2* in *C. elegans*. Thought to be the target of STAR proteins.

TGF See **transforming growth factor**.

TGFβ receptor type III A protein (betaglycan, 850aa) found ubiquitously on nearly all cell types in membrane-bound form and as soluble form in serum and in the extracellular matrix. It functions as a TGFβ superfamily co-receptor, by binding TGFβ1, TGFβ2, TGFβ3, inhibin, BMP-2, BMP-4, BMP-7, and GDF-5 and presenting these ligands to their respective signalling receptors. Entry in Atlas of Genetics and Cytogenetics in Oncology and Haematology: http://atlasgeneticsoncology.org/Genes/TGFBR3ID42541ch1p33.html

TGN Abbreviation for the **trans-Golgi network**.

TH domain Domain (Tec homology domain) that is characteristic of **Tec family** protein kinases, probably the ligand region for the **SH3** domain that follows downstream. The N-terminal 27aa of the TH domain are highly conserved (the **Btk** motif), and are followed by a proline-rich (PRR) region.

Th-POK A zinc-finger transcription factor (T-helper-inducing POZ/Kruppel-like factor; cKrox, 539aa) that acts as a key regulator of lineage commitment of immature T-cell precursors. Absence of Th-POK leads to complete conversion of CD4$^+$ cells to CD8$^+$ cells.

Th17 cells See **T-helper cells**.

thalassaemia *US.* thalassemia. A relatively common haemoglobinopathy, particularly in Mediterranean countries, in which there are mutations in globin genes that cause reduced haemoglobin production and anaemia. In alpha-thalassemias the mutation is in the globin alpha chain, in beta-thalassemias in the beta chain. Heterozygosity may provide some protection against malaria.

Thale cress See *Arabidopsis thalliana*.

thalidomide Sedative drug that was once extensively prescribed for nausea during pregancy until it became clear that it was a potent **teratogen**. Can produce a range of malformations of the fetus, in severe cases complete absence of limbs (amelia), or much reduced limb development (phocomelia). Thalidomide inhibits TNFα production and has anti-inflammatory effects and is used to treat leprosy and multiple myeloma.

thallic conidium See **conidium**.

thallus Simple plant body, not differentiated into stem, root, etc. Main form of the gametophyte generation of simpler plants such as liverworts.

thapsigargin Cell-permeable inhibitor (a sesquiterpene lactone extracted from *Thapsia garganica*) of calcium ATPase of endoplasmic reticulum (**SERCA**); leads to increase in cytoplasmic calcium ions. Acts independently of InsP3. A tumour promoter.

thaumatin Protein from the African plant *Thaumatococcus daniellii*. It tastes 105 times sweeter than sucrose.

thebaine *paramorphine* Minor natural constituent of opium, chemically similar to both morphine and codeine, but stimulatory rather than depressant. Thebaine can be converted into a variety of compounds including **codeine**, hydrocodone, **oxycodone**, naloxone. It is a partial agonist of the mu-opioid receptor.

Thellungiella halophila *Thellungiella salsuginea* A model halophyte with a small plant size, short life cycle, and small genome that shows tolerance to extreme salinity stress, chilling, freezing and ozone stress. *Thellungiella* genes exhibit high sequence identity (approximately 90% at the cDNA level) with *Arabidopsis* genes.

T-helper cells A major subset of CD4$^+$ **T-cells** further subdivided into Th1, Th2 and Th17 subclasses. Th1-cells are responsible for clearing intracellular pathogens and are involved in cell-mediated immunity. They produce IL-2, IFNγ and TNFα and are selectively activated by IFNγ and IL-12 and inhibited by IL-4 and IL-10. Th2-cells are involved with the humoral immune response, produce IL-4, IL-5 and IL-10 and promote antibody production. The two subclasses cross-inhibit so that one type becomes predominant and the immune response is of one kind or the other, rather than a mixture. Th17-cells produce **IL17** in response to autoimmune tissue damage but do also have an important anti-microbial role at epithelial or mucosal surfaces.

THEMIS Protein (thymocyte-expressed molecule involved in selection, 641aa) that controls both positive and negative T-cell selection. Interacts with phospholipase C-gamma-1, Itk (see **IL-2−inducible T-cell kinase**) and Grb2.

theobromine Principal alkaloid (3,7-dimethyl xanthine) of cacao bean; has similar properties to **theophylline** and **caffeine**.

theophylline A natural alkaloid (1,3-dimethyl-xanthine) that inhibits cAMP **phosphodiesterase**, and is often used experimentally in conjunction with exogenous dibutyryl cyclic-AMP to raise cellular cAMP levels. Other, less potent, methylxanthines are caffeine, theobromine and aminophylline.

therapeutic cloning Production of a cloned embryo by somatic cell nuclear transfer, with the intention of using stem cells from the early embryo for therapeutic purposes, e.g. in the treatment of Parkinson's disease. *Cf.* **reproductive cloning.**

therapeutic index The ratio of the lethal dose (LD_{50}) to the effective dose (ED_{50}) for a drug. Ideally the difference should be large.

thermal analysis Form of calorimetry in which the rate of heat flow (or some other property) to a solid is measured as a function of temperature. Detailed description: http://www.msm.cam.ac.uk/phasetrans/2002/Thermal1.pdf

thermal cycler *thermocycler* Instrument for automated **PCR** in which a sequence of temperature cycles is produced. See **Lightcycler.**

thermal melting profile In general a record of the phase state of a system over a temperature range. Phase changes can be detected by exothermy or endothermy. Valuable in studying lipid and DNA structures.

Thermococcus A spherical, motile hyperthermophile Archaean indigenous to anoxic thermal waters. An obligate anaerobic chemoheterotroph that uses sulphur as an electron acceptor, forming H_2S, at temperatures between $55°$ and $95°C$.

Thermodesulfobacteria A phylum of thermophilic sulphate-reducing bacteria. With a growth temperature optimum of $70°C$, the most thermophilic sulphate-reducing bacteria known.

thermodynamics The study of energy and energy flow in closed and open systems. The first law of thermodynamics states that energy can neither be created nor destroyed, the second law that energy systems have a tendency to increase their entropy (and that heat can flow from the hotter to the cooler, but not the converse; perpetual motion machines cannot exist), the third law of thermodynamics that as temperature approaches absolute zero, the entropy of a system approaches a constant minimum. Other laws have been proposed but these are probably the fundamental ones.

thermogenin See **uncoupling proteins.**

thermolysin Heat-stable zinc metallopeptidase (peptidase family M4) containing four calcium ions. Produced by various thermotolerant bacteria (EC 3.4.24.4, 548aa in *Bacillus thermoproteolyticus*) and retains 50% of its activity after 1 h at $80°C$.

Thermomicrobia See **Chloroflexni.**

thermonasty A **nastic movement** in response to temperature.

thermophile An organism that thrives at high temperature. The most extreme examples (hyperthermophiles) are from hot springs, deep-sea vents, and geysers that will tolerate temperatures above $80°C$. The archaebacterium *Pyrolobus fumarii* holds the current record, being able to grow at $113°C$.

thermotaxis A directed motile response to temperature. The grex of *Dictyostelium* shows a positive thermotaxis.

Thermotogae A phylum of the domain Bacteria containing the genus *Thermotoga*, an extreme thermophile capable of growth at temperatures as high as $90°C$. The genome sequence has revealed many genes with strong homology to genes of hyperthermophilic Archaea.

Thermus aquaticus Aerobic Gram-negative bacillus that was originally found in hot springs in Yellowstone National Park and was the source of **Taq polymerase**.

theta antigen See **thy1.**

theta replication The early replication method of bacteriophage lambda, involving the production of theta form intermediates. The circular phage DNA is replicated bidirectionally producing copies of the phage DNA (at an intermediate stage of the replication process the shape resembles the greek letter theta); at a later stage the phage shifts to using a rolling circle mechanism that produced concatemers with multiple copies of the genome that are later cut and processed to produce multiple individual phage genomes.

THG See **tetrahydrogestrinone.**

thiamine See **Vitamin B1.**

thiamine pyrophosphatase A nucleosidediphosphatase (TPPase; EC 3.6.1.6) that requires thiamine pyrophosphate as a cofactor. Has been used as a cytochemical marker for the trans-cisternae of the **Golgi apparatus.**

thiamine pyrophosphate *TPP* The coenzyme form of vitamin B1 (thiamine), a coenzyme for pyruvate dehydrogenase, α-ketoglutarate dehydrogenase and transketolase. Synthesized by thiamine pyrophosphokinase from free thiamine and MgATP.

thiamine pyrophosphokinase Enzyme (TPK, EC 2.7.6.2, 243aa) responsible for the synthesis of **thiamine pyrophosphate** (thiamine diphosphate) from thiamine and ATP.

thiazide diuretics Group of drugs with moderate diuretic effects. Work by inhibiting the sodium re-uptake transporter in the distal tubule of the kidney, thereby increasing sodium and water excretion but with concomitant loss of potassium ions (in contrast to the **potassium-sparing diuretics**). Used for long-term treatment of hypertension or oedema associated with congestive heart failure.

thick filaments Bipolar **myosin**-II filaments (12–14 nm diameter, 1.6 μm long) found in striated muscle. Myosin filaments elsewhere are often referred to as 'thick filaments', although their length may be considerably less. The myosin heads project from the thick filament in a regular fashion. There is a central 'bare' zone without projecting heads, the core being formed from antiparallel arrays of **LMM** regions of the myosin heavy chains. Thick filaments will self-assemble *in vitro* under the right ionic conditions.

thigmotaxis *Adj.* thigmotactic. A **taxis** in response to mechanical contact (touch). Although generally only used in the context of whole-organism behaviour it could be said that contact inhibition of locomotion is a cellular equivalent.

thigmotropism Tendency of an organism or part of an organism to turn towards or respond to a mechanical stimulus.

thin filaments Filaments 7–9 nm diameter attached to the **Z-discs** of striated muscle with opposite polarity in each half-sarcomere. Built of F-actin with associated **tropomyosin** and **troponin**.

thin layer chromatography *TLC* Chromatographic separation method in which a thin layer of the solid phase (often silica, aluminium oxide or cellulose) is fixed onto a glass or plastic sheet. Can be run one-dimensionally or in a second dimension with a different solvent system. Much used in separation of lipids. Visualization can be by staining or radioactive labelling and autoradiography.

thiobacillus Small rod-shaped proteobacteria living in sewage or soil and gaining energy from the oxidation of elemental sulphur and sulphur containing compounds. *Thiobacillus ferrooxidans* is generally considered to be important for accelerating the dissolution of metal sulphide from minerals.

thiocoraline A novel bioactive depsipeptide isolated from a marine actinomycete of the genus *Micromonospora*. Has potent cytotoxic activity and strong antimicrobial activity against Gram-positive microorganisms. Binds to supercoiled DNA and inhibits RNA synthesis.

thioctic acid See **lipoic acid**.

thioesters Compounds of the type. R-CO-S-R'. See **coenzyme A**, **palmitoylation**.

thioether The bond R-S-C, of which the best example is in methionine.

thiol endopeptidases *thiol proteinase* See **cysteine endopeptidases**.

thiol-activated haemolysins Cytolytic bacterial exotoxins (oxygen-labile haemolysins) that act by

binding to cholesterol in cell membranes and forming ring-like complexes that act as pores. SH-groups of these toxins must be in the reduced state for the toxin to function. Oxidation (to disulphide bridges) inactivates the toxin. Examples: **tetanolysin**, **streptolysin O**, θ-toxin (**perfringolysin**), **cereolysin**.

thionins Group of small, hydrophobic plant proteins of 45–48aa, of which 6–8 are cysteines; toxic to animals but probably a defense against pathogenic microorganisms, since they are up-regulated during fungal infections.

thiopentone *thiopental* A widely used intravenous anaesthetic. A rapid-onset, short-acting barbiturate.

thioredoxins Ubiquitous proteins (105aa in human) that act as antioxidants by facilitating the reduction of other proteins.They are secreted by a variety of cells, despite the lack of a **signal sequence**, in a manner resembling the alternative secretory pathway for IL1β, and a few other proteins. Mitochondrial thioredoxin (thioredoxin-2, 166aa) has an anti-apoptotic function and plays an important role in the regulation of mitochondrial membrane potential. Thioredoxin-related transmembrane protein 1 (280aa) and thioredoxin-like protein 1 (289aa) are also found in humans. Thioredoxins are kept in the reduced state by the flavoenzyme thioredoxin reductase. See **adult T-cell leukaemia-derived factor**, **DsbG**. Thioredoxin-interacting protein (thioredoxin-binding protein 2, TXNIP, 391aa) is overproduced by nutrient-sensing nerve cells when nutrient levels are high and this has been tentatively linked to obesity.

THO complex A subcomplex of the **TREX complex** containing, in *S. cerevisiae*, four proteins (Hpr1p, Tho2p, Mft1p, and Thp2p) that are essential for normal transcriptional elongation and genome stability. The THO sub-complex recruits two additional proteins, Sub2p and Yra1p, that are involved in nuclear RNA export and splicing.

Thomsen-Friedenreich-antigen *T-antigen* A tumour-associated epitope (Gal-GalNAc) on mucin-like glycoproteins (possibly a high molecular weight CD44 splice variant) that is recognised by the lectin, **jacalin**. Research article: http://www.ncbi.nlm.nih.gov/pmc/articles/PMC1568008/

thoracic duct The major efferent lymph duct into which lymph from most of the peripheral lymph nodes drains. Recirculating lymphocytes that have left the circulation in the lymph node return to the blood through the thoracic duct.

THP-1 Human monocytic cell line derived from peripheral blood of 1-year-old boy with acute monocytic leukaemia. Have Fc and C3b receptors and will differentiate into macrophage-like cells.

threonine The hydroxylated polar amino acid (Thr, T, 119 Da). See Table A1.

threonine peptidases A family of peptidases characterised by a threonine nucleophile at the N terminus of the mature enzyme. The family includes the **proteasome** peptidases, **polycystin-1** and glycosylasparaginase (EC 3.5.1.26, 346aa). The latter enzyme is deficient in **aspartylglycosaminuria**.

threose A four-carbon sugar in which the two central hydroxyl groups are in *trans* orientation (*cis* in erythrose).

threshold limit value-time-weighted average *TLV-TWA* The concentration of a substance to which it is believed nearly all workers may be repeatedly exposed without adverse effect. Based upon a time-weighted average concentration for a conventional eight hour workday and a 40-hour week (shorter than that worked by most scientists).

thrombasthenia Condition in which there is defective platelet aggregation, though adherence is normal. See **Glanzmann's thrombasthenia**.

thrombin Coagulation factor II, a serine endopeptidase (EC 3.4.21.5) derived from prothrombin (622aa) by proteolytic processing that generates the light chain (36aa) and heavy chain (259aa) which are linked by a disulphide bond. Thrombin cleaves **fibrinogen** to produce **fibrin** as the final stage in blood clotting and also activates platelets, leukocytes, and endothelium at sites of vascular injury through the **thrombin receptor**. Prothrombin can be cleaved either by the action of the extrinsic system (tissue factor+ phospholipid) or, more importantly, the intrinsic system (contact of blood with a foreign surface or connective tissue). Both extrinsic and intrinsic systems activate plasma Factor X to form Factor Xa which then, in conjunction with phospholipid (tissue derived or **platelet factor 3**) and Factor V, catalyses the conversion. See also Table F1. Mutations in the prothrombin gene can lead to hypoprothrombinaemia (deficient production) or abnormal protein that cannot be cleaved and activated (dysprothrombinaemia). High plasma prothrombin levels (hyperprothombinaemia) may increase the risk of venous thrombosis.

thrombin receptors A family of G protein-coupled receptors which are unusual in that cleavage of the receptor by thrombin produces a new N-terminus that functions as a tethered ligand. Proteinase-activated receptor 1 (PAR1, 425aa) and PAR4 (385aa) are high affinity receptors, PAR2 (397aa) responds to trypsin but not thrombin and is upregulated in chronic inflammation. PAR3 (374aa) does not signal but acts as a cofactor for the cleavage and activation of PAR4.

thrombocyte Archaic name for a blood **platelet**.

thrombocythaemia A condition in which there are high platelet numbers in the circulation (in excess of the normal level of 150,000–450,000 per mm^3).

Can be a result of a defect in the **thrombopoietin** (THPO)/THPO receptor system that regulates proliferation of megakaryocytes or a somatic mutation in the **JAK2** gene.

thrombocytopenia Gross deficiency in platelet number (below 100,000/mm^3), consequently a tendency to bleeding.

thrombocytopenic purpura A condition in which **thrombocytopenia** leads to bleeding into skin producing small petechial haemorrhages. The familial form of thrombotic thrombocytopenic purpura (TTP, microangiopathic haemolytic anaemia, Schulman-Upshaw syndrome) is caused by mutation in ADAMTS13 (a disintegrin-like and metalloprotease with thrombospondin type-1 motif-13), the peptidase that cleaves **von Willebrand factor**. In autoimmune thrombocytopenic purpura there is increased platelet destruction mediated by autoantibodies to platelet-membrane antigens. Secondary thrombocytopenic purpura can be associated with chronic disorders or with disturbed immune function due to chronic infections, lymphoproliferative and myeloproliferative disorders, pregnancy, or autoimmune disorders. Overview: http://emedicine.medscape.com/article/779969-overview

thromboglobulin A heparin-binding protein fragment (β thromboglobulin, 81aa) derived by proteolytic cleavage from **platelet basic protein**. It is found as a homotetramer.

thrombomodulin A thrombin receptor (575aa) found on the luminal surface of endothelial cells. The receptor-thrombin complex converts **protein C** to Ca, that in turn acts on Factors Va and VIIIa and reduced the production of thrombin. It is structurally similar to **coated pit** receptors.

thrombophilia Susceptibility to venous thrombosis, a condition that can be a result of mutation in genes for clotting, anticoagulant, or thrombolytic factors including **antithrombin 3**, **protein C**, **protein S**, **factor V**, **histidine-rich glycoprotein**, **plasminogen**, **plasminogen activator inhibitor**, **fibrinogen** and thrombomodulin. Environmental factors are also important.

thromboplastin The substance in plasma that converts prothrombin to thrombin, although now known not to be a single substance. 'Thromboplastin reagent' is used to test the blood-clotting time of patients receiving anti-coagulant treatment or showing signs of clotting dysfunction. It can be either an extract of mammalian tissue (lungs, heart or brain of animals) rich in **tissue factor**, or a recombinant preparation of human tissue factor in combination with phospholipids. See **International sensitivity index** and **International normalised ratio**.

thrombopoietin Growth factor (TPO, 353aa) that regulates the proliferation of megakaryocytes and the production of platelets (thrombopoiesis). Receptor is c-mpl (CD110, 635aa), a type I cytokine receptor that can cause phosphorylation of **STAT**3 and STAT5 through Jak3. The receptor interacts with **ataxin**-2-like protein, triggering phosphorylation on tyrosine residues.

thrombosis Formation of a solid mass (a **thrombus**) in the lumen of a blood vessel or the heart.

thrombospondin-related anonymous protein A type 1 membrane protein (TRAP, 559aa) that is an **invasin** of *Plasmodium falciparum* and an essential protein for sporozoite motility and for liver cell invasion. TRAP has multiple adhesive domains in its extracellular region and interacts, through aldolase, with microfilaments. Related proteins are found in *Babesia spp*.

thrombospondins A family of homotrimeric glycoproteins (thrombospondin-1, 1170aa; TSP2, 1172aa; TSP3, 956aa; TSP4, 961aa; **cartilage oligomeric matrix protein**) from α granules of **platelets**, and synthesized by various cell types in culture. Also found in extracellular matrix of cultured endothelial, smooth muscle, and fibroblastic cells. May have autocrine growth-regulatory properties: involved in platelet aggregation. Thrombospondins 1 and 2 promote synaptogenesis in the CNS. Thrombospondin 3 is a developmentally regulated heparin binding protein; thrombospondin 4 is similar to TSP3 but differs in being a potential integrin ligand.

thrombosthenin Obsolete name for platelet contractile protein: now known to be actomyosin (which makes up 15−20% of the total platelet protein).

thrombotic thrombocytopenic purpura See **thrombocytopenic purpura**.

thromboxanes Arachidonic acid metabolites produced by the action of thromboxane synthetase on prostaglandin cyclic endoperoxides. Thromboxane A2 (TxA2) is a potent inducer of platelet aggregation and release, and although unstable, the activation of platelets leads to the further production of TxA2. Also causes arteriolar constriction. Another endoperoxide product, **prostacyclin**, has the opposite effects.

thrombus Solid mass that forms in a blood vessel, usually as a result of damage to the wall. The first aggregate is of platelets and fibrin, but the thrombus may propagate by clotting in the stagnant downstream blood.

THUMP domain An ancient domain (thiouridine synthases, RNA methyltransferases, and pseudouridine synthases-domain, ~110aa) found in genes from Eukaryota and Archaea, but not eubacteria. The genes in question are for ThiI-like thiouridine synthases, conserved RNA methylases, archaeal pseudouridine synthases and several uncharacterized

proteins. It has been predicted that this domain is an RNA-binding domain and its restricted distribution reflects the distinct tRNA-processing strategies of Eukarya/Archaea and Eubacteria. SMART database entry: http://smart.embl-heidelberg.de/smart/do_annotation.pl?DOMAIN = THUMP&BLAST = DUMMY

thuringolysin O Cholesterol binding toxin from *Bacillus thuringiensis*.

thy1 A differentiation antigen (CD90, theta antigen, 161aa) on surface of T-cells, neurons, endothelial cells and fibroblasts. GPI-anchored and a member of the **immunoglobulin superfamily** with only one V-type (variable) domain.

thylakoids Membranous cisternae of the chloroplast, found as part of the grana and also as single cisternae interconnecting the grana. Contain the photosynthetic pigments, reaction centres and electron-transport chain. Each thylakoid consists of a flattened sac of membrane enclosing a narrow intra-thylakoid space.

thymectomy The excision of the thymus by operation, radiation or chemical means. Since the thymus is important in T-cell maturation, this has major effects on the immune system.

thymic aplasia A lack of T-lymphocytes, due to failure of the thymus to develop, resulting in very reduced cell-mediated immunity though serum immunoglobulin levels may be normal. See **thymic hypoplasia**.

thymic hypoplasia *di George syndrome* A deficiency in the development of the thymus and parathyroid as a result of a hemizygous deletion of 1.5 to 3.0-Mbp from chromosome 22q11.2 which encodes the TBX1 (T-box 1) transcription factor), or point mutations in the TBX1 gene. Haploinsufficiency of the factor causes hypoplasia of the thymus and secondarily of T-cells so that cell-mediated immunity is compromised. Hypoplasia of the parathyroid causes hypocalcemia. There are other developmental abnormalities associated with the syndrome, some caused by defects in cervical neural crest migration.

thymic stromal lymphopoietin A cytokine (TSLP, 159aa) that appears to induce a TH2−inducing signal in dendritic cells. It induces the release of T-cell chemoattractants from monocytes and enhances the maturation of CD11c$^+$ dendritic cells. The receptor is composed of CRLF2 (cytokine receptor-like factor 2, 371aa) and IL7R. Abstract of research article: http://stke.sciencemag.org/cgi/content/abstract/sigtrans;3/105/ra4?etoc

thymidine The name invariably used for the nucleoside thymine deoxyriboside; not the riboside which naming of the other nucleosides might lead one to expect.

thymidine block A method for synchronizing cells in culture. In the absence of thymidine, DNA synthesis cannot occur, so cells are blocked before S-phase; release of the block allows synchronous entry into cycle.

thymidine kinase Enzyme (TK, EC 2.7.1.21, 234aa) of the pyrimidine salvage pathway, catalysing phosphorylation of thymine deoxyriboside to form its 5′ phosphate, the nucleotide thymidylate. Animal cells lacking this enzyme can be selected by resistance to bromodeoxyuridine which they do not incorporate into DNA (which would be lethal) and can be used as parentals in somatic hybridization, since they are unable to grow in **HAT medium**.

thymidine phosphorylase See **PD-ECGF**.

thymine Pyrimidine base (2,6-di-hydroxy, 5-methyl-pyrimidine, 5-methyluracil) found in DNA (in place of uracil in RNA). Thymine dimers can be formed in DNA by covalent linkage between two adjacent (cis) thymidine residues, in response to ultraviolet irradiation. Although repair enzymes exist that can excise thymine dimers this is potentially mutagenic. See **xeroderma pigmentosum**.

thymine-DNA glycosylase A mismatch-specific DNA-binding glycosylase (TDG, EC 3.2.2.29, 410aa) that corrects G/T mismatches to G/C basepairs.

thymocyte Lymphocyte within the thymus; usually applied to an immature lymphocyte.

thymoma A tumour of thymic origin.

thymopentin Biologically active pentapeptide corresponding to residues 32−36 of thymopoietin. Will induce prothymocytes and activate peripheral T-cells.

thymopoietin A thymic hormone that induces differentiation of thymocytes, although the gene encodes a nuclear protein (lamina-associated polypeptide-2) present in three alternatively spliced isoforms that associate with **lamin A/C**. Thymopoietin-alpha (LAP2 α-isoform, 694aa) is proteolytically processed to produce thymopentin and thymopoietin (splenin, 49aa). LAP2 β and γ isoforms (454aa) are similarly processed to produce thymopentin and thymopoietin: the LAPs may help direct the assembly of the nuclear lamina and thereby help maintain the structural organization of the nuclear envelope. Mutation in the LAP2 gene can cause some forms of **dilated cardiomyopathy**.

thymosin Thymosin alpha1 is a naturally occurring thymic peptide (28aa) that primes dendritic cells for antifungal T helper type 1 resistance through toll-like receptor (TLR)-9 signalling. Beta-thymosins are polypeptides involved in the regulation of actin polymerization (see **thymosin β4**).

thymosin β4 Small protein (5 kDa, 43aa) found in large amounts in many vertebrate cells (approximately 0.2 mM in neutrophils) and that binds G-actin thereby inhibiting polymerization. Also has a range of cytokine-like activities. Identical to Fx peptide.

thymus The site of **T-cell** maturation, in mammals just anterior to the heart within the rib cage; in other vertebrates in rather undefined regions of the neck or within the gill chamber in teleost fish. The organ is composed of stroma (thymic epithelium) and lymphocytes, almost all of the T-cell lineage. The thymus regresses as the animal matures.

thymus-derived lymphocyte See **T-cell**.

thypedin See **pedin**.

thyroglobulin The glycoprotein precursor (2768aa) of the iodinated **thyroid hormones** thyroxine (T4) and triiodothyronine (T3). It is homodimeric. Variations are associated with susceptibility to autoimmune thyroid diseases (see **Graves' disease** and **Hashimoto's thyroiditis**). Formerly referred to as colloid. Not the same as **thyroxine-binding globulin**.

thyroid hormones Hormones produced by the thyroid gland that influence growth, metabolic rate and, in amphibians, metamorphosis. Tri-iodothyronine (T3) is more potent than thyroxine (T4; tetra-iodothyronine) from which it is derived. **Calcitonin** is also of thyroid origin but is not usually classed as a thyroid hormone. See also Tables H2 and H3.

thyroid-stimulating antibody Long-acting thyroid stimulator is an autoantibody found in many cases of primary thyrotoxicosis which causes hyperplasia of the thyroid by undetermined mechanisms. Human thyroid-stimulating immunoglobulin is a different antibody found in all or nearly all cases of primary thyrotoxicosis and may act by binding to the thyrotropin (TSH) receptor site, causing increased synthesis of **thyroglobulin**.

thyroid-stimulating hormone Polypeptide hormone (TSH, thyrotropin), secreted by the anterior pituitary gland, that activates cyclic AMP production in thyroid cells leading to production and release of the **thyroid hormones**. It is a heterodimer with a unique beta subunit (112aa) and an alpha subunit (116aa) shared with other pituitary hormones (chorionic gonadotropin, follicle-stimulating hormone and luteinizing hormone). The TSH receptor (TSHR, 764aa) is G-protein linked and is also a receptor for **thyrostimulin**. Various mutations in the receptor can cause hypothyroidism or hyperthyroidism; autoantibodies against TSHR are responsible for **Graves' disease**.

thyroiditis See **Graves' disease, Hashimoto's thyroiditis, hypothyroidism, Riedel's disease**.

thyroliberin See **thyrotropic releasing hormone**.

thyrostimulin A heterodimeric glycoprotein hormone with thyroid-stimulating activity that has a unique beta subunit (130aa) but shares the same alpha subunit as **thyroid-stimulating hormone** (TSH). Binds to the same receptor as TSH.

thyrotoxicosis Clinical syndrome caused by an excess of circulating free **thyroid hormones** (thyroxine (T4) and triiodothyronine (T3)). The most common cause is **Graves' disease**.

thyrotropic-releasing hormone Tripeptide (protirelin, TRH, thyroliberin, TRF, pyroGlu-His-Pro-NH$_2$) that releases thyrotropin from the anterior pituitary by stimulating adenyl cyclase. It is produced as 242aa precursor polypeptide with 6 copies of the sequence -Glu-His-Pro-Gly-. May also have neurotransmitter and paracrine functions.

thyrotropin See **thyroid-stimulating hormone**.

thyroxine See **thyroid hormones** and Table H3.

thyroxine-binding globulin The highest-affinity plasma carrier (TBG, 414aa) of **thyroid hormones**, a serpin (serpinA7), although it has no inhibitory function. **Transthyretin** and albumin also carry the hormones but TBG has the highest affinity and carries the majority of T4.

Ti plasmid Plasmid of *Agrobacterium tumefaciens*, transferred to higher plant cells in crown gall disease, carrying the **T-DNA** (about 200 kb) that is incorporated into the plant cell genome. Used as a vector to introduce foreign DNA into plant cells.

Tiam-1 The product of the T-lymphoma invasion and metastasis gene 1 (1591aa), a GDP-GTP exchange factor (GEF) for the small GTPase **rac** implicated in tumour invasion and metastasis. Tiam-2 (1701aa) is similar.

tiarin Olfactomedin-family member, a secreted scaffold protein (olfactomedin-noelin-tiarin factor 1, 492aa) that promotes dorsal neural specification in *Xenopus*. Also found in chickens.

Tic complex The translocating system of the chloroplast inner membrane that acts in concert with the **Toc complex** to transfer proteins from cytoplasm to the inside of the chloroplast or vice versa. Eight Tic subunits have been described so far, including two potential channel proteins (Tic110 and Tic20), the 'motor complex' (Tic40 associated with the stromal chaperone Hsp93) and the 'redox regulon' (Tic62, Tic55, and Tic32) involved in regulation of protein import via the metabolic redox status of the chloroplast. Regulation can additionally occur via thioredoxins (Tic110 and Tic55) or via the calcium/calmodulin network (Tic110 and Tic32). Abstract: http://www.ncbi.nlm.nih.gov/pubmed/20100520

TICAMs Adapter molecules (Toll/IL-1R domain-containing adapter-1, TICAM1, 792aa; TICAM2,

TRIF-related adapter molecule, 235aa) used by **toll-like receptors** 3 and 4 to mediate NFκB and interferon-regulatory factor (IRF) activation, and to induce apoptosis. Involved in immunity against invading pathogens. TRIF-induced cleavage of **TRAF1** is required for its inhibition of TRIF signalling.

Tie Endothelium-specific receptor tyrosine kinases (EC 2.7.10.1: Tie1, 1138aa; Tie2, Tek, CD202b, 1124aa) required for normal embryonic vascular development and tumour angiogenesis. Associates with p85 of PI3kinase. Tie2 is the receptor for **angiopoietin** and interacts with Tie1; the interaction is essential for response to VEGF but VEGF does not appear to be a ligand for Tie1. Abstract: http://www.ncbi.nlm.nih.gov/pubmed/19376222

tify domain A domain (36aa) originally identified in the ZIM protein (zinc-finger protein expressed in inflorescence meristem) and called the ZIM domain. It is called the tify domain after its most conserved amino acids (TIF[F/Y]XG) and is found in a variety of plant transcription factors that contain GATA domains. Group I is formed by proteins possessing a CCT (CONSTANS, CO-like, and TOC1) domain and a GATA-type zinc finger in addition to the tify domain. Group II contains proteins characterised by the tify domain but lacking a GATA-type zinc finger.

TIGAR A p53-inducible protein (Tp53-induced glycolysis and apoptosis regulator, 270aa), probably a fructose-2,6-bisphosphatase (EC 3.1.3.46) that functions to regulate glycolysis and to protect against oxidative stress. TIGAR expression is associated with a decrease in fructose-2,6-bisphosphate levels, inhibition of glycolysis and protection from ROS-associated apoptosis.

tight junction See **zonula occludens**.

tiling Another word for tessellation. 'Rep-tiles' can be joined together to make larger replicas of themselves. The aim in tiling is to cover the whole area completely, thus the term has come to be used in the context of the design of DNA arrays to give unbiased coverage, or tiling, of genomic DNA for the large-scale identification of transcribed sequences and regulatory elements. Increasingly sophisticated algorithms are being developed to assist in designing appropriate arrays to achieve this end. Technical report: http://medicine.tums.ac.ir/FA/Users/keramatipour/Cancer%20Genetics%20Articles/12394138.pdf

Tillman's reagent See **DCIP**.

TIM23 complex A multi-subunit complex (translocase of the inner membrane) that is responsible for transporting proteins across the inner mitochondrial membrane. The complex includes TIMM23 (209aa), which may form a pore, TIMM17 A or B (171aa) and TIMM50 (353aa). The complex interacts with the TIMM44 component of the PAM complex (presequence translocase-associated motor complex). See **TOM complex**. Diagram: http://www.nature.com/nrm/journal/v5/n7/fig_tab/nrm1426_F3.html

TIME FOR COFFEE A plant protein (1550aa) involved in regulating clock function. It acts in the mid to late night and may act on the transcriptional induction of **LATE ELONGATED HYPOCOTYL**.

time-lapse Technique applied to speed up the action in a film or videotape sequence. In filming by taking a frame every few seconds and projecting at conventional speed (16 or 24 frames per second) the movements of cells can be greatly speeded up, and then become conspicuous. With digital recording, which has now almost completely superceded the use of film, the recording is made at slow speed and replayed at full speed. The opposite of slow-motion.

time-resolved fluorescence Method to avoid interference by autofluorescence. Using an emitter fluorochrome that has slow decay characteristics coupled to the reagent of interest and temporally separating excitation and measurement, the signal can be arranged to derive almost entirely from the reporter fluorophore. (Autofluorescence decays very rapidly.) See **HTRF**.

time-weighted average A special averaging technique that takes accounts of variation in the basis of the individual measurements and adjusts for time irregularities from sample to sample.

timeless *Drosophila* gene (*tim*) essential for the production of **circadian rhythms**. The protein product, TIM, is a transcription factor (1208aa) which interacts with the *period* gene product. TIM and PER associate with one another and the regulated interaction seems to determine the entry of PER into the nucleus: both TIM and PER are produced in a circadian cycle.

Timothy syndrome A autosomal dominant disorder in which there is multiorgan dysfunction, including lethal arrhythmias (**long QT-8**), caused by a mutation in gene for the alpha-1C subunit of the voltage-dependent L-type calcium channel from cardiac muscle (CACNL1A1). Genetics Home Reference entry: http://ghr.nlm.nih.gov/condition = timothysyndrome

TIMP See **tissue inhibitors of metalloproteinases**.

TIN2 A protein (TRF1-interacting nuclear factor 2, 451aa) that forms part of the **shelterin** complex where it interacts with **telomeric repeat binding factor 1** (TRF1). May be a negative regulator of telomerase length. Mutations can be one cause of **dyskeratosis congenita** and of **Revesz's syndrome**. TIN2-interacting protein (POT1- and TIN2-organising protein, PTOP, adrenocortical dysplasia protein homologue, 544aa) blocks the interaction of **POT1** with telomeres and allows telomere extension.

tincar *Drosophila* gene, *tincar* (tinc), that encodes a protein with eight putative transmembrane domains that is involved in the regulation of eye photoreceptor cell development. There are four isoforms ranging from 986aa to 1513aa. The same authors originally reported it to be expressed specifically in four of the six pairs of cardioblasts in each segment, in a pattern identical to that of **tinman** (tin).

tinman Member of the NK homeobox family (416aa), essential for the specification of cardiac cells in *Drosophila*. Tinman and Bagpipe (Bap), another homeodomain protein, form homo- and heterodimeric complexes. See **tincar**.

TIP (1) An inhibitory regulator (272aa) of protein phosphatases 2A, 4 and 6, homologous to yeast Tip41, with a role within the ATM/ATR signalling pathway that controls DNA replication and repair. (2) A T-cell immunomodulatory protein (612aa) that induces the secretion of interferon gamma, TNF and IL10. (3) Tension induced/inhibited protein (methyltransferase-like protein 8, EC 2.1.1.-, 281aa), induced by stretch in murine mesenchymal cells. (4) Tax-1 interacting protein-3 (TIP1, 124aa) is involved in rho and wnt signalling and contains a single PDZ domain. TIP-1 is able to bind beta-**catenin** with high affinity and inhibit its transcriptional activity; also functions as a negative regulator of PDZ-based scaffolding. (5) TIP30 (Tat-interacting protein 30, oxidoreductase HTATIP2, EC 1.1.1.-, 242aa) is a proapoptotic factor that interacts with HIV-1 tat protein. (6) Tat-interacting protein of 110 kDa (TIP110, squamous cell carcinoma antigen recognized by T-cells 3, 963aa) regulates Tat transactivation. (7) TIP47 (tail-interacting protein of 47 kDa) is one of the **PAT family** of proteins now known to be identical to mannose 6-phosphate receptor-binding protein 1 (perilipin-3, placental protein 17, 434aa) that binds the cytoplasmic domain of the receptor and is required for its transport from endosomes to the trans-Golgi network. (8) Tuberoinfundibular peptide of 39aa, TIP39, **parathyroid hormone** 2; 100aa, processed to produce the active hormone of 39aa, a potent and selective agonist of PTH2Receptor. (9) In *Drosophila*, histone acetyltransferase Tip60 (EC 2.3.1.48, 541aa) the catalytic subunit of the Tip60 chromatin-remodeling complex which is involved in DNA repair. (10) In *Arabidopsis*, palmitoyltransferase TIP1 (EC 2.3.1.-, 620aa) involved in cell growth regulation. (11) **Tonoplast intrinsic protein** (aquaporin TIP1-1). (12) In *Schizosaccharomyces*, Tip elongation protein 1 (461aa) which stabilizes the tips of microtubules aligned with the long axis of the cell.

TIR domain Domain ($\sim$200aa) characteristic of **toll-like receptors** and the interleukin-1 receptor. The known human TIR-domain-containing adapter proteins include MyD88, TIR domain-containing adapter inducing IFN-beta (TRIF), TRIF-related adapter molecule (TRAM), and TIR-domain containing adapter protein (TIRAP). TIR domains also occur in some plant defence proteins. Structure: http://pawsonlab.mshri.on.ca/index.php?option = com_content& task = view&id = 182&Itemid = 64

TIRAP See **TIR domain**.

TIRF See **total internal reflection microscopy**.

tisB The protein toxin (29aa) in a type I toxin-antitoxin system of *E.coli* K12. Translation of the toxin is blocked by a small antitoxin RNA (**IstR-1**, 75nt) which blocks the ribosome-binding site on the tisB mRNA. Part of the programmed response to DNA damage which leads to increased levels of tisB which slows or stops bacterial growth until repair is complete. TisA is thought to be a pseudogene.

tissue Group of cells, often of mixed types and usually held together by extracellular matrix, that perform a particular function. Thus, tissues represent a level of organization between that of cells and of organs (which may be composed of several different tissues). Sometimes used in a more general sense – for example epithelial tissue, where the common factor is the pattern of organization, or connective tissue, where the common feature is the function.

tissue array Array of small sections of tissue (tissue microarray) on a microscope slide that can be used for immunocytochemical staining or *in situ* hybridisation. The tissue samples are, in some cases, selected from larger embedded tissue samples (e.g. of tumours) following histopathological examination.

tissue culture Originally the maintenance and growth of pieces of explanted tissue (plant or animal) in culture away from the source organism. Now usually refers to the (much more frequently used) technique of cell culture, using cells dispersed from tissues, or distant descendants of such cells.

tissue culture plastic Polystyrene that has been rendered wettable making it a surface to which proteins like vitronectin, and subsequently animal cells, will adhere and thus provides a suitable substratum for **anchorage-dependent** cells to grow. Can be achieved by oxidation or commercially by glow discharge. Article: http://www.springerlink.com/content/q2t02u4446476088/

tissue engineering The creation of new body parts for transplantation by *in vitro* culture of cells using an artificial matrix or support. There are considerable practical difficulties, particularly the absence of vascularisation, and few tissues have yet been successfully manufactured in this way.

tissue factor Integral membrane glycoprotein (factor III, CD142, 263aa), that initiates blood clotting by the **tissue factor pathway** after binding factors VII or VIIa. Tissue factor pathway inhibitor (lipoprotein-associated coagulation inhibitor, 304aa) is a protease

inhibitor that blocks the proteolytic activation of other clotting factors.

tissue factor pathway The mechanism for inducing blood-clotting by factors released from damaged tissue (extrinsic pathway), rather than as a result of contact with a foreign surface (the contact activation or intrinsic pathway). Tissue **thromboplastin** (Factor III) in conjunction with Factor VII (proconvertin) activates Factor X which then converts prothrombin to **thrombin**.

tissue inhibitors of metalloproteinases *TIMPs* Family of proteins (~200aa) that inhibit metallopeptidases that degrade extracellular matrix (MMPs) by forming stoichiometric 1:1 complexes. TIMP1 (fibroblast collagenase inhibitor, 207aa) acts on most MMPs except MMP14. Others have different ranges of inhibitory activity.

tissue plasminogen activator *TPA* See **plasminogen activator**.

tissue-typing The process of determining the allelic types of the antigens of the **major histocompatibility complex** (MHC) that determine whether a tissue graft will be accepted or rejected. At present carried out either by use of polyclonal or monoclonal antibodies against MHC antigens, or less usually by tests of MHC-restricted cell function or skin grafting (the latter not in humans).

titin *connectin* Family of enormous proteins (EC 2.7.11.1, 2000-3500 kDa, 34350aa) found in the sarcomere of striated muscle. Form a scaffolding of elastic fibres that may be important for correct assembly of the sarcomere and for resisting tension. There is a calmodulin-regulated ser/thr kinase domain. Each titin molecule spans from M-line to Z-disc. Mutations in titin cause various forms of **hypertrophic cardiomyopathy**, **myopathy** and **muscular dystrophy**.

TL antigens The mouse antigens (thymus leukemia antigens) coded for by the **TLa complex**; in normal animals only found on intrathymic lymphocytes, but also seen on leukaemic cells in certain forms of the disease in mice. The molecules have structures similar in some ways to Class I MHC products but are disulphide bonded tetramers of two 45 kDa chains and two 12 kDa chains of **beta-2-microglobulin** type.

TL1 (1) A secreted TNF-like ligand (tumour necrosis factor ligand superfamily member 15, TNFSF15, vascular endothelial growth inhibitor; VEGI, 251aa) that binds to **death receptor** DR3. TL1 is highly expressed in endothelial cells and, mediates activation of NFκB, and inhibits angiogenesis *in vitro*. Expression is upregulated in **Crohn's disease**. (2) In apples (*Malus domestica*) an important allergen, thaumatin-like protein 1a (pathogenesis-related protein 5a, Mal d 2, 246aa). Also found in other plant species.

TLa complex Genes coding for and controlling **TL antigens**; the complex is situated close to the H-2 complex on mouse chromosome 17 and resembles H-2 in several ways.

TLC See **thin layer chromatography**.

TLCK A protease inhibitor (tosyl lysyl chloromethylke-tone), particularly effective against trypsin and papain.

T-loop of RNA Thymine pseudo-uracil loop. The T-loop (Ty loop) of tRNA is the region of the molecule that is responsible for ribosome recognition.

TLR See **toll-like receptors**.

T-lymphocyte See **T-cell**.

TMB-8 An inhibitor (3,4,5-trimethoxybenzoic acid 8-(diethylamino)octyl ester) of the release of calcium from intracellular stores but also a potent, non-competitive, functional antagonist of various **nicotinic acetylcholine receptor** subtypes.

TMEFFs Proteins (TMEFF1, transmembrane protein with EGF-like and 2 follistatin-like domains, tomor-egulin-1, 380aa; TMEFF2; tomoregulin-2, 374aa) that may be tumour suppressors in brain and prostate cancers. In *Xenopus* development TMEFF1 may reg-ulate **nodal** and **BMP** signalling. TMEFF2 may be a survival factor for hippocampal and mesencephalic neurons and is mainly expressed in the prostate and brain; there are 3 splice variants, TRa, TRb, and TRc.

T-memory cells T-cells (Tm-cells) involved in **immunological memory** of specific antigens. The standard view is that they are effector T-cells that remain quiescent until re-exposed to antigen, but a second view is that there are a distinct population of central memory cells (TCM) with a capacity for self-renewal that act as memory stem cells. There do seem to be at least two distinct classes of Tm-cells. Memory Tsubsets: http://www.nature.com/nature/journal/v402/n6763supp/full/402d034a0.html

TMPD An easily oxidised compound (tetramethyl-1, 4-phenylendiamine; Wurster's reagent) that can be used as a reducing co-substrate for haem peroxidases. Following one-electron oxidation TMPD produces a highly coloured product that absorbs at 611 nm. Can be used for the detection of peroxidases on polyacryl-amide gels.

tmRNA See ssrA.

TMS See **transcranial magnetic stimulation**.

TMV See **tobacco mosaic virus**.

TNF Tumour necrosis factor. A pleiotropic cytokine, TNFα (cachectin), originally described as a tumour-inhibiting factor in the blood of animals exposed to bacterial **lipopolysaccharide** or **Bacille Calmette-Guerin** (BCG). Preferentially kills tumour cells *in vivo* and *in vitro*, causes necrosis of certain transplanted tumours in mice and inhibits experimental metastases. Human TNFα (157aa) is pro-inflammatory and the soluble form is released from the cell surface by the action of **TACE** (TNFα converting enzyme), a metalloproteinase of the **ADAM** type. TNFβ (lympho-toxin) has 35% structural and sequence homology with TNFα and binds to the same **TNF receptors** but is secreted in a conventional manner by activated T and B-cells. A superfamily of TNF-like cytokines are now recognised, many with specific receptors of the **TNF receptor** superfamily (see Table T1).

TNF receptors A superfamily of receptors (TNFSFRs) for **TNFα**, **FAS**, CD40, CD27, and **RANK**. The type I receptor (TNFRSF1A, p60, CD120a, 55 kDa, 455aa) is ubiquitous but the type II receptor (TNFRSF1B, p80, CD120b, 75 kDa, 461aa) is found mainly on haemato-poietic cells. The Type I and II receptors bind both TNFα and TNFβ (lymphotoxin) and are members of the NGF receptor family but have different signalling capacities. TNFSFR1A contains a **death domain** and forms a homotrimer when ligand binds; this stimulates interaction of various adaptor proteins (**TRADD, TRAF, FADD**) on the cytoplasmic face and triggers a range of cellular responses. TNFRSF3 is the receptor for heterotrimeric lymphotoxin (TNFβ) and TNFS14/LIGHT; it promotes apoptosis via TRAF3 and TRAF5. TNFRSF10B (death receptor 5, 440aa) is the receptor for TNFSF10/TRAIL. TNFRSF14 (Herpes virus entry mediator A, 283aa) is a receptor for BTLA, TNFSF14/LIGHT and homotrimeric TNFSF1, is involved in lymphocyte activation and plays an important role in HSV pathogenesis. See Table T1. TNFSF/ TNFRSF http://www.jbc.org/content/281/20/13964.long

TNP (1) Trinitrophenol (picric acid), a bright yellow compound often used as a hapten and an ingredient in Bouin's fixative. Can be explosive when dry. (2) TNP-470, a semisynthetic derivative of **fumagillin**, an angiogenesis inhibitor. (3) Tn5 transposase (Tnp), an enzyme involved in DNA transposition that causes genomic instability by mobilizing DNA elements. (4) See **TNP-AMP**.

TNP-AMP Trinitrophenol-adenosine monophosphate. Compound that can be used as an ATP analogue in studying ATPases. TNP nucleotides undergo an equi-librium transition to a semiquinoid structure when bound to the nucleotide-binding site of some proteins and are only fluorescent in this form.

TNT (1) Trinitrotoluene, a well-known explosive. (2) **Troponin** T (TnT).

tobacco mosaic virus *TMV* Plant RNA virus, the first to be isolated. Consists of a single central strand of RNA (a helix of 6500 nucleotides) enclosed within a coat consisting of 2130 identical capsomeres that, in the absence of the RNA, will self-assemble into a cylinder similar to the normal virus but of indeterminate length. Causes mottling of the leaves of the tobacco plant.

TABLE T1. TNF and TNF Receptors

Tumor Necrosis Factor Ligand Superfamily

Ligand	Synonyms	Size (aa)	Receptor(s)
TNFSF1	Lymphotoxin-α, LTA, TNFβ	205	As homotrimer: TNFRSF1A, TNFRSF1B and TNFRSF14; as heterotrimer with LTB, TNFRSF3
TNFSF2	TNFα , Cachectin	233	TNFRSF1A, TNFRSF1B
TNFSF3	TNF-C, lymphotoxin β	244	LTBR/TNFRSF3
TNFSF4	OX40 ligand, CD252, TAX transcriptionally-activated glycoprotein 1	183	TNFRSF4
TNFSF5	CD40 ligand, TRAP	261	CD40
TNFSF6	Fas ligand, CD178	281	Fas/TNFRSF6
TNFSF7	CD70	197	CD27/TNFRSF7
TNFSF8	CD153	234	CD30/TNFRSF8
TNFSF9	4-1BB ligand	254	TNFRSF9
TNFSF10	TNF-related apoptosis-inducing ligand, TRAIL	281	TNFRSF10A, TNFRSF10B, TNFRSF10C, TNFRSF10D
TNFSF11	Receptor activator of NFκB ligand, RANKL	317	TNFRSF11B/OPG and TNFRSF11A/RANK
TNFSF12	TNF-related weak inducer of apoptosis, TWEAK	249	FN14 and possibly TNRFSF12/APO3
TNFSF13	[APRIL]	250	TNFRSF13B/TACI and TNFRSF17/BCMA
TNFSF13B	B-cell-activating factor	285	TNFRSF13B and TNFRSF17
TNFSF14	Herpes virus entry mediator ligand, LIGHT	240	TNFRSF3/LTBR and D receptor TNFRSF6B
TNFLSF15	Vascular endothelial cell growth inhibitor, TNF ligand-related molecule 1	251	TNFRSF25 and TNFRSF6B
TNFSF18	Glucocorticoid-induced TNF-related ligand, GITRL, AITRL	177	TNFRSF18

TNF Receptor Superfamily

Receptor	Synonyms	Role	Size (aa)	Ligands
TNFRSF1A	TNF-R1, Type I TNF receptor, p55	S	455	TNFSF2/TNFα and homotrimeric TNFSF1/lymphotoxin-α
TNFRSF1B	TNF-R2, Type II TNF receptor, p75	S	461	TNFSF2/TNF-alpha, low affinity for homotrimeric TNFSF1
TNFRSF2	-			
TNFRSF3	TNF-R3, Lymphotoxin-β receptor	S	435	Heterotrimeric lymphotoxin and TNFSF14
TNFRSF4	OX40L receptor, CD134	S	277	TNFSF4
TNFRSF5	CD40L receptor	S	277	TNFSF5
TNFRSF6	Fas antigen, CD95	S	335	TNFSF6
TNFRSF6B	Decoy receptor for Fas ligand	D	300	TNFSF6, TNFS14
TNFRSF7	CD27	S	260	CD70
TNFRSF8	CD30L receptor, Lymphocyte activation antigen CD30	S	565	TNFSF8

TABLE T1. (Continued)

TNF Receptor Superfamily

Receptor	Synonyms	Role	Size (aa)	Ligands
TNFRSF9	4-1BB ligand receptor, CD137	S	255	TNFSF14
TNFRSF10A	Death receptor 4, TRAIL-R1, CD261	S		
TNFRSF10B	Death receptor 5, TRAIL-R2, CD262	S	440	TNFSF10/TRAIL.
TNFRSF10C	TRAIL-R3,	D	259	TNFSF10
TNFRSF10D	TRAIL-R4	D	386	TNFSF10
TNFRSF11A	Receptor activator of NFκB, RANK, CD265	S	616	TNFSF11/RANKL/TRANCE/OPGL
TNFRSF11B	OPG, [Osteoprotegerin]	D	401	TNFSF10
TNFRSF12A	Tweak-receptor, CD266	S	129	TNFSF12/TWEAK
TNFRSF13	Transmembrane activator and [CAML] interactor, CD267	S	293	TNFSF13 and TNFSF13B
TNFRSF13C	B-cell-activating factor receptor, CD268		184	TNFSF13B specifically
TNFRSF14	Herpes virus entry mediator A	S	283	BTLA, TNFSF14/LIGHT and homotrimeric TNFSF1
TNFRSF16	Low-affinity nerve growth factor receptor	S	427	[NGF]
TNFRSF17	B-cell maturation protein, CD269	S	184	TNFSF13B and TNFSF13
TNFRSF18	GITR, activation-inducible TNFR family member, [AITR]	S	241	TNFSF18
TNFRSF19	Toxicity and JNK inducer, TRADE, Troy	S	423	TNFSF1
TNFRSF19L	Receptor expressed in lymphoid tissues, [RELT]	S	430	Unknown
TNFRSF21	Death receptor 6	S	655	Unknown
TNFRSF25	Apoptosis-mediating receptor TRAMP, Death receptor 3	S	417	TNFSF12/APO3L/TWEAK
TNFRSF27	X-linked [ectodysplasin]-A2 receptor	S	297	EDA isoform A2 only
TNFRSF-EDAR	Ectodysplasin-A1 receptor	S	448	Ectodysplasin-A1 only

Role: 'S' = signalling. 'D' = Decoy

tobramycin Aminoglycoside **antibiotic** used for serious infections that are resistant to gentamicin. Derived from species of *Streptomyces* or produced synthetically, inhibits protein synthesis by binding with the 30S ribosomal subunit.

Toc complex Transport system of the outer membrane of the chloroplast. Analogous to **Tom complex** though proteins are not the same. Toc75 seems to form the pore, Toc159 and Toc 34 are thought to be GTP-regulated import receptors; Toc34 and Toc75 act sequentially to mediate docking and insertion of Toc159 resulting in assembly of the functional translocon.on the cytoplasmic side. See **Tic complex**. Doctoral thesis: http://edoc.ub.uni-muenchen.de/3952/1/Becker_Thomas.pdf

TOC1 An *Arabidopsis* gene (TIMING OF CHLOROPHYLL A/B BINDING PROTEIN 1) that, in association with CIRCADIAN CLOCK ASSOCIATED 1 (CCA1) and LATE ELONGATED HYPOCOTYL (LHY) is thought to constitute the central oscillator of the *Arabidopsis* circadian clock. Abstract: http://www.genetics.org/cgi/content/abstract/176/3/1501

tocopherol See **vitamin E**.

toeprinting A primer extension inhibition assay used to study the initiation step of protein synthesis. In a toeprinting assay, mRNA is translated using purified ribosomal complexes and cycloheximide is added to the reaction to inhibit elongation, thereby arresting the position of the ribosomes on the transcript. The mRNA complex is then copied into cDNA using a specific labeled primer and, where the reverse transcriptase meets the ribosome bound to the mRNA, polymerization is halted, and a 'toeprint' fragment is generated. Generally the P site of the stalled ribosome is 15−17 nucleotides upstream of the toeprint.

TOG domain A domain (tumour overexpressed gene domain, ~250aa) thought to bind tubulin dimers and promote microtubule polymerization, found the Dis1/xmap215 family of proteins in fungi, plants, and animals. All the xmap215 family are essential for correct microtubule function during cell division. It is composed of N-terminally-arrayed hexa-HEAT (huntingtin, elongation factor 3, the PR65/A subunit of protein phosphatase 2A and the lipid kinase, Tor) repeats. Abstract: http://www.ncbi.nlm.nih.gov/pubmed/19754440

Togaviridae A family of viruses with a single non-segmented positive strand RNA genome (Class IV). They have a bullet-shaped capsid, enveloped by a membrane formed from the host cell plasma-membrane; the budded membrane contains host lipids and viral 'spike' glycoproteins. The group can be divided into two main genera: Alphaviruses, that include **Semliki Forest virus**, **Chikungunya virus** and **Sindbis virus**, and Rubiviruses (of which the type species is **rubella** (German measles) virus). Many are transmitted by insects and were previously classified as **arboviruses**. ICTV database: http://www.ncbi.nlm.nih.gov/ICTVdb/ICTVdB/00.073.htm

tokogenetic Relationships between individuals within species in contrast to phylogenetic relationships between species or separate lineages.

Tol-Pal proteins Proteins (TolA, TolB, TolQ, TolR and Pal) of the cell envelope of *E. coli*, and other Gram-negative bacteria, that are required for maintaining outer membrane integrity. TolA bridges between the inner and outer membranes via its interaction with the Pal lipoprotein. TolQ, TolR and TolA form a complex in the inner membrane, whereas TolB is a periplasmic protein. The Pal lipoprotein interacts with many components, such as TolA, TolB, OmpA, the major lipoprotein and the murein layer.

tolbutamide A **sulphonylurea** that will bind to **SUR** and enhance insulin release from pancreatic **B cells**.

tolerance See **immunological tolerance**.

tolerogen Substance that will induce **immunological tolerance**.

toll A *Drosophila* gene required for dorsoventral polarity determination. The gene encodes a trans-membrane receptor (1097aa). Interacts downstream with **pelle** and **tube** and defines dorsoventral polarity in the embryo. Toll, which is present over the entire surface of the embryo, is activated ventrally by interaction with a spatially restricted, extracellular ligand (spaetzle).

toll-like receptors Family of receptors, first discovered in *Drosophila* (see **toll**), involved in innate immunity and that recognise and respond to different microbial components. Type I transmembrane proteins with significant homology in their cytoplasmic domains to the IL-1 receptor Type I. See Table T2.

tolloid A *Drosophila* gene encoding a metallopeptidase (EC 3.4.24.-, 1067aa); involved in the process of specifying dorsal-ventral polarity, probably by activating **decapentaplegic**. Member of the bone morphogenetic (BMP) family of proteins and the homologue of BMP-1. Tolloid-like protein 1 (1013aa) in humans processes procollagen C-propeptides, such as chordin, pro-biglycan and pro-lysyl oxidase and is required for dorso-ventral patterning and skeletogenesis.

toluidine blue A thiazine dye (CI Basic Blue 17) related to methylene blue and Azure A in structure; often used for staining thick resin sections. Typically exhibits metachromasia.

Tom complex Protein transport complex (translocase of outer membrane, general import pore) of the outer membrane of the mitochondrion. The complex contains eight different proteins: Tom40 (40 kDa, 362aa) forms the 2.2 nm hydrophilic pore and spans the outer membrane; Tom5, Tom6 and Tom 7 are embedded within the membrane adjacent to Tom40, Tom20, Tom 22, Tom37, and Tom70 are on the cytosolic face with Tom22 on the inner face as well. Tom70 and Tom20 are are involved in the recognition of preproteins. The comparable system in chloroplasts is the **Toc complex**. Research article: http://jcb.rupress.org/content/147/5/959.full.pdf

Tom1 One of a small family of proteins (target of Myb1, Tom1, 492aa; Tom1L1/Srcasm, 476aa; Tom1L2, 507aa) probably involved in intracellular trafficking and production of which is regulated by the **myb** transcription factor. Tom1L1 has potential binding sequences for Tsg101. Homologues are found in other animal species and in *Dictyostelium*. Cf. **Tom complex**.

Tombusviridae A family of single-stranded positive sense RNA plant viruses. The RNA is encapsulated in an icosahedral capsid, composed of 180 units of a single coat protein. Members of the family are responsible for economically important diseases such as **dieback**.

tomoregulin See **TMEFFs**.

TABLE T2. Toll like receptors

Name	Distribution	Function/Comments
TLR1 (TIL, Rsc786)	Low level in monocytes	Regulates TLR2 response
TLR2 (TIL4)	Monocytes, granulocytes; Upregulated on macrophages	Interacts with microbial lipoproteins, peptidoglycans, and LPS; NFκB pathway
TLR3	Low level fibroblasts	Interacts with dsRNA; NFκB pathway; induces production of type I interferons.
TLR4 (HToll, Ly87, Rasl2-8, Ran/M1)	Monocytes, upregulated on endothelium	Interacts with microbial lipoproteins, CD14-dependent response to LPS, NFκB pathway
TLR5 (TIL3)	mRNA in leukocytes, prostate, liver, lung	Interacts with microbial lipoproteins, NFκB, response to *Salmonella*
TLR6	mRNA: leukocytes, ovary, lung	Interacts with microbial lipoproteins, regulates TLR2 response
TLR7	mRNA: spleen, lung, placenta; upreg on macrophages	Low similarity to other TLRs
TLR8	mRNA: leukocytes, lung	
TLR9	DC*, B cells (intracellular, low)	Receptor for CpG bacterial DNA
TLR10	mRNA: lymphoid tissues	Related to TLR1 and TLR6
RP105 (CD180, Ly-78)	Protein: mature B cells	B cell activation, LPS recognition
MD-1 (Ly86)	Mature B cells	Regulates surface expression of RP105
MD-2 (Ly96)	Macrophages	Regulates surface expression of TLR4, signals LPS presence

A useful resource is the information provided by eBioscience at: http://www.ebioscience.com/ebioscience/whatsnew/tlr.htm. This table is based upon the one that they have on that site.
*DC = dendritic cells.

tonic See **adaptation**.

tonofilaments Cytoplasmic filaments (10 nm diameter: **intermediate filaments**) inserted into **desmosomes**.

tonoplast Membrane that surrounds the vacuole in a plant cell.

tonoplast intrinsic proteins Tonoplast intrinsic proteins (TIPs), a subfamily of **aquaporins**, are widely used as markers for vacuolar compartments in higher plants. Ten TIP isoforms are encoded by the *Arabidopsis* genome, further classified into five subgroups: three γ-TIPs (TIP1), three δ-TIPs (TIP2), the seed-specific α- and β-TIP (TIP3;1 and TIP3;2), one ε-TIP (TIP4;1) and one ζ-TIP (TIP5;1). Different TIP isoforms may define different vacuole functions.

TOO MANY MOUTHS A single-pass membrane protein (TMM, 496aa) found in epidermal cells of developing shoots that acts as a receptor for **EPIDERMAL PATTERNING FACTORs**.

tophus Mass of urate crystals surrounded by a chronic inflammatory reaction: characteristic of gout.

TOPO cloning vector A proprietary cloning methodology that uses DNA topoisomerase I, which functions both as a restriction enzyme and as a ligase. http://www.invitrogen.com/site/us/en/home/brands/topo.html

topographic map In general, a map that illustrates the surface features (topography) but used more specifically for the spatially ordered projection of neurons onto their target; e.g. in the retino-tectal projection, retinal ganglion cell axons project along the **optic nerve** to the contralateral tectum where they ramify to form terminal arbors. The target sites of the terminal arbors are ordered: neurons from a specific region of the retina consistently project to a specific region of the tectum, forming a map of the retina on the tectum.

topographical control Those phenomena of cell behaviour in which the shape of the local substratum affects behaviour, see for example **contact guidance**.

topoinhibition Term used to describe the inhibition of cell proliferation as the cells become closely packed on a culture dish: generally superseded by the term **density dependent inhibition**.

topoisomerases Enzymes that change the degree of supercoiling in DNA by cutting one or both strands and rejoining them. Type I topoisomerases (EC 5.99.1.2) cut only one strand of DNA; type I topoisomerase of *E. coli* (omega protein) relaxes negatively supercoiled DNA and does not act on positively supercoiled DNA. Type II topoisomerases (EC 5.99.1.3) cut both strands of DNA; type II topoisomerase of *E. coli* (DNA gyrase) increases the degree of negative supercoiling in DNA and requires ATP. It is inhibited by several antibiotics, including nalidixic acid and ovobiocin. A third type of eukaryotic topoisomerase is thought to be important in recombination and in the alternative lengthening of telomeres (ALT) pathway; it has homology with prokaryotic topoisomerase 1. Topoisomerase inhibitors are used in tumour chemotherapy. In humans DNA topoisomerase 1 (TOP1, 765aa) is inhibited by **camptothecin**; there are two isoforms of topoisomerase 2 (TOP2A, 1531aa; TOP2B, 1626). TOP2A is inhibited by amsacrine, TOP2B is associated with the **WINAC complex**. The type 3 topoisomerases (TOP3A, 1001aa; TOP3B, 862aa) are components of the **RMI complex**.

TOR (1) The target of rapamycin (TOR) part of the ras/MAP kinase signalling pathway, originally characterized in yeast. In *S. pombe*, tor1 and tor2 are phosphatidylinositol 3-kinases (EC 2.7.1.137, 2335aa and 2337aa). In *S. cerevisiae* TOR1 and TOR2 are serine/threonine-protein kinases (phosphatidylinositol kinase homologues, EC2.7.11.1, 2470aa and 2474aa). The TOR Complex 1 (**TORC 1**) is inhibited by rapamycin that is bound to **FKBP**; TORC2 is not. (2) CREB-regulated transcription coactivators (transducer of regulated cAMP response element-binding protein 1, TORC1, 634aa; TORC2, 693aa; TORC3, 619aa). (3) In *E. coli*, an anaerobic respiratory chain involving trimethylamine-N-oxide reductase (torA, EC 1.7.2.3, 848aa; torZ, 809aa) associated respectively with a cytochrome c-type protein (torC, 390aa or torY, 366aa). The tor operon encodes various of the components involved including the response regulator TorR, and the sensor TorS. (4) See **torsins**.

TORC (1) In *S. cerevisiae*, the TOR (target of rapamycin) complex. TORC1 contains **TOR1** or **TOR2**, KOG1 (Kontroller of growth protein 1, 1557aa), TCO89 (target of rapamycin complex 1 subunit 89 kDa, 799aa) and LST8 (lethal with SEC13 protein 8, 303aa) and regulates multiple cellular processes to control cell growth in response to environmental signals. TORC2 contains TOR2, AVO1 (adheres voraciously to TOR2 protein 1, 1176aa), AVO2 (426aa), AVO3 (1430aa), BIT61 (61 kDa binding partner of TOR2 protein, 543aa) and **LST8**; it regulates cell cycle-dependent polarization of the actin-cytoskeleton and cell wall integrity. **FKBP**-rapamycin binds TORC1, but not TORC2, and TORC2 disruption causes an actin defect,

suggesting that TORC2 mediates the rapamycin-insensitive, TOR2-unique pathway. mTOR is the equivalent mammalian system. (2) See **MECT**.

torpedo In *Drosophila*, the epidermal growth factor receptor (EC 2.7.10.1, gurken receptor, 1426aa). See **star**.

Torres body An intranuclear inclusion body of host protein seen in liver cells infected with yellow fever virus (**Togaviridae**) although not a reliable diagnostic feature. *Cf.* **Councilman bodies**.

torsins A sub-family of the AAA family of ATPases (torsin1A, tor1A, 332aa; tor1B, 289aa; tor2A, 321aa; tor3A, 397aa) that are associated with the endoplasmic reticulum and the nuclear envelope. The bioactive **salusin** peptides are derived from an isoform of torsin 2A. Mutations in torsin A are associated with early-onset torsion dystonia, a movement disorder characterised by twisting muscle contractures.

torso A maternally contributed receptor tyrosine kinase (EC. 2.7.1.112, 923aa) made by nurse cells during oogenesis in *Drosophila* and deposited into the oocyte. It is involved in specification and pattern formation in cells located in the anterior and posterior terminal regions of the embryo but is only present in the early stages of development and is not found after gastrulation. The ligand is thought to be **trunk** and activation of torso triggers expression of gap genes by antagonising Gro-mediated repression (see **groucho**). Article: http://www3.interscience. wiley.com/cgi-bin/fulltext/109923806/PDFSTART

torulosis An obsolete name for the infection caused by *Cryptococcus neoformans*, formerly known as *Torula histolytica*. See **cryptococcosis**.

torus (1) In general, a topological form, a cylinder distorted into a ring-shape with the ends fused (often referred to as doughnut-shaped) although a 1-torus is a line formed into a closed circle. (2) In anatomy, any rounded projection or swelling. (3) In botanical anatomy, a structure found at the centre of a bordered **pit**, especially in conifers, forming a thickened region of the pit membrane. When subjected to a pressure gradient, it seals the pit by pressing against the pit border.

total internal reflection fluorescence microscopy *TIRF* A form of fluorescence microscopy in which total internal reflection of light at a glass-liquid interface generates an evanescent wave that will excite fluorochromes that are close to the interface. Used, for example, to image points of contact of cells with the surface.

totipotent Capable of giving rise to all types of differentiated cell found in that organism. A single totipotent cell could, by division, reproduce the whole organism. See **embryonic stem cells**.

Townes-Brocks syndrome An autosomal dominant developmental disorder affecting a range of systems (anus, ears, kidney) caused by mutation in the gene encoding the **SALL1** putative transcription factor.

toxic shock syndrome Endotoxic shock caused by bacterial contamination of tampons; toxin responsible is produced by some strains of *Staphylococcus aureus*.

toxicity Formally, the capacity to cause injury to a living organism, usually defined with reference to the quantity of substance administered or absorbed, the mode of administration (how and how frequently), the type and severity of injury, the time needed to produce the injury, the nature of the organism(s) affected, and other relevant conditions. Toxicity is often expressed as the reciprocal of the absolute value of median lethal dose ($1/LD_{50}$) or lethal concentration ($1/LC_{50}$).

toxicodynamics By analogy with **pharmacodynamics**, study of the way in which potentially toxic substances interact with target sites, and the biochemical and physiological consequences that lead to adverse effects.

toxicogenetics Study of the genetic cause of individual variations in the response to potentially toxic substances. Tends to focus on single gene polymorphisms, rather than the overall genomic environment in which particular genes are being expressed, but is often used more or less interchangably with **toxicogenomics**. See **pharmacogenetics, pharmacogenomics**.

toxicogenomics Analysis of the genomic determinants of drug efficacy and toxicity; this encompasses not only genetic polymorphisms in, for example, **cytochrome P450**, but also the genomic environment in which individual genes are being expressed.

toxicokinetics Analysis of the uptake of potentially toxic substances by the body, the biotransformation they undergo, the distribution of the substances and their metabolites in the tissues, and the elimination of the substances and their metabolites from the body. By analogy with **pharmacokinetics**.

toxigenicity The ability of a pathogenic organism to produce injurious substances that damage the host.

toxin A naturally-produced poisonous substance that will damage or kill other cells. Bacterial toxins are frequently the major cause of the pathogenicity of the organism in question. See **endotoxins** and **exotoxins**.

toxin-antitoxin systems A set of closely linked genes encoding a protein toxin and a specific antitoxin often contained on a plasmid. Only daughter cells that inherit the plasmid survive after cell division because they have the anti-toxin. In type I toxin-antitoxin systems, the translation of messenger RNA for the toxin is inhibited by the binding of a small non-coding RNA antitoxin: an example is the tisAB / IstR system. In a type II system the protein toxin is post-translationally inhibited by a protein antitoxin (e.g. the mazF/mazE system in *Staphylococcus aureus*). Type III toxin-antitoxin systems rely on direct interaction between a toxic protein and an RNA antitoxin. The only known example is the ToxIN system from the bacterial plant pathogen *Erwinia carotovora*. The toxic ToxN protein (170aa) is inhibited by ToxI RNA.

Toxocara A genus of nematodes (roundworms) that cause intestinal infections, although larval stages may migrate into tissues. Human infections are usually acquired from dogs (*Toxocara canis*) or cats (*Toxocara cati*).

toxofilin An actin binding protein (245aa) that sequesters actin monomers and caps actin filaments in *Toxoplasma gondii*. Regulated by phosphorylation. Research article: http://www.molbiolcell.org/cgi/content/full/11/1/355

toxoid Non-toxic derivative of a bacterial exotoxin produced by formaldehyde or other chemical treatment: useful as a vaccine because it retains most antigenic properties of the toxin.

Toxoplasma A genus of parasitic protozoa. *T. gondii* is an intracellular parasite whose intermediate hosts includes humans, the final host being felines of many species. Causes toxoplasmosis in humans in which the parasite finally locates in tissues such as brain, heart, or eye causing serious and sometimes fatal lesions.

TP-1 See **trophoblast protein 1**.

tp53 The gene encoding the tumour suppressor **p53**.

TPA (1) See **plasminogen activator**. (2) A **phorbol ester** tumour promoter, 12-O-tetradecanoyl-phorbol-13-acetate also known as PMA, phorbol myristyl acetate.

TPCK Nonspecific protease inhibitor (tosyl phenyl chloromethyl ketone), interacts with histidine residues and inactivates by interfering with the active site.

TphiCG loop $T\phi CG$ See **T-loop of RNA**.

Tpn1 family See **En/Spm**.

TPO (1) Thyroid peroxidase (TPO, EC 1.11.1.8, 933aa), an important enzyme in the production of **thyroid hormones**. (2) A family of polyamine transporters (TPO1, 586aa; TPO2, 614aa; TPO3, 622aa; TPO4, 659aa, TPO5, 618aa), members of the major facilitator superfamily, involved in the detoxification of excess polyamines in the cytoplasm of *S. cerevisiae*. A related protein in mammals (tetracyline transporter-like protein, TETRAN, 455aa) confers cellular resistance to apoptosis induced by indometacin and diclofenac. It has 12 transmembrane domains and may act as an efflux pump. (3) Synonym for serine incorporator 5 (developmentally regulated proteinTpo1, serinc5, 460aa) that is expressed at high levels in the white matter and the oligodendroglial cells of the brain and may have a role in providing serine for the formation of myelin glycosphingolipids in oligodendrocytes.

TPR motif Degenerate consensus sequence (tetratricopeptide motif, $\sim$34aa) found in various proteins that are involved in the regulation of RNA synthesis, protein import and *Drosophila* development.

Tpr protein A conserved component of the nuclear pore complex (translocated promoter region protein, 2349aa) that may be involved in protein export.

tra-2 (1) Product of *C. elegans* gene *tra-2* that is required for female development and predicted to encode a large transmembrane protein (sex-determining transformer protein 2, 1475aa) that inhibits downstream male determinants. Translationally regulated by two elements, called TGEs (for tra-2 and GLI elements), located in the 3′ untranslated region of the mRNA, to which the **STAR protein GLD-1** binds. (2) A *Drosophila* protein (Transformer-2 sex-determining protein, tra-2, 272aa) required for female sex determination in somatic cells and for spermatogenesis in male germ cells. Human homologues (transformer-2 protein homologue alpha, 282aa; TRA2-beta, 288aa) are sequence-specific RNA-binding proteins involved in the control of pre-mRNA splicing. A homologue (326aa) is also found in *Dictyostelium*.

TRAAK One of the tandem-pore potassium channels (**TWIK**-related arachidonic acid-stimulated potassium channel protein, KCNK4, 393aa) that is voltage insensitive and outwardly rectifying. It is activated by polyunsaturated fatty acids.

trabecula A transverse structure across a cavity, e.g. strands of connective tissue projecting into an organ or the small interconnecting rods that make up cancellous bone.

trabecular bone See **cancellous bone**.

trabecular meshwork A specialized eye tissue essential in regulating intraocular pressure. See **myocilin**.

tracheid Water-conducting cell forming part of the plant **xylem**. Contains thick, lignified secondary cells walls, with no protoplast at maturity. Interconnects with neighbouring tracheids through pits; the end-walls are not perforated (*cf.* **vessel elements**).

Tracheophyta Vascular plants.

tracheophyte A vascular plant, a fern (pteridophyte) or a seed plant (gymnosperm or angiosperm).

trachoma A highly contagious form of bilateral keratoconjunctivitis caused by infection of the conjunctiva with *Chlamydia trachomatis*. Lacrimal glands and ducts are often affected as well and corneal abrasion can cause blindness.

TRADD A protein (TNF-receptor-1 associated death domain protein, 312aa) that binds through its **death domain** to the cytoplasmic tail of **TNF receptor-1**, and to **FADD** and **RIP**, though does not itself seem to have any catalytic activity.

TRAFs TNF receptor-associated proteins. Family of proteins that associate with the cytoplasmic domain of the TNF receptor and TNFR-family such as CD40. Act as cytoplasmic adaptor proteins in NFκB signalling; multiple isoforms are recognised.

TRAIL *Apo2L* TNF-related apoptosis-inducing ligand. One of the TNF ligand superfamily family that can be expressed either as a transmembrane protein (281aa) or in soluble form; induces apoptosis in a variety of tumour cell lines but not typically in normal or non-transformed cells. Structurally similar to CD95-ligand. There are four receptors, two of which seem to act as decoy receptors.

tram (1) Translocating chain-associated membrane protein 1 (TRAM1, 374aa in humans) required for the translocation of secretory proteins across the ER membrane. (2) Nuclear receptor coactivator 3 (EC 2.3.1.48, thyroid hormone receptor activator molecule 1, TRAM1, steroid receptor-associated coactivator 3, EC 2.3.1.48, 1424aa) a histone acetyltransferase that binds nuclear receptors and stimulates the transcriptional activities in a hormone-dependent fashion. (3) In *E. coli*, a transfer gene protein (traM, 127aa) that is involved in the bacterial conjugation. (4) In *Agrobacterium tumefaciens*, a transcriptional repressor (traM, 102aa) that negatively regulates conjugation and the expression of tra genes. (5) TRIF-related adaptor molecule, see **TIR-domain**.

TRAMP (1) Old name for one of the TNF receptor superfamily (TNFRSF25, death receptor 3, DR3, wsl-1; Apo-3, 417aa) abundantly expressed on thymocytes and lymphocytes, the receptor for TWEAK (TNFSF12). (2) A strain of mice (TRAMP mice) with transgenic adenocarcinoma of the mouse prostate, used as a model for human prostatic carcinoma. (3) Tyrosine-rich acidic matrix protein (TRAMP); see **dermatopontin**.

tramtrack In *Drosophila*, isoform beta of a repressor protein (tramtrack p69, 643aa) that binds to a number of sites in the transcriptional regulatory region of **fushi tarazu**.

trans-acting siRNAs *ta-siRNAs* A class of endogenous **siRNAs** that direct the cleavage of complementary mRNA targets, but the targets of these siRNAs are different from the transcripts that give rise to them (hence 'trans-acting'). In plants, TAS gene transcripts are cleaved by miRNAs; the cleavage products are protected against degradation by SGS3, copied into dsRNA by RNA-dependent RNA polymerase 6 (RDR6), and diced into ta-siRNAs by Dicer like-4 (DCL4).

trans-Golgi network A complex of membranous tubules and vesicles, near the trans-face of the **Golgi apparatus**, which is thought to be a major intersection for intracellular traffic of vesicles.

trans-splicing The splicing of two different pre-mRNA molecules together in contrast to the normal cis-splicing of conventional RNA molecules.

transactivation Experimental approach to control gene expression by the introduction of a transactivator gene and special promoter regions of DNA into a genomic region of interest. The transactivator-coded **transcription factor** will activate multiple genes with the appropriate upstream promoters (acts 'in trans') and by putting the transactivator under control of an inducible promoter, gene expression can be switched on or off. A reporter gene linked to the transactivator will show whether it is active or not. There are virus-encoded transactivators (e.g. **tat** of HIV), some of which may induce tumours by acting on proto-oncogenes.

transactivator See **transactivation**.

transacylase Any enzyme (EC 2.3 class) that transfers an acyl group. EC 2.3.1 enzymes transfer groups other than amino-acyl groups, EC 2.3.2 are aminoacyltransferases, EC 2.3.3 transfer acyl groups that are converted into alkyl on transfer.

transaldolase A conserved enzyme (EC 2.2.1.2, 337aa) that catalyzes the reversible transfer of a three-carbon ketol unit from sedoheptulose 7-phosphate to glyceraldehyde 3-phosphate to form erythrose 4-phosphate and fructose 6-phosphate. Together with transketolase, links the pentose phosphate pathway with glycolysis by converting pentoses to hexoses.

transaminases Class of enzymes (aminotransferases; EC 2.6.1. -) that convert amino acids to keto acids in a cyclic process using pyridoxal phosphate as cofactor; e.g. aspartate amino transferase catalyses the reaction: aspartate + α-ketoglutarate = oxaloacetate + glutamate.

transcranial magnetic stimulation *TMS* A neurophysiological technique that allows the induction of a current in the brain using a magnetic field generated by a current through a coil of copper wire that is encased in plastic and held over the subject's head. It is possible to apply TMS in trains of multiple stimuli per second (repetitive TMS) which may prove a valuable investigative tool.

transcriptase See **reverse transcriptase**.

transcription Synthesis of RNA by RNA polymerases using a DNA template.

transcription factors Eukaryotic proteins that bind to specific DNA sequences and regulate the binding of RNA polymerases and the production of mRNA. Some transcription factors activate mRNA production, others (transcriptional repressors) are inhibitory: many operate in conjunction with other proteins. So many are now known (possibly 10% of all known proteins) that it is no longer possible to provide a comprehensive listing but there is a web-based database (Transcription Factor Classification). See, *inter alia*, **AP2, GATA transcription factors, helix-loop-helix, helix-turn-helix, leucine zipper, TFII, TFIII, winged helix transcription factors, zinc finger**. See also Table T3. TRANSFAC database: http://www.gene-regulation.com/pub/databases/transfac/cl.html

transcription squelching Anomalous suppression of transcription of a gene by overexpression of a transcription factor that would be expected to raise transcription levels. Thought to be caused by sequestration of a limiting cofactor by the overexpressed transcription factor.

transcription unit A region of DNA that is transcribed to produce a single primary RNA transcript, i.e. a newly synthesized RNA molecule that has not been processed. Transcription units can be mapped by kinetic studies of RNA synthesis, and in some instances directly visualized by electron microscopy.

transcriptional control Control of gene expression by controlling the number of RNA transcripts of a region of DNA. A major regulatory mechanism for differential control of protein synthesis in both pro- and eukaryotic cells.

transcriptional insulators DNA sequences that subdivide the eukaryotic genome into regulatory domains by limiting the activities of enhancer and silencer elements. News & Views article: http://www.nature.com/ng/journal/v36/n10/full/ng1004-1036.html

transcriptional silencing A transcriptional control mechanism in which DNA is bundled into **heterochromatin** in order to make it permanently inaccessible for future transcription. Effectively, this allows for memory in the **determination** of cell fate in developing organisms. In *Drosophila*, **homeotic genes** are silenced by members of the **Polycomb** group of genes.

transcriptome The transcripts (mRNA) that are present in a cell. The content of the transcriptome varies as cells differentiate or respond to cues. It is analogous, but not identical, to the **proteome**.

transcriptosome Proposed model in which the RNA transcription machinery is preassembled as a unitary particle completely separate from the chromosomes, by analogy with the ribosome. *Cf.* **transcriptome**.

TABLE T3. Transcription factors

Many transcription factors have now been identified and an important resource is the TRANSFAC database at http://www.gene-regulation.com/pub/databases/transfac/cl.html. In this database, transcription factors are classified into Superclasses, Classes, Familes and sub-families: only the first three levels are listed here.

1. *Superclass*: Basic Domains

Class: Leucine zipper factors (bZIP).
- AP-1(-like) components
- CREB
- C/EBP-like factors
- bZIP/PAR
- Plant G-box binding factors
- ZIP only
- Other bZIP factors

Class: Helix-loop-helix factors (bHLH).
- Ubiquitous (class A) factors
- Myogenic transcription factors
- Achaete-Scute
- Tal/Twist/Atonal/Hen
- Hairy
- Factors with PAS domain
- INO
- HLH domain only
- Other bHLH factors

Class: Helix-loop-helix/leucine zipper factors (bHLH-ZIP).
- Ubiquitous bHLH-ZIP factors
- Cell-cycle controlling factors

Class: NF-1
- NF-1

Class: RF-X
- RF-X

Class: bHSH
- AP-2

2. *Superclass*: Zinc-coordinating DNA-binding domains

Class: Cys4 zinc finger of nuclear receptor type.
- Steroid hormone receptors
- Thyroid hormone receptor-like factors

Class: Diverse Cys4 zinc fingers.
- GATA-Factors
- Trithorax
- Other factors

Class: Cys2His2 zinc finger domain.
- Ubiquitous factors
- Developmental/cell cycle regulators
- Metabolic regulators in fungi
- Large factors with NF-6B-like binding properties
- Viral regulators

Class: Cys6 cysteine-zinc cluster.
- Metabolic regulators in fungi

Class: Zinc fingers of alternating composition
- Cx7Hx8Cx4C zinc fingers
- Cx2Hx4Hx4C zinc fingers

3. *Superclass*: Helix-turn-helix

Class: Homeo domain.
- Homeo domain only
- POU domain factors
- Homeo domain with LIM region
- Homeo domain plus zinc finger motifs

Class: Paired box.
- Paired plus homeo domain
- Paired domain only

Class: Fork head/winged helix.
- Developmental regulators
- Tissue-specific regulators
- Cell-cycle controlling factors
- Other regulators

Class: Heat shock factors
- HSF

Class: Tryptophan clusters.
- Myb
- Ets-type
- Interferon-regulating factors

Class: TEA domain.
- TEA

4. *Superclass*: beta-Scaffold Factors with Minor Groove Contacts

Class: RHR (Rel homology region).
- Rel/ankyrin
- Ankyrin only
- NF-AT

Class: STAT
- STAT

TABLE T3. (Continued)

Class: p53

- p53

Class: MADS box.

- Regulators of differentiation
- Responders to external signals
- Metabolic regulators

Class: beta-Barrel alpha-helix transcription factors

- E2

Class: TATA-binding proteins

- TBP

Class: HMG.

- SOX
- TCF-1
- HMG2-related
- UBF
- MATA
- Other HMG box factors

Class: Heteromeric CCAAT factors

- Heteromeric CCAAT factors

Class: Grainyhead

- Grainyhead

Class: Cold-shock domain factors.

- csd

Class: Runt.

- Runt

5. *Superclass*: Other Transcription Factors

Class: Copper fist proteins

- Fungal regulators

Class: HMGI(Y)

- HMGI(Y)

Class: Pocket domain

- Rb
- CBP

Class: E1A-like factors

- E1A

Class: AP2/EREBP-related factors

- AP2
- EREBP
- AP2/B3

transcytosis Process of transport of material across an epithelium by uptake on one face into a coated vesicle, which may then be sorted in the endosomal compartment, and then delivery to the opposite face of the cell, still within a vesicle.

transcytotic vesicle Membrane-bounded vesicle that shuttles fluid from one side of the endothelium to the other. There is some controversy as to whether or not they form pores.

transdetermination Change in determined state observed in experiments on *Drosophila* **imaginal discs**. These can be cultured for many generations in the abdomen of an adult, where they proliferate but do not differentiate. If transplanted into a larva, they differentiate after pupation according to the disc from which they were derived; they maintain their **determination**. Occasionally the disc will differentiate into a structure appropriate to another disc, showing transdetermination. It is a rare event, involves a population of cells, and certain changes are more common than others; e.g. leg to wing is more frequent than wing to leg.

transdifferentiation Change of a cell or tissue from one differentiated state to another. Rare, and has mainly been observed with cultured cells. In newts the pigmented cells of the iris transdifferentiate to form lens cells if the existing lens is removed.

transducin A **GTP-binding protein** found in the disc membrane of **retinal rods** and **cones** and involved in the cascade involved in transduction of light to a nervous impulse. A complex of three subunits; α (350aa), β (340aa) and γ (42aa); alpha subunits are encoded by different genes in rods and cones. Photoexcited rhodopsin interacts with transducin and promotes the exchange of GTP for GDP on the α subunit. The GTP-α subunit dissociates from the complex and activates a cGMP-phosphodiesterase by removing an inhibitory subunit. The a subunit of transducin can be ADP-ribosylated by cholera toxin and pertussis toxin.

transducisome *signalosome* A macromolecular complex of signalling molecules. Abstract of review: http://www.ncbi.nlm.nih.gov/pubmed/10712921

transduction (**1**) The transfer of a gene from one bacterium to another by a **bacteriophage**. In generalized transduction any gene may be transferred as a result of accidental incorporation during phage packaging. In specialized transduction only specific genes can be transferred, as a result of improper recombination out of the host chromosome of the **prophage** of a **lysogenic** phage. Transduction is an infrequent event but transducing phages have proved useful in the genetic analysis of bacteria. (**2**) See **signal transduction**.

transfection The introduction of DNA into a recipient eukaryote cell and its subsequent integration into the recipient cell's chromosomal DNA. Usually accomplished using DNA precipitated with calcium ions though a variety of other methods can be used (e.g. **electroporation**). Only about 1% of cultured cells are normally transfected. Transfection is analogous to bacterial transformation but in eukaryotes transformation is used to describe the changes in cultured cells caused by **tumour viruses**. Though originally used to describe the situation in which the transfected DNA is integrated, it is now frequently used just to mean introduction of DNA into a target cell, hence the necessity to specify **stable transfection**.

transfer cell In plants a cell specialized for transfer of water-soluble material to or from a neighbouring cell, usually a phloem sieve tube or a xylem tracheid. Elaborate wall ingrowths greatly increase the area of plasma membrane at the cell face across which transfer occurs. A type of **companion cell**.

transfer factor (**1**) A dialysable factor obtained from sensitised T-cells by freezing and thawing, that may possibly immunopotentiate animals. The transfer of specific immunity from one animal to another has been claimed and it is still being reported in the literature. (**2**) Transfer factor or carbon monoxide diffusing capacity (DL(CO)) is a test of the appropriateness of gas exchange across the alveolar membrane in the lung.

transfer RNA See **tRNA**.

transferase A suffix to the name of an enzyme indicating that it transfers a specific grouping from one molecule to another; e.g. acyl transferases transfer acyl groups.

transferrin The iron storage and transport protein (serotransferrin, siderophilin, 698aa) found in mammalian plasma; a β-globulin. Binds ferric iron with a K_{ass} of around 21 at pH 7.4, 18.1 at pH 6.6. An important constituent of growth media. **Transferrin receptors** on the cell surface bind transferrin as part of the transport route of iron into cells.

transferrin receptor The cell surface receptor (CD71, 760aa) a disulphide-linked homodimer, found ubiquitously on actively growing cells. It is responsible for internalisation of transferrin-bound iron and its intracellular release. In iron deficiency the amount of soluble transferrin receptor in plasma, produced by proteolysis of the membrane-bound form, increases.

transformasome Name proposed for a membranous extension responsible for binding and uptake of DNA; found on the surface of transformation-competent *Haemophilus influenzae* bacteria. Abstract of paper: http://www.ncbi.nlm.nih.gov/pubmed/19228200

transformation Any alteration in the properties of a cell that is stably inherited by its progeny. Classical example was the transformation of an avirulent strain of *Diplococcus* (*Streptococcus*) *pneumoniae* to virulence by DNA from a virulent strain, achieved in 1944 by Avery, MacLeod & McCarty. Now known that certain species of Bacteria and Archaea are naturally transformable. Currently usually refers to malignant transformation, but is used in other senses also, such as **blast transformation** of lymphocytes, so can only be distinguished by context. Malignant transformation is a change in animal cells in culture that usually greatly increases their ability to cause tumours when injected into animals. (It is assumed that parallel changes occur during carcinogenesis *in vivo*). Transformation can be recognized by changes in growth characteristics, particularly in requirements for macromolecular growth factors, and often also by changes in morphology. See **abortive transformation**, **genetic transformation**, **germ-line transformation**, **sol-gel transformation**, **spontaneous transformation**, **stable nuclear transformation**, **viral transformation**.

transformed cell See **transformation**.

transforming genes Genes, originally of tumour viruses, responsible for their ability to transform cells. The term now serves as an operational definition of oncogenes.

transforming growth factor *TGF* A family of growth factors secreted by transformed cells that can stimulate growth of normal cells. An unfortunate misnomer, since they induce aspects of transformed phenotype, such as growth in semi-solid agar, but do not actually transform. TGFα, a 50aa polypeptide originally isolated from viral-transformed rodent cells, contains **EGF-like domain**, binds to the EGF receptor, stimulates growth of microvascular endothelial cells and is angiogenic. TGFβ polypeptide, a homodimer of two 112aa chains, is secreted by many different cell types, stimulates wound healing but *in vitro* is also a growth inhibitor for certain cell types. The TGF family includes many of the bone morphogenetic proteins (BMPs) See also **activin**, **syndecan**. Mutations in TGFβ cause **Camurati-Engelmann disease**.

transforming virus Viruses capable of inducing malignant **transformation** of animal cells in culture. Among the **Oncovirinae**, non-defective viruses that lack oncogenes can induce tumours such as leukaemias in animals, but cannot transform *in vitro*. On acquisition of oncogenes they become (acute) transforming viruses.

transgelins Highly conserved actin binding proteins related to **calponin** that cause gelation. In humans transgelin 1 (smooth muscle protein 22-alpha, 201aa) is found in smooth muscle and fibroblasts. Transgelin 2 (199aa) is more widely distributed. Transgelin 3

(199aa) is mostly found in brain. Transgelin is also found in *S. cerevisiae* (calponin homologue 1, 200aa).

transgene A **gene** or DNA fragment from one organism that has been stably incorporated into the **genome** of another organism to produce a transgenic plant or animal.

transgenic mitigation strategy A strategy aimed at reducing the risk of transfer of genes from a transgenic organism to other (plant) species, particularly wild variants. It can involve 'mitigator genes' tandemly linked to the primary transgene, which will reduce the fitness of hybrids and their rare progeny. Article: http://www.isb.vt.edu/articles/feb0603.htm

transgenic organisms Organisms that have integrated foreign DNA into their germ line as a result of the experimental introduction of DNA.

transglutaminase An important extracellular enzyme (protein-glutamine γ-glutamyltransferase, Factor XIIIa, fibrinoligase, fibrin stabilizing factor, EC 2.3.2.13, 732aa) that catalyses the formation of an amide bond between side-chain glutamine and side-chain lysine residues in proteins with the elimination of ammonia. The linkage is stable and plays an important role in many extracellular assembly processes.

transglycosylation Transfer of a glycosidically bound sugar to another hydroxyl group. Various bacterially-derived transglycosylases, especially those from thermophiles, are being studied for use in biotechnological synthetic processes. Xyloglucan endotransglycosylases (XETs) cleave and re-ligate xyloglucan polymers in plant cell walls via a transglycosylation mechanism and are important in cell wall remodelling.

transhydrogenase The NAD(P) transhydrogenase (pyridine nucleotide transhydrogenase, 1086aa in humans) is a tetramer composed of 2 alpha (EC 1.6.1.1) and 2 beta (EC 1.6.1.2) subunits, an integral mitochondrial membrane protein that couples the proton transport across the membrane to the reversible transfer of hydride ion equivalents between NAD and NADP. *E. coli* contains both a soluble (466aa) and a membrane-bound (510aa) proton-translocating pyridine nucleotide transhydrogenase.

transient expression Transient transfection. When **transfecting** animal cells, cells in which the transgene has not been physically incorporated into the genome (**stable transfection**), but is carried as an episome that can be lost. This means that expression levels will not be constant over time, and will eventually fall away.

transin A peptidase (matrix metalloproteinase 3, procollagenase activator, proteoglycanase, stromelysin 1, EC3.4.24.17, 477aa) secreted by carcinoma cells: carboxy-terminal domain has haemopexin-like domains, and the N-terminal domain has the proteolytic activity. May be involved in digestion of extracellular matrix.

trans-inhibition A phenomenon observed with **Pak1** which forms homodimers *in vivo*, dimerization being regulated by the intracellular level of GTP-Cdc42 or GTP-Rac1. The homodimer is catalytically inactive because the N-terminal inhibitory portion of one Pak1 molecule in the dimer binds and inhibits the catalytic domain of the other. One GTPase interaction can result in activation of both partners.

transition probability model A model, now probably obsolete, that postulates that transition from G1 to S phase is probabilistic rather than depending on the accumulation of critical levels of particular proteins.

transition proteins In **spermatogenesis**, group of proteins that displace **histones** from nuclear DNA, and that are in turn displaced by **protamines** to produce the transcriptionally inactive nuclear DNA characteristic of the sperm nucleus.

transition temperature The temperature at which there is a transition in the organization of, for example, the phospholipids of a membrane where the transition temperature marks the shift from fluid to more crystalline. Usually determined by using an **Arrhenius plot** of activity against the reciprocal of absolute temperature, the transition temperature being that temperature at which there is an abrupt change in the slope of the plot. In membranes such phase-transitions tend to be inhibited by the presence of cholesterol.

transitional elements Region (transitional endoplasmic reticulum) at the boundary of the rough endoplasmic reticulum (RER) and the Golgi. **Transport vesicles** are responsible for the transfer of secretory proteins from this part of the RER to the Golgi system.

transitional epithelium An epithelial sheet made up of cells that change shape when the epithelium is stretched. Usually a **stratified epithelium**: best known example is in the bladder.

transketolase An enzyme (EC 2.2.1.1, 623aa in humans, 675aa in *Zea mays*) of both the pentose phosphate pathway in animals and the Calvin cycle of photosynthesis. In humans catalyses the reaction: sedoheptulose 7-phosphate + D-glyceraldehyde 3-phosphate = D-ribose 5-phosphate + D-xylulose 5-phosphate. Transketolase-like protein 1 (596aa in humans) is a second form of the enzyme. Defects in transketolase activity are implicated in Wernicke-Korsakoff syndrome. See **Wernicke's encephalopathy**.

translation The process whereby the information in mRNA is used to specify the sequence of amino acids in a polypeptide chain.

translational control Control of protein synthesis by regulation of the translation step, for example by selective usage of preformed mRNA or instability of the mRNA.

translational research Research that relates to the process of taking a scientific discovery into a clinical context; the whole process of moving a newly discovered compound from discovery research, through development and clinical trials and through the regulatory process so that it can be used on patients.

translationally controlled tumour protein See **TCTP**.

translesion synthesis The replication of DNA despite damage to the template by DNA polymerases of the Y family.

translin A DNA-binding protein (228aa) that recognizes single-stranded DNA ends generated by staggered breaks occuring at recombination hot spots. Translin-associated factor X (290aa) interacts with translin and may have a role in spermatogenesis.

translocase (1) A general term for an enzyme that is involved in moving molecules from one location to another. (2) A bacterial GTPase (elongation factor G; EF-G; EC 3.6.1.48, 704aa) that is a force-generating motor that causes peptidyl-tRNA to move from the A-site to the P-site in the **ribosome** and the mRNA to move so that the next codon is in position for usage. A mitochondrial enzyme (EC 3.6.5.3, 751aa in humans) carries out the same function and will work in bacteria, although the converse is not true, the bacterial enzyme will not work in mitochondria. (3) See **TOM complex**.

translocated intimin receptor Bacterial protein (Tir, ~550aa) secreted by attaching and effacing (A/E) pathogens that is translocated into target mammalian cells via a type III secretion system and forms a receptor for the adhesin **intimin**. The translocated Tir forms a tetrameric complex in the host membrane that binds two intimin molecules and triggers formation of host-cell signalling systems and actin polymerisation. The N-terminal region of Tir binds α-actinin. Article: http://www.cell.com/current-biology/abstract/S0960-9822%2800%2900543-1

translocating chain-associating membrane protein See **tram**.

translocation Rearrangement of a chromosome in which a segment is moved from one location to another, either within the same chromosome or to another chromosome. This is sometimes reciprocal, when one fragment is exchanged for another.

translocator protein In mammals, an 18-kDa peripheral-type benzodiazepine receptor (TSPO/PBR, 169aa) localized in the outer mitochondrial membrane

where it acts to transport haem, porphyrins, steroids and anions. The *Arabidopsis* homologue functions in the transport of protoporphyrinogen IX to the mitochondria where haem can be synthesized. In *Rhodobacter sphaeroides* TSPO is localized in the outer membrane and its expression is induced by oxygen: in conditions of high oxygen, TSPO negatively regulates the expression of photosynthetic genes. Role in plants: http://www.biomedcentral.com/1471-2229/11/108

translocon The complex of proteins associated with the translocation of nascent polypeptides into the cisternal space of the endoplasmic reticulum. The translocon is a multi-functional complex involved in regulating the interaction of ribosomes with the ER as well as regulating translocation and the integration of membrane proteins in the correct orientation. **Tram**, **signal peptidase** and signal recognition protein are among the proteins associated with the translocon.

transmembrane adaptor proteins See **TRAPS**.

transmembrane protein A protein with the polypeptide chain exposed on both sides of the membrane. The term does not apply when different subunits of a protein complex are exposed at opposite surfaces. Many integral membrane proteins are also transmembrane proteins. Types I and II are single pass molecules with the N-terminus to the exterior or in the cytoplasm respectively. Types II and IV are multi-pass proteins.

transmembrane transducer A system that transmits a chemical or electrical signal across a membrane. Usually involves a transmembrane receptor protein that is thought to undergo a conformational change that is expressed on the inner surface of the membrane. Many such transducing species are dimeric and the conformation change may involve interaction between the two components.

transmigration Migration of cells from one surface of a monolayer of cells to the other side; used particularly of the migration of leucocytes from the lumen of a blood vessel across vascular endothelium of the post-capillary venule and into tissue, and by extension *in vitro* models of this process.

transmissible mink encephalopathy A transmissible **spongiform encephalopathy**, though originally thought to be an unconventional **slow virus** infection.

transmission electron microscopy *TEM* See **electron microscopy**.

transpeptidases (1) Enzymes that create amide links from a free amino group and a carbonyl group on polypeptides. The bacterial transpeptidase (D-alanyl-D-alanine carboxypeptidase, beta-lactamase, EC 3.4.16.4, 403aa) cross-links peptidoglycan chains to form rigid cell walls and is the target of penicillin-type antibiotics. **(2)** A family of D-glutamyl transpeptidases (e.g. gamma-glutamyltranspeptidase-1, D-glutamyltransferase, EC 2.3.2.2, 569aa) are involved in breakdown of extracellular glutathione (GSH), provide cells with a local cysteine supply and contribute to maintaining intracellular GSH levels.

transpiration Loss of water-vapour from land-plants into the atmosphere, causing movement of water through the plant from the soil to the atmosphere via roots, shoot and leaves. Occurs mainly through the **stomata**.

transplantation antigen Any antigen that is antigenically active in graft rejection. In practice the **major histocompatibility complex** and the H-Y antigens, and to a lesser extent minor histocompatibility antigens.

transplantation reaction The set of cellular phenomena observed after an allogeneic (mismatched) graft is made to an organism that leads to destruction, detachment or isolation of the graft. In mammals this includes the invasion and destruction of the graft by cytotoxic lymphocytes, inhibition of **angiogenesis** and other processes.

transport diseases Single-gene defect diseases in which there is an inability to transport particular small molecules across membranes. Examples are aminoacidurias such as cystinuria, **iminoglycinuria**, **Hartnup disease**, **Fanconi syndrome**.

TRANSPORT INHIBITOR RESPONSE1 *TIR1* One of the proteins (TIR1, weak ethylene-insensitive protein 1, 594aa) that acts as a substrate-recognition component of a **SCF** E3 ubiquitin ligase complex that binds auxin and promotes the degradation of Aux/IAA repressor proteins.

transport protein A class of transmembrane protein that allows substances to cross plasma membranes far faster than would be possible by diffusion alone. A major class of transport proteins expend energy to move substances (**active transport**); these are transport ATPases. See **facilitated diffusion**, **symport**, **antiport**.

transport vesicle Vesicles that transfer material from the rough endoplasmic reticulum (RER) to the receiving face of the Golgi.

transporter See **transport protein**.

transportins A small family of proteins involved in transport from cytoplasm to the nucleus, serving as receptors for nuclear localization signals. Transportin 1 (TNPO1, importin beta2, karyopherin beta2, 898aa) and transportin 2 (importin beta2, karyopherin bet-2B, 897aa) recognise ribonucleoproteins (heterogeneous nuclear ribonucleoproteins, hnRNPs) that have a 38aa M9 domain. Transportin 3 (transportin-SR, importin12, 923aa), binds specifically

to the phosphorylated RS domains of various splicing factor proteins (SR proteins). Transportin activity is regulated by Ran-GTP. See **importins**.

transposable element See **transposon**.

transposase An enzyme that brings about the transposition of a sequence of DNA within a chromosome or between chromosomes. Transposase exists in dimer form with each monomer binding a separate transposon end. The DNA binding domain of the protein is responsible for recognising the appropriate sequence. See **transposon**.

transposition Movement form one location to another, particularly the movement of a DNA sequence (**transposon**) within the genome.

transposome Preformed **transposase-transposon** complexes that have been electroporated into bacterial cells.

transposon *transposable element* Small, mobile DNA sequences that can replicate and insert copies at random sites within chromosomes. They have nearly identical sequences at each end, oppositely oriented (inverted) repeats, and code for the enzyme, transposase, that catalyses their insertion. Bacteria have two types of transposon; simple transposons that have only the genes needed for insertion, and complex transposons that contain genes in addition to those needed for insertion. Eukaryotes contain three classes of mobile genetic elements; Class I are **retrotransposons** which move by producing RNA that is transcribed, by reverse transcriptase, into DNA which is then inserted at a new site, Class II are like bacterial transposons in that DNA sequences move directly, Class III are the miniature inverted-repeats transposable elements (MITEs). Gypsy database: http://gydb.uv.es/index.php/Intro

transposon tagging A widely used method for generating insertion mutants for gene functional analysis, particularly in plants (using **T-DNA).** Transposable elements create mutations at the site of insertion and genes tagged by a transposable element can be isolated by using the tag as a probe. See **activation tagging**.

transthyretin Plasma protein (thyroxine-binding prealbumin, 4.5 mM in plasma) that transports **thyroxine** and retinol (vitamin A). Tetrameric with 4 identical 127aa subunits. Transthyretin forms a complex under physiological conditions with retinol binding protein (2 mM in plasma) so that RBP is not lost by filtration in the kidney. Various defects can cause amyloidosis type 1 or 7, in which fibrils of transthyretin are deposited in the brain, or hyperthyroxinemia.

transudate Plasma-derived fluid that accumulates in tissue and causes **oedema**. A result of increased venous and capillary pressure, rather than altered vascular permeability (which leads to cellular exudate formation).

transvection An epigenetic phenomenon that depends upon an interaction between an allele on one chromosome and the corresponding allele on the homologous chromosome. It seems to depend upon chromosome pairing and the interaction does not occur if the alleles involved are on different chromosomes. The result can be either gene activation or repression. See **zeste**.

transverse tubule See **T tubule**.

transversion A point mutation in which a pyrimidine is replaced by a purine or *vice versa*.

TRAP See **transmembrane adaptor proteins, thrombospondin-related anonymous protein** and **Telomeric Repeat Amplification Protocol assay**.

TRAP assay See **Telomeric Repeat Amplification Protocol assay**.

TRAPS Transmembrane adaptor proteins. A family of proteins that include linker for activation of T-cells (**LAT**), the phosphoprotein associated with glycosphingolipid-enriched micro domains (PAG)/C-terminal Src kinase (csk) binding protein (Cbp), SHP2-interacting transmembrane adaptor protein (**SIT**), T-cell receptor interacting molecule (**TRIM**), non-T-cell activation linker (NTAL) and pp30. TRAPs share several common structural features, in particular multiple sites for tyrosine phosphorylation in their cytoplasmic tails. See also **adaptor proteins**.

traumatic acid A compound (9-hydroxy-traumatin, 9-hydroxy-12-oxo-10E-dodecenoic acid) produced in response to stress in plants, a product of the **lipoxygenases**. It acts as a wound healing agent by stimulating cell division near a trauma site to form a protective callus and to heal the damaged tissue. It may also act as a growth hormone, especially in inferior plants (e.g. algae). Traumatic acid is biosynthesized in plants by non-enzymatic oxidation of traumatin (12-oxo-trans-10-dodecenoic acid), another wound hormone.

TRB Mammalian homologues of the *Drosophila* protein tribbles. Tribbles homolog 1 (TRB1, 372aa) and TRB2 (343aa in mouse) interact with MAPK kinases and regulate activation of MAP kinases but may not have kinase activity (pseudokinases). Tribbles homologue 3 (neuronal cell death-inducible putative kinase, p65-interacting inhibitor of NFκB, 358aa) seems to ubiquitinate ACC1 (the rate-limiting enzyme in fatty acid synthesis) resulting in inactivation of β-oxidation. Impairs **Akt** activation induced by growth factors and insulin and seems to be the endogenous inhibitor of Akt/PKB.

TRE (1) Human oncogene (*tre*), isolated from NIH3T3 cells transfected with human Ewing's sarcoma DNA, that encodes a deubiquitinating enzyme (ubiquitin carboxyl-terminal hydrolase 6,

deubiquitinating enzyme 6, EC 3.1.2.15, 1406aa). *Drosophila* and yeast homologues have been discovered. (**2**) Thyroid hormone response element (TRE), a DNA sequence recognized by the thyroid hormone receptor. (**3**) **TPA** responsive element. (**4**) In *S. cerevisiae* a putative zinc metallopeptidase (TRE2, transferrin receptor-like protein 2, 809aa). (**5**) See **tre locus**.

tre locus A bacterial trehalose utilization locus, consisting of a transcriptional regulator, **treR**; a trehalose phosphoenolpyruvate transferase system (PTS) transporter, treB; and a trehalose-6-phosphate hydrolase, treC.

Treacher Collins syndrome See **treacle**.

treacle Product of the gene mutated in Treacher Collins syndrome (a disorder of craniofacial development) that encodes a protein (TCOF1, 1488aa) that may be involved in nucleolar-cytoplasmic transport. Haploinsufficiency may result in inadequate rRNA production during the early stages of embryogenesis.

treadmilling Name given to the proposed process in microtubules in which there is continual addition of subunits at one end, and disassembly at the other, so that the tubule stays of constant length, but individual subunits move along. Could in principle be used as a transport mechanism, although this is not currently favoured as a possibility. Has also been suggested for microfilaments.

treble clef A type of **zinc finger** motif found in a diverse group of proteins, the best characterized being the nuclear hormone receptors. Survey: http://www.ncbi.nlm.nih.gov/pmc/articles/PMC140525/?tool = pubmed

TreeView Software used for constructing phylogenetic trees. Article: http://taxonomy.zoology.gla.ac.uk/rod/treeview.html

trefoil factors Family of proteins (trefoil peptides) containing a trefoil motif, considered as scatter factors, proinvasive and angiogenic agents acting through cyclooxygenase-2 (COX-2)-dependent and thromboxane A2 receptor (TXA2-R)-dependent signalling pathways. Includes three members: TFF1 (pS2, 84aa), TFF2 or spasmolytic peptide (SP, 129aa) and TFF3 (intestinal trefoil factor, 80aa). TFFs are associated with mucin-secreting epithelial cells and play a crucial role in mucosal defense and healing. TFF1 is upregulated by estrogen in many breast carcinomas and is present in many other carcinomas.

trefoil motif Domain (P domain, 40aa) found in various secretory polypeptides that has highly conserved cysteine residues that are disulphide bonded in such a way as to generate a trefoil structure (bonded 1−5, 2−4, 3−6). There are also highly conserved A, G and W residues.

T-regulatory cells *T-suppressor cells* A class of T-cells the existence of which was disputed for a long time but is now fairly generally accepted. They suppress activation of the immune system in an antigen-specific manner, inducing and maintaining immune tolerance. Two subtypes are recognised, $CD8^+CD28^-$ and $CD4^+CD25^+$. The T-suppressor cell alloantigen (Tsud) maps near immunoglobulin allotype genes and may be a heavy chain constant-region marker on a T-cell receptor. Overview: http://www.clinicalmolecularallergy.com/content/7/1/5

trehalase Hydrolase (EC.3.2.1.28, alpha,alpha-trehalase, 583aa.) that converts trehalose into glucose. Deficiency of the intestinal enzyme in humans causes isolated trehalose intolerance.

trehalose A disaccharide sugar (342 Da) found widely in invertebrates, bacteria, algae, plants and fungi, formed by the dimerisation of glucose. Yields glucose on acid hydrolysis.

TREKs Tandem pore potassium channels (TWIK-related K^+ channels, TREK-1, 426aa; TREK-2, 538aa) that are outward rectifying. TREK-1 is activated by volatile general anesthetics. TREK-2 is activated by arachidonic acid and other unsaturated free fatty acids.

TREMs A family of receptors (triggering receptors expressed on myeloid cells) that regulate myeloid cell function. TREM1 (234aa) stimulates neutrophil and monocyte-mediated inflammatory responses and triggers release of pro-inflammatory chemokines and cytokines; a crucial mediator of septic shock. TREM-2 (230aa) is expressed on macrophages, microglia and pre-osteoclasts. It may have a role in chronic inflammation, stimulating production of constitutive chemokines and cytokines. Both signal through the adaptor DAP12. Defects of TREM-2 and DAP12 result in a rare syndrome characterized by presenile dementia and bone cysts. TREM3 (Trem-like transcript 3, 195aa) and TREM4 (200aa) are also known. TREM and TREM-like Receptors in Inflammation and Disease: http://www.ncbi.nlm.nih.gov/pmc/articles/PMC2723941/?tool = pubmed

Treponema Genus of bacteria of the spirochaete family (Spirochaetaceae) that are commensals or pathogens of humans and animals. *T. pallidum* causes syphilis. Cells are corkscrew-like, (6−15 µm long, 0.1−0.2 µm wide), motile, anaerobic, and with a **peptidoglycan** cell wall and a capsule of glycosaminoglycans similar to hyaluronic acid and chondroitin sulphate in composition. Membrane has **cardiolipin**.

treppe *staircase phenomenon* The gradual increase in the extent of muscular contraction following rapid repeated stimulation, particularly in cardiac muscle

that has received a number of stimuli of the same intensity following a quiescent period.

TREX complex A multiprotein complex (transcription/export complex), evolutionarily conserved from yeast to man, required for coupled transcription elongation and nuclear export of mRNAs. It specifically associates with spliced mRNA and not with unspliced pre-mRNA. Research article: http://www.biomedcentral.com/1741-7007/8/1

TRF (1) Thyrotropin-releasing factor, a synonym for **thyroid stimulating hormone**. (2) **Telomeric repeat binding factors** 1 & 2 (TRF-1, 439aa; TRF-2, 500aa). TRF2-interacting protein (399aa) is part of the **shelterin** complex and the homologue of the yeast telomeric RAP1 protein. (3) Topoisomerase-related function proteins. In humans TRF4-1 (DNA polymerase sigma, EC 2.7.7.7, 542aa) is probably involved in DNA repair.

triabody A trimeric antibody fragment with three Fv heads, each of which consists of a VH domain from one antibody paired with a VL domain from an unrelated antibody with a linker sequence between them that is too short to permit intramolecular pairing of the domains. See **diabody**.

triacyl glycerols See **triglycerides**.

triad The junction (triad junction) between the T tubules and the sarcoplasmic reticulum in striated muscle.

triadin A transmembrane protein (729aa) found in junctional terminal cisternae of sarcoplasmic reticulum where it co-localises with the **ryanodine receptor** and **junctin** and anchors **calsequestrin**. Suggested to have a central role in the mechanism of skeletal muscle excitation/contraction coupling.

tribbles A *Drosophila* protein (484aa) that regulates the cell cycle by specifically inducing proteosomal degradation of the CDC25 mitotic activators **string** and **twine**. There are three mammalian homologues: tribbles-homologue 1 (trib1, 372aa) inhibits **AP-1** activation and interacts with several mitogen-activated protein kinases. Trib2 (343aa) is a regulator of the inflammatory activation of monocytes. Trib3 (358aa) affects insulin signalling at the level of **Akt-2**. Although there is a protein kinase domain they do not appear to have enzyme activity.

tributyltin *TBT* Organic compound added to marine anti-fouling paint; now banned because it causes abnormalities in some marine creatures notably causes dog whelks to suffer from **imposex**.

tricarboxylic acid cycle *TCA cycle* Citric acid cycle, Krebs cycle. The central feature of oxidative metabolism. Cyclic reactions whereby acetyl CoA is oxidized to carbon dioxide providing reducing equivalents (NADH or $FADH_2$) to power the electron transport chain. Also provides intermediates for biosynthetic processes.

Trichinella spiralis See **trichiniasis**.

trichiniasis *trichinosis* Nematode infestation of the human intestine with the nematode worm *Trichinella* (or *Trichina*) *spiralis*, the larvae of which migrate from the gut and become encysted in muscle. Usually acquired as a result of eating raw or underdone pork.

Trichocephalida An Order containing three families: Cystoopsidae, Trichinellidae, and Trichuridae. See *Trichuris*.

trichocyst Small membrane-bounded vesicle lying below the pellicle of many ciliates. Fusion of the trichocyst with the plasma membrane occurs at a predictable site which can therefore be examined for membrane specialization.

Trichoderma viride A filamentous ascomycete fungus widely distributed in soil, plant material, decaying vegetation, and wood. *Trichoderma* may cause infections in immunocompromised individuals.

trichohyalin Major structural protein (THH, 1943aa) of inner root sheath cells and medulla of hair follicle, similar to **profilaggrin**, **involucrin** and **loricrin**, present in small amounts in other specialised epithelia. Trichohyalin is alpha-helix-rich and forms rigid structures as a result of postsynthetic modification by transglutaminases that cross-link the proteins and peptidyl-arginine deiminase that converts arginine to citrulline and modifies structure of the protein. Modified trichohyalin is thought to serve as a **keratin** intermediate filament-associated matrix protein, like **filaggrin**.

trichomes Epidermal outgrowths on plants, variously hair-like or scale-like. Plants with trichomes are hirsute, those without are glabrous although there is a range of terms according to the types and numbers of hairs or scales. Hairs may be unicellular or multicellular. *Arabidopsis* trichomes are used as a model of plant cell differentiation and cell biology, and the control of early trichome development is well-understood, involving transcription factors GLABRA3 and GLABRA1 and the WD-repeat protein TRANSPARENT TESTA GLABRA (TTG). See **GL1**.

Trichomonads Mastigophoran protozoa, pear-shaped and ranging in size from 4 μm to 30 μm. They possess three to five anterior flagella with a recurrent anterior flagellum which is attached to the body as an undulating membrane. Trichomonads reproduce by simple longitudinal binary fission although sexual reproduction may also occur in some species especially those parasitic on invertebrates. Many species are parasitic in vertebrates, including *Trichomonas vaginalis* which

is responsible for a sexually transmitted disease in man (trichomoniasis).

Trichonympha Genus of flagellated protozoans symbiotic in the intestine of some cockroaches and termites where they are responsible for the digestion of cellulose.

trichostatin A An antifungal antibiotic from *Streptomyces platensis* that is a potent reversible histone deacetylase (**HDAC**) inhibitor. Will activate the transcription of DNA methylation-mediated silenced genes in human cancer cells.

trichothecenes Mycotoxins (T-2 toxin, HT-2 toxin, diacetoxyscirpenol, deoxynivalenol) produced by certain species of fungi, e.g. *Fusarium*, that contaminate various agricultural products and are toxic for granulocytic and erythroblastic progenitor cells. Inhibit translation, although have a range of other effects.

Trichuris spp Nematodes (whip-worms). *Trichuris trichiura* is a common parasite of the human gut, especially in tropical Asia.

triclosan A chlorophenol used for its antibacterial properties, an ingredient in many detergents, cosmetics, lotions, insect repellants and an additive in various plastics and textiles. Probably a human health risk and environmental risk. Blocks the active site of the bacterial enzyme enoyl-acyl carrier-protein reductase that is essential for fatty acid synthesis.

tricyclic antidepressants A group of drugs formerly used in the treatment of moderate to severe depressive illness, probably working by inhibiting the re-uptake of norepinephrine and serotonin. Now superceded by **SSRIs**.

tricyclodecan-9-yl xanthate An inhibitor of sphingolipid biosynthesis that interferes with the trafficking of **prosaposin** to the lysosomal compartment. Also an inhibitor of phosphatidylcholine-specific phospholipase C.

TRIF See **TICAMs**.

triflavin See **disintegrin**.

trifluoperazine Antipsychotic drug (trifluperazine, Stellazine) that inhibits **calmodulin** at levels just below those at which it kills cells.

trigeminal system Neurons associated with the fifth or trigeminal nerve, the largest cranial nerve. The trigeminal system provides sensory innervation to the face and mucous membrane of the oral cavity, along with motor innervation to the muscles of mastication. It is called trigeminal because it has three major peripheral branches, the opthalmic, the maxillary and the mandibular nerves.

triglycerides Storage fats of animal adipose tissue where they are largely glycerol esters of saturated fatty acids. In plants they tend to be esters of unsaturated fatty acids (vegetable oils). Present as a minor component of cell membrane. Important energy supply in heart muscle.

trigramin See **disintegrin**.

triiodobenzoic acid An inhibitor of basipetal **auxin** transport in plants.

triiodothyronine See **thyroid hormones**.

TRIM (1) T-cell receptor interacting molecule (T-cell receptor-associated transmembrane adaptor 1, TRAT1, 186aa) is a **TRAP** found in lymphocytes. (2) A large family of E3 ubiquitin-protein ligases (EC 6.3.2.-, **tripartite motif**-containing proteins). TRIM22 (498aa) is an interferon-induced antiviral protein involved in cell innate immunity. TRIM25 (EC 6.3.2.19, 630aa) is involved in innate immune defense against viruses by mediating ubiquitination of **DDX58**. TRIM32 (653aa) ubiquitinates DTNBP1 (**dysbindin**), TRIM33 (1127aa) promotes **Smad4** ubiquitination, TRIM41 (630aa) catalyzes the ubiquitin-mediated degradation of protein kinase C, TRIM63 (353aa) is muscle-specific and regulates degradation of cardiac troponin I. Many others are known.

TriMEDB Triticeae Mapped EST Database. A database of mapped expressed sequence tags from Triticeae that provides annotated information on cDNA markers that are related to barley and their homologues in wheat. Link to database: http://trimedb.psc.riken.jp/

trinucleotide repeat Repetitive part of a genome that may form part of the coding sequence of a gene. The length of such repeats is frequently **polymorphic**, and unstably amplified repeats appear to be the major cause of such genetic diseases as **Huntington's chorea**, **fragile X syndrome**, spinobulbar muscular atrophy and myotonic dystrophy. Tripartite motif-containing protein-32 (653aa) has an E3 ubiquitin ligase activity and ubiquitinates **dysbindin**; defects cause limb-girdle **muscular dystrophy** type 2H and **Bardet-Biedl syndrome** type 11.

trio Multidomain protein (2861aa) that binds **LAR** transmembrane tyrosine phosphatase, has a serine/threonine protein kinase domain and separate rac- and rho-specific **GEF**-domains.

triolein A triglyceride that is the major component of olive oil.

TRIP6 Thyroid receptor-interacting protein 6 (zyxin-related protein 1, 476aa) of the zyxin/ajuba family, involved in relaying signals from the cell surface to the nucleus to weaken adherens junctions and promote actin cytoskeleton reorganization and cell invasiveness, including that induced by lysophosphatidic acid. It is also a transcriptional coactivator for NFκB and JUN. See **supervillin**.

tripartite motif A motif that is found in a large family of **TRIM** proteins that share a common function, they form homo-multimers and identify specific cell compartments. The motif consists of a RING finger, a B box finger and a coiled-coil domain. Various other domains in TRIM proteins give them their specific functions.

triple A syndrome An autosomal recessive neuroendocrinological disease (alachrima-achalasia-adrenal insufficiency syndrome (Addinsonianism)) caused by mutations in a gene that encodes aladin (546 aa), a **nucloporin**.

triple response The vascular changes in the skin in response to mild mechanical injury, an outward-spreading zone of reddening (flare) followed rapidly by a weal (swelling) at the site of injury. Redness, heat, and swelling, three of the 'cardinal signs' of inflammation, are present.

triple vaccine Vaccine for diphtheria, tetanus and whooping cough in infants. A sterile preparation of diphtheria and tetanus toxoids with acellular pertussis vaccine.

triploid Having three times the haploid number of chromosomes.

triptans General term for a family of drugs used for the acute treatment of migraine attacks, serotonin 5-HT1B/1D receptor agonists that act to induce vasoconstriction of extracerebral blood vessels and reduce neurogenic inflammation.

triskelion A three-legged structure assumed by **clathrin** isolated from **coated vesicles**. A trimer of clathrin with three light chains is probably the physiological subunit of clathrin coats in coated vesicles.

trisomy An additional copy of a chromosome so that there are three copies not two in a diploid organism. Best known example is trisomy 21 in **Down's syndrome**.

tristetraprolin An ARE (**AU-rich element**) binding protein (TTP, Poly(A)-specific ribonuclease, PARN, EC 3.1.13.4, 639aa) the only trans-acting factor shown to be capable of regulating AU-rich element-dependent mRNA turnover at the level of the intact animal.

tritanopia See **colour blindness**.

tritium A radioactive isotope of hydrogen (^{3}H) with a half-life of 12.25 years. Weak β-emitter, very suitable for autoradiography, and relatively easy to incorporate into complex molecules.

Triton X-100 Non-ionic detergent (iso-octylphenoxypolyethoxyethanol) used in isolating membrane proteins: the detergent replaces the phospholipids that normally surround such a protein. Other detergents of the Triton group are occasionally used so the full name should be quoted.

trituration Reduction in particle size by grinding in a mortar and pestle.

Triturus Genus of newts, much studied for their **lampbrush chromosomes**.

trk Oncogene, from human colon carcinoma, a chimeric gene formed through a somatic rearrangement involving the neighbouring genes for neurotrophic tyrosine kinase receptor type 1 (**NTRK1**) and tropomyosin-3. The *trkA* gene product (790aa) is a receptor for **NGF**, that of *trkB* the receptor (833aa) for **neurotrophin** 4 and **BDNF**, and of *trkC* the receptor (839aa) for NT-3. Other chimeric genes with trk tyrosine kinase linked to parts of other genes produce a constitutively active kinase and are associated with tumours. One example is the TRK-T3 oncogene which encodes a thyroid oncoprotein.

tRNA *transfer RNA* The low molecular weight RNAs (4S RNA) that specifically bind amino acids by amino-acylation to form **aminoacyl tRNA**, and that possess a special nucleotide triplet, the anticodon, sometimes containing the base inosine. They recognize codons on mRNA. By this recognition the appropriate tRNAs are brought into alignment in turn in the ribosome during protein synthesis (translation), there being at least one species of tRNA for each amino acid. In practice most cells possess about 30 types of tRNA. The amino acids are bound at the 3′ terminus that is always 3′-ACC. The anticodon is around 34−38 nucleotides from the 5′ end and the total length of the various tRNAs is 70−80 bases.

trochophore Free-living ciliated larval form of several different invertebrate phyla.

trophectoderm The extra-embryonic part of the ectoderm of mammalian embryos at the blastocyst stage before the mesoderm becomes associated with the ectoderm.

trophic Concerning food or nutrition. Not to be confused with tropic (stimulatory).

trophinin An integral membrane protein (749aa) that forms a cell adhesion molecule complex in which it binds **bystin** directly and **tastin** indirectly. Important for attachment of the blastocyst to uterine epithelium in implantation.

trophoblast Extra-embryonic layer of epithelium that forms around the mammalian blastocyst, and attaches the embryo to the uterus wall. Forms the outer layer of the chorion, and together with maternal tissue will form the placenta.

trophoblast protein 1 Protein (TP1, Interferon tau-1, trophoblastin, 195aa) secreted by the **trophoblast** in ruminants; prolongs the lifetime of the corpus luteum, thus signalling pregnancy.

trophophase In a culture system, for example in a bioreactor, the phase during which the cells or organisms are growing rapidly but not producing the metabolic products of interest. See **idiophase**.

trophosome An organ of dark green-brown spongy tissue, found in hydrothermal vent tubeworms (vestimentiferans) such as *Riftia pachyptila* (Siboglinidae, Polychaeta), in which there are endosymbiotic thiotrophic bacteria within host bacteriocytes that comprise 70% of the trophosome's volume.

trophozoite The feeding stage of a protozoan (as distinct from reproductive or encysted stages).

tropism A tropism is the orientation of an organism or part of an organism (e.g. leaf or stem) that involves turning or curving by differential growth in response to an environmental stimulus. Should be distinguished from a tactic response that involves movement of the whole cell or organism in response to a gradient of some sort. See **chemotropism**. In plants well-known examples include the response to light (phototropism) and to gravity (gravitropism).

tropocollagen Subunit from which collagen fibrils self-assemble: generated from **procollagen** by proteolytic cleavage of the extension peptides.

tropoelastin Soluble polypeptide (786aa), precursor to **elastin**, consisting mainly of repetitive elements of four, five, six and nine hydrophobic residues. Tropoelastin molecules, having been chaperoned out of the cell, are aligned on a scaffold of **fibrillin**-rich microfibrils and then stabilized by the formation of intermolecular cross-links (**desmosines**) to form elastin fibres.

tropomodulin Actin-capping protein that interacts with tropomyosin and **nebulin** at the pointed end of actin filaments. Tropomodulin-1 (359aa) is found in the erythrocyte membrane skeleton, tropomodulin-2 (351aa) is neuron-specific, tropomodulin-3 (352aa) is ubiquitous, tropomodulin-4 (354aa) is found in striated muscle. See **leiomodulin**.

tropomyosin Protein (284aa) associated with actin filaments both in cytoplasm and (in association with **troponin**) in the thin filament of striated muscle. Composed of two elongated α-helical chains 40 nm long, 2 nm diameter. Each chain has six or seven similar domains and interacts with as many G-actin molecules as there are domains. Not only does the binding of tropomyosin stabilize the F-actin, but the association with troponin in striated muscle is important in control by calcium ions. There are 4 genes in humans, expressed in different tissues.

troponin *TnC, TnI, TnT* Complex of three proteins, troponins C, I, and T, associated with **tropomyosin** and actin on the thin filament of striated muscle, upon which it confers calcium sensitivity. There is one troponin complex per tropomyosin. Troponin C (161aa) binds calcium ions reversibly, has a variable number of **EF-hand** motifs and is the least variable of the subunits. TnC binds TnI and TnT, but not actin. Troponin I (cardiac, 210aa; fast 182aa, slow, 187aa) binds to actin and at 1:1 stoichiometry can inhibit the actin-myosin interaction on its own. Troponin T (278aa) binds strongly to tropomyosin. Defects in troponin are associated with various myopathies.

trp (1) Tryptophan. (2) Operon encoding tryptophan metabolism genes in *E. coli*. (3) Transient receptor potential mutant in *Drosophila*, gene codes for a constitutively active calcium channel activated by **thapsigargin** (though trp1 is insensitive to thapsigargin). See **TRP channels**. (4) Human tyrosinase-related protein (Trp1, 537aa; Trp2, 517aa) found in **melanosomes** and involved in pigment formation.

TRP channels Non-selective cation channels (transient receptor potential channels) found in dorsal root ganglia neurons and skin cells. Six main subfamilies are recognised: the TRPC (canonical), TRPV (vanilloid), TRPM (melastatin), TRPP (polycystin), TRPML (mucolipin) and TRPA (ankyrin) groups. Some are responsible for temperature sensing and have unusually high temperature sensitivity ($Q_{10} > 10$). Six thermo-TRP channels have been cloned; TRPV1-4 are heat activated, whereas TRPM8 and TRPA1 are activated by cold. **Capsaicin** and resiniferatoxin are agonists for TRPV1, menthol for TRPM8 (cold receptor), and icilin for both TRPM8 and TRPA1. See **vanilloid receptor**. International Union of Pharmacology. XLIX. Nomenclature and Structure-Function Relationships of Transient Receptor Potential Channels http://pharmrev.aspetjournals.org/content/57/4/427.long

TRRAP An adaptor protein (transformation/transcription domain-associated protein, 3859aa) found in various multiprotein chromatin complexes that have histone acetyltransferase activity including the **SAGA** complex.

TRT Telomerase reverse transcriptase: see **telomerase** and **hTERT**.

Trypan blue Biological stain, an azo dye (diamine blue; Niagara blue 3B), often used to determine cell viability because it will not cross intact plasma membranes, and so only stains dead cells.

Trypanosoma Genus of Protozoa that causes serious infections in humans and domestic animals (trypanosomiasis). African trypanosomes, of the *T. brucei* group, are carried by Tsetse flies (*Glossina*) and, when they enter the bloodstream of the mammalian host go through a complex series of stages. Perhaps the most interesting feature is that there are recurrent bouts of parasitaemia as the parasite alters its surface antigens to evade the immune response of the host (see **antigenic variation**). The repertoire of antigenic variation is considerable. The S. American trypanosomes (of which *T. cruzi* is the best known) are carried by reduviid bugs, and cause a chronic and

incurable disease (Chagas' disease). Other interesting features of trypanosomes are the kinetoplast DNA and glycosomes (organelles containing enzymes of the glycolytic chain).

trypomastigote Any trypanosome-like stage in the life cycle of certain trypanosomatid protozoa, resembling the typical adult form of members of the genus *Trypanosoma*, slender elongate cells with a kinetoplast and basal body located at the posterior end and a flagellum running anteriorly along an undulating membrane. The trypomastigote stage of *Trypanosoma spp.* is the infectious stage carried by the insect vector.

trypsin Serine peptidase (EC 3.4.21.4, 247aa) from the pancreas of vertebrates. Cleaves peptide bonds involving the amino groups of lysine or arginine.

tryptophan One of the 20 amino acids (Trp, W, 204 Da) found in proteins, an essential dietary component in humans. Precursor of nicotinamide. See Table A1.

TSC-22 family A family of eukaryotic proteins involved in transcriptional regulation. Includes several TSC-22 domain proteins, the mammalian **delta sleep-inducing peptide**-immunoreactive peptide), the *Drosophila* protein bunched (shortsighted) a probable transcription factor required for peripheral nervous system morphogenesis, eye development and oogenesis and *C. elegans* hypothetical protein T18D3.7. UniProt entry: http://www.ebi.ac.uk/interpro/IEntry?ac = IPR000580

TSC1 See **hamartin**.

TSG-14 TNF-inducible gene (TNF-stimulated gene 14) of fibroblasts encoding a protein of the **pentraxin** family.

TSG101 Tumour susceptibility gene 101 protein (390aa), a component of the ESCRT-I complex involved in regulating vesicular trafficking. Binds to ubiquitinated cargo proteins.

TSH See **thyroid-stimulating hormone**.

TSH releasing factor A tripeptide produced by the hypothalamus that stimulates the anterior pituitary to release **thyroid-stimulating hormone** (TSH).

t-SNARE See **syntaxins**.

TSS See **toxic shock syndrome**.

Tst (1) Toxic shock syndrome toxin-1 (234aa) produced by some strains of *Staphylococcus aureus*. (2) A toxin (Tst1, 84aa) from *Tityus stigmurus* (Brazilian scorpion) that binds to mammalian sodiium channels and shifts the activation voltage toward more negative potentials, promoting spontaneous and repetitive firing. (3) Thiosulphate sulphurtransferase (EC 2.8.1.1, 297aa) that catalyses the formation of iron-sulphur complexes and is involved in cyanide detoxification.

tsu In *Drosphila* a protein (tsunagi, RNA-binding protein 8A, 165aa) involved in mRNA quality control via the **nonsense-mediated mRNA decay** pathway. Also involved in localization of **oskar** mRNA in the posterior pole of oocytes via its interaction with **mago nashi**.

T-suppressor cell See **T regulatory cells**.

TTC (1) 2,3,5-triphenyltetrazolium chloride (TTC), a compound used in spectrophotometric estimation of cellular respiration which reduces it to generate red formazan. (2) Threshold of toxicological concern (TTC), a pragmatic risk assessment tool for human exposure to chemicals. (3) **Tetanus toxin** C-fragment (TTC). (4) A trinucleotide repeat that is expanded in intron 1 of the frataxin gene, causing Friedreich's ataxia. (5) A stable prostacyclin analogue, TTC-909. (6) Prefix for a number of tetratricopeptide repeat proteins, for example E3 ubiquitin-protein ligase TTC3 (2025aa).

TTF (1) Transcription termination factors, proteins that regulates transcription of genes (TTF1, 905aa; TTF2, 1162aa) coupling ATP hydrolysis with removal of RNA polymerase II from the DNA template. (2) Thyroid transcription factor (TTF1, homeobox protein Nkx-2.1, 371aa) that activates thyroid specific genes such as thyroglobulin, thyroperoxidase and the thyrotropin receptor. Other TTFs are known.

T-toxin A series of linear β-polyketols (35−45 carbons) produced by the plant pathogenic fungus *Cochliobolus heterostrophus* that causes southern corn leaf blight. Mitochondria of T-cytoplasm corn are the main target for the toxin. Paper: http://www.ncbi.nlm.nih.gov/pmc/articles/PMC1056761/

ttp (1) Thrombotic thrombocytopenic purpura, a condition attributed to the presence of an autoantibody to ADAMTS13, the metallopreptidase that degrades ultra-large von Willebrand protein multimers. (2) See **tristetraprolin**. (3) See **SH3BP4**.

TTX See **tetrodotoxin**.

T-type channels Voltage-gated calcium channels of the low-voltage activated type (opening at quite negative potentials and showing voltage-dependent inactivation) which are transiently activated (hence 'T'). T-type channels are involved in pacemaker activity in neurons of the CNS, in the heart and in vascular smooth muscle and are also involved in stimulus-secretion coupling. The α1 subunit determines the channel's properties and there are multiple variants in humans including α1G (Cav3.1, CACNA1G, 2377aa), α1H (Cav3.2, CACNA1H, 2353aa) and α1I (Cav3.3, CACNA1I, 2188aa). The channels are modulated by

various hormones and neurotransmitters and are targets for some anti-epileptic drugs.

tubby Tubby and **tubby-like proteins** (TULPs) are encoded by members of a small gene family and have some features in common with **scramblases**. Tubby (506aa) is involved in signal transduction from heterotrimeric G protein-coupled receptors and binds to membranes containing phosphatidylinositol-4,5-bisphosphate. An autosomal recessive mutation in the mouse *tub* gene leads to progressive retinal degeneration, deafness, and maturity-onset obesity. In *C. elegans*, as in mammals, mutations in the tubby homologue, tub-1, promote increased fat deposition. There is a TUBBY-like protein gene family with 11 members in *Arabidopsis* where they may participate in the **abscisic acid** signalling pathway.

tubby-like proteins Proteins related to **tubby**. TULP-1 (542aa) is a photoreceptor-specific protein required for normal development of photoreceptor synapses, photoreceptor function and long-term survival. Interacts with cytoskeleton proteins and may play a role in protein transport Mutations in TULP1 can cause retinitis pigmentosa. TULP2 (520aa) is strongly expressed in testis, TULP3 (442aa) binds to phosphorylated inositide lipids and negatively regulates **sonic hedgehog** signalling. TULP4 (1543aa) may be a substrate-recognition component of an E3 ubiquitin ligase complex.

Tube A *Drosophila* mutant. Tube is a maternally encoded protein that, together with **pelle** transduces the signal from **toll**. Toll, Cactus and Dorsal, along with Tube and Pelle, participate in a common signal transduction pathway to specify the embryonic dorsal-ventral axis.

tubercle Chronic inflammatory focus, a **granuloma**, caused by *Mycobacterium tuberculosis*.

tuberculin skin test See **Mantoux test**, **Heaf test**.

tuberculosis A disease (TB, consumption) caused by *Mycobacterium tuberculosis* (or *Mycobacterium bovis* from cattle). The infection is usually of the lungs but can affect other tissues. The bacterium is intracellular and thus protected from immune responses. Drug-resistant TB, particularly in immunocompromised individuals, is one of the major causes of mortality and morbidity, especially in the developing countries. See **Bacille Calmette-Guerin**, **mycobacteria**.

tuberin Product (1784aa) of the tumour-suppressor gene *TSC2* that is involved in **tuberous sclerosis**. Hamartin (product of TSC1) and tuberin form a complex, of which tuberin is assumed to be the functional component. The TSC proteins have been implicated in the control of cell cycle by activating the cyclin-dependent kinase inhibitor p27 and in cell size regulation by inhibiting the mammalian target of rapamycin (mTOR)/p70S6K cascade. Tuberin has a GTPase activating domain that acts on Rheb, which in turn acts on the Ras/B-Raf/C-Raf/MEK signalling network.

tuberous sclerosis Autosomal dominant disorder caused by mutation in tumour suppressor genes TSC1 or TSC2. The disease is characterized by range of features including seizures, mental retardation, renal dysfunction and dermatological abnormalities. TSC1 encodes **hamartin**, TSC2 encodes **tuberin**. The complex of tuberin and hamartin (tumour suppressor complex) integrates inputs from multiple signalling cascades to inactivate the **small GTPase** rheb, and thereby inhibit **mTOR**-dependent cell growth.

TUBG1 See **tubulin**.

tubulin An abundant cytoplasmic protein, found mainly in two forms, α (451aa) and β (444aa). A tubulin **heterodimer** (one α, one β), constitutes the **protomer** for microtubule assembly. Multiple copies of tubulin genes are present (and are expressed) in most eukaryotic cells studied so far. The different tubulin isoforms seem, however, to be functionally equivalent. γ-tubulin (TUBG1) is localized in the **centrosome** and is involved in nucleation of microtubule assembly during the cell cycle. Highly conserved from yeast to mammals. See **gamma-tubulin complex**.

tubulin-folding cofactors Proteins (cofactors A–E) involved in regulating the formation and dissociation of tubulin heterodimers. Cofactors A and D stabilize tubulin in a quasi-native conformation. Cofactors B (cytoskeleton-associated protein 1, 244aa) and E can form a heterodimer which binds to alpha-tubulin. Cofactor E binds to the cofactor D-tubulin complex; interaction with cofactor C then causes the release of tubulin polypeptides that are committed to the native state.

tudor domain A conserved protein structural motif (~50aa) originally identified in the *Drosophila* protein, tudor. The proteins TP53BP1 (tumour suppressor p53-binding protein 1) and its fission yeast homolog Crb2 and JMJD2A (Jumonji domain containing 2A) contain either tandem or double Tudor domains and recognize methylated histones. Tudor domains have also been found to be involved in RNA binding. The structural basis for ligand binding by tudor domains is not understood. Other proteins with tudor domains are AKAP1 (A-kinase anchor protein 1) and ARID4A (AT rich interactive domain 4A) among others. Structural detail: http://pawsonlab.mshri. on.ca/index.php?option = com_content&task = view& Itemid = 64&id = 210

tularaemia A disease (deer-fly fever) primarily of a wide variety of wild mammals and birds but can be transmitted to humans by ticks and mosquitoes. It is an acute, febrile, granulomatous infection with the Gram-negative bacterium, *Francisella tularensis*

(formerly *Pasteurella tularense*). WHO information: http://www.who.int/csr/delibepidemics/tularaemia/en/

tumor Alternative (US) spelling of **tumour**.

tumorigenic Capable of causing tumours. Can refer either to a carcinogenic substance or agent such as radiation that affects cells, or to transformed cells themselves.

tumour Strictly, any abnormal swelling, but usually applied to a mass of neoplastic cells.

tumour angiogenesis factor *TAF* Substance(s) released from a tumour that promotes vascularization of the mass of neoplastic cells. Once a tumour becomes vascularized, it will grow more rapidly, and is more likely to metastasise. TAF is almost certainly more than one substance. See **angiogenin** and **VEGF**.

tumour cell A malignant (tumour-causing) cell derived from a tumour in an animal. Loosely, a transformed cell able to give rise to tumours. See **neoplasia**.

tumour initiation First stage of **tumour** development. See also **tumour progression**.

tumour necrosis factor See **TNF**.

tumour progression Changes that occur as a tumour develops, often an increase in malignancy, usually ascribed to a further change in the cells. Heterogeneity among tumour cells is probably a result of further mutations.

tumour promoter An agent that in classical studies of carcinogenesis in rodent skin increased the probability of tumour formation by a previously applied primary carcinogen, but did not induce tumours when used alone. The best example was croton oil, the active ingredients of which are phorbol esters (for example phorbol myristate acetate, PMA, which probably acts as an analogue of diacylglycerols, and may activate protein kinase C). Strictly speaking, not the same as a co-carcinogen, which is defined as being active when administered at the same time. Tumour promoters generally prove to be carcinogens when tested more stringently.

tumour specific antigen An antigen (tumour specific transplantation antigen, TSTA) on tumour cells detected by cell-mediated immunity. For virus-transformed cells TSTA (unlike **T-antigen**) is found to differ for different individual tumours induced by the same virus. May consist of fragments of T-antigens exposed at the cell surface.

tumour suppressor A gene (anti-oncogene, cancer susceptibility gene) that encodes a product that normally negatively regulates the cell cycle, and that must be mutated, inactivated or down-regulated (see **hdm2**) before a cell can escape normal growth control and proceed to rapid division (**neoplasia**).

Many examples are now known: see **adenomatous polyposis coli, BASC complex, bridging integrators, CTCF, death-associated protein, deleted in colonic cancer, deleted in malignant brain tumours-1, exostosin, GCIP, H19 gene, ING1, INK4, IRFs, menin, merlin, myeloid-derived suppressor cells, myopodin, p53, p107, parafibromin, PCGF, PINX1, PML protein, prohibitin, PTEN, retinoblastoma** (Rb), **SMCT, tes, transmembrane proteins with EGF-like and 2 follistatin-like domains, tuberous sclerosis, von Hippel-Lindau tumour suppressor protein, Wilms' tumour, WWOX, zf9.**

tumour virus A virus that will induce tumours. See **avian leukaemia, DNA tumour, Epstein-Barr, Friend murine leukaemia, Kirsten sarcoma, mammary tumour, Papillomaviridae, Polyomaviridae, Retroviridae, Shope fibroma** and **yaba** virus.

tumour-suppressor complex See **tuberous sclerosis**.

tumstatin The autoantigen in **Goodpasture's syndrome**, the N-terminal portion of an isoform of **collagen** type IV (COL4A3).

TUNEL method A method (transferase-mediated dUTP nick-end labelling) for identifying cells undergoing apoptosis by labelling the ends of their fragmented DNA. Full description: http://www.jhc.org/cgi/content/abstract/44/9/959 Technical tips: http://www.roche-applied-science.com/PROD_INF/MANUALS/CELL_MAN/apoptosis_113.pdf

tunicamycin A mixture of nucleoside antibiotics from *Streptomyces lysosuperificus* that inhibits N-glycosylation in eukaryotic cells by preventing the addition of N-acetyl glucosamine to dolichol phosphate.

tunicates A group of marine animals of the phylum Urochordata, closely related to the phylum Chordata that includes all vertebrates. Although the adult form is generally a sessile filter-feeder (sometimes colonial) the larva is free-living and tadpole-like. Sea squirts.

tunichromes Yellow, polyphenolic tripeptides prevalent in blood cells of tunicates.

turbidimetry Measurement of the turbidity, cloudiness, of a solution, often using an instrument called a **nephelometer**. The turbidity is caused by suspended particles; the relationship between particle size and number and the amount of light-scattering is complex.

Turcot's syndrome A disorder (mismatch repair cancer syndrome) in which there is colorectal polyposis and primary tumours of the central nervous system. It is caused by homozygous or compound heterozygous mutations in the **mismatch repair** genes MLH1, MSH2, MSH6 or PMS2.

turgor The pressure within cells, especially plant cells, derived from osmotic pressure differences between the inside and outside of the cell giving mechanical rigidity to the cells. Turgor drives cell expansion and certain movements such as the closing or opening of stomata.

Turner's syndrome Genetic defect (Ullrich-Turner syndrome) in humans in which there is only one X chromosome (affected individuals are therefore phenotypically female), probably as a result of meiotic non-disjunction. X-linked diseases normally restricted to males may manifest in such patients, further confusing the picture.

turnover number Equivalent to V_{max}, being the number of substrate molecules converted to product by one molecule of enzyme in unit time, when the substrate is saturating.

TW-240/260 An obsolete name for a **spectrin**-like protein (240/260 kDa) found in the **terminal web** of intestinal epithelial cells.

TWEAK TNF-related weak inducer of apoptosis. A member of the TNF family (TNFSF12, 249aa), the ligand for **TRAMP** (TNFRSF25). It is a weak inducer of apoptosis in some cell types, mediates NFκB activation, promotes angiogenesis and the proliferation of endothelial cells. Also involved in induction of inflammatory cytokines.

Tween Detergents used in various ways: for example, as blocking agents for membrane based immunoassays, for solubilizing membrane proteins, for lysing mammalian cells (at concentration of 0.05 to 0.5%). Tween variants (Tween-20, -40, -60, -80) have different chain-length fatty acid moieties.

TWIK One of the tandem pore potassium channels. TWIK1 (product of the KCNK1 gene; channel KCNO1, 336aa) is weakly inwardly-rectifying. TWIK2 (313aa) shows outward rectification in a physiological K^+ gradient. See **TASK, TRAAK** and **TREK**.

twine A *Drosophila* Cdc25-like protein phosphatase (EC 3.1.3.48, 426aa) that regulates the transition from G2 to the onset of the first meiotic division. Its activity is regulated by **tribbles**.

twinfilin Yeast actin-depolymerizing factor (332aa) containing two **ADF-H domains**. Localizes to the cortical actin cytoskeleton and is involved in motility and in endocytosis. Will sequester G-actin by forming tight 1:1 complex but does not seeem to cross-link filaments. Human homologues (twinfilin-1, 350aa; twinfilin-2, 349aa) have been identified and twinfilins are found in a range of animal phyla. Research article: http://www.nature.com/emboj/journal/v25/n6/full/7601019a.html

twinkle A mitochondrial protein (EC 3.6.1.-, 684aa) encoded by the C10ORF2 gene (chromosome 10 open reading frame-2) that may be an adenine nucleotide-dependent DNA **helicase** and may regulate mtDNA copy number in mammals. The protein colocalizes with mtDNA in a pattern fancifully likened to that of twinkling stars. A splice variant (twinky) missing residues 579–684 localizes in a similar way. Mutations are associated with a range of disorders.

twinky See **twinkle**.

twinstar The *Drosophila* homologue (148aa) of **cofilin**/ADF (actin depolymerization factor), is a component of the cytoskeleton that regulates actin dynamics. Mutations in twinstar result in defects in centrosome migration and cytokinesis.

twist A highly conserved bHLH transcription factor, known to promote epithelial-mesenchymal transition (**EMT**), induced by a cytokine signalling pathway that requires the dorsal-related protein RelA; Twist-1 (202aa) and twist-2 (160aa) repress cytokine gene expression through interaction with RelA, thus forming a negative feedback loop that represses the NFκB-dependent cytokine pathway. DNA-binding requires formation of a heterodimer with another bHLH protein. Twist proteins transiently inhibit **Runx2** function during skeletogenesis. Loss-of-function mutations of the TWIST 1 gene are responsible for **Saethre-Chotzen syndrome, Robinow-Sorauf syndrome** and **craniosynostosis** type 1. In *Drosophila* twist (490aa) is involved in dorsoventral patterning of embryonic germ layers and in *C. elegans* twist-related protein (178aa) is involved in postembryonic mesodermal cell fate specification.

twitch muscle Striated muscle innervated by a single motoneuron and having an electrically excitable membrane that exhibits an all-or-none response (*cf.* tonic muscle): in mammals almost all skeletal muscles are twitch muscles. Physiologists often divide muscles into fast- and slow-twitch types, the fast-twitch muscles being associated with fast motor units.

twitchin Large protein (6048aa) associated with myosin and important in muscle assembly in invertebrates. Has multiple fibronectin III-homology repeats. Product of *unc-22* gene in *C. elegans*, mutations in which produce animals that show twitching movements.

two-component systems Stimulus-response coupling mechanisms found in many bacteria (e.g. in bacterial chemotaxis) which usually involve a membrane-bound histidine kinase that senses a specific environmental stimulus and a corresponding response regulator that mediates the cellular response via differential expression of target genes. See **histidine-aspartate phosphorelay**.

two-dimensional gel electrophoresis A high resolution separation technique in which protein samples are separated by isoelectric focusing in one dimension and then laid on an SDS gel for size-determined separation in the second dimension. Can resolve hundreds of components on a single gel. Description: http://www.aesociety.org/areas/pdfs/AES_2d.pdf

two-hybrid system Screening system to identify genes encoding proteins which interact specifically with other proteins. One gene is expressed in yeast as a **fusion protein** with the DNA-binding site of the **GAL4** transcription factor, and the other gene co-expressed as a fusion with the transcriptional activator domain of GAL4. Only if the two proteins interact directly are the two GAL4 domains held in close enough proximity to trigger expression of a **reporter gene** (usually **LacZ**) downstream of the UASG promoter recognized by GAL4. False positives do occur and an independent verification of the interaction by another technique is advisable.

two-photon microscopy A method related to **confocal microscopy** but using very brief pulses of intense laser light to stimulate **fluorescence** emission that is restricted to the focal volume, the only place where there is sufficient absorption to stimulate emission. The method allows deeper imaging than conventional confocal methods.

two-pore channels Ion channels (two-pore segment channels) similar to voltage-sensitive calcium and sodium channels but with only two of the domains. They are related to **CATSPER** and **transient receptor potential channels**. Human two-pore channel 1 (TPC1, 816aa) and TPC2 (752aa) are voltage-sensitive calcium channels; variations in TPC2 affect skin **pigmentation**. In *Arabidopsis* TPC1 (733aa) functions as a voltage-gated inward-rectifying Ca^{2+} channel across the vacuole membrane and is inhibited by Al^{3+}.

TxA2 See **thromboxanes**.

Ty element Transposable element of *S. cerevisiae* that appears to resemble a primitive retrovirus. Each consists of a central region of around 5.6 kb flanked by direct repeats of around 330 bp. There are multiple Ty elements ($\sim$35) in each haploid genome. Web-page: http://www.ndsu.edu/pubweb/~mcclean/plsc431/transelem/trans2.htm

TY-5 See **macrophage inflammatory protein 1** α.

Tyk2 A non-receptor tyrosine kinase of the **JAK** family (EC 2.7.10.2, 1187aa) upstream of STAT3, and involved in IFN signalling. Deficiency of tyk2 causes autosomal recessive hyper-IgE syndrome.

tylose A parenchyma-cell outgrowth that wholly or partly blocks a xylem vessel. It grows out from an axial or ray parenchyma cell through a pit in the vessel wall.

Tymovirales A newly established order of viruses in the 2009 classification (see **virus taxonomy**) This group consists of viruses which have a positive-sense single stranded RNA genomes. There are four Families:

Alphaflexiviridae, Betaflexiviridae, Gammaflexiviridae and Tymoviridae.

Type I secretion system *T1SS, TOSS* See **secretion systems**.

Type II secretion system *T2SS* See **secretion systems**.

Type III secretion system *TTSS, T3SS* See **secretion systems**.

Type IV secretion system *T4SS, TFSS* See **secretion systems**.

typhoid An enteric fever due to infection with *Salmonella typhi*. There is prolonged fever, a rose rash and inflammation of the small intestine with ulceration. Fecal contamination of food or water is the standard cause of infection. *Cf.* **typhus**.

typhus Fever due to infection with *Rickettsia prowazekii*, transmitted by lice. See **scrub typhus**, a less serious infection caused by a related rickettsia.

tyrosinase A copper-containing protein (a monoxygenase, EC 1.14.18.1, 548aa) that catalyses the oxidation of tyrosine, and sets in train spontaneous reactions that yield melanin, the black pigment of skin, hair and eyes. The first intermediate is 3,4-dihydroxyphenylalanine (DOPA). Lack of tyrosinase activity is responsible for albinism.

tyrosine One of the twenty amino acids (Tyr, Y, 181 Da) directly coded in proteins. Non-essential in humans since it can be synthesized from phenylalanine. See Table A1.

tyrosine hydroxylase Enzyme (EC 1.14.16.2, with four splice variants, 497−528aa) required for the synthesis of the neurotransmitters **noradrenaline** and **dopamine**. Defects cause autosomal recessive DOPA-responsive dystonia (**Segawa syndrome**).

tyrosine kinases Kinases that phosphorylate a tyrosine residue in a protein. The phosphorylation site is generally characterized by a lysine or an arginine seven residues to the N-terminal side of the phosphorylated tyrosine, and an acidic residue (Asp or Glu) three or four residues to the N-terminal side of the tyrosine. There are however a number of exceptions to this rule, such as the tyrosine phosphorylation sites of enolase and lipocortin II. Tyrosine kinases play major roles in mitogenic signalling, and can be divided into two subfamilies: receptor tyrosine kinases (see **Eph-related receptor tyrosine kinases**, **flt**, insulin receptor family, **NTRK1**, c-ros receptor, sevenless, **stem cell-derived tyrosine kinase**), that have an extracellular ligand-binding domain, a single transmembrane domain, and an intracellular tyrosine kinase domain; and nonreceptor tyrosine kinases, which are soluble, cytoplasmic kinases (see **btk**, **lck**, **soluble tyrosine kinase**, **src family**, **zap70**). See also **dyrks**.

tyrosine phosphorylation See **tyrosine kinases**.

tyrosine recombinase retroelements A class of **retrotransposon** (YR retroelements) found in plants, protists, fungi, and a range of animals including vertebrates. The genome organization of YR retroelements is similar to that of LTR retrotransposons but they lack the PR region in the **pol** polyprotein and usually have a tyrosine recombinase rather than an integrase. YR retroelements can be divided into three families **DIRS**, **Ngaro** and **VIPER**.

tyrphostins Protein **tyrosine kinase** inhibitors derived from **erbstatin**; various types are available (tyrphostin A51, A25 etc.) with specificity for particular kinases. Review (2006) 'Tyrphostins and Other Tyrosine Kinase Inhibitors'. www.ncbi.nlm.nih.gov/pubmed/16756486. Older review (free). http://www.fasebj.org/cgi/reprint/6/14/3275

U

U Single letter abbreviation for uridine.

U protein (Obsolete) Hypothetical protein thought to regulate the transition of cells from G0 to G1 phase of the cell cycle, and thus inevitably into S-phase. The idea was that the concentration of this unstable (U) protein would have to exceed a threshold level for triggering progression through the cycle. This has never gained currency in the literature and has been superceded by **cyclins**.

U-box See **E3 ligase**.

U1 snRNP One of the classes of **small nuclear RNAs**. There are six U-types, all of which have a high uridylic acid content: U1-U5 are synthesized by RNA Polymerase II, U6 by RNA Polymerase III. See **U2 snRNP**.

U2 snRNP Small nuclear ribonucleoprotein particle forming part of the **spliceosome** that is targeted to the 3′ splice site of pre-mRNA by **U2AF**. U1 snRNP defines the 5′ splice site.

U2AF A heterodimeric protein (U2 snRNP auxiliary factor, 475aa and 240aa) that is a major determinant in 3′ splice site selection. Both subunits are conserved in other organisms, and homologues have been identified in *Drosophila, Schizosaccharomyces,* and *Caenorhabditis*. It binds site specifically to the intron pyrimidine tract between the branchpoint sequence and 3′ splice site at an early step in spliceosome assembly and recruits **U2 snRNP** to the branch site.

U2OS cells Human osteosarcoma cell line with epitheliod morphology, originally known as the 2T line. Time-lapse video: http://www.microscopyu.com/moviegallery/livecellimaging/u2/

U6 A small nuclear RNA (U6 snRNA) associates with the specific protein Prp24p, and a set of seven LSm2p-8p proteins, to form the U6 small nuclear ribonucleoprotein (snRNP). The U6 RNA intramolecular stem-loop (ISL) is a conserved component of the **spliceosome**. U6 promoters are widely used to drive expression of siRNA in cells.

U87 cells *U87 MG* A cell line derived from a human grade IV glioma. The line has been fully sequenced. http://www.cancer.ucla.edu/Index.aspx?page = 644&recordid = 320&returnURL = %2findex.aspx

U937 Human myelomonocytic cell line frequently used as a model for myeloid cells, although the cells are rather undifferentiated. Derived from a patient with histiocytic leukaemia.

UASG Upstream activation site G. Promoter sequence recognized by the GAL4 transcription factor. Used to control expression of a wide range gene products in transgenic plants and animals.

UBC proteins (1) Family of proteins involved in conjugating ubiquitin to proteins. UBC1, UBC4 and UBC5 have a role in targeting proteins for degradation, but others have more complex roles, including an involvement in cell cycle control (UBC3) and the secretory pathway (UBC6). (2) A synonym for ubiquitin C (polyubiquitin, 546aa) which is proteolytically cleaved to produced ubiquitin monomers. There are a variable number of ubiquitin repeats in polyubiquitin from different species.

ubinucleins A family of ubiquitously expressed nuclear proteins. Ubinuclein-1 (1134aa) is involved in the formation of senescence-associated heterochromatin foci (SAHF) and repression of proliferation-promoting genes.

ubiquinone *coenzyme Q* Small molecule with a hydrocarbon chain (usually of several isoprene units) that serves as an important electron carrier in the respiratory chain. The acquisition of an electron and a proton by ubiquinone produces ubisemiquinone (a free radical); a second proton and electron convert this to dihydro-ubiquinone. Plastoquinone, which is almost identical to ubiquinone, is the plant form.

ubiquitin A protein (76aa) found (ubiquitously) in all eukaryotic cells. Can be linked to the lysine side chains of proteins by formation of an amide bond to its C-terminal glycine in an ATP-requiring process involving ubiquitin conjugating enzymes. The protein/ubiquitin complex is then subject to rapid proteolysis. Ubiquitin has a role in the heat-shock response, is involved in quality control of nascent proteins, membrane trafficking, cell signalling, cell cycle control, X chromosome inactivation and the maintenance of chromosome structure. See **E3 ligase**.

ubiquitin carboxy-terminal hydrolases Cysteine endopeptidases that cleave polyubiquitin (ubiquitin c) and ubiquitin-protein linkages. UCH-L1 (EC 3.4.19.12, 223aa) is abundant in brain (up to 2% of total protein) and is found in **Lewy bodies**. Mutations in UCH-L1 have been associated with familial forms of Parkinsonism. UCHL3 (230aa) is similar but is more widely distributed and can also

The Dictionary of Cell and Molecular Biology. DOI: http://dx.doi.org/10.1016/B978-0-12-384931-1.00021-0

hydrolyse links with **nedd8**. UCHL5 (329aa) is associated with the **proteasome**.

ubiquitin conjugating enzymes A family of enzymes that catalyze the addition of ubiquitin to proteins. The E1 enzyme (ubiquitin-activating enzyme) adenylates ubiquitin and forms a thioester-linked complex that is transferred to the E2 enzyme which is then able to interact with an E3 ligase that transfers the ubiquitin to the target protein. There are multiple E3 ligases generally subdivided into four classes: **HECT**-type, RING-type, PHD-type, and U-box containing, each of which targets specific proteins. Ubiquitinylated proteins are rapidly hydrolysed by the **proteasome**. See also **N-end rule**. Webpages: http://www.ebi.ac.uk/interpro/potm/2004_12/Page2.htm

ubiquitin ligase See **ubiquitin conjugating enzymes**.

ubiquitinylation *ubiquitinoylation* The covalent addition of **ubiquitin** residues to proteins. Single or multiple residues can be added and bound ubiquitin can also be a site for further addition of ubiquitin residues.

ubisemiquinone See **ubiquinone**.

UDG An abundant and ubiquitous enzyme (uracil-DNA glycosylase, UNG1, EC 3.2.2.-, 313aa) that removes uracil residues from DNA (which can happen either by misincorporation of dUMP residues by DNA polymerase or deamination of cytosine. The mitochondrial and the nuclear isoforms are generated from the same gene by different promoter usage and alternative splicing. Defects in UDG1 are a cause of immunodeficiency with hyper-IgM type 5 syndrome. UDG2 (327aa) is cyclin O and the product of a different gene. Homologues or orthologues are found in all organisms. Diagrams and structures: http://www.biochem.ucl.ac.uk/bsm/xtal/teach/repair/udgase.html

UDP Uridine diphosphate.

UDP-galactose Uridine diphosphate-galactose. Sugar nucleotide, active form of galactose for galactosyl transfer reactions.

UDP-glucose Uridine diphosphate-glucose. Sugar nucleotide, active form of glucose for glucosyl transfer reactions.

UFD1L A component (ubiquitin fusion degradation protein 1 homologue, 307aa) of the pathway which degrades ubiquitin fusion proteins. It forms a complex with **valosin-containing protein** and **NPL4** that binds ubiquitinated proteins and is necessary for the export of misfolded proteins from the ER to the cytoplasm where they are degraded.

UK1 (**1**) Recombinant kringle domain (UK1) of the urokinase type **plasminogen activator** (uPA). Reported to have anti-angiogenic activity. (**2**) A benzoxazole antibiotic and antifungal natural product

isolated from *Streptomyces sp.* that is an inhibitor of topoisomerase II with a wide spectrum of potent anticancer activities. (**3**) A pathogenic strain of *Salmonella typhimurium* (UK1).

ulcerative colitis Inflammation of the colon and rectum: cause unclear, although there are often antibodies to colonic epithelium and *E. coli* strain 0119 B14. The pattern of inflammation differs from that in **Crohns disease** (the other form of inflammatory bowel disease) and there can be inflammation of other tissues (e.g uveitis). Susceptibility to ulcerative colitis and to other forms of inflammatory bowel disease is determined by multiple genes and in some cases there is variation in the interleukin-23 receptor.

Ulothrix A genus of filamentous green algae.

ultrabithorax *Ubx Drosophila* **homeotic** gene that is part of the **bithorax complex** and encodes a transcription factor (389aa) that provides positional information in the anterior-posterior axis. Mutations in Ubx affect parasegments 5–6, corresponding to the posterior thorax and anterior abdomen of the adult.

ultracentrifugation Centrifugation at very high g-forces: used to separate molecules e.g. mitochondrial from nuclear DNA on a caesium chloride gradient. In analytical ultracentrifugation the sample is optically monitored during centrifugation allowing characterization of the solution-state behaviour of macromolecules. The method, having almost disappeared, is coming back into use with the availability of new instruments. Review: http://www.abrf.org/JBT/1999/December99/dec99cole.html

ultradian Cycles of biological activity that occur with a frequency of less than 24 hours. *Cf.* **circadian**.

ultrafiltration Filtration under pressure. In the kidney, an ultrafiltrate is formed from plasma because the blood is at higher pressure than the lumen of the glomerulus. Also used experimentally to fractionate and concentrate solutions in the laboratory using selectively permeable artificial membranes.

ultrastructure General term to describe the level of organization that is below the level of resolution of the light microscope. In practice, a shorthand term for structure observed using the electron microscope, although other techniques could give information about structure in the sub-micrometre range.

ultraviolet *UV* Continuous spectrum beyond the violet end of the visible spectrum (wavelength less than 400 nm), and above the X-ray wavelengths (greater than 5 nm). Glass absorbs UV, so optical systems at these wavelengths have to be made of quartz. Nucleic acids absorb UV most strongly at around 260 nm, and this is the wavelength most likely to cause mutational damage (by the formation of thymine dimers). It is the UV component of sunlight that causes actinic keratoses to form in skin, but that is also required for Vitamin D synthesis. The common subdivisions are into UVA (400–315 nm), UVB (315–280 nm) and UVC (280–100 nm).

umami One of the five basic tastes (salt, sweet, sour, bitter and umami) generally described as that of monosodium glutamate.

umbelliferone Common plant compound (7-hydroxycoumarin) that can be used as a fluorescent pH indicator.

unc (1) Genes that produce uncoordinated behaviour in *C. elegans*. Their products are diverse, some encoding proteins involved in muscle structure and function, others important in neural developement and function, yet others in neurotransmitter metabolism or synapse formation. See Table U1. More information: http://www.mrc-lmb.cam.ac.uk/

genomes/FlyGee/Teichmann_Ig.pdf (2) A *Drosophila* gene (uncoordinated) is involved in ciliogenesis and when mutated the flies lack transduction in ciliated mechanosensory neurons and show uncoordinated movement; centriole function is unaffected. The gene product (1386aa) is an early marker for the conversion of a mitotic centriole into a ciliogenic basal body. More information: http://www.sdbonline.org/fly/genebrief/uncoordinated.htm

uncoupling agent Agents that uncouple electron transport from oxidative phosphorylation. Ionophores can do this by discharging the ion gradient across the mitochondrial membrane that is generated by electron transport. In general the term applies to any agent capable of dissociating two linked processes.

uncoupling proteins Members of the mitochondrial transporter superfamily that form dimers which are proton channels. UCP-1 (thermogenin, 307aa) is expressed exclusively in brown adipose tissue;

TABLE U1. UNC Proteins

Protein		Size*	Function**
UNC1	**membrane protein**	285	stomatin-like
UNC3	**transcription factor**	491	regulation of proteins needed for pathfinding?
UNC4	**homeobox protein**	252	regulates synaptic specificity
UNC5	netrin receptor	919	required for motor axon guidance
UNC6	netrin	612	extracellular matrix guidance cue
UNC7	innexin	522	component of gap junction
UNC8	**degenerin**	777	multi-pass membrane protein.
UNC9	**innexin**	386	gap junction component
UNC10	**rab-3-interacting molecule**	1563	synaptic vesicle release
UNC11	**PI-binding clathrin assembly protein**	586	recruits adaptor protein complex 2
UNC13	**phorbol ester/DAG-binding protein**	2155	may be involved in neurotransmitter release
UNC14	?	665	?
UNC15	paramyosin	882	structural component of thick filament
UNC16	JNK-interacting protein	1157	regulation of synaptic vesicle transport?
UNC17	**Vesicular acetylcholine transporter**	532	ACh transport to synaptic vesicles
UNC18	ACh regulator?	673	axonal transport?
UNC22	twitchin		unconfirmed
UNC24	?	414	?
UNC27	troponin I2	242	muscle control
UNC29	AChR beta subunit	493	ACh receptor
UNC30	homeobox protein	323	transcriptional regulator in Type D neurons
UNC31	**Ca-dependent secretion activator**	1396	vesicle exocytosis
UNC32	**V-type ATPase subunit**	905	proton pump assembly
UNC33	**cytoplasmic protein**	854	axonal guidance
UNC34		468	cell migration?

TABLE U1. (Continued)

Protein		Size*	Function**
UNC36	**voltage-dependent calcium channel subunit a2**	1249	involved in serotonin response
UNC37	transcription factor	612	motor neuron specificity determination
UNC38	**AChR alpha subunit**	511	ACh receptor
UNC40	**protein**	1415	receptor for UNC6?
UNC41	**stoned-b homologue?**	1657	adapter, involved in vesicle recycling?
UNC45		961	?
UNC47	**vesicular GABA transporter**	487	uptake of GABA into synaptic vesicles
UNC50	**multi-pass membrane protein**	301	required for expression of ACh receptors in muscle
UNC51	**serine/threonine-protein kinase**	856	important for axonal elongation and guidance
UNC52	perlecan homologue	3375	matrix component linked to myofibrils via integrin
UNC53	adapter protein	1654	migration and outgrowth of muscles & axons
UNC54	myosin 4	1966	muscle contraction
UNC58	multi-pass membrane protein	591	potassium transport in muscle?
UNC60	actin depolymerising factor-1	212	cytoskeletal reorganisation
UNC62	homeobox protein	564	postembryonic development of the ectoderm
UNC63	AChR alpha subunit	502	AChR
UNC64	syntaxin-1A homologue	291	various secretory processes
UNC73	trio homologue	?	cell migration and axon guidance
UNC78	actin-interacting protein-1	611	regulation of actin in myofibrils
UNC80	multi-pass membrane protein UNC80	3225	component of the nca-1 sodium channel
UNC83	**nuclear migration protein**	1041	involved in nuclear migration during development
UNC84	**nuclear migration and anchoring protein**	1111	interacts with anc-1
UNC86	transcription factor	467	involved in lineage specification
UNC87	calponin-family protein	565	localized to the I-band
UNC89	M-line assembly protein	8081	structural component of M-line
UNC93	**potassium channel regulatory protein**	705	coordination of muscle contraction?
UNC94	**tropomodulin**	392	microfilament pointed-end capping
UNC97	**LIM domain-containing protein**	348	component of adherens junction?
UNC98	zinc finger protein	310	transcription factor for muscle structure
UNC101	**API subunit**	422	component of clathrin-associated complex
UNC104	**kinesin-like protein**	1584	intracellular transport
UNC105	degenerin-like	887	monovalent cation channel
UNC112	Mitogen-inducible mig-2 protein-like	720	involved in cell-matrix adhesion?
UNC116	**kinesin heavy chain**	815	organelle transport
UNC119	developmental protein	219	necessary for nervous system development

A very diverse set of proteins that, when mutated, result in uncoordinated behaviour of *C. elegans*. The Table is based upon reviewed UniProt entries as at 25th Feb 2010.
ACh is acetylcholine, AChR is the acetylcholine receptor.
*Number of amino acid residues in *C. elegans*
**? indicates probable function

UCP-3 (312aa) predominantly in skeletal muscle. Mice over-expressing UCP-3 are hyperphagic but weigh less than normal littermates and have an increased glucose clearance rate, suggesting that UCP-3 has a role in influencing metabolic rate and glucose homeostasis. Genetic variation in human UCP2 (309aa) influences susceptibility to obesity.

underlapping A situation in which, following a collision between two cells in culture, particularly head-side collision, one cell crawls underneath the other, retaining contact with the substratum, and obtaining traction from contact with the rigid substratum (unlike **overlapping**, where traction must be gained on the dorsal surface of the other cell).

unequal crossing-over Crossing-over between homologous chromosomes that are not precisely paired, resulting in non-reciprocal exchange of material and chromosomes of unequal length. Tends to occur in regions containing tandemly repeated sequences.

unfolded protein response A cellular response to stress activated by an accumulation of unfolded or misfolded proteins in the lumen of the endoplasmic reticulum. The response involves reduction of translation, cell cycle arrest and an increase in the availability of chaperonins such as BiP/Grp78.

unigene (1) A unigene is a computationally derived sequence obtained by assembling several ESTs and each unigene cluster is considered to be a unique gene. It is effectively a database entry for a gene containing a group of corresponding ESTs. (2) Unigene is also the name of a database managed by **NCBI** in which each entry is a set of transcript sequences that appear to come from the same transcription locus (gene or expressed pseudogene), together with information on protein similarities, gene expression, cDNA clone reagents, and genomic location. It covers a wide range of organisms and can be used to do digital differential display – an approach to identifying genes that are differentially expressed in various tissues or in the same tissue under different conditions.

unineme hypothesis An hypothesis, now generally accepted as correct, that proposes that each chromosome (before S phase) consists of a single (double-helical) strand of DNA.

uniporter A class of transmembrane **transport proteins** that conveys a single species across the plasma membrane.

UniProt The Universal Protein Resource database, a comprehensive resource for protein sequence and annotation data. It is maintained through a collaboration between the European Bioinformatics Institute (EBI), the Swiss Institute of Bioinformatics (SIB) and the Protein Information Resource (PIR). The database has been invaluable for checking entries in this dictionary. Link: http://www.uniprot.org/uniprot

unit membrane The three-ply, approximately 7 nm-wide membrane structure found in all cells, composed of a fluid lipid bilayer with intercalated proteins. The unit membrane theory carries with it the presumption that all biological membranes have basically the same structure.

universal stress proteins A set of proteins originally described from *E. coli* but now known to occur in a range of organisms. The UspA family of proteins are found in the genomes of bacteria, archaea, fungi, protozoa, and plants, but the biological and biochemical functions of these proteins are not known, although they are required for resistance to DNA-damaging agents. They are within the adenine nucleotide alpha hydrolases-like superfamily. In *E.coli* the UspA family includes UspA, 144aa; UspC, 142aa, UspD, UspE, 316aa, UspF, 144aa, and UspG, 142aa). The UspB family is distinct (UspB, 111aa) and only reported from bacteria.

univoltine Producing only a single set of offspring during the breeding season. *Cf.* multivoltine.

unsaturated fatty acid A **fatty acid** with one or more double bonds. See Table L3.

untranslated region *UTR* The regions of a **cDNA** which are not translated to make a peptide. The 5′ UTR (the leader sequence that precedes the initiation (ATG) site) usually contains a ribosome binding site (in bacteria called the **Shine Dalgarno sequence**) and can be around one hundred or more nucleotides in length. The 3′ UTR is followed by the poly-A tail region. Both 3′ and 5′ UTRs can have sequences for binding of proteins or miRNAs that affect the stability or fate of the message.

uPA See **plasminogen activator**.

uPAR The **uPA** receptor (urokinase plasminogen activator (uPA) receptor, CD87, 335aa) that localizes and promotes plasmin production. It is found in secreted and GPI-anchored forms. The receptor is itself cleaved by uPA which acts as a negative feedback. The uPA/uPAR system controls matrix degradation in the processes of tissue remodeling, cell migration, and invasion and is involved in tumour progression and metastasis of a variety of cancers.

upstream Usually refers to nucleotide sequences that precede the codons specifying mRNA or that precede (are on the 5′ side of) the protein-coding sequence in mRNA or the early steps in a signalling cascade but can be more generally applied to early events in any process that involves sequential reactions.

uracil The pyrimidine base (2,6-dihydroxypyrimidine) from which uridine is derived.

uracil-DNA glycosylase A DNA glycosylase (EC 3.2.2.27, 313aa) that excises uracil residues from DNA (which can arise as a result of misincorporation of dUMP or deamination of cytosine).

uranyl acetate Uranium salt that is very electron-dense, and that is used as a stain in electron microscopy, usually for staining nucleic acid-containing structures in sections.

URB754 Potent, non-competitive inhibitor of rat brain **monoacyl-glycerol lipase** (MGL). Has only weak inhibitory activity against rat brain fatty acid amide hydrolase (FAAH) and binds only weakly to the rat cannabinoid receptor CB1. Reportedly does not inhibit human and mouse forms of the enzyme and the selectivity has been questioned.

ure2 In *S. cerevisiae* a **prion**-like protein that, if overexpressed, leads to the formation of increased amounts of the conformationally altered form, URE3 and of the misfolded forms of Sup35p (PSI$^+$) and Rnq1p (PIN$^+$). Paper: http://www.genetics.org/cgi/content/abstract/183/3/929

Urea $(NH_2)_2CO$ The final nitrogenous excretion product of many organisms (but see **uric acid**). The first organic compound to be artificially synthesized from inorganic starting materials, now produced commercially in huge amounts (as a fertilizer) from synthetic ammonia and carbon dioxide. Urea solutions are used experimentally to denature proteins.

urea cycle *ornithine cycle* The biochemical reactions that produce urea from ammonia, mostly in the liver in mammals. See **ornithine transcarbamylase**.

uredospore *urediniospore, urediospore* A binucleate spore which rapidly propagates; the dikaryotic phase of a rust fungus.

uric acid The final product of nitrogenous excretion in animals that require to conserve water, such as terrestrial insects, or have limited storage space, such as birds and their eggs. Uric acid has very low water-solubility, and crystals may be deposited in, for example, butterflies' wings to impart iridescence. See also **tophus**.

uricase Enzyme (urate oxidase, EC 1.7.3.3) that catalyzes the oxidation of urate to **allantoin**, the terminal reaction in purine degradation in most mammals (but not in primates and birds). There are four subunits of 303aa and one copper atom per molecule. The sequence is well conserved and in animals it is mainly localized in the liver, where it forms a large electron-dense paracrystalline core in many peroxisomes.

uridine *U* The ribonucleoside formed by the combination of ribose and uracil.

uridyl Chemical group formed by the loss of a hydroxyl from the ribose of uridine. Not the same as **uridylyl**.

uridylyl The uridine monophospho group derived from urydylic acid: not the same as **uridyl**. Can be post-translationally added to proteins, RNAs or sugar phosphates (uridylylation). The transfer of UDP from glucose to galactose in bacteria is catalysed by uridylyl transferase (EC 2.7.7.12, 890aa in *E. coli*). See **terminal uridylyl transferase**.

urinary plasminogen activator *uPA* See **plasminogen activator**.

urocanic acid An intermediate in degradation of L-histidine. Found as its trans isomer (t-UA, approximately 30 mg/cm2) in the uppermost layer of the skin (stratum corneum). Absorption of UV light caused isomerisation to the cis-form which is believed to inhibit various immune responses, including contact and delayed hypersensitivity.

Urochordata See **tunicates**.

urocortins Neuropeptides of the CRF (**corticotropin-releasing factor**) family in the mammalian brain. Urocortin I (40aa) is synthesized in human anterior pituitary cells and stimulates secretion of ACTH through the type 1 **corticotropin-releasing hormone** receptor. Urocortin II (urocortin-related peptide, stresscopin-related peptide, 43aa) and urocortin III (stresscopin, 40aa) are ligands for the type-2 CRH receptor which mediates stress-coping responses.

Urodela Amphibians of the Order Caudata that have tails: newts and salamanders.

urodilatin A natriuretic peptide (32aa) secreted from the distal tubules of the kidney that causes diuresis by increasing renal blood flow. Produced by post-translational processing of **atrial natriuretic peptide** prohormone.

urogastrone A peptide isolated from human urine that inhibits gastric acid secretion. Now known to be identical to **epidermal growth factor**.

uroguanylin See **guanylin**.

uroid Tail region of a moving amoeba.

urokinase See **plasminogen activator**.

uromodulin A GPI-linked cell surface-associated form of the Tamm-Horsfall glycoprotein (640aa) which is found as a high-molecular-weight polymer in urine. It has potent immunosuppressive activity. Contains a **ZP domain** that is important for polymerization. Mutations cause juvenile hyperuricemic nephropathy.

uronic acids Carboxylic acids related to hexose sugars etc. by oxidation of the primary alcohol group, e.g. glucuronic, galacturonic acid.

uroplakins Conserved integral membrane proteins, tetraspanins (UPK-1A & B, 260aa; UPK-2, 184aa; UPK-3A, 287aa; UPK-3B, 320aa) of the apical membrane of urothelium (an unusual asymmetric unit membrane). Mutations in UPK3A cause renal adysplasia. See **Potter's syndrome**.

uroporphyrinogen I synthetase An enzyme (porphobilinogen deaminase, hydroxymethylbilane synthase, EC.4.3.1.8) of haem biosynthesis that is defective in the inherited (autosomal dominant) disease, acute intermittent porphyria. There are two isoforms, one (317aa) active in all tissues, the other (334aa) restricted to erythrocytes. Uroporphyringogen I (UPI) is isomerised to UPIII by UPIII synthetase (EC 4.2.1.75, 265aa), defective in the autosomal recessive disease, congenital erythropoietic porphyria.

urotensin II A conserved neuropeptide (11aa), originally isolated from teleosts, that is the endogenous ligand for the orphan G-protein-coupled receptor, GPR14 (389aa) and a potent vasoconstrictor.

urticaria pigmentosa See **mastocytosis**.

Usher syndrome A rare, autosomal recessive syndrome in which there is congenital deafness and progressive blindness. Various subtypes are distinguished according to the defective protein. Affected genes include those for myosin-7A (expressed in hair cells), **harmonin, otocadherin, protocadherin 15, SANS**, usherin, GPR98 (a G-protein coupled receptor), **whirlin** and **clarin 1**.

usherin A type I transmembrane protein (5202aa) that interacts with collagen IV and fibronectin in the basal lamina and with **harmonin, whirlin** and ninein-like protein (NINL). It is found in many tissues and is a component of the interstereocilia linkages in inner ear sensory cells and the basal lamina just beneath the retinal pigment epithelial cells which presumably explains why defects can cause **Usher syndrome** type 2A and **retinitis pigmentosa** type 39.

uteroglobin **Progesterone**-binding protein originally found in lagomorphs (rabbits, hares etc.), which is also a potent inhibitor of **phospholipase A2**. Forms an antiparallel dimer, linked by disulphide bonds at either end. Later shown to be identical to Clara cell phospholipid-binding protein (CC16, secretoglobulin 1A1, 91aa) which inhibits inflammation in the lung.

UTR See **untranslated region**.

utricle *utriculus* Any small inflated bladder-like structures. In vertebrates, the upper chamber of the inner ear.

utrophin Autosomal homologue of **dystrophin** (dystrophin associated protein, 3433aa) localized near the neuromuscular junction in adult muscle, though in the absence of dystrophin (i.e. in **Duchenne muscular dystrophy**) utrophin is also located on the cytoplasmic face of the sarcolemma.

uveitis Inflammation of the iris, ciliary body and choroid of the eye.

uvomorulin Glycoprotein (882aa in humans) originally defined as the antigen responsible for eliciting antibodies capable of blocking compaction in early mouse embryos (at the morula stage), and inhibiting calcium-dependent aggregation of mouse teratocarcinoma cells. Now generally referred to as cadherin 1 (E-cadherin).

V

V (1) Single letter code for valine. (2) General prefix for the viral form of an oncogene, *v-onc*, *cf. c-onc*, the normal, cellular **proto-oncogene**.

V-gene See **variable gene**.

V-region Those regions in the amino acid sequence of both the heavy and the light chains of immunoglobulins where there is considerable sequence variability between one immunoglobulin and another of the same class, in contrast to constant sequence (C) regions. The V-regions are associated with the antigen-binding areas. They contain hypervariable regions of particularly high sequence diversity. Similar immunoglobulin-type V-regions are found in the **T-cell receptor**.

V-SNARE See **synaptobrevins**.

V-type ATPase Vacuolar ATPase. One of three major classes of ion transport ATPase, highly conserved and related to the **F-type ATPases**. They have a multi-subunit structure, and lack a phosphorylated intermediate. In the yeast V-type ATPase there are 13 protein subunits arranged in a cytoplasmic V1 complex and a membrane-embedded Vo complex with at least six types of subunits (a, d, c, c′, c″ and e). The V1 complex has three copies of the catalytic A and B subunits, three copies of the stator subunits E and G, one copy of the regulatory C and H subunits and subunits D and F forming a central axle for the rotor. The ATP hydrolysis-driven rotation of the ring is thought to pump protons across the membrane. V-type ATPases are found in membranes of various organelles (where the proton gradient may then be used to drive co-transport systems) and in the plasma membrane of some epithelia (e.g. intercalated cells of kidney) and cells such as osteoclasts. The ATPase is sensitive to **bafilomycin**. Mutation in the human a3 isoform cause infantile malignant osteopetrosis and mutations to the a4 isoform cause distal renal tubular acidosis. See also **P-type ATPase**.

V5 (1) Extra-striate visual area of the brain, (V5/MT). (2) An antiapoptotic pentapeptide V5, (VPMLK). (3) Vanadium^{5+}. (4) One of the catalytic domains of PKC beta. (5) The envelope (Env) gene V3-V5 regions of feline immunodeficiency virus that encodes the neutralizing epitopes. (6) The epithelial Ca^{2+} channel transient receptor potential cation channel V5 (TRPV5), involved in active Ca^{2+} reabsorption. (7) One of the splice variants of CD44.

V8 protease Serine peptidase (glutamyl endopeptidase; EC 3.4.21.19, 336aa) from *Staphylococcus aureus* strain V8. Cleaves peptide bonds on the carboxyl side of aspartic and glutamic acid residues. Used experimentally for selective cleavage of proteins for amino acid sequence determination or peptide mapping.

VAC Vascular anticoagulant (VAC)-alpha is now called **annexin** V, VAC-beta is annexin-8. See **VacA**.

VacA Vacuolating cytotoxin. A multi-subunit toxin (600-700 kDa: hexamers or heptamers of identical 1287aa monomers) released into culture supernatant by Type I *Helicobacter pylori*. The subunits (140 kDa) are cleaved to form a 95 kDa VacA monomer which is further cleaved to produce 37 and 58 kDa fragments that behave as an AB toxin and bind to receptor-like protein tyrosine phosphatases to form a hexameric, anion-selective pore in gastric epithelial cells. It stimulates the formation of large vacuoles in epithelial cells and causes ulceration and gastric lesions. Papers: http://jb.oxfordjournals. org/cgi/content/abstract/136/6/741, http://cmr.asm. org/cgi/reprint/19/3/449

vaccination The process of inducing immunity to a pathogenic organism by injecting either an antigenically related but non-pathogenic strain (attenuated strain) of the organism or related non-pathogenic species, or killed or chemically modified organisms of low pathogenicity (a vaccine). In all cases the aim is to expose the human or animal being vaccinated to an antigenic stimulus that leads to immune protection against disease, without inducing appreciable pathogenesis from the injection.

vaccine An antigen preparation that when injected will elicit the expansion of one or more clones of responding lymphocytes so that immune protection is provided against a disease. Often combined with some form of **adjuvant**.

vaccinia A dsDNA virus of the Orthopoxvirus family used in vaccination against smallpox. Related to, but not identical to, cowpox virus. Also used as a vector for introducing DNA into animal cells. The genome is large (190 kbp, ∼250 genes). Overview: http://emedicine.medscape.com/article/231773-overview

VACTERL syndrome A disorder in which there are vertebral defects, anal atresia, tracheoesophageal fistula with esophageal atresia, radial and renal dysplasia, cardiac and limb abnormalities. VACTERL associated with hydrocephalus may be caused by mutation in **PTEN**, other forms by mutation in the HOXD13 gene.

The Dictionary of Cell and Molecular Biology. DOI: http://dx.doi.org/10.1016/B978-0-12-384931-1.00022-2

vacuolar ATPase See **V-type ATPase**.

vacuole A membrane-bounded vesicle in a eukaryotic cell. Secretory, endocytotic, and phagocytotic vesicles can be termed vacuoles. Botanists tend to confine the term to the large vesicles found in plant cells that provide both storage and space-filling functions.

vacuolins Small molecules that induce rapid formation of large, swollen structures derived from endosomes and lysosomes by homotypic fusion. Vacuolin-1 (577 Da), the most potent compound, blocks the Ca^{2+}-dependent exocytosis of lysosomes induced by ionomycin or plasma membrane wounding.

valanimycin An antibiotic with antitumour activity, the product of a biosynthetic gene cluster in *Streptomyces viridifaciens*. Details: http://umbbd.msi.umn.edu/val/val_map.html

valine An essential amino acid (Val, V, 117 Da). See Table A1.

valinomycin A potassium **ionophore** antibiotic, produced by *Streptomyces fulvissimus*. Composed of 3 molecules (L-valine, D-α-hydroxyisovaleric acid, L-lactic acid) linked alternately to form a 36-membered ring, that folds to make a cage shaped like a tennis-ball seam. This wraps specifically around potassium ions, enclosing them in the hydrophilic interior, and providing a hydrophobic exterior. Potassium is thus free to diffuse through the lipid bilayer. Highly ion specific, valinomycin is used in **ion-selective electrodes**.

valosin-containing protein A structural protein of the **AAA family** (transitional endoplasmic reticulum ATPase, VCP, p97, 806aa) that forms a ternary complex with UFD1L (ubiquitin fusion degradation protein 1 homologue) and NPL4 and is involved in the formation of the transitional endoplasmic reticulum, the reassembly of Golgi after mitosis and the export of misfolded proteins from the ER to the cytoplasm, where they are degraded. Missense mutations in VCP are the cause of inclusion body myopathy with Paget disease of bone and frontotemporal dementia (**IBMPFD**).

valproate An anticonvulsant drug used to treat manic phase of bipolar disorder, epilepsy and migraine. May increase brain levels of gamma-aminobutyric acid (GABA).

VAMP See **synaptobrevins**.

van Buchems' disease A disorder (hyperostosis corticalis generalisata; hyperphosphatasemia tarda; leontiasis ossea) in which there is excessive bone growth. Type 1 is caused by the loss of downstream regulatory elements for the gene encoding **sclerostin**, a **bone morphogenic protein** (BMP) antagonist.

Type 2 is caused by mutation in the LRP5 gene that encodes the low density lipoprotein receptor-related protein 5 which interacts with **axin** in the **Wnt** signalling pathway.

Van Den Ende-Gupta syndrome A rare multiple congenital anomaly syndrome caused by mutation in **SCARF2**.

van der Waals attraction Attraction forces that arise from the interactions between atoms. The vibrations of molecules and assemblies of molecules cause electromagnetic interactions; these are attractive when the vibrational frequencies and absorptions are identical or similar, repulsive when non-identical. Other interactions originally proposed by van der Waals were included in this name, but these are usually separated into the Coulomb force, the Keesom force and the London force. Only the last is of electrodynamic nature. Probably important in holding lipid membranes into that structure and possibly in other interactions, e.g. cell adhesion. Electrodynamic forces between large-scale assemblies can be of relatively long-range nature.

van der Woude's syndrome An autosomal dominant developmental disorder that is the commonest cause of cleft lip and palate. Type 1 is caused by mutation in **IRF-6** (interferon regulatory factor-6) and Type 2 by mutation at a second locus (both loci on chromosome 1). A gene on chromosome 17 apparently enhances the probability of cleft palate in an individual carrying the two at-risk genes.

vanadate VO_4^{3-} Powerful inhibitor of many, but not all enzymes that cleave the terminal phosphate bond of ATP. The vanadate ion is believed to act as an analogue of the transition state of the cleavage reaction. **Dynein** is very sensitive to inhibition by vanadate, whereas **kinesin** is relatively insensitive. Similarly **tyrosine kinases** are sensitive to vanadate, but threonine/serine **protein kinases** are insensitive.

vancomycin A complex glycopeptide antibiotic produced by actinomycetes, commercially by *Amycolatopsis orientalis* (formerly *Nocardia orientalis*). Inhibits **peptidoglycan** synthesis by interacting with the substrate (unlike penicillins which interact with the enzyme that catalyses peptidoglycan addition). Active against many Gram-positive bacteria, although resistance to vancomycin has developed (vancomycin-resistant enterococci, VRE). Link: http://www.chm.bris.ac.uk/motm/vancomycin/text.htm

vang-like protein Multi-pass membrane proteins (Van Gogh-like protein 1, strabismus 2, VANGL1, 524aa; VANGL2, 521aa) involved in the control of early morphogenesis and patterning of axial midline structures including the neural plate. Defects in VANGL1 cause neural tube defects (e.g. **SDAM).** The vang-like proteins interact with proteins of the **dishevelled** family (see **planar cell polarity pathway**).

vanilloid receptors A sub-family of the transient receptor potential channels (**Trp channels**). Vanilloid receptor-1 (TrpV1, capsaicin receptor, 839aa) is widely expressed, particularly on sensory neurons. Binding of **capsaicin** (or **resiniferatoxin**) activates the receptor, which acts as a nonspecific calcium-permeant cation channel, and induces death of the cell. Heat will also activate the receptor which has response characteristics similar to thermal nociceptors and it may be involved in mediating inflammatory pain and hyperalgesia. Vanilloid receptor-like protein 1 (VRL1, TrpV2, 764aa) has similar channel properties but is not activated by vanilloids and acidic pH. May transduce physical stimuli in mast cells and is activated by temperatures higher than 52 degrees Celsius. TrpV3 (Vanilloid receptor-like 3, 790aa) is activated by warmth rather than heat and may interact with and modulate TrpV1. TrpV4 (vanilloid receptor-like 2, 871aa) is involved in osmotic sensitivity and mechanosensitivity; defects cause **brachyolmia type 3**, **spondylometaphyseal dysplasia** Kozlowski type and **metatropic dysplasia**. TrpV5 (epithelial calcium channel 1, calcium transport protein 2, 729aa) is a constitutively active calcium-selective cation channel thought to be involved in Ca^{2+} reabsorption in kidney and intestine. TrpV6 (725aa) is similar to TrpV5 with which it can form heteromultimeric channels that are voltage sensitive. Both *C. elegans* and *Drosophila* have TrpV-like genes. International Union of Pharmacology. XLIX. Nomenclature and Structure-Function Relationships of Transient Receptor Potential Channels: http://pharmrev.aspetjournals.org/content/57/4/427.long

vanins A family of small proteins of the **CN hydrolase** family, with similar functions. Vanin-1 (vascular non-inflammatory molecule 1, **pantetheinase**, EC 3.5.1.92, 513aa) is GPI-anchored and highly expressed in the gut and liver. It hydrolyzes pantetheine to pantothenic acid (vitamin B5) and the low-molecular-weight thiol cysteamine. Vanin-1 deficient mice have better-controlled inflammatory responses and transfection of thymic stromal cells with the Vanin-1 cDNA enhances thymocyte adhesion *in vitro*. There are at least two mouse (Vanin-1 and Vanin-3) and three human (VNN1, VNN2, VNN3) orthologous genes. There are vanin-like proteins in Drosophila.

Vaquez-Osler disease See **polycythaemia**.

variable antigens Term usually applied to the surface antigens of parasitic or pathogenic organisms that can alter their antigenic character to evade host immune responses. (See **antigenic variation**).

variable gene See **V-region**.

variable number tandem repeats See **satellite DNA**.

variable region See **V-region**.

Varicella zoster Member of the Alphaherpesvirinae: human herpes simplex virus type 3, causative agent of chickenpox and **shingles**.

variola virus Virus responsible for smallpox, now thought to have been completely eradicated. Large DNA orthopox virus (brick-like, $250-390$ nm $\times 20-260$ nm) with complex outer and inner membranes (not derived from plasma membrane of host cell).

varix (1) Ridges on the shells of gastropods representing periods of shell growth where thickened lips are formed. Old lips are previous varices and the current lip is the most recent varix. (2) An enlarged and dilated vein, usually tortuous, as in varicose veins.

vasa Drosophila gene (*vas*) involved in oogenesis and embryonic positional specification that encodes an ATP-dependent RNA helicase (661aa). The homologous human protein (DEAD box protein 4, 724aa) may be involved in germ cell development. Related proteins have been described in a range of animal species.

vascular anticoagulant See **annexin**.

vascular bundle Strand of vascular tissue in a plant; composed of xylem and phloem together with associated supporting cells.

vascular cell adhesion molecule See **VCAM**.

vascular cylinder See **stele**.

vascularization Growth of blood vessels into a tissue with the result that the oxygen and nutrient supply is improved. Vascularization of tumours is usually a prelude to more rapid growth and often to metastasis; excessive vascularization of the retina in diabetic retinopathy can lead indirectly to retinal detachment. Vascularization seems to be triggered by angiogenesis factors that stimulate endothelial cell proliferation and migration. See **angiogenin**, **tumour angiogenesis factor**, **VEGF**.

vasculitis Inflammation of the blood vessel wall. May be caused by immune complex deposition in or on the vessel wall.

vasoactive intestinal contractor Mouse homologue of **endothelin**-2, 175aa.

vasoactive intestinal peptide A peptide (VIP, 28aa), originally isolated from porcine intestine, but later found in the central nervous system where it acts as a neuropeptide, and is released by specific **interneurons**. May also affect behaviour of cells of the immune system. It is derived by proteolytic cleavage of a 170aa precursor from which two other vasoactive peptides (intestinal peptide PHV-42 and intestinal peptide PHM-27) are also generated. There are two G protein-coupled receptors for VIP (460 and 495aa).

vasoconstrictor An substance that reduces the luminal diameter of blood vessels. Most vasoconstrictors are alpha-**adrenergic receptor** agonists, e.g. **ephedrine**. *Cf.* **vasodilator**.

vasodilator Any drug or substance that causes expansion of blood vessels. Examples include **glyceryl nitrate**, **adrenomedullin**, **bradykinin**, **atrial natriuretic peptide**, **isoproterenol**, **nifedipine**, **prostacyclin**, **Substance P**. See **vasodilator stimulated phosphoprotein**. *Cf.* **vasoconstrictor**.

vasodilator stimulated phosphoprotein *VASP* A protein (380aa) that is a substrate for both cAMP- and cGMP-dependent protein kinases and that is associated with microfilament bundles in many tissue cells. Abundant in platelets; phosphorylation of VASP will inhibit platelet activation.

vasopressin A peptide hormone (antidiuretic hormone, ADH, 9aa) released from the posterior pituitary lobe but synthesized in the hypothalamus as a much larger precursor which includes the vasopressin-carrier, neurophysin and a glycoprotein. Has antidiuretic and **vasopressor** actions. There are 2 forms, differing only in the amino acid at position 8: arginine vasopressin (AVP) is widespread, while lysine vasopressin is found in pigs. There are three G protein-coupled AVP receptors, V1a, V1b, and V2. The V1a receptor (418aa) mediates cell contraction and proliferation, platelet aggregation, release of coagulation factor, and glycogenolysis and acts through phospholipase C. The V1b receptor (424aa) mediates calcium-triggered release of **ACTH**, beta-**endorphin**, and **prolactin** from the anterior pituitary. The V2 receptor (371aa) is found only in the kidney, where it is involved in water homeostasis; defects are one cause of **diabetes insipidus**.

vasopressor Any substance that causes constriction of blood vessels (vasoconstriction) thereby causing an increase in blood pressure.

vasostatins (1) Peptides (vasostatin 1, 76aa; vasostatin 2, 113aa) derived from chromogranin A (see **granins**) that inhibit vasoconstriction, promote fibroblast adhesion and have a range of other effects. Link: http://www.ncbi.nlm.nih.gov/pubmed/17566084?itool = EntrezSystem2.PEntrez.Pubmed.Pubmed_ResultsPanel.Pubmed_ RVDocSum&ordinalpos = 13 (2) The amino-terminal domain (180aa) of **calreticulin** that selectively inhibits the proliferation of endothelial cells; isolated from culture supernatants of an Epstein-Barr virus-immortalized cell line. Link: http://www.ncbi.nlm.nih.gov/pmc/articles/PMC2 746 267/?tool = pubmed

vasotocin Cyclic nonapeptide hormone (arginine vasotocin is CYIQNCPRG-NH$_2$), related to vasopressin and oxytocin, found in the neurohypophysis of

birds, reptiles and some amphibians and demonstrated by bioassay in the mammalian pineal gland, subcommissural organ and fetal pituitary gland.

vasp See **vasodilator stimulated phosphoprotein**.

vaspin A serine protease inhibitor (**serpin A12**, visceral adipose tissue-derived serpin) with approximately 40% homology to α-1-antitrypsin and an **adipokine**. Rat, mouse, and human vaspins are made up of 392, 394, and 395aa, respectively. Vaspin exerts an insulin-sensitizing effect on white adipose tissue in obese rats. Administration of vaspin to obese mice improves glucose tolerance, insulin sensitivity, and alters the expression of candidate genes for insulin resistance. Article: http://www.ncbi.nlm.nih.gov/pmc/articles/PMC1180799/?tool = pubmed

vault Large (40 × 67 nm) cytoplasmic ribonucleoprotein particle that has an eight-fold symmetry with a central pore and petal-like structures giving the appearance of an octagonal dome. The vault consists of a dimer of half-vaults, with each half-vault comprising 39 identical major vault protein (**MVP**) chains, **PARP4**, **telomerase-associated protein 1** and one or more vault RNAs (vRNAs). May be related to the central plug of the nuclear pore complex and may act as a scaffold for proteins involved in signal transduction.

vav A subfamily of the **dbl family** of guanine nucleotide exchange factors (**GEFs**) for the rho/rac GTPases. The *vav1* proto-oncogene product (p95vav, 845aa) is predominantly expressed in hematopoietic cells; vav2 (872aa) has a wider tissue distribution. There is enhanced phosphorylation of vav1 in Bcr-Abl-expressing cells, which activates vav and indirectly **rac-1**. Vav3 (847aa) is more highly expressed in glioblastomas and prostate cancer and is important in angiogenesis.

VCA region Verprolin-like, cofilin-like, acidic region. Motif found in **WASP** and **WAVE** that binds to G-actin and **Arp2/3** complex.

VCAM Vascular cell adhesion molecule. Cell adhesion molecule (CD106, 739aa) of the immunoglobulin superfamily expressed on endothelial cells, macrophages, dendritic cells, fibroblasts and myoblasts. Expression can be upregulated by inflammatory mediators (IL-1β, IL-4, TNFα, IFNγ) and it is the ligand for the integrin VLA4.

VDAC Voltage-dependent anion channel. See **porins**.

vector (1) A mathematical term to describe something that has both direction and magnitude. (2) Common term for a plasmid that can be used to transfer DNA sequences from one organism to another. See **transfection**. Different vectors may have properties particularly appropriate to give protein expression in the recipient, or for cloning, or may have different selectable markers. (3) An animal that transmits an infectious disease. Mosquitos are vectors of malaria.

vectorette method A method for **PCR** cloning an unknown sequence of DNA attached to a known sequence. To the end of the unknown sequence is attached a vectorette, a double-stranded sequence that includes a region of mismatch; primer for the known sequence drives the formation of a second primer site from the mismatch region and the unknown sequence is thus flanked with known sequences and can be specifically PCR amplified.

vectorial synthesis The term usually applied to the synthesis of proteins destined for export from the cell. As the protein is assembled it moves (vectorially) through the membrane of the rough endoplasmic reticulum, to which the ribosome is attached, and into the cisternal space. See **signal sequence**.

vectorial transport Transport of an ion or molecule across an epithelium in a certain direction (e.g. absorption of glucose by the gut). Vectorial transport implies a non-uniform distribution of **transport proteins** on the plasma membranes of two faces of the epithelium.

ved See **bozozok**.

vegetal pole The region of the egg opposite to the **animal pole**. The cytoplasm in this region is usually incorporated into future endoderm cells.

VEGF Growth factor (vascular endothelial growth factor, vascular permeability factor, VPF) of the **PDGF** family that stimulates mitosis in vascular endothelium and promotes angiogenesis. It also increases permeability of endothelial monolayers. Tissue-specific splice variants of VEGFA (232aa) are found (VEGF121, VEGF165, VEGF-C), VEGF165 having heparin-binding activity which VEGF121 lacks. VEGF-related factor (VEGF-B) is produced in two alternatively-spliced form (186- and 167-aa). Functional form is a dimer (or heterodimer of splice variants) and can form a heterodimer with placental growth factor (**PLGF**). The receptors are tyrosine kinases: flt-1 (VEGF-R1, fms-related tyrosine kinase 1, 1338aa), flt-4 (VEGF- R3, 1298aa, binds only VEGF-C) and flk-1 (VEGF-R2/KDR, 1356aa, binds only VEGF121). See **Tie** and **flt**. VEGF coregulated chemokine-1 (VCC1; CXCL17, dendritic cell and monocyte chemokine-like protein, DMC, 119aa) is upregulated in breast and colon tumours. See **neuropilin**.

veiled cell *dendritic cell* A cell type found in afferent lymph and defined (rather unsatisfactorily) on the basis of its morphology. Now generally referred to as dendritic cells. They are an accessory cell and migrate from the periphery (where they are referred to as **Langerhans cells** if in the skin) to the draining lymph node. In the lymph node they are also known as interdigitating cells and are found in the T-dependent areas of spleen or lymph nodes, involved in antigen presentation (Class II MHC positive). Have high levels of surface Ia antigens.

vein (1) A blood vessel that returns blood from the microvasculature to the heart; walls thinner and less elastic than those of artery. (2) In leaves, thickened portion containing a **vascular bundle**; the pattern, venation, is characteristic for each species. (3) See **vein protein**.

vein protein A ligand (epidermal growth factor-like protein, defective dorsal discs protein, 623aa) in *Drosophila* that binds the EGF receptor and is important in regulating proliferation of wing disc cells and in larval patterning.

veliger Free-living larval form of gastropod and bivalve molluscs; develop from the trochophore larva.

velocardiofacial syndrome A disorder with a highly variable phenotype, but with palatal anomalies as a major feature, caused by a deletion of 1.5 to 3.0-Mb on chromosome 22q11.2. As a result there is haploinsufficiency of the **T-box1** (TBX1) gene. Point mutations in TBX1 may have similar consequences. Another gene involved is for catechol-O-methyltransferase (COMT) but there are probably others. See **thymic hypoplasia** (DiGeorge syndrome), **goosecoid**. Article: http://www.genome.gov/25521139

velum A veil-like structure: (1) The posterior part of the soft palate in higher mammals. (2) In some Ciliophora, the delicate membrane bordering the oral cavity. (3) In sponges, a membrane constricting the lumen of an incurrent or excurrent canal. (4) In hydrozoan medusae, an annular shelf projecting inwards from the margin of the umbrella. (5) In rotifers, the trochal disk. (6) In molluscs, the ciliated locomotor organ of the veliger larva. (7) In Cephalochordata, the perforated membrane separating the buccal cavity from the pharynx.

venom A toxic secretion in animals that is actively delivered to the target organism, either to paralyse or incapacitate or else to cause pain as a defence mechanism. Most are protein or peptide toxins.

ventral nervous system defective *vnd* A *Drosophila* homeobox gene that encodes a protein (NK2, 723aa) that probably acts as a transcriptional regulator in neurogenesis. Flybase entry: http://www.sdbonline.org/fly/neural/vnd.htm

véraison A transitional phase of grape berry development, during which growth declines and berries start to change colour and soften.

verapamil A drug (454 Da) that blocks **L-type calcium channels**, used as a coronary vasodilator and anti-arrhythmic.

veratridine An alkaloid, a sodium channel activator that acts at neurotoxin receptor site 2 and preferentially binds to activated Na^+ channels causing persistent activation. Found in the seed of *Schoenocaulon officinale* (Cevadilla) and in the rhizome of *Veratrum album* (White hellebore).

vernalisation Treatment of plants with a period of low temperature to promote flowering earlier than they would otherwise.

Vero cells Cell line derived from kidney of African Green Monkey. Susceptible to a range of viruses.

verotoxin See **shiga toxin**.

verprolins Family of proteins involved in regulating cytoskeletal organisation. Verprolin was originally identified in yeast (817aa in *S. cerevisiae*, 309aa in *Schizosaccharomyces pombe*) and later shown to be needed for actin polymerization during polarized growth and during endocytosis. Verprolins regulate actin dynamics either by binding directly to actin, by binding the **WASP** family of proteins or by binding to other actin regulating proteins. Have been found in humans, insects and plants. See **WIP**, **WIRE**, **WAVE**.

Verrucomicrobia A bacterial phylum with only a few species described, mostly from fecally-contaminated soil or water (the type species is *Verrucomicrobia spinosum*).

versene Trivial name for **EDTA**.

versican A large chondroitin sulphate proteoglycan (large fibroblast glycoprotein, chondroitin sulphate core protein, core protein 3396aa) involved in cell signalling and connecting cells to extracellular matrix. The N-terminal region is similar to glial hyaluronic acid binding protein, centre has glycosaminoglycan attachment sites and the C-terminal region has EGF-like repeats. It binds hyaluronic acid and forms large multimolecular complexes in hyaline cartilage. Expression of certain versican isoforms has been implicated in migration and proliferation of cancer cells. Mutations in the versican gene cause **Wagner's syndrome** type 1.

versiconal An intermediate in the biosynthetic pathway that generates **aflatoxins**.

vertical transmission Transmission of an infectious agent from mother to offspring, transplacentally, through cross-infection during birth or through close contact during the neonatal period, cf. horizontal transmission between individuals.

Verticillium longisporum An important fungal pathogen of Brassicaceae that colonizes the xylem and causes wilting disease.

vesicle A closed membrane shell, derived from membranes either by a physiological process (budding) or mechanically by sonication. Vesicles of dimensions in excess of 50 nm are believed to be important in intracellular transport processes. See also coated vesicles.

vesicular monoamine transporters Transmembrane proteins (solute carriers SLC18A1, 525aa and SLC18A2, 514aa) that transport cytosolic monoamines (**dopamine**, **serotonin**) into synaptic vesicles, using the proton gradient across the vesicular membrane. Defects in these transport systems have been implicated in various neuropsychiatric disorders. **Reserpine** and **tetrabenazine** inhibit the transport.

vesicular stomatitis virus *VSV* A ssRNA virus of the **Rhabdoviridae** causing the disease soremouth in cattle. Widely used as a laboratory tool especially in studies on the spike glycoprotein as a model for the synthesis, post-translational modification and export of membrane proteins.

vesicular-arbuscular mycorrhiza Form of **mycorrhiza** in which the fungus invades the cortical cells to form vesicles and arbuscules (finely branched structures). Common among herbaceous plants and may significantly improve the mineral nutrition of the host.

vesiculin A highly acidic protein (10 kDa) found in **synaptic vesicles**. The name seems to be obsolete.

vessel Water-conducting system in the xylem, consisting of a column of cells (vessel elements) whose end-walls have been perforated or totally degraded, resulting in an uninterrupted tube.

Vg1 A TGFβ-family growth factor in *Xenopus* (DVR1, vegetal hemisphere VG1 protein, 360aa); the maternal mRNA is restricted to the vegetal pole of the mature oocyte and Vg1 is involved in induction of mesoderm. There are homologues of the gene in humans, chickens and zebrafish.

VH and VL genes/domains VH and VL genes define in part the sequences of the variable heavy and light regions of immunoglobulin molecules. VH and VL domains are the regions of amino acid sequence so defined. J genes and, in the case of the heavy chain, a D gene (D = diversity) also define these regions. Gene rearrangement plays a role in determining the sequences in which the genes are joined as the DNA of the immunoglobulin producing cell matures.

VHDL Very high density lipoprotein. See **lipoproteins**.

VHL The tumour suppressor protein (von Hippel-Lindau tumour suppressor protein, VHL30, 213aa) that is defective in the **von Hippel-Lindau syndrome**. A second protein, VHL19 is produced from an alternative start site. The proteins interact with **elongin** and **cullin-2** in the VCB (VHL-Elongin BC-CUL2) complex which acts as a ubiquitin-ligase

E3 that binds to hypoxia inducible factor-1 (HIF-1), targeting it for destruction when oxygen is present. In drosophila protein vhl (178aa) is involved in development of tracheal vasculature and has similar E3 ligase activity.

VHL tumour suppressor gene See **von Hippel-Lindau tumour suppressor protein**.

viability test Test to determine the proportion of living individuals, cells or organisms, in a sample (a viable count). Viability tests are most commonly performed on cultured cells and usually depend on the ability of living cells to exclude a dye, (an exclusion test), or to specifically take it up (inclusion test). For micro-organisms usually requires plating out of diluted samples to permit colony counting.

viable count See **viability test**.

vibratome A vibratory **microtome** that can be used to cut thin sections (down to 10 μm) from unfixed unfrozen tissue, usually embedded in agarose. Commercial website: www.vibratome.com

Vibrio cholerae A Gram-negative bacterium that causes cholera, the life-threatening aspects of which are caused by the exotoxin (see **cholera toxin**). Short, slightly curved rods, highly motile (single polar flagellum). Adhere to intestinal epithelium through an interaction between colonization factor **GbpA** and mucin, and produce enzymes (neuraminidase, proteases) that facilitate access of the bacterium to the epithelial surface.

VIC See **vasoactive intestinal contractor**.

Vicia faba Broad bean. Was often used in plant genetics because cells have only six large chromosomes. Fact sheet: http://www.hort.purdue.edu/newcrop/CropFactSheets/fababean.html

vicilin Seed storage protein of legumes. In *Pisum sativum* a trimer of 459aa subunits. High proportion of **beta pleated sheet** (40–50%) and only about 10% alpha-helix. See **cupins**.

victorin A host-specific toxin produced by the fungus *Cochliobolus victoriae* which causes Victoria blight of oats (*Avena sativa*). Sensitivity of oats to victorin, and to the fungus, is controlled by a single dominant gene that encodes a victorin binding protein (VBP, 1032aa), probably the P protein component of glycine decarboxylase.

vidarabine A nucleoside analogue (adenine arabinoside; 9-β-D-arabinofuranosyladenine; Ara-A) with antiviral properties that has been used to treat severe herpes virus infections.

vif The viral infectivity factor (192aa) of HIV that is expressed late during infection in a Rev-dependent manner and triggers degradation of the host antiviral cellular DNA deaminases (see **APOBECs**) acting through Cul5-ElonginB-ElonginC E3 ubiquitin ligase. Abstract; http://www.nature.com/mt/journal/v11/n1s/abs/mt2005556a.html

vigilin High density lipoprotein-binding protein. A protein (1268aa) with 14 **KH domains** that may protect cells from over-accumulation of cholesterol. It is, however, found in both the cytoplasm and the nucleus and does have RNA-binding properties. An apparently identical protein has been implicated in heterochromatin formation, chromosome segregation and the localization of mRNAs to actively translating ribosomes. Both roles for HBP/vigilin are discused in http://atvb.ahajournals.org/cgi/content/full/17/11/2350

Vik1 Non-motor protein (647aa) from *S. cerevisiae* that interacts with **KAR3** (a kinesin-14 protein) and has sequence and structural similarity to Cik1p. The Vik1 protein is detected in vegetatively growing cells but not in mating pheromone-treated cells. Vik1p physically associates with Kar3p in a complex separate from that of the Kar3p-Cik1p complex. Vik1p localizes to the spindle-pole body region in a Kar3p-dependent manner.

villin Microfilament-severing and -capping protein (827aa) from the microfilament bundle that provides mechanical support to the microvilli of intestinal epithelial cells. Severs at high calcium concentrations, caps at low. Villin-2 is **ezrin**.

vimentin Intermediate filament protein (466aa) found in mesodermally-derived cells. **Desmin** is the muscle homologue.

vinblastine See **vinca alkaloids**.

vinca alkaloids Alkaloids isolated from *Catharanthus roseus* (Madagascan periwinkle, formerly *Vinca rosea*) that bind to tubulin heterodimers and induce the formation of paracrystals rather than tubules. This depletes the cellular pool of heterodimers for assembly and as a result of treadmilling microtubules gradually disappear. Examples include vinblastine (818 Da), vincristine, vindesine and (semi-synthetic) vinorelbine are used in tumour chemotherapy.

vincristine See **vinca alkaloids**.

vinculin Protein (1134aa) isolated from muscle (cardiac and smooth), fibroblasts and epithelial cells. Associated with the cytoplasmic face of **focal adhesions**: may connect microfilaments to plasma membrane integral proteins through **talin**. Metavinculin, has an additional exon 19 that encodes 68aa. In cardiac muscle connects microfilaments to the intercalated disc. Mutation (R975W), in the alternatively spliced exon 19, is associated with hypertrophic cardiomyopathy.

vinorelbine See **vinca alkaloids**.

VIP (1) See **vasoactive intestinal peptide**. (2) VirE2-interacting protein 1 (VIP1) is a bZIP transcription factor that stimulates stress-dependent gene expression in plants by binding to VIP1 response elements (VREs). VREs are over-represented in promoters responding to activation of the mitogen-activated protein kinase (MPK3) pathway such as Trxh8 and MYB44. VIP1 plays an important role in the nuclear import of T-DNA during *Agrobacterium*-mediated plant transformation and becomes localized in the nucleus when phosphorylated by MPK3.

Viper A family of tyrosine recombinase retroelements found in the genome of some trypanosomes. The element (vestigial interposed retroelement) has an internal coding region flanked by a lineage of **small interspersed nuclear elements** (SINEs).

Vipera ammodytes Western sand viper. See **ammodytoxins**.

viperin The protein (virus inhibitory protein endoplasmic reticulum-associated interferon-inducible, 361aa) encoded by a gene that is induced in response to interferon, lipopolysaccharide, double-stranded RNA, poly(I-C) or Sendai virus. Viperin inhibits human cytomegalovirus replication possibly by an effect on virus budding by disrupting lipid rafts at the plasma membrane, by inactivating **FPPS**. It is a tightly regulated ISGF3 target gene, counter-regulated by **PRDI-BF1**.

VIR genes The vir region of the **Ti plasmid** of *Agrobacterium tumefaciens* contains six genes: *VirA* encodes a single protein which resembles a transmembrane chemoreceptor found in other bacteria and presumably binds acetosyringone, *VirB* may play a role in directing **T-DNA** transfer events at the bacterial cell surface, *VirC* and *VirD* encode a site-specific endonuclease that produces a T-strand that is the intermediate molecule that is transported to the plant cell, *VirE* encodes a single-stranded DNA binding protein that appears to coat the T-strand during transfer to the plant cell, *VirG* produces a positive regulatory protein which relays environmental information to other vir loci. Webpage: http://www.scribd.com/doc/510011/Mechanisms-of-Infection-Vir-Gene

viral antigens Those antigens specified by the viral genome (often coat proteins) that can be detected by a specific immunological response. Often of diagnostic importance.

viral haemorrhagic fevers A group of illnesses, some very severe and life-threatening, caused by several distinct families of viruses: **Arenaviridae**, **Filoviridae**, **Bunyaviridae** and **Flaviviridae**.

viral transformation Malignant transformation of a cell in culture, induced by a virus.

viremia Presence of virus in the blood.

virgin lymphocyte Outmoded term for a lymphocyte (CD45RA$^+$, CD45RO$^-$) that has not encountered the antigenic determinant for which it has receptors.

virino Hypothetical virus-like infectious agent — once suggested to be the cause of transmissible spongiform encephalopathies. Probably only of historical interest.

virion A single virus particle, complete with coat.

viroid Extremely small plant viruses. Their genome is a 240–350 nucleotide circular RNA strand, extensively base-paired with itself, so they resist RNAase attack. At one time the term was also used casually of self-replicative particles such as the **kappa particle** in *Paramecium*. Taxonomy of viroids: http://www.ncbi.nlm.nih.gov/ICTVdb/Ictv/fs_viroi.htm

virstatin Small molecule inhibitor of *Vibrio cholerae* virulence and intestinal colonization. It blocks the virulence transcriptional activator, ToxT, and prevents expression of the two major virulence factors, cholera toxin and the toxin coregulated pilus. Papers:http://www.ncbi.nlm.nih.gov/pubmed/17283330 http://www.ncbi.nlm.nih.gov/pubmed/16223984

virulence The degree of pathogenicity exhibited by a microbial or viral pathogen, usually a measure of the severity of disease that is caused by infection although sometimes referring to infectivity.

virus Viruses are obligate intracellular parasites of living but non-cellular nature, consisting of DNA or RNA and a protein coat. They range in diameter from 20–300 nm. Class I viruses (Baltimore classification) have double-stranded DNA as their genome; Class II have a single-stranded DNA genome; Class III have a double-stranded RNA genome; Class IV have a positive single-stranded RNA genome, the genome itself acting as mRNA; Class V have a negative single-stranded RNA genome used as a template for mRNA synthesis, and Class VI have a positive single-stranded RNA genome but with a DNA intermediate not only in replication but also in mRNA synthesis. The majority of viruses are recognized by the diseases they cause in plants, animals and prokaryotes. Viruses of prokaryotes are known as bacteriophages. See **virus taxonomy**.

virus taxonomy The taxonomy of viruses is organised in a similar fashion to that of other organisms, a hierarchical structure with levels of Order, Family, Subfamily, Genus, and Species. Standardization is done under the auspices of the International Committee on Taxonomy of Viruses (ICTV), a committee of the Virology Division of the International Union of Microbiological Societies. The current view (2011) is that there are six Orders, **Caudovirales**, **Herpesvirales**, **Mononegavirales**, **Nidovirales**, **Picornavirales** and **Tymovirales**, and a miscellaneous group of families not assigned to an order. The latter category includes 72 families amongst which are the **Adenoviridae, Arenaviridae, Baculoviridae, Bunyaviridae, Caliciviridae, Flaviviridae, Hepadnaviridae, Orthomyxoviridae, Papillomaviridae, Polyomaviridae, Poxviridae, Reoviridae, Retroviridae** and Togaviridae. Link: http://www.ictvonline.org/virusTaxonomy.asp?bhcp=1

viscoelastic Describing substances or structures showing non-Newtonian viscous behaviour, i.e. both elastic and viscous properties are demonstrable in response to mechanical shear.

viscous-mechanical coupling Method by which adjacent **cilia** are synchronized in a field. Coupling is through the transmission of mechanical forces, rather than of a synchronizing signal.

visfatin An enzyme (nicotinamide phosphoribosyl-transferase, NAmPRTase, EC 2.4.2.12, 491aa) that catalyzes an early step in biosynthesis of NAD, the condensation of nicotinamide with 5-phosphoribosyl-1-pyrophosphate to yield nicotinamide mononucleotide. Was formerly thought to be a cytokine (an **adipokine**) found in omental (visceral) adipose tissue but the publication reporting this was retracted.

visinin A cone-specific protein (192aa) first characterized in chicken retina, a homologue of **recoverin**. A human homologue (visinin-like-1, 191aa) is strongly expressed in granule cells of the cerebellum and as an intracellular EF-hand calcium sensor protein probably modulates calcium-mediated signalling.

Visna-maedi virus A **retrovirus** of sheep and goats. A member of the **Lentivirinae**, related to HIV. First identified in Iceland when it was introduced by sheep imported from Germany, and causes two diseases: maedi, the most common, is a pulmonary infection (maedi is Icelandic for shortness of breath); visna is due to infection of the nervous system, causing a paralysis similar to **multiple sclerosis** (visna is Icelandic for wasting).

visual purple See **rhodopsin**.

vital dye *vital stain* A dye used to label cells without affecting their normal function and that can be used to follow their fate during development (although the dye does eventually become diluted as cells divide). An example is the labelling of pre-migratory neural crest cells by injecting a solution of the fluorescent carbocyanine dye, **DiI** in order to follow their migration. An example: http://dev.biologists.org/content/111/4/857.full.pdf

vitamin A low-molecular weight organic compound of which small amounts are an essential component of the food supply for a particular animal or plant. For humans Vitamin A, the B series, C, D1 and D2, E, and K are required. See separate entries and Table V1.

vitamin B (1) Vitamin B1 (thiamine, aneurin) is water-soluble vitamin, and phosphate derivatives are important in many cellular processes. Thiamine pyrophosphate is a coenzyme for several enzymes, including pyruvate dehydrogenase, and is synthesized by thiamine pyrophosphokinase (EC 2.7.6.2). Thiamine deficiency causes beri-beri. (2) Vitamin B2 (riboflavin) is ribose attached to a flavin moiety and the core component of flavin nucleotides. Riboflavin deficiency (ariboflavinosis) can lead to iron-deficiency anaemia. (3) Vitamin B3 (niacin, nicotinic acid) is converted to nicotinamide and then to NAD and NADP. Deficiency causes pellagra. (4) Vitamin B5 (pantothenic acid) is needed to form coenzyme-A (CoA). (5) Vitamin B6 (pyridoxine, pyridoxal, and pyridoxamine) is a cofactor in many reactions of amino acid metabolism; the active form is pyridoxal phosphate. (6) Vitamin B7 (biotin, vitamin H) is important in fatty acid biosynthesis and catabolism and is an essential growth factor for many cells. Biotin is bound very strongly by **avidin** or **streptavidin**. (7) Vitamin B9 (folic acid, vitamin M) must be obtained in the diet or from intestinal microorganisms by humans. The active form is tetrahydrofolate (see **folate**). (8) Vitamin B12 is actually a class of related compounds, essential for various biological activities including the recycling of **folic acid**. The physiologically active forms are methylcobalamin and adenosylcobalamin which are generated from the common synthetic form, cyanocobalamin. See **intrinsic factor**.

vitamin C *ascorbic acid* An essential dietary vitamin for both humans and guinea pigs, involved in a range of metabolic processes but particularly in collagen synthesis. See **scurvy**.

vitamin D A group of fat-soluble prohormones (collectively calciferol). (1) Vitamin D2 (ergocalciferol) strongly absorbs UV radiation and may be protective in the skin. (2) Vitamin D3 (cholecalciferol) is produced in the skin in response to sunlight and deficiencies lead to rickets. (3) Calcitriol is the $1\alpha,25$-dihydroxy form of vitamin D3 and is important in calcium homeostasis, by increasing gastrointestinal and renal absorption, stimulating release from bone by osteoclasts and facilitating the effect of parathyroid hormone.

vitamin E The collective name for a set of 8 related fat-soluble vitamins with antioxidant properties (alpha-, beta-, gamma- and delta- tocopherols and tocotrienols), although most work has been on α-tocopherol. Vitamin E may protect against free-radical damage, particularly of membranes, but whether this is physiologically important is less clear.

vitamin H See **vitamin B7**.

vitamin K A group of 2-methilo-naphthoquinone derivatives involved in the carboxylation of certain

TABLE V1. Vitamins

Vitamin	Full name	Occurrence	Action	Deficiency disease
Fat soluble				
A	Retinol (11-cis retinal)	Vegetables	Phototransduction, morphogen	Night blindness, Xerophthalamia
D	1,25-dihydroxy-cholecalciferol	Action of sunlight on 7-dehydrocholesterol in skin	Ca^{++} regulation, Phosphate regulation	Rickets
E	α-tocopherol	Plants, esp. seeds, Wheatgerm	Antioxidant	Failure to grow to maturity, Infertility
K1		Higher green plants		
K2	Range of molecules	Intestinal bacteria		
Water soluble				
B1	Thiamine	Degradation of α-keto acids	Beriberi	
	Folic acid (Tetrahydofolic acid)	Plants	Purine biosynthesis	Anaemia
	Nicotinic acid (niacin)			Pellagra
	Pantothenic acid (CoA)	Can be made Plants & micro-organisms		
B2	Riboflavine	Plants & micro-organisms	Constituent of flavoproteins	
B6	Pyridoxine Pyridoxal Pyridoxamine		Transamination	Acrodynia in rats, Convulsions
B12	Cobalamine	Intestinal micro-organisms	Hydrogen transfer reactions	Pernicious anaemia
C	Ascorbic acid	Plants, esp. citrus fruits	Cofactor	Scurvy
H	Biotin	Intestinal bacteria	Protects against avidin toxicity, Intermediate CO_2 carrier	

glutamate residues in proteins to form gamma-carboxyglutamate residues (Gla-residues) that are usually involved in binding calcium. Several blood-clotting factors (prothrombin (factor II), factors VII, IX, X, protein C, protein S and protein Z) are Gla proteins (see **matrix Gla protein** and **osteocalcin**) and thus their synthesis is vitamin K-dependent. See **warfarin** and **coumarin**.

vitellin Most abundant protein in egg-yolk generated by proteolytic cleavage of **vitellogenin**.

vitelline membrane The membrane of protein fibres (not phospholipid), immediately outside the plasmalemma of the ovum and the earlier stages of the developing embryo. Its structure and composition vary in differing animal groups. In mammals, referred to as the **zona pellucida**. In hen's eggs, the vitelline membrane separates yolk from egg-white and is composed of various proteins including lyso-zyme C, **ovalbumin**, ovotransferrin (**conalbumin**), and **ovomucin** as well as eight **ZP domain**-containing proteins, **oviductin**, and two ATPases. Vitelline membrane outer layer protein I (VMO1, 163aa) binds to ovomucin fibrils; VMO II is beta-defensin-11. Paper: http://www.ncbi.nlm.nih.gov/pubmed/18452232

vitellogenic Giving rise to yolk of an egg.

vitellogenin A family of proteins that are the pre-cursor of several yolk proteins in non-mammalian eggs, especially **phosvitin** and **lipovitellin** in the eggs of various vertebrates. Vitellogenin is synthe-sized in the liver after oestrogen stimulation and in insects is synthesized and released from the fat-body during egg-formation. In mammals the large subunit of **microsomal triglyceride transfer protein** is vitellogenin-like.

vitiligo Patchy depigmentation (leucoderma) in which melanocytes in the skin, the mucous membranes, and the retina are destroyed, often with a sharp demarcation line. Associated with autoimmune disease. Loci for susceptibility to autoimmune disease, particularly vitiligo, have been mapped to chromosomes 1p31, 7, 8, and 4.

vitronectin Serum glycoprotein (478aa) also known as serum spreading factor from its activity in promoting adhesion and spreading of tissue cells in culture. Contains the cell-binding sequence Arg-Gly-Asp (RGD) first found in **fibronectin**. It is cleaved into three chains, one of which is somatomedin B.

vitrosin Obsolete name for collagen isolated from embryonic chick vitreous.

viviparous Organisms in which the young develop internally, not in eggs that are incubated externally.

VLA proteins Very late antigens. Proteins (VLA-1 and VLA-2) originally defined as antigens appearing on the surfaces of T-cells 2-4 weeks after *in vitro* activation; they are now know to be part of the β **integrin** family. Additional members of the subset are now known (VLA-3, VLA-4, VLA-5 and VLA-6), the β subunits all being identical. Some of the VLA proteins are receptors for collagen, laminin or fibronectin, and many are now known to be expressed on cells other than leucocytes.

VLDL See **lipoproteins**.

V_{max} The maximum initial velocity of an enzyme-catalyzed reaction, i.e. at saturating substrate levels.

vnx1 The gene encoding the vacuolar Na^+/H^+ exchanger (Low affinity vacuolar monovalent cation/proton antiporter, 908aa) in *S. cerevisiae*.

Vohwinkel's syndrome Skin disease caused by a defect in the gene for **connexin-26**; a variant form is caused by mutation in the **loricrin** gene.

voltage clamp A technique in electrophysiology, in which a microelectrode is inserted into a cell, and current injected through the electrode so as to hold the cell's membrane potential at some predefined level. The technique can be used with separate electrodes for voltage-sensing and current-passing; for small cells, the same electrode can be used for both. Voltage clamp is a powerful technique for the study of **ion channels**. See **patch clamp**.

voltage gradient Literally, the electric field in a region, defined as the potential difference between two points divided by the distance between them. Used more loosely, the potential difference across a plasma membrane.

voltage-gated ion channel A transmembrane **ion channel** whose permeability to ions is extremely sensitive to the transmembrane potential difference. These channels are essential for neuronal signal transmission, and for intracellular signal transduction. See **voltage-gated sodium channels**, **voltage-sensitive calcium channels**.

voltage-gated sodium channels See sodium channel.

voltage-sensitive calcium channels *VSCC* A variety of voltage-sensitive calcium channels are known and on the basis of electrophysiological and pharmacological criteria are grouped into six classes. The general function is to allow calcium influx into the cell as a result of membrane depolarization. The majority (**L-type, N-type, P-type** and **Q-type channels** require substantial depolarization and are sometimes collectively known as high voltage-activated types. The **R-type channels** activate after moderate depolarization and the **T-type channel** opens at relatively negative potentials.

volutin granules Metachromatic granules containing polyphosphate, a linear phosphate polymer found in bacteria, fungi, algae, and some higher eukaryotes that may serve as a stock of phosphate. They are particularly common in *Spirillum volutans*, corynebacteria and mycobacteria. Open access article: http://www.biology-direct.com/content/6/1/50

Volvox carteri A spherical alga that has around 2000 biflagellate somatic cells and 16 asexual reproductive cells (gonidia). Often used as a model for the early differentiation of function in multicellular organisms. Image: http://protist.i.hosei.ac.jp/pdb/Images/Chlorophyta/Volvox/carteri.html

vomeronasal organ Paired organs situated in the nasal area and connected by a narrow duct to the nasal cavity just inside the nostril. Mammalian vomeronasal sensory neurons detect specific chemicals, some of which may be **pheromones**. Called Jacobson's organ in snakes and lizards.

von Hippel-Lindau syndrome A dominant predisposition to various types of benign and malignant tumours caused by mutation in the gene encoding the von Hippel-Lindau tumour suppressor protein, **VHL**. Tumours are highly vascular and overproduce vascular endothelial growth factor (**VEGF**).

von Recklinghausen's disease See **neurofibromatosis** type 1.

von Willebrand factor *vWF* A multimeric plasma glycoprotein (2050aa) involved in platelet adhesion through an interaction with Factor VIII. Present in **Weibel-Palade bodies**, alpha-granules of platelets and in subendothelial connective tissue. The type A domain (vWF domain) is found in complement factors B, C2, CR3 and CR4; the integrins (I-domains); collagen types VI, VII, XII and XIV; and other extracellular proteins, all of which are involved in multi-protein complex formation. See **von Willebrand's disease**.

von Willebrand's disease An autosomal dominant platelet disorder in which adhesion to collagen, but not aggregation, is reduced. Both bleeding time and coagulation are increased. Factor VIII levels are secondarily reduced.

Vorticella Genus of ciliate **protozoa**. It has a bell-shaped body with a belt of cilia round the mouth of the bell, to sweep food particles towards the mouth and a long stalk, connecting it to the substratum, which contains the contractile **spasmoneme**.

VP16 (1) See **etoposide**. (2) Herpes simplex virion protein 16, a virus tegument phosphoprotein, contains two strong activation regions that can independently and co-operatively activate transcription in vivo and is a transcriptional activator of the viral immediate-early genes.

VP22 An abundant **tegument** protein (virion protein 22, 301aa) of alpha herpes simplex virus types 1 and 2 (HSV-1 and HSV-2) that may play a role in the nucleocapsid secondary envelopment at the trans-Golgi network during virus morphogenesis.

vpr A protein in HIV-1 (viral protein R, 96aa) expressed late during viral replication. Involved in the transport of the viral pre-integration (PIC) complex to the nucleus during the early stages of the infection and has a range of other effects including induction of cell cycle arrest in G2. Has cytotoxic effects on both infected cells and bystander cells, and exhibits both pro- and anti-apoptotic activity.

Vpu Viral protein U. A single-pass membrane protein (81aa) encoded by HIV-1 that targets CD4 and CD317 (**tetherin**) for proteasome degradation thereby enhancing virion release.

VRE (1) **Vancomycin**-resistant enterococci. (2) See VIP.

VSCC See **voltage-sensitive calcium channel**.

VSG Variant surface glycoprotein of trypanosomes. See **antigenic variation**. Article: http://www.biomedcentral.com/1471-2164/8/234

VSV-G tag Epitope tag (YTDIEMNRLGK) derived from the **vesicular stomatitis virus** G protein.

vWF See **von Willebrand factor**.

W

W Single letter code for tryptophan.

W locus Mouse coat colour locus, equivalent to the kit proto-oncogene, that encodes a **receptor tyrosine kinase** for stem cell factor, essential for development of haematopoietic and germ cells.

Waardenburg's syndrome Autosomal dominant disorders with deafness and pigmentary disturbances probably as a result of defects in function of **neural crest**. Various forms of the syndrome are recognized. Waardenburg's Syndrome 1 (WS1) and WS3 (also known as Klein-Waardenburg syndrome) are caused by mutation in Pax3 – an homologous defect to the mouse mutant Splotch that also has defective Pax-3. Waardenburg-Shah syndrome (WS4), in which Waardenburg's syndrome is associated with Hirschsprung's disease, is due to mutation in Sox10 and there is an homologous mutation in Dom mice (dominant megacolon), piebald-lethal and lethal spotting. WS2 is heterogeneous with mutation in the microphthalmia (MITF) gene or the gene for **slug**.

Waf1 An inhibitor (p21, cip1, cyclin-dependent kinase inhibitor 1A, melanoma differentiation-associated protein 6, 164aa) of **cdk** activity expressed in all adult tissues and found in a complex with cyclin D, cdk4 and **PCNA**. Can bind to and inhibit all members of the cdk family though affinity varies. Expression regulated by p53 tumour suppressor. Also found in active cyclin/cdk complexes – multiple copies of Waf1 may be necessary to produce inhibition.

Wagner's syndrome A disorder that affects the integrity of the vitreous humour of the eye caused by mutation in the gene encoding chondroitin sulphate proteoglycan-2 (**versican**); a similar disorder is Stickler's syndrome caused by defects in **collagen** COL2A1.

WAGR syndrome A contiguous gene syndrome in which there is deletion in chromosome 11 of a region containing the WT1 (**Wilms' tumour**) and PAX6 genes. The disorder is characterised by Wilms' tumour, susceptibility to aniridia, genitourinary abnormalities, and mental retardation. A larger deletion also removes the BDNF (brain derived neurotrophic factor) gene causing WAGRO syndrome in which there is obesity.

Waldenstrom's macroglobulinaemia A rare subtype of non-Hodgkin lymphoma in which there is a high level of monoclonal IgM in the blood. The IgM sometimes has detectable antibody activity,

e.g. rheumatoid factor. In some forms there is increased expression of 6 microRNAs, including MIR155 which acts through MAPK/ERK, PI3/AKT and NFκB pathways.

Walker motif Two separate sequences, A and B, which come together to form a nucleotide binding site. The Walker A motif (phosphate-binding loop, P-loop), is the sequence GXXXXGKT/S, and is probably the site for nucleotide binding in many proteins. The Walker B motif consensus is D(D/E)XX. The motif was originally described in the α- and β-subunits of F1-ATPase, myosin and other ATP-requiring enzymes by Walker and colleagues in 1982.

Walker-Warburg syndrome A disorder associated with several distinct congenital muscular dystrophies and developmental problems such as hydrocephalus, caused by mutation in the genes encoding protein O-mannosyltransferase-1 and -2 (POMT1 & 2) and in some cases mutations in genes for FKRP **fukutin**, fukutin-related protein and **LARGE** (large N-acetylglucosaminyltransferase).

wall appositions Localized modifications of the plant cell wall following mechanical stimulation by pathogens. They involve deposition of callose and site-directed secretion of other cell wall components and anti-microbial compounds including phenolics, silicon, H_2O_2 and pathogenesis-related proteins. They act as a physical and chemical barrier against invading pathogens, part of the basal defence response.

Wallerian degeneration The degenerative process that starts when a nerve fibre is cut or crushed. The distal portion of the axon, separated from the cell body degenerates (anterograde degeneration). The axoplasm is lost rapidly but the neurolemmal tube that tends to persist may act as a guidance cue for regenerating neurites in the peripheral nervous system.

WAP domain A conserved motif (whey acidic protein domain, ~50aa) containing eight cysteines in a characteristic 4-disulphide core arrangement. It has serine peptidase inhibitory activity in several proteins (e.g. WFIKKNs, elafin).

warfarin Synthetic derivative of **coumarin** that inhibits the effective synthesis of biologically active forms of the vitamin K-dependent clotting factors: used clinically as an antithrombotic. Was used as a rat poison though now superceded by more potent rodenticides.

The Dictionary of Cell and Molecular Biology. DOI: http://dx.doi.org/10.1016/B978-0-12-384931-1.00023-4

warm antibodies Antibodies, mostly IgG, that work best at body temperature, unlike the **cold agglutinins**. See **warm antibody haemolytic anaemia**.

warm antibody haemolytic anaemia An autoimmune disorder characterized by the premature destruction of red blood cells by autoantibody-induced lysis at temperatures above normal body temperature. Contrast **paroxysmal cold haemoglobinuria**.

wart (1) Benign tumour of basal cell of skin, the result of the infection of a single cell with wart virus (papillomavirus). Virus is undetectable in basal layer, but proliferates in keratinising cells of outer layers. (2) In *Drosophila* a ser/thr kinase, warts. See **Salvador-Warts-Hippo pathway**

WASH complex A multi-protein complex, important in endosome sorting, located on the endosomal surface where it recruits and activates the **Arp2/3** complex to induce actin polymerization. The complex is composed of F-actin-capping protein (**capZ**) alpha and beta subunits (various isoforms, depending upon the tissue), WASH (Wiskott-Aldrich Syndrome Protein and SCAR Homologues, WASH1, WASH2P, WASH3P, WASH4P, WASH5P or WASH6P, see **WASP family proteins**), FAM21 (FAM21A, 1341aa; FAM21B, 1253aa; FAM21C, 1318aa), KIAA1033 (1173aa), KIAA0196 (**strumpellin**) and CCDC53 (194aa). The complex is evolutionarily conserved through fungi, amoebae and metazoa. Research article: http://www.landesbioscience.com/journals/cib/article/DeriveryCIB3-3.pdf

WASL One of the **WASP** family proteins, (neural Wiskott-Aldrich syndrome protein, NWASP, 505aa) that regulates actin polymerization by stimulating the actin-nucleating activity of the **Arp2/3** complex, binds to SH3 domains of GRB2, to **WIP**, the PRPF40A gene product (**HIP-10**) and with **nostrin**.

WASP family proteins A family of proteins (Wiskott-Aldrich syndrome proteins) that includes WASP (502aa), N-WASP (WASP-like, WASL, 505aa), WAVE1, 2 & 3 (WASP family, verprolin homology domain-containg proteins) and the yeast homologue Las17p (633aa). WASP is an effector protein for rho-type GTPases and in conjunction with **cdc42** and the **arp2/3** complex regulates actin polymerisation. Defects in WASP are responsible for **Wiskott-Aldrich syndrome**. WASP is mainly found in myeloid cells whereas N-WASP is found in a larger range of tissues. See **verprolins**.

WASP homology domains The WASP-homology (WH1) domain (~115aa) is the amino-terminal region of WASP, binds proline-rich sequences and is involved in the formation of multicomponent assemblies. A subset of WH1 domains are termed **EVH1 domains** and appear to bind a polyproline motif. The WH2 domain (~35aa) is an actin monomer-binding motif found in many proteins (e.g. thymosin-β4, **ciboulot**, **WASP**, and **verprolin** /WIP) that are involved in regulation of the actin cytoskeleton. See also **beta-thymosin repeats**.

Wassermann reaction An obsolete test formerly used in the diagnosis of syphilis. Cardiolipin derived from ox heart was used as an antigen.

water potential The chemical potential (i.e. free energy per mole) of water in plants. Water moves within plants from regions of high water potential to regions of lower water potential, ie. down-gradient. Pure water has a potential of 0 MPa (mega Pascals). The symbol used is (Gk psi) Ψw. Water potential (Ψw) is the sum of solute potential (Ψs) and pressure potential (Ψp), thus $\Psi w = \Psi s + \Psi p$.

Watson's syndrome A disorder caused by mutation in the **neurofibromin** gene and probably allelic to neurofibromatosis (NF1).

WAVEs A family of proteins (WASP family, verprolin homology domain-containing proteins) with homology to the scar protein of *Dictyostelium* and formerly referred to as WAVE/SCAR. In humans wave1 (WASP family protein member 1, verprolin homology domain-containing protein 1, 559aa) and wave3 (502aa) are both strongly expressed in brain. Wave2 (498aa) has a wide tissue distribution. In *Arabidopsis* there are various WAVE-family proteins, referred to as SCAR proteins: SCAR1 (WAVE1, 821aa) regulates trichome branch positioning and expansion, SCAR2 (WAVE4, DISTORTED 3, IRREGULAR TRICHOME BRANCH 1, 1399aa) and various others all with similar roles.

waxes Complex mixtures of long-chain molecules that coat plant surfaces, minimizing water loss, possibly affording protection against pathogens.

WBC White blood cells: granulocytes (neutrophils, eosinophils and basophils), monocytes and lymphocytes.

WD-repeat proteins *WD40* A conserved structural motif (~40aa), usually ending with tryptophan and aspartic acid (WD), implicated in protein-protein interactions. Crystal structure of one WD-repeat protein (GTP-binding protein beta subunit) reveals that the seven repeat units form a circular propeller-like structure with seven blades. Boston University webpage: http://bmerc-www.bu.edu/projects/wdrepeat/

weal-and-flare See **triple response**.

weaver A strain of mice in which there is early apoptotic death during development of cells in testes, cerebellum and midbrain and that suffer from severe ataxia. The mutation is a base pair substitution in the G-protein-coupled inwardly-rectifying potassium channel 2 gene (GIRK2, 425aa in mouse). Up to 70% of the mesostriatal dopaminergic neurons are lost and major alterations of the dopaminergic dendrites of the substantia nigra have been described, a condition with similarities to Parkinson's disease. The defect does not seem to be due to reduced neurotrophin levels.

Weber's law See **Fechner's law**.

wee Cell cycle checkpoint genes that inhibit the entry into mitisis, found in *Schizosaccharomyces pombe*. Mutants in *wee-1* and *wee-2* have normal growth rate but divide earlier so the cells are smaller. In humans, Wee1-like protein kinase (EC 2.7.10.2, 646aa) acts as a negative regulator of the G2 to M transition by phosphorylating and inactivating cyclin B1-complexed **cdc2**.

Wegener's granulomatosis A systemic disease in which there are upper and lower respiratory tract granulomas and necrotising focal glomerulonephritis. There are antineutrophil cytoplasmatic autoantibodies (ANCAs) for which the antigen is the azurophil granule protease 3 (c-**ANCA**). Reference page: http://www3.niaid.nih.gov/topics/wegeners/

Wehi 3b cells A macrophage-like myelomonocytic cell line derived from Balb/c mouse: they constitutively secrete IL-3.

Weibel-Palade body Cytoplasmic organelle found in the vascular endothelial cells of some animals, though not in the endothelium of all vessels. Although markers for endothelium, their absence does not necessarily mean the cells are not of endothelial origin. They store and release **von Willebrand factor** and **P-selectin**. Image: http://www.pathologyimagesinc.com/emhandbook/diagn-organelles-section/organelle-pages/weibel-palade-body.html

Weil's disease See *Leptospira*.

Weil-Felix reaction An agglutination test used in the diagnosis of rickettsial infections (typhus etc.) which depends upon a carbohydrate cross-reacting antigen shared by ricckettsiae and Proteus group OX. It is relatively insensitive and is being superceded.

Weill-Marchesani syndrome A rare connective tissue disorder in which there are abnormalities of the lens of the eye, short stature, brachydactyly and joint stiffness. The autosomal recessive form is a result of mutations in the **ADAMTS10** gene, the autosomal dominant form by mutations in the **fibrillin 1** gene. NCBI information page: http://www.ncbi.nlm.nih.gov/bookshelf/br.fcgi?book = gene&part = weill-ms

Weismann's germ plasm theory The theory that organisms maintain genetic continuity from organism to offspring through the germ line cells (germ plasm) and that the other (somatic) cells play no part in the transmission of heritable factors. In some organisms somatic cells lose chromosomes at an early stage in development and are clearly no longer totipotent.

Weissenbacher-Zweymüller syndrome A rare autosomal dominant disorder of bone growth caused by defects in **collagen** type XI (COL11A2). Allelic with Stickler's syndrome type 3 and has features that are similar to otospondylomegaepiphyseal dysplasia (OSMED). Reference page: http://ghr.nlm.nih.gov/condition = weissenbacherzweymullersyndrome

Werdnig-Hoffman disease See **spinal muscular atrophy**.

Werner's syndrome A premature ageing syndrome (progeria) caused by mutations in the RECQL2 gene (WRN gene, predicted to encode a protein of 1432aa with similarities to DNA helicases). It is the human homologue of the *E. coli* RecQ **DNA helicase** that is involved in mismatch repair. There is chromosomal instability and a high incidence of tumours. Werner's syndrome cells in culture seem to be restricted to about 20 population doublings compared to the normal limit of around 60 (see **Hayflick limit**) unless they are experimentally induced to upregulate telomerase expression. Hutchinson-Gilford progeria syndrome is more severe and has an earlier onset caused by mutation in the *LMNA* gene that encodes both **lamin A** and lamin C. NCBI information page: http://www.ncbi.nlm.nih.gov/bookshelf/br.fcgi?book = gnd&part = wernersyndrome

Wernicke's encephalopathy *Wernicke-Korsakoff syndrome* Thiamine deficiency disease possibly caused by an abnormal isoform of transketolase, a thiamine-dependent enzyme. The deficiency can be dietary or due to alcohol abuse (which blocks thiamine absorption). Wernicke's encephalopathy is an acute syndrome which can cause death and neurologic morbidity; Korsakoff's syndrome (Korsakoff's psychosis) is a chronic neurologic condition that usually occurs subsequently.

West Nile virus An arthropod-borne Flavivirus that normally infects birds but can cause severe meningitis and encephalitis in humans.

Western blot See **blotting**.

wewakpeptins Depsipeptides (wewakpeptins A-D) isolated from the marine cyanobacterium *Lyngbya semiplena*. Have potential anti-tumour activity.

WFIKKNs Proteins (**WAP, follistatin, immunoglobulin, Kunitz** and **NTR domain**-containing proteins) with multiple peptidase-inhibitory domains that are thought to inhibit both serine peptidases and metallopeptidases. WFIKKN1 (548aa) and WFIKKN2 (WFIKKN-related protein, 576aa) have different tissue distributions. Research article: http://www.ncbi.nlm.nih.gov/pmc/articles/PMC31116/pdf/pq003705.pdf

WGHA See whole genome homozygosity association.

WHAMM A protein (**WASP** homologue-associated protein with actin, membranes and microtubules, WH2 domain-containing protein 1, 809aa) that is a nucleation-promoting factor that stimulates Arp2/3-mediated actin polymerization at the Golgi apparatus and along tubular membranes. It is involved in movement of vesicles from ER to Golgi.

Wharton's jelly Viscous hyaluronic acid-rich jelly found in the umbilical cord.

wheat germ The embryonic plant at the tip of the seed of wheat. Wheat germ has been used as the starting material for a cell-free translation system and is also the source of **wheat germ agglutinin**.

wheat germ agglutinin A plant lectin (WGA1, 212aa; WGA2, 213aa; WGA3, 186aa) that binds to N-acetylglucosaminyl and sialic acid residues. See **lectins**.

whirlin A **PDZ domain**-containing protein (**CASK**-interacting protein, CIP98, 98 kDa, 918aa in mouse) originally described from the mouse mutant, whirler, that is deaf and shows circling behaviour. Whirlin is necessary for the elongation and maintenance of inner and outer hair cell **stereocilia** in the organ of Corti. Mutations in human whirlin cause **Usher syndrome Type 2D** and autosomal recessive nonsyndromic deafness-31.

white An eye colour gene of *Drosophila*; the wild type product is essential for red eyes. The white locus is involved in the distribution of brown ommochrome and red pteridine pigments, found in the eyes of adult flies; the encoded protein is believed to be an integral membrane transporter protein for pigment precursors.The human homologue (ABCG1, 678aa) is involved in lipid homeostasis in macrophages and probably in other cells.

white adipose tissue Tissue composed largely of **adipocytes** but not specialised for production of heat (see **brown adipose tissue**). The major storage site for fat in the form of triglycerides, serves three functions: heat insulation, mechanical cushioning, and as an energy reserve. An increasing proportion of people in the so-called civilised world have an excess.

white blood cells *WBC* See **leucocytes** and specific classes (**basophils, coelomocytes, eosinophils, haemocytes, lymphocytes, neutrophils, monocytes**).

Whitmore's disease See **melioidosis**.

whole cell patch See **patch clamp**.

whole genome homozygosity association *WGHA* A method in which patterned clusters of SNPs demonstrating extended homozygosity are identified using arrays; these are tested for association with disease using genome-wide and regionally specific statistical tests. Research article: http://www.pnas.org/content/104/50/19942.full

Widal test A simple but unreliable diagnostic test for typhoid fever based on the presence of somatic (O) and flagellar (H) agglutinins to *Salmonella typhi* in serum.

Williams-Beuren syndrome A neurodevelopmental disorder (Williams' syndrome) with multisystem manifestations caused by heterozygosity for a deletion of Chromosome 7 band 7q11.23 spanning 1.5 million to 1.8 million base pairs and containing 26 to 28 genes. The deletion affects **elastin, LIM kinase-1**, replication factor C, and **MAGI** among others.

Wilms' tumour One of the most common solid tumours of childhood, a kidney tumour (nephroblastoma) thought to derive from renal stem cells which retain embryonic differentiation potential. Like **retinoblastoma**, both sporadic and inherited forms occur. In Wilm's tumour 1 (WT1) the mutation is in the gene for a zinc finger DNA-binding protein (429aa) that can act as a transcriptional activator or repressor depending on the cellular or chromosomal context. Wilm's tumour 2 (WT2) is associated with mutation of the imprinted H19 gene (also implicated in **Beckwith-Wiedemann syndrome**). Other forms of Wilm's tumour (WT3, 4) are associated with other loci. WT5 is caused by mutation in the POU6F2 transcription-factor gene. Wilms' tumour 1-associating protein (396aa) regulates the G2/M transition by stabilizing cyclin A2 mRNA. See **WAGR syndrome**.

Wilson's disease A rare autosomal recessive disease (hepatolenticular degeneration) caused by a defect in the copper-transporting ATPase (EC 3.6.3.4, 1465aa) coded by the *ATP7B* gene. There is excessive deposition of copper in liver, brain and kidney.

WINAC complex An ATP-dependent chromatin remodelling complex that directly interacts with the vitamin D receptor (VDR) through the **William's syndrome** transcription factor (WSTF). It contains SWI/SNF components and DNA replication-related factors such as **topoisomerase** TOP2B.

Winchester syndrome See **Mona 2**.

windbeutel A protein (257aa in *Drosophila*), thought to act as a chaperone by folding **pipe** and targeting it to the Golgi. Involved in dorsoventral axis patterning in early embryos. See **nudel**.

winged helix transcription factors Family of transcription factors (~100aa) characterized by a conserved DNA-binding domain found in *Drosophila* homeotic gene **forkhead** and rat hepatocyte nuclear factor-3 (HNF-3b). At least 80 genes with this motif are known and have been sub-divided on the basis of phylogenetic analysis into classes **FoxA** -FoxQ, many with developmentally-specific patterns of expression. FAST -1 is a member of this family. Nomenclature index: http://www.biology.pomona.edu/fox/

wingless In *Drosophila* a homologue of *int-1* (**Wnt-1**), a segment polarity gene that encodes a secreted signalling protein that functions in pattern formation. Wingless (468aa) secretion is dependent on **hedgehog** and it is a ligand for a family of **frizzled** receptors. Flybase entry: http://www.sdbonline.org/fly/segment/wingles1.htm

WIP (1) One of the **verprolin** family, (**WASP**-interacting protein, WAS/WASL-interacting protein family member 1, 503aa) that induces actin polymerization and redistribution and together with the adaptor protein **NCK1** and **GRB2** recruits, activates and regulates the subcellular localization of **WASL**. (2) In *Arabidopsis*, WPP domain-interacting proteins (WIP1, 489aa; WIP2, 509aa; WIP3, 459aa) that mediate docking of Ran GTPase-activating proteins with the nuclear envelope. (3) In *Zea mays*, Bowman-Birk type wound-induced proteinase inhibitor WIP1 (102aa).

WIRE One of the vertebrate **verprolins**, binds to **WASP** and N-WASP (**WASL**) and has a role in regulating actin dynamics downstream of the platelet-derived growth factor (PDGF) beta-receptor.

Wiskott-Aldrich syndrome Thrombocytopenia with severe immunodeficiency (both cell-mediated and IgM production). Associated with increased incidence of leukaemia. Caused by mutation in the gene that encodes **WASP**.

WISPs Members of the **CCN family** of regulatory proteins (**Wnt**-induced secreted proteins; WISP-1 (CCN4), 367aa; WISP-2 (CCN5), 250aa; WISP-3 (CCN6), 354aa). WISP-2 is thought to play a role in breast carcinoma and is overexpressed in response to estrogen. The other WISPs are expressed in various tumours.

Witkop's syndrome A condition (tooth-and-nail syndrome) in which there are defects in dentition and in nail growth caused by mutation in the *Msx1* gene.

WNKs Serine-threonine kinases (EC 2.7.11.1) of the STE20 family containing a cysteine but no lysine in the active site. Human WNK-1 (with no K (lysine) protein kinase-1, 1246aa) is expressed in most tissues, with different isoforms in the kidney, and in heart and skeletal muscle. Expression is high in polarized epithelia that transport chloride; WNK1 inhibits the activity of WNK4. WNK2 (2126aa) is widely expressed. WNK3 (1800aa) regulates transport mediated by Na-K-2Cl and Na-Cl cotransporters. WNK4 colocalizes with **ZO1** and regulates the thiazide-sensitive Na-Cl cotransporter. WNKs are activated by hyperosmotic stress and in turn activate

SPAK and **OSR-1**. Mutations can cause **pseudohypoaldosteronism**. WNK8 (563aa in *Arabidopsis*) regulates flowering time by modulating the photoperiod pathway and phosphorylates the vacuolar ATPase subunit C: WNK1 in *Arabidopsis* (700aa) has a similar role.

wnt Multigene family encoding various secreted signalling molecules important in morphogenesis. First member was *Drosophila* wingless, but many vertebrate homologues are now known. Wnt-1 (formerly int-1) binds to cell-surface **frizzled** family receptors which activate **dishevelled** family proteins. This pathway then changes beta-**catenin** and **plakoglobin** distribution and affects the association of APC **tumour suppressor** protein with catenin. Wnt proteins are believed to activate a transcription factor leukemia enhancer factor-1 (LEF-1) by inhibiting **glycogen synthase kinase**.

wobble hypothesis An hypothesis for why many m-RNA codon triplets translate to a single amino acid, why there are appreciably fewer t-RNAs than m-RNA codon types, and why the redundant nature of the genetic code translates into a precise set of 20 amino acids. Inosine in position 1 in the anticodon of tRNA can base pair with A,U or C in position 3 in the mRNA codon, so that for example UCU, UCC, UCA all code for serine using an inosine anticodon.

Wohlfart-Kugelberg-Welander disease See spinal muscular atrophy.

Wolbachia An obligate endosymbiotic bacterium of the filarial nematode *Brugia malayi*, and of a wide range of arthropods. Elimination of *Wolbachia* from *Brugia* leads to growth retardation, infertility and killing. *Wolbachia* infection can markedly affect reproductive behaviour in arthropods, making males sterile and inducing parthenogenesis in females.

Wolfram's syndrome An autosomal recessive disorder (diabetes insipidus, diabetes mellitus, optic atrophy, and deafness, DIMOAD) with growth retardation and various developmental and neurological defects. One form is caused by a mutation in **wolframin**, another by mutation in **CDGSH iron sulphur domain protein 2**.

wolframin A transmembrane protein (890aa) of the endoplasmic reticulum involved in cellular Ca^{2+} homeostasis. Mutations cause **Wolfram's syndrome** and non-syndromic sensorineural deafness autosomal dominant type 6 (DFNA6).

Wollaston prism Prism composed of two wedge-shaped prisms with optical axes at right angles. Light emerging has two beams of opposite polarization. Used in differential interference contrast microscopes.

Woodhouse-Sakati syndrome A rare autosomal recessive multisystemic disorder caused by a mutation in the C2ORF37 gene (chromosome 2 open reading frame 37) encoding a nucleolar protein of unknown function found in two alternatively spliced isoforms (240aa and 520aa).

woolsorter's disease Anthrax, infection with *Bacillus anthracis*, from infected wool or hair of animals.

WormBase Genomic database for *Caenorhabditis elegans* and related nematodes. Link: http://www.wormbase.org/

Woronin bodies A type of **peroxisome** specific to several genera of filamentous ascomycetes that seal the **septal pore** in response to cellular damage. The major protein (176aa) forms oligomers that self-assemble into hexagonal rods. Article: http://www.scielo.br/scielo.php?script = sci_arttext&pid = S1517-83822002000200005

wortmannin Fungal metabolite, isolated from *Penicillium wortmanni*, that is a fairly selective inhibitor of **PI-3-kinases** and possibly other points in the Ras/MAP kinase signalling pathway. Has been shown to inhibit **polo-like kinases.**

WPP domain A protein domain with a highly conserved Trp-Pro-Pro motif, involved in associating plant proteins with the nuclear envelope. The motif is found in **matrix attachment region-binding proteins** and in plant **ran**-GTPase activating proteins (RANGAPs). Article: http://www.ncbi.nlm.nih.gov/pmc/articles/PMC535872/

WRKY proteins Members of a large family of transcriptional regulators which contain the conserved amino acid sequence WRKYGQK together with a zinc-finger-like motif. Members of the WRKY transcription factor family are involved in transcriptional regulation associated with plant immune responses and development.

wssv A Group I (dsDNA) virus, white spot syndrome virus (WSSV), the type species of the genus *Whispovirus* in the family Nimaviridae, that infects penaeid shrimp and other crustaceans in commercial shrimp farms. It has a wide host range, is highly virulent and mortality can be very high.

WUSCHEL *WUS* A plant homeodomain protein involved in specifying stem cell fate in shoot and floral meristems; during embryogenesis promotes the vegetative-to-embryogenic transition. Abstract: http://www.ncbi.nlm.nih.gov/pubmed/12000682?dopt = AbstractPlus&holding = f1000, f1000m,isrctn

WW domain A semiconserved motif (38aa) that is a triple-stranded, antiparallel beta-sheet domain and binds proline-rich sequences. Found in diverse proteins (e.g. **WWOX**, **yes-associated protein** and **PIN-1**).

WWOX A tumour suppressor (WW domain-containing oxidoreductase, EC 1.1.1.-, 414aa) that is deleted or altered in several cancer types, plays a role in apoptosis and is required for normal bone development. May have a role in TGFβ1 and TNFα signalling and cell death, also in **wnt** signalling in *Drosophila*.

X

X Single letter code for an unspecified or unknown amino acid.

X chromosome A sex chromosome. In mammals paired in females. The **heterogametic sex** in birds, reptiles, some fish and amphibians is the female, not the male.

X chromosome inactivation centre *Xic* The region of the X chromosome responsible for the inactivation of that X chromosome in female mammals (see **Lyon hypothesis**). The gene involved, Xist, maps to this region and encodes a nuclear RNA (17 kb) that co-localizes with the inactivated X chromosome. Xist introduced on an autosome is capable of inactivation in cis and the Xist RNA becomes localized at that autosome. In many organisms, dosage compensation equalizes sex-linked gene expression in males and females; in flies and worms this is respectively by up- or down-regulation of X-linked expression.

X-Gal A noninducing chromogenic substrate for beta-galactosidase, which hydrolyzes X-Gal to form an intense blue precipitate. X-Gal is frequently used in conjunction with IPTG in **blue-white colour selection** to detect recombinants (white) from non-recombinants (blue) and for selection of beta-galactosidase reporter gene activity in transfection of eukaryotic cells.

X-inactivation The inactivation of one or other of each pair of X chromosomes to form the Barr body in female mammalian somatic cells (see **X chromosome inactivation centre**.) Thus tissues whose original zygote carried heterozygous X-borne genes should have individual cells expressing one or other but not both of the X-borne gene products. The inactivation is thought to occur early in development and leads to mosaicism of expression of such genes in the body. See also **Lyon hypothesis**.

X-linked disease Any inherited disease whose controlling gene or at least part of the relevant genome is carried on an X chromosome, e.g. haemophilia. Most known conditions are recessive and thus since males have only one X chromosome they will express any such recessive character. Females are only seriously affected if homozygous, which is rare, although there may be **haploinsufficiency**.

X-linked lymphoproliferative syndrome A set of disorders (XLP, Duncan disease, Purtilo syndrome) caused either (XLP1) by mutation in the SH2D1A gene (**SLAM-associated protein**) or (XLP2) by mutation in X-linked inhibitor of apoptosis,

XIAP. Affected individuals are extremely sensitive to infection with Epstein-Barr virus leading to severe or fatal mononucleosis, acquired hypogammaglobulinaemia, and malignant lymphoma.

X-ray diffraction Basis of powerful technique (X-ray crystallography) for determining the three-dimensional structure of molecules, including complex biological macromolecules such as proteins and nucleic acids, that form crystals or regular fibres. Low-angle X-ray diffraction is also used to investigate higher levels of ordered structure, as found in muscle fibres. Helpful website: http://www.panalytical.com/index.cfm?pid=135

X-ray microanalysis See **electron microprobe**.

Xaf1 A negative regulator of members of the IAP (inhibitor of apoptosis protein) family that inhibits the anti-caspase activity of **BIRC4**, (XIAP-associated factor 1, BIRC4-binding protein, 301aa). It binds **XIAP** and re-localizes it to the nucleus, inhibiting its activity and enhancing apoptosis. Expression is reduced or absent in tumours suggesting it may function as a tumour suppressor.

xanthine A purine base (2,6-dihydroxypurine). Its methylated derivatives (theophylline, theobromine, caffeine) are cAMP phosphodiesterase inhibitors.

xanthine dehydrogenase *xanthine oxidoreductase* A key enzyme (XDH, EC 1.17.1.4, 1333aa) in the purine degradation pathway that oxidises hypoxanthine to xanthine and xanthine to uric acid. Can be converted from the dehydrogenase (XDH) to the xanthine oxidase form (XO; EC 1.17.3.2) by proteolysis or reversibly through the oxidation of sulphydryl groups. Each subunit binds a molybdenum ion. Can generate reactive oxygen species which can cause cellular injury. Mutations in the *XDH* gene lead to **xanthinuria** type 1.

xanthine oxidase See **xanthine dehydrogenase**.

xanthinuria A disorder in which large amounts of xanthine are excreted in the urine. Type 1 is caused by mutation in **xanthine dehydrogenase**, in type 2 there is a deficiency of both xanthine dehydrogenase and aldehyde oxidase. Xanthinuria can also be caused by **molybdenum cofactor deficiency**.

xanthoma Localized lesion of subcutaneous tissues in which there is an accumulation of cholesterol-filled macrophages. Characteristic of primary biliary cirrhosis and various lipid-handling disorders.

The Dictionary of Cell and Molecular Biology. DOI: http://dx.doi.org/10.1016/B978-0-12-384931-1.00024-6

xanthophore A cell containing a yellow pigment found, for example, in the dermis of goldfish.

xanthophylls Yellow **carotenoid** pigments involved in photosynthesis. Consist of oxygenated carotenes, e.g. lutein, violaxanthin and neoxanthine. The yellow colour of the macula lutea in the eye and the yellow of egg yolk is from dietary xanthophylls, lutein and zeaxanthin.

xanthopterin Yellow pterin pigment found in some insect wings and in mammalian urine.

XBP-1 X-box-binding protein 1. A transcription factor (261aa) essential for hepatocyte growth, the differentiation of plasma cells, the secretion of immunoglobulin, and the unfolded protein response (see **inositol-requiring enzyme 1a**).

xenin A 25aa peptide hormone (xenopsin-related peptide) cleaved from the alpha- **coatomer** subunit. It is secreted into the blood after a meal, stimulates secretion from the exocrine pancreas and affects gut motility. Interacts with the **neurotensin** receptor subtype 1 of intestinal smooth muscle cells.

xenobiotic Any substance that does not occur naturally but that will affect living systems.

xenobiotic response element XRE A DNA regulatory sequence that binds transcription factors that regulate genes coding for enzymes involved in detoxification.

xenogeneic Literally, of foreign genetic stock; usually applied to tissue or cells from another species, as in xenogeneic transplantation.

xenograft A graft between individuals of unlike species, genus or family.

Xenopus The genus of African clawed frogs. *X. laevis* is widely used in developmental biology because the eggs are large and can be produced on demand by injecting the animals with human chorionic gonadotropin. *Xenopus* eggs are also used as an expression system for ion channnels because the eggs can be studied electrophysiologically. Biology and genomics resource: http://www.xenbase.org/common/

xenosome (1) A bacterial endosymbiont of certain marine protozoans. (2) Inorganic particles in various testate amoebae.

xenotropic Describing something that grows in a foreign environment. Used especially of endogenous retroviruses (see **xenotropic virus**).

xenotropic virus A virus that is benign in cells of one animal species and will only produce complete infective virus when it infects cells of a different species. Xenotropic murine leukemia viruses (a gamma retrovirus) cannot infect cells from laboratory mice because of the lack of a functional cell surface receptor required for virus entry but cells from many nonmurine species, including human cells, are fully permissive and the virus will infect and replicate. Xenotropic and polytropic (amphotropic) murine leukemia viruses (X-MLVs and P-MLVs) cross-interfere to various extents in non-mouse species and in wild Asian mice, suggesting that they might use a common receptor for infection. The possibility of a link between murine XMRV and human diseases (ME and both familial and sporadic prostate cancer) seems to have been disproved. See **xenotropic virus receptor**.

xenotropic virus receptor *X-receptor* Xenotropic and polytropic retrovirus receptor (XPR1, 696aa) has multiple membrane spanning domains, is expressed in a range of human tissues and may play a role in G protein-coupled signal transduction. It has been found in several mammals and in *Xenopus tropicalis*. A homologue (SPX and EXS domain-containing protein 3, 919aa) has been found in *Dictyostelium*.

xeroderma pigmentosum An autosomal recessive condition in humans associated with increased sensitivity to ultraviolet-induced mutagenesis, and thus to skin cancer. Sensitivity can be demonstrated in cultured cells, and appears to be due to deficiency in DNA repair, specifically in excision of ultraviolet-induced **thymine dimers**. There are 7 complementation groups (A-G) indicating that mutation at any one of at least 7 loci can cause defective DNA repair. The variant form of xeroderma pigmentosum (XPV) is caused by mutations in the DNA polymerase eta gene.

xerophile An organism that is able to live best in very dry environments.

XhoI Restriction enzyme from *Xanthomonas holcicola* now generally produced in *E. coli* into which the *XhoI* gene has been inserted.

XIAP X-linked inhibitor of apoptosis. An inhibitor of apoptosis, one of the **BIRC family** (BIRC4, 497aa) with three BIR repeats and **E3 ubiquitin-protein ligase activity**. Mediates the proteasomal degradation of target proteins, such as **caspase-3**, **smac** and **AIFM1**, also regulates the levels of **COMMD**. Binds tightly to caspase-9 in the apoptosome complex, and as a result caspase-7 processing is prevented. Mutations can cause **X-linked lymphoproliferative syndrome** type 2.

Xic See **X chromosome inactivation centre**.

XKCM1 A kinesin-like protein (KIF2C, kinesin central motor 1, 730aa) in *Xenopus* that promotes ATP-dependent removal of tubulin dimers from microtubules. It acts in opposition to **xmap215**.

XLA Synonym for the gene mutated in **Bruton's agammaglobulinemia**. See **btk**.

XLP X-linked lymphoproliferative disease. One of six X-linked immunodeficiencies, XLP (Duncan disease) is caused by a mutation in the gene for **SLAM-associated protein** which renders males unable to produce an effective immune response to **Epstein-Barr virus** (EBV). Proliferation of EBV-infected B-cells is apparently unregulated and invariably results in fatal mononucleosis, agammaglobulinemia, or malignant lymphoma. Review: http://emedicine.medscape.com/article/203780-overview

xmap215 *Xenopus* microtubule assembly protein of 215 kDa. A major microtubule-stabilizing factor (2065aa) in *Xenopus* egg extracts. It acts in opposition to the destabilizing factor **XKCM1**, a member of the kinesin superfamily, during interphase. Various homologues are known: human, ch-TOG (CKAP5); *C. elegans*, Zyg9; *Arabidopsis*, MOR1; yeast, Stu2p. See **TOG domain**. Abstract: http://www.ncbi.nlm.nih.gov/sites/entrez?db=pubmed&cmd=search&term=10620801

XMRV See **xenotropic virus**.

XRE See **xenobiotic response element**.

XRN1 A 5′ to 3′ exonuclease (EC 3.1.11.-, strand exchange protein 1, 1694aa) involved in mRNA degradation after the cap has been removed. Located in **GW bodies**.

XTT The XTT reduction method is used as a colorimetric assay for quantification of microbial respiratory activity. Product of reduction is an orange water-soluble formazan that has an absorption peak around 490 nm.

xylan Plant cell wall polysaccharide containing a backbone of β(1−4)-linked xylose residues. Sidechains of 4-O-methylglucuronic acid and arabinose are present in varying amounts (see **glucuronoxylan** and **arabinoxylan**), together with acetyl groups. Found in the **hemicellulose** fraction of the wall matrix.

xylanases Enzymes (e.g. in *Pseudomonas fluorescens*: xylanase A, EC 3.2.1.8, 611aa; xylanase B, EC 3.2.1.8, 592aa) that will hydrolyse xylan, a heteropolymer of the pentose sugar xylose found in wood. Potentially important in bioprocessing (see **Dictyoglomi**).

Xylella fastidiosa Bacterium responsible for citrus variegated chlorosis, a major disease of citrus fruit; the first plant pathogen whose genome was sequenced (July 2000). Bacterium is restricted to xylem where it causes deficiency in water transport and is transmitted by leafhoppers. Other strains of *X. fastidiosa* cause other economically important plant diseases. Xylella genome project: http://www.lbi.ic.unicamp.br/xf/

xylem Plant tissue responsible for the movement of water and inorganic solutes from the roots to the shoot and leaves. Contains **tracheids, vessels, fibre cells** and **parenchyma**. Also provides structural support for the plant, especially in wood.

xylene A mixture of the three isomers of dimethyl benzene used to clear fixed and dehydrated material before infiltrating with wax and sectioning.

xyloglucan Plant cell-wall polysaccharide containing a backbone of β(1−4)-linked glucose residues to most of which single xylose residues are attached as side chains. Galactose, fucose and arabinose may also be present in smaller amounts. It is the major **hemicellulose** of dicotyledonous primary walls, and non-graminaceous monocots and acts as a food reserve in some seeds. Informative website: http://www.ccrc.uga.edu/~mao/xyloglc/Xtext.htm

xyloglucan endo-transglycosylase Plant cell wall degrading enzyme (XET; EC 2.4.1.207, 295aa in cauliflower). Breaks a β-(1-4) bond in the backbone of a xyloglucan and transfers the xyloglucanyl segment on to O-4 of the non-reducing terminal glucose residue of an acceptor, which can be a xyloglucan or an oligosaccharide of xyloglucan.

xylose Monosaccharide (pentose) that is found in **xylans**, very abundant components of **hemicelluloses**.

xylulose A 5-carbon ketose sugar, whose 5-phosphate is an intermediate in the **pentose phosphate pathway** and the **Calvin-Benson cycle**. Found in the urine in patients with pentosuria, a disorder in which L-xylulose reductase is deficient.

XYY syndrome Condition in which the human male has an extra Y chromosome. They are normal males, except for minor growth and sometimes behavioural abnormalities.

Y

Y Single letter code for tyrosine.

Y chromosome Chromosome found only in the heterogametic sex. In mammals the male has one Y chromosome and one X chromosome. One region of the Y chromosome, the pseudoautosomal region, is homologous to and pairs with the X chromosome. The primary determinant of male sexual development (**sry**) is found on the unpaired, differentiated segment of the Y chromosome.

yaba virus A genus of double-stranded DNA viruses (in the **Poxviridae**) isolated from African monkeys that cause focal (benign) histiocytomas in primates. The yatapoxvirus genus contains three members: tanapox virus (TPV), yaba-like disease virus (YLDV) and yaba monkey tumour virus (YMTV). Tanapox virus (TPV) encodes and expresses a secreted TNFα-binding protein, TPV-2L or gp38, that has inhibitory properties. Yaba monkey tumour virus encodes a variant of the orthopoxvirus IL-18 binding proteins. Yaba-like disease virus secretes a glycoprotein, related to protein B18 from Vaccinia virus, that binds to and inhibits both type I and type III IFNs.

YAC See **yeast artificial chromosome**.

yaws *framboesia* A contagious tropical disease caused by infection with *Treponema pertenue*. There are characteristic raspberry-like papules on the skin.

yeast A colloquial and non-systematic name for fungi that tend to be unicellular for the greater part of their life cycle. They may be from various fungal families (Ascomycota, Basidiomycotina or Deuteromycotina). Commercially important yeasts include *Saccharomyces cerevisiae* (baker's yeast); pathogenic yeasts include the genus *Candida*. See also *Schizosaccharomyces pombe*.

yeast artificial chromosome *YAC* A vector system that has telomeric, centromeric, and replication origin sequences needed for replication and preservation in yeast and allows very large segments (>100 kbp−3000 kbp) of DNA to be cloned. Useful in chromosome mapping; contiguous YACs covering the whole *Drosophila* genome and certain human chromosomes are available. They are less stable than bacterial artificial chromosomes but proteins encoded by the inserts are post-translationally modified in a eukaryotic manner. Web resource: http://openwetware.org/wiki/Yeast_artificial_chromosomes

yeast elicitor A compound that mimics the effect of pathogens in eliciting immune responses in plants.

Purified elicitor consists of cell wall polysaccharides composed entirely of mannose. Abstract: http://www.ncbi.nlm.nih.gov/pubmed/15605242

yeast two-hybrid screening A method used to screen for proteins that interact. A cDNA library is constructed such that candidate proteins are expressed as translational fusions with part of (typically) the *GAL4* gene. Yeast cells are then co-transfected with a "bait" construct consisting of the cDNA of interest fused in-frame to the other part of the *GAL4* gene. Only if both expressed proteins physically interact will the two parts of the GAL4 protein come close enough to produce detectable beta-galactosidase activity. Similar systems have now been developed that tag bait and targets with heterodimeric proteins other than GAL4. Interactions identified in this way usually need to be confirmed by independent methods (e.g. pull-down assays).

yellow fever virus A positive-sense, single-stranded, encapsulated RNA virus of the Flaviviridae that causes yellow fever, the symptoms of which include fever and haemorrhage. Transmitted by the mosquitoes *Aedes aegypti* and *Haemagogus sp.* Only one antigenic type of the virus known.

Yersinia Genus of Gram-negative bacteria of the **Enterobacteriaceae**; all are parasites or pathogens. *Y. pestis* (formerly *Pasteurella pestis*) was probably the cause of plague (Black Death) a disease of rodents where humans are accidental hosts. *Yersinia enterocolitica* causes an enterocolitis that is most common in children and is self-limiting.

yes An oncogene originally identified in Yamaguchi avian sarcoma, encodes p62yes, a nonreceptor tyrosine kinase c-Yes (EC 2.7.10.2, 543aa), contained within EBP50 protein complexes by association with **yes-associated protein**.

Yes-associated protein A transcriptional coactivator (YAP65, 454aa) that interacts with and enhances p73-dependent apoptosis in response to DNA damage. Contains **WW domains** that bind to the PPPPY motif of **p73**.

YidC Translocase (548aa) found in both Gram-positive and Gram-negative bacteria. In association with SecYEG is involved in the insertion of proteins into the inner membrane of Gram-negative bacteria. Has sequence and functional homology with **Oxa1** in mitochondria and **Albino3** in chloroplasts.

YMRF-amide **FMRFamide**-like neuropeptide from *Hirudo medicinalis* (medicinal leech).

The Dictionary of Cell and Molecular Biology. DOI: http://dx.doi.org/10.1016/B978-0-12-384931-1.00025-8

yolk cells Cells lying between the endoderm and mesoderm, formed by cleavage of yolk-rich regions of embryos in which the yolk is restricted in distribution (telolecithal eggs). They probably give rise to the endothelium of vitelline vessels.

yolk sac One of the set of extra-embryonic membranes, growing out from the gut over the yolk surface, in birds formed from the splanchnopleure, an outer layer of splanchnic mesoderm and an inner layer of extra-embryonic endoderm. The yolk sac in mammals is the site of early haematopoiesis.

yorkie In *Drosophila* a transcriptional coactivator (418aa) which is the critical downstream regulatory target in the **Salvador-Warts-Hippo pathway**. Activity is regulated by multiple phosphorylation events. The human homologue (65 kDa Yes-associated protein, 504aa) is upregulated in some liver and prostate carcinomas.

YPD Yeast Proteome Database (YPDTM): contains information about the proteins of *Saccharomyces cerevisiae*. A subsection of a larger collection of curated databases. Link: http://www.proteome.com/YPDhome.html Not freely available without subscription.

YY transcription factor Yin Yang 1 (YY1, 414aa) is a multifunctional transcription factor involved in development, differentiation, cellular proliferation and apoptosis. It can act as a transcriptional repressor, an activator, or an initiator element binding protein that directs and initiates transcription of numerous cellular and viral genes. Yin Yang 2 (YY2, 372aa) may antagonize YY1 and function in development and differentiation. Full paper: http://www.jbc.org/content/279/24/25927.long

YY1 A multifunctional transcription factor (414aa) that exerts positive and negative control on a range of cellular and viral genes.

Z

zap70 A protein **tyrosine kinase** (zeta chain-associated protein, Syk-related tyrosine kinase, EC 2.7.10.2, 619aa) that associates with the zeta chain of the **T-cell receptor** (TCR) via tandem SH2-domains that interact with phosphorylated **ITAMs** of CD3-zeta following ligand binding. Zap-70 phosphorylates the **LAT adaptor** (linker of activated T-cells) and is essential for TCR-mediated IL-2 production. It interacts with several other enzymes involved in activation. Mutation in zap70 cause an autosomal recessive form of **severe combined immunodeficiency disease** (SCID) in which there is a selective absence of CD8$^+$ T-cells. Research article: http://www.ncbi.nlm.nih.gov/pmc/articles/PMC2191847/?tool = pubmed

Z-DEVD See DEVD.

Z-disc The region of the **sarcomere** into which **thin filaments** are inserted. **Alpha-actinin** in the Z-disc cross-links actin and **titin**.

Z-DNA The configuration adopted by DNA that has sequences of alternating **purines** and **pyrimidines**, a left-handed helix with the phosphate groups of the backbone zigzagged (hence Z) and a single deep groove. It is thought to provide torsional strain relief during transcription and the potential to form a Z-DNA structure correlates with regions of active transcription. See **A-DNA, B-DNA**. Predicting potential for Z-DNA formation: http://www.ncbi.nlm.nih.gov/pubmed/3780676

Zea mays Maize or Indian corn. In the US. 'corn' is taken to mean maize; in Europe 'corn' usually means wheat or barley.

zearalenone *ZEN* A non-steroidal estrogenic mycotoxin produced by several *Fusarium* spp. that has serious adverse effects in laboratory and domestic animals. The mechanism of ZEN toxicity involves binding to estrogen receptors. WHO report: http://www.inchem.org/documents/jecfa/jecmono/v44jec14.htm

zeatin A naturally-occurring **cytokinin** derived from adenine and originally isolated from maize seeds. Its riboside is also a cytokinin. High concentrations produce adventitious shoot formation. Website of supplier: http://www.zeatin.net/zeatin.htm

zeaxanthin A carotenoid pigment found in dark green, leafy vegetables (e.g. spinach, collard greens) and in the macula region of the retina. There is some (limited) evidence for its value as a nutritional supplement, particularly for macular degeneration.

ZEB Transcription factors, ZEB1 (δEF1, TCF8, AREB6, ZFHEP, NIL-2A, ZFHX1A, BZP, 1124aa) and ZEB2 (SIP1, SMADIP1, ZFHX1B, KIAA0569, 1214aa) that are involved in regulation of the epithelial-mesenchymal transition (**EMT**) during embryonic development and in neoplasia. They activate EMT by binding to E-box elements in the E-cadherin promoter. ZEB1 inhibits interleukin-2 (IL-2) gene expression. Defects in ZEB2 are the cause of Hirschsprung disease-mental retardation syndrome (**Mowat-Wilson syndrome**).

zebrafish A small freshwater fish (*Danio rerio*, formerly *Brachydanio rerio*), easily reared in an aquarium and a valuable laboratory animal. The embryo is transparent so that the progeny of single cells can be followed until quite late stages of development. Many mutant lines have been characterised and it is an important model system for **cell lineage** studies. The Zebrafish Model Organism Database: http://zfin.org/cgi-bin/webdriver?MIval = aa-ZDB_home.apg

zein Water-insoluble storage proteins (a broad class of **prolamine proteins**) in maize endosperm; can be extracted from corn gluten and used to form odourless, tasteless, clear, hard and almost invisible edible films. The solubility properties are unusual (insoluble in water and anhydrous alcohol, but soluble in a mixture of the two) and probably a consequence of having a preponderance of hydrophobic amino acids.

zeiosis Blebbing of the plasma membrane; sometimes referred to as 'cell boiling'.

zeitgeber Literally the 'time-giver'; the environmental agent or event that provides the cue for setting or resetting a biological clock.

Zellweger's syndrome A **peroxisome biogenesis disorder** (one of the **leukodystrophies**) that can be caused by mutations in several different genes involved in brain development and the formation of myelin. High levels of iron and copper build up in blood and tissue and cause the characteristic symptoms including an enlarged liver; facial deformities and neurological abnormalities. See **peroxins, mevalonate kinase, PAS genes**. NCBI information page: http://www.ninds.nih.gov/disorders/zellweger/zellweger.htm

Zenker's fluid A histochemical fixative that is good for preserving cytoplasmic structure but needs to be made up freshly and is hazardous because it contains potassium dichromate and mercuric chloride.

The Dictionary of Cell and Molecular Biology. DOI: http://dx.doi.org/10.1016/B978-0-12-384931-1.00026-X

zerknuellt 1 In *Drosophila*, a transcription factor (353aa) encoded by the *zen-1* gene, involved in dorsoventral axis determination. See **dorsal**.

zeste A regulatory protein (575aa) in *Drosophila* that is involved in **synapsis**-dependent gene expression (transvection). The synaptic pairing of chromosomes carrying genes with which zeste interacts influences the expression of these genes by binding to DNA and stimulating transcription from a nearby promoter. Zeste self-associates to form complexes of several hundred monomers.

zeta potential The electrostatic potential of a molecule or particle such as a cell, measured at the plane of hydrodynamic slippage outside the surface of the molecule or cell. Usually measured by electrophoretic mobility. Related to the surface potential and a measure of the electrostatic forces of repulsion the particle or molecule is likely to meet when encountering another of the same sign of charge. See **cell electrophoresis**.

zf9 A zinc finger transcription factor (283aa) from rat stellate cells (**Ito cells**) activated *in vitro* and may be an important signal in hepatic stellate cell activation *in vivo* after liver injury. The human homologue is KLF6 (core promoter element-binding protein, COPEB, **Kruppel**-like factor 6, 283aa) which may have a role in B-cell growth and development. Research article: http://www.pnas.org/content/95/16/9500.long

ZFIN Zebrafish Information Network, a database for *Danio rerio* covering genomics, anatomy, etc. Link: http://zfin.org/

Zfp57 Zinc finger protein 57. A mouse protein (452aa), the product of a maternal-zygotic effect gene, with homologues in other mammals. A transcription regulator that maintains maternal and paternal **genomic imprinting** by controlling DNA methylation during the earliest stages of zygotic development at multiple imprinting control regions. Mutation in the maternal Zfp57 gene causes loss of imprinting of maternal genes and is embryonically lethal. Abstract: http://www.cell.com/developmental-cell/abstract/S1534-5807%2808%2900338-9

Zic Zinc finger protein of cerebellum. A family of zinc-finger proteins that are crucial in neural development and are the vertebrate homologs of odd-paired in *Drosophila*. In *Xenopus*, Zic1 (443aa) induces expression of multiple genes and is important in the regulation of neural induction and neurogenesis. Zic2 (503aa) regulates anteroposterior patterning in early development by inhibiting expression of the **nodal** genes. Zic3 (441aa) is involved in early steps in determination of normal left-right asymmetry and the human form (467aa) is mutated in some cases of **heterotaxy**. Zic5 in *Xenopus* (515aa) is essential for neural crest development.

Abstract: http://www.ncbi.nlm.nih.gov/pubmed/15207846

Ziehl-Neelsen stain A stain that is used to identify acid fast bacteria (particularly *Mycobacterium tuberculosis*) which stain bright red.

Zigmond chamber See **orientation chamber**.

ZIM domain See **tify domain**.

zinc An essential 'trace' element that is a component of the active site of a variety of enzymes and has a structural role in zinc-finger transcription factors. Zn^{2+} has high affinity for the side chains of cysteine and histidine. Zinc is present in tissues at levels of ca 0.1 mM, but intracellular levels must be much lower. Zinc deficiency in soils is the most common micronutrient deficiency for crop plants.

zinc finger A motif (12aa) associated with many **DNA-binding proteins** stabilized by a zinc ion. A cysteine-histidine zinc finger has 2 cysteine and 2 histidine groups (a Cys2His2 type), or 4 cysteines (a Cys4 type), that directly coordinate the zinc. The loops (usually present in multiples) intercalate directly into the DNA helix. Originally identified in the RNA polymerase III transcription factor TFIIIA. A more systematic classification depends upon the folding pattern (e.g. the Cys2His2-like 'classic zinc fingers' and so-called **treble clef** forms). Classification: http://prodata.swmed.edu/zndb/zndb.php

zinc finger nucleases Synthetic proteins consisting of an engineered zinc finger DNA-binding domain (usually 3–6 fingers) fused to the cleavage domain of the FokI restriction endonuclease. ZFNs can be used to induce double-stranded breaks (DSBs) in specific DNA sequences and thereby promote site-specific homologous recombination. The zinc fingers can be designed to target specific genes. Website of zinc finger consortium: http://www.zinc-fingers.org/scientific-background.htm

zip (1) See **synaptonemal complex**. (2) In humans, a zinc transporter, ZIP1, (SLC39A1, 324aa).

zip kinase See **DAP kinases**.

zipper See **leucine zipper**.

zippering Process suggested to occur in phagocytosis in which the membrane of the phagocyte covers the particle by a progressive adhesive interaction. The evidence for such a mechanism comes from experiments in which capped B-cells are only partially internalized, whereas those with a uniform opsonizing coat of anti-IgG are fully engulfed.

zizimin A cdc42 guanine nucleotide exchange factor (zizimin-1, dedicator of cytokinesis protein 9, DOCK9, 2069aa; zizimin-2, DOCK11, 2073aa; zizimin-3, DOCK10, 2186aa).

ZJ-43 A potent inhibitor of NAAG peptidases (N-acetyl-aspartyl-glutamate peptidases), glutamate carboxypeptidase II and glutamate carboxypeptidase III that break down neurotransmitters. A Group II metabotropic glutamate receptor (mGluR) agonist.

ZO proteins *tight junction proteins* Peripheral membrane proteins of the **MAGUK** family associated with tight junctions (**zonula occludens**); there are tissue-specific isoforms. ZO-1 (TJP-1, 1748aa) is found on the cytoplasmic membrane surface, interacts with the other ZO proteins, and various other junction-associated proteins (**occludin**, **claudins**, CGN/**cingulin**, **CXADR**, GJA12 (**connexin**-47), GJD3 (gap junction delta-3 protein) and **ubinuclein-1**). ZO-2 (TJP2, 1190aa) occurs as a homodimer and as a heterodimer with ZO-1 and is also found in adherens junctions. Defects in ZO-2 are associated with familial **hypercholanemia**. ZO-3 (TJP3, 898aa) and ZO-4 (tight junction associated protein, TJAP1, TJP4, 557aa) also co-localise with ZO-1. ZO-4 is recruited to tight junctions late in the maturation of the complex. See **PDZ domains**. Research article: http://www.jbc.org/content/282/52/37710.full

Zollinger-Ellison syndrome A rare disorder in which tumours (gastrinomas) in the pancreas or duodenum overproduce **gastrin** which causes excess acid production leading to peptic ulcers. The inherited form (~25% of cases) is due to mutation in the MEN1 gene (see **multiple endocrine neoplasia**). NCBI Information page: http://digestive.niddk.nih.gov/ddiseases/pubs/zollinger/

zona pellucida A translucent layer of extracellular matrix material that surrounds the ovum and is responsible for species-specific sperm binding, induction of the acrosomal reaction and prevention of post-fertilization polyspermy. It is composed of three to four glycoproteins, ZP1 (638aa), ZP2 (745aa), ZP3 (424aa) and ZP4 (540aa). ZP3 is essential for sperm binding and zona matrix formation. Following fertilization ZP2 undergoes limited proteolysis by an oocyte-derived protease probably released during exocytosis of cortical granules and ZP3 ceases to be able to induce an acrosomal reaction. ZP2 and ZP4 may act as secondary sperm receptors. ZP1 is important for the structural integrity of the zona pellucida. All contain a **ZP domain** that is involved in polymerization: ZP2 and ZP3 are organized into long filaments cross-linked by ZP1 homodimers. Review: http://www.pubmedcentral.nih.gov/picrender.fcgi?artid=442960&blobtype=pdf

zone of polarizing activity *ZPA* The small group of mesenchymal cells located at the posterior margin of the developing avian limb bud that provides positional information to the developing limb. The ZPA expresses the morphogen **sonic hedgehog** expression of which can be induced in other cells by retinoic acid (thus explaining the identification of retinoic acid as the putative morphogen).

zonula adherens Specialized intercellular junction in which the membranes are separated by 15-25 nm, and into which are inserted microfilaments. Similar in structure to two apposed focal adhesions, though this may be misleading. Microfilaments inserted into the zonula adherens may interact (via myosin) with other microfilaments to generate contraction. Constitute mechanical coupling between cells.

zonula occludens *tight junction* A specialized intercellular junction in which the two plasma membranes are separated by only 1–2 nm. Found near the apical surface of cells in simple epithelia where they form a sealing 'gasket' around the cell that prevents fluid moving through the intercellular gap. The lateral diffusion of intrinsic membrane proteins between apical and basolateral domains of the plasma membrane is blocked by the zonula occludens.

zonula occludens toxin An enterotoxin (399aa) produced by *Vibrio cholerae*. Binds to a receptor on mature cells of small intestinal villi, but not in the colon, and activates a complex intracellular cascade of events that involve a dose- and time-dependent PKCa-related polymerization of actin filaments and redistribution of ZO-1 away from junctions making them more permeable. See **zonulin**. Research article: http://www.jbc.org/content/276/22/19160.full#sec-18

zonulin A protein (47 kDa) that was identified as being able to induce tight junction disassembly in non-human primate intestinal epithelia; it is an endogenous analogue of *Vibrio cholerae* **zonula occludens toxin** (Zot) and apparently operates competitively through the same signalling cascade. Its identity remained obscure until it was shown to be the precursor of haptoglobin 2. Zonulin is overexpressed in the intestinal mucosa of individuals with celiac disease. Research article: http://www.pnas.org/content/106/39/16799.full.pdf+html

zoo blot Blot with DNA or RNA from a wide range of species adsorbed onto a paper membrane (nylon or nitrocellulose) that can be probed with the sample of interest. Used, for example, to see the extent of conservation of a sequence between genera or even phyla.

zoochlorellae Intracellular symbiotic algae usually of the genus *Chlorella* found in some lamellibranch molluscs, protozoans, flatworms, sponges and corals.

zoonosis An infectious disease of humans whose natural reservoir is a non-human animal. Example: psittacosis, a viral disease of birds, occasionally infecting humans.

zootype Postulated pattern of gene expression shared by all animal phyla. The zootype hypothesis defines an animal by a specific spatial pattern of gene expression. and implies that six hox-type **homeobox**-containing genes should be present in all metazoa. Paper: http://www.ncbi.nlm.nih.gov/pmc/articles/PMC2222619/?tool = pubmed

zooxanthellae Intracellular photosynthetic symbiotic dinoflagellates found in a variety of marine invertebrates. Systematics uncertain.

zormin See **sallimus**.

ZP domain A domain (~260aa) found in various extracellular proteins, notably the **zona pellucida** glycoproteins, but also Tamm-Horsfall protein (THP, **uromodulin**), **glycoprotein-2**, **tectorins**, **TGF-beta receptor III** (betaglycan), **endoglin**, **DMBT-1** (deleted in malignant brain tumour-1), **NompA** (no-mechanoreceptor-potential-A), **dumpy** and **cuticlin-1**. It seems that the ZP domain is responsible for polymerization of these proteins into filaments of similar supramolecular structure. Article: http://www.nature.com/ncb/journal/v4/n6/full/ncb802.html Annotation: http://smart.embl-heidelberg.de/smart/do_annotation.pl?ACC = SM00241

ZPA See **zone of polarizing activity**.

Z-ring A filamentous ring structure that forms at the mid-point (equator) of a bacterial cell before division and contracts to separate daughter cells; analogous to the **constriction ring** of eukaryotic cell cytokinesis but composed of polymerized tubulin-like **FtsZ** protein. The mechanism of force generation is unclear. See **minD**. Article: http://www.pnas.org/content/104/41/16110.full

Z-scheme of photosynthesis A schematic representation of the **light-dependent reactions** of **photosynthesis**, in which the photosynthetic reaction centres and electron carriers are arranged according to their electrode potential (free energy) in one dimension and their reaction sequence in the second dimension. This gives a Z-shape, the two reaction centres (of photosystems I and II) being linked by the photosynthetic electron transport chain.

zwitterions Molecules that have a positive charge at one end and a negative charge at the other. Also known as ampholyte or dipolar ions.

zygomycetes A Class of the Eumycota or true fungi in which sexual reproduction is by the formation of a zygospore. Recent classifications subdivide them into eight Orders (Basidiobolales, Dimargaritales, Endogonales, Entomophthorales, Kickxellales, Mortierellales, Mucorales, Zoopagales, and Zygomycetes). Mostly saprophytes, although some are involved in mycorrhizas. Best known is probably *Mucor*. Zygomycetes website: http://www.zygomycetes.org/

zygonema See **zygotene**.

zygosporangium In the **zygomycetes**, a structure formed by the fusion of two different hyphal strands that contains a zygote.

zygospore Fungal spore produced by the fusion of two similar **gametes** or **hyphae**. Characteristic of the **Zygomycetes**.

zygote **Diploid** cell resulting from the fusion of male and female gametes at fertilization.

zygotene Classic term for the second stage of the **prophase** of **meiosis** I, during which the homologous chromosomes start to pair.

zygotic embryo A plant embryo formed following double fertilisation of the ovule, giving rise to the plant embryo and the endosperm. *Cf.* **somatic embryogenesis**.

zygotic-effect gene A gene encoded by the zygote and that is important in early development. Whether the maternally-derived or paternally-derived allele is expressed may depend upon **genomic imprinting**. Many early developmental activities are dependent upon mRNA already present in the egg and are maternally-derived. See **maternal-effect gene** and **Zfp57**.

zymogen Inactive precursor of an enzyme, particularly a proteolytic enzyme. Synthesized in the cell and secreted in this safe form, then converted to the active form by limited proteolytic cleavage.

zymogen granule A **secretory vesicle** containing an inactive precursor (zymogen). The contents are often very condensed.

zymogenic cell *gastric chief cell* Cells of the basal part of the gastric glands of the stomach. They contain extensive rough endoplasmic reticulum and zymogen granules and secrete pepsinogen, the inactive precursor of **pepsin**, and rennin.

zymogram *zymograph* Electrophoretic gel (or other separation) in which the position of an enzyme is revealed by a reaction that depends upon its enzymic activity with an appropriate substrate co-polymerised with the gel. The process is zymography.

zymosan Particulate yeast cell-wall polysaccharide (mannan-rich) that will activate **complement** in serum through the alternate pathway. Becomes coated with C3b/C3bi and is therefore a convenient opsonized particle; also leads to C5a production in the serum.

zyxin Protein (572aa) of the **zyxin/ajuba family** found at the **adherens junction** and focal adhesions. Interacts with **alpha-actinin** and **testin**. May be involved in transduction of signals that regulate gene expression in response to adhesion and enters the nucleus in the presence of HESX1 (homeobox protein ANF). See **zyxin-related protein 1**.

zyxin-related protein 1 One of the zyxin/ajuba family of proteins (ZRP-1, thyroid receptor-interacting protein 6, Trip6, 476aa) that interacts with the cytoplasmic domain of **endoglin** and the lysophosphatidic acid receptor (LPAR2). It is involved in feedback regulation of adherens junction by affecting gene expression and promotes the reorganization of the actin cytoskeleton and cell invasiveness. It is a transcriptional coactivator for NFκB and Jun.

zyxin/ajuba family A family of proteins characterized by three **LIM domains** that associate with the actin cytoskeleton, are components of both cell-cell junction adhesive complex and integrin-mediated adhesion complexes, and shuttle in and out of the nucleus where they indirectly influence gene expression. Members of the family include **ajuba**, **LIMD1**, LPP (**lipoma-preferred partner**), **zyxin-related protein 1** and **zyxin**.

Appendix

Prefixes for SI Units

Factor	Prefix	Symbol
10^{24}	yotta	Y
10^{21}	zetta	Z
10^{18}	exa	E
10^{15}	peta	P
10^{12}	tera	T
10^{9}	giga	G
10^{6}	mega	M
10^{3}	kilo	k
10^{2}	hecto	h
10^{1}	deca	da
10^{-1}	deci	d
10^{-2}	centi	c
10^{-3}	milli	m
10^{-6}	micro	μ
10^{-9}	nano	n
10^{-12}	pico	p
10^{-15}	femto	f
10^{-18}	atto	a
10^{-21}	zepto	z
10^{-24}	yocto	y

Greek Alphabet

A	α	alpha
B	β	beta
Γ	γ	gamma
Δ	δ	delta
E	ϵ	epsilon
Z	ζ	zeta
H	η	eta
Θ	θ	theta
I	τ	iota
K	κ	kappa
Λ	λ	lambda
M	μ	mu
N	ν	nu
Ξ	ξ	xi
O	o	omicron
Π	π	pi
P	ρ	rho
Σ	σ	sigma
T	τ	tau
Y	υ	upsilon
Φ	ϕ	phi
X	χ	chi
Ψ	ψ	psi
Ω	ω	omega

Useful constants

Avogadro's number (N)	6.022×10^{23} mol^{-1}
Boltzmann's constant (k)	1.318×10^{-23} J deg^{-1}
	3.298×10^{-24} cal deg^{-1}
Faraday constant (F)	9.649×10^{4} Coulomb mol^{-1}
Curie (Ci)	3.7×10^{10} disintegrations s^{-1}
Gas constant(R)	8.314 J mol^{-1} deg^{-1}
π	3.14159
e	2.71828

$\log_e x = 2.303 \log_{10} x$

Single-letter codes for amino acids

A	Ala	Alanine
R	Arg	Arginine
N	Asn	Asparagine
D	Asp	Aspartic acid
B		Asparagine or aspartic acid
C	Cys	Cysteine
Q	Gln	Glutamine
E	Glu	Glutamic acid
Z		Glutamine or glutamic acid
G	Gly	Glycine
H	His	Histidine
I	Ileu	Isoleucine
L	Leu	Leucine
K	Lys	Lysine
M	Met	Methionine
F	Phe	Phenylalanine
P	Pro	Proline
S	Ser	Serine
T	Thr	Threonine
W	Trp	Tryptophan
Y	Tyr	Tyrosine
V	Val	Valine

Printed in the United States
By Bookmasters